What Is an "Integrated Approach"?

One of the key themes in this book is that the body is an integrated set of systems. One of your tasks as you study will be to construct for yourself this global view of the body, its systems, and the many processes that keep the systems working.

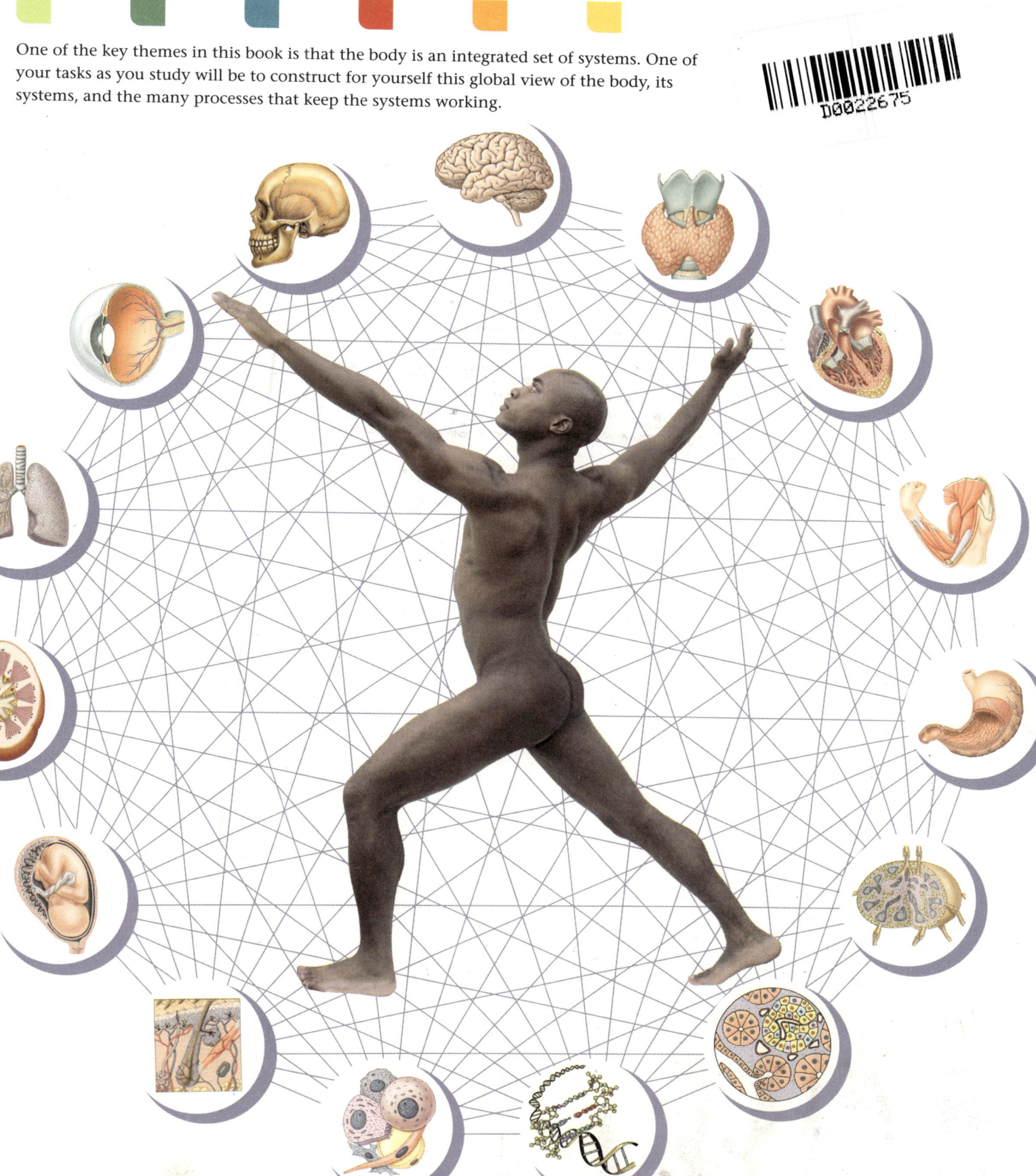

Top Ten Ways to Succeed in Classes that Use Active Learning

By Marilla Svinicki, Ph.D., Director
University of Texas Center for Teaching Effectiveness

1. Make the switch from an authority-based conception of learning to a self-regulated conception of learning. Recognize and accept your own responsibility for learning.

2. Be willing to take risks and go beyond what is presented in class or the text.

3. Be able to tolerate ambiguity and frustration in the interest of understanding.

4. See errors as opportunities to learn rather than failures. Be willing to make mistakes in class or in study groups so that you can learn from them.

5. Engage in active listening to what's happening in class.

6. Trust the instructor's experience in designing class activities and participate willingly if not enthusiastically.

7. Be willing to express an opinion or hazard a guess.

8. Accept feedback in the spirit of learning rather than as a reflection of you as a person.

9. Prepare for class physically, mentally, and materially (do the reading, work the problems, etc.).

10. Provide support for your classmate's attempts to learn. The best way to learn something well is to teach it to someone who doesn't understand.

Dr. Dee's Eleventh Rule:
DON'T PANIC! Pushing yourself beyond the comfort zone is scary, but you have to do it in order to improve.

Word Roots for Physiology

a- or **an-** without, absence
anti- against
ase signifies an enzyme
auto self
bi two
brady slow
cardio- heart
cephalo- head
cerebro- brain
contra- against
-crine a secretion
crypt- hidden
cutan- skin
-cyte or **cyto-** cell
de- without, lacking
di- two
dys- difficult, faulty
-elle small
-emia blood
endo- inside or within
epi- over
erythro- red
exo- outside
extra- outside
gastro- stomach
-gen, -genic produce
gluco-, glyco- sugar or sweet
hemi- half
hemo- blood
hepato- liver
homo- same
hydro- water
hyper- above or excess

hypo- beneath or deficient
inter- between
intra- within
-itis inflammation of
kali- potassium
leuko- white
lipo- fat
lumen inside of a hollow tube
-lysis split apart or rupture
macro- large
micro- small
mono- one
multi- many
myo- muscle
oligo- little, few
para- near, close
patho-, -pathy related to disease
peri- around
poly- many
post- after
pre- before
pro- before
pseudo- false
re- again
retro- backward or behind
semi- half
sub- below
super- above, beyond
supra- above, on top of
tachy- rapid
trans- across, through

HUMAN PHYSIOLOGY

AN INTEGRATED APPROACH FOURTH EDITION

Dee Unglaub Silverthorn, Ph.D.
University of Texas, Austin

WITH

William C. Ober, M.D.
Illustration Coordinator

Claire W. Garrison, R.N.
Illustrator

Andrew C. Silverthorn, M.D.
Clinical Consultant

WITH CONTRIBUTIONS BY

Bruce R. Johnson, Ph.D.
Cornell University

PEARSON

Benjamin Cummings

San Francisco Boston New York
Cape Town Hong Kong London Madrid Mexico City
Montreal Munich Paris Singapore Sydney Tokyo Toronto

Senior Acquisitions Editor: Deirdre Espinoza
Developmental Manager: Claire Alexander
Development Editor: Irene Nunes
Project Editor: Nicole George-O'Brien
Managing Editor: Wendy Earl
Editorial Assistants: Sabrina Larson and Amy Yu
Associate Media Editor: Sarah Young-Dualan
Editorial Media Producer: Alex Streczyn
Production Supervisor: Karen Gulliver
Art and Design Coordinator: Michele Mangelli
Photo Researcher: Maureen Spuhler
Text and Cover Design: Yvo Riebezos Design
Senior Manufacturing Buyer: Stacey Weinberger
Executive Marketing Manager: Lauren Harp
Compositor: Techbooks/GTS

Cover Image: Nikon Small World/Torsten Whitmann

Credits can be found on page Cr1

Library of Congress Cataloging-in-Publication Data

Silverthorn, Dee Unglaub, 1948–
  Human physiology: an integrated approach / Dee Unglaub Silverthorn; with
  William C. Ober, illustration coordinator; Claire W. Garrison, illustrator;
  Andrew C. Silverthorn, clinical consultant; with contributions by Bruce R.
  Johnson. —4th ed.
    p. cm.
  Includes index.
  ISBN 0-8053-6849-3
    1. Human physiology.   I. Title.
  QP34.5.S55 2006
  612—dc22   2005056517

www.aw-bc.com

ISBN 0-8053-6849-3

1 2 3 4 5 6 7 8 9 10—QWV—10 09 08 07 06

DEE UNGLAUB SILVERTHORN studied biology as an undergraduate at Tulane University and went on to earn a Ph.D. in marine science at the University of South Carolina. Her research interest is epithelial transport, and recent work in her laboratory has focused on transport properties of the chick allantoic membrane. She began her teaching career in the Physiology Department at the Medical University of South Carolina but over the years has taught a wide range of students, from medical and college students to those still preparing for higher education. At the University of Texas she lectures in physiology, coordinates student laboratories in physiology, and instructs graduate students in a course on developing teaching skills in the life sciences. She has received numerous teaching awards and honors, including the American Physiological Society's Claude Bernard Distinguished Lecturer and Arthur C. Guyton Physiology Educator of the Year, and UT–Austin's Texas Excellence Teaching Award, Texas Blazers Faculty Excellence Award, and College of Natural Sciences Teaching Excellence Award. The first edition of this textbook won the 1998 Robert W. Hamilton Author Award for best textbook published in 1997–98 by a University of Texas faculty member. Dee is currently editor-in-chief of *Advances in Physiology Education* and she works with members of the International Union of Physiological Sciences to improve physiology education in developing countries. She is also a member of the Human Anatomy and Physiology Society, the Society for Comparative and Integrative Biology, the Association for Biology Laboratory Education, and the Society for College Science Teaching. Her free time is spent creating multimedia fiber art and enjoying the Texas hill country with her husband, Andy, and their dogs.

ABOUT THE ILLUSTRATORS

DR. WILLIAM C. OBER, M.D. (art coordinator and illustrator) received his undergraduate degree from Washington and Lee University and his M.D. from the University of Virginia. While in medical school, he also studied in the Department of Art as Applied to Medicine at Johns Hopkins University. After graduation, Dr. Ober completed a residency in Family Practice and later was on the faculty at the University of Virginia in the Department of Family Medicine. He is currently an Instructor in the Division of Sports Medicine at UVA and is part of the Core Faculty at Shoals Marine Laboratory, where he teaches Biological Illustration every summer. The textbooks illustrated by Medical & Scientific Illustration have won numerous design and illustration awards.

CLAIRE W. GARRISON, R.N. (illustrator) practiced pediatric and obstetric nursing before turning to medical illustration as a full-time career. She returned to school at Mary Baldwin College where she received her degree with distinction in studio art. Following a 5-year apprenticeship, she has worked as Dr. Ober's partner in Medical and Scientific Illustration since 1986. She is on the Core Faculty at Shoals Marine Laboratory and co-teaches the Biological Illustration course.

ABOUT THE CLINICAL CONSULTANT

ANDREW C. SILVERTHORN, M.D. is a graduate of the United States Military Academy (West Point). He served in the infantry in Vietnam, and upon his return entered medical school at the Medical University of South Carolina in Charleston. He completed his family practice residency at the University of Texas Medical Branch, Galveston, where he was chief resident, and he is currently a family physician in solo private practice in Austin, Texas. When Andy is not busy seeing patients he may be found on the golf course or playing with his chocolate lab, Lady Godiva.

ABOUT THE CONTRIBUTOR

BRUCE JOHNSON is a Senior Research Associate in Neurobiology and Behavior at Cornell University. He earned a B.A. at Florida State University, an M.S. at Florida Atlantic University, and his Ph.D. in Biology through Boston University, working at the Marine Biological Laboratory in Woods Hole. At Cornell he teaches an undergraduate laboratory course entitled Principles of Neurophysiology. He is a coauthor of *Crawdad: A CD-ROM Lab Manual for Neurophysiology* and the *Laboratory Manual for Physiology*, and he continues development of model preparations for student neuroscience laboratories. He has taught neuroscience laboratories for NSF-sponsored Crawdad Workshops, in a Fulbright-sponsored course at the University of Copenhagen, for the Marine Biological Laboratory and the Shoals Marine Laboratory, and as a workshop leader for Project Kaleidoscope teaching conferences. He has received outstanding educator and distinguished teaching awards at Cornell University, and the Faculty for Neuroscience Educator of the Year Award. His research addresses the cellular and synaptic mechanisms of motor network plasticity. He also enjoys being a Soccer Dad, practicing karate, hiking, and traveling.

This edition is dedicated to my colleagues around the world who teach physiology.

Contents in Brief

Contents

UNIT 2

HOMEOSTASIS AND CONTROL

UNIT 3
INTEGRATION OF FUNCTION

UNIT 4
METABOLISM, GROWTH, AND AGING

Preface to the Student

WELCOME TO HUMAN PHYSIOLOGY!

As you begin your study of the human body, you should be prepared to make maximum use of the resources available to you, including your instructor, the library, the Internet, and your textbook. One of my goals in this book is to provide you not only with information about how the human body functions but also with tips for studying and problem solving. Many of these study aids have been developed with the input of my students, so I think you may find them particularly helpful.

In the Visual Introduction (p.xxx), I have put together a brief tour of the special features of the book, especially those that you may not have encountered previously in textbooks. Please take a few minutes to read this section so that you can make optimum use of the text as you study. If you take advantage of features such as the Concept Checks, the Figure and Graph Questions, and Running Problems, you will find that you begin to think about physiology in a different way.

One of your tasks as you study will be to construct for yourself a global view of the body, its systems, and the many processes that keep the systems working. This "big picture" is what physiologists call the integration of systems, and it is a key theme in the book. In order to integrate information, however, you must do more than simply memorize it. You need to truly understand it and be able to use it to solve problems that you have never encountered before. If you are headed for a career in the health professions, you will do this in the clinics. If you are headed for a career in biology, you will solve problems in the laboratory, field, or classroom. Analyzing, synthesizing, and evaluating information are skills that you need to develop while you are in school, and I hope that the features of this book will help you with this goal.

Good luck with your studies!

Warmest regards, Dr. Dee (as my students call me) *silverthorn@mail.utexas.edu.*

Preface to Instructors

In the three years since the third edition was published, physiology and medicine have continued to evolve rapidly. As scientists learn more about fundamental biological processes, their findings are translated into new medical treatments and suggestions for maintaining wellness. In this edition of the text we also have continued to update and focus on basic themes and concepts of physiology to help students establish a mental model of how the human body functions.

KEY THEMES OF THIS BOOK

In the fourth edition we kept the four approaches to learning physiology that have proved so popular since this book was first published in 1998.

1. A Focus on Problem Solving

One of the most valuable skills we can teach our students is the ability to think critically and use the information they learn to solve new problems. A number of features in the book, expanded in this edition, are designed to help students practice these skills. Concept Checks embedded in the text slow students down and ask them to think about what they've just read. Figure and Graph Questions challenge students to really think about what the illustrations are showing. The Running Problems provide real-life applications of the material in each chapter, and this edition features two new problems on blood doping and hyponatremia in athletes.

2. An Emphasis on Integration

The human body functions as a coordinated whole, not as isolated organ systems. Time and textbooks must follow a linear organization yet somehow students must come to appreciate that physiological functions form an intricate network of interlocking processes. The three chapters on Integrative Physiology (Chapters 13, 20, and 25) ask students to step back, look at the big picture, and discover for themselves how multiple organ systems communicate and cooperate to maintain homeostasis.

In this edition of the textbook, we have chosen metabolic syndrome—a clinical condition in which people exhibit the triad of type 2 diabetics, atherosclerosis, and high blood pressure—to illustrate an integrative disease that affects multiple body systems simultaneously. With obesity at epidemic proportions in the United States, it seemed appropriate to select a disease with immediate relevancy to our students. Examples of how the diseases that make up metabolic syndrome relate to normal physiology appear repeatedly throughout the book to underscore the importance of understanding functional integration in the body.

3. Integration of Physiology with Cellular and Molecular Physiology

Most physiology research today is being done at the cellular and molecular level, and this edition continues to integrate physiology with cellular and molecular biology. Many exciting developments in molecular medicine and physiology have occurred since the first edition. In this edition, we highlighted the basic principles of protein interactions by creating a new section with that title in Chapter 2. These principles—the importance of protein shape, protein isoforms, specificity and binding affinity, modulation of activity, competition for binding sites, and saturation of activity—appear repeatedly throughout physiology in discussions of enzymes, receptors, and transporters, but too often students fail to see the commonalities. We hope that by presenting protein interactions as a basic theme that runs throughout physiology, students will learn to recognize these principles when they encounter them in their studies.

4. Presentation of Physiology as a Dynamic Field

I have tried to present physiology as a dynamic discipline with numerous unanswered questions that merit further investigation and research. It is essential that students appreciate that many of the "facts" they are learning are really only our current theories, especially as new information emerges from the union of the human genome project and basic research. The demand for systems physiologists is increasing as research efforts focus on taking information uncovered by cellular and molecular research and applying it to clinical observations.

The physiology students of today are the next generation of scientists and health care providers—the people who will try to answer the unanswered questions. It is my hope that this book will provide an integrated view of physiology so that tomorrow's scientists and health care providers will enter their chosen professions with respect for the complexity of the human body and a clear vision of the potential of physiological and biomedical research.

NEW TO THIS EDITION

The fourth edition of *Human Physiology: An Integrated Approach* builds upon the thorough coverage of integrative and molecular physiology topics that have been the foundation of this book since its first edition. The text has been revised with additional figures and diagrams and extensive content updates.

Chapter-by-Chapter Content Updates

Chapter 1 Introduction to Physiology

- Expanded introduction to homeostasis
- New information on proteomics and moving beyond the genome
- New section on searching and reading the scientific literature
- New information on translational research (lab bench to bedside)
- New figure on control systems

Chapter 2 Molecular Interactions

- New chapter title to reflect new focus
- New section on protein interactions, including equilibrium constant and dissociation constant
- New figure on noncovalent bonds of proteins
- Revised discussion of and art for basic chemistry
- New discussion of trans fats
- Expanded discussion of molecular shape

Chapter 3 Compartmentation: Cells and Tissues

- New chapter title reflects new coverage and focus on theme of compartmentation
- New section on compartmentation and functional compartments of the body
- Moved cell membrane structure and description of body fluid compartments here from Ch. 5
- New section on membrane sphingolipids, lipid rafts, and lipid-anchored proteins
- Expanded information on proteasomes and protein degradation
- New figure of motor proteins
- Moved details of RER and Golgi function to Ch. 4
- New information on external lamina of nerve and muscle cells
- New photos of Golgi complex, sphingolipids, and lipid rafts
- New figure showing abdominopelvic, cranial, and thoracic cavities

Chapter 4 Energy and Cellular Metabolism

- New information on RNA interference and siRNA
- Expanded discussion of gene expression and protein synthesis
- New model of cisternal progression in Golgi complex function
- New table on properties of living organisms
- New summary figure of protein synthesis and processing

Chapter 5 Membrane Dynamics

- Introduced clearance by multiple organs as a mechanism for maintaining mass balance

- New table on GLUT transporter family
- Introduced acronyms for transporter families: SGLT name for Na^+-glucose transporter, OAT for organic acid transporter, NKCC for Na-K-2Cl transporter.
- Additional examples, new figure, and quantitative explanation of tonicity
- New information on the association of lipid rafts with caveolae
- New graph of competition for protein binding sites
- New figure on mass balance

Chapter 6 Communication, Integration, and Homeostasis

- Introduced desensitization as control mechanisms
- New information on programmed up-regulation

Chapter 7 Introduction to the Endocrine System

- Updated information on calcitonin gene-related peptide and hormone evolution

Chapter 8 Neurons: Cellular and Network Properties

- Updated information on neural stem cells
- Expanded information on conductance, ion current, excitability
- Expanded information on threshold for channel opening; channel activation, inactivation, deactivation
- Updated information on multiple isoforms of ion channels
- New information on tonically active and bursting pacemaker neurons
- Expanded discussion on differences between peripheral and central neurons and synapses
- Updated information on metabotropic glutaminergic receptors
- New information on long-term depression
- Added potentiators, additional receptor subtypes, and antagonists to Table 8-4
- New information on lipid neurocrines and cannabinoid receptors
- New figure on local current flow
- Revised figures for Na channel gating, refractory period, action potential conductance (w/figure question), saltatory conductance

Chapter 9 The Central Nervous System

- New information on brain-computer interface devices
- New figure question on meninges
- Expanded information on embryonic development
- Added information on why we sleep

Chapter 10 Sensory Physiology

- Expanded information on transient receptor potential V_1 ($TRPV_1$) channel, vanilloid receptors
- New information on nociceptive pain, inflammatory pain, neuropathic pain, pathological pain
- Updated information on olfaction, human pheromones, taste
- New information on role of gustducin in taste
- New information on stem cells and gene therapy to replace dead auditory hair cells

- Added information on innervation of lacrimal glands
- Expanded information on bipolar cells, horizontal cells, and amacrine cells
- Expanded information on types of ganglion cells in primate retina

Chapter 11 Efferent Division: Autonomic and Somatic Motor Control

- New information on sympathetic innervation of lymphoid tissue
- Added information on beta-3 receptors
- Updated information on ganglia as integrating centers
- Updated data for high school smoking question

Chapter 12 Muscles

- Introduced concept of torque in biomechanics discussion
- Updated information on acid production during exercise
- New EOC quantitative question on graph interpretation

Chapter 14 Cardiovascular Physiology

- New box on stem cells and heart disease
- New information on long QT syndrome
- Expanded discussion of contractility and afterload
- New information on cardiac glycosides and arrhythmias

Chapter 15 Blood Flow and the Control of Blood Pressure

- Expanded information on blood vessel anatomy
- Added information on collateral circulation in heart disease
- Expanded information on calcium channel blockers
- Added information on nitric oxide and its role in pathophysiology
- Expanded information on inflammatory markers (C-reactive protein and homocysteine) for cardiovascular disease
- Updated information on cardiovascular disease, and revised blood pressure guidelines for hypertension and prehypertension
- New graph on relationship between blood pressure and cardiovascular disease
- New table to review physics of blood flow
- New Level 3 question on patent ductus arteriosus

Chapter 16 Blood

- Updated information on umbilical cord blood and stem cells
- Added role of hypoxia-inducible factor 1 (HIF-1) in EPO synthesis
- New Running Problem on blood doping in athletes
- Two new Level 3 problems
- New photo of crenated and swollen RBCs

Chapter 17 Mechanics of Breathing

- Consolidated discussions of lung volumes and Boyle's law
- New box on congestive heart failure

Chapter 18 Gas Exchange and Transport

- Moved information on hemoglobin from Ch. 16 to discussion on oxygen transport
- New figure question on oxygen exchange

- New information on role of apoptosis in emphysema
- Updated information on neural control of ventilation

Chapter 19 The Kidneys

- New box on dialysis
- Updated information on tubuloglomerular feedback
- Two new Level 3 problems

Chapter 20 Integrative Physiology II: Fluid and Electrolyte Balance

- New discussion and figure on transport mechanisms in ascending loop of Henle
- New concept check questions on loop diuretics
- Added information on ENaC and ROMK channels
- New information on natriuretic peptides as markers for cardiovascular disease
- New Running Problem on hyponatremia

Chapter 21 Digestion

- Added information on receptive relaxation and amphipathic molecules
- Expanded information on intestinal absorption of iron and Ca^{2+}
- Added information on cytochrome P450 in the liver
- Updated information on incretin hormones and GI hormones
- Expended information on myenteric and submucosal plexuses, intrinsic and extrinsic neurons
- Consolidated information on digestion and absorption of nutrients
- Two new concept check questions, 1 new figure question
- Two new Level 3 Problem Solving questions
- New figure showing acini of pancreas

Chapter 22 Metabolism and Energy Balance

- New information on respiratory exchange ratio
- New section on lipid metabolism, coronary heart disease, and lipid-lowering drugs
- New section on metabolic syndrome
- Updated information and new table on diabetes drugs, including incretins, amylin, PPARs

Chapter 23 Endocrine Control of Growth and Metabolism

- Updated information on growth hormone treatment for short stature
- Updated information on androstenedione in sports
- New information on Ca-sensing receptor (CaSR) of parathyroid glands
- Added information on melanocytes and ACTH
- New information on use of the raloxifene and the PTH derivative *teriparatide* for treatment of osteoporosis
- New figure question on thyroid pathophysiology

Chapter 24 The Immune System

- Expanded information on eosinophils
- Updated information on steroid treatment of autoimmune disease
- New biotech box on humanized antibodies for treating cancer

Chapter 26 Reproduction and Development

- Updated information on embryonic sex determination and development
- Updated information on contraception, including new table of efficacy of various methods
- New information on link between erectile dysfunction and cardiovascular disease
- Updated information on drug therapy for female sexual dysfunction
- Revised figure for meiosis

Acknowledgments

Writing, editing, and publishing a textbook is a group project that requires the talents and expertise of many people. I would particularly like to acknowledge Bruce Johnson, Cornell University, Department of Neurobiology and Behavior, a superb neurobiologist and educator, who has contributed his expertise to ensure that the chapters on neurobiology are accurate and reflect the latest developments in that rapidly changing field.

Many other people devoted time and energy to making this book a reality, and I would like to thank them all, collectively and individually.

REVIEWERS

I am particularly grateful to the instructors who reviewed one or more chapters of the third edition. I appreciate the time and thought that went into their comments and I have included as many of their suggestions as I could. The reviewers for this edition include:

Patt Allen, Dixie College
Marvin H. Bernstein, New Mexico State University
Barrie P. Bode, St. Louis University
Howard Booth, Eastern Michigan University
Sunny Boyd, University of Notre Dame
N. Bryska, University of North Carolina, Charlotte
Shere Byrd, Fort Lewis College
Robert Canada, Howard University
James E. Davis, Alma College
P.A. George Fortes, University of California, San Diego
Nicholas Geist, Sonoma State University
Suzanne M. Gould, Ivy Tech, Anderson
Jean Grassman, Brooklyn College–CUNY
Albert Herrera, University of Southern California
Patrick K. Hidy, Central Texas College
James Hoffman, Diablo Valley College
Kelly Johnson, University of Kansas
John Lepri, University of North Carolina, Greensboro
Pat Lombardi, University of Oregon
Jeanne McLauchlin, University of Arkansas
Duncan R. Munro, College of Charleston
Elizabeth M. Rust, University of Michigan
Gail Sabbadini, San Diego State University
Curt Walker, Dixie College
Andrzej Wieraszko, College of Staten Island

Many other instructors and students took time to write or email queries or suggestions for clarification. I am always delighted to have input, and I apologize that I do not have room to acknowledge them individually.

SPECIALTY REVIEWS

No one can be an expert in every area of physiology, and I am deeply thankful for my friends and colleagues who reviewed entire chapters or answered specific questions. The specialty reviewers for this edition were:

Thomas Blevins, M.D., Austin TX
Cynthia Gill, University of Texas
Barbara Goodman, University of South Dakota
William Durham, Baylor College of Medicine
Susan L. Hamilton, Baylor College of Medicine
David Hood, York University
Steven C. King, Oregon Health & Science University
Michael Levitsky, Louisiana State University Health Sciences Center
Robert Linsenmeier, Northwestern University
Ron Markel, Pennsylvania State University
William Percy, University of South Dakota
Erik Swenson, University of Washington
Elizabeth Weiss, University of Texas–Austin

Specialty reviewers from previous editions include:

Beverly Bishop, SUNY Buffalo
Arthur Dalley, Vanderbilt University
Dan Hardy, M.D., Austin, Texas
Michael Joyner, Mayo Clinic
Carol C. Linder, The Jackson Laboratories
Henry S. Lucid, M.D., Austin, TX
Jeffrey Pommerville, Maricopa Community College
Withrow Gil Wier, University of Maryland, Baltimore School of Medicine
Richard Wilcox, University of Texas–Austin
Jack Wilmore, Texas A&M University
Robert G. Carroll, East Carolina University School of Medicine
Douglas Eaton, Emory University
Leona Young, Emory University
James Larimer, University of Texas
Paul Matsudaira, Massachusetts Institute of Technology
Joel Michael, Rush Medical College
Harold Modell, National Resource for Computers Life Science Education
James Stockand, University of Texas Health Science Center–San Antonio
Jeffrey Sands, Emory University
Virginia Scofield, University of Southern California

REVIEWERS FOR PREVIOUS EDITIONS

I would like to acknowledge again the contribution of the faculty who reviewed the previous editions.

Brenda Alston-Mills, North Carolina State University
Roland M. Bagby, University of Tennessee
Cynthia Beck, George Mason University
Patricia J. Berger, University of Utah
David Bernard, University of Texas–Arlington
Ronald Beumer, Armstrong State College
Sunny K. Boyd, University of Notre Dame
Timothy Bradley, University of California–Irvine
Alan P. Brockway, University of Colorado–Denver
David Byman, Pennsylvania State University Worthington–Scranton Campus
John Capeheart, University of Houston–Downtown
Annabelle Cohen, CUNY The College of Staten Island
Kim Cooper, Arizona State University
Charles L. Costa, Eastern Illinois University
Carlos Crocker, San Francisco State University
Catherine Dallemagne, Queensland University
Arianne Dantas, University of Melbourne
Gordon Garrett, Utah Valley State College
Lori K. Garrett, Danville Area Community College
Andrew Goliszek, North Carolina A&T State University
Margaret M. Gould, Georgia State University
Tammy Greco, Eastern Michigan University
David Hammerman, Long Island University
Janet L. Haynes, Long Island University–Brooklyn
Richard W. Heninger, Brigham Young University
Tracy M. Hodgson, University of Pittsburgh
Katja Hoehn, Mount Royal College
Carol Hoffman, San Jose City College
William C. Kleinelp, Middlesex Community College
John Lepri, University of North Carolina–Greensboro
Jennifer Lundmark, California State University–Sacramento
Daniel L. Mark, Penn Valley Community College
Kerry McDonald, University of Missouri
Roberta Meehan, Denver, Colorado
Wayne Meyer, Austin College
Alice C. Mills, Middle Tennessee State University
Ron Mobley, Wake Technical Community College
Stacia Moffett, Washington State University
John G. Moner, University of Massachusetts–Amherst
David Muehleisen, University of Utah
Tim Mullican, Dakota Wesleyan University
Patricia L. Munn, Longview Community College
Joanne Nosky, Okanagan Valley College of Massage Therapy
Scott Nunes, University of San Francisco
Linda Ogren, University of California–Santa Cruz
Dell Redding, Evergreen Valley College
Carol Rivard, Fanshawe College
Mary Anne Rokitka, SUNY Buffalo
Allen Rovick, Rush Medical College
Evelyn Schlenker, University of South Dakota
Norm Scott, University of Waterloo
Brian R. Shmaefsky, Kingwood College

Sabyasachi Sircar, Delhi, India
Amanda Starnes, Emory University
Judy Sullivan, Antelope Valley College
Roger Tanner-Theis, University of Oklahoma (Retired)
Deborah Taylor, Kansas City Kansas Community College
David Thomas, Fanshawe College
Benjamin Walcott, SUNY Stonybrook
Richard L. Walker, University of Calgary
Curt Walter, Dixie College
Donald Whitmore, University of Texas–Arlington
Ronald L. Wiley, Miami University of Ohio
Burl Williams, Jr., East Tennessee State University
David A. Woodman, University of Nebraska–Lincoln

PHOTOGRAPHS

I would like to thank the following colleagues who generously provided micrographs from their research:

Flora M. Love, University of Texas
Jane Lubisher, University of Texas
Carole L. Moncman, University of Texas
Young-Jin Son, University of Texas

SUPPLEMENTS

Several of my graduate students—Lynn Cialdella, Zara Oakes, and Alonna Beatty—revised this edition's test bank and web questions, originally crafted by Pat Munn at Longview Community College in Kansas and Alice Mills from University of Tennessee–Martin. Damian Hill worked with me once more to revise the Instructor's Resource Guide and Student Workbook.

THE DEVELOPMENT AND PRODUCTION TEAM

Writing a manuscript is only a first step in the long and complicated process that results in a bound book with all its ancillaries. In this edition I worked closely with Irene Nunes, my development editor, and Alan Titche, my copy editor, to refine and clarify the text. Irene was particularly helpful in the reorganization of the first unit of the book, and she brought a fresh eye to the project. Bill Ober and Claire Garrison, my art coauthors, used their considerable talents to make the art in this edition even clearer and more beautiful. Jim Gibson helped with art layout, and Anita Impagliazzo took the photos used in the art for icons and anatomy. Yvo Riezebos was the talented designer who created the cover and interior page designs.

The team at Benjamin Cummings worked tirelessly to see this edition move from manuscript to bound book. Nicole George-O'Brien was the project editor who pulled all the disparate elements of the project together and kept everyone on task, with advice from Deirdre McGill Espinoza, Senior Acquisitions Editor. Nicole masterfully coordinated the complexities of coordinating text, supplements, and associated projects. Early in the process Sabrina Larson arranged the reviews of the previous edition. Karen Gulliver, Production Editor, juggled schedules and guided the project through the transition from manuscript to pages. Maureen Sphuler tracked down new photographs for the book from sources around the world. Amy Yu coordinated the print supplements, and Alex Streczyn and Sarah Young-Dualan worked on the media supplements, including course management sites, Media Manager, and *The Physiology Place* companion website. Particular thanks go to Lauren Harp, Executive Marketing Manager, and the sales teams at Benjamin Cummings and Pearson International who provide an invaluable link between the author and physiology educators around the world.

SPECIAL THANKS

As always, I would like to thank my students who looked for errors and areas that needed improvement. In this edition, Matt Pahnke, a graduate student in exercise physiology at Texas, wrote the new running problem on hyponatremia that appears in Chapter 20.

My graduate teaching assistants have played a huge role in my teaching ever since I arrived at the University of Texas, and their input has helped shape how I teach. Many of them are now faculty members themselves. I would particularly like to thank:

Lynn Cialdella, M.S., M.B.A.
Patti Thorn, Ph.D.
Karina Loyo-Garcia, Ph.D.
Jan M. Machart, Ph.D.
Ari Berman, Ph.D.
Kurt Venator, Ph.D.
Peter English, Ph.D.
Kira Wenstrom, Ph.D.
Lawrence Brewer, Ph.D.
Carol C. Linder, Ph.D.

Finally, special thanks to my colleagues in the American Physiological Society and the Human Anatomy & Physiology Society, whose experiences in the classroom have enriched my own understanding of how to teach physiology. I would also like to recognize a special group of friends for their continuing support: Ruth Buskirk, Judy Edmiston, Jeanne Lagowski, Jan M. Machart, and Marilla Svinicki (University of Texas), Penelope Hansen (Memorial University, St. John's), Mary Anne Rokitka (SUNY Buffalo), Rob Carroll (East Carolina University School of Medicine), Cindy Gill (Hampshire College), and Joel Michael (Rush Medical College).

Once again, I thank my family and friends for their patience, understanding, and support during the chaos that seems inevitable with book revisions. And, as always, the biggest thank you to my husband Andy, who keeps me focused on what's really important in life.

A WORK IN PROGRESS

One of the most rewarding aspects of writing a textbook is the opportunity it has given me to meet or communicate with other instructors and students. In the years since the first edition was published, I have heard from people around the world, and have had the pleasure of hearing how the book has been incorporated into their teaching and learning.

Because science textbooks are revised every three or four years, they are always works in progress. I invite you to contact me or my publisher with any suggestions, corrections, or comments about this fourth edition. I am most reachable through e-mail at *silverthorn@mail.utexas.edu*. You can reach my editor at the following address:

Applied Sciences
Benjamin Cummings
1301 Sansome Street
San Francisco, CA 94111

Dee U. Silverthorn
University of Texas
Austin, Texas

VISUAL INTRODUCTION

TOOLS FOR DEVELOPING CRITICAL THINKING SKILLS

The Fourth Edition of *Human Physiology: An Integrated Approach* provides engaging opportunities for students to practice critical thinking and problem-solving skills that are essential for success in the Human Physiology course as well as in future coursework and careers in science or medicine.

FIGURE QUESTIONS AND GRAPH QUESTIONS

Many figures and graphs include questions that help students check their understanding of illustrated concepts, interpret scientific graphs, and apply information to new situations. The answers to the Figure and Graph questions appear at the end of each chapter.

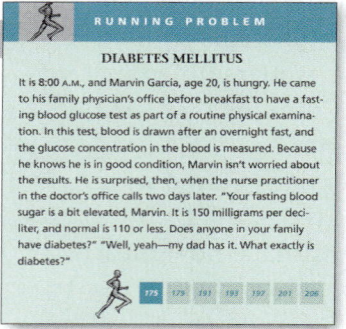

RUNNING PROBLEMS

Each chapter begins with a relevant, real-life problem involving a disease or disorder that unfolds in segments on subsequent pages. The numbered questions in the running problems require students to apply information that they have learned from reading the text. At the end of the chapter, students can check their problem-solving strategy against the Problem Conclusion.

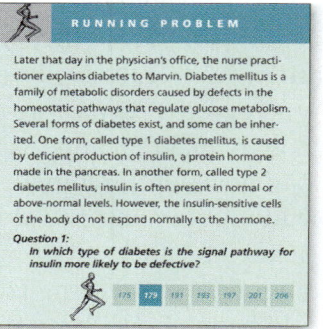

CONCEPT CHECKS

Concept Checks throughout each chapter provide a natural break for students to check their understanding of concepts in the preceding section. Students can review their answers to the Concept Check questions using the answer key at the end of the chapter.

CONCEPT CHECK

5. What are the three chemical classes of hormones?

6. The steroid hormone aldosterone has a short half-life for a steroid hormone—only about 20 minutes. What would you predict about the degree to which aldosterone is bound to blood proteins?   Answers: p. 242

REFLEX PATHWAY AND CONCEPT MAPS

The reflex pathways and concept maps organize physiological themes and details into a logical, consistent visual format. These figures use consistent colors and shapes to represent processes and will guide students to a better understanding of coordinated physiological function. Students will learn to create their own concept maps in the "Level Two" mapping questions at the end of the chapter.

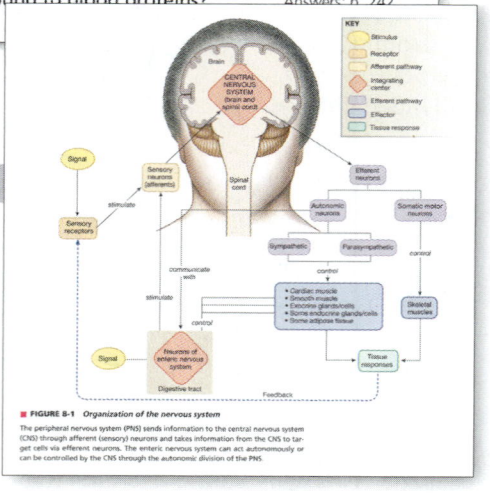

FIGURE 8-1 *Organization of the nervous system*

The peripheral nervous system (PNS) sends information to the central nervous system (CNS) through afferent (sensory) neurons and takes information from the CNS to target cells via efferent neurons. The enteric nervous system can act autonomously or can be controlled by the CNS through the autonomic division of the PNS.

Over many years of getting feedback from her Human Physiology students, author Dee Silverthorn has incorporated several study aids into her book.

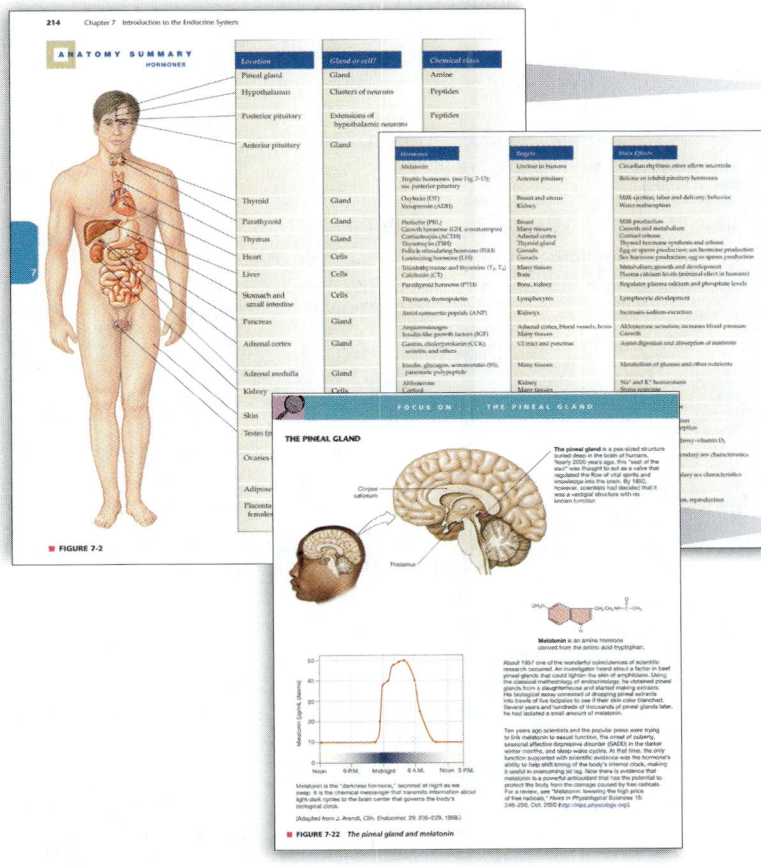

ANATOMY SUMMARIES

To understand physiology, students must also have a good understanding of basic anatomy. The Anatomy Summaries help to visualize the anatomy of a physiological system. Whether learning the anatomy for the first time or refreshing their memory from a previous course, these summaries allow students to see all of the essential features of each system in a single figure.

FOCUS ON…FIGURES

"Focus" figures highlight the anatomy and physiology of organs that are often overlooked in physiology texts, including the skin, the liver, and the pineal gland.

BACKGROUND BASICS

At the beginning of the chapter, this handy list reviews topics students will need to master for understanding the material that follows. The page references will save valuable study time.

BACKGROUND BASICS

Summation: **278** Second messenger systems: **180** Threshold: **196** G proteins: **183** Plasticity: **283**
Tonic control: **192** Membrane potential: **162** Graded potentials: **255** Neurotransmitter release: **271**

of these swellings, known as a **varicosity** [*varicosus,* abnormally enlarged or swollen], contains vesicles filled with neurotransmitter. Parasympathetic pathways end with axon terminals, or *terminal boutons* [from the French word for "knob" or "button"; Tbl. 8-1, p. 244], which also release neurotransmitter.

CONCEPT LINKS

These encircled "chain-links" connect concepts to topics that were discussed earlier in the text. Links will help students find material that they may have forgotten or that may be helpful for understanding new information.

The structure of an autonomic synapse differs from the model synapse shown in Figure 8-20 [p. 270]. Autonomic postganglionic axons end with a series of swollen areas at their distal ends, like beads spaced out along a string (Fig. 11-8 ■). Each

FIGURE REFERENCE LOCATORS

These colored squares function as place markers to make it easier to connect illustrations with the running narrative that describes the figure.

FOCUS BOXES

There are three kinds of boxes in this book, developed with the goal of helping students understand the role of physiology in science and medicine today:

BIOTECHNOLOGY boxes discuss physiology related applications and laboratory techniques from the fast-moving world of biotechnology.

CLINICAL boxes focus on clinical applications and pathologies. In addition to being interesting and applied to real life topics, these boxes will help with understanding homeostasis and normal function.

EMERGING CONCEPTS boxes offer a glimpse into new and exciting developments in physiological research. Some topics, such as stem cells, that were highlighted in Emerging Concepts boxes in earlier editions have become fundamental concepts in physiology and are now part of the main text.

An additional feature to all three of these focus boxes is a **DIABETES THEME**, which is used to illustrate the application of physiology to a current cultural topic. A diabetes theme box bears the icon of a Biotechnology box, a Clinical box, or an Emerging Concepts box, depending on the information presented. Because diabetes has such a widespread effect on the body, it makes a perfect example of integrated physiology.

BIOTECHNOLOGY

IMMUNOCYTOCHEMISTRY

Traditionally, physiologists thought that each hormone came from a single gland (even though some glands secrete multiple hormones), but the research technique called *immunocytochemistry* has turned that view of endocrinology upside down. Immunocytochemistry begins when an animal is injected with a foreign molecule (an antigen), such as a hormone. The animal's immune system recognizes the injected molecule as "not self" and produces specific antibodies against it. Th[...]
host animal and [...]
rescent, or enzy[...]
[...]
freshly exc[...]
microscope. If th[...]
tissue, the antib[...]
under the micro[...]
with the antibo[...]
investigators dis[...]
made in multipl[...]
which they were [...]

CLINICAL FOCUS

DIABETES

INSULIN AND METABOLISM

Why do people with diabetes have elevated levels of blood glucose ("blood sugar")? Part of the answer lies with the disruption of glucose metabolism when insulin is absent or when cells do not respond to it. In a normal person, insulin promotes cellular uptake and metabolism of glucose by stimulating the appropriate enzymes. Glucose absorbed from the digestive system goes through glycolysis to make ATP or is stored as glycogen. Fatty acids are stored as lipids, and amino acids are used to make proteins. In the absence of insulin, the enzymes associated with catabolic pathways are more active than those associated with anabolic pathways. In these overactive catabolic pathways, lipids undergo beta-oxidation, amino acids go through gluconeogenesis to become glucose, and [...] in-
[...]th
[...] the
[...] that

EMERGING CONCEPTS

MELANOPSIN—A NEW PHOTOSENSITIVE PIGMENT

Circadian rhythms in mammals are cued by light entering the eyes. For many years, scientists believed that pathways from the rods and cones of the retina led to the suprachiasmatic nucleus (SCN), the brain center for circadian rhythms. However, in 1999 researchers found that transgenic mice who lacked both rods and cones still had the ability to respond to changing light cues, suggesting that some other photoreceptor must exist in the retina. Now scientists believe they have found it: a subset of retinal ganglion cells that contain an opsin-like pigment called melanopsin. Axons from these cells project to the SCN, and it appears that these cells join rods and cones as the light-sensing cells of the mammalian retina.

Everything You Need to Teach Your Course Your Way

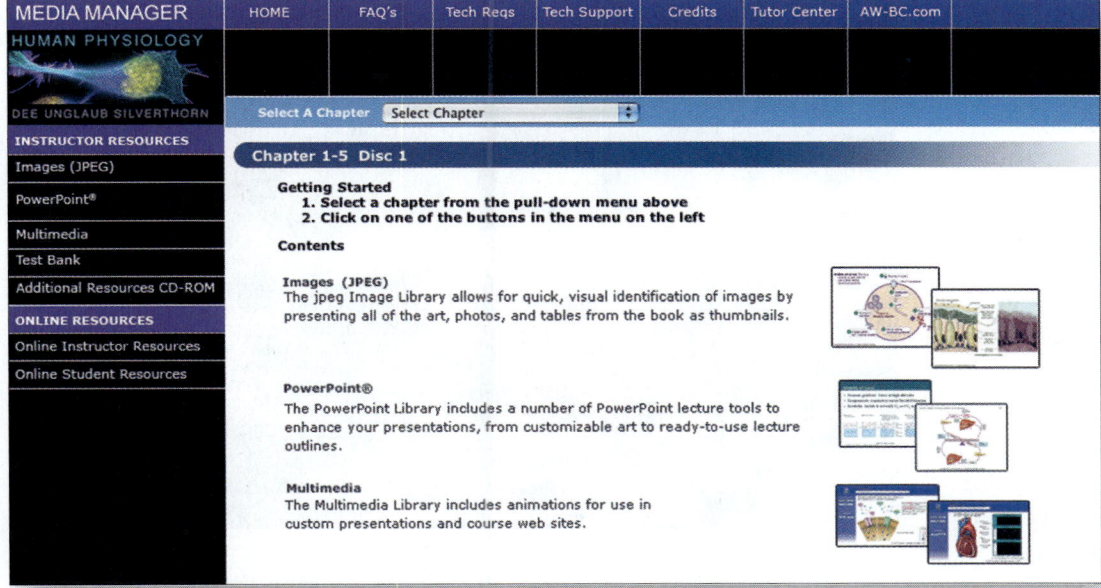

NEW! MEDIA MANAGER

Media Manager CD-ROMs include all of the art and photos from the book with customizable labels through the Text Edit feature, customizable PowerPoint® lecture slides, Step Edit Art selections featuring step-by-step "slide show" views of the book's complex art in PowerPoint, and links to online student activities and tutorials. All resources are listed and organized in a handy Quick Reference Guide.

INSTRUCTOR'S RESOURCE MANUAL

Developed and co-written by author Dee Silverthorn, this valuable teaching guide shows instructors how to incorporate active learning into their lectures and includes chapter summaries, outlines with helpful information and teaching tips, reading lists, discussion topics and demonstrations, extra quantitative problems for assignments, and more.

TRANSPARENCY ACETATES

New for this edition, all of the illustrations and photos from the text are included.

TEST BANK

Based on the textbook's four-level learning system, the test bank contains more than 3,000 questions, updated for this edition. They include multiple choice, matching, completion, short answer, essay, quantitative analysis, and graph questions. The computerized version comes on a cross-platform CD-ROM and utilizes TestGen® test software for easy editing, adding, or reorganizing of test questions.

COURSE MANAGEMENT

CourseCompass,™ WebCT, and Blackboard courses offer pre-loaded content including testing and assessment plus assets from The Physiology Place website. These three course management systems are available as pre-built, easy-to-use courses. Professors can use these powerful tools to create sophisticated web-based educational programs.

Learning Tools for Success

INTERACTIVEPHYSIOLOGY® 9-SYSTEM SUITE CD-ROM

Included with each new copy of the text, this award-winning, dynamic, and highly effective learning tool includes nine modules that help students master difficult concepts through detailed animations that bring complex physiological processes to life. Features a new module on the Digestive System.

THE PHYSIOLOGY PLACE

Including IP®Web and PhysioEx™ v6.0 for Human Physiology, this premium website features online access to three levels of quizzing, weblinks, muscle review, bone review, and new histology tutorials, chapter summaries, glossary, flashcards, and instructor's resources. Access codes are provided with each new copy of the text. (www.physiologyplace.com)

STUDENT WORKBOOK

Co-written by author Dee Silverthorn, this student workbook is adapted from materials the author developed for her own class. Each chapter features a "guided note-taking" series of questions that allow students to learn the basics as they read, plus other helpful study aids including try-it-yourself activities, quantitative and application-level review questions, chapter summaries, lab exercises, and reading lists.

PHYSIOEX™ VERSION 6.0 FOR HUMAN PHYSIOLOGY CD-ROM

This program consists of 13 modules containing 40 physiology lab simulations that allow students to perform physiological experiments that may be too costly, difficult, or time-consuming to conduct in a traditional wet lab setting. This easy-to-use software allows students to adjust the variables of each lab, perform experiments repeatedly, and even design their own labs. Available on both CD-ROM and the web, each experiment includes step-by-step instructions and review materials in the accompanying PhysioEx™ V6.0 for Human Physiology laboratory manual.

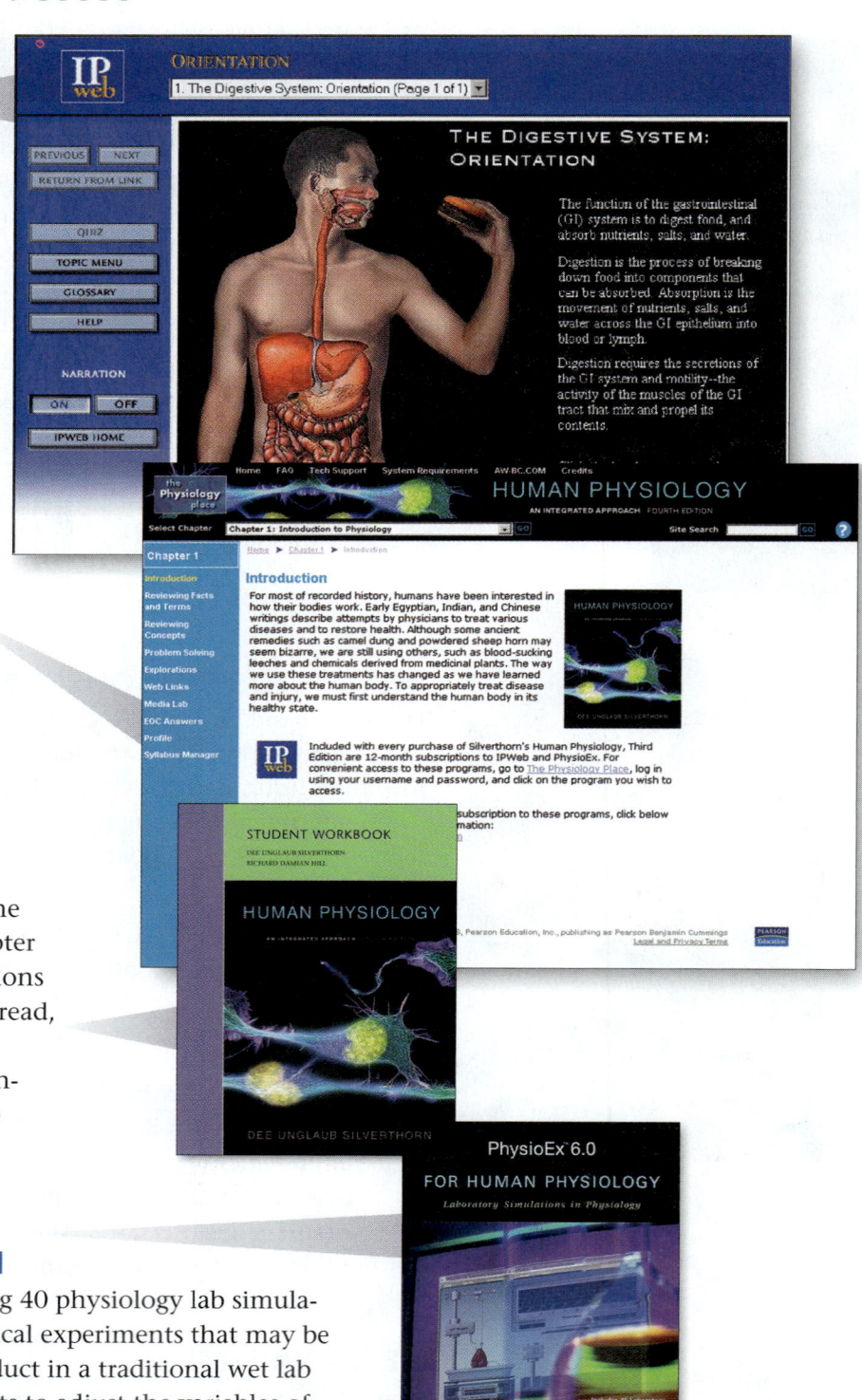

1

> *Physiology is not a science or a profession but a point of view.* —Ralph W. Gerard.

Introduction to Physiology

For most of recorded history, humans have been interested in how their bodies work. Early Egyptian, Indian, and Chinese writings describe attempts by physicians to treat various diseases and to restore health. Although some ancient remedies, such as camel dung and powdered sheep horn, may seem bizarre, we are still using others, such as blood-sucking leeches and chemicals derived from medicinal plants. The way we use these treatments has changed as we have learned more about the human body. To treat disease and injury appropriately, we must first understand the human body in its healthy state.

Physiology is the study of the normal functioning of a living organism and its component parts, including all its chemical and physical processes. The term *physiology* literally means "knowledge of nature." Aristotle (384–322 B.C.) used the word in this broad sense to describe the functioning of all living organisms, not just of the human body. However, Hippocrates (ca. 460–377 B.C.), considered the father of medicine, used the word *physiology* to mean "the healing power of nature," and thereafter the field became closely associated with medicine. By the sixteenth century in Europe, physiology had been formalized as the study of the vital functions of the human body, although today the term is again used to refer to the study of the functions of all animals and plants.

In contrast, **anatomy** is the study of structure, with much less emphasis on function. Despite this distinction, anatomy and physiology cannot truly be separated. The function of a tissue or organ is closely tied to its structure, and the structure of the tissue or organ presumably evolved to provide an efficient physical base for its function.

PHYSIOLOGICAL SYSTEMS

To understand how physiology relates to anatomy, we must first examine the component parts of the human body. Figure 1-1 ■ shows the different **levels of organization** of living organisms, ranging from atoms to groups of the same species (*populations*) to populations of different species living together in *ecosystems* and in the *biosphere*. The various sub-disciplines of

chemistry and biology related to the study of each organizational level are included in the figure. There is considerable overlap between the different fields, and these artificial divisions will vary according to who is defining them. One distinguishing feature of physiology is that it encompasses many levels of organization, from the molecular level all the way up to populations of a species.

At a fundamental level, atoms of elements link together to form molecules. The smallest unit of structure capable of carrying out all life processes is the **cell**. Cells are collections of molecules separated from the external environment by a barrier called the *cell* (or *plasma*) *membrane*. Simple organisms are composed of only one cell, but complex organisms have many cells with different structural and functional specializations.

Collections of cells that carry out related functions are known as **tissues** [*texere*, to weave]. Tissues form structural and functional units known as **organs** [*organon*, tool], and groups of organs integrate their functions to create **organ systems**.

Figure 1-2 ■ is a schematic diagram showing the interrelationships of most of the 10 physiological organ systems in the human body (Table 1-1 ■). The **integumentary system** [*integumentum*, covering], composed of the skin, forms a protective boundary that separates the body's internal environment from the external environment (the outside world). The **musculoskeletal system** provides support and body movement.

Four systems exchange materials between the internal and external environments. The **respiratory system** exchanges gases, the **digestive system** takes up nutrients and water and eliminates wastes, the **urinary system** removes excess water and waste material, and the **reproductive system** produces eggs or sperm.

The remaining four systems extend throughout the body. The **circulatory system** distributes materials by pumping blood through vessels. The **nervous** and **endocrine systems** coordinate body functions. Note that the figure shows them as a continuum rather than as two distinct systems. Why? Because as we have learned more about the integrative nature of

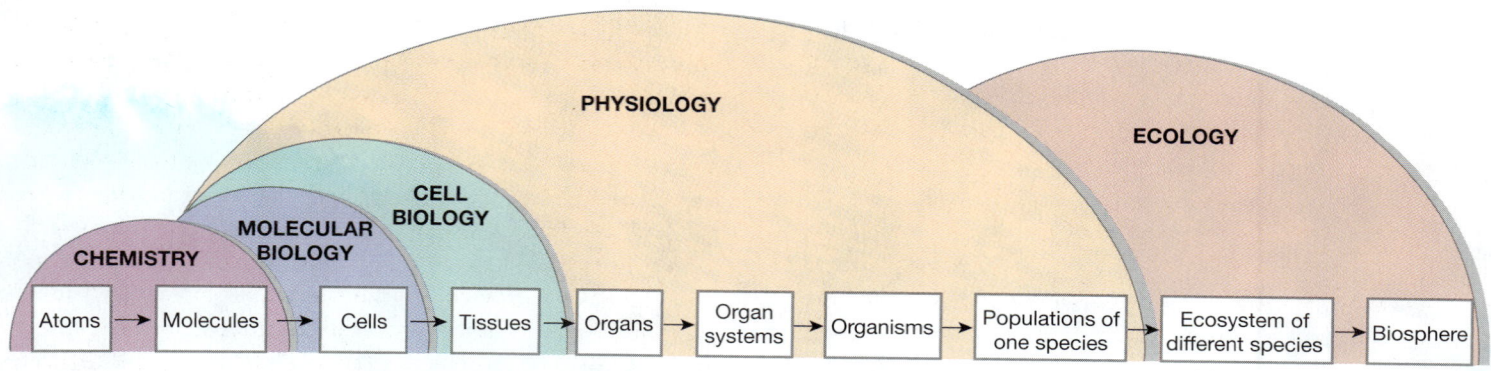

■ **FIGURE 1-1** *Levels of organization and the related fields of study*

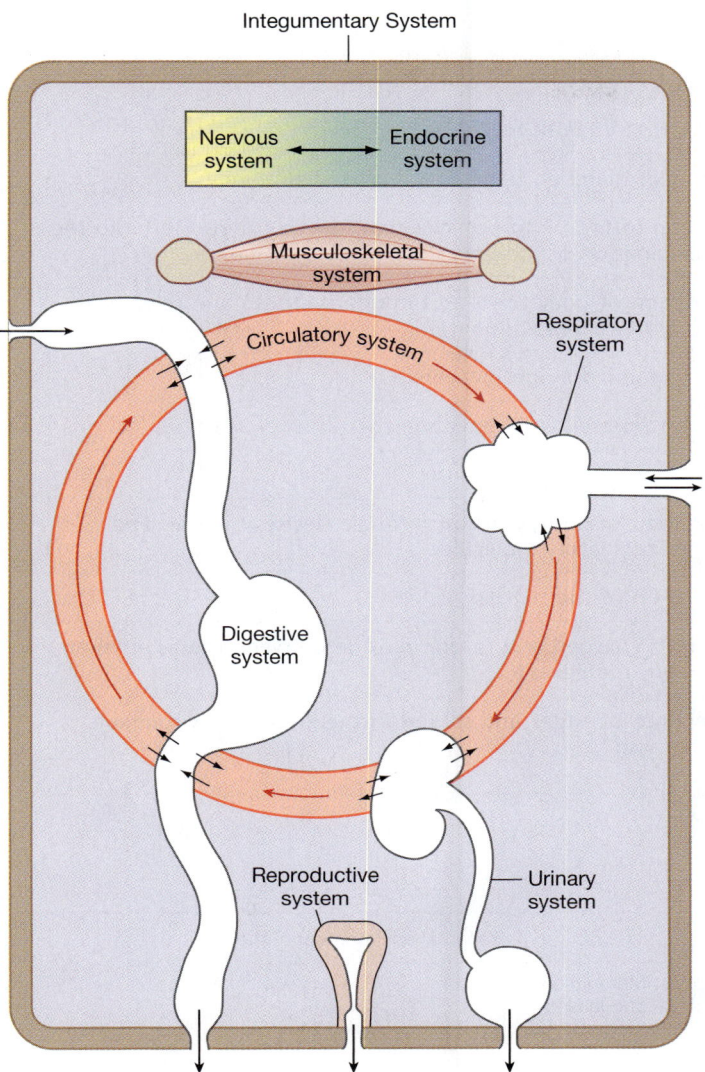

Integumentary System

Nervous system ↔ Endocrine system

Musculoskeletal system

Circulatory system

Respiratory system

Digestive system

Reproductive system

Urinary system

■ **FIGURE 1-2** *The integration between systems of the body*

This schematic figure indicates relationships between systems of the human body. The interiors of some hollow organs (shown in white) open to the external environment.

physiological function, the lines between these two systems have blurred. The one system not shown in Figure 1-2 is the diffuse **immune system.** Immune cells are positioned to intercept material that may enter through the exchange surfaces or through a break in the skin, and they protect the internal environment from foreign invaders. In addition, immune tissues are closely associated with the circulatory system.

FUNCTION AND PROCESS

Function and process are two related concepts in physiology. The **function** of a physiological system or event is the "why" of the system or event: why does the system exist and why

does the event happen? This way of thinking about a subject is called the **teleological approach** to science. For example, the teleological answer to the question of why red blood cells transport oxygen is "because cells need oxygen and red blood cells bring it to them." This answer explains the reason red blood cells transport oxygen but says nothing about *how* the cells transport oxygen.

In contrast, physiological **processes,** or **mechanisms,** are the "how" of a system. The **mechanistic approach** to physiology examines process. The mechanistic answer to the question "Why do red blood cells transport oxygen?" is "Oxygen binds to hemoglobin molecules contained in the red blood cells." This very concrete answer explains exactly how oxygen transport occurs but says nothing about the significance of oxygen transport to the intact animal.

Students often confuse these two approaches to thinking about physiology, and studies have shown that even medical students tend to answer questions with teleological explanations when the more appropriate response would be a mechanistic explanation. Often this is because the instructor asked "why" a physiological event occurs when he or she really wanted to know "how." Staying aware of the two approaches will help prevent confusion.

Although function and process seem to be two sides of the same coin, it is possible to study processes, particularly at the cellular and subcellular level, without understanding their function in the life of the organism. As biological knowledge becomes more complex, scientists sometimes become so involved in studying complex processes that they fail to step back and look at the significance of those processes to cells, organ systems, or the intact animal. One role of physiology is to integrate function and process into a cohesive picture.

HOMEOSTASIS

As we think about physiological functions, we often consider their *adaptive significance*—that is, why does a certain function help an animal survive in a particular situation? For example, humans are large, mobile, terrestrial animals whose bodies maintain a relatively constant water content despite living in a dry, highly variable external environment. What structures and mechanisms have evolved in our anatomy and physiology that enable us to survive in this hostile environment?

Most cells in our bodies are not very tolerant of changes in their surroundings. In this way they are similar to early organisms that lived in tropical seas, a stable environment where salinity, oxygen content, and pH vary little, and where light and temperature cycle in predictable ways. The internal composition of these ancient creatures was almost identical to that of seawater, and if environmental conditions changed, conditions inside the primitive organisms changed as well. Even today, marine invertebrates cannot tolerate significant changes

TABLE 1-1	Organ Systems of the Human Body	
SYSTEM NAME	**ORGANS OR TISSUES**	**REPRESENTATIVE FUNCTIONS**
Circulatory	Heart, blood vessels, blood	Transport of materials between all cells of the body
Digestive	Stomach, intestines, liver, pancreas	Conversion of food into particles that can be transported into the body; elimination of some wastes
Endocrine	Thyroid gland, adrenal gland	Coordination of body function through synthesis and release of regulatory molecules
Immune	Thymus, spleen, lymph nodes	Defense against foreign invaders
Integumentary	Skin	Protection from external environment
Musculoskeletal	Skeletal muscles, bones	Support and movement
Nervous	Brain, spinal cord	Coordination of body function through electrical signals and release of regulatory molecules
Reproductive	Ovaries and uterus, testes	Perpetuation of the species
Respiratory	Lungs, airways	Exchange of oxygen and carbon dioxide between the internal and external environments
Urinary	Kidneys, bladder	Maintenance of water and solutes in the internal environment; waste removal

in salinity and pH, as you know if you have ever maintained a saltwater aquarium. In both ancient and modern times, many marine organisms relied on the constancy of their external environment to keep their internal environment in balance.

In contrast, as organisms evolved and migrated from the ancient seas into estuaries, then into freshwater environments and onto the land, they encountered highly variable external environments. Rains dilute the salty water of estuaries, and organisms that live there must cope with the influx of water into their body fluids. Terrestrial organisms constantly lose internal water to the dry air around them. The organisms that survive in these challenging habitats cope with external variability by keeping their internal environment relatively stable, an ability known as **homeostasis** [*homeo-*, similar + *-stasis*, condition].

The watery internal environment of multicellular animals is called the **extracellular fluid** [*extra-*, outside of], a "sea within" the body that surrounds the cells (Fig. 1-3 ■). Extracellular fluid (ECF) serves as the transition between an organism's external environment and the **intracellular fluid** inside cells [*intra-*, within]. Because extracellular fluid is a buffer zone between the outside world and most cells of the body, elaborate physiological processes have evolved to keep its composition relatively stable.

When the extracellular fluid composition varies outside its normal range of values, compensatory mechanisms activate and return the fluid to the normal state. For example, when you

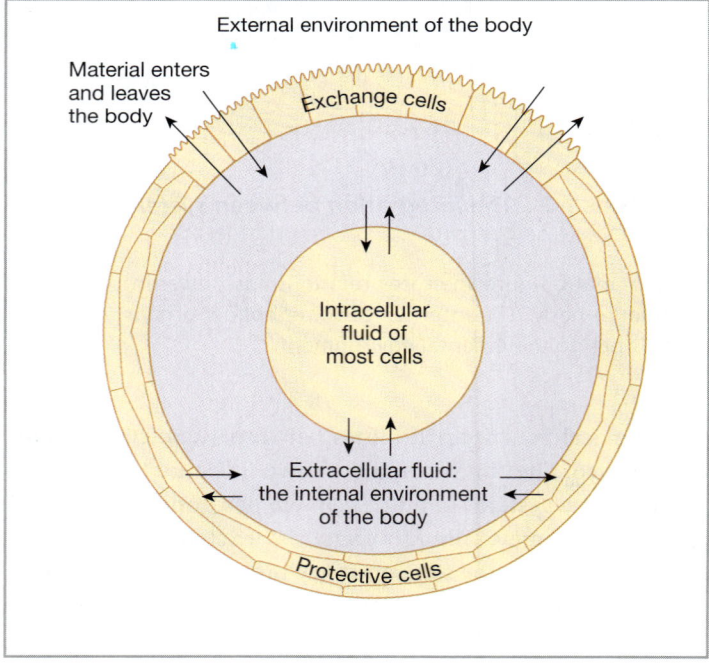

■ **FIGURE 1-3**   *Relationships between an organism's internal and external environments*

Few cells in the body are able to exchange material with the organism's external environment. Most cells are in contact with the body's internal environment, composed of the extracellular fluid.

drink a large volume of water, the dilution of your extracellular fluid triggers a mechanism that causes your kidneys to remove excess water and protect your cells from dilution. Most cells of multicellular animals do not tolerate much change and depend on the constancy of extracellular fluid to maintain normal function.

The concept of a relatively stable internal environment was developed by the French physician Claude Bernard in the mid-1800s. During his studies of experimental medicine, Bernard noted the stability of various physiological **parameters,*** such as body temperature, heart rate, and blood pressure. As the chair of physiology at the University of Paris, he wrote of "la fixité du milieu intérieur" (the constancy of the internal environment). This was an idea that applied to many of the experimental observations of his day, and it became the subject of discussion among physiologists and physicians.

In 1929, an American physiologist named Walter B. Cannon created the word *homeostasis* to describe the regulation of this internal environment. In his essay,[†] Cannon explained that he selected the prefix *homeo-* (meaning *like* or *similar*) rather than the prefix *homo-* (meaning *same*) because the internal environment is maintained within a range of values rather than at an exact fixed value. He also pointed out that the suffix *-stasis* in this instance means a *condition*, not a state that is static and unchanging. Thus, Cannon's homeostasis is a state of maintaining "a similar condition," also described as "a relatively constant internal environment."

Some physiologists contend that a literal interpretation of *stasis* [a state of standing] in the word *homeostasis* implies a static, unchanging state. They argue that we should use the word *homeodynamics* instead, to reflect the small changes constantly taking place in our internal environment [*dynamikos,* force or power]. Whether the process is called homeostasis or homeodynamics, the important concept to remember is that the body monitors its internal state and takes action to correct disruptions that threaten its normal function. Homeostasis and the regulation of the internal environment are central precepts of physiology and create an underlying theme of each chapter in this book.

Using observations made by numerous physiologists and physicians during the nineteenth and early twentieth centuries, Cannon proposed a list of parameters that are under homeostatic control. We now know that his list of physiological parameters was both accurate and complete. Cannon divided his parameters into what he described as environmental factors that affect cells (osmolarity, temperature, and pH) and "materials for cell needs" (nutrients, water, sodium, calcium, other inorganic ions, oxygen, and "internal secretions having general and continuous effects"). Cannon's "internal secretions"

are the hormones and other chemicals that our cells use to communicate with one another.

If the body fails to maintain homeostasis of these parameters, then normal function is disrupted and a disease state, or **pathological** condition [*pathos,* suffering], may result. Diseases can be divided into two general groups according to their origin: those in which the problem arises from internal failure of some normal physiological process, and those that originate from some outside source. Internal causes of disease include the abnormal growth of cells, which may cause cancer or benign tumors; the production of antibodies by the body against its own tissues (autoimmune diseases); and the premature death of cells or the failure of cell processes. Inherited disorders are also considered to have internal causes. External causes of disease include toxic chemicals, physical trauma, and foreign invaders such as viruses and bacteria. In both internally and externally caused diseases, when homeostasis is disturbed, the body attempts to compensate (Fig. 1-4 ■). If the compensation is successful, homeostasis is restored. If compensation fails, illness or disease may result. The study of body functions in a disease state is known as **pathophysiology.** You will encounter many examples of pathophysiology as we study the various systems of the body.

One very common pathological condition in the United States is **diabetes mellitus,** a metabolic disorder characterized by abnormally high blood glucose concentrations. Although we speak of diabetes as if it were a single disease, it is actually a

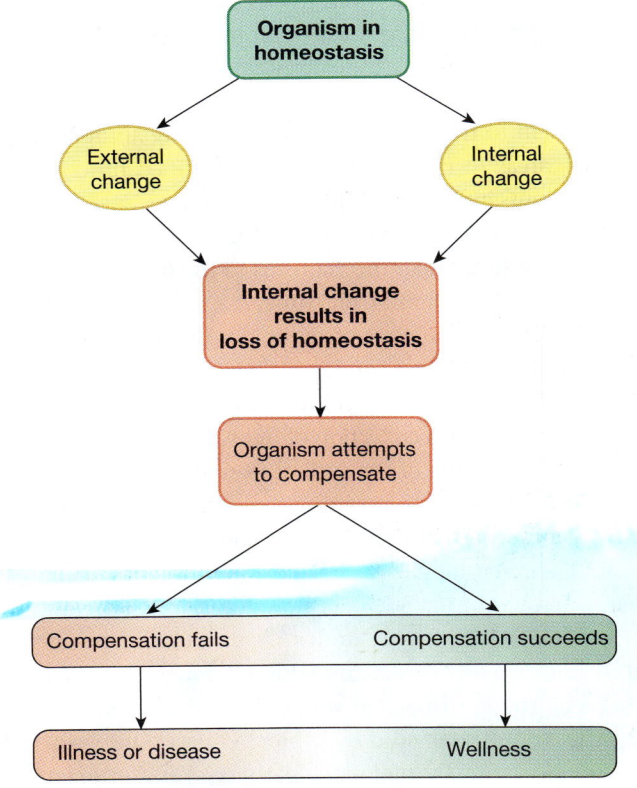

■ FIGURE 1-4 *Homeostasis*

---

*A parameter [*para-,* beside + *meter,* measure] is one of the variables in a system.
[†]Cannon WB. Organization for physiological homeostasis. *Physiol Rev 9:* 399–443, 1929.

1

whole family of diseases with various causes and manifestations. You will learn more about diabetes in the focus boxes scattered throughout the chapters of this book. The influence of this one disorder on many systems of the body makes it an excellent example of the integrative nature of physiology.

PHYSIOLOGY: MOVING BEYOND THE GENOME

There has never been a more exciting time to study human physiology. Today we benefit from centuries of work by physiologists who constructed a foundation of knowledge about how the human body functions. Since the 1970s, rapid advances in the fields of cellular and molecular biology have supplemented this work. A few decades ago we thought that we would find the key to the secret of life by sequencing the human *genome*, which is the collective term for all the genetic information contained in a species's DNA. However, this deconstructionist view of biology has proved to have its limitations because living organisms are much more than the simple sum of their parts.

The Human Genome Project (*www.ornl.gov/hgmis*) began in 1990 with the goal of identifying and sequencing all the genes in human DNA (the genome). However, as research advanced our understanding of DNA function, scientists' original idea that a given segment of DNA contained one gene that coded for one protein had to be revised when it became clear that one gene may code for many proteins. The Human Genome Project was completed in 2003, but before then researchers had moved beyond genomics to *proteomics*, the study of proteins in living organisms. Now scientists are coming to realize that knowing that a protein is made in a particular cell does not always tell us the significance of that protein to the cell, tissue, or functioning animal. The exciting new areas in biological research are called functional genomics, systems biology, and integrative biology, but fundamentally these are all fields of physiology that investigate integrated function, from subcellular mechanisms to the intact organism.

If you read the scientific literature, it appears that contemporary research, using the tools of molecular biology, has exploded into an era of "omes" and "omics." What is an "ome"? The term apparently derives from the Latin word for a mass or tumor, and it is now used to refer to a collection of items that make up a whole, such as a genome. One of the earliest uses of the "ome" suffix in biology is the term *biome*, meaning the entire community of organisms living in a major ecological region, such as the marine biome or the desert biome. A genome is all the genes in an organism, and a proteome includes all the proteins in that organism.

The related adjective "omics" describes the research related to studying an "ome," and adding "omics" to a root word has become the cutting-edge way to describe a research field. Thus, the traditional study of biochemistry now includes metabolomics (study of metabolic pathways) and interactomics (the study of protein-protein interactions). If you search the Internet, you will find numerous listings for the transcriptome

(RNA), lipidome (lipids), and glycome (carbohydrates). The "omics" craze has also given us such terms as cellome, enzymome, and unknome (proteins with unknown functions). There is even a journal named *OMICS*!

The Physiome Project (*www.physiome.org*) is an organized international effort to coordinate molecular, cellular, and physiological information about living organisms into an Internet database. Scientists around the world can access this information and apply it in their own research efforts to create better drugs or genetic therapies for curing and preventing disease. Some scientists are using the data to create mathematical models that explain how the body functions. The Physiome Project is an ambitious undertaking that promises to integrate information from diverse areas of research so that we can improve our understanding of the complex processes we call life.

PHYSIOLOGY IS AN INTEGRATIVE SCIENCE

Physiologists are trained to think about the **integration** of function across many levels of organization, from molecules to the living body. (To *integrate* means to bring varied elements together to create a unified whole.) One of the current challenges in physiology is integrating information from different body systems into a cohesive picture of the living human body. This process is complicated by the fact that many complex systems—including those of the human body—possess **emergent properties**, which are properties that cannot be predicted to exist based only on knowledge of the system's individual components. An emergent property is not a property of any single component of the system and is greater than the simple sum of the system's individual parts. Among the most complex emergent properties in humans are emotion, intelligence, and other aspects of brain function, none of which can be predicted from knowing the individual properties of nerve cells.

As previously noted, the integration of function among systems and across many levels of organization is a special focus of physiology. Traditionally, students study the cardiovascular system and regulation of blood pressure in one unit, then study the kidneys and control of body volume in a different unit. In the functioning body, however, the cardiovascular and renal systems communicate with each other, and a change in one is likely to cause a reaction in the other. For example, blood pressure is influenced by body volume, and changes in blood pressure may have a significant effect on kidney function. One of the most intriguing and challenging aspects of learning physiology is developing skills that help you understand how the different organ systems work together. One way physiologists do this is by using two types of maps. The first type of map, shown in Figure 1-5a ■, is a schematic representation of structure or function. The second type of map, shown in Figure 1-5b, diagrams a physiological process and is therefore referred to as a process map. (An alternative name is *flow chart*.)

(a) A map showing structure/function relationships

(b) A process map, or flow chart

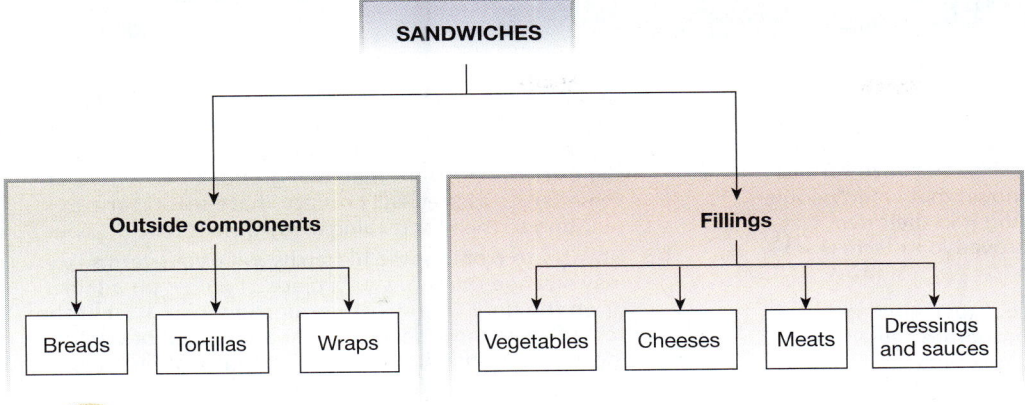

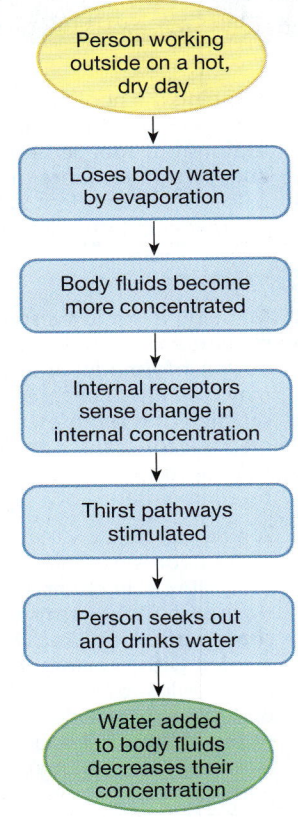

FIGURE QUESTIONS

1. Can you add more details and links to map **(a)**?

2. Here is an alphabetical list of terms for a map of the body. Use the steps in *Focus on Mapping* to create a map with them. Add additional terms to the map if you like. A sample answer can be found at the end of the chapter.

- bladder
- blood vessels
- brain
- cardiovascular system
- digestive system
- endocrine system
- heart
- immune system
- integumentary system
- intestine
- kidneys
- lungs
- lymph nodes
- mouth
- musculoskeletal system
- nervous system
- ovaries
- reproductive system
- respiratory system
- stomach
- testes
- the body
- thyroid gland
- urinary system
- uterus

■ **FIGURE 1-5** *Types of maps*

The concept of integrated function is the underlying principle in **translational research**, an approach sometimes described as "bench to bedside." Translational research applies the insights and results gained from basic biomedical research to treating and preventing human diseases. At the systems level, we know about most of the mechanics of body function. The unanswered questions involve integration and control of these mechanisms, particularly at the cellular and molecular levels. Nevertheless, explaining what happens in test tubes or isolated cells can only partially answer questions about function. For this reason, animal and human trials are essential steps in the process of applying basic research to treating or curing diseases.

THEMES IN PHYSIOLOGY

Recall the quotation that opens this chapter: "Physiology is not a science or a profession but a point of view."* Physiologists pride themselves on relating the processes they study to the functioning of the organism as a whole. Being able to think about how multiple body systems integrate their function is

one of the more difficult aspects of learning physiology, and if you are to develop expertise in physiology, you must do more than simply memorize facts and learn new terminology. Researchers have found that the ability to solve problems requires a conceptual framework, or "big picture," of the field.

This book will help you build a conceptual framework for physiology by explicitly describing basic biological concepts, or *themes* (Table 1-2 ■). Four major themes in this book are

TABLE 1-2	Key Concepts (Themes) in Physiology
1. Homeostasis and control systems	
2. Biological energy use	

3. Structure/function relationships
   Compartmentation
   Mechanical properties of cells, tissues, and organs
   Molecular interactions	

4. Communication
   Information flow
   Mass flow	

*Gerard RW. *Mirror to Physiology: A Self-Servey of Physiological Science*. Washington, D.C.: American Physiology Society, 1958.

1

FOCUS ON . . . MAPPING

Mapping is a nonlinear way of organizing material that has been shown to improve understanding and retention. It is a useful study tool because creating a map requires thinking about the importance of and relationships between various pieces of information. Studies have shown that when people interact with information by organizing it in their own way *before* they load it into memory, improved long-term learning results.

Mapping is not just a study technique. Experts in a field make maps when they are trying to integrate newly acquired information into their knowledge base, and they may create two or three versions of a map before they are satisfied that it represents their understanding. Scientists map out the steps in their experiments. Health care professionals create maps to guide them while diagnosing and treating patients.

A map can take a variety of forms but usually consists of terms (words or short phrases) linked by arrows to indicate associations. In general, there are three types of associations among the terms in a map: part/type, description/characteristic, and cause/effect. The linking arrows may be labeled with explanatory phrases or may be drawn in different colors to represent different types of links. Maps in physiology usually focus either on the relationships between anatomical structures and physiological processes (*structure/function maps*) or on normal homeostatic control pathways and responses to abnormal (pathophysiological) events (*process maps,* or *flow charts*). If appropriate, a map may also include graphs or pictures. Figures 1-5a is a structure map; Figures 3-9 and 3-22 are examples of structure/function maps. Figures 1-5b and 7-9 are examples of flow charts.

Many maps appear in this textbook, and they may serve as the starting point for your own maps. However, the real benefit from mapping comes from preparing maps yourself. By mapping information on your own, you think about the relationships between terms, organize concepts into a hierarchical structure, and look for similarities and differences between items. Interaction with the material in this way helps you process it into long-term memory instead of simply memorizing bits of information and forgetting them.

Getting Started on Mapping

1. First, **select the terms or concepts to map.** (In every chapter of this text, the end-of-chapter questions include at least one list of terms to map.) Sometimes it's helpful to write the terms on individual slips of paper or on 1-by-2-inch sticky notes so that you can rearrange the map more easily.

2. Usually the most difficult part of mapping is **deciding where to begin.** Start by grouping related terms in an organized fashion. You may find that you want to put some terms into more than one group. Make a note of these terms, as they will probably have several arrows pointing to them or leading away from them.

3. Now try to **create some hierarchy with your terms.** You may arrange the terms on a piece of paper, on a table, or on the floor. In a structure/function map, start at the top with the most general, most important, or overriding concept—the one from which all the others stem. In a process map, start with the first event to occur. Next, either break down the key idea into progressively more specific parts using the other concepts or else follow the event through its time course. Use arrows to point the direction of linkages and include horizontal links to tie related concepts together. The downward development of the map will generally mean either an increase in complexity or the passage of time.

   You may find that some of your arrows cross each other. Sometimes this can be avoided by rearranging the terms on the map. Labeling the linking arrows with explanatory words may be useful. For example,

   channel proteins ———*form*——➤ open channels.

   Color can be very effective on maps. You can use colors for different types of links or for different categories of terms. You may also add pictures and graphs that are associated with specific terms in your map.

4. Once you have created your map, **sit back and think about it.** Are all the items in the right place? You may want to move them around once you see the big picture. Revise your map to fill in the picture with new concepts or to correct wrong linkages. Review by recalling the main concept and then moving to the more specific details. Ask yourself questions like, "What is the cause? effect? parts involved? main characteristics?" to jog your memory.

5. A useful way to study with a map is to **trade maps with a classmate** and try to understand each other's maps. Your maps will almost certainly NOT look the same! That's okay. Remember that your map reflects the way you think about the subject, which may be different from the way someone else thinks about it. Did one of you put in something the other forgot? Did one of you have an incorrect relationship between two items?

6. **Practice making maps.** The study questions in each chapter will give you some ideas of what you should be mapping. Your instructor can help you get started.

---

homeostasis and control systems, biological energy use, structure-function relationships, and communication. As you work through the text, you will find that these concepts, with variations, occur repeatedly in the different organ systems.

Homeostasis and Control Systems   As discussed earlier in this chapter, a central precept of human physiology is homeo-

stasis, the maintenance of a relatively stable internal environment. The body has certain key parameters, such as blood pressure and blood glucose concentration, that must be kept within a particular operating range if the body is to remain healthy. These key parameters are *regulated variables* that are monitored and adjusted by physiological control systems. In its simplest

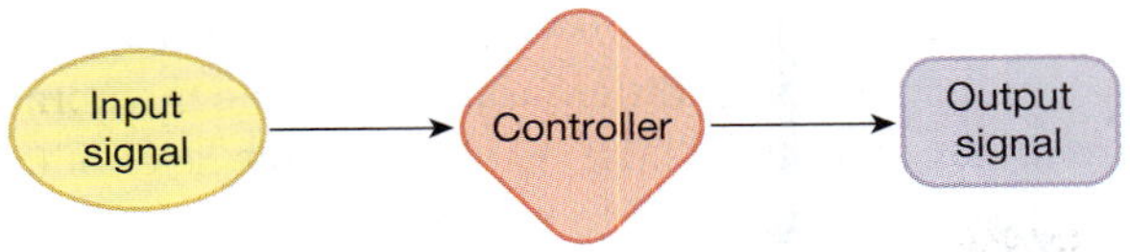

■ FIGURE 1-6 *A simple control system*

form, a **control system** has three components: (1) an input signal; (2) a controller, which is programmed to respond to certain input signals; and (3) an output signal (Fig. 1-6 ■). The different types of control systems will be discussed in Chapter 6.

Biological Energy Use The processes that take place in organisms require the continuous input of energy. Where does this energy come from, and how is it stored? Chapter 4 answers those questions and describes some of the ways that energy in the body is used for synthesis and breakdown of molecules. In later chapters you will learn how energy is used to transport molecules across cell membranes and to create movement.

Structure-Function Relationships The integration of structure and function is another physiological theme. Structural influences on function include molecular interactions, the division of the body into discrete compartments, and the mechanical properties of cells, tissues, and organs.

Molecular interactions The ability of individual molecules to bind to or react with other molecules is essential for biological function, and this ability is intimately related to each molecule's structure. Examples of molecular interactions include enzymes that speed up chemical reactions, or signal molecules that bind to membrane proteins. Molecular interactions will be discussed in Chapter 2.

Compartmentation of the body and of cells Compartmentation, or the presence of separate compartments, allows a cell, a tissue, or an organ to specialize and isolate functions. Membranes separate cells from one another and from the extracellular fluid, and create tiny compartments within the cell called *organelles*. At the macroscopic level, the tissues and organs of the body form discrete functional compartments. Compartmentation is the theme of Chapter 3.

Mechanical properties of cells, tissues, and organs The physical properties of cells and tissues are often a direct result of their anatomy. Some of the mechanical properties of cells and tissues that influence function include *compliance* (ability to stretch), *elastance* (ability to return to the unstretched state), strength, flexibility, and fluidity. In addition, the properties of specialized tissues and organs in the body enable them to function as biological pumps, filters, or motors.

Communication Both structure/function integration and homeostasis require that body cells communicate with one another rapidly and efficiently. *Information flow* in the body takes the form of either chemical signals or electrical signals.

Information may pass from one cell to its neighbors (local communication) or from one part of the body to another (long-distance communication). Information stored in the genetic code of DNA is passed from one cell to its daughters and from an organism to its offspring.

Communication between intracellular compartments and the extracellular fluid requires the transfer of information across the barrier of the cell membrane. The membrane is *selectively permeable,* which means that some molecules are able to pass through the barrier it presents but others are unable to cross. The details of how molecules cross biological membranes will be discussed in Chapter 5.

In some instances, specialized mechanisms allow extracellular signal molecules to pass their message across the cell membrane without entering the cell, a process known as **signal transduction.** Chemical communication and signal transduction are discussed in more detail in Chapter 6.

In physiology we are also concerned with *mass flow,* the movement of substances within and between compartments of the body. Blood flows through the circulatory system; air flows into and out of the lungs; gases, nutrients, and wastes move into and out of cells. The movement of materials in mass flow is made possible by a driving force, such as a pressure or concentration difference (a **gradient**). In some situations, mass flow is opposed by friction or other forces that create *resistance* to flow. Most mass flow in the body requires the input of energy at some point in the process. The concepts of gradients, flow, and resistance will be reintroduced in Chapter 5.

The themes just described will appear over and over again in the subsequent chapters of the book. Look for them in the summary material at the end of the chapters and in the end-of-chapter questions.

THE SCIENCE OF PHYSIOLOGY

How do we know what we know about the physiology of the human body? Observation and experimentation are the key elements of **scientific inquiry.** An investigator observes an event and, using prior knowledge, generates a **hypothesis** [*hypotithenai,* to assume], or logical guess, about how the event takes place. The next step is to test the hypothesis by designing an experiment in which the investigator manipulates some aspect of the event.

Good Scientific Experiments Must Be Carefully Designed

A common type of biological experiment either removes or alters some variable that the investigator thinks is an essential part of an observed phenomenon. That altered variable is the **independent variable.** For example, a biologist notices that birds at a feeder seem to eat more in the winter than in the summer. She generates a hypothesis that cold temperatures cause birds to increase their food intake. To test her hypothesis, she designs an experiment in which she will keep birds at different temperatures and monitor how much food they eat. In her experiment, temperature, the manipulated element, is the independent variable. Food intake, which is hypothesized to be dependent on temperature, becomes the **dependent variable.**

CONCEPT CHECK

1. Students in the laboratory run an experiment in which they drink different volumes of water and measure their urine output in the hour following drinking. What are the independent and dependent variables in this experiment?

Answers: p. 18

An essential feature of any experiment is a **control.** A control group is usually a duplicate of the experimental group in every respect except that the independent variable is not changed from its initial value. For example, in the bird-feeding experiment, the control group would be a set of birds maintained at a warm summer temperature but otherwise treated exactly like the birds held at cold temperatures. The purpose of the control is to ensure that any observed changes are due to the variable being manipulated and not to changes in some other variable. For example, suppose that in the bird-feeding experiment food intake increased after the investigator changed to a different food. Unless she had a control group that was also fed the new food, the investigator could not determine whether the increased food intake was due to temperature or to the fact that the new food was more palatable.

During an experiment, the investigator carefully collects information, or **data** [plural; singular *datum,* a thing given], about the effect that the manipulated (independent) variable has on the observed (dependent) variable. Once the investigator feels that she has sufficient information to draw a conclusion, she begins to analyze the data. Analysis can take many forms and usually includes statistical analysis to determine if apparent differences are statistically significant. A common format for presenting data is a graph (see Fig. 1-7 ■ in *Focus on Graphs*).

If one experiment supports the hypothesis that cold causes birds to eat more, then the experiment should be repeated to ensure that the results were not an unusual one-time event. This step is called **replication.** When the data support a hypothesis in multiple experiments, the hypothesis may become a **scientific theory.**

Most information presented in textbooks like this one is based on **models** that scientists have developed from the best available experimental evidence. On occasion, new experimental evidence that does not support a current model is published. In that case, the model must be revised to fit the available evidence. Thus, although you may learn a physiological "fact" while using this textbook, in 10 years that "fact" may be inaccurate because of what scientists have discovered in the interval. For example, in 1970 students learned that the cell membrane was a "butter sandwich," a structure composed of a layer of fats

sandwiched between two layers of proteins. In 1972, however, scientists presented a very different model of the membrane, in which globules of proteins float within a double layer of fats. As a result, textbook writers had to revise their descriptions of cell membranes, and students who had learned the butter sandwich model had to revise their mental model of the membrane.

Where do our scientific models for human physiology come from? We have learned much of what we know from experiments on animals ranging from squid to rats. In many instances, the physiological processes in such animals are either identical to those taking place in humans or else similar enough that we can extrapolate from the animal model to humans. It is important to use nonhuman models because experiments using human subjects can be difficult to perform.

However, not all studies done on animals can be applied to humans. For example, an antidepressant that had been used safely by Europeans for years was undergoing stringent testing required by the U.S. Food and Drug Administration before it could be sold in this country. When beagles were given the drug for a period of months, the dogs started dying from heart problems. Scientists were alarmed until further research showed that beagles have a unique genetic makeup that causes them to break down the drug into a more toxic substance. The drug was perfectly safe in other kinds of dogs and in humans, and it was subsequently approved for human use.

The Results of Human Experiments Can Be Difficult to Interpret

There are many reasons it is difficult to carry out physiological experiments in humans, including variability, psychological factors, and ethical considerations.

Variability Human populations have tremendous genetic and environmental **variability.** Although physiology books usually present *average* values for many physiological variables, such as blood pressure, these average values simply represent a number that falls somewhere near the middle of a wide range of values. Thus, to show significant differences between experimental and control groups in a human experiment, an investigator would, ideally, have to include a large number of identical subjects.

However, getting two groups of people who are *identical* in every respect is impossible. The researcher then must attempt to recruit subjects who are *similar* in as many aspects as possible. You may have seen newspaper advertisements requesting research volunteers: "Healthy males between 18 and 25, non-smokers, within 10% of ideal body weight, to participate in a study. . . ." The variability inherent in even a select group of humans must be taken into account when doing experiments with human subjects because variability may affect the researcher's ability to accurately interpret the significance of data collected on that group.

One way to reduce variability within a test population, whether human or animal, is to do a **crossover study.** In a crossover study, each individual acts both as experimental subject and as control. Thus, each individual's response to the treatment can be compared with his or her own control value. This method is particularly effective when there is wide variability within a population. For example, in a test of blood pressure medication, subjects might be divided into two groups. Group A takes an inactive substance (**placebo**, from the Latin for "I shall be pleasing") for the first half of the experiment, then changes to the experimental drug for the second half. Group B starts with the experimental drug, then changes to the placebo. This scheme enables the researcher to assess the effect of the drug on each individual. Statistically, the resulting data can be analyzed using various methods that look at the changes in the data within each individual rather than at changes in the collective group data.

Psychological Factors   Another significant variable in human studies is the psychological aspect of administering a treatment. If you give someone a pill and tell the person that it will help alleviate some problem, there is a strong possibility that the pill will have exactly that effect, even if it contains only sugar or an inert substance. This well-documented phenomenon is called the **placebo effect.** Similarly, if you warn people that a drug they are taking may have specific adverse side effects, those people will report a higher incidence of the side effects than a similar group of people who were not warned. This phenomenon is called the **nocebo effect**, from the Latin word *nocere*, to do harm. The placebo and nocebo effects show the ability of our minds to alter the physiological functioning of our bodies.

In setting up an experiment with human subjects, we must try to control for the placebo and nocebo effects. The simplest way to do this is with a **blind study**, in which the subjects do not know whether they are receiving the treatment or the placebo. Even this precaution can fail, however, if the researchers assessing the subjects know which type of treatment each subject is receiving. The researchers' expectations of what the treatment will or will not do may color their measurements or interpretations. To avoid this outcome, researchers often use **double-blind studies**, in which a third party, not involved in the experiment, is the only one who knows which group is receiving the experimental treatment and which group is receiving the control treatment. The most sophisticated experimental design for minimizing psychological effects is the **double-blind crossover study**, in which the control group in the first half of the experiment becomes the experimental group in the second half, and vice versa.

Ethical Considerations   Ethical questions arise when humans are used as experimental subjects, particularly when the subjects are people suffering from a disease or other illness. Is it ethical to withhold a new and promising treatment from the control group? A noteworthy example occurred a few years ago when researchers were testing the efficacy of a treatment for dissolving blood clots in heart attack victims. The survival rate among the treated patients was so much higher that testing was halted so that members of the control group could also be given the experimental drug.

In contrast, tests on some anticancer agents have shown that the experimental treatments were less effective in stopping the spread of cancer than were the standard treatments used by the controls. Was it ethical to undertreat patients in the experimental group by depriving them of the more-effective current medical practice? Most studies now are evaluated continually over the course of the study to minimize the possibility that subjects will be harmed by their participation.

In 2002 a trial on hormone replacement therapy in postmenopausal women was halted early when investigators realized that women taking a pill containing two hormones were developing cardiovascular disease and breast cancer at a higher rate than women on placebo pills. On the other hand, the women receiving hormones also had *lower* rates of colon cancer and bone fractures. The investigators decided that the risks associated with taking the hormones exceeded the potential benefits, and the study was stopped. To learn more about this clinical trial and the pros and cons of hormone replacement therapy, go to *www.nlm.nih.gov/medlineplus/hormonereplacementtherapy.html*, the website of the U.S. National Library of Medicine.

Human Studies Can Take Many Forms

Almost daily the newspapers carry articles about clinical trials studying the efficacy of drugs or other medical treatments. Many different aspects of experimental design can affect the validity and applicability of the results of these trials. For example, some trials are carried out for only a limited period of time on a limited number of people, such as studies being conducted for the U.S. Food and Drug Administration's drug approval process. In several instances in the past few years drugs approved as a result of such studies have later been withdrawn from the market when extended use of the drug uncovered adverse side effects, including deaths.

Longitudinal studies are designed to be carried out for a long period of time. One of the most famous longitudinal studies is the Framingham Heart Study (*www.framingham.com/heart*), started in 1948 and still ongoing. Framingham is a **prospective study** [*prospectus*, outlook, looking forward] that recruited healthy people and has been following them for years to identify factors that contribute to the development of cardiovascular disease. This study has already made important contributions to health care and is continuing with the adult children of the original participants.

Additional study designs you may encounter in the literature include cross-sectional and retrospective studies. **Cross-sectional studies** survey a population for the prevalence of a disease or condition. Data from cross-sectional studies identify trends to be investigated further, such as whether age group or socioeconomic status is associated with a higher risk of developing the condition being surveyed. **Retrospective studies** [*retro*, backwards + *spectare*, to look] match groups of people

FOCUS ON . . . GRAPHS

Graphs are pictorial representations of the relationship between two (or more) variables, plotted in a rectangular region (Fig. 1-7a ■). We use graphs to present a large amount of numerical data in a small space, to emphasize comparisons between variables, or to show trends over time. A viewer can extract information much more rapidly from a graph than from a table of numbers or from a written description. A well-constructed graph should contain (in very abbreviated form) everything the reader needs to know about the data, including the purpose of the experiment, how the experiment was conducted, and the results.

All scientific graphs have common features. The independent variable (the variable manipulated by the experimenter) is graphed on the horizontal *x*-axis. The dependent variable (the variable measured by the experimenter) is plotted on the vertical *y*-axis. If the experimental design is valid and the hypothesis is correct, changes in the independent variable (*x*-axis) will cause changes in the dependent variable (*y*-axis). In other words, *y* is a function of *x*. This relationship can be expressed mathematically, as $y = f(x)$.

Each axis of a graph is divided into units represented by evenly spaced tick marks on the axis. A label tells what variable the axis represents (time, temperature, amount of food consumed) and in what units it is marked (days, degrees Celsius, grams per day). The intersection of the two axes is called the *origin*. The origin usually, but not always, has a value of zero for both axes. A graph should have a title or legend that describes what the graph represents. If multiple groups are shown on one graph, the lines or bars representing the groups may have labels, or a key may show what group each symbol or color represents.

Most graphs you will encounter in physiology display data either as bars (bar graphs or histograms), as lines (line graphs), or as dots (scatter plots). Four typical types of

(b) Bar graph. Each bar shows a distinct variable. The bars are lined up side by side along one axis so that they can be easily compared with one another. Scientific bar graphs traditionally have the bars running vertically.

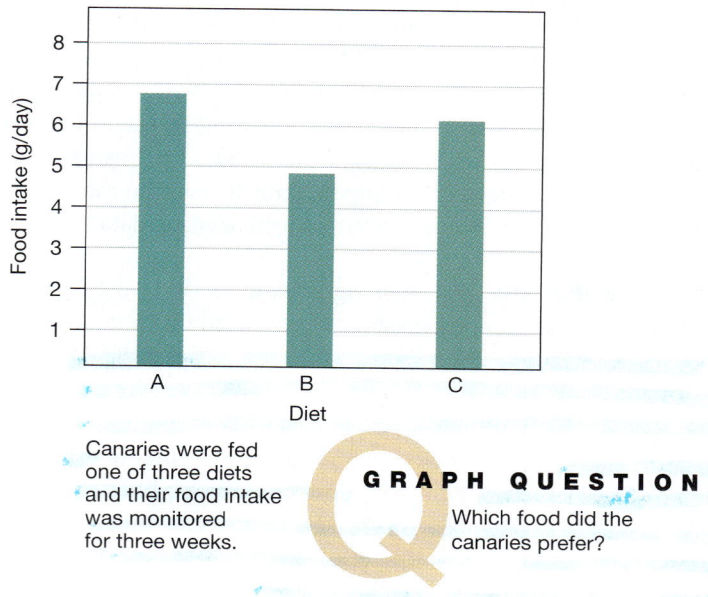

Canaries were fed one of three diets and their food intake was monitored for three weeks.

GRAPH QUESTION

Which food did the canaries prefer?

graphs are shown in Figure 1-7b–e. **Bar graphs** (Fig. 1-7b) are used when the independent variables are distinct entities. A **histogram** (Fig. 1-7c) is a specialized bar graph that shows the distribution of one variable over a range. The *x*-axis is divided into units (called "bins" in some computer graphing

(a) The standard features of a graph include units and labels on the axes, a key, and a figure legend.

(c) Histogram. A histogram quantifies the distribution of one variable over a range of values.

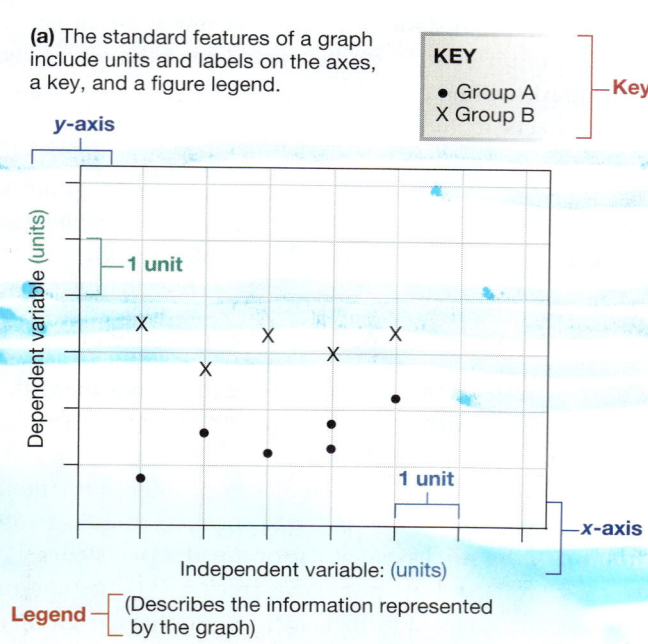

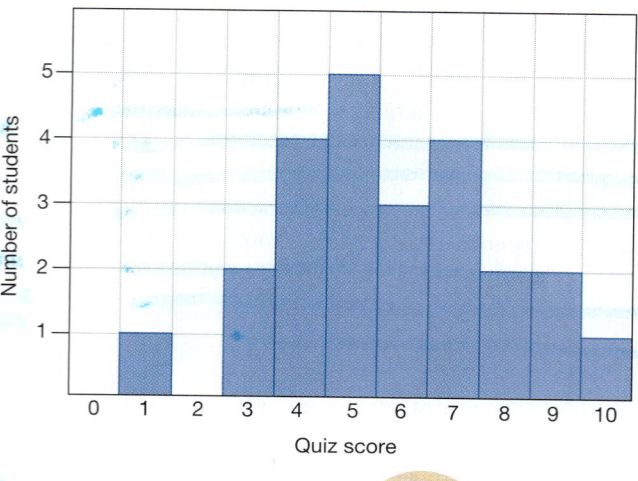

The distribution of student scores on a 10-point quiz is plotted on a histogram.

GRAPH QUESTION

How many students took the quiz?

■ **FIGURE 1-7**

(d) Line graph. The *x*-axis frequently represents time; the points represent average observations. The points may be connected by lines, in which case the slope of the line between two points shows the rate at which the variable changed.

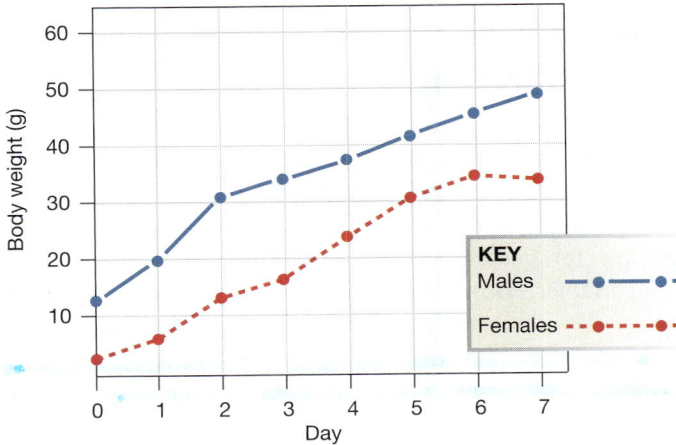

Male and female mice were fed a standard diet and weighed daily.

GRAPH QUESTION

When did male mice increase their body weight the fastest?

(e) Scatter plot. Each point represents one member of a test population. The individual points of a scatter plot are never connected by lines, but a best fit line may be estimated to show a trend in the data, or better yet, the line may be calculated by a mathematical equation.

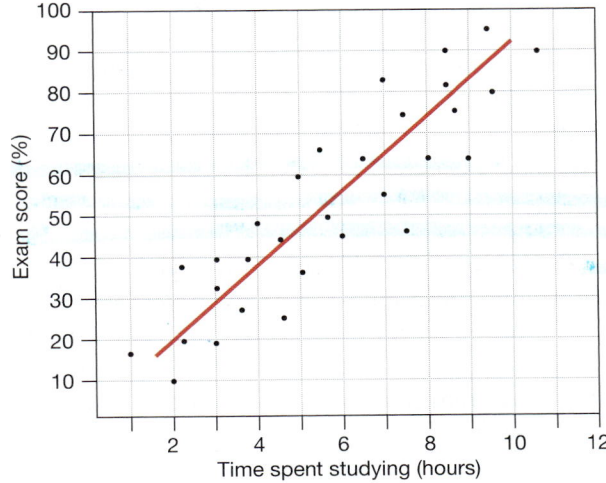

Student scores were directly related to the amount of time they spent studying.

GRAPH QUESTIONS

For graphs (d) and (e), answer the following:
- What was the investigator trying to determine?

- What are the independent and dependent variables?
- What are the results or trends indicated by the data?

programs), and the *y*-axis indicates how many pieces of data are associated with each bin.

   Line graphs (Fig. 1-7d) are commonly used when the independent variable on the *x*-axis is a continuous phenomenon, such as time, temperature, or weight. Each point on the graph may represent the average of a set of observations. Because the independent variable is a continuous function, the points on the graph can be connected with a line (point-to-point connections or a mathematically calculated "best fit" line or curve). Connecting the points allows the reader to **interpolate,** or estimate values between the measured values.

   Scatter plots (Fig. 1-7e) show the relationship between two variables, such as time spent studying for an exam and performance on that exam. Usually each point on the plot represents one member of a test population. Individual points on a scatter plot are never connected by a line, but a "best fit" line or curve may indicate a trend in the data.

   Here are some questions to ask when you are trying to extract information from a graph:

1. What variable does each axis represent?

2. What is the relationship between the variables represented by the axes? This relationship can usually be expressed by substituting the labels on the axes into the following statement: *y* varies with *x*. For example, in graph (b), the canaries' daily food intake varied with the type of diet.

3. Are any trends apparent in the graph? For line graphs and scatter plots, is the line horizontal (no change in the dependent variable when the independent variable changes), or does it have a slope? Is the line straight or curved? For bar graphs, are the bars the same height or different heights? If different heights, is there a trend in the direction of height change?

CONCEPT CHECK

2. Students in a physiology laboratory collected heart rate data on one another. In each case, heart rate was measured first for the subject at rest and again after the subject had exercised using a step test. Two findings from the experiment were (1) that heart rate was greater with exercise than at rest, and (2) that female subjects had higher resting heart rates than male subjects.

   (a) What was the independent variable in this experiment? What was the dependent variable?
   (b) Draw a graph and label each axis with the correct variable. Draw trend lines or bars that might approximate the data collected.

Answers: p. 18

who all have a particular disease to a similar but healthy control group to see whether development of the disease can be associated with a particular variable.

Often, the results of one or more published studies do not agree with the conclusions of other published studies. In some cases, the reason for the disagreement turns out to be a limitation of the experimental design, such as a small number of subjects who may not be representative of larger populations. In other cases, the disagreement may be due to small but potentially significant differences in the experimental designs of the different studies. One way scientists attempt to resolve contradictory results is to perform a **meta-analysis** of the data [*meta-*, at a higher level]. A meta-analysis combines all the data from a group of similar studies and uses sophisticated statistical techniques to extract significant trends or findings from the combined data. For example, multiple studies were done to assess whether glucosamine and chondroitin, two dietary supplements, can improve degenerative joint disease. However, the individual studies had small numbers of subjects (<50) and used different dosing regimens, making comparison of results difficult. A group of researchers then performed a meta-analysis using statistical methods and concluded that in most of the studies, glucosamine did have a beneficial effect on joint disease.*

The difficulty of using human subjects in experiments is one reason scientists use animals to develop many of our scientific models. Since the 1970s, physiological research has increasingly augmented animal experimentation with techniques developed by cellular biologists and molecular geneticists. As we have come to understand the fundamentals of how chemical signals in the body are received and interpreted by the cells, we have unlocked the mysteries of many processes. In doing so, we also have come closer to being able to treat many diseases by correcting their cause rather than simply alleviating their symptoms.

More and more, medicine is turning to therapies based on interventions at the molecular level. A classic example is the treatment of cystic fibrosis, an inherited disease in which the mucus of the lungs and digestive tract is unusually thick. For many years, patients with this condition had few treatment options, and most died at a young age. However, basic research into the mechanisms by which salt and water move across cell membranes provided clues to the underlying cause of cystic fibrosis: a defective protein in the membrane of certain cells. Once molecular geneticists found the gene that coded for that protein, the door was opened to the possibility of replacing the defective gene in cystic fibrosis patients with the gene for the normal protein. Without the basic research into how cells and tissues carry out their normal function, however, this treatment would never have been developed.

*Glucosamine and chondroitin for treatment of osteoarthritis: A systematic quality assessement and meta-analysis. *JAMA* 283(11): 1469–1475, 2000. (*www.jama.amaassn.org*)

As you read this book and learn what we currently know about how the human body works, keep in mind that many of the facts presented in it reflect models that represent our current understanding and are subject to change. There still are many questions in physiology waiting for investigators to find the answers.

SEARCHING AND READING THE SCIENTIFIC LITERATURE

One skill all physiology students should acquire is the ability to find information in the scientific literature. In today's world, the scientific literature can be found both in print, in the form of books and periodicals, and on the Web. Books are an excellent resource for general background information and learning about a subject. However, unless a book has a recent publication date, it may not be the most up-to-date source of information.

Scientific periodicals are called **journals**, and many are available now in both print and electronic format. Frequently journals are sponsored by scientific organizations, such as the American Physiological Society. A journal consists of contributed papers that describe the original scientific research of an individual or group. A scientist who speaks of writing "a paper" is usually referring to a scientific paper published in a journal. Most journal articles are **peer-reviewed**, which means that the research described in the paper has gone through a screening process in which the article is critiqued by an anonymous panel of two or three scientists whose credentials qualify them to judge the quality of the work. Peer review acts as a kind of quality control because a paper that does not meet the standards of the reviewers will be rejected by the editor of the journal.

Many journals publish **review articles**. A review article is a synopsis of recent research on a particular topic. When you are just beginning to research a topic, begin with recent review articles because they usually contain more-up-to-date information than a book on the same topic.

Many students begin their quest for information on a subject by searching the World Wide Web. Be cautious! Anyone can create a web page and publish information on the Web. There is no screening process comparable to peer review in journals, and so the reader of a web page must decide how valid the information is. Web sites published by recognized universities and not-for-profit organizations are likely to have good information, but an article about vitamins on the web page of a health food store should be viewed with a skeptical eye unless the article cites published peer-reviewed research.

Many people search the Web using Google (*www.google.com*), a search engine developed by Stanford University. When searching for scientific research articles, you might want to try Google Scholar (*www.scholar.google.com*). This new search engine locates more scholarly sources, such as peer-reviewed journals and academic publications.

Defining Search Terms When you decide to search for information, whether on the Web or through one of the many databases that catalog peer-reviewed and lay publications, your first step should be to define your search terms. You then do a **keyword search.** Databases such as MEDLINE, published by the U.S. National Library of Medicine, also have a standardized list of indexing terms. Suppose you want to learn more about channelopathies, the diseases caused by mutations in ion channels. If you go to PubMed, the free public access version of MEDLINE (*www.ncbi.nlm.nih.gov/PubMed/*), the basic search asks you to type in a term, such as CHANNELOPATHY. If you access MEDLINE through a search engine such as Ovid, the program will also try to match your keyword to the database's indexing terms, called MeSH headings for MEDLINE. For channelopathies, the related MeSH headings are various types of ion channels, "nervous system diseases," and pathologies related to ion channels, such as "arrhythmia" or "paralyses, familial periodic." To find review articles, either add REVIEW to your search terms or use the "limit" function to limit your search to reviews.

When you do a web search using a search engine such as Google, you must think more carefully about the search terms you use. Small wording changes can make a big difference. To see this for yourself, type CHANNELOPATHY into Google and see how many listings you get. Then, make the term plural—CHANNELOPATHIES—and see the difference in what and how many listings you get.

For an award-winning set of interactive tutorials on web searching, go to *http://tilt.lib.utsystem.edu/* (notice that there is no www in this address). These tutorials will help you learn how to search and evaluate web sites.

Citation Formats When you find an article either in print or on the Web, you should write down the full citation. Citation formats for papers vary slightly from source to source but will usually include the following elements (with the punctuation shown):

> Author(s). Article title. *Journal Name* volume(issue): inclusive pages, year of publication.

For example:

> Echevarria M and Ilundain AA. Aquaporins. *J Physiol Biochem* 54(2): 107–118, 1998.

In many citations, the name of a journal is abbreviated using standard abbreviations. For example, the *American Journal of Physiology* abbreviates to *Am J Physiol*. For each calendar year, all the issues of a given journal are assigned a **volume** number; the first issue of a given volume is designated **issue** 1, the second is issue 2, and so on.

Citing sources from the Web requires a different format. Here's one suggested format:

> Author/Editor (if known). Revision or copyright date (if available). Title of web page [Publication medium]. Publisher of web page. URL [Date accessed].

For example:

> Patton G (editor). 2005. Biological Journals and Abbreviations. [Online]. National Cancer Institute. *http://home.ncifcrf.gov/research/bja* [April 10, 2005].

Unlike print resources, web pages are not permanent and frequently disappear or move. If you access a print journal on the Web, you should give the print citation, not the URL.

> **CONCEPT CHECK**
>
> 3. Read the citation for the aquaporin paper mentioned above to answer these questions: (a) Who is the first author? (b) What is the full name of the journal in which this paper was published? (c) Assuming this journal is a monthly journal, in what month did this article appear? (d) On what page does the article begin?
>
> Answers: p. 18

Copying or paraphrasing material from another source without acknowledging that source is academic dishonesty. Word-for-word quotations placed within quotation marks are rarely used in scientific writing. Instead, we summarize the contents of the source paper and acknowledge the source, as follows:

> Some rare forms of epilepsy are known to be caused by mutations in ion channels (Mulley, *et al.*, 2003).

When a paper has three or more authors, we use the abbreviation *et al.*—from the Latin *et alii,* meaning "and others"—to save space in the body of the text. All authors' names are given in the full citation, which is usually included within a References section at the end of the paper. Reference lists are often arranged alphabetically by the last name of the paper's first author.

To learn more about searching and reading scientific literature, go to The Physiology Place (*www.physiologyplace.com*).

CHAPTER SUMMARY

1. **Physiology** is the study of the normal functioning of a living organism and its component parts. (p. 2)

Physiological Systems

2. Physiologists study the many **levels of organization** in living organisms, from molecules to populations of one species. (p. 2; Fig. 1-1)

3. The **cell** is the smallest unit of structure capable of carrying out all life processes. (p. 2)

4. Collections of cells that carry out related functions make up **tissues** and **organs.** (p. 2)

5. The human body has 10 physiological organ systems: **integumentary, musculoskeletal, respiratory, digestive, urinary, immune, circulatory, nervous, endocrine,** and **reproductive.** (pp. 2–3; Fig. 1-2)

Function and Process

6. The **function** of a physiological system or event is the "why" of the system; the **process** by which events occur is the "how" of a system. The **teleological approach** to physiology explains why events happen; the **mechanistic approach** explains how they happen. (p. 3)

Homeostasis

7. The human body as a whole is adapted to cope with a variable external environment, but most individual cells of the body can tolerate much less change. (p. 3)

8. **Homeostasis** is the maintenance of a relatively constant internal environment. **Parameters** that are homeostatically regulated include temperature, pH, ion concentrations, and water. (pp. 4–5)

9. The body's internal environment is the **extracellular fluid**. (p. 4; Fig. 1-3)

10. Failure to maintain homeostasis may result in illness or disease. (p. 5; Fig. 1-4)

Physiology: Moving Beyond the Genome

11. Contemporary research is moving from identifying the body's proteins to studying how those proteins and other molecules influence function in an organism. (p. 6)

Physiology Is an Integrative Science

12. Many complex functions are **emergent properties** that cannot be predicted from the properties of the individual component parts. (p. 6)

13. Translational research applies the results of basic physiological research to medical problems. (p. 7)

Themes in Physiology

14. The four key themes in physiology are homeostasis and control systems; biological energy use; structure-function relationships, such as **molecular interactions** and **compartmentation**; and communication among cells. (pp. 7–9)

The Science of Physiology

15. Observation and experimentation are the key elements of **scientific inquiry**. A **hypothesis** is a logical guess about how an event takes place. (p. 9)

16. In scientific experimentation, the factor manipulated by the investigator is the **independent variable**, and the observed factor is the **dependent variable**. All well-designed experiments have a **control** to ensure that observed changes are due to the experimental manipulation and not to some outside factor. (pp. 9–10)

17. **Data**, the information collected during an experiment, are analyzed and presented, often as a graph. (p. 10)

18. A **scientific theory** is a hypothesis that has been supported by data on multiple occasions. When new experimental evidence does not support a theory or a model, then the theory or model must be revised. (p. 10)

19. Animal experimentation is an important part of learning about human physiology because of the tremendous amount of **variability** within human populations, and because it is difficult to control human experiments. In addition, ethical questions arise when using humans as experimental animals. (p. 10)

20. In a **crossover study**, each subject acts as both experimental subject and control. For half the experiment, the subject takes an inactive substance known as a **placebo.** One difficulty in human experiments arises from the **placebo** and **nocebo effects**, in which changes take place even if the treatment is inactive. (pp. 10, 11)

21. In a **blind study**, the subjects do not know whether they are receiving the experimental treatment or a placebo. In a **double-blind study**, a third party removed from the experiment is the only one who knows which group is the experimental group and which is the control. In a **double-blind crossover study**, the control group in the first half of the experiment becomes the experimental group in the second half, and vice versa. (p. 11)

22. **Meta-analysis** of data combines data from many studies to look for trends. (p. 14)

Searching and Reading the Scientific Literature

23. Scientific literature can be found in print and in electronic formats. Readers should critically evaluate all sources for validity. (p. 14)

QUESTIONS

(Answers to the Review Questions begin on page A1.)

THE PHYSIOLOGY PLACE

Access more review material online at **The Physiology Place** website. There you'll find review questions, problem-solving activities, case studies, flashcards, and direct links to both *InterActive Physiology*® and *PhysioEx*™. To access the site, go to *www.physiologyplace.com* and select Human Physiology, Fourth Edition.

LEVEL ONE REVIEWING FACTS AND TERMS

1. Define physiology. Describe the relationship between physiology and anatomy.
2. Name the different levels of organization in the biosphere and the disciplines that study all the levels you have named.
3. Name the 10 systems of the body and give their major function(s).
4. What is meant by "Physiology is an integrative science"?

5. Define homeostasis. Name some physiological parameters that are maintained homeostatically.
6. Name four major themes in physiology.

LEVEL TWO REVIEWING CONCEPTS

7. **Mapping exercise:** Make a large map showing the organization of the human body. Show all levels of organization in the body (see Fig. 1-1) and all 10 organ systems. Try to include functions of all components on the map and remember that some structures may share functions. (Hint: Start with the human body as the most important term. You may also draw the outline of a body and make your map using it as the basis.)

8. Distinguish between the items in each group of terms.
   (a) tissues and organs
   (b) *x*-axis and *y*-axis on a graph
   (c) dependent and independent variables
   (d) teleological and mechanistic approaches

(e) the internal and external environments for a human

(f) blind, double-blind, and crossover studies

9. Name as many organs or body structures that connect directly with the external environment as you can.

10. Which organ systems are responsible for coordinating body function? For protecting the body from outside invaders? Which systems exchange material with the external environment, and what do they exchange?

LEVEL THREE PROBLEM SOLVING

11. A group of biology majors went to a mall and asked passersby, "Why does blood flow?" These are some of the answers they received. Which answers are teleological and which are mechanistic? (Not all answers are correct, but they can still be classified.)

(a) Because of gravity

(b) To bring oxygen and food to the cells

(c) Because if it didn't flow, we would die

(d) Because of the pumping action of the heart

12. Although dehydration is one of the most serious physiological obstacles that land animals must overcome, there are others. Think of as many as you can, and think of various strategies that different terrestrial animals have to overcome these obstacles. (Hint: Think of humans, insects, and amphibians; also think of as many different terrestrial habitats as you can.)

LEVEL FOUR QUANTITATIVE PROBLEMS

13. A group of students wanted to see what effect a diet deficient in vitamin D would have on the growth of baby guppies. They fed the guppies a diet low in vitamin D and measured fish body length every third day for three weeks. Their data looked like this:

Day	0	3	6	9	12	15	18	21
Average body length (mm)	6	7	9	12	14	16	18	21

(a) What was the dependent variable and what was the independent variable in this experiment?

(b) What was the control in this experiment?

(c) Make a fully labeled graph with a legend, using the data in the table.

(d) During what period of time was growth slowest? Most rapid? (Use your graph to answer this question.)

14. You performed an experiment in which you measured the volumes of nine slices of potato, then soaked the slices in solutions of different salinities for 30 minutes. At the end of 30 minutes, you again measured the volumes of the nine slices. The changes you found were:

% change in volume after 30 minutes

SOLUTION	SAMPLE 1	SAMPLE 2	SAMPLE 3
Distilled water	10%	8%	11%
1% salt (NaCl)	0%	−0.5%	1%
9% salt (NaCl)	−8%	−12%	−11%

(a) What was the independent variable in this experiment? What was the dependent variable?

(b) Can you tell from the information given whether or not there was a control in this experiment? If there was a control, what was it?

(c) Graph the results of the experiment using the most appropriate type of graph.

15. At the end of the semester, an intermediate-level class of 25 male weight lifters was measured for aerobic fitness and for midarm muscle circumference. The relationship between those two variables is graphed in Figure 1-8 ■.

(a) What kind of graph is this?

(b) What question were the investigators asking?

(c) In one sentence, summarize the relationship between the two variables plotted on the graph.

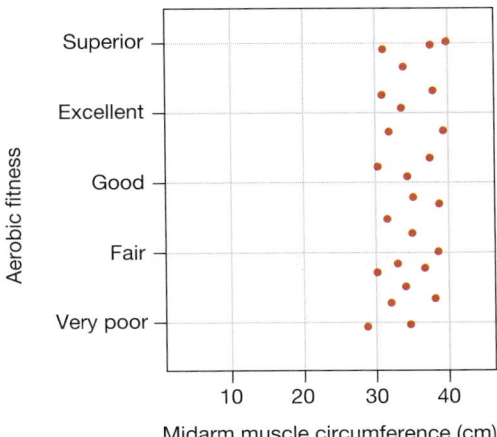

■ FIGURE 1-8

16. Answer the questions after the following article summary.

A recent study* was carried out on human volunteers to see whether two procedures performed during arthroscopic surgery [*arthro-*, joint + *scopium*, to look at] are effective in relieving knee pain associated with osteoarthritis, or degenerative joint disease [*osteon*, bone + *arthro-*, joint + *-itis*, inflammation]. The volunteers were up to 75 years old and were recruited from a Veterans Affairs Medical Center. They were 93% male and 60% white. One-third of the subjects had placebo operations; that is, they were given anesthesia and their knees were cut open, but the remainder of the treatment procedure was not done. The other two-thirds of the subjects had one of the two treatment procedures performed. Subjects were followed for two years. They answered questions about their knee pain and function and were given an objective walking and stair-climbing test. At the end of the study the results showed no significant difference in knee function or perception of pain between subjects getting one of the standard treatments and those getting the placebo operation.

(a) Do you think it is ethical to perform placebo surgeries on humans who are suffering from a painful condition, even if the subjects are informed that they might receive the placebo operation and not the standard treatment?

(b) Give two possible explanations for the decreased pain reported by the placebo operation subjects.

(c) Analyze and critique the experimental design of this study. Are the results of this study applicable to everyone with knee pain?

(d) Was this study a blind, double-blind, or double-blind crossover design?

(e) Why do you think the investigators felt it was necessary to include a placebo operation in this study?

---

*A controlled trial of arthroscopic surgery for osteoarthritis of the knee. *N Eng J Med* 347(2): 81–88, 2002.

ANSWERS

✓ Answers to Concept Check Questions

Page 10

1. The independent variable is the amount of water the students drink. The dependent variable is their urine output.

Page 13

2. (a) Activity level was the independent variable (*x*-axis), and heart rate was the dependent variable (*y*-axis). (b) A line graph would be appropriate for this data, but a bar graph could also be used if exercise intensity was the same in all subjects. (c) To show the difference between males and females, the graphs should have either separate lines or separate bars for males and females.

Page 15

3. (a) M. Echevarria (b) *Journal of Physiology and Biochemistry* (c) The (2) is issue number, which refers to February. (d) page 107

Q Answers to Figure and Graph Questions

Page 7

Fig. 1-5: (a) You might include different kinds of breads, meats, and so on, or add a category of sandwich characteristics, such as temperature (hot, cold) or layers (single, club).

Pages 12–13

Fig. 1-7b: The birds preferred diet A.

Fig. 1-7c: Twenty-four students took the quiz.

Fig. 1-7d: Male mice (blue line) increased their weight most between day 1 and day 2.

Fig. 1-7e: (a) In graph 1-7d the investigator was looking at changes of body weight over time in male and female mice. In graph 1-7e the investigator was trying to determine if there was a relationship between the amount of time spent studying for an examination and the student's score on that examination. (b) In graph 1-7d the independent variable was time and the dependent variable was body weight. In graph 1-7e the independent variable was number of hours spent studying and the dependent variable was the student score. (c) Graph 1-7d shows that male mice weigh more than female mice from the start of the experiment and that body weight increases with time. The rate of increase is about the same in males and females, indicated by the nearly parallel lines. Graph 1-7e shows that more hours spent studying resulted in higher exam scores.

Page 18

Fig. 1-9 ■. This sample map shows one possible way to map the terms. Notice how some items are linked to more than one other item. The map does not include many terms that could be included.

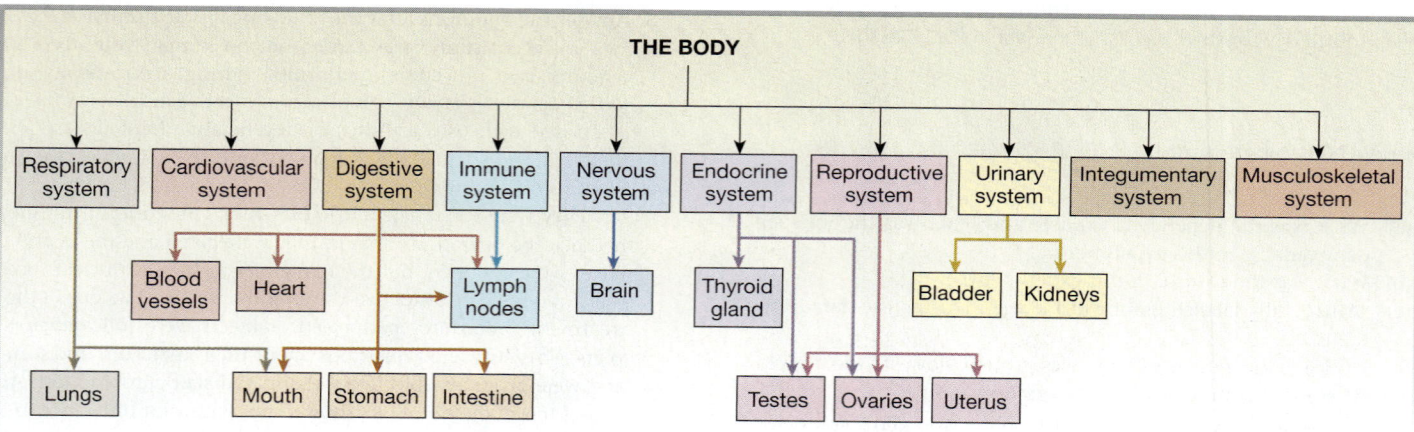

■ **FIGURE 1-9**

2

One of the things that distinguish ours from all earlier generations is this, that we have seen our atoms. —Karl Kelchner Darrow

Succinic acid and urea crystals.

Molecular Interactions

Chromium, zinc, copper, iron. . . . The list reads like the inventory of a metal company, but it actually comes from the label on a bottle of multivitamin-multimineral pills. Why are we supposed to ingest these metals, and what do they do in our bodies? A glance at a magazine article touting vitamins C and E as antioxidants may have you wondering about just what it is these vitamins are "against." What is an oxidant, and why do we need *anti*oxidants in our bodies?

To answer these questions, we must start at the lowest level of organization in the human body and look at the structure of atoms. We will then examine how atoms link together to form molecules, and how molecules, especially proteins, interact with one another.

Finally, because the human body is 60% water and most of its molecules are dissolved in this water, we will review how to make solutions and how to express the concentrations of molecules dissolved in them.

CHEMISTRY REVIEW

Atoms Are Composed of Protons, Neutrons, and Electrons

Atoms—the building blocks of all matter, including the human body—were once thought to be the smallest particles of matter [*atomos,* indivisible]. We now know, however, that an atom is composed of three types of even smaller particles: positively charged **protons**, uncharged **neutrons**, and negatively charged **electrons** (Fig. 2-1 ■). An atom contains an equal number of protons and electrons, giving it an overall electrical charge of zero.

The arrangement of protons, neutrons, and electrons in an atom is always the same. The protons and neutrons of the atom, which account for almost all of its mass, are clustered at the atom's center in a dense body called the **nucleus** [*nucleus,*

little nut; plural, nuclei]. The space around the nucleus—almost all of the atom's volume—contains the rapidly moving, lightweight, negatively charged electrons, held in their orbits by their attraction to the positively charged protons. Like the poles of a magnet, two oppositely charged atomic particles (+ and −) attract each other, while two particles with like charge (+ and +, or − and −) repel each other.

Atoms are very tiny, having diameters in the range of 1 to 5 angstroms (1 angstrom = 1 Å = 10^{-10} m). Most of an atom's volume is empty space. To get a feel for how little of that volume is occupied by the subatomic particles, imagine that the atom has a diameter equal to the length of a football field. The nucleus would then be the size of a large apple in the center of the field, and tiny electrons, each smaller than a pea, would move through the remainder of the space.

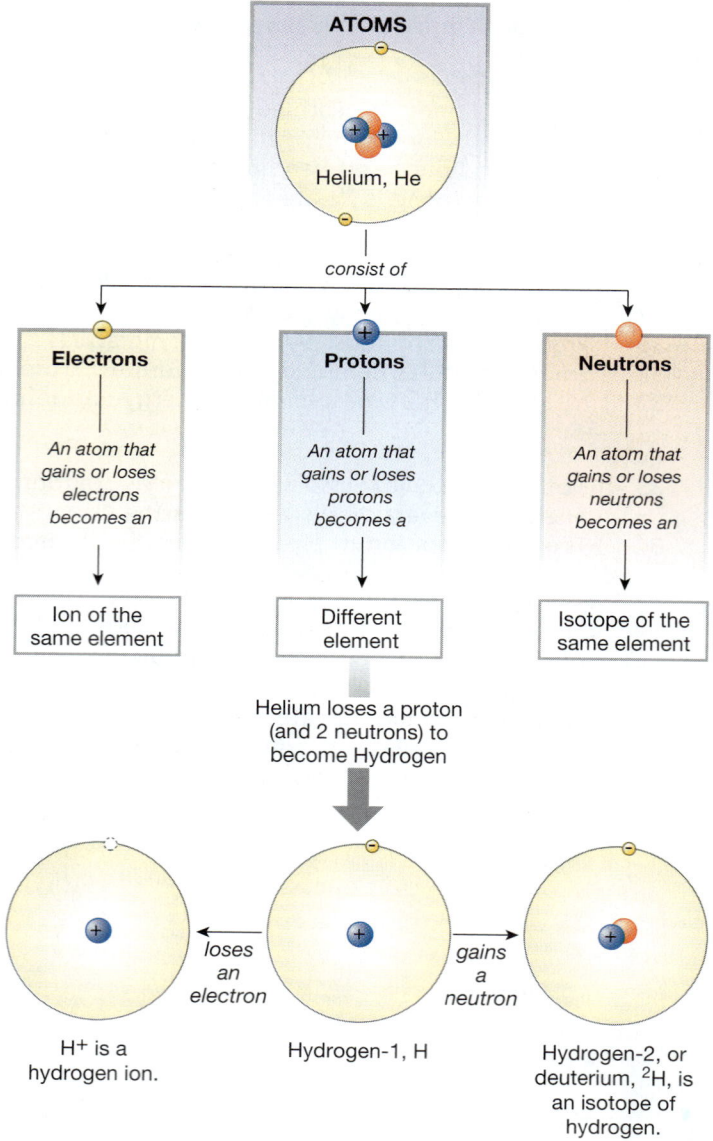

■ **FIGURE 2-1**   *A map showing the relationships among atoms, elements, ions, and isotopes*

The Number of Protons in the Nucleus Determines the Element

Atoms can be described two ways: by their atomic number and by their atomic mass. The **atomic number** of an atom is the number of protons in the nucleus. A hydrogen atom, for example, has one proton in its nucleus, so the atomic number of hydrogen is 1. A helium atom has two protons and thus an atomic number of 2, and so on.

It is an atom's atomic number that determines which *element* the atom is. An **element** is the simplest type of matter. Every atom having an atomic number of 1 (one proton in the nucleus) is an atom of the element *hydrogen,* for instance, and every atom having an atomic number of 2 is an atom of the element *helium* (Fig. 2-1). There are more than 100 known elements, and all of them are listed in the *periodic table of the elements,* shown on the inside back cover of this book. Each box in the table contains the name of an element and its one-, two-, or three-letter symbol. Some of the symbols come from the original Latin names for the elements and thus do not match our modern names. For example, the element sodium is abbreviated Na, from its Latin name *natrium.*

Three of the elements in the periodic table—oxygen, carbon, and hydrogen—make up more than 90% of the body's mass. Eight additional elements (nitrogen, phosphorus, sodium, potassium, calcium, magnesium, sulfur, and chlorine) round out the list of *major essential elements* of the body. In addition to these major essential elements, the body also requires trace amounts of other elements, such as selenium, chromium, manganese, and molybdenum. These nutrients are referred to as either *trace elements* or *minor essential elements.*

CONCEPT CHECK

1. Write the one- or two-letter symbol for each of the eleven major essential elements.

2. Look at the periodic table on the inside back cover, and find ten elements other than sodium whose symbols are not derived from their modern names.

Answers: p. 49

The **atomic mass** of an atom is the total mass of the protons and neutrons in the atom, expressed in **atomic mass units,** where 1 **amu** $= 1.6605 \times 10^{-27}$ kg. Protons and neutrons have nearly equal masses of about 1 amu. (Electrons are much smaller: it takes the mass of 1836 electrons to equal the mass of one neutron.) Thus, the atomic mass of the helium atom in Figure 2-1 is 4 amu, calculated by adding the masses of the two protons and two neutrons in the nucleus. Another unit used to express atomic mass is the *dalton* (Da), with 1 Da = 1 amu. The dalton unit honors John Dalton, who proposed the modern theory of the atom around 1803.

CONCEPT CHECK

3. What are the atomic masses of the two hydrogen atoms and one hydrogen ion shown at the bottom of Figure 2-1?

Answers: p. 49

RUNNING PROBLEM

What is chromium picolinate? Chromium (Cr) is an essential element that is linked to normal glucose metabolism. In the diet, chromium is found in brewer's yeast, broccoli, mushrooms, and apples. Because chromium in food and in chromium chloride supplements is poorly absorbed from the digestive tract, a scientist developed and patented the compound chromium picolinate. Picolinate, derived from amino acids, enhances chromium uptake at the intestine. The recommended daily allowance (RDA) of chromium is 50 micrograms. As we've seen, Stan takes several times this amount.

Question 1:
Locate chromium on the periodic table of the elements. What is chromium's atomic number? Atomic mass? How many electrons does one atom of chromium have? Which elements close to chromium are also essential elements?

| 20 | 21 | 23 | 26 | 35 | 38 | 44 |

Isotopes of an Element Contain Different Numbers of Neutrons

The number of protons in an element is constant, but the number of neutrons may vary. Atoms of an element that have different numbers of neutrons are known as **isotopes** of the element [*iso-,* same + *topos,* place (that is, the same place in the periodic table)]. Hydrogen, for instance, has three isotopes: hydrogen-1, hydrogen-2, and hydrogen-3. Hydrogen-1 contains only a single proton in the nucleus, hydrogen-2 contains one proton and one neutron, hydrogen-3 contains one proton and two neutrons (see the bottom of Fig. 2-1). Note that there is *always* only one proton in hydrogen but the atomic mass of hydrogen isotopes varies with the number of neutrons. The number after the name of the element indicates the isotope's atomic mass.

The symbol for an isotope indicates the atomic mass at the upper left of the chemical symbol for the element. For example, the symbol for hydrogen-1, the most common naturally occurring isotope of hydrogen, is ^{1}H. The symbol for hydrogen-2, also known as deuterium, is ^{2}H. The atomic mass shown below an element's symbol in the periodic table is a weighted average of the different isotopes of the element. All isotopes of an element have the same chemical properties because these properties are determined by the electron configuration of the atom and not by the number of neutrons.

Isotopes are important in biology and medicine because some, called **radioisotopes,** are unstable and emit energy called **radiation.** Radioisotopes emit three types of radiation: alpha (α), beta (β), and gamma (γ). Alpha radiation and beta radiation are composed of fast-moving particles (protons and neutrons in the case of alpha radiation, electrons in the case of beta radiation) ejected from an unstable atom. Gamma radiation is composed of high-energy waves rather than particles. Gamma radiation is able to penetrate matter much more deeply than α or β particles. X-rays are similar to gamma radiation and are used to create images of the body's interior.

Radioisotopes are widely used in medicine for diagnosis and treatment. In some instances, the radioisotope is used by itself. For example, the radioisotope iodine-131 is useful for the diagnosis and treatment of thyroid gland disorders. This radioisotope emits both α and β radiation. In small doses, this radiation does little damage, but larger doses can selectively destroy thyroid tissue when the gland is overactive. Another widely used radioisotope for diagnostic procedures is technetium-99. The use of radioisotopes in the diagnosis and treatment of disease is a specialty known as *nuclear medicine*.

✓ CONCEPT CHECK

4. Use the periodic table of the elements on the inside back cover to find the atomic number and average atomic mass of iodine and technetium. What is the letter symbol for technetium?

Answers: p. 49

Electrons Form Bonds Between Atoms and Capture Energy

The electrons in an atom do not move around the nucleus at random. Instead, they are arranged in a series of energy levels, or **shells**. The shell having the lowest energy is closest to the nucleus; as distance from the nucleus increases, shell energy levels increase. Each shell has a limit to the number of electrons it can hold. The first shell (the one closest to the nucleus) has a limit of two electrons, the next shell has a limit of eight, and so forth. The shells fill with electrons in sequence from inner to outer, a property that helps explain how atoms combine with each other. The arrangement of electrons in the outer shell of an atom determines its ability to bind with other atoms.

Electrons play four important roles in physiology: in the formation of covalent bonds, in the formation of ions, in the capture and transfer of energy, and in the formation of destructive free radicals.

1. **Covalent bonds.**   Shared electrons form strong covalent bonds to create molecules, as discussed in the next section.
2. **Ions.**   If an atom gains or loses one or more electrons, it acquires an electrical charge and becomes an **ion**. An atom that gains electrons acquires one negative charge (–1) for each electron added and is called an **anion**. An atom that loses electrons has one positive charge (+1) for each electron lost due to the protons that are left behind without matching electrons. Positively charged ions are called **cations**.

   By convention, ions are indicated by a superscript notation following the symbol of the element, with the number of charges first (if greater than 1) and the sign of the charge second. For example, a phosphate ion has two extra electrons, so we write this anion HPO_4^{2-}. Hydrogen loses

TABLE 2-1	Important Ions of the Body		
CATIONS		**ANIONS**	
Na^+	Sodium	Cl^-	Chloride
K^+	Potassium	HCO_3^-	Bicarbonate
Ca^{2+}	Calcium	HPO_4^{2-}	Phosphate
H^+	Hydrogen	SO_4^{2-}	Sulfate
Mg^{2+}	Magnesium		

its only electron and ends up as a single proton, H^+ (as shown at the left in the bottom row of Figure 2-1). Table 2-1 ■ lists the most important ions that you will encounter in physiology. Notice that some of the ions are single atoms, and others are combinations of atoms.

3. **High-energy electrons.**   The electrons in certain atoms can capture energy from their environment and transfer it to other atoms, so that the energy can be used for synthesis, movement, and other life processes. The released energy may also be emitted as radiation. For example, bioluminescence in fireflies is visible light emitted by high-energy electrons returning to their normal low-energy state.

4. **Free radicals.**   Our bodies are constantly exposed to radiation from natural sources, such as the sun, and from manufactured sources, such as microwave ovens and television sets. Some types of radiation alter the distribution of electrons in atoms, converting the atoms into unstable **free radicals**. Every free radical has at least one unpaired electron. One common free radical is the electrically neutral *hydroxy free radical*, OH, formed when a hydroxyl ion, OH^-, loses an electron. Another common free radical is the *superoxide* ion, $•O_2^-$. This free radical is constantly manufactured by the body during normal metabolism when a neutral oxygen molecule (O_2) gains an extra electron (•).

As you will learn in the section on molecular bonds, electrons are most stable when they occur in pairs. Thus, free radicals with their unpaired electron will try to "steal" an electron from another molecule. When this happens, the atomic species donating the electron may be left with an unpaired electron, converting it into a free radical. The newly created free radical then in effect looks for an electron it can "steal," creating a chain reaction of free radical production that can disrupt normal cell function. Free radicals are thought to contribute to aging and to the development of certain diseases, such as some cancers.

Antioxidants are substances that prevent damage to our cells by giving up electrons without becoming free radicals. Many antioxidants occur naturally in fruits and vegetables. The most common antioxidants promoted in vitamin supplements are vitamins C and E.

RUNNING PROBLEM

One advertising claim for chromium is that it improves the transfer of glucose—the simple sugar that cells use to fuel all their activities—from the bloodstream into cells. In diabetes mellitus, cells are unable to take up glucose from the blood efficiently. It seemed logical, therefore, to test whether addition of chromium to the diet would enhance glucose uptake in people with diabetes. In one Chinese study, diabetic patients receiving 500 micrograms of chromium picolinate twice a day showed significant improvement in their diabetes, but patients receiving 100 micrograms or a placebo did not.

Question 2:
If people have a chromium deficiency, would you predict that their blood glucose level would be lower or higher than normal? From the results of the Chinese study, can you conclude that all people with diabetes suffer from a chromium deficiency?

| 20 | 21 | **23** | 26 | 35 | 38 | 44 |

CONCEPT CHECK

Match each subatomic particle in the left column with all the phrases in the right column that describe it. Each phrase may be used more than once.

5. electrons
6. neutrons
7. protons

(a) each has atomic mass of 1 amu
(b) found in the nucleus
(c) negatively charged
(d) changing the number of these in an atom creates a new element
(e) adding or losing these makes an atom into an ion
(f) adding or losing these makes a different isotope of the same element

8. A magnesium ion, Mg^{2+}, has *gained/lost* two (choose one) *protons/neutrons/electrons*?

9. H^+ is also called a proton. Why is it given that name?

Answers: p. 49

MOLECULAR BONDS AND SHAPES

When two or more atoms link by sharing electrons, they make units known as **molecules**. Some molecules, such as those of oxygen gas (O_2) and nitrogen gas (N_2), are made of two atoms of the same element. Most molecules, however, contain more than one element, and therefore are known as **compounds** [*componere,* to put together]. For example, a glucose molecule ($C_6H_{12}O_6$) is composed of carbon, hydrogen, and oxygen atoms,

and so it is a compound. In contrast, oxygen gas (O_2) is a molecule but not a compound because it has only oxygen atoms.

The transfer of electrons from one atom to another or the sharing of electrons by two atoms is a critical part of forming **bonds**, the links between atoms. There are four common bond types, two strong and two weak. Covalent and ionic bonds are strong; hydrogen bonds and Van der Waals forces are weak bonds that contribute to molecular interactions.

Covalent Bonds Are Formed When Adjacent Atoms Share Electrons

Covalent bonds are strong bonds that result when two atoms share a pair of electrons, one electron from each atom. Most of the bonds that link atoms together into molecules are covalent bonds. Covalent bonds require the input of energy to break them apart.

It is possible to predict how many covalent bonds an atom can form by knowing how many unpaired electrons are in its outer shell, because an atom is most stable when all of its electrons are paired. Atoms therefore are likely to form molecules by sharing their unpaired electrons with other atoms that have unpaired electrons. Figure 2-2a ■ shows the electron configurations of two biologically important elements: hydrogen and oxygen. One atom of hydrogen, the lightest element, has one proton in its nucleus and one unpaired electron in its outer shell. Because hydrogen has only one electron to share, it always forms one covalent bond. Chemists represent the single electron by a single dot in the notation called *electron-dot shorthand*.

One atom of oxygen has two unpaired electrons in its outer shell (Fig. 2-2a). (Remember, as previously mentioned, the second electron shell can hold a maximum of eight electrons.) Oxygen therefore needs two additional electrons to complete its outer shell, and it can acquire them by sharing two electrons in covalent bonds. A water molecule (Fig. 2-2b) is a simple example of covalent bonds formed by oxygen. In a molecule of water, each of the two hydrogen atoms shares its electron with one of the unpaired electrons in the oxygen atom's outer shell. This sharing fills the outer shells of all three atoms with paired electrons and results in a stable molecule. Each shared electron pair represents a *single covalent bond*, indicated by the dash in the middle shorthand representation shown in Figure 2-2b.

CONCEPT CHECK

10. An atom of carbon has four unpaired electrons in its outer shell, which has space for eight electrons. How many covalent bonds will an atom of carbon form with other atoms? Use electron-dot shorthand to represent this atom.

11. One carbon atom will pair all its electrons by forming covalent bonds with how many hydrogen atoms? Draw this molecule using a dot or a line to represent each pair of shared electrons.

Answers: p. 49

Sometimes adjacent atoms share two pairs of electrons rather than just one pair. In this case the bond is called a

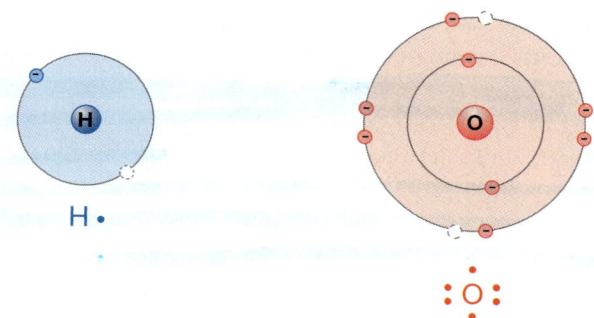

(a) Hydrogen has one proton in the nucleus and one unpaired electron. This is shown in shorthand as the symbol of the element with a dot for each electron in the atom's outer shell. **Oxygen** has eight protons and eight neutrons in the nucleus. Oxygen's outer shell has four paired electrons and two unpaired electrons.

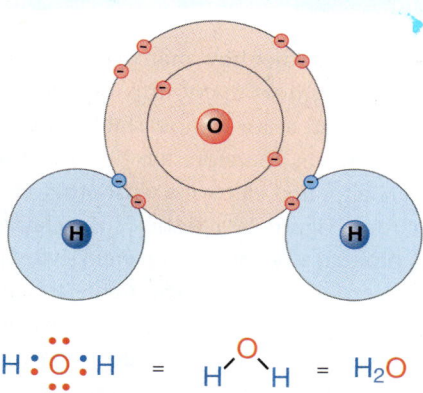

(b) A water molecule results when one oxygen atom forms covalent bonds with two hydrogen atoms. The bonds create paired electrons in the atoms' outer shells. Three different shorthand representations of the molecule are shown. In the middle shorthand, a dash replaces a pair of shared electrons. On the right is the chemical formula of water.

■ **FIGURE 2-2**   *Shared electrons in the outer shells of atoms form covalent bonds*

double bond and is represented by a double line (==). In the following example, two carbon atoms are sharing two pairs of electrons. As noted in Concept Check #10, each carbon has four unpaired electrons in its outer shell and therefore needs to share four additional electrons in covalent bonds:

$$4H\cdot\; +\; 2\;\cdot\overset{\cdot}{\underset{\cdot}{C}}\cdot \longrightarrow \quad \overset{H}{\underset{H}{}}\!\!:\!\overset{\cdot\cdot}{C}\!::\!\overset{\cdot\cdot}{C}\!\!:\overset{H}{\underset{H}{}} \quad or \quad \overset{H}{\underset{H}{}}\!\!>\!\!C\!=\!C\!<\!\overset{H}{\underset{H}{}}$$

Ethylene C_2H_4

The two adjacent carbon atoms create a double bond that shares four electrons, which leaves only two electrons per carbon to bond with hydrogen. If you count all the bonds in this drawing, you will find that each carbon forms a total of four bonds and each hydrogen forms one. The chemical formula for this molecule, called ethylene, is C_2H_4.

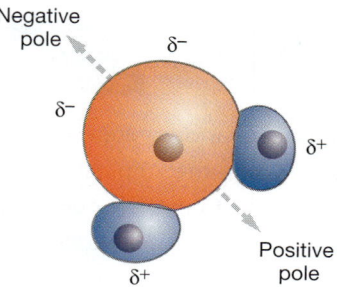

Water molecule

■ **FIGURE 2-3**   *Water is a polar molecule*

A water molecule develops separate regions, or poles, of partial positive (δ^+) and partial negative (δ^-) charge when the oxygen nucleus pulls the hydrogen electrons toward itself.

Polar and Nonpolar Molecules   Electron pairs in covalent bonds are not always evenly shared between the linked atoms. When the electrons are shared unevenly, the atom with the stronger attraction for electrons develops a slight negative charge (indicated by δ^-), and the atom with the weaker attraction for electrons develops a slight positive charge (δ^+). Molecules that develop these regions of partial positive and negative charge are called **polar molecules** because they can be said to have positive and negative poles. Atoms of certain elements, particularly nitrogen and oxygen, have a strong attraction for electrons and are often found in polar molecules.

A good example of a polar molecule is water (H_2O). The larger and stronger oxygen nucleus pulls the hydrogen electrons toward itself. This pull leaves the two hydrogen atoms of the molecule with a partial positive charge, and the single oxygen atom with a partial negative charge from the unevenly shared electrons (Fig. 2-3 ■). Note that the net charge for the entire water molecule is zero.

A **nonpolar molecule** is one whose shared electrons are distributed so evenly that there are no regions of partial positive or negative charge. For example, molecules composed mostly of carbon and hydrogen tend to be nonpolar. This is because carbon does not attract electrons as strongly as oxygen does.

The polarity of a molecule is important in determining whether the molecule will dissolve in water. Polar molecules generally dissolve easily and are said to be **hydrophilic** [*hydro-, water + -philic,* loving]. Nonpolar molecules do not dissolve well in water and are said to be **hydrophobic** [*-phobic,* hating].

Ionic Bonds Form When Atoms Gain or Lose Electrons

When an atom completely gains or loses one or more electrons, it becomes an ion. **Ionic bonds** result when an atom has such a strong attraction for electrons that it pulls one or more electrons completely away from another atom. Figure 2-4 ■ shows how sodium and chlorine atoms become ions and how

Step 1: Sodium gives up its one weakly held electron to chlorine, creating sodium and chloride ions, Na$^+$ and Cl$^-$.

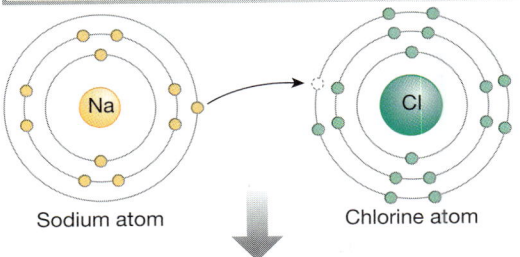

Sodium atom   Chlorine atom

Step 2: The sodium and chloride ions both have stable outer shells that are filled with electrons.

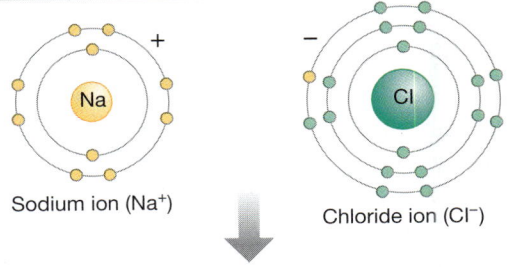

Sodium ion (Na$^+$)   Chloride ion (Cl$^-$)

Step 3: The Na$^+$ and Cl$^-$ ions are attracted to each other because of their opposite charges. In the solid state, ionic bonds between Na$^+$ and Cl$^-$ create a sodium chloride (NaCl) crystal.

Sodium ions   Chloride
(Na$^+$)   ions (Cl$^-$)

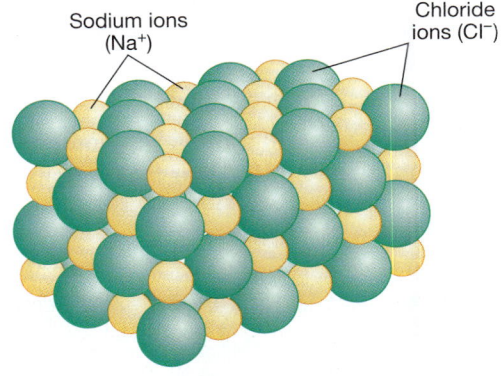

■ **FIGURE 2-4**   *Ions and ionic bonds*

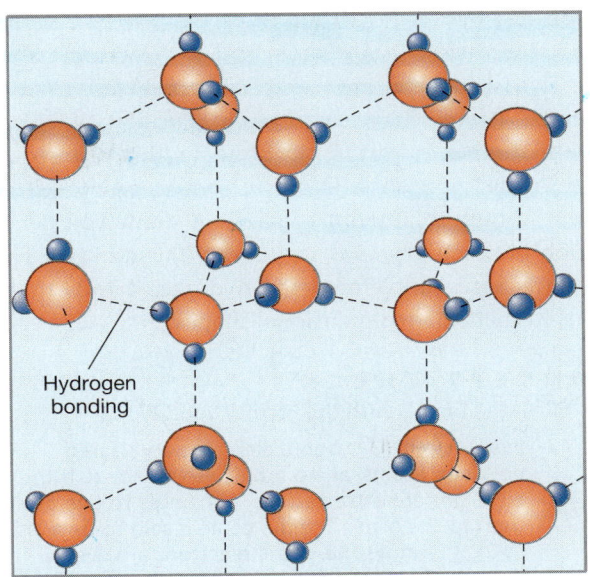

(a) The polar regions of adjacent water molecules allow them to form hydrogen bonds with one another.

Hydrogen bonding

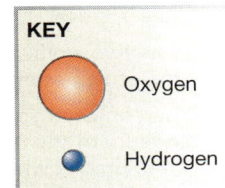

KEY

Oxygen

Hydrogen

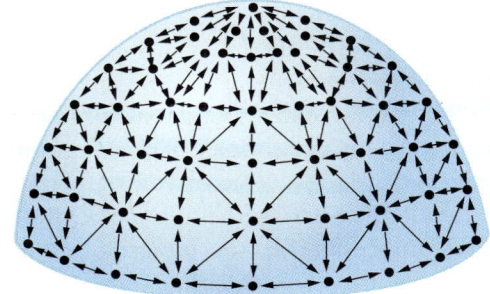

(b) Surface tension created by hydrogen bonds holds a drop of water on a flat surface in a hemispheric shape.

■ **FIGURE 2-5**   *Hydrogen bonds between water molecules*

those ions form ionic bonds. In table salt (the solid form of NaCl), the opposite charges attract, creating a neatly ordered crystal of alternating Na$^+$ and Cl$^-$ ions held together by ionic bonds. When NaCl crystals are placed in water, the ions interact with the polar regions of water molecules, which disrupts the ionic bonds and causes the crystals to dissolve (see Fig. 2-14 on p. 35).

Hydrogen Bonds and Van der Waals Forces Are Weak Interactions Between Atoms

A **hydrogen bond** is a weak attractive force between a hydrogen atom and a nearby oxygen, nitrogen, or fluorine atom. Hydrogen bonds may occur between atoms in neighboring molecules or between atoms in different parts of the same molecule. For example, one water molecule may hydrogen-bond with as many as four other water molecules. As a result, the molecules line up with their neighbors in a somewhat ordered fashion (Fig. 2-5a ■). This hydrogen bonding between molecules is responsible for the **surface tension** of water, the attractive force between water molecules that causes water to form spherical droplets when falling or to bead up when spilled onto a nonabsorbent surface (Fig. 2-5b). These weak forces of attraction also make it difficult to separate water molecules, as you may have noticed in trying to pick up a wet glass that is "stuck" to a slick table top by a thin film of water.

Van der Waals forces are weak, nonspecific attractions between the nucleus of any atom and the electrons of nearby atoms. Two atoms that are weakly attracted to each other by van der Waals forces move closer together until they are so close that their electrons begin to repel one another. Consequently, van der Waals forces allow atoms to pack closely together and occupy a minimum amount of space. A single van der Waals attraction between atoms is very weak, but multiple van der Waals attractions supplement the hydrogen bonds that hold proteins in their three-dimensional shapes.

✔ CONCEPT CHECK

Fill in the blanks with the correct bond type.

12. In a/an _covalent_ bond, electrons are shared between atoms, and a molecule is formed. If the electrons are attracted more strongly to one atom than to the other, the molecule is said to be a/an _polar_ molecule. If the electrons are evenly shared, the molecule is said to be a/an _nonpolar_ molecule.

13. A negatively charged ion is called a/an _anion_, and a positively charged ion is called a/an _cation_.

14. Name two elements whose presence contributes to a molecule becoming a polar molecule. _O & N_

15. Based on what you know from experience about the tendency of the following substances to dissolve in water, predict whether they are polar or nonpolar molecules: table sugar, vegetable oil.
_P_   _NP_
Answers: p. 49

Molecular Shape Is Related to Molecular Function

Molecular bonds—both covalent bonds and weak bonds—play a critical role in determining molecular shape. A molecule's shape, in turn, is closely related to its function. The chemical formula for a molecule, such as the formula $C_6H_{12}O_6$ for glucose, gives the number of atoms in a molecule, but doesn't tell you which atom is linked to which by covalent bonds. One form of chemical notation, shown in Figure 2-6 ■, does show which atoms are sharing electrons. In a glucose molecule (Fig. 2-6b), five carbon atoms and one oxygen join together to form a closed circle, or **ring**. The remaining atoms then attach to the atoms in the ring. In a leucine molecule (Fig. 2-6a), five carbons are linked by single covalent bonds into a straight chain, with additional atoms attached to the carbon chain.

The three-dimensional shape of a molecule is difficult to show on paper. The angles of covalent bonds between atoms may cause molecules to take on characteristic shapes. For example, the two hydrogen atoms of the water molecule shown in Figure 2-2b are attached to the oxygen with a bond angle of 104°. In addition, weak bonds between atoms create *noncovalent interactions* that influence molecular shape. The double helix of a DNA molecule (see Fig. 2-12, p. 33) is a complex shape that results from both covalent bonds between adjoining atoms, and hydrogen bonds between the two strands of the helix.

(a)

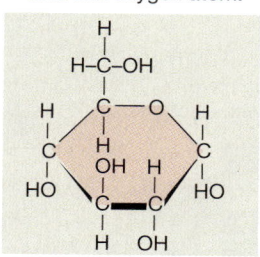

Leucine, an amino acid, is a straight chain of carbon atoms with hydrogens and other functional groups attached.

(b) Glucose, $C_6H_{12}O_6$, has a ring of five carbon atoms and one oxygen atom.

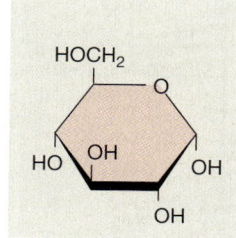

is the same as:

This glucose molecule is drawn with all of its carbons and hydrogens shown.

In the shorthand version, the carbons at each corner of the ring are omitted, as are the hydrogens attached to the carbons.

Q FIGURE QUESTIONS

- Write the chemical formula of leucine.
- What functional groups do leucine and glucose have?

■ **FIGURE 2-6** *Chemical structures and formulas of some biological molecules*

In the next section you will learn more about the characteristic properties and shapes of the four major groups of biologically important molecules.

RUNNING PROBLEM

Chromium is found in several ionic forms. The chromium usually found in biological systems and in dietary supplements is the cation Cr^{3+}. This ion is called trivalent because it has a net charge of +3. The hexavalent cation, Cr^{6+}, with a charge of +6, is used in industry, such as in the manufacturing of stainless steel and the chrome plating of metal parts.

Question 3:
How many electrons have been lost from the hexavalent ion of chromium? From the trivalent ion?

BIOMOLECULES

Most of the molecules of concern in human physiology have three elements in common: carbon, hydrogen, and oxygen. In addition, the elements phosphorus and nitrogen are also found in many biological molecules. Molecules that contain carbon are known as **organic molecules**, because it was once thought that they all existed in or were derived from plants and animals. Organic molecules associated with living organisms are also called **biomolecules**.

There are four major groups of biomolecules: carbohydrates, lipids, proteins, and nucleotides. The first three groups are used by the body for energy and as the building blocks for cellular components. The fourth group, the nucleotides, includes DNA and RNA, the structural components of genetic material; compounds that carry energy, such as adenosine triphosphate (ATP); and compounds that regulate metabolism, such as cyclic adenosine monophosphate (cAMP). Each group of biomolecules has a characteristic composition and molecular structure, and you will notice that many biomolecules are **polymers**, large molecules made up of repeating units [*poly-*, many + *-mer*, a part].

Several combinations of atoms, or **functional groups**, occur repeatedly in biological molecules. The atoms in functional groups tend to move from molecule to molecule as a single unit. For example, hydroxyl groups, —OH, common on many biological molecules, are added and removed as a group rather than as single hydrogen or oxygen atoms. Functional groups usually attach to molecules by single covalent bonds, represented as a single line such as the one on the left of the hydroxyl group above. The most common functional groups are listed in Table 2-2 ■.

TABLE 2-2 Common Functional Groups

Notice that oxygen, with two electrons to share, sometimes forms a double bond with another atom.

	SHORTHAND	BOND STRUCTURE
Carboxyl (acid)	—COOH	—C(=O)OH
Hydroxyl	—OH	—O—H
Amino	—NH$_2$	—N(H)(H)
Phosphate	—H$_2$PO$_4$	—O—P(=O)(OH)OH

Carbohydrates Are the Most Abundant Biomolecules

Carbohydrates get their name from their structure: literally, carbons [*carbo-*] with water [*hydro-*]. The general formula for a carbohydrate is (CH$_2$O)$_n$, showing that for each carbon there are two hydrogens and one oxygen, which is the same H:O ratio found in water. The *n* after the parentheses represents the number of repetitions of the CH$_2$O unit. To write the formula for a given carbohydrate, we multiply each subscript inside the parentheses by *n*: C$_n$H$_{2n}$O$_n$. For example, for glucose, C$_6$H$_{12}$O$_6$, *n* is 6.

Carbohydrates occur as simple sugars (**monosaccharides** and **disaccharides**) and as complex glucose polymers called **polysaccharides** (Fig. 2-7 ■). The most common monosaccharides are the building blocks of complex carbohydrates and have either five carbons, like **ribose**, or six carbons, like **glucose** (also known as **dextrose**). The major disaccharides are formed from the combination of glucose with another monosaccharide. You can recognize most mono- and disaccharides by the "-ose" ending on their names.

All living cells store glucose for energy in the form of a polysaccharide: **glycogen** in animals; **starch**, which humans can digest, in plants; and *dextran* in yeasts and bacteria. Some cells produce polysaccharides for structural purposes as well: **cellulose** in plants and *chitin* in invertebrate animals are two examples. It is unfortunate that humans are unable to digest cellulose and obtain its energy, because it is the most abundant organic molecule on earth.

CONCEPT CHECK

16. True or false? All organic molecules are biomolecules.
17. What is the general formula of a carbohydrate?

Answers: p. 49

Lipids Are Structurally the Most Diverse Biomolecules

Like carbohydrates, **lipids** are biomolecules made up of carbon, hydrogen, and oxygen, but as a rule lipids contain much less oxygen than carbohydrates do. An important characteristic of lipids is that they are nonpolar and therefore not very soluble in water. Technically, lipids are called fats if they are solid at room temperature and oils if they are liquid at room temperature. Most lipids derived from animal sources, such as lard and butter, are fats, whereas most plant lipids are oils. In addition to true lipids, this category includes three types of lipid-related molecules: phospholipids, steroids, and eicosanoids (Fig. 2-8 ■).

True lipids and phospholipids are most similar to each other in structure. Both groups contain a simple 3-carbon molecule known as **glycerol** and long molecules known as **fatty acids**. Phospholipids also include a phosphate group (—H$_2$PO$_4$).

Fatty acids are long chains of carbon atoms bound to hydrogens, with a carboxyl (—COOH) group at one end of the

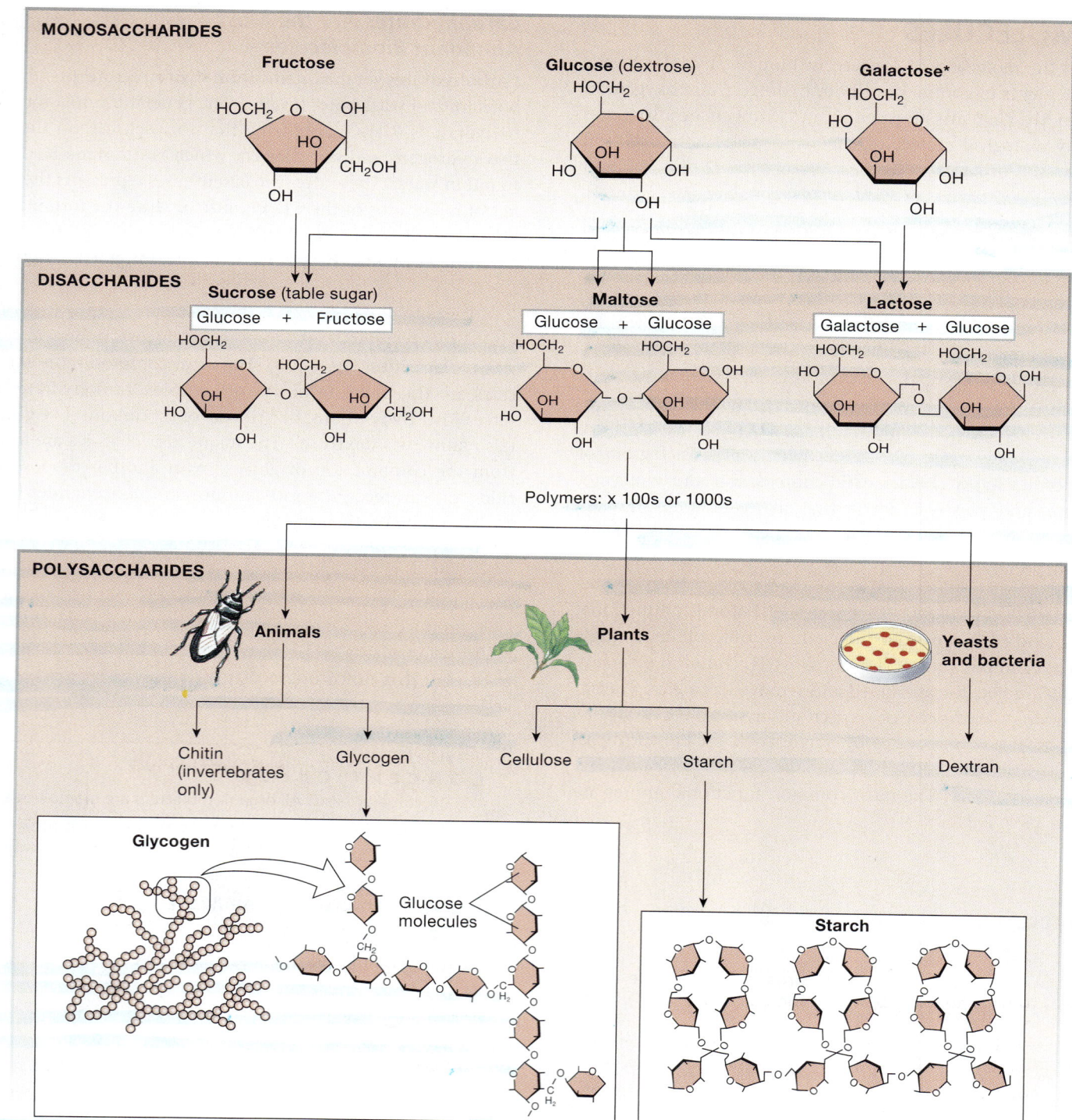

FIGURE 2-7 *Carbohydrates*

chain (Fig. 2-8). Fatty acids are described as **saturated** if there are no double bonds between carbons (palmitic acid); **mono-unsaturated** if there is one double bond in the molecule (oleic acid); or **polyunsaturated** (linolenic acid) if there are two or more double bonds in the molecule. For each double bond in a fatty acid, the molecule has two fewer hydrogen atoms attached

to the chain. The more saturated a fatty acid is, the more likely it is to be solid at room temperature.

Saturated fats, such as the fat that marbles a piece of choice beef, are generally considered to contribute to the development of artery-clogging fatty deposits. For a time, it was thought that the most unsaturated fats were least likely to be harmful, but

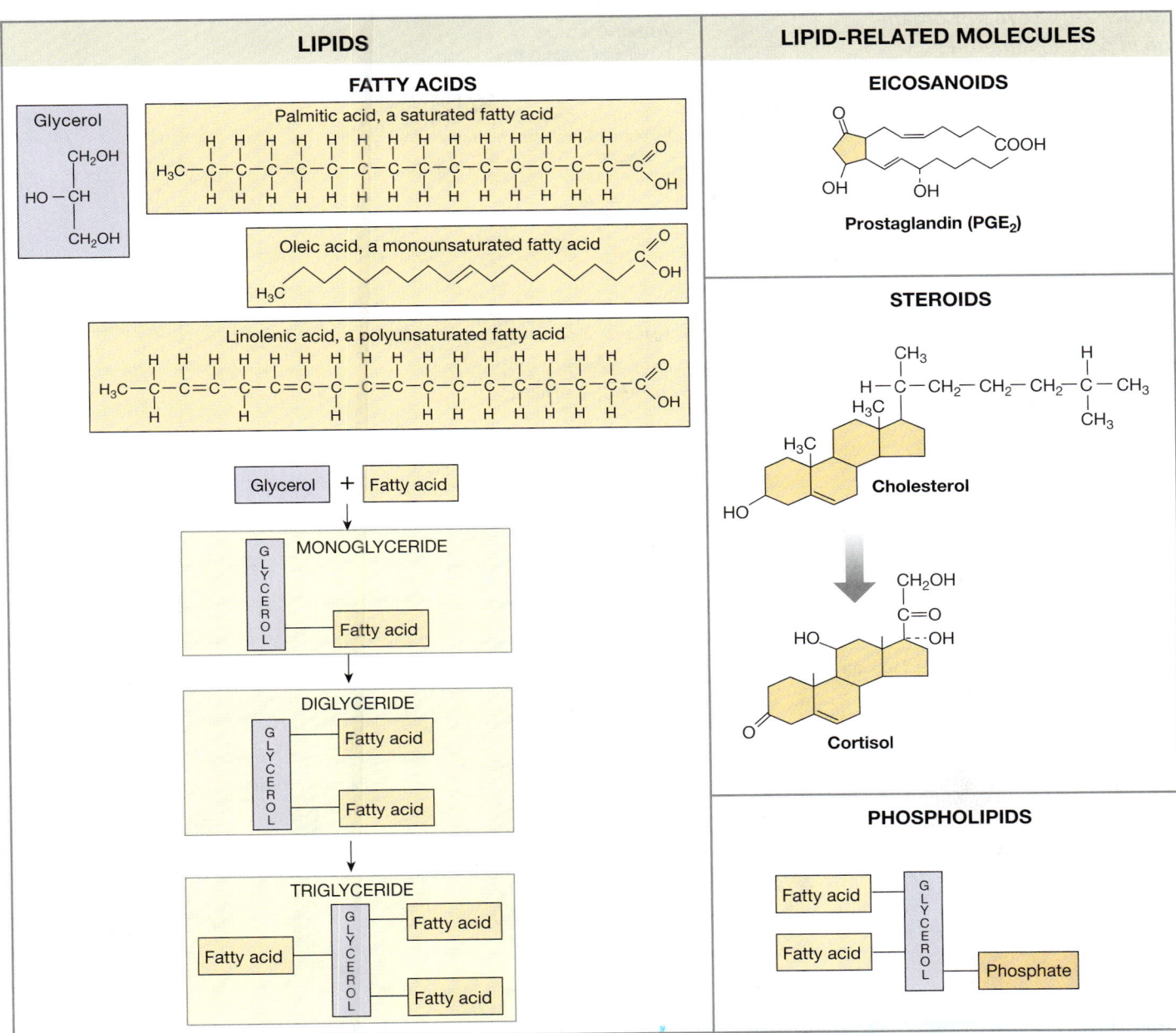

FIGURE 2-8 *Lipids and lipid-related molecules*

Palmitic and linolenic acids are shown with all bonds and atoms. Oleic acid, another fatty acid, is shown in a shorthand format that eliminates symbols for carbons and hydrogens in the middle of the molecule.

recent studies suggest that monounsaturated fats, such as olive oil, may be the most healthful. For years manufacturers of margarine and cooking oil liked to boast that their products were "high in polyunsaturates!" Now the new advertising buzzwords are "no trans fats!" *Trans fats* are unsaturated vegetable oils to which manufacturers attach hydrogens to make them more solid at room temperature. For many years it was assumed that these *hydrogenated* vegetable oils in margarine and vegetable shortening were more healthful than the saturated animal fats in butter and lard, but recent research indicates that

trans fats are just as likely to contribute to heart disease. By 2006 food manufacturers in the United States will be required to list the trans fat content of their products.

Glycerol links to one, two, or three fatty acids to form mono-, di-, or triglycerides. **Triglycerides** (more correctly known as **triacylglycerols**) are the most important form of lipid in the body; more than 90% of our lipids are in this form. Triglyceride levels in the blood are also important as predictors of artery disease; an elevated fasting triglyceride level is linked to a higher risk of disease.

FIGURE 2-9 *Levels of organization in protein molecules*

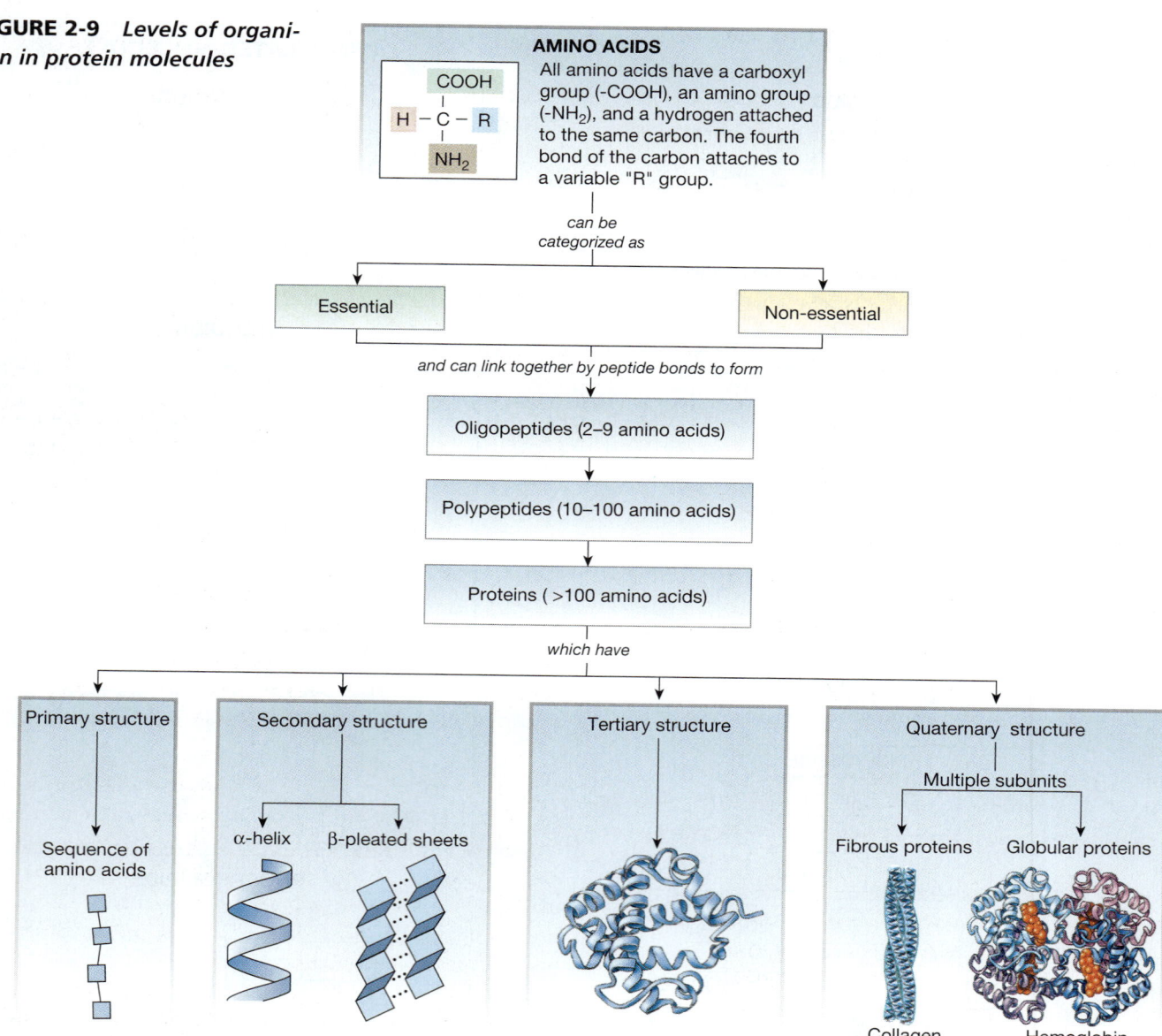

AMINO ACIDS
All amino acids have a carboxyl group (-COOH), an amino group (-NH$_2$), and a hydrogen attached to the same carbon. The fourth bond of the carbon attaches to a variable "R" group.

can be categorized as

Essential   Non-essential

and can link together by peptide bonds to form

Oligopeptides (2–9 amino acids)

Polypeptides (10–100 amino acids)

Proteins (>100 amino acids)

which have

Primary structure — Sequence of amino acids

Secondary structure — α-helix — β-pleated sheets

Tertiary structure

Quaternary structure — Multiple subunits — Fibrous proteins / Globular proteins

Collagen   Hemoglobin

Steroids are lipid-related molecules whose structure includes four linked carbon rings (Fig. 2-8). Cholesterol is the source of steroids in the human body and is the basis for a number of important hormones. It is also an important component of animal cell membranes.

The **eicosanoids** are modified 20-carbon fatty acids that are found in animals [*eikosi,* twenty]. These molecules all contain one open five- or six-carbon ring that makes the long carbon chain appear to double back on itself (Fig. 2-8). The main eicosanoids are thromboxanes, leukotrienes, and prostaglandins. As you will learn in later chapters, eicosanoids are regulators of various physiological functions.

CONCEPT CHECK

18. Use the chemical formulas given to decide which of the following fatty acids is most unsaturated:
    (a) $C_{18}H_{36}O_2$ (b) $C_{18}H_{34}O_2$ (c) $C_{18}H_{30}O_2$ Answers: p. 49

fewest H

Proteins Are the Most Versatile Biomolecules

Proteins are polymers of smaller building-block molecules called **amino acids** (Fig. 2-9 ■). Twenty different amino acids commonly occur in natural proteins, and the human body can synthesize all but nine of them. The nine that must be obtained from dietary proteins are called **essential amino acids**. A list of amino acids is in Appendix B.

There are also a few amino acids that do not occur in proteins but have important physiological functions. They include *homocysteine,* a sulfur-containing amino acid that occurs normally in the body but which in excess is associated with heart disease; γ-*amino butyric acid* (gamma-amino butyric acid, or *GABA*), a chemical made by nerve cells; and *creatine,* a molecule that stores energy when it binds to a phosphate group.

All amino acids have a similar core structure: a central carbon atom is linked to a hydrogen atom, a nitrogen-containing amino group (—NH$_2$), a carboxyl group (—COOH), and a group of atoms designated "R" that is different in each amino acid (Fig. 2-9). The nitrogen in the amino group makes proteins our major dietary source of nitrogen. The R groups are quite variable, differing in their size, shape, and ability to form hydrogen bonds or ions. Because of the different R groups, each amino acid reacts with other molecules in a unique way.

The twenty protein-forming amino acids assemble into polymers with an almost infinite number of combinations, just as the twenty-six letters of our alphabet combine to create different words. When two amino acids link together, the amino group of one is joined to the carboxyl group of the other, forming a **peptide bond**. The general name for any amino acid polymer is **peptide**, which refers to a polymer of any length, from two units to two million units. To introduce a degree of specificity, however, various modified terms are used. If a peptide consists of two to nine amino acids linked via peptide bonds, the molecule is called an **oligopeptide** [*oligo-*, few]. A chain of 10 to 100 amino acids is called a **polypeptide**. A chain of more than 100 amino acids is called a **protein**.

The sequence of amino acids in a peptide or protein chain is called the **primary structure** (Fig. 2-9). For example, the primary structure of thyrotropin-releasing hormone, a small peptide made in the brain, is the amino acid sequence glutamic acid-histidine-proline. The primary structure of a protein is genetically determined and is essential to proper function. For example, a substitution of only one amino acid causes sickle cell disease, a condition in which red blood cells cannot function normally.

As a polypeptide chain forms, it takes on a **secondary structure**, which is its spatial arrangement or shape (Fig. 2-9). The secondary structure is stabilized by hydrogen bonding between different parts of the molecule. The three most common shapes for polypeptide chains are a spiral called the **α-helix**, a **β-strand** whose bond angles create a zigzag shape rather than a spiral, and U-shaped **β-*turns***. In a polypeptide, β-strands often assemble into side-by-side *pleated sheets*. Proteins that are destined for structural uses may be composed almost entirely of pleated sheets because this configuration is very stable. Other proteins combine sheets with helices and turns.

The three-dimensional shape of a protein is its **tertiary structure**. Proteins are categorized into two large groups: fibrous and globular. **Fibrous proteins** are found as pleated sheets or in long chains of helices. The fibrous proteins are insoluble in water and form important structural components of cells and tissues. Examples include *collagen,* a fibrous protein found in many types of connective tissue, and *keratin,* a fibrous protein found in hair and nails.

Globular proteins have amino acid chains that fold back on themselves to create a complex tertiary structure containing pockets, channels, or protruding knobs. The tertiary structure of globular proteins arises partly from the angles of covalent bonds between amino acids, and partly from hydrogen bonds,

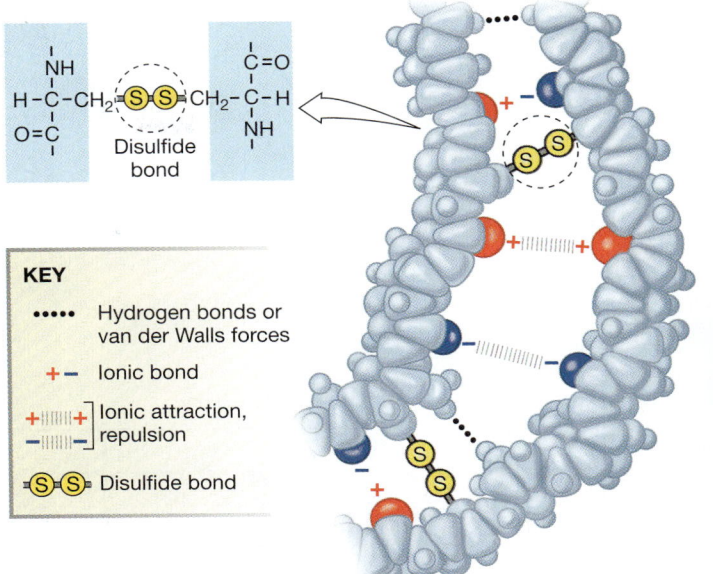

■ FIGURE 2-10 *Noncovalent bonds hold molecules or parts of molecules together*

Hydrogen bonds, disulfide bonds, van der Waals forces, and ionic bonds help create the tertiary structure of proteins. (Figure adapted from *Molecular Cell Biology 5/e* by H. Lodish *et al.*)

van der Waals forces, and ionic bonds that stabilize the tertiary structure (Fig. 2-10 ■). The amino acid cysteine also plays an important role in globular protein shape. Cysteine contains sulfur as part of a *sulfhydryl group* (—SH) (Fig. 2-10). Two cysteines in different parts of the polypeptide chain can bond covalently to each other in a **disulfide (S—S) bond**, pulling different sections of the chain together.

The globular proteins are soluble in water. They act as *carriers* for water-insoluble lipids in the blood, binding to the lipids and making them soluble. They also serve as *enzymes* that increase the rate of chemical reactions. Soluble proteins also act as cell-to-cell messengers in the form of hormones and neurotransmitters and as defense molecules to help fight foreign invaders.

If several protein chains associate with one another to form a functional protein, the protein is said to have **quaternary structure** (Fig. 2-9). Hemoglobin, an oxygen-carrying pigment in red blood cells, is the classic example of a protein with quaternary structure. One hemoglobin molecule has four subunits, making it a *tetramer.*

CONCEPT CHECK

19. What is the chemical formula of an amino group? Of a carboxyl group?
20. Why must we include essential amino acids in our diet but not the other amino acids?
21. What aspect of protein structure allows proteins to have more versatility than lipids or carbohydrates?

Answers: p. 49

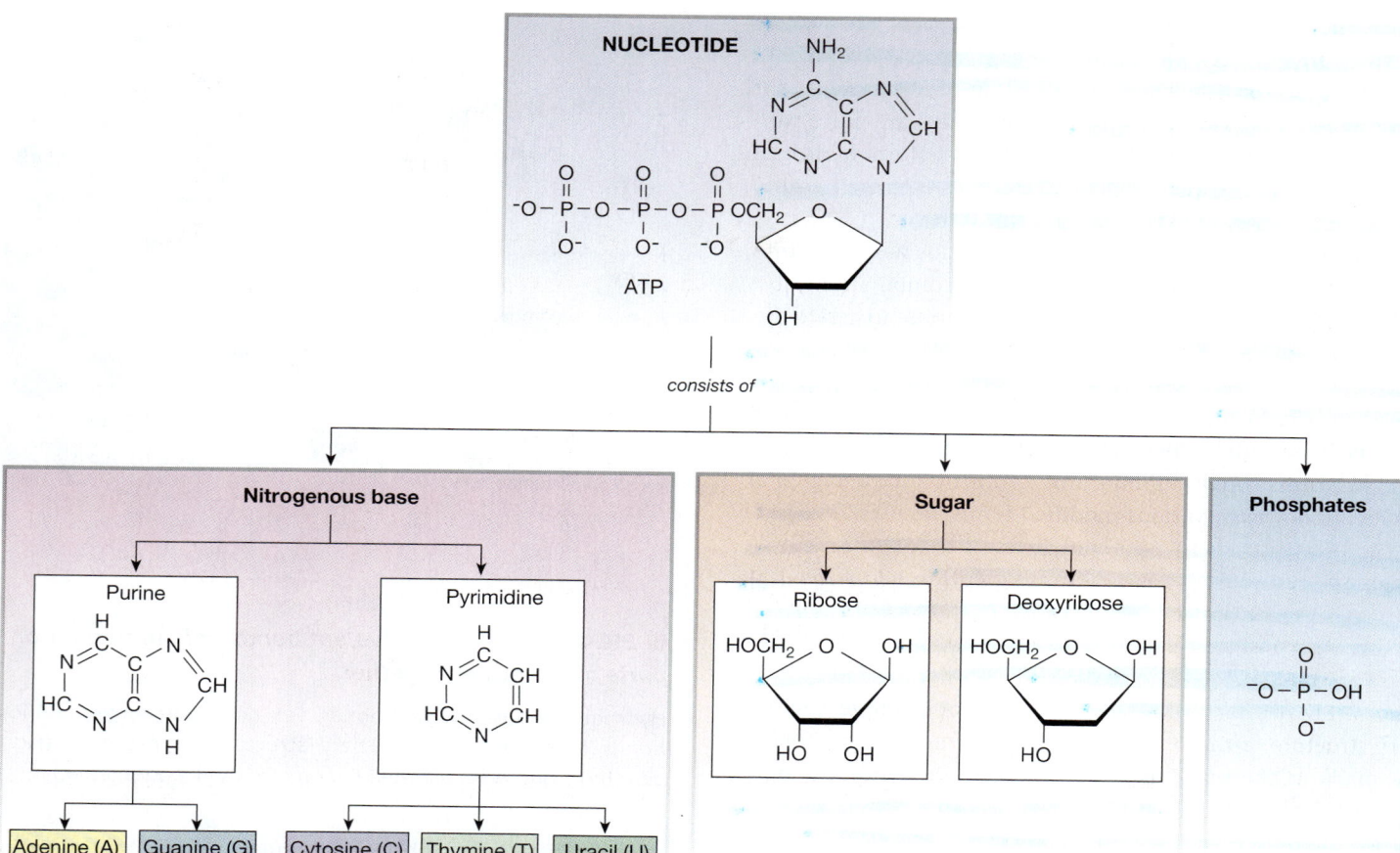

FIGURE 2-11   *Nucleotides and nucleic acids*

Some Molecules Combine Carbohydrates, Proteins, and Lipids

Not all biomolecules are pure protein, pure carbohydrate, or pure lipid. **Conjugated proteins** are protein molecules combined with another kind of biomolecule. Proteins combine with lipids to form **lipoproteins**. Lipoproteins are found in cell membranes, and they transport hydrophobic molecules, such as cholesterol, in the blood.

A *glycosylated* molecule is one to which a carbohydrate has been attached. Proteins combine with carbohydrates to form **glycoproteins**, and lipids bind to carbohydrates to become **glycolipids**. Glycoproteins and glycolipids, like

lipoproteins, are important components of cell membranes (see Chapter 3).

Nucleotides Transmit and Store Energy and Information

Nucleotides play an important role in many of the processes of the cell, including the storage and transmission of genetic information and energy. A **single nucleotide** is a three-part molecule composed of (1) one or more phosphate groups, (2) a five-carbon sugar, and (3) a carbon-nitrogen ring structure called a **nitrogenous base** (Fig. 2-11 ■). Every nucleotide contains one of two possible sugars: either the five-carbon sugar ribose or

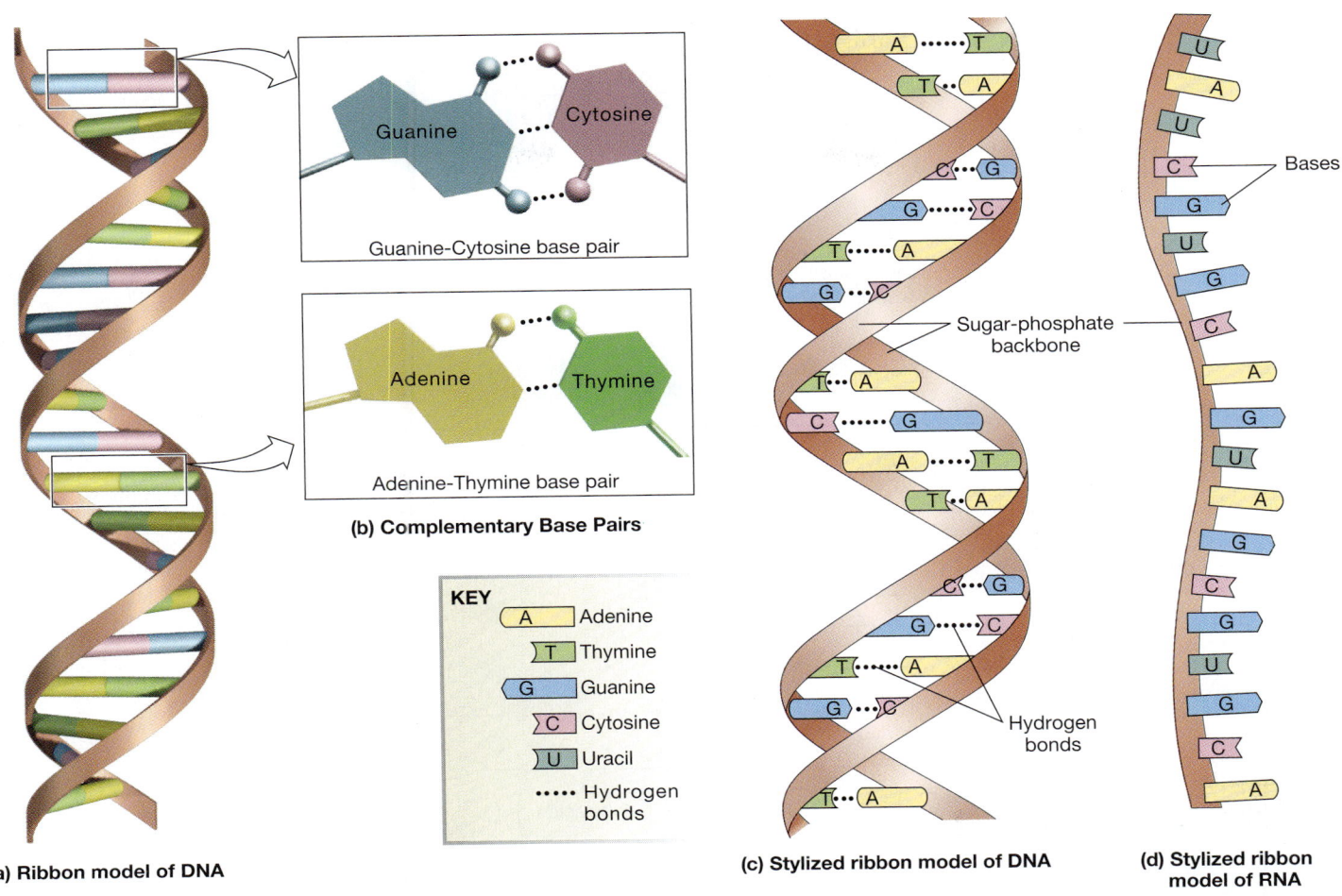

KEY

A	Adenine
T	Thymine
G	Guanine
C	Cytosine
U	Uracil
•••••	Hydrogen bonds

(a) Ribbon model of DNA

(b) Complementary Base Pairs

(c) Stylized ribbon model of DNA

(d) Stylized ribbon model of RNA

■ **FIGURE 2-12** *RNA and DNA*

deoxyribose [*de*-, without; -*oxy*-, oxygen], which is a ribose molecule minus one oxygen atom. There are also two types of nitrogenous bases: the purines and the pyrimidines. The **purines** have a double ring structure; the **pyrimidines** have a single ring. There are two purines, **adenine** and **guanine**, and three pyrimidines, **cytosine**, **thymine**, and **uracil**.

The smallest nucleotides include the energy-transferring compounds **ATP** (adenosine triphosphate) and **ADP** (adenosine diphosphate) (Fig. 2-11). A related nucleotide is **cyclic AMP** (cyclic adenosine monophosphate, or **cAMP**), a molecule important in the transfer of signals between the extracellular fluid and the cell. Other energy-transferring nucleotides are **NAD** (nicotinamide adenine dinucleotide) and **FAD** (flavin adenine dinucleotide).

DNA (deoxyribonucleic acid) and **RNA** (ribonucleic acid) are nucleotide polymers, or **nucleic acids**. They store genetic information within the cell and transmit it to future generations of cells. Both polymers are formed by linking nucleotides into long chains, similar to the long amino acid chains of proteins. The sugar of one nucleotide links to the phosphate of the next, creating a chain, or backbone, of alternating sugar-phosphate-sugar-phosphate groups. The nitrogenous bases extend to the side of the chain (Fig. 2-12 ■). Four nitrogenous bases—adenine, guanine, cytosine, and thymine—are found in DNA molecules. In RNA, the nitrogenous base *uracil* replaces the *thymine* found in DNA.

The RNA molecule usually takes the form of a single chain, although it temporarily binds to DNA while new RNA is being made. The DNA molecule is a *double helix*, a three-dimensional structure that forms when two DNA chains link through hydrogen bonds between complementary base pairs. Notice in Figure 2-12 that the purine bases are larger than the pyrimidine bases. Because of space limitations, base pairings in DNA are always one purine with one pyrimidine. A further restriction is that adenine (A) pairs only with thymine (T), and cytosine (C) pairs only with guanine (G). These base pairs (A-T and G-C) hold the double strand of DNA in its spiral shape. Base pairing and DNA replication are reviewed in Appendix B.

Figure 2-13 ■ summarizes the chemistry and biochemistry you will encounter as you continue your study of physiology.

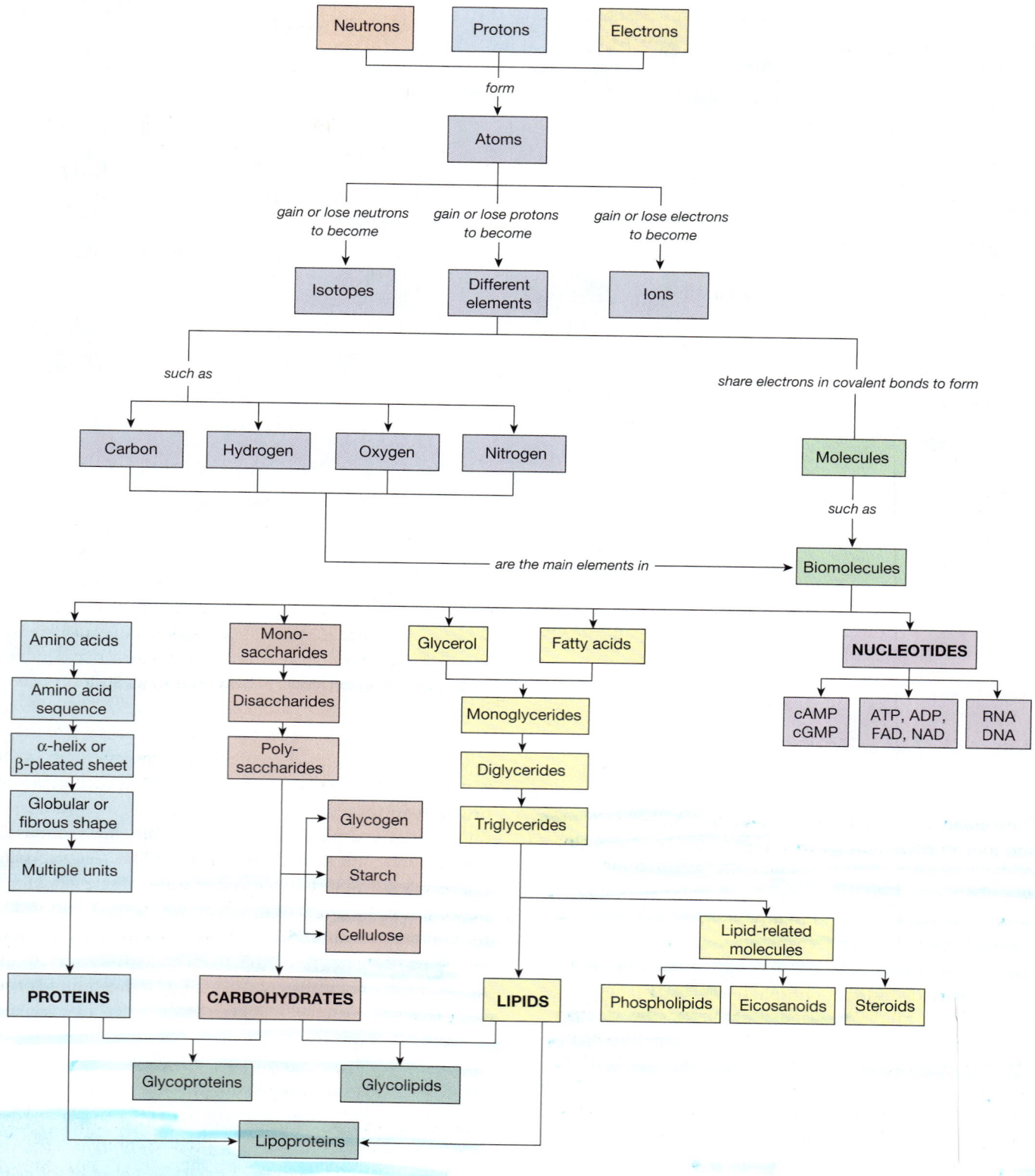

■ **FIGURE 2-13** *Chemistry summary*

The hexavalent form of chromium used in industry is known to be toxic to humans. In 1992, officials at California's Hazard Evaluation System and Information Service warned that inhaling chromium dust, mist, or fumes placed chrome and stainless steel workers at increased risk for lung cancer. Officials found no risk to the public from normal contact with chrome surfaces or stainless steel. The first link between the biological trivalent form of chromium (Cr^{3+}) and cancer came from *in vitro* studies [*vitrum,* glass; that is, a test tube] in which mammalian cells were kept alive in tissue culture. In these experiments, cells exposed to moderately high levels of chromium picolinate developed "potentially cancerous changes," possibly from free radicals formed inside the cells.

Question 4:
From this information, can you conclude that hexavalent and trivalent chromium are equally toxic?

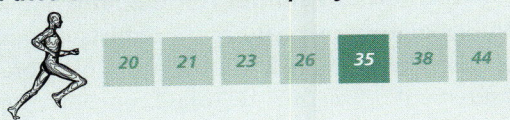

| 20 | 21 | 23 | 26 | **35** | 38 | 44 |

AQUEOUS SOLUTIONS, ACIDS, BASES, AND BUFFERS

Life as we know it is established on water-based, or *aqueous,* solutions that resemble dilute seawater in their ionic composition. The human body is about 60% water. Na^+, K^+, and Cl^- are the main ions in body fluids; other ions make up a lesser proportion. All molecules and cell components are either dissolved or suspended in these saline solutions. It is useful, therefore, to understand the properties of solutions.

Not All Molecules Dissolve in Aqueous Solutions

Substances that dissolve in liquid are collectively known as **solutes**. The liquids into which they dissolve are known as **solvents**. The combination of solutes dissolved in solvent is called a **solution**. In biological solutions, water is the universal solvent.

The degree to which a molecule is able to dissolve in a solvent is the molecule's **solubility**: the more easily it dissolves, the higher its solubility. Molecules that dissolve readily in water are said to be *hydrophilic.* Most are polar or ionic molecules, whose positive and negative regions readily interact with polar water molecules (Fig. 2-14 ■). Molecules that do not dissolve readily in water are said to be *hydrophobic.* They tend to be nonpolar molecules that are unable to make hydrogen bonds with water molecules.

The lipids (fats and oils) are the most hydrophobic group of biological molecules. They cannot interact with water molecules

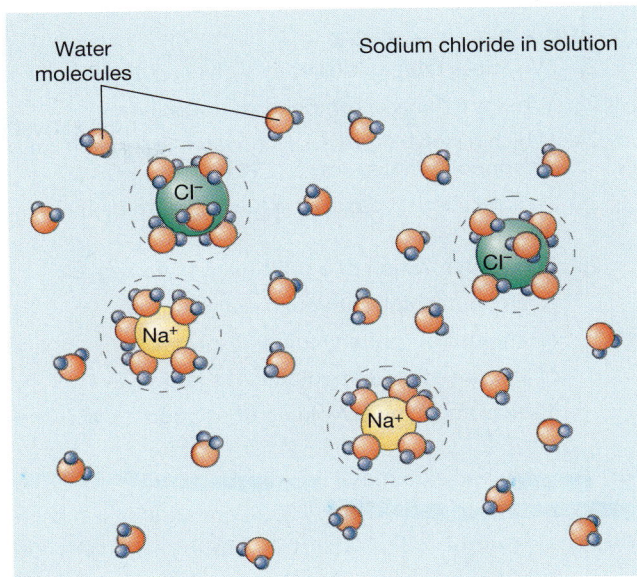

Water molecules   Sodium chloride in solution

■ FIGURE 2-14 *Sodium chloride dissolves in water*
Interaction with polar water molecules breaks ionic bonds that held the NaCl crystal together.

when placed in an aqueous solution and tend to form separate layers. One familiar example is salad oil floating on vinegar in a bottle of salad dressing. For hydrophobic molecules to dissolve in body fluids, they must be combined with a hydrophilic molecule that can effectively carry them into solution. For example, cholesterol, a common animal fat, is a hydrophobic molecule. Fat from a piece of meat dropped into a glass of warm water will float to the top, undissolved. Similarly, cholesterol in the blood will not dissolve unless it is combined with special water-soluble carrier molecules.

In addition to its role as a medium for dissolved substances, water has other properties that affect the way our bodies work. The surface tension of water, discussed on p. 25, is important in lung function, as you will see in Chapter 17. Water also has the ability to capture and store large amounts of heat. The heat-retaining property of water is significant in heat loss from the body through sweating. Temperature control in the body is discussed in Chapter 22.

There Are Several Ways to Express the Concentration of a Solution

The **concentration** of a solution is the amount of solute per unit volume of solution. There are many different ways to express concentration in physiology. The amount of solute can be described as the mass of the solute before it dissolves or as the number of molecules or ions of solute in the solution. Mass is usually given in grams (g) or milligrams (mg). Numbers of solute molecules are given in moles (mol), and numbers of solute ions are given in equivalents (eq). Volume is usually expressed as liters (L) or milliliters (mL).

22. Why does table salt (NaCl) dissolve in water?

23. What are the two components of a solution? *solute & solvent*

24. Which dissolve more easily in water, polar molecules or nonpolar molecules? *Polar*

25. A molecule that dissolves easily is said to be hydro____ *phill* ic.

26. The concentration of a solution is expressed as:

   a. amount of solvent/volume of solute

   b. amount of solute/volume of solvent

   c. amount of solvent/volume of solution

   d. amount of solute/volume of solution   Answers: p. 49

Moles

A mole [*moles*, a mass] is defined as 6.02×10^{23} atoms, ions, or molecules of a substance. Thus, one mole of a substance has the same number of particles as one mole of any other substance, just as a dozen eggs has the same number of items as a dozen roses. However, the mass of 1 mole of a substance depends on the mass of its individual particles.

Molecular Mass

From the chemical formula of a molecule, we can calculate its **molecular mass**. This is the mass of one molecule, expressed in atomic mass units or, more often, in daltons (where, as noted earlier in the chapter, 1 amu = 1 Da). To calculate molecular mass, multiply the atomic mass of each element in the molecule by the number of atoms of the element in the molecule, then add the results. For example, the molecular mass of glucose, $C_6H_{12}O_6$, would be calculated:

ELEMENT	NUMBER OF ATOMS	×	ATOMIC MASS OF THE ELEMENT	
Carbon	6		12.0 amu	= 72.0
Hydrogen	12		1.0 amu	= 12.0
Oxygen	6		16.0 amu	= 96.0
			SUM	180.0 amu
		Molecular Mass of Glucose	= 180 Da	

The mass of one mole of a substance is equal to the atomic or molecular mass of the substance expressed in grams. This value is called the **gram molecular mass**. For example, the molecular mass of glucose is 180 daltons. Therefore, one mole of glucose (in other words, 6.02×10^{23} glucose molecules) has a mass of 180 grams. One mole of NaCl, with the same number of molecules, has a mass of only 58.5 grams.

Molarity

Solution concentrations in biology are frequently given as **molarity**, the number of moles of solute in a liter of solution. Molarity is abbreviated as either mol/L or M. A one molar solution of glucose (1 mol/L, 1 M) contains 6.02×10^{23} mol-

ecules of glucose per liter. It is made by dissolving one mole (180 grams) of glucose in enough water to make one liter of solution.

Typical biological solutions are so dilute that solute concentrations are usually expressed as **millimoles** per liter (mmol/L or mM). A millimole is 1/1000 of a mole [*milli-*, 1/1000].

Equivalents

Concentrations of ions are sometimes expressed in equivalents per liter instead of moles per liter. An **equivalent** (eq) is equal to the molarity of an ion times the number of charges the ion carries. The sodium ion, with its charge of +1, has one equivalent per mole. The calcium ion, Ca^{2+}, has two equivalents per mole. The same rule is true for negatively charged ions. The chloride ion (Cl^-) has one equivalent per mole, and the hydrogen phosphate ion (HPO_4^{2-}) has two equivalents per mole.

A **milliequivalent** (meq) is 1/1000 of an equivalent. Milliequivalents are useful for expressing the concentration of dilute biological solutions. Many clinical measurements, such as the concentration of sodium ions in the blood, are reported in milliequivalents per liter (meq/L).

Weight/Volume, Volume/Volume, and Percent Solutions

In the laboratory or pharmacy, scientists cannot measure out solutes by the mole. Instead, they use the more conventional measurement of weight. The solute concentration in a solution prepared in this way may then be expressed as a percentage of the total solution, or **percent solution**. A 10% solution means 10 parts of solute per 100 parts *of total solution*.

If the solute is normally a solid at room temperature, the percent solution is expressed as *weight* of solute per *volume* of solution. A 5% glucose solution has 5 grams of glucose dissolved in water to make a final volume of 100 mL of solution. If the solute is a liquid at room temperature, such as concentrated hydrochloric acid, the solution is made using *volume/volume* measurements. To make 0.1% HCl, for instance, add 0.1 mL of concentrated acid to enough water to give a final volume of 100 mL.

Another weight/volume convention is common in medicine. The concentrations of drugs and other chemicals in the body are often expressed as milligrams of solute per deciliter of solution (mg/dL). A **deciliter** (dL) is 1/10 of a liter, or 100 mL. An out-of-date way of expressing *mg/dL* is mg%, where % means per 100 parts or 100 mL. A concentration of 20 mg/dL could also be expressed as 20 mg%.

27. Calculate the molecular mass of water, H_2O.

28. How much does a mole of KCl weigh?

29. Which solution is more concentrated: a 100 mM solution of glucose or a 0.1 M solution of glucose?

30. In making a 5% solution of glucose, why wouldn't you measure out 5 grams of glucose and add it to 100 mL of water?   Answers: p. 49

The Concentration of Hydrogen Ions in the Body Is Expressed in pH Units

One of the important ionic solutes in the body is the hydrogen ion, H^+. The concentration of H^+ in body fluids determines the body's *acidity*, a physiological parameter that is closely regulated because H^+ in solution can interfere with hydrogen bonding and van der Waals forces. These bonds are responsible for the shapes of many important molecules, and when the bonds are disrupted, the shape of a molecule can change, thereby destroying the molecule's ability to function.

Where do hydrogen ions in body fluids come from? Some of them come from the separation of water molecules (H_2O) into H^+ and OH^- ions. Others come from molecules that release H^+ when they dissolve in water. A molecule that contributes H^+ to a solution is called an **acid**. For example, carbonic acid is an acid made in the body from CO_2 (carbon dioxide) and water. In solution, carbonic acid separates into a bicarbonate ion and a hydrogen ion:

$$CO_2 + H_2O \rightleftharpoons H_2CO_3 \text{ (carbonic acid)} \rightleftharpoons H^+ + HCO_3^-$$

Note that when the H^+ is part of the intact carbonic acid molecule, it does not contribute to acidity. *It is only free H^+ that affects H^+ concentration and therefore acidity.*

Some molecules decrease the H^+ concentration of a solution by combining with free H^+. These molecules are called **bases**. Molecules that produce hydroxide ions, OH^-, in solution are bases because the hydroxide combines with H^+ to form water:

$$H^+ + OH^- \rightleftharpoons H_2O$$

Another molecule that acts as a base is ammonia, NH_3. It reacts with a free H^+ to form an ammonium ion:

$$NH_3 + H^+ \rightleftharpoons NH_4^+$$

The concentration of H^+ in body fluids is measured in terms of **pH**. The expression pH stands for "power of hydrogen" and is calculated from the negative logarithm of the hydrogen ion concentration: $pH = -\log [H^+]$ [Appendix B for logarithms]. The square brackets around the symbol for a solute are the shorthand notation for "concentration": $[H^+] =$ hydrogen ion concentration.

Another way to write the pH equation is $pH = \log (1/[H^+])$. This equation states that pH is inversely related to H^+ concentration. In other words, as the H^+ concentration goes up, the pH goes down. The pH of a solution is measured on a numeric scale between 0 and 14. Pure water has a pH value of 7.0, meaning its H^+ concentration is 1×10^{-7} M. Water is considered neutral. Solutions that have gained H^+ from an acid have H^+ concentrations higher than the 1×10^{-7} M concentration found in water. Because $[H^+]$ is inversely proportional to pH, **acidic** solutions have a pH less than 7. Solutions that have lost H^+ to a base have a H^+ concentration lower than that of pure water; such solutions are called either **basic** or **alkaline**. They

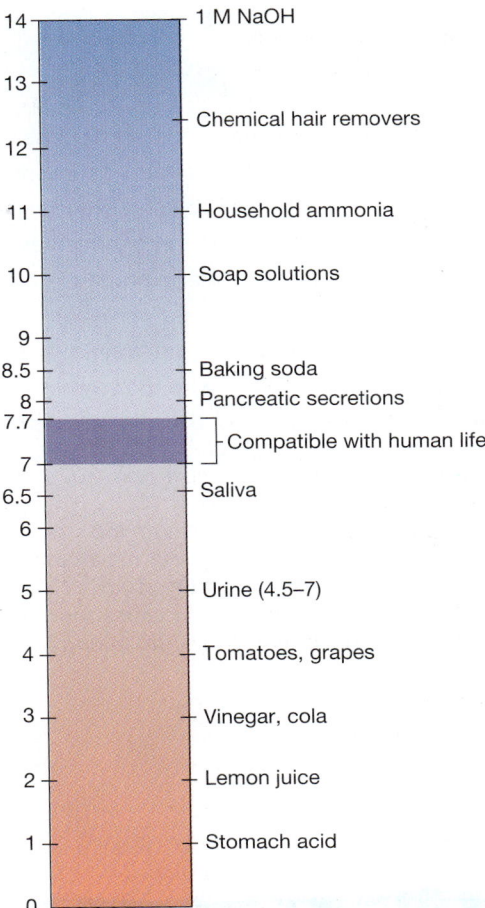

FIGURE QUESTIONS

- When the body becomes more acidic, does the pH increase or decrease?
- How can urine, stomach acid, and saliva have pH values outside the pH range that is compatible with life and yet be part of the living body?

■ **FIGURE 2-15**   *pH scale*

This logarithmic scale is a measure of the hydrogen ion concentration, $[H^+]$. For example, at a pH of 14, $[H^+] = 10^{-14}$ moles/liter.

have a pH value greater than 7. Remember, because pH is proportional to the reciprocal of the $[H^+]$, as the hydrogen ion concentration *increases*, the pH number *decreases*.

The pH scale is logarithmic, meaning that a change in pH value of 1 unit indicates a 10-fold increase or decrease in the hydrogen ion concentration. For example, if the pH of a solution is altered from a pH of 8 to a pH of 6, there has been a 100-fold (10^2 or 10×10) increase in the H^+ concentration.

Figure 2-15 ■ shows a pH scale and the pH values of various solutions. The normal pH of blood in the human body is 7.40, slightly alkaline. Regulation of the body's pH within a narrow range is critical because a blood pH more acidic than

RUNNING PROBLEM

Stan has been taking chromium picolinate because he heard that it would increase his strength and muscle mass. Then a friend told him that the Food and Drug Administration (FDA) said there was no evidence to show that chromium would help build muscle. In one study, a group of researchers gave high daily doses of chromium picolinate to football players during a two-month training period. By the end of the study, the players who took chromium supplements had not increased muscle mass or strength any more than players who did not take the supplement.

Question 5:

Based on the FDA announcement, this study (which did not support enhanced muscle development from chromium supplements), and the studies that suggest that chromium picolinate might cause cancer, do you think that Stan should continue taking chromium picolinate?

| 20 | 21 | 23 | 26 | 35 | **38** | 44 |

7.00 (pH < 7.00) or more alkaline than 7.70 (pH > 7.70) is incompatible with life.

Buffers are a key factor in the body's ability to maintain normal pH. A **buffer** is any substance that moderates changes in pH. Many buffers contain anions that have a strong affinity for H^+ molecules. When free H^+ is added to a buffer solution, these anions bond to the H^+, thereby minimizing any change in pH.

The bicarbonate anion, HCO_3^-, is an important buffer in the human body. The following equation shows how a sodium bicarbonate solution acts as a buffer when hydrochloric acid (HCl) is added. When placed in plain water, HCl dissociates into H^+ and Cl^- and creates a high H^+ concentration (low pH). When HCl dissociates in a sodium bicarbonate solution, however, some of the bicarbonate ions combine with some of the H^+ to form undissociated carbonic acid. "Tying up" the added H^+ in this way keeps the H^+ concentration of the solution from changing significantly, which minimizes the pH change.

$$H^+ + Cl^- + HCO_3^- + Na^+ \rightleftharpoons H_2CO_3 + Cl^- + Na^+$$

| Hydrochloric acid | Sodium bicarbonate | | Carbonic acid | Sodium chloride (table salt) |

CONCEPT CHECK

31. To be classified as an acid, a molecule must do what when dissolved in water?

32. pH is an expression of the concentration of what in a solution?

Answers: p. 49

PROTEIN INTERACTIONS

The history of the Human Genome Project (HGP; see Chapter 1) illustrates how our thinking about the role of proteins in biology has evolved. In 1990, when the HGP started, DNA was the molecule that would unlock all the secrets of life. As work progressed, however, scientists discovered that one nucleotide sequence in a strand of DNA could direct production of many different proteins. Consequently, the focus of contemporary research has shifted to the protein equivalent of the genome, the **proteome**. The Human Proteomics Initiative is an ambitious undertaking to catalog the structure and function of all proteins in the body (*http://us.expasy.org/sprot/hpi/*). One assumption of this project is that if scientists learn what proteins are in the body, they will be able to create models to predict the functions and interactions of those proteins.

As you learned earlier in this chapter, proteins are the "workhorses" of the body. Those that are not soluble in water are important structural components and will be discussed in more detail in the next chapter. Most soluble proteins fall into one of seven broad categories:

1. **Enzymes.** Some proteins act as enzymes to speed up chemical reactions. Enzymes play an important role in metabolism, the subject of Chapter 4.

2. **Membrane transporters.** Proteins in cell membranes help move substances back and forth between the intracellular and extracellular compartments. These proteins may form channels in the cell membrane, or they may bind to molecules and carry them through the membrane. Membrane transporters are discussed in detail in Chapter 5.

3. **Signal molecules.** Some proteins and smaller peptides act as hormones and other signal molecules. They are described in Chapter 6.

4. **Receptors.** Proteins that bind signal molecules and initiate cellular responses are called *receptors*. Receptors are discussed along with signal molecules in Chapter 6.

5. **Binding proteins.** These proteins, found mostly in the extracellular fluid, bind and transport molecules throughout the body. Examples you will encounter in this text include hemoglobin, which transports oxygen, and cholesterol-binding proteins such as LDL (low-density lipoprotein).

6. **Regulatory proteins.** Regulatory proteins turn cell processes on and off or up and down. The regulatory proteins known as *transcription factors* bind to DNA and alter gene expression and protein synthesis. The details of this process can be found in cell biology textbooks.

7. **Immunoglobulins.** These extracellular immune proteins, also called *antibodies*, help protect the body. Immune functions are discussed in Chapter 24.

Although soluble proteins are quite diverse, they do share some common features. They all bind to other molecules through noncovalent interactions similar to the ones illustrated in Figure 2-10. The binding, which takes place at a specific location on the protein molecule called a **binding site**, exhibits specificity, affinity, competition, and saturation, properties that will be discussed shortly. For proteins for which binding activates the protein and initiates a process (enzymes, membrane transporters, and receptors), we can consider the activity rate of the process and the factors that modulate, or alter, the rate.

Proteins Are Selective About the Molecules They Bind

Any molecule that binds to another molecule is called a **ligand** [*ligare*, to bind or tie]. Ligands that bind to enzymes and membrane transporters are also called **substrates** [*sub-* below + *stratum*, a layer]. Protein signal molecules and protein transcription factors are ligands. Immunoglobulins bind ligands, but the immunoglobulin-ligand complex itself then becomes a ligand, as you will learn in Chapter 24.

Ligand binding requires *molecular complementarity*; that is, the ligand and the binding site must be compatible. In protein binding, when the ligand and protein come close to each other, noncovalent interactions between the ligand and the protein's binding site allow the two molecules to bind. From studies of enzymes and other proteins, scientists have discovered that a protein's binding site and the shape of its ligand do not need to fit one another exactly. When the binding site and the ligand come close to each other, they begin to interact through hydrogen and ionic bonds and van der Waals forces. The protein's binding site then changes shape (*conformation*) to fit more closely to the ligand. This **induced-fit model** of protein-ligand interaction is shown in Figure 2-16 ■.

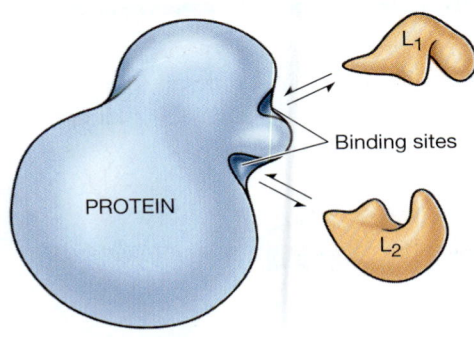

Induced-fit model

In this model of protein binding, the binding site shape is not an exact match to the ligands' (L) shape.

■ **FIGURE 2-16** *The induced-fit model of protein–ligand binding*

Protein Specificity The ability of a protein to bind to a certain ligand or a group of related ligands is called **specificity**. Some proteins are very specific about the ligands they bind, while others bind to whole groups of molecules. For example, the enzymes known as *peptidases* bind to polypeptides and break apart peptide bonds without regard for which two amino acids are joined by those bonds. Peptidases therefore are not considered to be very specific in their action. In contrast, *aminopeptidases* also break peptide bonds. However, they are more specific because they will bind only to one end of a protein chain (the end with an unbound amino group) and act on the terminal peptide bond.

Affinity The degree to which a protein is attracted to a ligand is called the protein's **affinity** for the ligand. A ligand for which a given protein has a high affinity is more likely to bind to the protein than is a ligand for which the protein has a low affinity. This can be shown in the same notation that we use to represent chemical reactions:

$$P + L \rightleftharpoons PL$$

where P is the protein, L is the ligand, and PL is the bound protein-ligand complex. The double arrow indicates that binding is reversible.

Reversible binding reactions go to a state of **equilibrium**, where the rate of binding ($P + L \rightarrow PL$) is exactly equal to the rate of unbinding, or *dissociation* ($P + L \leftarrow PL$). When a reaction is at equilibrium, the ratio of the unbound (*dissociated*) ligand and protein concentrations to the protein-ligand complex concentration is always the same. This ratio is called the *equilibrium constant* K_{eq}, and it applies to all reversible chemical reactions:

$$K_{eq} = \frac{[P][L]}{[PL]}$$

The square brackets [ ] around the letters indicate concentrations of the protein, ligand, and protein-ligand complex. For protein-binding reactions, where the equation is also a quantitative representation of the protein's binding affinity for the ligand, the equilibrium constant is also called the **dissociation constant** (K_d):

$$K_d = \frac{[P][L]}{[PL]}$$

Using algebra to rearrange the equation, this can also be expressed as

$$[PL] = \frac{[P][L]}{K_d}$$

From the rearranged equation you should be able to see that when K_d is large, the value of [PL] will be small. In other words, a large dissociation constant K_d means little binding of protein and ligand, and we can say the protein has a low affinity for the ligand. Conversely, a small K_d is a lower dissociation constant and means a higher value for [PL], indicating a higher affinity of the protein for the ligand.

If a protein binds to several related ligands, a comparison of their K_d values can tell us which ligand is more likely to bind to the protein. The related ligands will compete for the binding sites and are said to be **competitors**. Competition between ligands is a universal property of protein binding.

Ligands that mimic one another are called **agonists**. Agonists may occur in nature, such as *nicotine,* the chemical found in tobacco, which mimics the activity of the neurotransmitter *acetylcholine* by binding to the same receptor protein. Agonists can also be synthesized using what scientists learn from the study of protein-ligand binding sites. The ability of agonist molecules to mimic the activity of naturally occurring ligands has led to the development of many drugs.

CONCEPT CHECK

33. A researcher is trying to design a drug to bind to a particular cell receptor protein. Candidate molecule A has a K_d of 4.9 for the receptor. Molecule B has a K_d of 0.3. Which molecule has the most potential to be successful as the drug?

Answers: p. 49

Multiple Factors Can Alter Protein Binding

A protein's affinity for a ligand is not always constant. Chemical and physical factors can modulate binding affinity or can even totally eliminate it. Some proteins must be activated before they have a functional binding site. In this section we discuss some of the processes that have evolved to allow activation, modulation, and inactivation of protein binding.

Isoforms   Closely related proteins whose function is similar but whose affinity for ligands differs are called **isoforms** of one another. One example of protein isoforms that you will encounter later in this book is the oxygen-transporting protein *hemoglobin*. A hemoglobin molecule has a quaternary structure consisting of four subunits (see Fig. 2-9). In the hemoglobin isoform found in the developing fetus, the subunits are two α chains and two γ chains. Shortly after birth, the fetal hemoglobin molecules are broken down and replaced by adult hemoglobin, an isoform in which the subunits are two α chains and two β chains instead of γ chains. Both isoforms of hemoglobin bind oxygen, but the fetal isoform has a higher affinity for oxygen.

Activation   Some proteins are inactive when they are synthesized in the cell. Before they can become active, one or more portions of the molecule must be chopped off by enzymes (Fig. 2-17 ■). Protein hormones (a type of signal molecule) and enzymes are two groups that commonly undergo such *proteolytic activation* [*lysis,* to release]. The inactive forms of these proteins are often identified with the prefix *pro-* [before]: prohormone, proenzyme, proinsulin, for example. Some inactive enzymes have the suffix *-ogen* added to the name of the active enzyme instead, as in *chymotrypsinogen.*

The activation of some proteins requires the presence of a **cofactor**, which is an ion or small organic functional group.

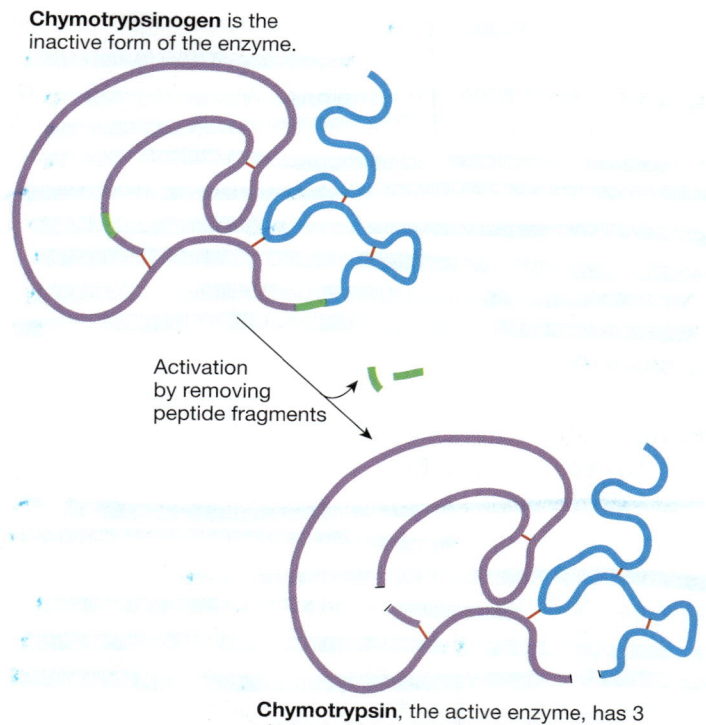

Chymotrypsinogen is the inactive form of the enzyme.

Activation by removing peptide fragments

Chymotrypsin, the active enzyme, has 3 chains held together by disulfide (S—S) bonds.

■ **FIGURE 2-17**   *Proteolytic activation of the enzyme chymotrypsin*

Cofactors must attach to the protein before the binding site will bind to ligand. Ionic cofactors include Ca^{2+}, Mg^{2+}, and Fe^{2+}. This process is shown in Figure 2-18 ■. Many enzymes will not function without cofactors.

Modulation Alters Protein Binding and Activity

The ability of a protein to bind a ligand and initiate a response can be altered by various factors, including temperature, pH, and molecules that interact with the protein. A factor that influences either protein binding or protein activity is known as a

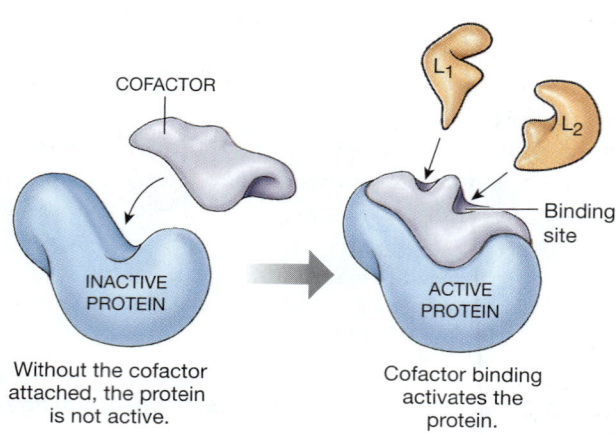

COFACTOR

L_1

L_2

Binding site

INACTIVE PROTEIN

ACTIVE PROTEIN

Without the cofactor attached, the protein is not active.

Cofactor binding activates the protein.

■ **FIGURE 2-18**   *Attachment of cofactors activates the protein*

TABLE 2-3 **Factors That Affect Protein Binding**

Essential for binding activity

Cofactors	Required for ligand binding at binding site
Proteolytic activation	Converts inactive to active form by removal of part of molecule. Examples: digestive enzymes, protein hormones

Modulators and factors that alter binding or activity

Competitive inhibitor	Competes directly with ligand by binding reversibly to active site
Irreversible inhibitor	Binds to binding site and cannot be displaced
Allosteric modulator	Binds to protein away from binding site and changes activity; may be inhibitors or activators
Covalent modulator	Binds covalently to protein and changes its activity. Example: phosphate groups
pH and temperature	Alter three-dimensional shape of enzyme by disrupting hydrogen or S—S bonds; may be irreversible if protein denatures

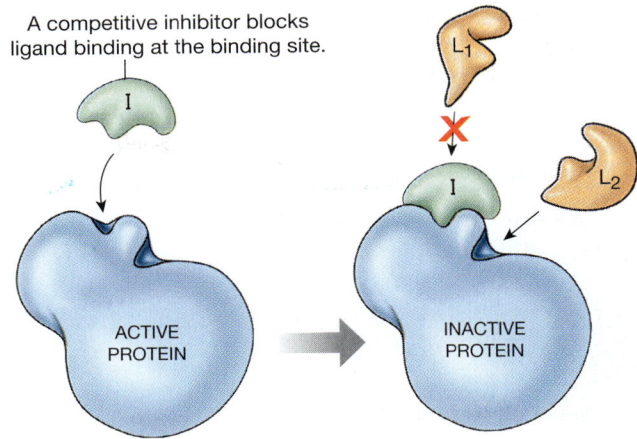

A competitive inhibitor blocks ligand binding at the binding site.

ACTIVE PROTEIN → INACTIVE PROTEIN

■ **FIGURE 2-19** *Competitive inhibition*

modulator. There are two basic mechanisms by which modulation takes place. The modulator either (1) changes the ability of the ligand to bind to the binding site or (2) changes either the protein's activity or its ability to create a response. The different types of modulation are discussed in the following sections and are summarized in Table 2-3 ■.

Chemical modulators are molecules that bind covalently or noncovalently to proteins and alter their binding ability or their activity. Chemical modulators may activate binding, enhance binding, inhibit binding, or completely inactivate the protein. Inactivation may be either reversible or irreversible.

Antagonists, also called *inhibitors,* are chemical modulators that bind to a protein and decrease its activity. Many are simply molecules that bind to the protein and block the binding site without causing a response. They are like the guy who slips into the front of the movie ticket line to chat with his girlfriend, the cashier. He has no interest in buying a ticket, but he prevents the people in line behind him from getting their tickets for the movie.

Competitive inhibitors are reversible antagonists that compete with the customary ligand for the binding site (Fig. 2-19 ■). The degree of inhibition depends on the relative concentrations of the competitive inhibitor and the customary ligand, as well as on the protein's affinities for the two. The binding of competitive inhibitors is reversible: increasing the concentration of the customary ligand can displace the competitive inhibitor and decrease the inhibition.

Irreversible antagonists, on the other hand, bind tightly to the protein and cannot be displaced by competition. Antagonist drugs have proven useful for treating many conditions. For example, tamoxifen, an antagonist to the estrogen receptor, is used in the treatment of hormone-dependent cancers of the breast.

Allosteric and covalent modulators may be either antagonists or activators. **Allosteric modulators** [*allos,* other + *stereos,* solid (as a shape)] bind to a protein away from the binding site and by doing so change the shape of the site (Fig. 2-20 ■). *Allosteric inhibitors* are antagonists that decrease the affinity of the binding site for the ligand and inhibit protein activity. *Allosteric activators* increase the probability of protein-ligand binding and enhance protein activity. In Chapter 18 you will learn how the oxygen-binding ability of hemoglobin changes with allosteric modulation by carbon dioxide, H^+, and several other factors.

Covalent modulators are atoms or functional groups that bind covalently to proteins and alter the proteins' properties. Like allosteric modulators, covalent modulators may either increase or decrease a protein's binding ability or its activity. One of the most common covalent modulators is the phosphate group. Many proteins in the cell can be activated or inactivated when a phosphate group forms a covalent bond with them, a process known as *phosphorylation.*

One of the best known chemical modulators is the antibiotic penicillin. Alexander Fleming discovered this compound in 1928, when he noticed that *Penicillium* mold inhibited bacterial growth. By 1938, researchers had extracted the active ingredient penicillin from the mold and used it to treat infections in humans. Yet it was not until 1965 that researchers figured out exactly how the antibiotic works. Penicillin is an antagonist that binds to a key bacterial protein by mimicking the normal ligand. Because penicillin forms unbreakable bonds with the

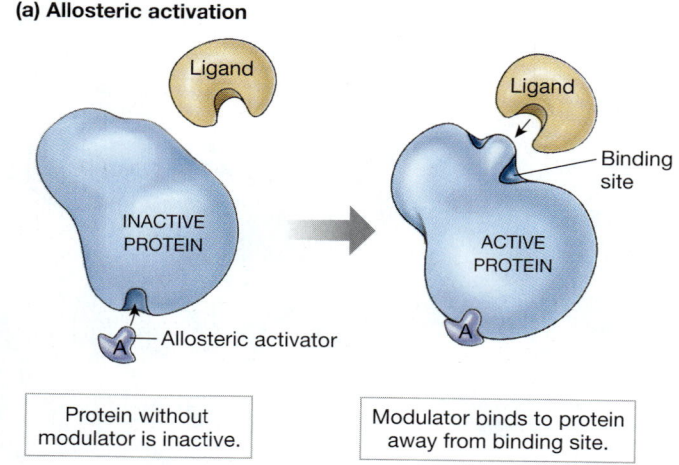

(a) Allosteric activation

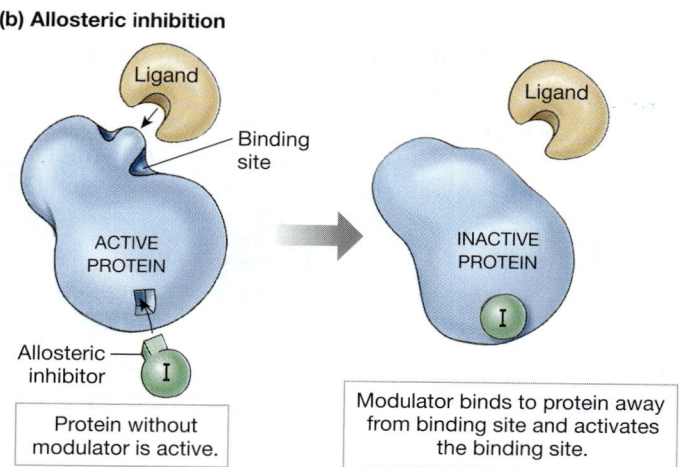

(b) Allosteric inhibition

■ **FIGURE 2-20**   *Allosteric modulation*

protein, the protein is irreversibly inhibited. Without the protein, the bacterium is unable to make a rigid cell wall. Without the rigid wall, the bacterium swells, ruptures, and dies.

Physical Factors Modulate or Inactivate Proteins

Physical conditions such as temperature and pH (acidity) can have dramatic effects on protein structure and function. Small changes in pH or temperature act as modulators to increase or decrease activity (Fig. 2-21 ■). However, once these factors exceed some critical value, they disrupt the noncovalent bonds holding the protein in its tertiary conformation. The protein

loses its shape and along with that, its activity. When the protein loses its conformation, it is said to be *denatured*.

If you have ever fried an egg, you have watched this transformation happen to the egg white protein *albumin* as it changes from a slithery clear state to a firm white state. Hydrogen ions in high enough concentration to be called acids have a similar effect on protein structure. During preparation of ceviche, the national dish of Ecuador, raw fish is marinated in lime juice. The acidic lime juice contains hydrogen ions that disrupt hydrogen bonds in the muscle proteins of the fish, causing the proteins to denature. As a result, the

■ **FIGURE 2-21**   *Effect of temperature on protein activity*

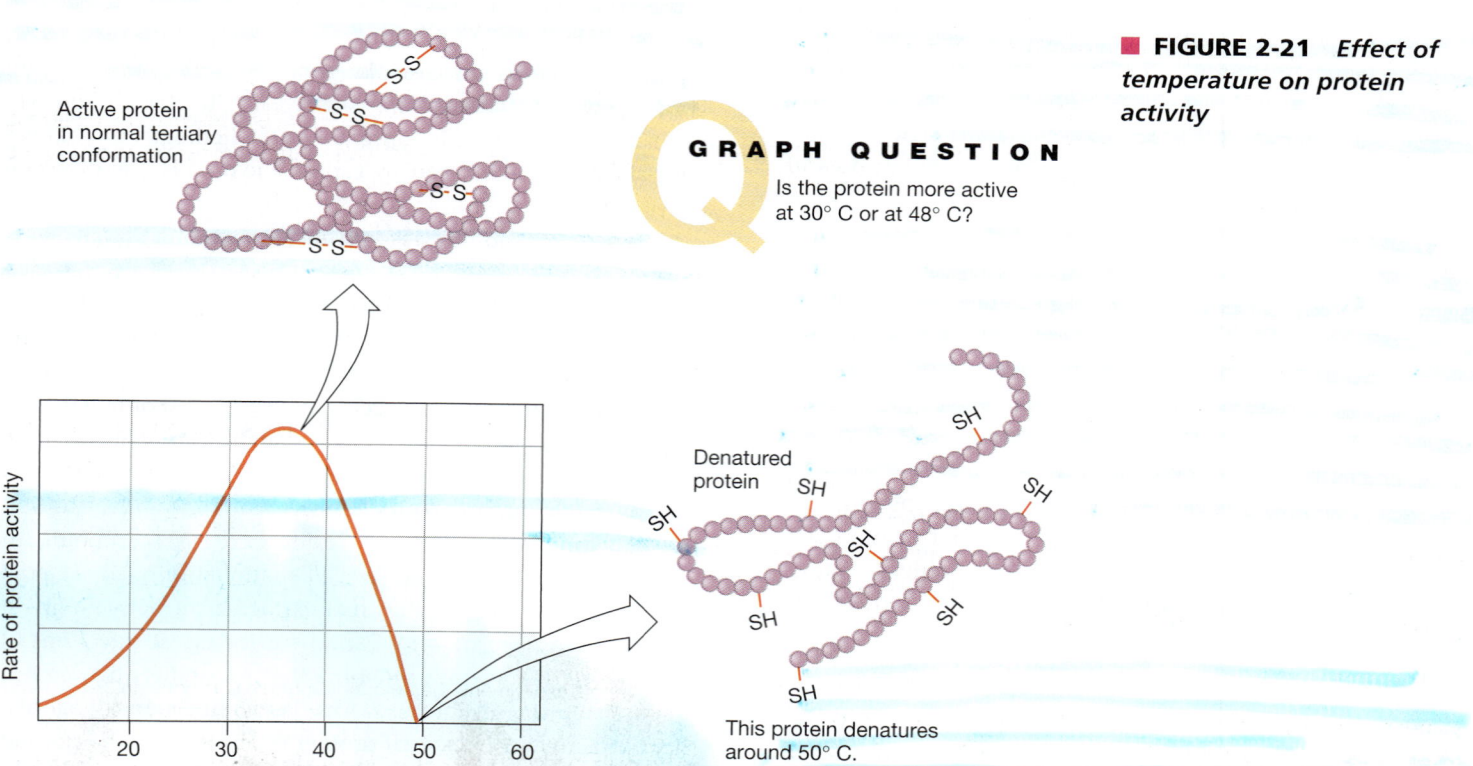

Active protein in normal tertiary conformation

GRAPH QUESTION

Is the protein more active at 30° C or at 48° C?

Denatured protein

This protein denatures around 50° C.

meat becomes firmer and opaque, just as it would if it were cooked with heat.

In a few cases, activity can be restored if the original temperature or pH return. The protein then resumes its original shape as if nothing had happened. Usually, however, denaturation produces a permanent loss of activity. There is certainly no way to unfry an egg or uncook a piece of fish. The potentially disastrous influence of temperature and pH on proteins is one reason these parameters are so closely regulated by the body.

CONCEPT CHECK

34. Match each chemical to its action(s).

(a) Allosteric modulator
(b) Competitive inhibitor
(c) Covalent modulator

c, a 1. Bind away from the binding site
b 2. Bind to the binding site
b 3. Inhibit activity only
c, a 4. Inhibit or enhance activity

Answers: p. 49

The Body Regulates the Amount of Protein Present in Cells

The final characteristic of proteins in the human body is that the amount of a given protein will vary over time, often in a regulated fashion. The body has mechanisms that enable it to monitor whether it needs more or less of particular proteins. Complex signaling pathways, many of which themselves involve proteins, then direct particular cells to make new proteins or to break down (*degrade*) existing proteins. The programmed production of new proteins (receptors, enzymes, and membrane transporters, in particular) is called **up-regulation**. Conversely, the programmed removal of proteins is called **down-regulation**. In both instances, the cell is directed to make or remove proteins to alter its response.

The amount of protein present in a cell has a direct influence on the magnitude of the cell's response. For example, the graph in Figure 2-22 ■ shows the results of an experiment in which the amount of ligand is held constant while the amount of protein is varied. As the graph shows, an increase in the amount of protein present causes an increase in the response.

As an analogy, think of the checkout lines in a supermarket. Imagine that each cashier is an enzyme, the waiting customers are ligand molecules, and those leaving the store with their purchases are products. One hundred people will get checked out faster when there are 25 lines open than when there are only 10 lines. Likewise, in an enzymatic reaction, the presence of more protein molecules means that more binding sites are available to interact with the ligand molecules. As a result, the ligands are converted to products more rapidly.

Regulating protein concentration is an important strategy that cells use to control their physiological processes. Cells alter the amount of a protein by influencing both its synthesis and its breakdown. If protein synthesis exceeds breakdown, protein

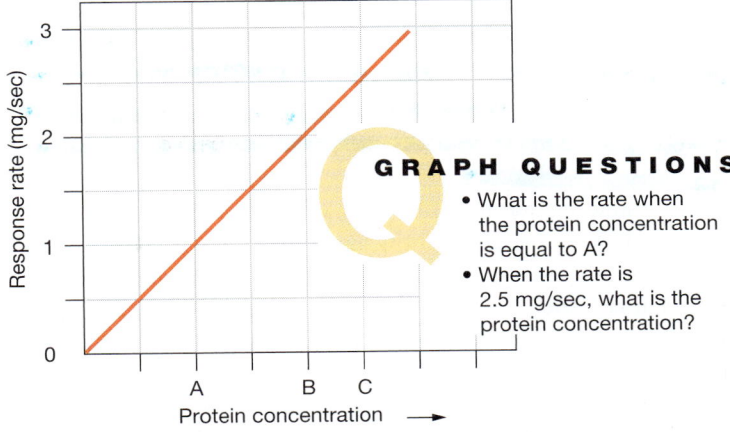

In this experiment, the ligand amount remains constant.

■ **FIGURE 2-22**　*Effect of changing protein concentration on rate*

accumulates and the reaction rate increases. If protein breakdown exceeds synthesis, the amount of protein decreases, as does the reaction rate. Even when the amount of protein is constant, there is still a steady turnover of protein molecules.

Reaction Rate Can Reach a Maximum

If the concentration of a protein in a cell is constant, then the concentration of the ligand determines the magnitude of the response. Fewer ligands activate fewer proteins and the response is low. As ligand concentrations increase, so does the magnitude of the response, up to a maximum where all protein binding sites are occupied.

Figure 2-23 ■ shows the results of a typical experiment in which the protein concentration is constant but the concentration of ligand varies. At low ligand concentrations, the response rate is directly proportional to the ligand concentration. Once

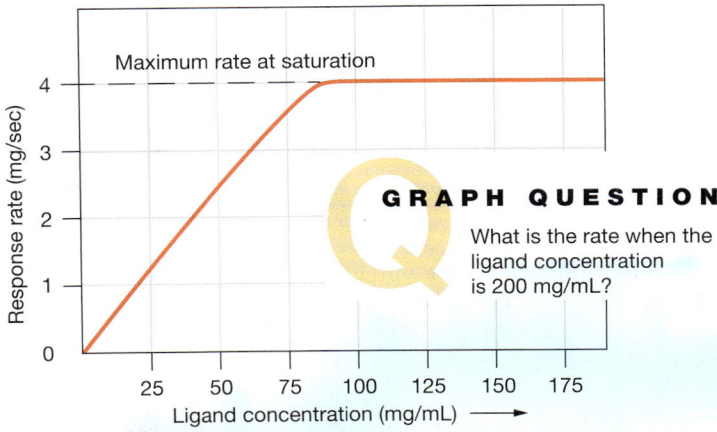

In this experiment, the amount of protein was constant. At the maximum rate the protein is said to be saturated.

■ **FIGURE 2-23**　*Effect of changing ligand concentration on rate* the concentration of ligand molecules exceeds a certain level, the protein molecules have no more free binding sites. The proteins are fully occupied, and the rate reaches a maximum value. This condition is known as **saturation**. Saturation applies to enzymes, membrane transporters, receptors, binding proteins, and immunoglobulins.

An analogy to saturation appeared in the early days of television on the *I Love Lucy* show. Lucille Ball was working at the conveyor belt of a candy factory, loading chocolates into the little paper cups of a candy box. Initially, the belt was moving slowly, and she had no difficulty picking up the candy and putting it into the box. Gradually, the belt brought candy to her more rapidly, and she had to increase her packing speed to keep up. Finally, the belt was bringing candy to her so fast that she couldn't pack it all in the boxes because she was working at her maximum rate. That was Lucy's saturation point. (Her

solution was to stuff the candy into her mouth as well as into the box!)

In summary, you have now learned about the important and nearly universal properties of soluble proteins. You will revisit these concepts many times as you work through the subsequent chapters of this book. In the next chapter you will learn more about the body's insoluble proteins, which are key structural components of cells and tissues.

CONCEPT CHECK

35. What happens to the rate of an enzymatic reaction as the amount of enzyme present decreases?

36. What happens to the rate of an enzymatic reaction when the enzyme has reached saturation?

Answers: p. 49

RUNNING PROBLEM CONCLUSION

CHROMIUM SUPPLEMENTS

In this running problem, you learned that claims of chromium picolinate's ability to enhance muscle mass have not been supported by evidence from controlled scientific experiments. You also learned that studies suggest that some forms of the biological trivalent form of chromium may be toxic. In 2003 the United Kingdom Joint Food Standards and Safety Group asked health supplement manufacturers to remove chromium picolinate from their products.

To learn more about current research, go to PubMed (*www.ncbi.nlm.nih.gov*) and search for *chromium picolinate*.

Compare what you find here with the results of a similar web search. Should you believe everything you read on the Web? For an award-winning set of interactive tutorials on web searching, go to *http://tilt.lib.utsystem.edu/* (notice that there is no www in this address). These tutorials will help you learn how to search and evaluate web sites.

Now compare your answers with those in the summary table.

	QUESTION	FACTS	INTEGRATION AND ANALYSIS
1a	Locate chromium on the periodic table of the elements.	The periodic table organizes the elements according to atomic number.	N/A*
1b	What is chromium's atomic number? Atomic mass?	Reading from the table, chromium (Cr) has an atomic number of 24 and an average atomic mass of 52.	N/A
1c	How many electrons does one atom of chromium have?	The atomic number of an element is equal to the number of protons in one atom. An atom has equal numbers of protons and electrons.	The atomic number of chromium is 24; therefore, one atom of chromium has 24 protons and 24 electrons.
1d	Which elements close to chromium are also essential elements?	Molybdenum, manganese, and iron.	N/A
2a	If people have a chromium deficiency, would you predict that their blood glucose level would be lower or higher than normal?	Chromium helps move glucose from the blood into cells.	If chromium is absent/lacking, less glucose will leave the blood and blood glucose would be higher than normal.
2b	From the results of the Chinese study, can you conclude that all people with diabetes suffer from chromium deficiency?	Higher doses of chromium supplements lowered elevated blood glucose levels, but lower doses had no effect. This is only one study, and no information is given about similar studies elsewhere.	We have insufficient evidence from the information presented to draw a conclusion about the role of chromium deficiency in diabetes.

	QUESTION	FACTS	INTEGRATION AND ANALYSIS
3	How many electrons have been lost from the hexavalent ion of chromium? From the trivalent ion?	For each electron lost from an ion, a positively charged proton is left behind in the nucleus of the ion.	The hexavalent ion of chromium, Cr^{6+}, has six unmatched protons and therefore has lost six electrons. The trivalent ion, Cr^{3+}, has lost three electrons.
4	From this information, can you conclude that hexavalent and trivalent chromium are equally toxic?	The hexavalent form is used in industry and, when inhaled, has been linked to an increased risk of lung cancer. Enough studies have shown an association that California's Hazard Evaluation System and Information Service has issued warnings to chromium workers. Evidence to date for the toxicity of trivalent chromium in chromium picolinate comes from studies done on isolated cells in tissue culture.	Although the toxicity of Cr^{6+} is well established, the toxicity of Cr^{3+} has not been conclusively determined. Studies performed on cells *in vitro* may not be applicable to humans. Additional studies need to be performed in which animals are given reasonable doses of chromium picolinate for an extended time period.
5	Based on the FTC announcement, this study (which did not support enhanced muscle development from chromium supplements), and the studies that suggest that chromium picolinate might cause cancer, do you think that Stan should continue taking chromium picolinate?	No research evidence supports a role for chromium picolinate in increasing muscle mass or strength. Other research suggests that chromium picolinate may cause cancerous changes in isolated cells.	The evidence presented suggests that for Stan, there is no benefit from taking chromium picolinate, and there may be risks. Using risk-benefit analysis, the evidence supports stopping the supplements. However, the decision is Stan's personal responsibility. He should remain informed of new developments that would change the risk-benefit analysis.

*Not applicable

20 21 23 26 35 38 **44**

CHAPTER SUMMARY

This chapter introduces biomolecules and ions, whose chemical properties underlie many of the key themes in physiology. To understand molecular interactions, communication, energy storage and transfer, and the mechanical properties of cells and tissues, you need a good background in basic chemistry.

Chemistry Review

1. **Atoms** are composed of protons, neutrons, and electrons. A **proton** has a mass of about 1 **atomic mass unit** (amu), or **dalton** (Da), and an electrical charge of +1. A **neutron** has a mass of about 1 amu and no electrical charge. An **electron** has negligible mass, compared with protons and neutrons, and a charge of −1. (p. 20; Fig. 2-1)

2. **Elements** are distinguished from each other by the **atomic number** of their atoms, which is the number of protons in the nucleus of one atom. (p. 21)

3. The **mass** of an atom is calculated by adding the number of protons and neutrons in its nucleus. (p. 21)

4. Atoms of one element that have different numbers of neutrons in their nuclei are known as **isotopes** of the element. Radioactive isotopes emit energy called **radiation**. (p. 21; Fig. 2-1)

5. The electrons around an atom are arranged in energy levels, or **shells**, centered on the nucleus. (p. 22)

6. Electrons are important for covalent and ionic bonds, energy capture and transfer, and free radical formation. (p. 22)

Molecular Bonds and Shapes

7. Two or more atoms linked together create a **molecule**. Molecules that include atoms of more than one element are called **compounds**. (p. 23)

8. An atom is most stable when all its outermost orbitals are occupied by pairs of electrons. (p. 23)

9. **Covalent bonds** are strong bonds formed when adjacent atoms share one or more pairs of electrons. Each pair of shared electrons forms a single covalent bond. (p. 23; Fig. 2-2)

10. Molecules whose atoms share electrons unevenly are called **polar molecules**. If electrons are shared evenly by the atoms in a molecule, the molecule is a **nonpolar molecule**. (p. 24; Fig. 2-3)

11. An atom that gains or loses electrons acquires an electrical charge and is called an **ion**. (p. 24; Fig. 2-4)

12. **Ionic bonds** are strong bonds formed when oppositely charged ions are attracted to each other. (p. 24)

13. Weak **hydrogen bonds** form when hydrogen atoms in polar molecules are attracted to oxygen, nitrogen, or fluorine atoms. Hydrogen bonding among water molecules is responsible for the **surface tension** of water. (p. 25; Fig. 2-5)

14. **Van der Waals forces** are weak bonds that form when molecules are attracted to each other. Van der Waals bonds may form within a molecule or between adjacent molecules. (p. 26)

15. Molecular shape is created by covalent bond angles and weak noncovalent interactions within a molecule. (p. 26)

Biomolecules

16. The four major groups of **biomolecules** are carbohydrates, lipids, proteins, and nucleotides. They all contain carbon, hydrogen, and oxygen. (p. 27)

17. **Carbohydrates** have the general formula $(CH_2O)_n$ and are found as **monosaccharides**, **disaccharides**, and **polysaccharides**. Glucose **polymers** store energy and serve as structural components of cells. (p. 27; Fig. 2-7)

18. **Lipids** contain carbon and hydrogen atoms but few oxygen atoms. Most lipids are nonpolar. **Fatty acids** are long chains of carbon atoms bound to hydrogens. **Triglycerides** are composed of glycerol and three fatty acids. (pp. 27–29; Fig. 2-8)

19. **Proteins** are the most versatile biomolecule because 20 different **amino acids** serve as their building blocks. (p. 30)

20. The **primary structure** of a protein is the sequence of its amino acids. The **secondary structure** is its spatial arrangement into either helixes or sheets. The **tertiary structure** is the three-dimensional shape of the folded amino acid chain, and the **quaternary structure** is the arrangement of multiple amino acid chains in a single protein. (p. 31; Fig. 2-9)

21. **Globular proteins** are soluble in water. They act as carriers, messengers, defense molecules, and enzymes. **Fibrous proteins** are not very soluble in water. They serve as structural components of cells and tissues. (p. 31)

22. Proteins, lipids, and carbohydrates combine to form **glycoproteins**, **glycolipids**, or **lipoproteins**. (p. 32)

23. Nucleotides are important in the transmission and storage of information and the transfer of energy. (p. 32)

24. A **nucleotide** is composed of one or more phosphate groups, a sugar, and either a purine or a pyrimidine nitrogenous base. Small nucleotides include ATP, ADP, and cyclic AMP. Nucleotide polymers are the **nucleic acids** DNA and RNA. (pp. 32–33; Figs. 2-11, 2-12)

Aqueous Solutions, Acids, Bases, and Buffers

IP **Fluids and Electrolytes: Acid-Base Homeostasis**

25. Molecules that dissolve in liquid are called **solutes**. The liquid in which they dissolve is called the **solvent**. The universal solvent for biological solutions is water. (p. 35)

26. The ease with which a molecule dissolves in a solvent is called its **solubility** in that solvent. **Hydrophilic molecules** dissolve easily in water, whereas **hydrophobic molecules** do not. (p. 35)

27. The **concentration** of a solution is the amount of solute per unit volume of solution. A one **molar** (1 M) solution has one mole of solute per liter of solution. The concentration of most biological solutions is expressed in **millimoles per liter** of solution (mM). (pp. 35–36)

28. A **mole** is 6.02×10^{23} particles of an element or compound. The **molecular mass** of a molecule is determined by adding the atomic masses of all its atoms. The mass in grams of 1 mole of any element or compound is known as the **gram molecular mass** of the element or compound. (p. 36)

29. Solution concentrations are described in moles/volume (**molarity**) or as **percent solutions**. Ion concentrations may be expressed in milliequivalents, where one **equivalent** equals the molarity of the ion times its electrical charge. (p. 36)

30. The **pH** of a solution is a measure of its hydrogen ion concentration. **Acids** are molecules that contribute H^+ to a solution. **Bases** are molecules that remove H^+ from solution. The more acidic the solution, the lower its pH. (p. 37; Fig. 2-15)

31. **Buffers** are solutions that moderate pH changes. (p. 38)

Protein Interactions

32. Most water-soluble proteins serve as enzymes, membrane transporters, signal molecules, receptors, binding proteins, transcription factors, or immunoglobulins. (pp. 38–39)

33. **Ligands** bind to proteins at a **binding site**. According to the **induced-fit model** of protein binding, the shapes of the ligand and binding site do not have to match exactly. (p. 39; Fig. 2-16)

34. Proteins are **specific** about the ligands they will bind. The attraction of a protein to its ligand is called the protein's **affinity** for the ligand. The **dissociation constant** (K_d) is a quantitative measure of a protein's affinity for a given ligand. (p. 39)

35. Ligands may compete for a protein's binding site. If competing ligands mimic each other's activity, they are **agonists**. (p. 40)

36. Closely related proteins having similar function but different affinities for ligands are called **isoforms** of one another. (p. 40)

37. Some proteins must be activated, either by proteolytic activation or by addition of **cofactors**. (p. 40; Figs. 2-17, 2-18)

38. **Competitive inhibitors** can be displaced from the binding site but irreversible antagonists cannot. (p. 41; Fig. 2-19)

39. **Allosteric modulators** bind to proteins at a location other than the binding site. **Covalent modulators** bind with covalent bonds. Both of these types of modulators may activate or inhibit the protein. (p. 41; Fig. 2-20)

40. Extremes of temperature or pH will denature proteins. (p. 42; Fig. 2-21)

41. Cells regulate their proteins by **up-regulation** or **down-regulation** of protein synthesis and destruction. The amount of protein is directly related to the magnitude of the cell's response. (p. 43; Fig. 2-22)

42. If the amount of protein is constant, the amount of ligand determines the cell's response. If all proteins become saturated with ligand, the response reaches its maximum. (p. 43; Fig. 2-23)

QUESTIONS

(Answers to the Review Questions begin on page A1.)

▸ **THE PHYSIOLOGY PLACE**

Access more review material online at **The Physiology Place** website. There you'll find review questions, problem-solving activities, case studies, flashcards, and direct links to both *InterActive Physiology*® and PhysioEx™. To access the site, go to *www.physiologyplace.com* and select Human Physiology, Fourth Edition.

LEVEL ONE   REVIEWING FACTS AND TERMS

1. List three major essential elements found in the human body.

2. When atoms of elements are bound to one another into compounds, such as H_2O or CO_2, one unit is called a _____.

3. (a) The subatomic particle with a mass of about 1 amu and a charge of $+1$ is called the_____. (b) The subatomic particle with a mass of about 1 amu and a neutral charge is the_____. (c) The remaining subatomic particle is the electron, with a negligible mass but a charge of _____.

4. Which subatomic particle in an element determines its atomic number?

5. Name the element associated with each of these symbols: Ca, C, O, Na, N, K, H, and P.

6. Isotopes of an element have the same number of _____ and _____, but differ in their number of _____.

7. Unstable isotopes emit energy called _____. Using this energy for the diagnosis or treatment of disease is a specialty called _____.

8. When electrons in an atom or molecule are stable, are they paired or unpaired?

9. When an atom of an element gains or loses one or more electrons, it is called a/an _____ of that element.

10. Match each type of bond with its description:
    (a) covalent bond
    (b) ionic bond
    (c) hydrogen bond
    (d) van der Waals force

    1. weak attractive force between hydrogen and oxygen or nitrogen
    2. formed when two atoms share one or more pairs of electrons
    3. weak attractive force between molecules
    4. formed when one atom loses one or more electrons to a second atom

11. In mixing compounds with water, the polarity of the compounds helps predict their solubility. When a compound's electrons are shared evenly, the compound is said to be _____; if shared unevenly, the compound is _____. Which group dissolves more readily in water?

12. The pH of a solution is an abbreviation conveying what information about that solution? A solution with a pH less than 7 is _____; a solution with a pH greater than 7 is _____.

13. A _____ is a solution that moderates changes in pH.

14. List the four kinds of biomolecules. Give an example of each kind that is relevant to physiology.

15. Match each carbohydrate with its description:
    (a) starch
    (b) chitin

    1. most abundant carbohydrate on earth
    2. disaccharide, found in milk
    (c) glucose
    (d) cellulose
    (e) glycogen

    3. storage form of glucose for animals
    4. storage form of glucose for plants
    5. structural polysaccharide of invertebrates
    6. monosaccharide

16. Proteins may be combined with other types of molecules. They are called _____ when combined with fatty components. They are called _____ when combined with carbohydrates.

17. Match each lipid with its description:
    (a) triglyceride
    (b) eicosanoid
    (c) steroid
    (d) oil
    (e) phospholipid

    1. most common form of lipid in the body
    2. liquid at room temperature, usually from plants
    3. important component of cell membrane
    4. structure composed of carbon rings
    5. modified 20-carbon fatty acid

18. Match these terms pertaining to proteins and amino acids:
    (a) the building blocks
    (b) must be included in our diet
    (c) proteins that speed the rate of chemical reactions
    (d) sequence of amino acids in a protein
    (e) protein chains folded into a ball-shaped structure

    1. essential amino acids
    2. primary structure
    3. amino acids
    4. globular proteins
    5. enzymes
    6. tertiary structure
    7. fibrous proteins

19. List the components of a nucleotide.

20. A molecule that binds to another molecule is called a _____.

21. Match these definitions with their terms (not all terms are used):
    (a) the ability of a protein to bind one molecule but not another
    (b) the part of a protein molecule that binds the ligand
    (c) the ability of a protein to alter its shape to fit more closely with that of the ligand

    1. irreversible inhibition
    2. induced fit
    3. binding site
    4. specificity
    5. saturation

22. An ion, such as Ca^{2+} or Mg^{2+}, that must be present in order for an enzyme to work is called a _____.

23. A protein whose structure is altered to the point that its activity is destroyed is said to be _____.

LEVEL TWO   REVIEWING CONCEPTS

24. **Mapping exercises:** Make each list of terms into a separate map (lists continue on page 48).

List 1: The atom	List 2: Solutions
anion	concentration
atomic mass	equivalent
atomic number	hydrogen bond
cation	hydrophilic
covalent bond	hydrophobic
electron	molarity
electron shell	mole
element	nonpolar molecule
ion	polar molecule

List 1: The atom (cont'd.)

isotope

neutron

nucleus

proton

List 2: Solutions (cont'd.)

solubility

solute

solvent

water

25. Use the periodic table of the elements on the inside back cover to answer the following questions: An atom of sodium has 11 protons in its nucleus.

    (a) How many electrons does the atom have?
    (b) What is the electrical charge of the atom?
    (c) How many neutrons does the average atom have?
    (d) If this atom loses one electron, it would be called a(n) _____?
    (e) What would be the electrical charge of the substance formed in (d)?
    (f) Write the chemical symbol for the ion referred to in (d).
    (g) What does the sodium atom become if it loses a proton from its nucleus?
    (h) Write the chemical symbol for the atom referred to in (g).

26. A solution in which $[H^+] = 10^{-3}$ M is _____ (acidic/basic), whereas a solution in which $[H^+] = 10^{-10}$ M is _____ (acidic/basic). Give the pH for each of these solutions.

27. Name three nucleotides, and tell why each one is important.

28. Explain the progression of protein structure from primary to quaternary.

29. Compare the structure of DNA with that of RNA.

30. Distinguish between purines and pyrimidines.

31. You know that two soluble proteins are isoforms of each other. What can you predict about their structures, functions, and affinities for ligands?

32. You have been asked to design some drugs for the purposes described below. Choose the desirable characteristics for each drug from the following list.

    (a) Drug A must bind to an enzyme and enhance its activity.
    (b) Drug B should mimic the activity of a normal nervous system signal molecule.
    (c) Drug C should block the activity of a membrane receptor protein.

    1. antagonist
    2. competitive inhibitor
    3. agonist
    4. allosteric activator
    5. covalent modulator

LEVEL THREE   PROBLEM SOLVING

33. You have been summoned to assist with the autopsy of an alien being whose remains have been brought to your lab. The chemical analysis returns with 33% C, 40% O, 4% H, 14% N, and 9% P. From this information you conclude that the cells do contain nucleotides, possibly even DNA or RNA. Your assistant is demanding that you tell him how you knew this. What do you tell him?

34. A cell carrying out work normally produces CO_2. The harder a cell works, the more CO_2 it produces. CO_2 is carried in the blood according to the following equation: $CO_2 + H_2O \rightleftharpoons H_2CO_3 \rightleftharpoons H^+ + HCO_3^-$. What effect does hard work by your muscle cells have on the pH of the blood?

35. When the radioactive isotope of iodine, I-131, emits beta radiation, it converts a neutron to a proton and an electron. What are the atomic mass, atomic number, and name of the resulting atom?

LEVEL FOUR   QUANTITATIVE PROBLEMS

36. Write the chemical formulas for each molecule depicted. Calculate the molecular weight of each molecule.

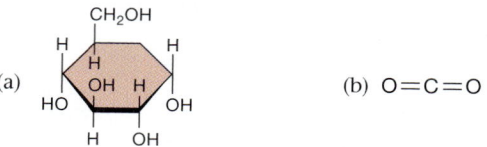

(a)

(b) O=C=O

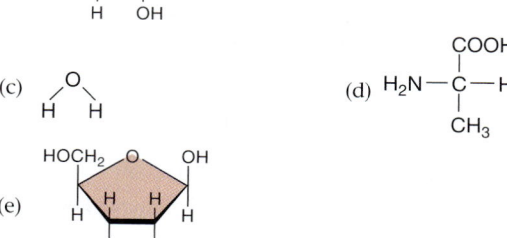

(c)

(d) $H_2N-\overset{\underset{|}{COOH}}{\underset{|}{C}}-H$
     CH_3

(e)

37. Calculate the amount of NaCl you would measure out to make one liter of 0.9% NaCl. Explain how you would make a liter of this solution.

38. A 1.0 M NaCl solution contains 58.5 g of salt per liter. (a) How many molecules of NaCl are present in this solution? (b) How many millimoles of NaCl are present? (c) How many equivalents of Na^+ are present? (d) Express 58.5 g of NaCl per liter as a percent solution.

39. How would you make 200 mL of a 5% glucose solution? Calculate the molarity of this solution. How many millimoles of glucose are present in 500 mL of this solution? (Hint: What is the molecular mass of glucose?)

40. The graph shown below represents the binding of oxygen molecules (O_2) to two different proteins, myoglobin and hemoglobin, over a range of oxygen concentrations. Based on the graph, which protein has the higher affinity for oxygen? Explain your reasoning.

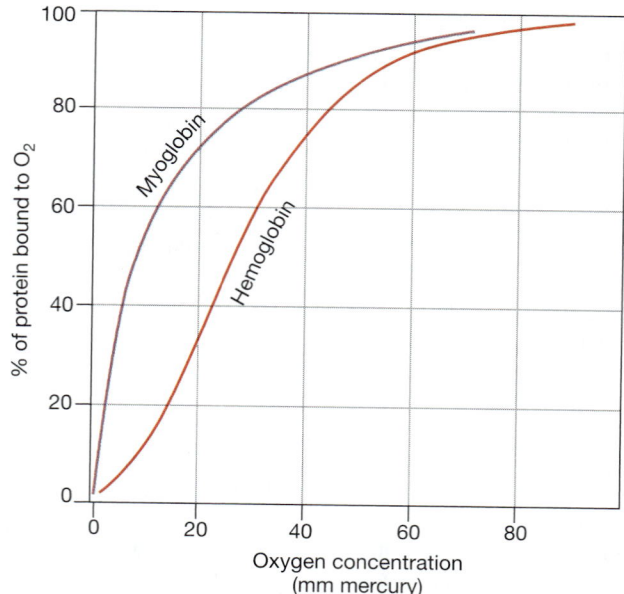

ANSWERS

✓ *Answers to Concept Check Questions*

Page 21

1. O, C, H, N, P, Na, K, Ca, Mg, S, Cl

2. Potassium (K), iron (Fe), copper (Cu), silver (Ag), tin (Sn), antimony (Sb), tungsten (W), gold (Au), mercury (Hg), and lead (Pb).

Page 21

3. H^+ and $H = 1$ amu; $^2H = 2$ amu.

Page 22

4. Iodine: atomic number 53, atomic mass 126.9; technetium: atomic number 43, atomic mass 98, symbol Tc.

Page 23

5. c, e

6. a, b, f

7. a, b, d

8. Mg^{2+} has *lost* two *electrons*.

9. The hydrogen ion is called a proton because loss of its one electron leaves behind one proton.

Page 23

10. One carbon atom needs to share four electrons to fill its outer shell; therefore it will form four covalent bonds. See electron-dot shorthand on left side of equation in next answer.

11. One carbon combines with four hydrogens to form a molecule of methane, an odorless gas that is a principal constituent of natural gas.

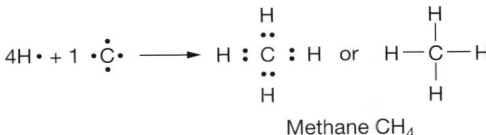

Methane CH_4

Page 26

12. (a) covalent, (b) polar, (c) nonpolar.

13. (a) anion, (b) cation.

14. Oxygen and nitrogen strongly attract electrons and tend to form polar bonds.

15. Table sugar dissolves easily in water, so it would be polar; vegetable oil does not dissolve in water, so it would be nonpolar.

Page 27

16. False. All biomolecules are organic molecules, but the reverse is not true.

17. $C_nH_{2n}O_n$

Page 30

18. Unsaturated fatty acids have double bonds between carbons. Each double bond removes two hydrogens from the molecule, therefore (c) $C_{18}H_{30}O_2$ is the most unsaturated because it has the fewest hydrogens.

Page 31

19. An amino group is $-NH_2$. A carboxyl group is $-COOH$.

20. Our bodies cannot make the essential amino acids. We must therefore obtain them through our diet. We can synthesize the other amino acids from nutrients.

21. Proteins are composed of 20 different amino acids that can be linked in an almost infinite number of sequences. These combinations allow proteins to take on different forms, such as the globular and fibrous proteins.

Page 36

22. Na^+ and Cl^- ions form hydrogen bonds with the polar water molecules. This disrupts the ionic bonds that held the NaCl crystal together.

23. A solution is composed of solute and solvent.

24. Polar molecules dissolve more easily in water than do nonpolar molecules.

25. Hydrophilic.

26. (d)

Page 36

27. 18 amu = 18 Da

28. 74.6 g

29. A 0.1 M solution is the same as a 100 mM solution, which means the concentrations are equal.

30. The 5 g of glucose will add volume to the solution, so if you begin with 100 mL of the solvent, you will end up with more than 100 mL of solution.

Page 38

31. An acid must dissociate into one or more H^+ plus anions.

32. pH is the concentration of H^+.

Page 40

33. Molecule B is a better candidate because its lower K_d means higher binding affinity.

Page 43

34. (a) 1, 4; (b) 2, 3; (c) 1, 4

Page 44

35. As the amount of protein decreases, the reaction rate decreases.

36. If a protein has reached saturation, the rate is at its maximum.

Q *Answers to Figure and Graph Questions*

Page 26

Figure 2-6: 1. $C_6H_{13}NO_2$ 2. Leucine has carboxyl and amino groups. Glucose has hydroxyl groups.

Page 37

Figure 2-15: 1. Increased acidity means that H^+ concentration increases and pH decreases. 2. Urine, stomach acid, and saliva are all contained within the lumen of hollow organs, where they are not part of the body's internal environment (see Fig. 1-2 on p. 4).

Page 42

Figure 2-21: The protein is more active at 30°C.

Page 43

Figure 2-22: 1. At protein concentration A, the rate is 1 mg/sec. 2. The rate is 2.5 mg/sec for protein concentration C.

Page 43

Figure 2-23: When the ligand concentration is 200 mg/mL, the rate is 4 mg/sec.

3

Cells are organisms, and entire animals and plants are aggregates of these organisms. —Theodor Schwann, 1839

Developing cell in tissue culture showing microtubules (red), actin (green), and DNA (blue).

Compartmentation: Cells and Tissues

BACKGROUND BASICS

Units of measure: inside back cover; Compartmentation: **9**; Extracellular fluid: **4**; Hydrophobic molecules: **24**; Proteins: **30**; Compound molecules: **32**; pH: **38**; Covalent and noncovalent interactions: **23, 25**

RUNNING PROBLEM

THE PAP SMEAR

Dr. George Papanicolaou has saved the lives of millions of women by popularizing the Pap smear, a screening method that detects the early signs of cervical cancer. In the past fifty years the Pap test has become widely used, and deaths from cervical cancer have dropped nearly 70%. If detected early, cervical cancer is one of the most treatable forms of cancer. Today, Jan Melton, who had an abnormal Pap smear several months ago, returns to her family physician for a repeat test. The results may determine whether she should undergo treatment for cervical cancer.

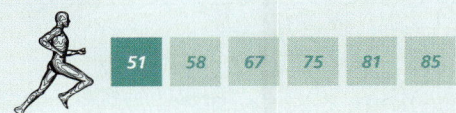

| 51 | 58 | 67 | 75 | 81 | 85 |

What makes a compartment? We may think of something totally enclosed, like a room or a box with a lid. But not all compartments are totally enclosed . . . think of the modular cubicles that make up many modern workplaces. And not all functional compartments have walls . . . think of a giant hotel lobby divided into conversational groupings by careful placement of rugs and furniture. Biological compartments come with the same type of anatomic variability, ranging from totally enclosed structures such as cells, to functional compartments without visible walls.

The first living compartment was probably a simple cell whose intracellular fluid was separated from the external environment by a wall made of phospholipids and proteins—the cell membrane. As we learned in Chapter 1, cells are the basic functional unit of living organisms, and an individual cell can carry out all the processes of life. As cells evolved, they acquired intracellular compartments separated from the intracellular fluid by membranes. Over time, groups of single-celled organisms began to cooperate and specialize their functions, eventually giving rise to multicellular organisms. As multicellular organisms evolved to become larger and more complex, their bodies became divided into various functional compartments.

Compartments are both an advantage and a disadvantage for organisms. On the advantage side, compartments separate biochemical processes that might otherwise conflict with one another. For example, protein synthesis takes place in one subcellular compartment while protein degradation is taking place in another. Barriers between compartments, whether inside a cell or inside a body, allow the composition of one compartment's contents to differ from the contents of adjacent compartments. An extreme example is the intracellular compartment called the lysosome, with an internal pH of 5

[Fig. 2-15, p. 37]. This pH is so acidic that if the lysosome ruptures, it severely damages or kills the cell that contains it.

The disadvantage to compartments is that barriers between them can make it difficult to move needed materials from one compartment to another. Living organisms overcome this problem with specialized mechanisms that transport selected substances across membranes. Membrane transport is the subject of Chapter 5.

In this chapter we first look at the various compartments that subdivide the human body. We then review the structure and function of individual cells before examining how groups of cells with similar functions join together to form the tissues and organs of the body.

FUNCTIONAL COMPARTMENTS OF THE BODY

As you learned in Chapter 1, the human body is a complex compartment separated from the outside world by layers of cells [p. 4]. Anatomically, the body is divided into three major body **cavities**: the *cranial cavity* (commonly referred to as the *skull*), the *thoracic cavity* (also called the *thorax*), and the *abdominopelvic cavity* (Fig. 3-1 ■). The cavities are separated from one another by bones and tissues, and they are lined with *tissue membranes*.

The cranial cavity [*cranium*, skull] contains the brain, our primary control center. The thoracic cavity is bounded by the spine and ribs on top and sides, with the muscular *diaphragm* forming the floor. This cavity surrounds the heart, which is enclosed in a membranous *pericardial sac* [*peri-*, around + *cardium*, heart] and the lungs, which are enclosed in two *pleural sacs*.

The *abdomen* and *pelvis* form one continuous cavity. A tissue lining called the *peritoneum* fully or partially surrounds most of the organs within the abdomen. The abdomen encloses the digestive organs (stomach, intestines, liver, pancreas, gallbladder), the spleen, and the kidneys. The pelvis contains reproductive organs, the urinary bladder, and the terminal portion of the large intestine.

The Lumens of Hollow Organs Are Not Part of the Internal Environment

The hollow organs, such as heart, lungs, blood vessels, and intestines, create another set of compartments within the body. The interior of a hollow organ is called the **lumen** [*lumin*, window]. A lumen may be wholly or partially filled with air or fluid. For example, the lumen of blood vessels is filled with the fluid we call blood.

For some organs, the lumen is essentially an extension of the external environment, and material in the lumen is not truly part of the body's internal environment until it crosses the wall of the organ. For example, we think of our digestive tract as being "inside" our body, but in reality its lumen is part of the body's external environment (see Fig. 3-24, p. 73). An analogy

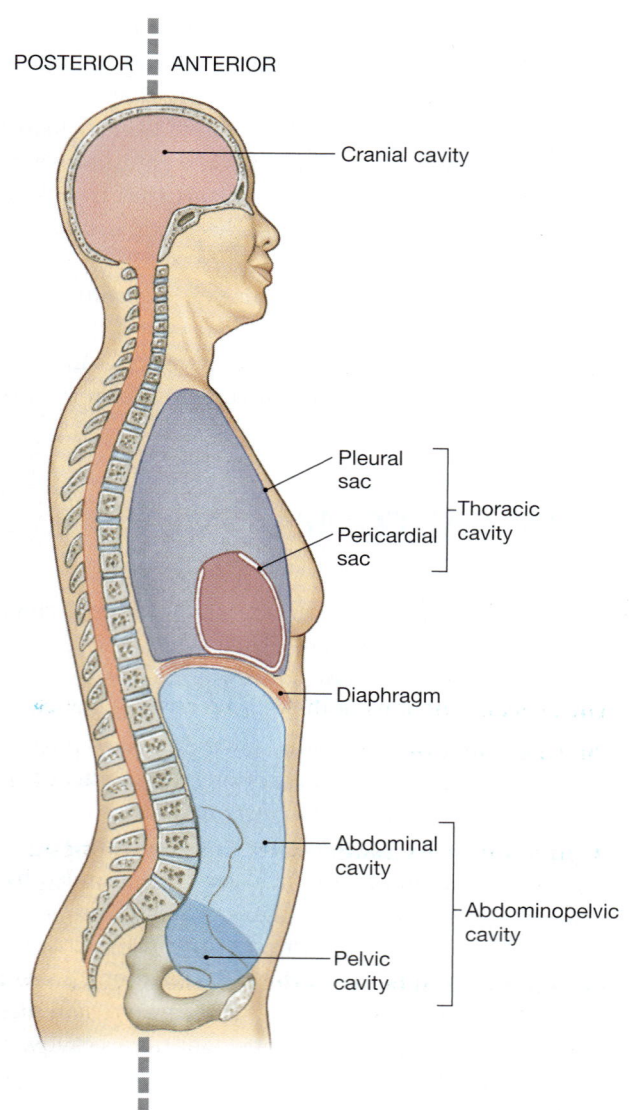

■ FIGURE 3-1   *Major cavities of the human body*

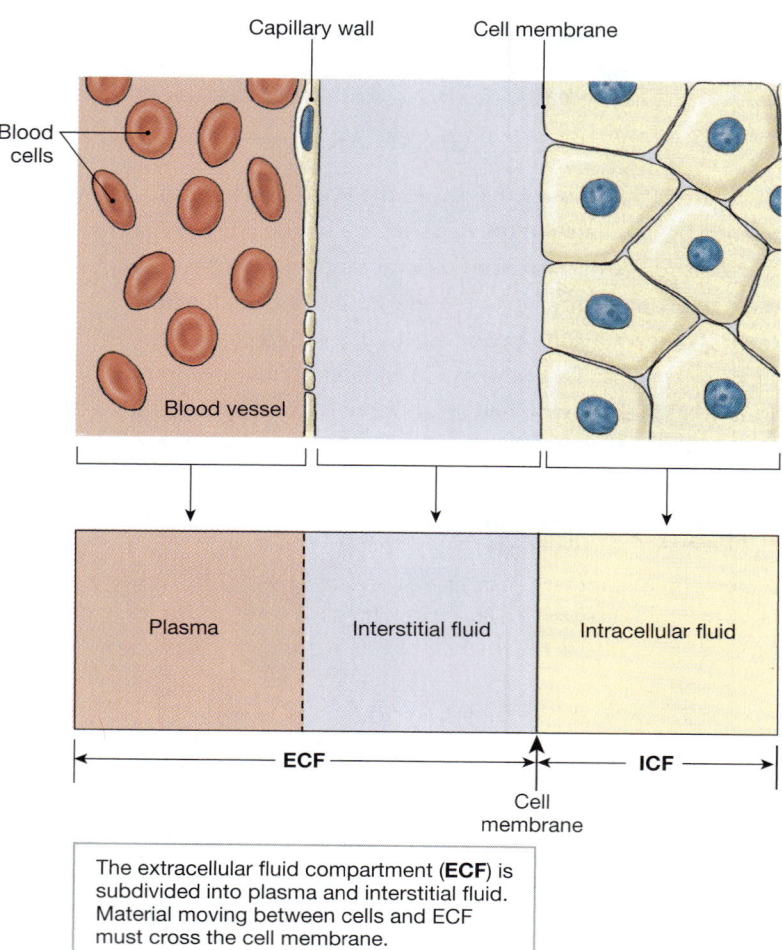

The extracellular fluid compartment (**ECF**) is subdivided into plasma and interstitial fluid. Material moving between cells and ECF must cross the cell membrane.

■ FIGURE 3-2   *Body fluid compartments*

The box in the bottom half schematically illustrates the relationships among the three fluid compartments of the body.

would be the hole through a bead. The hole passes through the bead but is not actually inside the bead.

An interesting illustration of this distinction between the internal environment and the external environment in a lumen involves the bacterium *Escherichia coli*. This organism normally lives and reproduces inside the large intestine, an internalized compartment whose lumen is continuous with the external environment. When *E. coli* is residing in this location, it does not harm the host. However, if the intestinal wall is punctured by disease or accident and *E. coli* enters the body's internal environment, a serious infection can result.

Functionally, the Body Has Three Fluid Compartments

In physiology we are often more interested in functional compartments than in anatomical compartments. As you learned in Chapter 1, most cells of the body are not in direct contact with

the outside world. Instead they are surrounded by extracellular fluid [⮂ Fig. 1-3, p. 4]. If we think of all the cells of the body together as one unit and the fluid surrounding them as a separate unit, we can divide the body into two main compartments: (1) the *intracellular fluid* (ICF) within the cells and (2) the *extracellular fluid* (ECF) outside the cells. These compartments are separated by the barrier of the cell membrane.

The extracellular fluid can be further subdivided (Fig. 3-2 ■). The dividing "wall" in this case is the wall of the circulatory system. **Plasma**, the fluid portion of the blood, lies within the circulatory system and forms one extracellular compartment. The other extracellular compartment is **interstitial fluid** [*inter-*, between + *stare*, to stand], which lies between the circulatory system and the cells. Figure 3-2 illustrates the standard box graphic that physiologists use to represent the three fluid compartments.

Although the body's cells can collectively be considered a single fluid compartment, in reality they are highly variable in their shape, size, and composition. In the next two sections we review the key features of cells and see how these fundamental units of life are themselves divided into compartments.

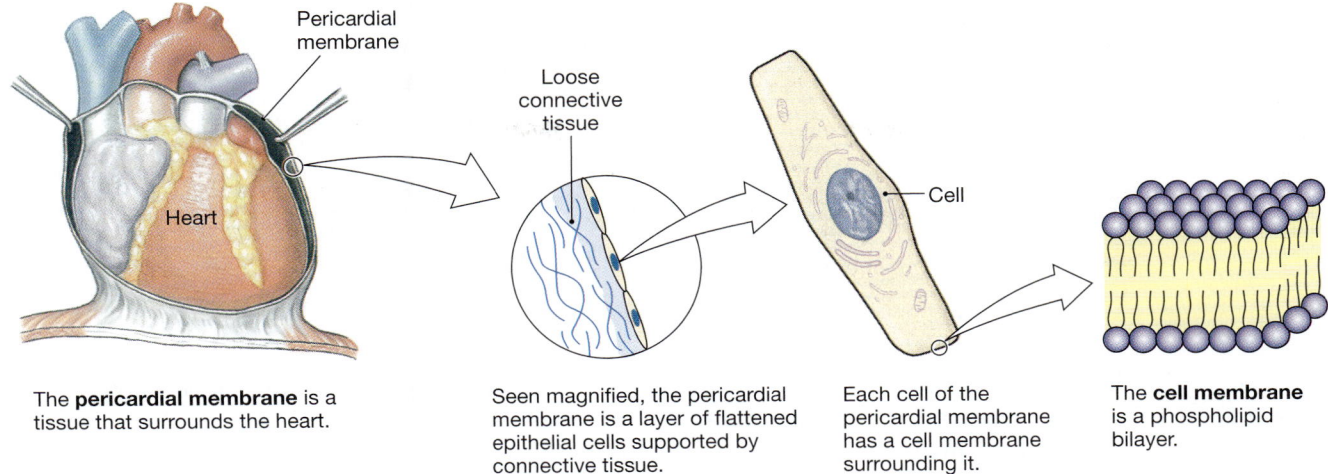

Pericardial membrane

Loose connective tissue

Heart

Cell

The **pericardial membrane** is a tissue that surrounds the heart.

Seen magnified, the pericardial membrane is a layer of flattened epithelial cells supported by connective tissue.

Each cell of the pericardial membrane has a cell membrane surrounding it.

The **cell membrane** is a phospholipid bilayer.

3

■ **FIGURE 3-3**   *Membranes in the body*

A membrane can refer either to a thin tissue layer, such as the pericardial membrane, or to the phospholipid bilayer surrounding every cell in the body (the cell membrane).

BIOLOGICAL MEMBRANES

The word *membrane* [*membrana,* a skin] has two meanings in biology. Before the invention of microscopes in the sixteenth century, a membrane always described a tissue that lined a cavity or separated two compartments. Even today, we speak of *mucous membranes* in the mouth and vagina, the *peritoneal membrane* that lines the inside of the abdomen, the *pleural membrane* that covers the surface of the lungs, and the *pericardial membrane* that surrounds the heart. These visible membranes are tissues: thin, translucent layers of cells.

Once scientists observed cells with a microscope, the nature of the barrier between the intracellular fluid and external environment became a matter of great interest. By the 1890s, scientists had concluded that the outer surface of cells, the **cell membrane,** was a thin layer of lipids that separated the aqueous fluids of the interior and outside environment. We now know that cell membranes consist of microscopic double layers (*bilayers*) of phospholipids with protein molecules inserted in them.

Thus, the word *membrane* may apply either to a tissue or to a phospholipid boundary layer (Fig. 3-3 ■). To add to the confusion, tissue membranes are often depicted in book illustrations as a single line, leading students to think of them as if they were similar in structure to the cell membrane. In this section you will learn more about these phospholipid membranes that create compartments for cells.

The Cell Membrane Separates the Cell from Its Environment

There are two synonyms for the term *cell membrane: plasma membrane* and *plasmalemma.* We will use the term *cell membrane* in this book rather than *plasma membrane* or *plasmalemma* to avoid confusion with the term *blood plasma.*

The general functions of the cell membrane include:

1. **Physical isolation:**   The cell membrane is a physical barrier that separates intracellular fluid inside the cell from the surrounding extracellular fluid.

2. **Regulation of exchange with the environment:**   The cell membrane controls the entry of ions and nutrients into the cell, the elimination of cellular wastes, and the release of products from the cell.

3. **Communication between the cell and its environment:**   The cell membrane contains proteins that enable the cell to recognize and respond to molecules or to changes in its external environment. Any alteration in the cell membrane may affect the cell's activities.

4. **Structural support:**   Proteins in the cell membrane hold the *cytoskeleton,* the cell's interior structural scaffolding, in place to maintain cell shape. Membrane proteins also create specialized junctions between adjacent cells or between cells and the *extracellular matrix* [*extra-,* outside], which is extracellular material that is synthesized and secreted by the cells. (**Secretion** is the process by which a cell releases a substance into the extracellular space.) Cell-cell and cell-matrix junctions stabilize the structure of tissues.

To understand how the cell membrane can carry out such diverse functions, we first examine the current model of cell membrane structure.

Membranes Are Mostly Lipid and Protein

In the early decades of the twentieth century, researchers trying to decipher membrane structure ground up cells and analyzed their composition. They discovered that all biological

TABLE 3-1	Composition of Selected Membranes		
MEMBRANE	PROTEIN	LIPID	CARBOHYDRATE
Red blood cell membrane	49%	43%	8%
Myelin membrane around nerve cells	18%	79%	3%
Inner mitochondrial membrane	76%	24%	0%

membranes consist of a combination of lipids and proteins plus a small amount of carbohydrate. However, a simple and uniform structure did not account for the highly variable properties of membranes found in different types of cells. How could water cross the cell membrane to enter a red blood cell but not be able to enter certain cells of the kidney tubule? The explanation had to lie in the molecular arrangement of the proteins and lipids in the various membranes.

The ratio of protein to lipid varies widely, depending on the source of the membrane (Table 3-1 ■). Generally, the more

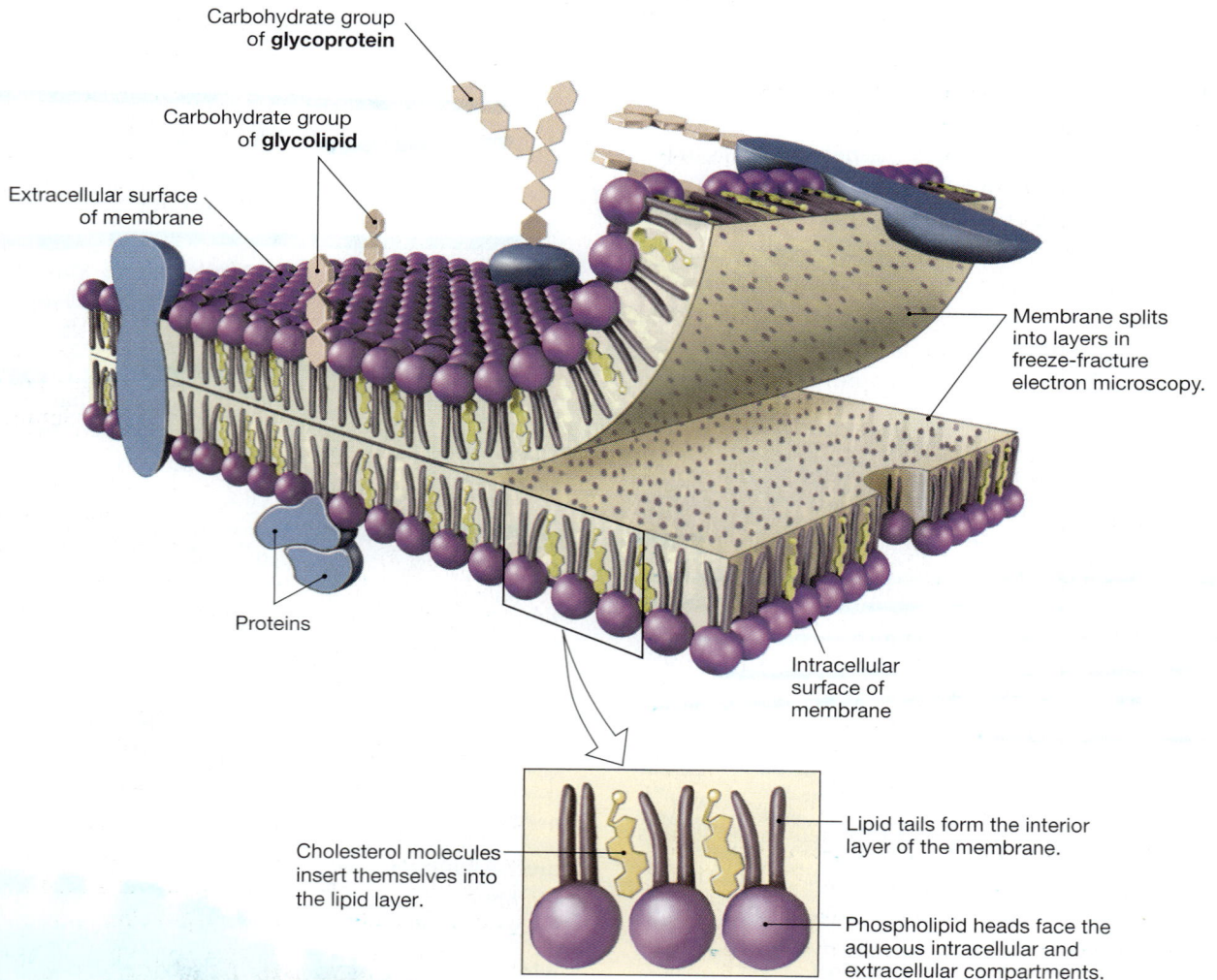

Carbohydrate group of **glycoprotein**

Carbohydrate group of **glycolipid**

Extracellular surface of membrane

Membrane splits into layers in freeze-fracture electron microscopy.

Proteins

Intracellular surface of membrane

Cholesterol molecules insert themselves into the lipid layer.

Lipid tails form the interior layer of the membrane.

Phospholipid heads face the aqueous intracellular and extracellular compartments.

■ **FIGURE 3-4** *The fluid mosaic model of a biological membrane*

In this model, the membrane is a double layer of phospholipid molecules. Proteins of various kinds are inserted into and through the phospholipid bilayer. Carbohydrates bind to proteins and lipids on the extracellular surface, creating glycoproteins and glycolipids.

metabolically active a membrane is, the more proteins it contains. For example, the inner mitochondrial membrane, which contains enzymes for ATP production, is three-quarters protein.

This chemical analysis of membranes was useful, but it did not explain how lipids and proteins were structurally arranged in a membrane. Studies in the 1920s suggested that there was enough lipid in a given area of membrane to create a double layer. The bilayer model was further modified in the 1930s to account for the presence of proteins. With the introduction of electron microscopy, scientists saw the cell membrane for the first time. The 1960s model of the membrane in electron micrographs was a "butter sandwich"—a clear layer of lipids sandwiched between two dark layers of protein.

By the early 1970s, freeze-fracture electron micrographs had revealed the actual three-dimensional arrangement of lipids and proteins within cell membranes. Because of what scientists learned from looking at freeze-fractured membranes, S. J. Singer and G. L. Nicolson in 1972 proposed the **fluid mosaic model** of the membrane. Figure 3-4 ■ highlights the major features of this contemporary model of membrane structure. The phospholipids are arranged in a bilayer so that the hydrophilic phosphate "heads" face the aqueous solutions inside and outside the cell, and the hydrophobic lipid "tails" are hidden in the center of the membrane. The membrane is studded with protein molecules, like raisins in a slice of bread, and the extracellular surface has glycoproteins and glycolipids. All cell membranes are of relatively uniform thickness, about 8 nm.

Membrane Lipids Form a Barrier Between the Cytoplasm and Extracellular Fluid

Three main types of lipids make up the cell membrane: phospholipids, sphingolipids, and cholesterol. As noted in Chapter 2, phospholipids are made of a glycerol backbone with two fatty acid chains extending to one side and a phosphate group extending to the other [p. 29]. The glycerol-phosphate head of the molecule is polar and thus hydrophilic (Fig. 3-5a ■). The fatty acid "tail" is nonpolar and thus hydrophobic.

When placed in an aqueous solution, phospholipids orient themselves so that the polar side of the molecule interacts with the water molecules while the nonpolar fatty acid tails "hide" by putting the polar heads between themselves and the water. This arrangement can be seen in three structures: the phospholipid bilayer, described in Figure 3-4; the micelle; and the liposome (Fig. 3-5b). **Micelles** are small droplets with a single layer of phospholipids arranged so that the interior of the micelle is filled with hydrophobic fatty acid tails. Micelles are important in the digestion and absorption of fats in the digestive tract.

Liposomes are larger spheres with bilayer phospholipid walls. This arrangement leaves a hollow center with an aqueous core that can be filled with water-soluble molecules. Biologists think that a liposome-like structure was the precursor of the first living cell. Today, liposomes are being used as a medium for the delivery of drugs through the skin.

Although phospholipids are the major lipid of membranes, some membranes have significant amounts of **sphingolipids**. Sphingolipids also have fatty acid tails, but their heads may be either phospholipids or glycolipids. Sphingolipids are slightly longer than phospholipids.

Cholesterol is also a significant part of many cell membranes. Cholesterol molecules, which are mostly hydrophobic, insert themselves between the hydrophilic heads of phospholipids (Fig. 3-4). Cholesterol helps make membranes impermeable to small water-soluble molecules and keeps membranes flexible over a wide range of temperatures.

Membrane Proteins May Be Loosely or Tightly Bound to the Membrane

According to some estimates, membrane proteins may be nearly one-third of all proteins coded in our DNA. Each cell has between 10 and 50 different types of proteins inserted into its membranes. Anatomically, membrane proteins are classified in three categories: integral proteins, peripheral proteins [*peripheria*, circumference], and lipid-anchored proteins (Fig. 3-6 ■).

Integral proteins, also called **transmembrane proteins** [*trans-* across], extend all the way across the cell membrane. They are tightly bound into the membrane because the 20–25 amino acids in the α-helix segments that pass through the bilayer are nonpolar. Being hydrophobic, these amino acids

(a)

Phospholipid molecules have polar heads and nonpolar tails.
The "R" group is a variable polar group.

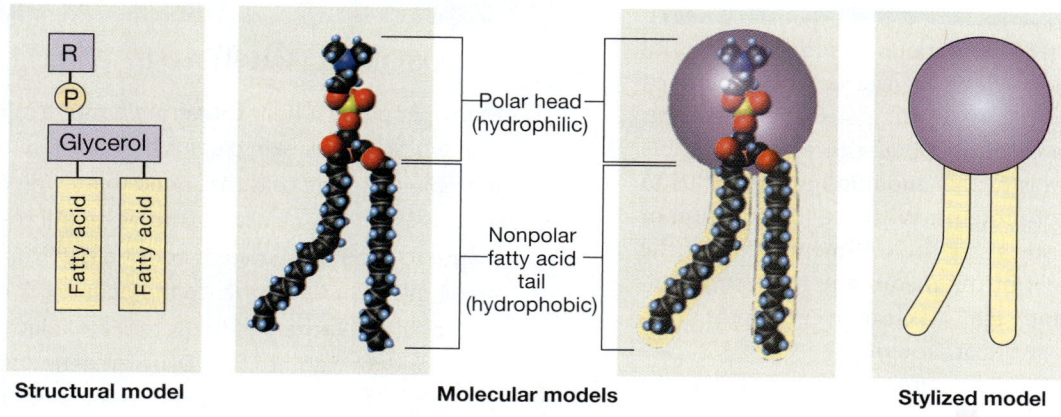

Structural model   Molecular models   Stylized model

Phospholipids arrange themselves so that their
nonpolar tails are not in contact with aqueous
solutions such as extracellular fluid.

(b)

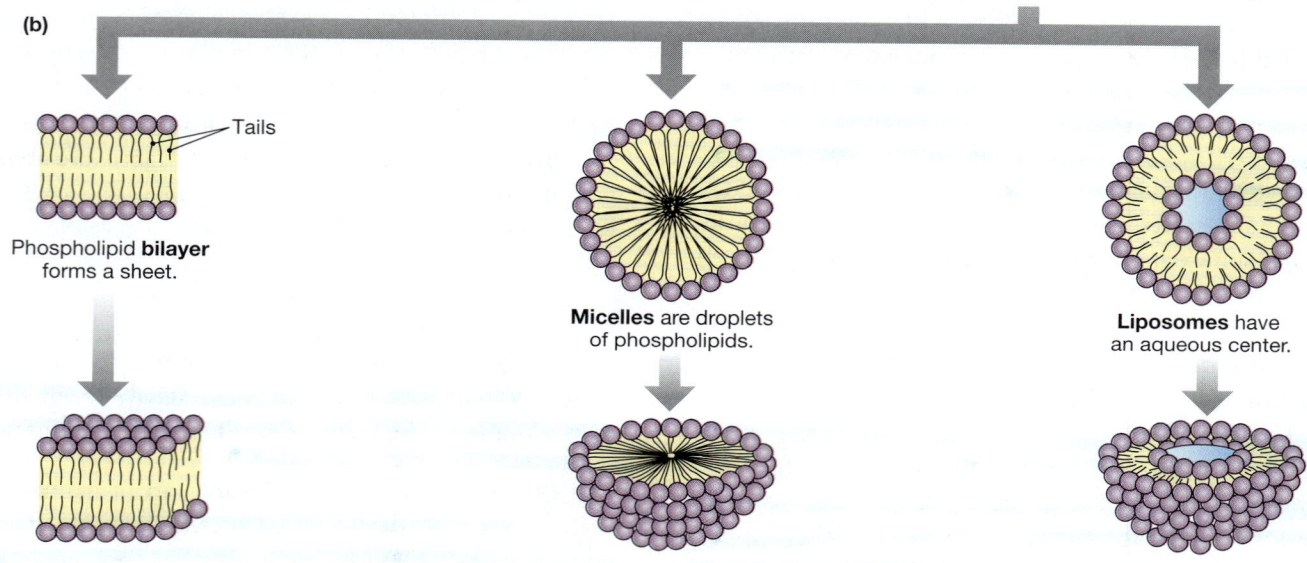

Tails

Phospholipid **bilayer**
forms a sheet.

Micelles are droplets
of phospholipids.

Liposomes have
an aqueous center.

■ **FIGURE 3-5**   *Membrane phospholipids form bilayers, micelles,
or liposomes*

Biological membranes are lipid bilayers. Micelles are important in lipid digestion.
Liposomes may resemble the first living cells.

create noncovalent interactions with the lipid tails of the phospholipids in the membrane bilayer [p. 25]. The only way transmembrane proteins can be removed is by disrupting the membrane structure with detergents or other harsh methods that destroy the membrane's integrity.

Transmembrane proteins are classified into families according to how many transmembrane segments they have. Many physiologically important membrane proteins have seven transmembrane segments (Fig. 3-7 ■), whereas others may have as few as one or as many as 12. When proteins cross the membrane more than once, the amino acid chains protrude into the cytoplasm and the extracellular fluid, as shown in

Figure 3-7. Carbohydrates may attach to the extracellular loops, and phosphate groups may attach to the intracellular loops. Phosphorylation of proteins is one regulatory method cells use to alter protein function [p. 41].

Peripheral proteins do not span the entire width of the cell membrane the way integral proteins do. Instead they attach themselves loosely to transmembrane proteins or to the polar heads of the phospholipids. They can be removed without disrupting the integrity of the membrane. Peripheral proteins include enzymes and some structural binding proteins that anchor the cytoskeleton to the membrane (Fig. 3-6).

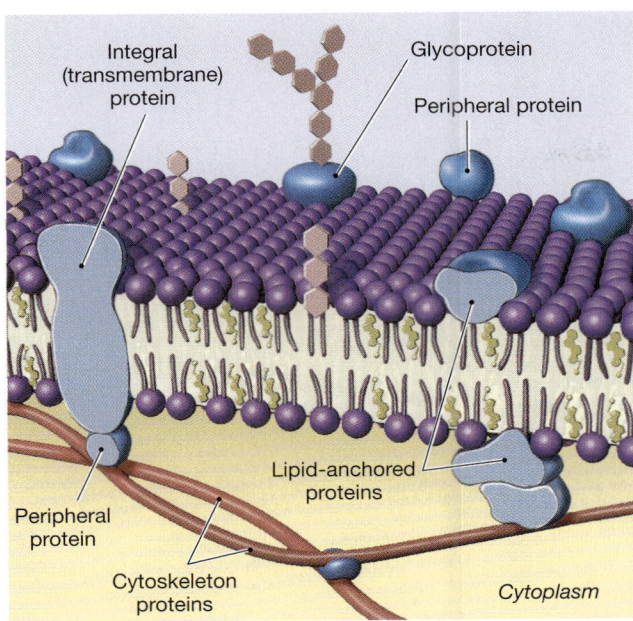

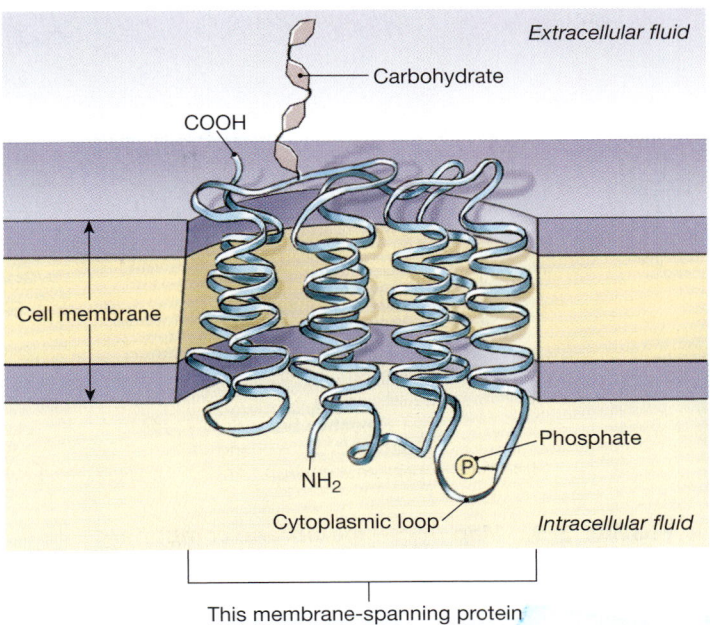

This membrane-spanning protein crosses the membrane seven times.

■ **FIGURE 3-6** *The three types of membrane proteins: integral, peripheral, and lipid-anchored*

Peripheral proteins can be removed without disrupting the integrity of the membrane.

Some membrane proteins that were previously thought to be peripheral proteins are now known to be **lipid-anchored proteins** (Fig. 3-6). These proteins are covalently bound to lipid tails that insert themselves into the bilayer. Many lipid-anchored proteins are found in association with membrane sphingolipids, leading to the formation of specialized patches of membrane

■ **FIGURE 3-7** *Integral proteins*

Membrane-spanning integral proteins have loops of peptide chains that extend into cytoplasm and extracellular fluid. Carbohydrates attach to the extracellular loops, and phosphate groups attach to the intracellular loops.

called *lipid rafts* (Fig. 3-8 ■). The longer tails of the sphingolipids elevate the lipid rafts over their phospholipid neighbors.

According to the original fluid mosaic model of the cell membrane, membrane proteins could move laterally from location to location, directed by protein fibers that run just under the

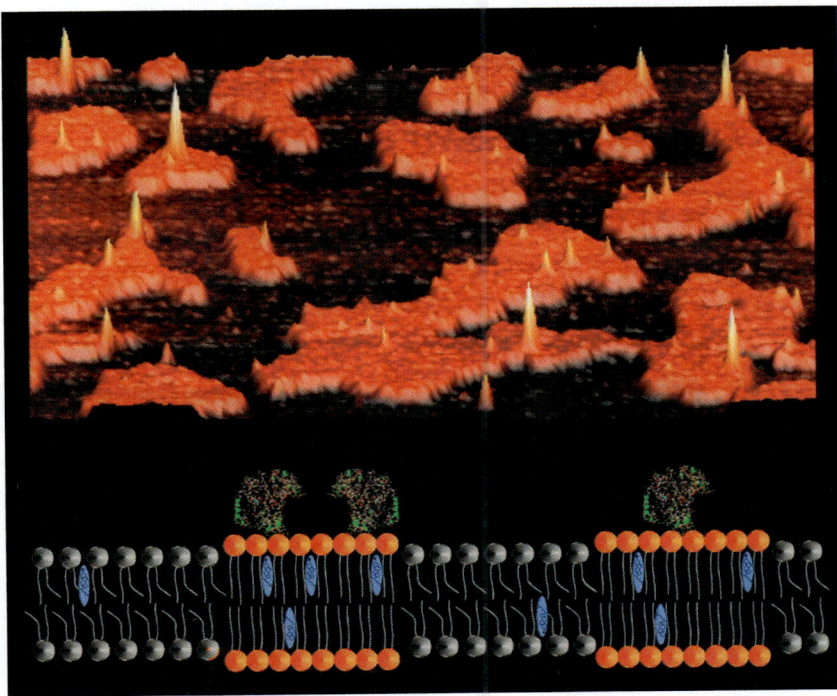

■ **FIGURE 3-8** *Lipid rafts are made of sphingolipids*

Sphingolipids (orange) are longer than phospholipids and stick up above the phospholipids of the membrane (black). A lipid-anchored enzyme, placental alkaline phosphatase (yellow), is almost always associated with a lipid raft. Image courtesy of DE Saslowsky, J Lawrence, X Ren, DA Brown, RM Henderson, and JM Edwardson. Placental alkaline phosphatase is efficiently targeted to rafts in supported lipid bilayers. *J. Biol. Chem.* 277: 26966–26970, 2002.

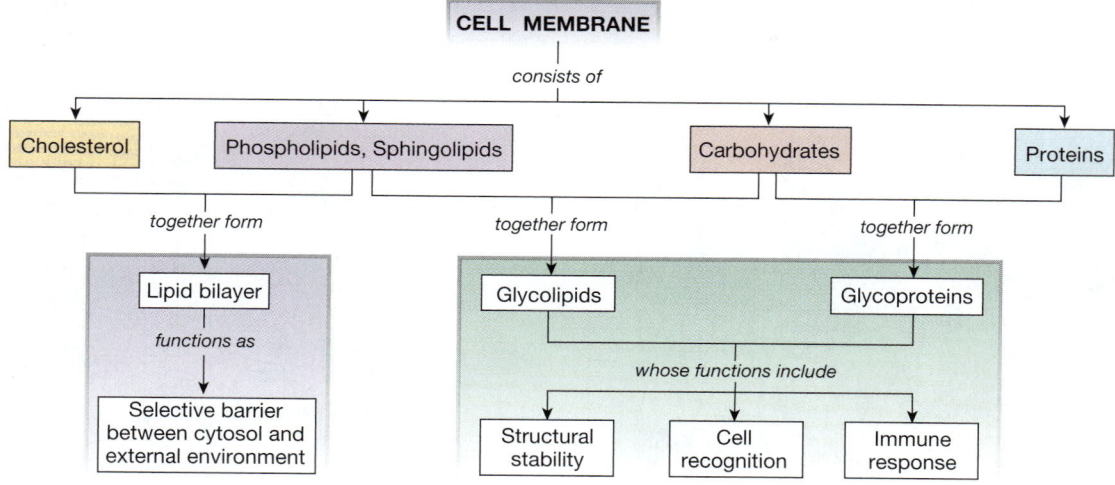

3

■ **FIGURE 3-9** *Map of cell membrane components*

membrane surface. However, researchers have learned that this is not true of all membrane proteins. Some integral proteins are anchored to cytoskeleton proteins (Fig. 3-6) and are therefore immobile. The ability of the cytoskeleton to restrict the movement of integral proteins allows cells to develop *polarity,* in which different faces of the cell have different proteins and therefore different properties. This is particularly important in the cells of the transporting epithelia, as you will learn in Chapter 5.

CONCEPT CHECK

1. Name three types of lipids found in cell membranes.
2. Describe three types of membrane proteins and how they are associated with the cell membrane.
3. Why do phospholipids in cell membranes form a bilayer instead of a single layer?
4. How many phospholipid bilayers does a substance passing through a mucous membrane cross?

Answers: p. 88

RUNNING PROBLEM

Cancer is a condition in which a small group of cells starts to divide uncontrollably and fails to differentiate into specialized cell types. Cancerous cells that originate in one tissue can escape from that tissue and spread to other organs through the circulatory system and the lymph vessels, a process known as *metastasis.*

Question 1:
   Why does the treatment of cancer focus on killing the cancerous cells?

51 58 67 75 81 85

Membrane Carbohydrates Attach to Both Lipids and Proteins

Most membrane carbohydrates are sugars attached either to membrane proteins (glycoproteins) or to membrane lipids (glycolipids). They are found exclusively on the external surface of the cell, where they form a protective layer known as the **glycocalyx** [*glyco-,* sweet + *kalyx,* husk or pod]. Glycoproteins on the cell surface play a key role in the body's immune response. An example with which you might be familiar is the ABO blood groups, which are determined by the number and composition of sugars attached to membrane sphingolipids.

Figure 3-9 ■ is a summary map organizing the structure of the cell membrane.

INTRACELLULAR COMPARTMENTS

Much of what we know about cells comes from studies of simple organisms that consist of one cell. But humans are much more complex, with trillions of cells in their bodies. It has been estimated that there are more than 200 different types of cells in the human body, each cell type with its own characteristic structure and function.

During development, cells specialize and take specific shapes and functions. Each cell in the body inherits identical genetic information in its DNA, but no one cell uses all of it. During **differentiation,** only selected genes are activated, transforming the cell into a specialized unit. In most cases, the final shape and size of a cell and its contents reflect its function. Figure 3-10 ■ shows some of the different cells in the human body. Although these mature cells look very different from one another, they all started out alike and they retain many features in common.

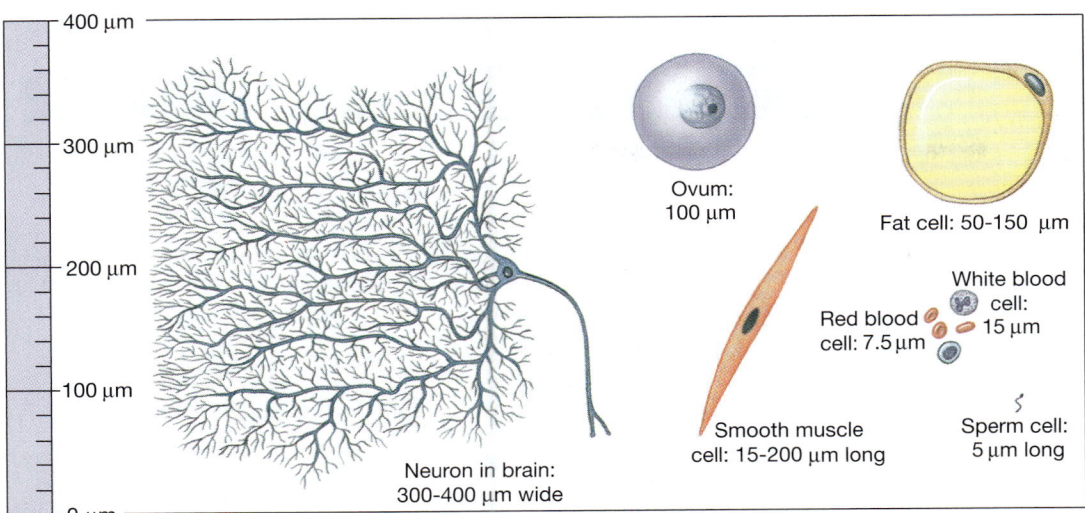

3

■ **FIGURE 3-10**

Representative cell types in the human body

Each cell has the genetic potential to become any type of cell, but selective activation of genes during differentiation determines the ultimate size, shape, and function of the cell.

Cells Are Divided into Compartments

The structural organization of a cell can be compared to that of a medieval walled city. The city is separated from the surrounding countryside by a high wall, with entry and exit strictly controlled through gates that can be opened and closed. The city inside the walls is divided into streets and a diverse collection of houses and shops with varied functions. Within the city, a ruler in the castle oversees the everyday comings and goings of the city's inhabitants. Because the city depends on food and raw material from outside the walls, the ruler negotiates with the farmers in the countryside. Foreign invaders are always a threat, so the city ruler communicates and cooperates with the rulers of neighboring cities.

In the cell, the outer boundary is the cell membrane. Like the city wall, it controls the movement of material between the cell interior and the outside by opening and closing "gates" made of protein. The inside of the cell is divided into compartments rather than into shops and houses. Each of these compartments has a specific purpose that contributes to the function of the cell as a whole. In the cell, DNA in the nucleus is the "ruler in the castle," controlling both the internal workings of the cell and its interaction with other cells. Like the city, the cell depends on supplies from its external environment. It must also communicate and cooperate with other cells in order to keep the body functioning in a coordinated fashion.

Figure 3-11 ■ is an overview map of cell structure. The cells of the body are surrounded by the dilute salt solution of the extracellular fluid. The cell membrane separates the inside environment of the cell (the intracellular fluid) from the extracellular fluid.

Internally the cell is divided into the *cytoplasm* and the *nucleus*. The cytoplasm consists of a fluid portion, called the *cytosol;* insoluble particles called *inclusions;* and membrane-bound structures collectively known as *organelles*. A typical cell from the lining of the small intestine is shown in Figure 3-12 ■. It has most of the structures found in animal cells.

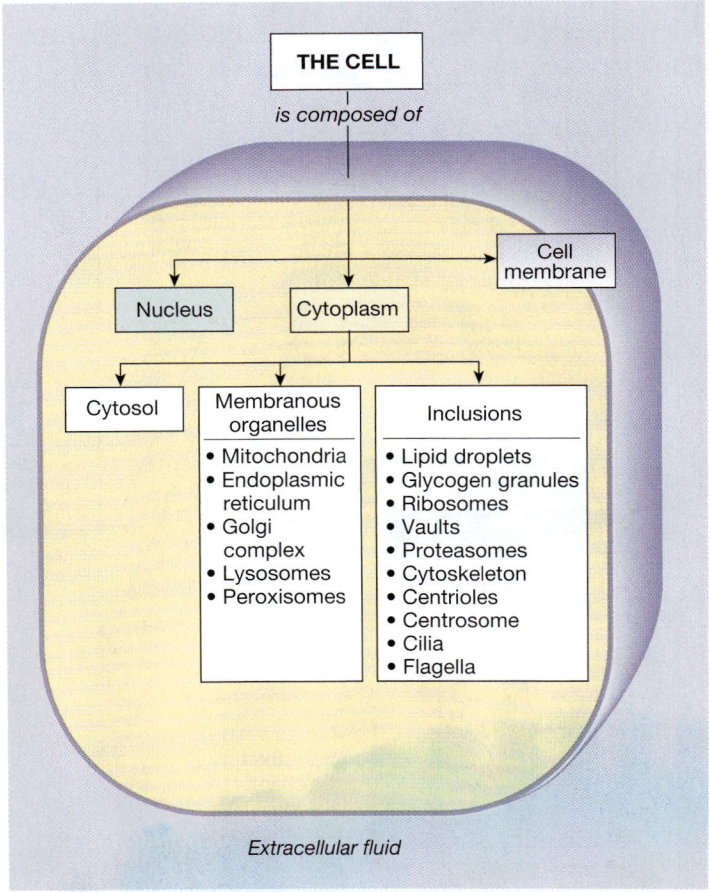

■ **FIGURE 3-11**   ***A map for the study of cell structure***

3

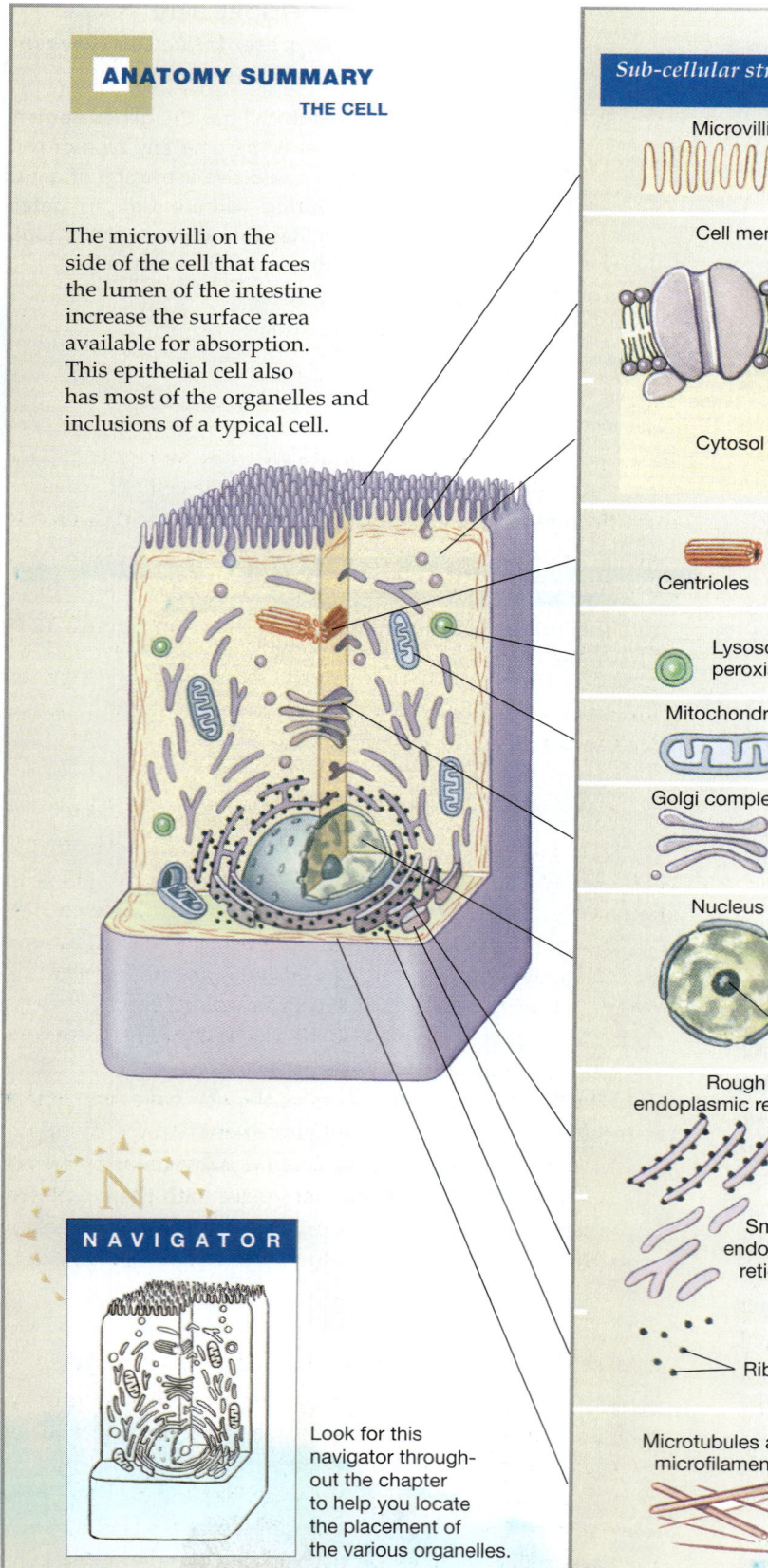

ANATOMY SUMMARY
THE CELL

The microvilli on the side of the cell that faces the lumen of the intestine increase the surface area available for absorption. This epithelial cell also has most of the organelles and inclusions of a typical cell.

NAVIGATOR

Look for this navigator throughout the chapter to help you locate the placement of the various organelles.

Sub-cellular structures	Characteristics	Functions
Microvilli	Extensive folding of the cell membrane	Increase surface area for absorption
Cell membrane	Bilayer of phospholipid molecules with inserted proteins	Acts as a gateway for and barrier to movement of material between the interior of the cell and the extracellular fluid
Cytosol	Semi-gelatinous substance	Contains dissolved nutrients, ions, wastes, insoluble inclusions; suspends the organelles
Centrioles	Bundles of microtubules	Direct movement of DNA during cell division
Lysosomes and peroxisomes	Membrane-bound vesicles filled with enzymes	Digest bacteria, old organelles; metabolize fatty acids
Mitochondrion	Double wall with central matrix	Produces most of cell's ATP
Golgi complex	Hollow membranous sacs	Modify and package proteins
Nucleus	Central lumen with a two-membrane outer envelope with pores	Contains DNA to direct all functions of the cell
Nucleolus	Region of DNA, RNA, and protein	Contains the genes that direct synthesis of ribosomal RNA
Rough endoplasmic reticulum	Membrane tubules that are continuous with the outer nuclear membrane; edged with ribosomes	Site of protein synthesis
Smooth endoplasmic reticulum	Same as rough ER without ribosomes	Synthesizes fatty acids, steroids, and lipids
Ribosomes	Granules of RNA and protein	Assemble amino acids into proteins
Microtubules and microfilaments	Protein fibers	Provide strength and support; enable cell motility; transport

■ **FIGURE 3-12**

TABLE 3-2	Diameter of Protein Fibers in the Cytoplasm		
	DIAMETER	**TYPE OF PROTEIN**	**FUNCTIONS**
Microfilaments	8–9 nm	Actin (globular)	Cytoskeleton; associates with myosin for muscle contraction
Intermediate filaments	10 nm	Myosin, neurofilament protein, keratin (filaments)	Cytoskeleton, hair and nails, protective barrier of skin; myosin forms thick filaments for muscle contraction
Microtubules	24–25 nm	Tubulin (globular)	Movement of cilia, flagella, and chromosomes; intracellular transport of organelles; cytoskeleton

The Cytoplasm Includes the Cytosol, Inclusions, and Organelles

The **cytoplasm** includes all material inside the cell membrane except for the nucleus. The cytoplasm has three components:

1. **Cytosol**, or intracellular fluid: The cytosol [*cyto-*, cell + *sol(uble)*] is a semi-gelatinous substance separated from the extracellular fluid by the cell membrane. The cytosol contains dissolved nutrients and proteins, ions, and waste products. The other two components of the cytoplasm—inclusions and organelles—are suspended in the cytosol.

2. **Inclusions** are particles of insoluble materials. Some are stored nutrients, but others are responsible for specific cell functions. The latter group is sometimes called the *nonmembranous organelles*.

3. **Organelles**—"little organs"—are like the organs of the body in that each has a specific role to play in the overall function of the cell. For example, the organelles called mitochondria (singular, *mitochondrion*) generate most of the cell's ATP, and the organelles called lysosomes act as the digestive system of the cell. The organelles work in an integrated manner, each organelle taking on one or more of the cell's functions.

Inclusions Are in Direct Contact with the Cytosol

The inclusions of cells do not have boundary membranes and so are in direct contact with the cytosol. Movement of material between inclusions and the cytosol does not require transport across a membrane. Nutrients are stored as glycogen granules and lipid droplets. Most inclusions with functions other than energy storage are made from protein or combinations of RNA and protein. They include ribosomes, proteasomes, vaults, and protein fibers.

Ribosomes (Fig. 3-12) are small, dense granules of RNA and protein that manufacture proteins under the direction of the cell's DNA (see Chapter 4). Ribosomes attached to the cytosolic face of organelles are called **fixed ribosomes**. Those suspended free in the cytosol are **free ribosomes**. Some free ribosomes form groups of 10 to 20 known as **polyribosomes**. A ribosome that is fixed one minute may release and become a free ribosome the next. Ribosomes are most numerous in cells that synthesize proteins for export out of the cell.

Proteasomes are hollow protein cylinders with a protein cap on each end. They have been described as "nanomachines" [*nanus*, dwarf] because they are the cell's site for targeted protein degradation. **Vaults**, first described in 1986, are made of RNA and proteins. Like proteasomes, vaults are hollow, barrel-shaped particles, but their function is still unknown. One vault protein is associated with tumor resistance to certain cancer-treating drugs.

Cytoplasmic Protein Fibers Come in Three Sizes

The three families of cytoplasmic protein fibers are classified by diameter and protein composition (Table 3-2 ■). All fibers are polymers of smaller proteins. The thinnest are **actin fibers**, also called *microfilaments*. Somewhat larger **intermediate filaments** may be made of different types of protein, including *myosin* in muscle, *keratin* in hair and skin, and *neurofilament* in nerve cells. The largest protein fibers are the hollow **microtubules**, made of a protein called **tubulin**.

The insoluble protein fibers of the cell have two general purposes: structural support and movement. Structural support comes primarily from the cytoskeleton. Movement of the cell or of elements within the cell takes place with the aid of protein fibers and a group of specialized enzymes called motor proteins. These functions are discussed in more detail in the sections that follow.

Microtubules Form Centrioles, Cilia, and Flagella

The largest cytoplasmic protein fibers, the microtubules, create the complex structures of centrioles, cilia, and flagella, which are all involved in some form of cell movement. The cell's *microtubule-organizing center*, the **centrosome**, assembles tubulin monomers into microtubules. The centrosome appears as a region of darkly staining material close to the cell nucleus. In most animal cells, the centrosome contains two **centrioles**, shown in the typical cell of Figure 3-12. Each centriole is a cylindrical bundle of 27 microtubules, arranged in nine triplets (Fig. 3-13a ■). In cell division,

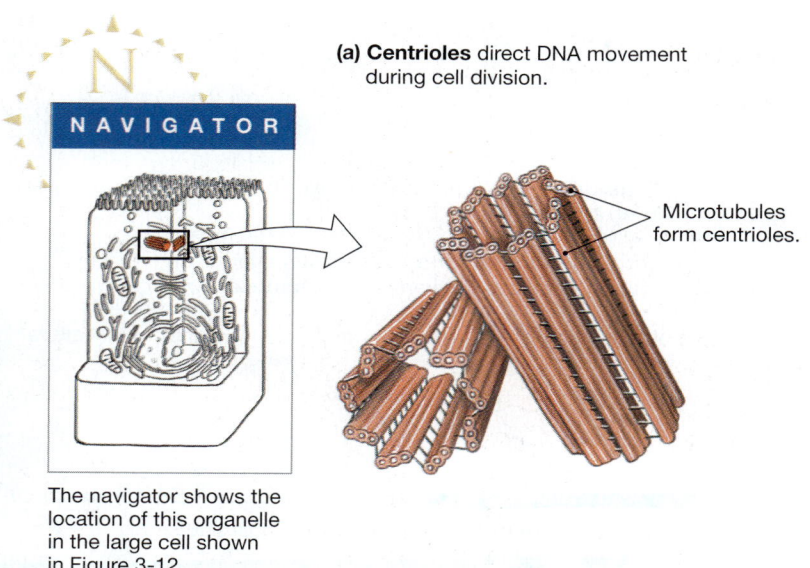

(a) Centrioles direct DNA movement during cell division.

NAVIGATOR

The navigator shows the location of this organelle in the large cell shown in Figure 3-12.

Microtubules form centrioles.

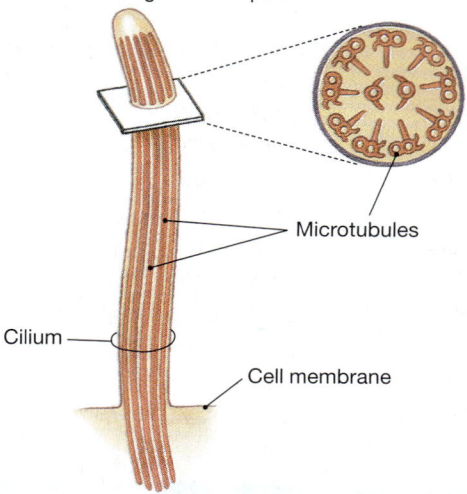

(c) Cilia and flagella have 9 pairs of microtubules surrounding a central pair.

Microtubules

Cilium

Cell membrane

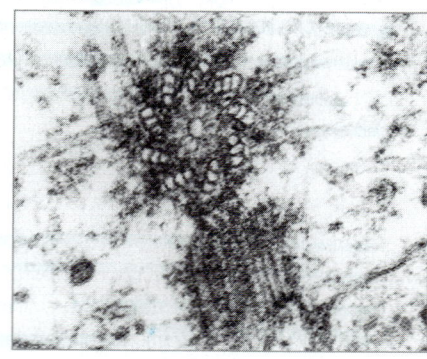

(b) A pair of centrioles

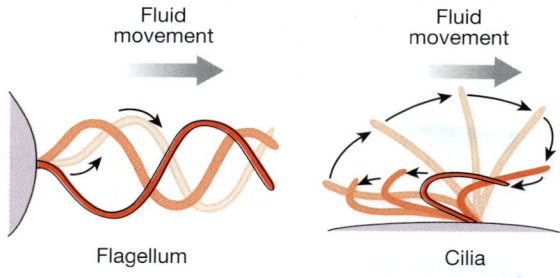

(d) The beating of cilia and flagella creates fluid movement.

Fluid movement

Fluid movement

Flagellum

Cilia

■ **FIGURE 3-13** *Centrioles, cilia, and flagella*

All three structures are formed from microtubules.

the centrioles direct the movement of DNA strands (see Appendix B). Cells that have lost their ability to undergo cell division, such as mature nerve cells, lack centrioles.

Cilia [singular, *cilium,* Latin for eyelash] are short, hairlike structures projecting from the cell surface like the bristles of a brush. The surface of a cilium is a continuation of the cell membrane, and its core contains nine pairs of microtubules surrounding a central pair (Fig. 3-13c). The microtubules terminate just inside the cell at the *basal body*. Cilia beat rhythmically back and forth when the microtubule pairs in their core slide past each other with the help of the motor protein dynein. Ciliary movement creates currents that sweep fluids or secretions across the cell surface.

Flagella [singular, *flagellum,* Latin for whip] have the same microtubule arrangement as cilia but are considerably longer. In addition, a flagellated animal cell usually has only one or two flagella, whereas ciliated cells may have one surface almost totally covered with cilia. Flagella are found on free-floating single cells; the only such cell in humans is the sperm cell of

males (see Fig. 3-10). The flagellum pushes the cell through fluid with its wavelike movements, just as undulating contractions of a snake's body push it headfirst through its environment. Flagella bend and move by the same basic mechanism as cilia.

The Cytoskeleton Is a Changeable Scaffold

The cytoskeleton is a flexible, changeable three-dimensional scaffolding of actin microfilaments, intermediate filaments, and microtubules that extends throughout the cytoplasm. Some cytoskeleton protein fibers are permanent, but most are synthesized or disassembled according to the cell's needs. Because of the cytoskeleton's changeable nature, its organizational details are complex and will not be discussed in detail here.

The cytoskeleton has at least five important functions.

1. **Cell shape.** The protein scaffolding of the cytoskeleton provides mechanical strength to the cell and in some cells plays an important role in determining the shape of the

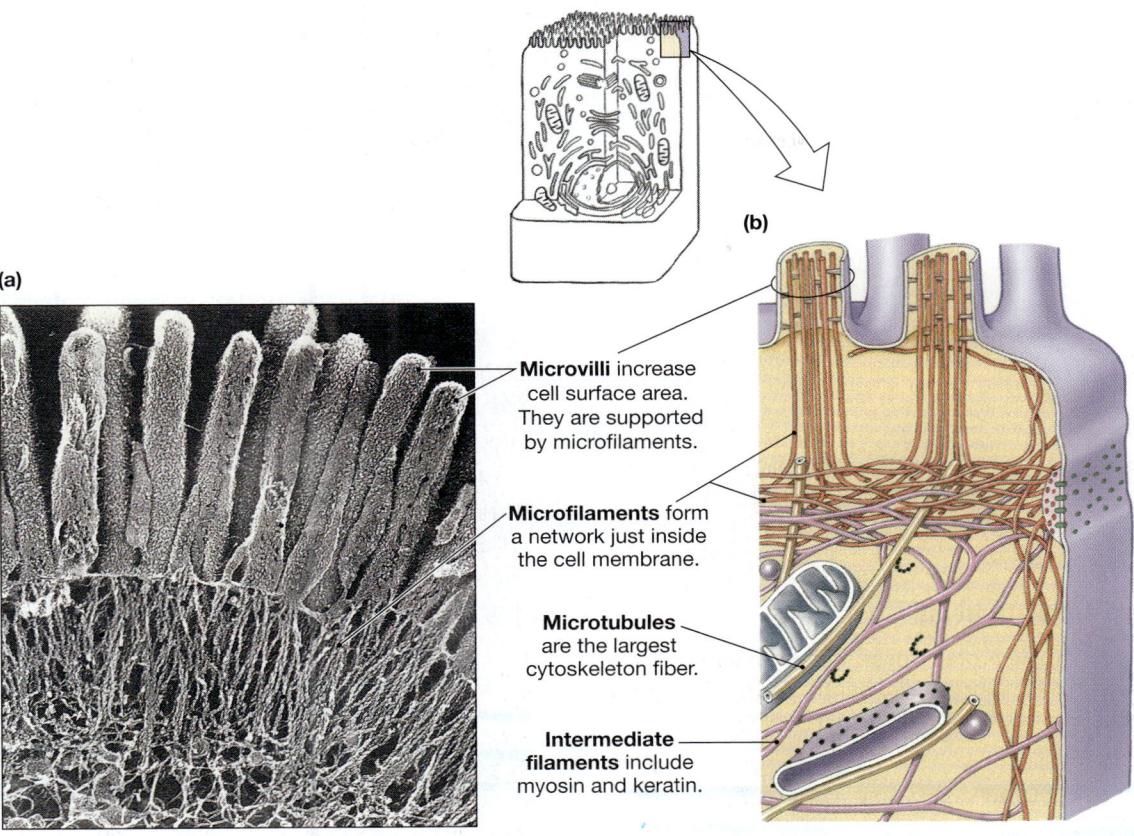

(a)

(b)

Microvilli increase cell surface area. They are supported by microfilaments.

Microfilaments form a network just inside the cell membrane.

Microtubules are the largest cytoskeleton fiber.

Intermediate filaments include myosin and keratin.

■ **FIGURE 3-14**   *The cytoskeleton and cytoplasmic protein fibers*

The cytoskeleton anchors some organelles in place and helps create cell shape.

cell. Figure 3-14a ■ shows how cytoskeletal fibers help support **microvilli** [*micro-*, small + *villus,* tuft of hair], fingerlike extensions of the cell membrane that increase the surface area for absorption of materials.

2. **Internal organization.**   Cytoskeletal fibers stabilize the positions of organelles. Figure 3-14b is a diagram of a typical cell with organelles held in place by the cytoskeleton. Note, however, that this figure is only a snapshot of one moment in the cell's life. The interior arrangement and composition of a cell are dynamic, changing from minute to minute in response to the needs of the cell, just as the inside of the walled city is always in motion. One disadvantage of the static illustrations in textbooks is that they are unable to accurately represent movement and the dynamic nature of many physiological processes.

3. **Intracellular transport.**   The cytoskeleton helps transport materials into the cell and within the cytoplasm, serving as an intracellular "railroad track" for moving organelles. This function is particularly important in cells of the nervous system, where material must be transported over intracellular distances as long as a meter.

4. **Assembly of cells into tissues.**   Protein fibers of the cytoskeleton connect with protein fibers in the extracellular space, linking cells to one another and to supporting material outside the cells. In addition to providing mechanical

strength to the tissue, these linkages allow the transfer of information from one cell to another.

5. **Movement.**   The cytoskeleton helps cells move, from white blood cells that squeeze out of blood vessels to growing nerve cells that send out long extensions as they elongate. Cilia and flagella on the cell membrane are able to move because of their microtubule cytoskeleton. Both movement and intracellular transport are facilitated by special motor proteins that use energy from ATP to slide or step along cytoskeletal fibers.

Motor Proteins Create Movement

Motor proteins are proteins that are able to convert stored energy into directed movement. There are three groups of motor proteins associated with the cytoskeleton: myosins, kinesins, and dyneins. All three groups use energy stored in ATP to propel themselves along cytoskeleton fibers. **Myosins** bind to actin fibers and are best known for their role in muscle contraction, described in detail in Chapter 12. **Kinesins** and **dyneins** are associated with movement along microtubules, and dyneins associated with the microtubule bundles of cilia and flagella help create their whiplike motion.

Most motor proteins are composed of multiple protein chains arranged into two heads that bind to the cytoskeleton

3

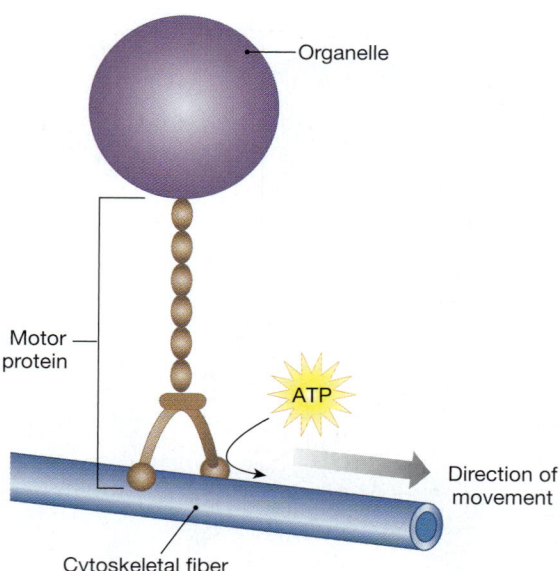

Organelle

Motor
protein

ATP

Direction of
movement

Cytoskeletal fiber

■ **FIGURE 3-15**   *Motor proteins*

Motor proteins have multiple protein chains that form two heads, a
neck, and a tail that can bind to organelles or other cargo. The heads
"walk" along cytoskeletal fibers with the help of energy from ATP.

fiber, a neck, and a tail region that is able to bind "cargo," such
as an organelle that needs to be transported through the cytoplasm (Fig. 3-15 ■). The heads alternately bind to the cytoskeleton fiber, then release and "step" forward. Our current
model of this process indicates that each step requires the energy stored in one ATP molecule.

CONCEPT CHECK

5. Name the three sizes of cytoplasmic protein fibers.
6. How would the absence of a flagellum affect a
   sperm cell?
7. What is the difference between cytoplasm and
   cytosol?
8. What is the difference between a cilium and a
   flagellum?
9. What is the function of motor proteins?   Answers: p. 88

Organelles Create Compartments for Specialized Functions

Organelles are separated from the cytosol by one or more phospholipid membranes similar in structure to the cell membrane.
These membranes create compartments that allow the cell to
isolate substances and separate functions. For example, an organelle might contain substances that could be harmful to the
cell, such as digestive enzymes. The process of protein modification and packaging is another example of separation of function. Proteins are modified in one organelle, packaged in
another, and stored in a third. Figure 3-12 shows the four major
groups of organelles: mitochondria, the endoplasmic reticulum, the Golgi complex, and lysosomes and peroxisomes.

Mitochondria   Mitochondria [singular, *mitochondrion; mitos,*
thread + *chondros,* granule] are small spherical to elliptical organelles with an unusual double wall (Fig. 3-16 ■). The outer
membrane of the wall gives the mitochondrion its shape. The

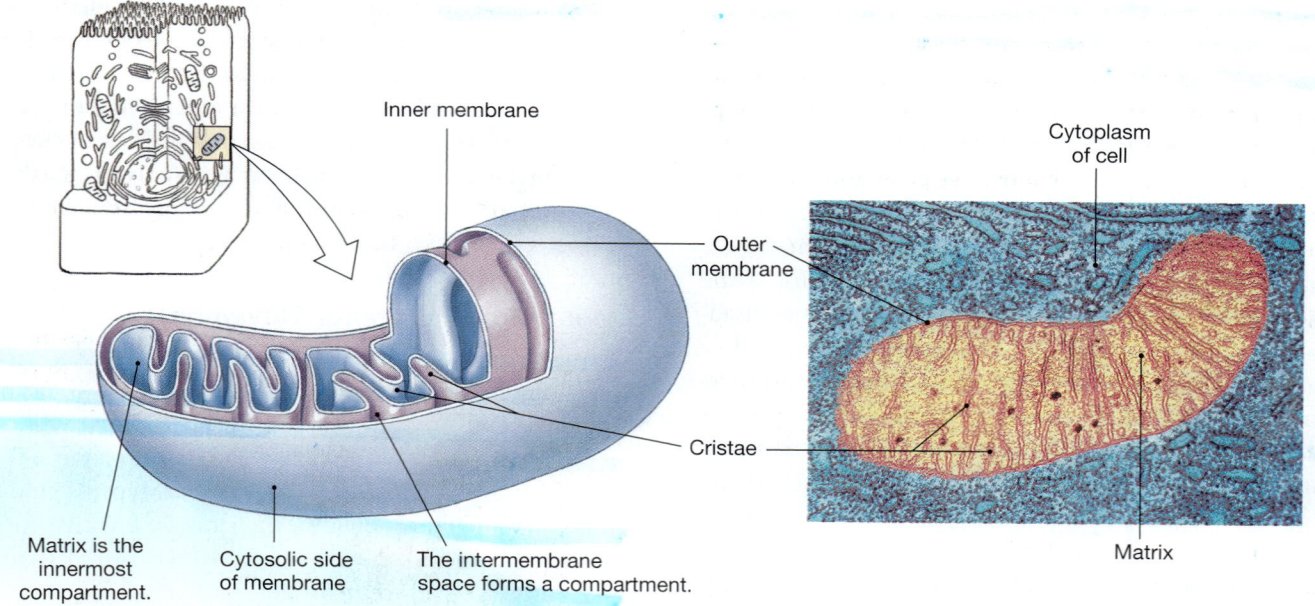

Inner membrane

Outer
membrane

Cytoplasm
of cell

Cristae

Matrix

Matrix is the
innermost
compartment.

Cytosolic side
of membrane

The intermembrane
space forms a compartment.

■ **FIGURE 3-16**   *Mitochondria*

Mitochondria have a double wall structure with a heavily folded inner membrane.
The double wall creates two compartments: the inner mitochondrial matrix and an intermembrane space.

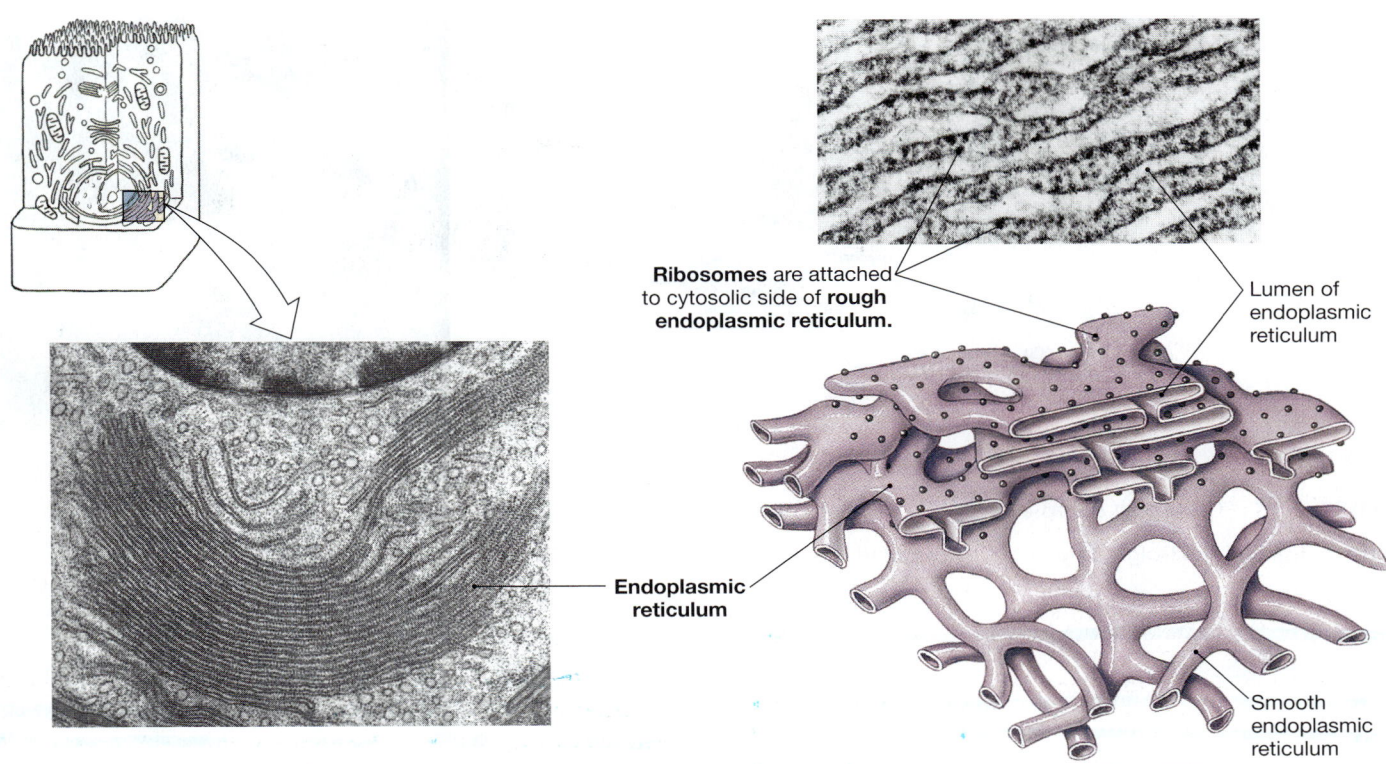

■ **FIGURE 3-17** *The endoplasmic reticulum*

Rough ER has attached ribosomes and is a major site of protein synthesis. Smooth ER
lacks ribosomes and is the site of lipid synthesis.

inner membrane is folded into leaflets or tubules called **cristae**. The double membrane creates two separate compartments within the mitochondrion. In the center, inside the inner membrane, is the compartment called the **mitochondrial matrix** [*matrix*, female animal for breeding]. The matrix contains enzymes, ribosomes, granules, and DNA. Between the outer and inner membranes is the **intermembrane space**, a compartment that plays an important role in the production of ATP by the mitochondria.

Mitochondria are the site of most ATP generation, which is why they are nicknamed the cell's "powerhouse." The number of mitochondria in a particular cell depends on the cell's energy needs. Cells such as skeletal muscle cells, which use a lot of energy, have many more mitochondria than less active cells, such as adipose (fat) cells.

Mitochondria are unusual organelles in two ways. First of all, in their matrix they have their own unique DNA. This **mitochondrial DNA**, along with matrix RNA, means that mitochondria can manufacture some of their own proteins. Why is this true of mitochondria and not other organelles? This question has been the subject of intense scrutiny. According to the *prokaryotic endosymbiont theory*, mitochondria are the descendants of bacteria that invaded cells millions of years ago. The bacteria developed a mutually beneficial relationship with their hosts and soon became an integral part of the host cells. Supporting evidence for this theory is the fact that our mitochon-

dria contain DNA, RNA, and enzymes similar to those found in bacteria but unlike those found in our own cell nuclei.

The second unusual characteristic of mitochondria is their ability to replicate themselves even when the cell to which they belong is not undergoing cell division. This process is aided by the presence of mitochondrial DNA, which allows the organelles to direct their own duplication. Mitochondrial replication takes place by budding, with small daughter mitochondria pinching off an enlarged parent. Cells such as exercising muscle cells, which are subjected to increased energy demands over a period of time, may meet the demand for more ATP by increasing the number of mitochondria in their cytoplasm.

The Endoplasmic Reticulum The **endoplasmic reticulum**, or **ER**, is a network of interconnected membrane tubes (Fig. 3-17 ■) that are a continuation of the outer membrane that surrounds the cell nucleus. The name *reticulum* comes from the Latin word for *net* and refers to the netlike appearance of the tubules. Electron micrographs reveal that there are two forms of endoplasmic reticulum: **rough endoplasmic reticulum** (RER) and **smooth endoplasmic reticulum** (SER). Rough endoplasmic reticulum has a granular, or rough, appearance from rows of ribosomes dotting its cytoplasmic surface. Smooth endoplasmic reticulum lacks the ribosomes and appears as smooth membrane tubes. Both types of endoplasmic reticulum

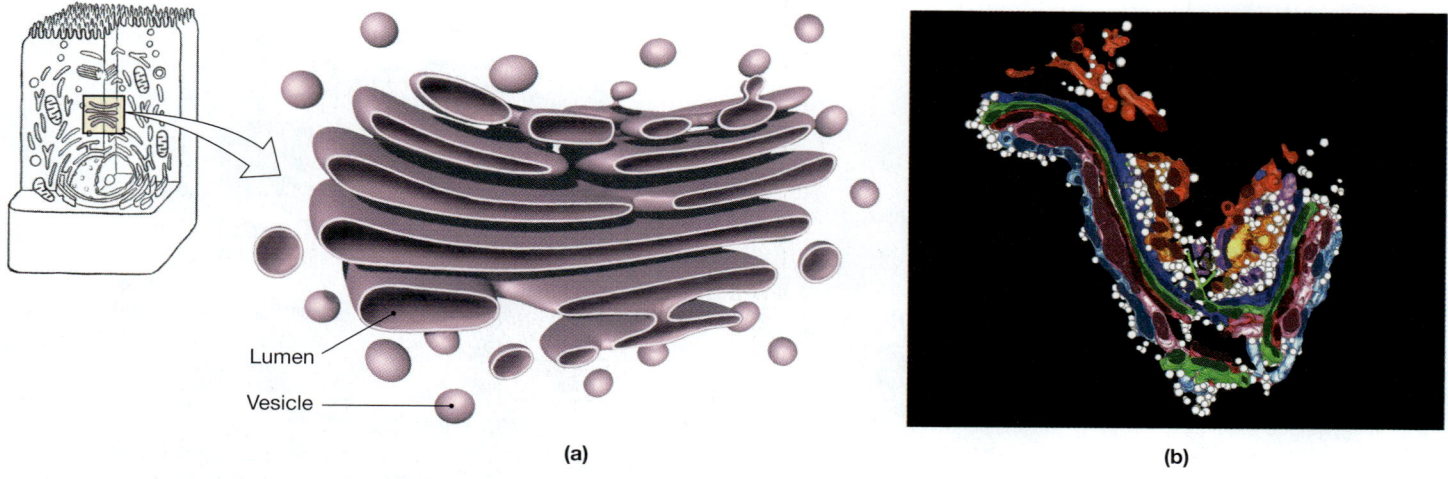

(a)   **(b)**

■ **FIGURE 3-18**   *The Golgi complex*

The flattened sacs of the Golgi complex are surrounded by vesicles.

have the same three major functions: synthesis, storage, and transport of biomolecules.

The smooth endoplasmic reticulum is the main site for the synthesis of fatty acids, steroids, and lipids [⊇ p. 29]. Phospholipids for the cell membrane are produced here, and cholesterol is modified into steroid hormones, such as the sex hormones estrogen and testosterone. The smooth endoplasmic reticulum of liver and kidney cells detoxifies or inactivates drugs. In skeletal muscle cells, a modified form of smooth endoplasmic reticulum stores calcium ions (Ca^{2+}) to be used in muscle contraction.

The rough endoplasmic reticulum is the main site for the synthesis of proteins. Proteins are assembled on ribosomes attached to the cytoplasmic surface of the rough endoplasmic reticulum, then inserted into the RER lumen, where they undergo chemical modification.

The Golgi Complex   The **Golgi complex** (Fig. 3-18 ■) was first described by Camillo Golgi in 1898. For years, some investigators thought that this organelle was just a result of the fixation process needed to prepare tissues for viewing under the light microscope. However, we now know from electron microscope studies that the Golgi complex is indeed a discrete organelle. It consists of a series of hollow curved sacs stacked on top of one another like a series of hot water bottles and surrounded by membrane-bound spheres called **vesicles** [*vesicula,* bladder]. The Golgi complex receives proteins made on the rough endoplasmic reticulum, modifies them, and packages them into vesicles. The details of protein synthesis, modification, and packaging are discussed in Chapter 4.

Cytoplasmic Vesicles   Membrane-bound cytoplasmic vesicles are of two kinds: secretory and storage. **Secretory vesicles** contain proteins that will be exported to other parts of the body. The contents of most **storage vesicles**, however, never leave the cytoplasm.

Lysosomes [*lysis,* dissolution + *soma,* body] are small, spherical storage vesicles that appear as membrane-bound granules in the cytoplasm (Fig. 3-19 ■). Lysosomes act as the digestive system of the cell. They use powerful enzymes to break down bacteria or old organelles, such as mitochondria, into their component molecules. Those molecules that can be

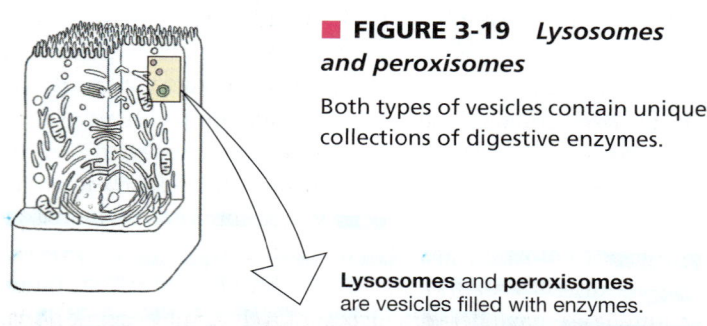

■ **FIGURE 3-19**   *Lysosomes and peroxisomes*

Both types of vesicles contain unique collections of digestive enzymes.

Lysosomes and **peroxisomes** are vesicles filled with enzymes.

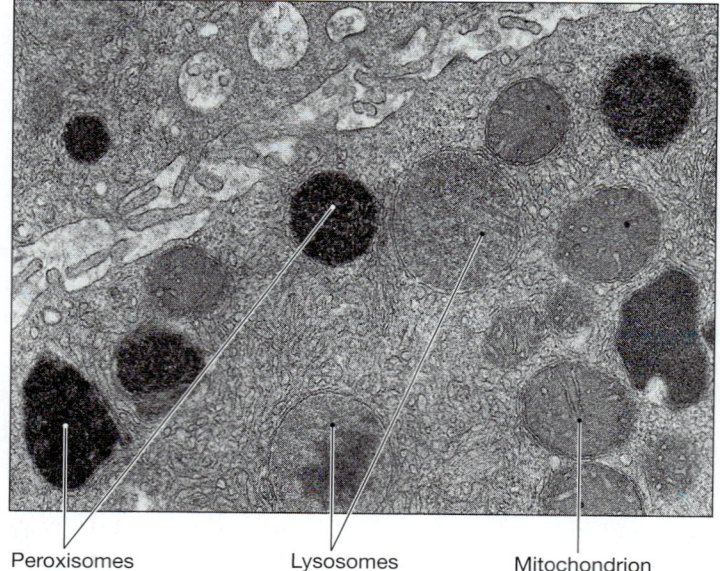

Peroxisomes   Lysosomes   Mitochondrion

reused are reabsorbed into the cytosol, while the rest are dumped out of the cell. As many as 50 types of enzymes have been identified from lysosomes of different cell types.

Because lysosomal enzymes are so powerful, one question puzzling early workers was why these enzymes do not normally destroy the cell that contains them. What scientists discovered was that lysosomal enzymes are activated only by very acidic conditions, 100 times more acidic than the normal acidity level in the cytoplasm. When lysosomes are first pinched off from the Golgi complex, their interior pH is about the same as that of the cytosol, 7.0–7.3. The enzymes are inactive at this pH. Their inactivity serves as a form of insurance. If the lysosome breaks or accidentally releases the enzyme, it will not harm the cell. However, as the lysosome sits in the cytoplasm, it accumulates H^+ in a process that uses energy. The pH inside the vesicle drops to 4.8–5.0, and the enzymes are activated. Once this has occurred, the enzymes are capable of breaking down various biomolecules in the lumen. The lysosomal membrane is not affected by the enzymes.

The digestive enzymes of lysosomes are not always kept isolated within the organelle. Occasionally, lysosomes release their enzymes outside the cell to dissolve extracellular support material, such as the hard calcium carbonate portion of bone. In other instances, cells allow the enzymes of their lysosomes to come in contact with the cytoplasm, leading to self-digestion of all or part of the cell. When muscles *atrophy* (shrink) from lack of use or the uterus diminishes in size after pregnancy, the loss of cell mass is due to the action of lysosomes.

The inappropriate release of lysosomal enzymes has been implicated in certain disease states, such as the inflammation and destruction of joint tissue in *rheumatoid arthritis*. In the inherited conditions known as *lysosomal storage diseases*, lysosomes are ineffective because they lack specific enzymes. One of the best-known lysosomal storage diseases is the fatal inherited condition known as *Tay-Sachs disease*. Infants with Tay-Sachs disease have defective lysosomes that fail to break down glycolipids. Accumulation of glycolipids in nerve cells causes nervous system dysfunction, including blindness and loss of coordination. Most infants afflicted with Tay-Sachs disease die in early childhood.

Peroxisomes are storage vesicles that are even smaller than lysosomes (Fig. 3-19). For years, they were thought to be a kind of lysosome, but we now know that they contain a different set of enzymes. Their main function appears to be to degrade long-chain fatty acids and potentially toxic foreign molecules.

Peroxisomes get their name from the fact that the reactions that take place inside them generate hydrogen peroxide (H_2O_2), a toxic molecule. The peroxisomes rapidly convert this peroxide to oxygen and water using the enzyme *catalase*. Peroxisomal disorders disrupt the normal processing of lipids and can severely disrupt neural function by altering the structure of nerve cell membranes.

CONCEPT CHECK

10. What distinguishes organelles from inclusions?
11. What is the anatomical difference between rough endoplasmic reticulum and smooth endoplasmic reticulum? What is the functional difference?
12. How do lysosomes differ from peroxisomes?
13. Apply the physiological theme of compartmentation to organelles in general and to mitochondria in particular.
14. Microscopic examination of a cell reveals many mitochondria. What does this observation imply about the cell's energy requirements?
15. Examining tissue from a previously unknown species of fish, you discover a tissue containing large amounts of smooth endoplasmic reticulum in its cells. What is one possible function of these cells?

Answers: p. 88

The Nucleus Is the Cell's Control Center

The nucleus of the cell contains DNA, the genetic material that ultimately controls all cell processes. How the nucleus receives information about conditions in the cell or elsewhere in the body and responds with fine-tuned control of the cell's protein synthesis is an active area of biological research. Figure 3-20 ■ illustrates the structure of a typical nucleus. Its boundary, or **nuclear envelope**, is a two-membrane structure that separates the nucleus from the cytoplasmic compartment. The outer membrane of the envelope is connected with the endoplasmic reticulum, and both membranes of the envelope are pierced here and there by round holes, or **pores**.

RUNNING PROBLEM

During a Pap test for cervical cancer, cell samples are swabbed from the cervix (neck) of the uterus. The cells are then smeared onto a glass slide and sent to a laboratory for examination by a trained cytologist. The cytologist looks for dysplasia, a change in the size and shape of the cells that is indicative of cancerous changes [*dys-*, abnormal + *-plasia*, growth or cell multiplication]. Cancer cells can usually be recognized by a large nucleus surrounded by a relatively small amount of cytoplasm. Jan's first Pap test showed all the hallmarks of dysplasia.

Question 2:
What property of cancer cells explains the large size of their nucleus and the relatively small amount of cytoplasm?

51 58 67 75 81 85

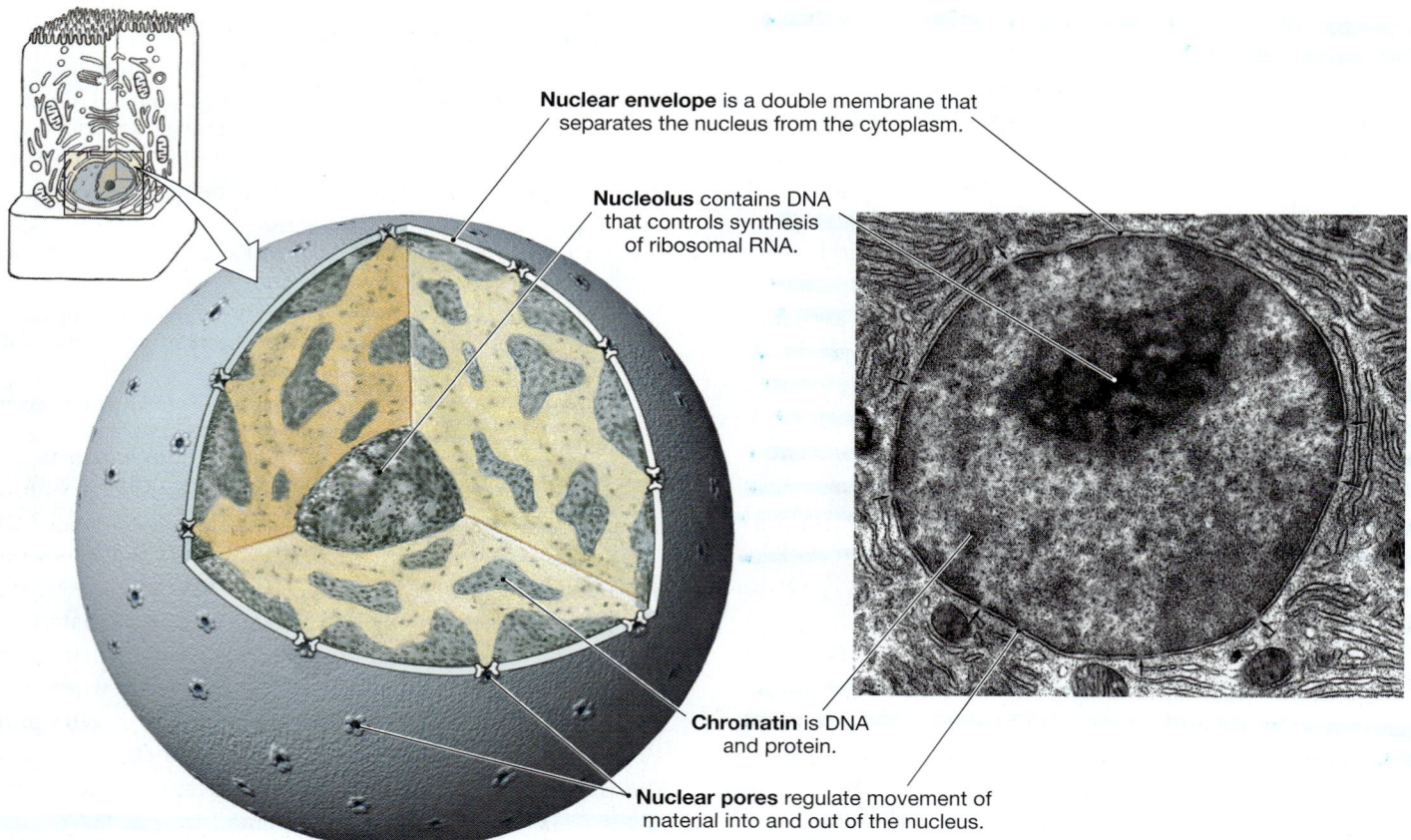

Nuclear envelope is a double membrane that separates the nucleus from the cytoplasm.

Nucleolus contains DNA that controls synthesis of ribosomal RNA.

Chromatin is DNA and protein.

Nuclear pores regulate movement of material into and out of the nucleus.

■ **FIGURE 3-20** *The nucleus*

Communication between the nucleus and cytosol occurs through the **nuclear pore complexes**, large protein complexes with a central channel. Ions and small molecules move freely through this channel when it is open, but large molecules such as proteins and RNA must be transported by a process that uses energy. This requirement allows the cell to restrict DNA to the nucleus and also to restrict various enzymes to either the cytoplasm or the nucleus.

In electron micrographs of cells that are not dividing, the nucleus appears filled with randomly scattered granular material, or **chromatin,** composed of DNA and associated proteins. Usually a nucleus also contains from one to four larger dark-staining bodies of DNA, RNA, and protein called **nucleoli** [singular, *nucleolus,* little nucleus]. Nucleoli contain the genes and proteins that control the synthesis of RNA for ribosomes (see Appendix B).

TISSUES OF THE BODY

Despite the amazing variety of intracellular structures, no single cell can carry out all the processes of the mature human body. Instead, cells assemble into the larger units we call tissues. The cells in any tissue are held together by specialized connections

called *cell junctions* and by other support structures. Tissues range in complexity from simple tissues containing only one cell type, such as the lining of blood vessels, to complex tissues containing many cell types and extensive extracellular material, such as connective tissue. The cells of most tissues work together to achieve a common purpose.

The study of tissue structure and function is known as **histology** [*histos,* tissue]. Histologists describe tissues by their physical features: (1) the shape and size of the cells, (2) how the cells are arranged in the tissue (in layers, scattered, and so on), (3) how the cells are connected to one another, and (4) the amount of extracellular material present in the tissue. There are four primary tissue types in the human body: epithelial, connective, muscle, and *neural,* or nerve. Before we consider each tissue type specifically, let's examine how cells link together to form tissues.

Extracellular Matrix Has Many Functions

Extracellular matrix (usually just called *matrix*) is extracellular material that is synthesized and secreted by the cells of a tissue. For years, scientists believed that matrix was an inert substance whose only function was to hold cells together. However,

TABLE 3-3	Major Cell Adhesion Molecules (CAMs)
NAME	**EXAMPLES**
Cadherins	Cell-cell junctions such as adherens junctions and desmosomes
Integrins	Primarily found in cell-matrix junctions
NCAMs (Nerve-cell adhesion molecules)	Nerve cell growth during development of the nervous system
Selectins	Temporary cell-cell adhesions

experimental evidence now shows that the extracellular matrix plays a vital role in many physiological processes, ranging from growth and development to cell death. A number of disease states are associated with overproduction or disruption of extracellular matrix, including chronic heart failure and the spread of cancerous cells throughout the body (*metastasis*).

The composition of extracellular matrix varies from tissue to tissue, but matrix always has two basic components: proteoglycans and insoluble protein fibers. **Proteoglycans** are glycoproteins, which means proteins covalently bound to polysaccharide chains [⊘ p. 32]. Insoluble protein fibers such as collagen, *fibronectin,* and *laminin* provide strength and anchor cells to the matrix. Attachments between the extracellular matrix and proteins in the cell membrane or the cytoskeleton are one means of communication between the cell and its external environment.

The amount of extracellular matrix in a tissue is highly variable. Nerve and muscle tissue have very little matrix, but the connective tissues, such as cartilage, bone, and blood, have extensive matrix that occupies as much volume as their cells. The consistency of extracellular matrix varies from watery (blood and lymph) to rigid (bone).

Cell Junctions Hold Cells Together to Form Tissues

During growth and development, cells form *cell-cell adhesions* that may be transient or that may develop into more permanent **cell junctions. Cell adhesion molecules,** or **CAMs,** are membrane-spanning proteins responsible both for cell junctions and for transient cell adhesions (Table 3-3 ■). Cell-cell or cell-matrix adhesions mediated by CAMs are essential for normal growth and development. For example, growing nerve cells creep across the extracellular matrix with the help of *nerve-cell adhesion molecules,* or *NCAMs.* Cell adhesion helps white blood cells escape from the circulation and move into infected tissues, and it allows clumps of platelets to cling to damaged blood vessels. Because cell adhesions are not permanent, the bond between CAMs and matrix is weak.

Cell junctions can be grouped into three categories: gap junctions, tight junctions, and anchoring junctions. Gap junctions allow direct cell-to-cell communication. Tight junctions are *occluding* [*occludere,* to close up] junctions that block movement of materials between cells. Anchoring junctions hold cells to one another and to the extracellular matrix.

1. **Gap junctions** are the simplest cell-cell junctions. They create cytoplasmic communication bridges between adjoining cells so that chemical and electrical signals pass rapidly from one cell to the next (Fig. 3-21c ■). Cylindrical proteins called *connexins* interlock to create passageways that look like hollow rivets with narrow channels through their centers. The channels are able to open and close, regulating the movement of small molecules and ions through them. Gap junctions were once thought to occur only in certain muscle and nerve cells, but we now know they are important in cell-to-cell communication in many tissues, including the liver, pancreas, ovary, and thyroid gland.

2. **Tight junctions** in humans and other vertebrates are occluding junctions designed to restrict the movement of material between the cells they link. In tight junctions the cell membranes of adjacent cells partly fuse together with the help of proteins called *claudins* and *occludins,* thereby making a barrier (Fig. 3-21a). Like many physiological processes, the barrier properties of tight junctions are dynamic and can be altered depending on the body's needs.

Tight junctions in the intestinal tract and kidney prevent most substances from moving freely between the external and internal environments and thus enable cells to regulate what enters and leaves the body. Tight junctions also create the so-called *blood-brain barrier* that prevents potentially harmful substances in the blood from reaching the extracellular fluid of the brain.

3. **Anchoring junctions** attach cells to each other (cell-cell anchoring junctions) or to the extracellular matrix (cell-matrix anchoring junctions). In vertebrates, cell-cell anchoring

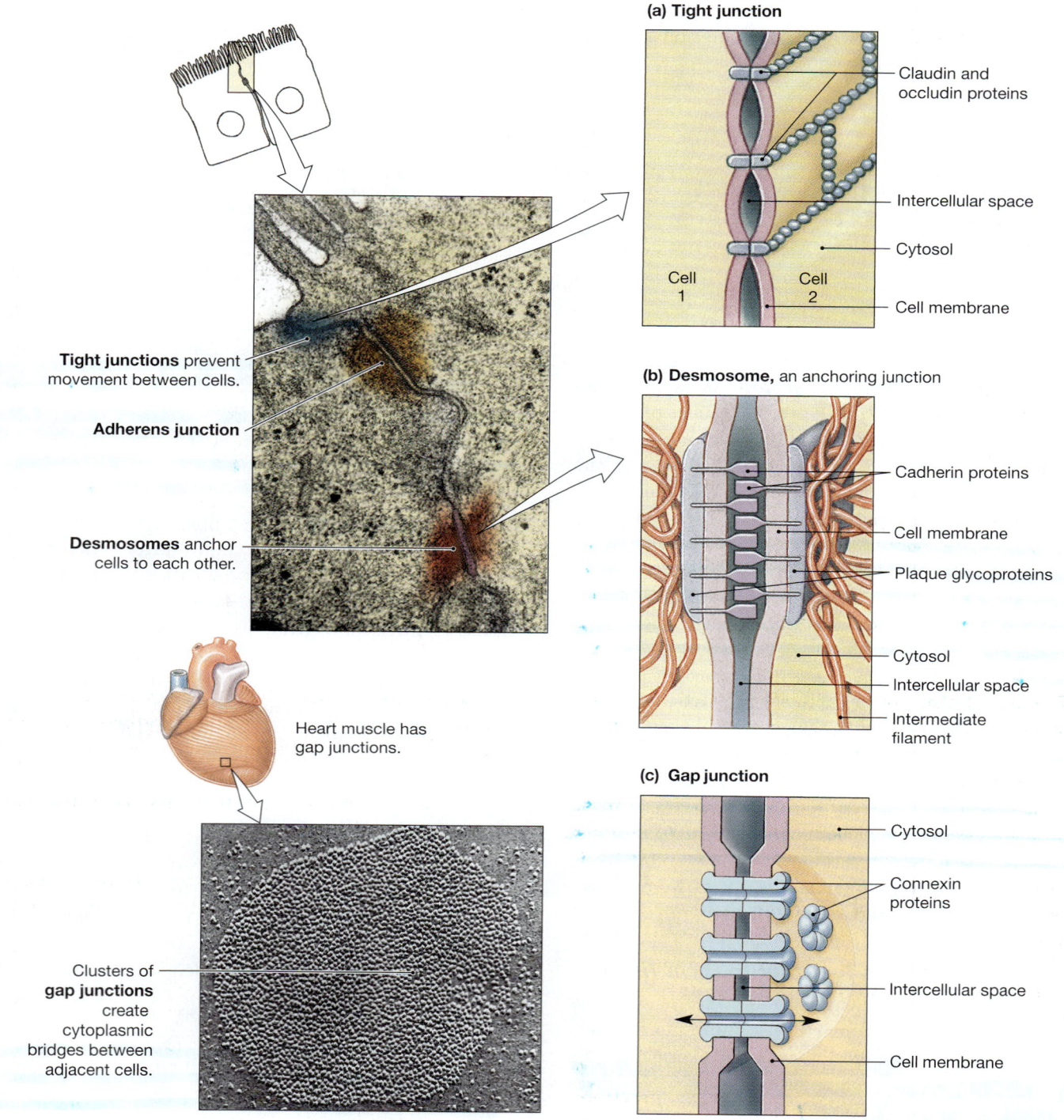

(a) Tight junction

Claudin and occludin proteins

Intercellular space

Cytosol

Cell 1   Cell 2

Cell membrane

Tight junctions prevent movement between cells.

Adherens junction

Desmosomes anchor cells to each other.

(b) Desmosome, an anchoring junction

Cadherin proteins

Cell membrane

Plaque glycoproteins

Cytosol

Intercellular space

Intermediate filament

Heart muscle has gap junctions.

(c) Gap junction

Cytosol

Connexin proteins

Intercellular space

Cell membrane

Clusters of **gap junctions** create cytoplasmic bridges between adjacent cells.

■ **FIGURE 3-21**   *Types of cell-cell junctions*

Tight junctions prevent movement of materials through the junction. Anchoring junctions hold cells together but allow materials to pass. Gap junctions create cytoplasmic bridges between adjacent cells.

junctions are created by CAMs called **cadherins,** which connect with one another across the intercellular space. Cell-matrix junctions use CAMs called **integrins**. Integrins are membrane proteins that can also bind to signal molecules in the cell's environment, transferring information carried by the signal across the cell membrane into the cytoplasm. Anchoring junctions have been compared to buttons or zippers that tie cells together and hold them in position within a tissue. Notice how the interlocking cadherin proteins in Figure 3.21b resemble the teeth of a zipper.

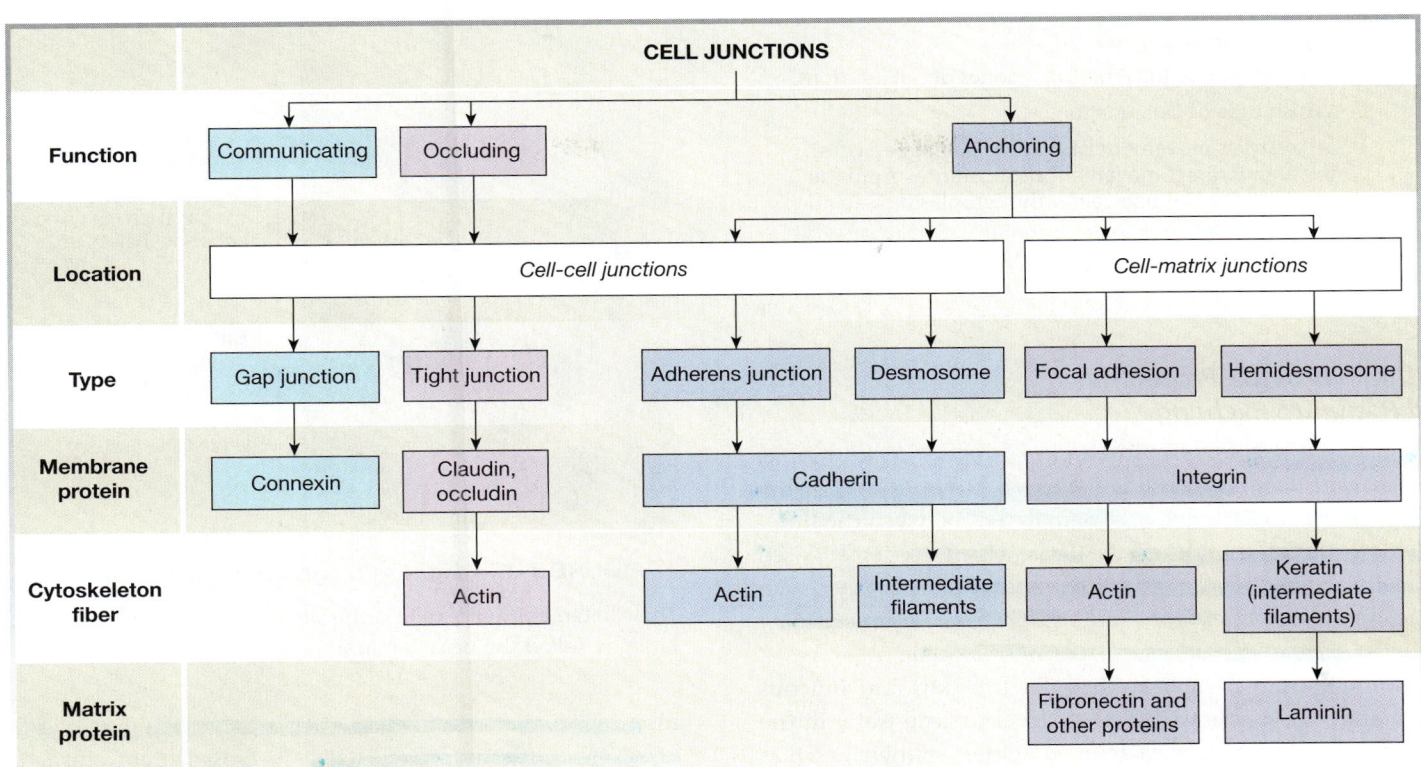

CELL JUNCTIONS

Function	Communicating	Occluding			Anchoring	
Location	Cell-cell junctions				Cell-matrix junctions	
Type	Gap junction	Tight junction	Adherens junction	Desmosome	Focal adhesion	Hemidesmosome
Membrane protein	Connexin	Claudin, occludin	Cadherin		Integrin	
Cytoskeleton fiber		Actin	Actin	Intermediate filaments	Actin	Keratin (intermediate filaments)
Matrix protein					Fibronectin and other proteins	Laminin

■ **FIGURE 3-22** *A map of cell junctions*

The protein linkage of anchoring cell junctions is very strong, allowing sheets of tissue in skin and lining body cavities to resist damage from stretching and twisting. Even the tough protein fibers of anchoring junctions can be broken, however. If you have shoes that rub against your skin, the stress can shear the proteins connecting the different skin layers. When fluid accumulates in the resulting space and the layers separate, a *blister* results.

Tissues held together with anchoring junctions are like a picket fence, where spaces between the connecting bars allow materials to pass from one side of the fence to the other. In contrast, tissues held together with tight junctions are more like a solid brick wall: very little can pass from one side of the wall to the other between the bricks.

Cell-cell anchoring junctions take the form of either adherens junctions or desmosomes. **Adherens junctions** (Fig. 3-21b) link actin fibers in adjacent cells together. **Desmosomes** [*desmos,* band + *soma,* body] attach to intermediate filaments of the cytoskeleton. Desmosomes are the strongest cell-cell junctions. In electron micrographs they can be recognized by the dense glycoprotein bodies, or *plaques,* that lie just inside the cell membranes in the region where the two cells connect (Fig. 3-21b). Desmosomes may be small points of contact between two cells (spot desmosomes) or bands that encircle the entire cell (belt desmosomes).

There are also two types of cell-matrix anchoring junctions. **Hemidesmosomes** [*hemi-,* half] are strong junctions

that anchor intermediate fibers of the cytoskeleton to fibrous matrix proteins such as laminin. **Focal adhesions** tie intracellular actin fibers to different matrix proteins such as fibronectin.

The loss of normal cell junctions plays a role in a number of diseases and in metastasis. Diseases in which cell junctions are destroyed or fail to form can have disfiguring and painful symptoms, such as blistering skin. One such disease is *pemphigus,* a condition in which the body attacks some of its own cell junction proteins (*www.pemphigus.org*).

The disappearance of anchoring junctions probably contributes to the metastasis of cancer cells throughout the body. Cancer cells lose their anchoring junctions because they have fewer cadherin molecules and are not bound as tightly to neighboring cells. Once a cancer cell is released from its moorings, it secretes protein-digesting enzymes known as *proteases.* These enzymes, especially those called *matrix metalloproteinases* (MMPs), dissolve the extracellular matrix so that escaping cancer cells can invade adjacent tissues or enter the bloodstream. Researchers are investigating ways to block MMP enzymes to see if they can prevent metastasis.

A summary map of cell junctions is shown in Figure 3-22 ■. Now that you understand how cells are held together into tissues, we will look at the four different tissue types in the body: (1) epithelial, (2) connective, (3) muscle, and (4) neural.

✓ **CONCEPT CHECK**

16. Name the three functional categories of cell junctions.

17. Which type of cell junction:
    (a) restricts movement of materials between cells?
    (b) allows direct movement of substances from the cytoplasm of one cell to the cytoplasm of an adjacent cell?
    (c) provides the strongest cell-cell junction?
    (d) anchors actin fibers in the cell to the extracellular matrix?

Answers: p. 88

Epithelia Provide Protection and Regulate Exchange

The **epithelial tissues**, or **epithelia** [*epi-*, upon + *thele-*, nipple; singular *epithelium*], protect the internal environment of the body and regulate the exchange of materials between the internal and external environments. These tissues cover exposed surfaces, such as the skin, and line internal passageways, such as the digestive tract. *Any substance that enters or leaves the internal environment of the body must cross an epithelium.*

Some epithelia, such as those of the skin and mucous membranes of the mouth, act as a barrier to keep water in the body and invaders such as bacteria out. Other epithelia, such as those in the kidney and intestinal tract, control the movement of materials between the external environment and the extracellular fluid of the body. Nutrients, gases, and wastes often must cross several different epithelia in their passage between cells and the outside world.

Another type of epithelium is specialized to manufacture and secrete chemicals into the blood or into the external environment. Sweat and saliva are examples of substances secreted by epithelia into the environment. Hormones are secreted into the blood.

Structure of Epithelia   Epithelia typically consist of one or more layers of cells connected to one another, with a thin layer of extracellular matrix lying between the epithelial cells and their underlying tissues (Fig. 3-23 ■). This matrix layer, called the **basal lamina** [*bassus*, low; *lamina*, a thin plate], or **basement membrane**, is composed of a network of collagen and laminin filaments embedded in proteoglycans. The protein filaments hold the epithelial cells to the underlying cell layers, just as cell junctions hold the individual cells in the epithelium to one another.

The cell junctions in epithelia are variable. Physiologists classify epithelia either as "leaky" or "tight," depending on how easily substances pass from one side of the epithelial layer to the other side. In a leaky epithelium, anchoring junctions allow molecules to cross the epithelium by passing through the gap between two adjacent epithelial cells. A typical leaky epithelium is the wall of capillaries (the smallest blood vessels), where all dissolved molecules except for large proteins can pass from the blood to the interstitial fluid by traveling through gaps between adjacent epithelial cells.

In a tight epithelium, such as that in the kidney, adjacent cells are bound to each other by tight junctions and infolded

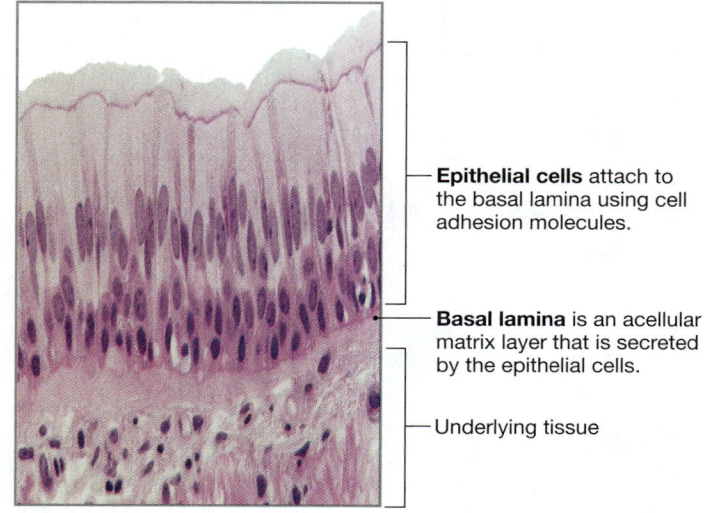

■ FIGURE 3-23   *Light micrograph of epithelial tissue*
The ciliated epithelial cells in this picture are attached to an acellular layer called the basal lamina.

Epithelial cells attach to the basal lamina using cell adhesion molecules.

Basal lamina is an acellular matrix layer that is secreted by the epithelial cells.

Underlying tissue

membranes that create a barrier that prevents substances from traveling between adjacent cells. To cross a tight epithelium, most substances must enter the epithelial cells and go *through* them. The tightness of an epithelium is directly related to how selective it is about what can move across it. Some epithelia, such as those of the intestine, have the ability to alter the tightness of their junctions according to the body's needs.

Types of Epithelia   Structurally, epithelial tissues can be divided into two general types: (1) sheets of tissue that lie on the surface of the body or that line the inside of tubes and hollow organs and (2) secretory epithelia that synthesize and release substances into the extracellular space. Histologists classify sheet epithelia by the number of cell layers in the tissue and by the shape of the cells in the surface layer. This classification scheme recognizes two types of layering—**simple** (one cell thick) and **stratified** (multiple cell layers) [*stratum*, layer + *facere*, to make]—and three cell shapes—**squamous** [*squama*, flattened plate or scale], **cuboidal**, and **columnar.** However, physiologists are more concerned with the functions of these tissues, so instead of using the histological descriptions, we will divide epithelia into five groups according to their function.

There are five functional types of epithelia: exchange, transporting, ciliated, protective, and secretory. *Exchange epithelia* permit rapid exchange of materials, especially gases. *Transporting epithelia* are selective about what can cross them and are found primarily in the intestinal tract and the kidney. *Ciliated epithelia* are located in the airways of the respiratory system and in the female reproductive tract. *Protective epithelia* are found on the surface of the body and just inside the openings of body cavities. *Secretory epithelia* synthesize and release secretory products into the external environment or into the blood.

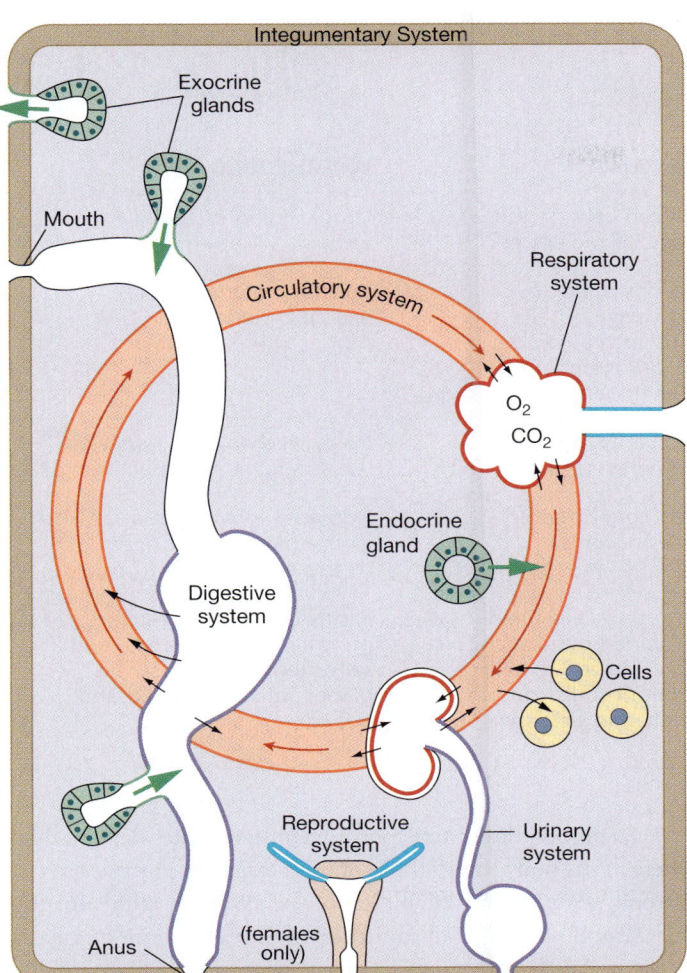

Integumentary System

Exocrine glands

Mouth

Respiratory system

Circulatory system

O_2
CO_2

Endocrine gland

Digestive system

Cells

Reproductive system

Urinary system

Anus

(females only)

■ **FIGURE 3-24**   *Distribution of epithelia in the body*

Q **FIGURE QUESTIONS**

• What systems have transporting epithelia? exchange epithelia? ciliated epithelia? protective epithelia?
• Where do secretions from endocrine glands go?
• Where do secretions from exocrine glands go?

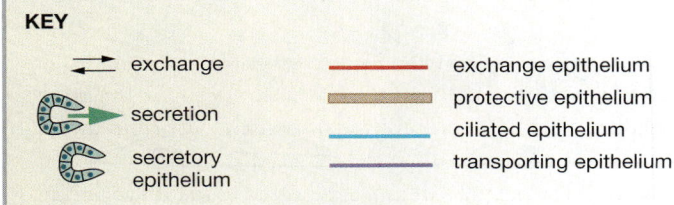

KEY

⇄	exchange	———	exchange epithelium
	secretion		protective epithelium
	secretory epithelium	———	ciliated epithelium
		———	transporting epithelium

Figure 3-24 ■ shows the distribution of these epithelia in the systems of the body. Notice that most epithelia face the external environment on one surface and the extracellular fluid on the other. The only exception is the endocrine glands. Table 3-4 ■ summarizes the types of epithelia.

Exchange epithelia   The **exchange epithelia** are composed of very thin, flattened cells that allow gases (CO_2 and O_2) to pass rapidly across the epithelium. This type of epithelium lines the blood vessels and the lungs, the two major sites of gas exchange in the body. In capillaries, gaps or pores in the epithelium also allow molecules smaller than proteins to pass *between* two adjacent epithelial cells, making this a leaky epithelium (Fig. 3-25a ■). Histologists classify thin exchange tissue as *simple squamous epithelium* because it has a single layer of thin, flattened cells. The simple squamous epithelium lining the heart and blood vessels is also called the **endothelium**.

Transporting epithelia   The **transporting epithelia** actively and selectively regulate the exchange of nongaseous materials, such as ions and nutrients, between the internal and external environments. These epithelia are found lining the hollow tubes of the digestive system and the kidney, where lumens open into the external environment [🔁 p. 51].

There are several characteristics of transporting epithelia (Fig. 3-25b):

1. **Cell shape.**   Cells of transporting epithelia are much thicker than cells of exchange epithelia, and they act as a barrier as well as an entry point. The cell layer is only one cell thick (a simple epithelium), but cells are cuboidal or columnar.

2. **Membrane modifications.**   The **apical membrane**, the surface of the epithelial cell that faces the lumen, has microvilli that increase the surface area available for transport. A cell with microvilli has at least 20 times the surface area of a cell without them. In addition, the **basolateral membrane**, the side of the epithelial cell facing the extracellular fluid, may also have folds that increase the cell's surface area.

3. **Cell junctions.**   The cells of transporting epithelia are firmly attached to adjacent cells by moderately tight to very tight junctions. This means that to cross the epithelium, material must move into an epithelial cell on one side of the tissue and out of the cell on the other side.

4. **Cell organelles.**   Most cells that transport materials have numerous mitochondria to provide energy for transport processes (discussed further in Chapter 5).

TABLE 3-4	Types of Epithelia			
	NUMBER OF CELL LAYERS	**CELL SHAPE**	**SPECIAL FEATURES**	**WHERE FOUND**
Exchange	One	Flattened	Pores between cells permit easy passage of molecules	Lungs, lining of blood vessels
Transporting	One	Columnar or cuboidal	Tight junctions prevent movement between cells; surface area increased by folding of cell membrane into villi, or fingers	Intestine, kidney, some exocrine glands
Ciliated	One	Cuboidal to columnar	One side covered with cilia to move fluid across surface	Nose, trachea, and upper airways; female reproductive tract
Protective	Many	Flattened in surface layers; polygonal in deeper layers	Cells tightly connected by many desmosomes	Skin and lining of cavities (such as the mouth) that open to the environment
Secretory	One to many	Columnar to polygonal	Protein-secreting cells filled with membrane-bound secretory granules and extensive RER; steroid-secreting cells contain lipid droplets and extensive SER	Exocrine glands, including pancreas, sweat glands, and salivary glands; endocrine glands, such as thyroid and gonads

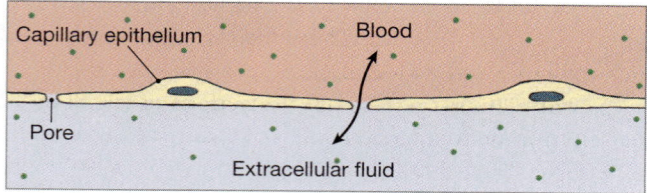

(a) Leaky exchange epithelium allows movement through gaps between the cells.

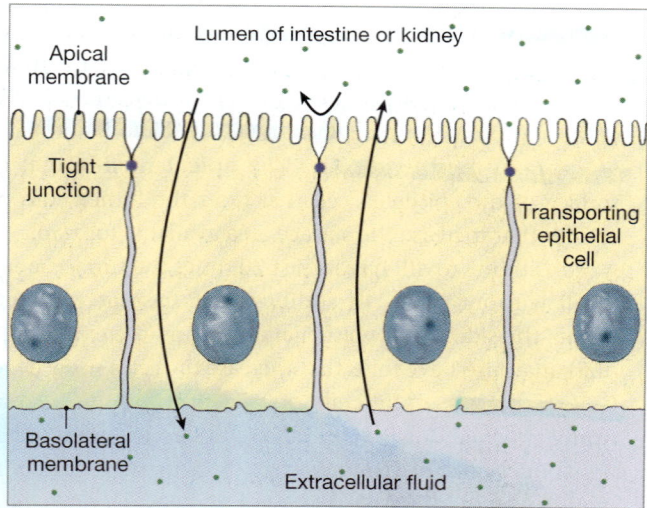

(b) Tight junctions in a transporting epithelium prevent movement between adjacent cells. Substances must instead pass through the epithelial cell, crossing two phospholipid cell membranes as they do so.

■ **FIGURE 3-25**   *Movement of substances across tight and leaky epithelia*

The properties of transporting epithelia differ depending on where in the body the epithelia are. For example, glucose can cross the epithelium lining the lumen of the small intestine and enter the extracellular fluid but cannot cross the epithelium of the large intestine. Furthermore, the transport properties of an epithelium can be regulated and modified in response to various stimuli. Hormones, for example, affect the transport of ions by the epithelium of the kidney. You will learn more about transporting epithelia when you study the kidney and digestive systems.

Ciliated epithelia   Ciliated epithelia are nontransporting tissues found lining the respiratory system and parts of the female reproductive tract. The surface of the tissue facing the lumen is covered with cilia that beat in a coordinated, rhythmic fashion, moving fluid and particles across the surface of the tissue. Injury to the cilia or to their epithelial cells can stop ciliary movement. Paralysis of the ciliated epithelium lining the respiratory tract is one effect of smoking. It is considered to contribute to the higher incidence of respiratory infection in smokers, because the mucus that traps bacteria is no longer swept up out of the lungs by the cilia. Figure 3-26 ■ shows the ciliated columnar epithelium of the trachea.

Protective epithelia   The **protective epithelia** prevent exchange between the internal and external environments and protect areas subject to mechanical or chemical stresses. They are stratified tissues, composed of many stacked layers of cells. Protective epithelia are toughened by the secretion of *keratin* [*keras*, horn], the same insoluble protein abundant in hair and nails. The *epidermis* [*epi*, upon + *derma*, skin] and linings of the mouth, pharynx, esophagus, urethra, and vagina are all protective epithelia.

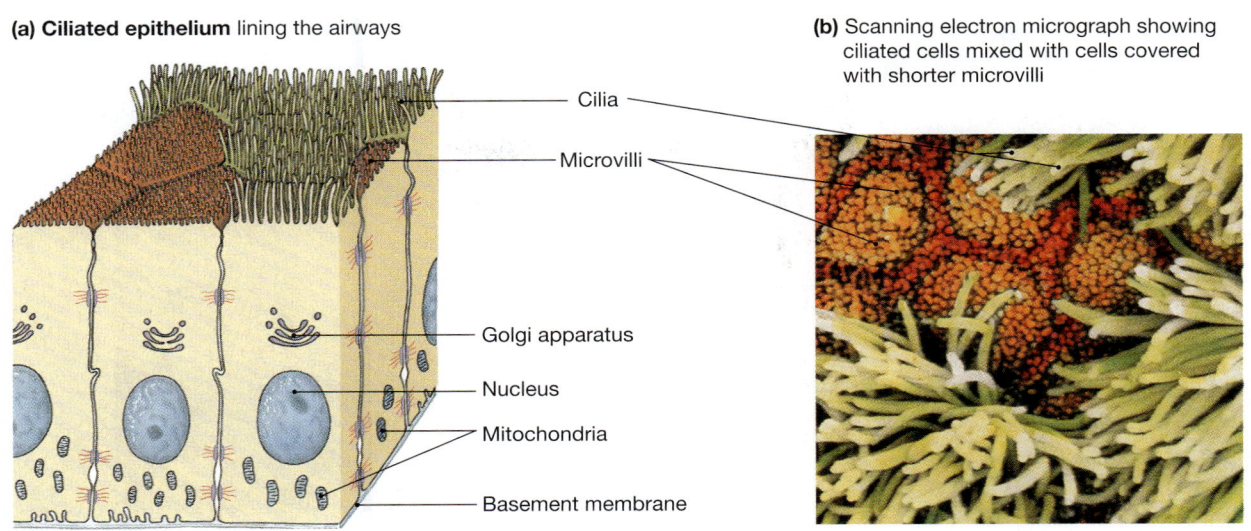

(a) Ciliated epithelium lining the airways

Cilia

Microvilli

Golgi apparatus

Nucleus

Mitochondria

Basement membrane

(b) Scanning electron micrograph showing ciliated cells mixed with cells covered with shorter microvilli

■ **FIGURE 3-26**   *Ciliated epithelium*

Some tissues have several types of epithelial cells. These cells have microvilli interspersed with ciliated cells.

Because protective epithelia are subjected to irritating chemicals, bacteria, and other destructive forces, the cells in them have a short life span. In deeper layers, new cells are produced continuously, displacing older cells at the surface. Each time you wash your face, you scrub off dead cells on the surface layer. As skin ages, the rate of cell turnover declines. Tretinoin (Retin-A®), a drug derived from vitamin A, speeds up cell division and surface shedding so treated skin develops a more youthful appearance.

RUNNING PROBLEM

Many kinds of cancer develop in epithelial cells, which are subject to damage or trauma. The cervix consists of two types of epithelia. Secretory epithelium with mucus-secreting glands lines the inside of the cervix, whereas a protective epithelium covers the outside of the cervix. At the opening of the cervix, these two types of epithelia come together. As Jan lies on the examining table, her physician uses a small spatula and a little brush to take samples of cells from both inside and outside the cervix. She then smears the cells on a slide and sprays them with fixative. The slides will now be sent to a lab for evaluation.

Question 3:
What kind of damage or trauma are cervical epithelial cells normally subjected to? Which of its two types of epithelia is more likely to be affected by trauma?

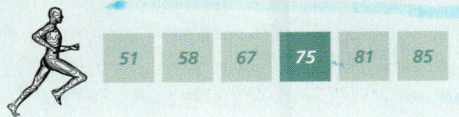

| 51 | 58 | 67 | **75** | 81 | 85 |

Secretory epithelia Secretory epithelia are composed of cells that produce a substance inside the cell and then secrete it into the extracellular space. Secretory cells may be scattered among other epithelial cells, or they may group together to form a multicellular **gland**. There are two types of secretory glands: exocrine and endocrine.

Exocrine glands release their secretions to the body's external environment [*exo-*, outside + *krinein*, to secrete]. This may be onto the surface of the skin or onto an epithelium lining one of the internal passageways, such as the airways of the lung or the lumen of the intestine. In effect, an exocrine secretion leaves the body. This explains how some exocrine secretions, like stomach acid, can have a pH that is incompatible with life [⊜ Fig. 2-15, p. 37]. Most exocrine glands release their products through open tubes known as **ducts**. Sweat glands, mammary glands found in the breast, salivary glands, the liver, and the pancreas are all exocrine glands.

Exocrine gland cells produce two types of secretions. **Serous secretions** are watery solutions, and many of them contain enzymes. Tears, sweat, and digestive enzyme solutions are all serous exocrine secretions. **Mucous secretions** (also called **mucus**) are sticky solutions containing glycoproteins and proteoglycans. **Goblet cells,** shown in Figure 3-27 ■, are single exocrine cells that produce mucus. Mucus acts as a lubricant for food to be swallowed, as a trap for foreign particles and microorganisms inhaled or ingested, and as a protective barrier between the epithelium and the environment. Some exocrine glands contain more than one type of secretory cell, and they produce both serous and mucous secretions. For example, the salivary glands release mixed secretions.

Unlike exocrine glands, **endocrine glands** are ductless and release their secretions, called **hormones**, into the body's extracellular compartment. Hormones enter the blood for distribution to

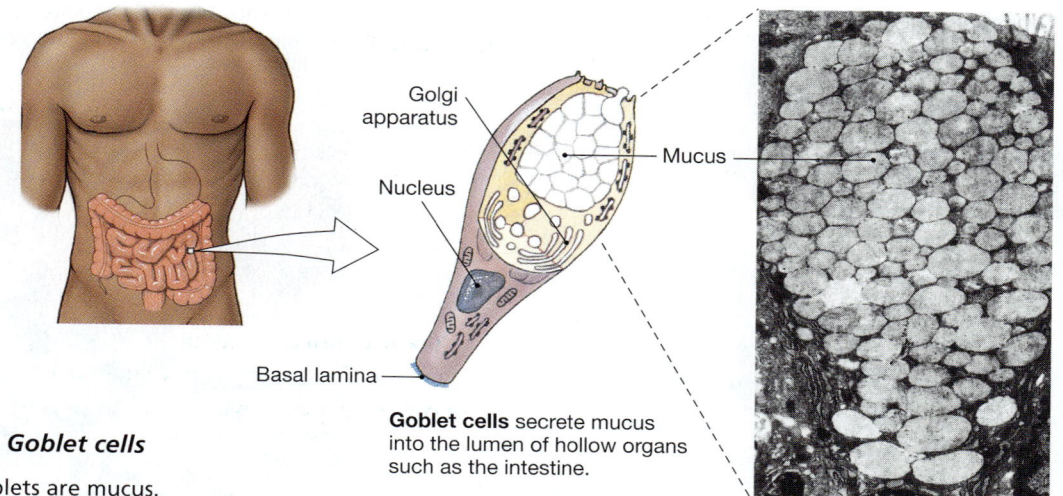

■ **FIGURE 3-27** *Goblet cells*

The large clear droplets are mucus.

Goblet cells secrete mucus into the lumen of hollow organs such as the intestine.

other parts of the body, where they regulate or coordinate the activities of various tissues, organs, and organ systems. Some of the best-known endocrine glands are the pancreas, the thyroid gland, the gonads, and the pituitary gland. For years, it was thought that all hormones were produced by cells grouped together into endocrine glands. We now know that isolated endocrine cells occur scattered in the epithelial lining of the digestive tract, in the tubules of the kidney, and in the walls of the heart.

Figure 3-28 ■ shows the epithelial origin of endocrine and exocrine glands. During development, epithelial cells grow downward into the supporting tissue. Exocrine glands retain a link to the parent epithelium in the form of a duct that transports the secretion to its destination (the external environment). Endocrine glands lose the connecting cells and secrete their hormones into the bloodstream.

CONCEPT CHECK

18. List the five functional types of epithelia.
19. Define secretion.
20. Name two properties that distinguish endocrine glands from exocrine glands.
21. The basal lamina of epithelium contains the protein fiber laminin. Are the overlying cells attached by focal adhesions or hemidesmosomes?
22. You look at a tissue under a microscope and see a simple squamous epithelium. Can it be a sample of the skin surface? Explain.
23. A cell of the intestinal epithelium secretes a substance into the extracellular fluid, where it is picked up by the blood and carried to the pancreas. Is the intestinal epithelium cell an endocrine or an exocrine cell?

Answers: p. 88

Connective Tissues Provide Support and Barriers

Connective tissues, the second major tissue type, provide structural support and sometimes a physical barrier that, along with specialized cells, helps defend the body from foreign

invaders such as bacteria. The distinguishing characteristic of connective tissues is the presence of extensive extracellular matrix containing widely scattered cells that secrete and modify the matrix. Connective tissues include blood, the support tissues for the skin and internal organs, and cartilage and bone.

Structure of Connective Tissue   The extracellular matrix of connective tissue is a **ground substance** of proteoglycans and water in which insoluble protein fibers are arranged, much like pieces of fruit suspended in a gelatin salad. The consistency of ground substance is highly variable, depending on the type of connective tissue. At one extreme is the watery matrix of blood, and at the other extreme is the hardened matrix of bone. In between are solutions of proteoglycans that vary in consistency from syrupy to gelatinous. The term *ground substance* is sometimes used interchangeably with *matrix*.

Connective tissue cells lie embedded in the extracellular matrix. These cells are described as *fixed* if they remain in one place and as *mobile* if they can move from place to place. **Fixed cells are responsible for local maintenance, tissue repair, and energy storage. Mobile cells are responsible mainly for defense.** The distinction between fixed and mobile cells is not absolute, because at least one cell type is found in both fixed and mobile forms.

Although extracellular matrix is nonliving, the connective tissue cells constantly modify it by adding, deleting, or rearranging molecules. The suffix *-blast* [*blastos*, sprout] on a connective tissue cell name often indicates a cell that is either growing or actively secreting extracellular matrix. **Fibroblasts**, for example, are connective tissue cells that secrete collagen-rich matrix. Cells that are actively breaking down matrix are identified by the suffix *-clast* [*klastos*, broken]. Cells that are neither growing, secreting matrix components, nor breaking down matrix may be given the suffix *-cyte*, meaning "cell." Remembering these suffixes should help you remember the functional differences between cells with similar names, such as the osteoblast, osteocyte, and osteoclast, three cell types found in bone.

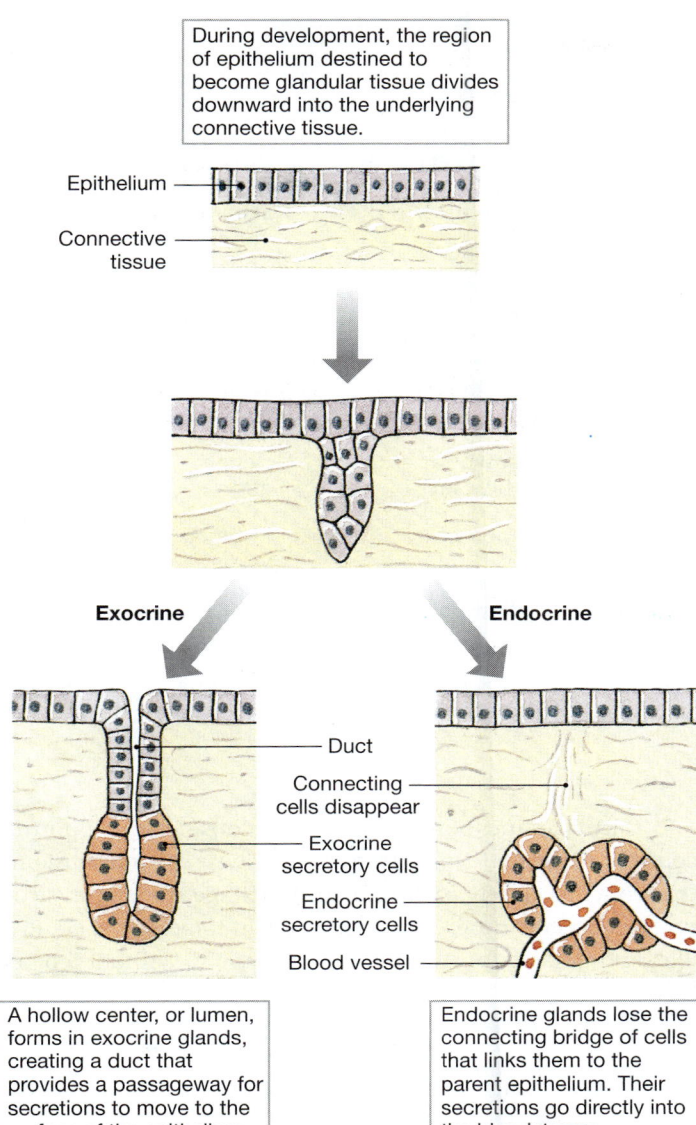

During development, the region of epithelium destined to become glandular tissue divides downward into the underlying connective tissue.

Epithelium

Connective tissue

Exocrine   **Endocrine**

Duct

Connecting cells disappear

Exocrine secretory cells

Endocrine secretory cells

Blood vessel

A hollow center, or lumen, forms in exocrine glands, creating a duct that provides a passageway for secretions to move to the surface of the epithelium.

Endocrine glands lose the connecting bridge of cells that links them to the parent epithelium. Their secretions go directly into the bloodstream.

■ **FIGURE 3-28** *Development of endocrine and exocrine glands from epithelium*

In addition to secreting proteoglycan ground substance, connective tissue cells produce matrix fibers. Four types of fiber proteins are found in matrix, aggregated into insoluble fibers. **Collagen** [*kolla*, glue + *-genes,* produced] is the most abundant protein in the human body, almost one-third of the body's dry weight. Collagen is also the most diverse of the four protein types, with at least 12 variations. It is found almost everywhere connective tissue is found, from the skin to muscles and bones. Individual collagen molecules pack together to form collagen fibers, flexible but inelastic fibers whose strength per unit weight exceeds that of steel. The amount and arrangement of collagen fibers are what distinguish one type of connective tissue from another.

Three other protein fibers in connective tissue are elastin, fibrillin, and fibronectin. **Elastin** is a coiled, wavy protein that returns to its original length after being stretched. This property is known

as *elastance*. Elastin combines with the very thin, straight fibers of **fibrillin** to form filaments and sheets of elastic fibers. These two fibers are important in elastic tissues such as the lungs, blood vessels, and skin. As mentioned earlier, **fibronectin** connects cells to extracellular matrix at focal adhesions. Fibronectins also play an important role in wound healing and in blood clotting.

Types of Connective Tissue   Table 3-5 ■ describes different types of connective tissue. The most common types are loose and dense connective tissue, adipose tissue, blood, cartilage, and bone.

Loose connective tissues (Fig. 3-29 ■) are the elastic tissues that underlie skin and provide support for small glands. **Dense connective tissues** provide strength or flexibility. Examples are tendons, ligaments, and the sheaths that surround muscles and nerves. In these dense tissues, collagen fibers are the dominant type. **Tendons** (Fig. 3-30 ■) attach skeletal muscles to bones. **Ligaments** connect one bone to another. Because ligaments contain elastic fibers in addition to collagen fibers, they have a limited ability to stretch, whereas tendons cannot stretch.

Adipose tissue is made up of **adipocytes**, or fat cells. An adipocyte of **white fat** typically contains a single enormous lipid droplet that occupies most of the volume of the cell (Fig. 3-31 ■). This is the most common form of adipose tissue in adults. **Brown fat** is composed of adipose cells that contain multiple lipid droplets rather than a single large droplet. This type of fat is almost completely absent in adults, but it plays an important role in temperature regulation in infants.

BIOTECHNOLOGY

GROWING NEW CARTILAGE

Have you torn the cartilage in your knee playing tennis? Maybe you won't have to have surgery to repair it. Replacing lost or damaged cartilage is moving from the realm of science fiction to the realm of reality. Researchers have developed a process in which they take a cartilage sample from a patient and put it into a tissue culture medium to reproduce. Once the culture has grown enough *chondrocytes*—the cells that synthesize the extracellular matrix of cartilage—the mixture is sent back to a physician, who surgically places the cells in the patient's knee at the site of cartilage damage. Once returned to the body, the chondrocytes secrete matrix and repair the damaged cartilage. Because the person's own cells are grown and reimplanted, there is no tissue rejection. This patented process, called Carticel®, has now been used on more than 10,000 patients.

TABLE 3-5   **Types of Connective Tissue**

TISSUE NAME	GROUND SUBSTANCE	FIBER TYPE AND ARRANGEMENT	MAIN CELL TYPES	WHERE FOUND
Loose connective tissue	Gel; more ground than fibers and cells	Collagen, elastic, reticular; random	Fibroblasts	Skin, around blood vessels and organs, under epithelia
Dense, irregular connective tissue	More fibers than ground	Mostly collagen; random	Fibroblasts	Muscle and nerve sheaths
Dense, regular connective tissue	More fibers than ground	Collagen; parallel	Fibroblasts	Tendons and ligaments
Adipose	Very little	None	Brown fat and white fat	Depends on age and sex
Blood	Aqueous	None	Blood cells	In blood and lymph vessels
Cartilage	Firm but flexible; hyaluronic acid	Collagen	Chondroblasts	Joint surfaces, spine, ear, nose, larynx
Bone	Rigid due to calcium salts	Collagen	Osteoblasts and osteoclasts	Bones

Blood is an unusual connective tissue that is characterized by its watery extracellular matrix, consisting of a dilute solution of ions and dissolved organic molecules. The matrix lacks insoluble protein fibers but contains a large variety of soluble proteins. We will discuss these in Chapter 16.

Cartilage and bone together are considered supporting connective tissues. These tissues have a dense ground substance that contains closely packed fibers. **Cartilage** is found in structures such as the nose, ears, knee, and windpipe. It is solid, flexible, and notable for its lack of blood supply. Without a blood

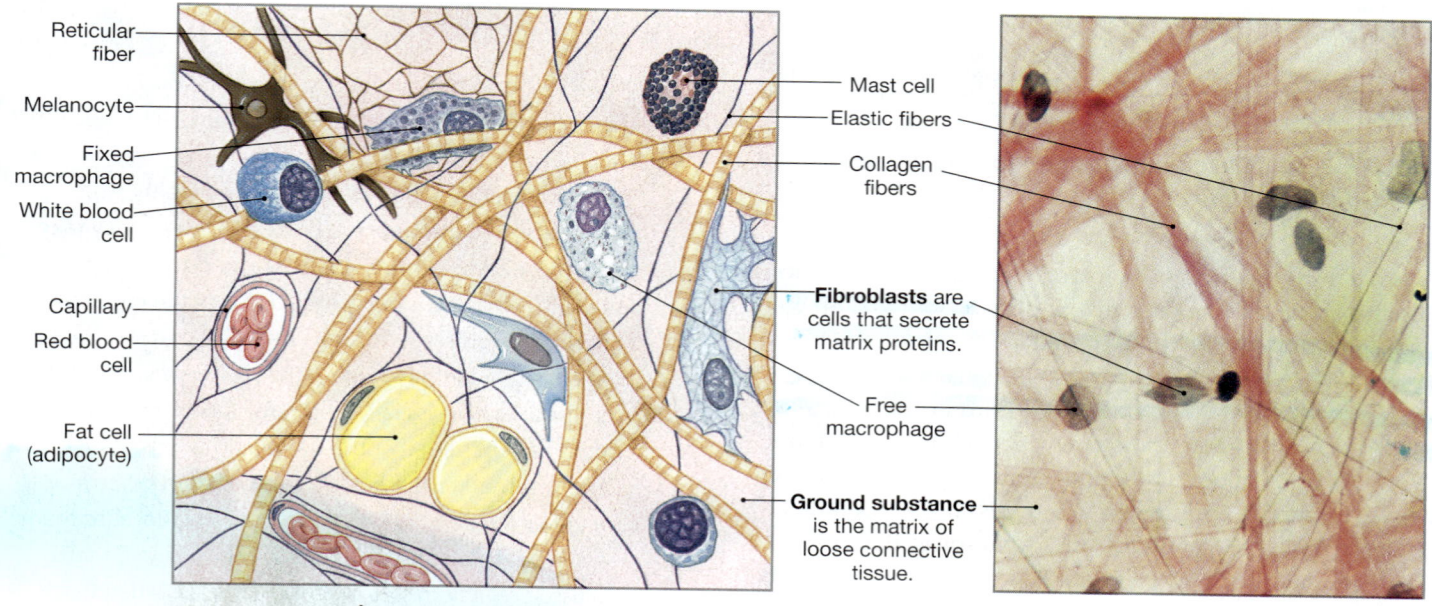

Reticular fiber
Melanocyte
Fixed macrophage
White blood cell
Capillary
Red blood cell
Fat cell (adipocyte)

Mast cell
Elastic fibers
Collagen fibers
Fibroblasts are cells that secrete matrix proteins.
Free macrophage
Ground substance is the matrix of loose connective tissue.

Loose connective tissue

Light micrograph of loose connective tissue

■ **FIGURE 3-29** *Cells and fibers of loose connective tissue*

Macrophages and mast cells are cells that defend against foreign invaders, such as bacteria.

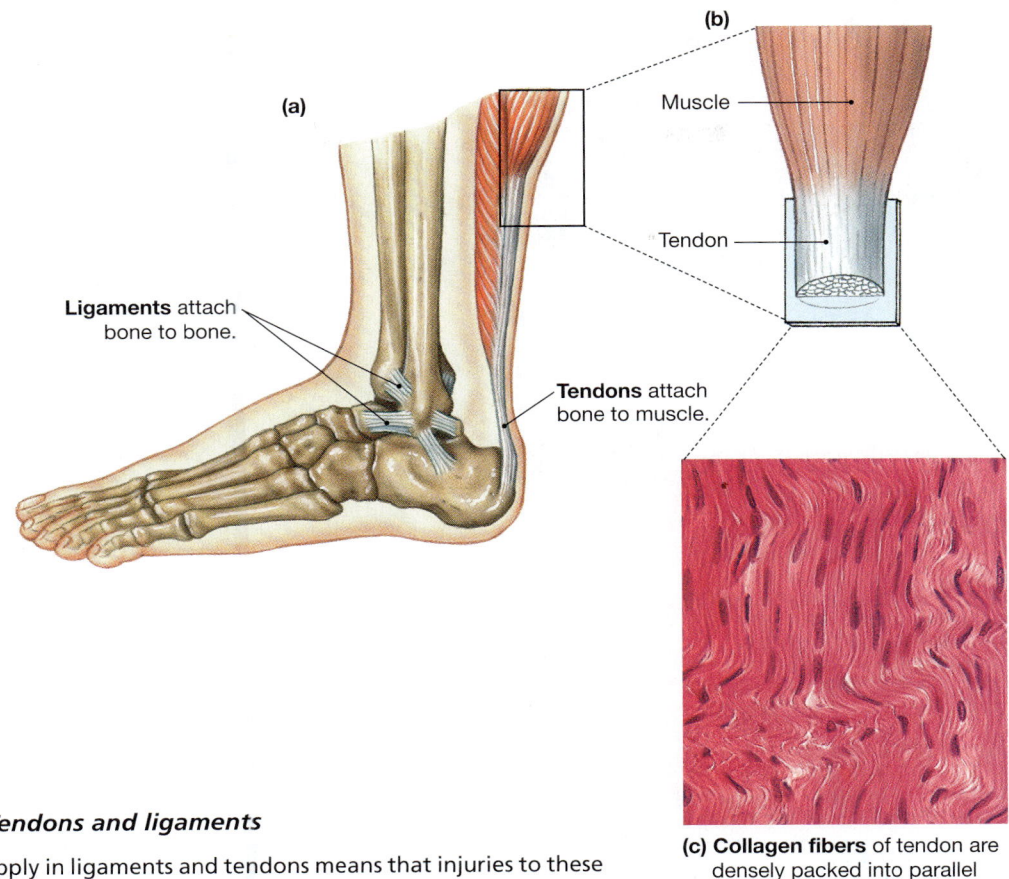

(a)

(b)

Muscle

Tendon

Ligaments attach bone to bone.

Tendons attach bone to muscle.

(c) **Collagen fibers** of tendon are densely packed into parallel bundles.

■ **FIGURE 3-30**   *Tendons and ligaments*

The minimal blood supply in ligaments and tendons means that injuries to these tissues heal very slowly.

supply, nutrients and oxygen must reach the cells of cartilage by diffusion. This is a slow process, which means that damaged cartilage heals slowly.

The fibrous extracellular matrix of **bone** is said to be *calcified* because it contains mineral deposits, primarily calcium salts, such as calcium phosphate. These minerals give the bone strength and rigidity. We will examine the structure and formation of bone along with calcium metabolism in Chapter 23.

Figure 3-32 ■ is a study map showing the components of connective tissues.

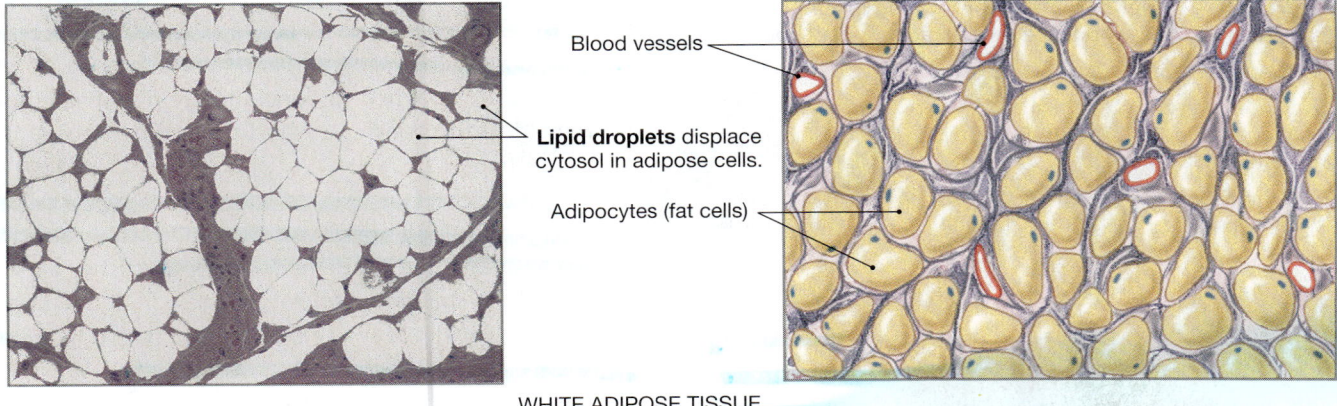

Blood vessels

Lipid droplets displace cytosol in adipose cells.

Adipocytes (fat cells)

WHITE ADIPOSE TISSUE

■ **FIGURE 3-31**   *Adipose tissue*

In the white adipose tissue shown here, a single lipid droplet almost fills each cell.

3

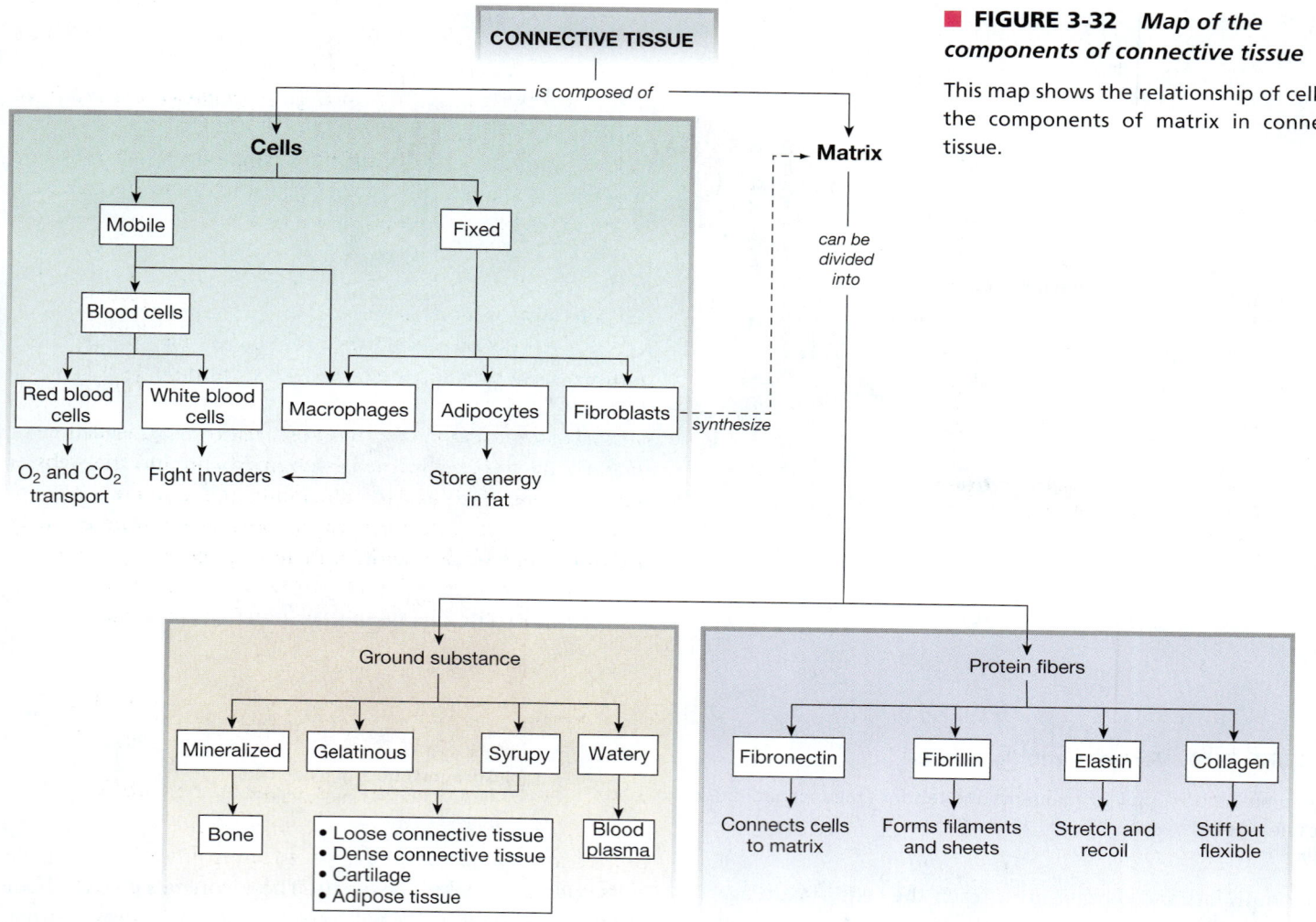

■ **FIGURE 3-32** *Map of the components of connective tissue*

This map shows the relationship of cells and the components of matrix in connective tissue.

CONCEPT CHECK

24. What is the distinguishing characteristic of connective tissues?

25. Name four types of protein fibers found in connective tissue matrix and give the characteristics of each.

26. Name six types of connective tissues.

27. Blood is a connective tissue with two components: plasma and cells. Which of these is the matrix in this connective tissue?

28. Why does torn cartilage heal more slowly than a cut in the skin?

Answers: p. 88

Muscle and Neural Tissues Are Excitable

The third and fourth of the body's four tissue types—muscle and neural—are collectively called the *excitable tissues* because of their ability to generate and propagate electrical signals called *action potentials*. Both of these tissue types have minimal extracellular matrix, usually limited to a supportive layer called the *external lamina*. Some types of muscle and nerve cells are also notable for their gap junctions, which play an important role in the rapid conduction of electrical signals from cell to cell.

Muscle tissue has the ability to contract and produce force and movement. There are three types of muscle tissue in the body: cardiac muscle, in the heart; smooth muscle, which makes up most internal organs; and skeletal muscle. Most skeletal muscles are attached to bone and are responsible for gross movement of the body. We will discuss muscle tissue in more detail in Chapter 12.

Neural tissue has two types of cells. **Neurons,** or nerve cells, carry information in the form of chemical and electrical signals from one part of the body to another. They are concentrated in the brain and spinal cord but also include a network of cells that extends to virtually every part of the body. **Glial cells,** or neuroglia, are the support cells for neurons. We will discuss the anatomy of neural tissue in Chapter 8.

A summary of the characteristics of the four tissue types can be found in Table 3-6 ■.

| 51 | 58 | 67 | 75 | **81** | 85 |

TISSUE REMODELING

Growth is a process that most people associate with the period from birth to adulthood. However, cell birth, growth, and death continue throughout a person's life. The tissues of the body are constantly changing as cells die and are replaced.

Apoptosis Is a Tidy Form of Cell Death

Cell death occurs two ways, one messy and one tidy. In **necrosis**, cells die from physical trauma, toxins, or lack of oxygen when their blood supply is cut off. Necrotic cells swell, their organelles deteriorate, and finally the cells rupture. The cell contents thus released include digestive enzymes that damage adjacent cells and trigger an inflammatory response. You see necrosis when you have a red area of skin surrounding a scab.

In contrast, cells that undergo *programmed cell death*, or **apoptosis** [ap-oh-TOE-sis or a-pop-TOE-sis; *apo-*, apart, away + *ptosis*, falling], do not disrupt their neighbors when they die. Apoptosis, also called cell suicide, is a complex process regulated by multiple chemical signals. Some signals keep apoptosis from occurring while other signals tell the cell to destruct. When the suicide signal wins out, chromatin in the nucleus condenses, the cell pulls away from its neighbors, shrinks, and finally breaks up into tidy membrane-bound *blebs* that are gobbled up by neighboring cells or by wandering cells of the immune system.

Apoptosis is a normal event in the life of an organism. During fetal development, apoptosis removes unneeded cells, such as half the cells in the developing brain and the webs of skin between fingers and toes. In adults, cells that are subject to wear and tear from exposure to the outside environment may live only a day or two before undergoing apoptosis. For example, it has been estimated that the intestinal epithelium is completely replaced with new cells every two to five days.

CONCEPT CHECK

29. What are some features of apoptosis that distinguish it from cell death due to injury?   Answers: p. 88

Stem Cells Can Create New Specialized Cells

If cells in the adult body are constantly dying, where do their replacements come from? This question is still being answered and is one of the hottest topics in biological research today. The following paragraphs describe what we currently know.

All cells in the body are derived from the single cell formed at conception. That cell and those that follow reproduce themselves by undergoing the cell division process known as **mitosis** (see Appendix B). The very earliest cells in the life of a human

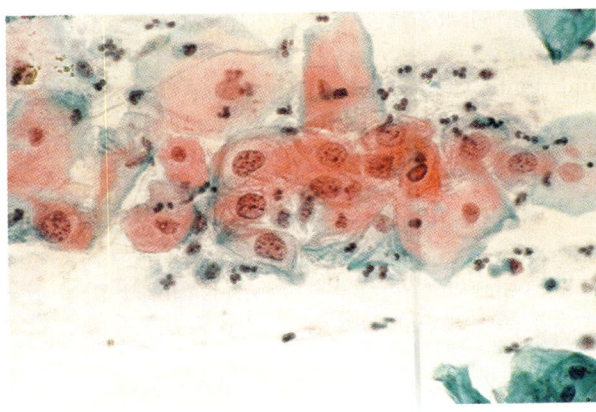

(a) Jan's abnormal Pap test.

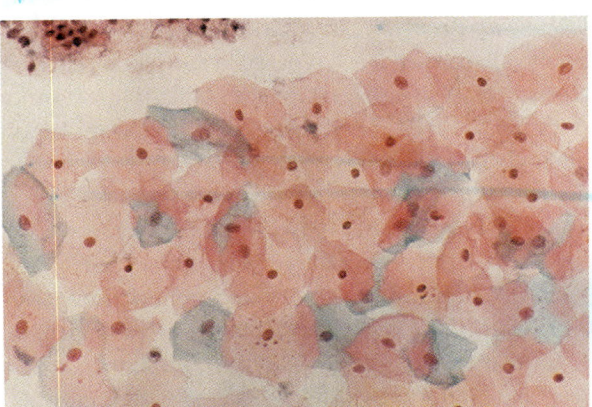

(b) Jan's second Pap test. Are these cells normal or abnormal?

■ **FIGURE 3-33** *Pap smears of cervical cells*

(a) Abnormal Pap smear. (b) Second Pap test. Are these cells normal or abnormal?

TABLE 3-6	Characteristics of the Four Tissue Types			
	EPITHELIAL	**CONNECTIVE**	**MUSCLE**	**NERVE**
Matrix amount	Minimal	Extensive	Minimal	Minimal
Matrix type	Basal lamina	Varied—protein fibers in ground substance that ranges from liquid to gelatinous to firm to calcified	External lamina	External lamina
Unique features	No direct blood supply	Cartilage has no blood supply	Able to generate electrical signals, force, and movement	Able to generate electrical signals
Surface features of cells	Microvilli, cilia	NA	NA	NA
Locations	Covers body surface; lines cavities and hollow organs and tubes; secretory glands	Supports skin and other organs; cartilage, bone, and blood	Makes up skeletal muscles, hollow organs, and tubes	Throughout body; concentrated in brain and spinal cord
Cell arrangement and shapes	Variable number of layers, from one to many; cells flattened, cuboidal, or columnar	Cells not in layers; usually randomly scattered in matrix; cell shape irregular to round	Cells linked in sheets or elongated bundles; cells shaped in elongated, thin cylinders; heart muscle cells may be branched	Cells isolated or networked; cell appendages highly branched and/or elongated

being are said to be **totipotent** [*totus,* entire] and have the ability to develop into any and all types of specialized cells. Any totipotent cell has the potential to become a functioning organism.

After about day 4 of development, the totipotent cells of the embryo begin to specialize, or *differentiate.* As they do so, they narrow their potential fates and become **pluripotent** [*plures,* many]. Pluripotent cells can develop into many different cell types but not all cell types. An isolated pluripotent cell cannot develop into an organism.

As differentiation continues, pluripotent cells develop into the various tissues of the body. As the cells specialize and mature, many lose the ability to reproduce themselves. Others, called **stem cells,** retain the ability to divide. In adults, these stem cells often can become only one particular cell type, such as a blood cell. Undifferentiated cells in a tissue that retain the ability to divide and develop into specialized cells of that tissue are said to be **multipotent** [*multi,* many]. Some of the most-studied multipotent adult stem cells are found in bone marrow and give rise to blood cells.

Biologists once believed that nerve and muscle cells, which are highly specialized in their mature forms, could not be replaced when they died. Now exciting new research indicates that stem cells for these tissues do exist in the body. However, these neural and muscle stem cells occur in very small numbers, are difficult to isolate, and do not thrive in the laboratory. Naturally occurring adult neural and muscle stem cells are unable to replace large masses of dead or dying tissue that occur as a result of diseases, such as strokes or heart attacks. Consequently, one goal of stem cell research is to find a source of pluripotent or multipotent

cells that could be grown in the laboratory and implanted to treat *degenerative* diseases, those in which cells degenerate and die.

One example of a degenerative disease is Parkinson's disease, in which certain types of nerve cells in the brain die. Embryos and fetal tissue are rich sources of stem cells, but the use of these two tissues is controversial and accompanied by many legal and ethical questions. Some researchers hope that adult stem cells will show **plasticity,** the ability to specialize into a cell of a type different from the type for which they were destined.

There are still many challenges facing us before stem cell therapy becomes a reality. One is finding a good source of stem cells. A second major challenge is determining the chemical signals that tell stem cells when to differentiate and what cell type to become. And even once these two challenges are overcome and donor stem cells are implanted, the body may recognize that the new cells are foreign tissue and try to reject them.

Stem cell research is an excellent example of the dynamic and often controversial nature of science. For the latest research findings as well as pending legislation and laws regulating stem cell research and use, you should check authoritative websites, such as that sponsored by the U.S. National Institutes of Health (*http://stemcells.nih.gov/*).

ORGANS

Groups of tissues that carry out related functions may form structures known as **organs.** The organs of the body contain the four types of tissue in various combinations. The skin is an

(a) The layers of the skin

Hair follicles secrete the nonliving keratin shaft of hair.

Sebaceous glands are exocrine glands that secrete a lipid mixture.

Arrector pili muscles pull hair follicles into a vertical position when the muscles contract, creating "goose bumps."

Sweat glands secrete a dilute salt fluid to cool the body.

Sensory receptors monitor external conditions.

Epidermis consists of multiple cell layers that create a protective barrier.

The **dermis** is loose connective tissue that contains exocrine glands, blood vessels, muscles, and nerve endings.

Hypodermis contains adipose tissue for insulation.

Artery   Vein

Blood vessels extend upward into the dermis.

Sensory nerve

Apocrine glands in the genitalia, anus, axillae (*axilla*, armpit), and eyelids release waxy or viscous milky secretions in response to fear or sexual excitement.

(b) Epidermis. The skin surface is a mat of linked keratin fibers left behind when old epithelial cells die.

Phospholipid matrix acts as the skin's main waterproofing agent.

Surface keratinocytes produce keratin fibers.

Desmosomes anchor epithelial cells to each other.

Epidermal cell

Basal lamina

(c) Connection between epidermis and dermis

Hemidesmosomes tie epidermal cells to fibers of the basal lamina.

Basal lamina or basement membrane is an acellular layer between epidermis and dermis.

3

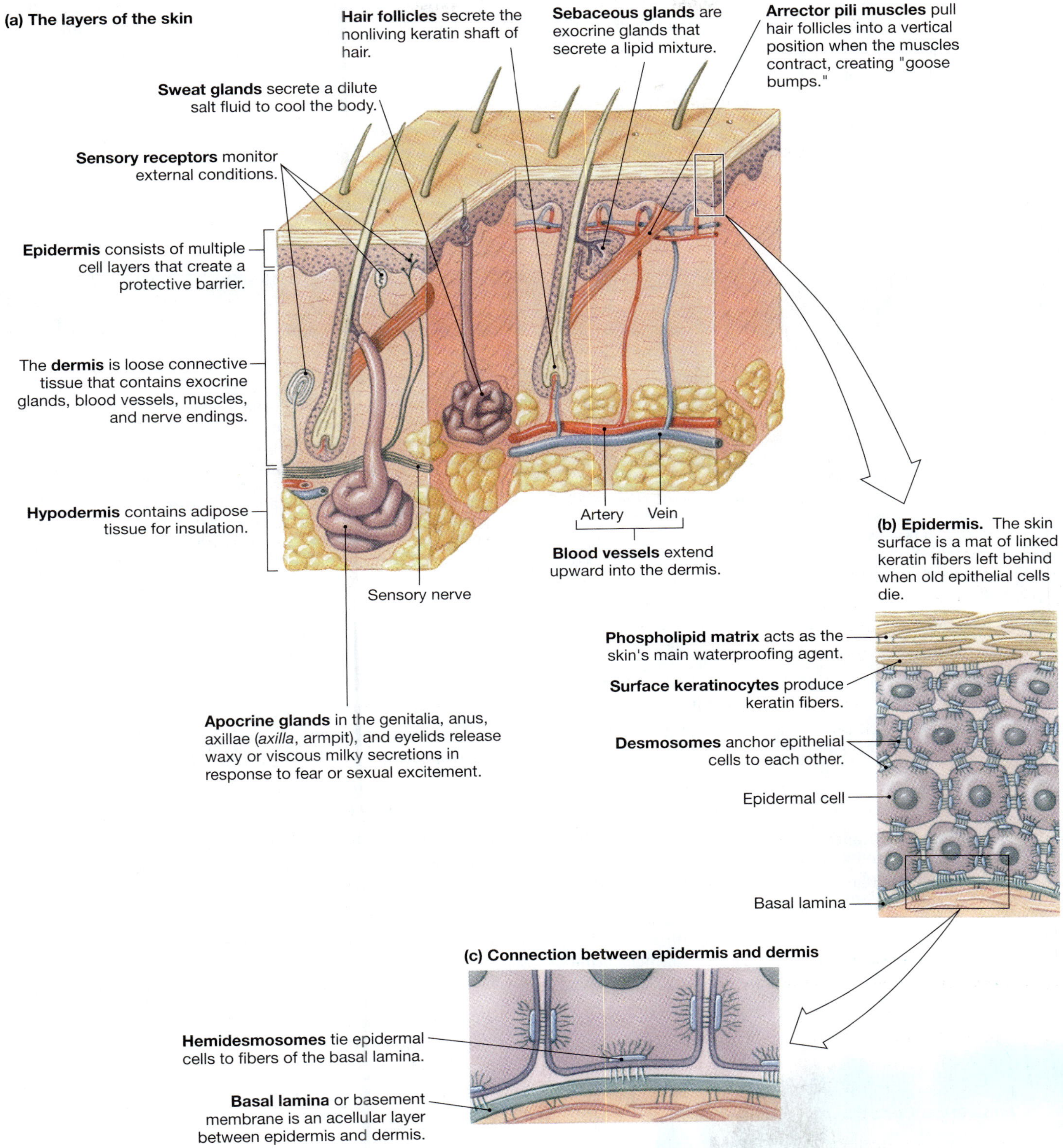

excellent example of an organ that incorporates all four types of tissue into an integrated whole. We think of skin as a thin layer that covers the external surfaces of the body, but in reality it is the heaviest single organ, at about 16% of an adult's total body weight! If it were flattened out, it would cover a surface area of between 1.2 and 2.3 square meters, about the size of a couple of card-table tops. Its size and weight make skin one of the most important organs of the body.

The functions of the skin do not fit neatly into any one chapter of this book, and this is true of other organs as well. We will highlight several of these organs in special *Organ Focus Features* throughout the book. These illustrated boxes discuss the structure and functions of these versatile organs so that you can gain an appreciation for the way different tissues combine for a united purpose. The first of these features, *Focus on the Skin*, appears on page 83.

As we consider the systems of the body in the succeeding chapters, you will see how diverse cells, tissues, and organs carry out the processes of the living body. Although the body's cells have different structures and different functions, they have one need in common: a continuous supply of energy. Without energy, cells cannot survive, let alone carry out all the other processes of daily living. The next chapter looks at cellular metabolism: how cells capture and harness the energy released by chemical reactions.

RUNNING PROBLEM CONCLUSION

THE PAP SMEAR

In this running problem, you learned that the Pap test can detect the early changes that herald cervical cancer. The diagnosis is not always simple because the change in cell cytology from normal to cancerous is a continuum and can be subject to individual interpretation. In 2001 a group of clinicians, cytotechnologists, pathologists, and patient advocates representing 45 professional societies from more than 20 countries met to review the evidence on which the guidelines for interpreting Pap smears are based. Their recommendations shaped the 2001 Bethesda System terminology guidelines (*http://bethesda2001.cancer.gov*).

As a result of the 2001 Bethesda System recommendations, abnormal Pap smears may now be tested for the presence of DNA from the human papilloma virus (HPV). Women with HPV have a higher risk of developing cervical cancer. To learn more about the association between HPV and cervical cancer, go to the National Cancer Institute homepage (*www.nci.nih.gov*) and search for "HPV." This site also contains information about cervical cancer. To check your understanding of the running problem, compare your answers with the information in the following summary table.

	QUESTION	FACTS	INTEGRATION AND ANALYSIS
1	Why does the treatment of cancer focus on killing the cancerous cells?	Cancerous cells divide uncontrollably and fail to coordinate with normal cells. Cancerous cells fail to differentiate into specialized cells.	Unless removed, cancerous cells will displace normal cells. This may cause destruction of normal tissues. In addition, because cancerous cells do not become specialized, they cannot carry out the same functions as the specialized cells they displace.
2	What property of cancer cells explains the large size of their nucleus and the relatively small amount of cytoplasm?	Cancerous cells divide uncontrollably. Dividing cells must duplicate their DNA prior to cell division, and this DNA duplication takes place in the nucleus, leading to the large size of that organelle [⬆ Appendix C]	Actively reproducing cells are likely to have more DNA in their nucleus as they prepare to divide, so their nuclei tend to be larger. Each cell division splits the cytoplasm between two daughter cells. If division is occurring rapidly, the daughter cells may not have time to synthesize new cytoplasm, so the amount of cytoplasm is less than in a normal cell.
3a	What kind of damage or trauma are cervical epithelial cells normally subjected to?	The cervix is the neck of the uterus.	The cervix is subject to trauma or damage during sexual intercourse and childbirth.

	QUESTION	FACTS	INTEGRATION AND ANALYSIS
3b	Which of its two types of epithelia is more likely to be affected by trauma?	The cervix consists of secretory epithelium with mucus-secreting glands lining the inside and protective epithelium covering the outside.	Protective epithelium is composed of multiple layers of cells and is designed to protect areas from mechanical and chemical stress [🔄 p. 72]. Therefore, the secretory epithelium with its single cell layer is more easily damaged.
4	Has Jan's dysplasia improved or worsened? What evidence do you have to support your answer?	The slide from Jan's first Pap smear shows abnormal cells with large nuclei and little cytoplasm. These abnormal cells do not appear in the second smear.	The disappearance of the abnormal cells indicates that Jan's dysplasia has resolved. She will return in six months for a repeat Pap smear. If it shows no dysplasia, her cervical cells have reverted to normal.

51 58 67 75 81 **85**

CHAPTER SUMMARY

Cell biology and histology illustrate one of the major themes in physiology: *compartmentation*. In this chapter you learned how a cell is subdivided into two main compartments, the nucleus and the cytoplasm. The cytoplasm is further broken down into three components: cytosol, organelles, and inclusions. You also learned how cells form tissues that create larger compartments within the body. A second theme in this chapter is the *mechanical properties* of cells and tissues that arise from the protein fibers of the cytoskeleton, from cell junctions, and from the molecules that make up the extracellular matrix, part of the "glue" that holds tissues together.

Functional Compartments of the Body

IP Fluids & Electrolytes: Introduction to Body Fluids

1. The **cell** is the functional unit of most living organisms. (p. 51)
2. The major human body cavities are the cranial cavity (skull), thoracic cavity (thorax), and abdominopelvic cavity. (p. 51; Fig. 3-1)
3. The **lumens** of some hollow organs are part of the body's external environment. (p. 51)
4. The body fluid compartments are the extracellular fluid (ECF) outside the cells and the intracellular fluid (ICF) inside the cells. The ECF can be subdivided into **interstitial fluid** bathing the cells and **plasma**, the fluid portion of the blood. (p. 52; Fig. 3-2)

Biological Membranes

5. The word **membrane** is used both for cell membranes and for epithelial tissues that line a cavity or separate two compartments. (p. 53; Fig. 3-3)
6. The **cell membrane** acts as a barrier between the intracellular and extracellular fluids, provides structural support, and regulates exchange and communication between the cell and its environment. (p. 53)

7. The **fluid mosaic model** of a biological membrane shows it as a **phospholipid bilayer** with proteins inserted into the bilayer. (p. 55; Fig. 3-4)
8. Membrane lipids include phospholipids, **sphingolipids**, and cholesterol. (p. 55)
9. Membrane-spanning **integral proteins** are tightly bound to the phospholipid bilayer. **Peripheral** and **lipid-anchored proteins** attach less tightly to either side of the membrane. (pp. 55–57; Fig. 3-6)
10. Carbohydrates attach to the extracellular surface of cell membranes. (p. 58)

Intracellular Compartments

11. The cytoplasm consists of semigelatinous **cytosol** with dissolved nutrients, ions, and waste products. Suspended in the cytosol are the two other components of the cytoplasm: insoluble **inclusions**, which have no enclosing membrane, and **organelles**, which are membrane-enclosed bodies that carry out specific functions. (p. 61; Fig. 3-11)
12. Important inclusions include **ribosomes**, which take part in protein synthesis; **proteasomes** for protein degradation; and insoluble protein fibers in three sizes: **actin fibers** (also called **microfilaments**), **intermediate filaments**, and **microtubules**. (p. 61; Tbl. 3-2)
13. **Centrosomes** are the microtubule-organizing center of the cell. (p. 61)
14. Three types of microtubules can be identified in the cell. **Centrioles** aid the movement of chromosomes during cell division, **cilia** move fluid or secretions across the cell surface, and **flagella** propel sperm through body fluids. (pp. 61–62; Fig. 3-13)
15. The **cytoskeleton** provides strength, support, and internal organization; aids transport of materials within the cell; links cells together; and enables motility in certain cells. (pp. 62–63; Fig. 3-14)

16. **Motor proteins** such as **myosins**, **kinesins**, and **dyneins** associate with cytoskeleton fibers to create movement. (p. 63; Fig. 3-15)

17. Membranes around organelles divide the cytosol into compartments whose functions can be separated. (p. 64)

18. **Mitochondria** generate most ATP for the cell. (p. 64; Fig. 3-16)

19. The **smooth endoplasmic reticulum** is the primary site of lipid synthesis. The **rough endoplasmic reticulum** is the primary site of protein synthesis. (pp. 65–66; Fig. 3-17)

20. The **Golgi complex** packages proteins into vesicles. **Secretory vesicles** release their contents into the extracellular fluid. (p. 66; Fig. 3-18)

21. **Lysosomes** and **peroxisomes** are **storage vesicles** that contain digestive enzymes. (pp. 66–67; Fig. 3-19)

22. The **nucleus** contains DNA, the genetic material that ultimately controls all cell processes, in the form of **chromatin**. The double-membrane **nuclear envelope** surrounding the nucleus has **nuclear pore complexes** that allow controlled chemical communication between the nucleus and cytosol. **Nucleoli** are nuclear areas that control the synthesis of RNA for ribosomes. (p. 67–68; Fig. 3-20)

Tissues of the Body

IP Muscular: Anatomy Review—Skeletal Muscle Tissue

23. There are four **primary tissue types** in the human body: epithelial, connective, muscle, and neural. (p. 68)

24. **Extracellular matrix** secreted by cells provides support and a means of cell-cell communication. It is composed of proteoglycans and insoluble protein fibers. (pp. 68–69)

25. Cell junctions fall into three categories. **Gap junctions** allow chemical and electrical signals to pass directly from cell to cell. **Tight junctions** prevent the movement of material between cells. **Anchoring junctions** hold cells to each other or to the extracellular matrix. (p. 69; Fig. 3-21)

26. Membrane proteins called **cell adhesion molecules (CAMs)** are essential in cell adhesion and in anchoring junctions. (p. 69; Tbl. 3-3)

27. **Desmosomes** and **adherens junctions** anchor cells to each other. **Focal adhesions** and **hemidesmosomes** anchor cells to matrix. (p. 71; Fig. 3-22)

28. **Epithelial tissues** protect the internal environment, regulate the exchange of material, or manufacture and secrete chemicals. There are five functional types found in the body: exchange, transporting, ciliated, protective, and secretory. (p. 72)

29. **Exchange epithelia** permit rapid exchange of materials, particularly gases. **Transporting epithelia** actively regulate the selective exchange of nongaseous materials between the internal and external environments. **Ciliated epithelia** move fluid and particles across the surface of the tissue. **Protective epithelia** help prevent exchange between the internal and external environments. The **secretory epithelia** release secretory products into the external environment or the blood. (pp. 72–74; Fig. 3-24)

30. **Exocrine glands** release their secretions into the external environment through **ducts**. **Endocrine glands** are ductless glands that release their secretions, called **hormones**, directly into the extracellular fluid. (p. 75; Fig. 3-28)

31. **Connective tissues** have extensive extracellular matrix that provides structural support and forms a physical barrier. (p. 76)

32. **Loose connective tissues** are the elastic tissues that underlie skin. **Dense connective tissues**, including **tendons** and **ligaments**, have strength or flexibility because they are made of collagen. **Adipose tissue** stores fat. The connective tissue we call **blood** is characterized by a watery matrix. **Cartilage** is solid and flexible and has no blood supply. The fibrous matrix of **bone** is hardened by deposits of calcium salts. (pp. 77–79)

33. Muscle and neural tissues are called excitable tissues because of their ability to generate and propagate electrical signals called action potentials. **Muscle tissue** has the ability to contract and produce force and movement. There are three types of muscle: cardiac, smooth, and skeletal. (p. 80)

34. **Neural tissue** includes **neurons**, which use electrical and chemical signals to transmit information from one part of the body to another, and support cells known as **glial cells** (neuroglia). (p. 80)

Tissue Remodeling

35. Cell death occurs by **necrosis**, which adversely affects neighboring cells, and by **apoptosis**, programmed cell death that does not disturb the tissue. (p. 81)

36. **Stem cells** are cells that are able to reproduce themselves and differentiate into specialized cells. Stem cells are most plentiful in embryos but are also found in the adult body. (pp. 81–82)

Organs

37. **Organs** are formed by groups of tissues that carry out related functions. The organs of the body contain the four types of tissues in various ratios. (p. 82)

QUESTIONS

(Answers to the Review Questions begin on page A1.)

The Physiology Place

Access more review material online at **The Physiology Place** website. There you'll find review questions, problem-solving activities, case studies, flashcards, and direct links to both *InterActive Physiology®* and *PhysioEx™*. To access the site, go to *www.physiologyplace.com* and select Human Physiology, Fourth Edition.

LEVEL ONE REVIEWING FACTS AND TERMS

1. List the four general functions of the cell membrane.

2. In 1972, Singer and Nicolson proposed a model of the cell membrane known as the _____ model. According to this model, the membrane is composed of a bilayer of _____ and a variety of embedded _____, with _____ on the extracellular surface.

3. What are the two primary types of biomolecules found in the cell membrane?

4. Define and distinguish between inclusions and organelles. Give an example of each.

5. Define cytoskeleton. List five functions of the cytoskeleton.

6. Match each term with the description that fits it best:

   (a) cilia

   (b) centriole

   (c) flagellum

   (d) centrosome

   1. in human cells, appears as single, long, whiplike tail
   2. short, hairlike structures that beat to produce currents in fluids
   3. a bundle of microtubules that aids in mitosis
   4. the microtubule-organizing center

7. Match each organelle with its function:

   (a) endoplasmic reticulum

   (b) Golgi complex

   (c) lysosome

   (d) mitochondria

   (e) peroxisome

   1. powerhouse of the cell where most ATP is produced
   2. degrades long-chain fatty acids and toxic foreign molecules
   3. network of membranous tubules that synthesize biomolecules
   4. digestive system of cell, degrading or recycling components
   5. modifies and packages proteins into vesicles

8. What process activates the enzymes inside lysosomes?

9. Exocrine glands produce watery secretions (such as tears or sweat) called _____, or stickier solutions called _____.

10. _____ glands release hormones, which enter the blood and regulate the activities of organs or systems.

11. List the four major tissue types. Give an example and location of each.

12. The largest and heaviest organ in the body is the _____.

13. Match each protein to its function. A function in the list may be used more than once.

    (a) cadherin
    (b) CAM
    (c) collagen
    (d) connexin
    (e) elastin
    (f) fibrillin
    (g) fibronectin
    (h) integrin
    (i) occludin

    1. membrane protein used to form cell junctions
    2. matrix glycoprotein used to anchor cells
    3. protein found in gap junctions
    4. matrix protein found in connective tissue

14. What types of glands can be found within the skin? Name the secretion of each type.

15. The term *matrix* can be used in reference to an organelle or to tissues. Compare the meanings of the term in these two contexts.

LEVEL TWO   REVIEWING CONCEPTS

16. **Mapping exercise:** Transform this list of terms into a map of cell structure. Add functions where appropriate.

actin	microtubule
cell membrane	microtubule-organizing center
centriole	mitochondria
centrosome	myosin
cilia	nonmembranous organelle
cytoplasm	nucleus
cytoskeleton	peroxisome
cytosol	proteasome
extracellular matrix	RER

flagella	ribosome
Golgi complex	secretory vesicle
intermediate filament	SER
keratin	storage vesicle
lysosome	thick filament
organelle	tubulin
microfilament	vault

17. List, compare, and contrast the three types of cell junctions and their subtypes. Give an example of where each type can be found in the body and describe its function in that location.

18. Which would have more rough endoplasmic reticulum: pancreatic cells that manufacture the protein hormone insulin, or adrenal cortex cells that synthesize the steroid hormone cortisol?

19. A number of organelles can be considered vesicles. Define *vesicle* and describe at least three examples.

20. Explain why a stratified epithelium offers more protection than a simple epithelium.

21. Sketch a short series of columnar epithelial cells. Label the apical and basolateral borders of the cells. Briefly explain the different kinds of junctions found on these cells.

22. Arrange the following compartments in the order a glucose molecule entering the body at the intestine would encounter them: interstitial fluid, plasma, intracellular fluid. Which of these fluids is/are considered to be extracellular fluid(s)?

23. Explain how inserting cholesterol into the phospholipid bilayer of the cell membrane decreases membrane permeability.

24. Compare and contrast the structure, locations, and functions of bone and cartilage.

25. Differentiate between the terms in each set below:
    (a) lumen/wall; **(b)** cytoplasm/cytosol; **(c)** myosin/keratin

26. When a tadpole turns into a frog, its tail shrinks and is reabsorbed. Is this an example of necrosis or apoptosis? Defend your answer.

27. Match the structures from the chapter to the basic physiological themes in the right column and give an example or explanation for each match. A structure may match with more than one theme.

    (a) cell junctions
    (b) cell membrane
    (c) cytoskeleton
    (d) organelles
    (e) cilia

    1. communication
    2. molecular interactions
    3. compartmentation
    4. mechanical properties
    5. biological energy use

28. In some instances, the extracellular matrix can be quite rigid. How might developing and expanding tissues cope with a rigid matrix to make space for themselves?

LEVEL THREE   PROBLEM SOLVING

29. One result of cigarette smoking is paralysis of the cilia that line the respiratory passageways. What function do these cilia serve? Based on what you have read in this chapter, why is it harmful when they no longer beat? What health problems would you expect to arise? How does this explain the hacking cough common among smokes?

30. Cancer is abnormal, uncontrolled cell division. What property of epithelial tissues might (and does) make them more prone to developing cancer?

31. What might happen to normal physiological function if matrix metalloproteinases are inhibited by drugs?

ANSWERS

✓ Answers to Concept Check Questions

Page 58

1. Three types of lipids found in cell membranes are phospholipid, sphingolipid, and cholesterol.

2. Integral proteins are tightly bound to the membrane. Peripheral proteins and lipid-anchored proteins are loosely bound to membrane components.

3. The tails of phospholipids are hydrophobic, and a single layer would put them in direct contact with aqueous body fluids.

4. A substance passing through a mucous membrane will cross two phospholipid bilayers.

Page 64

5. The cytoplasmic protein fibers are actin fibers (or microfilaments), intermediate filaments, and microtubules.

6. Without a flagellum, a sperm would be unable to swim to find an egg to fertilize.

7. Cytoplasm is everything inside the cell membrane except the nucleus. Cytosol is the semi-gelatinous substance in which organelles and inclusions are suspended.

8. Cilia are short, usually are very numerous on a cell, and move fluid or substances across the cell surface. Flagella are longer, usually occur singly on human sperm, and are used to propel a cell through a fluid.

9. Motor proteins use energy to create movement.

Page 67

10. Organelles are separated from the cytosol by a membrane; inclusions are not.

11. Rough ER has ribosomes attached to the cytoplasmic side of its membrane; smooth ER lacks ribosomes. Rough ER synthesizes proteins; smooth ER synthesizes lipids.

12. Lysosomes contain enzymes that break down bacteria and old organelles. Peroxisomes contain enzymes that break down fatty acids and foreign molecules.

13. The membranes of organelles create compartments that physically isolate their lumens from the cytosol. The double membrane of mitochondria creates two different compartments inside the organelle.

14. A large number of mitochondria suggests that the cell has a high energy requirement because mitochondria are the site of greatest energy production in the cell.

15. Large amounts of smooth endoplasmic reticulum suggest that the tissue synthesizes large amounts of lipids, fatty acids, or steroids, or that it detoxifies foreign molecules.

Page 72

16. Cell junctions are gap, tight, and anchoring.

17. (a) Tight, (b) gap, (c) anchoring (specifically, desmosome), (d) anchoring (specifically, focal adhesion).

Page 76

18. The five functional types of epithelia are protective, secretory, transporting, ciliated, and exchange.

19. Secretion is the process by which a cell releases a substance into the environment.

20. Endocrine glands do not have ducts, and they secrete into the blood. Exocrine glands have ducts and secrete into the external environment.

21. Hemidesmosomes attach to laminin (see Fig. 3-22).

22. No, skin has many layers of cells in order to protect the internal environment. A simple squamous epithelium (which is one cell thick with flattened cells) would not be a protective epithelium.

23. The cell is an endocrine cell because it secretes its product into the extracellular space for distribution in the blood.

Page 80

24. Connective tissues have extensive matrix.

25. Collagen provides strength and flexibility; elastin and fibrillin provide elastance; fibronectin helps anchor cells to matrix.

26. Connective tissues include bone, cartilage, blood, dense connective tissues (ligaments and tendons), loose connective tissue, and adipose.

27. The plasma, or liquid portion of blood, could be considered the extracellular matrix.

28. Cartilage lacks a blood supply, so oxygen and nutrients needed for repair must reach the cells by diffusion, a slow process.

Page 81

29. Apoptosis is a tidy form of cell death that removes cells without disrupting their neighbors. By contrast, necrosis releases digestive enzymes that damage neighboring cells.

Q Answers to Figure Questions

Page 73

Fig. 3-24: 1. Transporting epithelia (purple) are found primarily in the digestive tract and kidney, where they regulate the movement of substances into and out of the extracellular fluid. The lungs and the blood vessels are lined with exchange epithelium (red). Ciliated epithelia (blue) are found in the airways of the respiratory tract and in part of the female reproductive system. The outer surface of the body and the openings of the body cavities are covered with protective epithelium (brown). 2. Endocrine glands (without ducts) secrete their hormones into the blood. 3. Exocrine glands, with ducts, secrete their products outside the body—onto the surface of the skin or into the lumen of an organ that opens into the environment outside the body (see Fig. 1-2, p. 3).

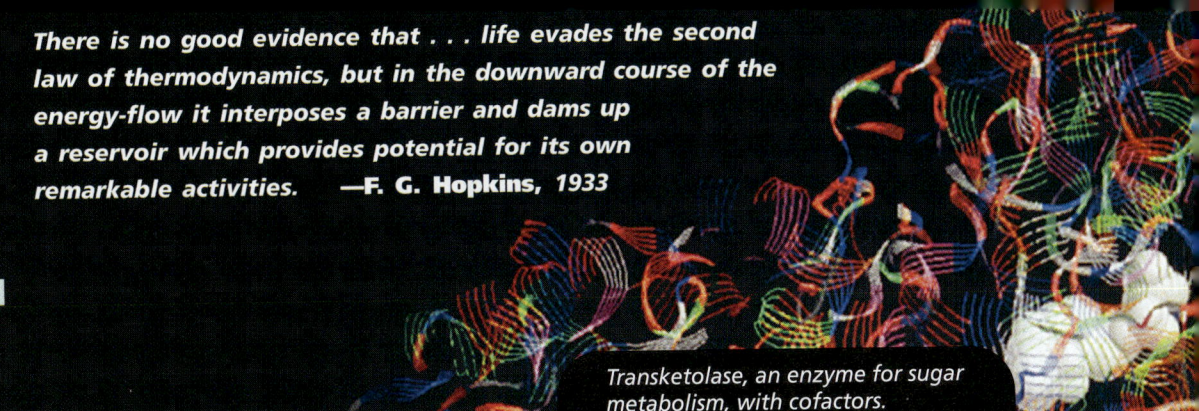

There is no good evidence that . . . life evades the second law of thermodynamics, but in the downward course of the energy-flow it interposes a barrier and dams up a reservoir which provides potential for its own remarkable activities. —F. G. Hopkins, 1933

Transketolase, an enzyme for sugar metabolism, with cofactors.

4

Energy and Cellular Metabolism

BACKGROUND BASICS

Forms of energy and work (Appendix B): **B1** DNA and RNA (Appendix C): **C1** Emergent properties: **6**
Hydrogen bonds: **25** Covalent bonds: **23** ATP: **33** Organelles: **61** Electrons: **22** Protein structure: **30**
Carbohydrates: **27** Nucleotides: **33** Graphing: **12** Lipids: **27**

RUNNING PROBLEM

TAY-SACHS DISEASE

In many American ultra-orthodox Jewish communities—in which arranged marriages are the norm—the rabbi is entrusted with an important, life-saving task. He keeps a confidential record of individuals known to carry the gene for Tay-Sachs disease, a fatal, inherited condition that strikes one in 3600 American Jews of Eastern European descent. Babies born with this disease rarely live beyond age 4, and there is no cure. Based on the family trees he constructs, the rabbi can avoid pairing two individuals who carry the deadly gene.

Sarah and David, who met while working on their college newspaper, are not orthodox Jews. Both are aware, however, that their Jewish ancestry might put any children they have at risk for Tay-Sachs disease. Six months before their wedding, they decide to see a genetic counselor to determine whether they are carriers of the gene for Tay-Sachs disease.

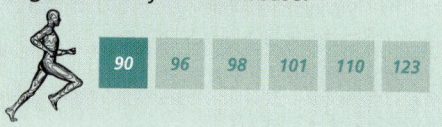

| 90 | 96 | 98 | 101 | 110 | 123 |

TABLE 4-1	Properties of Living Organisms

1. Have a highly organized, complex structure
2. Acquire, transform, store, and use energy
3. Sense and respond to internal and external environments
4. Maintain homeostasis through internal control systems with feedback
5. Store, use, and transmit information
6. Reproduce, develop, grow, and die
7. Have emergent properties that cannot be predicted from the simple sum of the parts
8. Species evolve

Christine Schmidt, Ph.D., and her graduate students seed isolated endothelial cells onto an engineered matrix and watch them grow. They know that if their work is successful, the tissue that results might someday help replace a blood vessel in the body. Just as a child playing with building blocks assembles them into a house, the bioengineer and her students create tissue from cells. In both examples, someone familiar with the starting components, building blocks or cells, can predict what the final product will be: blocks make buildings, cells make tissues.

Why then can't biologists, knowing the characteristics of nucleic acids, proteins, lipids, and carbohydrates, explain how combinations of these molecules acquire the remarkable attributes of a living cell? How can living cells carry out processes that far exceed what we would predict from understanding their individual components? The answer is *emergent properties* [🔁 p. 6], those distinctive traits that cannot be predicted from the simple sum of the component parts. For example, if you came across a collection of metal pieces and bolts from a disassembled car motor, could you predict (without prior knowledge) that, given an energy source, this collection properly arranged could create the power to move thousands of pounds?

The emergent properties of biological systems are of tremendous interest to scientists trying to explain how a simple compartment, such as a phospholipid liposome [🔁 p. 55], could have evolved into the first living cell. Pause for a moment and see if you can list the properties of life that characterize all living creatures. If you were a scientist looking at pictures and samples sent back from Mars, what would you look for to determine whether there is life in other parts of the universe?

Now compare your list with the one in Table 4-1 ■. Living organisms are highly organized and complex entities. Even a one-celled bacterium, although it appears simple under a microscope, has incredible complexity at the chemical level of organization. It uses intricately interconnected biochemical reactions to acquire, transform, store, and use energy and information. It senses and responds to changes in its internal and external environments so that it can maintain homeostasis. It reproduces, develops, grows, and dies, and over time, its species evolves.

Energy is essential for these processes we associate with living things. Without energy for growth, repair, and maintenance of the internal environment, a cell is like a ghost town filled with buildings that are slowly crumbling into ruin. Cells need energy to import raw materials, make new molecules, and repair or recycle aging parts. The ability of cells to extract energy from the external environment and use that energy to maintain themselves as organized, functioning units is one of their most outstanding characteristics. In this chapter, we look at the cell processes through which the human body obtains energy and maintains its ordered systems.

ENERGY IN BIOLOGICAL SYSTEMS

Energy cycling between the environment and living organisms is one of the fundamental concepts of biology. All cells use energy from their environment to grow, make new parts, and reproduce. Plants trap radiant energy from the sun and store it as chemical-bond energy through the process of photosynthesis (Fig. 4-1 ■). They extract carbon and oxygen from carbon dioxide, nitrogen from the soil, and hydrogen and

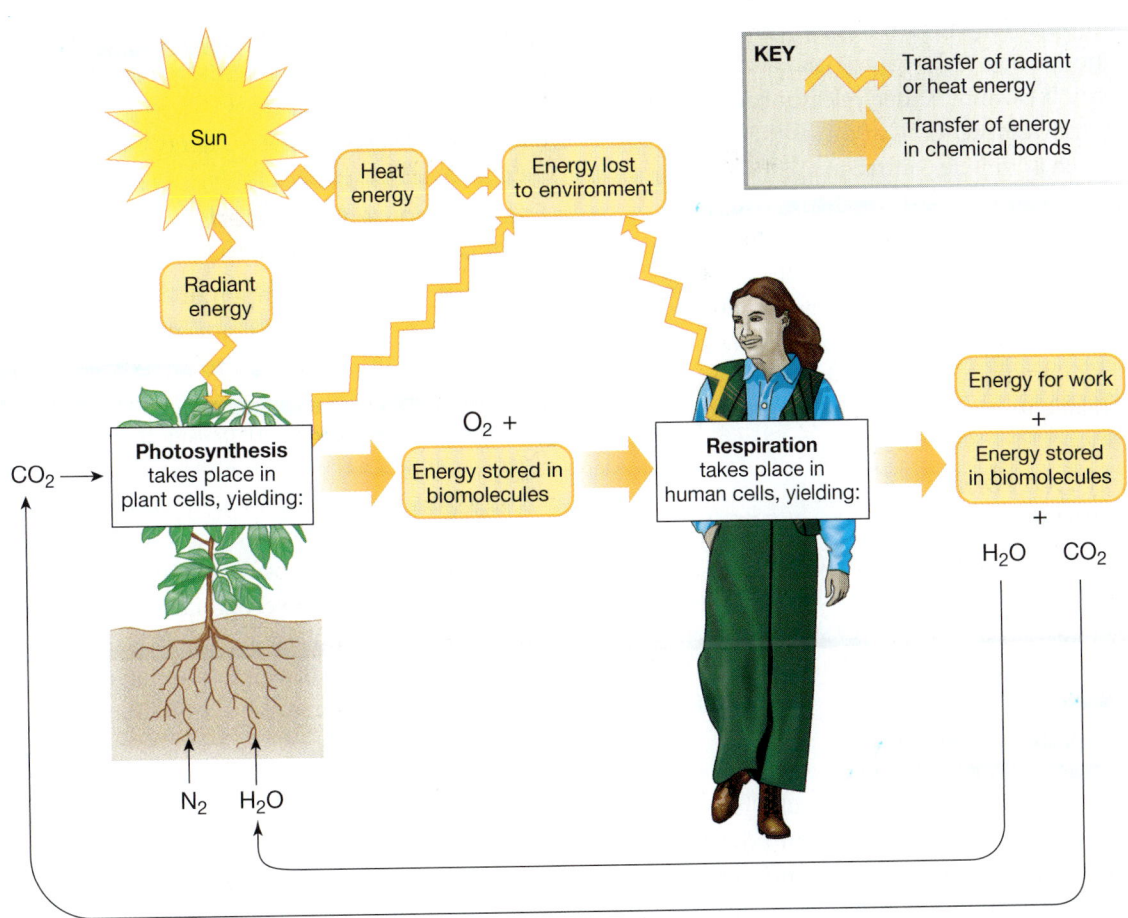

KEY
Transfer of radiant or heat energy
Transfer of energy in chemical bonds

Sun

Heat energy

Energy lost to environment

Radiant energy

Energy for work

CO_2 →

Photosynthesis
takes place in plant cells, yielding:

O_2 +
Energy stored in biomolecules

Respiration
takes place in human cells, yielding:

Energy stored in biomolecules

+

Energy stored in biomolecules

+

H_2O CO_2

N_2 H_2O

■ **FIGURE 4-1** *Energy transfer in the environment*

Plants trap radiant energy from the sun and use it to store energy in the chemical bonds of biomolecules. Animals eat the plants and either use the energy or store it.

oxygen from water to make biomolecules such as glucose and amino acids.

Animals, on the other hand, cannot trap energy from the sun or use carbon and nitrogen from the air and soil to synthesize biomolecules. They must import chemical-bond energy by ingesting the biomolecules of plants or other animals. Ultimately, however, energy from the sun that was trapped by photosynthesis is the energy source for all animals, including humans.

Animals extract energy from biomolecules through the process of *respiration,* which consumes oxygen and produces carbon dioxide and water. If animals ingest more energy than they need for immediate use, the excess energy is stored in chemical bonds, just as it is in plants. Glycogen (a glucose polymer) and lipid molecules are the main energy stores in animals [🔁 p. 27]. These storage molecules are available for use at times when an animal's energy needs exceed its food intake.

✔ CONCEPT CHECK

1. Which biomolecules always include nitrogen in their chemical makeup?
Answers: p. 126

Energy Is Used to Perform Work

All living organisms obtain, store, and use energy to fuel their activities. **Energy** can be defined as the capacity to do work, but what is *work*? We use this word in everyday life to mean various things, from hammering a nail to sitting at a desk writing a paper. In biological systems, however, the word means one of three specific things: chemical work, transport work, or mechanical work.

Chemical work, the making and breaking of chemical bonds, enables cells and organisms to grow, maintain a suitable internal environment, and store information needed for reproduction and other activities. Forming the chemical bonds of a protein the body will use for wound repair is an example of chemical work.

Transport work enables cells to move ions, molecules, and larger particles through the cell membrane and through the membranes of organelles in the cell. Transport work is particularly useful for creating **concentration gradients**, distributions of molecules in which the concentration is higher on one side of a membrane than on the other. For example, certain types of endoplasmic reticulum [🔁 p. 66] use energy to import calcium ions from the cytosol. This ion transport creates a high calcium concentration inside the organelle and a low concentration in the cytosol. If calcium is then released back into the cytosol, it creates a "calcium signal" that causes the cell to perform some action, such as muscle contraction.

Mechanical work in animals is used for movement. At the cellular level, movement includes organelles moving around in a cell, cells changing shape, and cilia and flagella beating [🔁 p. 63]. At the macroscopic level in animals, movement usually involves muscle contraction. Most mechanical work is mediated by motor proteins that make up certain intracellular fibers and filaments of the cytoskeleton [🔁 p. 62].

Energy Comes in Two Forms: Kinetic and Potential

Energy can be classified in various ways. We often think of energy in terms we deal with daily: thermal energy, electrical energy, mechanical energy. We speak of energy stored in chemical bonds. Each type of energy has its own characteristics. However, all types of energy share an ability to appear in two forms: as kinetic energy or as potential energy.

Kinetic energy is the energy of motion [*kinetikos,* motion]. A ball rolling down a hill, perfume molecules spreading through the air, electric charge flowing through power lines, heat warming a frying pan, and molecules moving across biological membranes are all examples of bodies that have kinetic energy.

Potential energy is stored energy. A ball poised at the top of a hill has potential energy because it has the potential to start moving down the hill. A molecule positioned on the high-concentration side of a concentration gradient stores potential energy because it has the potential energy to move down the gradient. In chemical bonds, potential energy is stored in the position of the electrons that form the bond [🔁 p. 23]. To learn more about kinetic and potential energy, see Appendix A.

A key feature of all types of energy is the ability of potential energy to become kinetic energy and vice versa.

Energy Can Be Converted from One Form to Another

Recall that a general definition of energy is the capacity to do work. Work always involves movement and therefore is associated with kinetic energy. Potential energy also can be used to perform work, but the potential energy must first be converted to kinetic energy. The conversion from potential energy to kinetic energy is never 100% efficient, and a certain amount of energy is lost to the environment, usually as heat.

The amount of energy lost in the transformation depends on the *efficiency* of the process. Many physiological processes in the human body are not very efficient. For example, 70% of the energy used in physical exercise is lost as heat rather than transformed into the work of muscle contraction.

Figure 4-2 ■ summarizes the relationship of kinetic energy and potential energy:

1. Kinetic energy of the moving ball is transformed into potential energy as work is used to push the ball up the ramp (Fig. 4-2a).
2. Potential energy is stored in the stationary ball at the top of the ramp (Fig. 4-2b). No work is being performed, but the capacity to do work is stored in the position of the ball.
3. The potential energy of the ball becomes kinetic energy when the ball rolls down the ramp (Fig. 4-2c). Some kinetic energy is lost to the environment as heat due to friction between the ball and the air and ramp.

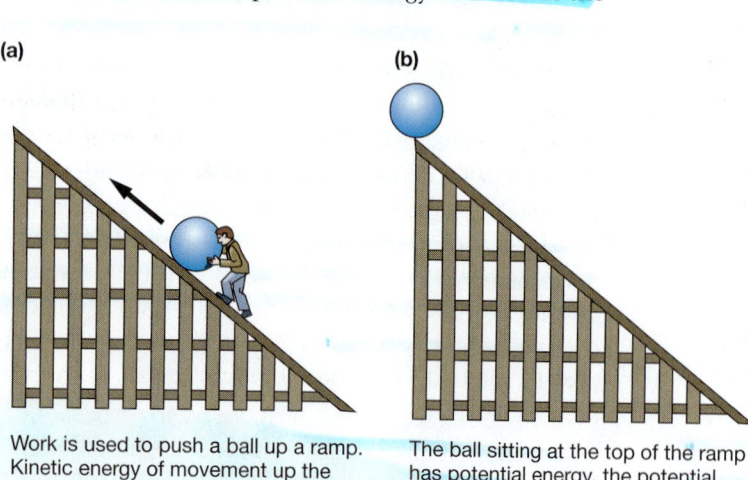
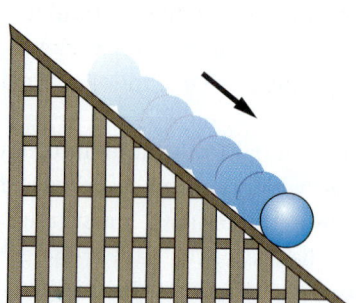

(a) Work is used to push a ball up a ramp. Kinetic energy of movement up the ramp is being stored in the potential energy of the ball's position.

(b) The ball sitting at the top of the ramp has potential energy, the potential to do work.

(c) The ball rolling down the ramp is converting the potential energy to kinetic energy. However, the conversion is not totally efficient, and some energy is lost as heat due to friction between the ball, ramp, and air.

■ **FIGURE 4-2**   *The relationship between kinetic energy and potential energy*

In biological systems, potential energy is stored in concentration gradients and chemical bonds. It is transformed into kinetic energy when needed to do chemical, transport, or mechanical work.

Thermodynamics Is the Study of Energy Use

Two basic rules govern the transfer of energy in biological systems and in the universe as a whole. The **first law of thermodynamics**, also known as the *law of conservation of energy*, states that the total amount of energy in the universe is constant. The universe is considered to be a *closed system*—nothing enters and nothing leaves. Energy can be converted from one type to another, but the total amount of energy in a closed system never changes.

The human body is not a closed system, however. As an *open system*, it exchanges materials and energy with its surroundings. Because our bodies cannot create energy, they import it from outside in the form of food. By the same token, our bodies lose energy, especially in the form of heat, to the environment. Energy that stays within the body can be changed from one type to another or can be used to do work.

The **second law of thermodynamics** states that natural spontaneous processes move from a state of order (nonrandomness) to a condition of randomness or disorder, also known as **entropy**. Creating and maintaining order in an open system such as the body requires the input of energy. Disorder occurs when open systems lose energy to their surroundings without regaining it. When this happens, we say that the entropy of the open system has increased.

The ghost-town analogy mentioned earlier illustrates the second law. When people put all their energy into activities away from town, the town slowly falls into disrepair and becomes less organized (its entropy increases). Similarly, without continual input of energy, a cell is unable to maintain its ordered internal environment. As the cell loses organization, its ability to carry out normal functions disappears, and it dies.

In the remainder of this chapter, you will learn how cells obtain energy from and store energy in the chemical bonds of biomolecules. Using chemical reactions, cells transform the potential energy of chemical bonds into kinetic energy for growth, maintenance, reproduction, and movement.

✓ CONCEPT CHECK

2. Name two ways animals store energy in their bodies.
3. What is the difference between potential energy and kinetic energy?
4. What is entropy?

Answers: p. 126

CHEMICAL REACTIONS

Living organisms are characterized by their ability to extract energy from the environment and use it to support life processes. The study of energy flow through biological systems is a field

REACTION TYPE	REACTANTS (SUBSTRATES)		PRODUCTS
Combination	A + B	→	C
Decomposition	C	→	A + B
Single displacement*	L + MX	→	LX + M
Double displacement*	LX + MY	→	LY + MX

TABLE 4-2 Types of Chemical Reactions

*X and Y represent atoms, ions, or chemical groups.

known as **bioenergetics** [*bios*, life + *en-*, in + *ergon*, work]. In a biological system, chemical reactions are a critical means of transferring energy from one part of the system to another.

Energy Is Transferred Between Molecules During Reactions

In a **chemical reaction**, a substance becomes a different substance, usually by the breaking and/or making of covalent bonds. A reaction begins with one or more molecules called **reactants** and ends with one or more molecules called **products** (Table 4-2 ■). In this discussion, we will consider a reaction that begins with two reactants and ends with two products:

$$A + B \longrightarrow C + D$$

The speed with which a reaction takes place, the **reaction rate**, is the disappearance rate of the reactants (A and B) or the appearance rate of the products (C and D). Reaction rate is measured as change in concentration during a certain time period and is often expressed as molarity per second (M/sec).

The purpose of chemical reactions in cells is either to transfer energy from one molecule to another or to use energy stored in reactant molecules to do work. The potential energy stored in the chemical bonds of a molecule is known as the **free energy** of the molecule. Generally, complex molecules have more chemical bonds and therefore higher free energies. For example, a large glycogen molecule has more free energy than a single glucose molecule, which in turn has more free energy than the carbon dioxide and water from which it was synthesized. The high free energy of complex molecules such as glycogen is the reason these molecules are used to store energy in cells.

To understand how chemical reactions transfer energy between molecules, we should answer two questions. First, how do reactions get started? The energy required to initiate a reaction is known as the *activation energy* for the reaction. Second, what happens to the free energy of the products and reactants during a reaction? The difference in free energy between reactants and products is known as the *net free energy change of the reaction*.

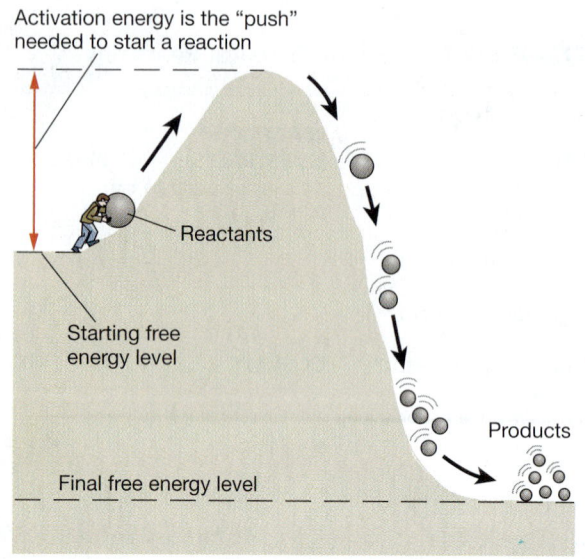

■ FIGURE 4-3 *Activation energy is the energy that must be put into reactants before a reaction can proceed*

Activation Energy Gets Reactions Started

Activation energy is the initial input of energy required to bring reactants into a position that allows them to react with one another. This "push" needed to start the reaction is shown in Figure 4-3 ■ as the little hill up which the ball must be pushed before it can roll by itself down the slope. A reaction with low activation energy will proceed spontaneously when the reactants are brought together. You can demonstrate a spontaneous reaction by pouring a little vinegar onto some baking soda and watching

the two react to form carbon dioxide. Reactions with high activation energies either will not proceed spontaneously or else will proceed too slowly to be useful. For example, if you pour vinegar over a pat of butter, no observable reaction takes place.

Energy Is Trapped or Released During Reactions

One characteristic property of any chemical reaction is the free energy change that occurs as the reaction proceeds. The products of a reaction will have either a lower free energy than the reactants or a higher free energy than the reactants. A change in free energy level means that the reaction has either released or trapped energy.

If the free energy of the products is lower than the free energy of the reactants, as in Figure 4-4a ■, the reaction releases energy and is called an **exergonic reaction** [*ex-*, out + *ergon*, work]. The energy released by an exergonic, or *energy-producing,* reaction may be used by other molecules to do work or may be given off as heat. In a few cases, the energy released in an exergonic reaction is stored as potential energy in a concentration gradient.

An important biological example of an exergonic reaction is the combination of ATP and water to form ADP, inorganic phosphate (P_i), and H^+. Energy is released during this reaction when the high-energy phosphate bond of the ATP molecule is broken:

$$ATP + H_2O \longrightarrow ADP + P_i + H^+ + energy$$

Now contrast the exergonic reaction of Figure 4-4a with the reaction represented in Figure 4-4b. In the latter, part of the activation energy added to the reactants is retained by the products, making their free energy greater than that of the reactants.

(a) **Exergonic reactions** release energy because the products have less energy than the reactants.

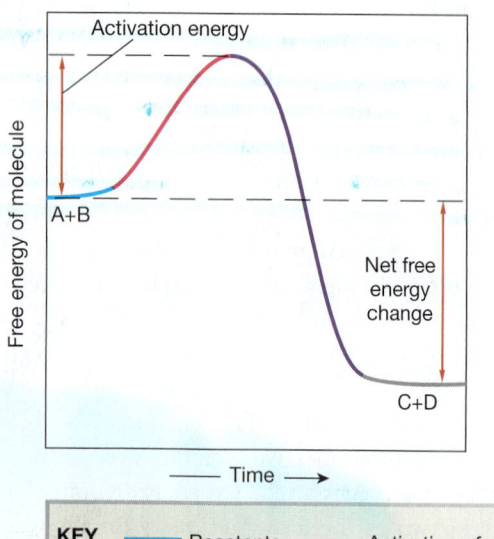

(b) **Endergonic reactions** trap some activation energy in the products, which then have more free energy than the reactants.

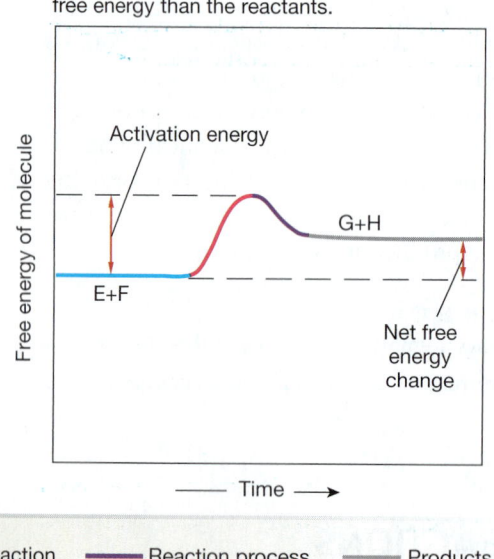

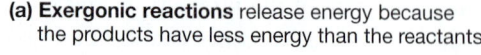

KEY — Reactants — Activation of reaction — Reaction process — Products

■ FIGURE 4-4 *Exergonic and endergonic reactions*

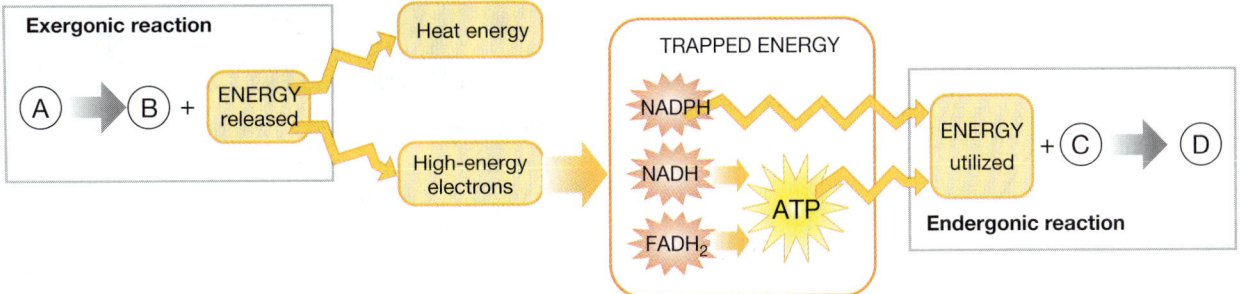

■ **FIGURE 4-5** *Energy transfer and storage in biological reactions*

Energy released by exergonic reactions can be trapped in the high-energy electrons of NADH, FADH$_2$, or NADPH. Energy that is not trapped is given off as heat.

These reactions that require a net input of energy are said to be **endergonic** [*end(o)*, within + *ergon*, work], or *energy-utilizing, reactions*.

Some of the energy added to an endergonic reaction remains trapped in the chemical bonds of the products. Thus, energy-utilizing reactions are often *synthesis* reactions, in which complex molecules are made from smaller molecules. For example, an endergonic reaction links many glucose molecules together to create the glucose polymer glycogen. The complex glycogen molecule has more free energy than the simple glucose molecules used to make it.

If a reaction traps energy as it proceeds in one direction (A + B → C + D), it will release energy as it proceeds in the reverse direction (C + D → A + B). For example, the energy trapped in the bonds of glycogen during its synthesis will be released when glycogen is broken back down into glucose.

Coupling Endergonic and Exergonic Reactions Where does the activation energy for metabolic reactions come from? The simplest way for a cell to acquire activation energy is to couple an exergonic reaction to an endergonic reaction. Some of the most familiar coupled reactions are those that use the energy released by breaking the high-energy bond of ATP to drive an endergonic reaction:

$$E + F \xrightarrow{\quad ATP \quad ADP + P_i \quad} G + H$$

In this type of coupled reaction, the two reactions take place simultaneously and in the same location, so that the energy from ATP can be used immediately to drive the endergonic reaction between reactants E and F.

However, it is not always practical for reactions to be directly coupled like this. Consequently, living cells have developed ways to trap the energy released by exergonic reactions and save it for later use. The most common method is to trap the energy in the form of high-energy electrons carried on

nucleotides [⮂ p. 22]. The nucleotide molecules NADH, FADH$_2$, and NADPH all capture energy in the electrons of their hydrogen atoms (Fig. 4-5 ■). NADH and FADH$_2$ usually transfer most of their energy to ATP, which can then be used to drive endergonic reactions.

Net Free Energy Change Determines Reaction Reversibility

The net free energy change of a reaction plays an important role in determining whether that reaction can be reversed, because the net free energy change of the forward reaction contributes to the activation energy of the reverse reaction. A chemical reaction that can proceed in both directions is called a **reversible reaction**. In a reversible reaction, the forward reaction A + B → C + D and its reverse reaction C + D → A + B are both likely to take place.* If a reaction proceeds in one direction but not the other, it is an **irreversible reaction**.

For example, look at the activation energy of the reaction C + D → A + B in Figure 4-6 ■. This reaction is the reverse of the reaction shown in Figure 4-4a. Because a large release of energy occurred in the forward reaction A + B → C + D, the reverse reaction (Fig. 4-6) has a large activation energy. As you will recall, the larger the activation energy, the less likely it is that the reaction will proceed spontaneously. Theoretically, all reactions can be reversed with enough energy input, but some reactions are so extremely exergonic that they are essentially irreversible.

In your study of physiology, you will encounter a few irreversible reactions. However, most biological reactions are reversible: if the reaction A + B → C + D is possible, then so is the reaction C + D → A + B. Reversible reactions are shown with arrows that point in both directions: A + B ⇌ C + D. One of the main reasons that many biological reactions are reversible is that they are aided by the specialized proteins known as enzymes.

---

*The naming of forward and reverse directions is arbitrary.

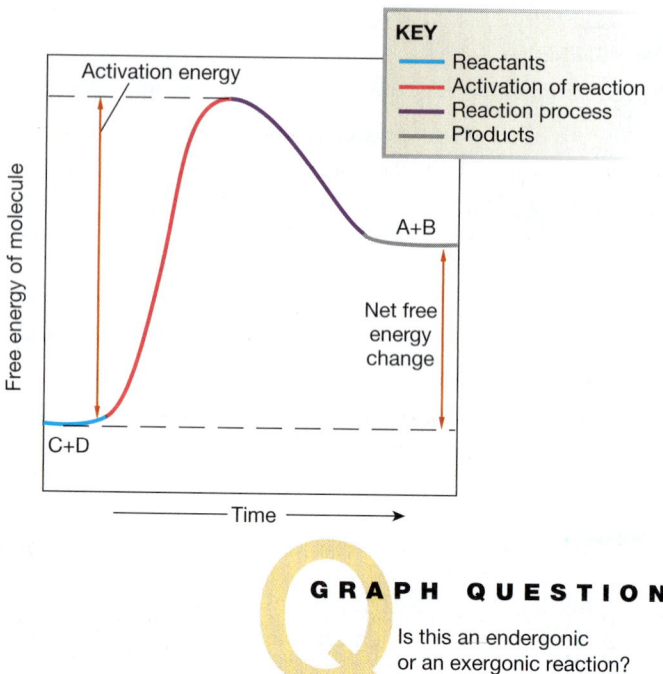

KEY
— Reactants
— Activation of reaction
— Reaction process
— Products

Activation energy

Free energy of molecule

Net free energy change

A+B

C+D

← Time →

GRAPH QUESTION

Is this an endergonic or an exergonic reaction?

■ **FIGURE 4-6** *Some reactions have large activation energies*

CONCEPT CHECK

5. What is the difference between endergonic and exergonic reactions?

6. If you mix baking soda and vinegar together in a bowl, the mixture reacts and foams up, releasing carbon dioxide gas. Name the reactant(s) and product(s) in this reaction.

7. Do you think the reaction of Question 6 is endergonic or exergonic? Do you think it is reversible? Defend your answers.

Answers: p. 126

ENZYMES

As noted in Chapter 2, **enzymes** are proteins that speed up the rate of chemical reactions. During these reactions, the enzyme molecules are not changed in any way, meaning they are biological *catalysts*. Without enzymes, most chemical reactions in a cell would go so slowly that the cell would be unable to live. Because an enzyme is not permanently changed or used up in the reaction it catalyzes, we might write it in a reaction equation this way:

$$A + B + enzyme \longrightarrow C + D + enzyme$$

This way of writing the reaction shows that the enzyme participates with reactants A and B but is unchanged at the end of the reaction. A more common shorthand for enzymatic reactions shows the name of the enzyme above the reaction arrow, like this:

$$A + B \xrightarrow{enzyme} C + D$$

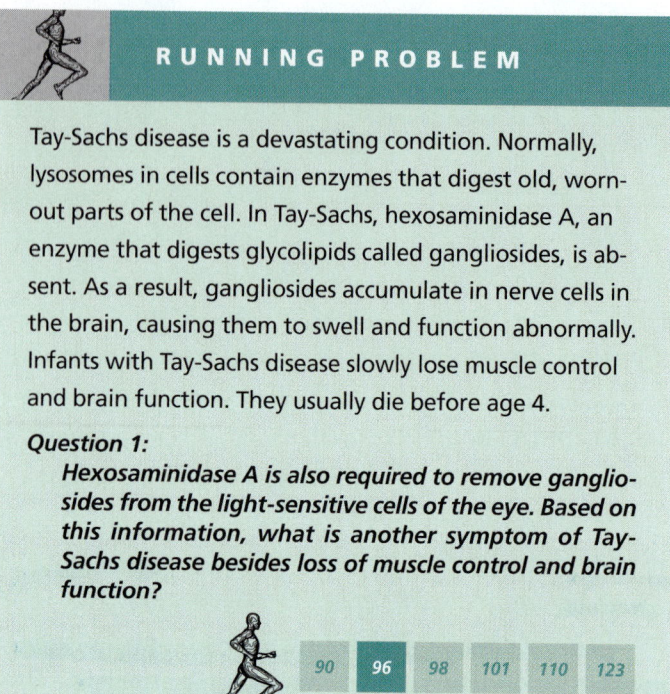

RUNNING PROBLEM

Tay-Sachs disease is a devastating condition. Normally, lysosomes in cells contain enzymes that digest old, worn-out parts of the cell. In Tay-Sachs, hexosaminidase A, an enzyme that digests glycolipids called gangliosides, is absent. As a result, gangliosides accumulate in nerve cells in the brain, causing them to swell and function abnormally. Infants with Tay-Sachs disease slowly lose muscle control and brain function. They usually die before age 4.

Question 1:
    Hexosaminidase A is also required to remove gangliosides from the light-sensitive cells of the eye. Based on this information, what is another symptom of Tay-Sachs disease besides loss of muscle control and brain function?

| 90 | 96 | 98 | 101 | 110 | 123 |

In enzymatically catalyzed reactions, the reactants are called **substrates**.

Enzymes Take Part in Typical Protein Interactions

Most enzymes are large proteins* with complex three-dimensional shapes. Like other proteins that bind to substrates, enzymes exhibit specificity, competition, and saturation [⇄ p. 39ff].

A few enzymes come in a variety of related forms (isoforms) and are known as **isozymes** [*iso-*, equal] of one another. Isozymes are enzymes that catalyze the same reaction but under different conditions or in different tissues. The structures of related isozymes are slightly different from one another, which causes the variability in their activity. Many isozymes have complex structures with multiple protein chains. For example, the enzyme *lactate dehydrogenase* (LDH) has two kinds of subunits, named H and M, that are assembled into *tetramers*—groups of four. LDH isozymes include H_4, H_2M_2, and M_4. The different LDH isozymes are tissue specific, including one found primarily in the heart and a second found in skeletal muscle and the liver.

Isozymes have an important role in the diagnosis of certain medical conditions. For example, in the hours following a heart attack, damaged heart muscle cells release enzymes into the blood. One way to determine whether a person's chest pain was indeed due to a heart attack is to look for elevated levels of heart isozymes in the blood. Some diagnostically important enzymes and the diseases of which they are suggestive are listed in Table 4-3 ■.

*Recently, researchers discovered that RNA can sometimes act as a catalyst.

TABLE 4-3   Diagnostically Important Enzymes

Elevated blood levels of these enzymes are suggestive of the pathologies listed.

ENZYME	RELATED DISEASES
Acid phosphatase*	Cancer of the prostate
Alkaline phosphatase	Diseases of bone or liver
Amylase	Pancreatic disease
Creatine kinase (CK)	Myocardial infarction (heart attack), muscle disease
Glutamate dehydrogenase (GDH)	Liver disease
Lactate dehydrogenase (LDH)	Myocardial infarction, liver disease, excess breakdown of red blood cells

*A newer test for a molecule called prostate specific antigen (PSA) has replaced the test for acid phosphatase in the diagnosis of prostate cancer.

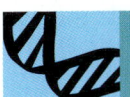

BIOTECHNOLOGY

SEPARATION OF ISOZYMES BY ELECTROPHORESIS

One way to determine which isozymes are present in a tissue sample is to use a technique known as **electrophoresis**. In this technique, a solution derived from the tissue sample is placed at one end of a container filled with a polyacrylamide polymer gel. An electric current is passed through the gel, causing the negatively charged proteins to move toward the positively charged end of the gel. The rate at which the proteins move depends on their size, their shape, and the electrical charge on their amino acids. As proteins move along the gel at different rates, they become separated from one another, so that they appear as individual bands of color when stained with a dye called Coomassie blue or with silver. Electrophoresis can be used to separate mixtures of charged macromolecules and is also used to identify DNA.

Enzymes May Be Activated, Inactivated, or Modulated

Enzyme activity, like that of other soluble proteins, can be altered by various factors. Some enzymes are synthesized as inactive molecules (*proenzymes* or *zymogens*) and activated on demand by proteolytic activation [p. 40]. Others require the binding of inorganic cofactors, such as Ca^{2+} or Mg^{2+}, before they become active.

Organic cofactors for enzymes are called **coenzymes**. Coenzymes do not alter the enzyme's binding site as inorganic cofactors do. Instead, coenzymes act as receptors and carriers for atoms or functional groups that are removed from the substrates during the reaction. Although coenzymes are needed for some metabolic reactions to take place, they are not required in large amounts.

Many of the substances that we call **vitamins** are the precursors of coenzymes. The water-soluble vitamins, such as the B vitamins, vitamin C, folic acid, biotin, and pantothenic acid, become coenzymes required for various metabolic reactions. For example, vitamin C is needed for adequate collagen synthesis.

Enzymes may be inactivated by inhibitors or by becoming denatured. Enzyme activity can be altered by chemical modulators or by changes in temperature and pH. Figure 4-7 ■ shows how enzyme activity can vary over a range of pH values. By turning reactions on and off or by increasing and decreasing the rate at which reactions take place, a cell can regulate the flow of biomolecules through different synthetic and energy-producing pathways.

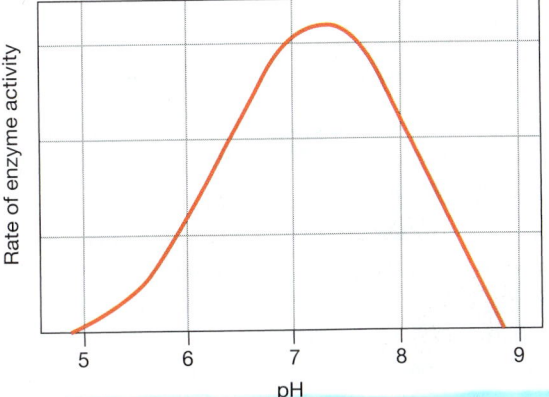

Most enzymes in humans have optimal activity near the body's internal pH of 7.4.

GRAPH QUESTION

If the pH falls from 8 to 7.4, what happens to the activity of the enzyme?

■ **FIGURE 4-7** *Effect of pH on enzyme activity*

RUNNING PROBLEM

Tay-Sachs disease is a *recessive* genetic disorder caused by a defect in the gene that directs synthesis of hexosaminidase A. *Recessive* means that for a baby to be born with Tay-Sachs disease, it must inherit two defective genes, one from each parent. People with one Tay-Sachs gene and one normal gene are called carriers of the disease. Carriers will not develop the disease but can pass the defective gene on to their children. About 1 in 27 people of Eastern European Jewish descent in the United States carries the Tay-Sachs gene. People who have two normal genes have normal amounts of hexosaminidase A in their blood. Carriers have lower-than-normal levels of the enzyme, but this amount is enough to prevent excessive accumulation of gangliosides in cells.

Question 2:

How could you test whether Sarah and David are carriers of the Tay-Sachs gene?

| 90 | 96 | **98** | 101 | 110 | 123 |

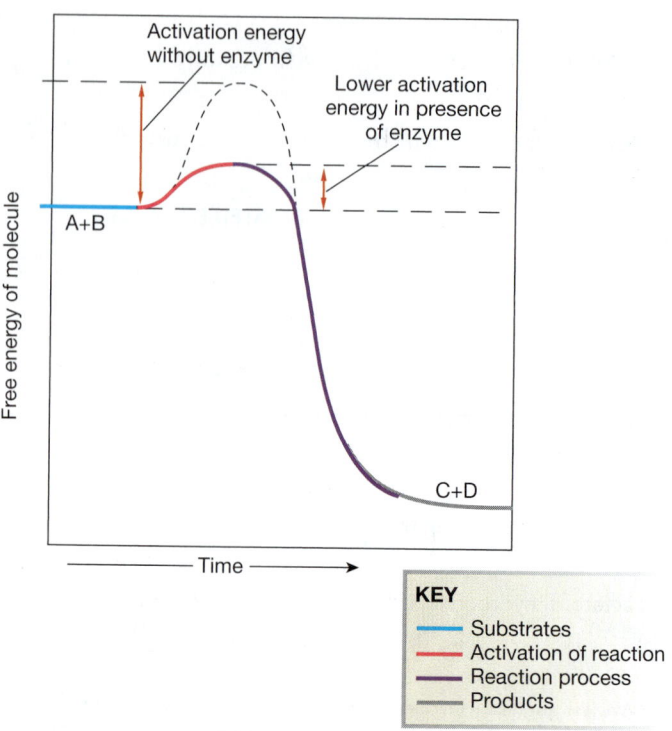

■ **FIGURE 4-8**   *Enzymes lower the activation energy of reactions*

In the absence of enzyme, the reaction (curved dashed line) would have a much greater activation energy.

✔ CONCEPT CHECK

8. What is a biological advantage of having multiple isozymes for a given reaction rather than only one form of the enzyme?

9. The four protein chains of an LDH isozyme are an example of what level of protein structure? a) primary b) secondary c) tertiary d) quaternary   Answers: p. 126

Enzymes Lower the Activation Energy of Reactions

How does an enzyme increase the rate of a reaction? In thermodynamic terms, it lowers the activation energy, making it more likely that the reaction will start (Fig. 4-8 ■). Enzymes accomplish this by binding to their substrates and bringing them into the best position for reacting with each other. Without enzymes, the reaction would depend on random collisions between substrate molecules to bring them into alignment.

The rate of a reaction catalyzed by an enzyme is much more rapid than the rate of the same reaction taking place in the absence of the enzyme. For example, consider the enzyme *carbonic anhydrase,* which converts CO_2 and water to carbonic acid. A single molecule of carbonic anhydrase takes one second to catalyze the conversion of 1 million molecules of CO_2 and

water to carbonic acid. In the absence of the enzyme, it takes nearly 100 seconds for only one molecule of CO_2 and water to be converted to carbonic acid. Without enzymes in cells, biological reactions would go so slowly that the cell would be unable to obtain enough energy to live.

Reaction Rates Are Variable

We measure the rate of an enzymatic reaction by monitoring either how fast the products are synthesized or how fast the substrates are consumed. Reaction rate can be altered by a number of factors, including changes in temperature, the amount of enzyme present, and substrate concentrations [⇄ p. 42]. In mammals we consider temperature to be essentially constant. This leaves enzyme amount and substrate concentration as the two main variables that affect reaction rate.

One strategy cells use to control reaction rates is to regulate the amount of enzyme in the cell. In the absence of appropriate enzyme, many biological reactions go very slowly or not at all. If enzyme is present, the rate of the reaction will be proportional to the amount of enzyme and the amount of substrate, unless there is so much substrate that all enzyme binding sites are saturated and working at maximum capacity [⇄ p. 43].

Reversible Reactions Obey the Law of Mass Action

You learned in the discussion of protein interactions in Chapter 2 that if the amount of enzyme is constant, the reaction rate is proportional to the substrate concentration [🔁 p. 43]. This seems simple until you consider a reversible reaction that can go in both

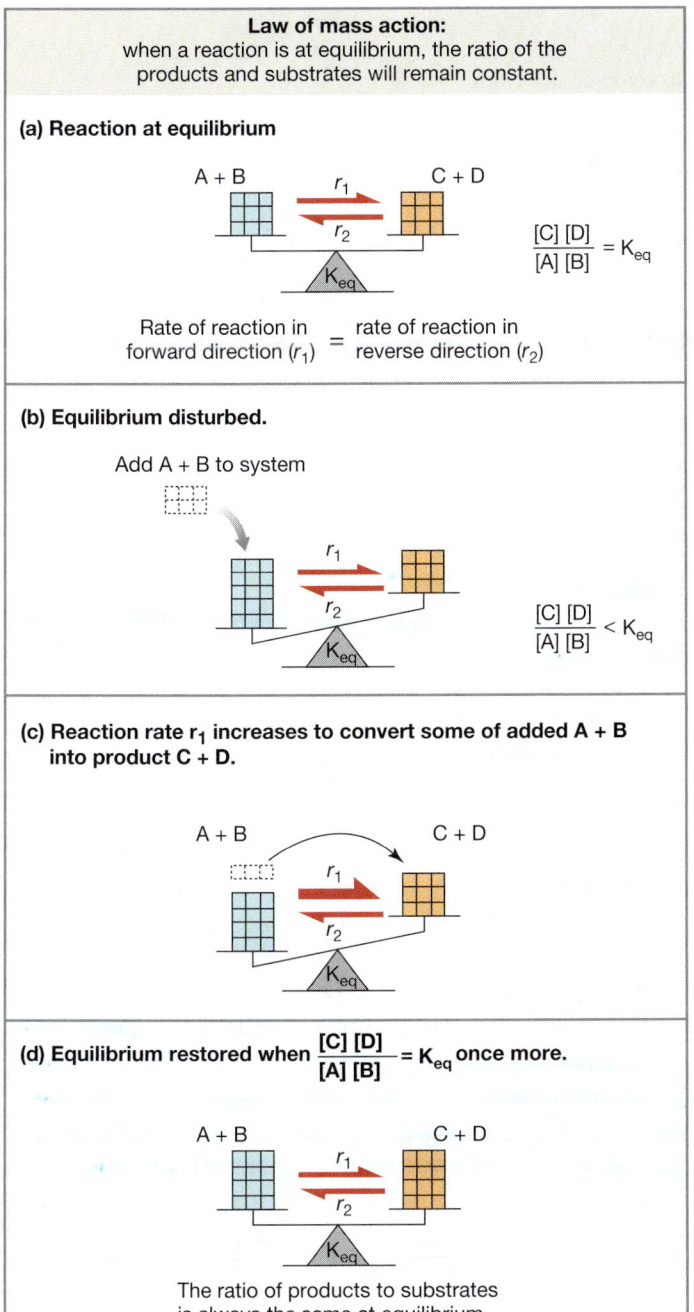

Law of mass action:
when a reaction is at equilibrium, the ratio of the products and substrates will remain constant.

(a) Reaction at equilibrium

A + B r_1 C + D
         r_2

$\dfrac{[C]\,[D]}{[A]\,[B]} = K_{eq}$

K_{eq}

$\begin{array}{c} \text{Rate of reaction in} \\ \text{forward direction } (r_1) \end{array} = \begin{array}{c} \text{rate of reaction in} \\ \text{reverse direction } (r_2) \end{array}$

(b) Equilibrium disturbed.

Add A + B to system

r_1
r_2

K_{eq}

$\dfrac{[C]\,[D]}{[A]\,[B]} < K_{eq}$

(c) Reaction rate r_1 increases to convert some of added A + B into product C + D.

A + B   C + D

r_1
r_2

K_{eq}

(d) Equilibrium restored when $\dfrac{[C]\,[D]}{[A]\,[B]} = K_{eq}$ once more.

A + B   C + D

r_1
r_2

K_{eq}

The ratio of products to substrates is always the same at equilibrium.

■ **FIGURE 4-9   Law of mass action**

This figure shows a reaction in which the ratio of products to substrates is 1:1, making the amount of A + B equal to the amount of C + D at equilibrium. The actual ratio of products to substrates depends on the reaction. For instance, the ratio could be 10:1 or 1:4.

directions. In that case, what determines in which direction the reaction is to go?

The answer, as noted in Chapter 2, is that reversible reactions go to a state of *equilibrium,* where the rate of the reaction in the forward direction (A + B → C + D) is equal to the rate of the reverse reaction (C + D → A + B), as shown in Figure 4-9a ■. At equilibrium, there is no net change in the amount of substrate or product. As fast as A and B convert to C and D, the reverse reaction takes place so that C and D turn back into A and B. At equilibrium, the ratio of the substrates and products is always the same and is equal to the reaction's *equilibrium constant,* K_{eq} [🔁 p. 39]. For the reaction given above,

$$K_{eq} = \frac{[C][D]}{[A][B]}$$

What happens if some factor disturbs this equilibrium by increasing the amount of either A or B in the system (Fig. 4-9b)? When the concentration of A or B goes up, the rate of the forward reaction increases to convert some of the added substrate into products C and D (Fig. 4-9c), thereby maintaining the original equilibrium product-to-substrate ratio, which in this example happens to be 1:1. As the ratio approaches its equilibrium value again, the rate of the forward reaction slows down until finally the system reaches equilibrium again (Fig. 4-9d).

The situation just described is an example of a reversible reaction obeying the **law of mass action,** a simple relationship that holds for chemical reactions whether in a test tube or in a cell. You may have learned this law in chemistry as *LeChâtelier's principle.* In very general terms, the law of mass action says that when a reaction is at equilibrium, the ratio of the products to the substrates is always the same [🔁 p. 39]. This law is important in physiology because adding or removing one of the participants in a chemical reaction has a chain-reaction effect that changes the concentrations of the other participants in the reaction.

✔ CONCEPT CHECK

10. Consider the previously described carbonic anhydrase reaction, which is reversible: $CO_2 + H_2O \rightleftharpoons H_2CO_3$. If the carbon dioxide concentration in the body increases, what happens to the concentration of carbonic acid (H_2CO_3)?   Answers: p. 126

Enzymatic Reactions Can Be Categorized

Most reactions catalyzed by enzymes can be classified into four categories: oxidation-reduction, hydrolysis-dehydration, exchange-addition-subtraction, and ligation reactions. Table 4-4 ■ summarizes these categories and gives the common enzyme names for different types of reactions.

An enzyme's name can provide important clues to the type of reaction the enzyme catalyzes. Most enzymes are instantly recognizable by the suffix *-ase.* The first part of the enzyme's name (everything that precedes the suffix) usually refers to the type of reaction, to the substrate upon which the enzyme

TABLE 4-4	**Types of Chemical Reactions**	

REACTION TYPE	WHAT HAPPENS	REPRESENTATIVE ENZYMES
1. Oxidation-reduction	Add or subtract electrons or H^+	
(a) Oxidation	Transfer electrons from donor to oxygen	Oxidase
	Remove electrons and H^+	Dehydrogenase
(b) Reduction	Gain electrons	Reductase
2. Hydrolysis-dehydration	Add or subtract a water	Hydrolase
(a) Hydrolysis	Split large molecules by adding water	Protease, lipase
(b) Dehydration	Remove water; used to make large molecules from several smaller ones	
3. Transfer chemical groups	Add, subtract, or exchange groups between molecules	
(a) Exchange reaction	Phosphate	Kinase
	Amino group	Transaminase
(b) Addition	Phosphate	Phosphorylase
	Amino group	Aminase
(c) Subtraction	Phosphate	Phosphatase
	Amino group	Deaminase
4. Ligation	Join two substrates using energy from ATP	Synthetase

acts, or to both. For example, *glucokinase* has glucose as its substrate, and as a *kinase* it will add a phosphate group [☰ p. 27] to the substrate. Addition of a phosphate group is called **phosphorylation**.

A few enzymes have two names. These enzymes were discovered before 1972, when the current standards for naming enzymes were first adopted. As a result, they have both a new name and a commonly used older name. Pepsin and trypsin, two digestive enzymes, are examples of older enzyme names.

Oxidation-Reduction Reactions

Oxidation-reduction reactions are the most important reactions in energy extraction and transfer in cells. These reactions transfer either electrons or protons (H^+) from one molecule to another. A molecule that gains electrons or loses H^+ is said to be **reduced**. One way to think of this is to remember that adding negatively charged electrons *reduces* the electric charge on the molecule. Likewise, the loss of a positively charged proton leaves behind an electron with a negative charge, again *reducing* the total charge on the molecule. Conversely, molecules that lose electrons or gain H^+ are said to be **oxidized**.

Hydrolysis-Dehydration Reactions

Hydrolysis and dehydration reactions are important in the breakdown and synthesis of large biomolecules. In **dehydration reactions** [*de-*, out + *hydr-*, water], a water molecule is one of the products. In many dehydration reactions, two molecules combine into one, losing water in the process. For example, the monosaccharides glucose and fructose join to make one sucrose molecule [☰ p. 27]. In the process, one substrate molecule loses a hydroxyl group (—OH) and the other substrate molecule loses a hydrogen to create water,

H_2O. When a dehydration reaction results in the synthesis of a new molecule, the process is known as *dehydration synthesis*.

In a **hydrolysis reaction** [*hydro*, water + *lysis*, to loosen or dissolve], a substrate changes into one or more products by adding water. In these reactions, the covalent bonds of the water molecule are broken ("lysed") so that the water reacts as a hydroxyl group (—OH) and a hydrogen (—H). For example, an amino acid can be removed from the end of a peptide with a hydrolysis reaction (see Fig. 4-19a, p. 111).

When an enzyme name consists of the substrate name plus the suffix *-ase*, the enzyme causes a hydrolysis reaction. One example is *lipase*, an enzyme that breaks up large lipids into smaller lipids by hydrolysis. A *peptidase* is an enzyme that removes an amino acid from a peptide.

Addition-Subtraction-Exchange Reactions

An **addition reaction** adds a functional group to one or more of the substrates. A **subtraction reaction** removes a functional group from one or more of the substrates. Functional groups are exchanged between or among substrates during **exchange reactions**.

For example, phosphate groups may be transferred from one molecule to another during addition, subtraction, or exchange reactions. The transfer of phosphate groups is an important means of covalent modulation [☰ p. 41], turning reactions on or off or increasing or decreasing their rates. Several types of enzymes catalyze reactions that transfer phosphate groups. Kinases transfer a phosphate group from a substrate to an ADP molecule to create ATP, or from an ATP molecule to a substrate. For example, creatine kinase transfers a phosphate group from creatine phosphate to ADP, forming ATP and leaving behind creatine.

RUNNING PROBLEM

In 1989, researchers discovered three genetic mutations responsible for Tay-Sachs disease. This discovery paved the way for a new, more accurate carrier screening test that detects the presence of the defective gene in blood cells rather than testing for lower-than-normal hexosaminidase A levels. David and Sarah will undergo this new genetic test.

Question 3:
Why is the new test for the Tay-Sachs gene more accurate than the old test, which detects decreased amounts of hexosaminidase A?

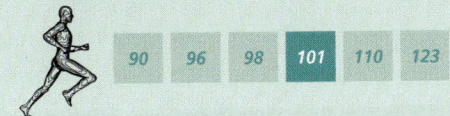

| 90 | 96 | 98 | 101 | 110 | 123 |

The addition, subtraction, and exchange of amino groups [🖥 p. 31] are also important in the body's use of amino acids. Removal of an amino group from an amino acid or peptide is a **deamination** reaction, illustrated in Figure 4-19b, p. 111. Addition of an amino group is **amination,** and the transfer of an amino group from one molecule to another is called **transamination**.

Ligation Reactions Ligation reactions join two molecules together using enzymes known as *synthetases* and energy from ATP. An example of a ligation reaction is the synthesis of *acetyl coenzyme A* (acetyl CoA) from fatty acids and coenzyme A. Acetyl CoA is an important molecule in the body, as you will learn in the next section.

✓ CONCEPT CHECK

11. Name the substrate for the enzymes lactase, peptidase, lipase, and sucrase.

12. Match the reaction type or enzyme in the left column to the group or particle involved.

    (a) Kinase   1. Amino group
    (b) Oxidation   2. Electrons
    (c) Hydrolysis   3. Phosphate group
    (d) Transaminase   4. Water   Answers: p. 126

METABOLISM

Metabolism refers to all chemical reactions that take place in an organism. These reactions (1) extract energy from nutrient biomolecules (such as proteins, carbohydrates, and lipids) and (2) either synthesize or break down molecules. Metabolism is often divided into **catabolism,** reactions that produce energy through the breakdown of large biomolecules, and **anabolism,** energy-utilizing reactions that result in the synthesis of large biomolecules. Anabolic and catabolic reactions take place simultaneously in cells throughout the body, so that at any given moment, some biomolecules are being synthesized while others are being broken down.

The energy released from or stored in the chemical bonds of biomolecules during metabolism is commonly measured in kilocalories (kcal). A **kilocalorie** is the amount of energy needed to raise the temperature of 1 liter of water by 1 degree Celsius. One kilocalorie is the same as a Calorie, with a capital C, used for quantifying the energy content of food. One kilocalorie is also equal to 1000 calories (small c).

Much of the energy released during catabolism is trapped in the high-energy phosphate bonds of ATP or in the high-energy electrons of NADH, $FADH_2$, or NADPH. Anabolic reactions then transfer energy from these temporary carriers to the covalent bonds of biomolecules.

Metabolism is a network of highly coordinated chemical reactions in which the activities taking place in a cell at any given moment are matched to the needs of the cell. Each step in a metabolic pathway is a different enzymatic reaction, and the reactions of a pathway proceed in sequence. Substrate A is changed into product B, which then becomes the substrate for the next reaction in the pathway, in which B is changed into C, and so forth:

$$A \longrightarrow B \longrightarrow C \longrightarrow D$$

We call the molecules of the pathway **intermediates** because the products of one reaction become the substrates for the next. You will sometimes hear metabolic pathways called *intermediary metabolism*. Certain intermediates, called *key intermediates*, participate in more than one pathway and act as the branch points for channeling substrate in one direction or another. Glucose, for instance, is a key intermediate in several metabolic pathways.

In many ways, a group of metabolic pathways is similar to a detailed road map (Fig. 4-10 ■). Just as a map shows a network of roads that connect various cities and towns, metabolism can be thought of as a network of chemical reactions connecting various intermediate products. Each city or town is a different chemical intermediate. One-way roads are irreversible reactions, and big cities with roads to several destinations are key intermediates. Just as there may be more than one way to get from one place to another, there can be several pathways between any given pair of chemical intermediates.

Cells Regulate Their Metabolic Pathways

How do cells regulate the flow of molecules through their metabolic pathways? They do so in five basic ways:

1. By controlling enzyme concentrations
2. By producing allosteric and covalent modulators
3. By using two different enzymes to catalyze reversible reactions

(a)

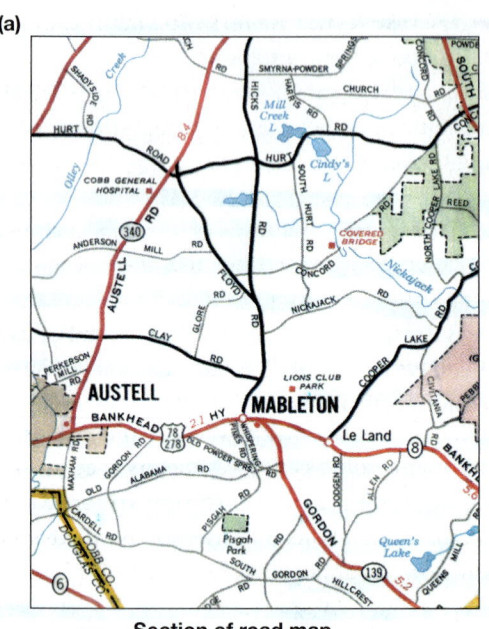

Section of road map

(b)

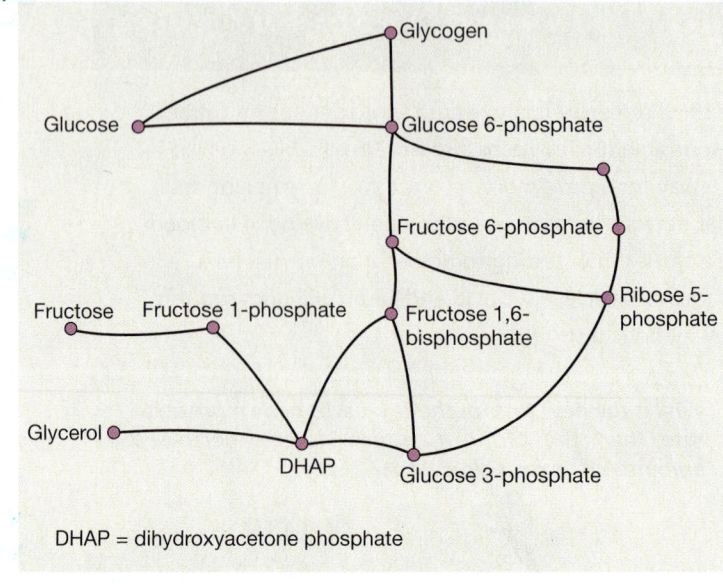

DHAP = dihydroxyacetone phosphate

Metabolic pathways drawn like a road map

■ **FIGURE 4-10**   *A group of metabolic pathways resembles a road map*

Cities on the map are equivalent to intermediates in metabolism. In metabolism, there may be more than one way to go from one intermediate to another, just as on the map there may be many ways to get from one city to another.

4. By isolating enzymes within intracellular organelles
5. By maintaining an optimum ratio of ATP to ADP

We discussed changing enzyme concentration in Chapter 2 [💾 p. 43]. The sections that follow examine the remaining four items on the list.

Modulation of Enzymes   Modulators, which alter the activity of a protein, were also introduced in Chapter 2 [💾 p. 40]. Enzyme modulation is frequently controlled by hormones and other signals coming from outside the cell. This type of outside regulation is a key element in the integrated control of the body's metabolism following a meal or during periods of fasting.

In addition, some metabolic pathways have their own built-in form of modulation, called **feedback inhibition**. In this form of modulation, the end product of a pathway, shown as Z in Figure 4-11 ■, acts as an inhibitory modulator of the pathway. As the pathway proceeds and Z accumulates, the

enzyme catalyzing the conversion of A to B is inhibited. Inhibition of the enzyme slows down production of Z until the cell can use it up. Once the levels of Z fall, feedback inhibition on enzyme 1 is removed and the pathway starts to run again. Because Z is the end product of the pathway, this type of feedback inhibition is sometimes called *end-product inhibition*.

Enzymes and Reversible Reactions   Cells can use reversible reactions to regulate the rate and direction of metabolism. If a single enzyme can catalyze the reaction in either direction (Fig. 4-12a ■), the reaction will go to a state of equilibrium as determined by the law of mass action. Such a reaction therefore cannot be closely regulated except by modulators and by control of the amount of enzyme. If a reversible reaction requires two different enzymes, however, one for the forward reaction and one for the reverse reaction, the cell can regulate the reaction more closely (Fig. 4-12b). If there is no enzyme for the reverse reaction present in the cell, the reaction is irreversible (Fig. 4-12c).

Compartmentation of Enzymes in the Cell   Many enzymes of metabolism are isolated in specific subcellular compartments. Some, like the enzymes of carbohydrate metabolism, are dissolved in the cytosol, whereas others are isolated within specific organelles. Mitochondria, endoplasmic reticulum, Golgi complex, and lysosomes all contain enzymes that are not found in the cytosol. This separation of enzymes means that the pathways controlled by the enzymes are also separated, which allows the cell to control metabolism by regulating the

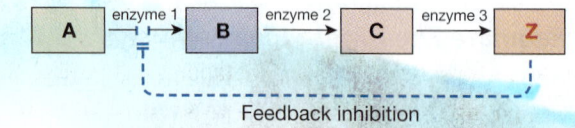

Feedback inhibition

■ **FIGURE 4-11**   *Feedback inhibition*

The accumulation of end product Z inhibits the first step of the pathway. As the cell consumes Z in another metabolic reaction, the inhibition is removed and the pathway resumes.

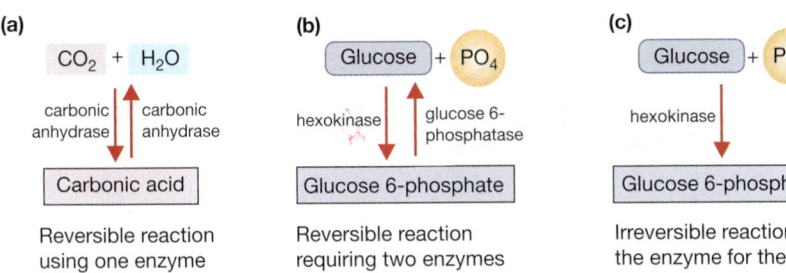

(a) Reversible reaction using one enzyme for both directions

(b) Reversible reaction requiring two enzymes

(c) Irreversible reaction lacks the enzyme for the reverse direction

■ **FIGURE 4-12** *Reversible and irreversible metabolic reactions*

(a) Some reversible reactions that take place in the body use one enzyme for both directions. (b) There can be more control over reversible reactions when the forward and reverse reactions require different enzymes. (c) Most cells of the body lack the enzyme glucose 6-phosphatase, and in these cells the reaction shown in (b) becomes irreversible.

movement of substrate from one cellular compartment to another. The isolation of enzymes within organelles is an example of structural and functional compartmentation [⟳ p. 9].

Ratio of ATP to ADP   The energy status of the cell is one final mechanism that can influence metabolic pathways. Through complex regulation, the ratio of ATP to ADP in the cell determines whether pathways that result in ATP synthesis are turned on or off. When ATP levels are high, production of ATP drops. When ATP levels are low, the cell sends substrates through pathways that result in more ATP synthesis. In the next section, we look further into the role of ATP in cellular metabolism.

ATP Transfers Energy Between Reactions

The usefulness of metabolic pathways as suppliers of energy is often measured in terms of the net amount of ATP the pathways can yield. Recall from Chapter 2 that ATP is a nucleotide containing three phosphate groups [⟳ p. 33]. One of the three phosphate groups is attached to ADP by a covalent bond in an energy-requiring reaction. Energy is stored in this **high-energy phosphate bond** and then released when the bond is broken during removal of the phosphate group. This relationship is shown by the following reaction:

$$ADP + P_i + energy \rightleftharpoons ADP\sim P \ (= ATP)$$

The squiggle ~ indicates a high-energy bond, and P_i is the abbreviation for an inorganic phosphate group. Estimates of the amount of free energy released when a high-energy phosphate bond is broken range from 7 to 12 kcal per mole of ATP.

ATP is more important as a carrier of energy than as an energy-storage molecule. For one thing, cells can contain only a limited amount of ATP. A resting adult human would need to store 40 kg (88 pounds) of ATP to supply the energy required to support one day's worth of metabolic activity, far more than our cells could store. Instead, the body acquires most of its daily energy requirement from the chemical-bond energy stored in complex biomolecules. Metabolic reactions transfer that energy to the high-energy bonds of ATP, or in a few cases, to the high-energy bonds of the related nucleotide *guanosine triphosphate*, **GTP**.

The metabolic pathways that yield the most ATP molecules are those that require oxygen—the **aerobic**, or *oxidative*, pathways. **Anaerobic** [*an-*, without + *aer*, air] pathways, which are those that can proceed without oxygen, also produce ATP molecules but in much smaller quantities. The lower ATP yield of anaerobic pathways means that most animals (including humans) are unable to survive for extended periods on anaerobic metabolism alone. In the next section we will consider how biomolecules are metabolized to transfer energy to ATP.

CONCEPT CHECK

13. Name five ways in which cells regulate the movement of substrates through metabolic pathways.

14. In which part of an ATP molecule is energy trapped and stored? In which part of a NADH molecule is energy stored?

15. What is the difference between aerobic and anaerobic pathways?

Answers: p. 126

ATP PRODUCTION

The catabolic pathways that extract energy from biomolecules and transfer it to ATP are summarized in Figure 4-13 ■. Aerobic production of ATP from glucose commonly follows two pathways: **glycolysis** [*glyco-*, sweet + *lysis*, dissolve] and the **citric acid cycle** (also known as the tricarboxylic acid cycle). Each of these pathways produces small amounts of ATP directly, but their most important contributions to ATP synthesis are high-energy electrons carried by NADH and $FADH_2$ to the electron transport system in the mitochondria. The electron transport system, in turn, transfers energy from electrons to the high-energy phosphate bond of ATP. At various points, the process produces carbon dioxide and water. The water can be used by the cell, but carbon dioxide is a waste product and must be removed from the body.

Carbohydrates enter glycolysis in the form of glucose. Lipids are broken down into glycerol and fatty acids [⟳ p. 27], which then enter the pathway at different points: glycerol feeds into glycolysis, and fatty acids are metabolized to acetyl CoA. Proteins are broken down into amino acids, which also enter at various points. Because glucose is the only molecule that follows both pathways in their entirety, we look first at glucose catabolism. We will then see how protein catabolism and lipid catabolism differ from glucose catabolism.

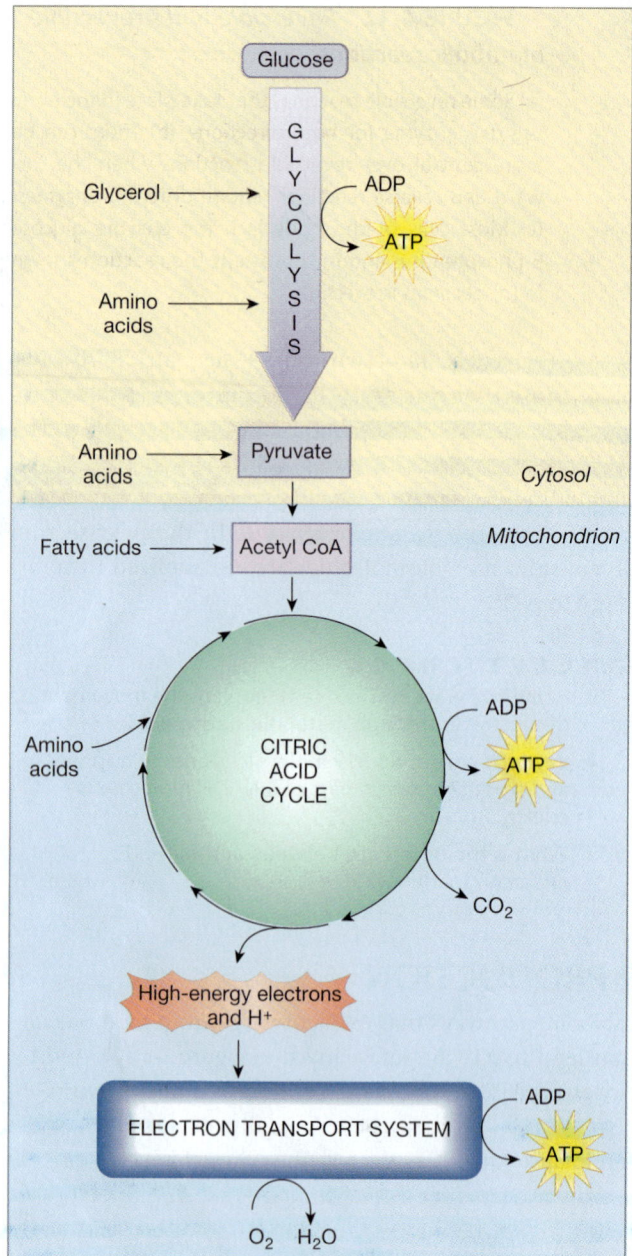

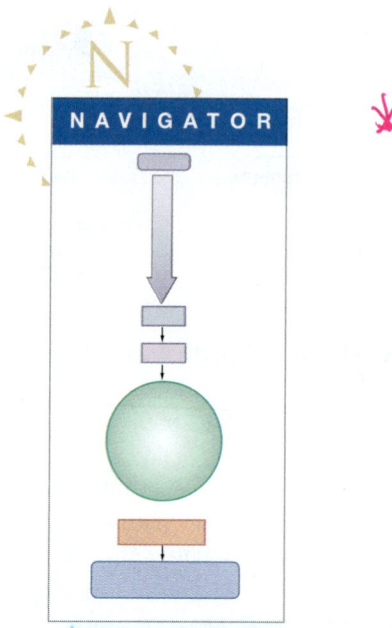

This icon represents the different steps in the metabolic summary figure. Look for it in the figures that follow to help you navigate your way through metabolism.

■ **FIGURE 4-13** *Overview of aerobic pathways for ATP production*

The body uses chemical-bond energy stored in biomolecules to make ATP. Glucose and glycerol enter the glycolysis pathway, amino acids enter at various points all along the pathway, and fatty acids enter at acetyl CoA.

The various portions of the aerobic pathway for ATP production are a good example of compartmentation. The enzymes of glycolysis are located in the cytosol, whereas the enzymes of the citric acid cycle are in the mitochondria.

CONCEPT CHECK ✓

16. Match each component on the left to the molecule(s) it is part of:

    (a) Amino acids   1. Carbohydrates
    (b) Fatty acids   2. Lipids
    (c) Glycerol   3. Polysaccharides
    (d) Glucose   4. Proteins
                       5. Triglycerides

    Answers: p. 127

Glycolysis Converts Glucose and Glycogen into Pyruvate

During glycolysis, one molecule of glucose is converted by a series of enzymatically catalyzed reactions into two pyruvate* molecules, producing a net release of energy. Thus, glycolysis is considered to be an exergonic pathway even though some steps of the pathway require energy input. A portion of the energy released during glycolysis is used to phosphorylate ADP

---

*Under normal cell conditions, many organic acids, such as pyruvic acid and lactic acid, are ionized and are therefore named as the corresponding anion: pyruvate and lactate. The suffix *-ate* identifies the anions of organic and some inorganic acids.

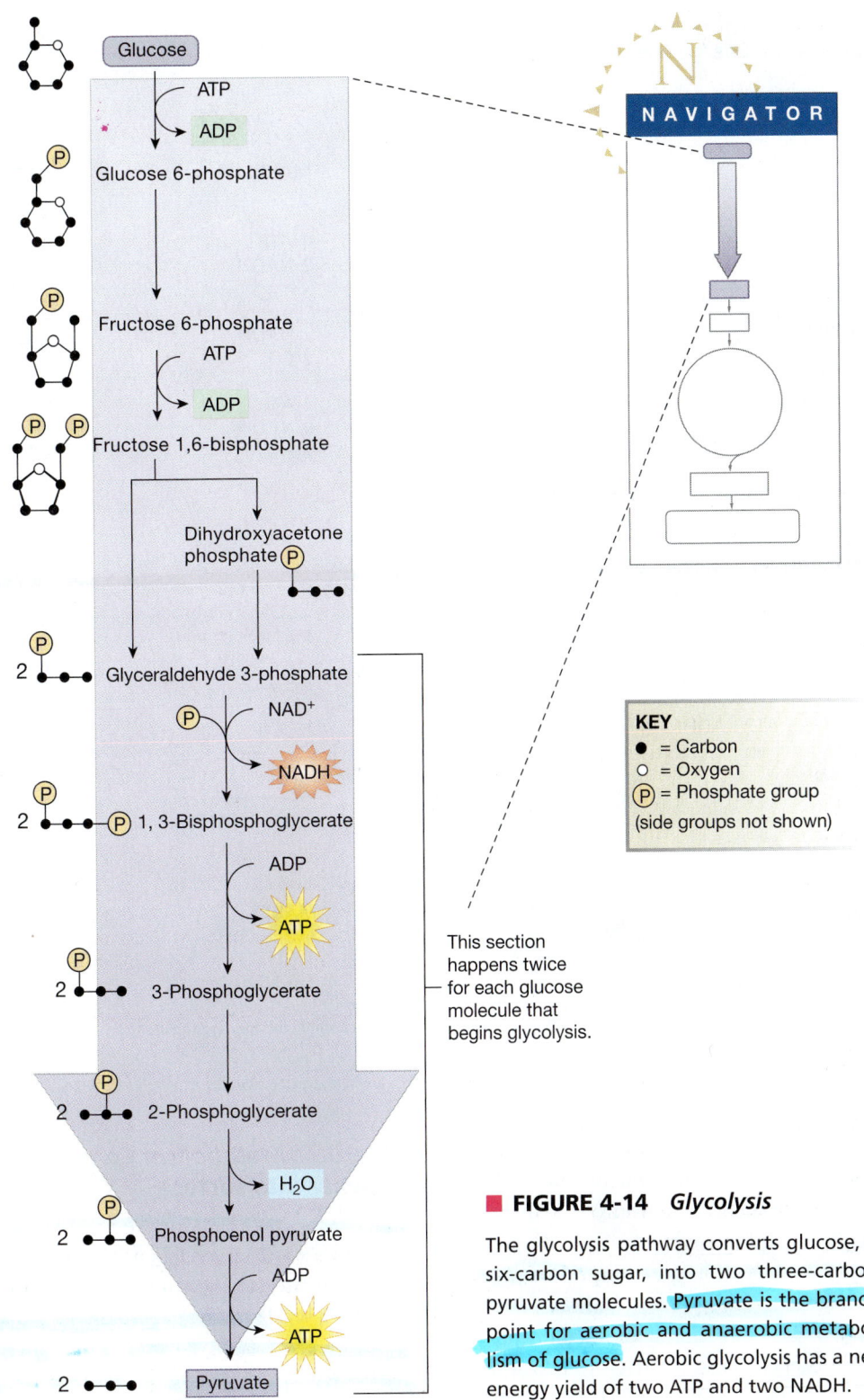

KEY
● = Carbon
○ = Oxygen
Ⓟ = Phosphate group
(side groups not shown)

This section
happens twice
for each glucose
molecule that
begins glycolysis.

■ **FIGURE 4-14** *Glycolysis*

The glycolysis pathway converts glucose, a
six-carbon sugar, into two three-carbon
pyruvate molecules. Pyruvate is the branch
point for aerobic and anaerobic metabolism of glucose. Aerobic glycolysis has a net
energy yield of two ATP and two NADH.

molecules, trapping the energy in ATP. The events of glycolysis
can be summarized as follows:

$$\text{Glucose} + 2\ \text{NAD}^+ + 2\ \text{ADP} + 2\ P_i \rightarrow$$
$$2\ \text{Pyruvate} + 2\ \text{ATP} + 2\ \text{NADH} + 2\ H^+ + 2\ H_2O$$

Notice that glycolysis does not require oxygen. It therefore
serves as a common pathway for aerobic and anaerobic catabolism of glucose.

Figure 4-14 ■ shows the steps of glycolysis. In the first
step, glucose is phosphorylated with the aid of ATP to glucose
6-phosphate.* Glucose 6-phosphate is a key intermediate that

---

*The "6" in glucose 6-phosphate tells you that the phosphate group has been at-
tached to carbon 6 of the glucose molecule.

TABLE 4-5	Summary of Energy Yield from Catabolic Pathways		

Aerobic metabolism

PATHWAY	USES	MAKES	NET ENERGY YIELD
Glycolysis glucose $\longrightarrow$ 2 pyruvate	2 ATP	4 ATP 2 NADH	+ 2 ATP + 2 NADH*
Aerobic Metabolism 2 pyruvate $\longrightarrow$ 2 acetyl CoA	—	2 NADH 2 CO_2	+ 2 NADH
2 acetyl CoA through citric acid cycle		2 ATP 6 NADH 2 $FADH_2$	+2 ATP +6 NADH +2 $FADH_2$
Total yield for 1 glucose through aerobic metabolism:			+4 ATP +2 NADH cytoplasm* +8 NADH mitochondria +2 $FADH_2$

Maximal ATP yield after mitochondrial oxidative phosphorylation:

4 ATP		4 ATP
2 cytoplasmic NADH $\times$ 1.5–2.5 ATP/NADH		3–5 ATP
8 mitochondrial NADH $\times$ 2.5 ATP/NADH		20 ATP
2 $FADH_2$ $\times$ 1.5 ATP/$FADH_2$		3 ATP
		= 30–32 ATP

Anaerobic metabolism

PATHWAY	USES	MAKES	NET ENERGY YIELD
Glucose $\rightarrow$ 2 lactate	2 ATP 2 NADH	4 ATP 2 NADH	+2 ATP 0 NADH

*Cytoplasmic NADH molecules cannot enter mitochondria. Their ATP yield depends on whether their electrons go to NADH or to $FADH_2$ in the mitochondria.

leads into glycolysis as well as into other pathways we will not consider. Note that by the time the six-carbon glucose molecule reaches the end of glycolysis, it has been split into two molecules of pyruvate, each containing three carbons.

Two steps of glycolysis are endergonic and require energy input from two ATP. Other steps are exergonic and produce four ATP and two NADH for each glucose that enters the pathway. The net energy yield from each glucose is therefore two ATP and two NADH. Table 4-5 ■ provides a summary of the energy yield of various catabolic pathways.

✓ CONCEPT CHECK

17. Do endergonic reactions release energy or trap it in the products?

18. Overall, is glycolysis an endergonic or exergonic pathway?

Answers: p. 127

Anaerobic Metabolism Converts Pyruvate into Lactate

Pyruvate is a branch point for metabolic pathways, like the hub cities on a road map. Depending on a cell's needs and condition, pyruvate can be shuttled off into one of two pathways (Fig. 4-15 ■). If the cell contains adequate oxygen, pyruvate continues into the citric acid cycle. If the cell lacks sufficient oxygen for aerobic pathways (a condition that may be caused by many factors, including strenuous exercise), pyruvate is converted into lactate with the assistance of the enzyme lactate dehydrogenase mentioned earlier:

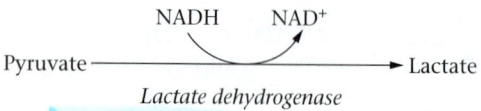

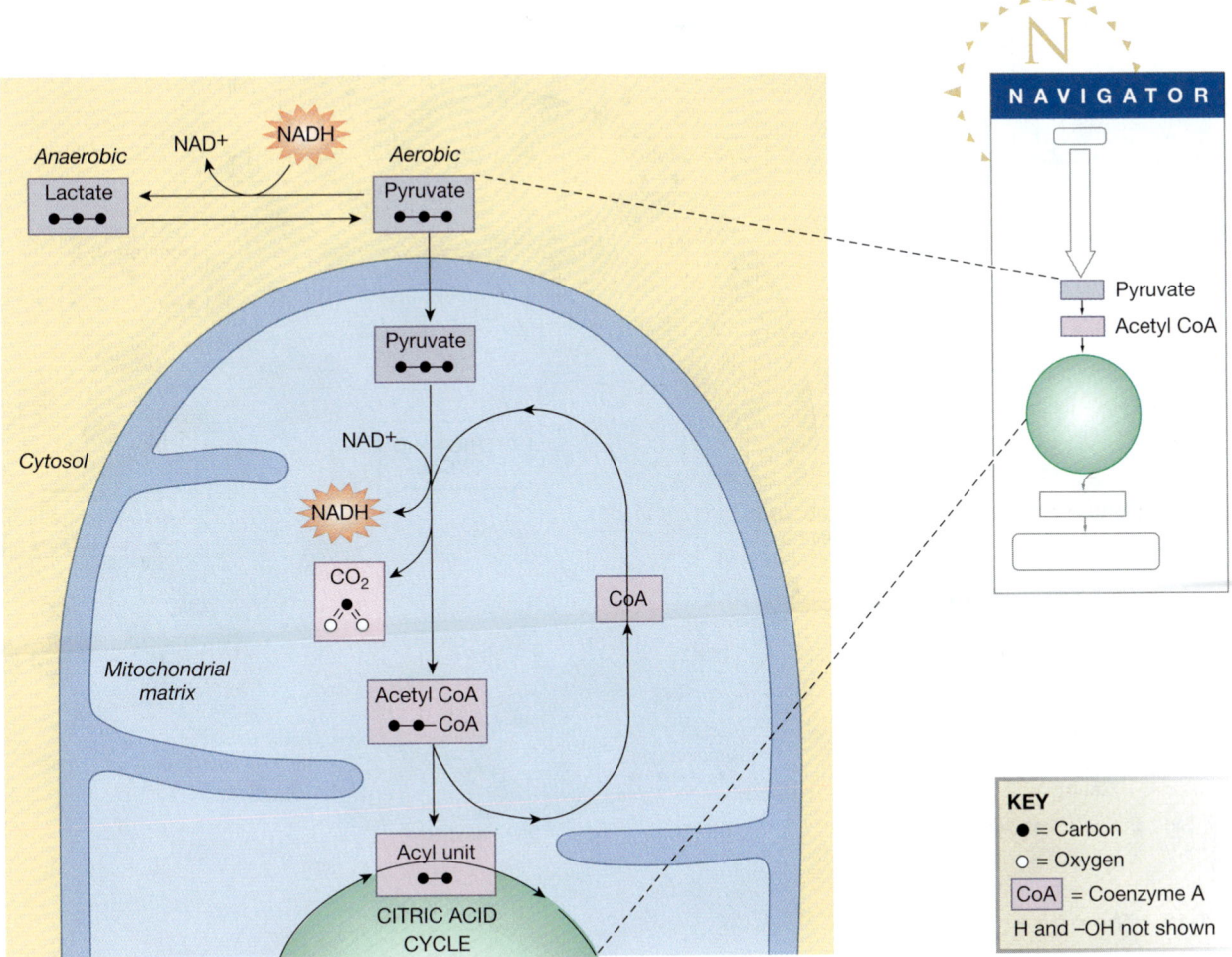

N   NAVIGATOR

KEY
- ● = Carbon
- ○ = Oxygen
- CoA = Coenzyme A
- H and –OH not shown

■ **FIGURE 4-15** *Pyruvate metabolism*

Anaerobic metabolism of pyruvate creates lactate (lactic acid). Under aerobic conditions, pyruvate is transported into mitochondria and converted to acetyl CoA. The acyl unit of acetyl CoA then feeds into the citric acid cycle.

The conversion of pyruvate to lactate changes one NADH back to NAD⁺ when a hydrogen atom and an electron are transferred to the lactate molecule. As a result, the net energy yield for the anaerobic metabolism of one glucose molecule is two ATP and no NADH (Table 4-5).

CONCEPT CHECK

19. Using Figure 4-14, identify the steps of glycolysis that are:
    (a) obviously exergonic
    (b) obviously endergonic
    (c) catalyzed by kinases
    (d) catalyzed by dehydrogenases (*Hint:* see Table 4-4.)
    (e) dehydration reactions

20. Lactate dehydrogenase acts on lactate by _____ (adding or removing?) a/an _____ and a/an _____. This process is called _____ (oxidation or reduction?).

Answers: p. 127

Pyruvate Enters the Citric Acid Cycle in Aerobic Metabolism

If the cell has adequate oxygen for aerobic metabolism, then pyruvate molecules formed from glucose during glycolysis are transported into the mitochondria (Fig. 4-15). Once in the mitochondrial matrix, pyruvate is converted into the key intermediate **acetyl CoA**. As its name suggests, this molecule has two parts: a two-carbon *acyl unit*, derived from pyruvate, and a coenzyme.

Coenzyme A is made from the vitamin *pantothenic acid*. Like enzymes, coenzyme A is not consumed during metabolic reactions and can be reused. The synthesis of acetyl CoA from pyruvate and coenzyme A is an exergonic reaction that produces one NADH. During the reaction, one of pyruvate's three carbon atoms combines with oxygen and is released as carbon dioxide (CO_2), leaving acetyl CoA with only two carbons.

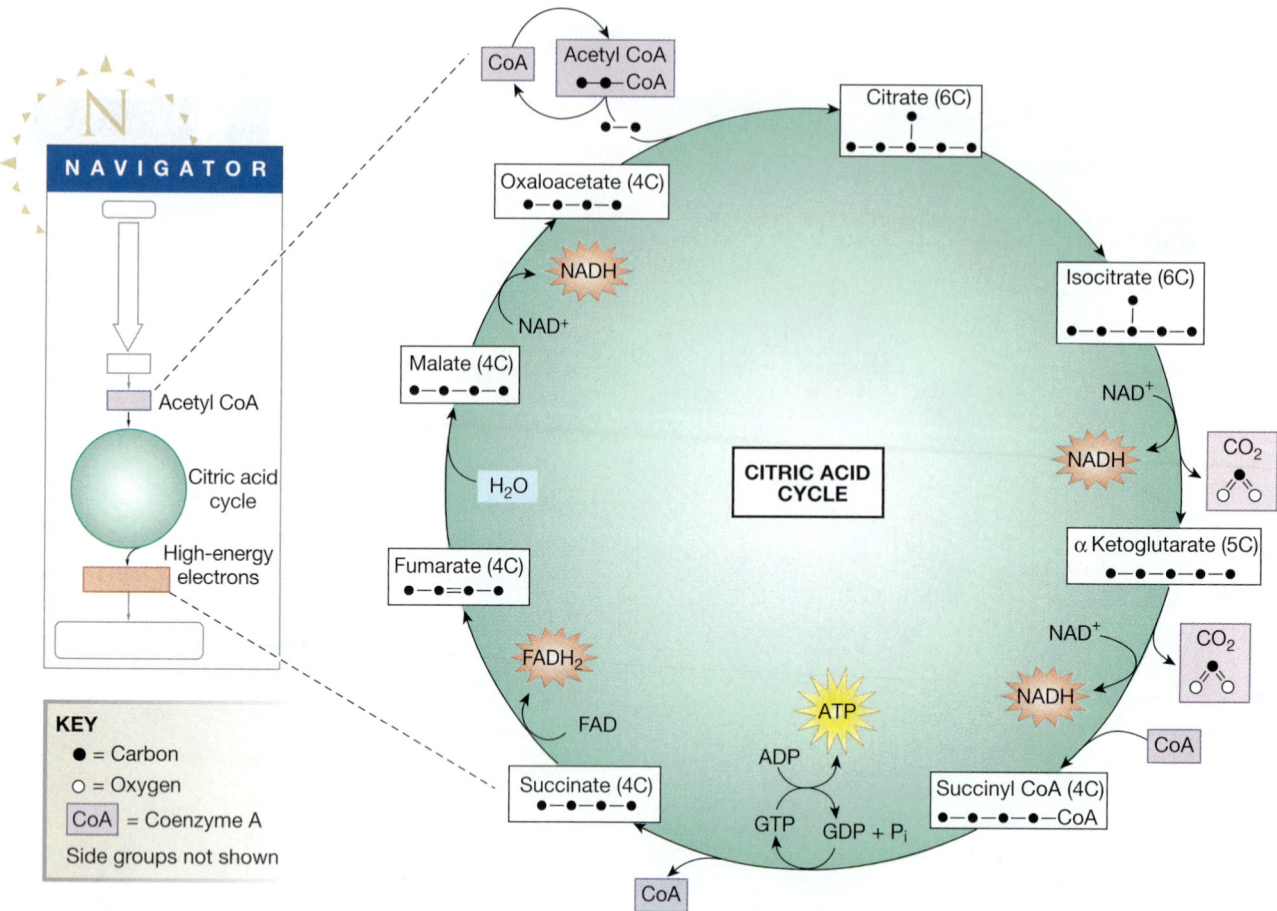

■ **FIGURE 4-16**   *The citric acid cycle*

A two-carbon acyl unit from acetyl CoA combines with a four-carbon oxaloacetate
molecule to make citrate. Each full turn of the cycle releases two carbons as carbon
dioxide and produces one ATP, three NADH, and one $FADH_2$.

Acetyl CoA releases its two-carbon acyl unit into the citric
acid cycle pathway (Fig. 4-16 ■). This pathway was first described by Hans A. Krebs, so it is also known as the *Krebs cycle*.
Because Krebs described other metabolic cycles, we will avoid
confusion by using the term citric acid cycle. The citric acid cycle makes a never-ending circle, adding carbons from an acetyl
CoA with each turn of the cycle and producing ATP, high-
energy electrons, and carbon dioxide.

The two-carbon acyl unit enters the citric acid cycle by
combining with a four-carbon oxaloacetate molecule that is the
last intermediate in the cycle. The resulting six-carbon citrate
molecule then goes through a series of reactions until it completes the cycle as another oxaloacetate molecule. Most of the
energy released by the reactions of the cycle is captured as high-
energy electrons on three NADH and one $FADH_2$. However,
some energy is used to make the high-energy phosphate bond
of one ATP, and the remainder is given off as heat. In two of the
reactions, carbon and oxygen are removed in the form of carbon dioxide. By the end of the cycle, the four-carbon oxaloacetate molecule that remains is ready to begin the cycle again.

In the next step of aerobic metabolism, NADH and $FADH_2$
transfer their high-energy electrons to the electron transport
system and return to the citric acid cycle as NAD^+ and FAD^+.

The Electron Transport System Transfers Energy from NADH and $FADH_2$ to ATP

The final step in aerobic ATP production is the transfer of energy from the high-energy electrons of NADH and $FADH_2$ to
ATP (Fig. 4-17 ■). This energy transfer is made possible by a
group of mitochondrial proteins known as the **electron transport system**, located in the inner mitochondrial membrane
[p. 64]. The protein complexes of the electron transport system include enzymes and the iron-containing proteins known
as **cytochromes**.

Movement of electrons through the electron transport system is described by a model called the **chemiosmotic theory**.
According to this model, as pairs of high-energy electrons pass
from complex to complex along the transport system, some of
the energy released by those reactions is used to pump H^+ from

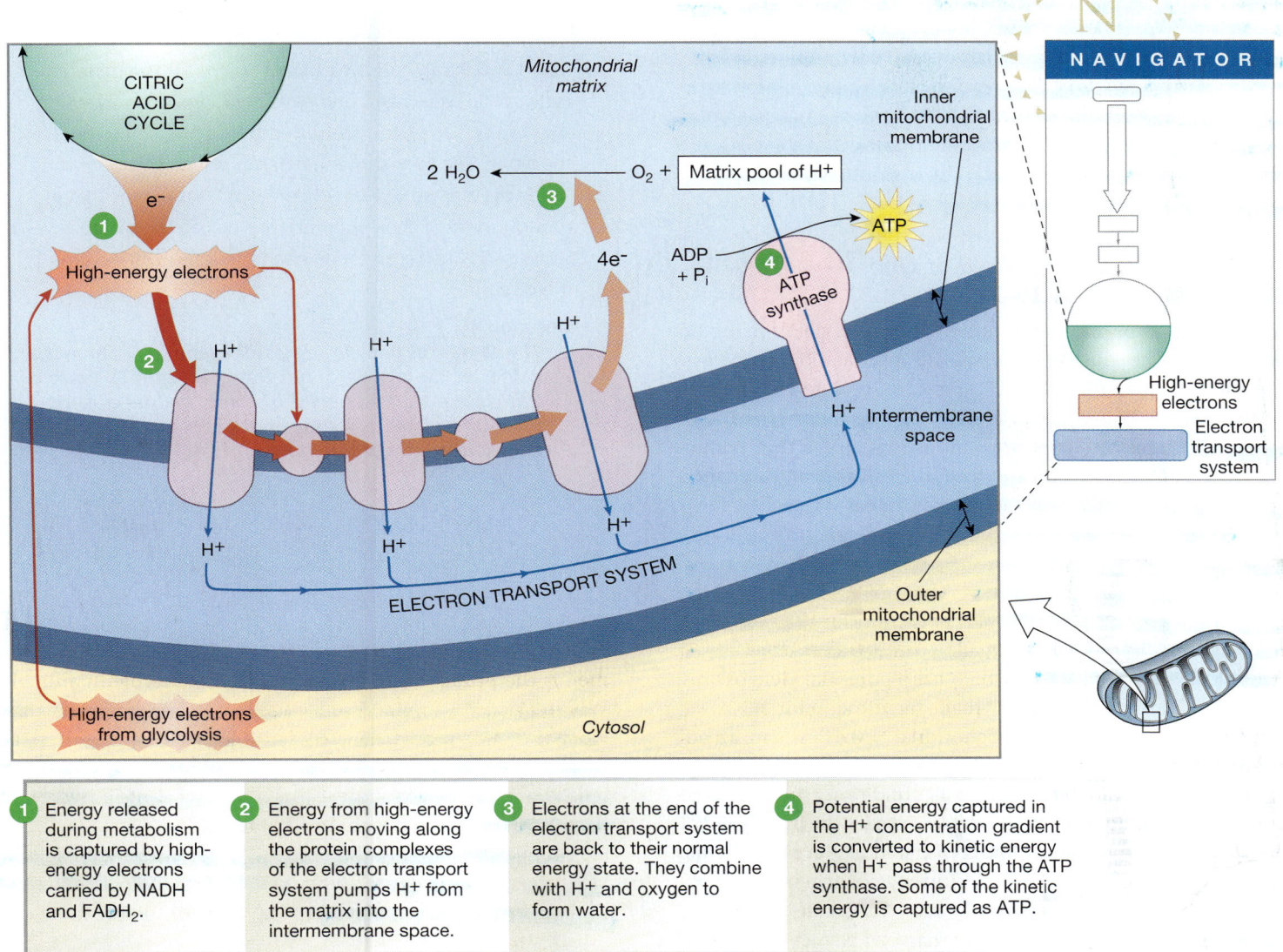

NAVIGATOR

High-energy electrons

Electron transport system

4

Mitochondrial matrix

$2 H_2O$ ← ← $O_2 +$ | Matrix pool of H^+ |

3

1

e⁻

High-energy electrons

2

$4e^-$

ADP + P_i

4

ATP

ATP synthase

CITRIC ACID CYCLE

H^+ H^+ H^+

H^+

Inner mitochondrial membrane

Intermembrane space

H^+ H^+ H^+

ELECTRON TRANSPORT SYSTEM

Outer mitochondrial membrane

High-energy electrons from glycolysis

Cytosol

1 Energy released during metabolism is captured by high-energy electrons carried by NADH and $FADH_2$.

2 Energy from high-energy electrons moving along the protein complexes of the electron transport system pumps H^+ from the matrix into the intermembrane space.

3 Electrons at the end of the electron transport system are back to their normal energy state. They combine with H^+ and oxygen to form water.

4 Potential energy captured in the H^+ concentration gradient is converted to kinetic energy when H^+ pass through the ATP synthase. Some of the kinetic energy is captured as ATP.

■ **FIGURE 4-17**   *The electron transport system and ATP synthesis via oxidative phosphorylation*

the mitochondrial matrix into the intermembrane space (step 2 of Fig. 4-17). This movement of H^+ into the intermembrane space creates an H^+ concentration gradient. Then, as the H^+ move back across the membrane (down their concentration gradient) into the mitochondrial matrix (step 4), potential energy stored in the concentration gradient is transferred to the high-energy bond of ATP. The synthesis of ATP using the electron transport system is called **oxidative phosphorylation** because the electron transport system requires oxygen to act as the final acceptor of electrons and H^+.

By the end of the electron transport chain, high-energy electrons have given up the usable portion of their stored energy. At that point, each pair of electrons passing through the transport system combines with two H^+ from the pool of H^+

ions in the matrix. The resulting hydrogen atoms then combine with an oxygen atom, creating a molecule of water, H_2O.

CONCEPT CHECK

21. What is phosphorylation? *the addition of a phosphate group*

22. Is the movement of electrons through the electron transport system endergonic or exergonic?
exergonic

Answers: p. 127

ATP Synthesis Is Coupled to Hydrogen Ion Movement

As we have seen, the energy released by electrons moving through the electron transport system is stored as potential energy by H^+ concentrated in the intermembrane space. This stored energy is converted into chemical-bond energy when the ions move back into the mitochondrial matrix through a protein known as $F_1F_OATPase$, or **ATP synthase** (Fig. 4-17, step 4). As H^+ move back into the mitochondrial matrix through a pore in the enzyme, the synthase transfers their kinetic energy to the high-energy phosphate bond of ATP. Because energy conversions are never completely efficient, a portion of the energy is released as heat. For each three H^+ that shuttle through the enzyme, a maximum of one ATP is formed.

The Maximum Energy Yield of One Glucose Molecule Is 30–32 ATP

If we tally up the maximum potential energy yield for the catabolism of one glucose molecule through aerobic pathways, the total comes to 30–32 ATP (Table 4-5). Note that we say *potential* yield. This is because often the mitochondria do not work up to capacity. There are various reasons for this, including the fact that a certain number of H^+ leak from the intermembrane space back into the mitochondrial matrix.

A second source of variability in the number of ATP produced per glucose comes from the two cytosolic NADH molecules produced during glycolysis. These NADH molecules are unable to enter mitochondria and must transfer their electrons through membrane carriers. Inside a mitochondrion, some of these electrons go to $FADH_2$, which has a potential yield of only 1.5 ATP rather than the 2.5 ATP made by mitochondrial NADH. If cytosolic electrons go to mitochondrial NADH instead, an additional two ATP molecules can be produced.

Despite the variability, under all conditions the ATP yield from aerobic metabolism of glucose far exceeds the puny yield of two ATP we see in the anaerobic conversion of glucose to lactate (Fig. 4-15 and Table 4-5). The low efficiency of anaerobic metabolism severely limits its usefulness in most vertebrate cells, whose metabolic energy demand is greater than can be met in this way. Some cells, such as exercising muscle cells, can tolerate anaerobic metabolism for a limited period of time. Eventually, however, they must shift back to aerobic metabolism.

CONCEPT CHECK

23. Describe two differences between aerobic and anaerobic metabolism of glucose.

24. What is the role of oxygen in oxidative phosphorylation?

25. How is the separation of mitochondria into two compartments essential to the chemiosmotic theory of ATP synthesis?

Answers: p. 127

Large Biomolecules Can Be Used to Make ATP

We have now examined how glucose is used for ATP production, but how are large biomolecules such as glycogen, proteins, and lipids used to make ATP? The answer is a series of pathways that convert these biomolecules into intermediates that then take part in glycolysis and the citric acid cycle (see overview in Figure 4-13, p. 104).

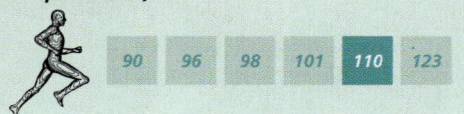

David and Sarah had their blood drawn for the genetic test several weeks ago and have been anxiously awaiting the results. Today, they returned to the hospital to hear the news. The tests show that Sarah carries the gene for Tay-Sachs disease but David does not. This means that although some of their children may be carriers of the Tay-Sachs gene like Sarah, none of the children will develop the disease.

Question 4:
The Tay-Sachs gene is a recessive gene (t). If Sarah is a carrier of the gene (Tt) but David is not (TT), what is the chance that any child of theirs will be a carrier? (Consult a general biology or genetics text if you need help solving this problem.)

| 90 | 96 | 98 | 101 | 110 | 123 |

Glycogen Converts to Glucose Glycogen, a glucose polymer, is the primary carbohydrate storage molecule in animals. When glucose supplies from outside the body are less than what the body's cells need to synthesize ATP, cells rely first on energy stored in the chemical bonds of glycogen. Glycogen breakdown, or **glycogenolysis**, is under the control of various hormones, as you will learn in Chapter 22.

In glycogenolysis, glycogen is broken down into glycolysis intermediates. Only about 10% of stored glycogen is hydrolyzed to glucose molecules (Fig. 4-18 ■). Before they can

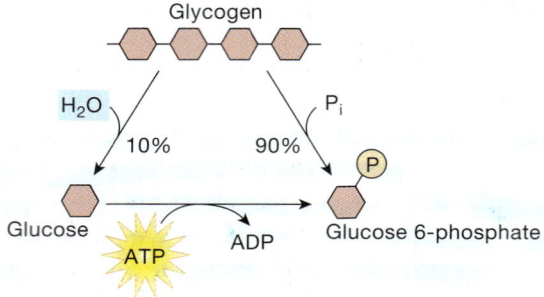

The direct conversion of glycogen to glucose 6-phosphate saves the cell one ATP per glucose.

■ **FIGURE 4-18** *Glycogenolysis, the breakdown of glycogen*

Glycogen can be converted directly to glucose 6-phosphate by the addition of phosphate. Glycogen that is broken down first to glucose and then phosphorylated "costs" the cell an extra ATP.

enter the glycolysis pathway, these glucose molecules must be phosphorylated to glucose 6-phosphate with the help of an ATP.

The majority of glycogen molecules that undergo glycogenolysis are converted directly to glucose 6-phosphate in a reaction that splits a glucose molecule from the polymer with the help of an inorganic phosphate from the cytosol. Because the direct phosphorylation of glycogen to glucose 6-phosphate saves one ATP, this pathway has a net yield of one additional ATP molecule per glucose (maximum 31–33 ATP for aerobic metabolism).

Proteins Can Be Catabolized to Produce ATP The first step in protein catabolism is the digestion of a protein into smaller polypeptides by enzymes called *proteases*. Once that has been accomplished, enzymes known as *peptidases* break the peptide bonds at the ends of the polypeptide, freeing individual amino acids (Fig. 4-19a ■). Most amino acids used for ATP synthesis come not from breaking down protein in cells, however, but from surplus amino acids in the diet.

Amino acids can be converted into intermediates for either glycolysis or the citric acid cycle. The first step in this conversion is deamination, creating an ammonia molecule and an organic acid (Fig. 4-19b). Among the organic acids created by deamination are pyruvate, acetyl CoA, and several intermediates of the citric acid cycle. The former amino acids thus enter the pathways of aerobic metabolism and become part of the ATP production process previously described.

The amino groups of amino acids are removed as ammonia (NH_3). They then rapidly pick up hydrogen ions (H^+) to become ammonium ions (NH_4^+), as indicated in Figure 4-19c. Because both ammonia and ammonium ions are toxic, liver cells rapidly convert them into urea (CH_4N_2O). Urea is the main nitrogenous waste of the body and is excreted by the kidneys.

Lipids Yield More Energy per Unit Weight than Glucose or Proteins Lipids are the primary fuel-storage molecule of the body because they have a higher energy content than proteins or carbohydrates. When the body needs to use stored energy, *lipases* break down lipids into glycerol and fatty acids through a series of reactions collectively known as **lipolysis** (Fig. 4-20 ■). Glycerol feeds into glycolysis about halfway through the pathway and goes through the same reactions as glucose from that point on.

The long carbon chains of fatty acids are not as easy to turn into ATP as glucose or amino acids. Most fatty acids must be transported from the cytosol into the mitochondrial matrix. There they are slowly disassembled as two-carbon units are chopped off the end of the chain, one unit at a time, in a process called **beta-oxidation**.

In most cells, the two-carbon units from fatty acids are converted into acetyl CoA, whose two-carbon acyl units feed di-

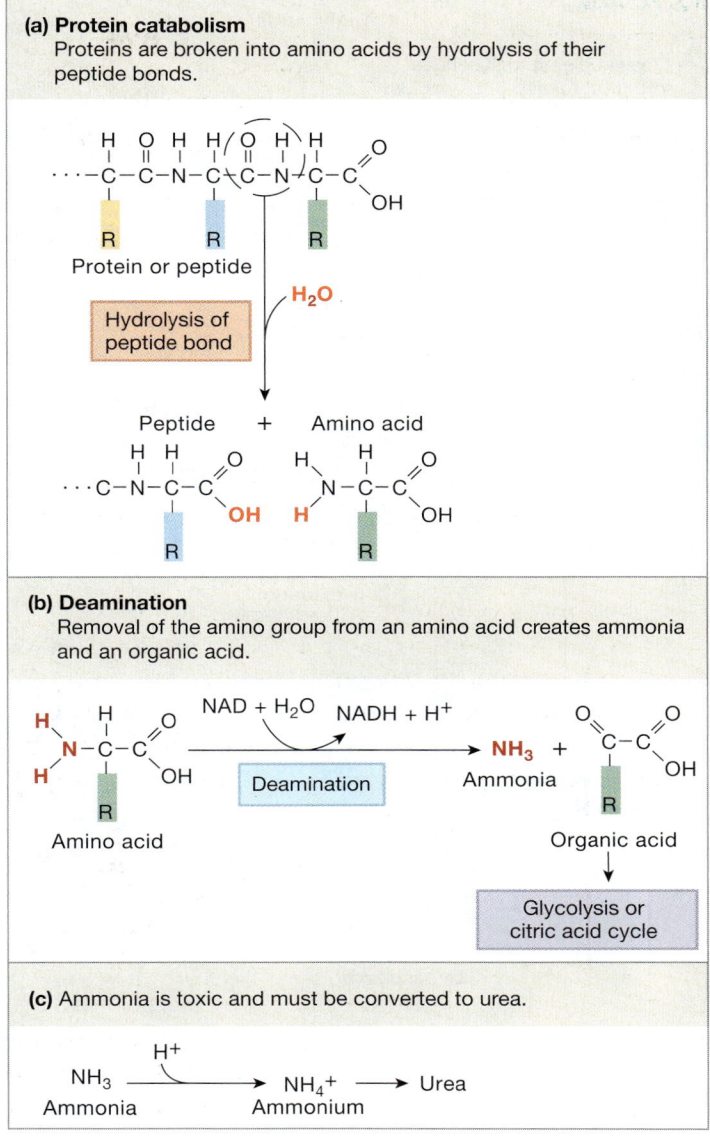

(a) Protein catabolism
Proteins are broken into amino acids by hydrolysis of their peptide bonds.

(b) Deamination
Removal of the amino group from an amino acid creates ammonia and an organic acid.

(c) Ammonia is toxic and must be converted to urea.

FIGURE QUESTIONS

- The use of water in reaction **(b)** makes the reaction:
  (1) hydration
  (2) hydrolysis
  (3) dehydration
- Name the family of enzymes that catalyze the reaction shown in **(a)**.

■ **FIGURE 4-19** *Protein catabolism and deamination*

rectly into the citric acid cycle. Because many acetyl CoA molecules can be produced from a single fatty acid, lipids contain 9 kcal of stored energy per gram, compared with 4 kcal per gram for proteins and carbohydrates. In the liver, if acetyl CoA production exceeds the capacity of the citric acid cycle to metabolize the acyl units, excess acetyl CoA is converted into **ketone bodies**, strong metabolic acids that can seriously disrupt the body's

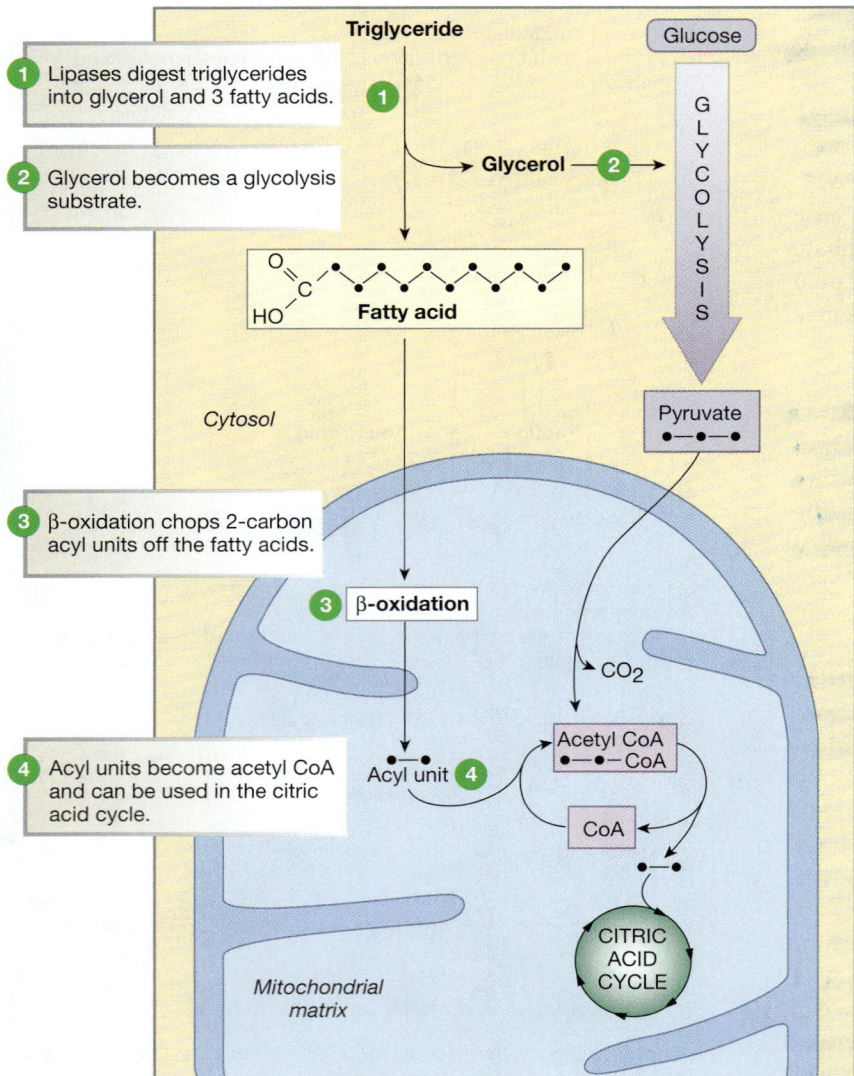

1. Lipases digest triglycerides into glycerol and 3 fatty acids.

2. Glycerol becomes a glycolysis substrate.

3. β-oxidation chops 2-carbon acyl units off the fatty acids.

4. Acyl units become acetyl CoA and can be used in the citric acid cycle.

■ **FIGURE 4-20** *Lipolysis*

pH balance. We will cover the details of fatty acid metabolism in Chapter 22.

CONCEPT CHECK

26. Is the digestion of proteins to polypeptides more likely to be a dehydration reaction or a hydrolysis reaction?

27. Based on what you have learned about enzyme names, what name would you give to enzymes that remove an amino acid from a molecule? Answers: p. 127

SYNTHETIC PATHWAYS

The metabolic reactions discussed to this point have been catabolic processes that break up large molecules into smaller ones to trap energy in the high-energy phosphate bond of ATP. We now examine anabolic reactions, those that synthe-size polysaccharides, lipids, and proteins. To carry out synthetic reactions, a cell must have an adequate supply of organic nutrients, as well as energy in the form of ATP.

Glycogen Can Be Made from Glucose

Glycogen, the main storage form of glucose in the body, is found in all cells, but the liver and skeletal muscles contain especially high concentrations. Glycogen in skeletal muscles provides a ready energy source for muscle contraction. Glycogen in the liver acts as the main source of glucose for the body in periods between meals. It is estimated that the liver keeps about a four-hour supply of glucose stored as glycogen. Once that glycogen is exhausted by catabolism, the cells must turn to lipids and proteins as their source of energy.

The synthesis of glycogen is essentially the reverse of glycogen breakdown (discussed in the previous section). Individual glucose molecules can be linked together into glycogen, and glucose 6-phosphate from glycolysis can be synthesized into glycogen by removal of the phosphate group (see Fig. 4-18). A single glycogen particle in the cytoplasm may contain as many as 55,000 linked glucose molecules!

Glucose Can Be Made from Glycerol or Amino Acids

Glucose is one of the key metabolic intermediates in the body for several reasons. As you have just learned, the aerobic metabolism of glucose is the most efficient way to make ATP, so glucose is the primary substrate for energy production in all cells. More important, glucose is normally the only substrate used for ATP synthesis in neural tissue. If the brain is deprived of glucose for any period of time, its cells begin to die. For this reason, the body has multiple metabolic pathways it can use to make glucose, ensuring a continuous supply for the brain.

The simplest source of glucose is the glycogenolysis pathway discussed earlier. But what happens when all the glycogen stores in the body are used up? In that case, cells can turn to lipids and proteins for the intermediates needed to make glucose.

The production of glucose from nonglucose precursors such as proteins or the glycerol portion of lipids is a process known as **gluconeogenesis**, literally "birth of new glucose" (Fig. 4-21 ■). The pathway is similar to glycolysis run in reverse but uses different enzymes for some steps. Gluconeogenesis can start with glycerol, with various amino acids, or with lactate. In all cells, the process can go as far as glucose 6-phosphate (G-6-P). However, only liver and kidney cells have significant amounts

CLINICAL FOCUS

ENERGY AND EXERCISE

Exercise is the major energy-consuming activity of the body, and lack of energy for muscle contraction is a major cause of fatigue in extended exercise sessions. Carbohydrate is the preferred fuel of skeletal muscle. At any given time, an adult human has approximately 4000 kcal of energy stored as glycogen (3000 kcal in liver and 1000 kcal in skeletal muscle), which provides enough energy to exercise at moderate intensity for about three hours. In comparison, the body stores about 10,000 kcal of energy in the lipids of adipose tissue, which provides enough energy to run 100 miles. So with those tremendous energy stores, why do athletes still experience fatigue? In fact, fatigue sets in when carbohydrate stores are depleted because lipids cannot be converted to ATP as quickly as carbohydrates. Therefore, as carbohydrate stores are depleted, the body relies on its lipid stores, which requires the athlete to exercise at a slower pace that matches the rate at which the energy in lipids is converted to energy in ATP.

of *glucose 6-phosphatase,* the enzyme that removes the phosphate group from G-6-P to make glucose. During periods of fasting, the liver is the primary source of glucose synthesis, producing about 90% of the glucose made from the breakdown of lipids and proteins.

CONCEPT CHECK

28. What is the difference between hexokinase and glucose 6-phosphatase? (*Hint*: see Table 4-4.)

Answers: p. 127

Acetyl CoA Is an Important Precursor for Lipid Synthesis

Lipids are so diverse that generalizing about their synthesis is difficult. Most lipids are synthesized by enzymes in the smooth endoplasmic reticulum and in the cytosol. Fatty acids are made in the cytosol by an enzyme called *fatty acid synthetase,* which links the two-carbon acyl groups of acetyl CoA together (Fig. 4-22 ■). The process also requires hydrogens and high-energy electrons from NADPH.

Glycerol can be made from glucose or from glycolysis intermediates. The combining of glycerol and fatty acids into triglycerides takes place in the smooth endoplasmic reticulum. The phosphorylation steps that turn triglyc-

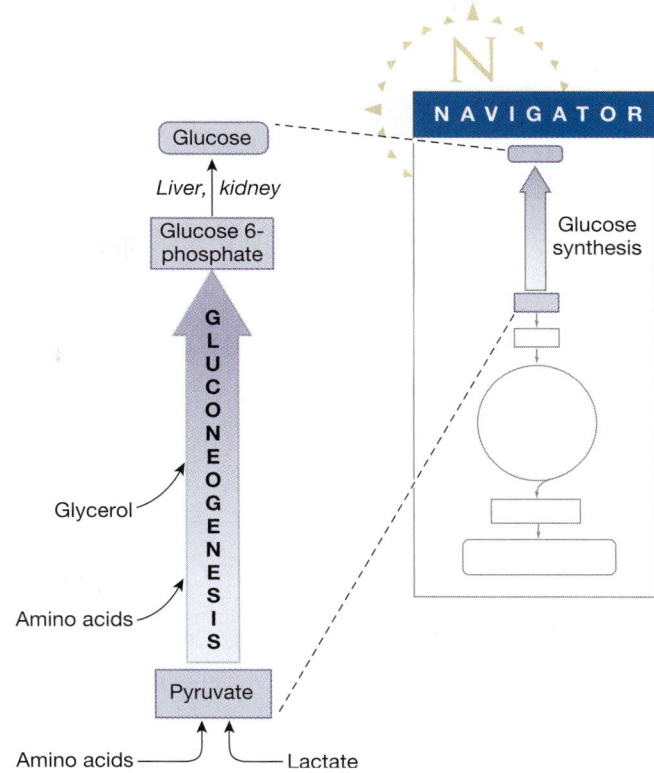

■ **FIGURE 4-21**   *Gluconeogenesis*

Glycerol, amino acids, and lactate can be used to make glucose by a series of reactions that are similar to the glycolysis pathway run in reverse. Only the liver and kidneys are able to convert glucose 6-phosphate to glucose.

erides into phospholipids also take place in the endoplasmic reticulum.

Cholesterol [⮂ Fig. 2-8, p. 29] is an important component of cell membranes, and it forms the skeleton of steroid hormones. Cholesterol is usually obtained from animal products in the diet, but it is such an important molecule for cells that the body will synthesize it if the diet is deficient. Even vegetarians who eat no animal products (vegans) have substantial amounts of cholesterol in their cells.

Cholesterol can be made from acetyl CoA. From there, it is a simple matter to add or subtract functional groups to the carbon rings to change cholesterol into various hormones and other steroids.

Proteins Are the Key to Cell Function

Proteins are the molecules that run a cell from day to day. Protein enzymes control the synthesis and breakdown of carbohydrates, lipids, structural proteins, and signal molecules. Protein transporters and pores in the cell membrane and in organelle membranes regulate the movement of molecules into and out of compartments. Other proteins form the structural skeleton of cells and tissues. Thus, protein synthesis is critical to cell function.

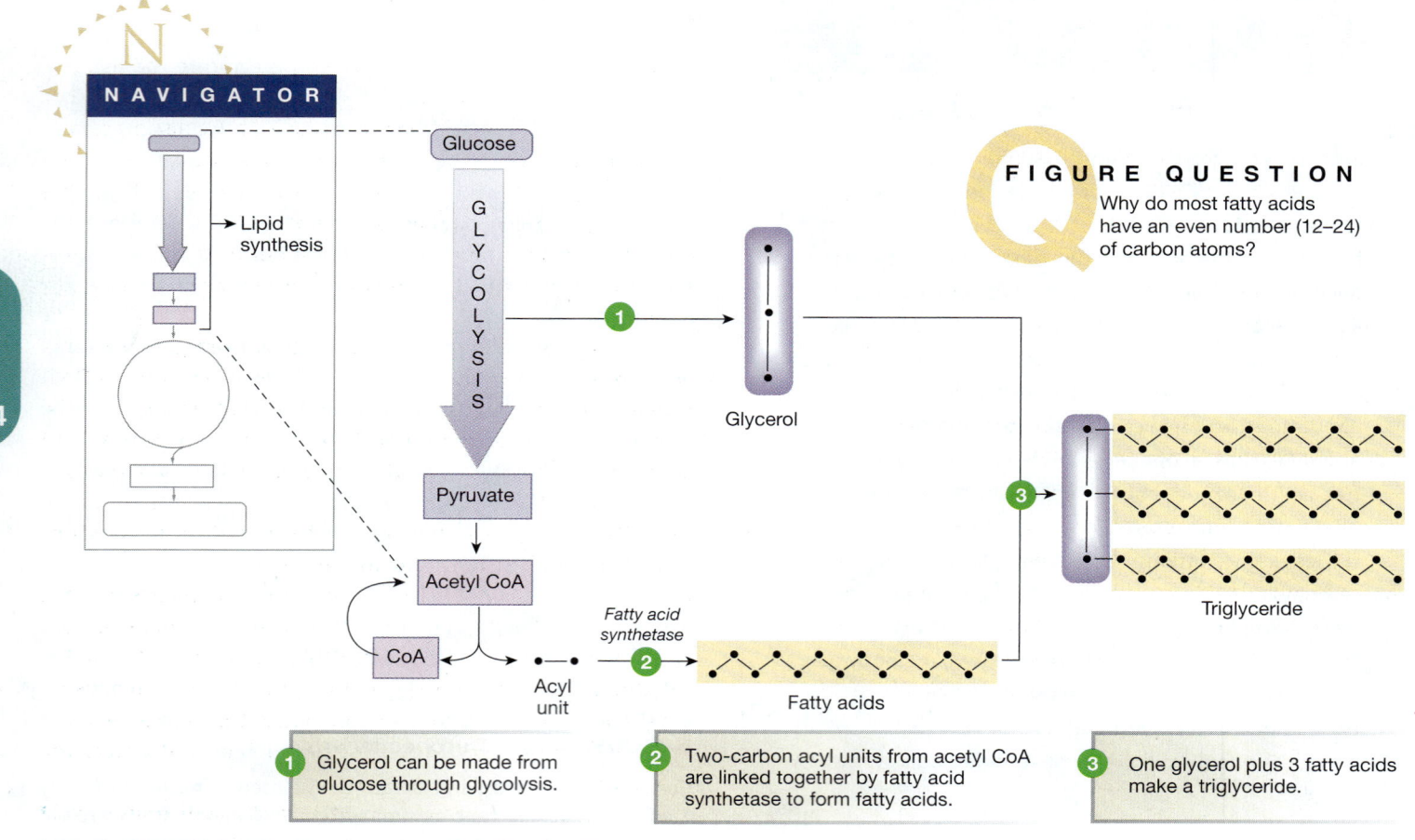

NAVIGATOR

Glucose

GLYCOLYSIS

Lipid synthesis

Pyruvate

Acetyl CoA

CoA

Acyl unit

Fatty acid synthetase

Glycerol

Fatty acids

Triglyceride

FIGURE QUESTION
Why do most fatty acids have an even number (12–24) of carbon atoms?

1 Glycerol can be made from glucose through glycolysis.

2 Two-carbon acyl units from acetyl CoA are linked together by fatty acid synthetase to form fatty acids.

3 One glycerol plus 3 fatty acids make a triglyceride.

■ **FIGURE 4-22** *Lipid synthesis*

The power of proteins arises from their tremendous variability and specificity. Protein synthesis using 20 amino acids can be compared to creating a language with an alphabet of 20 letters. The "words" of the protein "language" vary in length from three letters to hundreds of letters, resulting in thousands of different proteins with different functions. Like a language, a "spelling error" during protein synthesis can change the function of the protein, just as changing the last letter turns the word "foot" into "food."

In many inherited diseases, the substitution of a single amino acid alters the structure of a protein dramatically. One classic example is sickle cell anemia. In this inherited disease, a change of one amino acid alters the shape of hemoglobin enough so that red blood cells take on a crescent (sickle) shape, which causes them to get tangled up and block small blood vessels.

The Protein "Alphabet" Begins with DNA One of the mysteries of biology until the 1960s was the question of how only four nitrogenous bases in the DNA molecule—adenine (A), guanine (G), cytosine (C), and thymine (T)—could code for more than 20 different amino acids. If each base controlled the synthesis of one amino acid, a cell could make only four different amino acids. If *pairs* of bases represented different amino acids, the cell

could make $4^2 = 16$ different amino acids. Because we have 20 amino acids, this is still not satisfactory. If *triplets* of bases were the codes for different molecules, however, DNA could create $4^3 = 64$ different amino acids. These triplets, called **codons**, are indeed the way information is encoded in DNA and RNA. Figure 4-23 ■ shows the genetic code as it appears in one form of RNA. Remember that RNA substitutes the base uracil (U) for thymine [🔁 p. 33].

Of the 64 possible triplet combinations, one DNA codon (TAC) acts as the initiator or *"start codon"* that signifies the beginning of a coding sequence. Three codons are recognized as terminator or *"stop codons"* that show where the sequence ends. The remaining 60 triplets all code for amino acids. Methionine and tryptophan have only one codon each, but the other amino acids have between two and six different codons each. Thus, like letters spelling words, the DNA sequence determines the amino acid sequence of proteins.

How is information translated from DNA into a chain of amino acids? First, the base sequence of DNA is used to create a piece of **messenger RNA (mRNA)** in the process known as **transcription** [*trans,* over + *scribe,* to write]. The mRNA leaves the nucleus and enters the cytosol, where it directs **translation**, the assembly of amino acids into proteins.

DIABETES

INSULIN AND METABOLISM

Why do people with diabetes have elevated levels of blood glucose ("blood sugar")? Part of the answer lies with the disruption of glucose metabolism when insulin is absent or when cells do not respond to it. In a normal person, insulin promotes cellular uptake and metabolism of glucose by stimulating the appropriate enzymes. Glucose absorbed from the digestive system goes through glycolysis to make ATP or is stored as glycogen. Fatty acids are stored as lipids, and amino acids are used to make proteins. In the absence of insulin, the enzymes associated with catabolic pathways are more active than those associated with anabolic pathways. In these overactive catabolic pathways, lipids undergo beta-oxidation, amino acids go through gluconeogenesis to become glucose, and glycogen stores are turned back into glucose. The increased production of glucose by cells coupled with the inability of many cells to absorb glucose from the blood creates the elevated "blood sugar" levels that are a hallmark of diabetes.

Second base of codon

	U	C	A	G	
U	UUU ⎤ Phe UUC ⎦ UUA ⎤ Leu UUG ⎦	UCU ⎤ UCC ⎥ Ser UCA ⎥ UCG ⎦	UAU ⎤ Tyr UAC ⎦ UAA ⎤ **Stop** UAG ⎦	UGU ⎤ Cys UGC ⎦ UGA **Stop** UGG Trp	U C A G
C	CUU ⎤ CUC ⎥ Leu CUA ⎥ CUG ⎦	CCU ⎤ CCC ⎥ Pro CCA ⎥ CCG ⎦	CAU ⎤ His CAC ⎦ CAA ⎤ Gln CAG ⎦	CGU ⎤ CGC ⎥ Arg CGA ⎥ CGG ⎦	U C A G
A	AUU ⎤ AUC ⎥ Ile AUA ⎦ AUG Met **Start**	ACU ⎤ ACC ⎥ Thr ACA ⎥ ACG ⎦	AAU ⎤ Asn AAC ⎦ AAA ⎤ Lys AAG ⎦	AGU ⎤ Ser AGC ⎦ AGA ⎤ Arg AGG ⎦	U C A G
G	GUU ⎤ GUC ⎥ Val GUA ⎥ GUG ⎦	GCU ⎤ GCC ⎥ Ala GCA ⎥ GCG ⎦	GAU ⎤ Asp GAC ⎦ GAA ⎤ Glu GAG ⎦	GGU ⎤ GGC ⎥ Gly GGA ⎥ GGG ⎦	U C A G

First base of codon · Third base of codon

■ **FIGURE 4-23** *The genetic code as it appears in the codons of mRNA*

The three-letter abbreviations to the right of the brackets indicate the amino acid each codon represents. The start and stop codons are also marked.

Translating DNA to Protein Is a Complex Process

How does a cell know which of the thousands of bases present in its DNA to use in making a protein? It turns out that the information a cell needs to make a particular protein is contained in a segment of DNA known as a *gene*. What exactly is a gene? The definition keeps changing, but for this text we will say that a **gene** is a region of DNA that contains the information needed to make a functional piece of messenger RNA (mRNA). The mRNA in turn can be used to make a protein.

A summary map of the major steps from gene to functional protein is shown in Figure 4-24 ■. First, a section of DNA containing a gene must be activated so that its code can be read. Genes that are continuously being transcribed are said to be *constitutively* active. Usually these genes code for proteins that are essential to ongoing cell functions. Other genes are *regulated*; that is, their activity can be turned on (*induced*) or turned off (*repressed*) by regulatory proteins.

Once a gene is activated, transcription converts its DNA base sequence into a piece of mRNA. The newly made mRNA is processed in the nucleus; it either undergoes alternative splicing (discussed shortly) before leaving the nucleus or is "silenced" and destroyed by enzymes through *RNA interference*. Processed mRNA leaves the nucleus and enters the cytoplasm.

Once in the cytoplasm, mRNA works with *transfer RNA* (tRNA) and *ribosomal RNA* (rRNA) to assemble amino acids into protein chains. Newly synthesized proteins are then subject to **post-translational modification**. They fold into their complex shapes, are split by enzymes into smaller peptides, or have various chemical groups added to them. The remainder of this chapter explains these steps.

During Transcription, DNA Guides the Synthesis of a Complementary mRNA Molecule

The first steps in protein synthesis are compartmentalized within the nucleus because DNA is a very large molecule that cannot pass through the nuclear envelope. Transcription uses DNA as a template to create a small single strand of mRNA that can leave the nucleus. The synthesis of mRNA from the double-stranded DNA template requires an enzyme known as **RNA polymerase**, plus magnesium or manganese ions and energy in the form of high-energy phosphate bonds:

DNA template + nucleotides A, U, C, G

↓ RNA polymerase, Mg^{2+} or Mn^{2+}, and energy

mRNA + DNA template

4

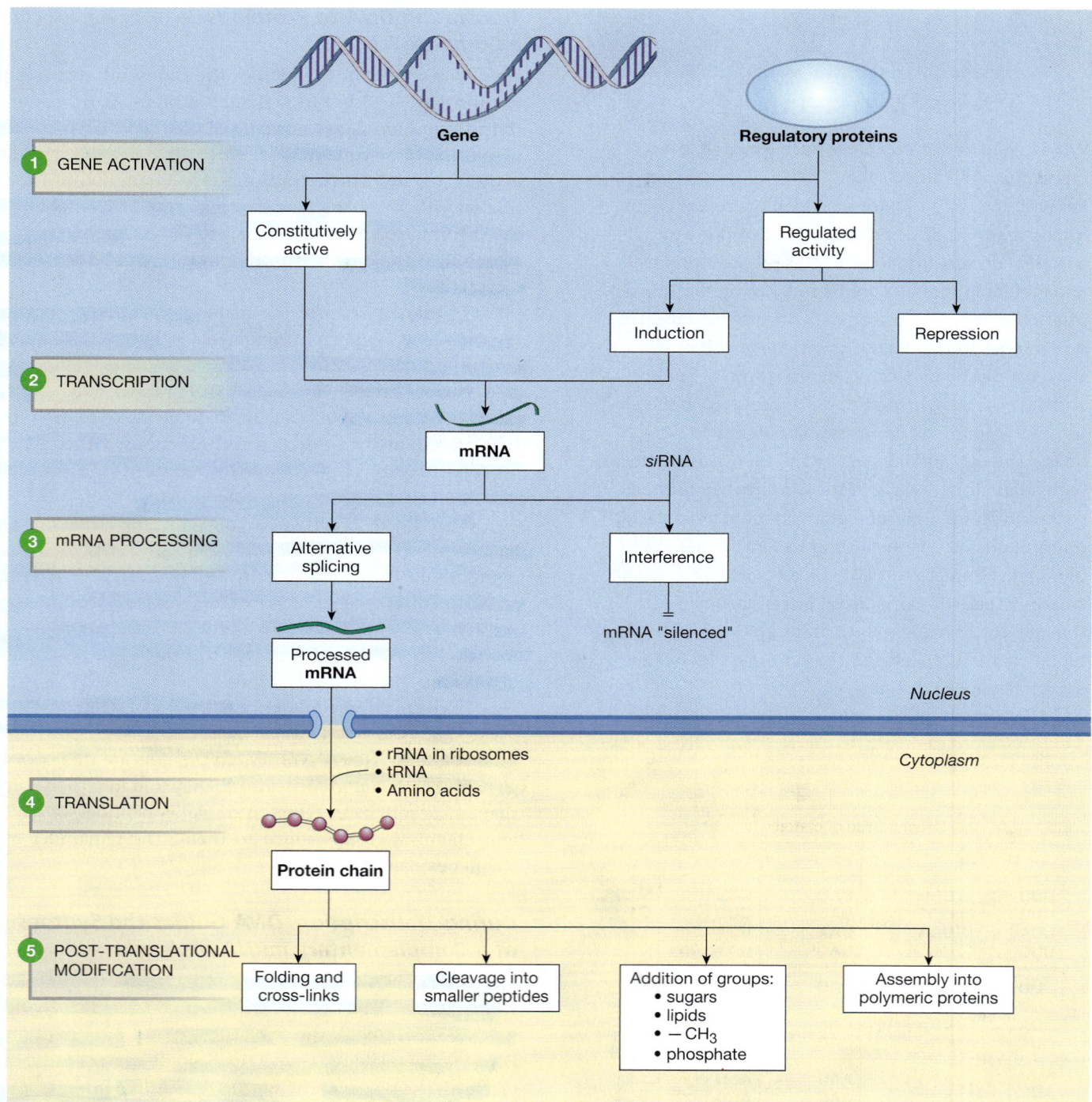

FIGURE 4-24 *The major steps required to convert the genetic code of DNA into a functional protein*

Each gene on DNA is preceded by a **promoter** region that must be activated before transcription can begin. Regulatory-protein **transcription factors** bind to DNA and activate the promoter, which tells the RNA polymerase where to bind to the DNA (Fig. 4-25 ■). The polymerase moves along the DNA molecule and "unwinds" the double strand by breaking the hydrogen

bonds between paired bases. One strand of DNA, called the *sense strand*, serves as the guide for mRNA synthesis, while the other strand, the *antisense strand*, sits idly by. The promoter region is not transcribed into mRNA.

During transcription, each base in the DNA sense strand pairs with the complementary mRNA base. This is similar to the

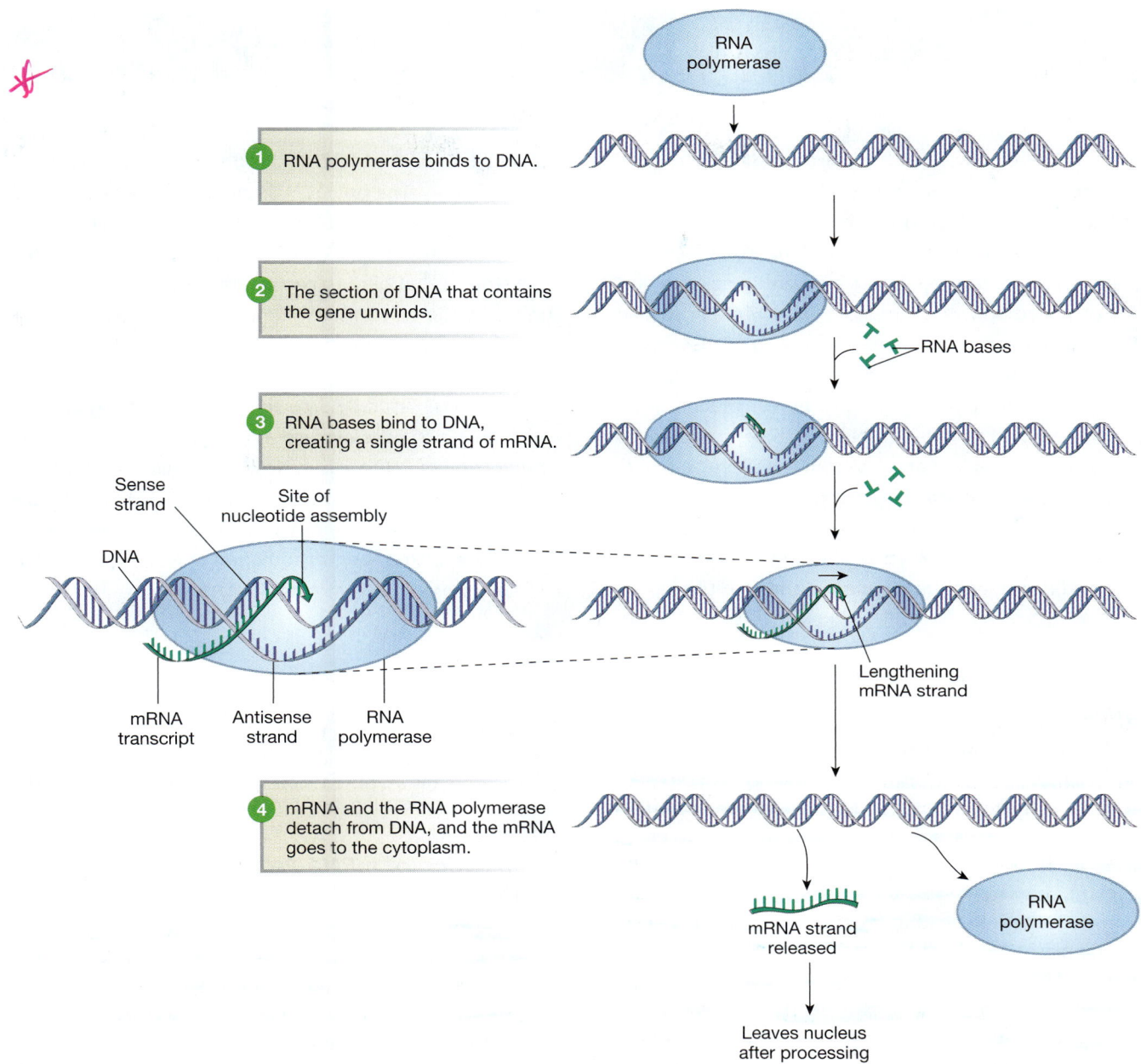

RNA polymerase

1 RNA polymerase binds to DNA.

2 The section of DNA that contains the gene unwinds.

RNA bases

3 RNA bases bind to DNA, creating a single strand of mRNA.

Sense strand
Site of nucleotide assembly
DNA
mRNA transcript
Antisense strand
RNA polymerase

Lengthening mRNA strand

4 mRNA and the RNA polymerase detach from DNA, and the mRNA goes to the cytoplasm.

mRNA strand released

RNA polymerase

Leaves nucleus after processing

■ **FIGURE 4-25** · *Transcription*

A gene is a segment of DNA that can produce a functional piece of mRNA. Base pairing is the same as in DNA synthesis, except that the base uracil (U) substitutes for thymine (T).

process by which a double strand of DNA forms (see Appendix B for a review of DNA synthesis). For example, a DNA segment containing the base sequence AGTAC is transcribed into the mRNA sequence UCAUG.

As the RNA bases bind to the DNA sense strand, they also bond with one another to create a single strand of mRNA. During transcription, bases are linked at an average rate of 40 per second. In humans, the largest mRNAs may contain as many as 5000 bases, and their transcription may take more than a

minute—a long time for a cellular process. When RNA polymerase reaches the stop codon, it stops adding bases to the growing mRNA strand and releases the strand.

CONCEPT CHECK

29. Use the genetic code in Figure 4-23 to write the DNA codons that correspond to the three mRNA stop codons.

30. What does its name tells you about the function of RNA polymerase?

Answers: p. 127

RNA INTERFERENCE

Who could have guessed that research to develop a more purple petunia would lead the way to one of the most exciting new areas of molecular biology research? RNA interference (*RNAi*) was first observed in 1990, when botanists who introduced purple pigment genes into petunias ended up with plants that were white or striped with white instead of the deeper purple color they expected. This observation did not attract attention until 1998, when scientists doing research in animal biology and medicine had a similar problem with gene expression in experiments on a nematode worm. Now RNAi is one of the hottest areas of biotechnology research.

Many details of RNAi are still not well understood. In very simple terms, RNA "silencing" of mRNA is a naturally occurring phenomenon accomplished through the production or introduction of double-stranded pieces of RNA (dsRNA). The dsRNA is chopped by an enzyme called *Dicer* into smaller (20–25 nucleotide) pieces called small interfering RNAs, or *siRNA*. The siRNAs separate into single strands that bind to complementary pieces of mRNA made by the cell. This attachment of siRNA blocks translation from the mRNA or even targets the mRNA for destruction.

RNAi is a naturally occurring RNA processing mechanism that may have evolved as a means of blocking the replication of RNA viruses. Now, however, researchers are hoping to learn how to use it to selectively block the production of single proteins within a cell. The scientists' ultimate goal is to create technologies that can be used for the diagnosis and treatment of disease.

Alternative Splicing Creates Multiple Proteins from One Sequence of DNA

The next step in the process is **mRNA processing,** which takes two forms (Fig. 4-24, step 3). In *RNA interference,* newly synthesized mRNA is inactivated or destroyed before it can be translated into proteins (see the Emerging Concepts box). In **alternative splicing,** enzymes clip segments out of the middle or off the ends of the mRNA strand. Other enzymes then splice the remaining pieces of the strand back together.

Alternative splicing is necessary because a gene contains both segments that encode proteins (**exons**) and noncoding segments called **introns** (Fig. 4-26 ■). That means the mRNA made from the gene's DNA also contains noncoding segments that must be removed before the mRNA leaves the nucleus. The result of alternative splicing is a piece of mRNA smaller than was made initially, a piece that now contains only the coding sequence for a specific protein.

One advantage of alternative splicing is that it allows a single base sequence on DNA to code for more than one protein. The designation of segments as coding or noncoding is not fixed for a given gene. Segments of mRNA that are removed one time can be left in the next time, producing a finished mRNA with a different sequence. The closely related forms of a single enzyme known as *isozymes* are probably made by alternative splicing of a single gene.

After mRNA has been processed, it exits the nucleus through nuclear pores and goes to ribosomes in the cytosol. There the mRNA directs the construction of protein.

CONCEPT CHECK

31. Explain in one or two sentences the relationship of mRNA, nitrogenous bases, introns, exons, mRNA processing, and proteins.

   Answers: p. 127

Translation of mRNA Produces a String of Amino Acids

Protein synthesis requires cooperation and coordination among three types of RNA: mRNA, rRNA, and tRNA. Upon arrival in the cytosol, processed mRNA binds to ribosomes, which are small particles of protein and several types of rRNA [≈ p. 61]. Each ribosome has two subunits, one large and one small, that come together when protein synthesis begins. The small ribosomal subunit binds the mRNA, then adds the large subunit so that the mRNA is sandwiched in the middle (Fig. 4-27 ■, step 3). Now the ribosome-mRNA complex is ready to begin translation.

During translation, the mRNA codons are matched to the proper amino acid. This is done with the assistance of a tRNA molecule (Fig. 4-27, step 4). One region of each tRNA contains a three-base sequence called an **anticodon** that is complementary to an mRNA codon. A different region of the tRNA molecule is bound to a specific amino acid.

As translation begins, the anticodons of tRNAs carrying amino acids bind to the complementary codons of mRNA on ribosomes. For example, a tRNA with anticodon sequence UUU carries the amino acid lysine. The UUU anticodon pairs with an AAA codon on mRNA, which Figure 4-23 indicates is one of

TAY-SACHS DISEASE

In this running problem you learned that Tay-Sachs disease is an incurable, recessive genetic disorder in which the enzyme that breaks down gangliosides in cells is missing. One in 27 Americans of Eastern European Jewish descent carries the gene for this disorder. You have also learned that a blood test can detect the presence of the deadly gene. Check your understanding of this running problem by comparing your answers to those in the summary table. To read more on Tay Sachs disease, go to the website of the National Tay Sachs & Allied Diseases Association (*www.ntsad.org*).

	QUESTION	FACTS	INTEGRATION AND ANALYSIS
1	What is another symptom of Tay-Sachs disease besides loss of muscle control and brain function?	Hexosaminidase A breaks down gangliosides. In Tay-Sachs disease, this enzyme is absent, and gangliosides accumulate in cells, including light-sensitive cells of the eye, and cause them to function abnormally.	Damage to light-sensitive cells of the eye could cause vision problems and even blindness.
2	How could you test whether Sarah and David are carriers of the Tay-Sachs gene?	Carriers of the gene have lower-than-normal levels of hexosaminidase A.	Run tests to determine the average enzyme levels in known carriers of the disease (i.e., people who are parents of children with Tay-Sachs disease) and in people who have little likelihood of being carriers. Compare the enzyme levels of suspected carriers with the averages for the known carriers and noncarriers.
3	Why is the new test for the Tay-Sachs disease more accurate than the old test?	The new test detects the defective gene. The old test analyzed levels of the enzyme produced by the gene.	The new test is a direct way to test if a person is a carrier. The old test was an indirect way. It is possible for factors other than a defective gene to alter a person's enzyme level. Can you think of some? (See answers at end of chapter.)
4	The Tay-Sachs gene is a recessive gene (t). What is the chance that any child of a carrier and a noncarrier will be a carrier?	Tt × TT TT Tt TT Tt	Each child has a 50% chance of being a carrier (Tt).

90   96   98   101   110   **123**

CHAPTER SUMMARY

The major theme of this chapter is *energy in biological systems* and how it is acquired, transferred, and used to do biological work. Energy is stored in large biomolecules such as fats and glycogen and is extracted from them through the processes of metabolism. Reactions and processes that require energy often extract it from the high-energy bond of ATP, a pattern you will see repeated as you learn more about the organ systems of the body.

Other themes in the chapter involve two kinds of *structure-function relationships:* molecular interactions and compartmentation. *Molecular interactions* are important in enzymes, where the function of an enzyme often depends on its folding into the proper shape so that it can bind to its substrate. *Compartmentation* of enzymes is essential for organizing and separating metabolic processes. Glycolysis takes place in the cytosol of the cell, for example, as does the anaerobic metabolism of pyruvate to lactate. The citric acid cycle occurs in the mitochondrial matrix, and the electron transport chain is located on the inner mitochondrial membrane.

Energy in Biological Systems

1. **Energy** is the capacity to do work. **Chemical work** enables cells and organisms to grow, reproduce, and carry out normal activities. **Transport work** enables cells to move molecules to create

concentration gradients. **Mechanical work** is used for movement. (p. 91–92)

2. **Kinetic energy** is the energy of motion. **Potential energy** is stored energy. (p. 92; Fig. 4-2)

Chemical Reactions

3. A **chemical reaction** begins with one or more **reactants** and ends with one or more **products** (Tbl. 4-2). **Reaction rate** is measured as the change in concentration of products with time. (p. 93)

4. The energy stored in the chemical bonds of a molecule and available to perform work is the **free energy** of the molecule. (p. 93)

5. **Activation energy** is the initial input of energy required to begin a reaction. (p. 94; Fig. 4-3)

6. **Exergonic reactions** are energy-producing. **Endergonic reactions** are energy-utilizing. (p. 94–95; Fig. 4-4)

7. Metabolic pathways couple exergonic reactions to endergonic reactions. (p. 95; Fig. 4-5)

8. Energy for driving endergonic reactions is stored in ATP. (p. 95)

9. A reaction that can proceed in both directions is called a **reversible reaction**. If a reaction can proceed in one direction but not the other, it is an **irreversible reaction**. The net free energy change of a reaction determines whether that reaction can be reversed. (p. 95)

Enzymes

10. **Enzymes** are biological catalysts that speed up the rate of chemical reactions without themselves being changed. In reactions catalyzed by enzymes, the reactants are called **substrates**. (p. 96)

11. Like other proteins that bind ligands, enzymes exhibit saturation, specificity, and competition. Related isozymes may have different activities. (p. 96)

12. Some enzymes are produced as inactive precursors and must be activated. This may require the presence of a **cofactor**. Organic cofactors are called **coenzymes**. (p. 97)

13. Enzyme activity is altered by temperature, pH, and modulator molecules. (p. 97)

14. Enzymes work by lowering the activation energy of a reaction. (p. 98; Fig. 4-8)

15. Reversible reactions go to a state of **equilibrium**, in which the rate of the reaction in the forward direction is equal to the rate of the reverse reaction. (p. 99; Fig. 4-9)

16. Reversible reactions obey the **law of mass action**: when a reaction is at equilibrium, the ratio of substrates to products is always the same. If the concentration of a substrate or product changes, the equilibrium will be disturbed. (p. 99; Fig. 4-9)

17. Most reactions can be classified as **oxidation-reduction**, **hydrolysis-dehydration**, **addition-subtraction-exchange**, or **ligation** reactions. (p. 100–101; Tbl. 4-4)

Metabolism

18. All the chemical reactions in the body are known collectively as **metabolism**. **Catabolic reactions** release energy and break down large biomolecules. **Anabolic reactions** require a net input of energy and synthesize large biomolecules. (p. 101)

19. Cells regulate the flow of molecules through their metabolic pathways by (1) controlling enzyme concentrations, (2) producing allosteric and covalent modulators, (3) using different enzymes to catalyze reversible reactions, (4) isolating enzymes in intracellular organelles, or (5) maintaining an optimum ratio of ATP to ADP. (p. 101–102)

20. **Aerobic pathways** require oxygen and yield the most ATP. **Anaerobic pathways** can proceed without oxygen but produce ATP in much smaller quantities. (p. 103)

ATP Production

IP Muscular: Muscle Metabolism

21. Through **glycolysis**, one molecule of glucose is converted into two pyruvate molecules, two ATP, two NADH, and two H^+. Glycolysis does not require the presence of oxygen. (p. 103–105; Fig. 4-14)

22. In **anaerobic metabolism**, pyruvate is converted into lactate, with a net yield of two ATP for each glucose molecule. (p. 106–107; Fig. 4-15)

23. **Aerobic metabolism** of pyruvate through the **citric acid cycle** yields ATP, carbon dioxide, water, and high-energy electrons captured by NADH and $FADH_2$. (p. 107–108; Fig. 4-16)

24. **High-energy electrons** from NADH and $FADH_2$ give up their energy as they pass through the **electron transport system**. Their energy is trapped in the high-energy bonds of ATP. (p. 108; Fig. 4-17)

25. Proteins can be broken down into amino acids, which are then deaminated to make intermediates that enter aerobic pathways for ATP production. (p. 111; Fig. 4-19)

27. Glycogen and lipids are the primary energy storage molecules in animals. Lipids are broken down for ATP production in the process of **beta-oxidation**. (p. 111; Fig. 4-20)

Synthetic Pathways

28. In **gluconeogenesis**, amino acids and glycerol can be converted to glucose. (p. 112; Fig. 4-21)

29. Lipid and cholesterol synthesis require acetyl CoA. (p. 113; Fig. 4-22)

30. Protein synthesis is controlled by nuclear **genes** made of DNA. The code represented by the base sequence in a gene is transcribed into a complementary base code on **messenger RNA**. **Alternative splicing** of mRNA in the nucleus allows one gene to code for multiple proteins. (pp. 114–118; Fig. 4-26)

31. mRNA leaves the nucleus and goes to the cytosol where, with the assistance of **transfer RNA** and **ribosomal RNA**, it assembles amino acids into a designated sequence. This process is called **translation**. (pp. 118–119; Fig. 4-27)

32. **Post-translational modification** converts the newly synthesized protein to its finished form. (p. 115)

33. Proteins may be modified in the rough endoplasmic reticulum or in the Golgi complex. In the Golgi complex, they are packaged into membrane-bound vesicles that become lysosomes, storage vesicles, or secretory vesicles. (p. 122; Fig. 4-28)

QUESTIONS

(Answers to the Review Questions begin on page A1.)

➤ **THE PHYSIOLOGY PLACE**

Access more review material online at **The Physiology Place** website. There you'll find review questions, problem-solving activities, case studies, flashcards, and direct links to both *InterActive Physiology®* and *PhysioEx™*. To access the site, go to *www.physiologyplace.com* and select *Human Physiology, Fourth Edition.*

LEVEL ONE REVIEWING FACTS AND TERMS

1. List the three basic forms of work and give a physiological example of each.

2. Explain the difference between potential energy and kinetic energy.

3. State the two laws of thermodynamics in your own words.

4. The sum of all chemical processes through which cells obtain and store energy is called ___Metabolism___.

5. In the reaction $CO_2 + H_2O \rightleftharpoons H_2CO_3$, water and carbon dioxide are the ___reactants___ and H_2CO_3 is the product. Because this reaction is catalyzed by an enzyme, it is also appropriate to call water and carbon dioxide ___substrate___. The speed at which this reaction occurs is called the reaction ___rate___, often expressed as molarity/second.

6. ___Enzymes___ are protein molecules that speed up chemical reactions by _____ (increasing or (decreasing?)) the activation energy of the reaction.

7. Match each definition in the left column with the correct term from the right column (you will not use all the terms):

   1. reaction that can run **D** in either direction
   2. reaction that releases energy **A**
   3. ability of an enzyme to **F** catalyze one reaction but not another
   4. boost of energy needed to get a reaction started **C**

   (a) exergonic
   (b) endergonic
   (c) activation energy
   (d) reversible
   (e) irreversible
   (f) specificity
   (g) free energy
   (h) saturation

8. Since 1972, enzymes have been designated by adding the suffix ___ase -___ to their name.

9. Organic molecules that must be present in order for an enzyme to function are called ___coenzymes___. The precursors of these organic molecules come from ___vitamins___ in our diet.

10. In an oxidation-reduction reaction, in which electrons are moved between molecules, the molecule that gains an electron is said to be ___reduced___, and the one that loses an electron is said to be ___oxidized___.

11. The removal of H_2O from reacting molecules is called ___dehydration___. Using H_2O to break down polymers, such as starch, is called ___hydrolysis___.

12. The removal of an amino group ($-NH_2$) from a molecule (such as an amino acid) is called ___deamination___. Transfer of an amino group from one molecule to the carbon skeleton of another molecule (to form a different amino acid) is called ___transamination___. What happens to the amino group removed from an amino acid?

13. In metabolism, ___catabolic___ reactions release energy and result in the breakdown of large biomolecules, and ___anabolic___ reactions require a net input of energy and result in the synthesis of large biomolecules. In what units do we measure the energy of metabolism?

14. Metabolic regulation in which the last product of a metabolic pathway (the end product) accumulates and slows or stops reactions earlier in the pathway is called ___inhibition___.

15. Explain how H^+ movement across the inner mitochondrial membrane results in ATP synthesis.

16. List the two carrier molecules that deliver high-energy electrons to the electron transport system.

17. The breakdown of lipids for energy is termed ___lipolysis___. The long chains of fatty acids are broken into two-carbon acetyl CoA molecules by a pathway called ___beta-oxidation___. Which pathway do these acetyl CoA molecules join in order to produce ATP?

18. How many kilocalories per gram are stored in carbohydrates, in proteins, and in lipids? ___carbs & proteins = 4 per gram Fat = 9___

19. Which cellular organelles are involved in lipid synthesis? ___smooth eR___

LEVEL TWO REVIEWING CONCEPTS

20. Create maps using the following terms.

Map 1: Metabolism	Map 2: Protein synthesis
acetyl CoA	alternative splicing
amino acids	base pairing
ATP	DNA
citric acid cycle	exon
CO_2	gene
cytosol	intron
electron transport system	mRNA
$FADH_2$	mRNA processing
fatty acids	bases (A, C, G, T, U)
gluconeogenesis	promoter
glucose	RNA polymerase
glycerol	sense strand
glycogen	start codon
glycolysis	stop codon
high-energy electrons	transcription
lactate	transcription factors
mitochondria	
NADH	
oxygen	
pyruvate	
water	

21. When bonds are broken during a chemical reaction, what are the three possible fates for the potential energy found in those bonds?

22. Match each metabolic process in the left column with the theme or themes from the right column that best describes the process:

   1. Glycolysis takes place in the cytosol; oxidative phosphorylation takes place in mitochondria.
   2. The electron transport system traps energy in a hydrogen ion concentration gradient.
   3. Proteins are modified in the endoplasmic reticulum.

   (a) Biological energy use
   (b) Compartmentation
   (c) Molecular interactions

(list continues next page)

4

4. Metabolic reactions are often coupled to the reaction ATP → ADP + P_i.

5. Some proteins have S—S bonds between nonadjacent amino acids.

6. Enzymes catalyze biological reactions.

23. Explain why it is advantageous for a cell to store or secrete an enzyme in an inactive form.

24. Compare the energy yields from the aerobic breakdown of one glucose to CO_2 and H_2O with the yield from one glucose going through anaerobic glycolysis ending with lactate. What are the advantages of each pathway?

25. What is the advantage of converting glycogen to glucose 6-phosphate rather than to glucose?

26. Briefly describe the processes of transcription and translation. Which organelles are involved in each process?

27. On what molecule does the anticodon appear? Explain the role of this molecule in protein synthesis.

28. Is the energy of ATP's phosphate bond an example of potential or kinetic energy?

29. If ATP releases energy to drive a chemical reaction, would you suspect the activation energy of that reaction to be large or small? Explain.

LEVEL THREE   PROBLEM SOLVING

30. Transporting epithelial cells in the intestine do not use glucose as their primary energy source. What molecules might they use instead?

31. Given the following strand of DNA, list the entire sequence of bases that would appear in the matching mRNA. For the underlined triplets, underline the corresponding mRNA codon. Then give the appropriate tRNA anticodons and the amino acids for which those four triplets are the code.

DNA: CGCTACAAGTCAGGTACCGTAACG

mRNA:

Anticodons:

Amino acids:

LEVEL FOUR   QUANTITATIVE PROBLEMS

32. The graph shows the free energy change for the reaction A + B → D. Is this an endergonic or exergonic reaction?

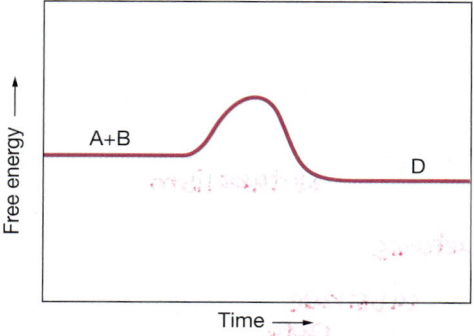

33. If the protein-coding portion of a piece of processed mRNA is 450 bases long, how many amino acids will be in the corresponding polypeptide? (*Hint:* the start codon is translated into an amino acid but the stop codon is not.)

ANSWERS

✓ Answers to Concept Check Questions

Page 91

1. Amino acids and nucleotides always contain nitrogen.

Page 93

2. Energy in the body is stored in chemical bonds and in concentration gradients.

3. Kinetic energy is the energy of motion: something is happening. Potential energy is stored energy: something is waiting to happen.

4. Entropy is a state of randomness or disorder.

Page 96

5. Endergonic reactions consume energy; exergonic reactions release energy.

6. The reactants are baking soda and vinegar; the product is carbon dioxide.

7. The foaming indicates that energy is being released, and so this is an exergonic reaction. The large amount of energy released indicates that the reaction is not readily reversible.

Page 98

8. The presence of isozymes enables one reaction to be catalyzed under a variety of conditions.

9. The four protein chains represent the quaternary level of protein structure.

Page 99

10. The carbonic acid concentration increases.

Page 101

11. The substrates are lactose (lactase), peptides (peptidase), lipids (lipase), and sucrose (sucrase).

12. (a) 3; (b) 2; (c) 4; (d) 1.

Page 103

13. Cells regulate substrate movement by (1) controlling the amount of enzyme, (2) producing allosteric and covalent modulators, (3) using two different enzymes to catalyze reversible reactions, (4) isolating enzymes within intracellular organelles, and (5) altering the ratio of ADP to ATP in the cell.

14. In ATP, energy is trapped and stored in one of the three phosphate bonds. In NADH, energy is stored in high-energy electrons.

15. Aerobic pathways require sufficient quantities of molecular oxygen (O_2) in the cell. Anaerobic pathways can proceed without oxygen.

Page 104

16. (a) 4; (b) 2, 5; (c) 2, 5; (d) 1, 3

Page 106

17. Endergonic reactions trap energy in the products.

18. Glycolysis is a net endergonic pathway.

Page 107

19. (a) Obviously exergonic reactions release energy that can be trapped by ATP or NADH: glyceraldehyde 3-phosphate to 1,3-bisphosphoglycerate; 1,3-bisphosphoglycerate to 3-phosphoglycerate; phosphoenol pyruvate to pyruvate.

(b) Obviously endergonic reactions require energy input from ATP: glucose to glucose 6-phosphate; fructose 6-phosphate to fructose 1,6-bisphosphate.

(c) Kinases add a phosphate group: glucose to glucose 6-phosphate; fructose 6-phosphate to fructose 1,6-bisphosphate; glyceraldehyde 3-phosphate to 1,3-bisphosphoglycerate.

(d) Dehydrogenases remove an electron and a hydrogen atom. In the step from glyceraldehyde 3-phosphate to 1,3-bisphosphoglycerate, NAD^+ acquires an electron and an H, suggesting that this step is catalyzed by a dehydrogenase.

(e) Dehydration reactions produce water as one of the products: 2-phosphoglycerate to phosphoenol pyruvate.

20. Lactate dehydrogenase acts on lactate by *removing* an *electron* and a *hydrogen atom*. This process is called *oxidation*.

Page 109

21. Phosphorylation is the addition of a phosphate group.

22. Electron transfer is exergonic.

Page 110

23. Anaerobic metabolism of glucose can proceed in the absence of oxygen; aerobic metabolism requires oxygen. Anaerobic metabolism produces two ATP per glucose; aerobic metabolism has a maximum yield of 30–32 ATP.

24. Oxygen acts as an acceptor of electrons and hydrogen ions at the termination point of the electron transport system.

25. According to the chemiosmotic theory, energy is trapped in a hydrogen ion concentration gradient and selectively released as the ions travel across the membrane separating the mitochondrial matrix from the intermembrane space.

Page 112

26. The digestion is more likely a hydrolysis reaction because large molecules are split into smaller molecules by the addition of water.

27. Deaminases remove amino acids.

Page 113

28. Hexokinase adds a phosphate group; glucose 6-phosphatase removes a phosphate group.

Page 117

29. The DNA triplets are ATT, ATC, and ACT.

30. RNA polymerase makes polymers of RNA.

Page 118

31. During mRNA processing, base sequences called introns are cut out of the mRNA. The remaining segments, the exons, are spliced back together and provide the code for a protein.

Page 122

32. Removal of a phosphate group is dephosphorylation.

33. Three types of post-translational modification are cleavage, addition of groups, and cross-linking.

34. Hemoglobin is a tetramer because it contains four protein chains.

Answers to Figure and Graph Questions

Page 96

Fig. 4-6: The graph shows an endergonic reaction.

Page 97

Fig. 4-7: When pH decreases from 8 to 7.4, enzyme activity increases.

Page 111

Fig. 4-19: 1. Reaction (b) is a hydrolysis reaction, as is reaction (a). 2. Hydrolases add water.

Page 114

Fig. 4-22: Most fatty acids have an even number of carbons because they are made from two-carbon acyl units. Some fatty acids with odd numbers of carbons are found in marine organisms.

Answer to Running Problem Conclusion

Page 123

Question 3. Factors other than a defective gene that could alter enzyme levels include decreased protein synthesis or increased protein breakdown in the cell. Such changes could occur even though the gene was normal.

4

5

Controversy surrounds the exact arrangements of molecules in cell membranes.

—**Robert D. DeVoe**, *in Mountcastle's* Medical Physiology, 1974

Cells (green actin and blue DNA) taking up red microspheres by endocytosis.

Membrane Dynamics

BACKGROUND BASICS

RUNNING PROBLEM

CYSTIC FIBROSIS

Over 100 years ago, midwives performed an unusual test on the infants they delivered: the midwife would lick the infant's forehead. A salty taste meant that the child was destined to die of a mysterious disease that withered the flesh and robbed the breath. Today, a similar "sweat test" will be performed in a major hospital—this time with state-of-the-art techniques—on Daniel Biller, a 2-year-old with a history of weight loss and respiratory problems. The name of the mysterious disease? Cystic fibrosis.

| 129 | 139 | 151 | 157 | 161 | 168 |

In the 1960s a group of conspiracy theorists obtained a lock of Napoleon Bonaparte's hair and sent it for chemical analysis in an attempt to show that he had been poisoned to death. Somewhere in Manhattan's Little Italy, a group of friends savor a delicious dinner and joke about the garlic odor on their breath that will result. At first glance these two scenarios appear to have little in common, but in fact, Napoleon's hair and garlic breath are both demonstrations of how the human body works to maintain the balance that we call homeostasis.

MASS BALANCE AND HOMEOSTASIS

You learned in the previous chapter that the body is an open system that exchanges heat and materials with the outside environment. To maintain a state of homeostasis—a relatively constant internal environment—the body utilizes the principle of mass balance. The **law of mass balance** says that if the amount of a substance in the body is to remain constant, any gain must be offset by an equal loss (Fig. 5-1■). For example, to maintain constant body concentration, water gain from the external environment and from metabolism must be offset by water loss back to the external environment. Other substances whose concentrations are maintained through mass balance include oxygen and carbon dioxide, salts, and hydrogen ions (pH). The law of mass balance is summarized by the following equation:

Total amount (or **load**) of substance x in the body
= intake + production − excretion − metabolism

Although most substances enter the body from the outside environment, some (such as carbon dioxide) can be produced internally through metabolism (Fig. 5-2■). In general, intake of water and nutrients into the body comes as food and drink

absorbed through the intestine. Oxygen and other gases and volatile molecules enter through the lungs, and a few lipid-soluble chemicals make their way to the internal environment by penetrating the barrier of the skin [p. 83].

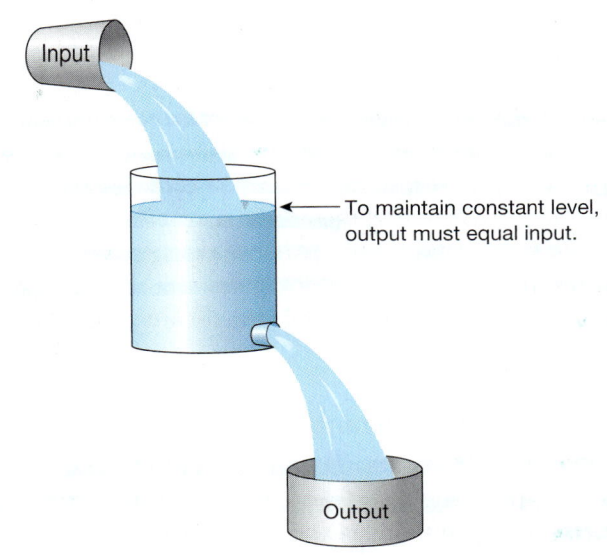

■ **FIGURE 5-1**   *Mass balance in an open system*

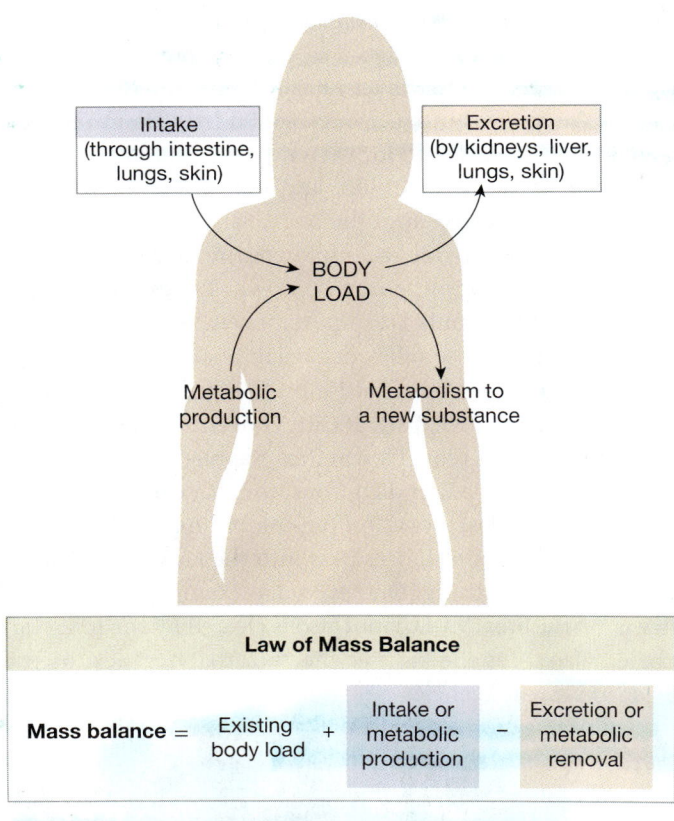

■ **FIGURE 5-2**   *Mass balance in the body*

Materials either enter the body by various routes or are produced through metabolism. Materials leave the body either by excretion or by metabolism.

To maintain mass balance, the body has two options for output. The simplest option is simply to excrete the material. **Excretion** is defined as the elimination of material from the body, and it usually takes place through the urine, feces, lungs, or skin. For example, carbon dioxide produced during metabolism is excreted by the lungs. Many *xenobiotics* [*xenos,* a foreigner] that enter the body, such as drugs or artificial food additives, are excreted by the liver and kidneys.

A second option to maintain mass balance is to metabolize the substance to a different substance. Nutrients that enter the body become the starting substrates in metabolic pathways, as you learned in Chapter 4. Metabolism converts the original nutrient to a different substance but in doing so creates a new mass balance disturbance by adding more of the *metabolite* to the body. (*Metabolite* is the general term for any product created in a metabolic pathway.)

Excretion Clears Substances from the Body

The rate at which a molecule disappears from the body by excretion, metabolism, or both is called the molecule's **clearance**. The quantitative expression of clearance will be introduced in Chapter 19, but it will be easier to understand that discussion if you already know that the kidney is only one organ of several that clear solutes from the body. The liver is the other major organ involved in clearing materials, especially xenobiotics. Hepatocytes [*hepaticus,* pertaining to the liver + *cyte,* cell], or liver cells, metabolize many different types of molecules, including hormones and drugs. The resulting metabolites may be secreted into the intestine for excretion in the feces or released into the blood for excretion by the kidneys.

Saliva, sweat, breast milk, and hair also contain solutes that have been cleared from the body. Salivary secretion of the hormone *cortisol* provides a simple noninvasive source of hormone for monitoring chronic stress. Drugs and alcohol excreted into breast milk are important because a breast-feeding infant will ingest these substances. Hair analysis can be used to test for arsenic in the body, and the 1960s analysis of Napoleon Bonaparte's hair showed significant concentrations of the poison. (The question remains whether Napoleon was murdered, accidentally poisoned, or died from stomach cancer.)

The lungs clear volatile lipid-soluble materials from the blood when these substances pass into the airways and are expelled during breathing. One everyday example of lung clearance is "garlic breath." Ethanol also is cleared by the lungs, and exhaled alcohol is the basis of the "breathalyzer" test used by law enforcement agencies.

Clearance is usually expressed as a volume of blood plasma *cleared* of substance *x* per unit of time. It therefore is only an indirect measure of the movement of substance *x*. A more direct measure is **mass flow**, defined as the rate of intake, production, or output of *x*:

$$\frac{\text{Mass flow}}{(\text{amount } x/\text{min})} = \frac{\text{concentration}}{(\text{amount } x/\text{vol})} \times \frac{\text{volume flow}}{(\text{vol/min})}$$

As an example, suppose a person is given an intravenous infusion of glucose solution that has a concentration of 50 grams of glucose per liter. If the infusion is given at a rate of 2 milliliters per minute, the mass flow of glucose into the body is:

50 g glucose/1000 mL solution $\times$ 2 mL solution/min

$$= 0.1 \text{ g glucose/min}$$

Mass flow applies not only to the entry, production, and removal of substances but also to the movement of substances from one compartment in the body to another. Recall from Chapter 3 [p. 52] that we can divide the body into two major compartments: the extracellular fluid and the intracellular fluid. Materials entering the body pass transiently through an epithelium, then become part of the extracellular fluid. How a substance is distributed after that depends on whether or not it crosses the barrier of the cell membrane and enters the cells.

CONCEPT CHECK

1. If a person eats 12 milligrams of salt in a day and excretes 11 milligrams of it in the urine, what happened to the remaining 1 milligram?

2. Glucose is aerobically metabolized to CO_2 and water. Explain the effect of this glucose metabolism on mass balance in the body.
   Answers: p. 172

Homeostasis Does Not Mean Equilibrium

When physiologists talk about homeostasis, they are often speaking of the stability of the body's *internal environment*—in other words, the stability of the extracellular fluid. One reason for this is that clinically we are able to monitor the composition of extracellular fluid by the simple act of withdrawing a blood sample and analyzing its fluid matrix, the plasma. It is much more difficult to follow what is taking place inside cells, although cells do maintain *cellular homeostasis*. However, although the composition of both body compartments is relative stable, individual solutes in the two compartments are not in equilibrium. Instead, the extracellular and intracellular fluid compartments usually exist in a state that might best be called a *dynamic disequilibrium*.

Water is essentially the only molecule that moves freely between most cells and the extracellular fluid. Because of this free movement of water, the extracellular and intracellular compartments can reach a state of **osmotic equilibrium** [*osmos,* push or thrust], in which the total amount of solute per volume of fluid is equal on the two sides of the cell membrane. At the same time, however, the body is in a state of **chemical disequilibrium**, in which the major solutes are more concentrated in one of the two body compartments than in the other (Figure 5-3 ■).

For example, sodium, chloride, and bicarbonate (HCO_3^-) ions are more concentrated in extracellular fluid than in intracellular fluid, whereas potassium ions are more concentrated inside the cell. Calcium (not shown in the figure) is more concentrated in the extracellular fluid than in the cytosol, although

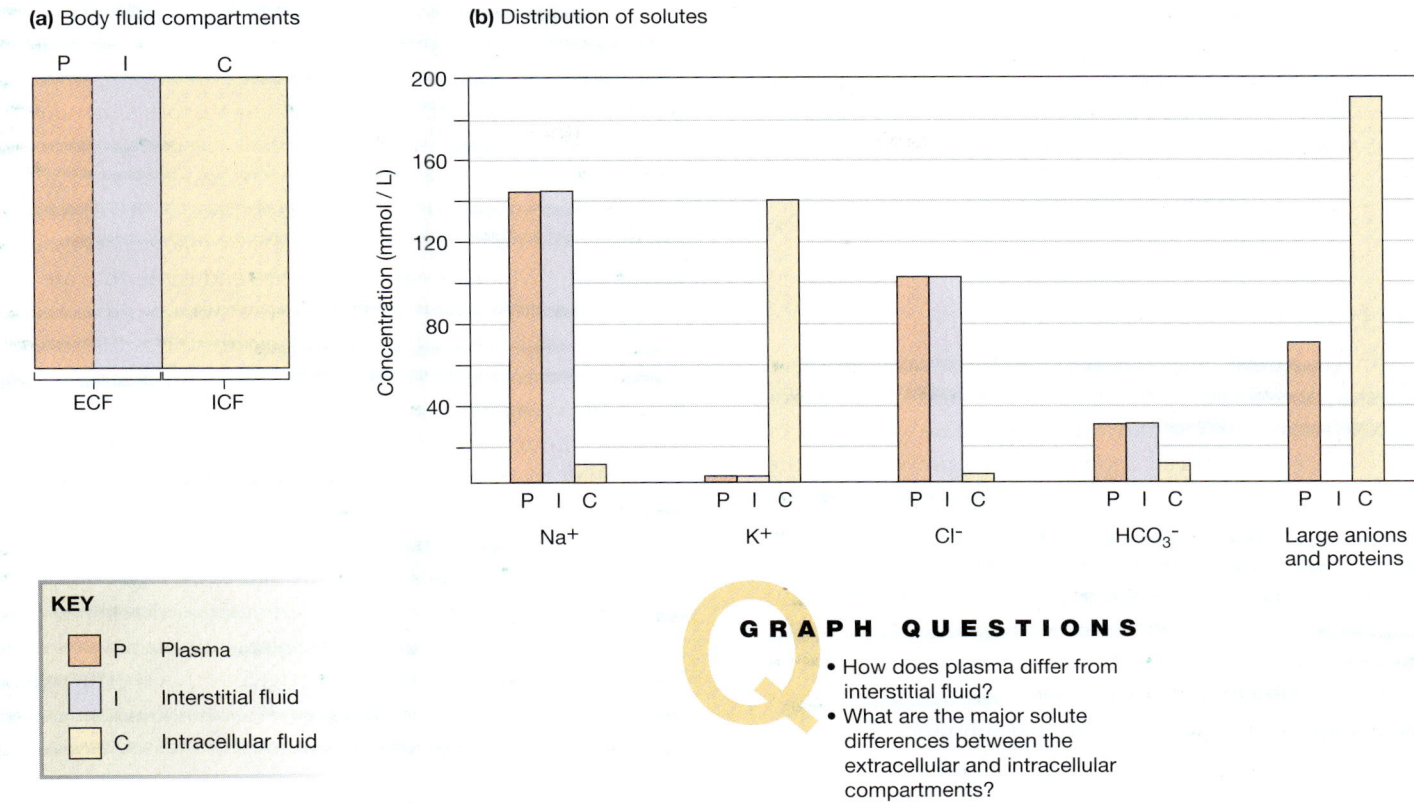

(a) Body fluid compartments

(b) Distribution of solutes

KEY

P	Plasma
I	Interstitial fluid
C	Intracellular fluid

GRAPH QUESTIONS

• How does plasma differ from interstitial fluid?
• What are the major solute differences between the extracellular and intracellular compartments?

■ **FIGURE 5-3** *Distribution of solutes in the body fluid compartments*

The extracellular fluid is composed of the blood plasma plus interstitial fluid. The compartments of the body are in a state of chemical disequilibrium.

many cells store Ca^{2+} inside organelles such as the endoplasmic reticulum and mitochondria.

Even the extracellular fluid is not at equilibrium between its two subcompartments [⇄ Fig. 3-2, p. 52]. A few solutes are more concentrated in the plasma than in the interstitial fluid. Figure 5-3 shows that proteins and other large anions are found both in intracellular fluid and in the plasma but are almost absent from the interstitial fluid.

The concentration differences of chemical disequilibrium are a hallmark of a living organism, as only the continual input of energy keeps the body in this state. If solutes leak across the cell membrane dividing the intracellular and extracellular compartments, energy is required to return them to the compartments they left. For example, K^+ that leak out of the cell and Na^+ that leak into the cell are returned to their original compartments by an energy-utilizing enzyme known as the Na^+-K^+-ATPase. When cells die and cannot use energy, they obey the second law of thermodynamics [⇄ p. 93] and return to a state of randomness that is marked by loss of chemical disequilibrium.

Many body solutes mentioned so far are ions, and therefore we must also consider the distribution of electrical charge between the intracellular and extracellular compartments. Although the body as a whole is electrically neutral, a few extra negative ions are found in the intracellular fluid, while their matching positive ions are located in the extracellular fluid. As a result, the inside of cells is slightly negative relative to the extracellular fluid. This electrical difference creates a state of **electrical disequilibrium** in the body, a topic discussed in more detail later in this chapter.

Thus, homeostasis is not the same as equilibrium. The intracellular and extracellular compartments of the body may be in osmotic equilibrium, but they are also in chemical and electrical disequilibrium. Furthermore, osmotic equilibrium and the two disequilibria are dynamic *steady states*. The modifier *dynamic* indicates that materials are constantly moving back and forth between the two compartments, but in a *steady state*, there is no *net* movement of materials between the compartments.

In the remainder of this chapter, we will discuss how the selective permeability of cell membranes is responsible for a body in which the intracellular and extracellular compartments are chemically and electrically different but have the same total concentration of solutes.

CONCEPT CHECK

3. Using what you learned about the naming conventions for enzymes [➢ p. 99], explain what the name *Na⁺-K⁺-ATPase* tells you about this enzyme's actions.

4. The intracellular fluid can be distinguished from the extracellular fluid by the ICF's high concentration of _____ ion and low concentration of _____, _____, and _____ ions.

Answers: p. 172

DIFFUSION

Although many materials move freely within a body compartment, exchange between the intracellular and extracellular compartments is restricted by the cell membrane. Whether or not a substance enters a cell depends on the properties of the cell membrane and those of the substance. Cell membranes are **selectively permeable**; that is, the lipid and protein composition of a given cell membrane determines which molecules will enter the cell and which will leave [➢ p. 55]. If a membrane allows a substance to pass through it, the membrane is said to be **permeable** to that substance [*permeare*, to pass through]. If a membrane does not allow a substance to pass, the membrane is said to be **impermeable** [*im-*, not] to that substance.

Membrane permeability is variable and can be changed by altering the proteins or lipids of the membrane. Some molecules, such as water, oxygen, carbon dioxide, and lipids, move easily across most cell membranes. On the other hand, ions, most polar molecules, and very large molecules (such as proteins), enter cells with more difficulty or may not enter at all.

Two properties of a molecule influence its movement across cell membranes: the size of the molecule and its lipid solubility [➢ p. 24]. Very small molecules and those that are lipid soluble can cross directly through the phospholipid bilayer. Larger or less lipid-soluble molecules are excluded from crossing the bilayer unless the cell has a specific mechanism for transporting them across. For these molecules, membrane proteins are the usual mediators cells use to transport molecules across their membranes. Very large lipophobic molecules must enter and leave cells in vesicles [➢ p. 66].

There are two ways to categorize how molecules move across membranes. One scheme, just described, separates movement according to physical requirements: whether it takes place through the phospholipid bilayer, with the aid of a membrane protein, or by using vesicles (Fig. 5-4 ■). A second scheme classifies movement according to its energy requirements. **Passive transport** does not require the input of energy. **Active**

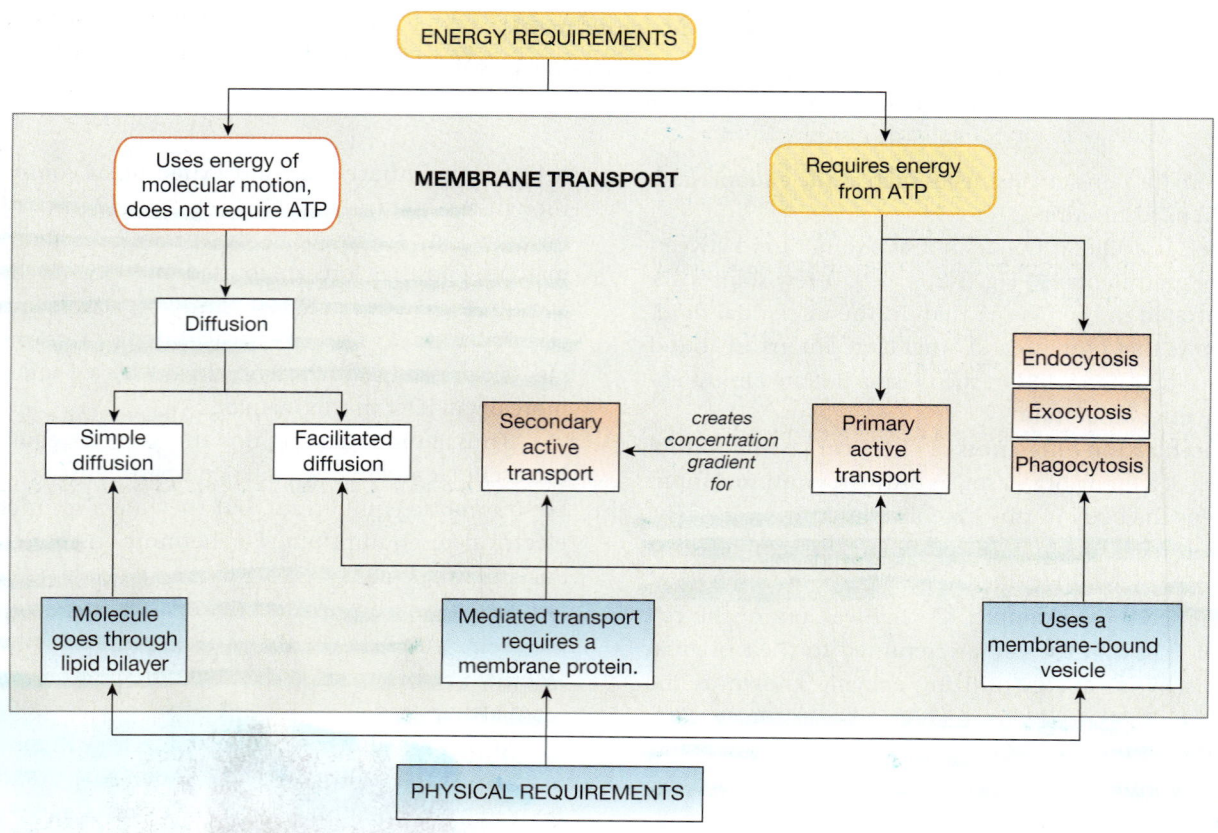

■ **FIGURE 5-4** *Map of membrane transport*

Movement of substances across cell membranes can be classified either by the energy requirements of transport (top part of map) or according to whether transport uses diffusion, a membrane protein, or a vesicle (bottom part of map).

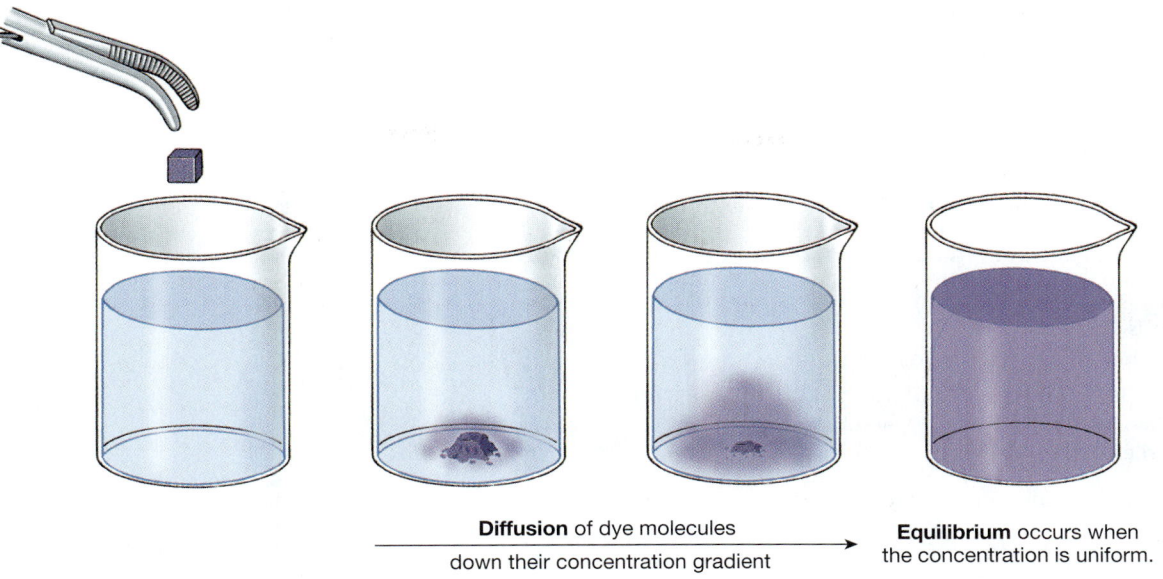

Diffusion of dye molecules →
down their concentration gradient

Equilibrium occurs when the concentration is uniform.

■ **FIGURE 5-5**   *Diffusion*

When a crystal of dye (such as potassium permanganate) is placed in water, the crystal dissolves, and dye molecules spread outward by diffusion.

transport requires the input of energy from some outside source, such as the high-energy phosphate bond of ATP.

Diffusion Uses Only the Energy of Molecular Movement

Passive transport across membranes uses the kinetic energy [🗲 p. 92] inherent in molecules. Gas molecules and molecules in solution constantly move from one place to another, bouncing off other molecules or off the sides of any container holding them. When molecules start out concentrated in one area of an enclosed space, their motion causes them to spread out gradually until they are evenly distributed throughout the available space. This process is known as diffusion.

Diffusion [*diffundere*, to pour out] may be defined as the movement of molecules from an area of higher concentration of the molecules to an area of lower concentration of the molecules.* If you leave a bottle of cologne open and later notice its fragrance across the room, it is because the aromatic molecules in the cologne have diffused from where they are more concentrated (in the bottle) to where they are less concentrated (across the room).

Diffusion has the following seven properties:

1. *Diffusion is a passive process.* By *passive,* we mean that diffusion does not require the input of energy from some

*Some texts use the term *diffusion* to mean any random movement of molecules. These texts call molecular movement along a concentration gradient *net diffusion.* To simplify matters, we will use the term *diffusion* to mean movement down a concentration gradient.

outside source. Diffusion uses only the kinetic energy possessed by all molecules.

2. *Molecules move from an area of higher concentration to an area of lower concentration.* A difference in the concentration of a substance between two places is called a concentration gradient [🗲 p. 91], also known as a **chemical gradient**. We say that molecules diffuse *down the gradient,* from higher concentration to lower concentration (Fig. 5-5 ■). The rate of diffusion depends on the magnitude of the concentration gradient. The larger the concentration difference, the faster diffusion takes place. For example, when you open a bottle of ammonia, the rate of diffusion is most rapid as the ammonia molecules first escape from the bottle into the air. Later, when the ammonia has spread evenly throughout the room, the rate of diffusion has dropped to zero because there is no longer a concentration gradient.

3. *Net movement of molecules occurs until the concentration is equal everywhere.* Once molecules of a given substance have distributed themselves evenly, the system reaches equilibrium and diffusion stops. Individual molecules are still moving at equilibrium, but for each molecule that exits an area, another one enters. The *dynamic* equilibrium state in diffusion means that the concentration has equalized throughout the system but molecules continue to move.

4. *Diffusion is rapid over short distances but much slower over long distances.* Albert Einstein studied the diffusion of molecules in solution and found that the time required for a molecule to diffuse from point A to point B is proportional to the square of the distance from A to B. In other words, if the distance doubles from 1 to 2, the time needed for diffusion increases from 1^2 to 2^2.

TABLE 5-1	Rules for Diffusion of Uncharged Molecules

General Properties of Diffusion

1. Diffusion uses the kinetic energy of molecular movement and does not require an outside energy source.

2. Molecules diffuse from an area of higher concentration to an area of lower concentration.

3. Diffusion continues until concentrations come to equilibrium. Molecular movement continues, however, after equilibrium has been reached.

4. Diffusion is faster
   - with higher concentration gradients.
   - over shorter distances.
   - at higher temperatures.
   - for smaller molecules.

5. Diffusion can take place in an open system or across a partition that separates two systems.

Simple Diffusion Across a Membrane

6. The rate of diffusion through a membrane is faster if
   - the membrane's surface area is larger.
   - the membrane is thinner.
   - the concentration gradient is larger.
   - the membrane is more permeable to the molecule.

7. Membrane permeability to a molecule depends on
   - the molecule's lipid solubility.
   - the molecule's size.
   - the lipid composition of the membrane.

What does the slow rate of diffusion over long distances mean for biological systems? In humans, nutrients take five seconds to diffuse from the blood to a cell that is 100 μm from the nearest capillary. At that rate, it would take years for nutrients to diffuse from the small intestine to cells in the big toe, and the cells would starve to death. To overcome the limitations of diffusion over distance, organisms have developed various transport mechanisms that speed up the movement of molecules. Most multicellular organisms have some form of circulatory system to bring oxygen and nutrients rapidly from the point at which they enter the body to the cells.

5. *Diffusion is directly related to temperature*. At higher temperatures, molecules move faster. Because diffusion is a result of molecular movement, the rate of diffusion increases as temperature increases. Generally, changes in temperature do not significantly affect diffusion rates in humans because we maintain a relatively constant body temperature.

6. *Diffusion rate is inversely related to molecular size*. Einstein showed that friction between the surface of a particle and the medium through which it diffuses is a source of resistance to movement. He calculated that diffusion is inversely proportional to the radius of the molecule: the larger the molecule, the slower its diffusion through a given medium.

7. *Diffusion can take place in an open system or across a partition that separates two systems*. Diffusion of ammonia or cologne within a room is an example of diffusion taking place in an open system. There are no barriers to molecular movement, and the molecules spread out to fill the entire system. Diffusion can also take place between two systems, such as the intracellular and extracellular compartments, but only if the partition dividing the two compartments allows the diffusing molecules to cross.

For example, if you close the top of an open bottle of ammonia, the ammonia molecules cannot diffuse out into the room because neither the bottle nor the cap is permeable to the ammonia. However, if you replace the metal cap with a plastic bag that has tiny holes in it, you will begin to smell the ammonia in the room because the bag is permeable to the ammonia. Similarly, if a cell membrane is permeable to a molecule, that molecule can enter or leave the cell by diffusion. If the membrane is not permeable to that particular molecule, the molecule cannot cross. Table 5-1 ■ summarizes these points.

An important point to note: ions do not move by diffusion, even though you will read and hear about ions *diffusing across membranes*. Diffusion is random molecular motion down a *concentration* gradient. Ion movement is influenced by *electrical* gradients because of the attraction of opposite charges and repulsion of like charges. Thus, ions move in response to combined electrical and concentration gradients, or *electrochemical gradients*. This electrochemical movement is a more complex process than diffusion resulting solely from a concentration gradient, and the two processes should not be confused. Ions and electrochemical gradients will be discussed in more detail at the end of this chapter.

In summary, diffusion is the passive movement of uncharged molecules down their concentration gradient due to random molecular movement. Diffusion is slower over long distances and slower for large molecules. When the concentration of the diffusing molecules is the same throughout a system, the system has come to chemical equilibrium, although the random movement of molecules continues.

CONCEPT CHECK

5. If the distance over which a molecule must diffuse doubles from 1 to 2, diffusion takes how many times as long?
   Answers: p. 172

Lipophilic Molecules Can Diffuse Through the Phospholipid Bilayer

Diffusion across membranes is a little more complicated than diffusion in an open system. Water is the primary solvent of the body, and many vital nutrients, ions, and other molecules dissolve in water because of its polar nature. However, substances that are hydrophilic and dissolve in water are lipo*phobic* as a rule: they do not readily dissolve in lipids. For this reason, the hydrophobic lipid core of the cell membrane acts as a barrier that prevents hydrophilic molecules from crossing.

Substances that can pass through the lipid center of a membrane move by diffusion. Diffusion directly across the phospholipid bilayer of a membrane is called **simple diffusion** and has the following properties in addition to the properties of diffusion listed earlier.

1. *The rate of diffusion depends on the ability of the diffusing molecule to dissolve in the lipid layer of the membrane.* Another way to say this is that the diffusion rate depends on how permeable the membrane is to the diffusing molecules. Most molecules in solution can mingle with the polar phosphate-glycerol heads of the bilayer, but only nonpolar molecules that are lipid-soluble (lipophilic) can traverse the central lipid core of the membrane. As a rule, only lipids, steroids, and small lipophilic molecules can move across the membrane by simple diffusion.

   One important exception to this statement concerns water. Water, although a polar molecule, may diffuse slowly across some phospholipid membranes. For years it

was thought that the polar nature of the water molecule prevented it from moving through the lipid center of the bilayer, but experiments done with artificial membranes have shown that the small size of the water molecule allows it to slip between the lipid tails in some membranes. How readily water passes through the membrane depends on the composition of the phospholipid bilayer. Membranes with a high cholesterol content are less permeable to water than those with a low cholesterol content, presumably because the cholesterol molecules fill the spaces between the fatty acid tails of the lipid bilayer and thus exclude water. For example, the cell membranes of some sections of the kidney are essentially impermeable to water unless the cells insert special water channel proteins into the phospholipid bilayer.

2. *The rate of diffusion across a membrane is directly proportional to the surface area of the membrane.* In other words, the larger the membrane's surface area, the more molecules can diffuse across per unit time. This fact may seem obvious, but it has important implications in physiology. One striking example of how a change in surface area affects diffusion is the lung disease emphysema. As lung tissue is destroyed, the surface area available for diffusion of oxygen decreases. Consequently, less oxygen can move into the body. In severe cases, the oxygen that reaches the cells is not enough to sustain any muscular activity and the patient is confined to bed.

3. *The rate of diffusion across a membrane is inversely proportional to the thickness of the membrane.* The thicker the membrane, the slower the rate at which diffusion takes place. For most biological membranes, thickness is essentially constant. However, diffusion distance comes into play in certain lung conditions in which the exchange epithelium of the lung is thickened with scar tissue. This slows diffusion so that the oxygen entering the body is not adequate to meet metabolic needs.

The rules for simple diffusion across membranes are summarized in Table 5-1. They can be combined mathematically into an equation known as **Fick's law of diffusion**, a relationship that involves the three factors just mentioned for membrane diffusion plus the factor of concentration gradient from our earlier general discussion of diffusion. In an abbreviated form, Fick's law says that:

$$\text{rate of diffusion} \propto \frac{\text{surface area} \times \text{concentration gradient} \times \text{membrane permeability}}{\text{membrane thickness}}$$

Figure 5-6 ■ illustrates the principles of Fick's law.

Membrane permeability is the most complex of the four terms in Fick's law because several factors influence it: (1) the size of the diffusing molecule, (2) the lipid-solubility of the molecule, and (3) the composition of the lipid bilayer across

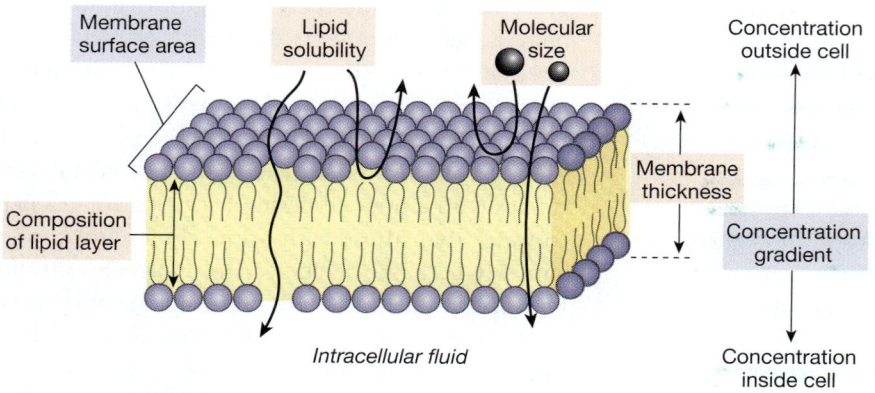

FIGURE 5-6 *Fick's law of diffusion*
This law mathematically relates factors that influence the rate of simple diffusion across a membrane.

which it is diffusing. As molecular size increases, membrane permeability decreases. As lipid solubility of the diffusing molecule increases, membrane permeability to the molecule increases. Alterations in lipid composition of the membrane change how easily diffusing molecules can slip between the individual phospholipids. For example, cholesterol molecules in membranes pack themselves into the spaces between the fatty acids tails and retard passage of molecules through those spaces [Fig. 3-4, p. 54], making the membrane less permeable.

In most physiological situations, membrane thickness is a constant. In that case, we can remove membrane thickness from our Fick's law equation and rearrange the equation to read:

$$\frac{\text{diffusion rate}}{\text{surface area}} = \frac{\text{concentration}}{\text{gradient}} \times \frac{\text{membrane}}{\text{permeability}}$$

This equation now describes the flux of a molecule across the membrane, because **flux** is defined as the diffusion rate per unit surface area of membrane:

flux = concentration gradient × membrane permeability

In other words, the flux of a molecule across a membrane depends on the concentration gradient and the membrane's permeability to the molecule.

One point to remember is that the principles of diffusion just discussed apply to all biological membranes, not just to the cell membrane. Movement of materials in and out of organelles follows the same rules.

CONCEPT CHECK

6. Where does the energy for diffusion come from?

7. Which is more likely to cross a cell membrane by simple diffusion: a fatty acid molecule or a glucose molecule?

8. What happens to the rate of diffusion in each of the following cases?

   (a) membrane thickness increases
   (b) concentration gradient increases
   (c) surface area decreases

9. Two compartments are separated by a membrane that is permeable only to water and to yellow dye molecules. Compartment A is filled with an aqueous solution of yellow dye, and compartment B is filled with an aqueous solution of an equal concentration of blue dye. If the system is left undisturbed for a long time, what color will compartment A be: yellow, blue, or green? (Remember, yellow plus blue makes green.) What color will compartment B be?

10. What keeps atmospheric oxygen from diffusing into our bodies across the skin? (*Hint:* what kind of epithelium is skin?)

   Answers: p. 172

PROTEIN-MEDIATED TRANSPORT

In the body, simple diffusion across membranes is limited to lipophilic molecules. The majority of molecules in the body are either lipophobic or electrically charged and therefore cannot

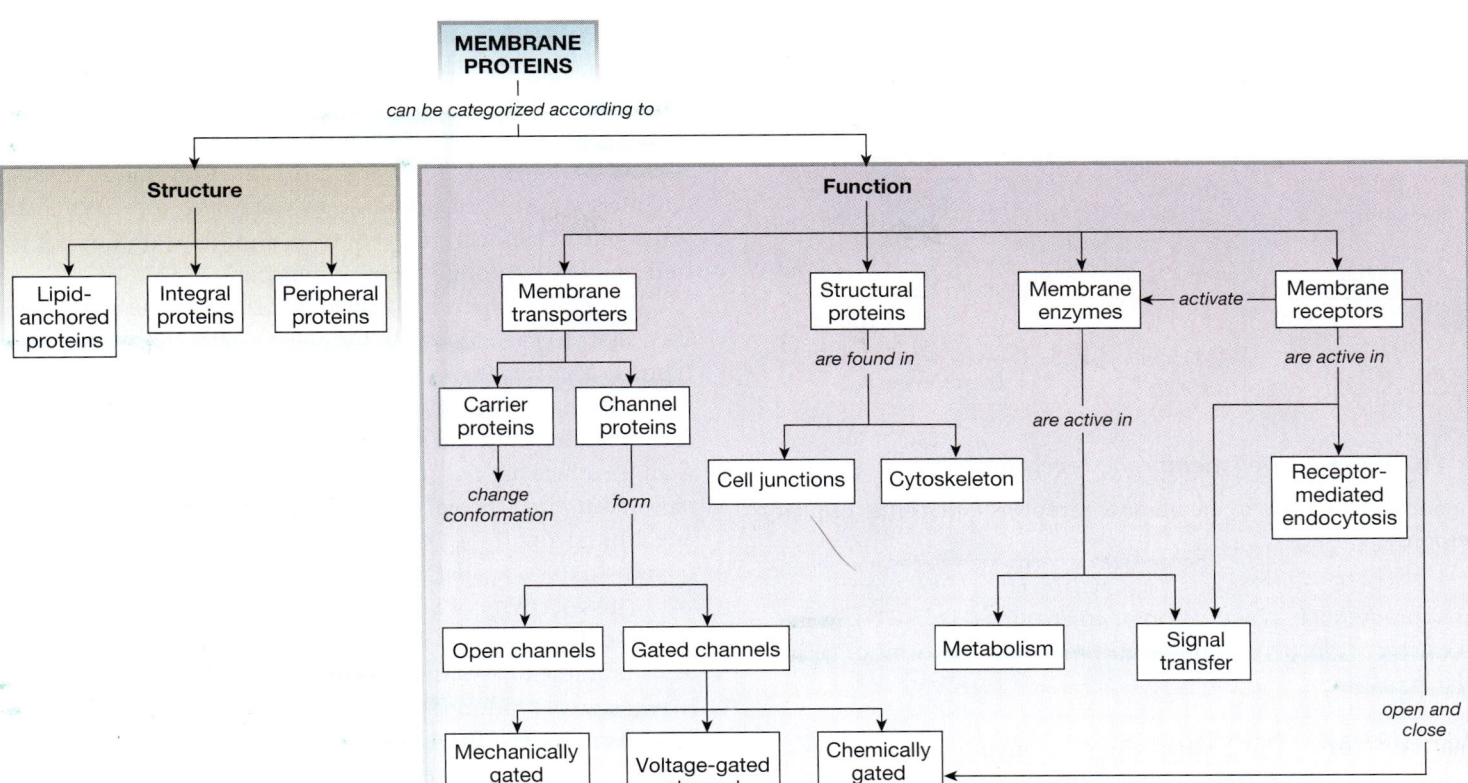

■ **FIGURE 5-7** *Map of membrane proteins*

Functional categories of membrane proteins include transporters, structural proteins, enzymes, and receptors.

cross membranes by simple diffusion. The vast majority of solutes cross membranes with the help of membrane proteins, a process we call **mediated transport**.

If mediated transport is passive and moves molecules down their concentration gradient, and if net transport stops when concentrations are equal on both sides of the membrane, the process is known as **facilitated diffusion**. If protein-mediated transport requires energy from ATP or another outside source and moves a substance against its concentration gradient, the process is known as **active transport**.

Membrane Proteins Function as Structural Proteins, Enzymes, Receptors, and Transporters

Protein-mediated transport across a membrane is carried out by two groups of membrane-spanning proteins: transporters and receptors. For physiologists, classifying membrane proteins by their function is more useful than classifying them by their structure. Our functional classification scheme recognizes four types of membrane proteins: (1) structural proteins, (2) enzymes, (3) receptors, and (4) transporters. Figure 5-7■ is a map comparing the structural and functional classifications of membrane proteins.

Structural Proteins The **structural proteins** have three major roles. The first is to connect the membrane to the cytoskeleton to maintain the shape of the cell. The microvilli of transporting epithelia [🔁 p. 73] are one example of membrane shaping by the cytoskeleton. The second role is to create cell junctions that hold tissues together, such as tight junctions and gap junctions [🔁 Fig. 3-21, p. 70]. Finally, the third role is to attach cells to the extracellular matrix by linking cytoskeleton fibers to extracellular collagen and other protein fibers [🔁 p. 69].

Enzymes Membrane enzymes catalyze chemical reactions that take place either on the cell's external surface or just inside the cell. For example, enzymes on the external surface of cells lining the small intestine are responsible for digesting peptides and carbohydrates. Enzymes attached to the intracellular surface of many cell membranes play an important role in transferring signals from the extracellular environment to the cytoplasm, as you will learn in Chapter 6.

Receptors Membrane receptor proteins are part of the body's chemical signaling system. The binding of a receptor with its ligand usually triggers another event at the

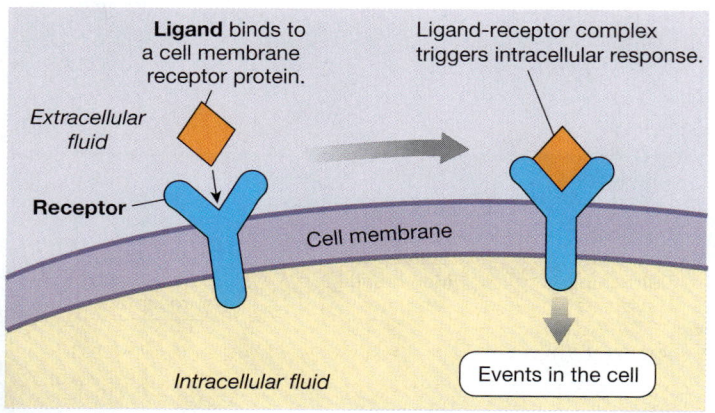

■ FIGURE 5-8 *Cell membrane receptors*

Ligand binding to a membrane receptor initiates a cellular response.

membrane, such as activation of an enzyme (Fig. 5-8 ■). Membrane receptors play an important role in some forms of vesicular transport.

Transporters   The fourth group of membrane proteins—**membrane transporters**—move molecules across membranes. We can further subdivide transporter proteins into two categories: channels and carriers. **Channel proteins** create water-filled passageways that directly link the intracellular and extracellular compartments (Fig. 5-9a ■). **Carrier proteins** bind to the substrates that they carry but never form a direct connection between the intracellular fluid and extracellular fluid. As Figure 5-9b shows, carriers are open to one side of the membrane or the other, but not to both at once the way channel proteins are.

Why do cells need both channels and carriers? The answer lies in the different properties of the two transporters. Channel proteins allow more rapid transport across the membrane but are not as selective about what they transport. Carriers, while slower, are better at discriminating between closely related molecules. Carriers can also move larger molecules than channels can.

Channel Proteins Form Open, Water-Filled Passageways

Channel proteins are made of membrane-spanning protein subunits that create a cluster of cylinders that surround a narrow, water-filled pore (Fig. 5-10 ■). Movement through channels is restricted primarily to water and ions. When the water-filled channels are open, tens of millions of ions per second can whisk through them unimpeded.

Channel proteins are named according to the substance(s) they allow to pass. Most cells have **water channels** made from a protein called *aquaporin*. In addition, more than 100 types of **ion channels** have been identified. Ion channels may be specific for one ion or may allow ions of similar size and charge to pass. For example, there are Na^+ channels, K^+ channels, and nonspecific *monovalent* ("one charge") cation channels that transport Na^+, K^+, and lithium ions (Li^+). Other ion channels you will encounter frequently in this text are Ca^{2+} channels and Cl^- channels.

The selectivity of a channel is determined by the diameter of its central pore and by the electrical charge of the amino acids that line the channel. If the channel amino acids are positively charged, positive ions will be repelled and negative ions will pass through the channel. On the other hand, a cation channel must have a negative charge that attracts cations but prevents the passage of Cl^- or other anions.

Channel proteins are like narrow doorways into the cell. If the door is closed, nothing can go through. If the door is open, there is a continuous passage between the two rooms connected by the doorway. The open or closed state of a channel is determined by regions of the protein molecule that act like swinging "gates."

According to current models, many channel gates are part of the cytoplasmic side of the membrane protein (Fig. 5-11 ■). Such a gate can be envisioned as a ball on a chain that swings up and blocks the mouth of the channel. A few channels have a gate in the middle of the protein, and one type of channel in nerve cells has two different gates.

Channels can be classified according to whether their gates are usually open or usually closed. **Open channels** spend most of their time with their gate open, allowing ions to move back

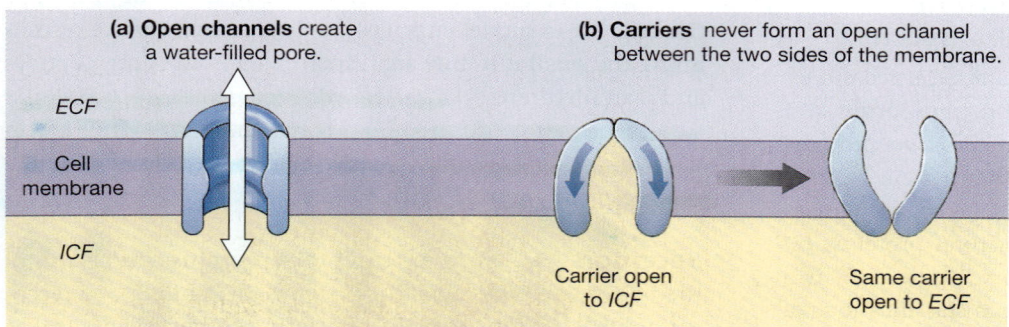

(a) **Open channels** create a water-filled pore.

ECF

Cell membrane

ICF

(b) **Carriers** never form an open channel between the two sides of the membrane.

Carrier open to *ICF*

Same carrier open to *ECF*

■ FIGURE 5-9 *Membrane transport proteins*

Many channels are made of multiple protein subunits that assemble in the membrane.

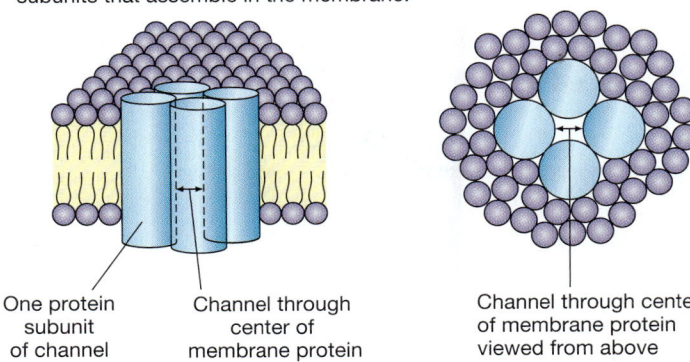

One protein subunit of channel   Channel through center of membrane protein   Channel through center of membrane protein viewed from above

■ **FIGURE 5-10**   *Structure of channel proteins*

Hydrophilic amino acids in the protein line the channel, creating a water-filled passage that allows ions and very small molecules, such as water, to pass through.

and forth across the membrane without regulation. These gates may occasionally flicker closed, but for the most part these channels behave as if they have no gates. Open channels are sometimes called either *leak channels* or *pores*, as in *water pores*.

Gated channels spend most of their time in a closed state, which allows these channels to regulate the movement of ions through them. When a gated channel opens, ions move through the channel just as they move through open channels. When a gated channel is closed, which it may be much of the time, it allows no ion movement between the intracellular and extracellular fluid.

What controls the opening and closing of gated channels? The gating can be controlled by intracellular messenger molecules or extracellular ligands (**chemically gated channels**), by the electrical state of the cell (**voltage-gated channels**), or by a physical change, such as increased temperature or a force that puts tension on the membrane and pops the channel open

Gated channels are usually closed. They open in response to chemical, mechanical, or electrical signals.

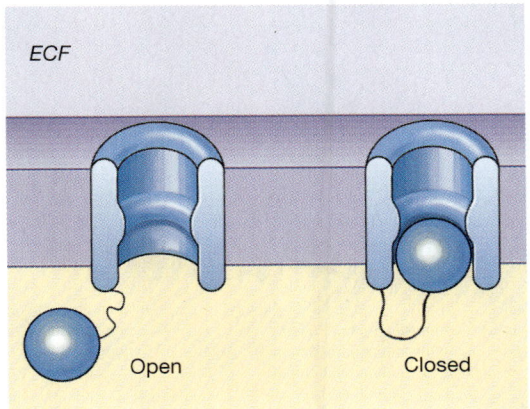

ECF

Open   Closed

■ **FIGURE 5-11**   *Gating of channel proteins*

RUNNING PROBLEM

Cystic fibrosis is a debilitating disease caused by a defect in a gated channel protein that normally transports chloride ions (Cl^-). The channel protein—called the cystic fibrosis transmembrane conductance regulator, or CFTR— is located in epithelia lining the airways, sweat glands, and pancreas. The opening and closing of this channel are regulated through the binding of nucleotides to specific regions of the channel. In people with cystic fibrosis, the CFTR is nonfunctional or absent. As a result, chloride transport across the epithelium is impaired.

Question 1:
   Is the CFTR a chemically gated, a voltage-gated, or a mechanically gated channel protein?

129 **139** 151 157 161 168

(**mechanically gated channels**). You will encounter many variations of these channel types as you study physiology.

CONCEPT CHECK

11. Positively charged ions are called _____, and negatively charged ions are called _____.

Answers: p. 172

Carrier Proteins Change Conformation to Move Molecules

The second type of transport protein is the carrier protein. Carrier proteins bind with specific substrates and carry them across the membrane by changing conformation. Small organic molecules (such as glucose and amino acids), which are too large to pass through channels, cross membranes using carriers. Ions such as Na^+ and K^+ may move by carriers as well as through channels.

Some carrier proteins move only one kind of molecule and are known as **uniport carriers** (Fig. 5-12a ■). However, it is common to find carriers that move two or even three kinds of molecules. A carrier that moves more than one kind of molecule at one time is called a **cotransporter**. If the molecules being transported are moving in the same direction, whether into or out of the cell, the carrier proteins are **symport carriers** [*sym-*, together + *portare*, to carry]. If the molecules are being carried in opposite directions, the carrier proteins are **antiport carriers** [*anti*, opposite + *portare*, to carry]. Symport and antiport carriers are shown in Figure 5-12b and c.

Carriers are large, complex proteins with multiple subunits. The conformation change required of a carrier protein makes this mode of transmembrane transport much slower than movement

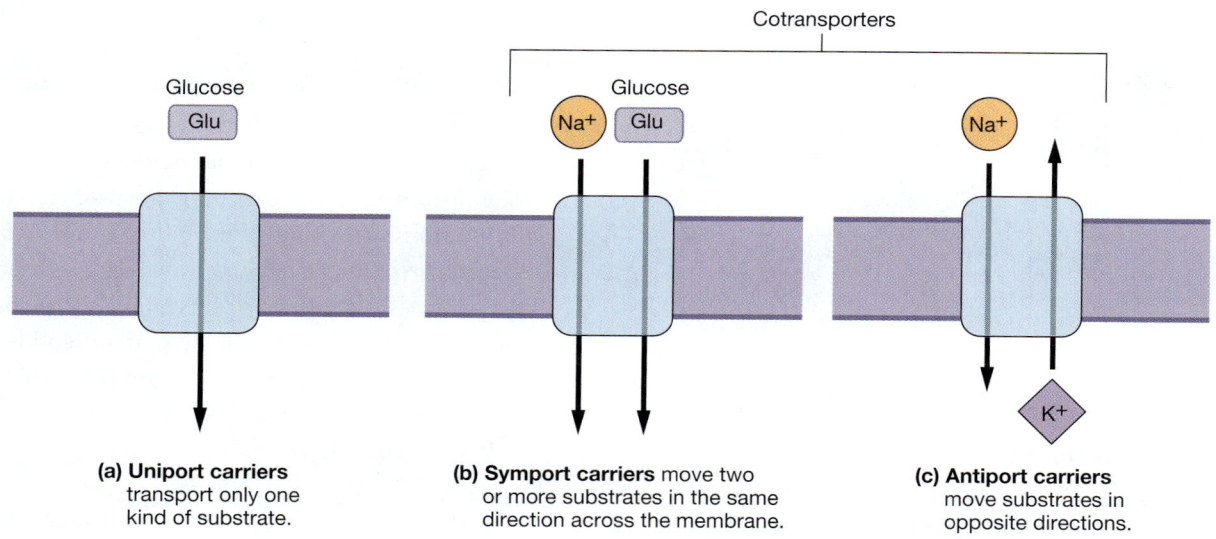

Cotransporters

Glucose

Glu

(a) Uniport carriers transport only one kind of substrate.

Glucose

Na⁺ Glu

(b) Symport carriers move two or more substrates in the same direction across the membrane.

Na⁺

K⁺

(c) Antiport carriers move substrates in opposite directions.

■ **FIGURE 5-12** *Types of carrier-mediated transport*

In the illustrations for this book, carrier proteins are represented by solid shapes in the cell membrane.

through channel proteins. Only 1000 to 1,000,000 molecules per second can be moved by a carrier protein, whereas tens of millions of ions per second move through a channel protein.

Carrier proteins differ from channel proteins in another way: carriers never create a continuous passage between the inside and outside of the cell. If channels are like doorways, then carriers are like revolving doors that allow movement between inside and outside without ever creating an open hole. Carrier proteins can transport molecules across a membrane in both directions, like a revolving door at a hotel, or they can restrict their transport to one direction, like the turnstile at an amusement park that allows you out of the park but not back in.

Carriers form passageways, but these openings are never open to both sides at the same time. One side is always closed, preventing free exchange across the membrane. In this respect, carrier proteins are like the Panama Canal (Fig. 5-13a ■). Picture

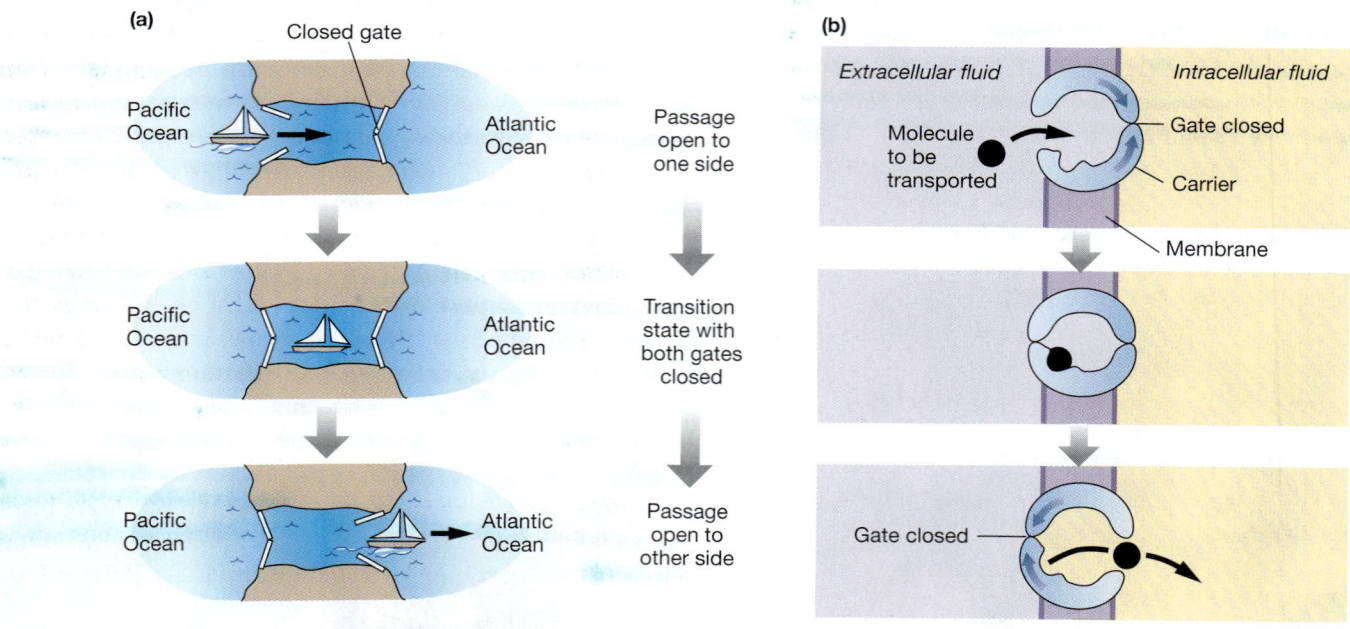

(a)

Closed gate

Pacific Ocean → Atlantic Ocean

Passage open to one side

Pacific Ocean — Atlantic Ocean

Transition state with both gates closed

Pacific Ocean → Atlantic Ocean

Passage open to other side

(b)

Extracellular fluid *Intracellular fluid*

Molecule to be transported

Gate closed

Carrier

Membrane

Gate closed

■ **FIGURE 5-13** *Facilitated diffusion by means of a carrier protein*

Carrier proteins, like the canal illustrated, never form a continuous passageway between the extracellular and intracellular fluids.

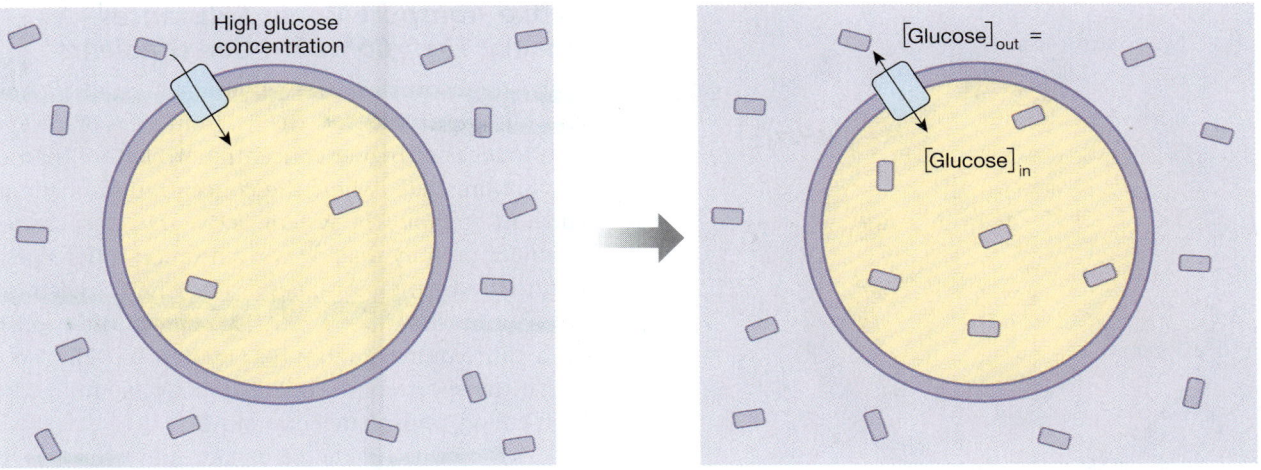

High glucose
concentration

Facilitated diffusion brings glucose into
the cell down its concentration gradient.

$[Glucose]_{out} =$

$[Glucose]_{in}$

Diffusion reaches equilibrium when the glucose
concentrations inside and outside the cell are equal.

■ **FIGURE 5-14**   *Net movement of molecules stops when facilitated diffusion reaches equilibrium*

This figure shows glucose transport using a GLUT carrier protein.

the canal with only two gates, one on the Atlantic side and one on the Pacific side. Only one gate at a time is open. When the Atlantic gate is closed, the canal opens into the Pacific. A ship enters the canal from the Pacific, and the gate closes behind it. Now the canal is isolated from both oceans with the ship trapped in the middle. Then the Atlantic gate opens, making the canal continuous with the Atlantic Ocean. The ship sails out of the gate and off into the Atlantic, having crossed the barrier of the land without the canal ever forming a continuous connection between the two oceans.

Movement across the membrane through a carrier protein is similar (Fig. 5-13b). The molecule to be transported binds to the carrier on one side of the membrane (the extracellular side in our example). This binding changes the conformation of the carrier so that the opening closes. After a brief transition in which both sides are closed, the opposite side of the carrier opens to the other side of the membrane. The carrier then releases the molecule being transported into the opposite compartment, having brought it through the membrane without creating a continuous connection between the extracellular and intracellular compartments.

CONCEPT CHECK

12. Name four functions of membrane proteins.

13. Which kinds of particles pass through open channels?

14. Name three ways channels differ from carriers.

15. If a channel is lined with amino acids that have a net positive charge, which of the following ions is/are likely to move freely through the channel? Na^+, Cl^-, K^+, Ca^{2+}.

16. Why doesn't glucose cross the cell membrane through open channels?

Answers: p. 172

Facilitated Diffusion Uses Carrier Proteins

As noted earlier, facilitated diffusion is protein-mediated transport in which no outside source of energy is needed to move molecules across the cell membrane, and active transport is protein-mediated transport that requires an outside energy source. Let us look first at facilitated diffusion.

Some polar molecules appear to move into and out of cells by diffusion, even though we know from their chemical properties that they are unable to pass easily through the lipid core of the cell membrane. The solution to this seeming contradiction is that these polar molecules cross the cell membrane by facilitated diffusion, with the aid of specific carriers. Sugars and amino acids are examples of molecules that enter or leave cells using facilitated diffusion. For example, a family of carrier proteins known as the GLUT transporters move glucose and related hexose sugars across membranes.

Facilitated diffusion has the same properties as simple diffusion (Table 5-1). The transported molecules move down their concentration gradient, the process requires no input of energy, and net movement stops at equilibrium, when the concentration inside the cell equals the concentration outside the cell (Fig. 5-14 ■):

$$[glucose]_{ECF} = [glucose]_{ICF}*$$

Cells in which facilitated diffusion takes place can avoid reaching equilibrium by keeping the concentration of substrate in the cell low. With glucose, for example, this is accomplished by

---

*In this book the presence of brackets around a solute's name indicates concentration.

5

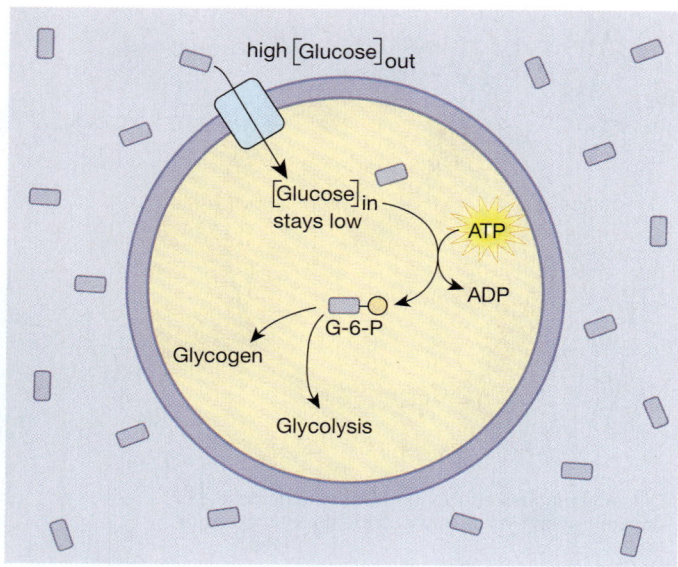

■ **FIGURE 5-15** *Diffusion of glucose into cells*

In many cells facilitated diffusion of glucose does not reach equilibrium because the cell converts glucose to glucose 6-phosphate (G-6-P), keeping the intracellular glucose concentration low.

phosphorylation (Fig. 5-15■). As soon as a glucose molecule enters the cell, it is phosphorylated to glucose 6-phosphate, the first step of glycolysis [🔁 p. 104]. Addition of the phosphate group prevents build-up of glucose inside the cell and also prevents glucose from leaving the cell.

CONCEPT CHECK ✓

17. Liver cells are able to convert glycogen to glucose, thereby making the intracellular glucose concentration higher than the extracellular glucose concentration. What do you think happens to facilitated diffusion of glucose when this occurs?   Answers: p. 172

Active Transport Moves Substances Against Their Concentration Gradients

Active transport is a process that moves molecules *against* their concentration gradient—that is, from areas of lower concentration to areas of higher concentration. Rather than creating an equilibrium state, where the concentration of the molecule is equal throughout the system, active transport creates a state of *dis*equilibrium by making concentration differences more pronounced. Moving molecules against their concentration gradient requires the input of outside energy, just as pushing a ball up a hill requires energy [🔁 Fig. 4-2, p. 92]. The energy for active transport comes either directly or indirectly from the high-energy phosphate bond of ATP.

Active transport can be divided into two types. In **primary (direct) active transport**, the energy to push molecules against their concentration gradient comes directly from the high-energy phosphate bond of ATP. **Secondary (indirect) active transport** uses potential energy [🔁 p. 92] stored in the concentration gradient of one molecule to push other molecules against their concentration gradient. All secondary active transport ultimately depends on primary active transport because the concentration gradients that drive secondary transport are created using energy from ATP.

The mechanism for both types of active transport appears to be similar to that for facilitated diffusion. A substrate to be transported binds to a membrane carrier and the carrier then changes conformation, releasing the substrate into the opposite compartment. Active transport differs from facilitated diffusion because the conformation change in the carrier protein requires energy input.

Primary Active Transport   Because primary active transport uses ATP as its energy source, many primary active transporters are known as **ATPases**. You may recall from Chapter 4 that the suffix *-ase* signifies an enzyme, and the stem (ATP) is the

■ **FIGURE 5-16**   *The sodium-potassium pump, Na⁺-K⁺-ATPase*

In this book, carrier proteins that hydrolyze ATP have the letters *ATP* written on the membrane protein.

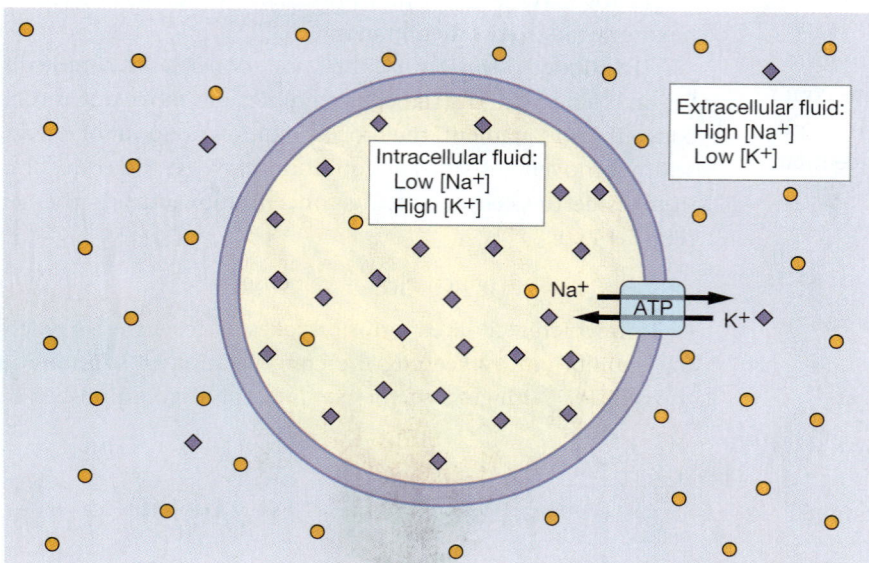

The Na⁺-K⁺-ATPase uses energy from ATP to pump Na⁺ out of the cell and K⁺ into the cell.

substrate upon which the enzyme is acting [🔁 p. 99]. These enzymes hydrolyze ATP to ADP and inorganic phosphate (P_i), releasing usable energy in the process. Most of the ATPases you will encounter in your study of physiology are listed in Table 5-2 ■. ATPases are sometimes called *pumps*, as in the sodium-potassium pump, Na^+-K^+-ATPase, mentioned earlier in this chapter.

The sodium-potassium pump is probably the single most important transport protein in animal cells because it maintains the concentration gradients of Na^+ and K^+ across the cell membrane (Fig. 5-16 ■). The transporter is arranged in the cell membrane so that it pumps 3 Na^+ out of the cell and 2 K^+ into the cell for each ATP consumed. In some cells, the energy needed to move these ions uses 30% of all the ATP produced by the cell. The current model of how the Na^+-K^+-ATPase works is illustrated in Figure 5-17 ■.

TABLE 5-2	Primary Active Transporters
NAMES	**TYPE OF TRANSPORT**
Na^+-K^+-ATPase or sodium-potassium pump	Antiport
Ca^{2+}-ATPase	Uniport
H^+-ATPase or proton pump	Uniport
H^+-K^+-ATPase	Antiport

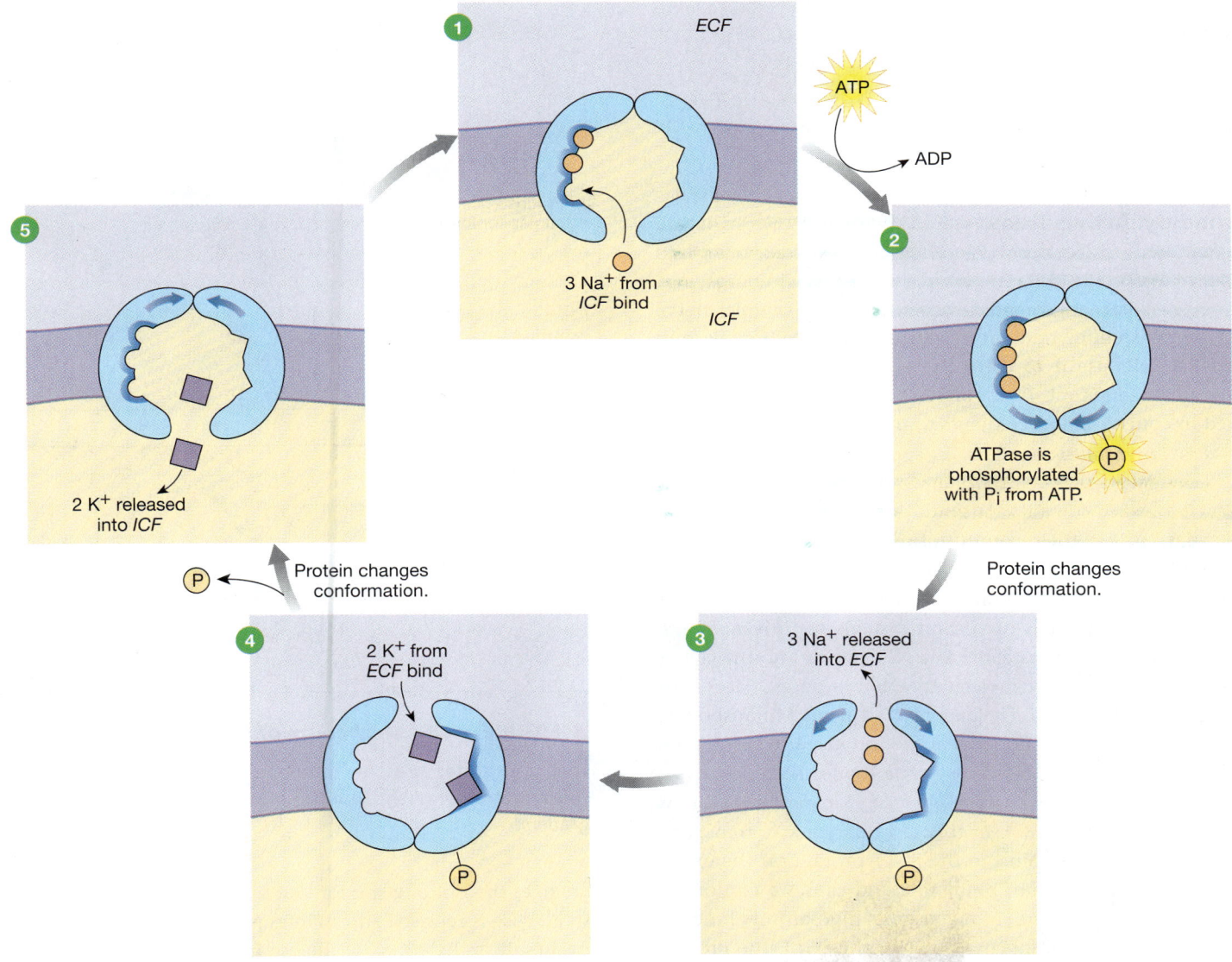

■ **FIGURE 5-17** *Mechanism of the Na^+-K^+-ATPase*

This figure presents one model of how the Na^+-K^+-ATPase uses energy and inorganic phosphate from ATP to move ions across a membrane.

TABLE 5-3	Common Secondary Active Transporters

SYMPORT CARRIERS	ANTIPORT CARRIERS
Sodium-dependent transporters	
Na^+-glucose (SGLT)	Na^+-H^+
Na^+-amino acids (several types)	Na^+-Ca^{2+}
Na^+-K^+-2 Cl^- (NKCC)	
Na^+-bile salts (small intestine)	
Na^+-choline uptake (nerve cells)	
Na^+-neurotransmitter uptake (nerve cells)	
Nonsodium-dependent transporters	
	HCO_3^--Cl^-
	H^+-K^+

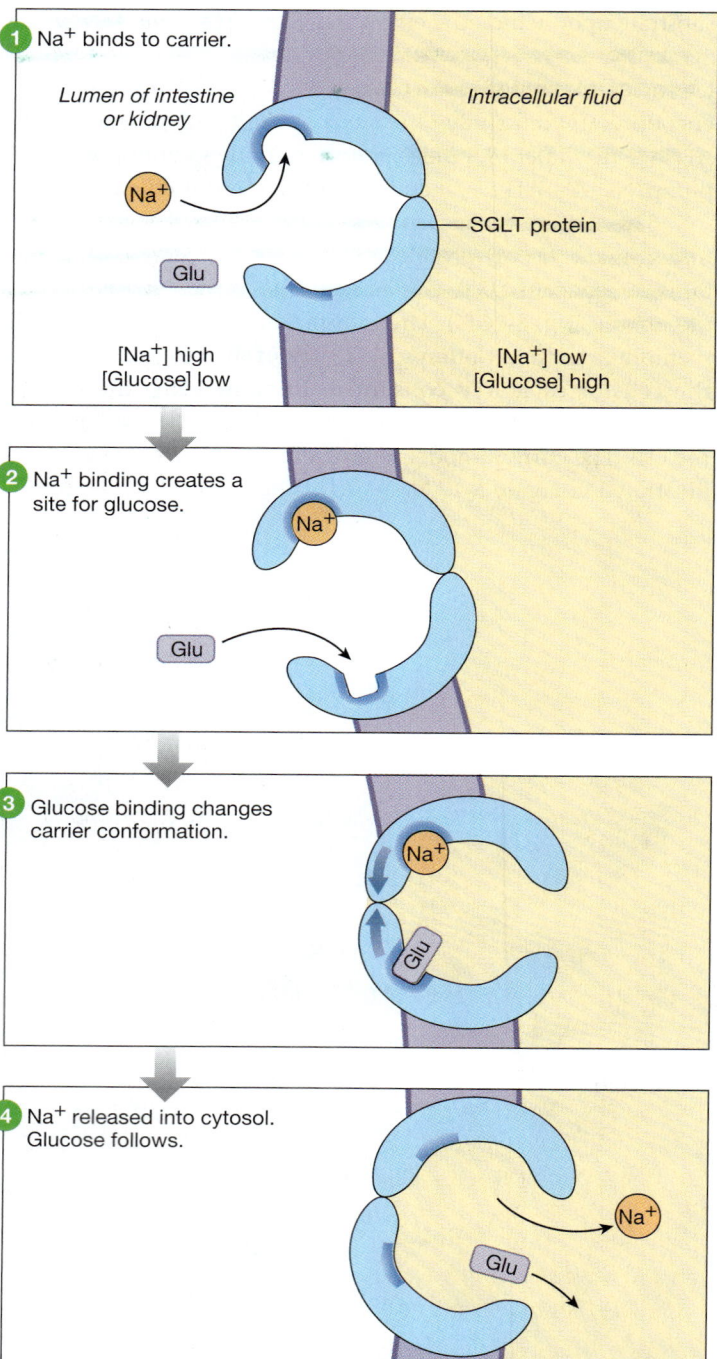

Secondary Active Transport   The sodium concentration gradient, with Na^+ concentration high in the extracellular fluid and low inside the cell, is a source of potential energy that the cell can harness for other functions. For example, nerve cells use the sodium gradient to transmit electrical signals, and epithelial cells use it to drive the uptake of nutrients, ions, and water. Membrane transporters that use potential energy stored in concentration gradients to move molecules are called *secondary active transporters*.

Secondary active transport uses the kinetic energy of one molecule moving down its concentration gradient to push other molecules against their concentration gradient. The cotransported molecules may go in the same direction across the membrane (symport) or in opposite directions (antiport). The most common secondary active transport systems are driven by the sodium concentration gradient. As a Na^+ moves into the cell, it either brings one or more molecules with it or trades places with molecules exiting the cell. The major Na^+-dependent transporters are listed in Table 5-3 ■. Notice that the cotransported molecules may be either other ions or uncharged molecules, such as glucose.

The mechanism of the Na^+-glucose transporter (SGLT) is illustrated in Figure 5-18 ■. In this secondary active transport, both Na^+ and glucose bind to the SGLT protein on the extracellular fluid side. Sodium binds first and causes a conformational change in the protein that creates a high-affinity binding site for glucose. When glucose binds to the SGLT, the protein changes conformation again and opens its channel to the intracellular fluid side. Sodium is released as it moves down its concentration gradient. The loss of Na^+ from the protein

FIGURE 5-18   *Mechanism of the SGLT transporter*

This transporter uses the potential energy stored in the Na^+ concentration gradient to move glucose against its concentration gradient.

changes the binding site for glucose back to a low-affinity site, so glucose is released and follows Na^+ into the cytoplasm. The net result is the entry of glucose into the cell against its concentration gradient, coupled to the movement of Na^+ into the cell down its concentration gradient.

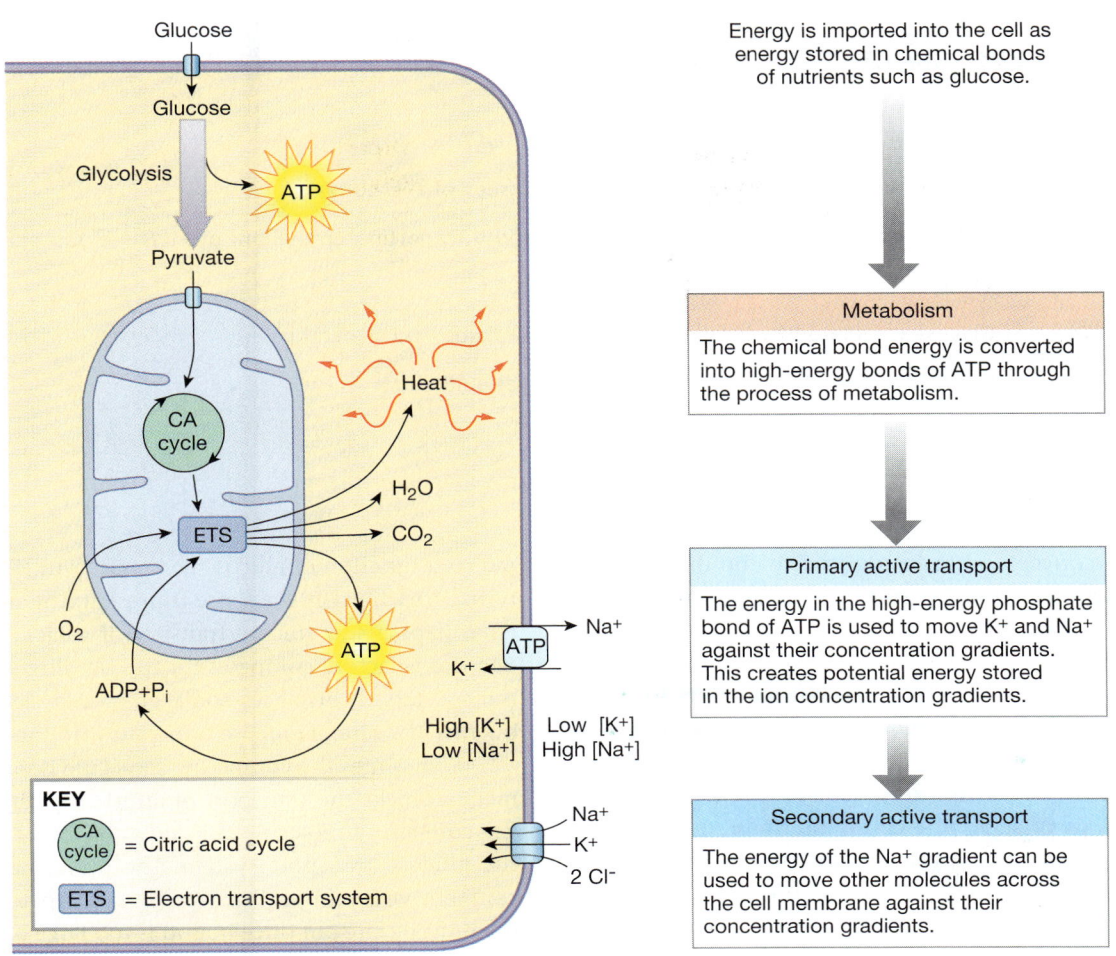

Glucose

Glucose

Glycolysis   ATP

Pyruvate

Heat

CA cycle

H_2O

ETS CO_2

O_2

ATP Na^+

$ADP+P_i$ ATP K^+

High $[K^+]$ Low $[K^+]$
Low $[Na^+]$ High $[Na^+]$

KEY

CA cycle = Citric acid cycle Na^+
K^+
$2 Cl^-$

ETS = Electron transport system

Energy is imported into the cell as energy stored in chemical bonds of nutrients such as glucose.

Metabolism
The chemical bond energy is converted into high-energy bonds of ATP through the process of metabolism.

Primary active transport
The energy in the high-energy phosphate bond of ATP is used to move K^+ and Na^+ against their concentration gradients. This creates potential energy stored in the ion concentration gradients.

Secondary active transport
The energy of the Na^+ gradient can be used to move other molecules across the cell membrane against their concentration gradients.

■ FIGURE 5-19   *Energy transfer in living cells*

Why does the body have both a facilitated diffusion GLUT carrier and a SGLT Na^+-glucose symporter? The simple answer is that GLUT carriers are found on all cells, but Na^+-glucose symporters are restricted to cells that are bringing glucose into the body from the external environment. GLUT transporters move glucose into or out of the cells depending on the concentration gradient. For example, when blood glucose levels are high, GLUT transporters on liver cells bring glucose into those cells. During times of fasting, when blood glucose levels fall, liver cells convert their glycogen stores to glucose. When the glucose concentration inside the liver cells builds up and exceeds the glucose concentration in the plasma, glucose leaves the cells on the reversible GLUT transporters. In contrast, the SGLT transporter can move substrate only into the cell because glucose must follow the Na^+ gradient. Consequently, SGLT transporters are found on epithelial cells that bring glucose into the body from the external environment.

The use of energy to drive active transport is summarized in Figure 5-19■. The organism imports energy from the environment in the chemical bonds of nutrients such as glucose. This energy is transferred to the high-energy bonds of ATP through oxidative phosphorylation in the mitochondria [p. 108]. The

energy in ATP is then used either to fuel primary active transport or to create ion concentration gradients for secondary active transport. For example, the Na^+-K^+-ATPase pushes sodium out of the cell and potassium into it. As this happens, the energy from ATP is transformed into potential energy stored in the ion concentration gradients. This potential energy can be harnessed by various sodium-dependent cotransporters, such as the Na^+-K^+-$2Cl^-$ transporter (NKCC). The kinetic energy of Na^+ moving into the cell down its concentration gradient is linked to the uphill movement of K^+ and Cl^- into the cell. As you study the different systems of the body, you will find these secondary active transporters taking part in many physiological processes.

CONCEPT CHECK

18. Name two ways active transport by the Na^+-K^+-ATPase (Fig. 5-17) differs from secondary transport by the SGLT (Fig. 5-18).   Answers: p. 172

Carrier-Mediated Transport Exhibits Specificity, Competition, and Saturation

Both passive and active forms of carrier-mediated transport demonstrate three properties: specificity, competition, and

	SUBSTANCES TRANSPORTED	**TISSUE LOCATIONS**
GLUT1	Glucose and other hexoses	Most tissues of the body
GLUT2	Glucose and other hexoses	Liver and transporting epithelium of intestine and kidney
GLUT3	Glucose and other hexoses	Neurons
GLUT4	Glucose (regulated by insulin)	Adipose tissue and skeletal muscle
GLUT5	Fructose	Intestinal epithelium

TABLE 5-4:   The GLUT Family* of Glucose Transporters

*GLUT6 through GLUT12 remain under investigation.

5

saturation. These concepts were introduced in the discussion on protein interactions [🔁 p. 43] and reflect the binding of a substrate to the protein.

Specificity   As noted in Chapter 2, specificity refers to the ability of a carrier to move only one molecule or only a group of closely related molecules. One example of specificity is found in the GLUT family of transporters, which move only six-carbon sugars (*hexoses*), such as glucose, mannose, galactose, and fructose [🔁 p. 27], across cell membranes. GLUT transporters have binding sites that recognize and transport hexoses, but they will not transport the disaccharide maltose or any form of glucose that is not normally found in nature. Thus we can say that GLUT transporters are specific for naturally occurring six-carbon monosaccharides.

For many years, scientists assumed that there must be different isoforms of the glucose carrier because they had observed that glucose transport was regulated by hormones in some cells but not in others. However, it was not until the 1980s that the first glucose transporter was isolated. To date, about 12 glucose facilitated diffusion carrier genes have been identified, and five GLUT proteins have been studied in detail (Table 5-4 ■). The proteins named GLUT6 through GLUT12 are still being investigated. The restriction of different GLUT transporters to different tissues is an important feature in the metabolism and homeostasis of glucose.

Competition   The property of competition is closely related to specificity. A transporter may move several members of a related group of substrates, but those substrates will compete with one another for the binding sites on the transporter. For example, GLUT transporters move the family of hexose sugars, but each different GLUT transporter has a "preference" for one or more hexoses, based on its binding affinity.

The results of an experiment demonstrating competition are shown in Figure 5-20 ■. The graph shows glucose transport rate as a function of glucose concentration. The top line shows transport when only glucose is present; the bottom line shows what happens to glucose transport if galactose is also present. Because galactose competes for the binding sites on the GLUT transporters, it displaces some glucose molecules, and consequently the rate of glucose transport into the cell decreases.

Sometimes the competing molecule is not transported but merely blocks the transport of another substrate. In this case, the competing molecule is a *competitive inhibitor* [🔁 p. 41]. In the glucose transport system, the disaccharide maltose is a competitive inhibitor (Fig. 5-21 ■). It competes with glucose for the binding site but once bound is too large to be moved across the membrane.

Competition between transported substrates has been put to good use in medicine. An example involves gout, a disease caused by elevated levels of uric acid in the plasma. One method of decreasing plasma uric acid is to enhance its excretion in the

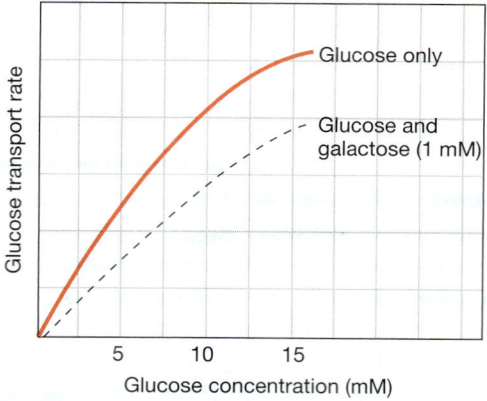

■ **FIGURE 5-20**   *Graph of transport competition*

This graph shows glucose transport rate as a function of glucose concentration. In one experiment, only glucose was present. In the second experiment a constant concentration of galactose was present.

(a)

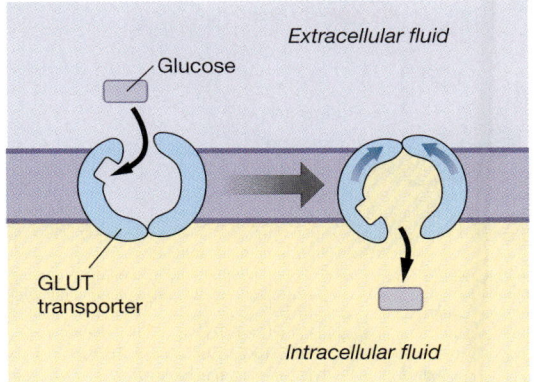

The **GLUT transporter** brings glucose across cell membranes.

(b)

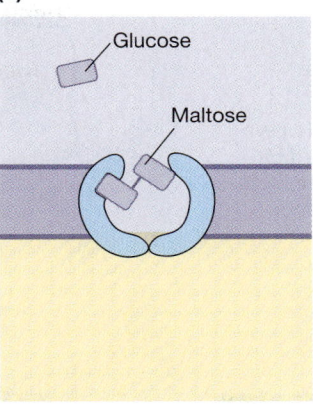

Maltose is a competitive inhibitor that binds to the GLUT transporter but is not itself carried across the membrane.

■ **FIGURE 5-21**   *Competitive inhibition of glucose transport*

The inhibitor in this case is the disaccharide maltose, a molecule much larger than glucose.

urine. Normally, the kidney's organic anion transporter (OAT) reclaims uric acid from the urine and returns the acid to the plasma. However, if an organic acid called probenecid is administered to the patient, the OAT binds to probenecid instead of to uric acid, preventing the reabsorption of uric acid. As a result, more uric acid leaves the body in the urine, lowering the uric acid concentration in the plasma.

Saturation   The rate of substrate transport depends on both the substrate concentration and the number of carrier molecules, a property that is shared by enzymes [🔁 p. 98]. For a fixed number of carriers, however, as substrate concentration increases, the transport rate increases up to a maximum, the point at which all carrier binding sites are filled with substrate. At this point, the carriers are said to have reached saturation. At saturation, the carriers are working at their maximum rate, and a further increase in substrate concentration will have no effect. Figure 5-22 ■ shows saturation represented graphically.

For an analogy, think of the carriers as doors into a concert hall. Each door has a maximum number of people that it can allow to enter the hall in a given period of time. Suppose all the doors together can allow a maximum of 100 people per minute to enter the hall. This is the maximum transport rate, also called the **transport maximum**. When the concert hall is empty, three maintenance people enter the doors every hour. The transport rate is 3 people/60 minutes, or 0.05 people/minute, well under the maximum. For a local dance recital, about 50 people per minute go through the doors, still well under the maximum. When the most popular rock group of the day appears in concert, however, thousands of people gather outside. When the doors open, thousands of people are clamoring to get in, but the doors will allow only 100 people/minute into the hall. The doors are working at the maximum rate, so it does not matter whether there are 1000 or 3000 people trying to get in. The transport rate is saturated at 100 people/minute.

How can cells increase their transport capacity and avoid saturation? One way is to increase the number of carriers in the membrane. This would be like opening more doors into the concert hall. Under some circumstances, cells are able to insert additional carriers into their membranes. Under other circumstances, a cell may withdraw carriers to decrease movement of a molecule into or out of the cell.

All forms of carrier-mediated transport show specificity, competition, and saturation, but as we learned earlier in the chapter, they also differ in one important way: passive mediated transport—better known as facilitated diffusion—requires no input of energy from an outside source. Active transport requires energy input from ATP, either directly or indirectly.

GRAPH QUESTION

On the *x*-axis, mark the substrate concentration at which the carriers first become saturated.

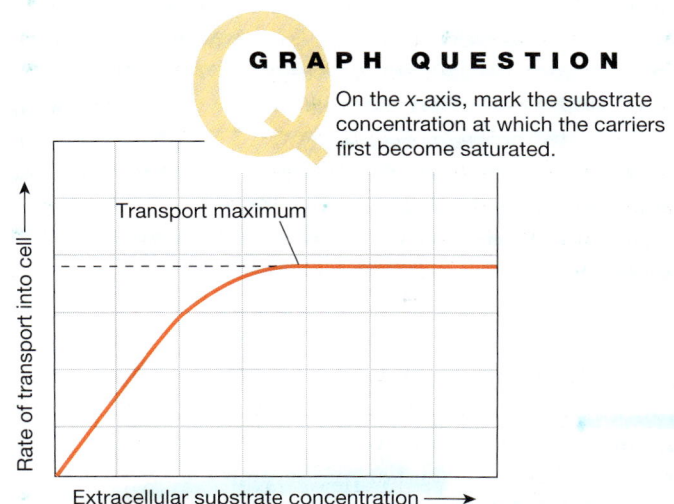

Transport rate is proportional to substrate concentration until the carriers are saturated.

■ **FIGURE 5-22**   *Graph showing saturation of carrier-mediated transport*

CONCEPT CHECK

19. What would you call a carrier that moves two substrates in opposite directions across a membrane?

20. In the concert-hall door analogy, we described how the maximum transport rate might be increased by increasing the number of doors leading into the hall. Using the same analogy, can you think of another way a cell might increase its maximum transport rate?

Answers: p. 172

VESICULAR TRANSPORT

What happens to the many macromolecules that are too large to enter or leave cells through protein channels or carriers? They move in and out of the cell with the aid of bubble-like vesicles created from the cell membrane. Cells use two basic mechanisms to import large molecules and particles: phagocytosis and endocytosis. Phagocytosis once was considered a type of endocytosis, but as scientists learned more about the mechanisms behind the two processes, they decided that phagocytosis was fundamentally different. Material leaves cells by the process known as exocytosis, a process that is similar to endocytosis run in reverse.

Phagocytosis Creates Vesicles Using the Cytoskeleton

If you studied *Amoeba* in your biology laboratory, you may have watched these one-cell creatures ingest their food by surrounding it and enclosing it within a vesicle that is brought into the cytoplasm. **Phagocytosis** [*phagein*, to eat + *cyte*, cell + *-sis*, process] is the actin-mediated process by which a cell engulfs a bacterium or other particle into a large membrane-bound vesicle called a **phagosome** [*soma*, body]. The phagosome pinches off from the cell membrane and moves to the interior of the cell, where it fuses with a lysosome [🔁 p. 66], whose digestive enzymes destroy the bacterium. Phagocytosis requires energy from ATP for the movement of the cytoskeleton and for the intracellular transport of the vesicles. In humans, phagocytosis occurs only in certain types of white blood cells called *phagocytes*, which specialize in "eating" bacteria and other foreign particles (Fig. 5-23 ■).

Endocytosis Creates Smaller Vesicles

Endocytosis, the second pathway by which large molecules or particles move into cells, differs from phagocytosis in that in endocytosis, the membrane surface indents rather than pushes out, and the vesicles formed are much smaller. In addition, some endocytosis is *constitutive;* that is, it is an essential function that is always taking place. In contrast, phagocytosis must be triggered by the presence of a substance to be ingested.

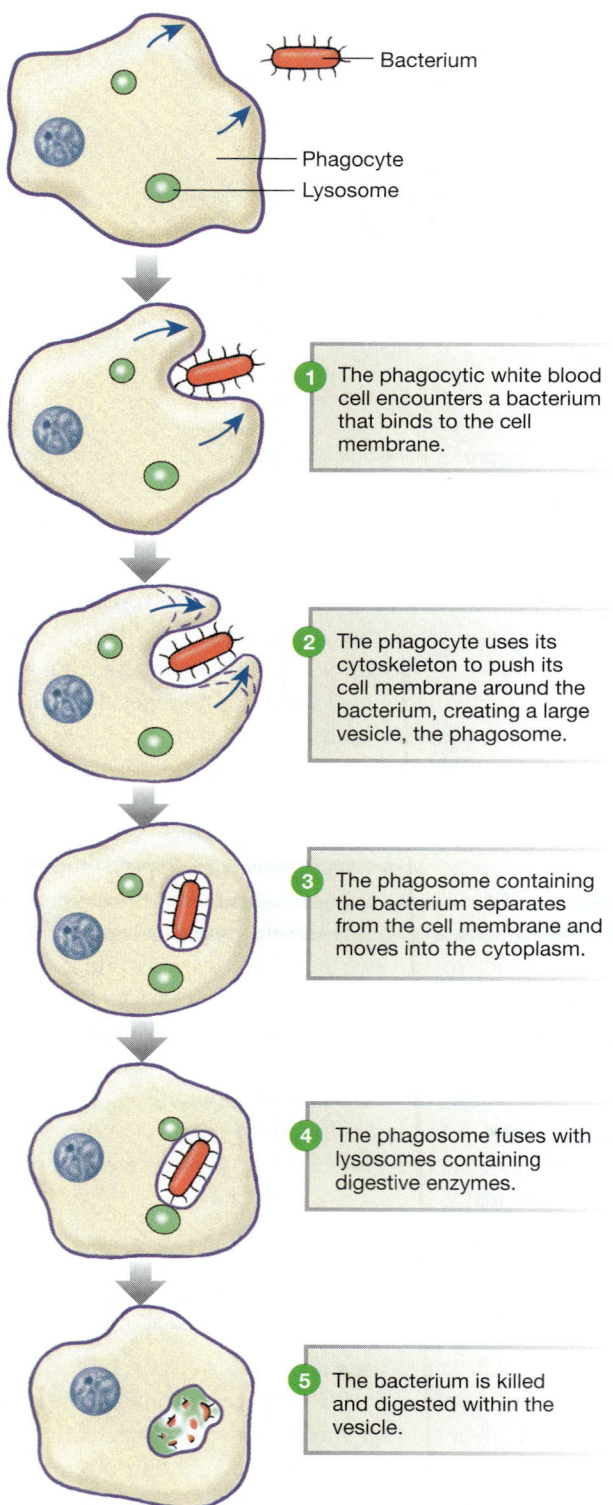

1   The phagocytic white blood cell encounters a bacterium that binds to the cell membrane.

2   The phagocyte uses its cytoskeleton to push its cell membrane around the bacterium, creating a large vesicle, the phagosome.

3   The phagosome containing the bacterium separates from the cell membrane and moves into the cytoplasm.

4   The phagosome fuses with lysosomes containing digestive enzymes.

5   The bacterium is killed and digested within the vesicle.

■ **FIGURE 5-23** *Phagocytosis*

Endocytosis is an active process that requires energy from ATP. It can be nonselective, allowing extracellular fluid to enter the cell—a process called **pinocytosis** [*pino-*, drink]—or it can be highly selective, allowing only specific molecules to enter the cell. Two types of endocytosis require a ligand to bind to a

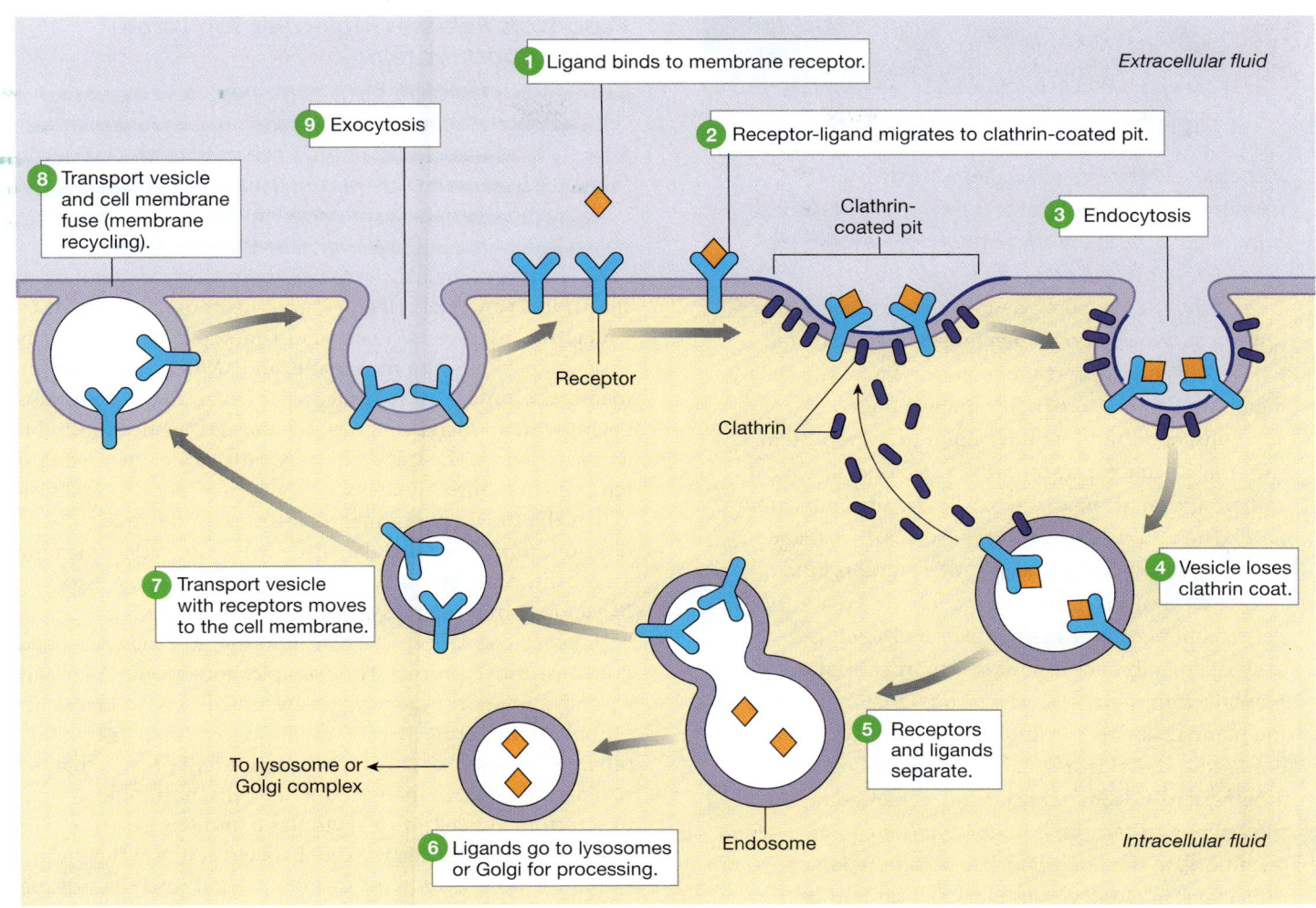

① Ligand binds to membrane receptor.

② Receptor-ligand migrates to clathrin-coated pit.

③ Endocytosis

④ Vesicle loses clathrin coat.

⑤ Receptors and ligands separate.

⑥ Ligands go to lysosomes or Golgi for processing.

⑦ Transport vesicle with receptors moves to the cell membrane.

⑧ Transport vesicle and cell membrane fuse (membrane recycling).

⑨ Exocytosis

Extracellular fluid

Clathrin-coated pit

Clathrin

Receptor

To lysosome or Golgi complex

Endosome

Intracellular fluid

■ **FIGURE 5-24** *Receptor-mediated endocytosis and exocytosis*

Steps 1–4 illustrate endocytosis; steps 7–9 show exocytosis.

membrane receptor protein: receptor-mediated endocytosis and potocytosis.

Receptor-Mediated Endocytosis Receptor-mediated endocytosis takes place in regions of the cell membrane known as clathrin-coated pits, indentations where the cytoplasmic side of the membrane has high concentrations of the protein *clathrin* (Fig. 5-24 ■). In the first step of the process, extracellular ligands that will be brought into the cell bind to their membrane receptors (step ① in Fig. 5-24). The receptor-ligand complex migrates along the cell surface until it encounters a coated pit ②. Once the receptor-ligand complex is in the coated pit, the membrane draws inward, or *invaginates* ③, then pinches off from the cell membrane and becomes a cytoplasmic vesicle. The clathrin molecules are released and recycle back to the membrane ④. In the vesicle, the receptor and ligand separate, leaving the ligand inside an *endosome* ⑤. The endosome moves to a lysosome if the ligand is to be destroyed, or to the Golgi complex if the ligand is to be processed ⑥.

Meanwhile, the ligand's receptors may be reused in a process known as **membrane recycling**. The vesicle with the receptors moves to the cell membrane ⑦ and fuses with it ⑧. The vesicle membrane then is incorporated back into the cell membrane by exocytosis ⑨. Notice in Figure 5-24 that the cytoplasmic face of the membrane remains the same throughout endocytosis and recycling. The extracellular surface of the cell membrane becomes the inside face of the vesicle membrane.

Receptor-mediated endocytosis transports a variety of substances into the cell, including protein hormones, growth factors, antibodies, and plasma proteins that serve as carriers for iron and cholesterol. Abnormalities in receptor-mediated removal of cholesterol from the blood are associated with elevated plasma cholesterol levels and cardiovascular disease.

Potocytosis and Caveolae The form of endocytosis known as **potocytosis** is distinguished from receptor-mediated endocytosis by the fact that potocytosis uses **caveolae** ("little caves")

THE LETHAL LIPOPROTEIN

Although cholesterol molecules are essential for membrane structure and for the synthesis of steroid hormones (such as the sex hormones), elevated cholesterol levels in the body can lead to heart disease. One of the reasons some people have too much cholesterol in their blood (*hypercholesterolemia*) is failure of cells to transport the molecule across the cell membrane. Cholesterol is insoluble in aqueous solutions, and therefore it is bound to a lipoprotein carrier molecule for transport in the blood. The most common form of the carrier is *low-density lipoprotein (LDL)*. The LDL-cholesterol complex (LDL-C) is taken into cells by endocytosis when LDL receptors bind to the LDL molecule. People who inherit a genetic defect that decreases the number of LDL receptors on their cell membranes cannot transport cholesterol normally into their cells. As a result, LDL-C remains in the plasma. Abnormally high blood levels of LDL-C predispose these people to the development of **atherosclerosis**, commonly known as hardening of the arteries [*atheroma*, a tumor + *skleros*, hard + *-sis*, condition]. In this condition, the accumulation of cholesterol in blood vessels blocks blood flow and contributes to heart attacks.

rather than clathrin-coated pits to concentrate and bring receptor-bound molecules into the cell. Caveolae are membrane regions with lipid rafts [🔁 p. 57], membrane receptor proteins, and a family of unique membrane proteins named *caveolins*. The receptors in caveolae are lipid-anchored proteins. In many cells, caveolae appear as small indented pockets on the cell membrane, which is how they acquired their name.

Caveolae have several functions: to concentrate and internalize small molecules, to help in the transfer of macromolecules across the capillary endothelium (see Fig. 5-27), and to participate in cell signaling. Caveolae appear to be involved in some disease processes, including viral and parasitic infections. Two forms of the disease *muscular dystrophy* are associated with abnormalities in the protein caveolin. Scientists are currently trying to discover more details about the role of caveolae in normal physiology and pathophysiology.

Exocytosis Releases Molecules Too Large for Transport Proteins

Exocytosis is the opposite of endocytosis. In exocytosis, intracellular vesicles move to the cell membrane, fuse with it (Fig. 5-24, ⑧), and then release their contents to the extracellular fluid ⑨. Cells use exocytosis to export large lipophobic molecules, such as proteins synthesized in the cell, and to get rid of wastes left in lysosomes from intracellular digestion.

The process by which the cell and vesicle membranes fuse is similar in a variety of cell types, from neurons to endocrine cells. Exocytosis involves two families of proteins: *Rabs,* which help vesicles dock onto the membrane, and *SNAREs,* which facilitate membrane fusion. In regulated exocytosis, the process usually begins with an increase in intracellular Ca^{2+} concentration that acts as a signal. The Ca^{2+} interacts with a calcium-sensing protein, which in turn initiates secretory vesicle docking and fusion. When the fused area of membrane opens, the vesicle contents diffuse into the extracellular space while the vesicle membrane stays behind and becomes part of the cell membrane. Exocytosis, like endocytosis, requires energy in the form of ATP.

Exocytosis takes place continuously in some cells, making it a constitutive process. For example, goblet cells [🔁 p. 75] in the intestine continuously release mucus by exocytosis, and fibroblasts in connective tissue release collagen [🔁 p. 77]. In other cell types, exocytosis is an intermittent process that is initiated by a signal. In many endocrine cells, hormones are stored in secretory vesicles in the cytoplasm and released in response to a signal from outside the cell. Exocytosis is also the means by which membrane proteins can be inserted into the cell membrane, as shown in Figure 5-24. You will encounter many examples of exocytosis in your study of physiology.

✓ CONCEPT CHECK

21. How does phagocytosis differ from endocytosis?
22. Name the two membrane protein families associated with endocytosis.
23. How do cells move large proteins into the cell? Out of the cell?

Answers: p. 172

TRANSEPITHELIAL TRANSPORT

All the transport processes described in the previous sections deal with the movement of molecules across a single membrane, that of the cell. However, any molecules entering and leaving the body across an epithelium must cross two cell membranes. Molecules cross the first membrane when they move into an epithelial cell from the external environment, and cross the second when they leave the epithelial cell to enter the extracellular fluid. Movement across epithelial cells, **transepithelial transport**, uses a combination of active and passive transport.

The transporting epithelia of the intestine and kidney are specialized to selectively transport molecules into and out of

Lumen of intestine
or kidney

Apical membrane
faces the lumen.

Tight junction prevents
movement of substances
between the cells.

Transporting
epithelial cell

Basolateral membrane
faces the ECF.

*Extracellular
fluid*

Transport proteins

Polarized epithelia have different transport proteins on apical and basolateral membranes. This allows selective directional transport across the epithelium. Transport from lumen to ECF is called **absorption**. Transport from ECF to lumen is called **secretion**.

■ **FIGURE 5-25** *Polarized cells of transporting epithelia*

The apical membrane and the basolateral membrane are the two poles of the cell.

5

the body. These epithelial cells are connected to one another by adhesive junctions and tight junctions [➿ p. 69] that act as barriers to minimize the unregulated diffusion of material between the cells. The tight junctions also mark the separation of the cell membrane into two regions, or poles. The surface of the epithelial cell that faces the lumen of an organ is called the *apical* [*apex,* the highest point] membrane (Fig. 5-25 ■; p. 51). It is often folded into microvilli that increase its surface area. The apical membrane is separated from the remainder of the cell membrane by tight junctions. The three surfaces of the cell that face the extracellular fluid are collectively called the *basolateral* membrane [*basal,* base + *latus,* side]. The apical membrane is also called the *mucosal* membrane; the corresponding term for the basolateral membrane is *serosal* membrane.

Transporting epithelial cells are said to be *polarized* because their apical and basolateral membranes have very different properties. Certain transport proteins, such as the Na^+-K^+-ATPase, are almost always found only on the basolateral membrane, whereas others, like the Na^+-glucose symporter, are localized to the apical membrane. This polarized distribution of transporters results in the one-way movement of certain molecules across the epithelium. Transport of material from the lumen of an organ to the extracellular fluid is called **absorption**. This process is the opposite of *secretion,* which is the movement of material from the ECF to the lumen.

The cells of transporting epithelia can alter their permeability by selectively inserting or withdrawing membrane proteins. Transporters pulled out of the membrane may be destroyed in lysosomes, or they may be stored in vesicles inside the cell, ready to be reinserted into the membrane in response to a signal (another example of membrane recycling).

The transepithelial movement of molecules requires that the molecules first enter an epithelial cell and then leave it. If the

molecule can be transported through a protein channel or on transport carriers, then the two-step process usually has one "uphill" step that requires energy and one "downhill" step in which the molecule moves passively down its gradient. Molecules that

RUNNING PROBLEM

The sweat test that Daniel will undergo analyzes levels of the salt NaCl in sweat. Sweat—a mixture of ions and water—is secreted into sweat ducts by the epithelial cells of sweat glands. As sweat moves toward the skin's surface through the sweat ducts, CFTR allows chloride ions to move out of the sweat and back into the epithelial cells. Sodium ions are also reabsorbed, following the Cl^-. This epithelium is not permeable to water, and so normal reabsorption of NaCl creates sweat with a low salt content. However, in the absence of CFTRs in the epithelium, salt is not reabsorbed. "Normally, sweat contains about 120 millimoles of salt per liter," says Beryl Rosenstein, M.D., of the Cystic Fibrosis Center at the Johns Hopkins Medical Institutions. "In cystic fibrosis, salt concentrations in the sweat can be four times the normal amount."

Question 2:
  Based on the information given, is CFTR on the apical or basolateral surface of the sweat gland epithelium?

129 139 **151** 157 161 168

are too large to be moved by membrane proteins can be transported across the cell in vesicles.

Transepithelial Transport of Glucose Uses Membrane Proteins

The movement of glucose from the lumen of the kidney tubule or intestine to the extracellular fluid is an important example of directional movement across a transporting epithelium. Transepithelial movement of glucose involves three transport systems: the secondary active transport of glucose with Na^+ from the lumen into the epithelial cell at the apical membrane, followed by the movement of both Na^+ and glucose out of the cell and into the extracellular fluid on the basolateral side of the cell. Sodium moves out by primary active transport via a Na^+-K^+-ATPase, and glucose leaves the cell by facilitated diffusion.

Figure 5-26 ■ shows the process in detail. The glucose concentration in the transporting epithelial cell is higher than the glucose concentration in either the extracellular fluid or the lumen of the kidney or intestine. Therefore, moving glucose from the lumen into the cell requires the input of energy—in this case, energy stored in the Na^+ concentration gradient. Sodium ions in the lumen bind to the symporter, as previously described (see Fig. 5-18), and bring glucose with them into the cell. The energy needed to move glucose against its concentration gradient comes from the kinetic energy of Na^+ moving down its concentration gradient.

Once glucose is in the epithelial cell, it leaves by moving down its concentration gradient on the facilitated diffusion GLUT transporter in the basolateral membrane (② Fig. 5-26). The Na^+ is pumped out of the cell on the basolateral side using the Na^+-K^+-ATPase ③. Sodium is more concentrated in the extracellular fluid than in the cell; therefore, this step requires energy provided by ATP.

The removal of Na^+ from the cell is essential if glucose is to continue to be absorbed from the lumen because the

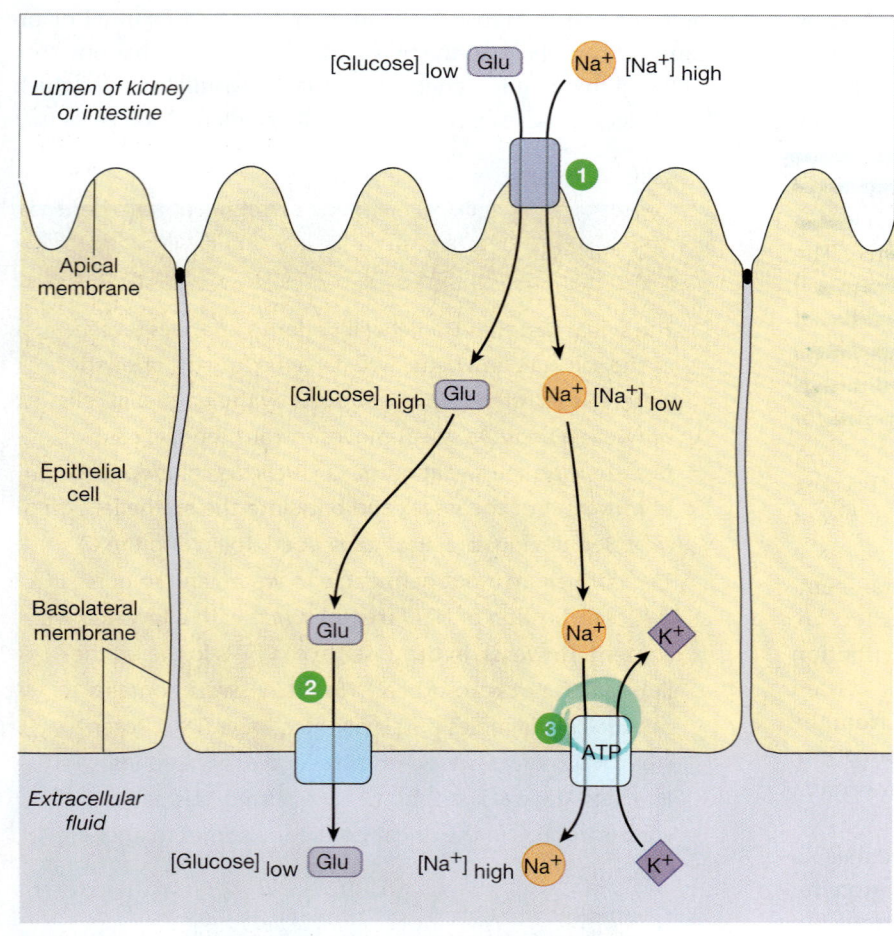

1 **Na^+- glucose symporter** brings glucose into cell against its gradient using energy stored in the Na^+ concentration gradient.

2 **GLUT transporter** transfers glucose to ECF by facilitated diffusion.

3 **Na^+-K^+- ATPase** pumps Na^+ out of the cell, keeping ICF Na^+ concentration low.

FIGURE QUESTIONS

- Match each transporter to its location.
  1. GLUT   a) apical membrane
  2. Na^+-glucose   b) basolateral membrane
     symporter
  3. Na^+-K^+-ATPase
- Is glucose movement across the basolateral membrane active or passive? Explain.
- Why doesn't Na^+ movement at the apical membrane require ATP?

■ **FIGURE 5-26**   *Transepithelial transport of glucose*

This process involves indirect (secondary) active transport of glucose across the apical membrane and glucose diffusion across the basolateral membrane.

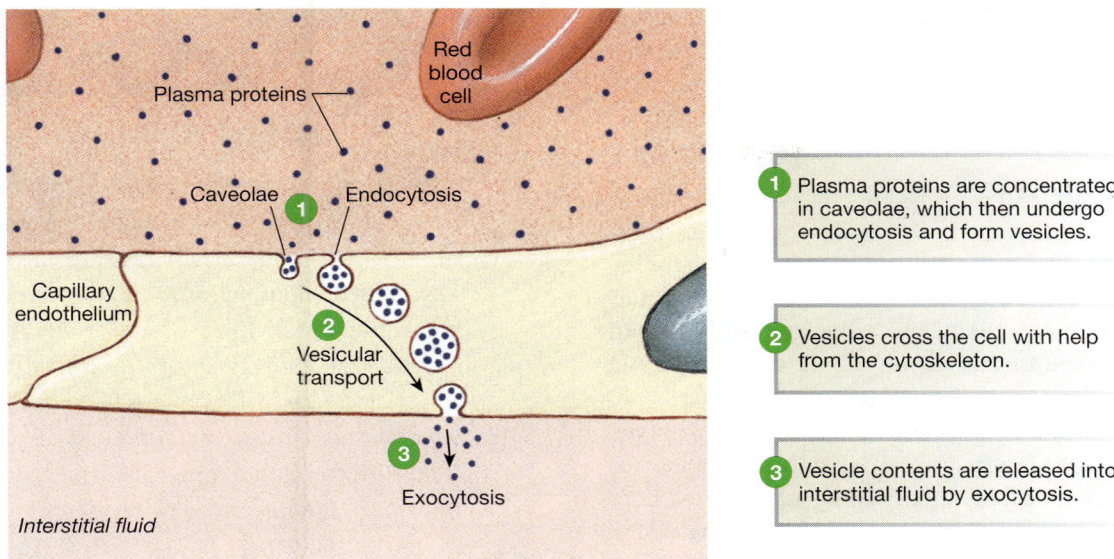

FIGURE 5-27 *Transcytosis across the capillary endothelium*

Na$^+$-glucose symporter depends on low intracellular concentrations of Na$^+$. If the basolateral Na$^+$-K$^+$-ATPase is poisoned with *ouabain* (pronounced wah-bane—a compound related to the heart drug digitalis), Na$^+$ that enters the cell cannot be pumped out. The Na$^+$ concentration inside the cell gradually increases until it is equal to that in the lumen. Without a sodium gradient, there is no energy source to run the Na$^+$-glucose symporter, and the movement of glucose across the epithelium stops.

Transepithelial transport can use ion movement through channels in addition to carrier-mediated transport. For example, the apical membrane of a transporting epithelium may use the Na$^+$-K$^+$-2 Cl$^-$ (NKCC) symporter to bring K$^+$ into the cell against its concentration gradient, using energy from the Na$^+$ gradient. Because the K$^+$ concentration inside the cell is higher than in the extracellular fluid, the K$^+$ can move out of the cell on the basolateral side through open K$^+$ leak channels. The Na$^+$ must be pumped out by the Na$^+$-K$^+$-ATPase. By this simple mechanism the body can absorb Na$^+$ and K$^+$ at the same time from the lumen of the intestine or the kidney.

CONCEPT CHECK

24. Why does Na$^+$ movement from the cytoplasm to the extracellular fluid require energy?

25. Ouabain, an inhibitor of the Na$^+$-K$^+$-ATPase, cannot pass through cell membranes. What would happen to the transepithelial glucose transport shown in Figure 5-26 if ouabain were applied to the apical side of the epithelium? To the basolateral side of the epithelium?

26. Which GLUT transporter is illustrated in Figure 5-26? (*Hint:* see Table 5-4.)

Answers: p. 173

Transcytosis Uses Vesicles to Cross an Epithelium

Some molecules, such as proteins, are too large to cross epithelia on membrane transporters. Instead they are moved across epithelia by **transcytosis**, which is a combination of endocytosis, vesicular transport across the cell, and exocytosis (Fig. 5-27 ■). In this process, the molecule is brought into the epithelial cell via receptor-mediated endocytosis or potocytosis. The resulting vesicle attaches to microtubules in the cell's cytoskeleton and is transported across the cell by a process known as **vesicular transport**. At the opposite side of the epithelium, the contents of the vesicle are expelled into the interstitial fluid by exocytosis.

Transcytosis makes it possible for large proteins to move across an epithelium and remain intact. It is the means by which infants absorb maternal antibodies in breast milk. The antibodies are absorbed on the apical surface of the infant's intestinal epithelium and then released into the extracellular fluid.

CONCEPT CHECK

27. If a poison that disassembles microtubules is applied to a capillary endothelial cell, what happens to transcytosis?

Answers: p. 173

Now that we have considered how solutes cross cell membranes, we will turn to the movement of water between the body's compartments.

OSMOSIS AND TONICITY

The distribution of solutes in the body depends on whether a substance can cross the cell membrane, either by simple diffusion, protein-mediated transport, or vesicular transport. Water, on the other hand, is able to move freely in and out of nearly every cell in the body by traversing water-filled ion channels and special water channels created by the protein aquaporin. In this section we examine the relationship between solute movement and water movement across cell membranes, a topic that provides the foundation for clinical use of intravenous (IV) fluid therapy.

The Body Is Mostly Water

Water is the most important molecule in the human body because it is the solvent for all living matter. As we look for life in distant parts of the solar system, one of the first questions scientists ask about a planet is, "Does it have water?" Without water, life as we know it cannot exist.

How much water is in the human body? Because one individual differs from the next, there is no single answer. However, in human physiology we often speak of standard values for physiological functions, based on "the 70-kg man." These standard values are derived from data obtained by studying young white males whose average weight was 70 kg. Thus, when we speak of standard or average values in physiology, remember that these numbers need to be adjusted for an individual's age, sex, weight, and ethnic origin.

The "standard" 70-kilogram (154-pound) male has 60% of his total body weight, or 42 kg (92.4 lb), in the form of water. Each kilogram of water has a volume of 1 liter, so his **total body water** is 42 liters. This is the equivalent of 21 two-liter soft drink bottles! Women have less water per kilogram of body mass than men because women have more adipose tissue. Look back at Figure 3-31 [🔁 p. 79] and note how the large fat droplets in adipose tissue occupy most of the cell, displacing the more aqueous cytoplasm. Age also influences body water content. Infants have relatively more water than adults, and water content decreases as people grow older than 60.

Table 5-5 ■ shows water content as a percentage of total body weight in people of various ages and both sexes. In clinical practice, it is necessary to allow for the variability of body

TABLE 5-5	Water Content as Percentage of Total Body Weight by Age and Sex	
AGE	MALE	FEMALE
Infant	65%	65%
1–9	62%	62%
10–16	59%	57%
17–39	61%	51%
40–59	55%	47%
60+	52%	46%

Adapted from Edelman and Leibman, *American Journal of Medicine* 27; 256–277, 1959.

CLINICAL FOCUS

ESTIMATING BODY WATER

Clinicians estimate a person's fluid loss in dehydration by equating weight loss to water loss. Because 1 liter of pure water weighs 1 kilogram, a decrease in body weight of 1 kilogram (or 2.2 lb) is considered equivalent to the loss of 1 liter of body fluid. A baby with diarrhea can easily be weighed to estimate its fluid loss. A decrease of 1.1 pounds (0.5 kg) of body weight is assumed to mean the loss of 500 mL of fluid. This calculation provides a quick estimate of how much fluid needs to be replaced.

water content when prescribing drugs. Because women and older people have less body water, they will have a higher concentration of a drug in the plasma than will young men if all are given an equal dose per kilogram of body mass.

The distribution of water among body compartments is less variable. When we look at the relative volumes of the body compartments, the intracellular compartment contains about two-thirds (67%) of the body's water (Fig. 5-28 ■). The remaining third (33%) is split between the interstitial fluid (which contains about 75% of the extracellular water) and the plasma (which contains about 25% of the extracellular water).

The Body Is in Osmotic Equilibrium

Water is able to move freely between cells and the extracellular fluid, and will distribute itself until water concentrations are equal throughout the body—in other words, until the body is in a state of osmotic equilibrium. The movement of water across a membrane in response to a solute concentration gradient is called **osmosis**. In osmosis, water moves to dilute the more concentrated solution. Once concentrations are equal, net movement of water stops.

Look at the example shown in Figure 5-29 ■, in which two compartments of equal volume are separated by a selectively permeable membrane that is permeable to water but that does not allow glucose to cross. In ①, compartments A and B contain equal volumes of glucose solution. Compartment B has more solute (glucose) per volume of solution and therefore is the more concentrated solution. A concentration gradient across the membrane exists for glucose, but the membrane is not permeable to glucose, so glucose cannot diffuse to equalize its distribution.

Water, on the other hand, can cross the membrane freely. It will move by osmosis from compartment A, which contains

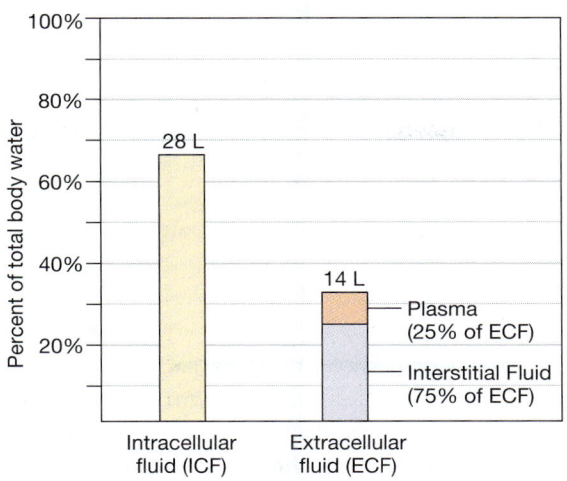

5

■ **FIGURE 5-28** *Distribution of water volume in the three body fluid compartments*

This figure shows the compartment volumes for the "standard" 70-kg man.

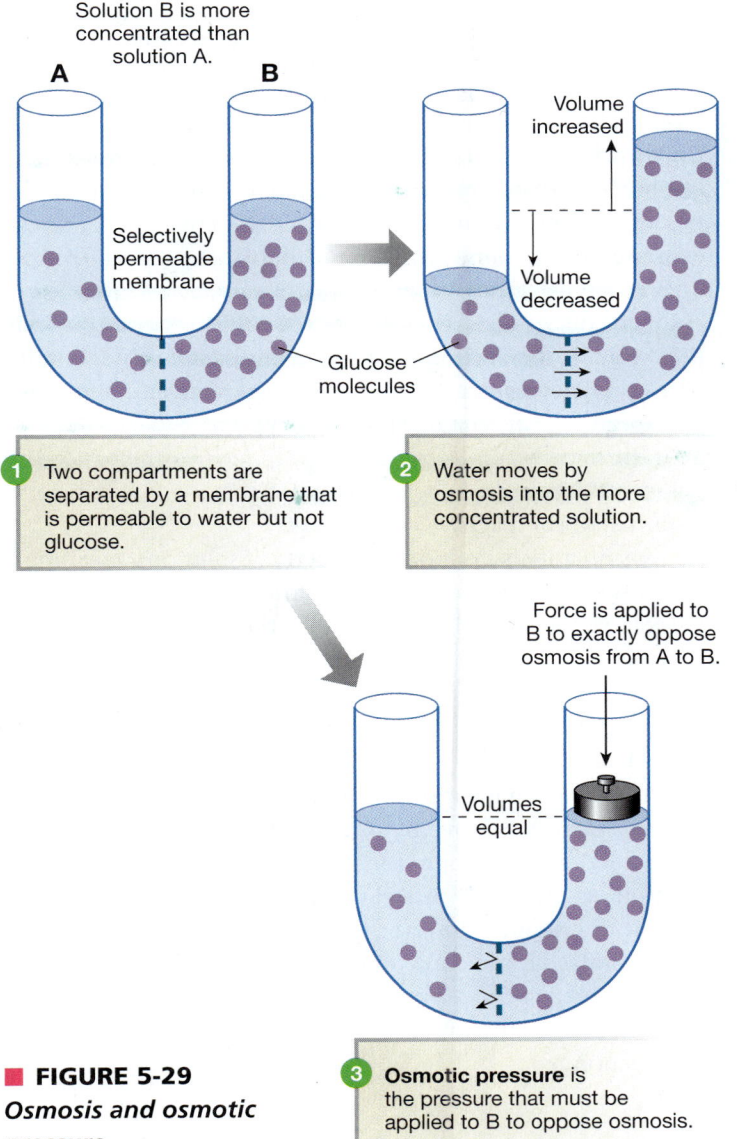

1 Two compartments are separated by a membrane that is permeable to water but not glucose.

2 Water moves by osmosis into the more concentrated solution.

3 **Osmotic pressure** is the pressure that must be applied to B to oppose osmosis.

■ **FIGURE 5-29**

Osmosis and osmotic pressure

the dilute glucose solution, to compartment B, which contains the more concentrated glucose solution. Thus, water moves to dilute the more concentrated solution (Fig. 5-29, ②).

How can we make quantitative measurements of osmosis? One method is shown in Figure 5-29, ③. Place a piston into compartment B, which has a higher solute concentration than compartment A. By pushing down on the piston, you can keep water from flowing from A to B. The pressure that must be applied to the piston to exactly oppose the osmotic movement of water into compartment B is known as the **osmotic pressure** of solution B. The units for osmotic pressure, just as with other pressures in physiology, are *atmospheres* (atm) or *millimeters of mercury* (mm Hg).*

Osmolarity Describes the Number of Particles in Solution

Another way to predict the osmotic movement of water quantitatively is to know the concentrations of the solutions with which we are dealing. In chemistry, concentrations are usually expressed as *molarity* (*M*), which is defined as number of moles of dissolved solute per liter of solution (mol/L). Recall from Chapter 2 that a *mole* is 6.02×10^{23} molecules [⮂ p. 36].

However, using molarity to describe biological concentrations can be misleading. The important factor for osmosis is the number of *particles* in a given volume of solution, not the number of molecules. Because some molecules dissociate into ions when they dissolve in a solution, the number of particles in solution is not always the same as the number of molecules. For example, one glucose molecule dissolved in

*A pressure of 1 mm Hg is equivalent to the hydrostatic pressure exerted on a 1-cm² area by a 1-mm-high column of mercury.

TABLE 5-6	**Comparing Osmolarities**	
SOLUTION A = 1 OsM GLUCOSE	**SOLUTION B = 2 OsM GLUCOSE**	**SOLUTION C = 1 OsM NaCl**
A is hyposmotic to B	B is hyperosmotic to A	C is isosmotic to A
A is isosmotic to C	B is hyperosmotic to C	C is hyposmotic to B

water yields one particle, but one NaCl dissolved in water yields two ions (particles): Na^+ and Cl^-. Water will move osmotically in response to the total concentration of *particles* in the solution. The particles may be ions, uncharged molecules, or a mixture of both.

Consequently, we express the concentration of biological solutions as **osmolarity**, the number of particles (ions or intact molecules) per liter of solution. Osmolarity is expressed in *osmoles* per liter (osmol/L or OsM) or, for very dilute physiological solutions, milliosmoles/liter (mOsM). To convert between molarity and osmolarity, use the following equation:

molarity (mol/L) × number of particles/molecule

= osmolarity (osmol/L)

Let us look at two examples, glucose and sodium chloride, and compare their molarity with their osmolarity.

One mole of glucose molecules dissolved in enough water to create 1 liter of solution yields a 1 molar solution (1 M). Because glucose does not dissociate in solution, the solution has only one mole of osmotically active particles:

1 M glucose × 1 particle per glucose molecule = 1 OsM glucose

Unlike glucose, sodium chloride dissociates into two ions when placed in solution.* Thus, one mole of NaCl dissociates in solution to yield two moles of particles: one mole of Na^+ and one mole of Cl^-. The result is a 2 OsM solution:

1 M NaCl × 2 ions per NaCl = 2 OsM NaCl

Osmolarity describes only the number of particles in the solution. It says nothing about the composition of the particles. A 1 OsM solution could be composed of pure glucose or pure Na^+ and Cl^- or a mixture of the three.

The normal osmolarity of the human body ranges from 280 to 296 milliosmoles per liter (mOsM). In this book, to simplify calculations we will round that number up slightly to 300 mOsM.

A term related to osmolarity that you may hear used is osmolality. **Osmolality** is concentration expressed as osmoles

of solute per kilogram of water. Because biological solutions are dilute and little of their weight comes from solute, physiologists sometimes use the terms *osmolarity* and *osmolality* interchangeably. Osmolality is usually used in clinical situations because it is easy to estimate people's body water content by weighing them.

CONCEPT CHECK

28. A mother brings her baby to the Emergency Room because he has had diarrhea and vomiting for two days. The staff weighs the baby and finds that he has lost 2 pounds. If you assume that all the weight loss is water loss, what volume of water has the baby lost? (2.2 pounds = 1 kilogram)   Answers: p. 173

Comparing Osmolarities of Two Solutions   Osmolarity is a property of every solution. You can compare the osmolarities of different solutions so long as the concentrations are expressed in the same units—for example, as milliosmoles per liter. If two solutions contain the same number of solute particles per unit volume, we say that the solutions are **isosmotic** [*iso-*, equal]. If solution A has a higher osmolarity (contains more particles per unit volume, is more concentrated) than solution B, we say that solution A is **hyperosmotic** to solution B. In the same example, solution B, with fewer osmoles per unit volume, is **hyposmotic** to solution A. Table 5-6 ■ shows some examples of comparative osmolarities.

Osmolarity is a *colligative* property of solutions, meaning it depends strictly on the *number* of particles per liter of solution. Osmolarity says nothing about what the particles are or how they behave. Before we can predict whether osmosis will take place between any two solutions divided by a membrane, we must know the properties of the membrane and of the solutes on each side of it.

If the membrane is permeable only to water and not to any solutes, water will move by osmosis from a less concentrated (hyposmotic) solution into a more concentrated (hyperosmotic) solution, as illustrated in Figure 5-29. Most biological systems are not this simple, however. Biological membranes are selectively permeable and allow some solutes to cross in addition to water. To predict the movement of water into and out of cells, you must know the *tonicity* of the solution, explained in the next section.

---

*For the purposes of this discussion, we will assume that all solutes that can dissociate do so completely (complete dissociation). The actual dissociation constant for NaCl is about 1.8.

129 139 151 **157** 161 168

Tonicity of a Solution Describes the Volume Change of a Cell Placed in That Solution

Tonicity [*tonikos*, pertaining to stretching] is a physiological term used to describe a solution and how that solution affects cell volume. If a cell placed in the solution gains water and swells, we say that the solution is **hypotonic** to the cell. If the cell loses water and shrinks when placed in the solution, the solution is said to be **hypertonic**. If the cell does not change size in the solution, the solution is **isotonic** (Table 5-7 ■). By convention, we always describe the tonicity of the solution relative to the cell. Tonicity describes the cell volume once the cell has come to equilibrium with the solution.

How, then, does tonicity differ from osmolarity?

1. Osmolarity describes the number of solute particles dissolved in a volume of solution. It has units, such as osmoles/liter. The osmolarity of a solution can be measured by a machine called an *osmometer*. Tonicity has no units; it is only a comparative term.

2. Osmolarity can be used to compare any two solutions, and the relationship is reciprocal (solution A is hyperosmotic to solution B; therefore, solution B is hyposmotic to solution A). Tonicity always compares a solution and a cell, and by convention, tonicity is used to describe only the solution—for example, "Solution A is hypotonic to red blood cells."

3. Osmolarity alone will not tell you what happens to a cell placed in a solution. Tonicity by definition tells you what happens to cell volume when the cell is placed in the solution.

This third point is the one that is most confusing to students. Why can't osmolarity be used to predict tonicity? The reason is that the tonicity of a solution depends not only on its concentration (osmolarity) but also on the *nature* of the solutes in the solution. By nature of the solutes, we mean whether the solute particles can cross the cell membrane. If the solute particles (ions or molecules) can enter the cell, we call them **penetrating solutes**. We call particles that cannot cross the cell membrane **nonpenetrating solutes**. Tonicity depends on the concentration of nonpenetrating solutes only. Let's see why this is true.

First, some preliminary information. The most important nonpenetrating solute in physiology is NaCl. If a cell is placed in a solution of NaCl, the Na^+ and Cl^- will not cross the membrane into the cell. (In reality, a few Na^+ ions may leak across, but they are immediately transported back to the extracellular fluid by the Na^+-K^+-ATPase. NaCl is therefore considered a *functionally* nonpenetrating solute.) By convention, we assume that cells are filled with other types of nonpenetrating solutes. In other words, the solutes inside the cell are unable to leave so long as the cell membrane remains intact.

Now we are ready to see why osmolarity alone cannot be used to predict tonicity. Suppose you know the composition and osmolarity of a solution. How can you figure out the tonicity of the solution without actually putting a cell in it? The key lies in knowing *the relative concentrations of nonpenetrating solutes in the cell and in the solution.*

Here are the rules for predicting tonicity:

1. *If the cell has a higher concentration of nonpenetrating solutes than the solution,* there will be net movement of water into the cell. The cell swells, and the solution is *hypotonic.*

	CELL BEHAVIOR WHEN PLACED IN THE SOLUTION	DESCRIPTION OF THE SOLUTION RELATIVE TO THE CELL
TABLE 5-7	**Tonicity of Solutions**	
SOLUTION		
A	Cell swells	Solution A is hypotonic
B	Cell doesn't change size	Solution B is isotonic
C	Cell shrinks	Solution C is hypertonic

(a)

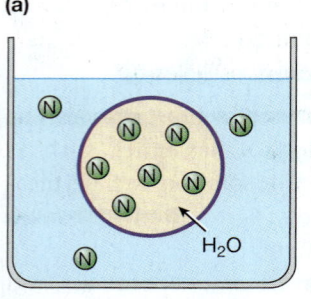

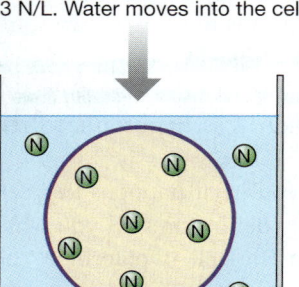

Cell has 6 N/L. Solution has
3 N/L. Water moves into the cell.

At equilibrium, the cell gained
volume. Therefore, the solution
was hypotonic.

(b)

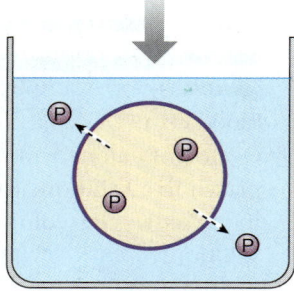

An artificial cell with 4 P/L is
placed in 1 L of pure water.

The penetrating solutes leave
the cell by diffusion, reaching
equilibrium at 2 P/L. No water
moves, so the solution is isotonic.

■ **FIGURE 5-30** *Tonicity depends on the relative concentrations of nonpenetrating solutes*

Water and penetrating solutes can cross the cell membrane and come to equilibrium, but nonpenetrating solutes cannot cross. The "cell" and the solution are each 1 liter in the top figures.

2. *If the cell has a lower concentration of nonpenetrating solutes than the solution,* there will be net movement of water out of the cell. The cell shrinks, and the solution is *hypertonic.*

3. *If the concentrations of nonpenetrating solutes are the same in the cell and the solution,* there is no net movement of water at equilibrium. The solution is *isotonic* to the cell.

Let's look at three examples of how this works. The first two examples will be simple because they use only nonpenetrating solutes or penetrating solutes. The third example will show how combining penetrating and nonpenetrating solutes can complicate the situation.

The artificial cell at the top of Figure 5-30a ■, contains six particles of nonpenetrating solute in 1 liter of cell volume. The 1 liter of solution contains three particles of nonpenetrating solute per liter. As in the tube example of Figure 5-29, there is a concentration gradient for the solute, but the solute cannot cross the cell membrane to equilibrate. There is also an osmotic gradient, however, and because water *can* cross the membrane, water moves into the cell until the solute concentrations have equilibrated (bottom of Fig. 5-30a). The cell gains volume at equilibrium, which means the solution was hypotonic to the cell. We could have predicted this by using rule 1: if the cell has

a higher concentration of nonpenetrating solutes than the solution, there will be net movement of water into the cell.

Now let's consider a situation in which an artificial cell with a volume of 1 liter has four particles of penetrating solute and is placed in 1 liter of pure water (top of Fig. 5-30b). Based on osmolarity alone, you might think that water would move into the cell because the cell is more concentrated. However, there is a concentration gradient for the solute, and the solute is able to cross the cell membrane. Solute therefore moves out of the cell by diffusion until solute concentrations in the cell and the solution reach equilibrium (bottom of Fig. 5-30b). Once the penetrating solute is at equilibrium, the cell and solution have equal osmolarities (two particles/liter), and the net movement of water is zero. The solution in this instance was isotonic. You could have predicted this from rule 3 above: if the concentrations of nonpenetrating solutes are the same in the cell and the solution, there is no net movement of water at equilibrium and the solution is isotonic. In this example, the concentrations of nonpenetrating solute in cell and solution were both zero. Remember, however, that we assume that real cells always contain only nonpenetrating solutes.

Now let's look at a complex example that mixes penetrating and nonpenetrating solutes. In Figure 5-31a ■, the artificial cell contains six particles of nonpenetrating solute in a 1 liter volume. The 1 liter of solution also contains six particles of solute: three nonpenetrating and three penetrating. Because cell and solution have the same concentrations (six particles per liter), they are isosmotic, and no water will move initially when the two are placed together (Fig. 5-31b).

Using the tonicity rules above, let us compare the concentrations of nonpenetrating solutes in the solution and cell. The nonpenetrating solutes are more concentrated in the cell than in the solution (six versus three). We therefore predict that water will move from the solution into the cell, increasing cell volume. However, the solution is isosmotic to the cell, and so what will cause osmosis?

You have learned that penetrating solutes move freely into and out of cells. In this example, there is a concentration gradient for the penetrating solute: 3/L in the solution versus 0/L in the cell. The penetrating solute therefore will diffuse down its concentration gradient and move into the cell. As soon as one particle moves in, we have a total of five solutes outside the cell but seven inside. This solute imbalance disturbs the osmotic balance and creates an osmotic gradient (Fig. 5-31c). Once this osmotic gradient exists, water moves into the cell and cell volume increases (solute movement has no significant effect on cell volume). Water movement continues until the concentrations of *nonpenetrating* solutes are equal, as in Figure 5-31d. Thus, in this example, an isosmotic solution is hypotonic because cell volume increased (rule 1).

If accepting the rules makes you uneasy, let's look at this example mathematically. We began with 1 L of solution in the cell and 1 L outside, a total of 2 L, and a total of 12 particles,

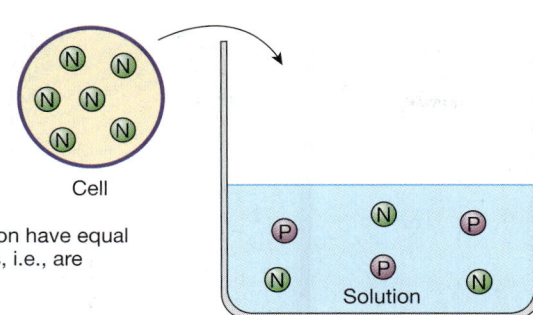

(a) Cell and solution have equal concentrations, i.e., are isosmotic.

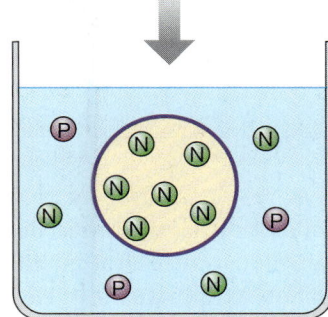

(b) When the cell is placed in the solution, no water moves initially because the cell and solution are in osmotic equilibrium.

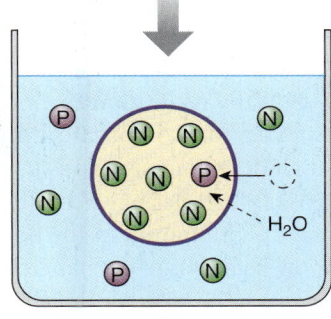

(c) However, there is a concentration gradient for the penetrating solute ⓟ, which diffuses into the cell. This disrupts the osmotic equilibrium, so water follows the solute into the cell.

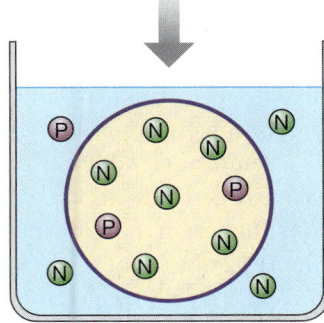

(d) Equilibrium is restored when the concentrations of nonpenetrating solutes Ⓝ are equal in the cell and the solution. The cell gained volume at equilibrium; therefore, the solution was hypotonic.

FIGURE QUESTION

Using the same cell, give the relative osmolarity and the tonicity of the following solutions, if the solution has
(a) 6 nonpenetrating particles
(b) 6 nonpenetrating and 3 penetrating particles or
(c) 3 nonpenetrating and 6 penetrating particles

■ **FIGURE 5-31**   *In a solution of mixed solutes, only the concentration of nonpenetrating solutes contributes to the tonicity of the solution*

TABLE 5-8	Rules for Osmolarity and Tonicity

1. Assume that all intracellular solutes are nonpenetrating.
2. Compare osmolarities before the cell and solution are put together, because at equilibrium, the cell contents and the solution outside the cell will be isosmotic.
3. Tonicity of a solution describes the volume change of a cell at equilibrium (Table 5-7).
4. Determine tonicity by comparing nonpenetrating solute concentrations in the cell and the solution. Net water movement will be into the compartment that has the higher concentration of nonpenetrating solutes.
5. A solution that is hyposmotic to a cell will always be hypotonic.

giving a concentration of 6 particles/L. The system will not gain or lose solute or water, so the overall concentration will remain 6 particles/L. When the cell is placed in the solution, the two penetrating solutes move into the cell, disturbing the osmotic balance. Water has to enter the cell so that the concentration remains at 6 particles/L:

$$8 \text{ particles/? liters} = 6 \text{ particles/L}$$

The cell volume at equilibrium will be 1.33 liters: 8 particles/1.33 L = 6 particles/L. The solution has lost 0.33 liter of water that moved into the cell:

$$4 \text{ particles}/(1 - 0.33) \text{ L} = 6 \text{ particles/L}$$

At equilibrium, the concentrations of the cell and the solution will always be the same.

Learning the rules rather than depending on mathematical calculations is important for clinical situations, when you will not know exact volumes for the person needing fluids. Table 5-8 ■ lists some rules to help you distinguish between osmolarity and tonicity.

Understanding the difference between the two properties is critical to making good clinical decisions about intravenous (IV) fluid therapy. In medicine, the tonicity of a solution is an important consideration. One purpose of IV fluids is to get water into dehydrated cells (in which case, a hypotonic IV solution is used) or to keep fluid in the extracellular fluid to replace blood loss (in which case, an isotonic IV solution is used). The choice of fluid depends on how the clinician wants the solutes and water to distribute between the extracellular and intracellular fluid compartments. Table 5-9 ■ lists the most common IV solutions and their approximate osmolarity and tonicity relative to the normal human cell.

TABLE 5-9	Intravenous Solutions		
SOLUTION	**ALSO KNOWN AS**	**OSMOLARITY**	**TONICITY**
0.9% saline*	Normal saline	Isosmotic	Isotonic
D_5—0.9% saline	5% dextrose** in normal saline	Hyperosmotic	Isotonic
D_5W	5% dextrose in water	Isosmotic	Hypotonic
0.45% saline	Half-normal saline	Hyposmotic	Hypotonic
D_5—0.45% saline	5% dextrose in half-normal saline	Hyperosmotic	Hypotonic

* Saline = NaCl.   **Dextrose = glucose.

CONCEPT CHECK

29. Which of the following solutions has/have the most water per unit volume: 1 M glucose, 1 M NaCl, or 1 OsM NaCl?

30. Two compartments are separated by a membrane that is permeable to water and urea but not to NaCl. Which way will water move when the following solutions are placed in the two compartments?

Compartment A	Membrane	Compartment B
(a) 1 M NaCl	|	1 OsM NaCl
(b) 1 M urea	|	2 M urea
(c) 1 OsM NaCl	|	1 OsM urea

31. You have a patient who lost 1 liter of blood, and you need to restore volume quickly while waiting for a blood transfusion to arrive from the blood bank.

    (a) Which would be better to administer: 5% dextrose (another name for glucose) in water or 0.9% NaCl in water? (Hint: think about how these solutes distribute in the body.) Defend your choice.
    (b) How much of your solution of choice would you have to administer to return blood volume to normal?

    Answers: p. 173

THE RESTING MEMBRANE POTENTIAL

Many of the body's solutes, including organic compounds such as pyruvate and lactate, are ions and therefore carry a net electrical charge. Potassium (K^+) is the major cation within cells, and sodium (Na^+) dominates the extracellular fluid (see Fig. 5-3, p. 131). Chloride ions (Cl^-) mostly remain with Na^+ in the extracellular fluid, whereas phosphate ions and negatively charged proteins are the major anions of the intracellular fluid.

However, the intracellular compartment is not electrically neutral: there are some protein anions inside cells that do not have matching cations, giving the cells a net negative charge. At the same time, the extracellular compartment has a net positive charge: some cations in the extracellular fluid do not have matching anions. One consequence of this uneven distribution of ions is that the intracellular and extracellular compartments are not in electrical equilibrium. Instead, the two compartments exist in a state of electrical disequilibrium [🔁 page 131].

The concept of electrical disequilibrium has traditionally been taught in chapters on nerve and muscle function because those tissues generate electrical signals known as action potentials. Yet one of the most exciting recent discoveries in physiology is the realization that other kinds of cells also use electrical signals for communication. In fact, all living organisms, including plants, use electrical signals! This section reviews the basic principles of electricity and discusses what creates electrical disequilibrium in the body. The chapter ends with a look at how beta cells of the pancreas use changes in the distribution of ions across cell membranes to trigger insulin secretion.

Electricity Review Atoms are electrically neutral [🔁 p. 20]. They are composed of positively charged protons, negatively charged electrons, and uncharged neutrons, but in balanced proportions, so that an atom is neither positive nor negative. The removal or addition of electrons to an atom creates the charged particles we know as ions. We have discussed several ions that are important in the human body, such as Na^+, K^+, and H^+. For each of these positive ions, somewhere in the body there is a matching electron, usually found as part of a negative ion. For example, when Na^+ in the body enters in the form of NaCl, the "missing" electron from Na^+ can be found on the Cl^-.

The following principles are important to remember when dealing with electricity in physiological systems:

1. The **law of conservation of electrical charge** states that the net amount of electrical charge produced in any process is zero. This means that for every positive charge on an ion, there is an electron on another ion. Overall, the human body is electrically neutral.

2. Opposite charges (+ and −) are attracted to each other, but two charges of the same type (+ and +, or − and −) repel each other. The protons and electrons in an atom exhibit this attraction.

3. Separating positive charges from negative charges requires energy. For example, energy is needed to separate the protons and electrons of an atom.

4. If separated positive and negative charges can move freely toward each other, the material through which they are moving is called a **conductor**. Water is a good conductor of electrical charge. If separated charges are unable to move through the material that separates them, the material is known as an **insulator**. The phospholipid bilayer of the cell membrane is a good insulator, as is the plastic coating on electrical wires.

The word *electricity* comes from the Greek word *elektron*, meaning "amber," the fossilized resin of trees. The Greeks discovered that if they rubbed a rod of amber with cloth, the amber acquired the ability to attract hair and dust. This attraction (called static electricity) arises from the separation of electrical charge that occurs when electrons move from the amber atoms to the cloth. To separate these charged particles, energy (work) must be put into the system. In the case of the amber, work was done by rubbing the rod. In the case of biological systems, the work is usually done by energy stored in ATP and other chemical bonds.

The Cell Membrane Enables Separation of Electrical Charge in the Body

In the body, separation of electrical charge takes place across the cell membrane. This process is shown in Figure 5-32 ■. The diagram shows an artificial cell filled with molecules that dissociate into positive and negative ions, represented by the plus and

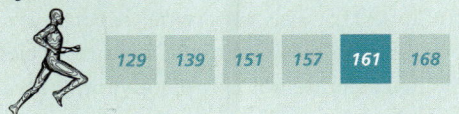
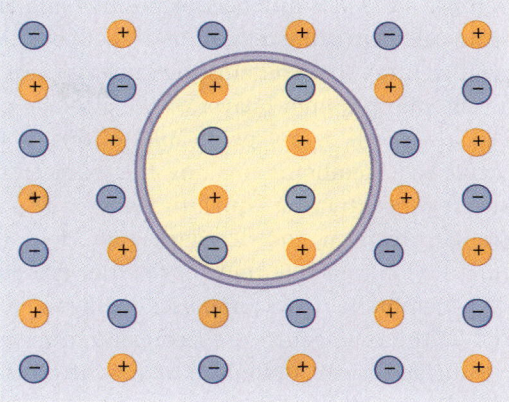

(a) Cell and solution are electrically and chemically at equilibrium.

(b) Cell and solution in chemical and electrical disequilbrium. Energy is used to pump one cation out of the cell, leaving a net charge of -1 in the cell and +1 outside the cell.

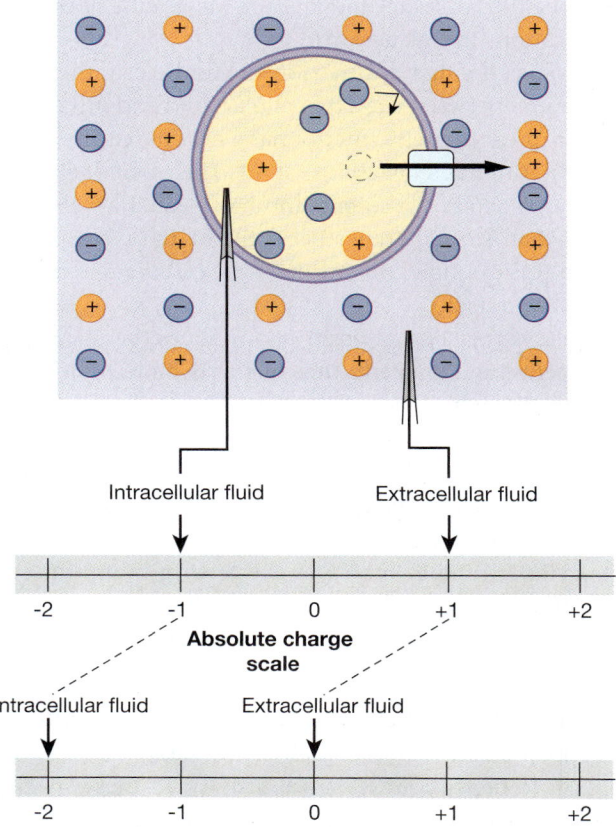

Intracellular fluid   Extracellular fluid

-2   -1   0   +1   +2
Absolute charge scale

Intracellular fluid   Extracellular fluid

-2   -1   0   +1   +2
Relative charge scale—extracellular fluid set to 0 (ground)

(c) On an absolute charge scale, the extracellular fluid (ECF) would be at +1 and the intracellular fluid (ICF) at -1. Physiological measurements, however, are always on a relative scale, on which the extracellular fluid is assigned a value of zero. This shifts the scale to the left and gives the inside of the cell a relative charge of -2.

■ **FIGURE 5-32** *Separation of electrical charge*

The cell membrane acts as an insulator to prevent free movement of ions between the intracellular and extracellular compartments.

minus signs. Because the molecules were electrically neutral to begin with, there are equal numbers of positive and negative ions inside the cell. The cell is placed in an aqueous solution, also electrically neutral, that contains the same types of cations and anions. The phospholipid bilayer of the artificial cell, like the bilayer of a real cell, is not permeable to ions. Water can freely cross this cell membrane, making the extracellular and intracellular ion concentrations equal. In Figure 5-32a, the system is at osmotic, chemical, and electrical equilibrium.

In Figure 5-32b, an active transport carrier protein is inserted into the membrane. This carrier uses energy to move positive ions out of the cell against their concentration gradient. The negative ions in the cell attempt to follow the positive ions because of the attraction of positive and negative charges. Because the membrane is impermeable to negative ions, however, they remain trapped in the cell. Positive ions outside the cell might try to move into the cell, attracted by the net negative charge of the intracellular fluid, but the membrane does not allow these cations to leak across it.

As soon as the first positive ion leaves the cell, the electrical equilibrium between the extracellular fluid and intracellular fluid is disrupted: the cell's interior has a net charge of -1 while the cell's exterior has a net charge of $+1$. The input of energy to transport ions across the membrane has created an **electrical gradient**—that is, a difference between the net charge in two regions. In this example, the inside of the cell became negative relative to the outside.

The active transport of positive ions out of the cell also creates a concentration gradient: there are now more positive ions outside the cell than inside. The combination of electrical and concentration gradients is called an **electrochemical gradient**. The cell remains in osmotic equilibrium because water can move freely across the membrane in response to solute movement.

An electrical gradient between the extracellular fluid and the intracellular fluid is known as the **resting membrane potential difference**, or **membrane potential** for short. Although the name sounds intimidating, we can break it apart to see what it means.

1. The *resting* part of the name comes from the fact that this electrical gradient is seen in all living cells, even those that appear to be without electrical activity. In these "resting" cells, the membrane potential has reached a steady state and is not changing.

2. The *potential* part of the name comes from the fact that the electrical gradient created by active transport of ions across the cell membrane is a form of stored, or potential, energy, just as concentration gradients are a form of potential energy. When oppositely charged molecules come back together, they release energy that can be used to do work, in the same way that molecules moving down their concentration gradient can do work. The work done by electrical energy includes opening voltage-gated membrane channels and sending electrical signals.

3. The *difference* part of the name is to remind you that the membrane potential represents a difference in the amount of electrical charge inside and outside the cell. The word *difference* is usually dropped from the name, as noted earlier.

In living systems, we measure electrical gradients on a relative scale rather than an absolute scale. Figure 5-32c compares the two scales. On the absolute scale, the extracellular fluid in our simple example has a net charge of $+1$ from the positive ion it gained, and the intracellular fluid has a net charge of -1 from the now-unbalanced negative ion that was left behind.

However, in real life we cannot measure the charges as numbers of electrons gained or lost. Instead we use a device that measures the *difference* in electrical charge between two points. This device artificially sets the net electrical charge of one side of the membrane at zero and measures the net charge of the second side relative to the first. In our example, resetting the extracellular fluid net charge to zero gives the intracellular fluid a net charge of -2, and we call this value the cell's resting membrane potential.

The equipment for measuring a cell's membrane potential is depicted in Figure 5-33 ■. *Electrodes* are created from hollow glass tubes drawn to very fine points. These *micropipets* are filled with a liquid that conducts electricity and then connected to a *voltmeter,* which measures the electrical difference between two points in units of either volts (V) or millivolts (mV). A *recording electrode* is inserted through the cell membrane into the cytoplasm of the cell. A *reference electrode* is placed in the external bath, which represents the extracellular fluid.

In living systems, by convention, the extracellular fluid is designated as the *ground* and assigned a charge of 0 mV (Fig. 5-32c). When the recording electrode is placed inside a living cell, the voltmeter measures the membrane potential—in other words, the electrical difference between the intracellular fluid and the extracellular fluid. A chart recorder connected to the voltmeter can make a recording of the membrane potential versus time.

For nerve and muscle cells, the voltmeter will record a resting membrane potential between -40 and -90 mV, indicating that the intracellular fluid is negative relative to the extracellular fluid (0 mV). (Throughout this discussion, remember that, as you saw in Figure 5-32c, the extracellular fluid is not really neutral because it has excess positive charges that exactly balance the excess negative charges inside the cell. The total body remains electrically neutral at all times.)

The Resting Membrane Potential Is Due Mostly to Potassium

Which ions create the resting membrane potential in animal cells? The artificial cell shown in Figure 5-32b used an active transport protein to move an unspecified positively charged ion across a membrane that was otherwise impermeable to ions. But what processes go on in living cells to create an electrical gradient?

Real cells are not completely impermeable to all ions. They have open channels and protein transporters that allow ions to

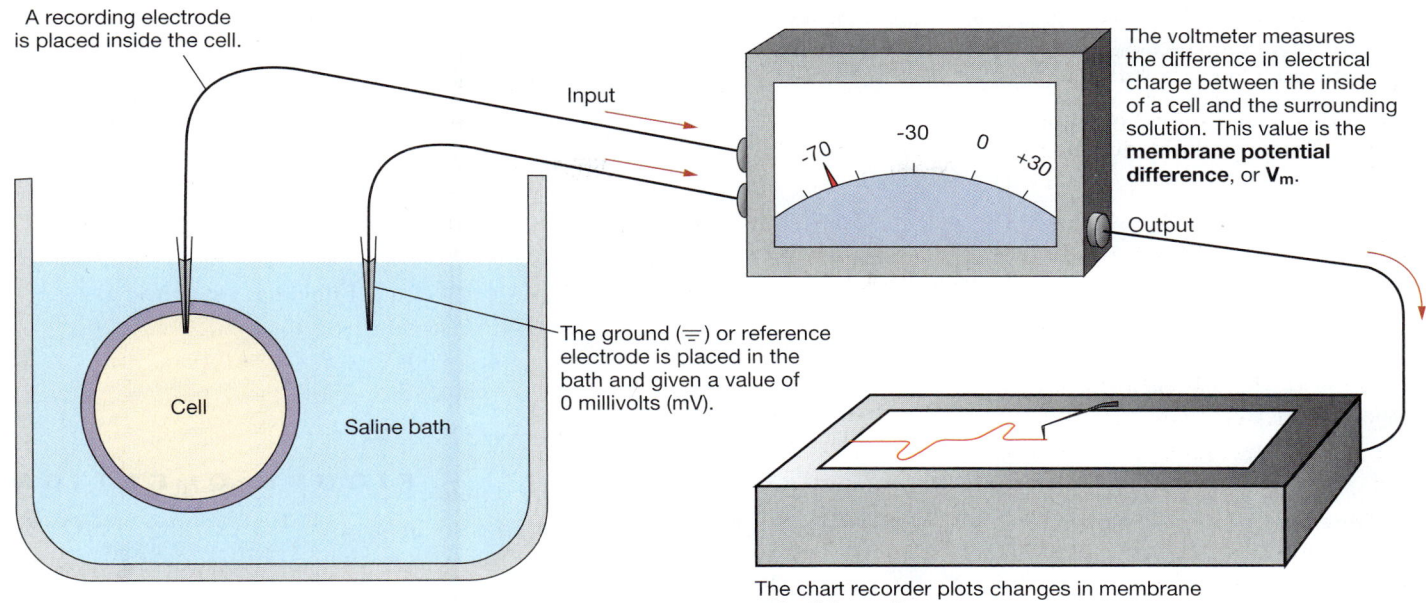

A recording electrode is placed inside the cell.

Input

The voltmeter measures the difference in electrical charge between the inside of a cell and the surrounding solution. This value is the **membrane potential difference**, or **V_m**.

-70 -30 0 +30

Output

The ground (⏚) or reference electrode is placed in the bath and given a value of 0 millivolts (mV).

Cell

Saline bath

The chart recorder plots changes in membrane potential over time.

■ FIGURE 5-33   *Measuring membrane potential difference*

In the laboratory, a voltmeter measures the difference in electrical charge between the inside of a cell and the surrounding solution.

move between the cytoplasm and the extracellular fluid. We can use a different artificial cell to show how the resting membrane potential arises in a typical living cell.

The artificial cell in Figure 5-34a ■ has a membrane that is impermeable to ions. The cell contains K^+ and large negatively charged proteins, represented by Pr^-. The cell is placed in a solution of Na^+ and Cl^-. Both the cell and the solution are electrically neutral, and the system is in electrical equilibrium. However, it is not in chemical equilibrium. There are concentration gradients for all four types of ions in the system, and they would all diffuse down their respective concentration gradients if they could cross the cell membrane.

In Figure 5-34b, a K^+ leak channel is inserted into the membrane, making it permeable only to K^+. Because there is no K^+ in the extracellular fluid initially, some K^+ will leak out of the cell, moving down their concentration gradient. As K^+ leaves the cell, the negatively charged proteins, Pr^-, are unable to follow because the cell membrane is not permeable to them. The proteins gradually build up a negative charge inside the cell as more and more K^+ diffuses out of the cell.

If the only force acting on K^+ were the concentration gradient, K^+ would leak out of the cell until the K^+ concentration inside the cell equaled the K^+ concentration outside. The loss of positive ions from the cell creates an electrical gradient, however. Because opposite charges attract each other, the negative Pr^- inside the cell try to pull K^+ back into the cell. At some point in this process, the electrical force attracting K^+ into the cell becomes equal in magnitude to the chemical concentration gradient driving K^+ out of the cell. At that point, net movement of K^+ across

the membrane stops (Fig. 5-34c). The rate at which K^+ move out of the cell down the concentration gradient is exactly equal to the rate at which K^+ move into the cell down the electrical gradient.

In a cell that is permeable to only one ion, such as the artificial cell just described, the membrane potential that exactly opposes the concentration gradient of the ion is known as the **equilibrium potential**, or **E_{ion}** (where the subscript *ion* is replaced by the symbol for whichever ion we are looking at). For example, when the concentration gradient is 150 mM K^+ inside and 5 mM K^+ outside the cell, the equilibrium potential for potassium, or E_K, is −90 mV. The equilibrium potential for any ion at 37° C (human body temperature) can be calculated using the Nernst equation:

$$E_{ion} = \frac{61}{z} \log \frac{[ion]_{out}}{[ion]_{in}}$$

where   61 is 2.303 RT/F at 37° C,*
        z is the electrical charge on the ion (+1 for K^+), and
        $[ion]_{out}$ and $[ion]_{in}$ are the ion concentration outside and inside the cell.

Now we will use the same artificial cell (K^+ and Pr^- inside, Na^+ and Cl^- outside), but this time we will make the membrane permeable only to Na^+ (Fig. 5-35■). Because Na^+ is more concentrated outside the cell, some Na^+ moves into the cell and accumulates there. Meanwhile, Cl^- left behind in the extracellular

---

*R is the ideal gas constant, T is absolute temperature, and F is the Faraday constant. For additional information, see Appendix A.

(a) An artificial cell whose membrane is impermeable to ions is filled with K$^+$ and large protein anions. It is placed in a solution of Na$^+$ and Cl$^-$. Both cell and solution are electrically neutral.

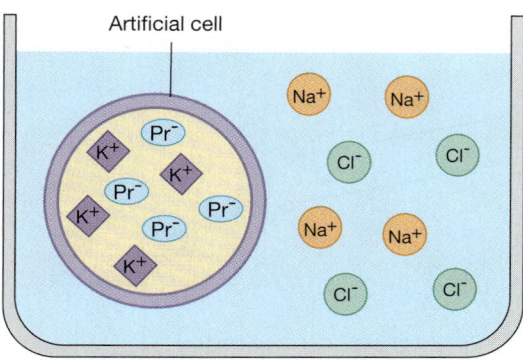

Artificial cell

(b) A K$^+$ leak channel is inserted into the membrane. K$^+$ leaks out of the cell because there is a K$^+$ concentration gradient.

K+ leak channel

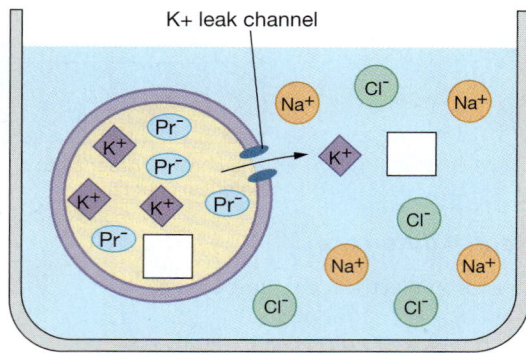

FIGURE QUESTION

In the white boxes write the net electrical charge of the intracellular and extracellular compartments as shown.

(c) The negative membrane potential attracts K$^+$ back into the cell. When the electrical gradient exactly opposes the K$^+$ concentration gradient, the resting membrane potential is the **equilibrium potential** for K$^+$ (E$_K$).

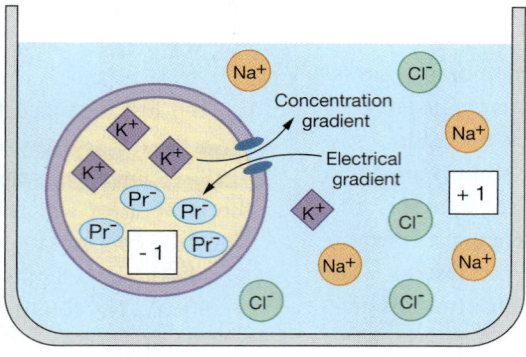

Concentration gradient

Electrical gradient

FIGURE QUESTION

Why don't Na$^+$, Cl$^-$, and the proteins (Pr$^-$) cross the membrane?

■ **FIGURE 5-34** *Potassium equilibrium potential*

The potassium equilibrium potential, or E$_K$, is the membrane potential at which the chemical and electrical gradients are equal in magnitude and opposite in direction, resulting in no net movement of K$^+$.

fluid gives that compartment a net negative charge. This imbalance creates an electrical gradient that tends to drive Na$^+$ back out of the cell. When the Na$^+$ concentration is 150 mM outside and 15 mM inside, the equilibrium potential for Na$^+$ (E$_{Na}$) is +60 mV. In other words, the concentration gradient moving Na$^+$ into the cell (150 mM outside, 15 mM inside) is exactly opposed by a positive membrane potential of +60 mV.

In reality, living cells are not permeable to only one ion. The situation in real cells is similar to a combination of the two artificial systems just described. If a cell is permeable to several ions, we cannot use the Nernst equation to calculate membrane potential. Instead we must use a related equation called the *Goldman equation* that considers concentration gradients of the permeable ions and the relative permeability of the cell to each ion. For more detail on the Goldman equation, see Chapter 8.

The cell illustrated in Figure 5-36 ■ has a resting membrane potential of −70 mV. Most cells are about 40 times more permeable to K$^+$ than to Na$^+$, and as a result a cell's resting membrane potential is closer to the E$_K$ of −90 mV than to the E$_{Na}$ of +60 mV. A small amount of Na$^+$ leaks into the cell, making the

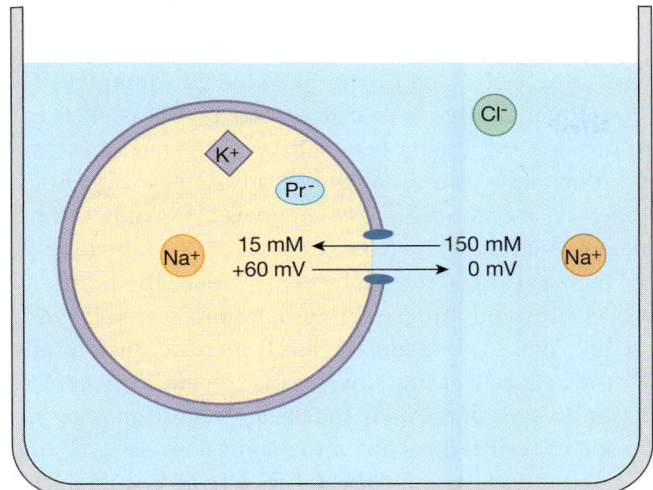

The Na$^+$ concentration gradient shown is exactly opposed by a membrane potential of +60 mV. Therefore, +60 mV is the E$_{Na}$ or Na$^+$ equilibrium potential.

■ **FIGURE 5-35** *Sodium equilibrium potential*

In this book, ion channels will be represented by open pores on the membrane.

inside of the cell less negative than it would be if Na$^+$ were totally excluded. Additional Na$^+$ that leaks in is promptly pumped out by the Na$^+$-K$^+$-ATPase, as described earlier. At the same time, K$^+$ ions that leak out of the cell are pumped back in. The pump contributes to the membrane potential by pumping 3 Na$^+$ out for every 2 K$^+$ pumped in. Because the Na$^+$-K$^+$-ATPase helps maintain the electrical gradient, it is called an *electrogenic* pump.

Not all ion transport creates an electrical gradient. Many transporters, like the Na$^+$-K$^+$-2 Cl$^-$ (NKCC) symporter, are electrically neutral. Some make an even exchange: for each charge that enters the cell, the same charge leaves. An example is the HCO$_3^-$-Cl$^-$ antiporter of red blood cells, which transports these ions in a one-for-one, electrically neutral exchange. Electrically neutral transporters have little effect on the resting membrane potential of the cell.

CONCEPT CHECK

32. Add a Cl$^-$ leak channel to the artificial cell shown in Figure 5-34a, and then figure out which way Cl$^-$ will move along the concentration and electrical gradients. Will the Cl$^-$ equilibrium potential be positive or negative?

33. What would happen to the resting membrane potential of a cell poisoned with ouabain (an inhibitor of the Na$^+$-K$^+$-ATPase)?   Answers: p. 173

Changes in Ion Permeability Change the Membrane Potential

As you have just learned, two factors influence a cell's membrane potential: (1) the concentration gradients of different ions across the membrane and (2) the permeability of the membrane to

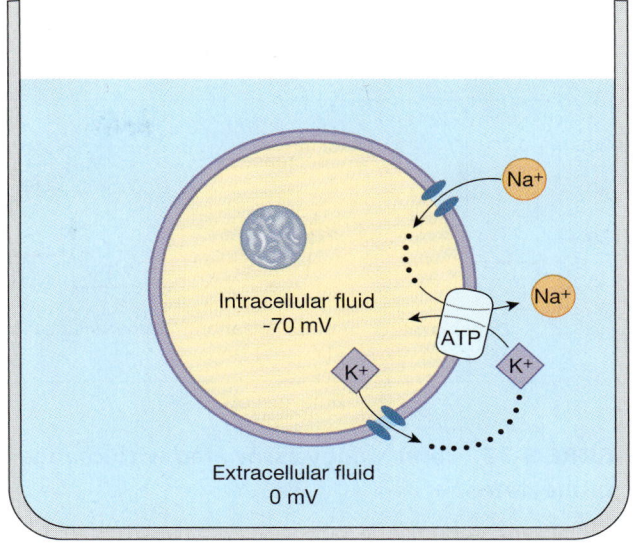

FIGURE QUESTIONS
- What force(s) promote(s) Na$^+$ leak into the cell?
- What force(s) promote(s) K$^+$ leak out of the cell?

■ **FIGURE 5-36** *Resting membrane potential in an actual cell*

Most cells in the human body are about 40 times more permeable to K$^+$ than to Na$^+$, and the resting membrane potential is about -70 mV.

those ions. If the cell's permeability to an ion changes, the cell's membrane potential changes. We monitor changes in membrane potential using the same intracellular recording electrodes that we use to record resting membrane potential (Fig. 5-33).

Figure 5-37 ■ shows a recording of membrane potential plotted against time. The membrane potential (V$_m$) begins at a steady resting value of -70 mV. When the trace moves upward (becomes less negative), the potential difference between the inside of the cell and the outside (set at 0 mV) decreases, and the cell is said to have *depolarized*. A return to the resting membrane potential is termed *repolarization*. If the resting potential moves away from 0 mV, the membrane potential becomes more negative, the potential difference has increased, and the cell has *hyperpolarized*.

A major point of confusion when talking about changes in membrane potential is the use of the phrases "the membrane potential decreased" or "the membrane potential increased." Normally, we associate "increase" with becoming more positive and "decrease" with becoming more negative—the opposite of what is happening in our cell discussion. One way to avoid confusion is to add the word *difference* after *membrane potential*. If the membrane potential *difference* is *increasing*, the value of V$_m$ must be moving away from the ground value of zero and becoming *more negative*. If the membrane potential *difference* is

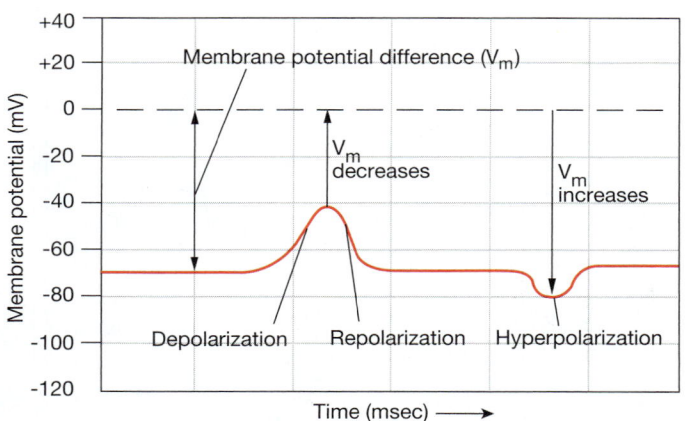

■ **FIGURE 5-37** *Terminology associated with changes in membrane potential*

If the membrane potential becomes less negative, the cell depolarizes. If the membrane potential becomes more negative than the resting potential, the cell hyperpolarizes.

decreasing, the value of V_m is moving closer to the ground value of 0 mV and is becoming *less negative.*

Four ions usually contribute the most to changes in membrane potential: Na^+, Ca^{2+}, Cl^-, and K^+. The first three are more concentrated in the extracellular fluid than in the cytosol, and the resting cell is minimally permeable to them. If a cell suddenly becomes more permeable to any one of these ions, then those ions will move across the membrane into the cell. Entry of Ca^{2+} or Na^+ will depolarize the cell (make the membrane potential more positive). Entry of Cl^- will hyperpolarize the cell (make the membrane potential more negative).

Most resting cells are fairly permeable to K^+, but making them more permeable will allow even more K^+ to leak out. The cell will hyperpolarize until it approaches the equilibrium potential for K^+. Making the cell *less* permeable to K^+ allows fewer K^+ to leak out of the cell. When the cell retains K^+, it becomes more positive and depolarizes. You will encounter instances of all these permeability changes as you study physiology.

It is important to remember that a significant change in membrane potential requires the movement of very few ions. *You need not reverse the concentration gradient to change the membrane potential.* For example, to change the membrane potential by 100 mV, only one of every 100,000 K^+ must enter or leave the cell. This is such a tiny fraction of the total number of K^+ in the cell that the concentration of K^+ remains essentially unchanged.

INTEGRATED MEMBRANE PROCESSES: INSULIN SECRETION

The movement of Na^+ and K^+ across cell membranes has been known to play a role in generating electrical signals in excitable tissues for many years. You will study these processes in detail in the chapters on the nervous and muscular systems. Recently,

however, we have come to understand that small changes in membrane potential act as signals in nonexcitable tissues, such as endocrine cells. One of the best-studied examples of this process involves the beta cell of the pancreas. Release of the hormone insulin by beta cells demonstrates how membrane processes—such as facilitated diffusion, exocytosis, and the opening and closing of ion channels by ligands and membrane potential—work together to regulate cell function.

The beta cells of the pancreas synthesize the protein hormone insulin and store it in cytoplasmic secretory vesicles [⊟ p. 66]. When blood glucose levels increase, such as after a meal, the beta cells release insulin by exocytosis. Insulin then directs other cells of the body to take up and use glucose, bringing blood concentrations down to pre-meal levels.

A key question about the process that went unanswered until recently was, "How does a beta cell 'know' that glucose levels have gone up and that it needs to release insulin?" The answer, we have now learned, links the beta cell's metabolism to its electrical activity.

Figure 5-38a ■ shows a beta cell at rest. Recall from earlier sections in this chapter that the gates of membrane channels can be opened or closed by chemical or electrical signals. The beta cell has two such channels that help control insulin release. One is a **voltage-gated Ca^{2+} channel**. This channel is closed at the cell's resting membrane potential (⑤ in Fig. 5-38a). The other is a K^+ leak channel (that is, the channel is usually open) that closes when ATP binds to it. It is called an **ATP-gated K^+ channel**, or K_{ATP} **channel**. In the resting cell, when glucose concentrations are low, the cell makes less ATP ① through ③. There is little ATP to bind to the K_{ATP} channel, and the channel remains open, allowing K^+ to leak out of the cell ④. At the resting membrane potential, the voltage–gated Ca^{2+} channels are closed, and there is no insulin secretion ⑤.

Figure 5-38b shows a beta cell secreting insulin. Following a meal, plasma glucose levels increase as glucose is absorbed from the intestine ①. Glucose reaching the beta cell diffuses into the cell with the aid of a GLUT transporter. Increased glucose in the cell stimulates the metabolic pathways of glycolysis and the citric acid cycle [⊟ p. 103], and ATP production increases ②, ③. When ATP binds to the K_{ATP} channel, the gate to the channel closes, preventing K^+ from leaking out of the cell ④. Retention of K^+ depolarizes the cell ⑤, which then causes the voltage-sensitive Ca^{2+} channels to open ⑥. Calcium ions enter the cell from the extracellular fluid, moving down their electrochemical gradient. The Ca^{2+} binds to proteins that initiate exocytosis of the insulin-containing vesicles, and insulin is released into the extracellular space ⑦.

The discovery that cells other than nerve and muscle cells use changes in membrane potential as signals for physiological responses changes our traditional thinking about the role of the resting membrane potential. In the next chapter, we will describe some other types of signals that the body uses for communication and coordination.

(a) Beta cell at rest. The K_{ATP} channel is open and the cell is at its resting membrane potential.

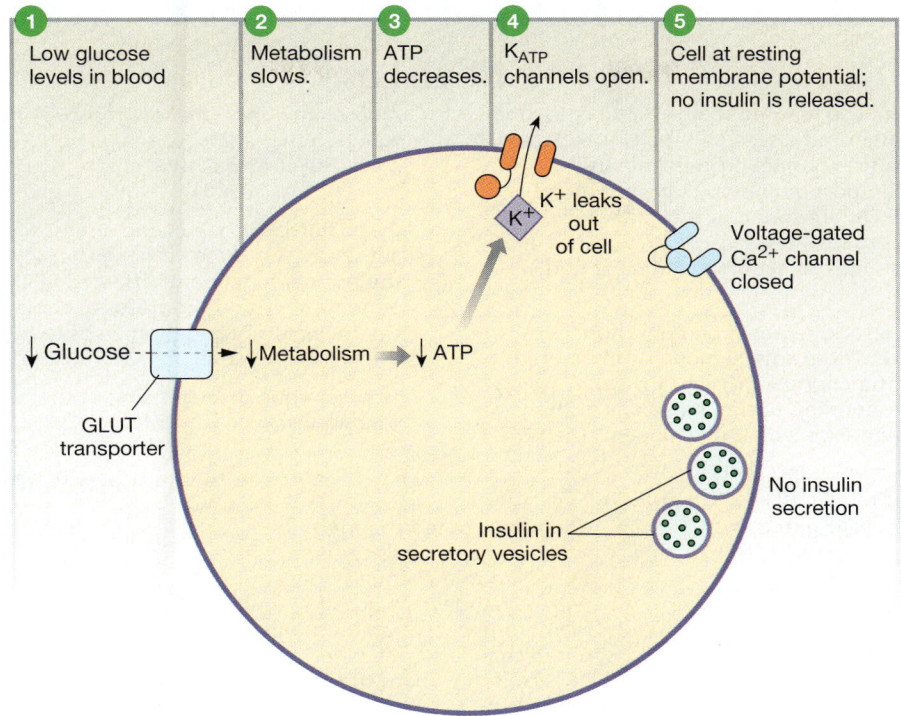

1	2	3	4	5
Low glucose levels in blood	Metabolism slows.	ATP decreases.	K_{ATP} channels open.	Cell at resting membrane potential; no insulin is released.

K^+ leaks out of cell

K^+

Voltage-gated Ca^{2+} channel closed

↓Glucose → ↓Metabolism ⇒ ↓ATP

GLUT transporter

Insulin in secretory vesicles

No insulin secretion

(b) Beta cell secreting insulin. Closure of the K_{ATP} channel depolarizes the cell, triggering exocytosis of insulin.

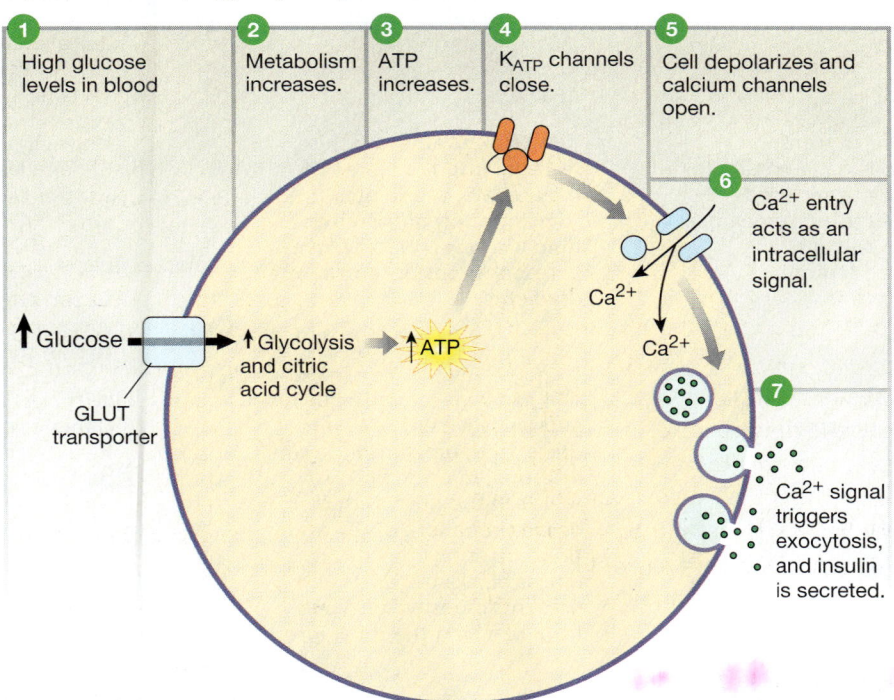

1	2	3	4	5
High glucose levels in blood	Metabolism increases.	ATP increases.	K_{ATP} channels close.	Cell depolarizes and calcium channels open.

6 Ca^{2+} entry acts as an intracellular signal.

Ca^{2+}

Ca^{2+}

↑Glucose → ↑Glycolysis and citric acid cycle ⇒ ↑ATP

GLUT transporter

7 Ca^{2+} signal triggers exocytosis, and insulin is secreted.

■ **FIGURE 5-38**   *Insulin secretion and membrane transport processes*

CYSTIC FIBROSIS

In this running problem, you learned that cystic fibrosis, one of the most common inherited diseases in the United States, is caused by a defect in the channel protein known as CFTR, which regulates the transport of chloride ions out of epithelial cells. Because CFTRs are found in the epithelial cell membranes of several organs—the sweat glands, lungs, and pancreas—cystic fibrosis may affect many different body processes.

Some of the most interesting research on cystic fibrosis uses genetically altered mice. These animal models can be bred to have either totally nonfunctional CFTRs or CFTRs that have altered functions corresponding to the mutations of the CFTR gene in humans. An excellent review paper, "The CF

mouse: an important tool for studying cystic fibrosis," can be found in the online journal *Expert Reviews in Molecular Medicine* at *http://www-ermm.cbcu.cam.ac.uk/ 01002551h. htm* (12 March 2001).

Cystic fibrosis is one of the diseases for which clinical trials are under way to see if gene therapy can successfully restore the normal function of affected CFTRs. To learn more about gene therapy and the status of current research, go to the Cystic Fibrosis Foundation website (*www.cff.org*) and click on the link to Research and Clinical Trials.

To check your understanding of the running problem, compare your answers with the information in the following table.

	QUESTION	FACTS	INTEGRATION AND ANALYSIS
1	Is the CFTR a chemically gated, a voltage-gated, or a mechanically gated channel protein?	Chemically gated channels open when a ligand binds to them. Voltage-gated channels open with a change in the cell's membrane potential. Mechanically gated channels open when a physical force opens the channel. CFTRs open in response to the binding of nucleotides.	Nucleotides are chemical ligands; therefore, CFTRs are chemically gated channel proteins.
2	Based on the information given, is the CFTR on the apical or basolateral surface of the sweat gland epithelium?	In normal people, the CFTRs move Cl^- from sweat into epithelial cells.	The epithelial surface that faces the lumen of the sweat gland, which contains sweat, is the apical membrane. Therefore, the CFTRs are on the apical surface.
3	Why would failure to transport chloride ions into the lumen of the airways cause the secreted mucus to be thick? (*Hint*: remember that water moves to dilute the more concentrated region.)	If Cl^- (followed by Na^+) is secreted into the lumen of the airways, the solute concentration of the airway fluid increases. Water moves in response to concentration gradients.	Normally, the movement of Cl^- and Na^+ creates an osmotic gradient so that water also enters the airway lumen, creating a saline solution that thins the thick mucus. If Cl^- cannot be secreted into the airways, there will be no fluid movement to thin the mucus.
4	Why will Daniel starve if he does not take artificial pancreatic enzymes?	The pancreas secretes mucus and digestive enzymes into ducts that empty into the small intestine. In cystic fibrosis, mucus in the ducts is thick because of a lack of Cl^- and fluid secretion. This thick mucus blocks the ducts and prevents digestive enzymes from reaching the small intestine.	Without digestive enzymes, Daniel cannot digest the food he eats. His weight loss over the past six months suggests that this has already became a problem. Taking artificial enzymes will enable him to digest his food.

CHAPTER SUMMARY

In this chapter you learned how the cell membrane acts as a barrier to create distinct intracellular and extracellular compartments, repeating the theme of *compartmentation* that was introduced in Chapters 3 and 4. Movement of materials from one compartment to another illustrates a new theme, *mass flow*. Mass flow across cell membranes occurs in response to osmotic, chemical (concentration), or electrical gradients. The selectively permeable cell membrane creates resistance to mass flow that can be overcome by changing the composition of the membrane lipids or by inserting membrane proteins that act as channels or transporters. *Movement of substances* in the body requires energy from different sources: molecular motion, concentration gradients, or chemical bonds. Finally, the theme of *protein interactions* is exemplified in the binding of substrates to transporters.

Mass Balance and Homeostasis

IP   Fluids and Electrolytes: Introduction to Body Fluids

1. The **law of mass balance** says that if the amount of a substance in the body is to remain constant, any input must be offset by an equal loss. (p. 129; Fig. 5-1)

2. Input of a substance into the body comes from metabolism or from the outside environment. Output occurs through metabolism or **excretion**. (p. 130; Fig. 5-2)

3. **Clearance** is the rate at which a material is removed from the blood by excretion, metabolism, or both. The liver, kidneys, lungs, and skin all clear substances from the blood. (p. 130)

4. The rate of intake, production, or output of a substance *x* is expressed as **mass flow**, where mass flow = concentration × volume flow. (p. 130)

5. Cells and the extracellular fluid both maintain homeostasis, but without being in equilibrium with each other. Most solutes are concentrated in either one compartment or the other, creating a state of **chemical disequilibrium**. (p. 130; Fig. 5-3)

6. Cations and anions are not distributed equally between the body compartments, creating a state of **electrical disequilibrium**. (p. 131)

7. Water moves freely between the cells and extracellular fluid, resulting in a state of **osmotic equilibrium**. (p. 130)

Diffusion

8. The cell membrane is a selectively permeable barrier that restricts free exchange between the cell and the interstitial fluid. The movement of a substance across the membrane depends on the **permeability** of the membrane to that substance. (p. 132)

9. Movement of molecules across membranes can be classified either by energy requirements or by the physical means the molecule uses to cross the membrane. (p. 132; Fig. 5-4)

10. Lipid-soluble substances can diffuse through the phospholipid bilayer. Less lipid-soluble molecules require the assistance of a membrane protein to cross the membrane. (p. 132)

11. **Passive transport** does not require the input of energy. (p. 132)

12. **Diffusion** is the passive movement of molecules down a chemical (concentration) gradient from an area of higher concentration to an area of lower concentration. Net movement stops when the sys-

tem reaches **equilibrium**, although molecular movement continues. (p. 133; Fig. 5-5)

13. Diffusion rate depends on the magnitude of the concentration gradient. Diffusion is slow over long distances, is directly related to temperature, and is inversely related to molecular size. (p. 135)

14. **Simple diffusion** across a membrane is directly proportional to membrane surface area, concentration gradient, and membrane permeability, and inversely proportional to membrane thickness. (p. 135; Fig. 5-6)

Protein-Mediated Transport

15. Most molecules cross membranes with the aid of membrane proteins. (p. 137)

16. Membrane proteins have four functional roles: **structural proteins** maintain cell shape and form cell junctions; **membrane-associated enzymes** catalyze chemical reactions and help transfer signals across the membrane; **receptor proteins** are part of the body's signaling system; and **transport proteins** move many molecules into or out of the cell. (pp. 137–138; Fig. 5-7)

17. **Channel proteins** form water-filled channels that link the intracellular and extracellular compartments. **Gated channels** regulate movement of substances through them by opening and closing. Gated channels are regulated by ligands, by the electrical state of the cell, or by physical changes such as pressure. (pp. 138–139; Fig. 5-11)

18. **Carrier proteins** never form a continuous connection between the intracellular and extracellular fluid. They bind to substrates, then change conformation. (p. 138; Fig. 5-9b)

19. Protein-mediated diffusion is called **facilitated diffusion**. It has the same properties as simple diffusion. (p. 141; Tbl. 5-1; Figs. 5-13, 5-14)

20. **Active transport** moves molecules against their concentration gradient and requires an outside source of energy. In **primary (direct) active transport**, the energy comes directly from ATP. **Secondary (indirect) active transport** uses the potential energy stored in a concentration gradient and is indirectly driven by energy from ATP. (p. 142)

21. The most important primary active transporter is the **sodium-potassium ATPase** (Na^+-K^+-ATPase), which pumps Na^+ out of the cell and K^+ into the cell. (pp. 142–143; Fig. 5-16)

22. Most secondary active transport systems are driven by the sodium concentration gradient. (p. 144; Tbl. 5-3; Fig. 5-17)

23. All carrier-mediated transport demonstrates **specificity**, **competition**, and **saturation**. Specificity refers to the ability of a transporter to move only one molecule or a group of closely related molecules. Related molecules may compete for a single transporter. Saturation occurs when a group of membrane transporters are working at their maximum rate. (pp. 145–147; Fig. 5-20, 5-22)

Vesicular Transport

24. Large macromolecules and particles are brought into cells by **phagocytosis** and **endocytosis**. Material leaves cells by **exocytosis**. When vesicles that come into the cytoplasm by endocytosis are returned to the cell membrane, the process is called **membrane recycling**. (p. 148; Figs. 5-23, 5-24)

25. In **receptor-mediated endocytosis**, ligands bind to membrane receptors that concentrate in **clathrin-coated pits**, the site of endocytosis. In **potocytosis**, receptors are located in **caveolae** that have a nonclathrin protein coating. (pp. 149–150; Fig. 5-24)

26. In exocytosis, the vesicle membrane fuses with the cell membrane before releasing its contents into the extracellular space. Exocytosis requires ATP. (p. 150)

Transepithelial Transport

27. Transporting epithelia in the intestine and kidney have different membrane proteins on their **apical** and **basolateral** surfaces. This polarization allows one-way movement of molecules across the epithelium. (pp. 150–151; Figs. 5-25, 5-26)

28. Larger molecules cross epithelia by **transcytosis**, which includes **vesicular transport**. (p. 153; Fig. 5-27)

Osmosis and Tonicity

29. The movement of water across a membrane in response to a concentration gradient is called **osmosis**. (p. 154; Fig. 5-29)

30. To compare solution concentrations, we express the concentration in terms of **osmolarity**, the number of particles (ions or intact molecules) per liter of solution, expressed as milliosmoles per liter (mOsM). (p. 156)

31. **Tonicity** of a solution describes the cell volume change that occurs when the cell is placed in that solution. Cells swell in **hypotonic solutions** and shrink in **hypertonic solutions**. If the cell does not change size at equilibrium, the solution is **isotonic**. (p. 157)

32. The osmolarity of a solution cannot be used to determine the tonicity of the solution. The relative concentrations of **nonpenetrating solutes** in the cell and in the solution determine tonicity.

Penetrating solutes contribute to the osmolarity of a solution but not to its tonicity. (p. 157; Fig. 5-30, 5-31)

The Resting Membrane Potential

IP   **Nervous I: The Membrane Potential**

33. Although the total body is electrically neutral, diffusion and active transport of ions across the cell membrane create an **electrical gradient**, with the inside of cells negative relative to the extracellular fluid. (p. 162; Fig. 5-33)

34. The electrical gradient between the extracellular fluid and the intracellular fluid is known as the **resting membrane potential difference.** (p. 162)

35. The movement of an ion across the cell membrane is influenced by the **electrochemical gradient** for that ion. (p. 162)

36. The membrane potential that exactly opposes the concentration gradient of an ion is known as the **equilibrium potential (E_{ion})**. The equilibrium potential for any ion can be calculated using the Nernst equation. (p. 163; Fig. 5-34)

37. In most living cells, K^+ is the primary ion that determines the resting membrane potential. (p. 163)

38. Changes in membrane permeability to ions such as K^+, Na^+, Ca^{2+}, or Cl^- will alter membrane potential and create electrical signals. (p. 166)

Integrated Membrane Processes: Insulin Secretion

39. The use of electrical signals to initiate a cellular response is a universal property of living cells. Pancreatic beta cells release insulin in response to a change in membrane potential. (p. 166; Fig. 5-38)

QUESTIONS

(Answers to the Review Questions begin on page A1.)

The Physiology Place

Access more review material online at **The Physiology Place** website. There you'll find review questions, problem-solving activities, case studies, flashcards, and direct links to both *InterActive Physiology®* and *PhysioEx™*. To access the site, go to *www.physiologyplace.com* and select Human Physiology, Fourth Edition.

LEVEL ONE   REVIEWING FACTS AND TERMS

1. List the four functions of membrane proteins, and give an example of each. 1.) structural proteins 2) transporter 3) receptors 4) enzymes

2. Distinguish between active transport and passive transport.

3. Which of the following processes are examples of active transport, and which are examples of passive transport? Simple diffusion, PT phagocytosis, facilitated diffusion, exocytosis, osmosis, endocytosis, potocytosis.

4. List four factors that will increase the rate of diffusion.

5. Match the membrane channels with the appropriate description(s). Answers may be used once, more than once, or not at all.

   (a) chemically gated channel

   (b) open pore 1, 6
   (c) voltage-gated channel 3
   (d) mechanically gated channel 5

   1. channel that spends most of its time in the open state
   2. channel that opens in response to a signal
   3. channel that opens when resting membrane potential changes
   4. channel that opens when a ligand binds to it
   5. channel that opens in response to membrane stretch
   6. channel through which water can pass

6. List the three physical methods by which materials enter cells.

7. A cotransporter is a protein that moves more than one molecule at a time. If the molecules are moved in the same direction, the cotransporters are called **symport** carriers; if the molecules are transported in opposite directions, the cotransporters are called **antiport** carriers. A transport protein that moves only one substrate is called a **uniport** carrier.

8. The two types of active transport are **direct**, which derives energy directly from ATP, and **indirect** which couples the kinetic energy of one molecule moving down its concentration gradient to the movement of another molecule against its concentration gradient.

9. A molecule that moves freely between the intracellular and extracellular compartments is said to be a _penetrating_ solute. A molecule that is not able to enter cells is called a _nonpenetrating_ solute.

10. Rank the following individuals in order of how much body water they contain, from highest to lowest: (a) a 25-year-old, 70-kg male; (b) a 25-year-old, 50-kg female; (c) a 65-year-old, 50-kg female; and (d) a 1-year-old male toddler. _d, a, b, c_

11. What determines the osmolarity of a solution? In what units is body osmolarity usually given?

12. What does it mean if we say that a solution is hypotonic to a cell? Hypertonic to the same cell? What determines the tonicity of a solution relative to a cell?

13. In your own words, state the four principles of electricity important in physiology.

14. Match each of the following items with its primary role in cellular activity.

    (a) Na^+-K^+-ATPase _7_
    (b) proteins _1, 4_
    (c) unit of measurement for membrane potential _6_
    (d) K^+ _5_
    (e) Cl^- _8_
    (f) ATP _3_
    (g) Na^+ _2_

    1. ion channel
    2. extracellular cation
    3. source of energy
    4. intracellular anion
    5. intracellular cation
    6. millivolts
    7. electrogenic pump
    8. extracellular anion
    9. milliosmoles

15. The membrane potential at which the electrical gradient exactly opposes the concentration gradient for an ion is known as the _equilibrium_

16. A material that allows free movement of electrical charges is called a(n) _conductors_, whereas one that prevents this movement is called a(n) _insulators_

LEVEL TWO   REVIEWING CONCEPTS   _5-4, 5-7, 5-24, 5-26_

17. Create a map of transport across cell membranes using the following terms. You may add additional terms if you wish.

    active transport
    carrier
    caveolae
    channel
    clathrin-coated pit
    concentration gradient
    electrochemical gradient
    exocytosis
    facilitated diffusion
    glucose
    GLUT transporter
    ion
    large polar molecule
    ligand
    osmosis

    Na^+-K^+-ATPase
    passive transport
    phospholipid bilayer
    potocytosis
    receptor
    receptor-mediated endocytosis
    secondary active transport
    simple diffusion
    small polar molecule
    transcytosis
    vesicle
    vesicular transport
    water

18. Draw a large rectangle to represent the total body volume. Using the information in Figure 5-28, divide the box proportionately into compartments to represent the different body compartments. Use the information in Figure 5-3 and add solutes to the compartments. Use large letters for solutes with higher concentrations, and small letters for solutes with low concentrations. Label the cell membranes and the endothelial membrane.

19. What factors influence the rate of diffusion across a membrane? Briefly explain each one.

20. Define the following terms and explain how they differ from one another: specificity, competition, saturation. Apply these terms in a short explanation of facilitated diffusion of glucose.

21. Red blood cells are suspended in a solution of NaCl. The cells have an osmolarity of 300 mOsM, and the solution has an osmolarity of 250 mOsM. (a) The solution is (hypertonic, isotonic, or hypotonic) to the cells. (b) Water would move (into the cells, out of the cells, or not at all).

22. Two compartments are separated by a membrane that is permeable to glucose. Each compartment is filled with 1 M glucose. After 6 hours, compartment A contains 1.5 M glucose and compartment B contains 0.5 M glucose. What kind of transport occurred?

23. A 2 M NaCl solution and a 2 M glucose solution are placed into two compartments separated by a membrane that is permeable to water but not to NaCl or glucose. Are the following statements true or false? Defend each answer with a short statement.

    (a) The salt solution is isosmotic to the glucose solution.
    (b) The salt solution is hyperosmotic to the glucose solution.
    (c) Water will move from the salt solution to the sugar solution.
    (d) The volume will increase on the glucose side of the membrane.

24. Referring to the previous question, change the molarity of the salt and sugar solutions so that each false statement would become true.

25. Explain the differences between a chemical gradient, an electrical gradient, and an electrochemical gradient.

LEVEL THREE   PROBLEM SOLVING

26. Sweat glands secrete into their lumen a fluid that is identical to interstitial fluid. As the fluid moves through the lumen on its way to the surface of the skin, the cells of the sweat gland's epithelium make the fluid hypotonic by removing Na^+ and leaving water behind. Design an epithelial cell that will reabsorb Na^+ but not water. You may place water pores, Na^+ leak channels, K^+ leak channels, and the Na^+-K^+-ATPase in the apical membrane, basolateral membrane, or both.

27. Insulin is a hormone that promotes the movement of glucose into many types of cells, thereby lowering blood glucose concentration. Propose a mechanism that explains how this occurs, using your knowledge of cell membrane transport.

28. The following terms have been applied to membrane transport molecules: specificity, competition, saturation. These terms can also be applied to enzymes. How does the application of these terms change in the two situations? What chemical characteristics do enzymes and transport molecules share that allow these terms to be applied to both?

29. NaCl is a nonpenetrating solute and urea is a penetrating solute for cells. Red blood cells are placed in each of the solutions below. The intracellular concentration of nonpenetrating solute is 300 mOsM. What will happen to the cell volume in each solution? Label each solution with all the terms that apply: hypertonic, isotonic, hypotonic, hyperosmotic, hyposmotic, isosmotic.

    (a) 150 mM NaCl plus 150 mM urea
    (b) 100 mM NaCl plus 50 mM urea
    (c) 100 mM NaCl plus 100 mM urea
    (d) 150 mM NaCl plus 100 mM urea
    (e) 100 mM NaCl plus 150 mM urea

30. Integral membrane glycoproteins have their sugars added as the proteins pass through the lumen of the endoplasmic reticulum and Golgi complex [p. 119]. Based on this information, where would

you predict finding the sugar "tails" of the proteins: on the cyto-plasmic side of the membrane, the extracellular side, or both? Explain your reasoning.

LEVEL FOUR   QUANTITATIVE PROBLEMS

31. The addition of dissolved solutes to water lowers the freezing point of water. A 1 OsM solution depresses the freezing point of water 1.86° C. If a patient's plasma shows a freezing-point depression of 0.550° C, what is her plasma osmolarity? (Assume that 1 kg water = 1 L.)

32. The patient in the previous question is found to have total body water volume of 42 L, ECF volume of 12.5 L, and plasma volume of 2.7 L.

    (a) What is her intracellular fluid (ICF) volume? Her interstitial fluid volume?

    (b) How much solute (osmoles) exists in her whole body? ECF? ICF? plasma? (*Hint*: concentration = solute amount / volume of solution)

33. What is the osmolarity of half-normal saline (= 0.45% NaCl)? [📰 p. 36] Assume that all NaCl molecules dissociate into two ions.

34. If you give 1 L of half-normal saline (see question 33) to the patient in question 32, what happens to each of the following at equilibrium? (*Hint:* NaCl is a nonpenetrating solute.)

    (a) her total body volume

    (b) her total body osmolarity

    (c) her ECF and ICF volumes

    (d) her ECF and ICF osmolarities

35. The following graph shows the results of an experiment in which a cell was placed in a solution of glucose. The cell had no glucose in it at the beginning, and its membrane can transport glucose. Which of the following processes is/are illustrated by this experiment?

    (a) simple diffusion

    (b) saturation

    (c) competition

    (d) active transport

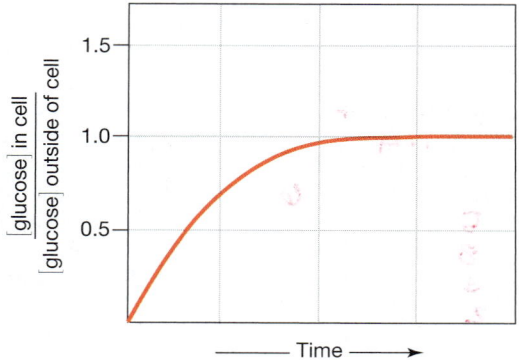

ANSWERS

✓ Answers to Concept Check Questions

Page 130

1. The remaining 1 milligram of salt remains in the body.

2. Glucose metabolism adds CO_2 and water to the body, disturbing the mass balance of these two substances. To maintain mass balance, both metabolites must be either excreted or further metabolized.

Page 132

3. The stem preceding the suffix –*ase* is the name of the substrate on which the enzyme acts; therefore ATP is a substrate for this enzyme.

4. Intracellular fluid has a high K^+ concentration and low Na^+, Cl^-, and Ca^{2+} concentrations.

Page 135

5. If distance doubles, diffusion takes four times as long.

Page 136

6. Energy for diffusion comes from molecular motion.

7. Because it is lipophilic, the fatty acid is more likely to cross by simple diffusion.

8. The diffusion rate (a) decreases, (b) increases, (c) decreases.

9. Compartment A remains yellow, and compartment B turns green.

10. The skin's thick extracellular matrix is generally impermeable to oxygen. Also, oxygen needs a moist surface to effectively diffuse across a membrane.

Page 139

11. Positive ions are cations, and negative ions are anions.

Page 141

12. Membrane proteins serve as structural proteins, receptors, enzymes, and transporters.

13. Ions and water molecules move through open channels.

14. Channel proteins form continuous connections between the two sides of a membrane, transport molecules more quickly, and bind less tightly to their substrates.

15. A channel lined with positive charges will attract anions, which in this instance means Cl^-.

16. Glucose is too large to pass through a channel.

Page 142

17. The direction of facilitated diffusion of glucose reverses, and glucose leaves the cell.

Page 143

18. The ATPase is an antiporter, but the SGLT is a symporter. The ATPase requires energy from ATP to change conformation, whereas the SGLT uses energy stored in the Na^+ concentration gradient.

Page 148

19. An antiporter moves substrates in opposite directions.

20. Larger doors could move more people. This would be analogous to a cell's synthesizing a new isoform of the transporter that would let the transporter move more substrate per second.

Page 150

21. In phagocytosis, the cytoskeleton pushes the membrane out to engulf a particle in a large vesicle. In endocytosis, the membrane surface indents and the vesicle is much smaller.

22. The proteins associated with endocytosis are clathrin and caveolin.

23. Proteins move into cells by endocytosis and out of cells by exocytosis.

Page 153

24. Sodium movement out of the cell requires energy because the direction of ion flow is against the concentration gradient.

25. Ouabain applied to the apical side would have no effect because there are no Na^+-K^+-ATPase molecules on that side. Ouabain applied to the basolateral side will stop the pump. Glucose transport will continue for a time until the Na^+ gradient between the cell and the lumen disappears due to Na^+ entry into the cell.

26. The GLUT2 transporter is illustrated.

Page 153

27. Transcytosis will stop because vesicular transport by the cytoskeleton depends on functioning microtubules.

Page 156

28. The baby has lost 0.91 kg of water, which is 0.91 liter.

Page 160

29. 1 M NaCl = 2 OsM NaCl. The 1 M (= 1 OsM) glucose and 1 OsM NaCl have the most water.

30. (a) Water moves into A because A is 2 OsM; (b) no net movement occurs because urea will diffuse across the membrane until it reaches equilibrium; (c) water moves into A, because A has a higher concentration of nonpenetrating solutes.

31. (a) The NaCl solution is better, even though both solutions are isosmotic to the body (Tbl. 5-9). Because, blood is lost from the extracellular compartment, the best replacement solution would remain in the ECF. Thus glucose is not as good a choice because it slowly enters cells, taking water with it. (b) If 1 L has been lost, you should replace at least 1 L.

Page 165

32. Cl^- will move into the cell down its concentration gradient, which would make the inside of the cell negative. The positive charges left outside would attract Cl^- back outside. The equilibrium potential would be negative.

33. Over time, Na^+ would leak into the cell, and the resting membrane potential would become more positive.

Q Answers to Figure and Graph Questions

Page 131

Fig. 5-3: 1. Plasma contains proteins and large anions not present in interstitial fluid. 2. The extracellular compartment contains more Na^+, Cl^-, and bicarbonate than the intracellular compartment, and fewer K^+.

Page 147

Fig. 5-22: You should mark the x-axis at the point where the curve levels off to a horizontal line.

Page 152

Fig. 5-26: (a) 1 = b; 2 = a; 3 = b. (b) Basolateral glucose transport is passive because the glucose moves down its concentration gradient. (c) Na^+ movement across the apical membrane does not require ATP because Na^+ is moving down its concentration gradient.

Page 155

Fig. 5-28: (1) 25% of 14 L = 3.5 L plasma. 75% = 10.5 L interstitial fluid. (2) Total body water = 42 L. (3) 3.5 L/42 L = 8.3% plasma; 10.5 L/42 L = 25% interstitial volume.

Page 159

Fig. 5-31: (a) isosmotic and isotonic; (b) hyperosmotic and isotonic; (c) hyperosmotic and hypotonic.

Page 164

Fig. 5-34: ICF = -1 and ECF = $+1$. The cell membrane is not permeable to Na^+, Cl^-, and proteins.

Page 165

Fig. 5-36: (1) Na^+ leak, into the cell is promoted by concentration and electrical gradients. (2) K^+ leaks out of the cell are promoted by the concentration gradient.

5

6

Future progress in medicine will require a quantitative understanding of the many interconnected networks of molecules that comprise our cells and tissues, their interactions, and their regulation.

—Overview of the NIH Roadmap, 2003

Communication, Integration, and Homeostasis

BACKGROUND BASICS

In 2003 the United States National Institutes of Health embarked on an ambitious project to promote translation of basic research into new medical treatments and strategies for disease prevention. Contributors to the NIH Roadmap (*http://nihroadmap.nih.gov*) are compiling information on biological pathways in an effort to understand how cells communicate with one another and maintain the body in a healthy state. This chapter examines the basic patterns of cell-to-cell communication and shows how the coordination of function resides in chemical and electrical signals. A combination of simple diffusion across small distances, widespread distribution of molecules through the circulatory system, and rapid, specific delivery of messages by the nervous system enables each cell in the body to communicate with most other cells and maintain homeostasis.

CELL-TO-CELL COMMUNICATION

In recent years the amount of information available about cell-to-cell communication has mushroomed as a result of advances in research technology. Signal pathways that once seemed fairly simple and direct are now known to be incredibly complex networks and webs of information transfer. In the sections that follow, we distill what is known about cell-to-cell communication into some basic patterns that you can learn and recognize when you encounter them again in your study of physiology.

By most estimates the human body is composed of about 75 *trillion* cells. Those cells face a daunting task—to communicate with one another in a manner that is rapid and yet conveys a tremendous amount of information. Surprisingly, there are only two basic types of physiological signals: electrical and chemical. **Electrical signals** are changes in a cell's membrane potential [p. 162]. **Chemical signals** are molecules secreted by cells into the extracellular fluid. Chemical signals are responsible for most communication within the body. The cells that receive electrical or chemical signals are called **target cells**, or **targets** for short.

Our bodies use four basic methods of cell-to-cell communication: (1) **gap junctions**, which allow direct cytoplasmic transfer of electrical and chemical signals between adjacent cells; (2) **contact-dependent signals**, which occur when surface molecules on one cell membrane bind to surface molecules on another cell membrane; (3) **local communication** by chemicals that diffuse through the extracellular fluid; and (4) **long-distance communication** through a combination of electrical signals carried by nerve cells and chemical signals transported in the blood. A given molecule can function as a signal by more than one method. For example, a molecule can act close to the cell that released it (local communication) as well as in distant parts of the body (long-distance communication).

Gap Junctions Create Cytoplasmic Bridges

The simplest form of cell-to-cell communication is the direct transfer of electrical and chemical signals through **gap junctions**, protein channels that create cytoplasmic bridges between adjacent cells (Fig. 6-1a) [p. 69]. A gap junction forms from

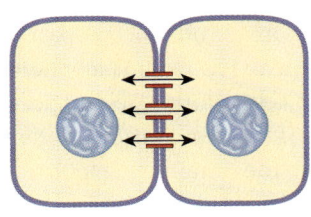

(a) Gap junctions form direct cytoplasmic connections between adjacent cells.

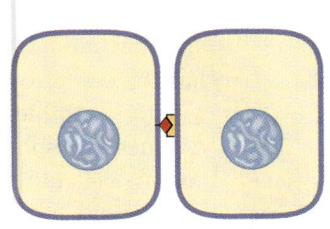

(b) Contact-dependent signals require interaction between membrane molecules on two cells.

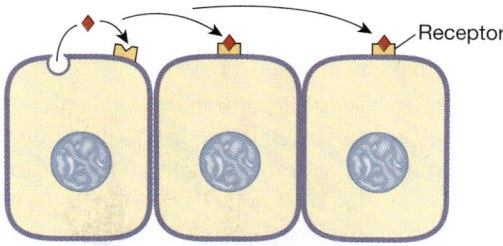

(c) Autocrine signals act on the same cell that secreted them. **Paracrine signals** are secreted by one cell and diffuse to adjacent cells.

Receptor

■ **FIGURE 6-1** *Direct contact and local cell-to-cell communication*

Local communication can be either by autocrine signals or by paracrine signals.

the union of membrane-spanning proteins, called *connexins,* on two adjacent cells. The united connexins create a protein channel (*connexon*) that can open and close. When the channel is open, the connected cells function like a single cell that contains multiple nuclei (a *syncytium*).

When gap junctions are open, ions and small molecules such as amino acids, ATP, and cyclic AMP diffuse directly from the cytoplasm of one cell to the cytoplasm of the next. As with other membrane channels, larger molecules cannot pass through gap junctions. In addition, gap junctions are the only means by which electrical signals can pass *directly* from cell to cell. Movement of molecules through gap junctions can be modulated or shut off completely.

Gap junctions are not all alike. Scientists have discovered more than 20 different isoforms of connexins that may mix or match to form gap junctions. The variety of connexin isoforms allows gap junction selectivity to vary from tissue to tissue. In mammals, gap junctions are found in almost every cell type, including heart muscle, some types of smooth muscle, lung, liver, and neurons of the brain.

Contact-Dependent Signals Require Cell-to-Cell Contact

Some cell-to-cell communication requires that surface molecules on one cell membrane bind to a membrane protein of another cell (Fig. 6-1b). Such *contact-dependent signaling* occurs in the immune system and during growth and development, such as when nerve cells send out long extensions that must grow from the central axis of the body to the *distal* (distant) ends of the limbs. **CAMs,** cell adhesion molecules first known for their role in cell-to-cell adhesion [⇄ p. 69], have now been shown to act as receptors in cell-to-cell signaling. CAMs are linked to the cytoskeleton and to intracellular enzymes. Through these linkages, CAMs transfer signals in both directions across cell membranes.

Paracrine and Autocrine Signals Carry Out Local Communication

Local communication is accomplished by paracrine and autocrine signaling. A **paracrine signal** [*para-*, beside + *krinen*, to secrete] is a chemical that acts on cells in the immediate vicinity of the cell that secreted the signal. If a chemical signal acts on the cell that secreted it, it is called an **autocrine signal** [*auto-*, self]. In some cases a molecule may act as both an autocrine signal and a paracrine signal.

Paracrine and autocrine signal molecules reach their target cells by diffusing through the interstitial fluid (Fig. 6-1c). Because distance is a limiting factor for diffusion, the effective range of paracrine signals is restricted to adjacent cells. A good example of a paracrine molecule is *histamine*, a chemical released from damaged cells. When you scratch yourself with a

pin, the red, raised *wheal* that results is due in part to the local release of histamine from the injured tissue. The histamine acts as a paracrine signal, diffusing to the capillaries in the immediate area of the injury and making them more permeable to white blood cells and antibodies in the plasma. Fluid also leaves the blood vessels and collects in the interstitial space, causing swelling around the area of injury.

Several important classes of molecules act as local signals. *Cytokines* are regulatory peptides that usually act close to the site where they are secreted. *Eicosanoids* [⇄ p. 29] are lipid-derived paracrine and autocrine signal molecules. Cytokines and eicosanoids will be discussed in more detail below.

Neural Signals, Hormones, and Neurohormones Carry Out Long-Distance Communication

All cells in the body can release paracrine signals, but most long-distance communication between cells is the responsibility of the nervous and endocrine systems. The endocrine system communicates by using **hormones** [*hormon,* to excite], chemical signals that are secreted into the blood and distributed all over the body by the circulation. Hormones come in contact with most cells of the body, but only those cells with receptors for the hormone are target cells (Fig. 6-2a ■).

The nervous system uses a combination of chemical signals and electrical signals to communicate over long distances. An electrical signal travels along a nerve cell (*neuron*) until it reaches the very end of the cell, where it is translated into a chemical signal secreted by the neuron (a **neurocrine**). If a neurocrine molecule diffuses from the neuron across a narrow extracellular space to a target cell and has a rapid effect, it is called a **neurotransmitter** (Fig. 6-2b). If a neurocrine acts more slowly as an autocrine or paracrine signal, it is called a **neuromodulator**. If a neurocrine released by a neuron diffuses into the blood for distribution, it is called a **neurohormone** (Fig. 6-2c). The similarities between neurohormones and classic hormones secreted by the endocrine system blur the distinction between the nervous and endocrine systems, making them a continuum rather than two distinct systems (see Fig. 6-31, p. 204).

Cytokines May Act as Both Local and Long-Distance Signals

Cytokines are among the most recently identified communication molecules. Initially the term *cytokine* referred only to proteins that modulate immune responses, but in the past few years the definition has been broadened to include a variety of regulatory peptides. All nucleated cells synthesize and secrete cytokines in response to stimuli. Cytokines control cell development, cell differentiation, and the immune response. In development and differentiation, cytokines function as autocrine or paracrine signals. In stress and inflammation, some

(a) Hormones are secreted by endocrine glands or cells into the blood. Only target cells with receptors for the hormone will respond to the signal.

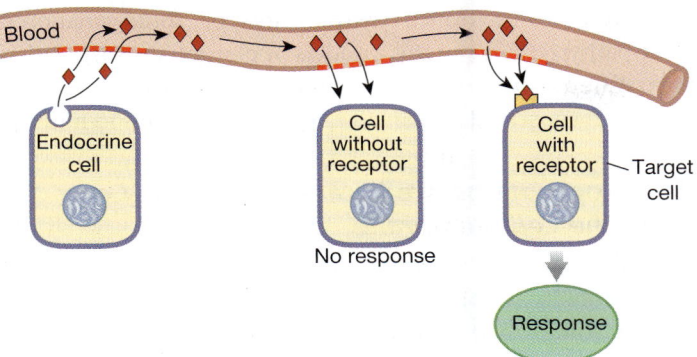

No response

Target cell

Response

(b) Neurotransmitters are chemicals secreted by neurons that diffuse across a small gap to the target cell. Neurons use electrical signals as well.

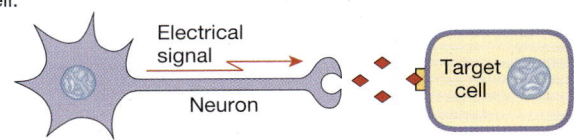

(c) Neurohormones are chemicals released by neurons into the blood for action at distant targets.

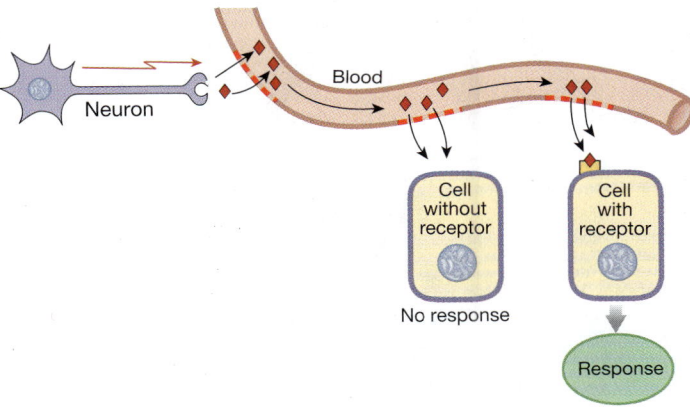

No response

Response

■ **FIGURE 6-2** *Long distance cell-to-cell communication*

cytokines may act on relatively distant targets and may be transported through the circulation just as hormones are.

How do cytokines differ from hormones? In general, cytokines act on a broader spectrum of target cells. In addition, cytokines are not produced by specialized glands the way hormones are, and they are made on demand. In contrast, most protein or peptide hormones are made in advance and stored in the endocrine cell until needed. However, the distinction between cytokines and hormones is sometimes blurry. For example, erythropoietin, the molecule that controls synthesis of red blood cells, is by tradition considered a hormone but functionally fits the definition of a cytokine.

CONCEPT CHECK

1. Match the communication method on the left with its property on the right.

   (a) autocrine
   (b) cytokine
   (c) gap junction
   (d) hormone
   (e) neurohormone
   (f) neurotransmitter
   (g) paracrine

   Communication occurs by:
   1. electrical signals
   2. chemical signals
   3. both electrical and chemical signals

2. Which signal molecules listed in the previous question are transported through the circulatory system? Which are associated with neurons?

3. A cat sees a mouse and pounces on it. Do you think the internal signal to pounce could have been transmitted by a paracrine signal? Give two reasons to explain why or why not. Answers: p. 209

SIGNAL PATHWAYS

Chemical signals in the form of paracrine and autocrine molecules and hormones are released from cells into the extracellular compartment. This is not a very specific way for these signals to find their targets because substances that travel through the blood reach nearly every cell in the body. Yet cells do not respond to every signal that reaches them. Why do some cells respond to a chemical signal while other cells ignore it? The answer lies in the target-cell **receptor proteins** to which chemical signals bind

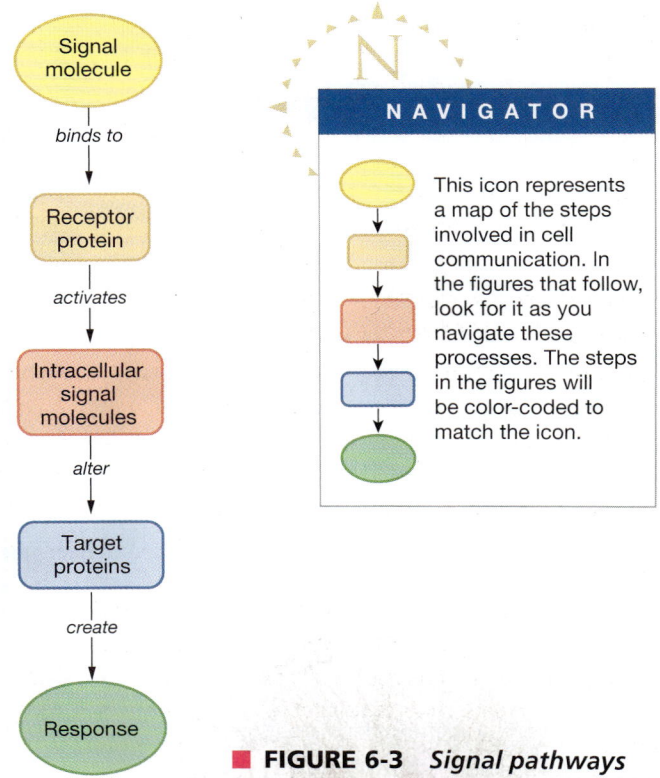

■ **FIGURE 6-3** *Signal pathways*

[⇄ p. 137]. *A cell cannot respond to a chemical signal if the cell lacks the appropriate receptor proteins for that signal* (Fig. 6-2a).

If a target cell has a receptor for a signal molecule, binding of the signal to the receptor protein will initiate a response. All signal pathways share the following common features (Fig. 6-3 ■):

1. The signal molecule is a *ligand* that binds to a receptor. The ligand is also known as a *first messenger* because it brings information to its target cell.
2. Ligand-receptor binding activates the receptor.
3. The receptor in turn activates one or more intracellular signal molecules.
4. The last signal molecule in the pathway initiates synthesis of target proteins or modifies existing target proteins to create a response.

In the following sections, we will describe some basic signal pathways. They may seem complex at first, but they follow patterns that you will encounter over and over as you study the systems of the body. Most physiological processes, from the beating of your heart to learning and memory, use some variation of these pathways. One of the wonders of physiology is the fundamental importance of these signal pathways and the way they have been conserved in animals ranging from worms to humans.

Receptor Proteins Are Located Inside the Cell or on the Cell Membrane

Chemical signals fall into two broad categories based on their lipid solubility: lipophilic or lipophobic. Target-cell receptors may be found in the nucleus, in the cytosol, or on the cell membrane as integral proteins. Where a chemical signal binds to its receptor largely depends on whether the signal molecule can enter the cell (Fig. 6-4 ■).

Lipophilic signal molecules can diffuse through the phospholipid bilayer of the cell membrane [⇄ p. 135] and bind to *cytosolic*

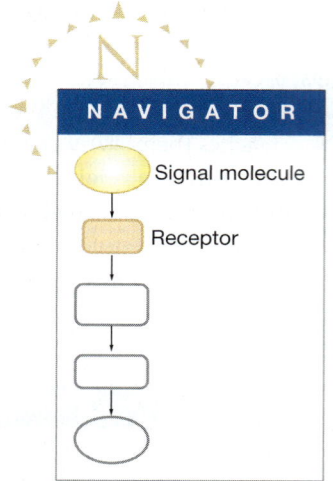

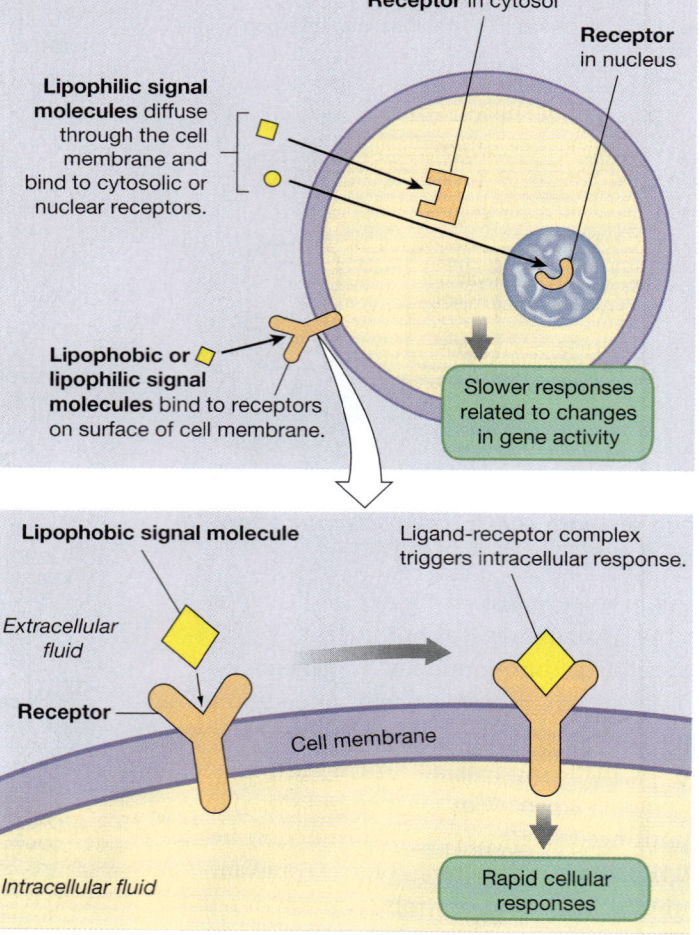

■ **FIGURE 6-4** *Target cell receptors*

Receptors are either membrane proteins or are located in the cytosol or nucleus.

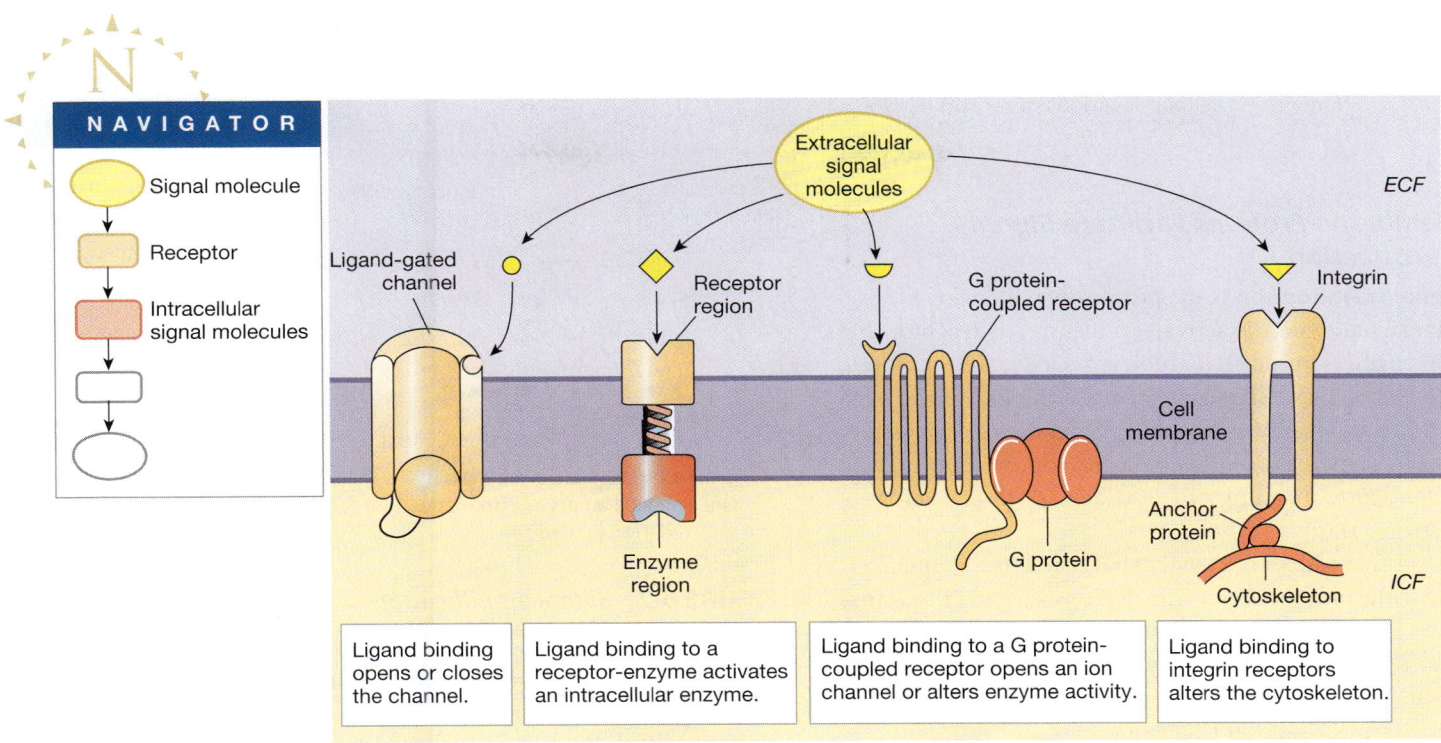

■ **FIGURE 6-5** *Four categories of membrane receptors*

receptors or *nuclear receptors*. In these cases, receptor activation often turns on a gene and directs the nucleus to make new mRNA (transcription, [ p. 115]). The mRNA then provides a template for synthesis of new proteins (translation, [ p. 118]). This is a relatively slow process, and the cell's response may not be noticeable for an hour or longer. In some instances the activated receptor can also turn off, or repress, gene activity. Many lipophilic signal molecules that follow this pattern are hormones, so we will defer further discussion of their action to the next chapter.

Lipophobic signal molecules are unable to diffuse through the phospholipid bilayer of the cell membrane. Instead, these signal molecules remain in the extracellular fluid and bind to receptor proteins on the cell membrane. (Some lipophilic signal molecules also bind to cell membrane receptors.) In general, the response time for pathways linked to membrane receptor proteins is very rapid, and responses can be seen within milliseconds to minutes.

Protein receptors for signal molecules play an important role in physiology and medicine. About half of all drugs currently in use act on receptor proteins. We can group membrane receptors into four major categories, illustrated in Figure 6-5 ■. The simplest receptors are chemically gated (*ligand-gated*) ion channels called *receptor-channels* [ p. 139]. Ligand binding opens or closes the channel and alters ion flow across the membrane.

Three other receptor types are shown in Figure 6-5: *receptor-enzymes, G protein-coupled receptors,* and *integrin receptors*. For all three, information from the signal molecule must be passed across the membrane to initiate an intracellular

response. This transmission of information from one side of a membrane to the other using membrane proteins is known as *signal transduction*. We will take a closer look at signal transduction before returning to the four receptor types that participate in it.

175   179   191   193   197   201   206

CONCEPT CHECK

4. List four components of signal pathways.
5. Name three cellular locations of receptors.

Answers: p. 209

Membrane Proteins Facilitate Signal Transduction

Signal transduction is the process by which an extracellular signal molecule activates a membrane receptor that in turn alters intracellular molecules to create a response. The extracellular signal molecule is the first messenger, and the intracellular molecules form a *second messenger system*. The term *signal transduction* comes from the verb *to transduce*, meaning "to lead across" [*trans*, across + *ducere*, to lead].

A **transducer** is a device that converts a signal from one form into a different form. For example, the transducer in a radio converts radio waves into sound waves (Fig. 6-6 ■). In biological systems, transducers convert the message of extracellular signal molecules into intracellular messages that trigger a response.

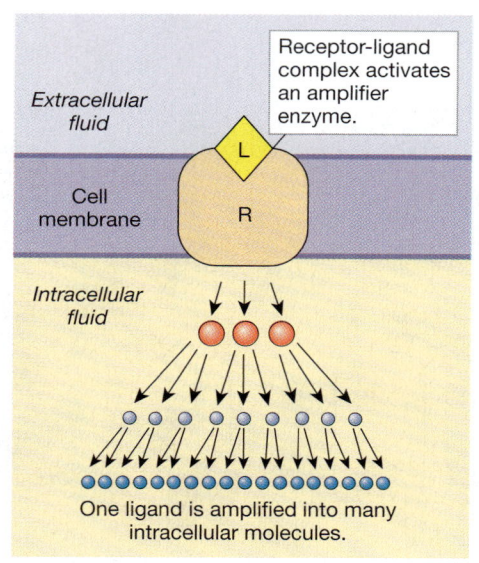

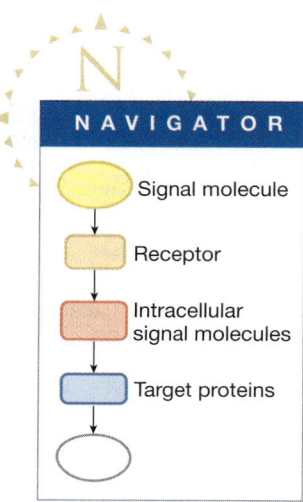

■ FIGURE 6-7 *Signal amplification*

In biological systems, as in a radio, the original signal is not only transformed but also amplified [*amplificare*, to make larger]. In cells, **signal amplification** turns one signal molecule into multiple second messenger molecules. The process begins when the ligand combines with its receptor (Fig. 6-7 ■). The receptor-ligand complex then turns on an **amplifier enzyme**, an enzyme that activates several more molecules. By the end of the process, the effects of the ligand have been amplified much more than if there were a 1:1 ratio between each step. Amplification gives the body "more bang for the buck" by enabling a small amount of ligand to create a large effect.

The basic pattern of a biological signal transduction pathway is shown in Figure 6-8 ■ and can be broken down into the following events.

1. An extracellular signal molecule binds to and activates a protein or glycoprotein membrane receptor.
2. The activated membrane receptor turns on its associated proteins. These proteins then may:
   (a) activate **protein kinases**, which are enzymes that transfer a phosphate group from ATP to a protein [≊ p. 100]. Phosphorylation is an important biochemical method of regulating cellular processes.
   (b) activate amplifier enzymes that create intracellular **second messengers**. The most common amplifier enzymes and second messengers are listed in Tables 6-1 ■ and 6-2 ■.
3. Second messenger molecules in turn
   (a) alter the gating of ion channels. Opening or closing ion channels creates electrical signals by altering the cell's membrane potential [≊ p. 162].
   (b) increase intracellular calcium. Calcium binding to proteins changes their function, creating a cellular response.

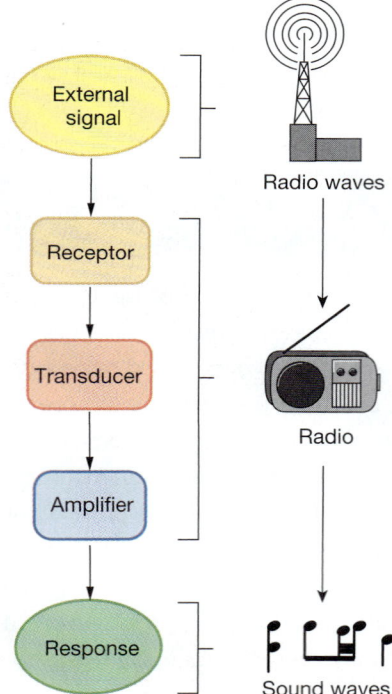

Signal transduction converts one form of signal into a different form.

■ FIGURE 6-6 *Signal transduction*

An example of signal transduction in the physical world is the process that takes place in a radio. The radio contains a transducer that converts radio waves into sound waves.

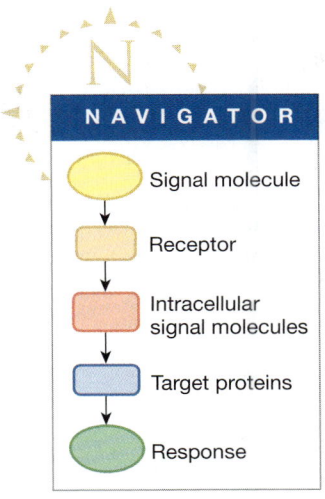

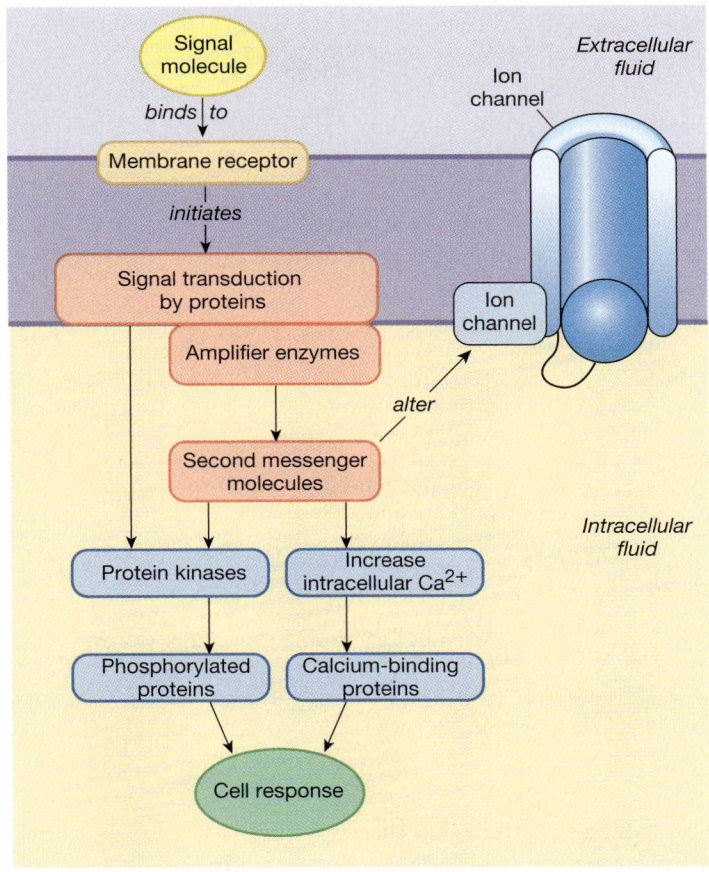

■ **FIGURE 6-8** *Biological signal transduction*

(c) change enzyme activity, especially of protein kinases or **protein phosphatases**, enzymes that remove a phosphate group. The phosphorylation or *dephosphorylation* of a protein can change its configuration and create a response. Examples of changes that occur with phosphorylation include increased or decreased enzyme activity and opening or closing gated ion channels.

4. The proteins modified by calcium binding and phosphorylation control one or more of the following:
   (a) metabolic enzymes,
   (b) motor proteins for muscle contraction and cytoskeletal movement,
   (c) proteins that regulate gene activity and protein synthesis, and
   (d) membrane transport and receptor proteins.

If you think this list includes almost everything a cell does, you're right!

TABLE 6-1	Amplifier Enzymes			
AMPLIFIER ENZYME	**CELLULAR LOCATION**	**ACTIVATED BY**	**CONVERTS**	**TO**
Adenylyl cyclase	Membrane	G protein-coupled receptor	ATP	cAMP
Guanylyl cyclase	Membrane Cytosol	Receptor-enzyme Nitric oxide (NO)	GTP	cGMP
Phospholipase C	Membrane	G protein-coupled receptor	Membrane phospholipids	IP_3 and DAG*

*IP_3 = Inositol trisphosphate; DAG = diacylglycerol

TABLE 6-2	Second Messenger Pathways	
SECOND MESSENGER	**ACTION**	**EFFECTS**
Ions		
Ca^{2+}	Binds to calmodulin	Alters enzyme activity
	Binds to other proteins	Exocytosis, muscle contraction, cytoskeleton movement, channel opening
Nucleotides		
cAMP	Activates protein kinases, especially protein kinase A	Phosphorylates proteins
	Binds to ion channels	Alters channel opening
cGMP	Activates protein kinases, especially protein kinase G	Phosphorylates proteins
	Binds to ion channels	Alters channel opening
Lipid-derived		
IP_3	Releases Ca^{2+} from intracellular stores	See Ca^{2+} effects above
DAG	Activates protein kinase C	Phosphorylates proteins

You can see from this basic pattern that the steps of a signal transduction pathway form a **cascade** (Fig. 6-9 ■) that starts when a stimulus (the signal molecule) converts inactive molecule A (the receptor) to an active form. Active A then converts inactive molecule B into active B, active molecule B in turn converts inactive molecule C into active C, and so on, until at the final step a substrate is converted into a product. Many intracellular signal pathways are cascades. Blood clotting is an important example of an extracellular cascade.

In the sections that follow, we will examine in more detail the four major types of membrane receptors (see Fig. 6-5). Keep in mind that these receptors may be responding to any of the different kinds of signal molecules—hormones, neurohormones, neurotransmitters, cytokines, paracrines, or autocrines.

CONCEPT CHECK

6. What are the four steps of signal transduction?

7. What happens during amplification?

8. Why do steroid hormones not require signal transduction and second messengers to exert their action? (*Hint:* are steroids lipophobic or lipophilic? [p. 17])

Answers: p. 209

Receptor-Enzymes Have Protein Kinase or Guanylyl Cyclase Activity

Receptor-enzymes have two regions: a receptor region on the extracellular side of the cell membrane, and an enzyme region

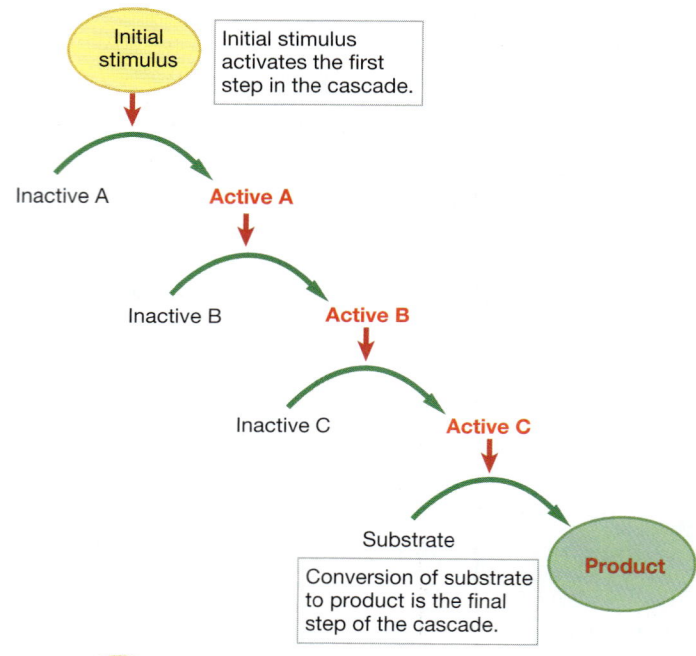

FIGURE QUESTION

Using Tables 6-1 and 6-2, create a cascade that includes ATP, cAMP, adenylyl cyclase, a phosphorylated protein, and protein kinase A. Match these molecules to the steps shown in the figure above.

■ **FIGURE 6-9**   *Steps of a cascade*

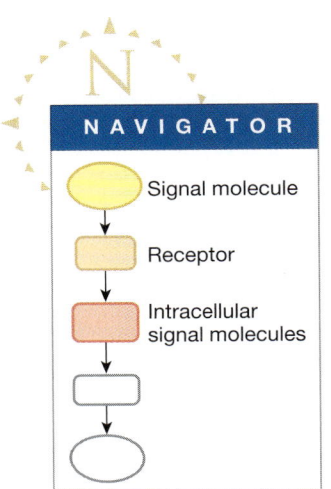

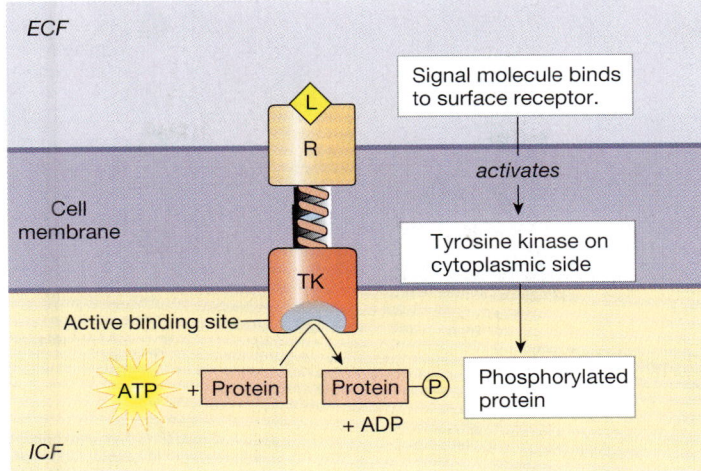

■ **FIGURE 6-10** *Tyrosine kinase, an example of a receptor-enzyme*

Tyrosine kinase (TK) transfers a phosphate group from ATP to a tyrosine (an amino acid) of a protein.

on the cytoplasmic side (see Fig. 6-5). In some instances, the receptor region and enzyme region are parts of the same protein molecule. In other cases, the enzyme region is a separate protein. Ligand binding to the receptor activates the enzyme. The enzymes of receptor-enzymes are either protein kinases, such as *tyrosine kinase* (Fig. 6-10 ■), or *guanylyl cyclase,* the amplifier enzyme that converts GTP to **cyclic GMP (cGMP)** [🔁 p. 32]. Ligands for receptor-enzymes include many growth factors and cytokines, as well as the hormone insulin.

Most Signal Transduction Uses G Proteins

The **G protein-coupled receptors (GPCR)** are a large and complex family of membrane-spanning proteins that cross the phospholipid bilayer seven times (see Fig. 6-5). The cytoplasmic tail of the receptor protein is linked to a three-part membrane transducer molecule known as a **G protein**. Hundreds of G protein-coupled receptors have been identified, and the list continues to grow. The types of ligands that bind to G protein-coupled receptors include hormones, growth factors, olfactory molecules, visual pigments, and neurotransmitters. In 1994 Alfred G. Gilman and Martin Rodbell received a Nobel prize for the discovery of G proteins and their role in cell signaling.

G proteins get their name from the fact that they bind guanosine nucleotides [🔁 p. 32]. Inactive G proteins are bound to guanosine diphosphate (GDP). Exchanging the GDP for guanosine triphosphate (GTP) activates the G protein. When G proteins are activated, they either (1) open an ion channel in the membrane or (2) alter enzyme activity on the cytoplasmic side of the membrane.

G proteins linked to amplifier enzymes make up the bulk of all known signal transduction mechanisms. The two most common amplifier enzymes for G protein-coupled receptors are adenylyl cyclase and phospholipase C. The pathways for these amplifier enzymes are described next.

Adenylyl Cyclase-cAMP Is the Signal Transduction System for Many Lipophobic Hormones

The **G protein-coupled adenylyl cyclase-cAMP system** was the first identified signal transduction pathway (Fig. 6-11 ■). It was discovered in the 1950s by Earl Sutherland when he was studying the effects of hormones on carbohydrate metabolism. This discovery proved so significant to our understanding of signal transduction that in 1971 Sutherland was awarded a Nobel prize for his work.

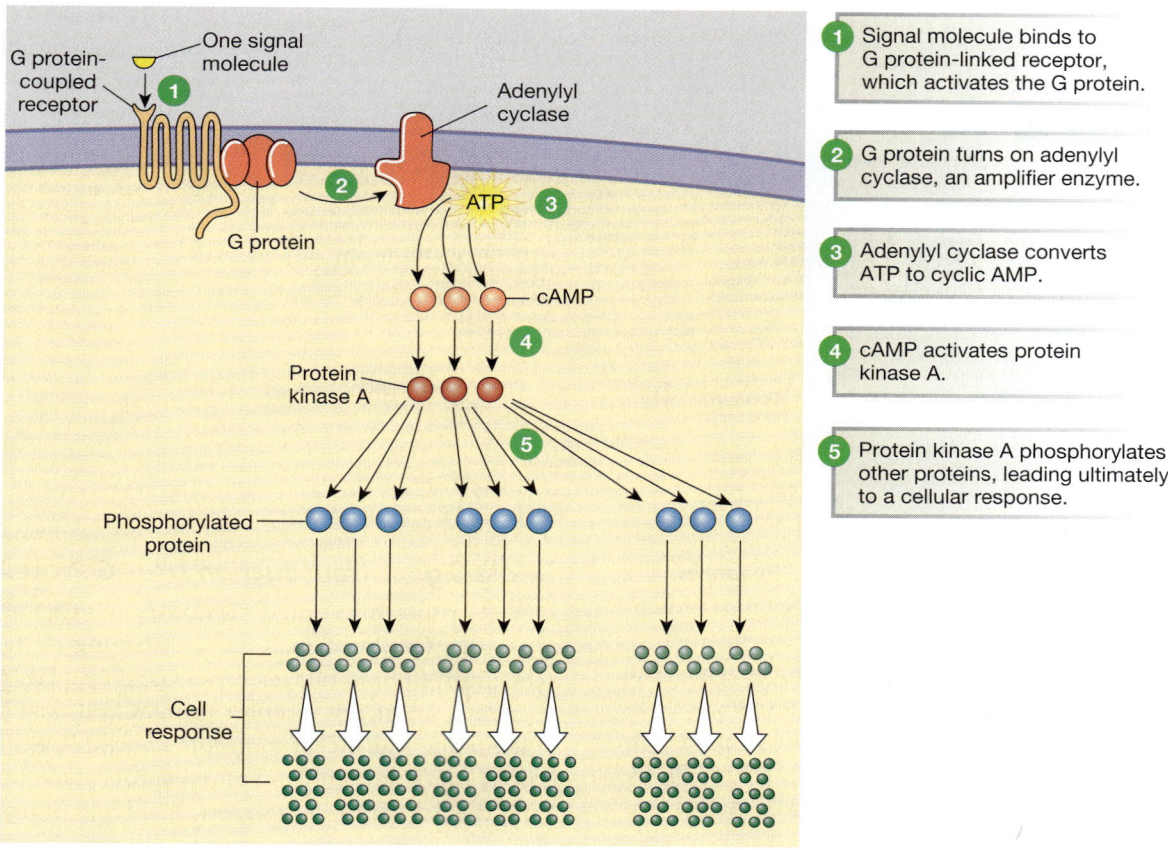

1. Signal molecule binds to G protein-linked receptor, which activates the G protein.

2. G protein turns on adenylyl cyclase, an amplifier enzyme.

3. Adenylyl cyclase converts ATP to cyclic AMP.

4. cAMP activates protein kinase A.

5. Protein kinase A phosphorylates other proteins, leading ultimately to a cellular response.

■ **FIGURE 6-11** *The G protein-coupled adenylyl cyclase-cAMP system*

This was the first second messenger system described.

In the adenylyl cyclase-cAMP system, *adenylyl cyclase* is the amplifier enzyme that converts ATP to the second messenger molecule *cyclic AMP* (cAMP). Cyclic AMP then activates *protein kinase A* (PK-A), which in turn phosphorylates other intracellular proteins as part of the signal cascade. The G protein-coupled adenylyl cyclase-cAMP system is the signal transduction system for many protein hormones.

G Protein-Coupled Receptors Also Use Lipid-Derived Second Messengers

Some G protein-coupled receptors are linked to a different amplifier enzyme: phospholipase C (Fig. 6-12 ■). When a signal molecule activates this G protein-coupled pathway, **phospholipase C** (**PL-C**) converts a membrane phospholipid (*phosphatidylinositol bisphosphate*) into two different second messenger molecules: diacylglycerol and inositol trisphosphate.

Diacylglycerol (DAG) is a nonpolar diglyceride that remains in the lipid portion of the membrane and interacts with **protein kinase C** (PK-C), a Ca^{2+}-activated enzyme associated with the cytoplasmic face of the cell membrane. Protein kinase C phosphorylates cytosolic proteins that continue the signal cascade.

Inositol trisphosphate (IP$_3$) is a water-soluble messenger molecule that enters the cytoplasm. There it binds to a calcium channel on the endoplasmic reticulum (ER). IP$_3$ binding opens the Ca^{2+} channel, allowing Ca^{2+} to diffuse out of the ER and into the cytosol. Calcium is itself an important signal molecule, as discussed below.

Integrin Receptors Transfer Information from the Extracellular Matrix

The membrane-spanning proteins called integrins [🔁 p. 70] mediate blood clotting, wound repair, cell adhesion and recognition in the immune response, and cell movement during development. On the extracellular side of the membrane, integrin receptors bind either to proteins of the extracellular matrix [🔁 p. 68] or to ligands such as antibodies and molecules involved in blood clotting. Inside the cell, integrins attach to the cytoskeleton via *anchor proteins* (Fig. 6-5). Ligand binding to the receptor causes integrins to activate intracellular enzymes or alter the organization of the cytoskeleton.

The importance of integrin receptors is illustrated by inherited conditions in which the receptor is absent. In one condition, platelets—cell fragments that play a key role in blood clotting—lack an integrin receptor. As a result, blood clotting is defective in these individuals.

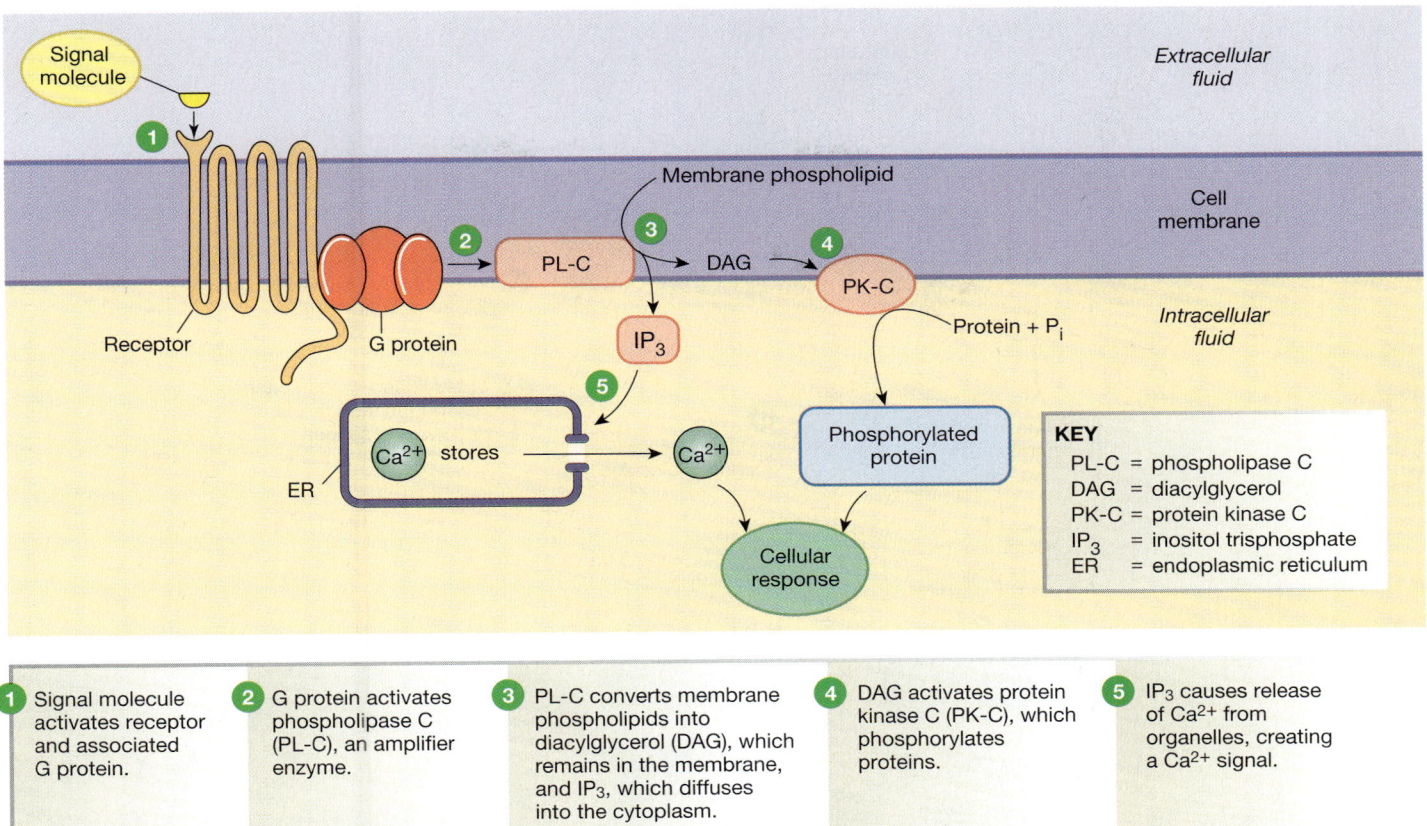

① Signal molecule activates receptor and associated G protein.

② G protein activates phospholipase C (PL-C), an amplifier enzyme.

③ PL-C converts membrane phospholipids into diacylglycerol (DAG), which remains in the membrane, and IP₃, which diffuses into the cytoplasm.

④ DAG activates protein kinase C (PK-C), which phosphorylates proteins.

⑤ IP₃ causes release of Ca²⁺ from organelles, creating a Ca²⁺ signal.

■ **FIGURE 6-12**   *The phospholipase C system*

In this system—another G protein-coupled second messenger system—phospholipase C (PL-C) converts membrane lipids into two second messengers: diacylglycerol (DAG) and inositol trisphosphate (IP₃).

The Most Rapid Signal Pathways Change Ion Flow Through Channels

Ligand-gated ion channels are the simplest receptors, and the activation of a **receptor-channel** initiates the most rapid intracellular responses. For that reason these receptors are often located in the excitable tissues of nerve and muscle. When an extracellular signal molecule binds to the receptor-channel protein, a channel gate opens or closes, changing the cell's permeability to an ion. One example of a receptor-channel is the acetylcholine-gated cation channel of skeletal muscle. The neurotransmitter *acetylcholine* from an adjacent neuron binds to the acetylcholine receptor and opens the channel, creating a cascade that initiates muscle contraction.

Increasing or decreasing the permeability of an ion channel changes the cell's membrane potential [⮂ p. 165]. For example, opening the acetylcholine-gated cation channel just described allows Na⁺ to enter the cytosol down that ion's electrochemical gradient. Net entry of cations depolarizes the cell, creating an electrical signal that alters voltage-sensitive proteins (Fig. 6-13 ■).

Note that not all ligand-gated ion channels are receptor-channels that are directly activated by extracellular signal molecules. Some ligand-gated channels are controlled by intracellular second messengers, such as cAMP. Others are indirectly linked to receptors by G proteins.

CONCEPT CHECK

9. Name the four categories of membrane receptors.

10. What is the difference between a first messenger and a second messenger?

11. Place the following terms in the correct order for a signal transduction pathway:

   (a) cell response, receptor, second messenger, ligand

   (b) amplifier enzyme, cell response, phosphorylated protein, protein kinase, second messenger

12. In each of the following situations, will a cell depolarize or hyperpolarize?

   (a) Cl⁻ channel opens
   (b) K⁺ channel opens
   (c) Na⁺ channel opens

Answers: p. 209

6

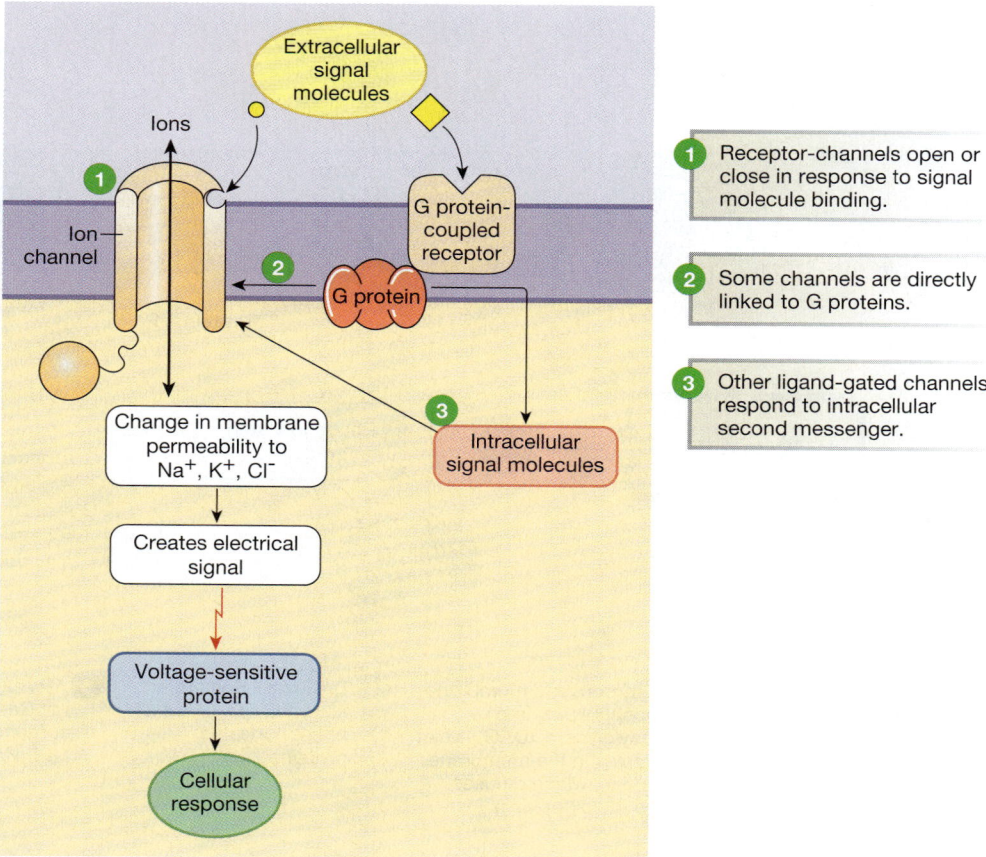

■ FIGURE 6-13 *How ions create electrical signals*

Figure 6-14 ■ is a summary map of basic signal transduction, showing the general relationships among first messengers, membrane receptors, second messengers, and cell responses.

NOVEL SIGNAL MOLECULES

In Chapters 7 and 8, you will learn more about hormones and neurotransmitters and their signal pathways. The following sections introduce you to some unusual signal molecules that are important in physiology and medicine. They include an ion (Ca^{2+}), two gases, and a family of lipid-derived messengers. The processes controlled by these signal molecules have been known for years, but the control signals themselves were discovered only relatively recently.

Calcium Is an Important Intracellular Signal

Calcium ions are the most versatile ionic messengers (Fig. 6-15 ■). Calcium enters the cytosol either through voltage-gated Ca^{2+} channels or through ligand-gated or mechanically gated channels. Calcium can also be released from intracellular compartments by second messengers, such as IP_3. Most intracellular

Ca^{2+} is stored in the endoplasmic reticulum [⮂ p. 67], where it is concentrated by active transport.

Release of Ca^{2+} into the cytosol (from any of the sources just mentioned) creates a Ca^{2+} signal, or Ca^{2+} "spark," that can be recorded using special Ca^{2+}-imaging techniques (see Biotechnology box: "Measuring Calcium Signals"). The calcium ions combine with cytoplasmic calcium-binding proteins to exert various effects. Several types of calcium-dependent events occur in the cell:

1. Ca^{2+} binds to the protein **calmodulin**, found in all cells, and alters enzyme or transporter activity or the gating of ion channels.

2. Calcium binds to other regulatory proteins and alters movement of contractile or cytoskeletal proteins such as microtubules. For example, Ca^{2+} binding to the regulatory protein *troponin* initiates muscle contraction in a skeletal muscle cell.

3. Ca^{2+} binds to regulatory proteins to trigger exocytosis of secretory vesicles [⮂ p. 148]. This was illustrated in Chapter 5 in the description of pancreatic beta cell release of insulin.

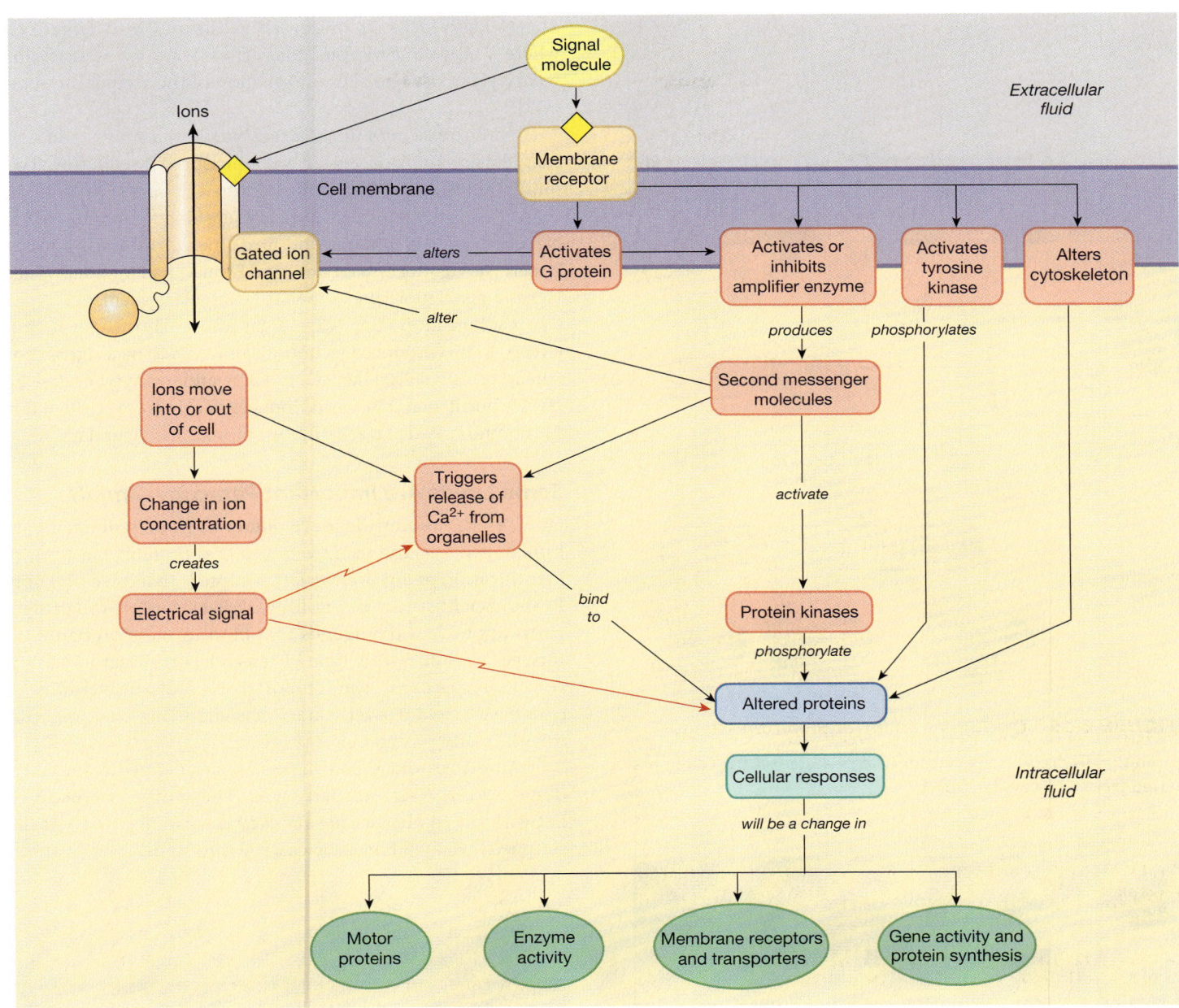

■ **FIGURE 6-14** *Summary map of signal transduction systems*

4. Ca^{2+} binds directly to ion channels to alter their gating state. An example of this target is a Ca^{2+}-activated K^+ channel found in nerve cells.

5. Ca^{2+} entry into a fertilized egg initiates development of the embryo.

✔ **C O N C E P T   C H E C K**

13. The concentration of extracellular Ca^{2+} averages 2.5 mmol/L. Free cytosolic Ca^{2+} concentration is about 0.001 mmol/L. If a cell is going to move calcium ions from its cytosol to the extracellular fluid, will it use passive or active transport? Explain.   Answers: p. 209

Gases Are Ephemeral Signal Molecules

Nitric oxide (NO), a soluble gas, is a novel short-acting paracrine/autocrine signal that acts close to where it is produced. Nitric oxide took years to identify because it is rapidly broken down, with a half-life of only 2 to 30 seconds. (*Half-life* is the time required for the signal to lose half of its activity.) In tissues, NO is synthesized by the action of the enzyme *nitric oxide synthase* (NOS) on the amino acid arginine:

$$\text{Arginine} + O_2 \xrightarrow{\textit{nitric oxide synthase}} \text{NO} + \text{citrulline (an amino acid)}$$

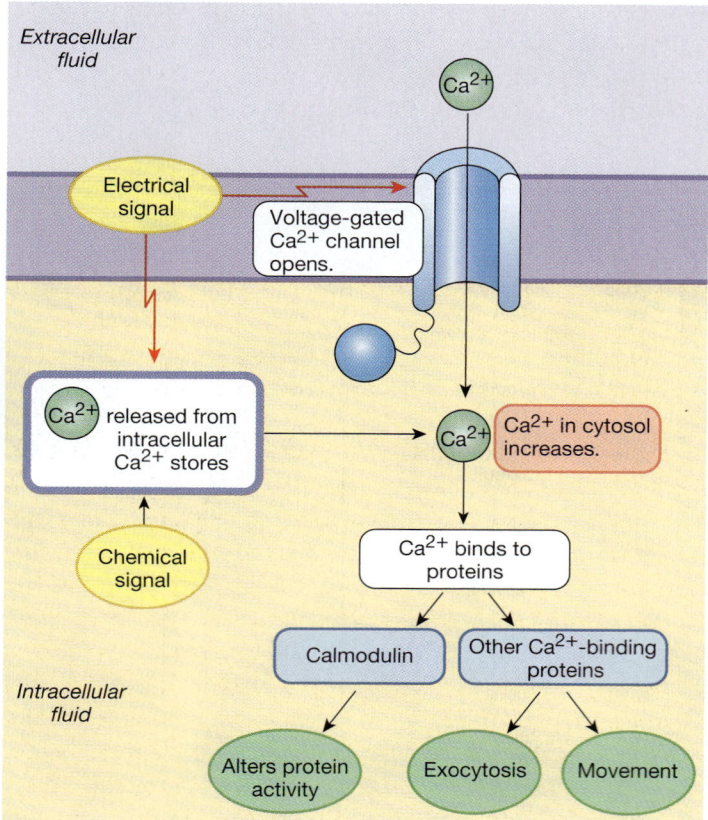

■ **FIGURE 6-15**   *Calcium as an intracellular messenger*

Calcium signals occur in the cytosol when Ca^{2+} enter the cell or are released from intracellular stores.

BIOTECHNOLOGY

MEASURING CALCIUM SIGNALS

If you've ever run your hand through a tropical ocean at night and seen the glow of bioluminescent jelly-fish, you've seen a calcium signal. Aequorin, a protein complex isolated from jellyfish, is one of the molecules that scientists use to monitor the presence of calcium ions during a cellular response. When aequorin combines with calcium, it releases light that can be measured by electronic detection systems. Since the first use of aequorin in 1967, researchers have been designing better and better indicators that allow them to follow calcium signals in cells. With the help of molecules called fura, Oregon green, BAPTA, and chameleons, we can now watch calcium ions diffuse through gap junctions and flow out of intracellular organelles.

The NO produced in this reaction diffuses into target cells, where it binds to a receptor that activates the cytosolic form of guanylyl cyclase and causes formation of the second messenger cGMP.

Nitric oxide in the brain acts as a neurotransmitter and a neuromodulator. In blood vessels, NO is produced by endothelial cells lining the vessels. It then diffuses into adjacent smooth muscle cells, causing them to relax and dilate the blood vessel. In 1998 the Nobel prize for physiology and medicine was awarded jointly to Robert Furchgott, Louis Ignarro, and Ferid Murad for their work on NO as a signal molecule in the cardiovascular system.

Carbon monoxide (CO), a gas known mostly for its toxic effects, is also produced in minute amounts to be a signal molecule in certain cells. Like NO, CO activates guanylyl cyclase and cGMP, but it may also work independently to exert its effects. Carbon monoxide targets smooth muscle and neural tissue.

Some Lipids Are Important Paracrine Signals

One of the interesting developments from sequencing the human genome and using genes to find proteins has been the identification of *orphan receptors,* receptors that have no known ligand. Scientists are trying to work backwards through signal pathways to find the ligands that bind to these orphan receptors. As a result of this type of research, investigators have recently recognized the importance and universality of *eicosanoids,* lipid-derived paracrine signals that play important roles in many physiological processes.

All eicosanoid signal molecules are derived from arachidonic acid, a 20-carbon fatty acid. The synthesis process is a network called the *arachidonic acid cascade* (Fig. 6-16 ■). For simplicity, we will break the cascade into steps.

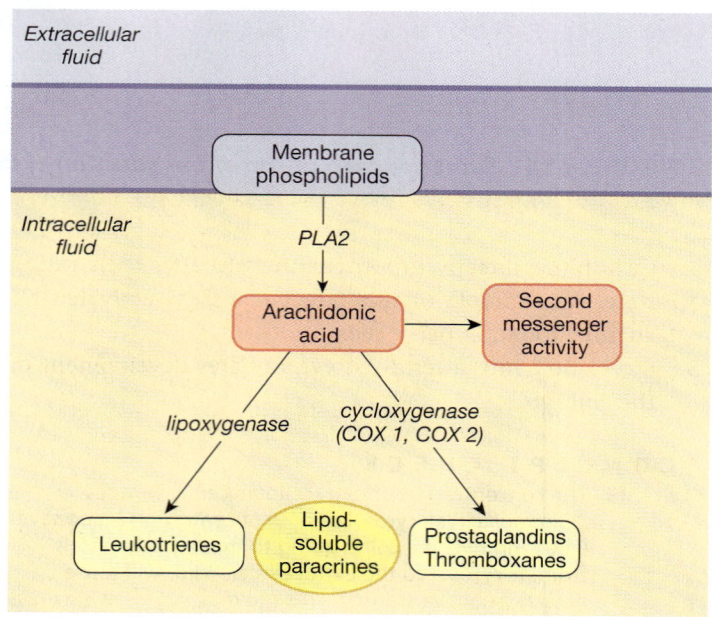

■ **FIGURE 6-16**   *The arachidonic acid cascade produces lipid messengers*

Arachidonic acid is produced from membrane phospholipids by the action of an enzyme, **phospholipase A₂** (PLA2). The activity of phospholipase A_2 is controlled by hormones and other signals. Arachidonic acid itself may act directly as a second messenger, altering ion channel activity and intracellular enzymes. It may also be converted into one of several classes of eicosanoid paracrines. These lipid-soluble molecules can diffuse out of the cell and combine with receptors on neighboring cells to exert their action.

There are two major groups of arachidonic acid-derived paracrines to be aware of:

1. **Leukotrienes** are molecules produced by the action of the enzyme *lipoxygenase* on arachidonic acid [*leuko-*, white + *triene*, a molecule with three double bonds between carbon atoms]. Leukotrienes are secreted by certain types of white blood cells. They play a significant role in asthma, a lung condition in which the smooth muscle of the airways constricts, making it difficult to breathe, and in the severe allergic reaction known as *anaphylaxis*. For this reason, pharmaceutical companies have been actively developing drugs to block leukotriene synthesis or action.

2. **Prostanoids** are molecules produced when the enzyme **cyclooxygenase** (COX) acts on arachidonic acid. Prostanoids include **prostaglandins** and **thromboxanes**. These eicosanoids act on many tissues of the body, including smooth muscle in various organs, platelets, kidney, and bone. In addition, prostaglandins are involved in sleep, inflammation, pain, and fever.

The nonsteroidal anti-inflammatory drugs (NSAIDs), such as aspirin and ibuprofen, help prevent inflammation by inhibiting COX enzymes and decreasing prostaglandin synthesis. However, NSAIDs are not specific and may have serious unwanted side effects, such as bleeding in the stomach. The discovery of two COX isozymes, COX1 and COX2, enabled the design of drugs that target a specific COX isozyme. By inhibiting only COX2, the enzyme that produces inflammatory prostaglandins, physicians hoped to treat inflammation with fewer side effects. However, recently studies showed that patients who took COX2 inhibitors for extended periods had increased risk of heart attacks and strokes, so whether use of those drugs should continue is still a matter of debate.

CONCEPT CHECK

14. Based on what you have learned about signal molecules, where might a drug that blocks leukotriene action act? How might a drug that blocks leukotriene synthesis act?

Answers: p. 209

MODULATION OF SIGNAL PATHWAYS

As you have just learned, signal pathways in the cell can be very complex. To complicate matters, different cells may respond differently to a given signal molecule. How can one molecule trigger response A in tissue 1 and response B in tissue 2? *For*

most signal molecules, the target cell response is determined by the receptor and its associated intracellular pathways, not by the ligand. Because of the importance of signal pathways, cells use receptors to maintain flexibility in their responses.

Receptors Exhibit Saturation, Specificity, and Competition

Receptors are proteins; therefore, receptor-ligand binding exhibits the protein-binding characteristics of specificity, competition, and saturation [p. 43]. You have learned about similar protein-binding properties in the enzymes discussed in Chapter 4 [p. 96] and the transporters discussed in Chapter 5 [p. 138].

Specificity and Competition: Multiple Ligands for One Receptor Receptors have binding sites for their ligands, just as enzymes and transporters do. As a result, different molecules with similar structures may be able to bind to the same receptor. A classic example of this principle involves two neurocrines responsible for the fight-or-flight response: the neurotransmitter *norepinephrine* and its cousin the neurohormone *epinephrine* (also called *adrenaline*). Both molecules bind to a class of receptors called *adrenergic receptors*. (*Adrenergic* is the adjective relating to adrenaline.) The ability of adrenergic receptors to bind these neurocrines, but not others, demonstrates specificity of the receptors.

Epinephrine and norepinephrine also compete for a single receptor type. Both neurocrines bind to subtypes of adrenergic receptors designated alpha (α) and beta (β). However, α-receptors have a higher binding affinity for norepinephrine, whereas the β_2-receptor subtype has a higher affinity for epinephrine.

Agonists and Antagonists When a ligand combines with a receptor, one of two events follows. Either the ligand turns the receptor on and elicits a response, or the ligand occupies the binding site and prevents the receptor from responding (Fig. 6-17 ■). Ligands that turn receptors on are known as *agonists*, and ligands that block receptor activity are called *antagonists*.

Pharmacologists use the principle of competing agonists [p. 40] to design drugs that are longer-acting and more resistant to enzymatic degradation than the endogenous ligand. One example is the family of modified estrogens (female sex hormones) in birth control pills. These drugs are agonists of naturally occurring estrogens but have chemical groups added to protect them from breakdown and extend their active life.

Multiple Receptors for One Ligand For many years physiologists were unable to explain the observation that a single signal molecule could have different effects in different tissues. For example, epinephrine, the neurohormone previously described, dilates blood vessels in skeletal muscle but constricts blood vessels in the intestine. How can one chemical have opposite effects? The answer became clear when scientists discovered that receptors, like other proteins, come as families of related isoforms [p. 40].

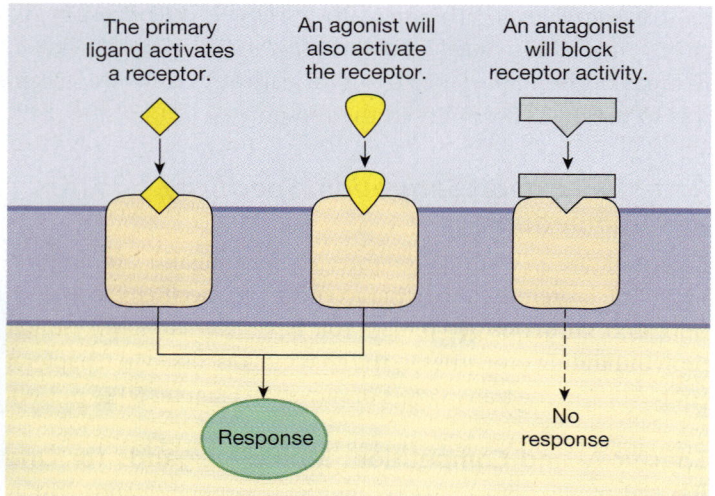

The primary ligand activates a receptor.

An agonist will also activate the receptor.

An antagonist will block receptor activity.

Response

No response

■ **FIGURE 6-17** *Agonists and antagonists*

The cellular response that follows binding of a signal molecule to a receptor depends on which isoform of the receptor is involved. For example, the α- and β-adrenergic receptors for epinephrine described earlier are isoforms of each other. When epinephrine binds to α-receptors on smooth muscle in intestinal blood vessels, the vessels constrict (Fig. 6-18 ■). When epinephrine binds to β-receptors on certain skeletal muscle blood vessels, the vessels dilate. In other words, the response of the blood vessel to epinephrine depends on the receptor isoform, not on the ligand that activates the receptor. Many drugs now are designed so that they are specific for only one receptor isoform.

CONCEPT CHECK

15. What common property of receptors, enzymes, and transporters explains why they all exhibit saturation, specificity, and competition?

16. Insulin increases the number of glucose transporters on a skeletal muscle cell but not on the membrane of a liver cell. List two possible mechanisms that could explain how this one hormone can have these two different effects.

Answers: p. 209

Up-Regulation and Down-Regulation Enable Cells to Modulate Responses

Saturation of proteins refers to the fact that protein activity reaches a maximum rate because cells contain limited numbers of protein molecules [p. 43]. This phenomenon can be observed with enzymes, transporters, and receptors. A cell's ability to respond to a chemical signal therefore can be limited by the finite number of receptors for that signal.

A single cell contains between 500 and 100,000 receptors on the surface of its cell membrane, with additional receptors in the cytosol and nucleus. In any given cell, the number of receptors may change over time. Old receptors are withdrawn from the membrane by endocytosis and are broken down in lysosomes. New receptors are inserted into the membrane by exocytosis. Intracellular receptors are also made and broken down. This flexibility permits a cell to vary its responses to chemical signals depending on the extracellular conditions and the internal needs of the cell.

What happens when a signal molecule is present in the body in abnormally high concentrations for a sustained period of time? Initially the increased signal level creates an enhanced response. As this enhanced response continues, the target cells may attempt to bring their response back to normal by *down-regulation* of the receptors for the signal [p. 43].

Down-regulation takes two forms: either a decrease in receptor number or a decrease in binding affinity. The cell can physically remove receptors from the membrane through endocytosis [Fig. 5-24, p. 149]. A quicker and more easily reversible type of down-regulation, called *desensitization,* can be achieved by binding a chemical modulator to the receptor protein. For example, the β-adrenergic receptors described in the previous section can be desensitized by phosphorylation of the receptor. The result of decreased receptor number or decreased binding affinity is a diminished response of the target cell even though the concentration of the signal molecule remains high.

Down-regulation is one explanation for the development of *drug tolerance,* a condition in which the response to a given dose

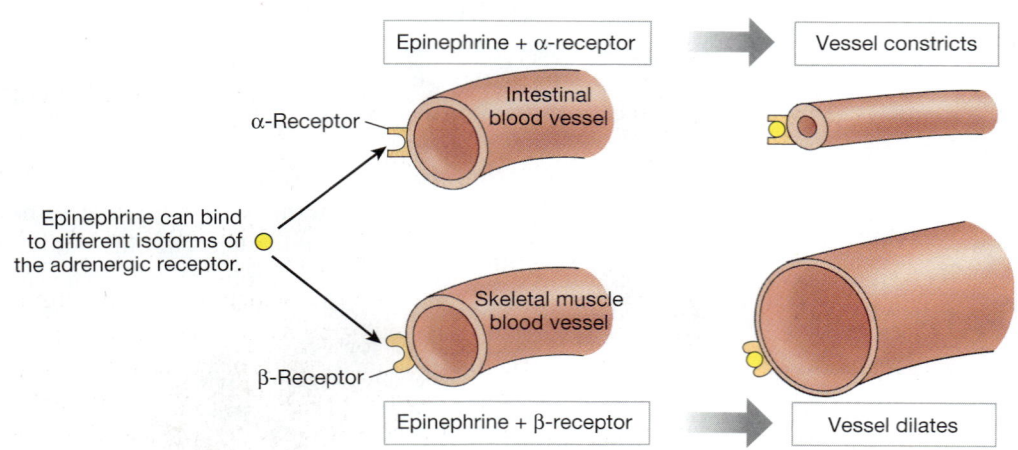

Epinephrine + α-receptor → Vessel constricts

α-Receptor

Intestinal blood vessel

Epinephrine can bind to different isoforms of the adrenergic receptor.

β-Receptor

Skeletal muscle blood vessel

Epinephrine + β-receptor → Vessel dilates

■ **FIGURE 6-18** *Target response depends on the target receptor*

In this example, epinephrine will either dilate or constrict blood vessels depending on the receptor found on the blood vessel.

RUNNING PROBLEM

"My dad takes insulin shots for his diabetes," Marvin says. "What does insulin do?" The nurse practitioner replies that normally insulin helps many cells take up and utilize glucose. In both types of diabetes, however, fasting blood glucose concentrations are elevated because the cells are not taking up and using glucose normally. If people with type 1 diabetes are given shots of insulin, their blood glucose levels decline. If people with type 2 diabetes are given insulin, blood glucose levels may change very little.

Question 2:

In which form of diabetes are the insulin receptors more likely to be up-regulated?

175 179 **191** 193 197 201 206

decreases despite continuous exposure to the drug. The development of tolerance to opiates, such as morphine and codeine, occurs when the receptors for these drugs down-regulate.

If the concentration of a ligand decreases, the target cell may use up-regulation to insert more receptors into the cell membrane in an attempt to keep its response at a normal level. For example, if a nerve cell is damaged and unable to release normal amounts of neurotransmitter, its target cell will up-regulate its receptors. This up-regulation makes the target cell more sensitive to whatever neurotransmitters are present. Up-regulation is also programmed during development as a mechanism that allows cells to vary their responsiveness to growth factors and other signal molecules.

CONCEPT CHECK

17. To down-regulate a receptor's binding affinity, a cell might (select all that apply):
    (a) synthesize a new isoform of the receptor
    (b) withdraw receptors from the membrane
    (c) insert new receptors into the membrane
    (d) use a covalent modulator (*Hint:* p. 41)

Answers: p. 209

Cells Must Be Able to Terminate Signal Pathways

In the body, signals turn on and off, so cells must be able to tell when a signal is over. This requires that signaling processes have built-in termination mechanisms. For example, to stop the response to a calcium signal, a cell removes Ca^{2+} from the cytosol by pumping it either back into the endoplasmic reticulum or out into the extracellular fluid.

Receptor activity can be stopped in a variety of ways. The extracellular first messenger can be degraded by enzymes in the extracellular space. An example of this is the breakdown of the neurotransmitter acetylcholine. Other first messengers, particularly neurotransmitters, can be removed from the extracellular fluid by being transported into neighboring cells. A widely used class of antidepressant drugs called *selective serotonin reuptake inhibitors,* or SSRIs, extends the active life of the neurotransmitter serotonin by slowing its removal from the extracellular fluid.

Once a ligand is bound to its receptor, activity can also be terminated by endocytosis of the receptor-ligand complex. This process was illustrated in Figure 5-24 [p. 149]. After the vesicle is in the cell, the ligand is removed, and the receptors can be returned to the membrane by exocytosis.

Many Diseases and Drugs Target the Proteins of Signal Transduction

As we learn more about cell signaling, scientists are realizing how many diseases are linked to problems with signal pathways. Diseases can be caused by alterations in receptors or by problems with G proteins or second messenger pathways (see Table 6-3 ■ for some examples). A single change in the amino acid sequence of a receptor protein can alter the shape of the receptor's binding site, thereby either destroying or modifying its activity.

Pharmacologists are using information about signaling mechanisms to design drugs to treat disease. Some of the alphabet soup of drugs in widespread use are ARBs (angiotensin receptor blockers), beta-adrenergic receptor blockers, and calcium-channel blockers for treating high blood pressure; SERMs (selective estrogen receptor modulators) for treating estrogen-dependent cancers; and H_1 (histamine type 1) receptor antagonists for decreasing acid secretion in the stomach. We will encounter many of these drugs again when we study the systems in which they are effective.

CONTROL PATHWAYS: RESPONSE AND FEEDBACK LOOPS

In Chapter 1 you learned that homeostasis is the ability of the body to maintain a relatively stable internal environment [p. 3]. Homeostasis is a continuous process that uses a **physiological control system** to monitor key functions, or **regulated variables**.

In its simplest form, any control system has three components: (1) an input signal; (2) a controller, which is programmed to respond to certain input signals; and (3) an output signal [Fig. 1-6, p. 9]. Physiological control systems are a little more complex. The input signal consists of the regulated variable and a specialized **sensor**. If the variable moves out of its desirable range, the sensor is activated and sends a signal to the

6

TABLE 6-3   Some Diseases or Conditions Linked to Abnormal Signaling Mechanisms

Genetically inherited abnormal receptors

RECEPTOR	PHYSIOLOGICAL ALTERATION	DISEASE OR CONDITION THAT RESULTS
Vasopressin receptor (X-linked defect)	Shortens half-life of the receptor	Congenital diabetes insipidus
Calcium sensor in parathyroid gland	Fails to respond to increase in plasma Ca^{2+}	Familial hypercalcemia
Rhodopsin receptor in retina of eye	Improper protein folding	Retinitis pigmentosa

Toxins affecting signal pathways

TOXIN	PHYSIOLOGICAL EFFECT	CONDITION THAT RESULTS
Bordetella pertussis toxin	Blocks inhibition of adenylate cyclase (i.e., keeps it active)	Whooping cough
Cholera toxin	Blocks enzyme activity of G proteins; cell keeps making cAMP	Ions secreted into lumen of intestine, causing massive diarrhea

controller (Fig. 6-19 ■). The controller acts as an **integrating center** [*integrare,* to restore] that evaluates information coming from the sensor and initiates a response that is designed to bring the regulated variable back into the desired range. The integrating center is often a nerve cell or an endocrine cell. The muscles and other tissues controlled by integrating centers are known as **effectors** [*effectus,* the carrying out of a task] because they effect a change.

Cannon's Postulates Describe Regulated Variables and Physiological Control Systems

Walter Cannon, the father of American physiology, described a number of properties of homeostatic control systems in the 1920s based on his observations of the body in health and disease states.* You will encounter these properties repeatedly as you study the various organ systems of the body. Cannon's four postulates are:

1. **The nervous system has a role in preserving the "fitness" of the internal environment.** *Fitness* in this instance means conditions that are compatible with normal function. The nervous system coordinates and integrates blood volume, blood osmolarity, blood pressure, and body temperature, among other regulated variables.

2. **Some systems of the body are under tonic control** [*tonos,* tone]. To quote Cannon, "An agent may exist which has a moderate activity which can be varied up and down." This type of control is like the volume control on a radio, which enables you to make the sound level louder or softer by turning a single knob. A physiological example of a tonically controlled system is the neural regulation of diameter in certain blood vessels, in which increased input from the nervous system decreases diameter, and decreased input from the nervous system increases diameter (Fig. 6-20 ■). *Tonic control* is one of the more difficult concepts in physiology because we have a tendency to think of responses stopping and starting when a controller turns off or on rather than as responses increasing and decreasing.

3. **Some systems of the body are under antagonistic control.** Cannon wrote, "When a factor is known which can shift a homeostatic state in one direction, it is reasonable to look for a factor or factors having an opposing effect." Systems that are not under tonic control are usually under *antagonistic control,* either by hormones or the nervous system. For example,

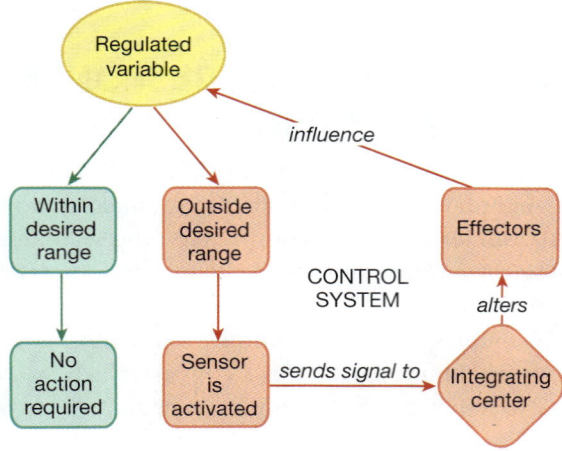

■ **FIGURE 6-19** *Physiological control systems keep regulated variables within a desired range during homeostasis*

*"Organization for Physiological Homeostasis," *Physiological Reviews* 9: 399–443, 1929.

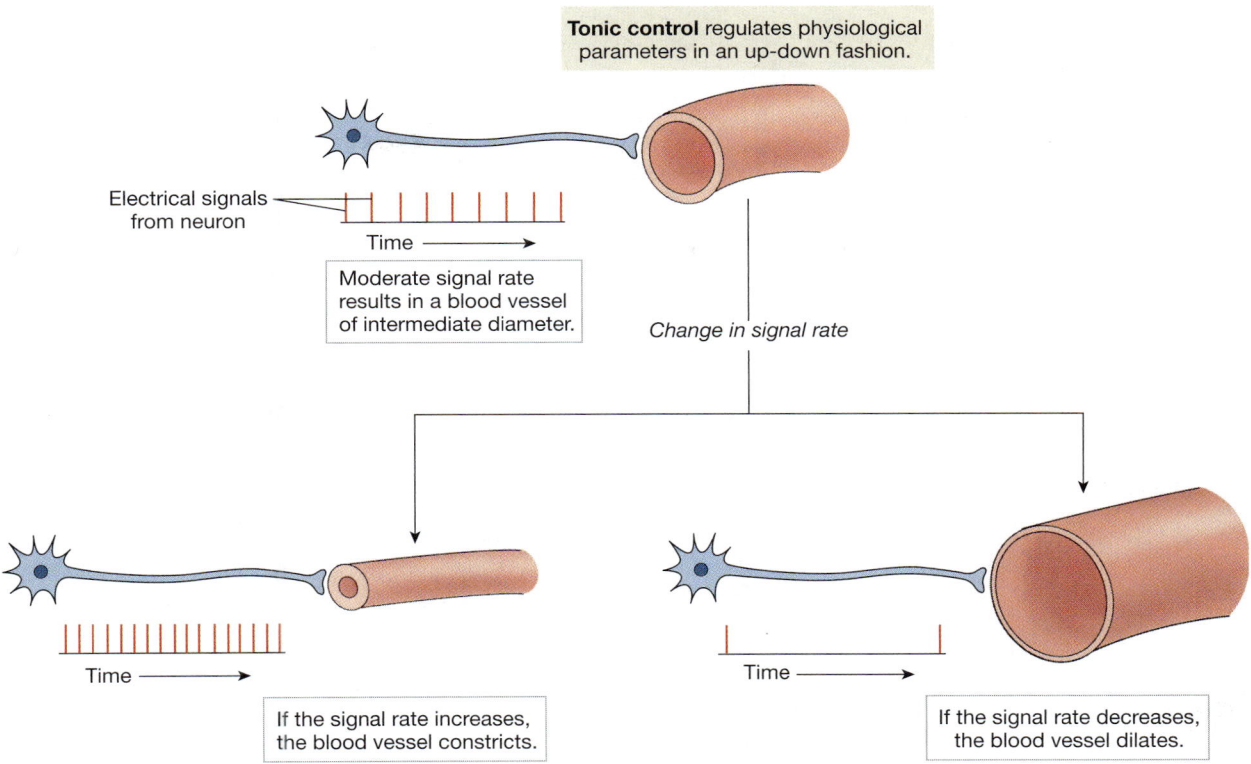

Tonic control regulates physiological parameters in an up-down fashion.

Electrical signals from neuron

Time ⟶

Moderate signal rate results in a blood vessel of intermediate diameter.

Change in signal rate

Time ⟶

If the signal rate increases, the blood vessel constricts.

Time ⟶

If the signal rate decreases, the blood vessel dilates.

■ **FIGURE 6-20** *Tonic control of blood vessel diameter*

insulin and glucagon are antagonistic hormones. Insulin decreases the concentration of glucose in the blood; glucagon increases it. In pathways controlled by the nervous system, the sympathetic and parasympathetic divisions often have opposing effects. For example, chemical signals from a sympathetic neuron increase heart rate, but chemical signals from a parasympathetic neuron decrease it (Fig. 6-21 ■).

4. **One chemical signal can have different effects in different tissues.** Cannon observed correctly that "homeostatic agents antagonistic in one region of the body may be cooperative in another region." However, it was not until we learned about cell receptors that the basis for the seemingly contradictory actions of some hormones or nerves became clear. As you learned in the first part of this chapter, a single chemical signal can have different effects depending on the receptor at the target cell. For example, epinephrine constricts or dilates blood vessels, depending on whether the vessel has α- or β-adrenergic receptors (see Fig. 6-18).

The remarkable accuracy of Cannon's postulates, now confirmed with cellular and molecular data, is a tribute to the observational skills of scientists in the nineteenth and early twentieth centuries.

 CONCEPT CHECK

18. What is the difference between tonic control and antagonistic control?

19. How can one chemical signal have opposite effects in two different tissues?   Answers: p. 209

RUNNING PROBLEM

"Why is an elevated blood glucose concentration bad?" Marvin asks. "The elevated blood glucose itself is not bad," says the nurse practitioner, "but when it is high after an overnight fast, it suggests that there is something wrong with the way your body is regulating its glucose metabolism." When a normal person absorbs a meal containing carbohydrates, blood glucose levels increase and stimulate insulin release. When cells have taken up the glucose from the meal and blood glucose levels fall, secretion of another pancreatic hormone, glucagon, increases. Glucagon raises blood glucose and helps keep the level within the homeostatic range.

Question 3:
The homeostatic regulation of blood glucose levels by the hormones insulin and glucagon is an example of which of Cannon's postulates?

175   179   191   **193**   197   201   206

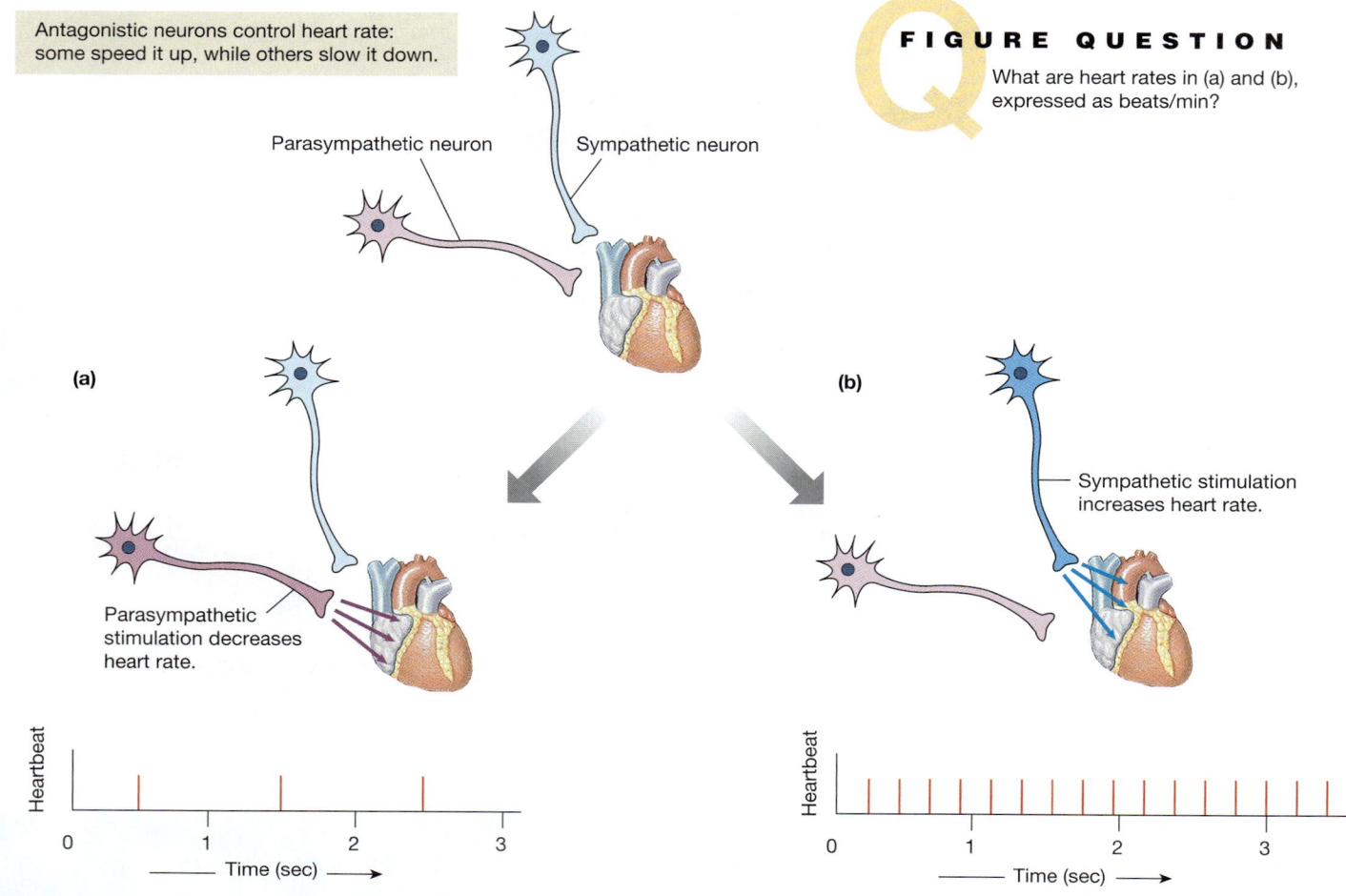

Antagonistic neurons control heart rate: some speed it up, while others slow it down.

Parasympathetic neuron   Sympathetic neuron

(a)

(b)

Parasympathetic stimulation decreases heart rate.

Sympathetic stimulation increases heart rate.

Heartbeat

0   1   2   3
Time (sec)

Heartbeat

0   1   2   3
Time (sec)

■ **FIGURE 6-21**   *Antagonistic control of heart rate*

The heart is controlled by antagonistic neurons. One set of neurons speeds up heart rate, and the other set slows it down.

Homeostasis May Be Maintained by Local or Long-Distance Pathways

The simplest control takes place at the tissue or cell involved. In **local control**, a relatively isolated change occurs in the vicinity of a cell or tissue and evokes a paracrine or autocrine response (Fig. 6-22 ■). More complicated **reflex control pathways** respond to changes that are widespread throughout the body or *systemic* in nature. In a reflex pathway, an integrating center located away from the affected cell or tissue receives information, evaluates it, and decides whether to send a chemical or electrical signal to initiate a response.

Long-distance reflex pathways are traditionally considered to involve two control systems: the nervous system and the endocrine system. However, cytokines [🔁 p. 176] are now known to be involved in some long-distance pathways. During stress and systemic inflammatory responses, cytokines work together with the nervous and endocrine systems to integrate information from all over the body into coordinated responses.

Local Control   Paracrine and autocrine signals are responsible for the simplest control systems. In local control, a cell or tissue senses a change in its immediate vicinity and responds. The response is restricted to the region where the change took place—hence the term *local control*.

One example of local control can be observed when oxygen concentration in a tissue decreases. The cells lining the small blood vessels bringing blood to that area sense the fall in oxygen concentration and respond by secreting a paracrine signal. The paracrine molecule relaxes muscles in the blood vessel wall, dilating the blood vessel and bringing more blood and therefore more oxygen to the area. Paracrine signal molecules involved in this response include carbon dioxide and metabolic products such as lactic acid.

Reflex Control   In a reflex control pathway, coordination of the reaction lies outside the organ that carries out the response. We will use the term *reflex* to mean any long-distance pathway that uses the nervous system, endocrine system, or both to receive

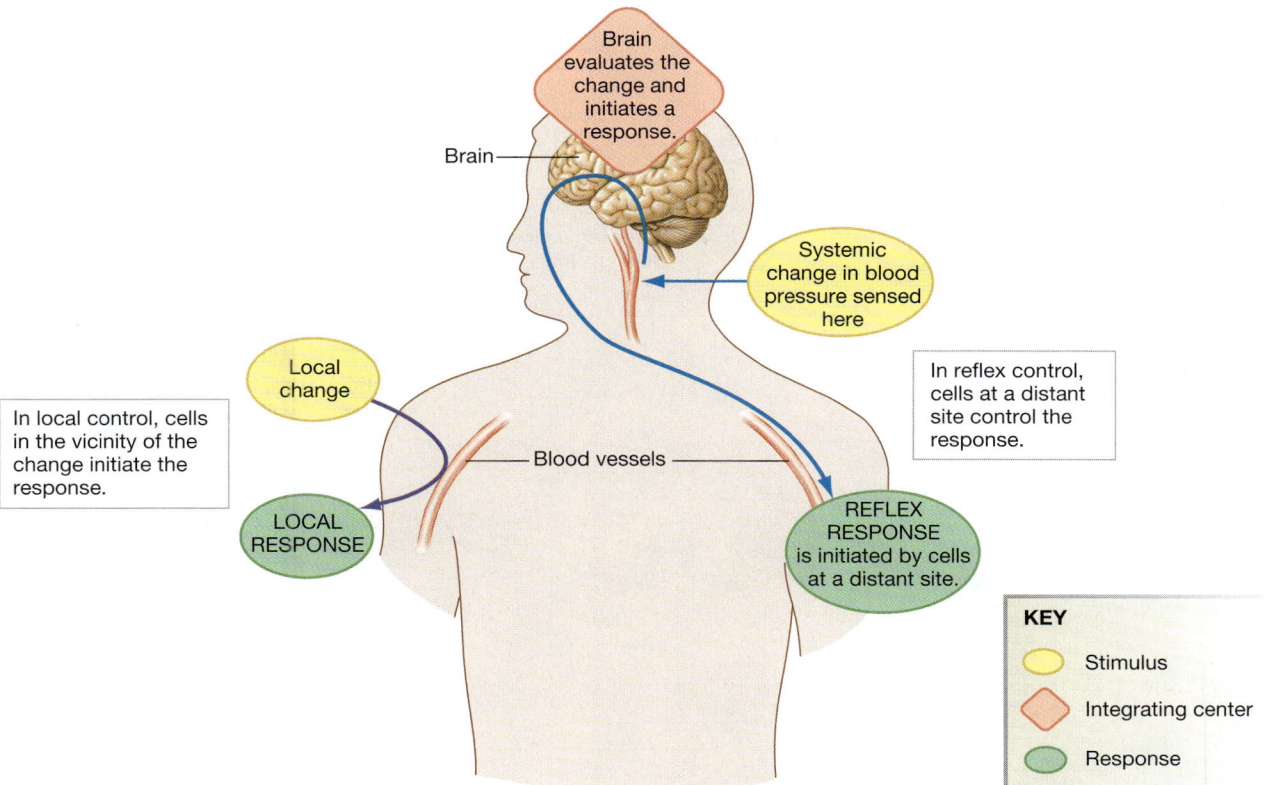

■ FIGURE 6-22 *Comparison of local and reflex control*

input about a change, integrate the information, and react appropriately. A reflex pathway can be broken down into two parts: a response loop and a feedback loop (Fig. 6-23 ■). The response loop begins with a stimulus and ends with the response of the target cell. We will discuss response loops first and then consider how they interact with feedback loops.

As with any other control system, a **response loop** has three primary components: an *input signal, integration of the signal,* and an *output signal.* These three components can be broken down into the following sequence of seven steps to form a pattern that is found with slight variations in all reflex pathways:

Stimulus → sensor or receptor → afferent pathway →
integrating center →
efferent pathway → target or effector → response

The input signal of a homeostatic reflex pathway consists of a stimulus, its sensory receptor, and an afferent (or incoming) pathway. (1) A **stimulus** is the disturbance or change that sets the pathway in motion. The stimulus may be a change in temperature, oxygen content, blood pressure, or any one of a myriad of other variables. The stimulus is sensed by (2) a **sensor** or sensory receptor that is continuously monitoring its environment. When alerted to a change, the receptor sends (3) a signal, or **afferent** (incoming) **pathway**, that links the receptor to (4) an integrating center. The integrating center then evaluates the incoming signal, compares it with the **setpoint**, or desired

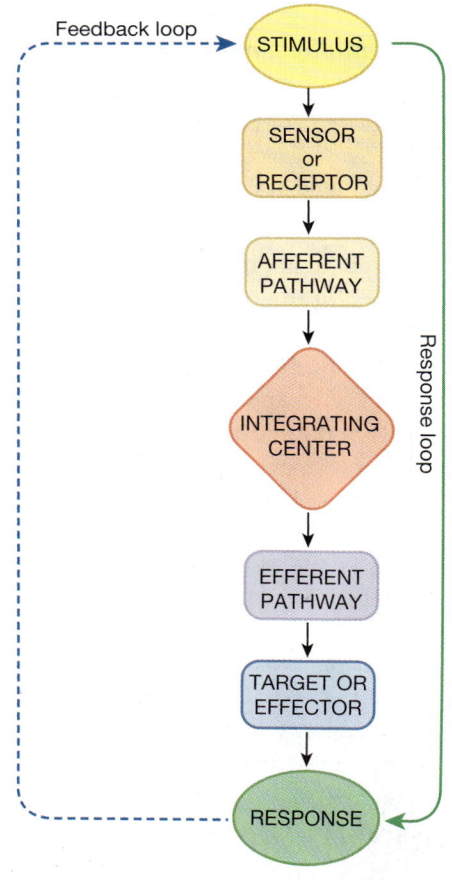

■ FIGURE 6-23 *Steps in a reflex control pathway*

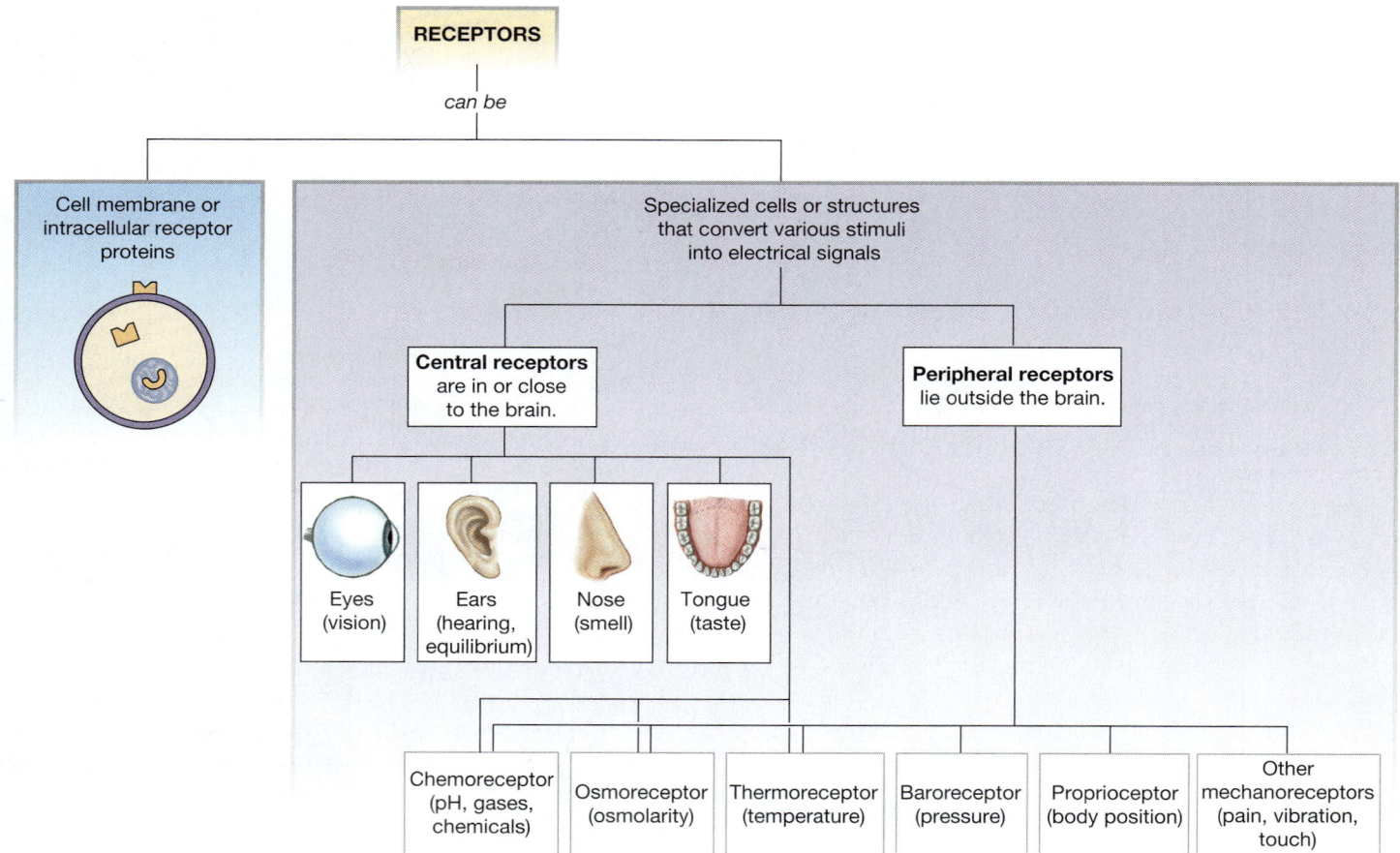

■ **FIGURE 6-24** *Multiple meanings of the word* receptor

The word *receptor* may mean a protein that binds to a ligand. Receptor can also mean a specialized cell or structure for transduction of stimuli into electrical signals (a *sensory receptor*). Sensory receptors are classified as central or peripheral, depending on whether they are found in the brain or outside the brain.

value, and decides on an appropriate response. The integrating center then initiates (5) the output signal, or **efferent** (outgoing) **pathway**. This is an electrical and/or chemical signal that travels to (6) the effector. The effector, or target, is the cell or tissue that carries out (7) the appropriate response to bring the situation back to within normal limits.

Receptor   The first step in a physiological response loop is activation of a sensor or receptor by the stimulus. NOTICE! This is a new and different application of the word *receptor*. Like many other terms in physiology, *receptor* can have different meanings (Fig. 6-24 ■). The sensory receptors of a neural reflex are not protein receptors that bind to signal molecules, like those involved in signal transduction. Rather, sensory receptors are specialized cells, parts of cells, or complex multicellular receptors such as the eye that respond to changes in the environment around them.

There are many sensory receptors in the body, each located where it is in the best position to monitor the variable it detects. The eyes, ears, and nose are receptors that sense light, sound and

motion, and odors, respectively. Your skin is covered with less complex receptors that sense touch, temperature, vibration, and pain. Other sensory receptors are internal: receptors in the joints of the skeleton that send information to the brain about body position, or blood pressure and oxygen receptors in blood vessels that monitor conditions in the circulatory system. The sensory receptors involved in neural reflexes are divided into *central receptors,* located in or closely linked to the brain, and *peripheral receptors,* which reside elsewhere in the body.

All sensory receptors have a **threshold**, a minimum stimulus that must be achieved to set the reflex response in motion. If a stimulus is below the threshold, no response loop will be initiated. You can demonstrate threshold easily by touching the back of your hand with a sharp, pointed object, such as a pin. If you touch the point to your skin lightly enough, you can see the contact between the point and your skin even though you do not feel anything. In this case, the stimulus (pressure from the point of the pin) is below threshold, and the pressure receptors of the skin are not responding. As you press harder, the stimulus reaches threshold, and the receptors respond by sending a signal through the afferent pathway, causing you to feel the pin.

Endocrine reflexes that are not associated with the nervous system do not use sensory receptors to initiate their pathways. Instead, endocrine cells act both as sensor and integrating center for the reflex. You were introduced to an example of this in Chapter 5, when we discussed how the pancreatic beta cells sense and respond to changes in blood glucose concentrations [ p. 166].

Afferent pathway   The afferent pathway in a reflex varies depending on the type of reflex. In a neural reflex, such as the pin touch above, the afferent pathway is the electrical and chemical signals carried by a nerve cell. In an endocrine reflex, there is no afferent pathway because the stimulus comes directly into the endocrine cell, which serves as both sensor and integrating center.

Integrating center   The integrating center in a reflex pathway is the cell that receives information about the change and is programmed to initiate an appropriate response. In endocrine reflexes, the integrating center is the endocrine cell. In neural reflexes, the integrating center usually lies within the central nervous system, which is composed of the brain and the spinal cord.

If information is coming from a single stimulus, it is a relatively simple task for an integrating center to compare that information with the setpoint and initiate a response (if necessary). Integrating centers really "earn their pay," however, when two or more conflicting signals come in from different sources. The center must evaluate each signal on the basis of its strength and importance and must come up with an appropriate response that integrates information from all contributing receptors. This is similar to the kind of decision making you must do when on one evening your parents want to take you to dinner, your friends are having a party, there is a television program you want to watch, and you have a major physiology test in three days. It is up to you to rank those items in order of importance and decide how you will act on them.

Efferent pathway   Efferent pathways are relatively simple. In the nervous system, the efferent pathway is always the electrical and chemical signals transmitted by an efferent neuron. Because all electrical signals traveling through the nervous system are identical, the distinguishing characteristic of the signal is the anatomical route taken by the nerve cell through which the signal goes. For example, the vagus nerve carries a neural signal to the heart, and the phrenic nerve carries one to the diaphragm. Because the nature of the electrical message is always the same and because there are relatively few types of neurotransmitters, nervous system efferent pathways are named using the anatomical description of the nerve that carries the signal.

In the endocrine system, the anatomical routing of the efferent pathway is always the same because all hormones travel in the blood to get to their target. Hormonal efferent pathways are distinguished by the chemical nature of the signal and are

therefore named for the hormone that carries the message. For example, the efferent pathway for a reflex integrated through the pancreas will be either the hormone insulin or the hormone glucagon, depending on the stimulus and the appropriate response.

Effectors   The effectors of reflex control pathways are the cells or tissues that carry out the response. The targets of neural pathways are muscles, glands, and some adipose tissue. The targets of endocrine pathways are any cells that have the proper receptor for the hormone.

Responses   There are two levels of response for any reflex control pathway. One is the very specific *cellular response* that takes place in the target cell. The more general *systemic response* describes what those specific cellular events mean either to the tissue or to the organism as a whole. For example, when the hormone epinephrine combines with β_2-adrenergic receptors on the walls of certain blood vessels, the cellular response is relaxation of the smooth muscle. The systemic response to relaxation of the blood vessel wall is increased blood flow through the vessel.

CONCEPT CHECK

20. What is the difference between local control and reflex control?

21. Name the seven steps in a reflex control pathway in their correct order.   Answers: p. 209

Response Loops Begin with a Stimulus and End with a Response

To illustrate response loops, we will now apply the concept to nonbiological and biological examples. A simple nonbiological analogy to a homeostatic reflex pathway is an aquarium whose

6

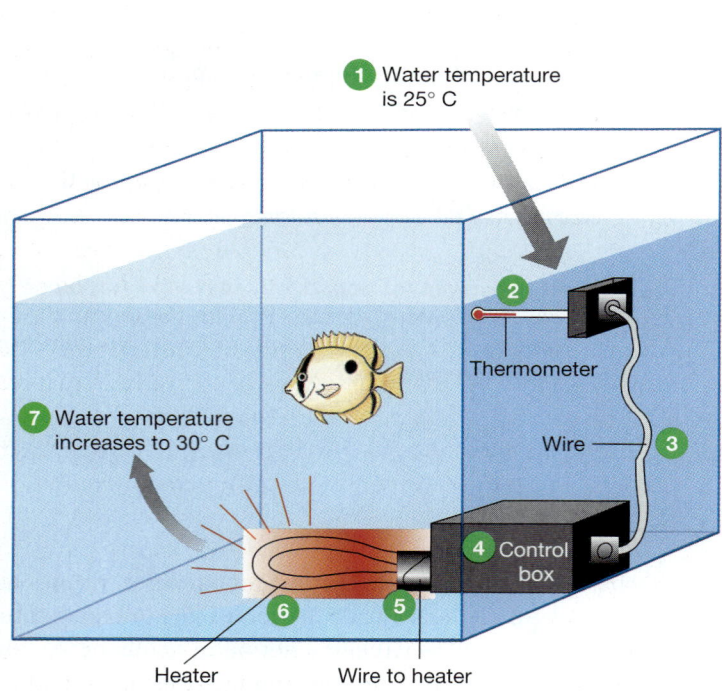

① Water temperature
is 25° C

② Thermometer

⑦ Water temperature
increases to 30° C

Wire ③

④ Control
box

⑥

⑤

Heater

Wire to heater

■ FIGURE 6-25   A nonbiological response loop

The control box of the aquarium is set to maintain a water temperature of 30° ± 1° C.

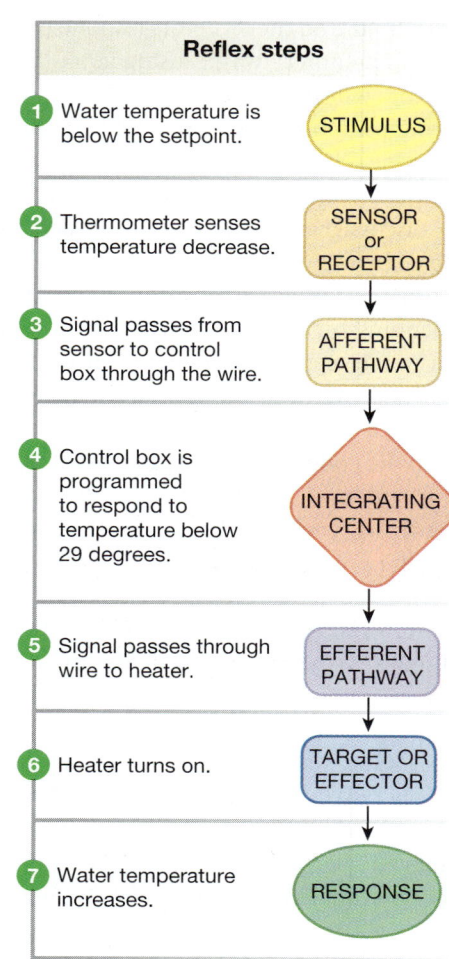

Reflex steps

① Water temperature is
below the setpoint.   **STIMULUS**

② Thermometer senses
temperature decrease.   **SENSOR
or
RECEPTOR**

③ Signal passes from
sensor to control
box through the wire.   **AFFERENT
PATHWAY**

④ Control box is
programmed
to respond to
temperature below
29 degrees.   **INTEGRATING
CENTER**

⑤ Signal passes through
wire to heater.   **EFFERENT
PATHWAY**

⑥ Heater turns on.   **TARGET OR
EFFECTOR**

⑦ Water temperature
increases.   **RESPONSE**

heater is programmed to maintain the water temperature at 30° C in a room whose temperature is 25° C (Fig. 6-25 ■). The desired water temperature (30° C) is the *setpoint* for the regulated variable.

Assume that initially the aquarium water is at room temperature, 25° C. When you turn the control box on, you set the response loop in motion. The thermometer (sensor) registers a temperature of 25° C. It sends this information via a wire (afferent path) to the control box (integrating center). The control box evaluates the incoming temperature signal, compares it with the setpoint for the system (30° C), and "decides" that a response is needed to bring the water temperature up to the setpoint. The control box sends a signal via another wire (efferent path) to the heater (effector), which turns on and starts heating the water (response). This sequence—from stimulus to response—is the response loop.

This aquarium example involves a variable (temperature) that is under *tonic control* (see p. 197) by a single control system (the heater). We can also describe a nonbiological analogy that illustrates Cannon's postulate of *antagonistic control*. For example, think of a house that has both heating and air-conditioning. The owner would like the house to remain at 70° F (about 21° C).

On chilly autumn mornings, the heater turns on to warm the house. Then, as the day warms up, the heater is no longer needed. When the sun heats the house above the setpoint, the air-conditioner turns on to cool the house back to 70° F. The heater and air-conditioner have antagonistic control over house temperature. A similar physiological example would be the hormones insulin and glucagon, which exert antagonistic control over glucose metabolism, as noted earlier.

CONCEPT CHECK

22. What is the drawback of having only a single control system (a heater) for maintaining aquarium water temperature in some desired range? Answers: p. 209

Setpoints Can Be Varied

In physiological systems, the setpoint for any given regulated variable can vary from person to person, or even for the same individual over a period of time. Factors that influence an individual's setpoint for a given variable include inheritance and the conditions to which the person has become accustomed. The adaptation of physiological processes to a given set of environmental conditions is known as **acclimatization** if it occurs

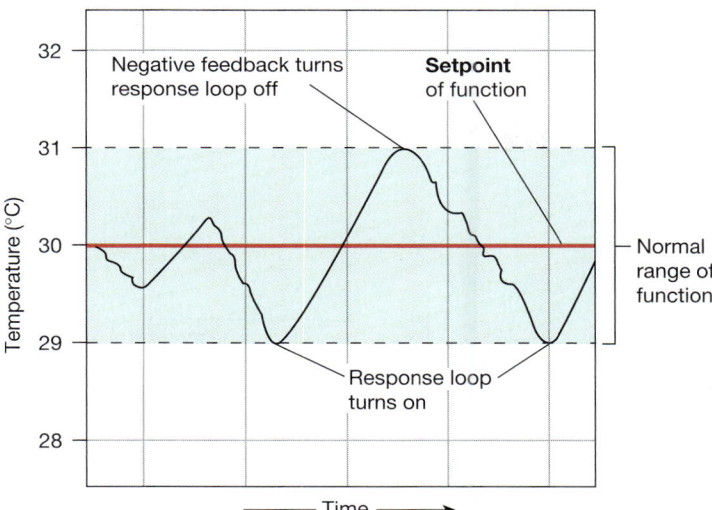

■ **FIGURE 6-26** *Oscillation around the setpoint*

Most functions that are controlled homeostatically have a setpoint, or normal value. The response loop that controls the function is activated when the function moves outside a predetermined normal range.

naturally, and as **acclimation** if the process is induced artificially in a laboratory setting. Each winter, northerners go south in February, hoping to escape the bitter subzero temperatures and snows of the northern climate. As they walk around in 40° F weather in short-sleeve shirts, the southerners, all bundled up in coats and gloves, think they are crazy: the weather is cold! The difference in behavior is due to different temperature acclimatization, a difference in the setpoint for body temperature regulation that is a result of prior conditioning.

Physiological setpoints also vary within individuals in response to external cues, such as the daily light-dark cycles and the seasons. These changing setpoints cause certain variables to vary in predictable ways over a period of time, forming patterns of change known as biorhythms (discussed later in this chapter).

Feedback Loops Modulate the Response Loop

The response loop is only part of a reflex. For example, in the aquarium just described, the sensor sends temperature information to the control box, which recognizes that the water is too cold. The control box responds by turning on the heater to warm the water. Once the response starts, though, what keeps the heater from sending the temperature up to, say, 50° C?

The answer is a **feedback loop**, where the response "feeds back" to influence the input portion of the pathway. In the aquarium example, turning on the heater increases the temperature of the water. The sensor continuously monitors the temperature and sends that information to the control box. When the temperature warms up to the maximum acceptable value, the control box shuts off the heater, thus ending the reflex response.

Negative Feedback Loops Are Homeostatic   For most reflexes, feedback loops are homeostatic—that is, designed to keep the system at or near a setpoint so that the variable being regulated will be relatively stable. How well an integrating center succeeds in maintaining stability depends on the *sensitivity* of the system. In the case of our aquarium, the control box is programmed to have a sensitivity of ±1° C. If the water temperature drops from 30° C to 29.5° C, it is still within the acceptable range, and no response is triggered. If the water temperature drops below 29° C (30° − 1°), the control box turns the heater on (Fig. 6-26 ■). As the water heats up, the control box constantly receives information about the water temperature from the sensor. When the water reaches 31° C (30° + 1°), the upper limit for the acceptable range, the feedback loop causes the control box to turn the heater off. The water then gradually cools off until the cycle starts all over again. The end result is a regulated variable that *oscillates* [*oscillare,* to swing] around the setpoint.

In physiological systems, some sensors are more sensitive than others. For example, the sensors for osmolarity trigger reflexes to conserve water when blood osmolarity increases only 3% above normal, but the sensors for low oxygen in the blood will not respond until oxygen has decreased by 40%.

A pathway in which the response opposes or removes the signal is known as **negative feedback** (Fig. 6-27a ■). Negative feedback loops *stabilize* the variable being regulated and thus aid the system in maintaining homeostasis. In the aquarium example, the heater warms the water (the response) and removes the stimulus (low water temperature). With loss of the stimulus for the pathway, the response loop shuts off. All homeostatic reflexes are controlled by negative feedback so that the variable being regulated will stay within a normal range. *Negative feedback loops can restore the normal state but cannot prevent the initial disturbance.*

Positive Feedback Loops Are Not Homeostatic   A few reflex pathways are not homeostatic. In a **positive feedback loop**, the response *reinforces* the stimulus rather than decreasing or removing it. In positive feedback, the response sends the variable being regulated even farther from its normal value, triggering a vicious cycle of ever-increasing response and sending the system temporarily out of control (Fig. 6-27b). Because positive feedback escalates the response, this type of feedback requires some intervention or event outside the loop to stop the response.

One example of a positive feedback loop involves the hormonal control of uterine contractions during childbirth (Fig. 6-28 ■). When the baby is ready to be delivered, it drops lower in the uterus and begins to put pressure on the *cervix,* the opening of the uterus. Sensory signals from the cervix to the brain cause release of the hormone *oxytocin,* which causes the uterus to contract and push the baby's head even harder against the cervix, further stretching it. The increased stretch causes more oxytocin release, which causes more contractions that push the baby harder against the cervix. This cycle continues until finally the baby is delivered, releasing the stretch on the cervix and stopping the positive feedback loop.

(a) Negative feedback: the response counteracts the stimulus, shutting off the response loop.

(b) Positive feedback: the response reinforces the stimulus, sending the varible farther from the setpoint.

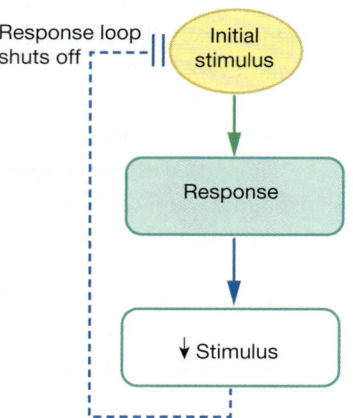

Response loop shuts off

Initial stimulus

↓

Response

↓

↓ Stimulus

Initial stimulus

↓

Response

⊕ Feedback loop

↑ Stimulus

An outside factor is required to shut off ⊕ feedback loop.

■ **FIGURE 6-27** *Negative and positive feedback*

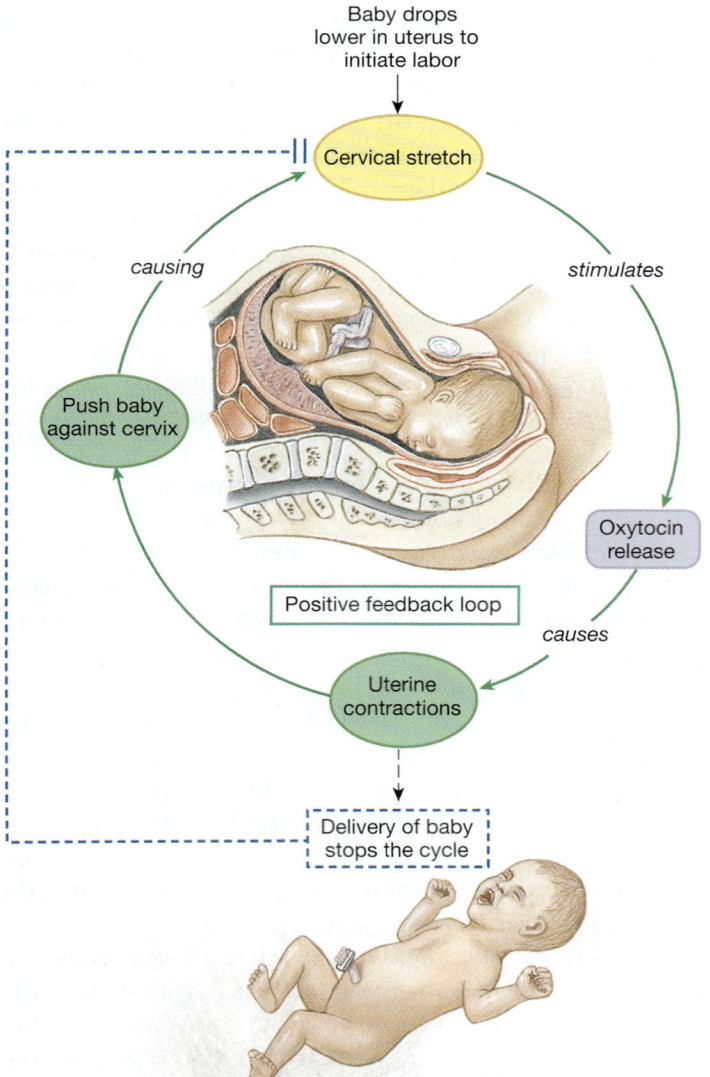

Baby drops lower in uterus to initiate labor

Cervical stretch

causing

stimulates

Push baby against cervix

Oxytocin release

Positive feedback loop

causes

Uterine contractions

Delivery of baby stops the cycle

■ **FIGURE 6-28** *A positive feedback loop*

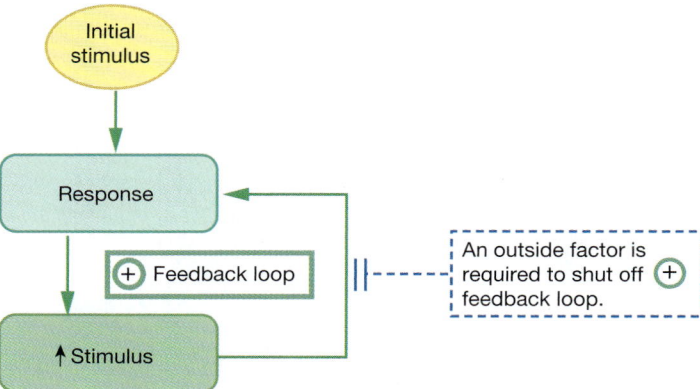

✔ **CONCEPT CHECK**

23. Does the aquarium heating system in Figure 6-25 function using positive feedback or negative feedback? Draw in the appropriate feedback loop on the reflex steps in the figure, using either a solid arrow accompanied by a plus sign inside a circle for positive feedback, or two short, parallel "stop" lines at the end of a dashed line for negative feedback, as illustrated in Figure 6-27. Answers: p. 210

Feedforward Control Allows the Body to Anticipate Change and Maintain Stability

Negative feedback loops can stabilize a function and maintain it within a normal range but are unable to prevent the change that triggered the reflex in the first place. A few reflexes have evolved that enable the body to predict that a change is about to occur and start the response loop in anticipation of the change. These anticipatory responses are called **feedforward control**.

An easily understood physiological example of feedforward control is the reflex of salivation. The sight, smell, or even the thought of food is enough to start our mouths watering. The saliva is present in expectation of the food that will soon be eaten. This reflex extends even further, because the same stimuli can start the secretion of hydrochloric acid as the stomach anticipates food on the way. One of the most complex feedforward reflexes appears to be the body's response to exercise, to be discussed in Chapter 25.

Biological Rhythms Result from Changes in a Setpoint

In many reflex pathways, the stimuli are obviously related to the function of the reflex. In the aquarium example, a change in temperature is the stimulus to ensure that temperature is maintained within the desired range. This is not true of all reflexes, however. Many hormones, for example, are secreted continuously, with

(a)

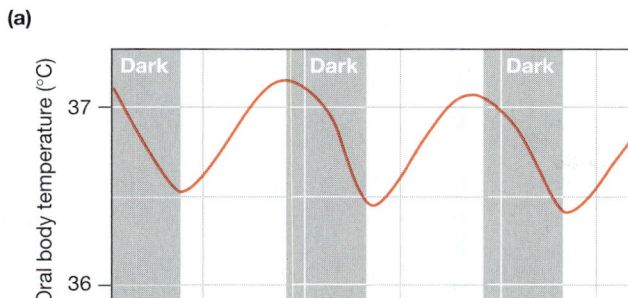

(b)

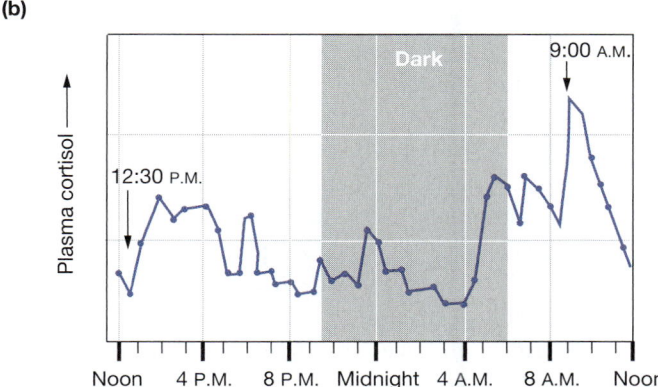

■ **FIGURE 6-29**   *Circadian rhythms*

levels that rise and fall throughout the day. Most examples of these apparently spontaneous reflexes occur in a predictable manner and are often timed to coincide with a predictable environmental change, such as light-dark cycles or the seasons.

All animals exhibit some form of daily biological rhythm, called a **circadian rhythm** [*circa,* about + *dies,* day]. Humans have circadian rhythms for many body functions, including blood pressure, body temperature, and metabolic processes. Body temperature peaks in the late afternoon and declines dramatically in the early hours of the morning (Fig. 6-29a ■). Have you ever been studying late at night and noticed that you feel cold? This is not because of a drop in environmental temperature but because your thermoregulatory reflex has turned down your internal thermostat.

Many hormones in humans are secreted so that their concentration in the blood fluctuates predictably through a 24-hour cycle as their setpoints change. Cortisol, growth hormone, and the sex hormones are among the most noted examples. If an abnormality in hormone secretion is suspected, it is important to know at what time of day the body fluid used for testing was taken from the patient. A cortisol value that is normal in a 9:00 A.M. sample would be abnormally high for a blood sample taken at noon (Fig. 6-29b). One strategy for avoiding this type of error uses a 24-hour collection period that results in an average value for the hormone over the course of a day. For example, cortisol secretion is monitored indirectly by measuring all urinary cortisol metabolites excreted in 24 hours.

What is the adaptive significance of functions that vary with a circadian rhythm? Our best answer is that biological rhythms create an anticipatory response to a predictable environmental variable. There are seasonal rhythms of reproduction in many mammalian and non-mammalian vertebrates and invertebrates, rhythms timed so that the offspring have food and other favorable conditions to maximize survival. Circadian rhythms cued by the light-dark cycle may correspond to our rest-activity cycles. These rhythms allow our bodies to anticipate behavior and coordinate body processes accordingly. You may hear someone who is accustomed to eating dinner at 6 P.M. say that he cannot digest his food if he waits until 10 P.M. to eat because his digestive system has "shut down" in anticipation of going to bed.

One of the interesting correlations between circadian rhythms and behavior involves body temperature. Researchers found that self-described "morning people" have temperature rhythms that cause body temperature to climb before they awaken in the morning, so that they get out of bed prepared to face the world. On the other hand, "night people" may be forced by school and work schedules to get out of bed while their body temperature is still at its lowest point, before their bodies are prepared for activity. These night people are still going strong and working productively in the early hours of the morning when the morning peoples' body temperatures are dropping and they are fast asleep.

Circadian rhythms arise from special groups of cells in the brain and are reinforced by information about the light-dark cycle that comes in through the eyes. Research in simpler animals such as flies is beginning to explain the molecular basis for biological rhythms. We will discuss the cellular and molecular basis for circadian rhythms in Chapter 9.

Now that you have been introduced to response loops and feedback loops, we turn to an analysis of the different control systems.

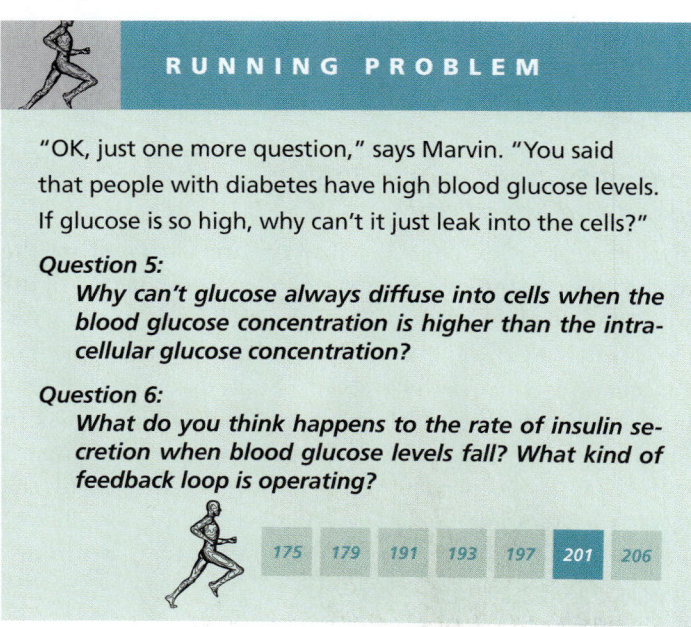

RUNNING PROBLEM

"OK, just one more question," says Marvin. "You said that people with diabetes have high blood glucose levels. If glucose is so high, why can't it just leak into the cells?"

Question 5:
Why can't glucose always diffuse into cells when the blood glucose concentration is higher than the intracellular glucose concentration?

Question 6:
What do you think happens to the rate of insulin secretion when blood glucose levels fall? What kind of feedback loop is operating?

175   179   191   193   197   **201**   206

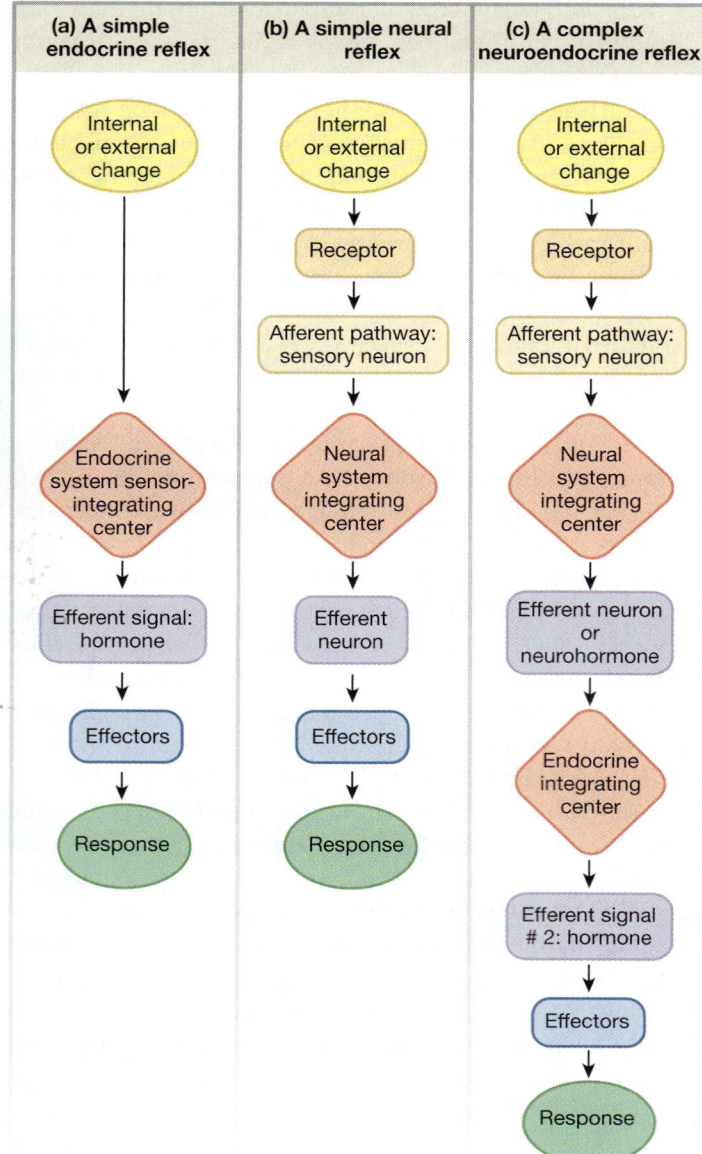

(a) A simple endocrine reflex	(b) A simple neural reflex	(c) A complex neuroendocrine reflex
Internal or external change	Internal or external change	Internal or external change
	Receptor	Receptor
	Afferent pathway: sensory neuron	Afferent pathway: sensory neuron
Endocrine system sensor-integrating center	Neural system integrating center	Neural system integrating center
Efferent signal: hormone	Efferent neuron	Efferent neuron or neurohormone
		Endocrine integrating center
Effectors	Effectors	Efferent signal #2: hormone
		Effectors
Response	Response	Response

■ **FIGURE 6-30** *Endocrine, neural, and neuroendocrine control pathways*

Control Systems Vary in Their Speed and Specificity

Physiological reflex control pathways are mediated by the nervous system, the endocrine system, or a combination of the two (Fig. 6-30 ■). A reflex mediated solely by the nervous system or solely by the endocrine system is relatively simple, but combination reflex pathways can be quite complex. In the most complex pathways, signals pass through three different integrating centers before finally reaching the target tissue. With so much overlap between pathways controlled by the nervous and endocrine systems, it makes sense to consider these systems as parts of a continuum rather than as two discrete systems.

Why does the body need different types of control systems? To answer that question, let us compare endocrine control with neural control to see what the differences are. Five major differences are summarized in Table 6-4 ■ and discussed next.

Specificity Neural control is very specific because each nerve cell has a specific target cell or cells to which it sends its message. Anatomically, we can isolate a neuron and trace it from its origin to where it terminates on its target cell(s). Endocrine control is more general because the chemical messenger is released into the blood and can reach virtually every cell in the body. As you learned in the first half of this chapter, the body's response to a specific hormone depends on which cells have receptors for that hormone. Multiple tissues in the body can respond to a hormone simultaneously.

Nature of the Signal The nervous system uses both electrical and chemical signals to send information throughout the body. Electrical signals travel long distances through nerve cells, releasing chemical signals (neurotransmitters) that diffuse across the small gap between the neuron and its target (Fig. 6-31, ① ■). In a limited number of instances, electrical signals pass directly from cell to cell through gap junctions.

The endocrine system uses only chemical signals: hormones secreted by endocrine glands or cells into the blood (Fig. 6-31, ⑥). The neurohormone pathway (Fig. 6-31, ②) represents a hybrid of the neural and endocrine reflexes. In a neurohormone pathway, a nerve cell creates an electrical signal, but the chemical the cell releases is a neurohormone that goes into the blood for general distribution.

✓ CONCEPT CHECK

24. (a) In the simple neural reflex shown in Figure 6-30b, which box or boxes represent(s) the brain and spinal cord? (b) Which box or boxes represent(s) the central and peripheral sense organs? (c) In Figure 6-30b, add a dashed line connecting boxes to show how a negative feedback loop would shut off the reflex.

Answers: p. 210

Speed Neural reflexes are much faster than endocrine reflexes. The electrical signals of the nervous system cover great distances very rapidly, with speeds of up to 120 m/sec. Neurotransmitters also create very rapid responses, on the order of milliseconds.

Hormones are much slower than neural reflexes. Their distribution through the circulatory system and diffusion from capillary to receptors take considerably longer than signals through nerve cells. In addition, hormones have a slower onset of action. In target tissues, the response may take minutes to hours before it can be measured.

Why do we need the speedy reflexes of the nervous system? Consider this example. A mouse ventures out of his hole and sees a cat ready to pounce on him and eat him. A signal must go from the mouse's eyes and brain down to his feet, telling him to run back into the hole. If his brain and feet were only 5 micrometers

TABLE 6-4	Comparison of Neural and Endocrine Control	
PROPERTY	**NEURAL REFLEX**	**ENDOCRINE REFLEX**
Specificity	Each neuron terminates on a single target cell or on a limited number of adjacent target cells.	Most cells of the body are exposed to a hormone. The response depends on which cells have receptors for the hormone.
Nature of the signal	Electrical signal passes through neuron, then chemical neurotransmitters pass the signal from cell to cell. In a few cases, cell-to-cell communication takes place through gap junctions.	Chemical signals are secreted in the blood for distribution throughout the body.
Speed	Very rapid.	Distribution of the signal and onset of action are much slower than in neural responses.
Duration of action	Usually very short. Responses of longer duration are mediated by neuromodulators.	Duration of action is usually much longer than in neural responses.
Coding for stimulus intensity	Each signal is identical in strength. Stimulus intensity is correlated with increased frequency of signaling.	Stimulus intensity is correlated with amount of hormone secreted.

(5 μm = 1/200 millimeter) apart, it would take a chemical signal 20 milliseconds (msec) to diffuse across the space; the mouse could escape. If the brain and feet were 50 μm (1/20 millimeter) apart, diffusion would take 2 seconds; the mouse might get caught. But because the head and tail of a mouse are *centimeters* apart, it would take a chemical signal *three weeks* to diffuse from the mouse's head to his feet. Poor mouse! Even if the distribution of the chemical were accelerated by help from the circulatory system, the chemical message would still take 10 seconds to get to the feet, and the mouse would become cat food. The moral of this tale is that reflexes requiring a speedy response are mediated by the nervous system because they are so much more rapid.

Duration of Action   Neural control is of shorter duration than endocrine control. The neurotransmitter released by a nerve cell combines with a receptor on the target cell and initiates a response. The response is usually very brief, however, because the neurotransmitter is rapidly removed from the vicinity of the receptor by various mechanisms. To get a sustained response, multiple repeating signals must be sent through the nerve cell.

Endocrine reflexes are slower to start, but they are of longer duration. This means that most of the ongoing, long-term functions of the body, such as metabolism and reproduction, fall under the control of the endocrine system.

Coding for Stimulus Intensity   As a stimulus increases in intensity, control systems must have a mechanism for conveying this information to the integrating center. The signal strength from any one neuron is constant in magnitude and therefore cannot reflect stimulus intensity. Instead, the frequency of signaling through the afferent neuron increases. In

the endocrine system, stimulus intensity is reflected by the amount of hormone released: the stronger the stimulus, the greater the amount of hormone released.

Complex Reflex Control Pathways Have Several Integrating Centers

Figure 6-31 summarizes variations in the neural, neuroendocrine, and endocrine reflex control pathways.

In a simple endocrine reflex pathway (Fig. 6-30a; Fig. 6-31, ⑥), the endocrine cell acts as both sensor and integrating center; there is no afferent pathway. The efferent pathway is the hormone, and the target is any cell having the appropriate receptor protein.

An example of a simple endocrine reflex is the secretion of the hormone insulin in response to changes in blood glucose level. The endocrine cells that secrete insulin monitor blood glucose concentrations by using ATP production in the cell as an indicator [⇄ Fig. 5-38, p. 167]. When blood glucose increases, intracellular ATP production exceeds the threshold level, and the endocrine cells respond by secreting insulin into the blood. Any target cell in the body that has insulin receptors will respond to the hormone and initiate processes that take glucose out of the blood. The removal of the stimulus acts in a negative feedback manner, and the response loop shuts off when blood glucose levels fall below a certain concentration.

In a simple neural reflex, all the steps of the pathway are present, from receptor to target (Fig. 6-30b, Fig. 6-31, ①). The neural reflex is represented in its simplest form by the knee jerk (or patellar tendon) reflex (see Fig. 13-7). A blow to the knee (the stimulus) activates a stretch receptor. An electrical and chemical signal travels through an afferent neuron to the

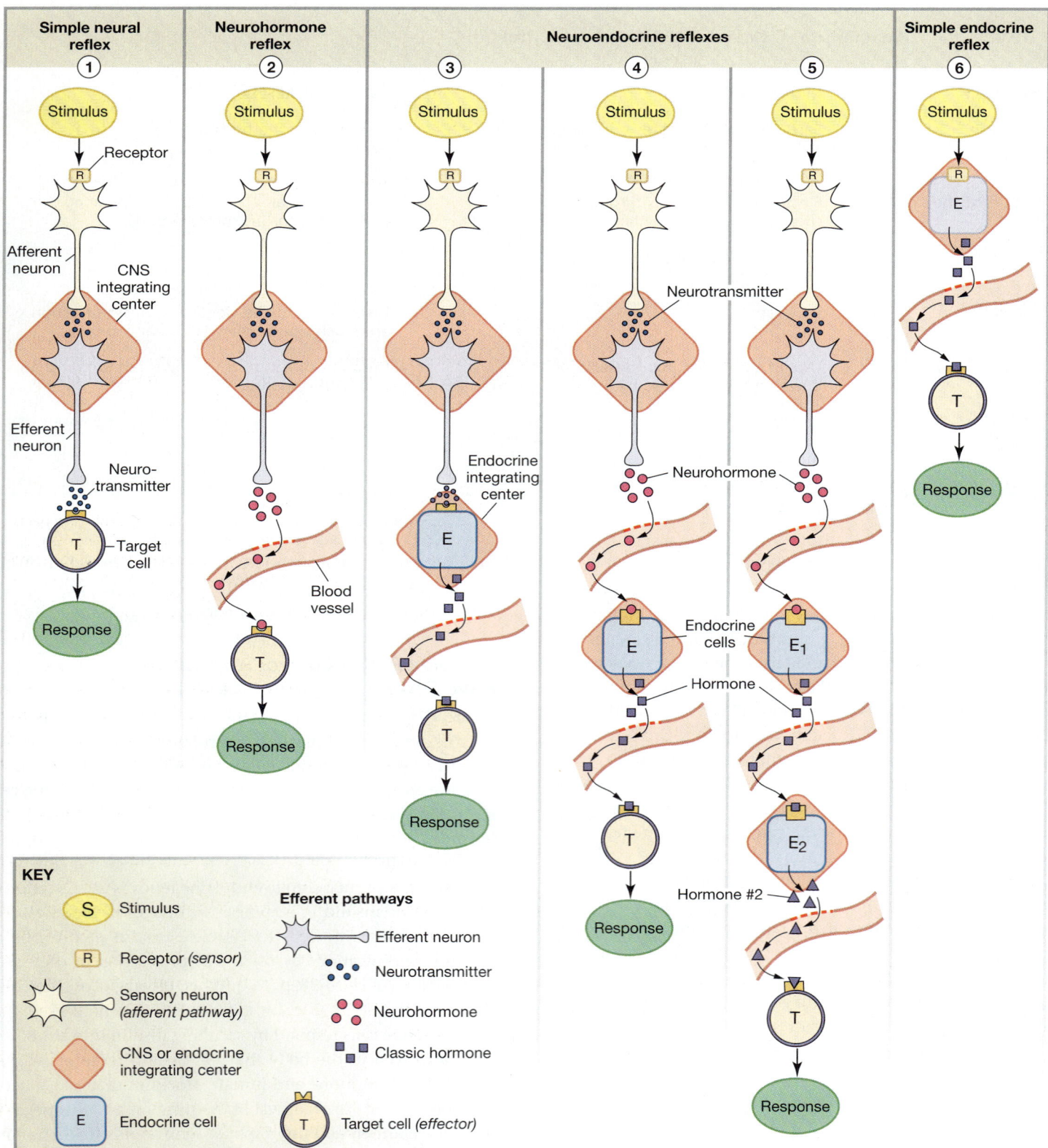

■ **FIGURE 6-31** *Some basic patterns of neural, endocrine, and neuroendocrine control pathways*

An example of pathway ① is the knee jerk reflex. An example of ② is secretion of oxytocin in response to suckling. An example of ③ is the secretion of insulin in response to a signal from the brain. ⑥ shows insulin release in response to an increase in blood glucose.

TABLE 6-5   Comparison of Neural, Neuroendocrine, and Endocrine Reflexes

	NEURAL	NEUROENDOCRINE	ENDOCRINE
Sensor or receptor	Special and somatic sensory receptors	Special and somatic sensory receptors	Endocrine cell
Afferent pathway	Afferent sensory neuron	Afferent sensory neuron	None
Integrating center	Brain or spinal cord	Brain or spinal cord	Endocrine cell
Efferent pathway	Efferent neuron (electrical signal and neurotransmitter)	Efferent neuron (electrical signal and neurohormone)	Hormone
Effector(s)	Muscles and glands, some adipose tissue	Most cells of the body	Most cells of the body
Response	Contraction and secretion primarily; may have some metabolic effects.	Change in enzymatic reactions, membrane transport, or cell proteins	Change in enzymatic reactions or membrane transport or cell proteins

spinal cord (the integrating center). If the blow is strong enough (exceeds threshold), a signal travels from the spinal cord through an efferent neuron to the muscles of the thigh (the target or effector). In response, the muscles contract, causing the lower leg to kick outward (the knee jerk).

CONCEPT CHECK

25. Match the following terms for parts of the knee jerk reflex to the parts of the simple neural reflex shown in Figure 6-30b: blow to knee, leg muscles, neuron to leg muscles, sensory neuron, brain and spinal cord, stretch receptor, muscle contraction.   Answers: p. 210

The neurohormone reflex, shown in Figure 6-31, ②, is identical to the neural reflex except that the neurohormone released by the neuron travels in the blood to its target, just like a hormone. A simple neurohormone reflex is the release of breast milk in response to a baby's suckling. The baby's mouth on the nipple stimulates sensory signals that travel through sensory neurons to the brain (integrating center). An electrical signal in the efferent neuron triggers the release of the neurohormone oxytocin from the brain into the circulation. Oxytocin is carried to the breast, where it causes contraction of smooth muscles in the breast (effectors), with the resultant ejection of milk.

In complex pathways, there may be more than one integrating center. Figure 6-31 shows three examples of complex neuroendocrine pathways. The simplest of these, pattern ③, combines a neural reflex with a classic endocrine reflex. The target of the neural reflex is an endocrine cell that releases a hormone. An example of this pattern can be found in the control of insulin release by the nervous system, in which both excitatory and inhibitory neurons terminate on the endocrine cells of the pancreas. The endocrine cells must integrate information from three sources: the two antagonistic inputs from the nervous system and their direct detection of blood glucose levels.

CONCEPT CHECK

26. Match the following terms with the appropriate parts of the simple neuroendocrine reflex in Figure 6-31, ③ (terms may be used more than once): food in stomach following a meal, brain and spinal cord, endocrine cells of pancreas, stretch receptors, efferent neuron to pancreas, insulin, adipose cell, blood, sensory neuron.   Answers: p. 210

Another complex reflex (Fig. 6-31, ④) uses a neurohormone to control the release of a classic hormone. The secretion of growth hormone is an example of this pathway. The most complex neuroendocrine pathways, shown as Figure 6-31, ⑤, include a neurohormone and two classic hormones. This pattern is typical of some hormones released by the anterior pituitary, an endocrine gland located just below the brain (see Chapter 7 for details).

In describing complex neuroendocrine reflex pathways, we identify only one receptor and afferent pathway, as indicated in Figure 6-31. In the three complex pathways shown, the brain is the first integrating center and the neurohormone is the first efferent pathway. In Figure 6-31, ⑤ the endocrine target (E_1) of the neurohormone is the second integrating center, and its hormone is the second efferent pathway. The second endocrine gland in the pathway (E_2) is the third integrating center, and its hormone is the third efferent pathway. The target of the last signal in the sequence is the effector.

Table 6-5 ■ compares the various steps in neural, neuroendocrine, and endocrine reflexes. In the remainder of the text, we will use the general patterns shown in Figure 6-31 as a tool for classifying complex reflex pathways. The next chapter looks in detail at some of the pathways of the endocrine system and the roles these pathways play in homeostasis.

RUNNING PROBLEM CONCLUSION

DIABETES MELLITUS

Marvin underwent further tests and was diagnosed with type 2 diabetes. With careful attention to his diet and with a regular exercise program, he has managed to keep his blood glucose levels under control. Diabetes is becoming a major health issue in the United States. To learn more about diabetes, see the American Diabetes Association website (*www.diabetes.org*) or the Centers for Disease Control and Prevention (*www.cdc.gov*).

In this running problem, you learned about glucose homeostasis and how it is maintained by insulin and glucagon. The disease diabetes mellitus is an indication that glucose homeostasis has been disrupted. Check your understanding of this running problem by comparing your answers to the information in the summary table.

	QUESTION	FACTS	INTEGRATION AND ANALYSIS
1	In which type of diabetes is the signal pathway for insulin more likely to be defective?	Insulin is a peptide hormone that uses membrane receptors linked to second messengers to transmit its signal to cells [🔁 p. 183]. People with type 1 diabetes lack insulin; people with type 2 diabetes have normal-to-elevated insulin levels.	Normal or high insulin levels suggest that the problem is not with amount of insulin but with the action of the insulin at the cell. The problem in type 2 diabetes could be a defective signal transduction mechanism.
2	In which form of diabetes are the insulin receptors more likely to be up-regulated?	Up-regulation of receptors usually occurs if a signal molecule is present in unusually low concentrations [🔁 p. 191]. In type 1 diabetes, insulin is not secreted by the pancreas.	In type 1 diabetes, insulin levels are low. Therefore, type 1 is more likely to cause up-regulation of the insulin receptors.
3	The homeostatic regulation of blood glucose levels by the hormones insulin and glucagon is an example of which of Cannon's postulates?	Cannon's postulates describe the role of the nervous system in maintaining homeostasis, and the concepts of tonic activity, antagonistic control, and different effects of signals in different tissues.	Insulin decreases blood glucose levels, and glucagon increases them. Therefore, the two hormones are an example of an antagonistic control.
4	In the insulin pathway that regulates blood glucose levels, what are the stimulus, the sensor, the integrating center, the efferent pathway, the effector(s), and the response(s)?	See the steps of reflex pathways [🔁 p. 195].	*Stimulus:* increase in blood glucose levels; *sensor:* beta cells of the pancreas that sense the change; *integrating center:* beta cells; *efferent pathway* (the signal): insulin; *effectors:* any tissues of the body that respond to insulin; *responses:* cellular uptake and use of glucose.
5	Why can't glucose always diffuse into cells when the blood glucose concentration is higher than the intracellular glucose concentration?	Glucose is lipophobic. Simple diffusion goes across the phospholipid bilayer. Facilitated diffusion uses protein carriers [🔁 p. 141].	Because glucose is lipophobic, it must cross the membrane by facilitated diffusion. If a cell lacks the necessary protein carriers, facilitated diffusion cannot take place.
6	What do you think happens to the rate of insulin secretion when blood glucose levels fall? What kind of feedback loop is operating?	The stimulus for insulin release is an increase in blood glucose levels. In negative feedback, the response offsets the stimulus. In positive feedback, the response enhances the stimulus.	An increase in blood glucose concentration stimulates insulin release; therefore, a decrease in blood glucose should decrease insulin release. In this example, the response (lower blood glucose) offsets the stimulus (increased blood glucose), so a negative feedback loop is operating.

175 179 191 193 197 201 **206**

CHAPTER SUMMARY

This chapter focuses on two of the major themes in physiology: *homeostasis and control systems,* and *communication.* The sensors, integrating centers, and effectors of physiological control systems are described in the context of reflex control pathways, which vary from simple to complex. Functional control systems require efficient communication that uses various combinations of chemical and electrical signals. Those signals that cannot cross cell membranes must use membrane receptor proteins and signal transduction to transfer their information into the cell. The interaction of signal molecules with protein receptors illustrates another fundamental theme of physiology, *molecular interactions.*

Cell-to-Cell Communication

1. There are two basic types of physiological signals: chemical and electrical. Chemical signals are the basis for most communication within the body. (p. 175)

2. There are four methods of cell-to-cell communication: (1) direct cytoplasmic transfer through gap junctions, (2) contact-dependent signaling, (3) local chemical communication, and (4) long-distance communication. (p. 175; Figs. 6-1 and 6-2)

3. **Gap junctions** are protein channels that connect two adjacent cells. When they are open, chemical and electrical signals pass directly from one cell to the next. (p. 175)

4. **Contact-dependent signals** require direct contact between surface molecules of two cells. (p. 176)

5. Local communication is accomplished by **paracrine signals**, chemicals that act on cells in the immediate vicinity of the cell that secreted the paracrine. A chemical that acts on the cell that secreted it is called an **autocrine signal**. The activity of paracrine and autocrine signal molecules is limited by diffusion distance. (p. 176)

6. Long-distance communication is accomplished by **neurocrine molecules** and electrical signals in the nervous system, and by **hormones** in the endocrine system. Only cells that possess receptors for a hormone will be **target cells.** (p. 176)

7. **Cytokines** are regulatory molecules that control cell development, differentiation, and the immune response. They function as both local and long-distance signals. (p. 176)

Signal Pathways

8. Chemical signals bind to **receptors** and change intracellular signal molecules that direct the response. (p. 177; Fig. 6-3)

9. Lipophilic signal molecules enter the cell and combine with cytoplasmic or nuclear receptors. Lipophobic signal molecules and some lipophilic molecules combine with membrane receptors. (p. 178; Fig. 6-4)

10. **Signal transduction** pathways use membrane receptor proteins and intracellular second messenger molecules to translate signal information into an intracellular response. (p. 179)

11. Some signal transduction pathways activate **protein kinases**. Others activate **amplifier enzymes** that create **second messenger** molecules. (p. 180)

12. Signal pathways create intracellular **cascades** that amplify the original signal. (p. 180; Figs. 6-7 and 6-9)

13. **Receptor-enzymes** activate protein kinases, such as **tyrosine kinase** (Fig. 6-10), or the amplifier enzyme **guanylyl cyclase**, which produces the second messenger **cGMP**. (p. 182)

14. **G proteins** linked to amplifier enzymes are the most prevalent signal transduction system. **G protein-coupled receptors** also alter ion channels. (p. 183)

15. The **G protein-coupled adenylyl cyclase-cAMP-protein kinase A** pathway is the most common pathway for protein and peptide hormones. (p. 183; Fig. 6-11)

16. The amplifier enzyme **phospholipase C** creates two second messengers: **IP$_3$** and **diacylglycerol**. IP$_3$ causes Ca^{2+} release from intracellular stores. Diacylglycerol activates **protein kinase C**. (p. 184; Fig. 6-12)

17. **Integrin** receptors link the extracellular matrix to the cytoskeleton. (p. 184; Fig. 6-5)

18. **Ligand-gated ion channels** open or close to create electrical signals. (p. 185; Fig. 6-13)

Novel Signal Molecules

19. Calcium is an important signal molecule that binds to **calmodulin** to alter enzyme activity. It also binds to other cell proteins to alter movement and initiate exocytosis. (p. 186; Fig. 6-15)

20. **Nitric oxide (NO)** and **carbon monoxide (CO)** are short-lived gaseous signal molecules. NO activates guanylyl cyclase directly. (p. 187)

21. The arachidonic acid cascade creates lipid paracrines, such as **leukotrienes**, **prostaglandins**, and **thromboxanes**. (p. 188; Fig. 6-16)

Modulation of Signal Pathways

22. The response of a cell to a signal molecule is determined by the cell's receptor for the signal. (p. 189)

23. Receptor proteins exhibit specificity, competition, and saturation. (p. 189)

24. A receptor may have multiple ligands. **Agonists** mimic the action of a signal molecule. **Antagonists** block the signal pathway. (p. 189; Fig. 6-17)

25. Receptors come in related forms called **isoforms**. One ligand may have different effects when binding to different isoforms. (p. 189; Fig. 6-18)

26. Cells exposed to abnormally high concentrations of a signal for a sustained period of time attempt to bring their response back to normal by decreasing the number of receptors or decreasing the binding affinity of the receptors. This is known as **down-regulation** of the receptors. **Up-regulation** is the opposite of down-regulation. (p. 190)

27. Cells have mechanisms for terminating signal pathways, such as removing the signal molecule or breaking down the receptor-ligand complex. (p. 191)

28. Many diseases have been linked to defects with various aspects of signal pathways, such as missing or defective receptors. (p. 191)

Control Pathways: Response and Feedback Loops

29. Walter Cannon first described four basic postulates of homeostasis: (1) The nervous system plays an important role in maintaining homeostasis. (2) Some parameters are under **tonic control**, which allows the parameter to be increased or decreased by a single signal

(Fig. 6-20). (3) Other parameters are under **antagonistic control**, in which one hormone or neuron increases the parameter while another decreases it (Fig. 6-21). (4) Chemical signals can have different effects in different tissues of the body, depending on the type of receptor present at the target cell. (p. 192)

30. The simplest homeostatic control takes place at the tissue or cell level and is known as **local control**. (p. 194)

31. In **reflex** control pathways, the decision that a response is needed is made away from the cell or tissue. A chemical or electrical signal to the cell or tissue then initiates the response. Long-distance reflex pathways involve the nervous and endocrine systems and cytokines. (p. 194; Fig. 6-22)

32. Reflex pathways can be broken down into **response loops** and **feedback loops**. A response loop begins when a **stimulus** is sensed by a sensor or **sensory receptor**. The sensor is linked by an **afferent pathway** to an **integrating center** that decides on an appropriate response. An **efferent pathway** travels from the integrating center to an **effector** that carries out the appropriate **response**. (p. 194; Fig. 6-23)

33. In **negative feedback**, a homeostatic response is turned off when the response of the system opposes or removes the original stimulus. (p. 199; Fig. 6-27a)

34. In **positive feedback** loops, the response reinforces the stimulus rather than decreasing or removing it. This destabilizes the system until some intervention or event outside the loop stops the response. (p. 199; Fig. 6-27b)

35. **Feedforward control** allows the body to predict that a change is about to occur and start the response loop in anticipation of the change. (p. 200)

36. Apparently spontaneous reflexes that occur in a predictable manner are called **biological rhythms**. Those that coincide with light-dark cycles are called **circadian rhythms**. (p. 200)

37. Neural control is faster and more specific than endocrine control but is usually of shorter duration. Endocrine control is less specific and slower to start but is longer lasting and is usually amplified. (p. 202; Tbl. 6-4)

38. Many reflex pathways are combinations of neural and endocrine control mechanisms. (p. 203; Fig. 6-31)

QUESTIONS

(Answers to the Review Questions begin on page A1.)

THE PHYSIOLOGY PLACE

Access more review material online at **The Physiology Place** website. There you'll find review questions, problem-solving activities, case studies, flashcards, and direct links to both *Inter Active Physiology*® and *PhysioEx*™. To access the site, go to **www.physiologyplace.com** and select Human Physiology, Fourth Edition.

LEVEL ONE   REVIEWING FACTS AND TERMS

1. What are the two routes for long-distance signal delivery in the body?

2. Which two body systems are charged with maintaining homeostasis by responding to changes in the environment?

3. What two types of physiological signals are used to send messages through the body? Of these two types, which is available to all cells?

4. The process of maintaining a relatively stable internal environment is called _____.

5. List at least three parameters maintained by homeostasis.

6. Distinguish between the "target" and the "receptor" in physiological systems.

7. In a signal pathway, the signal ligand, also called a _____ _____, binds to a _____, which activates and changes intracellular _____.

8. The three main amplifier enzymes are (a) _____, which forms cAMP; (b) _____, which forms cGMP; and (c) _____, which converts a phospholipid from the cell's membrane into two different second messenger molecules.

9. An enzyme known as protein kinase adds the functional group _____ to its substrate, by transferring it from a(n) _____ molecule.

10. Put the following parts of a reflex in the correct order for a physiological response loop: efferent pathway, afferent pathway, effector, stimulus, response, integrating center.

11. Distinguish between central and peripheral receptors.

12. Match each of the following terms with its description:

    (a) threshold
    (b) setpoint
    (c) effector
    (d) oscillation
    (e) sensitivity

    1. the desired target value for a parameter
    2. the distance allowed away from the setpoint before a response starts
    3. the minimum stimulus to trigger a response
    4. the organ or gland that performs the change
    5. movement of a parameter within the desired range

13. Receptors for signal pathways may be found in the _____, _____, or _____ of the cell.

14. The name for the daily fluctuations of body functions, including blood pressure, temperature, and metabolic processes, is _____ _____. These cycles arise in special cells in the _____.

15. Down-regulation results in a(n) _____ (increased or decreased?) sensitivity to a prolonged signal.

16. List the two ways that down-regulation may be achieved.

17. In a negative feedback loop, the effector moves the system in the _____ (same/opposite) direction as the stimulus.

LEVEL TWO   REVIEWING CONCEPTS

18. Explain the relationships of the terms in each of the following sets. Give a physiological example or location if applicable.

    (a) gap junctions, connexins, syncytium, connexon
    (b) autocrine, paracrine, cytokine, neurocrine, hormone
    (c) agonist, antagonist
    (d) transduction, amplification, cascade

19. List and compare the four classes of membrane receptors for signal pathways. Give an example of each.

20. Who was Walter Cannon? Restate his four postulates in your own words.

21. Briefly define the following terms and give an anatomical example when applicable: efferent pathway, afferent pathway, effector, stimulus, response, integrating center.

22. Explain the differences among positive feedback, negative feedback, and feedforward mechanisms. Under what circumstances would each be advantageous?

23. Compare and contrast the advantages and disadvantages of neural versus endocrine control mechanisms.

24. Label each of the following systems as positive or negative feedback.
    (a) glucagon secretion in response to declining blood glucose
    (b) increasing milk letdown and secretion in response to more suckling
    (c) urgency in emptying one's urinary bladder
    (d) sweating in response to rising body temperature

25. Identify the effector organ for each example in question 24.

26. Now identify the integrating center for each example in question 24.

LEVEL THREE PROBLEM SOLVING

27. In each of the following situations, identify the components of the reflex.
    (a) You are sitting quietly at your desk, studying, when you become aware of the bitterly cold winds blowing outside at 30 mph, and

you begin to feel a little chilly. You start to turn up the thermostat, remember last month's bill, and reach for an afghan to pull around you instead. Pretty soon you are toasty warm again.
    (b) While you are strolling through the shopping district, the aroma of cinnamon sticky buns reaches you. You inhale appreciatively, but remind yourself that you're not hungry, because you just had lunch an hour ago. You go about your business, but 20 minutes later you're back at the bakery, sticky bun in hand, ravenously devouring its sweetness, saliva moistening your mouth.

LEVEL FOUR QUANTITATIVE PROBLEMS

28. In a signal cascade for rhodopsin, a photoreceptor molecule, one rhodopsin activates 1000 molecules of transducin, the next molecule in the signal cascade. Each transducin activates one phosphodiesterase, and each phosphodiesterase converts 4000 cGMP to GMP.
    (a) What is the name of the phenomenon described in this paragraph?
    (b) Activation of one rhodopsin will result in the production of how many GMP molecules?

ANSWERS

✓ Answers to Concept Check Questions

Page 177

1. All the communication methods listed are chemical signals except for (c) gap junctions, which transfer both chemical and electrical signals. Neurohormones (e) and neurotransmitters (f) are associated with electrical signaling in neurons but are themselves chemicals.

2. Cytokines, hormones, and neurohormones travel through the blood. Neurohormones and neurotransmitters are released by neurons.

3. The signal to pounce could not have been a paracrine signal because the eyes are too far away from the legs and because the response was too rapid for it to have taken place by diffusion.

Page 180

4. The components of signal pathways are signal molecule, receptor, intracellular signal molecule(s), and target proteins.

5. The cellular locations of receptors are cell membrane, cytosol, and nucleus.

Page 182

6. The steps of signal transduction are (1) signal molecule binds to receptor that (2) activates a protein that (3) creates second messengers that (4) create a response.

7. Amplification turns one signal molecule (first messenger) into multiple second messenger molecules.

8. Steroids are lipophilic, so they can enter cells to bind to intracellular receptors.

Page 185

9. Receptors are either ligand-gated ion channels, receptor-enzymes, G protein-coupled receptors, or integrins.

10. First messengers are extracellular; second messengers are intracellular.

11. (a) ligand, receptor, second messenger, cell response; (b) amplifier enzyme, second messenger, protein kinase, phosphorylated protein, cell response

12. (a) Cl^- channel opens: cell hyperpolarizes; (b) K^+ channel opens: cell hyperpolarizes; (c) Na^+ channel opens: cell depolarizes.

Page 187

13. The cell must use active transport to move Ca^{2+} against their concentration gradient.

Page 189

14. A drug that blocks leukotriene action could act at the receptor or at any step downstream. A drug that blocks leukotriene synthesis might inhibit lipoxygenase.

Page 190

15. Receptors, enzymes, and transporters are all proteins.

16. Insulin could be using different second messenger systems or binding to different receptors.

Page 191

17. Choices (a) and (d) would down-regulate binding affinity. Changing receptor number would not affect binding affinity.

Page 193

18. Tonic control usually involves one control system, but antagonistic control uses two.

19. A signal can have opposite effects by using different receptors or different signal pathways.

Page 197

20. Local control takes place in or very close to the target cell. Reflex control is mediated by a distant integrating center.

21. Stimulus, sensor or receptor, afferent pathway, integrating center, efferent pathway, target or effector, response (tissue and systemic)

Page 198

22. If the aquarium water became overheated, there is no control mechanism for bringing it back into the desired range.

Page 200

23. Negative feedback shuts off the heater. This feedback loop is shown as a dashed line going from the response back to the stimulus, with parallel "stop" lines at the stimulus end.

Page 202

24. (a) The "neural system integrating center" is the brain and spinal cord. (b) "Receptor" represents the sense organs. (c) The dashed line indicating negative feedback runs from "Response" back to "Internal or external change."

Page 205

25. blow to knee = internal or external change; leg muscles = effectors; neuron to leg muscles = efferent neuron; sensory neuron = afferent pathway; brain and spinal cord = integrating center; stretch receptor = receptor; muscle contraction = response.

Page 205

26. food in stomach = stimulus; brain and spinal cord = CNS integrating center; endocrine cells of pancreas = E (integrating center); stretch receptors = receptor; efferent neuron to pancreas = efferent neuron; insulin = classic hormone; adipose cell = target cell; sensory neuron = afferent neuron.

Answers to Figure Questions

Page 182

Fig. 6-9: A (inactive and active) = adenylyl cyclase; inactive B = ATP; active B = cAMP; C (inactive and active) = protein kinase A; product = phosphorylated protein.

Page 194

Fig. 6-21: (a) 60 beats/min (b) 280 beats/min.

7

The separation of the endocrine system into isolated subsystems must be recognized as an artificial one, convenient from a pedagogical point of view but not accurately reflecting the interrelated nature of all these systems.

—Howard Rasmussen, *in* Williams' Textbook of Endocrinology, 1974

Radioactive scan of the thyroid gland.

Introduction to the Endocrine System

BACKGROUND BASICS

RUNNING PROBLEM

GRAVES' DISEASE

The ball slid by the hole and trickled off the green: another bogey. Ben Crenshaw's golf game was falling apart. The 33-year-old professional had won the Masters Tournament only a year ago, but now something was not right. He was tired and weak, had been losing weight, and felt hot all the time. He attributed his symptoms to stress, but his family thought otherwise. At their urging, he finally saw a physician. The diagnosis? Graves' disease, which results in an overactive thyroid gland.

| 212 | 222 | 228 | 231 | 235 | 236 | 238 |

David was seven years old when the symptoms first appeared. His appetite at meals increased, and he always seemed to be in the kitchen looking for food. Despite eating more, however, he was losing weight. When he started asking for water instead of soft drinks, David's mother became concerned, and when he wet the bed three nights in a row, she knew something was wrong. The doctor confirmed the suspected diagnosis after running tests to determine the concentration of glucose in David's blood and urine. David had diabetes mellitus. In his case, the disease was due to lack of insulin, a hormone produced by the pancreas. David was placed on insulin injections, a treatment he would continue for the rest of his life.

One hundred years ago, David would have died not long after the onset of symptoms. The field of **endocrinology**, the study of hormones, was then in its infancy. Most hormones had not been discovered, and the functions of known hormones were not well understood. There was no treatment for diabetes, no birth control pill for contraception. Babies born with inadequate secretion of thyroid hormone did not grow or develop normally.

Today, all that has changed. We have identified a long and growing list of hormones. The endocrine diseases that once killed or maimed can now be controlled by synthetic hormones and sophisticated medical procedures. Although physicians do not hesitate to use these treatments, we are still learning exactly how hormones act on their target cells. This chapter provides an introduction to the basic principles of hormone structure and function. You will learn more about individual hormones as you encounter them in your study of the various systems.

■ **FIGURE 7-1** *An endocrine disorder in ancient art*

This pre-Colombian stone carving of a woman shows a mass at her neck. This mass is an enlarged thyroid gland, a condition known as goiter. It was considered a sign of beauty among the people who lived high in the Andes mountains.

HORMONES

As you learned in Chapter 6, hormones are chemical messengers secreted into the blood by specialized cells. Hormones are responsible for many functions that we think of as long-term, ongoing functions of the body. Processes that fall mostly under hormonal control include growth and development, metabolism, regulation of the internal environment (temperature, water balance, ions), and reproduction. Hormones act on their target cells in one of three basic ways: (1) by controlling the rates of enzymatic reactions, (2) by controlling the transport of ions or molecules across cell membranes, or (3) by controlling gene expression and the synthesis of proteins.

Hormones Have Been Known Since Ancient Times

Although the scientific field of endocrinology is relatively young, diseases of the endocrine system have been documented for more than a thousand years. Evidence of endocrine abnormalities can even be seen in ancient art. For example, one pre-Colombian statue of a woman shows a mass on the front of her neck (Fig. 7-1 ■). The mass is an enlarged thyroid gland, or *goiter,* a common condition high in the Andes, where the dietary iodine needed to make thyroid hormones was lacking.

The first association of endocrine structure and function was probably the link between the testes and male sexuality. Castration of animals and men was a common practice in both Eastern and Western cultures because it was known to decrease the sex drive and render males infertile.

In 1849, A. A. Berthold performed the first classic experiment in endocrinology. He removed the testes from roosters and observed that the castrated birds had smaller combs, less aggressiveness, and less sex drive than uncastrated roosters. If the testes were surgically placed back into the donor cock or into another castrated bird, normal male behavior and comb development resumed. Because the reimplanted testes were not connected to nerves, Berthold concluded that the glands must be secreting something into the blood that affected the entire body.

Experimental endocrinology did not receive much attention, however, until 1889, when the 72-year-old French physician Charles Brown-Séquard made a dramatic announcement of his sexual rejuvenation after injecting himself with extracts made from bull testes ground up in water. An international uproar followed, and physicians on both sides of the Atlantic began to inject their patients with extracts of many different endocrine organs, a practice known as *organotherapy.*

We now know that the increased virility Brown-Séquard reported was most likely a placebo effect because testosterone is a hydrophobic steroid that cannot be extracted by an aqueous preparation. His research opened the door to hormone therapy, however, and in 1891 organotherapy had its first true success: a woman was treated for low thyroid hormone levels with glycerin extracts of sheep thyroid glands.

As the study of "internal secretions" grew, Berthold's experiments became a template for endocrine research. Once a gland or structure was suspected of secreting hormones, the classic steps for identifying an endocrine gland became:

1. Remove the suspected gland and monitor the animal for anatomical, behavioral, or physiological abnormalities. This is equivalent to inducing a state of *hormone deficiency*.
2. Either place the gland back in the animal or administer an extract of the gland and see if the abnormalities disappear. Such *replacement therapy* should eliminate the symptoms of hormone deficiency.
3. Either implant the gland in a normal animal or administer an extract from the gland to a normal animal, and see if symptoms characteristic of *hormone excess* appear.
4. Once a gland is identified as a potential source of hormones, purify extracts of the gland to isolate the active substance. The test for hormone activity is usually a biological assay in which an animal is injected with the purified extract and monitored for a response.

Hormones identified by this technique are sometimes called *classic hormones*. They include hormones of the pancreas, thyroid, adrenal glands, pituitary, and gonads, all discrete endocrine glands that could be easily identified and surgically removed. Not all hormones come from identifiable glands, however, and we have been slower to discover them.

The Anatomy Summary in Figure 7-2 ■ lists the major hormones of the body and the glands or cells that secrete them, along with the major properties of each hormone.

What Makes a Chemical a Hormone?

In 1905, the term *hormone* was coined from the Greek verb meaning "to excite or arouse." The traditional definition of a **hormone** is a chemical secreted by a cell or group of cells into the blood for transport to a distant target, where it exerts its effect at very low concentrations. However, as scientists learn more about chemical communication in the body, this definition is continually being challenged.

CLINICAL FOCUS

DIABETES

THE DISCOVERY OF INSULIN

Diabetes mellitus, the metabolic condition associated with pathologies of insulin function, has been known since ancient times. Detailed clinical descriptions of insulin-deficient diabetes were available to physicians, but they had no means of treating the disease, and patients invariably succumbed. However, in a series of classic experiments in endocrine physiology, researchers finally identified the cause of diabetes. In 1889, Oscar Minkowski at the University of Strasbourg (Germany) pinpointed the relationship between diabetes and the pancreas. Minkowski surgically removed the pancreas from dogs (*pancreatectomy*) and noticed that they developed symptoms that mimicked diabetes. He also found that implanting pieces of pancreas under the dogs' skin would prevent development of diabetes. Subsequently, in 1921 Fredrick G. Banting and Charles H. Best (Toronto, Canada) identified an antidiabetic substance in pancreas extracts. Banting and Best and others injected pancreatic extracts into diabetic animals and found that the extracts reversed the elevated blood glucose levels of diabetes. From there, it was a relatively short process until, in 1922, purified insulin was used in the first clinical trials.

ANATOMY SUMMARY

HORMONES

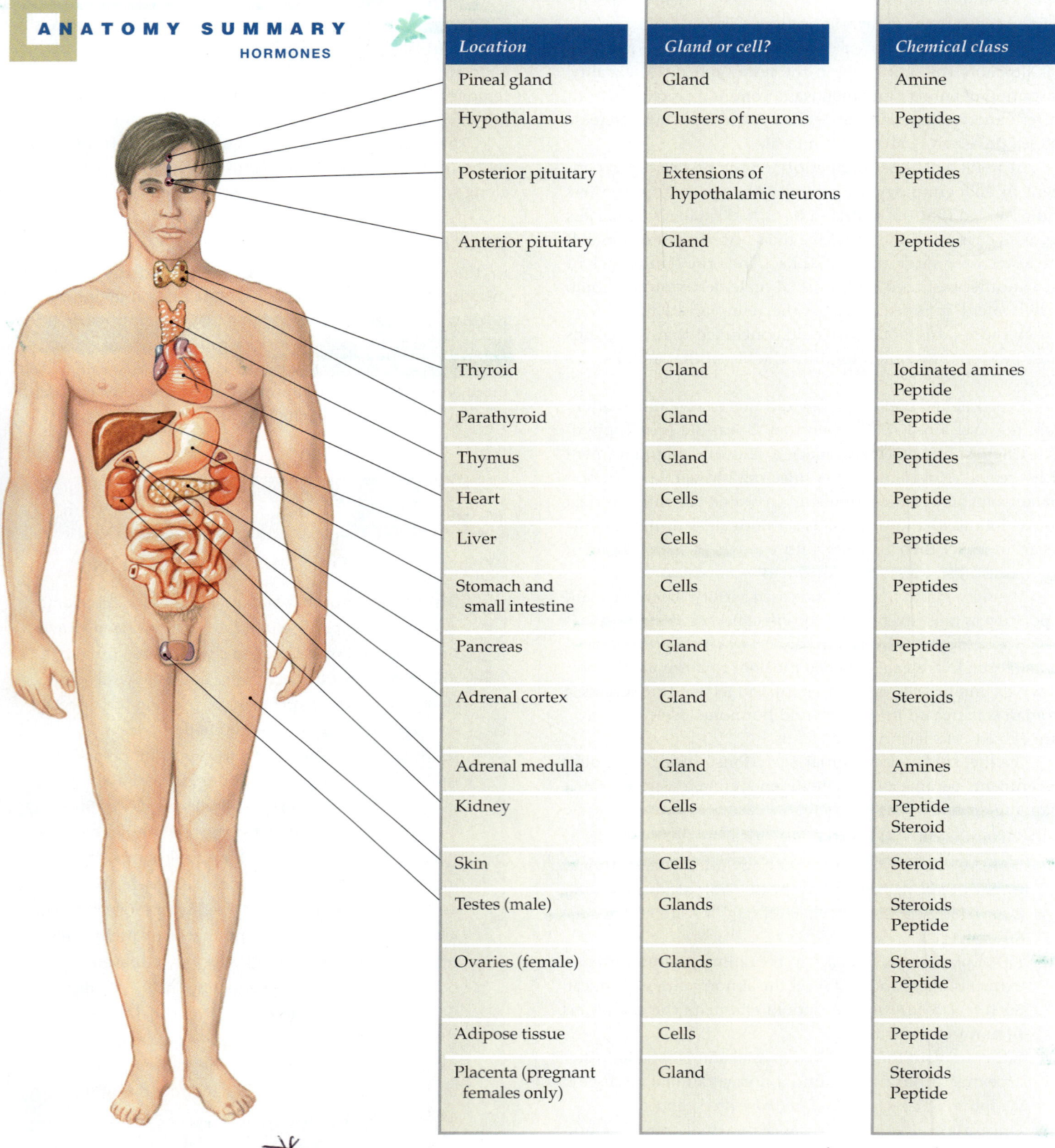

Location	Gland or cell?	Chemical class
Pineal gland	Gland	Amine
Hypothalamus	Clusters of neurons	Peptides
Posterior pituitary	Extensions of hypothalamic neurons	Peptides
Anterior pituitary	Gland	Peptides
Thyroid	Gland	Iodinated amines Peptide
Parathyroid	Gland	Peptide
Thymus	Gland	Peptides
Heart	Cells	Peptide
Liver	Cells	Peptides
Stomach and small intestine	Cells	Peptides
Pancreas	Gland	Peptide
Adrenal cortex	Gland	Steroids
Adrenal medulla	Gland	Amines
Kidney	Cells	Peptide Steroid
Skin	Cells	Steroid
Testes (male)	Glands	Steroids Peptide
Ovaries (female)	Glands	Steroids Peptide
Adipose tissue	Cells	Peptide
Placenta (pregnant females only)	Gland	Steroids Peptide

FIGURE 7-2

★ know location, hormones, Target & effect

Hormones	Targets	Main Effects
Melatonin	Unclear in humans	Circadian rhythms; other effects uncertain
Trophic hormones (see Fig. 7-13); see posterior pituitary	Anterior pituitary	Release or inhibit pituitary hormones
Oxytocin (OT)	Breast and uterus	Milk ejection; labor and delivery; behavior
Vasopressin (ADH)	Kidney	Water reabsorption
Prolactin (PRL)	Breast	Milk production
Growth hormone (GH, somatotropin)	Many tissues	Growth and metabolism
Corticotropin (ACTH)	Adrenal cortex	Cortisol release
Thyrotropin (TSH)	Thyroid gland	Thyroid hormone synthesis and release
Follicle stimulating hormone (FSH)	Gonads	Egg or sperm production; sex hormone production
Luteinizing hormone (LH)	Gonads	Sex hormone production; egg or sperm production
Triiodothyronine and thyroxine (T_3, T_4)	Many tissues	Metabolism; growth and development
Calcitonin (CT)	Bone	Plasma calcium levels (minimal effect in humans)
Parathyroid hormone (PTH)	Bone, kidney	Regulates plasma calcium and phosphate levels
Thymosin, thymopoietin	Lymphocytes	Lymphocyte development
Atrial natriuretic peptide (ANP)	Kidneys	Increases sodium excretion
Angiotensinogen	Adrenal cortex, blood vessels, brain	Aldosterone secretion; increases blood pressure
Insulin-like growth factors (IGF)	Many tissues	Growth
Gastrin, cholecystokinin (CCK), secretin, and others	GI tract and pancreas	Assist digestion and absorption of nutrients
Insulin, glucagon, somatostatin (SS), pancreatic polypeptide	Many tissues	Metabolism of glucose and other nutrients
Aldosterone	Kidney	Na^+ and K^+ homeostasis
Cortisol	Many tissues	Stress response
Androgens	Many tissues	Sex drive in females
Epinephrine, norepinephrine	Many tissues	Fight-or-flight response
Erythropoietin (EPO)	Bone marrow	Red blood cell production
1,25 Dihydroxy-vitamin D_3 (calciferol)	Intestine	Increases calcium absorption
Vitamin D_3	Intermediate form of hormone	Precursor of 1,25 dihydroxy-vitamin D_3
Androgen	Many tissues	Sperm production, secondary sex characteristics
Inhibin	Anterior pituitary	Inhibit FSH secretion
Estrogens and progesterone	Many tissues	Egg production; secondary sex characteristics
Ovarian inhibin	Anterior pituitary	Inhibits FSH secretion
Relaxin (pregnancy)	Uterine muscle	Relaxes muscle
Leptin and others	Hypothalamus, other tissues	Food intake, metabolism, reproduction
Estrogens and progesterone (P)	Many tissues	Fetal and maternal development
Chorionic somatomammotropin (CS)	Many tissues	Metabolism
Chorionic gonadotropin (CG)	Corpus luteum of ovary	Hormone secretion

7

Hormones Are Secreted by a Cell or Group of Cells

Traditionally, the field of endocrinology has focused on chemical messengers secreted by endocrine *glands,* the discrete and readily identifiable tissues derived from epithelial tissue [🔁p. 75]. However, we now know that molecules that act as hormones are secreted not only by classic endocrine glands but also by isolated endocrine cells (hormones of the *diffuse endocrine system*), by neurons *(neurohormones)*, and by cells of the immune system *(cytokines)*.

Hormones Are Secreted into the Blood

Secretion, first presented in Chapter 3 [🔁 p. 53], is the movement of a substance from the intracellular compartment either to the extracellular compartment or to the external environment. According to the traditional definition, hormones are secreted into the blood. However, the term *ectohormone* [*ektos,* outside] has been given to signal molecules secreted into the external environment.

Pheromones [*pherein,* to bring] are specialized ectohormones that act on other organisms of the same species to elicit a physiological or behavioral response. For example, sea anemones secrete alarm pheromones when danger threatens, and ants release trail pheromones to attract fellow workers to food sources. Pheromones are also used to attract members of the opposite sex for mating purposes. Sex pheromones can be found throughout the animal kingdom, in animals from fruit flies to dogs, but do humans have pheromones? This is still a matter of debate. Some studies have shown that human *axillary* (armpit) sweat glands secrete volatile steroids related to sex hormones that may serve as human sex pheromones. In one study, when female students were asked to rate the odors of T-shirts worn by male students, each woman preferred the odor of men who were genetically dissimilar from her. In another study, female axillary secretions rubbed on the upper lip of young women altered the timing of their menstrual cycles. How humans might sense pheromones will be discussed in Chapter 10.

Hormones Are Transported to a Distant Target

By the traditional definition, a hormone must be transported by the blood to a distant target cell. Experimentally, this property is sometimes difficult to demonstrate. Molecules that are suspected of being hormones but not fully accepted as such are called *candidate hormones*. They are usually identified by the word *factor*. For example, in the early 1970s, the hypothalamic regulating hormones were known as "releasing factors" and "inhibiting factors" rather than releasing and inhibiting hormones.

Currently, **growth factors**, a large group of substances that influence cell growth and division, are being studied to determine if they meet all the criteria for hormones. Although many growth factors have been shown to act locally as *autocrines* or *paracrines* [🔁 p. 176], evidence is lacking for their widespread distribution via the circulation. A similar situation exists with the lipid-derived signal molecules called *eicosanoids* [🔁 p. 29].

BIOTECHNOLOGY

IMMUNOCYTOCHEMISTRY

Traditionally, physiologists thought that each hormone came from a single gland (even though some glands secrete multiple hormones), but the research technique called *immunocytochemistry* has turned that view of endocrinology upside down. Immunocytochemistry begins when an animal is injected with a foreign molecule (an antigen), such as a hormone. The animal's immune system recognizes the injected molecule as "not self" and produces specific antibodies against it. The antibodies are extracted from the host animal and combined with a radioactive, fluorescent, or enzymatic marker. The tagged antibodies are applied to cells in tissue culture or to sections of freshly excised tissue, then examined under the microscope. If the original antigen is present in the tissue, the antibodies bind to it and become visible under the microscope. By screening multiple tissues with the antibodies against different hormones, investigators discovered that many hormones are made in multiple sites, often far from the tissue from which they were originally isolated.

Complicating the classification of signal molecules is the fact that a molecule may act as a hormone when secreted from one location but as a paracrine or autocrine signal when secreted from a different location. For example, in the 1920s scientists discovered that *cholecystokinin* (CCK) in extracts of intestine caused contraction of the gall bladder. For many years thereafter, CCK was known only as an intestinal hormone. Then in the mid-1970s, CCK was found in neurons of the brain, where it acts as a neurotransmitter or neuromodulator. In recent years, CCK has become famous for its possible role in controlling hunger.

Hormones Exert Their Effect at Very Low Concentrations

One hallmark of a hormone is its ability to act at concentrations in the nanomolar (10^{-9} M) to picomolar (10^{-12} M) range. Some chemical signals transported in the blood to distant targets are not considered hormones because they must be present in relatively high concentrations before an effect is noticed. For example, histamine released during allergic reactions may act on cells throughout the body, but its concentration exceeds the accepted range for a hormone.

As researchers discover new signal molecules and new receptors, the boundary between hormones and nonhormonal signal molecules continues to be challenged, just as the distinction between the nervous and endocrine systems has blurred. Many *cytokines* [🔁 p. 176] seem to meet the previously stated definition of a hormone. However, experts in cytokine research do not consider cytokines to be hormones because peptide cytokines are synthesized and released on demand, in contrast to classic peptide hormones, which are made in advance and stored in the parent endocrine cell. A few cytokines—for example, *erythropoietin,* the molecule that controls red blood cell production—were classified as hormones before the term *cytokine* was coined, contributing to the overlap between the two groups of signal molecules.

Hormones Act by Binding to Receptors

All hormones bind to target cell receptors and initiate biochemical responses. These responses are known as the **cellular mechanism of action** of the hormone. As you can see in Figure 7-2, one hormone may act on multiple tissues. To complicate matters, the effects may vary in different tissues or at different stages of development. Or a hormone may have no effect at all in a particular cell. Insulin is an example of a hormone with varied effects. In muscle and adipose tissues, it alters glucose transport proteins and enzymes for glucose metabolism. In the liver, it modulates enzyme activity but has no direct effect on glucose transport proteins. In the brain and certain other tissues, glucose metabolism is totally independent of insulin.

✔ CONCEPT CHECK

1. Name the membrane transport process by which glucose moves from the extracellular fluid into cells.

Answers: p. 241

The variable responsiveness of a cell to a hormone depends primarily on the cell's receptor and signal transduction pathways [🔁 p. 180]. If there are no hormone receptors in a tissue, its cells cannot respond. If tissues have different receptors and receptor-linked pathways for the same hormone, they will respond differently.

Hormone Action Must Be Terminated

Signal activity by hormones and other chemical signals must be of limited duration if the body is to respond to changes in its internal state. For example, insulin is secreted when blood glucose concentrations increase following a meal. As long as insulin is present, glucose leaves the blood and enters cells. However, if insulin activity continues too long, blood glucose levels could fall so low that the nervous system becomes unable to function properly—a potentially fatal situation. Normally the body avoids this situation in several ways: by limiting insulin secretion, by removing or inactivating insulin circulating in the blood, and by terminating insulin activity in target cells.

In general, hormones in the bloodstream are *degraded* (broken down) into inactive metabolites by enzymes found primarily in the liver and kidneys. The metabolites are then excreted in either the bile or the urine. The rate of hormone breakdown is indicated by a hormone's **half-life** in the circulation, the amount of time required to reduce the concentration of hormone by one-half. Half-life is one indicator of how long a hormone is active in the body.

Hormones bound to target membrane receptors have their activity terminated in several ways. Enzymes that are always present in the plasma can degrade peptide enzymes bound to cell membrane receptors. In some cases, the receptor-hormone complex is brought into the cell by endocytosis, and the hormone is then digested in lysosomes [🔁 Fig. 5-24, p. 149]. Intracellular enzymes metabolize hormones that enter cells.

✔ CONCEPT CHECK

2. What is the suffix in a chemical name that tells you a molecule is an enzyme? [*Hint:* 🔁 p. 99] Use that suffix to name an enzyme that digests peptides.

Answers: p. 241

THE CLASSIFICATION OF HORMONES

Hormones can be classified according to different schemes. The scheme used in Figure 7-2 groups them according to their source. A different scheme divides hormones into those whose release is controlled by the brain, and those whose release in not controlled by the brain. Another scheme groups hormones according to whether they bind to G protein-coupled receptors, tyrosine kinase-linked receptors, or intracellular receptors, and so on.

A final scheme divides hormones into three main chemical classes: peptide/protein hormones, steroid hormones, and amine hormones (Table 7-1 ■). The peptide hormones are composed of linked amino acids. The steroid hormones are all derived from cholesterol [🔁 p. 30]. The amine hormones are all derivatives of one of two amino acids: tryptophan or tyrosine.

✔ CONCEPT CHECK

3. What is the classic definition of a hormone?

4. Based on what you know about the organelles involved in protein and steroid synthesis [🔁 p. 66], what would be the major differences between the organelle composition of a steroid-producing cell and that of a protein-producing cell? Answers: p. 241

Most Hormones Are Peptides or Proteins

The peptide/protein hormones range from small peptides of only three amino acids to large proteins and glycoproteins. Despite the size variability among hormones in this group, all of them are usually called peptide hormones for the sake of simplicity. You can remember which hormones fall into this category by exclusion: if a hormone is not a steroid and not an amine, then it is probably a peptide.

TABLE 7-1	Comparison of Peptide, Steroid, and Amine Hormones			
	PEPTIDE HORMONES	**STEROID HORMONES**	**AMINES**	
			CATECHOLAMINES	**THYROID HORMONES**
Synthesis and storage	Made in advance; stored in secretory vesicles	Synthesized on demand from precursors	Made in advance; stored in secretory vesicles	Made in advance; precursor stored in secretory vesicles
Release from parent cell	Exocytosis	Simple diffusion	Exocytosis	Simple diffusion
Transport in blood	Dissolved in plasma	Bound to carrier proteins	Dissolved in plasma	Bound to carrier proteins
Half-life	Short	Long	Short	Long
Location of receptor	Cell membrane	Cytoplasm or nucleus; some have membrane receptors also	Cell membrane	Nucleus
Response to receptor-ligand binding	Activation of second messenger systems; may activate genes	Activation of genes for transcription and translation; may have nongenomic actions	Activation of second messenger systems	Activation of genes for transcription and translation
General target response	Modification of existing proteins and induction of new protein synthesis	Induction of new protein synthesis	Modification of existing proteins	Induction of new protein synthesis
Examples	Insulin, parathyroid hormone	Estrogen, androgens, cortisol	Epinephrine, norepinephrine	Thyroxine (T$_4$)

Peptide Hormone Synthesis, Storage, and Release

The synthesis and packaging of peptide hormones into membrane-bound secretory vesicles is similar to that of other proteins. The initial peptide that comes off the ribosome is a large inactive protein known as a preprohormone (Fig. 7-3 ■). Preprohormones contain one or more copies of a peptide hormone, a *signal sequence* that directs the protein into the lumen of the rough endoplasmic reticulum, and other peptide sequences that may or may not have biological activity.

As an inactive preprohormone moves through the endoplasmic reticulum and Golgi complex, the signal sequence is removed, creating a smaller, still-inactive molecule called a prohormone (Fig. 7-3). In the Golgi complex, the prohormone is packaged into secretory vesicles along with *proteolytic* [*proteo-*, protein + *lysis*, rupture] enzymes that chop the prohormone into active hormone and other fragments. This process is called *post-translational modification* [⊞ p. 120].

The secretory vesicles containing peptides are stored in the cytoplasm of the endocrine cell until the cell receives a signal for secretion. At that time, the vesicles move to the cell membrane and release their contents by calcium-dependent exocytosis

[⊞ p. 148]. All of the peptide fragments created from the prohormone are released together into the extracellular fluid, in a process known as *co-secretion*.

Post-Translational Modification of Prohormones   Studies of prohormone processing have led to some interesting discoveries. Some prohormones, such as that for *thyrotropin-releasing hormone* (TRH), contain multiple copies of the hormone (Fig. 7-4a ■). Another interesting prohormone is called *pro-opiomelanocortin* (Fig. 7-4b). This prohormone is split into three active peptides plus an inactive fragment. In some instances, even the fragments are clinically useful. For example, proinsulin is cleaved into active insulin and an inactive fragment known as *C-peptide* (Fig. 7-4c). Clinicians measure the levels of C-peptide in the blood of diabetics to monitor how much insulin the patient's pancreas is producing.

Transport in the Blood and Half-Life of Peptide Hormones   Peptide hormones are water soluble and therefore generally dissolve easily in the extracellular fluid for transport throughout the body. The half-life for peptide hormones is

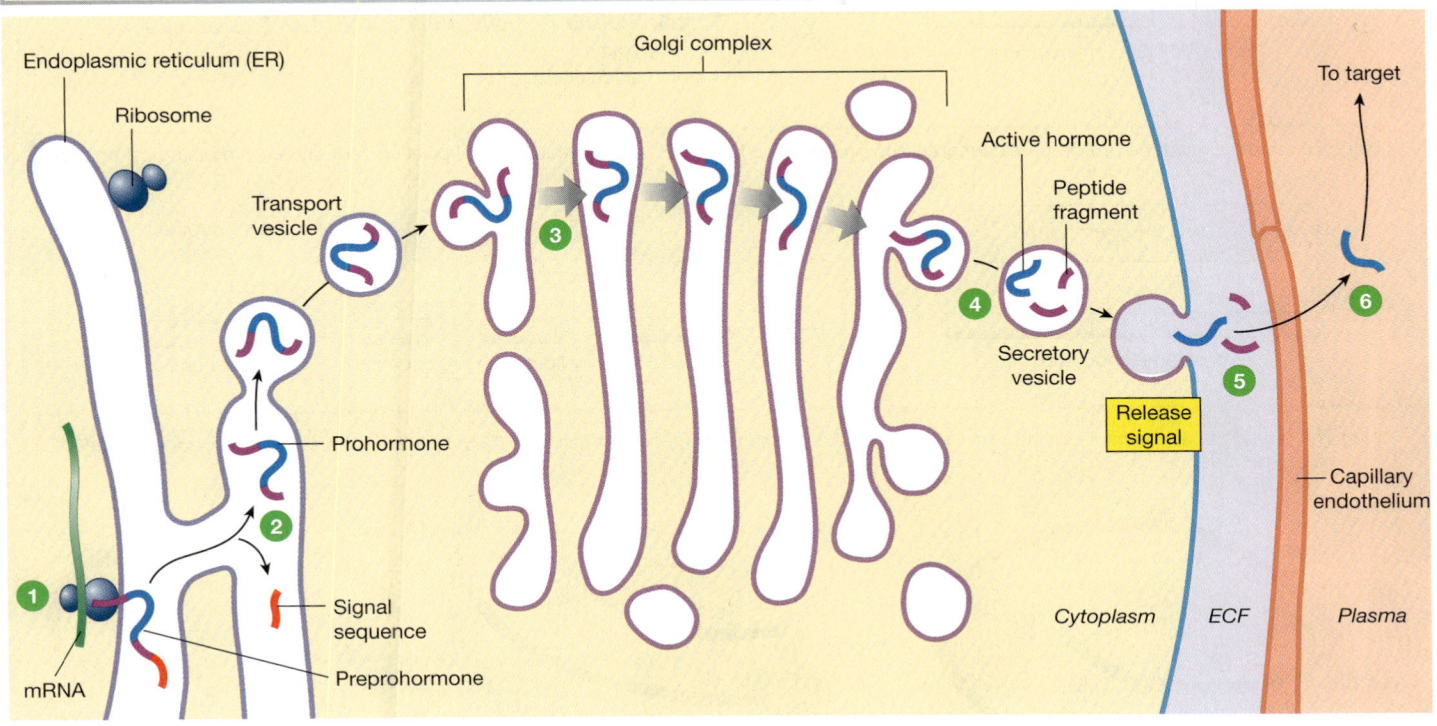

1 Messenger RNA on the ribosomes binds amino acids into a peptide chain called a **preprohormone**. The chain is directed into the ER lumen by a **signal sequence** of amino acids.

2 Enzymes in the ER chop off the signal sequence, creating an inactive **prohormone**.

3 The prohormone passes from the ER through the Golgi complex.

4 Secretory vesicles containing enzymes and prohormone bud off the Golgi. The enzymes chop the prohormone into one or more active peptides plus additional peptide fragments.

5 The secretory vesicle releases its contents by exocytosis into the extracellular space.

6 The hormone moves into the circulation for transport to its target.

FIGURE 7-3 *Peptide hormone synthesis, packaging, and release*

usually quite short, in the range of several minutes. If the response to a peptide hormone must be sustained for an extended period of time, the hormone must be secreted continually.

Cellular Mechanism of Action of Peptide Hormones

Because peptide hormones are lipophobic, they are usually unable to enter the target cell. Instead, they bind to surface membrane receptors. The hormone-receptor complex initiates the cellular response by means of a *signal transduction* system (Fig. 7-5 ■). Many peptide hormones work through cAMP second messenger systems [p. 183]. A few peptide hormone receptors, such as that of insulin, have tyrosine kinase activity [p. 183] or work through other signal transduction pathways.

The response of cells to peptide hormones is usually rapid because second messenger systems modify existing proteins. The changes triggered by peptide hormones include opening or closing membrane channels and modulating metabolic enzymes or transport proteins. Researchers have recently discovered that some peptide hormones also have longer-lasting effects when their second messenger systems activate genes and direct the synthesis of new proteins.

Steroid Hormones Are Derived from Cholesterol

The steroid hormones have a similar chemical structure because they are all derived from cholesterol (Fig. 7-6 ■). Unlike peptide hormones, which are made in tissues all over the body, steroid hormones are made in only a few organs. Three types of steroid hormones are made in the adrenal cortex, the outer portion of the adrenal glands [*cortex*, bark]. One adrenal gland sits atop each kidney [*ad-*, upon + *renal*, kidney]. The gonads produce the sex steroids (estrogens, progesterone, and androgens). In pregnant women, the placenta is also a source of steroid hormones.

Steroid Hormone Synthesis and Release Cells that secrete steroid hormones have unusually large amounts of smooth endoplasmic reticulum, the organelle in which steroids are synthesized. Steroids are lipophilic and diffuse easily across membranes, both out of their parent cell and into their target cell. This property also means that steroid-secreting cells cannot store hormones in secretory vesicles. Instead, they synthesize their hormone as it is needed. When a stimulus activates the endocrine cell, precursors in the cytoplasm are rapidly

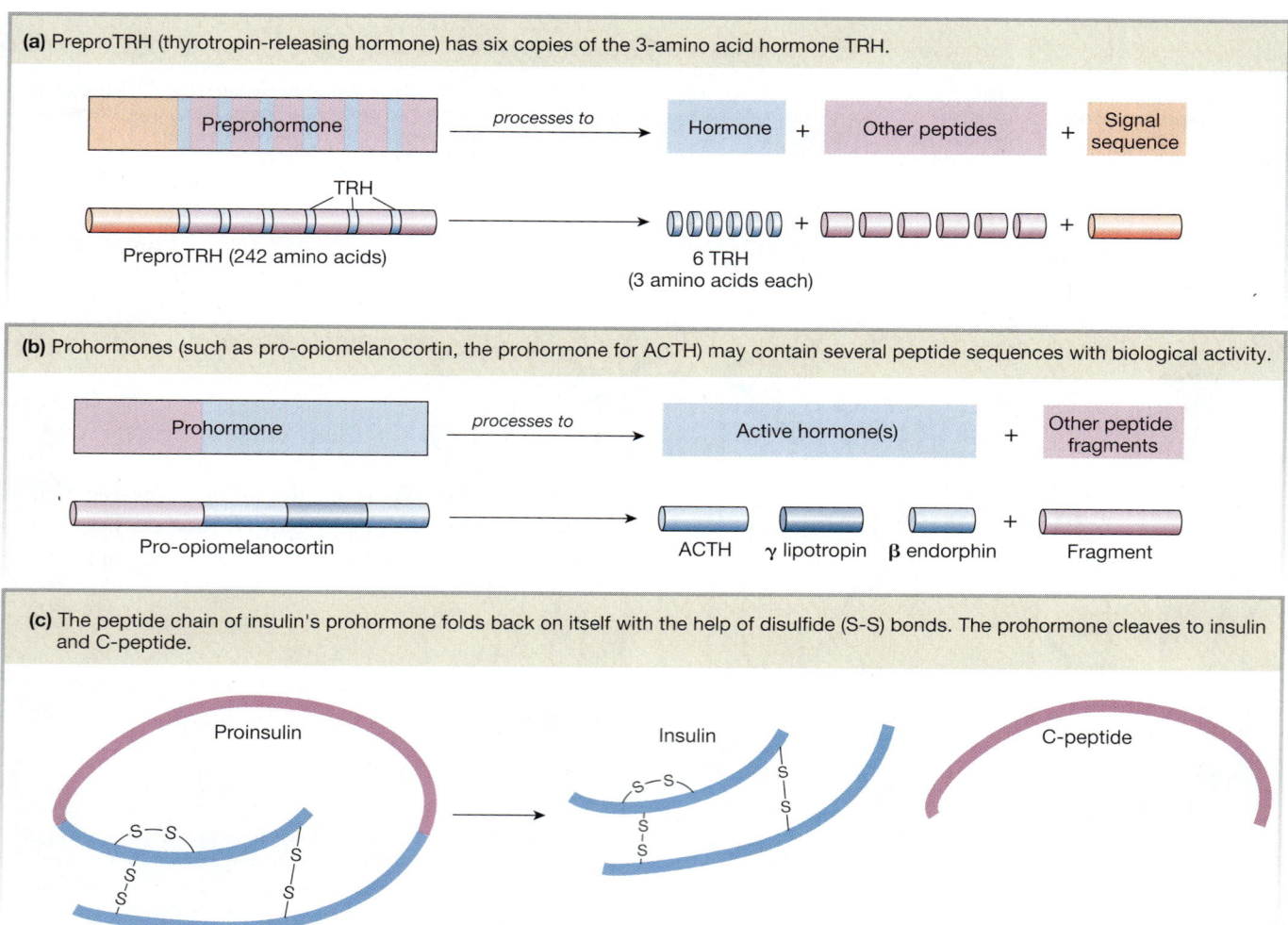

(a) PreproTRH (thyrotropin-releasing hormone) has six copies of the 3-amino acid hormone TRH.

PreproTRH (242 amino acids) → 6 TRH (3 amino acids each)

(b) Prohormones (such as pro-opiomelanocortin, the prohormone for ACTH) may contain several peptide sequences with biological activity.

Pro-opiomelanocortin → ACTH   γ lipotropin   β endorphin + Fragment

(c) The peptide chain of insulin's prohormone folds back on itself with the help of disulfide (S-S) bonds. The prohormone cleaves to insulin and C-peptide.

Proinsulin → Insulin   C-peptide

■ **FIGURE 7-4**   *Peptide hormone processing*

Peptide hormones are made as large, inactive preprohormones that include a signal sequence, one or more copies of the hormone, and additional peptide fragments.

converted to active hormone. The hormone concentration in the cytoplasm rises, and the hormones move out of the cell by simple diffusion.

Transport in the Blood and Half-Life of Steroid Hormones   Like their parent cholesterol, steroid hormones are not very soluble in plasma and other body fluids. For this reason, most of the steroid hormone molecules found in the blood are bound to protein carrier molecules. Some hormones have specific carriers, such as *corticosteroid-binding globulin*. Others simply bind to general plasma proteins, such as *albumin*.

The binding of a steroid hormone to a carrier protein protects the hormone from enzymatic degradation and results in an extended half-life. For example, **cortisol**, a hormone produced by the adrenal cortex, has a half-life of 60–90 minutes. (Compare this with epinephrine, an amine hormone whose half-life is measured in seconds.)

Although binding steroid hormones to protein carriers extends their half-life, it also blocks their entry into target cells. The carrier-steroid complex remains outside the cell because the carrier proteins are lipophobic and cannot diffuse through the membrane. Only unbound hormone molecule can diffuse into the target cell (Fig. 7-7 ■). As unbound hormone leaves the plasma, the carriers obey the law of mass action and release hormone so that the ratio of unbound to bound hormone in the plasma remains constant [the K_d; ≡ p. 39].

Fortunately, hormones are active in minute concentrations, and only a tiny amount of unbound steroid is enough to produce a response. As unbound hormone leaves the blood and enters cells, additional carriers release their bound steroid so that some unbound hormone is always in the blood and ready to enter a cell.

Peptide hormones (H) cannot enter their target cells and must combine with membrane receptors (R) that initiate signal transduction processes.

KEY

TK = Tyrosine kinase

AE = Amplifier enzyme

G = G protein

■ FIGURE 7-5 *Membrane receptors and signal transduction for peptide hormones*

Cellular Mechanism of Action of Steroid Hormones

The best-studied steroid hormone receptors are found within cells, either in the cytoplasm or in the nucleus. The ultimate destination of steroid receptor-hormone complexes is the nucleus, where the complex acts as a *transcription factor,* binding to DNA and either activating or *repressing* (turning off) one or more genes (Fig. 7-7,③). Activated genes create new mRNA that directs the synthesis of new proteins. Any hormone that alters gene activity is said to have a *genomic effect* on the target cell.

When steroid hormones activate genes to direct the production of new proteins, there is usually a lag time between hormone-receptor binding and the first measurable biological effects. This lag can be as much as 90 minutes. Consequently, steroid hormones do not mediate reflex pathways that require rapid responses.

In recent years researchers have discovered that several steroid hormones, including estrogens and aldosterone, have cell membrane receptors linked to signal transduction pathways, just as peptide hormones do. These receptors enable those steroid hormones to initiate rapid **nongenomic responses** in addition to their slower genomic effects. With the discovery of nongenomic effects of steroid hormones, the functional differences between steroid and peptide hormones seem almost to have disappeared.

7

CONCEPT CHECK

5. What are the three chemical classes of hormones?

6. The steroid hormone aldosterone has a short half-life for a steroid hormone—only about 20 minutes. What would you predict about the degree to which aldosterone is bound to blood proteins?

Answers: p. 242

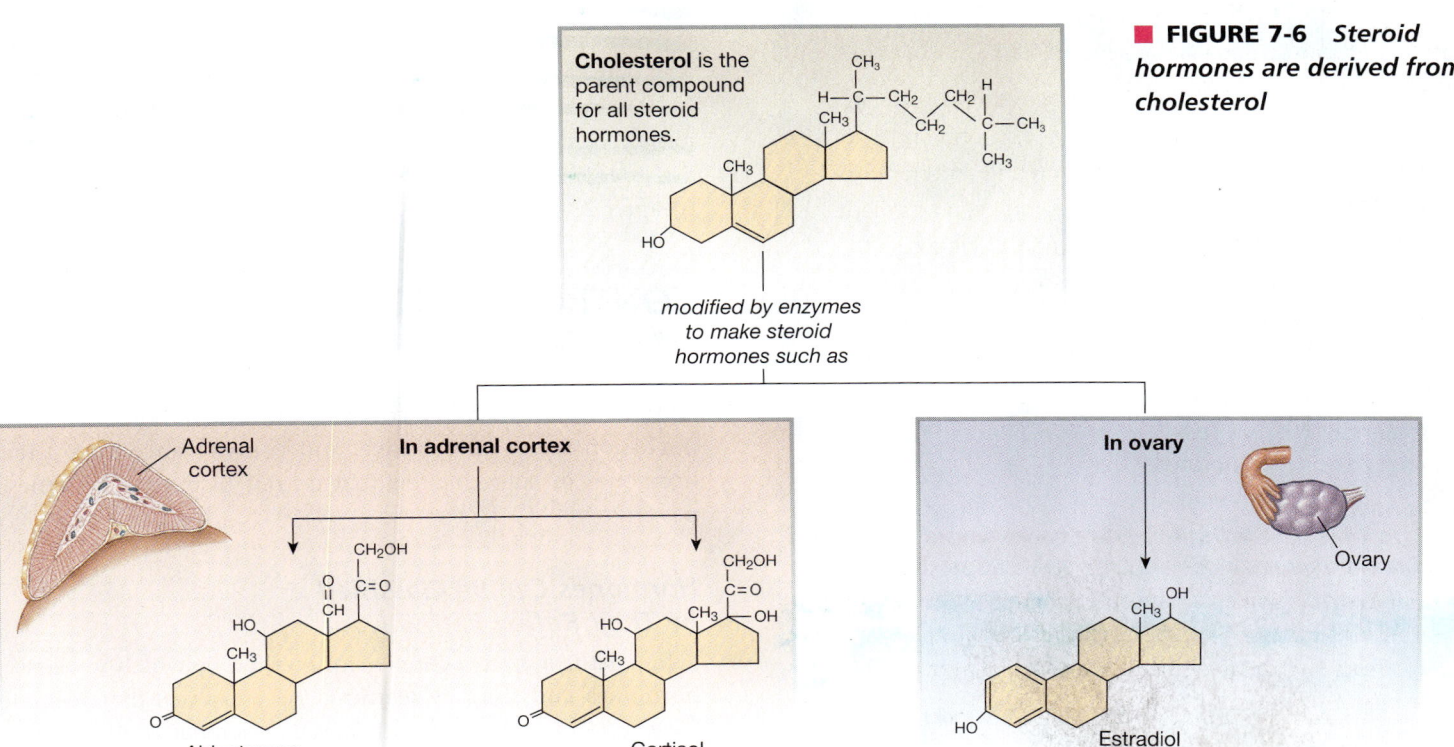

■ FIGURE 7-6 *Steroid hormones are derived from cholesterol*

Cholesterol is the parent compound for all steroid hormones.

modified by enzymes to make steroid hormones such as

Adrenal cortex

In adrenal cortex

Aldosterone

Cortisol

In ovary

Ovary

Estradiol (an estrogen)

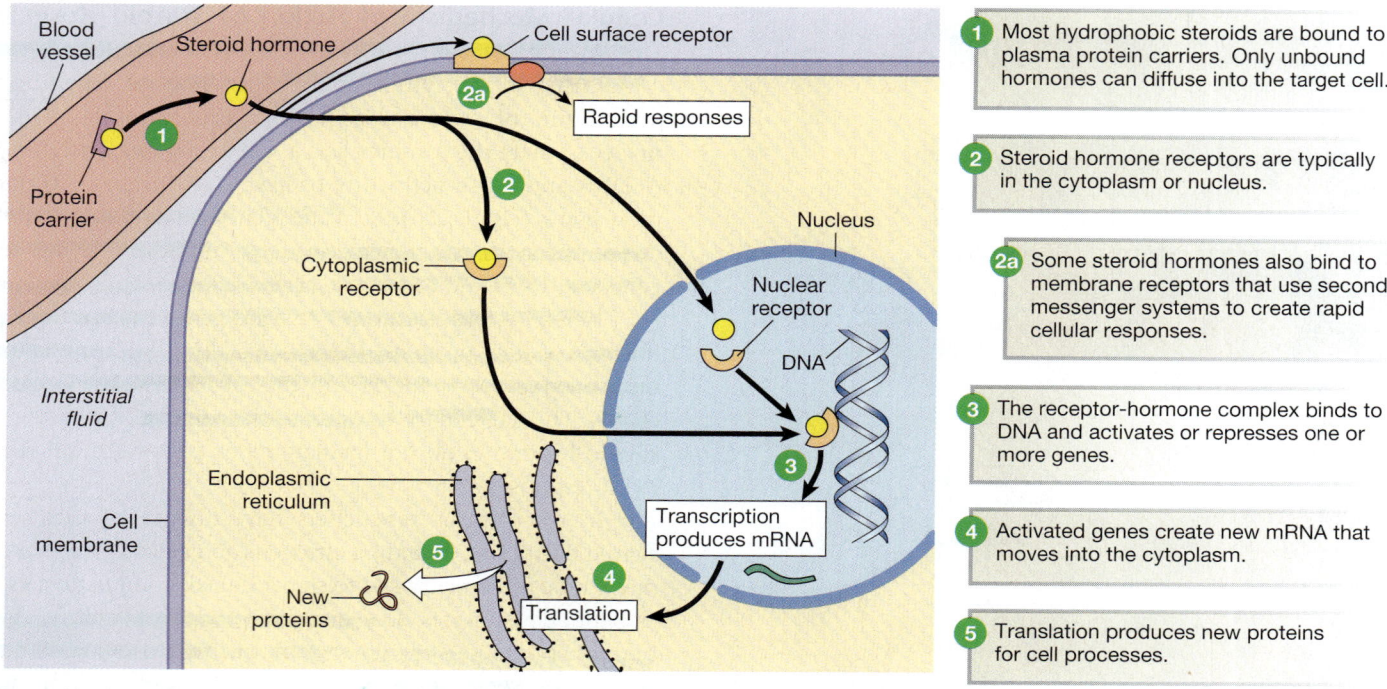

1. Most hydrophobic steroids are bound to plasma protein carriers. Only unbound hormones can diffuse into the target cell.

2. Steroid hormone receptors are typically in the cytoplasm or nucleus.

2a. Some steroid hormones also bind to membrane receptors that use second messenger systems to create rapid cellular responses.

3. The receptor-hormone complex binds to DNA and activates or represses one or more genes.

4. Activated genes create new mRNA that moves into the cytoplasm.

5. Translation produces new proteins for cell processes.

■ **FIGURE 7-7** *Steroid hormone action*

Amine Hormones Are Derived from One of Two Amino Acids

The amine hormones are small molecules created from either the amino acid tryptophan or the amino acid tyrosine. The amine hormone **melatonin** is derived from tryptophan (see *Focus on the Pineal Gland,* Fig. 7-22); all the other amine hormones—the catecholamines and thyroid hormones—are derived from tyrosine, notable for its ring structure (Fig. 7-8 ■). Catecholamines have one tyrosine molecule; the thyroid hormones have two tyrosines plus iodine atoms.

Despite a common precursor, the two groups of tyrosine-based amine hormones have little in common. The **catecholamines** (epinephrine, norepinephrine, and dopamine) are neurohormones that bind to cell membrane receptors the way peptide hormones do. The **thyroid hormones**, produced by the butterfly-shaped thyroid gland in the neck, behave more like steroid hormones, with intracellular receptors that activate genes. Thyroid hormones will be discussed in detail in Chapter 23.

CONTROL OF HORMONE RELEASE

Reflex pathways that help maintain homeostasis were introduced in Chapter 6. The sections that follow apply the basic patterns of reflexes to the control pathways for hormones. This discussion is not all-inclusive, and you will encounter a few hormones in later chapters that do not fit exactly into these patterns.

Hormones Can Be Classified by Their Reflex Pathways

Reflex pathways are a convenient method by which to classify hormones and simplify learning the pathways that regulate their secretion. All reflex pathways have similar components: a stimulus, an input signal, integration of the signal, an output

RUNNING PROBLEM

Shaped like a butterfly, the thyroid gland straddles the trachea just below the Adam's apple. Responding to hormonal signals from the hypothalamus and anterior pituitary, the thyroid gland concentrates iodine, an element found in food (most notably as an ingredient added to salt), and combines it with the amino acid tyrosine to make two thyroid hormones, thyroxine and triiodothyronine. These thyroid hormones perform many important functions in the body, including the regulation of growth and development, oxygen consumption, and the maintenance of body temperature.

Question 1:
a. To which of the three classes of hormones do the thyroid hormones belong?
b. If a person's diet is low in iodine, predict what happens to thyroxine production.

| 212 | **222** | 228 | 231 | 235 | 236 | 238 |

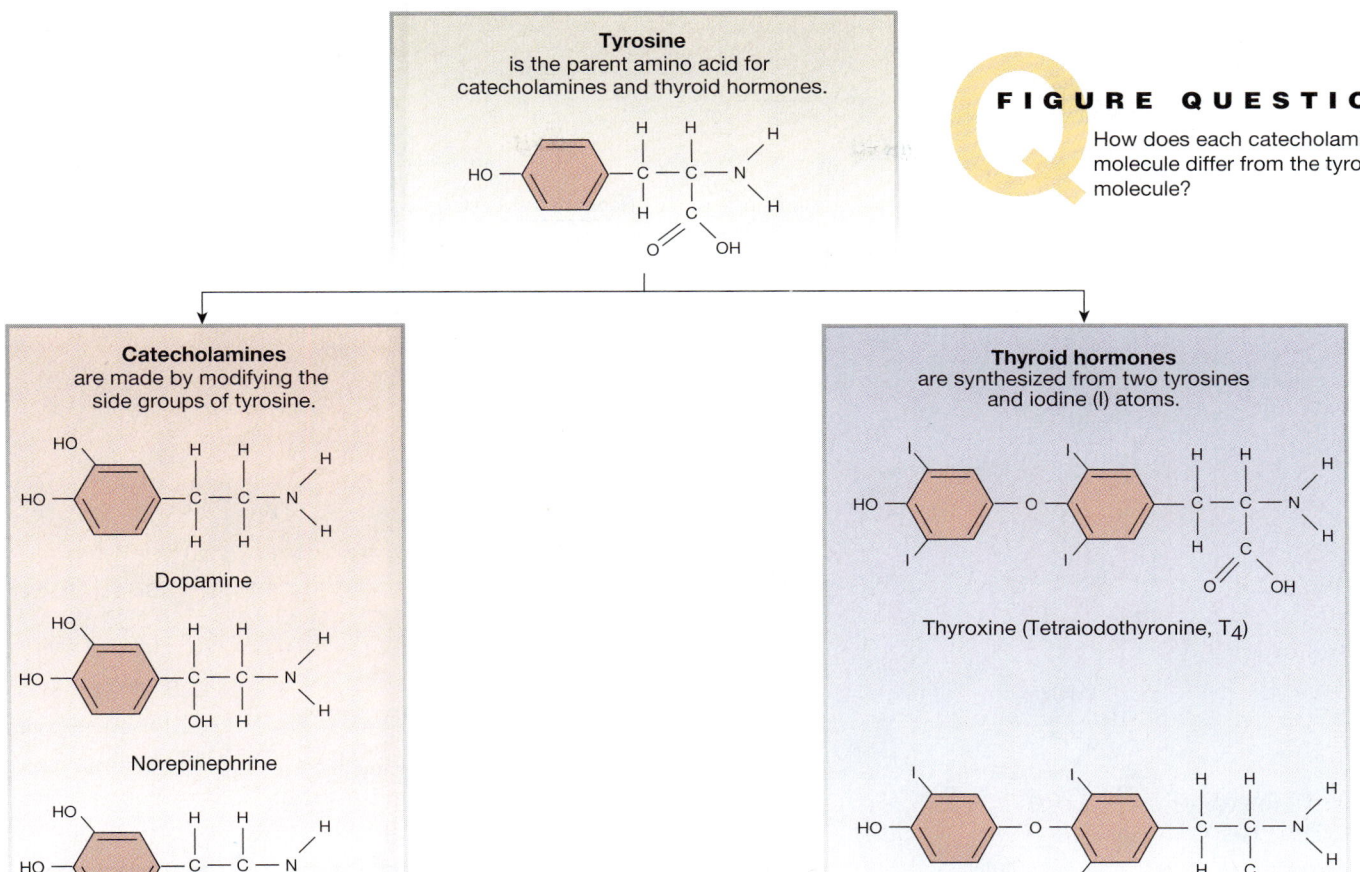

■ **FIGURE 7-8** *Tyrosine-derived amine hormones*

signal, and a response [⊇ Fig. 6-25, p. 198]. In endocrine and neuroendocrine reflexes, the output signal is a hormone or a neurohormone.

In a simple reflex pathway, the response of the pathway usually serves as the *negative feedback* signal that turns off the reflex [⊇ Fig. 6-27a, p. 200]. For example, an increase in blood glucose concentration initiates insulin secretion by the pancreas (Fig. 7-9 ■). Once released from the pancreas, insulin travels through the blood to its target tissues, which increase their glucose uptake and metabolism. The resultant decrease in blood glucose concentration acts as a negative feedback signal and turns off the reflex, ending release of insulin.

Hormones are not restricted to following only one reflex pathway pattern, however. For example, insulin secretion can also be triggered by input signals from the nervous system (Fig. 7-9) or by a hormone secreted from the digestive tract as a meal is eaten (not shown). The pancreatic endocrine cells—the integrating center for this reflex—thus must evaluate input signals from three different sources when "deciding" whether to secrete insulin.

CONCEPT CHECK

7. In Figure 6-23 (p. 195), which reflex step(s) represent(s) the input signal? Integration of the signal? The output signal?

8. In the blood glucose example, the increase in blood glucose corresponds to which step of a reflex pathway? Insulin secretion and the decrease in blood glucose correspond to which steps?

9. Which three reflex pathways in Figure 6-31 (p. 204) match insulin release by the three different mechanisms just described?   Answers: p. 242

The Endocrine Cell Is the Sensor in the Simplest Endocrine Reflexes

The simplest reflex control pathways in the endocrine system are those in which an endocrine cell directly senses a stimulus and responds by secreting its hormone [⊇ Fig. 6-31, pattern 6, p. 204]. In this type of pathway, the endocrine cell acts as both sensor (receptor) and integrating center. **Parathyroid hormone (PTH)** is an example of a hormone that operates via this simple

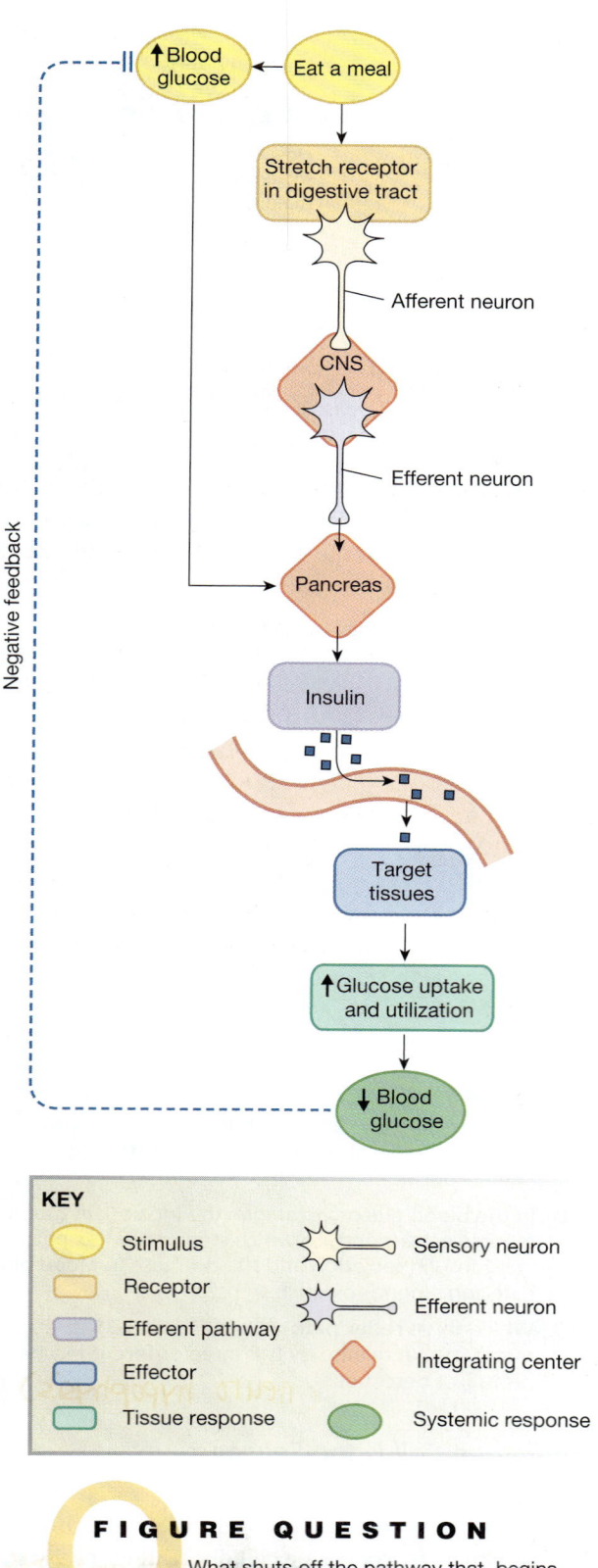

FIGURE QUESTION

What shuts off the pathway that begins with the stimulus of "eat a meal?"

■ **FIGURE 7-9** *Hormones may have multiple stimuli for their release*

Insulin can be released directly by an increase in blood glucose levels, or through nervous stimulation triggered by ingestion of a meal.

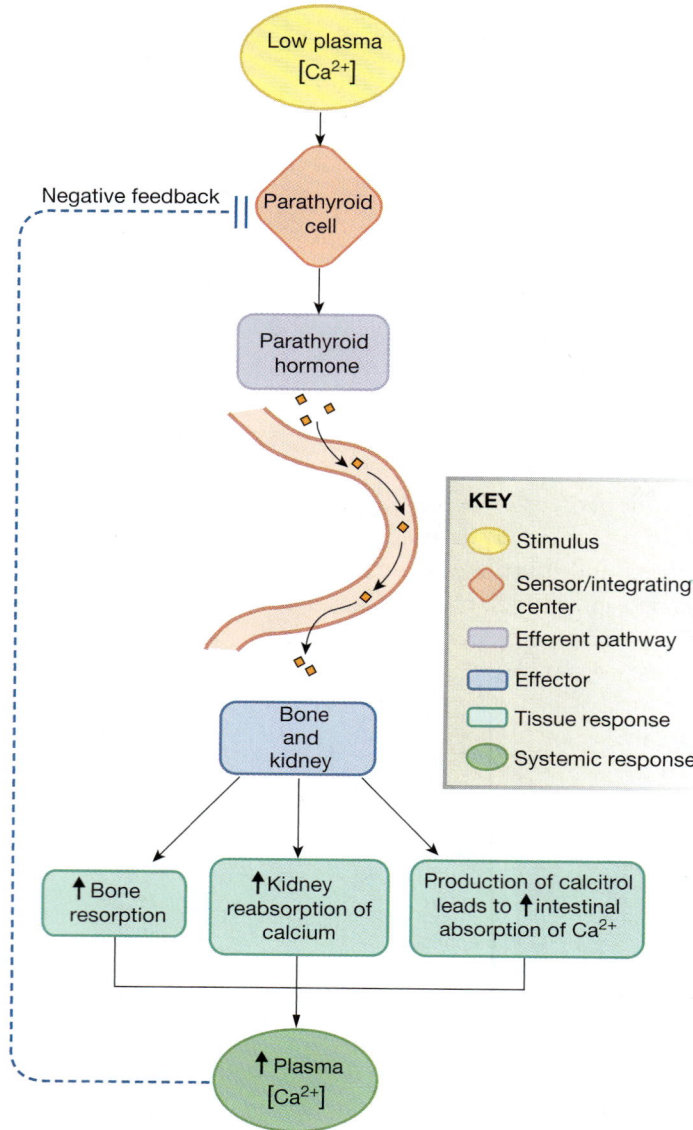

■ **FIGURE 7-10** *A simple endocrine reflex: parathyroid hormone*

endocrine reflex. Other hormones that follow a simple reflex pattern include the classic hormones insulin and glucagon, as well as some hormones of the diffuse endocrine system.

Parathyroid endocrine cells are clustered in four small glands that lie behind the thyroid gland. They monitor plasma Ca^{2+} concentration with the aid of G protein-coupled Ca^{2+} receptors on their cell membrane. When a certain number of receptors are bound to Ca^{2+}, PTH secretion is inhibited. If the plasma Ca^{2+} concentration falls below a certain level and fewer receptors are bound, inhibition ceases and the parathyroid cells secrete PTH (Fig. 7-10 ■). Parathyroid hormone travels through the blood to act on its target tissues, initiating responses that increase the concentration of Ca^{2+} in the plasma. The increase in plasma Ca^{2+} is a negative feedback signal that turns off the reflex, ending the release of parathyroid hormone.

CONCEPT CHECK
10. Draw a reflex pathway for parathyroid hormone as described in the previous paragraph.
Answers: p. 242

Many Endocrine Reflexes Involve the Nervous System

The nervous system and the endocrine system overlap in both structure and function. Stimuli that are integrated by the central nervous system influence the release of many hormones through efferent neurons, as previously described for insulin. In addition, specialized groups of neurons secrete neurohormones, and two endocrine structures are incorporated in the anatomy of the brain: the pineal gland (discussed in a *Focus* box on p. 237) and the pituitary gland.

One of the most fascinating links between the brain and the endocrine system is the influence of emotions over hormone secretion and function. Physicians for centuries have recorded instances in which emotional state has influenced health or normal physiological processes. Women today know that the timing of their menstrual periods may be altered by stressors such as travel or final exams. The condition known as "failure to thrive" in infants can often be linked to environmental or emotional stress that increases secretion of some pituitary hormones and decreases production of others. The interactions among stress, the endocrine system, and the immune system are receiving intense study, and we will discuss them further in Chapter 24.

Neurohormones Are Secreted into the Blood by Neurons

As noted in Chapter 6, neurohormones are chemical signals released into the blood by a neuron [⊜ Fig. 6-31, reflex 2]. The human nervous system produces three major groups of neurohormones: (1) catecholamines made by modified neurons in the adrenal medulla, (2) hypothalamic neurohormones secreted from the posterior pituitary, and (3) hypothalamic neurohormones that control hormone release from the anterior pituitary. Because the latter two groups of neurohormones are associated with the pituitary gland, we will first describe that important endocrine structure.

CONCEPT CHECK
11. Catecholamines belong to which chemical class of hormone?
Answers: p. 242

The Pituitary Gland Is Actually Two Fused Glands

The **pituitary gland** is a lima-bean-sized structure that extends downward from the brain, connected to it by a thin stalk and cradled in a protective pocket of bone (Fig. 7-11 ■). The first accurate description of the function of the pituitary gland came from Richard Lower (1631–1691), an experimental physiologist at Oxford University. Using observations and some experiments, he theorized that substances produced in the brain passed down the stalk into the gland and from there into the blood.

Lower did not realize that the pituitary gland is actually two different tissue types that merged during embryonic development. The **anterior pituitary** is a true endocrine gland of epithelial origin, derived from embryonic tissue that formed the roof of the mouth [⊜ Fig. 3-28, p. 77]. It is also called the *adenohypophysis* [*adeno-,* gland + *hypo-,* beneath + *phyein,* to grow], and its hormones are *adenohypophyseal* secretions. The **posterior pituitary**, or *neurohypophysis*, is an extension of the neural tissue of the brain. It secretes neurohormones made in the hypothalamus.

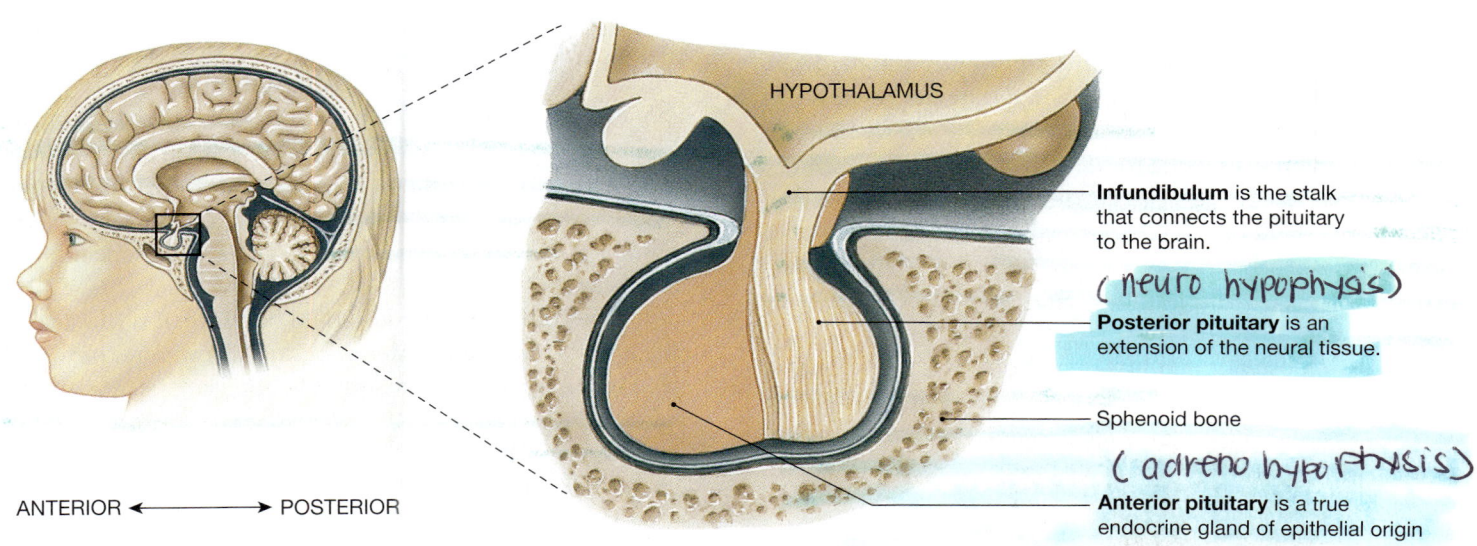

HYPOTHALAMUS

Infundibulum is the stalk that connects the pituitary to the brain.

(*neuro hypophysis*)
Posterior pituitary is an extension of the neural tissue.

Sphenoid bone

(*adreno hypophysis*)
Anterior pituitary is a true endocrine gland of epithelial origin

ANTERIOR ◄────► POSTERIOR

■ **FIGURE 7-11** *Pituitary gland anatomy*
The pituitary gland sits in a protected pocket of bone, connected to the brain by a thin stalk.

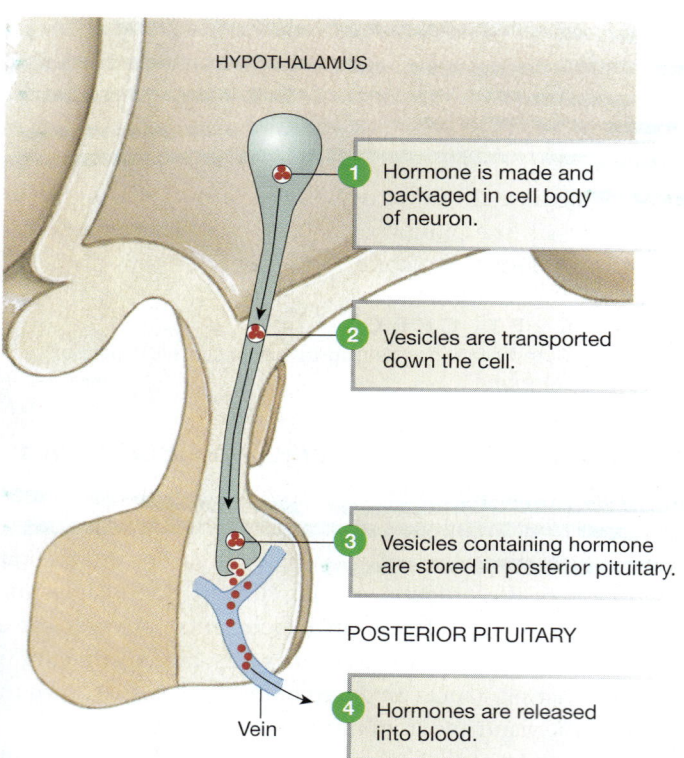

1. Hormone is made and packaged in cell body of neuron.

2. Vesicles are transported down the cell.

3. Vesicles containing hormone are stored in posterior pituitary.

POSTERIOR PITUITARY

4. Hormones are released into blood.

Vein

■ FIGURE 7-12 *Synthesis, storage, and release of posterior pituitary hormones*

The Posterior Pituitary Stores and Releases Two Neurohormones

The posterior pituitary is the storage and release site for two neurohormones: oxytocin and vasopressin. These small peptide hormones are synthesized in the cell bodies of neurons in the hypothalamus, a region of the brain that controls many homeostatic functions (Fig. 7-12 ■). Each hormone is made in a separate cell type.

Hypothalamic neurohormones follow the standard pattern for peptide synthesis, storage, and release described earlier in this chapter. However, secretory vesicles containing hormone are transported down long extensions of the neurons into the posterior pituitary, where they are stored in the cell terminals. When a stimulus reaches the hypothalamus, an electrical signal passes from the neuron cell body to the distal end of the cell in the posterior pituitary, and the vesicle contents are released into the circulation.

The two posterior pituitary neurohormones are composed of nine amino acids each. **Vasopressin** (also known as antidiuretic hormone) regulates water balance in the body. In women, **oxytocin** released from the posterior pituitary controls the ejection of milk during breast-feeding and contractions of the uterus during labor and delivery.

A few neurons release oxytocin as a neurotransmitter or neuromodulator onto neurons in other parts of the brain. A number of animal experiments plus a few human experiments suggest that oxytocin plays an important role in social, sexual,

and maternal behaviors. Some investigators postulate that *autism,* a developmental disorder in which patients are unable to form normal social relationships, may be related to defects in the normal oxytocin-modulated pathways of the brain.

CONCEPT CHECK

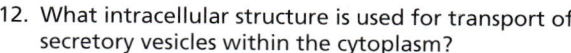

12. What intracellular structure is used for transport of secretory vesicles within the cytoplasm?

13. Name the membrane process by which the contents of secretory vesicles are released into the extracellular fluid.

Answers: p. 242

The Anterior Pituitary Secretes Six Hormones

As late as 1889, it was being said in reviews of physiological function that the pituitary was of little or no use to higher vertebrates! By the early 1900s, however, researchers had discovered that animals with their anterior pituitary glands surgically removed were unable to survive more than a day or two. This observation, combined with the clinical signs associated with pituitary tumors, made scientists realize that the anterior pituitary is a major endocrine gland that secretes not one but six physiologically significant hormones: prolactin, thyrotropin, adrenocorticotropin, growth hormone, follicle-stimulating hormone, and luteinizing hormone. Secretion of all the anterior pituitary hormones is controlled by hypothalamic neurohormones.

The anterior pituitary hormones, their associated hypothalamic neurohormones, and their targets are illustrated in Figure 7-13 ■. Notice that all but one of the anterior pituitary hormones have another endocrine gland or cell as one of their targets. A hormone that controls the secretion of another hormone is known as a **trophic hormone**.

The adjective *trophic* comes from the Greek word *trophikós,* which means "pertaining to food or nourishment" and refers to the manner in which the trophic hormone "nourishes" the target cell. Trophic hormones often have names that end with the suffix *-tropin,* as in *gonadotropin.** The root word to which the suffix is attached is the target tissue: the gonadotropins are hormones that are trophic to the gonads. The hypothalamic neurohormones that control release of the anterior pituitary hormones are also trophic hormones, but for historical reasons they are described as either *releasing hormones* (e.g., thyrotropin-releasing hormone) or *inhibiting hormones* (e.g., growth hormone-inhibiting hormone).

One complication you should be aware of is that many of the hypothalamic and anterior pituitary hormones have multiple names as well as standardized abbreviations. For example, **growth hormone-inhibiting hormone** (GHIH) is also called *somatostatin.* The abbreviations and alternate names are listed in the caption of Figure 7-13.

---

*A few hormones whose names end in *-tropin* do not have endocrine cells as their targets. For example, melanotropin acts on pigment-containing cells in many animals.

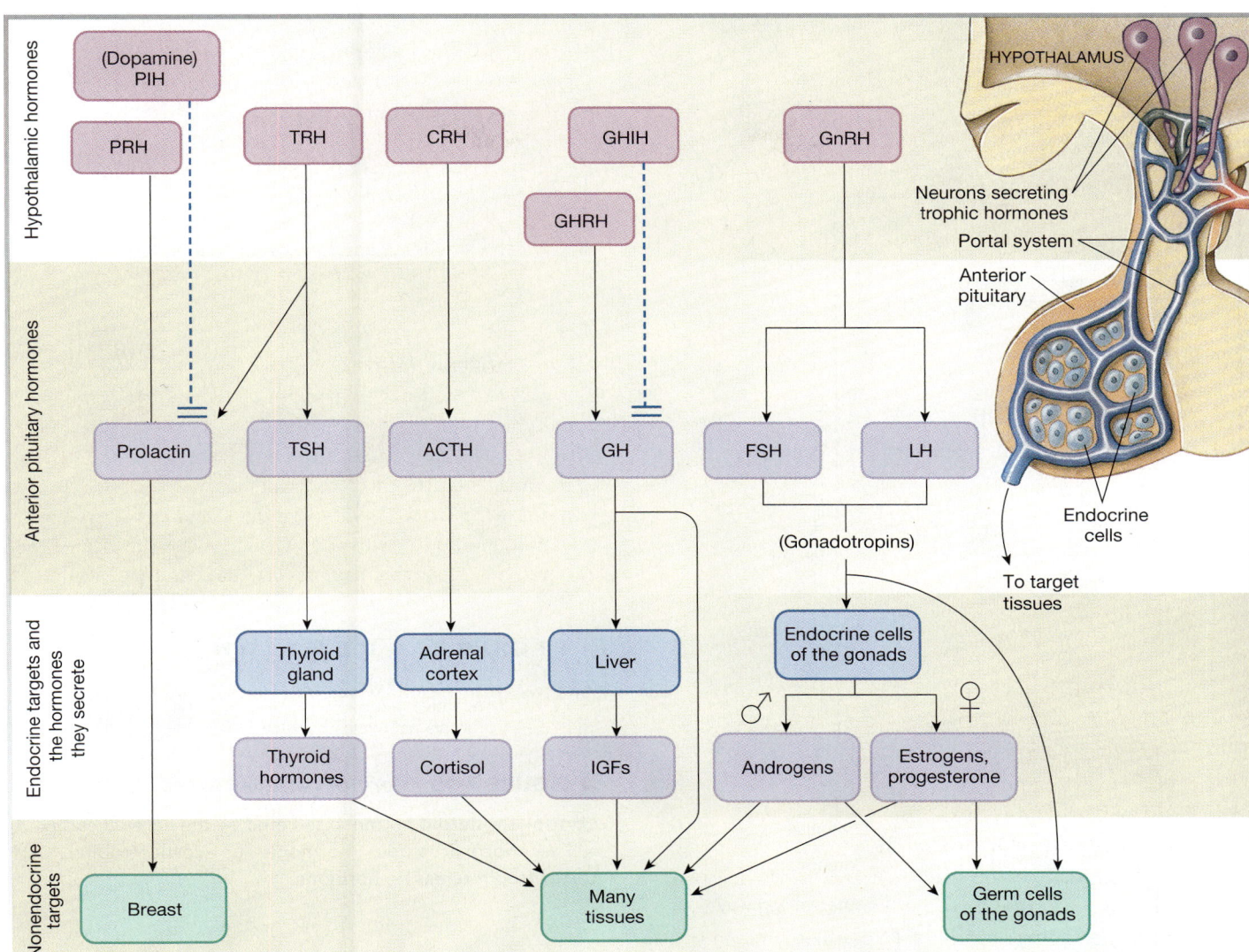

■ FIGURE 7-13 *Hormones of the hypothalamic-anterior pituitary pathway*

The hypothalamus secretes releasing hormones (-RH) and inhibiting hormones (-IH) that act on endocrine cells of the anterior pituitary to influence secretion of their hormones. Anterior pituitary hormones act either on additional endocrine glands or directly on target cells. ACTH = adrenocorticotropic hormone (corticotropin); CRH = corticotropin-releasing hormone; FSH = follicle-stimulating hormone; GH = growth hormone; GHIH = growth hormone-inhibiting hormone (somatostatin); GHRH = growth hormone-releasing hormone; GnRH = gonadotropin-releasing hormone; IGFs = insulin-like growth factors; LH = luteinizing hormone; PIH = prolactin-inhibiting hormone; PRH = prolactin-releasing hormone; TRH = thyrotropin-releasing hormone; TSH = thyroid-stimulating hormone (thyrotropin).

Feedback Loops Are Different in the Hypothalamic-Pituitary Pathway

The pathways in which anterior pituitary hormones act as trophic hormones are among the most complex endocrine reflexes because they involve three integrating centers: the hypothalamus, the anterior pituitary, and the endocrine target of the pituitary hormone (Fig. 7-14 ■). Feedback in these complex pathways follows a pattern that is different from the pattern described previously. Instead of the response acting as the negative feedback signal, the hormones themselves are the signal. Each hormone in the pathway feeds back to suppress hormone secretion by integrating centers earlier in the reflex pathway.

When secretion of one hormone in a complex pathway increases or decreases, the secretion of other hormones also changes because of the feedback loops that link the hormones. In pathways with two or three hormones in sequence, the "downstream" hormone usually feeds back to suppress the hormone(s) that controlled its secretion. (A major exception to this is feedback by ovarian hormones, as you will learn in Chapter 26.) For example, cortisol secreted from the adrenal cortex feeds back to suppress secretion of the trophic hormones corticotropin-releasing hormone and adrenocorticotropic hormone (Fig. 7-15 ■). This relationship is called **long-loop negative feedback**. In **short-loop**

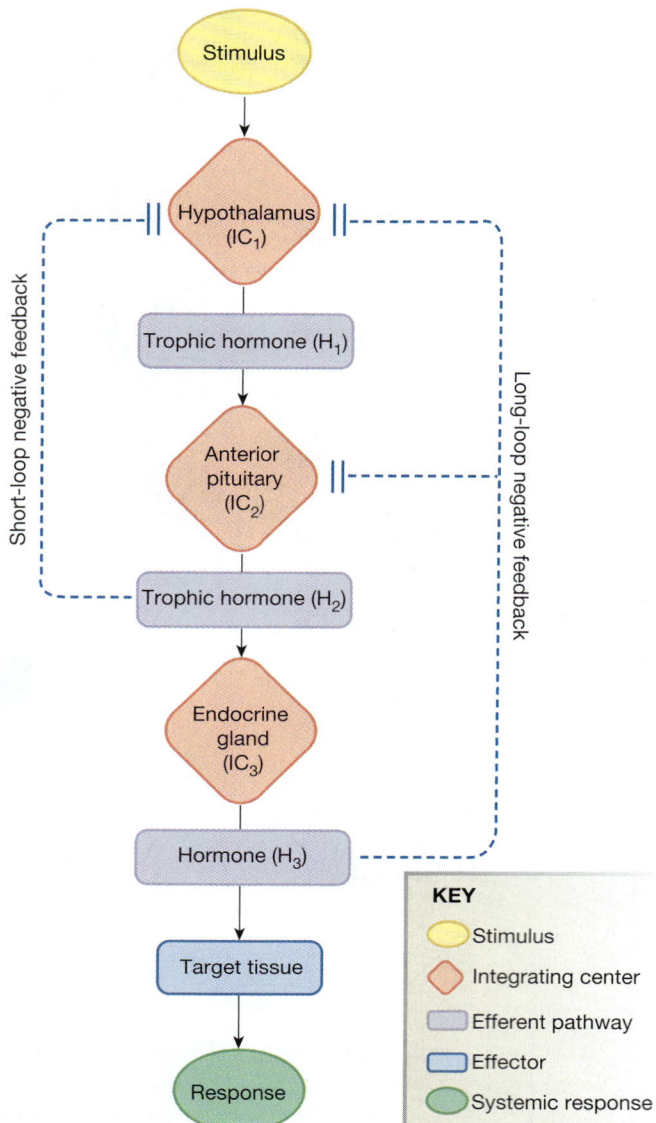

■ FIGURE 7-14 *Negative feedback loops in the hypothalamic-anterior pituitary pathway*

In complex endocrine pathways, the hormones of the pathway serve as negative feedback signals.

negative feedback, pituitary hormones feed back to decrease hormone secretion by the hypothalamus. We see this type of feedback in the cortisol secretion in Figure 7-15, where ACTH exerts short-loop negative feedback on the secretion of CRH. With this system of negative feedback, the hormones normally stay within the range needed for an appropriate response. Feedback patterns are important in the diagnosis of endocrine pathologies, to be discussed later in the chapter.

CONCEPT CHECK

14. Which pathway in Figure 6-31 (p. 204) fits the pituitary hormone pattern just described, in which there are three integrating centers?   Answers: p. 242

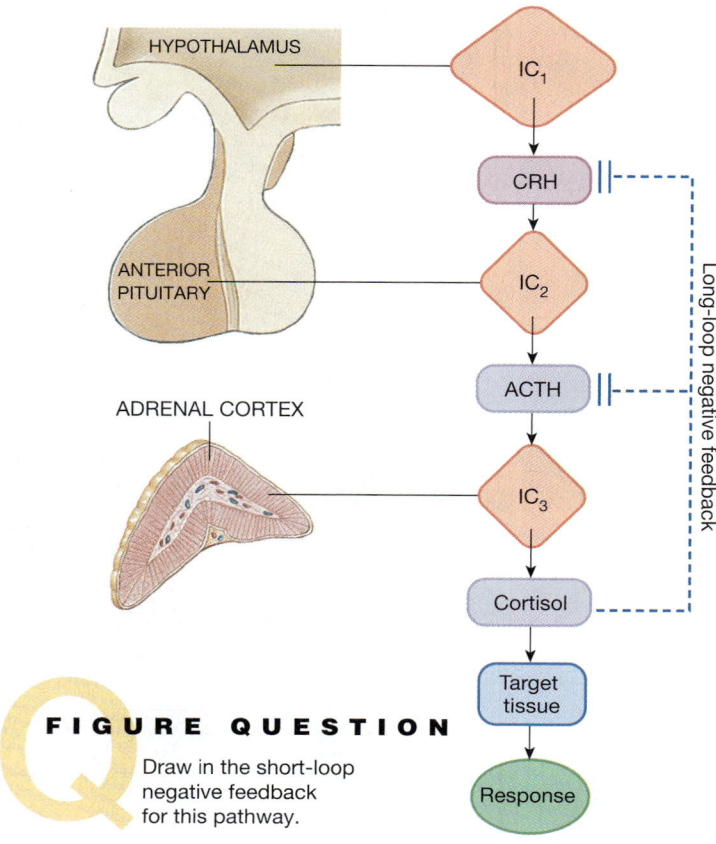

FIGURE QUESTION

Draw in the short-loop negative feedback for this pathway.

■ FIGURE 7-15 *Control pathway for cortisol secretion*

Cortisol is a steroid hormone secreted by the adrenal cortex. ACTH = adrenocorticotropic hormone or corticotropin; CRH = corticotropin-releasing hormone.

RUNNING PROBLEM

Thyroid hormone production is regulated by thyroid-stimulating hormone (TSH), a hormone secreted by the anterior pituitary. The production of TSH is in turn regulated by the neurohormone thyrotropin-releasing hormone (TRH) from the hypothalamus.

Question 2:
a. *In a normal person, when thyroid hormone levels in the blood increase, will negative feedback increase or decrease the secretion of TSH?*
b. *In a person with a hyperactive gland that is producing too much thyroid hormone, would you expect the level of TSH to be higher or lower than in a normal person?*

212   222   **228**   231   235   236   238

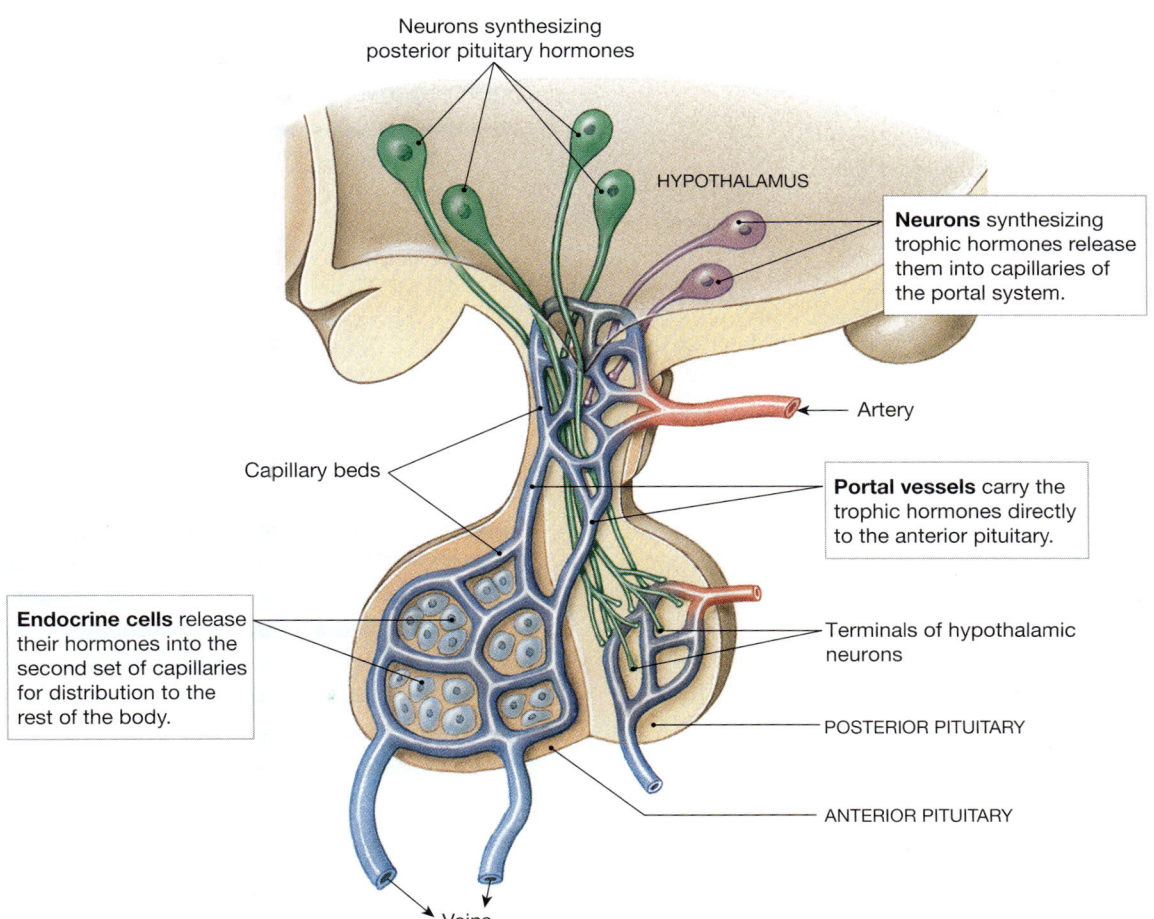

Neurons synthesizing
posterior pituitary hormones

HYPOTHALAMUS

Neurons synthesizing
trophic hormones release
them into capillaries of
the portal system.

← Artery

Capillary beds

Portal vessels carry the
trophic hormones directly
to the anterior pituitary.

Endocrine cells release
their hormones into the
second set of capillaries
for distribution to the
rest of the body.

Terminals of hypothalamic
neurons

POSTERIOR PITUITARY

ANTERIOR PITUITARY

7

Veins

■ **FIGURE 7-16**   *The hypothalamic-hypophyseal portal system*

The Hypothalamic-Hypophyseal Portal System Directs Trophic Hormone Delivery

The hypothalamic trophic hormones that regulate secretion of anterior pituitary hormones are transported directly to the pituitary through a special set of blood vessels known as the **hypothalamic-hypophyseal portal system** (Fig. 7-16 ■). A **portal system** is a specialized region of the circulation consisting of two sets of capillaries directly connected by a set of blood vessels. There are three portal systems in the body: one in the kidneys, one in the digestive system, and this one in the brain.

Hormones secreted into a portal system have a distinct advantage over hormones secreted into the general circulation because with a portal system, a much smaller amount of hormone can be secreted to elicit a given level of response. A dose of hormone secreted into the general circulation will be rapidly diluted by the total blood volume, which is typically more than 5 L. The same dose secreted into the tiny volume of blood flowing through the portal system remains concentrated while it is taken directly to its target. Thus, a small number of neurosecretory neurons in the hypothalamus can effectively control the anterior pituitary.

The minute amounts of hormone secreted into the hypothalamic-hypophyseal portal system posed a great challenge to the researchers who first isolated these hormones. Because such tiny quantities of hypothalamic-releasing hormones are secreted, Roger Guillemin and Andrew Shalley had to work with huge amounts of tissue to obtain enough hormone to analyze. Guillemin and his colleagues processed more than 50 tons of sheep hypothalami, and a major meat packer donated more than 1 million pig hypothalami to Shalley and his associates. For the final analysis, they needed 25,000 hypothalami to isolate and identify the amino acid sequence of just 1 mg of thyrotropin-releasing hormone, a tiny peptide made of three amino acids (see Fig. 7-4a). For their discovery, Guillemin and Shalley shared a Nobel prize in 1977.

Anterior Pituitary Hormones Control Growth, Metabolism, and Reproduction

The hormones of the anterior pituitary control so many vital functions that the pituitary is often called the master gland of the body. In general, we can say that the anterior pituitary hormones control metabolism, growth, and reproduction, all very complex processes.

■ **FIGURE 7-17** *A complex endocrine pathway*

Growth hormone acts directly on many body tissues but also influences liver production of insulin-like growth factors (IGFs or *somatomedins*), another group of hormones that regulate growth.

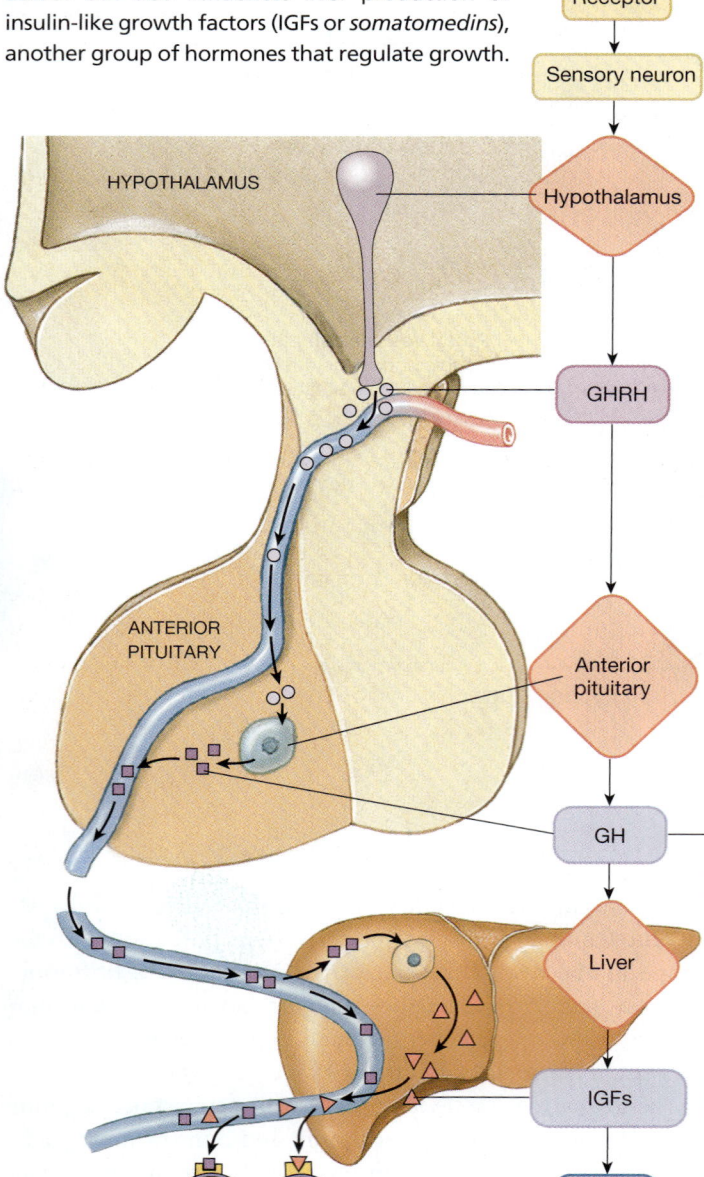

many tissues in addition to stimulating hormone production by the liver (Fig. 7-17 ■). Prolactin and growth hormone are the only two anterior pituitary hormones whose secretion is controlled by both releasing hormones and inhibiting hormones, as you can see in Figure 7-13 on page 227. We will discuss these hormones in detail in Chapters 26 and 23, respectively.

CONCEPT CHECK

15. Which pathway(s) in Figure 6-31 (p. 204) fit(s):
    (a) the hypothalamic trophic hormone-prolactin-breast pattern just described?
    (b) the growth hormone pathway shown in Figure 7-17?
    Answers: p. 242

The remaining four anterior pituitary hormones all have another endocrine gland as their primary target. **Follicle-stimulating hormone** (FSH) and **luteinizing hormone** (LH), known collectively as the **gonadotropins**, were originally named for their effects on the ovaries, but both hormones are trophic on male testes as well. **Thyroid-stimulating hormone** (TSH, or *thyrotropin*) controls hormone synthesis and secretion in the thyroid gland. **Adrenocorticotrophic hormone** (ACTH, or *corticotropin*) acts on certain cells of the adrenal cortex to control synthesis and release of the steroid hormone cortisol.

The hormones of the anterior and posterior pituitary will be discussed in more detail in later chapters.

CONCEPT CHECK

16. What is the target tissue of a hypothalamic hormone secreted into the hypothalamic-hypophyseal portal system?

17. Look at the pathway in Figure 7-9 for insulin release as the result of eating a meal. What event serves as the negative feedback signal to shut off insulin release?
    Answers: p. 242

HORMONE INTERACTIONS

One of the most complicated and confusing aspects of endocrinology is the way hormones interact at their target cells. It would be simple if each endocrine reflex were a separate entity and if each cell were under the influence of only a single hormone. In many instances, however, cells and tissues are controlled by multiple hormones that may be present at the same time. Complicating the picture is the fact that multiple hormones acting on a single cell can interact in ways that cannot be predicted by knowing the individual effects of the hormone. In this section, we examine three types of hormone interaction: synergism, permissiveness, and antagonism.

In Synergism, the Effect of Interacting Hormones Is More Than Additive

Sometimes different hormones have the same effect on the body, although they may accomplish that effect through different

One anterior pituitary hormone, **prolactin** (PRL), controls milk production in the female breast. In both sexes, prolactin appears to play a role in regulation of the immune system. **Growth hormone** (GH; also called *somatotropin*) affects metabolism of

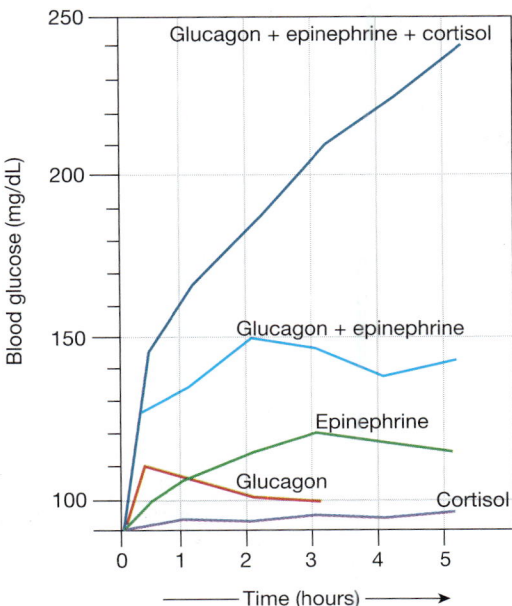

■ FIGURE 7-18 *Synergism*

This graph shows the effect of hormone infusions on blood glucose levels. The effects of combined hormones are greater than the summed effects of the individual hormones, indicating synergistic relationships. (Data adapted from Eigler et al., *J. Clin. Invest.* 63: 114, 1979.)

cellular mechanisms. One example is the hormonal control of blood glucose levels. Glucagon from the pancreas is the hormone primarily responsible for elevating blood glucose levels, but it is not the only hormone that has that effect. Cortisol raises blood glucose concentration, as does epinephrine.

What happens if two of these hormones are present in a target cell at the same time, or if all three hormones are secreted at the same time? You might expect their effects to be additive. In other words, if a given amount of epinephrine elevates blood glucose 5 mg/100 mL blood, and glucagon elevates blood glucose 10 mg/100 mL blood, you might expect both hormones acting at the same time to elevate blood glucose 15 mg/100 mL blood (5 + 10).

Frequently, however, two (or more) hormones interact at their targets so that the combination yields a result that is greater than additive. This type of interaction is called **synergism**. For our epinephrine/glucagon example, a synergistic reaction would be:

epinephrine	elevates blood glucose	5 mg/100 mL blood
glucagon	elevates blood glucose	10 mg/100 mL blood
epinephrine	elevate blood glucose	22 mg/100 mL blood

+ glucagon

That is, the combined effect of the two hormones is greater than the sum of the effects of the two hormones individually. Synergism is sometimes known as *potentiation,* as in "Epinephrine potentiates glucagon's effect on blood glucose."

An example of synergism involving epinephrine, glucagon, and cortisol is shown in Figure 7-18 ■. The cellular mechanisms that underlie synergistic effects are not always clear, but with peptide hormones, synergism is often linked to overlapping effects on second messenger systems.

Synergism is not limited to hormones. It can occur with any two (or more) chemicals in the body. Pharmacologists have developed drugs with synergistic components. For example, the effectiveness of the antibiotic penicillin is enhanced by the presence of clavulanic acid in the same pill.

A Permissive Hormone Allows Another Hormone to Exert Its Full Effect

In **permissiveness**, one hormone cannot fully exert its effects unless a second hormone is present. For example, maturation of the reproductive system is controlled by gonadotropin-releasing hormone from the hypothalamus, gonadotropins from the anterior pituitary, and steroid hormones from the gonads. However, if thyroid hormone is not present in sufficient amounts, maturation of the reproductive system is delayed. Because thyroid hormone by itself cannot stimulate maturation of the reproductive system, thyroid hormone is considered to have a permissive effect on sexual maturation.

RUNNING PROBLEM

Ben Crenshaw was diagnosed with Graves' disease, one form of hyperthyroidism. The goal of treatment is to reduce thyroid hormone activity, and Ben's physician offered him several alternatives. One treatment involves drugs that prevent the thyroid gland from using iodine. Another treatment is a single dose of radioactive iodine that destroys the thyroid tissue. A third treatment is surgical removal of all or part of the thyroid gland. Ben elected initially to use the thyroid-blocking drug. Several months later he was given radioactive iodine.

Question 3:
Why is radioactive iodine (rather than some other radioactive element, such as cobalt) used to destroy thyroid tissue?

The results of this interaction can be summarized as follows:

thyroid hormone alone	no development of reproductive system
reproductive hormones alone	delayed development of reproductive system
reproductive hormones with adequate thyroid hormone	normal development of reproductive system

The molecular mechanisms responsible for permissiveness are not well understood in most instances.

Antagonistic Hormones Have Opposing Effects

In some situations, two molecules work against each other, one diminishing the effectiveness of the other. This tendency of one substance to oppose the action of another is called *antagonism*. Recall from Chapter 6 that antagonism may result when two molecules compete for the same receptor [🔁 p. 44]. When one molecule binds to the receptor but does not activate it, that molecule acts as a competitive inhibitor, or antagonist, to the other molecule. This type of receptor antagonism has been put to use in the development of pharmaceutical compounds, such as the estrogen receptor antagonist *tamoxifen*.

In endocrinology, two hormones are considered functional antagonists if they have opposing physiological actions. For example, glucagon and growth hormone, both of which raise the concentration of glucose in the blood, are antagonistic to insulin, which lowers the concentration of glucose in the blood. Hormones with antagonistic actions do not necessarily compete for the same receptor. Instead, they may act through different metabolic pathways, or one hormone may decrease the number of receptors for the opposing hormone. For example, evidence suggests that growth hormone decreases the number of insulin receptors, providing part of its functional antagonistic effects on blood glucose concentration.

The synergistic, permissive, and antagonistic interactions of hormones make the study of endocrinology both challenging and intriguing. With this brief survey of hormone interactions, you have built a solid foundation for learning more about hormone interactions in later chapters.

ENDOCRINE PATHOLOGIES

As one endocrinologist said, "There are no good or bad hormones. A balance of hormones is important for a healthy life. . . . Unbalance leads to diseases."* We can learn much about the normal functions of a hormone by studying the diseases caused by hormone imbalances. There are three basic patterns of endocrine pathology: hormone excess, hormone deficiency, and abnormal responsiveness of target tissues to a hormone.

---

*W. König, preface to *Peptide and Protein Hormones* (New York: VCH Publishers, 1993).

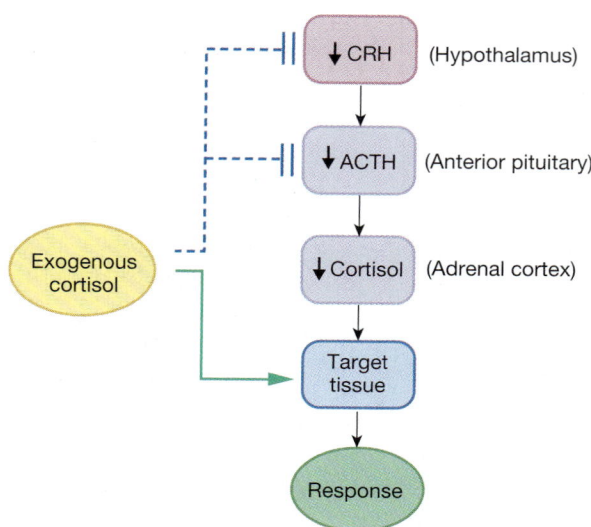

■ **FIGURE 7-19**   *Negative feedback by exogenous cortisol*
Hormone administered to a person for medical reasons will have the same negative feedback effect as the endogenous hormone.

To illustrate endocrine pathologies, we will use a single example, that of cortisol production by the adrenal cortex (see Fig. 7-15). This is a complex reflex pathway that starts with the secretion of corticotropin-releasing hormone (CRH) from the hypothalamus. CRH stimulates release of adrenocorticotropin (ACTH) from the anterior pituitary. ACTH in turn controls the synthesis and release of cortisol from the adrenal cortex. As in all other homeostatic reflex pathways, negative feedback shuts off the pathway. As cortisol increases, it acts as a negative feedback signal, causing the pituitary and hypothalamus to decrease their output of ACTH and CRH, respectively.

Hypersecretion Exaggerates a Hormone's Effects

If a hormone is present in excessive amounts, the normal effects of the hormone are exaggerated. Most instances of hormone excess are due to **hypersecretion**. There are numerous causes of hypersecretion, including benign tumors (*adenomas*) and cancerous tumors of the endocrine glands. Occasionally, nonendocrine tumors secrete hormones.

Any substance coming from outside the body is referred to as *exogenous* [*exo-*, outside + *-gen*, to be born], and sometimes a patient may exhibit signs of hypersecretion as the result of medical treatment with an exogenous hormone or agonist. In this case, the condition is said to be *iatrogenic*, or physician caused [*iatros*, healer + *-gen*, to be born]. It seems simple enough to correct the hormone imbalance by stopping treatment with the exogenous hormone, but this is not always the case.

In our example, exogenous cortisol in the body acts as a negative feedback signal, just as cortisol produced within the body would, shutting off the production of CRH and ACTH (Fig. 7-19 ■). Without the trophic "nourishing" influence of

ACTH, the body's own cortisol production shuts down. If the pituitary remains suppressed and the adrenal cortex is deprived of ACTH long enough, the cells of both glands shrink and lose their ability to manufacture ACTH and cortisol. The loss of cell mass is known as **atrophy** [*a-,* without + *trophikós,* nourishment].

If the cells of an endocrine gland atrophy because of exogenous hormone administration, they may be very slow or totally unable to regain normal function when the treatment with exogenous hormone is stopped. As you may know, steroid hormones can be used to treat poison ivy and severe allergies. However, when treatment is complete, the dosage must be tapered off gradually to allow the pituitary and adrenal gland to work back up to normal hormone production. As a result, packages of steroid pills direct patients ending treatment to take six pills one day, five the day after that, and so on. Low-dose, over-the-counter steroid creams usually do not pose a risk of feedback suppression when used as directed.

Hyposecretion Diminishes or Eliminates a Hormone's Effects

Symptoms of hormone deficiency occur when too little hormone is secreted (**hyposecretion**). Hyposecretion may occur anywhere along the endocrine control pathway, in the hypothalamus, pituitary, or other endocrine glands. The most common cause of hyposecretion syndromes is atrophy of the gland due to some disease process. For example, hyposecretion of thyroid hormone may occur if there is insufficient dietary iodine for the thyroid gland to manufacture the iodinated hormone.

Negative feedback pathways are affected in hyposecretion, but in the opposite direction from hypersecretion. The absence of negative feedback causes trophic hormone levels to rise as the trophic hormones attempt to make the defective gland increase its hormone output. For example, if the adrenal cortex atrophies as a result of tuberculosis, cortisol production diminishes. The hypothalamus and anterior pituitary sense that cortisol levels are below normal, so they increase secretion of CRH and ACTH, respectively, in an attempt to stimulate the adrenal gland into making more cortisol.

CONCEPT CHECK

18. Draw a reflex pathway similar to the one in Figure 7-19 to illustrate what happens to hormone levels and feedback when the adrenal cortex atrophies and cortisol secretion is below normal.
Answers: p. 242

Receptor or Second Messenger Problems Cause Abnormal Tissue Responsiveness

Endocrine diseases do not always arise from problems with endocrine glands. They may also be triggered by changes in the responsiveness of target tissues to the hormones. In these situations, the target tissues show abnormal responses even though the hormone levels may be within the normal range. Changes in the target tissue response are usually caused by abnormal interactions between the hormone and its receptor or by alterations in signal transduction pathways. These concepts were covered for signal molecules in general in Chapter 6 [p. 190], so we will restrict this discussion to some typical examples of abnormal tissue responsiveness in the endocrine system.

Down-Regulation If hormone secretion is abnormally high for an extended period of time, target cells may *down-regulate* (decrease the number of) their receptors in an effort to diminish their responsiveness to excess hormone. **Hyperinsulinemia** [*hyper-,* elevated + insulin + *-emia,* in the blood] is a classic example of down-regulation in the endocrine system. In this disorder, sustained high levels of insulin in the blood cause target cells to remove insulin receptors from the cell membrane. Patients suffering from hyperinsulinemia may show signs of diabetes despite their high blood insulin levels.

Receptor and Signal Transduction Abnormalities

Many forms of inherited endocrine pathologies can be traced to problems with hormone action in the target cell. Endocrinologists once believed that these problems were rare, but they are being recognized more and more as scientists increase their understanding of receptors and signal transduction mechanisms.

Some pathologies are due to problems with the hormone receptor. If a mutation alters the protein sequence of the receptor, the cellular response to receptor-hormone binding may be altered. In other mutations the receptors may be absent or completely nonfunctional. For example, in *testicular feminizing syndrome,* androgen receptors are nonfunctional in the male fetus because of a genetic mutation. As a result, androgens produced by the developing fetus are unable to influence development of the genitalia. The result is a child who appears to be female but lacks a uterus and ovaries.

Genetic alterations in signal transduction pathways can lead to symptoms of hormone excess or deficiency. In the disease called *pseudohypoparathyroidism* [*pseudo-,* false + *hypo-,* decreased + parathyroid + *-ism,* condition or state of being], patients show signs of low parathyroid hormone even though blood levels of the hormone are normal or elevated. These patients have inherited a defect in the G protein that links the hormone receptor to the cAMP amplifier enzyme, adenylyl cyclase. Because the signal transduction pathway does not function, the cells are unable to respond to parathyroid hormone, and signs of hormone deficiency appear.

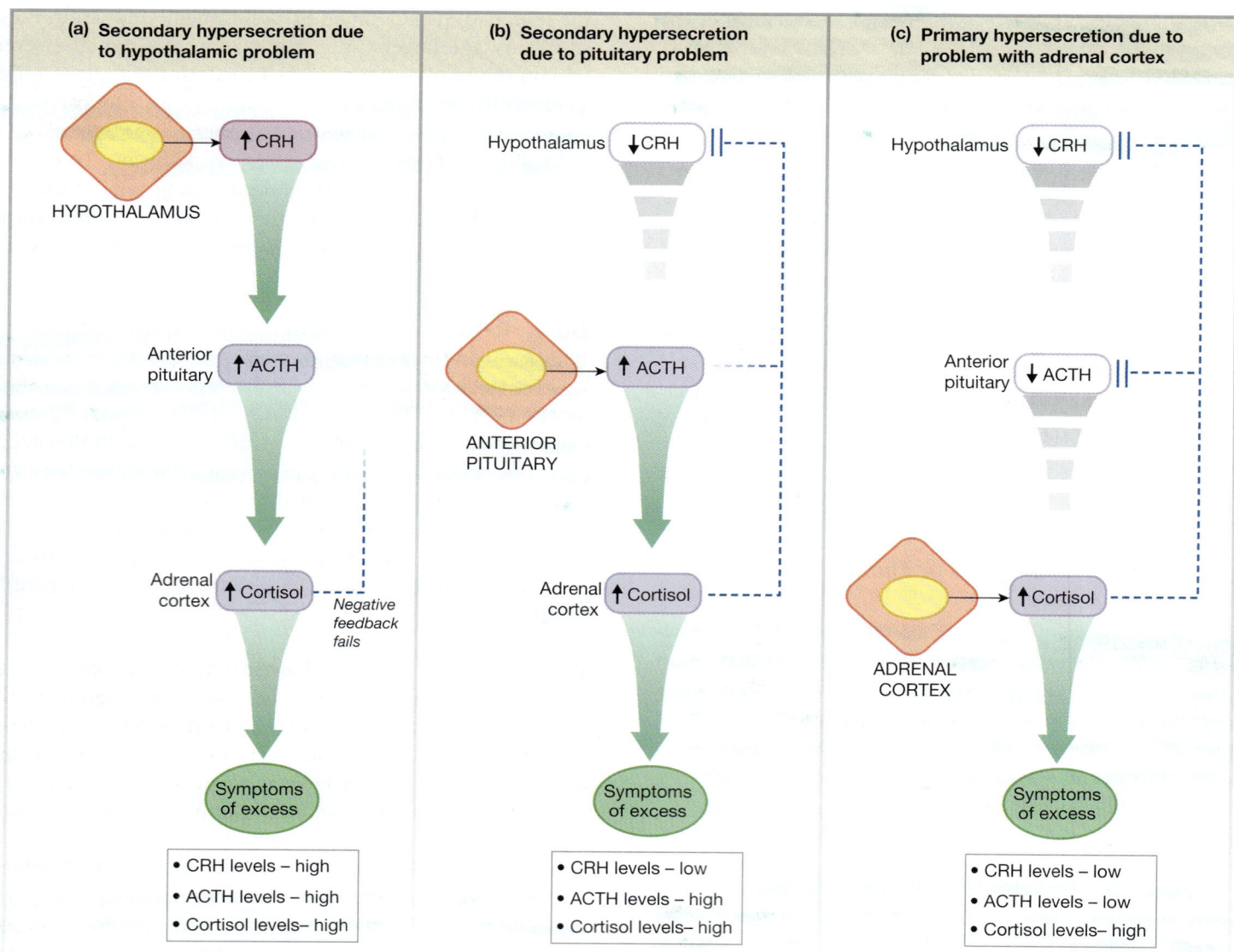

(a) Secondary hypersecretion due to hypothalamic problem

↑CRH

HYPOTHALAMUS

Anterior pituitary ↑ACTH

Adrenal cortex ↑Cortisol   *Negative feedback fails*

Symptoms of excess

- CRH levels – high
- ACTH levels – high
- Cortisol levels– high

(b) Secondary hypersecretion due to pituitary problem

Hypothalamus ↓CRH

↑ACTH

ANTERIOR PITUITARY

Adrenal cortex ↑Cortisol

Symptoms of excess

- CRH levels – low
- ACTH levels – high
- Cortisol levels– high

(c) Primary hypersecretion due to problem with adrenal cortex

Hypothalamus ↓CRH

Anterior pituitary ↓ACTH

↑Cortisol

ADRENAL CORTEX

Symptoms of excess

- CRH levels – low
- ACTH levels – low
- Cortisol levels– high

■ **FIGURE 7-20**   *Primary and secondary hypersecretion of cortisol*
When there is a pathology in an endocrine gland, negative feedback fails.

Diagnosis of Endocrine Pathologies Depends on the Complexity of the Reflex

Diagnosis of endocrine pathologies may be simple or complicated, depending on the complexity of the reflex. In the simplest endocrine reflex, such as that for parathyroid hormone, if there is too much or too little hormone, there is only one location where the problem can arise: the parathyroid glands (see Fig. 7-10). However, with complex hypothalamic-pituitary-endocrine gland reflexes, the diagnosis can be much more difficult.

If a pathology (deficiency or excess) arises in the last endocrine gland in a reflex, the problem is considered to be a **primary pathology**. For example, if a tumor in the adrenal cortex begins to produce excessive amounts of cortisol, the resulting condition is called *primary hypersecretion*. If dysfunction occurs in one of the tissues producing trophic hormones, the problem is a **secondary pathology**. For example, if the pituitary is damaged because of head trauma and ACTH secretion diminishes, the resulting cortisol deficiency is considered to be *secondary hyposecretion* of cortisol.

The diagnosis of pathologies in complex endocrine pathways depends on understanding negative feedback in the control pathway. Figure 7-20 ■ shows three possible causes of excess cortisol secretion. To determine which is the correct *etiology* (cause) of the disease in a particular patient, the clinician must assess the levels of the three hormones in the control pathway.

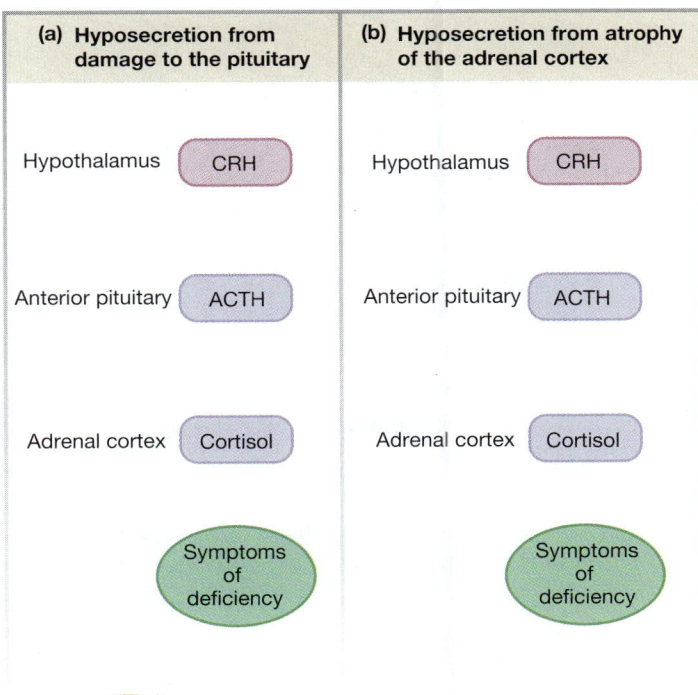

(a) Hyposecretion from damage to the pituitary

Hypothalamus — CRH

Anterior pituitary — ACTH

Adrenal cortex — Cortisol

Symptoms of deficiency

(b) Hyposecretion from atrophy of the adrenal cortex

Hypothalamus — CRH

Anterior pituitary — ACTH

Adrenal cortex — Cortisol

Symptoms of deficiency

FIGURE QUESTION

For each condition, use arrows to indicate whether levels of the three hormones in the pathway will be increased, decreased, or unchanged. Draw in negative feedback loops or indicate where feedback has failed.

■ **FIGURE 7-21** *Patterns of hormone secretion in hypocortisolism*

If the problem is overproduction of CRH by the hypothalamus (Fig. 7-20a), CRH levels will be higher than normal. High CRH in turn causes high ACTH, which in turn causes high cortisol. This is therefore secondary hypersecretion arising from a problem in the hypothalamus. In clinical practice, this type of pathology is very rare.

Figure 7-20b shows a secondary hypersecretion of cortisol due to an ACTH-secreting tumor of the pituitary. Once again, the high levels of ACTH cause high cortisol production, but in this example the high cortisol level has a negative feedback effect on the hypothalamus, decreasing production of CRH. The combination of low CRH and high ACTH isolates the problem to the pituitary. This pathology is responsible for about two-thirds of cortisol hypersecretion *syndromes* [*syn-,* together + *-drome,* running; a combination of symptoms characteristic of a particular pathology].

If cortisol levels are high but levels of both trophic hormones are low, the problem must be a primary disorder (Fig. 7-20c). There are two possible explanations: *endogenous* [*endo-,* within + *-gen,* to be born] cortisol hypersecretion or the exogenous administration of cortisol for therapeutic reasons (see Fig. 7-19).

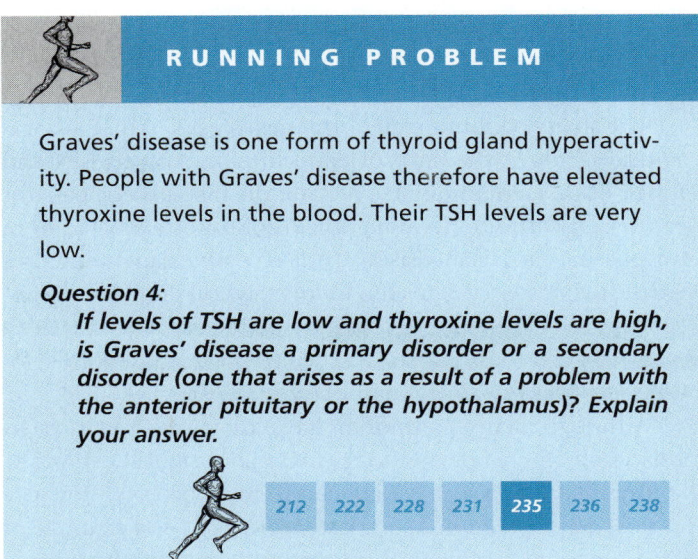

RUNNING PROBLEM

Graves' disease is one form of thyroid gland hyperactivity. People with Graves' disease therefore have elevated thyroxine levels in the blood. Their TSH levels are very low.

Question 4:
If levels of TSH are low and thyroxine levels are high, is Graves' disease a primary disorder or a secondary disorder (one that arises as a result of a problem with the anterior pituitary or the hypothalamus)? Explain your answer.

212   222   228   231   **235**   236   238

In either case, high levels of cortisol act as a negative feedback signal that shuts off production of CRH and ACTH. The pattern of high cortisol with low trophic hormone levels points to a primary disorder.

When the problem is an adrenal tumor that is secreting cortisol in an unregulated fashion, the normal control pathways are totally ineffective. Although negative feedback shuts off production of the trophic hormones, the tumor is not dependent on them for cortisol production, so cortisol secretion continues in their absence. The tumor must be removed or suppressed before cortisol secretion can be controlled.

Figure 7-21 ■ shows two possible etiologies for hyposecretion of cortisol. You can apply your understanding of negative feedback in the hypothalamic-pituitary control pathway to predict whether the levels of CRH, ACTH, and cortisol will be high or low in each case.

HORMONE EVOLUTION

Chemical signaling is an ancient method for communication and the maintenance of homeostasis. As scientists sequence the genomes of diverse species, they are discovering that in many cases hormone structure and function have changed amazingly little from the most primitive vertebrates through the mammals. In fact, hormone signaling pathways that were once considered exclusive to vertebrates, such as those for thyroid hormones and insulin, have now been shown to play physiological or developmental roles in invertebrates such as echinoderms and insects. This *evolutionary conservation* of hormone function is also demonstrated by the fact that some hormones from other organisms have biological activity when administered to humans. By studying which portions of a hormone molecule do not change from species to species, scientists have acquired important clues to aid in the design of agonist and antagonist drugs.

The ability of nonhuman hormones to work in humans was a critical factor in the birth of endocrinology. When Best and Banting discovered insulin in 1921 and the first diabetic patients were treated with the hormone, the insulin was extracted from cow, pig, or sheep pancreases. Until recently, slaughterhouses were the major source of insulin for the medical profession. Now, with genetic engineering, the human gene for insulin has been inserted into bacteria, which then synthesize the hormone, providing us with a plentiful source of human insulin.

Although many hormones have the same function in most vertebrates, a few hormones that play a significant role in the physiology of lower vertebrates seem to be evolutionarily "on their way out" in humans. Calcitonin is a good example of such a hormone. Although it plays a major role in calcium metabolism in fish, calcitonin apparently has no significant influence on daily calcium balance in adult humans. Neither calcitonin deficiency nor calcitonin excess is associated with any pathological condition or symptom.

Although calcitonin is not a significant hormone in humans, the calcitonin gene does code for a biologically active protein. Cells in the brain process calcitonin gene mRNA to make a peptide known as *calcitonin gene-related peptide* (CGRP), which apparently acts as a neurotransmitter. CGRP can act as a powerful dilator of blood vessels, and one recent study found that a CGRP receptor antagonist effectively treated migraine headaches, which are caused by cerebral blood vessel dilation (vasodilation). The ability of one gene to produce multiple peptides is one reason research is shifting from genomics to physiology and proteomics (the study of the role of proteins in physiological function) [⇄ p. 122].

Some endocrine structures that are important in lower vertebrates are *vestigial* [*vestigium,* trace] in humans, meaning that in humans these structures are present as minimally functional glands. For example, *melanocyte-stimulating hormone* (MSH) from the intermediate lobe of the pituitary controls pigmentation in reptiles and amphibians. However, adult humans have only a vestigial intermediate lobe and normally do not have measurable levels of MSH in their blood.

In the research arena, *comparative endocrinology*—the study of endocrinology in nonhuman organisms—has made significant contributions to our quest to understand the human body. Many of our models of human physiology are based on research carried out in fish or frogs or rats, to name a few. For example, the pineal gland hormone *melatonin* (Fig. 7-22 ■) was

RUNNING PROBLEM

Researchers have learned that Graves' disease is an autoimmune disorder in which the body fails to recognize its own tissue. In this condition, the body produces antibodies that mimic TSH and bind to the TSH receptor, turning it on. This false signal "fools" the thyroid gland into overproducing thyroid hormone. More women than men are diagnosed with Graves' disease, perhaps because of the influence of female hormones on thyroid function. Stress and other environmental factors have also been implicated in hyperthyroidism.

Question 5:
Antibodies are proteins that bind to the TSH receptor. From that information, what can you conclude about the cellular location of the TSH receptor?

Question 6:
In Graves' disease, why doesn't negative feedback shut off thyroid hormone production before it becomes excessive?

| 212 | 222 | 228 | 231 | 235 | **236** | 238 |

discovered through research using tadpoles. Many small non-human vertebrates have short life cycles that facilitate studying aging or reproductive physiology. Genetically altered mice (transgenic or knockout mice) have provided researchers valuable information about proteomics.

Opponents of animal research argue that scientists should not experiment with animals at all and should use only cell cultures and computer models. Although cultures and models are valuable tools and can be helpful in the initial stages of medical research, at some point new drugs and procedures must be tested on intact organisms prior to clinical trials in humans. Responsible scientists follow guidelines for appropriate animal use and limit the number of animals killed to the minimum needed to provide valid data.

In this chapter we have examined how the endocrine system with its hormones helps regulate the slower processes in the body. In the next chapter, you will learn how the nervous system can take care of the more rapid responses needed to maintain homeostasis.

THE PINEAL GLAND

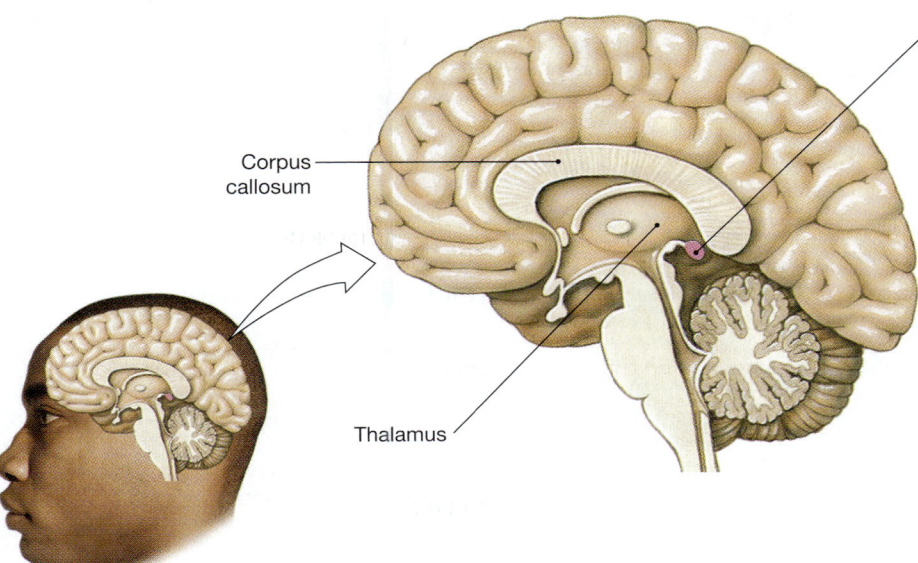

Corpus
callosum

Thalamus

The pineal gland is a pea-sized structure buried deep in the brain of humans. Nearly 2000 years ago, this "seat of the soul" was thought to act as a valve that regulated the flow of vital spirits and knowledge into the brain. By 1950, however, scientists had decided that it was a vestigial structure with no known function.

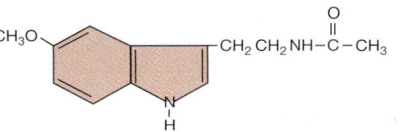

Melatonin is an amine hormone derived from the amino acid tryptophan.

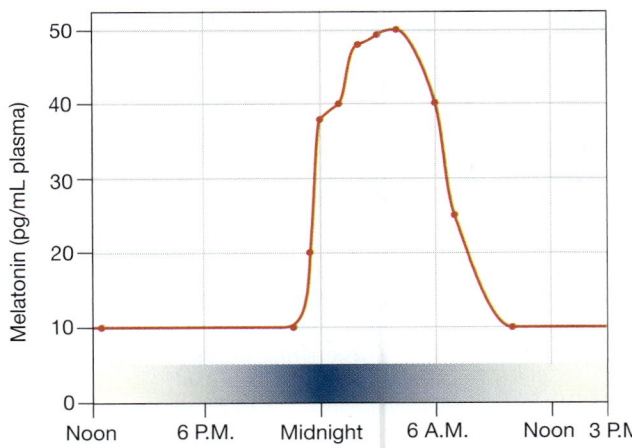

Melatonin is the "darkness hormone," secreted at night as we sleep. It is the chemical messenger that transmits information about light-dark cycles to the brain center that governs the body's biological clock.

(Adapted from J. Arendt, *Clin. Endocrinol.* 29: 205–229, 1988.)

About 1957 one of the wonderful coincidences of scientific research occurred. An investigator heard about a factor in beef pineal glands that could lighten the skin of amphibians. Using the classical methodology of endocrinology, he obtained pineal glands from a slaughterhouse and started making extracts. His biological assay consisted of dropping pineal extracts into bowls of live tadpoles to see if their skin color blanched. Several years and hundreds of thousands of pineal glands later, he had isolated a small amount of melatonin.

Ten years ago scientists and the popular press were trying to link melatonin to sexual function, the onset of puberty, seasonal affective depressive disorder (SADD) in the darker winter months, and sleep-wake cycles. At that time, the only function supported with scientific evidence was the hormone's ability to help shift timing of the body's internal clock, making it useful in overcoming jet lag. Now there is evidence that melatonin is a powerful antioxidant that has the potential to protect the body from the damage caused by free radicals. For a review, see "Melatonin: lowering the high price of free radicals," *News in Physiological Sciences* 15: 246–250, Oct. 2000 (*http://nips.physiology.org*).

■ **FIGURE 7-22** *The pineal gland and melatonin*

RUNNING PROBLEM CONCLUSION

GRAVES' DISEASE

In this running problem, you learned that in Graves' disease, thyroid hormone levels are high because an immune-system protein mimics TSH. You also learned that the thyroid gland concentrates iodine for synthesis of thyroid hormones and that radioactive iodine can concentrate in the gland and destroy the thyroid cells.

Graves' disease is the most common form of hyperthyroidism. Other famous people who have suffered from it include former U.S. President George H. W. Bush and First Lady Barbara Bush. To learn more about Graves' disease and other thyroid conditions, check out the Thyroid Foundation of America website at *www.tsh.org*. Check your answers to the problem questions by comparing them to the information in the summary table below.

	QUESTION	FACTS	INTEGRATION AND ANALYSIS
1a	To which of the three classes of hormones do thyroid hormones belong?	The three classes of hormones are peptides, steroids, and amines.	Thyroid hormones are made from the amino acid tyrosine; therefore, they are amines.
1b	If a person's diet is low in iodine, predict what happens to thyroxine production.	The thyroid gland concentrates iodine and combines it with the amino acid tyrosine to make thyroid hormones.	If iodine is lacking in the diet, a person will be unable to make thyroid hormones.
2a	In a normal person, when thyroid hormone levels in the blood increase, will negative feedback increase or decrease the secretion of TSH?	Negative feedback shuts off response loops.	Normally negative feedback will decrease TSH secretion.
2b	In a person with a hyperactive gland that is producing too much thyroid hormone, would you expect the level of TSH to be higher or lower than in a normal person?		If thyroid hormone is high, you would expect strong negative feedback and even lower levels of TSH.
3	Why is radioactive iodine (rather than some other radioactive element, such as cobalt) used to destroy thyroid tissue?	The thyroid gland concentrates iodine to make thyroid hormones.	Radioactive iodine will be concentrated in the thyroid gland and therefore will selectively destroy that tissue. Other radioactive elements would distribute more widely throughout the body and might harm normal tissues.
4	If levels of TSH are low and thyroxine levels are high, is Graves' disease a primary disorder or a secondary disorder (one that arises as a result of a problem with the anterior pituitary or the hypothalamus)? Explain your answer.	In secondary hypersecretion disorders, you would expect the levels of the hypothalamic and/or anterior pituitary trophic hormones to be elevated.	In Graves' disease, TSH from the anterior pituitary is very low. Therefore, the oversecretion of thyroid hormones is not the result of elevated TSH. This means that Graves' disease is a primary disorder that is caused by a problem in the thyroid gland itself.
5	Antibodies are proteins that bind to the TSH receptor. From that information, what can you conclude about the cellular location of the TSH receptor?	Receptors may be membrane receptors or intracellular receptors. Proteins cannot cross the cell membrane.	The TSH receptor is a membrane receptor. It uses the cAMP second messenger pathway for signal transduction.
6	In Graves' disease, why doesn't negative feedback shut off thyroid hormone production before it becomes excessive?	In normal negative feedback, increasing levels of thyroid hormone shut off TSH secretion. Without TSH stimulation, the thyroid stops producing thyroid hormone.	In Graves' disease, high levels of thyroid hormone have shut off TSH production. However, the thyroid gland still produces hormone in response to the binding of antibody to the TSH receptor. In this situation, negative feedback has failed.

212   222   228   231   235   236   238

CHAPTER SUMMARY

This chapter introduced you to the endocrine system and the role it plays in *communication* and *control* of physiological processes. As you've seen before, the fact that the body is *compartmentalized* into intracellular and extracellular compartments means that special mechanisms are required to enable signals to pass from one compartment to the other. The chapter also presented basic patterns that you will encounter again as you study various organ systems: differences among the three chemical classes of hormones, reflex pathways for hormones, types of hormone interaction, and endocrine pathologies.

Hormones

IP Endocrine System: Endocrine System Review

1. The specificity of a hormone depends on its receptors and their associated signal transduction pathways. (p. 216)
2. A **hormone** is a chemical secreted by a cell or group of cells into the blood for transport to a distant target, where it is effective at very low concentrations. (p. 212)
3. **Pheromones** are chemical signals secreted into the external environment. (p. 216)
4. Hormones bind to receptors to initiate responses known as the **cellular mechanism of action**. (p. 217)
5. Hormone activity is limited by terminating secretion, removing hormone from the blood, or terminating activity at the target cell. (p. 217)
6. The rate of hormone breakdown is indicated by a hormone's **half-life**. (p. 217)

The Classification of Hormones

IP Endocrine System: Biochemistry, Secretion and Transport of Hormones, and the Actions of Hormones on Target Cells

7. There are three types of hormones: **peptide/protein hormones**, composed of three or more amino acids; **steroid hormones**, derived from cholesterol; and **amine hormones**, derived from either tyrosine (e.g., melatonin) or tryptophan (e.g., catecholamines and thyroid hormones). (p. 217)
8. Peptide hormones are made as inactive **preprohormones** and processed to **prohormones**. Prohormones are chopped into active hormone and peptide fragments that are co-secreted. (p. 218; Figs. 7-3, 7-4)
9. Peptide hormones dissolve in the plasma and have a short half-life. They bind to surface receptors on their target cells and initiate rapid cellular responses through signal transduction. In some instances, peptide hormones also initiate synthesis of new proteins. (p. 218; Fig. 7-5)
10. Steroid hormones are synthesized as they are needed. They are hydrophobic, and most steroid hormones in the blood are bound to protein carriers. Steroids have an extended half-life. (p. 220)
11. Traditional steroid receptors are inside the target cell, where they turn genes on or off and direct the synthesis of new proteins. Cell response is slower than with peptide hormones. Steroid hormones may bind to membrane receptors and have nongenomic effects. (p. 221; Fig. 7-7)

12. Amine hormones may behave like typical peptide hormones or like a combination of a steroid hormone and a peptide hormone. (p. 222; Fig. 7-8)

Control of Hormone Release

IP Endocrine System: The Hypothalamic-Pituitary Axis

13. Classic endocrine cells act as both sensor and integrating center in the simple reflex pathway. (p. 223; Fig. 7-10)
14. Many endocrine reflexes involve the nervous system, either through **neurohormones** or through neurons that influence hormone release. (p. 225)
15. The pituitary gland is composed of the anterior pituitary (a true endocrine gland) and the posterior pituitary (an extension of the brain). (p. 225; Fig. 7-11)
16. The posterior pituitary releases two neurohormones, oxytocin and vasopressin, that are made in the hypothalamus. (p. 226; Fig. 7-12)
17. **Trophic hormones** control the secretion of other hormones. (p. 226)
18. Anterior pituitary hormones are controlled by releasing hormones and inhibiting hormones from the hypothalamus. (p. 226; Fig. 7-13)
19. In complex endocrine reflexes, hormones of the pathway act as negative feedback signals. (p. 227; Fig. 7-14)
20. The hypothalamic trophic hormones reach the pituitary through the **hypothalamic-hypophyseal portal system**. (p. 229; Fig. 7-16)
21. There are six anterior pituitary hormones: prolactin, growth hormone, follicle-stimulating hormone, luteinizing hormone, thyroid-stimulating hormone, and adrenocorticotrophic hormone. (p. 226; Fig. 7-13)

Hormone Interactions

22. If the combination of two or more hormones yields a result that is greater than additive, the interaction is **synergism**. (p. 231; Fig. 7-18)
23. If one hormone cannot exert its effects fully unless a second hormone is present, the second hormone is said to be **permissive** to the first. (p. 231)
24. If one hormone opposes the action of another, the two are **antagonistic** to each other. (p. 232)

Endocrine Pathologies

25. Diseases of hormone excess are usually due to **hypersecretion**. Symptoms of hormone deficiency occur when too little hormone is secreted (**hyposecretion**). **Abnormal tissue responsiveness** may result from problems with hormone receptors or signal transduction pathways. (p. 232)
26. **Primary pathologies** arise in the last endocrine gland in a reflex. A **secondary pathology** is a problem with one of the tissues producing trophic hormones. (p. 234; Fig. 7-20)

Hormone Evolution

27. Many human hormones are similar to hormones found in other vertebrate animals. (p. 235)

QUESTIONS

(Answers to the Review Questions begin on page A1.)

THE PHYSIOLOGY PLACE

Access more review material online at **The Physiology Place** website. There you'll find review questions, problem-solving activities, case studies, flashcards, and direct links to both *InterActive Physiology*® and *PhysioEx™*. To access the site, go to *www.physiologyplace.com* and select Human Physiology, Fourth Edition.

LEVEL ONE   REVIEWING FACTS AND TERMS

1. The study of hormones is called *endocrinology*

2. List the three basic ways hormones act on their target cells.

3. List five endocrine glands, and name one hormone secreted by each. Give one effect of each hormone you listed.

4. Match the following researchers with their experiments:

   (a) Lower
   (b) Berthold
   (c) Guillemin and Shalley
   (d) Brown-Séquard
   (e) Banting and Best

   1. isolated trophic hormones from the hypothalami of pigs and sheep
   2. claimed sexual rejuvenation after he injected himself with testicular extracts
   3. isolated insulin
   4. accurately described the function of the pituitary gland
   5. studied comb development in castrated roosters

5. Put the following steps for identifying an endocrine gland in order:

   (a) Purify the extracts and separate the active substances.
   (b) Perform replacement therapy with the gland or its extracts and see if the abnormalities disappear.
   (c) Implant the gland or administer the extract from the gland to a normal animal and see if symptoms characteristic of hormone excess appear.
   (d) Put the subject into a state of hormone deficiency by removing the suspected gland, and monitor the development of abnormalities.

6. For a chemical to be defined as a hormone, it must be secreted into the *blood* for transport to a *distant target* and take effect at *very low* concentrations.

7. What is meant by the term *half-life* in connection with the activity of hormone molecules?

8. Metabolites are inactivated hormone molecules, broken down by enzymes found primarily in the_____and_____, to be excreted in the_____and_____, respectively.

9. Candidate hormones often have the word_____as part of their name.

10. List and define the three chemical classes of hormones. Name one hormone in each class.

11. Decide if each of the following characteristics applies best to peptide hormones, steroid hormones, amine hormones, all of these, or none of these.

    (a) are lipophobic and must use a signal transduction system
    (b) have a short half-life, measured in minutes
    (c) often have a lag time of 90 minutes before effects are noticeable

    (d) are water-soluble, and thus easily dissolve in the extracellular fluid for transport
    (e) most hormones belong to this class
    (f) are all derived from cholesterol
    (g) consist of three or more amino acids linked together
    (h) are released into the blood to travel to a distant target organ
    (i) are transported in the blood bound to protein carrier molecules
    (j) are all lipophilic, so diffuse easily across membranes

12. Why do steroid hormones usually take so much longer to act than peptide hormones?

13. When steroid hormones act on a cell nucleus, the hormone-receptor complex acts as a/an *transcription* factor, binds to DNA, and activates one or more *genes*, which create mRNA to direct the synthesis of new *proteins*.

14. Researchers have discovered that some cells have additional steroid hormone receptors on their_____, enabling a faster response.

15. The amino acids that give rise to the amine hormones are_____, the basis for melatonin, and_____, from which the catecholamines and thyroid hormones are made.

16. A hormone that controls the secretion of another hormone is known as a *trophic* hormone.

17. In reflex control pathways involving trophic hormones and multiple integrating centers, the hormones themselves act as *feed-back* signals, suppressing trophic hormone secretion earlier in the reflex.

18. What characteristic defines neurohormones?

19. List the two hormones secreted by the posterior pituitary gland. To what chemical class do they belong?

20. What is the hypothalamic-hypophyseal portal system? Why is it important?

21. List the six hormones of the anterior pituitary gland; give an action of each. Which ones are trophic hormones?

22. How do long-loop negative feedback and short-loop negative feedback differ? Give an example of each type in the body's endocrine system.

23. When two hormones work together to create a result that is greater than additive, that interaction is called_____. When two hormones must both be present to achieve full expression of an effect, that interaction is called_____. When hormone activities oppose each other, that effect is called_____.

LEVEL TWO   REVIEWING CONCEPTS

24. Map the following groups of terms. Add additional terms if you like.

List 1	List 2
co-secretion	ACTH
endoplasmic reticulum	anterior pituitary
exocytosis	blood
Golgi complex	endocrine cell
hormone receptor	gonadotropins
peptide hormone	growth hormone
preprohormone	hypothalamus
prohormone	inhibiting hormone
secretory vesicle	neurohormone
signal sequence	neuron

synthesis

target cell response

oxytocin

peptide/protein

peripheral endocrine gland

portal system

posterior pituitary

prolactin

releasing hormone

trophic hormone

TSH

vasopressin

25. Compare and contrast the terms in each of the following sets:

    (a) paracrine, hormone, cytokine

    (b) primary and secondary endocrine pathologies

    (c) hypersecretion and hyposecretion

    (d) anterior and posterior pituitary

26. Compare and contrast the three chemical classes of hormones.

LEVEL THREE   PROBLEM SOLVING

27. You encountered the terms *specificity, receptors,* and *down-regulation* in other chapters in this text. Do their meanings change when applied to the endocrine system? What chemical and physical characteristics do hormones, enzymes, transport proteins, and receptors have in common that makes specificity important?

28. Dexamethasone is a drug used to suppress the secretion of adrenocorticotrophic hormone (ACTH) from the anterior pituitary. Two patients with hypersecretion of cortisol are given dexamethasone. Patient A's cortisol secretion level falls to normal levels as a result, but patient B's cortisol secretion level remains elevated. Draw maps of the reflex pathways for these two patients (see Fig. 7-15 for a template) and use the maps to determine which patient has primary hypercortisolism. Explain your reasoning.

29. Some early experiments for male birth control pills used drugs that suppressed gonadotropin (FSH and LH) release. However, men given these drugs stopped taking them because the drugs decreased testosterone secretion, which decreased the men's sex drive and caused impotence.

    (a) Use the information given in Figure 7-13 to draw the GnRH-FSH/LH-testosterone reflex pathway. Use the pathway to show how suppressing gonadotropins decreases sperm production and testosterone secretion.

    (b) Researchers subsequently suggested that a better treatment would be to give men extra testosterone. Draw another copy of the reflex pathway to show how testosterone could suppress sperm production without the side effect of impotence.

LEVEL FOUR   QUANTITATIVE PROBLEMS

30. The following graph represents the disappearance of a drug from the blood as the drug is metabolized and excreted. Based on the graph, what is the half-life of the drug?

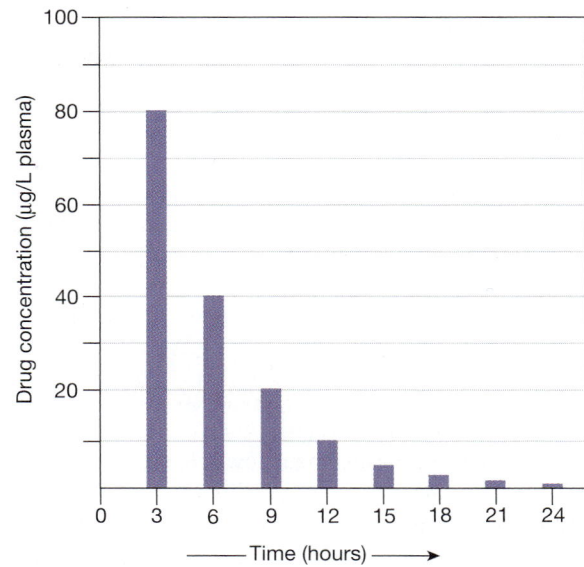

31. The following graph shows plasma TSH concentration in three groups of subjects. Pattern A is the normal subjects. Which pattern would be consistent with the following pathologies? Explain your reasoning.

    (a) primary hypothyroidism

    (b) primary hyperthyroidism

    (c) secondary hyperthyroidism

32. Based on what you have learned about the pathway for insulin secretion, draw and label a graph showing the effect of plasma glucose concentration on insulin secretion.

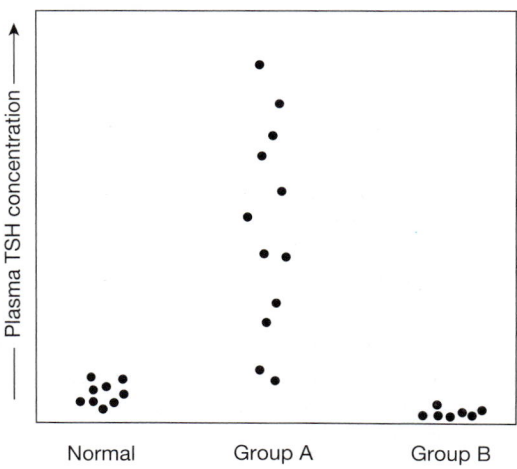

ANSWERS

✓ Answers to Concept Check Questions

Page 217

1. Glucose enters cells by facilitated diffusion (GLUT transporters).

Page 217

2. The suffix -*ase* indicates an enzyme. A *peptidase* digests peptides.

Page 217

3. A hormone is a chemical that is secreted into the blood and acts on a distant target in very low concentrations.

4. A steroid-producing cell would have extensive smooth endoplasmic reticulum; a protein-producing cell would have lots of rough endoplasmic reticulum.

Page 221

5. The three chemical classes of hormones are peptide, steroid, and amine.

6. The short half-life suggests that aldosterone is not bound to plasma proteins as much as other steroid hormones are.

Page 223

7. The input signal is the stimulus, receptor, and afferent pathway. Integration of the signal takes place in the integrating center. The output signal is the efferent pathway.

8. Increased blood glucose is the stimulus. Insulin secretion is the efferent pathway; decrease in blood glucose is the response.

9. Insulin release by blood glucose is reflex 6. Insulin release triggered by a neural signal following a meal is reflex 3. Insulin release in response to a digestive hormone is closest to reflex 4, although the question does not indicate whether the hormone is neurosecretory. If it is not, there is no matching pathway in Figure 6-31.

Page 225

10. Stimulus: decreased Ca^{2+}; sensor/integrating center: parathyroid cells; efferent path: PTH; effector: target tissues; response: increased Ca^{2+}.

Page 225

11. Catecholamines are amine hormones.

Page 226

12. Microtubules of the cytoskeleton move secretory vesicles.

13. Contents of secretory vesicles are released by exocytosis.

Page 228

14. Pathway 5 in Figure 6-31 fits the complex pattern described.

Page 230

15. (a) Pathway 4; (b) pathway 4 for GH acting directly on targets, and pathway 5 for GH acting on the liver.

Page 230

16. The target is endocrine cells of the anterior pituitary.

17. The pathway shuts off when food is no longer present in the intestine to create stretch. A decrease in blood glucose can also serve as a negative feedback signal.

Page 233

18. Cortisol level is low; low negative feedback makes CRH and ACTH levels high.

Q Answers to Figure Questions

Page 223

Fig. 7-8: The conversion of tyrosine to dopamine adds a hydroxyl (—OH) group to the 6-carbon ring and changes the carboxyl (—COOH) group to a hydrogen. Norepinephrine is made from dopamine by changing one hydrogen to a hydroxyl group. Epinephrine is made from norepinephrine by changing one hydrogen attached to the nitrogen to a methyl (—CH$_3$) group.

Page 224

Fig. 7-9: The pathway begun by eating a meal shuts off when the stretch stimulus disappears as the meal is digested and absorbed from the digestive tract.

Page 228

Fig. 7-15: In short-loop negative feedback, ACTH inhibits CRH.

Page 235

Fig. 7-21: (a) CRH high, ACTH low, cortisol low. No negative feedback loops are functioning. (b) CRH normal/high, ACTH high, cortisol low. Absence of negative feedback by cortisol increases trophic hormones. Short-loop negative feedback from ACTH may keep CRH within the normal range.

8

The future of clinical neurology and psychiatry is intimately tied to that of molecular neural science.

—**Eric R. Kandel**, **James H. Schwartz**, and **Thomas M. Jessell**, *in the preface to their book,* **Principles of Neural Science,** *2000*

Axons (purple) wrapped in Schwann cells (red). Green dots indicate a specific protein component of myelin.

Neurons: Cellular and Network Properties

BACKGROUND BASICS

RUNNING PROBLEM

MYSTERIOUS PARALYSIS

"Like a polio ward from the 1950s" is how Guy McKhann, M.D., a neurology specialist at the Johns Hopkins School of Medicine, describes a ward of Beijing Hospital that he visited on a trip to China in 1986. Dozens of paralyzed children— some attached to respirators to assist their breathing—filled the ward to overflowing. The Chinese doctors thought the children had Guillain-Barré syndrome (GBS), a rare paralytic condition, but Dr. McKhann wasn't convinced. There were simply too many stricken children for the illness to be the rare Guillain-Barré syndrome. Was it polio—as some of the Beijing staff feared? Or was it another illness, perhaps one that had not yet been discovered?

| 244 | 246 | 248 | 268 | 270 | 275 | 283 | 285 |

In an eerie scene from a science fiction movie, white-coated technicians move quietly through a room filled with bubbling cylindrical fish tanks. As the camera zooms in on one tank, no fish are seen darting through aquatic plants. The lone occupant of the tank is a gray mass with a convoluted surface like a walnut and a long tail that appears to be edged with beads. Floating off the beads are hundreds of fine fibers, waving softly as the oxygen bubbles weave through them. This is no sea creature. . . . It is a brain and spinal cord, removed from its original owner and awaiting transplantation into another body. Can this be real? Is this scenario possible? Or is it just the creation of an imaginative movie screenwriter?

The brain is regarded as the seat of the soul, the mysterious source of those traits that we think of as setting humans apart from animals. The brain and spinal cord are also integrating centers for homeostasis, movement, and many other body functions. They are the control center of the **nervous system**, a network of billions or trillions of nerve cells linked together in a highly organized manner to form the rapid control system of the body.

Nerve cells, or **neurons**, are designed to carry electrical signals rapidly and, in some cases, over long distances. They are uniquely shaped cells, and most have long, thin extensions, or **processes**, that can extend up to a meter in length. In most pathways, neurons release chemical signals, called **neurotransmitters**, into the extracellular fluid. In a few pathways, neurons are linked by *gap junctions* [🔁 p. 175], allowing electrical signals to pass directly from cell to cell.

Using electrical signals to release chemicals from a cell is not unique to neurons, as you learned in Chapter 5 [🔁 p. 166].

Single-celled protozoa and plants also employ electrical signaling mechanisms, in many cases using the same types of ion channels as vertebrates do. Scientists sequencing ion channel proteins have found that many of these proteins have been highly conserved during evolution, an indication of their fundamental importance.

What is unique to the nervous system is its sophisticated arrangement of interconnected networks. Reflex pathways in the nervous system do not necessarily follow a straight line from one neuron to the next. One neuron may influence multiple neurons, or many neurons may affect the function of a single neuron. The intricacy of neural networks underlies the **emergent properties** of the nervous system: complex processes, such as consciousness, intelligence, and emotion, that cannot be predicted from what we know about the properties of individual nerve cells. The search to explain emergent properties makes neuroscience one of the most active research areas in physiology today.

Neuroscience, like many other areas of science, has its own specialized language. In many instances, multiple terms describe a single structure or function, which potentially can lead to confusion. Table 8-1 ■ lists some neuroscience terms used in this book, along with their common synonyms.

TABLE 8-1	Synonyms in Neuroscience
TERM USED IN THIS BOOK	**SYNONYM(S)**
Action potential	Spike, nerve impulse, conduction signal
Autonomic nervous system	Visceral nervous system
Axon	Nerve fiber
Axonal transport	Axoplasmic flow
Axon terminal	Synaptic knob, synaptic bouton, presynaptic terminal
Axoplasm	Cytoplasm of an axon
Cell body	Cell soma, nerve cell body
Cell membrane of an axon	Axolemma
Glial cells	Neuroglia, glia
Interneuron	Association neuron
Rough endoplasmic reticulum	Nissl substance, Nissl body
Sensory neuron	Afferent neuron, afferent

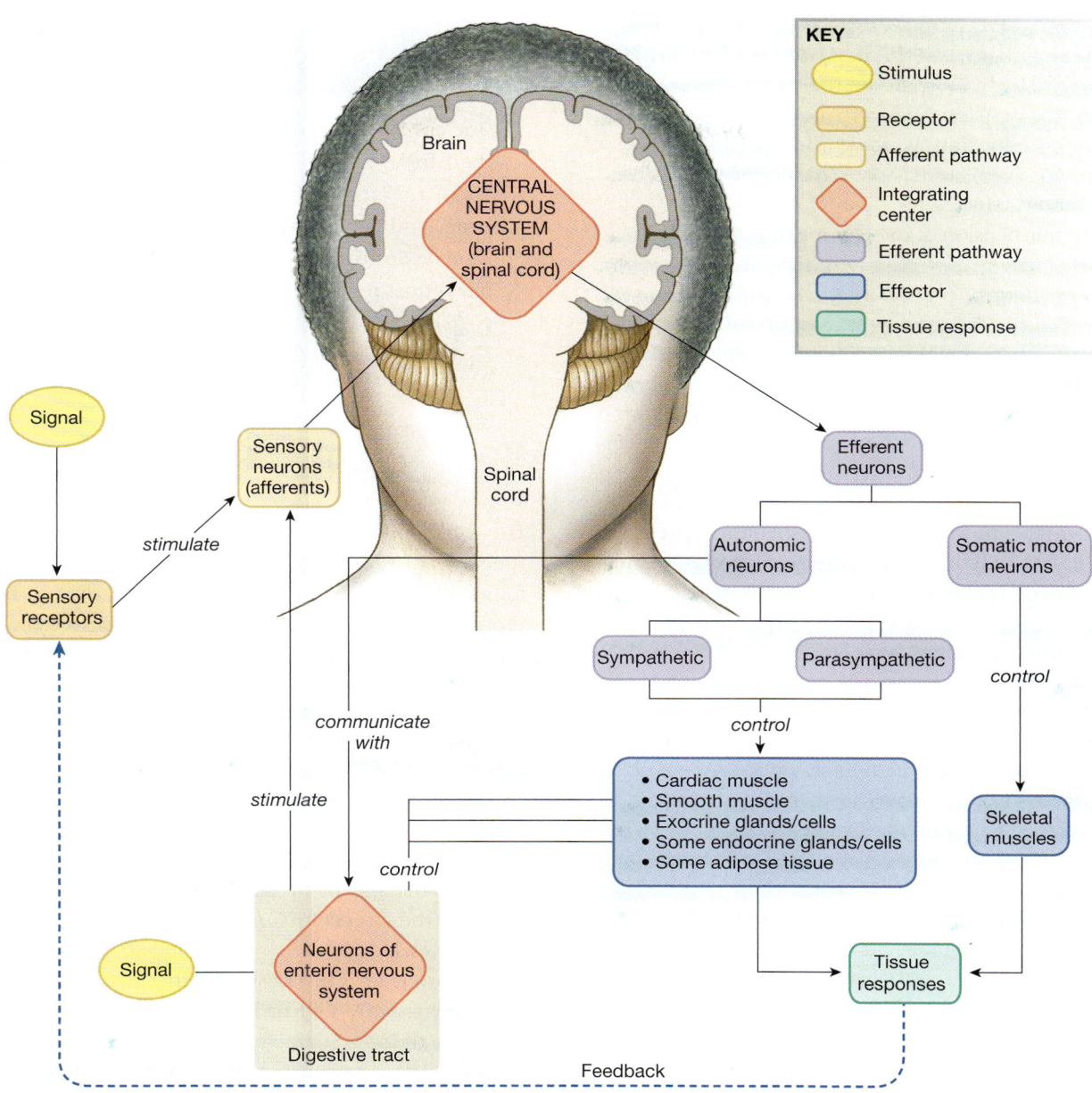

KEY

- Stimulus
- Receptor
- Afferent pathway
- Integrating center
- Efferent pathway
- Effector
- Tissue response

Brain

CENTRAL NERVOUS SYSTEM (brain and spinal cord)

Signal

Sensory neurons (afferents)

stimulate

Spinal cord

Sensory receptors

Efferent neurons

Autonomic neurons

Somatic motor neurons

Sympathetic

Parasympathetic

control

communicate with

control

stimulate

- Cardiac muscle
- Smooth muscle
- Exocrine glands/cells
- Some endocrine glands/cells
- Some adipose tissue

Skeletal muscles

control

Signal

Neurons of enteric nervous system

Digestive tract

Tissue responses

Feedback

■ **FIGURE 8-1** *Organization of the nervous system*

The peripheral nervous system (PNS) sends information to the central nervous system (CNS) through afferent (sensory) neurons and takes information from the CNS to target cells via efferent neurons. The enteric nervous system can act autonomously or can be controlled by the CNS through the autonomic division of the PNS.

ORGANIZATION OF THE NERVOUS SYSTEM

The nervous system can be divided into two parts. The **central nervous system (CNS)** consists of the **brain** and the **spinal cord**. The **peripheral nervous system (PNS)** consists of **afferent** (or **sensory**) **neurons** and **efferent neurons**. Information flow through the nervous system follows the basic reflex pattern described in Chapter 6 [⊞ p. 195].

Sensory receptors throughout the body continuously monitor conditions in both the internal and external environments (Fig. 8-1 ■). These receptors send information along afferent neurons to the central nervous system. Afferent pathways and their associated sensory receptors will be discussed in Chapter 10.

The CNS is the integrating center for neural reflexes. CNS neurons integrate information that arrives from the afferent branch of the PNS and determine whether a response is needed. Chapter 9 examines the complex structure and functions of the CNS.

The CNS then sends output signals directing an appropriate response (if any) that travel through efferent neurons to the effector cells of the body. Efferent neurons are subdivided into the **somatic motor division**,* which controls skeletal muscles, and the **autonomic division**, which controls smooth and cardiac muscles, exocrine glands, some endocrine glands, and some types of adipose tissue.

The autonomic division of the PNS is also called the *visceral nervous system* because it controls contraction and secretion in the various internal organs [*viscera*, internal organs]. Autonomic neurons are further divided into **sympathetic** and **parasympathetic branches**, which can be distinguished by their anatomical organization and by the chemicals they use to communicate with their target cells. Many internal organs receive innervation from both types of autonomic neurons, and it is a common pattern to find that the two divisions exert *antagonistic control* over a single target [p. 192]. The PNS will be discussed in Chapter 11.

In recent years, a third division of the nervous system has received considerable attention. The **enteric nervous system** is a network of neurons in the walls of the digestive tract. It is frequently controlled by the autonomic division of the nervous system, but it is also able to function autonomously as its own integrating center. We will discuss the enteric nervous system further in Chapter 21 on the digestive system.

Although most of this book will focus on the role of neural reflexes in communication, coordination, and homeostasis in the body, it is important to note that significant processes in the central nervous system can take place without input or output from the peripheral nervous system. The CNS has the ability to initiate activity without sensory input, and it need not create any measurable output. Two examples are thinking and dreaming, complex higher-brain functions that can take place totally within the CNS.

✓ CONCEPT CHECK

1. Organize the following terms describing functional types of neurons into a map or outline: afferent, autonomic, brain, central, efferent, enteric, parasympathetic, peripheral, sensory, somatic motor, spinal, sympathetic.

Answers: p. 289

CELLS OF THE NERVOUS SYSTEM

The nervous system is composed primarily of two cell types: neurons—the basic signaling units of the nervous system—and support cells known as glial cells (or glia or neuroglia).

Neurons Are Excitable Cells That Generate and Carry Electrical Signals

The neuron is the functional unit of the nervous system. (A *functional unit* is the smallest structure that can carry out the

---

*The expression *motor neuron* can be used to refer to all efferent neurons. However, clinically, the term *motor neuron* (or *motoneuron*) is usually used to describe somatic motor neurons that control skeletal muscles.

RUNNING PROBLEM

Guillain-Barré syndrome is a rare paralytic condition that strikes after a viral infection or an immunization. There is no cure, but usually the paralysis slowly disappears, and lost sensation slowly returns. In classic Guillain-Barré, patients can neither feel sensations nor move their muscles.

Question 1:
Which division(s) of the nervous system may be involved in Guillain-Barré syndrome (GBS)?

| 244 | **246** | 248 | 268 | 270 | 275 | 283 | 285 |

functions of a system.) Neurons are uniquely shaped cells with long processes that extend outward from the cell body. These processes are usually classified as either **dendrites** (which receive incoming signals) or **axons** (which carry outgoing information). The shape, number, and length of axons and dendrites vary from one neuron to the next, but these structures are an essential feature that allows neurons to communicate with one another and with other cells.

Traditionally we use efferent neurons similar to the model neurons shown in Figure 8-2 ■ and Figure 8-3e ■ to teach how a neuron functions. In other neuron types, the processes shown in the model may be missing or modified. However, if you understand these model neurons, you will understand how most neurons work.

Neurons may be classified either structurally or functionally. Structurally, neurons are classified by the number of processes that originate from the cell body. They may be described as *pseudounipolar* (axon and dendrites fuse during development to create one long process; Fig. 8-3a), *bipolar* (single axon and single dendrite; Fig. 8-3b), *multipolar* (many dendrites and branched axons; Fig. 8-3d), or *anaxonic* (lacking an identifiable axon; Fig. 8-3c). Because physiology is concerned chiefly with function, however, we will classify neurons not structurally but rather according to function: sensory (afferent) neurons, interneurons, and efferent (somatic motor and autonomic) neurons.

Sensory neurons carry information about temperature, pressure, light, and other stimuli from sensory receptors to the CNS. They vary from the model neuron in the length and organization of their processes. For example, peripheral sensory neurons have cell bodies that are found close to the CNS, with very long processes that extend to receptors in the limbs and internal organs. In these neurons, the cell body is out of the direct path of signals passing along the axon (Fig. 8-3a). In contrast, sensory neurons for smell and vision, with receptors located

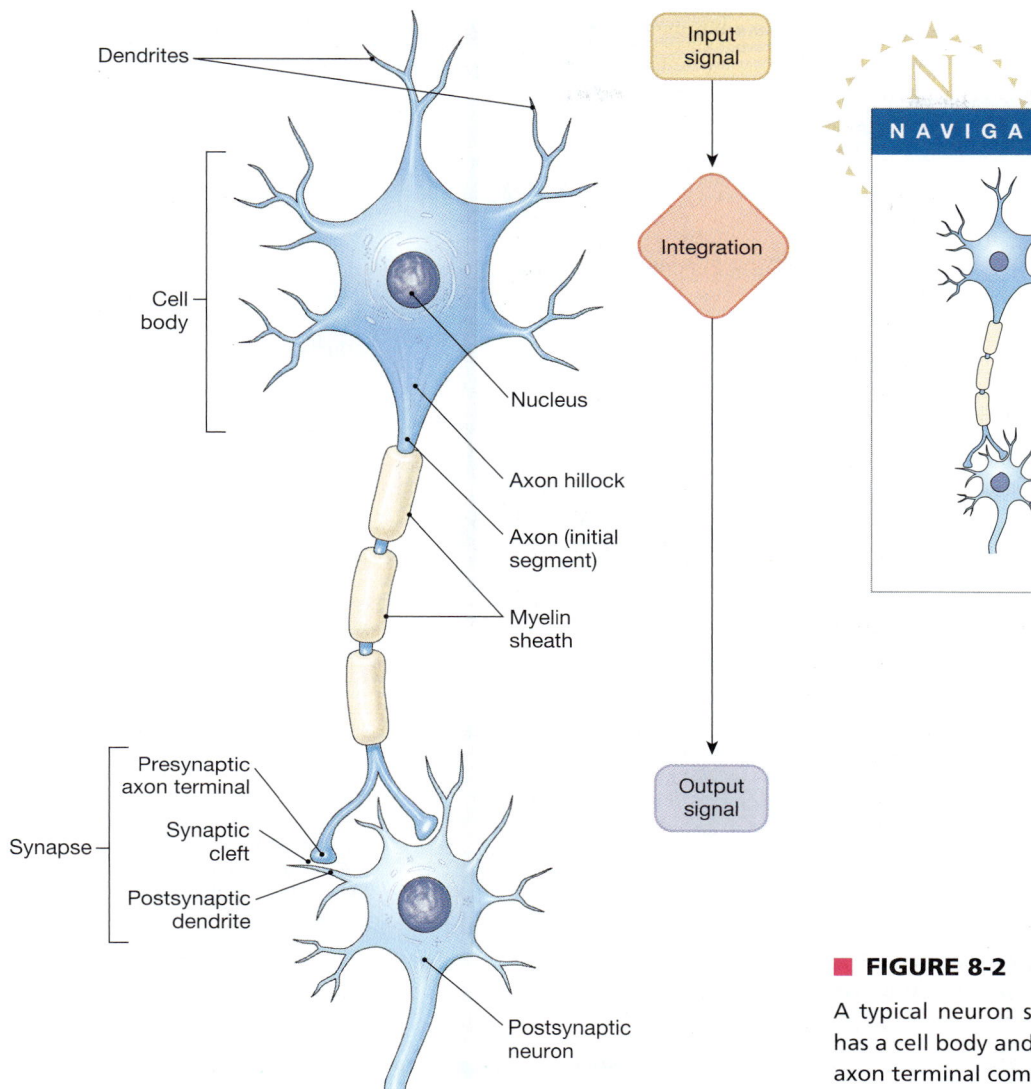

Input
signal

Integration

Output
signal

N

NAVIGATOR

This icon represents the different components of typical neurons. Look for it in the figures that follow to help you navigate your way through the structure and function of the nervous system.

8

■ **FIGURE 8-2**   *Model neuron*

A typical neuron such as the presynaptic neuron shown here has a cell body and numerous extensions. The region where an axon terminal communicates with its postsynaptic target cell is known as a synapse.

very close to the CNS, have two long processes so that signals that begin at the dendrites travel through the cell body to the axon (Fig. 8-3b).

Neurons that lie entirely within the CNS are known as **interneurons** (short for *interconnecting neurons*). They come in a variety of forms but often have quite complex branching processes that allow them to communicate with many other neurons (Fig. 8-3c). Some interneurons are quite small compared to the model neuron.

Efferent neurons, both somatic motor and autonomic, are generally very similar to the model neuron in Figure 8-2. In the autonomic division, some neurons have enlarged regions along the axon called **varicosities** (see Fig. 11-8). The varicosities store and release neurotransmitter.

The long axons of both afferent and efferent peripheral neurons are bundled together with connective tissue into cord-like fibers called **nerves** that extend from the CNS to the targets of the component neurons. Nerves may carry afferent signals

only (**sensory nerves**), efferent signals only (**motor nerves**), or signals in both directions (**mixed nerves**). Many nerves are large enough to be seen with the naked eye and have been given anatomical names. For example, the *phrenic nerve* runs from the spinal cord to the muscles of the diaphragm.

The Cell Body Is the Control Center of the Neuron   The **cell body** *(cell soma)* of a neuron resembles the typical cell described in Chapter 3, with a nucleus and all organelles needed to direct cellular activity [■ p. 59]. An extensive cytoskeleton extends outward into the axon and dendrites. The position of the cell body varies in different types of neurons, but in most neurons the cell body is small, generally making up one-tenth or less of the total cell volume.

Despite its small size, the cell body with its nucleus is essential to the well-being of the cell. If a neuron is cut apart, any sections separated from the cell body are likely to degenerate slowly and die because they lack the cellular machinery to

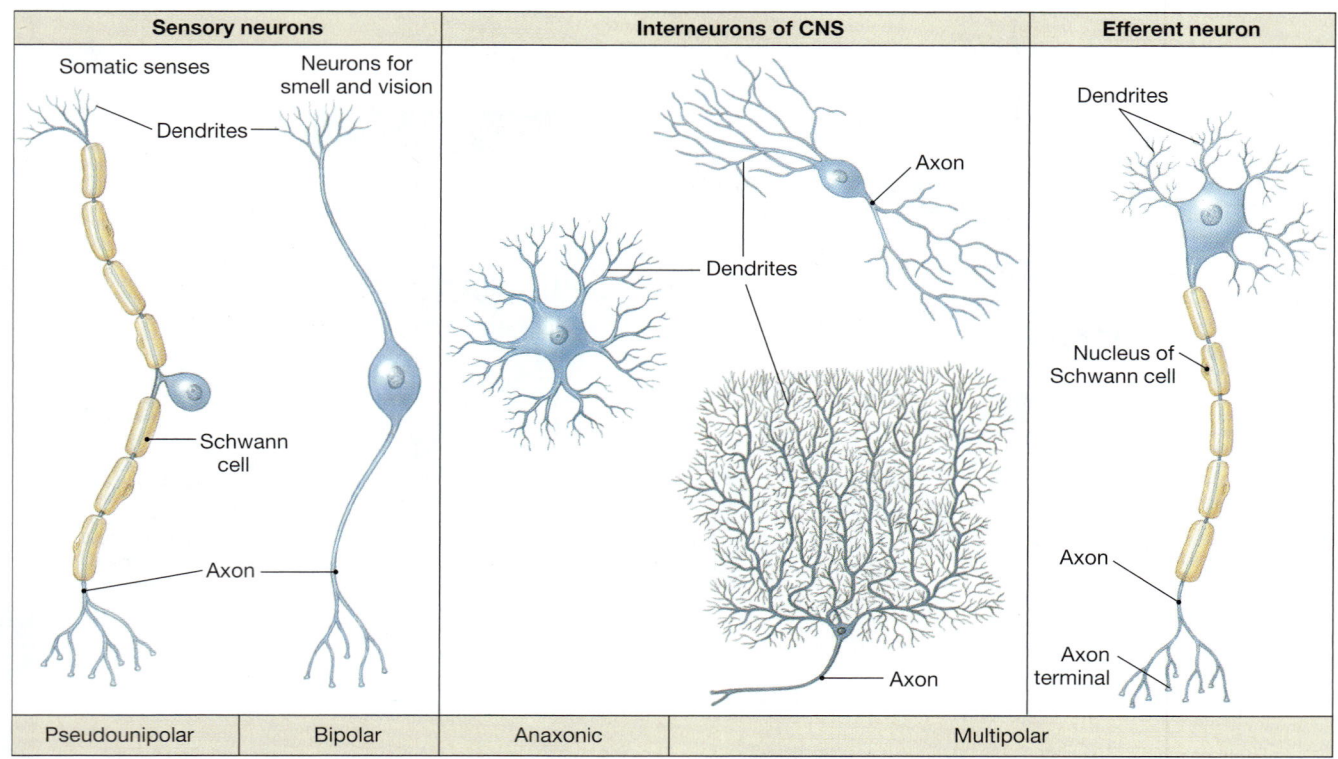

■ **FIGURE 8-3**   *Anatomic and functional categories of neurons*

Sensory neurons		Interneurons of CNS		Efferent neuron
Pseudounipolar	Bipolar	Anaxonic	Multipolar	

(a) Pseudounipolar neurons have a single process called the axon. During development, the dendrite fused with the axon.

(b) Bipolar neurons have two relatively equal fibers extending off the central cell body.

(c) Anaxonic CNS interneurons have no apparent axon.

(d) Multipolar CNS interneurons are highly branched but lack long extensions.

(e) A typical multipolar efferent neuron has five to seven dendrites, each branching four to six times. A single long axon may branch several times and end at enlarged axon terminals.

make essential proteins. In an injured somatic motor neuron whose axon has been severed, degeneration of the distal (distant) portions of the neuron will result in permanent paralysis of the muscles *innervated* by the neuron. (The term *innervated* means "controlled by a neuron.") If the damaged neuron is a sensory neuron, the person may experience loss of sensation (numbness or tingling) in the region previously innervated by the neuron. Scientists are trying to understand the cellular events that occur when neurons are damaged so that they can develop procedures to repair the functional loss that accompanies spinal cord and other neurological injuries.

Dendrites Receive Incoming Signals   Dendrites [*dendron,* tree] are thin, branched processes that receive incoming information from neighboring cells. Dendrites increase the surface area of a neuron, allowing it to communicate with multiple other neurons. The simplest neurons have only a single dendrite. At the other extreme, neurons in the brain may have multiple dendrites with incredibly complex branching (Fig. 8-3d). A dendrite's surface area can be expanded even more by the presence of **dendritic spines** that vary from thin spikes to mushroom-shaped knobs.

The primary function of dendrites in the peripheral nervous system is to receive incoming information and transfer it to an integrating region within the neuron. Within the CNS, dendrite function is more complex. Dendritic spines can function as independent compartments, sending signals back and

RUNNING PROBLEM

In classic Guillain-Barré syndrome, the disease affects both sensory and somatic motor neurons. Dr. McKhann observed that although the Beijing children could not move their muscles, they could feel a pin prick.

Question 2:
   Do you think the paralysis found in the Chinese children affected both sensory (afferent) and somatic motor neurons? Why or why not?

| 244 | 246 | **248** | 268 | 270 | 275 | 283 | 285 |

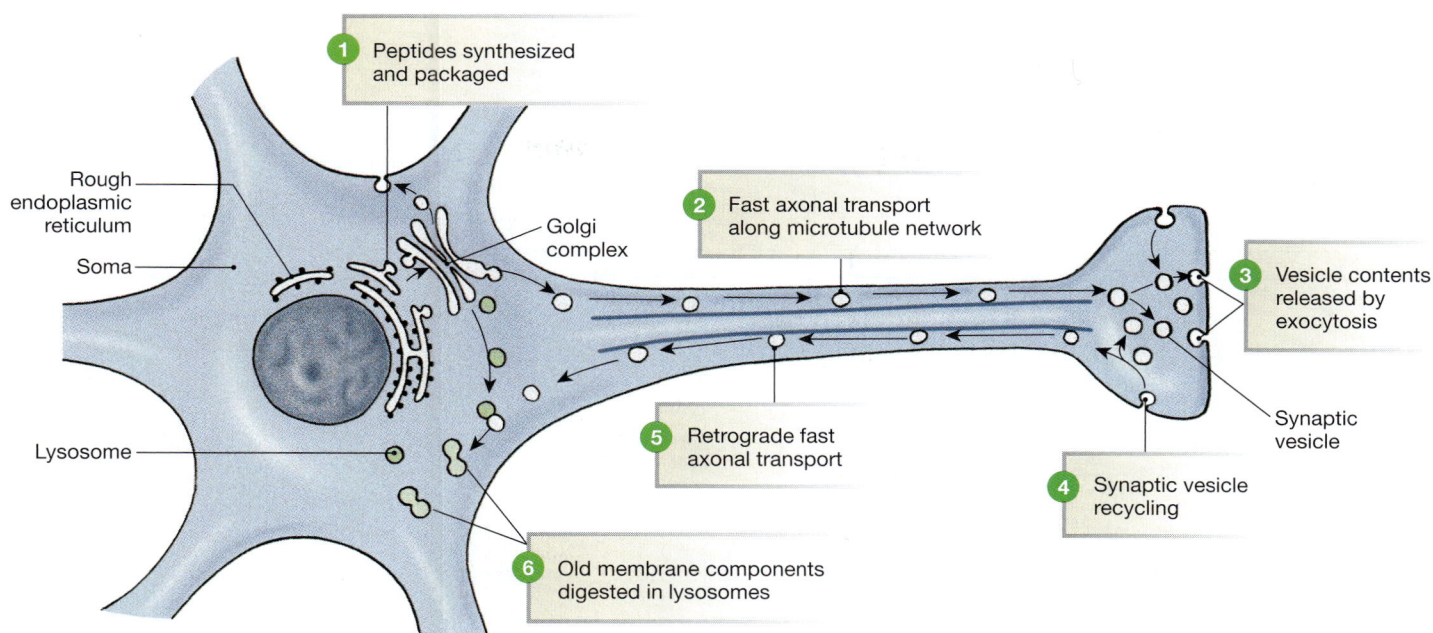

1. Peptides synthesized and packaged

Rough endoplasmic reticulum

Soma

Golgi complex

2. Fast axonal transport along microtubule network

3. Vesicle contents released by exocytosis

5. Retrograde fast axonal transport

Synaptic vesicle

4. Synaptic vesicle recycling

Lysosome

6. Old membrane components digested in lysosomes

■ **FIGURE 8-4** *Fast axonal transport of membranous organelles*

forth with other neurons in the brain. Many dendritic spines contain polyribosomes and can make their own proteins. Changes in spine morphology are associated with learning and memory as well as with various pathologies, such as genetic disorders that cause mental retardation and degenerative diseases such as Alzheimer's disease. Because of these associations, dendritic spines are a hot topic in neuroscience research.

Axons Carry Outgoing Signals to the Target

Most peripheral neurons have a single axon that originates from a specialized region of the cell body called the **axon hillock** (Fig. 8-2). Axons vary in length from more than a meter to only a few micrometers. They often branch sparsely along their length, forming **collaterals** [*col-*, with + *lateral*, something on the side]. In our model neuron, each collateral ends in a swelling called an **axon terminal**. The axon terminal contains mitochondria and membrane-bound vesicles filled with *neurocrine* molecules [🔲p. 176].

The primary function of an axon is to transmit outgoing electrical signals from the integrating center of the neuron to the end of the axon. At the distal end of the axon, the electrical signal is usually translated into a chemical message by secretion of a neurotransmitter, neuromodulator, or neurohormone. Neurons that secrete neurotransmitters and neuromodulators terminate near their target cells, which are usually other neurons, muscles, or glands.

The region where an axon terminal meets its target cell is called a **synapse** [*syn-*, together + *hapsis*, to join]. The neuron that delivers the signal to the synapse is known as the **presynaptic cell**, and the cell that receives the signal is called the **postsynaptic cell**. The narrow space between the two cells is called the **synaptic cleft**.

✓ CONCEPT CHECK
2. Where do neurons that secrete neurohormones terminate?
3. Draw a chain of three neurons that synapse on one another in sequence. Label the presynaptic and postsynaptic ends of each neuron, the cell bodies, dendrites, axons, and axon terminals.   Answers: p. 289

8

Axons are designed to convey chemical and electrical signals. Their cytoplasm is filled with many types of fibers and filaments but lacks ribosomes and endoplasmic reticulum. Therefore, any proteins destined for the axon or the axon terminal must be synthesized on the rough endoplasmic reticulum in the cell body. The proteins are then moved down the axon by a process known as **axonal transport**.

Slow axonal transport moves material by **axoplasmic** (cytoplasmic) **flow** from the cell body to the axon terminal. Material moves at a rate of only 0.2–2.5 mm/day; therefore, slow transport is used for components that are not consumed rapidly by the cell, such as enzymes and cytoskeleton proteins.

Fast axonal transport (Fig. 8-4 ■) moves organelles at rates of up to 400 mm (about 15.75 in.) per day. The neuron uses stationary microtubules as tracks along which transported vesicles and mitochondria "walk" with the aid of attached foot-like motor proteins. These motor proteins alternately bind and unbind to the microtubules with the help of ATP, stepping their organelles along the axon in a stop-and-go fashion. The role of motor proteins in axonal transport is similar to their role in muscle contraction and in the movement of chromosomes during cell division.

Fast axonal transport goes in two directions. Forward (or *anterograde*) transport moves synaptic and secretory vesicles

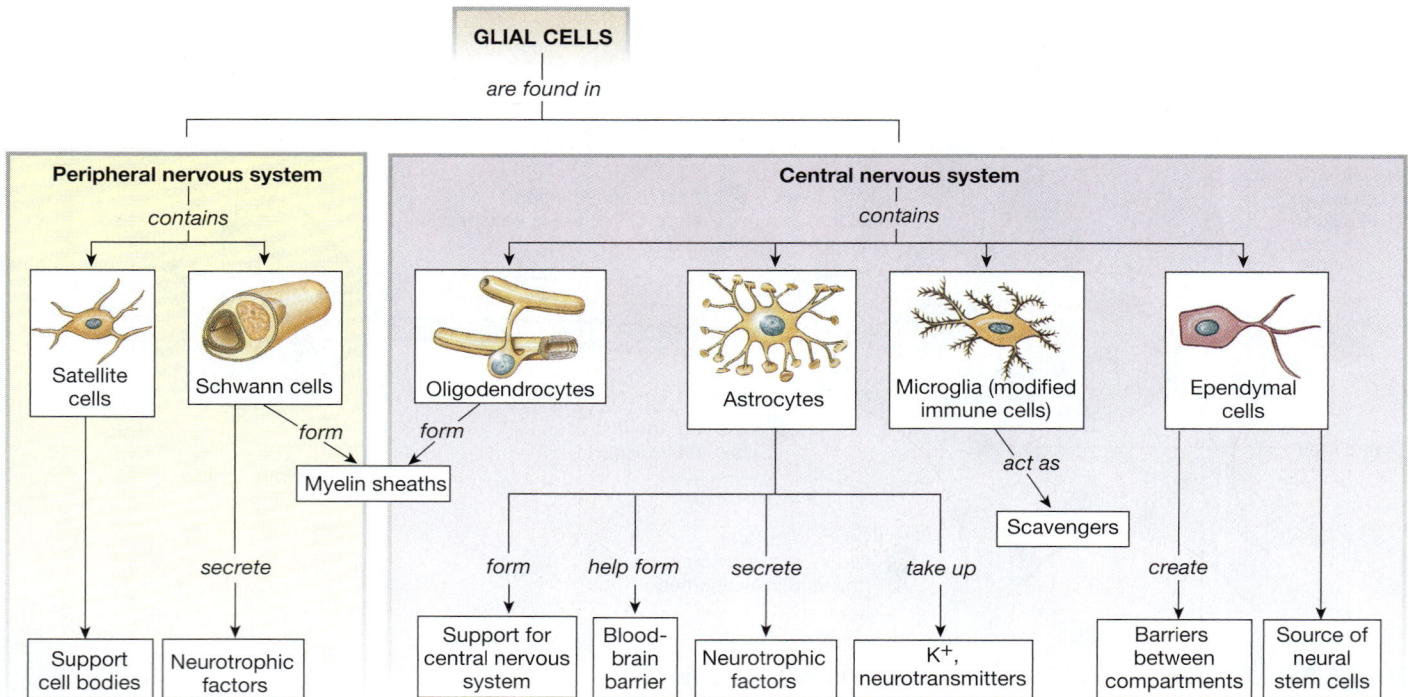

■ **FIGURE 8-5**   *Glial cells and their functions*

and mitochondria from the cell body to the axon terminal. Backward (or *retrograde*) transport returns old cellular components from the axon terminal to the cell body for recycling. There is evidence that nerve growth factors and some viruses also reach the cell body by fast retrograde transport.

Glial Cells Are the Support Cells of the Nervous System

Glial cells [*glia*, glue] are the unsung heroes of the nervous system, outnumbering neurons by 10–50 to 1. Although glial cells do not participate directly in the transmission of electrical signals over long distances, they provide important physical and biochemical support to neurons. Neural tissue secretes very little extracellular matrix [⇌ p. 68], and glial cells provide structural stability to neurons by wrapping around them. Some glial cells provide metabolic support to neurons, and they help maintain homeostasis of the brain's extracellular fluid by taking up excess metabolites and K^+.

Glial cells communicate with neurons and with one another primarily through chemical signals. Glial-derived growth and *trophic* (nourishing) factors help maintain neurons and guide them during repair and development. Research suggests that glial cells in turn respond to neurotransmitters and neuromodulators secreted by neurons. Glial cell function is an active area of neuroscience research, and we still do not fully understand all the roles these important cells play in the nervous system.

The peripheral nervous system has two types of glial cells—Schwann cells and satellite cells—and the CNS has four

types: oligodendrocytes, microglia, astrocytes, and ependymal cells (Fig. 8-5 ■). **Schwann cells** in the PNS and **oligodendrocytes** in the CNS support and insulate axons by forming **myelin,** a substance composed of multiple concentric layers of phospholipid membrane (Fig. 8-6 ■). Myelin forms when these glial cells wrap around an axon, squeezing out the glial cytoplasm so that each wrap becomes two membrane layers (Fig. 8-6b). As an analogy, think of wrapping a deflated balloon tightly around a pencil. Some neurons have as many as 150 wraps (300 membrane layers) in the myelin sheath that surrounds their axons. Gap junctions connect the membrane layers and allow flow of nutrients and information from layer to layer.

One difference between oligodendrocytes and Schwann cells is the number of axons each cell wraps around. In the CNS, one oligodendrocyte forms myelin around portions of several axons (Fig. 8-6a). In the peripheral nervous system, one Schwann cell associates with a single axon; a given axon may have as many as 500 Schwann cells, each wrapped around a 1–1.5 mm segment of the axon (Fig. 8-6c). Between the myelin-insulated areas, a tiny region of axon membrane remains in direct contact with the extracellular fluid. These gaps, called the **nodes of Ranvier,** play an important role in the transmission of electrical signals along the axon.

The second type of PNS glial cell, the **satellite cell,** is a nonmyelinating Schwann cell. Satellite cells form supportive capsules around nerve cell bodies located in ganglia. A **ganglion** [cluster or knot] is a cluster of nerve cell bodies found

(a) Glial cells of the central nervous system

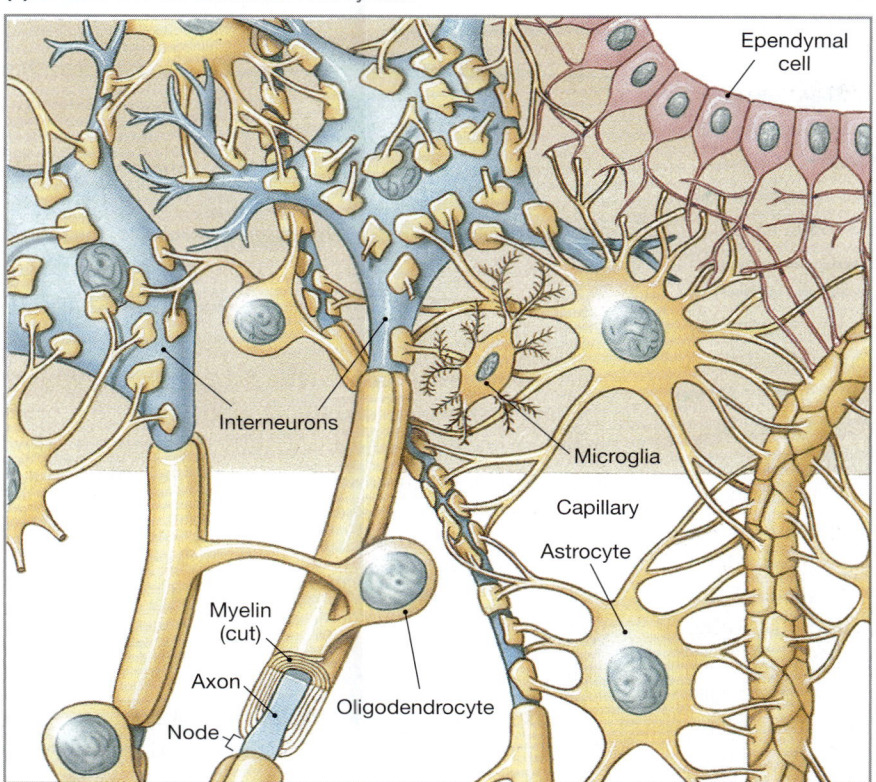

Ependymal cell

Interneurons

Microglia

Capillary

Astrocyte

Myelin (cut)

Axon

Oligodendrocyte

Node

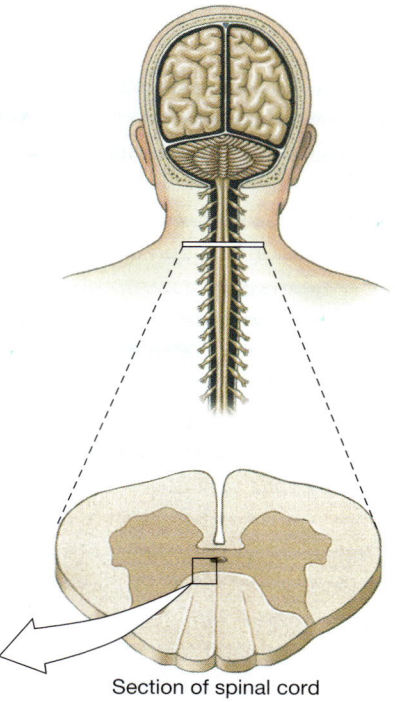

Section of spinal cord

8

(b) Myelin formation in the peripheral nervous system

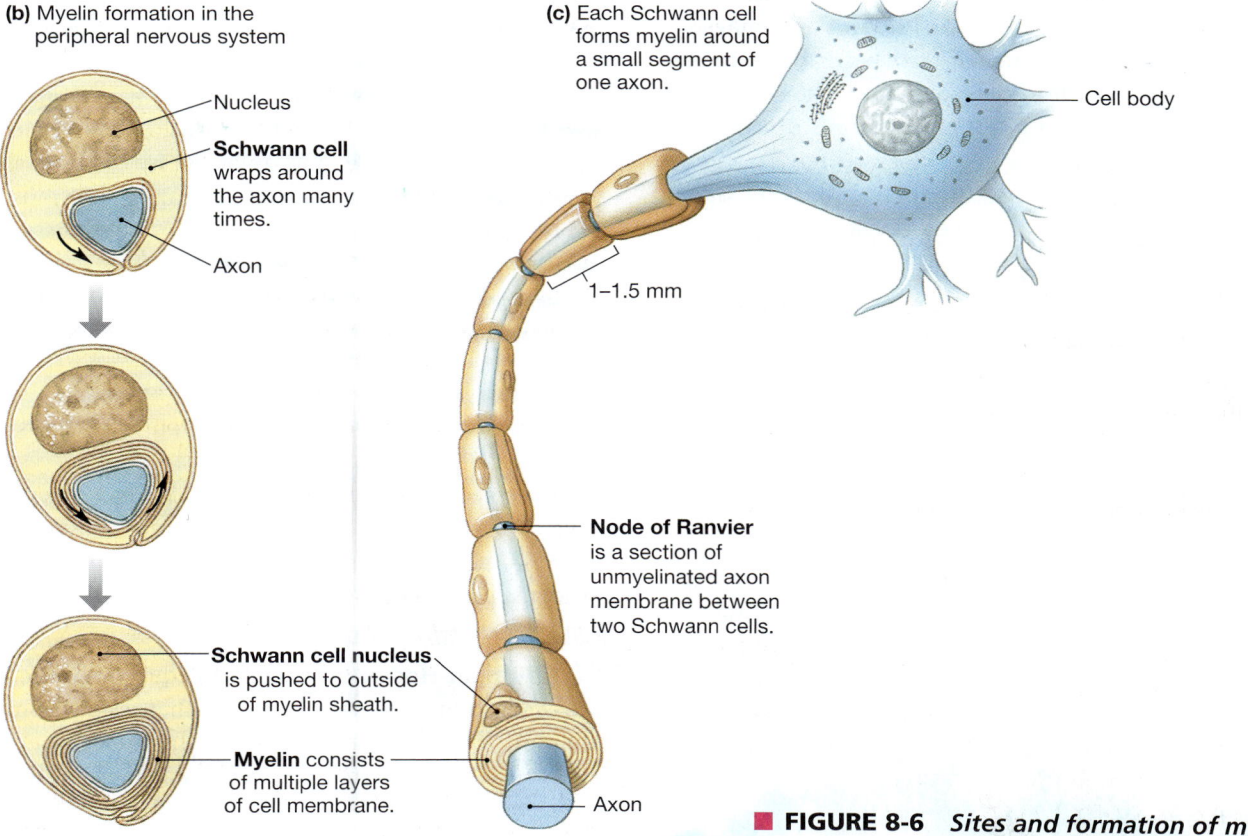

Nucleus

Schwann cell wraps around the axon many times.

Axon

Schwann cell nucleus is pushed to outside of myelin sheath.

Myelin consists of multiple layers of cell membrane.

(c) Each Schwann cell forms myelin around a small segment of one axon.

Cell body

1–1.5 mm

Node of Ranvier is a section of unmyelinated axon membrane between two Schwann cells.

Axon

■ **FIGURE 8-6**   *Sites and formation of myelin*

outside the CNS. Ganglia appear as knots or swellings along a nerve. The equivalent structure inside the CNS is known as a **nucleus** [plural, *nuclei*].

The glial cells known as **microglia** are specialized immune cells that reside permanently in the CNS. When activated, they remove damaged cells and foreign invaders. **Astrocytes** [*astron,* a star] are highly branched cells that contact neurons and blood vessels and may transfer nutrients between the two. In addition, they help maintain homeostasis in the extracellular fluid around CNS neurons by taking up K⁺ and neurotransmitters from the ECF. Astrocytes communicate with one another through gap junctions. **Ependymal cells** are specialized cells that create a selectively permeable epithelial layer, the *ependyma,* that separates the fluid compartments of the CNS.

The ependyma is a source of **neural stem cells** [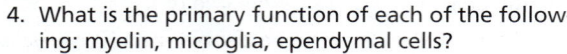 p. 81], immature cells that can differentiate into neurons and glial cells. Scientists once believed that if a neuron died, it could never be replaced. The discovery of neural stem cells changed that view. During early development, an undifferentiated cell layer called *neuroepithelium* lines the lumen of the neural tube, a structure that will later become the brain and spinal cord. As development proceeds, some cells migrate out of the neuroepithelium and differentiate into neurons. Other cells bordering the lumen of the neural tube specialize into the epithelium of the ependyma. However, among the ependymal cells and in the subependymal layer, some neural stem cells remain unspecialized, waiting until they are called upon to replace damaged cells. In some exciting work, researchers have been able to coax neural stem cells to produce dopamine, a neurotransmitter that is deficient in patients suffering from Parkinson's disease. It now appears possible that in the coming years we may see stem cell transplants designed to reverse the loss of function that comes with a number of degenerative neurological diseases.

✓ CONCEPT CHECK

4. What is the primary function of each of the following: myelin, microglia, ependymal cells?

5. Name the two glial cell types that form myelin. How do they differ from each other?

Answers: p. 289

ELECTRICAL SIGNALS IN NEURONS

The property of nerve and muscle cells that characterizes them as *excitable tissues* is their ability to propagate electrical signals rapidly in response to a stimulus. We now know that many other cell types generate electrical signals to initiate intracellular processes (see insulin secretion, p. 166), but the ability of nerve and muscle cells to send a constant electrical signal over long distance is characteristic of electrical signaling in these tissues.

The Nernst Equation Predicts Membrane Potential for a Single Ion

Recall from Chapter 5 that all living cells have a resting membrane potential (V_m) [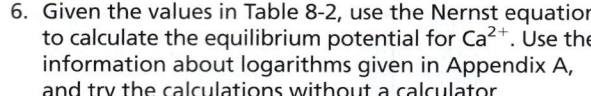 p. 162] that results from the uneven distribu-

tion of ions across the cell membrane. Two factors influence the membrane potential:

1. *concentration gradients of ions* across the membrane. Normally, sodium (Na^+), chloride (Cl^-), and calcium (Ca^{2+}) are more concentrated in the extracellular fluid than in the cytosol. Potassium (K^+) is more concentrated in the cytosol than in the extracellular fluid.

2. *membrane permeability to those ions.* The resting cell membrane is much more permeable to K^+ than to Na^+ or Ca^{2+}. This makes K^+ the major ion contributing to the resting membrane potential.

Chapter 5 introduced the *Nernst equation,* which describes the membrane potential that a single ion would produce if the membrane were permeable to only that one ion [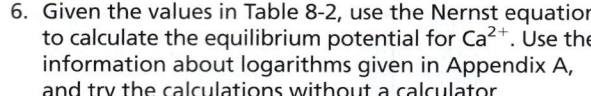 p. 163]. For any one ion, this membrane potential is called the E_{ion}, or *equilibrium potential* of the ion:

$$E_{ion} = \frac{61}{z} \log \frac{[ion]_{out}}{[ion]_{in}}$$

where:

61 is 2.303 RT/F at 37° C*

z is the electrical charge on the ion (+1 for K^+)

$[ion]_{out}$ and $[ion]_{in}$ are the ion concentrations outside and inside the cell

When we use the estimated intracellular and extracellular concentrations for K^+ (Table 8-2 ■) in the Nernst equation, the equation predicts a potassium equilibrium potential of −90 mV. However, an average value for the resting membrane potential of neurons is −70 mV (inside the cell relative to outside), more positive than predicted by the potassium equilibrium potential. This means that other ions must be contributing to the membrane potential. Neurons at rest are slightly permeable to Na^+, and the leak of positive Na^+ into the cell makes the resting membrane potential slightly more positive than it would be if the cell were permeable only to K^+.

✓ CONCEPT CHECK

6. Given the values in Table 8-2, use the Nernst equation to calculate the equilibrium potential for Ca^{2+}. Use the information about logarithms given in Appendix A, and try the calculations without a calculator.

Answers: p. 289

The GHK Equation Predicts Membrane Potential Using Multiple Ions

The **Goldman-Hodgkin-Katz (GHK) equation** is used to calculate the resting membrane potential that results from the contribution of all ions that can cross the membrane. The GHK equation includes membrane permeability values because the permeability of an ion will influence its contribution to the

---

*R is the ideal gas constant, T is absolute temperature, and F is the Faraday constant. For additional information, see Appendix B.

TABLE 8-2	**Ion Concentrations and Equilibrium Potentials**		
ION	**EXTRACELLULAR FLUID (mM)**	**INTRACELLULAR FLUID (mM)**	E_{ion} **AT 37° C**
K^+	5 mM (normal range: 3.5–5)	150 mM	−90 mV
Na^+	145 mM (normal range: 135–145)	15 mM	+60 mV
Cl^-	108 mM (normal range: 100–108)	10 mM (range: 5–15)	−63 mV
Ca^{2+}	1 mM	0.0001 mM	see Concept Check question 6

membrane potential. If the membrane is not permeable to an ion, that ion will not affect the membrane potential.

For mammalian cells, we assume that Na^+, K^+, and Cl^- are the three ions that influence membrane potential in resting cells. Each ion's contribution to the membrane potential is proportional to its ability to cross the membrane. The GHK equation for cells that are permeable to Na^+, K^+, and Cl^- is

$$V_m = 61 \log \frac{P_K[K^+]_{out} + P_{Na}[Na^+]_{out} + P_{Cl}[Cl^-]_{in}}{P_K[K^+]_{in} + P_{Na}[Na^+]_{in} + P_{Cl}[Cl^-]_{out}}$$

where:

V_m is the resting membrane potential at 37° C
61 is 2.303 RT/F at 37° C
P is the relative permeability of the membrane to the ion shown in the subscript
$[ion]_{out}$ and $[ion]_{in}$ are the ion concentrations outside and inside the cell

Although this equation looks quite intimidating, it can be simplified into words to say: Resting membrane potential is determined by the combined contributions of the (concentration gradient) × (membrane permeability) for each ion.

If the membrane is not permeable to an ion, the permeability term for that ion will be zero, and the ion drops out of the equation. For example, cells at rest normally are not permeable to Ca^{2+}, and therefore Ca^{2+} is not part of the GHK equation.

The GHK equation predicts resting membrane potentials based on given ion concentrations and membrane permeabilities, and it explains how the cell's slight permeability to Na^+ makes the resting membrane potential more positive than the E_K determined with the Nernst equation. The GHK equation can also be used to predict what happens to the membrane potential when ion concentrations or membrane permeabilities change. This makes the equation a valuable tool for studying electrical signaling in the body.

Ion Movement Across the Cell Membrane Creates Electrical Signals

The resting membrane potential of living cells is determined primarily by the K^+ concentration gradient and the cell's resting permeability to K^+, Na^+, and Cl^-. A change in either the K^+ concentration gradient or ion permeabilities will change the membrane potential. For example, at rest, the cell membrane of a neuron is only slightly permeable to Na^+. However, if the membrane suddenly increases its Na^+ permeability, Na^+ will enter the cell, moving down its electrochemical gradient [🔁 p. 135]. The addition of positive Na^+ to the intracellular fluid *depolarizes* the cell membrane [🔁 Fig. 5-37, p. 166] and creates an electrical signal.

The movement of ions across the membrane can also *hyperpolarize* a cell. If the cell membrane suddenly becomes more permeable to K^+, positive charge is lost from inside the cell and the cell becomes more negative (hyperpolarizes). A cell may also hyperpolarize if negatively charged ions, such as Cl^-, enter the cell from the extracellular fluid.

CONCEPT CHECK

7. Would a cell with a resting membrane potential of −70 mV depolarize or hyperpolarize in the following cases? (You must consider both the concentration gradient and the electrical gradient of the ion to determine net ion movement.)

   (a) Cell becomes more permeable to Ca^{2+}.
   (b) Cell becomes less permeable to K^+.

8. Would the cell membrane depolarize or hyperpolarize if a small amount of Na^+ leaked into the cell?

   Answers: p. 289

It is important to understand that a change in membrane potential from −70 mV to a positive value, such as +30 mV, *does not mean that the ion concentration gradients have reversed!* A significant change in membrane potential occurs with the movement of very few ions. For example, to change the membrane potential by 100 mV, only 1 of every 100,000 K^+ must enter or leave the cell. This is such a tiny fraction of the total number of K^+ in the cell that the intracellular concentration of K^+ remains essentially unchanged even though the membrane potential has changed by 100 mV.

An analogous situation is moving one grain of sand from a beach into your eye. There are so many grains of sand on the beach that the loss of one grain is not significant, just as the movement of one K^+ across the cell membrane does not significantly alter the concentration of K^+. However, the electrical signal created by moving a few K^+ across the membrane has a significant effect on the cell's membrane potential, just as getting that one grain of sand in your eye will create significant discomfort.

Gated Channels Control the Ion Permeability of the Neuron

How does a cell change its ion permeability? The simplest way is to open or close existing channels in the membrane. Neurons contain a variety of gated ion channels that alternate between open and closed states, depending on the intracellular and extracellular conditions [ p. 139]. A slower method for changing membrane permeability is for the cell to insert new channels into the membrane or remove some existing channels.

Ion channels are usually named according to the primary ion(s) they allow to pass through them. There are four major types of selective ion channels in the neuron: (1) Na^+ channels, (2) K^+ channels, (3) Ca^{2+} channels, and (4) Cl^- channels. Other channels are less selective, such as the monovalent cation channels that allow both Na^+ and K^+ to pass. The ease with which ions flow through a channel is called the channel's **conductance** [*conductus,* escort]. Channel conductance will vary with the gating state of the channel and with the channel protein isoform. As described in Chapter 5, ion channels may spend most of their time in an open state (*leak channels*), or they may have gates that open or close in response to particular stimuli:

1. **Mechanically gated ion channels** are found in sensory neurons and open in response to physical forces such as pressure or stretch.
2. **Chemically gated ion channels** in most neurons respond to a variety of ligands, such as extracellular neurotransmitters and neuromodulators or intracellular signal molecules.
3. **Voltage-gated ion channels** respond to changes in the cell's membrane potential. These channels play an important role in the initiation and conduction of electrical signals.

Not all channels behave in exactly the same way. The **threshold voltage**, or minimum stimulus [ p. 196], for channel opening varies from one channel type to another. For example, some channels we think of as leak channels are actually voltage-gated channels that remain open in the voltage range of the resting membrane potential.

The speed with which a gated channel opens and closes also differs among different types of channels. Channel opening is called channel *activation*. Channel closing is called *inactivation*. Activated channels allow ion flow; inactivated channels do not. For example, axons contain Na^+ channels and K^+ channels that are both activated by cell depolarization. The Na^+ channels open very rapidly, but the K^+ channels are slower to open. The result is an initial flow of Na^+ across the membrane, followed later by a flow of K^+.

Each major channel type has multiple subtypes with varying properties. Neurons contain at least five subtypes of voltage-gated Ca^{2+} channels, four subtypes of voltage-gated K^+ channels, and two subtypes of voltage-gated Na^+ channels. Within each subtype there are also multiple isoforms that may express different opening and closing properties.

Changes in Channel Permeability Create Electrical Signals

When ion channels open, ions may move into or out of the cell. The flow of electrical charge carried by an ion is called the ion's **current**, abbreviated I_{ion}. The direction of ion movement depends on the *electrochemical* (combined concentration and electrical) gradient for the ion. Potassium ions usually move out of the cell. Na^+, Cl^-, and Ca^{2+} usually flow into the cell. The net flow of ions across the membrane depolarizes or hyperpolarizes the cell, creating an electrical signal.

CLINICAL FOCUS

CHANNELOPATHIES

Ion channels are proteins, and like other proteins they may lose or change function if their amino acid sequence is altered. **Channelopathies** [*pathos,* suffering] are inherited diseases caused by mutations in ion channel proteins. Because ion channels are so intimately linked to the electrical activity of cells, many channelopathies manifest themselves as disorders of the excitable tissues (nerve and muscle). One significant contribution of molecular biology to medicine was the discovery that what the medical community considers to be one disease can actually be a family of related diseases with different causes but similar symptoms. For example, the condition known as *long Q-T syndrome* (LQTS; named for changes in the electrocardiogram test) is a cardiac problem characterized by an irregular heart beat (*arrhythmia*; *a-,* without), fainting, and sometimes sudden death. Research studies determined that LQTS is actually three separate conditions, two caused by mutations in a K^+ channel in heart muscle and one caused by a change in a Na^+ channel. Other well-known channelopathies include some forms of epilepsy and malignant hyperthermia. The most common channelopathy is cystic fibrosis, which results from defects in chloride channel function (see Chapter 5 Running Problem).

TABLE 8-3	Comparison of Graded Potential and Action Potential in Neurons	
	GRADED POTENTIAL	**ACTION POTENTIAL**
Type of signal	Input signal	Conduction signal
Where occurs	Usually dendrites and cell body	Trigger zone through axon
Types of gated ion channels involved	Mechanically, chemically, or voltage-gated channels	Voltage-gated channels
Ions involved	Usually Na^+, Cl^-, Ca^{2+}	Na^+ and K^+
Type of signal	Depolarizing (e.g., Na^+) or hyperpolarizing (e.g., Cl^-)	Depolarizing
Strength of signal	Depends on initial stimulus; can be summed	Is always the same; (all-or-none phenomenon); cannot be summed
What initiates the signal	Entry of ions through channels	Above-threshold graded potential at the trigger zone
Unique characteristics	No minimum level required to initiate	Threshold stimulus required to initiate
	Two signals coming close together in time will sum	Refractory period: two signals too close together in time cannot sum
	Initial stimulus strength is indicated by frequency of a series of action potentials	

Electrical signals can be classified into two basic types: graded potentials and action potentials (Table 8-3 ■). **Graded potentials** are variable-strength signals that travel over short distances and lose strength as they travel through the cell. They are used for short-distance communication. If a depolarizing graded potential is strong enough when it reaches an integrating region within a neuron, the graded potential will initiate an action potential. **Action potentials** are large, constant-strength depolarizations that can travel for long distances through a neuron without losing strength. Their function is rapid signaling over long distances.

Graded Potentials Reflect the Strength of the Stimulus That Initiates Them

Graded potentials in neurons are depolarizations or hyperpolarizations that occur in the dendrites and cell body or, less frequently, near the axon terminals. These changes in membrane potential are called "graded" because their size, or *amplitude* [*amplitudo,* large], is directly proportional to the strength of the triggering event. A large stimulus will cause a strong graded potential, and a small stimulus will result in a weak graded potential.

In neurons of the CNS and the efferent division, graded potentials occur when chemical signals from other neurons open chemically gated ion channels, allowing ions to enter or leave the neuron. Mechanical stimuli (such as stretch) open ion channels in some sensory neurons. Graded potentials may also occur when an open channel closes, decreasing the movement of ions through the cell membrane. For example, if K^+ channels close, fewer K^+ will leave the cell, and the retention of K^+ will depolarize the cell.

CONCEPT CHECK

9. Match each ion's movement with the type of graded potential it will create.

   (a) Na^+ entry
   (b) Cl^- entry
   (c) K^+ exit
   (d) Ca^{2+} entry

   1. depolarizing
   2. hyperpolarizing

   Answers: p. 289

Figure 8-7 ■ shows a graded potential that begins when a stimulus opens monovalent cation channels on the cell body of a neuron. Sodium ions move into the neuron, bringing in electrical energy. The positive charge carried in by the Na^+ spreads as a wave of depolarization through the cytoplasm, just as a stone thrown into water creates ripples or waves that spread outward from the point of entry. The wave of depolarization that moves through the cell is known as **local current flow**. By convention, current in biological systems is the net movement of *positive* electrical charge.

The strength of the initial depolarization in a graded potential is determined by how much charge enters the cell, just as the size of waves caused by a stone tossed in water is determined by the size of the stone. If more Na^+ channels open, more Na^+ enters, and the graded potential has a higher initial

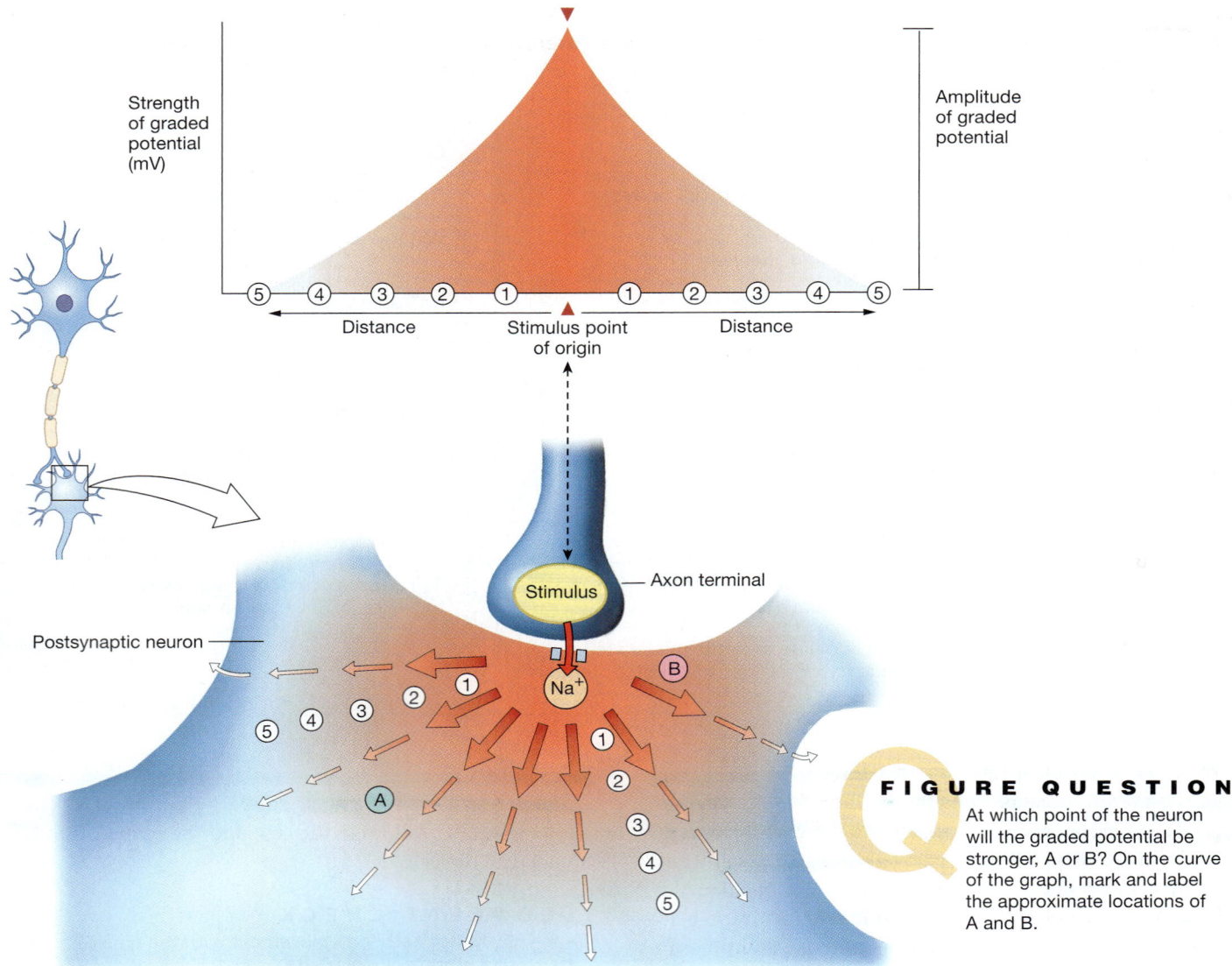

FIGURE QUESTION

At which point of the neuron will the graded potential be stronger, A or B? On the curve of the graph, mark and label the approximate locations of A and B.

■ **FIGURE 8-7**   *Graded potentials decrease in strength as they spread out from the point of origin*

amplitude. The stronger the initial amplitude, the farther the graded potential can spread through the neuron before dying out.

Why do graded potentials lose strength as they move through the cytoplasm? There are two reasons:

1. *Current leak*. Some of the positive charges leak back across the membrane as the depolarization wave moves through the cell. The membrane in the neuron cell body is not a good insulator and has open leak channels that allow positive charge to flow out into the extracellular fluid.

2. *Cytoplasmic resistance*. The cytoplasm itself provides resistance to the flow of electricity, just as water creates resistance that diminishes the waves from the stone. The combination of current leak and cytoplasmic resistance means that the strength of the signal inside the cell decreases over distance.

Graded potentials that are strong enough eventually reach the region of the neuron known as the **trigger zone**. In efferent neurons and interneurons, the trigger zone is the axon hillock and the very first part of the axon, a region known as the **initial segment** (see Fig. 8-2). In sensory neurons, the trigger zone is immediately adjacent to the receptor, where the dendrites join the axon.

CONCEPT CHECK

10. Identify the trigger zones of the neurons illustrated in Figure 8-3, if possible.   Answers: p. 289

The trigger zone is the integrating center of the neuron and contains a high concentration of voltage-gated Na$^+$ channels in its membrane. If graded potentials reaching the trigger zone depolarize the membrane to the threshold voltage, voltage-gated

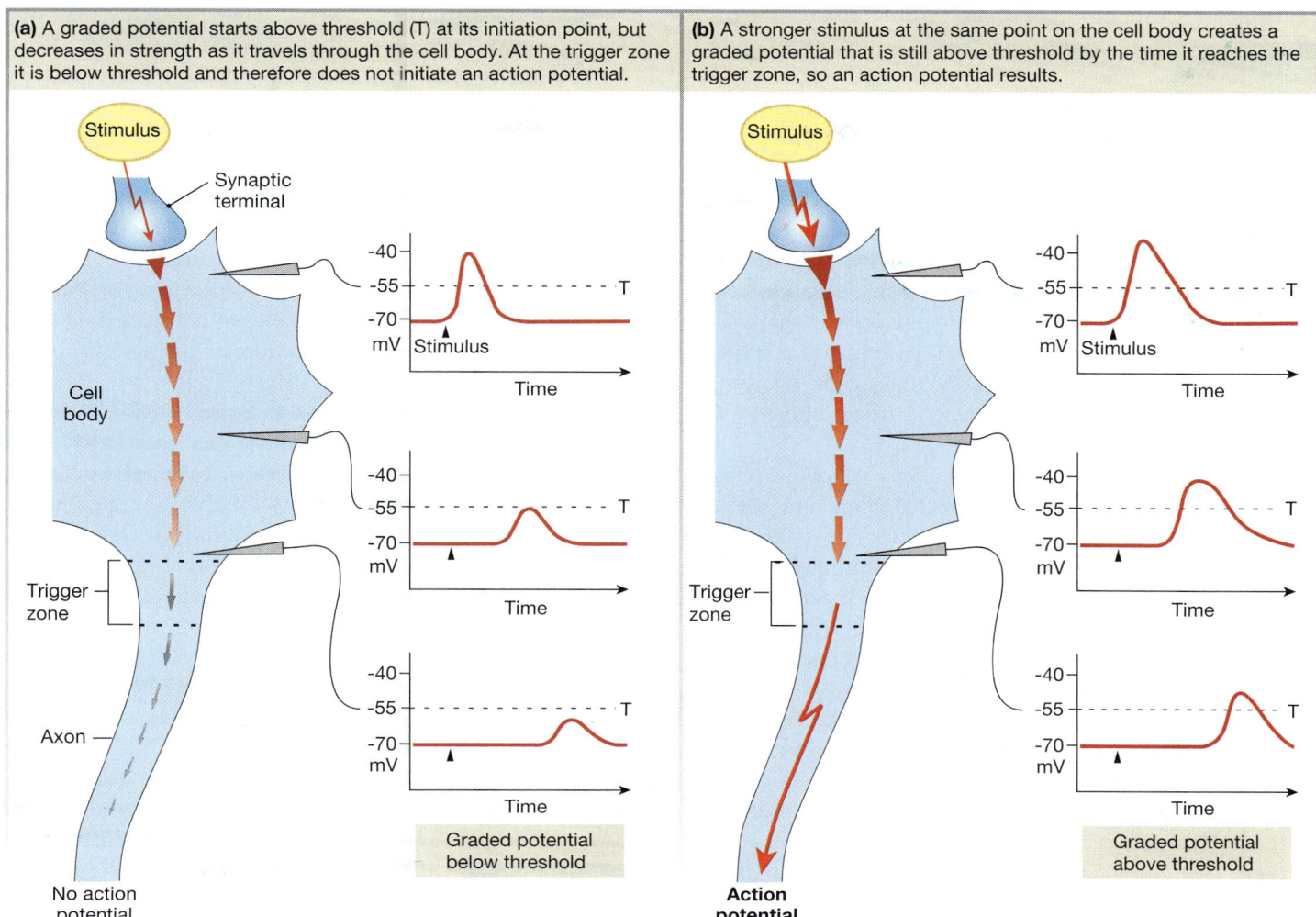

(a) A graded potential starts above threshold (T) at its initiation point, but decreases in strength as it travels through the cell body. At the trigger zone it is below threshold and therefore does not initiate an action potential.

(b) A stronger stimulus at the same point on the cell body creates a graded potential that is still above threshold by the time it reaches the trigger zone, so an action potential results.

■ **FIGURE 8-8**   *Subthreshold and suprathreshold graded potentials in a neuron*

In (a), the graded potential is below the threshold value of −55 mV as it reaches the trigger zone. In (b), the graded potential is above the threshold value when it reaches the trigger zone.

Na^+ channels open, and an action potential is initiated. If the depolarization does not reach threshold, the graded potential simply dies out as it moves into the axon.

Because depolarization makes a neuron more likely to fire an action potential, depolarizing graded potentials are considered to be *excitatory*. A hyperpolarizing graded potential moves the membrane potential farther from the threshold value and makes the neuron less likely to fire an action potential. Consequently, hyperpolarizing graded potentials are considered to be *inhibitory*.

Figure 8-8 ■ shows a neuron with three recording electrodes placed at intervals along the cell body and trigger zone. In Figure 8-8a, a single stimulus triggers a *subthreshold* excitatory postsynaptic potential, one that is below threshold by the time it reaches the trigger zone. Although the cell is depolarized to −40 mV at the site where the graded potential begins, the current decreases as it travels through the cell body. As a result,

the graded potential is below threshold by the time it reaches the trigger zone. (For the typical mammalian neuron, threshold is about −55 mV.) The stimulus is not strong enough to depolarize the cell to threshold at the trigger zone, and the graded potential dies out without triggering an action potential.

Figure 8-8b shows a stronger initial stimulus that initiates a stronger depolarization and ultimately causes an action potential. Although this excitatory graded potential also diminishes with distance through the neuron, its higher initial strength ensures that it is above threshold at the trigger zone. In this example, an action potential is triggered.

Action Potentials Travel Long Distances Without Losing Strength

Action potentials, also known as *spikes*, differ from graded potentials in that they do not diminish in strength as they travel through the neuron. The ability of a neuron to respond rapidly to a stimulus and fire an action potential is called the cell's **excitability**.

Recordings of neuronal action potentials show them to be depolarizations of about 100 mV amplitude. The strength of the graded potential that initiates an action potential has no influence on the amplitude of the action potential. Action potentials are sometimes called **all-or-none** phenomena because they either occur as a maximal depolarization (if the stimulus reaches threshold) or do not occur at all (if the stimulus is below threshold). An action potential measured at the distal end of an axon is identical to the action potential that started at the trigger zone. This property is essential for the transmission of signals over long distances, such as from a fingertip to the spinal cord.

The explanation of action potential firing that follows is typical of what goes on in PNS neurons. In this simple but elegant mechanism, for whose description A. L. Hodgkin and A. F. Huxley won a 1963 Nobel prize, a *suprathreshold* (above-threshold) stimulus causes an action potential, as you saw in Figure 8-8b. These action potentials require only two types of gated ion channels: a voltage-gated Na^+ channel and a voltage-gated K^+ channel, plus some leak channels.

The process of electrical signaling in the CNS can be much more complex, however. Brain neurons show different electrical personalities by firing action potentials in a variety of patterns, and sometimes without requiring an external stimulus to bring them to threshold. For example, neurons can be *tonically active* [🔁 p. 192], firing regular trains of action potentials (beating pacemakers), or they can exhibit *bursting,* bursts of action potentials rhythmically alternating with intervals of quiet (rhythmic pacemakers). These different firing patterns are created by ion channel variants whose activation and inactivation voltages, opening and closing speeds, and sensitivity to neuromodulators differ. This variability makes brain neurons more dynamic and complicated than the simple somatic motor neuron we will use as our model in the discussion that follows.

Action Potentials Represent Movement of Na^+ and K^+ Across the Membrane

Action potentials occur when voltage-gated ion channels open, altering membrane permeability to Na^+ and K^+. Figure 8-9 ■ shows the voltage and ion permeability changes that take place in one section of membrane during an action potential. The graph can be divided into three phases: the rising phase of the action potential, the falling phase, and the after-hyperpolarization phase. Before and after the action potential, at ① and ⑨, the neuron is at its resting membrane potential of −70 mV.

The rising phase of the action potential is due to a sudden temporary increase in the cell's permeability to Na^+. An action potential begins when a graded potential reaching the trigger zone ② depolarizes the membrane to threshold (−55 mV) ③.

As the cell depolarizes, voltage-gated Na^+ channels open, making the membrane much more permeable to Na^+. Because Na^+ is more concentrated outside the cell and because the negative membrane potential inside the cell attracts these positively charged ions, Na^+ flows into the cell.

The addition of positive charge to the intracellular fluid depolarizes the cell membrane, making it progressively more positive (shown by the steep rising phase on the graph ④). In the top third of the rising phase, the membrane potential has reversed polarity; that is, the inside of the cell has become more positive than the outside. This reversal is represented on the graph by the *overshoot,* that portion of the action potential above 0 mV.

As soon as the cell membrane potential becomes positive, the electrical driving force moving Na^+ into the cell disappears. However, the Na^+ concentration gradient remains, so Na^+ continues to move into the cell. As long as Na^+ permeability remains high, the membrane potential moves toward the Na^+ *equilibrium potential* (E_{Na}) of +60 mV. (Recall from Chapter 5 that E_{Na} is the membrane potential at which the movement of Na^+ into the cell down its concentration gradient is exactly opposed by the positive membrane potential [🔁 p. 163].) However, before the E_{Na} is reached, the Na^+ channels in the axon close. Sodium permeability decreases dramatically, and the action potential peaks at +30 mV ⑤.

The falling phase of the action potential corresponds to an increase in K^+ permeability. Voltage-gated K^+ channels, like Na^+ channels, start to open in response to depolarization. The K^+ channel gates are much slower to open, however, and peak K^+ permeability occurs later than peak Na^+ permeability (Fig. 8-9, lower graph). By the time the K^+ channels are open, the membrane potential of the cell has reached +30 mV because of Na^+ influx through faster-opening Na^+ channels.

When the Na^+ channels close at the peak of the action potential, the K^+ channels have just finished opening, making the membrane very permeable to K^+. At a positive membrane potential, the concentration and electrical gradients for K^+ favor movement of K^+ out of the cell. As K^+ moves out of the cell, the membrane potential rapidly becomes more negative, creating the falling phase of the action potential ⑥ and sending the cell toward its resting potential.

When the falling membrane potential reaches −70 mV, the voltage-gated K^+ channels have not yet closed. Potassium continues to leave the cell through both voltage-gated and K^+ leak channels, and the membrane hyperpolarizes, approaching the E_K of −90 mV. This after-hyperpolarization ⑦ is also called the *undershoot.* Once the slow voltage-gated K^+ channels finally close, some of the outward K^+ leak stops ⑧. Retention of K^+ and Na^+ leak inward bring the membrane potential back to −70 mV ⑨, the value that reflects the cell's resting permeability to K^+, Cl^-, and Na^+.

To summarize, the action potential is a change in membrane potential that occurs when voltage-gated ion channels in

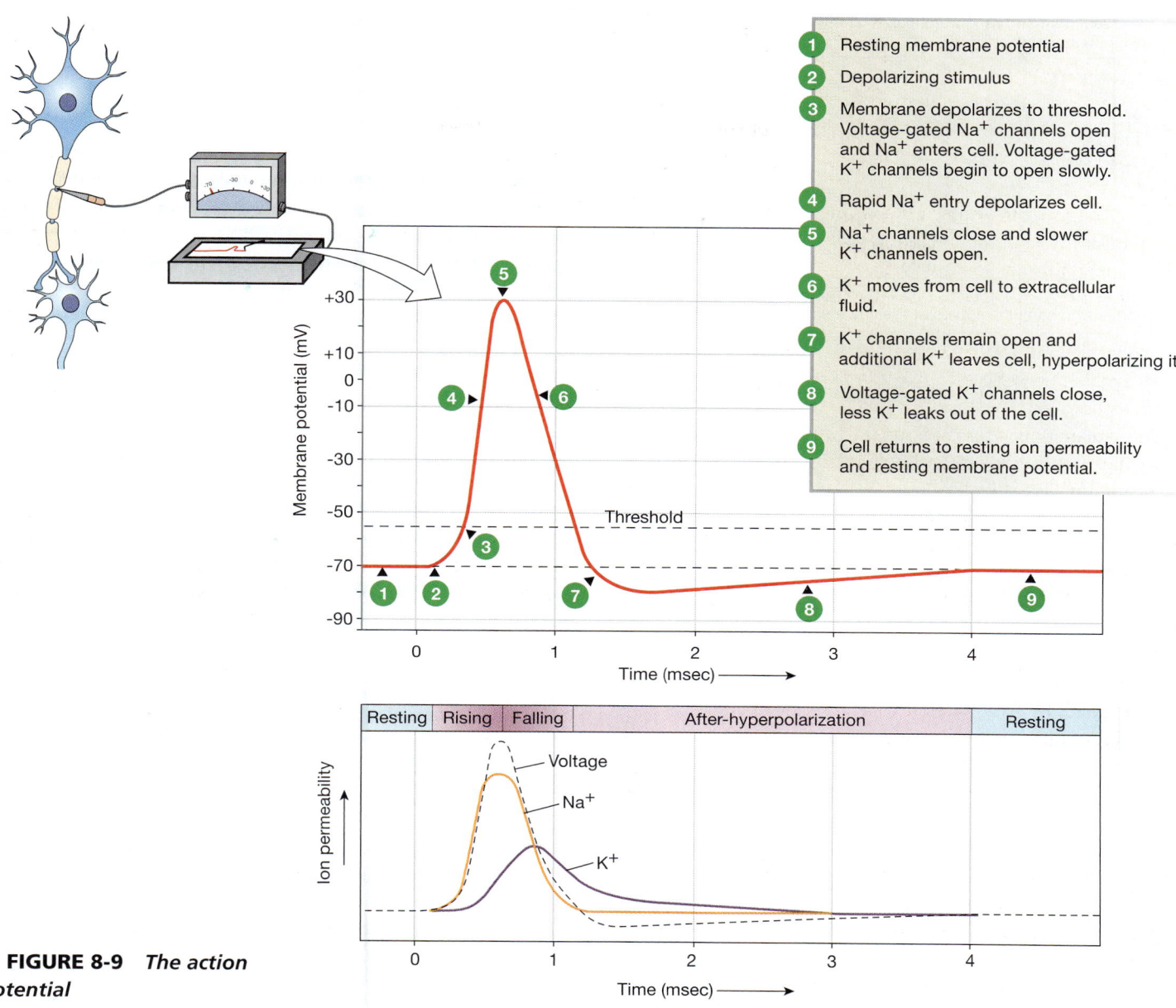

1 Resting membrane potential

2 Depolarizing stimulus

3 Membrane depolarizes to threshold. Voltage-gated Na⁺ channels open and Na⁺ enters cell. Voltage-gated K⁺ channels begin to open slowly.

4 Rapid Na⁺ entry depolarizes cell.

5 Na⁺ channels close and slower K⁺ channels open.

6 K⁺ moves from cell to extracellular fluid.

7 K⁺ channels remain open and additional K⁺ leaves cell, hyperpolarizing it.

8 Voltage-gated K⁺ channels close, less K⁺ leaks out of the cell.

9 Cell returns to resting ion permeability and resting membrane potential.

■ **FIGURE 8-9**   *The action potential*

the membrane open, increasing the cell's permeability first to Na⁺ and then to K⁺. The *influx* (movement into the cell) of Na⁺ depolarizes the cell. This depolarization is followed by K⁺ *efflux* (movement out of the cell), which restores the cell to the resting membrane potential.

Na⁺ Channels in the Axon Have Two Gates

One question that puzzled scientists for many years was how the voltage-gated Na⁺ channels could close when the cell was depolarized. Why should these channels *close* when depolarization was the stimulus for Na⁺ channel *opening*? After many years of study, they found the answer. These voltage-gated Na⁺ channels have two gates to regulate ion movement rather than a single gate. The two gates, known as **activation** and **inactivation gates**, flip-flop back and forth to open and shut the Na⁺ channel.

When a neuron is at its resting membrane potential, the activation gate of the Na⁺ channel is closed and no Na⁺ moves through the channel (Fig. 8-10a ■). The inactivation gate, apparently an amino acid sequence resembling a ball and chain on the cytoplasmic side of the channel, is open. When the cell membrane near the channel depolarizes, the activation gate swings open (Fig. 8-10b). This opens the channel pore and allows Na⁺ to move into the cell down its electrochemical gradient (Fig. 8-10c). The addition of positive charge further depolarizes the inside of the cell and starts a positive feedback loop [🔁p. 199] (Fig. 8-11 ■). More Na⁺ channels open, and more Na⁺ enters, further depolarizing the cell. As long as the cell remains depolarized, activation gates in Na⁺ channels remain open.

(a) At the resting membrane potential, the activation gate closes the channel.

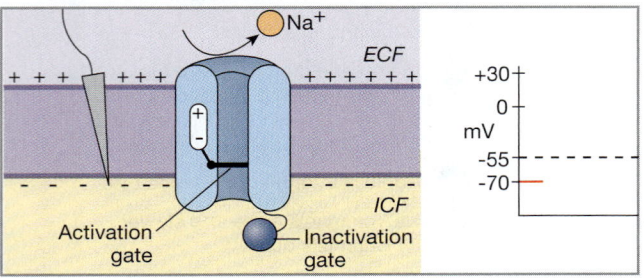

(b) Depolarizing stimulus arrives at the channel.

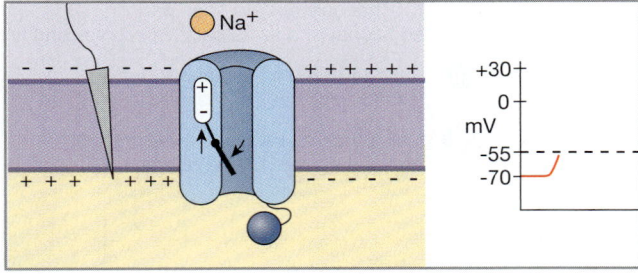

(c) With activation gate open, Na⁺ enters the cell.

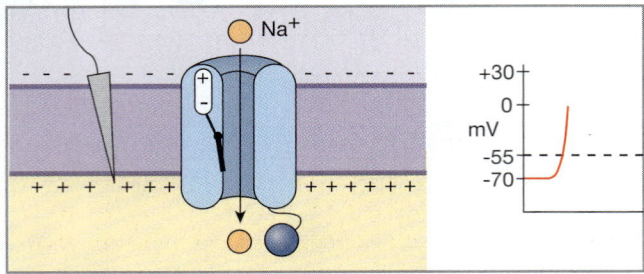

(d) Inactivation gate closes and Na⁺ entry stops.

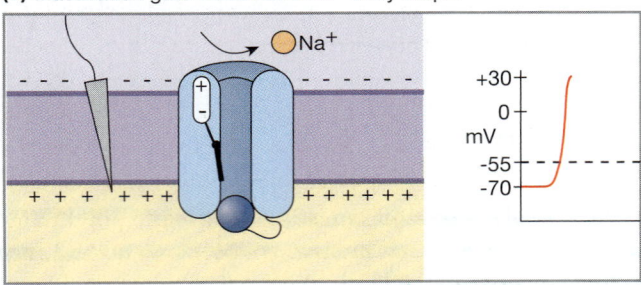

(e) During repolarization caused by K⁺ leaving the cell, the two gates reset to their original positions.

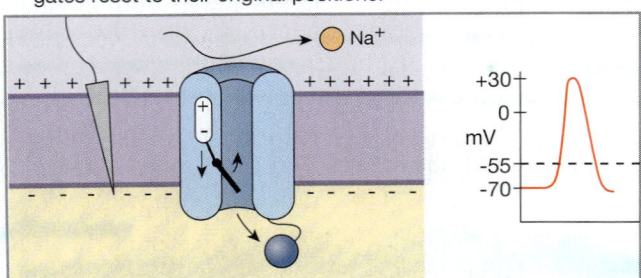

■ **FIGURE 8-10** *Model of the voltage-gated Na⁺ channel*

The distinguishing feature of this channel is the presence of two gates: an activation gate and an inactivation gate.

As happens in all positive feedback loops, outside intervention is needed to stop the escalating depolarization of the cell. This outside intervention is the role of the inactivation gates in the Na⁺ channels. Both activation and inactivation gates move in response to depolarization, but the inactivation gate delays its movement for 0.5 msec. During that delay the Na⁺ channel is open, allowing enough Na⁺ influx to create the rising phase of the action potential. When the slower inactivation gate finally closes (Fig. 8-10d), Na⁺ influx stops, and the action potential peaks.

While the neuron repolarizes during K⁺ efflux, the Na⁺ channel gates reset to their original positions so they can respond to the next depolarization (Fig. 8-10e). Thus, voltage-gated Na⁺ channels use a two-step process for opening and closing rather than a single gate that swings back and forth. This important property allows electrical signals along the axon to be conducted in only one direction, as you will see in the next section.

CONCEPT CHECK

11. The pyrethrin insecticides, derived from chrysanthemums, disable inactivation gates of Na⁺ channels so that the channels remain open. In neurons poisoned with pyrethrins, what would you predict would happen to the membrane potential? Explain your answer.

12. When Na⁺ channel gates are resetting, is the activation gate opening or closing? Is the inactivation gate opening or closing?

Answers: p. 290

Action Potentials Will Not Fire During the Absolute Refractory Period

The double gating of Na⁺ channels plays a major role in the phenomenon known as the **refractory period**. The adjective *refractory* comes from a Latin word meaning "stubborn." The "stubbornness" of the neuron refers to the fact that once an action potential has begun, a second action potential cannot be triggered for about 2 msec, no matter how large the stimulus. This period is called the **absolute refractory period** (Fig. 8-12 ■) and represents the time required for the Na⁺ channel gates to reset to their resting positions. The absolute refractory period ensures that a second action potential will not occur before the first has finished. *Action potentials cannot overlap and cannot travel backward because of their refractory periods.*

A **relative refractory period** follows the absolute refractory period. During the relative refractory period, a stronger-than-normal depolarizing graded potential is needed to bring the

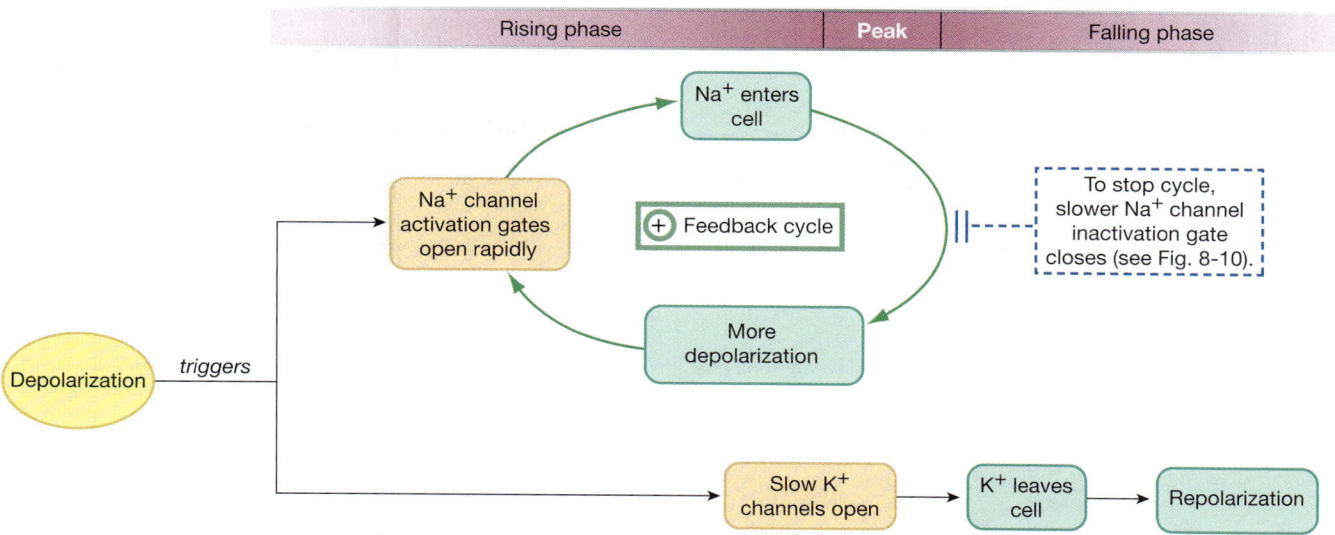

■ **FIGURE 8-11** *Ion movement during an action potential*

The entry of Na$^+$ into the cell creates a positive feedback loop that stops when the Na$^+$ channel inactivation gates close.

cell up to threshold, and the action potential will be smaller than normal. During this time, many but not all Na$^+$ channel gates have reset to their original positions and a threshold-level depolarization will open them. Those Na$^+$ channels that have not quite returned to their resting position can be opened by a higher-than-normal graded potential.

In addition, during the relative refractory period, K$^+$ channels are still open. Although Na$^+$ can enter through newly re-opened Na$^+$ channels, depolarization due to Na$^+$ entry will be offset by K$^+$ loss. As a result, any action potentials that fire will have a smaller amplitude than normal.

The refractory period is a key characteristic that distinguishes action potentials from graded potentials. If two stimuli reach the dendrites of a neuron within a short time, the successive graded potentials created by those stimuli can be added to one another. If, however, two suprathreshold graded potentials reach the action potential trigger zone within the absolute refractory period, the second graded potential will be ignored because the Na$^+$ channels are inactivated and are incapable of being opened again so soon.

Refractory periods limit the rate at which signals can be transmitted down a neuron. The absolute refractory period also ensures one-way travel of an action potential from cell body to axon terminal by preventing the action potential from traveling backward.

Stimulus Intensity Is Coded by the Frequency of Action Potentials

One distinguishing characteristic of action potentials is that every action potential in a given neuron will be identical to every other action potential in that neuron. If this is so, how does the neuron transmit information about the strength and

duration of the stimulus that started the action potential? The answer lies not in the amplitude of the action potential but in the frequency of action potential propagation.

A graded potential reaching the trigger zone does not usually trigger a single action potential. Instead, even a minimal graded potential that is above threshold will trigger a burst of action potentials (Fig. 8-13a ■). If the graded potential increases in strength (amplitude), the frequency of action potentials fired (in action potentials per second) increases (Fig. 8-13b).

The amount of neurotransmitter released at the axon terminal is directly related to the total number of action potentials that arrive at the terminal per unit time. Often a burst of action potentials arriving at the terminal results in increased neurotransmitter release, as shown in Figure 8-13. However, in some cases of sustained activity, neurotransmitter release may decrease because the axon cannot replenish its neurotransmitter supply rapidly enough.

One Action Potential Does Not Alter Ion Concentration Gradients

As you just learned, an action potential results from ion movements across the neuron membrane. First Na$^+$ moves into the cell, and then K$^+$ moves out. However, it is important to understand that very few ions move across the membrane in a single action potential, so that *the relative Na$^+$ and K$^+$ concentrations inside and outside the cell remain essentially unchanged.* For example, only 1 every 100,000 K$^+$ must leave the cell to shift the membrane potential from +30 to −70 mV, equivalent to the falling phase of the action potential. The tiny number of ions that cross the membrane during an action potential does not disrupt the Na$^+$ and K$^+$ concentration gradients.

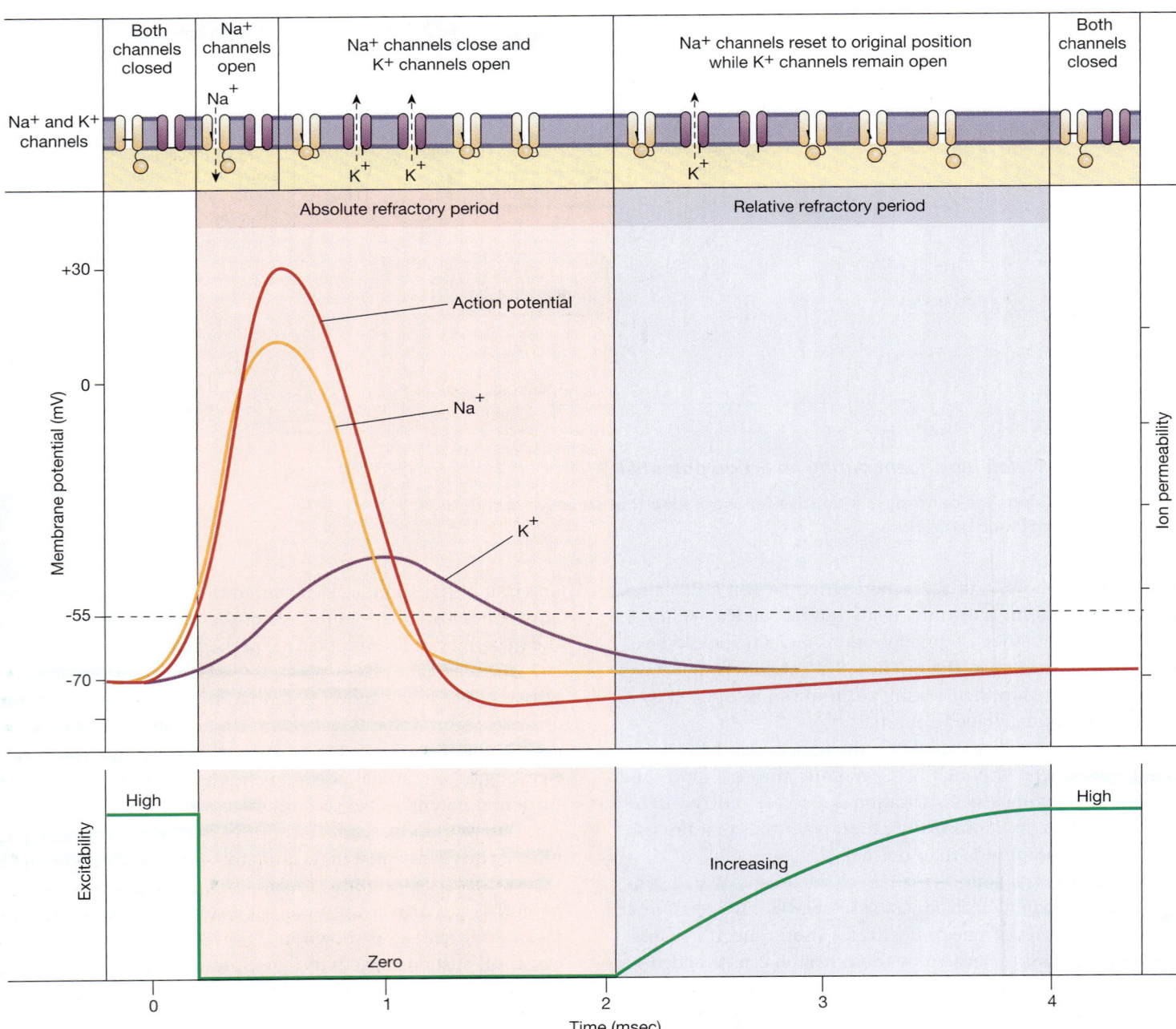

■ FIGURE 8-12 *Refractory periods*

During the absolute refractory period, no stimulus can trigger another action potential. During the relative refractory period, only a larger-than-normal stimulus can initiate a new action potential. A single channel shown during a phase means that the majority of channels are in this state. Where more than one channel of a particular type is shown, the population is split between the states.

Normally, the ions that do move into or out of the cell during action potentials are rapidly restored to their original compartments by the Na^+-K^+-ATPase (also known as the Na^+-K^+ pump). The pump uses energy from ATP to exchange Na^+ that enters the cell for K^+ that leaked out of it [p. 143]. *This exchange does not need to happen before the next action potential fires, however, because the ion concentration gradient was not significantly altered by one action potential!* A neuron without a functional Na^+-K^+ pump may fire a thousand or more action potentials before a significant change in the ion gradients occurs.

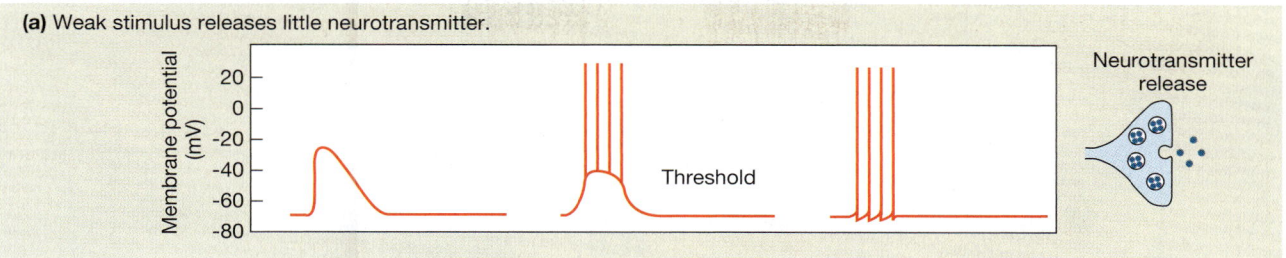

(a) Weak stimulus releases little neurotransmitter.

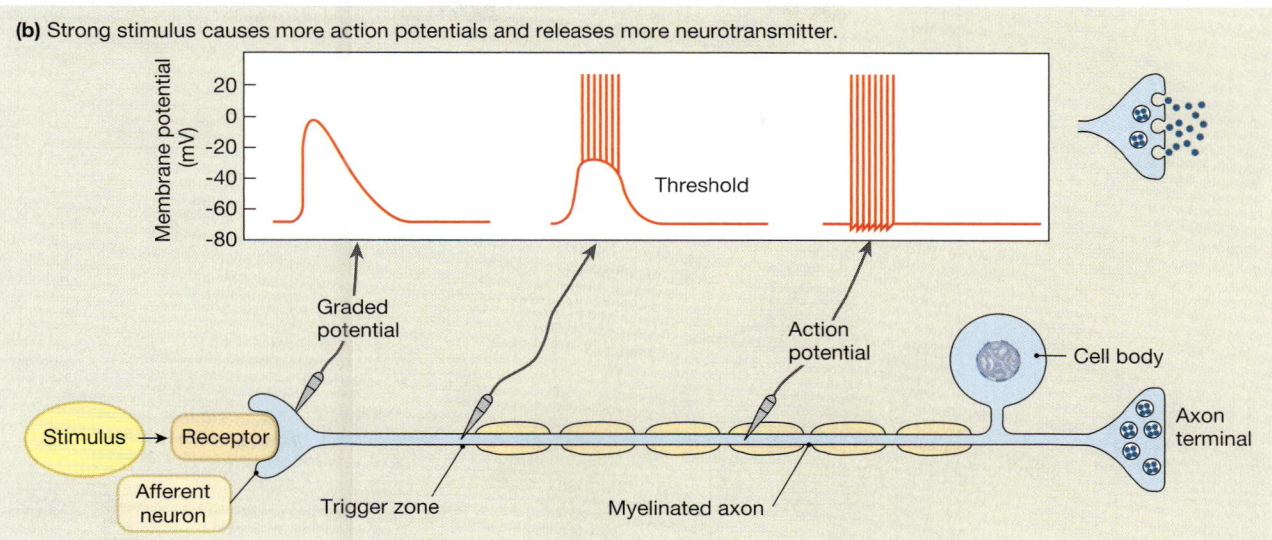

(b) Strong stimulus causes more action potentials and releases more neurotransmitter.

■ FIGURE 8-13   *Coding for stimulus intensity*

All action potentials in a neuron are identical; therefore, the strength of the stimulus is indicated by the frequency of action potential firing. Stronger stimuli release more neurotransmitter into the synapse.

CONCEPT CHECK ✓

13. If you put ouabain, a inhibitor of the Na⁺-K⁺ pump, on a neuron and then stimulate the neuron repeatedly, what do you expect to see happen to action potentials generated by that neuron?

    a. They cease immediately.
    b. There is no immediate effect, but they diminish over time and eventually disappear.
    c. They get smaller immediately, then stabilize with smaller amplitude.
    d. Ouabain has no effect on action potentials.

Answers: p. 290

Action Potentials Are Conducted from the Trigger Zone to the Axon Terminal

The high-speed movement of an action potential through the axon is called **conduction**. Conduction represents the flow of electrical energy from one part of the cell to another in a process that constantly replenishes lost energy. Thus, an action potential does not lose strength over distance, as a graded potential does. To see why the action potential that reaches the end of an axon is identical to the action potential that started at the trigger zone, we must examine conduction at the cellular level.

The depolarization of a section of axon causes positive ions to spread through the cytoplasm in all directions by local current flow (Fig. 8-14 ■). Simultaneously, on the outside of the axon membrane, current flows back toward the depolarized section. The local current flow in the cytoplasm will diminish over distance

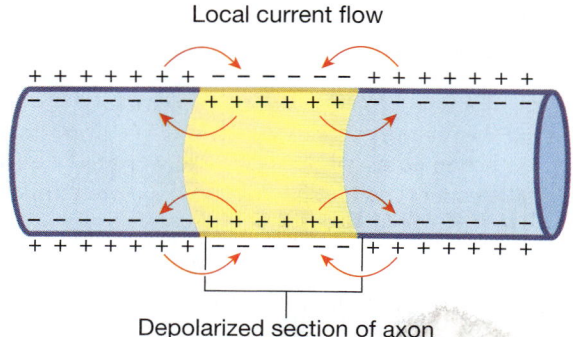

Local current flow

Depolarized section of axon

■ FIGURE 8-14   *Local current flow*

When a section of axon depolarizes, positive charges move by local current flow into adjacent sections of the cytoplasm. On the extracellular surface, current flows toward the depolarized region.

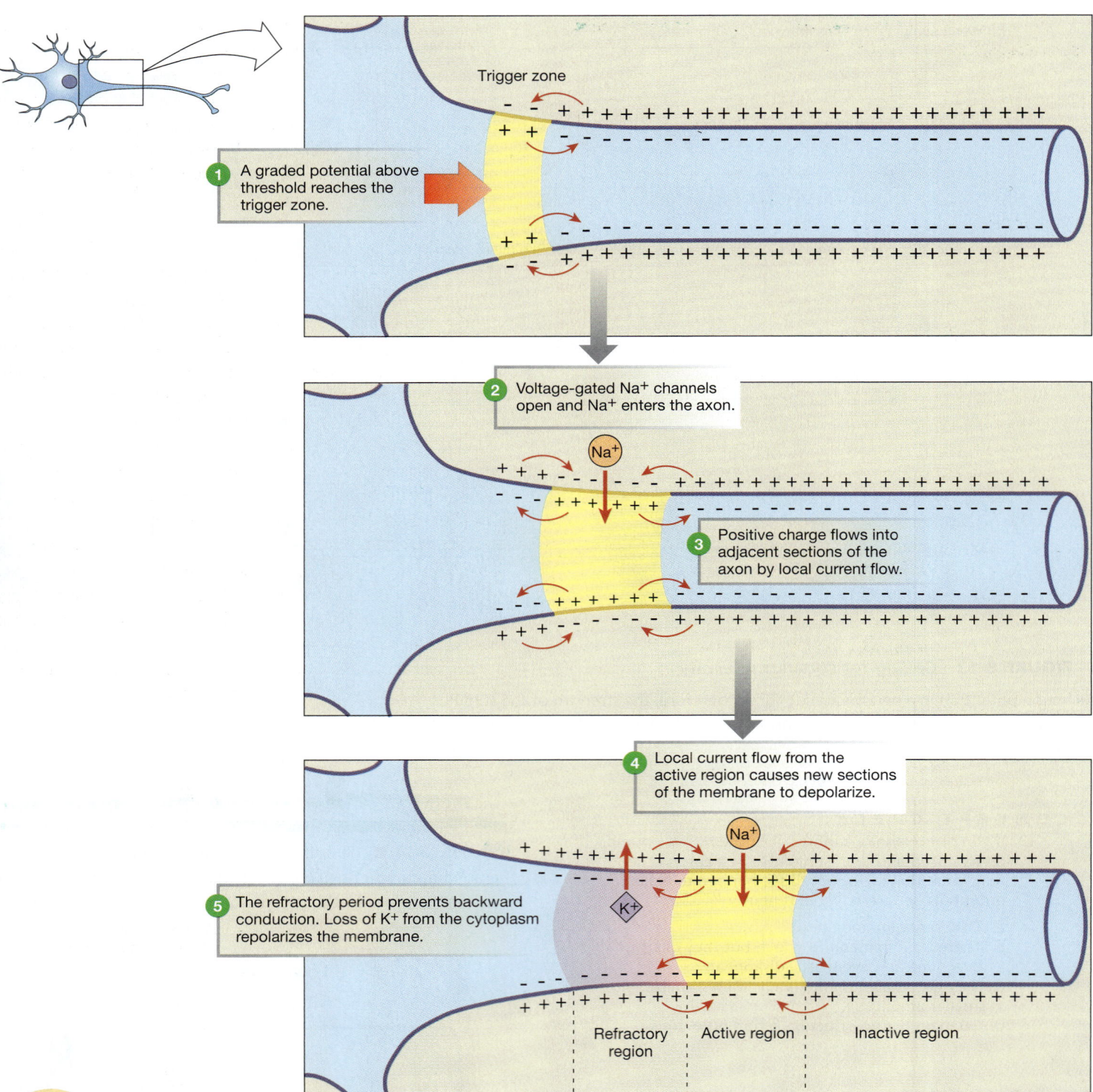

1 A graded potential above threshold reaches the trigger zone.

Trigger zone

2 Voltage-gated Na⁺ channels open and Na⁺ enters the axon.

3 Positive charge flows into adjacent sections of the axon by local current flow.

4 Local current flow from the active region causes new sections of the membrane to depolarize.

5 The refractory period prevents backward conduction. Loss of K⁺ from the cytoplasm repolarizes the membrane.

Refractory region   Active region   Inactive region

FIGURE QUESTION

Match the segments of the neuron in the bottom frame of the figure with the corresponding phrase(s):

a) proximal axon (blue)
b) absolute refractory period (pink)
c) active region (yellow)
d) relative refractory period (purple)
e) distal inactive region (blue)

1. rising phase of action potential
2. falling phase of action potential
3. after-hyperpolarization
4. resting potential

■ **FIGURE 8-15** *Conduction of action potentials*

and would eventually die out. The axon, however, is well supplied with voltage-gated Na⁺ channels. Whenever the depolarization reaches those channels, they open, allowing Na⁺ to enter the cell and reinforcing the depolarization. This mechanism initiates the positive feedback loop shown in Figure 8-11. Let's see how this works when an action potential begins at the axon's trigger zone.

The stimulus is a graded potential above threshold that enters the trigger zone (Fig. 8-15 ■, ①). The depolarization opens voltage-gated Na⁺ channels, Na⁺ enters the axon, and the initial segment of axon depolarizes ②. Positive charge from the depolarized trigger zone spreads to adjacent sections of membrane ③, repelled by the Na⁺ that entered the cytoplasm and attracted by the negative charge of the resting membrane potential.

The flow of local current toward the axon terminal (to the right in Figure 8-15) begins conduction of the action potential. When the membrane distal to the trigger zone depolarizes, its Na⁺ channels open, allowing Na⁺ into the cell ④. This starts the positive feedback loop: depolarization opens Na⁺ channels, Na⁺ enters, causing more depolarization and opening more Na⁺ channels in the adjacent membrane. The continuous entry of Na⁺ down the axon toward the axon terminal means that the strength of the signal does not diminish as the action potential propagates itself. (Contrast this with graded potentials in Fig. 8-7, in which Na⁺ enters only at the point of stimulus, resulting in a membrane potential change that loses strength over distance.)

As each segment of axon reaches the peak of the action potential, its Na⁺ channels inactivate. During the action potential's falling phase, K⁺ channels are open, allowing K⁺ to leave the cytoplasm. Finally, the K⁺ channels close and the membrane in that segment of axon returns to its resting potential.

Although positive charge from a depolarized segment of membrane may flow backward toward the cell body ⑤, depolarization in that direction has no effect. The section of axon proximal to the active region is in the absolute refractory period, with its Na⁺ channels inactivated; therefore, the action potential cannot move backwards.

What happens to current flow backward from the trigger zone into the cell body? We used to believe that there were few voltage-gated ion channels in the cell body, so that retrograde current flow could be ignored. However, we now know that the cell body and dendrites do have voltage-gated ion channels and may respond to the retrograde flow of charge from action potentials. These retrograde signals are able to influence and modify the next signal that reaches the cell.

When we talk about action potentials, it is important to realize that there is no single action potential that moves through the cell. The action potential that occurs at the trigger zone is like the movement in the first domino of a series of dominos standing on end. As the first domino falls, it strikes the next, passing on its kinetic energy. As the second domino falls, it passes kinetic energy to the third domino, and so on. If you could take a snapshot of the line of falling dominos, you would see that as the first domino is coming to rest in the fallen position, the next one is almost down, the third one most of the way down, and so forth, until you reach the domino that has just been hit and is starting to fall (Fig. 8-16a ■).

(a) The transmission of action potentials can be compared to a "snapshot" of dominos falling, where each domino is in a different position.

(b) Simultaneous recordings show that each section of axon is experiencing a different phase of the action potential.

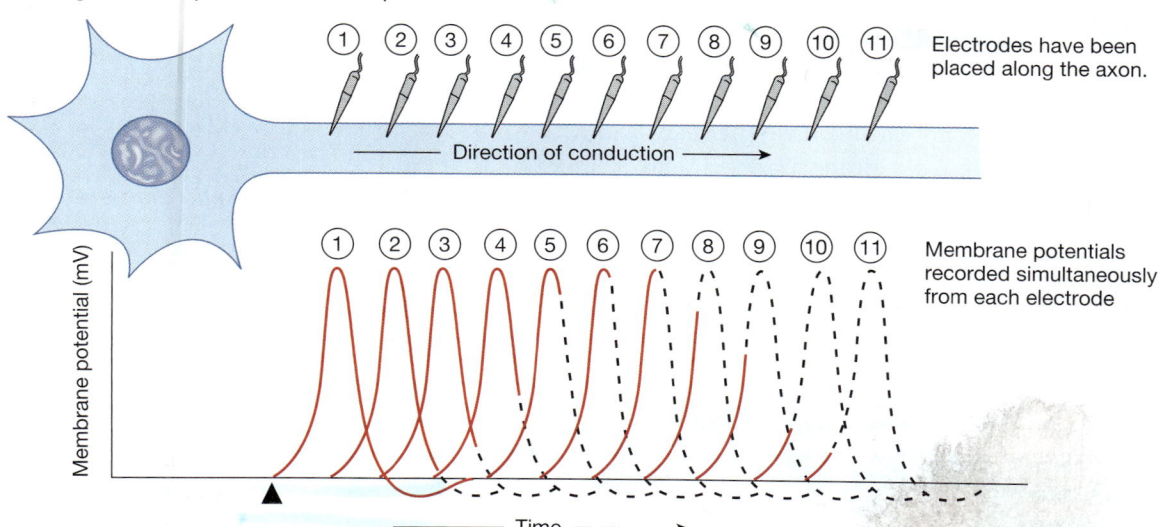

■ **FIGURE 8-16** *Action potentials along an axon*

In the same way, a wave of action potentials moves down the axon. An action potential is simply a representation of the membrane potential in a given segment of cell membrane at any given moment. As the electrical energy of the action potential passes from one part of the axon to the next, the energy state is reflected in the membrane potential of that region. If we were to insert a series of recording electrodes along the length of an axon and start an electrical signal at the trigger zone, we would see a series of overlapping action potentials at different parts of the wave form, just like the dominos that are frozen in different positions (Fig. 8-16b).

CONCEPT CHECK

14. If you place an electrode in the middle of an axon and artificially depolarize the cell above threshold, in which direction will an action potential travel: to the axon terminal, to the cell body, or to both? Explain your answer.
    Answers: p. 290

Larger Neurons Conduct Action Potentials Faster

Two key physical parameters influence the speed of action potential conduction in a mammalian neuron: (1) the diameter of the axon and (2) the resistance of the axon membrane to ion leakage out of the cell. The larger the diameter of the axon or the more leak-resistant the membrane, the faster an action potential will move.

To understand the relationship between diameter and conduction, think of a water pipe with water flowing through it. The water that touches the walls of the pipe encounters resistance due to friction between the flowing water molecules and the stationary walls. The water in the center of the pipe meets no direct resistance from the walls and therefore flows faster. In a large-diameter pipe, a smaller fraction of the water flowing through the pipe is in contact with the walls, making the total resistance lower. In the same way, charges flowing inside an axon meet resistance from the membrane. Thus, the larger the diameter of the axon, the lower its resistance to ion flow.

The connection between axon diameter and speed of conduction is especially evident in the nerve cells of invertebrate animals, such as squid and earthworms. The giant axons of these organisms, used for rapid escape responses, may be up to 1 mm in diameter (Fig. 8-17a ■). Their large diameter makes these axons easily punctured with electrodes, and studies of these species have been very important in the development of our understanding of electrical signaling. Larger-diameter axons for rapid conduction are found mostly in animals with small nervous systems, however.

If you compare a cross section of a squid giant axon with a cross section of a mammalian nerve, you will find that the mammalian nerve contains about 200 neurons in the same cross-sectional area (Fig. 8-17b). If each of those 200 axons were as large as a squid giant axon, that nerve would have to be 16 cm

(a) Large axons offer less resistance to current flow, but occupy space.

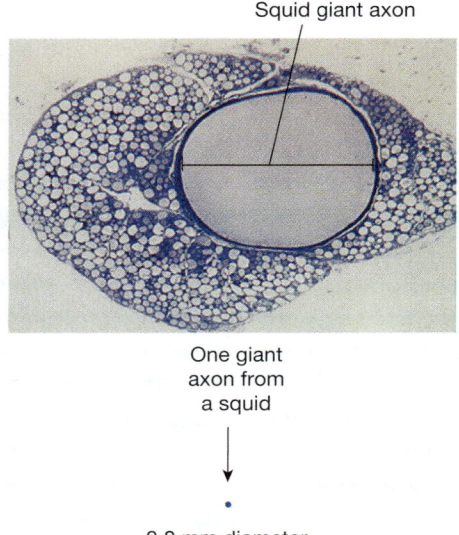

Squid giant axon

One giant axon from a squid

0.8 mm diameter

(b) Small myelinated axons conduct action potentials as rapidly as large unmyelinated axons.

Nerve   Axons

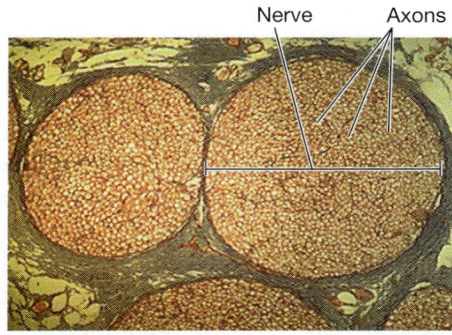

If 20 myelinated mammalian axons in the nerve above were each the size of a squid giant axon, the nerve would have to be this size.

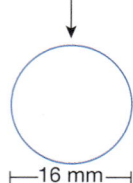

|—16 mm—|

■ **FIGURE 8-17** *Axon diameter and speed of conduction*

in diameter, much too large for our bodies! Complex nervous systems pack more axons into a small nerve by trading large-diameter axons for smaller-diameter axons wrapped in insulating membranes of myelin.

Conduction Is Faster in Myelinated Axons

The conduction of action potentials down an axon is faster in high-resistance axons, in which current leak out of the cell is minimized. The unmyelinated axon depicted in Figure 8-15 has

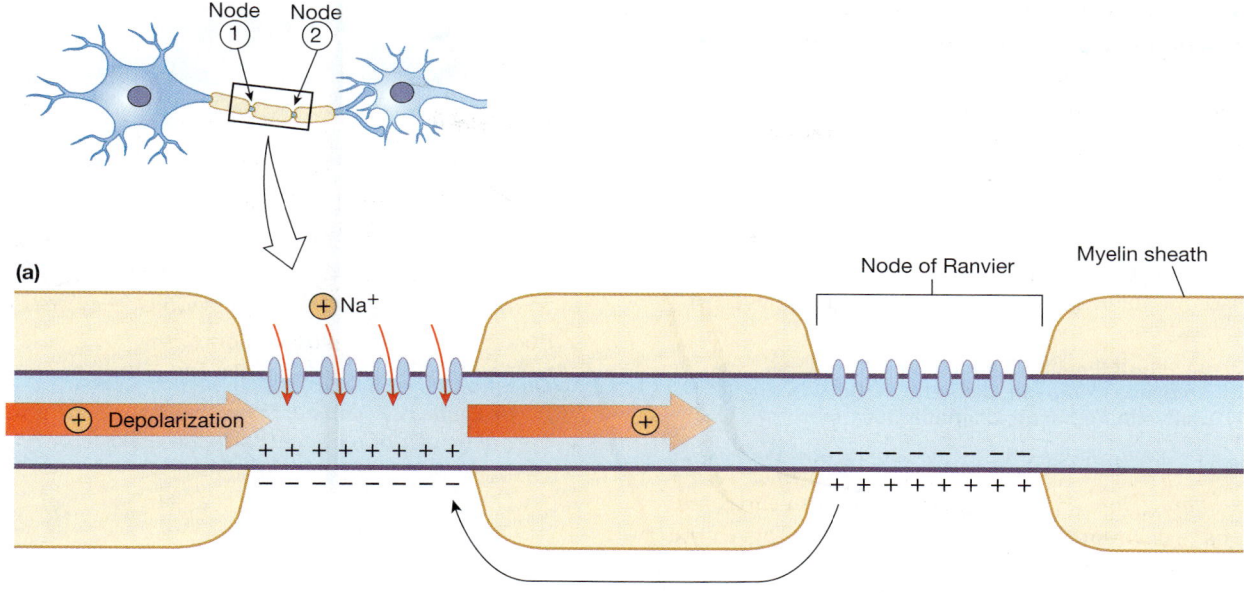

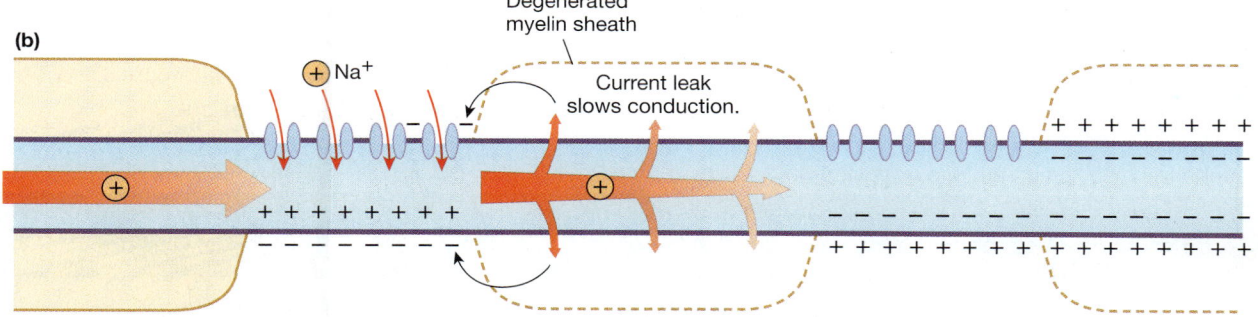

■ FIGURE 8-18 *Saltatory conduction*

(a) Action potentials appear to jump from one node of Ranvier to the next. (b) In demyelinating diseases, conduction slows due to current leakage out of the previously insulated regions between the nodes.

low resistance to current leak because the entire axon membrane is in contact with the extracellular fluid and has ion channels through which current can leak. In contrast, myelinated axons limit the amount of membrane in contact with the extracellular fluid. In these axons, small sections of bare membrane—the nodes of Ranvier—alternate with longer segments wrapped in multiple layers of membrane (the myelin sheath). The myelin sheath creates a high-resistance wall that prevents ion flow out of the cytoplasm. The myelin membranes are analogous to heavy coats of plastic surrounding electrical wires, as they increase the effective thickness of the axon membrane by as much as 100-fold.

As an action potential passes down the axon from trigger zone to axon terminal, it passes through alternating regions of myelinated axon and nodes of Ranvier (Fig. 8-18a ■). The conduction process is identical to that described previously for the unmyelinated axon, except that it occurs only at the nodes in myelinated axons. Each node has a high concentration of voltage-gated Na^+ channels, which open with depolarization and allow Na^+ into the axon. Sodium ions entering at a node reinforce the depolarization and keep the amplitude of the action potential constant as it passes from node to node. The apparent jump of the action potential from node to node is called **saltatory conduction**, from the Latin word *saltare*, meaning "to leap."

What makes conduction more rapid in myelinated axons? The answer is that channel opening slows conduction slightly. In unmyelinated axons, channels must open all the way down the axon membrane to maintain the amplitude of the action potential. In myelinated axons, only the nodes need Na^+ channels because of the insulating properties of the myelin membrane. As the action potential passes along myelinated segments, conduction is not slowed by channel opening.

Saltatory conduction thus is an effective alternative to large-diameter axons and allows rapid action potentials through small axons. A myelinated frog axon 10 μm in diameter conducts action potentials at the same speed as an unmyelinated 500-μm squid axon. Thus, electrical current flows through different axons at different rates, depending on the two parameters of axon diameter and myelination.

✓ CONCEPT CHECK

15. Place the following neurons in order of their speed of conduction, from fastest to slowest:

    (a) myelinated axon, diameter 20 μm
    (b) unmyelinated axon, diameter 20 μm
    (c) unmyelinated axon, diameter 200 μm

    Answers: p. 290

In *demyelinating diseases,* the loss of myelin from vertebrate neurons can have devastating effects on neural signaling. In the central and peripheral nervous systems, the loss of myelin slows the conduction of action potentials. In addition, when ions leak out of the now-uninsulated regions of membrane between the channel-rich nodes of Ranvier (Fig. 8-18b), the depolarization that reaches a node may not be above threshold and conduction may fail.

Multiple sclerosis is the most common and best-known demyelinating disease. It is characterized by a variety of neurological

RUNNING PROBLEM

Like multiple sclerosis, Guillain-Barré syndrome is an illness in which the myelin that insulates axons is destroyed. One way that multiple sclerosis and other demyelinating illnesses are diagnosed is through the use of a nerve conduction test. This test measures the combined strength of action potentials and the rate at which they are conducted as they travel down neurons.

Question 3:
In Guillain-Barré syndrome, what would you expect the results of a nerve conduction test to be?

| 244 | 246 | 248 | 268 | 270 | 275 | 283 | 285 |

complaints, including fatigue, muscle weakness, difficulty walking, and loss of vision. At this time, we can treat some of the symptoms but not the causes of demyelinating diseases, which are mostly either inherited or autoimmune disorders. Currently, researchers are using recombinant DNA technology to study demyelinating disorders in mice.

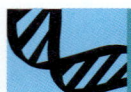

BIOTECHNOLOGY

MUTANT MOUSE MODELS

The use of animal models to study human diseases has become a valuable part of biomedical research. The mouse genome sequencing effort tells us that 99% of the approximately 30,000 mouse genes have direct human homologs (equivalents). This means that we can use the mouse to understand what genes, and the proteins they encode, do in healthy and sick humans. Sometimes, natural mutations produce animal diseases that resemble human diseases. Two examples of such mutants are the twitcher mouse, in which normal myelin degenerates owing to an inherited metabolic problem, and the wobbler mouse, in which somatic motor neurons controlling the limbs die. In other cases, scientists have used biotechnology techniques to create mice that lack specific genes (**knock-out mice**) or to breed mice that contain extra genes that were inserted artificially (**transgenic mice**). The mouse is the ideal organism to conduct these experiments. It is small, relatively inexpensive, and has a short lifespan. Additionally, its biological processes are similar to those of humans, and the mouse genome can be easily manipulated by genetic engineering technologies.

To learn more, read the articles in a special Mouse Genome issue of the journal *Nature* (*Nature* 420: 510–590, 2002). For an introduction, see the article written by the Mouse Genome Sequencing Consortium: "Initial sequencing and comparative analysis of the mouse genome," *Nature* 420: 520–562, 2002. You can obtain additional information from the Mouse Genome Informatics databases at The Jackson Laboratory (*www.informatics.jax.org*). Click on the *JAX® Mice and Services* button. In the *JAX® Mice* database you will find descriptions and photos of some of the different knock-out and transgenic mice being used for research.

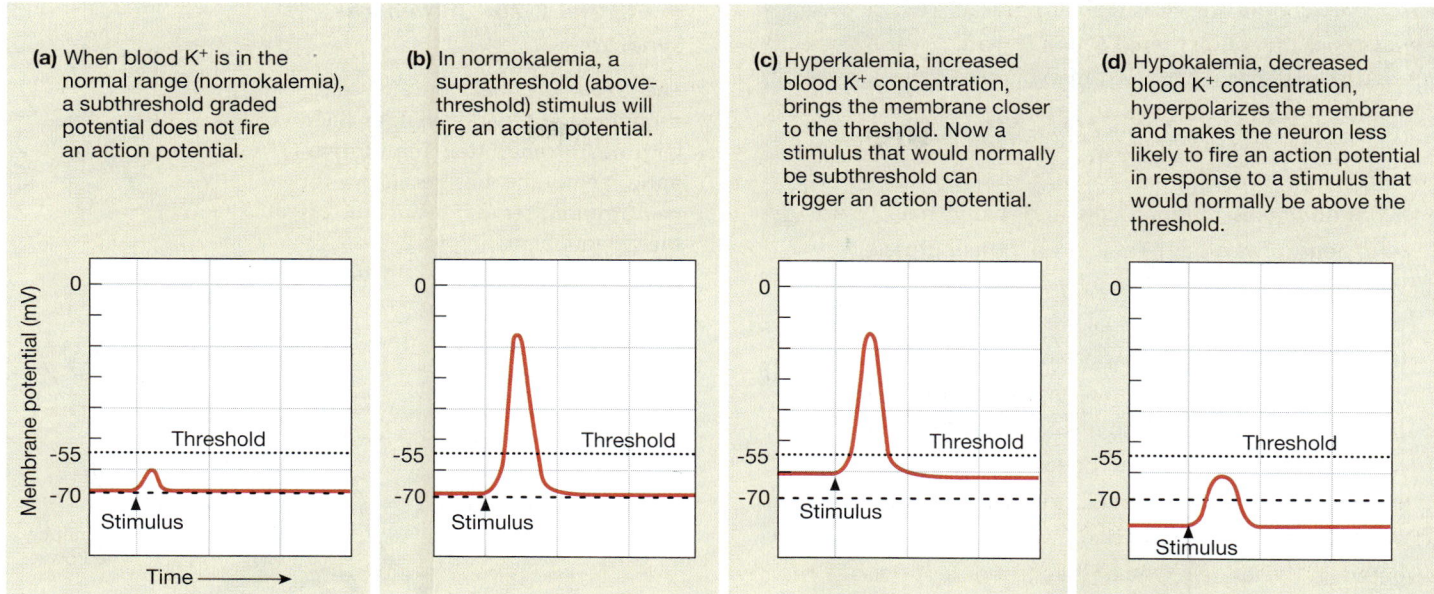

(a) When blood K$^+$ is in the normal range (normokalemia), a subthreshold graded potential does not fire an action potential.

(b) In normokalemia, a suprathreshold (above-threshold) stimulus will fire an action potential.

(c) Hyperkalemia, increased blood K$^+$ concentration, brings the membrane closer to the threshold. Now a stimulus that would normally be subthreshold can trigger an action potential.

(d) Hypokalemia, decreased blood K$^+$ concentration, hyperpolarizes the membrane and makes the neuron less likely to fire an action potential in response to a stimulus that would normally be above the threshold.

■ **FIGURE 8-19**   *Effect of extracellular potassium concentration on the excitability of neurons*

In hyperkalemia (c), the membrane depolarizes and the cell becomes more excitable. With hypokalemia (d), the membrane hyperpolarizes and the cell becomes less excitable.

Electrical Activity Can Be Altered by a Variety of Chemical Factors

A large variety of chemicals alter the conduction of action potentials by binding to Na$^+$, K$^+$, or Ca^{2+} channels in the neuron membrane. For example, some *neurotoxins* bind to and block Na$^+$ channels. Local anesthetics such as procaine, which block sensation, function the same way. If Na$^+$ channels are not functional, Na$^+$ cannot enter the axon. As a result, a depolarization that begins at the trigger zone loses strength as it moves down the axon, much like a normal graded potential. If the wave of depolarization manages to reach the axon terminal, it may be too weak to release neurotransmitter. As a result, the message of the presynaptic neuron is not passed on to the postsynaptic cell.

Alterations in the extracellular fluid concentrations of K$^+$ and Ca^{2+} are also associated with abnormal electrical activity in the nervous system. The relationship between extracellular fluid K$^+$ levels and the conduction of action potentials is the most straightforward and easiest to understand, as well as being one of the most clinically significant.

The concentration of K$^+$ in the blood and interstitial fluid is the major determinant of the resting potential of all cells. If K$^+$ concentration in the blood moves out of the normal range of 3.5–5 mmol/L, the result will be a change in the resting membrane potential of cells (Fig. 8-19 ■). This change is not important to most cells, but it can have serious consequences to the body as a whole because of the relationship between resting potential and the excitability of nervous and muscle tissue.

At normal K$^+$ levels, subthreshold graded potentials do not trigger action potentials, and suprathreshold graded potentials do (Fig. 8-19a, b). An increase in blood K$^+$ concentration—**hyperkalemia** [*hyper-*, above + *kalium*, potassium + *-emia*, in the blood]—will shift the resting membrane potential of a neuron closer to threshold and cause the cells to fire action potentials in response to smaller graded potentials (Fig. 8-19c). If blood K$^+$ concentration falls too low (**hypokalemia**), the resting membrane potential of the cells hyperpolarizes, moving farther from threshold. In this case, a stimulus strong enough to trigger an action potential when the resting potential is the normal -70 mV does not reach the threshold value (Fig. 8-19d). This condition shows up as muscle weakness because the neurons that control skeletal muscles are not firing normally.

Hypokalemia and its resultant muscle weakness are one reason that sport drinks supplemented with Na$^+$ and K$^+$, such as Gatorade™, were developed. When people sweat excessively, they lose both salts and water. If they replace this fluid loss with pure water, the K$^+$ remaining in the blood will be diluted, causing hypokalemia. By replacing sweat loss with a dilute salt solution, a person can prevent potentially dangerous drops in blood K$^+$ levels. Because of the importance of K$^+$ to normal function of the nervous system, the body regulates blood K$^+$ levels within a narrow range. We will discuss the important role of the kidneys in maintaining ion balance in Chapter 19.

8

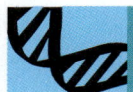

BIOTECHNOLOGY

OF SNAKES, SNAILS, SPIDERS, AND SUSHI

What do snakes, marine snails, and spiders have to do with neurophysiology? They all provide neuroscientists with compounds for studying synaptic transmission, extracted from the neurotoxic venoms these creatures use to kill their prey. The Asian snake *Bungarus multicinctus* provides us with α-bungarotoxin, a long-lasting poison that binds tightly to nicotinic acetylcholine receptors. The fish-hunting cone snail, *Conus geographus*, and the funnel web spider, *Agelenopsis aperta*, use toxins that block different types of voltage-gated Ca^{2+} channels. One of the most potent poisons known, however, comes from the Japanese puffer fish, a highly prized delicacy whose flesh is consumed as sushi. The puffer has tetrodotoxin (TTX) in its gonads. This neurotoxin blocks Na^+ channels on axons and prevents the transmission of action potentials, so ingestion of only a tiny amount can be fatal. The chefs who prepare the puffer fish, or *fugu*, for consumption are carefully trained to avoid contaminating the fish's flesh as they remove the toxic gonads. There's always some risk involved in eating *fugu*, though—one reason that the youngest person at the table is the first to sample the dish.

CELL-TO-CELL COMMUNICATION IN THE NERVOUS SYSTEM

Information flow through the nervous system using electrical and chemical signals is one of the most active areas of neuroscience research today. The specificity of neural communication depends on several factors: the signal molecules secreted by neurons, the target cell receptors for these chemicals, and the anatomical connections between neurons and their targets, which occur in regions known as synapses.

Information Passes from Cell to Cell at the Synapse

Each synapse has two parts: (1) the axon terminal of the *presynaptic cell* and (2) the membrane of the *postsynaptic cell* (Fig. 8-20 ■). In a neural reflex, information moves from presynaptic cell to postsynaptic cell. The postsynaptic cells may be neurons or non-neuronal cells. In most neuron-to-neuron synapses, the presynaptic axon terminals are next to either the dendrites or the cell body of the postsynaptic neuron.

■ **FIGURE 8-20**   *A chemical synapse*

The axon terminal contains mitochondria and synaptic vesicles filled with neurotransmitter. The postsynaptic membrane has receptors for neurotransmitter that diffuses across the synaptic cleft.

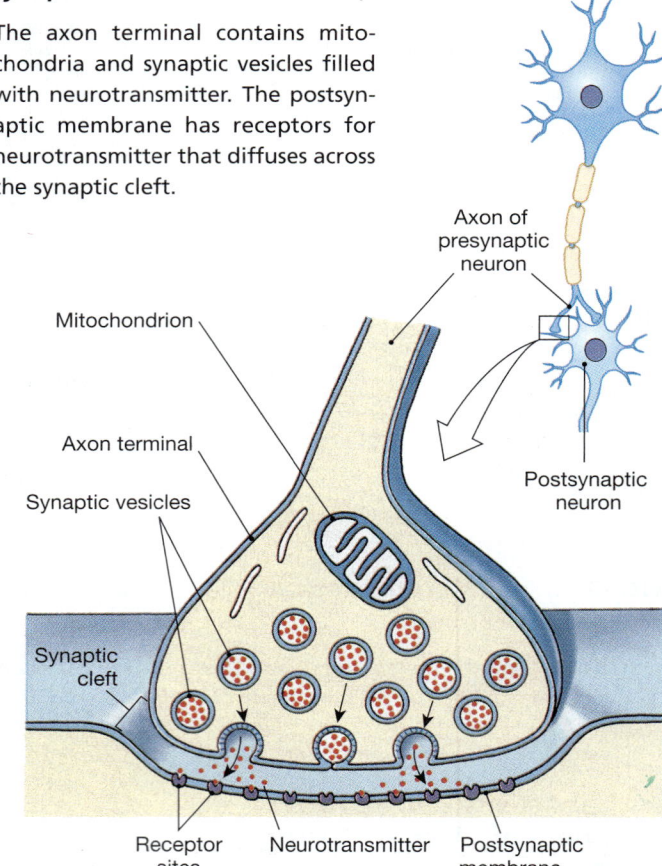

In general, postsynaptic neurons with many dendrites will also have many synapses. A moderate number of synapses is 10,000, but some cells in the brain are estimated to have 150,000 or more synapses on their dendrites! Synapses can also occur on the axon and even at the axon terminal of the postsynaptic cell. Synapses are classified as electrical or chemical depending on the

RUNNING PROBLEM

Dr. McKhann decided to perform nerve conduction tests on some of the paralyzed children in Beijing Hospital. He found that although the rate of conduction along the children's neurons was normal, the action potentials were greatly diminished.

Question 4:
   Is the paralytic illness that affected the Chinese children a demyelinating condition? Why or why not?

244 246 248 268 **270** 275 283 285

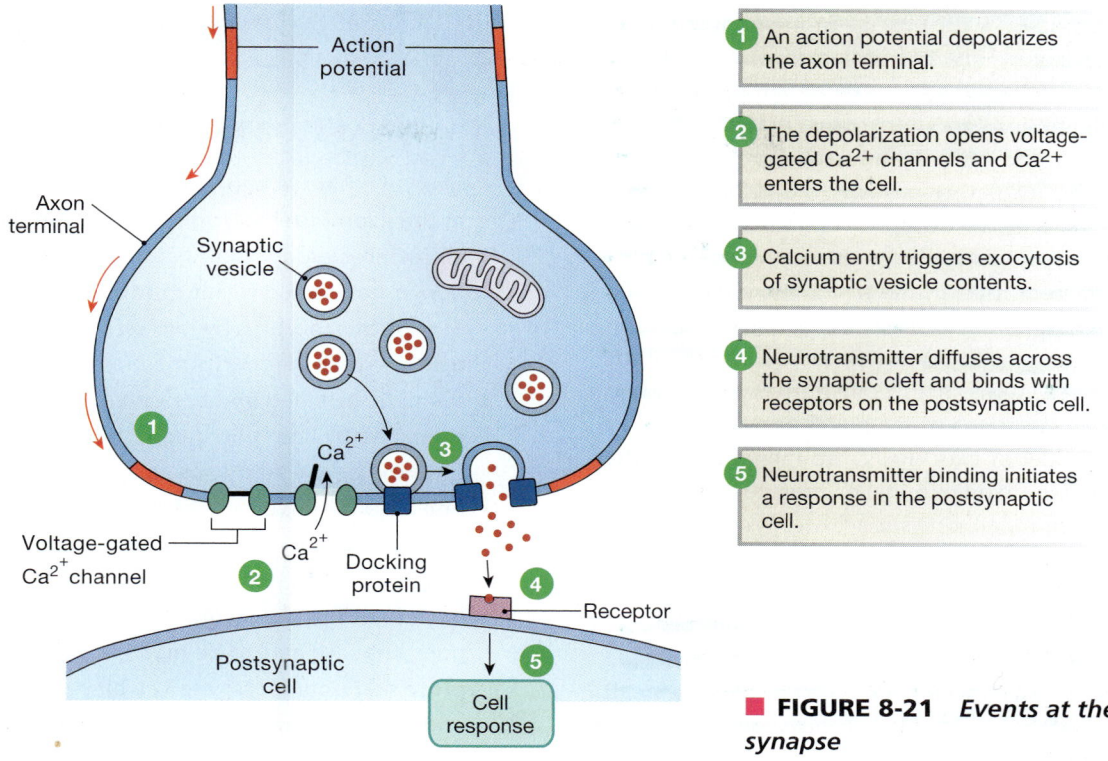

1 An action potential depolarizes the axon terminal.

2 The depolarization opens voltage-gated Ca^{2+} channels and Ca^{2+} enters the cell.

3 Calcium entry triggers exocytosis of synaptic vesicle contents.

4 Neurotransmitter diffuses across the synaptic cleft and binds with receptors on the postsynaptic cell.

5 Neurotransmitter binding initiates a response in the postsynaptic cell.

■ **FIGURE 8-21**   *Events at the synapse*

type of signal that passes from the presynaptic cell to the postsynaptic one.

Electrical Synapses   **Electrical synapses** pass an electrical signal, or current, directly from the cytoplasm of one cell to another through gap junctions. Information can flow in both directions through gap junctions. Electrical synapses occur mainly in neurons of the CNS. They are also found in glial cells, in cardiac and smooth muscle, and in nonexcitable cells that use electrical signals, such as the pancreatic beta cell. The primary advantage of electrical synapses is rapid conduction of signals from cell to cell that synchronizes activity within a network of cells. Gap junctions also allow chemical signal molecules to diffuse between adjacent cells.

Chemical Synapses   The vast majority of synapses in the nervous system are **chemical synapses**, which use neurotransmitters to carry information from one cell to the next. At chemical synapses, the electrical signal of the presynaptic cell is converted into a chemical signal that crosses the synaptic cleft between the presynaptic neuron and its target. Combination of the neurotransmitter with its receptor on the postsynaptic cell either initiates an electrical response or activates a second messenger pathway [🔁 p. 180].

Neurotransmitter synthesis can take place either in the nerve cell body or in the axon terminal. However, axon terminals do not have the organelles needed for protein synthesis. Consequently, polypeptide neurotransmitters and protein enzymes needed for metabolism in the axon terminal must be made in the cell body. The dissolved enzymes are brought to

axon terminals by slow axonal transport, but the neurotransmitters, which are consumed more rapidly than enzymes, are moved in vesicles by fast axonal transport.

When we examine the axon terminal of a presynaptic cell with an electron microscope, we find many small synaptic vesicles and large mitochondria in the cytoplasm (Fig. 8-20). Some vesicles are "docked" at active zones along the membrane closest to the synaptic cleft, waiting for a signal to release their contents. Other vesicles act as a reserve pool, clustering close to the docking sites. Each vesicle contains neurotransmitter that is released on demand.

CONCEPT CHECK

16. Which organelles are needed to synthesize proteins and package them into vesicles?

17. What is the function of mitochondria in a cell?

18. How do mitochondria get to the axon terminals?

Answers: p. 290

Calcium Is the Signal for Neurotransmitter Release at the Synapse

The release of neurotransmitters into the synaptic cleft takes place by exocytosis. From what we can tell, exocytosis in neurons is similar to exocytosis in other types of cells, such as the pancreatic beta cell described in Chapter 5 [🔁 p. 166]. Neurotoxins that block neurotransmitter release, including tetanus and botulinum toxins, exert their action by inhibiting specific proteins of the cell's exocytotic apparatus.

Figure 8-21 ■ shows how neurotransmitters are released by exocytosis. When the depolarization of an action potential reaches the axon terminal①, the change in membrane potential sets off a sequence of events. The axon terminal membrane has voltage-gated Ca^{2+} channels that open in response to depolarization②. Calcium ions are more concentrated in the extracellular fluid than in the cytosol, and so they move into the cell. The Ca^{2+} binds to regulatory proteins and initiates exocytosis③. The membrane of the synaptic vesicle fuses with the cell membrane, aided by multiple membrane proteins. The fused area opens and neurotransmitter inside the synaptic vesicle moves into the synaptic cleft ④. The neurotransmitter molecules diffuse across the gap to bind with membrane receptors on the postsynaptic cell. When neurotransmitters bind to their receptors, a response is initiated in the postsynaptic cell ⑤.

In the classic model of exocytosis, the membrane of the vesicle becomes part of the axon terminal membrane [🔁 Fig. 5-24, p. 149]. This increase in membrane surface area is countered by endocytotic recycling of the vesicles at regions away from the active sites (see Fig. 8-4, p. 249). However, a second model of secretion is emerging. In this model, called the **kiss-and-run pathway,** synaptic vesicles fuse to the presynaptic membrane at a complex called the **fusion pore.** This fusion opens a small channel that is just large enough for neurotransmitter to pass through. Then, instead of opening the fused area wider and incorporating the vesicle membrane into the cell membrane, the vesicle pulls back from the fusion pore and returns to the pool of vesicles in the cytoplasm.

✔ CONCEPT CHECK

19. In an experiment on synaptic transmission, a synapse was bathed in a Ca^{2+}-free medium that was otherwise equivalent to extracellular fluid. An action potential was triggered in the presynaptic neuron. Although the action potential reached the axon terminal at the synapse, the usual response of the postsynaptic cell did not occur. What conclusion did the researchers make on the basis of these results?

Answers: p. 290

Neurocrines Convey Information from Neurons to Other Cells

What are the neurocrine signal molecules that neurons release? Their chemical composition is varied, and they may function as neurotransmitters, neuromodulators, or neurohormones [🔁, page 176]. Neurotransmitters and neuromodulators act as *paracrine signals,* with target cells located close to the neuron that secretes them. Neurohormones, on the other hand, are secreted into the blood and distributed throughout the body [🔁 p. 213].

Generally, neurotransmitters act at a synapse and elicit a rapid response. Neuromodulators act at both synaptic and non-synaptic sites and are slower acting. Some neuromodulators and neurotransmitters also act on the cell that secretes them, making them *autocrine* signals as well as paracrines. The number of molecules identified as neurotransmitters and neuromodulators is large and growing daily.

EMERGING CONCEPTS

SYNAPTIC VESICLES KISS AND RUN

Most of what we know about synaptic vesicle release at the axon terminal comes from a combination of electrophysiology studies and microscopic examination. It has been difficult to follow the dynamic process of exocytosis by microscopy, however. Electron microscopy is performed in a vacuum, so the tissue must be killed and preserved. Living cells can be examined by light microscopy, but the resolution of even the best light microscope is too low to show fine details. However, a new technique called **atomic force microscopy** (AFM) is enabling scientists to watch secretion in live cells. With AFM, researchers have found a new membrane structure, the fusion pore, that appears to be made of actin and docking proteins. Fusion pores are key players in the kiss-and-run pathway, in which synaptic vesicles join the cell membrane only long enough to release their contents. Fusion pores have been observed in endocrine and exocrine cells as well as in neural tissue, suggesting that their presence may be essential in all forms of secretion. To learn more about fusion pores and to see computerized images of them from AFM, read "Fusion pore in live cells," *News in Physiological Sciences* 17: 219–222, December 2002 (*www.nips.org*).

The Nervous System Secretes a Variety of Neurocrines

The array of neurocrines in the body is truly staggering (Table 8-4 ■). They can be informally grouped into seven classes according to their structure: (1) acetylcholine, (2) amines, (3) amino acids, (4) purines, (5) gases, (6) peptides, and (7) lipids. CNS neurons release many different neurocrines, including some polypeptides known mostly for their hormonal activity. In contrast, the PNS secretes only three major neurocrines: the neurotransmitters acetylcholine and norepinephrine, and the neurohormone epinephrine.

Acetylcholine **Acetylcholine** (ACh), in a chemical class by itself, is synthesized from choline and acetyl coenzyme A (acetyl CoA). Choline is a small molecule also found in membrane phospholipids. Acetyl CoA is the metabolic intermediate that links glycolysis to the citric acid cycle [🔁 p. 107]. The synthesis of ACh from these two precursors is a simple enzymatic reaction that takes

TABLE 8-4	Major Neurocrines			
CHEMICAL	**RECEPTOR**	**TYPE**	**RECEPTOR LOCATION**	**KEY AGONISTS, ANTAGONISTS, AND POTENTIATORS***
Acetylcholine (ACh)	Cholinergic			
	Nicotinic	ICR** (Na^+, K^+)	Skeletal muscles, autonomic neurons, CNS	Nicotine: agonist; curare, α-bungarotoxin: antagonists
	Muscarinic	GPCR	Smooth and cardiac muscle, endocrine and exocrine glands, CNS	Muscarine: agonist; atropine: antagonist
Amines				
Norepinephrine (NE)	Adrenergic (α, β)	GPCR	Smooth and cardiac muscle, endocrine and exocrine glands, CNS	α: Prazosin (Minipress); β: propranolol
Dopamine (DA)	Dopamine (D)	GPCR	CNS	Antipsychotic drugs: antagonists; bromocriptine: agonist
Serotonin (5-hydroxytryptamine, 5-HT)	Serotonergic (5-HT)	ICR (Na^+, K^+) GPCR	CNS	Sumatriptan: agonist LSD: antagonist
Histamine	Histamine (H)	GPCR	CNS	Ranitidine (Zantac®) and cimetidine (Tagamet®): antagonists
Amino acids				
Glutamate	Glutaminergic ionotropic			
	AMPA	ICR (Na^+, K^+)	CNS	
	NMDA	ICR (Na^+, K^+, Ca^{2+})	CNS	
	Glutaminergic metabotropic	GPCR	CNS	Glycine: potentiator
GABA (γ-aminobutyric acid)	GABA ($GABA_A$, $GABA_B$)	ICR (Cl^-) GPCR	CNS	Picrotoxin: antagonist; alcohol, barbiturates: potentiators
Glycine	Glycine	ICR (Cl^-)	CNS	Strychnine: antagonist
Purines				
Adenosine	Purine (P)	GPCR	CNS	
Gases				
Nitric Oxide (NO)	None	N/A	N/A	

*This list does not include many chemicals that are used as agonists and antagonists in physiological research. To review potentiation, see p. 231.

**ICR = ion channel-receptor; GPCR = G protein-coupled receptor; AMPA = α-amino-3-hydroxy-5-methyl-4 isoxazole proprionic acid; NMDA = N-methyl-D-aspartate; LSD = lysergic acid diethylamine; N/A = not applicable.

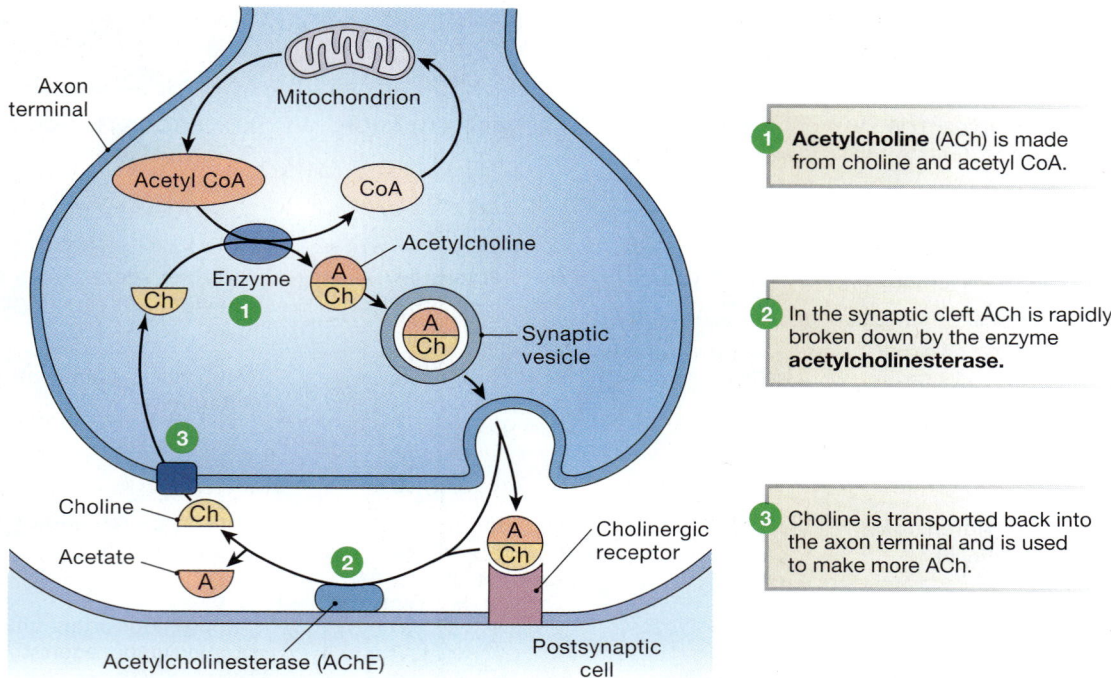

Axon terminal

Mitochondrion

Acetyl CoA

CoA

Enzyme

Acetylcholine

Ch

A Ch

1

A Ch

Synaptic vesicle

3

Choline

Ch

Acetate

A

2

A Ch

Cholinergic receptor

Acetylcholinesterase (AChE)

Postsynaptic cell

1 **Acetylcholine** (ACh) is made from choline and acetyl CoA.

2 In the synaptic cleft ACh is rapidly broken down by the enzyme **acetylcholinesterase.**

3 Choline is transported back into the axon terminal and is used to make more ACh.

■ **FIGURE 8-22**   *Synthesis and recycling of acetylcholine at a synapse*

8

place in the axon terminal (Fig. 8-22 ■). Neurons that secrete ACh and receptors that bind ACh are described as **cholinergic**.

Amines   Amine neurotransmitters, like the amine hormones [ p. 222], are derived from single amino acids. The amino acid tyrosine is converted to **dopamine, norepinephrine**, and **epinephrine**. All three of these neurocrines also function as neurohormones when secreted by the adrenal medulla [ Fig. 11-10, p. 385].

Neurons that secrete norepinephrine are called **adrenergic neurons**, or, more properly, **noradrenergic neurons**. The adjective *adrenergic* does not have the same obvious link to its neurotransmitter as *cholinergic* does to *acetylcholine*. Instead, the adjective derives from the British name for epinephrine, *adrenaline*. In the early part of the twentieth century, British researchers thought that sympathetic neurons secreted adrenaline (epinephrine), hence the modifier *adrenergic*. Although our understanding has changed, the name persists. Whenever you see reference to "adrenergic control" of a function, you must make the connection to a neuron secreting norepinephrine.

Other amine neurotransmitters include **serotonin** (also called *5-hydroxytryptamine* or **5-HT**), made from the amino acid tryptophan, and histamine, made from histidine. The amine neurotransmitters are all active in the CNS. In addition, norepinephrine is the major neurotransmitter of the peripheral autonomic sympathetic division.

Amino Acids   At least four amino acids function as neurotransmitters in the CNS. **Glutamate** is the primary excitatory neurotransmitter of the CNS, and **aspartate** serves the same

CLINICAL FOCUS

MYASTHENIA GRAVIS

What would you think was wrong if suddenly your eyelids started drooping, you had difficulty watching moving objects, and it became difficult to chew, swallow, and talk? What disease attacks these skeletal muscles but leaves the larger muscles of the arms and legs alone? The answer is **myasthenia gravis** [*myo-*, muscle + *asthenes*, weak + *gravis*, severe], an autoimmune disease in which the body fails to recognize the acetylcholine (ACh) receptors on skeletal muscle as part of "self." The immune system then produces antibodies to attack the receptors. The antibodies bind to the ACh receptor protein and change it in some way that causes the muscle cell to withdraw the receptors from the membrane and destroy them. This destruction leaves the muscle with fewer ACh receptors in the membrane. Even though neurotransmitter release is normal, the muscle target has a diminished response that is exhibited as muscle weakness. Currently medical science does not have an effective treatment for myasthenia gravis, although various drugs can help control its symptoms. To learn more about this disease, visit the website for the Myasthenia Gravis Foundation of America at *www.myasthenia.org*.

function in selected regions of the brain. The main inhibitory neurotransmitter in the brain is **gamma-aminobutyric acid (GABA)**. The amino acid **glycine** is also considered inhibitory, although it can bind to one type of glutamate receptor and be excitatory. Glycine is the primary inhibitory neurotransmitter of the spinal cord.

Peptides   The nervous system secretes a variety of peptides that act as neurotransmitters and neuromodulators in addition to functioning as neurohormones. These peptides include **substance P**, involved in some pain pathways, and the **opioid peptides** (**enkephalins** and **endorphins**) that mediate *analgesia* [*an-*, without + *algos*, pain]. Peptides that function as both neurohormones and neurotransmitters include *cholecystokinin (CCK)*, *vasopressin,* and *atrial natriuretic peptide.* Many peptide neurotransmitters are cosecreted with other neurotransmitters.

Purines   *Adenosine, adenosine monophosphate* (AMP), and *adenosine triphosphate* (ATP) can all act as neurotransmitters. These molecules, known collectively as *purines* [p. 33], bind to *purinergic* receptors in the CNS and on other excitable tissues such as the heart.

Gases   One of the most interesting neurotransmitters is *nitric oxide* (NO), an unstable gas synthesized from oxygen and the amino acid arginine. Nitric oxide acting as a neurotransmitter diffuses freely into a target cell rather than binding to a membrane receptor [p. 187]. Once inside the target cell, nitric oxide binds to proteins. With a half-life of only 2–30 seconds, nitric oxide is elusive and difficult to study. It is also released from cells other than neurons and often acts as a paracrine.

Recent work suggests that *carbon monoxide* (CO), best known as a toxic gas, is produced by the body in tiny amounts to serve as a neurotransmitter.

Lipids   Lipid neurocrines include several eicosanoids [p. 29] that are the endogenous ligands for *cannabinoid receptors.* The CB_1 cannabinoid receptor is found in the brain; the CB_2 receptor is found on immune cells. The receptors were named for one of their exogenous ligands, Δ^9-tetrahydrocannabinoid (THC), which comes from the plant *Cannabis sativa,* more commonly known as marijuana.

Multiple Receptor Types Amplify the Effects of Neurotransmitters

All neurotransmitters except nitric oxide have one or more receptor types with which they bind. Each receptor type may have multiple subtypes, a property that allows one neurotransmitter to have different effects in different tissues. Receptor subtypes are distinguished by combinations of letter and number subscripts. For example, serotonin (5-HT) has at least 20 receptor subtypes that have been identified, including 5-HT$_{1A}$ and 5-HT$_4$.

RUNNING PROBLEM

Dr. McKhann then asked to see autopsy reports on some of the children who had died of their paralysis at Beijing Hospital. In the reports, pathologists noted that the patients had normal myelin but damaged axons. In some cases, the axon had been completely destroyed, leaving only a hollow shell of myelin.

Question 5:
   Do the results of Dr. McKhann's investigation suggest that the Chinese children had Guillain-Barré syndrome? Why or why not?

Neurotransmitter receptors fall into two of the membrane receptor categories that we discussed in Chapter 6 [p. 179]: ligand-gated ion channels and G protein–coupled receptors (GPCR). Receptors that alter ion channel function are called *ionotropic* receptors, whereas those that exert their actions through second messenger systems are called *metabotropic* receptors. Some metabotropic GPCRs regulate the opening or closing of ion channels.

The study of neurotransmitters and their receptors has been greatly simplified by two advances in molecular biology. The genes for many receptor subtypes have been cloned, allowing researchers to create mutant receptors and study their properties. In addition, researchers have discovered or synthesized a variety of agonists and antagonist molecules that mimic or inhibit neurotransmitter activity by binding to the receptors [p. 189].

Table 8-4 includes descriptions of receptor types and some of their agonists or antagonists.

Cholinergic Receptors   **Cholinergic receptors** come in two main subtypes: **nicotinic**, named because *nicotine* is an agonist, and **muscarinic**, for which *muscarine*, a compound found in some fungi, is an agonist. Cholinergic nicotinic receptors are found on skeletal muscle, in the autonomic division of the PNS, and in the CNS. Nicotinic receptors are monovalent cation channels through which both Na^+ and K^+ can pass. Sodium entry into cells exceeds K^+ exit because the electrochemical gradient for Na^+ is stronger. As a result, net Na^+ entry depolarizes the postsynaptic cell and makes it more likely to fire an action potential.

Cholinergic muscarinic receptors come in five related subtypes. They are all coupled to G proteins and are linked to second messenger systems. The tissue response to activation of

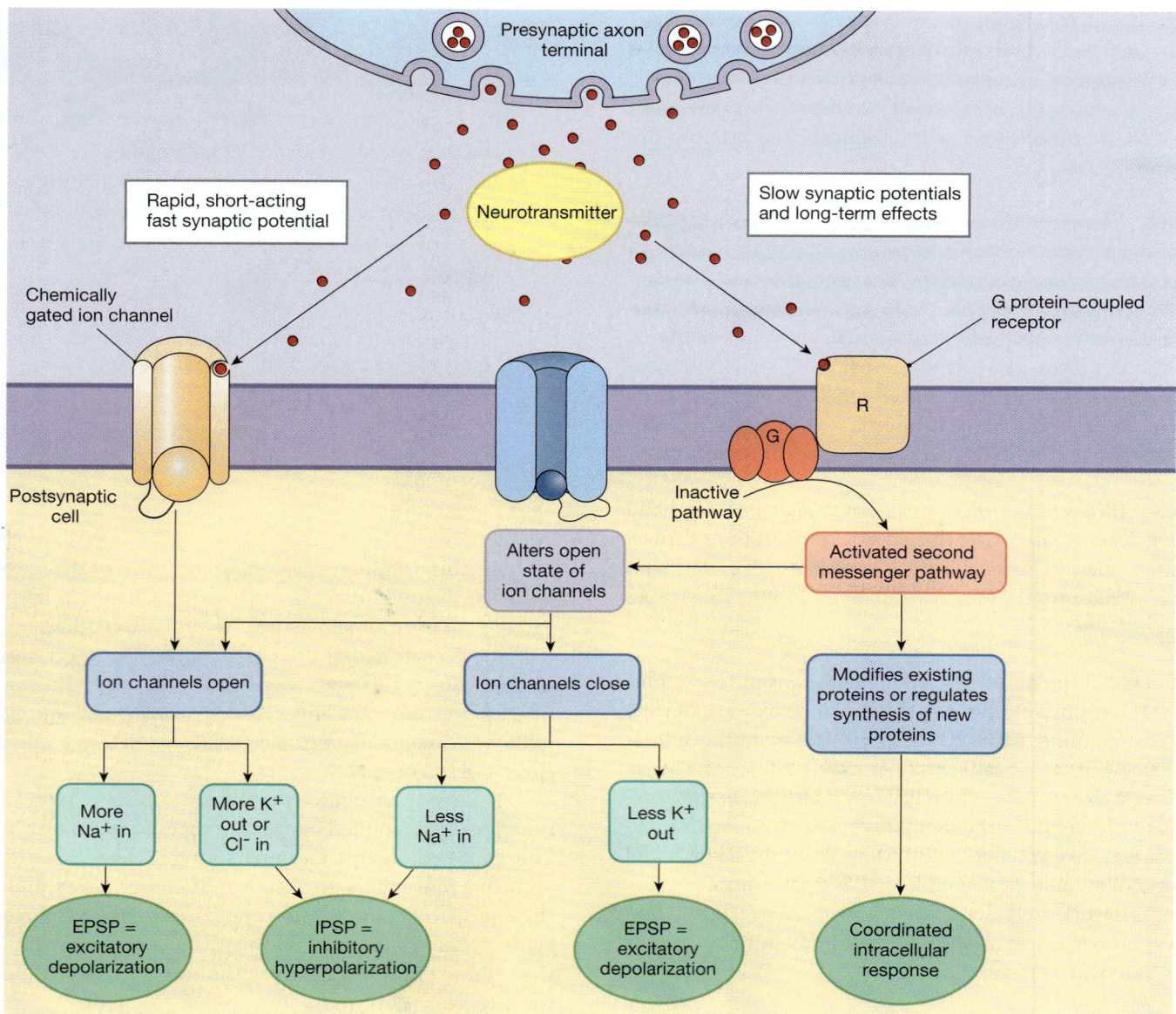

■ FIGURE 8-23 *Fast and slow responses in postsynaptic cells*

Some neurotransmitters create rapid, short-acting responses by directly opening ion channels. Other neurotransmitters create slower, longer-lasting responses by activating second messenger systems.

a muscarinic receptor varies with the receptor subtype. These receptors occur in the CNS and in the autonomic parasympathetic division of the PNS.

Adrenergic Receptors Adrenergic receptors are divided into two classes: α (alpha) and β (beta), with multiple subtypes of each. Like cholinergic muscarinic receptors, adrenergic receptors are linked to G proteins and initiate second messenger cascades. Indeed, it was the action of epinephrine on β-receptors in dog liver that led E. W. Sutherland to the discovery of cyclic AMP and the concept of second messenger systems as transducers

of extracellular messengers [🔁 p. 183]. The two classes of adrenergic receptors work through different second messenger pathways.

✓ CONCEPT CHECK

20. When pharmaceutical companies design drugs, they try to make a given drug as specific as possible for the particular receptor subtype they are targeting. For example, a drug might target adrenergic β_1- receptors rather than all adrenergic α- and β-receptors. What would be the advantage to this?

Answers: p. 290

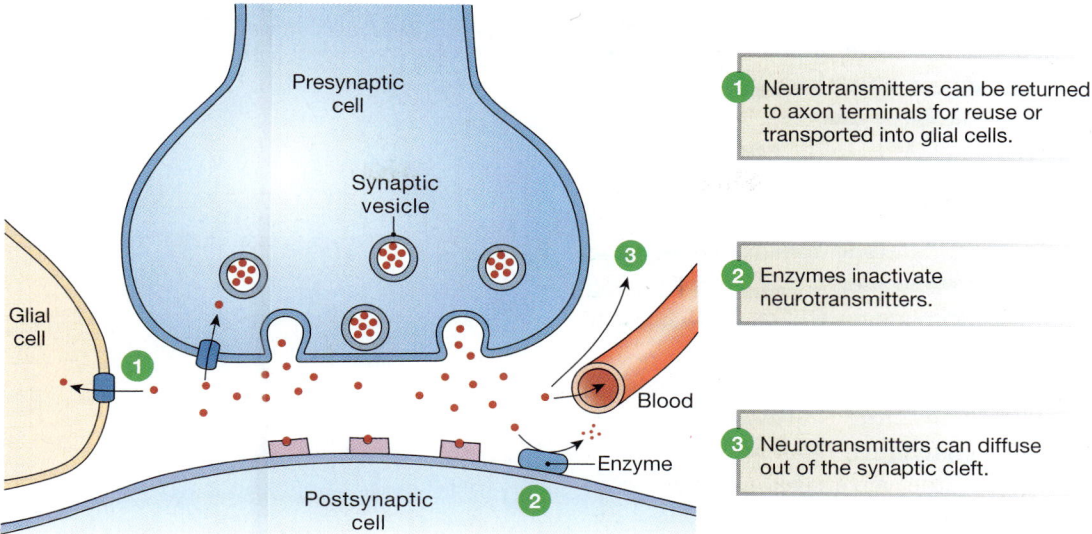

Presynaptic cell

Synaptic vesicle

Glial cell

Postsynaptic cell

Blood

Enzyme

1 Neurotransmitters can be returned to axon terminals for reuse or transported into glial cells.

2 Enzymes inactivate neurotransmitters.

3 Neurotransmitters can diffuse out of the synaptic cleft.

■ **FIGURE 8-24**   *Inactivation of neurotransmitters*

Glutaminergic Receptors   Glutamate is the main excitatory neurotransmitter in the CNS and also acts as a neuromodulator. The action of glutamate at a particular synapse depends on which of its receptor types occurs on the target cell. Metabotropic glutaminergic receptors act through GPCRs. Two ionotropic glutamate receptors are receptor-channels. NMDA receptors are named for the glutamate agonist N-*methyl*-D-*aspartate,* and AMPA receptors are named for their agonist α-*amino-3-hydroxy-5-methylisoxazole-4-proprionic acid.*

AMPA receptors are ligand-gated monovalent cation channels similar to nicotinic acetylcholine channels. Glutamate binding opens the channel, and the cell depolarizes due to net Na$^+$ influx.

NMDA receptors are unusual for several reasons. First, they are cation channels that allow Na$^+$, K$^+$, and Ca^{2+} to pass through the channel. Second, channel opening requires both glutamate binding and a change in membrane potential. The NMDA receptor channel is blocked by a magnesium ion (Mg^{2+}) at resting membrane potentials. Glutamate binding opens the ligand-activated gate, but ions cannot flow past the Mg^{2+}. However, if the cell depolarizes, the Mg^{2+} blocking the channel is expelled, and then ions flow through the pore (see Fig. 8-30).

Not All Postsynaptic Responses Are Rapid and of Short Duration

A neurotransmitter combining with its receptor sets in motion a series of responses in the postsynaptic cell (Fig. 8-23 ■). In the simplest response, the neurotransmitter binds to and opens a receptor-channel on the postsynaptic cell, leading to ion movement between the postsynaptic cell and the extracellular fluid. The resulting change in membrane potential is called a **fast synaptic potential** because it begins quickly and lasts only a few milliseconds. If the synaptic potential is depolarizing, it is called an **excitatory postsynaptic potential** (EPSP) because it makes the cell more likely to fire an action potential. If the synaptic potential is hyperpolarizing, it is called an **inhibitory postsynaptic potential** (IPSP) because hyperpolarization moves the membrane potential farther from threshold and makes the cell less likely to fire an action potential.

In slow postsynaptic responses, neurotransmitters bind to G protein-coupled receptors linked to second messenger systems. The second messengers may act from the cytoplasmic side of the cell membrane to open or close ion channels. (Fast synaptic potentials always open ion channels.) Membrane potentials resulting from this process are called **slow synaptic potentials** because the second messenger method takes longer to create a response. In addition, the response itself lasts longer, usually seconds to minutes.

Slow postsynaptic responses are not limited to altering the open state of ion channels. Neurotransmitter activation of second messenger systems may also modify existing cell proteins or regulate the production of new cell proteins. This type of slow response has been linked to the growth and development of neurons and to the mechanisms underlying long-term memory.

Neurotransmitter Activity Is Rapidly Terminated

A key feature of neural signaling is its short duration, which is achieved by the rapid removal or inactivation of neurotransmitter in the synaptic cleft. This can be accomplished in various ways (Fig. 8-24 ■). Some neurotransmitter molecules simply diffuse away from the synapse, separating them from their receptors. Other neurotransmitters are inactivated by enzymes in the synaptic cleft. Many neurotransmitters are removed from the extracellular fluid by transport either back into the presynaptic cell or into adjacent neurons or glial cells.

For example, acetylcholine (ACh) in the extracellular fluid is rapidly broken down by the enzyme **acetylcholinesterase** (AChE) in the extracellular matrix and in the membrane of the postsynaptic cell (see Fig. 8-22). Choline from degraded ACh is actively transported back into the presynaptic axon terminal and used to make new acetylcholine to refill recycled synaptic vesicles.

In contrast, norepinephrine action at the target tissue is terminated when the neurotransmitter is actively transported back into the presynaptic axon terminal. Once back in the axon terminal, norepinephrine is either repackaged into vesicles or broken down by intracellular enzymes such as *monoamine oxidase (MAO)*, found in mitochondria.

✓ CONCEPT CHECK

21. One class of antidepressant drugs is called selective serotonin reuptake inhibitors (SSRIs). What do these drugs do to serotonin activity at the synapse?

Answers: p. 290

INTEGRATION OF NEURAL INFORMATION TRANSFER

Traditionally, scientists viewed chemical synapses as sites of one-way communication, with all messages moving from a presynaptic cell to a postsynaptic cell. We now know that this is not always the case. In the brain, there are some synapses where cells on both sides of the synaptic cleft release neurotransmitter that acts on the opposite cell. Perhaps more important, we have learned that many postsynaptic cells "talk back" to their presynaptic neurons by sending neuromodulators that bind to presynaptic receptors. Let's examine some of the ways that complex neural pathways communicate.

Neural Pathways May Involve Many Neurons Simultaneously

Communication between neurons is not always a one-to-one event. Sometimes a single presynaptic neuron branches, and its collaterals synapse on multiple target neurons. This pattern is known as **divergence** (Fig. 8-25a ■). If a larger number of presynaptic neurons provides input to a smaller number of postsynaptic neurons, the pattern is known as **convergence** (Fig. 8-25b). Combination of convergence and divergence in the CNS may result in one postsynaptic neuron with synapses from as many as 10,000 presynaptic neurons (Fig. 8-26 ■).

One advantage of convergence in the nervous system is that input from multiple sources can influence the output of a single postsynaptic cell. When two or more presynaptic neurons converge on the dendrites or cell body of a single postsynaptic

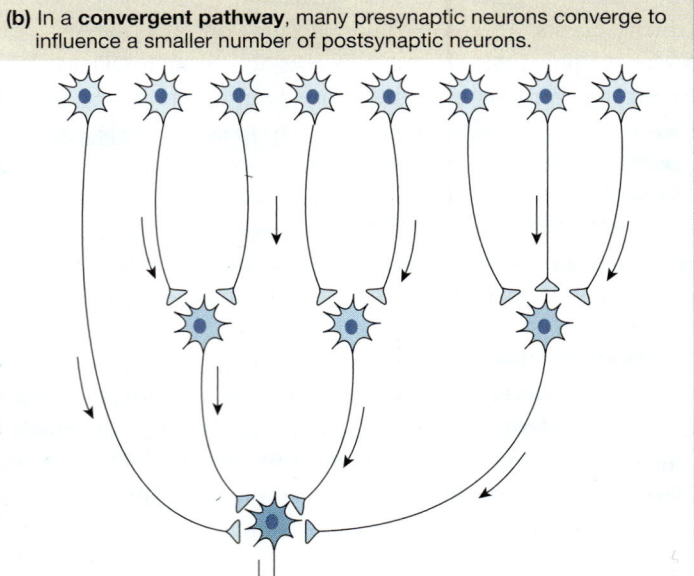

(a) In a **divergent pathway**, one presynaptic neuron branches to affect a larger number of postsynaptic neurons.

(b) In a **convergent pathway**, many presynaptic neurons converge to influence a smaller number of postsynaptic neurons.

■ **FIGURE 8-25** *Convergence and divergence*

cell, the response of the postsynaptic cell will be determined by the summed input from the presynaptic neurons. If the stimuli all create subthreshold excitatory postsynaptic potentials (EPSPs), the EPSPs can add together to create a suprathreshold potential at the trigger zone. The initiation of an action potential from several simultaneous graded potentials is an example of **spatial summation**. The word *spatial* [*spatium*, space] refers to the fact that the graded potentials originate at different locations (spaces) on the neuron.

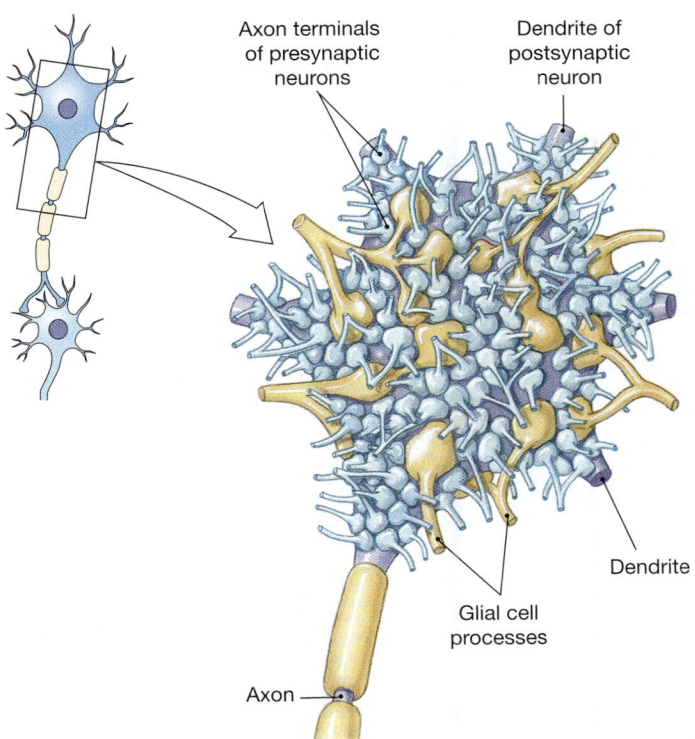

Axon terminals of presynaptic neurons

Dendrite of postsynaptic neuron

Dendrite

Glial cell processes

Axon

■ FIGURE 8-26 *The abundance of synapses on a postsynaptic neuron*

The cell body and dendrites of a somatic motor neuron are nearly covered by hundreds of axon terminals from other neurons.

Figure 8-27a ■ illustrates spatial summation when three excitatory presynaptic neurons converge on one postsynaptic neuron. Each EPSP is too weak to trigger an action potential by itself, but if the three presynaptic neurons fire simultaneously, the sum of the three EPSPs is suprathreshold and creates an action potential.

Postsynaptic inhibition may occur when a presynaptic neuron releases an inhibitory neurotransmitter onto a postsynaptic cell and alters its response. Figure 8-27b shows three neurons, two excitatory and one inhibitory, converging on a postsynaptic cell. The neurons fire, creating one inhibitory postsynaptic potential (IPSP) and two excitatory graded potentials that sum as they reach the trigger zone. The IPSP counteracts the two EPSPs, creating an integrated signal that is below threshold. As a result, no action potential leaves the trigger zone.

Summation of graded potentials does not always require input from more than one presynaptic neuron. Two subthreshold graded potentials from the same presynaptic neuron can be summed if they arrive at the trigger zone close enough together in time. Summation that occurs from graded potentials overlapping in time is called **temporal summation** [*tempus*, time]. Let's see how this can happen.

Figure 8-28a ■ shows recordings from an electrode placed in the trigger zone of a neuron. A stimulus (X_1) starts a subthreshold graded potential on the cell body at the time marked on the x-axis. The graded potential reaches the trigger zone and depolarizes it, as shown on the graph (A_1). A second stimulus (X_2) occurs later, and its subthreshold graded potential (A_2) reaches the trigger zone some time after the first. The interval between the two stimuli is so long that the two graded potentials do not overlap. Neither potential by itself is above threshold, so no action potential is triggered.

In Figure 8-28b, the two stimuli are given closer together in time. As a result, the two subthreshold graded potentials arrive at the trigger zone at almost the same time. The second graded potential adds its depolarization to that of the first, causing the trigger zone to depolarize to threshold.

In many situations, graded potentials in a neuron will incorporate both temporal and spatial summation. The summation of graded potentials demonstrates a key property of neurons: *postsynaptic integration*. When multiple signals reach a neuron, postsynaptic integration allows the neuron to evaluate the strength and duration of the signals. If the resultant integrated signal is above threshold, the neuron fires an action potential.

CONCEPT CHECK

22. In Figure 8-27b, assume the postsynaptic neuron has a resting membrane potential of −70 mV and a threshold of −55 mV. If the inhibitory presynaptic neuron creates an IPSP of 5 mV and the two excitatory presynaptic neurons have EPSPs of 10 and 12 mV, will the postsynaptic neuron fire an action potential?

23. In the graphs of Figure 8-28, why doesn't the membrane potential change at the same time as the stimulus?

Answers: p. 290

Synaptic Activity Can Also Be Modulated at the Axon Terminal

The examples of modulation just discussed all took place on the postsynaptic side of a synapse, but the activity of presynaptic cells can also be altered. When a modulatory neuron (inhibitory or excitatory) terminates on or close to an axon terminal of a presynaptic cell, summation of its EPSPs or IPSPs with the action potential reaching the terminal creates presynaptic modulation. If activity in the modulatory neuron decreases neurotransmitter release, the modulation is called *presynaptic inhibition* (Fig. 8-29a ■). In *presynaptic facilitation*, modulatory input increases neurotransmitter release by the presynaptic cell.

Presynaptic modulation provides a more precise means of control than postsynaptic modulation. If the responsiveness of a neuron is altered at the dendrites and cell body, all target cells of the neuron will be affected equally (Fig. 8-29b). In contrast, presynaptic inhibition on a divergent neuron

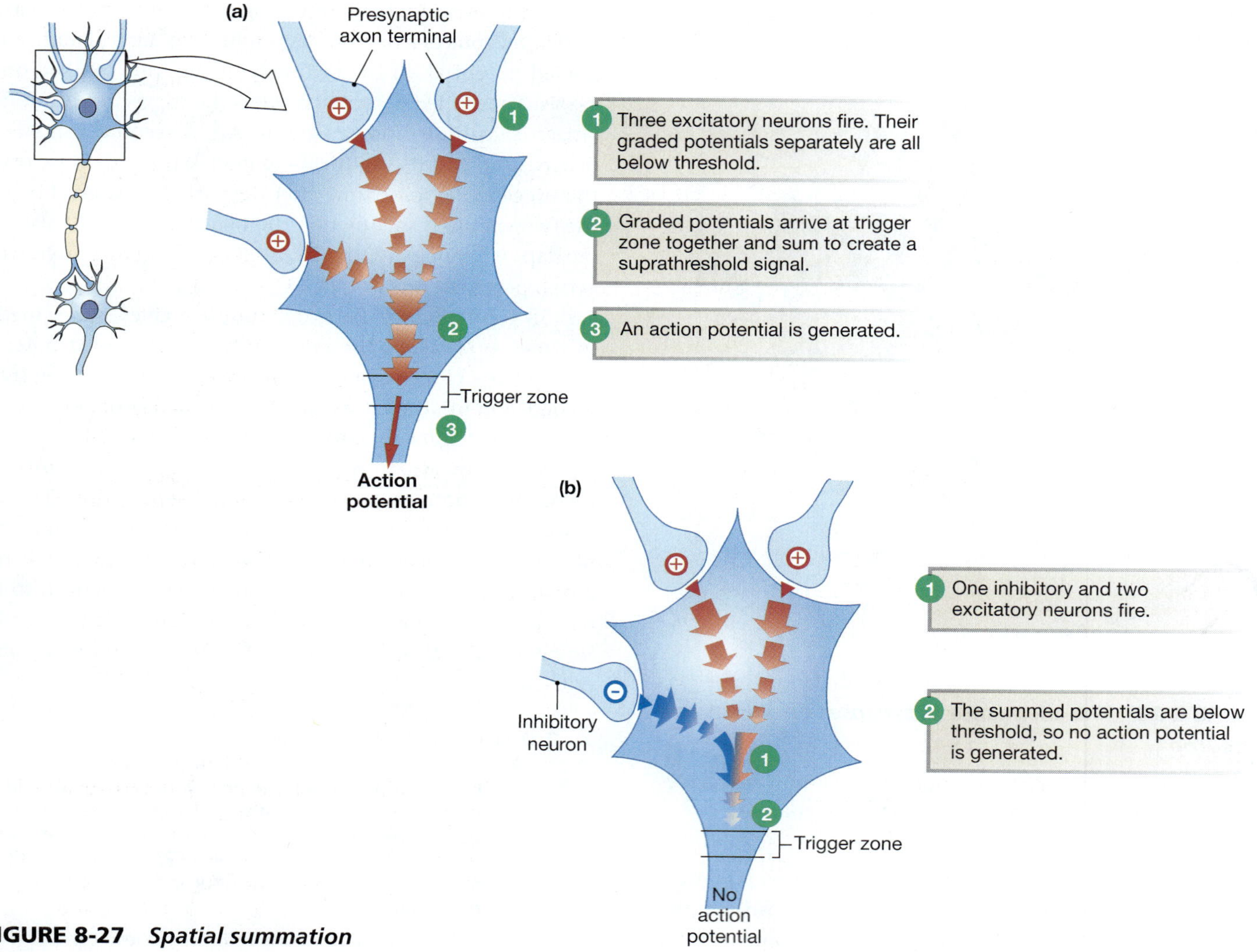

(a)
Presynaptic axon terminal
1 Three excitatory neurons fire. Their graded potentials separately are all below threshold.
2 Graded potentials arrive at trigger zone together and sum to create a suprathreshold signal.
3 An action potential is generated.
Trigger zone
Action potential
(b)
1 One inhibitory and two excitatory neurons fire.
2 The summed potentials are below threshold, so no action potential is generated.
Inhibitory neuron
Trigger zone
No action potential

FIGURE 8-27 ***Spatial summation***

Stimuli (X_1 & X_2)
Membrane potential (mV)
-55
-70
Threshold
A_1
A_2
X_1
X_2
Time (msec)
+30
0

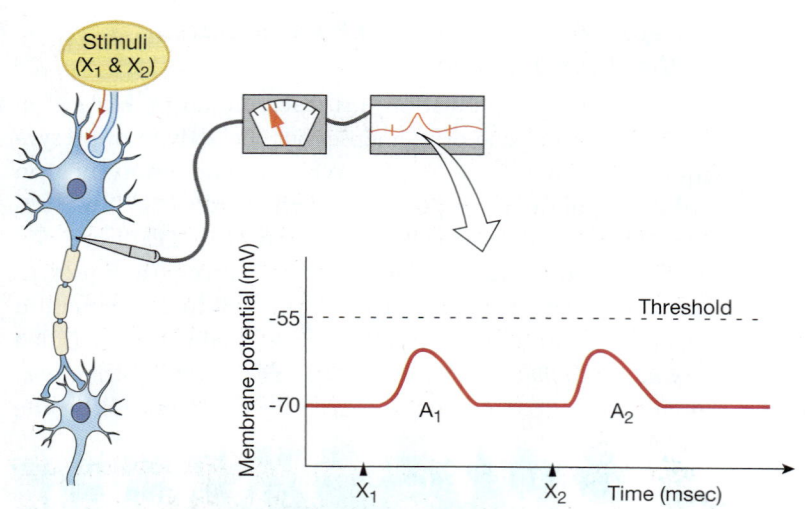

(a) No summation. Two graded potentials will not cause an action potential if they are far apart in time.

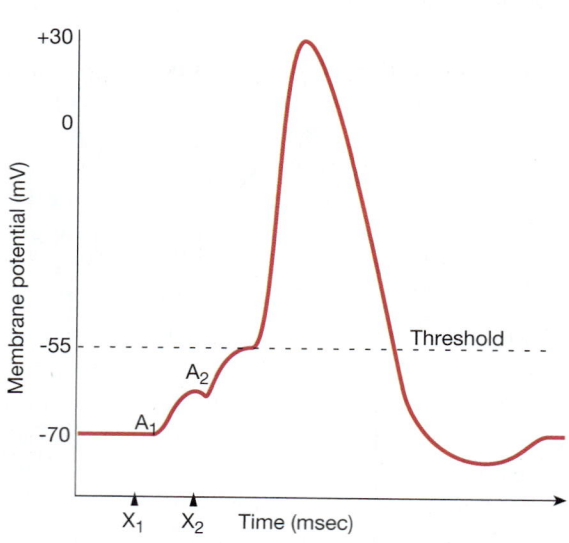

(b) Summation causing action potential. If two subthreshold potentials arrive at the trigger zone within a short period of time, they may sum and create an action potential.

FIGURE 8-28 ***Temporal summation***

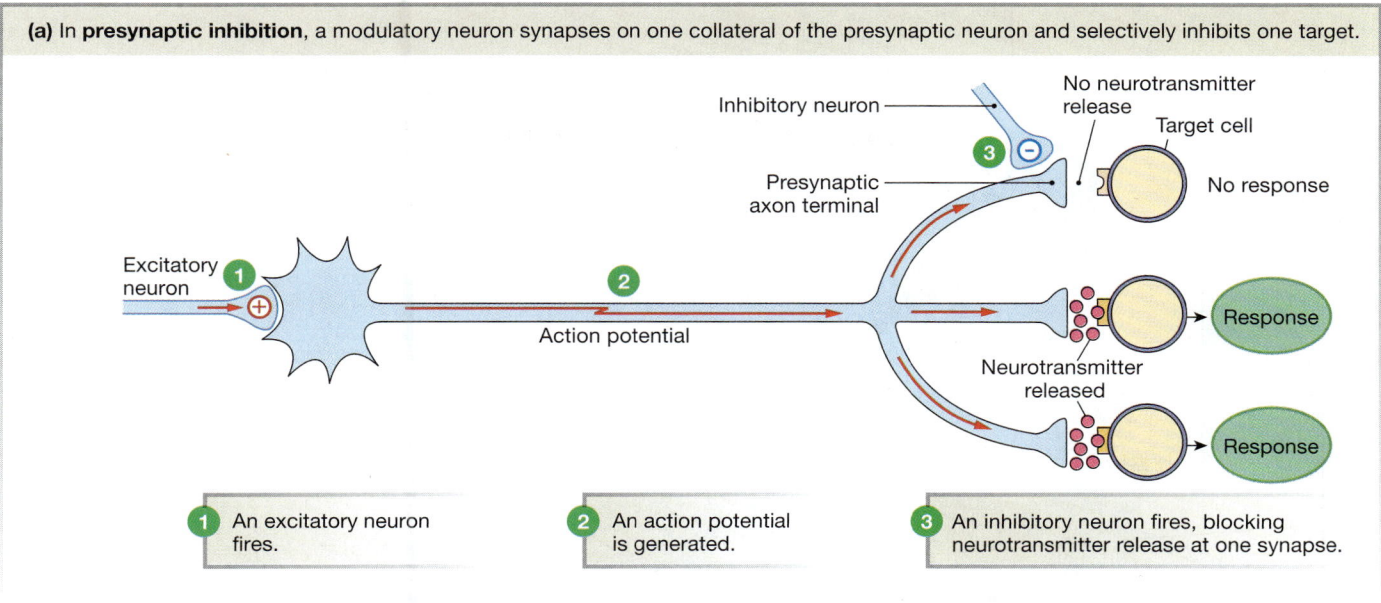

(a) In **presynaptic inhibition**, a modulatory neuron synapses on one collateral of the presynaptic neuron and selectively inhibits one target.

Inhibitory neuron

No neurotransmitter release

Target cell

No response

Presynaptic axon terminal

Excitatory neuron

Action potential

Neurotransmitter released

Response

Response

1. An excitatory neuron fires.

2. An action potential is generated.

3. An inhibitory neuron fires, blocking neurotransmitter release at one synapse.

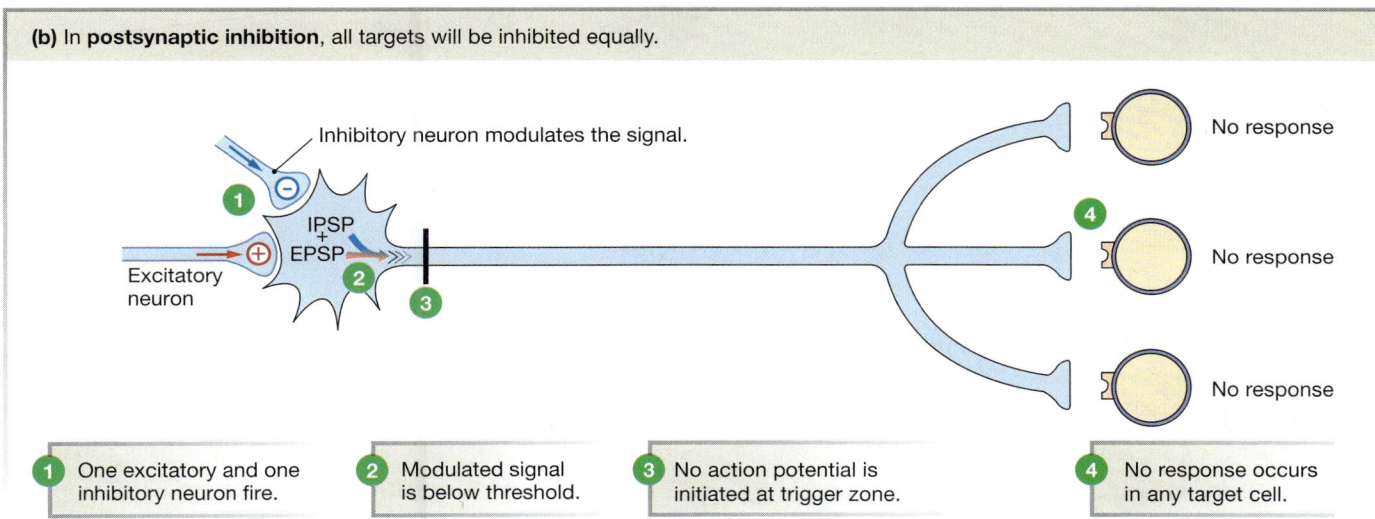

(b) In **postsynaptic inhibition**, all targets will be inhibited equally.

Inhibitory neuron modulates the signal.

IPSP + EPSP

Excitatory neuron

No response

No response

No response

1. One excitatory and one inhibitory neuron fire.

2. Modulated signal is below threshold.

3. No action potential is initiated at trigger zone.

4. No response occurs in any target cell.

■ **FIGURE 8-29** *Presynaptic and postsynaptic inhibition*

allows selective modulation of collaterals and their targets (Fig. 8-29a). One collateral can be inhibited while others remain unaffected.

Synaptic activity can also be altered by changing the responsiveness of the target cell to neurotransmitter. This may be accomplished by changing the identity, affinity, or number of neurotransmitter receptors. Modulators can alter all of these parameters by influencing the synthesis of enzymes, membrane transporters, and receptors. Most neuromodulators act through second messenger systems that alter existing proteins, and their effects last much longer than do those of neurotransmitters.

CONCEPT CHECK

24. Why are axon terminals sometimes called "biological transducers"?

Answers: p. 290

Long-Term Potentiation Alters Synaptic Communication

Two of the "hot topics" in neurobiology today are **long-term potentiation** (LTP) (*potentia*, power) and *long-term depression*, processes in which activity at a synapse induces sustained changes in the quality or quantity of synaptic connections. Long-term potentiation and depression are believed to be related to the neural processes for learning and memory. Many scientists are working to discover the molecular details of the mechanisms.

A key element in long-term potentiation is the amino acid glutamate, the main excitatory neurotransmitter in the CNS. As you learned previously, glutamate has two types of receptors: NMDA receptors and AMPA receptors. The NMDA channel has an unusual property: the channel is blocked both by a gate and

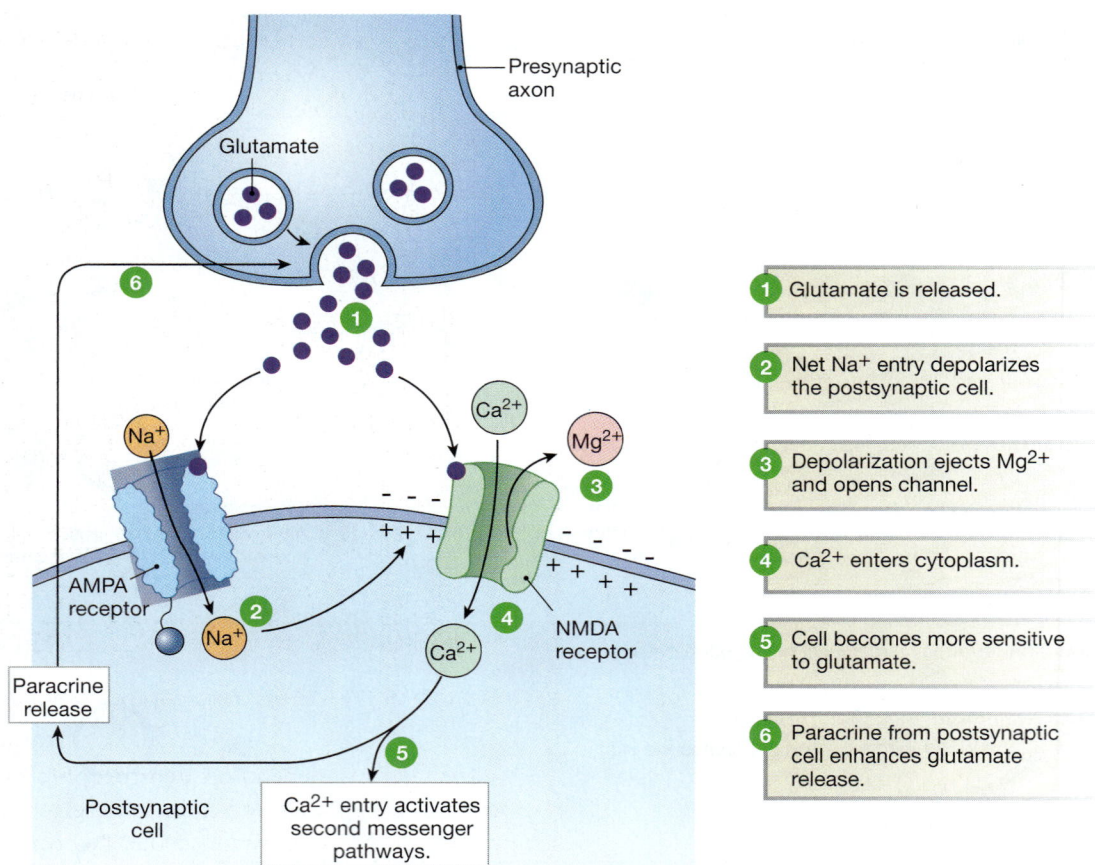

■ FIGURE 8-30   *Long-term potentiation*

The NMDA receptors for glutamate are responsible for the changes that lead to long-term potentiation. With long-term potentiation, the presynaptic cell increases its glutamate release, and the postsynaptic cell becomes more sensitive to glutamate.

1 Glutamate is released.

2 Net Na^+ entry depolarizes the postsynaptic cell.

3 Depolarization ejects Mg^{2+} and opens channel.

4 Ca^{2+} enters cytoplasm.

5 Cell becomes more sensitive to glutamate.

6 Paracrine from postsynaptic cell enhances glutamate release.

by a Mg^{2+} ion. Channel opening requires both neurotransmitter binding and depolarization of the cell.

When presynaptic neurons release glutamate, the chemical binds to both types of receptors on the postsynaptic cell (Fig. 8-30 ■). Activation of the AMPA receptor opens a cation channel, and net Na^+ entry depolarizes the cell. The gate of the NMDA receptor opens with glutamate binding, but the channel remains blocked by the Mg^{2+} until the cell depolarizes. Electrical repulsion then knocks the Mg^{2+} out of the channel, allowing Ca^{2+} to enter the cytosol.

The Ca^{2+} signal initiates second messenger pathways. As a result of these pathways, the postsynaptic cell releases a paracrine that acts on the presynaptic cell to enhance neurotransmitter release. The postsynaptic cell also becomes more sensitive to glutamate, possibly by inserting more glutamate receptors in the postsynaptic membrane.

✓ CONCEPT CHECK

25. Why would depolarization of the membrane repulse Mg^{2+} into the extracellular fluid?

Answers: p. 290

Disorders of Synaptic Transmission Are Responsible for Many Diseases

Synaptic transmission is the most vulnerable step in the process of signaling through the nervous system. It is the point at which many things go wrong, leading to disruption of normal function. Yet, at the same time, the receptors at synapses are exposed to the extracellular fluid, making them more accessible to drugs than intracellular receptors are. In recent years we have discovered that a variety of nervous system disorders are related to problems with synaptic transmission. These include Parkinson's disease, schizophrenia, and depression. The best understood diseases of the synapse are those that involve the neuromuscular junction. Diseases resulting from synaptic transmission problems within the CNS have proved more difficult to study because they are more difficult to isolate anatomically.

Drugs that act on synaptic activity, particularly synapses in the CNS, are the oldest known and most widely used of all pharmacological agents. Caffeine, nicotine, and alcohol are common drugs in many cultures. Some of the drugs we use to

RUNNING PROBLEM

Currently, Dr. McKhann believes that the disease afflicting the Chinese children—which he has named acute motor axonal polyneuropathy (AMAN)—may be caused by a bacterial infection. He also believes that the disease initiates its damage of axons at the neuromuscular junctions.

Question 6:
Based on information provided in this chapter, what other diseases involve altered synaptic transmission?

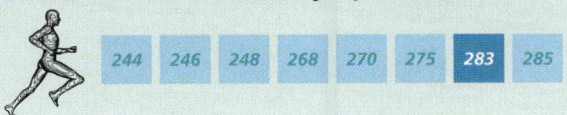

244 246 248 268 270 275 **283** 285

treat conditions such as schizophrenia, anxiety, and epilepsy act by influencing events at the synapse. In many disorders arising in the CNS, we do not yet fully understand either the cause of the disorder or the drug's mechanism of action. This subject is one major area of pharmacological research, and new classes of drugs are being formulated and approved every year.

Development of the Nervous System Depends on Chemical Signals

During the development of a system as complex as the nervous system, how can more than 100 billion neurons in the brain find their correct targets and make synapses among more than 10 times that many glial cells? How can a somatic motor neuron in the spinal cord find the correct pathway to form a synapse with its target muscle in the big toe? The answer is found in the chemical signals used by the developing embryo, ranging from factors that control differentiation of stem cells to those that direct an elongating axon to its target.

The axons of embryonic nerve cells send out special tips called **growth cones** that extend through the extracellular compartment until they find their target cell (Fig. 8-31 ■). In experiments in which target cells are moved to an unusual location in the embryo, the axons in many instances are still able to find their targets by "sniffing out" their chemical scent. Growth cones depend on many different types of signals to find their way: growth factors, molecules in the extracellular matrix, and membrane proteins on the growth cones and on cells along the path. For example, integrins [⟳ p. 70] on the growth cone membrane bind to laminins, protein fibers in the extracellular matrix. Nerve-cell adhesion molecules (NCAMs) [⟳ p. 69] interact with membrane proteins of other cells.

Once an axon reaches its target cell, a synapse forms. However, synapse formation must be followed by electrical and chemical activity, or the synapse will disappear. The survival of

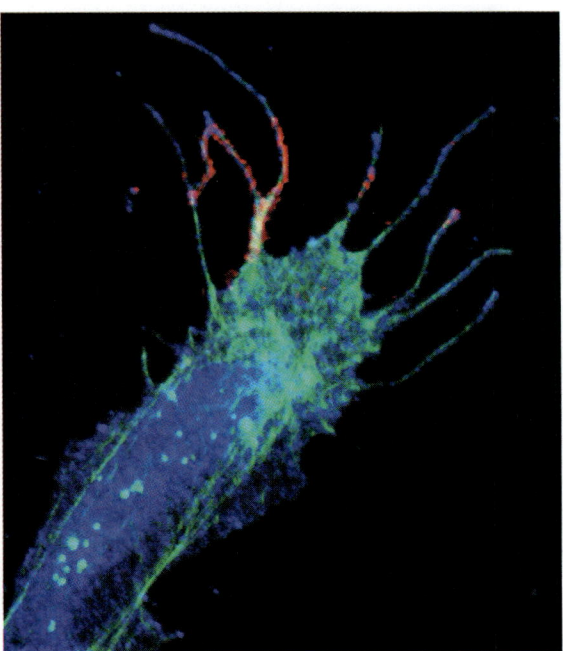

■ **FIGURE 8-31** *Growth cones of a developing axon*

neurons seems to depend on **neurotrophic factors** [*trophikos,* nourishment] secreted by neurons and glial cells. There is still much we do not understand about this complicated process, and it is an active area of physiological research.

This "use it or lose it" scenario is most dramatically reflected by the fact that the infant brain is only about one-fourth the size of the adult brain. Further brain growth is due not to an increase in cell number but to an increase in size and number of axons, dendrites, and synapses. Development depends on electrical activity in the brain—in other words, on action potentials moving through sensory pathways and interneurons.

Babies who are neglected or deprived of sensory input may experience delayed development because of the lack of nervous system stimulation. On the other hand, there is no evidence that extra stimulation in infancy will enhance intellectual development, despite a popular movement to expose babies to art, music, and foreign languages before they can even walk. Once synapses form, they are not fixed for life. Variations in electrical activity can cause rearrangement of the connections. This process, known as the **plasticity** (changeability) of synapses, continues throughout life. It is one reason that older adults are urged to keep learning new skills and information.

When Neurons Are Injured, Segments Separated from the Cell Body Die

We can grow neurons when young, but what happens when neurons are injured? The responses of mature neurons to injury are similar in many ways to the growth of neurons during

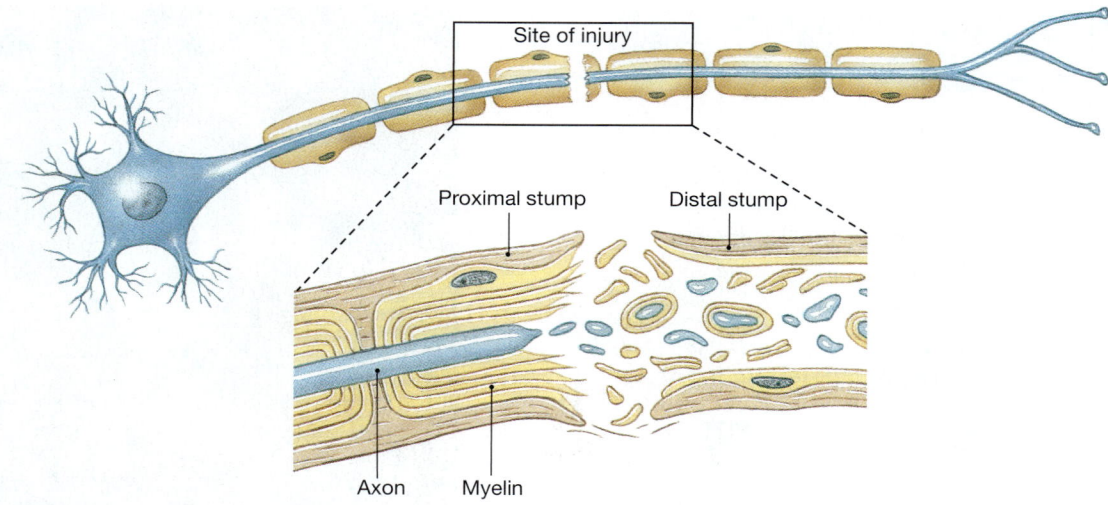

FIGURE 8-32 *Injury to neurons*

When an axon is cut, the section attached to the cell body (the proximal stump) continues to live, but the section of the axon distal to the cut (the distal stump) begins to disintegrate and die. Under some circumstances, the proximal stump may regrow through the existing sheath of Schwann cells and reform a synapse with the proper target.

development, relying on a combination of chemical and electrical signals. This is another area of active research because spinal cord injuries and brain damage from illnesses and accidents disable many people each year. The loss of one neuron from a reflex pathway can have drastic consequences for the entire pathway.

If the cell body dies when a neuron is injured, the entire neuron dies. If the cell body is intact and only the axon is severed, most of the neuron will survive (Fig. 8-32 ■). At the site of injury, the cytoplasm leaks out until membrane is recruited to seal the opening. The proximal stump (the segment closer to the cell body) swells as organelles and filaments carried into the axon by axonal transport accumulate. Chemical factors produced by Schwann cells near the injury site move by retrograde transport to the cell body, telling it that an injury has occurred. The distal segment of axon (the distal stump), deprived of its source of protein, begins to degenerate slowly. The death of this part of the neuron may take a month or longer, although synaptic transmission ceases almost immediately. The myelin sheath around the distal axon begins to unravel, and the axon itself collapses. The fragments are cleared

away by scavenger microglia or phagocytes that ingest and digest the debris.

Under some conditions, axons in the peripheral nervous system can regenerate and reestablish their synaptic connections. The Schwann cells of the damaged neuron secrete certain neurotrophic factors that keep the cell body alive and stimulate regrowth of the axon. The growing tip of a regenerating axon behaves much like the growth cone of a developing axon, following chemical signals in the extracellular matrix along its former path until the axon rejoins its target cell. Sometimes the loss of the distal axon is permanent, however, and the pathway is destroyed.

Regeneration of axons in the central nervous system is less likely to occur naturally. CNS glial cells tend to seal off and scar the damaged region, and damaged CNS cells secrete factors that inhibit axon regrowth. Many scientists are studying these mechanisms of axon growth and inhibition in the hopes of finding treatments that will restore function to victims of spinal cord injury and degenerative neurological disorders.

MYSTERIOUS PARALYSIS

In this running problem you learned about a baffling paralytic illness that appears to be a new disease. Although it bears some resemblance to Guillain-Barré syndrome, upon closer investigation the illness has some significantly different features. To read Dr. McKhann's original report, see "Clinical and electrophysiological aspects of acute paralytic disease of children and young adults in northern China," *Lancet* 338(8767): 593–597, 1991 Sep 7.

Check your understanding of this running problem by comparing your answers to the information in the summary table below.

	QUESTION	FACTS	INTEGRATION AND ANALYSIS
1	Which division(s) of the nervous system may be involved in Guillain-Barré syndrome (GBS)?	The nervous system is divided into peripheral nervous system, which has afferent (sensory) and efferent subdivisions, and central nervous system (CNS). Efferent neurons are either somatic motor neurons, which control skeletal muscles, or autonomic neurons, which control glands and smooth and cardiac muscle.	Patients with GBS can neither feel sensations nor move their muscles. This suggests a problem in both afferent and somatic motor neurons. However, it is also possible that there is a problem in the CNS integrating center. You do not have enough information to determine which division is affected.
2	Do you think the paralysis found in the Chinese children affected both sensory (afferent) and somatic motor neurons? Why or why not?	The Chinese children can feel a pin prick but cannot move their muscles.	Sensory (afferent) function is normal if they can feel the pin prick. Paralysis of the muscles suggests a problem with somatic motor neurons, with the CNS centers controlling movement, or with the muscles themselves.
3	In Guillain-Barré syndrome, what would you expect the results of a nerve conduction test to be?	Nerve conduction tests measure conduction speed and strength of action potentials. In GBS, myelin around neurons is destroyed.	Myelin insulates axons and increases conduction speed. Without myelin, ions leak out of the axon. Thus, in GBS you would expect decreased conduction speed and decreased strength of action potentials.
4	Is the paralytic illness that affected the Chinese children a demyelinating condition? Why or why not?	Nerve conduction tests showed normal conduction speed but decreased strength of the action potentials.	Myelin loss should decrease conduction speed as well as action potential strength. Therefore, this illness is probably not a demyelinating disease.
5	Do the results of Dr. McKhann's investigation suggest that the Chinese children had Guillain-Barré syndrome? Why or why not?	Autopsy reports on children who died from the disease showed that the axons were damaged but the myelin was normal.	GBS is a demyelinating disease that affects both sensory and motor neurons. The Chinese children had normal sensory function, and nerve conduction tests and histological studies indicated normal myelin. Therefore, it is reasonable to conclude that the disease is not GBS.
6	Based on information provided in this chapter, what other diseases involve altered synaptic transmission?	Synaptic transmission can be altered by blocking neurotransmitter release from the presynaptic cell, by interfering with the action of neurotransmitter on the target cell, or by removing neurotransmitter from the synapse.	Parkinson's disease, depression, schizophrenia, and myasthenia gravis are related to problems with synaptic transmission (p. 282).

244 246 248 268 270 275 283 285

CHAPTER SUMMARY

This chapter introduces the nervous system, one of the major control systems responsible for maintaining *homeostasis*. The divisions of the nervous system correlate with the steps in a reflex pathway. Sensory receptors monitor regulated variables and send input signals to the central nervous system through sensory (afferent) neurons. Output signals, both electrical and chemical, travel through the efferent divisions (somatic motor and autonomic) to their targets throughout the body. Information transfer and *communication* depend on electrical signals that pass along neurons, on *molecular interactions* between signal molecules and their receptors, and on signal transduction in the target cells. In Chapters 9 through 11 and 13 you will learn more details about the basic processes introduced here.

1. The **nervous system** is a complex network of neurons that form the rapid control system of the body. (p. 244)

2. **Emergent properties** of the nervous system include consciousness, intelligence, and emotion. (p. 244)

Organization of the Nervous System

3. The nervous system is divided into the **central nervous system (CNS)**, composed of the **brain** and **spinal cord**, and the **peripheral nervous system.** (p. 245; Fig. 8-1)

4. The peripheral nervous system has **afferent (sensory) neurons** that bring information into the CNS, and **efferent neurons** that carry information away from the CNS back to various parts of the body. (p. 245)

5. The efferent neurons include **somatic motor neurons,** which control skeletal muscles, and **autonomic neurons,** which control smooth and cardiac muscles, glands, and some adipose tissue. (p. 246)

6. Autonomic neurons are subdivided into **sympathetic** and **parasympathetic** branches. (p. 246)

Cells of the Nervous System

IP Nervous System I: Anatomy Review

7. Neurons have a **cell body** with a nucleus and organelles to direct cellular activity, **dendrites** to receive incoming signals, and an **axon** to transmit electrical signals from the cell body to the **axon terminal.** (p. 246; Fig. 8-2)

8. **Interneurons** are neurons that lie entirely within the CNS. (p. 247; Fig. 8-3)

9. The region where an axon terminal meets its target cell is called a **synapse.** The target cell is called the **postsynaptic cell,** and the neuron that releases the chemical signal is known as the **presynaptic cell.** The region between these two cells is the **synaptic cleft.** (p. 249)

10. Material is transported between the cell body and axon terminal by **axonal transport.** (p. 249; Fig. 8-4)

11. **Glial cells** provide physical support and direct the growth of neurons during repair and development. **Schwann cells** and **satellite cells** are glial cells associated with the peripheral nervous system. **Microglia, oligodendrocytes, astrocytes,** and **ependymal cells** are glial cells found in the CNS. Microglia are modified immune cells that act as scavengers. (p. 250; Fig. 8-5)

12. Schwann cells and oligodendrocytes form insulating **myelin sheaths** around neurons. The **nodes of Ranvier** are sections of uninsulated membrane occurring at intervals along the length of an axon. (p. 250; Fig. 8-6)

13. **Neural stem cells** are found in the ependymal layer. (p. 252)

Electrical Signals in Neurons

IP Nervous System I: The Membrane Potential; Ion Channels; The Action Potential

14. Membrane potential is influenced by the concentration gradients of ions across the membrane and by the permeability of the membrane to those ions. (p. 252)

15. The **Goldman-Hodgkin-Katz (GHK) equation** predicts membrane potential based on ion concentration gradients and membrane permeability. (p. 252)

16. The permeability of a cell to ions changes when ion channels in the membrane open and close. Movement of only a few ions significantly changes the membrane potential. (p. 253)

17. Gated ion channels in neurons open or close in response to chemical or mechanical signals or in response to depolarization of the cell membrane. (p. 254)

18. **Graded potentials** are depolarizations or hyperpolarizations whose strength is directly proportional to the strength of the triggering event. Graded potentials lose strength as they move through the cell. (p. 255; Tbl. 8-3; Fig. 8-7)

19. The wave of depolarization that moves through the cell with a graded potential is known as **local current flow.** (p. 255)

20. **Action potentials** are rapid electrical signals that travel undiminished in amplitude from the cell body to the axon terminals. (p. 255)

21. Action potentials begin in the **trigger zone** if either a single graded potential or the sum of multiple graded potentials exceeds a minimum depolarization known as the **threshold.** (p. 256; Fig. 8-8)

22. Depolarizing graded potentials make a neuron more likely to fire an action potential. Hyperpolarizing graded potentials make a neuron less likely to fire an action potential. (p. 257)

23. Action potentials are uniform, **all-or-none** depolarizations. (p. 258)

24. The rising phase of the action potential is due to increased Na^+ permeability. The falling phase of the action potential is due to increased K^+ permeability. (p. 258; Fig. 8-9)

25. The voltage-gated Na^+ channels of the axon have a fast **activation gate** and a slower **inactivation gate.** (p. 259; Fig. 8-10)

26. Once an action potential has begun, there is a brief period of time known as the **absolute refractory period** during which a second action potential cannot be triggered, no matter how large the stimulus. Because of this, action potentials cannot be summed. (p. 260; Fig. 8-12)

27. During the **relative refractory period**, a higher-than-normal graded potential is required to trigger an action potential. (p. 260)

28. Information about the strength and duration of a stimulus is conveyed by the frequency of action potential propagation. (p. 261; Fig. 8-13)

29. Very few ions cross the membrane during an action potential. The Na^+-K^+-ATPase eventually restores Na^+ and K^+ to their original compartments. (p. 261)

30. The movement of an action potential through the axon at high speed is called **conduction**. (p. 263; Fig. 8-15)

31. Larger axon diameter and increased membrane resistance speed up action potential conduction. The myelin sheath speeds up conduction by increasing membrane resistance and decreasing current leakage. (p. 266)

32. The apparent jumping of action potentials from node to node is called **saltatory conduction**. (p. 267; Fig. 8-18)

33. Changes in blood K^+ concentration affect resting membrane potential and the conduction of action potentials. (p. 269; Fig. 8-19)

Cell-to-Cell Communication in the Nervous System

IP **Nervous System II: Anatomy Review; Synaptic Transmission; Ion Channels**

34. In **electrical synapses**, an electrical signal passes directly from the cytoplasm of one cell to another through gap junctions. **Chemical synapses** use neurotransmitters to carry information from one cell to the next, with the neurotransmitters diffusing across the synaptic cleft. (p. 271)

35. Neurotransmitters are synthesized in the cell body or in the axon terminal. They are stored in **synaptic vesicles** and are released by exocytosis when an action potential reaches the axon terminal. Neurotransmitters combine with receptors on target cells. (p. 271; Fig. 8-21)

36. Neurotransmitters come in a variety of forms. **Cholinergic** neurons secrete **acetylcholine**. **Adrenergic** neurons secrete **norepinephrine**. **Glutamate**, **GABA**, **serotonin**, **adenosine**, and **nitric oxide** are other major neurotransmitters. (p. 274; Tbl. 8-4)

37. Neurotransmitter receptors are either ligand-gated ion channels or G protein–coupled receptors. Ion channels create **fast synaptic potentials**. G protein–coupled receptors either create **slow synaptic potentials** or modify cell metabolism. (p. 277; Fig. 8-23)

38. Neurotransmitter action is rapidly terminated by reuptake into cells, diffusion away from the synapse, or enzymatic breakdown. (p. 277; Fig. 8-24)

Integration of Neural Information Transfer

IP **Nervous System II: Synaptic Potentials & Cellular Integration**

39. If a presynaptic neuron synapses on a larger number of postsynaptic neurons, the pattern is known as **divergence**. If several presynaptic neurons provide input to a smaller number of postsynaptic neurons, the pattern is known as **convergence**. (p. 278; Fig. 8-25)

40. The summation of simultaneous graded potentials from different neurons is known as **spatial summation**. The summation of graded potentials that closely follow each other sequentially is called **temporal summation**. (p. 278; Figs. 8-27, 8-28)

41. **Presynaptic modulation** of an axon terminal allows selective modulation of collaterals and their targets. **Postsynaptic modulation** occurs when a modulatory neuron, usually inhibitory, synapses on a postsynaptic cell. (p. 279; Fig. 8-29)

42. **Long-term potentiation** is one mechanism by which neurons change the quality or quantity of their synaptic connections. (p. 281)

43. Developing neurons find their way to their targets by using chemical signals. (p. 283)

QUESTIONS

(Answers to the Review Questions begin on page A1.)

LEVEL ONE REVIEWING FACTS AND TERMS

1. List the three functional classes of neurons, and explain how they differ structurally and functionally.

2. Somatic motor neurons control ___skeletal muscles___ Autonomic neurons control smooth and cardiac muscles, exocrine and some endocrine glands, and some types of adipose tissue.

3. Autonomic neurons are classified as either ___sympathetic___ or ___para-sympathetic___ neurons.

4. Match each term with its description:

    (a) axon (3)
    (b) dendrite (1)
    (c) afferents (2)
    (d) efferents (5)
    (e) trigger zone (4)

    1. process of a neuron that receives incoming signals
    2. sensory neurons, transmit information to CNS
    3. long process that transmits signals to the target cell
    4. region of neuron where action potential begins
    5. neurons that transmit information from CNS to the rest of the body

5. Name the two primary cell types found in the nervous system.

6. Draw a typical neuron and label the cell body, axon, dendrites, nucleus, trigger zone, axon hillock, collaterals, and axon terminals. Draw mitochondria, rough endoplasmic reticulum, Golgi complex, and vesicles in the appropriate sections of the neuron.

7. Axonal transport refers to

    (a) the release of neurotransmitter molecules into the synaptic cleft.
    (b) the use of microtubules to send secretions from the cell body to the axon terminal.
    (c) the transport of vesicles containing proteins down the length of the axon.

(continued)

(d) the movement of the axon terminal to synapse with a new post-synaptic cell.

(e) none of these

8. Match the numbers of the appropriate characteristics with the two types of potentials. Characteristics may apply to one or both types.

(a) action potential 1, 4
(b) graded potential
   2, 3, 5, 6

1. all-or-none
2. can be summed
3. amplitude decreases with distance
4. exhibits a refractory period
5. amplitude depends on strength of stimulus
6. has no threshold

9. A typical neuron synapses with how many other neurons?

(a) 1
(b) fewer than 10
(c) 1000
(d) 10,000
(e) more than 100,000

10. List the four major types of ion channels found in neurons. Are they chemically gated, mechanically gated, or voltage-gated?

11. Arrange the following events in the proper sequence: e, b, d, a, c

(a) Efferent neuron reaches threshold and fires an action potential.
(b) Afferent neuron reaches threshold and fires an action potential.
(c) Effector organ responds by performing output.
(d) Integrating center reaches decision about response.
(e) Sensory organ detects change in the environment.

12. An action potential is (circle all correct answers)

(a) a reversal of the Na^+ and K^+ concentrations inside and outside the neuron.
(b) the same size and shape at the beginning and end of the axon.
(c) initiated by inhibitory postsynaptic graded potentials.
(d) transmitted to the distal end of a neuron and causes release of neurotransmitter.

In questions 13–17, choose from the following ions to fill in the blanks correctly: Na^+, K^+, Ca^{2+}, Cl^-.

13. The resting cell membrane is more permeable to ___K^+___ than to ___Na^+___. Although ___Na^+___ contribute little to the resting membrane potential, they play a key role in generating electrical signals in excitable tissues.

14. The concentration of ___Na^+___ is 12 times greater outside the cell than inside.

15. The concentration of ___K^+___ is 30 times greater inside the cell than outside.

16. An action potential occurs when ___Na^+___ enter the cell.

17. The resting membrane potential is due to the high ___K^+___ permeability of the cell.

18. What is the myelin sheath?

19. List two factors that enhance conduction speed.

20. List three ways neurotransmitters are removed from the synapse.

21. Draw and label a graph of an action potential. Below the graph, draw the positioning of the K^+ and Na^+ channel gates during each phase.

LEVEL TWO   REVIEWING CONCEPTS

22. Create a map showing the organization of the nervous system using the following terms, plus any terms you choose to add:

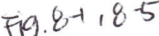
FIg. 8.1 8.5

afferent signals	integration	satellite cell
astrocyte	interneuron	Schwann cell
autonomic division	microglia	sensory division
brain	muscles	somatic motor
CNS	neuron	division
efferent neuron	neurotransmitter	spinal cord
efferent signals	oligodendrocyte	stimulus
ependymal cell	parasympathetic	sympathetic division
glands	division	
glial cells	peripheral division	

23. What causes the depolarization phase of an action potential? (Circle all that apply.)

(a) K^+ leaving the cell through voltage-gated channels
(b) K^+ being pumped into the cell by the Na^+-K^+-ATPase
(c) Na^+ being pumped into the cell by the Na^+-K^+-ATPase
(d) Na^+ entering the cell through voltage-gated channels
(e) opening of the Na^+ channel inactivation gate

24. List four types of glial cells, and briefly explain the role of each.

25. Arrange the following terms to describe the sequence of events after a neurotransmitter binds to a receptor on a postsynaptic neuron. Terms may be used more than once or not at all.

(a) action potential fires at axon hillock   f, j, g, e, b, k, c
(b) trigger zone reaches threshold   a, h, i, d
(c) cell depolarizes
(d) exocytosis
(e) graded potential occurs
(f) ligand-gated ion channel opens
(g) local current flow occurs
(h) saltatory conduction occurs
(i) voltage-gated Ca^{2+} channels open
(j) voltage-gated K^+ channels open
(k) voltage-gated Na^+ channels open

26. Match the best term (hyperpolarize, depolarize, repolarize) to the following events. The cell in question has a resting membrane potential of −70 mV. Fig. 5-37 pg. 166

(a) membrane potential changes from −70 mV to −50 mV D
(b) membrane potential changes from −70 mV to −90 mV H
(c) membrane potential changes from +20 mV to −60 mV R
(d) membrane potential changes from −80 mV to −70 mV D

27. A neuron has a resting membrane potential of −70 mV. Will the neuron hyperpolarize or depolarize when each of the following events occurs? (More than one answer may apply; list all those that are correct.)

(a) Na^+ enter the cell D
(b) K^+ leave the cell H
(c) Cl^- enter the cell H
(d) Ca^{2+} enter the cell D

28. Define, compare, and contrast the following concepts:

(a) threshold, subthreshold, suprathreshold, all-or-none, overshoot, undershoot
(b) graded potential, EPSP, IPSP, absolute refractory period, relative refractory period
(c) afferent neuron, efferent neuron, interneuron
(d) sensory neuron, somatic motor neuron, sympathetic neuron, autonomic neuron, parasympathetic neuron
(e) fast synaptic potential, slow synaptic potential
(f) temporal summation, spatial summation
(g) convergence, divergence

29. If all action potentials within a given neuron are identical, how does the neuron transmit information about the strength and duration of the stimulus?

30. The presence of myelin allows an axon to
    (a) produce more frequent action potentials.
    (b) conduct impulses more rapidly.
    (c) produce action potentials of larger amplitude.
    (d) produce action potentials of longer duration.

LEVEL THREE   PROBLEM SOLVING

31. If human babies' muscles and neurons are fully developed and functional at birth, why can't they focus their eyes, sit up, or learn to crawl within hours of being born? (*Hint:* muscle strength is not the problem.)

32. The voltage-gated Na^+ channels of a neuron open when the neuron depolarizes. If depolarization opens the channels, why do they close when the neuron is maximally depolarized?

33. One of the pills that Jim takes for high blood pressure caused his blood K^+ level to decrease from 4.5 mM to 2.5 mM. What happens to the resting membrane potential of his liver cells? (Circle all that are correct.)
    (a) decreases
    (b) increases
    (c) does not change
    (d) becomes more negative
    (e) becomes less negative
    (f) fires an action potential
    (g) depolarizes
    (h) hyperpolarizes
    (i) repolarizes

34. Characterize each of the following stimuli as being mechanical, chemical, or thermal:
    (a) bath water at 106° F
    (b) acetylcholine
    (c) a hint of perfume
    (d) epinephrine
    (e) lemon juice
    (f) a punch on the arm

LEVEL FOUR   QUANTITATIVE PROBLEMS

35. The GHK equation is sometimes abbreviated to exclude chloride, which plays a minimal role in membrane potential for most cells. In addition, because it is difficult to determine absolute membrane permeability values for Na^+ and K^+, the equation is revised to use the ratio of the two ion permeabilities as $\alpha = P_{Na}/P_K$:

$$V_m = 61 \log \frac{[K^+]_{out} + \alpha[Na^+]_{out}}{[K^+]_{in} + \alpha[Na^+]_{in}}$$

Thus, if you know the relative membrane permeabilities of the two ions and their intracellular and extracellular concentrations, you can predict the membrane potential for a cell.

(a) Looking at the previous equation, can you predict what will happen to V_m if the only change is an increase in sodium permeability?

(b) A resting cell has an alpha value of 0.025 and the following ion concentrations:

	INTRACELLULAR	EXTRACELLULAR
Na^+	5 mM	135 mM
K^+	150 mM	4 mM

What is its membrane potential?

36. In each of the following scenarios, will an action potential be produced? The postsynaptic neuron has a resting membrane potential of −70 mV.

(a) Fifteen neurons synapse on one postsynaptic neuron. At the trigger zone, 12 of the neurons produce EPSPs of 2 mV each, and the other three produce IPSPs of 3 mV each. The threshold for the postsynaptic cell is −50 mV.

(b) Fourteen neurons synapse on one postsynaptic neuron. At the trigger zone, 11 of the neurons produce EPSPs of 2 mV each, and the other three produce IPSPs of 3 mV each. The threshold for the postsynaptic cell is −60 mV.

(c) Fifteen neurons synapse on one postsynaptic neuron. At the trigger zone, 14 of the neurons produce EPSPs of 2 mV each, and the other one produces an IPSP of 9 mV. The threshold for the postsynaptic cell is −50 mV.

8

ANSWERS

✓ Answers to Concept Check Questions

Page 246

1. Compare your answer to the map in Figure 8-1, p. 246.

Page 249

2. Neurons that secrete neurohormones terminate close to blood vessels so that the neurohormones can enter the circulation.

3. See Figure 8-2.

Page 252

4. Myelin insulates axon membranes. Microglia are scavenger cells in the CNS. Ependymal cells form epithelial barriers between fluid compartments of the CNS.

5. Schwann cells are in the PNS, and each Schwann cell forms myelin around a small portion of one axon. Oligodendrocytes are in the CNS, and one oligodendrocyte forms myelin around axons of several neurons.

Page 252

6. For Ca^{2+}, the electrical charge z is +2; the ratio of ion concentrations is $1/0.0001 = 10,000$ or 10^4. Log of 10^4 is 4 (see Appendix B). Thus $E_{ion} = (61 \times 4)/(+2) = 122$ mV.

Page 253

7. (a) depolarizes (b) depolarizes

8. depolarize

Page 255

9. (a) 1, (b) 2, (c) 2, (d) 1

Page 256

10. The trigger zone for the sensory neurons will be close to where the dendrites converge. You cannot tell where the trigger zone is for the anaxonic neuron. For multipolar neurons, the trigger zone will be at the junction of the cell body and the axon.

Page 260

11. The membrane potential will depolarize and stay depolarized.

12. During resetting, the activation gate is closing, and the inactivation gate is opening.

Page 263

13. (b)

Page 266

14. The action potential will go in both directions because the Na^+ channels around the stimulation site have not been inactivated by a previous depolarization.

Page 268

15. a, c, b

Page 271

16. Proteins are synthesized on the ribosomes of the rough endoplasmic reticulum; then the proteins are directed into the Golgi complex to be packaged into vesicles.

17. Mitochondria are the primary sites of ATP synthesis.

18. Mitochondria reach the axon terminal by fast axonal transport along microtubules.

Page 272

19. The researchers concluded that some event between arrival of the action potential at the axon terminal and depolarization of the postsynaptic cell is dependent on extracellular Ca^{2+}. We now know that this event is neurotransmitter release.

Page 276

20. Because different receptor subtypes work through different signal transduction pathways, targeting drugs to specific receptor subtypes decreases the likelihood of unwanted side effects.

Page 278

21. SSRIs decrease reuptake of serotonin into the axon terminal, thereby increasing the time serotonin is active in the synapse.

Page 279

22. The postsynaptic neuron will fire an active potential, because the net effect would be a 17 mV depolarization to $-70 - (-17) = -53$ mV, which is just above the threshold of -55 mV.

23. The membrane potential does not change at the same time as the stimulus because the depolarization must travel from the point of the stimulus to the recording point.

Page 281

24. Axon terminals convert (transduce) the electrical action potential signal into a chemical neurotransmitter signal.

Page 282

25. Membrane depolarization makes the inside of the membrane more positive with respect to the outside. Like charges repel one another, so the more positive membrane potential tends to repel Mg^{2+}.

Q *Answers to Figure Questions*

Page 256

Figure 8-7: The graded potential is stronger at B. On the graph, A is between 3 and 4, and B is between the point of origin and 1.

Page 264

Figure 8-15: a) 4; b) 2, 3; c) 1; d) 3; e) 4

Neuronal assemblies have important properties that cannot be explained by the additive qualities of individual neurons.

—**O. Hechter,** *in* Biology and Medicine into the 21st Century, *1991*

PET scan of brain.

The Central Nervous System

BACKGROUND BASICS

RUNNING PROBLEM

INFANTILE SPASMS

At four months of age, Ben could roll over, hold up his head, and reach for things. At seven months, he was nearly paralyzed and lay listlessly in his crib. He had lost his abilities so gradually that it was hard to remember when each one had slipped away, but his mother could remember exactly when it began. She was preparing to feed him lunch one day when she heard a cry from the highchair where Ben was sitting. As she watched, Ben's head dropped to his chest, came back up, then went hurtling toward his lap, smacking into his highchair table. Ben's mother snatched him up into her arms, and she could feel him still convulsing against her shoulder. This was the first of many such spells that came with increasing frequency and duration.

292   300   315   317   321   322

Ed Lalor of Media Lab Europe sat immobile in a chair, his head clamped into a vise-like headpiece that arced over his skull. On the computer screen in front of him, a strange, frog-like creature balanced on a tightrope suspended over a futuristic city. As Ed shifted his gaze back and forth between two sets of flashing chequered boxes, the creature walked along the tightrope, controlled only by externally recorded electrical activity from Ed's brain. Meanwhile, in Canada and the United States, researchers demonstrated a brain-implanted device that enabled a quadriplegic to control a computer merely by thinking about what he wanted to do. These stories sound like science fiction, but the brain-computer interface devices described are real. The scientists who developed them are using what we know about the human brain to harness its electrical signals and create wireless bridges to external machines.

EMERGENT PROPERTIES OF NEURAL NETWORKS

Neurons in the nervous system link together to form circuits that have specific functions. The most complex circuits are those of the brain, in which billions of neurons are linked into intricate networks that converge and diverge, creating an infinite number of possible pathways. Signaling within these pathways creates thinking, language, feeling, learning, and memory—the complex behaviors that make us human. Some neuroscientists have proposed that the functional unit of the nervous

system be changed from the individual neuron to neural networks because even the most basic functions require circuits of neurons.

How is it that combinations of neurons linked together into chains or networks collectively possess emergent properties not found in any single neuron? We do not yet have an answer to this question. Some scientists seek to answer it by looking for parallels between the nervous system and the integrated circuits of computers.

Computer programs have been written that attempt to mimic the thought processes of humans. This field of study, called *artificial intelligence*, has created some interesting programs, such as the "psychiatrist" programmed to respond to typed complaints with appropriate comments and suggestions. We are nowhere near creating a brain as complex as that of a human, however, or even one as complex as that of Hal, the computer in the classic movie *2001: A Space Odyssey.*

Probably one reason computers cannot yet accurately model brain function is that computers lack *plasticity,* the ability to change circuit connections and function in response to sensory input and past experience [⮌ p. 283]. Although some computer programs can change their output under specialized conditions, they cannot begin to approximate the plasticity of human brain networks, which easily restructure themselves as the result of sensory input, learning, emotion, and creativity. In addition, we now know that the brain can add new connections when neural stem cells differentiate. Computers cannot add new circuits to themselves.

How can simply linking neurons together create **affective behaviors**, which are related to feeling and emotion, and **cognitive behaviors** [*cognoscere,* to get to know] related to thinking? In their search for the organizational principles that lead to these behaviors, scientists seek clues in the simplest animal nervous systems.

EVOLUTION OF NERVOUS SYSTEMS

All animals have the ability to sense and respond to changes in their environment. Even single-cell organisms such as *Paramecium* are able to carry out the basic tasks of life: finding food, avoiding becoming food, finding a mate. Yet these unicellular organisms have no obvious brain or integrating center, merely the resting membrane potential that exists in living cells.

The first multicellular animals to develop neurons were members of the phylum Cnidaria, the jellyfish and sea anemones. Their nervous system is a *nerve net* (Fig. 9-1a ■) composed of sensory neurons, connective interneurons, and motor neurons that innervate muscles and glands. These animals respond to stimuli with complex behaviors, yet without input from an identifiable control center. If you watch a jellyfish swim or a sea anemone stuff a piece of shrimp into its mouth, it is hard to imagine how a diffuse network of neurons can create such complex coordinated movements. However, the same basic principles

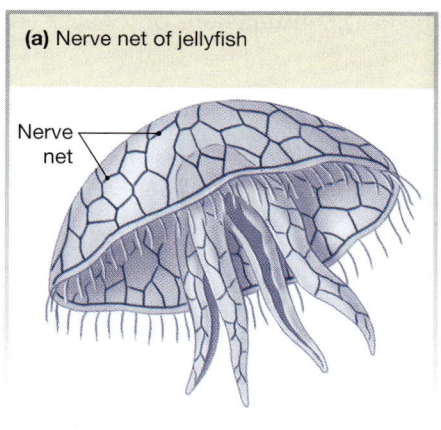

(a) Nerve net of jellyfish

Nerve net

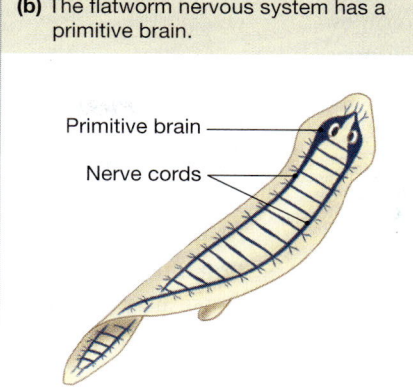

(b) The flatworm nervous system has a primitive brain.

Primitive brain

Nerve cords

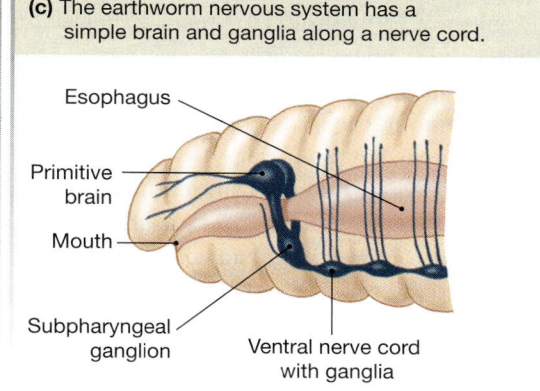

(c) The earthworm nervous system has a simple brain and ganglia along a nerve cord.

Esophagus

Primitive brain

Mouth

Subpharyngeal ganglion

Ventral nerve cord with ganglia

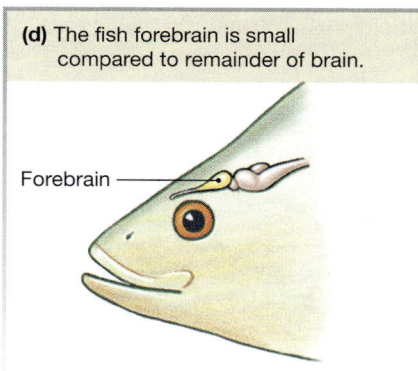

(d) The fish forebrain is small compared to remainder of brain.

Forebrain

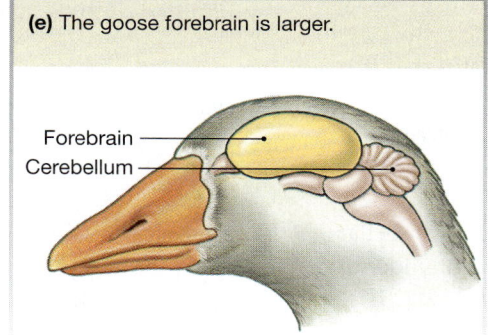

(e) The goose forebrain is larger.

Forebrain
Cerebellum

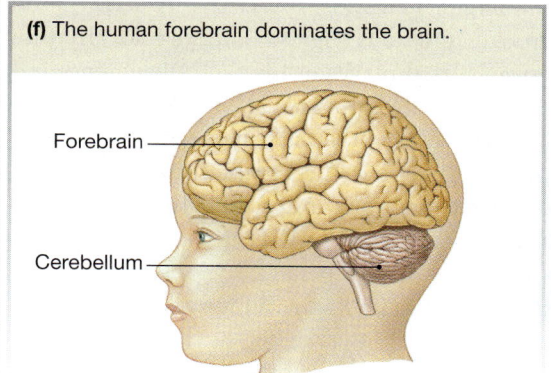

(f) The human forebrain dominates the brain.

Forebrain

Cerebellum

9

■ **FIGURE 9-1** *Evolution of the nervous system*

that you learned in Chapter 8 apply to jellyfish and humans. Electrical signals in the form of action potentials, and chemical signals passing across synapses, are the same in all animals. It is only in the number and organization of the neurons that one species differs from another.

CONCEPT CHECK

1. Match each of the following terms with the appropriate neuron type(s).

   (a) Afferent neuron
   (b) Efferent signal
   (c) Integrating center
   (d) Input signal
   (e) Output signal

   1. Interneuron
   2. Motor neuron
   3. Sensory neuron

   Answers: p. 326

In the primitive flatworms, we see the beginnings of a nervous system as we know it in higher animals, although in flatworms the distinction between central nervous system and peripheral nervous system is not clear. Flatworms have a rudimentary brain consisting of a cluster of nerve cell bodies concentrated in the head (*cephalic*) region. Two large nerves called *nerve cords* come off the primitive brain and lead to a nerve network that innervates distal regions of the flatworm body (Fig. 9-1b).

The segmented worms, or annelids, such as the earthworm, have a more advanced central nervous system (Fig. 9-1c). Clusters of cell bodies are no longer restricted to the head region, as

they are in flatworms, but also occur in fused pairs, called *ganglia* (singular *ganglion*) [p. 250], along a nerve cord. Because each segment of the worm contains a ganglion, simple reflexes can be integrated within a segment without input from the brain. Reflexes that do not require integration in the brain also occur in higher animals and are called **spinal reflexes** in humans and other vertebrates.

Annelids and higher invertebrates have complex reflexes controlled through neural networks. Researchers use leeches (a type of annelid) and *Aplysia,* a type of shell-less mollusk, to study neural networks and synapse formation because the neurons in these species are 10 times larger than human brain neurons, and because the networks have identical numbers of neurons from animal to animal. The neural function of these invertebrates provides a simple model that we can apply to more complex vertebrate networks.

Nerve cell bodies clustered into brains persist throughout the more advanced phyla and become increasingly more complex. One advantage to cephalic brains is that in most animals, the head is the part of the body that first contacts the environment as the animal moves. Thus, as brains evolved, they became associated with specialized cephalic receptors, such as eyes for vision and chemoreceptors for smell and taste.

In the higher arthropods, such as insects, specific regions of the brain are associated with particular functions. More complex brains are associated with complex behaviors, such as the ability of social insects to organize themselves into colonies, divide labor, and communicate with one another. The octopus (a cephalopod mollusk) has the most sophisticated brain development among the invertebrates, as well as the most sophisticated behavior.

In vertebrate brain evolution, the most dramatic change is seen in the *forebrain* region [*fore,* in front], which includes the **cerebrum** [*cerebrum,* brain; adjective *cerebral*]. In fish, the forebrain is a small bulge (Fig. 9-1d) dedicated mainly to processing olfactory information about odors in the environment. In birds and rodents, part of the forebrain has enlarged into a cerebrum with a smooth surface (Fig. 9-1e). In humans, the cerebrum is the largest and most distinctive part of the brain, with deep grooves and folds (Fig. 9-1f). More than anything else, the cerebrum is what makes us human. All evidence indicates that it is the part of the brain that allows reasoning and cognition.

The other brain structure whose evolution is obvious in the vertebrates is the **cerebellum**, a region of the *hindbrain* devoted to coordinating movement and balance. Birds (Fig. 9-1e) and humans (Fig. 9-1f) both have well-developed cerebellar structures. The cerebellum, like the cerebrum, is readily identifiable in these phyla by its grooves and folds.

In this chapter we begin with an overview of central nervous system anatomy and functions. We then look at how neu-

ral networks create the higher brain functions of thought and emotion.

ANATOMY OF THE CENTRAL NERVOUS SYSTEM

The vertebrate central nervous system (CNS) consists of the brain and the spinal cord. As you learned in the previous section, brains increase in complexity and degree of specialization as we move up the phylogenetic tree from fish to humans. However, if we look at the vertebrate nervous system during development, a basic anatomical pattern emerges. In all vertebrates, the CNS consists of layers of neural tissue surrounding a fluid-filled central cavity lined with epithelium.

(a) 20-day embryo (dorsal view)

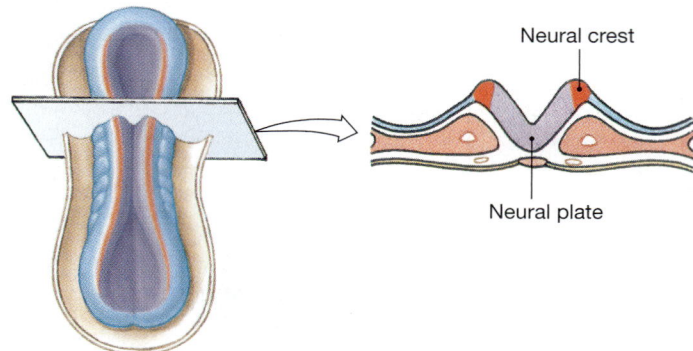

(b) 23-day embryo (dorsal view)

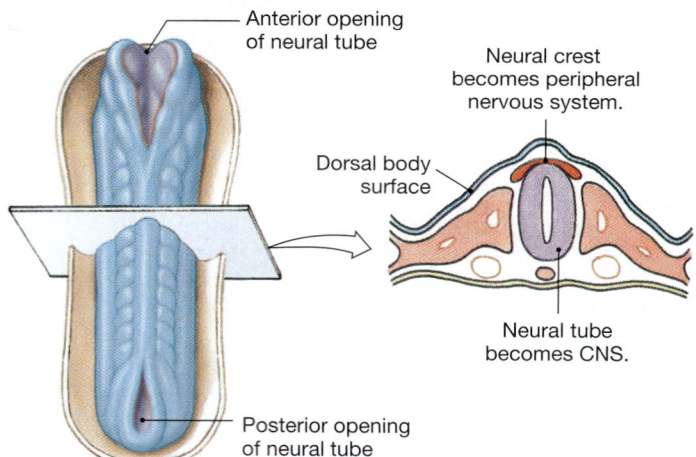

■ **FIGURE 9-2**   *The embryonic nervous system*

(a) In the 20-day embryo, neural plate cells (purple) begin to migrate toward the midline, where they will fuse to form the neural tube. Neural crest cells migrate with the neural plate cells. (b) By day 23, neural tube formation is almost complete. The neural tube will become the central nervous system. Neural crest cells on the dorsal side of the embryo will develop into the sensory and motor neurons of the peripheral nervous system.

BIOTECHNOLOGY

TRACING NEURONS IN A NETWORK

One of the challenges facing neurophysiologists and neuroanatomists is tracing the networks of neurons that control specific functions, a task that can be likened to following one tiny thread through a tangled mass the size of a beach ball. In 1971, K. Kristensson and colleagues introduced the use of *horseradish peroxidase* (HRP), an enzyme that acts in the presence of its substrate to produce a visible product. When injected into the extracellular fluid near axon terminals, HRP is brought into the neuron by endocytosis. The vesicles containing HRP are then transported by fast retrograde axonal transport to the cell body and dendrites [🔁 p. 249]. Once the enzyme-substrate reaction is completed, the neuron becomes visible, allowing the researcher to trace the entire neuron from its target back to its origin. Now, with the help of fluorescent antibodies, scientists can even tell which ion channels, neurotransmitters, and receptors the neuron uses.

(a) A 4-week human embryo showing the anterior end of the neural tube, which has specialized into three brain regions

(b) At 6 weeks, the neural tube has differentiated into the brain regions present at birth. The central cavity (lumen) shown in the cross section will become the ventricles of the brain. (see Fig. 9-5)

(c) By 11 weeks of embryonic development, the growth of the cerebrum is noticeably more rapid than that of the other divisions of the brain.

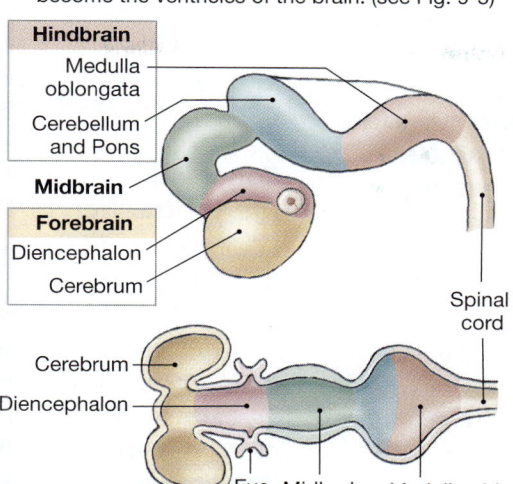

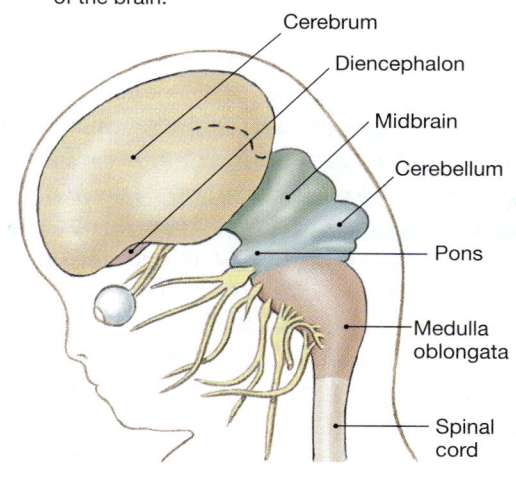

(d) At birth, the cerebrum has covered most of the other brain regions. Its rapid growth within the rigid confines of the cranium forces it to develop a convoluted, furrowed surface.

(e) The directions "dorsal" and "ventral" are different in the brain because of flexion in the neural tube during development.

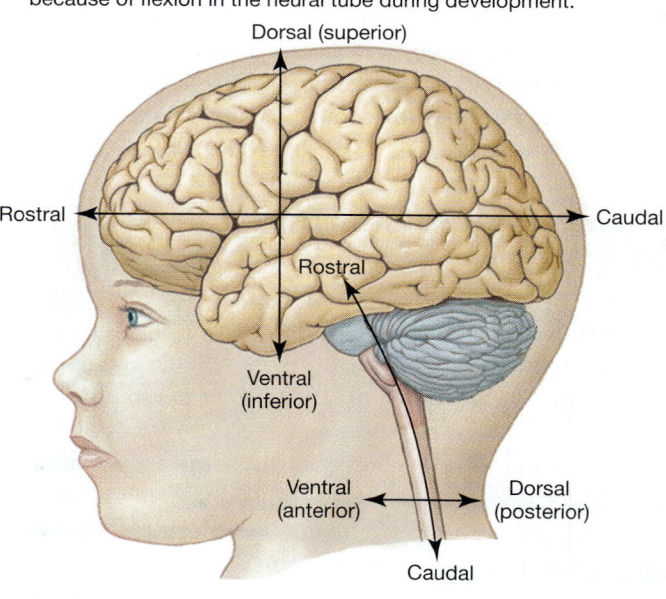

■ **FIGURE 9-3** *The neural tube specializes into the seven major regions of the nervous system*

The Central Nervous System Develops from a Hollow Tube

In the very early embryo, cells that will become the nervous system lie in a flattened region called the **neural plate**. As development proceeds (at about day 20 of human development), neural plate cells along the edge migrate toward the midline (Fig. 9-2a ■). By about day 23 of human development, the neural plate cells have fused with each other, creating a **neural tube** (Fig. 9-2b.). *Neural crest cells* from the lateral edges of the neural plate now lie dorsal to the neural tube.

The lumen of the neural tube will remain hollow and become the central cavity of the CNS. The cells lining the neural tube will either differentiate into the epithelial *ependyma* [⊋ p. 252] or remain as undifferentiated *neural stem cells*. The outer cell layers of the neural tube will become the neurons and glia of the CNS. Neural crest cells will become the sensory and motor neurons of the peripheral nervous system.

By week 4 of human development, the anterior portion of the neural tube has begun to specialize into the regions of the brain (Fig. 9-3a ■). Three divisions are obvious: a **forebrain**, a midbrain, and a **hindbrain**. The tube posterior to the hindbrain will become the spinal cord. At this stage, the portion of the forebrain that will become the cerebrum is not much larger than the other regions of the brain.

As development proceeds, the growth of the cerebrum begins to outpace that of the other regions (Fig. 9-3b). By week 6, the CNS has formed the seven major divisions that are present at birth. Six of these regions are in the brain—(1) the cerebrum, (2) the **diencephalon**, (3) the midbrain, (4) and (5) the cerebellum and **pons**, (6) the **medulla oblongata**—and the seventh is the spinal cord. The cerebrum and diencephalon develop from the forebrain. The cerebellum, pons, and medulla oblongata are divisions of the hindbrain.

By week 6 the central cavity (lumen) of the neural tube has begun to enlarge into the hollow **ventricles** [*ventriculus*, belly] of the brain. There are two *lateral ventricles* (the first and second) and two *descending ventricles* (the third and fourth). The central cavity of the neural tube also becomes the *central canal* of the spinal cord.

By week 11 the cerebrum is noticeably enlarged (Fig. 9-3c), and at birth the cerebrum is the largest and most obvious structure we see when looking at a human brain (Fig. 9-3d). The fully developed cerebrum surrounds the diencephalon, midbrain, and pons, leaving only the cerebellum and medulla oblongata visible below it. Because of the flexion (bending) of the neural tube early in development (see Fig. 9-3b), some directional terms have different meanings when applied to the brain (Fig. 9-3e).

The Central Nervous System Is Divided into Gray Matter and White Matter

The central nervous system, like the peripheral nervous system, is composed of neurons and supportive glial cells. As noted in Chapter 8, interneurons are neurons completely contained within the CNS. Sensory (afferent) and efferent neurons link interneurons to peripheral receptors and effectors.

When viewed on a macroscopic level, the tissues of the CNS are divided into gray matter and white matter (Fig. 9-4c ■). **Gray matter** consists of unmyelinated nerve cell bodies, dendrites, and axon terminals. The cell bodies are assembled in an organized fashion in both the brain and the spinal cord. They form layers in some parts of the brain and in other parts cluster into groups of neurons that have similar functions. Clusters of cell bodies in the brain and spinal cord are known as *nuclei*, and they are usually identified by specific names, such as the *lateral geniculate nucleus*, where visual information is processed.

White matter is made up mostly of myelinated axons and contains very few cell bodies. Its pale color comes from the myelin sheaths that surround the axons. Bundles of axons that connect different regions of the CNS are known as **tracts**. Tracts in the central nervous system are equivalent to nerves in the peripheral nervous system.

The consistency of the brain and spinal cord is soft and jellylike. Although individual neurons and glial cells have highly organized internal cytoskeletons that maintain cell shape and orientation, neural tissue has minimal extracellular matrix and must rely on external support for protection from trauma. This support comes in the form of an outer casing of bone, three layers of connective tissue membrane, and fluid between the membranes.

CONCEPT CHECK
2. Name the four kinds of glial cells found in the CNS, and describe the function(s) of each.   Answers: p. 326

Bone and Connective Tissue Support the Central Nervous System

In vertebrates, the brain is encased in a bony **skull**, or **cranium** (Fig. 9-4a), and the spinal cord runs through a canal in the **vertebral column**. The body segmentation that is characteristic of many invertebrates can still be seen in the bony **vertebrae** (singular *vertebra*), which are stacked on top of one another and separated by disks of connective tissue. Nerves of the peripheral nervous system enter and leave the spinal cord by passing through notches between the stacked vertebrae (Fig. 9-4c).

Three layers of membrane, collectively called the **meninges** [singular *meninx*, membrane], lie between the bones and tissues of the central nervous system. These membranes help stabilize the neural tissue and protect it from bruising against the bones of the skeleton. Starting from the bones and moving toward the neural tissue, the membranes are (1) the dura mater, (2) the arachnoid membrane, and (3) the pia mater (Fig. 9-4b, c). The **pia mater** [*pius*, pious + *mater*, mother] is a thin membrane that adheres to the surface of the brain and spinal cord. Arteries that supply blood to the brain are associated with this layer. The **arachnoid** [*arachnoides*, cobweblike] **membrane** is loosely tied to the pia, leaving a *subarachnoid space* between the two layers. The **dura mater** [*durare*, to last] is thick (think *durable*) and associated with veins that drain blood from the brain through vessels or cavities called *sinuses*.

The final protective component of the CNS is extracellular fluid, which helps cushion the delicate neural tissue. The cranium has an internal volume of 1.4 L, of which about 1 L is occupied by the cells. The remaining volume is divided into two distinct extracellular compartments: the blood (100–150 mL) and the cerebrospinal fluid and interstitial fluid (250–300 mL). The cerebrospinal fluid and interstitial fluid together form the extracellular environment for neurons. Interstitial fluid lies inside the pia mater; cerebrospinal fluid is found in the ventricles and in the space between the pia mater and the arachnoid membrane. The two fluids communicate with each other across the leaky junctions of the pial membrane and ependymal cell layer lining the ventricles.

CONCEPT CHECK
3. What is a ganglion? What is the equivalent structure in the CNS?

4. Peripheral nerves are equivalent to what organizational structure in the CNS?   Answers: p. 326

The Brain Floats in Cerebrospinal Fluid

Cerebrospinal fluid, or CSF, is a salty solution that is continuously secreted by the **choroid plexus**, a specialized region on the

ANATOMY SUMMARY
CENTRAL NERVOUS SYSTEM

CENTRAL NERVOUS SYSTEM, POSTERIOR VIEW

(a)

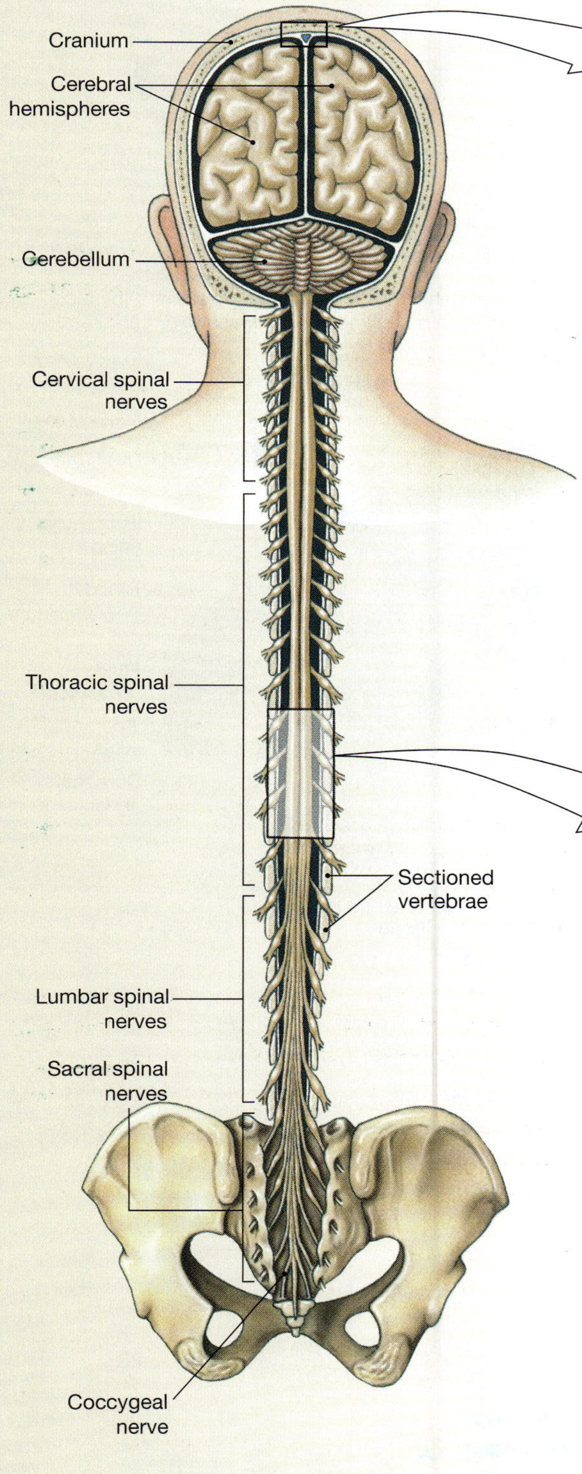

- Cranium
- Cerebral hemispheres
- Cerebellum
- Cervical spinal nerves
- Thoracic spinal nerves
- Sectioned vertebrae
- Lumbar spinal nerves
- Sacral spinal nerves
- Coccygeal nerve

SECTIONAL VIEWS OF THE CNS

(b) Meningeal layers of the brain cushion and protect delicate neural tissue.

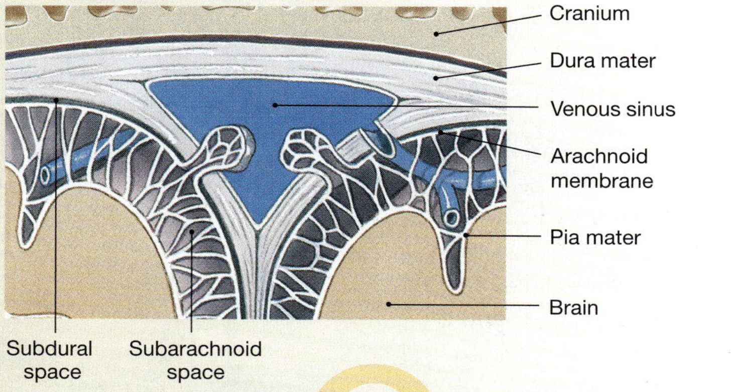

- Cranium
- Dura mater
- Venous sinus
- Arachnoid membrane
- Pia mater
- Brain
- Subdural space
- Subarachnoid space

FIGURE QUESTION

Moving from the cranium in, name the meninges that form the boundaries of the venous sinus and the subdural and subarachnoid spaces.

9

(c) Posterior view of spinal cord and vertebra

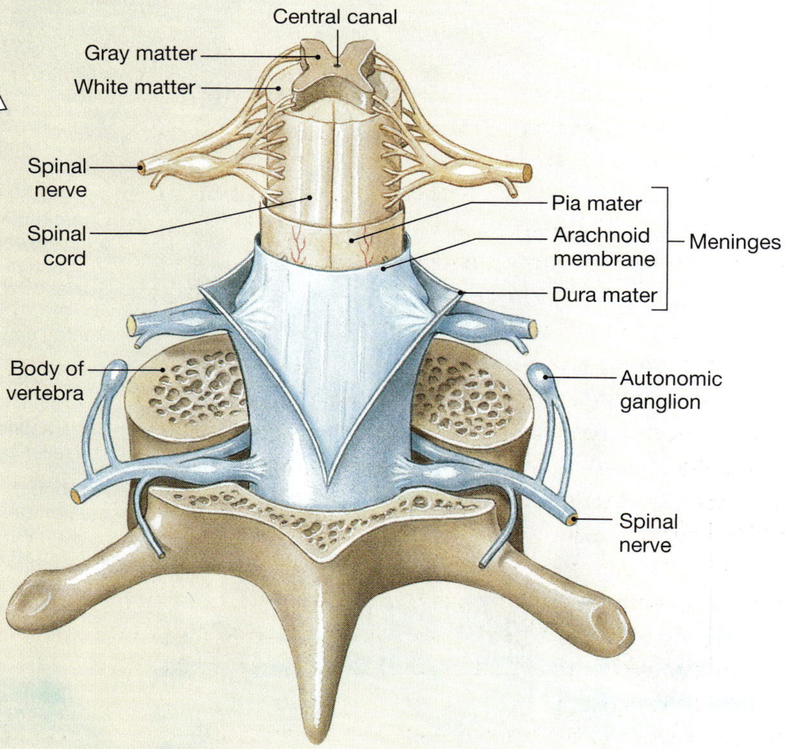

- Central canal
- Gray matter
- White matter
- Spinal nerve
- Spinal cord
- Body of vertebra
- Pia mater
- Arachnoid membrane
- Dura mater
- Meninges
- Autonomic ganglion
- Spinal nerve

FIGURE 9-4

(a) Ventricles of the brain
The lateral ventricles consist of the first and second ventricles. The third and fourth ventricles extend through the brain stem and connect to the central canal that runs through the spinal cord. Compare the frontal view to the cross-section in Fig. 9-3b.

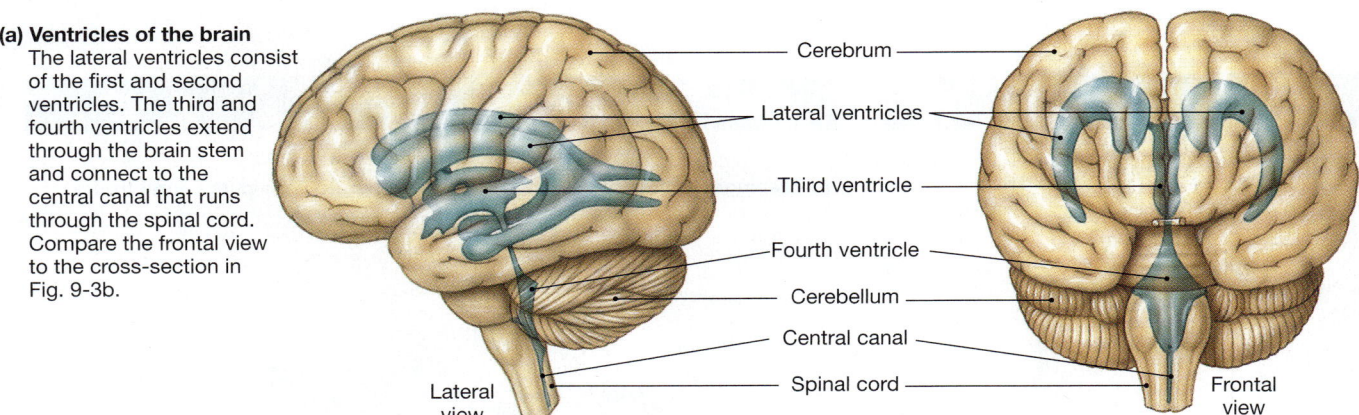

Cerebrum

Lateral ventricles

Third ventricle

Fourth ventricle

Cerebellum

Central canal

Spinal cord

Lateral view

Frontal view

(b) Cerebrospinal fluid is secreted into the ventricles and flows throughout the subarachnoid space where it cushions the central nervous system.

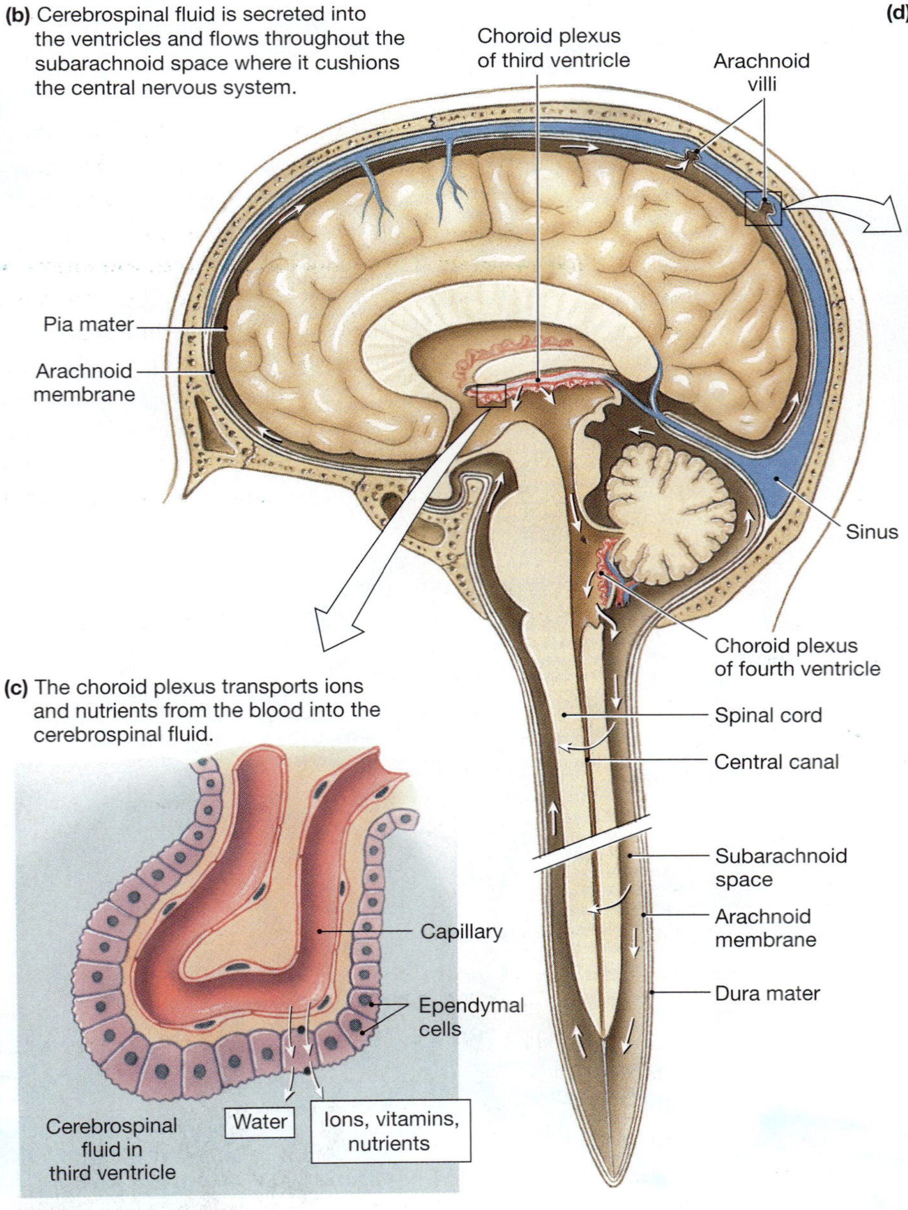

Choroid plexus of third ventricle

Arachnoid villi

Pia mater

Arachnoid membrane

Sinus

Choroid plexus of fourth ventricle

Spinal cord

Central canal

Subarachnoid space

Arachnoid membrane

Dura mater

(c) The choroid plexus transports ions and nutrients from the blood into the cerebrospinal fluid.

Capillary

Ependymal cells

Cerebrospinal fluid in third ventricle

Water

Ions, vitamins, nutrients

(d) Cerebrospinal fluid is reabsorbed back into the blood at fingerlike projections of the arachnoid membrane called villi.

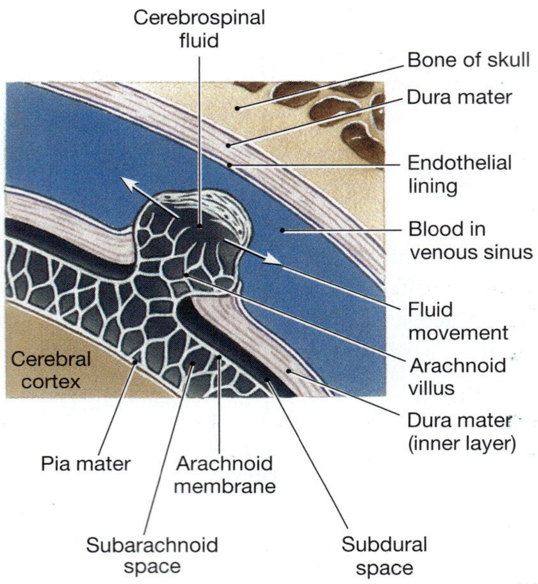

Cerebrospinal fluid

Bone of skull

Dura mater

Endothelial lining

Blood in venous sinus

Fluid movement

Arachnoid villus

Dura mater (inner layer)

Cerebral cortex

Pia mater

Arachnoid membrane

Subarachnoid space

Subdural space

9

FIGURE 9-5

(a)

(b)

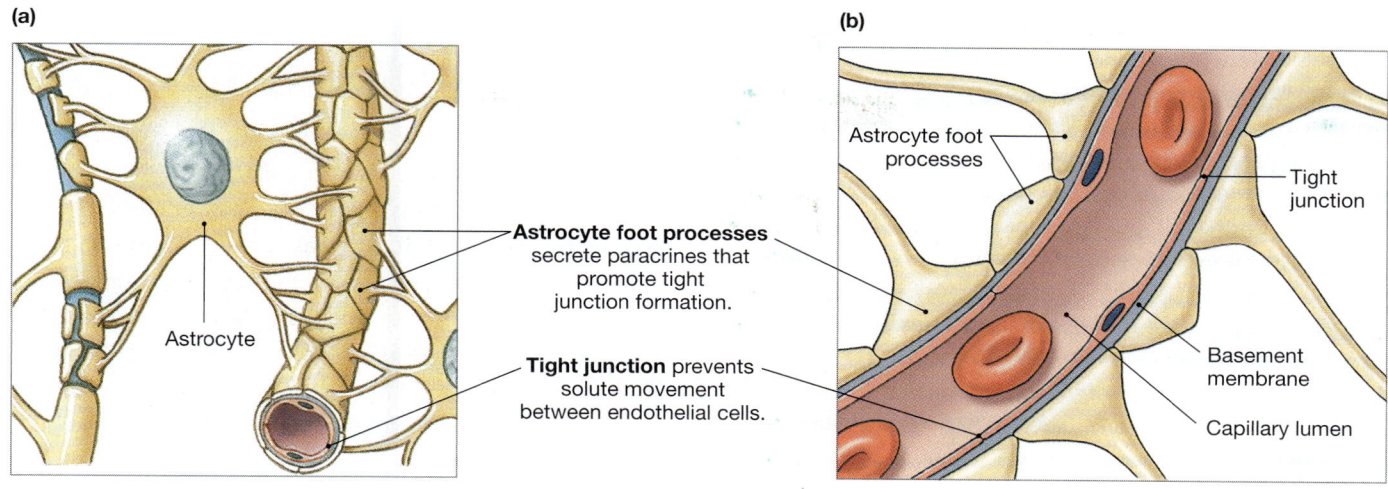

Astrocyte foot processes

Tight junction

Astrocyte foot processes secrete paracrines that promote tight junction formation.

Astrocyte

Tight junction prevents solute movement between endothelial cells.

Basement membrane

Capillary lumen

■ **FIGURE 9-6**   *The blood-brain barrier*

The movement of material across the blood-brain barrier is regulated to protect neurons from potentially harmful substances in the blood.

walls of the ventricles (Fig. 9-5b ■). The choroid plexus is remarkably similar to kidney tissue and consists of capillaries and a transporting epithelium [⟲ p. 73] derived from the ependyma. The choroid plexus cells selectively pump sodium and other solutes from plasma into the ventricles, creating an osmotic gradient that draws water along with the solutes (Fig. 9-5c).

From the ventricles, cerebrospinal fluid flows into the **subarachnoid space** between the pia mater and the arachnoid membrane, surrounding the entire brain and spinal cord in fluid (Fig. 9-5b). The cerebrospinal fluid flows around the neural tissue and is finally absorbed back into the blood by special **villi** [singular *villus*, shaggy hair] on the arachnoid membrane in the cranium (Fig. 9-5d). The rate of fluid flow through the central nervous system is sufficient to replenish the entire volume of cerebrospinal fluid about three times a day.

Cerebrospinal fluid serves two purposes: physical protection and chemical protection. The brain and spinal cord float in the thin layer of fluid between the membranes. The buoyancy of cerebrospinal fluid reduces the weight of the brain nearly 30-fold. Lighter weight translates into less pressure on blood vessels and nerves attached to the CNS.

The cerebrospinal fluid also provides protective padding. When there is a blow to the head, the fluid must be compressed before the brain can hit the inside of the cranium. For a dramatic example of the protective power of cerebrospinal fluid, shake a block of tofu (representing the brain) in an empty jar. Then shake another block of tofu in a jar completely filled with water to see how cerebrospinal fluid safeguards the brain.

In addition to physically protecting the delicate tissues of the CNS, cerebrospinal fluid creates a closely regulated extracellular environment for the neurons. The choroid plexus is selective about which substances it transports into the ventricles, and as a result, the composition of cerebrospinal fluid is different from that of the plasma. The concentrations of K^+, Ca^{2+}, HCO_3^-, and

glucose are lower in the cerebrospinal fluid, and the concentration of H^+ is higher. Only the concentration of Na^+ in cerebrospinal fluid is similar to that in the blood. In addition, cerebrospinal fluid contains very little protein and no blood cells.

Cerebrospinal fluid exchanges solutes with the interstitial fluid of the CNS and provides a route by which wastes can be removed. Clinically, a sample of cerebrospinal fluid is presumed to be an indicator of the chemical environment in the brain. This sampling procedure, known as a *spinal tap* or *lumbar puncture,* is generally done by withdrawing fluid from the subarachnoid space between vertebrae at the lower end of the spinal cord. The presence of proteins or blood cells in cerebrospinal fluid suggests an infection.

C O N C E P T   C H E C K

5. If the concentration of H^+ in cerebrospinal fluid is higher than that in the blood, what can you say about the pH of the CSF?

6. If a blood vessel running between the meninges ruptures as a result of a blow to the cranium, what do you predict might happen?

7. Is cerebrospinal fluid more like plasma or more like interstitial fluid? Defend your answer.   Answers: p. 326

The Blood-Brain Barrier Protects the Brain from Harmful Substances in the Blood

The final layer of protection for the brain is a functional barrier between the interstitial fluid and the blood. This barrier is necessary to isolate the body's main control center from potentially harmful substances in the blood and from blood-borne pathogens such as bacteria. To achieve this protection, most of the 400 miles of brain capillaries create a functional **blood-brain barrier** (Fig. 9-6 ■). Although not a literal barrier, the highly selective permeability of brain capillaries shelters the brain from toxins and from fluctuations in hormones, ions, and neuroactive substances such as neurotransmitters in the blood.

Why are brain capillaries so much less permeable than other capillaries? In most capillaries, leaky cell-cell junctions and pores allow free exchange of solutes between the plasma and interstitial fluid [🔁 p. 72]. In brain capillaries, however, the endothelial cells form tight junctions with one another, junctions that prevent solute movement between the cells. Tight junction formation apparently is induced by paracrine signals from adjacent astrocytes whose foot processes surround the capillary. Thus, it is the brain tissue itself that creates the blood-brain barrier.

The selective permeability of the blood-brain barrier can be attributed to its transport properties. The capillary endothelium uses selected membrane carriers and channels to move nutrients and other useful materials from the blood into the brain interstitial fluid. Other transporters move wastes from the interstitial fluid into the plasma. Any water-soluble molecule that is not transported on one of these carriers cannot cross the blood-brain barrier.

One interesting illustration of how the blood-brain barrier works is seen in *Parkinson's disease,* a neurological disorder in which brain levels of the neurotransmitter dopamine are too low because dopaminergic neurons are either damaged or dead. Dopamine administered in a pill or injection is ineffective because it is unable to cross the blood-brain barrier. The dopamine precursor L-*dopa,* however, is transported across the cells of the blood-brain barrier on an amino acid transporter [🔁 p. 144]. Once neurons have access to L-dopa in the interstitial fluid, they metabolize it to dopamine, thereby correcting the deficiency.

Although the blood-brain barrier excludes many water-soluble substances, smaller lipid-soluble molecules can diffuse through the cell membranes [🔁 p. 135]. This is one reason some antihistamines make you sleepy but others do not. Older antihistamines were lipid-soluble amines that readily crossed the blood-brain barrier and acted on brain centers controlling alertness. The newer drugs are much less lipid soluble and as a result do not have the same sedative effect.

A few areas of the brain lack a functional blood-brain barrier because their capillaries have leaky endothelium like most of the rest of the body. In these areas of the brain, the function of adjacent neurons depends in some way on direct contact with the blood. For instance, the posterior pituitary releases neurosecretory hormones that must pass into the capillaries for distribution throughout the body [🔁 p. 226]. Another region that lacks the blood-brain barrier is the vomiting center in the medulla oblongata. These neurons monitor the blood for possibly toxic foreign substances, such as drugs. If they sense something harmful, they initiate a vomiting reflex. Vomiting removes the contents of the digestive system and eliminates ingested toxins.

Neural Tissue Has Special Metabolic Requirements

A unique property of the central nervous system is its specialized metabolism. Neurons require a constant supply of oxygen and glucose to make ATP for active transport of ions and

RUNNING PROBLEM

Ben was diagnosed with *infantile spasms,* a form of epilepsy characterized by the onset of head-drop seizures at four to seven months and by arrested or deteriorating mental development. Ben was started on a month-long regimen of adrenocorticotropin (ACTH) [🔁 p. 228] shots to control the seizures. Scientists are unsure why this hormone is so effective in controlling this type of seizure, but they have found that among its effects, it increases myelin formation, increases blood-brain barrier integrity, and enhances binding of the neurotransmitter GABA at synapses. As expected, Ben's seizures disappeared completely before the month of treatment ended, and his development began to return to a normal level.

Question 1:
   How might a leaky blood-brain barrier lead to a cascade of action potentials that trigger a seizure?

Question 2:
   GABA opens Cl⁻ channels on the postsynaptic cell. What does this do to the cell's membrane potential? Does GABA make the cell more or less likely to fire action potentials?

Question 3:
   Why is it important to limit the duration of ACTH therapy, particularly in very young patients? [🔁 p. 232]

| 292 | **300** | 315 | 317 | 321 | 322 |

neurotransmitters. Oxygen passes freely across the blood-brain barrier, and membrane transporters move glucose from the plasma to the brain's interstitial fluid. Unusually low levels of either substrate can have devastating results on brain function.

Because of its high demand for oxygen, the brain receives about 15% of the blood pumped by the heart. If blood flow to the brain is interrupted, brain damage occurs after only a few minutes without oxygen. Neurons are equally sensitive to lack of glucose.

Under normal circumstances the only energy source for neurons is glucose. (In starvation situations, the brain can metabolize ketones produced by fat breakdown [🔁 p. 111].) By some estimates, the brain is responsible for about half of the body's glucose consumption. Consequently, the body uses several homeostatic pathways to ensure that glucose concentrations in the blood always remain adequate to meet the brain's demand. If homeostasis fails, progressive **hypoglycemia** (low blood glucose levels) leads to confusion, unconsciousness, and eventually death.

Now that you have a broad overview of the central nervous system, we will examine the structure and function of the spinal cord and brain in more detail.

HYPOGLYCEMIA AND THE BRAIN

Neurons are picky about their food. Under most circumstances, the only biomolecule that neurons use for energy is glucose. Surprisingly, this can present a problem for diabetic patients, whose problem is too much glucose in the blood. In the face of sustained hyperglycemia (elevated blood glucose), the cells of the blood-brain barrier down-regulate [🔁 p. 43] their glucose transporters. Then, if the patient's blood glucose level falls below normal due to insulin or other therapy, the neurons of the brain may be out of luck. They will be unable to take up glucose fast enough to sustain their electrical activity. The result is impaired CNS function, including confusion, irritability, and slurred speech. Unless the neurons are provided with glucose promptly, they can sustain permanent damage. In extreme cases, hypoglycemia can cause coma or even death.

CONCEPT CHECK

8. Oxidative phosphorylation takes place in which organelle?

9. Name the two metabolic pathways for aerobic metabolism of glucose. What happens to NADH produced in these pathways?

10. In the late 1800s the scientist Paul Ehrlich injected blue dye into the bloodstream of animals. He noticed that all tissues except the brain stained blue. He was not aware of the blood-brain barrier and drew a different conclusion from his results. What other explanation might Ehrlich have proposed?

11. In a subsequent experiment, a student of Ehrlich's injected the dye into the cerebrospinal fluid of the same animals. What do you think he observed about staining in the brain and in other body tissues?

Answers: p. 326

THE SPINAL CORD

The spinal cord is the major pathway for information flowing back and forth between the brain and the skin, joints, and muscles of the body. In addition, the spinal cord contains neural networks responsible for locomotion. If the spinal cord is severed, there is loss of sensation from the skin and muscles as well as *paralysis,* loss of the ability to voluntarily control muscles.

The spinal cord is divided into four regions (*cervical, thoracic, lumbar,* and *sacral*), named to correspond to the adjacent vertebrae (see Fig. 9-4a). Each spinal region is subdivided into segments, and each segment gives rise to a bilateral pair of **spinal nerves**. Just before a spinal nerve joins the spinal cord, it divides into two branches called **roots** (Fig. 9-7a ■). The **dorsal root** of each spinal nerve is specialized to carry incoming sensory information. The **dorsal root ganglia**, swellings found on the dorsal roots just before they enter the cord (Fig. 9-7b), contain cell bodies of sensory neurons. The **ventral root** carries information from the central nervous system to muscles and glands.

In cross section, the spinal cord has a butterfly- or H-shaped core of gray matter and a surrounding rim of white matter. Sensory fibers from the dorsal roots synapse with interneurons in the **dorsal horns** of the gray matter. The dorsal horn cell bodies are organized into two distinct nuclei, one for somatic information and one for visceral information (Fig. 9-7b).

The **ventral horns** of the gray matter contain cell bodies of motor neurons that carry efferent signals to muscles and glands. The ventral horns are organized into somatic motor and autonomic nuclei. Efferent fibers leave the spinal cord via the ventral root.

The white matter of the spinal cord is the biological equivalent of fiber-optic cables that telephone companies use to carry our communications systems. White matter can be divided into a number of **columns** composed of tracts of axons that transfer information up and down the cord. **Ascending tracts** take sensory information to the brain. They occupy the dorsal and external lateral portions of the spinal cord (Fig. 9-7c). **Descending tracts** carry mostly efferent (motor) signals from the brain to the cord. They occupy the ventral and interior lateral portions of the white matter. **Propriospinal tracts** [*proprius,* one's own] are those that remain within the cord.

The spinal cord can function as a self-contained integrating center for simple *spinal reflexes,* with signals passing from a sensory neuron through the gray matter to an efferent neuron (Fig. 9-8 ■). In addition, spinal interneurons may route sensory information to the brain through ascending tracts or bring commands from the brain to motor neurons. In many cases, the interneurons also modify information as it passes through them. Reflexes play a critical role in the coordination of movement and will be discussed in more detail in Chapter 13.

CONCEPT CHECK

12. What are the differences between horns, roots, tracts, and columns of the spinal cord?

13. If a dorsal root of the spinal cord is cut, what function will be disrupted?

Answers: p. 326

(a) One segment of spinal cord, ventral view, showing its pair of nerves

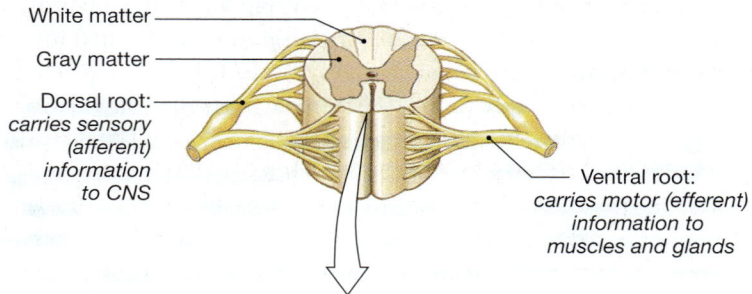

White matter

Gray matter

Dorsal root: *carries sensory (afferent) information to CNS*

Ventral root: *carries motor (efferent) information to muscles and glands*

(b) Gray matter consists of sensory and motor nuclei

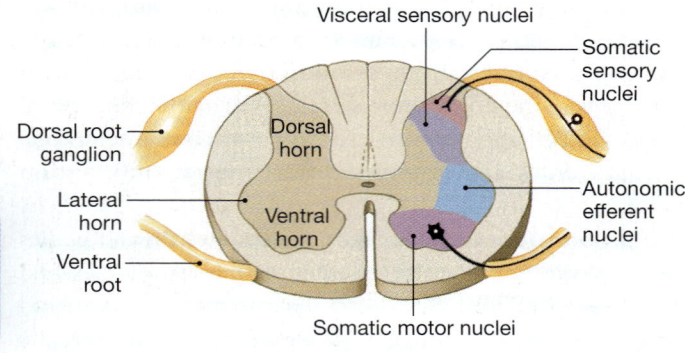

Visceral sensory nuclei

Somatic sensory nuclei

Dorsal root ganglion

Dorsal horn

Lateral horn

Ventral horn

Autonomic efferent nuclei

Ventral root

Somatic motor nuclei

(c) White matter in the spinal cord consists of axons carrying information to and from the brain.

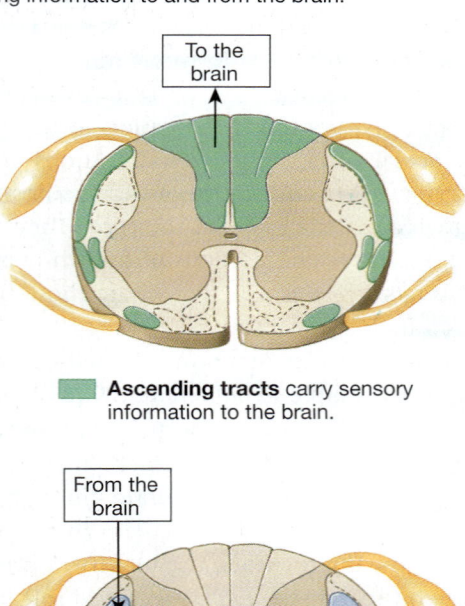

To the brain

Ascending tracts carry sensory information to the brain.

From the brain

Descending tracts carry commands to motor neurons.

■ **FIGURE 9-7** *Specialization in the spinal cord*

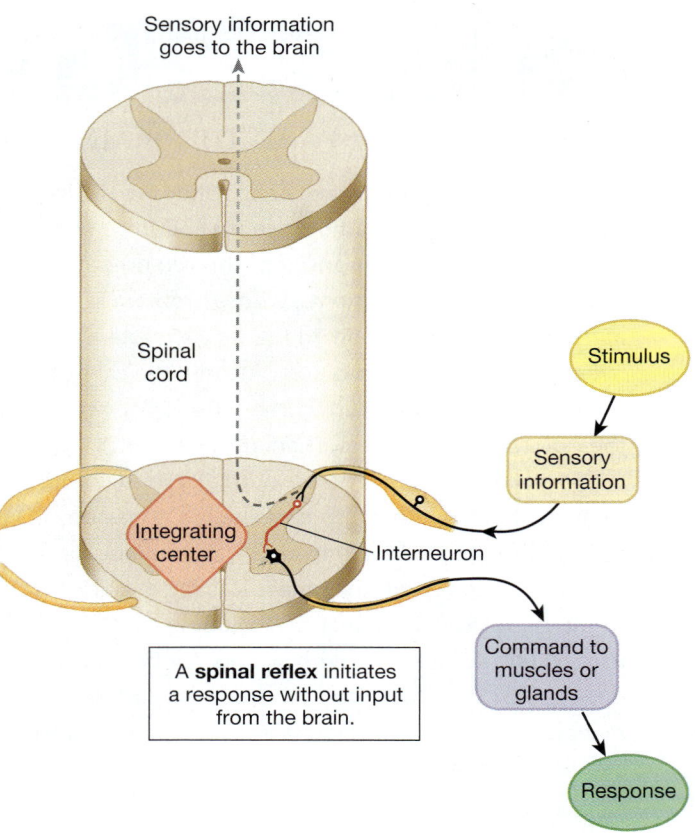

Sensory information goes to the brain

Spinal cord

Integrating center

Interneuron

Stimulus

Sensory information

Command to muscles or glands

Response

A **spinal reflex** initiates a response without input from the brain.

■ **FIGURE 9-8** *The spinal cord is an integrating center*

In a spinal reflex, sensory information entering the spinal cord is acted on without input from the brain. However, sensory information about the stimulus may be sent to the brain.

THE BRAIN

Thousands of years ago, Aristotle declared that the heart was the seat of the soul. However, most people now agree that the brain is the organ that gives the human species its unique attributes. The challenge facing today's scientists is to understand how circuits formed by millions of neurons result in complex behaviors such as speaking, writing a symphony, or creating imaginary worlds for an interactive computer game. Brain function may be the ultimate emergent property [🔁 p. 6]. The question remains whether we will ever be able to decipher how emotions such as happiness and love arise from the chemical and electrical signals passing along circuits of neurons.

It is possible to study the brain at many levels of organization. The most reductionist view looks at the individual neurons and at what happens to them in response to chemical or electrical signals. A more integrative study might look at groups of neurons and how they interact with one another in *circuits,* or *pathways.* The most complicated approach starts with a behavior or physiological response and works backward to dissect the neural circuits that create the behavior or response.

TABLE 9-1	The Cranial Nerves		
NUMBER	NAME	TYPE	FUNCTION
I	Olfactory	Sensory	Olfactory (smell) information from nose
II	Optic	Sensory	Visual information from eyes
III	Oculomotor	Motor	Eye movement, pupil constriction, lens shape
IV	Trochlear	Motor	Eye movement
V	Trigeminal	Mixed	Sensory information from face, mouth; motor signals for chewing
VI	Abducens	Motor	Eye movement
VII	Facial	Mixed	Sensory for taste; efferent signals for tear and salivary glands, facial expression
VIII	Vestibulocochlear	Sensory	Hearing and equilibrium
IX	Glossopharyngeal	Mixed	Sensory from oral cavity, baro- and chemoreceptors in blood vessels; efferent for swallowing, parotid salivary gland secretion
X	Vagus	Mixed	Sensory and efferents to many internal organs, muscles, and glands
XI	Spinal accessory	Motor	Muscles of oral cavity, some muscles in neck and shoulder
XII	Hypoglossal	Motor	Tongue muscles

For centuries, studies of brain function were restricted to anatomical descriptions. However, there is no tidy 1:1 relationship between structure and function when looking at the brain. An adult human brain has a mass of about 1400 g and contains an estimated 10^{12} neurons. When you consider that each one of these millions of neurons may receive as many as 200,000 synapses, the number of possible neuronal connections is mind boggling.

A basic principle to remember when studying the brain is that one function, even an apparently simple one such as bending your finger, will involve multiple brain regions (as well as the spinal cord). Conversely, one brain region may be involved in several functions at the same time. In other words, understanding the brain is not simple and straightforward.

Figure 9-9 ■ is an anatomy summary to follow as we discuss major brain regions, moving from the most primitive to the most complex. Of the six major divisions of the brain present at birth (see Fig. 9-3b), only the medulla, cerebellum, and cerebrum are visible when the intact brain is viewed in profile. The remaining three divisions (diencephalon, midbrain, and pons) are covered by the cerebrum.

The Brain Stem Is the Transition Between Spinal Cord and Midbrain

The **brain stem** is the oldest and most primitive region of the brain and consists of structures that derive from the embryonic midbrain and hindbrain. The brain stem can be divided into white matter and gray matter, and in some ways its anatomy is similar to that of the spinal cord. Some ascending tracts from the spinal cord pass through the brain stem, whereas others synapse there. Descending tracts from higher brain centers also travel through the brain stem on their way to the spinal cord.

Pairs of peripheral nerves branch off the brain stem, similar to spinal nerves along the spinal cord (Fig. 9-9d). Eleven of the 12 **cranial nerves** (numbers II–XII) originate along the brain stem. (The first cranial nerve, the olfactory nerve (I), enters the forebrain.) Cranial nerves carry sensory and motor information for the head and neck (Table 9-1 ■) and are described according to whether they include sensory fibers, efferent fibers, or both (mixed nerves). For example, cranial nerve X, the **vagus nerve** [*vagus*, wandering], is a mixed nerve that carries both sensory and motor fibers for many internal organs. An important component of a clinical neurological examination is testing the functions controlled by these nerves.

Located throughout the brain stem are discrete groups of nerve cell bodies (*nuclei*). Many of these nuclei are associated with the **reticular formation**, a diffuse collection of neurons that extends throughout the brain stem. The name *reticular* means "network" and comes from the crisscrossed axons that branch profusely up into superior sections of the brain and down into the spinal cord. Nuclei in the brain stem are involved

LATERAL VIEW OF THE CNS

(a)

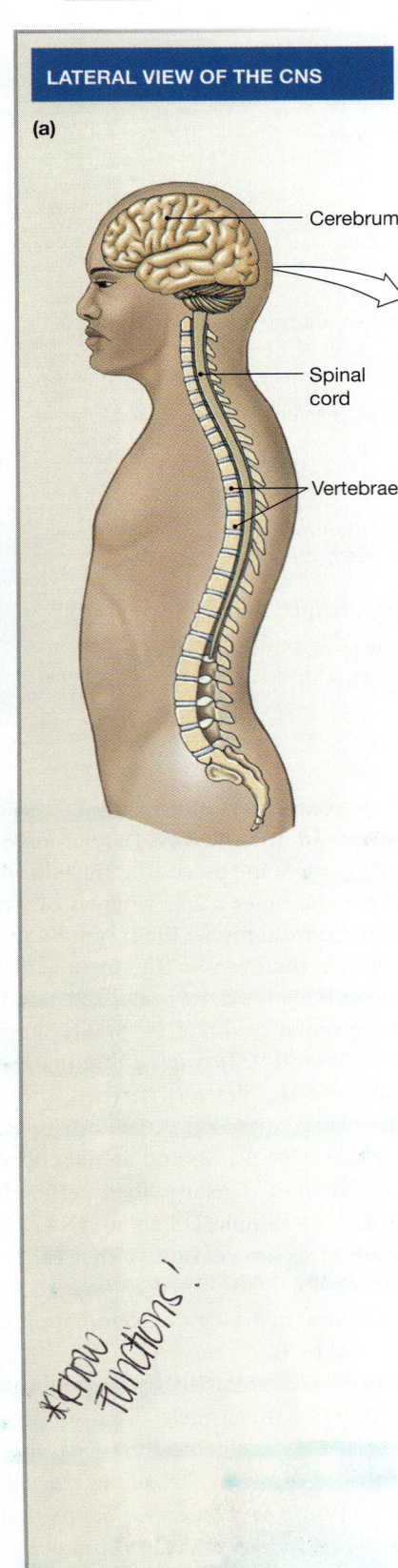

Cerebrum

Spinal cord

Vertebrae

*know functions!

ANATOMY OF THE BRAIN

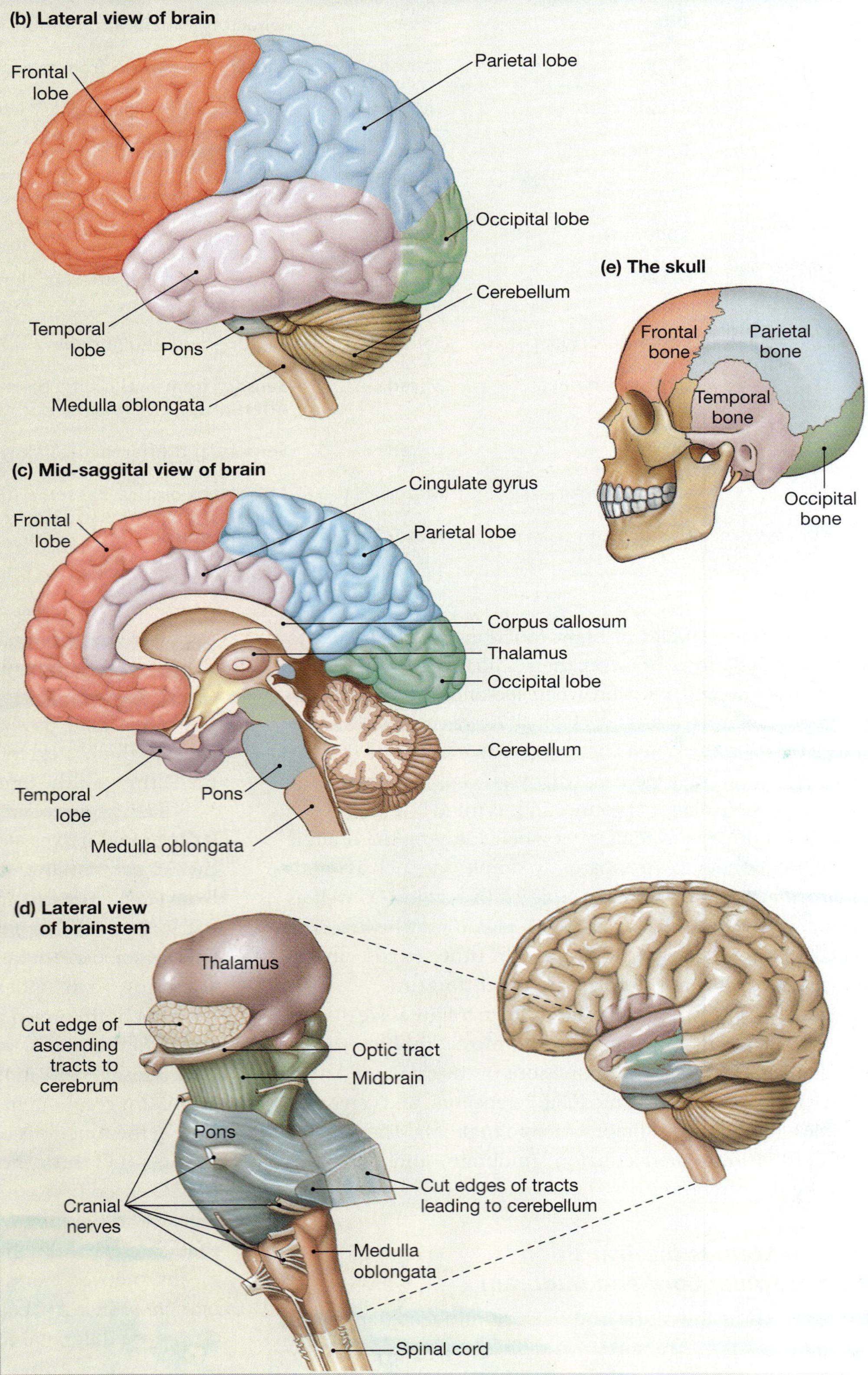

(b) **Lateral view of brain**

Frontal lobe

Parietal lobe

Occipital lobe

Cerebellum

Temporal lobe

Pons

Medulla oblongata

(e) **The skull**

Frontal bone

Parietal bone

Temporal bone

Occipital bone

(c) **Mid-saggital view of brain**

Frontal lobe

Cingulate gyrus

Parietal lobe

Corpus callosum

Thalamus

Occipital lobe

Cerebellum

Temporal lobe

Pons

Medulla oblongata

(d) **Lateral view of brainstem**

Thalamus

Cut edge of ascending tracts to cerebrum

Optic tract

Midbrain

Pons

Cranial nerves

Cut edges of tracts leading to cerebellum

Medulla oblongata

Spinal cord

FIGURE 9-9

FUNCTIONS OF THE BRAIN	
REGION	FUNCTION
Cerebrum (Frontal ●, Parietal ●, Occipital ●, and Temporal ● lobes)	
• Cerebral cortex (See Fig. 9-15)	
Sensory fields	Perception
Motor areas	Skeletal muscle movement
Association areas	Integration of information and direction of voluntary movement
● Basal ganglia (See Fig. 9-11)	Movement
● Limbic system (See Fig. 9-13)	
● Amygdala	Emotion and memory
● Hippocampus	Learning and memory
● **Diencephalon** (See Fig. 9-10)	
● Thalamus	Integrating center and relay station for sensory and motor information
● Hypothalamus	Homeostasis and behavioral drives (See Table 9-2)
● Pituitary	Hormone secretion
● Pineal gland	Melatonin secretion
● **Cerebellum**	Movement coordination
● **Brain stem**	
● Midbrain	Eye movement
● Pons	Relay station between cerebrum and cerebellum; coordination of breathing
● Medulla oblongata	Control of involuntary functions
Reticular formation (See Fig. 9-19)	Arousal, sleep, muscle tone, pain modulation

(continued)

in many basic processes, including arousal and sleep, muscle tone and stretch reflexes, coordination of breathing, blood pressure regulation, and modulation of pain.

CONCEPT CHECK

14. Are the following white matter or gray matter: (a) ascending tracts, (b) reticular formation, (c) descending tracts?

15. Using the information from Table 9-1, describe the types of activities you might ask a patient to perform if you wished to test the function of each cranial nerve.

16. In anatomical directional terminology, the cerebrum, which is located next to the top of the skull, is said to be _____ to the brain stem. Answers: p. 326

The Brain Stem Consists of Medulla, Pons, and Midbrain

Starting at the spinal cord and moving toward the top of the skull, the brain stem includes the medulla oblongata, the pons, and the midbrain (Fig. 9-9d). Some authorities include the cerebellum as part of the brain stem. The diamond-shaped fourth ventricle, visible in the hindbrain of the early embryo (see Fig. 9-3b), runs through the interior and connects to the central canal of the spinal cord (see Fig. 9-5).

The medulla oblongata, frequently just called the *medulla* [*medulla*, marrow; adjective *medullary*], is the transition from the spinal cord into the brain proper (Fig. 9-9d). Its white matter includes ascending **somatosensory tracts** [*soma*, body] that bring sensory information to the brain, and descending **corticospinal tracts** that convey information from the cerebrum to the spinal cord. About 90% of corticospinal tracts cross the midline to the opposite side of the body in a region of the medulla known as the **pyramids**. As a result of this crossover, each side of the brain controls the opposite side of the body. Gray matter in the medulla includes nuclei that control many involuntary functions, such as blood pressure, breathing, swallowing, and vomiting.

The pons [*pons*, bridge; adjective *pontine*] is a bulbous protrusion on the ventral side of the brain stem above the medulla and below the midbrain. Because its primary function is to act as a relay station for information transfer between the cerebellum and cerebrum, the pons is often grouped with the cerebellum. The pons also coordinates the control of breathing along with centers in the medulla.

The third region of the brain stem, the midbrain, or *mesencephalon* [*mesos*, middle], is a relatively small area that lies between the lower brain stem and the diencephalon. The primary function of the midbrain is control of eye movement, but it also relays signals for auditory and visual reflexes.

The Cerebellum Coordinates Movement

The cerebellum is the second largest structure in the brain (Fig. 9-9a–c). It is located inside the base of the skull, just above the nape of the neck. The name *cerebellum* [adjective *cerebellar*]

9

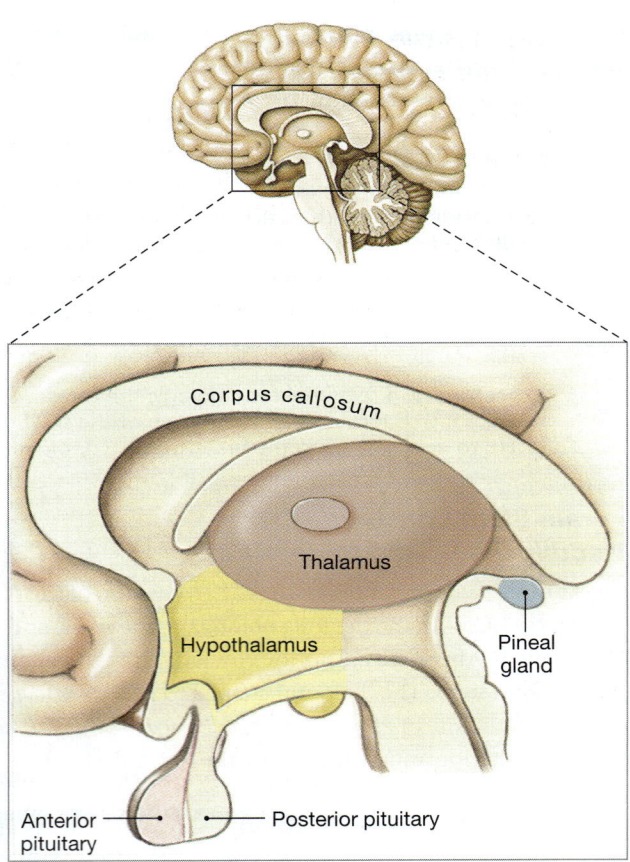

■ FIGURE 9-10 *The diencephalon*

The diencephalon lies between the brain stem and the cerebrum. It consists of thalamus, hypothalamus, pineal gland, and pituitary gland.

TABLE 9-2	Functions of the Hypothalamus

1. Activates sympathetic nervous system
   - Controls catecholamine release from adrenal medulla (as in fight-or-flight reaction)
   - Helps maintain blood glucose concentrations through effects on endocrine pancreas

2. Maintains body temperature
   - Stimulates shivering and sweating

3. Controls body osmolarity
   - Motivates thirst and drinking behavior
   - Stimulates secretion of vasopressin [⮂ p.226]

4. Controls reproductive functions
   - Directs secretion of oxytocin (for uterine contractions and milk release)
   - Directs trophic hormone control of anterior pituitary hormones FSH and LH [⮂ p.227]

5. Controls food intake
   - Stimulates satiety center
   - Stimulates feeding center

6. Interacts with limbic system to influence behavior and emotions

7. Influences cardiovascular control center in medulla oblongata

8. Secretes trophic hormones that control release of hormones from anterior pituitary gland

means "little brain," and, indeed, most of the nerve cells in the brain are in the cerebellum. The specialized function of the cerebellum is to process sensory information and coordinate the execution of movement. Sensory input into the cerebellum comes from somatic receptors in the periphery of the body and from receptors for equilibrium and balance located in the inner ear. The cerebellum also receives motor input from neurons in the cerebrum. Cerebellar function is described in more detail in Chapter 13.

The Diencephalon Contains the Centers for Homeostasis

The diencephalon, or "between-brain," lies between the brain stem and the cerebrum. It is composed of two main sections, the thalamus and the hypothalamus, and two endocrine structures, the pituitary and pineal glands (Fig. 9-10 ■).

Most of the diencephalon is occupied by many small nuclei that make up the **thalamus** [*thalamus,* bedroom; adjective *thalamic*]. The thalamus receives sensory fibers from the optic tract, ears, and spinal cord as well as motor information from the cerebellum. It projects fibers to the cerebrum, where the information is processed. The thalamus is often described as a relay station because almost all sensory information from lower

parts of the CNS passes through it. Like the spinal cord, the thalamus can modify information passing through it, making it an integrating center as well as a relay station.

The **hypothalamus** lies beneath the thalamus. Although the hypothalamus occupies less than 1% of total brain volume, it is the center for homeostasis and contains centers for various behavioral drives, such as hunger and thirst. Output from the hypothalamus also influences many functions of the autonomic division of the nervous system, as well as a variety of endocrine functions (Table 9-2 ■). The hypothalamus receives input from multiple sources, including the cerebrum, the reticular formation, and various sensory receptors. Output from the hypothalamus goes first to the thalamus and eventually to multiple effector pathways.

Two important endocrine structures are located in the diencephalon: the pituitary gland and the pineal gland. The pituitary gland was described in Chapter 7 [⮂ p. 225]. The posterior pituitary (neurohypophysis) is a downgrowth of the hypothalamus and secretes neurohormones that are synthesized in hypothalamic nuclei. The anterior pituitary (adenohypophysis) is a true endocrine gland. Its hormones are regulated by hypothalamic neurohormones secreted into the *hypothalamic-hypophyseal portal system*. The pineal gland, which secretes the hormone melatonin, was also introduced in Chapter 7 [⮂ p. 237] and will be discussed later in this chapter.

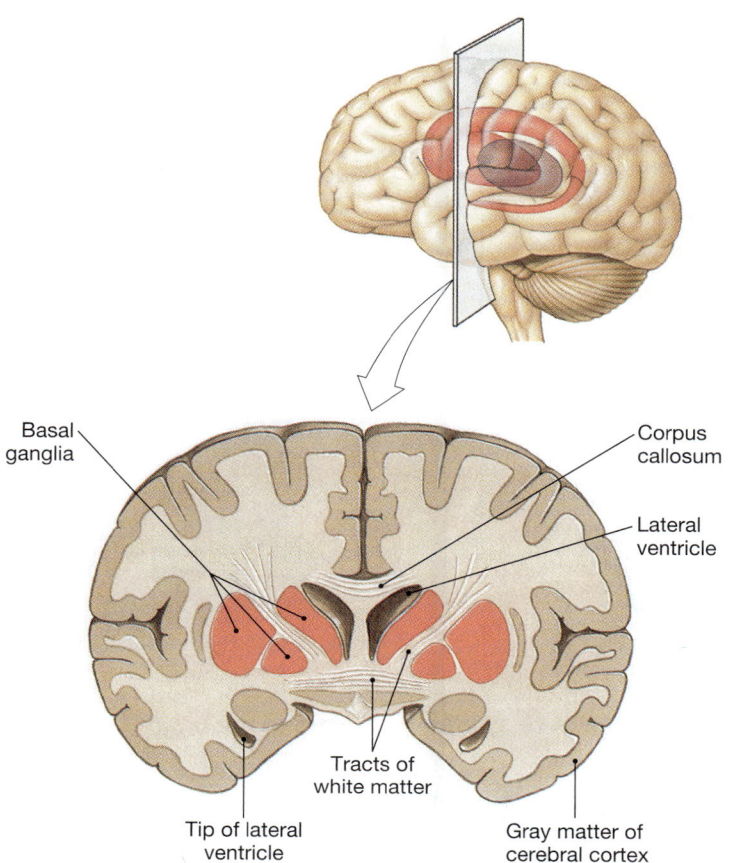

Basal ganglia

Corpus callosum

Lateral ventricle

Tracts of white matter

Tip of lateral ventricle

Gray matter of cerebral cortex

■ **FIGURE 9-11** *Gray matter of the cerebrum*

The cerebral cortex and basal ganglia are two of the three regions of gray matter in the cerebrum. The third region, the limbic system, is detailed in Figure 9-13. The frontal view shown here is similar to the sectional view obtained using modern diagnostic imaging techniques.

CONCEPT CHECK

17. Starting at the spinal cord and moving up, name the subdivisions of the brain stem.

18. What are the four primary structures of the diencephalon?

Answers: p. 326

The Cerebrum Is the Site of Higher Brain Functions

As noted earlier in the chapter, the cerebrum is the largest and most distinctive part of the human brain and fills most of the cranial cavity. It is composed of two hemispheres connected primarily at the **corpus callosum** (Figs. 9-9c and 9-10), a distinct structure formed by axons passing from one side of the brain to the other. This connection ensures that the two hemispheres communicate and cooperate with each other. Each cerebral hemisphere is divided into four lobes, named for the bones of the skull under which they are located: *frontal, parietal, temporal,* and *occipital* (Fig. 9-9b, c, e).

The surface of the cerebrum in humans and other primates has a furrowed, walnut-like appearance, with grooves called *sulci* [singular *sulcus,* a furrow] dividing convolutions called *gyri* [singular *gyrus,* a ring or circle]. During development, the cerebrum grows faster than the surrounding cranium, causing the tissue to fold back on itself to fit into a smaller volume. The degree of folding is directly related to the level of processing of which the brain is capable; less-advanced mammals, such as rodents, have brains with a relatively smooth surface. The human brain is so convoluted that if it were inflated enough to smooth the surfaces, it would be three times as large and would need a head the size of a beach ball.

The Cerebrum Has Distinct Regions of Gray Matter and White Matter

Cerebral gray matter can be into three major regions: the cerebral cortex, the basal ganglia, and the limbic system. The **cerebral cortex** [*cortex,* bark or rind; adjective *cortical,* plural *cortices*] is the outer layer of the cerebrum, only a few millimeters thick (Fig. 9-11 ■). Neurons of the cerebral cortex are arranged in anatomically distinct vertical columns and horizontal layers (Fig. 9-12 ■). It is within these layers that our higher brain functions arise.

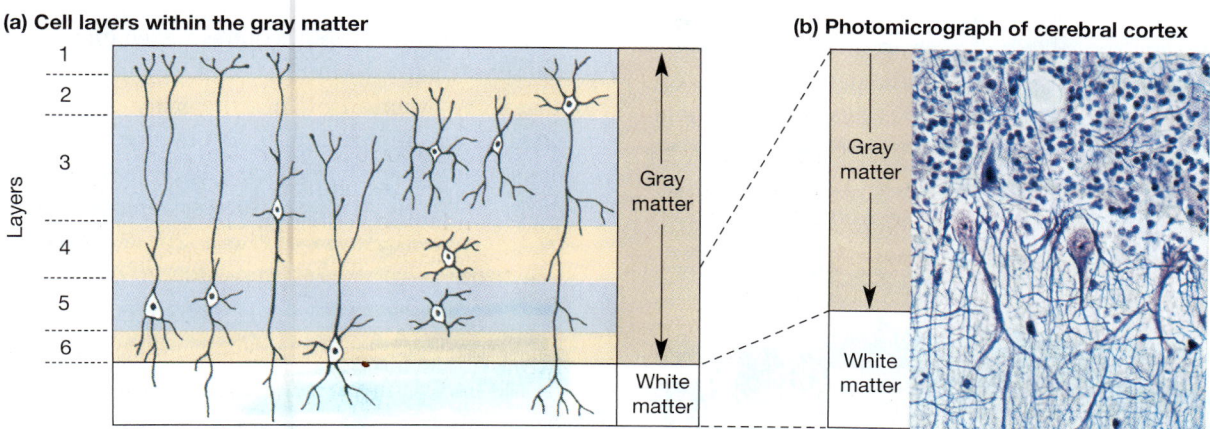

(a) Cell layers within the gray matter

Layers

1
2
3
4
5
6

Gray matter

White matter

(b) Photomicrograph of cerebral cortex

Gray matter

White matter

■ **FIGURE 9-12** *Cell bodies in the cerebral cortex form distinct layers*

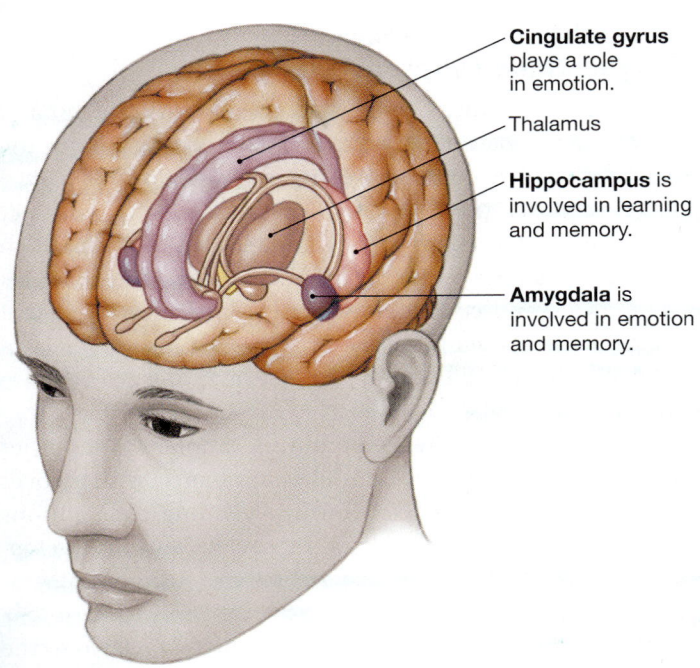

■ FIGURE 9-13   *The limbic system*

Anatomically, the limbic system is part of the gray matter of the cerebrum.

The second region of cerebral gray matter consists of the **basal ganglia*** (Fig. 9-11), which are involved in the control of movement and will be discussed in Chapter 13. The third region is the **limbic system** [*limbus*, a border], which surrounds the brain stem (Fig. 9-13 ■). The limbic system represents probably the most primitive region of the cerebrum. It acts as the link between higher cognitive functions, such as reasoning, and more primitive emotional responses, such as fear. The major areas of the limbic system are the **amygdala** and **cingulate gyrus**, which are linked to emotion and memory, and the **hippocampus**, which is associated with learning and memory.

White matter in the cerebrum is found mostly in the interior (Fig. 9-11). Bundles of fibers allow different regions of the cortex to communicate with one another and transfer information from one hemisphere to the other, primarily through the corpus callosum. According to some estimates, the corpus callosum may have as many as 200 million axons passing through it! Information entering and leaving the cerebrum goes along tracts that pass through the thalamus (with the exception of olfactory information, which goes directly from olfactory receptors to the cerebrum).

---

*The basal ganglia are also called the basal nuclei. Neuroanatomists prefer to reserve the term *ganglia* for clusters of nerve cell bodies outside the CNS, but the term *basal ganglia* is commonly used in clinical settings.

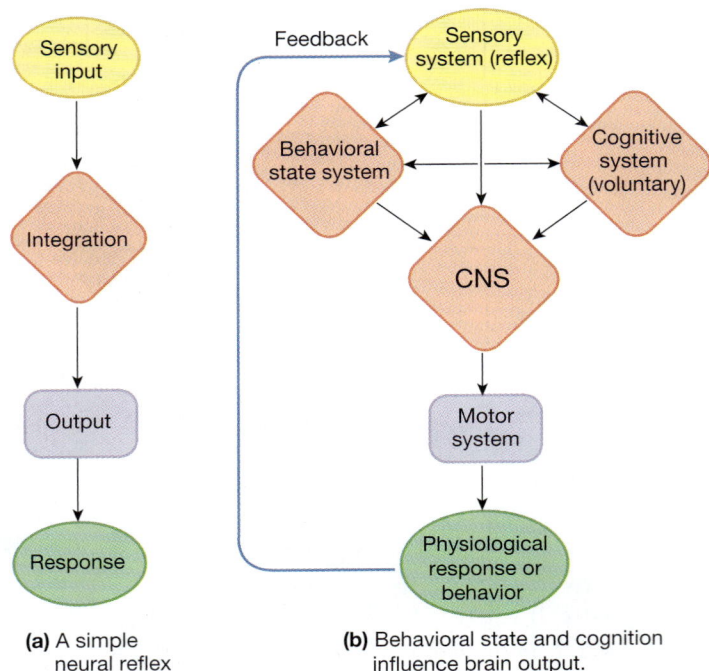

(a) A simple neural reflex

(b) Behavioral state and cognition influence brain output.

■ FIGURE 9-14   *Reflex pathways in the brain*

CONCEPT CHECK

19. Name the anatomical location in the brain where neurons from one side of the body cross to the opposite side.

20. Name the divisions of the brain in anatomical order, starting from the spinal cord.   Answers: p. 326

BRAIN FUNCTION

From a simplistic view, the brain is an information processor much like a computer. It receives sensory input from the internal and external environments, integrates and processes the information, and, if appropriate, creates a response (Fig. 9-14a ■). This is the basic reflex pathway introduced in Chapter 6 [☐ p. 195]. What makes the brain more complicated than this simple reflex pathway, however, is its ability to generate information and output signals *in the absence of external input*. Modeling this intrinsic input requires a diagram more complex than the simple pathway illustrated in Figure 9-14a.

Larry Swanson of the University of Southern California presents one approach to modeling brain function in his book *Brain Architecture: Understanding the Basic Plan* (Oxford University Press, 2003). He describes three systems that influence output by the motor systems of the body: (1) the **sensory system**, which monitors the internal and external environments and initiates reflex responses; (2) a **cognitive system** that resides in the cerebral cortex and is able to initiate voluntary responses; and (3) a **behavioral state system**, which also resides in the brain and governs sleep-wake cycles and other intrinsic behaviors.

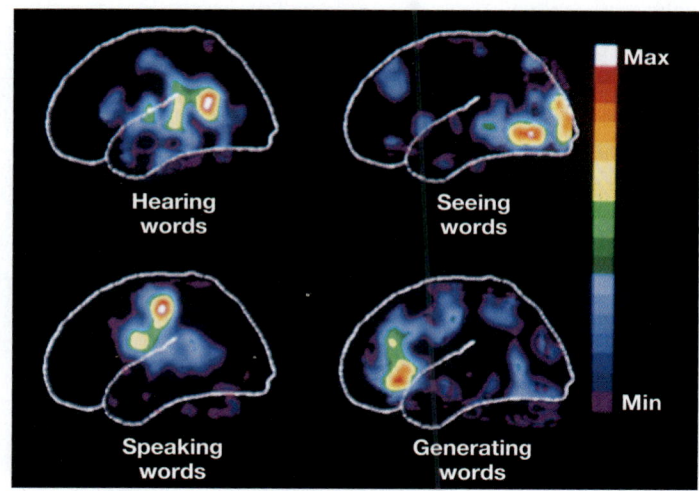

■ FIGURE 9-17   *PET scan of the brain at work*

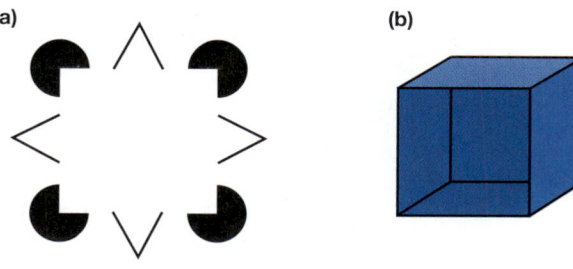

■ FIGURE 9-18   *Perception*

The brain has the ability to interpret sensory information to create the perception of (a) shapes or (b) three-dimensional objects.

and viscera. The somatosensory pathways carry information about touch, temperature, pain, itch, and body position. Damage to this part of the brain leads to reduced sensitivity of the skin on the opposite side of the body because sensory fibers cross to the opposite side of the midline as they ascend through the spine or medulla.

The special senses of vision, hearing, taste, and olfaction (smell) each have different brain regions devoted to processing their sensory input (see Fig. 9-15). The **visual cortex**, located in the occipital lobe, receives information from the eyes. The **auditory cortex**, located in the temporal lobe, receives information from the ears. The **olfactory cortex**, a small region in the temporal lobe, receives input from chemoreceptors in the nose. The **gustatory cortex**, deeper in the brain near the edge of the frontal lobe, receives sensory information from the taste buds. The somatic and special senses are described in detail in the next chapter.

Sensory Information Is Processed into Perception

Once sensory information reaches the appropriate cortical area, information processing has just begun. Neural pathways extend from sensory areas to appropriate association areas, which integrate somatic, visual, auditory, and other stimuli into perception, the brain's interpretation of sensory stimuli.

Often the perceived stimulus is very different from the actual stimulus. For instance, photoreceptors in the eye receive light waves of different frequencies, but we perceive the different wave energies as different colors. Similarly, the brain translates pressure waves hitting the ear into sound and interprets chemicals binding to chemoreceptors as taste or smell.

One interesting aspect of perception is the way our brain fills in missing information to create a complete picture, or translates a two-dimensional drawing into a three-dimensional shape (Fig. 9-18 ■). Thus, we sometimes perceive what our brain expects to perceive. Our perceptual translation of sensory stimuli allows the information to be acted upon and used in voluntary motor control or in complex cognitive functions such as language.

The Motor System Governs Output from the Central Nervous System

The motor output component of the nervous system is associated with the efferent division, shown in Figure 8-1 [p. 245]. Motor output can be divided into three major types: (1) skeletal muscle movement, controlled by the somatic motor division; (2) neuroendocrine signals, neurohormones secreted into the blood by neurons located primarily in the hypothalamus and adrenal medulla; and (3) *visceral* responses, the actions of smooth and cardiac muscle or endocrine and exocrine glands. Visceral responses are governed by the autonomic division of the nervous system.

Information about skeletal muscle movement is processed in several regions of the CNS. Simple stimulus-response pathways, such as the knee jerk reflex, are processed either in the spinal cord or in the brain stem. Although these reflexes do not require integration in the cerebral cortex, they can be modified or overridden by input from the cognitive system.

Voluntary movements are initiated by the cognitive system and originate in the **primary motor cortex** and **motor association area** in the frontal lobes of the cerebrum (see Fig. 9-15). These regions receive input from sensory areas as well as from the cerebellum and basal ganglia. Long output neurons called *pyramidal cells* project axons from the motor areas through the brain stem to the spinal cord. Other pathways go from the cortex to the basal ganglia and lower brain regions. Descending motor pathways cross to the opposite side of the body; therefore, damage to a motor area will be manifested by paralysis or function loss on the opposite side of the body. A detailed discussion of the coordination of body movement can be found in Chapter 13.

Neuroendocrine and visceral responses are coordinated primarily in the hypothalamus and medulla. The brain stem is responsible for maintaining many of the automatic life functions we take for granted, such as breathing, blood pressure, and control of body movement. It receives sensory information from

9

(a) ● Norepinephrine

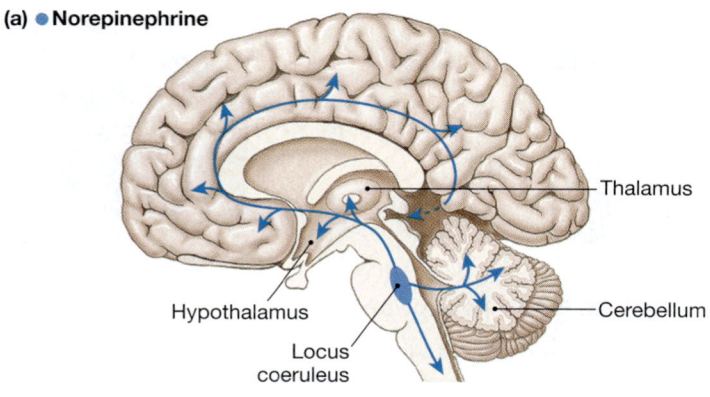

Thalamus

Hypothalamus

Cerebellum

Locus
coeruleus

(b) ● Serotonin

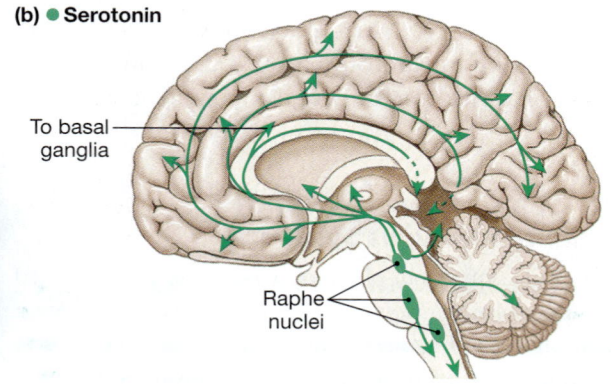

To basal
ganglia

Raphe
nuclei

(c) ● Dopamine

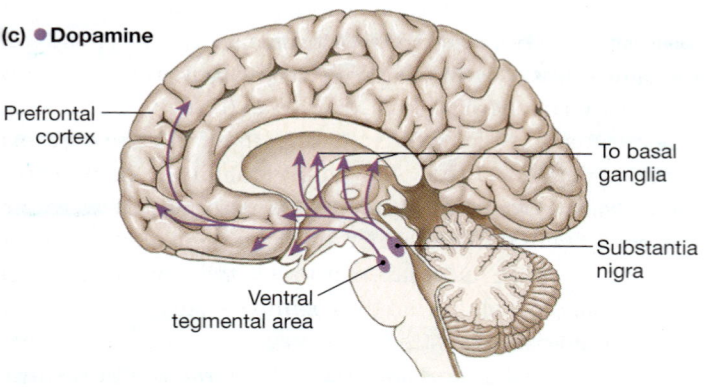

Prefrontal
cortex

To basal
ganglia

Substantia
nigra

Ventral
tegmental area

(d) ● Acetylcholine

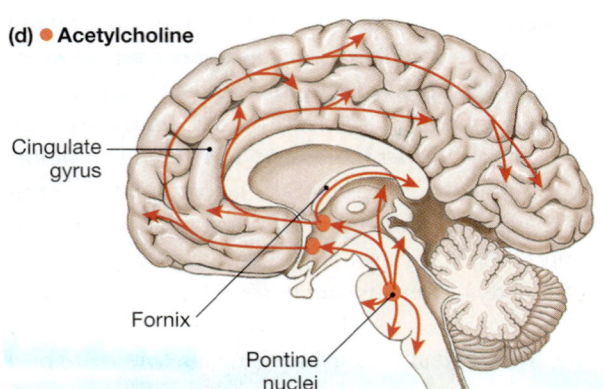

Cingulate
gyrus

Fornix

Pontine
nuclei

the body and relays motor commands to peripheral muscles and glands. The hypothalamus contains centers for temperature regulation, eating, and control of body osmolarity, among others. The responses to stimulation of these centers may be neural or hormonal reflexes or a behavioral response. Stress, reproduction, and growth are also mediated by the hypothalamus by way of multiple hormones. You will learn more about these reflexes in later chapters as we discuss the various systems of the body.

Sensory input is not the only factor determining motor output by the brain. The behavioral state system can modulate reflex pathways, and the cognitive system exerts both voluntary and involuntary control over motor functions.

The Behavioral State System Modulates Motor Output

The behavioral state system is an important modulator of sensory and cognitive processing. Many neurons in the behavioral state system are found in regions of the brain outside the cerebral cortex, including parts of the reticular formation in the brain stem, the hypothalamus, and the limbic system.

The neurons collectively known as the **diffuse modulatory systems** originate in the reticular formation in the brain stem and project their axons to large areas of the brain (Fig. 9-19 ■, Table 9-3 ■). There are four modulatory systems that are generally classified according to the neurotransmitter they secrete: *noradrenergic* (norepinephrine), *serotonergic* (serotonin), *dopaminergic* (dopamine), and *cholinergic* (acetylcholine). The diffuse modulatory systems regulate brain function by influencing attention, motivation, wakefulness, memory, motor control, mood, and metabolic homeostasis.

The Reticular Activating System Influences States of Arousal

One function of the behavioral state system is control of levels of consciousness and sleep-wake cycles. **Consciousness** is the body's state of arousal or awareness of self and environment.

Experimental evidence shows that the **reticular activating system**, a diffuse collection of neurons in the reticular formation, plays an essential role in keeping the "conscious brain" awake. If connections between the reticular formation and the cerebral cortex are disrupted surgically, an animal becomes comatose for long periods of time. Other evidence for the importance of the reticular formation in states of arousal comes from studies showing that general anesthetics depress synaptic transmission in that region of the brain. Presumably, blocking

■ **FIGURE 9-19** *The diffuse modulatory systems modulate brain function*

The neurons collectively known as the diffuse modulatory systems originate in the reticular formation of the brain stem and project their axons to large areas of the brain. The four systems are named for their neurotransmitters.

TABLE 9-3	The Diffuse Modulatory Systems		
SYSTEM (NEUROMODULATOR)	**SITE WHERE NEURONS ORIGINATE**	**STRUCTURES THAT NEURONS INNERVATE**	**FUNCTIONS MODULATED BY THE SYSTEM**
Noradrenergic (norepinephrine)	Locus coeruleus of the pons	Cerebral cortex, thalamus, hypothalamus, olfactory bulb, cerebellum, midbrain, spinal cord	Attention, arousal, sleep-wake cycles, learning, memory, anxiety, pain, and mood
Serotonergic (serotonin)	Raphe nuclei along brain stem midline	Lower nuclei project to spinal cord	Pain; locomotion
		Upper nuclei project to most of brain	Sleep-wake cycle, mood and emotional behaviors such as aggression and depression
Dopaminergic (dopamine)	Substantia nigra in midbrain Ventral tegumentum in midbrain	Cortex Cortex and parts of limbic system	Motor control "Reward" centers linked to addictive behaviors
Cholinergic (acetylcholine)	Base of cerebrum; pons and midbrain	Cerebrum, hippocampus, thalamus	Sleep-wake cycles, arousal, learning, memory, sensory information passing through thalamus

ascending pathways between the reticular formation and the cerebral cortex creates a state of unconsciousness.

Physiologically, what distinguishes being awake from various stages of sleep? One way to define arousal states is by the pattern of electrical activity created by the cortical neurons. The measurement of brain activity is recorded by a procedure known as **electroencephalography**. Surface electrodes placed on or in the scalp detect depolarizations of the cortical neurons in the region just under the electrode.

In awake states, many neurons are firing but not in a coordinated fashion. Presumably the desynchronization of electrical activity in waking states is produced by ascending signals from the reticular formation. An *electroencephalogram,* or **EEG,** of the waking-alert (eyes open) state shows a rapid, irregular pattern with no dominant waves. In awake-resting (eyes closed) states, sleep, or coma, electrical activity of the neurons begins to synchronize into waves with characteristic patterns (Fig. 9-20 ■). As the person's state of arousal lessens, the frequency of the waves decreases. The more synchronous the firing of cortical neurons, the larger the amplitude of the waves. Thus, the awake-resting state is characterized by low-amplitude, high-frequency waves, and deep sleep is marked by high-amplitude, low-frequency waves. The complete cessation of brain waves is one of the clinical criteria for determining death.

Why Do We Sleep?

In humans, our major rest period is marked by a behavior known as **sleep,** defined as an easily reversible state of inactivity characterized by lack of interaction with the external environment.

Why we need to sleep is one of the unsolved mysteries in neurophysiology. Some explanations that have been proposed include to conserve energy, to avoid predators, to allow the body to repair itself, and to process memories. Most mammals and birds show the same stages of sleep as humans, telling us that sleep is a very ancient property of vertebrate brains.

Until the 1960s, sleep was thought to be a passive state that resulted from withdrawal of stimuli to the brain. Then experiments showed that neuronal activity in ascending tracts from the brain stem to the cerebral cortex was required for sleep. From other studies, we know that the sleeping brain consumes as much oxygen as the awake brain, and sometimes even more. As a result, we now consider sleep to be an active state.

Sleep is divided into four stages, each marked by identifiable, predictable events associated with characteristic somatic changes and EEG patterns (Fig. 9-20a). The two major sleep phases are **slow-wave sleep** (also called **deep sleep** or **non-REM sleep**, stage 4) and **REM (rapid eye movement) sleep** (stage 1). Slow-wave sleep is apparent on the EEG by the presence of *delta waves,* high-amplitude, low-frequency waves of long duration that sweep across the cerebral cortex. During this phase of the sleep cycle, sleepers adjust body position without conscious commands from the brain to do so.

In contrast, REM sleep is marked by an EEG pattern closer to that of an awake person, with low-amplitude, high-frequency waves. During REM sleep, brain activity inhibits motor neurons to skeletal muscles, paralyzing them. Exceptions to this pattern are the muscles that move the eyes and those that control breathing. The control of homeostatic functions is depressed

(a)

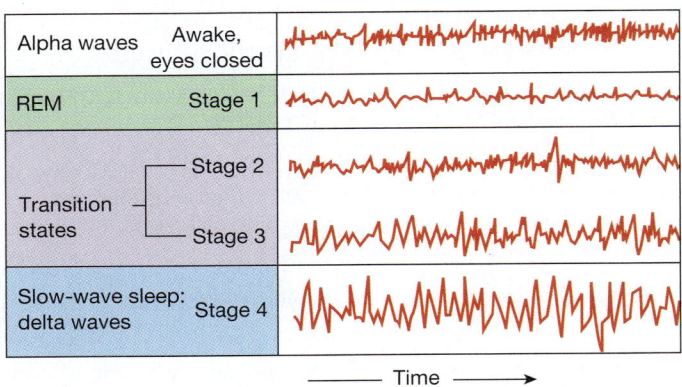

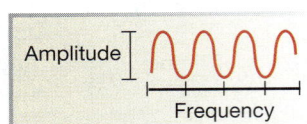

(b)

Which EEG pattern has the highest frequency? The greatest amplitude?

GRAPH QUESTIONS

■ **FIGURE 9-20**   *Electroencephalograms (EEGs) and the sleep cycle*

Recordings of electrical activity in the brain during awake-resting and sleep periods show characteristic patterns.

during REM sleep, and body temperature falls toward ambient temperature.

REM sleep is the period during which most dreaming takes place. The eyes move behind closed lids, as if following the action of the dream. Sleepers are most likely to wake up spontaneously from periods of REM sleep.

A typical eight-hour sleep consists of repeating cycles, as shown in Figure 9-20b. In the first hour, the person moves from wakefulness into a deep sleep (stage 4; first blue area in Figure 9-20b). The sleeper then cycles between deep sleep and REM sleep (stage 1), with transition stages (stages 2–3) occurring in between. Near the end of an eight-hour sleep period a sleeper spends the most time in stage 2 and REM sleep, until finally awakening for the day.

If sleep is a neurologically active process, what is it that makes us sleepy? The possibility of a sleep-inducing factor was first proposed in 1913, when scientists found that cerebrospinal fluid from sleep-deprived dogs could induce sleep in normal animals. Since then, a variety of sleep-inducing factors have been identified. Curiously, many of them are also substances that enhance the immune response, such as interleukin-1, interferon, serotonin, and tumor necrosis factor. As a result of this finding, some investigators have suggested that one answer to the puzzle of the biological reason for sleep is that we need to sleep to enhance our immune response. Whether or not that is a reason for why we sleep, the link between the immune system and sleep induction may help explain why we tend to sleep more when we are sick.

Sleep disorders are relatively common, as you can tell by looking at the variety of sleep-promoting agents available over

the counter in drugstores. Among the more common sleep disorders are *insomnia* (the inability to go to sleep or remain asleep long enough to awake refreshed), sleep apnea, and sleepwalking. *Sleep apnea* [*apnoos,* breathless] is a condition in which the sleeper awakes when the airway muscles relax to the point of obstructing normal breathing.

Sleepwalking, or *somnambulism* [*somnus,* sleep + *ambulare,* to walk], is a sleep behavior disorder that for many years was thought to represent the acting out of dreams. However, most dreaming occurs during REM sleep (stage 1), whereas sleepwalking takes place during deep sleep (stage 4). During sleepwalking episodes, which may last from 30 seconds to 30 minutes, the subject's eyes are open and registering the surroundings. The subject is able to avoid bumping into objects, can negotiate stairs, and in some cases is reported to perform such tasks as preparing food or folding clothes. The subject usually has little if any conscious recall of the sleepwalking episode upon awakening. Sleepwalking is most common in children, and the frequency of episodes declines with age. There is also a genetic component, as the tendency to sleepwalk runs in families. To learn more about the different sleep disorders, see the U.S. National Institutes of Health website for the National Center for Sleep Disorder Research (*www.nhlbi.nih.gov/about/ncsdr*).

✓ CONCEPT CHECK

21. During sleep, relay neurons in the thalamus reduce information reaching the cerebrum by altering their membrane potential. Are these neurons more likely to have depolarized or hyperpolarized? Explain your reasoning.

Answers: p. 326

EMERGING CONCEPTS

ADENOSINE AND THAT "JAVA JOLT"

Caffeine and its methylxanthine cousins *theobromine* and *theophylline* (found in chocolate and tea) are probably the most widely consumed psychoactive drugs, known since ancient times for their stimulant effect. Molecular research has revealed that the methylxanthines are receptor antagonists for *adenosine,* a molecule composed of the nitrogenous base adenine plus the sugar ribose [⊇ p. 32]. Adenosine acts as an important neuromodulator in the central nervous system. Four subtypes of adenosine receptor have been identified, and they are all G protein-coupled, cAMP-dependent membrane proteins. The discovery that the stimulant effect of caffeine comes from its blockade of adenosine receptors has led scientists to investigate adenosine's role in sleep-wake cycles. Evidence suggests that adenosine accumulates in the extracellular fluid during waking hours, increasingly suppressing activity of the neurons that promote wakefulness. Other roles for adenosine in the brain include its possible involvement in the addiction/reward system and in the development of depression.

Physiological Functions Exhibit Circadian Rhythms

All organisms (even plants) have alternating daily patterns of rest and activity. Sleep-wake rhythms, like many other biological cycles, generally follow a 24-hour light-dark cycle and are known as *circadian rhythms* [⊇, p. 201]. When an organism is placed in conditions of constant light or darkness, these activity rhythms persist, apparently cued by an internal clock.

In mammals, the "clock" resides in networks of neurons located in the **suprachiasmatic nucleus** of the hypothalamus. A very simple interpretation of recent experiments on the molecular basis of the clock is that clock cycling is the result of a complex feedback loop in which specific genes turn on and direct protein synthesis. The proteins accumulate, turn off the genes, and then are themselves degraded. In the absence of the proteins, the genes turn back on and the cycle begins again. The clock has intrinsic activity that is synchronized with the external environment by sensory information about light cycles received through the eyes.

Circadian rhythms in humans can be found in most physiological functions and usually correspond to the phases of our

RUNNING PROBLEM

About six months after the start of ACTH treatment, Ben's head-drop seizures returned and his development began to decline once again. An EEG following Ben's relapse did not demonstrate the erratic wave patterns specific to infantile spasms but did show abnormal activity in the right cortex. A neurologist ordered a positron emission tomography (PET) scan to determine the focus of Ben's seizure activity.

Ben received an injection of radioactively labeled glucose. He was then placed in the center of a PET machine lined with radiation detectors that created a map of his brain showing areas of high and low radioactivity. Those parts of his brain that were more active absorbed more glucose and thus emitted more radiation when the radioactive compound began to decay.

Question 4:
What is the rationale for using radioactively labeled glucose (and not some other nutrient) for the PET scan?

| 292 | 300 | **315** | 317 | 321 | 322 |

sleep-wake cycles. For example, body temperature and cortisol secretion both cycle on a daily basis [⊇ Fig. 6-29, p. 201]. Melatonin secretion by the pineal gland also is strongly linked to light-dark cycling and appears to feed back to the suprachiasmatic nucleus to modulate clock cycling.

Emotion and Motivation Involve Complex Neural Pathways

Emotion and motivation are two aspects of brain function that probably represent an overlap of the behavioral state system and cognitive system. The pathways involved are complex and form closed circuits that cycle information among various parts of the brain, including the hypothalamus, limbic system, and cerebral cortex. We still do not understand the underlying neural mechanisms, and this is a large and active area of neuroscience research.

Emotions are difficult to define. We know what they are and can name them, but in many ways they defy description. One characteristic of emotions is that they are difficult to voluntarily turn on or off. The most commonly described emotions, which arise in different parts of the brain, are anger, aggression, sexual feelings, fear, pleasure, contentment, and happiness.

The limbic system, particularly the region known as the *amygdala,* is the center of emotion in the human brain. Scientists

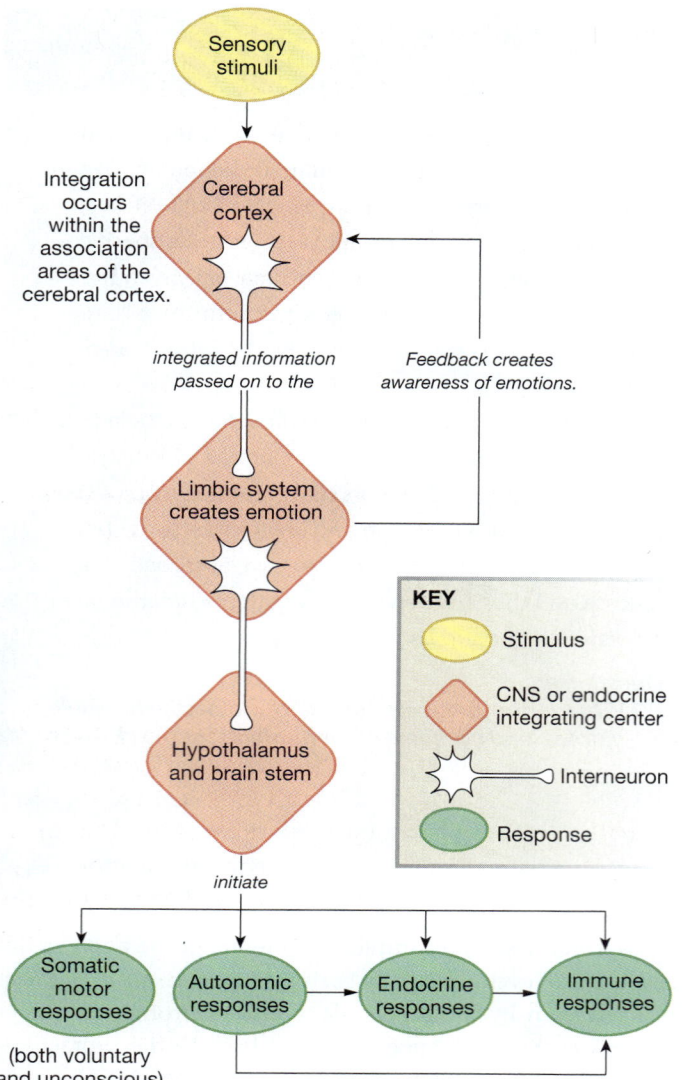

■ **FIGURE 9-21**   *The link between emotions and physiological functions*

The association between stress and increased susceptibility to viruses is an example of an emotionally linked immune response.

have learned about the role of this brain region through experiments in humans and animals. When the amygdala is artificially stimulated in humans, as it might be during surgery for epilepsy, patients report experiencing feelings of fear and anxiety. Experimental lesions that destroy the amygdala in animals cause the animals to become tamer and to display hypersexuality. As a result, neurobiologists believe that the amygdala is the center for basic instincts such as fear and aggression.

The pathways for emotions are complex (Fig. 9-21 ■). Sensory stimuli feeding into the cerebral cortex are constructed in the brain to create a representation (perception) of the world. After information is integrated by the association areas, it is passed on to the limbic system. Feedback from the limbic system to the cerebral cortex creates awareness of the emotion,

while descending pathways to the hypothalamus and brain stem initiate voluntary behaviors and unconscious responses mediated by autonomic, endocrine, immune, and somatic motor systems. The physical result of emotions can be as dramatic as the pounding heart of a fight-or-flight reaction or as insidious as the development of an irregular heartbeat. The links between mind and body are difficult to study and will take many years of research to understand.

Motivation is defined as internal signals that shape voluntary behaviors. Some of these behaviors, such as eating, drinking, and having sex, are related to survival; others, such as curiosity and having sex (again), are linked to emotions. Some motivational states are known as **drives** and generally have three properties in common: (1) they create an increased state of CNS arousal or alertness, (2) they create goal-oriented behavior, and (3) they are capable of coordinating disparate behaviors to achieve that goal.

Motivated behaviors often work in parallel with autonomic and endocrine responses in the body, as you might expect with behaviors originating in the hypothalamus. For example, if you eat salty popcorn, your body osmolarity increases. This stimulus acts on the thirst center of the hypothalamus, motivating you to seek something to drink. Increased osmolarity also acts on an endocrine center in the hypothalamus, releasing a hormone that increases water retention by the kidneys. Thus, one stimulus triggers both a motivated behavior and a homeostatic endocrine response.

Some motivated behaviors can be activated by internal stimuli that may not even be obvious to the person in whom they are occurring. Eating, curiosity, and sex drive are three examples of behaviors with complex stimuli underlying their onset. We may eat, for example, because we are hungry or because the food looks good or because we do not want to hurt someone's feelings. Many motivated behaviors stop when the person has reached a certain level of satisfaction, or **satiety**, but they may also continue *despite* feeling satiated.

Pleasure is a motivational state that is being intensely studied because of its relationship to *addictive behaviors,* such as drug use. Animal studies have shown that pleasure is a physiological state that is accompanied by increased activity of the neurotransmitter dopamine in certain parts of the brain. Drugs that are addictive, such as cocaine and possibly nicotine, act by enhancing the effectiveness of dopamine, thereby increasing the pleasurable sensations perceived by the brain. As a result, use of these drugs rapidly becomes a learned behavior. Interestingly, not all behaviors that are addictive are pleasurable. For example, there are a variety of compulsive behaviors that involve self-mutilation, such as pulling out hair by the roots. Fortunately, many behaviors can be modulated, given motivation.

Moods Are Long-Lasting Emotional States

Moods are similar to emotions but are longer-lasting, relatively stable subjective feelings related to one's sense of well-being.

Moods are difficult to define at a neurobiological level, but evidence obtained in studying and treating mood disorders indicates that what was once thought to be purely psychological in origin is actually a CNS function. Depression and other mood disorders result from abnormal neurotransmitter release or reception in different brain regions.

Mood disorders are estimated to be the fourth leading cause of illness in the world today. **Depression** is a mood disturbance that affects nearly 10% of the United States population each year. It is characterized by sleep and appetite disturbances and alterations of mood and libido that may seriously impact the person's ability to function at school or work or in personal relationships. Many people do not realize that depression is not a sign of mental or moral weakness, or that it can be treated successfully with drugs and psychotherapy. (For detailed information about depression, go to *www.nlm.nih.gov/medlineplus/depression.html*).

The drug therapy for depression has changed in recent years, but all the major categories of antidepressant drugs alter some aspect of synaptic transmission. The older *tricyclic antidepressants,* such as amitriptyline, block reuptake of norepinephrine into the presynaptic neuron, thus extending the active life of the neurotransmitter. The antidepressants known as *selective serotonin reuptake inhibitors,* or SSRIs, slow down the removal of serotonin (and possibly also norepinephrine) from the synapse. As a result of uptake inhibition, serotonin lingers in the synaptic cleft longer than usual, increasing serotonin-dependent activity in the postsynaptic neuron. Other antidepressant drugs alter brain levels of dopamine, indicating that norepinephrine, serotonin, and dopamine are all involved in brain pathways for mood and emotion.

Interestingly, patients need to take antidepressant drugs for several weeks before they experience their full effect. This suggests that the changes taking place in the brain are long-term modulation of pathways rather than simply enhanced fast synaptic responses. The search to uncover the biological basis of disturbed brain function is a major focus of neuroscience research today.

Some research into brain function has become quite controversial, particularly that dealing with sexuality and the degree to which behavior in general is genetically determined in humans. We will not delve deeply into any of these subjects because they are complex and would require lengthy explanations to do them justice. Instead, we will look briefly at some of the recent models proposed to explain the mechanisms that are the basis for higher cognitive functions.

Learning and Memory Change Synaptic Connections in the Brain

For many years, motivation, learning, and memory (all of which are aspects of the cognitive state) were considered to be in the realm of psychology rather than biology. Neurobiologists in decades past were more concerned with the network and

RUNNING PROBLEM

Ben's halted development is a feature unique to infantile spasms. The abnormal portions of the brain send out continuous action potentials during frequent seizures and ultimately change the interconnections of brain neurons. The damaged portions of the brain harm normal portions to such an extent that medication or surgery should be started as soon as possible. If intervention is not begun early, the brain can be permanently damaged and developmental recovery will never occur.

Question 5:
   The brain's ability to change its synaptic connections as a result of neuronal activity is called_____.

292   300   315   **317**   321   322

cellular aspects of neuronal function. In recent years, however, the two fields have overlapped more and more. Scientists have discovered that the underlying basis for cognitive function seems to be explainable in terms of cellular events that influence plasticity—events such as long-term potentiation [p. 281]. The ability of neuronal connections to change with experience is fundamental to the two cognitive processes of learning and memory.

Learning Is the Acquisition of Knowledge

How do you know when you have learned something? Learning can be demonstrated by behavioral changes, but behavioral changes are not required in order for learning to occur. Learning can be internalized and is not always reflected by overt behavior while the learning is taking place. Would someone watching you read your textbook be able to tell whether you had learned anything?

Learning can be classified into two broad types: associative and nonassociative. **Associative learning** occurs when two stimuli are associated with each other, such as Pavlov's classic experiment in which he simultaneously presented dogs with food and rang a bell. After a period of time, the dogs came to associate the sound of the bell with food and began to salivate in anticipation of food whenever the bell was rung. Another form of associative learning occurs when an animal associates a stimulus with a given behavior. An example would be a mouse that gets a shock each time it touches a certain part of its cage. It soon associates that part of the cage with an unpleasant experience and avoids the area.

Nonassociative learning is a change in behavior that takes place after repeated exposure to a single stimulus. This type of learning includes habituation and sensitization, two adaptive behaviors that allow us to filter out and ignore background stimuli while responding more sensitively to potentially disruptive stimuli. In **habituation**, an animal shows a decreased response to an irrelevant stimulus that is repeated over and over. For example, a sudden loud noise may startle you, but if the noise is repeated over and over again, your brain begins to ignore it. Habituated responses allow us to filter out stimuli that we have evaluated and found to be insignificant.

Sensitization is the opposite of habituation, and the two behaviors combined help increase an organism's chances for survival. In sensitization learning, exposure to a noxious or intense stimulus causes an enhanced response upon subsequent exposure. For example, people who become ill while eating a certain food may find that they lose their desire to eat that food again. Sensitization is adaptive because it helps us avoid potentially harmful stimuli.

Memory Is the Ability to Retain and Recall Information

Memory is the ability to retain and recall information. We have several types of memory: short-term and long-term, reflexive and declarative. Processing for different types of memory appears to take place through different pathways. With noninvasive imaging techniques such as MRI and PET scans, researchers have been able to track brain activity as individuals learned to perform tasks.

Memories are stored throughout the cerebral cortex in pathways known as **memory traces**. Some components of memories are stored in the sensory cortices where they are processed. For example, pictures are stored in the visual cortex, and sounds in the auditory cortex.

Learning a task or recalling a task already learned may involve multiple brain circuits that work in parallel. This *parallel processing* helps provide backup in case one of the circuits is damaged. It is also believed to be the means by which specific memories are generalized, allowing new information to be matched to stored information. For example, a person who has never seen a volleyball will recognize it as a ball because the volleyball has the same general characteristics as all other balls the person has seen.

In humans, the hippocampal region of the limbic system seems to be an important structure in both learning and memory. For example, patients who had part of the hippocampus destroyed to relieve a certain type of epilepsy also had trouble remembering new information. When given a list of words to repeat, they could remember the words as long as their attention stayed focused on the task. If they were distracted, however, the memory of the words disappeared, and they had to learn the list again. Information stored in long-term memory before the operation was not affected. This inability to remember newly acquired information is a defect known as **anterograde amnesia** [*amnesia,* oblivion].

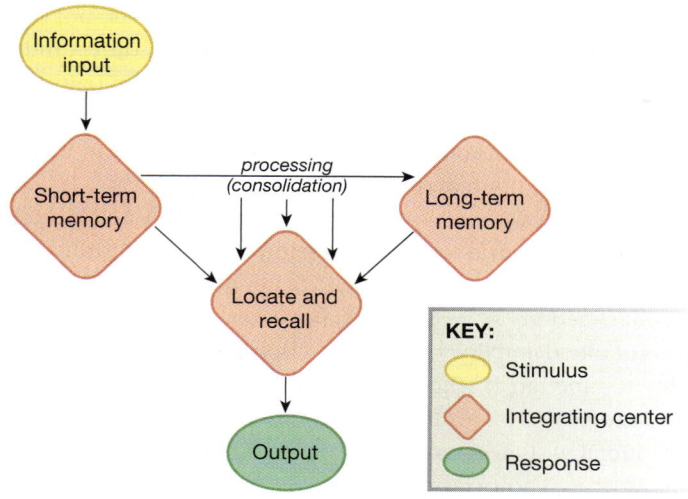

■ **FIGURE 9-22** *Memory processing*
New information goes into short-term memory but will be lost unless it is processed and stored in long-term memory.

Memory has multiple levels of storage, and our memory bank is constantly changing (Fig. 9-22 ■). When a stimulus comes into the CNS, it first goes into **short-term memory**, a limited storage area that can hold only about 7 to 12 pieces of information at a time. Items in short-term memory will disappear unless an effort, such as repetition, is made to put them into a more permanent form.

Working memory is a special form of short-term memory processed in the prefrontal lobes. This region of the cerebral cortex is devoted to keeping track of bits of information long enough to put them to use in a task that takes place after the information has been acquired. Working memory in these regions is linked to long-term memory stores, so that newly acquired information can be integrated with stored information and acted on. For example, suppose you are trying to cross a busy road. You look to the left and see that there are no cars coming for several blocks. You then look to the right and see that there are no cars coming from that direction either. Working memory has stored the information that the road to the left is clear, and so using stored knowledge about safety, you are able to conclude that there is no traffic from either direction and it is safe to cross the road.

In people with damage to the prefrontal lobes of the brain, this task becomes more difficult because they are unable to recall whether the road is clear from the left once they have looked away to assess traffic coming from the right. Working memory allows us to collect a series of facts from short- and long-term memory and connect them in a logical order to solve problems or plan actions.

Long-term memory is a storage area capable of holding vast amounts of information. The processing of information that converts short-term memory into long-term memory is known

TABLE 9-4	Types of Long-Term Memory

REFLEXIVE (IMPLICIT) MEMORY	DECLARATIVE (EXPLICIT) MEMORY
Recall is automatic and does not require conscious attention	Recall requires conscious attention
Acquired slowly through repetition	Depends on higher-level thinking skills such as inference, comparison, and evaluation
Includes motor skills and rules and procedures	Memories can be reported verbally
Procedural memories can be demonstrated	

as **consolidation** (Fig. 9-22). Consolidation can take varying periods of time, from seconds to minutes. Information passes through many intermediate levels of memory during consolidation, and in each of these stages, the information can be located and recalled.

As scientists studied the consolidation of short-term memory into long-term memory, they discovered that the process involves changes in synaptic connections of the circuits involved in learning. In some cases, new synapses form; in others, the effectiveness of synaptic transmission is altered either through long-term potentiation or through long-term depression. These changes are evidence of plasticity and show us that the brain is not "hard-wired."

Long-term memory has been divided into two types that are consolidated and stored using different neuronal pathways (Table 9-4 ■). **Reflexive (implicit) memory,** which is automatic and does not require conscious processes for either creation or recall, involves the amygdala region of the limbic system and the cerebellum. Information stored in reflexive memory is acquired slowly through repetition. Motor skills fall into this category, as do procedures and rules. For example, you do not need to think about putting a period at the end of each sentence or about how to pick up a fork. Reflexive memory has also been called *procedural memory* because it generally concerns how to do things. Reflexive memories can be acquired through either associative or nonassociative learning processes, and these memories are stored.

Declarative (explicit) memory, on the other hand, requires conscious attention for its recall. Its creation generally depends on the use of higher-level cognitive skills such as inference, comparison, and evaluation. The neuronal pathways involved in this type of memory are in the temporal lobes. Declarative memories deal with knowledge about ourselves and the world around us that can be reported or described verbally.

Sometimes information can be transferred from declarative memory to reflexive memory. The quarterback on a football team is a good example. When he learned to throw the football as a small boy, he had to pay close attention to gripping the ball and coordinating his muscles to throw the ball accurately. At that point of learning to throw the ball, the process was in declarative memory and required conscious effort as the boy analyzed his movements. With repetition, however, the mechanics of throwing the ball were transferred to reflexive memory: they became a reflex that could be executed without conscious thought. That transfer allowed the quarterback to use his conscious mind to analyze the path and timing of the pass while the mechanics of the pass became automatic. Athletes often refer to this automaticity of learned body movements as *muscle memory*.

Memory is an individual thing. We process information on the basis of our experiences and perception of the world. Because people have widely different experiences throughout their lives, it follows that no two people will process a given piece of information in the same way. If you ask a group of people about what happened during a particular event such as a lecture or an automobile accident, no two descriptions will be identical. Each person will have processed the event according to her or his own perceptions and experiences. Experiential processing is important to remember when studying in a group situation, because it is unlikely that all group members will learn or recall information the same way.

Memory loss and the inability to process and store new memories are devastating medical conditions. In younger people, memory problems are usually associated with trauma to the brain from accidents. In older people, strokes and progressive *dementia* [*demens,* out of one's mind] are the main causes of memory loss. **Alzheimer's disease** is a progressive neurodegenerative disease of cognitive impairment that accounts for about half the cases of dementia in the elderly. By one estimate, it affects about 4 million Americans, with the number expected to rise as Baby Boomers age. Alzheimer's is characterized by memory loss that progresses to a point where the patient does not recognize family members. Over time, even the personality changes, and in the final stages, other cognitive functions fail so that patients cannot communicate with caregivers.

Diagnosis of Alzheimer's is usually made through the patient's declining performance on mental status examinations. The only definitive diagnosis comes after death, when brain tissue can be examined for neuronal degeneration, extracellular plaques made of β-*amyloid protein,* and intracellular tangles of *tau,* a protein that is normally associated with microtubules. Although the presence of plaques and tangles is diagnostic, the underlying cause of Alzheimer's is unclear. There is a known genetic component, and other theories include oxidative stress and chronic inflammation. Currently there is no proven prevention or treatment, although drugs that are acetylcholine

9

(a) Speaking a written word

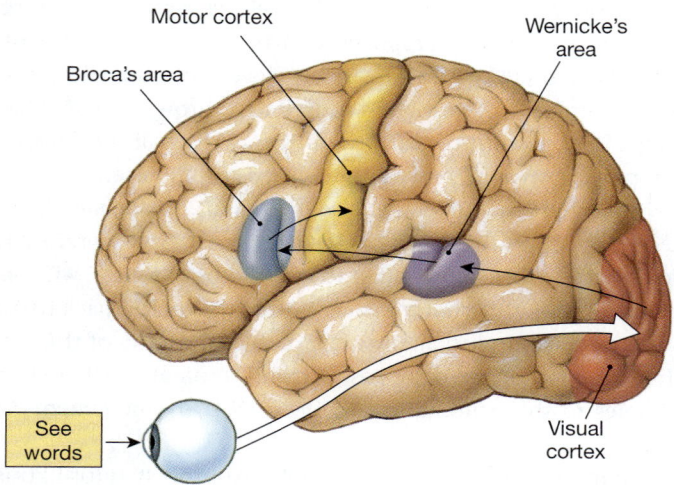

(b) Speaking a heard word

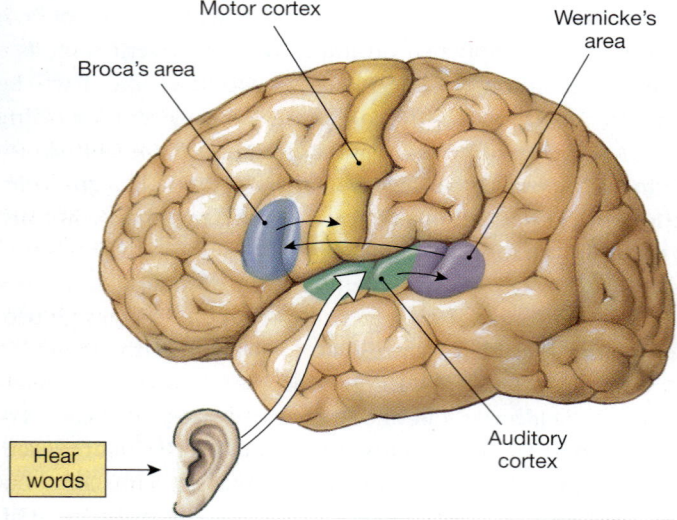

■ **FIGURE 9-23** *Cerebral processing of spoken and visual language*

People with damage to Wernicke's area do not understand spoken or written communication. Those with damage to Broca's area understand but are unable to respond appropriately.

agonists or acetylcholinesterase inhibitors slow the progression of the disease. The forecast of 14 million people with Alzheimer's by the year 2030 has put this disease in the forefront of neurobiological research.

Language Is the Most Elaborate Cognitive Behavior

One of the hallmarks of an advanced nervous system is the ability of one member of a species to exchange complex information with other members of the same species. Although found predominantly in birds and mammals, this ability also occurs in certain insects that convey amazingly detailed information by means of sound (crickets), touch and sight (bees), and odor (ants). In humans, the exchange of complex information takes place primarily through spoken and written language. Because language is considered the most elaborate cognitive behavior, it has received considerable attention from neurobiologists.

Language skills require the input of sensory information (primarily from hearing and vision), processing in various centers in the cerebral cortex, and the coordination of motor output for vocalization and writing. In most people the centers for language ability are found in the left hemisphere of the cerebrum. Even 70% of people who are either left-handed (right-brain dominant) or ambidextrous use their left brain for speech. The ability to communicate through speech has been divided into two processes: the combination of different sounds to form words (vocalization) and the combination of words into grammatically correct and meaningful sentences.

The integration of spoken language in the human brain involves two regions in the cerebral cortex: **Wernicke's area** in the temporal lobe and **Broca's area** in the frontal lobe close to the motor cortex (Fig. 9-23 ■). Most of what we know about these areas comes from studies of people with brain lesions (because nonhuman animals are not capable of speech). Even primates that communicate on the level of a small child through sign language and other visual means do not have the physical ability to vocalize the sounds of human language.

Input into the language areas comes from either the visual cortex (reading) or the auditory cortex (listening). Sensory input from either cortex goes first to Wernicke's area, then to Broca's area. After integration and processing, output from Broca's area to the motor cortex initiates a spoken or written action. If damage occurs to Wernicke's area, a person is unable to understand any spoken or visual information. The person's own speech, as a result, is nonsense, because the person is unaware of his or her own errors. This condition is known as **receptive aphasia** [*a-*, not + *phatos*, spoken] because the person is unable to understand sensory input.

On the other hand, people with damage to Broca's area understand spoken and written language but are unable to speak or write in normal syntax. Their response to a question often consists of appropriate words strung together in random order. These patients may have a difficult time dealing with their disability because they are aware of their mistakes but are powerless to correct them. Damage to Broca's area causes an **expressive aphasia**.

Mechanical forms of aphasia occur as a result of damage to the motor cortex. Patients with this type of damage find themselves unable to physically shape the sounds that make up words, or unable to coordinate the muscles of their arm and hand to write.

RUNNING PROBLEM

The PET scan revealed two abnormal spots, or *loci* (plural of *locus*), on Ben's right hemisphere, one on the parietal lobe and one overlapping a portion of the primary motor cortex. Because the loci triggering Ben's seizures were located on the same hemisphere and were in the cortex, Ben was a candidate for a *hemispherectomy,* removal of the cortex of the affected hemisphere. Surgeons removed 80 percent of his right cerebral cortex, sparing areas crucial to vision, hearing, and sensory processing. Normally the motor cortex would be spared as well, but in Ben's case a seizure locus overlapped much of the region.

Question 6:
In which lobes are the centers for vision, hearing, and sensory processing located?

Question 7:
Which of Ben's abilities might have suffered if his left hemisphere had been removed instead?

Question 8:
By taking only the cortex of the right hemisphere, what parts of the cerebrum did surgeons leave behind?

Question 9:
Why were the surgeons careful to spare Ben's right lateral ventricle?

292   300   315   317   **321**   322

Personality Is a Combination of Experience and Inheritance

One of the most difficult aspects of brain function to translate from the abstract realm of psychology into the physical circuits of neurobiology is the combination of attributes we call **personality**. What is it that makes us individuals? The parents of more than one child will tell you that their offspring were different from birth, and even before that in the womb. If we all have the same brain structure, what makes us different?

This question fascinates many people. The answer that is evolving from neurobiology research is that we are a combination of our experiences and the genetic constraints we inherit. One complicating factor is the developmental aspect of "experience," as scientists are showing that exposure of developing embryos to hormones while still in the womb can alter brain pathways.

What we learn or experience and what we store in memory create a unique pattern of neuronal connections in our brains. Sometimes these circuits malfunction, creating depression, schizophrenia, or any number of other personality disturbances. Psychiatrists for many years attempted to treat these disorders as if they were due solely to events in the person's life, but now we know that there is a genetic component to many of these disorders.

Schizophrenia [*schizein,* to split + *phren,* the mind] is an example of a brain disorder that has both a genetic and an environmental basis. In the American population as a whole, the risk of developing schizophrenia is about 1%. However, if one parent has schizophrenia, the risk increases to 10%, indicating that people can inherit a susceptibility to developing the disease. The cause of schizophrenia is not currently known. However, as with many other conditions involving altered mental states, schizophrenia can be treated with drugs that influence neurotransmitter release and activity in the brain. To learn more about diagnosis and treatment of schizophrenia, see the National Institutes of Health website *www.nlm.nih.gov/ medlineplus/schizophrenia.html.*

We still have much to learn about repairing damage to the CNS. One of the biggest tragedies in life is the personality change that sometimes accompanies head trauma. Physical damage to the delicate circuits of the brain not only can alter intelligence and memory storage but also can create a new personality. The person who exists after the injury may not seem to be the same person who inhabited that body before the injury. Although the change may not be noticeable to the injured person, it can be devastating to the victim's family and friends. Perhaps as we learn more about how neurons link to one another, we will be able to find a means of restoring damaged networks and preventing the lasting effects of head trauma and brain disorders.

INFANTILE SPASMS

Ben has remained seizure-free since the surgery and shows normal development in all areas except motor skills. He remains somewhat weaker and less coordinated on his left side, the side opposite (*contralateral*) to the surgery, but over time the weakness should subside with the aid of physical therapy. Ben's recovery stands as a testament to the incredible plasticity of the brain.

Apart from the physical damage caused to the brain, a number of epileptics have developmental delays that stem from the social aspects of their disorder. Young children with frequent seizures often have difficulty socializing with their peers due to overprotective parents, missed school days, and the fear of people who do not understand epilepsy. Their problems can extend into adulthood, when epileptics may have difficulty finding employment or driving if their seizures are not controlled. There are numerous examples of adults who undergo successful epilepsy surgery but are still unable to fully enter society due to their lack of social and employment skills. Not surprisingly, the rate of depression is much higher among epileptics. To learn more about epilepsy, start with the Epilepsy Foundation (*www.epilepsyfoundation.org*).

This Running Problem was written by Susan E. Johnson while she was an undergraduate student at the University of Texas at Austin studying for a career in the biomedical sciences.

	QUESTION	FACTS	INTEGRATION AND ANALYSIS
1	How might a leaky blood-brain barrier lead to action potentials that trigger a seizure?	Neurotransmitters and other chemicals circulating freely in the blood are normally separated from brain tissue by the blood-brain barrier.	Negatively charged molecules and neurotransmitters entering the brain might depolarize neurons and trigger action potentials.
2	What does GABA do to the cell's membrane potential? Does GABA make the cell more or less likely to fire action potentials?	GABA opens Cl^- channels.	Cl^- entering a neuron hyperpolarizes the cell and makes it less likely to fire action potentials.
3	Why is it important to limit the duration of ACTH therapy?	Exogenous ACTH acts in a short negative feedback loop, decreasing the output of CRH from the hypothalamus and ACTH production by the anterior pituitary. See Figure 7-15, p. 228.	Long-term suppression of endogenous hormone secretion by ACTH can cause CRH- and ACTH-secreting neurons to atrophy, resulting in a lifelong cortisol deficiency.
4	What is the rationale for using radioactively labeled glucose (and not some other nutrient) for the PET scan?	Glucose is the primary energy source for the brain.	Glucose usage is more closely correlated to brain activity than any nutrient in the body. Areas of abnormally high glucose usage are suggestive of overactive cells.
5	The brain's ability to change its synaptic connections as a result of neuronal activity is called_____.	Changes in synaptic connections as a result of neuronal activity are an example of plasticity.	N/A*
6	In which lobes are the centers for vision, hearing, and sensory processing located?	Vision is processed in the occipital lobe, hearing in the temporal lobe, and sensory information in the parietal lobe.	N/A
7	Which of Ben's abilities might have suffered if his left hemisphere had been removed instead?	In most people, the left hemisphere contains Wernicke's area and Broca's area, two centers vital to speech. The left brain controls right-sided sensory and motor functions.	Patients who have undergone left hemispherectomies have difficulty with speech (abstract words, grammar, and phonetics). They show loss of right-side sensory and motor functions.
8	By taking only the cortex of the right hemisphere, what parts of the cerebrum did surgeons leave behind?	The cerebrum consists of gray matter in the cortex and interior nuclei, white matter, and the ventricles.	The surgeons left behind the white matter, interior nuclei, and ventricles.
9	Why were the surgeons careful to spare Ben's right lateral ventricle?	The walls of the ventricles contain the choroid plexus, which secretes cerebrospinal fluid (CSF). CSF plays a vital protective role by cushioning the brain.	CSF protection is particularly important following removal of portions of brain tissue because the potential damage from jarring of the head is much greater.

292 300 315 317 321 **322**

*N/A—not applicable

CHAPTER SUMMARY

The brain is the primary control center of the body, and (as you will learn in later chapters), homeostatic responses in many organ systems are designed to maintain brain function. The ability of the brain to create complex thoughts and emotions in the absence of external stimuli is one of its *emergent properties*.

Emergent Properties of Neural Networks

1. Neural networks create **affective** and **cognitive behaviors**. (p. 292)
2. The brain exhibits **plasticity**, the ability to change connections as a result of experience. (p. 292)

Evolution of Nervous Systems

3. Nervous systems evolved from a simple network of neurons to complex brains. (p. 292; Fig. 9-1)
4. The **cerebrum** is responsible for thought and emotion. (p. 294)

Anatomy of the Central Nervous System

5. The central nervous system consists of layers of cells around a fluid-filled central cavity and develops from the **neural tube** of the embryo. (p. 295; Fig. 9-2)
6. The **gray matter** of the CNS consists of unmyelinated nerve cell bodies, dendrites, and axon terminals. The cell bodies either form layers in parts of the brain or else cluster into groups known as **nuclei**. (p. 296)
7. Myelinated axons form the **white matter** of the CNS and run in bundles called **tracts**. (p. 296)
8. The brain and spinal cord are encased in the **meninges** and the bones of the **cranium** and vertebrae. The meninges are the **pia mater**, the **arachnoid membrane**, and the **dura mater**. (p. 296; Fig. 9-4)
9. The **choroid plexus** secretes **cerebrospinal fluid** (CSF) into the **ventricles** of the brain. Cerebrospinal fluid cushions the tissue and creates a chemically controlled environment. (p. 296; Fig. 9-5)
10. Tight junctions in brain capillaries create a **blood-brain barrier** that prevents possibly harmful substances in the blood from entering the interstitial fluid. (p. 299; Fig. 9-6)
11. The normal fuel source for neurons is glucose, and the body closely regulates blood glucose concentrations. (p. 300)

The Spinal Cord

12. Each segment of the spinal cord is associated with a pair of **spinal nerves**. (p. 301)
13. The **dorsal root** of each spinal nerve carries incoming sensory information. The **dorsal root ganglia** contain the nerve cell bodies of sensory neurons. (p. 301; Fig. 9-7)
14. The **ventral roots** carry information from the central nervous system to muscles and glands. (p. 301)
15. **Ascending tracts** of white matter carry sensory information to the brain, and **descending tracts** carry efferent signals from the brain. **Propriospinal tracts** remain within the spinal cord. (p. 301)
16. **Spinal reflexes** are integrated in the spinal cord. (p. 301; Fig. 9-8)

The Brain

17. The brain has six major divisions: cerebrum, diencephalon, midbrain, cerebellum, pons, and medulla oblongata. (p. 303; Fig. 9-9)
18. The **brain stem** is divided into medulla oblongata, pons, and midbrain (mesencephalon). **Cranial nerves** II to XII originate here. (p. 303; Tbl. 9-1)
19. The **reticular formation** is a diffuse collection of neurons that play a role in many basic processes. (p. 303)
20. The **medulla oblongata** contains **somatosensory** and **corticospinal tracts** that convey information between the cerebrum and spinal cord. Most tracts cross the midline in the **pyramid** region. The medulla contains control centers for many involuntary functions. (p. 305)
21. The **pons** acts as a relay station for information between the cerebellum and cerebrum. (p. 305)
22. The **midbrain** controls eye movement and relays signals for auditory and visual reflexes. (p. 305)
23. The **cerebellum** processes sensory information and coordinates the execution of movement. (p. 305)
24. The **diencephalon** is composed of the thalamus and hypothalamus. The **thalamus** relays and modifies sensory and motor information going to and from the cerebral cortex. (p. 306; Fig. 9-10)
25. The **hypothalamus** contains centers for behavioral drives and plays a key role in homeostasis by its control over endocrine and autonomic function. (p. 306; Tbl. 9-2)
26. The **pituitary gland** and **pineal gland** are endocrine glands located in the diencephalon. (p. 306)
27. The cerebrum is composed of two hemispheres connected at the **corpus callosum**. Each cerebral hemisphere is divided into **frontal**, **parietal**, **temporal**, and **occipital lobes**. (p. 307)
28. Cerebral gray matter includes the **cerebral cortex**, basal ganglia, and limbic system. (p. 307; Fig. 9-11)
29. The **basal ganglia** help control movement. (p. 308)
30. The **limbic system** acts as the link between cognitive functions and emotional responses. It includes the **amygdala** and **cingulate gyrus**, linked to emotion and memory, and the **hippocampus**, associated with learning and memory. (p. 308; Fig. 9-13)

Brain Function

31. Three brain systems influence motor output: a **sensory system**, a **cognitive system**, and a **behavioral state system**. (p. 308; Fig. 9-14)
32. Higher brain functions, such as reasoning, arise in the cerebral cortex. The cerebral cortex contains three functional specializations: **sensory areas**, **motor areas**, and **association areas**. (p. 309; Fig. 9-15)
33. Each hemisphere of the cerebrum has developed functions not shared by the other hemisphere, a specialization known as **cerebral lateralization**. (p. 309; Fig. 9-16)
34. Sensory areas receive information from sensory receptors. The **primary somatic sensory cortex** processes information about touch, temperature, and other somatic senses. The **visual cortex**, **auditory cortex**, **gustatory cortex**, and **olfactory cortex** receive information about vision, sound, taste, and odors, respectively. (pp. 310–311)

35. **Association areas** integrate sensory information into perception. **Perception** is the brain's interpretation of sensory stimuli. (p. 311)

36. Motor output includes skeletal muscle movement, neuroendocrine secretion, and visceral responses. (p. 311)

37. Motor areas direct skeletal muscle movement. Each cerebral hemisphere contains a **primary motor cortex** and **motor association area**. (p. 311)

38. The **behavioral state system** controls states of arousal and modulates the sensory and cognitive systems. (p. 312)

39. The **diffuse modulatory systems** of the reticular formation influence attention, motivation, wakefulness, memory, motor control, mood, and metabolic homeostasis. (p. 312; Fig. 9-19, Tbl. 9-3)

40. The **reticular activating system** keeps the brain **conscious**, or aware of self and environment. Electrical activity in the brain varies with levels of arousal and can be recorded by **electroencephalography**. (p. 312; Fig. 9-20)

41. **Circadian rhythms** are controlled by an internal clock in the **suprachiasmatic nucleus**. (p. 315)

42. **Sleep** is an easily reversible state of inactivity with characteristic stages. The two major phases of sleep are **REM (rapid eye movement) sleep** and **slow-wave sleep** (non-REM sleep). The physiological reason for sleep is unknown. (p. 313)

43. The limbic system is the center of **emotion** in the human brain. Emotional events influence physiological functions. (p. 315; Fig. 9-21)

44. **Motivation** arises from internal signals that shape voluntary behaviors related to survival or emotions. Motivational **drives** create goal-oriented behaviors. (p. 316)

45. **Moods** are long-lasting emotional states. Many mood disorders can be treated by altering neurotransmission in the brain. (p. 316

46. **Learning** is the acquisition of knowledge about the world around us. **Associative learning** occurs when two stimuli are associated with each other. **Nonassociative learning** includes imitative behaviors, such as learning a language. (p. 317)

47. In **habituation**, an animal shows a decreased response to a stimulus that is repeated over and over. In **sensitization**, exposure to a noxious or intense stimulus creates an enhanced response on subsequent exposure. (p. 318)

48. **Memory** has multiple levels of storage and is constantly changing. Information is first stored in **short-term memory** but will disappear unless consolidated into long-term memory. (p. 318; Fig. 9-22)

49. **Long-term memory** includes **reflexive memory**, which does not require conscious processes for its creation or recall, and **declarative memory**, which uses higher-level cognitive skills for formation and requires conscious attention for its recall. (p. 318; Tbl. 9-4)

50. The **consolidation** of short-term memory into long-term memory appears to involve changes in the synaptic connections of the circuits involved in learning. (p. 319)

51. Language is considered the most elaborate cognitive behavior. The integration of spoken language in the human brain involves information processing in **Wernicke's area** and **Broca's area**. (p. 320; Fig. 9-23)

QUESTIONS

(Answers to the Review Questions begin on page A1.)

THE PHYSIOLOGY PLACE

Access more review material online at **The Physiology Place** website. There you'll find review questions, problem-solving activities, case studies, flashcards, and direct links to both *InterActive Physiology*® and *PhysioEx*™. To access the site, go to *www.physiologyplace.com* and select Human Physiology, Fourth Edition.

LEVEL ONE REVIEWING FACTS AND TERMS

1. The ability of human brains to change circuit connections and function in response to sensory input and past experience is known as _plasticity_.

2. _____ behaviors are related to feeling and emotion. _____ behaviors are related to thinking.

3. The part of the brain called the _cerebrum_ is what makes us human, allowing human reasoning and cognition.

4. In vertebrates, the central nervous system is protected by the bones of the _skull (brain)_ and _vertebral column (spinal cord)_.

5. Name the meninges, beginning with the layer next to the bones. _dura mater, arachnoid mm., pia mater_

6. List and explain the purposes of cerebrospinal fluid (CSF). Where is CSF made?

7. Compare the CSF concentration of each of the following substances with its concentration in the blood plasma.

    (a) HCO_3^- (b) Ca^{2+}
    (c) glucose   (d) H^+
    (e) Na^+ (f) K^+

8. The only fuel source for neurons under normal circumstances is _glucose_. Low concentration of this fuel in the blood is termed _hypoglycemia_. To synthesize enough ATP to continually transport ions, the neurons consume large quantities of _O_2_. To supply these needs, about _15_ % of the blood pumped by the heart goes to the brain.

9. Match each of the following areas with its function.

    (a) medulla oblongata _5_
    (b) pons _7_
    (c) midbrain _9_
    (d) reticular formation _3_
    (e) cerebellum _1_
    (f) diencephalon _2_
    (g) thalamus _6_
    (h) hypothalamus _8_
    (i) cerebrum _4_

    1. coordinates execution of movement
    2. is composed of the thalamus and hypothalamus
    3. controls arousal and sleep
    4. is composed of two hemispheres and fills most of the cranium
    5. contains control centers for blood pressure and breathing
    6. relays and modifies information going to and from the cerebrum
    7. transfers information to the cerebellum
    8. contains integrating centers for homeostasis
    9. relays signals and visual reflexes, plus eye movement

10. What is the blood-brain barrier, and what is its function?

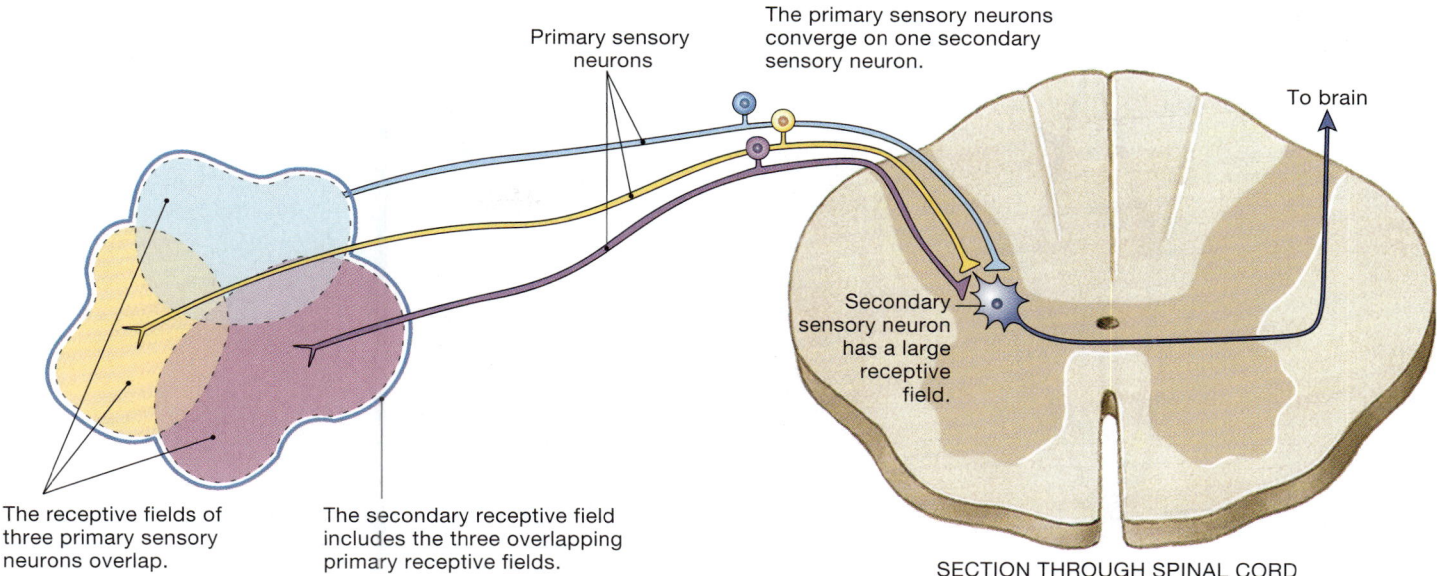

The primary sensory neurons converge on one secondary sensory neuron.

Primary sensory neurons

To brain

Secondary sensory neuron has a large receptive field.

The receptive fields of three primary sensory neurons overlap.

The secondary receptive field includes the three overlapping primary receptive fields.

SECTION THROUGH SPINAL CORD

■ **FIGURE 10-2** *Receptive fields of sensory neurons*

Convergence of primary sensory neurons allows simultaneous subthreshold stimuli to sum at the secondary sensory neuron and initiate an action potential. (In this illustration, only part of the secondary sensory neuron is shown.)

In addition, sensory neurons of neighboring receptive fields may exhibit *convergence* [⇄ p. 278], in which multiple presynaptic neurons provide input to a smaller number of postsynaptic neurons. Convergence allows multiple simultaneous subthreshold stimuli to sum at the postsynaptic (secondary) neuron. When multiple primary sensory neurons converge on a single secondary sensory neuron, their individual receptive fields merge into a single, large *secondary receptive field,* as indicated in Figure 10-2.

The size of secondary receptive fields determines how sensitive a given area is to a stimulus. For example, sensitivity to touch is demonstrated by a **two-point discrimination test** (Fig. 10-3a ■). In some regions of skin, such as that on the arms and legs, two pins placed within 20 mm of each other will be interpreted by the brain as a single pinprick. In these areas, many primary neurons converge on a single secondary neuron, so the secondary receptive field is very large. In contrast, more sensitive areas of skin have smaller receptive fields, with as little as a 1:1 relationship between primary and secondary sensory neurons. In these regions, two pins separated by as little as 2 mm can be perceived as two separate touches (Fig. 10-3b).

The Central Nervous System Integrates Sensory Information

Sensory information from much of the body enters the spinal cord and travels through ascending pathways to the brain. Some sensory information goes directly into the brain stem via the cranial nerves [⇄ p. 303]. Sensory information that initiates visceral reflexes is integrated in the brain stem or spinal cord and usually does not reach conscious perception. An example of an unconscious visceral reflex is the control of blood pressure by centers in the brain stem.

Each major division of the brain processes one or more types of sensory information (Fig. 10-4 ■). For example, the midbrain receives visual information, and the medulla oblongata receives input for sound and taste. Information about balance and equilibrium is processed primarily in the cerebellum. These pathways, along with those carrying somatosensory information, project to the thalamus, which acts as a relay and processing station before passing the information on to the cerebrum.

Only *olfactory* [*olfacere,* to sniff] information is not routed through the thalamus. The sense of smell, a type of chemoreception, is considered to be the one of the oldest senses; even the most primitive vertebrate brains have well-developed regions for processing olfactory information. Information about odors travels from the nose through the *olfactory bulb* and first cranial nerve [⇄ p. 303] to the olfactory cortex in the cerebrum. Perhaps it is because of this direct input to the cerebrum that odors are so closely linked to memory and emotion. Most people have experienced encountering a smell that suddenly brings back a flood of memories of places or people from the past.

One interesting aspect of CNS processing of sensory information is the **perceptual threshold**, the level of stimulus intensity necessary for you to be aware of a particular sensation. Stimuli bombard your sensory receptors constantly, but your

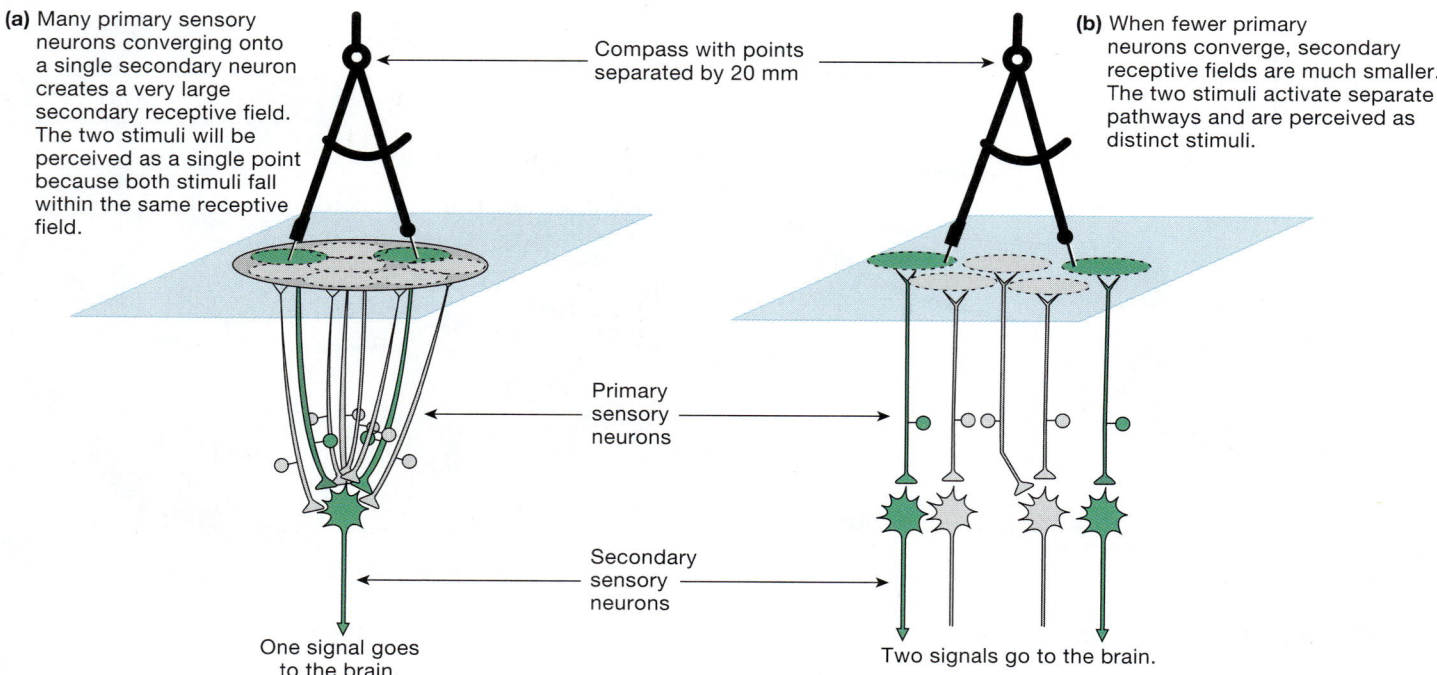

(a) Many primary sensory neurons converging onto a single secondary neuron creates a very large secondary receptive field. The two stimuli will be perceived as a single point because both stimuli fall within the same receptive field.

Compass with points separated by 20 mm

(b) When fewer primary neurons converge, secondary receptive fields are much smaller. The two stimuli activate separate pathways and are perceived as distinct stimuli.

Primary sensory neurons

Secondary sensory neurons

One signal goes to the brain.

Two signals go to the brain.

■ **FIGURE 10-3** *Two-point discrimination*

(a) Less-sensitive regions of the skin are found on the arms and legs. In these regions, two stimuli 20 mm apart cannot be felt separately. (b) In more sensitive regions, such as the fingertips, two stimuli separated by as little as 2 mm will be perceived as two distinct stimuli.

brain can "turn off" some stimuli to avoid being overwhelmed with information. You experience a change in perceptual threshold when you "tune out" the radio while studying or when you "zone out" during a lecture. In both cases, the noise is adequate to stimulate sensory neurons in the ear, but neurons higher in the pathway dampen the perceived signal so that it does not reach the conscious brain.

Decreased perception of a stimulus is accomplished by *inhibitory modulation* [🔁 p. 41], which diminishes a suprathreshold stimulus until it is below the perceptual threshold. Inhibitory modulation often occurs in the secondary and tertiary neurons of a sensory pathway. If the modulated stimulus suddenly becomes important, such as when the professor asks you a question, you can consciously focus your attention and overcome the inhibitory modulation. At that point, your conscious brain seeks to retrieve and recall recent sound input from your subconscious so that you can answer the question.

Coding and Processing Distinguish Stimulus Modality, Location, Intensity, and Duration

If all stimuli are converted to action potentials in the sensory neurons and all action potentials are identical, how can the central nervous system tell the difference between, say, heat and pressure, or between a pinprick to the toe and one to the hand? The attributes of the stimulus must somehow be preserved once the stimulus enters the nervous system for processing. This means that the CNS must distinguish four properties of a stimulus: (1) its nature, or **modality**, (2) its location, (3) its intensity, and (4) its duration.

Sensory Modality The modality of a stimulus is indicated by which sensory neurons are activated and by where the pathways of the activated neurons terminate in the brain. Each

RUNNING PROBLEM

Ménière's disease—named for its discoverer, the nineteenth-century French physician Prosper Ménière—is caused by a buildup of fluid in the inner ear. Symptoms of Ménière's disease include episodic attacks of vertigo, nausea, and tinnitus, accompanied by hearing loss. *Vertigo* is a false sensation of movement that patients often describe as dizziness.

Question 1:
  In which part of the brain is sensory information about equilibrium processed?

328 332 354 357 361 363 371

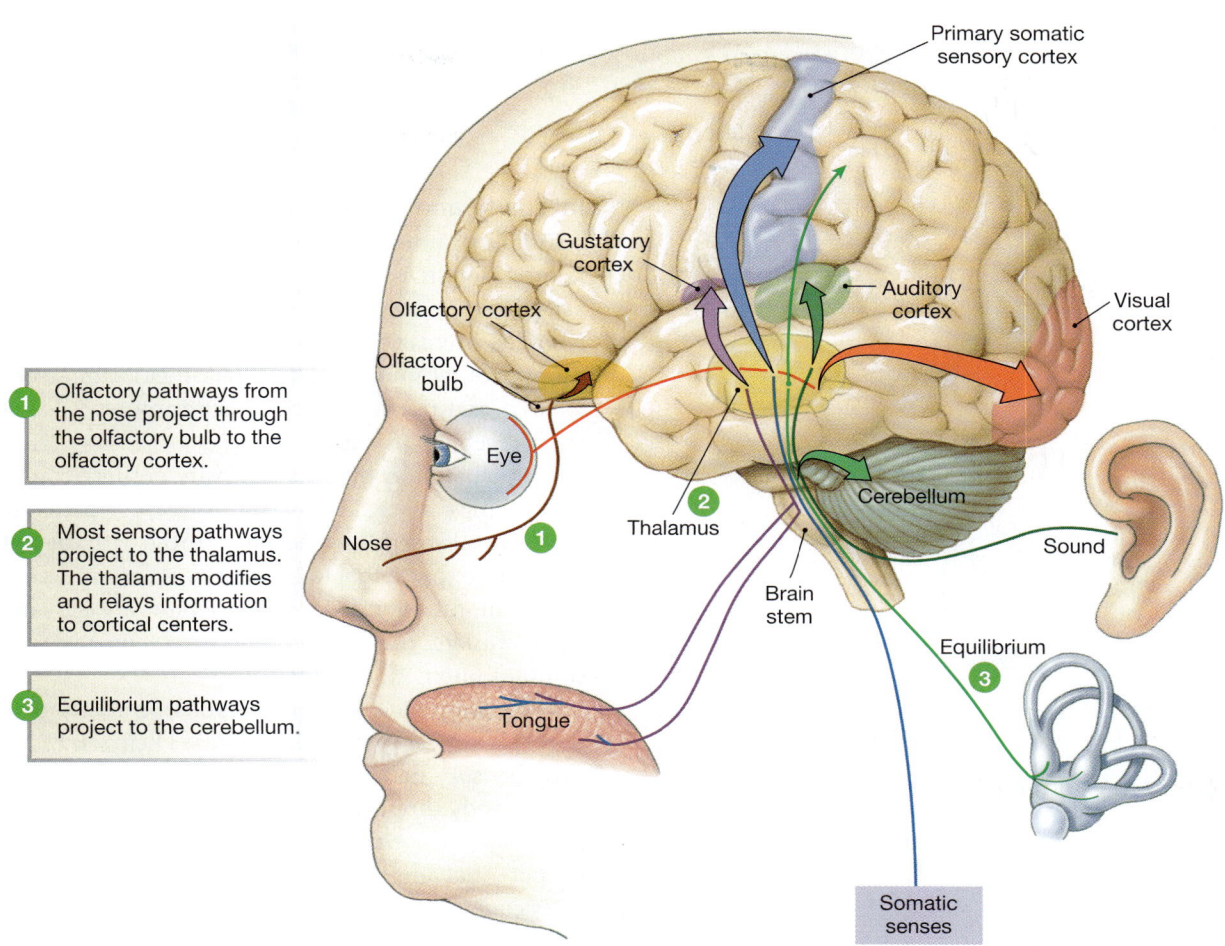

Primary somatic
sensory cortex

Gustatory
cortex

Auditory
cortex

Visual
cortex

Olfactory cortex

Olfactory
bulb

Eye

Cerebellum

Sound

Nose

Thalamus

Brain
stem

Equilibrium

Tongue

Somatic
senses

1 Olfactory pathways from
the nose project through
the olfactory bulb to the
olfactory cortex.

2 Most sensory pathways
project to the thalamus.
The thalamus modifies
and relays information
to cortical centers.

3 Equilibrium pathways
project to the cerebellum.

10

■ **FIGURE 10-4** *Sensory pathways in the brain*

Most pathways except the olfactory pathway pass through the thalamus on their way
to the cerebral cortex. A few equilibrium pathways go directly to the cerebellum.

receptor type is most sensitive to a particular modality of stimulus. For example, some neurons respond most strongly to touch; others respond to changes in temperature. Each sensory modality can be subdivided. For instance, color vision is divided into red, blue, and green according to the wavelengths that most strongly stimulate the different visual receptors.

In addition, the brain associates a signal coming from a specific group of receptors with a specific modality. This 1:1 association of a receptor with a sensation is called **labeled line coding**. Stimulation of a cold receptor will always be perceived as cold, whether the actual stimulus was cold or an artificial depolarization of the receptor. The blow to the eye that causes us to "see" a flash of light is another example of labeled line coding.

Location of the Stimulus   The location of a stimulus is also coded according to which receptive fields are activated. The sensory regions of the cerebrum are highly organized with respect to incoming signals, and input from adjacent sensory receptors is processed in adjacent regions of the cortex. This arrangement preserves the topographical organization of receptors on the skin, eye, or other regions in the processing centers of the brain.

For example, touch receptors in the hand project to a specific area of the cerebral cortex. Experimental stimulation of that area of the cortex during brain surgery is interpreted as a touch to the hand, even though there may have been no actual contact. Similarly, the phantom limb pain reported by some amputees occurs when secondary sensory neurons in the spinal cord become hyperactive, resulting in the sensation of pain in a limb that is no longer there.

Auditory information is an exception to the localization rule, however. Neurons in the ears are sensitive to different frequencies of sound, but they have no receptive fields and their activation provides no information about the location of the sound. Instead, the brain uses the timing of receptor activation to compute a location, as shown in Figure 10-5 ■. A sound coming from directly in front of a person will reach both ears simultaneously. A sound placed off to one side will reach the closer ear several milliseconds before it reaches the other ear.

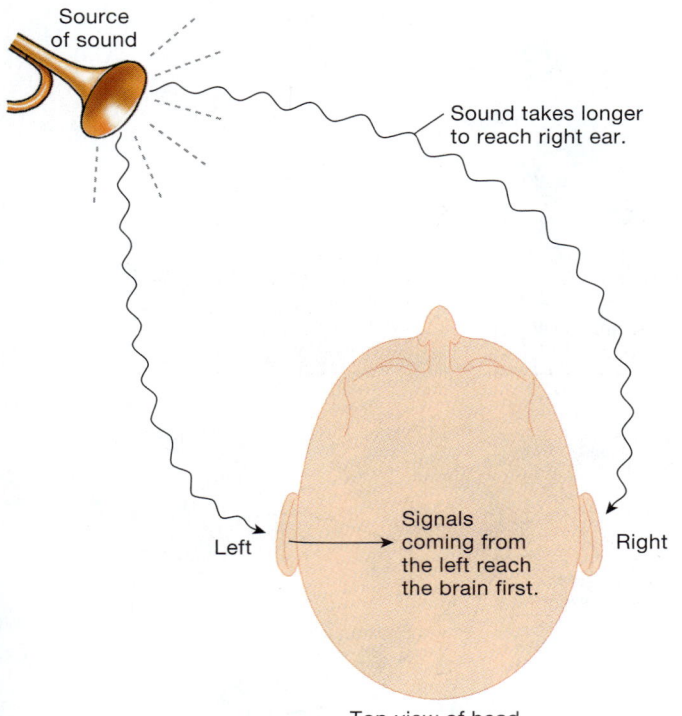

Source of sound

Sound takes longer to reach right ear.

Left

Signals coming from the left reach the brain first.

Right

Top view of head

■ **FIGURE 10-5** *Localization of sound*

The brain uses timing differences rather than specific neurons to localize sound.

The difference in the time it takes for the sound stimuli to reach the two sides of the auditory cortex is registered by the brain and used to compute the sound's source.

Lateral inhibition, which increases the contrast between activated receptive fields and their inactive neighbors, is another way the location of a stimulus can be isolated. Figure 10-6 ■ shows this process for a pressure stimulus to the skin. The pin pushing on the skin activates three primary sensory neurons, each of which releases neurotransmitters onto its corresponding secondary neuron. However, the three secondary neurons do not all respond in the same fashion. The secondary neuron closest to the stimulus (neuron B) suppresses the response of the secondary neurons lateral to it (that is, on either side), where the stimulus is weaker, and simultaneously allows its own pathway to proceed without interference. The inhibition of neurons farther from the stimulus enhances the contrast between the center and the sides of the receptive field, making the sensation more easily localized. Lateral inhibition is also used in the visual system to sharpen our perception of visual edges.

The pathway in Figure 10-6 also demonstrates **population coding**, the way multiple receptors can function together to send the CNS more information than would be possible from a single receptor. By comparing the input from multiple receptors, the CNS can make complex calculations about the spatial and temporal characteristics of a stimulus.

CONCEPT CHECK

4. In Figure 10-6, what kind(s) of ion channel might open in neurons A and C that would depress their responsiveness: Na^+, K^+, Cl^-, or Ca^{2+}? Answers: p. 374

Intensity and Duration of the Stimulus The intensity of a stimulus cannot be directly calculated from a single sensory neuron action potential because a single action potential is "all-or-none," with constant amplitude and duration. Instead, stimulus intensity is coded in two types of information: the number of receptors activated (another example of population coding) and the frequency of action potentials coming from those receptors (*frequency coding*).

Population coding for intensity occurs because not all receptors have the same threshold to their preferred stimulus. Only the most sensitive receptors (those with the lowest thresholds) will respond to a low-intensity stimulus. As a stimulus increases in intensity, additional receptors are activated. The CNS then translates the number of active receptors into a measure of stimulus intensity.

For individual sensory neurons, intensity discrimination begins at the receptor. If a stimulus is below threshold, the primary sensory neuron does not respond. Once stimulus intensity exceeds threshold, the primary sensory neuron begins to fire action potentials. As stimulus intensity increases, the receptor potential amplitude (strength) increases in proportion, and the frequency of action potentials in the primary sensory neuron increases, up to a maximum rate (Fig. 10-7 ■).

Similarly, the duration of a stimulus is coded by the duration of action potentials in the sensory neuron. In general, a longer stimulus generates a longer series of action potentials in the primary sensory neuron. However, if a stimulus persists, some receptors **adapt**, which means they cease to respond. Receptors fall into one of two classes, depending on how they adapt to continuous stimulation.

Tonic receptors are slowly adapting receptors that fire rapidly when first activated, then slow and maintain their firing as long as the stimulus is present (Fig. 10-8a ■). Pressure-sensitive baroreceptors, irritant receptors, and some tactile receptors and proprioceptors fall into this category. In general, the stimuli that activate tonic receptors are parameters that must be monitored continuously by the body.

In contrast, **phasic receptors** are rapidly adapting receptors that fire when they first receive a stimulus but cease firing if the strength of the stimulus remains constant (Fig. 10-8b). Phasic receptors are attuned specifically to *changes* in a parameter. Once a stimulus reaches a steady intensity, phasic receptors adapt to the new steady state and turn off. This type of response allows the body to ignore information that has been evaluated and found not to threaten homeostasis or well-being.

Our sense of smell is an example of a sense that uses phasic receptors. For example, you can smell your cologne when you put it on in the morning, but as the day goes on your

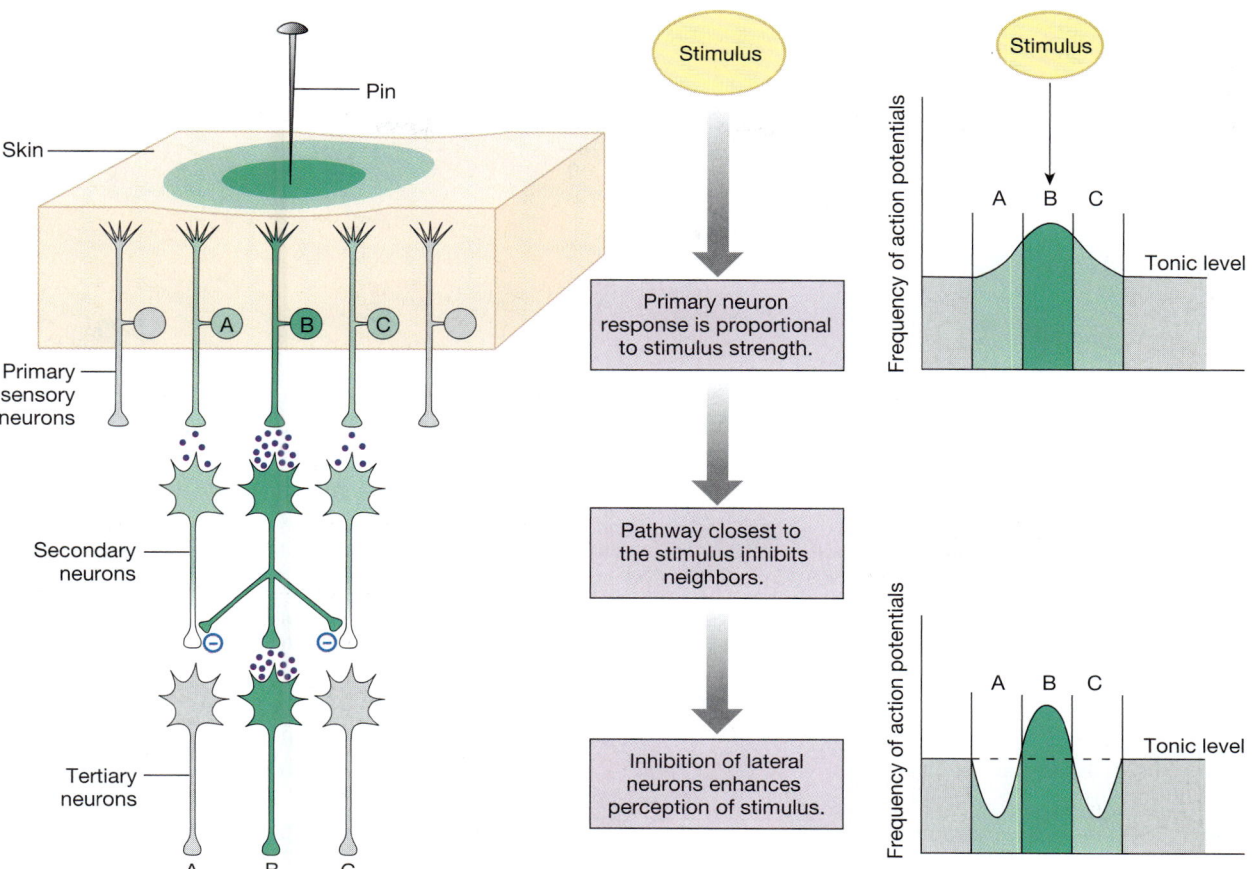

■ **FIGURE 10-6** *Lateral inhibition*

The responses of primary sensory neurons A, B, and C are proportional to the intensity of the stimulus in each receptive field. Secondary sensory neurons A and C are inhibited by secondary sensory neuron B, creating greater contrast between B and its neighbors.

olfactory receptors adapt and are no longer stimulated by the cologne molecules. You no longer smell the fragrance, yet others may comment on it. Adaptation of phasic receptors allows us to filter out extraneous sensory information and concentrate on what is new, different, or essential.

What happens at a receptor during adaptation depends on the receptor type. In some receptors, K^+ channels in the receptor membrane open, causing the membrane to repolarize and stopping the signal. In other receptors, Na^+ channels quickly inactivate. In yet other receptors, biochemical pathways alter the receptor's responsiveness.

Accessory structures may also decrease the amount of stimulus reaching the receptor. In the ear, for example, tiny muscles contract and dampen the vibration of small bones in response to loud noises, thus decreasing the sound signal before it reaches auditory receptors. In general, once adaptation of a phasic receptor has occurred, the only way to create a new signal is to either increase the intensity of the excitatory stimulus or remove the stimulus entirely and allow the receptor to reset.

To summarize, the specificity of sensory pathways is established in several ways:

1. Each receptor is most sensitive to a particular type of stimulus.
2. A stimulus above threshold initiates action potentials in a sensory neuron that projects to the CNS.
3. Stimulus intensity and duration are coded in the pattern of action potentials reaching the CNS.
4. Stimulus location and modality are coded according to which receptors are activated or (in the case of sound and smell) by the timing of receptor activation.
5. Each sensory pathway projects to a specific region of the cerebral cortex dedicated to a particular receptive field. The brain can then tell the origin of each incoming signal.

CONCEPT CHECK

5. How do sensory receptors tell the central nervous system about the intensity of a stimulus?
6. What is the adaptive significance of having irritant receptors that are tonic instead of phasic?

Answers: p. 375

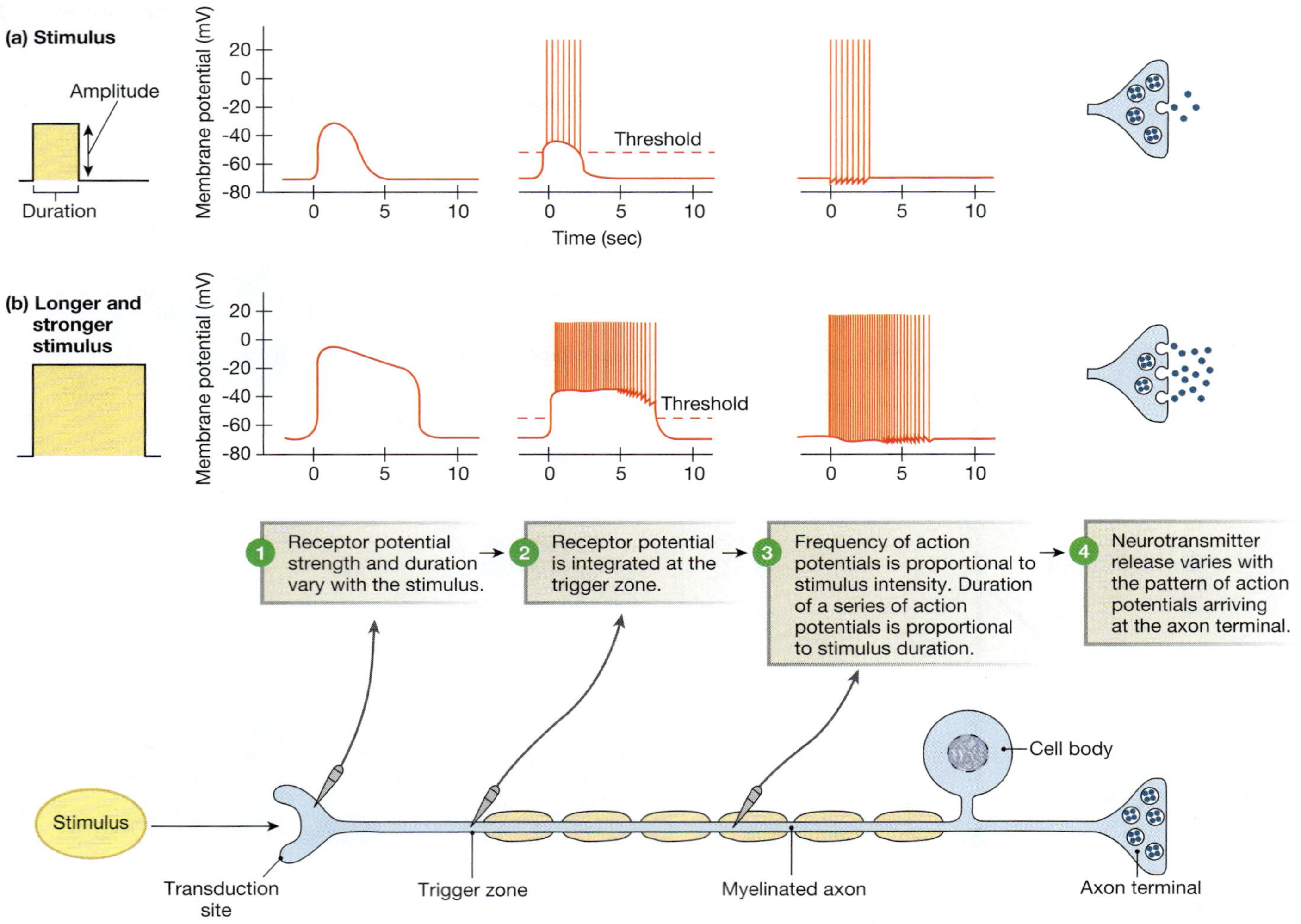

(a) Stimulus

Amplitude

Duration

Membrane potential (mV)

20
0
-20
-40
-60
-80

0   5   10

Threshold

Time (sec)

(b) Longer and stronger stimulus

Membrane potential (mV)

20
0
-20
-40
-60
-80

0   5   10

Threshold

① Receptor potential strength and duration vary with the stimulus.

② Receptor potential is integrated at the trigger zone.

③ Frequency of action potentials is proportional to stimulus intensity. Duration of a series of action potentials is proportional to stimulus duration.

④ Neurotransmitter release varies with the pattern of action potentials arriving at the axon terminal.

Cell body

Stimulus

Transduction site

Trigger zone

Myelinated axon

Axon terminal

■ **FIGURE 10-7**   *Sensory coding for stimulus intensity and duration*

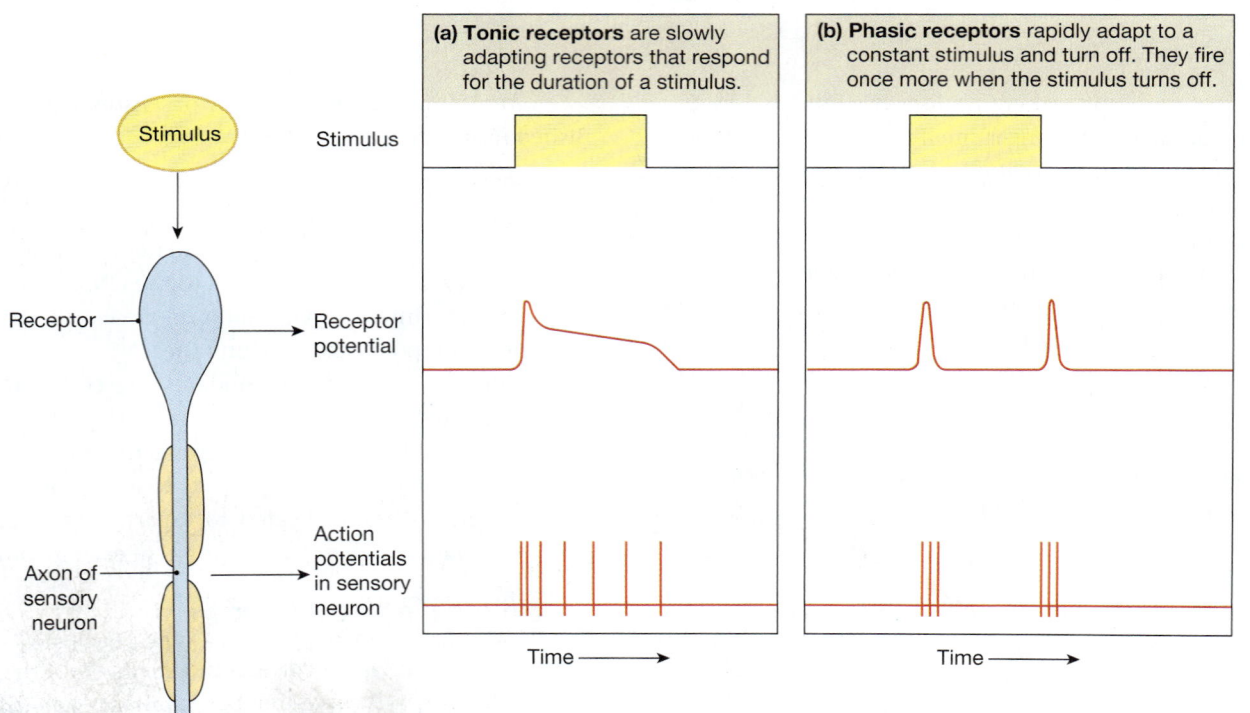

(a) Tonic receptors are slowly adapting receptors that respond for the duration of a stimulus.

(b) Phasic receptors rapidly adapt to a constant stimulus and turn off. They fire once more when the stimulus turns off.

Stimulus

Stimulus

Receptor

Receptor potential

Axon of sensory neuron

Action potentials in sensory neuron

Time →

Time →

■ **FIGURE 10-8**   *Tonic and phasic receptors*

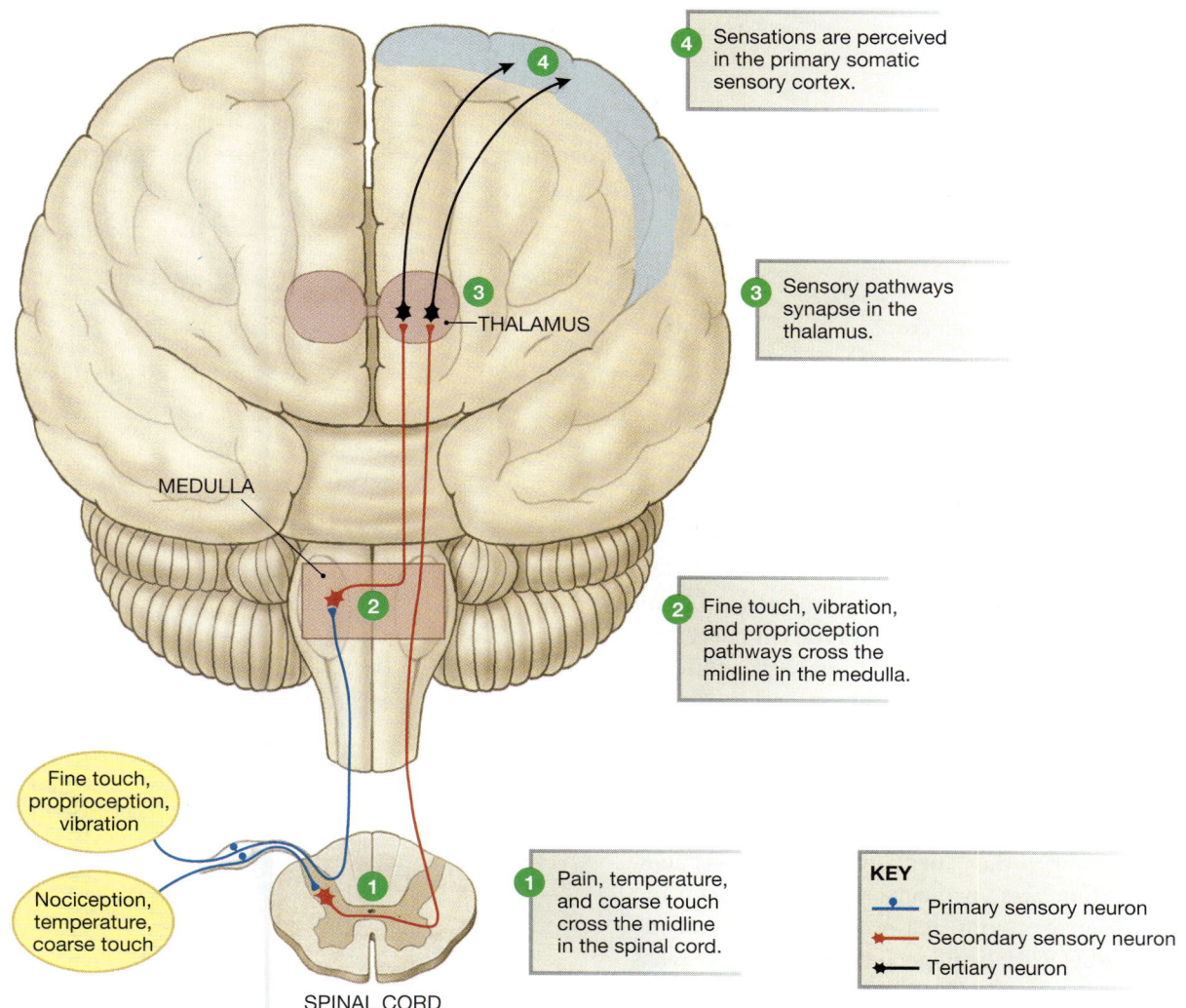

4 Sensations are perceived in the primary somatic sensory cortex.

3 Sensory pathways synapse in the thalamus.

THALAMUS

MEDULLA

2 Fine touch, vibration, and proprioception pathways cross the midline in the medulla.

Fine touch, proprioception, vibration

Nociception, temperature, coarse touch

1 Pain, temperature, and coarse touch cross the midline in the spinal cord.

KEY
— Primary sensory neuron
— Secondary sensory neuron
— Tertiary neuron

SPINAL CORD

■ **FIGURE 10-9** *Sensory pathways cross the body's midline*

SOMATIC SENSES

There are four somatosensory modalities: touch, proprioception, temperature, and nociception, which includes pain and itch. We will discuss details of proprioception in Chapter 13.

Pathways for Somatic Perception Project to the Somatosensory Cortex and Cerebellum

Receptors for the somatic senses are found both in the skin and in the viscera. Receptor activation triggers action potentials in the associated primary sensory neuron. In the spinal cord, many primary sensory neurons synapse onto interneurons that serve as the secondary sensory neurons. The location of the synapse between a primary neuron and a secondary neuron varies according to the type of receptor (Fig. 10-9 ■). Neurons associated with receptors for nociception, temperature, and coarse touch synapse onto their secondary neurons shortly after entering the spinal cord. In contrast, most fine touch, vibration, and proprioceptive neurons have very long axons

that project up the spinal cord all the way to the medulla (Table 10-3 ■).

All secondary sensory neurons cross the midline of the body at some point, so that sensations from the left side of the body are processed in the right hemisphere of the brain, and vice versa. The secondary neurons for nociception, temperature, and coarse touch cross the midline in the spinal cord, then ascend to the brain. Fine touch, vibration, and proprioceptive neurons cross the midline in the medulla.

In the thalamus, secondary sensory neurons synapse onto **tertiary sensory neurons**, which in turn project to the somatosensory region of the cerebral cortex. In addition, most sensory pathways send branches to the cerebellum so that it can use the information to coordinate balance and movement.

The **somatosensory cortex** [p. 310] is the part of the brain that recognizes where ascending sensory tracts originate. Each sensory tract has a corresponding region of the cortex, so that all sensory pathways for the left hand terminate in one area, all pathways for the left foot terminate in another area,

TABLE 10-3	Sensory Pathways	
	STIMULUS	
	FINE TOUCH, PROPRIOCEPTION, VIBRATION	IRRITANTS, TEMPERATURE, COARSE TOUCH
Primary sensory neuron terminates in:	Medulla	Dorsal horn of spinal cord
	Path crosses midline of body.	Path crosses midline of body.
Secondary sensory neuron terminates in:	Thalamus	Thalamus
Tertiary sensory neuron terminates in:	Somatosensory cortex	Somatosensory cortex

and so on (Fig. 10-10 ■). Within the cortical region for a particular body part, columns of neurons are devoted to particular types of receptors. For example, a cortical column activated by cold receptors in the left hand may be found next to a column activated by pressure receptors in the skin of the left hand. This columnar arrangement creates a highly organized structure that maintains the association between specific receptors and the sensory modality they transmit.

Some of the most interesting research about the somatosensory cortex has been done on patients during brain surgery for epilepsy. Because brain tissue has no pain fibers, this type of surgery can be performed with the patient awake under local anesthesia. The surgeon stimulates a particular region of the brain and asks the patient about sensations that occur. The ability of the patient to communicate with the surgeon during this process has expanded our knowledge of brain regions tremendously.

Experiments can also be done on nonhuman animals by stimulating peripheral receptors and monitoring electrical activity in the cortex. We have learned from these experiments that the more sensitive a region of the body is to touch and other stimuli, the larger the corresponding region in the cortex. Interestingly, the size of the regions is not fixed. If a particular body part is used more extensively, its topographical region in the cortex will expand. For example, people who are visually handicapped and learn to read Braille with their fingertips develop an enlarged region of the somatosensory cortex devoted to the fingertips. This plasticity [🔁 p. 283] shows the amazing versatility of the brain.

Touch Receptors Respond to Many Different Stimuli

Touch receptors are among the most common receptors in the body. These receptors respond to many forms of physical contact, such as stretch, steady pressure, fluttering or stroking movement, vibration, and texture. They are found both in the skin (Fig. 10-11■) and in deeper regions of the body. Touch receptors associated with muscles, joints, or internal organs will be discussed in later chapters.

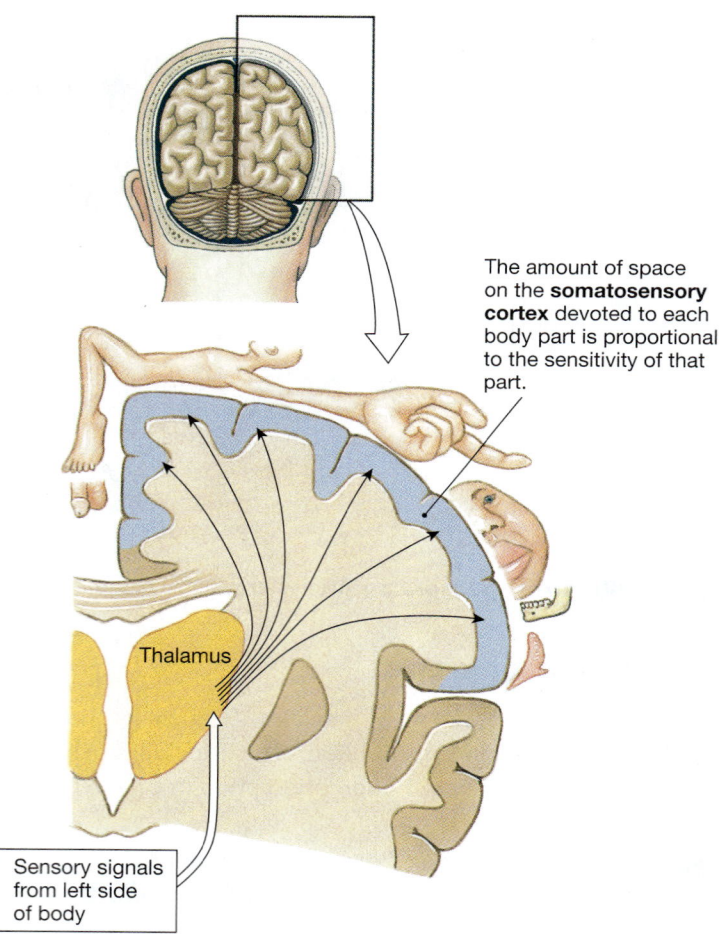

The amount of space on the **somatosensory cortex** devoted to each body part is proportional to the sensitivity of that part.

Thalamus

Sensory signals from left side of body

Cross section of the right cerebral hemisphere and sensory areas of the cerebral cortex

■ FIGURE 10-10   *The somatosensory cortex*

Each representation of a body part is shown adjacent to the area of the sensory cortex that processes stimuli for that body part.

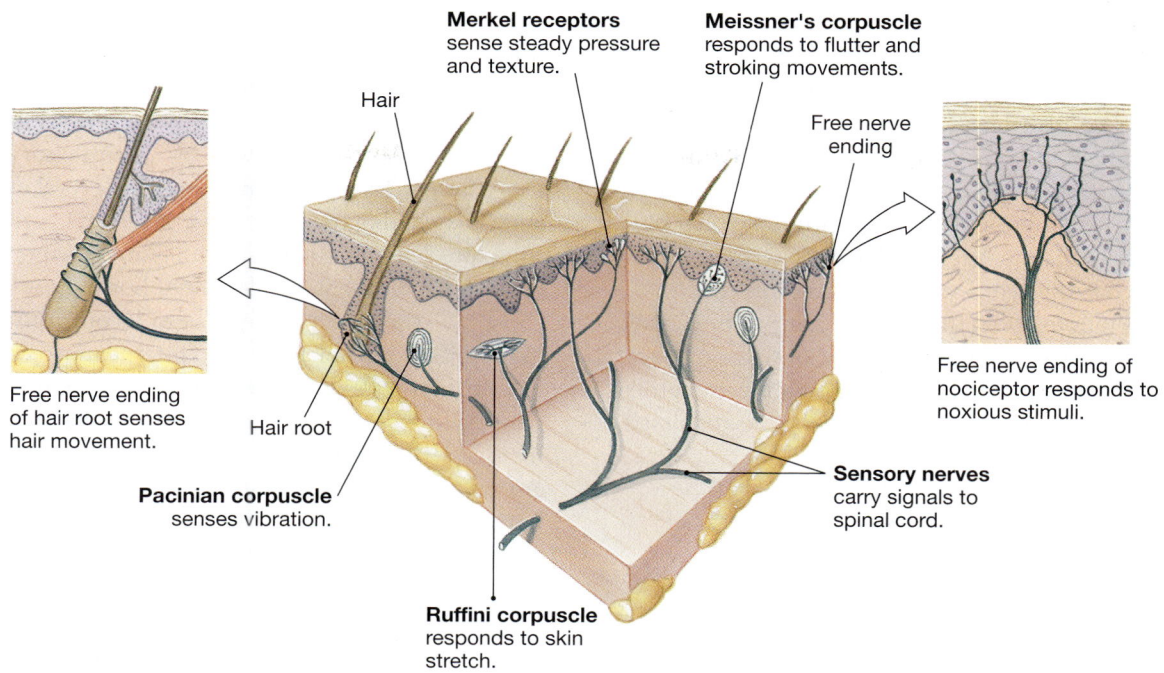

Merkel receptors
sense steady pressure
and texture.

Meissner's corpuscle
responds to flutter and
stroking movements.

Hair

Free nerve
ending

Free nerve ending
of hair root senses
hair movement.

Hair root

Free nerve ending of
nociceptor responds to
noxious stimuli.

Pacinian corpuscle
senses vibration.

Sensory nerves
carry signals to
spinal cord.

Ruffini corpuscle
responds to skin
stretch.

■ **FIGURE 10-11** *Touch receptors in the skin*

Different types of touch receptors are scattered throughout the superficial and deep
layers of the skin.

Touch receptors come in many forms (Table 10-4 ■). Some are free nerve endings, such as those that wrap around the base of hairs; others are more complex. Most touch receptors are difficult to study because of their small size. However, **Pacinian corpuscles**, which respond to vibration, are some of the largest receptors in the body, and much of what we know about somatosensory receptors comes from studies on these structures.

Pacinian corpuscles are composed of nerve endings encapsulated in layers of connective tissue (see Fig. 10-1b). They are found in the subcutaneous layers of skin and in muscles, joints, and internal organs. The concentric layers of connective tissue in the corpuscles create large receptive fields. Pacinian corpuscles respond best to high-frequency vibrations, whose energy is

transferred through the connective tissue capsule to the nerve ending, where the energy opens mechanically gated ion channels [⟳ p. 139]. Pacinian corpuscles are rapidly adapting phasic receptors, and this property allows them to respond to a change in touch but then ignore it.

Properties of the remaining touch receptors depicted in Figure 10-11—Meissner's corpuscles, Ruffini corpuscles, and Merkel receptors—are summarized in Table 10-4.

Temperature Receptors Are Free Nerve Endings

Temperature receptors are free nerve endings that terminate in the subcutaneous layers of the skin. **Cold receptors** are sensitive primarily to temperatures lower than body temperature.

TABLE 10-4	Touch Receptors in Skin			
RECEPTOR	**STIMULUS**	**LOCATION**	**STRUCTURE**	**ADAPTATION**
Free nerve endings	Various touch and pressure stimuli	Around hair roots and under surface of skin	Unmyelinated nerve endings	Variable
Meissner's corpuscles	Flutter, stroking	Superficial layers of skin	Encapsulated in connective tissue	Rapid
Pacinian corpuscles	Vibration	Deep layers of skin	Encapsulated in connective tissue	Rapid
Ruffini corpuscles	Stretch of skin	Deep layers of skin	Enlarged nerve endings	Slow
Merkel receptors	Steady pressure, texture	Superficial layers of skin	Enlarged nerve endings	Slow

Warm receptors are stimulated by temperatures in the range extending from normal body temperature (37° C) to about 45° C. Above that temperature, pain receptors are activated, creating a sensation of painful heat.

The receptive field for a thermoreceptor is about 1 mm in diameter, and the receptors are scattered across the body. There are considerably more cold receptors than warm ones. Temperature receptors slowly adapt between 20° and 40° C: their initial response tells us that the temperature is changing, and their sustained response tells us about the ambient temperature. Outside the 20–40° C range, where the likelihood of tissue damage is greater, the receptors do not adapt. Temperature receptors play an important role in thermoregulation, which is discussed in Chapter 22.

Nociceptors Initiate Protective Responses

Nociceptors [*nocere,* to injure] are receptors that respond to a variety of strong noxious stimuli (chemical, mechanical, or thermal) that cause or have the potential to cause tissue damage. Nociceptors are sometimes called pain receptors, even though pain is a perceived sensation rather than a stimulus. Sensory input from nociceptors initiates adaptive, protective responses, such as the reflexive action of pulling a hand away from a hot stove touched accidentally. Nociceptors are not limited to the skin. Discomfort from overuse of muscles and joints helps us avoid additional damage to these structures.

Nociceptive pain is mediated by free nerve endings that respond to a variety of chemical, mechanical, and thermal stimuli with the help of membrane ion channels. For example, the membrane channels called *vanilloid receptors* (also called *transient receptor potential* V_1 or $TRPV_1$ channels) respond to damaging heat from a stove or other source, as well as to *capsaicin,* the chemical that makes hot chili peppers burn your mouth. At the opposite end of the temperature spectrum, researchers recently identified a membrane protein that responds both to cold and to menthol, one reason mint-flavored foods feel cool. Nociceptor activation is modulated by local chemicals that are released upon tissue injury, such as K^+, histamine, and prostaglandins released from damaged cells; serotonin released from platelets activated by tissue damage; and a peptide known as **substance P** secreted by primary sensory neurons. These

chemicals, which also mediate the inflammatory response at the site of injury, either activate nociceptors or sensitize them by lowering their activation threshold. Increased sensitivity to pain at sites of tissue damage is called **inflammatory pain**.

Nociceptors may activate two pathways: (1) reflexive protective responses that are integrated at the level of the spinal cord and (2) ascending pathways to the cerebral cortex that become conscious sensation (pain or itch). Primary sensory neurons from nociceptors terminate in the dorsal horn of the spinal cord (see Fig. 10-9). There they synapse onto secondary sensory neurons that project to the brain or onto interneurons for local circuits.

Irritant responses that are integrated in the spinal cord initiate rapid unconscious protective reflexes that automatically remove a stimulated area from the source of the stimulus. For example, if you accidentally touch a hot stove, an automatic **withdrawal reflex** will cause you to pull back your hand even before you are aware of the heat. This is one example of the spinal reflexes discussed in Chapter 9 [➡ p. 301].

The lack of brain involvement in many protective reflexes has been demonstrated in the classic "spinal frog" preparation, in which the animal's brain has been destroyed. If the frog's foot is placed in a beaker of hot water or strong acid, the withdrawal reflex causes the leg to contract and move the foot away from the stimulus. The frog is unable to feel pain because the brain, which translates sensory input into perception, is not functional.

Pain and Itching Are Mediated by Nociceptors

Afferent signals from nociceptors are carried to the CNS in three types of primary sensory fibers: *Aβ* (A-beta) *fibers, AΔ* (A-delta) *fibers,* and *C fibers* (Table 10-5 ■). Two sensations may be perceived when nociceptors are activated: pain and itch. The most common sensation is pain, but whenever histamine or some other stimulus activates a subtype of C fiber, we perceive the sensation we call **itch** (*pruritis*). Itch comes only from nociceptors in the skin and is characteristic of many rashes and other skin conditions. However, itch can also be a symptom of a number of systemic diseases, including multiple sclerosis, hyperparathyroidism, and diabetes mellitus. The higher pathways for itch are not as well understood as the pathways for pain.

Pain is a subjective perception, the brain's interpretation of sensory information transmitted along pathways that begin

TABLE 10-5	Classes of Somatosensory Nerve Fibers		
FIBER TYPE	**FIBER CHARACTERISTICS**	**SPEED OF CONDUCTION**	**ASSOCIATED WITH**
Aβ (beta)	Large, myelinated	30–70 m/sec	Mechanical stimuli
Aδ (delta)	Small, myelinated	12–30 m/sec	Cold, fast pain, mechanical stimuli
C	Small, unmyelinated	0.5–2 m/sec	Slow pain, heat, cold, mechanical stimuli

at nociceptors. Pain is also highly individual and may vary with a person's emotional state. The discussion here is limited to the sensory experience of pain.

Fast pain, described as sharp and localized, is rapidly transmitted to the CNS by small, myelinated Aδ fibers. **Slow pain**, described as duller and more diffuse, is carried on small, unmyelinated C fibers. The timing distinction between the two is most obvious when the stimulus originates far from the CNS, such as when you stub your toe. You first experience a quick stabbing sensation (fast pain), followed shortly by a dull throbbing (slow pain).

The ascending pathways for pain cross the body's midline in the spinal cord and ascend to the thalamus and sensory areas of the cortex. The pathways also send branches to the limbic system and hypothalamus. As a result, pain may be accompanied by emotional distress and a variety of autonomic reactions, such as nausea, vomiting, or sweating.

Our perception of pain is subject to modulation at several levels in the nervous system. It can be magnified by past experiences or suppressed in emergencies when survival depends on ignoring injury. In the latter situation, descending pathways travel through the thalamus and inhibit nociceptor neurons in the spinal cord. Artificial stimulation of these inhibitory pathways is one of the newer techniques being used to control chronic pain.

Pain can also be suppressed in the dorsal horn of the spinal cord, before the stimuli are sent to ascending spinal tracts. Normally, tonically active inhibitory interneurons in the spinal cord inhibit ascending pathways for pain (Fig. 10-12a ■). C fibers from nociceptors synapse on these inhibitory interneurons. When activated by a painful stimulus, the C fibers simultaneously excite the ascending path and block the tonic inhibition (Fig. 10-12b). This action allows the pain signal from the C fiber to travel unimpeded to the brain.

In the **gate control theory** of pain modulation (Fig. 10-12c), Aβ fibers carrying sensory information about mechanical stimuli help block pain transmission. The Aβ fibers synapse on the inhibitory interneurons and *enhance* the interneuron's inhibitory activity. If simultaneous stimuli reach the inhibitory neuron from the Aβ and C fibers, the integrated response is partial inhibition of the ascending pain pathway so that pain perceived by the brain is lessened. The gate control theory explains why rubbing a bumped elbow or shin lessens your pain: the tactile stimulus of rubbing activates Aβ fibers and helps decrease the sensation of pain.

Pain can be felt in skeletal muscles (*deep somatic pain*) as well as in the skin. Muscle pain during exercise is associated with the onset of anaerobic metabolism. It is often perceived as a burning sensation in the muscle (as in "go for the burn!"), a sensation apparently created by a metabolite released during anaerobic metabolism. Some investigators have suggested that the metabolite is K^+, known to enhance the pain response. Muscle pain from **ischemia** (lack of adequate blood flow that

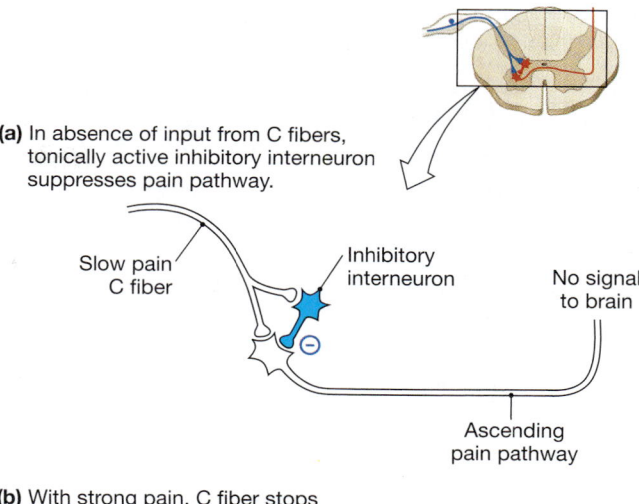

(a) In absence of input from C fibers, tonically active inhibitory interneuron suppresses pain pathway.

Slow pain C fiber

Inhibitory interneuron

No signal to brain

Ascending pain pathway

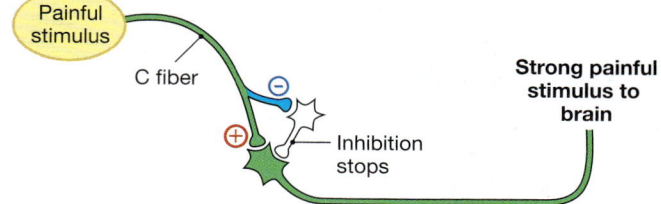

(b) With strong pain, C fiber stops inhibition of the pathway, allowing a strong signal to be sent to the brain.

Painful stimulus

C fiber

Strong painful stimulus to brain

Inhibition stops

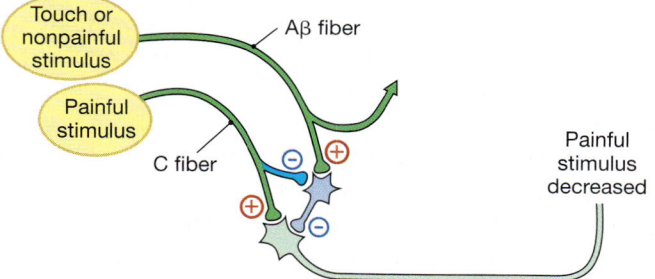

(c) Pain can be modulated by simultaneous somatosensory input.

Touch or nonpainful stimulus

Aβ fiber

Painful stimulus

C fiber

Painful stimulus decreased

■ **FIGURE 10-12** *The gate-control theory of pain modulation*

In the gate-control model, nonpainful stimuli can diminish the pain signal.

reduces oxygen supply) also occurs during *myocardial infarction* (heart attack).

Pain in the heart and other internal organs (*visceral pain*) is often poorly localized and may be felt in areas far removed from the site of the stimulus (Fig. 10-13a ■). For example, the pain of cardiac ischemia may be felt in the neck and down the left shoulder and arm. This **referred pain** apparently occurs because multiple primary sensory neurons converge on a single ascending tract (Fig. 10-13b). According to this model, when painful stimuli arise in visceral receptors, the brain is unable to

10

(a) Pain in internal organs is often sensed on the surface of the body, a sensation known as **referred pain**.

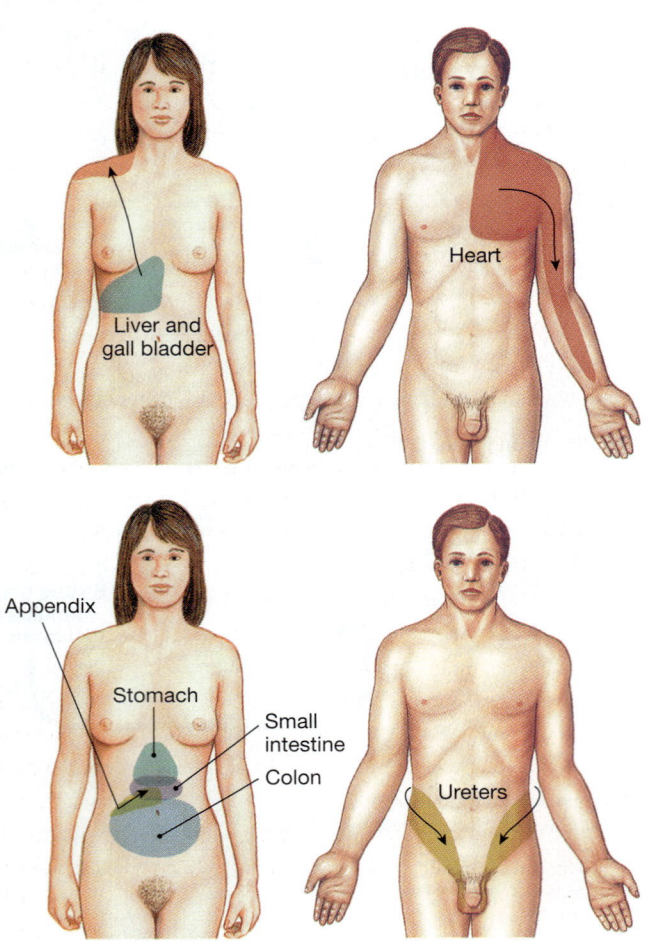

(b) One theory of referred pain says that nociceptors from several locations converge on a single ascending tract in the spinal cord. Pain signals from the skin are more common than pain from internal organs, and the brain associates activation of the pathway with pain in the skin. Adapted from H.L. Fields, *Pain* (McGraw Hill, 1987).

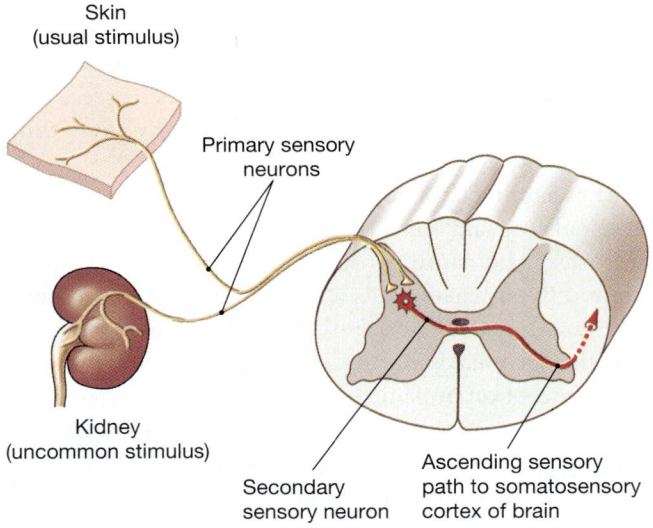

■ **FIGURE 10-13**   *Referred pain*

distinguish visceral signals from the more common signals arising from somatic receptors. As a result, it interprets the pain as coming from the somatic regions rather than the viscera.

Chronic pain of one sort or another affects millions of people in this country every year. This type of pain is often much greater than nociceptor activation would indicate and reflects damage to or long-term changes in the nervous system. Chronic pain is a **pathological pain** and is also called *neuropathic pain*. Scientists do not yet fully understand what causes neuropathic pain, which makes its treatment difficult.

The alleviation of pain is of considerable interest to health professionals. **Analgesic drugs** [*analgesia*, painlessness] range from aspirin to potent opiates such as morphine. Aspirin inhibits prostaglandins and presumably slows the transmission of pain signals from the site of injury. The opiate drugs act directly on CNS *opioid receptors* that are part of an analgesic system that responds to endogenous opiate molecules [🔁 p. 275]. The *enkephalins* and *dynorphin* are secreted by neurons associated with pain pathways, and the endogenous opiate β-*endorphin* is produced from the same prohormone as ACTH (adreno-corticotropin) in neuroendocrine cells of the hypothalamus [🔁 Fig. 7-4b, p. 220]. Activation of opioid receptors blocks pain

perception by decreasing neurotransmitter release from primary sensory neurons and through postsynaptic inhibition of the secondary sensory neurons.

Although opiates are effective at relieving pain, they are also highly addictive, and scientists are exploring alternative strategies for pain relief. In cases of severe chronic pain, it is sometimes possible to get surgical relief by severing afferent nerves at the dorsal root or by stimulating the inhibitor pathways of the brain. Acupuncture can also be effective, although the physiological reason for its effectiveness is not clear. The leading theory on how acupuncture works proposes that properly placed acupuncture needles trigger the release of endorphins by the brain.

✔ **CONCEPT CHECK**

7. What is the adaptive advantage of a spinal reflex?

8. Rank the speed of signal transmission through the following fiber types, from fastest to slowest: (a) small diameter, myelinated fiber; (b) large diameter, myelinated fiber; (c) small diameter, unmyelinated fiber.

9. Your sense of smell uses phasic receptors. What other receptors (senses) adapt to ongoing stimuli?

Answers: p. 375

CLINICAL FOCUS

NATURAL PAIN-KILLERS

Many drugs we use today for pain relief are derivatives of plant or animal molecules. One of the newest pain-killers in this group is *ziconotide,* a synthetic compound related to the poison used by South Pacific cone snails to kill fish. This drug works by blocking calcium channels on nociceptive neurons. Ziconotide, approved in 2004 for the treatment of severe chronic pain, is highly toxic. To minimize systemic side effects, it must be injected directly into the cerebrospinal fluid surrounding the spinal cord. Ziconotide relieves pain but may also cause hallucinations and other psychiatric symptoms, so it is a last-resort treatment. Other pain-killing drugs obtained from biological sources include aspirin, derived from the bark of willow trees (genus *Salix*), and opiate drugs such as morphine and codeine that come from the opium poppy, *Papver somniferum*. These drugs have been used in western and Chinese medicine for centuries, and even today you can purchase willow bark as an herbal remedy.

CHEMORECEPTION: SMELL AND TASTE

The five special senses—smell, taste, hearing, equilibrium, and vision—are concentrated in the head region. Like somatic senses, the special senses rely on receptors to transform information about the environment into patterns of action potentials that can be interpreted by the brain. Smell and taste are both forms of *chemoreception,* one of the oldest senses from an evolutionary perspective. Unicellular bacteria use chemoreception to sense their environment, and primitive animals without formalized nervous systems use chemoreception to locate food and mates. It has been hypothesized that chemoreception evolved into chemical synaptic communication in animals.

Olfaction Is One of the Oldest Senses

Imagine waking up one morning and discovering a whole new world around you, a world filled with odors that you never dreamed existed—scents that told you more about your surroundings than you ever imagined from looking at them. This is exactly what happened to a young patient of Dr. Oliver Sacks (an account is in *The Man Who Mistook His Wife for a Hat and Other Clinical Tales*). Or imagine skating along the sidewalk without a helmet, only to fall and hit your head. When you regain consciousness, the world has lost all odor: no smell of grass or perfume or garbage. Even your food has lost much of its taste, and you now eat only to survive because eating has lost its pleasure.

We do not realize the essential role that our sense of smell plays in our lives until a head cold or injury robs us of the ability to smell. **Olfaction** [*olfacere,* to sniff] allows us to discriminate among thousands of different odors. Even so, our noses are not nearly as sensitive as those of many other animals whose survival depends on olfactory cues. The **olfactory bulb** (Fig. 10-14 ■), the extension of the forebrain that receives input from the primary olfactory neurons, is much better developed in vertebrates whose survival is more closely linked to chemical monitoring of their environment (see Fig. 9-1, p. 293).

The human olfactory system consists of primary sensory neurons (**olfactory cells**) that synapse with secondary sensory neurons in the olfactory bulb, which then processes the incoming information (Figure 10-14b). Secondary and higher-order neurons project from the olfactory bulb through the first cranial nerve to the *olfactory cortex*. In addition, olfactory pathways lead to the amygdala and hippocampus, parts of the limbic system involved with emotion and memory. This arrangement seems quite simple, but complex processing takes place in the olfactory bulb before signals pass on to the cortex. Some descending modulatory pathways from the cortex terminate in the olfactory bulb, and there are reciprocal modulatory connections within and between the two branches of the olfactory bulb.

One amazing aspect of olfaction is the link between smell, memory, and emotion. A special cologne or the aroma of food can trigger memories and create a wave of nostalgia for the time, place, or people with whom the aroma is associated. In some way that we do not understand, the processing of odors through the limbic system creates deeply buried olfactory memories. Particular combinations of olfactory receptors become linked to other patterns of sensory experience so that stimulating one pathway stimulates them all.

In rodents, an accessory olfactory structure in the nasal cavity, the **vomeronasal organ** (VNO), is known to be involved in behavioral responses to sex pheromones [🔁 p. 216].) Anatomical studies in humans have not provided clear evidence for or against a functional VNO, but experiments with compounds believed to act as human pheromones suggest that humans do communicate with chemical signals.

Olfactory cells in humans are concentrated in a 3-cm^2 patch of **olfactory epithelium** high in the nasal cavity (Fig. 10-14a). The olfactory cells are neurons whose processes extend from one side of the cell body to form dendrites on the surface of the olfactory epithelium, and extend from the other to the olfactory bulb, located on the underside of the frontal lobe. Olfactory cells, unlike other neurons in the body, are continuously dividing cells, with a turnover time of about two months (Fig. 10-14c).

10

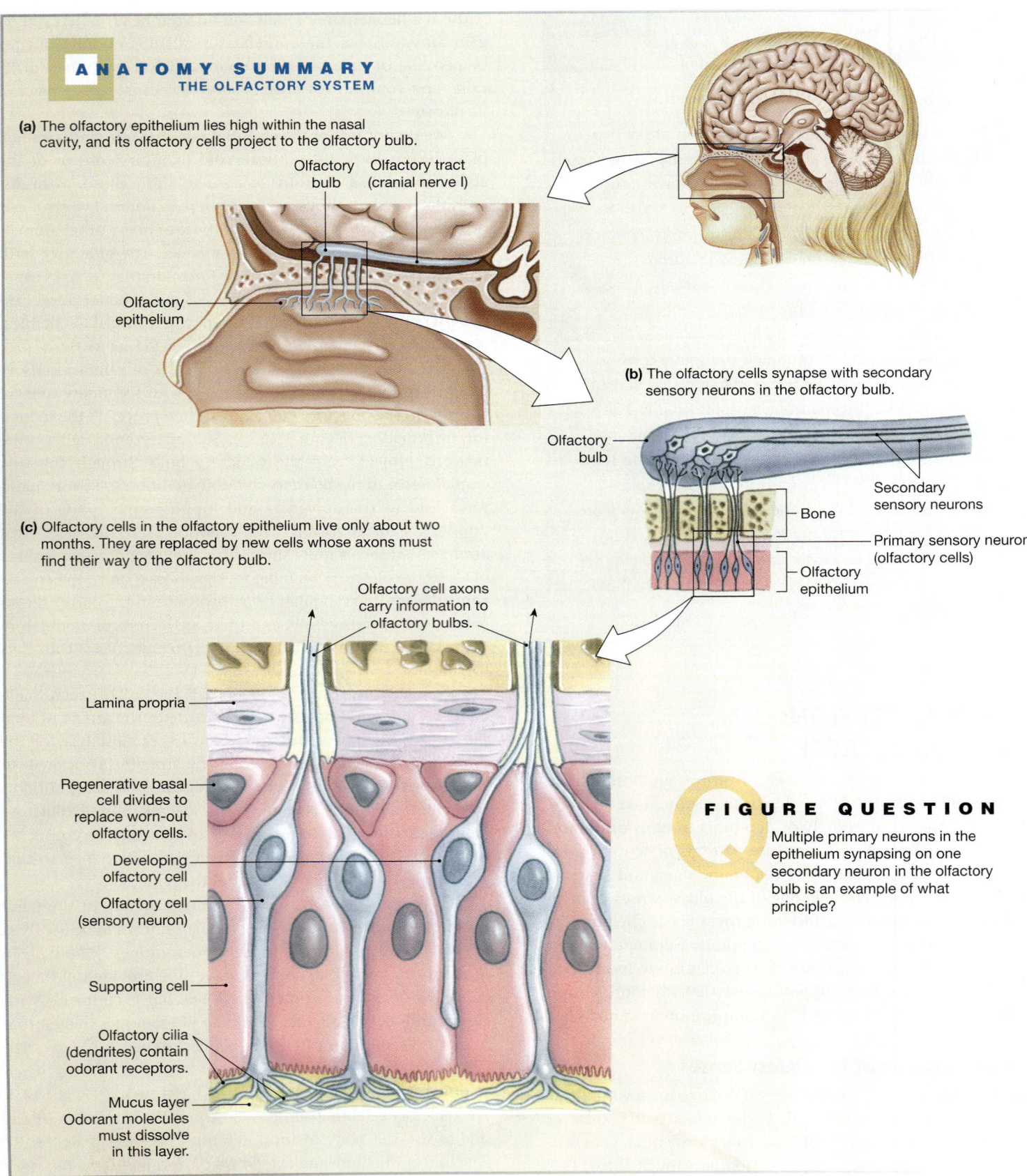

ANATOMY SUMMARY
THE OLFACTORY SYSTEM

(a) The olfactory epithelium lies high within the nasal cavity, and its olfactory cells project to the olfactory bulb.

Olfactory bulb

Olfactory tract (cranial nerve I)

Olfactory epithelium

(b) The olfactory cells synapse with secondary sensory neurons in the olfactory bulb.

Olfactory bulb

Secondary sensory neurons

Bone

Primary sensory neurons (olfactory cells)

Olfactory epithelium

(c) Olfactory cells in the olfactory epithelium live only about two months. They are replaced by new cells whose axons must find their way to the olfactory bulb.

Olfactory cell axons carry information to olfactory bulbs.

Lamina propria

Regenerative basal cell divides to replace worn-out olfactory cells.

Developing olfactory cell

Olfactory cell (sensory neuron)

Supporting cell

Olfactory cilia (dendrites) contain odorant receptors.

Mucus layer
Odorant molecules must dissolve in this layer.

Q FIGURE QUESTION

Multiple primary neurons in the epithelium synapsing on one secondary neuron in the olfactory bulb is an example of what principle?

■ **FIGURE 10-14**

This means that the axon of each new neuron must find its way to the olfactory bulb and make the proper synaptic connection. Discovering how these neurons manage to repeat the same connection each time will give us insight into how developing neurons find their targets.

The surface of the olfactory epithelium is composed of the knobby terminals of the olfactory cells, each knob sprouting multiple nonmobile cilia (Fig. 10-14c). The cilia are embedded in a layer of mucus, and odorant molecules must first dissolve in and penetrate the mucus before they can bind to an **odorant receptor** protein. Each odorant receptor is sensitive to a variety of substances.

Odorant receptors are G protein-cAMP-linked membrane receptor proteins [p. 183]. Odorant receptor genes form the largest known gene family in vertebrates (about 1000 genes, or 3–5% of the genome), but only about 400 odorant receptor proteins are expressed in humans. The combination of an odorant molecule with its odorant receptor activates a special G protein, G_{olf}, which in turn increases intracellular cAMP. The increase in cAMP concentration opens cAMP-gated cation channels, depolarizing the cell and triggering a signal that travels along the olfactory cell axon to the olfactory bulb.

What is occurring at the cellular and molecular levels that allows us to discriminate between thousands of different odors? Current research suggests that each individual olfactory cell contains a single type of odorant receptor. The axons of cells with the same receptors converge on a few secondary neurons in the olfactory bulb, which then can modify the information before sending it on to the olfactory cortex. The brain uses information from hundreds of olfactory cells in different combinations to create the perception of many different smells, just as combinations of letters create different words. This is another example of population coding in the nervous system [p. 334].

CONCEPT CHECK

10. Create a map or diagram of the olfactory pathway from an olfactory cell to the olfactory cortex.
11. Create a map or diagram that starts with a molecule from the environment binding to its odorant receptor in the nose, and ends with neurotransmitter release from the primary olfactory neuron.
12. Which part of an olfactory cell is the dendrites?
13. Are olfactory neurons pseudounipolar, bipolar, or multipolar? (*Hint:* see Fig. 8-3, p. 248) Answers: p. 375

Taste Is a Combination of Five Basic Sensations

Our sense of taste (**gustation**) is closely linked to olfaction. Indeed, much of what we call the taste of food is actually the aroma, as you know if you have ever had a bad cold. Although smell is sensed by hundreds of types of receptors, taste is a combination of only five sensations: sweet, sour, salty, bitter, and **umami**, a taste associated with the amino acid glutamate and some nucleotides. Umami, a name derived from the Japanese

word for "deliciousness," is a basic taste that enhances the flavor of foods. It is the reason that monosodium glutamate (MSG) is used as a food additive in some countries.

Interestingly, although we have only a limited number of taste sensations, each one is associated with an essential body function. Sour taste is triggered by the presence of H^+ and salty by the presence of Na^+, two ions whose concentrations in body fluids are closely regulated. The other three taste sensations result from organic molecules. Sweet and umami are associated with nutritious food. Bitter taste is recognized by the body as a warning of possibly toxic components. If something tastes bitter, our first reaction is often to spit it out.

The receptors for taste are located primarily on **taste buds** clustered together on the surface of the tongue (Fig. 10-15). One taste bud is composed of 50–150 **taste cells**, along with support cells and regenerative basal cells. Taste receptors are also scattered through other regions of the oral cavity, such as the palate.

Each taste cell is a non-neural polarized epithelial cell [p. 151] tucked down into the epithelium so that only a tiny tip protrudes into the oral cavity through a *taste pore*. In a given bud, tight junctions link the apical ends of adjacent cells together, limiting movement of molecules between the cells. The apical membrane of a taste cell is modified into microvilli to increase the amount of surface area in contact with the environment (Fig. 10-15c). The basal side of the cell forms a synapse with a primary sensory neuron.

For a substance to be tasted, it must first dissolve in the saliva and mucus of the mouth. Dissolved taste ligands then interact with an apical membrane protein on a taste cell (Fig. 10-15c). Although the details differ for the five taste sensations, interaction of a taste ligand with a membrane protein initiates a signal transduction cascade that ends with an intracellular calcium signal, due to Ca^{2+} influx from the extracellular fluid or Ca^{2+} release from intracellular stores. The Ca^{2+} signal triggers exocytosis of neurotransmitter [p. 150] into the synapse and initiates a series of action potentials in the primary sensory neuron.

The mechanisms of taste transduction are a good example of how our models of physiological function must periodically be revised as new research data are published. For many years the widely held view of taste transduction was that an individual taste cell could sense more than one taste, with cells differing in their sensitivities. However, recently published research using molecular biology and knockout mice [p. 268] suggests that each taste cell is sensitive to only one taste.

The details of taste cell signal transduction, which once appeared to be relatively straightforward, are more complex than scientists initially thought (Fig. 10-16). The ligands for bitter, sweet, and umami tastes bind to different G protein-coupled receptors, including about 30 variants of bitter receptors. Bitter and sweet (and perhaps umami) taste transduction

(a) Taste buds are located on the dorsal surface of the tongue. The umami receptors (not shown) are in the back of the pharynx.

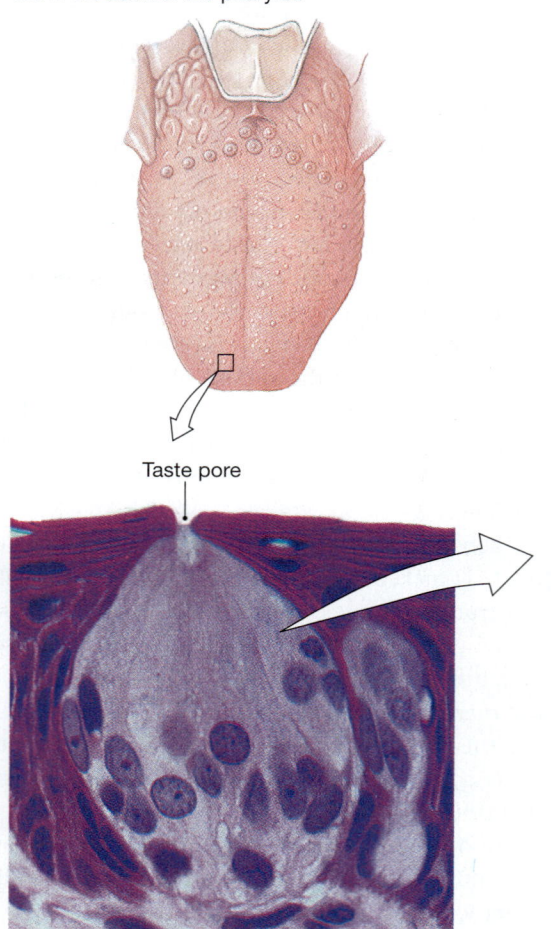

Taste pore

(b) A light micrograph of a taste bud. Each taste bud is composed of taste cells and support cells, joined near the apical surface with tight junctions.

(c) Taste ligands bind to receptors and create Ca^{2+} signals that release neurotransmitters onto primary sensory neurons.

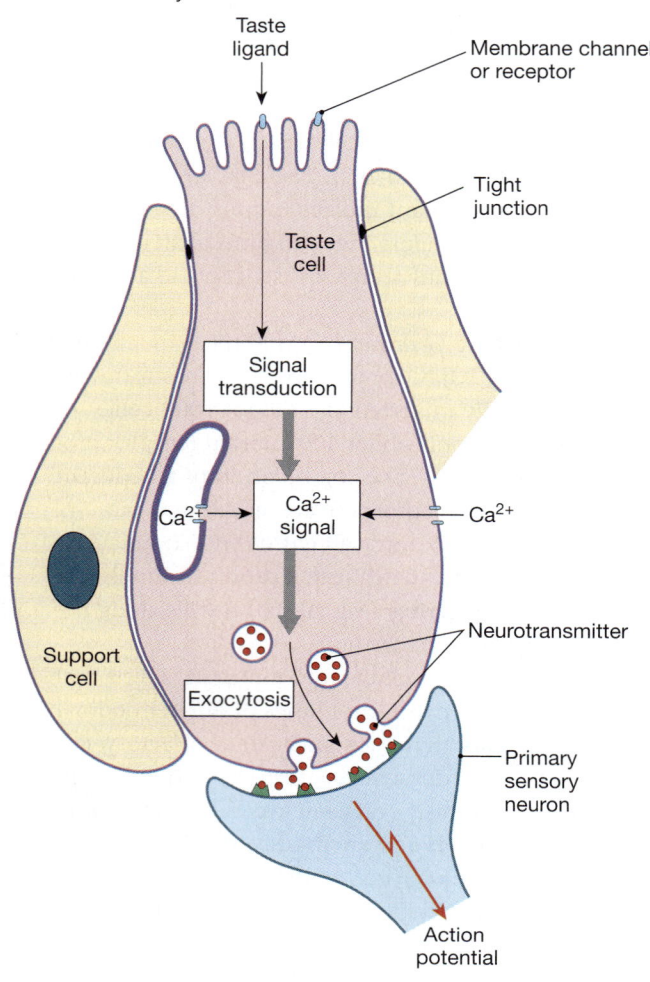

■ **FIGURE 10-15**   *Taste buds*

involve a special G protein called **gustducin**. The receptor proteins appear to be linked to several signal transduction pathways. Some release Ca^{2+} from intracellular stores, while others ultimately open cation channels and allow Ca^{2+} to enter the cell. In any case, the final step in all the pathways is a Ca^{2+} signal that causes exocytosis of neurotransmitter, just as occurs in an axon terminal.

Salty and sour transduction are mediated by ion channels rather than by G protein-coupled receptors. The channels allow Na^+ or H^+ to enter and depolarize the taste cell. The sour ligand H^+ also binds to and closes K^+ channels, which depolarizes the cell. Depolarization opens a voltage-gated Ca^{2+} channel and creates a Ca^{2+} signal as previously described.

Neurotransmitters from taste cells activate primary sensory neurons whose axons run through cranial nerves VII, IX,

and X to the medulla, where they synapse. Sensory information then passes through the thalamus to the gustatory cortex (see Fig. 10-4). Central processing of sensory information compares the input from multiple taste cells and interprets the taste sensation based on which populations of neurons are responding most strongly.

An interesting psychological aspect of taste is the phenomenon named **specific hunger**. Humans and other animals that are lacking a particular nutrient may develop a craving for that substance. **Salt appetite**, representing a lack of Na^+ in the body, has been recognized for years. Hunters have used their knowledge of this specific hunger to stake out salt licks because they know that animals will seek them out. Salt appetite is directly related to Na^+ concentration in the body and cannot be assuaged by ingestion of other cations, such as Ca^{2+} or K^+. Other appetites, such as cravings for chocolate, are more difficult

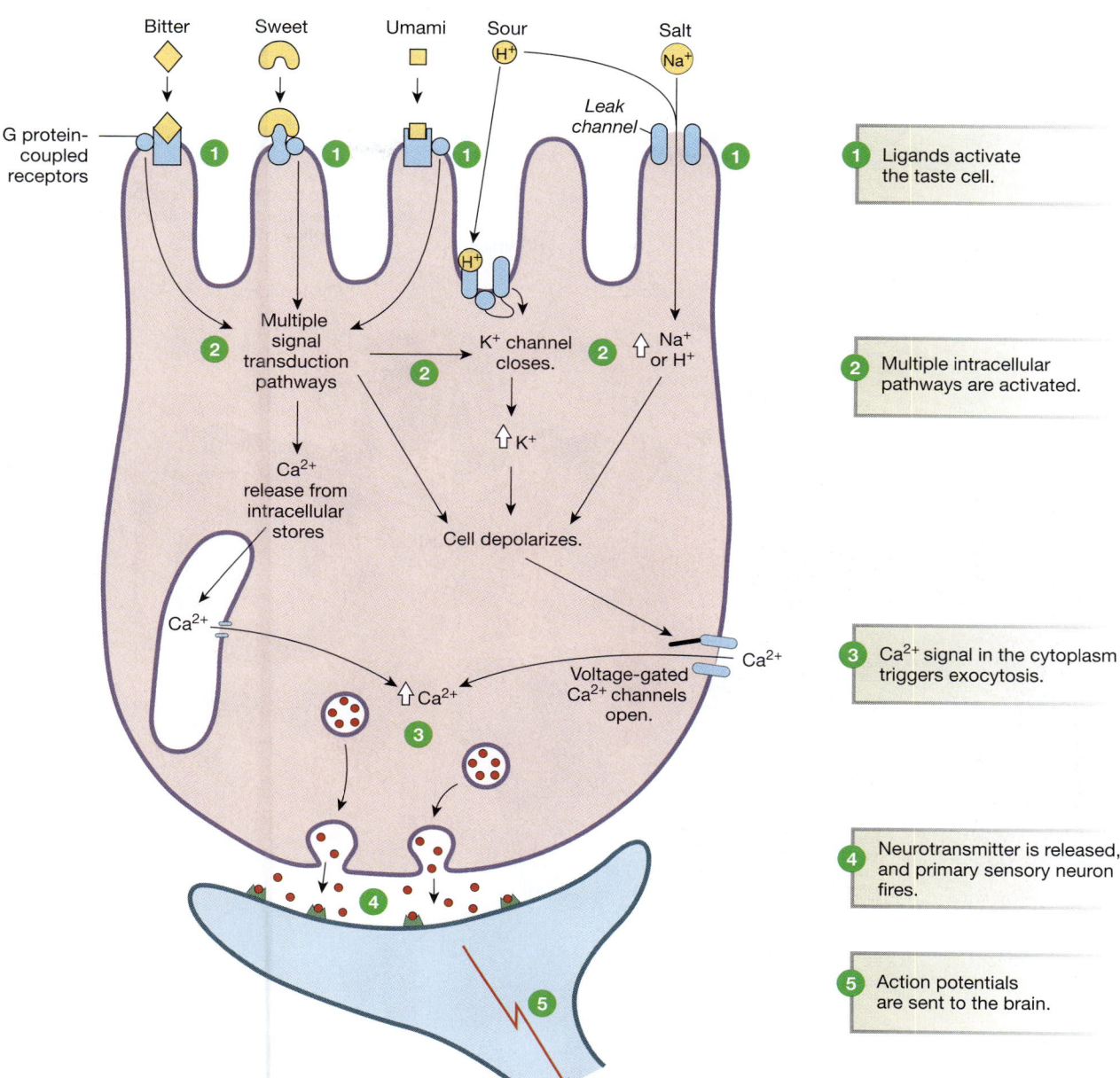

Bitter   Sweet   Umami   Sour   Salt

G protein-coupled receptors

1. Ligands activate the taste cell.

Multiple signal transduction pathways

K⁺ channel closes.

Na⁺ or H⁺

2. Multiple intracellular pathways are activated.

Ca²⁺ release from intracellular stores

⇧ K⁺

Cell depolarizes.

Ca²⁺

⇧ Ca²⁺

Voltage-gated Ca²⁺ channels open.

Ca²⁺

3. Ca²⁺ signal in the cytoplasm triggers exocytosis.

4. Neurotransmitter is released, and primary sensory neuron fires.

5. Action potentials are sent to the brain.

■ **FIGURE 10-16** *Summary of taste transduction*

Bitter, sweet, and umami ligands bind to G protein–coupled membrane receptors. Ionic ligands for salt and sour taste enter through ion channels. Each taste cell senses only one type of ligand, not all five as shown in this figure.

to relate to specific nutrient needs and probably reflect complex mixtures of physical, psychological, environmental, and cultural influences.

CONCEPT CHECK

14. With what essential nutrient is the umami taste sensation associated?

15. Map or diagram the neural pathway from a taste cell to the gustatory cortex.

Answers: p. 375

THE EAR: HEARING

The ear is a sense organ that is specialized for two distinct functions: hearing and equilibrium. It can be divided into external, middle, and inner sections, with the neurological elements housed in and protected by structures in the inner ear. The vestibular complex of the inner ear is the primary sensor for equilibrium. The remainder of the ear is used for hearing.

The *external ear* consists of the outer ear, or **pinna**, and the **ear canal** (Fig. 10-17 ■). The pinna is another example of an

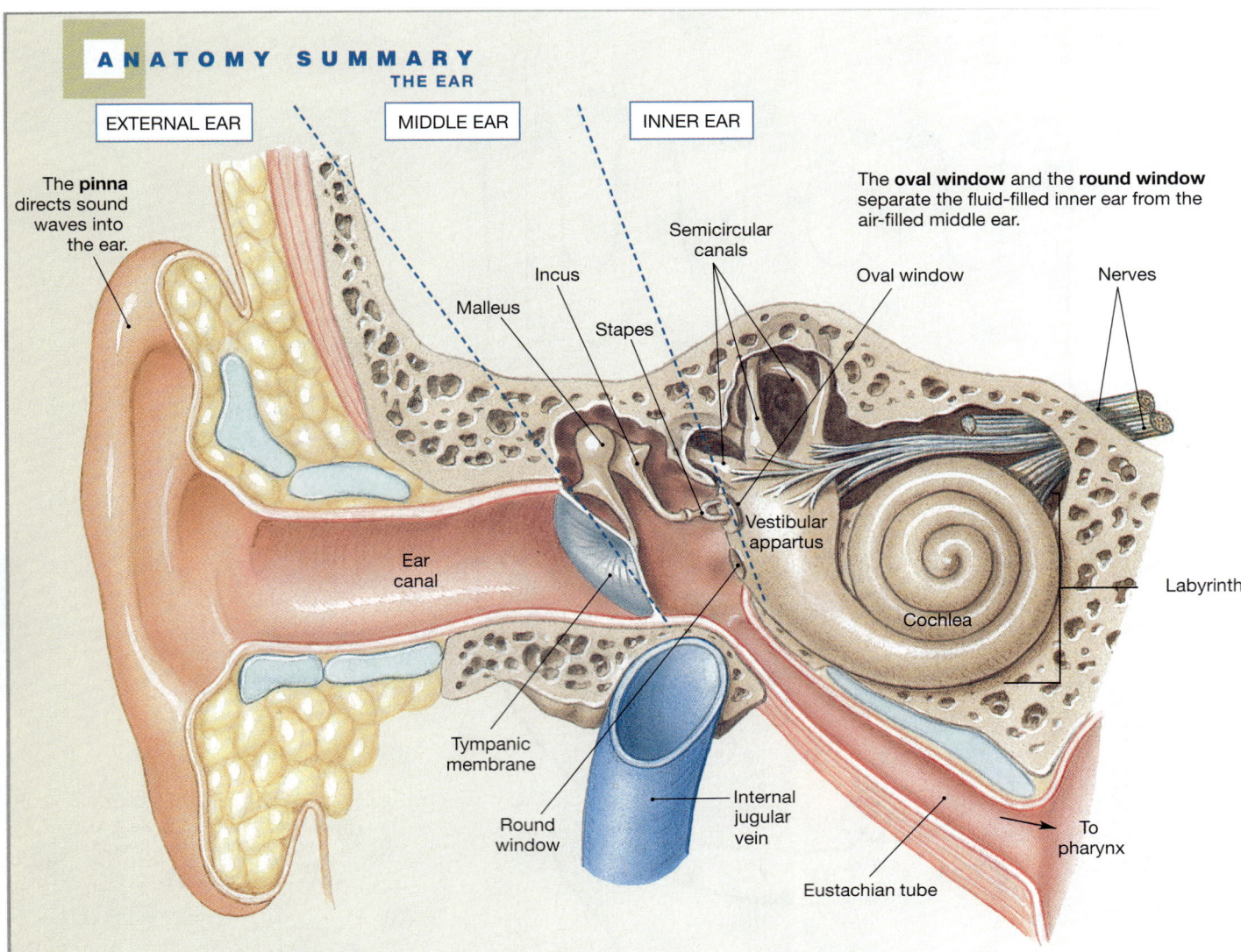

ANATOMY SUMMARY
THE EAR

EXTERNAL EAR   MIDDLE EAR   INNER EAR

The **pinna** directs sound waves into the ear.

The **oval window** and the **round window** separate the fluid-filled inner ear from the air-filled middle ear.

Semicircular canals

Incus

Malleus

Stapes

Oval window

Nerves

Vestibular appartus

Ear canal

Labyrinth

Cochlea

Tympanic membrane

Round window

Internal jugular vein

To pharynx

Eustachian tube

■ FIGURE 10-17

important accessory structure to a sensory system, and it varies in shape and location from species to species, depending on the animals' survival needs. The ear canal is sealed at its internal end by a thin membranous sheet of tissue called the **tympanic membrane**, or *eardrum.*

The tympanic membrane separates the external ear from the *middle ear,* an air-filled cavity that connects with the pharynx through the **eustachian tube.** The eustachian tube is normally collapsed, sealing off the middle ear, but it opens transiently to allow middle ear pressure to equilibrate with atmospheric pressure during chewing, swallowing, and yawning. Colds or other infections that cause swelling can block the eustachian tube and result in fluid buildup in the middle ear. If bacteria are trapped in the middle ear fluid, the ear infection known as *otitis media* [*oto-,* ear + *-itis,* inflammation + *media,* middle] results.

Three small bones of the middle ear conduct sound from the external environment to the inner ear: the **malleus** [ham-

mer], the **incus** [anvil], and the **stapes** [stirrup]. The three bones are connected to one another with the biological equivalent of hinges. One end of the malleus is attached to the tympanic membrane, and the stirrup end of the stapes is attached to a thin membrane that separates the middle ear from the inner ear.

The inner ear consists of two major sensory structures. The *vestibular apparatus* with its *semicircular canals* is the sensory transducer for our sense of equilibrium. It will be described in the following section. The **cochlea** of the inner ear contains sensory receptors for hearing. On external view the cochlea is a membranous tube that lies coiled like a snail shell within a bony cavity called the **labyrinth.** Two membranous disks, the **oval window** (to which the stapes is attached) and the **round window,** separate the liquid-filled cochlea from the air-filled middle ear. Branches of cranial nerve VIII, the *vestibulocochlear nerve,* lead from the inner ear to the brain.

(a) Sound waves are pressure waves with alternating peaks of compressed air and valleys where molecules are farther apart.

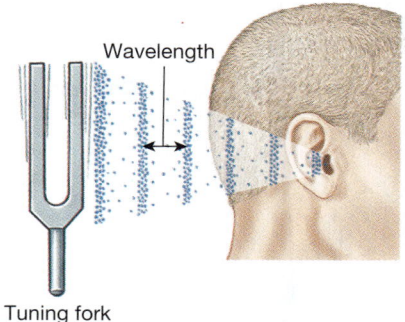

Wavelength

Tuning fork

(b) Sound waves are distinguished by their amplitude, measured in decibels (dB), and frequency, measured in hertz (Hz).

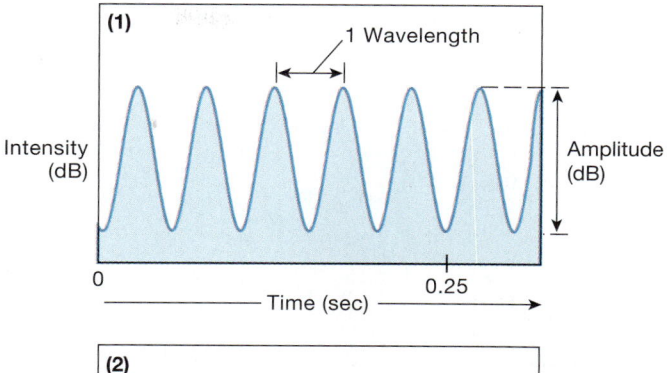

(1)

1 Wavelength

Intensity (dB)

Amplitude (dB)

0

0.25

Time (sec)

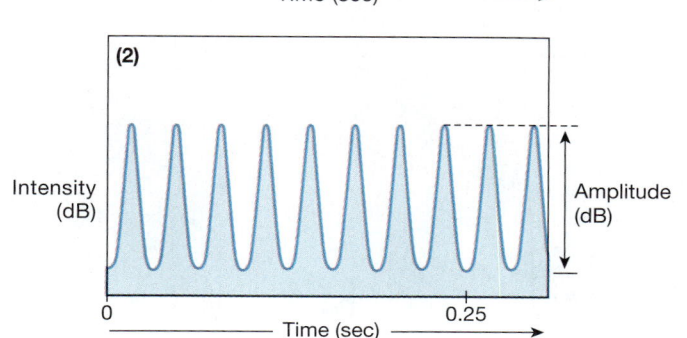

(2)

Intensity (dB)

Amplitude (dB)

0

0.25

Time (sec)

GRAPH QUESTIONS

- What are the frequencies of the sound waves in graphs (1) and (2) in Hz (waves/second)?
- Which set of sound waves would be interpreted as having lower pitch?

■ **FIGURE 10-18**   *Sound waves*

Hearing Is Our Perception of Sound

Hearing is our perception of the energy carried by *sound waves,* which are pressure waves with alternating peaks of compressed air and valleys in which the air molecules are farther apart (Fig. 10-18a ■). The classic question about hearing is, "If a tree falls in the forest with no one to hear, does it make a noise?" The physiological answer is no. The falling tree will emit sound waves, but there is no noise unless someone or something is present to process and perceive the wave energy as sound.

Sound is our interpretation of the frequency, amplitude, and duration of the sound waves that reach our ears. Our brains translate **frequency** of sound waves (the number of wave peaks that pass a given point each second) into the **pitch** of a sound. Low-frequency waves are perceived as low-pitched sounds, such as the rumble of distant thunder. High-frequency waves create high-pitched sounds, such as the screech of fingernails on a blackboard.

Sound wave frequency (Fig. 10-18b) is measured in waves per second, or **hertz (Hz)**. The average human ear can hear sounds over the frequency range of 20–20,000 Hz, with the most acute hearing between 1000–3000 Hz. Our hearing is not as acute as that of many other animals, just as our sense of smell is less acute. Bats listen for ultra-high-frequency sound waves (in the kilohertz range) that bounce off objects in the dark. Elephants and some birds can hear sounds in the infrasound (very low frequency) range.

Loudness is our interpretation of sound intensity and is influenced by the sensitivity of an individual's ear. The intensity

of a sound wave is a function of the wave **amplitude** (Fig. 10-18b). Intensity is measured on a logarithmic scale in units called **decibels (dB)**. Each 10-dB increase represents a 10-fold increase in intensity. Normal conversation has a typical noise level of about 60 dB. Sounds of 80 dB or more can damage the sensitive hearing receptors of the ear, resulting in hearing loss. A typical heavy metal rock concert has noise levels around 120 dB, an intensity that puts listeners in immediate danger of damage to their hearing. The amount of damage depends on the duration and frequency of the noise as well as its intensity.

✓ **CONCEPT CHECK**

16. What is a kilohertz?

Answers: p. 375

Sound Transduction Is a Multistep Process

Hearing is a complex sense that involves four distinct transductions. Energy from sound waves in the air first becomes mechanical vibrations, then fluid waves, then chemical signals, and finally action potentials. This section summarizes the process. We will then look in detail at signal transduction in the cochlea.

The four transduction steps and sound transmission through the ear are shown in Figure 10-19 ■. Sound waves striking the outer ear are directed down the ear canal until they hit the tympanic membrane and are converted into vibrations of the membrane (first transduction). These vibrations are transferred to the malleus, the incus, and the stapes, in that order.

10

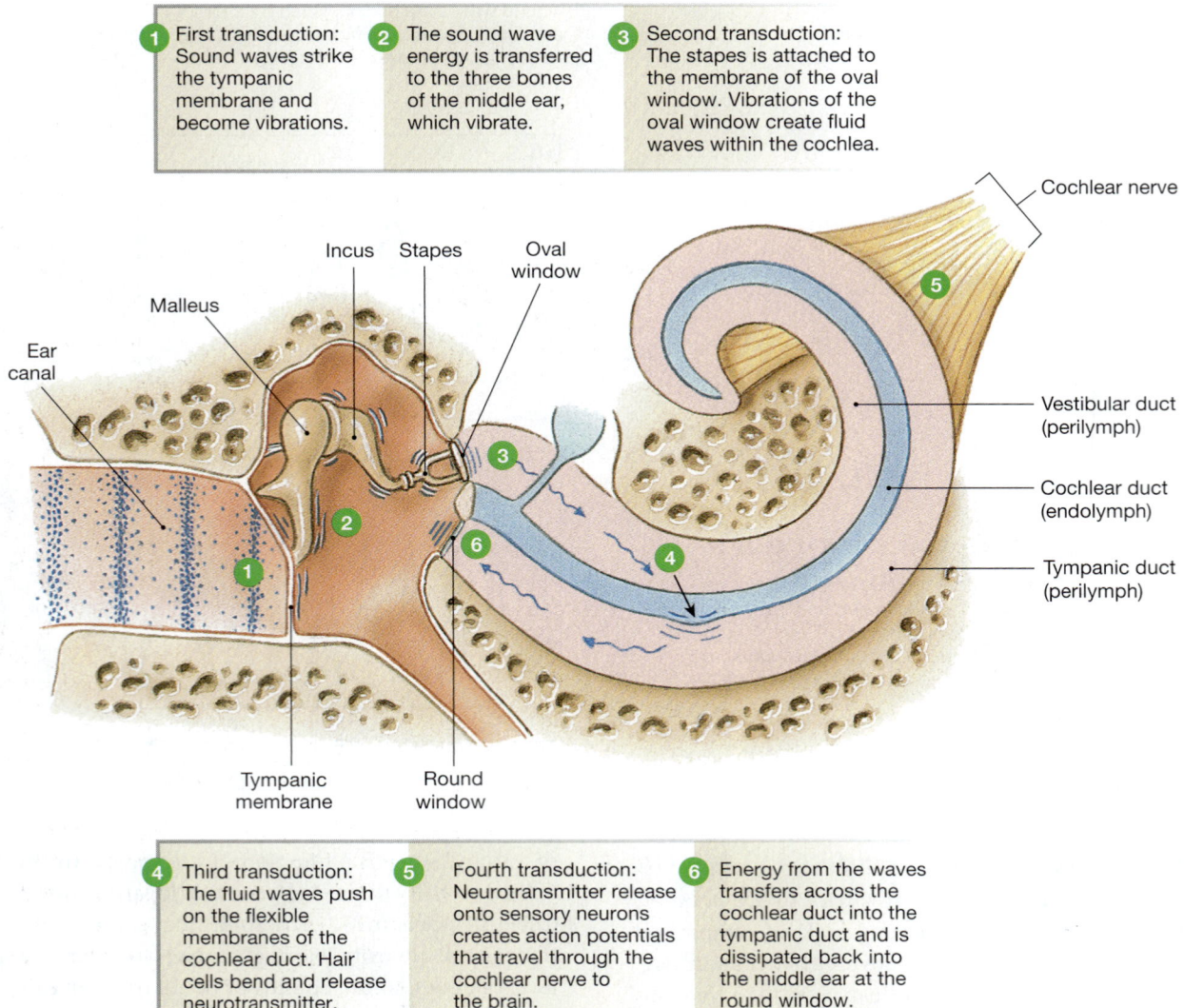

| ① First transduction: Sound waves strike the tympanic membrane and become vibrations. | ② The sound wave energy is transferred to the three bones of the middle ear, which vibrate. | ③ Second transduction: The stapes is attached to the membrane of the oval window. Vibrations of the oval window create fluid waves within the cochlea. |

| ④ Third transduction: The fluid waves push on the flexible membranes of the cochlear duct. Hair cells bend and release neurotransmitter. | ⑤ Fourth transduction: Neurotransmitter release onto sensory neurons creates action potentials that travel through the cochlear nerve to the brain. | ⑥ Energy from the waves transfers across the cochlear duct into the tympanic duct and is dissipated back into the middle ear at the round window. |

■ **FIGURE 10-19** *Sound transmission through the ear*

The arrangement of the three connected middle ear bones creates a "lever" that multiplies the force of the vibration (*amplification*) so that very little sound energy is lost due to friction. If noise levels are so high that there is danger of damage to the inner ear, small muscles in the middle ear can pull on the bones to decrease their movement and thereby dampen sound transmission to some degree.

As the stapes vibrates, it pulls and pushes on the thin tissue of the oval window, to which it is attached. Vibrations at the oval window create waves in the fluid-filled channels of the cochlea (second transduction), but because water is not compressible, the wave energy dissipates back into the air of the middle ear at the round window.

As waves move through the cochlea, they push on the flexible membranes of the *cochlear duct* and bend sensory **hair cells** inside the duct. This movement causes the hair cells to release neurotransmitter onto primary sensory neurons (the chemical signal, third transduction). Neurotransmitter binding to the sensory neurons initiates action potentials (fourth transduction) that send coded information about sound through the *cochlear nerve* to cranial nerve VIII and the brain.

The Cochlea Is Filled with Fluid

The transduction of wave energy into action potentials takes place in the cochlea of the inner ear. Uncoiled, the cochlea can be seen to be composed of three parallel, fluid-filled channels: (1) the **vestibular duct**, or *scala vestibuli* [*scala*, stairway; *vestibulum*, entrance]; (2) the central **cochlear duct**, or *scala media* [*media*, middle]; and (3) the **tympanic duct**, or *scala tympani* [*tympanon*, drum] (Fig. 10-20 ■). The vestibular and tympanic ducts are continuous with each other, and they connect at the tip of the cochlea through a small opening known as the **helicotrema** [*helix*, a spiral + *trema*, hole]. The cochlear duct is a dead-end tube but it connects to the vestibular apparatus through a small opening.

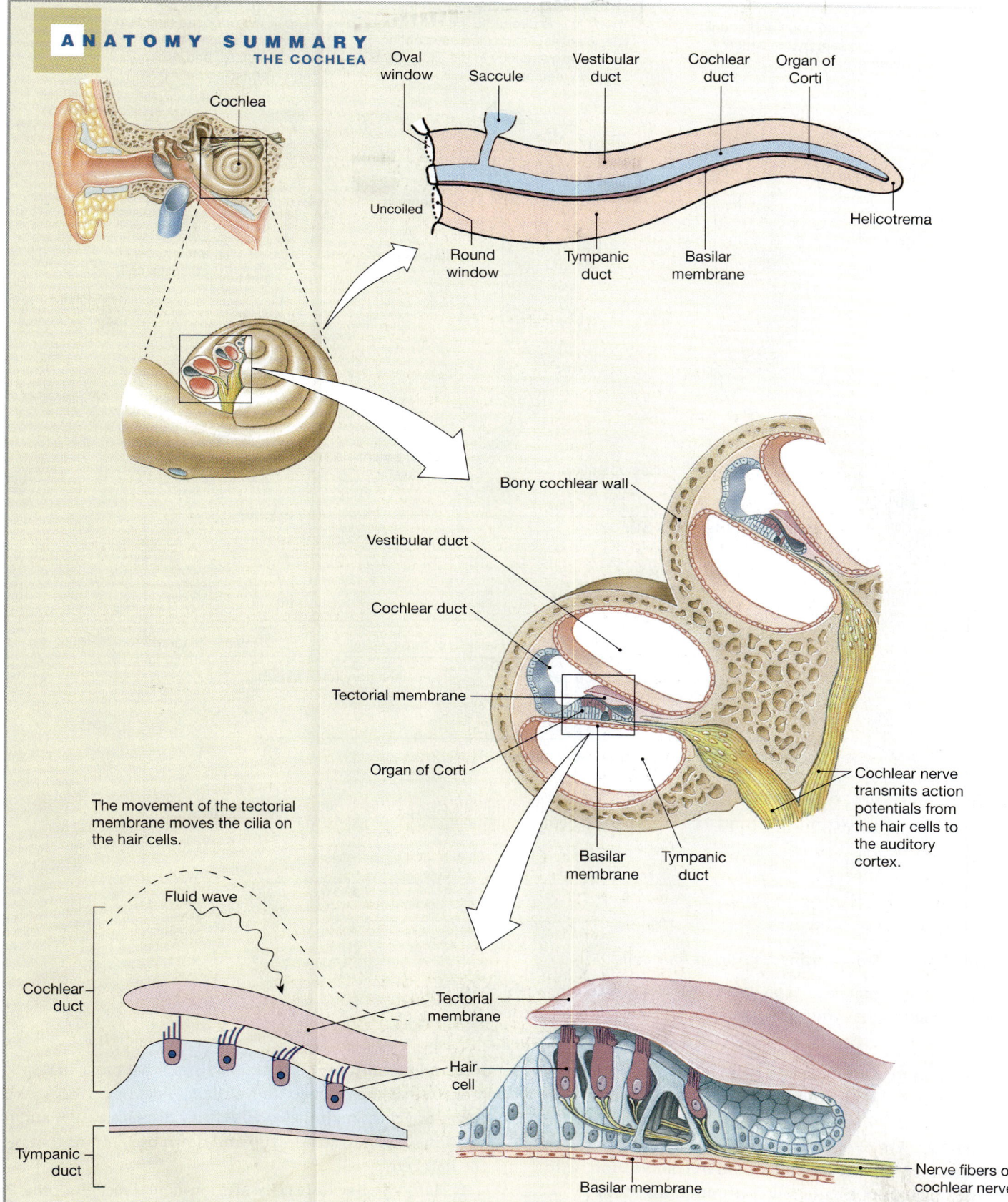

ANATOMY SUMMARY
THE COCHLEA

Cochlea

Oval window

Saccule

Vestibular duct

Cochlear duct

Organ of Corti

Uncoiled

Helicotrema

Round window

Tympanic duct

Basilar membrane

Bony cochlear wall

Vestibular duct

Cochlear duct

Tectorial membrane

Organ of Corti

Basilar membrane

Tympanic duct

Cochlear nerve transmits action potentials from the hair cells to the auditory cortex.

The movement of the tectorial membrane moves the cilia on the hair cells.

Fluid wave

Cochlear duct

Tympanic duct

Tectorial membrane

Hair cell

Basilar membrane

Nerve fibers of cochlear nerve

10

■ FIGURE 10-20

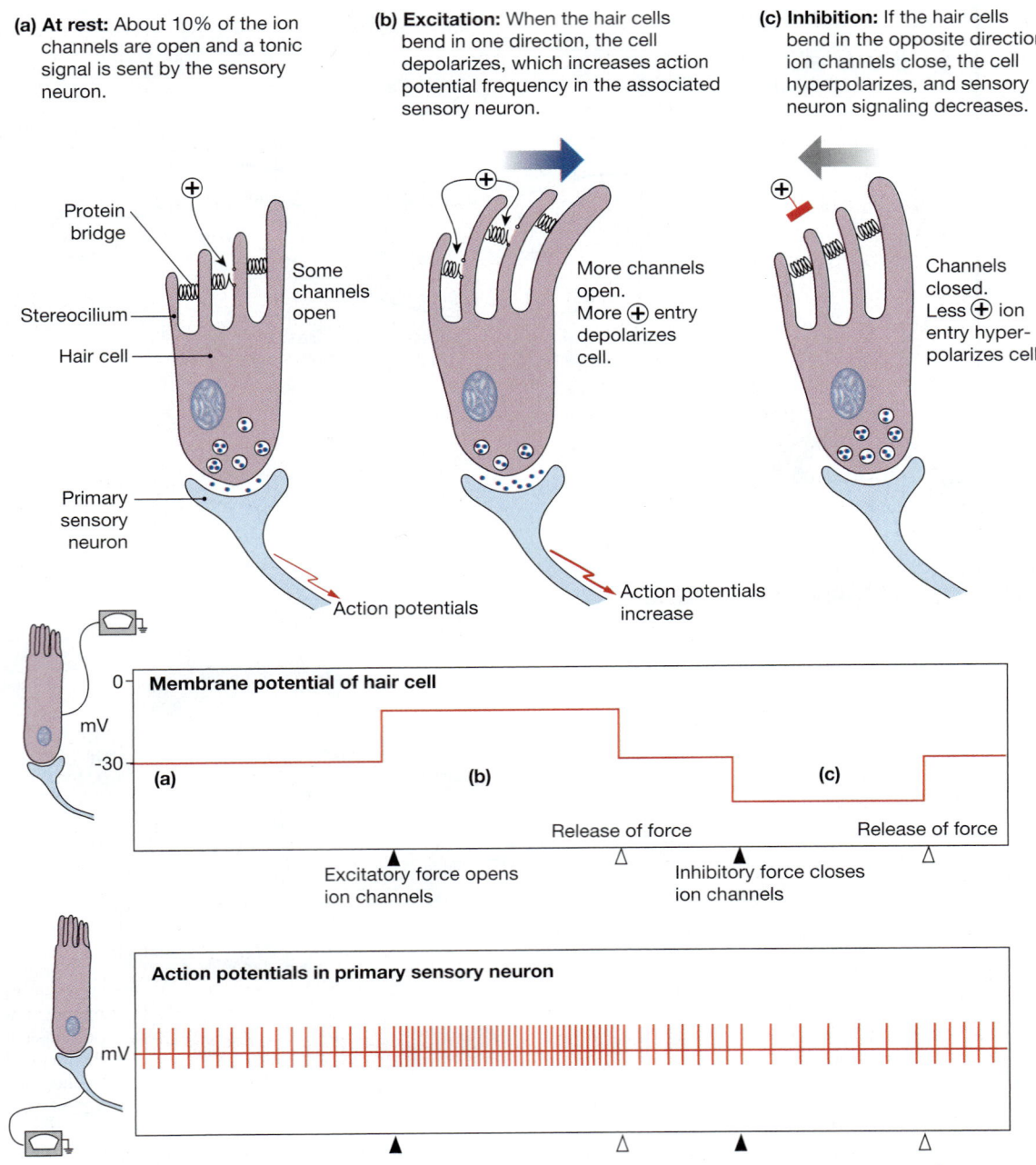

(a) At rest: About 10% of the ion channels are open and a tonic signal is sent by the sensory neuron.

(b) Excitation: When the hair cells bend in one direction, the cell depolarizes, which increases action potential frequency in the associated sensory neuron.

(c) Inhibition: If the hair cells bend in the opposite direction, ion channels close, the cell hyperpolarizes, and sensory neuron signaling decreases.

Protein bridge

Stereocilium

Hair cell

Some channels open

More channels open. More ⊕ entry depolarizes cell.

Channels closed. Less ⊕ ion entry hyper-polarizes cell.

Primary sensory neuron

Action potentials

Action potentials increase

Membrane potential of hair cell

0

mV

-30

(a)

(b)

(c)

Release of force

Release of force

Excitatory force opens ion channels

Inhibitory force closes ion channels

Action potentials in primary sensory neuron

mV

Time

■ **FIGURE 10-21**   *Signal transduction in hair cells*

The stereocilia of hair cells have "trap doors" that close off ion channels. These openings are controlled by the protein-bridge "springs" connecting adjacent cilia.

The fluid in the vestibular and tympanic ducts is similar in ion composition to plasma and is known as **perilymph**. The cochlear duct is filled with **endolymph** secreted by epithelial cells in the duct. Endolymph is unusual because it is more like intracellular fluid than extracellular fluid in composition, with high concentrations of K^+ and low concentrations of Na^+.

The cochlear duct contains the **organ of Corti**, composed of hair cell receptors and support cells. The organ of Corti sits on the **basilar membrane** and is partially covered by the **tectorial membrane**, both flexible tissues that move in response to waves passing through the vestibular duct (Fig. 10-20). As the waves travel through the cochlea, they displace basilar and tectorial membranes, creating up-and-down oscillations that bend the hair cells.

Hair cells, like taste cells, are non-neural receptor cells. The apical surface of each hair cell is modified into 50–100 stiffened cilia known as **stereocilia**, arranged in ascending height (Fig. 10-21a ■). The longest cilium of each hair cell, called a kinocilium [*kinein,* to move] is embedded in the overlying tectorial membrane. If the tectorial membrane moves, the embedded kinocilia do also. This movement is transmitted to the stereocilia of the hair cell.

When hair cells move in response to sound waves, their stereocilia flex, first one way, then the other. The stereocilia are attached to each other by protein bridges. These bridges act like little springs and are connected to "trap doors" that open and close ion channels in the cilia membrane. When the hair cells and cilia are in a neutral position, about 10% of the ion channels are open, and there is a low level of tonic neurotransmitter released onto the primary sensory neuron.

When waves deflect the tectorial membrane so that cilia bend toward the tallest members of a bundle, the bridges pop more channels open, so cations enter the cell, which then depolarizes (Fig. 10-21b). Voltage-gated Ca^{2+} channels open, neurotransmitter release increases, and the sensory neuron increases its firing rate. When the tectorial membrane pushes the cilia away from the tallest members, the springy bridges relax and all the ion channels close. Cation influx slows, the membrane hyperpolarizes, less transmitter is released, and sensory neuron firing decreases (Fig. 10-21c).

The vibration pattern of waves reaching the inner ear is thus transduced into a pattern of action potentials going to the CNS. Because tectorial membrane vibrations reflect the frequency of the incoming sound wave, the hair cells and sensory neurons must be able to respond to sounds of nearly 20,000 waves per second, the highest frequency audible by a human ear.

Sounds Are Processed First in the Cochlea

The auditory system processes sound waves so that they can be discriminated by location, pitch, and loudness. Localization of sound is a complex process that requires sensory input from both ears coupled with sophisticated computation by the brain (see Fig. 10-5). In contrast, the initial processing for pitch and loudness takes place in the cochlea of each ear.

Coding sound for pitch is primarily a function of the basilar membrane. This membrane is stiff and narrow near its attachment between the round and oval windows but widens and becomes more flexible near its distal end (Fig. 10-22a ■). High-frequency waves entering the vestibular duct create maximum displacement of the portion of basilar membrane close to the oval window and consequently are not transmitted very far along the cochlea. Low-frequency waves travel along the length of the basilar membrane and create their maximum displacement near the flexible distal end. This differential response to frequency transforms the temporal aspect of frequency (number of sound waves per second) into spatial coding by location along the basilar membrane (Fig. 10-22b). The spatial coding is preserved in the auditory cortex as neurons project from hair cells along the basilar membrane to corresponding regions in the brain.

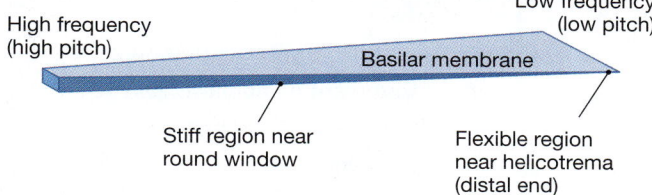

(a) The basilar membrane has variable sensitivity to sound wave frequency along its length.

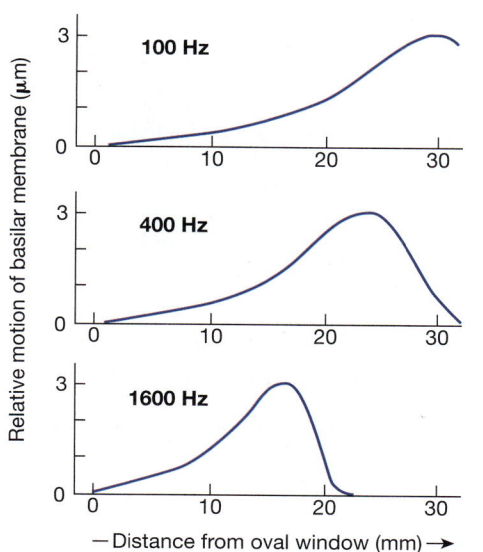

(b) The frequency of sound waves determines the displacement of the basilar membrane. The location of active hair cells creates a code that the brain translates as information about the pitch of sound.

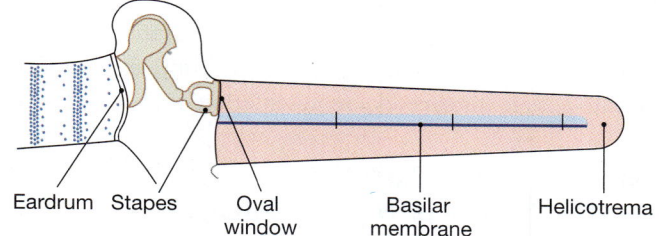

■ **FIGURE 10-22   *Sensory coding for pitch***

Loudness is coded by the ear in the same way that signal strength is coded in somatic receptors. The louder the noise, the more rapidly action potentials fire in the sensory neuron.

Auditory Pathways Project to the Auditory Cortex

Once the cochlea transforms sound waves into electrical signals, primary sensory neurons transfer this information to the brain. The cochlear nerve becomes part of cranial nerve VIII, and its neurons project to nuclei in the medulla oblongata. At that point, sound from each ear is projected to both *ipsilateral* (same side) and *contralateral* (opposite side) nuclei. As a result, each side of the brain gets information from both ears. Ascending tracts from the medulla divide and go in three directions.

Anant reports to the otolaryngologist that his attacks of dizziness strike without warning and last from 10 minutes to an hour. They often cause him to vomit. He also reports that he has a persistent low buzzing sound in one ear, and that he does not seem to hear low tones as well as he could before the attacks started. The buzzing sound (tinnitus) often gets worse during his dizzy attacks.

Question 2:
> *Subjective tinnitus occurs when an abnormality somewhere along the anatomical pathway for hearing causes the brain to perceive a sound that does not exist outside the auditory system. Starting from the ear canal, name the auditory structures in which problems may arise.*

| 328 | 332 | **354** | 357 | 361 | 363 | 371 |

The main auditory pathway synapses in nuclei in the midbrain and thalamus before projecting to the auditory cortex (see Fig. 10-4). Collateral pathways take information to the reticular formation and the cerebellum.

The localization of a sound source is an integrative task that requires simultaneous input from both ears. Unless sound is coming from directly in front of a person, it will not reach both ears at the same time (see Fig. 10-5). The brain records the time differential for sound arriving at the ears and uses a complex computation to create a three-dimensional representation of the sound source.

Hearing Loss May Result from Mechanical or Neural Damage

There are three forms of hearing loss: conductive, central, and sensorineural. In *conductive hearing loss,* sound cannot be transmitted through either the external ear or the middle ear. The causes of conductive hearing loss range from an ear canal plugged with ear wax (*cerumen*), to fluid in the middle ear from an infection, to diseases or trauma that impede vibration of the malleus, incus, or stapes. Correction of conductive hearing loss includes microsurgical techniques in which the bones of the middle ear can be reconstructed.

Central hearing loss results either from damage to the neural pathways between the ear and cerebral cortex or from damage to the cortex itself, as might occur from a stroke. This form of hearing loss is relatively uncommon.

Sensorineural hearing loss arises from damage to the structures of the inner ear, including death of hair cells as a result of loud noises. The loss of hair cells in mammals is currently irreversible. Birds and lower vertebrates, however, are able to regenerate hair

COCHLEAR IMPLANTS

One technique used to treat sensorineural hearing loss is the cochlear implant. The newest generation of cochlear implants has three components: a computerized speech processor, a transmitter, and 8–24 electrodes. The speech processor is a transducer that converts sound into electrical impulses. The smallest speech processors now fit behind the ear like a conventional hearing aid, thanks to microchip technology. The transmitter converts the processor's electrical impulses into radio waves that activate the electrodes, which are surgically placed under the skin and run directly into the cochlea, where they stimulate the sensory nerves. Cochlear implants have been remarkably successful for many profoundly deaf people, allowing them to hear loud noises and modulate their own voices. In the most successful cases, individuals can even use the telephone.

cells to replace those that die. This discovery has researchers exploring strategies to duplicate the process in mammals, including transplantation of neural stem cells and gene therapy to induce nonsensory cells to differentiate into hair cells.

A therapy that replaces hair cells would be an important advance because the incidence of hearing loss in younger people is increasing due to prolonged exposure to rock music and environmental noises. Ninety percent of hearing loss in the elderly—called *presbycusis* [*presbys,* old man + *akoustikos,* able to be heard]—is sensorineural. Currently the primary treatment for sensorineural hearing loss is the use of hearing aids, but amazing results have been obtained with cochlear implants attached to tiny computers (see Biotechnology box).

Hearing is probably our most important social sense. Suicide rates are higher among deaf people than among those who have lost their sight. More than any other sense, hearing connects us to other people and to the world around us.

CONCEPT CHECK

17. Map or diagram the pathways followed by a sound wave entering the ear, starting in the air at the outer ear and ending on the auditory cortex.

18. Why is somatosensory information projected to only one hemisphere of the brain but auditory information is projected to both hemispheres? (*Hint:* see Figs. 10-5 and 10-9.)

19. Would a cochlear implant help a person who suffers from nerve deafness? From conductive hearing loss?

Answers: p. 375

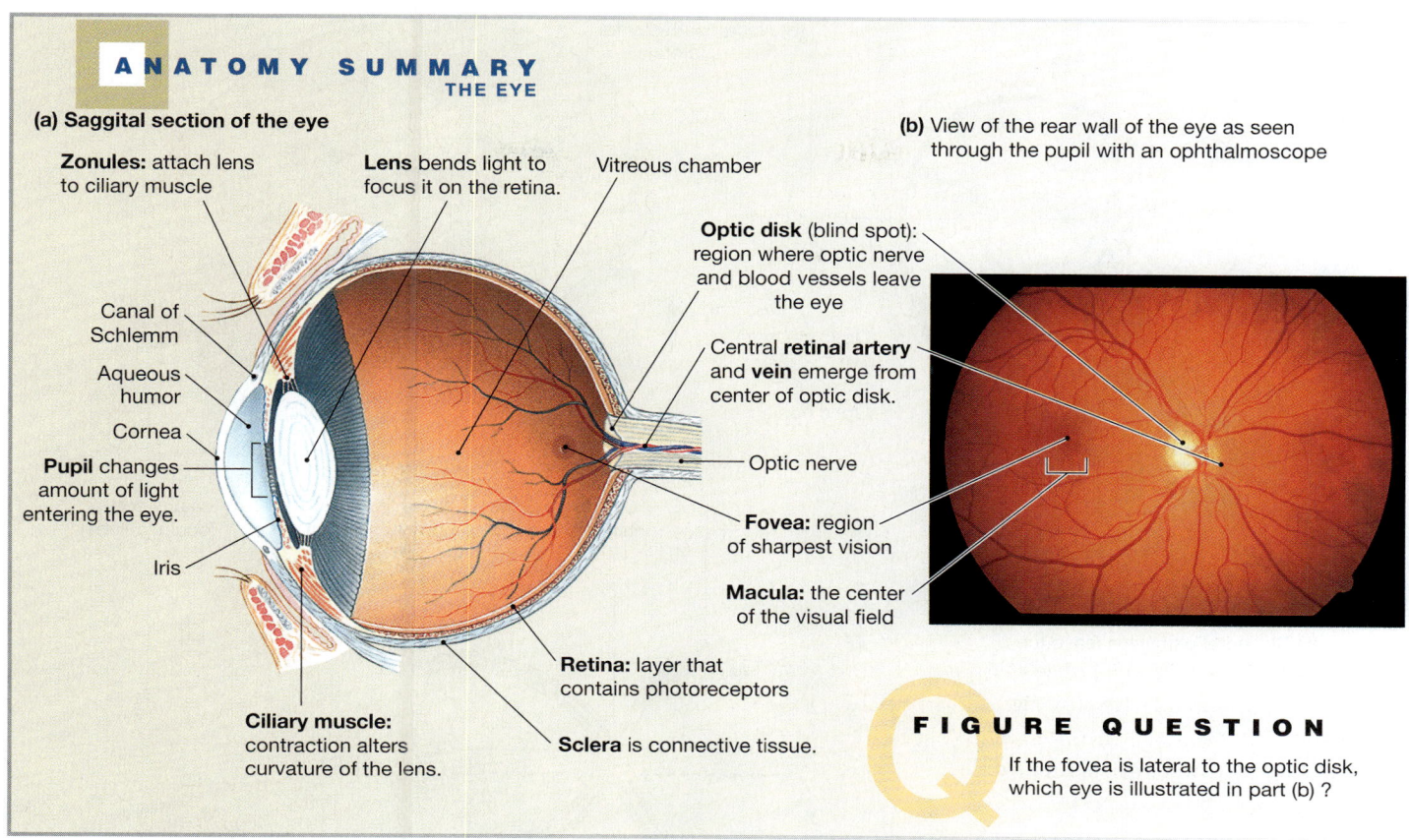

ANATOMY SUMMARY
THE EYE

(a) Saggital section of the eye

Zonules: attach lens to ciliary muscle

Lens bends light to focus it on the retina.

Vitreous chamber

Canal of Schlemm

Aqueous humor

Cornea

Pupil changes amount of light entering the eye.

Iris

Ciliary muscle: contraction alters curvature of the lens.

(b) View of the rear wall of the eye as seen through the pupil with an ophthalmoscope

Optic disk (blind spot): region where optic nerve and blood vessels leave the eye

Central **retinal artery** and **vein** emerge from center of optic disk.

Optic nerve

Fovea: region of sharpest vision

Macula: the center of the visual field

Retina: layer that contains photoreceptors

Sclera is connective tissue.

FIGURE QUESTION

If the fovea is lateral to the optic disk, which eye is illustrated in part (b) ?

10

■ **FIGURE 10-28**

extrinsic eye muscles, skeletal muscles that attach to the outer surface of the eyeball and control eye movements. Cranial nerves III, IV, and VI innervate these muscles. The upper and lower *eyelids* close over the anterior surface of the eye, and the *lacrimal apparatus,* a system of glands and ducts, keeps a continuous flow of tears washing across the exposed surface so that it remains moist and free of debris. Tear secretion is stimulated by parasympathetic neurons from cranial nerve VII.

The eye itself is a hollow sphere divided into two compartments (chambers) separated by a lens (Fig. 10-28a ■). The **lens,** suspended by ligaments called **zonules,** is a transparent disk that focuses light. The anterior chamber in front of the lens is filled with **aqueous humor,** a low-protein plasma-like fluid secreted by the ciliary epithelium supporting the lens. Behind the lens is a much larger chamber, the **vitreous chamber,** filled mostly with the **vitreous body** [*vitrum,* glass], a clear, gelatinous matrix that helps maintain the shape of the eyeball. The outer wall of the eyeball, the **sclera,** is composed of connective tissue.

Light enters the eye through the **cornea,** a transparent disk of tissue on the anterior surface that is a continuation of the sclera. After crossing the aqueous humor, light passes through the opening of the pupil and strikes the lens, which has two convex surfaces. The cornea and lens together bend

incoming light rays so that they focus on the **retina,** the light-sensitive lining of the eye that contains the photoreceptors.

When viewed through the pupil with an ophthalmoscope [*ophthalmos,* eye], the retina is seen to be crisscrossed with small arteries and veins that radiate out from one spot, the **optic disk** (Fig. 10-28b). The optic disk is the location where neurons of the visual pathway form the **optic nerve** (cranial nerve II) and exit the eye. Lateral to the optic disk is a small dark spot, the *fovea.* The fovea and a narrow ring of tissue surrounding it, the *macula,* are the regions of the retina with the most acute vision.

Neural pathways for the eyes are illustrated in Figure 10-29 ■. The optic nerves from the eyes go to the **optic chiasm** in the brain, where some of the fibers cross to the opposite side. After synapsing in the **lateral geniculate body** (*lateral geniculate nucleus*) of the thalamus, the vision neurons of the tract terminate in the occipital lobe at the **visual cortex.** Collateral pathways go from the thalamus to the midbrain, where they synapse with efferent neurons that control the diameter of the pupils.

CONCEPT CHECK

23. What functions does the aqueous humor serve?

Answers: p. 375

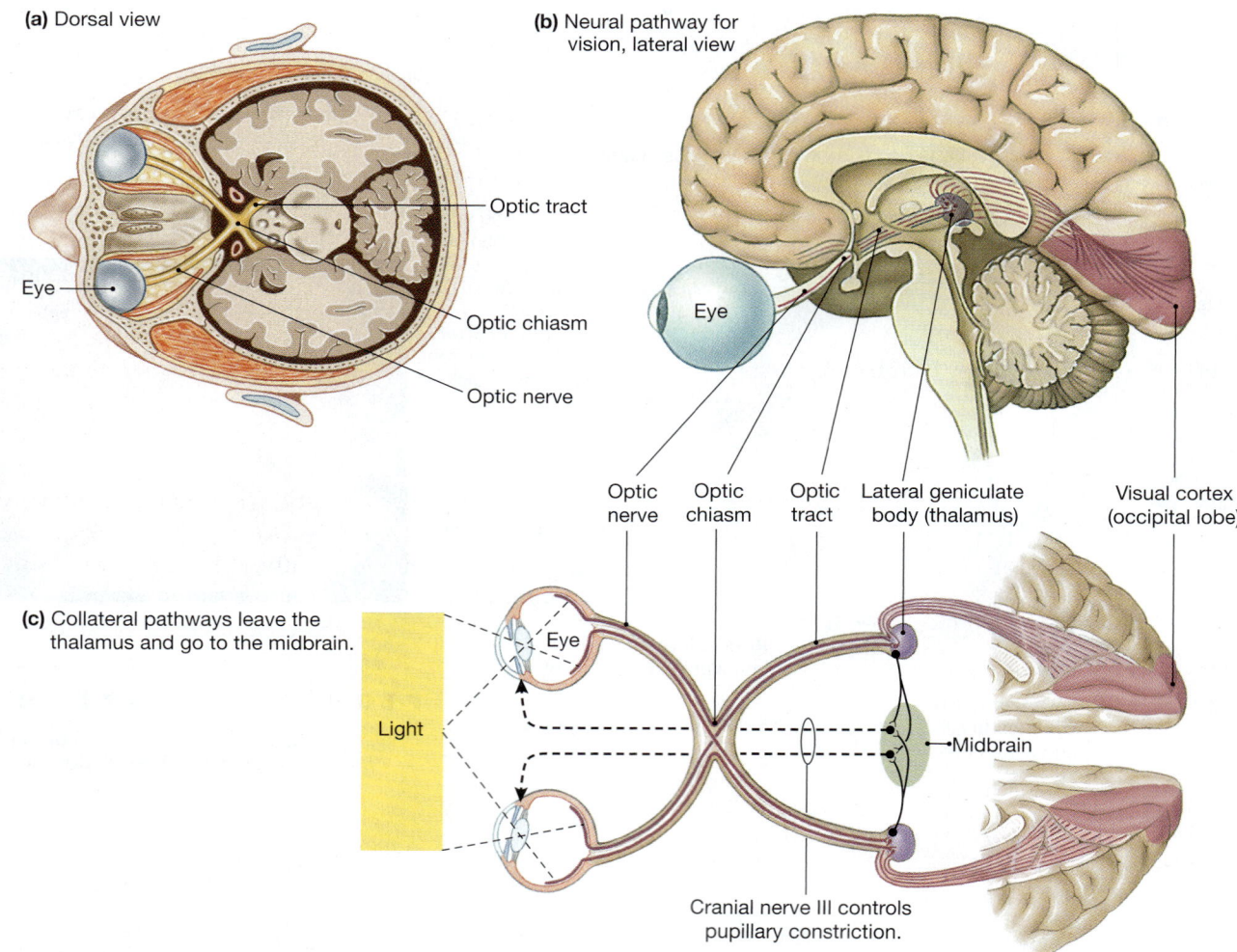

(a) Dorsal view

Eye

Optic tract

Optic chiasm

Optic nerve

(b) Neural pathway for vision, lateral view

Eye

Optic nerve

Optic chiasm

Optic tract

Lateral geniculate body (thalamus)

Visual cortex (occipital lobe)

(c) Collateral pathways leave the thalamus and go to the midbrain.

Light

Eye

Midbrain

Cranial nerve III controls pupillary constriction.

■ **FIGURE 10-29** *Neural pathways for vision and the pupillary reflex*

Collateral pathways synapsing in the midbrain control constriction of the pupils.

Light Enters the Eye Through the Pupil

In the first step of the visual pathway, light from the environment enters the eye. Before it strikes the retina, however, the light is modified two ways. First, the amount of light that reaches photoreceptors is modulated by changes in the size of the pupil. Second, the light is focused by changes in the shape of the lens.

The human eye functions over a 100,000-fold range of light intensity. Most of this ability comes from the sensitivity of the photoreceptors, but the pupils assist by regulating the amount of light that falls on the retina. In bright sunlight, the pupils narrow to about 1.5 mm diameter when a parasympathetic pathway constricts the circular pupillary muscles. In the dark, the opening of the pupil dilates to 8 mm, a 28-fold increase in pupil area. Dilation occurs when radial muscles lying perpendicular to the circular muscles contract under the influence of sympathetic neurons.

Testing **pupillary reflexes** is a standard part of a neurological examination. Light hitting the retina in one eye activates

the reflex. Signals travel through the optic nerve to the thalamus, then to the midbrain, where efferent neurons constrict the pupils in *both* eyes (Fig. 10-29c). This response is known as the **consensual reflex** and is mediated by parasympathetic fibers running through cranial nerve III.

CONCEPT CHECK

24. Use the neural pathways in Figure 10-29c to answer the following questions.

    (a) Why does shining light into one eye cause pupillary constriction in both eyes?

    (b) If you shine a light in the left eye and get pupillary constriction in the right eye but not in the left eye, what can you conclude about the afferent path from the left eye to the brain? About the efferent pathways to the pupils?

25. Parasympathetic fibers constrict the pupils, and sympathetic fibers dilate them. This is an example of what kind of control?

Answers: p. 375

In addition to regulating the amount of light that hits the retina, the pupils create what is known as **depth of field**. A

simple example comes from photography. Imagine a picture of a puppy sitting in the foreground amid a field of wildflowers. If only the puppy and the flowers immediately around her are in focus, the picture is said to have a shallow depth of field. If instead the puppy and the wildflowers all the way back to the horizon are in focus, the picture has full depth of field. Depth of field is created by constricting the pupil (or the diaphragm on a camera) so that only a narrow beam of light enters the eye. In this way, a greater depth of the image is focused on the retina.

The Lens Focuses Light on the Retina

When light enters the eye, it passes through the cornea and lens prior to striking the retina. When light rays pass from air into a medium of different density, such as glass or water, they bend, or **refract**. Light entering the eye is refracted twice: about

two-thirds of the refraction occurs when light passes through the cornea, and the remaining one-third occurs when it passes through the lens. To simplify this discussion, we will consider only the refraction that occurs as light passes through the lens because the lens is capable of changing its shape to focus light.

When light passes from one medium into another, the angle of refraction (how much the light rays bend) is influenced by two factors: (1) the difference in density of the two media and (2) the angle at which the light rays meet the surface of the medium into which it is passing. For light passing through the lens of the eye, we will assume that the density of the lens is the same as the density of the air so that this factor can be ignored. The angle at which light meets the face of the lens depends on the curvature of the lens surface and the direction of the light beam.

Imagine parallel light rays striking the surface of a transparent lens. If the lens surface is perpendicular to the rays, the light will pass through without bending. If the surface is not perpendicular, however, the light rays will bend. Parallel light rays striking a *concave* lens, such as that shown in Figure 10-30a ■, will be refracted into a wider beam. Parallel rays striking a *convex* lens will bend inward and focus to a point—*convex* lenses *converge* light waves (Fig. 10-30b). You can demonstrate the properties of a convex lens by using a magnifying glass to focus sunlight onto a piece of paper or other surface.

When parallel light rays pass through a convex lens, the single point where the rays converge is called the **focal point** (Fig. 10-30b). The distance from the center of a lens to its focal point is known as the **focal length** (or *focal distance*). For any given lens, the focal length and focal point are fixed. To change the focal length, the shape of the lens must change.

When light from an object passes through the lens of the eye, the focal point must fall precisely on the retina if the object is to be seen in focus. In Figure 10-31a ■, parallel light rays strike a lens whose surface is relatively flat. For this lens, the focal point falls on the retina. The object is therefore in focus. For the

10

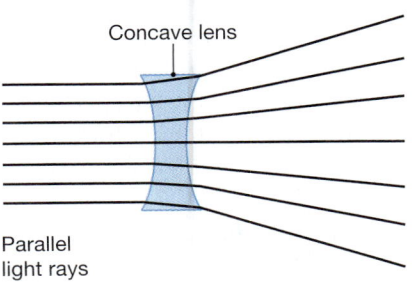

(a) A **concave lens** scatters light rays.

Concave lens

Parallel
light rays

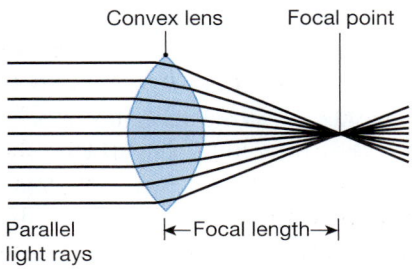

(b) A **convex lens** causes light rays to converge.

Convex lens   Focal point

Parallel
light rays |←Focal length→|

The focal length is the distance from
the center of the lens to the focal point.

■ **FIGURE 10-30**   *Refraction of light*

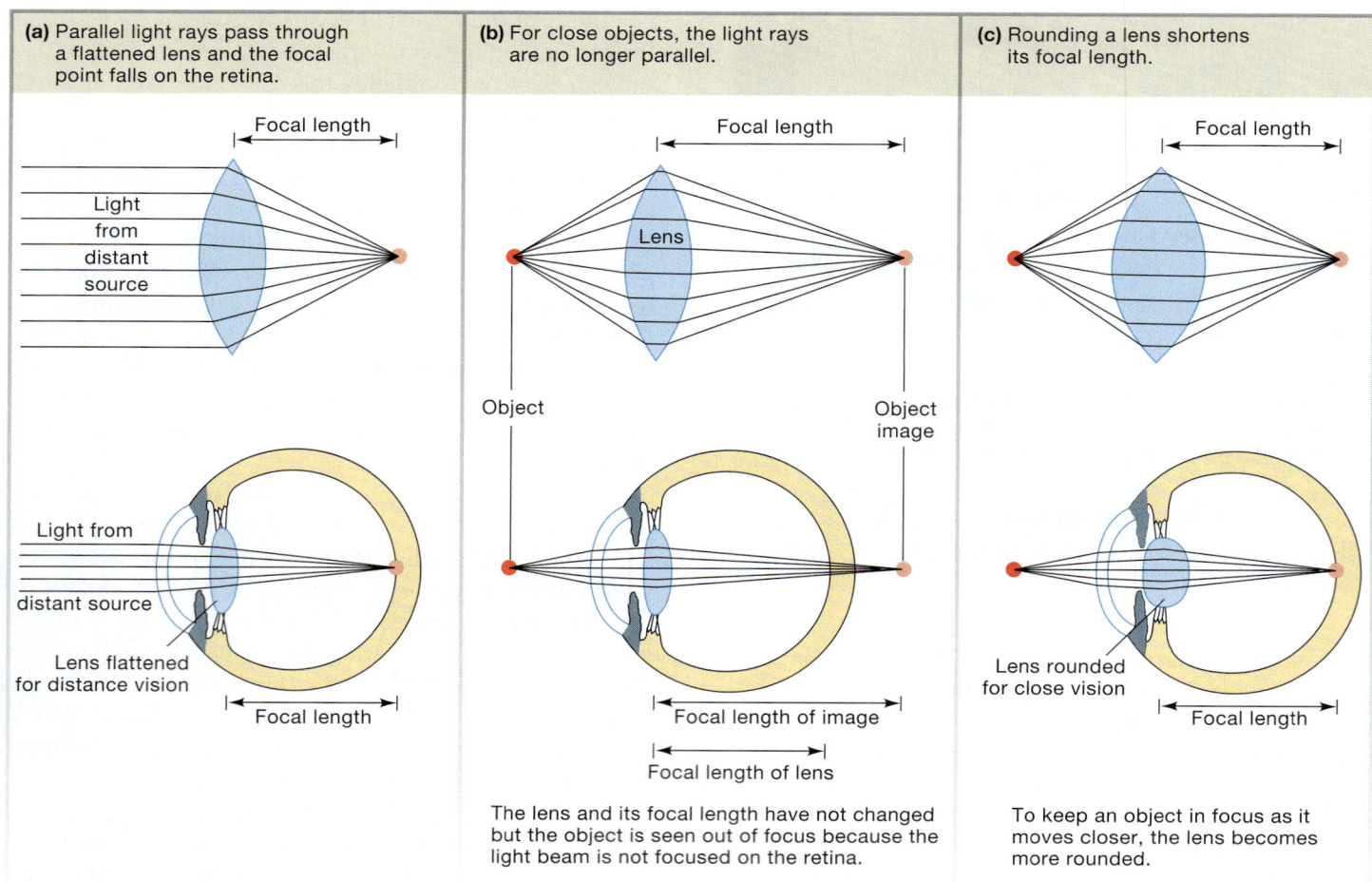

(a) Parallel light rays pass through a flattened lens and the focal point falls on the retina.

Focal length

Light from distant source

Light from distant source

Lens flattened for distance vision

Focal length

(b) For close objects, the light rays are no longer parallel.

Focal length

Lens

Object

Object image

Focal length of image

Focal length of lens

The lens and its focal length have not changed but the object is seen out of focus because the light beam is not focused on the retina.

(c) Rounding a lens shortens its focal length.

Focal length

Lens rounded for close vision

Focal length

To keep an object in focus as it moves closer, the lens becomes more rounded.

■ FIGURE 10-31   *Optics*

human eye, any object that is 20 feet or more from the eye will create parallel light rays.

What happens, though, when an object is closer than 20 feet to the lens? In that case, the light rays from the object are not parallel and strike the lens at an oblique angle (Fig. 10-31b). The light rays falling on the retina no longer converge to a point, and the object appears fuzzy and out of focus.

To keep a near object in focus, the lens must be rounded in order to increase the angle of refraction (Fig. 10-31c). Making the lens more convex causes the light rays to converge on the retina instead of behind it, and the object comes into focus. The process by which the eye adjusts the shape of the lens to keep objects in focus is known as **accommodation**. If an object is so close to the lens that it falls within the focal length, no amount of accommodation can bring the object into focus. The closest distance at which you can focus on an object is known as the **near point of accommodation**.

You can demonstrate the accommodation reflex easily by closing one eye and holding your hand up about 8 inches in front of your open eye, fingers spread apart. Focus your eye on some object in the distance that is visible between your fingers. Notice that when you do so, your fingers remain visible

but out of focus. Your lens is flattened for distance vision, so the focal point for near objects falls behind the retina. Those objects appear out of focus. Now shift your gaze to your fingers and notice that they come into focus. The light rays coming off your fingers have not changed their angle, but your lens has become more rounded, and the light rays now converge on the retina.

How can the lens, which is clear and does not have any muscle fibers in it, change shape? The answer lies in the **ciliary muscle**, a ring of smooth muscle that surrounds the lens and is attached to it by the inelastic ligaments called zonules (Fig. 10-32 ■). If no tension is placed on the lens by the ligaments, the lens assumes its natural rounded shape because of the elasticity of its capsule. If the ligaments pull on the lens, it flattens out and assumes the shape required for distance vision.

Tension on the ligaments is controlled by the ciliary muscle. When this circular muscle contracts, the muscle ring gets smaller, releasing tension on the ligaments so that the lens rounds. When the ciliary muscle is relaxed, the ring is more open and the lens is pulled into a flatter shape.

(a) The lens is attached to the ciliary muscle by inelastic ligaments (zonules).

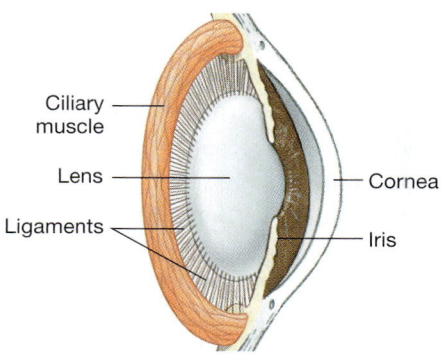

- Ciliary muscle
- Lens
- Ligaments
- Cornea
- Iris

(b) When ciliary muscle is relaxed, the ligaments pull on and flatten the lens.

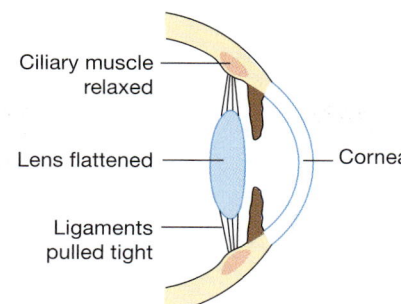

- Ciliary muscle relaxed
- Lens flattened
- Ligaments pulled tight
- Cornea

(c) When ciliary muscle contracts, it releases tension on the ligaments and the lens becomes more rounded.

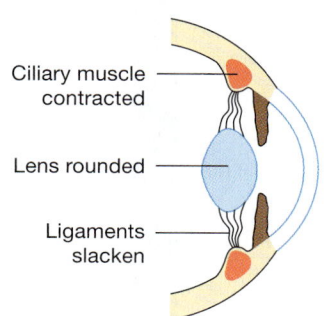

- Ciliary muscle contracted
- Lens rounded
- Ligaments slacken

■ **FIGURE 10-32**   *Mechanism of accommodation*

Young people can focus on items as close as 8 cm, but the accommodation reflex diminishes from the age of 10 on. By age 40, accommodation is only about half of what it was at age 10, and by age 60, many people lose the reflex completely because the lens has lost flexibility. The loss of accommodation, **presbyopia**, is the reason most people begin to wear reading glasses in their 40s.

Two other common vision problems, near-sightedness (**myopia**) and far-sightedness (**hyperopia**), occur when the focal point falls either in front of the retina or behind the retina, respectively (Fig. 10-33 ■). These conditions can be corrected by placement of a lens having the appropriate curvature in front of the eye to change the focal length. A third common vision problem, **astigmatism**, is usually caused by a cornea that is not a perfectly shaped dome, resulting in distorted images.

CONCEPT CHECK

26. If a person's cornea, which helps focus light, is more rounded than normal (has a greater curvature), is this person more likely to be hyperopic or myopic? (*Hint:* see Figure 10-33.)

Answers: p. 375

(a) Hyperopia, or far-sightedness, occurs when the focal point falls behind the retina.

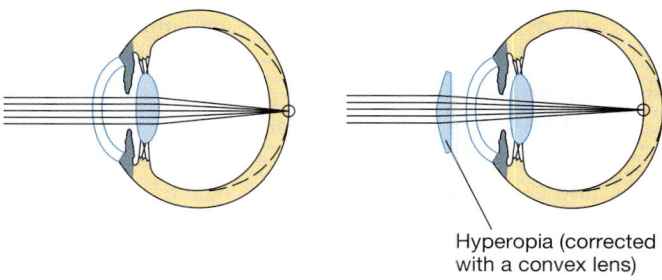

Hyperopia (corrected with a convex lens)

(b) Myopia, or near-sightedness, occurs when the focal point falls in front of the retina.

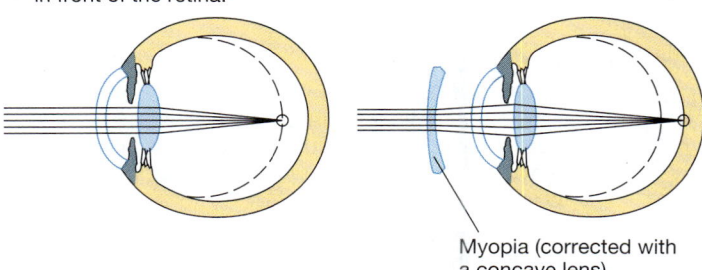

Myopia (corrected with a concave lens)

FIGURE QUESTION

Explain (1) what the corrective lenses do to the refraction of light, and (2) why each lens type corrects the visual defect.

■ **FIGURE 10-33**   *Common visual defects*

RUNNING PROBLEM

Anant's condition does not improve with the low-salt diet and diuretics, and he continues to suffer from disabling attacks of vertigo with vomiting. In severe cases of Ménière's disease, surgery is sometimes performed as a last resort. In one surgical procedure for the disease, the vestibular nerve is severed. This surgery is difficult to perform, as the vestibular nerve lies near many other important nerves, including facial nerves and the auditory nerve. Patients who undergo this procedure are advised that the surgery can result in deafness if the cochlear nerve is inadvertently severed.

Question 6:
Why would severing the vestibular nerve alleviate Ménière's disease?

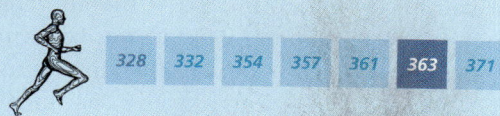

328   332   354   357   361   363   371

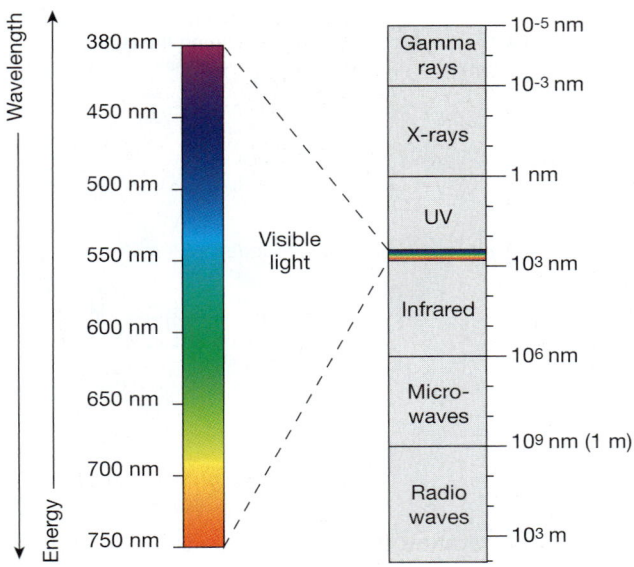

■ **FIGURE 10-34**   *The electromagnetic spectrum*

Phototransduction Occurs at the Retina

The transduction of light energy into action potentials takes place when light hits the retina, the sensory organ of the eye. Actually, only a small portion of the light that enters our eyes is perceived by our brain. **Visible light** is that electromagnetic energy whose waves have a frequency of $4.0–7.5 \times 10^{14}$ waves per second (Hz) and a wavelength of 400–750 nanometers (nm).

The full electromagnetic spectrum extends from high-energy, very-short-wavelength waves such as X-rays and gamma rays to low-energy, lower-frequency microwaves and radio waves (Fig. 10-34 ■). Ultraviolet and infrared light border the ends of the visible light spectrum. Although our eyes are unable to perceive ultraviolet and infrared wavelengths, the eyes of some other animals can see these wavelengths. For example, bees use ultraviolet "runways" on flowers to guide them to pollen and nectar.

In the human eye, **photoreceptors** in the retina transduce light energy into electrical signals. The retina develops from the same embryonic tissue as the brain, and (as in the cortex of the brain) neurons in the retina are organized into layers (Fig. 10-35d ■). Backing the photosensitive portion of the human retina is a dark **pigment epithelium** layer. Its function is to absorb any light rays that escape the photoreceptors, preventing distracting light from reflecting inside the eye and distorting the visual image. The black color of these epithelial cells comes from granules of the pigment *melanin.*

The layers of retinal neurons are in reverse order from what you might expect. Instead of the photoreceptors facing the pupil, where photons of light will strike them first, they are at the back, with their photosensitive tips against the pigment epithelium. Most light entering the eye must pass through

several relatively transparent layers of neurons before striking the photoreceptors.

One exception to this pattern occurs in a small region of the retina known as the **fovea** [pit]. In this area, photoreceptors receive light directly because the intervening neurons are pushed off to the side (Fig. 10-35c). The fovea is also free of blood vessels that would block light reception. As noted earlier, the fovea and the **macula** immediately surrounding it are the areas of most acute vision, and they form the center of the visual field.

The fovea is the point on which light focuses when you look at an object. For example, look at Figure 10-36 ■. The eye is focused on the green-yellow border of the color bar. Light from that section of the visual field falls on the fovea and is in sharp focus. Notice that the image falling on the retina is upside down. Subsequent visual processing by the brain reverses the image again so that we perceive it in the correct orientation.

Sensory information about light passes from the photoreceptors to **bipolar neurons**, then to a layer of **ganglion cells**. The axons of ganglion cells form the optic nerve, which leaves the eye at the optic disk. Because the optic disk has no photoreceptors, images projected onto this region cannot be seen, creating what is called the eye's **blind spot.**

One hallmark of signal processing in the retina is convergence [🔁 p. 278], in which multiple neurons synapse onto a single postsynaptic cell (Fig. 10-35e). As many as 15 to 45 photoreceptors may synapse onto one bipolar neuron. Multiple bipolar neurons in turn innervate a single ganglion cell, so that the information from hundreds of millions of photoreceptors is condensed down to a mere 1 million axons leaving the eye in each optic nerve. Convergence is minimal in the fovea, where some photoreceptors have a 1:1 relationship with their bipolar neurons, and greatest at the outer edges of the retina.

Signal processing in the retina is modulated by input from two additional sets of cells (Fig. 10-35d). **Horizontal cells** synapse with photoreceptors and bipolar cells. **Amacrine cells** modulate information flowing between bipolar cells and ganglion cells.

✓ **CONCEPT CHECK**

27. Animals that see well in very low light lack a pigment epithelium and instead have a layer called the *tapetum lucidum* behind the retina. What property might this layer have that would enhance vision in low light?

28. How is the difference in visual acuity between the fovea and the edge of the visual field similar to the difference in touch discrimination between the fingertips and the skin of the arm?

29. Macular degeneration is the leading cause of blindness in Americans over the age of 55. Impaired function in the macula of the retina causes vision loss in which part of the visual field?

Answers: p. 375

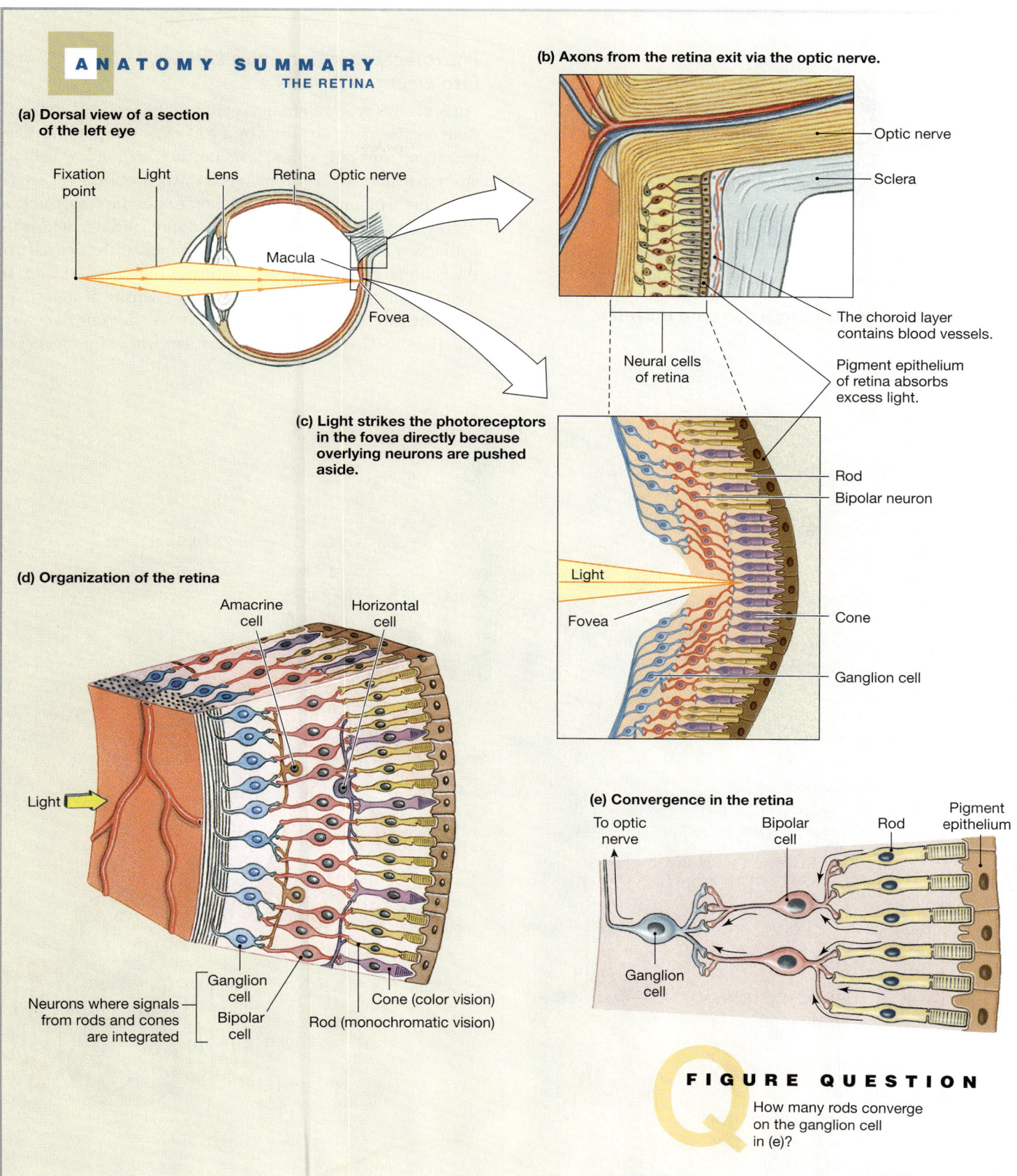

ANATOMY SUMMARY
THE RETINA

(a) Dorsal view of a section of the left eye

Fixation point
Light
Lens
Retina
Optic nerve
Macula
Fovea

(b) Axons from the retina exit via the optic nerve.

Optic nerve
Sclera
The choroid layer contains blood vessels.
Pigment epithelium of retina absorbs excess light.
Neural cells of retina

(c) Light strikes the photoreceptors in the fovea directly because overlying neurons are pushed aside.

Light
Fovea
Rod
Bipolar neuron
Cone
Ganglion cell

(d) Organization of the retina

Amacrine cell
Horizontal cell
Light
Neurons where signals from rods and cones are integrated
Ganglion cell
Bipolar cell
Cone (color vision)
Rod (monochromatic vision)

(e) Convergence in the retina

To optic nerve
Bipolar cell
Rod
Pigment epithelium
Ganglion cell

FIGURE QUESTION

How many rods converge on the ganglion cell in (e)?

10

■ **FIGURE 10-35**

Drawing of photoreceptors in the fovea adapted from E. R. Kandel et al., *Principles of Neural Science,* 3rd edition (New York: McGraw Hill, 2000).

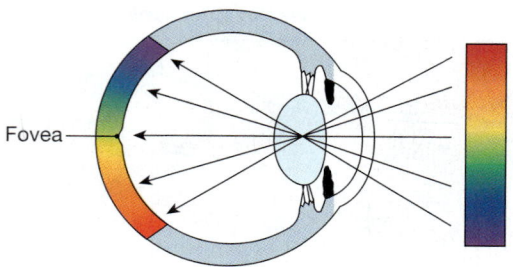

The projected image is upside down on the retina.
Visual processing in the brain reverses the image.

■ **FIGURE 10-36** *Image projection onto the retina*

Photoreceptors Transduce Light into Electrical Signals

There are two types of photoreceptors in the eye: rods and cones. **Rods** are the more numerous by a 20:1 ratio, except in the fovea, which contains only cones. Rods function well in low light and are responsible for nighttime vision, when objects are seen in black and white rather than in color. **Cones** are responsible for *high-acuity* vision and color vision during the daytime, when light levels are higher. *Acuity* means keenness and is derived from the Latin *acuere,* meaning "to sharpen." The fovea, which is the region of sharpest vision, has a very high density of cones.

The two types of photoreceptors have the same basic structure (Fig. 10-37 ■): (1) an outer segment whose tip touches the

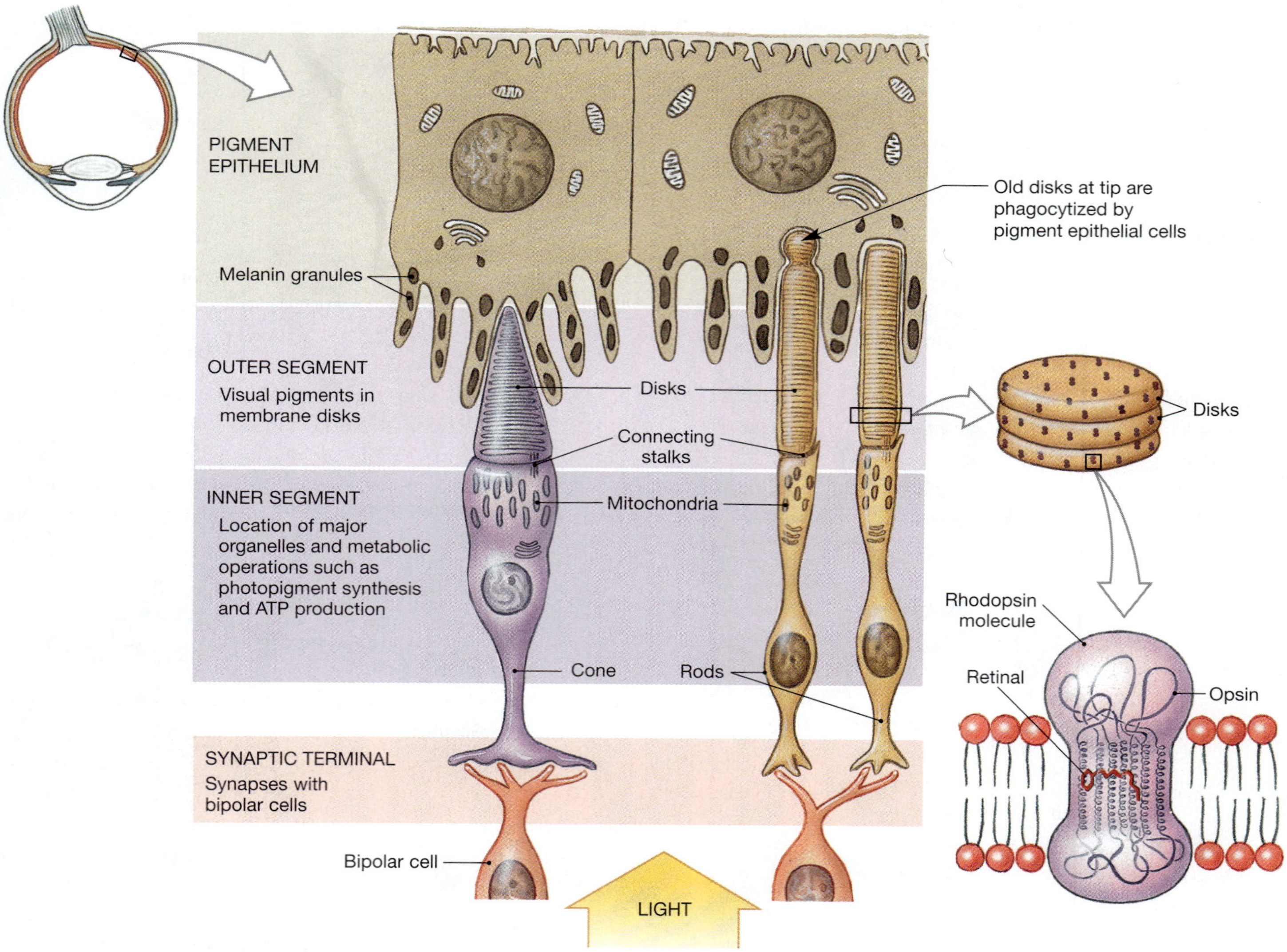

Fovea

PIGMENT EPITHELIUM

Melanin granules

Old disks at tip are phagocytized by pigment epithelial cells

OUTER SEGMENT
Visual pigments in membrane disks

Disks

Connecting stalks

Disks

INNER SEGMENT
Location of major organelles and metabolic operations such as photopigment synthesis and ATP production

Mitochondria

Cone

Rods

Rhodopsin molecule

Retinal

Opsin

SYNAPTIC TERMINAL
Synapses with bipolar cells

Bipolar cell

LIGHT

10

■ **FIGURE 10-37** *Photoreceptors: rods and cones*

Light transduction takes place in the outer segment of the photoreceptor. Changes in photoreceptor membrane potential alter neurotransmitter release onto bipolar cells. The dark pigment epithelium absorbs extra light and prevents that light from reflecting back and distorting vision.

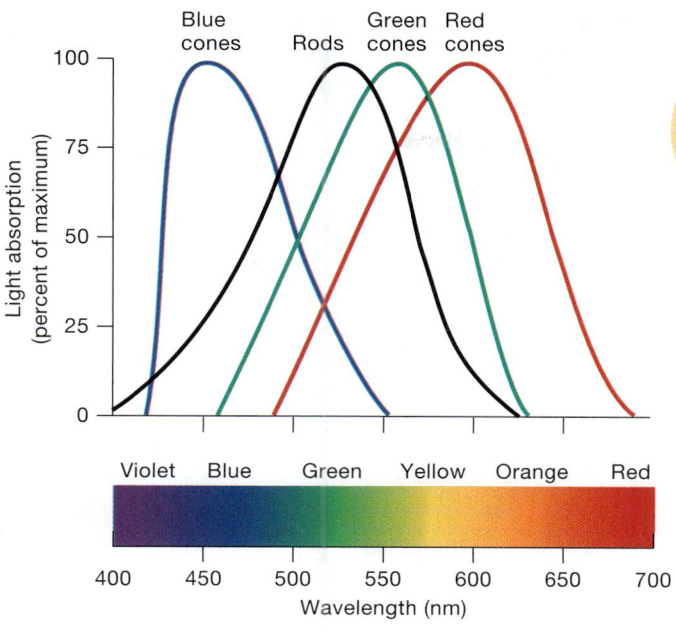

GRAPH QUESTIONS
- Which color pigment absorbs light over the broadest spectrum of wavelengths?
- Over the narrowest?
- Which color pigment absorbs the most light at 500 nm?

■ **FIGURE 10-38** *Light absorption of visual pigments*

There are three types of cone pigment, each with a characteristic light absorption spectrum.

pigment epithelium of the retina, (2) an inner segment that contains the cell nucleus and organelles for ATP and protein synthesis, and (3) a base segment that synapses with the bipolar cells. In the outer segment, the cell membrane has deep folds that form disklike layers, like candy mints stacked in a wrapper. Toward the tip of the outer segments in rods, these layers actually separate from the cell membrane and form free-floating disks. In the cones, they stay attached.

Light-sensitive **visual pigments** are bound to cell membranes in the outer segments of photoreceptors. These pigments are the transducers that convert light energy into a change in membrane potential. Rods have one type of visual pigment, **rhodopsin**. Cones have three different pigments that are closely related to rhodopsin.

Cones are responsible for color vision and contain visual pigments that are excited by different wavelengths of light. White light is a combination of colors, as you see when you separate white light by passing it through a prism. The eye contains cones for red, green, and blue light. Each cone type is stimulated by a range of light wavelengths but is most sensitive to a particular wavelength (Fig. 10-38 ■). Red, green, and blue are the three primary colors for light, just as red, blue, and yellow are the three primary colors for paints.

The color of any object we are looking at depends on the wavelengths of light reflected by the object. Green leaves reflect green light, and bananas reflect yellow light. White objects reflect most wavelengths. Black objects absorb most wavelengths, which is one reason they heat up in sunlight while white objects stay cool.

Our brain recognizes the color of an object by interpreting the combination of signals coming to it from the three different color cones. This seems like a simple process. However, the details of color vision are still not well understood, and there is some controversy about how color is processed in the cerebral cortex. **Color-blindness** is a condition in which a person inherits a defect in one or more of the three types of cones and has difficulty distinguishing certain colors. Probably the best-known form of color-blindness is red-green, in which people cannot tell red and green apart.

✓ **CONCEPT CHECK**
30. Why is our vision in the dark in black and white rather than in color?

Answers: p. 375

Phototransduction The process of phototransduction is similar for rhodopsin (in rods) and the three color pigments (in cones). Rhodopsin is composed of two molecules: **opsin**, a protein embedded in the membrane of the rod disks, and **retinal**, a vitamin A derivative that is the light-absorbing portion of the pigment (see Fig. 10-37). In the absence of light, retinal binds snugly into a binding site on the opsin (Fig. 10-39a ■). When activated by as little as one photon of light, retinal changes shape to a new configuration. The activated retinal no longer binds to opsin and is released from the pigment in the process known as **bleaching** (Fig. 10-39b).

How does rhodopsin bleaching lead to action potentials traveling through the optical pathway? In order to understand the link, we must look at other properties of the rods. As you learned in Chapters 5 and 8, electrical signals in cells occur as a

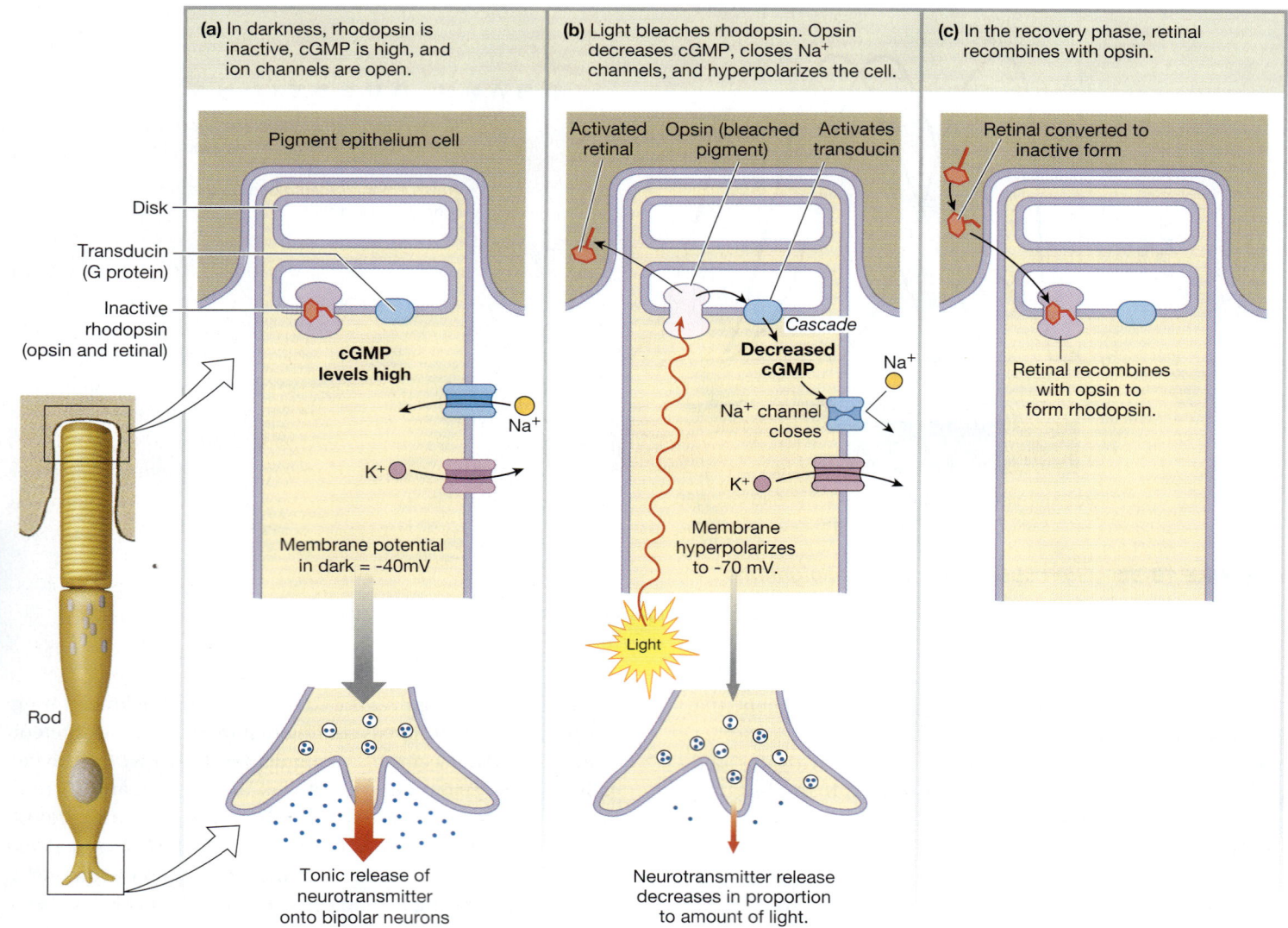

(a) In darkness, rhodopsin is inactive, cGMP is high, and ion channels are open.

Pigment epithelium cell

Disk

Transducin (G protein)

Inactive rhodopsin (opsin and retinal)

cGMP levels high

Na⁺

K⁺

Membrane potential in dark = -40mV

Rod

Tonic release of neurotransmitter onto bipolar neurons

(b) Light bleaches rhodopsin. Opsin decreases cGMP, closes Na⁺ channels, and hyperpolarizes the cell.

Activated retinal   Opsin (bleached pigment)   Activates transducin

Cascade

Decreased cGMP

Na⁺

Na⁺ channel closes

K⁺

Membrane hyperpolarizes to -70 mV.

Light

Neurotransmitter release decreases in proportion to amount of light.

(c) In the recovery phase, retinal recombines with opsin.

Retinal converted to inactive form

Retinal recombines with opsin to form rhodopsin.

■ **FIGURE 10-39** *Phototransduction in rods*

result of the movement of ions between the intracellular and extracellular compartments. A rod contains many cation channels that allow Na⁺ to enter the rod, and K⁺ channels that allow K⁺ to leak out of the rod.

When a rod is in darkness and rhodopsin is not active, cyclic GMP levels in the rod are high and both sets of channels are open. Sodium ion influx is greater than K⁺ efflux, so the rod stays depolarized to an average membrane potential of −40 mV, instead of the more usual −70 mV. At this less-negative membrane potential, there is tonic (continuous) release of neurotransmitter from the synaptic portion of the rod onto the adjacent bipolar cell (Fig. 10-39a).

When light activates rhodopsin, a second-messenger cascade is initiated through **transducin** (Fig. 10-39b), a molecule closely related to gustducin, the G protein found in bitter taste receptors. The transducin second-messenger cascade decreases the concentration of cGMP, which closes the cation channels.

As a result, Na⁺ influx slows or stops. Less Na⁺ influx but continued K⁺ efflux causes the inside of the cell to hyperpolarize. Neurotransmitter release onto the bipolar neurons consequently decreases. Bright light will close all Na⁺ channels and stop all neurotransmitter release. Dimmer light will cause a response that is graded in proportion to the light intensity.

After activation, retinal diffuses out of the rod and is transported into the pigment epithelium. There it reverts to its inactive form before moving back into the rod and being reunited with opsin (Fig. 10-39c). The recovery of rhodopsin from bleaching can take some time and is a major factor in the slow adaptation of the eyes when moving from bright light into the dark.

We now move from the cellular mechanism of light transduction to the processing of light signals in the bipolar and ganglion cells of the retina, the third and final step in our vision pathway.

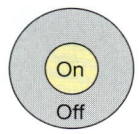 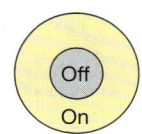

	On Off	Off On
Light shone onto center	Excites ganglion cell	Inhibits ganglion cell
Light shone onto surround	Inhibits ganglion cell	Excites ganglion cell
Diffuse light on both center and surround	Weak response from ganglion cell	Weak response from ganglion cell

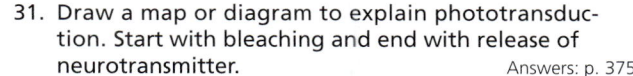

The retina uses contrast rather than absolute light intensity for better detection of weak stimuli.

■ **FIGURE 10-40** *Ganglion cell receptive fields*

CONCEPT CHECK

31. Draw a map or diagram to explain phototransduction. Start with bleaching and end with release of neurotransmitter.

Answers: p. 375

Signal Processing Begins in the Retina

After the neurotransmitter glutamate is released from photoreceptors onto bipolar neurons, signal processing in the retina becomes quite complex. For one thing, there are two types of bipolar cells, *light-on* and *light-off*. Light-on bipolar cells are inhibited by glutamate release in the dark and are activated (released from inhibition) in the light. Light-off bipolar cells are excited by glutamate release in the dark and inhibited in the light. Whether glutamate is excitatory or inhibitory depends on the type of glutamate receptor on the bipolar neuron. In this fashion, one stimulus (light) creates two responses with a single neurotransmitter.

Another point of signal processing takes place at the synapse between the bipolar neurons and the ganglion cells. These connections can also be either excitatory or inhibitory. We know more about ganglion cells because they lie on the surface of the retina, and thus their axons are the most accessible to researchers. Extensive studies have been done in which researchers stimulated the retina with carefully placed light and evaluated the response of the ganglion cells.

Each ganglion cell receives signals from a particular area of the retina. These areas, known as **visual fields**, are similar to receptive fields in the somatic sensory system [🖳 p. 330]. The visual field of a ganglion cell near the fovea is quite small. Only a few photoreceptors are associated with each ganglion, and so

visual acuity is greatest in these areas. At the edge of the visual field, multiple photoreceptors converging onto a single ganglion cell results in vision that is not as sharp.

An analogy for this arrangement is to think of pixels on your computer screen. Assume that two screens have the same number of "photoreceptors," as indicated by a maximal screen resolution of 1280×1024 pixels. If screen A has one photoreceptor becoming one "ganglion cell" pixel, the actual screen resolution is 1280×1024, and the image will be very clear. If eight photoreceptors on screen B converge into one ganglion cell pixel, then the actual screen resolution falls to 160×128, resulting in a very blurry and perhaps indistinguishable image.

Visual fields in the retina have two interesting properties. First, they are circular, unlike the irregular shape of somatic sensory receptive fields. Second, they are divided into two sections: a round center and its doughnut-shaped **surround** (Fig. 10-40 ■). There are two basic types of visual fields. In an *on-center/off-surround field,* the associated ganglion cell responds most strongly with a series of action potentials when light is brightest in the center of the field. If light is brightest in the off-surround region of the field, the ganglion cell will be inhibited and stop firing action potentials.

The reverse happens with *off-center/on-surround fields*. If light is uniform across the visual field, the ganglion cell responds weakly. Thus, the retina uses *contrast* rather than absolute light intensity to recognize objects in the environment. One advantage of using contrast is that it allows better detection of weak stimuli.

Scientists have now identified multiple types of ganglion cells in the primate retina. The two predominant types, which account for 80% of retinal ganglion cells, are M cells and P cells. Large *magnocellular* ganglion cells, or **M cells**, are more sensitive to information about movement. Smaller *parvocellular* ganglion cells, or **P cells**, are more sensitive to signals that pertain to form and fine detail, such as the texture of objects in the visual field. A recently discovered type of ganglion cell contains an opsin-like protein called *melanopsin* (see the Emerging Concepts box).

Once action potentials leave ganglion cell bodies, they travel along the optic nerves to the CNS for further processing. As noted earlier, the optic nerves enter the brain at the optic chiasm. At this point, some nerve fibers from each eye cross to the other side of the brain for processing. Figure 10-41 ■ shows how information from the right side of the visual field is processed on the left side of the brain, and information from the left side of the field is processed on the right side of the brain.

The central portion of the visual field, where left and right visual fields overlap, is the **binocular zone**. The two eyes have slightly different views of objects in this region, and the brain processes and integrates the two views to create three-dimensional representations of the objects. Our sense of depth perception—that is, whether one object is in front of or behind

10

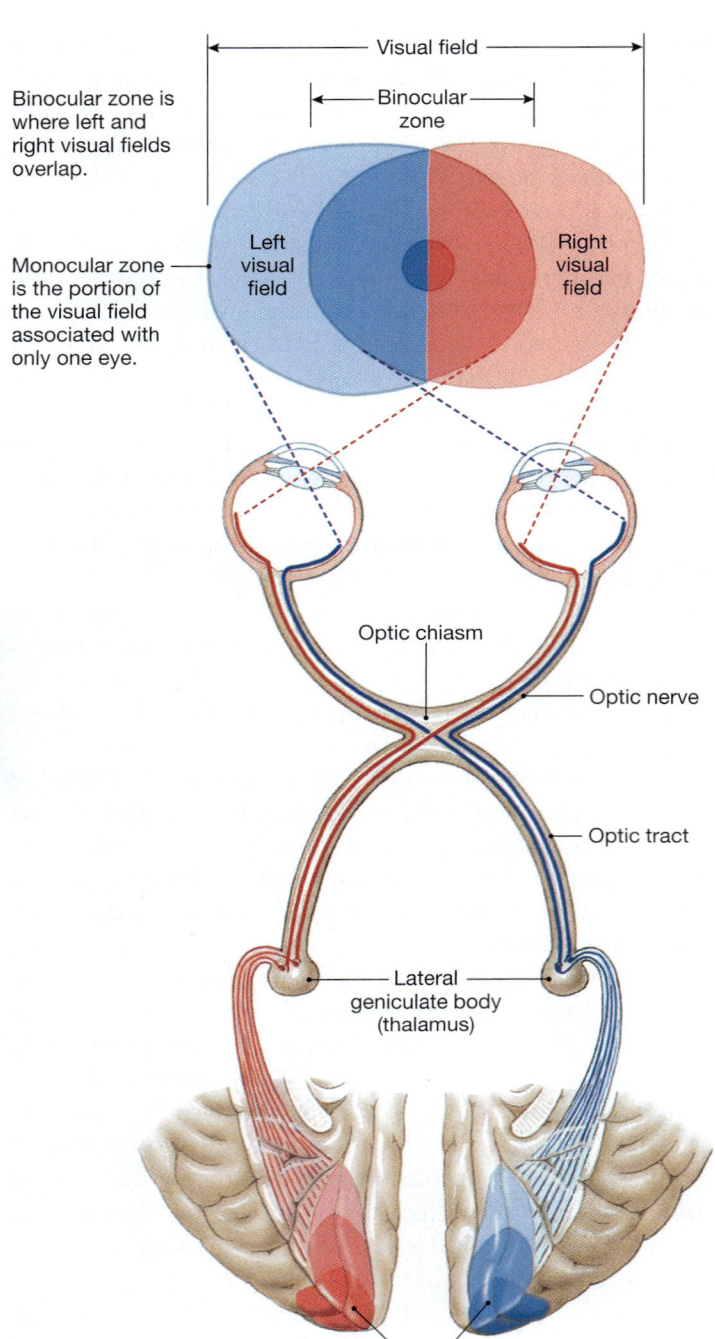

Binocular zone is where left and right visual fields overlap.

Monocular zone is the portion of the visual field associated with only one eye.

Visual field

Binocular zone

Left visual field

Right visual field

Optic chiasm

Optic nerve

Optic tract

Lateral geniculate body (thalamus)

Visual cortex

■ **FIGURE 10-41** *Visual fields and binocular vision*

The left side of the visual field in each eye is projected to the visual cortex on the right side of the brain, and the right side of each field is projected to the left visual cortex. Objects in the binocular zone are perceived in three dimensions. Objects falling outside the binocular zone are perceived in only two dimensions.

another—depends on binocular vision. Objects that fall within the visual field of only one eye are in the **monocular zone** and are viewed in two dimensions.

Once axons leave the optic chiasm, a few fibers project to the midbrain, where they participate in control of eye movement

EMERGING CONCEPTS

MELANOPSIN—A NEW PHOTOSENSITIVE PIGMENT

Circadian rhythms in mammals are cued by light entering the eyes. For many years, scientists believed that pathways from the rods and cones of the retina led to the suprachiasmatic nucleus (SCN), the brain center for circadian rhythms. However, in 1999 researchers found that transgenic mice who lacked both rods and cones still had the ability to respond to changing light cues, suggesting that some other photoreceptor must exist in the retina. Now scientists believe they have found it: a subset of retinal ganglion cells that contain an opsin-like pigment called melanopsin. Axons from these cells project to the SCN, and it appears that these cells join rods and cones as the light-sensing cells of the mammalian retina.

or coordinate with somatosensory and auditory information for balance and movement. Most axons, however, project to the lateral geniculate body of the thalamus, where the optic fibers synapse onto neurons leading to the visual cortex in the occipital lobe.

The lateral geniculate body is organized in layers that correspond to the different parts of the visual field. Thus, information from adjacent objects is processed together. This **topographical organization** is maintained in the visual cortex, with the six layers of neurons grouped into vertical columns. Within each portion of the visual field, information is further sorted by form, color, and movement.

Visual processing and our perception of the world around us are exceedingly complex subjects whose details are beyond the scope of this book. The cortex merges monocular information from the two eyes to give us a binocular view of our surroundings. Information from on/off combinations of ganglion cells is translated into sensitivity to line orientation in the simplest pathways, or into color, movement, and detailed structure in the most complex. Each of these attributes of visual stimuli is processed through a separate pathway, creating a network whose complexity we are just beginning to unravel.

RUNNING PROBLEM CONCLUSION

MÉNIÈRE'S DISEASE

Anant was told about the surgical options but elected to continue medical treatment rather than risk deafness. Over the next two months, his Ménière's disease gradually resolved. Its cause was never pinpointed.

In this running problem, you learned that vestibular problems can result in severe dizziness. You also learned about some treatments that are available to alleviate Ménière's disease. Now check your understanding of this running problem by comparing your answers to those in the summary table.

	QUESTION	FACTS	INTEGRATION AND ANALYSIS
1	In which part of the brain is sensory information about equilibrium processed?	The major equilibrium pathways project to the cerebellum. Some information is also processed in the cerebrum.	[Not applicable]
2	Subjective tinnitus occurs when an abnormality somewhere along the anatomical pathway for hearing causes the brain to perceive a sound that does not exist outside the auditory system. Starting from the ear canal, name the auditory structures in which problems may arise.	The middle ear consists of malleus, incus, and stapes, bones that vibrate with sound. The hearing portion of the inner ear consists of hair cells in the fluid-filled cochlea. The cochlear nerve leads to the brain.	Subjective tinnitus could arise from a problem with any of the structures named. Abnormal bone growth can affect the middle ear bones. Excessive fluid accumulation in the inner ear will affect the hair cells. Neural defects may cause the cochlear nerve to fire spontaneously, creating the perception of sound.
3	When a person with positioning vertigo changes position, the displaced crystals float toward the semicircular canals. Why would this cause dizziness?	The ends of the semicircular canals contain sensory cristae, each crista consisting of a cupula with embedded hair cells. Displacement of the cupula creates a sensation of rotational movement.	If the floating crystals displace the cupula, the brain will perceive movement that is not matched to sensory information coming from the eyes. The result is vertigo, an illusion of movement.
4	Compare the symptoms of positioning vertigo and Ménière's disease. On the basis of Anant's symptoms, which condition do you think he has?	The primary symptom of positioning vertigo is brief dizziness following a change in position. Ménière's disease combines vertigo with tinnitus and hearing loss.	Anant complains of dizzy attacks typically lasting up to an hour. The dizziness caused by positioning vertigo is very brief because it is caused by a change in position. It is more likely that Anant has Ménière's disease.
5	Why is limiting salt (NaCl) intake suggested as a treatment for Ménière's disease?	Ménière's disease is characterized by too much endolymph in the inner ear. Endolymph is an extracellular fluid.	Reducing salt intake should also reduce the amount of fluid in the extracellular compartment because the body will not retain water. Reduction of ECF volume may decrease fluid accumulation in the inner ear.
6	Why would severing the vestibular nerve alleviate Ménière's disease?	The vestibular nerve transmits information about balance and rotational movement from the vestibular apparatus to the brain.	Severing the vestibular nerve prevents false information about body rotation from reaching the brain, thus alleviating the vertigo of Ménière's disease.

10

328 332 354 357 361 363 **371**

CHAPTER SUMMARY

We all live in the same world, but different animals perceive the world differently. Dogs hear sounds we can't, for instance, and nocturnal animals have better night vision than we do. An animal can perceive only those stimuli for which it has sensory receptors. In this chapter you explored sensory receptors in the human body and learned how each type is designed to enable us to perceive different aspects of the world around us.

Despite the unique characteristics of each sense, basic patterns emerge for sensory transduction and perception. Membrane proteins in the form of ion channels or G protein-coupled receptors convert chemical, mechanical, thermal, and light energy into electrical signals that pass along sensory neurons to CNS integrating centers. The brain processes and filters incoming signals, sometimes acting on sensory information without that information ever reaching conscious awareness. Many of the visceral reflexes you will study in later chapters are unconscious responses to sensory input.

1. Sensory stimuli are divided into the **special senses** of vision, hearing, taste, smell, and equilibrium, and the **somatic senses** of touch, temperature, pain, itch, and proprioception. (p. 328; Tbl. 10-1)

General Properties of Sensory Systems

2. Sensory pathways begin with a stimulus that is converted by a receptor into an electrical potential. (p. 328)

3. If the stimulus is above threshold, action potentials pass along a sensory neuron to the central nervous system. We become aware of some stimuli but are never conscious of others. (p. 329)

4. Sensory receptors vary from free nerve endings to encapsulated nerve endings to specialized receptor cells. (p. 329; Fig. 10-1)

5. There are four types of sensory receptors, based on the stimulus to which they are most sensitive: **chemoreceptors, mechanoreceptors, thermoreceptors,** and **photoreceptors**. (p. 330; Tbl. 10-2)

6. Each receptor type has an **adequate stimulus**, a particular form of energy to which it is most responsive. (p. 330)

7. A stimulus that is above **threshold** will create a graded potential in the receptor. (p. 330)

8. Multiple sensory neurons may converge on one secondary neuron and create a single large **receptive field**. (p. 330; Figs. 10-2, 10-3)

9. Sensory information from the spinal cord projects to the thalamus, then on to the sensory areas of the cerebral cortex. Only olfactory information is not first routed through the thalamus. (p. 331; Fig. 10-4)

10. The central nervous system is able to change our level of awareness of sensory input. The **perceptual threshold** is the level of stimulus intensity necessary for us to be aware of a particular sensation. (p. 331)

11. The **modality** of a signal and its location are indicated by which sensory neurons are activated. The association of a receptor with a specific sensation is called **labeled line coding**. (pp. 332–333)

12. Localization of auditory information depends on the timing of receptor activation in each ear. (p. 333; Fig. 10-5)

13. **Lateral inhibition** enhances the contrast between the center of the receptive field and the edges of the field. In **population coding** the brain uses input from multiple receptors to calculate location and timing of a stimulus. (p. 334; Fig. 10-6)

14. Stimulus intensity is coded by the number of receptors activated and by the frequency of their action potentials. (p. 334; Fig. 10-7)

15. For **tonic receptors**, the sensory neuron fires action potentials as long as the **receptor potential** is above threshold. **Phasic receptors** respond to a change in stimulus intensity but adapt if the strength of the stimulus remains constant. (pp. 334–335; Fig. 10-8)

Somatic Senses

16. There are four somatosensory modalities: touch, proprioception, temperature, and nociception. (p. 337)

17. **Secondary sensory neurons** cross the midline so that one side of the brain processes information from the opposite side of the body. Ascending sensory tracts terminate in the **somatosensory cortex**. (p. 337; Fig. 10-9, Tbl. 10-3)

18. Touch receptors come in many varieties. Temperature receptors sense heat and cold. (p. 339; Fig. 10-11, Tbl. 10-4)

19. **Nociceptors** are free nerve endings that respond to chemical, mechanical, or thermal stimuli that are perceived as pain and itch. (p. 340)

20. Some responses to irritants are protective **spinal reflexes**. (p. 340)

21. **Fast pain** is transmitted rapidly by small, myelinated fibers. **Slow pain** is carried on small, unmyelinated fibers. Pain may be modulated either by descending pathways from the brain or by **gating** mechanisms in the spinal cord. (p. 341; Fig. 10-12, Tbl. 10-5)

22. **Referred pain** from internal organs occurs when multiple primary sensory neurons converge onto a single ascending tract. (p. 341; Fig. 10-13)

Chemoreception: Smell and Taste

23. Chemoreception is divided into the special senses of smell (**olfaction**) and taste (**gustation**). (p. 343)

24. **Olfactory cells** in the nasal cavity are bipolar neurons whose pathways project directly to the olfactory cortex. They are the only neurons in the body that divide continually. (p. 343; Fig. 10-14)

25. **Odorant receptors** are G protein-coupled membrane proteins. (p. 345)

26. Taste is a combination of five sensations: sweet, sour, salty, bitter, and **umami**. (p. 345)

27. **Taste cells** are non-neural cells with membrane proteins that interact with taste ligands. This interaction creates an intracellular Ca^{2+} signal that triggers exocytosis of neurotransmitter onto the primary sensory neuron. (p. 345; Figs. 10-15, 10-16)

The Ear: Hearing

28. **Hearing** is our perception of the energy carried by sound waves. Sound transduction turns air waves into mechanical vibrations, then fluid waves, chemical signals, and finally action potentials. (p. 349; Fig. 10-19)

29. The **cochlea** of the inner ear contains three parallel, fluid-filled ducts. The **cochlear duct** contains the **organ of Corti**, which contains **hair cell** receptors. (pp. 350, 352; Fig. 10-20)

30. When sound bends the hair cell cilia, the cells depolarize and release neurotransmitter onto the sensory neurons. (p. 353; Fig. 10-21)

31. The initial processing for pitch, loudness, and duration of sound takes place in the cochlea. Localization of sound is a higher function that requires sensory input from both ears and sophisticated computation by the brain. (p. 353; Fig. 10-22)

The Ear: Equilibrium

32. **Equilibrium** is mediated through hair cells in the **vestibular apparatus** and **semicircular canals** of the inner ear. Gravity and acceleration provide the force that moves the cilia. (p. 355; Figs. 10-23, 10-24, 10-25)

The Eye and Vision

33. **Vision** is the translation of reflected light into a mental image. Photoreceptors of the **retina** transduce light energy into an electrical signal that passes to the visual cortex for processing. (pp. 358–359; Figs. 10-28, 10-29)

34. The amount of light entering the eye is altered by changing the size of the pupil. (p. 360)

35. Light waves are focused by the lens, whose shape is adjusted by contracting or relaxing the **ciliary muscle**. (p. 362; Figs. 10-31, 10-32)

36. Light is converted into electrical energy by the **photoreceptors** of the retina. Signals pass through bipolar neurons to ganglion cells, whose axons form the optic nerve. (p. 364; Fig. 10-35)

37. The **fovea** has the most acute vision because it has the smallest receptive fields. (p. 364; Fig. 10-36)

38. **Rods** are responsible for monochromatic nighttime vision. **Cones** are responsible for high-acuity vision and color vision during the daytime. (p. 366; Figs. 10-37, 10-38)

39. Light-sensitive **visual pigments** in photoreceptors convert light energy into a change in membrane potential. The visual pigment in rods is **rhodopsin**. Cones have three different visual pigments. (p. 367; Fig. 10-38)

40. Rhodopsin is composed of **opsin** and **retinal**. In the absence of light, retinal binds snugly to opsin. (p. 367; Figs. 10-37, 10-39)

41. When light bleaches rhodopsin, **retinal** is released and **transducin** begins a second-messenger cascade that hyperpolarizes the rod and releases less neurotransmitter onto the bipolar neurons. (p. 368; Fig. 10-39)

42. Signals pass from photoreceptors through bipolar neurons to ganglion cells, with modulation by horizontal and amacrine cells. (p. 365; Fig. 10-35)

43. Ganglion cells called **M cells** convey information about movement. Ganglionic **P cells** transmit signals that pertain to the form and texture of objects in the visual field. (p. 369)

44. Information from one side of the visual field is processed on the opposite side of the brain. Three-dimensional objects fall within the visual fields of both eyes. (pp. 369–370; Fig. 10-41)

QUESTIONS

(Answers to the Review Questions begin on page A1.)

THE PHYSIOLOGY PLACE

Access more review material online at **The Physiology Place** website. There you'll find review questions, problem-solving activities, case studies, flashcards, and direct links to both *InterActive Physiology®* and *PhysioEx™*. To access the site, go to *www.physiologyplace.com* and select Human Physiology, Fourth Edition.

LEVEL ONE   REVIEWING FACTS AND TERMS

1. What is the role of the afferent division of the nervous system?
2. Define proprioception.
3. What are the common elements of all sensory pathways?
4. List and briefly describe the four major types of somatic receptors based on the type of stimulus to which they are most sensitive.
5. The receptors of each primary sensory neuron pick up information from a specific area, known as the _____.
6. Match the brain area with the sensory information processed there:
   (a) sounds
   (b) odors
   (c) visual information
   (d) taste
   (e) equilibrium
   1. midbrain
   2. cerebrum
   3. medulla
   4. cerebellum
   5. none of the above
7. The conversion of stimulus energy into a change in membrane potential is called _____. The form of energy to which a receptor responds is called its _____. The minimum stimulus required to activate a receptor is known as the _____.
8. When a sensory receptor membrane depolarizes (or hyperpolarizes in a few cases), the change in membrane potential is called the _____ potential. Is this a graded potential or an all-or-none potential?

9. Explain what is meant by adequate stimulus to a receptor.
10. The organization of sensory regions in the _____ of the brain preserves the topographical organization of receptors on the skin, eye, or other regions. However, there are exceptions to this rule. In which two senses does the brain rely on the timing of receptor activation to determine the location of the initial stimulus?
11. What is lateral inhibition?
12. Define tonic receptors and list some examples. Define phasic receptors and give some examples. Which type adapts? How is adaptation useful?
13. When heart pain is perceived as coming from the neck and down the left arm, this is an example of _____ pain.
14. What are the five basic tastes? What is the adaptive significance of each taste sensation?
15. The unit of sound wave measurement is _____, which measures the frequency of sound waves per second. The loudness, or intensity, of a sound is a function of the _____ of the sound waves and is measured in _____. The range of hearing for the average human ear is from _____ to _____ [units], with the most acute hearing in the range of _____ to _____ [units].
16. Which structure of the inner ear codes sound for pitch? Define spatial coding.
17. Loud noises cause action potentials to: (choose all correct answers)
   (a) fire more frequently.
   (b) have higher amplitudes.
   (c) have longer refractory periods.
18. Once sound waves have been transformed into electrical signals in the cochlea, sensory neurons transfer information to the _____, from which collaterals then take the information to the _____ and _____. The main auditory pathway synapses in the _____ and _____ before finally projecting to the _____ in the _____.

19. The parts of the vestibular apparatus that tell our brain about our movements through space are the _____, which sense rotation, and the _____, which responds to linear forces. Movement is the _____ component, whereas the otolith organs also function in telling us about head position while standing, the _____ component.

20. List the following structures in the sequence in which a beam of light entering the eye will encounter them:
    (a) aqueous humor
    (b) cornea
    (c) lens
    (d) pupil
    (e) retina

21. The three primary colors of vision are _____, _____, and _____. White light containing these colors stimulates photoreceptors called _____. Lack of the ability to distinguish some colors is called _____.

22. List six types of cells found in the retina, and briefly describe their functions.

LEVEL TWO   REVIEWING CONCEPTS

23. Compare and contrast the following:
    (a) the special senses with the somatic senses
    (b) different types of touch receptors with respect to structure, size, and location
    (c) transmission of sharp localized pain with the transmission of dull and diffuse pain (include the particular fiber types involved as well as the presence or absence of myelin in your discussion)
    (d) the forms of hearing loss

24. Draw three touch receptors having overlapping receptive fields (see Fig. 10-2). Draw a primary and secondary sensory neuron for each receptor so that they have separate ascending pathways to the cortex. Use the information in your drawing to answer this question: How many different regions of the skin can the brain distinguish using input from these three receptors?

25. Describe the neural pathways that link pain with emotional distress, nausea, and vomiting.

26. Trace the neural pathways involved in olfaction. What is G_{olf}?

27. Compare the process of signal transduction in taste buds for salty/sour ligands with the process for sweet/bitter/umami ligands.

28. Put the following structures in the order in which a sound wave would encounter them: (a) pinna, (b) cochlear duct, (c) stapes, (d) ion channels, (e) oval window, (f) hair cells/stereocilia, (g) tympanic membrane, (h) incus, (i) vestibular duct, (j) malleus.

29. Sketch the structures and receptors of the vestibular apparatus for equilibrium. Label the components. Briefly describe how they function to notify the brain of movement.

30. Explain how accommodation by the eye occurs. What is the loss of accommodation called?

31. List four common visual problems, and explain how they occur.

32. Explain how the intensity and duration of a stimulus are coded so that the stimulus can be interpreted by the brain. (Remember, action potentials are all-or-none phenomena.)

33. Map the following terms related to vision. Add additional terms if you wish.

accommodation reflex	lens
binocular vision	melanin
bipolar cells	opsin
bleaching	optic chiasm
blind spot	phototransduction
ciliary muscle	pupillary reflex
cones	retina
depth of field	retinal
field of vision	rhodopsin
focal point	rods
fovea	transducin
ganglion cells	visual field
lateral geniculate	zonules

34. Make a table of the special senses. In the first row, write these stimuli: sound, standing on the deck of a rocking boat, light, a taste, an aroma. In row 2, describe the location of the receptor for each sense. In row 3, describe the structure or properties of each receptor. In a final row, name the cranial nerve(s) that convey(s) each sensation to the brain. [⮂ p. 303]

LEVEL THREE   PROBLEM SOLVING

35. You are prodding your blindfolded lab partner's arm with two needle probes (with her permission). Sometimes she can tell you are using two probes, but when you probe less-sensitive areas, she thinks there is just one probe. Which sense are you testing? Which receptors are being stimulated? Explain why she sometimes feels only one probe.

36. Consuming alcohol depresses the nervous system and vestibular apparatus. In a sobriety check, police officers use this information to determine if an individual is inebriated. What kinds of tests can you suggest that would show evidence of this inhibition?

37. Often, children are brought to medical attention because of speech difficulties. If you were a clinician, which sense would you test first in such patients, and why?

38. A clinician shines a light into a patient's left eye, and neither pupil constricts. Shining the light into the right eye elicits a normal consensual reflex. What problem in the reflex pathway could explain these observations?

ANSWERS

✓ Answers to Concept Check Questions

Page 330

1. Myelinated axons have a faster conduction velocity than unmyelinated axons.

2. The pinna funnels sound into the ear canal.

3. Muscle length/tension, proprioception = mechanoreceptor. Pressure, inflation, distension = mechanoreception. Osmolarity = mechanoreception. Temperature = thermoreception. Oxygen, glucose, pH = chemoreception.

Page 334

4. K^+ and Cl^- channels in neurons A and C are probably opening and causing hyperpolarization.

Page 335

5. Sensory neurons signal intensity of a stimulus by the rate at which they fire action potentials.

6. Nociceptors warn the body of danger. If possible, the body should respond in some way that stops the harmful stimulus. Therefore, it is important that signals continue as long as the stimulus is present, meaning the receptors should be tonic rather than phasic.

Page 342

7. The adaptive advantage of a spinal reflex is a rapid reaction.

8. b, a, c (see Table 10-5).

9. There are many examples, including taste receptors and touch receptors that adapt to the pressure of clothes against skin.

Page 345

10. Olfactory cell (primary neuron) → secondary neuron in olfactory bulb → cranial nerve I → olfactory cortex in temporal lobe.

11. If you need help, use Figure 10-16 as the basic pattern for creating this map.

12. The knobby terminals of olfactory cells function as dendrites.

13. Olfactory neurons are bipolar neurons.

Page 347

14. Umami is associated with ingestion of the amino acid glutamate.

15. Taste cell → primary sensory neuron through cranial nerves VII, IX, or X → medulla (synapse with secondary neuron) → thalamus → gustatory cortex in parietal lobe.

Page 349

16. A kilohertz is 1000 Hz, which means 1000 waves per second.

Page 354

17. Follow the pathway shown in Figure 10-19, beginning with sound waves in the ear canal. For the portion not shown in the illustration, your map should show the cochlear nerve going to cranial nerve VIII to the medulla. The main pathway goes to midbrain, thalamus, and auditory cortex in the temporal lobe. Collateral pathways go to the reticular formation in the brain stem and to the cerebellum.

18. Somatosensory information projects to the hemisphere of the brain opposite to the side of the body on which the signal originates. The location of sound is coded by the time a stimulus arrives in each hemisphere, so a signal to both hemispheres is necessary.

19. A cochlear implant would not help people with nerve deafness or conductive hearing loss. It can only help people with sensorineural hearing loss.

Page 357

20. K^+ entry into hair cells causes depolarization.

21. When fluid builds up in the middle ear, the eardrum is unable to move freely, so it cannot transmit sound through the bones of the middle ear as efficiently.

22. When a dancer spots, the endolymph in the ampulla moves with each head rotation but then stops as the dancer holds his/her head still. This results in less inertia than if the head were continuously turning.

Page 359

23. The aqueous humor supports the cornea and lens. It also brings nutrients to and removes wastes from the epithelial layer of the cornea, which has no blood supply.

Page 360

24. (a) The sensory pathway from one eye diverges to activate motor pathways for both pupils. (b) The afferent path and its integration must be functioning because there is an appropriate response on the right side. The motor (efferent) path to the left eye must not be functioning.

25. This is antagonistic control.

Page 363

26. A more curved cornea causes light rays to converge more sharply. This causes the focal point to fall in front of the retina, so the person will be myopic.

Page 364

27. The tapetum lucidum reflects light, which enhances the amount of light hitting the photoreceptors.

28. In both the retina and skin, the finest discrimination occurs in the region with the smallest visual or receptive fields.

29. Damage to the macula, which surrounds the fovea, results in vision loss in the central portion of the visual field. Peripheral vision remains unaffected.

Page 367

30. Our dark vision is in black and white because only rods (black and white vision), not cones (color vision), are sensitive enough to be stimulated by such low levels of light.

Page 369

31. Use the information in Figure 10-39 to create your map.

Answers to Figure and Graph Questions

Page 344

Figure 10-14: Multiple neurons synapsing on a single neuron is an example of convergence.

Page 349

Figure 10-18: Graph (1) shows 20 Hz waves (5 waves in the 0.25-sec interval shown means 20 waves in 1 minute). Graph (2) shows 32 Hz waves. The waves in (1) have the lower pitch because they have the lower frequency.

Page 359

Figure 10-28: The right eye is shown in this photograph.

Page 363

Figure 10-33: (1) Convex lenses focus a beam of light, and concave lenses scatter a beam of light passing through them. (2) In hyperopia, the focal point lies behind the retina; a convex lens shortens the focal length, so that the focal point is moved onto the retina. In myopia, the focal point lies in front of the retina; a concave lens increases the focal length and moves the focal point onto the retina.

Page 365

Figure 10-35: Six rods converge on the ganglion cell.

Page 367

Figure 10-38: The pigment in red cones absorbs light over the broadest spectrum, blue cones over the narrowest. At 500 nm, the pigments in blue and green cones absorb light equally.

10

11

Because a number of cells in the autonomic nervous system act in conjunction, they have relinquished their independence to function as a coherent whole.

—Otto Appenzeller and Emilio Oribe,
in The Autonomic Nervous System, 1997

Somatic motor neurons (green) terminate at the neuromuscular junction, where axon terminals (red) release ACh onto nicotinic cholinergic receptors of the muscle (blue).

Efferent Division: Autonomic and Somatic Motor Control

BACKGROUND BASICS

RUNNING PROBLEM

A POWERFUL ADDICTION

Every day, more than 50 million Americans intentionally absorb a toxin that has been linked to the deaths of more than 6 million people since 1964. Why would people knowingly poison themselves? If you've guessed that the toxin is nicotine, you already know part of the answer. One of more than 4000 chemicals found in tobacco, nicotine is highly addictive. So powerful is this addiction that fewer than 10% of tobacco users are able to quit smoking the first time they try. Shanika, a smoker for six years, is attempting for the second time to quit smoking. The odds are in her favor this time, however, because she has sought help. She has decided to try the nicotine patch.

| 377 | 378 | 382 | 387 | 390 | 392 |

The picnic lunch was wonderful. You are now dozing on the grass in the warm spring sunlight as you let the meal digest. Suddenly you feel something moving across your lower leg. You open your eyes, and as they adjust to the bright light, you see a four-foot-long snake slithering over your foot. More by instinct than reason, you fling the snake into the grass while scrambling to a safe perch on top of the nearby picnic table. You are breathing heavily, and your heart is pounding.

In less than a second, your body has gone from a state of quiet rest and digestion to a state of panic and frantic activity. This reflex reaction is integrated and coordinated through the central nervous system (CNS), then carried out by the efferent division of the peripheral nervous system (PNS). The fibers of efferent neurons are bundled together into nerves that carry commands from the CNS to the muscles and glands of the body. Some nerves, called *mixed nerves,* also carry sensory information through afferent fibers [p. 328].

The efferent division of the peripheral nervous system can be subdivided into **somatic motor neurons,** which control skeletal muscles, and **autonomic neurons,** which control smooth muscle, cardiac muscle, many glands, and some adipose tissue. The somatic and autonomic divisions are sometimes called the voluntary and involuntary divisions of the nervous system, respectively. However, although it is true that movement controlled by somatic pathways usually requires conscious thought and that autonomic reflexes are mainly involuntary, this distinction does not always hold true. For example, some skeletal muscle reflexes, such as swallowing and the knee jerk reflex, are involuntary. And a person can use biofeedback training to learn to modulate some involuntary autonomic functions, such as heart rate and blood pressure.

We begin our study of the efferent division of the PNS by looking at the autonomic division. Then we will consider the somatic motor division, as preparation for learning about muscles in Chapter 12.

THE AUTONOMIC DIVISION

The autonomic division of the efferent nervous system (or *autonomic nervous system* for short) is also known in older writings as the *vegetative nervous system,* reflecting the observation that its functions are not under voluntary control. The word *autonomic* comes from the same roots as *autonomous,* meaning *self-governing.* Another name for the autonomic division is *visceral nervous system* because of its control over internal organs.

The autonomic division is subdivided into **sympathetic** and **parasympathetic branches** (often called the *sympathetic* and *parasympathetic nervous systems*). Some parts of the sympathetic branch were first described by the Greek physician Claudius Galen (ca. A.D. 130–200), who is famous for his compilation of anatomy, physiology, and medicine as they were known during his time. As a result of his dissections, Galen proposed that "animal spirits" flowed from the brain to the tissues through hollow nerves, creating "sympathy" between the different parts of the body. Galen's "sympathy" later gave rise to the name for the sympathetic branch. The prefix *para-,* for the parasympathetic branch, means *beside* or *alongside.*

Although the sympathetic and parasympathetic branches can be distinguished anatomically, there is no simple way to separate the actions of the two branches on their targets. The best generalization is to characterize them according to the type of situation in which they are most active. The picnic scene that began the chapter illustrates the two extremes at which the sympathetic and parasympathetic branches function. If you are resting quietly after a meal, the parasympathetic branch is dominant, taking command of the routine, quiet activities of day-to-day living, such as digestion. Consequently, parasympathetic neurons are sometimes said to control "rest and digest" functions.

In contrast, the sympathetic branch is dominant in stressful situations, such as the potential threat from the snake. One of the most dramatic examples of sympathetic action is the **fight-or-flight** response, in which the brain triggers massive simultaneous sympathetic discharge throughout the body. As the body prepares to fight or flee, the heart speeds up; blood vessels to muscles of the arms, legs, and heart dilate; and the liver starts to produce glucose to provide energy for muscle contraction. Digestion becomes a low priority when life and limb are threatened, and so blood is diverted from the gastrointestinal tract to skeletal muscles.

The massive sympathetic discharge that occurs in fight-or-flight situations is mediated through the hypothalamus and

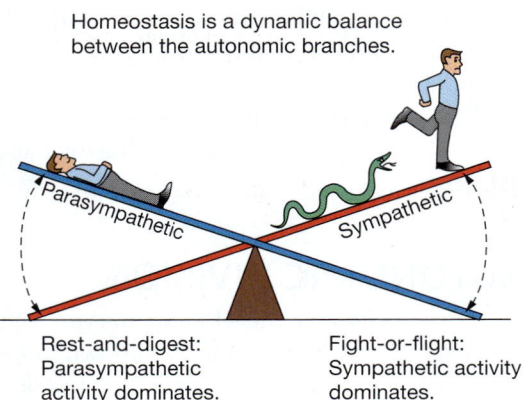

Homeostasis is a dynamic balance between the autonomic branches.

Rest-and-digest: Parasympathetic activity dominates.

Fight-or-flight: Sympathetic activity dominates.

■ **FIGURE 11-1** *Role of the autonomic division in homeostasis*

The parasympathetic branch is dominant during most rest-and-digest functions. The sympathetic branch dominates in fight-or-flight situations. Most normal activities reflect a balance between the two branches.

is a total-body response to a crisis. If you have ever been scared by the squealing of brakes or a sudden sound in the dark, you know how rapidly the nervous system can influence multiple body systems. Most sympathetic responses are not the all-out response of a fight-or-flight reflex, however, and more important, activating one sympathetic pathway does not automatically activate them all.

The role of the sympathetic nervous system in mundane daily activities is as important as a fight-or-flight response. For example, one key function of the sympathetic branch is control of blood flow to the tissues, as you will learn in Chapter 15. Most of the time, autonomic control of body function "seesaws" back and forth between the sympathetic and parasympathetic branches (Figure 11-1 ■) as they cooperate to fine-tune various processes. Only occasionally, as in the fight-or-flight example, does the seesaw move to one extreme or the other.

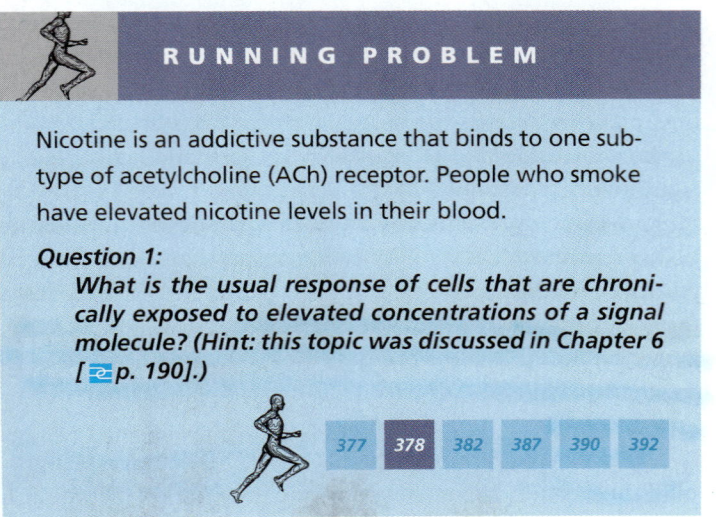

RUNNING PROBLEM

Nicotine is an addictive substance that binds to one subtype of acetylcholine (ACh) receptor. People who smoke have elevated nicotine levels in their blood.

Question 1:
What is the usual response of cells that are chronically exposed to elevated concentrations of a signal molecule? (Hint: this topic was discussed in Chapter 6 [p. 190].)

| 377 | **378** | 382 | 387 | 390 | 392 |

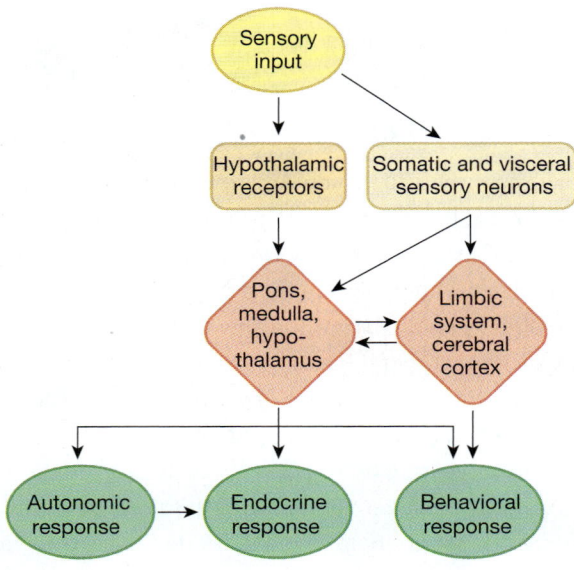

Sensory input

Hypothalamic receptors

Somatic and visceral sensory neurons

Pons, medulla, hypo-thalamus

Limbic system, cerebral cortex

Autonomic response → Endocrine response

Behavioral response

■ **FIGURE 11-2** *The hypothalamus, pons, and medulla initiate autonomic, endocrine, and behavioral responses*

Hypothalamic trophic hormones control secretion of anterior pituitary hormones [p. 226]. The hypothalamus secretes neurohormones from the posterior pituitary [p. 226]. Autonomic neurons also control hormone release from peripheral endocrine cells, such as pancreatic beta cells [p. 224]. Emotional responses mediated through the limbic system can influence autonomic responses [p. 308].

CONCEPT CHECK

1. The afferent division of the nervous system has what two components?
2. The central nervous system consists of the _____ and the _____.

Answers: p. 395

Autonomic Reflexes Are Important for Homeostasis

The autonomic nervous system works closely with the endocrine system and the behavioral state system [p. 312] to maintain homeostasis in the body. Sensory information from somatosensory and visceral receptors goes to homeostatic control centers in the hypothalamus, pons, and medulla (Fig. 11-2 ■). These centers monitor and regulate important functions such as blood pressure, temperature regulation, and water balance (Fig. 11-3 ■). The hypothalamus also contains neurons that act as sensors, such as osmoreceptors, which monitor osmolarity, and thermoreceptors, which monitor body temperature. Motor output from the hypothalamus and brain stem creates autonomic responses, endocrine responses, and behavioral responses such as drinking, food-seeking, and temperature regulation (getting out of the heat, putting on a sweater). The behavioral responses are integrated in brain centers responsible for motivated behaviors and control of movement.

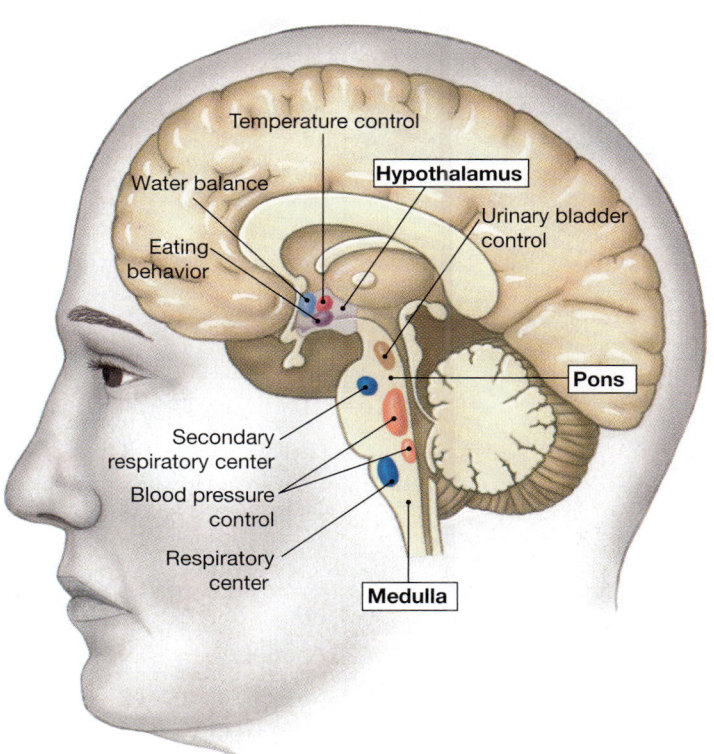

FIGURE 11-3 *Autonomic control centers in the brain*

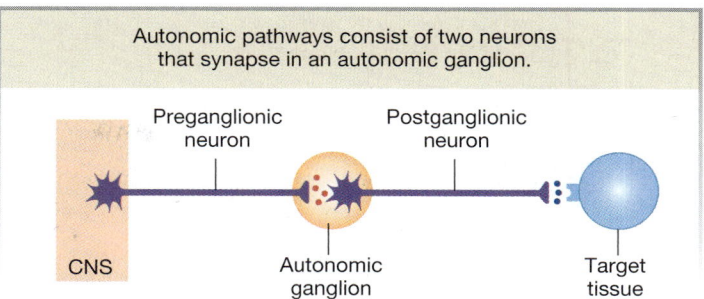

Autonomic pathways consist of two neurons that synapse in an autonomic ganglion.

Preganglionic neuron   Postganglionic neuron

CNS   Autonomic ganglion   Target tissue

FIGURE 11-4 *Autonomic pathways*

In addition, sensory information integrated in the cerebral cortex and limbic system can create emotions that influence autonomic output, as Figure 11-2 illustrates. Blushing, fainting at the sight of a hypodermic needle, and "butterflies in the stomach" are all examples of emotional influences on autonomic functions. Understanding the autonomic and hormonal control of organ systems is the key to understanding the maintenance of homeostasis in virtually every system of the body.

Some autonomic reflexes are capable of taking place without input from the brain. These *spinal reflexes* [🔁 Fig. 9-8, p. 301] include urination, defecation, and penile erection—bodily functions that can be influenced by descending pathways from the brain but do not require this input. For example, people with spinal cord injuries that disrupt communication between the brain and spinal cord may retain some spinal reflexes but lose the ability to sense or control them.

Antagonistic Control Is a Hallmark of the Autonomic Division

The sympathetic and parasympathetic branches of the autonomic nervous system display all four of Walter Cannon's properties of homeostasis: (1) preservation of the fitness of the internal environment, (2) up-down regulation by tonic control, (3) antagonistic control, and (4) chemical signals with different effects in different tissues [🔁 p. 142].

Most internal organs are under *antagonistic control*, in which one autonomic branch is excitatory and the other

branch is inhibitory (see the table in Fig. 11-5). For example, sympathetic innervation increases heart rate while parasympathetic stimulation decreases it. Consequently, heart rate can be regulated by altering the relative proportions of sympathetic and parasympathetic control.

Exceptions to dual antagonistic innervation include the sweat glands and the smooth muscle in most blood vessels. These tissues are innervated only by the sympathetic branch and rely strictly on tonic (up-down) control.

Although the two autonomic branches are usually antagonistic in their control of a given target tissue, they sometimes work cooperatively on different tissues to achieve a common goal. For example, blood flow for penile erection is under control of the parasympathetic branch, and muscle contraction for sperm ejaculation is directed by the sympathetic branch.

In some autonomic pathways, the neurotransmitter receptor determines the response of the target tissue. For instance, most blood vessels contain one type of *adrenergic receptor* [🔁 p. 276] that causes smooth muscle contraction (vasoconstriction). However, some blood vessels also contain a second type of adrenergic receptor that causes smooth muscle to relax (vasodilation). Both receptors are activated by catecholamines [🔁 p. 222]. In this example the receptor, not the chemical signal, determines the response.

CONCEPT CHECK
3. Define homeostasis.

Answers: p. 395

Autonomic Pathways Have Two Efferent Neurons in Series

All autonomic pathways (sympathetic and parasympathetic) consist of two neurons in series (Fig. 11-4 ■). The first neuron, called the **preganglionic neuron**, originates in the central nervous system and projects to an **autonomic ganglion** outside the CNS. There the preganglionic neuron synapses with the second neuron in the pathway, the **postganglionic neuron**. This neuron has its cell body in the ganglion and projects its axon to the target tissue. (A *ganglion* is a cluster of nerve cell bodies that lie outside the CNS. The equivalent in the CNS is a *nucleus* [🔁 p. 250].)

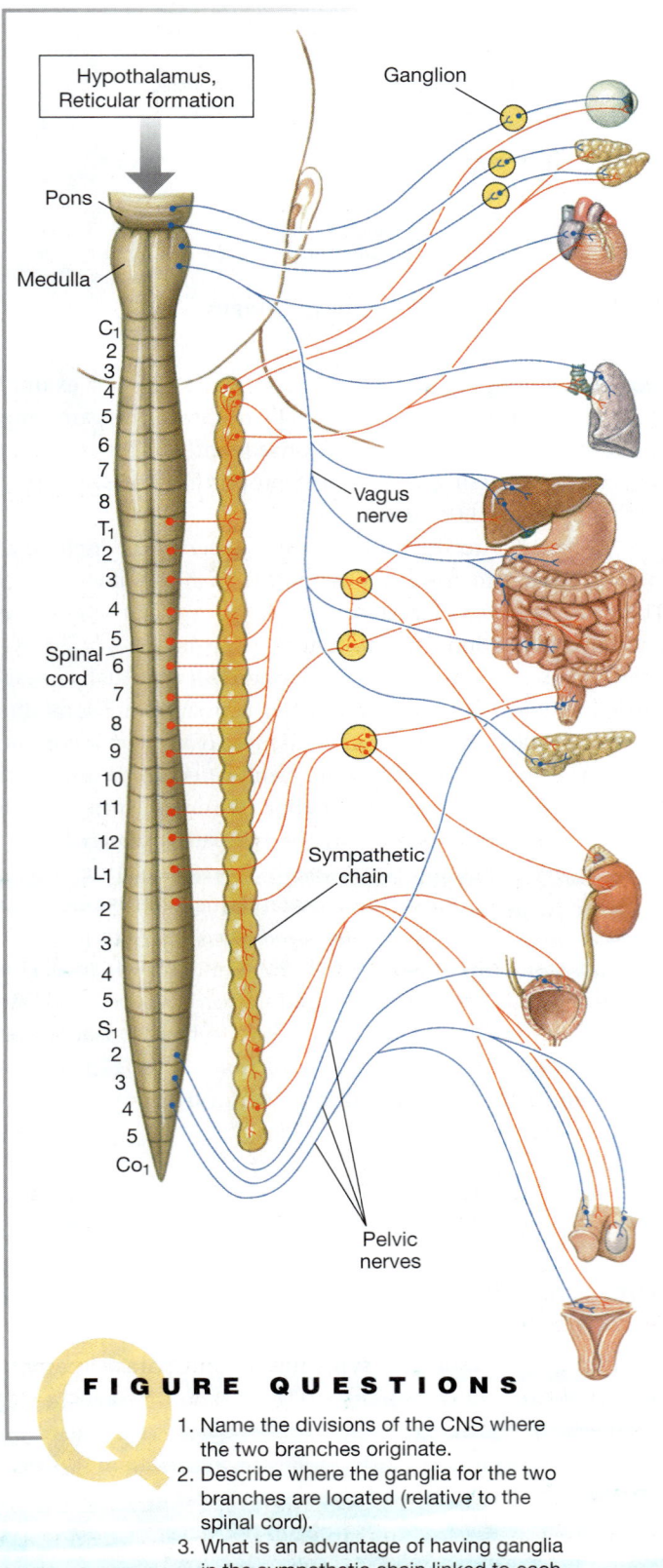

Effector Organ	Parasympathetic Response *	Sympathetic Response	Adrenergic Receptor
Pupil of eye	Constricts	Dilates	α
Salivary glands	Watery secretion	Mucus, enzymes	α and β_2
Heart	Slows rate	Increases rate and force of contraction	β_1
Arterioles and veins	——	Constricts Dilates	α β_2
Lungs	Bronchioles constrict	Bronchioles dilate	β_2**
Digestive tract	Increases motility and secretion	Decreases motility and secretion	α, β_2
Exocrine pancreas	Increases enzyme secretion	Decreases enzyme secretion	α
Endocrine pancreas	Stimulates insulin secretion	Inhibits insulin secretion	α
Adrenal medulla	——	Secretes catecholamines	——
Kidney	——	Increases renin secretion	β_1
Urinary bladder	Release of urine	Urinary retention	α, β_2
Adipose tissue	——	Fat breakdown	β
Sweat glands		Localized sweating	α
Male and female sex organs	Erection	Ejaculation (male)	α
Uterus	Depends on stage of cycle	Depends on stage of cycle	α, β_2
Lymphoid tissue (not illustrated)	——	Generally inhibitory	α, β_2

*All parasympathetic responses are mediated by muscarinic receptors.

**Hormonal epinephrine only

FIGURE QUESTIONS

1. Name the divisions of the CNS where the two branches originate.
2. Describe where the ganglia for the two branches are located (relative to the spinal cord).
3. What is an advantage of having ganglia in the sympathetic chain linked to each other?

KEY

●—< Parasympathetic
●—< Sympathetic

■ **FIGURE 11-5** *Autonomic sympathetic and parasympathetic pathways*

There are actually two sympathetic ganglion chains, one on either side of the spinal cord.

Divergence [🔁 p. 278] is an important feature of autonomic pathways. On average, one preganglionic neuron entering a ganglion synapses with eight or nine postganglionic neurons. Some synapse on as many as 32 neurons! Each postganglionic neuron may then innervate a different target, meaning that a single signal from the CNS can affect a large number of target cells simultaneously.

In the traditional view of the autonomic division, autonomic ganglia were simply a way-station for the transfer of signals from preganglionic neurons to postganglionic neurons. We now know, however, that ganglia are more than a simple collection of axon terminals and nerve cell bodies: they also contain interneurons that lie completely within them. These interneurons enable the autonomic ganglia to act as mini-integrating centers, receiving sensory input from the periphery of the body and modulating outgoing autonomic signals to target tissues. Presumably this arrangement means that a reflex could be integrated totally within a ganglion, with no involvement of the CNS. That pattern of control is known to exist in the enteric nervous system [🔁 p. 246], which will be discussed in Chapter 21.

Sympathetic and Parasympathetic Branches Exit the Spinal Cord in Different Regions

How, then, do the two autonomic branches differ anatomically? The main anatomical differences are (1) where the pathways originate in the CNS and (2) the location of the autonomic ganglia. As Figure 11-5 ■ shows, most sympathetic pathways (red) originate in the thoracic and lumbar regions of the spinal cord. *Sympathetic ganglia* are found primarily in two chains that run along either side of the spinal column, with additional ganglia along the descending aorta. Long nerves (axons of postganglionic neurons) project from the ganglia to the target tissues. Because most sympathetic ganglia lie close to the spinal cord, sympathetic pathways generally have short preganglionic neurons and long postganglionic neurons.

Many parasympathetic pathways (shown in blue in Figure 11-5) originate in the brain stem, and their axons leave the brain in several cranial nerves [🔁 p. 303]. Other parasympathetic pathways originate in the sacral region (near the lower end of the spinal cord) and control pelvic organs. In general, parasympathetic ganglia are located either on or near their target organs. Consequently, parasympathetic preganglionic neurons have long axons, and parasympathetic postganglionic neurons have short axons.

Parasympathetic innervation goes primarily to the head, neck, and internal organs. The major parasympathetic tract is the **vagus nerve** (cranial nerve X), which contains about 75% of all parasympathetic fibers. This nerve carries both sensory information from internal organs to the brain, and parasympathetic output from the brain to organs (Fig. 11-6 ■). *Vagotomy,* a procedure in which the vagus nerve is surgically cut, was an experimental technique used in the nineteenth and early twentieth centuries to study the effects of the autonomic nervous

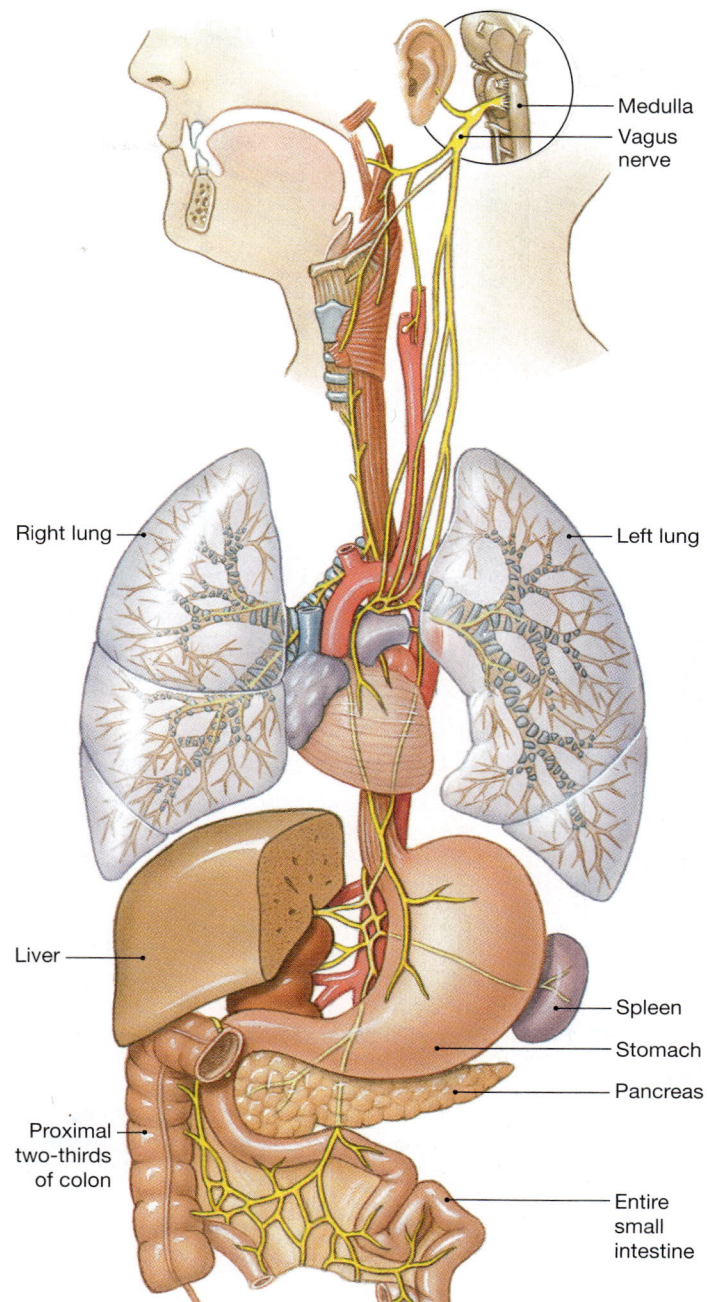

■ FIGURE 11-6 *The vagus nerve*

This nerve carries parasympathetic fibers to most of the internal organs. It also carries sensory (afferent) neurons from the organs to the brain.

system on various organs. For a time, vagotomy was the preferred treatment for stomach ulcers because removal of parasympathetic innervation to the stomach decreased the secretion of stomach acid. However, this procedure had many unwanted side effects and has been abandoned in favor of drug therapies that treat the problem more specifically.

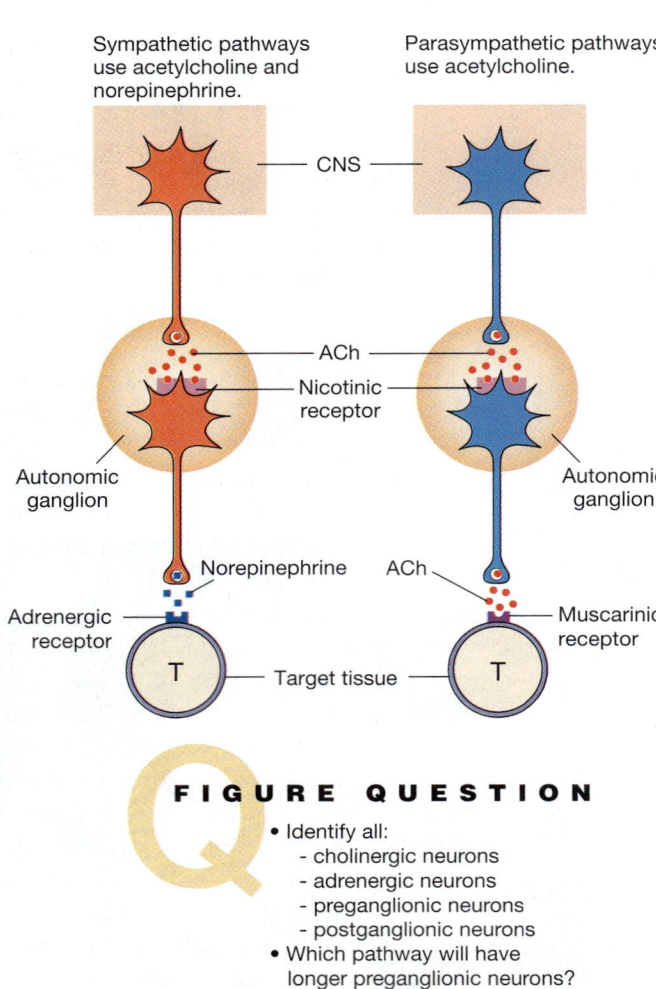

Sympathetic pathways use acetylcholine and norepinephrine.

Parasympathetic pathways use acetylcholine.

CNS

ACh

Nicotinic receptor

Autonomic ganglion

Autonomic ganglion

Norepinephrine

ACh

Adrenergic receptor

Muscarinic receptor

T

T

Target tissue

FIGURE QUESTION

- Identify all:
  - cholinergic neurons
  - adrenergic neurons
  - preganglionic neurons
  - postganglionic neurons
- Which pathway will have longer preganglionic neurons? (*Hint:* See Fig. 11-5)

■ **FIGURE 11-7** *Sympathetic and parasympathetic pathways*

All autonomic pathways have two neurons in series. Sympathetic and parasympathetic pathways differ in their neurotransmitters and receptors.

CONCEPT CHECK

4. A nerve that carries both sensory and motor information is called a _____ nerve.

5. Name the four regions of the spinal cord in order, starting from the brain stem.

Answers: p. 395

The Autonomic Nervous System Uses a Variety of Neurotransmitters and Modulators

Chemically, the sympathetic and parasympathetic branches can be distinguished by their neurotransmitters and receptors, using the following rules Figure 11-7 ■:

1. Both sympathetic and parasympathetic preganglionic neurons release acetylcholine (ACh) onto *nicotinic cholinergic receptors* on the postganglionic cell [🔁 p. 275].

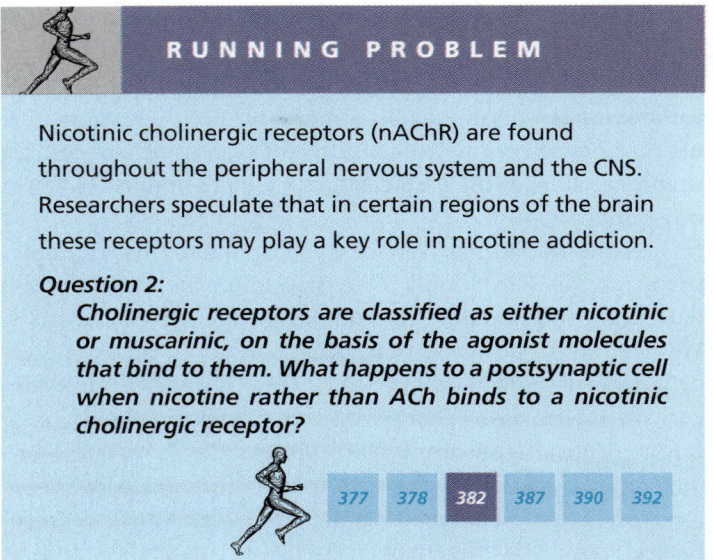

RUNNING PROBLEM

Nicotinic cholinergic receptors (nAChR) are found throughout the peripheral nervous system and the CNS. Researchers speculate that in certain regions of the brain these receptors may play a key role in nicotine addiction.

Question 2:
Cholinergic receptors are classified as either nicotinic or muscarinic, on the basis of the agonist molecules that bind to them. What happens to a postsynaptic cell when nicotine rather than ACh binds to a nicotinic cholinergic receptor?

| 377 | 378 | 382 | 387 | 390 | 392 |

2. Most postganglionic sympathetic neurons secrete norepinephrine onto *adrenergic receptors* on the target cell.

3. Most postganglionic parasympathetic neurons secrete acetylcholine onto *muscarinic cholinergic receptors* on the target cell.

However, there are some exceptions to these rules. A few sympathetic postganglionic neurons, such as those that terminate on sweat glands, secrete ACh rather than norepinephrine. These neurons are therefore called *sympathetic cholinergic neurons*.

A small number of autonomic neurons secrete neither norepinephrine nor acetylcholine and are known as *non-adrenergic, non-cholinergic neurons*. Some of the chemicals they use as neurotransmitters include substance P, somatostatin, vasoactive intestinal peptide (VIP), adenosine, nitric oxide, and ATP. We will mention these unusual neurotransmitters in later chapters when they play a significant role. The non-adrenergic, non-cholinergic neurons are assigned to either the sympathetic or parasympathetic branch according to the location where their preganglionic fibers leave the nerve cord.

Autonomic Pathways Control Smooth and Cardiac Muscle, Glands, and Lymphoid and Adipose Tissues

The targets of autonomic neurons are smooth muscle, cardiac muscle, many exocrine glands, a few endocrine glands, lymphoid tissues, and some adipose tissue. The synapse between a postganglionic autonomic neuron and its target cell is called the **neuroeffector junction**.

The structure of an autonomic synapse differs from the model synapse shown in Figure 8-20 [🔁 p. 270]. Autonomic postganglionic axons end with a series of swollen areas at their distal ends, like beads spaced out along a string (Fig. 11-8 ■). Each

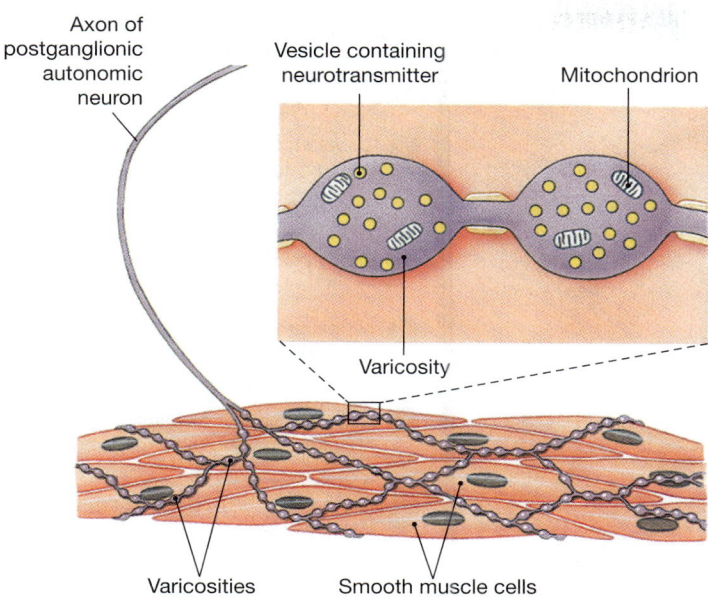

Axon of postganglionic autonomic neuron

Vesicle containing neurotransmitter

Mitochondrion

Varicosity

Varicosities   Smooth muscle cells

■ **FIGURE 11-8**   *Varicosities in autonomic neurons*
These swellings, which are found along the distal end of postganglionic axons, release neurotransmitter over the surface of target cells.

of these swellings, known as a **varicosity** [*varicosus,* abnormally enlarged or swollen], contains vesicles filled with neurotransmitter. Parasympathetic pathways end with axon terminals, or *terminal boutons* [from the French word for "knob" or "button"; ⇄ Tbl. 8-1, p. 244], which also release neurotransmitter.

The branched ends of the axon lie across the surface of the target tissue, but the underlying target cell membrane does not possess clusters of neurotransmitter receptors in specific sites. Instead, the neurotransmitter is simply released into the interstitial fluid to diffuse to wherever the receptors are located. The result is a less-directed form of communication than that which occurs between a somatic motor neuron and a skeletal muscle. The diffuse release of autonomic neurotransmitter means that a single postganglionic neuron can affect a large area of target tissue.

The release of autonomic neurotransmitters is subject to modulation from a variety of sources. For example, sympathetic varicosities contain receptors for hormones and for paracrines such as histamine. These modulators may either facilitate or inhibit neurotransmitter release. Some preganglionic neurons co-secrete neuropeptides along with acetylcholine. The peptides act as neuromodulators, producing slow synaptic potentials that modify the activity of postganglionic neurons [⇄ p. 277].

Autonomic Neurotransmitters Are Synthesized in the Axon

In the autonomic division, neurotransmitter synthesis takes place in the axon varicosities. The primary autonomic neurotransmitters are acetylcholine (ACh) and norepinephrine, both small molecules easily synthesized by cytoplasmic enzymes.

Neurotransmitter made in the varicosities is packaged into synaptic vesicles for storage.

The process of neurotransmitter release follows the same pattern you learned in Chapter 8 [⇄ p. 271]. When an action potential arrives at the varicosity, voltage-gated Ca^{2+} channels open, Ca^{2+} enters the neuron, and the synaptic vesicle contents are released by exocytosis. Once neurotransmitters are released into the synapse, they either diffuse through the interstitial fluid until they encounter a receptor on the target cell or drift away from the synapse.

The concentration of neurotransmitter in the synapse is a major factor in the control that an autonomic neuron exerts on its target: the more neurotransmitter, the longer or stronger the response. The concentration of neurotransmitter in a synapse is influenced by its rate of breakdown or removal. Neurotransmitter activation of its receptor terminates when the neurotransmitter either (1) diffuses away, (2) is metabolized by enzymes in the extracellular fluid, or (3) is actively transported into cells around the synapse. The uptake of neurotransmitter by varicosities allows neurons to reuse the chemicals.

These steps were illustrated for acetylcholine in Figure 8-22 [⇄ p. 274] and are shown for norepinephrine in Figure 11-9 ■. Norepinephrine is synthesized in the varicosity from the amino acid tyrosine. Once released into the synapse, norepinephrine may combine with an adrenergic receptor on the target cell, diffuse away, or be transported back into the varicosity. Inside the neuron, recycled norepinephrine is either repackaged into vesicles or broken down by **monoamine oxidase** (MAO), the main enzyme responsible for degradation of catecholamines.

Table 11-1 ■ compares the characteristics of the two primary autonomic neurotransmitters.

Most Sympathetic Pathways Secrete Norepinephrine onto Adrenergic Receptors

Sympathetic pathways secrete catecholamines that bind to adrenergic receptors on their target cells. Adrenergic receptors come in two varieties: α (alpha) and β (beta), with several subtypes of each. Alpha receptors—the most common sympathetic receptor—respond strongly to norepinephrine and only weakly to epinephrine (Table 11-2 ■).

The three main subtypes of beta receptors differ in their affinity for catecholamines. **β₁-receptors** respond equally strongly to norepinephrine and epinephrine. **β₂-receptors** are more sensitive to epinephrine than to norepinephrine. Interestingly, the β₂-receptors are not innervated (no sympathetic neurons terminate near them), which limits their exposure to the neurotransmitter norepinephrine. **β₃-receptors**, which are found primarily on adipose tissue, are innervated and more sensitive to norepinephrine than to epinephrine.

All adrenergic receptors are G protein-coupled receptors rather than ion channels [⇄ p. 183]. This means that the target cell response is slower to start and usually lasts longer. The different adrenergic receptor subtypes use different second messenger

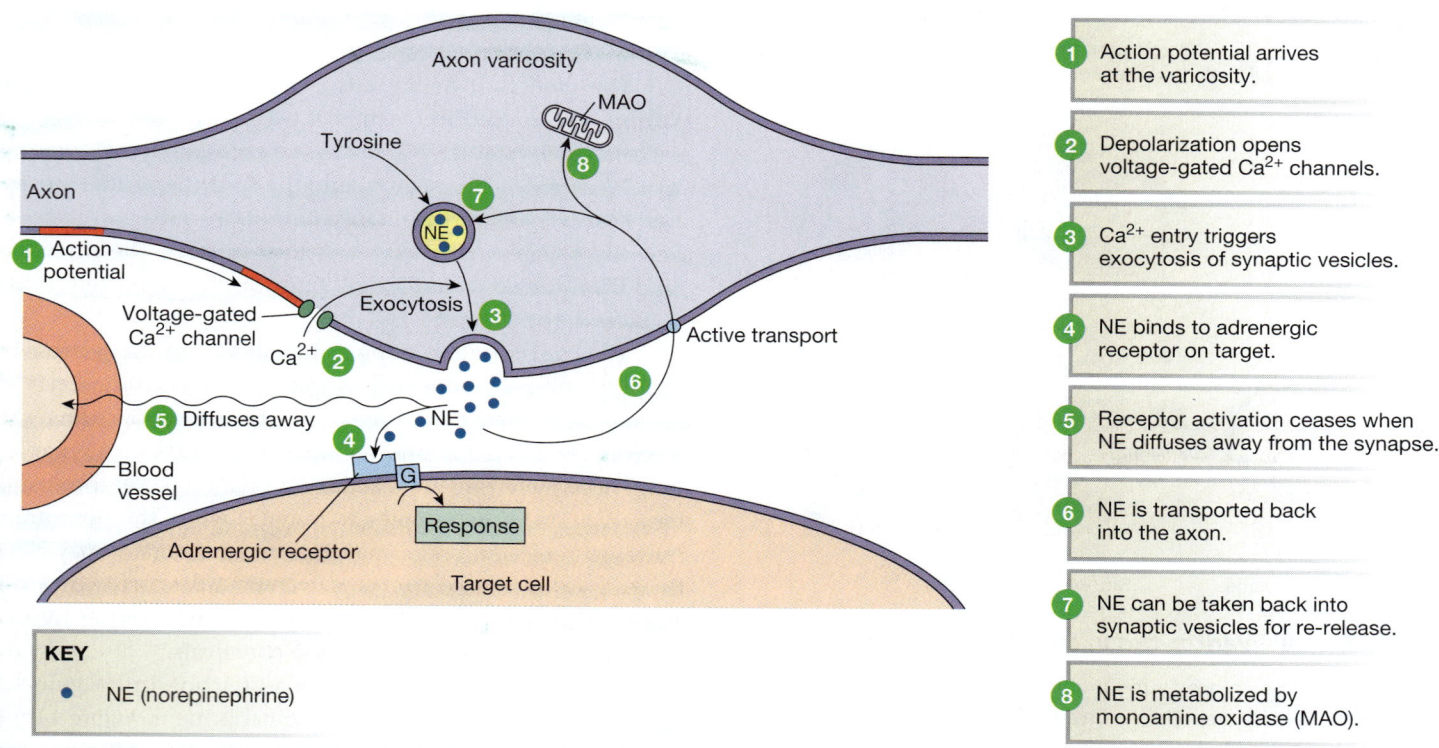

KEY
- NE (norepinephrine)

1. Action potential arrives at the varicosity.
2. Depolarization opens voltage-gated Ca^{2+} channels.
3. Ca^{2+} entry triggers exocytosis of synaptic vesicles.
4. NE binds to adrenergic receptor on target.
5. Receptor activation ceases when NE diffuses away from the synapse.
6. NE is transported back into the axon.
7. NE can be taken back into synaptic vesicles for re-release.
8. NE is metabolized by monoamine oxidase (MAO).

■ **FIGURE 11-9** *Norepinephrine release at a varicosity of a sympathetic neuron*

pathways (Table 11-2). Catecholamine binding to β-receptors increases cyclic AMP and triggers the phosphorylation of intracellular proteins. The target cell response then depends on the specific downstream pathway. For example, activation of β$_1$-receptors enhances cardiac muscle contraction, but activation of β$_2$-receptors relaxes smooth muscle in many tissues.

α$_1$-**receptors** activate phospholipase C, creating IP$_3$ and DAG [🔁 Fig. 6-12, p. 185]. DAG initiates a cascade that phosphorylates proteins. IP$_3$ opens Ca^{2+} channels, creating intracellular Ca^{2+} signals. In general, activation of α$_1$-receptors causes

muscle contraction or secretion by exocytosis. α$_2$-**receptors** decrease intracellular cyclic AMP and cause smooth muscle relaxation (gastrointestinal tract) or decreased secretion (pancreas).

For all adrenergic receptors, second messenger activity in the target tissue can persist for a longer time than is usually associated with the rapid action of the nervous system. The long-lasting metabolic effects of some autonomic pathways result from modification of existing proteins or from the synthesis of new proteins. We will discuss the specific effects of catecholamines on various tissues in subsequent chapters.

TABLE 11-1	Postganglionic Autonomic Neurotransmitters	
	SYMPATHETIC DIVISION	**PARASYMPATHETIC DIVISION**
Neurotransmitter	Norepinephrine	Acetylcholine (ACh)
Receptor types	α- and β-adrenergic	Nicotinic and muscarinic cholinergic
Synthesized from	Tyrosine	Acetyl CoA + choline
Inactivation enzyme	Monoamine oxidase (MAO) in mitochondria of varicosity	Acetylcholinesterase (AChE) in synaptic cleft
Varicosity membrane transports	Norepinephrine	Choline

TABLE 11-2   Properties of Adrenergic Receptors

RECEPTOR	FOUND IN	SENSITIVITY	EFFECT ON SECOND MESSENGER
α_1	Most sympathetic target tissues	NE > E*	Activates phospholipase C
α_2	Gastrointestinal tract and pancreas	NE > E	Decreases cAMP
β_1	Heart muscle, kidney	NE = E	Increases cAMP
β_2	Certain blood vessels and smooth muscle of some organs	E > NE	Decrease cAMP
β_3	Adipose tissue	NE > E	Increases cAMP

*NE = norepinephrine, E = epinephrine.

CONCEPT CHECK

6. In what organelle is most intracellular Ca^{2+} stored?

7. What enzyme (a) converts ATP to cAMP? (b) does cAMP activate? [⊇ Fig. 6-11, p. 184]   Answers: p. 395

The Adrenal Medulla Secretes Catecholamines

The **adrenal medulla** [*ad-*, upon + *renal*, kidney; *medulla*, marrow] is a specialized neuroendocrine tissue associated with the sympathetic nervous system. During development, the neural tissue destined to secrete the catecholamines norepinephrine and epinephrine splits into two functional entities: the sympathetic branch of the nervous system, which secretes norepinephrine, and the adrenal medulla, which secretes epinephrine primarily.

The adrenal medulla forms the core of the *adrenal glands,* which sit atop the kidneys (Fig. 11-10a ■). Like the pituitary gland, each adrenal gland is actually two glands of different embryological origin that fused during development (Fig. 11-10b). The outer portion, the *adrenal cortex,* is a true endocrine gland

11

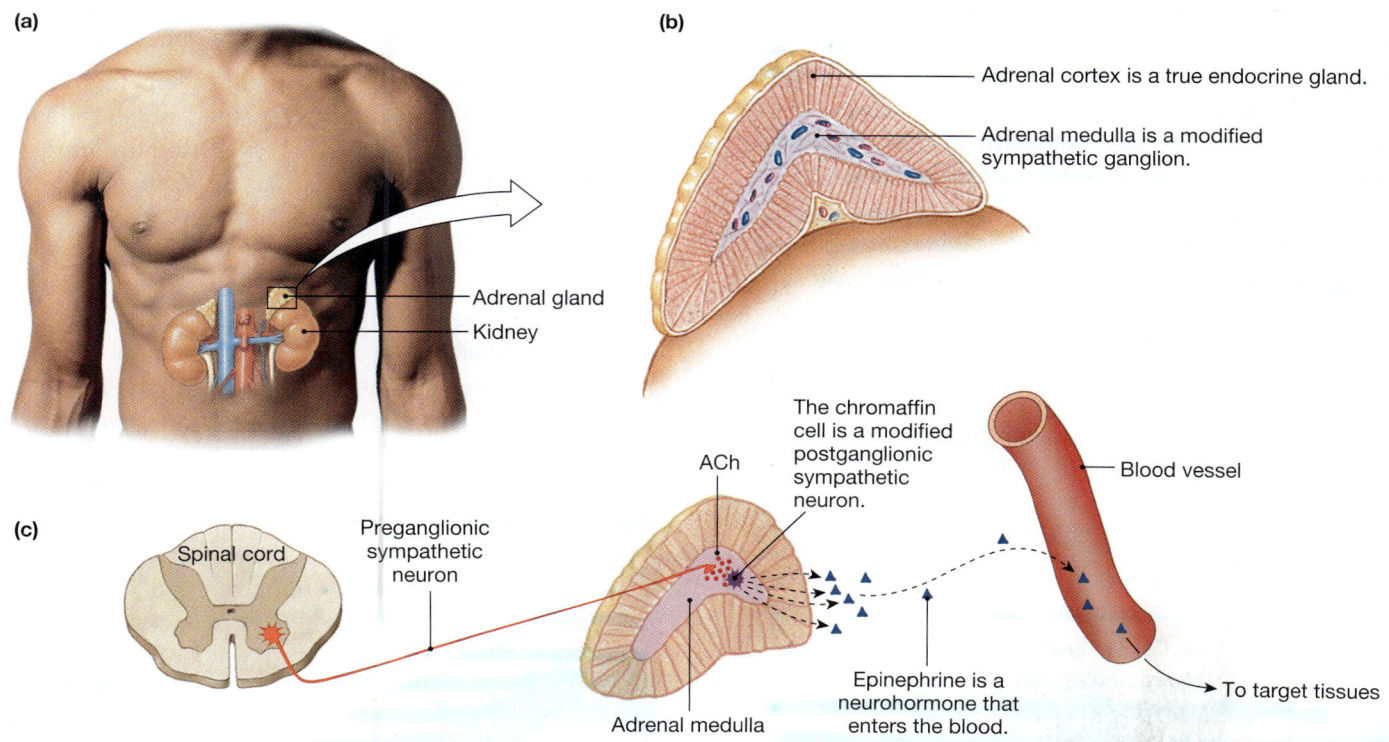

(a)

(b)

Adrenal cortex is a true endocrine gland.

Adrenal medulla is a modified sympathetic ganglion.

Adrenal gland
Kidney

The chromaffin cell is a modified postganglionic sympathetic neuron.

ACh

Blood vessel

(c)
Spinal cord
Preganglionic sympathetic neuron

Adrenal medulla

Epinephrine is a neurohormone that enters the blood.

To target tissues

■ **FIGURE 11-10** *The adrenal medulla*

The adrenal medulla secretes epinephrine into the blood.

TABLE 11-3   **Agonists and Antagonists of Neurotransmitter Receptors**

RECEPTOR TYPE	NEUROTRANSMITTER	AGONIST	ANTAGONISTS	INDIRECT AGONISTS/ANTAGONISTS
Cholinergic	Acetylcholine			AChE* *inhibitors:* neostigmine,
Muscarinic		Muscarine	Atropine, scopolamine	
Nicotinic		Nicotine	α-bungarotoxin (muscle only), TEA (tetraethylammonium; ganglia only), curare	
Adrenergic	Norepinephrine (NE), epinephrine			*Stimulate NE release:* ephedrine, amphetamines *Prevents NE uptake:* cocaine
Alpha		Phenylephrine	"Alpha-blockers"	
Beta		Isoproterenol	"Beta-blockers": propranolol (β_1 and β_2), metoprolol (β_1 only)	

*AChE = acetylcholinesterase.

of epidermal origin that secretes steroid hormones [🔁 p. 221]. The adrenal medulla, which forms the small core of the gland, develops from the same embryonic tissue as sympathetic neurons and is a neurosecretory structure.

The adrenal medulla is often described as a *modified sympathetic ganglion.* Preganglionic sympathetic neurons project from the spinal cord to the adrenal medulla, where they synapse (Fig. 11-10c). However, the postganglionic neurons lack the axons that would normally project to target cells. Instead, the axonless cell bodies, called *chromaffin cells,* secrete the neurohormone epinephrine directly into the blood. In response to alarm signals from the CNS, the adrenal medulla releases large amounts of epinephrine for general distribution throughout the body as part of a fight-or-flight response.

✓ CONCEPT CHECK

8. Is the adrenal medulla most like the anterior pituitary or the posterior pituitary? Explain.

9. Which type of ACh receptors do you suppose chromaffin cells have, nicotinic or muscarinic?   Answers: p. 395

Parasympathetic Pathways Secrete Acetylcholine onto Muscarinic Receptors

As a rule, parasympathetic neurons release ACh at their targets. As noted earlier, muscarinic cholinergic receptors [🔁 p. 275] are found at the neuroeffector junctions of the parasympathetic branch. Muscarinic receptors are all G protein-coupled receptors. Receptor activation initiates second messenger pathways, some of which open K^+ or Ca^{2+} channels. The tissue response to activation of a muscarinic receptor varies with the receptor subtype, of which there are at least five.

Autonomic Agonists and Antagonists Are Important Tools in Research and Medicine

The study of the two autonomic branches has been greatly simplified by advances in molecular biology. The genes for many autonomic receptors and their subtypes have been cloned, allowing researchers to create mutant receptors and study their properties. In addition, researchers have either discovered or synthesized a variety of agonist and antagonist molecules (Table 11-3 ■). Direct agonists and antagonists combine with the target receptor to mimic or block neurotransmitter action. Indirect agonists and antagonists act by altering secretion, reuptake, or degradation of neurotransmitters.

For example, cocaine is an indirect agonist that blocks the reuptake of norepinephrine into adrenergic nerve terminals, thereby extending norepinephrine's stimulatory effect on the target. *Anticholinesterases* (cholinesterase inhibitors) are indirect agonists that block ACh degradation and extend the active life of each ACh molecule. The toxic *organophosphate insecticides,* such as parathion and malathion, are anticholinesterases.

Many drugs used to treat depression are indirect agonists that act either on membrane transporters for neurotransmitters (tricyclic antidepressants and selective serotonin reuptake inhibitors) or on their metabolism (monoamine oxidase inhibitors). The older antidepressant drugs that act on norepinephrine transport and metabolism (tricyclics and MAO inhibitors) may have side effects related to their actions in the autonomic nervous system, including cardiovascular problems, constipation, urinary difficulty, and sexual dysfunction [*dys-*, abnormal or ill]. The newer serotonin reuptake inhibitors have fewer autonomic side effects because serotonin is primarily a CNS neurotransmitter.

One research study that examined brains at autopsy found that smokers have a greater number of neurotransmitter receptors that bind nicotine than do nonsmokers. Normally, chronic exposure to an agonist causes cells to down-regulate their receptors. Up-regulation, an increase in receptor numbers, is usually seen when cells are chronically exposed to receptor antagonists. Molecular studies of the nicotinic cholinergic receptor show that initial binding of ACh or agonists opens the receptor's channel, but continued exposure to ACh closes the channel.

Question 3:
Although nicotine has been shown in short-term studies to be an agonist at nicotinic receptors, chronic exposure to nicotine causes up-regulation of the receptors. Speculate on why up-regulation might occur.

Question 4:
Name another ion channel you have studied that opens in response to a stimulus but inactivates and closes shortly thereafter.

| 377 | 378 | 382 | **387** | 390 | 392 |

DIABETES

AUTONOMIC NEUROPATHY

Primary disorders of the autonomic division are rare, but the condition known as **diabetic autonomic neuropathy** is quite common. This complication of diabetes often begins as a sensory neuropathy, with tingling and loss of sensation in the hands and feet. In some patients, pain is the primary symptom [🔁 p. 340]. About 30% of diabetic patients go on to develop autonomic neuropathies, manifested by dysfunction of the cardiovascular, gastrointestinal, urinary, and reproductive systems (abnormal heart rate, constipation, incontinence, impotence). The cause of diabetic neuropathies is not clear. Patients who have chronically elevated blood glucose levels are more likely to develop neuropathies, but the underlying metabolic pathway has not been identified. Other contributing factors for neuropathy include decreased blood flow and autoimmune reactions. Currently there is no prevention for diabetic neuropathies other than controlling blood glucose levels, and no cure. The only recourse for patients is taking drugs that treat the symptoms.

Many new drugs have been developed from our studies of agonists and antagonists. The discovery of α- and β-adrenergic receptors led to the development of drugs that block only one of the two receptor types. Antagonists of α-receptors proved to have little clinical significance, but the drugs known as beta-blockers have given physicians a powerful tool for treating high blood pressure, one of the most common disorders in the United States today.

Primary Disorders of the Autonomic Nervous System Are Relatively Uncommon

Diseases and malfunction of the autonomic nervous system are relatively rare. Direct damage (trauma) to hypothalamic control centers may disrupt the body's ability to regulate water balance or temperature. Generalized sympathetic dysfunction may result from systemic diseases such as cancer and diabetes mellitus. There are also some conditions, such as *multiple system atrophy,* in which the CNS control centers for autonomic functions degenerate. In many cases of sympathetic dysfunction, the symptoms are manifested most strongly in the cardiovascular system, when diminished sympathetic input to blood vessels results in abnormally low blood pressure. Other prominent symptoms of sympathetic pathology include urinary *incontinence* [*in-,* unable + *continere,* to contain], which is the loss of bladder control, and *impotence,* which is the inability to obtain or sustain a penile erection.

Occasionally, patients suffer from primary autonomic failure when sympathetic neurons degenerate. In the face of continuing diminished sympathetic input, target tissues up-regulate [🔁 p. 190], putting more receptors into the cell membrane to maximize the cell's response to available norepinephrine. This increase in receptor abundance leads to *denervation hypersensitivity,* a state in which the administration of exogenous adrenergic agonists causes a greater-than-expected response.

Summary of Sympathetic and Parasympathetic Branches

As you have seen in this discussion, the branches of the autonomic nervous system share some features but are distinguished by others. Many of these features are summarized in Figure 11-11 ■ and compared in Table 11-4 ■.

1. Both sympathetic and parasympathetic pathways consist of two neurons (preganglionic and postganglionic) in series. One exception to this rule is the adrenal medulla, in which postganglionic sympathetic neurons have been modified into a neuroendocrine organ.

2. All preganglionic autonomic neurons secrete acetylcholine onto nicotinic receptors. Most sympathetic neurons secrete norepinephrine onto adrenergic receptors.

11

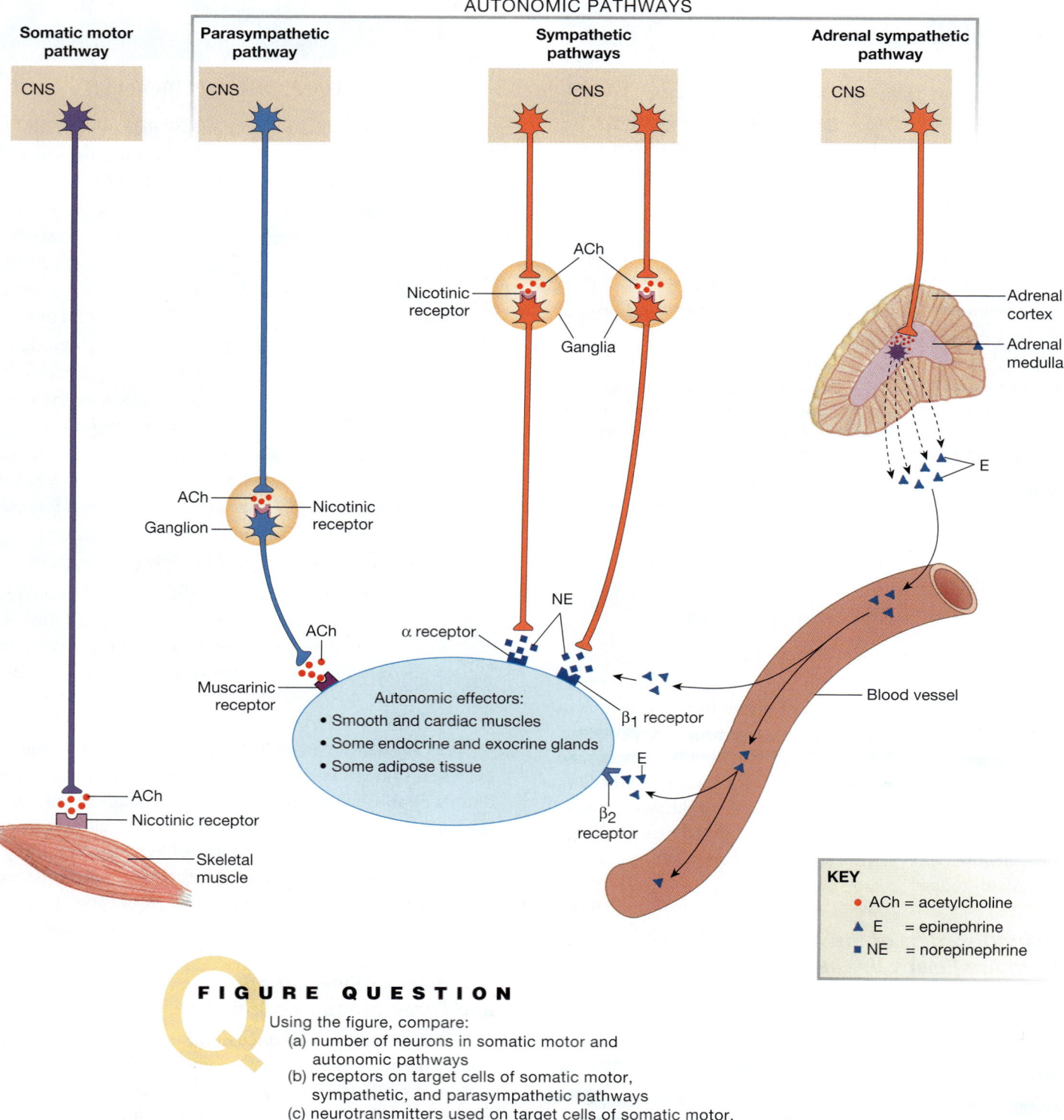

FIGURE QUESTION

Using the figure, compare:
(a) number of neurons in somatic motor and autonomic pathways
(b) receptors on target cells of somatic motor, sympathetic, and parasympathetic pathways
(c) neurotransmitters used on target cells of somatic motor, sympathetic, and parasympathetic pathways
(d) receptor subtypes for epinephrine to subtypes for norepinephrine
(e) ganglion location for sympathetic and parasympathetic pathways.

■ **FIGURE 11-11** *Summary of the efferent pathways of the peripheral nervous system*

TABLE 11-4 **Comparison of Sympathetic and Parasympathetic Branches**

	SYMPATHETIC	PARASYMPATHETIC
Point of CNS origin	1st thoracic to 2nd lumbar segments	Midbrain, medulla, and 2nd–4th sacral segments
Location of peripheral ganglia	Primarily in paravertebral sympathetic chain; 3 outlying ganglia located alongside descending aorta	On or near target organs
Structure of region from which neurotransmitter is released	Varicosities	Varicosities and axon terminals
Neurotransmitter at target synapse	Norepinephrine (adrenergic neurons)	ACh (cholinergic neurons)
Inactivation of neurotransmitter at synapse	Uptake into varicosity, diffusion	Enzymatic breakdown, diffusion
Neurotransmitter receptors on target cells	α and β	Muscarinic
Ganglionic synapse	ACh on nicotinic receptor	ACh on nicotinic receptor
Neuron-target synapse	NE on α- or β-receptor	ACh on muscarinic receptor

Most parasympathetic neurons secrete acetylcholine onto muscarinic receptors.

3. Sympathetic pathways originate in the thoracic and lumbar regions of the spinal cord. Most sympathetic ganglia are located close to the spinal cord (are *paravertebral*). Parasympathetic pathways leave the CNS at the brain stem and in the sacral region of the spinal cord. Parasympathetic ganglia are located close to or in the target tissue.

4. The sympathetic branch controls functions that are useful in stress or emergencies (fight-or-flight). The parasympathetic branch is dominant during rest-and-digest activities.

THE SOMATIC MOTOR DIVISION

Somatic motor pathways, which control skeletal muscles, differ from autonomic pathways both anatomically and functionally (Table 11-5 ■). Somatic motor pathways have a single neuron that originates in the CNS and projects its axon to the target tissue, which is always a skeletal muscle. Unlike autonomic pathways, which may be either excitatory or inhibitory, somatic pathways are always excitatory.

TABLE 11-5 **Comparison of Somatic and Autonomic Divisions**

	SOMATIC	AUTONOMIC
Number of neurons in efferent path	1	2
Neurotransmitter/receptor at neuron-target synapse	ACh/nicotinic	ACh/muscarinic or NE/α or β
Target tissue	Skeletal muscle	Smooth and cardiac muscle; some endocrine and exocrine glands; some adipose tissue
Neurotransmitter released from	Axon terminals	Varicosities and axon terminals
Effects on target tissue	Excitatory only: muscle contracts	Excitatory or inhibitory
Peripheral components found outside the CNS	Axons only	Preganglionic axons, ganglia, postganglionic neurons
Summary of function	Posture and movement	Visceral function, including movement in internal organs and secretion; control of metabolism

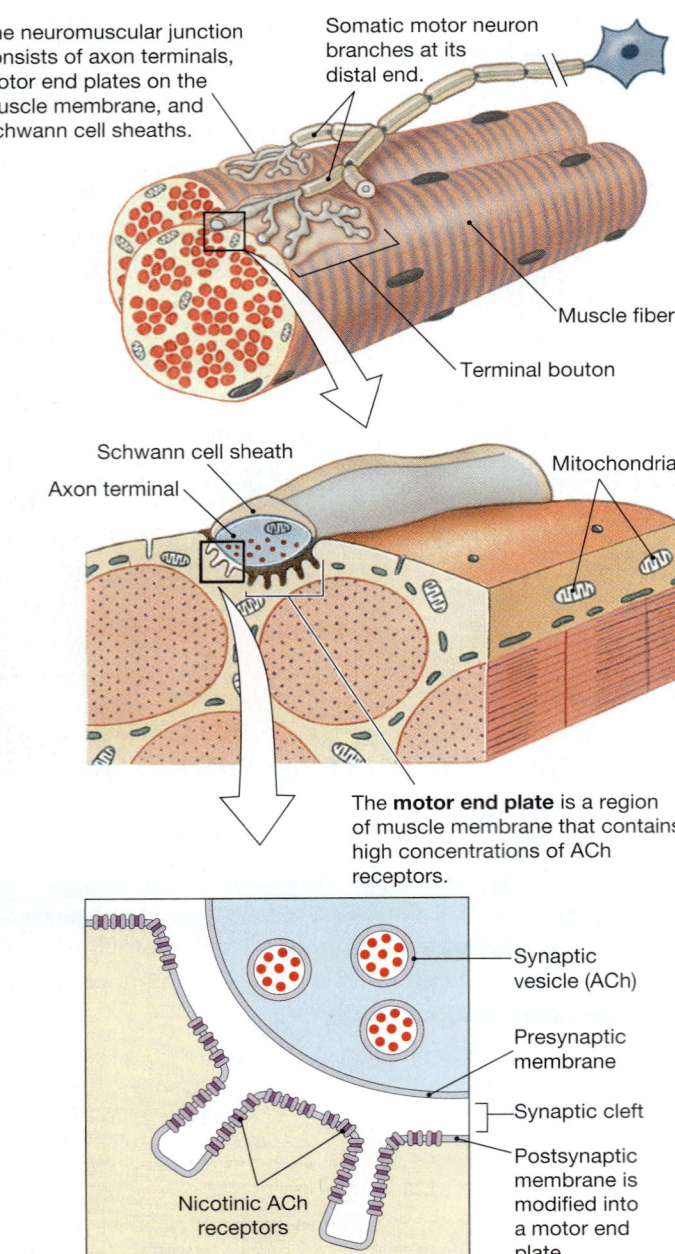

The neuromuscular junction consists of axon terminals, motor end plates on the muscle membrane, and Schwann cell sheaths.

Somatic motor neuron branches at its distal end.

Muscle fiber

Terminal bouton

Schwann cell sheath

Axon terminal

Mitochondria

The **motor end plate** is a region of muscle membrane that contains high concentrations of ACh receptors.

Synaptic vesicle (ACh)

Presynaptic membrane

Synaptic cleft

Postsynaptic membrane is modified into a motor end plate.

Nicotinic ACh receptors

■ **FIGURE 11-12**   *Anatomy of the neuromuscular junction*

A Somatic Motor Pathway Consists of One Neuron

The cell bodies of somatic motor neurons are located either in the ventral horn of the spinal cord [p. 301] or in the brain, with a long single axon projecting to the skeletal muscle target (Fig. 11-11). These myelinated axons may be a meter or more in length, like the somatic motor neurons that innervate the muscles of the foot and hand.

Somatic motor neurons branch close to their targets. Each branch divides into a cluster of enlarged axon terminals that lie on the surface of the skeletal muscle fiber (Fig. 11-12 ■). This branching structure allows a single motor neuron to control many muscle fibers at one time.

The synapse of a somatic motor neuron on a muscle fiber is called the **neuromuscular junction** (Fig. 11-12). Like all other synapses, this region has three components: (1) the motor neuron's presynaptic axon terminal filled with synaptic vesicles and mitochondria, (2) the synaptic cleft, and (3) the postsynaptic membrane of the skeletal muscle fiber.

In addition, the neuromuscular junction includes extensions of Schwann cells that form a thin layer covering the top of the axon terminals. For years it was thought that this cell layer simply provided insulation to speed up the conduction of the action potential, but it is now known that Schwann cells play a critical role in the formation of neuromuscular junctions during development, and that they continue to secrete supportive growth factors throughout their life.

On the postsynaptic side of the neuromuscular junction, the muscle cell membrane that lies opposite the axon terminal is modified into a **motor end plate**, a series of folds that look like shallow gutters. Along the upper edge of each gutter, nicotinic ACh receptor channels cluster together in an active zone. Between the axon and the muscle, the synaptic cleft is filled with a fibrous matrix whose collagen fibers hold the axon terminal and the motor end plate in the proper alignment. The matrix also

(a) Acetylcholine (ACh) combines with nicotinic receptors or is metabolized by acetylcholinesterase (AChE).

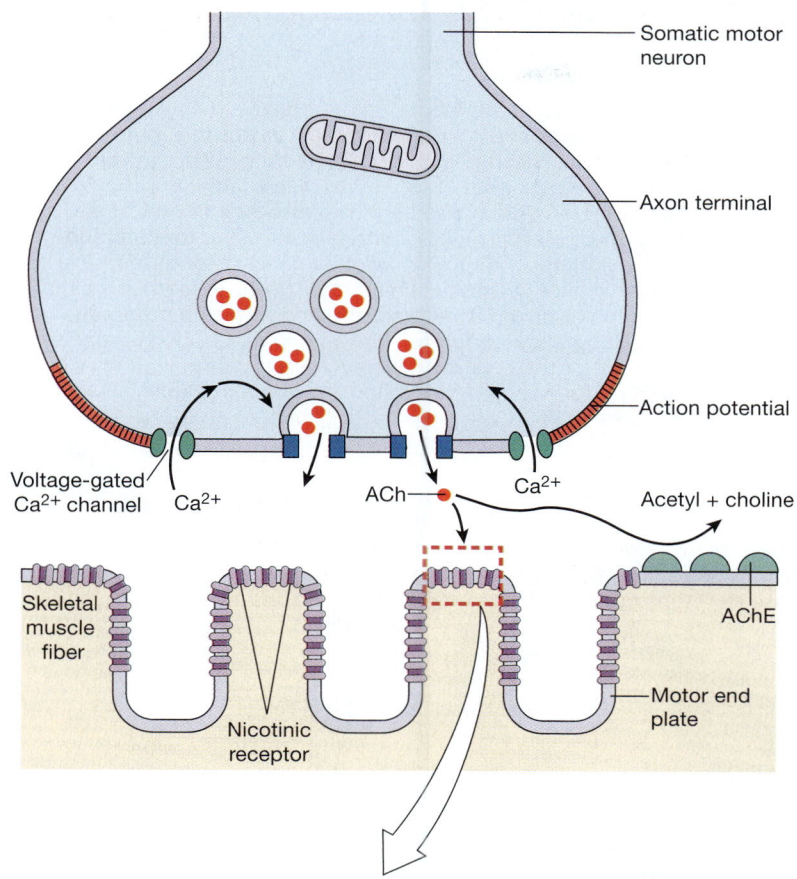

(b) The nicotinic cholinergic receptor binds two ACh molecules, opening a nonspecific monovalent cation channel.

Closed channel

Open channel: ACh bound to nicotinic receptor allows Na^+ and K^+ to pass.

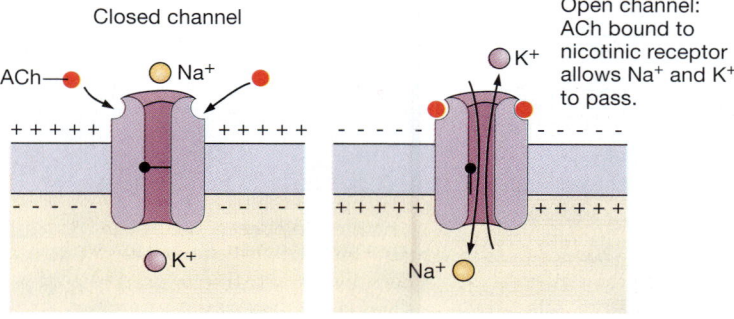

■ **FIGURE 11-13** *Events at the neuromuscular junction*

The Neuromuscular Junction Contains Nicotinic Receptors

As in all neurons, action potentials arriving at the axon terminal open voltage-gated Ca^{2+} channels in the membrane. Calcium diffuses into the cell down its electrochemical gradient, triggering the release of ACh-containing synaptic vesicles. Acetylcholine diffuses across the synaptic cleft and combines with nicotinic receptor channels on the skeletal muscle membrane (Fig. 11-13 ■). These receptors are similar but not identical to the nicotinic ACh receptors found in autonomic ganglia (and those found in the CNS). This difference is evidenced by the fact that the snake toxin α-bungarotoxin binds to nicotinic skeletal muscle receptors but not to those in autonomic ganglia.

Nicotinic cholinergic receptors (nAChR) are chemically gated ion channels with two binding sites for ACh (Fig. 11-13b). When ACh binds to the receptor, the channel gate opens and allows monovalent cations to flow through. Net Na^+ entry into the muscle fiber depolarizes it, triggering an action potential that causes contraction of the skeletal muscle cell.

Acetylcholine acting on a skeletal muscle's motor end plate is always excitatory and creates muscle contraction. There is no antagonistic innervation to relax skeletal muscles. Instead, relaxation occurs when the somatic motor neurons are inhibited in the CNS, preventing ACh release. In Chapter 13 you will learn more about how inhibition of somatic motor pathways controls body movement.

CONCEPT CHECK

12. A nonsmoker who chews nicotine-containing gum might notice an increase in heart rate, a function controlled by sympathetic neurons. If postganglionic sympathetic neurons secrete norepinephrine, why would nicotine affect heart rate?

13. Patients with myasthenia gravis have a deficiency of ACh receptors on their skeletal muscles and have weak muscle function as a result. Why would administration of an anticholinesterase drug (one that inhibits acetylcholinesterase) improve muscle function in these patients?
    Answers: p. 395

Somatic motor neurons do more than simply create contractions: they are necessary for muscle health. "Use it or lose it" is a cliché that is very appropriate to the dynamics of muscle mass because disrupting synaptic transmission at the neuromuscular junction has devastating effects on the entire body. Without communication between the motor neuron and the muscle, the skeletal muscles for movement and posture weaken, as do the skeletal muscles for breathing. In the severest cases, loss of respiratory function can be fatal unless the patient is placed on artificial ventilation. Myasthenia gravis, the loss of ACh receptors discussed earlier, is the most common disorder of the neuromuscular junction.

contains **acetylcholinesterase** (AChE), the enzyme that rapidly deactivates ACh by degrading it into acetyl and choline [🔁p. 272].

CONCEPT CHECK

10. Compare gating and ion selectivity of ion channels in the motor end plate with that of ion channels along the axon of a somatic motor neuron.

11. Is the ventral horn of the spinal cord, which contains the cell bodies of somatic motor neurons, gray matter or white matter?
    Answers: p. 395

A POWERFUL ADDICTION

Shanika is determined to stop smoking this time because her grandfather, a smoker for many years, was just diagnosed with lung cancer. Finding that the patch alone does not stop her craving for a cigarette, she attends behavioral modification classes as well. After six months, she proudly informs her family that she thinks she has kicked the habit.

As you learned in this problem, chronic exposure to nicotine causes target cells to up-regulate their receptor number. When smokers stop smoking, the removal of nicotine from the system allows receptors to return to an active state. The nicotine patch allows the former smoker to gradually decrease nicotine levels in the body, preventing withdrawal symptoms during the time the cells are down-regulating their receptors back to the normal number.

Neuroscientists have learned that addictive behaviors develop because addictive chemicals act as *positive reinforcers* in the brain. Nicotine enhances dopamine release in the brain's reward centers and creates pleasurable sensations. Over time, the brain also begins to associate the social aspects of cigarette smoking with pleasure. The combination of physical dependence on nicotine and the conditioned response makes quitting even more difficult. To learn more about nicotine addiction and smoking cessation programs, try a Google search for *nicotine addiction*. Check your understanding of this running problem by comparing your answers to the information in the following summary table.

	QUESTION	FACTS	INTEGRATION AND ANALYSIS
1	What is the usual response of cells that are chronically exposed to elevated concentrations of a signal molecule?	A cell exposed to elevated concentrations of a signal molecule will decrease its receptors for that molecule.	Down-regulation of receptors allows a cell to respond normally even if the concentration of ligand is elevated.
2	Cholinergic receptors are classified as either nicotinic or muscarinic, on the basis of the agonist molecules that bind to them. What happens to a postsynaptic cell when nicotine rather than ACh binds to a nicotinic cholinergic receptor?	Nicotine is an agonist of ACh. Agonists mimic the activity of a ligand.	Nicotine binding to a nAChR will open ion channels in the postsynaptic cell, and the cell will depolarize. This is the same effect that ACh binding creates.
3	Although nicotine has been shown in short-term studies to be an agonist at nicotinic receptors, chronic exposure to nicotine causes up-regulation of the receptors. Speculate on why up-regulation might occur.	Chronic exposure to an agonist usually causes down-regulation. Chronic exposure to an antagonist usually causes up-regulation. nAChR channels open with initial exposure to agonists but close with continued exposure.	Although nicotine is a short-term agonist, it appears to be acting as an antagonist during long-term exposure. Cells up-regulate their receptors when exposed to antagonists.
4	Name another ion channel you have studied that opens in response to a stimulus but inactivates and closes shortly thereafter.	The voltage-gated Na⁺ channel of the axon first opens, then closes when the inactivation gate shuts.	[Not applicable]
5	Why might excessive levels of nicotine cause respiratory paralysis?	Nicotinic receptors are found at the neuromuscular junction that controls skeletal muscle contraction. The diaphragm and chest wall muscles that regulate breathing are skeletal muscles.	The nicotinic receptors of the neuromuscular junction are not as sensitive to nicotine as are those of the CNS and autonomic ganglia. However, excessively high amounts of nicotine will activate the nAChR of the motor end plate, causing the muscle fiber to depolarize and contract. The continued presence of nicotine keeps these ion channels open, and the muscle remains depolarized. In this state, the muscle is unable to contract again, resulting in paralysis.

377 378 382 387 390 392

CHAPTER SUMMARY

This chapter completes our survey of the nervous system: the central nervous system, which acts as the integrating center for nervous reflexes; the sensory (afferent) system, which is the input division of the peripheral nervous system; and the autonomic and somatic motor divisions, which are the output pathways of the PNS. *Communication* among the three divisions depends primarily on chemical signaling and *molecular interactions* between neurotransmitters and their receptors. *Homeostasis* requires constant surveillance of body parameters by the nervous system, working in conjunction with the endocrine and immune systems. As you learn about the function of other body systems, you will continue to revisit the principles of communication and coordination introduced in Chapters 6 through 11.

The Autonomic Division

1. The efferent division of the peripheral nervous system is divided into **somatic motor neurons**, which control skeletal muscles, and **autonomic neurons**, which control smooth muscle, cardiac muscle, many glands, lymphoid tissue, and some adipose tissue. (p. 377)

2. The autonomic division is subdivided into a **sympathetic branch** and a **parasympathetic branch**. (p. 377; Tbl. 11-4)

3. The maintenance of homeostasis within the body is a balance of autonomic control, endocrine control, and behavioral responses. (p. 378)

4. The autonomic division is controlled by centers in the hypothalamus, pons, and medulla. Some autonomic reflexes are spinal reflexes. Many of these can be modulated by input from the brain. (p. 378; Figs. 11-2, 11-3)

5. The two autonomic branches demonstrate Cannon's properties of homeostasis: maintenance of the internal environment, tonic control, antagonistic control, and variable tissue responses. (p. 379)

6. An autonomic pathway is composed of a **preganglionic neuron** from the CNS that synapses with a **postganglionic neuron** in an **autonomic ganglion**. Autonomic ganglia can modulate and integrate information passing through them. (p. 379; Fig. 11-4)

7. Most sympathetic pathways originate in the thoracic and lumbar regions of the spinal cord. Most sympathetic ganglia lie either close to the spinal cord or along the descending aorta. (p. 381; Fig. 11-5)

8. Parasympathetic pathways originate in the brain stem or the sacral region of the spinal cord. Parasympathetic ganglia are located on or near their target organs. (p. 381; Fig. 11-5)

9. The primary autonomic neurotransmitters are **acetylcholine** and **norepinephrine**. All preganglionic neurons secrete ACh onto **nicotinic cholinergic receptors**. As a rule, postganglionic sympathetic neurons secrete norepinephrine onto **adrenergic receptors**, and postganglionic parasympathetic neurons secrete ACh onto **muscarinic cholinergic receptors**. (p. 383; Fig. 11-7, Tbl. 11-1)

10. The synapse between an autonomic neuron and its target cells is called the **neuroeffector junction**. (p. 382)

11. Autonomic axons end with **varicosities** from which neurotransmitter is released. (p. 383; Figs. 11-8, 11-9)

12. The **adrenal medulla** secretes epinephrine and is controlled by sympathetic preganglionic neurons. (p. 385; Fig. 11-10)

13. Adrenergic receptors are G protein-coupled receptors. Alpha receptors respond most strongly to norepinephrine. β_1**-receptors** respond equally to norepinephrine and epinephrine. β_2**-receptors** are not associated with sympathetic neurons and respond most strongly to epinephrine. β_3**-receptors** respond most strongly to norepinephrine. (p. 383; Fig. 11-11, Tbl. 11-2)

14. Muscarinic receptors are also G protein-coupled receptors. (p. 386)

The Somatic Motor Division

15. Somatic motor pathways, which control skeletal muscles, have a single neuron that originates in the CNS and terminates on a skeletal muscle. Somatic motor neurons are always excitatory and cause muscle contraction. (p. 389; Fig. 11-11)

16. A single **somatic motor neuron** controls many muscle fibers at one time. (p. 390)

17. The synapse of a somatic motor neuron on a muscle fiber is called the **neuromuscular junction**. The muscle cell membrane is modified into a **motor end plate** that contains a high concentration of nicotinic ACh receptors. (p. 390; Fig. 11-12)

18. ACh binding to nicotinic receptor opens cation channels. Net Na^+ entry into the muscle fiber depolarizes the fiber. Acetylcholine in the synapse is broken down by the enzyme **acetylcholinesterase**. (p. 391; Fig. 11-13)

11

QUESTIONS

(Answer to the Review Questions begin on page A1.)

▶ THE PHYSIOLOGY PLACE

Access more review material online at **The Physiology Place** website. There you'll find review questions, problem-solving activities, case studies, flashcards, and direct links to both *InterActive Physiology*® and *PhysioEx*™. To access the site, go to *www.physiologyplace.com* and select Human Physiology, Fourth Edition.

LEVEL ONE   REVIEWING FACTS AND TERMS

1. Name the two efferent divisions of the peripheral nervous system. What type of effectors does each control?

2. The autonomic nervous system is sometimes called the _____ nervous system. Why is this an appropriate name? List some functions controlled by the autonomic nervous system.

3. What are the two branches of the autonomic nervous system? How are these branches distinguished from each other anatomically and physiologically?

4. Which neurosecretory endocrine gland is closely allied to the sympathetic branch?

5. Neurons that secrete acetylcholine are described as _____ neurons, whereas those that secrete norepinephrine are called either _____ or _____ neurons.

6. List four things that can happen to autonomic neurotransmitters after they are released into a synapse.

7. The main enzyme responsible for catecholamine degradation is _____, abbreviated _____.

8. Somatic motor pathways
   (a) are excitatory or inhibitory?
   (b) are composed of a single neuron or a preganglionic and a post-ganglionic neuron?
   (c) synapse with glands or with smooth, cardiac, or skeletal muscle?

9. What is acetylcholinesterase? Describe its action.

10. What kind of receptor is found on the postsynaptic cell in a neuro-muscular junction?

LEVEL TWO   REVIEWING CONCEPTS

11. Concept map: Use the following terms to make a map comparing the somatic motor division and the sympathetic and parasympathetic branches of the autonomic division. You may add additional terms.

acetylcholine	muscarinic receptor
adipose tissue	nicotinic receptor
alpha receptor	norepinephrine
autonomic division	one-neuron pathway
beta receptor	parasympathetic branch
cardiac muscle	skeletal muscle
cholinergic receptor	smooth muscle
efferent division	somatic motor division
endocrine gland	sympathetic branch
exocrine gland	two-neuron pathway
ganglion	

12. What is the advantage of divergence of neural pathways in the autonomic nervous system?

13. Compare and contrast
   (a) neuroeffector junctions and neuromuscular junctions.
   (b) alpha, beta, muscarinic, and nicotinic receptors. Describe where each is found and the ligands that bind to them.

14. Compare and contrast
   (a) autonomic ganglia and CNS nuclei.
   (b) the adrenal medulla and the posterior pituitary gland.
   (c) boutons and varicosities.

15. If a target cell's receptor is _____ (use items in left column), the neuron(s) releasing neurotransmitter onto the receptor must be _____ (use all appropriate items in right column).
   (a) nicotinic cholinergic   1. somatic motor neuron
   (b) adrenergic α   2. autonomic preganglionic neuron
   (c) muscarinic cholinergic   3. sympathetic postganglionic neuron
   (d) adrenergic β   4. parasympathetic postganglionic neuron

16. Ganglia contain the cell bodies of (choose all that apply)
   (a) somatic motor neurons.
   (b) preganglionic autonomic neurons.
   (c) interneurons.
   (d) postganglionic autonomic neurons.
   (e) sensory neurons.

LEVEL THREE   PROBLEM SOLVING

17. If nicotinic receptor channels allow both Na^+ and K^+ to flow through, why does Na^+ influx exceed K^+ efflux? (Hint: p. 135)

18. You have discovered a neuron that innervates an endocrine cell in the intestine. To learn more about this neuron, you place a marker substance at the endocrine cell synapse. The marker is taken into the neuron and transported in a vesicle by retrograde axonal transport to the nerve cell body.
   (a) By what process is the marker probably taken into the axon terminal?
   (b) The nerve cell body is found in a ganglion very close to the endocrine cell. To which branch of the peripheral nervous system does the neuron probably belong? (Be as specific as you can.)
   (c) Which neurotransmitter do you predict will be secreted by the neuron onto the endocrine cell?

19. The Huaorani Indians of South America use blowguns to shoot darts poisoned with curare at monkeys. Curare is a plant toxin that binds to and inactivates nicotinic ACh receptors. What happens to a monkey struck by one of these darts?

LEVEL FOUR   QUANTITATIVE PROBLEMS

20. The U.S. Centers for Disease Control and Prevention (CDC) conduct biennial Youth Risk Behavior Surveys (YRBS) in which they ask high school students to self-report risky behaviors such as alcohol consumption and smoking. The graphs that follow were created from data in the latest report on cigarette smoking among American high school students (www.cdc.gov/mmwr/preview/mmwrhtml/mm5323a1.htm). *Current smoking* is defined as smoking cigarettes on at least one day in the 30 days preceding the survey.
   (a) What can you say about cigarette smoking among high school students in the ten-year period from 1991 to 2003?
   (b) Which high school students are most likely to be smokers? Least likely to be smokers?

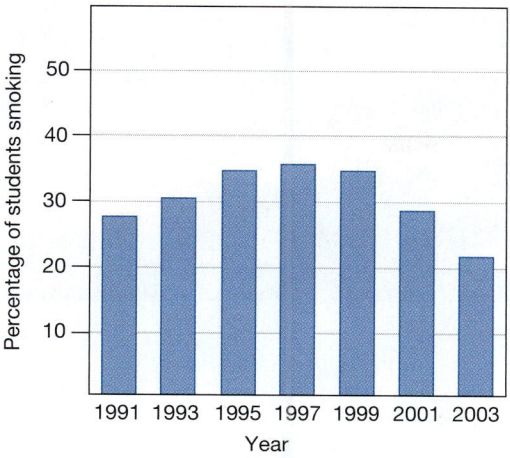

Percentage of students who reported
current smoking (1991-2003)

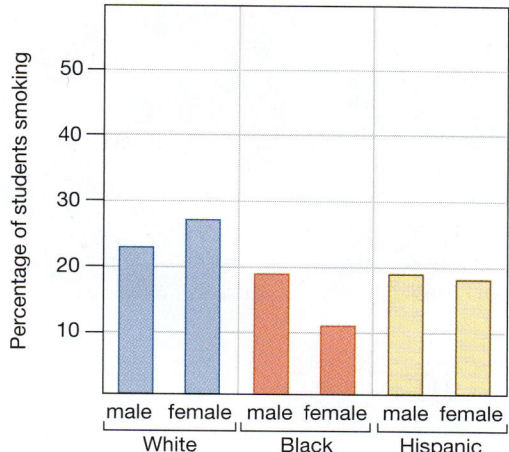

Percentage of students in 2003 who reported
current smoking, separated by sex and race/
ethnicity*

*Other race/ethnic groups are not shown because the
numbers of students reporting were too small for meaningful
statistical analysis.

ANSWERS

✓ *Answers to Concept Check Questions*

Page 378

1. The afferent division consists of sensory receptors and sensory neurons.

2. The CNS consists of the brain and spinal cord.

Page 395

3. Homeostasis is the maintenance of a relatively stable internal environment.

Page 382

4. Mixed nerves carry sensory and motor signals.

5. The regions of the spinal cord are cervical, thoracic, lumbar, and sacral.

Page 385

6. Ca^{2+} is stored in the endoplasmic reticulum.

7. (a) Adenylyl cyclase converts ATP to cAMP; (b) cAMP activates protein kinase A.

Page 386

8. The adrenal medulla is neurosecretory and therefore like the posterior pituitary.

9. Chromaffin cells are modified postganglionic neurons, so they have nicotinic receptors.

Page 391

10. The nAChR of the motor end plate are ligand-gated monovalent cation (Na^+ and K^+) channels. The axon contains voltage-gated channels, with separate channels for Na^+ and K^+ [⮌ p. 254].

11. The ventral horn is gray matter.

Page 391

12. The postganglionic sympathetic neurons are activated by acetylcholine acting on nicotinic receptors. This means nicotine will also excite sympathetic neurons, such as those that increase heart rate.

13. Anticholinesterase drugs decrease the rate at which ACh is broken down at the motor end plate. The slower breakdown rate allows ACh to remain active at the motor end plate for a longer time and helps offset the decrease in active receptors.

Q *Answers to Figure Questions*

Page 380

Figure 11-5: 1. (a) Sympathetic pathways originate in the thoracic and lumbar regions of the spinal cord; parasympathetic pathways originate in the brain stem or sacral region. (b) Sympathetic ganglia are located close to the spinal column or along the descending aorta (not shown); parasympathetic ganglia are located on or near their target organs. 2. Connections between the sympathetic ganglia allow rapid communication within the sympathetic branch.

Page 382

Figure 11-7: (a) The three neurons that secrete ACh are cholinergic. The one neuron that secretes norepinephrine is adrenergic. The cell bodies of preganglionic neurons are in the CNS; the cell bodies of postganglionic neurons are in a ganglion. (b) Parasympathetic pathways have the longer preganglionic neurons.

Page 388

Figure 11-11: (a) Somatic has one neuron, autonomic has two. (b) Somatic motor targets have nicotinic ACh receptors, parasympathetic targets have muscarinic ACh receptors, and sympathetic targets have adrenergic receptors. (c) Somatic motor and parasympathetic pathways use ACh; sympathetic uses norepinephrine. (d) Epinephrine is most active on β_1- and β_2-receptors; norepinephrine is most active on β_1- and α-receptors. (e) Sympathetic ganglia are close to the CNS; parasympathetic ganglia are closer to their target tissues.

11

12

A muscle is...an engine, capable of converting chemical energy into mechanical energy. It is quite unique in nature, for there has been no artificial engine devised with the great versatility of living muscle.

—**Ralph W. Stacy and John A. Santolucito,**
in Modern College Physiology, 1966

One somatic motor neuron branches to innervate several skeletal muscle fibers.

Muscles

BACKGROUND BASICS

RUNNING PROBLEM

PERIODIC PARALYSIS

This morning, Paul Leong, age 6, gave his mother the fright of her life. One minute he was happily playing in the backyard with his new beagle puppy. The next minute, after sitting down to rest, he could not move his legs. In answer to his screams, his mother came running and found her little boy unable to walk. Panic-stricken, she scooped him up, brought him into the house, and dialed 9-1-1. But as she hung up the phone and prepared to wait for the paramedics, Paul got to his feet and walked over to her. "I'm OK now, Mom," he announced. "I'm going outside."

397 410 412 419 423 430

It was his first time to be the starting pitcher. As he ran from the bullpen onto the field, his heart was pounding and his stomach felt as if it were tied in knots. He stepped onto the mound and gathered his thoughts before throwing his first practice pitch. Gradually, as he went through the familiar routine of throwing and catching the baseball, his heart slowed and his stomach relaxed. It was going to be a good game.

The pitcher's pounding heart, queasy stomach, and movements as he runs and throws are all the results of muscle contraction. Our muscles have two common functions: to generate motion and to generate force. Our skeletal muscles also generate heat and contribute significantly to the homeostasis of body temperature. When homeostasis is threatened by cold conditions, the brain may direct our muscles to shiver, creating additional heat.

Three types of muscle tissue occur in the human body: skeletal muscle, cardiac muscle, and smooth muscle. Most **skeletal muscles** are attached to the bones of the skeleton, enabling these muscles to control body movement. **Cardiac muscle** [*kardia,* heart] is found only in the heart and is responsible for moving blood through the circulatory system. Skeletal and cardiac muscles are classified as **striated muscles** [*stria,* groove] because of their alternating light and dark bands under the light microscope (Fig. 12-1a, 1b ■).

Smooth muscle is the primary muscle of internal organs and tubes, such as the stomach, urinary bladder, and blood vessels. Its primary function is to influence the movement of material into, out of, and within the body. An example is the passage of food through the gastrointestinal tract. Under the microscope, smooth muscle lacks the obvious cross-bands of striated muscles (Fig. 12-1c). Its lack of banding results from the less organized arrangement of contractile fibers within the muscle cells.

(a) Skeletal muscle

- Nucleus
- Muscle fiber (cell)
- Striations

(b) Cardiac muscle

- Striations
- Muscle fiber
- Intercalated disk
- Nucleus

(c) Smooth muscle

- Muscle fiber
- Nucleus

■ **FIGURE 12-1**   *The three types of muscles*

Skeletal muscles are often described as voluntary muscles, and smooth and cardiac muscle as involuntary. However, this is not a precise classification. Skeletal muscles can contract without conscious direction, and we can learn a certain degree of conscious control over some smooth and cardiac muscle [p. 377].

Skeletal muscles are unique in that they contract only in response to a signal from a somatic motor neuron. They cannot initiate their own contraction, nor is their contraction influenced directly by hormones. In contrast, cardiac and smooth muscle have multiple levels of control. Although their primary extrinsic control arises through autonomic innervation, some types of smooth and cardiac muscle can contract spontaneously, without signals from the central nervous system. In addition, the activity of cardiac and some smooth muscle is subject to modulation by the endocrine system. Despite these differences, smooth and cardiac muscle share many properties with skeletal muscle.

This chapter discusses skeletal and smooth muscle anatomy and contraction, and then concludes by comparing the properties of skeletal muscle, smooth muscle, and cardiac muscle. We will look at the details of cardiac muscle in Chapter 14 when we study the heart. The metabolism and endocrinology of skeletal muscle will be covered in Chapters 22, 23, and 25.

12

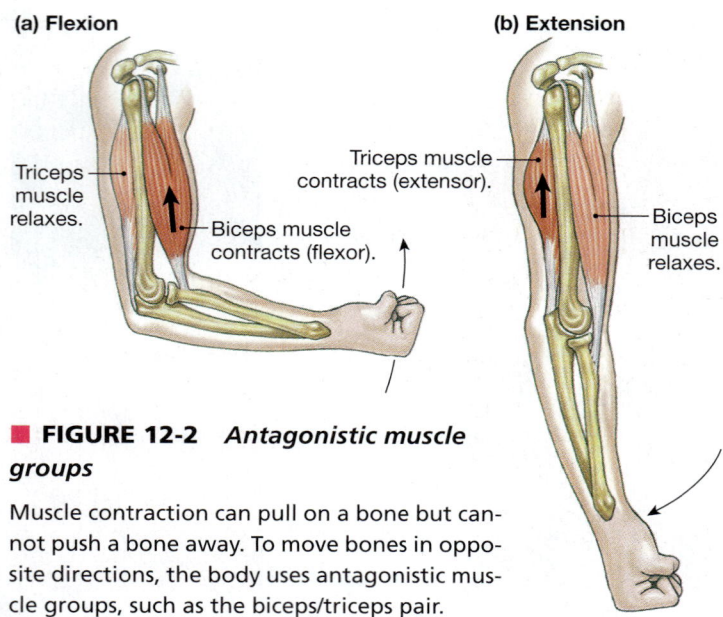

(a) Flexion

Triceps muscle relaxes.

Biceps muscle contracts (flexor).

(b) Extension

Triceps muscle contracts (extensor).

Biceps muscle relaxes.

■ **FIGURE 12-2** *Antagonistic muscle groups*

Muscle contraction can pull on a bone but cannot push a bone away. To move bones in opposite directions, the body uses antagonistic muscle groups, such as the biceps/triceps pair.

SKELETAL MUSCLE

Skeletal muscles make up the bulk of muscle in the body and constitute about 40% of total body weight. They are responsible for positioning and moving the skeleton, as their name suggests. Skeletal muscles are usually attached to bones by **tendons** made of collagen [⇄ p. 79]. The **origin** of a muscle is the end of the muscle that is attached closest to the trunk or to the more stationary bone. The **insertion** of the muscle is the more *distal* [*distantia,* distant] or more mobile attachment.

When the bones attached to a muscle are connected by a flexible joint, contraction of the muscle moves the skeleton. If the centers of the connected bones are brought closer together when the muscle contracts, the muscle is called a **flexor**, and the movement is called *flexion*. If the bones move away from each other when the muscle contracts, the muscle is called an **extensor**, and the movement is called *extension*.

Most joints in the body have both flexor and extensor muscles, because a contracting muscle can pull a bone in one direction but cannot push it back. Flexor-extensor pairs are called **antagonistic muscle groups** because they exert opposite effects. Figure 12-2 ■ shows a pair of antagonistic muscles in the arm: the *biceps brachii* [*brachion,* arm], which acts as the flexor, and the *triceps brachii,* which acts as the extensor. When the biceps muscle contracts, the hand and forearm move toward the shoulder. When the triceps contracts, the flexed forearm moves away from the shoulder. In each case, when one muscle contracts and shortens, the antagonistic muscle must relax and lengthen.

CONCEPT CHECK

1. Identify as many pairs of antagonistic muscle groups in the body as you can. If you cannot name them, point out the probable location of the flexor and extensor of each group.

Answers: p. 434

Skeletal Muscles Are Composed of Muscle Fibers

Muscles function together as a unit. A skeletal muscle is a collection of muscle cells, or **muscle fibers** (see Anatomy Summary, Fig. 12-3 ■), just as a nerve is a collection of neurons. Each skeletal muscle fiber is a long, cylindrical cell with up to several hundred nuclei on the surface of the fiber. Skeletal muscle fibers are the largest cells in the body, created by the fusion of many individual embryonic muscle cells.

The fibers in a given muscle are arranged with their long axes in parallel (Fig. 12-3a), and each skeletal muscle fiber is sheathed in connective tissue. Groups of adjacent fibers are bundled together into units called **fascicles**. Collagen, elastic fibers, nerves, and blood vessels are found between the fascicles. The entire muscle is enclosed in a connective tissue sheath that is continuous with the connective tissue around the muscle fibers and fascicles and with the tendons holding the muscle to underlying bones.

Muscle Fiber Anatomy   Muscle physiologists, like neurobiologists, have adopted a specialized vocabulary (Table 12-1 ■). The cell membrane of a muscle fiber is called the **sarcolemma** [*sarkos,* flesh + *lemma,* shell], and the cytoplasm is called the **sarcoplasm**. The main intracellular structures in striated muscles are **myofibrils** [*myo-,* muscle], highly organized bundles of contractile and elastic proteins that carry out the work of contraction. Skeletal muscles also contain extensive **sarcoplasmic reticulum** (SR), a form of modified endoplasmic reticulum that wraps around each myofibril like a piece of lace (Figs. 12-3b, 12-4 ■). The sarcoplasmic reticulum consists of longitudinal tubules, which release Ca^{2+} ions, and the **terminal cisternae**, which concentrate and sequester Ca^{2+} [*sequestrare,* to put in the hands of a trustee].

A branching network of **transverse tubules**, also known as **t-tubules**, is closely associated with the terminal cisternae. One t-tubule with its two flanking terminal cisternae is known as a *triad* (Fig. 12-4). The membranes of t-tubules are a continuation of the muscle fiber membrane. This makes the lumen of the t-tubules continuous with the extracellular fluid.

TABLE 12-1	Muscle Terminology
GENERAL TERM	**MUSCLE EQUIVALENT**
Muscle cell	Muscle fiber
Cell membrane	Sarcolemma
Cytoplasm	Sarcoplasm
Modified endoplasmic reticulum	Sarcoplasmic reticulum

To understand how this network of tubules in the heart of the muscle fiber communicates with the outside, take a lump of soft clay and poke your finger into the middle of it. Notice how the outside surface of the clay (analogous to the cell membrane of the muscle fiber) is now continuous with the sides of the hole that you poked in the clay (the membrane of the t-tubule).

T-tubules rapidly move action potentials that originate at the neuromuscular junction on the cell surface into the interior of the fiber. Without t-tubules, the action potential could reach the center of the fiber only by the diffusion of positive charge through the cytosol, a slower process that would delay the response time of the muscle fiber.

The cytosol between the myofibrils contains many glycogen granules and mitochondria. Glycogen, the storage form of glucose found in animals, is a reserve source of energy. Mitochondria provide much of the ATP for muscle contraction through oxidative phosphorylation of glucose and other biomolecules.

Myofibrils Are the Contractile Structures of a Muscle Fiber

One muscle fiber contains a thousand or more myofibrils that occupy most of the intracellular volume, leaving little space for cytosol and organelles (Fig. 12-3b). Each myofibril is composed of several types of proteins: the contractile proteins *myosin* and *actin*, the regulatory proteins *tropomyosin* and *troponin*, and the giant accessory proteins *titin* and *nebulin*.

Myosin [*myo-*, muscle] is the motor protein of the myofibril. Various isoforms of myosin occur in different types of muscle and help determine the muscle's speed of contraction. Each myosin molecule is composed of protein chains that intertwine to form a long tail and a pair of tadpolelike heads (Fig. 12-3e). In skeletal muscle, about 250 myosin molecules join to create a **thick filament**. Each thick filament is arranged so that the myosin heads are clustered at the ends of the filament, and the central region of the filament is a bundle of myosin tails. The rodlike core of the thick filament is stiff, but the protruding myosin heads have an elastic hinge region where the heads join the rods. This hinge region allows the heads to swivel around their point of attachment.

Actin [*actum*, to do] is a protein that makes up the **thin filaments** of the muscle fiber. One actin molecule is a globular protein (*G-actin*), represented in Figure 12-3f by a round ball. Usually, multiple G-actin molecules polymerize to form long chains or filaments, called *F-actin*. In skeletal muscle, two F-actin polymers twist together like a double strand of beads, creating the thin filaments of the myofibril.

Most of the time, the parallel thick and thin filaments of the myofibril are connected by **crossbridges** that span the space between the filaments. These crossbridges form when myosin heads of the thick filaments bind loosely to actin in the thin filaments (Fig. 12-3d). Each G-actin molecule has a single binding site for a myosin head.

Under a light microscope, the arrangement of thick and thin filaments in a myofibril creates a repeating pattern of alternating light and dark bands (Figs. 12-1a, 12-3c). One repeat of the pattern forms a **sarcomere** [*sarkos*, flesh + *-mere*, a unit or segment], which has the following elements (Fig. 12-5 ■):

1. **Z disks**: One sarcomere is composed of two Z disks and the filaments found between them. Z disks are zigzag protein structures that serve as the attachment site for thin filaments. The abbreviation *Z* comes from *zwischen*, the German word for "between."

2. **I band**: These are the lightest color bands of the sarcomere and represent a region occupied only by thin filaments. The abbreviation *I* comes from *isotropic*, a description from early microscopists meaning that this region reflects light uniformly under a polarizing microscope. A Z disk runs through the middle of every I band, so each half of an I band belongs to a different sarcomere.

3. **A band**: This is the darkest of the sarcomere's bands and encompasses the entire length of a thick filament. At the outer edges of the A band, the thick and thin filaments overlap. The center of the A band is occupied by thick filaments only. The abbreviation *A* comes from *anisotropic* [*an-*, not], meaning that the protein fibers in this region scatter light unevenly.

4. **H zone**: This central region of the A band is lighter than the outer edges of the A band because the H zone is occupied by thick filaments only. The *H* comes from *helles*, the German word for "clear."

5. **M line**: This band represents proteins that form the attachment site for thick filaments, equivalent to the Z disk for the thin filaments. Each M line divides an A band in half. *M* is the abbreviation for *mittel*, the German word for "middle."

In three-dimensional array, the actin and myosin molecules form a lattice of parallel, overlapping thin and thick filaments, held in place by their attachments to the Z-disk and M-line proteins, respectively (Fig. 12-5b). When viewed end-on, each thin filament is surrounded by three thick filaments, and six thin filaments encircle each thick filament (Fig. 12-5c, rightmost circle).

The proper alignment of filaments within a sarcomere is ensured by two types of proteins: titin and nebulin (Fig. 12-6 ■). **Titin** is a huge elastic molecule and the largest known protein, composed of more than 25,000 amino acids. A single titin molecule stretches from one Z disk to the neighboring M line. To get an idea of the immense size of titin, imagine that one titin molecule is an 8-foot-long piece of the very thick rope used to tie ships to a wharf. By comparison, a single actin molecule would be about the length and weight of a single eyelash.

Titin has two functions: (1) it stabilizes the position of the contractile filaments and (2) its elasticity returns stretched muscles to their resting length. Titin is helped by **nebulin**, an inelastic

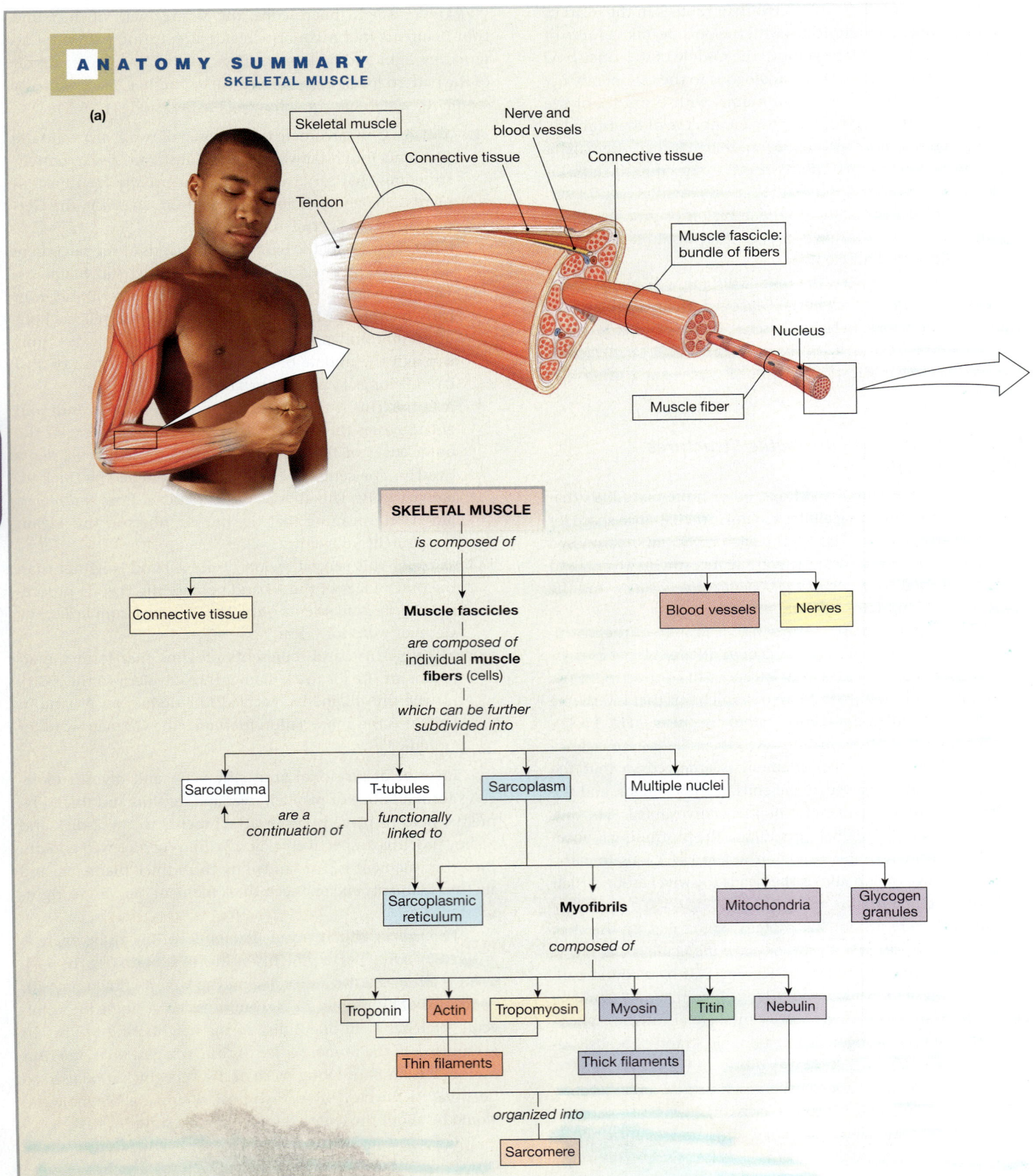

ANATOMY SUMMARY
SKELETAL MUSCLE

(a)

Skeletal muscle

Nerve and blood vessels

Connective tissue

Connective tissue

Tendon

Muscle fascicle: bundle of fibers

Nucleus

Muscle fiber

SKELETAL MUSCLE

is composed of

| Connective tissue | Muscle fascicles | Blood vessels | Nerves |

are composed of individual **muscle fibers** (cells)

which can be further subdivided into

| Sarcolemma | T-tubules | Sarcoplasm | Multiple nuclei |

are a continuation of

functionally linked to

Sarcoplasmic reticulum

Myofibrils

Mitochondria

Glycogen granules

composed of

| Troponin | Actin | Tropomyosin | Myosin | Titin | Nebulin |

Thin filaments

Thick filaments

organized into

Sarcomere

■ **FIGURE 12-3**

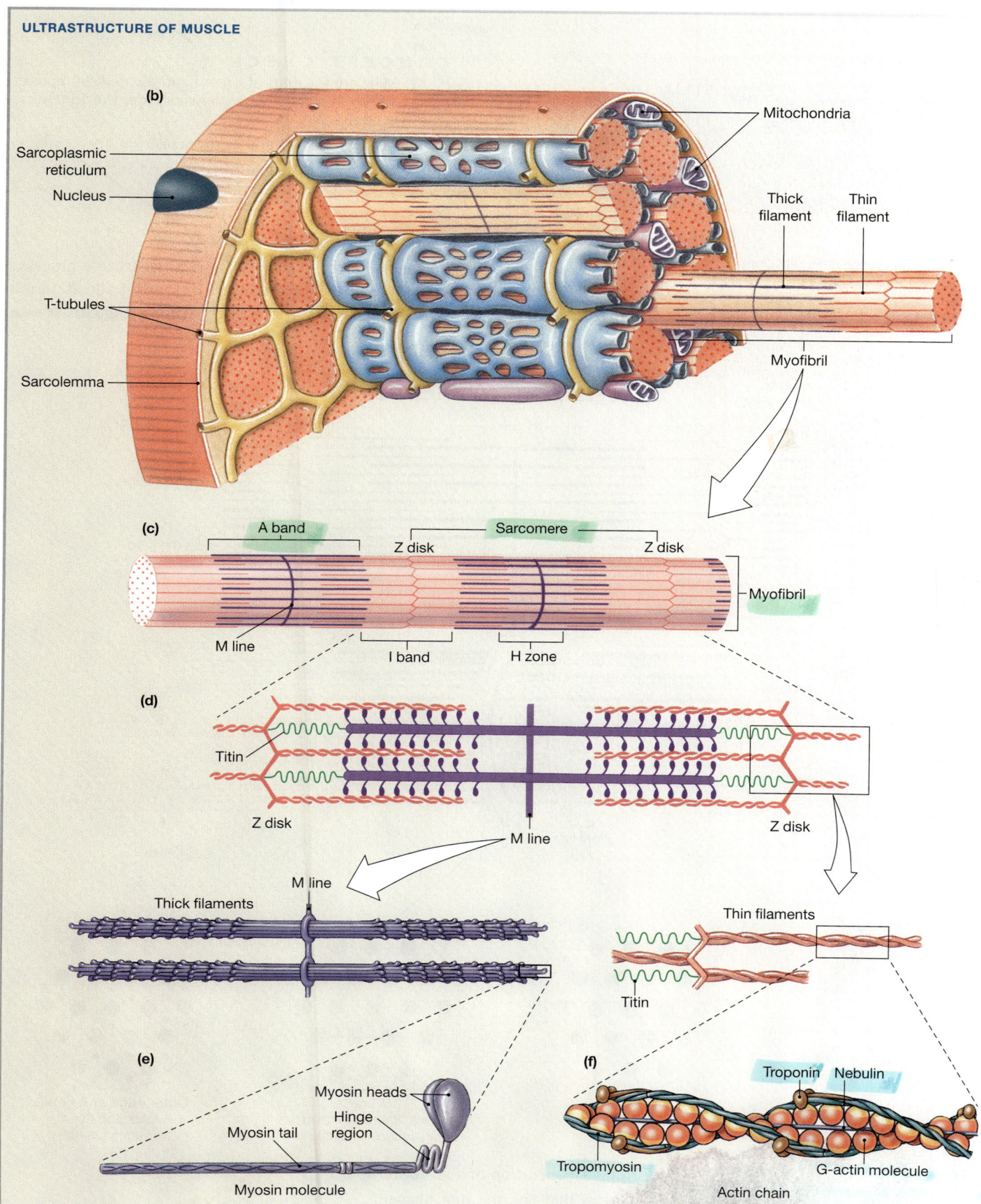

ULTRASTRUCTURE OF MUSCLE

FIGURE 12-3 *(continued)*

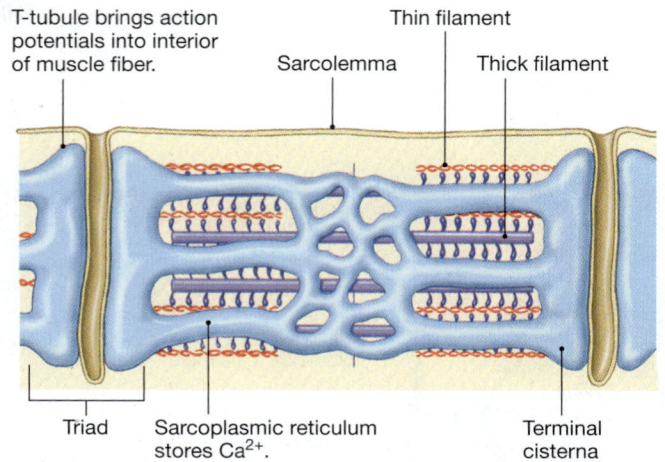

T-tubule brings action potentials into interior of muscle fiber.

Thin filament

Sarcolemma

Thick filament

Triad

Sarcoplasmic reticulum stores Ca^{2+}.

Terminal cisterna

■ **FIGURE 12-4** *T-tubules and the sarcoplasmic reticulum*

giant protein that lies alongside thin filaments and attaches to the Z disk. Nebulin helps align the actin filaments of the sarcomere.

CONCEPT CHECK

2. Why are the ends of the A band the darkest region of the sarcomere when viewed under the light microscope?

3. What is the function of t-tubules?

4. Why are skeletal muscles described as striated?

Answers: p. 434

Muscle Contraction Creates Force

The contraction of muscle fibers is a remarkable process that enables us to create force to move or to resist a load. In muscle physiology, the force created by the contracting muscle is called

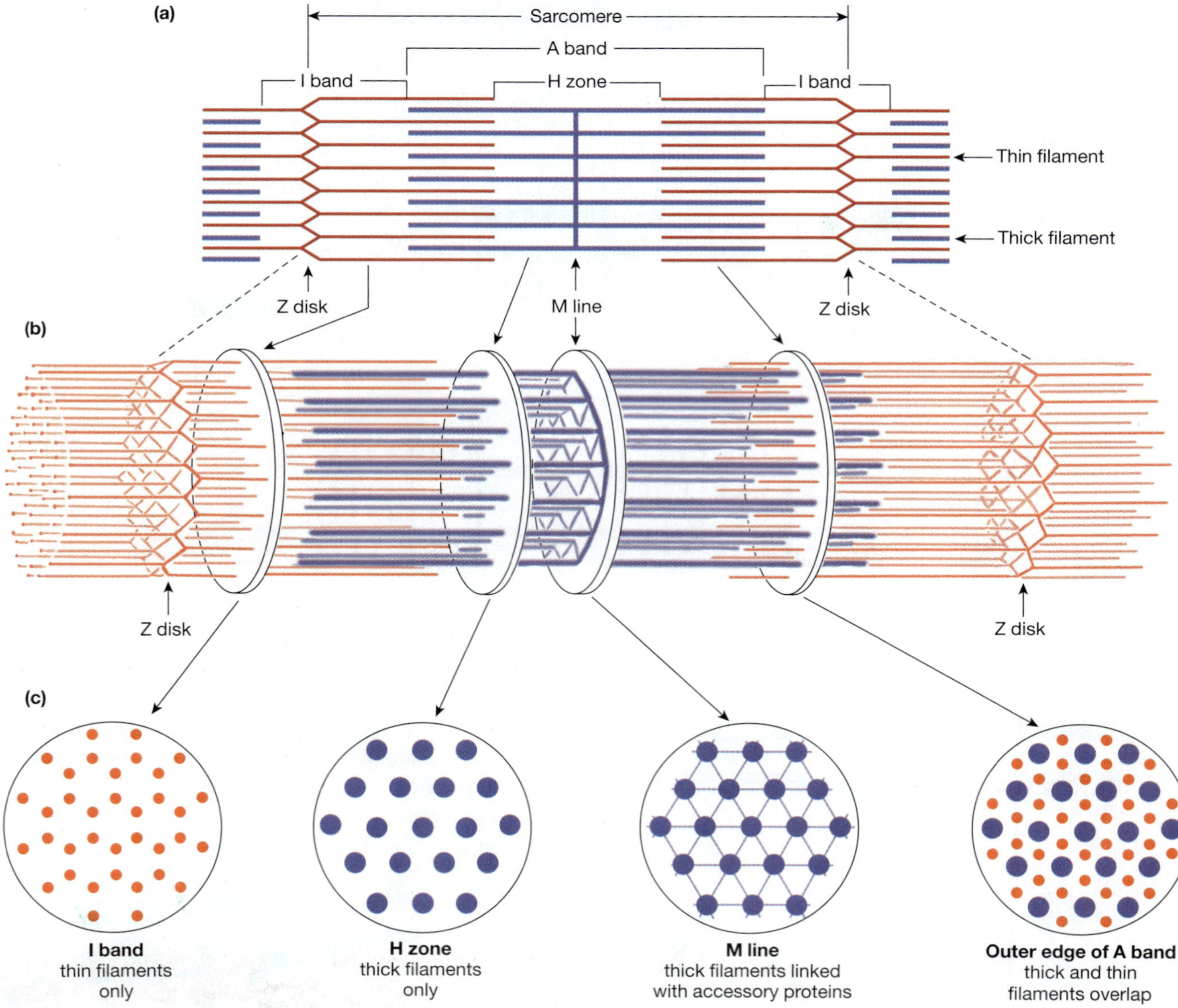

(a)

Sarcomere

A band

I band

H zone

I band

Thin filament

Thick filament

Z disk

M line

Z disk

(b)

Z disk

Z disk

(c)

I band
thin filaments only

H zone
thick filaments only

M line
thick filaments linked with accessory proteins

Outer edge of A band
thick and thin filaments overlap

■ **FIGURE 12-5** *The two- and three-dimensional organization of a sarcomere*

The Z disk (not shown in part c) has accessory proteins that link the thin filaments together, similar to the accessory proteins shown for the M line.

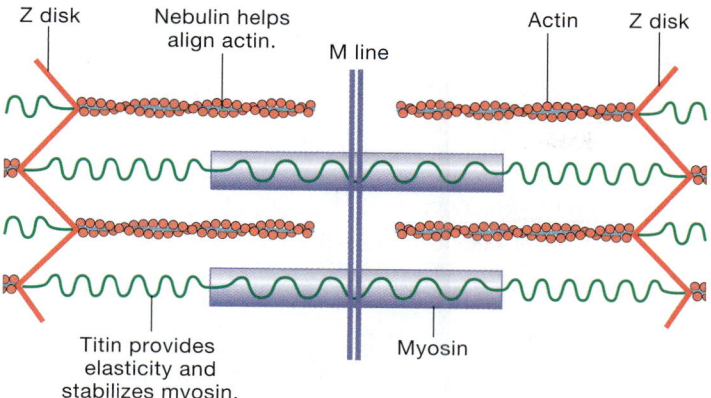

Z disk | Nebulin helps align actin. | M line | Actin | Z disk

Titin provides elasticity and stabilizes myosin. | Myosin

■ **FIGURE 12-6** *Titin and nebulin*

Titin spans the distance from one Z disk to the neighboring M line. Nebulin, lying along the thin filaments, attaches to a Z disk but does not extend to the M line.

the **muscle tension**. The **load** is a weight or force that opposes contraction of a muscle. **Contraction**, the creation of tension in a muscle, is an active process that requires energy input from ATP. **Relaxation** is the release of tension created by a contraction.

Figure 12-7 ■ summarizes the major steps leading up to skeletal muscle contraction.

1. **Events at the neuromuscular junction** convert a chemical signal from a somatic motor neuron into an electrical signal in the muscle fiber. These events were described in Chapter 11 [💧 p. 391].
2. **Excitation-contraction coupling** is the process in which muscle action potentials initiate calcium signals that in turn activate a contraction-relaxation cycle.
3. At the molecular level, a **contraction-relaxation cycle** can be explained by the *sliding filament theory of contraction*. In intact muscles, one contraction-relaxation cycle is called a muscle *twitch*.

In the sections that follow, we will examine each of these steps. We start with the molecular basis for muscle contraction that is the heart of the process. From there, we go to the integrated function of a muscle fiber as it undergoes excitation-contraction coupling. The skeletal muscle section ends with a discussion of the organization of intact muscles and how they move bones around joints.

✔ CONCEPT CHECK

5. What are the three anatomical elements of a neuromuscular junction?
6. What is the chemical signal at a neuromuscular junction?

Answers: p. 434

Muscles Shorten When They Contract

In previous centuries, scientists observed that when muscles move a load, they shorten. This observation led to early theories

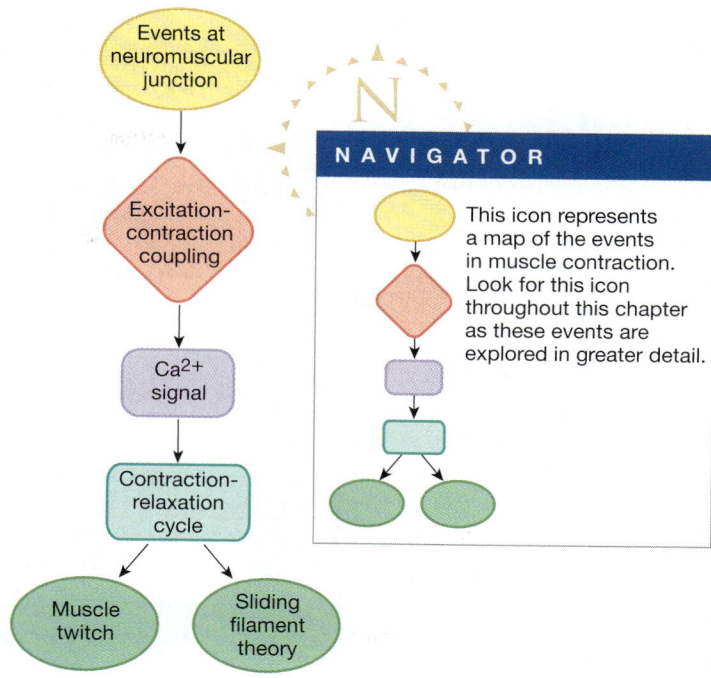

■ **FIGURE 12-7** *Summary of muscle contraction*

of contraction, which proposed that muscles were made of molecules that curled up and shortened when active, then relaxed and stretched at rest, like elastic in reverse. The theory received support when myosin was found to be a helical molecule that shortened upon heating (the reason meat shrinks when you cook it).

In 1954, however, scientists Andrew Huxley and Rolf Niedeigerke discovered that the length of the A band of a myofibril remains constant during contraction. Because the A band represents the myosin filament, Huxley and Niedeigerke realized that shortening of the myosin molecule could not be responsible for contraction. Subsequently, they proposed an alternative model, **the sliding filament theory of contraction.** In this model, overlapping actin and myosin filaments of fixed length slide past one another in an energy-requiring process, resulting in muscle contraction.

The sliding filament theory explains how a muscle can contract and create force without creating movement. For example, if you push on a wall, you are creating tension in many muscles of your body without moving the wall. According to the sliding filament theory, tension generated in a muscle fiber is directly proportional to interaction between the thick and thin filaments.

Sliding Filament Theory of Contraction If you examined a myofibril at its resting length, you would see that within each sarcomere, the ends of the thick and thin filaments overlap slightly. As the muscle contracts, the thick and thin filaments slide past each other, moving the Z disks of the sarcomere closer together. This phenomenon can be seen in light micrographs of

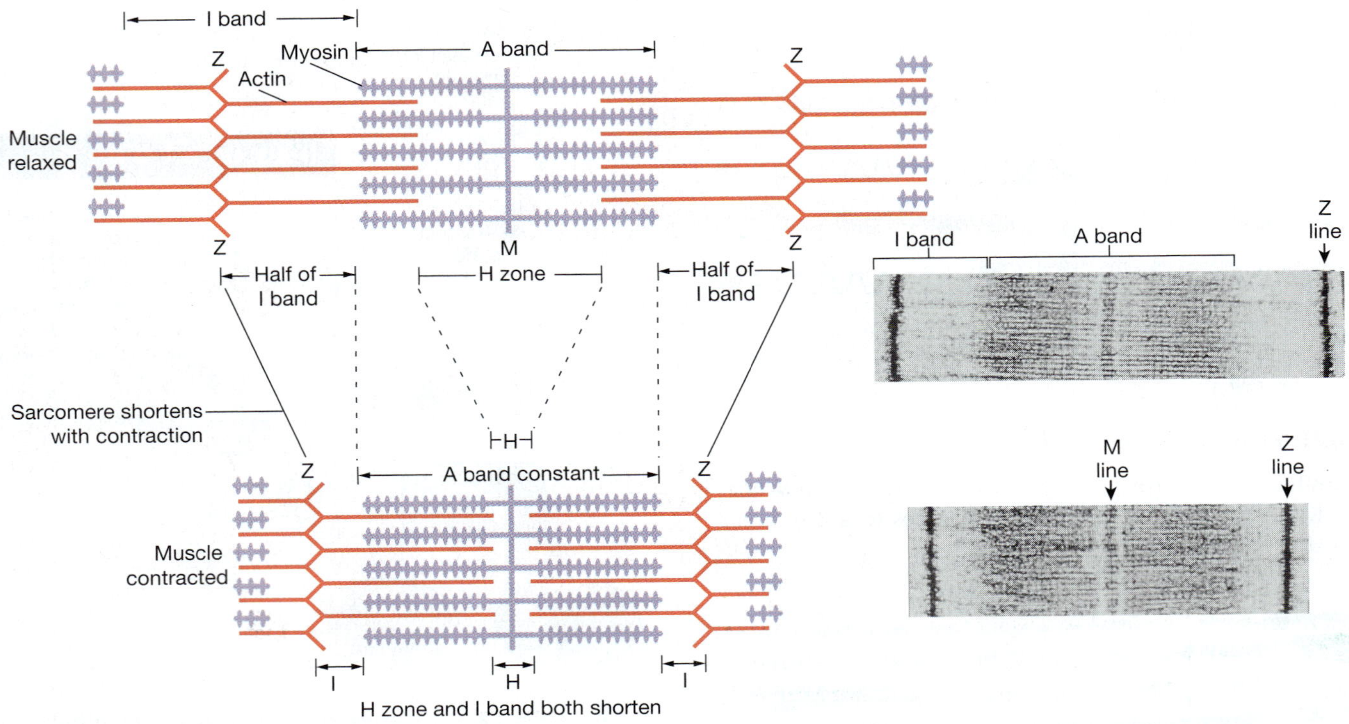

■ **FIGURE 12-8** ***Changes in sarcomere length during contraction***

As contraction takes place, the thick and thin filaments do not change length but instead slide past one another.

relaxed and contracted muscle (Fig. 12-8 ■). In the relaxed state, a sarcomere has a large I band (thin filaments only) and an A band whose length is the length of the thick filament. As contraction occurs, the sarcomere shortens. The two Z disks at each end move closer together while the I band and H zone—regions where actin and myosin do not overlap in resting muscle—almost disappear.

Despite the shortening of the sarcomere, the length of the A band remains constant. These changes are consistent with the sliding of thin actin filaments along the thick myosin filaments as the actin filaments move toward the M line in the center of the sarcomere. It is from this process that the sliding filament theory of contraction derives its name.

The force that pushes the actin filament is the movement of myosin crossbridges that link actin and myosin. Each myosin head has two binding sites on it: one for an ATP molecule and one for actin. The actin molecules serve as the "rope" to which the myosin heads bind.

During the **power stroke** that is the basis for muscle contraction, movement of the flexible myosin crossbridges pushes actin filaments toward the center of the sarcomere. At the end of a power stroke, each myosin head releases its bound actin, then swings back and binds to a new actin molecule, ready to start another cycle. This process repeats many times as a muscle fiber contracts. The myosin heads repeatedly bind and release actin molecules as myosin pushes the thin filaments toward the center of the sarcomere. An analogy that may help you visualize this is to think of a tug-of-war team holding onto one end of a long rope. When the order comes, each person begins pulling on the rope, hand over hand, grabbing, pulling, and releasing as the rope moves past.

In a muscle fiber, what causes movement of the myosin molecules? The answer is energy from ATP. Myosin is a motor protein with the ability to create movement [p. 63]. It accomplishes this task by converting the chemical bond energy of ATP into the mechanical energy of motion. Each myosin molecule is an ATPase (*myosin ATPase*) that binds ATP and hydrolyzes it to ADP and inorganic phosphate (P_i), releasing energy. The released energy is trapped and stored as potential energy in the angle between the myosin head and the long axis of the myosin filament. This potential energy then fuels the power stroke that moves actin.

Figure 12-9 ■ shows the molecular events of a contractile cycle.

1. **The rigor state**. Myosin heads create crossbridges by tightly binding to G-actin molecules. In this state, no nucleotide (ATP or ADP) occupies the second binding site on the myosin head. In living muscle, the rigor state [*rigere,* to be stiff] occurs for only a very brief period.
2. **ATP binds and myosin detaches**. An ATP molecule binds to the myosin head. This changes the actin-binding affinity of myosin, and the head releases from the G-actin molecule.

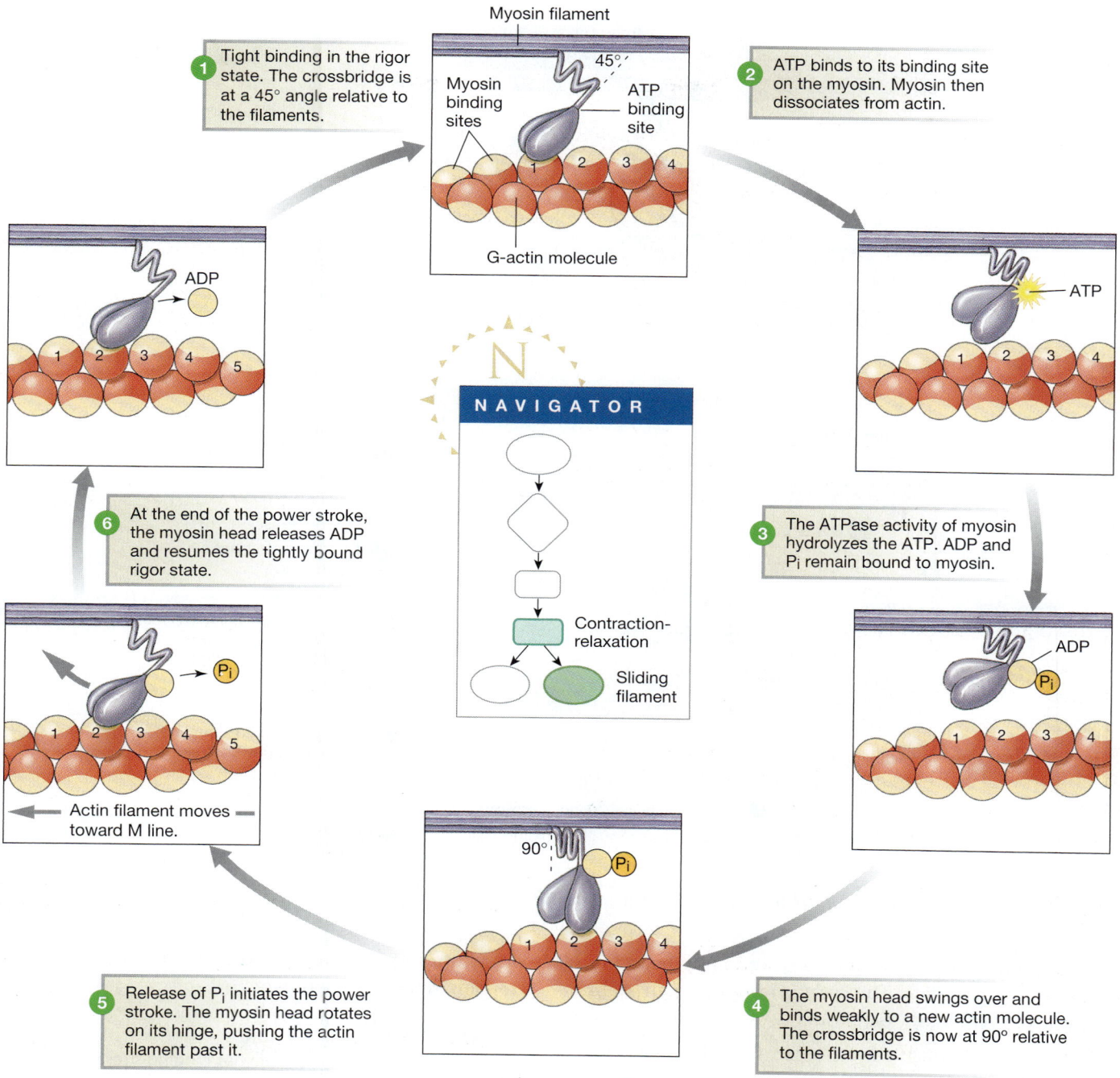

Tight binding in the rigor state. The crossbridge is at a 45° angle relative to the filaments. **1**

Myosin filament

Myosin binding sites

ATP binding site

45°

G-actin molecule

ATP binds to its binding site on the myosin. Myosin then dissociates from actin. **2**

ATP

At the end of the power stroke, the myosin head releases ADP and resumes the tightly bound rigor state. **6**

ADP

NAVIGATOR

Contraction-relaxation

Sliding filament

The ATPase activity of myosin hydrolyzes the ATP. ADP and P_i remain bound to myosin. **3**

ADP

P_i

Actin filament moves toward M line.

P_i

90°

P_i

Release of P_i initiates the power stroke. The myosin head rotates on its hinge, pushing the actin filament past it. **5**

The myosin head swings over and binds weakly to a new actin molecule. The crossbridge is now at 90° relative to the filaments. **4**

■ **FIGURE 12-9** *The molecular basis of contraction*

To review actin and myosin structure, see Figure 12-3(d–f).

3. **ATP hydrolysis**. The ATP-binding site on the myosin head closes around ATP and hydrolyzes it to ADP and inorganic phosphate (P_i). Both products remain bound to the head.

4. **Myosin reattaches: weak binding**. The energy released from ATP causes the myosin head to swing and bind weakly to a new G-actin molecule, one or two positions away from the G-actin to which it was previously bound. At this point, the myosin has potential energy, like a stretched spring, and is ready to execute the power stroke that will move the actin filament past it. ADP and P_i are still bound to myosin.

5. **P_i release and the power stroke**. The power stroke begins when inorganic phosphate is released from its myosin binding site. As the myosin head swings toward the M line, it pushes the attached actin filament in the same direction. The power stroke is also called *crossbridge tilting* because the myosin head and hinge region tilt from a 90° angle relative to the thick and thin filaments to a 45° angle.

6. **Release of ADP**. In the last step of the contractile cycle, myosin releases ADP, the second product of ATP hydrolysis.

12

(a) Relaxed state

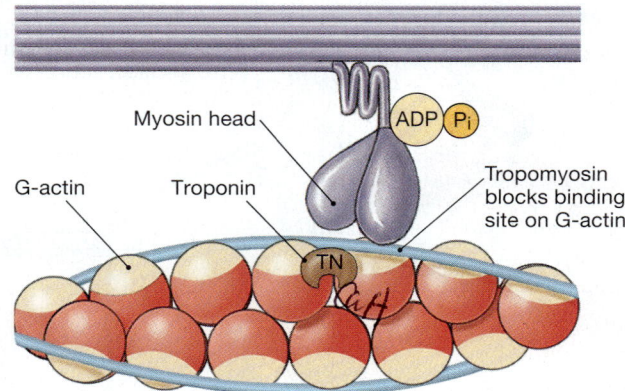

Myosin head

G-actin   Troponin

ADP Pi

Tropomyosin blocks binding site on G-actin

TN

(b) Initiation of contraction

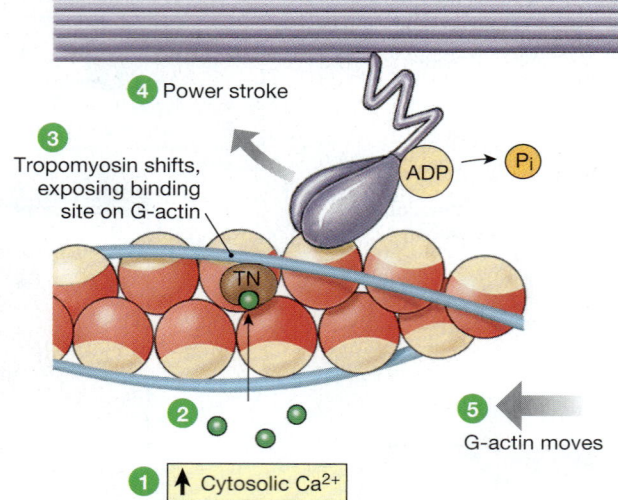

4 Power stroke

3 Tropomyosin shifts, exposing binding site on G-actin

ADP → Pi

TN

2 ●●●

1 ↑ Cytosolic Ca²⁺

5 G-actin moves

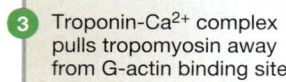
1 Ca^{2+} levels increase in cytosol.

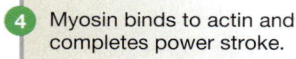
2 Ca^{2+} binds to troponin.

3 Troponin-Ca^{2+} complex pulls tropomyosin away from G-actin binding site.

4 Myosin binds to actin and completes power stroke.

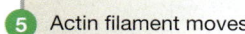
5 Actin filament moves.

■ **FIGURE 12-10** *Regulatory role of tropomyosin and troponin*

At this point, the myosin head is again tightly bound to actin in the rigor state. The cycle is ready to begin again when a new ATP binds to myosin.

Although the contractile cycle began with the rigor state in which no ATP or ADP was bound to myosin, relaxed muscle fibers remain mostly in step 4. The rigor state in living muscle is normally brief because the muscle fiber has a sufficient supply of ATP that binds to myosin when ADP is released in step 6.

After death, however, when metabolism stops and ATP supplies are exhausted, muscles are unable to bind more ATP, so they remain in the tightly bound state described in step 1. In the condition known as *rigor mortis,* the muscles "freeze" owing to immovable crossbridges. The tight binding of actin and myosin persists for a day or so after death, until enzymes within the decaying fiber begin to break down the muscle proteins.

Actin-myosin binding and crossbridge tilting explain how actin filaments slide past myosin filaments during contraction. But what regulates this process? If ATP is always available in the living muscle fiber, what keeps the filaments from continuously interacting? The answer lies with the two regulatory proteins tropomyosin and troponin.

CONCEPT CHECK

7. Each myosin molecule has binding sites for what molecules?

8. What is the difference between F-actin and G-actin?

Answers: p. 434

Contraction Is Regulated by Troponin and Tropomyosin

In skeletal muscle, the thin actin filaments of the myofibril are associated with two regulatory proteins that prevent myosin heads from completing their power stroke, much as the safety latch on a gun keeps the cocked trigger from being pulled. **Tropomyosin** [*tropos,* to turn] is an elongated protein polymer that wraps around the actin filament and partially blocks the myosin-binding sites (Fig. 12-10a ■). **Troponin** (TN) is a calcium-binding protein that controls the position of tropomyosin.

When tropomyosin is in its blocking (or "off") position, weak actin-myosin binding can take place, but myosin is blocked from completing its power stroke. For contraction to occur, tropomyosin must be shifted to an "on" position that uncovers the remainder of the binding site. This allows the power stroke to take place.

The off-and-on positioning of tropomyosin is regulated by troponin, a complex of three proteins associated with tropomyosin. As contraction begins in response to a calcium signal (① in Fig. 12-10b), one protein of the complex—**troponin C**—binds reversibly to Ca^{2+} (②). Calcium binding pulls tropomyosin toward the groove of the actin filament and unblocks the myosin-binding sites (③). This "on" position enables the myosin heads to carry out their power strokes and move the actin filament (④ and ⑤). Contractile cycles repeat as long as the binding sites are uncovered.

For relaxation to occur, Ca^{2+} concentrations in the cytosol must decrease so that Ca^{2+} unbinds from troponin. Without Ca^{2+}, the troponin-tropomyosin complex returns to its "off" position, covering most of the myosin-binding site. During the portion of the relaxation phase when actin and myosin are not bound to each other, the filaments of the sarcomere slide back to their original positions with the aid of titin and elastic connective tissues within the muscle.

Although the preceding discussion sounds as if we know everything there is to know about the molecular basis of muscle contraction, in reality this is simply our current model. The process is more complex than presented here, and it now appears that myosin influences Ca^{2+}-troponin binding, depending on whether the myosin is bound to actin in a strong (rigor) state, bound to actin in a weak state, or not bound at all. The details of this influence are still being worked out.

Studying contraction and the movement of molecules in a myofibril has proved very difficult. Many research techniques rely on crystallized molecules, electron microscopy, and other tools that cannot be used with living tissues. Often we can see the thick and thin filaments only at the beginning and end of contraction. Progress is being made, however, and perhaps in the next decade you will see a "movie" of muscle contraction, constructed from photographs of sliding filaments.

✓ CONCEPT CHECK

9. Name an elastic fiber in the sarcomere that aids relaxation.

10. In the sliding filament theory of contraction, what prevents the filaments from sliding back to their original position each time a myosin head releases to bind to the next actin binding site? (*Hint:* what would happen if all crossbridges released simultaneously?)

Answers: p. 434

Acetylcholine Initiates Excitation-Contraction Coupling

Now that you have been introduced to the molecular basis of muscle contraction, we will start at the neuromuscular junction

BIOTECHNOLOGY

THE *IN VITRO* MOTILITY ASSAY

One big step forward in understanding the power stroke of myosin was the development of the *in vitro* motility assay in the 1980s. In this assay, isolated myosin molecules are randomly bonded to a specially coated glass coverslip. A fluorescently labeled actin molecule is placed on top of the myosin molecules. With ATP as a source of energy, the myosin heads bind to the actin and move it across the coverslip, marked by a fluorescent trail as it goes. In even more ingenious experiments, developed in 1995, a single myosin molecule is bound to a tiny bead that elevates it above the surface of the cover slip. An actin molecule is placed on top of the myosin molecule, like the balancing pole of a tightrope walker. As the myosin "motor" moves the actin molecule, lasers measure the nanometer movements and piconewton forces created with each cycle of the myosin head. Because of this technique, researchers can now measure the mechanical work being done by a single myosin molecule!

and follow the contraction process as it begins at that point. It can be broken down into the following basic steps:

1. Acetylcholine (ACh) is released from the somatic motor neuron.
2. ACh initiates an action potential in the muscle fiber.
3. The muscle action potential triggers calcium release from the sarcoplasmic reticulum.
4. Calcium combines with troponin and initiates contraction.

As you learned earlier in the chapter, this combination of electrical and mechanical events in a muscle fiber is called excitation-contraction coupling. Now let's look at the process in detail.

Acetylcholine released into the synapse at a neuromuscular junction binds to ACh receptor-channels on the motor end plate of the muscle fiber (Fig. 12-11 ■) [🔁 p. 390]. When these channels open, they allow both Na^+ and K^+ to cross the membrane. However, Na^+ influx exceeds K^+ efflux because the electrochemical driving force is greater for Na^+ [🔁 p. 135]. The addition of net positive charge to the muscle fiber depolarizes the membrane, creating an **end-plate potential** (EPP). Normally, end-plate potentials always reach threshold and initiate a muscle action potential.

The action potential is conducted across the surface of the muscle fiber and into the t-tubules by the opening of voltage-gated

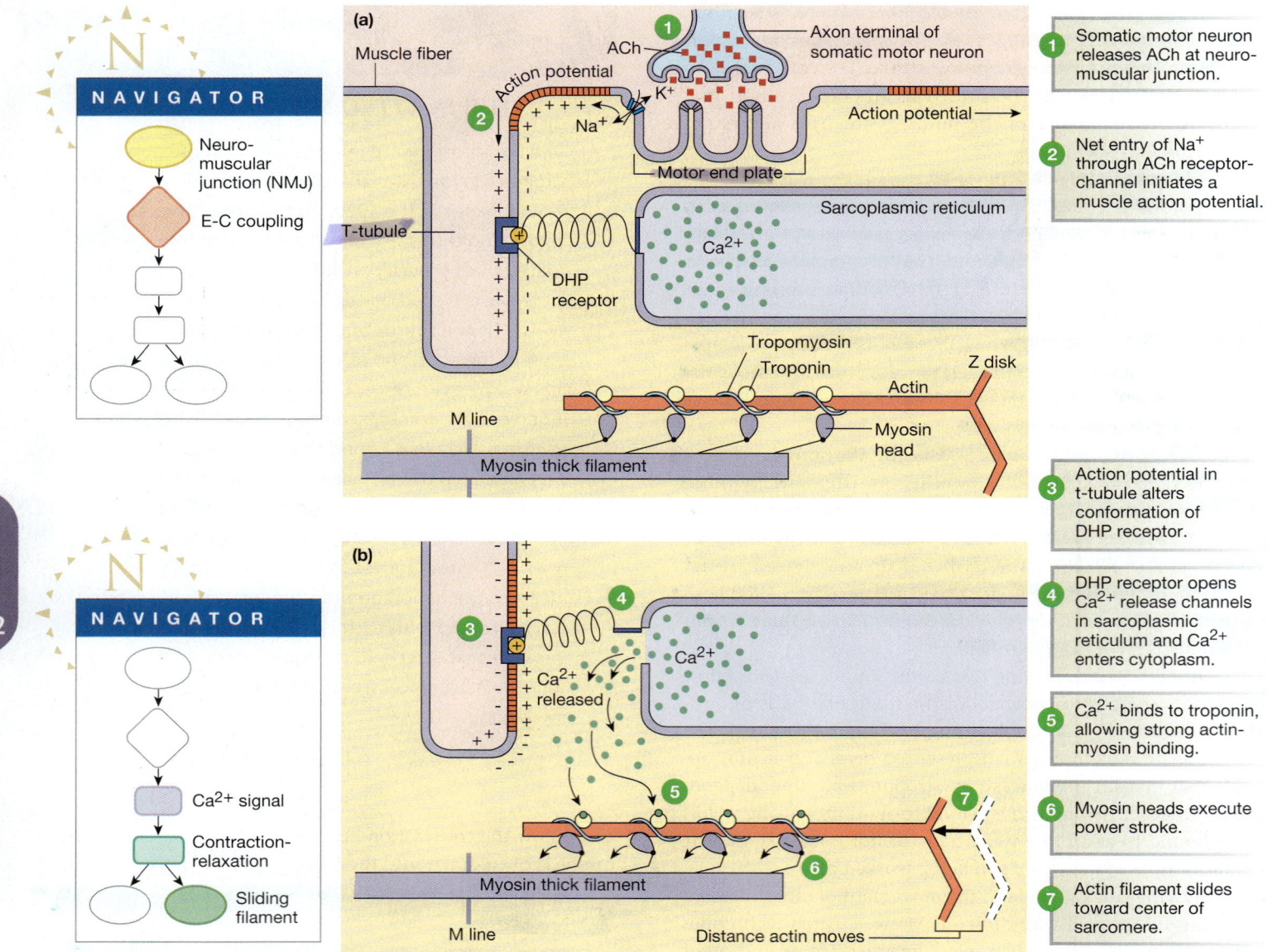

FIGURE 12-11 *Excitation-contraction coupling*

Na⁺ channels. The process is similar to the conduction of action potentials in axons, although action potentials in skeletal muscle are conducted more slowly than action potentials in neurons [⮀ p. 257].

The action potential that moves across the membrane and down the t-tubules causes Ca^{2+} release from the sarcoplasmic reticulum (Fig. 12-11b). The t-tubule membrane contains voltage-sensing receptors (**dihydropyridine**, or **DHP, receptors**) that are mechanically linked to Ca^{2+} **release channels** in the adjacent sarcoplasmic reticulum. (These channels are also called **ryanodine receptors**, or **RyR**.) When a wave of depolarization reaches a DHP receptor, its conformation changes, opening Ca^{2+} release channels in the sarcoplasmic reticulum. Stored Ca^{2+} then moves down its electrochemical gradient into the cytosol, where it initiates contraction as depicted in Figure 12-10b.

Free cytosolic Ca^{2+} levels in a resting muscle are normally quite low, but after an action potential, they increase about 100-fold. When cytosolic Ca^{2+} levels are high, Ca^{2+} binds to troponin, tropomyosin moves to the "on" position, and contraction occurs.

Relaxation occurs when the sarcoplasmic reticulum pumps Ca^{2+} back into its lumen using **Ca^{2+}-ATPase** [⮀ p. 142]. As the free cytosolic Ca^{2+} concentration decreases, Ca^{2+} releases from troponin, tropomyosin slides back to block the myosin-binding site, and the fiber relaxes.

The discovery that Ca^{2+}, not the action potential, is the signal for contraction was the first piece of evidence suggesting that calcium acts as a messenger inside cells. For years it was thought that calcium signals occurred only in muscles, but we now know that calcium is an almost universal second messenger [⮀ p. 186].

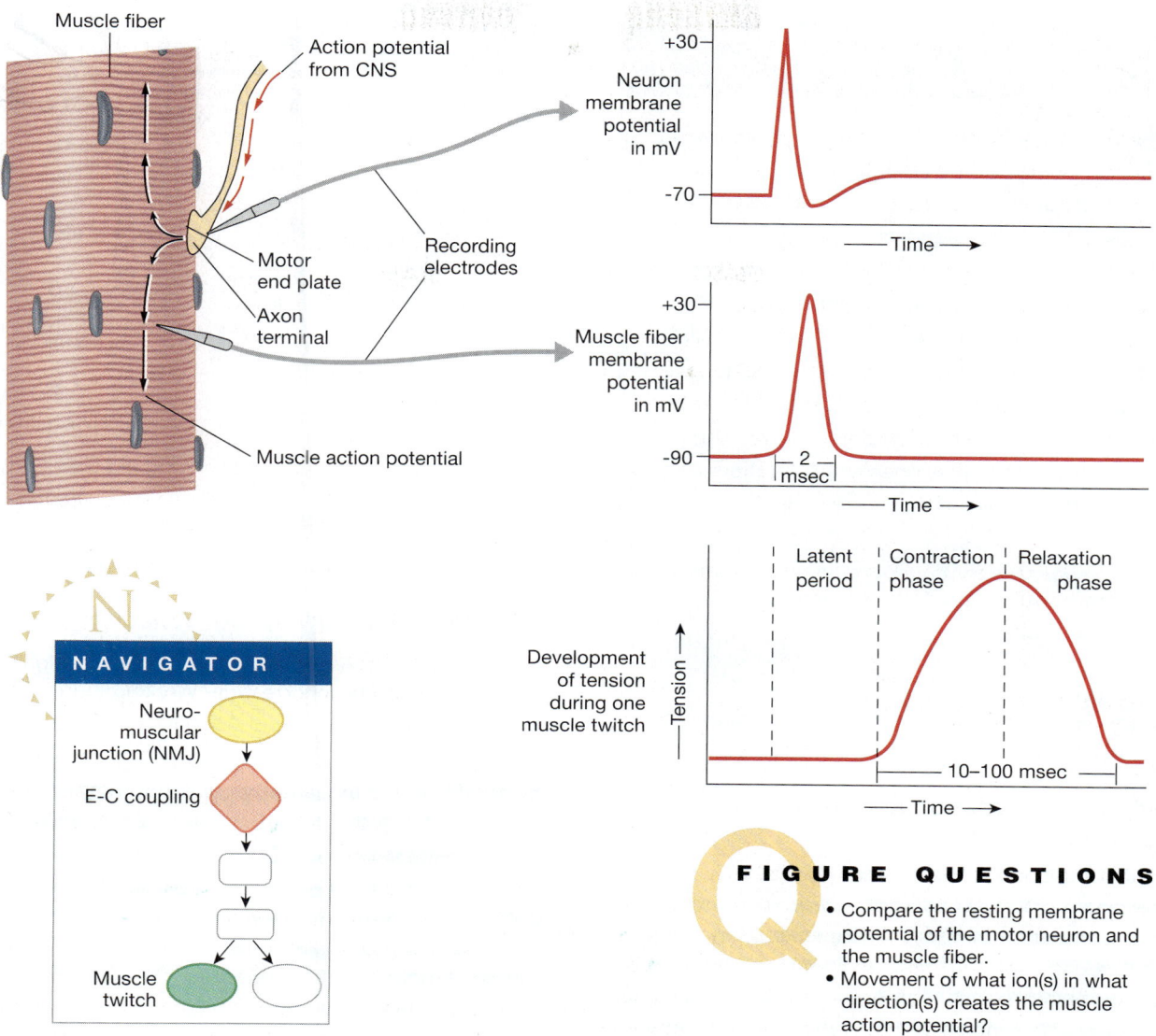

■ FIGURE 12-12 *Electrical and mechanical events in muscle contraction*

Action potentials in the axon terminal (top graph) and in the muscle fiber (middle graph) are followed by a muscle twitch (bottom graph).

Figure 12-12 ■ shows the timing of electrical and mechanical events during excitation-contraction coupling. The somatic motor neuron action potential is followed by the skeletal muscle action potential, which in turn is followed by contraction. A single contraction-relaxation cycle in a skeletal muscle fiber is known as a **twitch**. Notice that there is a short delay—the **latent period**—between the muscle action potential and the beginning of muscle tension development. This delay represents the time required for excitation-contraction coupling to take place.

Once contraction begins, muscle tension increases steadily to a maximum value as crossbridge interaction increases. Tension then decreases in the relaxation phase of the twitch. During relaxation, elastic elements of the muscle return the sarcomeres to their resting length.

A single action potential in a muscle fiber evokes a single twitch (Fig. 12-12, bottom graph). However, muscle twitches vary from fiber to fiber in the speed with which they develop tension (the rising slope of the twitch curve), the maximum tension they achieve (the height of the twitch curve), and the duration of the twitch (the width of the twitch curve). We will discuss factors that affect these parameters in upcoming sections. First we will discuss how muscles produce ATP to provide energy for contraction and relaxation.

✔ **CONCEPT CHECK**

11. Which part of contraction requires ATP? Does relaxation require ATP?

12. What events are taking place during the latent period before contraction begins?

Answers: p. 434

Skeletal Muscle Contraction Requires a Steady Supply of ATP

The muscle fiber's use of ATP is a key feature of muscle physiology. Muscles require energy constantly: during contraction for crossbridge movement and release, during relaxation to pump Ca^{2+} back into the sarcoplasmic reticulum, and after excitation-contraction coupling to restore Na^+ and K^+ to the extracellular and intracellular compartments, respectively. Where do muscles get the ATP they need for this work?

The amount of ATP in a muscle fiber at any one time is sufficient for only about eight twitches. As a backup energy source, muscles contain **phosphocreatine**, a molecule whose high-energy phosphate bonds are created from creatine and ATP when muscles are at rest (Fig. 12-13 ■). When muscles become active, such as during exercise, the high-energy phosphate group of phosphocreatine is transferred to ADP, creating more ATP to power the muscles.

The enzyme responsible for transferring the phosphate group from phosphocreatine to ADP is **creatine kinase** (CK), also known as *creatine phosphokinase* (CPK). Muscle cells contain large amounts of this enzyme. Consequently, elevated blood levels of creatine kinase usually indicate damage to skeletal or cardiac muscle. Because the two muscle types contain different isozymes [🔁 p. 96], clinicians can distinguish cardiac tissue damage during a heart attack from skeletal muscle damage.

Because energy stored in high-energy phosphate bonds is very limited, muscle fibers must use metabolism to transfer energy from the chemical bonds of nutrients to ATP. Carbohydrates,

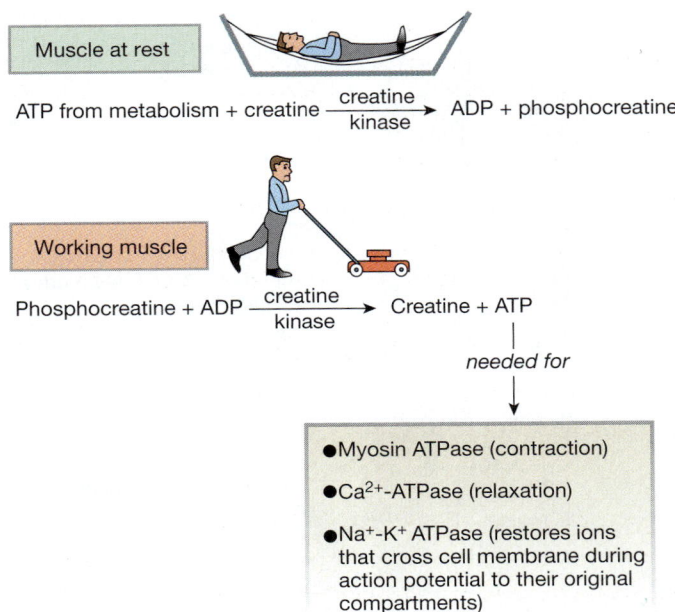

■ FIGURE 12-13   *Phosphocreatine*

Resting muscle stores energy from ATP in the high-energy phosphate bonds of phosphocreatine. Working muscle then uses that stored energy.

particularly glucose, are the most rapid and efficient source of energy for ATP production. Glucose is metabolized through glycolysis to pyruvate [🔁 p. 104]. In the presence of adequate oxygen, pyruvate goes into the citric acid cycle, producing about 30 ATP for each molecule of glucose.

When oxygen concentrations are too low to maintain aerobic metabolism, the muscle fiber shifts to *anaerobic glycolysis*. In this pathway, glucose is metabolized to lactic acid with a yield of only 2 ATP per glucose. Anaerobic metabolism of glucose is a quicker source of ATP but produces many fewer ATP per glucose. When muscle energy demands outpace the amount of ATP that can be produced through anaerobic metabolism of glucose, muscles can function for only a short time without fatiguing.

Muscle fibers also obtain energy from fatty acids, although this process always requires oxygen [🔁 p. 111]. During rest and light exercise, skeletal muscles burn fatty acids along with glucose, one reason that modest exercise programs of brisk walking are an effective way to reduce body fat. However, *beta-oxidation*—the process by which fatty acids are converted to acetyl CoA—is a slow process and cannot produce ATP rapidly enough to meet the energy needs of muscle fibers during heavy exercise. Under these conditions, muscle fibers rely more on glucose.

Proteins normally are not a source of energy for muscle contraction. Most amino acids found in muscle fibers are used to synthesize proteins rather than to produce ATP.

Do muscles ever run out of ATP? You might think so if you have ever exercised to the point of fatigue, the point at which you feel that you cannot continue or your limbs refuse to obey commands from your brain. Most studies show, however, that even intense exercise uses only 30% of the ATP in a muscle fiber. The condition we call fatigue must come from other changes in the exercising muscle.

✓ CONCEPT CHECK

13. According to the convention for naming enzymes, what does the name *creatine kinase* tell you about this enzyme's function? [*Hint:* 🔁 p. 99]

14. The reactions in Figure 12-13 show that creatine kinase catalyzes the creatine-phosphocreatine reaction in both directions. What then determines the direction that the reaction goes at any given moment? [*Hint:* 🔁 p. 39]

Answers: p. 434

Muscle Fatigue Has Multiple Causes

The physiological term **fatigue** describes a condition in which a muscle is no longer able to generate or sustain the expected power output. Fatigue is highly variable. It is influenced by the intensity and duration of the contractile activity, by whether the muscle fiber is using aerobic or anaerobic metabolism, by the composition of the muscle, and by the fitness level of the individual.

Multiple factors have been proposed as playing a role in fatigue (Fig. 12-14 ■). They can be classified into **central fatigue** mechanisms, which arise in the central nervous system, and **peripheral fatigue** mechanisms, which arise anywhere between the neuromuscular junction and the contractile elements of the muscle. Most experimental evidence suggests that muscle fatigue arises from excitation-contraction failure in the muscle fiber rather than from failure of control neurons or neuromuscular transmission.

Central fatigue includes subjective feelings of tiredness and a desire to cease activity. Several studies have shown that this psychological fatigue precedes physiological fatigue in the muscles and therefore may be a protective mechanism. Low pH from acid production during ATP hydrolysis is often mentioned as a possible cause of fatigue, and some evidence suggests that acidosis may influence the sensation of fatigue perceived by the brain. However, homeostatic mechanisms for pH balance maintain blood pH at normal levels until exertion is nearly maximal, so pH as a factor in central fatigue probably applies only in cases of maximal exertion.

Neural causes of fatigue could arise either from communication failure at the neuromuscular junction or from failure of the CNS command neurons. For example, if ACh is not synthesized in the axon terminal fast enough to keep up with neuron firing rate, neurotransmitter release at the synapse will decrease. Consequently, the muscle end-plate potential may fail to reach the threshold value needed to trigger a muscle fiber action potential, resulting in contraction failure. This type of fatigue is associated with some neuromuscular diseases, but it is probably not a factor in normal exercise.

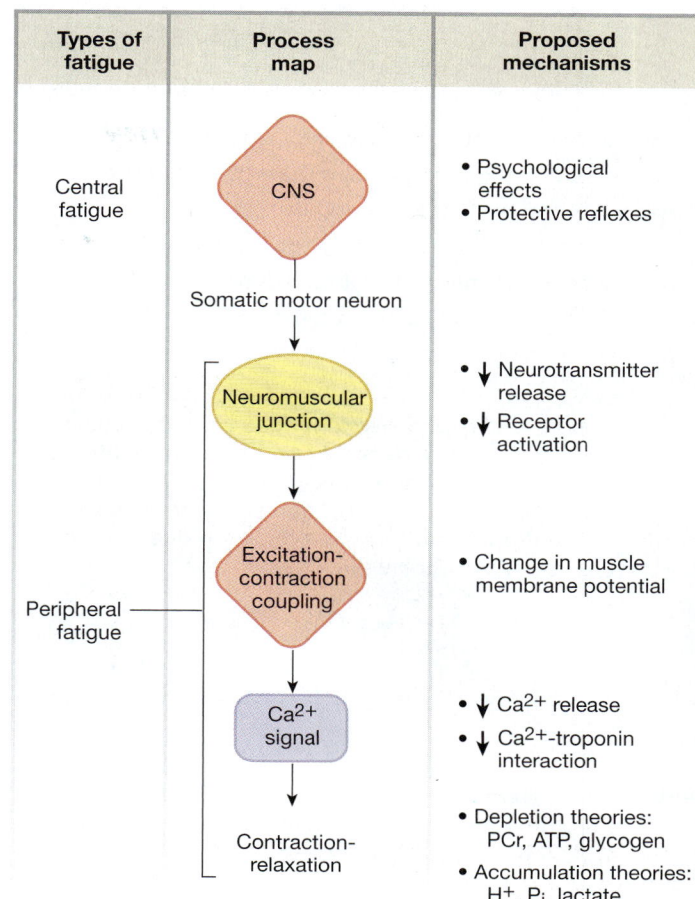

Types of fatigue	Process map	Proposed mechanisms
Central fatigue	CNS	• Psychological effects • Protective reflexes
	Somatic motor neuron	
	Neuromuscular junction	• ↓ Neurotransmitter release • ↓ Receptor activation
Peripheral fatigue	Excitation-contraction coupling	• Change in muscle membrane potential
	Ca²⁺ signal	• ↓ Ca²⁺ release • ↓ Ca²⁺-troponin interaction
	Contraction-relaxation	• Depletion theories: PCr, ATP, glycogen • Accumulation theories: H⁺, Pᵢ, lactate

■ **FIGURE 12-14** *Locations and possible causes of muscle fatigue*

There appears to be no single cause for muscle fatigue during exercise, and the study of this phenomenon is quite complex. Fatigue within the muscle fiber can occur in any of several sites. In extended submaximal exertion, fatigue is associated with the depletion of muscle glycogen stores. Because most studies show that lack of ATP is not a limiting factor, glycogen depletion may be affecting some other aspect of contraction, such as the release of Ca^{2+} from the sarcoplasmic reticulum.

The cause of fatigue in short-duration maximal exertion seems to be different. It apparently results from increased levels of inorganic phosphate (P_i) produced when ATP and phosphocreatine are used for energy in the muscle fiber. Elevated inorganic phosphate levels may slow P_i release from myosin and thereby alter the power stroke (Fig. 12-9, ⑤). Another theory suggests that elevated phosphate levels decrease Ca^{2+} release because the phosphate combines with Ca^{2+} to become calcium phosphate. Some investigators feel that phosphate alters the dynamics of Ca^{2+} release from the sarcoplasmic reticulum.

Potassium is another factor implicated in fatigue. During maximal exercise, K^+ leaves the muscle fiber with each action potential, and as a result K^+ concentrations rise in the extracellular

RUNNING PROBLEM

Two forms of periodic paralysis exist. One form, called *hypokalemic periodic paralysis,* is characterized by decreased blood levels of K$^+$ during paralytic episodes. The other form, *hyperkalemic periodic paralysis,* is characterized by either normal or increased blood levels of K$^+$ during episodes. Results of a blood test revealed that Paul has the hyperkalemic form.

Question 2:
In people with hyperkalemic periodic paralysis, attacks tend to occur after a period of exercise (that is, after a period of repeated muscle contractions). What ion is responsible for the repolarization phase of the muscle action potential, and in which direction does this ion move across the muscle fiber membrane?

| 397 | 410 | **412** | 419 | 423 | 430 |

fluid of the t-tubules. The shift in K$^+$ alters the membrane potential of the muscle fiber and is believed to decrease Ca^{2+} release from the sarcoplasmic reticulum.

CONCEPT CHECK

15. If K$^+$ concentration increases in the extracellular fluid surrounding a cell but does not change significantly in the cell's cytoplasm, the cell membrane will _____ (*depolarize/hyperpolarize*) and become _____ (*more/less*) negative. *Answers: p. 434*

Skeletal Muscle Fibers Are Classified by Contraction Speed and Resistance to Fatigue

Skeletal muscle fibers can be classified on the basis of their speed of contraction and their resistance to fatigue with repeated stimulation. The groups include slow-twitch fibers (also called ST or type I), fast-twitch oxidative-glycolytic fibers (FOG or type IIA), and fast-twitch glycolytic fibers (FG or type IIB).

Fast-twitch muscle fibers (type II) develop tension two to three times faster than **slow-twitch fibers** (type I). The speed with which a muscle fiber contracts is determined by the isoform of myosin ATPase present in the fiber's thick filaments. Fast-twitch fibers split ATP more rapidly and can therefore complete multiple contractile cycles more rapidly than slow-twitch fibers. This speed translates into faster tension development in the fast-twitch fibers.

The duration of contraction also varies according to fiber type. Twitch duration is determined largely by how fast the sarcoplasmic reticulum removes Ca^{2+} from the cytosol. As cytosolic Ca^{2+} concentrations fall, Ca^{2+} unbinds from troponin, allowing

tropomyosin to move into position to partially block the myosin-binding sites. With the power stroke inhibited in this way, the muscle fiber relaxes.

Fast-twitch fibers pump Ca^{2+} into their sarcoplasmic reticulum more rapidly than slow-twitch fibers do, so fast-twitch fibers have quicker twitches. The twitches in fast-twitch fibers last only about 7.5 msec, making these muscles useful for fine, quick movements, such as playing the piano. Contractions in slow-twitch muscle fibers may last more than 10 times as long. Fast-twitch fibers are used occasionally, but slow-twitch fibers are used almost constantly for maintaining posture, standing, or walking.

The second major difference between muscle fiber types is their ability to resist fatigue. Glycolytic fibers (fast-twitch type IIB) rely primarily on anaerobic glycolysis to produce ATP. However, the accumulation of H$^+$ from ATP hydrolysis contributes to acidosis, a condition implicated in the development of fatigue, as noted previously. As a result, glycolytic fibers fatigue more easily than do oxidative fibers, which do not depend on anaerobic metabolism. Oxidative fibers rely primarily on oxidative phosphorylation [💫 p. 108] for production of ATP—hence their descriptive name. These fibers, which include slow-twitch fibers and fast-twitch oxidative-glycolytic fibers, have more mitochondria (the site of enzymes for the citric acid cycle and oxidative phosphorylation) than glycolytic fibers. They also have more blood vessels in their connective tissue to bring oxygen to the cells (Fig. 12-15 ■).

The efficiency with which muscle fibers obtain oxygen is a factor in their preferred method of glucose metabolism. Oxygen in the blood must diffuse into the interior of muscle fibers in order to reach the mitochondria. This process is facilitated by the presence of **myoglobin**, a red oxygen-binding pigment with a high affinity for oxygen. This affinity allows myoglobin to act as a transfer molecule, bringing oxygen more rapidly to the interior of the fibers. Because oxidative fibers contain more myoglobin, oxygen diffusion is faster than in glycolytic fibers. Oxidative fibers are described as *red muscle* because large amounts of myoglobin give them their characteristic color.

In addition to myoglobin, oxidative fibers have smaller diameters, so the distance through which oxygen must diffuse before reaching the mitochondria is shorter. Because oxidative fibers have more myoglobin and more capillaries to bring blood to the cells and are smaller in diameter, they maintain a better supply of oxygen and are able to use oxidative phosphorylation for ATP production.

Glycolytic fibers, in contrast, are described as *white muscle* because of their lower myoglobin content. These muscle fibers are also larger in diameter than slow-twitch fibers. The combination of larger size, less myoglobin, and fewer blood vessels means that glycolytic fibers are more likely to run out of oxygen after repeated contractions. Glycolytic fibers therefore rely primarily on anaerobic glycolysis for ATP synthesis and fatigue most rapidly.

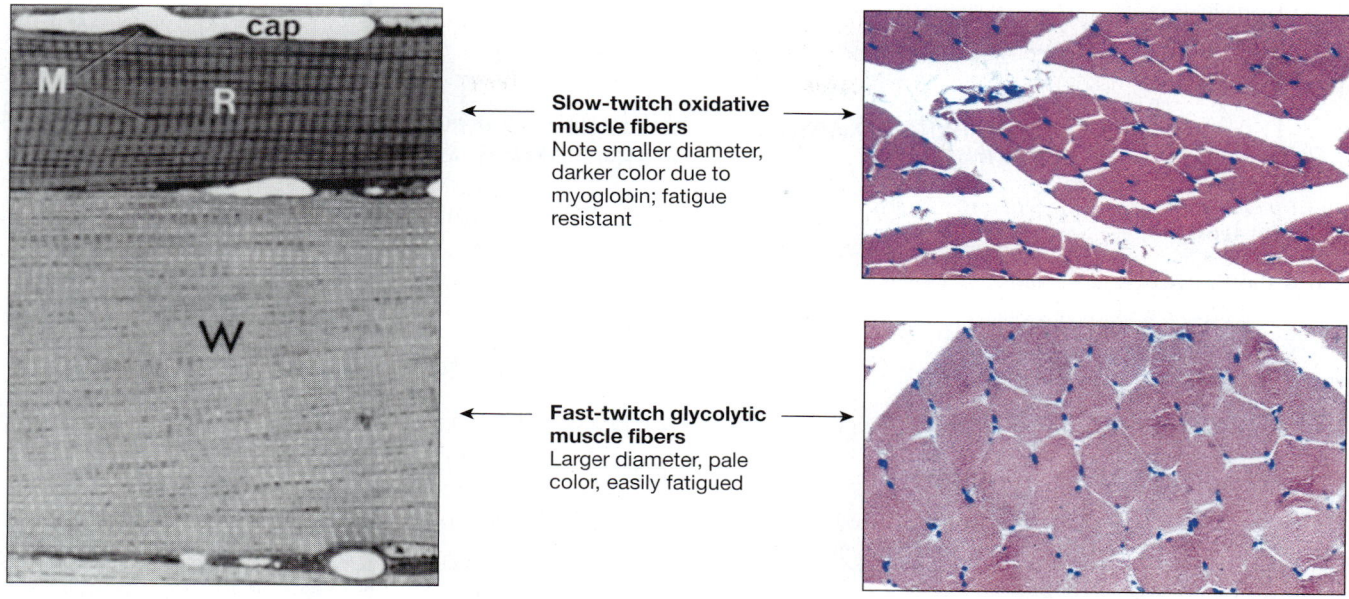

■ FIGURE 12-15 *Fast-twitch glycolytic and slow-twitch oxidative muscle fibers*

Large amounts of red myoglobin, numerous mitochondria (M), and extensive capillary blood supply (cap) distinguish slow-twitch oxidative muscle (labeled R here for red muscle) from fast-twitch glycolytic muscle (labeled W for white muscle).

Fast-twitch oxidative-glycolytic fibers exhibit properties of both oxidative and glycolytic fibers. They are smaller than fast-twitch glycolytic fibers and use a combination of oxidative and glycolytic metabolism to produce ATP. Because of their intermediate size and the use of oxidative phosphorylation for ATP synthesis, fast-twitch oxidative-glycolytic fibers are more fatigue resistant than their fast-twitch glycolytic cousins. Fast-twitch oxidative-glycolytic fibers, like slow-twitch fibers, are classified as red muscle because of their myoglobin content.

Human muscles are a mixture of fiber types, with the ratio of types varying from muscle to muscle and from one individual to another. For example, who would have more fast-twitch fibers in leg muscles, a marathon runner or a high-jumper? Characteristics of the three muscle fiber types are compared in Table 12-2 ■.

Tension Developed by Individual Muscle Fibers Is a Function of Fiber Length

In a muscle fiber, the tension developed during a twitch is a direct reflection of the length of individual sarcomeres before contraction begins (Fig. 12-16 ■). Each sarcomere will contract with optimum force if it is at optimum length (neither too long nor too short) before the contraction begins. Fortunately, the normal resting length of skeletal muscles usually ensures that sarcomeres are at optimum length when they begin a contraction.

At the molecular level, sarcomere length reflects the overlap between the thick and thin filaments. The sliding filament theory predicts that *the tension a muscle fiber can generate is*

directly proportional to the number of crossbridges formed between the thick and thin filaments. If the fibers start a contraction at a very long sarcomere length, the thick and thin filaments are barely overlapping, forming few crossbridges (Fig. 12-16e). This means that in the initial part of the contraction, the sliding filaments can interact only minimally and therefore cannot generate much force.

At the optimum sarcomere length (Fig. 12-16c), the filaments begin contracting with numerous crossbridges between the thick and thin filaments, allowing the fiber to generate optimum force in that twitch. If the sarcomere is shorter than optimum length at the beginning of the contraction (Fig. 12-16b), the thick and thin fibers have too much overlap before the contraction begins. Consequently, the thick filaments can move the thin filaments only a short distance before the thin actin filaments from opposite ends of the sarcomere start to overlap. This overlap prevents crossbridge formation. If the sarcomere is so short that the thick filaments run into the Z disks (Fig. 12-16a), myosin is unable to find new binding sites for crossbridge formation, and tension decreases rapidly. Thus the development of single-twitch tension in a muscle fiber is a passive property that depends on filament overlap and sarcomere length.

Force of Contraction Increases with Summation of Muscle Twitches

Although we have just seen that single-twitch tension is determined by the length of the sarcomere, it is important to note that a single twitch does not represent the maximum force that

TABLE 12-2	Characteristics of Muscle Fiber Types		
	SLOW-TWITCH OXIDATIVE; RED MUSCLE	FAST-TWITCH OXIDATIVE-GLYCOLYTIC; RED MUSCLE	FAST-TWITCH GLYCOLYTIC; WHITE MUSCLE
Speed of development of maximum tension	Slowest	Intermediate	Fastest
Myosin ATPase activity	Slow	Fast	Fast
Diameter	Small	Medium	Large
Contraction duration	Longest	Short	Short
Ca^{2+}-ATPase activity in SR	Moderate	High	High
Endurance	Fatigue resistant	Fatigue resistant	Easily fatigued
Use	Most used: posture	Standing, walking	Least used: jumping
Metabolism	Oxidative; aerobic;	Glycolytic but becomes more oxidative with endurance training	Glycolytic; more anaerobic than fast-twitch oxidative-glycolytic type
Capillary density	High	Medium	Low
Mitochondria	Numerous	Moderate	Few
Color	Dark red (myoglobin)	Red	Pale

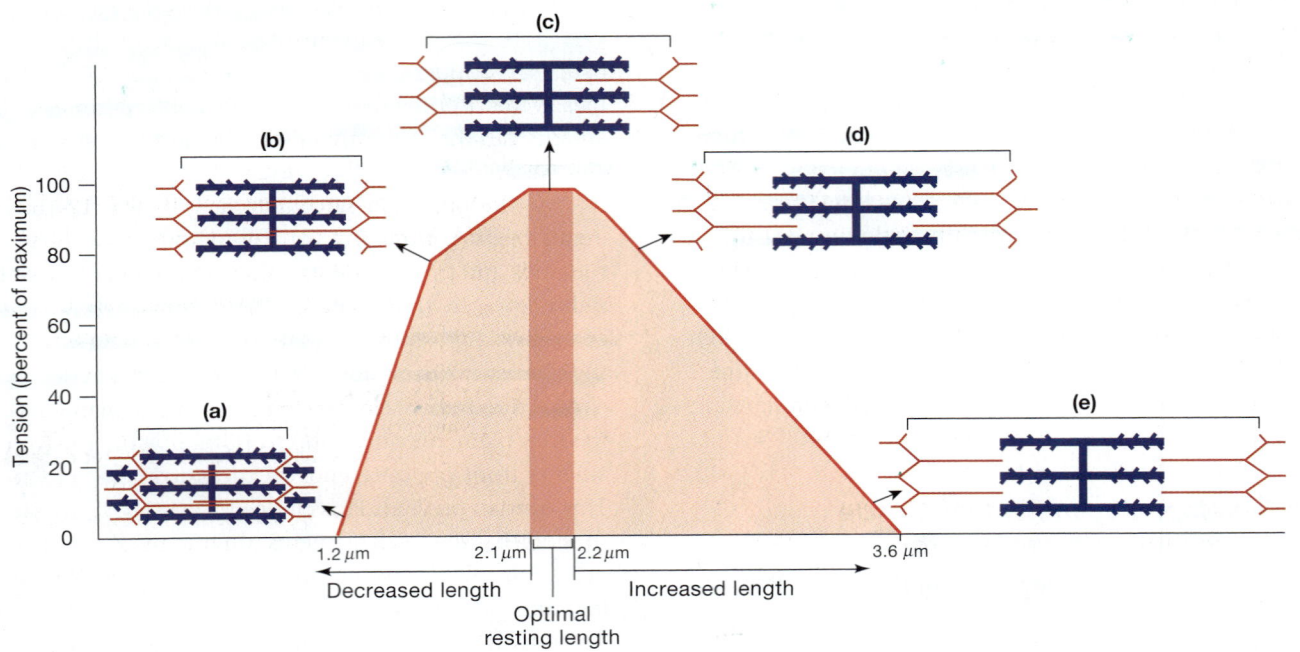

■ **FIGURE 12-16** *Length-tension relationships in contracting skeletal muscle*

The graph plots tension generated by a skeletal muscle at various resting lengths before contraction starts.

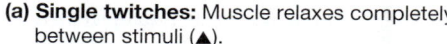
(a) Single twitches: Muscle relaxes completely between stimuli (▲).

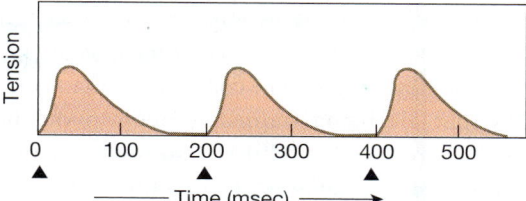

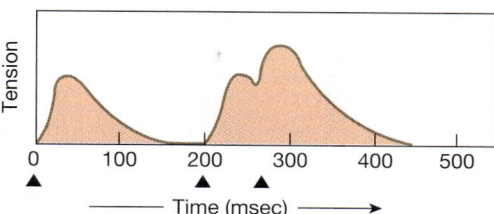
(b) Summation: Stimuli closer together do not allow muscle to relax fully.

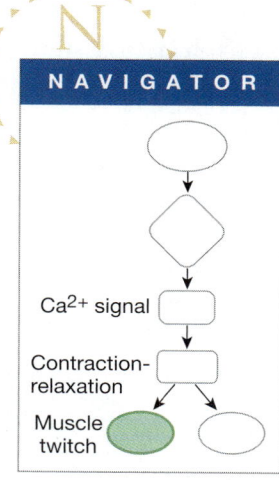

(c) Summation leading to unfused tetanus: Stimuli are far enough apart to allow muscle to relax slightly between stimuli.

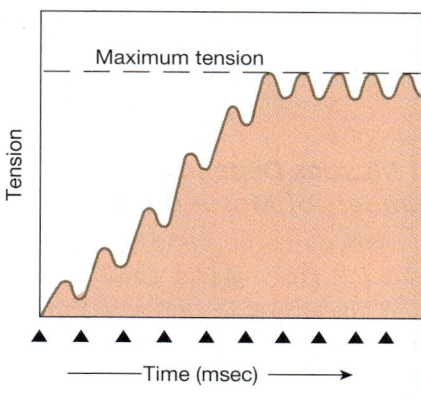

(d) Summation leading to complete tetanus: Muscle reaches steady tension.

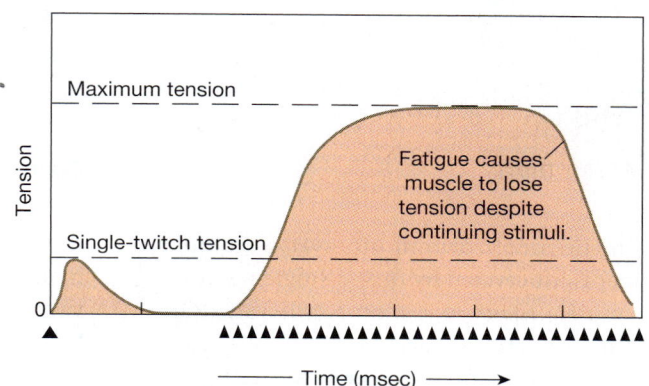

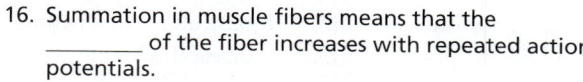

■ **FIGURE 12-17** *Summation of contractions*

a muscle fiber can develop. The force generated by the contraction of a single muscle fiber can be increased by increasing the rate (frequency) at which muscle action potentials stimulate the muscle fiber.

A typical muscle action potential lasts between 1 and 3 msec, while the muscle contraction may last 100 msec (see Fig. 12-12). If repeated action potentials are separated by long intervals of time, the muscle fiber has time to relax completely between stimuli (Fig. 12-17a ■). If the interval of time between action potentials is shortened, the muscle fiber does not have time to relax completely between two stimuli, resulting in a more forceful contraction (Fig. 12-17b). This process is known as **summation** and is similar to the temporal summation of graded potentials that takes place in neurons [p. 278].

If action potentials continue to stimulate the muscle fiber repeatedly at short intervals (high frequency), relaxation between contractions diminishes until the muscle fiber achieves a state of maximal contraction known as **tetanus**. There are two types of tetanus. In *incomplete* (or *unfused*) *tetanus*, the stimulation rate of the muscle fiber is not at a maximum value, and consequently the fiber relaxes slightly between stimuli (Fig. 12-17c). In *complete* (or *fused*) *tetanus*, the stimulation rate is fast enough

that the muscle fiber does not have time to relax. Instead, it reaches maximum tension and remains there (Fig. 12-17d).

Thus it is possible to increase the tension developed in a single muscle fiber by changing the rate at which action potentials occur in the fiber. Muscle action potentials are initiated by the somatic motor neuron that controls the muscle fiber.

CONCEPT CHECK

16. Summation in muscle fibers means that the _____ of the fiber increases with repeated action potentials.

17. Temporal summation in neurons means that the _____ of the neuron increases when two depolarizing stimuli occur close together in time.

Answers: p. 434

A Motor Unit Is One Somatic Motor Neuron and the Muscle Fibers It Innervates

The basic unit of contraction in an intact skeletal muscle is a **motor unit**, composed of a group of muscle fibers that function together and the somatic motor neuron that controls them (Fig. 12-18 ■). When the somatic motor neuron fires an action potential, all muscle fibers in the motor unit contract. Note

One muscle may have many motor units of different fiber types.

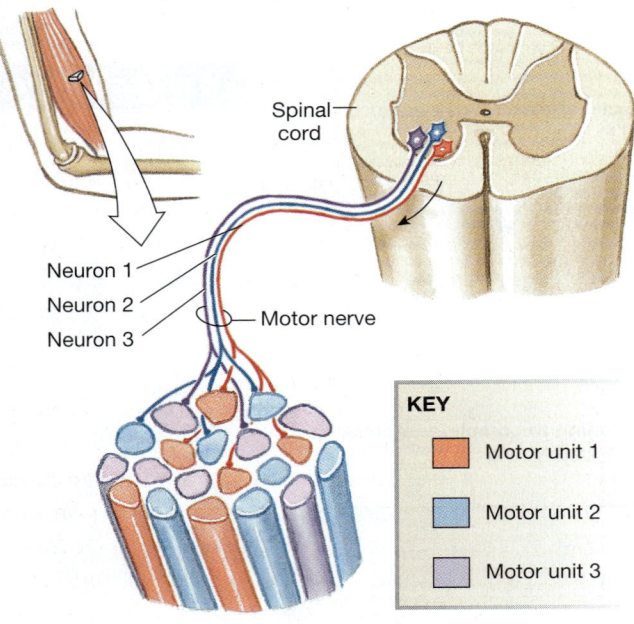

Spinal cord

Neuron 1
Neuron 2
Neuron 3

Motor nerve

KEY

Motor unit 1

Motor unit 2

Motor unit 3

■ **FIGURE 12-18** *Motor units*

that although one somatic motor neuron innervates multiple fibers, each muscle fiber is innervated by only a single neuron.

The number of muscle fibers in a motor unit varies. In muscles used for fine motor actions, such as the muscles that move the eyes or the muscles of the hand, a motor unit contains as few as three to five muscle fibers. If one such motor unit is activated, only a few fibers contract, and the muscle response is quite small. If additional motor units are activated, the response increases by small increments because only a few more muscle fibers contract with the addition of each motor unit. This arrangement allows fine gradations of movement.

In muscles used for gross motor actions such as standing or walking, each motor unit may contain hundreds or even thousands of muscle fibers. The gastrocnemius muscle in the calf of the leg, for example, has about 2000 muscle fibers in each motor unit. Each time an additional motor unit is activated in these muscles, many more muscle fibers contract, and the muscle response jumps by correspondingly greater increments.

All muscle fibers in a single motor unit are of the same fiber type. Thus, there are fast-twitch motor units and slow-twitch motor units. Which kind of muscle fiber associates with a particular neuron appears to be a function of the neuron. During embryological development, each somatic motor neuron secretes a growth factor that directs the differentiation of all muscle fibers in its motor unit so that they develop into the same fiber type.

Intuitively, it would seem that people who inherit a predominance of one fiber type over another would excel in certain sports. They do, to some extent. Endurance athletes, such

as distance runners and cross-country skiers, have a predominance of slow-twitch fibers, whereas sprinters, ice hockey players, and weight lifters tend to have larger percentages of fast-twitch fibers. Inheritance is not the only determining factor for fiber composition in the body, however, because the metabolic characteristics of muscle fibers can be changed to some extent. With endurance training, the aerobic capacity of some fast-twitch fibers can be enhanced until they are almost as fatigue-resistant as slow-twitch fibers. Since the conversion occurs only in those muscles that are being trained, a neuromodulator chemical is probably involved. In addition, endurance training increases the number of capillaries and mitochondria in the muscle tissue, allowing more oxygen-carrying blood to reach the contracting muscle and contributing to the increased aerobic capacity of the muscle fibers.

CONCEPT CHECK

18. Which type of runner would you expect to have more slow-twitch fibers, a sprinter or a marathoner?

Answers: p. 434

Contraction in Intact Muscles Depends on the Types and Numbers of Motor Units

In a skeletal muscle, each motor unit contracts in an all-or-none manner. How then can muscles create graded contractions of varying force and duration? The answer lies in the fact that intact muscles are composed of multiple motor units of different types (Fig. 12-18). This diversity allows the muscle to vary contraction by (1) changing the types of motor units that are active or (2) changing the number of motor units that are responding at any one time.

The force of contraction in a skeletal muscle can be increased by recruiting additional motor units. **Recruitment** is controlled by the nervous system and proceeds in a standardized sequence. A weak stimulus directed onto a pool of somatic motor neurons in the central nervous system activates only the neurons with the lowest thresholds [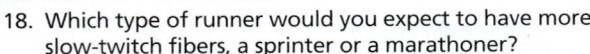 p. 254]. Studies have shown that these low-threshold neurons control fatigue-resistant slow-twitch fibers, which generate minimal force.

As the stimulus onto the motor neuron pool increases in strength, additional motor neurons with higher thresholds begin to fire. These neurons in turn stimulate motor units composed of fatigue-resistant fast-twitch oxidative-glycolytic fibers. Because more motor units (and thus more muscle fibers) are participating in the contraction, greater force is generated in the muscle.

As the stimulus increases to even higher levels, somatic motor neurons with the highest thresholds begin to fire. These neurons stimulate motor units composed of glycolytic fast-twitch fibers. At this point, the muscle contraction is approaching its maximum force. Because of differences in myosin and crossbridge formation, fast-twitch fibers can generate more force than slow-twitch fibers. However, because fast-twitch fibers fatigue more rapidly, it is impossible to hold a muscle contraction at maximum force for an extended period of time. You can demonstrate this by clenching your fist as hard as you can: how long can you hold it before some of the muscle fibers begin to fatigue?

Sustained contractions in a muscle require a continuous train of action potentials from the central nervous system to the muscle. As you learned earlier, however, increasing the stimulation rate of a muscle fiber results in summation of its contractions. If the muscle fiber is easily fatigued, summation will lead to fatigue and diminished tension (Fig. 12-17d).

One way the nervous system avoids fatigue in sustained contractions is by **asynchronous recruitment** of motor units. The nervous system modulates the firing rates of the motor neurons so that different motor units take turns maintaining muscle tension. The alternation of active motor units allows some of the motor units to rest between contractions, preventing fatigue.

Asynchronous recruitment will prevent fatigue only in submaximal contractions, however. In high-tension, sustained contractions, the individual motor units may reach a state of unfused tetanus, in which the muscle fibers cycle between contraction and partial relaxation. In general, we do not notice this cycling because the different motor units in the muscle are contracting and relaxing at slightly different times. As a result, the contractions and relaxations of the motor units average out and appear to be one smooth contraction. But as different motor units fatigue, we are unable to maintain the same amount of tension in the muscle, and the force of the contraction gradually decreases.

CONCEPT CHECK

19. What is the response of a muscle fiber to an increase in the firing rate of the somatic motor neuron?

20. How does the nervous system increase the force of contraction in a muscle composed of many motor units?

Answers: p. 434

MECHANICS OF BODY MOVEMENT

Because one main role of skeletal muscles is to move the body, we now turn to the mechanics of body movement. The term *mechanics* refers to how muscles move loads and how the anatomical relationship between muscles and bones maximizes the work the muscles can do.

Isotonic Contractions Move Loads, but Isometric Contractions Create Force Without Movement

When we described the function of muscles earlier in this chapter, we noted that they can create force to generate movement but can also create force without generating movement. You can demonstrate both properties with a pair of heavy weights. Pick up one weight in each hand and then bend your elbows so that the weights touch your shoulders. You have just performed an **isotonic**

contraction [*iso*, equal + *teinein*, to stretch]. Any contraction that creates force and moves a load is an isotonic contraction.

When you bent your arms at the elbows and brought the weights to your shoulders, the biceps muscles shortened in a **concentric action**. If you now slowly extend your arms, resisting the tendency of the weights to pull them down, you are performing another type of isotonic contraction known as an **eccentric** (or lengthening) **action** [*ec-*, out of + *centrum*, center]. Eccentric motion is thought to contribute most to cellular damage after exercise and to lead to delayed muscle soreness.

If you pick up the weights and hold them stationary in front of you, the muscles of your arms are creating tension (force) to overcome the load of the weights but are not creating movement. Contractions that create force without moving a load are called **isometric** (static) **contractions** [*iso*, equal + *metric*, measurement]. Isotonic and isometric contractions are illustrated in Figure 12-19 ■. To demonstrate an isotonic contraction experimentally, we hang a weight (the load) from the muscle in Figure 12-19a and electrically stimulate the muscle to contract. The muscle contracts, lifting the weight. The graph on the right shows the development of force throughout the contraction.

To demonstrate an isometric contraction experimentally, we attach a heavier weight to the muscle, as shown in Figure 12-19b. When the muscle is stimulated, it develops tension, but the force created is not enough to move the load. In isometric contractions, muscles create force without shortening significantly. For example, when your exercise instructor yells at you to "tighten those glutes," your response is isometric contraction of the gluteal muscles in your buttocks.

How can an isometric contraction create force if the length of the muscle does not change significantly? The elastic elements of the muscle provide the answer. All muscles contain elastic fibers in the tendons and other connective tissues that attach muscles to bone, and in the connective tissue between muscle fibers. In muscle fibers, elastic cytoskeletal proteins occur between the myofibrils and as part of the sarcomere. All of these elastic components behave collectively as if they were connected in series (one after the other) to the contractile elements of the muscle. Consequently, they are often called the **series elastic elements** of the muscle (Fig. 12-20 ■).

When the sarcomeres shorten in an isometric contraction, the elastic elements stretch. This stretching of the elastic elements allows the fibers to maintain a relatively constant length even though the sarcomeres are shortening and creating tension (Fig. 12-20, ②). Once the elastic elements have been stretched and the force generated by the sarcomeres equals the load, the muscle shortens in an isotonic contraction and lifts the load.

Bones and Muscles Around Joints Form Levers and Fulcrums

The anatomical arrangement of muscles and bones in the body is directly related to how muscles work. The body uses its bones and joints as levers and fulcrums on which muscles exert force

(a) Isotonic contraction: muscle contracts, shortens, and creates enough force to move the load.

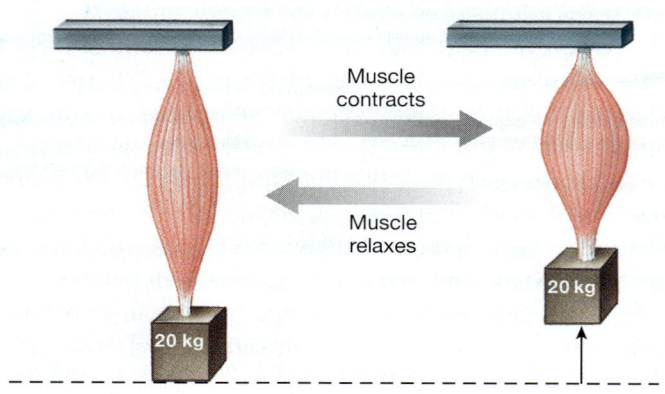

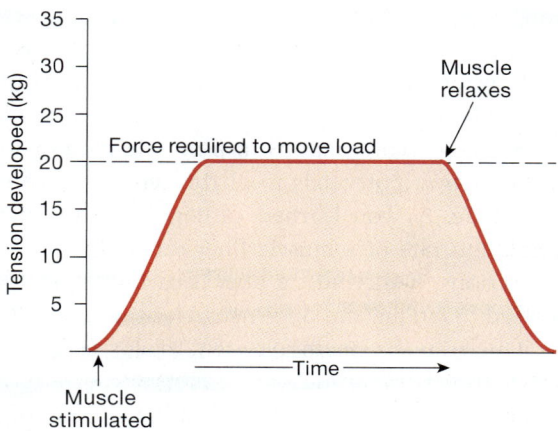

(b) Isometric contraction: muscle contracts but does not shorten. Force cannot move the load.

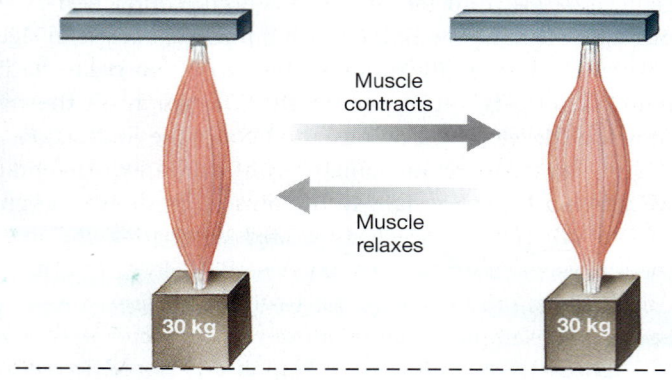

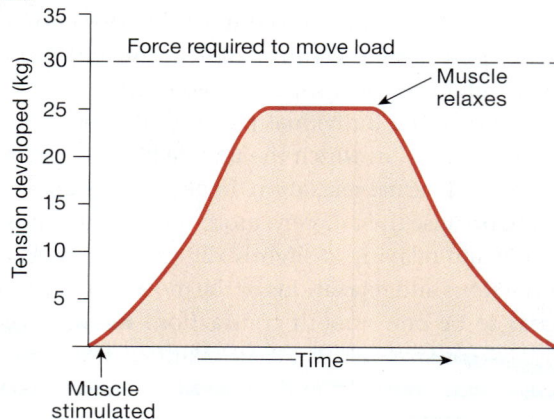

■ **FIGURE 12-19**   *Isotonic and isometric contractions*

to move or resist a load. A **lever** is a rigid bar that pivots around a point known as the **fulcrum** (Fig. 12-21a ■). In the body, bones form levers, flexible joints form the fulcrums, and muscles attached to bones create force by contracting. Most lever systems in the body are similar to the one shown in Figure 12-21a, where the fulcrum is located at one end of the lever, the load is near the other end of the lever, and the muscle attaches between the fulcrum and the load. This arrangement maximizes the distance and speed with which the lever can move the load but also requires that the muscles do more work. We will use the flexion of the forearm to illustrate how this lever system functions.

In the lever system of the forearm, the elbow joint acts as the fulcrum around which rotational movement of the forearm (the lever) takes place (Fig. 12-21a). The biceps muscle is attached at its origin at the shoulder and inserts onto the radius bone of the forearm a few centimeters away from the elbow joint. When the biceps contracts, it creates the upward force F_1 (Fig. 12-21b) as it pulls on the bone. The total rotational force* created by the

biceps depends on the force of muscle contraction and on the distance between the fulcrum and the point at which the muscle inserts onto the radius.

If the biceps is to hold the forearm stationary and flexed at a 90° angle, the muscle must exert enough upward rotational force to exactly oppose the downward rotational force exerted by gravity on the forearm (Fig. 12-21b). The downward rotational force on the forearm is proportional to the weight of the forearm (F_2) times the distance from the fulcrum to the forearm's center of gravity (the point along the lever at which the forearm load exerts its force). For the arm illustrated in Figure 12-21b, the biceps must exert 6 kg of force to hold the arm at a 90° angle. Because the muscle is not shortening, this is an isometric contraction.

Now what happens if a 7-kg weight is placed in the hand? This weight places an additional load on the lever that is farther from the fulcrum than the forearm's center of gravity (Fig. 12-21c). Unless the biceps can create additional upward force to offset the downward force created by the weight, the hand will fall. If you know the force exerted by the added weight and its distance from the elbow, you can calculate the additional muscle force needed to keep the arm from dropping the 7-kg weight.

---

*In physics, rotational force is expressed as *torque*, and the force of contraction is expressed in newtons (mass × acceleration due to gravity). For simplicity, we will ignore the contribution of gravity in this discussion and use the mass unit "kilograms" for force of contraction.

Schematic of the series elastic elements

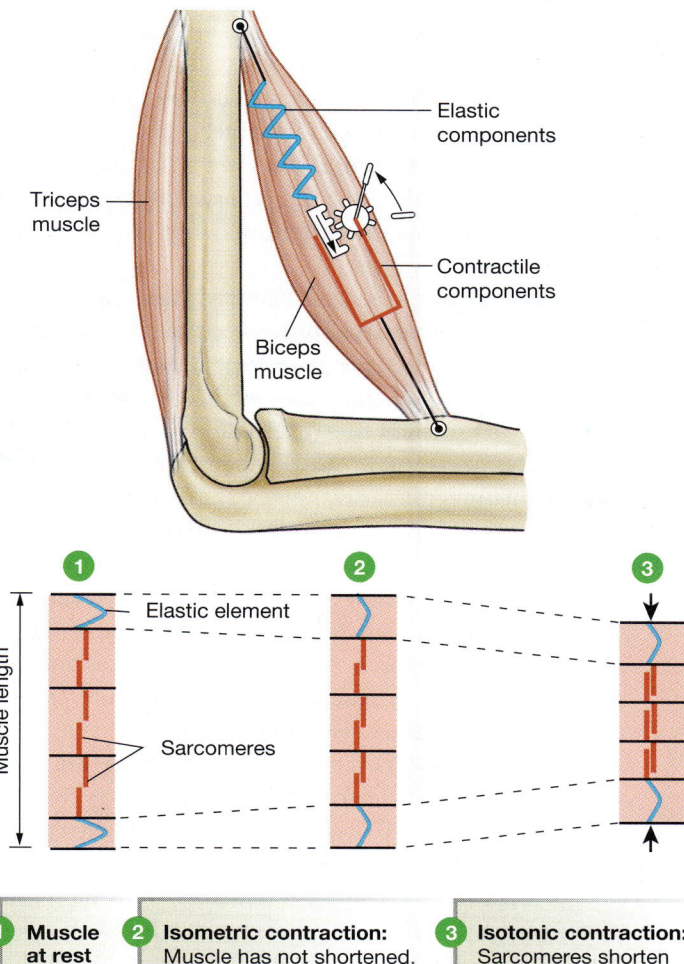

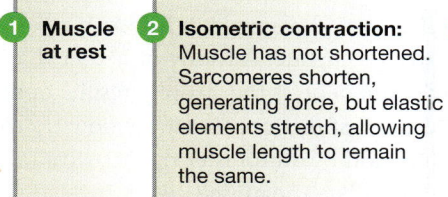

1 **Muscle at rest**	2 **Isometric contraction:** Muscle has not shortened. Sarcomeres shorten, generating force, but elastic elements stretch, allowing muscle length to remain the same.	3 **Isotonic contraction:** Sarcomeres shorten more but, because elastic elements are already stretched, the entire muscle must shorten.

■ **FIGURE 12-20** *Series elastic elements in muscle*

A muscle has both contractile components (sarcomeres, shown here as a gear and ratchet) and elastic components (shown here as a spring).

What happens to the force required of the biceps to support a weight if the distance between the fulcrum and the muscle insertion point changes? Genetic variability in the insertion point can have a dramatic effect on the force required to move or resist a load. For example, if the biceps in Figure 12-21b inserted 6 cm from the fulcrum instead of 5 cm, it would only need to generate 5 kg of force to offset the weight of the arm. Some studies have shown a correlation between muscle insertion points and success in certain athletic events.

In the example so far, we have assumed that the load is stationary and that the muscle is contracting isometrically. What happens if we want to flex the arm and lift the load? To move the load from its position, the biceps must exert a force that exceeds the force created by the stationary load.

The disadvantage of a lever system in which the fulcrum is positioned near one end of the lever is that the muscle is required to create large amounts of force to move or resist a small load, as we just saw. However, the advantage of this type of lever-fulcrum system is that it maximizes speed and mobility. A small movement of the forearm at the point where the muscle inserts becomes a much larger movement at the hand (Fig. 12-22 ■). In addition, the two movements occur in the same amount of time, and so the speed of contraction at the insertion point is amplified at the hand. Thus, the lever-fulcrum system of the arm amplifies both the distance the load is moved and the speed at which this movement takes place.

In muscle physiology, the speed with which a muscle contracts depends on the type of muscle fiber (fast-twitch or slow-twitch) and on the load that is being moved. Intuitively, you can see that you can flex your arm much faster with nothing in your hand than you can while holding a 7-kg weight in your hand. The relationship between load and velocity of contraction in a muscle fiber, determined experimentally, is graphed in Figure 12-23 ■. Contraction is fastest when the load on the muscle is zero. When the load on the muscle equals the ability of the muscle to create force, the muscle is unable to move the load and the velocity drops to zero. The muscle can still contract, but the contraction becomes isometric instead of isotonic. Because speed is a function of load and muscle fiber type, it cannot be regulated by the body except through recruitment of faster muscle fiber types. However, the arrangement of muscles, bones, and joints allows the body to amplify speed so that regulation at the cellular level becomes less important.

RUNNING PROBLEM

Paul's doctor explained to Mrs. Leong that the paralytic attacks associated with hyperkalemic periodic paralysis last only a few minutes to a few hours and generally involve only the muscles of the extremities. "Is there any treatment?" asked Mrs. Leong. The doctor replied that although the condition is incurable, attacks may be prevented with drugs. Diuretics, for example, increase the rate at which the body excretes water and ions (including Na^+ and K^+), and these medications have been shown to prevent attacks of paralysis in people with this condition.

Question 3:
Why does a Na^+ channel that will not inactivate cause paralysis?

397 410 412 **419** 423 430

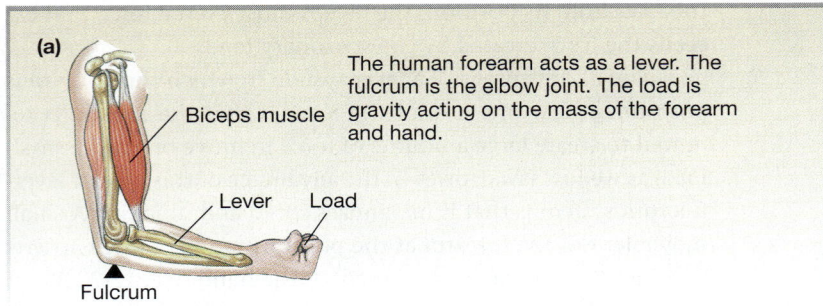

(a)

Biceps muscle

The human forearm acts as a lever. The fulcrum is the elbow joint. The load is gravity acting on the mass of the forearm and hand.

Lever   Load

Fulcrum

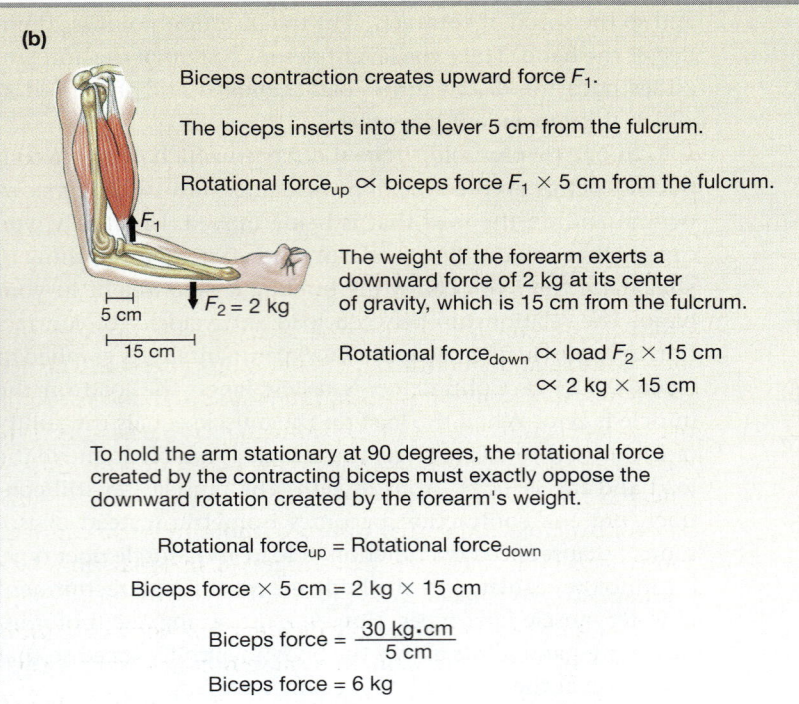

(b)

Biceps contraction creates upward force F_1.

The biceps inserts into the lever 5 cm from the fulcrum.

Rotational force$_{up}$ $\propto$ biceps force $F_1 \times$ 5 cm from the fulcrum.

F_1

5 cm

$F_2 = 2$ kg

15 cm

The weight of the forearm exerts a downward force of 2 kg at its center of gravity, which is 15 cm from the fulcrum.

Rotational force$_{down}$ $\propto$ load $F_2 \times$ 15 cm

$\propto$ 2 kg $\times$ 15 cm

To hold the arm stationary at 90 degrees, the rotational force created by the contracting biceps must exactly oppose the downward rotation created by the forearm's weight.

Rotational force$_{up}$ = Rotational force$_{down}$

Biceps force $\times$ 5 cm = 2 kg $\times$ 15 cm

$$\text{Biceps force} = \frac{30 \text{ kg·cm}}{5 \text{ cm}}$$

Biceps force = 6 kg

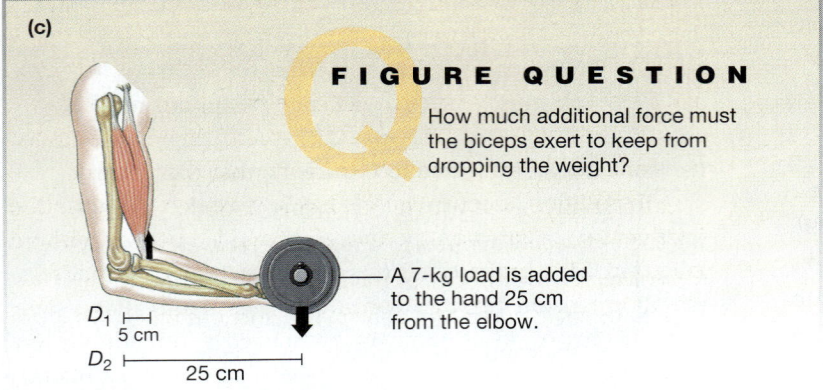

(c)

FIGURE QUESTION

How much additional force must the biceps exert to keep from dropping the weight?

A 7-kg load is added to the hand 25 cm from the elbow.

D_1

5 cm

D_2

25 cm

■ **FIGURE 12-21**   *The arm is a lever and fulcrum system*

CONCEPT CHECK

21. One study found that many world-class athletes have muscle insertions that are farther from the joint than in the average person. Why would this trait translate into an advantage for a weight lifter?

Answers: p. 434

Muscle Disorders Have Multiple Causes

Dysfunction in skeletal muscles can arise from a problem with the signal from the nervous system, from miscommunication at the neuromuscular junction, or from defects in the muscle. Unfortunately, in many muscle conditions, even the simple ones, we do not fully understand the mechanism of the primary defect. As a result, we can treat the symptoms but may not be able to cure the problem.

One common muscle disorder is a "charlie horse," or *muscle cramp*—a sustained painful contraction of skeletal muscles. Many muscle cramps are caused by hyperexcitability of the somatic motor neurons controlling the muscle. As the neuron fires repeatedly, the muscle fibers of its motor unit go into a state of painful sustained contraction. Sometimes muscle cramps can be relieved by forcibly stretching the muscle. Apparently, stretching sends sensory information to the central nervous system that inhibits the somatic motor neuron, relieving the cramp.

The simplest muscle disorders arise from overuse. Most of us have exercised too long or too hard and suffered from muscle fatigue or soreness as a result. Trauma to muscles can cause muscle fibers, the connective tissue sheath, or the union of muscle and tendon to tear.

Disuse of muscles can be as traumatic as overuse. With prolonged inactivity, such as may occur when a limb is immobilized in a cast, the skeletal muscles will atrophy. Blood supply to the muscle diminishes, and the muscle fibers get smaller. If activity is resumed in less than a year, the fibers usually will regenerate. Atrophy of longer than one year is usually permanent. If the atrophy results from somatic motor neuron dysfunction, therapists now try to maintain muscle function by administering electrical impulses that directly stimulate the muscle fibers.

Acquired disorders that affect the skeletal muscle system include infectious diseases, such as influenza, that lead to weakness and achiness, and poisoning by toxins, such as those produced in botulism (*Clostridium botulinus*) and tetanus (*Clostridium tetani*). Botulinum toxin acts by decreasing the release of acetylcholine from the somatic motor neuron. Clinical investigators have successfully used injections of botulinum toxin as a treatment for writer's cramp, a disabling cramp of the hand that apparently arises as a result of hyperexcitability in the distal portion of the somatic motor neuron. You may also have heard of Botox® injections for cosmetic wrinkle reduction. Botulinum toxin injected under the skin temporarily paralyzes facial muscles that pull the skin into wrinkles.

Inherited muscular disorders are the most difficult to treat. These conditions include various forms of muscular dystrophy as well as biochemical defects in glycogen and lipid storage. In **Duchenne's muscular dystrophy**, the structural protein **dystrophin**, which links actin to proteins in the cell membrane, is

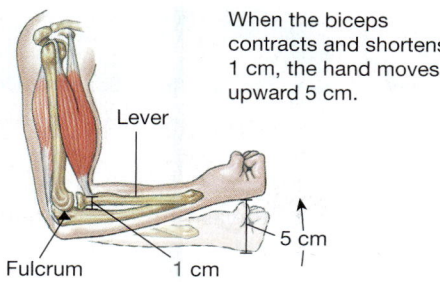

Because the insertion of the biceps is close to the fulcrum, a small movement of the biceps becomes a much larger movement of the hand.

When the biceps contracts and shortens 1 cm, the hand moves upward 5 cm.

Lever

Fulcrum   1 cm

5 cm

FIGURE QUESTION

If the biceps shortens 1 cm in 1 second, how fast does the hand move upward?

■ **FIGURE 12-22**   *The arm amplifies speed of movement of the load.*

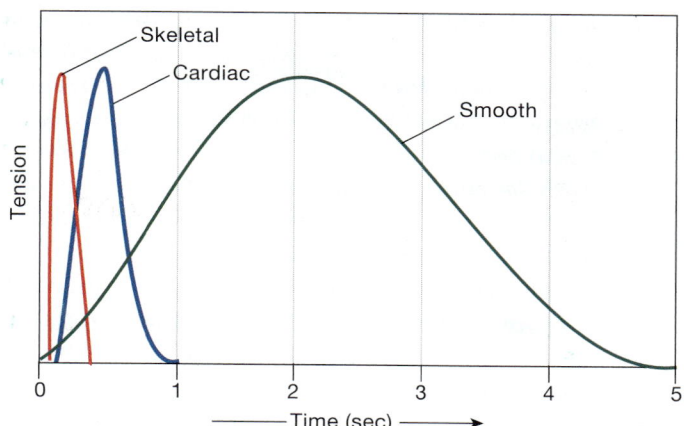

■ **FIGURE 12-24**   *Duration of muscle contraction in the three types of muscle*

absent. In muscle fibers that lack dystrophin, tiny tears in the membrane allow extracellular Ca^{2+} to enter the fiber. Consequently, intracellular enzymes are activated, resulting in breakdown of the fiber components. The major symptom of Duchenne dystrophy is progressive muscle weakness, and patients usually die before age 30 from failure of the respiratory muscles.

McArdle's disease, also known as *myophosphorylase deficiency glycogenosis,* is a condition in which the enzyme that converts glycogen to glucose 6-phosphate is absent in muscles. As a result, muscles lack a usable glycogen energy supply, and exercise tolerance is limited.

One way we are trying to learn more about muscle diseases is by using animal models, such as genetically engineered mice that lack the genes for certain muscle proteins. Researchers are trying to correlate the absence of protein with particular disruptions in function.

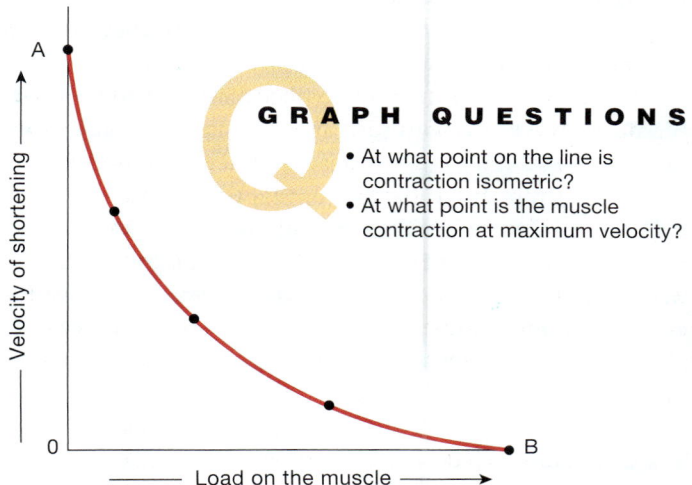

GRAPH QUESTIONS

• At what point on the line is contraction isometric?
• At what point is the muscle contraction at maximum velocity?

■ **FIGURE 12-23**   *Load-velocity relationship in skeletal muscle*

SMOOTH MUSCLE

Although skeletal muscle has the most muscle mass in the body, cardiac and smooth muscle are more important in the maintenance of homeostasis. Smooth muscle is found predominantly in the walls of hollow organs and tubes, where its contraction changes the shape of the organ. Often smooth muscle generates force to move material through the lumen of the organ. For example, sequential waves of smooth muscle contraction in the intestinal tract move ingested material from the esophagus to the colon.

Smooth muscle is noticeably different from striated muscle in the way it develops tension. In a smooth muscle twitch, contraction and relaxation occur much more slowly than in either skeletal or cardiac muscle (Fig. 12-24 ■). At the same time, smooth muscle uses less energy to generate a given amount of force, and it can maintain its force for long periods. By one estimate, for example, a smooth muscle cell can generate maximum tension with only 25–30% of its crossbridges active. We still do not fully understand the mechanisms through which this is accomplished.

In addition, smooth muscle has low oxygen consumption rates yet can sustain contractions for extended periods without fatiguing. This property allows organs such as the bladder to maintain tension despite a continued load. It also allows some smooth muscles to be tonically contracted and maintain tension most of the time. The esophageal and urinary bladder **sphincters** [*sphingein,* to close] are examples of tonically contracted muscles whose function is to close off the opening to a hollow organ. These sphincters relax when it is necessary to allow material to enter or leave the organ.

Until recently, smooth muscle had not been studied as extensively as skeletal muscle for many reasons:

1. **Smooth muscle has more variety**. Many types of smooth muscle with widely differing properties are found throughout the animal kingdom, making a single model of smooth muscle function impossible. In humans, smooth muscle

can be divided into six major groups: *vascular* (blood vessel walls), *gastrointestinal* (walls of digestive tract and associated organs, such as the gall bladder), *urinary* (walls of bladder and ureters), *respiratory* (airway passages), *reproductive* (uterus in females and other reproductive structures in both females and males), and *ocular* (eye). These muscles have different functions in the body, and their physiology is a reflection of their specialized functions. In contrast, skeletal muscle is relatively uniform throughout the body.

2. **Smooth muscle anatomy makes functional studies difficult.** The contractile fibers of smooth muscle are arranged in oblique bundles rather than in parallel sarcomeres. Consequently, a contraction pulls on the cell membrane in many directions at once. In addition, within an organ the layers of smooth muscle may run in several directions. For example, the intestine has one layer that encircles the lumen and a perpendicular layer that runs the length of the intestine. It is difficult to measure tension developing in both layers at once.

3. **Smooth muscle contraction is controlled by hormones and paracrines in addition to neurotransmitters.** Unlike skeletal muscle, which is controlled only by acetylcholine from somatic motor neurons, smooth muscle activity may be controlled by acetylcholine, norepinephrine, and a variety of other neurotransmitters, hormones, and paracrines.

4. **Smooth muscle has variable electrical properties.** Normal skeletal muscles always respond to an action potential with a twitch, but smooth muscles exhibit a variety of electrical behaviors. They may hyperpolarize as well as depolarize, and they can depolarize without firing action potentials. Contraction may take place after an action potential, after a subthreshold graded potential, or without any change in membrane potential.

5. **Multiple pathways influence contraction and relaxation of smooth muscle.** Skeletal muscles contract in response to acetylcholine from a somatic motor neuron, and relax when the stimulus for contraction ceases. In marked contrast, multiple neurotransmitters, hormones, and paracrines acting on a smooth muscle fiber can inhibit contraction as well as stimulate it. And because several different signals might reach the muscle fiber simultaneously, smooth muscle fibers must act as integrating centers. For example, sometimes blood vessels receive contradictory messages from two sources: one message signals for contraction, the other for relaxation. The smooth muscle fibers must integrate the two signals and execute an appropriate response. The complexity of overlapping regulatory pathways influencing smooth muscle tone makes the tissue difficult to work with in the laboratory.

Because of the variability in smooth muscle types, we will introduce only their general features in this chapter. Properties that are specific to a certain type will be dealt with when you learn about the different muscles in later chapters.

CLINICAL FOCUS

SMOOTH MUSCLE AND ATHEROSCLEROSIS

For years, it was believed that mature smooth muscle cells had lost the ability to undergo mitosis. Recent studies, however, have shown otherwise. Not only can smooth muscle cells reproduce themselves, but their proliferation may also play a significant role in *atherosclerosis,* a disease of blood vessels in which the vessel lumen narrows as the vessel wall thickens. Calcified fatty deposits are found in the extracellular matrix, leading to the hardened state that gives atherosclerosis its popular name; hardening of the arteries. As part of the disease process, smooth muscle cells in the blood vessel wall convert from a contractile state to a proliferative state. During proliferation, the muscle cells migrate toward the lumen, divide, and accumulate cholesterol. This change in smooth muscle function is regulated by growth factors and cytokines released by white blood cells in the region of the atherosclerotic lesion. Researchers are trying to unravel the complex story behind the smooth muscle changes in atherosclerosis so that they can develop drugs to help prevent the disease.

Smooth Muscles Are Much Smaller than Skeletal Muscle Fibers

Smooth muscles are small, spindle-shaped cells with a single nucleus, in contrast to the large multinucleated fibers of skeletal muscles. In neurally controlled smooth muscle, neurotransmitter is released from autonomic neuron varicosities [⮂ p. 383] close to the surface of the muscle fibers. Smooth muscle lacks specialized receptor regions such as the motor end plates found in skeletal muscle synapses. Instead, the neurotransmitter simply diffuses across the cell surface until it finds a receptor.

Most smooth muscle is **single-unit smooth muscle** (*unitary smooth muscle*), so called because the individual muscle cells contract as a single unit. Single-unit smooth muscle is also called **visceral smooth muscle** because it forms the walls of internal organs (viscera), such as blood vessels and the intestinal tract. All the fibers of single-unit smooth muscle are electrically connected to one another, so an action potential in one cell will spread rapidly through gap junctions to make the entire sheet of tissue contract (Fig. 12-25a ■). Because all fibers contract every time, no reserve units are left to be recruited to increase contraction force. Instead, the amount of Ca^{2+} that enters the cell determines the force of contraction, as you will learn in the discussion that follows.

(a) **Single-unit smooth muscle cells** are connected by gap junctions, and the cells contract as a single unit.

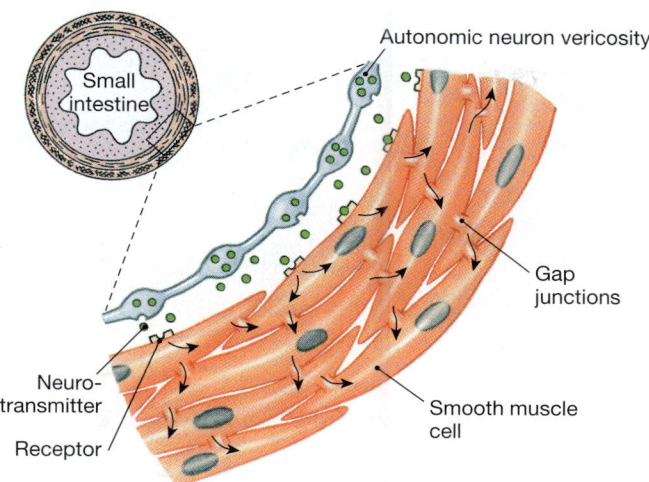

(b) **Multi-unit smooth muscle cells** are not electrically linked, and each cell must be stimulated independently.

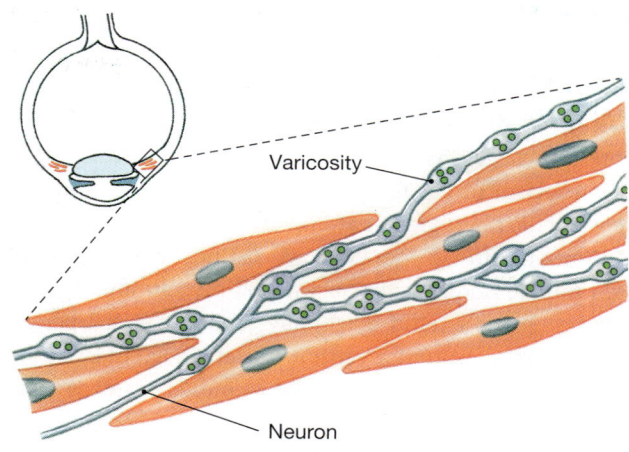

■ **FIGURE 12-25**   *Types of smooth muscle*

Multi-unit smooth muscle consists of cells that are not linked electrically. Consequently, each individual muscle cell must be closely associated with an axon terminal or varicosity and stimulated independently (Fig. 12-25b). This arrangement allows fine control of contractions in these muscles through selective activation of individual muscle cells. As in skeletal muscle, increasing the force of contraction requires recruitment of additional fibers.

Multi-unit smooth muscle is found in the iris and ciliary body of the eye, in part of the male reproductive tract, and in the uterus except just prior to labor and delivery. Interestingly, the multi-unit smooth muscle of the uterus changes and becomes single-unit during the final stages of pregnancy. Genes for synthesis of gap junction connexin proteins turn on, apparently under the influence of pregnancy hormones. The addition of gap junctions to the uterine muscle cells synchronizes electrical signals, allowing the uterine muscle to contract more effectively while expelling the baby.

 CONCEPT CHECK

22. What is the difference in how contraction force is varied in multi-unit and single-unit smooth muscle?

Answers: p. 434

Smooth Muscle Has Longer Actin and Myosin Filaments

Smooth muscle uses many of the same contractile elements as skeletal muscle: actin-myosin crossbridges, sarcoplasmic reticulum with Ca^{2+} release channels, and a Ca^{2+} signal that initiates the process. However, details of the structural elements and the contraction process differ in the two muscle types.

Smooth muscles have longer actin and myosin filaments than skeletal muscles, and the myosin isoform in smooth muscle is different from that in skeletal muscle. Smooth muscle myosin ATPase activity is much slower, decreasing the rate of crossbridge cycling and lengthening the contraction phase. In addition, one

of the smaller protein chains in the myosin head plays a regulatory role in controlling contraction and relaxation. This small regulatory protein chain is called a **myosin light chain**.

Actin is more plentiful in smooth muscle than in striated muscle, with an actin-to-myosin ratio of 10–15 to 1, compared with 2–4 to 1 in striated muscle. Smooth muscle actin is associated with tropomyosin, as in skeletal muscle. However, unlike skeletal muscle, smooth muscle lacks troponin.

Smooth muscle has less sarcoplasmic reticulum than skeletal muscle, although the amount varies from one type of smooth muscle to another. The primary Ca^{2+} release channel in smooth muscle sarcoplasmic reticulum is an **IP_3-receptor channel**. Inositol trisphosphate (IP_3) is a second messenger created in the phospholipase C pathway [🔁 p. 184]. The calcium-storage

---

RUNNING PROBLEM

Three weeks later, Paul had another attack of paralysis, this time at kindergarten after a game of tag. He was rushed to the hospital and given glucose by mouth. Within minutes, he was able to move his legs and arms and asked for his mother.

Question 4:

Explain why oral glucose might help bring Paul out of his paralysis. (Hint: cells use glucose to produce ATP for active transport. The Na^+-K^+-ATPase actively exchanges K^+ and Na^+ across the cell membrane. What happens to the extracellular K^+ level when Paul is given glucose?)

| 397 | 410 | 412 | 419 | **423** | 430 |

function of the sarcoplasmic reticulum may be supplemented by *caveolae* [⊇ p. 149], small vesicles that cluster close to the cell membrane (Fig. 12-26 ■).

Smooth Muscle Contractile Filaments Are Not Arranged in Sarcomeres

Smooth muscle gets its name from the homogeneous appearance of its cytoplasm under the microscope (Figs. 12-1c and 12-26). The contractile fibers are not arranged in sarcomeres, which is the reason smooth muscle does not have distinct banding patterns as striated muscle does. Instead, actin and myosin are arranged in long bundles that extend diagonally around the cell periphery, forming a lattice around a central nucleus (Fig. 12-27a ■). The oblique arrangement of contractile elements beneath the cell membrane causes smooth muscle fibers to become globular when they contract (Fig. 12-27b), rather than simply shortening as skeletal muscles do.

The long actin filaments of smooth muscle attach to **dense bodies** of protein in the cytoplasm and terminate at protein *attachment plaques* in the cell membrane (Fig. 12-27a, c). The less numerous myosin filaments lie bundled between the long actin filaments and are arranged so that their entire surface is covered by myosin heads (Fig. 12-27d). (Recall that in

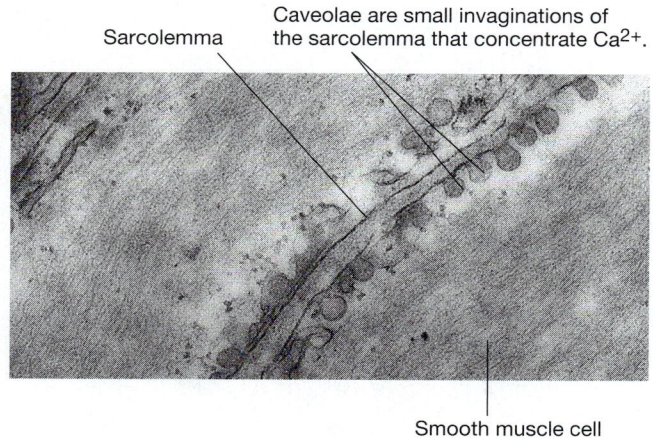

the sarcomere of skeletal muscles, the center of the myosin filament lacks myosin heads.)

The continuous line of myosin heads allows actin to slide along the myosin for longer distances. This unique organization enables smooth muscle to be stretched more while still maintaining enough overlap to create optimum tension. This is

■ **FIGURE 12-26**   *Caveolae in smooth muscle*

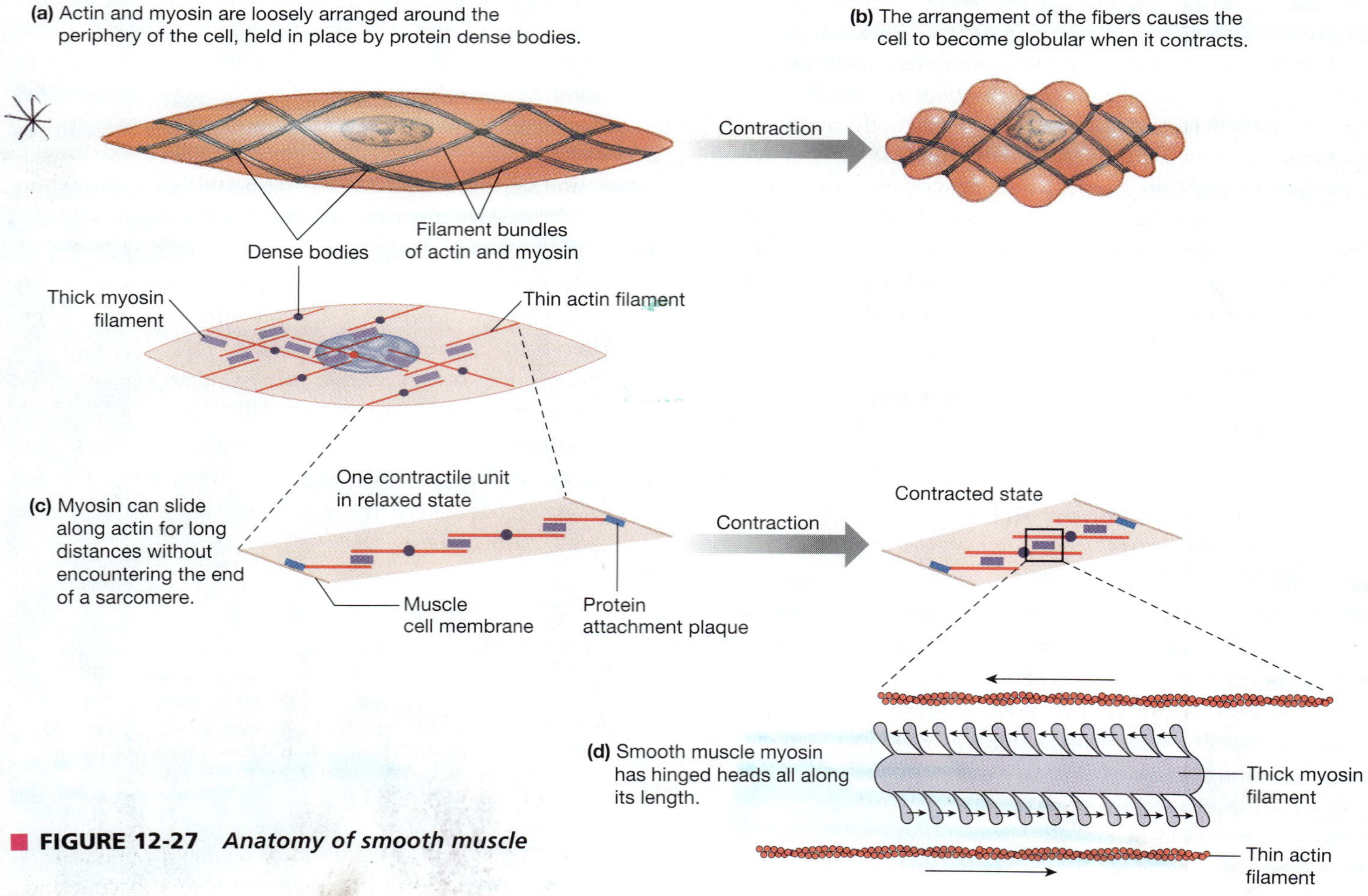

(a) Actin and myosin are loosely arranged around the periphery of the cell, held in place by protein dense bodies.

(b) The arrangement of the fibers causes the cell to become globular when it contracts.

Contraction

Filament bundles of actin and myosin

Dense bodies

Thick myosin filament

Thin actin filament

One contractile unit in relaxed state

(c) Myosin can slide along actin for long distances without encountering the end of a sarcomere.

Muscle cell membrane

Protein attachment plaque

Contraction

Contracted state

(d) Smooth muscle myosin has hinged heads all along its length.

Thick myosin filament

Thin actin filament

■ **FIGURE 12-27**   *Anatomy of smooth muscle*

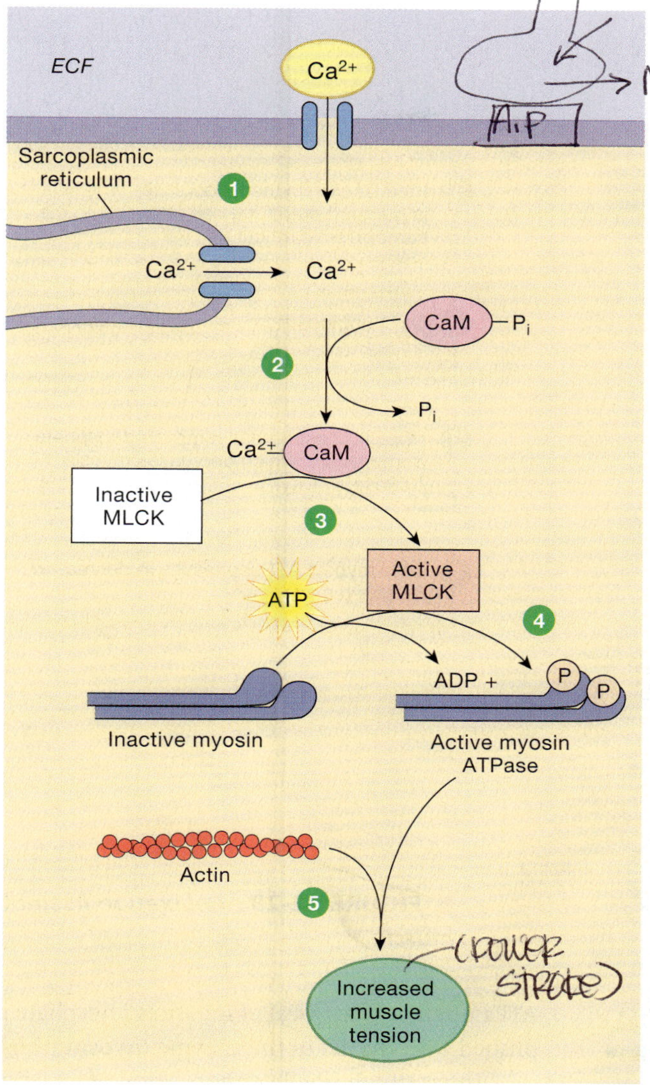

ECF

Ca²⁺

→ Nor, epi.

A.P

Sarcoplasmic
reticulum

1

Ca²⁺ ⟶ Ca²⁺

CaM — Pᵢ

2

⟶ Pᵢ

Ca²⁺ — CaM

Inactive
MLCK

3

ATP

Active
MLCK

4

ADP +

P P

Inactive myosin

Active myosin
ATPase

Actin

5

Increased
muscle
tension

(POWER STROKE)

1 Intracellular Ca²⁺ concentrations increase
when Ca²⁺ enters cell and is released from
sarcoplasmic reticulum.

2 Ca²⁺ binds to calmodulin (CaM).

3 Ca²⁺–calmodulin activates myosin light
chain kinase (MLCK).

4 MLCK phosphorylates light chains in myosin
heads and increases myosin ATPase activity.

5 Active myosin crossbridges slide along actin
and create muscle tension.

12

■ **FIGURE 12-28** *Smooth muscle
contraction*

an important property for internal organs, such as the bladder, whose volume varies as it alternately fills and empties.

CONCEPT CHECK

23. The dense bodies that anchor smooth muscle actin are analogous to what structure in a sarcomere? (*Hint*: see Fig. 12-5.)

24. Name three ways smooth muscle myosin differs from skeletal muscle myosin.

25. Name one way actin and its associated proteins differ in skeletal and smooth muscle.
    Answers: p. 434

Phosphorylation of Proteins Plays a Key Role in Smooth Muscle Contraction

The molecular events of smooth muscle contraction are similar in many ways to those in skeletal muscle, but some important differences exist. The primary difference is the role of phosphorylation in the regulation of contraction. Here is a summary of our current understanding of the key points of smooth muscle contraction. In smooth muscle:

1. An increase in cytosolic Ca²⁺ initiates contraction. Ca²⁺ is released from the sarcoplasmic reticulum and also enters from the extracellular fluid.

2. Ca²⁺ binds to **calmodulin**, a binding protein found in the cytosol. (In skeletal muscle, Ca²⁺ binds to troponin, which smooth muscle lacks.)

3. Ca²⁺ binding to calmodulin is the first step in a cascade that ends in contraction. (In skeletal muscle, Ca²⁺ binding initiates contraction immediately.)

4. Phosphorylation of proteins is an essential step in contraction.

5. The primary control of contraction resides in the regulation of myosin ATPase activity.

CONCEPT CHECK

26. Compare list items 1, 3, 4, and 5 above with what happens during skeletal muscle contraction. (*Hint*: see Fig. 12-11.)
    Answers: p. 434

Figure 12-28 ■ illustrates the steps of smooth muscle contraction. Contraction begins when cytosolic Ca²⁺ concentrations

12

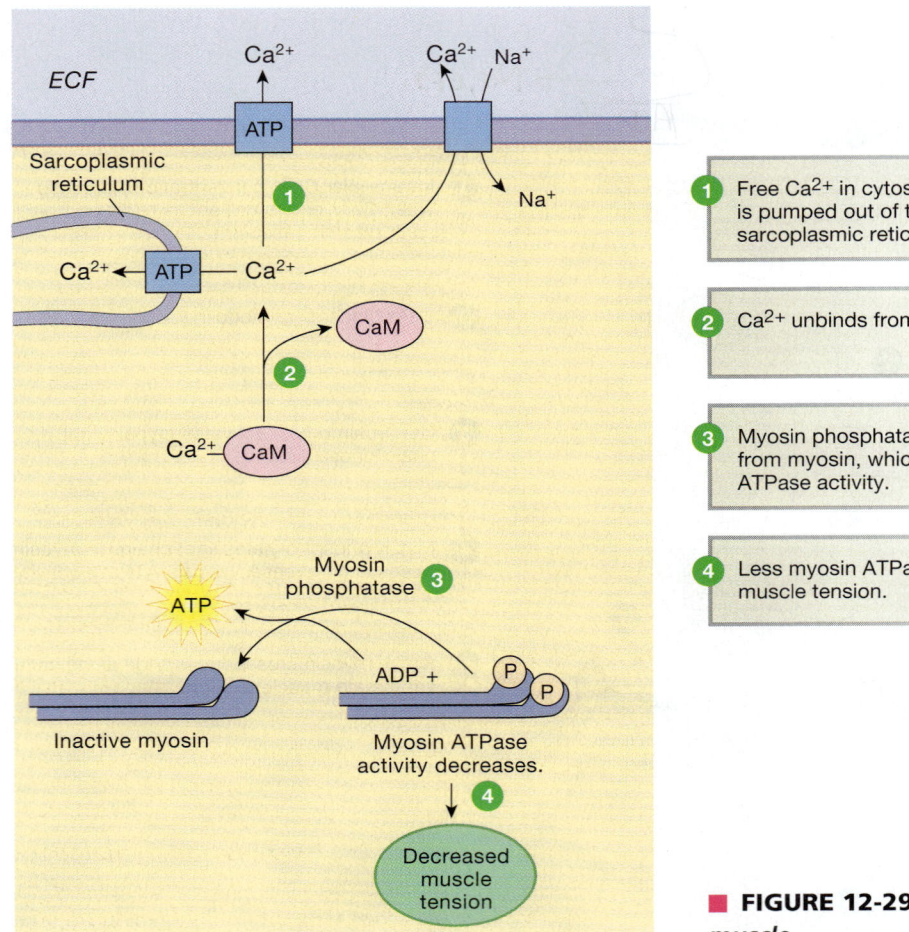

1   Free Ca²⁺ in cytosol decreases when Ca²⁺ is pumped out of the cell or back into the sarcoplasmic reticulum.

2   Ca²⁺ unbinds from calmodulin (CaM).

3   Myosin phosphatase removes phosphate from myosin, which decreases myosin ATPase activity.

4   Less myosin ATPase results in decreased muscle tension.

■ **FIGURE 12-29** *Relaxation in smooth muscle*

increase following Ca²⁺ entry from the extracellular fluid and Ca²⁺ release from the sarcoplasmic reticulum. The calcium ions bind to calmodulin (CaM), obeying the law of mass action [🔁 p. 39]. The Ca²⁺-calmodulin complex then activates an enzyme called **myosin light chain kinase (MLCK)**. This enzyme enhances myosin ATPase activity by phosphorylating light protein chains in the myosin head. When myosin ATPase activity is high, actin binding and crossbridge cycling increase tension in the muscle. Thus, smooth muscle contraction is primarily controlled through myosin-linked regulatory processes rather than the actin-linked regulation of skeletal muscle.

Relaxation in Smooth Muscle Has Several Steps

Relaxation in a smooth muscle fiber is a multistep process (Fig. 12-29 ■). As in skeletal muscle, free Ca²⁺ is removed from the cytosol when Ca²⁺-ATPase pumps it back into the sarcoplasmic reticulum. In addition, some of the Ca²⁺ is pumped out of the cell with the help of a Ca²⁺-Na⁺ antiport exchanger [🔁 p. 139] and Ca²⁺-ATPase. By the law of mass action, a decrease in free cytosolic Ca²⁺ causes Ca²⁺ to unbind from calmodulin. In the absence of Ca²⁺-calmodulin, myosin light chain kinase inactivates.

The additional step in smooth muscle relaxation is dephosphorylation of the myosin light chain, which decreases myosin

ATPase activity. Removal of myosin's phosphate group is accomplished with the aid of the enzyme **myosin phosphatase**.

Interestingly, dephosphorylation of myosin does not automatically result in relaxation. Under conditions that we do not fully understand, dephosphorylated myosin may remain attached to actin for a period of time in what is known as a **latch state**. This condition maintains tension in the muscle fiber without consuming ATP. It is a significant factor in the ability of smooth muscle to sustain contraction without fatiguing. The hinge muscles of certain bivalve mollusks such as oysters can enter a similar latch state that allows them to remain tightly closed under anaerobic conditions.

CONCEPT CHECK

27. What happens to contraction if a smooth muscle is placed in a saline bath from which all calcium has been removed?
     Answers: p. 434

Calcium Entry Is the Signal for Smooth Muscle Contraction

In smooth muscle, an increase in cytosolic Ca²⁺ concentration is the signal to initiate contraction. As you saw in Figure 12-28, Ca²⁺ enters the cell from the extracellular fluid and is released from the sarcoplasmic reticulum. However, because Ca²⁺ stores

The increase in cytosolic Ca^{2+} initiates contraction through the regulation of myosin ATPase as discussed in the previous section. In some types of smooth muscle, myosin regulation of crossbridge cycling is supplemented by regulation of actin. Several actin-associated regulatory proteins have been identified, including one called *caldesmon* and another called *calponin*. Second messengers that modulate contraction may act on myosin light chain kinase, on myosin phosphatase, or on the actin-associated regulatory proteins. The details of this modulation are still being worked out.

CONCEPT CHECK

28. Compare Ca^{2+} release channels in skeletal and smooth muscle sarcoplasmic reticulum.   Answers: p. 434

Muscle Stretch Opens Ca^{2+} Channels

Smooth muscle cells contain stretch-activated Ca^{2+} channels that open when pressure or other force distorts the cell membrane. Because contraction in this instance originates from a property of the muscle fiber itself, it is known as a **myogenic contraction**. Myogenic contractions are common in blood vessels that maintain a certain amount of tone at all times.

Some smooth muscles adapt if the muscle cells are stretched for an extended period of time. As the stretch stimulus continues, the Ca^{2+} channels begin to close in a time-dependent fashion. Then, as Ca^{2+} is pumped out of the cell, the muscle relaxes. This adaptation response explains why the bladder develops tension as it fills, then relaxes as it adjusts to the increased volume. (There is a limit to the amount of stretch the muscle can endure, however, and once a critical volume is reached, the urination reflex empties the bladder.)

Some Smooth Muscles Have Unstable Membrane Potentials

A major route of Ca^{2+} influx for smooth muscle is entry through voltage-gated Ca^{2+} channels in the cell membrane. However, the role of membrane potentials in smooth muscle contraction is much more complex than in skeletal muscle, where contraction always begins in response to an action potential. In smooth muscle, an action potential is not required to open voltage-gated Ca^{2+} channels. Graded potentials open a few Ca^{2+} channels, allowing small amounts of Ca^{2+} into the cell. This ion entry depolarizes the cell and may open additional Ca^{2+} channels.

In addition, many types of smooth muscle display unstable resting membrane potentials that vary between -40 and -80 mV. Cells that exhibit cyclic depolarization and repolarization of their membrane potential are said to have **slow wave potentials** (Fig. 12-31a ■). Sometimes the cell simply cycles through a series of subthreshold slow waves. However, if the peak of the depolarization reaches threshold, action potentials fire, followed by contraction of the muscle.

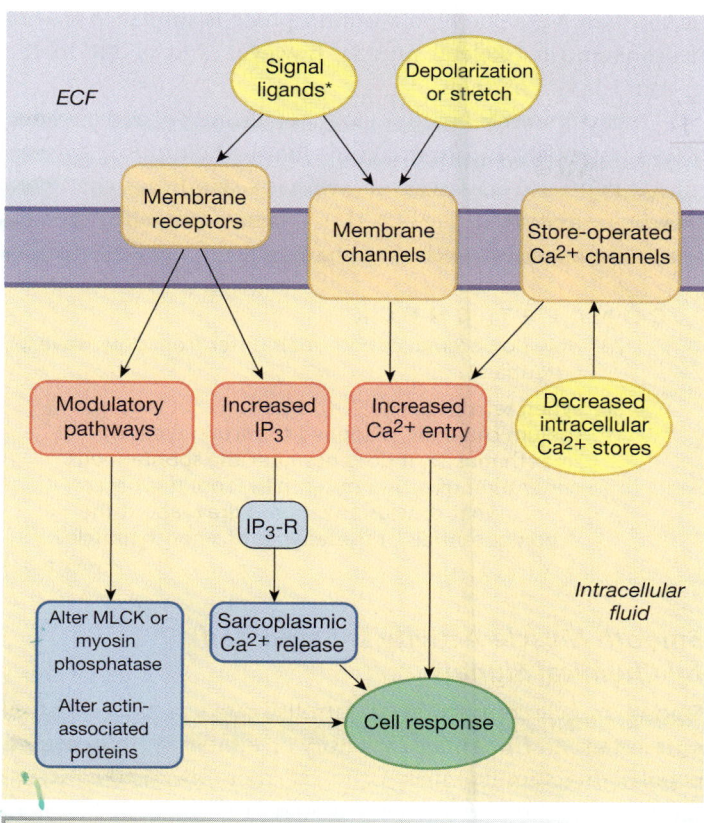

■ **FIGURE 12-30** *Control of smooth muscle contraction*

KEY

IP_3-R = IP_3–activated receptor–channel

* Ligands include norepinephrine, ACh, other neurotransmitters, hormones, and paracrines.

in smooth muscle are limited, sustained contractions depend on continued influx of Ca^{2+} from the extracellular fluid. The entry of variable amounts of Ca^{2+} into the muscle fiber creates contractions whose force is graded according to the strength of the Ca^{2+} signal.

Recall from the introduction to this section that smooth muscle contraction can be initiated by neurotransmitters, hormones, or paracrines. This is where our model becomes complex. Figure 12-30 ■ is a generalized summary of the multiple pathways that can initiate or modulate contraction in smooth muscle.

Ca^{2+} enters a smooth muscle cell through membrane channels opened by depolarization, membrane stretch, or chemical signals. One recently identified family of Ca^{2+} channels has been named **store-operated Ca^{2+} channels** because these channels open (by an as-yet-unidentified mechanism) in response to depleted intracellular stores of Ca^{2+}. Voltage-gated and stretch-activated channels are discussed in more detail below.

Sarcoplasmic reticulum Ca^{2+} release is mediated primarily by an IP_3-activated receptor-channel. The IP_3 receptor-channel opens as a result of signal transduction pathways begun by extracellular signal molecules.

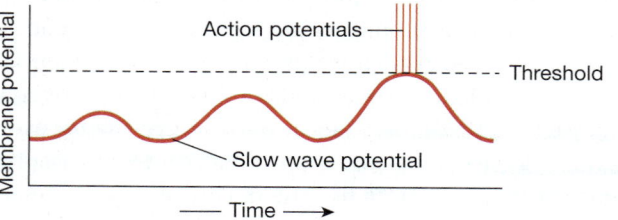

(a) Slow wave potentials fire action potentials when they reach threshold.

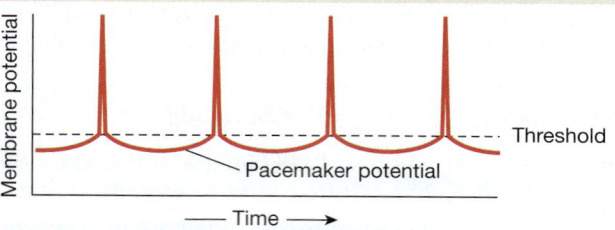

(b) Pacemaker potentials always depolarize to threshold.

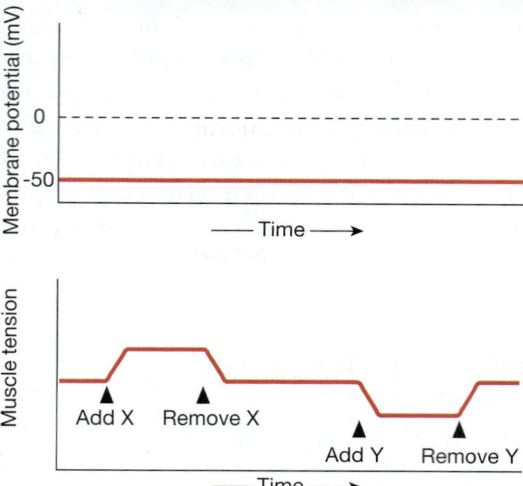

(c) Pharmacomechanical coupling occurs when chemical signals change muscle tension without a change in membrane potential.

■ **FIGURE 12-31** *Membrane potentials vary in smooth muscle*

Other types of smooth muscle with unstable membrane potentials have regular depolarizations that always reach threshold and fire an action potential (Fig. 12-31b). These depolarizations are called **pacemaker potentials** because they create regular rhythms of contraction. Pacemaker potentials are found in some cardiac muscles as well as in smooth muscle. Both slow wave and pacemaker potentials are due to ion channels in the cell membrane that spontaneously open and close.

Some smooth muscles have action potentials that look very much like action potentials in neurons. At the cellular level, however, smooth muscle action potentials are different from those in neurons and skeletal muscle because in smooth muscle, the depolarization phase is due to the entry of Ca^{2+}

rather than Na^+. The repolarization phase is similar to that of neurons and in skeletal muscle, however, and is due to the opening of K^+ channels.

Smooth muscle function does not always depend on firing action potentials. In **pharmacomechanical coupling**, smooth muscle contraction may occur without a significant change in membrane potential (Fig. 12-31c). Chemical signals may also relax muscle tension without a change in membrane potential.

CONCEPT CHECK

29. How do pacemaker potentials differ from slow wave potentials?

30. When tetrodotoxin (TTX), a poison that blocks sodium channels, is applied to certain types of smooth muscle, it does not alter the spontaneous generation of action potentials. From this observation, what conclusion can you draw about the action potentials of these types of smooth muscle?

Answers: p. 434

Smooth Muscle Activity Is Regulated by Chemical Signals

Smooth muscle contraction is controlled by a variety of chemical signals that may be either excitatory or inhibitory. In general, a depolarizing stimulus makes the cell more likely to contract. Hyperpolarization of the cell decreases the likelihood of contraction. Contraction can be modulated by second messenger action on myosin or by actin regulation.

Autonomic Neurotransmitters Many smooth muscles have dual innervation and are controlled by both sympathetic and parasympathetic neurons. However, other smooth muscles, such as those found in blood vessels, are controlled by only one of the two autonomic branches. In this type of *tonic control*, the response is graded by increasing or decreasing the amount of neurotransmitter released onto the muscle.

As we have seen, a neurotransmitter can have different effects in different tissues, depending on the receptors to which it binds. Thus, both the neurotransmitter and its receptor determine the response of a smooth muscle to nervous stimulation. Adrenergic and cholinergic muscarinic receptors act via IP_3 and cAMP second messenger systems. The exact response of the muscle depends on the adrenergic or cholinergic receptor subtype that is activated.

In addition to norepinephrine and acetylcholine, an amazing variety of other neurotransmitters are active in smooth muscle. In many instances we know how these chemicals affect smooth muscle contraction, but we do not understand the reflex pathways that trigger their release.

CONCEPT CHECK

31. How can a neuron alter the amount of neurotransmitter it releases? (*Hint:* see Fig. 8-13, ⧉ p. 263)

32. Explain how hyperpolarization decreases the likelihood of contraction in smooth muscle.

Answers: p. 434

TABLE 12-3	Comparison of the Three Muscle Types		
	SKELETAL	**SMOOTH**	**CARDIAC**
Appearance under light microscope	Striated	Smooth	Striated
Fiber arrangement	Sarcomeres	Oblique bundles	Sarcomeres
Fiber proteins	Actin, myosin; troponin and tropomyosin	Actin, myosin, tropomyosin	Actin, myosin; troponin and tropomyosin
Control	• Voluntary • Ca^{2+} and troponin • Fibers independent of one another	• Involuntary • Ca^{2+} and calmodulin • Fibers electrically linked via gap junctions	• Involuntary • Ca^{2+} and troponin • Fibers electrically linked via gap junctions
Nervous control	Somatic motor neuron	Autonomic neurons	Autonomic neurons
Hormonal influence	None	Multiple hormones	Epinephrine
Location	Attached to bones; a few sphincters close off hollow organs	Forms the walls of hollow organs and tubes; some sphincters	Heart muscle
Morphology	Multinucleate; large, cylindrical fibers	Uninucleate; small spindle-shaped fibers	Uninucleate; shorter branching fibers
Internal structure	T-tubule and sarcoplasmic reticulum	No t-tubules; sarcoplasmic reticulum reduced or absent	T-tubule and sarcoplasmic reticulum
Contraction speed	Fastest	Slowest	Intermediate
Contraction force of single fiber twitch	All-or-none	Graded	Graded
Initiation of contraction	Requires input from motor neuron	Can be autorhythmic	Autorhythmic

Hormones and Paracrines Hormones and paracrines also control smooth muscle contraction—unlike skeletal muscle, whose contraction is controlled only by the nervous system. Smooth muscles in the cardiovascular, gastrointestinal, urinary, respiratory, and reproductive systems respond either to blood-borne or to locally released chemicals. For example, asthma is a condition in which smooth muscle of the airways constricts in response to histamine release. This constriction can be reversed by the administration of epinephrine, a neurohormone that relaxes smooth muscle and dilates the airway. Note from this example that not all physiological responses are adaptive or favorable to the body: constriction of the airways triggered during an asthma attack, if left untreated, can be fatal.

Another important paracrine that affects smooth muscle contraction is *nitric oxide* [p. 187]. This gas is synthesized by the endothelial lining of blood vessels and relaxes adjacent smooth muscle that regulates the diameter of the blood vessels. For many years, the identity of this *endothelium-derived relaxing factor*, or EDRF, eluded scientists even though its presence could be demonstrated experimentally. We know now that EDRF is nitric oxide, an important paracrine in many systems of the body.

Although smooth muscles do not have nearly the mass of skeletal muscles, they play a critical role in the function of most organ systems. You will learn more about smooth muscle physiology in the chapters to come.

CARDIAC MUSCLE

Cardiac muscle, the specialized muscle of the heart, shares features with both smooth and skeletal muscle (Table 12-3 ■). Like skeletal muscle fibers, cardiac muscle fibers are striated and have a sarcomere structure. However, cardiac muscle fibers are shorter than skeletal muscle fibers, may be branched, and have a single nucleus (unlike multinucleate skeletal muscle fibers).

As in single-unit smooth muscle, cardiac muscle fibers are electrically linked to one another. The gap junctions are contained in specialized cell junctions known as *intercalated disks*. Some cardiac muscle, like some smooth muscle, exhibits pacemaker potentials. In addition, cardiac muscle is under sympathetic and parasympathetic control as well as hormonal control. In Chapter 14 you will learn more about cardiac muscle and how it functions within the heart.

PERIODIC PARALYSIS

In this running problem, you were introduced to hyperkalemic periodic paralysis, a condition caused by a genetic defect in the Na^+ channels on muscle cell membranes. To learn more about hyperkalemic periodic paralysis, go to the web and search for Medlineplus, the illustrated medical encyclopedia of the U.S. National Library of Medicine. Once you read the information about periodic paralysis, compare the hyperkalemic and hypokalemic forms of the disease.

Now check your understanding of this running problem by comparing your answers with the information in the following summary table.

	QUESTION	FACTS	INTEGRATION AND ANALYSIS
1	What effect does the continuous influx of Na^+ have on the membrane potential of Paul's muscle fibers?	The resting membrane potential of cells is negative relative to the extracellular fluid.	The influx of positive charge will depolarize the membrane potential.
2	What ion is responsible for the repolarization phase of the muscle action potential, and in which direction does this ion move across the muscle fiber membrane?	Each muscle twitch results from an action potential in the muscle fiber. In the repolarization phase of the action potential, K^+ leaves the cell.	During the repeated contractions of exercise, K^+ leaves the muscle fiber and accumulates in the t-tubules. This increase in K^+ concentration in the t-tubules affects the Na^+ channels and triggers an attack.
3	Why does a Na^+ channel that will not inactivate cause paralysis?	During an attack, the Na^+ channels remain open and continuously admit Na^+, and the muscle fiber remains depolarized.	If the muscle fiber is unable to repolarize, it cannot fire additional action potentials. The first action potential causes a twitch, but the muscle then goes into a state of flaccid (uncontracted) paralysis.
4	Explain why oral glucose might help bring Paul out of his paralysis.	Muscle fibers use glucose to produce ATP to run the Na^+-K^+-ATPase. This transporter actively moves K^+ and Na^+ across the cell membrane.	Providing glucose to cells may increase the amount of ATP available to run the Na^+-K^+-ATPase, which removes K^+ from the extracellular fluid. Insulin, a hormone that enhances glucose uptake into cells, also helps relieve the paralysis.

397 410 412 419 423 **430**

CHAPTER SUMMARY

Muscles exhibit many of the physiological properties introduced in Chapter 1. They provide an excellent system for studying *structure-function* relationships at all levels, from actin, myosin, and sliding filaments in the cell to muscles pulling on bones and joints. *Mechanical properties* of muscles that influence contraction include elastic components, such as the protein titin and the series elastic elements of the intact muscle. *Compartmentation* is essential to muscle function, as demonstrated by the concentration of Ca^{2+} in the sarcoplasmic reticulum and the key role of Ca^{2+} signals in initiating contraction. The *law of mass action* is at work in the dynamics of Ca^{2+}-calmodulin and Ca^{2+}-troponin binding and unbinding. Muscles also show how *biological energy use* transforms stored energy in ATP's chemical bonds to the movement of motor proteins.

Muscles provide many examples of *communication* and *control* in the body. Communication occurs on a scale as small as electrical signals spreading among smooth muscle cells via gap junctions, or as large as a somatic motor neuron innervating multiple muscle fibers. Skeletal muscles are controlled only by somatic motor neurons, but smooth and cardiac muscle have complex regulation that ranges from neurotransmitters to hormones and paracrines.

1. Muscles generate motion, force, and heat. (p. 397)
2. The three types of muscle are **skeletal muscle**, **cardiac muscle**, and **smooth muscle**. Skeletal and cardiac muscles are **striated muscles**. (p. 397; Fig. 12-1)
3. Skeletal muscles are controlled by somatic motor neurons. Cardiac and smooth muscle are controlled by autonomic innervation, paracrines, and hormones. Some smooth and cardiac muscles are autorhythmic and contract spontaneously. (p. 397)

Skeletal Muscle

IP Muscular Physiology

4. Skeletal muscles are usually attached to bones by tendons. The **origin** is the end of the muscle attached closest to the trunk or to the more stationary bone. The **insertion** is the more distal or mobile attachment. (p. 398)

5. At a flexible joint, muscle contraction moves the skeleton. **Flexors** bring bones closer together; **extensors** move bones away from each other. Flexor-extensor pairs are examples of **antagonistic muscle groups**. (p. 398; Fig. 12-2)

6. A skeletal muscle is a collection of **muscle fibers**, large cells with many nuclei. (p. 398; Fig. 12-3, Tbl. 12-1)

7. **T-tubules** allow action potentials to move rapidly into the interior of the fiber and release calcium from the **sarcoplasmic reticulum**. (p. 398; Fig. 12-4)

8. **Myofibrils** are intracellular bundles of contractile and elastic proteins. **Thick filaments** are made of **myosin**. **Thin filaments** are made mostly of **actin**. **Titin** and **nebulin** hold thick and thin filaments in position. (pp. 399, 402; Figs. 12-3, 12-6)

9. Myosin binds to actin, creating **crossbridges** between the thick and thin filaments. (p. 399; Fig. 12-3)

10. One **sarcomere** is composed of two **Z disks** and the filaments between them. A sarcomere is divided into **I bands** (thin filaments only), an **A band** that runs the length of a thick filament, and a central **H zone** occupied by thick filaments only. The **M line** and Z disks represent attachment sites for myosin and actin, respectively. (p. 399; Fig. 12-5)

11. The force created by a contracting muscle is called **muscle tension**. The **load** is a weight or force that opposes contraction of a muscle. (p. 403)

12. The **sliding filament theory of contraction** states that during contraction, overlapping thick and thin filaments slide past each other in an energy-dependent manner as a result of actin-myosin crossbridge movement. (p. 403; Fig. 12-8)

13. Myosin converts energy from ATP into motion. **Myosin ATPase** hydrolyzes ATP to ADP and P_i. (p. 404; Fig. 12-9)

14. When myosin releases P_i, the myosin molecule moves in the **power stroke**. At the end of the power stroke, myosin releases ADP. The cycle ends in the **rigor state**, with myosin tightly bound to actin. (pp. 404–405; Fig. 12-9)

15. **Tropomyosin** blocks the myosin-binding site on actin. As contraction begins, **troponin** binds to Ca^{2+}. This unblocks the myosin-binding sites and allows myosin to complete its power stroke. (p. 406; Fig. 12-10)

16. During relaxation, the sarcoplasmic reticulum uses a Ca^{2+}-ATPase to pump Ca^{2+} back into its lumen. (p. 408)

17. In **excitation-contraction coupling**, a somatic motor neuron releases ACh, which initiates a skeletal muscle action potential that leads to contraction. (p. 407; Fig. 12-11a)

18. Voltage-sensing **DHP receptors** in the t-tubules open Ca^{2+} **release channels** in the sarcoplasmic reticulum. (p. 408; Fig. 12-11a)

19. A single contraction-relaxation cycle is known as a **twitch**. The **latent period** between the end of the muscle action potential and the beginning of muscle tension development represents the time required for Ca^{2+} release and binding to troponin. (p. 409; Fig. 12-12)

20. Muscle fibers store energy for contraction in **phosphocreatine**. Anaerobic metabolism of glucose is a rapid source of ATP but is not efficient. Aerobic metabolism is very efficient but requires an adequate supply of oxygen to the muscles. (p. 410; Fig. 12-13)

21. **Muscle fatigue** is a condition in which a muscle is no longer able to generate or sustain the expected power output. Fatigue has multiple causes. (p. 411; Fig. 12-14)

22. Skeletal muscle fibers can be classified on the basis of their speed of contraction and resistance to fatigue into **fast-twitch glycolytic fibers**, **fast-twitch oxidative-glycolytic fibers**, and **slow-twitch (oxidative) fibers**. Oxidative fibers are the most fatigue resistant. (p. 412–413; Fig. 12-15; Tbl. 12-2)

23. **Myoglobin** is an oxygen-binding pigment that transfers oxygen to the interior of the muscle fiber. (p. 412)

24. The tension of a skeletal muscle contraction is determined by the length of the sarcomeres before contraction begins. (p. 413; Fig. 12-16)

25. Increasing the stimulus frequency causes summation of twitches with an increase of tension. A state of maximal contraction is known as **tetanus**. (p. 415; Fig. 12-17)

26. A **motor unit** is composed of a group of muscle fibers and the somatic motor neuron that controls them. The number of muscle fibers in a motor unit varies, but all fibers in a single motor unit are of the same fiber type. (p. 415; Fig. 12-18)

27. The force of contraction within a skeletal muscle can be increased by **recruitment** of additional motor units. (p. 416)

Mechanics of Body Movement

28. An **isotonic contraction** creates force and moves a load. An **isometric contraction** creates force without moving a load. **Concentric actions** are shortening contractions. **Eccentric actions** are lengthening contractions. (p. 417; Fig. 12-19)

29. Isometric contractions occur because **series elastic elements** allow the fibers to maintain constant length even though the sarcomeres are shortening and creating tension. (p. 417; Fig. 12-20)

30. The body uses its bones and joints as **levers** and **fulcrums**. Most lever-fulcrum systems in the body maximize distance and speed but also require that muscles do more work. (p. 418; Figs. 12-21, 12-22)

31. Contraction speed is a function of muscle fiber type and load. Contraction is fastest when the load on the muscle is zero. (p. 419; Fig. 12-23)

Smooth Muscle

32. Smooth muscle is slower than skeletal muscle but can sustain contractions for longer without fatiguing. (p. 421; Fig. 12-24)

33. **Single-unit smooth muscle** contracts as a single unit when depolarizations pass from cell to cell through gap junctions. In **multi-unit smooth muscle**, individual muscle fibers are stimulated independently. (pp. 422–423; Fig. 12-25)

34. Actin and myosin are arranged along the periphery of a smooth muscle cell. Smooth muscle actin lacks troponin. (p. 423; Fig. 12-27)

35. Smooth muscle has relatively little sarcoplasmic reticulum, and the primary Ca^{2+} release channel is an **IP$_3$-receptor channel**. (p. 423)

36. In smooth muscle contraction, Ca^{2+} binds to **calmodulin** and activates **myosin light chain kinase** (MLCK). (pp. 425–426; Fig. 12-28)

37. MLCK phosphorylates **myosin light protein chains**, which activates myosin ATPase. This allows crossbridge power strokes. (p. 426; Fig. 12-28)

12

38. During relaxation, Ca^{2+} is pumped out of the cytosol, and myosin light chains are dephosphorylated by **myosin phosphatase**. (p. 426; Fig. 12-29)

39. Ca^{2+} for contraction enters the cell through Ca^{2+} channels in the cell membrane. Additional Ca^{2+} is released from the sarcoplasmic reticulum. (p. 426; Fig. 12-30)

40. In **myogenic contraction**, stretch opens membrane Ca^{2+} channels. (p. 427)

41. Unstable membrane potentials in smooth muscle take the form of either **slow wave potentials** or **pacemaker potentials**. (pp. 427–428; Fig. 12-31a, b)

42. The rising phase of smooth muscle action potentials is due to Ca^{2+} entry rather than Na^+ entry. (p. 428)

43. In **pharmacomechanical coupling**, Ca^{2+} entry causes smooth muscle contraction without a significant change in membrane potential. (p. 428; Fig. 12-31c)

44. Smooth muscle is controlled by sympathetic and parasympathetic neurons and a variety of chemical signals. (p. 428)

Cardiac Muscle

45. Cardiac muscle fibers are striated, have a single nucleus, and are electrically linked through gap junctions. Cardiac muscle shares features with both skeletal and smooth muscle. (p. 429; Tbl. 12-3)

QUESTIONS

(Answers to the Review Questions begin on page A1.)

THE PHYSIOLOGY PLACE

Access more review material online at **The Physiology Place** website. There you'll find review questions, problem-solving activities, case studies, flashcards, and direct links to both *InterActive Physiology*® and *PhysioEx™*. To access the site, go to *www.physiologyplace.com* and select Human Physiology, Fourth Edition.

LEVEL ONE   REVIEWING FACTS AND TERMS

1. The three types of muscle tissue found in the human body are skeletal , smooth , and cardiac . Which type is attached to the bones, enabling it to control body movement?

2. Which two muscle types are striated? skeletal & cardiac

3. Which type of muscle tissue is controlled only by somatic motor neurons? skeletal

4. Which of the following statement(s) is(are) true about skeletal muscles?
   F (a) They constitute about 60% of a person's total body weight.
   T (b) They position and move the skeleton.
   T (c) The insertion of the muscle is more distal or mobile than the origin.
   T (d) They are often paired into antagonistic muscle groups called flexors and extensors.

5. Arrange the following skeletal muscle components in order from outermost to innermost: sarcolemma①, connective tissue sheath②, thick and thin filaments④, myofibrils③.   *myofilaments*

6. The modified endoplasmic reticulum of skeletal muscle is called the sarcoplasmic reticulum . Its role is to sequester Ca2+ ions.

7. T-tubules allow a.p. to move to the interior of the muscle fiber.

8. List six proteins that make up the myofibrils. Which protein creates the power stroke for contraction? actin, myosin, troponin, tropomyosin, titin, nebulin

9. List the letters used to label the elements of a sarcomere. Which band has a Z disk in the middle? Which is the darkest band? Why? Which element forms the boundaries of a sarcomere? Name the line that divides the A band in half. What is the function of this line? where thick filaments link each other   I Band   A-Band   M-line   Zdisk

10. Briefly explain the functions of titin and nebulin.

11. During contraction, the A band remains a constant length. This band is composed primarily of myosin molecules. Which components approach each other during contraction? Z DISKS

12. Explain the sliding filament theory.

13. Match the following characteristics with the appropriate type(s) of muscle.
    (a) has the largest diameter a,b,e   1. fast-twitch glycolytic fibers
    (b) uses anaerobic metabolism a,f,d, thus fatigues quickly e   2. fast-twitch oxidative-glycolytic fibers
    (c) has the most blood vessels   3. slow-twitch oxidative fibers   c,d,f,h
    (d) has some myoglobin
    (e) is used for quick, fine movements
    (f) is also called red muscle
    (g) uses a combination of oxidative and glycolytic metabolism
    (h) has the most mitochondria

14. Explain the roles of troponin, tropomyosin, and Ca^{2+} in skeletal muscle contraction.

15. Which neurotransmitter is released by somatic motor neurons? acetylcholine

16. What is the motor end plate, and what kinds of receptors are found there? Explain how neurotransmitter binding to these receptors creates an action potential.

17. A single contraction-relaxation cycle in a skeletal muscle fiber is known as a twitch .

18. List the steps of skeletal muscle contraction that require ATP.

19. The basic unit of contraction in an intact skeletal muscle is the motor unit . The force of contraction within a skeletal muscle is increased by recruitment additional motor units.

20. The two functional types of smooth muscle are single-unit (visceral) and multi-unit .

LEVEL TWO   REVIEWING CONCEPTS

21. Make a map of muscle fiber structure using the following terms. Add additional terms if you like (continued page 433).

    actin   myosin
    Ca2+   nucleus
    cell   regulatory protein
    cell membrane   sarcolemma

12-3 to 12-6

#d T band shortens during

contractile protein

crossbridges

cytoplasm

elastic protein

glycogen

mitochondria

muscle fiber

sarcoplasm

sarcoplasmic reticulum

titin

tropomyosin

troponin

t-tubule

22. Arrange the following terms to create a map of skeletal muscle excitation, contraction, and relaxation. Terms may be used more than once. Add any additional terms you like. 12–9 to 12–11

acetylcholine

ACh receptor

actin

action potential

ADP

ATP

axon terminal

Ca^{2+}

Ca^{2+}-ATPase

calcium-release channels

contraction

crossbridge

DHP receptor

end-plate potential

exocytosis

motor end plate

myosin

Na^+

neuromuscular junction

P_i

power stroke

relaxation

rigor state

sarcoplasmic reticulum

somatic motor neuron

tropomyosin

troponin

t-tubules

voltage-gated Ca^{2+} channels

23. How does an action potential in a muscle fiber trigger a Ca^{2+} signal inside the fiber?

24. Muscle fibers depend on a continuous supply of ATP. How do the fibers in the different types of muscle generate ATP? What is used for a backup energy source?

25. Define muscle fatigue. Summarize factors that could play a role in its development. How can muscle fibers adapt to resist fatigue?

26. Explain how you vary the strength and effort made by your muscles in picking up a pencil versus picking up a full gallon container of milk.

27. Compare and contrast the cellular anatomy and neural and chemical control of contraction in skeletal and smooth muscle.

28. What is the role of the sarcoplasmic reticulum in muscular contraction? How can smooth muscle contract when it has so little sarcoplasmic reticulum?

29. Compare and contrast:
    (a) fast-twitch oxidative-glycolytic, fast-twitch glycolytic, and slow-twitch muscle fibers
    (b) a twitch and tetanus
    (c) action potentials in motor neurons and action potentials in skeletal muscles
    (d) temporal summation in motor neurons and summation in skeletal muscles
    (e) isotonic contraction, isometric contraction, concentric action, and eccentric action
    (f) slow-wave and pacemaker potentials
    (g) the source and role of Ca^{2+} in skeletal and smooth muscle contraction

30. Explain the different factors that influence Ca^{2+} entry and release in smooth muscle fibers.

LEVEL THREE   PROBLEM SOLVING

31. One way that scientists study muscles is to put them into a state of rigor by removing ATP. In this condition, actin and myosin are strongly linked but unable to move. On the basis of what you know about muscle contraction, predict what would happen to these muscles in a state of rigor if you (a) added ATP but no free calcium ions; (b) added ATP with a substantial concentration of calcium ions.

32. When curare, a South American Indian arrow poison, is placed on a nerve-muscle preparation, the muscle will not contract when the nerve is stimulated, even though neurotransmitter is still being released from the nerve. Give all possible explanations for the action of curare that you can think of.

33. On the basis of what you have learned about muscle fiber types and metabolism, predict what variations in structure you would find among these athletes:
    (a) a 7 foot, 2 inch tall, 325-pound basketball player
    (b) a 5 foot, 10 inch tall, 180-pound steer wrestler
    (c) a 5 foot, 7 inch tall, 130-pound female figure skater
    (d) a 4 foot, 11 inch tall, 89-pound female gymnast

LEVEL FOUR   QUANTITATIVE PROBLEMS

34. Look at the following graph, created from data published in "Effect of ambient temperature on human skeletal muscle metabolism during fatiguing submaximal exercise," *Journal of Applied Physiology* 86(3):902–908, 1999. What hypotheses might you develop about the cause(s) of muscle fatigue based on these data?

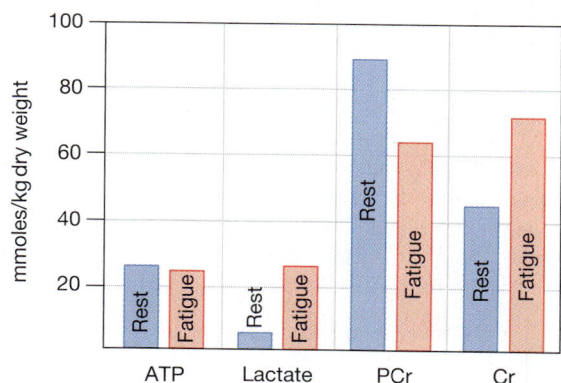

Muscle metabolites in resting muscle and after cycling exercise to fatigue

35. Use the arm in Figure 12-21b to answer the following questions.
    (a) How much force would a biceps muscle inserted 4 cm from the fulcrum need to exert to hold the arm stationary at a 90° angle? How does this force compare with the force needed when the insertion point is 5 cm from the fulcrum?
    (b) If a 7-kg weight band is placed around the wrist 20 cm from the fulcrum, how much force does the biceps inserted 5 cm from the fulcrum need to exert to hold the arm stationary at a 90° angle? How does this force compare with the force needed to keep the arm horizontal in the situation shown in Figure 12.21c, with the same weight in the hand (25 cm from the fulcrum)?

12

ANSWERS

✓ Answers to Concept Check Questions

Page 398

1. Some examples are biceps/triceps in the upper arm; hamstring (flexor)/quadriceps (extensor) in the upper leg; tibialis anterior (flexor)/gastrocnemius (extensor) for foot movement at the ankle.

Page 402

2. Ends of the A bands are darkest because they are where the thick and thin filaments overlap.

3. T-tubules allow action potentials to travel from the surface of the muscle fiber to its interior.

4. The banding pattern of organized filaments in the sarcomere forms striations in the muscle.

Page 403

5. A neuromuscular junction consists of axon terminals from one somatic motor neuron, the synaptic cleft, and the motor end plate on the muscle fiber.

6. The chemical signal at a neuromuscular junction is acetylcholine.

Page 406

7. Each myosin molecule has binding sites for ATP and actin.

8. F-actin is a polymer filament of actin made from globular G-actin molecules.

Page 407

9. Titin is an elastic fiber in the sarcomere.

10. The crossbridges do not all unlink at one time, so while some myosin heads are free and swiveling, others are still tightly bound.

Page 409

11. The release of myosin heads from actin requires ATP binding. Energy from ATP is required for the power stroke. Relaxation does not directly require ATP, but relaxation will not occur unless Ca^{2+} is pumped back into the sarcoplasmic reticulum using a Ca^{2+}-ATPase.

12. The events of the latent period include creation of the muscle action potential, release of Ca^{2+} from the sarcoplasmic reticulum, and diffusion of Ca^{2+} to the contractile filaments.

Page 411

13. *Creatine* is the substrate, and *kinase* tells you that this enzyme phosphorylates the substrate.

14. Because creatine kinase catalyzes the reaction in both directions, the relative concentrations of the reactants and products determine the direction of the reaction. The reaction obeys the law of mass action and goes to equilibrium.

Page 412

15. Increasing extracellular K^+ will cause the cell to depolarize and become less negative.

Page 415

16. Tension.

17. Strength of the graded potential.

Page 416

18. A marathoner probably has more slow-twitch muscle fibers, whereas a sprinter probably has more fast-twitch muscle fibers.

Page 417

19. Increased motor neuron firing rate causes summation in a muscle fiber, which increases the force of contraction.

20. The nervous system increases the force of contraction by recruiting additional motor units.

Page 420

21. If the muscle insertion point is farther from the joint, the leverage is better and a contraction creates more rotational force.

Page 423

22. Multi-unit smooth muscle increases force by recruiting additional muscle fibers; single-unit smooth muscle increases force by increasing Ca^{2+} entry.

Page 425

23. Dense bodies are analogous to Z disks.

24. Smooth muscle myosin is longer, has heads the entire length of the filament, and has slower ATPase activity.

25. Smooth muscle actin is longer than skeletal muscle actin, and it lacks troponin.

Page 425

26. Item 1: In skeletal muscles, all Ca^{2+} comes from the sarcoplasmic reticulum (④ in Fig. 12-11). Item 3: Ca^{2+} binds directly to troponin in skeletal muscle. Item 4: Phosphorylation does not regulate skeletal muscle myosin. Item 5: The primary control point in skeletal muscle is Ca^{2+} binding to troponin.

Page 426

27. Without ECF Ca^{2+}, contraction either decreases or stops altogether because little or no Ca^{2+} is available to initiate the process.

Page 427

28. Skeletal muscle Ca^{2+} release channels are linked to DHP receptors and open upon depolarization. Smooth muscle Ca^{2+} release channels are activated by IP_3.

Page 428

29. Pacemaker potentials always reach threshold and create regular rhythms of contraction. Slow wave potentials are variable in magnitude and may not reach threshold each time.

30. The depolarization phase of the action potentials must not be due to Na^+ entry.

Page 428

31. More action potentials in the neuron increase neurotransmitter release.

32. Many Ca^{2+} channels open with depolarization; therefore, hyperpolarization decreases the likelihood that these channels open. The presence of Ca^{2+} is necessary for contraction.

Q Answers to Figure and Graph Questions

Page 409

Fig. 12-12: (a) Muscle V_m is -90 mV, and neuron V_m is -70 mV. This illustrates the fact that V_m is not the same in all cells. (b) Muscle action potential is due to Na^+ entering the fiber during depolarization, and K^+ leaving during repolarization.

Page 420

Fig. 12-21: Biceps force $\times$ 5 cm = 7 kg $\times$ 25 cm = 35 kg (additional force).

Page 421

Fig. 12-22: The hand moves upward at a speed of 5 cm/sec.

Page 421

Fig. 12-23: Contraction is isometric at B because at this point muscle does not shorten. Maximum velocity is at A, where the load on the muscle is zero.

Extracting signals directly from the brain to directly control robotic devices has been a science fiction theme that seems destined to become fact.

—**Dr. Eberhard E. Fetz,** *Science News* 156: 142, 8/28/99

13

Integrative Physiology I: Control of Body Movement

BACKGROUND BASICS

RUNNING PROBLEM

TETANUS

"She has not been able to talk to us. We're afraid she may have had a stroke." That is how her neighbors described 77-year-old Cecile Evans when they brought her to the emergency room. But when a neurological examination revealed no problems other than Mrs. Evans's inability to open her mouth and stiffness in her neck, emergency room physician Dr. Doris Ling began to consider other diagnoses. She noticed some scratches healing on Mrs. Evans's arms and legs and asked the neighbors if they knew what had caused them. "Oh, yes. She told us a few days ago that her dog jumped up and knocked her against the barbed wire fence." At that point, Dr. Ling realized she was probably dealing with her first case of tetanus.

| 436 | 439 | 443 | 447 | 451 | 452 |

Think back to the baseball pitcher in the last chapter. As he stands on the mound, looking in at the first batter, he receives sensory information from multiple sources: the sound of the crowd, the sight of the batter and the catcher, the feel of the ball in his hand, the alignment of his body as he begins his windup. Sensory receptors code this information and send it to the central nervous system (CNS), where it is integrated.

The pitcher acts consciously on some of the information: he decides to throw a fastball. But he processes other information at the subconscious level and acts on it without conscious thought. As he thinks about starting his motion, for instance, he shifts his weight to offset the impending movement of his arm. The integration of sensory information into an involuntary response is the hallmark of a *reflex* [🔁 p. 195].

NEURAL REFLEXES

All neural reflexes begin with a stimulus that activates a sensory receptor. The receptor sends information in the form of action potentials through sensory neurons to the CNS [🔁 p. 328]. The CNS is the integrating center that evaluates all incoming information and selects an appropriate response. It then initiates action potentials in efferent neurons to direct the response of muscles and glands—the *effectors*.

A key feature of many reflex pathways is *negative feedback*, a concept introduced in Chapter 6 [🔁 p. 199]. Feedback signals from muscle and joint receptors keep the CNS continuously informed of changing body position. Some reflexes have a

feedforward component that allows the body to anticipate a stimulus and begin the response [🔁 p. 200]. Bracing yourself in anticipation of a collision is an example of a feedforward response.

Neural Reflex Pathways Can Be Classified in Different Ways

Reflex pathways in the nervous system consist of chains or networks of neurons that link sensory receptors to muscles or glands. Neural reflexes can be classified in several ways (Table 13-1 ■):

1. *By the efferent division of the nervous system that controls the response.* Reflexes that involve somatic motor neurons and skeletal muscles are known as **somatic reflexes**. Reflexes whose responses are controlled by autonomic neurons are called **autonomic reflexes**.

2. *By the CNS location where the reflex is integrated.* **Spinal reflexes** are integrated in the spinal cord. These reflexes may be modulated by higher input from the brain, but they can occur without that input. Reflexes integrated in the brain are called **cranial reflexes**.

3. *By whether the reflex is innate or learned.* Many reflexes are **innate**; that is, we are born with them, and they are genetically determined. One example is the knee jerk reflex, in which the lower leg kicks out when the patellar tendon at the lower edge of the kneecap is tapped. Other reflexes are acquired through experience [🔁 p. 317]. The example of

TABLE 13-1	Classification of Neural Reflexes

Neural reflexes can be classified by:

1. **Efferent division that controls the effector**
   a. Somatic motor neurons control skeletal muscles.
   b. Autonomic neurons control smooth and cardiac muscle, glands, and adipose tissue.

2. **Integrating region within the central nervous system**
   a. Spinal reflexes do not require input from the brain.
   b. Cranial reflexes are integrated within the brain.

3. **Time at which the reflex develops**
   a. Innate (inborn) reflexes are genetically determined.
   b. Learned (conditioned) reflexes are acquired through experience.

4. **The number of neurons in the reflex loop**
   a. Monosynaptic reflexes have only two neurons: one afferent (sensory) and one efferent. Only somatic motor reflexes can be monosynaptic.
   b. Polysynaptic reflexes include one or more interneurons between the afferent and efferent neurons. All autonomic reflexes are polysynaptic because they have three neurons: one afferent and two efferent.

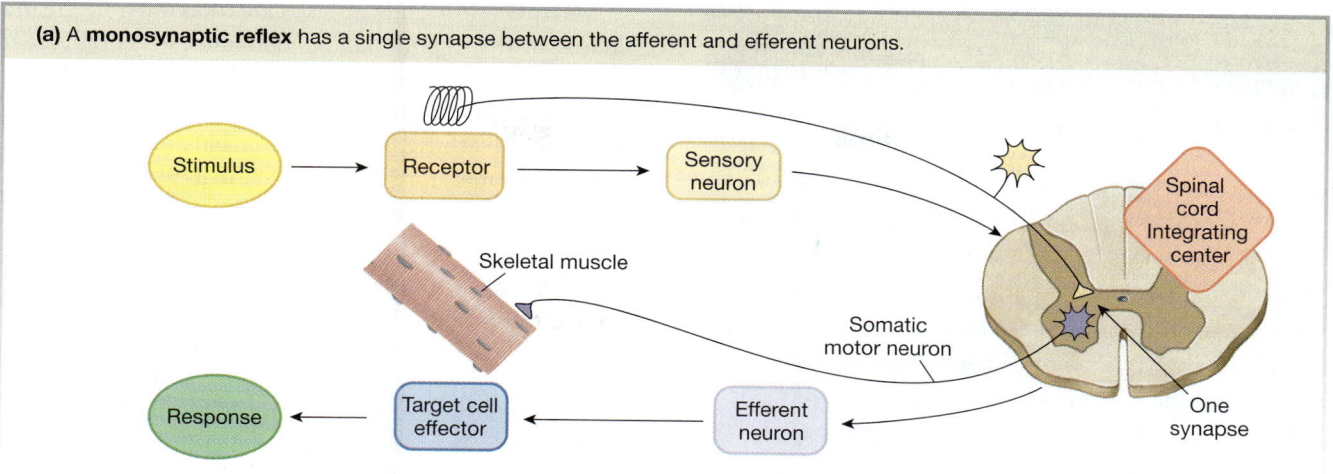

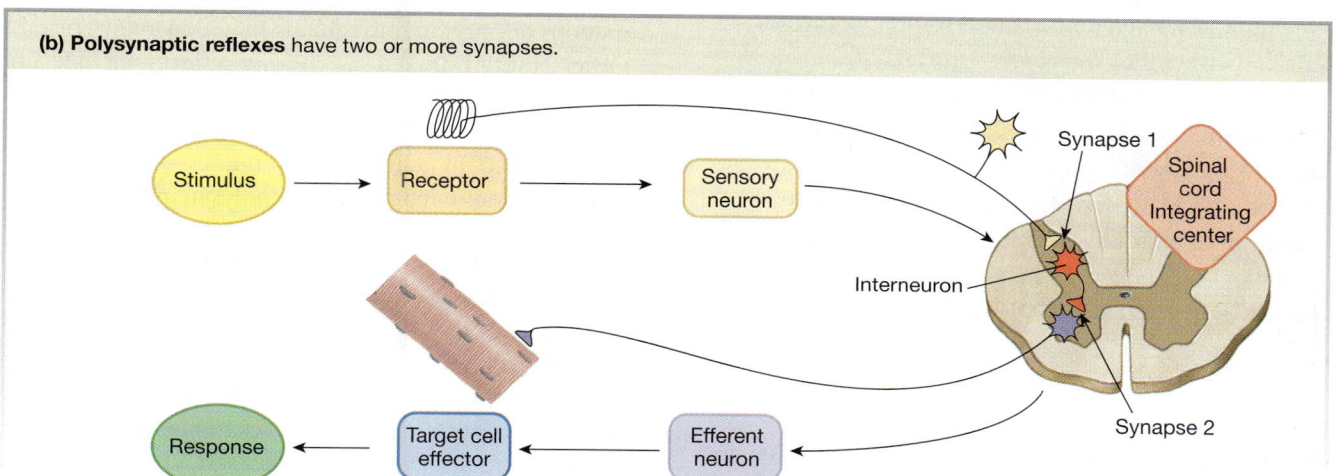

■ FIGURE 13-1   *Monosynaptic and polysynaptic somatic motor reflexes*

All synapses for somatic motor reflexes occur in the central nervous system, as illustrated.

Pavlov's dogs salivating upon hearing a bell is the classic example of a **learned reflex**, also referred to as a **conditioned reflex**.

4. *By the number of neurons in the reflex pathway.* The simplest reflex is a **monosynaptic reflex**, named for the single synapse between the two neurons in the pathway: an afferent sensory neuron and an efferent somatic motor neuron. These two neurons synapse in the spinal cord, allowing a signal initiated at the receptor to go directly from the sensory neuron to the motor neuron (Fig. 13-1a ■). (The synapse between the somatic motor neuron and its muscle target is ignored.)

Most reflexes have three or more neurons in the pathway (and at least two synapses), leading to their designation as **polysynaptic reflexes** (Fig. 13-1b). Polysynaptic reflexes may be quite complex, with extensive branching in the CNS to form networks involving multiple interneurons. *Divergence* of pathways allows a single stimulus to affect multiple targets [⇄ p. 278]. *Convergence* integrates the input from multiple sources to modulate the response. Recall from Chapter 8 that modulation in polysynaptic pathways may involve excitation or inhibition [⇄ p. 279].

AUTONOMIC REFLEXES

Autonomic reflexes are also known as *visceral reflexes* because they often involve the internal organs of the body. Some visceral reflexes, such as urination and defecation, are spinal reflexes that can take place without input from the brain. However, spinal reflexes are often modulated by excitatory or inhibitory signals from the brain, carried by descending tracts from higher brain centers. For example, urination may be voluntarily initiated by conscious thought, or it may be inhibited by emotion or a stressful situation, such as the presence of

VISUALIZATION TECHNIQUES IN SPORTS

Presynaptic facilitation, in which modulatory input increases neurotransmitter release, is now believed to be the physiological mechanism that underlies the success of visualization techniques in sports. Visualization, also known as *guided imagery,* enables athletes to maximize their performance by "psyching" themselves, picturing in their minds the perfect vault or the perfect leaping catch for a touchdown. By pathways that we still do not understand, the mental image conjured up by the cerebral cortex is translated into signals that find their way to the muscles. Guided imagery is also being used in medicine as an *adjunct* (supplementary) therapy for cancer treatment and pain management. The ability of the conscious brain to alter physiological function is only one example of the many fascinating connections between the higher brain and the body.

other people (a syndrome known as "bashful bladder"). Often, the higher control of a spinal reflex is a learned response. The toilet training that we master as toddlers is an example of a learned reflex that the CNS uses to modulate the simple spinal reflex of urination.

Other autonomic reflexes are integrated in the brain, primarily in the hypothalamus, thalamus, and brain stem. These regions contain centers that coordinate body functions needed to maintain homeostasis, such as heart rate, blood pressure, breathing, eating, water balance, and maintenance of body temperature [⇄ Fig. 11-3, p. 379]. The brain stem also contains the integrating centers for autonomic reflexes such as salivating, vomiting, sneezing, coughing, swallowing, and gagging.

An interesting type of autonomic reflex is the conversion of emotional stimuli into visceral responses. The limbic system [⇄ p. 308]—the site of primitive drives such as sex, fear, rage, aggression, and hunger—has been called the "visceral brain" because of its role in these emotionally driven reflexes. We speak of "gut feelings" and "butterflies in the stomach"—all transformations of emotion into somatic sensation and visceral function. Other emotion-linked autonomic reflexes include urination, defecation, blushing, blanching, and *piloerection,* in which tiny muscles in the hair follicles pull the shaft of the hair erect ("I was so scared my hair stood on end!").

Autonomic reflexes are all polysynaptic, with at least one synapse in the CNS between the sensory neuron and the preganglionic autonomic neuron, and an additional synapse in the ganglion between the preganglionic and postganglionic neurons (Fig. 13-2 ■).

Many autonomic reflexes are characterized by tonic activity, a continuous stream of action potentials that creates ongoing activity in the effector. For example, the tonic control of blood vessels discussed in Chapter 6 is an example of a continuously active autonomic reflex [⇄ p. 190]. You will encounter many autonomic reflexes as you continue your study of the systems of the body.

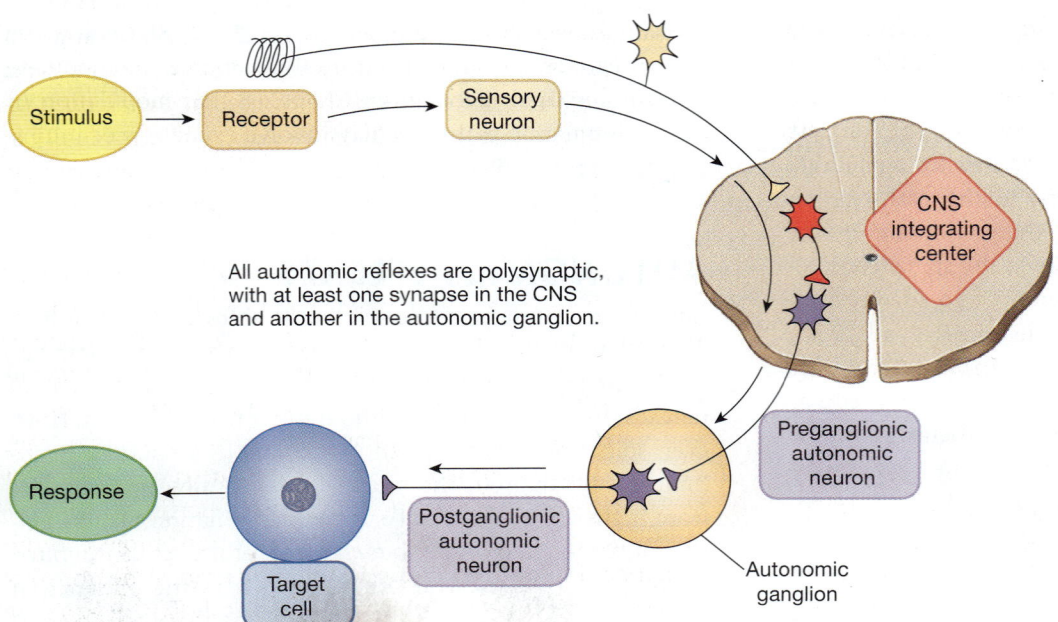

■ **FIGURE 13-2** *Autonomic reflexes*

All autonomic reflexes are polysynaptic, with at least one synapse in the CNS and another in the autonomic ganglion.

SKELETAL MUSCLE REFLEXES

Although we are not always aware of them, skeletal muscle reflexes are involved in almost everything we do. Receptors that sense changes in joint movements, muscle tension, and muscle length feed this information to the CNS, which responds one of two ways. If muscle contraction is the appropriate response, the CNS activates somatic motor neurons to the muscle fibers. If a muscle needs to be relaxed to achieve the response, sensory input activates inhibitory interneurons in the CNS, and these interneurons *inhibit* activity in somatic motor neurons controlling the muscle.

Recall that somatic motor neurons always cause contraction in skeletal muscle [p. 391]. There is no inhibitory neuron that synapses on skeletal muscles to cause them to relax. Instead, *relaxation results from the absence of excitatory input by the somatic motor neuron.* Inhibition and excitation of somatic motor neurons and their associated skeletal muscles must occur at synapses within the CNS.

Skeletal muscle reflexes have the following components:

1. *Sensory receptors,* known as **proprioceptors**, are located in skeletal muscles, joint capsules, and ligaments. They monitor the position of our limbs in space, our movements, and the effort we exert in lifting objects.

2. *Sensory neurons* carry the input signal from proprioceptors to the CNS.

3. *The central nervous system* integrates the input signal using networks and pathways of excitatory and inhibitory interneurons. In a reflex, sensory information is integrated and acted on subconsciously. However, some sensory information may be integrated in the cerebral cortex and become perception, and some reflexes can be modulated by conscious input.

4. *Somatic motor neurons* carry the output signal. The somatic motor neurons that innervate skeletal muscle contractile fibers are called **alpha motor neurons**.

5. The effectors are contractile skeletal muscle fibers, also known as extrafusal muscle fibers. Action potentials in alpha motor neurons cause contraction in extrafusal fibers.

Three types of proprioceptors are found in the body: muscle spindles, Golgi tendon organs, and joint receptors. *Joint receptors* are found in the capsules and ligaments around joints in the body. They are stimulated by mechanical distortion that

accompanies changes in the relative positioning of bones linked by flexible joints. Sensory information from joint receptors is integrated primarily in the cerebellum.

In the next two sections we examine the function of muscle spindles and Golgi tendon organs, two interesting and unique receptors. These receptors lie inside skeletal muscles and sense changes in muscle length and tension. Their sensory output activates muscle reflexes.

Muscle Spindles Respond to Muscle Stretch

Muscle spindles are stretch receptors that send information to the spinal cord and brain about muscle length and changes in muscle length. They are small, elongated structures scattered among and arranged parallel to the contractile extrafusal muscle fibers (Fig. 13-3a ■). With the exception of one muscle in the jaw, every skeletal muscle in the body has many muscle spindles. For example, a cat's hindlimb muscle has 50 or more spindles.

Each muscle spindle consists of a connective tissue *capsule* that encloses a group of small muscle fibers known as **intrafusal fibers** [*intra-,* within + *fusus,* spindle]. Intrafusal muscle fibers are modified so that the ends are contractile but the central region lacks myofibrils (Fig. 13-3b). The contractile ends of the intrafusal fibers have their own innervation from **gamma motor neurons**. The noncontractile central region of each intrafusal fiber is wrapped by sensory nerve endings that are stimulated by stretch. The sensory neurons project to the spinal cord and synapse directly on alpha motor neurons innervating the muscle in which the spindles lie.

13

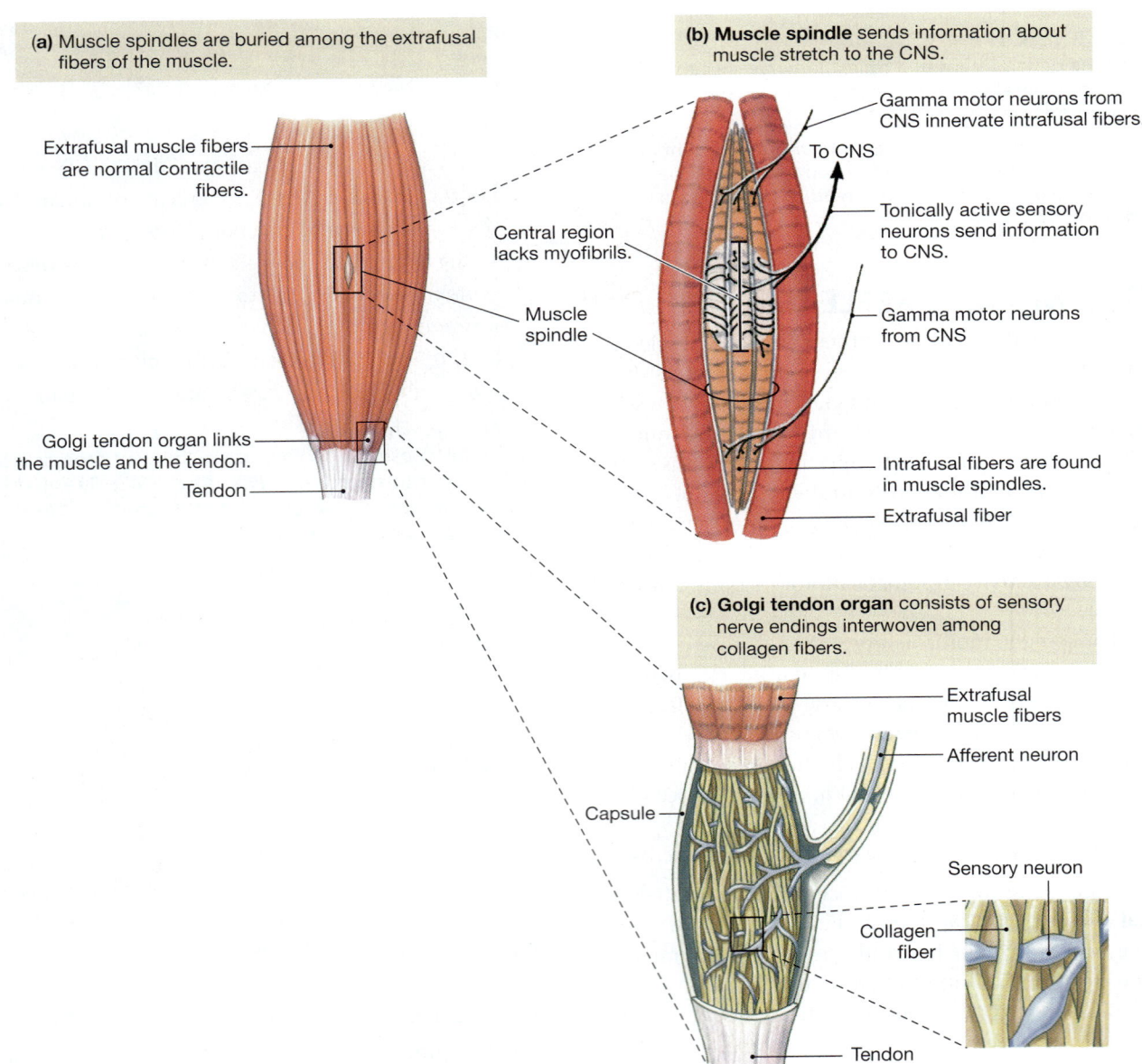

(a) Muscle spindles are buried among the extrafusal fibers of the muscle.

Extrafusal muscle fibers are normal contractile fibers.

Golgi tendon organ links the muscle and the tendon.

Tendon

Muscle spindle

Central region lacks myofibrils.

(b) Muscle spindle sends information about muscle stretch to the CNS.

Gamma motor neurons from CNS innervate intrafusal fibers.

To CNS

Tonically active sensory neurons send information to CNS.

Gamma motor neurons from CNS

Intrafusal fibers are found in muscle spindles.

Extrafusal fiber

(c) Golgi tendon organ consists of sensory nerve endings interwoven among collagen fibers.

Extrafusal muscle fibers

Afferent neuron

Capsule

Sensory neuron

Collagen fiber

Tendon

■ **FIGURE 13-3** *Muscle spindles and Golgi tendon organs are sensory receptors in muscle*

When a muscle is at its resting length, the central region of each muscle spindle is stretched enough to activate the sensory fibers (Fig. 13-4a ■). Thus, sensory neurons from the spindles are tonically active, sending a steady stream of action potentials to the CNS. Because of this tonic activity, even a muscle at rest maintains a certain level of tension, known as **muscle tone**.

Muscle spindles are anchored in parallel to the extrafusal muscle fibers. Any movement that increases muscle length also stretches the muscle spindles and causes their sensory fibers to fire more rapidly (Fig. 13-4b). This creates a reflex contraction of the muscle, which prevents damage from overstretching. The reflex pathway in which muscle stretch initiates a contraction response is known as a **stretch reflex**.

CONCEPT CHECK

3. Using the standard steps of a reflex pathway (stimulus, receptor, and so forth), draw a reflex map of the stretch reflex. Answers: p. 455

If stretch activates muscle spindles, what happens to spindle activity when a resting muscle contracts and shortens? You might predict that the release of tension on the center of the intrafusal fibers causes the afferent neurons to slow their firing rate or stop firing altogether (Fig. 13-5a ■). However, in a normal muscle this does not happen because of the process known as **alpha-gamma coactivation**.

The role of gamma motor neurons is to adjust the stretch sensitivity of the muscle spindles so that the spindles are active

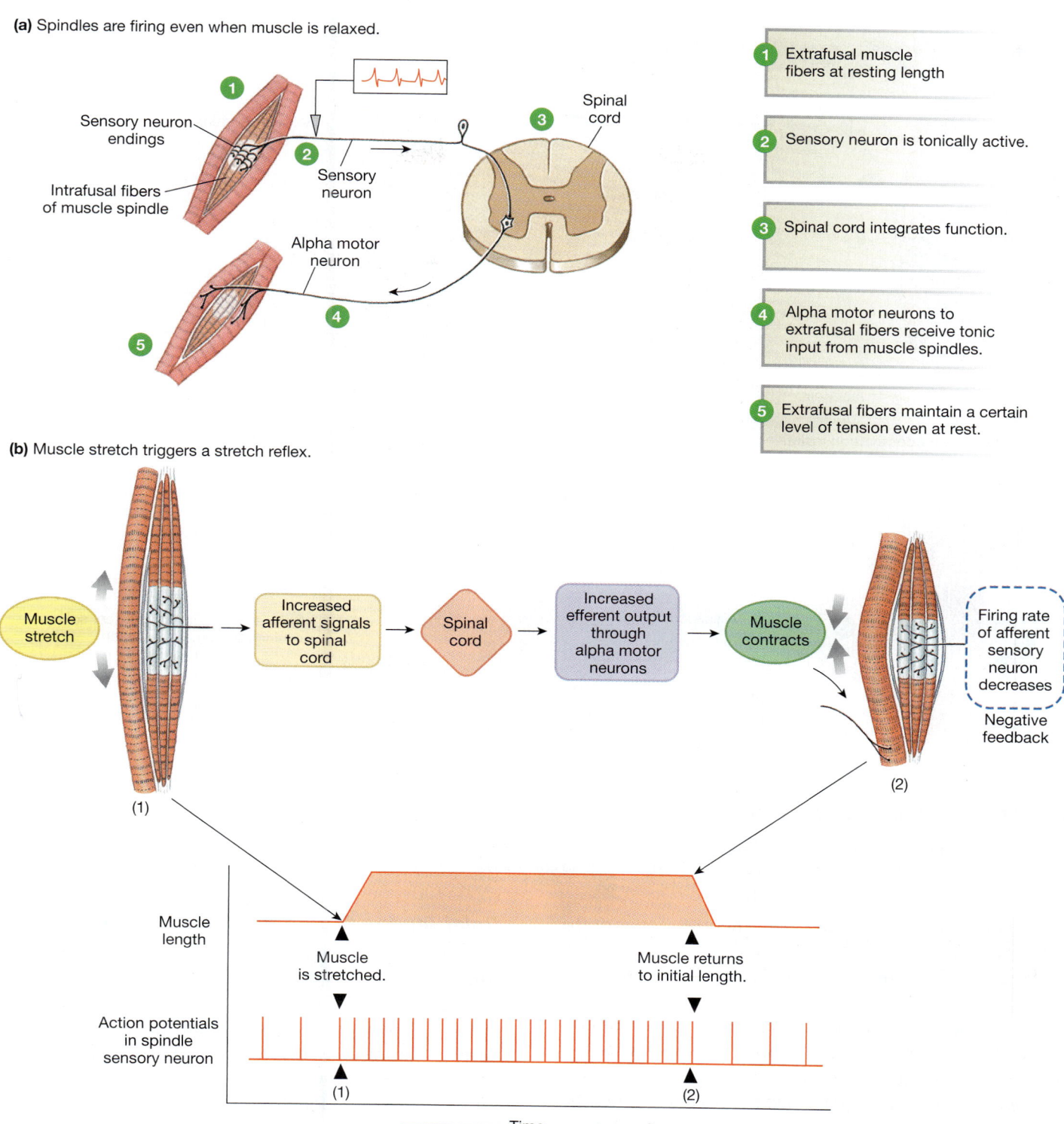

FIGURE 13-4 *Muscle spindles monitor muscle length and prevent overstretching*

no matter what the muscle length is. When the gamma motor neurons innervating the contractile ends of the intrafusal fibers fire, the ends of the fibers contract and shorten (Fig. 13-5b). This contraction of the spindle ends lengthens the central region and maintains stretch on the sensory nerve endings.

In alpha-gamma coactivation, when alpha motor neurons to the extrafusal fibers of a muscle fire, the gamma motor neurons to the spindles of that muscle are activated at the same time. The extrafusal fibers contract and the muscle shortens, releasing tension on the muscle spindle capsule. However, gamma

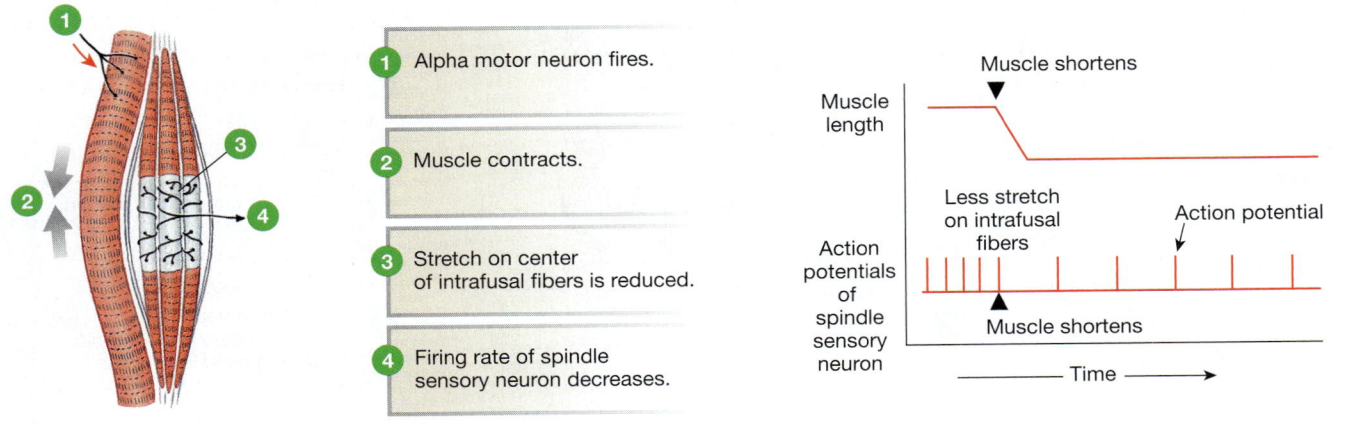

(a) If gamma motor axons are cut, the spindle loses activity when muscle contracts.

1 Alpha motor neuron fires.

2 Muscle contracts.

3 Stretch on center of intrafusal fibers is reduced.

4 Firing rate of spindle sensory neuron decreases.

Muscle length

Muscle shortens

Action potentials of spindle sensory neuron

Less stretch on intrafusal fibers

Action potential

Muscle shortens

Time

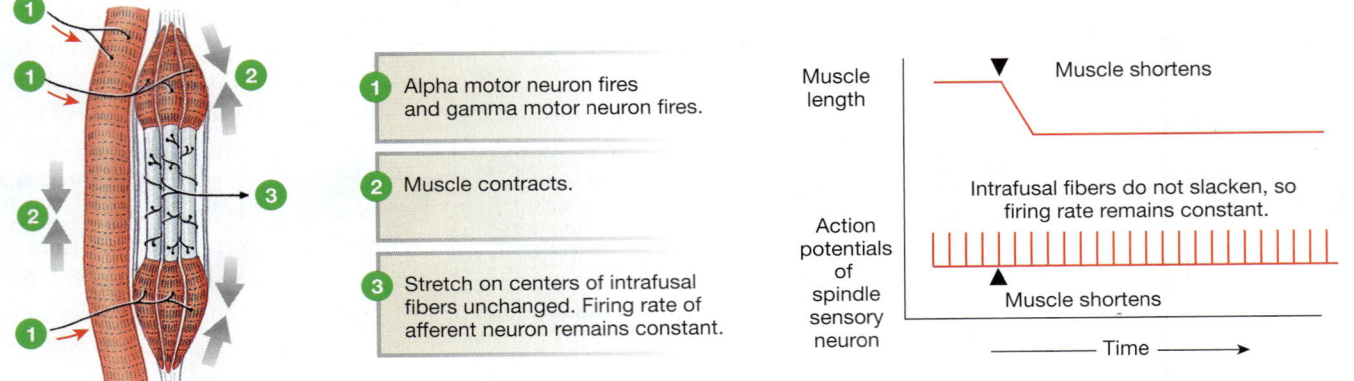

(b) Alpha-gamma coactivation maintains spindle function when muscle contracts.

1 Alpha motor neuron fires and gamma motor neuron fires.

2 Muscle contracts.

3 Stretch on centers of intrafusal fibers unchanged. Firing rate of afferent neuron remains constant.

Muscle length

Muscle shortens

Action potentials of spindle sensory neuron

Intrafusal fibers do not slacken, so firing rate remains constant.

Muscle shortens

Time

■ **FIGURE 13-5**   *Alpha-gamma coactivation*

Gamma motor neurons innervate muscle fibers at the ends of muscle spindles. Alpha-gamma coactivation keeps the spindles stretched when the muscle contracts.

coactivation of the intrafusal fibers stretches the centers of the spindle fibers and counteracts the release of tension on the capsule. Thus, the spindle remains active even during muscle contraction.

An example of how muscle spindles work during a stretch reflex is shown in Figure 13-6a–c ■. You can demonstrate this yourself with an unsuspecting friend. Have your friend stand with eyes closed, one arm extended with the elbow at 90°, and the hand palm up. Place a small book or other flat weight in the outstretched hand and watch the arm muscles contract to compensate for the added weight. Then suddenly drop a heavier load, such as another book, onto the subject's hand. The added weight will send the hand downward, stretching the biceps muscle and activating its muscle spindles. Sensory input into the spinal cord then activates the alpha motor neurons of the biceps muscle. The biceps will contract, bringing the arm back to its original position.

Golgi Tendon Organs Respond to Muscle Tension

A second type of muscle proprioceptor is the **Golgi tendon organ**. These receptors are found at the junction of tendons and muscle fibers, placing them in series with the muscle fibers. Golgi tendon organs respond primarily to the tension a muscle develops during an isometric contraction, and they cause a relaxation reflex. This is the opposite of muscle spindles, which cause reflex contraction. In contrast to the muscle spindle, the Golgi tendon organ is relatively insensitive to muscle stretch.

Golgi tendon organs are composed of free nerve endings that wind between collagen fibers inside a connective tissue capsule (Fig. 13-3c). When a muscle contracts, its tendons act as an elastic component during the isometric phase of the contraction [➲ p. 419]. Contraction pulls collagen fibers within the Golgi tendon organ tight, pinching sensory endings of the afferent neurons and causing them to fire.

Afferent input from activation of the Golgi tendon organ excites *inhibitory* interneurons in the spinal cord. The interneurons

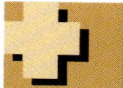

CLINICAL FOCUS

REFLEXES AND MUSCLE TONE

Clinicians use reflexes to investigate the condition of the nervous system and the muscles. For a reflex to be normal, there must be normal conduction through all neurons in the pathway, normal synaptic transmission at the neuromuscular junction, and normal muscle contraction. A reflex that is absent, abnormally slow, or greater than normal (hyperactive) suggests the presence of disease. Interestingly, not all abnormal reflexes suggest neuromuscular disorders. For example, slowed relaxation of the ankle flexion reflex suggests hypothyroidism. (The cellular mechanism linking low thyroid to slow reflexes is not known.) Besides testing reflexes, clinicians assess muscle tone. Even when relaxed and at rest, muscles have a certain resistance to stretch that is the result of continuous (tonic) output by alpha motor neurons. The absence of muscle tone or a muscle's resistance to being passively stretched by the examiner (increased tone) indicates a problem with the pathways that control muscle contraction.

inhibit alpha motor neurons innervating the muscle, and muscle contraction decreases or ceases. Under most circumstances, this reflex slows muscle contraction as the force of contraction increases. In other instances, the Golgi tendon organs prevent excessive contraction that might injure the muscle.

In the example of the books placed on the outstretched hand, if supporting the added weight requires more tension than the muscle can develop, the Golgi tendon organ will respond as muscle tension nears its maximum. The Golgi tendon organ triggers reflex *inhibition* of the biceps motor neurons, causing the biceps to relax and the arm to fall. Thus, the person will drop the added weight before the muscle fibers can be damaged (Figs. 13-6d, e). Golgi tendon organ input is an important source of inhibition to alpha motor neurons.

✓ CONCEPT CHECK

4. Using the standard steps of a reflex pathway, create a map showing alpha-gamma coactivation and the Golgi tendon reflex. Begin with the stimulus "Alpha motor neuron fires."
Answers: p. 455

Stretch Reflexes and Reciprocal Inhibition Control Movement Around a Joint

Movement around most flexible joints in the body is controlled by groups of synergistic and antagonistic muscles that act in a coor-

dinated fashion. Sensory neurons from muscle receptors and efferent motor neurons that control the muscle are linked together by diverging and converging pathways of interneurons within the spinal cord. The collection of pathways controlling a single joint is known as a **myotatic unit** [*myo-*, muscle + *tasis*, stretching].

The simplest reflex in a myotatic unit is the **monosynaptic stretch reflex**, which involves only two neurons: the sensory neuron from the muscle spindle and the somatic motor neuron to the muscle. The knee jerk reflex is an example of a monosynaptic stretch reflex (Fig. 13-7 ■). To demonstrate the knee jerk reflex, the subject sits on the edge of a table so that the lower leg hangs relaxed. When the patellar tendon below the kneecap is tapped with a small rubber hammer, the tap stretches the quadriceps muscle, which runs up the front of the thigh. This stretching activates muscle spindles and sends action potentials via the sensory fibers to the spinal cord. The sensory neurons synapse directly onto the motor neurons that control contraction of the quadriceps muscle (a monosynaptic reflex). The action potentials in the motor neurons cause motor units in the quadriceps to contract, and the lower leg swings forward.

For muscle contraction to extend the leg, the antagonistic flexor muscles must relax (**reciprocal inhibition**). In the leg, this requires relaxation of the hamstring muscles running up the back of the thigh. The single stimulus of the tap to the tendon accomplishes both contraction of the quadriceps muscle and reciprocal inhibition of the hamstrings. The sensory fibers branch upon entering the spinal cord. Some of the branches activate motor neurons innervating the quadriceps, while the other branches synapse on inhibitory interneurons. The inhibitory interneurons suppress activity in the motor neurons controlling the hamstrings (a polysynaptic reflex). The result is a relaxation of the hamstrings that allows contraction of the quadriceps to proceed unopposed.

RUNNING PROBLEM

Once in the spinal cord, tetanospasmin is released from the motor neuron. It then selectively blocks neurotransmitter release at inhibitory synapses. Patients with tetanus experience muscle spasms that begin in the jaw and may eventually affect the entire body. When the extremities become involved, the arms and legs may go into painful, rigid spasms.

Question 2:
Using the reflex pathways diagrammed in Figures 13-7 and 13-8, explain why inhibition of inhibitory interneurons might result in uncontrollable muscle spasms.

| 436 | 439 | **443** | 447 | 451 | 452 |

13

Muscle spindle reflex: the addition of a load stretches the muscle and the spindles, creating a reflex contraction.

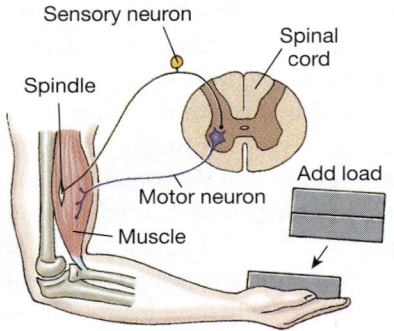

(a) Add load to muscle.

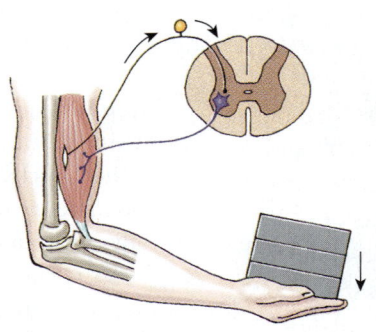

(b) Muscle and muscle spindle stretch as arm falls.

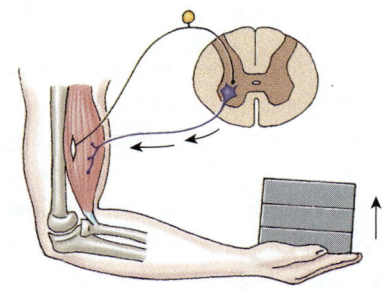

(c) Reflex contraction initiated by muscle spindle restores arm position.

Golgi tendon reflex protects the muscle from excessively heavy loads by causing the muscle to relax and drop the load.

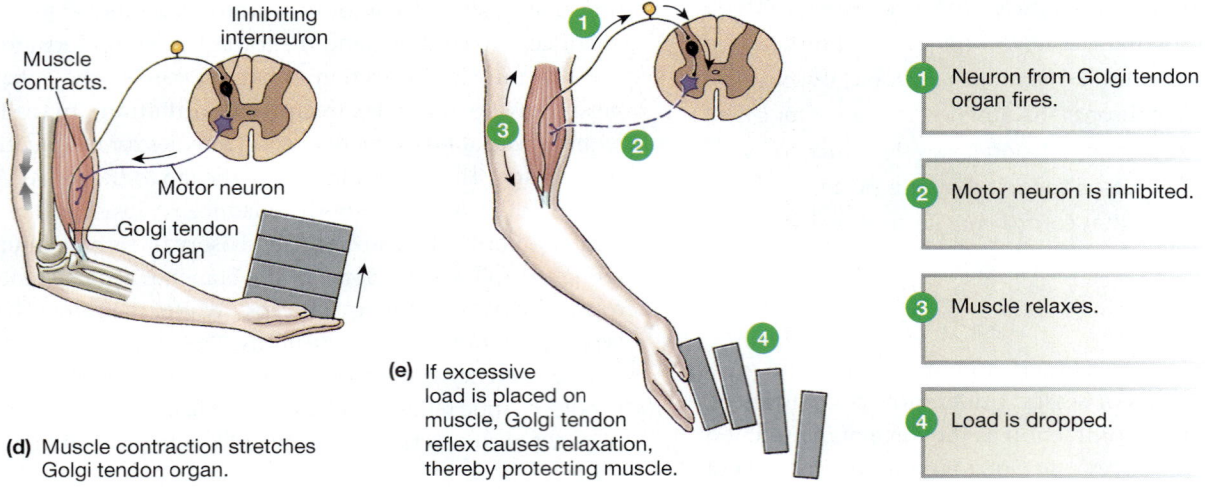

(d) Muscle contraction stretches Golgi tendon organ.

(e) If excessive load is placed on muscle, Golgi tendon reflex causes relaxation, thereby protecting muscle.

1 Neuron from Golgi tendon organ fires.

2 Motor neuron is inhibited.

3 Muscle relaxes.

4 Load is dropped.

■ **FIGURE 13-6** *Muscle reflexes*

Flexion Reflexes Pull Limbs Away from Painful Stimuli

Flexion reflexes are polysynaptic reflex pathways that cause an arm or leg to be pulled away from a painful stimulus, such as a pinprick or a hot stove. These reflexes, like the reciprocal inhibition reflex just described, rely on divergent pathways in the spinal cord. Figure 13-8 ■ uses the example of stepping on a tack to illustrate a flexion reflex.

When the foot contacts the point of the tack, nociceptors (pain receptors) in the foot send sensory information to the spinal cord. Here the signal diverges, activating multiple excitatory interneurons. Some of these interneurons excite alpha motor neurons, leading to contraction of the flexor muscles of the stimulated limb. Other interneurons simultaneously activate inhibitory interneurons that cause relaxation of the antagonistic muscle groups. Because of this reciprocal inhibition, the

limb is flexed, withdrawing it from the painful stimulus. This type of reflex requires more time than a stretch reflex (such as the knee jerk reflex) because it is a polysynaptic rather than a monosynaptic reflex.

CONCEPT CHECK

5. Draw a reflex map of the flexion reflex initiated by a painful stimulus to the sole of a foot. (Leave space to add additional information later.)
   Answers: p. 455

Flexion reflexes, particularly in the legs, are usually accompanied by the **crossed extensor reflex**, a postural reflex that helps maintain balance when one foot is lifted from the ground. In this reflex, also shown in Figure 13-8, the quick withdrawal of the right foot from a painful stimulus (a tack) is matched by extension of the left leg so that it can support the sudden shift in weight. The extensors contract in the supporting

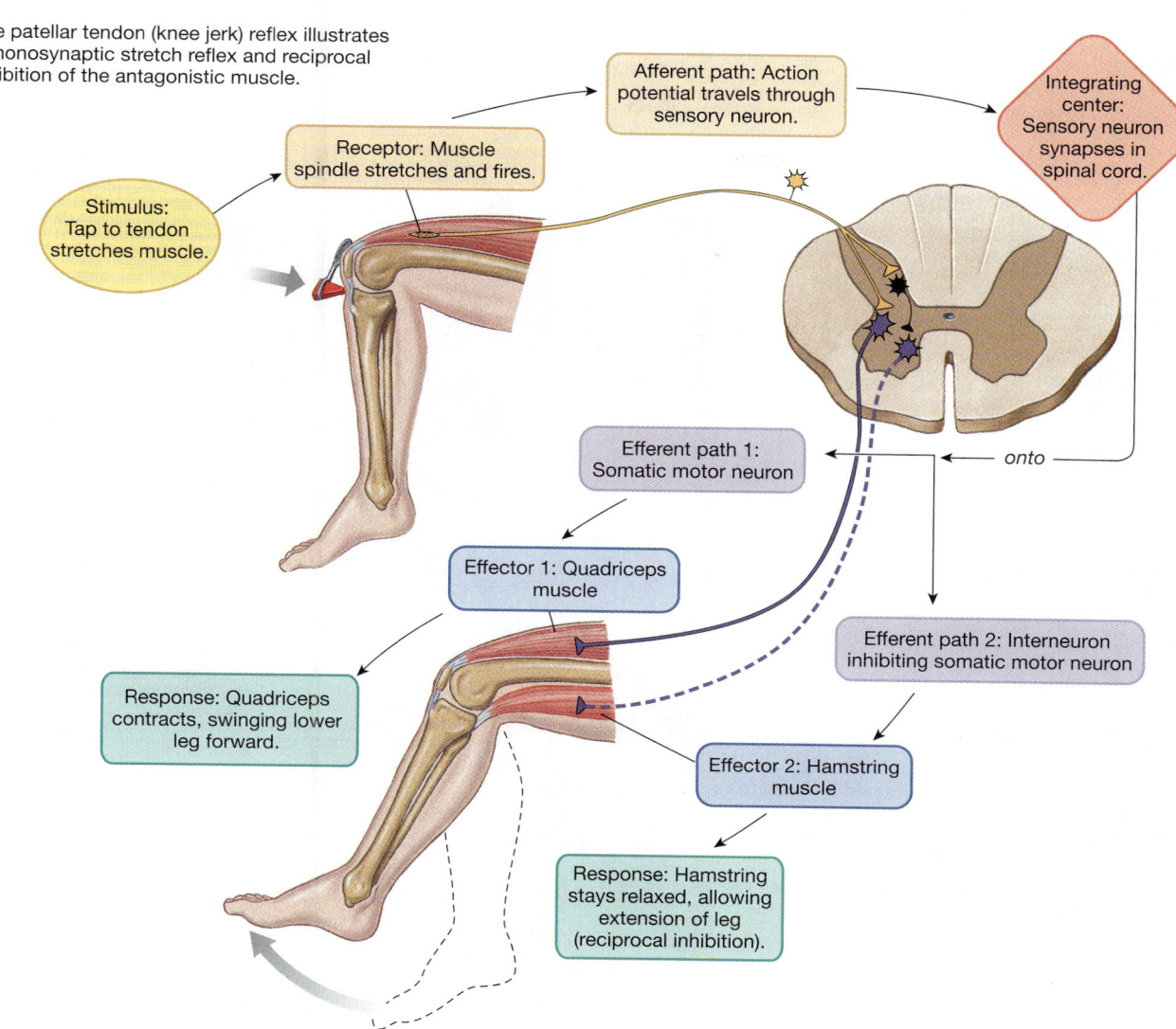

The patellar tendon (knee jerk) reflex illustrates a monosynaptic stretch reflex and reciprocal inhibition of the antagonistic muscle.

Stimulus: Tap to tendon stretches muscle.

Receptor: Muscle spindle stretches and fires.

Afferent path: Action potential travels through sensory neuron.

Integrating center: Sensory neuron synapses in spinal cord.

onto

Efferent path 1: Somatic motor neuron

Efferent path 2: Interneuron inhibiting somatic motor neuron

Effector 1: Quadriceps muscle

Response: Quadriceps contracts, swinging lower leg forward.

Effector 2: Hamstring muscle

Response: Hamstring stays relaxed, allowing extension of leg (reciprocal inhibition).

■ **FIGURE 13-7** *The patellar tendon (knee jerk) reflex*

left leg and relax in the withdrawing right leg, while the opposite occurs in the flexor muscles.

Note in Figure 13-8 how the one sensory neuron synapses on multiple interneurons. Divergence of the sensory signal permits a single stimulus to control two sets of antagonistic muscle groups as well as to send sensory information to the brain. This type of complex reflex with multiple neuron interactions is more typical of our reflexes than the simple monosynaptic knee jerk stretch reflex.

The most complicated spinal reflex pathways are controlled by networks of neurons in the CNS called **central pattern generators**. Once activated, central pattern generators create spontaneous repetitive movement. In humans, rhythmic movements controlled by central pattern generators include locomotion and the unconscious rhythmicity of quiet breathing.

In the next section we look at how the CNS controls movements that range from involuntary reflexes to complex,

voluntary movement patterns such as dancing, throwing a ball, or playing a musical instrument.

✓ **CONCEPT CHECK**

6. Add the crossed extensor reflex in the supporting leg to the map you created in Concept Check 5.

7. As you pick up a heavy weight, which of the following are active in your biceps muscle: alpha motor neuron, gamma motor neuron, muscle spindle afferent neurons, Golgi tendon organ afferent neurons?

8. What distinguishes a stretch reflex from a crossed extensor reflex?

Answers: p. 455

THE INTEGRATED CONTROL OF BODY MOVEMENT

Most of us never think about how our body translates thoughts into action. Even the simplest movement requires proper timing so that antagonistic and synergistic muscle groups contract in

13

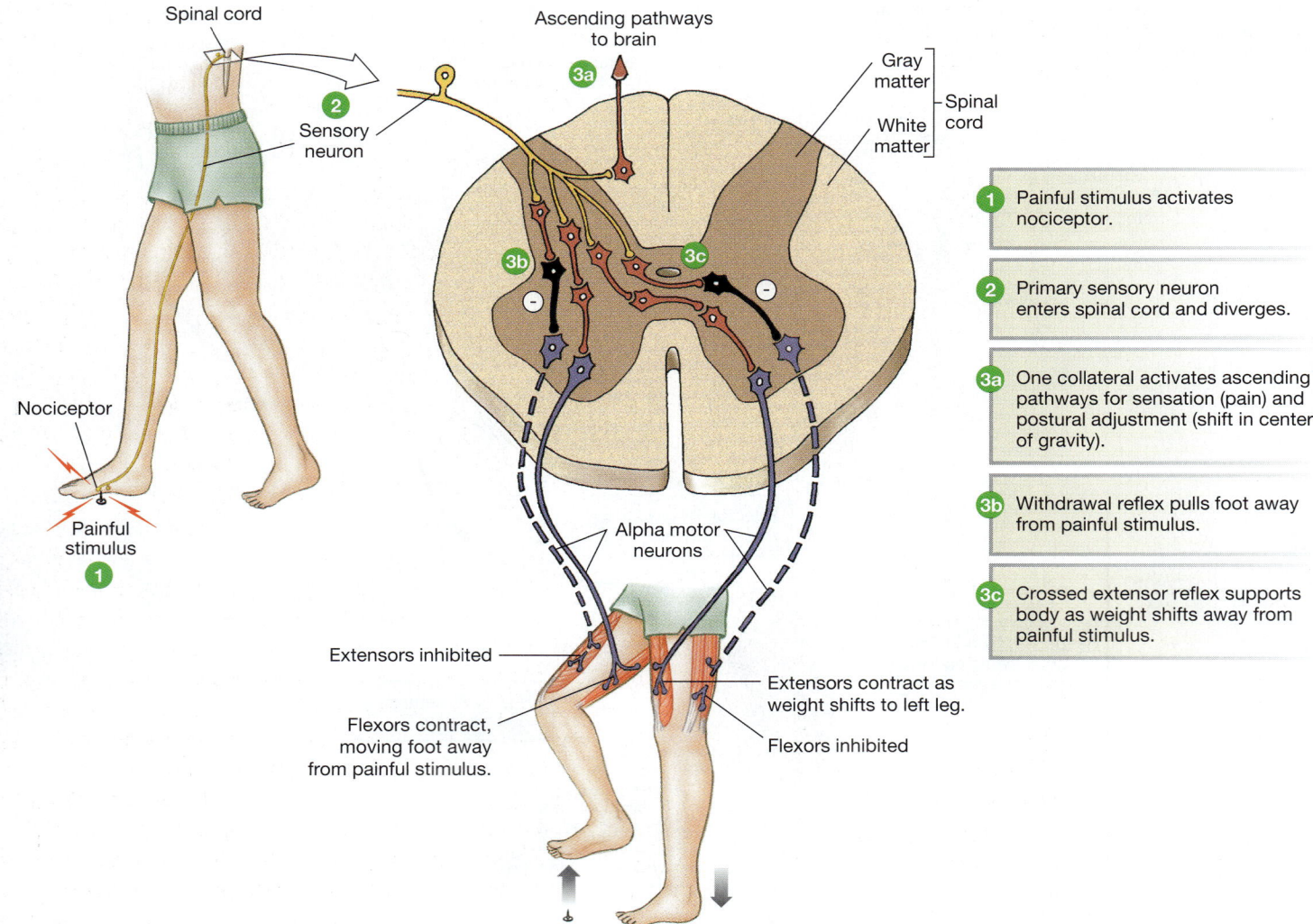

Spinal cord

Ascending pathways to brain

3a

Gray matter — Spinal cord
White matter —

2

Sensory neuron

Nociceptor

Painful stimulus

1

3b

3c

Alpha motor neurons

Extensors inhibited

Flexors contract, moving foot away from painful stimulus.

Extensors contract as weight shifts to left leg.

Flexors inhibited

1. Painful stimulus activates nociceptor.

2. Primary sensory neuron enters spinal cord and diverges.

3a. One collateral activates ascending pathways for sensation (pain) and postural adjustment (shift in center of gravity).

3b. Withdrawal reflex pulls foot away from painful stimulus.

3c. Crossed extensor reflex supports body as weight shifts away from painful stimulus.

■ **FIGURE 13-8** *Flexion reflex and the crossed extensor reflex*

Flexion in one limb simultaneously causes extension in the opposite limb. Coordination of reflexes with postural adjustments is essential for maintaining balance.

the appropriate sequence and to the appropriate degree. In addition, the body must continuously adjust its position to compensate for differences between the intended movement and the actual one. For example, the baseball pitcher steps off the mound to field a ground ball but in doing so slips on a wet patch of grass. His brain quickly compensates for the unexpected change in position through reflex muscle activity, and he stays on his feet to intercept the ball.

Skeletal muscles cannot communicate with one another directly, and so they send messages to the CNS, allowing the integrating centers to take charge and direct movement. Most body movements are highly integrated, coordinated responses that require input from multiple regions of the brain. In this section we examine a few of the CNS integrating centers that are responsible for control of body movement.

Movement Can Be Classified as Reflex, Voluntary, or Rhythmic

Movement can be loosely classified into three categories: reflex movement, voluntary movement, and rhythmic movement (Table 13-2 ■). **Reflex movements** are the least complex and are integrated primarily in the spinal cord (for example, see the knee jerk reflex in Fig. 13-7). However, like other spinal reflexes, reflex movements can be modulated by input from higher brain centers. In addition, the sensory input that initiates reflex movements, such as the input from muscle spindles and Golgi tendon organs, goes to the brain and participates in the coordination of voluntary movements and postural reflexes.

Postural reflexes that help us maintain body position as we stand or move through space are integrated in the brain stem. They require continuous sensory input from visual and vestibular (inner ear) sensory systems and from the muscles themselves [🔁 p. 355]. Muscle, tendon, and joint receptors

RUNNING PROBLEM

Dr. Ling admits Mrs. Evans to the intensive care unit. There Mrs. Evans is given tetanus antitoxin to deactivate any toxin that has not yet entered motor neurons. She also receives penicillin, an antibiotic that kills the bacteria, and drugs to help relax her muscles. Despite these treatments, by the third day Mrs. Evans is having difficulty breathing because of spasms in her chest muscles. Dr. Ling calls in the chief of anesthesiology to administer metocurine, a drug similar to curare. Curare and metocurine induce temporary paralysis of muscles by binding to ACh receptors on the motor end plate. Patients receiving metocurine must be placed on respirators that breathe for them. For people with tetanus, however, metocurine can temporarily halt the muscle spasms and allow the body to recover.

Question 3:
 Why does the binding of metocurine to ACh receptors on the motor end plate induce muscle paralysis? (Hint: what is the function of ACh in synaptic transmission?) Is metocurine an agonist or an antagonist of ACh?

| 436 | 439 | 443 | **447** | 451 | 452 |

provide information about proprioception, the positions of various body parts relative to one another. You can tell if your arm is bent even when your eyes are closed because these receptors provide information about body position to the brain.

Information from the vestibular apparatus of the ear and visual cues help us maintain our position in space. For example, we use the horizon to tell us our spatial orientation relative to the ground. In the absence of visual cues, we rely on tactile input. People trying to move in a dark room instinctively reach for a wall or piece of furniture to help orient themselves. Without visual and tactile cues, our orientation skills may fail. The lack of cues is what makes flying airplanes in clouds or fog impossible without instruments. The effect of gravity on the vestibular system is such a weak input when compared with visual or tactile cues that pilots may find themselves flying upside down relative to the ground.

Voluntary movements are the most complex type of movement. They require integration at the cerebral cortex, and they can be initiated at will without external stimuli. Learned voluntary movements improve with practice, and some even become involuntary, like reflexes. Think about how difficult it was to learn to ride a bicycle. However, once you learned to pedal smoothly and to keep your balance, the movements became automatic. "Muscle memory" is the name dancers and athletes give the ability of the unconscious brain to reproduce voluntary, learned movements and positions.

Rhythmic movements, such as walking or running, are a combination of reflex movements and voluntary movements. Rhythmic movements are initiated and terminated by input from the cerebral cortex, but once initiated, they can be sus-

13

TABLE 13-2	Types of Movement		
	REFLEX	**VOLUNTARY**	**RHYTHMIC**
Stimulus that initiates movement	Primarily external via sensory receptors; minimally voluntary	External stimuli or at will	Initiation and termination voluntary
Example	Knee jerk, cough, postural reflexes	Playing piano	Walking, running
Complexity	Least complex; integrated at level of spinal cord or brain stem with higher center modulation	Most complex; integrated in cerebral cortex	Intermediate complexity; integrated in spinal cord with higher center input required
Comments	Inherent, rapid	Learned movements that improve with practice; once learned, may become subconscious ("muscle memory")	Spinal circuits act as pattern generators; activation of these pathways requires input from brain stem

EMERGING CONCEPTS

CENTRAL PATTERN GENERATORS AND SPINAL CORD INJURIES

The ability of central pattern generators to sustain rhythmic movement without continued sensory input has proved important for research on spinal cord injuries. Scientists have artificially stimulated rhythmic movements in experimental animals when the spinal cord between the brain and the motor neurons has been severed. For example, an animal paralyzed by a spinal cord injury that blocks the "start walking" signal from the brain will walk if supported on a moving treadmill and given an electrical stimulus to activate the spinal pattern generator governing that motion. As the treadmill moves the animal's legs, the pattern generator, reinforced by sensory signals from muscle spindles, drives contraction of the leg muscles. Researchers are trying to take advantage of these rhythmic reflexes in people with spinal cord injuries. Some of the newest therapeutic techniques involve artificially stimulating portions of the spinal cord so that movement is restored to formerly paralyzed limbs.

tained without further input from the brain. The rhythmic activity of skeletal muscles in the limbs is maintained by networks of spinal interneurons. These interneurons are central pattern generators, causing alternate contraction and relaxation of muscles in a repetitive, rhythmic fashion until instructed to stop by signals from the brain. As an analogy, think of a battery-operated bunny. When the switch is thrown to "on," the bunny begins to hop. It continues its repetitive hopping until someone turns it off (or until the battery runs down).

The distinctions among reflex, voluntary, and rhythmic movements are not always clear-cut. The precision of voluntary movements improves with practice, but so does that of some reflexes. Voluntary movements, once learned, can become reflexive. In addition, most voluntary movements require continuous input from postural reflexes. **Feedforward reflexes** allow the body to prepare for a voluntary movement, and feedback mechanisms are used to create a smooth, continuous motion. Coordination of movement thus requires cooperation from many parts of the brain.

The CNS Integrates Movement

Three levels of the nervous system control movement: (1) the spinal cord, which integrates spinal reflexes and contains central pattern generators; (2) the brain stem and cerebellum, which control postural reflexes and hand and eye movements; and (3) the cerebral cortex and basal ganglia [🔁 p. 308], which are responsible for voluntary movements. The thalamus relays and modifies signals being sent from the spinal cord, basal ganglia, and cerebellum to the cerebral cortex (Table 13-3 ■).

TABLE 13-3   Neural Control of Movement

LOCATION	ROLE	RECEIVES INPUT FROM:	SENDS INTEGRATIVE OUTPUT TO:
Spinal cord	Spinal reflexes; locomotor pattern generators	Sensory receptors	Brain stem, cerebellum, thalamus/cerebral cortex
Brain stem	Posture, hand and eye movements	Cerebellum, visual and vestibular sensory receptors	Spinal cord
Motor areas of cerebral cortex	Planning and coordinating complex movement	Thalamus	Brain stem, spinal cord (corticospinal tract), cerebellum, basal ganglia
Cerebellum	Monitors output signals from motor areas and adjusts movements	Spinal cord (sensory), cerebral cortex (commands)	Brain stem, cerebral cortex (Note: All output is inhibitory.)
Thalamus	Contains relay nuclei that modulate and pass messages to cerebral cortex	Basal ganglia, cerebellum, spinal cord	Cerebral cortex
Basal nuclei	Motor planning	Cerebral cortex	Cerebral cortex, brain stem

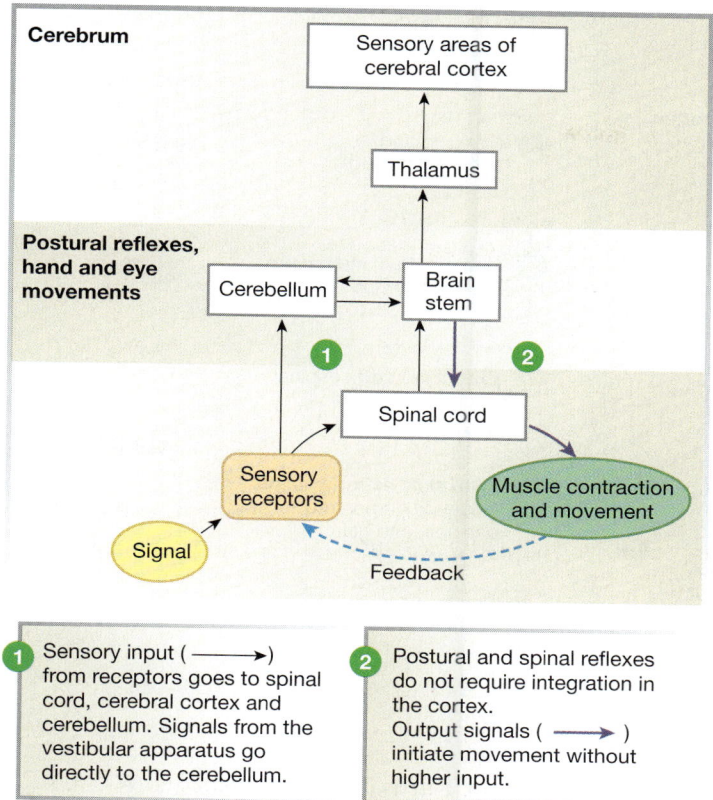

1 Sensory input (———→) from receptors goes to spinal cord, cerebral cortex and cerebellum. Signals from the vestibular apparatus go directly to the cerebellum.

2 Postural and spinal reflexes do not require integration in the cortex. Output signals (———→) initiate movement without higher input.

■ **FIGURE 13-9** *Integration of muscle reflexes*

Reflex movements do not require input from the cerebral cortex. Proprioceptors such as muscle spindles, Golgi tendon organs, and joint capsule receptors provide information to the spinal cord, brain stem, and cerebellum (Fig. 13-9 ■). The brain stem is in charge of postural reflexes and hand and eye movements. It also gets commands from the cerebellum, the part of the brain responsible for "fine-tuning" movement. The result is reflex movement. However, some sensory information is sent

through ascending pathways to sensory areas of the cortex, where it can be used to plan voluntary movements.

Voluntary movements require coordination between the cerebral cortex, cerebellum, and basal ganglia. The control of voluntary movement can be divided into three steps: (1) decision making and planning, (2) initiating the movement, and (3) executing the movement (Fig. 13-10 ■). The cerebral cortex plays a key role in the first two steps. Behaviors such as movement require knowledge of the body's position in space (where am I?), a decision on what movement should be executed (what shall I do?), a plan for executing the movement (how shall I do it?), and the ability to hold the plan in memory long enough to carry it out (now, what was I just doing?). As with reflex movements, sensory feedback is used to continuously refine the process.

Let's return to our baseball pitcher and trace the process as he decides whether to throw a fastball or a slow curve. Standing out on the mound, the pitcher is acutely aware of his surroundings: the other players on the field, the batter in the box, and the dirt beneath his feet. With the help of visual and somatosensory input to the sensory areas of the cortex, he is aware of his body position as he steadies himself for the pitch (Fig. 13-11 ■). Deciding which type of pitch to throw and anticipating the consequences occupy many pathways in his prefrontal cortex and association areas. These pathways loop down through the basal ganglia and thalamus for modulation before cycling back to the cortex.

Once the pitcher makes the decision to throw a fastball, the motor cortex takes charge of organizing the execution of this complex movement. To initiate the movement, descending information travels from the motor association areas and motor cortex to the brain stem, the spinal cord, and the cerebellum. The cerebellum assists in making postural adjustments by integrating feedback from peripheral sensory receptors. The basal ganglia, which assisted the cortical motor areas in planning the pitch, also provide information about posture, balance, and gait to the brain stem.

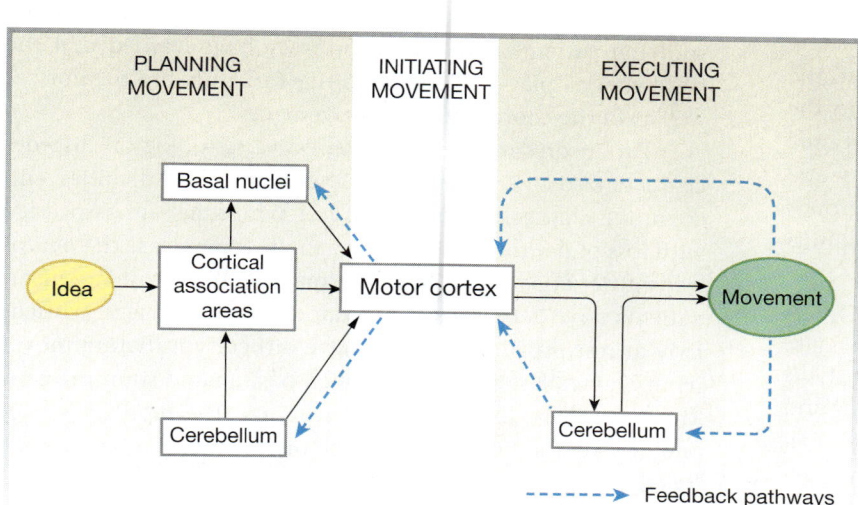

----→ Feedback pathways

■ **FIGURE 13-10**   *CNS control of voluntary movement*

Voluntary movements can be divided into three phases: planning, initiation, and execution. Sensory feedback allows the brain to correct for any deviation between the planned movement and the actual movement.

13

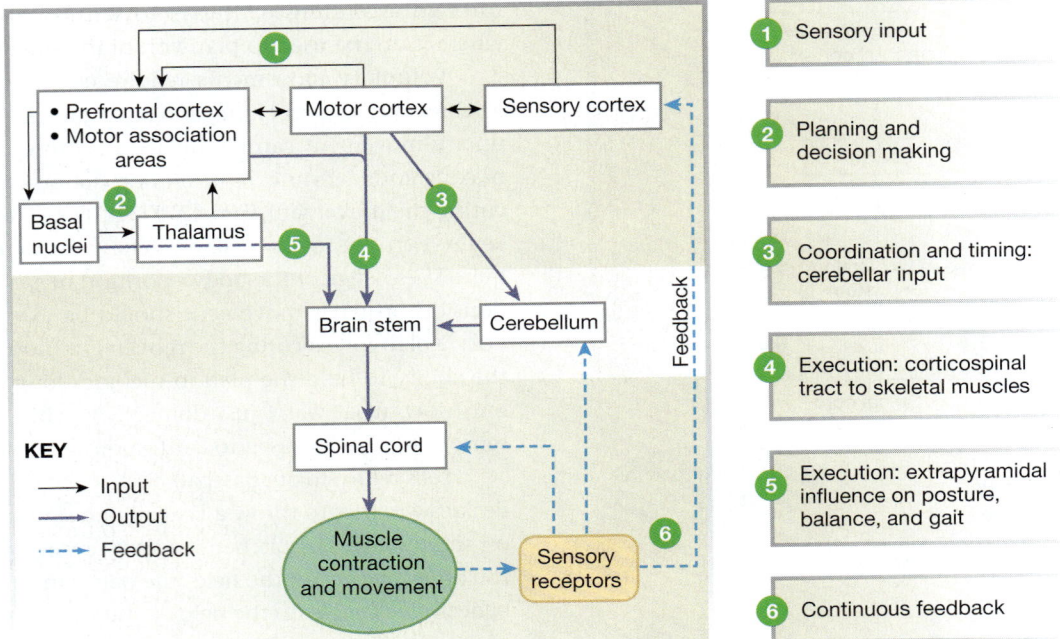

■ **FIGURE 13-11**   *Control of voluntary movements*

The pitcher's decision to throw a fastball now is translated into action potentials that travel down through the **corticospinal tract**, a group of interneurons that run from the motor cortex to the spinal cord (Fig. 13-12 ■), where they synapse directly onto somatic motor neurons. As noted in Chapter 9, most of these descending pathways cross to the opposite side of the body in a region of the medulla known as the *pyramids*. Consequently, this pathway is sometimes called the *pyramidal tract*. Descending motor pathways with multiple synapses in the CNS make up what is sometimes called the *extrapyramidal tract* or the *extrapyramidal system*. However, we now know that the pyramidal and extrapyramidal pathways interact and are not as distinct in their function as was once believed.

As the pitcher begins the pitch, *feedforward postural reflexes* adjust the body position, shifting weight slightly in anticipation of the changes about to occur (Fig. 13-13 ■). Through the appropriate divergent pathways, action potentials race to the pools of somatic motor neurons that control the muscles used for pitching: some are excited, others are inhibited. The neural circuitry allows precise control over antagonistic muscle groups as the pitcher flexes and retracts his right arm. His weight shifts onto his right foot as his right arm moves back.

Each of these movements activates sensory receptors that feed information back to the spinal cord, brain stem, and cerebellum, activating postural reflexes. These reflexes adjust his body position so that the pitcher does not lose his balance and fall over backward. Finally, he releases the ball, catching his balance on the follow-through—another example of postural reflexes mediated through sensory feedback. His head stays erect,

and his eyes track the ball as it reaches the batter. WHACK . . . home run. As the pitcher's eyes follow the ball and he evaluates the result of his pitch, his brain is preparing for the next batter, hoping to use what it has learned from these pitches to improve those to come.

Symptoms of Parkinson's Disease Reflect the Functions of the Basal Ganglia

Our understanding of the role of the basal ganglia in the control of movement has been slow to develop because, for many years, animal experiments yielded little information. Randomly destroying portions of the basal ganglia did not appear to affect research animals. However, research focusing on **Parkinson's disease** (Parkinsonism) in humans has been more fruitful. From studying patients with Parkinson's, we have learned that the basal ganglia play a role in cognitive function and memory as well as in the coordination of movement.

Parkinson's disease is a progressive neurological disorder characterized by abnormal movements, speech difficulties, and cognitive changes. These signs and symptoms are associated with loss of neurons in the basal ganglia that release the neurotransmitter dopamine. One abnormal sign that most Parkinson patients have is tremors in the hands, arms, and legs, particularly at rest. In addition, they have difficulty initiating movement and walk slowly with stooped posture and shuffling gait. They lose facial expression, fail to blink (the reptilian stare), and may develop depression, sleep disturbances, and personality changes.

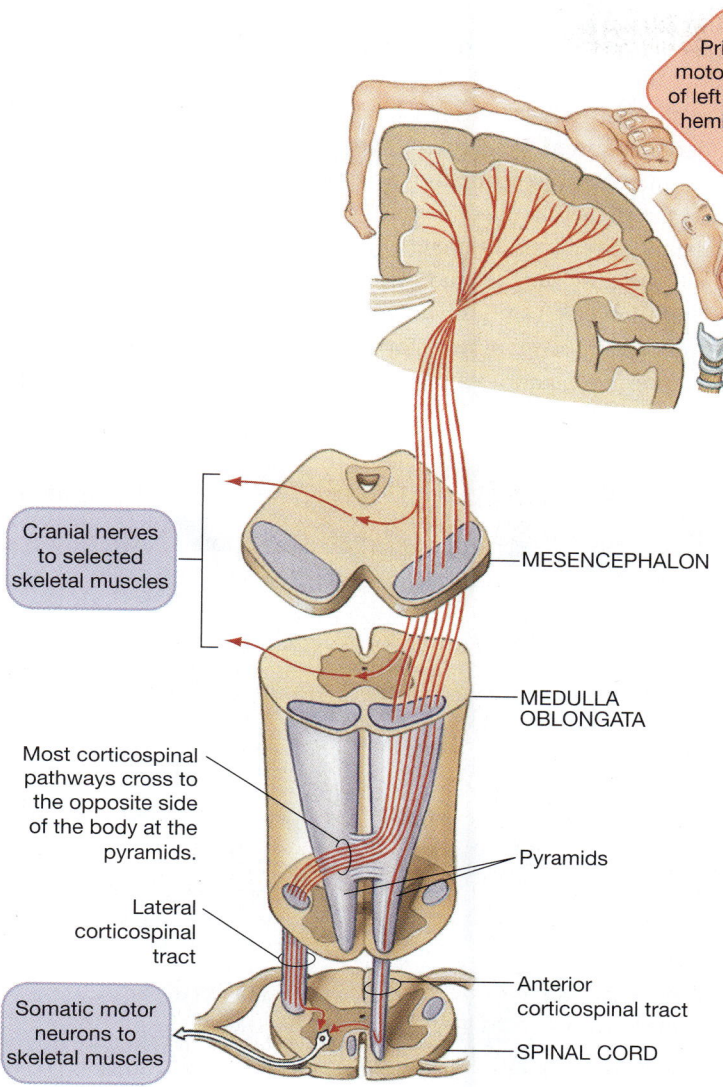

Primary motor cortex of left cerebral hemisphere

Cranial nerves to selected skeletal muscles

MESENCEPHALON

MEDULLA OBLONGATA

Most corticospinal pathways cross to the opposite side of the body at the pyramids.

Pyramids

Lateral corticospinal tract

Anterior corticospinal tract

Somatic motor neurons to skeletal muscles

SPINAL CORD

■ **FIGURE 13-12** *The corticospinal tract*

These interneurons run directly from the motor cortex to their synapses with somatic motor neurons. Most corticospinal neurons cross the midline at the pyraminds.

The cause of Parkinson's disease is usually not known and appears to be a combination of environmental factors and genetic susceptibility. However, a few years ago, a number of young drug users were diagnosed with Parkinsonism. Their disease was traced to the use of homemade heroin containing a toxic contaminant that destroyed *dopaminergic* (dopamine-secreting) neurons. This contaminant has been isolated and now enables researchers to induce Parkinson's disease in experimental animals so that we have an animal model on which to test new treatments.

The primary current treatment for Parkinson's is administration of L-dopa, a precursor of dopamine that can cross the blood-brain barrier. Experimental treatments include transplants of dopamine-secreting neurons. Proponents of stem cell

research feel that Parkinson's may be one of the conditions that would benefit from the transplant of stem cells into affected brains.

CONTROL OF MOVEMENT IN VISCERAL MUSCLES

Movement created by contracting smooth and cardiac muscles is very different from that created by skeletal muscles, in large part because smooth and cardiac muscle are not attached to bone. In the internal organs, or viscera, muscle contraction usually changes the shape of an organ, narrowing the lumen of a hollow organ or shortening the length of a tube. In many hollow internal organs, muscle contraction pushes material through the lumen of the organ: the heart pumps blood, the digestive tract moves food, the uterus expels a baby.

Visceral muscle contraction is often reflexively controlled by the autonomic nervous system, but not always. Some types of smooth and cardiac muscle are capable of generating their own action potentials, independent of an external signal. Both the heart and digestive tract have

13

RUNNING PROBLEM

Four weeks later, Mrs. Evans is ready to go home, completely recovered and showing no signs of lingering effects. Once she could talk, Mrs. Evans, who was born on the farm where she still lived, was able to tell Dr. Ling that she had never had immunization shots for tetanus or any other diseases. "Well, that made you one of only a handful of people in the United States who will develop tetanus this year," Dr. Ling told her. "You've been given your first two tetanus shots here in the hospital. Be sure to come back in six months for the last one so that this won't happen again." Because of national immunization programs begun in the 1950s, tetanus is now a rare disease in the United States. However, in developing countries without immunization programs, tetanus is still a common and serious condition.

Question 4:
On the basis of what you know about who receives immunization shots in the United States, predict the age and background of people who are most likely to develop tetanus this year.

  436 439 443 447 **451** 452

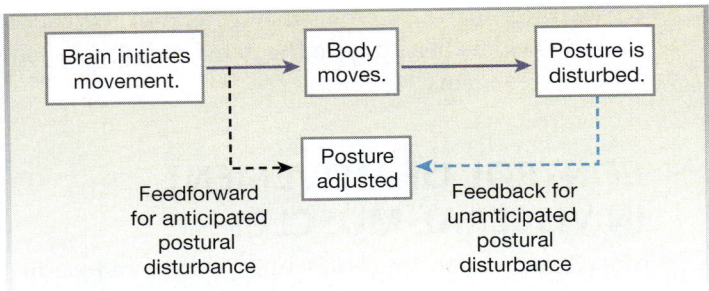

■ **FIGURE 13-13** *Feedforward reflexes and feedback of information during movement*

spontaneously depolarizing muscle fibers (often called *pacemakers*) that give rise to regular, rhythmic contractions.

Reflex control of visceral smooth muscle varies from that of skeletal muscle. Skeletal muscles are controlled only by the nervous system, but in many types of visceral muscle, hormones are important in regulating contraction. In addition, some visceral muscle cells are connected to one another by gap junctions that allow electrical signals to pass directly from cell to cell.

Because smooth and cardiac muscle have such a variety of control mechanisms, we will discuss their control as we cover the appropriate organ system for each type of muscle. In the next chapter, we examine cardiac muscle and its function in the heart.

RUNNING PROBLEM CONCLUSION

TETANUS

In this running problem, you learned about the tetanus toxin tetanospasmin, a potent poison made by the bacterium *Clostridium tetani*. As little as 175 billionths of a gram (175 nanograms) can be fatal to a 70-kg human. Both tetanus toxin and botulinum toxin cause paralysis, but tetanus is a rigid (contracted muscle) paralysis, while botulism is a flaccid (relaxed muscle) paralysis. To learn more about tetanus, visit the web site of the U.S. Centers for Disease Control and Prevention (*www.cdc.gov*). Now check your understanding of this running problem by comparing your answers with the information in the summary table.

	QUESTION	FACTS	INTEGRATION AND ANALYSIS
1a	By what process is tetanospasmin taken up into neurons?	Tetanospasmin is a protein.	Proteins are too large to cross cell membranes by mediated transport. Therefore, tetanospasmin must be taken up by endocytosis [🔁 p. 148].
1b	By what process does tetanospasmin travel up the axon to the nerve cell body?	Movement of substances from the axon terminal to the cell body is called retrograde axonal transport [🔁 p. 249].	Tetanospasmin is taken up by endocytosis, so it will be contained in endocytotic vesicles. These vesicles "walk" along microtubules through retrograde axonal transport.
2	Using the reflex pathways diagrammed in Figures 13-7 and 13-8, explain why inhibition of inhibitory interneurons might result in uncontrollable muscle spasms.	Muscles often occur in antagonistic pairs. When one muscle is contracting, its antagonist must be inhibited.	If the inhibitory interneurons are not functioning, both sets of antagonistic muscles can contract at the same time. This would lead to muscle spasms and rigidity because the bones attached to the muscles would be unable to move in any direction.
3a	Why does the binding of metocurine to ACh receptors on the motor end plate induce muscle paralysis?	ACh is the somatic motor neuron neurotransmitter that initiates skeletal muscle contraction.	If metocurine binds to ACh receptors, it prevents ACh from binding. Without ACh binding, the muscle fiber will not depolarize and cannot contract, resulting in paralysis.
3b	Is metocurine an agonist or an antagonist of ACh?	Agonists mimic the effects of a substance; antagonists block the effects of a substance.	Metocurine blocks ACh action; therefore, it is an antagonist.

QUESTION	FACTS	INTEGRATION AND ANALYSIS
4 On the basis of what you know about who receives immunization shots in the United States, predict the age and background of people who are most likely to develop tetanus this year.	Immunizations are required for all children of school age. This practice has been in effect since about the 1950s. In addition, most people who suffer puncture wounds or dirty wounds are given tetanus booster shots when they are treated for those wounds.	Most cases of tetanus in the United States will occur in people over the age of 60 who have never been immunized, in immigrants (particularly migrant workers), and in newborn infants. Another source of the disease is contaminated heroin; injection of the drug under the skin may cause tetanus.

436   439   443   447   451   **452**

CHAPTER SUMMARY

How many times have you heard people say, "I did it without thinking"? In effect, they were telling you that their action was a reflex response. There are many ways to control the functions of muscles and glands of the body, but a neural reflex is the simplest and the fastest.

This chapter integrates the material you learned in Chapters 8–12 and shows how the nervous *control system* controls body movement.

Neural Reflexes

1. A neural reflex consists of the following elements: stimulus, receptor, sensory neurons, efferent neurons, effectors (muscles and glands), and response. (p. 436)

2. Neural reflexes can be classified in several ways. **Somatic reflexes** involve somatic motor neurons and skeletal muscles. **Autonomic (or visceral) reflexes** are controlled by autonomic neurons. (p. 436; Tbl. 13-1)

3. **Spinal reflexes** are integrated in the spinal cord. **Cranial reflexes** are integrated in the brain. (p. 436)

4. Many reflexes are innate. Others are acquired through experience. (p. 436)

5. The simplest reflex pathway is a **monosynaptic reflex** with only two neurons. **Polysynaptic reflexes** have three or more neurons in the pathway. (p. 437; Fig. 13-1)

Autonomic Reflexes

6. Some autonomic reflexes are spinal reflexes that are modulated by input from the brain. Other reflexes needed to maintain homeostasis are integrated in the brain, primarily in the hypothalamus, thalamus, and brain stem. (p. 438)

7. Autonomic reflexes are all polysynaptic, and many are characterized by tonic activity. (p. 438; Fig. 13-2)

Skeletal Muscle Reflexes

8. Skeletal muscle relaxation must be controlled by the CNS because somatic motor neurons always cause contraction in skeletal muscle. (p. 439)

Postural and spinal reflexes follow the basic pattern of a reflex, with sensory input being integrated in the CNS and acted on when an output signal goes to skeletal muscles. Voluntary movements do not require sensory input to be initiated, but they integrate sensory feedback to ensure smooth execution.

9. The normal contractile fibers of a muscle are called **extrafusal muscle fibers**. Their contraction is controlled by **alpha motor neurons**. (p. 439; Fig. 13-3)

10. **Muscle spindles** send information about muscle length to the CNS. These receptors consist of **intrafusal fibers** with sensory neurons wrapped around the noncontractile center. **Gamma motor neurons** innervate the contractile ends of the intrafusal fibers. (p. 439; Fig. 13-3)

11. Muscle spindles are tonically active stretch receptors. Their output creates tonic contraction of extrafusal muscle fibers. Because of this tonic activity, a muscle at rest maintains a certain level of tension, known as **muscle tone**. (p. 440; Fig. 13-4a)

12. If an intact muscle stretches, the intrafusal fibers of the spindle stretch and initiate reflex contraction of the muscle. The contraction prevents damage from overstretching. This reflex pathway is known as a **stretch reflex**. (p. 440; Fig. 13-4b)

13. When a muscle contracts, **alpha-gamma coactivation** ensures that the muscle spindle remains active. Activation of gamma motor neurons causes contraction of the ends of the intrafusal fibers. This contraction lengthens the central region of the intrafusal fiber and maintains stretch on the sensory nerve endings. (p. 440; Fig. 13-5)

14. **Golgi tendon organs** are found at the junction of the tendons and muscle fibers. They consist of free nerve endings that wind between collagen fibers. Golgi tendon organs respond to both stretch and contraction of the muscle by causing a reflexive relaxation. (p. 442; Figs. 13-3, 13-6)

15. The synergistic and antagonistic muscles that control a single joint are known as a **myotatic unit**. When one set of muscles in a

13

myotatic unit contracts, the antagonistic muscles must relax through a reflex known as **reciprocal inhibition**. (p. 443; Fig. 13-7)

16. **Flexion reflexes** are polysynaptic reflexes that cause an arm or leg to be pulled away from a painful stimulus. Flexion reflexes that occur in the legs are usually accompanied by the **crossed extensor reflex**, a postural reflex that helps maintain balance when one foot is lifted from the ground. (p. 444; Fig. 13-8)

17. **Central pattern generators** are networks of neurons in the CNS that function spontaneously to control certain rhythmic muscle movements. (p. 445)

The Integrated Control of Body Movement

18. Movement can be loosely classified into three categories: reflex movement, voluntary movement, and rhythmic movement. (p. 446; Tbl. 13-2)

19. **Reflex movements** are integrated primarily in the spinal cord. **Postural reflexes** are integrated in the brain stem. (p. 446; Fig. 13-9, Tbl. 13-3)

20. **Voluntary movements** are integrated in the cerebral cortex and can be initiated at will. Learned voluntary movements improve with practice and may even become involuntary, like reflexes. (p. 447; Fig. 13-11)

21. **Rhythmic movements**, such as walking, are a combination of reflexes and voluntary movements. Rhythmic movements can be sustained by central pattern generators. (p. 447)

22. Motor commands travel from cortex to spinal cord through the **corticospinal tract** and from the **basal ganglia** through extrapyramidal pathways. (p. 448; Fig. 13-12)

23. **Feedforward reflexes** allow the body to prepare for a voluntary movement; feedback mechanisms are used to create a smooth, continuous motion. (p. 448; Fig. 13-13)

Control of Movement in Visceral Muscles

24. Contraction in smooth and cardiac muscles may occur spontaneously or may be controlled by hormones or by the autonomic division of the nervous system. (p. 451)

QUESTIONS

(Answers to the Review Questions begin on page A1.)

THE PHYSIOLOGY PLACE

Access more review material online at **The Physiology Place** website. There you'll find review questions, problem-solving activities, case studies, flashcards, and direct links to both *InterActive Physiology*® and *PhysioEx*™. To access the site, go to *www.physiologyplace.com* and select Human Physiology, Fourth Edition.

LEVEL ONE REVIEWING FACTS AND TERMS

1. All neural reflexes begin with a _____ that activates a receptor.
2. Somatic reflexes involve _____ muscles; _____ (or visceral) reflexes are controlled by autonomic neurons.
3. The pathway pattern that brings information from many neurons into a smaller number of neurons is known as _____.
4. When the axon terminal of a modulatory neuron (cell M) terminates close to the axon terminal of a presynaptic cell (cell P) and decreases the amount of neurotransmitter released by cell P, the resulting type of modulation is called _____. (*Hint:* see p. 279.)
5. Autonomic reflexes are also called _____ reflexes. Why?
6. Some autonomic reflexes are spinal reflexes; others are integrated in the brain. List examples of each.
7. Which part of the brain transforms emotions into somatic sensation and visceral function? List three autonomic reflexes that are linked to emotions.
8. How many synapses occur in the simplest autonomic reflexes? Where do the synapses occur?
9. List the three types of sensory receptors that convey information for muscle reflexes.
10. Because of tonic activity in neurons, a resting muscle maintains a low level of tension known as _____.
11. Stretching an intact skeletal muscle causes sensory neurons to (increase/decrease) their rate of firing, causing the muscle to contract, thereby relieving the stretch. Why is this a useful reflex?

12. Match the structure to all correct statements about it.
    (a) muscle spindle
    (b) Golgi tendon organ
    (c) joint capsule mechanoreceptor

    1. is strictly a sensory receptor
    2. has afferent neurons that bring information to the CNS
    3. is associated with two types of efferent neurons
    4. conveys information about the relative positioning of bones
    5. is innervated by gamma motor neurons
    6. modulates activity in alpha motor neurons

13. The Golgi tendon organ responds to both _____ and _____, although _____ elicits the stronger response. Activation (increases/decreases) muscle contraction via the _____ neuron.
14. The simplest reflex requires a minimum of how many neurons? How many synapses? Give an example.
15. List and differentiate the three categories of movement. Give an example of each.

LEVEL TWO REVIEWING CONCEPTS

16. What is the purpose of alpha-gamma coactivation? Explain how it occurs.
17. Modulatory neuron M synapses on the axon terminal of neuron P, just before P synapses with the effector organ. If M is an inhibitory neuron, what happens to neurotransmitter release by P? What effect does M's neurotransmitter have on the postsynaptic membrane potential of P? (*Hint:* draw this pathway.)
18. At your last physical, your physician checked your patellar tendon reflex by tapping just below your knee while you sat quietly on the edge of the table. (a) What was she checking when she did this test? (b) What would happen if you were worried about falling off the table and were very tense? Where does this additional input to the efferent motor neurons originate? Are these modulatory neurons causing EPSPs or IPSPs at the spinal motor neuron? (c) Your physician notices that you are tense and asks you to count backward from

100 by 3's while she repeats the test. Why would carrying out this counting task enhance your reflex?

LEVEL THREE PROBLEM SOLVING

19. There are several theories about how presynaptic inhibition works at the cellular level. Use what you have learned about membrane potentials and synaptic transmission to explain how each of the following mechanisms could result in presynaptic inhibition:

 (a) Voltage-gated Ca^{2+} channels in axon terminal are inhibited.
 (b) Cl^- channels in axon terminal open.
 (c) K^+ channels in axon terminal open.

20. Andy is working on improving his golf swing. He must watch the ball, swing the club back and then forward, twist his hips, straighten his left arm, then complete the follow-through, where the club arcs in front of him. Which parts of the brain are involved in adjusting how hard he hits the ball, keeping all his body parts moving correctly, watching the ball, and then repeating these actions once he has verified that this swing is successful?

21. It's Halloween, and you are walking through the scariest Haunted House around. As you turn a corner and enter the dungeon, a skeleton reaches out and grabs your arm. You let out a scream. Your heart rate quickens, and you feel the hairs on your arm stand on end. (a) What has just happened to you? (b) Where in the brain is fear processed? What are the functions of this part of the brain? Which branch (somatic or autonomic) of the motor output does it control? What are the target organs for this response? (c) How is it possible for your hair to stand on end when hair is made of proteins that do not contract? (*Hint:* see Fig. 10-11.) Given that the autonomic nervous system is mediating this reflex response, which type of tissue do you expect to find attached to hair follicles?

22. Using what you've learned about tetanus and botulinum toxins, make a table to compare the two. In what ways are tetanus and botulinum toxin similar to each other? How are they different from each other?

ANSWERS

✓ Answers to Concept Check Questions

Page 439

1. Sensor (sensory receptor), afferent pathway (sensory neuron), integrating center (central nervous system), efferent pathway (autonomic or somatic motor neuron), effector (muscles, glands, some adipose tissue).
2. Upon hyperpolarization, the membrane potential becomes more negative and moves farther from threshold.

Page 440

3. Your map of a stretch reflex should match the components shown in Figure 13-4b.

Page 455

4. Your map of alpha-gamma coactivation should match the steps in Figure 13-5b. The stimulus of muscle contraction is the same for the Golgi tendon reflex, but your map should then branch to show the steps in Figure 13-6d and e.

Page 444

5. Your flexion reflex map should match the steps shown for the knee jerk in Figure 13-7, with the added contraction of hip flexor muscles in addition to the quadriceps.

Page 445

6. The initial steps of the crossed extensor reflex are the same as those of the flexion reflex until the CNS. There the crossed extensor reflex follows the diagram shown in Figure 13-8, step 3c.
7. When you pick up a weight, alpha and gamma neurons, spindle afferents, and Golgi tendon organ afferents are all active.
8. A stretch reflex is initiated by stretch and causes a reflex contraction. A crossed extensor reflex is a postural reflex initiated by withdrawal from a painful stimulus; the extensor muscles contract, but the corresponding flexors are inhibited.

13

Only in the 17th century did the brain displace the heart as the controller of our actions.

—Mary A. B. Brazier, *A History of Neurophysiology in the 19th Century, 1988*

Immunofluorescent staining of actin (pink) and nebulin (green) in cultured cardiac muscle cells.

Cardiovascular Physiology

BACKGROUND BASICS

In the classic movie *Indiana Jones and the Temple of Doom,* the evil priest reaches into the chest of a sacrificial victim and pulls out his heart, still beating. This act was not dreamed up by some Hollywood scriptwriter—it was taken from rituals of the ancient Mayans, who documented this grisly practice in their carvings and paintings. The heart has been an object of fascination for centuries, but how can this workhorse muscle, which pumps 7200 liters of blood a day, keep beating outside the body? Before we can answer that question, we must first consider the role of hearts in cardiovascular systems.

As life evolved, simple one-celled organisms began to band together, first into cooperative colonies and then into multicelled organisms. In most multicellular animals, only the surface layer of cells is in direct contact with the environment. This body plan presents a problem because diffusion slows as distance increases [⇄ p. 133]. For example, oxygen consumption in the interior cells of larger animals exceeds the rate at which oxygen can diffuse from the body surface.

One solution to overcome slow diffusion was the evolutionary development of circulatory systems that move fluid between the body's surface and its deepest parts. In simple animals, muscular activity creates fluid flow when the animal moves. More complex animals have muscular pumps called hearts to circulate internal fluid.

In the most efficient circulatory systems, the heart pumps blood through a closed system of vessels. This one-way circuit steers the blood along a specific route and ensures systematic distribution of gases, nutrients, signal molecules, and wastes. A

circulatory system comprising a heart, blood vessels, and blood is known as a **cardiovascular system** [*kardia,* heart + *vasculum,* little vessel].

Although the idea of a closed cardiovascular system that cycles blood in an endless loop seems intuitive to us today, it has not always been so. **Capillaries**, the microscopic vessels in which blood exchanges material with the interstitial fluid, were not discovered until Marcello Malpighi, an Italian anatomist, observed them through a microscope in the middle of the seventeenth century. At that time European medicine was still heavily influenced by the ancient belief that the cardiovascular system distributed both blood and air.

Blood was thought to be made in the liver and distributed throughout the body in the veins. Air went from the lungs to the heart, where it was digested and picked up "vital spirits." From the heart, air was distributed to the tissues through vessels called arteries. Anomalies—such as the fact that a cut artery squirted blood rather than air—were ingeniously explained by unseen links between arteries and veins that opened upon injury.

According to this model of the circulatory system, the tissues consumed all blood delivered to them, and the liver had to synthesize new blood continuously. It took the calculations of William Harvey (1578–1657), court physician to King Charles I of England, to show that the weight of blood pumped by the heart in a single hour exceeds the weight of the entire body! Once it became obvious that the liver could not make blood as rapidly as the heart pumped it, Harvey looked for an anatomical route that would allow the blood to recirculate rather than be consumed in the tissues. He showed that valves in the heart and veins created a one-way flow of blood, and that veins carried blood back to the heart, not out to the limbs. He also showed that blood entering the right side of the heart had to go to the lungs before it could go to the left side of the heart.

The results of these studies created a furor among Harvey's contemporaries, leading Harvey to say in a huff that no one under the age of 40 could understand his conclusions. Ultimately, Harvey's work became the foundation of modern cardiovascular physiology. Today, we understand the structure of the cardiovascular system at microscopic and molecular levels that Harvey never dreamed existed. Yet some things have not changed. Even now, with our sophisticated technology, we are searching for "spirits" in the blood, although today we call them by such names as *hormone* and *cytokine.*

OVERVIEW OF THE CARDIOVASCULAR SYSTEM

In the simplest terms, a cardiovascular system is a series of tubes (the blood vessels) filled with fluid (blood) and connected to a pump (the heart). Pressure generated in the heart propels blood through the system continuously. The blood picks up oxygen at

14

the lungs and nutrients in the intestine and then delivers these substances to the body's cells while simultaneously removing cellular wastes for excretion. In addition, the cardiovascular system plays an important role in cell-to-cell communication and in defending the body against foreign invaders. This chapter focuses on an overview of the cardiovascular system and on the heart as a pump. We will examine the properties of the blood vessels and the homeostatic controls that regulate blood flow and blood pressure in Chapter 15.

The Cardiovascular System Transports Materials Throughout the Body

The primary function of the cardiovascular system is the transport of materials to and from all parts of the body. Substances transported by the cardiovascular system can be divided into (1) nutrients, water, and gases that enter the body from the external environment, (2) materials that move from cell to cell within the body, and (3) wastes that the cells eliminate (Table 14-1 ■).

Oxygen enters the body at the exchange surface of the lungs. Nutrients and water are absorbed across the intestinal epithelium. Once in the blood, all these materials are distributed by the circulation. A steady supply of oxygen for the cells is particularly important because many cells deprived of oxygen become irreparably damaged within a short period of time. For example, about 5–10 seconds after blood flow to the brain is stopped, a person loses consciousness. If oxygen delivery stops for 5–10 minutes, permanent brain damage results. Neurons of the brain have a very high rate of oxygen consumption and are unable to meet their metabolic need for ATP by using anaerobic pathways, which have low yields of ATP/glucose [🌐 p. 106]. Because of the brain's sensitivity to *hypoxia* [*hypo-,* low + *-oxia,* oxygen], homeostatic controls do everything possible to maintain cerebral blood flow, even if it means depriving other cells of oxygen.

Cell-to-cell communication is a key function of the cardiovascular system. For example, hormones secreted by endocrine glands are carried in the blood to their targets. Nutrients, such as glucose from the liver and fatty acids from adipose tissue, are transported to metabolically active cells. Finally, the defense team of white blood cells and antibodies patrols the circulation to intercept foreign invaders.

The cardiovascular system also picks up carbon dioxide and metabolic wastes released by cells and transports them to the lungs and kidneys for excretion. Some waste products are transported to the liver for processing before they are excreted in the urine or feces. Heat also circulates through the blood, moving from the body core to the surface, where it dissipates.

The Cardiovascular System Consists of the Heart, Blood Vessels, and Blood

The cardiovascular system is composed of the heart, the blood vessels (also known as the *vasculature*), and the cells and plasma of the blood. Blood vessels that carry blood away from the heart are called **arteries**. Blood vessels that return blood to the heart are called **veins**.

TABLE 14-1   Transport in the Cardiovascular System

SUBSTANCE MOVED	FROM	TO
Materials entering the body		
Oxygen	Lungs	All cells
Nutrients and water	Intestinal tract	All cells
Materials moved from cell to cell		
Wastes	Some cells	Liver for processing
Immune cells, antibodies, clotting proteins	Present in blood continuously	Available for any cell that needs them
Hormones	Endocrine cells	Target cells
Stored nutrients	Liver and adipose tissue	All cells
Materials leaving the body		
Metabolic wastes	All cells	Kidneys
Heat	All cells	Skin
Carbon dioxide	All cells	Lungs

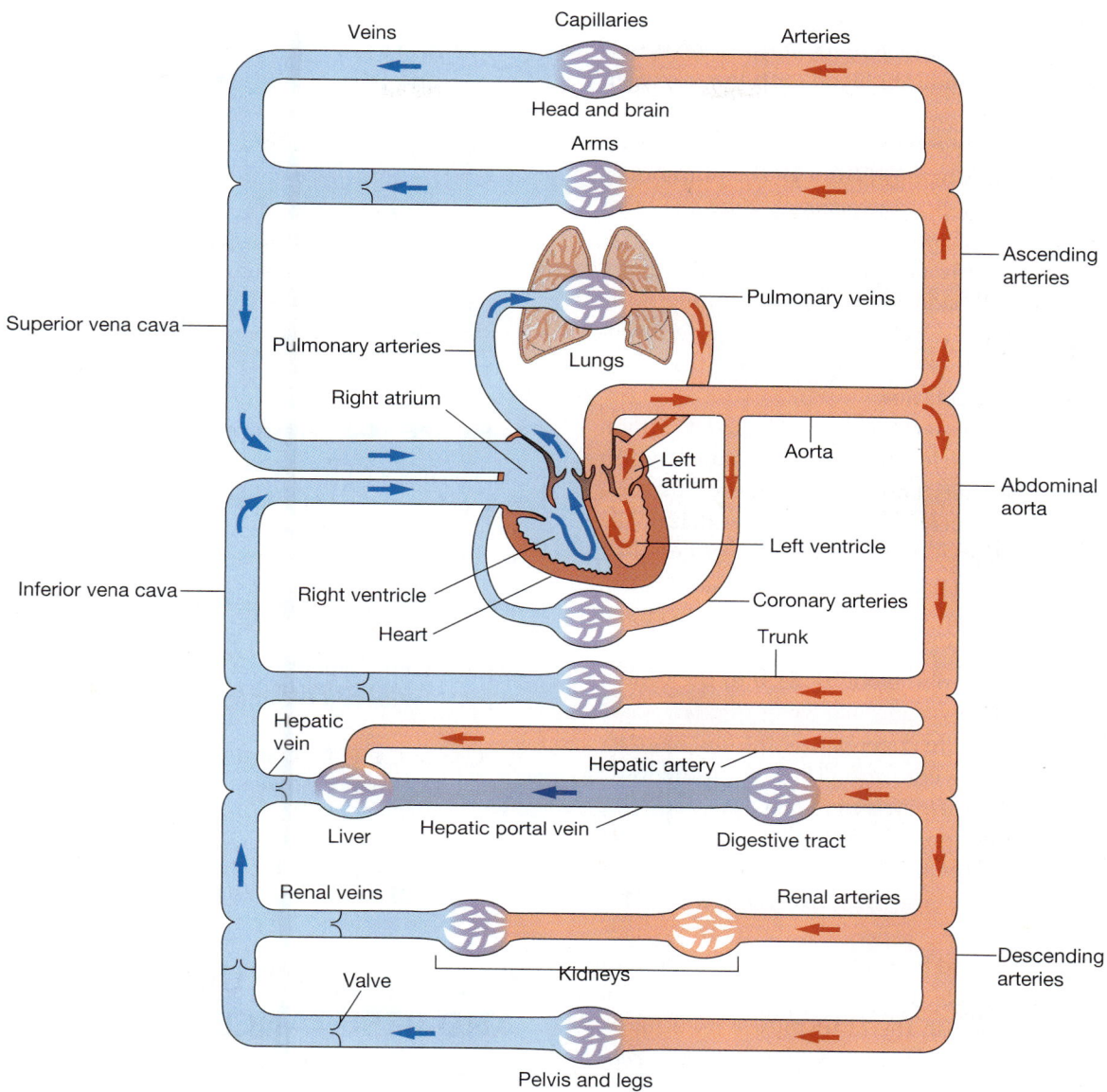

Veins   Capillaries   Arteries

Head and brain

Arms

Superior vena cava

Pulmonary arteries

Right atrium

Pulmonary veins

Lungs

Left atrium

Aorta

Ascending arteries

Abdominal aorta

Left ventricle

Inferior vena cava

Right ventricle

Heart

Coronary arteries

Trunk

Hepatic vein

Hepatic artery

Liver   Hepatic portal vein   Digestive tract

Renal veins   Renal arteries

Valve   Kidneys   Descending arteries

Pelvis and legs

■ **FIGURE 14-1** *Overview of cardiovascular system anatomy*

As blood moves through the cardiovascular system, a system of valves in the heart and veins ensures that the blood flows in one direction only. Like the turnstiles at an amusement park, the valves keep the blood from reversing its direction of flow. Figure 14-1 ■ is a schematic diagram that shows these components and the route that blood follows through the body. Notice in this illustration, as well as in most other diagrams of the heart, that the right side of the heart is on the left side of the page; that is, the heart is labeled as if you were viewing the heart of a person facing you.

The heart is divided by a central wall, or **septum**, into left and right halves. Each half functions as an independent pump that consists of an **atrium** [*atrium*, central room; plural *atria*] and a **ventricle** [*ventriculus*, belly]. The atrium receives blood returning to the heart from the blood vessels; the ventricle pumps

blood out into the blood vessels. The right side of the heart receives blood from the tissues and sends it to the lungs for oxygenation. The left side of the heart receives newly oxygenated blood from the lungs and pumps it to tissues throughout the body.

Starting in the right atrium in Figure 14-1, trace the path taken by blood as it flows through the cardiovascular system. Note that blood in the right side of the heart is colored blue. This is a convention used to show blood from which the tissues have extracted oxygen. Although this blood is often described as *deoxygenated*, it is not completely devoid of oxygen. It simply has less oxygen than blood going from the lungs to the tissues.

In living people, well-oxygenated blood is bright red, and low-oxygen blood is a darker red. Under some conditions, low-oxygen blood can impart a bluish color to certain areas of the skin, such as around the mouth and under the fingernails. This condition, known as *cyanosis* [*kyanos,* dark blue], is the reason blue is used in drawings to indicate blood with lower oxygen content.

From the right atrium, blood flows into the right ventricle of the heart. From there it is pumped through the **pulmonary arteries** to the lungs, where it is oxygenated. Note the color change from blue to red in Figure 14-1, indicating higher oxygen content after the blood leaves the lungs. From the lungs, blood travels to the left side of the heart through the **pulmonary veins**. The blood vessels that go from the right ventricle to the lungs and back to the left atrium are known collectively as the **pulmonary circulation**.

Blood from the lungs enters the heart at the left atrium and passes into the left ventricle. Blood pumped out of the left ventricle enters the large artery known as the **aorta**. The aorta branches into a series of smaller and smaller arteries that finally lead into networks of capillaries. Notice at the top of Figure 14-1 the color change from red to blue as the blood passes through the capillaries, indicating that oxygen has left the blood and diffused into the tissues.

After leaving the capillaries, blood flows into the venous side of the circulation, moving from small veins into larger and larger veins. The veins from the upper part of the body join together to form the **superior vena cava**. Those from the lower part of the body form the **inferior vena cava**. The two *venae cavae* empty into the right atrium. The blood vessels that carry blood from the left side of the heart to the tissues and back to the right side of the heart are collectively known as the **systemic circulation**.

Return to Figure 14-1 and follow the divisions of the aorta after it leaves the left ventricle. The first branch represents the *coronary arteries,* which nourish the heart muscle itself. Blood from these arteries flows into capillaries, then into the *coronary veins,* which empty directly into the right atrium at the *coronary sinus.* Ascending branches of the aorta go to the arms, head, and brain. The abdominal aorta supplies blood to the trunk, the legs, and the internal organs such as the kidneys (*renal arteries*), liver (*hepatic artery*), and digestive tract.

Notice two special arrangements of the circulation. One is the blood supply to the digestive tract and liver. Both regions receive well-oxygenated blood through their own arteries, but, in addition, blood leaving the digestive tract goes directly to the liver by means of the *hepatic portal vein.* The liver is an important site for nutrient processing and plays a major role in the detoxification of foreign substances. Most nutrients absorbed in the intestine are routed directly to the liver, allowing that organ to process material before it is released into the general circulation. The two capillary beds of the digestive tract and liver, joined by the hepatic portal vein, are examples of a *portal system.*

A second portal system occurs in the kidneys, where two capillary beds are connected in series. You will learn more

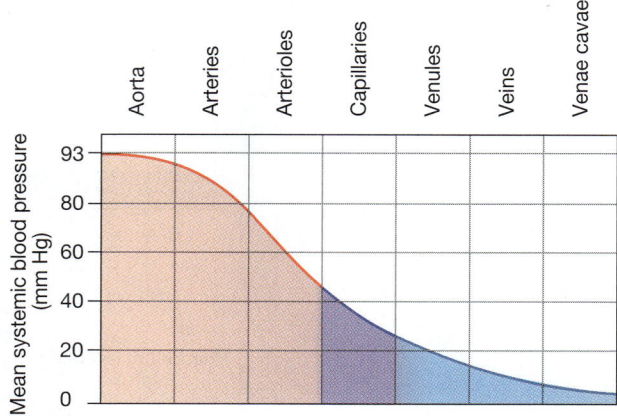

■ FIGURE 14-2 *Pressure gradient in the blood vessels*

The mean blood pressure of the systemic circulation ranges from a high of 93 mm Hg (millimeters of mercury) in the aorta to a low of a few mm Hg in the venae cavae.

about this special vascular arrangement in Chapter 19. A third portal system, discussed earlier but not shown here, is the hypothalamic-hypophyseal portal system, which connects the hypothalamus and the anterior pituitary [🔄 p. 229].

✔ CONCEPT CHECK

1. A cardiovascular system has what three major components?

2. What is the difference between (a) the pulmonary and systemic circulations, (b) an artery and a vein, (c) an atrium and a ventricle?

Answers: p. 498

PRESSURE, VOLUME, FLOW, AND RESISTANCE

If you ask people why blood flows through the cardiovascular system, many of them respond, "So that oxygen and nutrients can get to all parts of the body." Although this is true, it is a teleological answer, one that describes the purpose of blood flow. In physiology, we are also concerned with how blood flows—in other words, with the mechanisms or forces that create blood flow.

A simple mechanistic answer to "Why does blood flow?" is that liquids and gases flow down **pressure gradients** (ΔP) from regions of higher pressure to regions of lower pressure. Therefore, blood can flow in the cardiovascular system only if one region develops higher pressure than other regions.

In humans, high pressure is created in the chambers of the heart when it contracts. Blood flows out of the heart (the region of highest pressure) into the closed loop of blood vessels (a region of lower pressure). As blood moves through the system, pressure is lost because of friction between the fluid and the blood vessel walls. Consequently, pressure falls continuously as blood moves farther from the heart (Fig. 14-2 ■). The highest

(a)

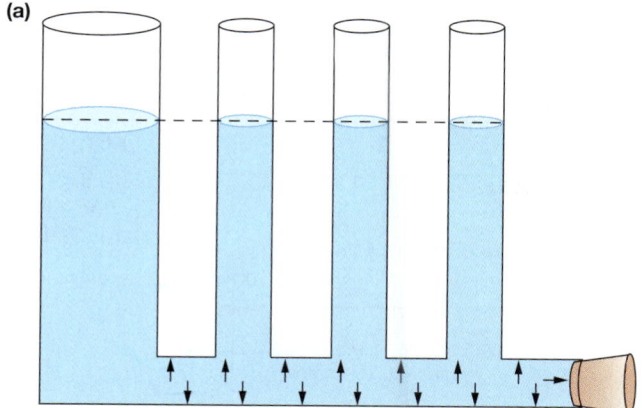

Hydrostatic pressure is the pressure exerted on the walls of the container by the fluid within the container. Hydrostatic pressure is proportional to the height of the water column.

(b)

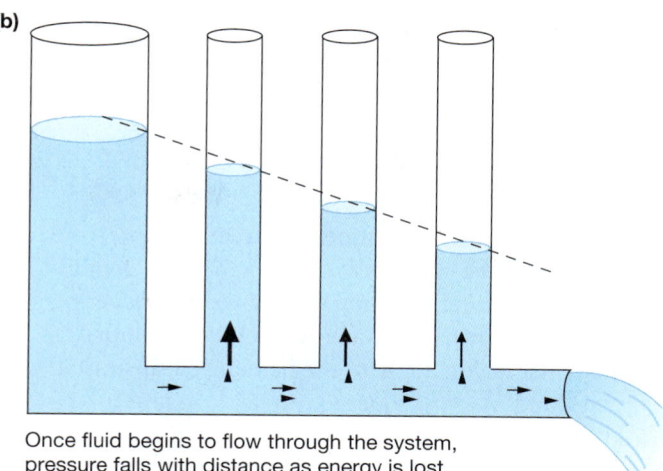

Once fluid begins to flow through the system, pressure falls with distance as energy is lost because of friction. This is the situation in the cardiovascular system.

■ **FIGURE 14-3** *Pressure differences in static and flowing fluids*

pressure in the vessels of the cardiovascular system is found in the aorta and systemic arteries as they receive blood from the left ventricle. The lowest pressure is in the venae cavae, just before they empty into the right atrium.

In this section we review the laws of physics that explain the interaction of pressure, volume, flow, and resistance in the cardiovascular system. Many of these principles apply broadly to the flow of all types of liquids and gases, including the flow of air in the respiratory system. However, in this chapter we will focus on the flow of blood and its relevance to the function of the heart.

The Pressure of Fluid in Motion Decreases over Distance

Pressure in a fluid is the force exerted by the fluid on its container. If the fluid is not moving, the pressure it exerts is called **hydrostatic pressure** (Fig. 14-3a ■), and force is exerted equally in all directions. For example, a column of fluid in a tube exerts hydrostatic pressure on the floor and sides of the tube. In the heart and blood vessels, pressure is commonly measured in millimeters of mercury (mm Hg).* One millimeter of mercury is equivalent to the hydrostatic pressure exerted by a 1-mm-high column of mercury on an area of 1 cm^2.

In a system in which fluid is flowing, pressure falls over distance as energy is lost because of friction (Fig. 14-3b). In addition, the pressure exerted by moving fluid has two components: a dynamic, flowing component that represents the kinetic energy of the system, and a lateral component that represents the hydrostatic pressure (potential energy) exerted on the walls of the system. Pressure within our cardiovascular system is usually called hydrostatic pressure even though it is a

---

*Some physiological literature reports pressures in torr (1 torr = 1 mm Hg) or in centimeters of water: 1 cm H_2O = 0.74 mm Hg.

When people speak of a "heart attack," they are referring to a clot that stops the blood supply to part of the heart, creating a condition known as *ischemia* [*ischien,* to suppress + *-emia,* blood]. In medical terms, a heart attack is called a *myocardial infarction,* referring to an area of heart muscle that is dying because of a lack of blood supply. The clot in Walter's coronary artery had restricted blood flow to his heart, and its cells were beginning to die from lack of oxygen. When someone has a heart attack, immediate medical intervention is critical. While waiting for the ambulance, the medic gave Walter oxygen, hooked him to a heart monitor, and started an intravenous (IV) injection of normal (isotonic) saline. With an intravenous injection line in place, other drugs could be given rapidly if Walter's condition should suddenly worsen.

Question 1:
Why did the medic give Walter oxygen?

Question 2:
What effect would the injection of isotonic saline have on Walter's extracellular fluid volume? On his intracellular fluid volume? On his total body osmolarity? [Hint: 🔁 p. 157]

14

457 **461** 471 479 484 489 494

system in which fluid is in motion. Some textbooks are beginning to replace the term *hydrostatic pressure* with the term *hydraulic pressure*. Hydraulics is the study of fluid in motion.

Pressure Changes in Liquids Without a Change in Volume

If the walls of a fluid-filled container contract, the pressure exerted on the fluid in the container increases. You can demonstrate this principle by filling a balloon with water and squeezing the water balloon in your hand. Water is minimally compressible, and so the pressure you apply to the balloon is transmitted throughout the fluid. As you squeeze, higher pressure in the fluid causes parts of the balloon to bulge. If the pressure becomes high enough, the stress on the balloon will cause it to pop. The water volume inside the balloon did not change, but the pressure in the fluid increased.

In the human heart, contraction of the blood-filled ventricles is similar to squeezing a water balloon: pressure created by the contracting muscle is transferred to the blood. This high-pressure blood then flows out of the ventricle and into the blood vessels, displacing lower-pressure blood already in the vessels. The pressure created in the ventricles is called the **driving pressure** because it is the force that drives blood through the blood vessels.

When the walls of a fluid-filled container expand, the pressure exerted on the fluid decreases. Thus, when the heart relaxes and expands, pressure in the fluid-filled chambers falls.

Pressure changes can also take place in the blood vessels. If blood vessels dilate, blood pressure inside them falls. If blood vessels constrict, blood pressure increases. Volume changes of the blood vessels and heart are major factors that influence blood pressure in the cardiovascular system.

Blood Flows from an Area of Higher Pressure to One of Lower Pressure

As stated earlier, blood flow through the cardiovascular system requires a pressure gradient. This pressure gradient is analogous to the difference in pressure between two ends of a tube through which fluid flows (Fig. 14-4a ■). Flow through the tube is directly proportional to ($\propto$) the pressure gradient (ΔP):

$$\text{Flow} \propto \Delta P \qquad (1)$$

where $\Delta P = P_1 - P_2$. This relationship says that the higher the pressure gradient, the greater the fluid flow.

A pressure gradient is not the same thing as the absolute pressure in the system. For example, the tube in Figure 14-4b has an absolute pressure of 100 mm Hg at each end. However, because there is no pressure gradient between the two ends of the tube, there is no flow through the tube.

On the other hand, two identical tubes can have very different absolute pressures but the same flow. The top tube in Figure 14-4c has a hydrostatic pressure of 100 mm Hg at one

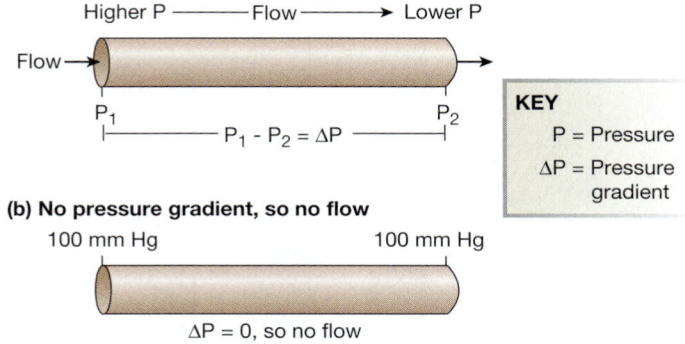

(a) Fluid flows only if there is a positive pressure gradient.

Higher P ——— Flow ——→ Lower P

Flow→

P_1 P_2

$P_1 - P_2 = \Delta P$

KEY
P = Pressure
ΔP = Pressure gradient

(b) No pressure gradient, so no flow

100 mm Hg   100 mm Hg

$\Delta P = 0$, so no flow

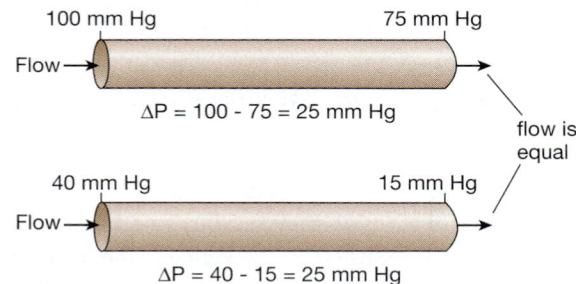

(c) Flow depends on ΔP, not absolute P.

100 mm Hg   75 mm Hg

Flow→

$\Delta P = 100 - 75 = 25$ mm Hg

flow is equal

40 mm Hg   15 mm Hg

Flow→

$\Delta P = 40 - 15 = 25$ mm Hg

■ **FIGURE 14-4** *Fluid flow through a tube*

end and 75 mm Hg at the other end, which means that the pressure gradient between the ends of the tube equals 25 mm Hg. The identical bottom tube has a hydrostatic pressure of 40 mm Hg at one end and 15 mm Hg at the other end. This tube has lower absolute pressure all along its length but the same pressure gradient as the top tube—25 mm Hg. Because the pressure difference in the two tubes is identical, the fluid flow through the tubes is the same.

Resistance Opposes Flow

In an ideal system, a substance in motion would remain in motion. However, no system is ideal because all movement creates friction. Just as a ball rolled across the ground loses energy to friction, blood flowing through blood vessels encounters friction from the walls of the vessels and from cells within the blood rubbing against one another as they flow.

The tendency of the cardiovascular system to oppose blood flow is called the system's **resistance** to flow. Resistance (R) is a term that most of us understand from everyday life. We speak of people being resistant to change or taking the path of least resistance. This concept translates well to the cardiovascular system because blood flow also takes the path of least resistance. An increase in the resistance of a blood vessel results in a decrease in the flow through that vessel. We can express that relationship as

$$\text{Flow} \propto 1/R \qquad (2)$$

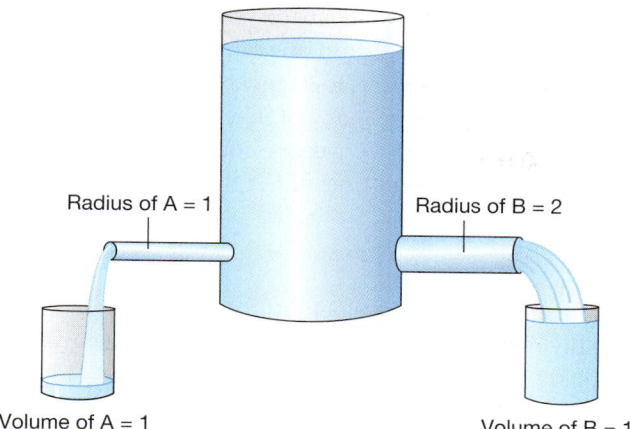

Resistance $\propto \dfrac{1}{radius^4}$		Flow $\propto \dfrac{1}{resistance}$	
Tube A	Tube B	Tube A	Tube B
$R \propto \dfrac{1}{1^4}$	$R \propto \dfrac{1}{2^4}$	Flow $\propto \dfrac{1}{1}$	Flow $\propto \dfrac{1}{\frac{1}{16}}$
$R \propto 1$	$R \propto \dfrac{1}{16}$	Flow $\propto 1$	Flow $\propto 16$

■ **FIGURE 14-5** *The role of radius in determining resistance to flow*

This expression says that flow is inversely proportional to resistance: if resistance increases, flow decreases; if resistance decreases, flow increases.

What parameters determine resistance? For fluid flowing through a tube, resistance is influenced by three components: the radius of the tube (r), the length of the tube (L), and the viscosity (thickness) of the fluid (η, the Greek letter eta). The following equation, derived by the French physician Jean Leonard Marie Poiseuille and known as **Poiseuille's law**, shows the relationship of these factors:

$$R = 8L\eta/\pi r^4 \qquad (3)$$

Because the value of $8/\pi$ is a constant, this factor can be removed from the equation, and the relationship can be rewritten as

$$R \propto L\eta/r^4 \qquad (4)$$

This expression says that (1) the resistance to fluid flow offered by a tube increases as the length of the tube increases, (2) resistance increases as the viscosity of the fluid increases, but (3) resistance decreases as the tube's radius increases.

To remember these relationships, think of drinking through a straw. You do not need to suck as hard on a short straw as on a long one (the resistance offered by the straw increases with length). Drinking water through a straw is easier than drinking a thick milkshake (resistance increases with viscosity). And drinking the milkshake through a big fat straw is much easier than through a skinny cocktail straw (resistance increases as radius decreases).

How significant are tube length, fluid viscosity, and tube radius to blood flow in a normal individual? The length of the systemic circulation is determined by the anatomy of the system and is essentially constant. Blood viscosity is determined by the ratio of red blood cells to plasma and by how much protein is in the plasma. Normally, viscosity is constant, and small changes in either length or viscosity have little effect on resistance. This leaves changes in the radius of the blood vessels as the main variable that affects resistance in the systemic circulation.

Let's return to the example of the straw and the milkshake to illustrate how changes in radius affect resistance. If we assume that the length of the straw and the viscosity of the milkshake do not change, this system is similar to the cardiovascular system—the radius of the tube has the greatest effect on resistance. If we consider only resistance (R) and radius (r) from equation 4, the relationship between resistance and radius can be expressed as

$$R \propto 1/r^4 \qquad (5)$$

If the skinny straw has a radius of 1, its resistance is proportional to $1/1^4$, or 1. If the fat straw has a radius of 2, the resistance it offers is $1/2^4$, or 1/16th, that of the skinny straw (Fig. 14-5 ■). Because flow is inversely proportional to resistance, flow increases 16-fold when the radius doubles.

As you can see from this example, a small change in the radius of a tube has a large effect on the flow of a fluid through that tube. Thus a small change in the radius of a blood vessel will have a large effect on the resistance to blood flow offered by that vessel. A decrease in blood vessel diameter is known as **vasoconstriction** [*vas*, a vessel or duct]. An increase in blood vessel diameter is called **vasodilation**. Vasoconstriction decreases blood flow through a vessel; vasodilation increases blood flow through a vessel.

In summary, by combining equations 1 and 2, we get the equation

$$\text{Flow} \propto \Delta P/R \qquad (6)$$

which, translated into words, says that the flow of blood in the cardiovascular system is directly proportional to the pressure gradient in the system, and inversely proportional to the resistance of the system to flow. If the pressure gradient remains constant, then flow will vary inversely with resistance.

14

CONCEPT CHECK

3. Which is more important for determining flow through a tube: absolute pressure or the pressure gradient?

4. The following two identical tubes have the pressures shown at each end. Which tube has the greater flow? Defend your choice.

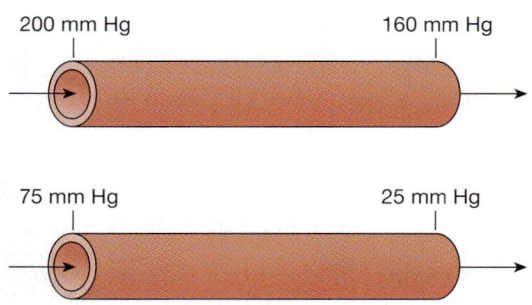

5. The following four tubes have the same driving pressure. Which tube has the highest flow? Which has the lowest flow? Defend your choices.

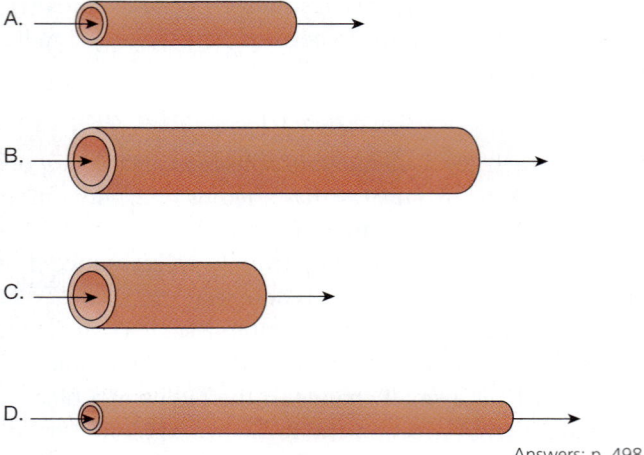

Answers: p. 498

Velocity of Flow Depends on the Flow Rate and the Cross-Sectional Area

The word *flow* is sometimes used imprecisely in cardiovascular physiology, leading to confusion. Flow usually means **flow rate**, the volume of blood that passes a given point in the system per unit time. In the circulation, flow is expressed in either liters per minute (L/min) or milliliters per minute (mL/min). For instance, blood flow through the aorta of a 70-kg man at rest is about 5 L/min.

Flow rate should not be confused with **velocity of flow**, the distance a fixed volume of blood travels in a given period of time. Velocity of flow is a measure of *how fast* blood flows past a point. Flow rate measures *how much* (volume) blood flows past a point in a given period of time. For example, look through the

open door at the hallway outside your classroom. The number of people passing the door in one minute is the flow rate of people through the hallway. How quickly those people are walking past the door is their velocity of flow.

The relationship between velocity of flow (v), flow rate (Q), and cross-sectional area of the tube (A) is expressed by the equation

$$v = Q/A \qquad (7)$$

which shows that the velocity through a tube equals the flow rate divided by the tube's cross-sectional area. In a tube of fixed diameter (and thus fixed cross-sectional area), velocity of flow is directly related to flow rate. In a tube of variable diameter, if the flow rate is constant, velocity of flow varies inversely with the diameter. In other words, velocity is faster in narrow sections, and slower in wider sections.

Figure 14-6 ■ shows how the velocity of flow varies as the cross-sectional area of the tube changes. The vessel in the figure has two widths: narrow, with a cross-sectional area of 1 cm², and wide, with a cross-sectional area of 12 cm². The flow rate is identical in both parts of the vessel: 12 cm³ per minute.* This flow rate means that in one minute, 12 cm³ of fluid flows past point X in the narrow section, and 12 cm³ of fluid flows past point Y in the wide section.

But *how fast* does the fluid need to flow to accomplish that rate? According to equation 7, the velocity of flow at point X is 12 cm/min, but at point Y it is only 1 cm/min. Thus, fluid flows more rapidly through narrow sections of a tube than through wide sections.

To see this principle in action, watch a leaf as it floats down a stream. Where the stream is narrow, the leaf moves rapidly, carried by the fast velocity of the water. In sections where the stream widens into a pool, the velocity of the water decreases and the leaf meanders more slowly.

In this chapter and the next, we will apply the physics of fluid flow to the cardiovascular system. The heart generates pressure when it contracts and pumps blood into the arterial side of the circulation. Arteries act as a pressure reservoir during the heart's relaxation phase, maintaining the *mean arterial pressure* (MAP) that is the primary driving force for blood flow. Mean arterial pressure is influenced by two parameters: *cardiac output* (the volume of blood the heart pumps per minute) and *peripheral resistance* (the resistance of the blood vessels to blood flow through them):

Mean arterial pressure ∝ cardiac output ×
peripheral resistance

Chapter 15 discusses peripheral resistance and blood flow. The remainder of this chapter examines heart function and the parameters that influence cardiac output.

---

*1 cm³ = 1 cubic centimeter (cc) = 1 mL.

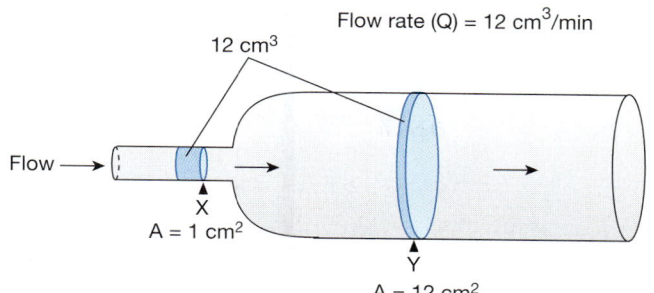

Flow rate (Q) = 12 cm³/min

12 cm³

Flow →

X
A = 1 cm²

Y
A = 12 cm²

Velocity (v) = $\dfrac{\text{Flow rate (Q)}}{\text{Cross-sectional area (A)}}$	
At point X	**At point Y**
$v = \dfrac{12 \text{ cm}^3/\text{min}}{1 \text{ cm}^2}$	$v = \dfrac{12 \text{ cm}^3/\text{min}}{12 \text{ cm}^2}$
v = 12 cm/min	v = 1 cm/min

The narrower the vessel, the faster the velocity of flow.

■ **FIGURE 14-6** *Flow rate versus velocity of flow*

CONCEPT CHECK

6. Two canals in Amsterdam are identical in size, but the water flows faster through one than through the other. Which canal has the higher flow rate?

Answers: p. 498

CARDIAC MUSCLE AND THE HEART

To ancient civilizations, the heart was more than a pump—it was *the seat of the mind*. When ancient Egyptians mummified their dead, they removed most of the viscera but left the heart in place so that the gods could weigh it as an indicator of the owner's worthiness. Aristotle characterized the heart as the most important organ of the body, as well as *the seat of intelligence*. We can still find evidence of these ancient beliefs in modern expressions such as "heartfelt emotions." The link between the heart and mind is one that is still explored today.

The heart is the workhorse of the body, a muscle that contracts continually, resting only in the milliseconds-long pause between beats. By one estimate, in one minute the heart performs work equivalent to lifting a 5-pound weight up 1 foot. The energy demands of this work require a continuous supply of nutrients and oxygen to the heart muscle.

The Heart Has Four Chambers

The heart is a muscular organ, about the size of a fist, that lies in the center of the *thoracic cavity* (see Anatomy Summary, Fig. 14-7a, b, d ■). The pointed *apex* of the heart angles down to the left side of the body, while the broader *base* lies just behind the breastbone, or *sternum*. Because we usually associate the word *base* with the bottom, remember that the base of a cone is the broad end, and the apex is the pointed end. The heart can be thought of as an inverted cone with apex down and base up. Within the thoracic cavity, the heart lies on the ventral side, sandwiched between the two lungs, with its apex resting on the diaphragm (Fig. 14-7b).

The heart is encased in a tough membranous sac, the **pericardium** [*peri*, around + *kardia*, heart] (Fig. 14-7d, e). A thin

layer of clear pericardial fluid inside the pericardium lubricates the external surface of the heart as it beats within the sac. Inflammation of the pericardium (*pericarditis*) may reduce this lubrication to the point that the heart rubs against the pericardium, creating a sound known as a *friction rub*.

The heart itself is composed mostly of cardiac muscle, or **myocardium** [*myo*, muscle + *kardia*, heart], covered by thin outer and inner layers of epithelium and connective tissue. Seen from the outside, the bulk of the heart is the thick muscular walls of the ventricles, the two lower chambers (Fig. 14-7f). The thinner-walled atria lie above the ventricles.

The major blood vessels all emerge from the base of the heart. The aorta and *pulmonary trunk* (artery) direct blood from the heart to the tissues and lungs, respectively. The venae cavae and pulmonary veins return blood to the heart (Table 14-2 ■). When the heart is viewed from the front (anterior view), as in Figure 14-7f, the pulmonary veins are hidden behind the other major blood vessels. Running across the surface of the ventricles are shallow grooves containing the **coronary arteries** and **coronary veins**, which supply blood to the heart muscle.

The relationship between the atria and ventricles can be seen in a cross-sectional view of the heart (Fig. 14-7g). As noted earlier, the left and right sides of the heart are separated by the interventricular septum, so that blood on one side does not mix with blood on the other side. Although blood flow in the left heart is separated from flow in the right heart, the two sides contract in a coordinated fashion. First the atria contract together, then the ventricles contract together.

Blood flows from veins into the atria and from there through one-way valves into the ventricles, the pumping chambers. Blood leaves the heart via the pulmonary trunk from the right ventricle and via the aorta from the left ventricle. A second set of valves guards the exits of the ventricles so that blood cannot flow back into the heart once it has been ejected.

Notice in Figure 14-7g that blood enters each ventricle at the top of the chamber but also leaves at the top. This is because during development, the tubular embryonic heart twists back on itself (Fig. 14-8b ■). This twisting puts the arteries (through which blood leaves) close to the top of the ventricles. Functionally, this

14

ANATOMY SUMMARY
THE HEART

ANATOMY OF THE THORACIC CAVITY

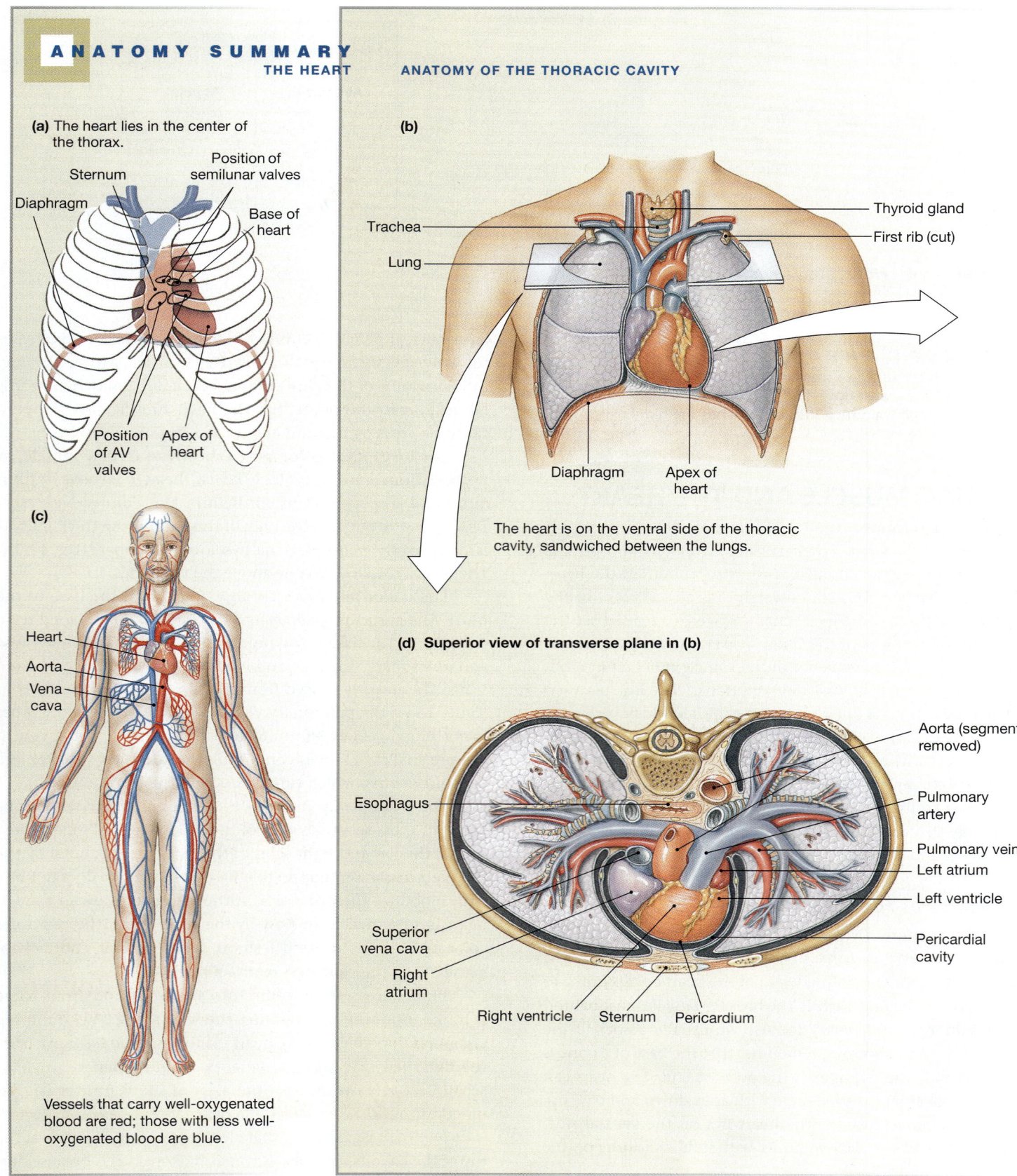

(a) The heart lies in the center of the thorax.

- Sternum
- Position of semilunar valves
- Diaphragm
- Base of heart
- Position of AV valves
- Apex of heart

(b)

- Trachea
- Lung
- Thyroid gland
- First rib (cut)
- Diaphragm
- Apex of heart

The heart is on the ventral side of the thoracic cavity, sandwiched between the lungs.

(c)

- Heart
- Aorta
- Vena cava

Vessels that carry well-oxygenated blood are red; those with less well-oxygenated blood are blue.

(d) Superior view of transverse plane in (b)

- Esophagus
- Aorta (segment removed)
- Pulmonary artery
- Pulmonary vein
- Left atrium
- Left ventricle
- Pericardial cavity
- Superior vena cava
- Right atrium
- Right ventricle
- Sternum
- Pericardium

■ FIGURE 14-7

STRUCTURE OF THE HEART

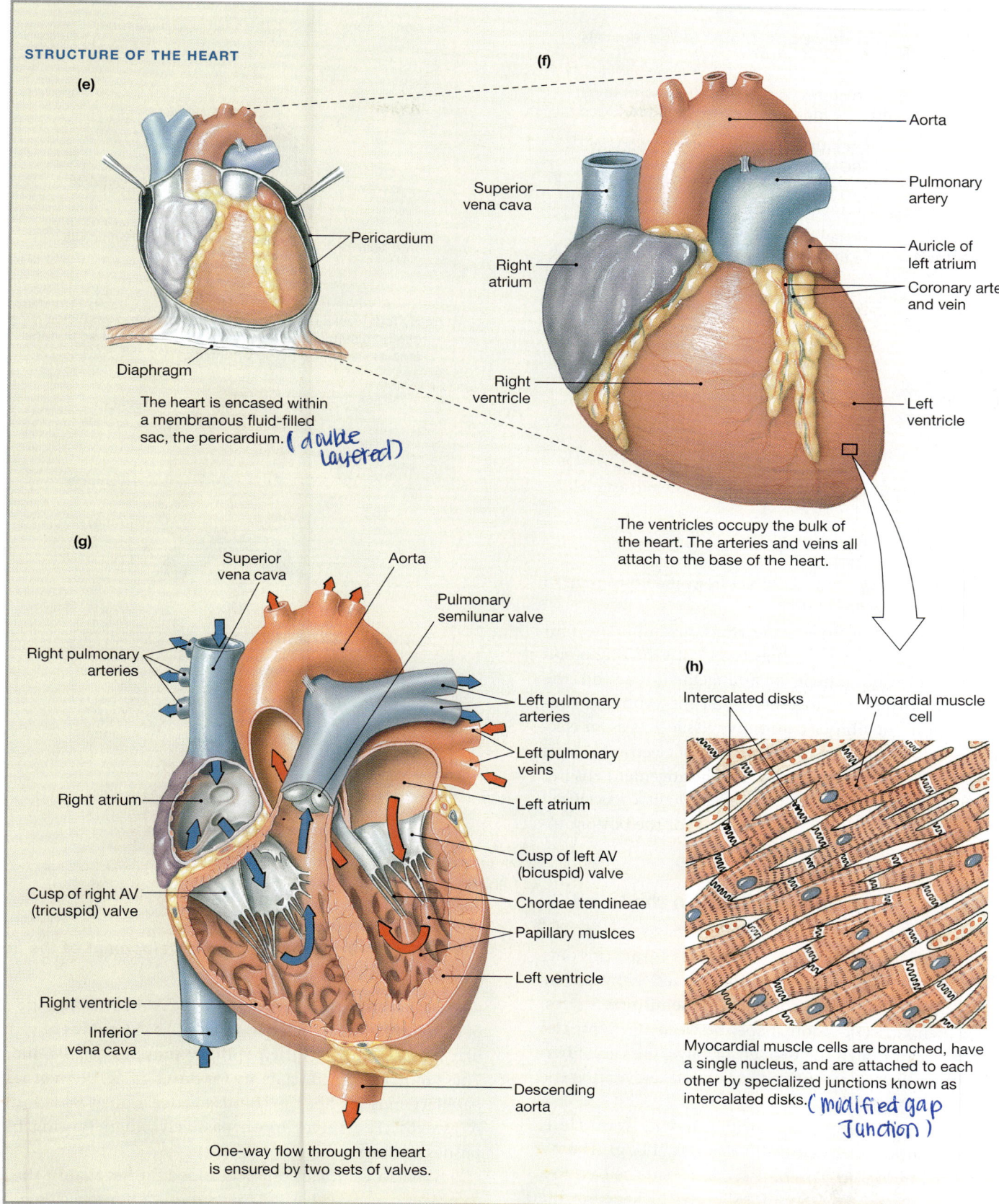

(e)

Pericardium

Diaphragm

The heart is encased within a membranous fluid-filled sac, the pericardium. (double layered)

(f)

Aorta

Superior vena cava

Pulmonary artery

Right atrium

Auricle of left atrium

Coronary artery and vein

Right ventricle

Left ventricle

The ventricles occupy the bulk of the heart. The arteries and veins all attach to the base of the heart.

(g)

Superior vena cava

Aorta

Pulmonary semilunar valve

Right pulmonary arteries

Left pulmonary arteries

Left pulmonary veins

Left atrium

Right atrium

Cusp of left AV (bicuspid) valve

Chordae tendineae

Cusp of right AV (tricuspid) valve

Papillary muslces

Left ventricle

Right ventricle

Inferior vena cava

Descending aorta

One-way flow through the heart is ensured by two sets of valves.

(h)

Intercalated disks

Myocardial muscle cell

Myocardial muscle cells are branched, have a single nucleus, and are attached to each other by specialized junctions known as intercalated disks. (modified gap junction)

14

■ **FIGURE 14-7** *(continued)*

TABLE 14-2	The Heart and Major Blood Vessels	
Blue type indicates structures containing blood with lower oxygen content; red type indicates well-oxygenated blood.		
	RECEIVES BLOOD FROM	**SENDS BLOOD TO**
Heart		
Right atrium	Venae cavae	Right ventricle
Right ventricle	Right atrium	Lungs
Left atrium	Pulmonary veins	Left ventricle
Left ventricle	Left atrium	Body except for lungs
Vessels		
Venae cavae	Systemic veins	Right atrium
Pulmonary trunk (artery)	Right ventricle	Lungs
Pulmonary vein	Veins of the lungs	Left atrium
Aorta	Left ventricle	Systemic arteries

means that the ventricles must contract from the bottom up so that blood is squeezed out of the top.

Four fibrous connective tissue rings surround the four heart valves (Fig. 14-7a). These rings form both the origin and insertion for the cardiac muscle, an arrangement that pulls the apex and base of the heart together when the ventricles contract. In addition, the fibrous connective tissue acts as an electrical insulator, blocking most transmission of electrical signals between the atria and the ventricles. This arrangement ensures that the electrical signals can be directed through a specialized conduction system to the apex of the heart for the bottom-to-top contraction.

Heart Valves Ensure One-Way Flow in the Heart

As the arrows in Figure 14-7g indicate, blood flows through the heart in one direction. Two sets of heart valves ensure this one-way flow: one set (the **atrioventricular valves**) between the atria and ventricles, and the second set (the **semilunar valves**, named for their crescent-moon shape) between the ventricles and the arteries. Although the two sets of valves are very different in structure, they serve the same function: preventing the backward flow of blood.

The opening between each atrium and its ventricle is guarded by an atrioventricular (AV) valve (Fig. 14-7g). The AV valve is formed from thin flaps of tissue joined at the base to a connective tissue ring. The flaps are slightly thickened at the edge and connect on the ventricular side to collagenous tendons, the **chordae tendineae** (Fig. 14-9b, d ■). Most of the

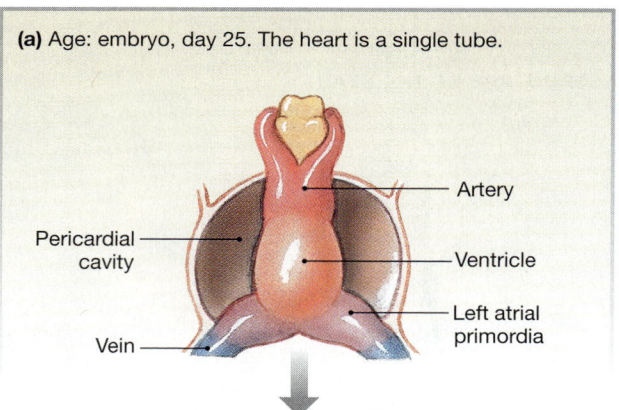

(a) Age: embryo, day 25. The heart is a single tube.

Artery
Pericardial cavity
Ventricle
Left atrial primordia
Vein

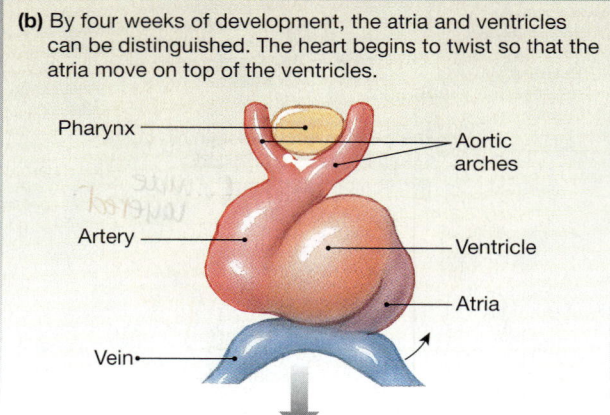

(b) By four weeks of development, the atria and ventricles can be distinguished. The heart begins to twist so that the atria move on top of the ventricles.

Pharynx
Aortic arches
Artery
Ventricle
Atria
Vein

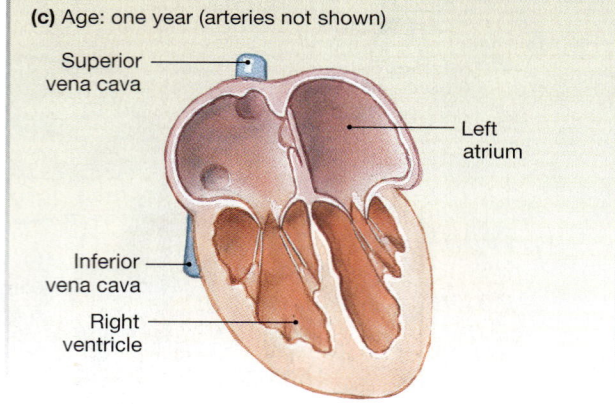

(c) Age: one year (arteries not shown)

Superior vena cava
Left atrium
Inferior vena cava
Right ventricle

■ **FIGURE 14-8** *Embryological development of the heart*

chordae fasten to the edges of the valve flaps. The opposite ends of the chordae are tethered to moundlike extensions of ventricular muscle known as the **papillary muscles** [*papilla,* nipple]. These muscles provide stability for the chordae, but neither the papillary muscles nor the chordae actively open and close the AV valves. The valves move passively when flowing blood pushes on them.

When a ventricle contracts, blood pushes against the bottom side of its AV valve and forces it upward into a closed position (Fig. 14-9b). The chordae tendineae prevent the valve from being pushed back into the atrium, just as the struts on an

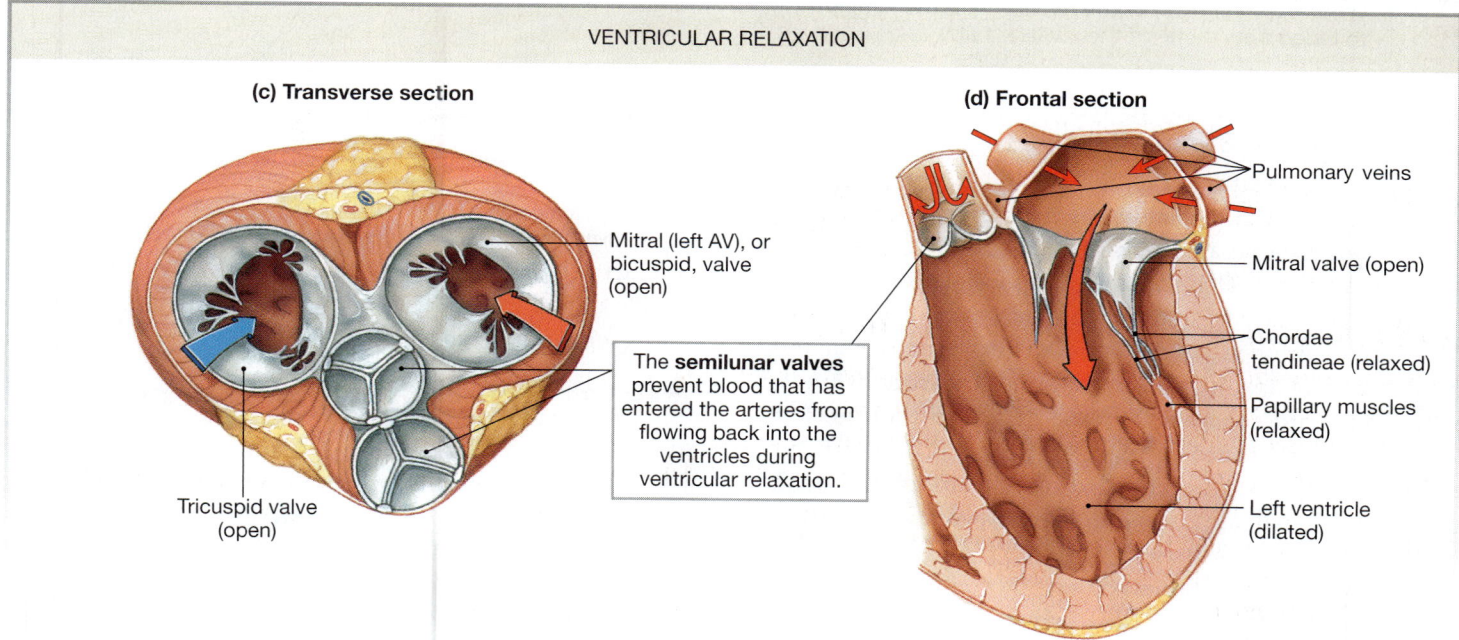

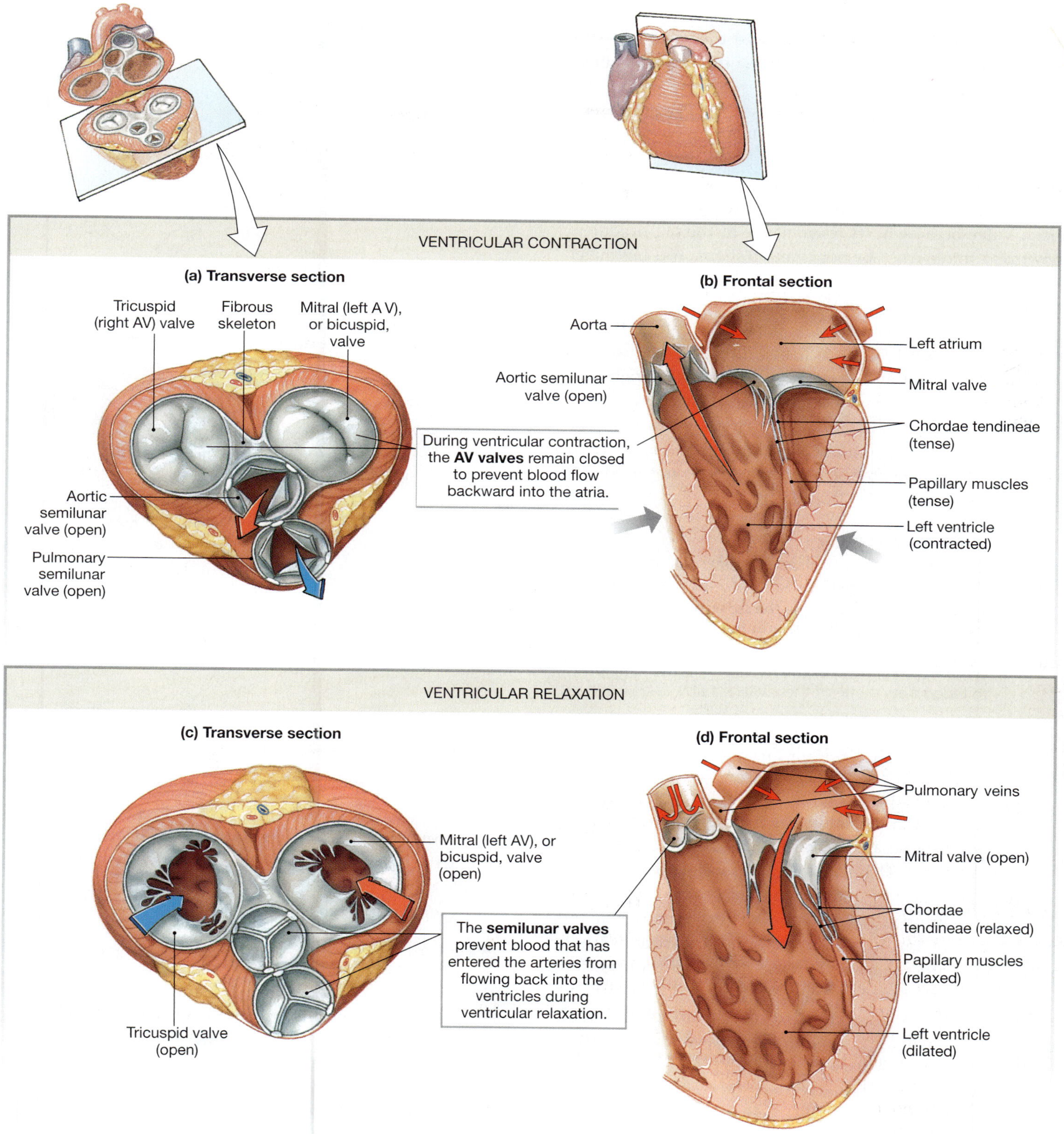

VENTRICULAR CONTRACTION

(a) Transverse section

Tricuspid (right AV) valve
Fibrous skeleton
Mitral (left AV), or bicuspid, valve
Aortic semilunar valve (open)
Pulmonary semilunar valve (open)

During ventricular contraction, the **AV valves** remain closed to prevent blood flow backward into the atria.

(b) Frontal section

Aorta
Aortic semilunar valve (open)
Left atrium
Mitral valve
Chordae tendineae (tense)
Papillary muscles (tense)
Left ventricle (contracted)

VENTRICULAR RELAXATION

(c) Transverse section

Mitral (left AV), or bicuspid, valve (open)
Tricuspid valve (open)

The **semilunar valves** prevent blood that has entered the arteries from flowing back into the ventricles during ventricular relaxation.

(d) Frontal section

Pulmonary veins
Mitral valve (open)
Chordae tendineae (relaxed)
Papillary muscles (relaxed)
Left ventricle (dilated)

■ **FIGURE 14-9** *Heart valves*

Views **(a)** and **(c)** show the AV valves as viewed from the atria and the semilunar valves as viewed from inside the arteries.

14

umbrella keep the umbrella from turning inside out in a high wind. Occasionally, the chordae fail, and the valve is pushed back into the atrium during ventricular contraction, an abnormal condition known as *prolapse*.

The two AV valves are not identical. The valve that separates the right atrium and right ventricle has three flaps and is called the **tricuspid valve** [*cuspis*, point] (Fig. 14-9a). The valve between the left atrium and left ventricle has only two flaps and is called the **bicuspid valve**. The bicuspid is also called the **mitral valve** because of its resemblance to the tall headdress, known as a miter, worn by popes and bishops. You can match AV valves to the proper side of the heart by remembering that the Right Side has the Tricuspid (R-S-T).

The semilunar valves separate the ventricles from the major arteries. The **aortic valve** is between the left ventricle and the aorta, and the **pulmonary valve** lies between the right ventricle and the pulmonary trunk. Both sets of semilunar valves have three cuplike leaflets that snap closed when blood attempting to flow back into the ventricles fills them (Fig. 14-9c, d). Because of their shape, the semilunar valves do not need connective tendons as the AV valves do.

✔ CONCEPT CHECK

7. What prevents electrical signals from passing through the connective tissue in the heart?

8. Trace a drop of blood from the superior vena cava to the aorta, naming all structures the drop encounters along its route.

9. What is the function of the AV valves? What happens to blood flow if one of these valves fails?   Answers: p. 498

Cardiac Muscle Cells Contract Without Nervous Stimulation

The bulk of the heart is composed of cardiac muscle cells, or myocardium. Most cardiac muscle is contractile, but about 1% of the myocardial cells are specialized to generate action potentials spontaneously. These cells are responsible for a unique property of the heart: its ability to contract without any outside signal. As mentioned in the introduction to this chapter, records tell us of Spanish explorers in the New World witnessing human sacrifices in which hearts torn from the chests of living victims continued to beat for minutes. The heart can contract without a connection to other parts of the body because the signal for contraction is *myogenic*, originating within the heart muscle itself.

The signal for myocardial contraction comes not from the nervous system but from specialized myocardial cells known as **autorhythmic cells**. The autorhythmic cells are also called **pacemakers** because they set the rate of the heartbeat. Myocardial autorhythmic cells are anatomically distinct from contractile cells: autorhythmic cells are smaller and contain few contractile fibers. Because they do not have organized sarcomeres, autorhythmic cells do not contribute to the contractile force of the heart.

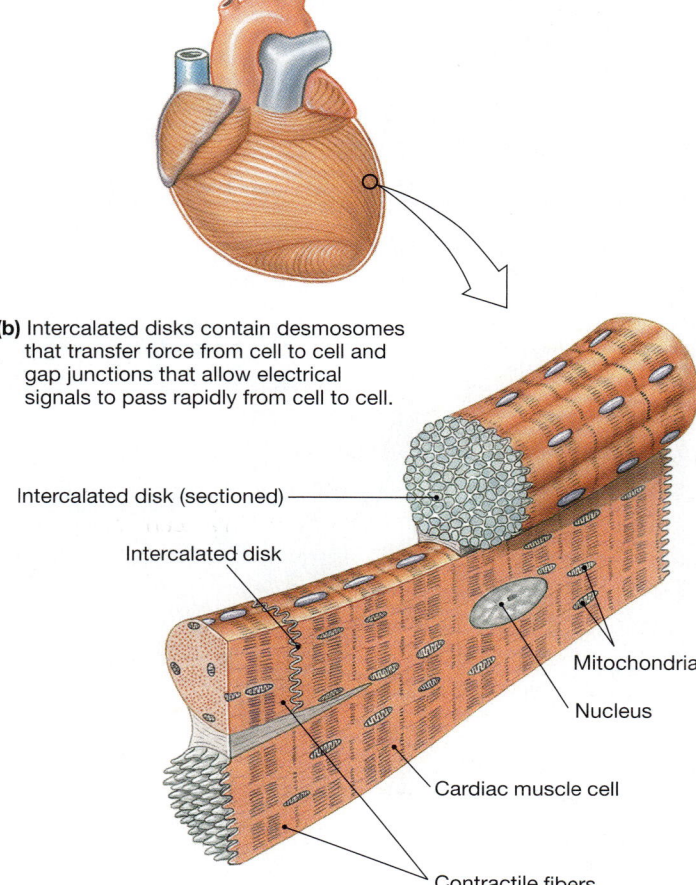

(a) The spiral arrangement of ventricular muscle allows ventricular contraction to squeeze the blood upward from the apex of the heart.

(b) Intercalated disks contain desmosomes that transfer force from cell to cell and gap junctions that allow electrical signals to pass rapidly from cell to cell.

Intercalated disk (sectioned)

Intercalated disk

Mitochondria

Nucleus

Cardiac muscle cell

Contractile fibers

■ **FIGURE 14-10**   *Cardiac muscle*

For an electron micrograph of gap junctions, see Figure 3-21c on p. 70.

Contractile cells are typical striated muscle, however, with contractile fibers organized into sarcomeres [🔁 p. 399]. Cardiac muscle differs in significant ways from skeletal muscle and shares some properties with smooth muscle:

1. Cardiac muscle fibers are much smaller than skeletal muscle fibers and usually have a single nucleus per fiber.

2. Individual cardiac muscle cells branch and join neighboring cells end-to-end to create a complex network (Fig. 14-10b ■). The cell junctions, known as **intercalated disks** [*inter-*, between + *calare*, to proclaim], consist of interdigitated membranes. Intercalated disks have two components: *desmosomes* [🔁 p. 71] and gap junctions [🔁 p. 175]. Desmosomes are strong connections that tie adjacent cells together, thereby allowing force created in one cell to be transferred to the adjacent cell.

3. *Gap junctions* in the intercalated disks electrically connect cardiac muscle cells to one another. Waves of depolarization thus can spread rapidly from cell to cell, allowing all

the heart muscle cells to contract almost simultaneously. In this respect, cardiac muscle resembles single-unit smooth muscle.

4. The t-tubules of myocardial cells are larger than those of skeletal muscle, and they branch inside the myocardial cells.

5. Myocardial sarcoplasmic reticulum is smaller than that of skeletal muscle, reflecting the fact that cardiac muscle depends in part on extracellular Ca^{2+} to initiate contraction. In this respect, cardiac muscle resembles smooth muscle.

6. Mitochondria occupy about one-third the cell volume of a cardiac contractile fiber, a reflection of the high energy demand of these cells. By one estimate, cardiac muscle consumes 70–80% of the oxygen delivered to it by the blood, more than twice the amount extracted by other cells in the body.

During periods of increased activity, the heart uses almost all the oxygen brought to it by the coronary arteries. Thus, the only way to get more oxygen to exercising heart muscle is to increase the blood flow. Reduced myocardial blood flow from narrowing of a coronary vessel by a clot or fatty deposit can damage or even kill myocardial cells.

See Table 12-3, p. 429, for a summary comparison of the three muscle types.

Cardiac EC Coupling Combines Features of Skeletal and Smooth Muscle

In Chapters 11 and 12 you learned how acetylcholine from a somatic motor neuron causes a skeletal muscle action potential to begin excitation-contraction coupling (EC coupling) [p. 403]. In cardiac muscle, an action potential also initiates EC coupling, but the action potential originates spontaneously in the heart's pacemaker cells and spreads into the contractile cells through gap junctions. (Neurotransmitters modulate the pacemaker rate, as you will learn later in this chapter.)

An action potential that enters a contractile cell moves across the sarcolemma and into the t-tubules, where it opens voltage-gated Ca^{2+} channels in the cell membrane. Ca^{2+} enters the cell and opens *ryanodine receptor-channels (RyR)* in the sarcoplasmic reticulum (Fig. 14-11 ■). Note that these ryanodine receptors are operated by Ca^{2+} binding, not by mechanical linkage as the RyR channels in skeletal muscle are.

The ryanodine receptors are Ca^{2+} channels, and opening them causes **Ca^{2+}-induced Ca^{2+} release**. Stored Ca^{2+} flows out of the sarcoplasmic reticulum and into the cytosol, creating a Ca^{2+} "spark" that can be seen using special biochemical methods [p. 188]. Multiple sparks from different RyR channels sum to create a Ca^{2+} signal.

Calcium released from the sarcoplasmic reticulum provides about 90% of the Ca^{2+} needed for muscle contraction. Calcium diffuses through the cytosol to the contractile elements, where the ions bind to troponin and initiate the cycle of crossbridge formation and movement. Contraction takes place

RUNNING PROBLEM

When Walter arrived at the University of Texas Southwestern Medical Center emergency room, one of the first tasks was to determine whether he had actually had a heart attack. Walter's vital signs (pulse and breathing rates, blood pressure, and temperature) were taken. He was given aspirin and heparin to decrease blood clotting, and nitroglycerin to dilate coronary blood vessels. A technician drew blood for enzyme assays to determine the level of cardiac creatine kinase in Walter's blood. When heart muscle cells die, they release various enzymes that serve as markers of a heart attack. A second tube of blood was sent for an assay of its troponin I level. Troponin I is a good indicator of heart damage following a heart attack.

Question 3:
A related form of creatine kinase is found in skeletal muscle. What are related forms of an enzyme called? [Hint: ⮂ p. 96]

Question 4:
What is troponin, and why would elevated blood levels of troponin indicate heart damage? [Hint: ⮂ p. 407]

by the same type of sliding filament movement that occurs in skeletal muscle [p. 403].

Relaxation in cardiac muscle is generally similar to that in skeletal muscle. As cytoplasmic Ca^{2+} concentrations decrease, Ca^{2+} unbinds from troponin, myosin releases actin, and the contractile filaments slide back to their relaxed position. As in skeletal muscle, Ca^{2+} is transported back into the sarcoplasmic reticulum with the help of a Ca^{2+}-ATPase. However, in cardiac muscle Ca^{2+} is also removed from the cell in exchange for Na^+ via a Na^+-Ca^{2+} antiport protein. Each Ca^{2+} moves out of the cell against its electrochemical gradient in exchange for 3 Na^+ entering the cell down their electrochemical gradient. Sodium that enters the cell during this transfer is removed by the Na^+-K^+-ATPase.

CONCEPT CHECK

10. If a myocardial contractile cell is placed in interstitial fluid and depolarized, the cell will contract. If Ca^{2+} is removed from the fluid surrounding the myocardial cell and the cell is depolarized, it will not contract. If the experiment is repeated with a skeletal muscle fiber, the skeletal muscle will contract when depolarized, whether or not Ca^{2+} is present in the surrounding fluid. What conclusion can you draw from the results of this experiment?

Answers: p. 498

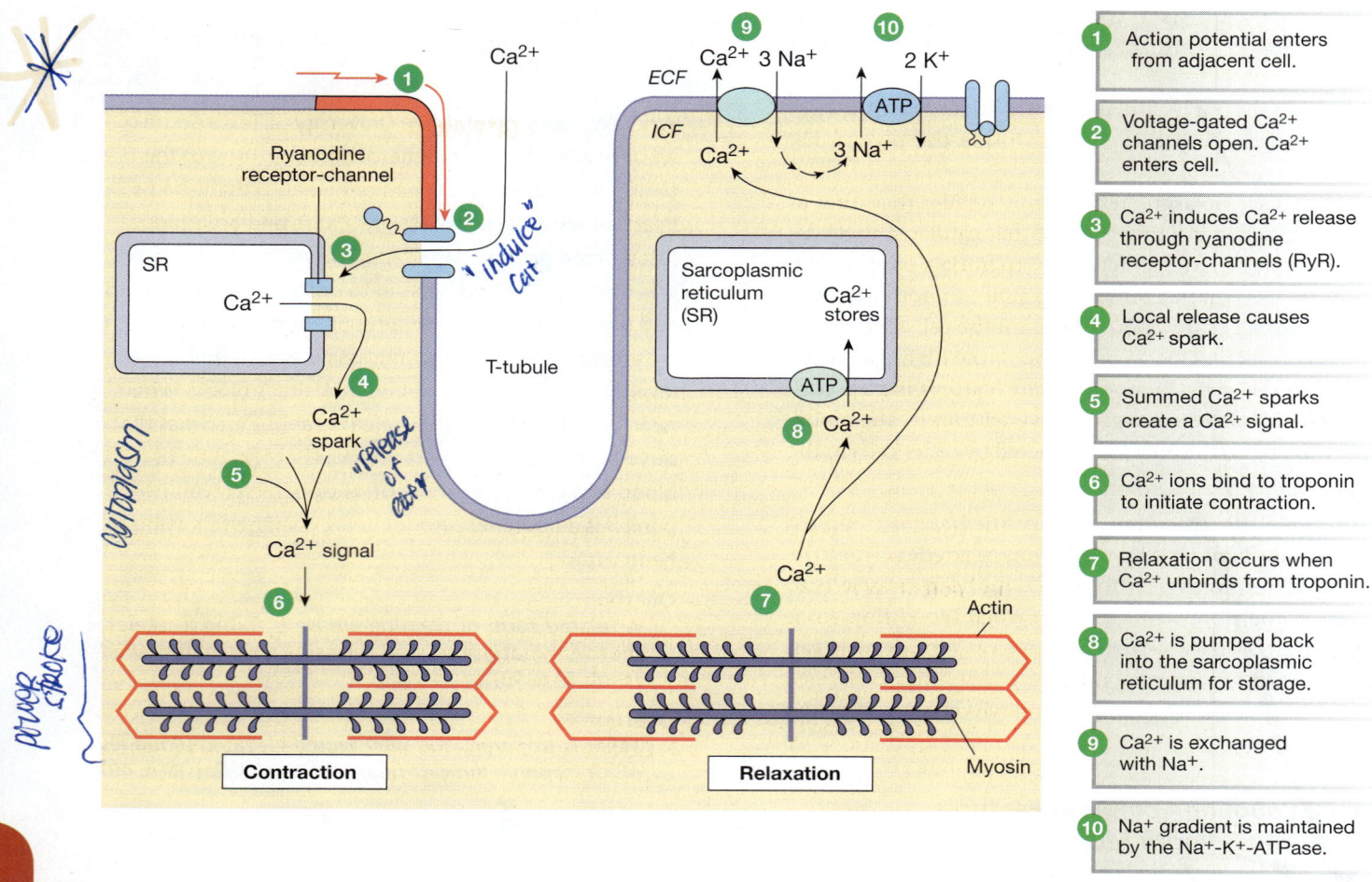

1. Action potential enters from adjacent cell.

2. Voltage-gated Ca^{2+} channels open. Ca^{2+} enters cell.

3. Ca^{2+} induces Ca^{2+} release through ryanodine receptor-channels (RyR).

4. Local release causes Ca^{2+} spark.

5. Summed Ca^{2+} sparks create a Ca^{2+} signal.

6. Ca^{2+} ions bind to troponin to initiate contraction.

7. Relaxation occurs when Ca^{2+} unbinds from troponin.

8. Ca^{2+} is pumped back into the sarcoplasmic reticulum for storage.

9. Ca^{2+} is exchanged with Na^+.

10. Na^+ gradient is maintained by the Na^+-K^+-ATPase.

■ **FIGURE 14-11** *Excitation-contraction coupling and relaxation in cardiac muscle*

Cardiac Muscle Contraction Can Be Graded

A key property of cardiac muscle cells is the ability of a single muscle fiber to execute *graded contractions,* in which the fiber varies the amount of force it generates. (Recall that in skeletal muscle, contraction in a single fiber is all-or-none at any given fiber length.) The force generated by cardiac muscle is proportional to the number of crossbridges that are active. The number of active crossbridges is determined by how much Ca^{2+} is bound to troponin.

If cytosolic Ca^{2+} concentrations are low, some crossbridges will not be activated and contraction force will be small. If additional Ca^{2+} enters the cell from the extracellular fluid, more Ca^{2+} is released from the sarcoplasmic reticulum. This additional Ca^{2+} binds to troponin, enhancing the ability of myosin to form crossbridges with actin and creating additional force.

When Cardiac Muscle Is Stretched, It Contracts More Forcefully

Another factor that affects the force of contraction in cardiac muscle is the sarcomere length at the beginning of contraction. For both cardiac and skeletal muscle, the tension generated is directly proportional to the initial length of the muscle fiber

[📖 p. 413]. As muscle fiber length and sarcomere length increase, tension increases, up to a maximum (Fig. 14-12 ■).

In the intact heart, stretch on the individual fibers is a function of how much blood is in the chambers of the heart. The relationship between force and ventricular volume is an important property of cardiac function and is discussed in detail later in this chapter.

CONCEPT CHECK

11. A drug that blocks all Ca^{2+} channels in the myocardial cell membrane is placed in the solution around the cell. What will happen to the force of contraction in that cell?

Answers: p. 499

Action Potentials in Myocardial Cells Vary According to Cell Type

Cardiac muscle, like skeletal muscle and neurons, is an excitable tissue with the ability to generate action potentials. Each of the two types of cardiac muscle cells has a distinctive action potential. In both types, Ca^{2+} plays an important role in the action potential, in contrast to the action potentials of skeletal muscle and neurons.

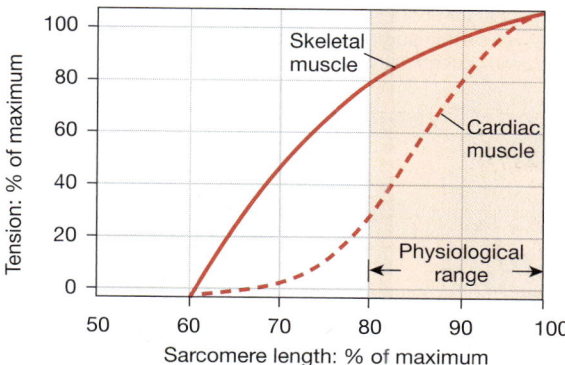

■ **FIGURE 14-12** *Length-tension relationships in skeletal and cardiac muscle*

These data represent tension developed during isometric contraction. The physiological range is the sarcomere length in which the muscle normally functions.

Myocardial Contractile Cells

The action potentials of myocardial contractile cells are similar in several ways to those of neurons and skeletal muscle [🖳 p. 257]. The rapid depolarization phase of the action potential is the result of Na^+ entry, and the steep repolarization phase is due to K^+ leaving the cell. The main difference between the action potential of the myocardial contractile cell and that of a skeletal muscle fiber or a neuron is that in the myocardial cell, there is a lengthening of the action potential caused by Ca^{2+} entry. Let's take a look at these longer action potentials (Fig. 14-13 ■).

Phase 4: resting membrane potential. Myocardial contractile cells have a stable resting potential of about –90 mV.

Phase 0: depolarization. When a wave of depolarization moves into a contractile cell through gap junctions, the membrane potential becomes more positive. Voltage-gated Na^+ channels open, allowing Na^+ to enter the cell and rapidly depolarize it. The membrane potential reaches about +20 mV before the Na^+ channels close. These are double-gated Na^+ channels, similar to the voltage-gated Na^+ channels of the axon [🖳 p. 260].

Phase 1: initial repolarization. When the Na^+ channels close, the cell begins to repolarize as K^+ leaves through open K^+ channels.

Phase 2: the plateau. The initial repolarization is very brief. The action potential then flattens into a plateau as the result of two events: a decrease in K^+ permeability and an increase in Ca^{2+} permeability. Voltage-gated Ca^{2+} channels activated by depolarization have been slowly opening during phases 0 and 1. When they finally open, Ca^{2+} enters the cell. At the same time, some K^+ channels close. The combination of Ca^{2+} influx and decreased K^+ efflux causes the action potential to flatten out into a plateau.

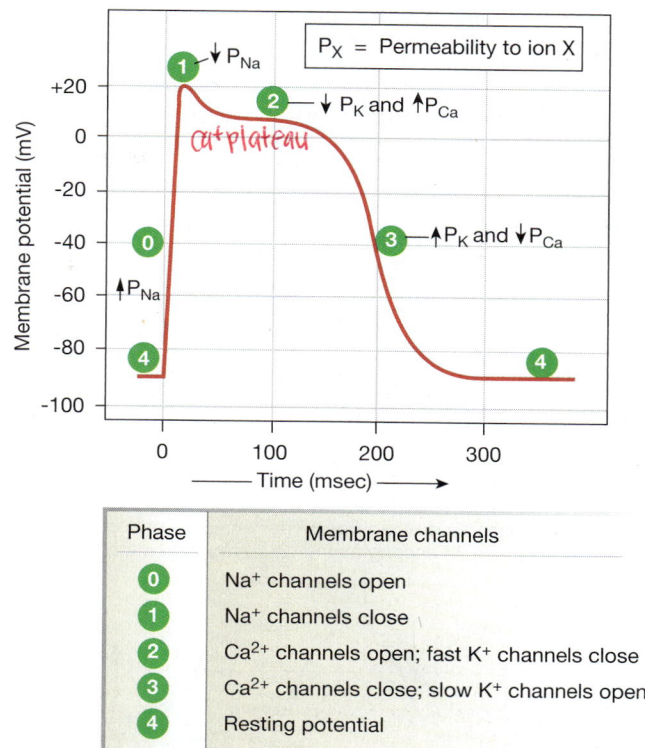

Phase	Membrane channels
0	Na^+ channels open
1	Na^+ channels close
2	Ca^{2+} channels open; fast K^+ channels close
3	Ca^{2+} channels close; slow K^+ channels open
4	Resting potential

■ **FIGURE 14-13** *Action potential of a cardiac contractile cell*

The phase numbers are a convention.

Phase 3: rapid repolarization. The plateau ends when Ca^{2+} channels close and K^+ permeability increases once more. The K^+ channels responsible for this phase are similar to those in the neuron: they are activated by depolarization but are slow to open. When the delayed K^+ channels open, K^+ exits rapidly, returning the cell to its resting potential (phase 4).

The influx of Ca^{2+} during phase 2 lengthens the total duration of a myocardial action potential. A typical action potential in a neuron or skeletal muscle fiber lasts between 1 and 5 msec. In a contractile myocardial cell, the action potential typically lasts 200 msec or more. The longer myocardial action potential helps prevent the sustained contraction called tetanus. Prevention of tetanus in the heart is important because cardiac muscles must relax between contractions so the ventricles can fill with blood.

To understand why a longer action potential prevents tetanus, compare the relationship between action potentials and contraction in skeletal and cardiac muscle cells (Fig. 14-14 ■). As you may recall from Chapter 12, the skeletal muscle action potential (red curve) is ending as contraction (blue curve) begins. Thus, a second action potential fired immediately after the refractory period will cause summation of the contractions. If a series of action potentials occurs in rapid succession, the sustained

14

(a) Skeletal muscle fast-twitch fiber: The refractory period (yellow) is very short compared with the amount of time required for the development of tension.

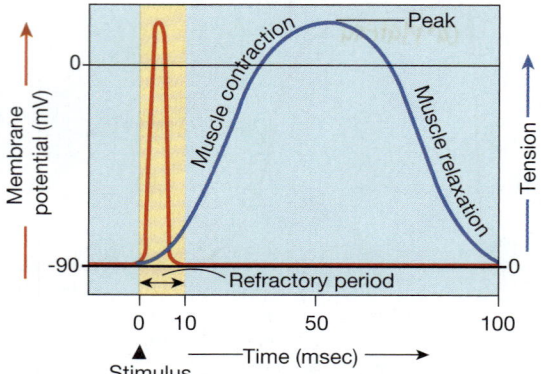

(b) Skeletal muscles that are stimulated repeatedly will exhibit summation and tetanus (action potentials not shown).

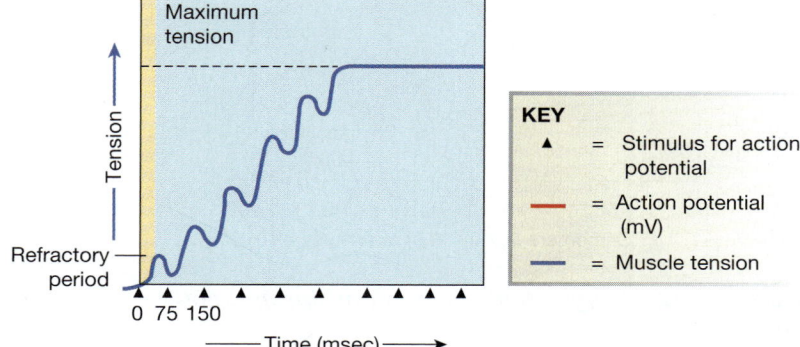

KEY

▲ = Stimulus for action potential

— = Action potential (mV)

— = Muscle tension

(c) Cardiac muscle fiber: The refractory period lasts almost as long as the entire muscle twitch.

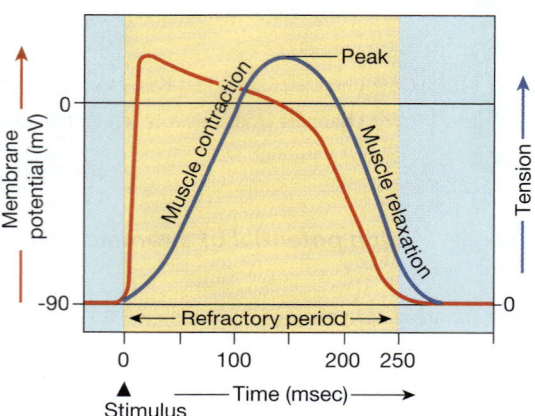

(d) Long refractory period in a cardiac muscle prevents tetanus.

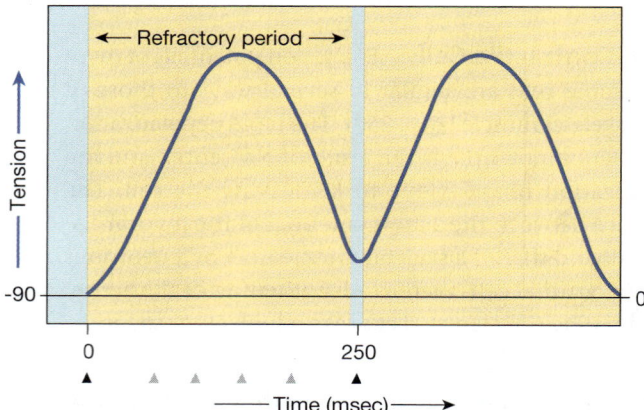

■ **FIGURE 14-14** *Refractory periods and summation in skeletal and cardiac muscle*

14

contraction known as tetanus results (Fig. 14-14b). Tetanus cannot occur in cardiac muscle because the longer action potential means the refractory period and the contraction end almost simultaneously (Fig. 14-14c). By the time a second action potential takes place, the myocardial cell has almost completely relaxed. Consequently, no summation occurs (Fig. 14-14d).

CONCEPT CHECK

12. Which ions moving in what directions cause the depolarization and repolarization phases of a neuronal action potential?

13. At the molecular level, what is happening during the refractory period in neurons and muscle fibers?

14. Lidocaine is a molecule that blocks the action of voltage-gated cardiac Na$^+$ channels. What will happen to the action potential of a myocardial contractile cell if lidocaine is applied to the cell? Answers: p. 499

Myocardial Autorhythmic Cells What gives myocardial autorhythmic cells their unique ability to generate action potentials spontaneously in the absence of input from the nervous

system? This property results from their unstable membrane potential, which starts at −60 mV and slowly drifts upward toward threshold (Fig. 14-15a ■). Because the membrane potential never "rests" at a constant value, it is called a **pacemaker potential** rather than a resting membrane potential. Whenever the pacemaker potential depolarizes to threshold, the autorhythmic cell fires an action potential.

What causes the membrane potential of these cells to be unstable? Our current understanding is that the autorhythmic cells contain channels that are different from the channels of other excitable tissues. When the cell membrane potential is −60 mV, I$_f$ channels that are permeable to both K$^+$ and Na$^+$ open (Fig. 14-15c). These channels are called I$_f$ channels because they allow current (I) to flow and because of their unusual properties. The researchers who first described the ion current through these channels initially did not understand its behavior and named it *funny* current—hence the subscript *f*.

When I$_f$ channels open at negative membrane potentials, Na$^+$ influx exceeds K$^+$ efflux (Fig. 14-15b). (This is similar to

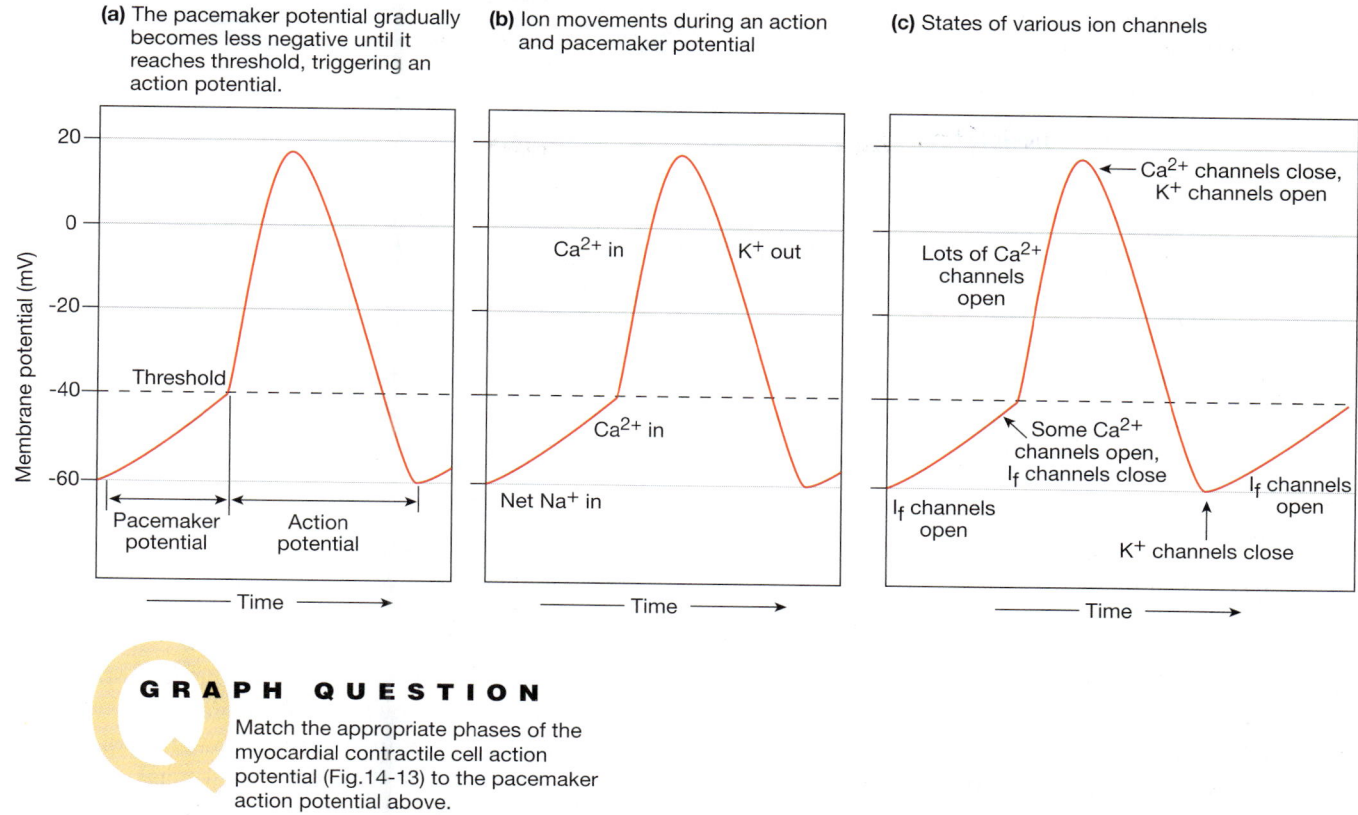

(a) The pacemaker potential gradually becomes less negative until it reaches threshold, triggering an action potential.

(b) Ion movements during an action and pacemaker potential

(c) States of various ion channels

GRAPH QUESTION

Match the appropriate phases of the myocardial contractile cell action potential (Fig.14-13) to the pacemaker action potential above.

■ **FIGURE 14-15**   *Action potentials in cardiac autorhythmic cells*

what happens at the neuromuscular junction when nonspecific cation channels open [p. 390].) The net influx of positive charge slowly depolarizes the autorhythmic cell. As the membrane potential becomes more positive, the I_f channels gradually close and some Ca^{2+} channels open. The subsequent influx of Ca^{2+} continues the depolarization, and the membrane potential moves steadily toward threshold.

When the membrane potential reaches threshold, additional Ca^{2+} channels open. Calcium rushes into the cell, creating the steep depolarization phase of the action potential. Note that this process is different from that in other excitable cells, in which the depolarization phase is due to the opening of voltage-gated Na^+ channels.

When the Ca^{2+} channels close at the peak of the action potential, slow K^+ channels have opened. The repolarization phase of the autorhythmic action potential is due to the resultant efflux of K^+. This phase is similar to repolarization in other types of excitable cells.

Autonomic Neurotransmitters Modulate Heart Rate

The speed with which pacemaker cells depolarize determines the rate at which the heart contracts (the heart rate). The interval between action potentials can be modified by altering the permeability of the autorhythmic cells to different ions. Increased permeability to Na^+ and Ca^{2+} during the pacemaker potential phase speeds up depolarization and heart rate. Decreased Ca^{2+} permeability or increased K^+ permeability slows depolarization and thus slows the heart rate.

CONCEPT CHECK

15. What would increasing K^+ permeability do to the membrane potential of the cell?   Answers: p. 499

Sympathetic stimulation of pacemaker cells speeds up heart rate. The catecholamines norepinephrine (from sympathetic neurons) and epinephrine (from the adrenal medulla) increase ion flow through both I_f and Ca^{2+} channels. More rapid cation entry speeds up the rate of the pacemaker depolarization, causing the cell to reach threshold faster and increasing the rate of action potential firing (Fig. 14-16a ■). When the pacemaker fires action potentials more rapidly, heart rate increases.

Catecholamines exert their effect by binding to and activating β_1-adrenergic receptors on the autorhythmic cells. The β_1-receptors use a cAMP second messenger system to alter the transport properties of the ion channels. In the case of the I_f channels, cAMP itself is the messenger. When cAMP binds to open I_f channels, they remain open longer. The I_f channels

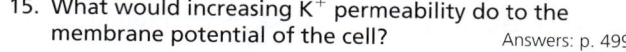

14

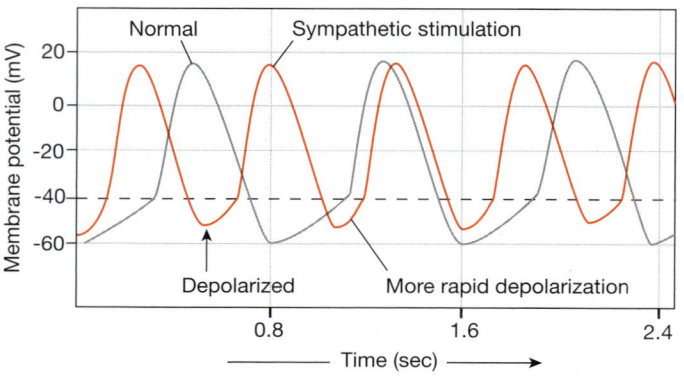

(a) Sympathetic stimulation and epinephrine depolarize the autorhythmic cell and speed up the depolarization rate, increasing the heart rate.

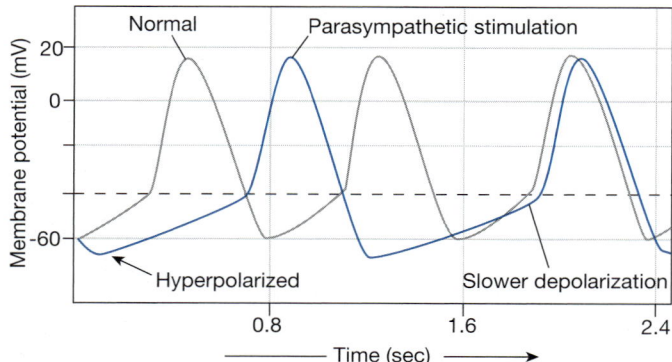

(b) Parasympathetic stimulation hyperpolarizes the membrane potential of the autorhythmic cell and slows depolarization, decreasing the heart rate.

■ **FIGURE 14-16**   *Modulation of heart rate by the nervous system*

belong to a family called the *HCN channels*, or *hyperpolarization-activated cyclic nucleotide-gated channels*. Other members of the HCN family are found in neurons.

The parasympathetic neurotransmitter acetylcholine (ACh) slows heart rate. Acetylcholine activates muscarinic cholinergic receptors that influence K^+ and Ca^{2+} channels in the pacemaker cell. Potassium permeability increases, hyperpolarizing the cell so that the pacemaker potential begins at a more negative value (Fig. 14-16b). At the same time, Ca^{2+} permeability

of the pacemaker decreases. Decreased Ca^{2+} permeability slows the rate at which the pacemaker potential depolarizes. The combination of the two effects causes the cell to take longer to reach threshold, delaying the onset of the action potential in the pacemaker and slowing the heart rate.

Table 14-3 ■ compares action potentials of the two types of myocardial muscle with those of skeletal muscle. Next we will see how action potentials of autorhythmic cells spread throughout the heart to coordinate contraction.

TABLE 14-3   **Comparison of Action Potentials in Cardiac and Skeletal Muscle**

	SKELETAL MUSCLE	CONTRACTILE MYOCARDIUM	AUTORHYTHMIC MYOCARDIUM
Membrane potential	Stable at −70 mV	Stable at −90 mV	Unstable pacemaker potential; usually starts at −60 mV
Events leading to threshold potential	Net Na^+ entry through ACh-operated channels	Depolarization enters via gap junctions	Net Na^+ entry through I_f channels; reinforced by Ca^{2+} entry
Rising phase of action potential	Na^+ entry	Na^+ entry	Ca^{2+} entry
Repolarization phase	Rapid; caused by K^+ efflux	Extended plateau caused by Ca^{2+} entry; rapid phase caused by K^+ efflux	Rapid; caused by K^+ efflux
Hyperpolarization	Due to excessive K^+ efflux at high K^+ permeability when K^+ channels close; leak of K^+ and Na^+ restores potential to resting state	None; resting potential is −90 mV, the equilibrium potential for K^+	None; when repolarization hits −60 mV, the I_f channels open again
Duration of action potential	Short: 1–2 msec	Extended: 200+ msec	Variable; generally 150+ msec
Refractory period	Generally brief	Long because resetting of Na^+ channel gates delayed until end of action potential	None

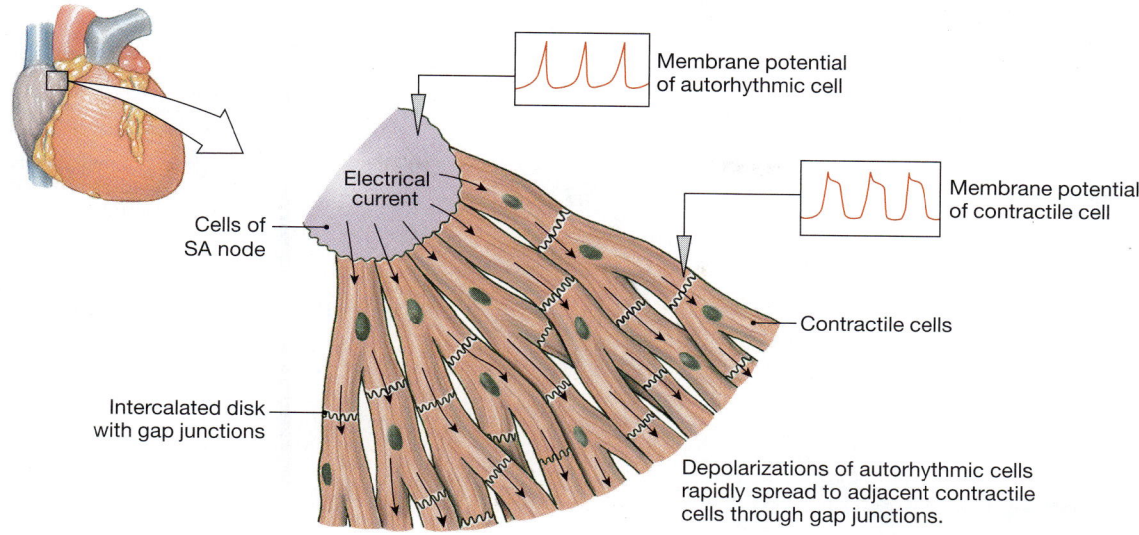

Membrane potential of autorhythmic cell

Membrane potential of contractile cell

Electrical current

Cells of SA node

Contractile cells

Intercalated disk with gap junctions

Depolarizations of autorhythmic cells rapidly spread to adjacent contractile cells through gap junctions.

■ **FIGURE 14-17**   *Electrical conduction in myocardial cells*

CONCEPT CHECK

16. Do you think that the Ca^{2+} channels in autorhythmic cells are the same as the Ca^{2+} channels in contractile cells? Defend your answer.

17. What will happen to the action potential of a myocardial autorhythmic cell if tetrodotoxin, which blocks voltage-gated Na^+ channels, is applied to the cell?

18. In an experiment, the vagus nerve [🔁 p. 381] to the heart was cut. The investigators noticed that heart rate increased. What can you conclude about the vagal neurons that innervate the heart?   Answers: p. 499

THE HEART AS A PUMP

We now turn from single myocardial cells to the intact heart. How can one tiny noncontractile autorhythmic cell cause the entire heart to beat? And why do those doctors on TV shows shock patients with electric paddles when their hearts malfunction? You're about to learn the answers to these questions.

Electrical Conduction in the Heart Coordinates Contraction

The heart is like a group of people around a stalled car. One person can push on the car, but it's not likely to move very far unless everyone pushes together. In the same way, individual myocardial cells must depolarize and contract in a coordinated fashion if the heart is to create enough force to circulate blood.

Electrical communication in the heart begins with an action potential in an autorhythmic cell. The depolarization spreads rapidly to adjacent cells through gap junctions in the intercalated disks (Fig. 14-17 ■). The depolarization wave is followed by a wave of contraction that passes across the atria, then moves into the ventricles.

The depolarization begins in the **sinoatrial node (SA node)**, autorhythmic cells in the right atrium that serve as the main pacemaker of the heart (Fig. 14-18 ■). The depolarization wave then spreads rapidly through a specialized conducting system of noncontractile autorhythmic fibers. A branched **internodal pathway** connects the SA node to the **atrioventricular node (AV node)**, a group of autorhythmic cells near the floor of the right atrium. From the AV node, the depolarization moves into **Purkinje fibers** in the **atrioventricular bundle (AV bundle)*** in the septum between the ventricles. (Purkinje fibers are specialized conducting cells that transmit electrical signals very rapidly.) A short way down the septum, the AV bundle fibers divide into left and right **bundle branches**. The bundle branch fibers continue downward to the apex of the heart, where they divide into smaller Purkinje fibers that spread outward among the contractile cells.

The electrical signal for contraction begins when the SA node fires an action potential and the depolarization spreads to adjacent cells through gap junctions (Fig. 14-18 ①). Electrical conduction is rapid through the internodal conducting pathways ② but slower through the contractile cells of the atria ③.

As action potentials spread across the atria, they encounter the fibrous skeleton of the heart at the junction of the atria and ventricles. This barricade prevents the transfer of electrical signals from the atria to the ventricles. Consequently, the AV node is the only pathway through which action potentials can reach the contractile fibers of the ventricles.

The electrical signal passes from the AV node through the AV bundle and bundle branches to the apex of the heart (Fig. 14-18, ④). The Purkinje fibers transmit impulses very

14

---

*The AV bundle is also called the **bundle of His** ("hiss").

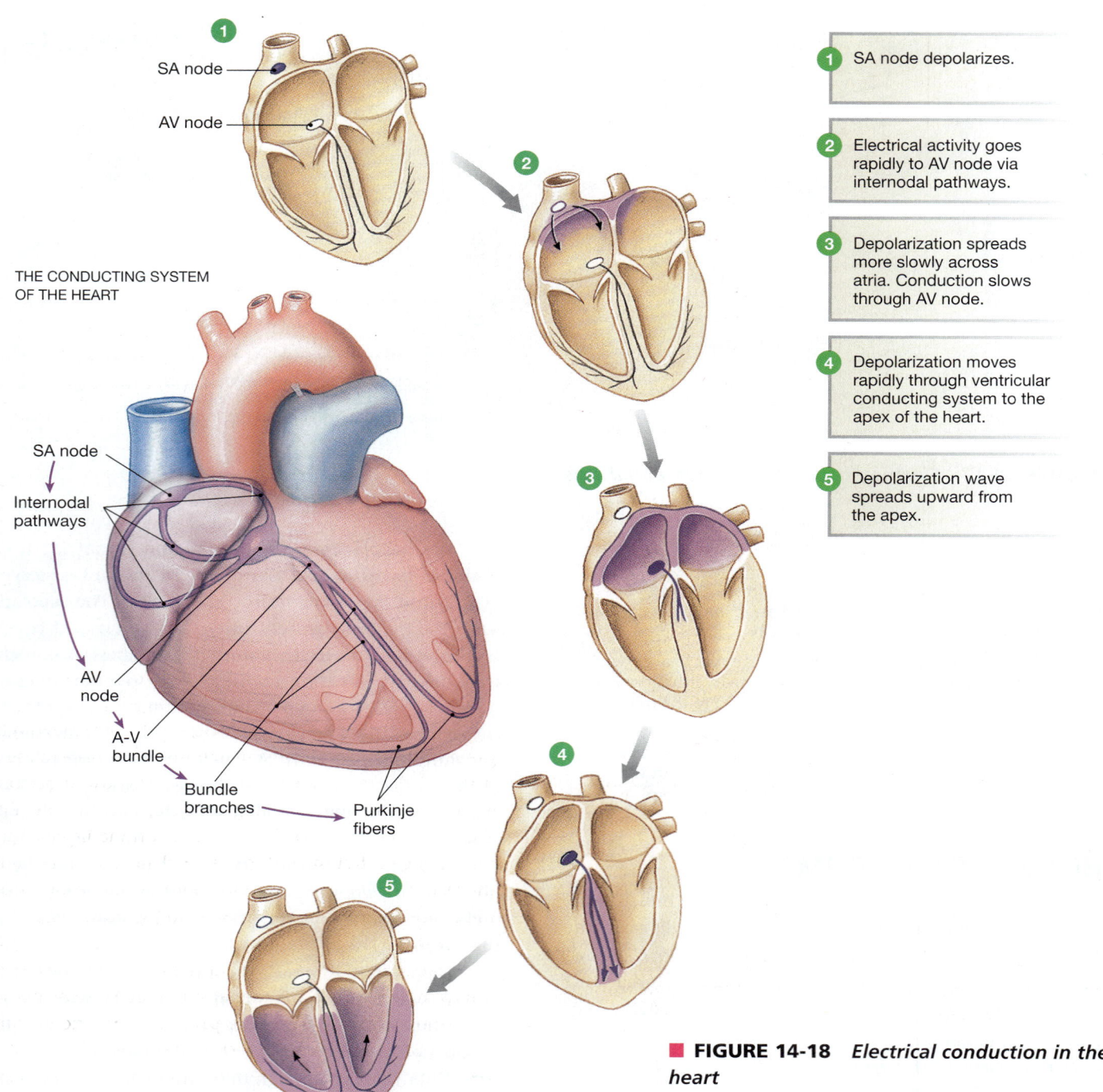

THE CONDUCTING SYSTEM OF THE HEART

1 SA node depolarizes.

2 Electrical activity goes rapidly to AV node via internodal pathways.

3 Depolarization spreads more slowly across atria. Conduction slows through AV node.

4 Depolarization moves rapidly through ventricular conducting system to the apex of the heart.

5 Depolarization wave spreads upward from the apex.

■ **FIGURE 14-18**   *Electrical conduction in the heart*

Purple shading in steps 2–5 represents depolarization.

rapidly, with speeds up to 4 m/sec, so that all contractile cells in the apex contract nearly simultaneously ⑤.

Why is it necessary to direct the electrical signals through the AV node? Why not allow them to spread downward from the atria? The answer lies in the fact that blood is pumped out of the ventricles through openings at the top of the chambers (see Fig. 14-9b). If electrical signals from the atria were conducted directly into the ventricles, the ventricles would start contracting at the top. Then blood would be squeezed downward and would become trapped in the bottom of the ventricles

(think of squeezing a toothpaste tube at the top). The apex-to-base contraction squeezes blood toward the arterial openings at the base of the heart.

The ejection of blood from the ventricles is aided by the spiral arrangement of the muscles in the walls (see Fig. 14-10a). As these muscles contract, they pull the apex and base of the heart closer together, squeezing blood out the openings at the top of the ventricles.

A second function of the AV node is to delay the transmission of action potentials slightly, allowing the atria to complete

RUNNING PROBLEM

The results of the creatine kinase and troponin I assays do not come back for several hours. If a coronary artery were blocked, damage to the heart muscle could be severe by that time. In Walter's case, an electrocardiogram (ECG) showed an abnormal pattern of electrical activity. "He's definitely had an MI," said the ER physician, referring to a myocardial infarction, or heart attack. "Let's start him on a beta-blocker and t-PA." t-PA (short for *tissue plasminogen activator*) activates plasminogen, a substance that is produced in the body and dissolves blood clots. Given within 1–3 hours of a heart attack, t-PA can help dissolve blood clots that are blocking blood flow to the heart muscle. This will help limit the extent of ischemic damage.

Question 5:
 How do electrical signals move from cell to cell in the myocardium?

Question 6:
 What happens to contraction in a myocardial contractile cell if a wave of depolarization passing through the heart bypasses it?

Question 7:
 A beta-blocker is an antagonist to β_1-adrenergic receptors. What will this drug do to Walter's heart rate? Why is that response helpful following a heart attack?

| 457 | 461 | 471 | **479** | 484 | 489 | 494 |

CLINICAL FOCUS

FIBRILLATION

Coordination of myocardial contraction is essential for normal cardiac function. In extreme cases in which the myocardial cells contract in a disorganized manner, a condition known as *fibrillation* results. Ventricular fibrillation is a life-threatening emergency because without coordinated contraction of the muscle fibers, the ventricles cannot pump enough blood to supply adequate oxygen to the brain. One way to correct this problem is to administer an electrical shock to the heart. The shock creates a depolarization that triggers action potentials in all cells simultaneously, coordinating them again. You have probably seen this procedure on television hospital shows, when a doctor places flat paddles on the patient's chest and tells everyone to stand back ("Clear!") while the paddles pass an electrical current through the body.

---

their contraction before ventricular contraction begins. The **AV node delay** is accomplished by slowing conduction through the nodal cells. Action potentials here move at only 1/20 the rate of action potentials in the atrial internodal pathway.

Pacemakers Set the Heart Rate

The cells of the SA node set the pace of the heartbeat. Other cells in the conducting system, such as the AV node and the Purkinje fibers, have unstable resting potentials and can also act as pacemakers under some conditions. However, because their rhythm is slower than that of the SA node, they do not usually have a chance to set the heartbeat. The Purkinje fibers, for example, can spontaneously fire action potentials, but their firing rate is very slow, between 25 and 40 beats per minute.

Why does the fastest pacemaker determine the pace of the heartbeat? Consider the following analogy. A group of people are playing "follow the leader" as they walk. Initially, everyone

is walking at a different pace—some fast, some slow. When the game starts, everyone must match his or her pace to the pace of the person who is walking the fastest. The fastest person in the group is the SA node, walking at 70 steps per minute. Everyone else in the group (autorhythmic and contractile cells) sees that the SA node is fastest, and so they pick up their pace and follow the leader. In the heart, the cue to follow the leader is the electrical signal sent from the SA node to the other cells.

Now suppose the SA node gets tired and drops out of the group. The role of leader defaults to the next fastest person, the AV node, who is walking at a rate of 50 steps per minute. The group slows to match the pace of the AV node, but everyone is still following the fastest walker.

What happens if the group divides? Suppose that when they reach a corner, the AV node leader goes left but a renegade Purkinje fiber decides to go right. Those people who follow the AV node continue to walk at 50 steps per minute, but the people who follow the Purkinje fiber slow down to match his pace of 35 steps per minute. Now there are two leaders, each walking at a different pace.

In the heart, the SA node is the fastest pacemaker and normally sets the heart rate. If this node is damaged and cannot function, one of the slower pacemakers in the heart takes over. Heart rate then matches the rate of the new pacemaker. It is even possible for different parts of the heart to follow different pacemakers, just as the walking group split at the corner.

In a condition known as *complete heart block,* the conduction of electrical signals from the atria to the ventricles through the AV node is disrupted. The SA node fires at its rate of 70 beats per minute, but those signals never reach the ventricles. So the ventricles coordinate with their fastest pacemaker. Because ventricular autorhythmic cells discharge only about 35 times a minute, the rate at which the ventricles contract is much slower than the rate at which the atria contract. If ventricular contraction is too slow to maintain adequate blood flow, it may be necessary for the heart's rhythm to be set artificially by a surgically implanted mechanical pacemaker. These battery-powered devices artificially stimulate the heart at a predetermined rate.

CONCEPT CHECK

19. Name two functions of the AV node. What is the purpose of AV node delay?

20. Where is the SA node located?

21. Occasionally an ectopic pacemaker [*ektopos,* out of place] will develop in part of the heart's conducting system. What happens to heart rate if an ectopic atrial pacemaker depolarizes at a rate of 120 times per minute?

Answers: p. 499

The Electrocardiogram Reflects the Electrical Activity of the Heart

At the end of the nineteenth century, physiologists discovered that they could place electrodes on the skin's surface and record the electrical activity of the heart. These recordings, called **electrocardiograms,** or ECGs,* provide indirect information about heart function. It is possible to use surface electrodes to record internal electrical activity because salt solutions, such as our NaCl-based extracellular fluid, are good conductors of electricity.

The first human electrocardiogram was recorded in 1887, but the procedure was not refined for clinical use until the first years of the twentieth century. The father of the modern ECG was a Dutch physiologist named Walter Einthoven. He named the parts of the ECG as we know them today and created "Einthoven's triangle," a hypothetical triangle created around the heart when electrodes are placed on both arms and the left leg (Fig. 14-19 ■). The sides of the triangle are numbered to correspond with the three leads, or pairs of electrodes, used for a recording. An ECG is recorded from one lead at a time. One electrode acts as the positive electrode of a lead and a second electrode acts as the negative electrode of the lead. (The third electrode is inactive.) For example, in lead I, the left arm electrode is designated as positive and the right arm electrode is designated as negative.

An ECG tracing shows the summed electrical potentials generated by all cells of the heart. Different components of the ECG reflect depolarization or repolarization of the atria and ventricles. Because depolarization initiates muscle contraction,

---

*The abbreviation EKG—from the Greek word *kardia,* meaning *heart*—is sometimes used.

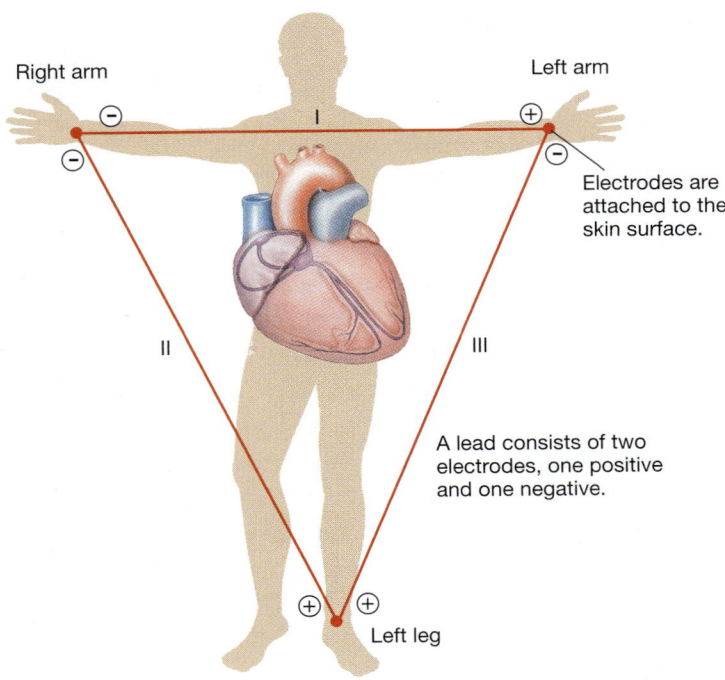

■ FIGURE 14-19 *Einthoven's triangle*

Electrodes attached to both arms and the left leg form a triangle. Two electrodes are active at any one time. Each two-electrode pair constitutes one *lead* (pronounced "leed"). Lead I, for instance, has the negative electrode attached to the right arm and the positive electrode attached to the left arm.

these *electrical events* (waves) of an ECG can be associated with contraction or relaxation (collectively referred to as the *mechanical events* in the heart). Let's follow an ECG through a single contraction-relaxation cycle, otherwise known as a **cardiac cycle.**

There are two major components of an ECG: waves and segments. *Waves* appear as deflections above or below the baseline. *Segments* are sections of baseline between two waves. *Intervals* are combinations of waves and segments.

Three major waves can be seen on a normal ECG recorded from lead I (Fig. 14-20 ■). The first wave is the **P wave,** which corresponds to depolarization of the atria (Fig. 14-21 ■). The next trio of waves, the **QRS complex,** represents the progressive wave of ventricular depolarization. The final wave, the **T wave,** represents the repolarization of the ventricles. Atrial repolarization is not represented by a special wave but is incorporated into the QRS complex.

The mechanical events of the cardiac cycle lag slightly behind the electrical signals, just as the contraction of a single cardiac muscle cell follows its action potential (see Fig. 14-14c). Atrial contraction begins during the latter part of the P wave and continues during the PR segment. Ventricular contraction begins just after the Q wave and continues through the T wave.

One thing many people find confusing is that you cannot tell if an ECG recording represents depolarization or repolarization

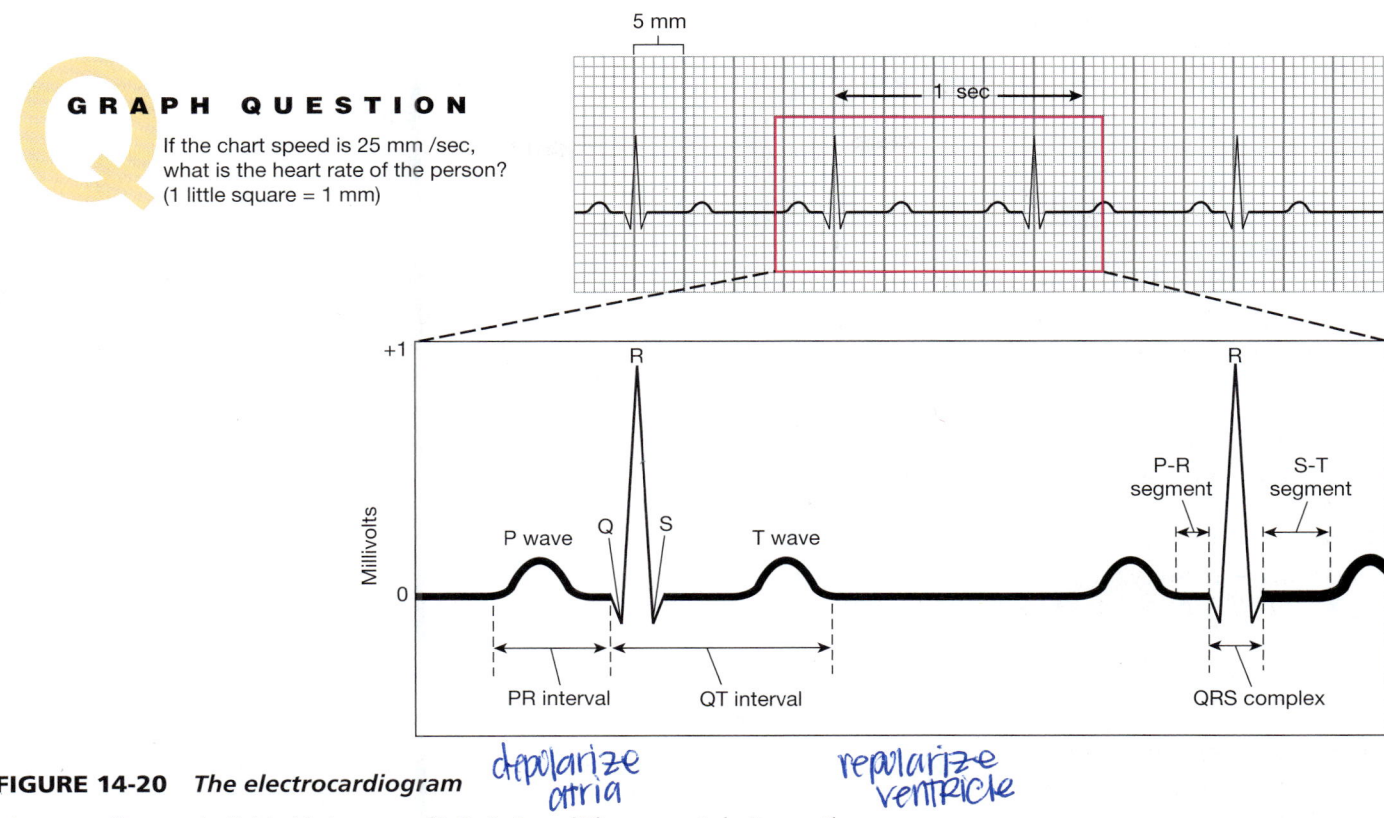

GRAPH QUESTION

If the chart speed is 25 mm /sec, what is the heart rate of the person? (1 little square = 1 mm)

■ **FIGURE 14-20**   *The electrocardiogram*

An electrocardiogram is divided into waves (P, Q, R, S, and T), segments between the waves (the P-R and S-T segments, for example), and intervals consisting of a combination of waves and segments (such as the PR and QT intervals). This ECG tracing was recorded from lead I.

by looking at the shape of the waves relative to the baseline. For example, the P wave represents atrial depolarization and the T wave represents ventricular repolarization, but both the P wave and the T wave are deflections above the baseline. This is very different from the action potential recordings of neurons and muscle fibers, in which an upward deflection always represents depolarization [🔁 p. 257].

An ECG is not the same as a single action potential (Fig. 14-22 ■). An action potential is one electrical event in a single cell, recorded using an intracellular electrode. The ECG is an extracellular recording that represents the sum of multiple action potentials taking place in many heart muscle cells. When an electrical wave moving through the heart is directed toward the positive electrode, the ECG wave goes up from the baseline. If net charge movement through the heart is toward the negative electrode, the wave points downward. In addition, the amplitudes of action potential and ECG recordings are very different. A ventricular action potential has a voltage change of 110 mV, for example, but the ECG signal has an amplitude of only 1 mV by the time it reaches the surface of the body.

An important point to remember is that an ECG is an electrical "view" of a three-dimensional object. This is one reason we use multiple leads to assess heart function. Think of looking at an automobile. From the air, it looks like a rectangle, but from the side and front it has different shapes. Not everything that you see from the front of the car can be seen from its side, and vice versa. In the same way, the leads of an ECG provide different electrical "views" and give information about different regions of the heart. A 12-lead ECG (the three limb electrodes plus nine more electrodes placed on the chest and trunk) is the standard for clinical use. The additional leads provide detailed information about electrical conduction in the heart. Electrocardiograms are important diagnostic tools in medicine because they are quick, painless, and noninvasive (that is, do not puncture the skin).

An ECG provides information on heart rate and rhythm, conduction velocity, and even the condition of tissues in the heart. Thus, although obtaining an ECG is simple, interpreting some of its subtleties can be quite complicated. The interpretation of an ECG begins with the following questions (Fig. 14-23 ■).

1. What is the heart rate? Heart rate is normally timed either from the beginning of one P wave to the beginning of the next P wave or from the peak of one R wave to the peak of the next R wave. A normal resting heart rate is 60–100 beats per minute, although trained athletes often have slower heart rates at rest. A faster-than-normal rate is known as *tachycardia*, and a slower-than-normal rate is called *bradycardia* [*tachys*, swift; *bradys*, slow].

14

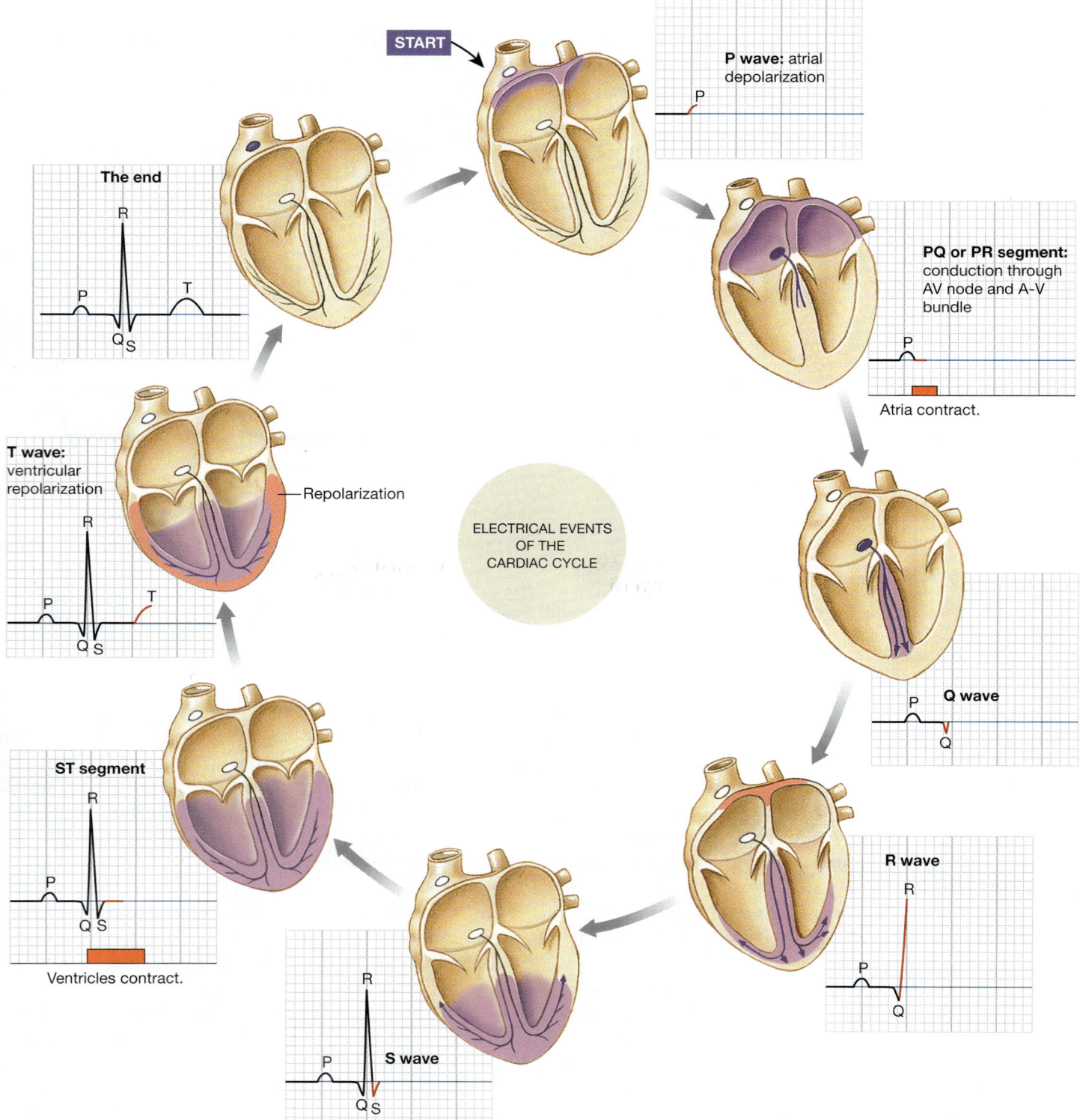

START

P wave: atrial depolarization

P

PQ or PR segment: conduction through AV node and A-V bundle

P

Atria contract.

The end

R

P   T

Q S

T wave: ventricular repolarization

Repolarization

ELECTRICAL EVENTS OF THE CARDIAC CYCLE

R

P   T

Q S

Q wave

P

Q

ST segment

R

P

Q S

Ventricles contract.

R

P   **S wave**

Q S

R wave

R

P

Q

14

■ **FIGURE 14-21** *Correlation between an ECG and electrical events in the heart*

The figure shows the correspondence between electrical events in the ECG and depolarizing (purple) and repolarizing (peach) regions of the heart.

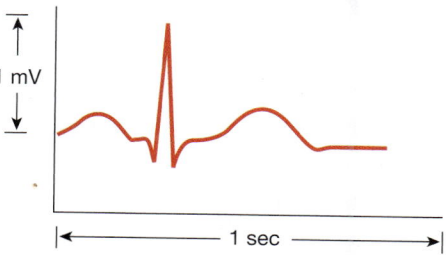

(a) The electrocardiogram represents the summed electrical activity of all cells recorded from the surface of the body.

1 mV

1 sec

(b) The ventricular action potential is recorded from a single cell using an intracellular electrode. Notice that the voltage change is much greater when recorded intracellularly.

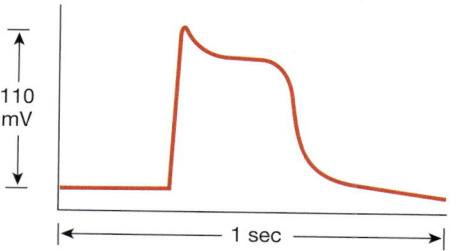

110 mV

1 sec

■ **FIGURE 14-22** *Comparison of an ECG and a myocardial action potential*

2. Is the rhythm of the heartbeat regular (that is, occurs at regular intervals) or irregular? An irregular rhythm, or *arrhythmia* [*a-*, without + rhythm], can result from a benign extra beat or from more serious conditions such as atrial fibrillation, in which the SA node has lost control of the pacemaking.

   After determining heart rate and rhythm, the next step in analyzing an ECG is to look at the relationship of the various waves. To help your analysis, you might want to write the letters above the P, R, and T waves.

3. Does a QRS complex follow each P wave, and is the P-R segment constant in length? If not, a problem with conduction of signals through the AV node may exist. In heart block (the conduction problem mentioned earlier), action potentials from the SA node sometimes fail to be transmitted through the AV node to the ventricles. In these conditions, one or more P waves may occur without initiating a QRS complex. In the most severe form of heart block (third-degree), the atria depolarize regularly at one pace while the ventricles contract at a much slower pace (Fig. 14-23b).

   The more difficult aspects of interpreting an ECG include looking for subtle changes, such as alterations in the shape or duration of various waves or segments. An experienced clinician can find signs pointing to changes in conduction velocity,

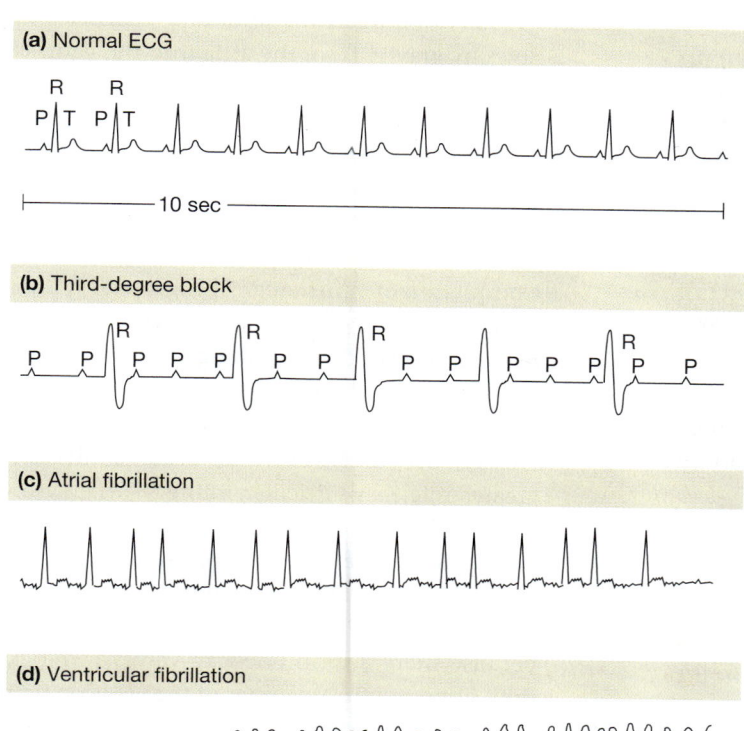

(a) Normal ECG

R R
P T P T

10 sec

(b) Third-degree block

P P R P P P R P P R P P P R P P

(c) Atrial fibrillation

(d) Ventricular fibrillation

■ **FIGURE 14-23** *Normal and abnormal electrocardiograms*

All tracings represent 10-sec recordings.

Questions to ask when analyzing ECG tracings:

1. What is the rate? Is it within the normal range of 60-100 beats per minute?

2. Is the rhythm regular?

3. Are all normal waves present in recognizable form?

4. Is there one QRS complex for each P wave? If yes, is the P-R segment constant in length?

5. If there is not one QRS complex for each P wave, count the heart rate using the P waves, then count it according to the R waves. Are the rates the same? Which wave would agree with the pulse felt at the wrist?

14

enlargement of the heart, or tissue damage resulting from periods of ischemia. An amazing number of conclusions can be drawn about heart function simply by looking at alterations in the heart's electrical activity as recorded on an ECG.

✓ **CONCEPT CHECK**

22. Three abnormal ECGs are shown in Figure 14-23b–d. Study them and see if you can relate the ECG changes to disruption of the normal electrical conduction pattern in the heart.

23. Identify the waves on the following 10-second ECG recording. Look at the pattern of their occurrence and describe what has happened to electrical conduction in the heart.

Answers: p. 499

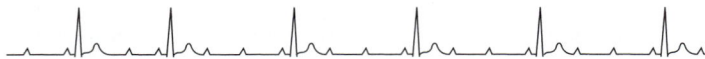

Cardiac arrhythmias are a family of cardiac pathologies that range from benign to those with potentially fatal consequences. Arrhythmias are electrical problems that arise during the generation or conduction of action potentials through the heart, and they can usually be seen on an ECG. Some arrhythmias are "dropped beats" that result when the ventricles do not get their usual signal to contract. Other arrhythmias, such as *premature ventricular contractions* (PVCs), are extra beats that occur when an autorhythmic cell other than the SA node jumps in and fires an action potential out of sequence.

One interesting heart condition that can be observed on an ECG is *long QT syndrome* (LQTS), named for the change in the QT interval. LQTS has several forms. Some are inherited channelopathies, in which mutations occur in myocardial Na^+ or K^+ channels [⟳ p. 254]. In another form of LQTS, the ion channels are normal but the protein *ankyrin-B* that anchors the channels to the cell membrane is defective. *Iatrogenic* (physician-caused) forms of LQTS can occur as a side effect of taking certain medications. One well-publicized incident occurred in the 1990s when patients took a non-sedating antihistamine called terfenadine (Seldane®) that binds to K^+ repolarization channels. After at least eight deaths were attributed to the drug, the U.S. Food and Drug Administration removed Seldane from the market.

The Heart Contracts and Relaxes Once During a Cardiac Cycle

Each cardiac cycle has two phases: **diastole**, the time during which cardiac muscle relaxes, and **systole**, the time during which the muscle is contracting [*diastole*, dilation; *systole*, contraction]. Because the atria and ventricles do not contract and relax at the same time, we will discuss atrial and ventricular events separately.

In thinking about blood flow during the cardiac cycle, remember that blood flows from an area of higher pressure to one

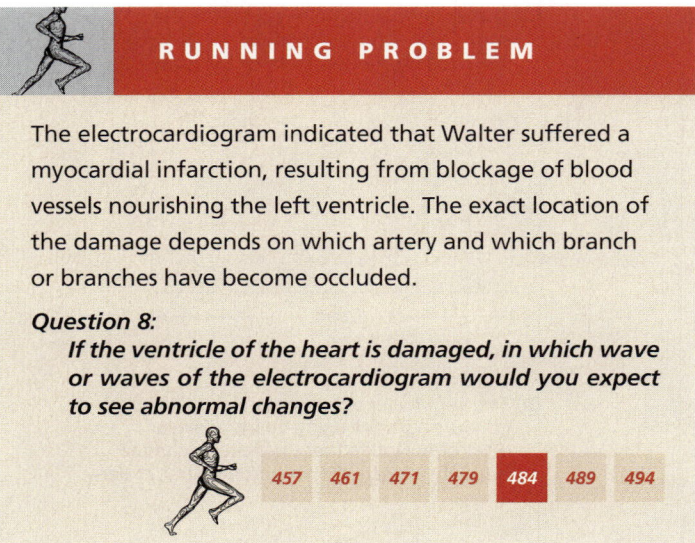

RUNNING PROBLEM

The electrocardiogram indicated that Walter suffered a myocardial infarction, resulting from blockage of blood vessels nourishing the left ventricle. The exact location of the damage depends on which artery and which branch or branches have become occluded.

Question 8:
If the ventricle of the heart is damaged, in which wave or waves of the electrocardiogram would you expect to see abnormal changes?

457   461   471   479   **484**   489   494

of lower pressure, and that contraction increases pressure while relaxation decreases pressure. In this discussion, we divide the cardiac cycle into the five phases shown in Figure 14-24 ■:

1. **The heart at rest: atrial and ventricular diastole.** We enter the cardiac cycle at the brief moment during which both the atria and the ventricles are relaxing. The atria are filling with blood from the veins, and the ventricles have just completed a contraction. As the ventricles relax, the AV valves between the atria and ventricles open. Blood flows by gravity from the atria into the ventricles. The relaxing ventricles expand to accommodate the entering blood.

✓ **CONCEPT CHECK**

24. During atrial filling, is pressure in the atrium higher or lower than pressure in the venae cavae?   Answers: p. 499

2. **Completion of ventricular filling: atrial systole.** Although most blood enters the ventricles while the atria are relaxed, the last 20% of filling is accomplished when the atria contract and push blood into the ventricles. (This applies to a normal person at rest. When heart rate increases, as during exercise, atrial contraction plays a greater role in ventricular filling.) Atrial systole, or contraction, begins following the wave of depolarization that sweeps across the atria. The pressure increase that accompanies contraction pushes blood into the ventricles. A small amount of blood is forced backward into the veins because there are no one-way valves to block backward flow, although the openings of the veins do narrow during contraction. This retrograde movement of blood back into the veins may be observed as a pulse in the jugular vein of a normal person who is lying with the head and chest elevated about 30°. (Look in the hollow formed where the sternocleidomastoid muscle runs under the clavicle.) An

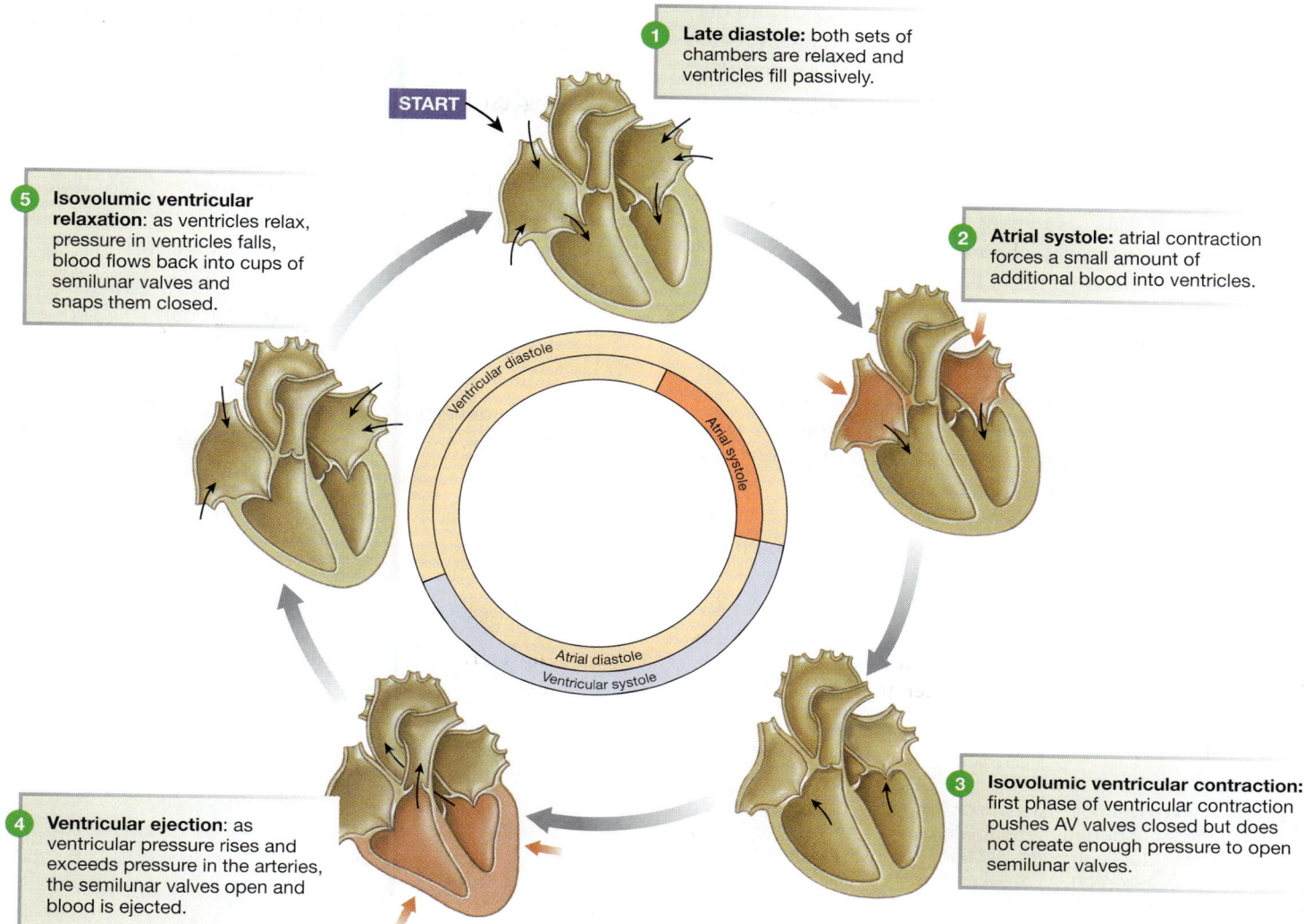

1 **Late diastole:** both sets of chambers are relaxed and ventricles fill passively.

START

5 **Isovolumic ventricular relaxation:** as ventricles relax, pressure in ventricles falls, blood flows back into cups of semilunar valves and snaps them closed.

2 **Atrial systole:** atrial contraction forces a small amount of additional blood into ventricles.

Ventricular diastole

Atrial systole

Atrial diastole
Ventricular systole

4 **Ventricular ejection:** as ventricular pressure rises and exceeds pressure in the arteries, the semilunar valves open and blood is ejected.

3 **Isovolumic ventricular contraction:** first phase of ventricular contraction pushes AV valves closed but does not create enough pressure to open semilunar valves.

■ **FIGURE 14-24** *Mechanical events of the cardiac cycle*

observable jugular pulse higher on the neck of a person sitting upright is a sign that pressure in the right atrium is higher than normal.

3. **Early ventricular contraction and the first heart sound.** As the atria are contracting, the depolarization wave is moving slowly through the conducting cells of the AV node, then rapidly down the Purkinje fibers to the apex of the heart. Ventricular systole begins there as spiral bands of muscle squeeze the blood upward toward the base. Blood pushing against the underside of the AV valves forces them closed so that blood cannot flow back into the atria. Vibrations following closure of the AV valves create the **first heart sound**, S_1, the "lub" of "lub-dup."

With both sets of AV and semilunar valves closed, blood in the ventricles has nowhere to go. Nevertheless, the ventricles continue to contract, squeezing on the blood in the same way that you might squeeze a water balloon in your hand. This is similar to an isometric contraction, in which

muscle fibers create force without movement [p. 417]. To return to the toothpaste tube analogy, it is like squeezing the tube with the cap on: high pressure develops within the tube, but the toothpaste has nowhere to go. This phase is called **isovolumic ventricular contraction** [*iso-*, equal], to underscore the fact that the volume of blood in the ventricle is not changing.

While the ventricles begin to contract, the atrial muscle fibers are repolarizing and relaxing. When atrial pressure falls below that in the veins, blood flows from the veins into the atria again. Closure of the AV valves isolates the upper and lower cardiac chambers, meaning that atrial filling is independent of events taking place in the ventricles.

4. **The heart pumps: ventricular ejection.** As the ventricles contract, they generate enough pressure to open the semilunar valves and push blood into the arteries. The pressure created by ventricular contraction becomes the driving

force for blood flow. High-pressure blood is forced into the arteries, displacing the low-pressure blood that fills them and pushing it farther into the vasculature. During this phase, the AV valves remain closed and the atria continue to fill.

5. **Ventricular relaxation and the second heart sound.** At the end of ventricular ejection, the ventricles begin to repolarize and relax. As they do so, ventricular pressure decreases. Once ventricular pressure falls below the pressure in the arteries, blood starts to flow backward into the heart. This backflow of blood fills the cuplike cusps of the semilunar valves, forcing them together into the closed position. The vibrations created by semilunar valve closure are the **second heart sound, S_2,** the "dup" of "lub-dup."

Once the semilunar valves close, the ventricles again become sealed chambers. The AV valves remain closed because ventricular pressure, although falling, is still higher than atrial pressure. This period is called **isovolumic ventricular relaxation** because the volume of blood in the ventricles is not changing.

When ventricular relaxation causes ventricular pressure to become less than atrial pressure, the AV valves open. Blood that has been accumulating in the atria during ventricular contraction rushes into the ventricles. The cardiac cycle has begun again.

✔ CONCEPT CHECK

25. Which chamber—atrium or ventricle—has higher pressure during the following phases of the cardiac cycle? (a) ventricular ejection, (b) isovolumic ventricular relaxation, (c) atrial and ventricular diastole, (d) isovolumic ventricular contraction

26. *Murmurs* are abnormal heart sounds caused either by blood forced through a narrowed valve opening or by backward flow (regurgitation) through a valve that has not closed completely. *Valvular stenosis* [*stenos,* narrow] may be an inherited condition or may result from inflammation or other disease processes. At which step(s) in the cardiac cycle (Fig. 14-24) would you expect to hear a murmur caused by the following pathologies? (a) aortic valvular stenosis, (b) mitral valve regurgitation, (c) aortic valve regurgitation   Answers: p. 499

Pressure-Volume Curves Represent One Cardiac Cycle

Another way to describe the cardiac cycle is with a pressure-volume graph, shown in Figure 14-25 ■. This figure represents the changes in volume (*x*-axis) and pressure (*y*-axis) that occur during one cardiac cycle. The flow of blood through the heart is governed by the same principle that governs the flow of all liquids and gases: flow proceeds from areas of higher pressure to areas of lower pressure. When the heart contracts, the pressure increases and blood flows out of the heart into areas of lower pressure. Figure 14-25 represents pressure and volume changes in the left ventricle, which sends blood into the systemic circu-

CLINICAL FOCUS

GALLOPS, CLICKS, AND MURMURS

The simplest direct assessment of heart function consists of listening to the heart through the chest wall, a process known as **auscultation** [*auscultare,* to listen to] that has been practiced since ancient times. In its simplest form, auscultation is done by placing an ear against the chest. Today, however, it is usually performed by listening through a stethoscope placed against the chest and the back. Normally, there are two audible heart sounds. The first ("lub") is associated with closure of the AV valves. The second ("dup") is associated with closure of the semilunar valves.

Two additional heart sounds can be recorded with very sensitive electronic stethoscopes. The third heart sound is caused by turbulent blood flow into the ventricles during ventricular filling, and the fourth sound is associated with turbulence during atrial contraction. In certain abnormal conditions, these latter two sounds may become audible through a regular stethoscope. They are called gallops because their timing puts them close to one of the normal heart sounds: "lub—dup-dup," or "lub-lub—dup." Other abnormal heart sounds include clicking, caused by abnormal movement of one of the valves, and murmurs, caused by the "whoosh" of blood leaking through an incompletely closed or excessively narrowed (*stenosed*) valve.

lation. The left side of the heart creates higher pressures than the right side, which sends blood through the shorter pulmonary circuit.

The cycle begins at point A. The ventricle has completed a contraction and contains the minimum amount of blood that it will hold during the cycle. It has relaxed, and its pressure is also at its minimum value. Blood is flowing into the atrium from the pulmonary veins.

Once pressure in the atrium exceeds pressure in the ventricle, the mitral valve between the atrium and ventricle opens (Fig. 14-25, point A). Atrial blood now flows into the ventricle, increasing its volume (point A to point B). As blood flows in, the relaxing ventricle expands to accommodate the entering blood. Consequently, the volume of the ventricle increases but the pressure in the ventricle goes up very little.

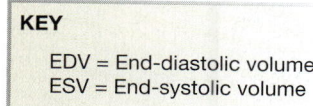

KEY

EDV = End-diastolic volume
ESV = End-systolic volume

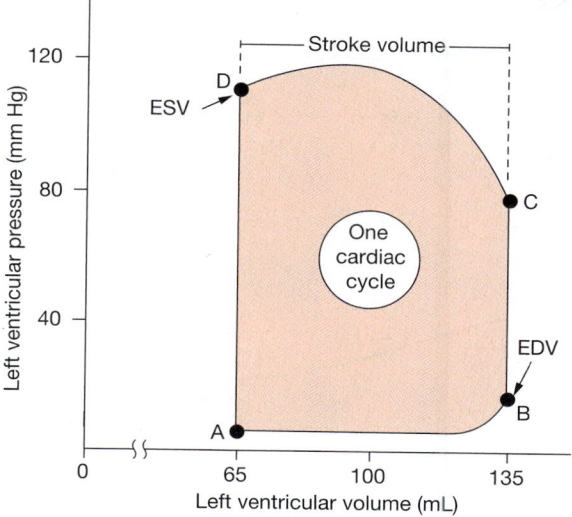

GRAPH QUESTIONS

Match the following segments
to the corresponding ventricular events:

A ⟶ B:	(a) Ejection of blood into aorta
B ⟶ C:	(b) Isovolumic contraction
C ⟶ D:	(c) Isovolumic relaxation
D ⟶ A:	(d) Passive filling and atrial contraction

Match the following events to points A – D:
(a) aortic valve opens
(b) mitral valve opens
(c) aortic valve closes
(d) mitral valve closes

■ **FIGURE 14-25** *Left ventricular pressure-volume
changes during one cardiac cycle*

The last portion of ventricular filling is completed by atrial contraction. The ventricle now contains the maximum volume of blood that it will hold during this cardiac cycle (point B). Because maximum filling occurs at the end of ventricular relaxation (diastole), this volume is called the **end-diastolic volume (EDV)**. In a 70-kg man at rest, end-diastolic volume is about 135 mL, but this value will vary under different conditions. During periods of very high heart rate, for instance, when the ventricle does not have time to fill completely between beats, the end-diastolic value may be less than 135 mL.

When ventricular contraction begins, the mitral valve closes. With both the AV valve and the semilunar valve closed, blood in the ventricle has nowhere to go. Nevertheless, the ventricle continues to contract, causing the pressure in this chamber to increase rapidly during isovolumic contraction (B → C in Fig. 14-25). Once ventricular pressure exceeds the pressure in

the aorta, the aortic valve opens (point C). Pressure continues to increase as the ventricle contracts further, but ventricular volume decreases as blood is pushed out into the aorta (C → D).

The heart does not empty itself completely of blood each time the ventricle contracts. The amount of blood left in the ventricle at the end of contraction is known as the **end-systolic volume (ESV)**. The ESV (point D) is the minimum amount of blood the ventricle will contain during one cycle. An average ESV value in a person at rest is 65 mL, meaning that nearly half of the 135 mL that was in the ventricle at the start of the contraction is still there at the end of the contraction.

At the end of each ventricular contraction, the ventricle begins to relax. As it does so, ventricular pressure decreases. Once pressure in the ventricle falls below aortic pressure, the semilunar valve closes, and the ventricle again becomes a sealed chamber. The remainder of relaxation occurs without a change in blood volume, and so this phase is called *isovolumic relaxation* (Fig. 14-25, D → A). When ventricular pressure finally falls to the point at which atrial pressure exceeds ventricular pressure, the mitral valve opens and the cycle begins again.

The electrical and mechanical events of the cardiac cycle are summarized together in Figure 14-26 ■, known as a Wiggers diagram after the physiologist who first created it.

CONCEPT CHECK

27. In Figure 14-24, at what points in the cycle do EDV and ESV occur?

28. On the Wiggers diagram in Figure 14-26, match the following events to the lettered boxes: (a) end diastolic volume, (b) aortic valve opens, (c) mitral valve opens, (d) aortic valve closes, (e) mitral valve closes, (f) end systolic volume

29. Why does atrial pressure increase just to the right of point C in Figure 14-26? Why does it decrease during the initial part of ventricular systole, then increase? Why does it decrease to the right of point D?

30. Why does ventricular pressure shoot up suddenly at point C in Figure 14-26?

Answers: p. 499

Stroke Volume Is the Volume of Blood Pumped by One Ventricle in One Contraction

What is the purpose of blood remaining in the ventricles at the end of each contraction? For one thing, the resting end-systolic volume of 65 mL provides a safety margin. With a more forceful contraction, the heart can decrease its ESV, sending additional blood to the tissues. Like many organs of the body, the heart does not usually work "all out."

The amount of blood pumped by one ventricle during a contraction is known as the **stroke volume**. It is measured in milliliters per beat and can be calculated as follows:

Volume of blood before contraction − Volume of
blood after contraction = stroke volume

EDV − ESV = stroke volume

14

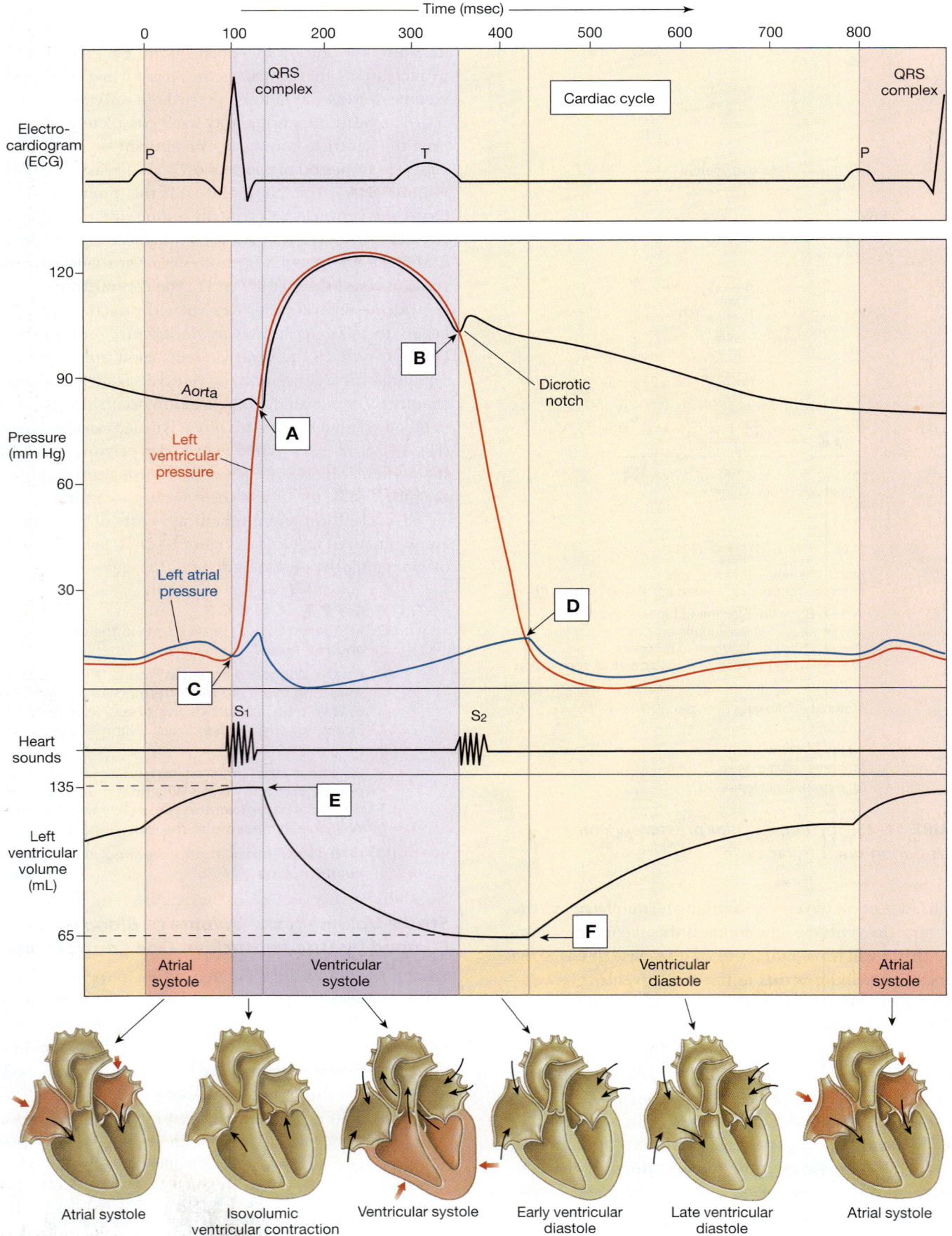

Time (msec)

| 0 | 100 | 200 | 300 | 400 | 500 | 600 | 700 | 800 |

Electro-cardiogram (ECG)

QRS complex

Cardiac cycle

QRS complex

P

T

P

Pressure (mm Hg)

120

90

Aorta

Left ventricular pressure

B

Dicrotic notch

A

60

30

Left atrial pressure

D

C

Heart sounds

S₁

S₂

Left ventricular volume (mL)

135

E

65

F

Atrial systole | Ventricular systole | Ventricular diastole | Atrial systole

Atrial systole | Isovolumic ventricular contraction | Ventricular systole | Early ventricular diastole | Late ventricular diastole | Atrial systole

■ **FIGURE 14-26** *The Wiggers diagram*

The boxed letters refer to Concept Checks 28–30.

For the average contraction in a person at rest:

135 mL − 65 mL = 70 mL, the normal stroke volume

Stroke volume is not constant and can increase to as much as 100 mL during exercise. Stroke volume, like heart rate, is homeostatically regulated by mechanisms discussed later in this chapter.

Cardiac Output Is a Measure of Cardiac Performance

How can we assess the effectiveness of the heart as a pump? One way is to measure **cardiac output**, the volume of blood pumped by one ventricle in a given period of time. Because all blood that leaves the heart flows through the tissues, cardiac output is an indicator of total blood flow through the body. However, cardiac output does not tell us how blood is distributed to various tissues. That aspect of blood flow is regulated at the tissue level.

Cardiac output (CO) can be calculated by multiplying heart rate (beats per minute) by stroke volume (mL per beat, or per contraction):

$$\text{cardiac output} = \text{heart rate} \times \text{stroke volume}$$

For an average resting heart rate of 72 beats per minute and a stroke volume of 70 mL per beat, we have

CO = 72 beats/min × 70 mL/beat
= 5040 mL/min   (or approx. 5 L/min)

Average total blood volume is about 5 liters. This means that, at rest, one side of the heart pumps all the blood in the body through it in only one minute!

Normally, cardiac output is the same for both ventricles. However, if one side of the heart begins to fail for some reason and is unable to pump efficiently, cardiac output becomes mismatched. In that situation, blood will pool in the circulation behind the weaker side of the heart.

During exercise, cardiac output may increase to 30–35 L/min. Homeostatic changes in cardiac output are accomplished by varying the heart rate, the stroke volume, or both. Both local and reflex mechanisms can alter cardiac output, as we will see in the sections that follow.

✔ CONCEPT CHECK

31. If the stroke volume of the left ventricle is 250 mL/beat and the stroke volume of the right ventricle is 251 mL/beat, what will happen to the relative distribution of blood between the systemic and pulmonary circulation after 10 beats?   Answers: p. 499

Heart Rate Is Modulated by Autonomic Neurons and Catecholamines

An average resting heart rate in an adult is about 70 beats/minute (bpm). The normal range is highly variable, however. Trained athletes may have resting heart rates of 50 bpm or less, while

RUNNING PROBLEM

Walter was in the cardiac care unit by 1 P.M., where the cardiologist visited him. "We need to keep an eye on you here for the next week. There is a good chance the damage from your heart attack could cause an irregular heartbeat." Once Walter was stable, he would have a coronary angiogram, a procedure in which an opaque dye visible on X-rays shows where coronary artery lumens have narrowed from atherosclerotic plaques. Depending on the results of that test, the physician might recommend either balloon angioplasty, in which a tube passed into the coronary artery is inflated to open up the blockage, or coronary bypass surgery, in which veins from other parts of the body are grafted onto the heart arteries to provide bypass channels around blocked regions.

Question 9:
If Walter's heart attack has damaged the muscle of his left ventricle, what do you predict will happen to his left cardiac output?

457 461 471 479 484 **489** 494

someone who is excited or anxious may have a rate of 125 bpm or higher. Children have higher average heart rates than adults. Although heart rate is initiated by autorhythmic cells in the SA node, it is modulated by neural and hormonal input.

The sympathetic and parasympathetic branches of the autonomic division influence heart rate through antagonistic control (Fig. 14-27 ■) [🔁 Fig. 6-21, p. 194]. Parasympathetic activity slows heart rate, while sympathetic activity speeds it up. Normally, tonic control of heart rate is dominated by the parasympathetic branch. This can be shown experimentally by blocking all autonomic input to the heart. When all sympathetic and parasympathetic input is blocked, the spontaneous depolarization rate of the SA node is 90–100 times per minute. To achieve a resting heart rate of 70 beats per minute, tonic parasympathetic activity must slow the intrinsic rate down from 90 beats per minute.

An increase in heart rate can be achieved in two ways. The simplest method for increasing rate is to decrease parasympathetic activity. As parasympathetic influence is withdrawn from the autorhythmic cells, they resume their intrinsic rate of depolarization, and heart rate increases to 90–100 beats per minute. Alternatively, sympathetic input is required to increase heart rate above the intrinsic rate. As you learned earlier, norepinephrine (or epinephrine) on β_1-receptors speeds up the

14

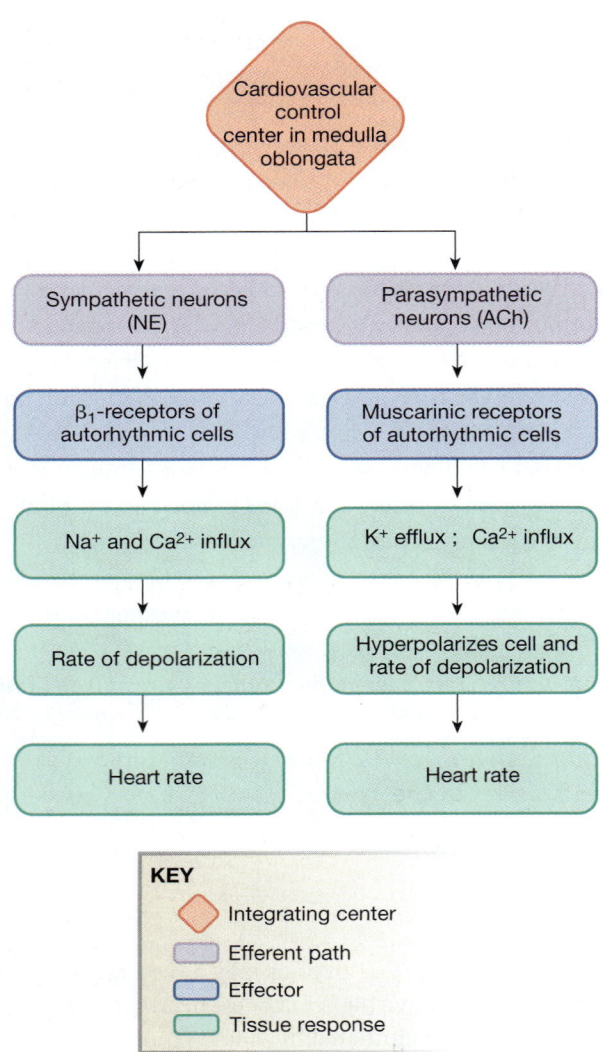

FIGURE 14-27 *Reflex control of heart rate*

depolarization rate of the autorhythmic cells and increases heart rate.

Both autonomic branches also alter the rate of conduction through the AV node. Acetylcholine secreted by parasympathetic neurons slows the conduction of action potentials through the AV node, thereby increasing AV node delay. In contrast, the catecholamines epinephrine and norepinephrine enhance conduction of action potentials through the AV node and through the conducting system.

Multiple Factors Influence Stroke Volume

Stroke volume, the volume of blood pumped per ventricle per contraction, is directly related to the force generated by cardiac muscle during a contraction. Normally, the greater the force of contraction, the greater the stroke volume. In the isolated heart, the force of ventricular contraction is affected by two parameters: the length of muscle fibers at the beginning of contraction and the contractility of the heart. The volume of blood in the

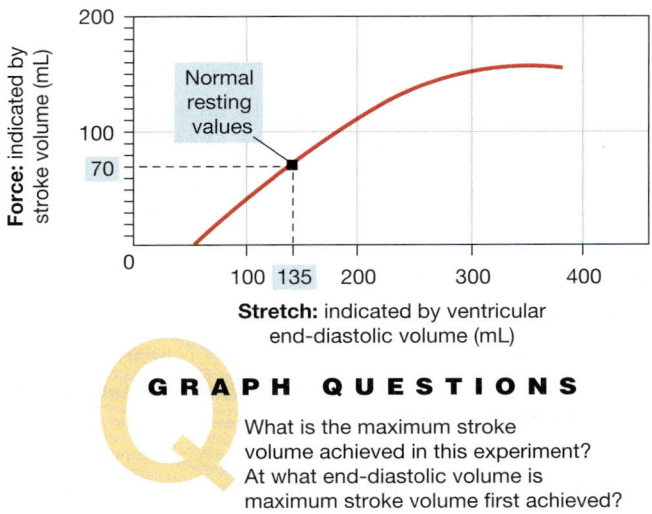

GRAPH QUESTIONS

What is the maximum stroke volume achieved in this experiment? At what end-diastolic volume is maximum stroke volume first achieved?

■ **FIGURE 14-28** *Length-force relationships in the intact heart: a Starling curve*

In the intact heart, stroke volume is used as an indicator of contractile force.

ventricle at the beginning of contraction (the end-diastolic volume) determines the length of the muscle. **Contractility** is the intrinsic ability of a cardiac muscle fiber to contract at any given fiber length and is a function of Ca^{2+} interaction with the contractile filaments.

Length-Tension Relationships and the Frank-Starling Law of the Heart As you learned earlier in this chapter, the force created by a myocardial muscle fiber is directly related to the length of the sarcomere. As sarcomere length increases (up to an optimum length), contraction force increases. In the intact heart, as stretch of the ventricular wall increases, so does the stroke volume. If additional blood flows into the ventricles, the muscle fibers stretch, then contract more forcefully, ejecting more blood. The degree of myocardial stretch before contraction begins is called the **preload** on the heart because this stretch represents the load placed on cardiac muscles before they contract.

This relationship between stretch and force in the intact heart was first described by a German physiologist, Otto Frank. A British physiologist, Ernest Starling, then expanded on Frank's work. Starling attached an isolated heart-lung preparation from a dog to a reservoir so that he could regulate the amount of blood returning to the heart. He found that in the absence of any nervous or hormonal control, the heart pumped all the blood that returned to it.

The relationship between stretch and force in the intact heart is plotted on a *Starling curve* (Fig. 14-28 ■). The *x*-axis represents the end-diastolic volume. This volume is a measure of stretch in the ventricles, which in turn determines sarcomere length. The *y*-axis of the Starling curve represents the stroke volume and is an indicator of the force of contraction.

The graph shows that stroke volume is proportional to EDV. As additional blood enters the heart, the heart contracts more forcefully and ejects more blood. This relationship is known as the **Frank-Starling law of the heart**. It means that within physiological limits, the heart pumps all the blood that returns to it.

Stroke Volume and Venous Return According to the Frank-Starling law, stroke volume increases as end-diastolic volume increases. End-diastolic volume is normally determined by **venous return**, the amount of blood that enters the heart from the venous circulation. Three factors affect venous return: (1) contraction or compression of veins returning blood to the heart (the skeletal muscle pump), (2) pressure changes in the abdomen and thorax during breathing (the respiratory pump), and (3) sympathetic innervation of veins.

Skeletal muscle pump is the name given to skeletal muscle contractions that squeeze veins (particularly in the legs), compressing them and pushing blood toward the heart. During exercise that involves the lower extremities, the skeletal muscle pump helps return blood to the heart. During periods of sitting or standing motionless, the skeletal muscle pump does not assist venous return.

The **respiratory pump** is created by movement of the thorax during inspiration (breathing in). As the chest expands and the diaphragm moves toward the abdomen, the thoracic cavity enlarges and develops a subatmospheric pressure. This low pressure decreases pressure in the inferior vena cava as it passes through the thorax, which helps draw more blood into the vena cava from veins in the abdomen. The respiratory pump is aided by the higher pressure placed on the outside of abdominal veins when the abdominal contents are compressed during inspiration. The combination of increased pressure in the abdominal veins and decreased pressure in thoracic veins enhances venous return during inspiration.

Constriction of veins by sympathetic activity is the third factor that affects venous return. When the veins constrict, their volume decreases, squeezing more blood out of them and into the heart. With a larger ventricular volume at the beginning of the next contraction, the ventricle contracts more forcefully, sending the blood out into the arterial side of the circulation. In this manner, sympathetic innervation of veins allows the body to redistribute some venous blood to the arterial side of the circulation.

Contractility Is Controlled by the Nervous and Endocrine Systems

Any chemical that affects contractility is called an **inotropic agent** [*ino*, fiber], and its influence is called an **inotropic effect.** If a chemical increases the force of contraction, it is said to have

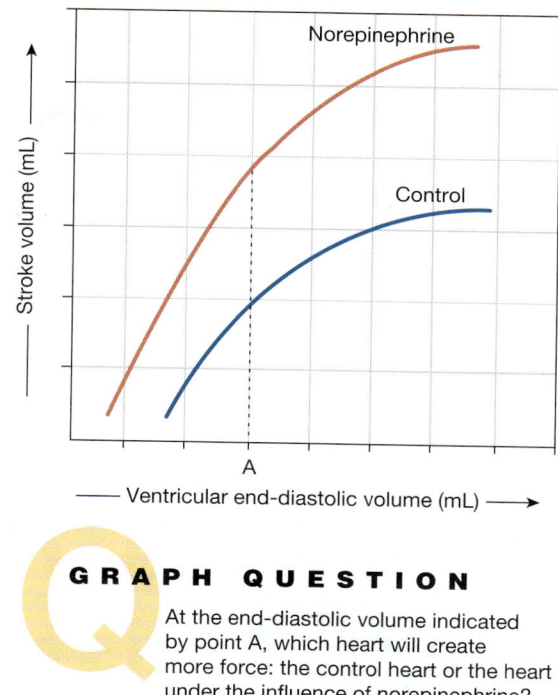

GRAPH QUESTION

At the end-diastolic volume indicated by point A, which heart will create more force: the control heart or the heart under the influence of norepinephrine?

■ **FIGURE 14-29** *The effect of norepinephrine on contractility of the heart*

Norepinephrine is a positive inotropic agent.

a positive inotropic effect. For example, the catecholamines epinephrine and norepinephrine and drugs such as digitalis enhance contractility and are therefore considered to have a positive inotropic effect. Chemicals with negative inotropic effects decrease contractility.

Figure 14-29 ■ shows a normal Starling curve (the control curve) along with a curve showing how the stroke volume changes with increased contractility due to norepinephrine. Note that contractility is distinct from the length-tension relationship. A muscle can remain at one length (for example, the end-diastolic volume marked A in Figure 14-29) but show increased contractility. Contractility increases as the amount of calcium available for contraction increases. Contractility was once considered to be distinct from changes in force resulting from variations in muscle (sarcomere) length. However, it now appears that increasing sarcomere length also makes cardiac muscle more sensitive to Ca^{2+}, thus linking contractility to muscle length.

The mechanism by which catecholamines increase Ca^{2+} entry and storage and exert their positive inotropic effect is mapped in Figure 14-30 ■. The signal molecules bind to and activate β_1-adrenergic receptors [🔁 p. 270] on the contractile myocardial cell membrane. Activated β_1-receptors use a cyclic AMP second messenger system to phosphorylate specific intracellular proteins [🔁 p. 183]. Phosphorylation of voltage-gated Ca^{2+}

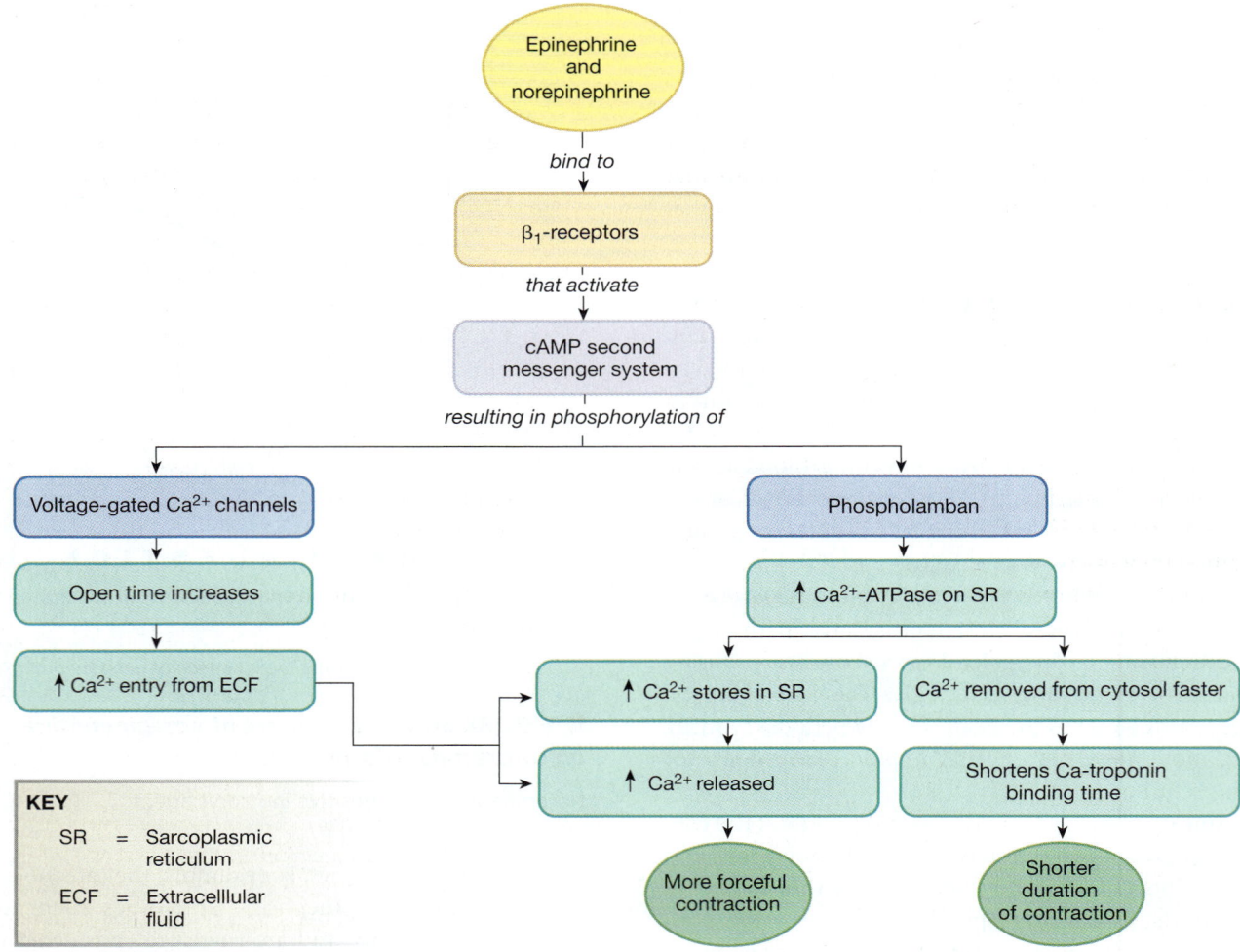

■ FIGURE 14-30   *Modulation of cardiac contraction by catecholamines*

channels increases the probability that they will open. More open channels allow more Ca^{2+} to enter the cell.

The catecholamines increase Ca^{2+} storage through the use of a regulatory protein called **phospholamban** (Fig. 14-30). Phosphorylation of phospholamban enhances Ca^{2+}-ATPase activity in the sarcoplasmic reticulum. The ATPase concentrates Ca^{2+} in the sarcoplasmic reticulum, making more Ca^{2+} available for calcium-induced calcium release. Because more cytosolic Ca^{2+} means more active crossbridges, and because the force of contraction is proportional to the number of active crossbridges, the net result of catecholamine stimulation is a stronger contraction.

In addition to increasing the force of cardiac contraction, catecholamines also shorten the duration of contraction. The enhanced Ca^{2+}-ATPase speeds up removal of Ca^{2+} from the cytosol. This in turn shortens the time that Ca^{2+} is bound to troponin and decreases the active time of the myosin crossbridges. The muscle twitch is therefore briefer.

A different mechanism that enhances contractility can be triggered by the administration of cardiac glycosides, a class of

molecules first discovered in the plant *Digitalis purpurea* (purple foxglove). Cardiac glycosides include digitoxin and the related compound *ouabain*, a molecule used to inhibit sodium transport in research studies. Glycosides increase contractility by slowing Ca^{2+} removal from the cytosol (in contrast to the catecholamines just discussed, which speed up Ca^{2+} removal). This mechanism is a pharmacological effect and does not occur in the absence of the drug.

Cardiac glycosides have been used since the eighteenth century as a remedy for *heart failure,* a pathological condition in which the heart is unable to contract forcefully. These highly toxic drugs depress Na^{+}-K^{+}-ATPase activity in all cells, not just those of the heart. With depressed Na^{+}-K^{+}-ATPase activity, Na^{+} builds up in the cytosol, and the concentration gradient for Na^{+} across the cell membrane diminishes. This in turn decreases the potential energy available for indirect active transport [🔁 p. 144]. In the myocardial cell, cardiac glycosides decrease the cell's ability to remove Ca^{2+} by means of the Na^{+}-Ca^{2+} exchanger. The resultant increase in cytosolic Ca^{2+} causes more forceful myocardial contractions.

EMERGING CONCEPTS

STEM CELLS FOR HEART DISEASE

One of the interesting (and scary) aspects of translating basic scientific research into medicine is that sometimes therapies work but no one knows why. An example is the use of bone marrow stem cells to treat heart disease. After a heart attack, portions of the myocardium may be so damaged from lack of oxygen that they can no longer contract and contribute to cardiac function. A therapy that could replace dead and damaged cells and restore function would be a dream come true. In 2001 a group of researchers reported that bone marrow stem cells injected into mice with damaged hearts differentiated into new myocardial cells. This dramatic result prompted rapid translation of the basic research into human clinical trials. By 2004 there were more than 12 trials in which bone marrow cells were being injected into patients with impaired cardiac function. Early results indicated that some patients were exhibiting functional improvement. At the same time, however, scientists were reporting that they had been unable to duplicate the 2001 findings that bone marrow stem cells differentiate into myocardial cells. The reason for the discrepancy between the clinical improvement of patients injected with stem cells and the apparent failure of stem cells to form new myocardial cells remains unclear. One possible explanation is that the bone marrow extracts contain growth factors that promoted endogenous myocardial cell repair. Both basic research and clinical trials on this subject are continuing.

CONCEPT CHECK

32. Using the myocardial cell in Figure 14-11 as a model, draw a contractile cell and show how catecholamines increase myocardial contractility.

Answers: p. 499

EDV and Arterial Blood Pressure Determine Afterload

Many of the experiments that uncovered the relationship between myocardial stretch and contractile force were conducted using isolated hearts. In the intact animal, ventricular force must be used to overcome the resistance created by blood filling the arterial system. The combined load of EDV and arterial resistance during ventricular contraction is known as **afterload**.

As an analogy, think of waiters carrying trays of food through a swinging door. A tray is a load equivalent to blood in the ventricles at the beginning of contraction. The door is an additional load that the waiter must push against to leave the kitchen. Normally this additional load is relatively minor. If someone decides to play a prank, however, and piles furniture against the dining room side of the door (increased afterload), the waiter must expend considerably more force to push through the door. Similarly, ventricular contraction must push a load of blood through a semilunar valve and out into the blood-filled arteries.

Increased afterload is found in several pathological situations, including elevated arterial blood pressure and loss of stretchability (compliance) in the aorta. To maintain constant stroke volume when afterload increases, the ventricle must increase its force of contraction, which then increases the muscle's need for oxygen and ATP production. If increased afterload becomes a chronic situation, the myocardial cells hypertrophy, resulting in increased thickness of the ventricular wall.

Clinically, arterial blood pressure is often used as an indirect indicator of afterload. Other aspects of ventricular function can be assessed noninvasively by echocardiography, an ultrasound procedure in which sound waves are reflected off heart tissue. A common functional index derived from this procedure is the **ejection fraction**, or percentage of EDV ejected with one contraction (stroke volume/EDV). Using our standard values for the 70-kg man, ejection fraction at rest is 70 mL/135 mL, or 52%. If stroke volume increases to 100 mL with exercise, the ejection fraction increases to 74%.

CONCEPT CHECK

33. A person's aortic valve opening has become constricted, creating a condition known as *aortic stenosis*. Which ventricle is affected by this change? What happens to the afterload on this ventricle?

Answers: p. 499

14

The factors that determine cardiac output are summarized in Figure 14-31 ■. Cardiac output varies with both heart rate and stroke volume. Heart rate is modulated by the autonomic division of the nervous system and by epinephrine. The determination of stroke volume is more complex because stroke volume is a function of an intrinsic myocardial response to stretch (the length-tension relationship of the Frank-Starling law) interacting with adrenergically mediated changes in contractility. Venous return is a major determinant of end-diastolic volume and stretch.

The heart is a complex organ with many parts that can malfunction. In the next chapter we will examine how cardiac output plays a key role in blood flow through the circulation. You will learn about high blood pressure and atherosclerosis, and how these conditions can cause the heart to fail in its role as a pump.

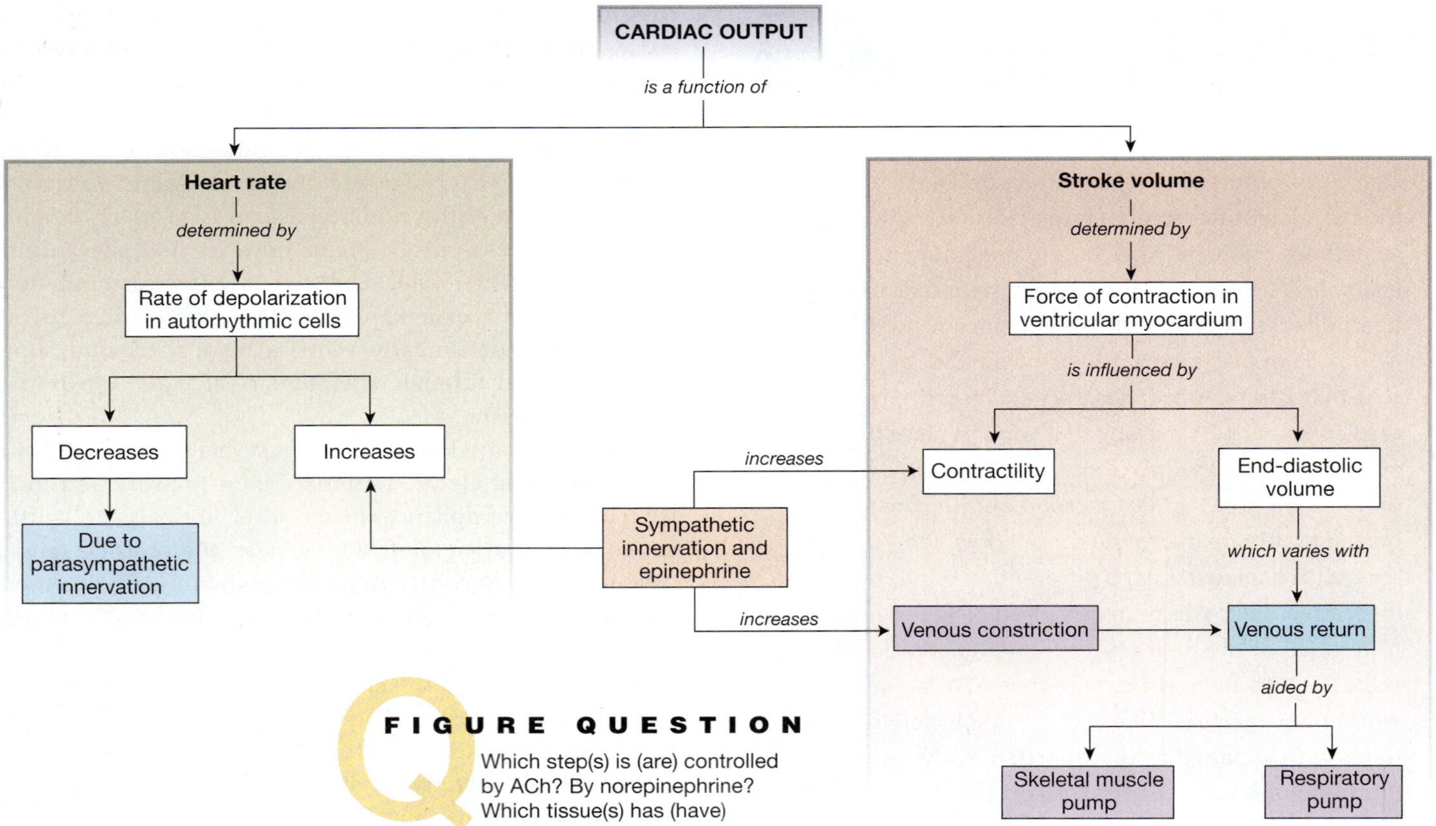

```
                          ┌─────────────────────┐
                          │   CARDIAC OUTPUT    │
                          └─────────────────────┘
                                is a function of
                    ┌───────────────────────────────────┐
```

Heart rate

determined by

Rate of depolarization in autorhythmic cells

Decreases | Increases

Due to parasympathetic innervation

Sympathetic innervation and epinephrine

increases → Contractility

increases → Venous constriction

Stroke volume

determined by

Force of contraction in ventricular myocardium

is influenced by

Contractility | End-diastolic volume

which varies with

Venous return

aided by

Skeletal muscle pump | Respiratory pump

FIGURE QUESTION

Which step(s) is (are) controlled by ACh? By norepinephrine? Which tissue(s) has (have) muscarinic receptors? β_1-receptors?

■ **FIGURE 14-31** *Factors that affect cardiac output*

14

RUNNING PROBLEM CONCLUSION

MYOCARDIAL INFARCTION

Walter's angiogram showed two blocked arteries, which were opened by balloon angioplasty. He returned home with instructions from his doctor for modifying his lifestyle to include a better diet, regular exercise, and no cigarette smoking. As part of his follow-up, Walter had a *myocardial perfusion imaging* test, in which he was administered radioactive thallium. The distribution of thallium throughout the heart is an indicator of blood flow to the heart muscle.

In this running problem, you learned about some current techniques for diagnosing and treating heart attacks. You also learned that many of these treatments depend on speed to work effectively. Check your understanding of the physiology covered in this problem by comparing your answers with the information in the summary table.

	QUESTION	FACTS	INTEGRATION AND ANALYSIS
1	Why did the medic give Walter oxygen?	The medic suspects that Walter has had a heart attack. Blood flow and oxygen supply to the heart muscle may be blocked.	If the heart is not pumping effectively, the brain may not receive adequate oxygen. Administration of oxygen will raise the amount of oxygen that reaches both the heart and the brain.
2	What effect would the injection of isotonic saline have on Walter's extracellular fluid volume? On his intracellular fluid volume? On his total body osmolarity?	An isotonic solution is one that does not change cell volume [🔁 p. 157]. Isotonic saline (NaCl) is isosmotic to the body.	The extracellular volume will increase because all of the saline administered will remain in that compartment. Intracellular volume and total body osmolarity will not change.

	QUESTION	FACTS	INTEGRATION AND ANALYSIS
3	A related form of creatine kinase is found in skeletal muscle. What are related forms of an enzyme called?	Related forms of an enzyme are called isozymes [⊇ p. 96].	Although isozymes are variants of the same enzymes, their activity may vary under different conditions, and their structures are slightly different. Cardiac and skeletal muscle isozymes can be distinguished by their different structures.
4	What is troponin, and why would elevated blood levels of troponin indicate heart damage?	Troponin is the regulatory protein bound to tropomyosin [⊇ p. 407]. Ca^{2+} binding to troponin uncovers the myosin-binding site of actin to allow contraction.	Troponin is part of the contractile apparatus of the muscle cell. If troponin escapes from the cell and enters the blood, this is an indication that the cell either has been damaged or is dead.
5	How do electrical signals move from cell to cell in the myocardium?	Electrical signals pass through gap junctions in intercalated disks [⊇ p. 175].	The cells of the heart are electrically linked by gap junctions.
6	What happens to contraction in a myocardial contractile cell if a wave of depolarization passing through the heart bypasses it?	Depolarization in a muscle cell is the signal for contraction.	If a myocardial cell is not depolarized, it will not contract. Failure to contract creates a nonfunctioning region of heart muscle and impairs the pumping function of the heart.
7	What will a beta-blocker do to Walter's heart rate? Why is that response helpful following a heart attack?	A beta-blocker is an antagonist to β_1-adrenergic receptors. Activation of β_1-receptors increases heart rate.	A beta-blocker therefore decreases heart rate and lowers oxygen demand. Cells that need less oxygen are less likely to die if their blood supply is diminished.
8	If the ventricle of the heart is damaged, in which wave or waves of the electrocardiogram would you expect to see abnormal changes?	The P wave represents atrial depolarization. The QRS complex and T wave represent ventricular depolarization and repolarization, respectively.	The QRS complex and the T wave are most likely to show changes after a heart attack. Changes indicative of myocardial damage include enlargement of the Q wave, shifting of the S-T segment off the base-line (elevated or depressed), and inversion of the T wave.
9	If Walter's heart attack has damaged the muscle of his left ventricle, what do you predict will happen to his left cardiac output?	Cardiac output equals stroke volume times heart rate.	If the ventricular myocardium has been weakened, stroke volume may decrease. Decreased stroke volume will in turn decrease cardiac output.

457   461   471   479   484   489   **494**

CHAPTER SUMMARY

The cardiovascular system exemplifies many of the basic themes in physiology. Blood flows through vessels as a result of high pressure created during ventricular contraction (*mass flow*). The circulation of blood provides an essential route for *cell-to-cell communication*, particularly for hormones and other chemical signals. Myocardial contraction, like contraction in skeletal and smooth muscle, demonstrates the importance of *molecular interactions, biological energy use,* and the *mechanical properties* of cells and tissues. This chapter also introduced the *control systems* for cardiovascular physiology, a theme that will be expanded in the next chapter.

Overview of the Cardiovascular System

IP **Cardiovascular—Anatomy Review: The Heart**

1. The human **cardiovascular system** consists of a **heart** that pumps **blood** through a closed system of **blood vessels**. (p. 457; Fig. 14-1)

2. The primary function of the cardiovascular system is the transport of nutrients, water, gases, wastes, and chemical signals to and from all parts of the body. (p. 458; Tbl. 14-1)

3. Blood vessels that carry blood away from the heart are called **arteries**. Blood vessels that return blood to the heart are called **veins**. **Valves** in the heart and veins ensure unidirectional blood flow. (p. 458; Fig. 14-1)

4. The heart is divided into two **atria** and two **ventricles**. (p. 459; Fig. 14-1)

5. The **pulmonary circulation** goes from the right side of the heart to the lungs and back to the heart. The **systemic circulation** goes from the left side of the heart to the tissues and back to the heart. (p. 460; Fig. 14-1)

Pressure, Volume, Flow, and Resistance

6. Blood flows down a **pressure gradient** (ΔP), from the highest pressure in the **aorta** and arteries to the lowest pressure in the **venae cavae** and **pulmonary veins**. (p. 460; Fig. 14-2)

7. In a system in which fluid is flowing, pressure decreases over distance. (p. 461; Fig. 14-3)

8. The pressure created when the ventricles contract is called the **driving pressure** for blood flow. (p. 462)

9. **Resistance** of a fluid flowing through a tube increases as the length of the tube and the **viscosity** (thickness) of the fluid increase, and as the radius of the tube decreases. Of these three factors, radius has the greatest effect on resistance. (pp. 462–463)

10. If resistance increases, flow rate decreases. If resistance decreases, flow rate increases. (p. 462; Fig. 14-5)

11. Fluid flow through a tube is proportional to the pressure gradient (ΔP). A pressure gradient is not the same thing as the absolute pressure in the system. (p. 462; Fig. 14-4)

12. **Flow rate** is the volume of blood that passes one point in the system per unit time. (p. 464)

13. **Velocity of flow** is the distance a volume of blood travels in a given period of time. At a constant flow rate, the velocity of flow through a small tube is faster than the velocity of flow through a larger tube. (p. 464; Fig. 14-6)

Cardiac Muscle and the Heart

IP **Cardiovascular—Cardiac Action Potential**

14. The heart is composed mostly of cardiac muscle, or **myocardium**. Most cardiac muscle is typical striated muscle. (p. 465; Fig. 14-7h)

15. The signal for contraction originates in **autorhythmic cells** in the heart. Autorhythmic cells are noncontractile myocardium. (p. 470)

16. Myocardial cells are linked to one another by **intercalated disks** that contain gap junctions. The junctions allow depolarization to spread rapidly from cell to cell. (p. 470; Fig. 14-10)

17. In contractile cell excitation-contraction coupling, an action potential opens Ca^{2+} channels. Ca^{2+} entry into the cell triggers the release of additional Ca^{2+} from the sarcoplasmic reticulum through **calcium-induced calcium release**. (p. 471; Fig. 14-11)

18. The force of cardiac muscle contraction can be graded according to how much Ca^{2+} enters the cell. (p. 471)

19. As initial muscle fiber length increases, force of contraction increases. (p. 472; Fig. 14-12)

20. The action potentials of myocardial contractile cells have a rapid depolarization phase created by Na^+ influx, and a steep repolarization phase due to K^+ efflux. The action potential also has a plateau phase created by Ca^{2+} influx. (p. 473; Fig. 14-13)

21. Autorhythmic myocardial cells have an unstable membrane potential called a **pacemaker potential**. The pacemaker potential is due to I_f **channels** that allow net influx of positive charge. (p. 474; Fig. 14-15)

22. The steep depolarization phase of the autorhythmic cell action potential is caused by Ca^{2+} influx. The repolarization phase is due to K^+ efflux. (p. 474; Fig. 14-15)

23. Norepinephrine and epinephrine act on β_1-receptors to speed up the rate of the pacemaker depolarization and increase heart rate. Acetylcholine activates muscarinic receptors and slows down heart rate. (p. 475; Fig. 14-16)

The Heart as a Pump

IP **Cardiovascular—Intrinsic Conduction System**

24. Action potentials originate at the **sinoatrial node** (SA node) and spread rapidly from cell to cell in the heart. Action potentials are followed by a wave of contraction. (p. 477; Fig. 14-17)

25. The electrical signal moves from the SA node through the **internodal pathway** to the **atrioventricular node** (AV node), then into the **AV bundle**, **bundle branches**, terminal **Purkinje fibers**, and myocardial contractile cells. (p. 477; Fig. 14-18)

26. The SA node sets the pace of the heartbeat. If the SA node malfunctions, other autorhythmic cells in the AV node or ventricles will take control of heart rate. (p. 479)

27. An **electrocardiogram** (ECG) is a surface recording of the electrical activity of the heart. The **P wave** represents atrial depolarization. The **QRS complex** represents ventricular depolarization. The **T wave** represents ventricular repolarization. Atrial repolarization is incorporated in the QRS complex. (p. 480; Figs. 14-20, 14-21)

28. An ECG provides information on heart rate and rhythm, conduction velocity, and the condition of cardiac tissues. (p. 481)

IP **Cardiovascular—The Cardiac Cycle**

29. One **cardiac cycle** includes one cycle of contraction and relaxation. **Systole** is the contraction phase; **diastole** is the relaxation phase. (p. 484; Fig. 14-24)

30. Most blood enters the ventricles while the atria are relaxed. Only 20% of ventricular filling at rest is due to atrial contraction. (p. 484)

31. The **AV valves** prevent backflow of blood into the atria. Vibrations following closure of the AV valves create the **first heart sound**. (p. 485; Figs. 14-9, 14-24)

32. During **isovolumic ventricular contraction**, the ventricular blood volume does not change, but pressure rises. When ventricular pressure exceeds arterial pressure, the **semilunar valves** open, and blood is ejected into the arteries. (pp. 485–486; Figs. 14-9, 14-24)

33. When the ventricles relax and ventricular pressure falls, the semilunar valves close, creating the **second heart sound**. (p. 486; Fig. 14-9, 14-24)

34. The amount of blood pumped by one ventricle during one contraction is known as the **stroke volume**. (p. 487; Fig. 14-25)

IP **Cardiovascular—Cardiac Output**

35. **Cardiac output** is the volume of blood pumped per ventricle per unit time. It is equal to heart rate times stroke volume. The average cardiac output at rest is 5 L/min. (p. 489)

36. Homeostatic changes in cardiac output are accomplished by vary-ing heart rate, stroke volume, or both. (p. 493; Fig. 14-31)

37. Parasympathetic activity slows heart rate; sympathetic activity speeds it up. (p. 489; Fig. 14-27)

38. The **Frank-Starling law of the heart** says that an increase in **end-diastolic volume** results in a greater stroke volume. (p. 491; Fig. 14-28)

39. Epinephrine and norepinephrine increase the force of myocardial contraction when they bind to β_1-adrenergic receptors. They also shorten the duration of cardiac contraction. (p. 491; Fig. 14-30)

40. End-diastolic volume and **preload** are determined by **venous return**. Venous return is affected by skeletal muscle contractions,

the respiratory pump, and constriction of veins by sympathetic ac-tivity. (p. 491)

41. **Contractility** of the heart is enhanced by catecholamines and cer-tain drugs. Chemicals that alter contractility are said to have an **inotropic effect**. (p. 491; Fig. 14-29)

42. **Afterload** is the load placed on the ventricle as it contracts. After-load reflects the preload and the effort required to push the blood out into the arterial system. Mean arterial pressure is a clinical indi-cator of afterload. (p. 493)

43. **Ejection fraction**, the percent of EDV ejected with one contraction (stroke volume/EDV), is one measure for evaluating ventricular function. (p. 493)

QUESTIONS

(Answers to the Review Questions begin on page A1.)

THE PHYSIOLOGY PLACE

Access more review material online at **The Physiology Place** website. There you'll find review questions, problem-solving activities, case studies, flashcards, and direct links to both *InterActive Physiology*® and *PhysioEx*™. To access the site, go to *www.physiologyplace.com* and select Human Physiology, Fourth Edition.

LEVEL ONE   REVIEWING FACTS AND TERMS

1. What contributions to understanding the cardiovascular system did each of the following people make?
   - (a) William Harvey
   - (b) Otto Frank and Ernest Starling
   - (c) Marcello Malpighi

2. List three functions of the cardiovascular system.

3. Put the following structures in the order in which blood passes through them, starting and ending with the left ventricle:
   - (a) left ventricle
   - (b) systemic veins
   - (c) pulmonary circulation
   - (d) systemic arteries
   - (e) aorta
   - (f) right ventricle

4. The primary factor causing blood to flow through the body is a _____ gradient. In humans, the value of this gradient is high-est at the _____ and in the _____. It is lowest in the _____. In a system in which fluid is flowing, pressure de-creases over distance because _____.

5. If vasodilation occurs in a blood vessel, pressure (increases/decreases).

6. The specialized cell junctions between myocardial cells are called _____. These areas contain _____ that allow rapid con-duction of electrical signals.

7. Trace an action potential from the SA node through the conducting system of the heart.

8. Distinguish between the two members of each of the following pairs:
   - (a) end-systolic volume and end-diastolic volume
   - (b) sympathetic and parasympathetic control of heart rate
   - (c) diastole and systole
   - (d) systemic and pulmonary circulation
   - (e) AV node and SA node

9. Match the descriptions with the correct anatomic terms. Not all terms are used. Give a definition for the unused terms.
   - (a) tough membranous sac that encases the heart
   - (b) valves between ventricles and the main arteries
   - (c) a vessel that carries blood away from the heart
   - (d) lower chamber of the heart
   - (e) valve between left atrium and left ventricle
   - (f) primary artery of the systemic circulation
   - (g) muscular layer of the heart
   - (h) narrow end of the heart; points downward
   - (i) valve with papillary muscles
   - (j) the upper chambers of the heart

   1. aorta
   2. apex
   3. artery
   4. atria
   5. atrium
   6. AV valve
   7. base
   8. bicuspid valve
   9. endothelium
   10. myocardium
   11. pericardium
   12. semilunar valve
   13. tricuspid valve
   14. ventricle

10. What events cause the two principal heart sounds?

11. What is the proper term for each of the following?
    - (a) number of heart contractions per minute
    - (b) volume of blood in the ventricle before the heart contracts
    - (c) volume of blood that enters the aorta with each contraction
    - (d) volume of blood that leaves the heart in one minute
    - (e) volume of blood in the entire body

LEVEL TWO   REVIEWING CONCEPTS

12. Concept maps:
    - (a) Create a map showing blood flow through the heart and body. Label as many structures as you can.
    - (b) Create a map for control of cardiac output using the following terms. You may add additional terms.

ACh	heart rate
adrenal medulla	length-tension relationship
autorhythmic cells	muscarinic receptor
β_1-receptor	norepinephrine
Ca^{2+}	parasympathetic neurons
Ca^{2+}-induced Ca^{2+} release	respiratory pump
cardiac output	skeletal muscle pump
contractile myocardium	stroke volume
contractility	sympathetic neurons
force of contraction	venous return

13. List the events of the cardiac cycle in sequence, beginning with atrial and ventricular diastole. Note when valves open and close. Describe what happens to pressure and blood flow in each chamber at each step of the cycle.

14. Compare and contrast the structure of a cardiac muscle cell with that of a skeletal muscle cell. What unique properties of cardiac muscle are essential to its function?

15. Explain why contractions in cardiac muscle cannot sum or exhibit tetanus.

16. Correlate the waves of an ECG with mechanical events in the atria and ventricles. Why are there only three electrical events but four mechanical events?

17. Match the following ion movements with the appropriate phrase. More than one ion movement may apply to a single phrase. Some choices may not be used.

   (a) slow rising phase of autorhythmic cells
   (b) plateau phase of contractile cells
   (c) rapid rising phase of contractile cells
   (d) rapid rising phase of autorhythmic cells
   (e) rapid falling phase of contractile cells
   (f) falling phase of autorhythmic cells
   (g) cardiac muscle contraction
   (h) cardiac muscle relaxation

   1. K^+ from ECF to ICF
   2. K^+ from ICF to ECF
   3. Na^+ from ECF to ICF
   4. Na^+ from ICF to ECF
   5. Ca^{2+} from ECF to ICF
   6. Ca^{2+} from ICF to ECF

18. List and briefly explain four types of information that an ECG provides about the heart.

19. Define inotropic effect. Name two drugs that have a positive inotropic effect on the heart.

LEVEL THREE PROBLEM SOLVING

20. Two drugs used to reduce cardiac output are calcium channel blockers and beta (receptor) blockers. What effect do these drugs have on the heart that explains how they decrease cardiac output?

21. Police Captain Jeffers has suffered a myocardial infarction.

   (a) Explain to his (nonmedically oriented) family what has happened to his heart.

(b) When you analyzed his ECG, you referred to several different leads, such as lead I and lead III. What are leads?

(c) Why is it possible to record an ECG on the body surface without direct access to the heart?

22. What might cause a longer-than-normal PR interval in an ECG?

23. The following paragraph is a summary of a newspaper article:

   A new treatment for atrial fibrillation due to an excessively rapid rate at the SA node involves a high-voltage electrical pulse administered to the AV node to destroy its autorhythmic cells. A ventricular pacemaker is then implanted in the patient.

   Briefly explain the physiological rationale for this treatment. Why is a rapid atrial depolarization rate dangerous? Why is the AV node destroyed in this procedure? Why must a pacemaker be implanted?

LEVEL FOUR QUANTITATIVE PROBLEMS

24. Police Captain Jeffers in question 21 has an ejection fraction (SV divided by EDV) of only 25%. His stroke volume is 40 mL/beat, and his heart rate is 100 beats/min. What are his EDV, ESV, and CO? Show your calculations.

25. If 1 cm water = 0.74 mm Hg:

   (a) Convert a pressure of 120 mm Hg to cm H_2O.
   (b) Convert a pressure of 90 cm H_2O to mm Hg.

26. Calculate cardiac output if stroke volume is 65 mL/beat and heart rate is 80 beats/min.

27. Calculate end-systolic volume if end-diastolic volume is 150 mL and stroke volume is 65 mL/beat.

28. A person has a total blood volume of 5 L. Of this total, assume that 4 L is contained in the systemic circulation and 1 L is in the pulmonary circulation. If the person has a cardiac output of 5 L/min, how long will it take (a) for a drop of blood leaving the left ventricle to return to the left ventricle and (b) for a drop of blood to go from the right ventricle to the left ventricle?

14

ANSWERS

✓ Answers to Concept Check Questions

Page 460

1. A cardiovascular system has tubes (vessels), fluid (blood), and a pump (heart).

2. (a) The pulmonary circulation takes blood to and from the lungs; the systemic circulation takes blood to and from the rest of the body. (b) An artery carries blood away from the heart; a vein carries blood to the heart. (c) An atrium is an upper heart chamber that receives blood entering the heart; a ventricle is a lower heart chamber that pumps blood out of the heart.

Page 464

3. The pressure gradient is more important.

4. The bottom tube has the greater flow because it has the larger pressure gradient (50 mm Hg versus 40 mm Hg for the top tube).

5. Tube C has the highest flow because it has the largest radius of the four tubes (less resistance) and the shorter length (less resistance). (Tube B has the same radius as tube C but a longer length and there-

fore offers greater resistance to flow.) Tube D, with the greatest resistance due to longer length and narrow radius, has the lowest flow.

Page 465

6. If the canals are identical in size and therefore in cross-sectional area A, the canal with the higher velocity of flow v has the higher flow rate Q. (From equation 7, $Q = v \times A$.)

Page 470

7. Connective tissue is not excitable and is therefore unable to conduct action potentials.

8. Superior vena cava → right atrium → tricuspid (right AV) valve → right ventricle → pulmonary (right semilunar) valve → pulmonary trunk → pulmonary vein → left atrium → mitral (bicuspid, left AV) valve → left ventricle → aortic (left semilunar) valve → aorta

9. The AV valves prevent backward flow of blood. If one fails, blood leaks back into the atrium.

Page 471

10. From this experiment, it is possible to conclude that myocardial cells require extracellular calcium for contraction but skeletal muscle cells do not.

Page 472

11. If all calcium channels in the muscle cell membrane are blocked, there will be no contraction. If only some are blocked, the force of contraction will be smaller than the force created with all channels open.

Page 474

12. Na^+ influx causes neuronal depolarization, and K^+ efflux causes neuronal repolarization.

13. The refractory period represents the time required for the Na^+ channel gates to reset (activation gate closes, inactivation gate opens).

14. If cardiac Na^+ channels are blocked with lidocaine, the cell will not depolarize and therefore will not contract.

Page 475

15. Increasing K^+ permeability hyperpolarizes the membrane potential.

Page 477

16. The Ca^{2+} channels in autorhythmic cells are not the same as those in contractile cells. Autorhythmic Ca^{2+} channels open rapidly when the membrane potential reaches about -50 mV and close when it reaches about $+20$ mV. The Ca^{2+} channels in contractile cells are slower and do not open until the membrane has depolarized fully.

17. If tetrodotoxin is applied to a myocardial autorhythmic cell, nothing will happen because there are no voltage-gated Na^+ channels in the cell.

18. Cutting the vagus nerve caused heart rate to increase, so the nerve must contain parasympathetic fibers that slow heart rate.

Page 480

19. The AV node conducts action potentials from atria to ventricles. It also slows down the speed at which those action potentials are conducted, allowing atrial contraction to end before ventricular contraction begins.

20. The SA node is in the upper right atrium.

21. The fastest pacemaker sets the heart rate, so the heart rate increases to 120 beats/min.

Page 484

22. In Figure 14-23b, notice that there is no regular association between the P waves and the QRS complexes (the P-R segment varies in length). Notice also that not every P wave has an associated QRS complex. Both P waves and QRS complexes appear at regular intervals, but the atrial rate (P waves) is faster than the ventricular rate (QRS complexes). In (c), there are identifiable R waves but no P waves. In (d), there are no recognizable waves at all, indicating that the depolarizations are not following the normal conduction path.

23. Starting at left, the waves are P, P, QRS, T, P, P, QRS, T, P, P, P, and so on. Each P wave that is not followed by a QRS wave suggests an intermittent conduction block at the AV node. See also Figure 14-23b.

Page 484

24. The atrium has lower pressure than the venae cavae.

Page 486

25. (a) ventricle, (b) ventricle, (c) atrium, (d) ventricle

26. (a) ventricular ejection, (b) isovolumic ventricular contraction and ventricular ejection (c) from isovolumic ventricular relaxation until ventricular contraction begins again

Page 487

27. EDV occurs in step 3, and ESV occurs in step 5.

28. (a) E, (b) A, (c) D, (d) B, (e) C, (f) F

29. Atrial pressure increases because pressure on the mitral valve pushes the valve back into the atrium, decreasing atrial volume. Atrial pressure decreases during the initial part of ventricular systole as the atrium relaxes. The pressure then increases as the atrium fills with blood. Atrial pressure begins to decrease at point D, when the mitral valve opens and blood flows down into the ventricles.

30. Ventricular pressure shoots up when the ventricles contract on a fixed volume of blood.

Page 489

31. After 10 beats, the pulmonary circulation will have gained 10 mL of blood and the systemic circulation will have lost 10 mL.

Page 493

32. Your drawing should show a β_1-receptor on the cell membrane activating intracellular cAMP, which should have an arrow drawn to Ca^{2+} channels on the sarcoplasmic reticulum. Open channels should be shown as increasing cytoplasmic Ca^{2+}. A second arrow should go from cAMP to Ca^{2+}-ATPase on the SR and the cell membrane, showing increased uptake in the SR and increased removal of Ca^{2+} from the cell.

Page 493

33. The aortic valve is found in the left ventricle. A stenotic aortic valve would increase the afterload on the ventricle.

Answers to Figure and Graph Questions

Page 475

Figure 14-15: Phase 2 (the plateau) of the contractile cell action potential has no equivalent in the autorhythmic cell action potential. Phase 4 is approximately equivalent to the pacemaker potential. Both action potentials have rising phases, peaks, and falling phases.

Page 481

Figure 14-20: The heart rate is either 75 beats/min or 80 beats/min, depending on how you calculate it. If you use the data from one R peak to the next, the time interval between the two peaks is 0.8 sec; therefore,

$$\frac{1 \text{ beat}}{0.8 \text{ sec}} \times \frac{60 \text{ sec}}{1 \text{ min}} = 75 \text{ beats/min}$$

However, it is more accurate to estimate rate by using several seconds of the ECG tracing rather than one RR interval because beat-to-beat intervals may vary. If you start counting at the first R wave on the top graph and go right for 3 sec, there are 4 beats in that time period, which means

$$\frac{4 \text{ beats}}{3 \text{ sec}} \times \frac{60 \text{ sec}}{1 \text{ min}} = 80 \text{ beats/min}$$

Page 487

Figure 14-25: Match segments to events: (a) C → D, (b) B → C, (c) D → A, (d) A → B. Match events to points: (a) C, (b) A, (c) D, (d) B.

Page 490

Figure 14-28: Maximum stroke volume is about 160 mL/beat, first achieved when end-diastolic volume is about 330 mL.

Page 491

Fig. 14-29: At point A, the heart under the influence of norepinephrine has a larger stroke volume and is therefore creating more force.

Page 494

Fig. 14-31: Heart rate is the only parameter controlled by ACh. Heart rate and contractility are both controlled by norepinephrine. The SA node has muscarinic receptors. The SA node and contractile myocardium have β_1-receptors.

14

15

Since 1900, CVD (cardiovascular disease) has been the No. 1 killer in the United States every year but 1918.

—American Heart Association, Heart Disease and Stroke Statistics—2003 Update

Endothelial cell of the microcirculation (green actin, red microtubules, blue DNA).

Blood Flow and the Control of Blood Pressure

BACKGROUND BASICS

RUNNING PROBLEM

ESSENTIAL HYPERTENSION

"Doc, I'm as healthy as a horse," says Kurt English, age 56, during his annual physical examination. "I don't want to waste your time. Let's get this over with." But to Dr. Arthur Cortez, Kurt does not appear to be the picture of health: he is about 30 pounds overweight. When Dr. Cortez asks about his diet, Kurt replies, "Well, I like to eat." Exercise? "Who has the time? " replies Kurt. Dr. Cortez wraps a blood pressure cuff around Kurt's arm and takes a reading. "Your blood pressure is 164 over 100," says Dr. Cortez. "We'll take it again in 15 minutes. If it's still high, we'll need to discuss it further." Kurt stares at his doctor, flabbergasted. "But how can my blood pressure be too high? I feel fine!" he protests.

501 508 510 513 521 529

Anthony was sure he was going to be a physician, until the day in physiology laboratory they studied blood types. When the lancet pierced his fingertip and he saw the drop of bright red blood well up, the room started to spin, then everything went black. He awoke, much embarrassed, to the sight of his classmates and the teacher bending over him.

Anthony suffered an attack of *vasovagal syncope* (fainting), a benign and common emotional reaction to blood, hypodermic needles, or other upsetting sights. Normally, homeostatic regulation of the cardiovascular system maintains adequate blood flow (*perfusion*) to the heart and brain. In vasovagal syncope, signals from the nervous system cause a sudden drop in blood pressure, and the individual faints from lack of oxygen to the brain. In this chapter you will learn how the heart and blood vessels work together most of the time to prevent such problems.

A simplified model of the cardiovascular system (Fig. 15-1 ■) illustrates the key points we will discuss in this chapter. This model shows the heart as two separate pumps that work in series (one after the other), with the right heart pumping blood first to the lungs and then to the left heart. The left heart then pumps blood through the rest of the body and back to the right heart.

Blood leaving the left heart enters systemic arteries, shown here as an expandable, elastic region. Pressure produced by

15

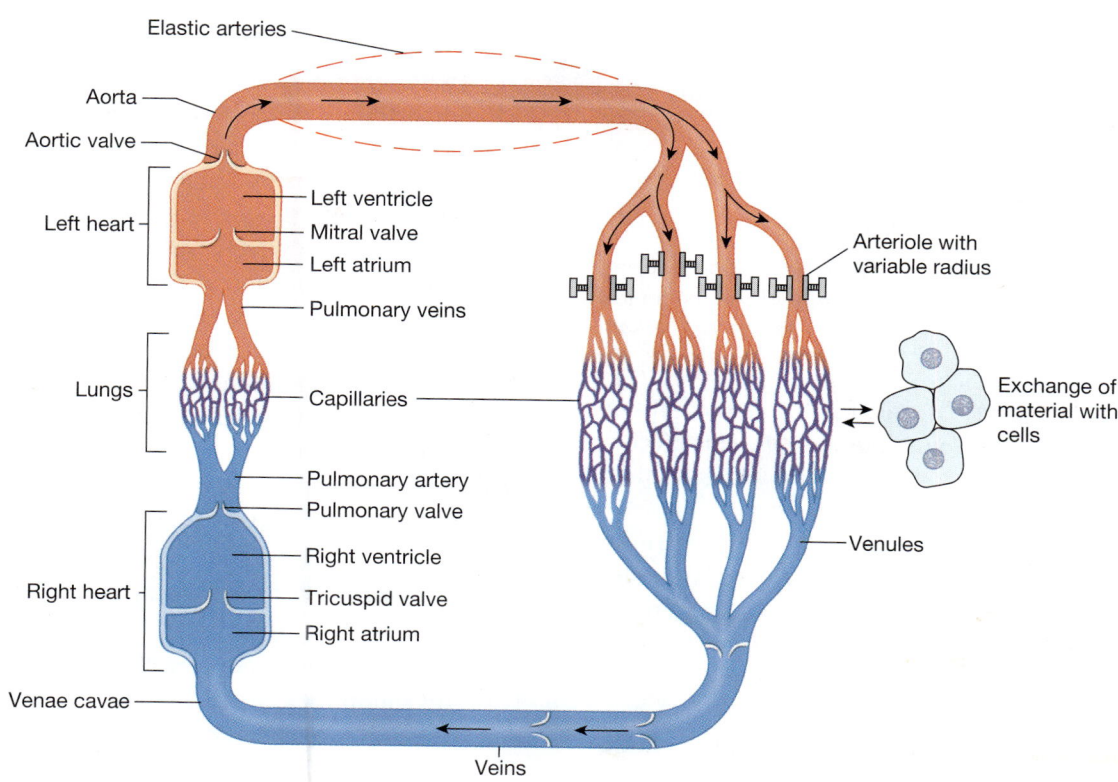

■ **FIGURE 15-1** *Functional model of the cardiovascular system*

The heart functions as two pumps working in series. The systemic arteries are a pressure reservoir that maintains blood flow during ventricular relaxation. The arterioles, shown with adjustable screws representing variable radii, are the site of variable resistance. Exchange between the blood and cells takes place only at the capillaries.

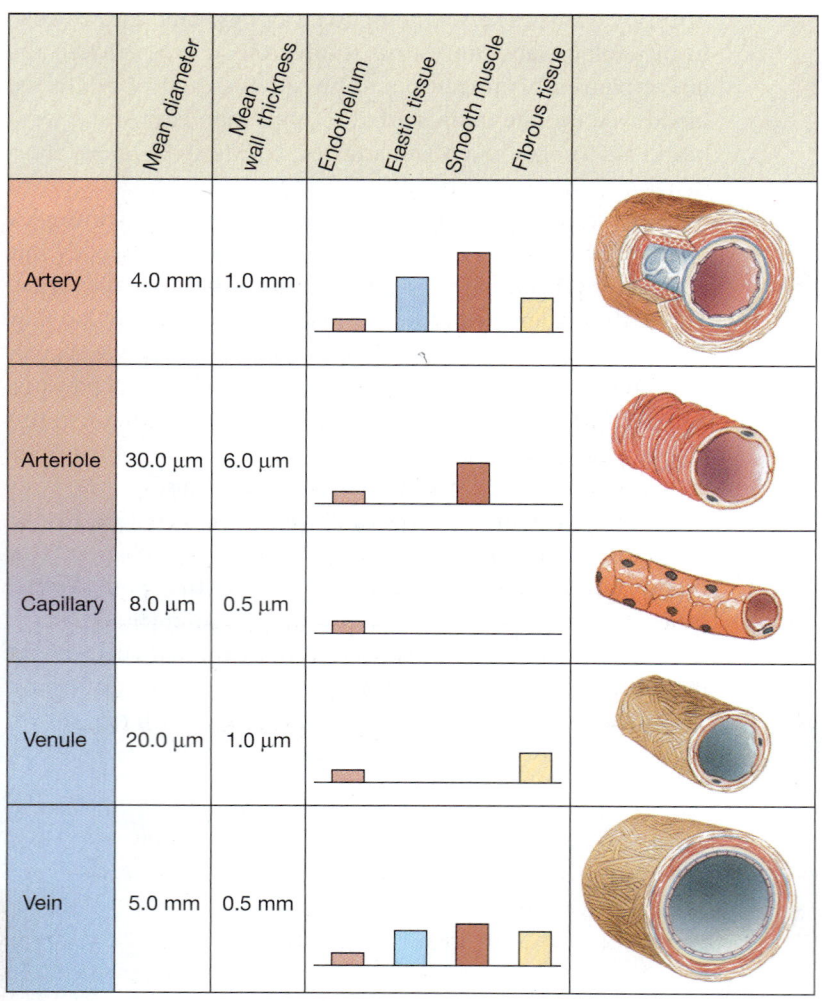

	Mean diameter	Mean wall thickness	Endothelium	Elastic tissue	Smooth muscle	Fibrous tissue	
Artery	4.0 mm	1.0 mm					
Arteriole	30.0 μm	6.0 μm					
Capillary	8.0 μm	0.5 μm					
Venule	20.0 μm	1.0 μm					
Vein	5.0 mm	0.5 mm					

■ **FIGURE 15-2** *Blood vessel structure*

The walls of blood vessels vary in diameter and composition. The endothelium and its underlying elastic tissue together form the tunica intima. (Adapted from A. C. Burton, *Physiol. Rev.* 34:619–642, 1954.)

contraction of the left ventricle is stored in the elastic walls of arteries and slowly released through *elastic recoil*. This mechanism maintains a continuous driving pressure for blood flow during the time when the ventricles are relaxing.

Downstream from the arteries, small vessels called **arterioles** create a high-resistance outlet for arterial blood flow. Arterioles direct distribution of blood flow to individual tissues by selectively constricting and dilating. Arteriolar diameter is regulated both by local factors, such as tissue oxygen concentrations, and by homeostatic control.

Once blood flows into the capillaries, a leaky epithelium allows exchange of materials between the plasma, the interstitial fluid, and the cells of the body. At the distal end of the capillaries, blood flows into the venous side of the circulation and from there back to the right heart.

Total blood flow through any level of the circulation is equal to cardiac output. For example, if cardiac output is 5 L/min, blood flow through all the systemic capillaries is 5 L/min. In the same manner, blood flow through the pulmonary side of the circulation is equal to blood flow through the systemic circulation.

THE BLOOD VESSELS

The walls of blood vessels are composed of layers of smooth muscle, elastic connective tissue, and fibrous connective tissue (Fig. 15-2 ■). The inner lining of all blood vessels is a thin layer of **endothelium**, a type of epithelium. For years, the endothelium was thought to be simply a passive barrier. However, we now know that endothelial cells secrete many paracrines and play important roles in the regulation of blood pressure, blood vessel growth, and absorption of materials.

Surrounding the endothelium are layers of connective tissue and smooth muscle. The endothelium and its adjacent elastic connective tissue together make up the *tunica intima*, usually called simply the *intima*. The layers of smooth muscle and connective tissue surrounding the intima vary in thickness in different vessels. The descriptions that follow apply to the vessels of the systemic circulation, although those of the pulmonary circulation are generally similar.

Blood Vessels Contain Vascular Smooth Muscle

The smooth muscle of blood vessels is known as **vascular smooth muscle**. Most blood vessels contain smooth muscle, arranged in either circular or spiral layers. *Vasoconstriction* narrows the diameter of the vessel lumen, whereas *vasodilation* widens it.

In most blood vessels, smooth muscle cells maintain a state of partial contraction at all times, creating the condition known as *muscle tone* [🔁 p. 440]. Contraction of smooth muscle, like that of cardiac muscle, depends on the entry of Ca^{2+} from the extracellular fluid through Ca^{2+} channels [🔁 p. 426]. A variety of chemicals influences vascular smooth muscle tone, including neurotransmitters, hormones, and paracrines. Many vasoactive paracrines are secreted either by endothelial cells lining blood vessels or by tissues surrounding the vessels.

Arteries and Arterioles Carry Blood Away from the Heart

The aorta and major arteries are characterized by walls that are both stiff and springy. Arteries have a thick smooth muscle layer and large amounts of elastic and fibrous connective tissue (Fig. 15-2). Because of the stiffness of the fibrous tissue, substantial amounts of energy are required to stretch the walls of an artery outward. This energy comes in the form of high-pressure blood ejected from the left ventricle. Once the artery is distended with blood, energy stored by stretched elastic fibers is released through elastic recoil.

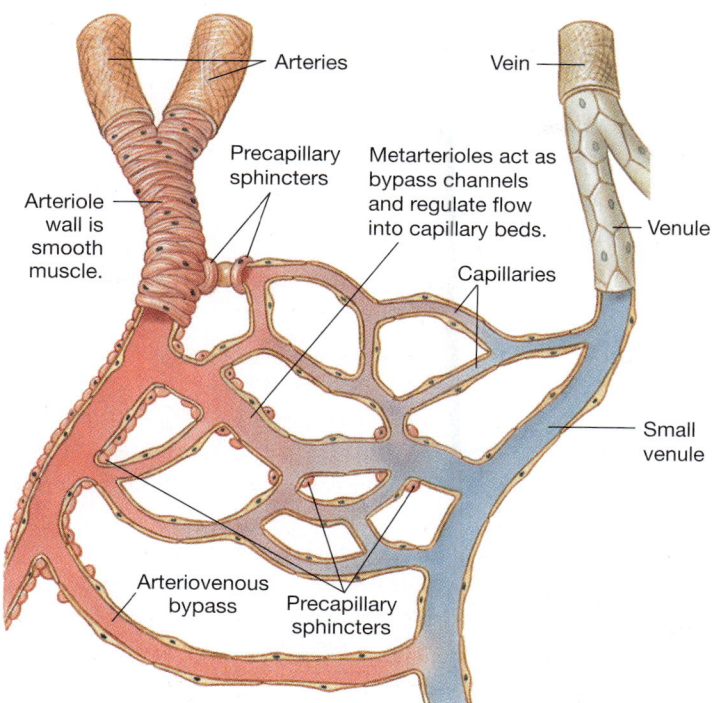

■ **FIGURE 15-3**   *Metarterioles*

The arteries and arterioles are characterized by a divergent [*divergere,* bend apart] pattern of blood flow. As major arteries divide into smaller and smaller arteries, the character of the wall changes, becoming less elastic and more muscular. The walls of arterioles contain several layers of smooth muscle that contract and relax under the influence of various chemical signals.

Some arterioles branch into vessels known as **metarterioles** [*meta-,* beyond] (Fig. 15-3 ■). Blood flowing through metarterioles can either be directed into adjoining capillary beds or can bypass the capillaries and go directly to the venous circulation if muscle rings called **precapillary sphincters** [*sphingein,* to hold tight] are contracted. True arterioles have a continuous smooth muscle layer, but only part of the wall of metarterioles is surrounded by smooth muscle.

In addition to regulating blood flow through the capillaries, metarterioles allow white blood cells to go directly from the arterial to the venous circulation. Capillaries are barely large enough to let red blood cells through, much less white blood cells, which are twice as large.

Arterioles, along with capillaries and small postcapillary vessels called venules, form the *microcirculation*. Regulation of blood flow through the microcirculation is an active area of physiological research.

Exchange Between the Blood and Interstitial Fluid Takes Place in the Capillaries

Capillaries are the smallest vessels in the cardiovascular system. They and the postcapillary venules are the site of exchange between the blood and the interstitial fluid. To facilitate exchange of materials, capillaries lack smooth muscle and elastic or fibrous tissue reinforcement (Fig. 15-2). Instead, their walls consist of a flat layer of endothelium, one cell thick, supported on an acellular matrix called the *basal lamina* (basement membrane) [🔁 p. 72].

Many capillaries are closely associated with cells known as **pericytes** [*peri-,* around]. These highly branched contractile cells surround the capillaries, forming a meshlike outer layer between the capillary endothelium and the interstitial fluid. Pericytes contribute to the "tightness" of capillary permeability: the more pericytes, the less leaky the capillary endothelium. Cerebral capillaries, for example, are surrounded by pericytes and glial cells and have tight junctions that create the *blood-brain barrier* [🔁 p. 299].

Pericytes secrete factors that influence capillary growth, and they can differentiate to become new endothelial or smooth muscle cells. Loss of pericytes around capillaries of the retina is a hallmark of the disease *diabetic retinopathy,* a leading cause of blindness. Scientists are now trying to determine whether pericyte loss is a cause or consequence of the retinopathy.

Blood Flow Converges in the Venules and Veins

Blood flows from the capillaries into small vessels called venules. The very smallest venules are similar to capillaries, with a thin exchange epithelium and little connective tissue (Fig. 15-2). They are distinguished from capillaries by their convergent pattern of flow.

Smooth muscle begins to appear in the walls of larger venules. From venules, blood flows into veins that become larger in diameter as they travel toward the heart. Finally, the largest veins, the venae cavae, empty into the right atrium.

Veins are more numerous than arteries and have a larger diameter. As a result of their large volume, the veins hold more than half of the blood in the circulatory system. Veins lie closer to the surface of the body than arteries, forming the bluish blood vessels that you see running just under the skin. Veins have thinner walls than arteries, with less elastic tissue. As a result, they expand easily when they fill with blood.

When you have blood drawn from your arm (*venipuncture*), the technician uses a tourniquet to exert pressure on the blood vessels. Blood flow into the arm through deep high-pressure arteries is not affected, but pressure exerted by the tourniquet stops outflow through the low-pressure veins. As a result, blood collects in the surface veins, making them stand out against the underlying muscle tissue.

Angiogenesis Creates New Blood Vessels

One topic of tremendous interest to researchers is **angiogenesis**, [*angeion,* vessel + *gignesthai,* to beget], the process by which new blood vessels develop, especially after birth. In children, blood vessel growth is necessary for normal development. In adults, angiogenesis takes place as wounds heal and as the uterine

(a) Ventricular contraction

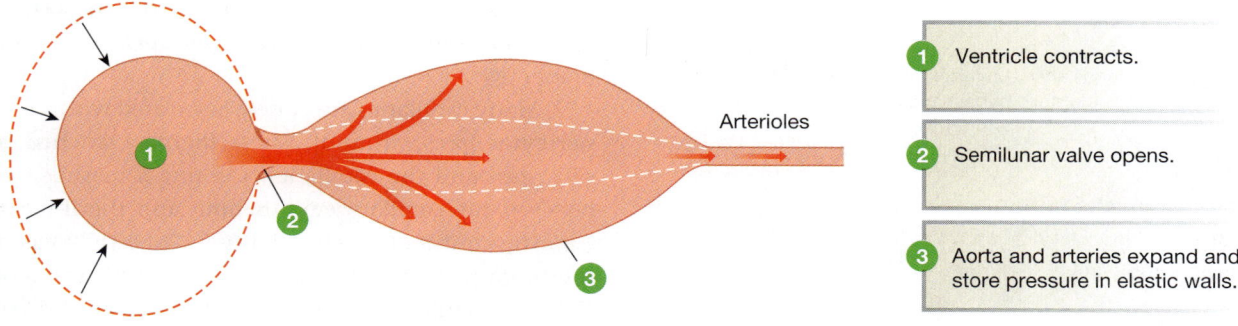

Arterioles

1 Ventricle contracts.

2 Semilunar valve opens.

3 Aorta and arteries expand and store pressure in elastic walls.

(b) Ventricular relaxation

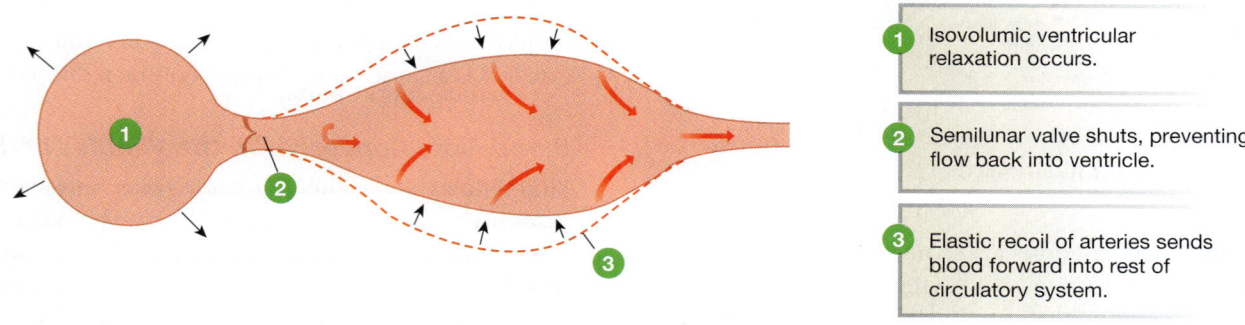

1 Isovolumic ventricular relaxation occurs.

2 Semilunar valve shuts, preventing flow back into ventricle.

3 Elastic recoil of arteries sends blood forward into rest of circulatory system.

■ **FIGURE 15-4** *Elastic recoil in the arteries*

The arteries maintain driving pressure when the ventricles are in diastole.

lining grows after menstruation. Angiogenesis also occurs with endurance exercise training, enhancing blood flow to the heart muscle and to skeletal muscles.

From studies of normal blood vessels and tumor cells, scientists learned that angiogenesis is controlled by a balance of angiogenic and antiangiogenic cytokines. A number of related growth factors, including *vascular endothelial growth factor* (VEGF) and *fibroblast growth factor* (FGF), promote angiogenesis. These growth factors are *mitogens,* meaning they promote mitosis, or cell division. They are normally produced by smooth muscle cells and pericytes.

Cytokines that inhibit angiogenesis include *angiostatin,* made from the blood protein plasminogen, and *endostatin* [*stasis,* a state of standing still]. Scientists are currently using these cytokines to develop new treatments for two major illnesses: cancer and coronary heart disease. As cancer cells invade tissues and multiply, they instruct the host tissue to develop new blood vessels to feed the growing tumor. Without these new vessels, the interior cells of a cancerous mass would be unable to get adequate oxygen and nutrients, and would die. Angiostatin and endostatin are being used in clinical trials to

see if they can block angiogenesis and literally starve tumors to death.

In contrast, **coronary heart disease**, also known as *coronary artery disease,* is a condition in which we would like to be able to selectively induce angiogenesis. In coronary heart disease, blood flow to the myocardium is decreased by fatty deposits that narrow the lumen of the coronary arteries. In some individuals, new blood vessels develop spontaneously and form *collateral circulation* that supplements flow through the partially blocked artery. Researchers are looking for way to duplicate this natural process and induce angiogenesis to replace occluded vessels [*occludere,* to close up].

BLOOD PRESSURE

The pressure created by ventricular contraction is the driving force for blood flow through the cardiovascular system. As blood is ejected from the left ventricle, the aorta and arteries expand to accommodate it (Fig. 15-4 ■). When the ventricle relaxes and the semilunar valve closes, the elastic arterial walls recoil, propelling the blood forward into smaller arteries and

TABLE 15-1	Pressure, Flow, and Resistance in the Cardiovascular System

Flow ∝ ΔP/R

1. Blood flows if a pressure gradient (ΔP) is present.
2. Blood flows from areas of higher pressure to areas of lower pressure.
3. Blood flow is opposed by the resistance R of the system.
4. Three factors affecting resistance are radius of the blood vessels, viscosity of the blood, and length of the system [p. 462].
5. Flow is usually expressed in either liters or milliliters per minute (L/min or mL/min).
6. Velocity of flow is usually expressed in either centimeters per minute (cm/min) or millimeters per second (mm/sec).
7. The primary determinant of velocity of flow (when flow rate is constant) is the total cross-sectional area of the vessel(s).

arterioles (Fig. 15-4b). By sustaining the driving pressure for blood flow during ventricular relaxation, the arteries create continuous blood flow through the blood vessels.

Blood flow obeys the rules of fluid flow that were introduced in Chapter 14 [p. 462]. Flow is directly proportional to the pressure gradient between any two points, and inversely proportional to the resistance of the vessels to flow (Table 15-1 ■). Unless otherwise noted, the discussion that follows is restricted to the events that take place in the systemic circuit. Pulmonary blood flow will be discussed in Chapter 17.

Systemic Blood Pressure Is Highest in Arteries and Lowest in Veins

Blood pressure is highest in the arteries and decreases continuously as blood flows through the circulatory system (Fig. 15-5 ■). The decrease in pressure occurs because energy is lost as a result of the resistance to flow offered by the vessels. Resistance to blood flow also results from friction between the blood cells.

In the systemic circulation, the highest pressure occurs in the aorta and reflects pressure created by the left ventricle. Aortic pressure reaches an average high of 120 mm Hg during ventricular systole (**systolic pressure**), then falls steadily to a low of 80 mm Hg during ventricular diastole (**diastolic pressure**). Notice that although pressure in the ventricle falls to nearly 0 mm Hg as the ventricle relaxes, diastolic pressure in the large arteries remains relatively high. The high diastolic pressure in arteries reflects the ability of those vessels to capture and store energy in their elastic walls.

The rapid pressure increase that occurs when the left ventricle pushes blood into the aorta can be felt as a **pulse**, or pressure wave, transmitted through the fluid-filled arteries. The pressure wave travels about 10 times faster than the blood itself. Even so, a pulse felt in the arm is occurring slightly after the ventricular contraction that created the wave.

The amplitude of the pressure wave decreases over distance because of friction, and the wave finally disappears at the

capillaries (Fig. 15-5). **Pulse pressure,** a measure of the strength of the pressure wave, is defined as systolic pressure minus diastolic pressure:

Systolic pressure − diastolic pressure = pulse pressure

For example, in the aorta:

120 mm Hg − 80 mm Hg = 40 mm Hg pulse pressure

By the time blood reaches the veins, pressure has fallen because of friction, and a pressure wave no longer exists. Low-pressure blood in veins below the heart must flow "uphill," or against gravity, to return to the heart. Try holding your arm straight down without moving for several minutes and notice how the veins in the back of your hand begin to stand out as they fill with blood. (This effect may be more evident in older people, whose subcutaneous connective tissue has lost elasticity.)

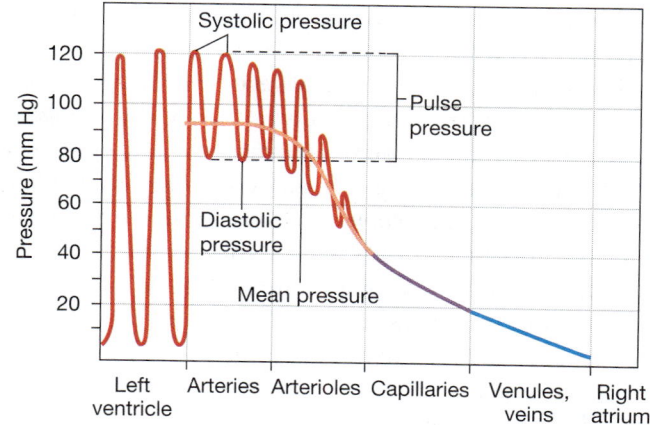

■ **FIGURE 15-5** *Pressure throughout the systemic circulation*

Pressure waves created by ventricular contraction are reflected into the blood vessels. They diminish in amplitude with distance and disappear at the capillaries.

Valves in the veins
prevent backflow
of blood.

When the skeletal muscles compress
the veins, they force blood toward the
heart (the skeletal muscle pump).

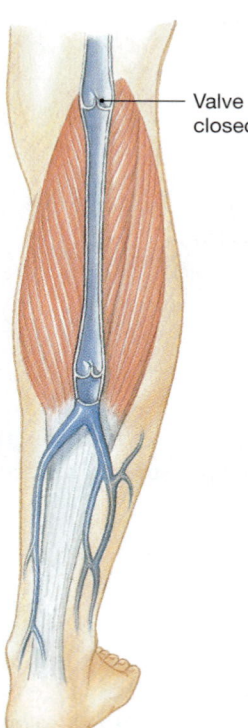
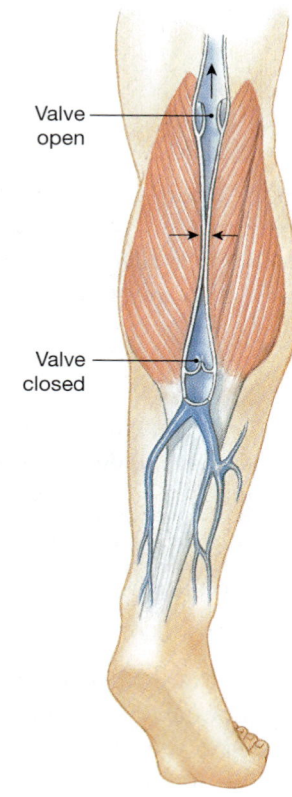

Valve
closed

Valve
open

Valve
closed

■ **FIGURE 15-6**   *Valves ensure one-way flow in veins*

Then raise your hand so that gravity assists the venous flow and watch the bulging veins disappear.

To assist venous flow, some veins have internal one-way valves (Fig. 15-6 ■). These valves, like those in the heart, ensure that blood passing the valve cannot flow backward. Once blood reaches the vena cava, there are no valves. Venous blood flow is steady rather than pulsatile, pushed along by the continuous movement of blood out of the capillaries.

Venous return to the heart is aided by the *skeletal muscle pump* and the *respiratory pump* [↩ p. 491]. When muscles such as those in the calf of the leg contract, they compress the veins, forcing the blood upward past the valves. While your hand is hanging down, try clenching and unclenching your fist to see the effect muscle contraction has on distention of the veins.

CONCEPT CHECK

1. Would you expect to find valves in the veins leading from the brain to the heart? Defend your answer.

2. If you checked the pulse in a person's carotid artery and left wrist at the same time, would the pressure waves occur simultaneously? Explain.

3. Who has the higher pulse pressure, someone with blood pressure of 90/60 or someone with blood pressure of 130/95?

Answers: p. 533

Arterial Blood Pressure Reflects the Driving Pressure for Blood Flow

Arterial blood pressure, or simply "blood pressure," reflects the driving pressure created by the pumping action of the heart. Because ventricular pressure is difficult to measure, it is customary to assume that arterial blood pressure reflects ventricular pressure. Because arterial pressure is pulsatile, we use a single value—the **mean arterial pressure** (MAP)—to represent driving pressure. Mean arterial pressure is estimated as diastolic pressure plus one-third of pulse pressure:

$$\text{MAP} = \text{diastolic P} + 1/3 \,(\text{systolic P} - \text{diastolic P})$$

For a person whose systolic pressure is 120 and diastolic pressure is 80:

$$\text{MAP} = 80 \text{ mm Hg} + 1/3 \,(120 - 80 \text{ mm Hg})$$

$$\text{MAP} = 93 \text{ mm Hg}$$

Mean arterial pressure is closer to diastolic pressure than to systolic pressure because diastole lasts twice as long as systole.

Abnormally high or low arterial blood pressure can be indicative of a problem in the cardiovascular system. If blood pressure falls too low (*hypotension*), the driving force for blood flow will be unable to overcome opposition by gravity. In this instance, blood flow and oxygen supply to the brain are impaired, and the subject may become dizzy or faint.

On the other hand, if blood pressure is chronically elevated (a condition known as *hypertension*, or high blood pressure), high pressure on the walls of blood vessels may cause weakened areas to rupture and bleed into the tissues. If a rupture occurs in the brain, it is called a *cerebral hemorrhage* and may cause the loss of neurological function commonly called a *stroke*. If a weakened area ruptures in a major artery, such as the descending aorta, rapid blood loss into the abdominal cavity will cause blood pressure to fall below the critical minimum. Without prompt treatment, rupture of a major artery is fatal.

CONCEPT CHECK

4. The formula given for calculating MAP applies to a typical resting heart rate of 60–80 beats/min. If heart rate increases, would the contribution of systolic pressure to mean arterial pressure decrease or increase, and would MAP decrease or increase?

5. Peter's systolic pressure is 112 mm Hg, and his diastolic pressure is 68 mm Hg (written 112/68). What is his pulse pressure? His mean arterial pressure?

Answers: p. 533

Blood Pressure Is Estimated by Sphygmomanometry

We estimate arterial blood pressure in the radial artery of the arm using a *sphygmomanometer*, an instrument consisting of an inflatable cuff and a pressure gauge [*sphygmus*, pulse + *manometer*, an instrument for measuring pressure of a fluid]. The cuff encircles the upper arm and is inflated until it exerts

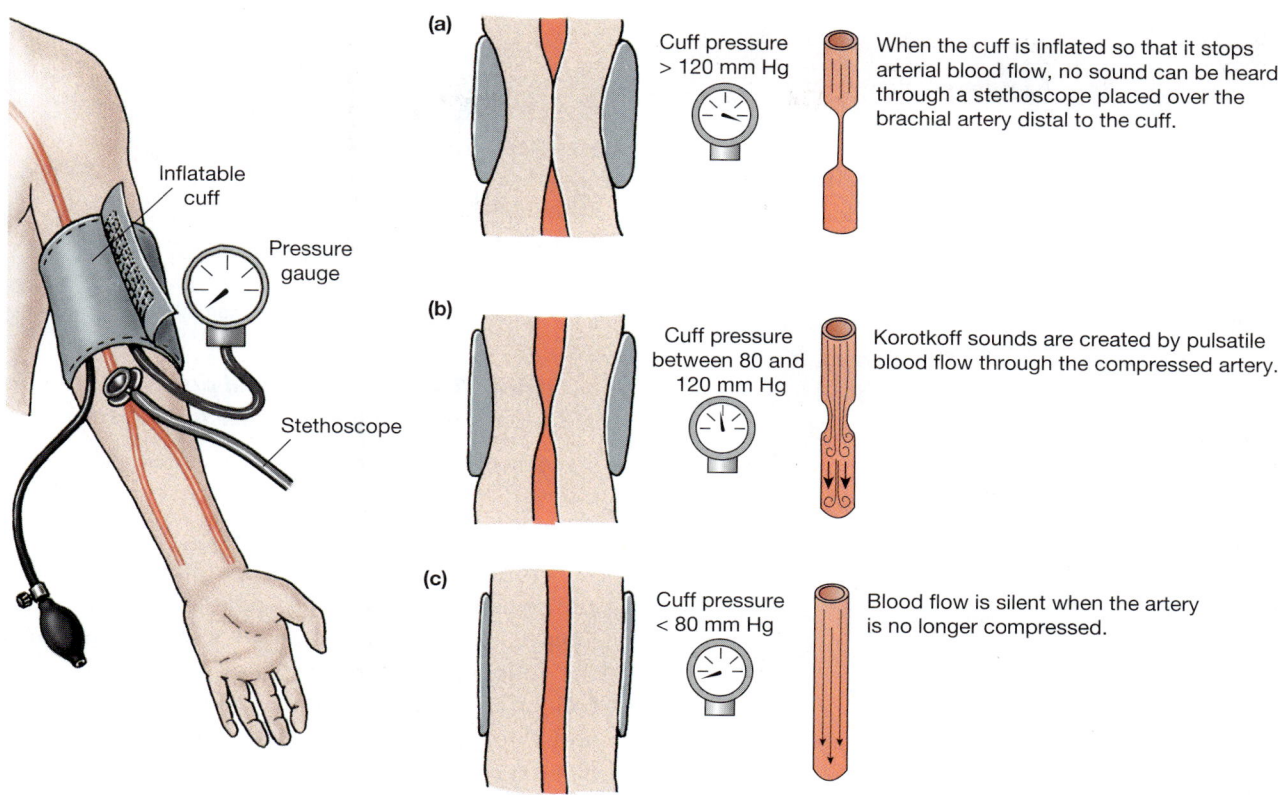

(a) Cuff pressure > 120 mm Hg

When the cuff is inflated so that it stops arterial blood flow, no sound can be heard through a stethoscope placed over the brachial artery distal to the cuff.

(b) Cuff pressure between 80 and 120 mm Hg

Korotkoff sounds are created by pulsatile blood flow through the compressed artery.

(c) Cuff pressure < 80 mm Hg

Blood flow is silent when the artery is no longer compressed.

Inflatable cuff

Pressure gauge

Stethoscope

■ **FIGURE 15-7**   *Measurement of arterial blood pressure*

Arterial blood pressure is measured with a sphygmomanometer (an inflatable cuff plus a pressure gauge) and a stethoscope. The inflation pressure shown is for a person whose blood pressure is 120/80.

pressure higher than the systolic pressure driving arterial blood. When cuff pressure exceeds arterial pressure, blood flow into the lower arm stops (Fig. 15-7a ■).

Now pressure on the cuff is gradually released. When cuff pressure falls below systolic arterial blood pressure, blood begins to flow again. As blood squeezes through the still-compressed artery, a thumping noise called a **Korotkoff sound** can be heard with each pressure wave (Fig. 15-7b). Once the cuff pressure no longer compresses the artery, the sounds disappear (Fig. 15-7c).

The pressure at which a Korotkoff sound is first heard represents the highest pressure in the artery and is recorded as the systolic pressure. The point at which the Korotkoff sounds disappear is the lowest pressure in the artery and is recorded as the diastolic pressure. By convention, blood pressure is written as systolic pressure over diastolic pressure.

For years the "average" value for blood pressure has been stated as 120/80. Like many average physiological values, however, these numbers are subject to wide variability, both from one person to another and within a single individual from moment to moment. A systolic pressure that is consistently over 140 mm Hg at rest, or a diastolic pressure that is chronically over 90 mm Hg, is considered a sign of hypertension in an otherwise healthy person. Furthermore, the guidelines published

in the 2003 JNC 7 Report* recommend that individuals maintain their blood pressure *below* 120/80. Persons whose systolic pressure is consistently in the range of 120–139 or whose diastolic pressure is in the range of 80–89 now are considered to be prehypertensive and should be counseled on lifestyle modification strategies to reduce their blood pressure.

Cardiac Output and Peripheral Resistance Determine Mean Arterial Pressure

Mean arterial pressure is the driving force for blood flow, but what determines mean arterial pressure? Arterial pressure is a balance between blood flow into the arteries and blood flow out of the arteries. If flow in exceeds flow out, blood collects in the arteries, and mean arterial pressure increases. If flow out exceeds flow in, mean arterial pressure falls.

Blood flow into the aorta is equal to the cardiac output of the left ventricle. Blood flow out of the arteries is influenced primarily by **peripheral resistance**, defined as the resistance to

---

*Seventh Report of the Joint National Committee on Prevention, Detection, Evaluation, and Treatment of High Blood Pressure, National Institutes of Health. *www.nhlbi.nih.gov/guidelines/hypertension*

RUNNING PROBLEM

Kurt's second blood pressure reading is 158/98. Dr. Cortez asks him to take his blood pressure at home daily for two weeks and then return to the doctor's office. When Kurt comes back with his diary, the story is the same: his blood pressure continues to average 160/100. After running some tests, Dr. Cortez concludes that Kurt is one of approximately 50 million adult Americans with high blood pressure, also called hypertension. If not controlled, hypertension can lead to heart failure, stroke, and kidney failure.

Question 1:

Why are people with high blood pressure at greater risk for having a hemorrhagic (or bleeding) stroke?

| 501 | **508** | 510 | 513 | 521 | 529 |

flow offered by the arterioles (Fig. 15-8 ■). We can express the relationship between cardiac output, peripheral resistance, and mean arterial pressure as

Mean arterial pressure ∝
cardiac output × resistance of arterioles

$$MAP \propto CO \times R_{arterioles}$$

Let's consider how this works. If cardiac output increases, the heart pumps more blood into the arteries per unit time. If resistance to blood flow out of the arteries does not change, flow into the arteries is greater than flow out, blood volume in the arteries increases, and arterial blood pressure increases.

In another example, suppose cardiac output remains unchanged but peripheral resistance increases. Flow in is unchanged, but flow out is less. Blood again accumulates in the arteries, and the arterial pressure again increases. Most cases of hypertension are believed to be caused by increased peripheral resistance without changes in cardiac output.

Two additional factors can influence arterial blood pressure: total blood volume and the distribution of blood in the systemic circulation.

Changes in Blood Volume Affect Blood Pressure

Although the volume of the blood in the circulation is usually relatively constant, changes in blood volume can affect arterial blood pressure. If blood volume increases, blood pressure increases. When blood volume decreases, blood pressure decreases.

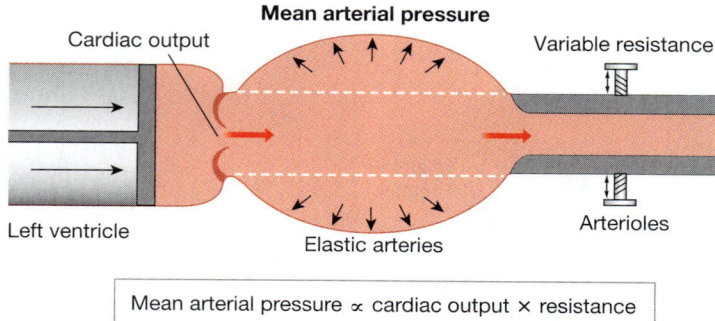

Mean arterial pressure ∝ cardiac output × resistance

■ **FIGURE 15-8**   *Mean arterial pressure is a function of cardiac output and resistance in the arterioles*

In this model, the ventricle is represented by a syringe. The variable diameter of the arterioles is represented by adjustable screws.

To understand the relationship between blood volume and pressure, think of the circulatory system as an elastic balloon filled with water. If only a small amount of water is in the balloon, little pressure is exerted on the walls, and the balloon is soft and flabby. As more water is added to the balloon, more pressure is exerted on the elastic walls. If you fill a balloon close to the bursting point, you risk popping the balloon. The best way to reduce this pressure is to remove some of the water.

Small increases in blood volume occur throughout the day due to ingestion of food and liquids, but these increases usually do not create long-lasting changes in blood pressure because of homeostatic compensations. Adjustments for increased blood volume are primarily the responsibility of the kidneys. If blood volume increases, the kidneys restore normal volume by excreting excess water in the urine (Fig. 15-9 ■).

Compensation for decreased blood volume is more difficult and requires an integrated response from the kidneys and the cardiovascular system. If blood volume decreases, *the kidneys cannot restore the lost fluid.* The kidneys can only conserve blood volume and thereby prevent further decreases in blood pressure. The only way to restore lost fluid volume is through drinking or intravenous infusions. This is an example of mass balance: volume lost to the external environment must be replaced from the external environment.

Cardiovascular compensation for decreased blood volume includes vasoconstriction and increased sympathetic stimulation of the heart [⟳ Fig. 14-31, p. 494]. However, there are limits to the effectiveness of cardiovascular compensation, and if the fluid loss is too great, adequate blood pressure cannot be maintained. Typical events that might cause significant changes in blood volume include dehydration, hemorrhage, and ingestion of a large quantity of fluid. The integrated compensation for these events will be discussed in Chapter 20.

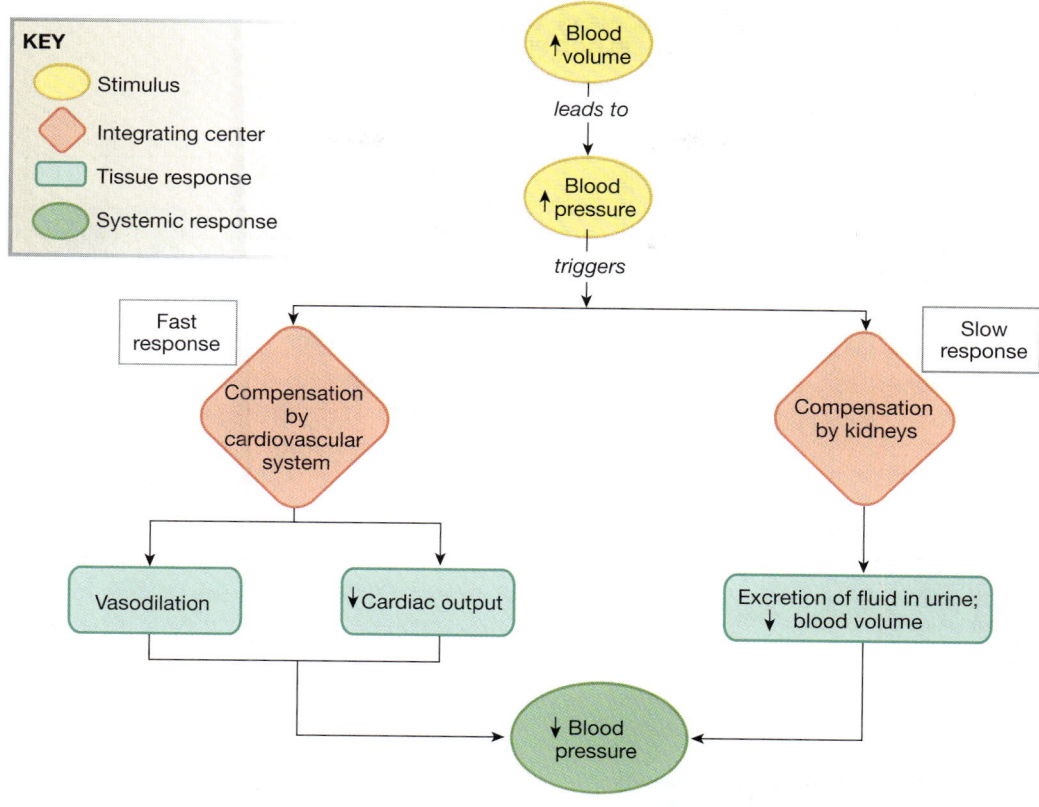

KEY
- ◯ Stimulus
- ◇ Integrating center
- ▭ Tissue response
- ◯ Systemic response

↑ Blood volume

leads to

↑ Blood pressure

triggers

Fast response

Compensation by cardiovascular system

Slow response

Compensation by kidneys

Vasodilation

↓ Cardiac output

Excretion of fluid in urine; ↓ blood volume

↓ Blood pressure

■ **FIGURE15-9** *Blood pressure control involves both the cardiovascular system and the renal system*

The cardiovascular system carries out rapid responses, while the renal response is slower.

In addition to the absolute volume of blood in the cardiovascular system, the relative distribution of blood between the arterial and venous sides of the circulation can be an important factor in maintaining arterial blood pressure. Arteries are low-volume vessels that usually contain only about 11% of total blood volume at any one time. Veins, in contrast, are high-volume vessels that hold about 60% of the circulating blood volume at any one time. The veins act as a volume reservoir, holding blood that can be redistributed to the arteries if needed. When arterial blood pressure falls, increased sympathetic activity constricts veins, decreasing their holding capacity and redistributing blood to the arterial side of the circulation. Figure. 15-10 ■ summarizes the four key factors that influence mean arterial blood pressure.

RESISTANCE IN THE ARTERIOLES

As we saw in Chapter 14 [≥p. 462], resistance to blood flow (R) is directly proportional to the length of the tubing through which the fluid flows (L) and to the viscosity (η) of the fluid, and inversely proportional to the fourth power of the tubing radius (r):

$$R \propto L\eta/r^4$$

Normally the length of the systemic circulation and the blood's viscosity are relatively constant. That leaves only the radius of the blood vessels as the primary resistance to blood flow:

$$R \propto 1/r^4$$

The arterioles are the main site of variable resistance in the systemic circulation and contribute more than 60% of the total resistance to flow in the system. Resistance in arterioles is variable because of the large amounts of smooth muscle in the arteriolar walls. When the smooth muscle contracts and relaxes, the radius of the arterioles changes.

Arteriolar resistance is influenced by both systemic and local control mechanisms:

1. *Sympathetic reflexes* mediated by the central nervous system maintain mean arterial pressure and govern blood distribution for certain homeostatic needs, such as temperature regulation.

2. *Local control of arteriolar resistance* matches tissue blood flow to the metabolic needs of the tissue. In the heart and skeletal muscle, these local controls take precedence over reflex control by the central nervous system.

3. *Hormones*—particularly those that regulate salt and water excretion by the kidneys—influence blood pressure by acting directly on the arterioles and by altering autonomic reflex control.

15

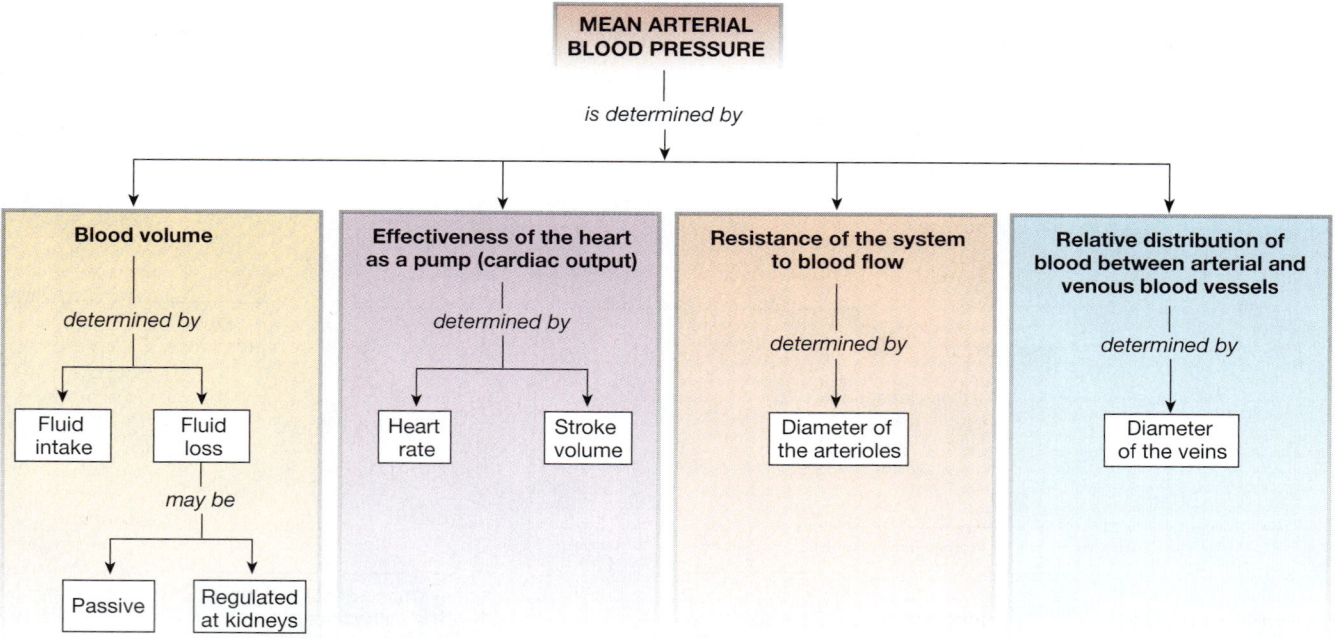

■ **FIGURE 15-10** *Factors that influence mean arterial pressure*

15

SHOCK

Shock is a broad term that refers to generalized, severe circulatory failure. Shock can arise from multiple causes: failure of the heart to maintain normal cardiac output (*cardiogenic shock*), decreased circulating blood volume (*hypovolemic shock*), bacterial toxins (*septic shock*), and miscellaneous causes, such as the massive immune reactions that cause *anaphylactic shock*. No matter what the cause, the results are similar: low cardiac output and falling peripheral blood pressure. When tissue perfusion can no longer keep up with tissue oxygen demand, the cells begin to sustain damage from inadequate oxygen and from the buildup of metabolic wastes. Once this damage occurs, a positive feedback cycle begins. The shock becomes progressively worse until it becomes irreversible, and the patient dies. The management of shock includes administration of oxygen, fluids, and norepinephrine, which stimulates vasoconstriction and increases cardiac output. If the shock arises from a cause that is treatable, such as a bacterial infection, measures must also be taken to remove the precipitating cause.

Table 15-2 ■ lists the chemicals that mediate arteriolar resistance by producing vasoconstriction or vasodilation. The following sections look at the factors that influence blood flow at the tissue level.

Myogenic Autoregulation Automatically Adjusts Blood Flow

Vascular smooth muscle has the ability to regulate its own state of contraction, a process called **myogenic autoregulation**. In the absence of autoregulation, an increase in blood pressure increases

RUNNING PROBLEM

Most hypertension is *essential hypertension,* which means high blood pressure that cannot be attributed to any particular cause. "Since your blood pressure is only mildly elevated," Dr. Cortez tells Kurt, "let's see if we can control it with lifestyle changes. You need to reduce salt and fat in your diet, get some exercise, and lose some weight." "Looks like you're asking me to turn over a whole new leaf," says Kurt.

Question 2:
What is the rationale for reducing salt intake to control hypertension? (Hint: salt causes water retention.)

501 508 **510** 513 521 529

TABLE 15-2   **Chemicals Mediating Vasoconstriction and Vasodilation**

CHEMICAL	PHYSIOLOGICAL ROLE	SOURCE	TYPE
Vasoconstriction			
Norepinephrine (α-receptors)	Baroreceptor reflex	Sympathetic neurons	Neurotransmitter
Serotonin	Platelet aggregation, smooth muscle contraction	Neurons, digestive tract, platelets	Paracrine, neurotransmitter
Substance P	Pain, increase capillary permeability	Neurons, digestive tract	Paracrine, neurotransmitter
Endothelin	Paracrine mediator	Vascular endothelium	Paracrine
Vasopressin	Increase blood pressure in hemorrhage	Posterior pituitary	Neurohormone
Angiotensin II	Increase blood pressure	Plasma hormone	Hormone
Vasodilation			
Epinephrine (β$_2$-receptors)	Increase blood flow to skeletal muscle, heart, liver	Adrenal medulla	Neurohormone
Acetylcholine (via NO)	Erection of clitoris or penis	Parasympathetic neurons	Neurotransmitter
Vasoactive intestinal peptide	Digestive secretion, relax smooth muscle	Neurons	Neurotransmitter, neurohormone
Nitric oxide (NO)	Paracrine mediator	Endothelium	Paracrine
Bradykinin (via NO)	Increase blood flow	Multiple tissues	Paracrine
Adenosine	Increase blood flow to match metabolism	Hypoxic cells	Paracrine
$\downarrow O_2, \uparrow CO_2, \uparrow H^+, \uparrow K^+$	Increase blood flow to match metabolism	Cell metabolism	Paracrine
Histamine	Increase blood flow	Mast cells	Paracrine
Natriuretic peptides (example—ANP)	Reduce blood pressure	Atrial myocardium, brain	Hormone, neurotransmitter

blood flow through an arteriole. However, when smooth muscle fibers in the wall of the arteriole stretch because of increased blood pressure, the arteriole constricts. This vasoconstriction increases the resistance offered by the arteriole, automatically decreasing blood flow through the vessel. With this simple and direct response to pressure, arterioles regulate their own blood flow.

The mechanism responsible for the intrinsic response of vascular smooth muscle is stretch that opens mechanically gated Ca^{2+} channels in the muscle membrane. Calcium entering the smooth muscle cell combines with calmodulin and activates myosin light chain kinase, which in turn increases myosin ATPase activity and crossbridge activity [⎘ p. 426].

Paracrines Alter Vascular Smooth Muscle Contraction

Local control of arteriolar resistance is an important method by which individual tissues regulate their own blood supply.

Local regulation is accomplished by paracrines (including the gases O_2, CO_2, and NO) secreted by the vascular endothelium or by cells to which the arterioles are supplying blood (Table 15-2).

The concentrations of many paracrines change as cells become more or less metabolically active. For example, if aerobic metabolism increases, tissue O_2 levels decrease while CO_2 production goes up. Both low O_2 and high CO_2 dilate arterioles. This vasodilation increases blood flow into the tissue, bringing additional O_2 to meet the increased metabolic demand and removing waste CO_2 (Fig. 15-11a ■). The process in which an increase in blood flow accompanies an increase in metabolic activity is known as **active hyperemia** [*hyper-*, above normal + *(h)aimia*, blood].

If blood flow to a tissue is occluded [*occludere*, to close up] for a few seconds to a few minutes, O_2 levels fall and metabolically produced paracrines such as CO_2 and H^+ accumulate in the interstitial fluid. Local *hypoxia* [*hypo-*, low + *oxia*, oxygen]

15

(a) Active hyperemia

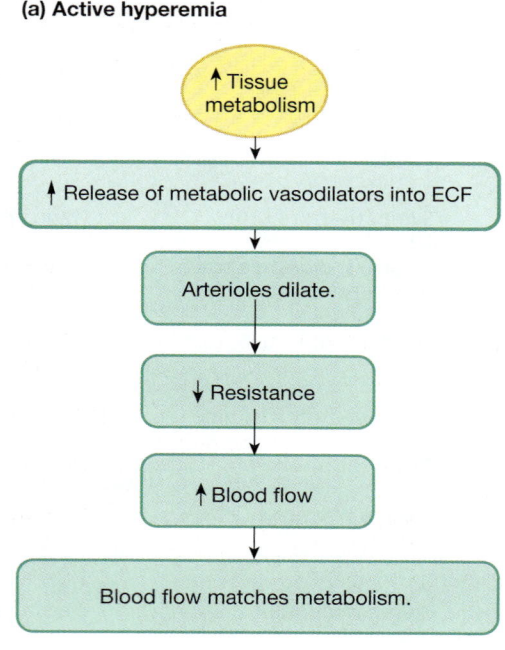

(b) Reactive hyperemia

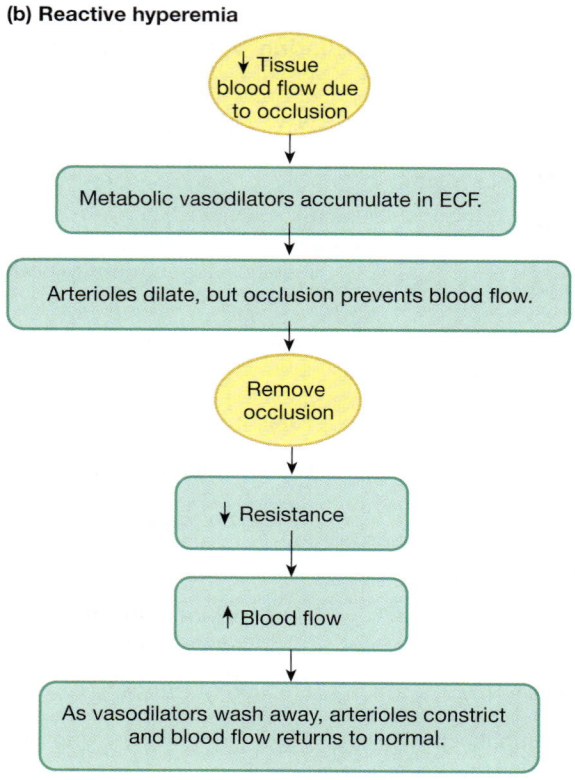

■ **FIGURE 15-11**   *Hyperemia*

causes endothelial cells to synthesize the vasodilator nitric ox-ide. When blood flow to the tissue resumes, the increased concentrations of NO, CO_2, and other paracrines immediately trigger significant vasodilation. As the vasodilators are metabolized or washed away by the restored tissue blood flow, the radius of the arteriole gradually returns to normal. An increase in tissue blood flow following a period of low perfusion is known as **reactive hyperemia** (Fig. 15-11b).

Our knowledge concerning the importance of nitric oxide as a vasodilator is growing as additional research studies are published. NO plays an important physiological role in the male erection reflex (see Chapter 26), and drugs used to treat erectile dysfunction prolong NO activity. Decreases in endogenous NO activity are suspected to play a role in several significant conditions, including hypertension and the elevated blood pressure that sometimes occurs during pregnancy.

Another vasodilator paracrine is the nucleotide **adenosine**. If oxygen consumption in heart muscle exceeds the rate at which oxygen is supplied by the blood, myocardial hypoxia results. In response to low tissue oxygen, the myocardial cells release adenosine. Adenosine dilates coronary arterioles in an attempt to bring additional blood flow into the muscle.

Not all vasoactive paracrines reflect changes in metabolism. For example, *kinins* and *histamine* are potent vasodilators that play a role in inflammation. *Serotonin* (5-HT), previously mentioned as a CNS neurotransmitter [p. 274], is also a vaso-constricting paracrine released by activated platelets. When

damaged blood vessels activate platelets, the subsequent serotonin-mediated vasoconstriction helps slow blood loss. Serotonin agonists called triptans (for example, *sumatriptan*) are drugs that bind to 5-HT$_1$ receptors and cause vasoconstriction. These drugs are used to treat migraine headaches, which are caused by inappropriate cerebral vasodilation.

CONCEPT CHECK

6. Resistance to blood flow is determined *primarily* by (a) blood viscosity, (b) blood volume, (c) cardiac output, (d) blood vessel diameter, or (e) blood pressure gradient (ΔP).

7. The extracellular fluid concentration of K^+ increases in exercising skeletal muscles. What effect will this increase in K^+ have on blood flow in the muscles?

Answers: p. 533

The Sympathetic Branch Controls Most Vascular Smooth Muscle

Smooth muscle contraction in arterioles is regulated by neural and hormonal signals in addition to locally produced paracrines. Among the hormones with significant vasoactive properties are *atrial natriuretic peptide* and *angiotensin II*. These hormones also have significant effects on the kidney's excretion of ions and water, as you will learn in Chapter 20.

Most systemic arterioles are innervated by sympathetic neurons. A notable exception is arterioles involved in the erection reflex of the penis and clitoris. They are controlled

EMERGING CONCEPTS

FROM DYNAMITE TO VASODILATION

In 1998 the Nobel prize in physiology and medicine was awarded to Robert Furchgott, Louis Ignarro, and Ferid Murad for their research on the role of nitric oxide as a signal molecule in the cardiovascular system (*www.nobel.se/medicine*). Who would have thought that this component of smog and derivative of dynamite would turn out to be a biological messenger? Certainly not the referees who initially rejected Louis Ignarro's attempts to publish his research findings on the elusive gas. However, the ability of nitrate-containing compounds to relax blood vessels has been known for more than 100 years, ever since workers in Alfred Nobel's dynamite factory complained of headaches caused by nitrate-induced vasodilation. Ever since the 1860s, physicians have used nitroglycerin to relieve *angina*—heart pain that results from constricted blood vessels—and even today heart patients carry little nitroglycerin tablets to slide under their tongues when angina strikes. Still, it took years of work to isolate the short-lived gas that is the biologically active molecule derived from nitroglycerin. Despite all our twenty-first-century technology, direct research on NO is still difficult, and many studies investigate its influence indirectly by studying the location and activity of nitric oxide synthase (NOS), the enzyme that produces NO.

RUNNING PROBLEM

After a month, Kurt returns to the doctor's office for a checkup. He has lost five pounds and is walking at least a mile daily, but his blood pressure has not changed. "I swear, I'm trying to do better," says Kurt, "but it's difficult." Because lifestyle changes alone have not lowered Kurt's blood pressure, Dr. Cortez prescribes an antihypertensive drug. "This drug, called an ACE inhibitor, blocks production of a chemical called angiotensin II, a powerful vasoconstrictor. This medication should bring your blood pressure back to a normal value."

Question 3:
Why would blocking the action of a vasoconstrictor lower blood pressure?

501	508	510	**513**	521	529

indirectly by parasympathetic innervation that causes paracrine release of nitric oxide, resulting in vasodilation.

Tonic discharge of norepinephrine from sympathetic neurons helps maintain myogenic tone of arterioles (Fig. 15-12 ■). Norepinephrine binding to α-receptors on vascular smooth muscle causes vasoconstriction. If sympathetic release of norepinephrine decreases, the arterioles dilate. If sympathetic stimulation increases, the arterioles constrict.

Epinephrine from the adrenal medulla travels through the blood and binds with α-receptors, reinforcing vasoconstriction. However, α-receptors have a lower affinity for epinephrine and do not respond as strongly to it as they do to norepinephrine [📘 p. 276].

Epinephrine also binds to β_2-receptors, found on vascular smooth muscle of heart, liver, and skeletal muscle arterioles.

These receptors are not innervated and therefore respond primarily to circulating epinephrine [📘 p. 276]. Activation of vascular β_2-receptors by epinephrine causes vasodilation.

One way to remember which arterioles have β_2-receptors is to think of a fight-or-flight response to a stressful event [📘 p. 377]. This response includes a generalized increase in sympathetic activity, along with the release of epinephrine. Blood vessels that have β_2-receptors respond to epinephrine by vasodilating. Such β_2-mediated vasodilation enhances blood flow to the heart, skeletal muscles, and liver, tissues that are active during the fight-or-flight response. (The liver produces glucose for muscle contraction.)

During fight or flight, increased sympathetic activity at arteriolar α-receptors causes vasoconstriction. The increase in resistance diverts blood from nonessential organs, such as the gastrointestinal tract, to the skeletal muscles, liver, and heart. The nervous system's ability to selectively alter blood flow to organs is an important aspect of cardiovascular regulation.

✓ CONCEPT CHECK

8. What happens when epinephrine combines with β_1-receptors in the heart? With β_2-receptors in the heart? (*Hint:* "in the heart" is vague. The heart has multiple tissue types. Which heart tissues possess the different types of β-receptors? [📘 p. 491])

9. Skeletal muscle arterioles have both α- and β-receptors on their smooth muscle. Epinephrine can bind to both. Will the arterioles constrict or dilate in response to epinephrine? Explain.

Answers: p. 533

■ **FIGURE 15-12** *Tonic control of arteriolar diameter*

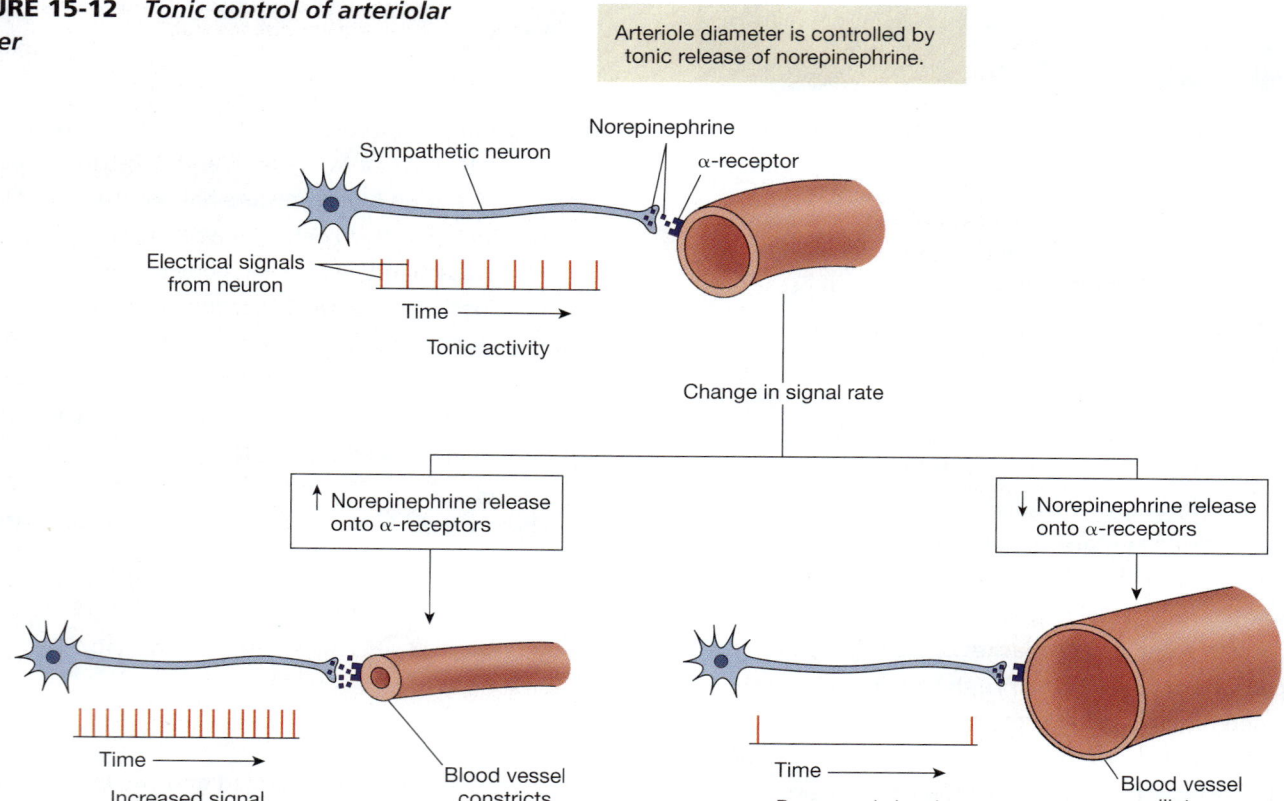

DISTRIBUTION OF BLOOD TO THE TISSUES

The distribution of systemic blood varies according to the metabolic needs of individual organs. Distribution is governed by a combination of local control mechanisms and homeostatic reflexes. For example, skeletal muscles at rest receive about 20% of cardiac output. During exercise, when the muscles use more oxygen and nutrients, they receive as much as 85%.

Blood flow to individual organs is set to some degree by the number and size of arteries feeding the organ. Figure 15-13 ■ shows how blood is distributed to various organs when the body is at rest. Usually, more than two-thirds of the cardiac output is routed to the digestive tract, liver, muscles, and kidneys.

Variations in blood flow to individual tissues are possible because the arterioles in the body are arranged in parallel. That is, all arterioles receive blood at the same time from the aorta (Fig. 15-1). Total blood flow through *all* the arterioles of the body always equals the cardiac output.

However, the flow through individual arterioles depends on their resistance. The higher the resistance in the arteriole, the lower the blood flow through it. If an arteriole constricts and resistance goes up, blood flow through that arteriole goes down (Fig. 15-14 ■):

$$\text{Flow}_{\text{arteriole}} \propto 1/\text{resistance}_{\text{arteriole}}$$

In other words, blood is diverted from high-resistance arterioles to lower-resistance arterioles. You might say that blood traveling through the arterioles takes the path of least resistance.

In a tissue, blood flow into individual capillaries can be regulated by the precapillary sphincters described earlier in the chapter. When these small bands of smooth muscle at metarteriole-capillary junctions constrict, they restrict blood flow into the capillaries (Fig. 15-15 ■). When the sphincters dilate, blood flow into the capillaries increases. This mechanism provides an additional site for local control of blood flow.

CONCEPT CHECK
10. Use Figure 15-13 to answer these questions. (a) Which tissue has the highest blood flow per unit weight? (b) Which tissue has the least blood flow, regardless of weight?
    Answers: p. 533

EXCHANGE AT THE CAPILLARIES

The transport of materials around the body is only part of the function of the cardiovascular system. Once blood reaches the capillaries, the plasma and the cells exchange materials across the thin capillary walls. Most cells are located within 0.1 mm of the nearest capillary, and diffusion over this short distance proceeds rapidly.

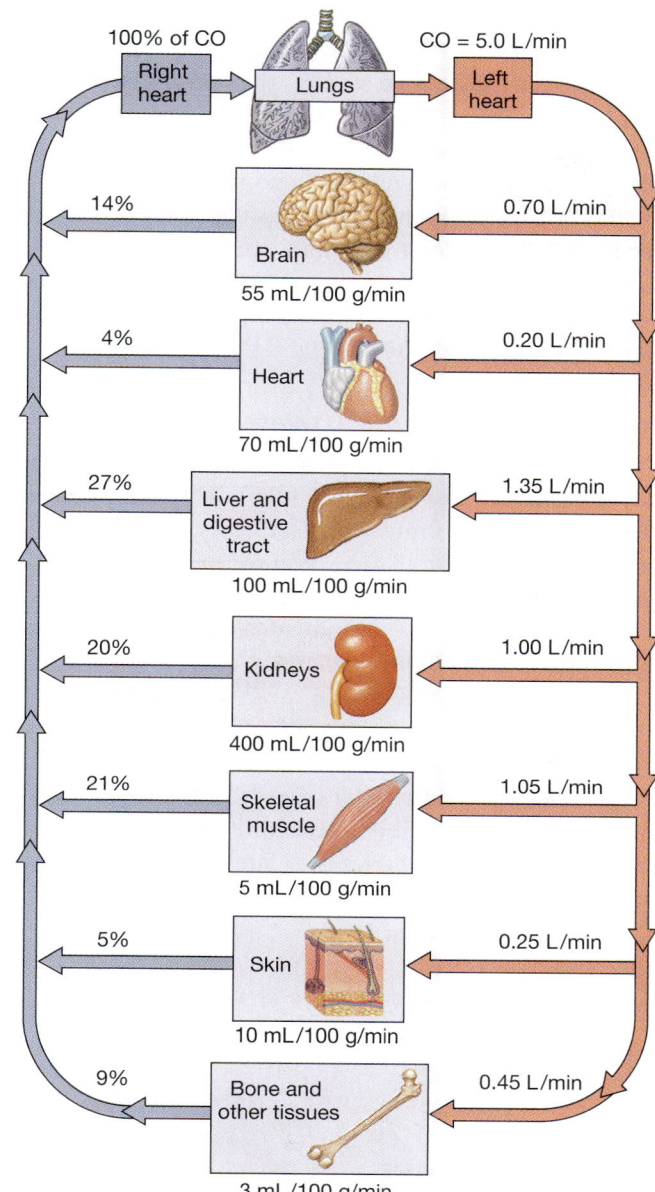

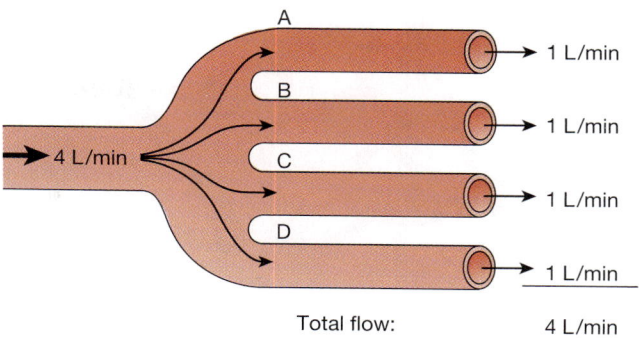

(a) Blood flow through four identical vessels (A–D) is equal. Total flow into vessels equals total flow out.

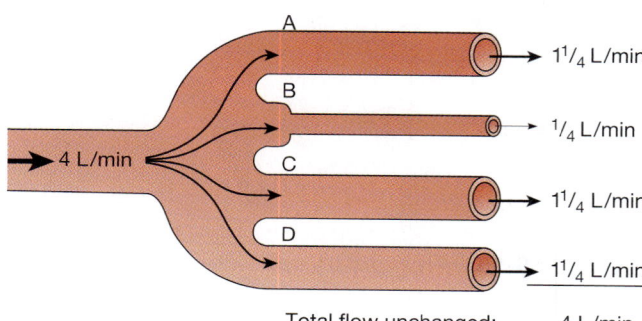

(b) When vessel B constricts, resistance of B increases and flow through B decreases. Flow diverted from B is divided among the lower-resistance vessels A, C, and D.

■ **FIGURE 15-14** *Blood flow through individual blood vessels is determined by the vessel's resistance to flow*

and glands have the highest. By one estimate, the adult human body has about 50,000 miles of capillaries, with a total exchange surface area of more than 6300 m^2, nearly the surface area of two football fields.

Capillaries have the thinnest walls of all the blood vessels, composed of a single layer of flattened endothelial cells supported on a basal lamina (Figure 15-2). The diameter of a capillary is just larger than that of a red blood cell, forcing the blood cells to pass through single file. Cell junctions between the endothelial cells vary from tissue to tissue and help determine the "leakiness" of the capillary.

Capillaries can be divided into two types based on structure. The most common capillaries are **continuous capillaries,** whose endothelial cells are joined to one another with leaky junctions (Fig. 15-16a ■). These capillaries are found in muscle, connective tissue, and neural tissue. The continuous capillaries of the brain have evolved to form the blood-brain barrier, with tight junctions that protect neural tissue from toxins that may be present in the bloodstream [🔁 p. 299].

Fenestrated capillaries [*fenestra,* window] have large pores (*fenestrae*) that allow high volumes of fluid to pass rapidly between the plasma and interstitial fluid (Fig. 15-16b). These capillaries

FIGURE QUESTION

What is the rate of blood flow through the lungs?

■ **FIGURE 15-13** *Distribution of blood in the body at rest*

Blood flow to the major organs is represented in three ways: as a percentage of total flow, as volume per 100 grams of tissue per minute, and as an absolute rate of flow (in L/min).

The capillary density in any given tissue is directly related to the metabolic activity of the tissue's cells. Tissues with a higher metabolic rate require more oxygen and nutrients. Those tissues have more capillaries per unit area. Subcutaneous tissue and cartilage have the lowest capillary density. Muscles

15

(a) When precapillary sphincters are relaxed, blood flows through all capillaries in the bed.

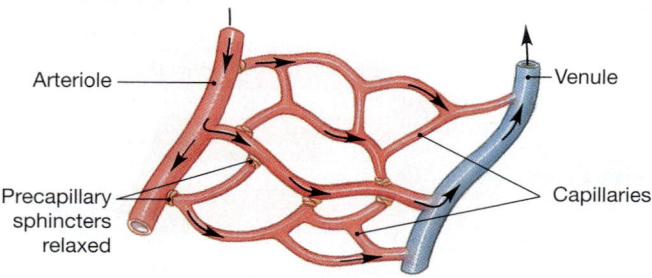

(b) If precapillary sphincters constrict, blood flow bypasses capillaries completely and flows through metarterioles.

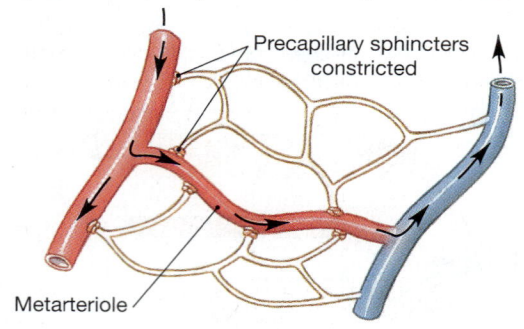

■ **FIGURE 15-15** *Precapillary sphincters*

are found primarily in the kidney and the intestine, where they are associated with absorptive transporting epithelia.

Three tissues—the bone marrow, the liver, and the spleen—do not have typical capillaries. Instead they have modified vessels called **sinusoids** that are as much as five times wider than a capillary. The sinusoid endothelium has fenestrations, and there may be gaps between the cells as well. Sinusoids are found in locations where blood cells and plasma proteins need to cross the endothelium to enter the blood. Figure 16-5, Focus on Bone Marrow, shows blood cells leaving the bone marrow by squeezing between endothelial cells.

Velocity of Blood Flow is Lowest in the Capillaries

In Chapter 14 you learned that at a constant flow rate, velocity of flow is higher in a smaller vessel than in a larger vessel [🔁 p. 464]. From this, you might conclude that blood moves very rapidly through the capillaries because they are the smallest blood vessels. However, the primary determinant for velocity of flow is not the diameter of an individual capillary but the *total cross-sectional area* of *all* the capillaries.

What is total cross-sectional area? Imagine circles representing cross sections of all the capillaries placed edge to edge, and you have it. For the capillaries, those circles would cover an area much larger than the total cross-sectional areas of all the arteries and veins combined. Therefore, because total cross-sectional area of the capillaries is so large, the velocity of flow through them is low.

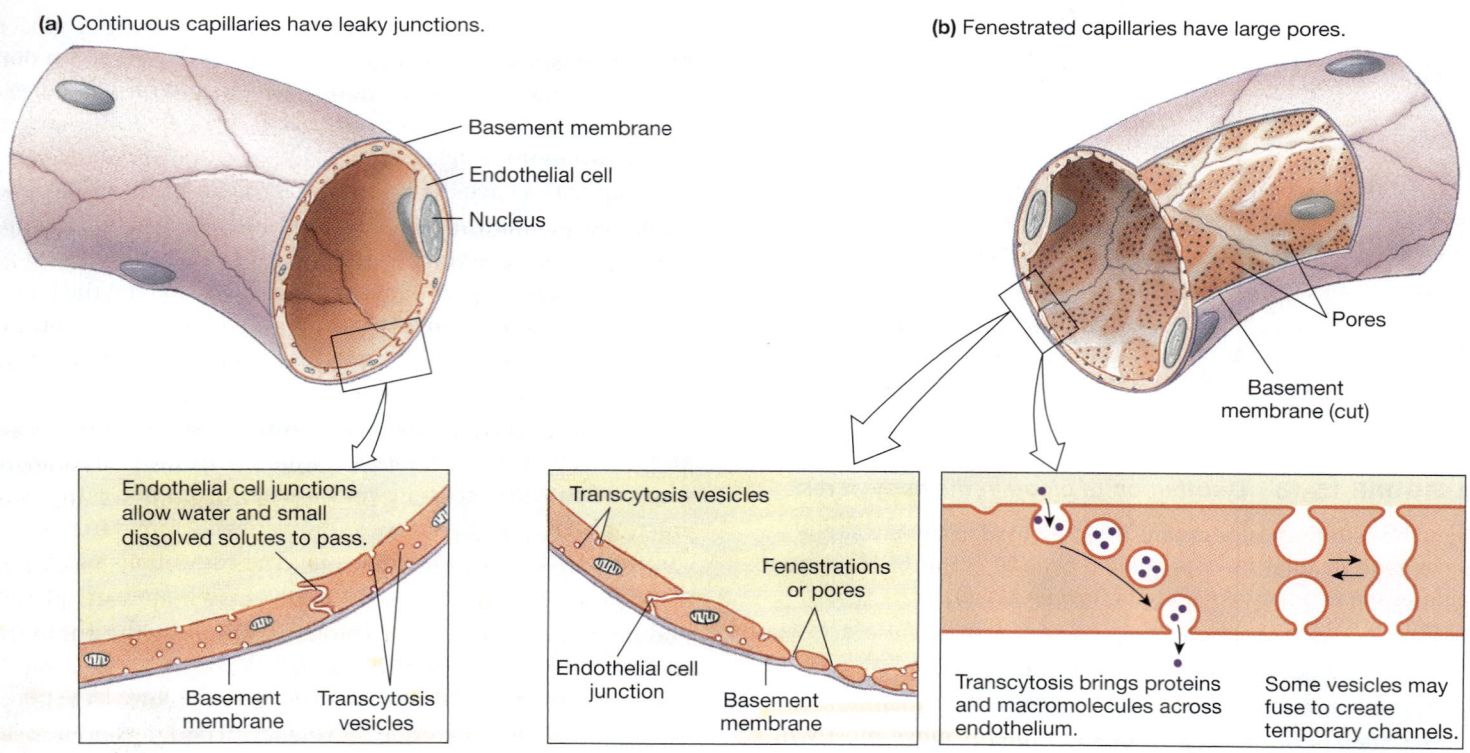

■ **FIGURE 15-16** *The two types of capillaries*

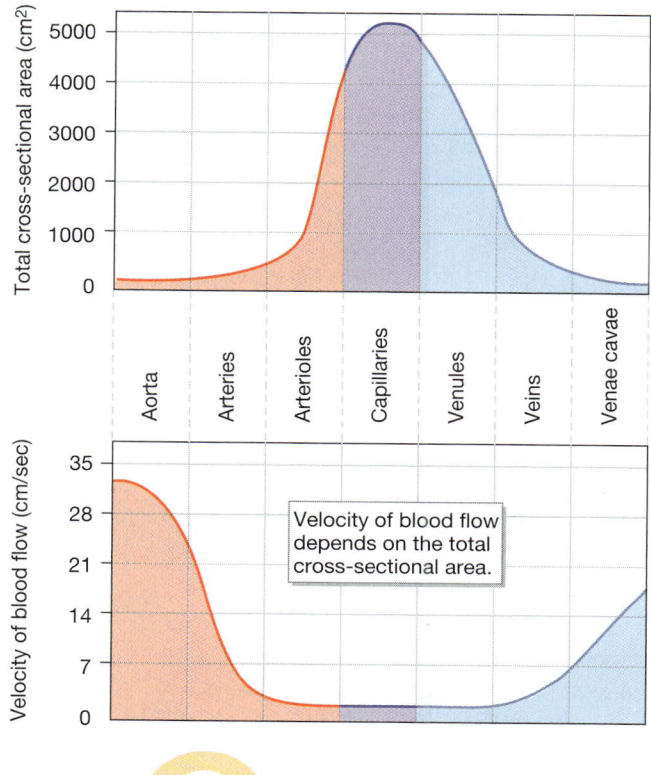

Velocity of blood flow depends on the total cross-sectional area.

GRAPH QUESTION

(a) Is velocity of flow directly proportional to or inversely proportional to cross-sectional area?

(b) What effect does changing only the cross-sectional area have on flow rate?

■ **FIGURE 15-17**   *Velocity of flow depends on total cross-sectional area of the vessels*

Figure 15-17 ■ compares cross-sectional areas of different parts of the systemic circulation with the velocity of blood flow in each part. The fastest flow is in the relatively small-diameter arterial system. The slowest flow is in the capillaries and venules, which collectively have the largest cross-sectional area. The low velocity of flow through capillaries is a useful characteristic that allows diffusion enough time to go to equilibrium [p. 133].

Most Capillary Exchange Takes Place by Diffusion and Transcytosis

Exchange between the plasma and interstitial fluid takes place either by movement between endothelial cells (the *paracellular pathway*) or by movement through the cells (*transendothelial transport*). Smaller dissolved solutes and gases move by diffusion between or through the cells, depending on their lipid solubility [p. 24]. Larger solutes and proteins move mostly by vesicular transport [p. 148].

The diffusion rate for dissolved solutes is determined primarily by the concentration gradient between the plasma and the interstitial fluid. Oxygen and carbon dioxide diffuse freely across the thin endothelium. Their concentrations reach equilibrium with the interstitial fluid and cells by the time blood reaches the venous end of the capillary. In capillaries with leaky cell junctions, most small dissolved solutes can diffuse freely between the cells or through the fenestrae.

Blood cells and most plasma proteins are unable to pass through the junctions between capillary endothelial cells. However, we know that proteins do move from plasma to interstitial fluid and vice versa. In most capillaries, larger molecules (including selected proteins) are transported across the endothelium by *transcytosis* [p. 153]. The endothelial cell surface appears dotted with numerous *caveolae* and noncoated pits that become vesicles for transcytosis. It appears that in some capillaries, chains of vesicles fuse to create open channels that extend across the endothelial cell (Fig. 15-16b).

Capillary Filtration and Absorption Take Place by Bulk Flow

A third form of capillary exchange is bulk flow into and out of the capillary. **Bulk flow** refers to the mass movement of fluid between the blood and the interstitial fluid as the result of hydrostatic or osmotic pressure gradients. If the direction of bulk flow is into the capillary, the fluid movement is called **absorption**. If the direction of flow is out of the capillary, the fluid movement is known as **filtration**. Capillary filtration is caused by hydrostatic pressure that forces fluid out of the capillary through leaky cell junctions. As an analogy, think of garden "soaker" hoses whose perforated walls allow water to ooze out.

Most capillaries show a transition from net filtration at the arterial end to net absorption at the venous end. There are some exceptions to this rule, though. Capillaries in part of the kidney filter fluid along their entire length, for instance, and some capillaries in the intestine are only absorptive, picking up digested nutrients that have been transported into the interstitial fluid from the lumen of the intestine.

Two forces regulate bulk flow in the capillaries. One is hydrostatic pressure, the lateral pressure component of blood flow that pushes fluid out through the capillary pores [p. 461], and the other is osmotic pressure [p. 155]. These forces are sometimes called *Starling forces*, after the English physiologist E. H. Starling, who first described them (the same Starling as in the Frank-Starling law of the heart).

Osmotic pressure is determined by solute concentration of a compartment. The main solute difference between plasma and interstitial fluid is due to proteins, which are present in the plasma but mostly absent from interstitial fluid. The osmotic pressure created by the presence of these proteins is known as **colloid osmotic pressure** (π). Note that colloid osmotic pressure is *not* equivalent to the total osmotic pressure in a capillary.

15

(a) Filtration in systemic capillaries

Net pressure = hydrostatic pressure − colloid osmotic pressure

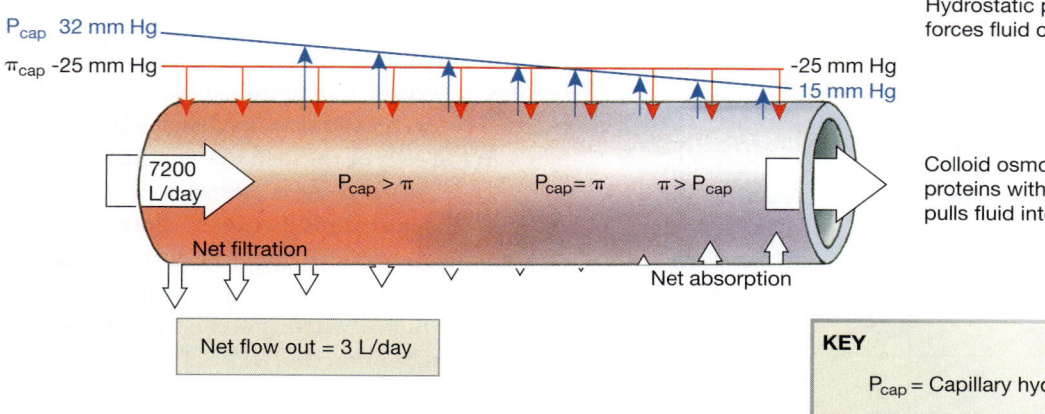

Hydrostatic pressure P_{cap} forces fluid out of the capillary.

P_{cap} 32 mm Hg

π_{cap} -25 mm Hg

-25 mm Hg

15 mm Hg

7200 L/day

$P_{cap} > \pi$ $P_{cap} = \pi$ $\pi > P_{cap}$

Net filtration

Net absorption

Colloid osmotic pressure of proteins within the capillary pulls fluid into the capillary.

Net flow out = 3 L/day

KEY

P_{cap} = Capillary hydrostatic pressure

π = Colloid osmotic pressure

(b) Relationship between capillaries and lymph vessels

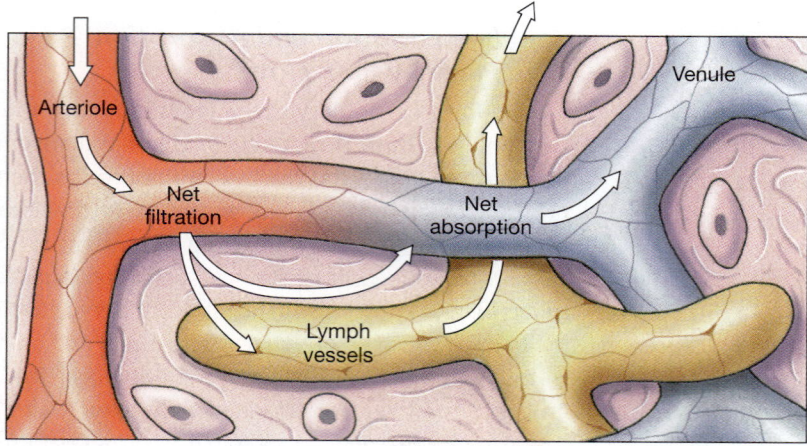

Venule

Arteriole

Net filtration

Net absorption

Lymph vessels

The excess water and solutes that filter out of the capillary are picked up by the lymph vessels and returned to the circulation.

■ **FIGURE 15-18**   *Fluid exchange at a capillary*

A net average of 3 L/day of fluid filters out of the capillaries.

It is simply a measure of the osmotic pressure created by proteins. Because the capillary endothelium is freely permeable to ions and other solutes in the plasma and interstitial fluid, these other solutes do not contribute to the osmotic gradient.

Colloid osmotic pressure is higher in the plasma (π_{cap} = 25mm Hg) than in the interstitial fluid (π_{IF} = 0 mm Hg). Therefore, the osmotic gradient favors water movement by osmosis from the interstitial fluid into the plasma, represented by the red vertical arrows in Figure 15-18a ■. For the purposes of our discussion, we will consider colloid osmotic pressure to be constant along the length of the capillary, at π_{cap} = 25 mm Hg.

Capillary hydrostatic pressure (P_{cap}), on the other hand, decreases along the length of the capillary as energy is lost to friction. Average values for capillary hydrostatic pressure, shown in Figure 15-18a, are 32 mm Hg at the arterial end of a capillary and 15 mm Hg at the venous end. The hydrostatic

pressure of the interstitial fluid (P_{IF}) is very low, and so we will consider it to be essentially zero. This means that water movement due to hydrostatic pressure will always be directed out of the capillary, as denoted by the blue vertical arrows in Figure 15-18a, with the pressure gradient decreasing from the arterial end to the venous end.

Net fluid flow across the capillary is determined by the difference between the hydrostatic pressure gradient favoring filtration and the colloid osmotic pressure favoring absorption:

Filtration (P_{out}) = hydrostatic pressure gradient = $P_{cap} - P_{IF}$

Absorption (π_{in}) = colloid osmotic pressure gradient

$$= \pi_{IF} - \pi_{cap}$$

Net pressure = hydrostatic pressure gradient

+ colloid osmotic pressure gradient

$$= P_{out} + \pi_{in}$$

If we assume that the interstitial hydrostatic and colloid osmotic pressures are zero, as discussed above, then we get the following values at the arterial end of a capillary:

$$\text{Net pressure}_{\text{arterial end}} = (32 \text{ mm Hg} - 0) + (0 - 25 \text{ mm Hg})$$
$$= 32 - 25 \text{ mm Hg} = 7 \text{ mm Hg}$$

Because at the arterial end P_{out} is greater than π_{in}, the net pressure is 7 mm Hg of filtration pressure. At the venous end, where capillary hydrostatic pressure is less:

$$\text{Net P}_{\text{venous end}} = (15 \text{ mm Hg} - 0) + (0 - 25 \text{ mm Hg})$$
$$= 15 - 25 \text{ mm Hg} = -10 \text{ mm Hg}$$

Here π_{in} is greater than P_{out}, and therefore the net pressure is 10 mm Hg favoring absorption. (A negative net pressure indicates absorption.)

Fluid movement down the length of a capillary is shown in Figure 15-18a. At the arterial end there is net filtration, and at the venous end there is net absorption. If the point at which filtration equals absorption occurred in the middle of the capillary, there would be no net movement of fluid. All volume that was filtered at the arterial end would be absorbed at the venous end. However, filtration is usually greater than absorption, resulting in bulk flow of fluid out of the capillary into the interstitial space. By most estimates, that bulk flow amounts to about 3 liters per day, which is the equivalent of the entire plasma volume! Unless this filtered fluid is returned to the plasma, the blood will turn into a sludge of blood cells and proteins. Restoring fluid lost from the capillaries to the circulatory system is one of the functions of the lymphatic system, which we discuss next.

CONCEPT CHECK

11. Suppose that the hydrostatic pressure P_{cap} at the arterial end of a capillary increases from the 32 mm Hg shown in Figure 15-18a to 35 mm Hg. If P_{cap} remains 15 mm Hg at the venous end, will net filtration in this capillary decrease, increase, or stay the same?

12. A person with liver disease may lose the ability to synthesize plasma proteins. What will happen to the colloid osmotic pressure of his blood? What will happen to the balance between filtration and absorption in his capillaries?

13. Why did this discussion refer to the osmotic pressure of the plasma rather than the osmolarity of the plasma?

Answers: p. 533

THE LYMPHATIC SYSTEM

The vessels of the lymphatic system interact with three other physiological systems: the cardiovascular system, the digestive system, and the immune system. Functions of the lymphatic system include (1) returning fluid and proteins filtered out of the capillaries to the circulatory system, (2) picking up fat absorbed at the small intestine and transferring it to the circulatory system, and (3) serving as a filter to help capture and destroy foreign pathogens. In this discussion we focus on the role of the lymphatic system in fluid transport. The other two functions will be discussed in connection with digestion (Chapter 21) and immunity (Chapter 24).

The lymphatic system is designed for the one-way movement of interstitial fluid from the tissues into the circulation. Blind-end lymph vessels (*lymph capillaries*) lie close to all blood capillaries except those in the kidney and central nervous system (Fig. 15-18b). The smallest lymph vessels are composed of a single layer of flattened endothelium that is even thinner than the capillary endothelium.

The walls of these tiny lymph vessels are anchored to the surrounding connective tissue by fibers that hold the thin-walled vessels open. Large gaps between cells allow fluid, interstitial proteins, and particulate matter such as bacteria to be swept into the lymph vessels by bulk flow. Once inside the lymphatics, this clear fluid is called simply **lymph**.

Lymph vessels in the tissues join one another to form larger lymphatic vessels that progressively increase in size (Fig. 15-19 ■). These vessels have a system of semilunar valves, similar to valves in the venous circulation. The largest lymph ducts empty into the venous circulation just under the collarbones, where the left and right subclavian veins join the internal jugular veins. At intervals along the way, vessels enter **lymph nodes**, bean-shaped nodules of tissue with a fibrous outer capsule and an internal collection of immunologically active cells, including lymphocytes and macrophages.

The lymphatic system has no single pump like the heart. Lymph flow depends primarily on waves of contraction of smooth muscle in the walls of the larger lymph vessels. Flow is aided by contractile fibers in the endothelial cells, by the one-way valves, and by external compression created by skeletal muscles.

The skeletal muscle pump plays a significant role in lymph flow, as you know if you have ever injured a wrist or ankle. An immobilized limb frequently swells from the accumulation of fluid in the interstitial space, a condition known as **edema** [*oidema*, swelling]. Patients with edema in an injured limb are told to elevate the limb above the level of the heart so gravity will assist lymph flow back to the blood.

An important reason for returning filtered fluid to the circulation is the recycling of plasma proteins. The body must maintain a low protein concentration in the interstitial fluid because colloid osmotic pressure is the only significant force that opposes capillary hydrostatic pressure. If proteins move from the plasma to the interstitial fluid, the osmotic pressure gradient that opposes filtration decreases. With less opposition to capillary hydrostatic pressure, additional fluid moves into the interstitial space.

Inflammation is an example of a situation in which the balance of colloid osmotic and hydrostatic pressures is disrupted.

15

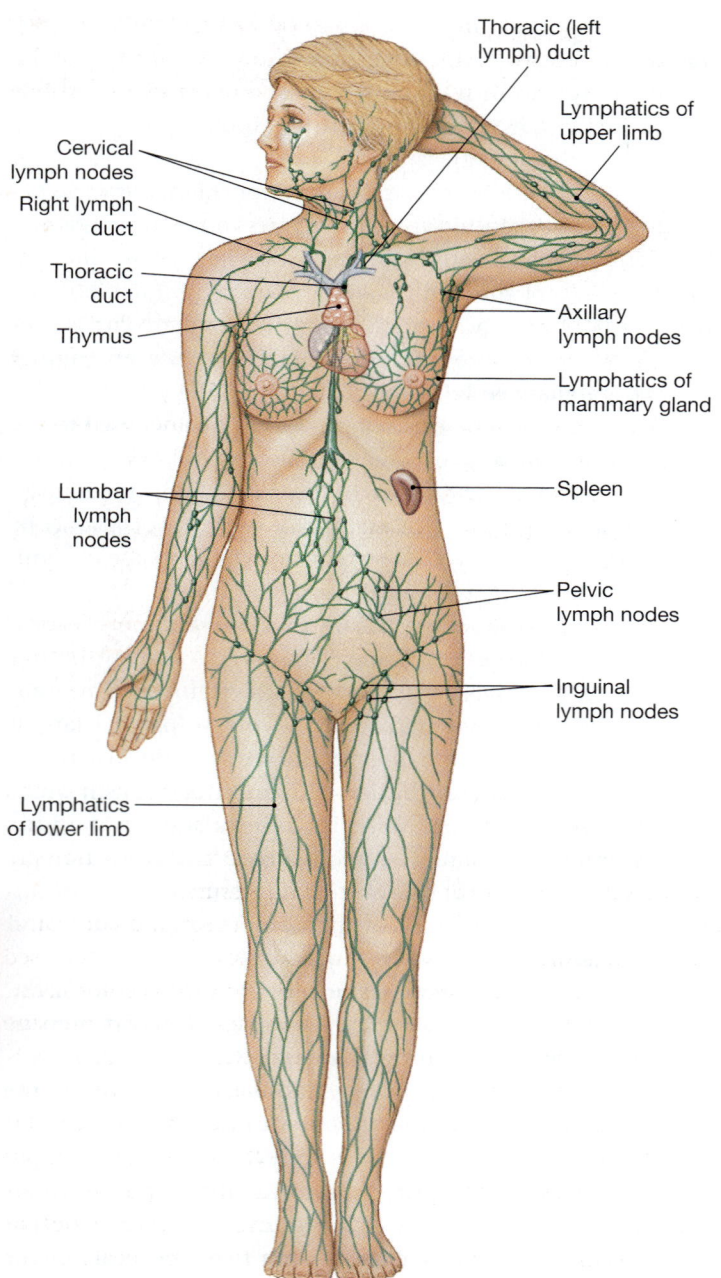

Lymph empties into the venous circulation.

■ **FIGURE 15-19** *The lymphatic system*

Blind-end lymph vessels in the tissues remove fluid and filtered proteins.

Histamine released in the inflammatory response makes capillary walls leakier and allows proteins to escape from the plasma into the interstitial fluid. Thus, the local swelling that accompanies a region of inflammation is an example of edema caused by redistribution of proteins from the plasma to the interstitial fluid.

Edema Is the Result of Alterations in Capillary Exchange

Edema is a sign that normal exchange between the circulatory system and the lymphatics has been disrupted. Edema usually arises from one of two causes: (1) inadequate drainage of lymph or (2) blood capillary filtration that greatly exceeds capillary absorption.

Inadequate lymph drainage occurs with obstruction of the lymphatic system, particularly at the lymph nodes. Parasites, cancer, or fibrotic tissue growth caused by therapeutic radiation can block the movement of lymph through the system. For example, *elephantiasis* is a chronic condition marked by gross enlargement of the legs and lower appendages when parasites block the lymph vessels. Lymph drainage may also be impaired if lymph nodes are removed during surgery, a common procedure in the diagnosis and treatment of cancer.

Factors that disrupt the normal balance between capillary filtration and absorption include:

1. *An increase in capillary hydrostatic pressure.* Increased hydrostatic pressure is usually indicative of elevated venous pressure. An increase in arterial pressure is generally not noticeable at the capillaries because of autoregulation of pressure in the arterioles.

   One common cause of increased venous pressure is *heart failure,* a condition in which one ventricle loses pumping power and can no longer pump all the blood sent to it by the other ventricle (see Ch. 14 Concept Check #31 on p. 489). For example, if the right ventricle begins to fail but the left ventricle maintains its cardiac output, blood accumulates in the systemic circulation. Blood pressure rises first in the right atrium, then in the veins and capillaries draining into the right side of the heart. When capillary hydrostatic pressure increases, filtration greatly exceeds absorption, leading to edema.

2. *A decrease in plasma protein concentration.* Plasma protein concentrations may decrease as a result of severe malnutrition or liver failure. The liver is the main site for plasma protein synthesis.

3. *An increase in interstitial proteins.* As discussed earlier, excessive leakage of proteins out of the blood will decrease plasma colloid osmotic pressure and will increase net capillary filtration.

On occasion, changes in the balance between filtration and absorption help the body maintain homeostasis. For example, if arterial blood pressure falls, capillary hydrostatic pressure also decreases. This change increases fluid absorption. If blood pressure falls low enough, there will be net absorption in the capillaries rather than net filtration. This passive mechanism helps maintain blood volume in situations in which blood pressure is very low, such as hemorrhage or severe dehydration.

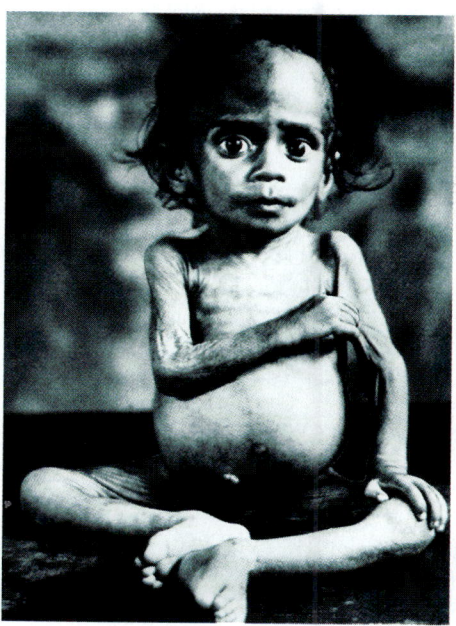

FIGURE 15-20 *Ascites (abdominal edema) in a child with protein malnutrition*

The African word for protein malnutrition is *kwashiorkor.*

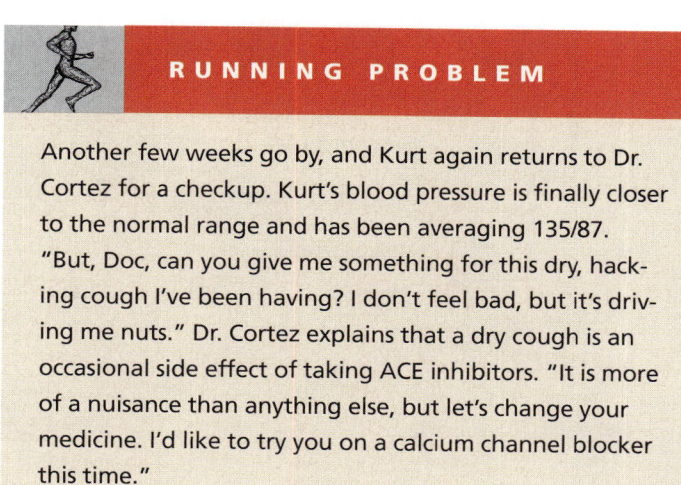

Another few weeks go by, and Kurt again returns to Dr. Cortez for a checkup. Kurt's blood pressure is finally closer to the normal range and has been averaging 135/87. "But, Doc, can you give me something for this dry, hacking cough I've been having? I don't feel bad, but it's driving me nuts." Dr. Cortez explains that a dry cough is an occasional side effect of taking ACE inhibitors. "It is more of a nuisance than anything else, but let's change your medicine. I'd like to try you on a calcium channel blocker this time."

Question 4:
  How do calcium channel blockers lower blood pressure?

501   508   510   513   **521**   529

CONCEPT CHECK

14. If the left ventricle fails to pump normally, blood will back up into what set of blood vessels? Where would you expect edema to occur?

15. Malnourished children who have inadequate protein in their diet often have grotesquely swollen bellies; this condition, which can be described as edema of the abdomen, is called *ascites* (Fig. 15-20 ■). Use the information you have just learned about capillary filtration to explain why malnutrition causes ascites.

Answers: p. 533

REGULATION OF BLOOD PRESSURE

The central nervous system coordinates the reflex control of blood pressure. The main integrating center is in the medulla oblongata. Because of the difficulty of studying neural networks in the brain, however, we still know relatively little about the nuclei, neurotransmitters, and interneurons of the **medullary cardiovascular control centers**.

The Baroreceptor Reflex Is the Primary Homeostatic Control for Blood Pressure

The primary function of the cardiovascular control center is to maintain adequate blood flow to the brain and heart. Sensory input to this integrating center comes from a variety of peripheral sensory receptors. Stretch-sensitive mechanoreceptors known as **baroreceptors** are located in the walls of the carotid arteries and aorta (Fig. 15-21 ■), where they monitor the pres-

sure of blood flowing to the brain (carotid baroreceptors) and to the body (aortic baroreceptors). The carotid and aortic baroreceptors are tonically active stretch receptors that fire action potentials continuously at normal blood pressures.

The primary reflex pathway for homeostatic control of blood pressure is the **baroreceptor reflex**. When increased blood pressure in the arteries stretches the baroreceptor membrane, firing rate of the receptor increases. If blood pressure falls, the firing rate of the receptor decreases.

Action potentials from the baroreceptors travel to the medullary cardiovascular control center via sensory neurons. The cardiovascular control center integrates the sensory input and initiates an appropriate response. The response of the baroreceptor reflex is quite rapid: changes in cardiac output and peripheral resistance occur within two heartbeats of the stimulus.

Efferent output from the cardiovascular control center is carried via both sympathetic and parasympathetic autonomic neurons. Peripheral resistance is under tonic sympathetic control, with increased sympathetic discharge causing vasoconstriction.

Heart function is regulated by antagonistic control. Increased sympathetic activity increases heart rate at the SA node, shortens conduction time through the AV node, and enhances the force of myocardial contraction. Increased parasympathetic activity slows heart rate but has only a small effect on ventricular contraction.

The baroreceptor reflex is summarized in Figure 15-22 ■. Baroreceptors increase their firing rate as blood pressure increases, activating the medullary cardiovascular control center.

15

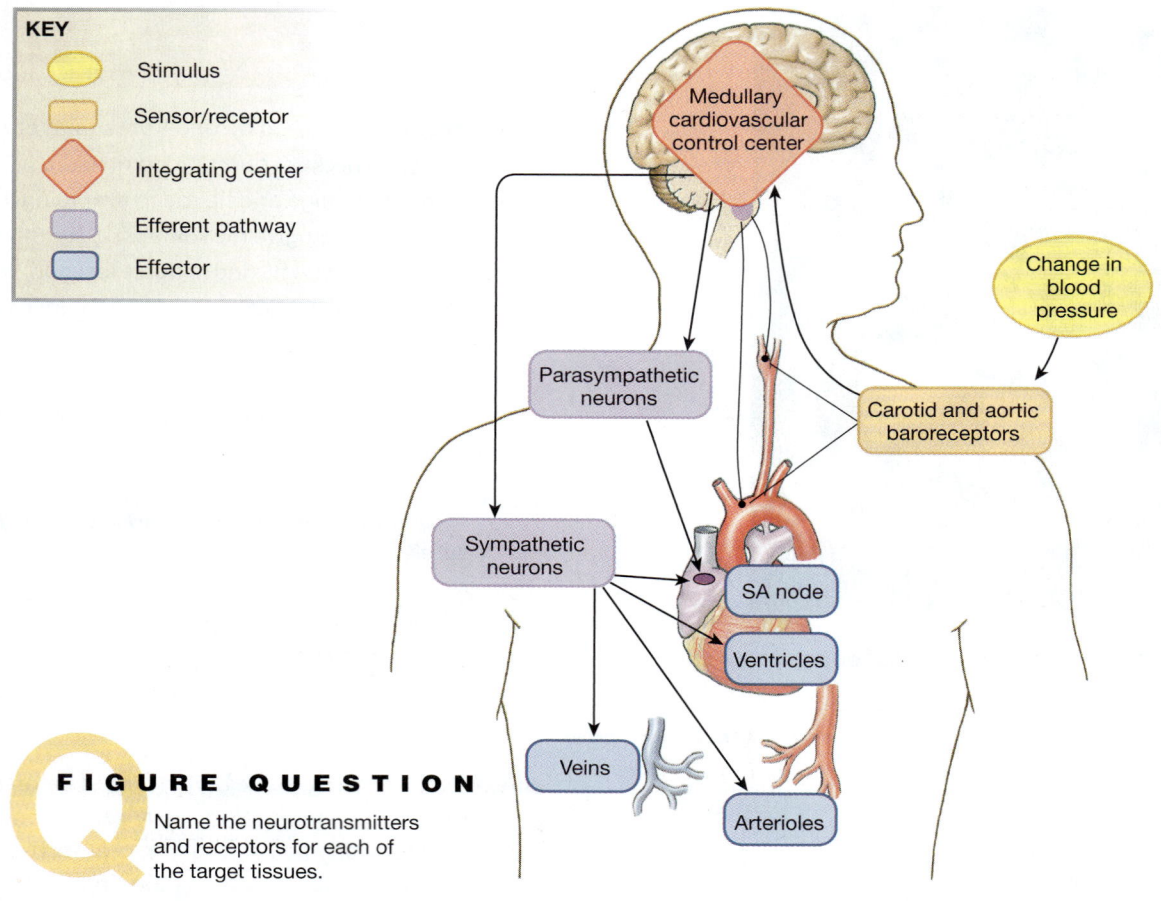

KEY

- 🟡 Stimulus
- 🟧 Sensor/receptor
- 🔶 Integrating center
- 🟪 Efferent pathway
- 🟦 Effector

Q **FIGURE QUESTION**

Name the neurotransmitters and receptors for each of the target tissues.

■ **FIGURE 15-21**   *Components of the baroreceptor reflex*

In response, the cardiovascular control center increases parasympathetic activity and decreases sympathetic activity to slow down the heart. When heart rate falls, cardiac output falls. In the circulation, decreased sympathetic activity causes dilation of the arterioles, allowing more blood to flow out of the arteries. The combination of reduced cardiac output and decreased peripheral resistance lowers the mean arterial blood pressure.

Cardiovascular function can be modulated by input from peripheral receptors other than the baroreceptors. For example, arterial chemoreceptors activated by low blood oxygen levels increase cardiac output. The cardiovascular control center also has reciprocal communication with centers in the medulla that control breathing.

The integration of function between the respiratory and circulatory systems is adaptive. If tissues require more oxygen, it is supplied by the cardiovascular system working in tandem with the respiratory system. Consequently, increases in breathing rate are usually accompanied by increases in cardiac output.

Blood pressure is also subject to modulation by higher brain centers, such as the hypothalamus and cerebral cortex.

The hypothalamus is responsible for vascular responses involved in body temperature regulation (see Chapter 22) and for the fight-or-flight response. Learned and emotional responses may originate in the cerebral cortex and be expressed by cardiovascular responses such as blushing and fainting.

One such reflex is the *vasovagal response,* which may be triggered in some people by the sight of blood or a hypodermic needle. (Recall Anthony's experience at the beginning of this chapter.) In this pathway, increased parasympathetic activity and decreased sympathetic activity slow heart rate and cause widespread vasodilation. Cardiac output and peripheral resistance both fall, triggering a precipitous fall in blood pressure. With insufficient blood to the brain, the individual faints.

Regulation of blood pressure in the cardiovascular system is closely tied to regulation of body fluid balance by the kidneys. Thus, certain hormones secreted from the heart act on the kidneys, while hormones secreted from the kidneys act on the heart and blood vessels. Together, the heart and kidneys play a major role in maintaining homeostasis of body fluids, the subject of Chapter 20. The overlap of these systems is an excellent example of the integration of organ system function.

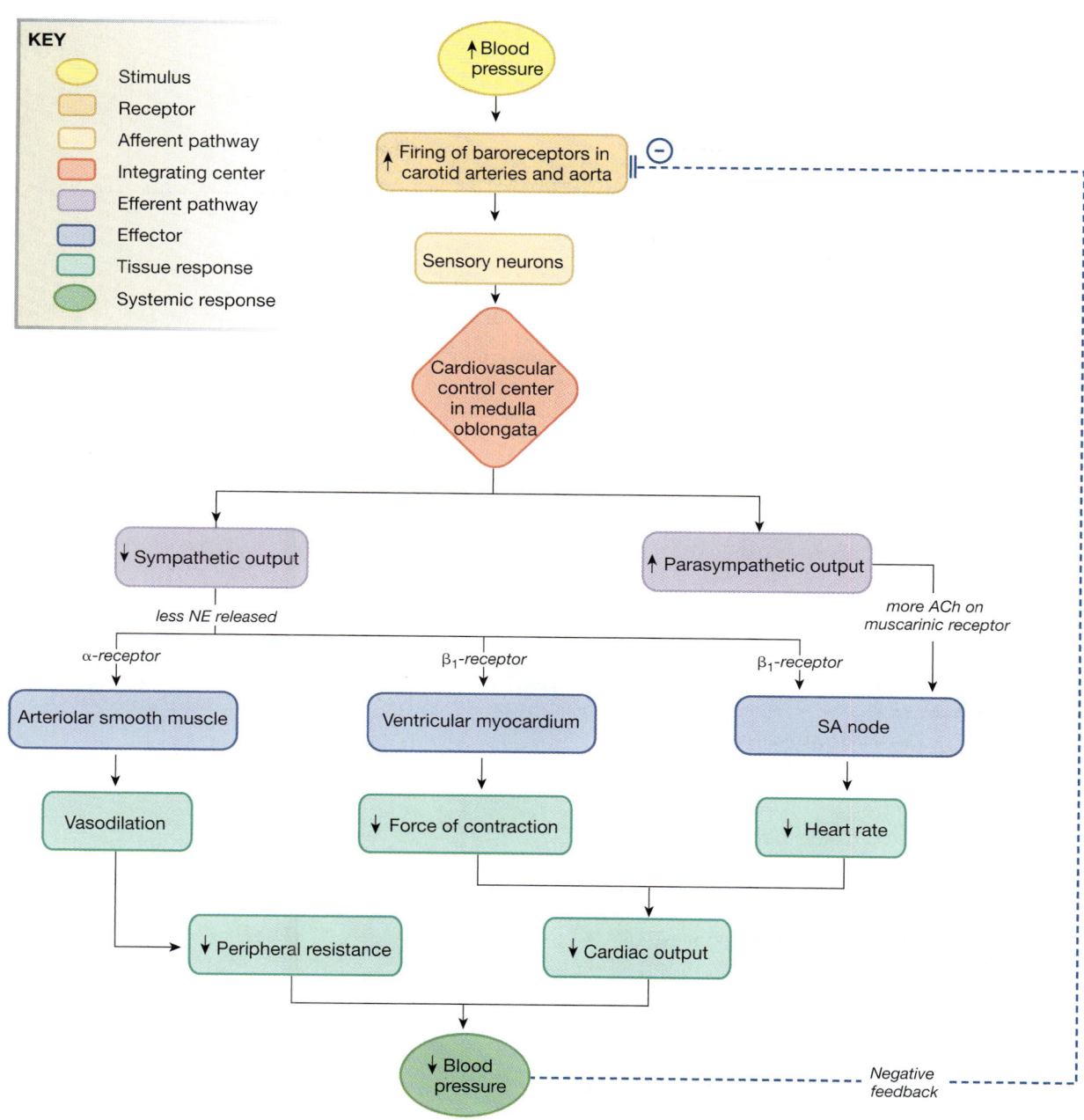

KEY

- Stimulus
- Receptor
- Afferent pathway
- Integrating center
- Efferent pathway
- Effector
- Tissue response
- Systemic response

↑Blood pressure

↑ Firing of baroreceptors in carotid arteries and aorta

Sensory neurons

Cardiovascular control center in medulla oblongata

↓ Sympathetic output — *less NE released*

↑ Parasympathetic output — *more ACh on muscarinic receptor*

α-receptor

Arteriolar smooth muscle

Vasodilation

β₁-receptor

Ventricular myocardium

↓ Force of contraction

β₁-receptor

SA node

↓ Heart rate

↓ Peripheral resistance

↓ Cardiac output

↓ Blood pressure

Negative feedback

■ **FIGURE 15-22** *The baroreceptor reflex: the response to increased blood pressure*

CONCEPT CHECK

16. Baroreceptors have stretch-sensitive ion channels in their cell membrane. Increased pressure stretches the receptor cell membrane, opens the channels, and initiates action potentials. What ion probably flows through these channels and in which direction (into or out of the cell)?

Answers: p. 533

Orthostatic Hypotension Triggers the Baroreceptor Reflex

The baroreceptor reflex functions every morning when you get out of bed. When you are lying flat, gravitational forces are distributed evenly up and down the length of your body, and blood is distributed evenly throughout the circulation. When you stand up, gravity causes blood to pool in the lower extremities. This pooling creates an instantaneous decrease in venous return. As a result, less blood is in the ventricles at the beginning of the next contraction. Cardiac output falls from 5 L/min to 3 L/min, causing arterial blood pressure to decrease. This decrease in blood pressure upon standing is known as *orthostatic hypotension* [*orthos,* upright + *statikos,* to stand].

Orthostatic hypotension in a normal person triggers the baroreceptor reflex. The carotid and aortic baroreceptors respond to the fall in arterial blood pressure by decreasing their firing rate (Fig. 15-23■). Diminished sensory input into the cardiovascular control center increases sympathetic activity and

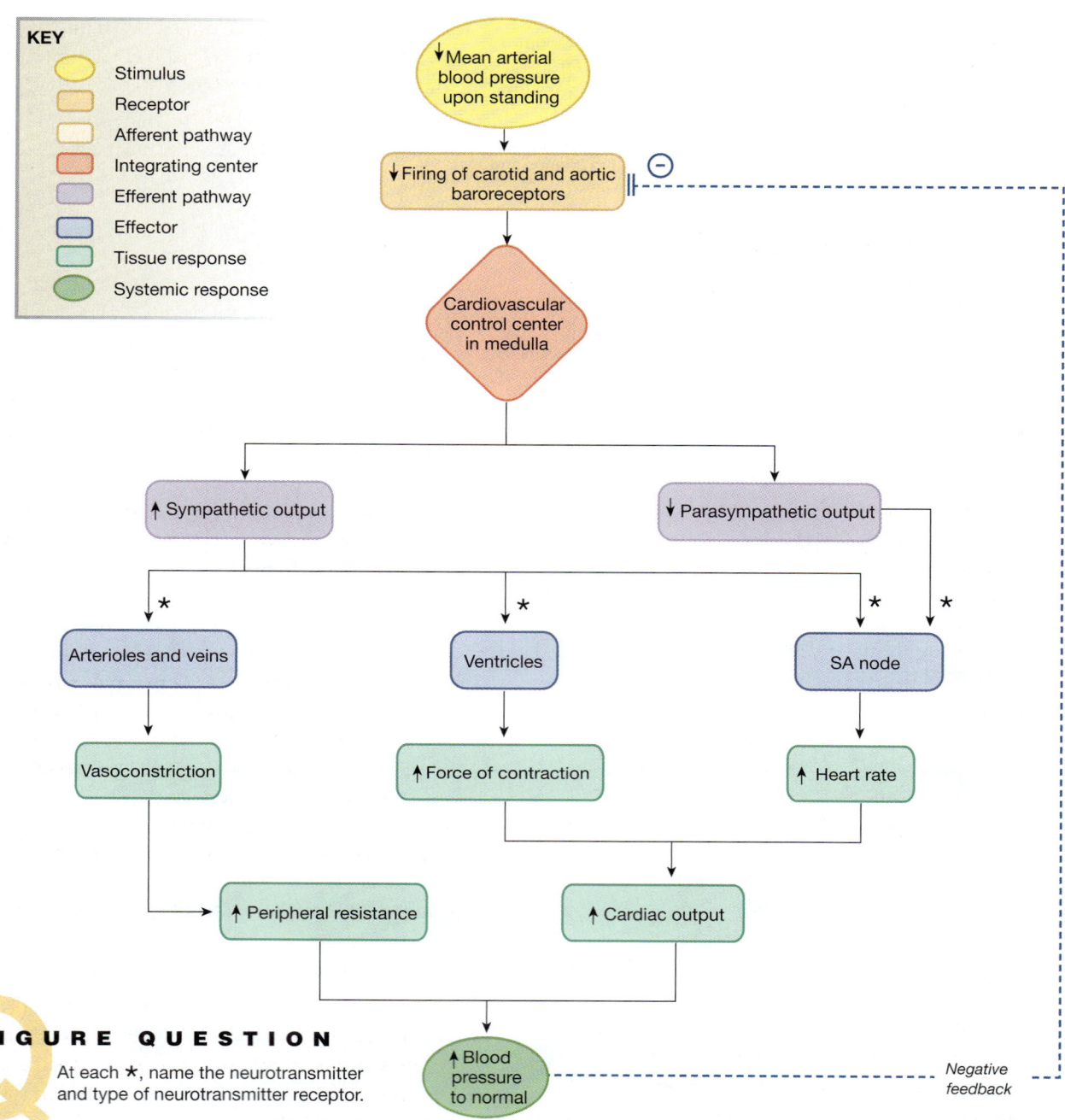

KEY

- Stimulus
- Receptor
- Afferent pathway
- Integrating center
- Efferent pathway
- Effector
- Tissue response
- Systemic response

↓ Mean arterial blood pressure upon standing

↓ Firing of carotid and aortic baroreceptors

⊖

Cardiovascular control center in medulla

↑ Sympathetic output

↓ Parasympathetic output

Arterioles and veins *

Ventricles *

SA node * *

Vasoconstriction

↑ Force of contraction

↑ Heart rate

↑ Peripheral resistance

↑ Cardiac output

↑ Blood pressure to normal

Negative feedback

FIGURE QUESTION

Q At each ★, name the neurotransmitter and type of neurotransmitter receptor.

■ **FIGURE 15-23** *The baroreceptor reflex: the response to orthostatic hypotension*

decreases parasympathetic activity. As a result of autonomic changes, heart rate and force of contraction increase while arterioles and veins constrict. The combination of increased cardiac output and increased peripheral resistance increases mean arterial pressure and restores it to normal within two heartbeats. The skeletal muscle pump also contributes to the recovery by enhancing venous return when abdominal and leg muscles contract to maintain an upright position.

The baroreceptor reflex is not always effective, however. For example, during extended bed rest or in the zero-gravity conditions of space flights, blood from the lower extremities is distributed evenly throughout the body rather than pooling in the lower extremities. This even distribution raises arterial pressure, triggering the kidneys to excrete what is perceived as excess fluid. Over the course of three days, excretion of water leads to a 12% decrease in blood volume. When the person finally gets out of bed or returns to earth, gravity again causes blood to pool in the legs. Orthostatic hypotension occurs, and the baroreceptors attempt to compensate. In this instance, however, the cardiovascular system is unable to restore normal pressure because of the loss of blood volume. As a result, the subject may become dizzy or even faint from reduced delivery of oxygen to the brain.

17. In the movie *Jurassic Park,* Dr. Ian Malcolm must flee from the *T. rex.* Draw a reflex map showing the cardiovascular response to his fight-or-flight situation. (*Hints:* what is the stimulus? Fear is integrated in the limbic system.)

Answers: p. 533

CARDIOVASCULAR DISEASE

Diseases of the heart and blood vessels, such as heart attacks and strokes, play a role in more than half of all deaths in the United States. According to the American Heart Association, in 2005 cardiovascular diseases will cost people in the United States more than $393 billion in medical expenses and lost wages. The prevalence of cardiovascular disease is reflected in the tremendous amount of research being done worldwide. The scientific investigations range from large-scale clinical studies that track cardiovascular disease in thousands of people, such as the Framingham (Massachusetts) Heart Study, to experiments at the cellular and molecular levels.

Much of the research at the cellular and molecular levels is designed to expand our understanding of both normal and abnormal function in the heart and blood vessels. Scientists are studying a virtual alphabet soup of transporters and regulators. Some of these molecules, such as adenosine, endothelin, vascular endothelial growth factor (VEGF), phospholamban, and nitric oxide, you have studied here and in Chapter 14.

As we increase our knowledge of cardiovascular function, we also begin to understand the actions of drugs that have been used for centuries. A classic example is the cardiac glycoside *digitalis* [📖 p. 492], whose mechanism of action was explained when scientists discovered the role of Na^+-K^+-ATPase. It is a sobering thought to realize that for many therapeutic drugs, we know *what* they do without fully understanding *how* they do it.

Risk Factors for Cardiovascular Disease Include Smoking, Obesity, and Inheritable Factors

As you learned in Chapter 1, conducting and interpreting research on humans is a complicated endeavor in part because of the difficulty of designing well-controlled experiments [📖 p. 10]. The economic and social importance of cardiovascular disease (CVD) make it the focus of many studies each year as researchers try to improve treatments and prediction algorithms. (An *algorithm* is a set of rules or a sequence of steps used to solve a problem.) We can predict the likelihood that a person will develop cardiovascular disease during his or her lifetime by examining the various risk factors that the person possesses. The list of risk factors described here is the result of following the medical histories of thousands of people for many years in studies such as the Framingham Heart Study. As more data become available, additional risk factors may be added.

Risk factors are generally divided into those over which the person has no control and those that can be controlled.

Medical intervention is aimed at reducing risk from the controllable factors. The risk factors that cannot be controlled include gender, age, and a family history of early cardiovascular disease. As noted earlier in the chapter, *coronary heart disease* (CHD) is a form of cardiovascular disease in which the coronary arteries become blocked by cholesterol deposits and blood clots. Up until middle age, men have a 3–4 times higher risk of developing CHD than do women. After age 55, when most women have entered menopause, the death rate from CHD equalizes in men and women. In general, the risk of coronary heart disease increases as people age. Heredity also plays an important role. If a person has one or more close relatives with this condition, his or her risk is elevated.

Risk factors that can be controlled include cigarette smoking, obesity, sedentary lifestyle, and untreated hypertension. In the United States, smoking-related illnesses are the primary preventable cause of death, followed by conditions related to overweight and obesity. Physical inactivity and obesity have been steadily increasing in the United States since 1991, and currently 70% of U.S. adults are either overweight or obese.

Two risk factors for cardiovascular disease—blood lipids and diabetes mellitus—have both an uncontrollable genetic component and a modifiable lifestyle component. Diabetes mellitus is a metabolic disorder that puts a person at risk for developing coronary heart disease by contributing to the development of **atherosclerosis** ("hardening of the arteries"), in which fatty deposits form inside arterial blood vessels. Elevated serum cholesterol and triglycerides also lead to atherosclerosis. (Lipid metabolism is discussed in more detail in Chapter 22.)

The increasing prevalence of these risk factors has created an epidemic in the United States, with more than 1.4 million deaths in 2002 attributed to all forms of cardiovascular disease.

15

Atherosclerosis Is an Inflammatory Process

Coronary heart disease accounts for the majority of cardiovascular disease deaths and is the single largest killer of Americans. Let's look at the underlying cause of this disease; atherosclerosis.

The role of elevated blood cholesterol in the development of atherosclerosis is well established. Cholesterol, like other lipids, is not very soluble in aqueous solutions, such as the plasma. Therefore, when cholesterol in the diet is absorbed from the digestive tract, it combines with lipoproteins to make it more soluble. (The details of cholesterol digestion, absorption, and metabolism are discussed in Chapters 21 and 22.) Multiple lipoproteins associate with cholesterol, but clinicians generally are concerned with two: those found in **high-density lipoprotein-cholesterol (HDL-C)** complexes and those found in **low-density lipoprotein-cholesterol (LDL-C)** complexes. HDL-C is the more desirable form of blood cholesterol because high levels of HDL are associated with lower risk of heart attacks. (Memory aid: let the "H" in HDL stand for "healthy.")

LDL-C is sometimes called "bad" cholesterol because elevated plasma LDL-C levels are associated with coronary heart

CLINICAL FOCUS

DIABETES AND CARDIOVASCULAR DISEASE

Having diabetes is one of the major risk factors for developing cardiovascular disease, and almost two-thirds of people with diabetes will die from cardiovascular problems. In diabetes, cells that cannot use glucose turn to fats and proteins for their energy. The body breaks down fat into fatty acids [⊜ p. 112] and dumps them into the blood. Plasma cholesterol levels are also elevated. When LDL-C remains in the blood, the excess is ingested by macrophages, starting a series of events that lead to atherosclerosis. Because of the pivotal role that LDL-C plays in atherosclerosis, many forms of therapy, ranging from dietary modification and exercise to drugs, are aimed at lowering LDL-C levels. Left untreated, blockage of small and medium-sized blood vessels in the lower extremities can lead to loss of sensation and *gangrene* (tissue death) in the feet. Atherosclerosis in larger vessels causes heart attacks and strokes. To learn more about diabetes and the increased risk of cardiovascular disease, visit the websites of the American Diabetes Association (*www.diabetes.org*) and the American Heart Association (*www.americanheart.org*).

disease. (Remember this by associating "L" with "lethal.") Normal levels of LDL-C are not bad, however, because LDL is necessary for cholesterol transport into cells. LDL-C's binding site—a protein called **apoB**—combines with an LDL receptor found in clathrin-coated pits on the cell membrane, and the receptor-LDL-C complex is brought into the cell by endocytosis [⊜ Fig. 5-24, p. 149]. The LDL receptor recycles to the cell membrane, and the endosome fuses with a lysosome. LDL-C's proteins are digested to amino acids, and the freed cholesterol is used to make cell membranes or steroid hormones.

Although LDL is needed for cellular uptake of cholesterol, excess levels of plasma LDL-C lead to atherosclerosis. Endothelial cells lining the arteries transport LDL-C into the extracellular space so that it accumulates just under the intima. There, macrophages ingest cholesterol and other lipids to become lipid-filled *foam cells* (Fig. 15-24b ■). Cytokines released by the macrophages promote smooth muscle cell division. This early-stage *lesion* [*laesio*, injury] is called a *fatty streak*. As the condition progresses, the lipid core grows, and smooth muscle cells reproduce, forming bulging *plaques* that protrude into the lumen

of the artery. In the advanced stages of atherosclerosis, the plaques develop hard, calcified regions and fibrous collagen caps (Fig. 15-24c). The mechanism by which calcium carbonate is deposited is still being investigated.

We once believed that the occlusion (blockage) of coronary blood vessels by large plaques was the primary cause of heart attacks, but that model has been revised. The new model indicates that blood clot formation on plaques is more dependent on the structure of a plaque than on its size. *Vulnerable plaques* have thin fibrous caps that are more likely to rupture, exposing collagen and activating platelets that initiate blood clots (*thrombi*) (Fig. 15-24d). *Stable plaques* have thick fibrous caps that separate the lipid core from the blood and do not activate platelets. Atherosclerosis is now considered to be an inflammatory process in which macrophages release enzymes that convert stable plaques to vulnerable plaques.

EMERGING CONCEPTS

INFLAMMATORY MARKERS FOR CARDIOVASCULAR DISEASE

In clinical studies, it is sometimes difficult to determine whether a factor that has a positive correlation with a disease functions in a cause-effect relationship or represents a simple association. For example, two factors associated with higher incidence of heart disease are C-reactive protein and homocysteine. *C-reactive protein* (CRP) is a molecule involved in the body's response to inflammation, and in one study, women who had elevated blood CRP levels were more than twice as likely to have a serious cardiovascular problem as women with low CRP. Does that mean that CRP is causing cardiovascular disease? Or could it simply be a marker that can be used clinically to predict who is more likely to develop cardiovascular complications, such as a heart attack or stroke?

Similarly, elevated homocysteine levels are associated with an increased incidence of CVD. (*Homocysteine* is an amino acid that takes part in a complicated metabolic pathway that also requires folate and vitamin B_{12} as cofactors.) Should physicians routinely measure homocysteine along with cholesterol? Currently there is little clinical evidence to show that reducing either CRP or homocysteine decreases a person's risk of developing CVD. If these two markers are not indicators for *modifiable* risk factors, should a patient's insurance be asked to pay for the tests used to detect them?

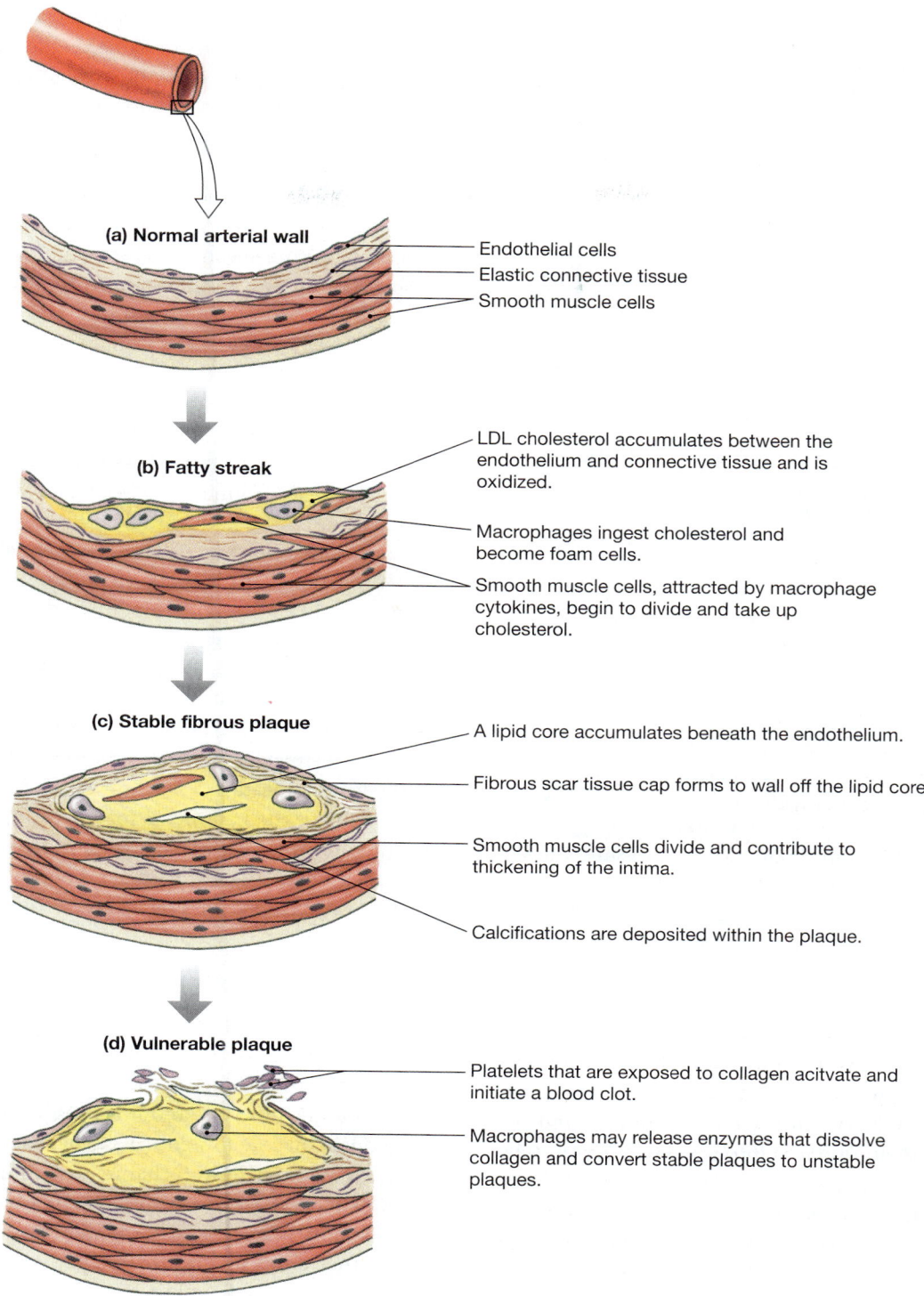

(a) **Normal arterial wall**
— Endothelial cells
— Elastic connective tissue
— Smooth muscle cells

(b) **Fatty streak**
— LDL cholesterol accumulates between the endothelium and connective tissue and is oxidized.
— Macrophages ingest cholesterol and become foam cells.
— Smooth muscle cells, attracted by macrophage cytokines, begin to divide and take up cholesterol.

(c) **Stable fibrous plaque**
— A lipid core accumulates beneath the endothelium.
— Fibrous scar tissue cap forms to wall off the lipid core.
— Smooth muscle cells divide and contribute to thickening of the intima.
— Calcifications are deposited within the plaque.

(d) **Vulnerable plaque**
— Platelets that are exposed to collagen acitvate and initiate a blood clot.
— Macrophages may release enzymes that dissolve collagen and convert stable plaques to unstable plaques.

■ **FIGURE 15-24** *The development of atherosclerotic plaques*

If a clot blocks blood flow to the heart muscle, a heart attack (*myocardial infarction*) results. Blocked blood flow in a coronary artery cuts off the oxygen supply to myocardial cells supplied by that artery. The oxygen-starved cells must then rely on anaerobic metabolism [🔁 p. 106], which produces lactic acid (H^+). As ATP production declines, the contractile cells are unable to pump Ca^{2+} out of the cell. The combination of unusually high Ca^{2+} and H^+ concentrations in the cytosol closes gap junctions in the damaged cells. Closure electrically isolates the damaged cells so that they no longer contract, and it forces action potentials to find an alternate route from cell to cell. If the damaged area of myocardium is large, the disruption can lead

15

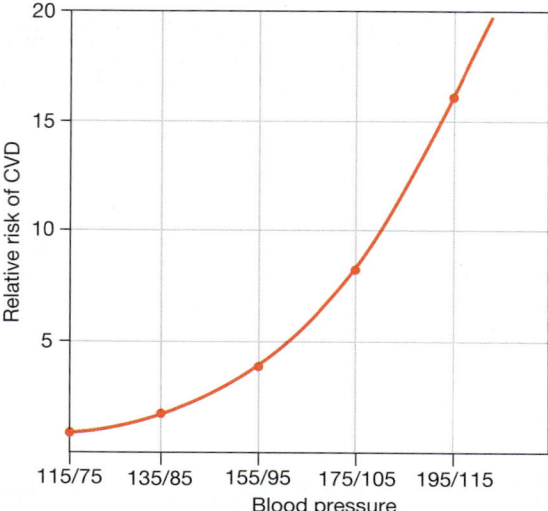

■ FIGURE 15-25   *The relationship between increasing blood pressure and the risk of developing cardiovascular disease*

to an irregular heartbeat (*arrhythmia*) and potentially result in cardiac arrest and/or death.

Hypertension Represents a Failure of Homeostasis

One controllable risk factor for cardiovascular disease is hypertension—chronically elevated blood pressure, with systolic pressures greater than 140 mm Hg or diastolic pressures greater than 90 mm Hg. Hypertension is a common disease in the United States and is one of the most common reasons for visits to physicians and for the use of prescription drugs. High blood pressure is associated with increasing risk of CVD: the risk doubles for each 20/10 mm Hg increase in blood pressure over a baseline value of 115/75 (Fig. 15-25 ■).

More than 90% of all patients with hypertension are considered to have *essential* (or *primary*) *hypertension*, with no clear-cut cause other than heredity. Cardiac output is usually normal in these people, and their elevated blood pressure appears to be associated with increased peripheral resistance. Some investigators have speculated that the increased resistance may be due to a lack of nitric oxide, the locally produced vasodilator formed by endothelial cells in the arterioles. In the remaining 5–10% of hypertensive cases, the cause is known, and the hypertension is considered to be secondary to an underlying pathology. For instance, the cause might be an endocrine disorder that causes fluid retention.

A key feature of hypertension from all causes is adaptation of the carotid and aortic baroreceptors to higher pressure, with subsequent down-regulation of their activity. Without input from the baroreceptors, the cardiovascular control center interprets the high blood pressure as "normal," and no reflex reduction of pressure occurs.

Hypertension is a risk factor for atherosclerosis because high pressure in the arteries damages the endothelial lining of the vessels and promotes the formation of atherosclerotic plaques. In addition, high arterial blood pressure puts additional strain on the heart by increasing afterload [↺ p. 493]. When resistance in the arterioles is high, the myocardium must work harder to push the blood into the arteries.

Amazingly, stroke volume in hypertensive patients remains constant up to a mean blood pressure of about 200 mm Hg, despite the increasing amount of work that the ventricle must perform as blood pressure increases. The cardiac muscle of the left ventricle responds to chronic high systemic resistance in the same way that skeletal muscle responds to a weight-lifting routine. The heart muscle *hypertrophies*, increasing the size and strength of the muscle fibers.

However, if resistance remains high over time, the heart muscle is unable to meet the work load and begins to fail: cardiac output by the left ventricle decreases. If cardiac output of the right heart remains normal while the output from the left side decreases, fluid collects in the lungs, creating *pulmonary edema*. At this point, a detrimental positive feedback loop begins. Oxygen exchange in the lungs diminishes because of the pulmonary edema, leading to less oxygen in the blood. Lack of oxygen for aerobic metabolism further weakens the heart, and its pumping effectiveness diminishes even more. Unless treated, this condition, known as *congestive heart failure*, eventually leads to death.

Many of the treatments for hypertension have their basis in the cardiovascular physiology you have learned. For example, calcium entry into vascular smooth muscle and cardiac muscle can be decreased by a class of drugs known as *calcium channel blockers*. These drugs bind to Ca^{2+} channel proteins, making it less likely that the channels will open in response to depolarization. With less Ca^{2+} entry, vascular smooth muscle dilates, while in the heart the depolarization rate of the SA node and the force of contraction decrease. Vascular smooth muscle is more sensitive than cardiac muscle to certain classes of calcium channel blockers, and it is possible to get vasodilation at drug doses that are low enough to have no effect on heart rate. Other tissues with Ca^{2+} channels, such as neurons, are only minimally affected by calcium channel blockers because their Ca^{2+} channels are of a different subtype.

Other drugs used to treat hypertension include diuretics, which decrease blood volume, and beta-blocking drugs that target β_1-receptors and decrease catecholamine stimulation of cardiac output. Two other groups of antihypertensive drugs, the ACE inhibitors and the angiotensin receptor blockers, act by decreasing the activity of angiotensin, a powerful vasoconstrictor substance. You will learn more about angiotensin later in the book, when you study the integrated control of blood pressure by the cardiovascular and renal systems. In the future, we may be seeing new treatments for hypertension that are based on the molecular physiology of the heart and blood vessels.

RUNNING PROBLEM CONCLUSION

ESSENTIAL HYPERTENSION

Kurt remained on the calcium channel blocker, and after several months his blood pressure stabilized at 130/85—a significant improvement. Kurt's new diet also brought his total blood cholesterol down below 200 mg/dL plasma. By improving two of his controllable risk factors, Kurt decreased his chances of having a heart attack. To learn more about hyper-

tension and some of the therapies currently used to treat it, visit the website of the American Heart Association (*www.americanheart.org*).

Check your understanding of this running problem by comparing your answers with the information in the summary table.

	QUESTION	FACTS	INTEGRATION AND ANALYSIS
1	Why are people with high blood pressure at greater risk for having a hemorrhagic (or bleeding) stroke?	High blood pressure exerts force on the walls of the blood vessels.	If an area of blood vessel wall is weakened or damaged, high blood pressure may cause that area to rupture, allowing blood to leak out of the vessel into the surrounding tissues.
2	What is the rationale for reducing salt intake to control hypertension?	Salt causes water retention.	Blood pressure will increase if the circulating blood volume increases. By restricting salt in the diet, a person can decrease retention of fluid in the extracellular compartment, which includes the plasma.
3	Why would blocking the action of a vasoconstrictor lower blood pressure?	Peripheral blood pressure is determined by cardiac output and peripheral resistance.	Resistance is inversely proportional to the radius of the blood vessels. Therefore, if blood vessels dilate as a result of blocking a vasoconstrictor, resistance and blood pressure will decrease.
4	How do calcium channel blockers lower blood pressure?	Calcium entry from the extracellular fluid plays an important role in both smooth muscle and cardiac muscle contraction.	Blocking Ca^{2+} entry through Ca^{2+} channels will decrease the force of cardiac contraction and decrease the contractility of vascular smooth muscle. Both of these effects will lower blood pressure.

501   508   510   513   521   **529**

15

CHAPTER SUMMARY

Blood flow through the cardiovascular system is an excellent example of *mass flow* in the body. Cardiac contraction creates high pressure in the ventricles, and this pressure drives blood through the vessels of the systemic and pulmonary circuits, speeding up cell-to-cell *communication*. Resistance to flow is regulated by *local and reflex control mechanisms* that act on arteriolar smooth muscle and help match tissue perfusion to tissue needs. The *homeostatic* baroreceptor reflex monitors arterial pressure to ensure adequate perfusion of the brain and heart. Capillary *exchange of material* between the plasma and interstitial fluid *compartments* uses several transport mechanisms, including diffusion, transcytosis, and bulk flow.

1. Homeostatic regulation of the cardiovascular system is aimed at maintaining adequate blood flow to the brain and heart. (p. 501)

2. Total blood flow at any level of the circulation is equal to the cardiac output. (p. 502)

The Blood Vessels

IP Cardiovascular—Anatomy Review: Blood Vessel Structure & Function

3. Blood vessels are composed of layers of smooth muscle, elastic and fibrous connective tissue, and **endothelium**. (p. 502; Fig. 15-2)

4. **Vascular smooth muscle** maintains a state of muscle tone. (p. 502)

5. The walls of the aorta and major arteries are both stiff and springy. This property allows them to absorb energy and release it through elastic recoil. (p. 502)

6. **Metarterioles** regulate blood flow through capillaries and allow white blood cells to go directly from arterioles to the venous circulation. (p. 503; Fig. 15-3)

7. Capillaries and postcapillary **venules** are the site of exchange between blood and interstitial fluid. (p. 503)

8. Veins hold more than half of the blood in the circulatory system. Veins have thinner walls with less elastic tissue than arteries, so veins expand easily when they fill with blood. (p. 503)

9. **Angiogenesis** is the process by which new blood vessels grow and develop, especially after birth. (p. 503)

Blood Pressure

IP Cardiovascular—Measuring Blood Pressure

10. The ventricles create high pressure that is the driving force for blood flow. The aorta and arteries act as a pressure reservoir during ventricular relaxation. (p. 504; Fig. 15-4)

11. Blood pressure is highest in the arteries and decreases as blood flows through the circulatory system. At rest, average **systolic pressure** is 120 mm Hg, and average **diastolic pressure** is 80 mm Hg. (p. 505; Fig. 15-5)

12. Pressure created by the ventricles can be felt as a **pulse** in the arteries. **Pulse pressure** equals systolic pressure minus diastolic pressure. (p. 505)

13. Blood flow against gravity in the veins is assisted by one-way valves and by the respiratory and skeletal muscle pumps. (p. 505; Fig. 15-6)

14. Arterial blood pressure is indicative of the driving pressure for blood flow. **Mean arterial pressure** (MAP) is defined as diastolic pressure + 1/3 (systolic pressure−diastolic pressure). (p. 506)

15. Arterial blood pressure is usually measured with a sphygmomanometer. Blood squeezing through a compressed brachial artery makes **Korotkoff sounds**. (p. 507; Fig. 15-7)

16. Arterial pressure is a balance between cardiac output and the resistance to blood flow offered by the arterioles (**peripheral resistance**). (p. 507; Fig. 15-8)

17. If blood volume increases, blood pressure increases. If blood volume decreases, blood pressure decreases. (p. 508; Fig. 15-9)

18. Venous blood volume can be shifted to the arteries if arterial blood pressure falls. (p. 509; Fig. 15-10)

Resistance in the Arterioles

IP Cardiovascular—Factors that Affect Blood Pressure

19. The arterioles are the main site of variable resistance in the systemic circulation. A small change in the radius of an arteriole creates a large change in resistance: $R \propto 1/r^4$. (p. 509)

20. Arterioles regulate their own blood flow through **myogenic autoregulation**. Vasoconstriction increases the resistance offered by an arteriole and decreases the blood flow through the arteriole. (p. 510)

21. Arteriolar resistance is influenced by local control mechanisms that match tissue blood flow to the metabolic needs of the tissue. Vasodilator paracrines include nitric oxide, H^+, K^+, CO_2, prostaglandins, adenosine, and histamine. Low O_2 causes vasodilation. Endothelins are powerful vasoconstrictors. (p. 511; Tbl. 15-2)

22. **Active hyperemia** is a process in which increased blood flow accompanies increased metabolic activity. **Reactive hyperemia** is an increase in tissue blood flow following a period of low perfusion. (p. 511; Fig. 15-11)

23. Most systemic arterioles are under tonic sympathetic control. Norepinephrine causes vasoconstriction. Decreased sympathetic stimulation causes vasodilation. (p. 512)

24. Epinephrine binds to arteriolar α-receptors and causes vasoconstriction. Epinephrine on β_2-receptors, found in the arterioles of the heart, liver, and skeletal muscle, causes vasodilation. (p. 513)

Distribution of Blood to the Tissues

25. Changing the resistance of the arterioles affects mean arterial pressure and alters blood flow through the arteriole. (p. 514; Fig. 15-14)

26. The flow through individual arterioles depends on their resistance. The higher the resistance in an arteriole, the lower the blood flow in that arteriole: $Flow_{arteriole} \propto 1/R_{arteriole}$. (p. 514)

27. Blood flow into individual capillaries can be regulated by **precapillary sphincters**. (p. 514; Fig. 15-15)

Exchange at the Capillaries

IP Cardiovascular—Autoregulation and Capillary Dynamics

28. Exchange of materials between the blood and the interstitial fluid occurs primarily by diffusion. (p. 514)

29. **Continuous capillaries** have leaky junctions between cells but also transport material using transcytosis. Continuous capillaries with tight junctions form the blood-brain barrier. (p. 515; Fig. 15-16)

30. **Fenestrated capillaries** have pores that allow large volumes of fluid to pass rapidly. (p. 515; Fig. 15-17)

31. The velocity of blood flow through the capillaries is slow, allowing diffusion to go to equilibrium. (p. 516; Fig. 15-17)

32. **Bulk flow** refers to the mass movement of fluid between the blood and the interstitial fluid. Fluid movement is called **filtration** if the direction of flow is out of the capillary, and **absorption** if the flow is directed into the capillary. (p. 517; Fig. 15-18)

33. The osmotic pressure difference between plasma and interstitial fluid due to the presence of plasma proteins is the **colloid osmotic pressure**. (p. 517)

The Lymphatic System

IP Fluids & Electrolytes—Electrolyte Homeostasis, Edema

34. About 3 liters of fluid filter out of the capillaries each day. The lymphatic system returns this fluid to the circulatory system. (p. 519; Fig. 15-19)

35. Lymph capillaries accumulate fluid, interstitial proteins, and particulate matter by bulk flow. Lymph flow depends on smooth muscle in vessel walls, one-way valves, and the skeletal muscle pump. (p. 519)

36. The condition in which excess fluid accumulates in the interstitial space is called **edema**. Factors that disrupt the normal balance between capillary filtration and absorption cause edema. (p. 519; Fig. 15-20)

Regulation of Blood Pressure

IP Cardiovascular—Blood Pressure Regulation

37. The reflex control of blood pressure resides in the medulla oblongata. **Baroreceptors** in the carotid artery and the aorta monitor

arterial blood pressure and trigger the **baroreceptor reflex**. (p. 521; Figs. 15-21, 15-22)

38. Efferent output from the medullary **cardiovascular control center** goes to the heart and arterioles. Increased sympathetic activity increases heart rate and force of contraction. Increased parasympathetic activity slows heart rate. Increased sympathetic discharge at the arterioles causes vasoconstriction. There is no significant parasympathetic control of arterioles. (p. 522)

39. Cardiovascular function can be modulated by input from higher brain centers and from the respiratory control center of the medulla. (p. 522)

40. The baroreceptor reflex functions each time a person stands up. The decrease in blood pressure upon standing is known as orthostatic hypotension. (p. 523; Fig. 15-23)

Cardiovascular Disease

41. **Cardiovascular disease** is the leading cause of death in the United States. Risk factors predict the likelihood that a person will develop cardiovascular disease during her or his lifetime. (p. 525)

42. **Atherosclerosis** is a condition in which fatty deposits called plaques develop in arteries. If plaques are unstable, they may block the arteries by triggering blood clots. (p. 525; Fig. 15-24)

44. Hypertension is a significant risk factor for the development of cardiovascular disease. (p. 528; Fig. 15-25)

QUESTIONS

(Answers to the Review Questions begin on page A1.)

THE PHYSIOLOGY PLACE

Access more review material online at **The Physiology Place** website. There you'll find review questions, problem-solving activities, case studies, flashcards, and direct links to both *InterActive Physiology*® and *PhysioEx*™ To access the site, go to *www.physiologyplace.com* and select Human Physiology, Fourth Edition.

LEVEL ONE REVIEWING FACTS AND TERMS

1. The first priority of blood pressure homeostasis is to maintain adequate perfusion to which two organs?

2. Match the types of blood vessels with the terms that describe them. Each vessel type may have more than one match, and matching items may be used more than once.
   - (a) arterioles
   - (b) arteries
   - (c) capillaries
   - (d) veins
   - (e) venules
   1. store pressure generated by the heart
   2. have walls that are both stiff and elastic
   3. carry low-oxygen blood
   4. have thin walls of exchange epithelium
   5. act as a volume reservoir
   6. their diameter can be altered by neural input
   7. blood flow slowest through these vessels
   8. have lowest blood pressure
   9. are the main site of variable resistance

3. List the four tissue components of blood vessel walls, in order from inner lining to outer covering. Briefly describe the importance of each tissue.

4. Blood flow to individual tissues is regulated by selective vasoconstriction and vasodilation of which vessels?

5. Aortic pressure reaches a typical high value of _____ (give both numeric value and units) during _____, or contraction of the heart. As the heart relaxes during the event called _____, aortic pressure declines to a typical low value of _____. This blood pressure reading would be written as _____/_____.

6. The rapid pressure increase that occurs when the left ventricle pushes blood into the aorta can be felt as a pressure wave, or _____. What is the equation used to calculate the strength of this pressure wave?

7. List the factors that aid venous return to the heart.

8. What is hypertension, and why is it a threat to a person's health?

9. When measuring a person's blood pressure, at what point in the procedure are you likely to hear Korotkoff sounds?

10. List three paracrines that cause vasodilation. What is the source of each one? In addition to paracrines, list two other ways to control smooth muscle contraction in arterioles.

11. What is hyperemia? How does active hyperemia differ from reactive hyperemia?

12. Most systemic arterioles are innervated by the _____ branch of the nervous system. Increased sympathetic input will have what effect on arteriole diameter?

13. Match each event in the left column with all appropriate neurotransmitter(s) *and* receptor(s) from the list on the right.
   - (a) vasoconstriction of intestinal arterioles
   - (b) vasodilation of coronary arterioles
   - (c) increased heart rate
   - (d) decreased heart rate
   - (e) vasoconstriction of coronary arterioles
   1. norepinephrine
   2. epinephrine
   3. acetylcholine
   4. β_1-receptor
   5. α-receptor
   6. β_2-receptor
   7. nicotinic receptor
   8. muscarinic receptor

14. Which organs receive more than two-thirds of the cardiac output at rest? Which organs have the highest flow of blood on a per unit weight basis?

15. By looking at the density of capillaries in a tissue, you can make assumptions about what property of the tissue? Which tissue has the lowest capillary density? Which tissue has the highest?

16. What type of transport is used to move each of the following substances across the capillary endothelium?
   - (a) oxygen
   - (b) proteins
   - (c) glucose
   - (d) water

17. With which three physiological systems do the vessels of the lymphatic system interact?

18. Define edema. List some ways in which it can arise.

19. Define the following terms and explain their significance to cardiovascular physiology.
   - (a) perfusion
   - (b) colloid osmotic pressure
   - (c) vasoconstriction
   - (d) angiogenesis
   - (e) metarterioles
   - (f) pericytes

20. The two major lipoprotein carriers of cholesterol are _____ and _____. Which type is bad when present in the body in elevated amounts?

15

LEVEL TWO REVIEWING CONCEPTS

21. **Concept map:** Map all the following factors that influence mean arterial pressure. You may add additional terms.

aorta	parasympathetic neuron
arteriole	peripheral resistance
baroreceptor	SA node
blood volume	sensory neuron
cardiac output	stroke volume
carotid artery	sympathetic neuron
contractility	vein
heart rate	venous return
medulla oblongata	ventricle

22. Compare and contrast the following sets of terms:
    (a) lymphatic capillaries and systemic capillaries
    (b) roles of the sympathetic and parasympathetic branches in blood pressure control
    (c) lymph and blood
    (d) continuous capillaries and fenestrated capillaries
    (e) hydrostatic pressure and colloid osmotic pressure in systemic capillaries

23. Calcium channel blockers prevent Ca^{2+} movement through Ca^{2+} channels. Explain two ways this action lowers blood pressure. Why are neurons and other cells unaffected by these drugs?

24. Define myogenic autoregulation. What mechanisms have been proposed to explain it?

25. Left ventricular failure may be accompanied by edema, shortness of breath, and increased venous pressure. Explain how these signs and symptoms develop.

LEVEL THREE PROBLEM SOLVING

26. Draw a reflex map that explains Anthony's vasovagal syncope at the sight of blood. Include all the steps of the reflex, and explain whether pathways are being stimulated or inhibited.

27. A physiologist placed a section of excised arteriole in a perfusion chamber containing saline. When the oxygen content of the saline perfusing (flowing through) the arteriole was reduced, the arteriole dilated. In a follow-up experiment, she used an isolated piece of arteriolar smooth muscle that had been stripped away from the other layers of the arteriole wall. When the oxygen content of the saline was reduced as in the first experiment, the isolated muscle showed no response. What do these two experiments suggest about how low oxygen exerts local control over arterioles?

28. Robert is a 52-year-old nonsmoker. He weighs 180 lbs and stands 5´9˝ tall, and his blood pressure averaged 145/95 on three successive visits to his doctor's office. His father, grandfather, and uncle all had heart attacks in their early 50s, and his mother died of a stroke at the age of 71.
    (a) Identify Robert's risk factors for coronary heart disease.
    (b) Does Robert have hypertension? Explain.
    (c) Robert's doctor prescribes a drug called a beta blocker. Explain the mechanism by which a beta-receptor-blocking drug may help lower blood pressure.

29. The following figure is a schematic representation of the systemic circulation. Use it to help answer the following questions. (CO = cardiac output, MAP = mean arterial pressure.)
    (a) If resistance in vessels 1 and 2 increases because of the presence of local paracrines but cardiac output is unchanged, what happens to MAP? What happens to flow through vessels 1 and 2? Through vessels 3 and 4?

    (b) Homeostatic compensation occurs within seconds. Draw a reflex map to explain the compensation (stimulus, receptor, and so on).
    (c) When vessel 1 constricts, what happens to the filtration pressure in the capillaries downstream from that arteriole?

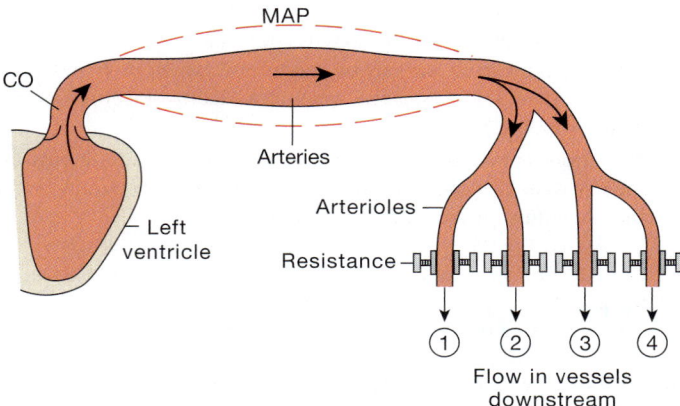

Flow in vessels downstream

30. The following graphs are recordings of contractions in an isolated frog heart. The intact frog heart is innervated by sympathetic neurons that increase heart rate, and by parasympathetic neurons that decrease heart rate. Based on these four graphs, what conclusion can you draw about the mechanism of action of atropine? (Atropine does not cross the cell membrane.)

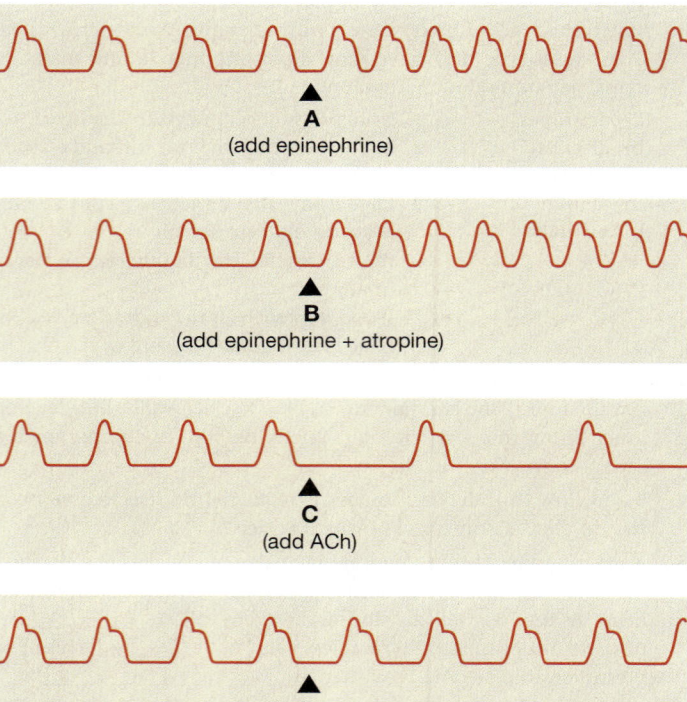

A
(add epinephrine)

B
(add epinephrine + atropine)

C
(add ACh)

D
(add ACh + atropine)

31. In advanced atherosclerosis, calcified plaques cause the normally elastic aorta and arteries to become stiff and noncompliant. (a) What effect does this change in the aorta have on afterload? (b) If cardiac output remains unchanged, what happens to peripheral resistance and mean arterial pressure?

32. During fetal development, most blood in the pulmonary artery bypasses the lungs and goes into the aorta by way of a channel called the *ductus arteriosus*. Normally this fetal bypass channel closes during the first day after birth, but each year about 4000 babies in the United States maintain a *patent* (open) ductus arteriosus and require surgery to close the channel.

    (a) Use this information to draw an anatomical diagram showing blood flow in an infant with a patent ductus arteriosus.
    (b) In the fetus, why does most blood bypass the lungs?
    (c) If the systemic side of the circulatory system is longer than the pulmonary side, which circuit has the greater resistance?
    (d) If flow is equal in the pulmonary and systemic circulations, which side of the heart must generate more pressure to overcome resistance?
    (e) Use your answer to (d) to figure out which way blood will flow through a patent ductus arteriosus.

LEVEL FOUR   QUANTITATIVE PROBLEMS

33. Using the appropriate equation, mathematically explain what happens to blood flow if the *diameter* of a blood vessel increases from 2 mm to 4 mm.

34. Duplicate the calculations that led William Harvey to believe that blood circulated in a closed loop:
    (a) Take your resting pulse.
    (b) Assume that your heart at rest pumps 70 mL/beat, and that 1 mL of blood weighs one gram. Calculate how long it would take your heart to pump your weight in blood. (2.2 pounds = 1 kilogram)

35. Calculate the mean arterial pressure (MAP) and pulse pressure for a person with a blood pressure of 115/73.

36. According to the Fick principle, the oxygen consumption rate of an organ is equal to the blood flow through that organ times the amount of oxygen extracted from the blood as it flows through the organ:

    oxygen consumption rate = blood flow
    × (arterial oxygen content − venous oxygen content)
    mL O_2 consumed/min = mL blood/min × mL O_2/mL blood

    A person has a total body oxygen consumption rate of 250 mL/min. The oxygen content of blood in his aorta is 200 mL O_2/L blood, the oxygen content of his pulmonary artery blood is 160 mL O_2/L blood. What is his cardiac output?

ANSWERS

✓ Answers to Concept Check Questions

Page 506

1. Veins from the brain do not require valves because blood flow is aided by gravity.

2. The carotid wave would arrive slightly ahead of the wrist wave because the distance from heart to carotid artery is shorter.

3. Pressure of 130/95 has the higher pulse pressure (35 mm Hg).

Page 506

4. If heart rate increases, the relative time spent in diastole decreases. In that case, the contribution of systolic pressure to mean arterial pressure increases, and MAP increases.

5. Pulse pressure is 112 − 68 = 44 mm Hg. MAP is 68 + 1/3(44) = 82.7 mm Hg.

Page 512

6. (d)

7. Extracellular K^+ dilates arterioles, which increases blood flow.

Page 513

8. Epinephrine binding to myocardial β_1-receptors increases heart rate and force of contraction. Epinephrine binding to β_2-receptors on heart arterioles causes vasodilation.

9. α_1-Receptors have lower affinity for epinephrine than β_2-receptors, so the β_2-receptors dominate and arterioles dilate.

Page 514

10. (a) The kidney has the highest blood flow per unit weight. (b) The heart has the lowest total blood flow.

Pages 519

11. Net filtration will increase as a result of the increased hydrostatic pressure.

12. Loss of plasma proteins will decrease colloid osmotic pressure. As a result, hydrostatic pressure will have a greater effect in the filtration-absorption balance, and filtration will increase.

13. Using osmotic pressure rather than osmolarity allows a direct comparison between absorption pressure and filtration pressure, both of which are expressed in mm Hg.

Page 521

14. If the left ventricle fails, blood will back up into the left atrium and pulmonary veins, and then into lung capillaries. Edema in the lungs is known as *pulmonary edema*.

15. Low-protein diets result in a low concentration of plasma proteins. Capillary absorption is reduced while filtration remains constant resulting in edema and as cites.

Page 523

16. The most likely ion is Na^+ moving into the receptor cell.

Page 525

17. Stimulus: sight, sound, and smell of the *T. rex*. Receptors: eyes, ears, and nose. Integrating center: cerebral cortex, with descending pathways through the limbic system. Divergent pathways go to the cardiovascular control center, which increases sympathetic output to heart and arterioles. A second descending spinal pathway goes to the adrenal medulla, which releases epinephrine. Epinephrine on β_2-receptors of liver, heart, and skeletal muscle arterioles causes vasodilation of those arterioles. Norepinephrine onto α-receptors in other arterioles causes vasoconstriction. Both catecholamines increase heart rate and force of contraction.

15

Q Answers to Figure and Graph Questions

Page 515

Figure 15-13: Blood flow through the lungs is 5 L/min.

Page 517

Figure 15-17: (a) Velocity of flow is inversely proportional to area: as area increases, velocity decreases. (b) Changing only cross-sectional area has no effect on flow rate because flow rate is determined by cardiac output.

Page 522

Figure 15-21: SA node: muscarinic cholinergic receptors for ACh and β_1-receptors for catecholamines. Ventricles: β_1-receptors for catecholamines. Arterioles and veins: α-receptors for norepinephrine.

Page 524

Figure 15-23: Arterioles and veins: norepinephrine, α-receptors. Ventricles: norepinephrine, β_1-receptors. SA node: norepinephrine, β_1-receptors; and ACh, muscarinic receptors.

15

16

Who would have thought the old man to have had so much blood in him?

—Macbeth, V, i, 42, *by William Shakespeare*

Bovine pulmonary artery endothelium

Blood

Plasma and the Cellular Elements of Blood

Blood Cell Production

Red Blood Cells

Platelets and Coagulation

BACKGROUND BASICS

BLOOD DOPING IN ATHLETES

Athletes spend hundreds of hours training, trying to build their endurance. For Johann Muehlegg, a cross-country skier at the 2002 Salt Lake City Winter Olympics, it appeared that his training had paid off when he captured three gold medals. On the last day of the Games, however, Olympics officials expelled Muehlegg and stripped him of his gold medal in the 50-kilometer classical race. The reason? Muehlegg had tested positive for a performance-enhancing chemical that increased the oxygen-carrying capacity of his blood. Officials claimed Muehlegg's endurance in the grueling race was the result of blood doping, not training.

536   542   547   551   554

Blood, the fluid that circulates in the cardiovascular system, has occupied a prominent place throughout history as an almost mystical fluid. Humans undoubtedly had made the association between blood and life by the time they began to fashion tools and hunt animals. A wounded animal that lost blood would weaken and die if the blood loss was severe enough. The logical conclusion was that blood was necessary for existence. This observation eventually led to the term *lifeblood*, meaning anything essential for existence.

Ancient Chinese physicians linked blood to energy flow in the body. They wrote about the circulation of blood through the heart and blood vessels long before William Harvey described it in seventeenth-century Europe. In China, changes in blood flow were used as diagnostic clues to illness. Chinese physicians were expected to recognize some 50 variations in the pulse. Because blood was considered a vital fluid to be conserved and maintained, bleeding patients to cure disease was not a standard form of treatment.

In contrast, Western civilizations came to believe that disease-causing evil spirits circulated in the blood. The way to remove these spirits was to remove the blood containing them. Because it was recognized that blood was an essential fluid, however, bloodletting had to be done judiciously. Veins were opened with knives or sharp instruments (*venesection*), or blood-sucking leeches were applied to the skin. In ancient India, people believed that leeches could distinguish between healthy and infected blood.

There is no written evidence that venesection was practiced in ancient Egypt, but the work carried out by Galen of Pergamum in the second century A.D. influenced Western medicine for nearly 2000 years. This early Greek physician advocated bleeding as treatment for many disorders. The location, timing, and frequency of the bleeding depended on the condition, and the physician was instructed to remove enough blood to bring the patient to the point of fainting. Over the years, this practice undoubtedly killed more people than it cured.

What is even more remarkable is the fact that as late as 1923, an American medical textbook advocated bleeding for treating certain infectious diseases, such as pneumonia! Now that we better understand the importance of blood in the immune response, it is doubtful that modern medicine will ever again turn to blood removal as a nonspecific means of treating disease. It is still used, however, for selected *hematological disorders* [*haima,* blood].

PLASMA AND THE CELLULAR ELEMENTS OF BLOOD

What is this remarkable fluid that flows through the circulatory system? Blood makes up one-fourth of the extracellular fluid, the internal environment that bathes cells and acts as a buffer between cells and the external environment. Blood is the circulating portion of the extracellular fluid, responsible for carrying material from one part of the body to another. Total blood volume in a 70-kg man is equal to about 7% of his total body weight, or $0.07 \times 70\,\text{kg} = 4.9\,\text{kg}$. Thus, if we assume that 1 kg of blood occupies a volume of 1 liter, a 70-kg man has about 5 liters of blood. Of this volume, about 2 liters is composed of blood cells, while the remaining 3 liters is composed of plasma, the fluid portion of the blood.

This chapter presents an overview of the components of blood. We will consider the functions of plasma, red blood cells, and platelets in this chapter but will reserve detailed discussion of hemoglobin for Chapter 18, and of white blood cells and blood types for Chapter 24.

Plasma Is Composed of Water, Ions, Organic Molecules, and Dissolved Gases

Plasma is the fluid portion of the blood, within which cellular elements are suspended (Fig. 16-1 ■). Water is the main component of plasma, accounting for about 92% of its weight. Proteins account for another 7%. The remaining 1% is dissolved organic molecules (amino acids, glucose, lipids, and nitrogenous wastes), ions (Na^+, K^+, Cl^-, H^+, Ca^{2+}, and HCO_3^-), trace elements and vitamins, and dissolved oxygen (O_2) and carbon dioxide (CO_2).

Plasma is identical in composition to interstitial fluid except for the presence of **plasma proteins**. **Albumins** are the most prevalent type of protein in the plasma, making up about 60% of the total. Albumins and nine other proteins—including *globulins,* the clotting protein *fibrinogen,* and the iron-transporting protein *transferrin*—make up more than 90% of all plasma proteins. The liver makes most plasma proteins and secretes them

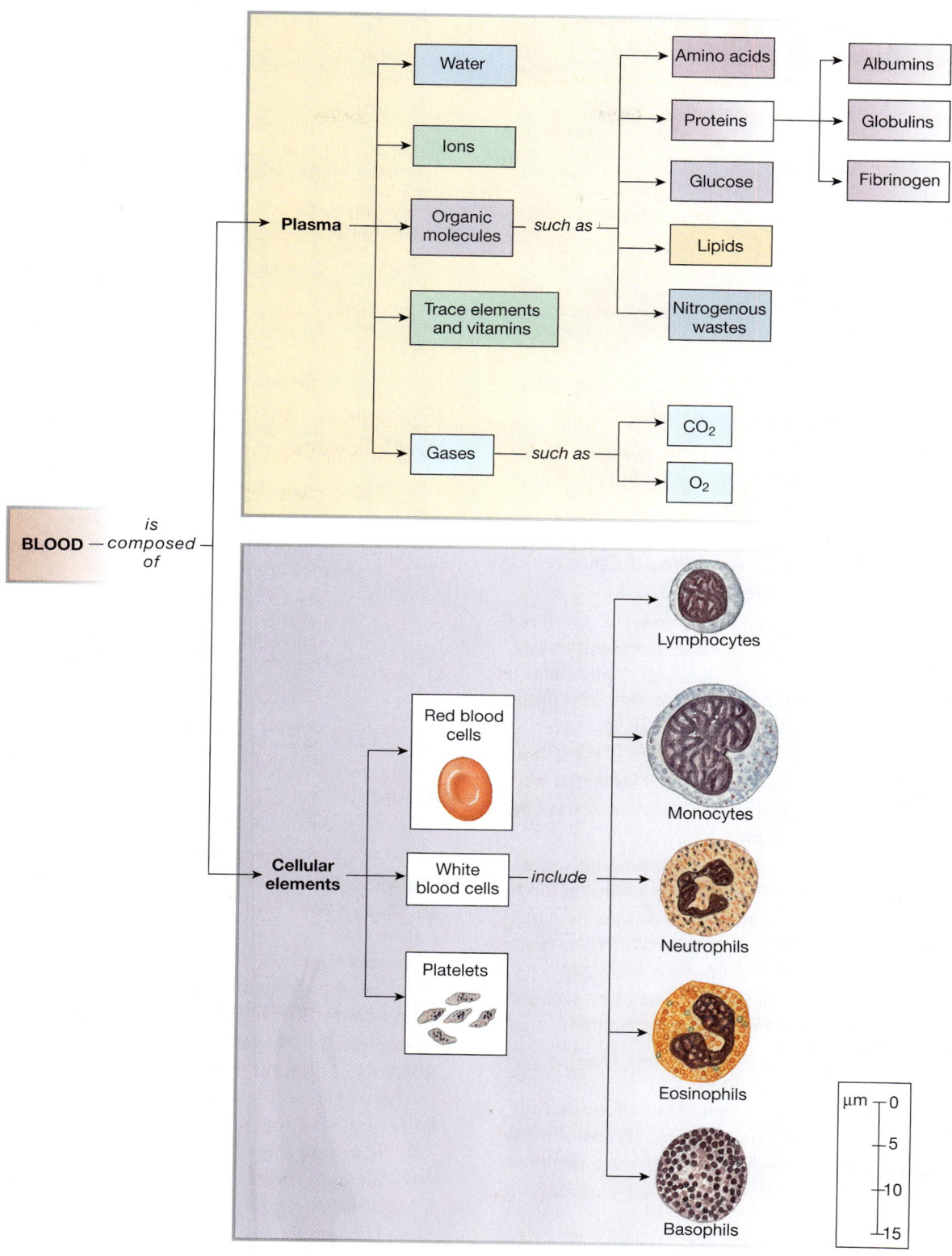

■ FIGURE 16-1   *Composition of blood*

into the blood. Some globulins, known as *immunoglobulins* or *antibodies,* are synthesized and secreted by specialized cells rather than by the liver.

The presence of proteins in the plasma makes the osmotic pressure of the blood higher than that of the interstitial fluid. This osmotic gradient tends to pull water from the interstitial fluid into the capillaries, as you learned in Chapter 15 [🔁 p. 518].

Plasma proteins participate in many functions, including blood clotting and defense against foreign invaders. In addition,

16

TABLE 16-1	Functions of Plasma Proteins	
NAME	**SOURCE**	**FUNCTION**
Albumins (multiple types)	Liver	Major contributors to colloid osmotic pressure of plasma; carriers for various substances
Globulins (multiple types)	Liver and lymphoid tissue	Clotting factors, enzymes, antibodies, carriers for various substances
Fibrinogen	Liver	Forms fibrin threads essential to blood clotting
Transferrin	Liver and other tissues	Iron transport

they act as carriers for steroid hormones, cholesterol, drugs, and certain ions such as iron (Fe^{2+}). Finally, some plasma proteins act as hormones or as extracellular enzymes. Table 16-1 ■ summarizes the functions of plasma proteins.

The Cellular Elements Include Red Blood Cells, White Blood Cells, and Platelets

Three main cellular elements are found in blood (Fig. 16-1): **red blood cells**, also called **erythrocytes** [*erythros*, red]; **white blood cells**, also called **leukocytes** [*leukos*, white]; and **platelets** or *thrombocytes* [*thrombo-*, lump, clot]. Only white blood cells are fully functional cells in the circulation. Red blood cells have lost their nuclei by the time they enter the bloodstream, and platelets, which also lack a nucleus, are cell fragments that have split off a relatively large parent cell known as a **megakaryocyte** [*mega*, extremely large + *karyon*, kernel + *-cyte*, cell].

Red blood cells play a key role in transporting oxygen from lungs to tissues, and carbon dioxide from tissues to lungs. Platelets are instrumental in *coagulation*, the process by which blood clots prevent blood loss in damaged vessels. White blood cells play a key role in the body's immune responses, defending the body against foreign invaders, such as parasites, bacteria, and viruses. Although most white blood cells circulate through the body in the blood, their work is usually carried out in the tissues rather than in the circulatory system.

Blood contains five types of mature white blood cells: (1) **lymphocytes**, (2) **monocytes**, (3) **neutrophils**, (4) **eosinophils**, and (5) **basophils**. Monocytes that leave the circulation and enter the tissues develop into **macrophages**. Tissue basophils are called **mast cells**.

The types of white blood cells may be grouped according to common morphological or functional characteristics. Neutrophils, monocytes, and macrophages are collectively known as **phagocytes** because they can engulf and ingest foreign particles such as bacteria (phagocytosis) [e p. 148]. Lymphocytes are sometimes called **immunocytes** because they are responsible for specific immune responses directed against invaders. Basophils, eosinophils, and neutrophils are called **granulocytes**

because they contain cytoplasmic inclusions that give them a granular appearance.

✔ CONCEPT CHECK

1. Name the five types of leukocytes.
2. Why do we say that erythrocytes and platelets are not fully functional cells?
3. On the basis of what you have learned about the origin and role of plasma proteins, explain why patients with advanced liver degeneration frequently suffer from edema [e p. 520].   Answers: p. 557

BLOOD CELL PRODUCTION

Where do these different blood cells come from? They are all descendants of a single precursor cell type known as the *pluripotent hematopoietic stem cell* (Fig. 16-2 ■). This cell type is found primarily in **bone marrow**, a soft tissue that fills the hollow center of bones. Pluripotent stem cells have the remarkable ability to develop into many different cell types. As they specialize, they narrow their possible fates; first they become *uncommitted stem cells*, then *progenitor cells* that are committed to developing into one or perhaps two cell types. Progenitor cells differentiate into red blood cells, lymphocytes, other white blood cells, and megakaryocytes, the parent cells of platelets. It is estimated that only about one out of every 100,000 cells in the bone marrow is an uncommitted stem cell, making the isolation and study of these cells very difficult.

In recent years scientists have been working to isolate and grow uncommitted hematopoietic stem cells to use as replacements for patients whose own stem cells have been killed by cancer chemotherapy. These stem cells were first obtained from either bone marrow or peripheral blood. Now umbilical cord blood, collected at birth, has been found to be a rich source of hematopoietic stem cells. The potential for a source of easily collected stem cells has led to the development of public and private cord blood banking programs in the United States and Europe. Some research suggests that stem cells from umbilical cords may be able to form new liver cells and even new neurons, making this a hot topic in research.

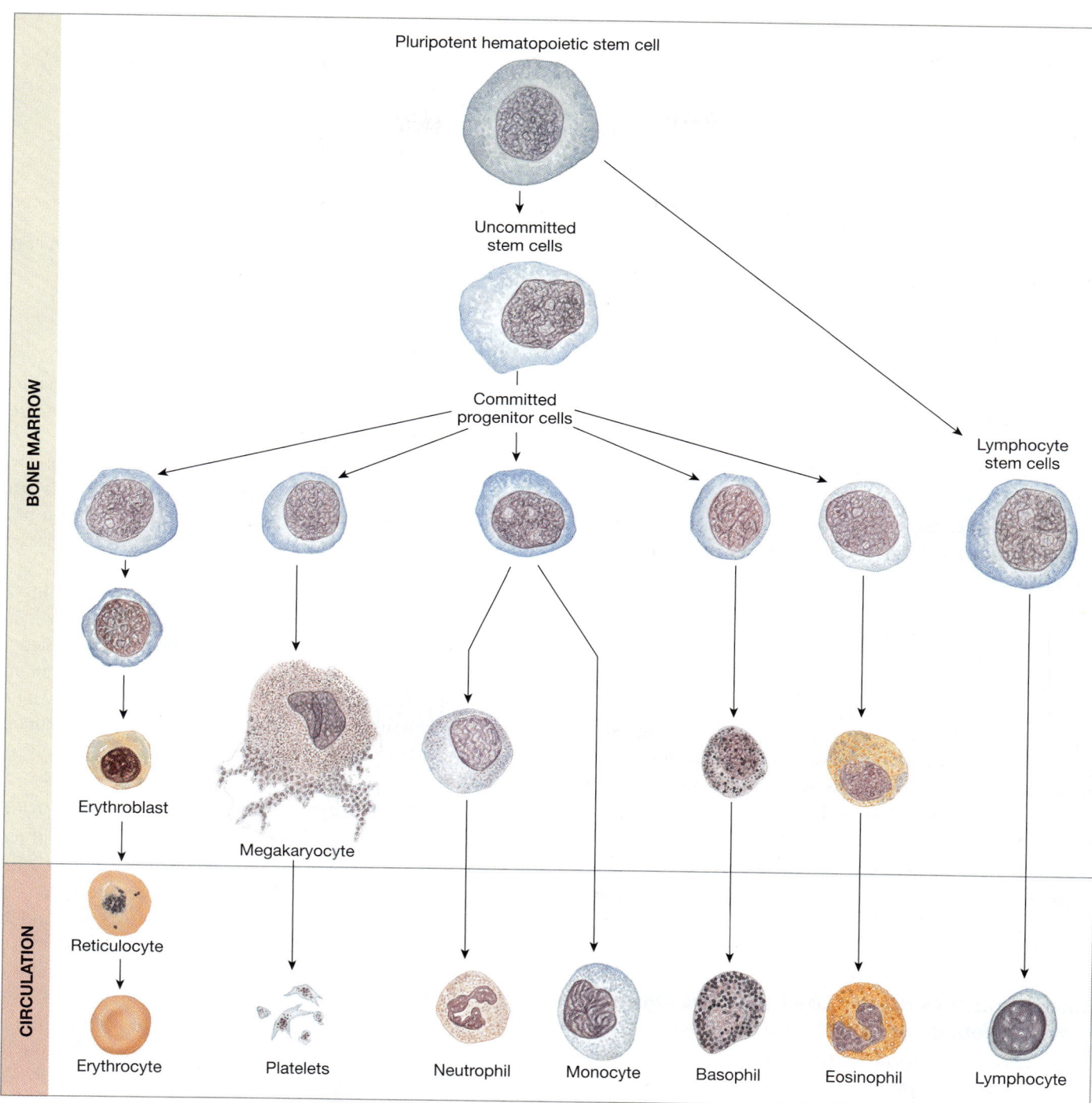

BONE MARROW

CIRCULATION

Pluripotent hematopoietic stem cell

Uncommitted
stem cells

Committed
progenitor cells

Lymphocyte
stem cells

Erythroblast

Megakaryocyte

Reticulocyte

Erythrocyte

Platelets

Neutrophil

Monocyte

Basophil

Eosinophil

Lymphocyte

■ FIGURE 16-2 *Hematopoiesis*

Cells below the horizontal line are the predominant forms found circulating in the
blood. Cells above the line are found mostly in the bone marrow.

Blood Cells Are Produced in the Bone Marrow

Hematopoiesis [*haima*, blood + *poiesis*, formation], the synthesis of blood cells, begins early in embryonic development and continues throughout a person's life. In about the third week of fetal development, specialized cells in the yolk sac of the embryo form clusters. Some of these cell clusters are destined to become the endothelial lining of blood vessels, while others become blood cells. The common embryological origin of the endothelium and blood cells perhaps explains why many cytokines that control hematopoiesis are released by the vascular endothelium.

As the embryo develops, blood cell production spreads from the yolk sac to the liver, spleen, and bone marrow. By

16

TABLE 16-2	Cytokines Involved in Hematopoiesis	
NAME	SITES OF PRODUCTION	INFLUENCES GROWTH OR DIFFERENTIATION OF
Erythropoietin (EPO)	Kidney cells	Red blood cells
Thrombopoietin (TPO)	Liver primarily	Megakaryocytes
Colony-stimulating factors, interleukins, stem cell factor	Endothelium and fibroblasts of bone marrow, leukocytes	All types of blood cells; mobilizes hematopoietic stem cells

birth, the liver and spleen no longer produce blood cells. Hematopoiesis continues in the marrow of all the bones of the skeleton until age five. As the child continues to age, the active regions of marrow decrease, so that in adults, the pelvis, spine, ribs, cranium, and proximal ends of long bones are the only areas producing blood cells.

Active bone marrow is red because of the presence of **hemoglobin**, the oxygen-binding protein of red blood cells. Inactive marrow is yellow because of an abundance of adipocytes (fat cells). (You can see the difference between red and yellow marrow the next time you look at bony cuts of meat in the grocery store.) Although blood synthesis in adults is limited, the liver, spleen, and inactive (yellow) regions of marrow can resume blood cell production in times of need.

In the regions of marrow that are actively producing blood cells, about 25% of the developing cells are red blood cells, while 75% are destined to become white blood cells. The life span of white blood cells is considerably shorter than that of red blood cells, and so the former must be replaced more frequently. For example, neutrophils have a six-hour half-life, and the body must make *100 billion* neutrophils each day in order to replace those that die. Red blood cells, on the other hand, live for nearly four months in the circulation.

Hematopoiesis Is Controlled by Colony-Stimulating Factors, Interleukins, and Other Cytokines

What controls the production and development of blood cells? The chemical factors known as cytokines are responsible. Cytokines are peptides or proteins released from one cell that affect the growth or activity of another cell [p. 176]. Chapter 7 described how newly discovered hormones are identified by the word *factor,* and the same applies to cytokines. When first discovered, they are often called *factors* and given a modifier that describes their actions: growth factor, differentiating factor, trophic factor.

Some of the best-known cytokines in hematopoiesis are the *colony-stimulating factors,* molecules made by endothelial cells and white blood cells. Others are the **interleukins** [*inter-*, between + *leuko,* white], such as IL-3. The name *interleukin* was first given to cytokines released by one white blood cell to act

on another white blood cell. Numbered interleukin names are given to cytokines once their amino acid sequences have been identified. Interleukins also play important roles in the immune system, as we will discuss in Chapter 24.

Another hematopoietic cytokine is *erythropoietin,* which controls red blood cell synthesis. Erythropoietin is usually called a hormone, but technically it fits the definition of a cytokine because it is made on demand rather than stored in vesicles.

Table 16-2 ■ lists a few of the many cytokines linked to hematopoiesis. The role cytokines play in blood cell production is so complicated that one review on this topic was titled "Regulation of hematopoiesis in a sea of chemokine family members with a plethora of redundant activities"!* Because of the complexity of the subject, we will not attempt to discuss hematopoietic cytokines in detail.

Colony-Stimulating Factors Regulate Leukopoiesis

Colony-stimulating factors (CSFs) were identified and named for their ability to stimulate the growth of leukocyte colonies in culture. These cytokines, made by endothelial cells, marrow fibroblasts, and white blood cells, regulate leukocyte production and development (**leukopoiesis**). They are required to induce both cell division (mitosis) and cell maturation in stem cells. Once a leukocyte matures, it loses its ability to undergo mitosis.

One fascinating aspect of leukopoiesis is that production of new white blood cells is regulated in part by existing white blood cells. This form of control allows leukocyte development to be very specific and tailored to the body's needs. When the body's defense system is called on to fight off foreign invaders, both the absolute number of white blood cells and the relative proportions of the different types of white blood cells in the circulation change. Clinicians often rely on a *differential white cell count* to help them arrive at a diagnosis (Fig. 16-3 ■). For example, a person with a bacterial infection will usually have a high total number of white blood cells in the blood, with an increased percentage that are neutrophils. Cytokines released by

*H. E. Broxmeyer, and C. H. Kim, *Experimental Hematology* 27(7): 1113–1123, 1999, July.

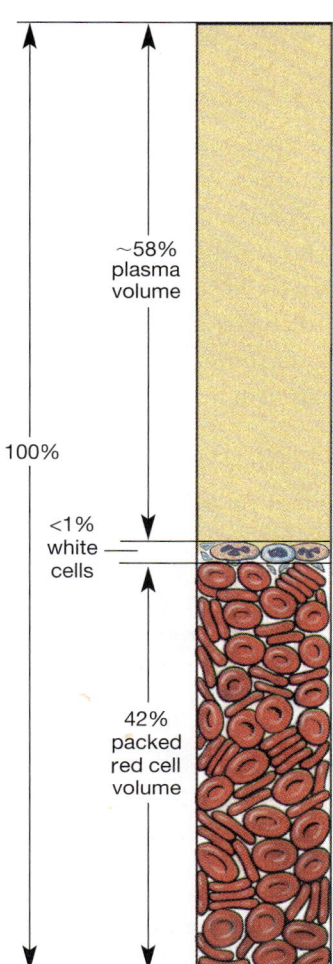

	Males	Females
Hematocrit	40–54%	37–47%
Hemoglobin (g Hb/dL* blood)	14–17	12–16
Red cell count (cells/μL)	$4.5–6.5 \times 10^6$	$3.9–5.6 \times 10^6$
Total white cell count (cells/μL)	$4–11 \times 10^3$	$4–11 \times 10^3$
Differential white cell count		
Neutrophils	50-70%	50-70%
Eosinophils	1-4%	1-4%
Basophils	<1%	<1%
Lymphocytes	20–40%	20–40%
Monocytes	2–8%	2–8%
Platelets (per μL)	$200–500 \times 10^3$	$200–500 \times 10^3$

* 1 deciliter (dL) = 100 mL

■ **FIGURE 16-3**   *The blood count*

The table lists the normal ranges of values. **Hematocrit** is the percentage of total blood volume that is occupied by packed (centrifuged) red blood cells. The **hemoglobin** value reflects the oxygen-carrying capacity of red blood cells. In a **red cell count**, a machine counts erythrocytes as they steam through a beam of light. **Total white cell count** includes all types of leukocytes but does not distinguish between them. The **differential white cell count** presents estimates of the relative proportions of the five types of leukocytes in a thin blood smear stained with biological dyes. **Platelet count** is suggestive of the blood's ability to clot.

active white blood cells fighting the bacterial infection stimulate the production of additional neutrophils and monocytes. A person with a viral infection may have a high, normal, or low total white cell count but often shows an increase in the percentage of lymphocytes. The complex process by which leukocyte production is matched to need is still not well understood and is an active area of research.

Scientists are working to create a model for the control of leukopoiesis so that they can develop effective treatments for diseases characterized by either a lack or an excess of white blood cells. The *leukemias* are a group of diseases characterized by the abnormal growth and development of white blood cells. In *neutropenias* [*penia,* poverty], patients have too few white blood cells and are unable to fight off bacterial and viral infections. Researchers hope to find better treatments for both leukemias and neutropenias by unlocking the secrets of how the body regulates cell growth and division.

Thrombopoietin Regulates Platelet Production

Thrombopoietin (**TPO**) is a glycoprotein that regulates the growth and maturation of megakaryocytes, the parent cells of platelets. (Recall that *thrombocyte* is an alternative name for

platelet.) TPO is produced primarily in the liver but is also present in the kidney. The gene for this cytokine, the activity of which was first described in 1958, was cloned in 1994. Within a year, TPO was widely available to researchers through the use of recombinant DNA techniques, and there has been an explosion of papers describing its effects on megakaryocytes and platelet production. Although we still do not understand everything about the basic biology of the process, thrombopoietin synthesis is a huge step forward in the search for answers.

Erythropoietin Regulates Red Blood Cell Production

Red blood cell production (**erythropoiesis**) is controlled by the glycoprotein **erythropoietin** (**EPO**), assisted by several cytokines. Erythropoietin is made primarily in the kidneys of adults. The stimulus for EPO synthesis and release is *hypoxia,* low oxygen levels in the tissues. Hypoxia stimulates production of a transcription factor called *hypoxia-inducible factor 1* (HIF-1), which turns on the EPO gene to increase EPO synthesis. This pathway, like other endocrine pathways, helps the body maintain homeostasis. By stimulating red blood cell synthesis, EPO puts more hemoglobin into the circulation to carry oxygen.

16

The existence of a hormone controlling red blood cell production was first suggested in the 1950s, but two decades passed before scientists succeeded in purifying the substance. One reason for the delay is that EPO is made on demand and not stored, as in an endocrine cell. It took scientists another nine years to identify the amino acid sequence of EPO and to isolate and clone the gene for it. However, an incredible leap was made after the EPO gene was isolated: only two years later, the hormone was produced by recombinant DNA technology and put into clinical use.

Today, physicians can prescribe not only genetically engineered EPO (epoetin alfa) but also several colony-stimulating factors (sargramostim and filgrastim) that stimulate white blood cell synthesis. Other hematopoietic cytokines are being tested in clinical trials. Cancer patients in whom hematopoiesis has been suppressed by chemotherapy have benefited from injections of these hematopoietic hormones.

✓ CONCEPT CHECK

4. Name the cytokine(s) that regulates growth and maturation in (a) erythrocytes, (b) leukocytes, and (c) megakaryocytes.

Answers: p. 557

RED BLOOD CELLS

Erythrocytes are the most abundant cell type in the blood. A microliter of blood contains about 5 million red blood cells, compared with only 4000–11,000 white blood cells and 200,000–500,000 platelets. The primary function of red blood cells is to facilitate oxygen transport from the lungs to cells, and carbon dioxide transport from cells to lungs.

The ratio of red blood cells to plasma is indicated clinically by the **hematocrit** and is expressed as a percentage of the total blood volume (Fig. 16-3). Hematocrit is determined by drawing a blood sample into a narrow capillary tube and spinning it in a centrifuge so that the heavier red blood cells go to the bottom of the sealed tube, leaving the thin "buffy layer" of lighter white blood cells and platelets in the middle, and plasma on top.

The column of *packed red cells* is measured, and the hematocrit value is reported as a percentage of the total sample volume. The normal range of hematocrit is 40–54% for a man and 37–47% for a woman. This test provides a rapid and inexpensive way to estimate a person's red cell count because blood for a hematocrit can be collected by simply sticking a finger.

Mature Red Blood Cells Lack a Nucleus

In the bone marrow, committed progenitor cells differentiate through several stages into large, nucleated *erythroblasts*. As erythroblasts mature, the nucleus condenses and the cell shrinks in diameter from 20 μm to about 7 μm. In the last stage before maturation, the nucleus is pinched off and phagocytosed by bone marrow macrophages. At the same time, other membra-

RUNNING PROBLEM

Blood doping to increase the oxygen-carrying capacity of the blood has been a problem in endurance sports for more than 30 years. The first sign that Muehlegg might be cheating by this method was the result of a simple blood test for hemoglobin and a hematocrit, taken several hours before his 50-kilometer race. Muehlegg's blood hemoglobin level registered above 17.5 g/dL. A repeat test was within acceptable limits, however, and Muehlegg was allowed to race.

Question 1:
   What is the normal range for Muehlegg's hemoglobin (Fig. 16-3)?

Question 2:
   Olympic officials also tested Muehlegg's hematocrit. With blood doping, would you expect a hematocrit value to be lower or higher than normal?

| 536 | **542** | 547 | 551 | 554 |

nous organelles (such as mitochondria) break down and disappear. The final immature cell form, called a *reticulocyte,* leaves the marrow and enters the circulation, where it matures into an erythrocyte in about 24 hours (Fig. 16-4 ■).

Mature mammalian red blood cells are biconcave disks, shaped much like jelly doughnuts with the filling squeezed out of the middle (Fig. 16-5a ■). They are simple membranous "bags" filled with enzymes and hemoglobin. Because red blood cells contain no mitochondria, they cannot carry out aerobic metabolism. Glycolysis is their primary source of ATP. Without a nucleus and endoplasmic reticulum to carry out protein synthesis, erythrocytes are unable to make new enzymes or to renew membrane components. This inability leads to an increasing loss of membrane flexibility, making older cells more fragile and likely to rupture.

The biconcave shape of the red blood cell is one of its most distinguishing features. The membrane is held in place by a complex cytoskeleton composed of filaments linked to transmembrane attachment proteins (Fig. 16-5c). Despite the cytoskeleton, red cells are remarkably flexible, like a partially filled water balloon that can compress into various shapes. This flexibility allows erythrocytes to change shape as they squeeze through the narrow capillaries of the circulation.

The disklike structure of red blood cells also allows them to modify their shape in response to osmotic changes in the blood. An erythrocyte placed in a slightly hypotonic medium [p. 157] will swell and form a sphere without disruption of its

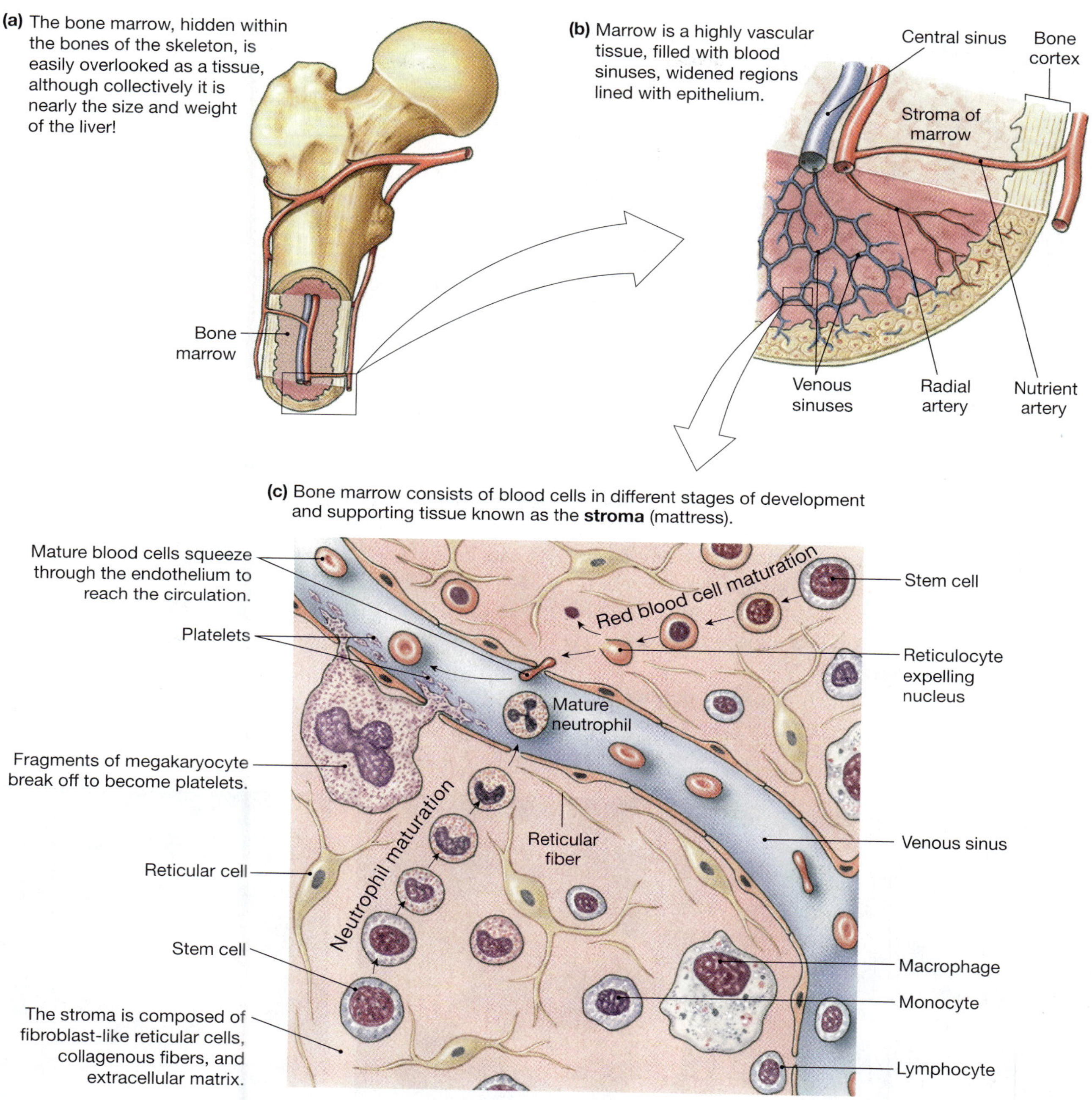

(a) The bone marrow, hidden within the bones of the skeleton, is easily overlooked as a tissue, although collectively it is nearly the size and weight of the liver!

Bone marrow

(b) Marrow is a highly vascular tissue, filled with blood sinuses, widened regions lined with epithelium.

Central sinus

Bone cortex

Stroma of marrow

Venous sinuses

Radial artery

Nutrient artery

(c) Bone marrow consists of blood cells in different stages of development and supporting tissue known as the **stroma** (mattress).

Mature blood cells squeeze through the endothelium to reach the circulation.

Platelets

Fragments of megakaryocyte break off to become platelets.

Reticular cell

Stem cell

The stroma is composed of fibroblast-like reticular cells, collagenous fibers, and extracellular matrix.

Neutrophil maturation

Reticular fiber

Mature neutrophil

Red blood cell maturation

Stem cell

Reticulocyte expelling nucleus

Venous sinus

Macrophage

Monocyte

Lymphocyte

■ **FIGURE 16-4**

16

membrane integrity (Fig. 16-6a ■). In hypertonic media, red blood cells shrink up and develop a spiky surface when the membrane pulls tight against the cytoskeleton (Fig. 16-6c).

The **morphology** [*morphe,* form] of red blood cells can provide clues to the presence of disease. Sometimes the cells lose their flattened disk shape and become spherical (*spherocytosis*). In sickle cell anemia, the cells are shaped like a sickle or crescent moon. In some disease states, the size of red blood cells—the **mean red cell volume** (MCV)—may be either abnor-

mally large or abnormally small. For example, red blood cells can be abnormally small in iron-deficiency anemia.

Hemoglobin Synthesis Requires Iron

Hemoglobin (Hb) is a tetramer comprising four globular protein chains (*globins*), each centered around an iron-containing *heme* group. The details of how hemoglobin transports oxygen will be discussed in Chapter 18, but in this section we look at the process of hemoglobin synthesis.

(a) SEM shows biconcave disk shape of RBCs.

(b) Cross-section of RBC

(c) The cytoskeleton creates the unique shape of RBCs.

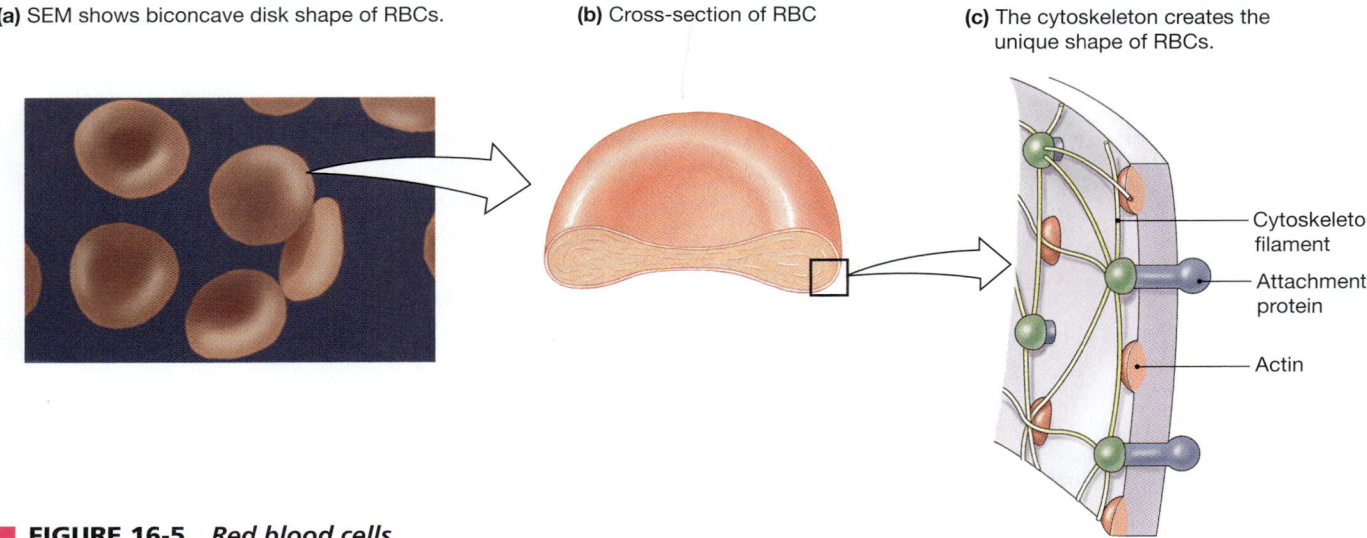

Cytoskeleton filament

Attachment protein

Actin

■ **FIGURE 16-5**   *Red blood cells*

Hemoglobin synthesis requires an adequate supply of iron in the diet (Fig. 16-7 ■), from sources such as red meat, beans, spinach, and iron-fortified bread. Iron is absorbed in the small intestine by active transport. It is transported in the blood by a carrier protein called **transferrin**. Developing red blood cells in the bone marrow use iron to make the heme group of hemoglobin. Excess iron in the body is stored, mostly in the liver, in the protein **ferritin** and its derivatives.

It is possible to ingest too much iron, and poisoning in children sometimes occurs when children ingest too many vitamin pills containing iron. Initial symptoms are gastrointestinal pain, cramping, and internal bleeding, which occurs as iron corrodes the digestive epithelium. Subsequent problems include liver failure, which can be fatal.

Red Blood Cells Live About Four Months

Red blood cells in the circulation live for about 120 days (plus or minus 20 days). Increasingly fragile older erythrocytes may rupture as they try to squeeze through narrow capillaries, or

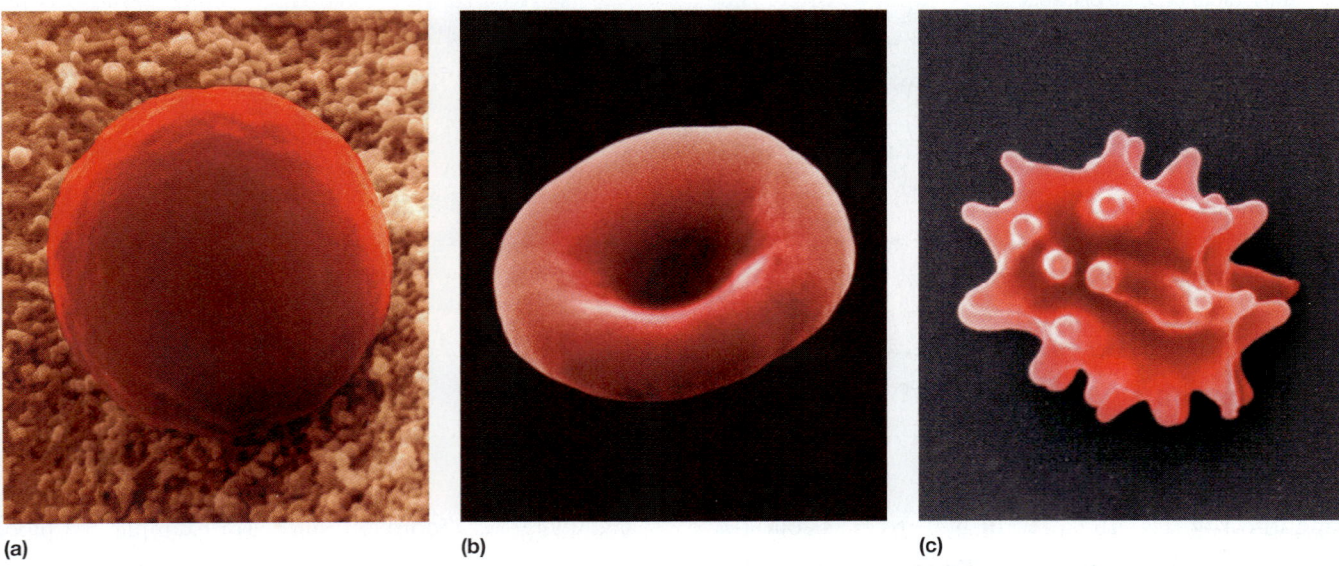

(a)

(b)

(c)

■ **FIGURE 16-6**   *Osmotic changes to red blood cell shape*

(a) Erythrocytes placed in a hypotonic medium swell and lose their characteristic biconcave disk shape, shown in **(b)**. **(c)** Erythrocytes placed in a hypertonic medium shrink, but the rigid cytoskeleton remains intact, creating a spiky surface. These cells are said to be *crenated* [*crenatus,* a notch].

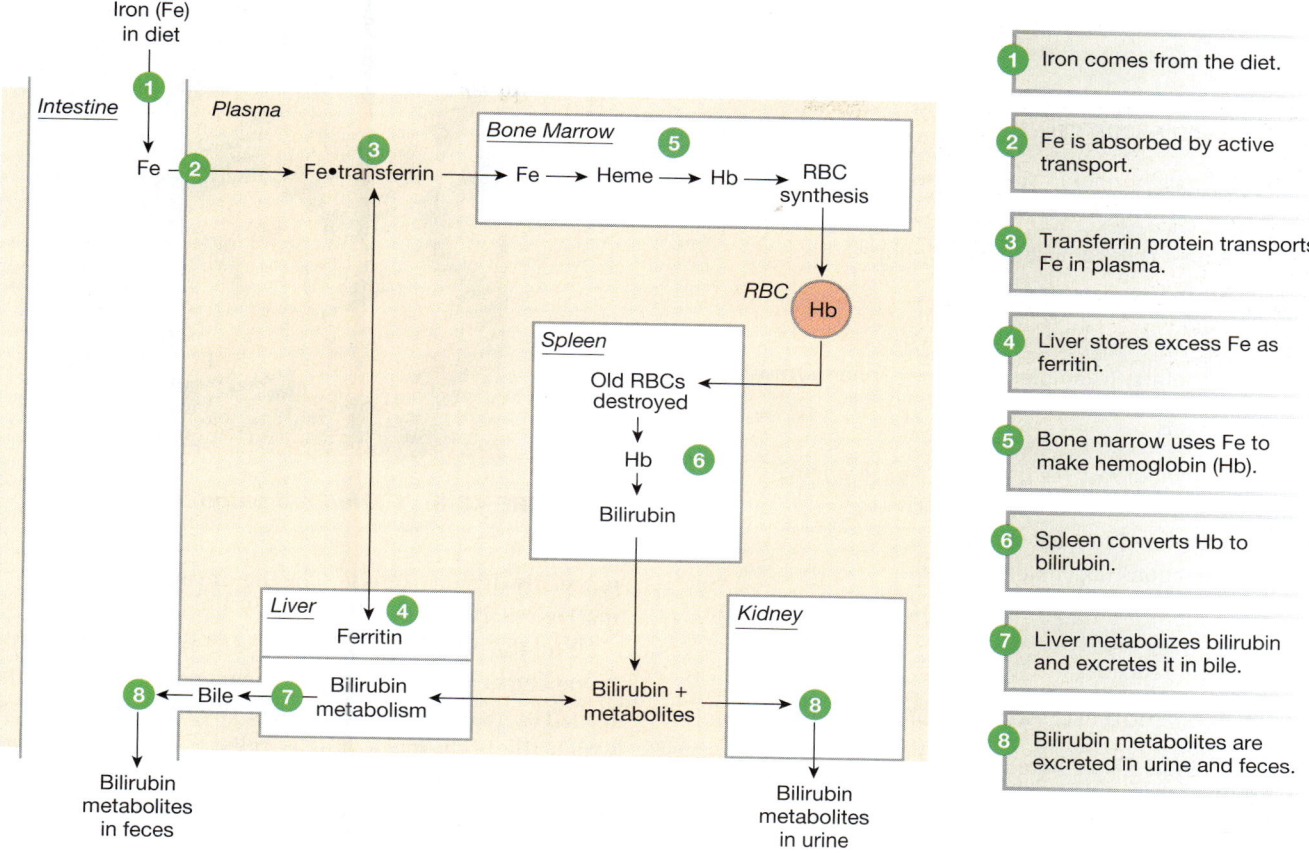

■ **FIGURE 16-7** *Iron metabolism*

1. Iron comes from the diet.
2. Fe is absorbed by active transport.
3. Transferrin protein transports Fe in plasma.
4. Liver stores excess Fe as ferritin.
5. Bone marrow uses Fe to make hemoglobin (Hb).
6. Spleen converts Hb to bilirubin.
7. Liver metabolizes bilirubin and excretes it in bile.
8. Bilirubin metabolites are excreted in urine and feces.

CLINICAL FOCUS

DIABETES

HEMOGLOBIN AND HYPERGLYCEMIA

One of the goals of diabetes treatment is to keep blood glucose concentrations as close to normal as possible, but how can a clinician tell if a patient has been doing this? One way is to measure the patient's hemoglobin. Glucose in the plasma will bind covalently to hemoglobin, producing a glycohemoglobin known as **hemoglobin A_{1c}** ("A-one-C"). The amount of hemoglobin A_{1c} in the plasma is directly related to hemoglobin's exposure to glucose over the preceding 8–12 weeks. By using this assay, a clinician can monitor long-term fluctuations in blood glucose levels and adjust a diabetic patient's therapy appropriately.

they may be engulfed by scavenging macrophages as they pass through the spleen.

Many components of hemoglobin are recycled. Amino acids from the globin chains are incorporated into new proteins, and some iron from the heme groups is reused in new heme groups. The remnants of the heme groups are converted by cells of the spleen and liver to a colored pigment called **bilirubin**. Bilirubin is carried by plasma albumin to the liver, where it is metabolized and incorporated into a secretion called **bile** (Fig. 16-7). Bile is secreted into the digestive tract, and the bilirubin metabolites leave the body in the feces. Small amounts of other bilirubin metabolites are filtered from the blood in the kidneys, where they contribute to the yellow color of urine.

In some circumstances, bilirubin levels in the blood become elevated (*hyperbilirubinemia*). This condition, known as **jaundice**, causes the skin and whites of the eyes to take on a yellow cast. The accumulation of bilirubin can occur from multiple causes. Newborns whose fetal hemoglobin is being broken down and replaced with adult hemoglobin are particularly susceptible to bilirubin toxicity, so doctors monitor babies for jaundice in the first weeks of life. Another common cause of jaundice is liver disease, in which the liver is unable to process or excrete bilirubin.

16

TABLE 16-3	Causes of Anemia
ACCELERATED RED BLOOD CELL LOSS	
Blood loss: cells are normal in size and hemoglobin content but low in number	
Hemolytic anemias: cells rupture at an abnormally high rate	
Hereditary	
Membrane defects (example: hereditary spherocytosis)	
Enzyme defects	
Abnormal hemoglobin (example: sickle cell anemia)	
Acquired	
Parasitic infections (example: malaria)	
Drugs	
Autoimmune reactions	
DECREASED RED BLOOD CELL PRODUCTION	
Defective red blood cell or hemoglobin synthesis in the bone marrow	
Aplastic anemia: can be caused by certain drugs or radiation	
Inadequate dietary intake of essential nutrients	
Iron deficiency (iron is required for heme production)	
Folic acid deficiency (folic acid is required for DNA synthesis)	
Vitamin B_{12} deficiency (B_{12} is required for DNA synthesis): may be due to lack of intrinsic factor for B_{12} absorption.	
Inadequate production of erythropoietin	

16

CONCEPT CHECK

5. Distinguish between (a) heme and hemoglobin, and (b) ferritin and transferrin.

6. Is bile an endocrine secretion or an exocrine secretion?

Answers: p. 557

Red Blood Cell Disorders Decrease Oxygen Transport

Because hemoglobin plays a critical role in oxygen transport, the red blood cell count and hemoglobin content of the body are important. If hemoglobin content is too low—a condition known as **anemia**—the blood cannot transport enough oxygen to the tissues. People with anemia are usually tired and weak,

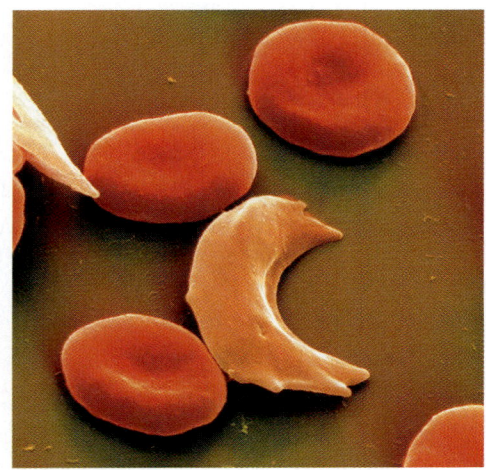

■ **FIGURE 16-8**   *Sickled red blood cells*

especially during exercise. The major causes of anemia are summarized in Table 16-3 ■.

In the *hemolytic anemias* [*lysis,* rupture], the rate of red blood cell destruction exceeds the rate of red blood cell production. The hemolytic anemias are usually hereditary defects in which the body makes fragile cells. For example, in *hereditary spherocytosis,* the erythrocyte cytoskeleton does not link properly because of defective or deficient cytoskeletal proteins. Consequently, the cells are shaped more like spheres than like biconcave disks. This disruption in the cytoskeleton results in red blood cells that rupture easily and are unable to withstand osmotic changes as well as normal cells. Several of the hemolytic anemias are acquired diseases, as indicated in Table 16-3.

Some anemias are the result of abnormal hemoglobin molecules. *Sickle cell disease* is a genetic defect in which glutamate, the sixth amino acid in the 146-amino acid beta chain of hemoglobin, is replaced by valine. The result is abnormal hemoglobin (a form referred to as *HbS*) that crystallizes when it gives up its oxygen. This crystallization pulls the red blood cells into a sickle shape, like a crescent moon (Fig. 16-8 ■). The sickled cells become tangled with other sickled cells as they pass through the smallest blood vessels, causing the cells to jam up and block blood flow to the tissues. This blockage creates tissue damage and pain from hypoxia. One treatment for sickle cell disease is the administration of *hydroxyurea,* a compound that inhibits DNA synthesis. Hydroxyurea alters bone marrow function so that immature red blood cells produce the fetal form of hemoglobin (*HbF*) instead of adult hemoglobin. HbF interferes with the crystallization of hemoglobin, so HbS no longer forms and the red blood cells no longer sickle. In addition, some studies show improvements in sickle cell symptoms before HbF levels increase. One theory of why this happens is based on the finding that hydroxyurea is metabolized to nitric oxide (NO), which causes vasodilation. Inhaled nitric oxide is now being tested as a treatment for sickle cell disease symptoms.

Other anemias result from the failure of the bone marrow to make adequate amounts of hemoglobin. One of the most common examples of an anemia that results from insufficient hemoglobin synthesis is *iron-deficiency anemia*. If iron loss by the body exceeds iron intake, the marrow does not have adequate iron to make heme groups, and hemoglobin synthesis slows. People with iron-deficiency anemia have either a low red blood cell count (reflected in a low hematocrit) or a low hemoglobin content in their blood. Their red blood cells are often smaller than usual (*microcytic* red blood cells), and the lower hemoglobin content may cause the cells to be paler than normal, in which case they are described as being *hypochromic* [*hypo-*, below normal; *chrom-*, color]. Women who menstruate are likely to suffer from iron-deficiency anemia because of iron loss in menstrual blood.

Although the anemias are common, it is also possible to have too many red blood cells. *Polycythemia vera* [*vera*, true] is a stem cell dysfunction that produces too many blood cells, white as well as red. These patients may have hematocrits as high as 60–70% (normal is 37–54%). The increased number of cells causes the blood to become more viscous and thus more resistant to flow through the circulatory system [🔁 p. 463].

In *relative polycythemia*, the person's red blood cell number is normal but the hematocrit is elevated due to loss of plasma volume. This might occur with dehydration, for example. The opposite problem can also occur. If an athlete overhydrates, the hematocrit may decrease temporarily because of increased plasma volume. In both of these situations, there is no actual pathology involving the red blood cells.

✔ CONCEPT CHECK

7. A person who goes from sea level to a city that is 5000 feet above sea level will show an increased hematocrit within two or three days. Draw the reflex pathway that links the hypoxia of high altitude to increased red blood cell production.   Answers: p. 557

PLATELETS AND COAGULATION

Because of its fluid nature, blood flows freely throughout the circulatory system. However, if there is a break in the "piping" of the system, blood will be lost unless steps are taken. One of the challenges for the body is to plug holes in damaged blood vessels while still maintaining blood flow through the vessel.

It would be simple to block off a damaged blood vessel completely, like putting a barricade across a street full of potholes. However, just as shopkeepers on that street lose business if traffic is blocked, cells downstream from the point of injury die from lack of oxygen and nutrients if the vessel is completely blocked. The body's task is to allow blood flow through the vessel while simultaneously repairing the damaged wall.

RUNNING PROBLEM

In its earliest form, blood doping was accomplished by blood transfusions, which increased the athlete's oxygen-carrying capacity. One hallmark of a recent blood transfusion is elevated hemoglobin and hematocrit levels. Muehlegg claimed that his elevated hemoglobin was a result of his special diet and of dehydration from diarrhea he had suffered the night before.

Question 3:
   Explain how diarrhea could cause a temporarily elevated hematocrit.

Question 4:
   How might Muehlegg quickly reduce his hematocrit without removing red blood cells?

| 536 | 542 | 547 | 551 | 554 |

This challenge is complicated by the fact that blood in the system is under pressure. If the repair "patch" is too weak, it will simply be blown out by the blood pressure. Thus, stopping blood loss involves several steps. First, the pressure in the vessel must be decreased long enough to create a secure mechanical seal in the form of a blood clot. Once the clot is in place and blood loss has been stopped, the body's repair mechanisms can take over. Then, as the wound heals, enzymes gradually dissolve the clot while scavenger white blood cells ingest and destroy the debris.

Platelets Are Small Fragments of Cells

As noted earlier, platelets are cell fragments produced in the bone marrow from huge cells called megakaryocytes. Megakaryocytes develop their formidable size by undergoing mitosis up to seven times without undergoing nuclear or cytoplasmic division. The result is a *polyploid* cell with a lobed nucleus (Fig. 16-9a ■).

The outer edges of marrow megakaryocytes extend through the endothelium into the lumen of marrow blood sinuses, where the cytoplasmic extensions fragment into disklike platelets (Fig. 16-4). Platelets are smaller than red blood cells, are colorless, and have no nucleus. Their cytoplasm contains mitochondria, smooth endoplasmic reticulum, and many granules filled with clotting proteins and cytokines.

The typical life span of a platelet is about 10 days. Platelets are always present in the blood, but they are not active unless damage has occurred to the walls of the circulatory system.

16

(a)

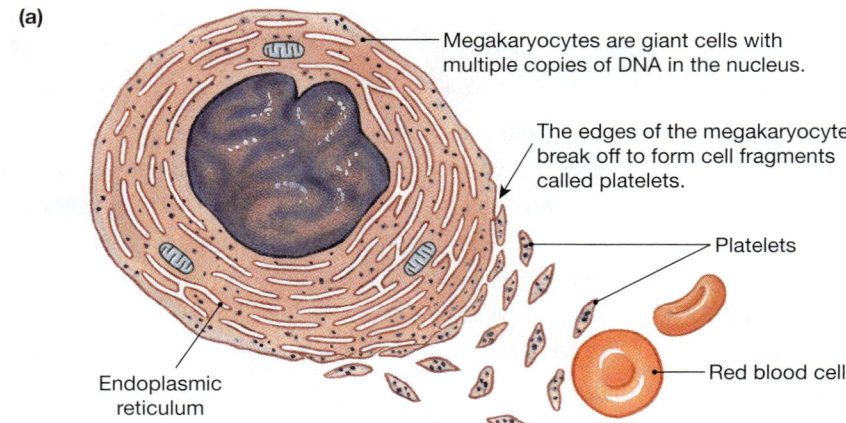

Megakaryocytes are giant cells with multiple copies of DNA in the nucleus.

The edges of the megakaryocyte break off to form cell fragments called platelets.

Platelets

Red blood cell

Endoplasmic reticulum

(b)

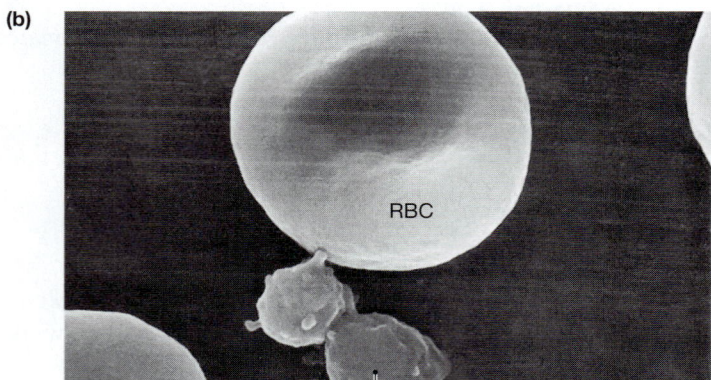

RBC

Inactive platelets are small disklike cell fragments.

(c)

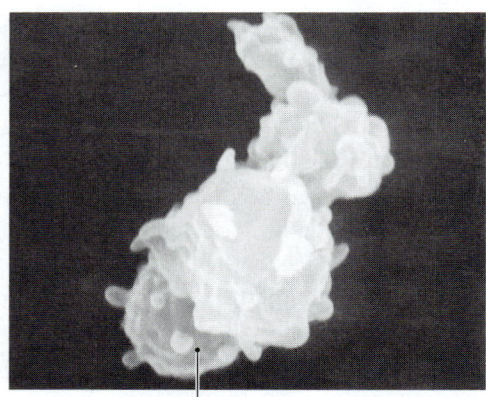

Activated platelets (shown enlarged) develop a spiky outer surface and adhere to each other.

■ **FIGURE 16-9**   *Platelets and megakaryocytes*

Hemostasis Prevents Blood Loss from Damaged Vessels

Hemostasis [*haima,* blood + *stasis,* stoppage] is the process of keeping blood within a damaged blood vessel. (The opposite of hemostasis is *hemorrhage* [*-rrhagia,* abnormal flow].) Hemostasis has three major steps: (1) vasoconstriction, (2) temporary blockage of a break by a platelet plug, and (3) blood coagulation, or formation of a clot that seals the hole until tissues are repaired.

Figure 16-10 ■ summarizes this process. The first step in hemostasis is immediate constriction of damaged vessels caused by vasoconstrictive paracrines released by the endothelium. Vasoconstriction temporarily decreases blood flow and pressure within the vessel. When you put pressure on a bleeding wound, you also decrease flow within the damaged vessel.

Vasoconstriction is rapidly followed by mechanical blockage of the hole by a **platelet plug**. The plug forms as platelets stick to the exposed collagen (**platelet adhesion**) and become activated, releasing cytokines into the area around the injury. Platelet factors reinforce local vasoconstriction and activate more platelets, which stick to one another (**platelet aggregation**) to form a loose platelet plug.

Simultaneously, exposed collagen and **tissue factor** (a protein-phospholipid mixture) initiate a series of reactions known as the **coagulation cascade**. The cascade is a series of enzymatic reactions that ends in the formation of a *fibrin* protein fiber mesh that stabilizes the platelet plug. The reinforced platelet plug is called a **clot**. Some chemical factors involved in the coagulation cascade also promote platelet adhesion and aggregation in the damaged region. Eventually, as cell growth and division repair the damaged vessel, the clot retracts and is slowly dissolved by the enzyme **plasmin**.

The body must maintain the proper balance during hemostasis. Too little hemostasis allows excessive bleeding; too much creates a **thrombus**, a blood clot that adheres to the undamaged wall of a blood vessel [*thrombos,* a clot or lump]. A large thrombus can block the lumen of the vessel and stop blood flow.

Although hemostasis seems straightforward, unanswered questions remain at the cellular and molecular levels. Because inappropriate blood clotting plays an important role in strokes and heart attacks, this area of research is very active. Research has led to the development and use of "clot busters," enzymes that can dissolve clots in arteries after heart attacks and strokes.

16

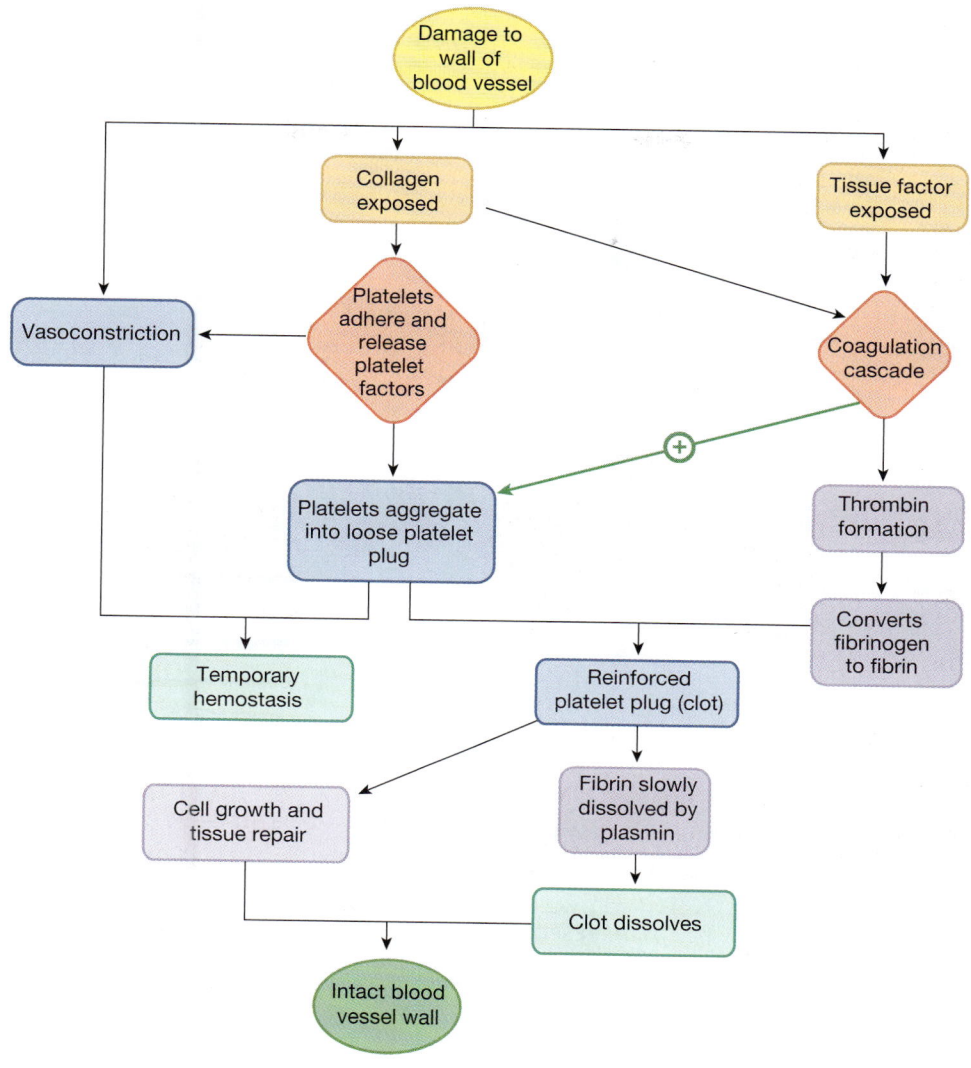

■ **FIGURE 16-10** *Overview of hemostasis and tissue repair*

A detailed study of hemostasis involves many chemical factors, some of which play multiple roles and have multiple names. Thus learning about hemostasis can be especially challenging. For example, some factors participate in both platelet plug formation and coagulation, and one factor in the cascade activates enzymes for both clot formation and clot dissolution. Because of the complexity of the coagulation cascade, we will discuss only a few aspects of hemostasis in additional detail.

Platelet Activation Begins the Clotting Process

When a blood vessel wall is first damaged, exposed collagen and chemicals from endothelial cells activate platelets (Fig. 16-11 ■; Table 16-4 ■). Normally, the blood vessel's endothelium separates the collagenous matrix fibers from the circulating blood. But when the vessel is damaged, collagen is exposed, and platelets rapidly begin to adhere to it.

Platelets adhere to collagen with the help of *integrins*, membrane receptor proteins that are linked to the cytoskeleton

[🔁 p. 70]. Binding activates platelets so that they release the contents of their intracellular granules, including **serotonin** (5-hydroxytryptamine), ADP, and **platelet-activating factor (PAF)**. PAF sets up a positive feedback loop by activating more platelets. It also initiates pathways that convert platelet membrane phospholipids into **thromboxane** A_2 [🔁 p. 189]. Serotonin and thromboxane A_2 are vasoconstrictors. They also contribute to platelet aggregation, along with ADP and PAF. The net result is a growing platelet plug that seals the damaged vessel wall.

If platelet aggregation is a positive feedback event, what prevents the platelet plug from continuing to form and thus spreading beyond the site of injury to other areas of the vessel wall? The answer lies in the fact that platelets will not adhere to normal endothelium. Intact vascular endothelial cells convert their membrane lipids into **prostacyclin**, an eicosanoid [🔁 p. 29] that blocks platelet adhesion and aggregation. Nitric oxide released by normal, intact endothelium also inhibits platelets

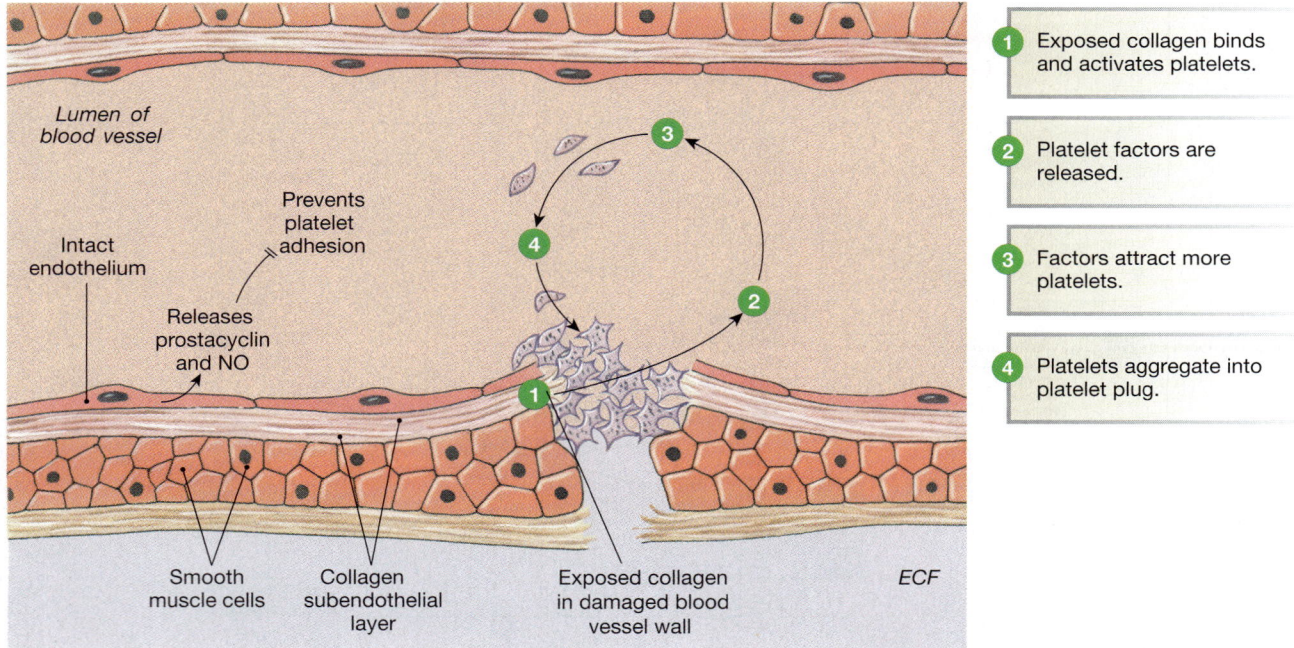

Lumen of blood vessel

Intact endothelium

Prevents platelet adhesion

Releases prostacyclin and NO

Smooth muscle cells

Collagen subendothelial layer

Exposed collagen in damaged blood vessel wall

ECF

1 Exposed collagen binds and activates platelets.

2 Platelet factors are released.

3 Factors attract more platelets.

4 Platelets aggregate into platelet plug.

■ **FIGURE 16-11** *Platelet plug formation*

TABLE 16-4   Factors Involved in Platelet Function

CHEMICAL FACTOR	SOURCE	ACTIVATED BY OR RELEASED IN RESPONSE TO	ROLE IN PLATELET PLUG FORMATION	OTHER ROLES AND COMMENTS
Collagen	Subendothelial extracellular matrix	Injury exposes platelets to collagen	Binds platelets to begin platelet plug	——
von Willebrand factor (vWF)	Endothelium, megakaryocytes	Exposure to collagen	Links platelets to collagen	Deficiency or defect causes prolonged bleeding
Serotonin	Secretory vesicles of platelets	Platelet activation	Platelet aggregation	Vasoconstrictor
Adenosine diphosphate (ADP)	Platelet mitochondria	Platelet activation, thrombin	Platelet aggregation	——
Platelet-activating factor (PAF)	Platelets, neutrophils, monocytes	Platelet activation	Platelet aggregation	Plays role in inflammation; increases capillary permeability
Thromboxane A_2	Phospholipids in platelet membranes	Platelet activating factor	Platelet aggregation	Vasoconstrictor; eicosanoid
Platelet-derived growth factor (PDGF)	Platelets	Platelet activation	——	Promotes wound healing by attracting fibroblasts and smooth muscle cells

16

RUNNING PROBLEM

After the 1984 Olympics Games, where U.S. cyclists reportedly suffered bad side effects following blood transfusions, the International Olympic Committee and other organizations banned blood doping. Then recombinant human EPO (rhEPO) became available in the late 1980s, and athletes started injecting the drug to increase their body's red blood cell production. Subsequently the biotechnology firm Amgen created a longer-acting derivative of EPO named darbepoietin. Athletes using rhEPO and darbepoietin hoped to escape detection by using these natural hormones, but sports organizations worked with scientists to develop methods for detection.

Question 5:
Endogenous EPO, rhEPO, and darbepoietin all induce red blood cell synthesis, but they can be distinguished from one another when a urine sample is tested by electrophoresis [⟲ p. 97]. Explain how three hormones made from the same gene can all be active yet different enough from one another to be detectable in the laboratory.

Question 6:
One hallmark of illegal EPO use is elevated reticulocytes in the blood. Why would this suggest greater-than-normal EPO activity?

| 536 | 542 | 547 | **551** | 554 |

---

from adhering. The combination of platelet attraction to the injury site and repulsion from the normal vessel wall creates a localized response that limits the platelet plug to the area of damage.

Coagulation Converts the Platelet Plug into a More Stable Clot

The third major step in hemostasis, coagulation, is a complex process in which fluid blood forms a gelatinous clot. Coagulation is divided into two pathways (Fig. 16-12 ■). An **intrinsic pathway** (yellow) begins with collagen exposure and involves proteins already present in the plasma. Collagen activates the first enzyme, factor XII, to begin the cascade. An **extrinsic pathway** (blue) starts when damaged tissues expose tissue factor, also called *tissue thromboplastin*. Tissue factor activates factor VII to begin the extrinsic pathway. The two pathways unite at the **common pathway** (green) to create **thrombin**, the enzyme that converts **fibrinogen** into insoluble **fibrin** polymers. These fibrin fibers become part of the clot.

Coagulation was initially regarded as a cascade similar to the second messenger cascades diagrammed in Chapter 6 [⟲ p. 182]. At each step an enzyme converts an inactive precursor into an active enzyme, often with the help of Ca^{2+}, membrane phospholipids, or additional factors. We now know, however, that the process is more than a simple cascade. Factors in the intrinsic and extrinsic pathways interact with each other, making coagulation a network rather than a simple cascade. In addition, several positive feedback loops sustain the cascade until one or more of the participating plasma proteins is completely consumed.

The end result of coagulation is the conversion of fibrinogen into fibrin, a reaction catalyzed by the enzyme thrombin (Fig. 16-13 ■). The fibrin fibers weave through the platelet plug and trap red blood cells within their mesh (see the chapter opening photo on p. 535). Active factor XIII converts fibrin into a cross-linked polymer that stabilizes the clot.

As the clot forms, it incorporates plasmin molecules, the seeds of its own destruction. **Plasmin** is an enzyme created from plasminogen by thrombin and **tissue plasminogen activator (t-PA)**. Plasmin breaks down fibrin polymers into fibrin fragments (Fig. 16-13). The dissolution of fibrin by plasmin is known as **fibrinolysis**. As the damaged vessel wall slowly repairs itself, fibrinolysis helps remove the clot.

The large number of factors involved in coagulation and the fact that a single factor may have many different names can be confusing (Table 16-5 ■). Scientists assigned numbers to the coagulation factors, but the factors are not numbered in the order in which they participate in the coagulation cascade. Instead, they are numbered according to the order in which they were discovered.

CONCEPT CHECK

8. In Figure 16-10, which box corresponds to the beginning of the intrinsic pathway of coagulation? Which corresponds to the beginning of the extrinsic pathway? To the beginning of the common pathway?

Answers: p. 557

Anticoagulants Prevent Coagulation

Once coagulation begins, what keeps it from continuing until the entire circulation has clotted? Two mechanisms limit the extent of blood clotting within a vessel: (1) inhibition of platelet adhesion and (2) inhibition of the coagulation cascade and fibrin production (Table 16-6 ■). As mentioned earlier, factors such as prostacyclin in the blood vessel endothelium and plasma ensure that the platelet plug is restricted to the area of damage.

In addition, endothelial cells release chemicals known as **anticoagulants**, which prevent coagulation from taking place. Most act by blocking one or more of the reactions in the coagulation cascade. The body produces two anticoagulants, **heparin** and **antithrombin III**, which work together to block

16

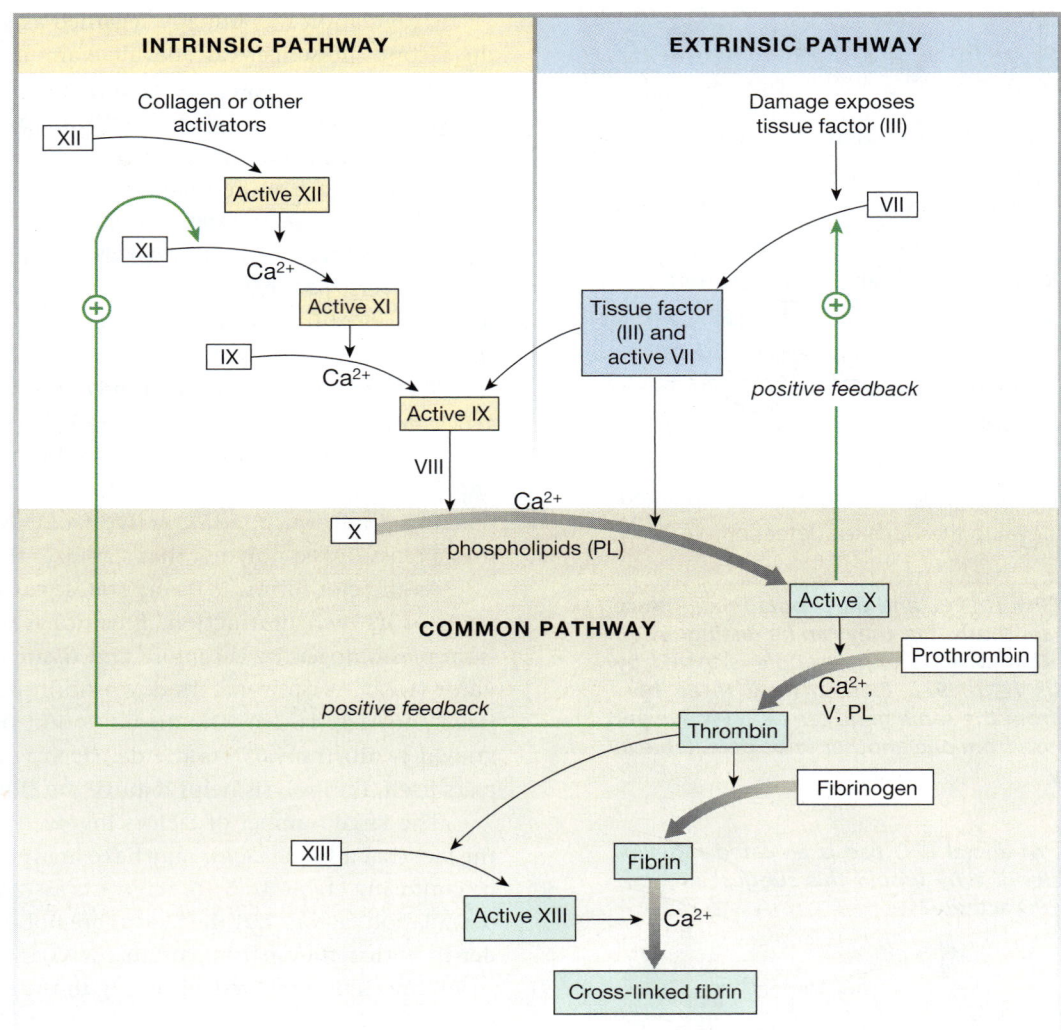

■ FIGURE 16-12   *The coagulation cascade*

Inactive plasma proteins are converted into active enzymes in each step of the pathway.

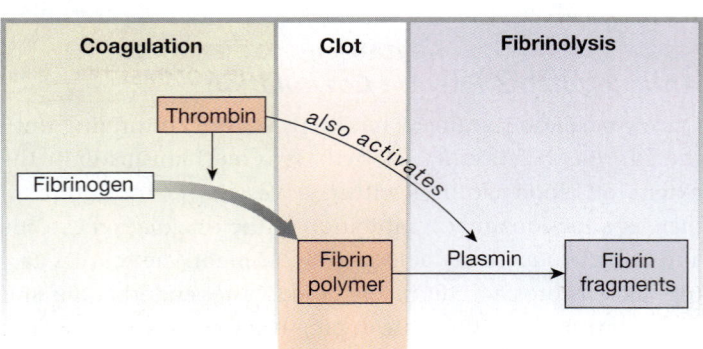

■ FIGURE 16-13   *Coagulation and fibrinolysis*

active factors IX, X, XI, and XII. **Protein C,** another anticoagulant in the body, inhibits clotting factors V and VIII.

Anticoagulant drugs may be prescribed for people who are in danger of forming small blood clots that could block off critical vessels in the brain, heart, or lungs. The family of *coumarin anticoagulants,* such as *warfarin* (Coumadin®), block the action of vitamin K, a cofactor [🔁 p. 40] in the synthesis of clotting factors II, VII, IX, and X. These anticoagulants were discovered when cattle that developed severe bleeding problems were found to have been eating spoiled sweet clover.

When blood samples are drawn into glass tubes, clotting takes place very rapidly unless the tube contains an anticoagulant. Several of the anticoagulants used for this purpose remove free Ca^{2+} from the plasma. Calcium is an essential clotting factor, so with no Ca^{2+}, no coagulation can occur. In the living body, however, plasma Ca^{2+} levels never decrease to levels that interfere with coagulation.

TABLE 16-5   **Factors Involved in Coagulation**

CHEMICAL FACTOR	SOURCE	ACTIVATED BY OR RELEASED IN RESPONSE TO	ROLE IN COAGULATION	OTHER ROLES AND COMMENTS
Collagen	Subendothelial extracellular matrix	Injury that exposes collagen to plasma clotting factors	Starts intrinsic pathway	NA
von Willebrand factor (vWF)	Endothelium, megakaryocytes	Exposure to collagen	Regulates level of factor VIII	Deficiency or defect causes prolonged bleeding
Kininogen and kallikrein	Liver and plasma	Cofactors normally present in plasma pathway	Cofactors for contact activation of intrinsic pathway	Mediate inflammatory response; enhance fibrinolysis
Tissue factor (tissue thromboplastin or factor III)	Most cells except platelets	Damage to tissue	Starts extrinsic pathway	NA
Prothrombin and thrombin (factor II)	Liver and plasma	Platelet lipids, Ca^{2+}, and factor V	Fibrin production	NA
Fibrinogen and fiblin (factor I)	Liver and plasma	Thrombin	Form insoluble fibers that stabilize platelet plug	NA
Fibrin-stabilizing factor (XIII)	Liver, megakaryocytes	Platelets	Cross-links fibrin polymers to make stable mesh	NA
Ca^{2+} (factor IV)	Plasma ions	NA	Required for several steps of coagulation cascade	Never a limiting factor
Vitamin K	Diet	NA	Needed for synthesis of factors II, VII, IX, X	NA

16

TABLE 16-6   **Endogenous Factors Involved in Fibrinolysis and Anticoagulation**

CHEMICAL FACTOR	SOURCE	ACTIVATED BY OR RELEASED IN RESPONSE TO	ROLE IN ANTICOAGULATION OR FIBRINOLYSIS	OTHER ROLES AND COMMENTS
Plasminogen and plasmin	Liver and plasma	t-PA and thrombin	Dissolves fibrin and fibrinogen	NA
Tissue plasminogen activator (t-PA)	Many tissues	Normally present; levels increase with stress, protein C	Activates plasminogen	Recombinant t-PA used clinically to dissolve clots
Antithrombin III	Liver and plasma	NA	Anticoagulant; blocks factors IX, X, XI, XII, thrombin, kallikrein	Facilitated by heparin; no effect on thrombin despite name
Prostacyclin (prostaglandin I, or PGI_2)	Endothelial cells	NA	Blocks platelet aggregation	Vasodilator

EMERGING CONCEPTS

CLOT BUSTERS AND ANTIPLATELET AGENTS

The discovery of the factors controlling coagulation and fibrinolysis was an important step in developing treatments for heart attacks, more properly called *myocardial infarctions* (MIs). An **infarct** is an area of tissue deprived of its blood supply (and consequently its oxygen) because the vessel is blocked either by a clot or by an atherosclerotic plaque. Unless the blockage is removed promptly, the tissue will die or be severely damaged. If the blockage occurs in a coronary artery, the heart may sustain so much damage that it no longer functions properly. Many clinical trials around the world are studying the best way to treat myocardial infarctions. One option is to use fibrinolytic drugs—such as streptokinase (from bacteria) and tissue plasminogen activator—to dissolve the clots. These drugs are now being combined with other agents that prevent further platelet plug and clot formation. Radio advertisements for aspirin remind consumers that this old standby can help prevent heart attack damage. The newest drugs for treating heart attacks are antiplatelet agents, including some that are antagonists to platelet integrin receptors.

Acetylsalicylic acid (aspirin) is an agent that prevents platelet plug formation. It acts by inhibiting the COX enzymes [p. 00] that promote synthesis of the platelet activator thromboxane A_2. People who are at risk of developing small blood clots are sometimes told to take one aspirin every other day "to thin the blood." The aspirin does not actually dilute the blood, but it does prevent clots from forming by blocking platelet aggregation. Aspirin is now given routinely as emergency treatment for a suspected heart attack (see the box on Clot Busters).

Several inherited diseases affect the coagulation process. Patients with coagulation disorders bruise easily. In severe forms, spontaneous bleeding may occur throughout the body. Bleeding into the joints and muscles can be painful and disabling. If bleeding occurs in the brain, it can be fatal.

The best-known coagulation disorder is **hemophilia**, a name given to several diseases in which one of the factors in the coagulation cascade is either defective or lacking. Hemophilia A, a factor VIII deficiency, is the most common form, occurring in about 80% of all cases. This disease is a recessive sex-linked trait that usually affects only males.

One exciting development in the treatment of hemophilia was a report on the first patients to be given gene therapy for hemophilia B, a deficiency in clotting factor IX. Patients injected with a virus engineered to carry the gene for factor IX started to produce some of the factor on their own, reducing their need for expensive injections of artificial factor IX. To learn more about these clinical trials and the latest treatments for hemophilia, visit the National Hemophilia Foundation website at *www.hemophilia.org*.

RUNNING PROBLEM CONCLUSION

BLOOD DOPING IN ATHLETES

Johann Muehlegg's elevated hemoglobin and hematocrit prior to his 50-km race meant an automatic urine drug test following the race. At the time of the 2002 Olympics, athletes knew that there was a urine test for EPO, but they were not aware that the same test could detect darbepoietin. Both of Muehlegg's urine samples tested positive for darbepoietin, and he was stripped of his 50-km gold metal. The International Olympic Committee tested other athletes for rhEPO at the 2002 Salt Lake City Winter Olympics and had more than 100 positive results. Despite official prohibitions, blood doping in endurance sports remains a major problem.

Now check your understanding of the physiology behind blood doping by comparing your answers with the information in the following table.

	QUESTION	FACTS	INTEGRATION AND ANALYSIS
1	What is the normal range for Muehlegg's hemoglobin?	Normal hemoglobin range for males is 14–17 g/dL whole blood.	N/A
2	With blood doping, would you expect a hematocrit value to be lower or higher than normal?	Hematocrit is the percent of a blood sample volume that is packed red blood cells. A primary function of red blood cells is to carry oxygen.	Blood doping is done to increase oxygen-carrying capacity; therefore, the athlete would want more blood cells. This would mean a higher hematocrit.

QUESTION	FACTS	INTEGRATION AND ANALYSIS
3 Explain how diarrhea could cause a temporarily elevated hematocrit.	Diarrhea causes dehydration, which is loss of fluid volume. Plasma is the fluid component of blood.	If the total volume of red cells is unchanged but plasma volume decreases with dehydration, the hematocrit will increase.
4 How might Muehlegg quickly reduce his hematocrit without removing red blood cells?	Hematocrit = $$\frac{\text{red cell volume}}{\text{red cell volume + plasma volume}}$$	If plasma volume increases, hematocrit will decrease even though red cell volume does not change. By drinking fluids, Muehlegg could increase his plasma volume quickly.
5 Endogenous EPO, rhEPO, and darbepoietin all induce red blood cell synthesis but can be distinguished from one another by electrophoresis. Explain how three hormones made from the same gene can all be active yet different enough from one another to be detectable in the laboratory.	Activity depends on the protein binding to the receptor's binding site. Post-translational modification allows proteins from the same gene to be altered so that they are different from one another.	The three hormones have sites that bind to and activate the EPO receptor. They have different sizes or charges that cause them to separate during electrophoresis. For example, the glycosylation pattern in rhEPO is different from the pattern in endogenous EPO.
6 One hallmark of illegal EPO use is elevated reticulocytes in the blood. Why would this suggest greater-than-normal EPO activity?	Reticulocytes are the final immature stage of red blood cell development.	If red blood cell development becomes more rapid, reticulocytes may be pushed out of the bone marrow and into the blood before they have time to mature.

536   542   547   551   **554**

CHAPTER SUMMARY

Blood is an interesting tissue, with blood cells and cell fragments suspended in a liquid matrix—the plasma—which forms one of the two extracellular *compartments*. *Exchange* between the plasma and interstitial fluid takes place only in the capillaries. *Mass flow* of blood through the body depends on the pressure gradient created by the heart. At the same time, high pressure in the blood vessels poses a danger should the wall of the vessel rupture. Collectively, the cellular and protein components of blood serve as a functional unit that provides protection against hemorrhage. Blood cells are also essential for oxygen transport and defense, as you will learn in later chapters.

Plasma and the Cellular Elements of Blood

1. Blood is the circulating portion of the extracellular fluid. (p. 536)
2. **Plasma** is composed mostly of water, with dissolved proteins, organic molecules, ions, and dissolved gases. (p. 536; Fig. 16-1)
3. The **plasma proteins** include **albumins, globulins**, and the clotting protein **fibrinogen**. They function in blood clotting, defense, and as hormones, enzymes, or carriers for different substances. (p. 536)
4. The cellular elements of blood are **red blood cells (erythrocytes)**, **white blood cells (leukocytes)**, and **platelets**. Platelets are fragments of cells called **megakaryocytes**. (p. 538; Fig. 16-1)

5. Blood contains five types of white blood cells: (1) **lymphocytes**, (2) **monocytes**, (3) **neutrophils**, (4) **eosinophils**, and (5) **basophils**. (p. 538; Fig. 16-1)

Blood Cell Production

6. All blood cells develop from a pluripotent hematopoietic stem cell. (p. 538; Fig. 16-2)
7. **Hematopoiesis** begins early in embryonic development and continues throughout a person's life. Most hematopoiesis takes place in the bone marrow. (p. 539; Fig. 16-4)
8. **Colony-stimulating factors** and other cytokines control white blood cell production. **Thrombopoietin** regulates the growth and maturation of megakaryocytes. Red blood cell production is regulated primarily by **erythropoietin**. (pp 540–542)

Red Blood Cells

9. Mature mammalian red blood cells are biconcave disks lacking a nucleus. They contain hemoglobin, a red oxygen-carrying pigment. (p. 542; Fig. 16-5)
10. Hemoglobin synthesis requires iron in the diet. Iron is transported in the blood on **transferrin** and stored mostly in the liver, on **ferritin**. (p. 544; Fig. 16-7)

16

11. When hemoglobin is broken down, some heme groups are converted into **bilirubin**, which is incorporated into **bile** and excreted. Elevated bilirubin concentrations in the blood cause **jaundice**. (p. 545; Fig. 16-7)

Platelets and Coagulation

12. Platelets are cell fragments filled with granules containing clotting proteins and cytokines. Platelets are activated by damage to vascular endothelium. (p. 547; Fig. 16-9)

13. **Hemostasis** begins with vasoconstriction and the formation of a **platelet plug**. (p. 548; Fig. 16-10)

14. Exposed collagen triggers **platelet adhesion** and **platelet aggregation**. The platelet plug is converted into a clot when reinforced by **fibrin**. (p. 548; Fig. 16-10)

15. In the last step of the **coagulation cascade**, fibrin is made from **fibrinogen** through the action of **thrombin**. (p. 551; Fig. 16-12)

16. As the damaged vessel is repaired, **plasmin** trapped in the platelet plug dissolves fibrin (**fibrinolysis**) and breaks down the clot. (p. 551; Fig. 16-13)

17. Platelet plugs are restricted to the site of injury by **prostacyclin** in the membrane of intact vascular endothelium. **Anticoagulants** limit the extent of blood clotting within a vessel. (p. 551; Fig. 16-11)

QUESTIONS

(Answers to the Review Questions begin on page A1.)

THE PHYSIOLOGY PLACE

Access more review material online at **The Physiology Place** website. There you'll find review questions, problem-solving activities, case studies, flashcards, and direct links to both *InterActive Physiology*® and *PhysioEx*™. To access the site, go to *www.physiologyplace.com* and select Human Physiology, Fourth Edition.

LEVEL ONE   REVIEWING FACTS AND TERMS

1. The fluid portion of the blood, called _____, is composed mainly of _____.

2. List the three families of plasma proteins. Name at least one function of each type. Which type is most prevalent in the body?

3. List the cellular elements found in blood, and name at least one function of each.

4. Blood cell production is called _____. When and where does it occur?

5. What role do colony-stimulating factors, cytokines, and interleukins play in blood cell production? How are these chemical signal molecules different? Give two examples of each.

6. List the technical terms for production of red blood cells, production of platelets, and production of white blood cells.

7. The hormone that directs red blood cell synthesis is called _____. Where is it produced, and what is the stimulus for its production?

8. How are the terms *hematocrit* and *packed red cells* related? What are normal hematocrit values for men and women?

9. Distinguish between an erythroblast and an erythrocyte. Give three distinct characteristics of erythrocytes.

10. Which chemical element in the diet is important for hemoglobin synthesis?

11. Define the following terms and explain their significance in hematology.
    (a) jaundice
    (b) anemia
    (c) transferrin
    (d) hemophilia

12. Chemicals that prevent blood clotting from occurring are called _____.

LEVEL TWO   REVIEWING CONCEPTS

13. **Concept maps:** Combine each list of terms into a map. You may add other terms.

    List 1

ADP	integrins
collagen	membrane
phospholipids	platelet-activating factor
platelet activation	positive feedback
platelet adhesion	serotonin
platelet aggregation	thromboxane A_2
platelet plug	vasoconstriction

    List 2

clot	infarct
coagulation	plasmin
fibrin	plasminogen
fibrinogen	polymer
fibrinolysis	thrombin

    List 3

bile	hemoglobin
bilirubin	intestine
bone marrow	iron
erythropoietin	liver
ferritin	porphyrin
globin	reticulocyte
hematocrit	transferrin
heme	

14. Distinguish between the intrinsic, extrinsic, and common pathways of the coagulation cascade.

15. Once platelets are activated to aggregate, what factors halt their activity?

LEVEL THREE   PROBLEM SOLVING

16. Rachel is undergoing chemotherapy for breast cancer. She has blood cell counts at regular intervals, with these results:

	Normal Count	Patient Count 10 Days Post-Chemotherapy	Patient Count 20 Days Post-Chemotherapy
WBC ($\times 10^3$)	4.6–10.2	2.6	4.9
RBC ($\times 10^6$)	4.04–5.48	3.85	4.2
Platelets	142–424	133	151

At the time of the 10-day test, Jen (the nurse) notes that Rachel, although pale and complaining of being tired, does not have any bruises on her skin. Jen tells Rachel to eat foods high in protein, take a multivitamin tablet containing iron, and stay home and away from crowds as much as possible. How are Jen's observations and recommendations related to the results of the 10-day and 20-day blood tests?

17. Hemochromatosis is an inherited condition in which the body over-absorbs iron, resulting in an elevated total body load of iron.

    (a) What plasma protein would you expect to be elevated in this disease?

    (b) Which organ(s) would you expect to show damage in this disease?

    (c) Can you think of a simple treatment that could decrease the body's overload of iron in hemochromatosis?

18. Erythropoietin (EPO) was first isolated from the urine of anemic patients who had high circulating levels of the hormone. Although these patients had high concentrations of EPO, they were unable to produce adequate amounts of hemoglobin or red cells. Give some possible reasons why the patients' EPO was unable to correct their anemia.

LEVEL FOUR   QUANTITATIVE PROBLEMS

19. If we estimate that total blood volume is 7% of body weight, calculate the total blood volume in a 200-lb man and in a 130-lb woman (2.2 lb/kg). What are their plasma volumes if the man's hematocrit is 52% and the woman's hematocrit is 41%?

20. The total blood volume of an average person is 7% of total body weight. Using this figure and the fact that 1 kg of blood occupies a volume of about 1 liter, figure the total erythrocyte volume of a 50-kg woman with a hematocrit of 40%.

ANSWERS

 Answers to Concept Check Questions

Page 538

1. The five types of leukocytes are lymphocytes, monocytes/macrophages, basophils/mast cells, neutrophils, and eosinophils.

2. Erythrocytes and platelets lack nuclei and therefore are unable to carry out protein synthesis.

3. Liver degeneration reduces the total plasma protein concentration, which reduces the osmotic pressure in the capillaries. This decease in osmotic pressure increases net capillary filtration, and edema results.

Page 542

4. (a) erythropoietin, (b) colony-stimulating factors and interleukins, (c) thrombopoietin.

Page 546

5. (a) Heme is an iron-containing subunit of a hemoglobin molecule. (b) Ferritin is the liver protein that stores iron. Transferrin is the plasma protein that transports iron in the blood.

6. Bile is an exocrine secretion because it is secreted into the intestine.

Page 547

7. Low atmospheric oxygen at high altitude → low arterial oxygen → sensed by kidney cells → secrete erythropoietin → acts on bone marrow → increased production of red blood cells.

Page 551

8. The intrinsic pathway starts at the gold "Collagen exposed" box, the extrinsic pathway starts at the gold "Tissue factor exposed" box, and the common pathway begins at the red "Coagulation cascade" diamond.

16

17

This being of mine, whatever it really is, consists of a little flesh, a little breath, and the part which governs.

—Marcus Aurelius Antoninus (A.D. 121–180)

Mouse fibroblasts secrete collagen-rich matrix.

Mechanics of Breathing

BACKGROUND BASICS

EMPHYSEMA

"Diagnosis: COPD (blue bloater)," reads Edna Wilson's patient chart. COPD—chronic obstructive pulmonary disease—is a name given to diseases in which air exchange is impaired by narrowing of the airways. Most people with COPD have emphysema or chronic bronchitis or a combination of the two. Individuals in whom chronic bronchitis predominates are nicknamed "blue bloaters," owing to the bluish tinge of their skin and a tendency to be overweight. "Pink puffers" suffer more from emphysema. They tend to be thin, have normal (pink) skin coloration, and breathe shallow, rapid breaths. Because COPD is usually caused by smoking, most people can avoid the disease simply by not smoking. Unfortunately, Edna has been a heavy smoker for 35 of her 47 years.

559   561   569   574   581   582

Imagine covering the playing surface of a racquetball court (about 75 m²) with thin plastic wrap, then crumpling up the wrap and stuffing it into a 3-liter soft drink bottle. Impossible? Maybe so, if you use plastic wrap and a drink bottle. But the lungs of a 70-kg man have a gas exchange surface the size of that plastic wrap, compressed into a volume that is less than that of the bottle. This tremendous surface area for gas exchange is needed to supply the trillions of cells in the body with adequate amounts of oxygen.

Aerobic metabolism in cells depends on a steady supply of oxygen and nutrients from the environment, coupled with the removal of carbon dioxide. In very small aquatic animals, simple diffusion across the body surface meets these needs. Distance limits diffusion rate, however, so most multicelled animals require specialized respiratory organs associated with a circulatory system. Respiratory organs take a variety of forms, but all possess a large surface area compressed into a small space.

Besides needing a large exchange surface, humans and other terrestrial animals face an additional physiological challenge: dehydration. The exchange surface must be thin and moist to allow gases to pass from air into solution, and yet at the same time it must be protected from drying out as a result of exposure to air. Some terrestrial animals, such as the slug (a shell-less snail), meet the challenge of dehydration with behavioral adaptations that restrict them to humid environments and nighttime activities.

However, a more common solution is anatomical: an internalized respiratory epithelium. Human lungs are enclosed in the chest cavity to control their contact with the outside air. Internalization creates a humid environment for the exchange of gases with the blood and protects the delicate exchange surface from damage.

Internalized lungs create another problem, however: how to move air between the atmosphere and the exchange surface deep within the body. Air flow requires a muscular pump to create pressure gradients. Thus, in more complex body plans, the respiratory system consists of two separate components: a muscle-driven pump and a thin, moist exchange surface. In humans, the pump is the musculoskeletal structure of the thorax [p. 51]. The lungs themselves consist of the exchange epithelium and associated blood vessels.

The four primary functions of the respiratory system are:

1. **Exchange of gases between the atmosphere and the blood**. The body brings in O_2 for distribution to the tissues and eliminates CO_2 waste produced by metabolism.
2. **Homeostatic regulation of body pH**. The lungs can alter body pH by selectively retaining or excreting CO_2.
3. **Protection from inhaled pathogens and irritating substances**. Like all other epithelia that contact the external environment, the respiratory epithelium is well supplied with mechanisms that trap and destroy potentially harmful substances before they can enter the body.
4. **Vocalization**. Air moving across the vocal cords creates vibrations used for speech, singing, and other forms of communication.

In addition to serving these functions, the respiratory system is also a significant source of water loss and heat loss from the body. These losses must be balanced using homeostatic compensations.

In this chapter you will learn how the respiratory system carries out these functions by exchanging air between the environment and the interior air spaces of the lungs. This exchange is the *bulk flow* of air, and it follows many of the same principles that govern the bulk flow of blood through the cardiovascular system:

1. Flow takes place from regions of higher pressure to regions of lower pressure.
2. A muscular pump creates pressure gradients.
3. Resistance to air flow is influenced primarily by the diameter of the tubes through which the air is flowing.

The primary difference between air flow in the respiratory system and blood flow in the circulatory system is that air is a compressible mixture of gases while blood is a noncompressible liquid.

THE RESPIRATORY SYSTEM

The word *respiration* has several meanings in physiology (Fig. 17-1 ■). **Cellular respiration** refers to the intracellular reaction of oxygen with organic molecules to produce carbon dioxide,

17

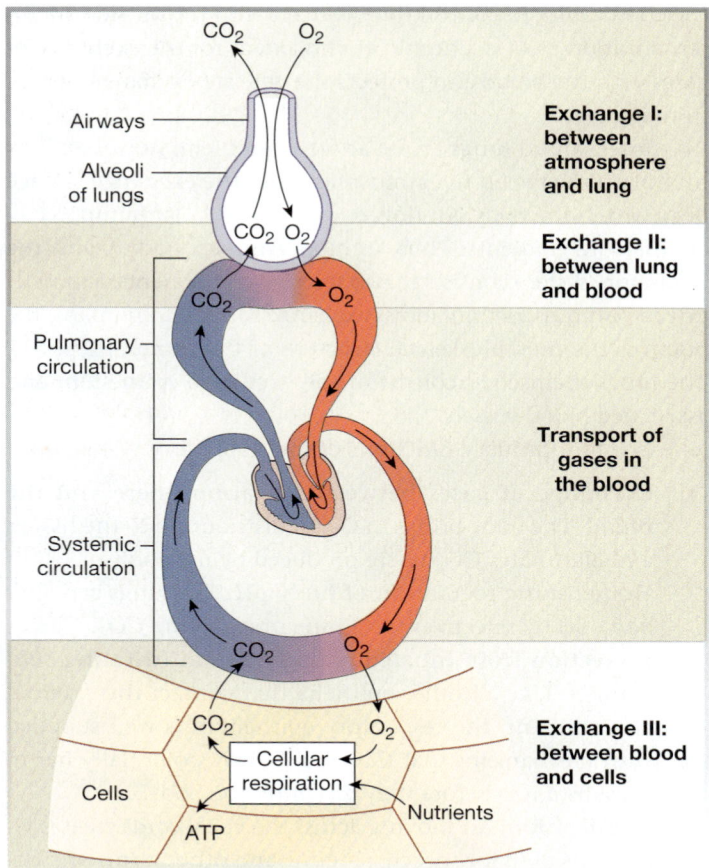

Exchange I: between atmosphere and lung

Exchange II: between lung and blood

Transport of gases in the blood

Exchange III: between blood and cells

Airways

Alveoli of lungs

Pulmonary circulation

Systemic circulation

Cells

Cellular respiration

Nutrients

ATP

■ FIGURE 17-1 *Overview of external and cellular respiration*

water, and energy in the form of ATP [🌐 p. 107]. **External respiration,** the topic of this chapter and the next, is the movement of gases between the environment and the body's cells. External respiration can be subdivided into four integrated processes, illustrated in Figure 17-1:

1. *The exchange of air between the atmosphere and the lungs.* This process is known as **ventilation,** or breathing. **Inspiration** (inhalation) is the movement of air into the lungs. **Expiration** (exhalation) is the movement of air out of the lungs. The mechanisms by which ventilation takes place are collectively called the *mechanics of breathing.*
2. *The exchange of O_2 and CO_2 between the lungs and the blood.*
3. *The transport of O_2 and CO_2 by the blood.*
4. *The exchange of gases between blood and the cells.*

External respiration requires the coordinated functioning of the respiratory and cardiovascular systems. The **respiratory system** consists of structures involved in ventilation and gas exchange (Fig. 17-2 ■):

1. The **conducting system** of passages, or **airways,** that lead from the external environment to the exchange surface of the lungs.
2. The **alveoli** (singular **alveolus**) [*alveus,* a concave vessel], a series of interconnected sacs that collectively form the ex-

change surface, where oxygen moves from inhaled air to the blood, and carbon dioxide moves from the blood to air that is about to be exhaled.

3. The bones and muscles of the thorax (chest cavity) and abdomen that assist in ventilation.

The respiratory system can be divided into two parts. The **upper respiratory tract** consists of the mouth, nasal cavity, pharynx, and larynx. The **lower respiratory tract** consists of the trachea, two primary bronchi, their branches, and the lungs. The lower tract is also known as the *thoracic portion* of the respiratory system because it is enclosed in the thorax.

Bones and Muscles of the Thorax Surround the Lungs

The thorax is bounded by the bones of the spine and rib cage and their associated muscles. Together the bones and muscles are called the *thoracic cage.* The ribs and spine (the *chest wall*) form the sides and top of the cage. A dome-shaped sheet of skeletal muscle, the **diaphragm,** forms the floor (Fig. 17-2a).

Two sets of **intercostal muscles,** internal and external, connect the 12 pairs of ribs (Fig. 17-2b). Additional muscles, the **sternocleidomastoids** and the **scalenes,** run from the head and neck to the sternum and first two ribs.

Functionally, the thorax is a sealed container filled with three membranous bags, or sacs. One, the *pericardial sac,* contains the heart. The other two bags, the **pleural sacs,** contain the lungs [*pleura,* rib or side]. The esophagus and thoracic blood vessels and nerves pass between the pleural sacs (Fig. 17-2d).

Pleural Sacs Enclose the Lungs

The **lungs** (Fig. 17-2a, c) consist of light, spongy tissue whose volume is occupied mostly by air-filled spaces. These irregular cone-shaped organs nearly fill the thoracic cavity, with their bases resting on the curved diaphragm. Rigid conducting airways—the bronchi—connect the lungs to the main airway, the trachea.

Each lung is surrounded by a double-walled pleural sac whose membranes line the inside of the thorax and cover the outer surface of the lungs (Fig. 17-3 ■). Each *pleural membrane,* or **pleura,** contains several layers of elastic connective tissue and numerous capillaries. The opposing layers of pleural membrane are held together by a thin film of **pleural fluid** whose total volume is only a few milliliters. The result is similar to an air-filled balloon (the lung) surrounded by a water-filled balloon (the pleural sac). Most illustrations exaggerate the volume of the pleural fluid, but you can appreciate its thinness if you imagine spreading 3 mL of water evenly over the surface of a 3-liter soft drink bottle.

Pleural fluid serves several purposes. First, it creates a moist, slippery surface so that the opposing membranes can slide across one another as the lungs move within the thorax. Second, it holds the lungs tight against the thoracic wall. To visualize this arrangement, think of two panes of glass stuck together by a thin film of water. You can slide the panes back and forth across each other, but you cannot pull them apart because of the cohesiveness of the water [🔁 p. 25]. A similar fluid bond between the two pleural membranes makes the lungs "stick" to the thoracic cage and holds them stretched in a partially inflated state, even at rest.

Airways Connect Lungs to the External Environment

Air enters the upper respiratory tract through the mouth and nose and passes into the **pharynx**, a common passageway for food, liquids, and air [*pharynx,* throat]. From the pharynx, air flows through the **larynx** into the **trachea**, or windpipe (Fig. 17-2a). The larynx contains the **vocal cords**, connective tissue bands that tighten to create sound when air moves past them.

The trachea is a semiflexible tube held open by 15 to 20 C-shaped cartilage rings (Fig. 17-2e). It extends down into the thorax, where it branches (division 1) into a pair of **primary bronchi**, one *bronchus* to each lung. Within the lungs, the bronchi branch repeatedly (divisions 2–11) into progressively smaller bronchi (Fig. 17-2a, e). Like the trachea, the bronchi are semirigid tubes supported by cartilage.

In the lungs, the smallest bronchi branch to become **bronchioles**, small collapsible passageways with walls of smooth muscle. The bronchioles continue branching (divisions 12–23) until the *respiratory bronchioles* form a transition between the airways and the exchange epithelium of the lung.

The diameter of the airways becomes progressively smaller from the trachea to the bronchioles, but as the individual airways get narrower, their numbers increase (Fig. 17-4 ■). As a result, the total cross-sectional area increases with each division of the airways. Total cross-sectional area is lowest in the upper respiratory tract and greatest in the bronchioles, analogous to the increase in cross-sectional area that occurs from the aorta to the capillaries in the circulatory system [🔁 p. 517].

✔ CONCEPT CHECK

1. What is the difference between cellular respiration and external respiration?
2. Name the components of the upper respiratory tract, and those of the lower respiratory tract.
3. Based on the total cross-sectional area of different airways, where is the velocity of air flow highest and lowest?
4. Give two functions of pleural fluid.
5. Name the components (including muscles) of the thoracic cage. List the contents of the thorax.
6. Which air passages of the respiratory system are collapsible?

Answers: p. 586

Alveoli Are the Site of Gas Exchange

The alveoli, clustered at the ends of terminal bronchioles, make up the bulk of lung tissue (Fig. 17-2f, g). Their primary function is the exchange of gases between themselves and the blood.

Each tiny alveolus is composed of a single layer of epithelium (Fig. 17-2g). Two types of epithelial cells are found in the alveoli, and they occur in roughly equal numbers. The smaller but thicker **type II alveolar cells** synthesize and secrete a chemical known as **surfactant.** Surfactant mixes with the thin fluid lining of the alveoli to aid lungs as they expand during breathing, as we will see later in this chapter. Type II cells also help minimize the amount of fluid present in the alveoli by transporting solutes, followed by water, out of the alveolar air space.

The larger **type I alveolar cells** are very thin so that gases can diffuse rapidly through them (Fig. 17-2h). In much of the exchange area, a layer of basement membrane fuses the alveolar epithelium to the capillary endothelium, and only a small amount of interstitial fluid is present.

The thin walls of the alveoli do not contain muscle because muscle fibers would block rapid gas exchange. As a result, lung tissue itself cannot contract. However, connective tissue between the alveolar epithelial cells contains many elastin fibers that create elastic recoil when lung tissue is stretched.

The close association of the alveoli with an extensive network of capillaries demonstrates the intimate link between the respiratory and cardiovascular systems. Blood vessels cover 80–90% of the alveolar surface, forming an almost continuous "sheet" of blood in close contact with the air-filled alveoli. The proximity of capillary blood to alveolar air is essential for the rapid exchange of gases.

The Pulmonary Circulation Is a High-Flow, Low-Pressure System

The pulmonary circulation begins with the pulmonary trunk, which receives low-oxygen blood from the right ventricle. The pulmonary trunk divides into two pulmonary arteries, one to each lung [🔁 Fig. 14-1, p. 459]. Oxygenated blood from the lungs returns to the left atrium via the pulmonary veins.

RUNNING PROBLEM

Patients with chronic bronchitis have excessive mucus production and general inflammation of the entire respiratory tract. The mucus narrows the airways and makes breathing difficult.

Question 1:
What does narrowing of the airways do to the resistance airways offer to air flow? (Hint: the relationship between radius and resistance is the same for air flow in the respiratory system as it is for blood flow in the circulatory system. [🔁 p. 509])

559   561   569   574   581   582

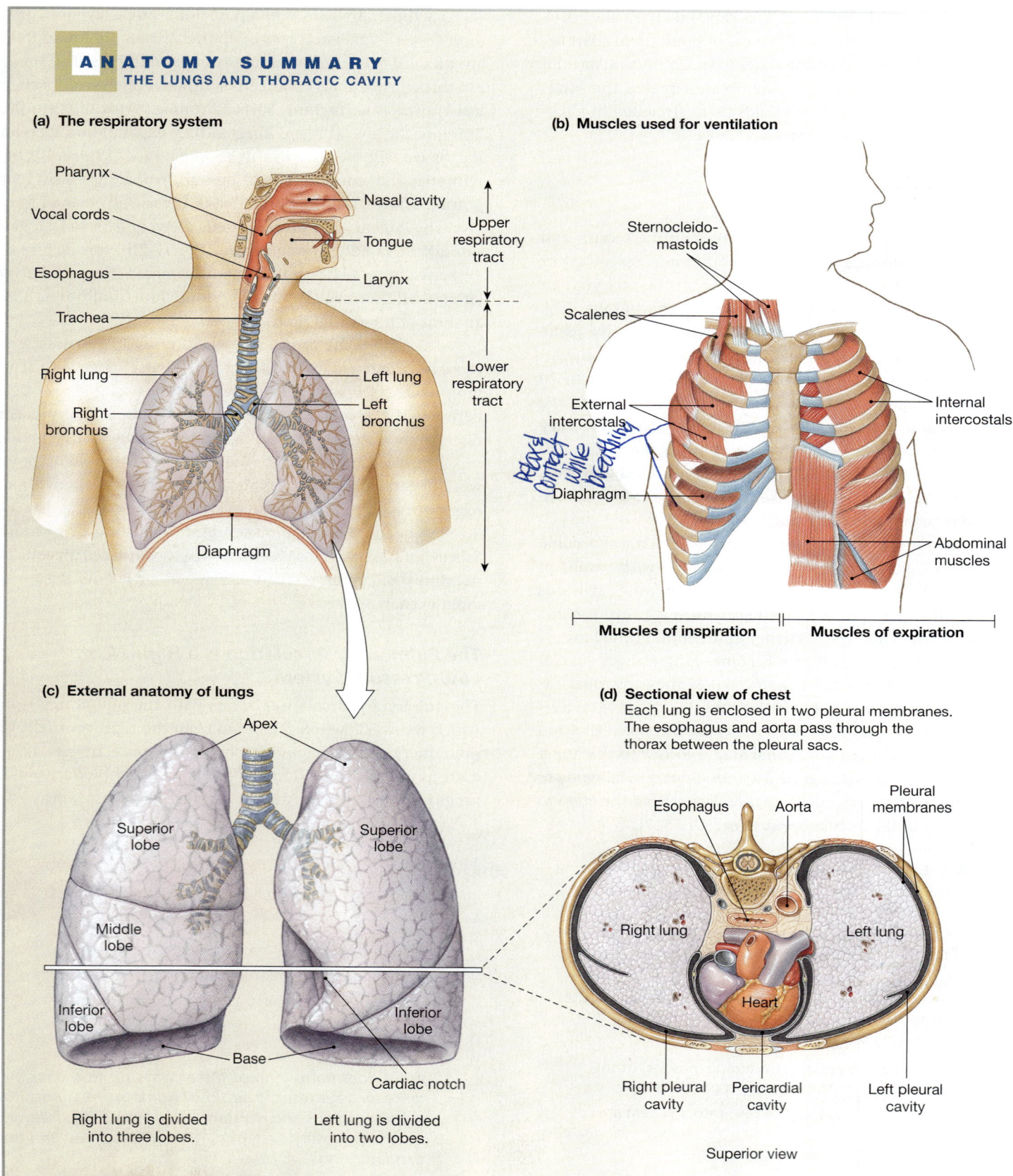

ANATOMY SUMMARY
THE LUNGS AND THORACIC CAVITY

(a) The respiratory system

Pharynx

Vocal cords

Esophagus

Trachea

Right lung

Right bronchus

Diaphragm

Nasal cavity

Tongue

Larynx

Left lung

Left bronchus

Upper respiratory tract

Lower respiratory tract

(b) Muscles used for ventilation

Sternocleido-mastoids

Scalenes

External intercostals

Diaphragm

Internal intercostals

Abdominal muscles

Relax & Contract while breathing

Muscles of inspiration   **Muscles of expiration**

(c) External anatomy of lungs

Apex

Superior lobe

Middle lobe

Inferior lobe

Base

Superior lobe

Inferior lobe

Cardiac notch

Right lung is divided into three lobes.

Left lung is divided into two lobes.

(d) Sectional view of chest
Each lung is enclosed in two pleural membranes. The esophagus and aorta pass through the thorax between the pleural sacs.

Esophagus   Aorta   Pleural membranes

Right lung   Left lung

Heart

Right pleural cavity   Pericardial cavity   Left pleural cavity

Superior view

FIGURE 17-2

THE BRONCHI AND ALVEOLI

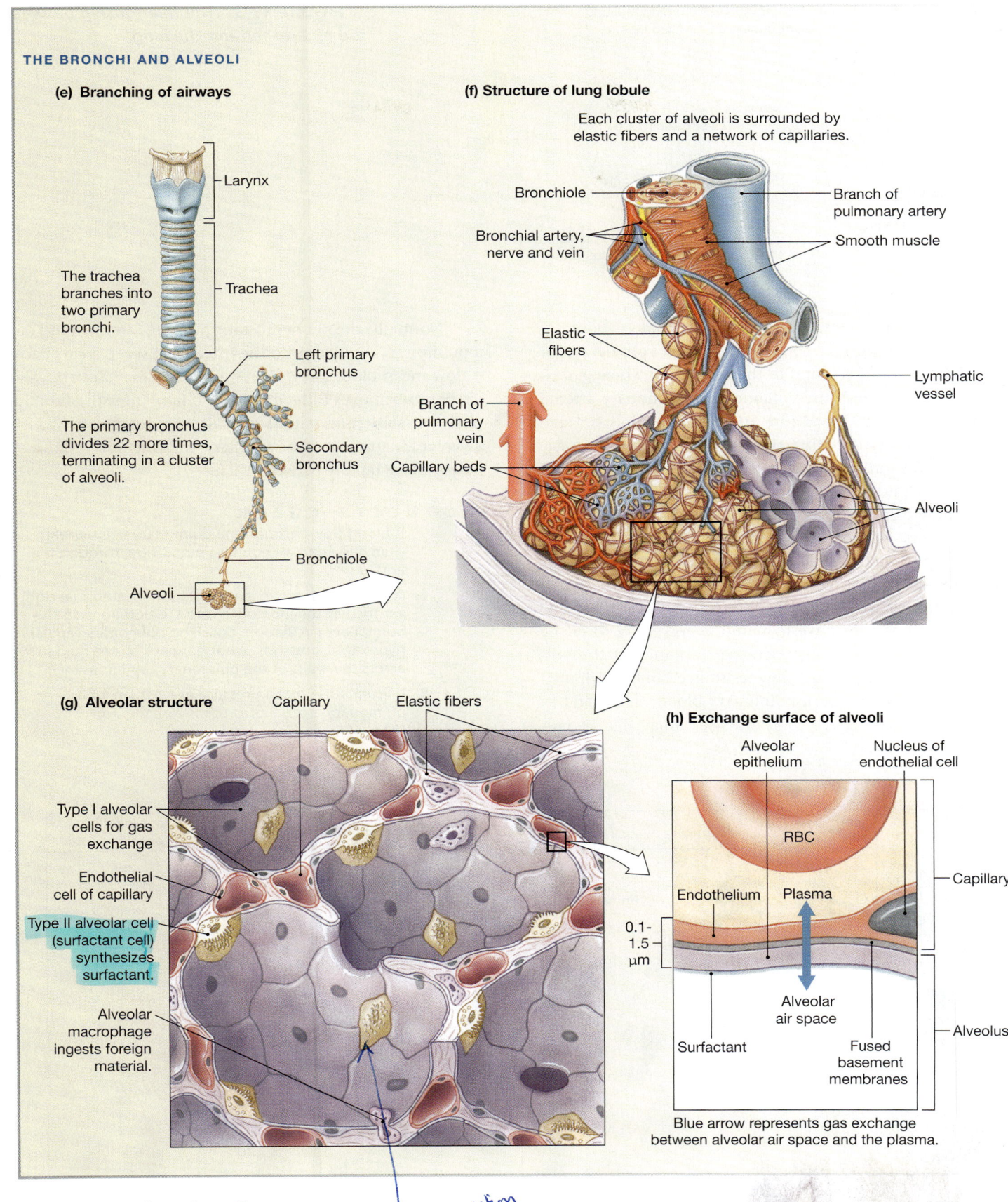

(e) Branching of airways

Larynx

The trachea branches into two primary bronchi.

Trachea

The primary bronchus divides 22 more times, terminating in a cluster of alveoli.

Left primary bronchus

Secondary bronchus

Bronchiole

Alveoli

(f) Structure of lung lobule

Each cluster of alveoli is surrounded by elastic fibers and a network of capillaries.

Bronchiole

Branch of pulmonary artery

Bronchial artery, nerve and vein

Smooth muscle

Elastic fibers

Lymphatic vessel

Branch of pulmonary vein

Capillary beds

Alveoli

(g) Alveolar structure

Capillary

Elastic fibers

Type I alveolar cells for gas exchange

Endothelial cell of capillary

Type II alveolar cell (surfactant cell) synthesizes surfactant.

Alveolar macrophage ingests foreign material.

For secretion (prevent lungs from collapsing)

(h) Exchange surface of alveoli

Alveolar epithelium

Nucleus of endothelial cell

RBC

Capillary

Endothelium

Plasma

0.1–1.5 μm

Alveolar air space

Endothelium

Surfactant

Fused basement membranes

Alveolus

Blue arrow represents gas exchange between alveolar air space and the plasma.

■ FIGURE 17-2   (continued)

17

The pleural sac forms a double membrane surrounding the lung, similar to a fluid-filled balloon surrounding an air-filled balloon.

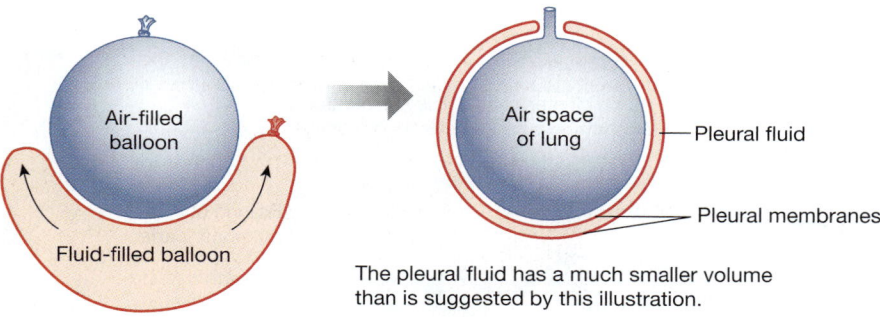

The pleural fluid has a much smaller volume than is suggested by this illustration.

At any given moment, the pulmonary circulation contains about 0.5 liter of blood, or 10% of total blood volume. About 75 mL of this amount is found in the capillaries, where gas exchange takes place, with the remainder in pulmonary arteries and veins. The rate of blood flow through the lungs is much higher than the rate in other tissues [⟳ p. 515] because the lungs receive the entire cardiac output of the right ventricle: 5 L/min. This means that as much blood flows through the lungs in one minute as flows through the entire rest of the body in the same amount of time!

Despite the high flow rate, pulmonary blood pressure is low. Pulmonary arterial pressure averages 25/8 mm Hg, much lower than the average systemic pressure of 120/80 mm Hg. The right ventricle does not have to pump as forcefully to create blood flow through the lungs because resistance of the pulmonary circulation is low. This low resistance can be attributed to the shorter total length of pulmonary blood vessels and to the distensibility and large total cross-sectional area of pulmonary arterioles.

Normally, the net hydrostatic pressure filtering fluid out of a pulmonary capillary into the interstitial space is low because of low mean blood pressure [⟳ p. 518]. The lymphatic system efficiently removes filtered fluid, and lung interstitial fluid volume is usually minimal. As a result, the distance between the alveolar air space and the capillary endothelium is short, and gases diffuse rapidly between them.

CONCEPT CHECK

7. Is blood flow through the pulmonary trunk greater than, less than, or equal to blood flow through the aorta?

8. A person has left ventricular failure but normal right ventricular function. As a result, blood pools in the pulmonary circulation, doubling pulmonary capillary hydrostatic pressure. What happens to net fluid flow across the walls of the pulmonary capillaries?

9. Calculate the mean pressure in a person whose pulmonary arterial pressure is 25/8 mm Hg. [⟳ p. 506]

Answers: p. 586

	Name	Division	Diameter (mm)	How many?	Cross-sectional area (cm²)
Conducting system	Trachea	0	15–22	1	2.5
	Primary bronchi	1	10–15	2	
	Smaller bronchi	2		4	
		3			
		4	1–10		
		5			
		6–11		1 x 10⁴	
				2 x 10⁴	100
	Bronchioles	1–23	0.5–1		
Exchange surface				8 x 10⁷	5 x 10³
	Alveoli	24	0.3	3–6 x 10⁸	>1 x 10⁶

■ **FIGURE 17-4** *Branching of the airways*

CONGESTIVE HEART FAILURE

When is a lung problem not a lung problem? The answer: when it's really a heart problem. Congestive heart failure (CHF) is an excellent example of the interrelationships among body systems, and of how disruptions in one system can have a domino effect in the others. The primary symptoms of heart failure are shortness of breath (*dyspnea*), wheezing during breathing, and sometimes a productive cough whose *phlegm* (pronounced "flem") may be pinkish due to the presence of blood. Congestive heart failure arises when the right heart is a more effective pump than the left heart (see Ch. 14, Concept Check 31, p. 489). When blood accumulates in the pulmonary circulation, increased volume increases pulmonary blood pressure and capillary hydrostatic pressure. Capillary filtration exceeds the ability of the lymph system to drain interstitial fluid, resulting in pulmonary edema. Treatment of CHF includes increasing urinary output of fluid, which brings yet another organ system into the picture. By current estimates, nearly 5 million Americans suffer from CHF. To learn more about this condition, visit the American Heart Association website (*www.americanheart.org*) or see the Health Information section for the National Heart, Lung, and Blood Institute of the National Institutes of Health (*www.nhlbi.nih.gov*).

GAS LAWS

Respiratory air flow is very similar in many respects to blood flow in the cardiovascular system, even though blood is a noncompressible liquid and air is a compressible mixture of gases. Blood pressure and environmental air pressure (**atmospheric pressure**) are both reported in millimeters of mercury (mm Hg).*

At sea level, normal atmospheric pressure is 760 mm Hg. However, in this book we will follow the convention of designating atmospheric pressure as 0 mm Hg. Because atmospheric pressure varies with altitude and because very few people live exactly at sea level, this convention allows us to compare pressure differences that occur during ventilation without correct-

---

*Respiratory physiologists sometimes report gas pressures in units of centimeters of water: 1 mm Hg = 1.36 cm H_2O.

TABLE 17-1	**Gas Laws**
1. The total pressure of a mixture of gases is the sum of the pressures of the individual gases (Dalton's law).	
2. Gases, singly or in a mixture, move from areas of higher pressure to areas of lower pressure.	
3. If the volume of a container of gas changes, the pressure of the gas will change in an inverse manner (Boyle's law).	

ing for altitude. Negative numbers designate subatmospheric pressures, and positive numbers denote higher-than-atmospheric pressures.

Table 17-1 ■ summarizes the rules that govern the behavior of gases in air. These rules provide the basis for the exchange of air between the external environment and the alveoli. Gas laws that govern the solubility of gases in solution will be considered in Chapter 18.

Air Is a Mixture of Gases

The atmosphere surrounding the earth is a mixture of gases and water vapor. **Dalton's law** states that the total pressure exerted by a mixture of gases is the sum of the pressures exerted by the individual gases. Thus, in dry air at an atmospheric pressure of 760 mm Hg, 78% of the total pressure is due to N_2, 21% to O_2, and so on (Table 17-2 ■).

In respiratory physiology, we are concerned not only with total atmospheric pressure but also with the individual pressures of oxygen and carbon dioxide. The pressure of a single gas in a mixture is known as its **partial pressure** (P_{gas}). To find the partial pressure of any one gas in a sample of air, multiply the atmospheric pressure (P_{atm}) by the gas's relative contribution (%) to P_{atm}:

Partial pressure of an atmospheric gas =
$P_{atm} \times$ % of gas in atmosphere

Partial pressure of oxygen = 760 mm Hg $\times$ 21%

$P_{O_2} = 760 \times 0.21 = 160$ mm Hg

Thus, the partial pressure of oxygen (P_{O_2}) in dry air at sea level is 160 mm Hg. The pressure exerted by an individual gas is determined only by its relative abundance in the mixture and is independent of the molecular size or mass of the gas.

The partial pressures of gases in air vary slightly depending on how much water vapor is in the air because the pressure of water vapor "dilutes" the contribution of other gases to the total pressure. To calculate the partial pressure of a gas in humid air, you must first subtract the water vapor pressure from the total pressure. Table 17-2 compares the partial pressures of some gases in dry air and at 100% humidity.

TABLE 17-2 Partial Pressures (P_{gas}) of Atmospheric Gases at 760 mm Hg

GAS AND ITS PERCENTAGE IN AIR	P_{gas} IN DRY, 25° C AIR	P_{gas} IN 25° C AIR, 100% HUMIDITY	P_{gas} IN 37° C AIR, 100% HUMIDITY
Nitrogen (N_2) 78%	593 mm Hg	574 mm Hg	556 mm Hg
Oxygen (O_2) 21%	160 mm Hg	155 mm Hg	150 mm Hg
Carbon dioxide (CO_2) 0.033%	0.25 mm Hg	0.24 mm Hg	0.235 mm Hg
Water vapor	0 mm Hg	24 mm Hg	47 mm Hg

CONCEPT CHECK

10. If nitrogen is 78% of atmospheric air, what is the partial pressure of nitrogen (P_{N_2}) in a sample of dry air that has an atmospheric pressure of 720 mm Hg?

Answers: p. 586

Gases Move from Areas of Higher Pressure to Areas of Lower Pressure

Air flow occurs whenever there is a pressure gradient. Air flow, like blood flow, is directed from areas of higher pressure to areas of lower pressure. Meteorologists predict the weather by knowing that areas of high atmospheric pressure move in to replace areas of low pressure. In ventilation, bulk flow of air down pressure gradients explains how air exchanges between the external environment and the lungs. Movement of the thorax during breathing creates alternating conditions of high and low pressure in the lungs.

Movement down pressure gradients also applies to single gases. For example, oxygen moves from areas of higher oxygen partial pressure to areas of lower oxygen partial pressure. Diffusion of individual gases is important in the alveoli-blood and blood-cell gas exchanges discussed in Chapter 18.

Boyle's Law Describes Pressure-Volume Relationships of Gases

The pressure exerted by a gas or mixture of gases in a sealed container is created by the collisions of moving gas molecules with the walls of the container and with each other. If the size of the container is reduced, the collisions between the gas molecules and the walls become more frequent, and the pressure rises. This relationship can be expressed by the equation

$$P_1V_1 = P_2V_2$$

where P represents pressure and V represents volume.

For example, start with a 1-liter container (V_1) of a gas whose pressure is 100 mm Hg (P_1), as shown in Figure 17-5 ∎. What happens to the pressure of the gas when the lid of the container moves in to decrease the volume to 0.5 L? According to our equation,

$$P_1V_1 = P_2V_2$$

$$100 \text{ mm Hg} \times 1 \text{ L} = P_2 \times 0.5 \text{ L}$$

$$P_2 = 200 \text{ mm Hg}$$

This calculation tells us that if the volume is reduced by one-half, the pressure doubles. If the volume were to double, the pressure would be reduced by one-half. This relationship between pressure and volume was first noted by Robert Boyle in the 1600s and has been called **Boyle's law** of gases.

In the respiratory system, changes in the volume of the chest cavity during ventilation cause pressure gradients that create air flow. When the chest volume increases, the alveolar pressure falls, and air flows into the respiratory system. When the chest volume decreases, the alveolar pressure rises, and air flows out into the atmosphere. This movement of air is bulk flow because the entire gas mixture is moving rather than merely one or two of the gas species contained in the air.

Boyle's Law: $P_1V_1 = P_2V_2$

Decreasing volume increases collisions and increases pressure.

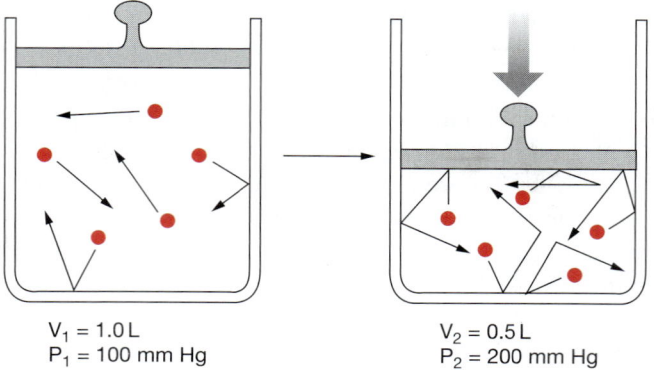

$V_1 = 1.0 \text{ L}$
$P_1 = 100 \text{ mm Hg}$

$V_2 = 0.5 \text{ L}$
$P_2 = 200 \text{ mm Hg}$

∎ **FIGURE 17-5** *Boyle's law*

Boyle's law ($P_1V_1 = P_2V_2$) assumes that temperature and the number of gas molecules remain constant.

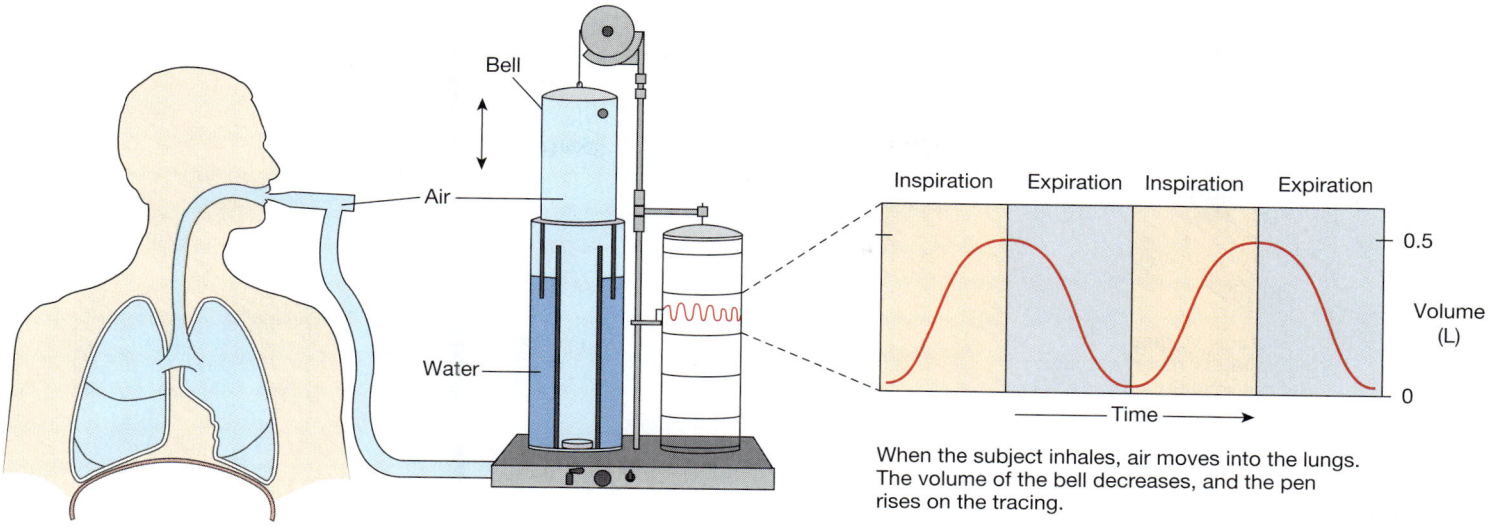

Bell

Air

Water

Inspiration Expiration Inspiration Expiration

Volume
(L)

—— Time ——→

0.5

0

When the subject inhales, air moves into the lungs.
The volume of the bell decreases, and the pen
rises on the tracing.

■ **FIGURE 17-6** *A spirometer*

The subject inserts a mouthpiece that is attached to an inverted bell filled with air or
oxygen. The volume of the bell and the volume of the subject's respiratory tract create a closed system because the bell is suspended in water.

VENTILATION

The first exchange in respiratory physiology is ventilation, or
breathing, the bulk flow exchange of air between the atmosphere and the alveoli (Fig. 17-1).

Lung Volumes Change During Ventilation

Physiologists and clinicians assess a person's pulmonary function by measuring how much air the person moves during quiet
breathing, then with maximum effort. These **pulmonary function tests** use a **spirometer**, an instrument that measures the
volume of air moved with each breath (Fig. 17-6 ■). (Most
spirometers in clinical use today are small computerized machines rather than the traditional spirometer illustrated here.)

When a subject is attached to the traditional spirometer
through a mouthpiece and the subject's nose is clipped closed,
the subject's respiratory tract and the spirometer form a closed
system. When the subject breathes in, air moves from the
spirometer into the lungs, and the recording pen, which traces
a graph on a rotating cylinder, moves up. When the subject exhales, air moves from the lungs back into the spirometer, and
the pen moves down.

Lung Volumes The air moved during breathing can be divided into four lung volumes: (1) tidal volume, (2) inspiratory
reserve volume, (3) expiratory reserve volume, and (4) residual
volume. The numerical values given in Figure 17-7 ■ represent
average volumes for a 70-kg man. The volumes for women are
typically about 20–25% less. Lung volumes vary considerably
with age, sex, and height. Each of the following paragraphs be-

gins with the instructions you would be given if you were being
tested for these volumes.

"Breathe quietly." The volume of air that moves during a
single inspiration or expiration is known as the **tidal volume**
(V_T). Average tidal volume during quiet breathing is about
500 mL.

"Now, at the end of a quiet inspiration, take in as much
additional air as you possibly can." The additional volume you
inspire above the tidal volume represents your **inspiratory reserve volume** (IRV). In a 70-kg man, this volume is about
3000 mL, a sixfold increase over the normal tidal volume.

"Now stop at the end of a normal exhalation, then exhale
as much air as you possibly can." The amount of air forcefully
exhaled after the end of a normal expiration is the **expiratory
reserve volume** (ERV), which averages about 1100 mL.

The fourth volume cannot be measured directly. Even if you
blow out as much air as you can, air still remains in the lungs and
the airways. The volume of air in the respiratory system after
maximal exhalation—about 1200 mL—is called the **residual volume** (RV). Most of this residual volume exists because the lungs
are held stretched against the thoracic wall by the pleural fluid.

Lung Capacities The sum of two or more lung volumes is
called a **capacity**. The **vital capacity** (VC) is the sum of the inspiratory reserve volume, expiratory reserve volume, and tidal
volume. Vital capacity represents the maximum amount of air
that can be voluntarily moved into or out of the respiratory system with one breath. To measure vital capacity, you would instruct the person to take in as much air as possible, then blow it
all out. Vital capacity decreases with age.

17

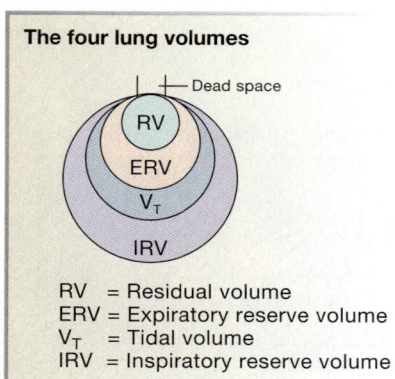

The four lung volumes

Dead space

RV
ERV
V_T
IRV

RV = Residual volume
ERV = Expiratory reserve volume
V_T = Tidal volume
IRV = Inspiratory reserve volume

Pulmonary volumes

		Males	Females	
Vital capacity	IRV	3000	1900	Inspiratory capacity
	V_T	500	500	
	ERV	1100	700	Functional residual capacity
Residual volume		1200	1100	
		5800 mL	4200 mL	

A spirometer tracing showing lung volumes and capacities

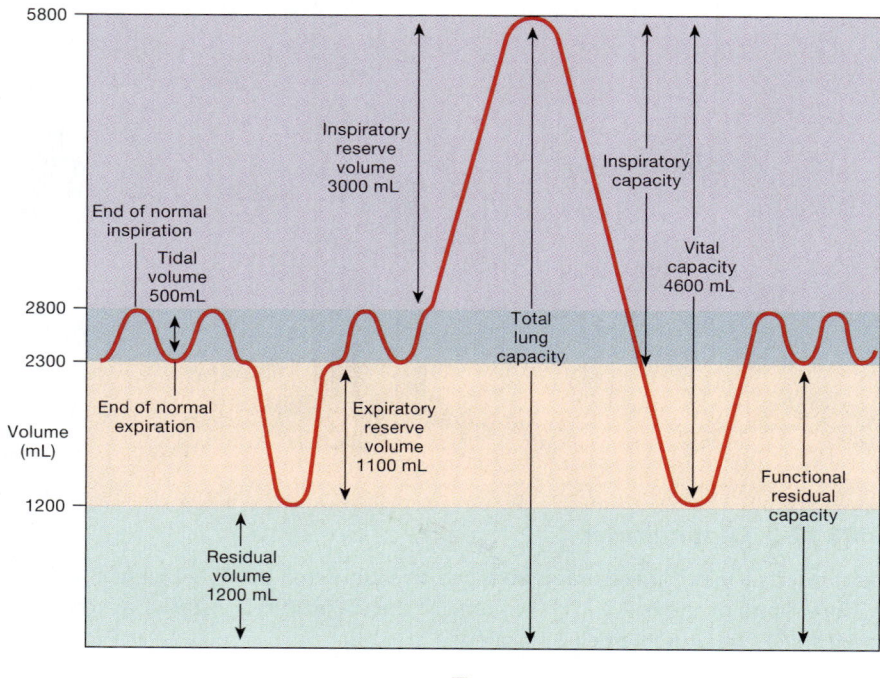

Capacities are sums of two or more volumes.

■ **FIGURE 17-7**   *Lung volumes and capacities*

The vital capacity plus the residual volume yields the **total lung capacity** (TLC). Other capacities of importance in pulmonary medicine include the **inspiratory capacity** (tidal volume + inspiratory reserve volume) and the **functional residual capacity** (expiratory reserve volume + residual volume).

CONCEPT CHECK

11. How are lung volumes related to lung capacities?
12. Which lung volume cannot be measured directly?
13. If vital capacity decreases with age but total lung capacity does not change, which lung volume must be changing? In which direction?   *Answers: p. 586*

The Airways Warm, Humidify, and Filter Inspired Air

During breathing, the upper airways and the bronchi do more than simply serve as passageways for air. They play an important role in conditioning air before it reaches the alveoli. Conditioning has three components:

1. *Warming* air to body temperature (37° C), so that core body temperature will not change and alveoli will not be damaged by cold air;
2. *Adding water vapor* until the air reaches 100% humidity, so that the moist exchange epithelium will not dry out; and
3. *Filtering out foreign material,* so that viruses, bacteria, and inorganic particles will not reach the alveoli.

Inhaled air is warmed by the body's heat and moistened by water evaporating from the mucosal lining of the airways. Under normal circumstances, by the time air reaches the trachea, it has been conditioned to 100% humidity and 37° C.

Breathing through the mouth is not nearly as effective at warming and moistening air as breathing through the nose. If you exercise outdoors in very cold weather, you may be familiar with the ache in your chest that results from breathing cold air through your mouth.

Filtration of air takes place both in the trachea and in the bronchi. These airways are lined with ciliated epithelium, that secretes both mucus and a dilute saline solution. The cilia are bathed in a watery saline layer (Fig. 17-8 ■). On top of them lies a sticky layer of mucus that traps most inhaled particles larger than 2 μm.

The mucus layer is secreted by *goblet cells* in the epithelium (Fig. 17-8). The cilia beat with an upward motion that moves the mucus continuously toward the pharynx, creating what is called the *mucociliary escalator.* Mucus contains *immunoglobulins* that can disable many pathogens, and once the mucus reaches the pharynx and is swallowed, stomach acid and enzymes destroy any remaining microorganisms.

Secretion of the watery saline layer beneath the mucus is essential for a functional mucociliary escalator. In the disease *cystic fibrosis,* for example, inadequate ion secretion decreases fluid movement in the airways. Without the saline layer, cilia

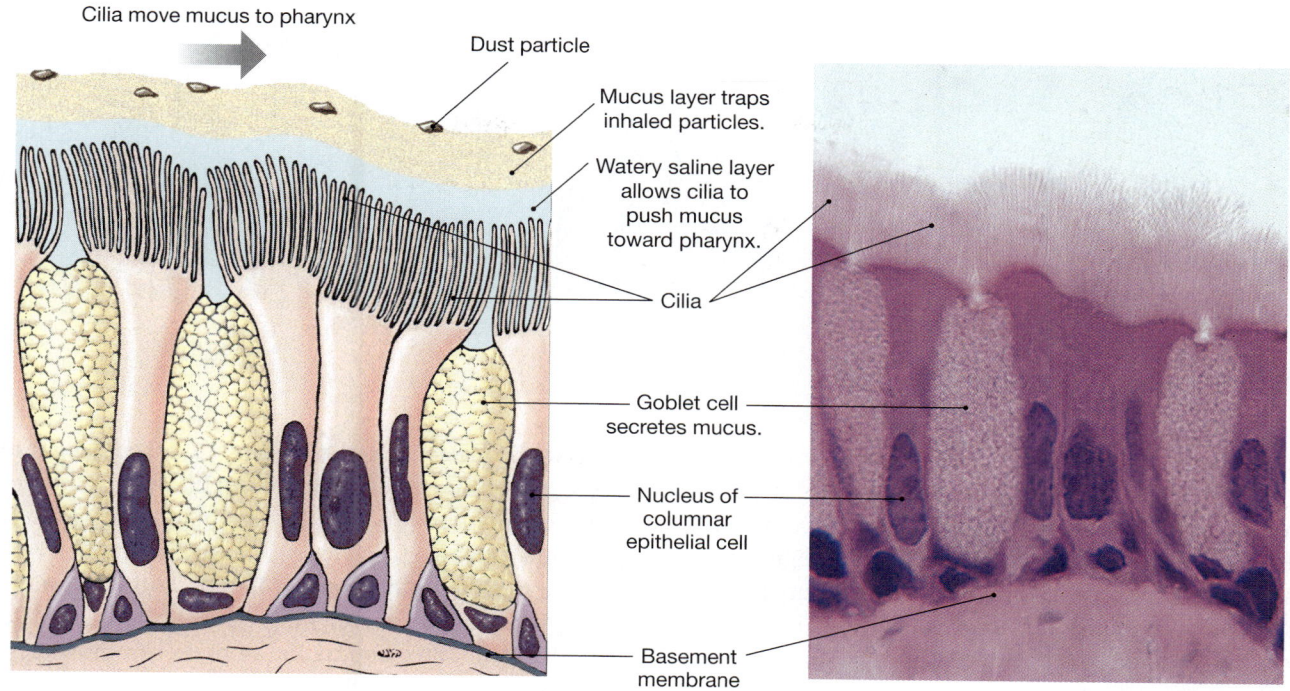

Cilia move mucus to pharynx

Dust particle

Mucus layer traps inhaled particles.

Watery saline layer allows cilia to push mucus toward pharynx.

Cilia

Goblet cell secretes mucus.

Nucleus of columnar epithelial cell

Basement membrane

Ciliated epithelium of the trachea

■ **FIGURE 17-8** *Ciliated respiratory epithelium*

Ciliary movement of the mucus layer toward the pharynx removes inhaled pathogens and particulate matter.

become trapped in thick, sticky mucus. Mucus cannot be cleared, and bacteria colonize the airways, resulting in recurrent lung infections.

CONCEPT CHECK

14. As inhaled air becomes humidified passing down the airways, what happens to the P_{O_2} of the air?

15. Cigarette smoking paralyzes cilia in the airways. Why would paralysis of the cilia cause smokers to develop a cough?

Answers: p. 586

During Ventilation, Air Flows Because of Pressure Gradients

Air flows into the lungs because of pressure gradients created by a pump, just as blood flows because of the pumping action of the heart. In the respiratory system, most lung tissue is thin exchange epithelium, so muscles of the thoracic cage and diaphragm must function as the pump. When these muscles contract, the lungs expand, held to the inside of the chest wall by the pleural fluid.

Breathing is an active process that uses muscle contraction to create a pressure gradient. The primary muscles involved in quiet breathing (breathing at rest) are the diaphragm, the external intercostals, and the scalenes. During forced breathing, other muscles of the chest and abdomen may be recruited to assist. Examples of physiological situations in which breathing is

forced include exercise, playing a wind instrument, and blowing up a balloon.

As noted earlier in the chapter, air flow in the respiratory tract obeys the same rule as blood flow:

$$\text{Flow} \propto \Delta P/R$$

RUNNING PROBLEM

Smokers usually develop chronic bronchitis before they develop emphysema. Cigarette smoke paralyzes the cilia that sweep debris and mucus out of the airways. Without the action of cilia, mucus and debris pool in the airways, leading to a chronic cough. Eventually, breathing becomes difficult.

Question 2:

Why do people with chronic bronchitis have a higher-than-normal rate of respiratory infections?

| 559 | 561 | **569** | 574 | 581 | 582 |

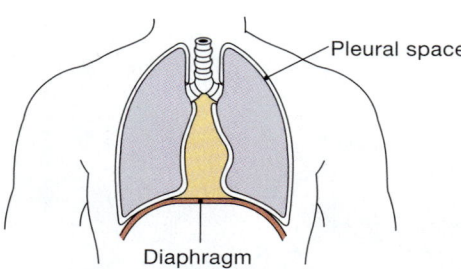

(a) At rest, diaphragm is relaxed.

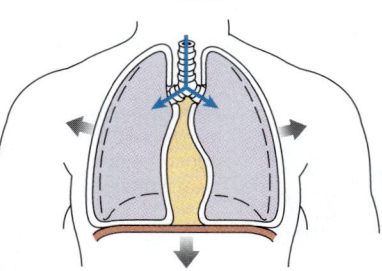

(b) Diaphragm contracts, thoracic volume increases.

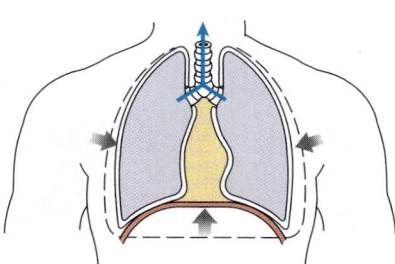

(c) Diaphragm relaxes, thoracic volume decreases.

Pleural space

Diaphragm

■ **FIGURE 17-9**   *Movement of the diaphragm*

This equation means that (1) air flows in response to a pressure gradient (ΔP) and (2) flow decreases as the resistance (R) of the system to flow increases. Before we discuss resistance, we will consider how the respiratory system creates a pressure gradient.

Pressures in the respiratory system can be measured either in the air spaces of the lungs (**alveolar pressure**, P_A) or in the pleural fluid (**intrapleural pressure**). Because atmospheric pressure is relatively constant, pressure in the lungs must be higher or lower than atmospheric pressure for air to flow between the atmosphere and the alveoli.

Air moves into the lungs when you inhale and out of the lungs when you exhale. A single **respiratory cycle** consists of an inspiration followed by an expiration. Because the respiratory system ends in a dead end, the direction of air flow must reverse. The pressure-volume relationships of Boyle's law provide the basis for pulmonary ventilation.

CONCEPT CHECK

16. Compare the direction of air movement during one respiratory cycle with the direction of blood flow during one cardiac cycle.

17. Explain the relationship between the lungs, the pleural membranes, the pleural fluid, and the thoracic cage.

Answers: p. 586

Inspiration Occurs When Alveolar Pressure Decreases

For air to move into the lungs, pressure inside the lungs must become lower than atmospheric pressure. According to Boyle's law, an increase in volume will create a decrease in pressure. During inspiration, thoracic volume increases when certain skeletal muscles of the rib cage and diaphragm contract.

When the diaphragm contracts, it loses its dome shape and drops down toward the abdomen. In quiet breathing, the diaphragm moves about 1.5 cm. This movement increases thoracic volume by flattening its floor (Fig. 17-9 ■). Contraction of the diaphragm causes between 60% and 75% of the inspiratory volume change during normal quiet breathing.

Movement of the rib cage creates the remaining 25–40% of the volume change. During inhalation, the external inter-

costal and scalene muscles (see Fig. 17-2b) contract and pull the ribs upward and out. Rib movement during inspiration has been likened to a pump handle lifting up and away from the pump (the ribs moving up and away from the spine; Fig. 17-10a ■) and to the movement of a bucket handle as it lifts away from the side of a bucket (ribs moving outward in a lateral direction; Fig. 17-10b). The combination of these two movements broadens the rib cage in all directions. As thoracic volume increases, pressure decreases, and air flows into the lungs.

(a) "Pump handle" motion increases anterior-posterior dimension of rib cage.

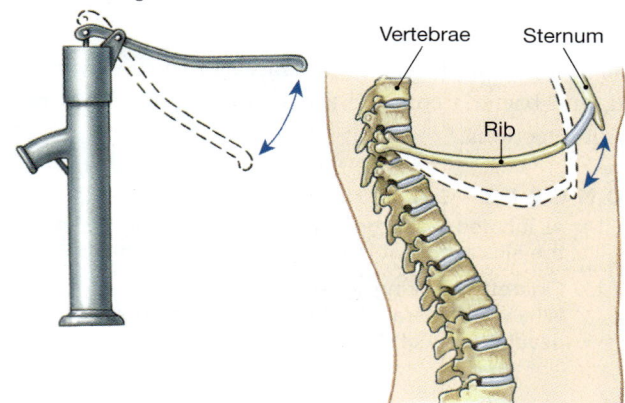

Vertebrae   Sternum   Rib

(b) "Bucket handle" motion increases lateral dimension of rib cage.

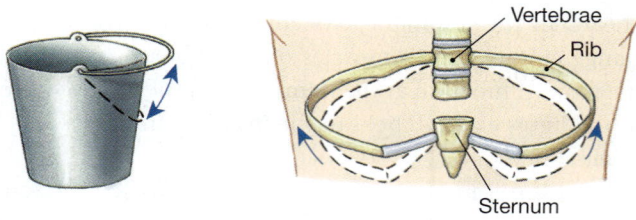

Vertebrae   Rib   Sternum

■ **FIGURE 17-10**   *Movement of the rib cage during inspiration*

Handles on a hand pump and a bucket serve as good models for rib movement during inspiration.

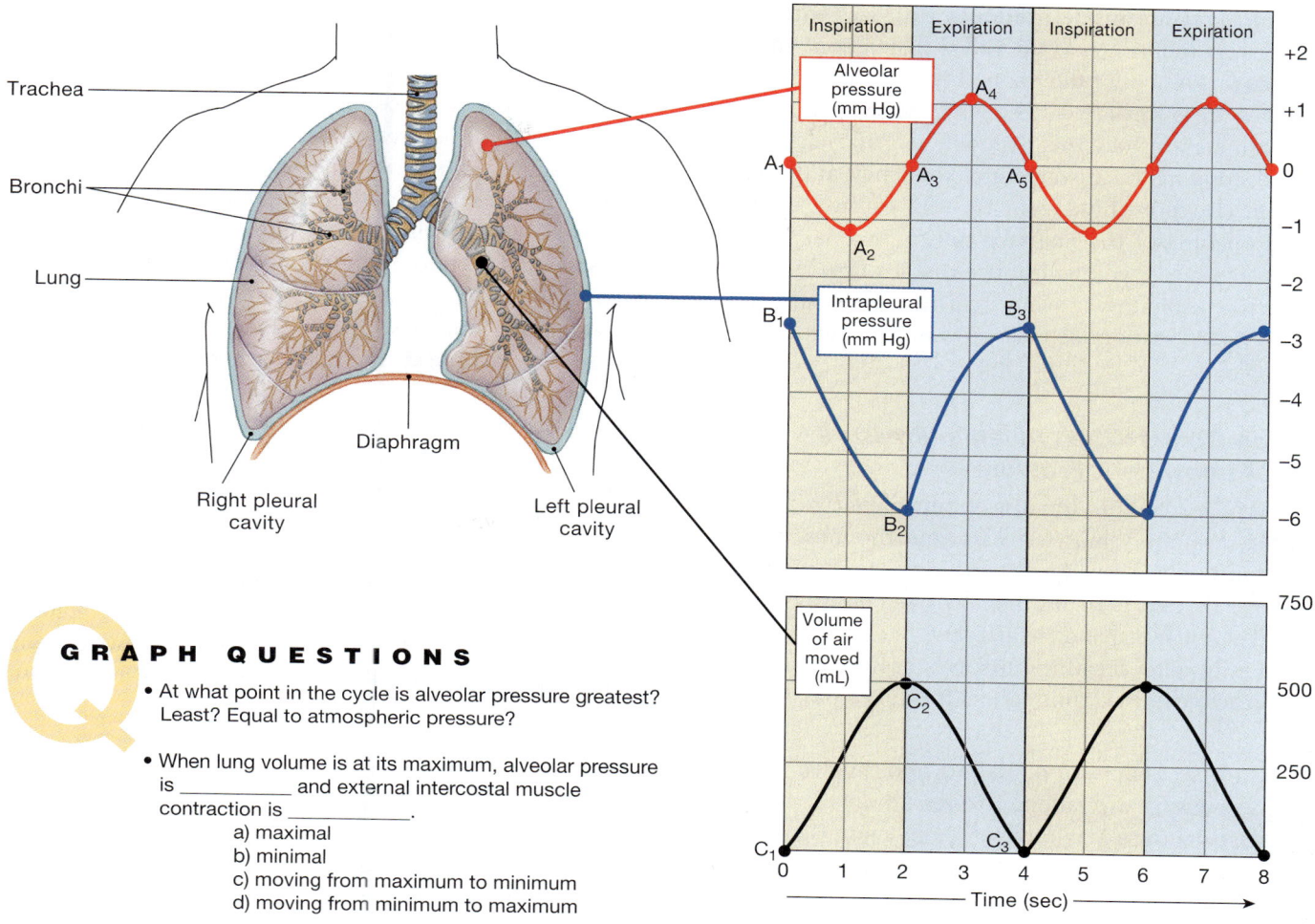

GRAPH QUESTIONS

- At what point in the cycle is alveolar pressure greatest? Least? Equal to atmospheric pressure?

- When lung volume is at its maximum, alveolar pressure is _____ and external intercostal muscle contraction is _____.
    - a) maximal
    - b) minimal
    - c) moving from maximum to minimum
    - d) moving from minimum to maximum

■ FIGURE 17-11 *Pressure changes during quiet breathing*

For many years, quiet breathing was attributed solely to the action of the diaphragm and the external intercostal muscles. It was thought that the scalenes and sternocleidomastoid muscles were active only during deep breathing. In recent years, however, studies have changed our understanding of how these accessory muscles contribute to quiet breathing. If an individual's scalenes are paralyzed, inspiration is achieved primarily by contraction of the diaphragm. Observation of patients with neuromuscular disorders has revealed that although the contracting diaphragm increases thoracic volume by moving toward the abdominal cavity, it also tends to pull the lower ribs inward, working against inspiration. In normal individuals, we know that the lower ribs move up and out during inspiration rather than in. The fact that there is no up-and-out rib motion in patients with paralyzed scalenes tells us that normally the scalenes must be contributing to inspiration by lifting the sternum and upper ribs.

New evidence also downplays the role of the external intercostal muscles during quiet breathing. However, the external intercostals play an increasingly important role as respiratory activity increases. Because the exact contribution of external

intercostals and scalenes varies depending on the type of breathing, we will group these muscles together and call them the *inspiratory muscles.*

Now let's see how alveolar pressure changes during a single inspiration. Follow the graphs in Figure 17-11 ■ as you read through the process.

Time 0. In the brief pause between breaths, alveolar pressure is equal to atmospheric pressure (0 mm Hg at point A_1). When pressures are equal, there is no air flow.

Time 0–2 sec: Inspiration. As inspiration begins, inspiratory muscles contract, and thoracic volume increases. With the increase in volume, alveolar pressure falls about 1 mm Hg below atmospheric pressure (-1 mm Hg, point A_2), and air flows into the alveoli (point C_1 to point C_2). Because the thoracic volume changes faster than air can flow, alveolar pressure reaches its lowest value about halfway through inspiration (point A_2).

As air continues to flow into the alveoli, pressure increases until the thoracic cage stops expanding, just before the end of inspiration. Air movement continues for a fraction of a second longer, until pressure inside the lungs equalizes with atmospheric pressure (point A_3). At the end of inspiration, lung volume is at its maximum for the respiratory cycle (point C_2), and alveolar pressure is equal to atmospheric pressure.

You can demonstrate this phenomenon by taking a deep breath and stopping the movement of your chest at the end of inspiration. (Do not "hold your breath" because doing so closes the opening of the pharynx and prevents air flow.) If you do this correctly, you will notice that air flow stops after you freeze the inspiratory movement. This exercise shows that at the end of inspiration, alveolar pressure is equal to atmospheric pressure.

Expiration Occurs When Alveolar Pressure Exceeds Atmospheric Pressure

At the end of inspiration, impulses from somatic motor neurons to the inspiratory muscles cease, and the muscles relax. Elastic recoil of the lungs and thoracic cage returns the diaphragm and rib cage to their original relaxed positions, just as a stretched elastic waistband recoils when released. Because expiration during quiet breathing involves passive elastic recoil rather than active muscle contraction, it is called **passive expiration**.

Time 2–4 sec: expiration. As lung and thoracic volumes decrease during expiration, air pressure in the lungs increases, reaching a maximum of about 1 mm Hg above atmospheric pressure (Fig. 17-11, point A_4). Alveolar pressure is now higher than atmospheric pressure, so air flow reverses and air moves out of the lungs.

Time 4 sec. At the end of expiration, air movement ceases when alveolar pressure is again equal to atmospheric pressure (point A_5). Lung volume reaches its minimum for the respiratory cycle (point C_3). At this point, the respiratory cycle has ended and is ready to begin again with the next breath.

The pressure differences shown in Figure 17-11 apply to quiet breathing. During exercise or forced heavy breathing, these values will become proportionately larger. **Active expiration** occurs during voluntary exhalations and when ventilation exceeds 30–40 breaths per minute. (Normal resting ventilation rate is 12–20 breaths per minute for an adult.) Active expiration uses the internal intercostal muscles and the abdominal muscles (see Fig. 17-2b), which are not used during inspiration. These muscles are collectively called the *expiratory muscles*.

The internal intercostal muscles line the inside of the rib cage. When they contract, they pull the ribs inward, reducing the volume of the thoracic cavity. To feel this action, place your hands on your rib cage. Forcefully blow as much air out of your lungs as you can, noting the movement of your hands as you do so.

The internal and external intercostals function as antagonistic muscle groups [🔁 p. 398] to alter the position and volume

of the rib cage during ventilation. The diaphragm, however, has no antagonistic muscles. Instead, abdominal muscles contract during active expiration to supplement the activity of the internal intercostals. Abdominal contraction pulls the lower rib cage inward and decreases abdominal volume, actions that displace the intestines and liver upward. The displaced viscera push the diaphragm up into the thoracic cavity and passively decrease chest volume even more. The action of abdominal muscles during forced expiration is why aerobics instructors tell you to blow air out as you lift your head and shoulders during abdominal "crunches." The active process of blowing air out helps contract the abdominals, the very muscles you are trying to strengthen.

Any neuromuscular disease that weakens skeletal muscles or damages their motor neurons can adversely affect ventilation. With decreased ventilation, less fresh air enters the lungs. In addition, loss of the ability to cough increases the risk of pneumonia and other infections. Examples of diseases that affect the motor control of ventilation include *myasthenia gravis* [🔁 p. 274], an illness in which acetylcholine receptors of the motor end plates of skeletal muscles are destroyed, and *polio* (poliomyelitis), a viral illness that paralyzes skeletal muscles.

✓ CONCEPT CHECK

18. Scarlett O'Hara is trying to squeeze herself into a corset with an 18-inch waist. Will she be more successful by taking a deep breath and holding it or by blowing all the air out of her lungs? Why?

19. Why would loss of the ability to cough increase the risk of respiratory infections? (*Hint:* what does coughing do to mucus in the airways?) Answers: p. 586

Intrapleural Pressure Changes During Ventilation

Ventilation requires that the lungs, which are unable to expand and contract on their own, move in association with the contraction and relaxation of the thorax. As we noted earlier in this chapter, the lungs are "stuck" to the thoracic cage by cohesive forces exerted by the fluid between the two pleural membranes. Thus, if the thoracic cage moves, the lungs move with it.

The intrapleural pressure in the fluid between the pleural membranes is normally subatmospheric. This subatmospheric pressure arises during fetal development, when the thoracic cage with its associated pleural membrane grows more rapidly than the lung with its associated pleural membrane. The two pleural membranes are held together by the pleural fluid bond, so the elastic lungs are forced to stretch to conform to the larger volume of the thoracic cavity. At the same time, however, elastic recoil of the lungs creates an inwardly directed force that tends to pull the lungs away from the chest wall (Fig. 17-12a ■). The combination of the outward pull of the thoracic cage and inward recoil of the elastic lungs creates a subatmospheric intrapleural pressure of about −3 mm Hg.

You can create a similar situation by half-filling a syringe with water and capping it with a plugged-up needle. At this

(a) Normal lung at rest

(b) Pneumothorax

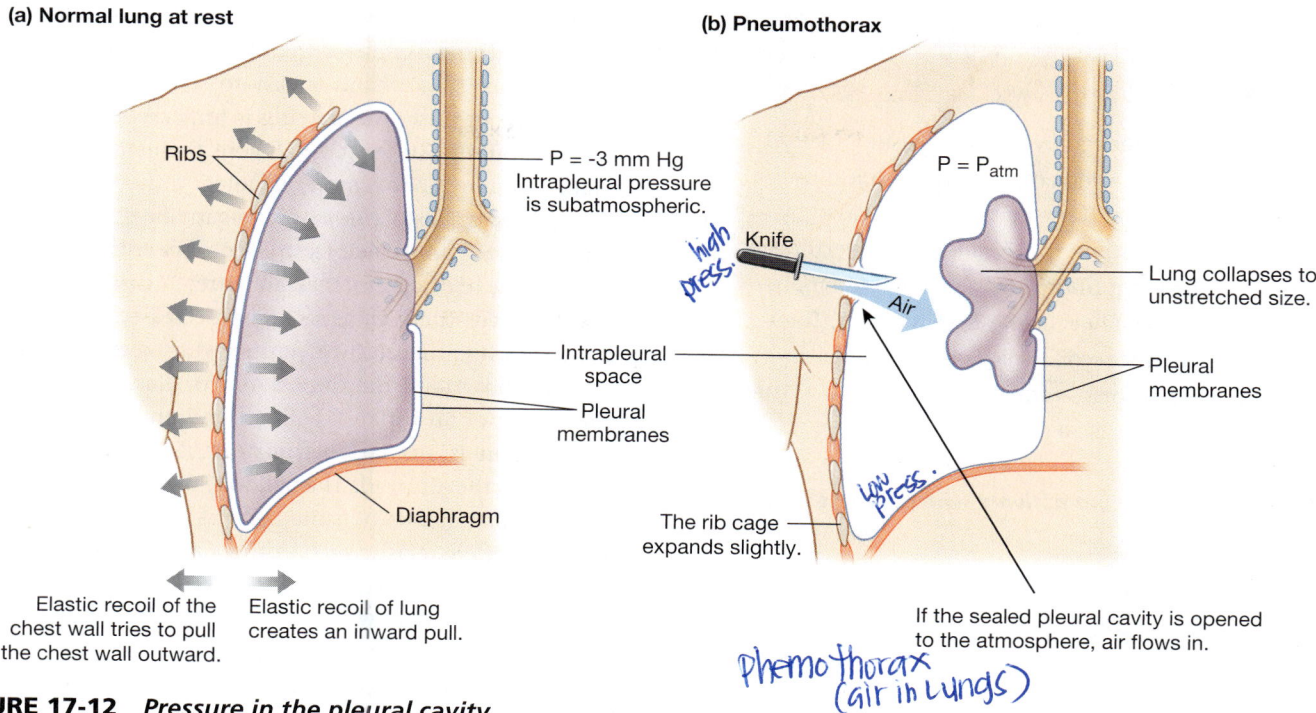

- Ribs
- P = -3 mm Hg Intrapleural pressure is subatmospheric.
- *high press.*
- Knife
- Air
- Intrapleural space
- Pleural membranes
- Diaphragm
- *low press.*
- The rib cage expands slightly.

Elastic recoil of the chest wall tries to pull the chest wall outward.

Elastic recoil of lung creates an inward pull.

P = P_atm

Lung collapses to unstretched size.

Pleural membranes

If the sealed pleural cavity is opened to the atmosphere, air flows in.

Pneumothorax (air in lungs)

■ **FIGURE 17-12** *Pressure in the pleural cavity*

point, the pressure inside the barrel is equal to atmospheric pressure. Now hold the syringe barrel (the chest wall) in one hand while you try to withdraw the plunger (the elastic lung pulling away from the chest wall). As you pull on the plunger, the volume inside the barrel increases slightly, but the cohesive forces between the water molecules cause the water to resist expansion. The pressure in the barrel, which was initially equal to atmospheric pressure, decreases slightly as you pull on the plunger. If you release the plunger, it snaps back to its resting position, restoring atmospheric pressure inside the syringe.

What happens to subatmospheric intrapleural pressure if an opening is made between the sealed pleural cavity and the atmosphere? A knife thrust between the ribs, a broken rib that punctures the pleural membrane, or any other event that opens the pleural cavity to the atmosphere will allow air to flow in down its pressure gradient, just as air enters when you break the seal on a vacuum-packed can.

Air in the pleural cavity breaks the fluid bond holding the lung to the chest wall. The chest wall expands outward while the elastic lung collapses to an unstretched state, like a deflated balloon (Fig. 17-12b). This condition, called **pneumothorax** [*pneuma*, air + *thorax*, chest], results in a collapsed lung that is unable to function normally. Pneumothorax can also occur spontaneously if a congenital *bleb* (or weakened section of lung tissue) ruptures, allowing air from inside the lung to enter the pleural cavity.

Correction of a pneumothorax has two components: removing as much air from the pleural cavity as possible with a suction pump, and sealing the hole to prevent more air from entering. Any air remaining in the cavity will gradually be absorbed

into the blood, restoring the pleural fluid bond and reinflating the lung.

Pressures in the pleural fluid vary during a respiratory cycle. At the beginning of inspiration, intrapleural pressure is about -3 mm Hg (Fig. 17-11, point B_1). As inspiration proceeds, the pleural membranes and lungs follow the expanding thoracic cage because of the pleural fluid bond, but the elastic lung tissue resists being stretched. The lungs attempt to pull farther away from the chest wall, causing the intrapleural pressure to become even more negative (Fig. 17-11, point B_2).

Because this process is difficult to visualize, return to the analogy of the water-filled syringe with the plugged-up needle. You can pull the plunger out a small distance without much effort, but the cohesiveness of the water makes it difficult to pull the plunger out any farther. The increased amount of work you do trying to pull the plunger out is paralleled by the work your inspiratory muscles must do when they contract during inspiration. The bigger the breath, the more work is required to stretch the elastic lung.

By the end of a quiet inspiration, when the lungs are fully expanded, intrapleural pressure falls to around -6 mm Hg (Fig. 17-11, point B_2). During exercise or other powerful inspirations, intrapleural pressure may reach -8 mm Hg.

During expiration, the thoracic cage returns to its resting position. The lungs are released from their stretched position, and the intrapleural pressure returns to its normal value of about -3 mm Hg (point B_3). Notice that intrapleural pressure never equilibrates with atmospheric pressure because the pleural cavity is a closed compartment.

17

RUNNING PROBLEM

Emphysema is characterized by a loss of elastin, the elastic fibers that help the alveoli recoil during expiration. Elastin is destroyed by elastase, an enzyme released by cells of the immune system, which must work overtime in smokers to rid the lungs of irritants. People with emphysema have more difficulty exhaling than inhaling. Their alveoli have lost elastic recoil, which makes expiration—normally a passive process—require conscious effort. They literally must work to push air out of their lungs.

Question 3:
Name the muscles that patients with emphysema use to exhale actively.

| 559 | 561 | 569 | **574** | 581 | 582 |

Pressure gradients required for air flow are created by the work of skeletal muscle contraction. Normally, about 3–5% of the body's energy expenditure is used for quiet breathing. During exercise, the energy required for breathing increases substantially. The two factors that have the greatest influence on the amount of work needed for breathing are the stretchability of the lungs and the resistance of the airways to air flow.

CONCEPT CHECK

20. A person has periodic spastic contractions of the diaphragm, otherwise known as hiccups. What happens to intrapleural and alveolar pressures when a person hiccups?

21. A stabbing victim is brought to the emergency room with a knife wound between the ribs on the left side of his chest. What has probably happened to his left lung? To his right lung? Why does the left side of his rib cage seem larger than the right side?   Answers: p. 586

Lung Compliance and Elastance May Change in Disease States

Adequate ventilation depends on the ability of the lungs to expand normally. Most of the work of breathing goes into overcoming the resistance of the elastic lungs and the thoracic cage to stretching. Clinically, the ability of the lung to stretch is called **compliance**. A high-compliance lung stretches easily, just as a compliant person is easy to persuade. A low-compliance lung requires more force from the inspiratory muscles to stretch it.

Compliance is different from **elastance** (elasticity). The fact that a lung stretches easily (high compliance) does not necessarily mean that it will return to its resting volume when the stretching force is released (elastance). You may have experienced something like this with old gym shorts. After many washings the elastic waistband is easy to stretch (high compliance) but lacking in elastance, making it impossible for the shorts to stay up around your waist. Analogous problems occur in the respiratory system. For example, as noted in the Running Problem, emphysema is a disease in which elastin fibers normally found in lung tissue are destroyed. Destruction of elastin results in lungs that exhibit high compliance and stretch easily during inspiration. However, these lungs also have decreased elastance, so they do not recoil to their resting position during expiration.

To understand the importance of elastic recoil to expiration, think of an inflated balloon and an inflated plastic bag. The balloon is similar to the normal lung. Its elastic walls squeeze on the air inside the balloon, thereby increasing the internal air pressure. When the neck of the balloon is opened to the atmosphere, elastic recoil causes air to flow out of the balloon. The inflated plastic bag, on the other hand, is like the lung of an individual with emphysema. It has high compliance and is easily inflated, but it has little elastic recoil. If the inflated plastic bag is opened to the atmosphere, most of the air remains inside the bag.

CLINICAL FOCUS

FIBROTIC LUNG DISEASE

One type of lung disease results from chronic inhalation of fine particulate matter that escapes the mucus lining the airways and reaches the exchange epithelium of the alveoli. The only protective mechanism in that region of the respiratory system is removal by wandering alveolar macrophages (see Fig. 17-2g). These phagocytic cells patrol the alveoli, engulfing any airborne particles that reach them. If the particles are organic, the macrophages digest them with lysosomal enzymes. However, if the particles cannot be digested or if they accumulate in large numbers, an inflammatory process ensues. Intracellular accumulation of particles causes the macrophage to secrete growth factors that stimulate fibroblasts in the lung's connective tissue. These fibroblasts produce collagen that forms inelastic, fibrous scar tissue. Large amounts of scar tissue reduce the compliance of the lung and result in **fibrotic lung disease**, or **fibrosis**. Particles that can trigger fibrosis include asbestos, coal dust, silicon, and even dust, paper particles, and industrial pollutants.

(a) Pressure is greater in the smaller bubble.

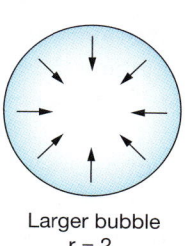

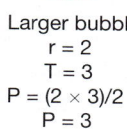

Larger bubble
r = 2
T = 3
P = (2 × 3)/2
P = 3

Smaller bubble
r = 1
T = 3
P = (2 × 3)/1
P = 6

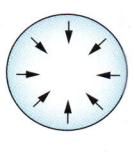

Law of LaPlace: P = 2T/r

P = pressure
T = surface tension
r = radius

According to the law of LaPlace, if two bubbles have the same surface tension, the smaller bubble will have higher pressure.

(b) Surfactant reduces surface tension (T). Pressure is equalized in the large and small bubbles.

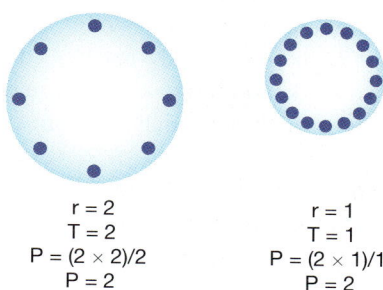

r = 2
T = 2
P = (2 × 2)/2
P = 2

r = 1
T = 1
P = (2 × 1)/1
P = 2

■ **FIGURE 17-13**   *The law of LaPlace*

(a) Because r is in the denominator in the LaPlace equation, the pressure inside a bubble increases as bubble radius decreases. **(b)** The presence of surfactant lowers the surface tension in a film. In this way, surfactant helps equalize the interior pressure in bubbles of different sizes.

A decrease in lung compliance affects ventilation because more work must be expended to stretch a stiff lung. Pathological conditions in which compliance is reduced are called **restrictive lung diseases.** In these conditions, the energy expenditure required to stretch less-compliant lungs can far exceed the normal work of breathing. Two common causes of decreased compliance are inelastic scar tissue formed in *fibrotic lung diseases,* and inadequate alveolar production of surfactant, a chemical that facilitates lung expansion.

Surfactant Decreases the Work of Breathing

For years, physiologists assumed that elastin and other elastic fibers were the primary source of resistance to stretch in the lung. However, studies comparing the work required to expand air-filled and saline-filled lungs showed that air-filled lungs are much harder to inflate. From this result, researchers concluded that lung tissue itself contributes less to resistance than once thought. Some other property of the normal air-filled lung, a property not present in the saline-filled lung, must create most of the resistance to stretch.

This property is the surface tension [🔗 p. 25] created by the thin fluid layer between the alveolar cells and the air. At any air-fluid interface, the surface of the fluid is under tension, like a thin membrane being stretched. When the fluid is water, surface tension arises because of the hydrogen bonds between water molecules. The water molecules on the fluid's surface are attracted to other water molecules beside and beneath them but are not attracted to gases in the air at the air-fluid interface.

Alveolar surface tension is similar to that which exists in a spherical bubble. The surface tension created by the thin film of fluid is directed toward the center of the bubble and creates pressure in the interior (Fig. 17-13a ■). The **law of LaPlace** is an expression of this pressure. It states that the pressure (P) inside

a bubble formed by a fluid film is a function of two factors: the surface tension of the fluid (T) and the radius of the bubble (r). This relationship is expressed by the equation

$$P = 2T/r$$

If two bubbles have different diameters but are formed by fluids that have the same surface tension, the pressure inside the smaller bubble is greater than that inside the larger bubble (Fig. 17-13a).

How does this apply to the lung? In physiology, we equate the bubble to a fluid-lined alveolus (although alveoli are not perfect spheres). The fluid lining all the alveoli creates surface tension. If the surface tension (T) of the fluid were the same in small and large alveoli, small alveoli would have higher inwardly-directed pressure than larger alveoli, and increased resistance to stretch. As a result, more work would be needed to expand smaller alveoli.

Normally, however, our lungs secrete a surfactant that reduces surface tension. Surfactants ("*surf*ace *act*ive age*nts*") are molecules that disrupt cohesive forces between water molecules by substituting themselves for water at the surface. For example, that product you add to your dishwasher to aid in the rinse cycle is a surfactant that keeps the rinse water from beading up on the dishes (and forming "spots" when the water beads dry). In the lungs, surfactant decreases surface tension of the alveolar fluid and thereby decreases resistance of the lung to stretch.

Surfactant is more concentrated in smaller alveoli, making their surface tension less than that in larger alveoli (Fig. 17-13b). Lower surface tension helps equalize the pressure among alveoli of different sizes and makes it easier to inflate the smaller alveoli. With lower surface tension, the work needed to expand the alveoli with each breath is greatly reduced. Human surfactant is a mixture containing proteins and phospholipids, such as

17

dipalmitoylphosphatidylcholine, which are secreted into the alveolar air space by type II alveolar cells (see Fig. 17-2g).

Normally, surfactant synthesis begins about the twenty-fifth week of fetal development under the influence of various hormones. Production usually reaches adequate levels by the thirty-fourth week (about six weeks before normal delivery). Babies who are born prematurely without adequate concentrations of surfactant in their alveoli develop *newborn respiratory distress syndrome (RDS).* In addition to "stiff" (low-compliance) lungs, RDS babies also have alveoli that collapse each time they exhale. These infants must use a tremendous amount of energy to expand their collapsed lungs with each breath. Unless treatment is initiated rapidly, about 50% of these infants die. In the past, all physicians could do for RDS babies was administer oxygen. Today, however, the prognosis for RDS babies is much better. Amniotic fluid can be sampled to assess whether or not the fetal lungs are producing adequate amounts of surfactant. If they are not, and if delivery cannot be delayed, RDS babies can be treated with aerosol administration of artificial surfactant until the lungs mature enough to produce their own. The current treatment also includes artificial ventilation that forces air into the lungs and keeps the alveoli open.

Airway Diameter Is the Primary Determinant of Airway Resistance

The other factor besides compliance that influences the work of breathing is the resistance of the respiratory system to air flow. Resistance in the respiratory system is similar in many ways to resistance in the cardiovascular system [🔁 p. 462]. Three parameters contribute to resistance (R): the system's length (L), the viscosity of the substance flowing through the system (η), and the radius of the tubes in the system (r). As with flow in the cardiovascular system, Poiseuille's law relates these factors to one another:

$$R \propto L\eta/r^4$$

Because the length of the respiratory system is constant, we can ignore L in the equation. The viscosity of air is almost constant, although you may have noticed that it feels harder to breathe in a sauna filled with steam than in a room with normal humidity. Water droplets in the steam increase the viscosity of the steamy air, thereby increasing its resistance to flow. Viscosity also changes slightly with atmospheric pressure, decreasing as pressure decreases. A person at high altitude may feel less resistance to air flow than a person at sea level. Despite these exceptions, viscosity plays a very small role in resistance to air flow.

Because length and viscosity are essentially constant for the respiratory system, the radius (or diameter) of the airways becomes the primary determinant of airway resistance. Normally, however, the work needed to overcome resistance of the airways to air flow is much less than the work needed to overcome the resistance of the lungs and thoracic cage to stretch.

Nearly 90% of airway resistance normally can be attributed to the trachea and bronchi, rigid structures with the smallest

total cross-sectional area. Because these structures are supported by cartilage and bone, their diameters normally do not change, and their resistance to air flow is constant. However, mucus accumulation from allergies or infections can dramatically increase resistance. If you have ever tried breathing through your nose when you have a cold, you can appreciate how the narrowing of an upper airway limits air flow!

The bronchioles normally do not contribute significantly to airway resistance because their total cross-sectional area is about 2000 times that of the trachea. Because the bronchioles are collapsible tubes, however, a decrease in their diameter can suddenly turn them into a significant source of airway resistance. **Bronchoconstriction** increases resistance to air flow and decreases the amount of fresh air that reaches the alveoli.

Bronchioles, like arterioles, are subject to reflex control by the nervous system and by hormones. However, most minute-to-minute changes in bronchiolar diameter occur in response to paracrines. Carbon dioxide in the airways is the primary paracrine that affects bronchiolar diameter. Increased CO_2 in expired air relaxes bronchiolar smooth muscle and causes **bronchodilation**.

Histamine is a paracrine that acts as a powerful bronchoconstrictor. This chemical is released by *mast cells* [🔁 p. 538] in response to either tissue damage or allergic reactions. In severe allergic reactions, large amounts of histamine may lead to widespread bronchoconstriction and difficult breathing. Immediate medical treatment in these patients is imperative.

The primary neural control of bronchioles comes from parasympathetic neurons that cause bronchoconstriction, a reflex designed to protect the lower respiratory tract from inhaled irritants. There is no significant sympathetic innervation of the bronchioles in humans. However, smooth muscle in the bronchioles is well supplied with β_2-receptors that respond to epinephrine. Stimulation of β_2-receptors relaxes airway smooth muscle and results in bronchodilation. This reflex is used therapeutically in the treatment of asthma and various allergic reactions characterized by histamine release and bronchoconstriction. Table 17-3 ■ summarizes the factors that alter airway resistance.

✔ CONCEPT CHECK

22. In a normal person, which contributes more to the work of breathing: airway resistance or lung and chest wall compliance?

23. Coal miners who spend years inhaling fine coal dust have much of their alveolar surface area covered with scarlike tissue. What happens to their lung compliance as a result?

24. How does the work required for breathing change when surfactant is not present in the lungs?

25. A cancerous lung tumor has grown into the walls of a group of bronchioles, narrowing their lumens. What has happened to the resistance to air flow in these bronchioles?

26. Name the neurotransmitter and receptor for parasympathetic bronchoconstriction.   *Answers: p. 586*

TABLE 17-3	**Factors That Affect Airway Resistance**	
FACTOR	**AFFECTED BY**	**MEDIATED BY**
Length of the system	Constant; not a factor	
Viscosity of air	Usually constant; humidity and altitude may alter slightly	
Diameter of airways		
Upper airways	Physical obstruction	Mucus and other factors
Bronchioles	Bronchoconstriction	Parasympathetic neurons (muscarinic receptors), histamine, leukotrienes
	Bronchodilation	Carbon dioxide, epinephrine (β_2-receptors)

Rate and Depth of Breathing Determine the Efficiency of Breathing

You may recall that the efficiency of the heart is measured by the cardiac output, which is calculated by multiplying heart rate by stroke volume. Likewise, we can estimate the effectiveness of ventilation by calculating **total pulmonary ventilation**, the volume of air moved into and out of the lungs each minute. Total pulmonary ventilation, also known as the *minute volume*, is calculated as follows:

total pulmonary ventilation = ventilation rate × tidal volume

The normal ventilation rate for an adult is 12–20 breaths per minute. Using the average tidal volume (500 mL) and the slowest ventilation rate, we get:

total pulmonary ventilation =

12 breaths/min × 500 mL/breath
= 6000 mL/min = 6 L/min

Total pulmonary ventilation represents the physical movement of air into and out of the respiratory tract, but is it a good indicator of how much fresh air reaches the alveolar exchange surface? Not necessarily.

Some air that enters the respiratory system does not reach the alveoli because part of every breath remains in the conducting airways, such as the trachea and bronchi. Because the conducting airways do not exchange gases with the blood, they are known as the **anatomic dead space**. Anatomic dead space averages about 150 mL.

To illustrate the difference between the total volume of air that enters the airways and the volume of fresh air that reaches the alveoli, let's consider a typical breath that moves 500 mL of air during a respiratory cycle (Fig. 17-14 ■).

1. At the end of an inspiration, lung volume is maximal, and fresh air fills the dead space.

2. The tidal volume of 500 mL is exhaled. However, the first portion of this 500 mL to exit the airways is the 150 mL of fresh air that had been in the dead space, followed by 350 mL of "stale" air from the alveoli. Thus even though 500 mL of air exited the alveoli, only 350 mL of that volume left the body. The remaining 150 mL of "stale" alveolar air stays in the dead space.

3. At the end of expiration, lung volume is at its minimum, and stale air from the most recent expiration fills the anatomic dead space.

4. With the next inspiration, 500 mL of fresh air enters the airways. The entering air returns the 150 mL of stale air in the anatomic dead space to the alveoli, followed by the first 350 mL of the fresh air. The last 150 mL of inspired fresh air again remains in the dead space and never reaches the alveoli.

Thus, although 500 mL of air entered the alveoli, only 350 mL of that volume was fresh air. The fresh air entering the alveoli equals the tidal volume minus the dead space volume.

Because a significant portion of inspired air never reaches an exchange surface, a more accurate indicator of ventilation efficiency is **alveolar ventilation**, the amount of fresh air that reaches the alveoli each minute. Alveolar ventilation is calculated by multiplying ventilation rate by the volume of fresh air that reaches the alveoli:

alveolar ventilation =

ventilation rate × (tidal volume − dead space)

Using the same ventilation rate and tidal volume as before, and a dead space of 150 mL, then

alveolar ventilation =

12 breaths/min × (500 − 150 mL/breath)
= 4200 mL/min

Thus, at 12 breaths per minute, the alveolar ventilation is 4.2 L/min. Although 6 L/min of fresh air enters the respiratory system, only 4.2 L reaches the alveoli.

Alveolar ventilation can be drastically affected by changes in the rate or depth of breathing. Table 17-4 ■ shows that three people can have the same total pulmonary ventilation but dramatically different alveolar ventilation. **Maximum voluntary ventilation**, which involves breathing as deeply and quickly as possible, may increase total pulmonary ventilation to as much as 170 L/min. Table 17-5 ■ describes various patterns of ventilation, and Table 17-6 ■ gives normal ventilation values.

17

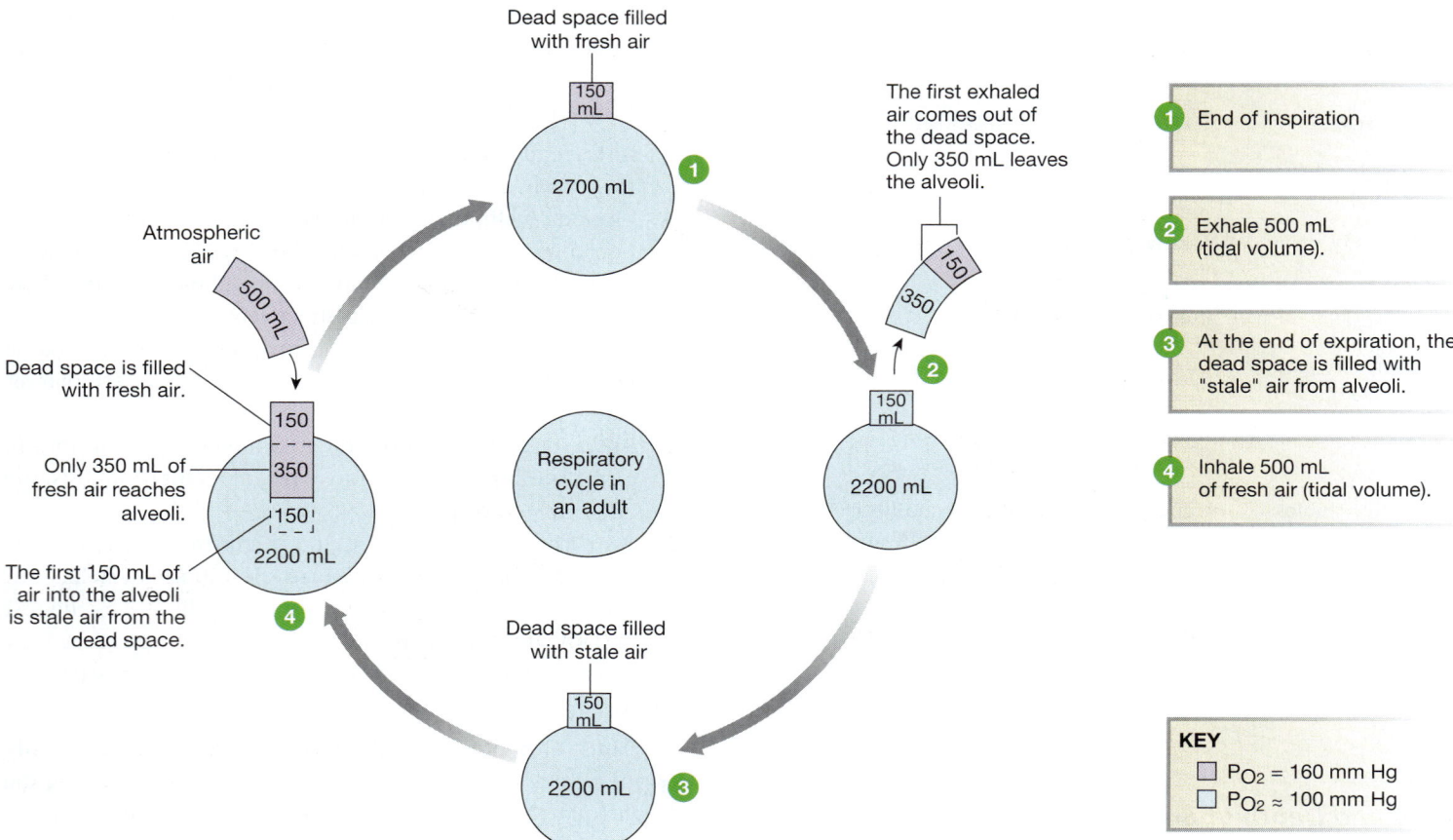

■ **FIGURE 17-14** *Total pulmonary ventilation and alveolar ventilation*

In this example, total pulmonary ventilation is 500 mL/breath × ventilation rate but alveolar ventilation is only 350 mL/breath × ventilation rate.

Gas Composition in the Alveoli Varies Little During Normal Breathing

How much can a change in alveolar ventilation affect the amount of fresh air and oxygen that reach the alveoli? Fig. 17-15 ■ shows how the partial pressures P_{O_2} and P_{CO_2} in the alveoli vary with hyper- and hypoventilation. As alveolar ventilation increases above normal levels (**hyperventilation**), alveolar P_{O_2} rises to about 120 mm Hg, and alveolar P_{CO_2} falls to around 20 mm Hg. During **hypoventilation**, when less fresh air enters the alveoli, alveolar P_{O_2} decreases and alveolar P_{CO_2} increases.

Although a dramatic change in alveolar ventilation pattern will affect gas partial pressures in the alveoli, the P_{O_2} and

TABLE 17-4 Effects of Breathing Pattern on Alveolar Ventilation

TIDAL VOLUME (ML)	VENTILATION RATE (BREATHS/MIN)	TOTAL PULMONARY VENTILATION (ML/MIN)	FRESH AIR TO ALVEOLI (ML) (TIDAL VOLUME–DEAD SPACE VOLUME*)	ALVEOLAR VENTILATION (ML/MIN)
500 (normal)	12 (normal)	6000	350	4200
300 (shallow)	20 (rapid)	6000	150	3000
750 (deep)	8 (slow)	6000	600	4800

*Dead space volume is assumed to be 150 mL.

TABLE 17-5   Types and Patterns of Ventilation

NAME	DESCRIPTION	EXAMPLES
Eupnea	Normal quiet breathing	
Hyperpnea	Increased respiratory rate and/or volume in response to increased metabolism	Exercise
Hyperventilation	Increased respiratory rate and/or volume without increased metabolism	Emotional hyperventilation; blowing up a balloon
Hypoventilation	Decreased alveolar ventilation	Shallow breathing; asthma; restrictive lung disease
Tachypnea	Rapid breathing; usually increased respiratory rate with decreased depth	Panting
Dyspnea	Difficulty breathing (a subjective feeling sometimes described as "air hunger")	Various pathologies or hard exercise
Apnea	Cessation of breathing	Voluntary breath-holding; depression of CNS control centers

P_{CO_2} in the alveoli change surprisingly little during normal quiet breathing. Alveolar P_{O_2} is fairly constant at 100 mm Hg, and alveolar P_{CO_2} stays close to 40 mm Hg.

Intuitively, you might think that P_{O_2} would increase when fresh air first enters the alveoli, then decrease steadily as oxygen leaves to enter the blood. Instead, we find only very small swings in P_{O_2}. Why? The reasons are that (1) the amount of oxygen that enters the alveoli with each breath is roughly equal to the amount of oxygen that enters the blood, and (2) the amount of fresh air that enters the lungs with each breath is only a little more than 10% of the total lung volume at the end of inspiration.

CONCEPT CHECK

27. If a person increased his tidal volume, what would happen to his alveolar P_{O_2}?

28. If his breathing rate increased, what would happen to his alveolar P_{O_2}?

Answers: p. 586

TABLE 17-6   Normal Ventilation Values in Pulmonary Medicine

Total pulmonary ventilation	6 L/min
Total alveolar ventilation	4.2 L/min
Maximum voluntary ventilation	125–170 L/min
Respiration rate	12–20 breaths/min

Ventilation and Alveolar Blood Flow Are Matched

Moving oxygen from the atmosphere to the alveolar exchange surface is only the first step in external respiration. Next, gas exchange must occur across the alveolar-capillary interface.

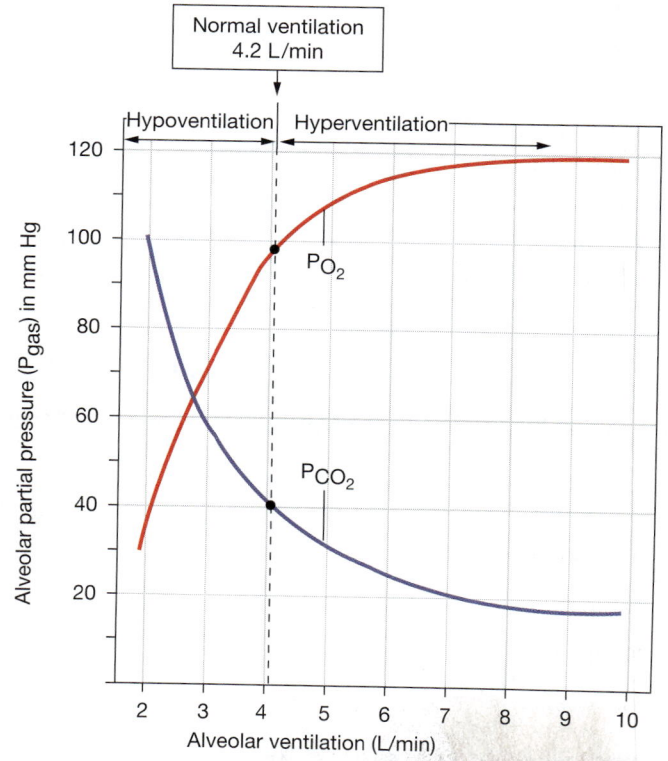

FIGURE 17-15 *Effect of changing alveolar ventilation on P_{O_2} and P_{CO_2} in the alveoli*

Finally, blood flow (*perfusion*) past the alveoli must be high enough to pick up the available oxygen. Matching the ventilation rate into groups of alveoli with blood flow past those alveoli is a two-part process involving local regulation of both air flow and blood flow.

Alterations in pulmonary blood flow depend almost exclusively on properties of the capillaries and on such local factors as the concentrations of oxygen and carbon dioxide in the lung tissue. Capillaries in the lungs are unusual because they are collapsible. If the pressure of blood flowing through the capillaries falls below a certain point, the capillaries close off, diverting blood to pulmonary capillary beds in which blood pressure is higher.

In a person at rest, some capillary beds in the apex (top) of the lung are closed off because of low hydrostatic pressure. Capillary beds at the base of the lung have higher hydrostatic pressure because of gravity and thus remain open. Consequently, blood flow is diverted toward the base of the lung. During exercise, when blood pressure rises, the closed apical capillary beds open, ensuring that the increased cardiac output will be fully oxygenated as it passes through the lungs. The ability of the lungs to recruit additional capillary beds during exercise is an example of the reserve capacity of the body.

At the local level, the body attempts to match air flow and blood flow in each section of the lung by regulating the diameters of the arterioles and bronchioles. Bronchiolar diameter is mediated primarily by CO_2 levels in exhaled air passing through them (Table 17-7 ■). An increase in the P_{CO_2} of expired air causes bronchioles to dilate. A decrease in the P_{CO_2} of expired air causes bronchioles to constrict.

Although there is some autonomic innervation of pulmonary arterioles, there is apparently little neural control of pulmonary blood flow. The resistance of pulmonary arterioles to blood flow is regulated primarily by the oxygen content of the interstitial fluid around the arteriole (Fig. 17-16 ■). If ventilation of alveoli in one area of the lung is diminished, the P_{O_2} in that area decreases, and the arterioles respond by constricting (Fig. 17-16c). This local vasoconstriction is adaptive because it diverts blood away from the underventilated region to better-ventilated parts of the lung.

On the other hand, if P_{O_2} in some region of the lung becomes higher than normal, the alveoli in that region are being overventilated relative to the blood flow past them. The arterioles supplying those alveoli dilate to bring in additional blood to pick up the extra oxygen.

Note that constriction of pulmonary arterioles in response to low P_{O_2} is the opposite of what occurs in the systemic circulation [🔁 p. 511]. In the systemic circulation, a decrease in the P_{O_2} of a tissue causes local arterioles to dilate, bringing more blood to those tissues that are consuming oxygen.

Another important point must be noted here. Local control mechanisms are not effective regulators of air and blood flow under all circumstances. If blood flow is blocked in one pulmonary artery, or if air flow is blocked at the level of the larger airways, local responses that shunt air or blood to other parts of the lung are ineffective because in these cases no part of the lung has normal ventilation or perfusion.

CONCEPT CHECK

29. If a lung tumor decreases blood flow in one small section of the lung to a minimum, what happens to P_{O_2} in the alveoli in that section and in the surrounding interstitial fluid? What happens to P_{CO_2} in that section? What is the compensatory response of the bronchioles in the affected section? Will the compensation bring ventilation in the affected section of the lung back to normal? Explain. *Answers: p. 586*

Auscultation and Spirometry Assess Pulmonary Function

Most pulmonary function tests are relatively simple to perform. Auscultation of breath sounds is an important diagnostic technique in pulmonary medicine, just as auscultation of heart sounds is an important technique in cardiovascular diagnosis [🔁 p. 486]. Breath sounds are more complicated to interpret than heart sounds, however, because breath sounds have a wider range of normal variation.

TABLE 17-7   Local Control of Arterioles and Bronchioles by Oxygen and Carbon Dioxide

GAS COMPOSITION	BRONCHIOLES	PULMONARY ARTERIOLES	SYSTEMIC ARTERIOLES
P_{CO_2} increases	Dilate	(Constrict)*	Dilate
P_{CO_2} decreases	Constrict	(Dilate)	Constrict
P_{O_2} increases	(Constrict)	Dilate	Constrict
P_{O_2} decreases	(Dilate)	Constrict	Dilate

*Responses in parentheses indicate weak responses.

(a) Ventilation in alveoli is matched to perfusion through pulmonary capillaries.

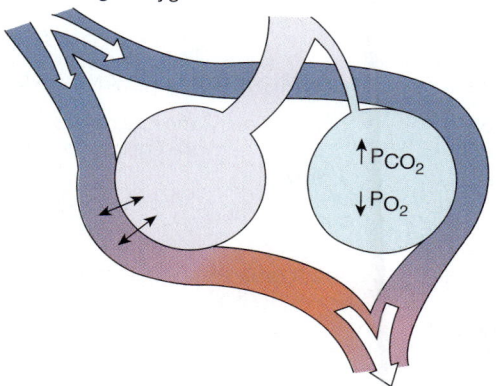

(b) Ventilation-perfusion mismatch.

If ventilation decreases in a group of alveoli (blue), P_{CO_2} increases and P_{O_2} decreases. Blood flowing past those alveoli does not get oxygenated.

(c) Local control mechanisms try to keep ventilation and perfusion matched.

Decreased tissue P_{O_2} around underventilated alveoli constricts their arterioles, diverting blood to better-ventilated alveoli.

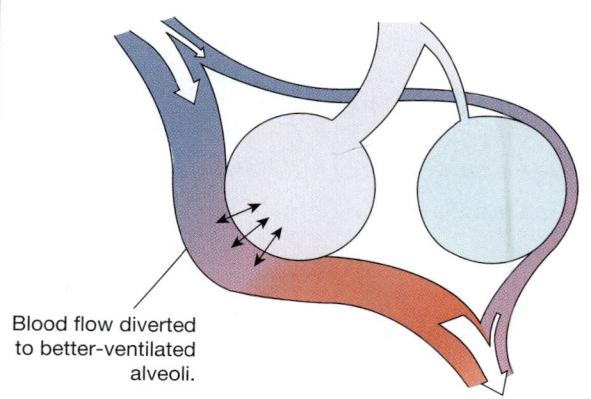

Blood flow diverted to better-ventilated alveoli.

■ **FIGURE 17-16** *Local control matches ventilation and perfusion*

Normally, breath sounds are distributed evenly over the lungs and resemble a quiet "whoosh" made by flowing air. When air flow is reduced, such as in pneumothorax, breath sounds may be either diminished or absent. Abnormal sounds include various squeaks, pops, wheezes, and bubbling sounds caused by fluid and secretions in the airways or alveoli. Inflammation of the pleural membrane results in a crackling or grating sound known as a *friction rub*. It is caused by swollen, inflamed pleural membranes rubbing against each other, and it disappears when fluid again separates them.

Diseases in which air flow during expiration is diminished as a result of narrowing of the bronchioles are known as **obstructive lung diseases**. When patients with obstructive lung diseases are asked to exhale forcefully, air whistling through the narrowed lower airways creates a wheezing sound that can be heard even without a stethoscope. Depending on the severity of the disease, the bronchioles may even collapse and close off before a forced expiration is completed, reducing both the amount and rate of air flow as measured by a spirometer.

Obstructive lung diseases include asthma, emphysema, and chronic bronchitis. The latter two are sometimes called *chronic obstructive pulmonary disease* (COPD) because of their ongoing, or chronic, nature. Asthma is an inflammatory condition, often associated with allergies, that is characterized by bronchoconstriction and airway edema. Asthma can be triggered by exercise (exercise-induced asthma) and by rapid changes in the temperature or humidity of inspired air. Asthmatic patients complain of "air hunger" and difficulty breathing (dyspnea). The severity of asthma attacks ranges from mild to life threatening.

Studies of asthma at the cellular level show that a variety of chemical signals may be responsible for inducing asthmatic bronchoconstriction, including acetylcholine, histamine, substance P (a neuropeptide), and leukotrienes secreted by mast cells, macrophages, and eosinophils. *Leukotrienes* are lipid-like bronchoconstrictors that are released during the inflammatory response. Asthma is treated with inhaled and oral medications that include β_2-adrenergic agonists, anti-inflammatory drugs, and leukotriene antagonists.

CONCEPT CHECK

30. Restrictive lung diseases [ p. 575] decrease lung compliance. How will inspiratory reserve volume change in patients with a restrictive lung disease?

31. Chronic obstructive lung disease causes patients to lose the ability to exhale fully. How does residual volume change in these patients? Answers: p. 586

This completes our discussion of the mechanics of ventilation. In the next chapter, we shift from bulk flow of air to the diffusion and transport of oxygen and carbon dioxide as they travel between the air spaces of the alveoli and the cells of the body.

RUNNING PROBLEM CONCLUSION

EMPHYSEMA

Edna is discharged from the hospital after three days and given prescriptions for bronchodilator and anti-inflammatory drugs to keep her airways as open as possible. Unfortunately, the lung changes that take place with COPD are not reversible, and she will require treatment for the rest of her life. According to the American Lung Association (*www.lungusa.org*), COPD is the fourth leading cause of death in the United States and costs more than $30 billion per year in direct medical costs and indirect costs such as lost wages.

In this running problem you learned about chronic obstructive pulmonary disease. Now check your understanding of the physiology in the problem by comparing your answers with those in the following table.

	QUESTION	FACTS	INTEGRATION AND ANALYSIS
1	What does narrowing of the airways do to the resistance airways offer to air flow?	The relationship between tube radius and resistance is the same for air flow as for blood flow: as radius decreases, resistance increases [p. 462].	When resistance increases, the body must use more energy to create air flow.
2	Why do people with chronic bronchitis have a higher-than-normal rate of respiratory infections?	Cigarette smoke paralyzes the cilia that sweep debris and mucus out of the airways. Without the action of cilia, mucus and trapped particles pool in the airways.	Bacteria trapped in the mucus can multiply and cause respiratory infections.
3	Name the muscles that patients with emphysema use to exhale actively.	Normal expiration depends on elastic recoil of muscles and elastic tissue in the lungs.	Forceful expiration involves the internal intercostal muscles and the abdominal muscles.
4	Why does Edna have an increased hematocrit?	Because of Edna's COPD, her arterial P_{O_2} is low. The major stimulus for red blood cell synthesis is hypoxia.	Low arterial oxygen levels trigger the synthesis of additional red blood cells. This provides more binding sites for oxygen transport.

559 561 569 574 581 582

CHAPTER SUMMARY

The process by which air flows into and out of the lungs is another example of the principle of *mass flow*. Like blood flow, air flow requires a pump to create a pressure gradient and encounters resistance primarily from changes in the diameter of the tubes through which it flows. The *mechanical properties* of the pleural sacs and elastic recoil in the chest wall and lung tissue are essential for normal ventilation.

1. Aerobic metabolism in living cells consumes oxygen and produces carbon dioxide. (p. 559)

2. Gas exchange requires a large, thin, moist exchange surface; a pump to move air; and a circulatory system to transport gases to the cells. (p. 559)

3. Respiratory system functions include gas exchange, pH regulation, vocalization, and protection from foreign substances. (p. 559)

Respiratory System

IP **Respiratory System: Anatomy Review**

4. **Cellular respiration** refers to cellular metabolism that consumes oxygen. **External respiration** is the exchange of gases between the atmosphere and cells of the body. It includes ventilation, gas exchange at the lung and cells, and transport of gases in the blood. **Ventilation** is the movement of air into and out of the lungs. (pp. 559–560; Fig. 17-1)

5. The **respiratory system** consists of anatomical structures involved in ventilation and gas exchange. (p. 560)

6. The **upper respiratory tract** includes the mouth, nasal cavity, **pharynx**, and **larynx**. The **lower respiratory tract** includes the **trachea**, **bronchi**, **bronchioles**, and exchange surfaces of the **alveoli**. (p. 560; Fig. 17-2a)

7. The thoracic cage is bounded by the ribs, spine, and **diaphragm**. Two sets of **intercostal muscles** connect the ribs. (p. 560; Fig. 17-2b)

8. Each **lung** is contained within a double-walled **pleural sac** that contains a small quantity of **pleural fluid**. (p. 560; Figs. 17-2d, 17-3)

9. The two **primary bronchi** enter the lungs. Each primary bronchus divides into progressively smaller bronchi and finally into collapsible **bronchioles**. (p. 561; Figs. 17-2e, 17-4)

10. The alveoli consist mostly of thin-walled **type I alveolar cells** for gas exchange. **Type II alveolar cells** produce surfactant. A network of capillaries surrounds each alveolus. (p. 561; Fig. 17-2f, g)

11. Blood flow through the lungs equals cardiac output. Resistance to blood flow in the pulmonary circulation is low. Pulmonary arterial pressure averages 25/8 mm Hg. (p. 564)

Gas Laws

IP **Respiratory System: Pulmonary Ventilation**

12. The total pressure of a mixture of gases is the sum of the pressures of the individual gases in the mixture (**Dalton's law**). **Partial pressure** is the pressure contributed by a single gas in a mixture. (p. 565; Tbls. 17-1, 17-2)

13. Bulk flow of air occurs down pressure gradients, as does the movement of any individual gas making up the air. (p. 566)

14. **Boyle's law** states that as the volume available to a gas increases, the gas pressure decreases. The body creates pressure gradients by changing thoracic volume. (p. 566; Fig. 17-5)

Ventilation

IP **Respiratory System: Pulmonary Ventilation**

15. **Tidal volume** is the amount of air taken in during a single normal inspiration. **Vital capacity** is tidal volume plus **expiratory** and **inspiratory reserve volumes**. Air volume in the lungs at the end of maximal expiration is the **residual volume**. (p. 567; Fig. 17-7)

16. The upper respiratory system filters, warms, and humidifies inhaled air. (p. 568)

17. Air flow in the respiratory system is directly proportional to the pressure gradient, and inversely related to the resistance to air flow offered by the airways. (p. 569)

18. A single **respiratory cycle** consists of an inspiration and an expiration. (p. 570)

19. During **inspiration**, **alveolar pressure** decreases, and air flows into the lungs. Inspiration requires contraction of the inspiratory muscles and the diaphragm. (p. 570; Fig. 17-9)

20. **Expiration** is usually passive, resulting from elastic recoil of the lungs. (p. 572)

21. **Active expiration** requires contraction of the internal intercostal and abdominal muscles. (p. 572)

22. **Intrapleural pressures** are always subatmospheric because the pleural cavity is a sealed compartment. (p. 572; Fig. 17-11)

23. **Compliance** is the ease with which the chest wall and lungs expand. Loss of compliance increases the work of breathing. **Elastance** is the ability of a stretched lung to return to its normal volume. (p. 574)

24. **Surfactant** decreases surface tension in the fluid lining the alveoli. This prevents smaller alveoli from collapsing. (p. 575; Fig. 17-13)

25. The diameter of the bronchioles determines how much resistance they offer to air flow. (p. 576)

26. Increased CO_2 in expired air dilates bronchioles. Parasympathetic neurons cause **bronchoconstriction** in response to irritant stimuli. There is no significant sympathetic innervation of bronchioles, but epinephrine causes **bronchodilation**. (p. 576; Tbl. 17-3)

27. **Total pulmonary ventilation** = tidal volume $\times$ ventilation rate. **Alveolar ventilation** = ventilation rate $\times$ (tidal volume − dead space volume). (p. 577; Fig. 17-14)

28. Alveolar gas composition changes very little during a normal respiratory cycle. **Hyperventilation** increases alveolar P_{O_2} and decreases alveolar P_{CO_2}. **Hypoventilation** has the opposite effect. (p. 578; Fig. 17-15)

29. Air flow is matched to blood flow around the alveoli by local mechanisms. Increased levels of CO_2 dilate bronchioles, and decreased O_2 constricts pulmonary arterioles. (p. 579; Fig. 17-16, Table 17-6)

17

QUESTIONS

(Answers to the Review Questions begin on page A1.)

LEVEL ONE   REVIEWING FACTS AND TERMS

1. List four functions of the respiratory system.

2. Give two definitions for the word *respiration*.

3. Which sets of muscles are used for normal quiet inspiration? For
    normal quiet expiration? For active expiration? What kind(s) of
    muscles are the different respiratory muscles (skeletal, cardiac, or
    smooth)?

4. What is the function of pleural fluid?

5. Name the anatomical structures that an oxygen molecule passes on
    its way from the atmosphere to the blood.

6. Diagram the structure of an alveolus, and state the function of each
    part. How are capillaries associated with an alveolus?

7. Trace the path of the pulmonary circulation. About how much
    blood is found here at any given moment? What is a typical arterial
    blood pressure for the pulmonary circuit, and how does this pres-
    sure compare with that of the systemic circulation?

8. What happens to inspired air as it is conditioned during its passage
    through the airways?

9. During inspiration, most of the thoracic volume change is the result
    of movement of the _____.

10. Describe the changes in alveolar and intrapleural pressure during
    one respiratory cycle.

11. What is the function of surfactants in general? In the respiratory sys-
    tem?

12. Of the three factors that contribute to the resistance of air flow
    through a tube, which plays the largest role in changing resistance
    in the human respiratory system?

13. Match the following items with their correct effect on the bronchi-
    oles:

(a) histamine	1. bronchoconstriction
(b) epinephrine	2. bronchodilation
(c) acetylcholine	3. no effect
(d) increased P_{CO_2}	

14. On the spirogram, label:

    (a) tidal volume (V_T), inspiratory and expiratory reserve volumes
        (IRV and ERV), residual volume (RV), vital capacity (VC), total
        lung capacity (TLC).

    (b) What is the value of each of the volumes and capacities you
        labeled?

    (c) What is this person's ventilation rate?

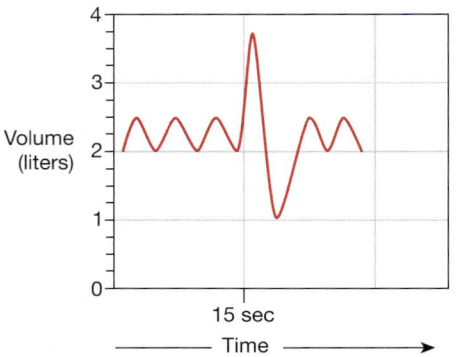

LEVEL TWO   REVIEWING CONCEPTS

15. Compile the following terms into a map of ventilation. Use up ar-
    rows, down arrows, greater than symbols (>), and less than symbols
    (>) as modifiers. You may add additional terms.

abdominal muscles	inspiratory muscles
air flow	internal intercostals
contract	P_A
diaphragm	P_{atm}
expiratory muscles	$P_{intrapleural}$
external intercostals	quiet breathing
forced breathing	relax
in, out, from, to	scalenes

16. Compare and contrast the terms in each of the following sets:

    (a) compliance and elastance
    (b) inspiration, expiration, and ventilation
    (c) intrapleural pressure and alveolar pressure
    (d) total pulmonary ventilation and alveolar ventilation
    (e) type I and type II alveolar cells
    (f) pulmonary circulation and systemic circulation

17. List the major paracrines and neurotransmitters that cause bron-
    choconstriction and bronchodilation. What receptors do they act
    through? (muscarinic, nicotinic, α, β_1, β_2)

18. Decide whether each of the following parameters will increase, de-
    crease, or not change in the situations given.

    (a) airway resistance with bronchodilation
    (b) intrapleural pressure during inspiration
    (c) air flow with bronchoconstriction
    (d) bronchiolar diameter with increased P_{CO_2}
    (e) tidal volume with decreased compliance
    (f) alveolar pressure during expiration

19. Define the following terms: pneumothorax, spirometer, ausculta-
    tion, hypoventilation, bronchoconstriction, minute volume, partial
    pressure of a gas.

20. The cartoon coyote is blowing up a balloon as part of an attempt to
    once more catch the roadrunner. He first breathes in as much air as
    he can, then blows out all he can into the balloon.

    (a) The volume of air in the balloon is equal to the _____
        _____ of the coyote's lungs. This volume can be measured
        directly by measuring the balloon volume or by adding which
        respiratory volumes together?

(b) In 10 years, when the coyote is still chasing the roadrunner, will he still be able to put as much air into the balloon in one breath? Explain.

21. Match the descriptions on the right to the appropriate phase(s) of ventilation:

(a) inspiration
(b) expiration
(c) both inspiration and expiration
(d) neither

_____ usually depend(s) on elastic recoil
_____ is/are easier when lung compliance decreases
_____ is/are driven mainly by positive intrapleural pressure generated by muscular contraction
_____ is usually an active process requiring smooth muscle contraction

22. Draw and label a graph showing the P_{O_2} of air in the primary bronchi during one respiratory cycle. (*Hint:* what parameter goes on each axis?)

23. Lung compliance decreases as we age. In the absence of other changes, would the following parameters increase, decrease, or not change as compliance decreases?

(a) work required for breathing
(b) ease with which lungs inflate
(c) lung elastance
(d) airway resistance during inspiration

24. Will pulmonary surfactant increase, decrease, or not change the following?

(a) work required for breathing
(b) lung compliance
(c) surface tension in the alveoli

LEVEL THREE PROBLEM SOLVING

25. Assume a normal female has a resting tidal volume of 400 mL, a respiratory rate of 13 breaths/min, and an anatomic dead space of 125 mL. When she exercises, which of the following scenarios would be most efficient for increasing her oxygen delivery to the lungs?

(a) increase respiratory rate to 20 breaths/min but have no change in tidal volume
(b) increase tidal volume to 550 mL but have no change in respiratory rate
(c) increase tidal volume to 500 mL and respiratory rate to 15 breaths/min

Which of these scenarios is most likely to occur during exercise in real life?

26. A 30-year-old computer programmer has had asthma for 15 years. When she lies down at night, she has spells of wheezing and coughing. Over the years, she has found that she can breathe better if she sleeps sitting nearly upright. Upon examination, her doctor finds that she has an enlarged thorax. Her lungs are overinflated on X-ray. Here are the results of her examination and pulmonary function tests. Use the normal values and abbreviations in Figure 17-7 to help answer the questions.

Ventilation rate:	16 breaths/min
Tidal volume:	600 mL
ERV:	1000 mL
RV:	3500 mL

Inspiratory capacity:	1800 mL
Vital capacity:	2800 mL
Functional residual capacity:	4500 mL
TLC:	6300 mL

After she is given a bronchodilator, her vital capacity increased to 3650 mL.

(a) What is her minute volume?
(b) Explain the change in vital capacity with bronchodilators.
(c) Which other values are abnormal? Can you explain why they might be, given her history and findings?

LEVEL FOUR QUANTITATIVE PROBLEMS

27. A container of gas with a movable piston has a volume of 500 mL and a pressure of 60 mm Hg. The piston is moved, and the new pressure is 150 mm Hg. What is the new volume of the container?

28. You have a mixture of gases in dry air, with an atmospheric pressure of 760 mm Hg. Calculate the partial pressure of each gas if the composition of the air is:

(a) 21% oxygen, 78% nitrogen, 0.3% carbon dioxide
(b) 40% oxygen, 13% nitrogen, 45% carbon dioxide, 2% hydrogen
(c) 10% oxygen, 15% nitrogen, 1% argon, 25% carbon dioxide

29. Li is a tiny woman, with a tidal volume of 400 mL and a respiratory rate of 12 breaths per minute at rest. What is her total pulmonary ventilation? Just before a physiology exam, her ventilation increases to 18 breaths per minute from nervousness. Now what is her total pulmonary ventilation? Assuming her anatomic dead space is 120 mL, what is her alveolar ventilation in each case?

30. You collected the following data on your classmate Neelesh.

Minute volume = 5004 mL/min
Respiratory rate = 3 breaths/15 sec
Vital capacity = 4800 mL
Expiratory reserve volume = 1000 mL

What are Neelesh's tidal volume and inspiratory reserve volume?

31. Use the illustrations below to help solve this problem. A spirometer with a volume of 1 liter (V_1) is filled with a mixture of oxygen and helium, with the helium concentration being 4 g/L (C_1). Helium does not move from the lungs into the blood or from the blood into the lungs. A subject is told to blow out all the air he possibly can. Once he finishes that exhalation, his lung volume is V_2. He then puts the spirometer tube in his mouth and breathes quietly for several breaths. At the end of that time, the helium is evenly dispersed in the spirometer and the subject's lungs. A measurement shows the new concentration of helium is 1.9 g/L. What was the subject's lung volume at the start of the experiment? (*Hint:* $C_1V_1 = C_2V_2$)

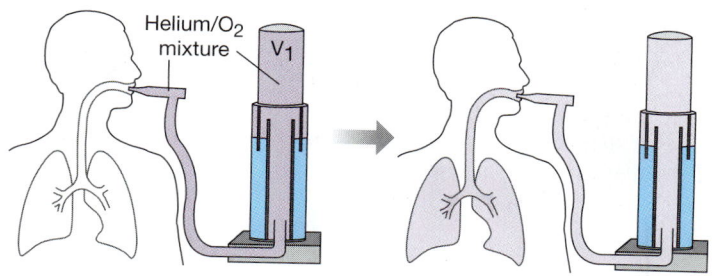

Helium/O₂ mixture V_1

17

ANSWERS

 Answers to Concept Check Questions

Page 561

1. Cellular respiration is intracellular and uses O_2 and organic substrates to produce ATP. External respiration is exchange and transport of gases between the atmosphere and cells.

2. Upper respiratory tract includes mouth, nasal cavity, pharynx, and larynx. Lower respiratory tract includes trachea, bronchi, bronchioles, and exchange surface of lungs.

3. Velocity is highest in the trachea and lowest in bronchioles.

4. Pleural fluid reduces friction and holds lungs tight against the chest wall.

5. The thoracic cage consists of rib cage with intercostal muscles, spinal (vertebral) column, and diaphragm. The thorax contains two lungs in pleural sacs, the heart and pericardial sac, esophagus, and major blood vessels.

6. The bronchioles are collapsible.

Page 564

7. Blood flow is equal in pulmonary trunk and aorta.

8. Increased hydrostatic pressure causes greater net filtration out of capillaries and may result in pulmonary edema.

9. Mean pressure = 8 mm Hg + 1/3(25 − 8) mm Hg = 8 + 17/3 mm Hg = 13.7 mm Hg.

Page 566

10. 720 mm Hg × 0.78 = 562 mm Hg

Page 568

11. Lung capacities are the sum of two or more lung volumes.

12. Residual volume cannot be measured directly.

13. If aging individuals have reduced vital capacity while total lung capacity does not change, then residual volume must increase.

Page 569

14. As air becomes humidified, the P_{O_2} decreases.

15. If cilia cannot move mucus, the collected mucus will trigger a cough reflex to clear the mucus out.

Page 570

16. Air flow reverses direction during a respiratory cycle, but blood flows in a loop and never reverses direction.

17. See Figures 17-2d and 17-3. The lungs are enclosed in a pleural sac. One pleural membrane attaches to the lung, and the other lines the thoracic cage. Pleural fluid fills the pleural sac.

Page 572

18. Scarlett will be more successful if she exhales deeply, as this will decrease her thoracic volume and will pull her lower rib cage inward.

19. Inability to cough decreases the ability to expel the potentially harmful material trapped in airway mucus.

Page 574

20. A hiccup causes a rapid decrease in both intrapleural pressure and alveolar pressure.

21. The knife wound would collapse the left lung if the knife punctured the pleural membrane. Loss of adhesion between the lung and chest wall would release the inward pressure exerted on the chest wall, and the rib cage would expand outward. The right side would be unaffected as the right lung is contained in its own pleural sac.

Page 576

22. Normally, lung and chest wall compliance contribute more to the work of breathing.

23. Scar tissue reduces lung compliance.

24. Without surfactant, the work of breathing increases.

25. When bronchiolar diameter decreases, resistance increases.

26. Neurotransmitter is acetylcholine, and receptor is muscarinic.

Page 579

27. Increased tidal volume increases alveolar P_{O_2}.

28. Increased breathing rate increases alveolar P_{O_2}. Increasing breathing rate or tidal volume increases alveolar ventilation.

Page 580

29. P_{O_2} in alveoli in the affected section will increase because O_2 is not leaving the alveoli. P_{CO_2} will decrease because new CO_2 is not entering the alveoli from the blood. Bronchioles constrict when P_{CO_2} decreases (see Table 17-6), shunting air to areas of the lung with better blood flow. This compensation cannot restore normal ventilation in this section of lung, and local control is insufficient to maintain homeostasis.

Page 582

30. Inspiratory reserve volume decreases.

31. Residual volume increases in patients who cannot fully exhale.

 Answers to Graph Questions

Page 571

Fig. 17-11 Alveolar pressure is greatest in the middle of expiration and least in the middle of inspiration. It is equal to atmospheric pressure at the beginning and end of inspiration and expiration. When lung volume is at its minimum, alveolar pressure is (c) moving from maximum to minimum and external intercostal muscle contraction is (b) minimal.

18

The successful ascent of Everest without supplementary oxygen is one of the great sagas of the 20th century.

—John B. West, *Climbing with O's*, NOVA Online (www.pbs.org)

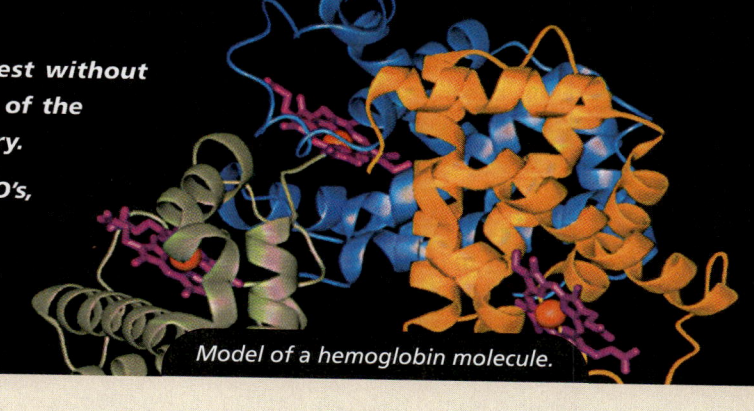

Model of a hemoglobin molecule.

Gas Exchange and Transport

BACKGROUND BASICS

RUNNING PROBLEM

HIGH ALTITUDE

In 1981 a group of 20 physiologists, physicians, and climbers, supported by 42 Sherpa assistants, formed the American Medical Research Expedition to Mt. Everest. The purpose of the expedition was to study human physiology at extreme altitudes, starting with the base camp at 5400 m (18,000 ft) and continuing on to the summit at 8850 m (over 29,000 ft). From the work of these scientists and others, we now have a good picture of the physiology of high-altitude acclimatization.

| 588 | 592 | 593 | 596 | 601 | 606 | 608 |

The book *Into Thin Air* by Jon Krakauer chronicles an ill-fated trek to the top of Mt. Everest. To reach the summit of Mt. Everest, climbers must pass through the "death zone" located at about 8000 meters (over 26,000 ft). Of the thousands of people who have attempted the summit, only about 2000 have been successful, and more than 185 have died. What are the physiological challenges of climbing Mt. Everest (8850 m or 29,035 ft), and why did it take so many years before humans successfully reached the top? The lack of oxygen at high altitude is part of the answer.

In the previous chapter we looked at the mechanics of breathing, the events that create the bulk flow of air into and out of the lungs. In this chapter we focus on the two gases that are most significant to physiology, oxygen and carbon dioxide, and look at how they move between alveolar air spaces and the cells of the body. The process can be divided into two components: the exchange of gases between compartments, which requires diffusion across cell membranes, and the transport of gases in the blood (Fig. 18-1 ■).

DIFFUSION AND SOLUBILITY OF GASES

The simple diffusion of oxygen and carbon dioxide across cell layers (between alveoli and pulmonary capillaries, or between systemic capillaries and cells) obeys the rules for simple diffusion across a membrane that are summarized in *Fick's law of diffusion* [🔁 p. 135]:

diffusion rate ∝

$$\frac{\text{surface area} \times \text{concentration gradient} \times \text{membrane permeability}}{\text{membrane thickness}}$$

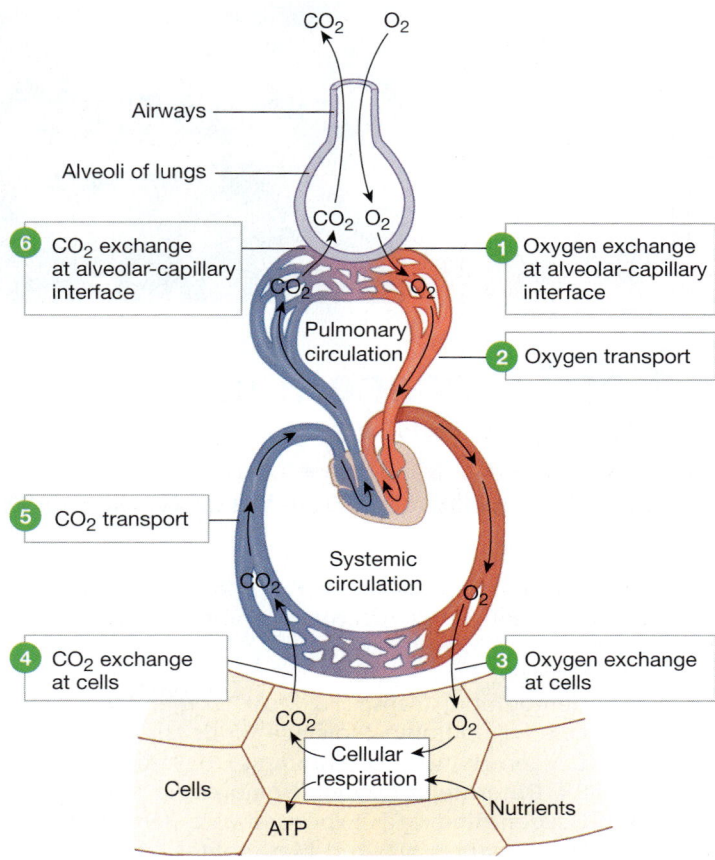

■ FIGURE 18-1 *Overview of oxygen and carbon dioxide exchange and transport*

Gases move into and out of the blood both at the pulmonary capillaries (steps 1 and 6) and at the systemic capillaries (steps 3 and 4).

If we assume that membrane permeability is constant, then three factors influence diffusion in the lungs:

1. *Surface area.* The rate of diffusion is directly proportional to the available surface area.
2. *Concentration gradient.* The rate of diffusion is directly proportional to the concentration gradient of the diffusing substance.
3. *Membrane thickness.* The rate of diffusion is inversely proportional to the thickness of the membrane.

From the general rules for diffusion, we can add a fourth influence: *diffusion distance*—diffusion is most rapid over short distances.

Under most circumstances, diffusion distance, surface area, and membrane thickness are constants in the body and are maximized to facilitate diffusion. Thus the most important factor for gas exchange in normal physiology is the concentration gradient.

When we think of concentrations in physiology, units such as moles per liter and milliosmoles per liter come to mind. However, respiratory physiologists commonly use *partial pressures* to express gas concentrations in solution. This measure allows direct comparison with partial pressures of the gases in air,

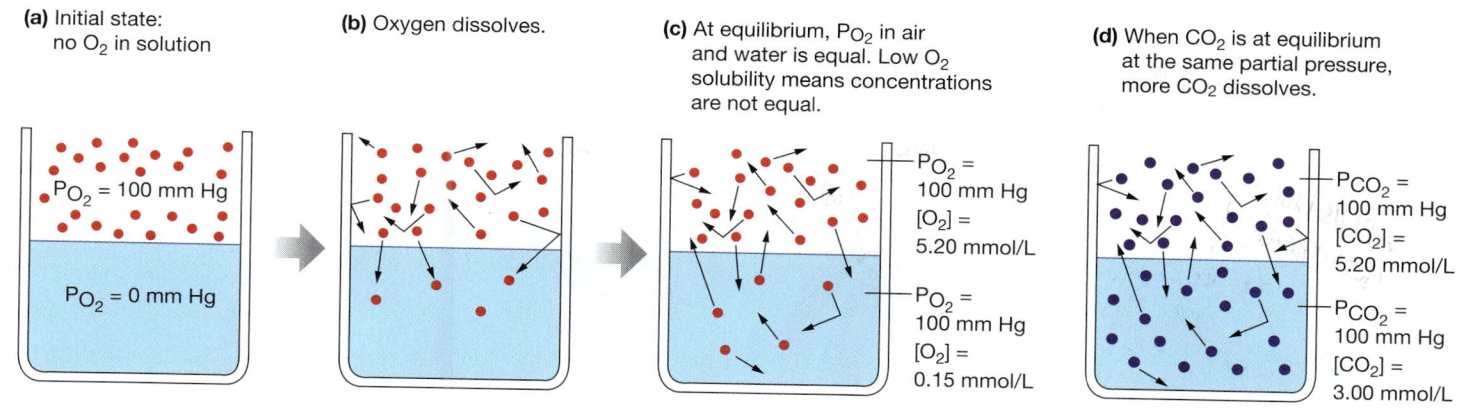

(a) Initial state:
no O_2 in solution

P_{O_2} = 100 mm Hg

P_{O_2} = 0 mm Hg

(b) Oxygen dissolves.

(c) At equilibrium, P_{O_2} in air and water is equal. Low O_2 solubility means concentrations are not equal.

P_{O_2} = 100 mm Hg
$[O_2]$ = 5.20 mmol/L

P_{O_2} = 100 mm Hg
$[O_2]$ = 0.15 mmol/L

(d) When CO_2 is at equilibrium at the same partial pressure, more CO_2 dissolves.

P_{CO_2} = 100 mm Hg
$[CO_2]$ = 5.20 mmol/L

P_{CO_2} = 100 mm Hg
$[CO_2]$ = 3.00 mmol/L

■ **FIGURE 18-2** *Behavior of gases in solution*

When temperature remains constant, the amount of a gas that dissolves in a liquid depends on both the solubility of the gas in the liquid and the partial pressure of the gas.

which is important for establishing whether there is a concentration gradient between the alveoli and the blood.

The Solubility of Gases in Liquids Depends on Pressure, Solubility, and Temperature

When a gas is placed in contact with water and there is a pressure gradient, the gas molecules move from one phase to the other. If gas pressure is higher in the water than in the gaseous phase, then gas molecules will leave the water. If the gas pressure is higher in the gaseous phase than in the water, then the gas will dissolve into the water.

The movement of gas molecules from air into a liquid is directly proportional to three factors: (1) the pressure gradient of the gas, (2) the solubility of the gas in the liquid, and (3) temperature. Because temperature is relatively constant in mammals, we will ignore its contribution in this discussion.

The ease with which a gas dissolves in a liquid is the **solubility** of the gas in that liquid. If a gas is very soluble, large numbers of gas molecules will go into solution at a low gas partial pressure. With less soluble gases, even a high partial pressure may cause only a few molecules of the gas to dissolve in the liquid. For example, consider a container of water exposed to air with a P_{O_2} of 100 mm Hg (Fig. 18-2a ■). Initially, the water has no oxygen dissolved in it (water P_{O_2} = 0 mm Hg). As the air stays in contact with the water, some of the moving oxygen molecules in the air diffuse into the water and dissolve (Fig. 18-2b). This process continues until equilibrium is reached. At equilibrium (Fig. 18-2c), the movement of oxygen from the air into the water is equal to the movement of oxygen from the water back into the air.

We refer to the amount of oxygen that dissolves in the water at any given P_{O_2} as the *partial pressure of the gas in solution*. In our example, therefore, if the air has a P_{O_2} of 100 mm Hg, at equilibrium the water will also have a P_{O_2} of 100 mm Hg.

Note that this does *not* mean that the concentration of oxygen is the same in the air and in the water! The concentration of dissolved oxygen also depends on the *solubility* of oxygen in water. For example, when P_{O_2} is 100 mm Hg in both the air and the water, the oxygen concentration in the air is 5.2 mmol O_2/L air, but the oxygen concentration in the water is only 0.15 mmol O_2/L water (Fig. 18-2c). As you can see, oxygen is not very soluble in water and, by extension, in any aqueous solution. Its low solubility is one reason for the evolution of oxygen-carrying molecules in the aqueous solution we call blood.

Now compare oxygen solubility with CO_2 solubility (Fig. 18-2d). Carbon dioxide is 20 times as soluble in water as oxygen is. At a P_{CO_2} of 100 mm Hg, the CO_2 concentration in air is 5.2 mmol CO_2/L air, and its concentration in water is 3.0 mmol/L water. So although P_{O_2} and P_{CO_2} are both 100 mm Hg in the water, the amount of each gas that dissolves in the water is very different.

✓ CONCEPT CHECK

1. A saline solution is exposed to a mixture of nitrogen gas and hydrogen gas in which $P_{H_2} = P_{N_2}$. What information do you need to predict whether equal amounts of H_2 and N_2 will dissolve in the solution?

2. If nitrogen is 78% of atmospheric air, what is the partial pressure of this gas when the dry atmospheric pressure is 720 mm Hg?

3. True or false? Plasma with a P_{O_2} of 40 mm Hg and a P_{CO_2} of 40 mm Hg has the same concentrations of oxygen and carbon dioxide.
Answers: p. 611

GAS EXCHANGE IN THE LUNGS AND TISSUES

The gas laws state that individual gases flow from regions of higher partial pressure to regions of lower partial pressure, and this rule governs the exchange of oxygen and carbon dioxide in the lungs and tissues.

18

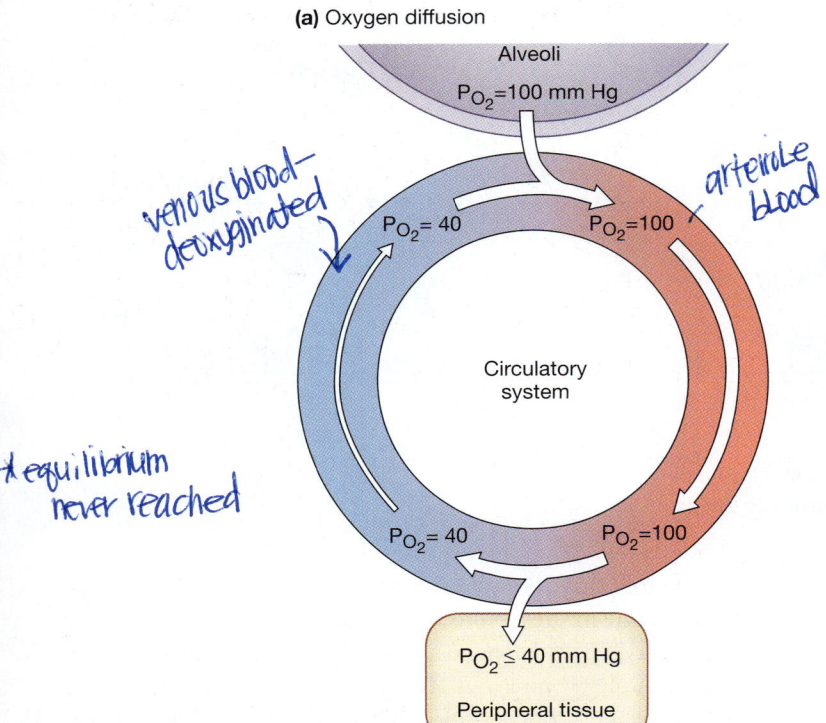

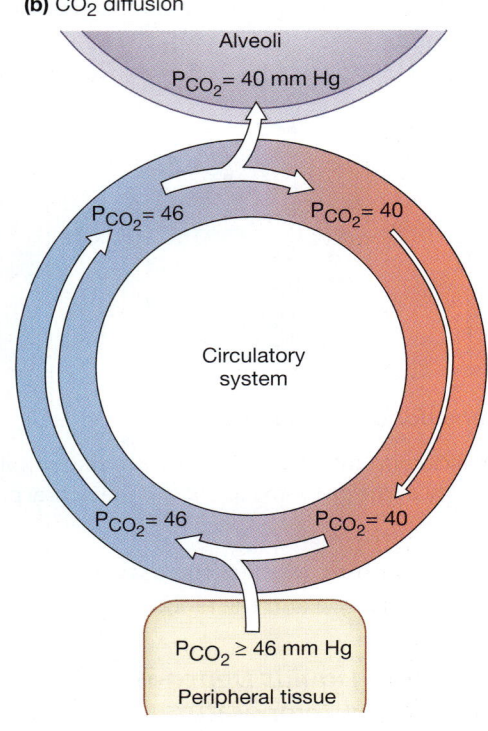

■ **FIGURE 18-3** *Gas exchange at the alveoli and cells*

Gases diffuse down their partial pressure (concentration) gradient.

Normal alveolar P_{O_2} is about 100 mm Hg (Fig. 18-3a ■). The P_{O_2} of systemic venous blood arriving at the lungs is 40 mm Hg. Oxygen therefore moves down its partial pressure (concentration) gradient from the alveoli into the capillaries. Diffusion goes to equilibrium, and the P_{O_2} of arterial blood leaving the lungs is the same as in the alveoli: 100 mm Hg.

When arterial blood reaches tissue capillaries, the gradient is reversed. Cells are continuously using oxygen for oxidative phosphorylation. In the cells of a person at rest, intracellular P_{O_2} averages 40 mm Hg. Arterial blood arriving at the cells has a P_{O_2} of 100 mm Hg. Because P_{O_2} is lower in the cells, oxygen diffuses down its partial pressure gradient from plasma into cells. Once again, diffusion goes to equilibrium, and as a result venous blood has the same P_{O_2} as the cells it just passed.

Conversely, P_{CO_2} is higher in tissues than in systemic capillary blood because of CO_2 production during metabolism (Fig. 18-3b). Cellular P_{CO_2} in a person at rest is about 46 mm Hg, compared to an arterial plasma P_{CO_2} of 40 mm Hg. The gradient causes CO_2 to diffuse out of cells into the capillaries. Diffusion goes to equilibrium, and systemic venous blood averages a P_{CO_2} of 46 mm Hg.

At the pulmonary capillaries, the process reverses. Venous blood bringing waste CO_2 from the cells has a P_{CO_2} of 46 mm Hg. Alveolar P_{CO_2} is 40 mm Hg. Because P_{CO_2} is higher in the plasma, CO_2 moves from the capillaries into the alveoli. By the time blood leaves the alveoli, it has a P_{CO_2} of 40 mm Hg, identical to the P_{CO_2} of the alveoli (Fig. 18-3b). Table 18-1 ■ summarizes the arterial and venous partial pressures just discussed.

If the diffusion of gases between alveoli and blood is significantly impaired, *hypoxia* (a state of too little oxygen) results. Hypoxia frequently (but not always!) goes hand in hand with **hypercapnia**, elevated concentrations of carbon dioxide. These two conditions are clinical signs, not diseases, and clinicians must gather additional information to pinpoint their cause. Table 18-2 ■ lists several types of hypoxia and some typical causes.

Three categories of problems result in low arterial oxygen content: (1) inadequate oxygen reaching the alveoli, (2) problems with oxygen exchange between alveoli and pulmonary capillaries, and (3) inadequate transport of oxygen in the blood. We will consider these issues in the sections that follow.

TABLE 18-1	Normal Blood Values in Pulmonary Medicine	
	ARTERIAL	**VENOUS**
P_{O_2}	95 mm Hg (85–100)	40 mm Hg
P_{CO_2}	40 mm Hg (35–45)	46 mm Hg
pH	7.4 (7.38–7.42)	7.37

TABLE 18-2   Classification of Hypoxias

TYPE	DEFINITION	TYPICAL CAUSES
Hypoxic hypoxia	Low arterial P_{O_2}	High altitude; alveolar hypoventilation; decreased lung diffusion capacity; abnormal ventilation-perfusion ratio
Anemic hypoxia	Decreased total amount of O_2 bound to hemoglobin	Blood loss; anemia (low [Hb] or altered HbO_2 binding); carbon monoxide poisoning
Ischemic hypoxia	Reduced blood flow	Heart failure (whole-body hypoxia); shock (peripheral hypoxia); thrombosis (hypoxia in a single organ)
Histotoxic hypoxia	Failure of cells to use O_2 because cells have been poisoned	Cyanide and other metabolic poisons

CONCEPT CHECK ✔

4. Cellular metabolism review: which of the following three metabolic pathways—glycolysis, the citric acid cycle, and the electron transport system—is *directly* associated with (a) O_2 consumption and with (b) CO_2 production?

5. Why doesn't the movement of oxygen from the alveoli to the plasma decrease the P_{O_2} of the alveoli?

Answers: p. 611

A Decrease in Alveolar P_{O_2} Decreases Oxygen Uptake at the Lungs

The first requirement for adequate oxygen delivery to the tissues is adequate oxygen intake from the atmosphere, as reflected by the P_{O_2} of the alveoli. A decrease in alveolar P_{O_2} will result in less oxygen entering the blood. There are two possible causes of low alveolar P_{O_2}: either (1) the inspired air has abnormally low oxygen content or (2) alveolar ventilation is inadequate.

The main factor that affects the oxygen content of inspired air is altitude. The partial pressure of oxygen in air decreases along with total atmospheric pressure as you move from sea level (where normal atmospheric pressure is 760 mm Hg) to higher altitudes. For example, Denver, 1609 m above sea level, has an atmospheric pressure of about 628 mm Hg. The P_{O_2} of dry air in Denver is 132 mm Hg, down from 160 mm Hg at sea level.

Unless a person is traveling, however, altitude remains constant. If alveolar P_{O_2} is low but the composition of inspired air is normal, then the problem lies with alveolar ventilation. Low alveolar ventilation is also known as *hypoventilation* and is characterized by lower-than-normal volumes of fresh air entering the alveoli. Pathological factors that cause alveolar hypoventilation (Fig. 18-4 ■) include decreased lung compliance (fibrosis; Fig. 18-4c), increased airway resistance (asthma;

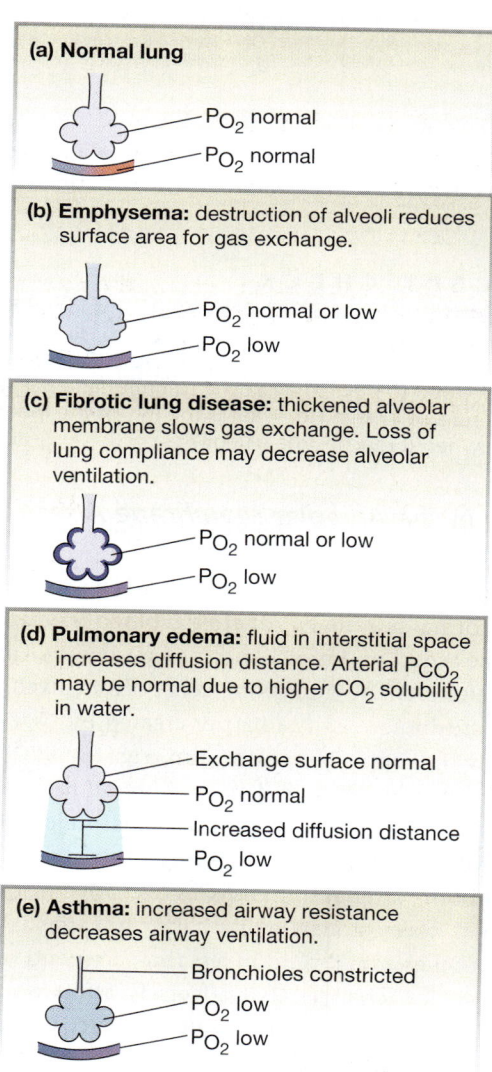

(a) Normal lung
P_{O_2} normal
P_{O_2} normal

(b) Emphysema: destruction of alveoli reduces surface area for gas exchange.
P_{O_2} normal or low
P_{O_2} low

(c) Fibrotic lung disease: thickened alveolar membrane slows gas exchange. Loss of lung compliance may decrease alveolar ventilation.
P_{O_2} normal or low
P_{O_2} low

(d) Pulmonary edema: fluid in interstitial space increases diffusion distance. Arterial P_{CO_2} may be normal due to higher CO_2 solubility in water.
Exchange surface normal
P_{O_2} normal
Increased diffusion distance
P_{O_2} low

(e) Asthma: increased airway resistance decreases airway ventilation.
Bronchioles constricted
P_{O_2} low
P_{O_2} low

■ **FIGURE 18-4** *Pathological conditions that reduce alveolar ventilation and gas exchange*

18

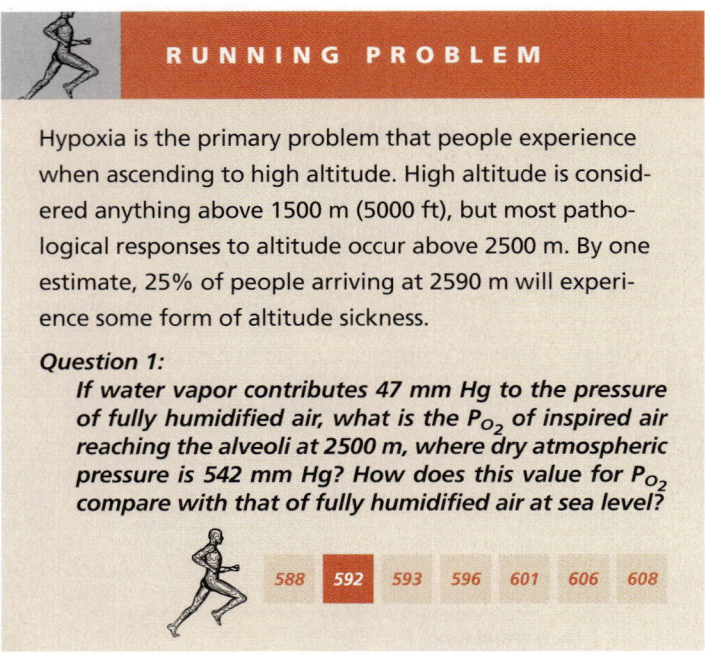

RUNNING PROBLEM

Hypoxia is the primary problem that people experience when ascending to high altitude. High altitude is considered anything above 1500 m (5000 ft), but most pathological responses to altitude occur above 2500 m. By one estimate, 25% of people arriving at 2590 m will experience some form of altitude sickness.

Question 1:

If water vapor contributes 47 mm Hg to the pressure of fully humidified air, what is the P_{O_2} of inspired air reaching the alveoli at 2500 m, where dry atmospheric pressure is 542 mm Hg? How does this value for P_{O_2} compare with that of fully humidified air at sea level?

| 588 | **592** | 593 | 596 | 601 | 606 | 608 |

Fig. 18-4e), and overdoses of drugs (including alcohol) that depress the central nervous system and slow ventilation rate and depth.

CONCEPT CHECK

6. At the summit of Mt. Everest, an altitude of 8850 m, atmospheric pressure is only 250 mm Hg. What is the P_{O_2} of dry atmospheric air atop Everest? If water vapor added to inhaled air at the summit has a partial pressure of 47 mm Hg, what is the P_{O_2} of the inhaled air when it reaches the alveoli?

Answers: p. 611

Changes in the Alveolar Membrane Alter Gas Exchange

In situations in which the composition of the air reaching the alveoli is normal but the P_{O_2} of arterial blood leaving the lungs is low, some aspect of the exchange process between alveoli and blood is defective. The transfer of oxygen from alveoli to blood requires diffusion across the barrier created by type I alveolar cells and by the capillary endothelium (Fig. 18-5 ■).

Normally, the diffusion distance is small because the cells are thin and there is little or no interstitial fluid between the two cell layers. In addition, both oxygen and carbon dioxide are soluble in both water and lipids. Gas exchange in the lungs is rapid, blood flow through pulmonary capillaries is slow, and diffusion reaches equilibrium in less than 1 second.

Pathological changes that adversely affect gas exchange include (1) a decrease in the amount of alveolar surface area available for gas exchange, (2) an increase in the thickness of the alveolar membrane, and (3) an increase in the diffusion distance between the alveoli and the blood.

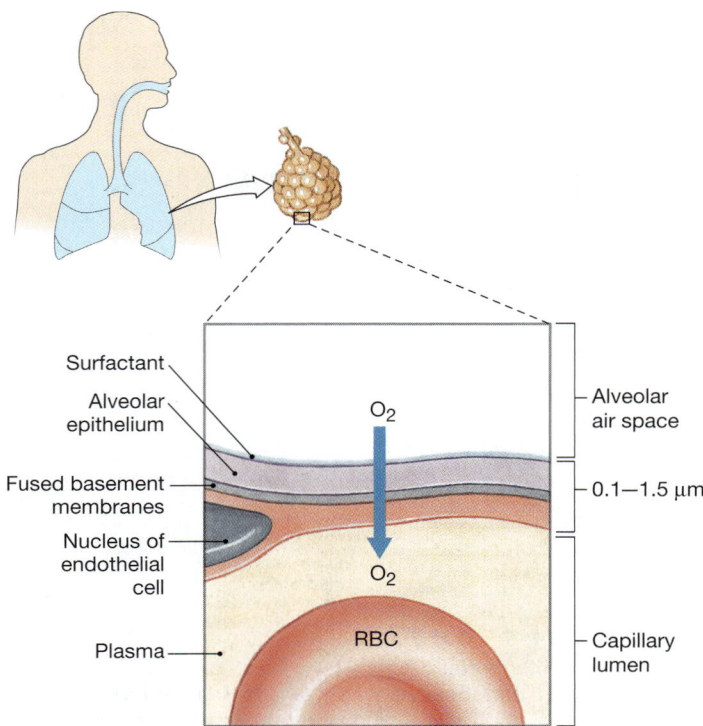

■ **FIGURE 18-5**   *Oxygen diffuses across alveolar epithelial cells and capillary endothelial cells to enter the plasma*

Physical loss of alveolar surface area is dramatic in *emphysema*, a degenerative lung disease most often caused by cigarette smoking (Fig. 18-4b). The irritating effect of smoke in the alveoli activates alveolar macrophages that release proteolytic enzymes. These enzymes destroy the elastic fibers of the lung [p. 77] and induce apoptosis of cells, breaking down the walls of the alveoli. The result is a high-compliance/low-elastic recoil lung with fewer and larger alveoli and less surface area for gas exchange.

Pathological changes in the alveolar membrane that alter its properties slow gas exchange. For example, in fibrotic lung diseases, deposition of scar tissue thickens the alveolar membrane (Fig. 18-4c). Diffusion of gases through this scar tissue is much slower than normal. However, because the lungs have a built-in reserve capacity, one-third of the exchange epithelium must be incapacitated before arterial P_{O_2} falls significantly.

One pathological condition in which the diffusion distance between alveoli and blood increases is **pulmonary edema**, characterized by excessive interstitial fluid volume in the lungs (Fig. 18-4d). Normally, only small amounts of interstitial fluid are present in the lungs, the result of low pulmonary blood pressure and effective lymph drainage. However, if pulmonary blood pressure rises for some reason, such as left ventricular failure or mitral valve dysfunction, the normal filtration/reabsorption balance at the capillary is disrupted [Fig. 15-18, p. 518]. When capillary hydrostatic pressure increases, more fluid filters out of the capillary. If filtration increases too much,

THE PULSE OXIMETER

One important clinical indicator of the effectiveness of gas exchange in the lungs is the amount of oxygen in arterial blood. Obtaining an arterial blood sample is difficult and painful because it means finding an accessible artery. (Most blood is drawn from superficial veins rather than from arteries, which lie deeper within the body.) Over the years, however, scientists have developed instruments that quickly and painlessly measure blood oxygen levels through the surface of the skin on a finger or earlobe. One such instrument, the *pulse oximeter*, clips onto the skin and in seconds gives a digital reading of arterial hemoglobin saturation. The oximeter works by measuring light absorbance of the tissue at two wavelengths. Another instrument, the *transcutaneous oxygen sensor*, measures dissolved oxygen using a variant of traditional gas-measuring electrodes. Both methods have limitations but are popular because they provide a rapid, noninvasive means of estimating arterial oxygen concentration.

RUNNING PROBLEM

Acute mountain sickness is the mildest illness caused by altitude hypoxia. The primary symptom is a headache that may be accompanied by dizziness, nausea, fatigue, or confusion. More severe illnesses are *high-altitude pulmonary edema* (HAPE) and *high-altitude cerebral edema*. HAPE is the major cause of death from altitude sickness. It is characterized by high pulmonary arterial pressure, extreme shortness of breath, and sometimes a productive cough yielding a pink, frothy fluid. Treatment is immediate relocation to lower altitude and administration of oxygen.

Question 2:
  Why would someone with HAPE be short of breath?

Question 3:
  Based on what you learned about the mechanisms for matching ventilation and perfusion in the lung [⬛ p. 579], can you explain why patients with HAPE have elevated pulmonary arterial blood pressure?

588   592   **593**   596   601   606   608

---

the lymphatics are unable to remove all the fluid, and excess accumulates in the pulmonary interstitial space, creating pulmonary edema. In severe cases, fluid even leaks across the alveolar membrane, collecting inside the alveoli.

Oxygen has low solubility in body fluids and takes longer to cross the increased diffusion distance present in pulmonary edema, resulting in decreased arterial P_{O_2}. Carbon dioxide, in contrast, is relatively soluble in body fluids, so the increased diffusion distance may not significantly affect carbon dioxide exchange. In some cases of pulmonary edema, arterial P_{O_2} is low but arterial P_{CO_2} is normal because of the different solubilities of the two gases.

CONCEPT CHECK

  7. Why does left ventricular failure or mitral valve dysfunction cause elevated pulmonary blood pressure?

  8. If alveolar ventilation increases, what will happen to arterial P_{O_2}? To arterial P_{CO_2}? To venous P_{O_2} and P_{CO_2}? Explain your answers.
  Answers: p. 611

GAS TRANSPORT IN THE BLOOD

Now that we have described gas exchange, we turn our attention to how oxygen and carbon dioxide are transported in the blood. The *law of mass action* [⬛ p. 39] plays an important role in this process. Changes in oxygen or CO_2 concentration disturb the equilibrium of reactions, shifting the balance between substrates and products.

Hemoglobin Transports Most Oxygen to the Tissues

Oxygen is transported two ways in the blood: dissolved in the plasma and bound to hemoglobin (Hb). This fact can be summarized as follows:

Total blood oxygen content = amount dissolved in plasma
                                   + amount bound to hemoglobin

Hemoglobin, the oxygen-binding protein in red blood cells, binds reversibly to oxygen, as summarized in the equation

$$Hb + O_2 \rightleftharpoons HbO_2$$

In the pulmonary capillaries, where plasma P_{O_2} increases as oxygen diffuses in from the alveoli, hemoglobin binds oxygen. At the cells, where oxygen is being used and plasma P_{O_2} falls, hemoglobin gives up its oxygen. Because oxygen is only slightly soluble in aqueous solutions, we must have adequate amounts of hemoglobin in our blood to survive.

The importance of hemoglobin in oxygen transport is summarized in Figure 18-6 ⬛. More than 98% of the oxygen in a given volume of blood is bound to hemoglobin and transported inside red blood cells. The balance remains dissolved in plasma.

To understand just how low the solubility of oxygen is, consider the following example. In a situation in which blood has no hemoglobin, only 3 mL of O_2 will dissolve in the plasma

18

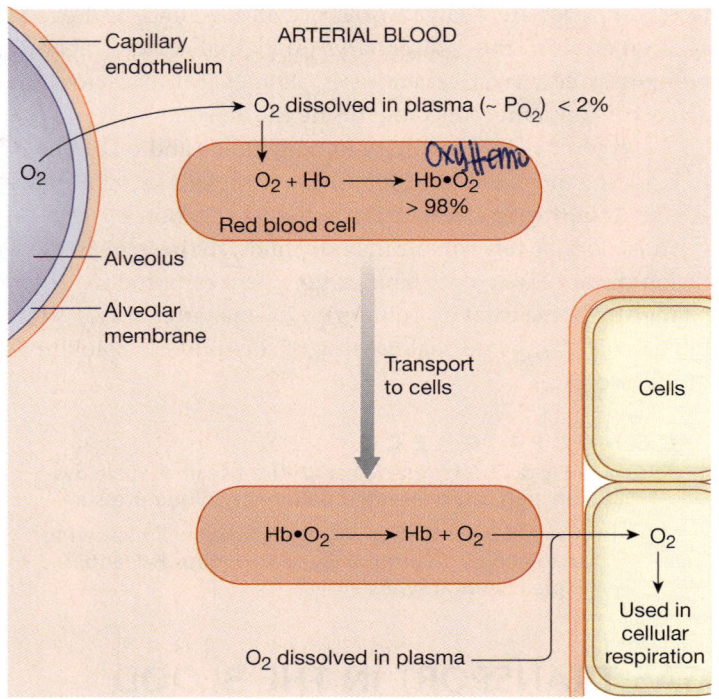

ARTERIAL BLOOD

Capillary endothelium

O_2 dissolved in plasma (~ P_{O_2}) < 2%

O_2

O_2 + Hb ⟶ Hb•O_2 OxyHemo

Red blood cell   > 98%

Alveolus

Alveolar membrane

Transport to cells

Cells

Hb•O_2 ⟶ Hb + O_2 O_2

O_2 dissolved in plasma

Used in cellular respiration

FIGURE QUESTION

How many cell membranes will O_2 cross in its passage between the airspace of the alveolus and binding to hemoglobin?

■ **FIGURE 18-6** *Summary of oxygen transport in the blood*

More than 98% of the oxygen in blood is bound to hemoglobin in red blood cells; less than 2% is dissolved in plasma.

18

fraction of 1 liter of arterial blood (Fig. 18-7a ■). Thus, with a typical cardiac output of 5 L blood/min, about 15 mL of dissolved oxygen reaches the systemic tissues each minute. This small amount cannot begin to meet the needs of the tissues, because oxygen consumption at rest is about 250 mL O_2 per minute, and that figure increases dramatically with exercise. Thus the oxygen brought to cells dissolved in plasma provides less than 10% of what the cells consume, so the body is heavily dependent on oxygen carried by hemoglobin.

At normal hemoglobin levels, red blood cells carry about 197 mL O_2/L blood (Fig. 18-7b). Thus:

total arterial O_2 carrying capacity = 3 mL dissolved O_2/L blood + 197 mL HbO_2/L blood = 200 mL O_2/L blood

If cardiac output remains 5 L/min, hemoglobin-assisted oxygen delivery to cells is almost 1000 mL/min, four times the oxygen consumption needed by the tissues at rest.

The amount of oxygen that binds to hemoglobin depends on two factors: (1) the P_{O_2} in the plasma surrounding the red

blood cells and (2) the number of potential binding sites available in the red blood cells.

Plasma P_{O_2} is the primary factor determining how many of the available hemoglobin binding sites are occupied by oxygen. Figure 18-7c shows what happens to oxygen transport when alveolar and arterial P_{O_2} decrease. As you learned in previous sections, arterial P_{O_2} is established by (1) the composition of inspired air, (2) the alveolar ventilation rate, and (3) the efficiency of gas exchange from alveoli to blood.

The total number of oxygen-binding sites depends on the number of hemoglobin molecules in red blood cells. Clinically, this number can be estimated either by counting the red blood cells and quantifying the amount of hemoglobin per red blood cell (*mean corpuscular hemoglobin*) or by determining the blood hemoglobin content (g Hb/dL whole blood). Any pathological condition that decreases the amount of hemoglobin in the cells or the number of red blood cells will adversely affect the blood's oxygen-transporting capacity.

People who have lost large amounts of blood need to replace hemoglobin for oxygen transport. A blood transfusion is the ideal replacement for blood loss, but in emergencies this is

EMERGING CONCEPTS

BLOOD SUBSTITUTES

Physiologists have been attempting to find a substitute for blood ever since 1878, when an intrepid physician named T. Gaillard Thomas transfused a patient with whole milk in place of blood. Although milk seems an unlikely replacement for blood, it has two important properties: proteins to provide colloid osmotic pressure and molecules (emulsified lipids) capable of binding to oxygen. In the development of hemoglobin substitutes, oxygen transport is the most difficult property to mimic. A hemoglobin solution would seem to be the obvious answer, but hemoglobin that is not compartmentalized in red blood cells behaves differently than hemoglobin that is compartmentalized. Investigators are making progress by polymerizing hemoglobin into larger, more stable molecules and loading these hemoglobin polymers into phospholipid liposomes [ p. 55]. Perfluorocarbon emulsions are also being tested as oxygen carriers. To learn more about this research, read "Physiological properties of blood substitutes," in *News in Physiological Sciences* 16(1):38–41, 2001 Feb. (*http://nips.physiology.org*).

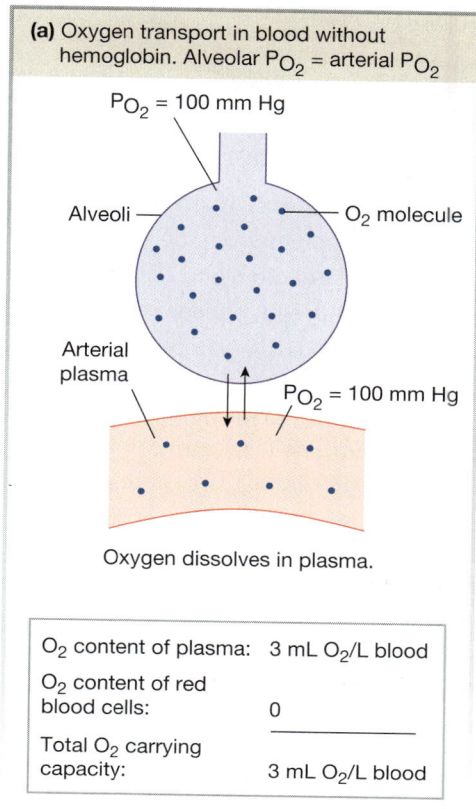

(a) Oxygen transport in blood without hemoglobin. Alveolar P_{O_2} = arterial P_{O_2}

P_{O_2} = 100 mm Hg

Alveoli

O_2 molecule

Arterial plasma

P_{O_2} = 100 mm Hg

Oxygen dissolves in plasma.

O_2 content of plasma:	3 mL O_2/L blood
O_2 content of red blood cells:	0
Total O_2 carrying capacity:	3 mL O_2/L blood

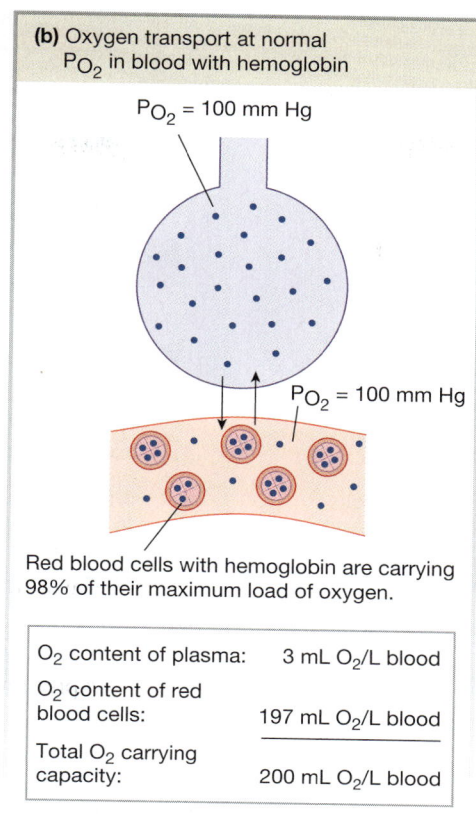

(b) Oxygen transport at normal P_{O_2} in blood with hemoglobin

P_{O_2} = 100 mm Hg

P_{O_2} = 100 mm Hg

Red blood cells with hemoglobin are carrying 98% of their maximum load of oxygen.

O_2 content of plasma:	3 mL O_2/L blood
O_2 content of red blood cells:	197 mL O_2/L blood
Total O_2 carrying capacity:	200 mL O_2/L blood

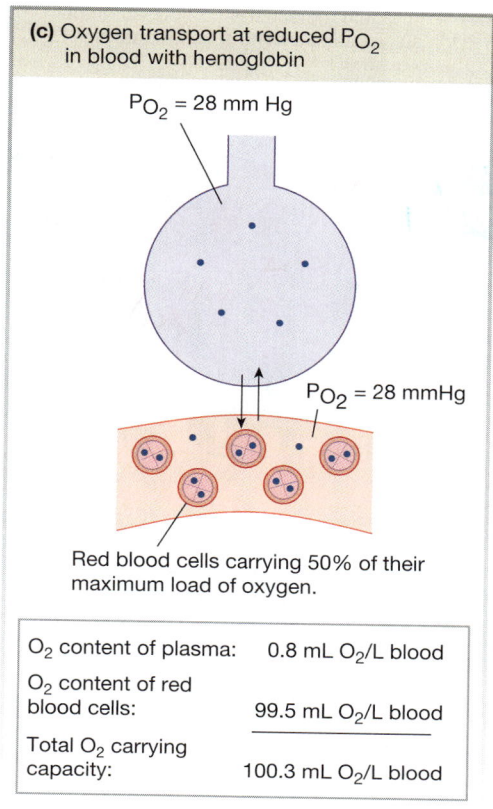

(c) Oxygen transport at reduced P_{O_2} in blood with hemoglobin

P_{O_2} = 28 mm Hg

P_{O_2} = 28 mmHg

Red blood cells carrying 50% of their maximum load of oxygen.

O_2 content of plasma:	0.8 mL O_2/L blood
O_2 content of red blood cells:	99.5 mL O_2/L blood
Total O_2 carrying capacity:	100.3 mL O_2/L blood

■ **FIGURE 18-7**   *The role of hemoglobin in oxygen transport*

not always possible. Saline infusions can replace lost blood volume, but saline (like plasma) cannot transport sufficient quantities of oxygen to support cellular respiration. Faced with this problem, researchers are currently testing artificial oxygen carriers to replace hemoglobin. In times of large-scale disasters, these hemoglobin substitutes would eliminate the need to identify a patient's blood type before giving transfusions.

One Hemoglobin Molecule Binds Up to Four Oxygen Molecules

Why is hemoglobin an effective oxygen carrier? The answer lies in its molecular structure. Hemoglobin is a large, complex protein whose quaternary structure has four globular protein chains, each of which is wrapped around an iron-containing **heme group** (Fig. 18-8a ■). The four heme groups in a hemoglobin molecule are identical. Each consists of a carbon-hydrogen-nitrogen *porphyrin ring* with an iron atom (Fe) in the center. About 70% of the iron in the body is found in the heme groups of hemoglobin.

The central iron atom of each heme group can bind reversibly with one oxygen molecule. Because there are four iron atoms per hemoglobin, each hemoglobin molecule has the potential to bind four oxygen molecules. The iron-oxygen interaction is a weak bond that can be easily broken without altering either the hemoglobin or the oxygen.

There are several forms of **globin** protein chains in hemoglobin. The most common forms are designated *alpha* (α), *beta* (β), *gamma* (γ), and *delta* (δ), depending on the structure of the chain. Most adult hemoglobin (designated *HbA*) has two alpha chains and two beta chains. However, a small portion of adult hemoglobin (about 2.5%) has two alpha chains and two delta chains (HbA_2).

The human fetus has a different isoform of hemoglobin that is adapted to attract oxygen from maternal blood in the placenta. *Fetal hemoglobin (HbF)* has two gamma chains in place of the two beta chains found in adult hemoglobin. Shortly after birth, fetal hemoglobin is replaced with the adult form as new red blood cells are made. The different binding properties of adult and fetal hemoglobin are discussed in the sections that follow.

Oxygen-Hemoglobin Binding Obeys the Law of Mass Action

Hemoglobin bound to oxygen is known as **oxyhemoglobin**, abbreviated HbO_2. (It would be more accurate to show the number of oxygen molecules carried on each hemoglobin molecule—$Hb(O_2)_{1-4}$—but we will use the simpler abbreviation because the number of bound oxygen molecules varies from one hemoglobin molecule to another.)

18

(a) A hemoglobin molecule is composed of four protein globin chains, each surrounding a central heme group.

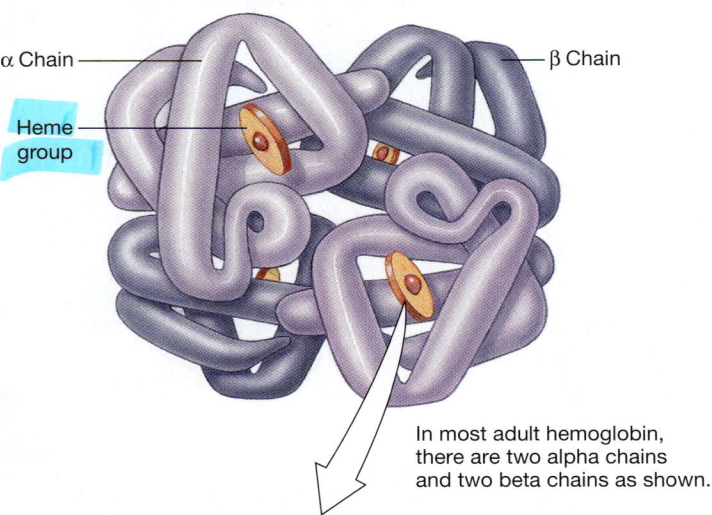

α Chain

β Chain

Heme group

In most adult hemoglobin, there are two alpha chains and two beta chains as shown.

(b) Each heme group consists of a porphyrin ring with an iron atom in the center.

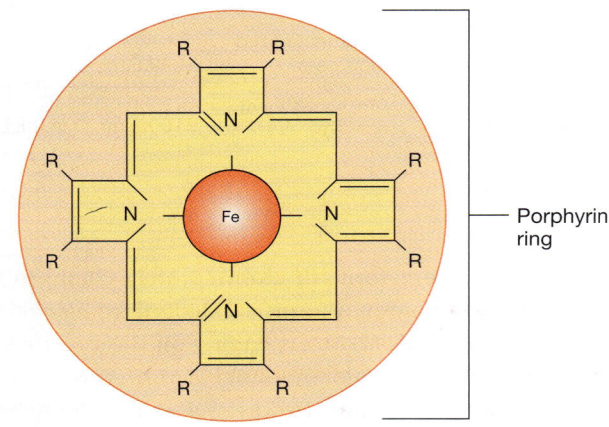

Porphyrin ring

R = additional C, H, O groups

■ **FIGURE 18-8** *The hemoglobin molecule*

Oxygen-hemoglobin binding obeys the law of mass action:

$$Hb + O_2 \rightleftharpoons HbO_2$$

If oxygen concentration increases, this *oxygen-hemoglobin binding reaction* shifts to the right and more oxygen binds to hemoglobin. If the concentration of oxygen decreases, the reaction shifts to the left and hemoglobin releases some of its bound oxygen.

P_{O_2} Determines Oxygen-Hemoglobin Binding

Because of the law of mass action, the amount of oxygen bound to hemoglobin depends primarily on the P_{O_2} of plasma surrounding the red blood cells (Fig. 18-6). In the pulmonary capillaries, O_2 dissolved in the plasma diffuses into red blood cells, where it binds to hemoglobin. This removes dissolved O_2 from the plasma, causing more oxygen to diffuse in from the alveoli.

RUNNING PROBLEM

In most people arriving at high altitude, normal physiological responses kick in to help acclimatize the body to the chronic hypoxia. Within two hours of arrival, hypoxia triggers the release of erythropoietin from the kidneys and liver. This hormone stimulates red blood cell production, and as a result, new erythrocytes appear in the blood within four days.

Question 4:
How does adding erythrocytes to the blood help a person acclimatize to high altitude?

Question 5:
What does adding erythrocytes to the blood do to the viscosity of the blood? What effect will that change in viscosity have on blood flow?

588 592 593 **596** 601 606 608

The transfer of oxygen from alveolar air to plasma to red blood cells and onto hemoglobin occurs so rapidly that blood in the pulmonary capillaries normally picks up as much oxygen as the P_{O_2} of the plasma and the number of red blood cells permit.

Once arterial blood reaches the tissues, the exchange process that took place in the lungs reverses. Dissolved oxygen diffuses out of systemic capillaries into cells, and the resultant decrease in plasma P_{O_2} disturbs the equilibrium of the oxygen-hemoglobin binding reaction by removing oxygen from the left side of the equation. The equilibrium shifts to the left according to the law of mass action, and the hemoglobin molecules release their oxygen stores, as represented in the bottom half of Figure 18-6.

Like oxygen loading at the lungs, this process of transferring oxygen to the body's cells takes place very rapidly and goes to equilibrium. The P_{O_2} of the cells determines how much oxygen unloads from hemoglobin. As cells increase their metabolic activity, their P_{O_2} decreases, and hemoglobin releases more oxygen to them.

Oxygen Binding Is Expressed as a Percentage

The amount of oxygen bound to hemoglobin at any given P_{O_2} is expressed as a percentage:

(amount of O_2 bound/maximum that could be bound) × 100
= percent saturation of hemoglobin

The **percent saturation of hemoglobin** refers to the percentage of available binding sites that are bound to oxygen. If all binding sites of all hemoglobin molecules are occupied by oxygen molecules, the blood is 100% oxygenated, or *saturated* with oxygen. If half the available binding sites are carrying oxygen, the hemoglobin is 50% saturated, and so on.

The relationship between plasma P_{O_2} and percent saturation of hemoglobin can be explained with the following analogy. The hemoglobin molecules carrying oxygen are like students moving books from an old library to a new one. Each student (a hemoglobin molecule) can carry a maximum of four books (100% saturation). The librarian in charge controls how many books (O_2 molecules) each student will carry, just as plasma P_{O_2} determines the amount of oxygen that binds to hemoglobin.

At the same time, the total number of books being carried depends on the number of available students, just as the amount of oxygen delivered to the tissues depends on the number of available hemoglobin molecules. For example, if there are 100 students, and the librarian gives each of them four books (100% saturation), then 400 books will be carried to the new library. If the librarian gives out only three books to each student (decreased plasma P_{O_2}), then only 300 books will go to the new library, even though each student could carry four. (Students carrying only three out of a possible four books corresponds to 75% saturation of hemoglobin). If the librarian is handing out four books per student but only 50 students show up (fewer hemoglobin molecules), then only 200 books will get to the new library, even though the students will be carrying the maximum number of books they can carry.

The physical relationship between P_{O_2} and how much oxygen binds to hemoglobin can be studied *in vitro*. Researchers expose samples of hemoglobin to various P_{O_2} levels and quantitatively determine the amount of oxygen that binds. **Oxyhemoglobin dissociation curves**, such as the one shown in Figure 18-9 ■, are the result of these *in vitro* binding studies.

The shape of the HbO_2 dissociation curve reflects the properties of the hemoglobin molecule and its affinity for oxygen. If you look at the curve, you find that at normal alveolar and arterial P_{O_2} (100 mm Hg), 98% of the hemoglobin is bound to oxygen. In other words, as blood passes through the lungs under normal conditions, hemoglobin picks up nearly the maximum amount of oxygen that it can carry.

Notice that the curve is nearly flat at P_{O_2} levels higher than 100 mm Hg (that is, the slope approaches zero). At P_{O_2} above 100 mm Hg, even large changes in P_{O_2} cause only minor changes in percent saturation. In fact, hemoglobin will not be 100% saturated until the P_{O_2} reaches nearly 650 mm Hg, a partial pressure far higher than anything we encounter in everyday life.

The flattening of the dissociation curve at higher P_{O_2} also means that alveolar P_{O_2} can fall a good bit below 100 mm Hg without significantly lowering hemoglobin saturation. As long as P_{O_2} in the alveoli (and thus in the pulmonary capillaries) stays above 60 mm Hg, hemoglobin will be more than 90% saturated and will maintain near-normal levels of oxygen

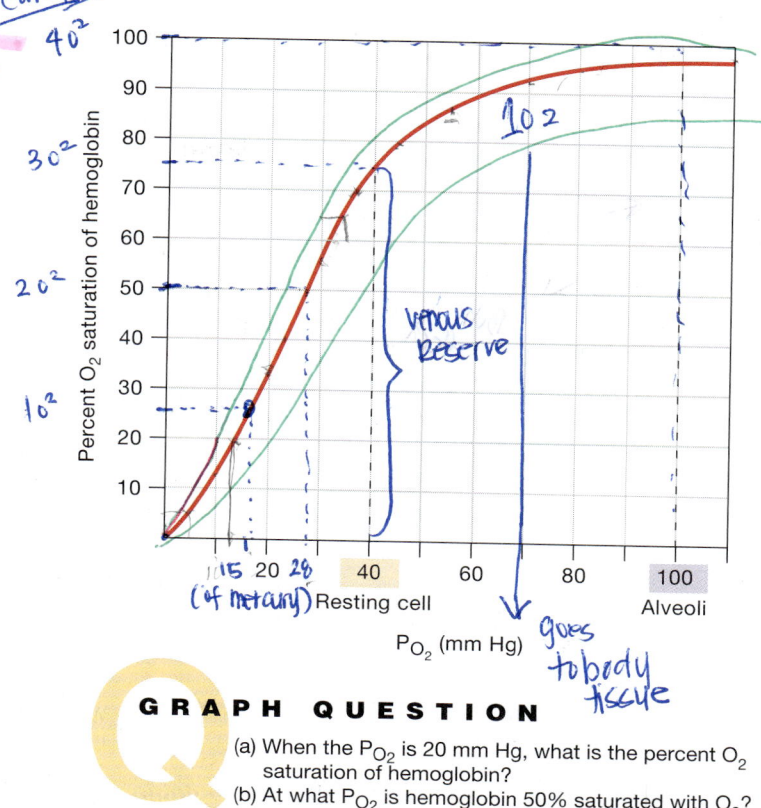

GRAPH QUESTION

(a) When the P_{O_2} is 20 mm Hg, what is the percent O_2 saturation of hemoglobin?
(b) At what P_{O_2} is hemoglobin 50% saturated with O_2?

■ **FIGURE 18-9**   *An oxygen-hemoglobin dissociation curve*

transport. However, once P_{O_2} falls below 60 mm Hg, the curve becomes steeper. The steep slope means that a small decrease in P_{O_2} causes a relatively large release of oxygen.

For example, if P_{O_2} falls from 100 mm Hg to 60 mm Hg, the percent saturation of hemoglobin goes from 98% to about 90%, a decrease of 8%. This is equivalent to a saturation change of 2% for each 10 mm Hg change. If P_{O_2} falls further, from 60 to 40 mm Hg, the percent saturation goes from 90% to 75%, a decrease of 7.5% for each 10 mm Hg. In the 40–20 mm Hg range, the curve is even steeper. Hemoglobin saturation declines from 75% to 35%, a change of 20% for each 10 mm Hg change.

What is the physiological significance of the shape of the dissociation curve? In blood leaving systemic capillaries with a P_{O_2} of 40 mm Hg (the normal partial pressure of resting cells), hemoglobin is still 75% saturated, which means that at the cells it released only one-fourth of the oxygen it is capable of carrying. The oxygen that remains bound serves as a reservoir that cells can draw on if metabolism increases.

When metabolically active tissues use additional oxygen, their cellular P_{O_2} decreases, and additional oxygen is released by hemoglobin at the cells. At a P_{O_2} of 20 mm Hg (an average value for exercising muscle), hemoglobin saturation falls to about 35%. With this 20 mm Hg decrease in P_{O_2} (40 mm Hg to 20 mm Hg), hemoglobin releases an additional 40% of the

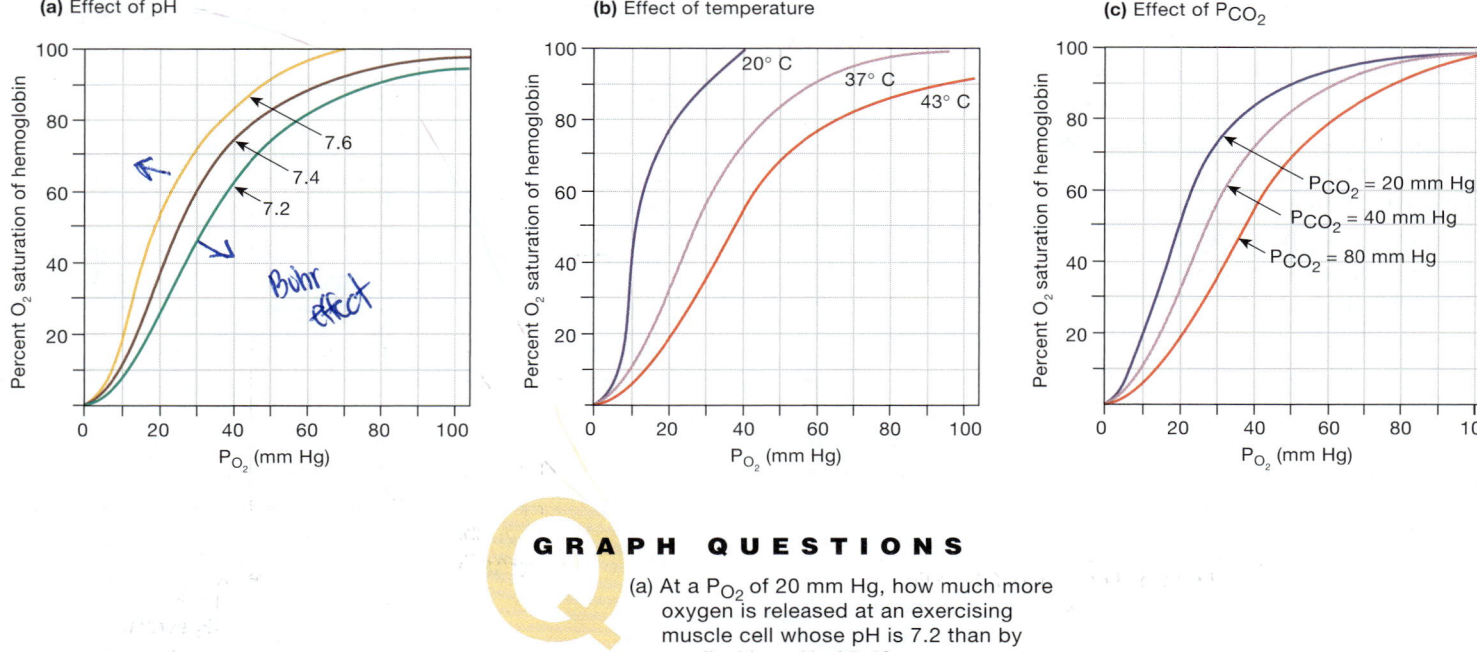

(a) Effect of pH

(b) Effect of temperature

(c) Effect of P_{CO_2}

GRAPH QUESTIONS

(a) At a P_{O_2} of 20 mm Hg, how much more oxygen is released at an exercising muscle cell whose pH is 7.2 than by a cell with a pH of 7.4?

(b) What happens to oxygen release when the exercising muscle cell warms up?

■ **FIGURE 18-10** *Physical factors alter hemoglobin's affinity for oxygen*

oxygen it is capable of carrying. This is another example of the built-in reserve capacity of the body.

CONCEPT CHECK

9. Can a person breathing 100% oxygen at sea level achieve 100% saturation of her hemoglobin?

10. What effect does hyperventilation have on the percent saturation of arterial hemoglobin? (*Hint:* see Fig. 17-15.)

Answers: p. 611

Temperature, pH, and Metabolites Affect Oxygen-Hemoglobin Binding

Any factor that changes the conformation of the hemoglobin protein may affect its ability to bind oxygen. In humans, physiological changes in plasma pH, P_{CO_2}, and temperature all alter the oxygen-binding affinity of hemoglobin. Changes in binding affinity are reflected by changes in the shape of the HbO_2 dissociation curve.

Increased temperature, increased P_{CO_2}, or decreased pH will decrease the affinity of hemoglobin for oxygen and shift the oxygen-hemoglobin dissociation curve to the right (Fig. 18-10 ■). When these factors change in the opposite direction, binding affinity increases, and the curve shifts to the left. Notice that when the curve shifts in either direction, the changes are much more pronounced in the steep part of the curve. Physiologically, this means that oxygen binding at the lungs (in the 90–100 mm Hg P_{O_2} range) is not greatly affected, but oxygen

delivery at the tissues (in the 20–40 mm Hg range) will be significantly altered.

Let's examine one example, the affinity shift that takes place when pH decreases from 7.4 (normal) to 7.2 (more acidic). (The normal range for blood pH is 7.38–7.42, but a pH of 7.2 is compatible with life.) Look at the graph in Figure 18-10a. At a P_{O_2} of 40 mm Hg (equivalent to a resting cell) and pH of 7.4, hemoglobin is about 75% saturated. At the same P_{O_2}, if the pH falls to 7.2, the percent saturation decreases to about 62%. This means that hemoglobin molecules release 13% more oxygen at pH 7.2 than they do at pH 7.4.

When does the body undergo shifts in blood pH? One situation is with maximal exertion that pushes cells into anaerobic metabolism. Anaerobic metabolism in exercising muscle fibers produces lactic acid, which in turn releases H^+ into the cytoplasm and extracellular fluid. As H^+ concentrations increase, pH falls, the affinity of hemoglobin for oxygen decreases, and the HbO_2 dissociation curve shifts to the right. More oxygen is released at the tissues as the blood becomes more acidic (pH decreases). A shift in the hemoglobin saturation curve that results from a change in pH is called the **Bohr effect**.

An additional factor that affects oxygen-hemoglobin binding is **2,3-diphosphoglycerate** (2,3-DPG; also called *2,3-bisphosphoglycerate* or *2,3-BPG*), a compound made from an intermediate of the glycolysis pathway. **Chronic hypoxia** (extended periods of low oxygen) triggers an increase in 2,3-DPG

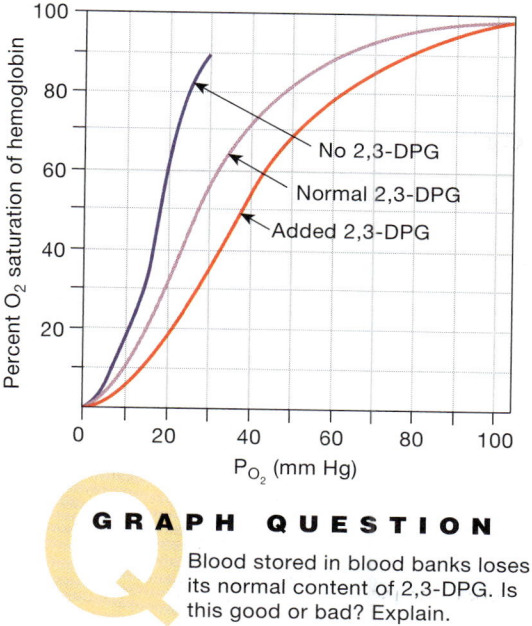

GRAPH QUESTION

Blood stored in blood banks loses its normal content of 2,3-DPG. Is this good or bad? Explain.

■ **FIGURE 18-11**   *2,3-DPG alters hemoglobin's affinity for oxygen*

production in red blood cells. Increased levels of 2,3-DPG lower the binding affinity of hemoglobin and shift the HbO_2 dissociation curve to the right (Fig. 18-11 ■). Ascent to high altitude and anemia are two situations that increase 2,3-DPG production.

Changes in hemoglobin's structure also change its oxygen-binding affinity. For example, fetal hemoglobin has gamma chain isoforms for two of its subunits. The presence of gamma chains enhances the ability of fetal hemoglobin to bind oxygen in the low-oxygen environment of the placenta. The altered binding affinity is reflected by the shape of the fetal HbO_2 dissociation curve, which differs from the shape of the adult curve (Fig. 18-12 ■).

Figure 18-13 ■ summarizes the factors that influence the total oxygen content of arterial blood.

CONCEPT CHECK

11. A muscle that is actively contracting may have a cellular P_{O_2} of 25 mm Hg. What happens to oxygen binding to hemoglobin at this low P_{O_2}? What is the P_{O_2} of the venous blood leaving the active muscle?

Answers: p. 611

Carbon Dioxide Is Transported in Three Ways

Gas transport in the blood includes carbon dioxide removal from the cells as well as oxygen delivery to cells. The hemoglobin molecule, as we will see, also plays an important role in CO_2 transport. Carbon dioxide is a by-product of cellular respiration [🔁 p. 107]. It is more soluble in body fluids than oxygen is, but the cells produce far more CO_2 than can dissolve in the plasma.

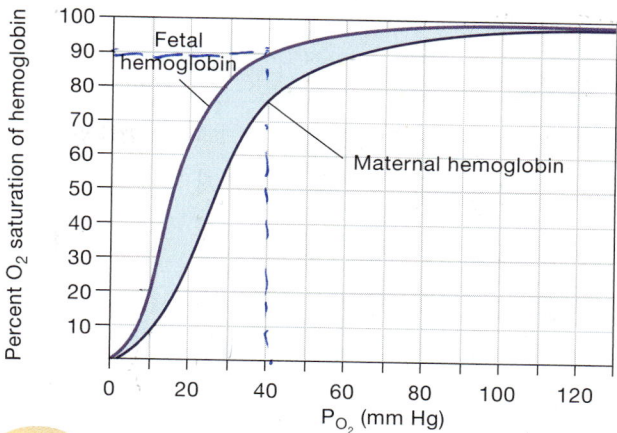

GRAPH QUESTIONS

(a) Because of incomplete gas exchange across the thick membranes of the placenta, hemoglobin in fetal blood leaving the placenta is 80% saturated with oxygen. What is the P_{O_2} of that placental blood?

(b) Blood in the vena cava of the fetus has a P_{O_2} around 10 mm Hg. What is the percent O_2 saturation of maternal hemoglobin at the same P_{O_2}?

■ **FIGURE 18-12**   *Differences in the oxygen-binding properties of maternal and fetal hemoglobin*

Only about 7% of the CO_2 carried by venous blood is dissolved in the blood. The remaining 93% diffuses into red blood cells, where 70% is converted to bicarbonate ion, as explained below, and 23% binds to hemoglobin (Hb–CO_2). Figure 18-14 ■ summarizes these three mechanisms of carbon dioxide transport in the blood.

Why is removing CO_2 from the body so important? The reason is that elevated P_{CO_2} (*hypercapnia*) causes the pH disturbance known as *acidosis*. Extremes of pH interfere with hydrogen bonding of molecules and can denature proteins [🔁 p. 37]. Abnormally high P_{CO_2} levels also depress central nervous system function, causing confusion, coma, or even death. Thus, CO_2 is a potentially toxic waste product that must be removed by the lungs.

CO_2 and Bicarbonate Ions   As just noted, about 70% of the CO_2 that enters the blood is transported to the lungs as bicarbonate ions (HCO_3^-) dissolved in the plasma. The conversion of CO_2 to HCO_3^- serves two purposes: (1) it provides an additional means by which CO_2 can be transported from cells to lungs, and (2) HCO_3^- is available to act as a buffer for metabolic acids [🔁 p. 38], thereby helping stabilize the body's pH.

How does CO_2 turn into HCO_3^-? The rapid conversion depends on the presence of **carbonic anhydrase (CA)**, an enzyme found concentrated in red blood cells. Let's see how this happens. Dissolved CO_2 in the plasma diffuses into red blood cells, where it may react with water in the presence of carbonic

18

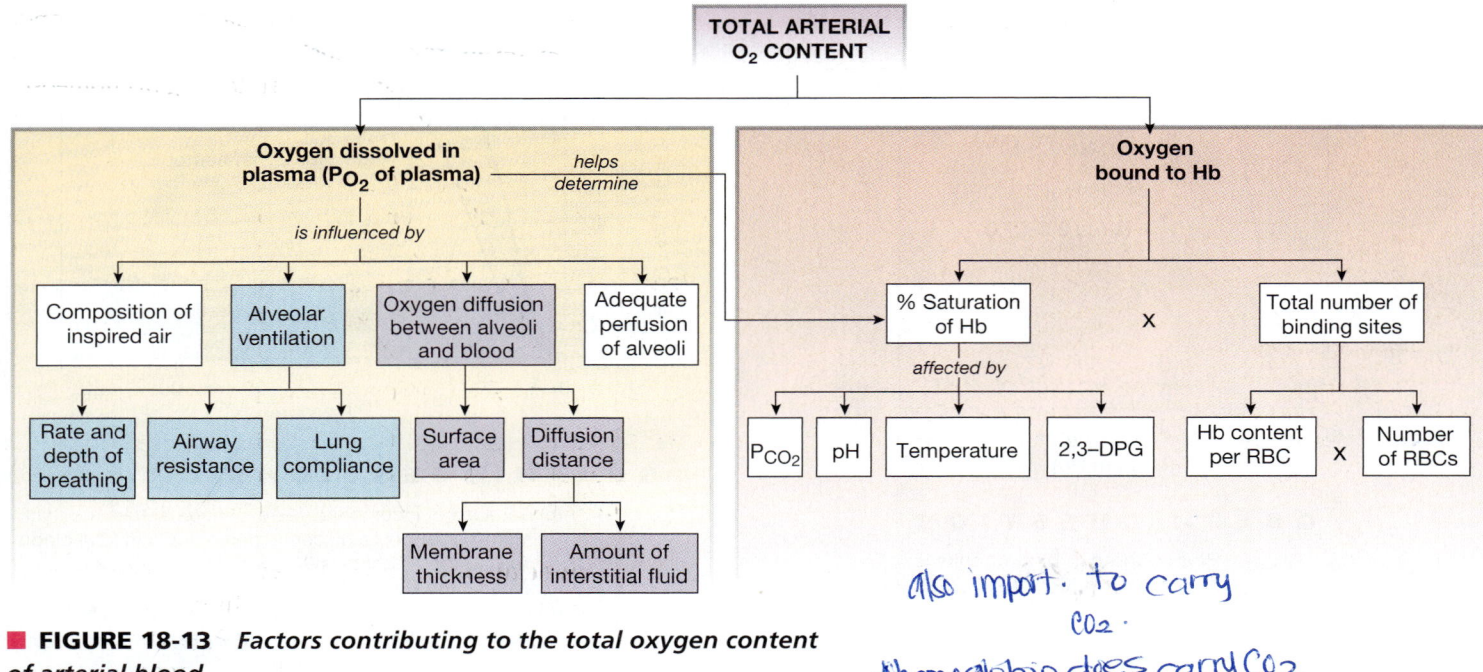

FIGURE 18-13 *Factors contributing to the total oxygen content of arterial blood*

also import. to carry CO2.

Hemoglobin does carry CO2

anhydrase to form *carbonic acid* (top portion of Fig. 18-14). Carbonic acid then dissociates into a hydrogen ion and a bicarbonate ion:

$$CO_2 + H_2O \rightleftharpoons \underset{\substack{\text{carbonic} \\ \text{acid}}}{H_2CO_3} \rightleftharpoons H^+ + HCO_3^-$$

Because the carbonic acid dissociates readily, we sometimes ignore the intermediate step and summarize the reaction as:

$$CO_2 + H_2O \rightleftharpoons H^+ + HCO_3^-$$

This reaction is reversible. The rate in either direction depends on the relative concentrations of the substrates and obeys the law of mass action.

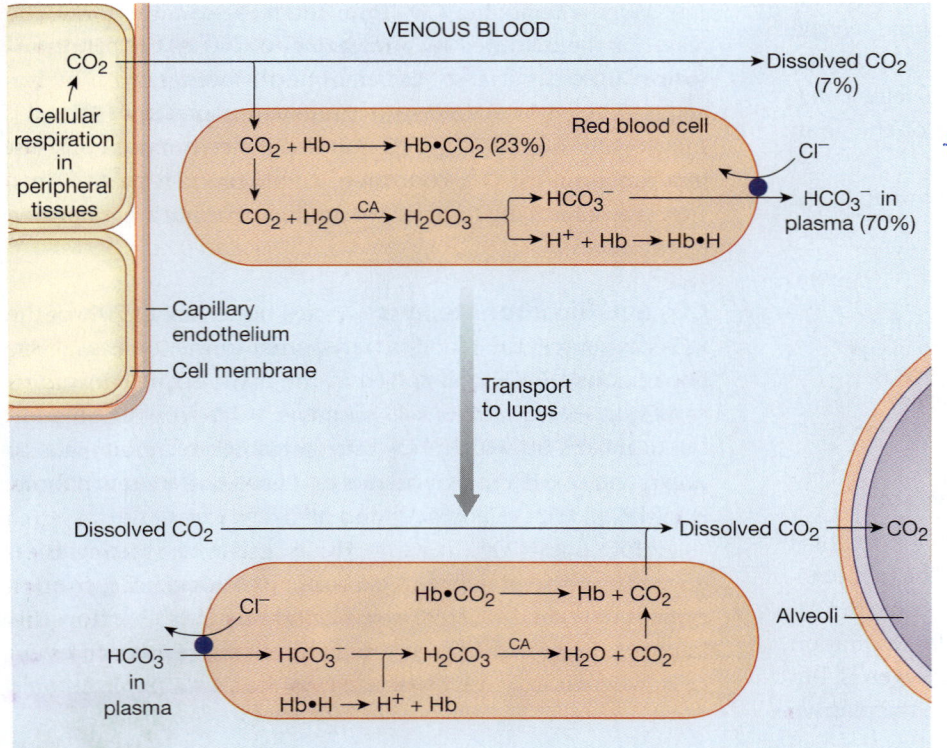

FIGURE 18-14 *Carbon dioxide transport in the blood*

CA = carbonic anhydrase.

★ if CO2 levels are high, hemoglobin can't carry all of H2. So it goes the plasma, plasma then drops.

RUNNING PROBLEM

The usual homeostatic response to high-altitude hypoxia is hyperventilation, which begins on arrival. Hyperventilation enhances alveolar ventilation, but this may not help elevate arterial P_{O_2} levels significantly when atmospheric P_{O_2} is low. However, hyperventilation does lower plasma P_{CO_2}.

Question 6:
What happens to plasma pH during hyperventilation? (Hint: apply the law of mass action to figure out what happens to the balance between CO_2 and $H^+ + HCO_3^-$.)

Question 7:
How does this change in pH affect oxygen binding at the lungs when P_{O_2} is decreased? How does it affect unloading of oxygen at the cells?

| 588 | 592 | 593 | 596 | **601** | 606 | 608 |

The conversion of carbon dioxide to H^+ and HCO_3^- will continue until equilibrium is reached. (Water is always in excess in the body, so water concentration plays no role in the dynamic equilibrium of this reaction.) To keep the reaction going, the products (H^+ and HCO_3^-) must be removed from the cytoplasm of the red blood cell. If the product concentrations are kept low, the reaction cannot reach equilibrium. Carbon dioxide continues to move out of plasma into the red blood cells, which in turn allows more CO_2 to diffuse out of tissues into the blood.

Two separate mechanisms remove free H^+ and HCO_3^-. In the first, bicarbonate leaves the red blood cell on an antiport protein [⟳p. 139]. This transport process, known as the **chloride shift**, exchanges one HCO_3^- for one Cl^-. The one-for-one exchange maintains electrical neutrality so that the cell's membrane potential is not affected. The transfer of HCO_3^- into the plasma makes this buffer [⟳p. 38] available to moderate pH changes caused by the production of metabolic acids. Bicarbonate is the most important extracellular buffer in the body.

Hemoglobin and H^+   The second mechanism removes free H^+ from the red blood cell cytoplasm. Hemoglobin within the red blood cell acts as a buffer and binds hydrogen ions in the reaction

$$H^+ + Hb \rightarrow Hb \cdot H$$

Hemoglobin's buffering of H^+ is an important step that prevents large changes in the body's pH. If blood P_{CO_2} is elevated much above normal, the hemoglobin buffer cannot soak up all the H^+ produced from the reaction of CO_2 and water. In those

cases, excess H^+ accumulates in the plasma, causing the condition known as **respiratory acidosis**. Further information on the role of the respiratory system in maintaining pH homeostasis is found in Chapter 20.

Hemoglobin and CO_2   Although most carbon dioxide that enters red blood cells is converted to bicarbonate ions, about 23% of the CO_2 in venous blood binds directly to hemoglobin. When oxygen leaves its binding sites on the hemoglobin molecule, CO_2 binds with free hemoglobin at exposed amino groups ($-NH_2$), forming **carbaminohemoglobin**:

$$CO_2 + Hb \rightleftharpoons Hb \cdot CO_2 \text{ (carbaminohemoglobin)}$$

The formation of carbaminohemoglobin is facilitated by the presence of CO_2 and H^+ because both these factors decrease hemoglobin's binding affinity for oxygen (see Fig. 18-10).

CO_2 Removal at the Lungs   When venous blood reaches the lungs, the processes that took place in the systemic capillaries reverse (bottom portion of Fig. 18-14). The P_{CO_2} of the alveoli is lower than that of venous blood in the pulmonary capillaries. Therefore, CO_2 diffuses down its pressure gradient—in other words, out of plasma into the alveoli—and the plasma P_{CO_2} begins to fall.

The decrease in plasma P_{CO_2} allows dissolved CO_2 to diffuse out of the red blood cells. As CO_2 levels in the red blood cells decrease, the equilibrium of the CO_2-HCO_3^- reaction is disturbed, shifting toward production of more CO_2. Removal of CO_2 causes H^+ to leave the hemoglobin molecules, and the chloride shift reverses: Cl^- returns to the plasma in exchange for HCO_3^- moving back into the red blood cells. The HCO_3^- and newly released H^+ re-form into carbonic acid, which is then converted into water and CO_2. This CO_2 is then free to diffuse out of the red blood cell and into the alveoli.

Figure 18-15 ■ shows the combined transport of CO_2 and O_2 in the blood. At the alveoli, O_2 diffuses down its pressure gradient, moving from the alveoli into the plasma and then from the plasma into the red blood cells. Hemoglobin binds to O_2, increasing the amount of oxygen that can be transported to the cells.

At the cells, the process reverses. Because P_{O_2} is lower in cells than in the arterial blood, O_2 diffuses from the plasma into the cells. The decrease in plasma P_{O_2} causes hemoglobin to release O_2, making additional oxygen available to enter cells.

Carbon dioxide from aerobic metabolism simultaneously leaves cells and enters the blood, dissolving in the plasma. From there, CO_2 enters red blood cells, where most is converted to HCO_3^- and H^+. The HCO_3^- is returned to the plasma in exchange for a Cl^- while the H^+ binds to hemoglobin. A fraction of the CO_2 that enters red blood cells also binds directly to hemoglobin. At the lungs, the process reverses as CO_2 diffuses out of the pulmonary capillaries and into the alveoli.

18

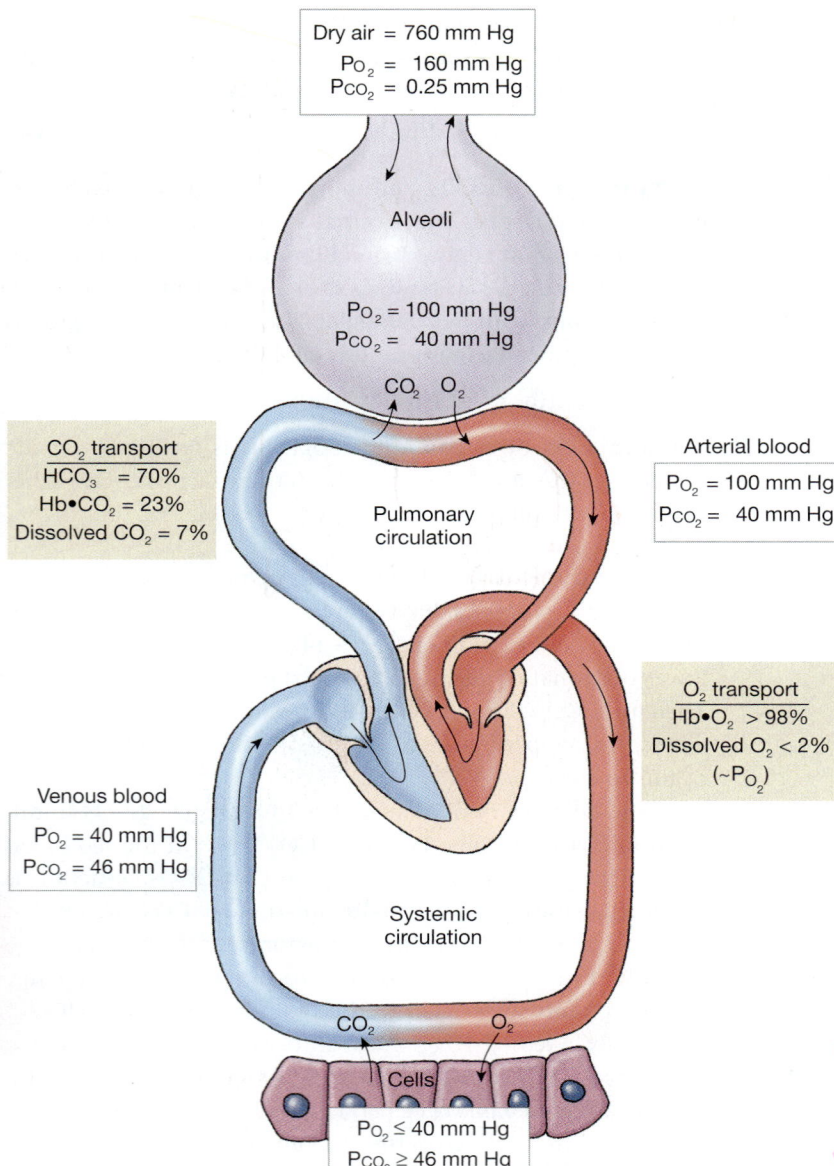

CO₂ transport
$HCO_3^- = 70\%$
$Hb \cdot CO_2 = 23\%$
Dissolved $CO_2 = 7\%$

Dry air = 760 mm Hg
P_{O_2} = 160 mm Hg
P_{CO_2} = 0.25 mm Hg

Alveoli

P_{O_2} = 100 mm Hg
P_{CO_2} = 40 mm Hg

CO_2 O_2

Arterial blood
P_{O_2} = 100 mm Hg
P_{CO_2} = 40 mm Hg

Pulmonary
circulation

O_2 transport
$Hb \cdot O_2 > 98\%$
Dissolved $O_2 < 2\%$
$(\sim P_{O_2})$

Venous blood
P_{O_2} = 40 mm Hg
P_{CO_2} = 46 mm Hg

Systemic
circulation

CO_2 O_2

Cells

$P_{O_2} \leq 40$ mm Hg
$P_{CO_2} \geq 46$ mm Hg

■ **FIGURE 18-15**   *Summary of gas transport*

To understand fully how the respiratory system coordinates delivery of oxygen to the lungs with transport of oxygen in the circulation, we will now consider the central nervous system control of ventilation.

CONCEPT CHECK

12. How would an obstruction of the airways affect alveolar ventilation, arterial P_{CO_2}, and the body's pH?

Answers: p. 611

REGULATION OF VENTILATION

Breathing is a rhythmic process that usually occurs without conscious thought or awareness. In that respect, it resembles the rhythmic beating of the heart. However, skeletal muscles, unlike autorhythmic cardiac muscles, are not able to contract

spontaneously. Instead, skeletal muscle contraction must be initiated by somatic motor neurons, which in turn are controlled by the central nervous system. The intrinsic rhythmicity of ventilation is subject to modulation by a variety of influences, including chemoreceptor reflexes, interaction with the cardiovascular control center, emotions, and conscious and unconscious control by higher brain centers.

In the respiratory system, contraction of the diaphragm and intercostals is initiated by groups of neurons in the pons and medulla of the brain stem (Fig. 18-16 ■). These neurons take the form of a network with a **central pattern generator** that has intrinsic rhythmic activity [🔁 p. 445]. Research suggests the rhythmic activity arises from pacemaker neurons that have unstable membrane potentials.

Direct study of the brain centers controlling ventilation is difficult because of the complexity of the neuronal network and its anatomical location. Consequently, some of our understanding of how ventilation is controlled has come from observing patients with brain damage. Other information has come from animal experiments in which the neural connections between major parts of the brain stem were severed.

From these observations and the hypotheses formed from them, the following model for the control of ventilation has been proposed. Although some parts of the model are well supported with experimental evidence, other aspects are still under investigation. The model states that:

1. Respiratory neurons in the medulla control inspiration and expiration.
2. Neurons in the pons modulate ventilation.
3. The rhythmic pattern of breathing arises from a network of spontaneously discharging neurons.
4. Ventilation is subject to modulation by various chemoreceptor-linked reflexes and by higher brain centers.

Neurons in the Medulla Control Breathing

Classic descriptions of how the brain controls ventilation divided groups of neurons into various control centers. More recent descriptions, however, are less specific about assigning function to particular centers and simply refer to the network of neurons in the brain stem as the central pattern generator.

The central pattern generator functions automatically throughout a person's life but can also be controlled voluntarily, up to a point. Complicated synaptic interactions between neurons in the network create the rhythmic cycles of inspiration and expiration, influenced continuously by sensory input from chemoreceptors for CO_2, O_2, and H^+. Ventilation pattern depends in large part on the levels of these three substances in the extracellular fluid.

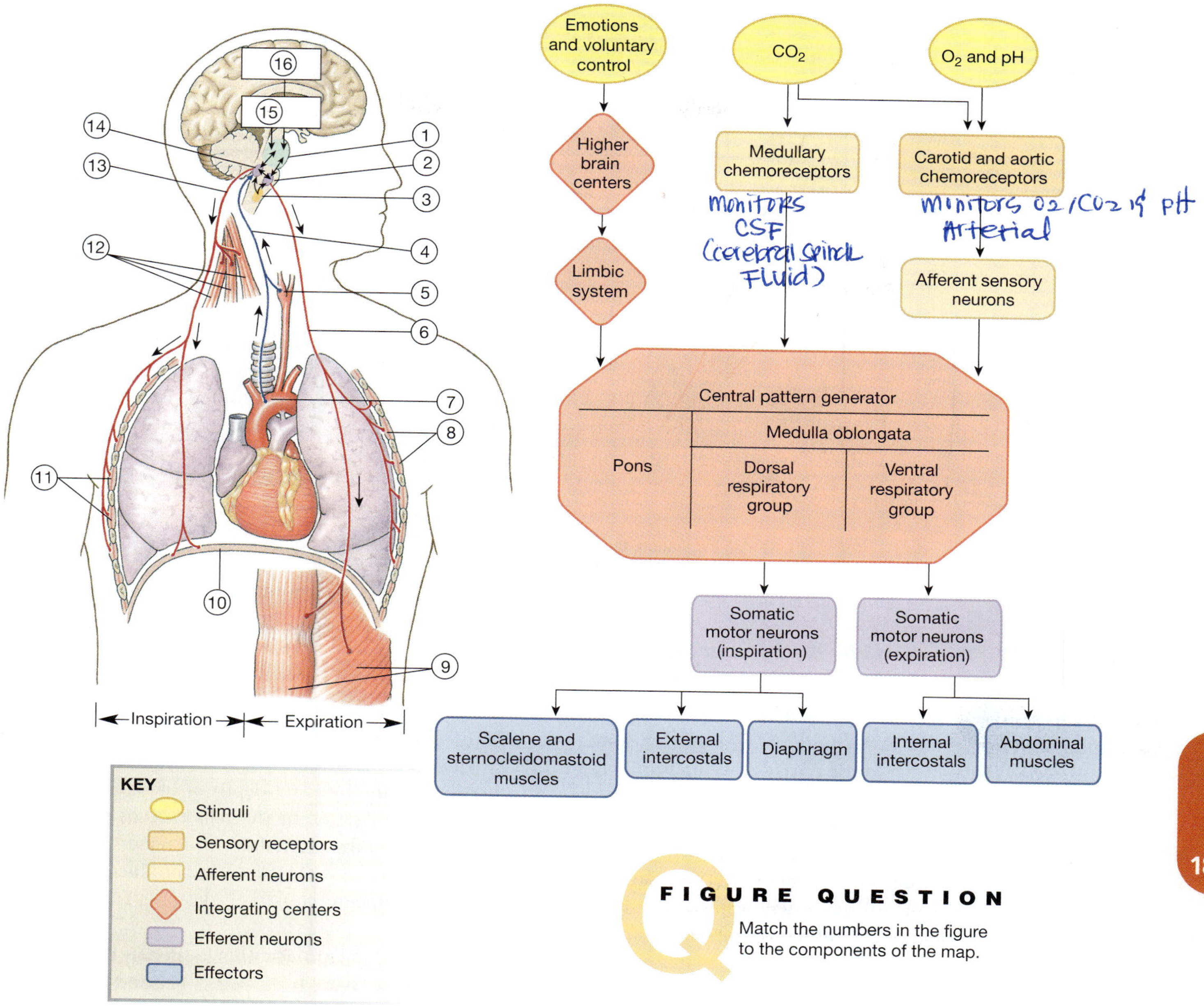
16
15
14
13
12
11
10
9
8
7
6
5
4
3
2
1
Inspiration
Expiration
Emotions and voluntary control
CO_2
O_2 and pH
Higher brain centers
Limbic system
Medullary chemoreceptors
monitors CSF (cerebral spinal fluid)
Carotid and aortic chemoreceptors
monitors O_2/CO_2 & pH Arterial
Afferent sensory neurons
Central pattern generator
Pons
Medulla oblongata
Dorsal respiratory group
Ventral respiratory group
Somatic motor neurons (inspiration)
Somatic motor neurons (expiration)
Scalene and sternocleidomastoid muscles
External intercostals
Diaphragm
Internal intercostals
Abdominal muscles
KEY
Stimuli
Sensory receptors
Afferent neurons
Integrating centers
Efferent neurons
Effectors

18

FIGURE QUESTION

Match the numbers in the figure to the components of the map.

■ **FIGURE 18-16** ***Reflex control of ventilation***

Chemoreceptors monitor blood gases and pH. Control centers in the brain stem regulate activity in somatic motor neurons leading to respiratory muscles.

Although there is still much to be learned about the central pattern generator, we do know that respiratory neurons are concentrated in two centers in the medulla oblongata. The **dorsal respiratory group** (DRG) of neurons contains mostly *inspiratory neurons* that control the diaphragm (Fig. 18-16). The **ventral respiratory group** (VRG) of neurons controls muscles used for active expiration and some inspiratory muscles, particularly those active during greater-than-normal inspiration, such as occurs during vigorous exercise. Fibers from the VRG also innervate muscles of the larynx, pharynx, and tongue to keep the upper airways open during breathing.

During quiet respiration at 12 breaths/min, the inspiratory neurons gradually increase stimulation of the inspiratory muscles for 2 seconds. This increase is sometimes called *ramping* because of the shape of the graph of inspiratory neuron activity (Fig. 18-17 ■). A few inspiratory neurons fire to begin the ramp. The firing of these neurons recruits other inspiratory neurons to fire in an apparent positive feedback loop. As more neurons fire, more skeletal muscle fibers are recruited. The rib cage expands smoothly as the diaphragm contracts.

At the end of 2 seconds, the inspiratory neurons abruptly stop firing, and the respiratory muscles relax. Over the next

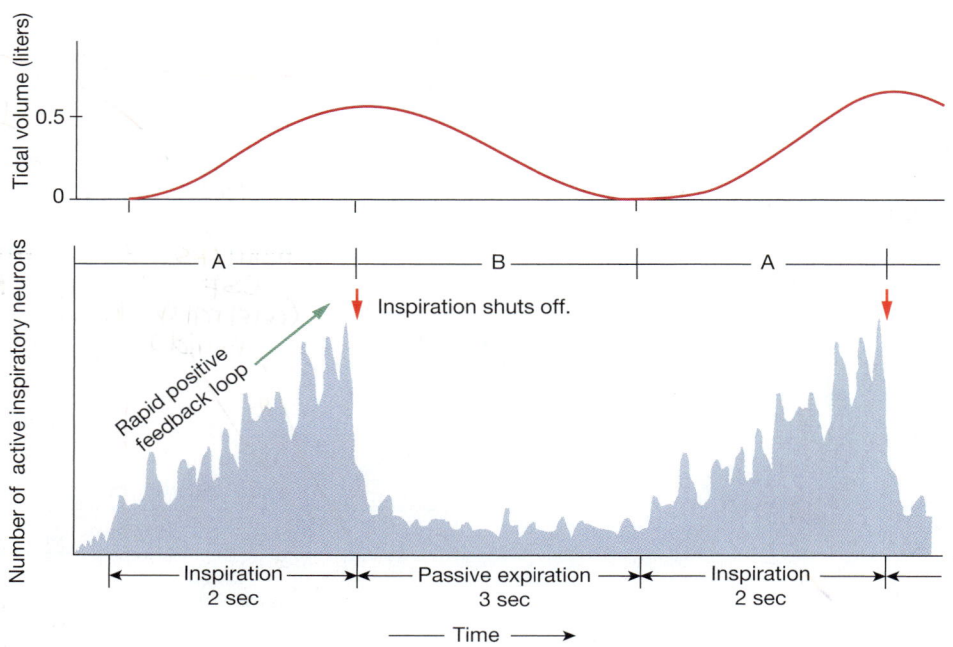

During inspiration, the activity of inspiratory neurons increases steadily, apparently through a positive feedback mechanism. At the end of inspiration, the activity shuts off abruptly and expiration takes place through recoil of elastic lung tissue.

■ **FIGURE 18-17**   *Neural activity during quiet breathing*

3 seconds, passive expiration occurs because of elastic recoil of the inspiratory muscles and elastic lung tissue. However, some motor neuron activity occurs during passive expiration, suggesting that perhaps muscles in the upper airways contract to slow the flow of air out of the respiratory system.

The expiratory neurons of the ventral respiratory group remain mostly inactive during quiet respiration. They function primarily during forced breathing, when inspiratory movements are exaggerated, and during active expiration.

In forced breathing, increased activity of inspiratory neurons stimulates accessory inspiratory muscles, such as the sternocleidomastoids. Contraction of the accessory inspiratory muscles enhances expansion of the thorax by raising the sternum and upper ribs.

With active expiration, expiratory neurons from the ventral respiratory group activate the internal intercostal and abdominal muscles. There seems to be some reciprocal inhibition between inspiratory and expiratory neurons, as inspiratory neurons are inhibited during active expiration.

Carbon Dioxide, Oxygen, and pH Influence Ventilation

Sensory input from central and peripheral chemoreceptors [🔁 p. 330] modifies the rhythmicity of the central pattern generator. Carbon dioxide is the primary stimulus for changes in ventilation. Oxygen and plasma pH play lesser roles.

The chemoreceptors for oxygen and carbon dioxide are strategically associated with the arterial circulation. If too little oxygen is present in arterial blood destined for the brain and other tissues, the rate and depth of breathing increase. If the rate of CO_2 production by the cells exceeds the rate of CO_2 removal by the lungs, arterial P_{CO_2} increases, and ventilation is intensified to match CO_2 removal to production. These homeostatic reflexes operate constantly, keeping arterial P_{O_2} and P_{CO_2} within a narrow range.

Peripheral chemoreceptors located in the carotid and aortic arteries sense changes in the P_{O_2}, pH, and P_{CO_2} of the plasma (Fig. 18-16). These **carotid** and **aortic bodies** are close to the locations of the baroreceptors involved in reflex control of blood pressure [🔁 p. 521]. **Central chemoreceptors** in the brain respond to changes in the concentration of CO_2 in the cerebrospinal fluid. These central receptors lie on the ventral surface of the medulla, close to neurons involved in respiratory control.

Peripheral Chemoreceptors   When specialized **glomus cells** [*glomus*, a ball-shaped mass] in the carotid and aortic bodies are activated by a decrease in P_{O_2} or pH or by an increase in P_{CO_2}, they trigger a reflex increase in ventilation. The details of glomus cell function remain to be worked out, but the basic mechanism by which these chemoreceptors respond to low oxygen is similar to the mechanism you learned for insulin release by pancreatic beta cells [🔁 p. 166] or taste transduction in taste buds [🔁 p. 345].

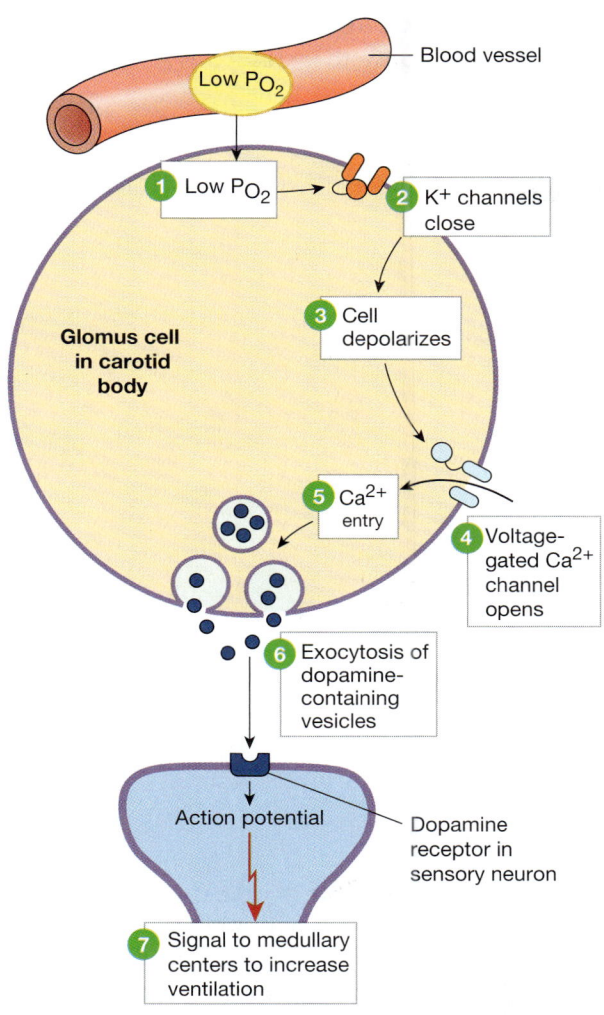

■ FIGURE 18-18   *Carotid body oxygen sensor releases neurotransmitter when P_{O_2} decreases*

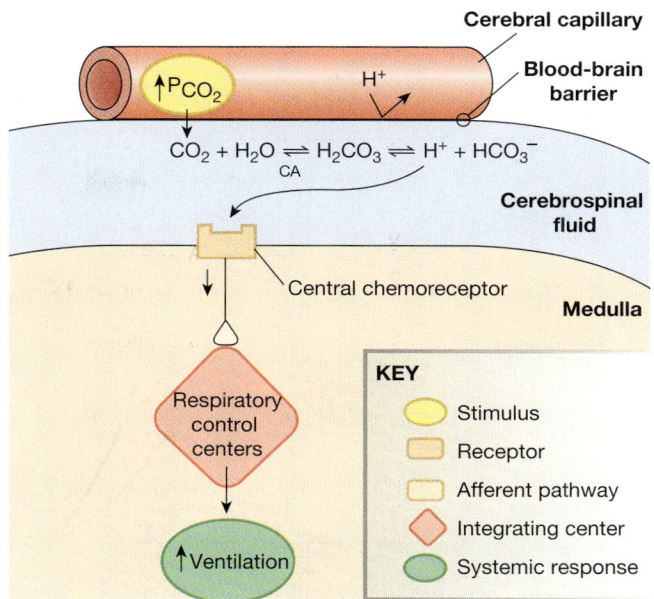

■ FIGURE 18-19   *Central chemoreceptors monitor CO_2 in cerebrospinal fluid*

CA = carbonic anhydrase.

In all three examples, a stimulus inactivates K^+ channels, causing the receptor cell to depolarize (Fig. 18-18 ■). Depolarization opens voltage-gated Ca^{2+} channels, and Ca^{2+} entry causes exocytosis of neurotransmitter (probably dopamine in glomus cells) onto the sensory neuron. In the carotid and aortic bodies, dopamine initiates action potentials in sensory neurons leading to the brain stem respiratory centers, signaling them to increase ventilation.

Under most circumstances, oxygen is not an important factor in modulating ventilation because arterial P_{O_2} must fall to less than 60 mm Hg before ventilation is stimulated. This large decrease in P_{O_2} is equivalent to ascending to an altitude of 3000 m. (For reference, Denver is located at an altitude of 1609 m.)

Because the peripheral chemoreceptors respond only to dramatic changes in arterial P_{O_2}, they do not play a role in the everyday regulation of ventilation. However, unusual physiological conditions, such as ascending to high altitude, and pathological conditions, such as chronic obstructive pulmonary disease

(COPD), can reduce arterial P_{O_2} to levels that are low enough to activate the peripheral chemoreceptors.

The peripheral chemoreceptors in the carotid and aortic bodies are more responsive to changes in P_{O_2} than to changes in plasma pH and P_{CO_2}. However, any condition that reduces plasma pH or increases P_{CO_2} stimulates ventilation by way of the peripheral chemoreceptors.

Central Chemoreceptors   The most important chemical controller of ventilation is carbon dioxide, mediated through central chemoreceptors located in the medulla (Fig. 18-19 ■). These receptors set the respiratory pace, providing continuous input into the central pattern generator. When arterial P_{CO_2} increases, CO_2 crosses the blood-brain barrier quite rapidly and activates the central chemoreceptors. These receptors signal the central pattern generator to increase the rate and depth of ventilation, thereby increasing alveolar ventilation and removing CO_2 from the blood (Fig. 18-20 ■).

Although we say that the central chemoreceptors monitor CO_2, they actually respond to pH changes in the cerebrospinal fluid. Carbon dioxide that diffuses across the blood-brain barrier into the cerebrospinal fluid is converted to bicarbonate and H^+:

$$CO_2 + H_2O \rightleftharpoons H_2CO_3^- \rightleftharpoons H^+ + HCO_3^-$$

Experiments indicate that the H^+ produced by this reaction is what actually initiates the chemoreceptor reflex, rather than the increased level of CO_2.

Note, however, that pH changes in the plasma *do not* usually influence the central chemoreceptors directly. Free H^+ in

18

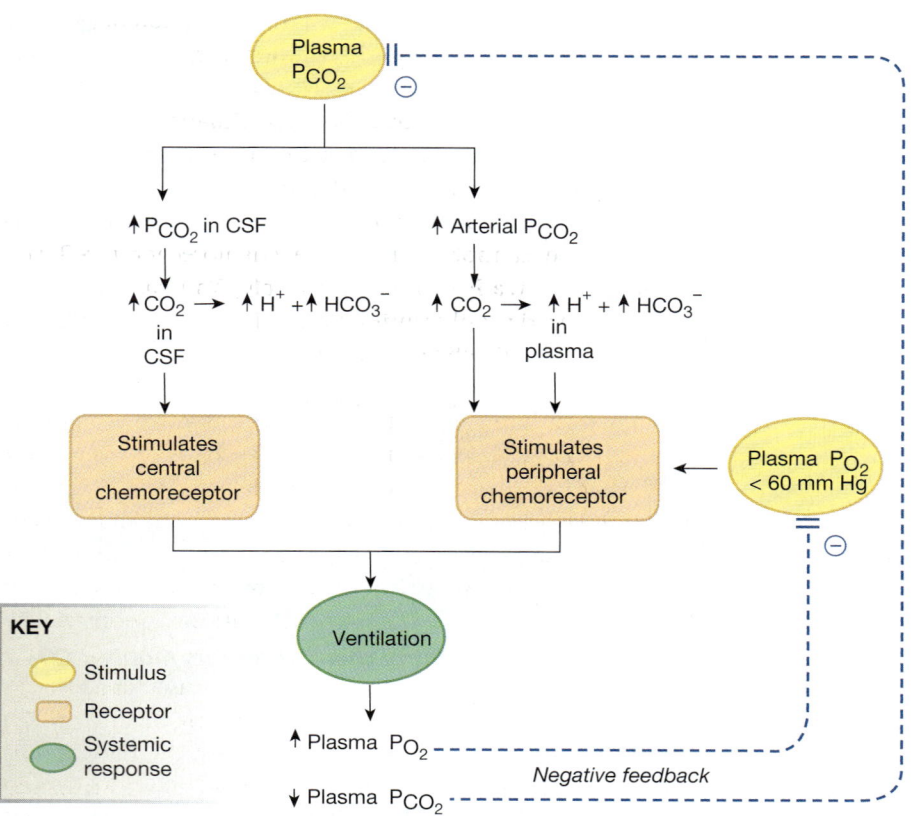

■ **FIGURE 18-20** *Chemoreceptor response to increased* P_{CO_2}

the plasma crosses the blood-brain barrier very slowly and therefore has little effect on the central chemoreceptors.

When plasma P_{CO_2} increases, the central chemoreceptors initially respond strongly by increasing ventilation. However, if P_{CO_2} remains elevated for several days, ventilation falls back toward normal rates as the chemoreceptors adapt. This adaptation occurs because over time the choroid plexus begins to transport HCO_3^- into the cerebrospinal fluid. These ions act as a buffer, removing H^+ from the cerebrospinal fluid and decreasing H^+ stimulation of the central pattern generator.

Fortunately for people with chronic lung diseases, the response of peripheral chemoreceptors to low arterial P_{O_2} remains intact over time, even though central chemoreceptors adapt to high P_{CO_2}. In some situations, low P_{O_2} becomes the primary chemical stimulus for ventilation. For example, patients with severe chronic lung disease, such as emphysema, have chronic hypercapnia and hypoxia. Their arterial P_{CO_2} may rise to 50–55 mm Hg (normal is 35–45) while their P_{O_2} falls to 45–50 mm Hg (normal 75–100). Because these levels are chronic, the central chemoreceptors adapt to the elevated P_{CO_2}. Most of the chemical stimulus for ventilation in this situation then comes from low P_{O_2}, sensed by the carotid and aortic chemoreceptors. If these patients are given too much oxygen, they may stop breathing because their chemical stimulus for ventilation is eliminated.

The central chemoreceptors respond to decreased arterial P_{CO_2} as well as to increased P_{CO_2}. If alveolar P_{CO_2} falls, as it

might during hyperventilation, plasma P_{CO_2} and cerebrospinal fluid P_{CO_2} follow suit. As a result, central chemoreceptor activity declines, and the central pattern generator slows the ventilation rate. When ventilation decreases, carbon dioxide begins to accumulate in alveoli and the plasma. Eventually, the arterial P_{CO_2} rises above the threshold level for the chemoreceptors. At that point, the receptors fire, and the central pattern generator again increases ventilation.

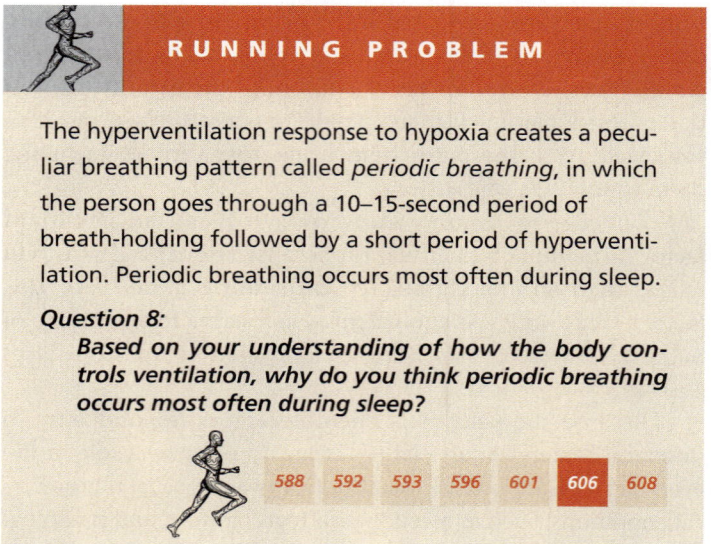

RUNNING PROBLEM

The hyperventilation response to hypoxia creates a peculiar breathing pattern called *periodic breathing*, in which the person goes through a 10–15-second period of breath-holding followed by a short period of hyperventilation. Periodic breathing occurs most often during sleep.

Question 8:
Based on your understanding of how the body controls ventilation, why do you think periodic breathing occurs most often during sleep?

| 588 | 592 | 593 | 596 | 601 | **606** | 608 |

Protective Reflexes Guard the Lungs

In addition to the chemoreceptor reflexes that help regulate ventilation, the body has protective reflexes that respond to physical injury or irritation of the respiratory tract and to over-inflation of the lungs. The major protective reflex is *bronchoconstriction,* mediated through parasympathetic neurons that innervate bronchiolar smooth muscle. Inhaled particles or noxious gases stimulate **irritant receptors** in the airway mucosa. The irritant receptors send signals through sensory neurons to control centers in the central nervous system that trigger bronchoconstriction. Protective reflex responses also include coughing and sneezing.

The **Hering-Breuer inflation reflex** prevents overexpansion of the lungs during strenuous exercise. If tidal volume exceeds 1 liter, stretch receptors in the lung signal the brain stem to terminate inspiration. However, because normal tidal volume is only 500 mL, this reflex does not operate during quiet breathing and mild exertion.

Higher Brain Centers Affect Patterns of Ventilation

Conscious and unconscious thought processes also affect respiratory activities. Higher centers in the hypothalamus and cerebrum can alter the activity of the central pattern generator and change ventilation rate and depth. Voluntary control of ventilation falls into this category. Higher brain center control is not a *requirement* for ventilation, however. Even if the brain stem above the pons is severely damaged, essentially normal respiratory cycles continue.

Respiration can also be affected by stimulation of portions of the limbic system. As a result, emotional and autonomic activities such as fear and excitement may affect the pace and depth of respiration. In some of these situations, the neural pathway goes directly to the somatic motor neurons, bypassing the central pattern generator in the brain stem.

Although we can temporarily alter our respiratory performance, we cannot override the chemoreceptor reflexes. Holding your breath is a good example. You can hold your breath voluntarily only until elevated P_{CO_2} in the blood and cerebrospinal fluid activates the chemoreceptor reflex, forcing you to inhale.

Small children having temper tantrums sometimes attempt to manipulate parents by threatening to hold their breath until they die. However, the chemoreceptor reflexes make it impossible for the children to carry out that threat. Extremely strong-willed children can continue holding their breath until they turn blue and pass out from hypoxia, but once they are unconscious, normal breathing automatically resumes.

Breathing is intimately linked to cardiovascular function. The integrating centers for both functions are located in the brain stem, and interneurons project between the two centers, allowing signaling back and forth. In Chapter 20 we will examine the integration of cardiovascular, respiratory, and renal function as the three systems work together to maintain fluid and acid-base homeostasis. First, however, we examine the physiology of the kidneys in Chapter 19.

18

HIGH ALTITUDE

On May 29, 1953, Edmund Hillary and Tenzing Norgay of the British Everest Expedition were the first humans to reach the summit of Mt. Everest. They carried supplemental oxygen with them, as it was believed that this feat was impossible without it. In 1978, however, Reinhold Messner and Peter Hebeler achieved the "impossible." On May 8, they struggled to the summit using sheer willpower and no extra oxygen. In Messner's words, "I am nothing more than a single narrow gasping lung, floating over the mists and summits." Learn more about these Everest expeditions by doing a Google search for *Hillary Everest* or *Messner Everest*.

To learn more about different types of mountain sickness, see "All about altitude illness," *www.high-altitude-medicine. com;* "High altitude medicine," *American Family Physician* 1998 Apr. 15, *www.aafp.org/afp/980415ap/harris.html;* and "High-altitude pulmonary edema," *New England Journal of Medicine* 346(21):1606–1607, 2002 May 23, *www.nejm.org.*

In this running problem you learned about normal and abnormal responses to high altitude. Check your understanding of the physiology behind this respiratory challenge by comparing your answers with the information in the following table.

	QUESTION	FACTS	INTEGRATION AND ANALYSIS
1	What is the P_{O_2} of inspired air reaching the alveoli at 2500 m, where dry atmospheric pressure is 542 mm Hg? How does this value for P_{O_2} compare with the P_{O_2} value for fully-humidified air at sea level?	Water vapor contributes a partial pressure of 47 mm Hg to fully humidified air. Oxygen is 21% of dry air. Normal atmospheric pressure at sea level is 760 mm Hg.	Correction for water vapor: 542 − 47 = 495 mm Hg × 21% = P_{O_2} = 104 mm Hg. In humidified air at sea level, P_{O_2} = 150 mm Hg.
2	Why would someone with HAPE be short of breath?	Pulmonary edema increases the diffusion distance for oxygen.	The increased diffusion distance worsens the normal hypoxia of altitude.
3	Based on what you learned about mechanisms for matching ventilation and perfusion in the lung, can you explain why patients with HAPE have elevated pulmonary arterial blood pressure?	Low oxygen levels constrict pulmonary arterioles.	Constriction of pulmonary arterioles causes blood to collect in the pulmonary arteries behind the constriction. This increases arterial blood pressure.
4	How does adding erythrocytes to the blood help a person acclimatize to high altitude?	98% of arterial oxygen is carried bound to hemoglobin.	Additional hemoglobin increases the oxygen-carrying capacity of the blood.
5	What does adding erythrocytes to the blood do to the viscosity of the blood? What effect will that change in viscosity have on blood flow?	Adding cells increases blood viscosity.	According to Poiseuille's law, increased viscosity increases resistance to flow, so blood flow will decrease.
6	What happens to plasma pH during hyperventilation?	Apply the law of mass action to the equation $CO_2 + H_2O \rightleftharpoons H^+ + HCO_3^-$.	The amount of CO_2 in the plasma decreases during hyperventilation, which means the equation shifts to the left. This shift decreases H^+, which increases pH (alkalosis).
7	How does this change in pH affect oxygen binding at the lungs when P_{O_2} is decreased? How does it affect unloading of oxygen at the cells?	See Figure 18-10a.	The left shift of the curve means that, at any given P_{O_2}, more O_2 binds to hemoglobin. Less O_2 will unbind at the tissues for a given P_{O_2}, but P_{O_2} in the cells is probably lower than normal, and consequently there may be no change in unloading.
8	Why do you think periodic breathing occurs most often during sleep?	Periodic breathing alternates periods of breath-holding (apnea) and hyperventilation.	An awake person is more likely to make a conscious effort to breathe during the breath-holding spells, eliminating the cycle of periodic breathing.

588   592   593   596   601   606   **608**

CHAPTER SUMMARY

In this chapter, you learned why climbing Mt. Everest is such a respiratory challenge for the human body, and why people with emphysema experience the same respiratory challenges at sea level. The exchange and transport of oxygen and carbon dioxide in the body illustrate the *mass flow* of gases along concentration gradients. The *homeostatic bal-* ance of these blood gases demonstrates *mass balance*: the concentration in the blood varies according to what enters or leaves at the lungs and tissues. The *law of mass action* governs the chemical reactions through which hemoglobin binds oxygen, and carbonic anhydrase catalyzes the conversion of CO_2 and water to carbonic acid.

Diffusion and Solubility of Gases

1. Diffusion of oxygen and CO_2 is influenced by partial pressure gradients, surface area, membrane thickness, and diffusion distance. (p. 588)
2. The amount of a gas that dissolves in a liquid is proportional to the partial pressure of the gas and to the **solubility** of the gas in the liquid. Carbon dioxide is 20 times more soluble in aqueous solutions than oxygen is. (p. 589; Fig. 18-2)

Gas Exchange in the Lungs and Tissues

IP Respiratory System: Gas Exchange

3. Normal alveolar and arterial P_{O_2} is about 100 mm Hg. Normal arterial P_{CO_2} is about 40 mm Hg. (p. 590; Fig. 18-3)
4. Normal venous P_{O_2} is 40 mm Hg, and normal venous P_{CO_2} is 46 mm Hg. (p. 590; Fig. 18-3)
5. Both the composition of inspired air and the effectiveness of alveolar ventilation affect alveolar P_{O_2}. (p. 591)
6. Changes in alveolar surface area, in alveolar membrane thickness, and in interstitial distance between alveoli and pulmonary capillaries all affect gas exchange efficiency and arterial P_{O_2}. (p. 592; Fig. 18-4)

Gas Transport in the Blood

IP Respiratory System: Gas Transport

7. Oxygen is transported either dissolved in plasma (<2%) or bound to hemoglobin (>98%). (p. 593; Fig. 18-6)
8. The P_{O_2} of plasma determines how much oxygen will bind to hemoglobin. (p. 596; Fig. 18-9)
9. Oxygen-hemoglobin binding is influenced by pH, temperature, and **2,3-diphosphoglycerate**. (p. 598; Figs. 18-10, 18-11)

10. Venous blood carries 7% of its carbon dioxide dissolved in plasma, 23% as **carbaminohemoglobin**, and 70% as bicarbonate ion in the plasma. (p. 601; Fig. 18-14)
11. **Carbonic anhydrase** in red blood cells converts CO_2 to carbonic acid, which dissociates into H^+ and HCO_3^-. The H^+ then binds to hemoglobin, and HCO_3^- enters the plasma using the **chloride shift**. (p. 599)

Regulation of Ventilation

IP Respiratory System: Control of Respiration

12. Respiratory control resides in a **central pattern generator**, a network of neurons in the brain stem. (p. 602)
13. The medullary **dorsal respiratory group** contains inspiratory neurons that control somatic motor neurons to the diaphragm. The **ventral respiratory group** of neurons assists in inspiration and active expiration. (p. 603; Fig. 18-16)
14. **Peripheral chemoreceptors** in the carotid and aortic bodies monitor P_{O_2}, P_{CO_2}, and pH. Ventilation increases when P_{O_2} falls below 60 mm Hg. (p. 604; Fig. 18-18)
15. Carbon dioxide is the primary stimulus for changes in ventilation. **Central chemoreceptors** in the medulla respond to changes in P_{CO_2}. (p. 604; Fig. 18-19)
16. Protective reflexes monitored by peripheral receptors prevent injury to the lungs from overinflation or irritants. (p. 607)
17. Conscious and unconscious thought processes can affect respiratory activity. (p. 607)

18

QUESTIONS

(Answers to the Review Questions begin on page A1.)

THE PHYSIOLOGY PLACE

Access more review material online at **The Physiology Place** website. There you'll find review questions, problem-solving activities, case studies, flashcards, and direct links to both *InterActive Physiology*® and PhysioEx™. To access the site, go to *www.physiologyplace.com* and select Human Physiology, Fourth Edition.

LEVEL ONE REVIEWING FACTS AND TERMS

1. List five factors that influence the diffusion of gases between alveolus to blood.

2. More than _____% of the oxygen in arterial blood is transported bound to hemoglobin. How is the remaining oxygen transported to the cells?
3. Name four factors that influence the amount of oxygen that binds to hemoglobin. Which of these four factors is the most important?
4. Describe the structure of a hemoglobin molecule. What chemical element is essential for hemoglobin synthesis?
5. The centers for control of ventilation are found in the _____ and _____ of the brain. What do the dorsal and ventral respiratory groups of neurons control? What is a central pattern generator?

6. Name the chemoreceptors that influence ventilation, and explain how they do so. What chemical is the most important controller of ventilation?

7. Describe the protective reflexes of the respiratory system. What does the Hering-Breuer reflex prevent? How is this reflex initiated?

8. What causes the exchange of oxygen and carbon dioxide between alveoli and blood or between blood and cells?

9. List five possible physical changes that could result in less oxygen reaching the arterial blood.

LEVEL TWO   REVIEWING CONCEPTS

10. **Concept map:** Construct a map of gas transport using the following terms. You may add additional terms.

P_{CO_2}	carbonic anhydrase
P_{O_2}	chloride shift
alveoli	dissolved CO_2
arterial blood	dissolved O_2
carbaminohemoglobin	hemoglobin
hemoglobin saturation	pressure gradient
oxyhemoglobin	red blood cell
plasma	venous blood

11. In respiratory physiology, it is customary to talk of the P_{O_2} of the plasma. Why is this not the most accurate way to describe the oxygen content of blood?

12. Compare and contrast the concepts in the following pairs:
    (a) transport of oxygen and of carbon dioxide in arterial blood
    (b) partial pressure and concentration of a gas dissolved in a liquid

13. Will HbO_2 binding increase, decrease, or not change with decreased pH?

14. Define hypoxia, COPD, and hypercapnia.

15. Why did oxygen-transporting pigments need to evolve in animals?

16. Draw and label the following graphs:
    (a) The effect of ventilation on arterial P_{O_2}
    (b) The effect of arterial P_{CO_2} on ventilation

17. As the P_{O_2} of plasma increases:
    (a) what happens to the amount of oxygen that dissolves in plasma?
    (b) what happens to the amount of oxygen that binds to hemoglobin?

18. If a person is anemic and has a lower-than-normal level of hemoglobin in her red blood cells, what will her arterial P_{O_2} be compared to normal?

19. Create reflex pathways (stimulus, receptor, afferent path, and so on) for the chemical control of ventilation, starting with the following stimuli:
    (a) Increased arterial P_{CO_2}
    (b) Arterial $P_{O_2} = 55$ mm Hg
    Be as specific as possible regarding anatomical locations. Where known, include neurotransmitters and their receptors.

LEVEL THREE   PROBLEM SOLVING

20. Marco tries to hide at the bottom of a swimming hole by breathing in and out through a garden hose, which greatly increases his anatomic dead space. What happens to the following parameters in his arterial blood, and why?
    (a) P_{CO_2} (b) P_{O_2}
    (c) bicarbonate ion   (d) pH

21. Which person will carry more oxygen in his blood:
    (a) one with a normal hemoglobin level of 15 g/dL and an arterial P_{O_2} of 80 mm Hg
    (b) one with a reduced hemoglobin level of 12 g/dL and a normal arterial P_{O_2} of 100 mm Hg

22. What would happen to each of the following parameters in a person suffering from an acute asthma attack (bronchoconstriction)?
    (a) arterial P_{O_2}
    (b) arterial hemoglobin saturation
    (c) alveolar ventilation

23. In early research on the control of rhythmic breathing, scientists made the following observations. What hypotheses might the researchers have formulated from each observation?
    (a) *Observation.* If the brain stem is severed below the medulla, all respiratory movement ceases.
    (b) *Observation.* If the brain stem is severed above the level of the pons, ventilation is normal.
    (c) *Observation.* If the medulla is completely separated from the pons and higher brain centers, ventilation becomes gasping but the respiratory rhythm remains.

24. A hospitalized patient with severe chronic obstructive lung disease has a P_{CO_2} of 55 mm Hg and a P_{O_2} of 50 mm Hg. To elevate his blood oxygen, he is given pure oxygen through a nasal tube. The patient immediately stops breathing. Explain why this might occur.

25. You are a physiologist on a space flight to a distant planet. You find intelligent humanoid creatures inhabiting the planet, and they willingly submit to your tests. Some of the data you have collected are described below. The first graph shows the oxygen dissociation curve for the oxygen-carrying pigment in the blood of the humanoid named Bzork. Bzork's normal alveolar P_{O_2} is 85 mm Hg. His normal cell P_{O_2} is 20 mm Hg, but it drops to 10 mm Hg with exercise.

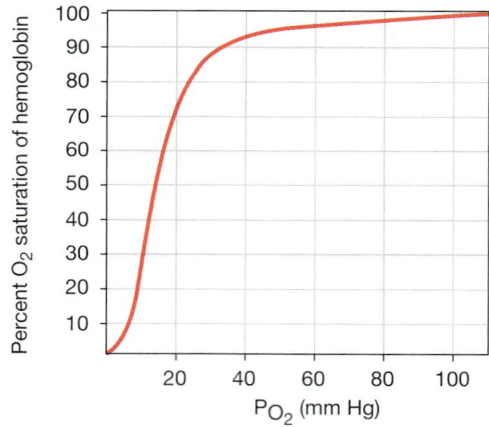

(a) What is the percent saturation for Bzork's oxygen-carrying pigment in blood at the alveoli? In blood at an exercising cell?
(b) Based on the graph, what conclusions can you draw about Bzork's oxygen requirements during normal activity and during exercise?

Answers **611**

26. The next experiment on Bzork involves his ventilatory response to different conditions. The data from that experiment are graphed below. Interpret the results of experiments A and C.

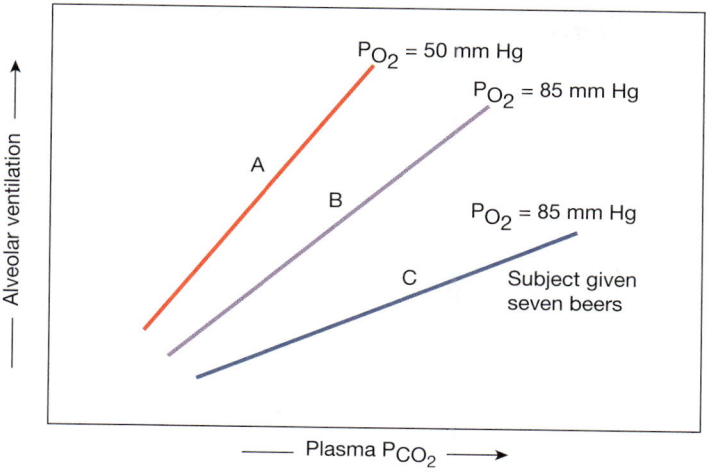

Alveolar ventilation

P_{O_2} = 50 mm Hg

P_{O_2} = 85 mm Hg

A

B

P_{O_2} = 85 mm Hg

C   Subject given seven beers

Plasma P_{CO_2}

LEVEL FOUR QUANTITATIVE PROBLEMS

27. You are given the following information on a patient.

    Blood volume = 5.2 liters

    Hematocrit = 47%

    Hemoglobin concentration = 12 g/dL whole blood

    Total amount of oxygen carried in blood = 1015 mL

    Arterial plasma P_{O_2} = 100 mm Hg

    You know that at a plasma P_{O_2} of 100 mm Hg, plasma contains 0.3 mL oxygen/dL, and that hemoglobin is 98% saturated. Each hemoglobin molecule can bind to a maximum of four molecules of oxygen. Using this information, calculate the maximum oxygen-carrying capacity of hemoglobin (100% saturated). Units will be mL oxygen/g hemoglobin.

28. Adolph Fick, the nineteenth-century physiologist who derived Fick's law of diffusion, also developed the Fick equation that relates oxygen consumption, cardiac output, and blood P_{O_2}:

$$O_2 \text{ consumption} = \frac{\text{cardiac output}}{\text{arterial } P_{O_2} - \text{venous } P_{O_2}}$$

A person has a cardiac output of 4.5 L/min, an arterial oxygen content of 105 mL O_2/L blood, and a vena cava oxygen content of 60 mL O_2/L blood. What is this person's oxygen consumption?

29. Describe what happens to the oxygen-hemoglobin dissociation curve in Figure 18-9 when blood hemoglobin falls from 15 g/dL blood to 10 g/dL blood.

ANSWERS

✓ Answers to Concept Check Questions

Page 589

1. The other factor that affects how much of each gas will dissolve in the saline solution is the solubility of the gas in that solution.

2. 720 mm Hg × 0.78 N_2 = 561.6 mm Hg

3. False. Plasma is essentially water, and Figure 18-2 shows that CO_2 is more soluble in water than O_2.

Page 591

4. (a) electron transport system (b) citric acid cycle

5. The P_{O_2} of the alveoli does not decrease because it is constantly being replenished by fresh air being inhaled from the external environment. [↩ p. 578]

Page 592

6. Air is 21% oxygen. Therefore, for dry air on Everest, P_{O_2} = 0.21 × 250 mm Hg = 53 mm Hg. Correction for P_{H_2O}: P_{O_2} = (250 mm Hg − 47 mm Hg) × 21% = (203 mm Hg) × 0.21 = 43 mm Hg.

Page 593

7. These pathological conditions cause blood to pool in the lungs because the left heart is unable to pump all the blood coming into it from the pulmonary circulation. Increased blood volume in the lungs increases pulmonary blood pressure.

8. When alveolar ventilation increases, arterial P_{O_2} increases because more fresh air is entering the alveoli. Arterial P_{CO_2} decreases because the low P_{CO_2} of fresh air dilutes alveolar P_{CO_2} below its normal value (40 mm Hg). The CO_2 pressure gradient between the venous blood (46 mm Hg) and the alveoli increases, causing more CO_2 to leave the blood. Venous P_{O_2} and P_{CO_2} do not change because these pressures

are determined by levels of oxygen consumption and CO_2 production in the cells.

Page 598

9. Yes. Hemoglobin reaches 100% saturation at 650 mm Hg. At sea level, atmospheric pressure is 760 mm Hg, and if the "atmosphere" is 100% oxygen, then P_{O_2} is 760 mm Hg.

10. The flatness at the top of the P_{O_2} curve in Figure 17-15 tells you that hyperventilation causes only a minimal increase in percent saturation of arterial hemoglobin.

Page 599

11. As the P_{O_2} of the exercising muscle falls, more oxygen is released by hemoglobin. The P_{O_2} of the venous blood leaving the muscle will be 25 mm Hg, the same as the P_{O_2} of the muscle.

Page 602

12. An airway obstruction would decrease alveolar ventilation and increase arterial P_{CO_2}. Elevated arterial P_{CO_2} would increase the H^+ concentration in the arterial blood and decrease pH.

Q Answers to Figure and Graph Questions

Page 594

Fig. 18-6: O_2 will cross five cell membranes: two of the alveolar cell, two of the capillary endothelium, and one of the red blood cell.

Page 597

Fig. 18-9: (a) When P_{O_2} is 20 mm Hg, hemoglobin is about 34% saturated with oxygen. (b) Hemoglobin is 50% saturated with oxygen at a P_{O_2} of 28 mm Hg.

18

Page 598

Fig. 18-10: (a) When pH falls from 7.4 to 7.2, hemoglobin saturation decreases by 13%, from about 37% saturation to 24%. (b) When an exercising muscle cell warms up, it releases more oxygen at any given P_{O_2}.

Page 599

Fig. 18-11: Loss of 2,3-DPG is not good because then hemoglobin binds more tightly to oxygen at the P_{O_2} values found in cells.

Page 599

Fig. 18-12: (a) The P_{O_2} of placental blood is about 28 mm Hg. (b) At a P_{O_2} of 10 mm Hg, maternal blood is only about 8% saturated with oxygen.

Page 603

Fig. 18-16: 1. pons; 2. ventral respiratory group; 3. medullary chemoreceptor; 4. sensory neuron; 5. carotid chemoreceptor; 6. somatic motor neuron (expiration); 7. aortic chemoreceptor; 8. internal intercostals; 9. abdominal muscles; 10. diaphragm; 11. external intercostals; 12. scalenes and sternocleiodmastoids; 13. somatic motor neuron (inspiration); 14. dorsal respiratory group; 15. limbic system; 16. higher brain centers (emotions and voluntary control)

18

19

*Plasma undergoes modification
to urine in the nephron.*

—**Arthur Grollman,** *in Clinical Physiology:
The Functional Pathology of Disease, 1957*

Glomerular capillaries (green) of a living rat.

The Kidneys

BACKGROUND BASICS

GOUT

Michael Moustakakis, 43, has spent the last two days on the sofa, suffering from a relentless throbbing pain in his left big toe. When the pain began, Michael thought he had a mild sprain or perhaps the beginnings of arthritis. Then the pain intensified, and the toe joint became hot and red. Michael finally hobbled into his doctor's office, feeling a little silly about his problem. On hearing his symptoms, the doctor seems to know instantly what is wrong. "Sounds to me like you have gout," says Dr. Garcia.

| 614 | 618 | 630 | 631 | 633 | 636 |

About A.D. 100, Aretaeus the Cappadocian wrote, "Diabetes is a wonderful affection, not very frequent among men, being a melting down of the flesh and limbs into urine. . . . The patients never stop making water [urinating], but the flow is incessant, as if from the opening of aqueducts." Physicians have known for centuries that **urine**, the fluid waste produced by the kidneys, reflects the functioning of the body. To aid them in their diagnosis of illness, they even carried special flasks for the collection and inspection of patients' urine.

The first step in examining a urine sample was to determine its color. Was it dark yellow (concentrated), pale straw (dilute), red (indicating the presence of blood), or black (indicating the presence of hemoglobin metabolites)? One form of malaria was called *blackwater fever* because metabolized hemoglobin from the abnormal breakdown of red blood cells turned victims' urine black or dark red.

Physicians also inspected urine samples for clarity, froth (indicating abnormal presence of proteins), smell, and even taste. Physicians who did not want to taste the urine themselves would allow their students the "privilege" of tasting it for them. A physician without students might expose insects to the urine and study their reaction.

Probably the most famous example of using urine for diagnosis was the taste test for diabetes mellitus, historically known as the *honey-urine disease*. Diabetes is an endocrine disorder characterized by the presence of glucose in the urine. The urine of diabetics tasted sweet and attracted insects, making the diagnosis clear.

Although today we have much more sophisticated tests for glucose in the urine, the first step of a *urinalysis* is still to examine the color, clarity, and odor of the urine. In this chapter you will learn why we can tell so much about how the body is functioning by what is present in the urine.

FUNCTIONS OF THE KIDNEYS

If you asked people on the street, "What is the most important function of the kidney?" they would probably say, "The removal of wastes." Actually, the most important function of the kidney is the homeostatic regulation of the water and ion content of the blood, also called *salt and water balance* or *fluid and electrolyte balance*. Waste removal is important, but disturbances in blood volume or ion levels will cause serious medical problems before the accumulation of metabolic wastes reaches toxic levels.

The kidneys maintain normal blood concentrations of ions and water by balancing intake of those substances with their excretion in the urine, obeying the principle of mass balance [p. 129]. We can divide kidney function into six general areas:

1. **Regulation of extracellular fluid volume and blood pressure.** When extracellular fluid volume decreases, blood pressure also decreases [p. 508]. If ECF volume and blood pressure fall too low, the body cannot maintain adequate blood flow to the brain and other essential organs. The kidneys work in an integrated fashion with the cardiovascular system to ensure that blood pressure and tissue perfusion remain within an acceptable range.

2. **Regulation of osmolarity.** The body integrates kidney function with behavioral drives, such as thirst, to maintain blood osmolarity at a value close to 290 mOsM. We will examine the reflex pathways for regulation of ECF volume and osmolarity in Chapter 20.

3. **Maintenance of ion balance.** The kidneys keep concentrations of key ions within a normal range by balancing dietary intake with urinary loss. Sodium (Na^+) is the major ion involved in the regulation of extracellular fluid volume and osmolarity. Potassium (K^+) and calcium (Ca^{2+}) concentrations are also closely regulated. We will discuss renal control of sodium and potassium balance in Chapter 20 but defer the discussion of calcium until Chapter 22, when we look at all aspects of calcium homeostasis.

4. **Homeostatic regulation of pH.** The pH of plasma is normally kept within a narrow range [p. 37]. If extracellular fluid becomes too acidic, the kidneys remove H^+ and conserve bicarbonate ions (HCO_3^-), which act as a buffer [p. 38]. Conversely, when extracellular fluid becomes too alkaline, the kidneys remove HCO_3^- and conserve H^+. The kidneys play a significant role in pH homeostasis, but they do not correct pH disturbances as rapidly as the lungs do, as you will learn in Chapter 20.

5. **Excretion of wastes.** The kidneys remove metabolic waste products and foreign substances, such as drugs and environmental toxins. Metabolic wastes include creatinine from muscle metabolism [p. 410] and the nitrogenous wastes *urea* [p. 111] and *uric acid*. A metabolite of hemoglobin called *urobilinogen* gives urine its characteristic yellow color. Hormones are another endogenous substance the kidneys clear from the blood.

Examples of foreign substances that the kidneys actively remove include the artificial sweetener *saccharin* and the anion *benzoate,* part of the preservative *potassium benzoate* that you ingest each time you drink a diet soft drink.

6. **Production of hormones.**   Although the kidneys are not endocrine glands, they play important roles in three endocrine pathways. Kidney cells synthesize *erythropoietin,* the cytokine/hormone that regulates red blood cell synthesis [🔁 p. 541]. They also release *renin,* an enzyme that regulates the production of hormones involved in sodium balance and blood pressure homeostasis. Renal enzymes help convert vitamin D$_3$ into a hormone that regulates Ca^{2+} balance.

The kidneys, like many other organs in the body, have a tremendous reserve capacity. By most estimates, you must lose nearly three-fourths of your kidney function before homeostasis begins to be affected. Many people function perfectly normally with only one kidney, including the one person in 1000 born with only one kidney (the other fails to develop during gestation) or those people who donate a kidney for transplantation.

CONCEPT CHECK

1. Ion regulation is a key feature of kidney function. What happens to the resting membrane potential of a neuron if extracellular K$^+$ levels decrease? [🔁 p. 269]
2. What happens to the force of cardiac contraction if plasma Ca^{2+} levels decrease substantially? [🔁 p. 471]

Answers: p. 639

ANATOMY OF THE URINARY SYSTEM

The **urinary system** is composed of the kidneys and accessory structures (Fig. 19-1a ■). The study of kidney function is called **renal physiology**, from the Latin word *renes,* meaning "kidneys."

The Urinary System Consists of Kidneys, Ureters, Bladder, and Urethra

Let's begin by following the route a drop of water takes on its way from plasma to excretion in the urine. In the first step of urine production, water and solutes move from plasma into the hollow tubules (*nephrons*) that make up the bulk of the paired kidneys. These tubules modify the composition of the fluid as it passes through. The modified fluid leaves the kidney and passes into a hollow tube called a **ureter**. There are two ureters, one leading from each kidney to the **urinary bladder**. The bladder expands and fills with urine until, by reflex action, it contracts and expels urine through a single tube, the **urethra**.

The urethra in males exits the body through the shaft of the penis. In females, the urethral opening is found anterior to the openings of the vagina and anus. *Micturition,* or urination, is the process by which urine is excreted.

The kidneys are the site of urine formation. They lie on either side of the spine at the level of the eleventh and twelfth ribs, just above the waist (Fig. 19-1b). Although they are below the diaphragm, they are technically outside the abdominal cavity,

sandwiched between the membranous **peritoneum**, which lines the abdomen, and the bones and muscles of the back. Because of their location behind the peritoneal cavity, the kidneys are sometimes described as being *retroperitoneal* [*retro-*, behind].

The concave surface of each kidney faces the spine. The renal blood vessels, nerves, lymphatics, and ureters all emerge from this surface. **Renal arteries**, which branch off the abdominal aorta, supply blood to the kidneys. **Renal veins** carry blood from the kidneys to the inferior vena cava.

At any given time, the kidneys receive 20–25% of the cardiac output, even though they constitute only 0.4% of total body weight (4.5–6 ounces each). This high rate of blood flow through the kidneys is critical to renal function.

The Nephron Is the Functional Unit of the Kidney

A cross section through a kidney shows that the interior is arranged in two layers: an outer **cortex** and inner **medulla** (Fig. 19-1c). The layers are formed by the organized arrangement of microscopic tubules called **nephrons**. About 80% of the nephrons in a kidney are almost completely contained within the cortex (*cortical* nephrons), but the other 20%—called *juxtamedullary* nephrons [*juxta-*, beside]—dip down into the medulla (Fig. 19-1i).

The nephron is the functional unit of the kidney. (A *functional unit* is the smallest structure that can perform all the functions of an organ.) Each of the 1 million nephrons in a kidney is divided into sections (Fig. 19-1j), and each section is closely associated with specialized blood vessels (Fig. 19-1g, h).

19

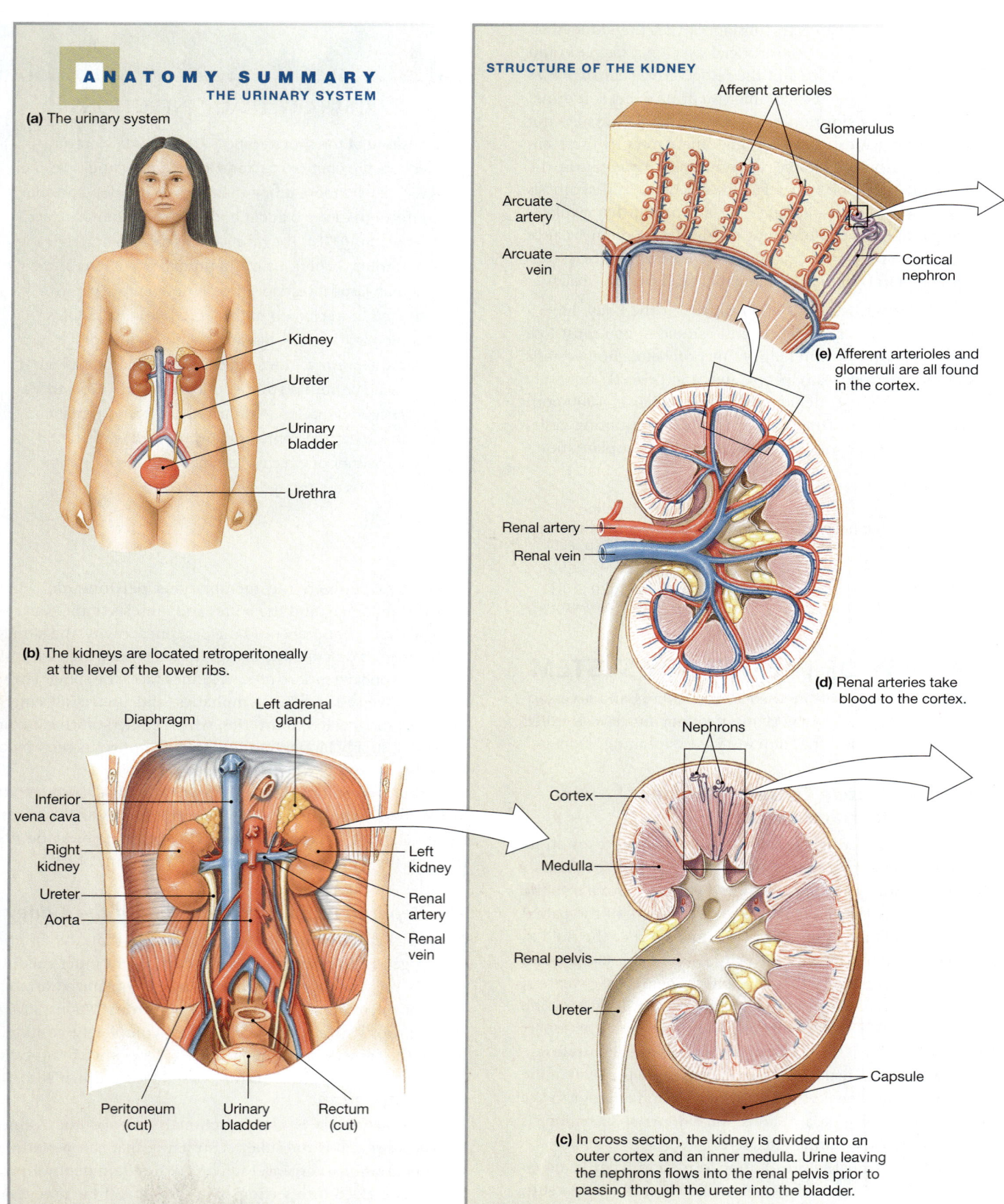

A N A T O M Y   S U M M A R Y
THE URINARY SYSTEM

(a) The urinary system

Kidney

Ureter

Urinary
bladder

Urethra

(b) The kidneys are located retroperitoneally
at the level of the lower ribs.

Diaphragm

Left adrenal
gland

Inferior
vena cava

Right
kidney

Ureter

Aorta

Left
kidney

Renal
artery

Renal
vein

Peritoneum
(cut)

Urinary
bladder

Rectum
(cut)

STRUCTURE OF THE KIDNEY

Afferent arterioles

Glomerulus

Arcuate
artery

Arcuate
vein

Cortical
nephron

(e) Afferent arterioles and
glomeruli are all found
in the cortex.

Renal artery

Renal vein

(d) Renal arteries take
blood to the cortex.

Nephrons

Cortex

Medulla

Renal pelvis

Ureter

Capsule

(c) In cross section, the kidney is divided into an
outer cortex and an inner medulla. Urine leaving
the nephrons flows into the renal pelvis prior to
passing through the ureter into the bladder.

■ FIGURE 19-1

19

26. The next experiment on Bzork involves his ventilatory response to different conditions. The data from that experiment are graphed below. Interpret the results of experiments A and C.

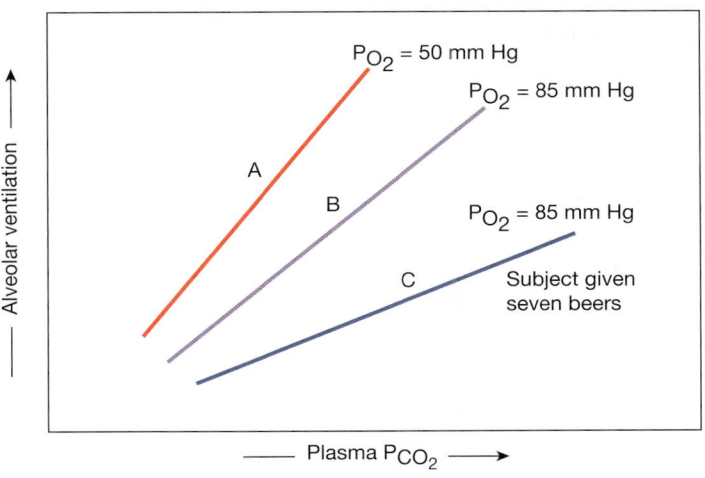

- P_{O_2} = 50 mm Hg
- P_{O_2} = 85 mm Hg
- A
- B
- P_{O_2} = 85 mm Hg
- C
- Subject given seven beers

Alveolar ventilation

Plasma P_{CO_2}

LEVEL FOUR QUANTITATIVE PROBLEMS

27. You are given the following information on a patient.

Blood volume = 5.2 liters

Hematocrit = 47%

Hemoglobin concentration = 12 g/dL whole blood

Total amount of oxygen carried in blood = 1015 mL

Arterial plasma P_{O_2} = 100 mm Hg

You know that at a plasma P_{O_2} of 100 mm Hg, plasma contains 0.3 mL oxygen/dL, and that hemoglobin is 98% saturated. Each hemoglobin molecule can bind to a maximum of four molecules of oxygen. Using this information, calculate the maximum oxygen-carrying capacity of hemoglobin (100% saturated). Units will be mL oxygen/g hemoglobin.

28. Adolph Fick, the nineteenth-century physiologist who derived Fick's law of diffusion, also developed the Fick equation that relates oxygen consumption, cardiac output, and blood P_{O_2}:

$$O_2 \text{ consumption} = \frac{\text{cardiac output}}{\text{arterial } P_{O_2} - \text{venous } P_{O_2}}$$

A person has a cardiac output of 4.5 L/min, an arterial oxygen content of 105 mL O_2/L blood, and a vena cava oxygen content of 60 mL O_2/L blood. What is this person's oxygen consumption?

29. Describe what happens to the oxygen-hemoglobin dissociation curve in Figure 18-9 when blood hemoglobin falls from 15 g/dL blood to 10 g/dL blood.

ANSWERS

✓ Answers to Concept Check Questions

Page 589

1. The other factor that affects how much of each gas will dissolve in the saline solution is the solubility of the gas in that solution.

2. 720 mm Hg $\times$ 0.78 N_2 = 561.6 mm Hg

3. False. Plasma is essentially water, and Figure 18-2 shows that CO_2 is more soluble in water than O_2.

Page 591

4. (a) electron transport system (b) citric acid cycle

5. The P_{O_2} of the alveoli does not decrease because it is constantly being replenished by fresh air being inhaled from the external environment. [⊇ p. 578]

Page 592

6. Air is 21% oxygen. Therefore, for dry air on Everest, P_{O_2} = 0.21 $\times$ 250 mm Hg = 53 mm Hg. Correction for P_{H_2O}: P_{O_2} = (250 mm Hg – 47 mm Hg) $\times$ 21% = (203 mm Hg) $\times$ 0.21 = 43 mm Hg.

Page 593

7. These pathological conditions cause blood to pool in the lungs because the left heart is unable to pump all the blood coming into it from the pulmonary circulation. Increased blood volume in the lungs increases pulmonary blood pressure.

8. When alveolar ventilation increases, arterial P_{O_2} increases because more fresh air is entering the alveoli. Arterial P_{CO_2} decreases because the low P_{CO_2} of fresh air dilutes alveolar P_{CO_2} below its normal value (40 mm Hg). The CO_2 pressure gradient between the venous blood (46 mm Hg) and the alveoli increases, causing more CO_2 to leave the blood. Venous P_{O_2} and P_{CO_2} do not change because these pressures are determined by levels of oxygen consumption and CO_2 production in the cells.

Page 598

9. Yes. Hemoglobin reaches 100% saturation at 650 mm Hg. At sea level, atmospheric pressure is 760 mm Hg, and if the "atmosphere" is 100% oxygen, then P_{O_2} is 760 mm Hg.

10. The flatness at the top of the P_{O_2} curve in Figure 17-15 tells you that hyperventilation causes only a minimal increase in percent saturation of arterial hemoglobin.

Page 599

11. As the P_{O_2} of the exercising muscle falls, more oxygen is released by hemoglobin. The P_{O_2} of the venous blood leaving the muscle will be 25 mm Hg, the same as the P_{O_2} of the muscle.

Page 602

12. An airway obstruction would decrease alveolar ventilation and increase arterial P_{CO_2}. Elevated arterial P_{CO_2} would increase the H^+ concentration in the arterial blood and decrease pH.

Q Answers to Figure and Graph Questions

Page 594

Fig. 18-6: O_2 will cross five cell membranes: two of the alveolar cell, two of the capillary endothelium, and one of the red blood cell.

Page 597

Fig. 18-9: (a) When P_{O_2} is 20 mm Hg, hemoglobin is about 34% saturated with oxygen. (b) Hemoglobin is 50% saturated with oxygen at a P_{O_2} of 28 mm Hg.

18

Page 598

Fig. 18-10: (a) When pH falls from 7.4 to 7.2, hemoglobin saturation decreases by 13%, from about 37% saturation to 24%. (b) When an exercising muscle cell warms up, it releases more oxygen at any given P_{O_2}.

Page 599

Fig. 18-11: Loss of 2,3-DPG is not good because then hemoglobin binds more tightly to oxygen at the P_{O_2} values found in cells.

Page 599

Fig. 18-12: (a) The P_{O_2} of placental blood is about 28 mm Hg. (b) At a P_{O_2} of 10 mm Hg, maternal blood is only about 8% saturated with oxygen.

Page 603

Fig. 18-16: 1. pons; 2. ventral respiratory group; 3. medullary chemoreceptor; 4. sensory neuron; 5. carotid chemoreceptor; 6. somatic motor neuron (expiration); 7. aortic chemoreceptor; 8. internal intercostals; 9. abdominal muscles; 10. diaphragm; 11. external intercostals; 12. scalenes and sternocleiodmastoids; 13. somatic motor neuron (inspiration); 14. dorsal respiratory group; 15. limbic system; 16. higher brain centers (emotions and voluntary control)

19

Plasma undergoes modification to urine in the nephron.

—*Arthur Grollman, in Clinical Physiology: The Functional Pathology of Disease, 1957*

Glomerular capillaries (green) of a living rat.

The Kidneys

Functions of the Kidneys

Anatomy of the Urinary System

Overview of Kidney Function

Filtration

Reabsorption

Secretion

Excretion

Micturition

BACKGROUND BASICS

GOUT

Michael Moustakakis, 43, has spent the last two days on the sofa, suffering from a relentless throbbing pain in his left big toe. When the pain began, Michael thought he had a mild sprain or perhaps the beginnings of arthritis. Then the pain intensified, and the toe joint became hot and red. Michael finally hobbled into his doctor's office, feeling a little silly about his problem. On hearing his symptoms, the doctor seems to know instantly what is wrong. "Sounds to me like you have gout," says Dr. Garcia.

| 614 | 618 | 630 | 631 | 633 | 636 |

About A.D. 100, Aretaeus the Cappadocian wrote, "Diabetes is a wonderful affection, not very frequent among men, being a melting down of the flesh and limbs into urine. . . . The patients never stop making water [urinating], but the flow is incessant, as if from the opening of aqueducts." Physicians have known for centuries that **urine**, the fluid waste produced by the kidneys, reflects the functioning of the body. To aid them in their diagnosis of illness, they even carried special flasks for the collection and inspection of patients' urine.

The first step in examining a urine sample was to determine its color. Was it dark yellow (concentrated), pale straw (dilute), red (indicating the presence of blood), or black (indicating the presence of hemoglobin metabolites)? One form of malaria was called *blackwater fever* because metabolized hemoglobin from the abnormal breakdown of red blood cells turned victims' urine black or dark red.

Physicians also inspected urine samples for clarity, froth (indicating abnormal presence of proteins), smell, and even taste. Physicians who did not want to taste the urine themselves would allow their students the "privilege" of tasting it for them. A physician without students might expose insects to the urine and study their reaction.

Probably the most famous example of using urine for diagnosis was the taste test for diabetes mellitus, historically known as the *honey-urine disease*. Diabetes is an endocrine disorder characterized by the presence of glucose in the urine. The urine of diabetics tasted sweet and attracted insects, making the diagnosis clear.

Although today we have much more sophisticated tests for glucose in the urine, the first step of a *urinalysis* is still to examine the color, clarity, and odor of the urine. In this chapter you will learn why we can tell so much about how the body is functioning by what is present in the urine.

FUNCTIONS OF THE KIDNEYS

If you asked people on the street, "What is the most important function of the kidney?" they would probably say, "The removal of wastes." Actually, the most important function of the kidney is the homeostatic regulation of the water and ion content of the blood, also called *salt and water balance* or *fluid and electrolyte balance*. Waste removal is important, but disturbances in blood volume or ion levels will cause serious medical problems before the accumulation of metabolic wastes reaches toxic levels.

The kidneys maintain normal blood concentrations of ions and water by balancing intake of those substances with their excretion in the urine, obeying the principle of mass balance [⊇ p. 129]. We can divide kidney function into six general areas:

1. **Regulation of extracellular fluid volume and blood pressure.** When extracellular fluid volume decreases, blood pressure also decreases [⊇ p. 508]. If ECF volume and blood pressure fall too low, the body cannot maintain adequate blood flow to the brain and other essential organs. The kidneys work in an integrated fashion with the cardiovascular system to ensure that blood pressure and tissue perfusion remain within an acceptable range.

2. **Regulation of osmolarity.** The body integrates kidney function with behavioral drives, such as thirst, to maintain blood osmolarity at a value close to 290 mOsM. We will examine the reflex pathways for regulation of ECF volume and osmolarity in Chapter 20.

3. **Maintenance of ion balance.** The kidneys keep concentrations of key ions within a normal range by balancing dietary intake with urinary loss. Sodium (Na^+) is the major ion involved in the regulation of extracellular fluid volume and osmolarity. Potassium (K^+) and calcium (Ca^{2+}) concentrations are also closely regulated. We will discuss renal control of sodium and potassium balance in Chapter 20 but defer the discussion of calcium until Chapter 22, when we look at all aspects of calcium homeostasis.

4. **Homeostatic regulation of pH.** The pH of plasma is normally kept within a narrow range [⊇ p. 37]. If extracellular fluid becomes too acidic, the kidneys remove H^+ and conserve bicarbonate ions (HCO_3^-), which act as a buffer [⊇ p. 38]. Conversely, when extracellular fluid becomes too alkaline, the kidneys remove HCO_3^- and conserve H^+. The kidneys play a significant role in pH homeostasis, but they do not correct pH disturbances as rapidly as the lungs do, as you will learn in Chapter 20.

5. **Excretion of wastes.** The kidneys remove metabolic waste products and foreign substances, such as drugs and environmental toxins. Metabolic wastes include creatinine from muscle metabolism [⊇ p. 410] and the nitrogenous wastes *urea* [⊇ p. 111] and *uric acid*. A metabolite of hemoglobin called *urobilinogen* gives urine its characteristic yellow color. Hormones are another endogenous substance the kidneys clear from the blood.

Examples of foreign substances that the kidneys actively remove include the artificial sweetener *saccharin* and the anion *benzoate*, part of the preservative *potassium benzoate* that you ingest each time you drink a diet soft drink.

6. **Production of hormones.**   Although the kidneys are not endocrine glands, they play important roles in three endocrine pathways. Kidney cells synthesize *erythropoietin,* the cytokine/hormone that regulates red blood cell synthesis [🔁 p. 541]. They also release *renin,* an enzyme that regulates the production of hormones involved in sodium balance and blood pressure homeostasis. Renal enzymes help convert vitamin D_3 into a hormone that regulates Ca^{2+} balance.

The kidneys, like many other organs in the body, have a tremendous reserve capacity. By most estimates, you must lose nearly three-fourths of your kidney function before homeostasis begins to be affected. Many people function perfectly normally with only one kidney, including the one person in 1000 born with only one kidney (the other fails to develop during gestation) or those people who donate a kidney for transplantation.

CONCEPT CHECK

1. Ion regulation is a key feature of kidney function. What happens to the resting membrane potential of a neuron if extracellular K^+ levels decrease? [🔁 p. 269]
2. What happens to the force of cardiac contraction if plasma Ca^{2+} levels decrease substantially? [🔁 p. 471]

Answers: p. 639

ANATOMY OF THE URINARY SYSTEM

The **urinary system** is composed of the kidneys and accessory structures (Fig. 19-1a ■). The study of kidney function is called **renal physiology**, from the Latin word *renes,* meaning "kidneys."

The Urinary System Consists of Kidneys, Ureters, Bladder, and Urethra

Let's begin by following the route a drop of water takes on its way from plasma to excretion in the urine. In the first step of urine production, water and solutes move from plasma into the hollow tubules (*nephrons*) that make up the bulk of the paired kidneys. These tubules modify the composition of the fluid as it passes through. The modified fluid leaves the kidney and passes into a hollow tube called a **ureter**. There are two ureters, one leading from each kidney to the **urinary bladder**. The bladder expands and fills with urine until, by reflex action, it contracts and expels urine through a single tube, the **urethra**.

The urethra in males exits the body through the shaft of the penis. In females, the urethral opening is found anterior to the openings of the vagina and anus. *Micturition,* or urination, is the process by which urine is excreted.

The kidneys are the site of urine formation. They lie on either side of the spine at the level of the eleventh and twelfth ribs, just above the waist (Fig. 19-1b). Although they are below the diaphragm, they are technically outside the abdominal cavity,

CLINICAL FOCUS

URINARY TRACT INFECTIONS

Because of the shorter length of the female urethra and its proximity to bacteria leaving the large intestine, women are more prone than men to develop bacterial infections of the bladder and kidneys, or **urinary tract infections** (UTIs). The most common cause of UTIs is the bacterium *Escherichia coli,* a normal inhabitant of the human large intestine. *E. coli* is not harmful while restricted to the lumen of the large intestine, but it is pathogenic [*patho-*, disease + *-genic*, causing] if it gets into the urethra. The most common symptoms of a UTI are pain or burning during urination and increased frequency of urination. A urine sample from a patient with a UTI often contains many red and white blood cells, neither of which is commonly found in normal urine. UTIs are treated with antibiotics.

sandwiched between the membranous **peritoneum**, which lines the abdomen, and the bones and muscles of the back. Because of their location behind the peritoneal cavity, the kidneys are sometimes described as being *retroperitoneal* [*retro-*, behind].

The concave surface of each kidney faces the spine. The renal blood vessels, nerves, lymphatics, and ureters all emerge from this surface. **Renal arteries**, which branch off the abdominal aorta, supply blood to the kidneys. **Renal veins** carry blood from the kidneys to the inferior vena cava.

At any given time, the kidneys receive 20–25% of the cardiac output, even though they constitute only 0.4% of total body weight (4.5–6 ounces each). This high rate of blood flow through the kidneys is critical to renal function.

19

The Nephron Is the Functional Unit of the Kidney

A cross section through a kidney shows that the interior is arranged in two layers: an outer **cortex** and inner **medulla** (Fig. 19-1c). The layers are formed by the organized arrangement of microscopic tubules called **nephrons**. About 80% of the nephrons in a kidney are almost completely contained within the cortex (*cortical* nephrons), but the other 20%—called *juxtamedullary* nephrons [*juxta-*, beside]—dip down into the medulla (Fig. 19-1i).

The nephron is the functional unit of the kidney. (A *functional unit* is the smallest structure that can perform all the functions of an organ.) Each of the 1 million nephrons in a kidney is divided into sections (Fig. 19-1j), and each section is closely associated with specialized blood vessels (Fig. 19-1g, h).

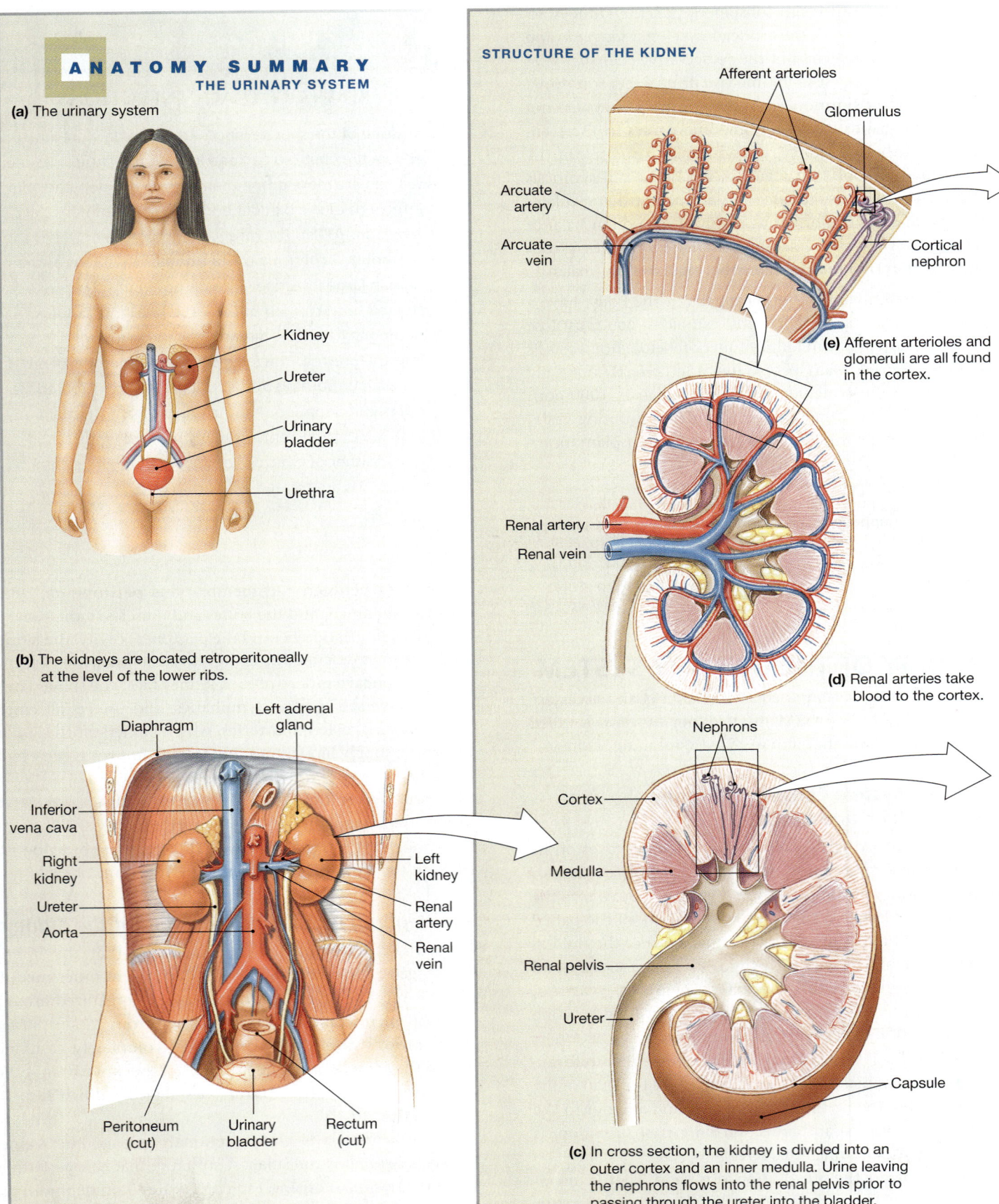

(a) The urinary system

Kidney

Ureter

Urinary bladder

Urethra

(b) The kidneys are located retroperitoneally at the level of the lower ribs.

Diaphragm

Left adrenal gland

Inferior vena cava

Right kidney

Ureter

Aorta

Left kidney

Renal artery

Renal vein

Peritoneum (cut)

Urinary bladder

Rectum (cut)

STRUCTURE OF THE KIDNEY

Afferent arterioles

Glomerulus

Arcuate artery

Arcuate vein

Cortical nephron

(e) Afferent arterioles and glomeruli are all found in the cortex.

Renal artery

Renal vein

(d) Renal arteries take blood to the cortex.

Nephrons

Cortex

Medulla

Renal pelvis

Ureter

Capsule

(c) In cross section, the kidney is divided into an outer cortex and an inner medulla. Urine leaving the nephrons flows into the renal pelvis prior to passing through the ureter into the bladder.

■ FIGURE 19-1

STRUCTURE OF THE NEPHRON

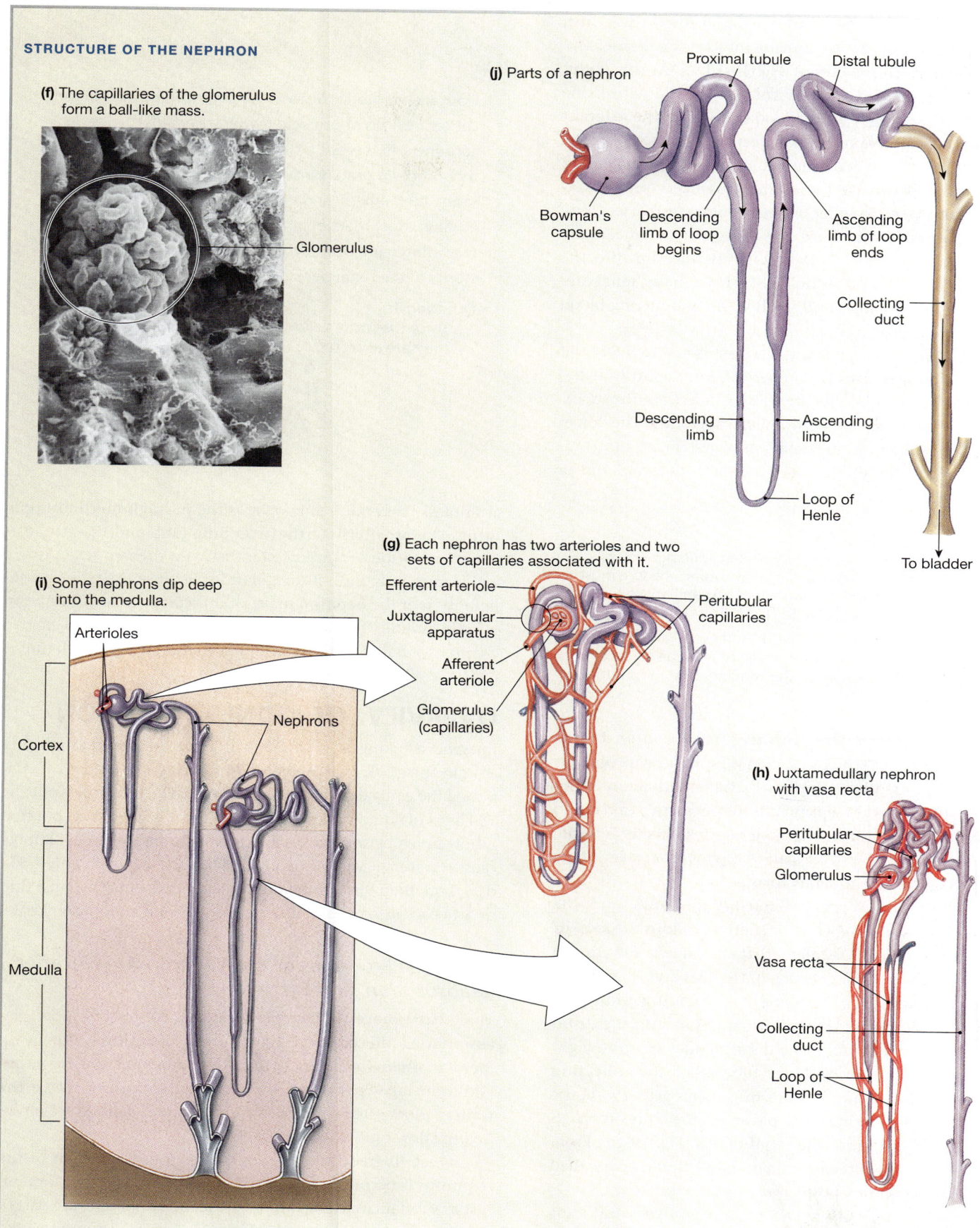

(f) The capillaries of the glomerulus form a ball-like mass.

Glomerulus

(j) Parts of a nephron

Proximal tubule

Distal tubule

Bowman's capsule

Descending limb of loop begins

Ascending limb of loop ends

Collecting duct

Descending limb

Ascending limb

Loop of Henle

To bladder

(i) Some nephrons dip deep into the medulla.

Arterioles

Nephrons

Cortex

Medulla

(g) Each nephron has two arterioles and two sets of capillaries associated with it.

Efferent arteriole

Juxtaglomerular apparatus

Afferent arteriole

Glomerulus (capillaries)

Peritubular capillaries

(h) Juxtamedullary nephron with vasa recta

Peritubular capillaries

Glomerulus

Vasa recta

Collecting duct

Loop of Henle

19

FIGURE 19-1 *(continued)*

Vascular Elements of the Kidney Blood enters the kidney through the renal artery before flowing into smaller arteries and then into arterioles in the cortex (Fig. 19-1d, e). At this point, the arrangement of blood vessels turns into a *portal system,* one of three in the body [p. 229]. Blood flows from the **afferent arteriole** into a ball-like network of capillaries known as the **glomerulus** [*glomus,* a ball-shaped mass; plural *glomeruli*] (Fig. 19-1f). Blood leaving the glomerulus flows into an **efferent arteriole**, then into a set of **peritubular capillaries** [*peri-,* around] that surround the tubule (Fig. 19-1g). In juxtamedullary nephrons, the long peritubular capillaries that dip into the medulla are called the **vasa recta** (Fig. 19-1h). Finally, renal capillaries join to form venules and small veins, conducting blood out of the kidney through the renal vein.

The function of the renal portal system is first to filter fluid out of the blood and into the lumen of the nephron at the glomerular capillaries, and then to *reabsorb* fluid from the tubule back into the blood at the peritubular capillaries. The forces behind fluid movement in the renal portal system are similar to those that govern the filtration of water and molecules out of systemic capillaries in other tissues.

CONCEPT CHECK

3. If net filtration out of glomerular capillaries occurs, then you know that capillary hydrostatic pressure must be (*greater than/less than/equal to*) capillary colloid osmotic pressure. [p. 517]

4. If net reabsorption into peritubular capillaries occurs, then capillary hydrostatic pressure must be (*greater than/less than/equal to*) the capillary colloid osmotic pressure.

Answers: p. 639

Tubular Elements of the Kidney The nephron begins with a hollow, ball-like structure called **Bowman's capsule** that surrounds the glomerulus (Fig. 19-1j). The endothelium of the glomerulus is fused to the epithelium of Bowman's capsule so that fluid filtering out of the capillaries passes directly into the lumen of the tubule. The combination of glomerulus and Bowman's capsule is called the **renal corpuscle**.

From Bowman's capsule, filtered fluid flows into the **proximal tubule** [*proximal,* close or near], then into the **loop of Henle,** a hairpin-shaped segment that dips down toward the medulla and then back up. The loop of Henle is divided into two limbs, a thin **descending limb** and an **ascending limb** with thin and thick segments. The fluid then passes into the **distal tubule** [*distal,* distant or far]. The distal tubules of up to eight nephrons drain into a single larger tube called the **collecting duct**. (The distal tubule and its collecting duct together form the **distal nephron.**) Collecting ducts pass from the cortex through the medulla and drain into the **renal pelvis** (Fig. 19-1c). From the renal pelvis, the filtered and modified fluid, now called **urine,** flows into the ureter on its way to excretion.

Notice in Figure 19-1g how the nephron twists and folds back on itself so that the final part of the ascending limb of the loop of Henle passes between the afferent and efferent

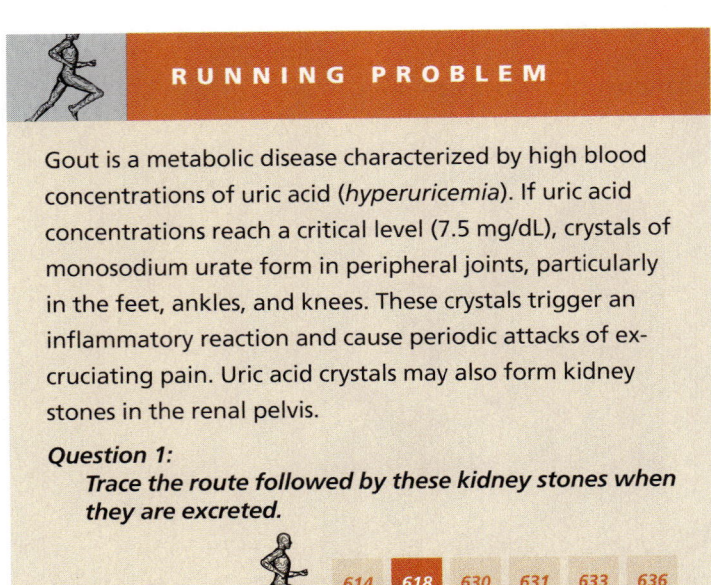

RUNNING PROBLEM

Gout is a metabolic disease characterized by high blood concentrations of uric acid (*hyperuricemia*). If uric acid concentrations reach a critical level (7.5 mg/dL), crystals of monosodium urate form in peripheral joints, particularly in the feet, ankles, and knees. These crystals trigger an inflammatory reaction and cause periodic attacks of excruciating pain. Uric acid crystals may also form kidney stones in the renal pelvis.

Question 1:
Trace the route followed by these kidney stones when they are excreted.

614   618   630   631   633   636

arterioles. This region is known as the **juxtaglomerular apparatus**. The proximity of the ascending limb and the arterioles allows paracrine communication between the two structures, a key feature of kidney autoregulation. Because the twisted configuration of the nephron makes it difficult to follow fluid flow, we will unfold the nephron in the remaining figures in this chapter so that fluid flows from left to right across the figure.

OVERVIEW OF KIDNEY FUNCTION

Imagine drinking a 12-ounce soft drink every three minutes around the clock: by the end of 24 hours, you would have consumed the equivalent of 90 2-liter bottles. The thought of putting 180 liters of liquid into your intestinal tract is staggering, but that is how much plasma moves into the nephrons every day! But because the average volume of urine leaving the kidneys is only 1.5 L/day, more than 99% of the fluid that enters nephrons must find its way back into the blood, or the body would rapidly dehydrate.

The Three Processes of the Nephron Are Filtration, Reabsorption, and Secretion

Three basic processes take place in the nephron: filtration, reabsorption, and secretion (Fig. 19-2 ■). **Filtration** is the movement of fluid from blood into the lumen of the nephron. Filtration takes place only in the renal corpuscle, where the walls of glomerular capillaries and Bowman's capsule are modified to allow bulk flow of fluid.

Once filtered fluid (*filtrate*) passes into the lumen of the nephron, it becomes part of the body's external environment, just as substances in the lumen of the intestinal tract are part of the external environment [Figure 1-2, p. 4]. Thus, anything that filters into the nephron is destined for removal in the urine unless it is reabsorbed into the body.

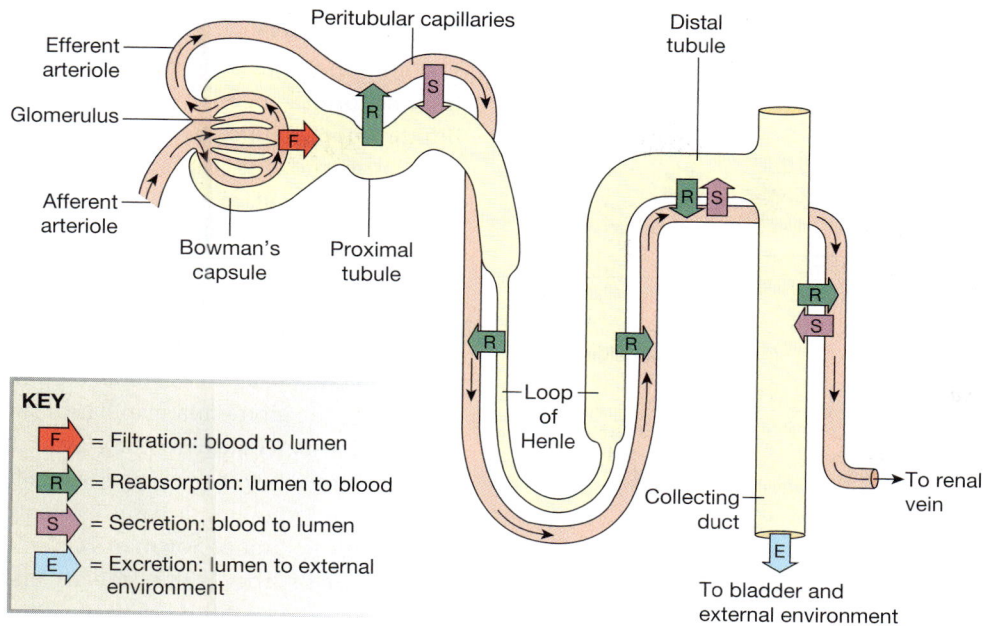

■ **FIGURE 19-2**   *Filtration, reabsorption, secretion, and excretion*

After filtrate leaves Bowman's capsule, it is modified by reabsorption and secretion. **Reabsorption** is the process of moving substances in the filtrate from the lumen of the tubule back into the blood flowing through peritubular capillaries. **Secretion** removes selected molecules from the blood and adds them to the filtrate in the tubule lumen. Although secretion and glomerular filtration both move substances from blood into the tubule, secretion is a more selective process that usually uses membrane proteins to move molecules across the tubule epithelium.

Volume and Osmolarity Change as Fluid Flows Through the Nephron

Now let's follow some filtrate through the nephron to learn what happens to it in the various segments (Table 19-1 ■). The 180 liters of fluid that filters into Bowman's capsule each day are almost identical in composition to plasma and nearly isosmotic—about 300 mOsM [🔁 p. 155]. As this filtrate flows through the proximal tubule, about 70% of its volume is reabsorbed, leaving 54 liters in the lumen. Reabsorption occurs when proximal tubule cells transport solutes out of the lumen, and water follows by osmosis. Filtrate leaving the proximal tubule has the same osmolarity as filtrate that entered. Thus, we say that the primary function of the proximal tubule is the *bulk reabsorption of isosmotic fluid.*

After leaving the proximal tubule, filtrate passes into the loop of Henle, the primary site for creating dilute urine. As the filtrate passes through the loop, proportionately more solute is reabsorbed than water, and the filtrate becomes hyposmotic relative to the plasma. By the time filtrate flows out of the loop, it averages 100 mOsM, and its volume has fallen from 54 L/day to about 18 L/day. Now 90% of the volume originally filtered into Bowman's capsule has been reabsorbed into the capillaries.

From the loop of Henle, filtrate passes into the distal tubule and the collecting duct. In these two segments, the fine regulation of salt and water balance takes place under the control of several hormones. Reabsorption and (to a lesser extent) secretion determine the final composition of the filtrate. By the end of the collecting duct, the filtrate has a volume of 1.5 L/day and an osmolarity that can range from 50 mOsM to 1200 mOsM. The final volume and osmolarity of urine depend on the body's need to conserve or excrete water and solute.

A word of caution here: it is very easy to confuse *secretion* with *excretion*. Try to remember the origins of the two prefixes. *Se-* means *apart,* as in to separate something from its source. In the nephron, secreted solutes are moved from plasma to tubule lumen. *Ex-* means *out,* or *away,* as in out of or away from the body. Excretion refers to the removal of a substance from the body.

TABLE 19-1	Changes in Filtrate Volume and Osmolarity Along the Nephron	
LOCATION IN NEPHRON	**VOLUME OF FLUID**	**OSMOLARITY OF FLUID**
Bowman's capsule	180 L/day	300 mOsM
End of proximal tubule	54 L/day	300 mOsM
End of loop of Henle	18 L/day	100 mOsM
End of collecting duct (final urine)	1.5 L/day (average)	50–1200 mOsM

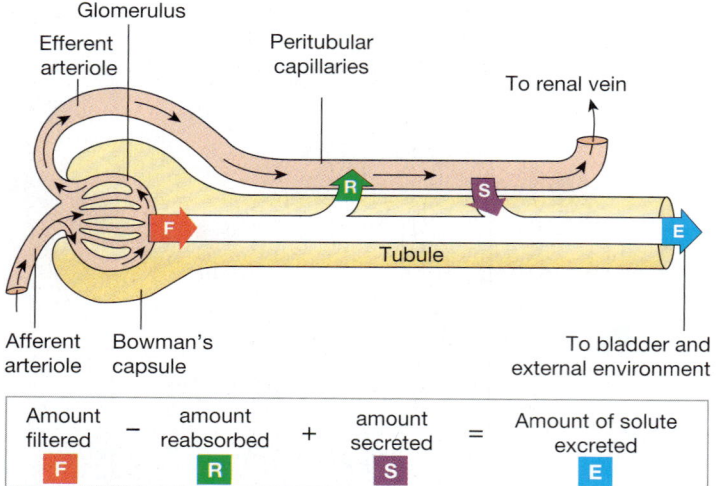

Amount filtered **F** − amount reabsorbed **R** + amount secreted **S** = Amount of solute excreted **E**

■ **FIGURE 19-3** *The urinary excretion of a substance depends on its filtration, reabsorption, and secretion*

Besides the kidneys, other organs that carry out excretory processes include the lungs (CO_2) and intestines (undigested food, bilirubin).

Figure 19-2 summarizes filtration, reabsorption, secretion, and excretion. Filtration takes place in the renal corpuscle as fluid moves from the capillaries of the glomerulus into Bowman's capsule. Reabsorption and secretion occur along the remainder of the tubule, transferring materials between the lumen and the peritubular capillaries. The quantity and composition of the substances being reabsorbed and secreted vary in different segments of the nephron. Filtrate that remains in the lumen at the end of the nephron is excreted as urine.

The amount of any substance excreted in the urine reflects how that substance was handled during its passage through the nephron (Fig. 19-3 ■). The amount excreted is equal to the amount filtered into the tubule, minus the amount reabsorbed into the blood, plus the amount secreted into the tubule lumen:

amount excreted =

amount filtered − amount reabsorbed + amount secreted

This equation is a useful way to think about renal handling of solutes. In the following sections, we will look in more detail at the important processes of filtration, reabsorption, secretion, and excretion.

✔ CONCEPT CHECK

5. Name one way in which filtration and secretion are alike. Name one way in which they differ.

6. A water molecule enters the renal corpuscle from the blood and ends up in the urine. Name all the anatomical structures that the molecule passes through on its trip to the outside world.

7. What would happen to the body if filtration continued at a normal rate but reabsorption dropped to half the normal rate?

Answers: p. 639

FILTRATION

The filtration of plasma into the kidney tubule is the first step in urine formation. This relatively nonspecific process creates a filtrate whose composition is like that of plasma minus most of the plasma proteins. Under normal conditions, blood cells remain in the capillary, so that the filtrate is composed only of water and dissolved solutes.

The Renal Corpuscle Contains Three Filtration Barriers

Filtration takes place in the renal corpuscle (Fig. 19-4 ■), which consists of the glomerular capillaries surrounded by Bowman's capsule. Substances leaving the plasma must pass through three *filtration barriers* before entering the tubule lumen: the glomerular capillary endothelium, a basal lamina (basement membrane), and the epithelium of Bowman's capsule (Fig. 19-4d).

The first barrier is the capillary endothelium. Glomerular capillaries are *fenestrated capillaries* [🔁 p. 515] with large pores that allow most components of the plasma to filter through the endothelium. The pores are small enough, however, to prevent blood cells from leaving the capillary. The negatively charged proteins on the pore surfaces also help repel negatively charged plasma proteins.

Glomerular **mesangial cells** lie between and around the glomerular capillaries (Fig. 19-4c). Mesangial cells have cytoplasmic bundles of actin-like filaments that enable them to contract and alter blood flow through the capillaries. In addition, mesangial cells secrete cytokines associated with immune and inflammatory processes. Disruptions of mesangial cell function have been linked to several disease processes in the kidney.

The second filtration barrier is an acellular layer of extracellular matrix called the **basal lamina**, which separates the capillary endothelium from the epithelial lining of Bowman's capsule (Fig. 19-4d). The basal lamina consists of negatively charged glycoproteins and a collagen-like material that act like a coarse sieve, excluding most plasma proteins from the fluid that filters through it.

The third filtration barrier is the epithelium of Bowman's capsule. The portion of the capsule epithelium that surrounds each glomerular capillary consists of specialized cells called **podocytes** [*podos*, foot] (Fig. 19-4c). Podocytes have long cytoplasmic extensions called **foot processes** that extend from the main cell body (Fig. 19-4a, b). These processes wrap around the glomerular capillaries and interlace with one another, leaving narrow **filtration slits** closed by a membrane. Podocytes, like mesangial cells, contain contractile fibers. These are connected to the basal lamina by integrins.

When you visualize plasma filtering out of the glomerular capillaries, it is easy to imagine that all the plasma in the capillary moves into Bowman's capsule. However, filtration of all the plasma would leave behind a sludge of blood cells and proteins that could not flow out of the glomerulus. Instead, only about one-fifth of the plasma that flows through the kidneys filters into

(a) The epithelium around glomerular capillaries is modified into podocytes.

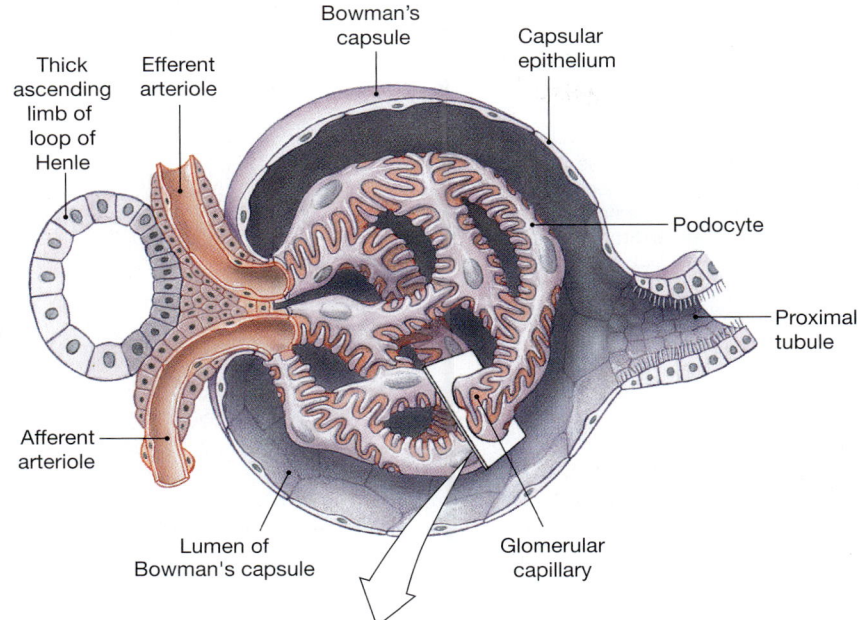

(b) Micrograph showing podocyte foot processes around glomerular capillary.

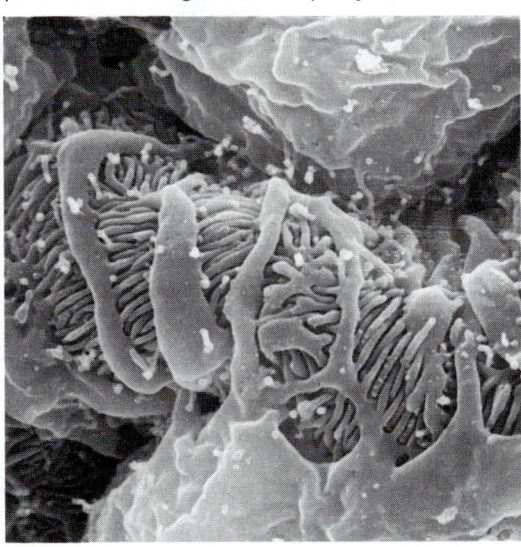

(c) Podocyte foot processes surround each capillary, leaving slits through which filtration takes place.

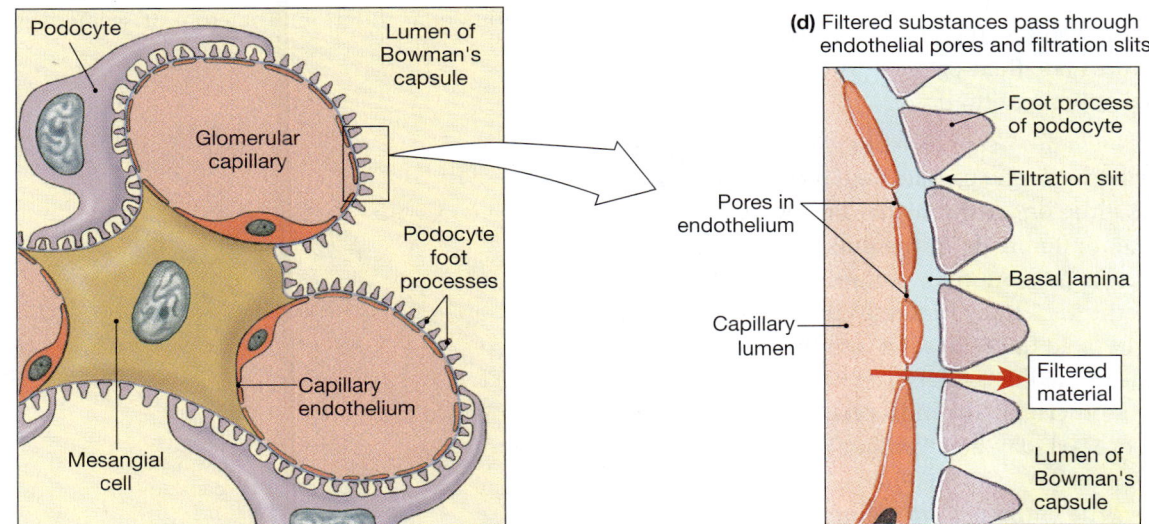

(d) Filtered substances pass through endothelial pores and filtration slits.

■ **FIGURE 19-4** *The renal corpuscle*

The glomerular capillary endothelium, basal lamina, and Bowman's capsule epithelium create a three-layer filtration barrier.

the nephrons. The remaining four-fifths of the plasma, along with most plasma proteins and blood cells, flows into the peritubular capillaries (Fig. 19-5 ■). The percentage of total plasma volume that filters into the tubule is called the **filtration fraction**.

Filtration Occurs Because of Hydrostatic Pressure in the Capillaries

What drives filtration across the walls of the glomerular capillaries? The process is similar in many ways to filtration of fluid

out of systemic capillaries [🔁 p. 517] and is influenced by the following forces:

1. The *hydrostatic pressure* (P_H) of blood flowing through the glomerular capillaries forces fluid through the leaky endothelium. Capillary blood pressure averages 55 mm Hg and favors filtration into Bowman's capsule. Although pressure declines along the length of the capillaries, it remains higher than the opposing pressures. Consequently, filtration takes place along nearly the entire length of the glomerular capillaries.

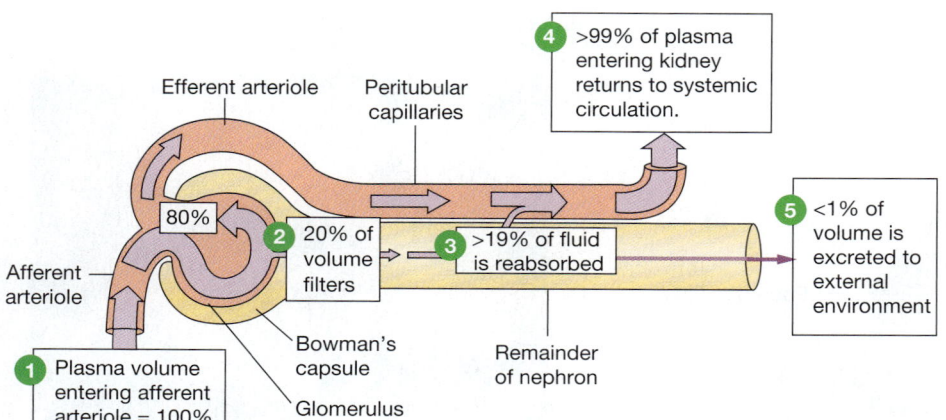

Only 20% of the plasma that passes through the glomerulus is filtered. Less than 1% of filtered fluid is eventually excreted.

2. The *colloid osmotic pressure* (π) inside glomerular capillaries is higher than that of the fluid in Bowman's capsule. This pressure gradient is due to the presence of proteins in the plasma. The osmotic pressure gradient averages 30 mm Hg and favors fluid movement back into the capillaries.

3. Bowman's capsule is an enclosed space (unlike the interstitial fluid), and so the presence of fluid in the capsule creates a hydrostatic *fluid pressure* (P_{fluid}) that opposes fluid movement into the capsule. Fluid filtering out of the capillaries must displace the fluid already in the capsule lumen. Hydrostatic fluid pressure in the capsule averages 15 mm Hg, opposing filtration.

The three pressures—capillary blood pressure, capillary colloid osmotic pressure, and capsule fluid pressure—that influence glomerular filtration are summarized in Figure 19-6 ■. The net driving force is 10 mm Hg in the direction favoring filtration. Although this pressure may not seem very high, when combined with the very leaky nature of the fenestrated capillaries, it results in rapid fluid filtration into the tubules.

The volume of fluid that filters into Bowman's capsule per unit time is the **glomerular filtration rate (GFR)**. Average GFR is 125 mL/min, or 180 L/day, an incredible rate considering that total plasma volume is only about 3 liters. This rate means that the kidneys filter the entire plasma volume 60 times a day, or 2.5 times every hour. If most of the filtrate were not reabsorbed during its passage through the nephron, we would run out of plasma in only 24 minutes of filtration!

GFR is influenced by two factors: the net filtration pressure just described and the filtration coefficient. Filtration pressure is determined primarily by renal blood flow and blood pressure. The **filtration coefficient** has two components: the surface area of the glomerular capillaries available for filtration and the permeability of the capillary-Bowman's capsule interface. In this respect, glomerular filtration is similar to gas exchange at the alveoli, where the rate of gas exchange depends on partial pressure differences, the surface area of the alveoli, and the permeability of the alveolar membrane.

CONCEPT CHECK

8. Why is the osmotic pressure of plasma in efferent arterioles higher than that in afferent arterioles?

Answers: p. 640

Blood Pressure and Renal Blood Flow Influence GFR

Blood pressure provides the hydrostatic pressure that drives glomerular filtration. Therefore, it seems reasonable to assume

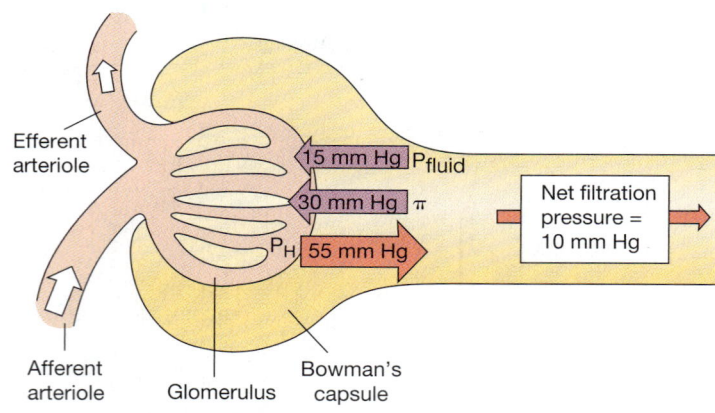

P_H	−	π	−	P_{fluid}	= net filtration pressure
55 mm Hg	−	30 mm Hg	−	15 mm Hg	= 10 mm Hg

KEY

P_H = Hydrostatic pressure (blood pressure)
π = Colloid osmotic pressure gradient due to proteins in plasma but not in Bowman's capsule
P_{fluid} = Fluid pressure created by fluid in Bowman's capsule

■ **FIGURE 19-6**   *Filtration pressure in the renal corpuscle depends on hydrostatic pressure, colloid osmotic pressure, and fluid pressure*

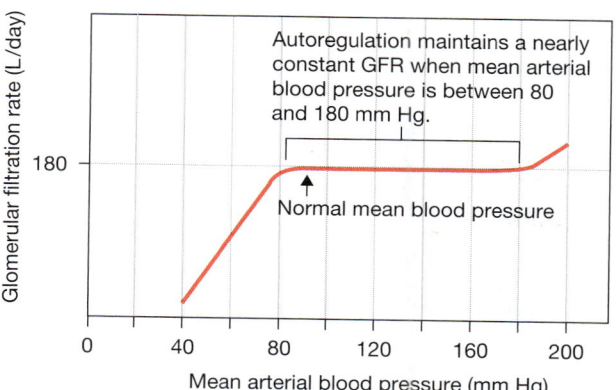

FIGURE 19-7 *Autoregulation of glomerular filtration rate takes place over a wide range of blood pressures*

that if blood pressure increases, GFR goes up, and if blood pressure falls, GFR goes down. That is not usually the case, however. Instead, GFR is remarkably constant over a wide range of blood pressures. As long as mean arterial blood pressure remains between 80 mm Hg and 180 mm Hg, GFR averages 180 L/day (Fig. 19-7 ■).

GFR is controlled primarily by regulation of blood flow through the renal arterioles (Fig. 19-8 ■). If the overall resistance of the renal arterioles increases, renal blood flow decreases, and

EMERGING CONCEPTS

DIABETES

DIABETIC NEPHROPATHY

End-stage renal failure, in which kidney function has deteriorated beyond recovery, is a life-threatening complication in 30–40% of people with type 1 diabetes and in 10–20% of those with type 2 diabetes. As with many other complications of diabetes, the exact causes of renal failure are not clear. Diabetic nephropathy usually begins with an increase in glomerular filtration. This is followed by the appearance of proteins in the urine (*proteinuria*), an indication that the normal filtration barrier has been altered. In later stages, filtration rates decline. This stage is associated with thickening of the glomerular basal lamina and changes in both podocytes and mesangial cells. Abnormal proliferation of mesangial cells compresses the glomerular capillaries and impedes blood flow, contributing to the decrease in glomerular filtration. At this point, patients must have their kidney function supplemented by dialysis, and eventually they may need a kidney transplant.

(a) Renal blood flow and GFR change if resistance in the arterioles changes.

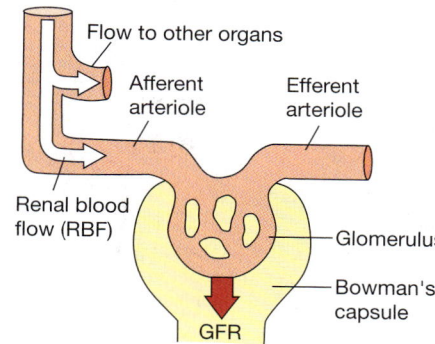

(b) Vasoconstriction of the afferent arteriole increases resistance and decreases renal blood flow, capillary blood pressure (P_H), and GFR.

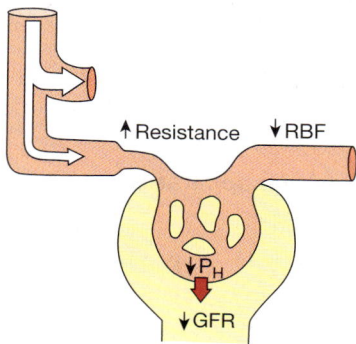

(c) Increased resistance of efferent arteriole decreases renal blood flow but increases P_H and GFR.

(d)

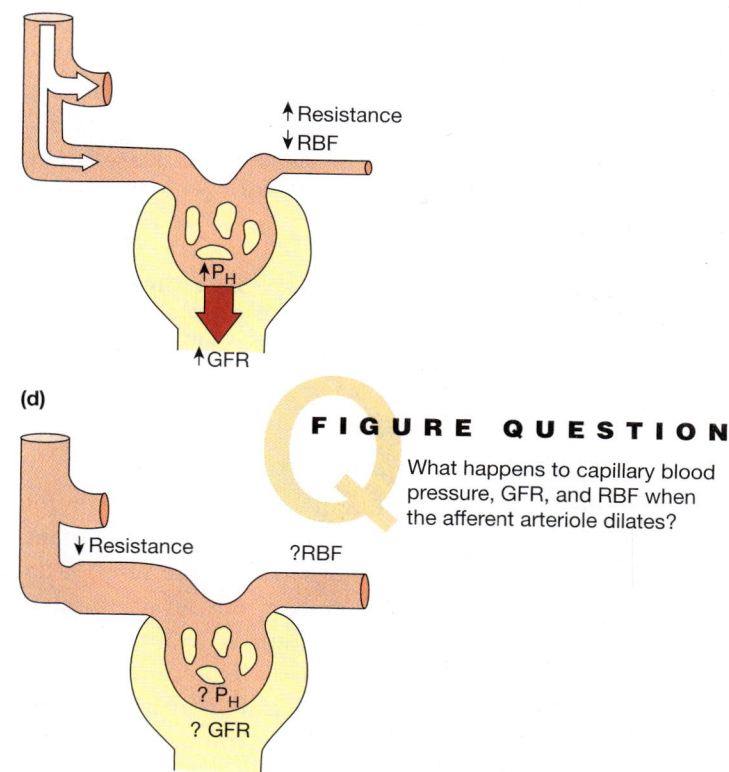

FIGURE QUESTION

What happens to capillary blood pressure, GFR, and RBF when the afferent arteriole dilates?

FIGURE 19-8 *Resistance changes in renal arterioles alter GFR and renal blood flow*

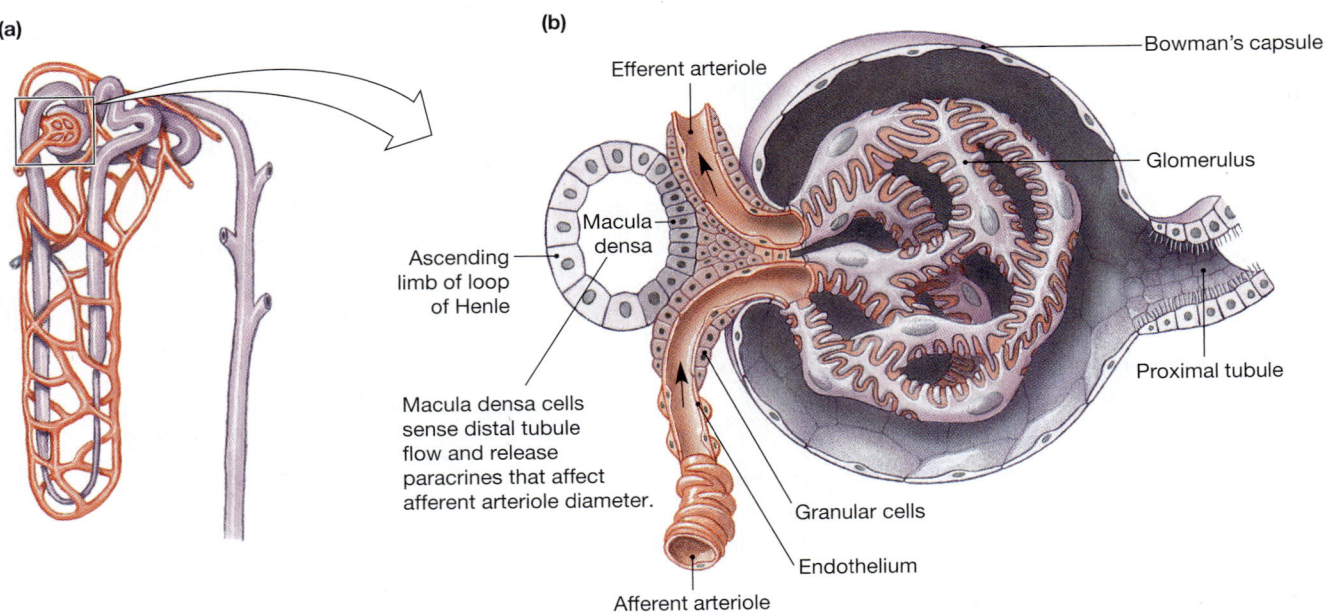

(a)

(b)

Efferent arteriole

Bowman's capsule

Glomerulus

Macula densa

Ascending
limb of loop
of Henle

Macula densa cells
sense distal tubule
flow and release
paracrines that affect
afferent arteriole diameter.

Proximal tubule

Granular cells

Endothelium

Afferent arteriole

■ **FIGURE 19-9** *The juxtaglomerular apparatus*

blood is diverted to other organs [Fig. 15-14, p. 515]. The effect of increased resistance on GFR, however, depends on *where* the resistance change takes place.

If resistance increases in the *afferent* arteriole (Fig. 19-8b), hydrostatic pressure decreases on the glomerular side of the constriction. This translates into a decrease in GFR. If resistance increases in the *efferent* arteriole, blood "dams up" in front of the constriction, and hydrostatic pressure in the glomerular capillaries increases (Fig. 19-8c). Increased glomerular pressure increases GFR. The opposite changes occur with decreased resistance in the afferent or efferent arterioles. Most regulation occurs at the afferent arteriole.

CONCEPT CHECK

9. If a hypertensive person's blood pressure is 143/107 mm Hg and mean arterial pressure is diastolic pressure + 1/3 the pulse pressure, what is this person's mean arterial pressure? What is this person's GFR according to Figure 19-7? Answers: p. 640

GFR Is Subject to Autoregulation

Autoregulation of glomerular filtration rate is a local control process in which the kidney maintains a relatively constant GFR in the face of normal fluctuations in blood pressure. We do not completely understand the process, but several mechanisms are at work. The **myogenic response** is the intrinsic ability of vascular smooth muscle to respond to pressure changes. **Tubuloglomerular feedback** is a paracrine signaling mechanism through which changes in fluid flow through the loop of Henle influence GFR.

Myogenic Response The myogenic response of afferent arterioles is similar to autoregulation in other systemic arterioles.

When smooth muscle in the arteriole wall stretches because of increased blood pressure, stretch-sensitive ion channels open, and the muscle cells first depolarize, then contract [p. 472]. Vasoconstriction increases resistance to flow, and so blood flow through the arteriole diminishes. The decrease in blood flow decreases filtration pressure in the glomerulus.

If blood pressure decreases, the tonic level of arteriolar contraction disappears, and the arteriole becomes maximally dilated. However, vasodilation is not as effective at maintaining GFR as vasoconstriction because normally the afferent arteriole is fairly relaxed. Consequently, when mean blood pressure drops below 80 mm Hg, GFR decreases. This decrease is adaptive in the sense that if less plasma is filtered, the potential for fluid loss in the urine is decreased. In other words, a decrease in GFR helps the body conserve blood volume.

Tubuloglomerular Feedback Tubuloglomerular feedback is a local control pathway in which fluid flow through the tubule influences GFR. The twisted configuration of the nephron causes the final portion of the ascending limb of the loop of Henle to pass between the afferent and efferent arterioles (Fig. 19-9a ■). The tubule and arteriolar walls are modified in the regions where they contact each other and together form the *juxtaglomerular apparatus*.

The modified portion of the tubule epithelium is a plaque of cells called the **macula densa** (Fig. 19-9b). The adjacent wall of the afferent arteriole has specialized smooth muscle cells called **granular cells** (also known as *juxtaglomerular cells* or *JG cells*). The granular cells secrete *renin*, an enzyme involved in salt and water balance. When NaCl delivery past the macula densa increases as a result of increased GFR, the macula densa cells send a paracrine message to the neighboring afferent arteriole

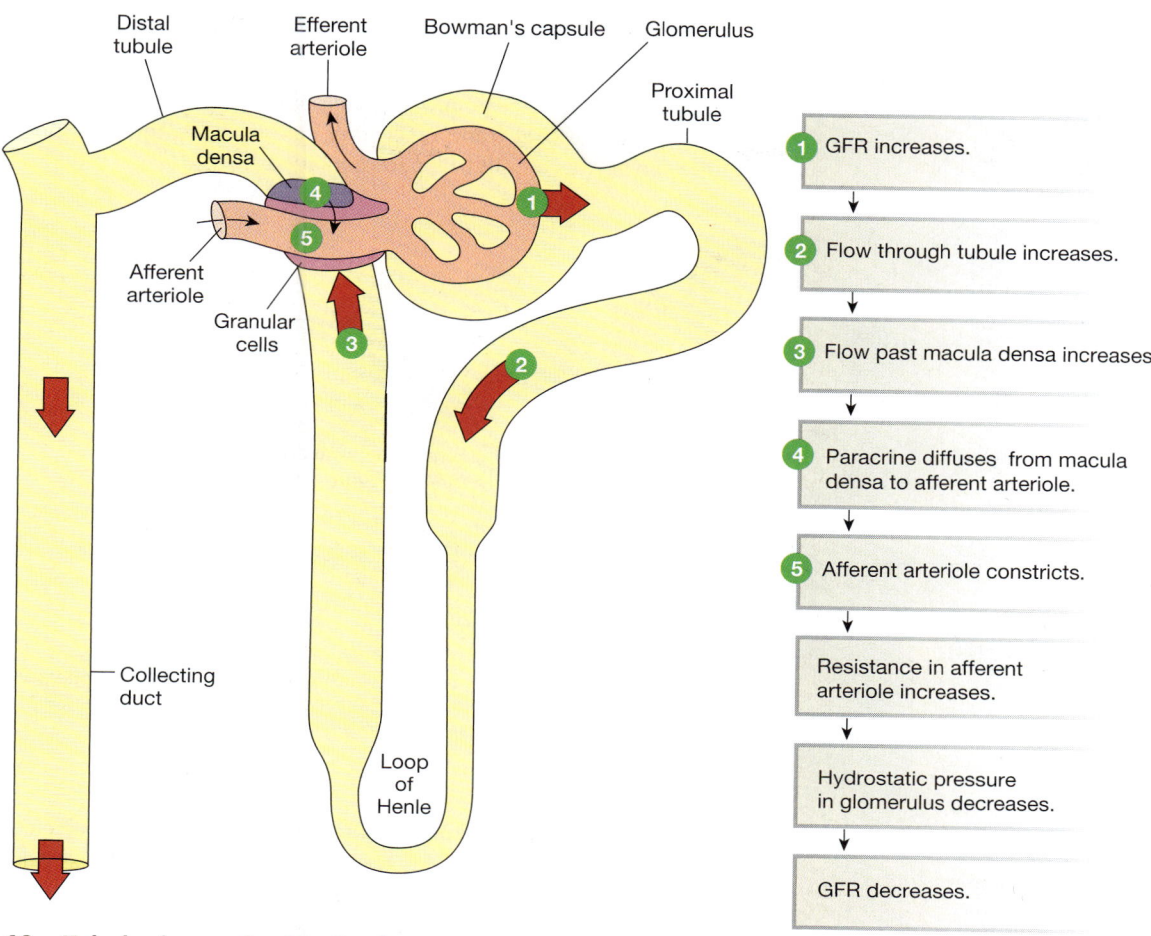

Distal tubule — Efferent arteriole — Bowman's capsule — Glomerulus — Proximal tubule

Macula densa

Afferent arteriole

Granular cells

Collecting duct

Loop of Henle

1. GFR increases.
2. Flow through tubule increases.
3. Flow past macula densa increases.
4. Paracrine diffuses from macula densa to afferent arteriole.
5. Afferent arteriole constricts.

Resistance in afferent arteriole increases.

Hydrostatic pressure in glomerulus decreases.

GFR decreases.

■ **FIGURE 19-10** *Tubuloglomerular feedback*

(Fig. 19-10 ■). The afferent arteriole constricts, increasing resistance and decreasing GFR.

Experimental evidence indicates that the macula densa cells transport NaCl, and that changes in salt transport initiate tubuloglomerular feedback. The paracrine signaling between the macula densa and the afferent arteriole is complex, and the details are not clear. Experiments show that several paracrine signals, including ATP, adenosine, and nitric oxide, pass from the macula densa to the arteriole.

Hormones and Autonomic Neurons Also Influence GFR

One of the major functions of the kidneys is the integrated control of systemic blood pressure. Consequently, integrating centers outside the kidney may use hormones and the autonomic nervous system to alter glomerular filtration rate. These neural and hormonal signals work in two ways: by changing resistance in the arterioles and by altering the filtration coefficient.

Neural control of GFR is mediated by sympathetic neurons that innervate both the afferent and efferent arterioles. Sympathetic innervation of α-receptors on vascular smooth muscle causes vasoconstriction [p. 512]. If sympathetic activity is moderate, there is little effect on GFR. If systemic blood pressure

drops sharply, however, as occurs with hemorrhage or severe dehydration, sympathetically induced vasoconstriction of the arterioles decreases GFR and renal blood flow. This is an adaptive response that helps conserve fluid volume.

A variety of hormones also influence arteriolar resistance. Among the most important are *angiotensin II,* a potent vasoconstrictor, and *prostaglandins,* which act as vasodilators. These same hormones may affect the filtration coefficient by acting on podocytes or mesangial cells. Podocytes change the size of the glomerular filtration slits. If the slits widen, more surface area is available for filtration, and GFR increases. Contraction of mesangial cells apparently changes the glomerular capillary surface area available for filtration. We still have much to learn about these processes, and physiologists are actively investigating them.

CONCEPT CHECK

10. If systemic blood pressure remains constant but the afferent arteriole of a nephron constricts, what happens to renal blood flow and GFR in that nephron?

11. A person with cirrhosis of the liver has lower-than-normal levels of plasma proteins and consequently a higher-than-normal GFR. Explain why a decrease in plasma protein concentration causes an increase in GFR.

Answers: p. 640

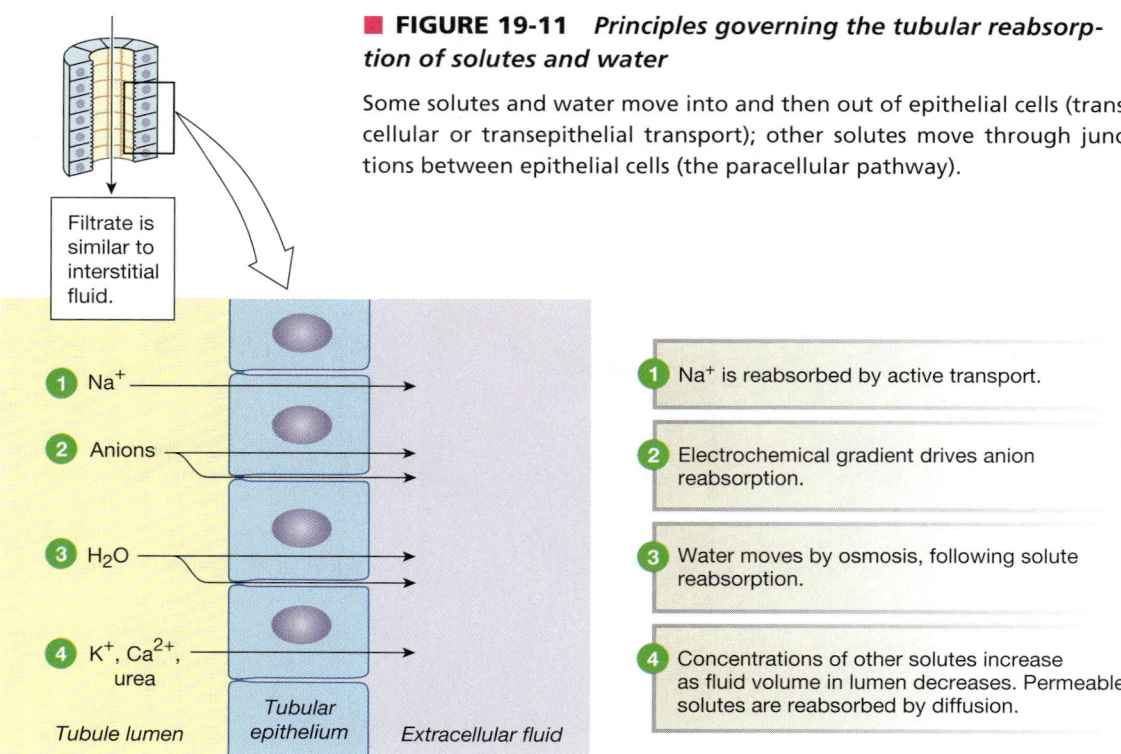

■ **FIGURE 19-11** *Principles governing the tubular reabsorption of solutes and water*

Some solutes and water move into and then out of epithelial cells (transcellular or transepithelial transport); other solutes move through junctions between epithelial cells (the paracellular pathway).

Filtrate is similar to interstitial fluid.

1 Na^+

2 Anions

3 H_2O

4 K^+, Ca^{2+}, urea

Tubule lumen *Tubular epithelium* *Extracellular fluid*

1 Na^+ is reabsorbed by active transport.

2 Electrochemical gradient drives anion reabsorption.

3 Water moves by osmosis, following solute reabsorption.

4 Concentrations of other solutes increase as fluid volume in lumen decreases. Permeable solutes are reabsorbed by diffusion.

REABSORPTION

Each day, 180 liters of filtered fluid pass from the glomerular capillaries into the tubules, yet only about 1.5 liters are excreted in the urine. Thus more than 99% of the fluid entering the tubules must be reabsorbed into the blood as filtrate moves through the nephrons. Most of this reabsorption takes place in the proximal tubule, with a smaller amount of reabsorption in the distal segments of the nephrons. Regulated reabsorption in the distal nephron allows the kidneys to return ions and water to the plasma selectively—as needed to maintain homeostasis.

One question you might be asking is, "Why bother to filter 180 L/day and then reabsorb 99% of it? Why not simply filter and excrete the 1% that needs to be eliminated?" There are two reasons. First, many foreign substances are filtered into the tubule but not reabsorbed into the blood. The high daily filtration rate helps clear such substances from the plasma very rapidly.

Second, filtering ions and water into the tubule simplifies their regulation. If a portion of filtrate that reaches the distal nephron is not needed to maintain homeostasis, it passes into the urine. With a high GFR, this excretion can occur quite rapidly. However, if the ions and water are needed, they are reabsorbed.

Reabsorption May Be Active or Passive

Reabsorption of water and solutes from the tubule lumen to the extracellular fluid depends on active transport. The filtrate flowing out of Bowman's capsule into the proximal tubule has the same solute concentrations as extracellular fluid. To move solute out of the lumen, the tubule cells must therefore use active transport to create concentration or electrochemical gradients. Water osmotically follows solutes as they are reabsorbed.

Figure 19-11 ■ is a general overview of reabsorption. Active transport of Na^+ from the tubule lumen to the extracellular fluid creates a transepithelial electrical gradient in which the lumen is more negative than the ECF. Anions then follow the positively charged Na^+ out of the lumen. The net movement of Na^+ and anions from lumen to extracellular fluid causes water to follow by osmosis. The loss of volume from the lumen increases the concentration of solutes (including K^+, Ca^{2+}, and urea) left behind in the filtrate: an unchanged amount of solute in a smaller volume produces an increased concentration. Once luminal solute concentrations are higher than solute concentrations in the extracellular fluid, the solutes are free to diffuse out of the lumen if the epithelium of the tubule is permeable to them.

Reabsorption involves both **transepithelial transport** (also called *transcellular transport*), in which substances cross both the apical and basolateral membranes of the tubule epithelial cell [🔁 p. 150], and the **paracellular pathway**, in which substances pass through the junction between two adjacent cells. Which route a solute takes depends on the permeability of the epithelial junctions and on the electrochemical gradient for the solute. For solutes that move through the epithelial cells, their concentration gradients determine their transport mechanisms. Solutes moving down their concentration (or electrochemical) gradient use open leak channels or facilitated diffusion carriers to cross the cell membrane. Molecules that need to be pushed against their concentration gradient are moved by either primary or secondary active transport. Sodium is directly or indirectly involved in many instances of both passive and active transport.

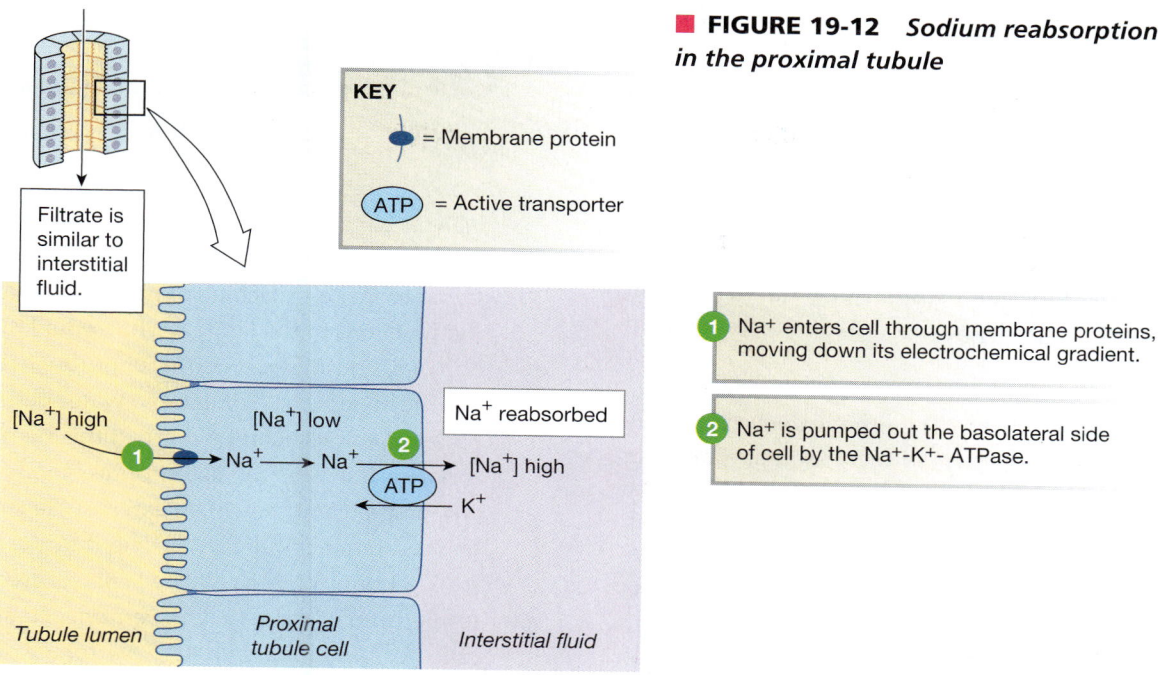

■ **FIGURE 19-12**   *Sodium reabsorption in the proximal tubule*

KEY

●   = Membrane protein

(ATP)   = Active transporter

1   Na⁺ enters cell through membrane proteins, moving down its electrochemical gradient.

2   Na⁺ is pumped out the basolateral side of cell by the Na⁺-K⁺- ATPase.

Filtrate is similar to interstitial fluid.

[Na⁺] high

[Na⁺] low

Na⁺ reabsorbed

1   Na⁺ → Na⁺   2   [Na⁺] high

ATP

K⁺

Tubule lumen   Proximal tubule cell   Interstitial fluid

Active Transport of Sodium   The active transport of Na⁺ is the primary driving force for most renal reabsorption. As noted earlier, filtrate entering the proximal tubule is similar in ion composition to plasma, with a higher Na⁺ concentration than is found in cells. Thus Na⁺ in the filtrate can enter tubule cells passively by moving down its electrochemical gradient (Fig. 19-12 ■). Apical movement of Na⁺ uses a variety of symport and antiport transport proteins [⟲ p. 139] or open leak channels. In the proximal tubule, a **Na⁺-H⁺ antiporter** plays a major role in Na⁺ reab-

sorption. Once inside a tubule cell, Na⁺ is actively transported out across the basolateral membrane by the Na⁺-K⁺-ATPase. The end result is Na⁺ reabsorption across the tubule epithelium.

Secondary Active Transport: Symport with Sodium
Sodium-linked secondary active transport in the nephron is responsible for the reabsorption of many substances, including glucose, amino acids, ions, and various organic metabolites. Figure 19-13 ■ shows one example: Na⁺-dependent

19

■ **FIGURE 19-13**   *Sodium-linked glucose reabsorption in the proximal tubule*

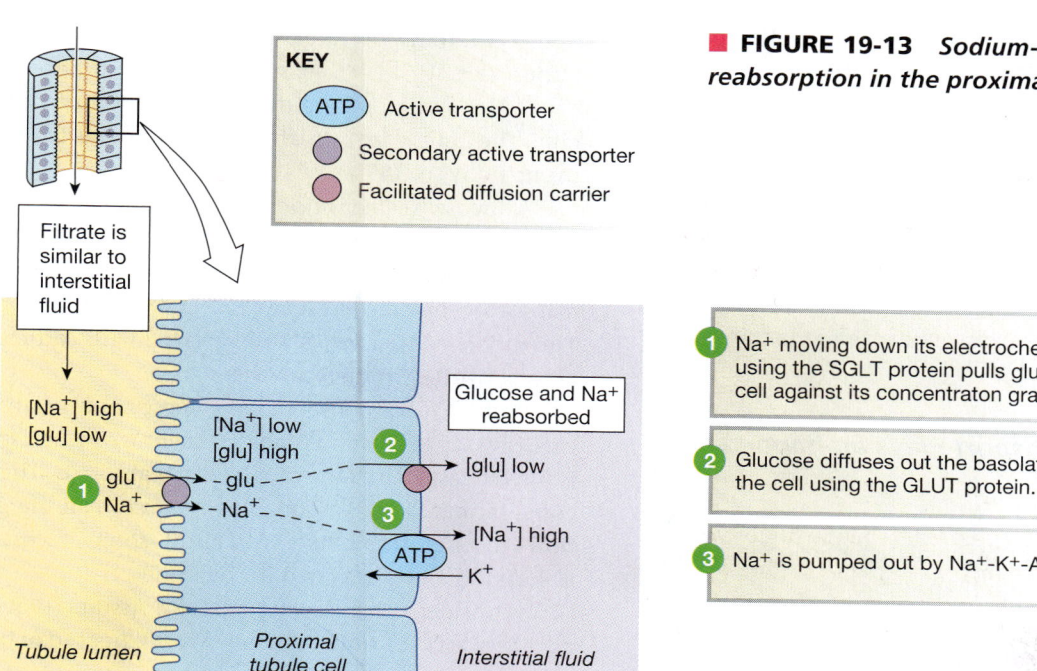

KEY

(ATP)   Active transporter

●   Secondary active transporter

●   Facilitated diffusion carrier

Filtrate is similar to interstitial fluid

[Na⁺] high
[glu] low

[Na⁺] low
[glu] high

Glucose and Na⁺ reabsorbed

1   glu — glu   2   [glu] low
Na⁺ — Na⁺

3   [Na⁺] high

ATP

K⁺

Tubule lumen   Proximal tubule cell   Interstitial fluid

1   Na⁺ moving down its electrochemical gradient using the SGLT protein pulls glucose into the cell against its concentraton gradient.

2   Glucose diffuses out the basolateral side of the cell using the GLUT protein.

3   Na⁺ is pumped out by Na⁺-K⁺-ATPase.

reabsorption across the proximal tubule epithelium [, Fig. 5-26, p. 152]. The apical membrane contains the Na$^+$-glucose co-transporter (SGLT) that brings glucose into the cytoplasm against its concentration gradient by harnessing the energy of Na$^+$ moving down its electrochemical gradient. On the basolateral side of the cell, Na$^+$ is pumped out by the Na$^+$-K$^+$-ATPase while glucose diffuses out with the aid of a facilitated diffusion GLUT transporter. The same basic pattern holds for many other molecules absorbed by Na$^+$-dependent transport: an apical symport protein and a basolateral facilitated diffusion carrier.

Passive Reabsorption: Urea The nitrogenous waste product urea has no active transporters in the proximal tubule but will move across the epithelium by diffusion if there is a urea concentration gradient. Initially, urea concentrations in the filtrate and extracellular fluid are equal. However, the active transport of Na$^+$ and other solutes in the proximal tubule creates a urea concentration gradient by the following process.

When Na$^+$ and other solutes are reabsorbed from the proximal tubule, the transfer of osmotically active particles makes the extracellular fluid more concentrated than the filtrate remaining in the lumen (see Fig. 19-11). In response to the osmotic gradient, water moves by osmosis across the epithelium. Up to this point, no urea molecules have moved out of the lumen because there has been no urea concentration gradient. Now, however, when water leaves the lumen, the filtrate concentration of urea increases because the same amount of urea is contained in a smaller volume. Once a concentration gradient for urea exists, urea diffuses out of the lumen into the extracellular fluid.

Transcytosis: Plasma Proteins Filtration of plasma at the glomerulus normally leaves most plasma proteins in the blood, but some smaller protein hormones and enzymes can pass through the filtration barrier. Most filtered proteins are reabsorbed in the proximal tubule, with the result that normally only trace amounts of protein appear in urine.

Small as they are, filtered proteins are too large to be reabsorbed by carriers or through channels. Instead they enter proximal tubule cells by endocytosis at the apical membrane. Once in the cells, the proteins may be digested and released as amino acids or delivered intact to the extracellular fluid via transcytosis [p. 153].

Saturation of Renal Transport Plays an Important Role in Kidney Function

Most transport in the nephron uses membrane proteins and exhibits the three characteristics of mediated transport: saturation, specificity, and competition [p. 145].

Saturation refers to the maximum rate of transport that occurs when all available carriers are occupied by (are saturated with) substrate. At substrate concentrations below the satura-

tion point, transport rate is directly related to substrate concentration (Fig. 19-14 ■). At substrate concentrations equal to or above the saturation point, transport occurs at a maximum rate. The transport rate at saturation is the **transport maximum T$_m$** [p. 147].

Glucose reabsorption in the nephron is an excellent example of the consequences of saturation. At normal plasma glucose concentrations, all glucose that enters the nephron is reabsorbed before it reaches the end of the proximal tubule. The tubule epithelium is well supplied with carriers to capture glucose as the filtrate flows past.

But what happens if blood glucose concentrations become excessive, as they do in diabetes mellitus? In that case, glucose is filtered faster than the carriers can reabsorb it. The carriers become saturated and are unable to reabsorb all the glucose that flows through the tubule. As a result, some glucose escapes reabsorption and is excreted in the urine.

Consider the following analogy. Assume that the carriers are like seats on a train at Disney World. Instead of boarding the stationary train from a stationary platform, passengers step

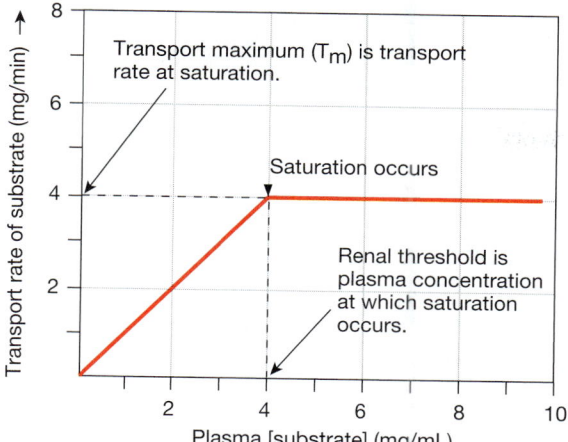

GRAPH QUESTION

What is the transport rate at the following plasma substrate concentrations: 3 mg/mL, 5 mg/mL, 8 mg/mL? At what plasma substrate concentration is the transport rate 4 mg/min?

■ **FIGURE 19-14** *Saturation of mediated transport*

The transport rate of a substance is proportional to the plasma concentration of the substance, up to the point at which transporters become saturated. Once saturation occurs, transport rate reaches a maximum, known as the transport maximum. The plasma concentration of substrate at which the transport maximum occurs is called the renal threshold.

onto a moving sidewalk that rolls them past the train. As the passengers see an open seat, they grab it. However, if more people are allowed onto the moving sidewalk than there are seats in the train, some people will not find seats. And because the sidewalk is moving people past the train toward an exit, they cannot wait for the next train. Instead, they end up being transported out the exit.

In the same fashion, glucose molecules entering Bowman's capsule in the filtrate are like passengers stepping onto the moving sidewalk. In order to be reabsorbed, each glucose molecule must bind to a transporter as the filtrate flows through the proximal tubule. If only a few glucose molecules enter the tubule at a time, each one will find a free transporter and be reabsorbed, just as a small number of people on the moving sidewalk all find seats on the train. However, if glucose molecules filter into the tubule faster than the glucose carriers can transport them, some glucose remains in the lumen and is excreted in the urine.

Figure 19-15 ■ is a graphic representation of glucose handling by the kidney. Figure 19-15a shows that the filtration rate of glucose from plasma into Bowman's capsule is proportional to the plasma concentration of glucose. Because filtration does not exhibit saturation, the graph continues infinitely in a

straight line: the filtrate glucose concentration is always equal to the plasma glucose concentration.

Figure 19-15b plots the reabsorption rate of glucose in the proximal tubule against the plasma concentration of glucose. Reabsorption exhibits a maximum transport rate (T_m) when the carriers reach saturation.

Figure 19-15c plots the excretion rate of glucose in relation to the plasma concentration of glucose. When plasma glucose concentrations are low enough that 100% of the filtered glucose is reabsorbed, no glucose is excreted. Once the carriers reach saturation, glucose excretion begins. The plasma concentration at which glucose first appears in the urine is called the **renal threshold** for glucose.

Figure 19-15d is a composite graph that compares filtration, reabsorption, and excretion of glucose. Recall from our earlier discussion that

amount excreted =
  amount filtered − amount reabsorbed + amount secreted

For glucose, which is not secreted, the equation can be rewritten as

glucose excreted = glucose filtered − glucose reabsorbed

Under normal conditions, all filtered glucose is reabsorbed. In other words, filtration is equal to reabsorption.

Notice in Figure 19-15d that the lines representing filtration and reabsorption are identical up to the plasma glucose concentration that equals the renal threshold. If filtration equals reabsorption, the algebraic difference between the two is zero, and there is no excretion. Once the renal threshold is reached, filtration begins to exceed reabsorption. Notice on the graph that the filtration and reabsorption lines diverge at this point. The difference between the filtration line and the reabsorption line represents the excretion rate:

$$\text{excretion} = \underset{(increasing)}{\text{filtration}} - \underset{(constant)}{\text{reabsorption}}$$

Excretion of glucose in the urine is called **glucosuria** or **glycosuria** [-*uria*, in the urine] and usually indicates an elevated blood glucose concentration. Rarely, glucose appears in the urine even though the blood glucose concentrations are normal. This situation is due to a genetic disorder in which the nephron does not make enough carriers.

Peritubular Capillary Pressures Favor Reabsorption

The reabsorption we have just discussed refers to the movement of solutes and water from the tubule lumen to the interstitial fluid. How does that reabsorbed fluid then get into the capillary? The answer is that the driving force for reabsorption from the interstitial fluid into the capillaries is the low hydrostatic pressure that exists along the entire length of the peritubular capillaries. This low pressure favors reabsorption.

19

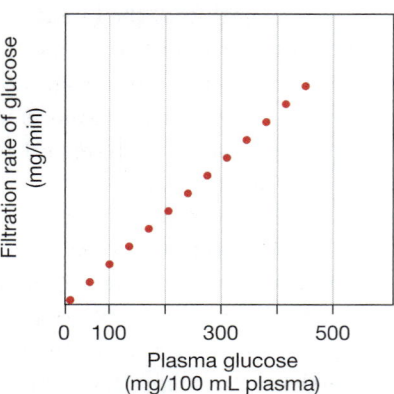

(a) Filtration of glucose is proportional to the plasma concentration.

Filtration rate of glucose (mg/min)

Plasma glucose (mg/100 mL plasma)

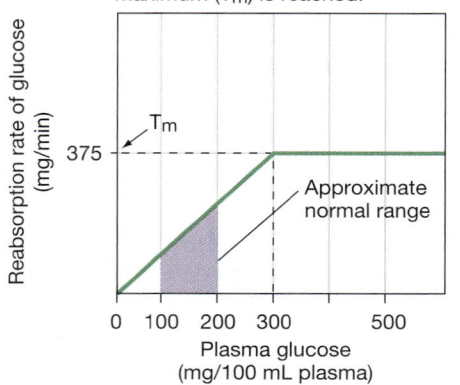

(b) Reabsorption of glucose is proportional to plasma concentration until the transport maximum (T_m) is reached.

Reabsorption rate of glucose (mg/min)

T_m

375

Approximate normal range

Plasma glucose (mg/100 mL plasma)

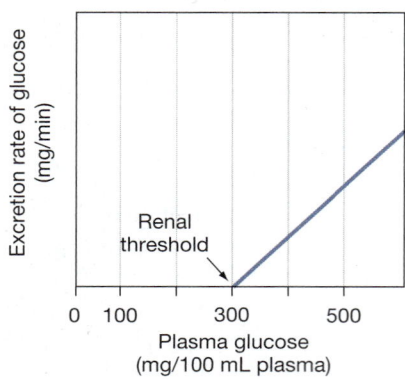

(c) Glucose excretion is zero until the renal threshold is reached.

Excretion rate of glucose (mg/min)

Renal threshold

Plasma glucose (mg/100 mL plasma)

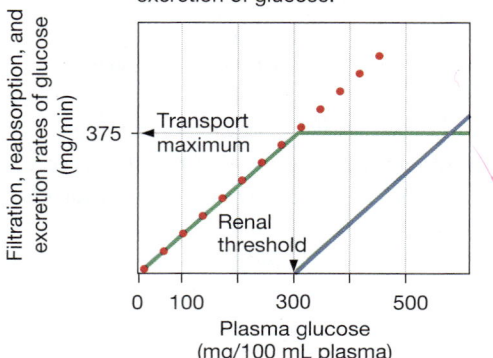

(d) Composite graph shows the relationship between filtration, reabsorption, and excretion of glucose.

Filtration, reabsorption, and excretion rates of glucose (mg/min)

375

Transport maximum

Renal threshold

Plasma glucose (mg/100 mL plasma)

■ **FIGURE 19-15** *Glucose handling by the nephron*

19

RUNNING PROBLEM

Uric acid, the molecule that causes gout, is a normal product of purine metabolism. Increased uric acid production may be associated with cell and tissue breakdown, or it may occur as a result of inherited enzyme defects. Uric acid in the blood filters freely into Bowman's capsule but is almost totally reabsorbed in the proximal tubule. Some uric acid is secreted into the proximal tubule.

Question 2:
   Why would uric acid levels in the blood go up when cell breakdown increases? (Hint: purines are part of which category of biomolecules?)

Question 3:
   Based on what you have learned about uric acid, predict two ways a person may develop hyperuricemia.

614   618   **630**   631   633   636

The peritubular capillaries have an average hydrostatic pressure of 10 mm Hg (in contrast to the glomerular capillaries, where hydrostatic pressure averages 55 mm Hg). Colloid osmotic pressure, which favors movement of fluid into the capillaries, is 30 mm Hg. As a result, the pressure gradient in peritubular capillaries is 20 mm Hg, favoring the absorption of fluid into the capillaries. Fluid that is reabsorbed passes from the capillaries to the venous circulation and returns to the heart.

SECRETION

Secretion is the transfer of molecules from extracellular fluid into the lumen of the nephron (see Fig. 19-2). Secretion, like reabsorption, depends mostly on membrane transport systems. The secretion of K^+ and H^+ by the nephron is important in the homeostatic regulation of those ions (see Chapter 20). In addition, many organic compounds are secreted. These compounds include both metabolites produced in the body and substances brought into the body.

Secretion enables the nephron to enhance excretion of a substance. If a substance is filtered and not reabsorbed, it will be excreted very effectively. If, however, the substance is filtered into the tubule, not reabsorbed, *and* then more of it is secreted into the tubule from the peritubular capillaries, excretion will be even more effective.

Secretion is an active process because it requires moving substrates against their concentration gradients. Most organic compounds are transported across the tubule epithelium into the lumen by secondary active transport.

Competition Decreases Penicillin Secretion

An interesting and important example of an organic molecule secreted by the nephron is the antibiotic *penicillin*. Many people today take antibiotics for granted, but until the early decades of the twentieth century, infections were a leading cause of death.

In 1928, Alexander Fleming discovered a substance in the bread mold *Penicillium* that retarded the growth of bacteria. But the antibiotic was difficult to isolate, so it did not become available for clinical use until the late 1930s. During World War II, penicillin made a major difference in the number of deaths and amputations caused by infected wounds. The only means of producing penicillin, however, was to isolate it from bread mold, and supplies were limited.

Demand for the drug was heightened by the fact that kidney tubules secrete penicillin. Renal secretion is so efficient at clearing foreign molecules from the blood that within three to four hours after a dose of penicillin has been administered, about 80% has been excreted in the urine. During the war, the drug was in such short supply that it was common procedure to collect the urine from patients being treated with penicillin so that the antibiotic could be isolated and reused.

This was not a satisfactory solution, however, and so researchers looked for a way to slow penicillin secretion. They hoped to find a molecule that could compete with penicillin for the organic anion transporter responsible for secretion [🔁 p. 146]. That way, when presented with both drugs, the carrier would bind preferentially to the competitor and secrete it, leaving penicillin behind in the blood. A synthetic compound named *probenecid* was the answer. When probenecid is administered concurrently with penicillin, the transporter removes probenecid preferentially, prolonging the activity of penicillin in the body. Once mass-produced synthetic penicillin became available and supply was no longer a problem, the medical use of probenecid declined.

EXCRETION

Urine output is the result of all the processes that take place in the kidney. By the time fluid reaches the end of the nephron, it bears little resemblance to the filtrate that started in Bowman's capsule. Glucose, amino acids, and useful metabolites are gone, having been reabsorbed into the blood, and organic wastes are more concentrated. The concentrations of ions and water in the urine are highly variable, depending on the state of the body.

Although excretion tells us what the body is eliminating, excretion by itself cannot tell us the details of renal function. Recall that for any substance,

excretion = filtration − reabsorption + secretion

Thus simply looking at the excretion rate of a substance tells us nothing about how the kidney handled that substance. The excretion rate of a substance depends on (1) its filtration rate and (2) whether the substance is reabsorbed, secreted, or both as it passes through the tubule.

Renal handling of a substance and GFR are often of clinical interest. For example, clinicians use information about a person's glomerular filtration rate as an indicator of overall kidney function. And pharmaceutical companies developing drugs must provide the Food and Drug Administration with complete information on how the human kidney handles each new compound.

But how can investigators dealing with living humans assess filtration, reabsorption, and secretion at the level of the individual nephron? They have no way to do this directly because the kidneys are not easily accessible and the nephrons are microscopic. Scientists therefore had to develop a technique that would allow them to assess renal function using only analysis of the urine and blood. From their work came the concept of clearance.

RUNNING PROBLEM

Michael finds it amazing that a metabolic problem could lead to pain in his big toe. "There's a way to treat gout, right?" he asks. Dr. Garcia explains that the treatment includes anti-inflammatory agents, lots of water, and avoidance of alcohol, which can trigger gout attacks. "In addition, I would like to put you on a *uricosuric agent*, like probenecid, which will enhance renal excretion of uric acid," replies Dr. Garcia. "By enhancing excretion, we can reduce uric acid levels in your blood and thus provide relief." Michael agrees to try these measures and on his way home stops off at a convenience store for a case of his favorite bottled water.

Question 4:
Uric acid is reabsorbed from the proximal tubule on a membrane transporter. Uricosuric agents are organic acids. Given these two facts, explain how uricosuric agents might enhance excretion of uric acid.

| 614 | 618 | 630 | **631** | 633 | 636 |

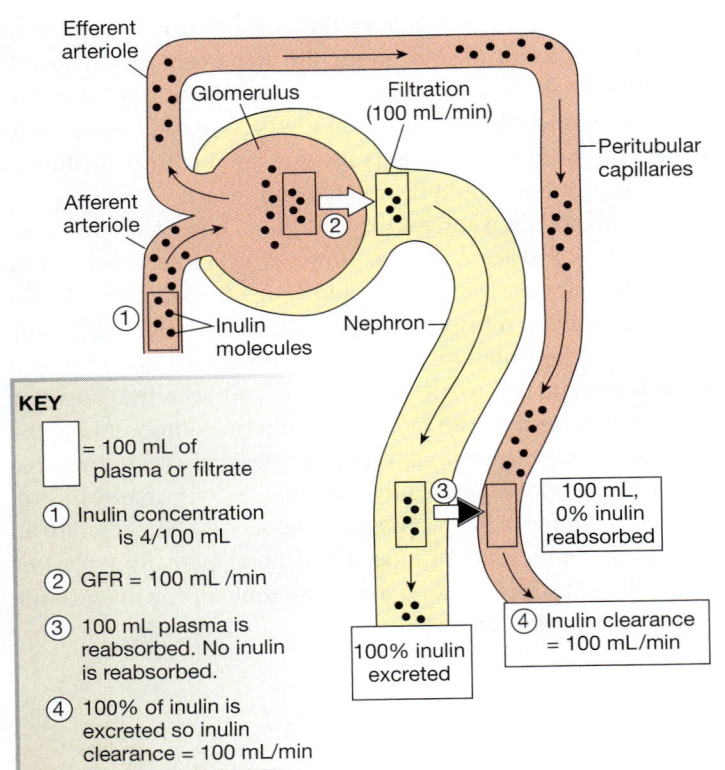

KEY

▢ = 100 mL of plasma or filtrate

① Inulin concentration is 4/100 mL

② GFR = 100 mL /min

③ 100 mL plasma is reabsorbed. No inulin is reabsorbed.

④ 100% of inulin is excreted so inulin clearance = 100 mL/min

100 mL, 0% inulin reabsorbed

100% inulin excreted

④ Inulin clearance = 100 mL/min

■ **FIGURE 19-16**   *Inulin clearance*

Clearance Is a Noninvasive Way to Measure GFR

Clearance of a solute is the rate at which that solute disappears from the body by excretion or by metabolism [🔁 p. 130]. For any solute that is cleared only by renal excretion, clearance is expressed as the volume of plasma passing through the kidneys that has been totally cleared of that solute in a given period of time. Because this is such an indirect way to think of excretion (how much blood has been cleared of X rather than how much X has been excreted), clearance is often a very difficult concept for students.

Before we jump into the mathematical expression of clearance, let's look at an example that shows how clearance relates to kidney function. For our example, we will use **inulin**, a polysaccharide isolated from the tuberous roots of a variety of plants. (Inulin is not the same as *insulin*, the protein hormone that regulates glucose metabolism.) Scientists discovered from experiments with isolated nephrons that inulin injected into the plasma filters freely into the nephron. As it passes through the kidney tubule, inulin is neither reabsorbed nor secreted. Thus, 100% of the inulin that filters into the tubule is excreted.

How does this relate to clearance? To answer this question, take a look at Figure 19-16 ■, which assumes that 100% of a filtered volume of plasma is reabsorbed. (This is not too far off the actual value, which is more than 99%.) Inulin has been injected so that its plasma concentration is 4 inulin molecules per

100 mL plasma. If GFR is 100 mL plasma filtered per minute, we can calculate the filtration rate, or *filtered load*, of inulin using the equation

$$\text{filtered load of X} = [X]_{plasma} \times \text{GFR}$$

$$\begin{aligned}\text{filtered load of inulin} &= (4 \text{ inulin}/100 \text{ mL plasma}) \\ &\quad \times 100 \text{ mL plasma filtered/min} \\ &= 4 \text{ inulin/min}\end{aligned}$$

As the filtered inulin and the filtered plasma pass along the nephron, all the plasma is reabsorbed but all the inulin remains in the tubule. The reabsorbed plasma contains no inulin, so we say it has been totally *cleared* of inulin. The *inulin clearance* therefore is 100 mL of plasma cleared/min. At the same time, the excretion rate of inulin is 4 inulin molecules excreted per minute.

What good is this information? For one thing, we can use it to calculate the glomerular filtration rate. Notice from Figure 19-16 that inulin clearance (100 mL plasma cleared/min) is equal to the GFR (100 mL plasma filtered/min). Thus, *for any substance that is freely filtered but neither reabsorbed nor secreted, its clearance is equal to GFR.*

Now let's show mathematically that inulin clearance is equal to GFR. We already know that

$$\text{filtered load of inulin} = [\text{inulin}]_{plasma} \times \text{GFR} \quad (1)$$

We also know that 100% of the inulin that filters into the tubule is excreted. In other words:

$$\text{filtered load of inulin} = \text{excretion rate of inulin} \quad (2)$$

Because of this equality, we can substitute excretion rate for filtered load in equation (1) by using algebra (if A = B and A = C, then B = C):

$$\text{excretion rate of inulin} = [\text{inulin}]_{plasma} \times \text{GFR} \quad (3)$$

This equation can be rearranged to read

$$\text{GFR} = \frac{\text{excretion rate of inulin}}{[\text{inulin}]_{plasma}} \quad (4)$$

It turns out that the right side of this equation is identical to the clearance equation for inulin. Thus the general equation for the clearance of any substance X (mL plasma cleared/min) is

$$\text{clearance of X} = \frac{\text{excretion rate of X (mg/min)}}{[X]_{plasma} \text{ (mg/mL plasma)}} \quad (5)$$

For inulin:

$$\text{inulin clearance} = \frac{\text{excretion rate of inulin}}{[\text{inulin}]_{plasma}} \quad (6)$$

The right sides of equations (4) and (6) are identical, so by using algebra again, we can say that:

$$\text{GFR} = \text{inulin clearance} \quad (7)$$

So why is this important? For one thing, you have just learned how we can measure GFR in a living human by taking only blood and urine samples. Try the example in Concept Check 12 to see if you understand the preceding discussion.

Inulin is not practical for routine clinical applications because it does not occur naturally in the body and must be administered by continuous intravenous infusion. As a result, inulin use is restricted to research. Unfortunately, no substance that occurs naturally in the human body is handled by the kidney exactly the way inulin is handled.

In clinical settings, physicians use creatinine to estimate GFR. **Creatinine** is a breakdown product of phosphocreatine, an energy-storage compound found primarily in muscles [🔁 p. 410]. It is constantly produced by the body and need not be administered. Normally, the production and breakdown rates of phosphocreatine are relatively constant, and the plasma concentration of creatinine does not vary much.

Although creatinine is always present in the plasma and is easy to measure, it is not the perfect molecule for estimating GFR because a small amount is secreted into the urine. However, the amount secreted is small enough that, in most people, *creatinine clearance* is routinely used to estimate GFR.

CONCEPT CHECK

12. If plasma creatinine = 1.8 mg/100 mL plasma, urine creatinine = 1.5 mg/mL urine, and urine volume is 1100 mL in 24 hours, what is the creatinine clearance? What is GFR?

Answers: p. 640

Clearance and GFR Help Us Determine Renal Handling of Solutes

Once we know a person's GFR, we can determine how the kidney handles any solute by measuring the solute's plasma concentration and its excretion rate. If we assume that the solute is freely filtered at the glomerulus, we know from equation (1) that

$$\text{Filtered load of X} = [X]_{plasma} \times GFR$$

By comparing the filtered load of the solute with its excretion rate, we can tell how the nephron handled that substance

(Table 19-2 ■). For example, if less of the substance appears in the urine than was filtered, net reabsorption occurred (excreted = filtered − reabsorbed). If more of the substance appears in the urine than was filtered, there must have been net secretion of the substance into the lumen (excreted = filtered + secreted). If the same amount of the substance is filtered and excreted, then the substance is handled like inulin—neither reabsorbed nor secreted.

TABLE 19-2   **Renal Handling of Solutes**

For any molecule X that is freely filtered at the glomerulus:

If filtration rate is greater than excretion rate,	there is net reabsorption of X.
If excretion rate is greater than filtration rate,	there is net secretion of X.
If filtration and excretion rate are the same,	X passes through the nephron without net reabsorption or secretion.
If the clearance of X is less than inulin clearance,	there is net reabsorption of X.
If the clearance of X is equal to inulin clearance,	X is neither reabsorbed nor secreted.
If the clearance of X is greater than inulin clearance,	there is net secretion of X.

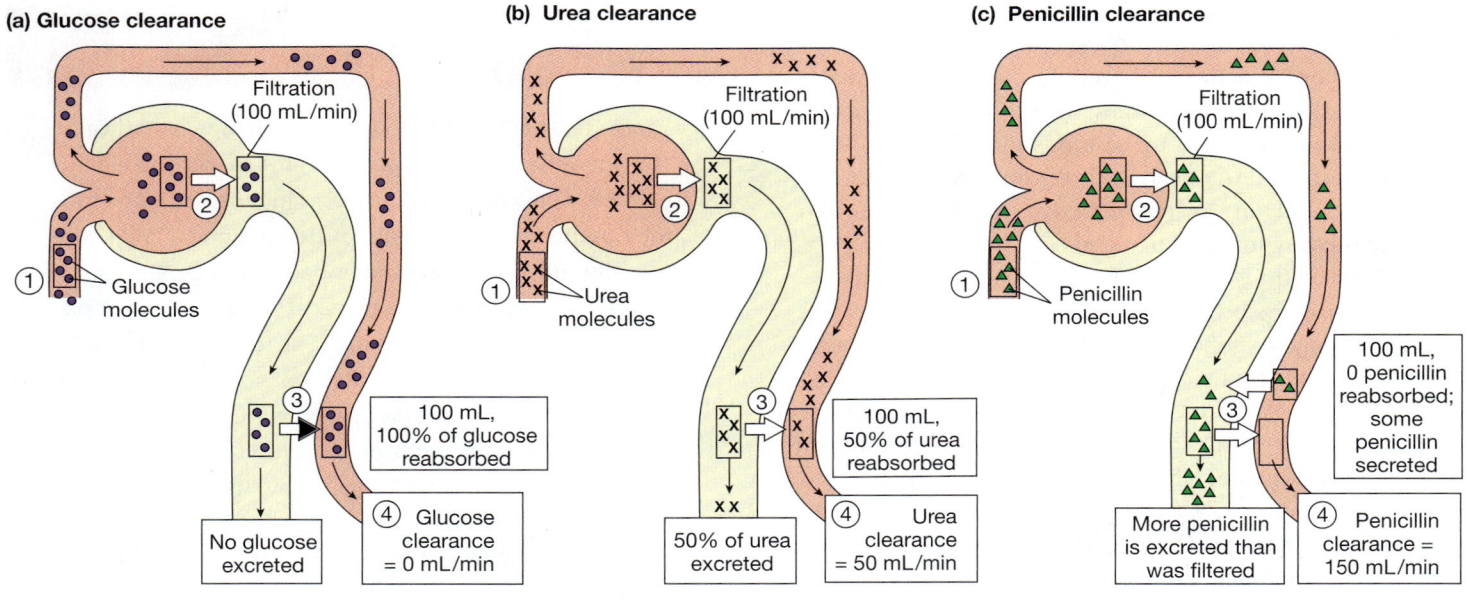

(a) Glucose clearance

(b) Urea clearance

(c) Penicillin clearance

■ FIGURE 19-17 *The relationship between clearance and excretion*

The figure represents the events taking place in one minute. For simplicity, 100% of the filtered volume is assumed to be reabsorbed.

KEY

☐ = 100 mL of plasma or filtrate

① Plasma concentration is 4/100 mL

② GFR = 100 mL /min

③ 100 mL plasma is reabsorbed.

④ Clearance depends on renal handling of solute

Suppose that glucose is present in the plasma at 100 mg glucose/dL plasma, and GFR is calculated from creatinine clearance to be 125 mL plasma/min. For these values, equation (1) tells us that

Filtered load of glucose = (100 mg glucose/100 mL plasma)
$\times$ 125 mL plasma/min

Filtered load of glucose = 125 mg glucose/min

There is no glucose in this person's urine, however: glucose excretion is zero. Because glucose was filtered at a rate of 125 mg/min but excreted at a rate of 0 mg/min, it must have been totally reabsorbed.

Clearance values can also be used to determine how the nephron handles a filtered solute. In this method, researchers calculate creatinine or inulin clearance, then compare the clearance of the solute being investigated with the creatinine or inulin clearance (Table 19-2). If clearance of the solute is less than the inulin clearance, the solute has been reabsorbed. If the clearance of the solute is higher than the inulin clearance, additional solute has been secreted into the urine. More plasma was cleared of the solute than was filtered, so the additional solute must have been removed from the plasma by secretion.

Figure 19-17 ■ shows clearance of three molecules: glucose, urea, and penicillin. All solutes have the same concentration in the blood entering the glomerulus: 4 molecules/100 mL plasma. GFR is 100 mL/min, and we assume for simplicity that the entire 100 mL of plasma filtered into the tubule is reabsorbed.

For any solute, its clearance reflects how the kidney tubule handles it. For example, 100% of the glucose that filters is reabsorbed, and glucose clearance is zero (Fig. 19-17a). Urea is partially reabsorbed; four molecules filter, but only two are reabsorbed (Fig. 19-17b, ③). Consequently, urea clearance is 50 mL plasma per minute. Urea and glucose clearance are both less than the inulin clearance of 100 mL/min, which tells you that urea and glucose have been reabsorbed.

Penicillin (Fig. 19-17c) is filtered but not reabsorbed, and additional penicillin molecules are secreted from plasma in the peritubular capillaries. In this example, an extra 50 mL of plasma have been cleared of penicillin in addition to the original 100 mL that filtered. The penicillin clearance therefore is 150 mL plasma per minute. Penicillin clearance is greater than the inulin clearance of 100 mL/min, which tells you that net secretion of penicillin occurs.

Note that a comparison of clearance values tells you only the *net* handling of the solute. It does not tell you if a molecule is both reabsorbed and secreted. For example, nearly all K^+ filtered is reabsorbed in the proximal tubule and loop of Henle, and then a small amount is secreted back into the tubule lumen at the distal nephron. On the basis of K^+ clearance, it appears that only reabsorption occurred.

Clearance calculations are relatively simple because all you need to know are the urine excretion rates and the plasma concentrations for any solute of interest; both values are easily

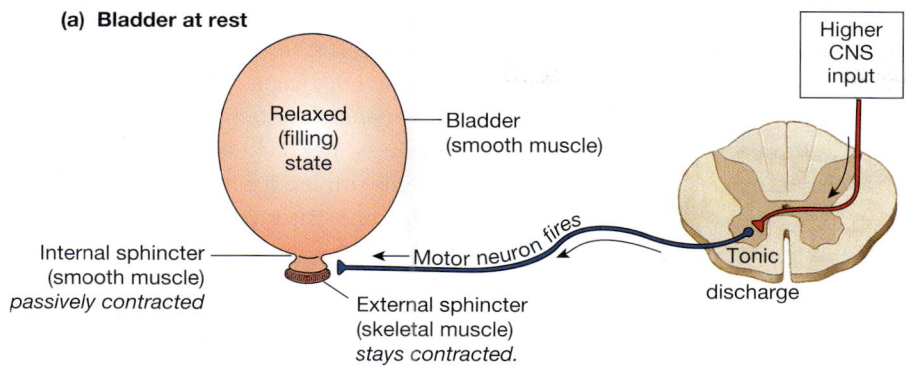

(a) Bladder at rest

Relaxed (filling) state

Bladder (smooth muscle)

Internal sphincter (smooth muscle) *passively contracted*

Motor neuron fires

External sphincter (skeletal muscle) *stays contracted.*

Higher CNS input

Tonic discharge

(b) Micturition

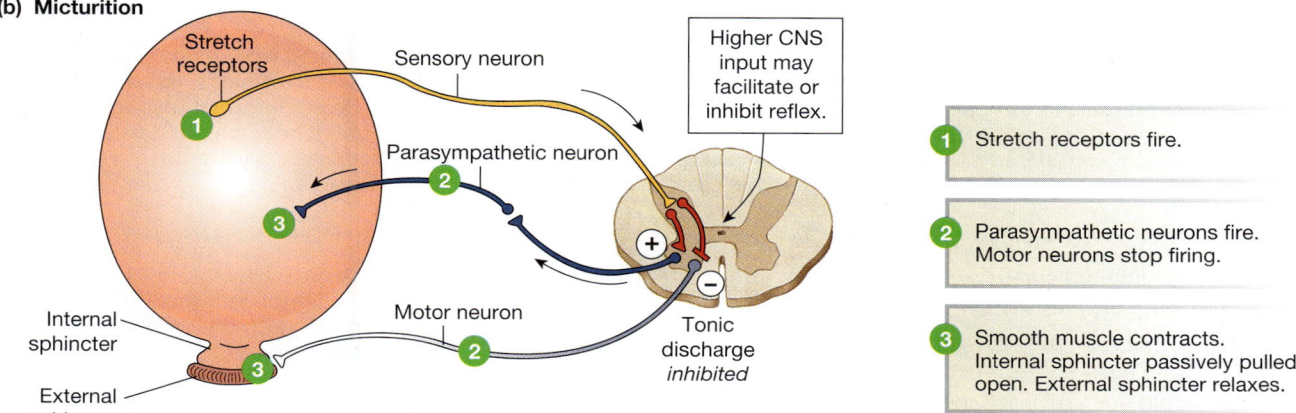

Stretch receptors

Sensory neuron

Higher CNS input may facilitate or inhibit reflex.

Parasympathetic neuron

Internal sphincter

Motor neuron

External sphincter

Tonic discharge *inhibited*

1 Stretch receptors fire.

2 Parasympathetic neurons fire. Motor neurons stop firing.

3 Smooth muscle contracts. Internal sphincter passively pulled open. External sphincter relaxes.

■ **FIGURE 19-18** *The storage of urine and the micturition reflex*

obtained. If you also know either inulin or creatinine clearance, then you can determine the renal handling of any compound.

MICTURITION

Once filtrate leaves the collecting ducts, it can no longer be modified, and its composition does not change. The filtrate, now called urine, flows into the renal pelvis and then down the *ureter* to the bladder with the help of rhythmic smooth muscle contractions. The bladder is a hollow organ whose walls contain well-developed layers of smooth muscle. In the bladder, urine is stored until released in the process known as urination, voiding, or more formally, **micturition** [*micturire*, to desire to urinate].

The bladder can expand to hold a volume of about 500 mL. The neck of the bladder is continuous with the *urethra*, a single tube through which urine passes to reach the external environment. The opening between the bladder and urethra is closed by two rings of muscle called *sphincters* (Fig. 19-18a ■).

The **internal sphincter** is a continuation of the bladder wall and consists of smooth muscle. Its normal tone keeps it contracted. The **external sphincter** is a ring of skeletal muscle

controlled by somatic motor neurons. Tonic stimulation from the central nervous system maintains contraction of the external sphincter except during urination.

Micturition is a simple spinal reflex that is subject to both conscious and unconscious control from higher brain centers. As the bladder fills with urine and its walls expand, stretch receptors send signals via sensory neurons to the spinal cord (Fig. 19-18b). There the information is integrated and transferred to two sets of neurons. The stimulus of a full bladder excites parasympathetic neurons leading to the smooth muscle in the bladder wall. The smooth muscle contracts, increasing the pressure on the bladder contents. Simultaneously, somatic motor neurons leading to the external sphincter are inhibited.

Contraction of the bladder occurs in a wave that pushes urine downward toward the urethra. Pressure exerted by the urine forces the internal sphincter open while the external sphincter relaxes. Urine passes into the urethra and out of the body, aided by gravity.

This simple micturition reflex occurs primarily in infants who have not yet been toilet trained. A person who has been toilet trained acquires a learned reflex that keeps the micturition reflex inhibited until she or he consciously desires to urinate. The learned reflex involves additional sensory fibers in the bladder that signal the degree of fullness. Centers in the brain stem and cerebral cortex receive that information and override the basic micturition reflex by directly inhibiting the parasympathetic fibers and by reinforcing contraction of the external sphincter. When an appropriate time to urinate arrives, those same centers remove the inhibition and facilitate the reflex by inhibiting the external sphincter.

In addition to conscious control of urination, various subconscious factors can affect the micturition reflex. "Bashful bladder" is a condition in which a person is unable to urinate in the presence of other people despite the conscious intent to do so. The sound of running water facilitates micturition and is often used to help patients urinate if the urethra is irritated from insertion of a *catheter,* a tube inserted into the bladder to drain it passively.

RUNNING PROBLEM CONCLUSION

GOUT

In this running problem, you learned that gout, which often presents as a debilitating pain in the big toe, is a metabolic problem whose cause and treatment may be linked to kidney function. Gout is one of the oldest known diseases and for many years was considered a "rich man's" disease caused by too much rich food and drink. Thomas Jefferson and Benjamin Franklin both suffered from gout. To learn more about its causes, symptoms, and treatments, go to the U.S. National Library of Medicine's Health Information page (*http://www.nlm.nih.gov/hinfo.html*). Check your understanding of this running problem by comparing your answers against the information in the summary table.

	QUESTION	FACTS	INTEGRATION AND ANALYSIS
1	Trace the route followed by kidney stones when they are excreted.	Kidney stones often form in the renal pelvis.	From the renal pelvis, a stone passes down the ureter, into the urinary bladder, then into the urethra and out of the body.
2	Why would uric acid levels in the blood go up when cell breakdown increases?	Purines include adenine and guanine, which are components of DNA, RNA, and ATP [🔁 p. 32].	When a cell dies, the nucleus and other components are broken down into biomolecules. Degradation of the cell's DNA, RNA, and ATP increases purine production, which in turn increases uric acid production.
3	Based on what you have learned about uric acid, predict two ways a person may develop hyperuricemia.	Hyperuricemia is a condition of elevated uric acid levels in the blood. Uric acid is made from purines. It is filtered, reabsorbed, and secreted in the kidneys.	Hyperuricemia results either from overproduction of uric acid or from failure of the kidneys to excrete uric acid. Because uric acid normally gets into the urine by secretion, a defect in the renal secretory mechanism would lead to hyperuricemia.
4	Uric acid is reabsorbed on a membrane transporter. Uricosuric agents are organic acids. Given these two facts, explain how uricosuric agents might enhance excretion of uric acid.	Mediated transport exhibits competition, in which related molecules compete for one transporter. Usually, one molecule binds preferentially and therefore inhibits transport of the second molecule.	Uricosuric agents may compete with uric acid for the proximal tubule transporter. Preferential binding of the uricosuric agents would block uric acid access to the transporter, leaving it in the lumen and increasing the amount of it excreted.
5	Explain why not drinking enough water while taking uricosuric agents may cause uric acid stones to form in the urinary tract.	Uric acid stones form when uric acid concentrations exceed a critical level.	If a person drinks large volumes of water, the excess water will be excreted by the kidneys. Large amounts of water dilute the urine, thereby preventing the high concentrations of uric acid needed for stone formation.

614 618 630 631 633 **636**

19

CHAPTER SUMMARY

The urinary system, like the lungs, uses the principle of *mass balance* to maintain homeostasis. The components of urine are constantly changing and reflect the kidney's functions of regulating ions and water and removing wastes.

One of the body's three *portal systems*—each of which includes two capillary beds—is found in the kidney. Filtration occurs in the first capillary bed, and reabsorption in the second. The *pressure-flow-resistance*

relationship you encountered in the cardiovascular and pulmonary systems also plays a role in glomerular filtration and urinary excretion.

Compartmentation is illustrated by the movement of water and solutes between the internal and external environments as filtrate is modified along the nephron. Reabsorption and secretion of solutes depend on *molecular interactions* and on the *movement of molecules across membranes* of the tubule cells.

Functions of the Kidneys

1. The kidneys regulate extracellular fluid volume, blood pressure, and osmolarity; maintain ion balance; regulate pH; excrete wastes and foreign substances; and participate in endocrine pathways. (p. 614)

Anatomy of the Urinary System

IP Urinary System: Glomerular Filtration

2. The **urinary system** is composed of two kidneys, two ureters, a bladder, and a urethra. (p. 615; Fig. 19-1a)

3. Each **kidney** has about 1 million **nephrons**. In cross-section, a kidney is arranged into an outer **cortex** and inner **medulla**. (p. 615; Fig. 19-1c)

4. Renal blood flow goes from **afferent arteriole** to **glomerulus** to **efferent arteriole** to **peritubular capillaries**. The **vasa recta** capillaries dip into the medulla. (p. 618; Fig. 19-1h, i)

5. Fluid filters from the glomerulus into **Bowman's capsule**. From there, it flows through the **proximal tubule**, **loop of Henle**, **distal tubule**, and **collecting duct**, then drains into the **renal pelvis**. **Urine** flows through the **ureter** to the **urinary bladder**. (p. 618; Fig. 19-1c, j)

Overview of Kidney Function

6. **Filtration** is the movement of fluid from plasma into Bowman's capsule. **Reabsorption** is the movement of filtered materials from tubule to blood. **Secretion** is the movement of selected molecules from blood to tubule. (pp. 618–619; Fig. 19-2)

7. Average urine volume is 1.5 L/day. Osmolarity varies between 50 and 1200 mOsM. (p. 619; Tbl. 19-1)

8. The amount of a solute excreted equals the amount filtered minus the amount reabsorbed plus the amount secreted. (p. 620; Fig. 19-3)

Filtration

IP Urinary System: Glomerular Filtration

9. Filtered solutes must pass first through glomerular capillary endothelium, then through a **basal lamina**, and finally through Bowman's capsule epithelium before reaching the lumen of Bowman's capsule. (p. 620; Fig. 19-4)

10. Filtration allows most components of plasma to enter the tubule but excludes blood cells and most plasma proteins. (p. 620)

11. The Bowman's capsule epithelium has specialized cells called **podocytes** that wrap around the glomerular capillaries and create **filtration slits**. **Mesangial cells** are associated with the glomerular capillaries. (p. 620; Fig. 19-4)

12. One-fifth of renal plasma flow filters into the tubule lumen. The percentage of total plasma volume that filters is called the **filtration fraction**. (p. 621; Fig. 19-5)

13. Hydrostatic pressure in glomerular capillaries averages 55 mm Hg, favoring filtration. Opposing filtration are colloid osmotic pressure of 30 mm Hg and hydrostatic capsule fluid pressure averaging 15 mm Hg. The net driving force is 10 mm Hg, favoring filtration. (pp. 621–622; Fig. 19-6)

14. The **glomerular filtration rate** (**GFR**) is the amount of fluid that filters into Bowman's capsule per unit time. Average GFR is 125 mL/min, or 180 L/day. (p. 622)

15. Hydrostatic pressure in glomerular capillaries can be altered by changing resistance in the afferent and efferent arterioles. (p. 624; Fig. 19-8)

16. Autoregulation of glomerular filtration is accomplished by a **myogenic response** of vascular smooth muscle in response to pressure changes and by **tubuloglomerular feedback**. When fluid flow through the distal tubule increases, the **macula densa** cells send a paracrine signal to the afferent arteriole, which constricts. (p. 624; Fig. 19-10)

17. Reflex control of GFR is mediated through systemic signals, such as hormones, and through the autonomic nervous system. (p. 625)

Reabsorption

IP Urinary System: Early Filtrate Processing

18. Most reabsorption takes place in the proximal tubule. Finely regulated reabsorption takes place in the more distal segments of the nephron. (p. 626)

19. The active transport of Na^+ and other solutes creates concentration gradients for passive reabsorption of urea and other solutes. (p. 626; Fig. 19-11)

20. Most reabsorption involves transepithelial transport, but some solutes and water are reabsorbed by the paracellular pathway. (p. 626)

21. Glucose, amino acids, ions, and various organic metabolites are reabsorbed by Na^+-linked secondary active transport. (p. 627; Fig. 19-13)

22. Most renal transport is mediated by membrane proteins and exhibits saturation, specificity, and competition. The **transport maximum** T_m is the transport rate at saturation. (p. 628; Fig. 19-14)

23. The **renal threshold** is the plasma concentration at which a substance first appears in the urine. (p. 629; Fig. 19-14)

24. Peritubular capillaries reabsorb fluid along their entire length. (p. 629)

Secretion

25. Secretion enhances excretion by removing solutes from the peritubular capillaries. K^+, H^+, and a variety of organic compounds are secreted. (p. 631)

26. Molecules that compete for renal carriers slow the secretion of a molecule. (p. 631)

Excretion

27. The excretion rate of a solute depends on (1) its filtered load and (2) whether it is reabsorbed or secreted as it passes through the nephron. (p. 631)

28. **Clearance** describes how many milliliters of plasma passing through the kidneys have been totally cleared of a solute in a given period of time. (p. 632)

29. **Inulin** clearance is equal to GFR. In clinical settings, **creatinine** is used to measure GFR. (p. 632; Fig. 19-16)

30. Clearance can be used to determine how the nephron handles a solute filtered into it. (p. 633; Fig. 19-17)

Micturition

31. The external sphincter of the bladder is skeletal muscle that is tonically contracted except during urination. (p. 635; Fig. 19-18)

32. Micturition is a simple spinal reflex subject to conscious and unconscious control. Parasympathetic neurons cause contraction of the smooth muscle in the bladder wall. Somatic motor neurons leading to the external sphincter are simultaneously inhibited. (p. 635)

QUESTIONS

(Answers to the Review Questions begin on page A1.)

THE PHYSIOLOGY PLACE

Access more review material online at **The Physiology Place** website. There you'll find review questions, problem-solving activities, case studies, flashcards, and direct links to both *InterActive Physiology*® and *PhysioEx*™. To access the site, go to *www.physiologyplace.com* and select Human Physiology, Fourth Edition.

LEVEL ONE REVIEWING FACTS AND TERMS

1. List and explain the significance of the five characteristics of urine that can be found by physical examination.

2. List and explain the six major kidney functions.

3. At any given time, what percentage of cardiac output goes to the kidneys?

4. List the major structures of the urinary system in their anatomical sequence, from the kidneys to the urine leaving the body. Describe the function of each structure.

5. Arrange the following structures in the order that a drop of water entering the nephron would encounter them:
   (a) afferent arteriole
   (b) Bowman's capsule
   (c) collecting duct
   (d) distal tubule
   (e) glomerulus
   (f) loop of Henle
   (g) proximal tubule
   (h) renal pelvis

6. Name the three filtration barriers that solutes must cross as they move from plasma to the lumen of Bowman's capsule. What components of blood are usually excluded by these layers?

7. What force(s) promotes glomerular filtration? What force(s) opposes it? What is meant by the term *net driving force*?

8. What does the abbreviation GFR stand for? What is a typical numerical value for GFR in milliliters per minute? In liters per day?

9. Identify the following structures, then explain their significance in renal physiology:
   (a) juxtaglomerular apparatus
   (b) macula densa
   (c) mesangial cell
   (d) podocyte
   (e) sphincters in the bladder
   (f) renal cortex

10. In which segment of the nephron does most reabsorption take place? When a molecule or ion is reabsorbed from the lumen of the nephron, where does it go? If a solute is filtered and not reabsorbed from the tubule, where does it go?

11. Match each of the following ions or molecules with its primary mode(s) of transport across the kidney epithelium.
   (a) Na^+
   (b) glucose
   (c) urea
   (d) plasma proteins
   (e) water

   1. transcytosis
   2. primary active transport
   3. secondary active transport
   4. facilitated diffusion
   5. movement through open channels
   6. simple diffusion through the phospholipid bilayer

12. List three solutes secreted into the tubule lumen.

13. What solute that is normally present in the body is used to estimate GFR in humans?

14. What is micturition?

LEVEL TWO REVIEWING CONCEPTS

15. Map the following terms. You may add additional terms if you like.

   α-receptor
   afferent arteriole
   autoregulation
   basal lamina
   Bowman's capsule
   capillary blood pressure
   capsule fluid pressure
   colloid osmotic pressure
   efferent arteriole
   endothelium
   epithelium
   GFR

   glomerulus
   JG cells
   macula densa
   mesangial cell
   myogenic autoregulation
   norepinephrine
   paracrine
   plasma proteins
   podocyte
   resistance
   vasoconstriction

16. Define, compare, and contrast the items in the following sets of terms:
   (a) filtration, secretion, and excretion
   (b) saturation, transport maximum, and renal threshold
   (c) probenecid, creatinine, inulin, and penicillin
   (d) clearance, excretion, and glomerular filtration rate

17. What are the advantages of a kidney that filters a large volume of fluid and then reabsorbs 99% of it?

18. If the afferent arteriole of a nephron constricts, what happens to GFR in that nephron? If the efferent arteriole of a nephron constricts, what happens to GFR in that nephron? Assume that no autoregulation takes place.

19. Diagram the micturition reflex. How is this reflex altered by toilet training? How do higher brain centers influence micturition?

20. Antimuscarinic drugs are the accepted treatment for an overactive bladder. Explain why they work for this condition.

LEVEL THREE   PROBLEM SOLVING

21. You have been asked to study kidney function in a new species of rodent found in the Amazonian jungle. You isolate some nephrons and expose them to inulin. The following graph shows the results of your studies. (a) How is the rodent nephron handling inulin? Is inulin filtered? Is it excreted? Is there net inulin reabsorption? Is there net secretion? (b) On the graph, accurately draw a line indicating the net reabsorption or secretion. (*Hint:* excretion = filtration − reabsorption + secretion)

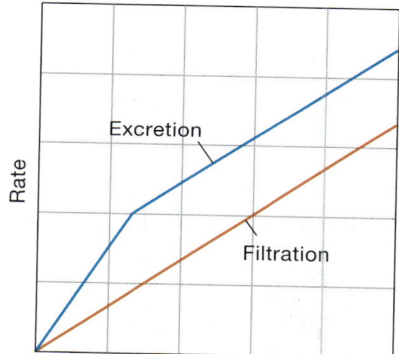

22. Draw a section of renal tubule epithelium showing three cells joined by cell junctions. Label the apical and basolateral membranes, the tubule lumen, and the extracellular fluid. Use the following written description of proximal tubule processes to draw a model cell.

   The proximal tubule cells contain carbonic anhydrase, which promotes the conversion of CO_2 and water to carbonic acid. Car-

bonic acid then dissociates to H^+ and HCO_3^-. Sodium is reabsorbed by an apical Na^+-H^+ antiporter and a basolateral Na^+-K^+ATPase. Chloride is passively reabsorbed by movement through the paracellular pathway. Bicarbonate produced in the cytoplasm leaves the cell on a basolateral Na^+-HCO_3^- symporter.

23. Read the box on hemodialysis on p. 628 and see if you can create a model system that would work for dialysis. Draw two compartments (one to represent blood and one to represent dialysis fluid) separated by a semipermeable membrane. In the blood compartment, list normal extracellular fluid solutes and their concentrations (see the table with normal values of blood components inside the back cover of this book). What will happen to the concentrations of these solutes during kidney failure? Which of these solutes should you put in the dialysis fluid, and what should their concentrations be? (*Hint:* do you want diffusion into the dialysis fluid, out of the dialysis fluid, or no net movement?) How would you change the dialysis fluid if the patient was retaining too much water?

LEVEL FOUR   QUANTITATIVE PROBLEMS

24. Darlene weighs 50 kg. Assume that her total blood volume is 8% of her body weight, that her heart pumps her total blood volume once a minute, and that her renal blood flow is 25% of her cardiac output. Calculate the volume of blood that flows through Darlene's kidneys each minute.

25. Dwight was competing for a spot on the Olympic equestrian team. As his horse, Nitro, cleared a jump, the footing gave way, causing the horse to somersault, landing on Dwight and crushing him. The doctors feared kidney damage and ran several tests. Dwight's serum creatinine level was 2 mg/100 mL. His 24-hour urine specimen had a volume of 1 L and a creatinine concentration of 20 mg/mL. A second specimen taken over the next 24 hours had the same serum creatinine value and urine volume, but a urine creatinine concentration of 4 mg/ml. How many milligrams of creatinine are in each specimen? What is Dwight's creatinine clearance in each test? What is his GFR? Evaluate these results and comment on Dwight's kidney function.

26. You are a physiologist taking part in an archeological expedition to search for Atlantis. One of the deep-sea submersibles has come back with a mermaid, and you are taking a series of samples from her. You have determined that her GFR is 250 mL/min and that her kidneys reabsorb glucose with a transport maximum of 50 mg/min. What is her renal threshold for glucose? When her plasma concentration of glucose is 15 mg/mL, what is its glucose clearance?

ANSWERS

✓ Answers to Concept Check Questions

Page 615

1. If extracellular K^+ decreases, more K^+ leaves the neuron, and the membrane potential hyperpolarizes (becomes more negative, increases).

2. If plasma Ca^{2+} decreases, the force of contraction decreases.

Page 618

3. When net filtration out of the glomerular capillaries occurs, the capillary hydrostatic pressure must be greater than the capillary colloid osmotic pressure.

4. When net reabsorption into the peritubular capillaries occurs, the capillary hydrostatic pressure must be less than the capillary colloid osmotic pressure.

Page 620

5. Filtration and secretion both represent movement from the extracellular fluid into the lumen. Filtration takes place only at Bowman's capsule; secretion takes place all along the rest of the tubule.

6. Glomerulus → Bowman's capsule → proximal tubule → loop of Henle → distal tubule → collecting duct → renal pelvis → ureter → urinary bladder → urethra.

7. If reabsorption decreases to half the normal rate, the body would run out of plasma in under an hour.

Page 622

8. Osmotic pressure is higher in efferent arterioles because fluid volume is decreased there, leaving the same amount of protein in a smaller volume.

Page 624

9. This person's mean arterial pressure is 119 mm Hg. This person's GFR is 180 L/day.

Page 625

10. If the afferent arteriole constricts, the resistance in that arteriole increases, and blood flow through that arteriole is diverted to lower-resistance arterioles. GFR will decrease in the nephron whose arteriole constricted.

11. The primary driving force for GFR is blood pressure opposed by fluid pressure in Bowman's capsule and colloid osmotic pressure due to plasma proteins (Fig. 19-6). With fewer plasma proteins, the plasma has lower-than-normal colloid osmotic pressure. With less colloid osmotic pressure opposing GFR, GFR increases.

Page 633

12. Creatinine clearance = creatinine excretion rate/$[creatinine]_{plasma}$ = (1.5 mg creatinine/mL urine $\times$ 1.1 L urine/day)/1.8 mg creatinine/100 mL plasma. Creatinine clearance is about 92 L/day, and GFR is equal to creatinine clearance.

Answers to Figure and Graph Questions

Page 623

Figure 19-8: Capillary blood pressure, GFR, and renal blood flow all increase.

Page 629

Figure 19-14: The transport rate at 3 mg/mL is 3 mg/min; at 5 and 8 mg/mL, it is 4 mg/min. The transport rate is 4 mg/min at any plasma concentration equal to or greater than 4 mg/mL, the renal threshold for this substance.

20

At a 10% loss of body fluid, the patient will show signs of confusion, distress, and hallucinations and at 20%, death will occur.

—**Poul Astrup**, *in Salt and Water in Culture and Medicine, 1993*

Integrative Physiology II: Fluid and Electrolyte Balance

BACKGROUND BASICS

RUNNING PROBLEM

HYPONATREMIA

Lauren was competing in her first ironman-distance triathlon, a 140.6-mile race consisting of 2.4 miles of swimming, 112 miles of cycling, and 26.2 miles of running. At mile 22 of the run, approximately 16 hours after starting the race, she collapsed. After being admitted to the medical tent, Lauren complained of nausea, a headache, and general fatigue. The medical staff noted that Lauren's face and clothing were covered in white crystals. When they weighed her and compared that value with her pre-race weight recorded at registration, they realized Lauren had gained 2 kg during the race.

642   645   658   661   665   671

The American businesswoman in Tokyo finished her workout and stopped at the snack bar of the fitness club to ask for a sports drink. The attendant handed her a bottle labeled "Pocari Sweat®." Although the thought of drinking sweat is not very appealing, the physiological basis for the name is sound.

During exercise, the body secretes sweat, a dilute solution of water and ions, particularly Na^+, K^+, and Cl^-. To maintain homeostasis, the body must replace any substances it has lost to the external environment. Therefore, the replacement fluid a person consumes after exercise should resemble sweat.

In this chapter, we explore how humans maintain salt and water balance, also known as fluid and electrolyte balance. The homeostatic control mechanisms for fluid and electrolyte balance in the body are aimed at maintaining four parameters: fluid volume, osmolarity, the concentrations of individual ions, and pH.

FLUID AND ELECTROLYTE HOMEOSTASIS

The human body is in a state of constant flux. Over the course of a day we ingest about 2 liters of food and drink that contain 6–15 grams of NaCl. In addition, we bring in varying amounts of other electrolytes, including K^+, H^+, Ca^{2+}, HCO_3^-, and phosphate ions (HPO_4^{2-}). The body's task is to maintain *mass balance* [p. 129]: what comes in must be excreted if the body does not need it.

The body has several routes for excreting ions and water. The kidneys are the primary route for water loss and for removal of many ions. Under normal conditions, small amounts of both water and ions are lost in the feces as well. In addition, the lungs help remove H^+ and HCO_3^- by excreting CO_2.

Although physiological mechanisms that maintain fluid

also play an essential role. *Thirst* is critical because drinking is the only normal way to replace lost water. *Salt appetite* is a behavior that leads people and animals to seek and ingest salt (sodium chloride, NaCl).

Why are we concerned with homeostasis of these substances? Water and Na^+ are associated with extracellular fluid volume and osmolarity. Disturbances in K^+ balance can cause serious problems with cardiac and muscle function by disrupting the membrane potential of excitable cells. Ca^{2+} is involved in a variety of body processes, from exocytosis and muscle contraction to bone formation and blood clotting, and H^+ and HCO_3^- are the ions whose balance determines body pH.

ECF Osmolarity Affects Cell Volume

Why is maintaining osmolarity so important to the body? The answer lies in the fact that water crosses most cell membranes freely. If the osmolarity of the extracellular fluid changes, water moves into or out of cells and changes intracellular volume. If ECF osmolarity decreases as a result of excess water intake, water moves into the cells and they swell. If ECF osmolarity increases as a result of salt intake, water moves out of the cells and they shrink.

Changes in cell volume—either shrinking or swelling—can impair cell function. When cells swell, for example, ion channels in the membrane open, disrupting membrane potential and cell signaling. The brain, encased in the rigid skull, is particularly vulnerable to damage from swelling. Some cells can regulate their volume in response to swelling or shrinking, but this capability is limited. In general, maintenance of ECF osmolarity within a normal range is essential to maintaining homeostasis.

EMERGING CONCEPTS

REGULATION OF CELL VOLUME

The regulation of cell volume is so important that many cells have independent mechanisms for regulating volume. Renal tubule cells in the medulla of the kidney, for example, are constantly exposed to high extracellular fluid osmolarity, yet these cells maintain normal cell volume. They do so by synthesizing organic solutes as needed so that their intracellular osmolarity matches that of the extracellular fluid. The organic solutes used to raise intracellular osmolarity include sugar alcohols and certain amino acids. In addition, cells can regulate their volume by changing their ionic composition. In some instances, changes in cell volume are believed to act as signals that initiate certain cellular responses. For example, swelling of liver cells activates protein and glycogen synthesis, whereas shrinkage of liver cells causes breakdown of protein and glycogen.

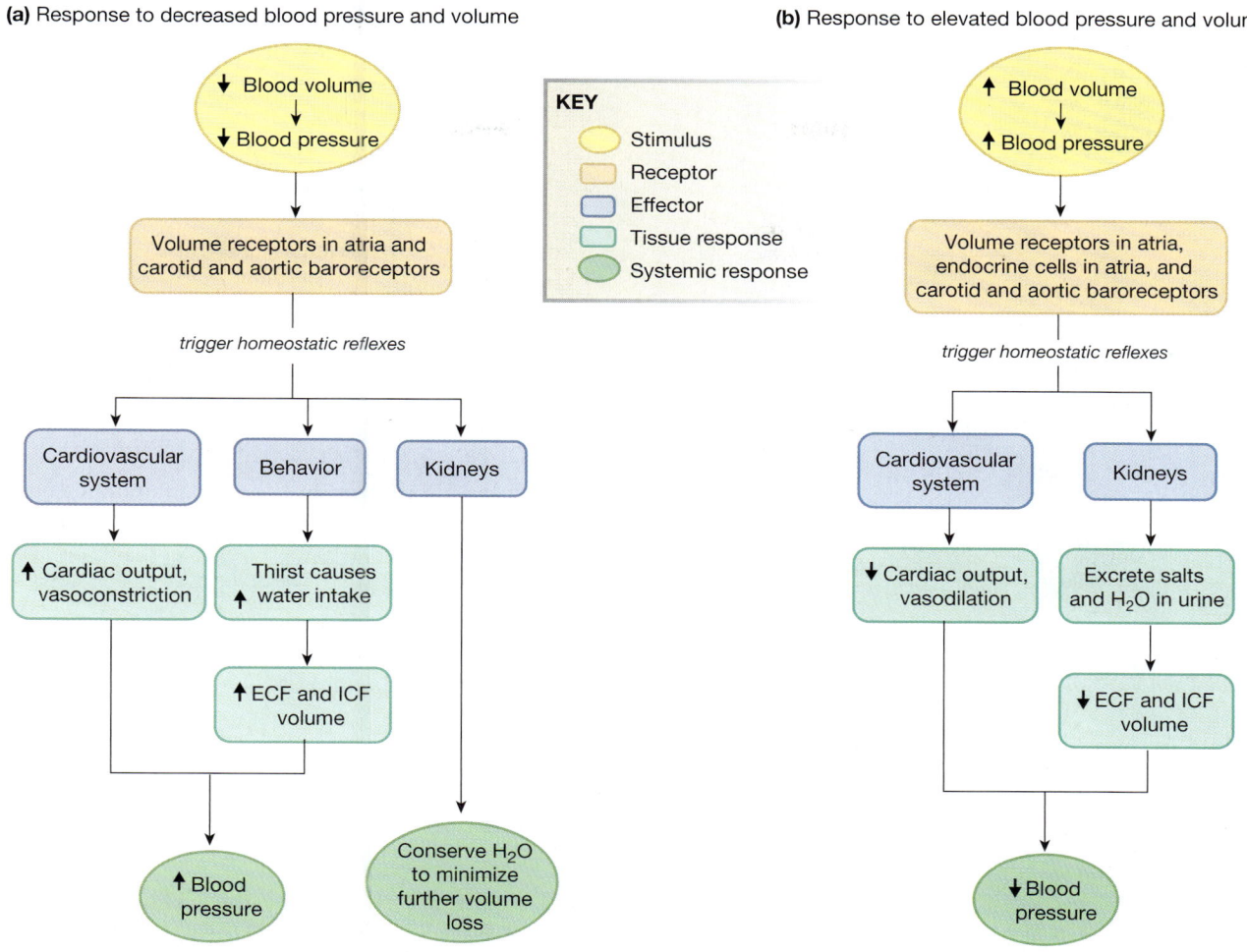

(a) Response to decreased blood pressure and volume

(b) Response to elevated blood pressure and volume

KEY
- Stimulus
- Receptor
- Effector
- Tissue response
- Systemic response

(a)

↓ Blood volume
↓
↓ Blood pressure

Volume receptors in atria and carotid and aortic baroreceptors

trigger homeostatic reflexes

- Cardiovascular system
- Behavior
- Kidneys

↑ Cardiac output, vasoconstriction

Thirst causes ↑ water intake

↑ ECF and ICF volume

↑ Blood pressure

Conserve H_2O to minimize further volume loss

(b)

↑ Blood volume
↓
↑ Blood pressure

Volume receptors in atria, endocrine cells in atria, and carotid and aortic baroreceptors

trigger homeostatic reflexes

- Cardiovascular system
- Kidneys

↓ Cardiac output, vasodilation

Excrete salts and H_2O in urine

↓ ECF and ICF volume

↓ Blood pressure

■ **FIGURE 20-1** *The body's integrated responses to changes in blood volume and blood pressure*

Fluid and Electrolyte Balance Requires Integration of Multiple Systems

The process of fluid and electrolyte balance is truly integrative because it involves the respiratory and cardiovascular systems in addition to renal and behavioral responses. Adjustments made by the lungs and cardiovascular system are primarily under neural control and can be made quite rapidly. Homeostatic compensation by the kidneys occurs more slowly because the kidneys are primarily under endocrine and neuroendocrine control. For example, small changes in blood pressure that result from increases or decreases in blood volume are quickly corrected by the cardiovascular control center in the brain [⊜ p. 521]. If volume changes are persistent or of large magnitude, the kidneys step in to help maintain homeostasis.

Figure 20-1 ■ summarizes the integrated response of the body to changes in blood volume and blood pressure. Signals from carotid and aortic baroreceptors and atrial volume receptors initiate a quick neural response mediated through the cardiovascular control center, and a slower response elicited from the kidneys. In addition, low blood pressure stimulates thirst. In both situations, renal function integrates with the cardiovascular system to keep blood pressure within a normal range.

Because of the overlap in their functions, a change made by one system—whether renal or cardiovascular—is likely to have consequences that affect the other. Endocrine pathways initiated by the kidneys have direct effects on the cardiovascular system, for instance, and hormones released by myocardial cells act on the kidneys. Sympathetic pathways from the cardiovascular control center affect not only cardiac output and vasoconstriction but also glomerular filtration and hormone release by the kidneys.

Thus the maintenance of blood pressure, blood volume, and ECF osmolarity forms a network of interwoven control pathways. This integration of function in multiple systems is one of the more difficult concepts in physiology; it is also one of the most exciting areas of physiological research.

WATER BALANCE

Water is the most abundant molecule in the body, constituting about 50% of total body weight in females ages 17 to 39, and 60% of total body weight in males of the same age group.

20

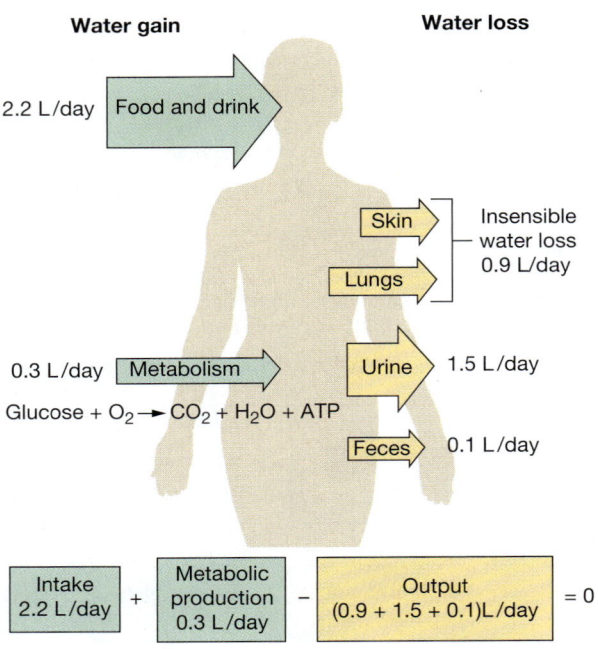

Water gain

2.2 L/day Food and drink

0.3 L/day Metabolism
Glucose + $O_2 \rightarrow CO_2 + H_2O$ + ATP

Water loss

Skin
Lungs
— Insensible water loss 0.9 L/day

Urine 1.5 L/day

Feces 0.1 L/day

| Intake 2.2 L/day | + | Metabolic production 0.3 L/day | − | Output (0.9 + 1.5 + 0.1)L/day | = 0 |

■ **FIGURE 20-2** *Water balance in the body*

A 60-kg (132-lb) woman therefore contains about 30 liters of body water, and the "standard" 70-kg man contains about 42 liters. Two-thirds of his water (about 28 liters) is inside the cells, about 3 liters are in the plasma, and the remaining 11 liters are in the interstitial fluid [⬆ Fig. 5-28, p. 155].

Daily Water Intake and Excretion Are Balanced

To maintain a constant volume of water in the body, we must take in the same amount of water that we excrete: intake must equal output. There are multiple sources for water gain and loss during a day (Fig. 20-2 ■). On average, an adult ingests a little more than 2 liters of water in food and drink in a day. Normal cellular respiration (glucose + $O_2 \rightarrow CO_2 + H_2O$) adds about 0.3 liter of water, bringing the total daily intake to approximately 2.5 liters.

Notice that the only means by which water normally enters the body from the external environment is by absorption through the digestive tract. Unlike some animals, we cannot absorb significant amounts of water directly through our skin. If fluids must be rapidly replaced or an individual is unable to eat and drink, fluid can be added directly to the plasma by means of **intravenous (IV) injection**, a medical procedure.

The major route of water loss from the body is the urine, which has a daily volume of about 1.5 liters (Fig. 20-2). A small volume of water (about 100 mL) is lost in the feces. Additionally, water leaves the body through **insensible water loss**. This water loss, called *insensible* because we are not normally aware of it, occurs across the skin surface and during the exhalation of humidified air. Even though the human epidermis is modified with an outer layer of keratin to reduce evaporative water loss in a terrestrial environment [⬆ p. 83], we still lose about 900 mL of

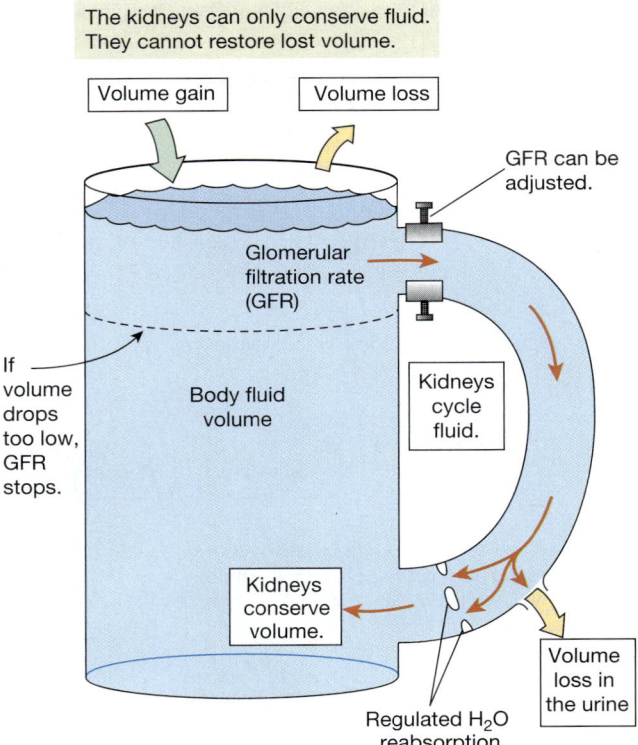

The kidneys can only conserve fluid. They cannot restore lost volume.

Volume gain
Volume loss

GFR can be adjusted.

Glomerular filtration rate (GFR)

If volume drops too low, GFR stops.

Body fluid volume

Kidneys cycle fluid.

Kidneys conserve volume.

Regulated H_2O reabsorption

Volume loss in the urine

■ **FIGURE 20-3** *A model of the role of the kidneys in water balance*

water insensibly each day. Thus the 2.5 liters of water we take in are balanced by the 2.5 liters that leave the body. Only water loss in the urine can be regulated.

Although urine is normally the major route of water loss, in certain situations other routes of water loss can become significant. Excessive sweating is one example. Another way in which water is lost is through diarrhea, a condition that can pose a major threat to the maintenance of water balance, particularly in infants.

Pathological water loss disrupts homeostasis in two ways. Volume depletion of the extracellular compartment decreases blood pressure. If blood pressure cannot be maintained, the tissues do not get adequate oxygen. If the fluid lost is hyposmotic to the body (as is the case in excessive sweating), the solutes left behind in the body raise osmolarity, potentially disrupting cell function.

Normally, water balance takes place automatically. Salty food makes us thirsty. Drinking 42 ounces of a soft drink means an extra trip to the bathroom. Salt and water balance is a subtle process that we are only peripherally aware of, like breathing and the beating of the heart.

Now that we have seen *why* regulation of osmolarity is important, let's see *how* the body accomplishes that goal.

The Kidneys Conserve Water

Figure 20-3 ■ summarizes the role of the kidneys in water balance. The mug represents the body, and its hollow handle

RUNNING PROBLEM

The medical staff was concerned with Lauren's large weight increase during the race. They asked her to recall what she ate and drank during the race. Lauren reported that to avoid getting dehydrated in the warm weather, she had drunk large quantities of water in addition to sports gel and sports drinks containing carbohydrates and electrolytes.

Question 1:
What are the names of the fluids found in the two major body compartments and the major ions in each fluid?

Question 2:
Based on Lauren's history, give a reason for why her weight increased during the race.

642   **645**   658   661   665   671

represents the kidneys, where body fluid filters into the nephrons and then may or may not be reabsorbed into the body. Some solutes and water leave the body in the urine, but the volume that leaves can be regulated, as indicated at the bottom of the handle.

The normal range for fluid volume in the mug lies between the dashed line and the open top. Fluid in the mug enters the handle (equivalent to being filtered into the kidney) and cycles back into the body of the mug to maintain the mug's volume. If fluid is added to the mug and threatens to overflow, the extra fluid is allowed to drain out of the handle (comparable to excess water excreted in urine). If a small volume is lost from the mug, fluid still flows through the handle, but all fluid loss from the handle is turned off to prevent additional fluid loss. The only way to replace the lost fluid is to add water from a source outside the mug. Translating this model to the body underscores the fact that the kidneys cannot replenish lost water: all they can do is conserve it. And as shown in the mug model, if fluid loss is severe and volume falls below the dashed line, fluid no longer flows into the handle, just as a major decrease in blood volume and blood pressure shuts down renal filtration.

Urine Concentration Is Determined in the Loop of Henle and Collecting Duct

The concentration, or osmolarity, of urine is a measure of how much water is excreted by the kidneys. When the body needs to eliminate excess water, the kidneys put out copious amounts of dilute urine with an osmolarity as low as 50 mOsM. Removal of excess water in urine is known as **diuresis** [*diourein,* to pass in urine]. (This is why drugs that promote the excretion of urine are called *diuretics.*) When the kidneys are conserving water, the urine becomes quite concentrated, up to four times as concentrated as the blood (1200 mOsM versus the blood's 300 mOsM).

The kidneys alter urine concentration by varying the amounts of water and Na^+ reabsorbed in the distal nephron. (For a review of renal anatomy, see Figure 19-1, p. 616.) To produce dilute urine, the kidney must reabsorb solute without allowing water to follow by osmosis. This means that the cell membranes across which solute is transported must not be permeable to water. To produce concentrated urine, the nephron must reabsorb water while leaving solute in the tubule lumen. At one time, scientists believed that water was actively transported on carriers, just as Na^+ and other ions are. However, once micropuncture techniques for sampling fluid inside kidney tubules were developed, scientists discovered that water is reabsorbed only by osmosis.

Osmosis will not take place unless a concentration gradient exists. Through an unusual arrangement of blood vessels and renal tubules, to be discussed later, the renal medulla maintains a high osmotic concentration in its interstitial fluid. This high *medullary interstitial osmolarity* allows urine to be concentrated.

Let's follow some filtered fluid through a nephron to see where these changes in osmolarity take place (Fig. 20-4 ■). Recall from Chapter 19 that reabsorption in the proximal tubule is isosmotic [p. 619]. Filtrate entering the loop of Henle has an osmolarity of about 300 mOsM. Fluid leaving the loop of Henle, however, is hyposmotic, with an osmolarity of around 100 mOsM. This hyposmotic fluid is created when cells in the thick portion of the ascending limb of the loop transport Na^+, K^+, and Cl^- out of the tubule lumen. These cells are unusual because their apical surface (facing the tubule lumen) is impermeable to water. Thus when these cells transport solute out of the lumen, water cannot follow, creating a hyposmotic solution.

Once the hyposmotic fluid leaves the loop of Henle and passes through the distal tubule and into a collecting duct, its concentration is determined by the water permeability of epithelial cells in the distal nephron. If the apical membrane of these cells is not permeable to water, water remains in the tubule and the filtrate remains dilute. A small amount of additional solute can be reabsorbed as fluid passes along the collecting duct, making the filtrate even more dilute. When this happens, the concentration of urine can be as low as 50 mOsM.

Conversely, if the urine that is excreted is to be more concentrated than 100 mOsM, the tubule epithelium in the distal nephron must become permeable to water through the insertion of water pores. When the collecting duct membrane is permeable to water, osmosis draws water out of the lumen and into the interstitial fluid. At maximal water permeability, removal of

20

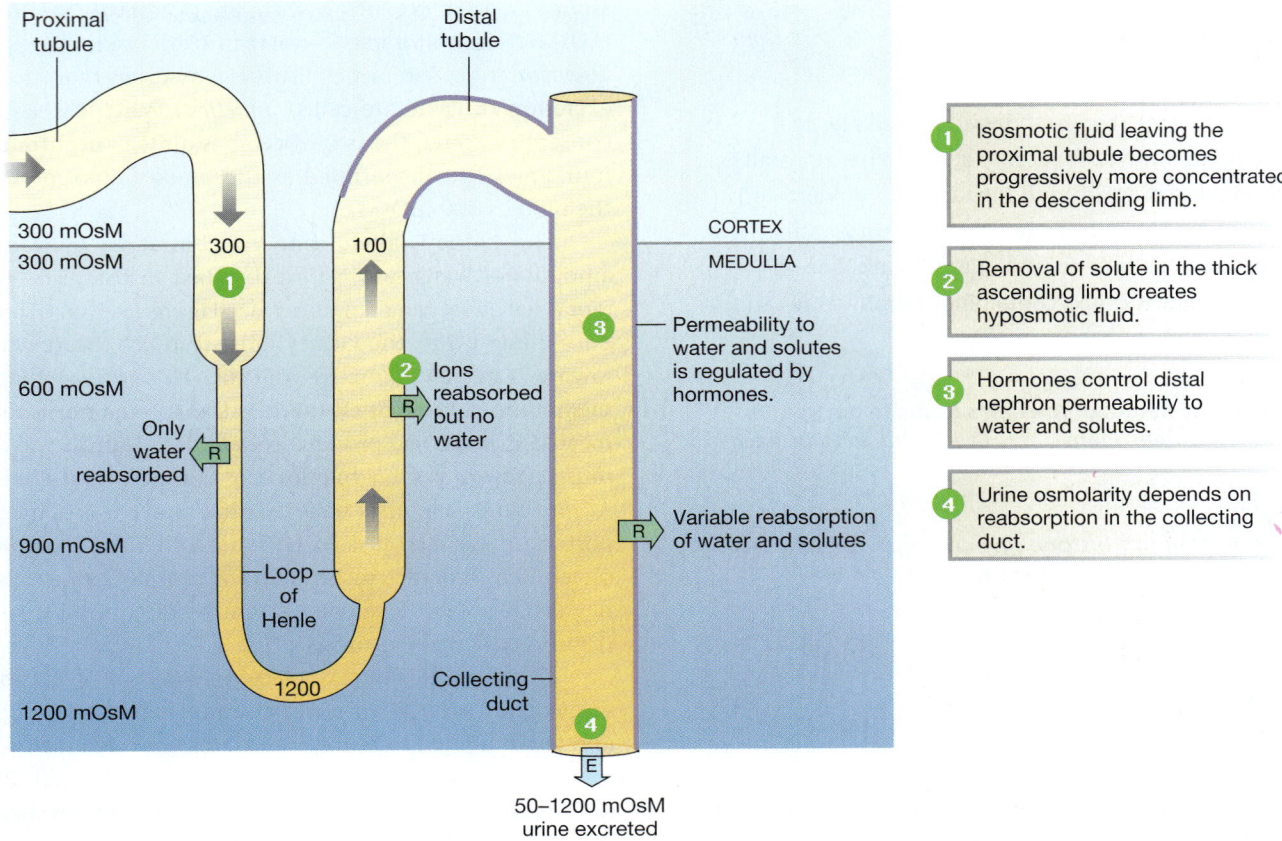

Proximal tubule

Distal tubule

300 mOsM
300 mOsM

300

100

CORTEX

MEDULLA

600 mOsM

Only
water
reabsorbed

R

R

Ions
reabsorbed
but no
water

900 mOsM

Loop
of
Henle

R

Permeability to
water and solutes
is regulated by
hormones.

R

Variable reabsorption
of water and solutes

1200

Collecting
duct

1200 mOsM

50–1200 mOsM
urine excreted

1. Isosmotic fluid leaving the proximal tubule becomes progressively more concentrated in the descending limb.

2. Removal of solute in the thick ascending limb creates hyposmotic fluid.

3. Hormones control distal nephron permeability to water and solutes.

4. Urine osmolarity depends on reabsorption in the collecting duct.

■ **FIGURE 20-4** *Osmolarity changes as filtrate flows through the nephron*

water from the tubule leaves behind a concentrated urine with an osmolarity that can be as high as 1200 mOsM.

Water reabsorption in the kidneys conserves water and can decrease body osmolarity to some degree when coupled with excretion of solute in the urine. However, the kidney's homeostatic mechanisms can do nothing to restore lost fluid volume. Only the ingestion or infusion of water can replace water that has been lost.

Vasopressin Controls Water Reabsorption

How do the distal tubule and collecting duct cells alter their permeability to water? The process involves adding or removing water pores in the apical membrane under the direction of the posterior pituitary hormone vasopressin [p. 226]. Because vasopressin causes the body to retain water, it is also known as **antidiuretic hormone (ADH).**

When vasopressin acts on target cells, water pores are present in the apical membrane, allowing water to move out of the lumen by osmosis (Fig. 20-5a ■). The water moves by osmosis because solute concentration in the cells and interstitial fluid of the renal medulla is higher than that of fluid in the tubule. In the absence of vasopressin, the collecting duct is impermeable to water (Fig. 20-5b). Although a concentration gradient is

---

DIABETES

CLINICAL FOCUS

OSMOTIC DIURESIS

The primary sign of diabetes mellitus is an elevated blood glucose concentration. In untreated diabetics, if blood glucose levels exceed the renal threshold for glucose reabsorption [p. 629], glucose will be excreted in the urine. This may not seem like a big deal, but any additional nonreabsorbable solute that remains in the lumen forces additional water to be excreted, causing *osmotic diuresis*. Suppose, for example, that the nephrons must excrete 300 milliosmoles of NaCl. If the urine is maximally concentrated at 1200 mOsM, the NaCl will be excreted in a volume of 0.25 L. However, if the NaCl is joined by 300 milliosmoles of glucose that must be excreted, the volume of urine doubles, to 0.5 L. Osmotic diuresis in untreated diabetics (primarily type 1) causes *polyuria* (excessive urination) and *polydipsia* (excessive thirst) [*dipsios*, thirsty] as a result of dehydration and high plasma osmolarity.

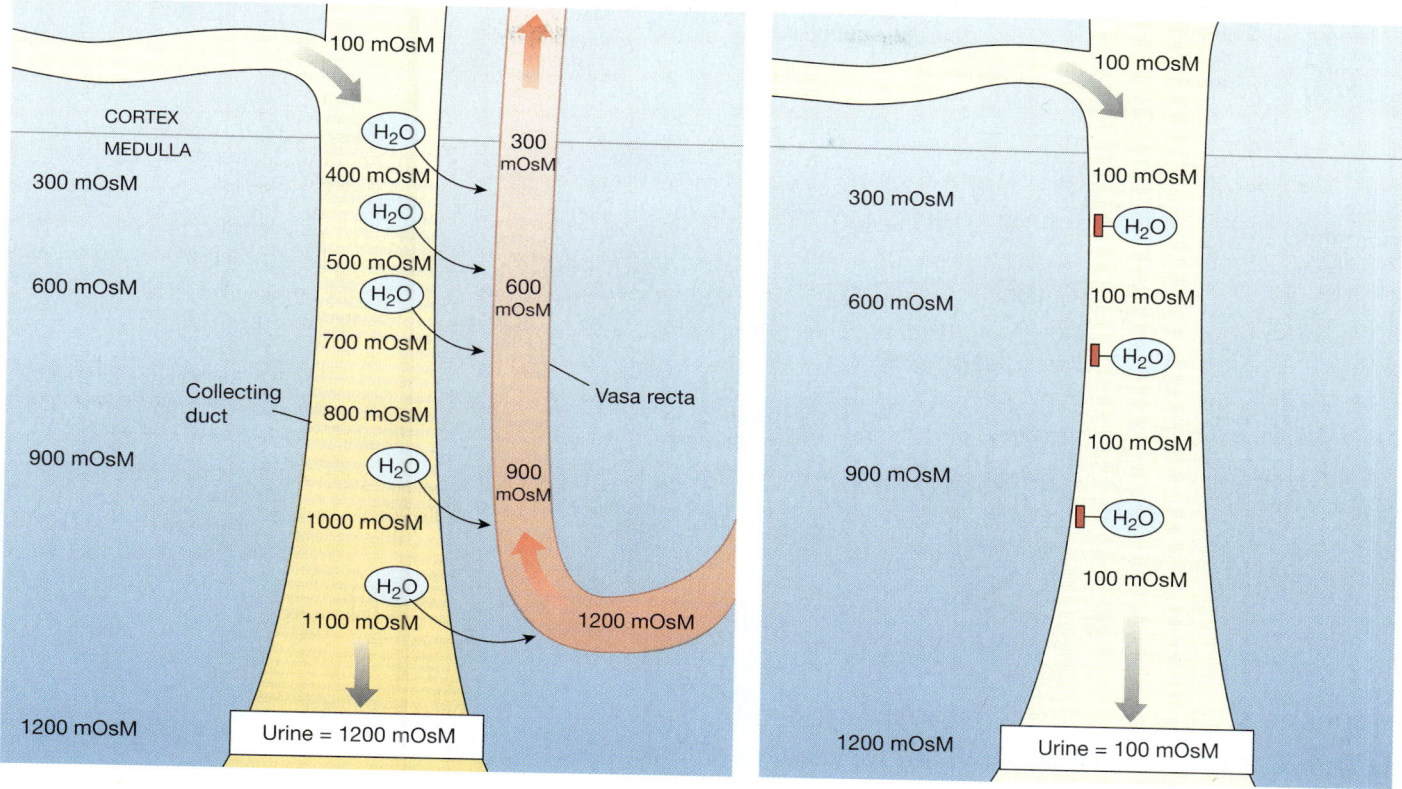

(a) With maximal vasopressin, the collecting duct is freely permeable to water. Water leaves by osmosis and is carried away by the vasa recta capillaries. Urine is concentrated.

(b) In the absence of vasopressin, the collecting duct is impermeable to water and the urine is dilute.

■ **FIGURE 20-5** *Water movement in the collecting duct in the presence and absence of vasopressin*

present across the epithelium, water remains in the tubule, producing dilute urine.

The water permeability of the collecting duct is not an all-or-none phenomenon, as the previous paragraph might suggest. Permeability is variable, depending on how much vasopressin is present. The graded effect of vasopressin allows the body to match urine concentration closely to the body's needs.

Vasopressin and Aquaporins Most membranes in the body are freely permeable to water. What makes the cells of the distal nephron different? The answer lies with the *water pores* found in these cells. Water pores are **aquaporins**, a family of membrane channels with at least 10 different isoforms that occur in mammalian tissues. The kidney has multiple isoforms of aquaporins, including *aquaporin-2* (AQP2), the water channel regulated by vasopressin.

AQP2 in a collecting duct cell may be found in two locations: on the apical membrane facing the tubule lumen and in the membrane of cytoplasmic storage vesicles (Fig. 20-6 ■). (Two other isoforms of aquaporins are present in the basolateral membrane, but they are not regulated by vasopressin.) When vasopressin levels (and, consequently, collecting duct water permeability) are low, the collecting duct cell has few water

pores in its apical membrane and stores its AQP2 water pores in cytoplasmic storage vesicles.

When vasopressin from the posterior pituitary arrives at its target, it binds to its receptor on the basolateral side of the cell (step 1 in Fig. 20-6). Binding activates a G-protein/cAMP second messenger system [p. 183]. Subsequent phosphorylation of intracellular proteins causes the AQP2 vesicles to move to the apical membrane and fuse with it. Exocytosis inserts the AQP2 water pores into the apical membrane. Now the cell is permeable to water. This process, in which parts of the cell membrane are alternately added by exocytosis and withdrawn by endocytosis, is known as **membrane recycling** [Fig. 5-24, p. 149).

CONCEPT CHECK

1. Will the apical membrane of a collecting duct cell have more water pores when vasopressin is present or when it is absent?

2. The interstitial fluid in contact with the basolateral side of collecting duct cells has an extremely high osmolarity, and yet the cells do not shrivel up. How can they maintain normal cell volume in the face of such high ECF osmolarity? (*Hint:* read the box on regulation of cell volume, p. 642.)

Answers: p. 675

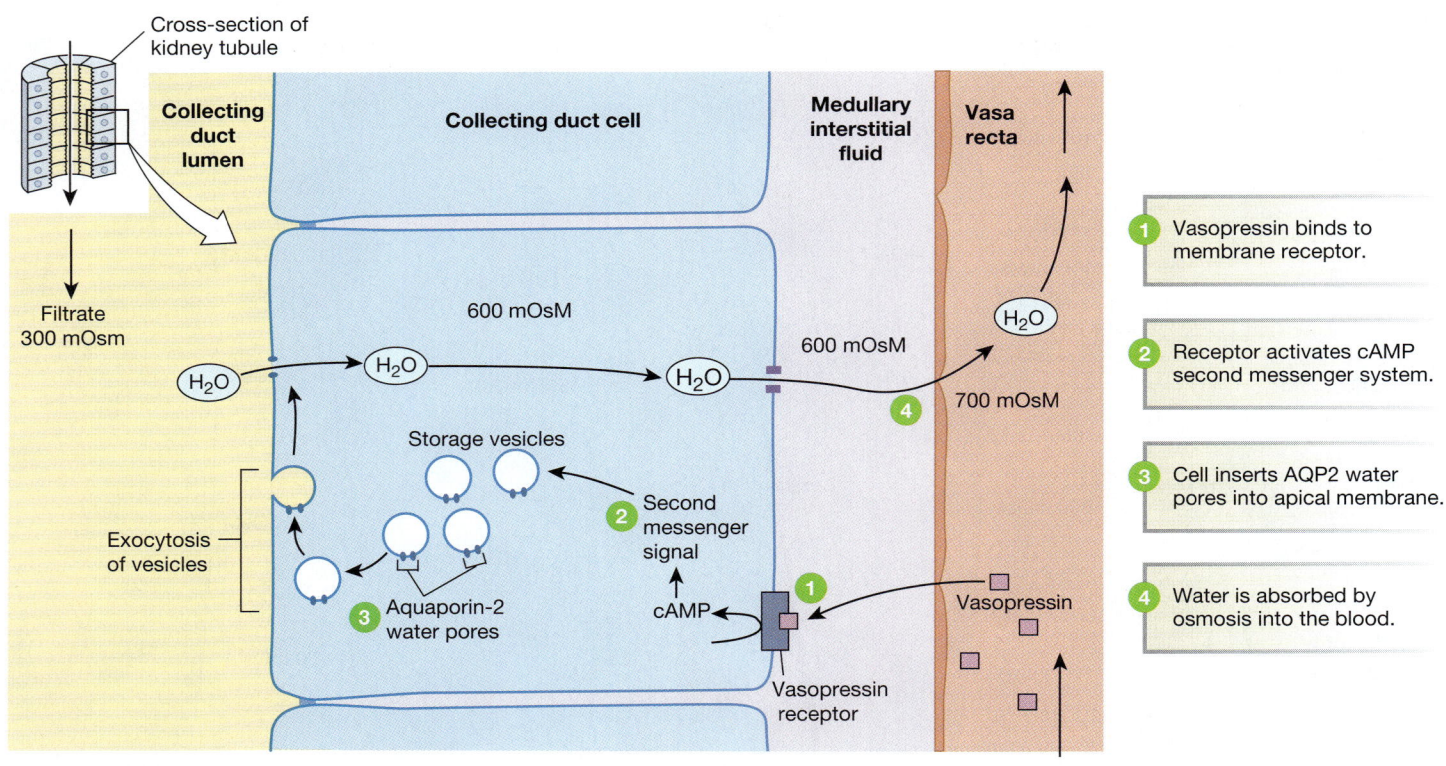

Cross-section of kidney tubule

| Collecting duct lumen | Collecting duct cell | Medullary interstitial fluid | Vasa recta |

Filtrate 300 mOsm

600 mOsM

H_2O H_2O H_2O 600 mOsM

700 mOsM   **4**

Storage vesicles

Second messenger signal   **2**

Exocytosis of vesicles

3 Aquaporin-2 water pores

cAMP   **1**

Vasopressin

Vasopressin receptor

1 Vasopressin binds to membrane receptor.

2 Receptor activates cAMP second messenger system.

3 Cell inserts AQP2 water pores into apical membrane.

4 Water is absorbed by osmosis into the blood.

■ **FIGURE 20-6**   *The mechanism of vasopressin action*

In the absence of vasopressin, the water pores are withdrawn from the apical membrane by endocytosis and stored on cytoplasmic vesicles.

Changes in Blood Pressure, Volume, and Osmolarity Trigger Water Balance Reflexes

What stimuli control vasopressin secretion? There are three: plasma osmolarity, blood volume, and blood pressure (Fig. 20-7 ■). The most potent stimulus for vasopressin release is an increase in plasma osmolarity. Osmolarity is monitored by **osmoreceptors**, stretch-sensitive cells that activate sensory neurons when osmolarity increases above threshold.

The primary osmoreceptors for vasopressin release are found in the hypothalamus. When plasma osmolarity is below the threshold value of 280 mOsM, the osmoreceptors do not fire, and vasopressin release from the pituitary ceases (Fig. 20-8 ■). If plasma osmolarity rises above 280 mOsM, the osmoreceptors stimulate release of vasopressin.

Decreases in blood pressure and blood volume are less powerful stimuli for vasopressin release. The primary receptors for decreased volume are stretch-sensitive receptors in the atria. Blood pressure is monitored by the same carotid and aortic baroreceptors that initiate cardiovascular responses [ p. 521]. When blood pressure or blood volume is low, these receptors signal the hypothalamus to secrete vasopressin and conserve fluid.

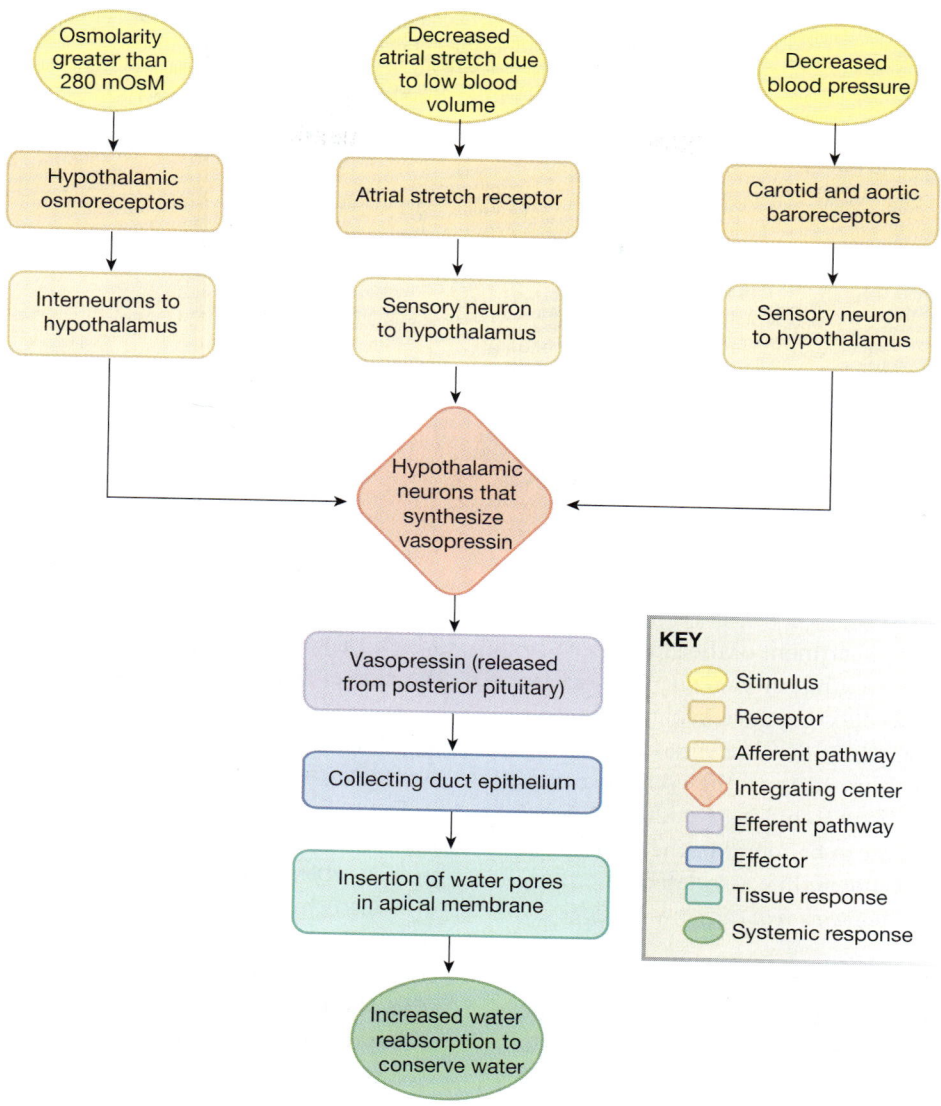

FIGURE 20-7   *Factors affecting vasopressin release*

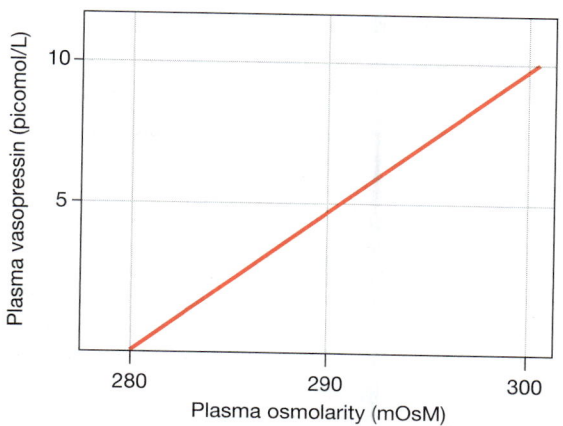

FIGURE 20-8   *The effect of plasma osmolarity on vasopressin secretion by the posterior pituitary*

CONCEPT CHECK

3. A scientist monitoring the activity of osmoreceptors notices that infusion of hyperosmotic saline (NaCl) causes increased firing of the osmoreceptors. Infusion of hyperosmotic urea (a penetrating solute) [📄 p. 157] had no effect on the firing rate. If osmoreceptors fire only when cell volume decreases, explain why hyperosmotic urea did not affect them.

4. If vasopressin increases water reabsorption by the blood vessels of a nephron, would vasopressin secretion be increased or decreased with dehydration?

5. If vasopressin secretion is suppressed, will urine be dilute or concentrated?

Answers: p. 675

The Loop of Henle Is a Countercurrent Multiplier

Vasopressin is the signal for water reabsorption out of the nephron tubule, but the key to the kidney's ability to produce concentrated urine is the high osmolarity of the medullary

(a) If blood vessels are not close to each other, heat is dissipated to the external environment.

(b) Countercurrent heat exchanger allows warm blood entering the limb to transfer heat directly to blood flowing back into the body.

■ **FIGURE 20-9**   *A countercurrent heat exchanger*

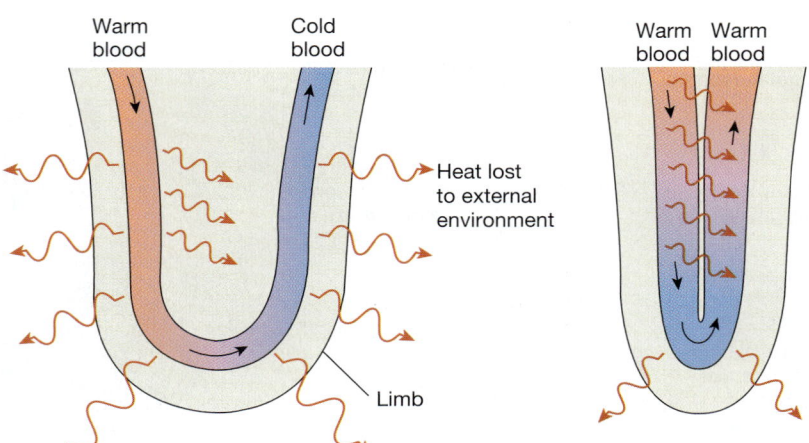

interstitium (interstitial fluid compartment of the kidney). Without it, there would be no concentration gradient for osmotic movement of water out of the collecting duct. What creates this high ECF osmolarity? And why isn't the interstitial fluid osmolarity reduced as water is reabsorbed from the collecting duct and descending limb of the loop of Henle (see Fig. 20-4, p. 646)? The answers to these questions can be found in the anatomical arrangement of the loop of Henle and its associated blood vessels, the vasa recta, to form a *countercurrent exchange system*.

Countercurrent Exchange Systems   Countercurrent exchange systems require arterial and venous blood vessels that pass very close to each other, with their fluid flow moving in opposite directions (the name *countercurrent* comes from the fact that the two flows run *counter to* each other). This anatomical arrangement allows the transfer of heat or molecules from one vessel to the other. Because the countercurrent heat exchanger is easier to understand, we first examine how it works and then translate the same principle to the kidney.

The countercurrent heat exchanger in mammals and birds evolved to reduce heat loss from flippers, tails, and other limbs that are poorly insulated and have a high surface-area-to-volume ratio. Without a heat exchanger, warm blood flowing from the body core into the limb would easily lose heat to the surrounding environment (Fig. 20-9a ■). With a countercurrent heat exchanger, warm arterial blood entering the limb transfers its heat to cooler venous blood flowing from the tip of the limb back into the body (Fig. 20-9b). This arrangement reduces the amount of heat lost to the external environment.

The countercurrent exchange system of the kidney—the loop of Henle—works on the same principle, except that it transfers solutes instead of heat. However, because the kidney forms a closed system, the solutes are not lost to the environment. Instead, the solutes concentrate in the interstitium. This process is aided by active transport of solutes out of the ascending limb,

which makes the ECF osmolarity even greater. For this reason, the loop of Henle is known as a **countercurrent multiplier**.

The Renal Countercurrent Multiplier   An overview of the countercurrent multiplier system in the renal medulla is shown in Figure 20-10 ■. Filtrate from the proximal tubule flows into the descending limb of the loop of Henle. The descending limb is permeable to water but does not transport ions. As the loop dips into the medulla, water moves by osmosis from the descending limb into the progressively more concentrated interstitial fluid, leaving solutes behind in the tubule lumen.

The filtrate becomes progressively more concentrated as it moves deeper into the medulla. At the tips of the longest loops of Henle, the filtrate reaches a concentration of 1200 mOsM. Filtrate in shorter loops (which do not extend into the most concentrated regions of the medulla) does not reach such a high concentration.

When the fluid flow reverses direction and enters the ascending limb of the loop, the properties of the tubule epithelium change. The tubule epithelium in this segment of the nephron is impermeable to water while actively transporting Na^+, K^+ and Cl^- out of the tubule into the interstitial fluid. The loss of solute from the lumen causes the filtrate osmolarity to decrease steadily, from 1200 mOsM at the bottom of the loop to 100 mOsM at the point where the ascending limb leaves the medulla and enters the cortex. The net result of the countercurrent multiplier in the kidney is to produce hyperosmotic interstitial fluid in the medulla, and hyposmotic filtrate leaving the loop of Henle.

Normally, about 25% of all Na^+ and K^+ reabsorption takes place in the ascending limb of the loop. Some transporters responsible for active ion reabsorption in the thick portion of the ascending limb are shown in Figure 20-11 ■. The *NKCC symporter* uses energy stored in the Na^+ concentration gradient to transport Na^+, K^+ and 2 Cl^- from the lumen into the epithelial cells of the ascending limb. The Na^+-K^+-ATPase removes Na^+ from

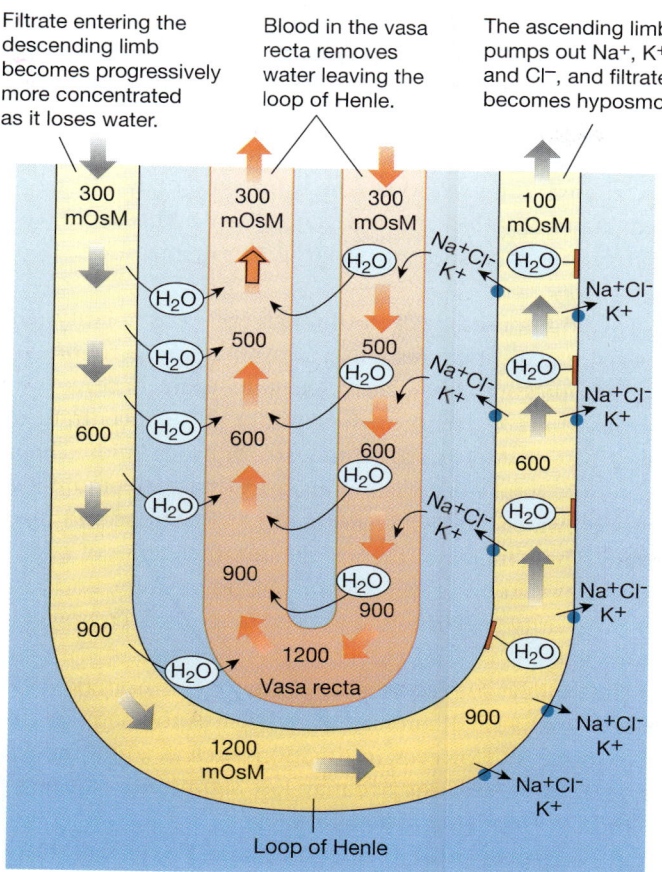

Filtrate entering the descending limb becomes progressively more concentrated as it loses water.

Blood in the vasa recta removes water leaving the loop of Henle.

The ascending limb pumps out Na⁺, K⁺, and Cl⁻, and filtrate becomes hyposmotic.

■ **FIGURE 20-10** *Countercurrent exchange in the medulla of the kidney*

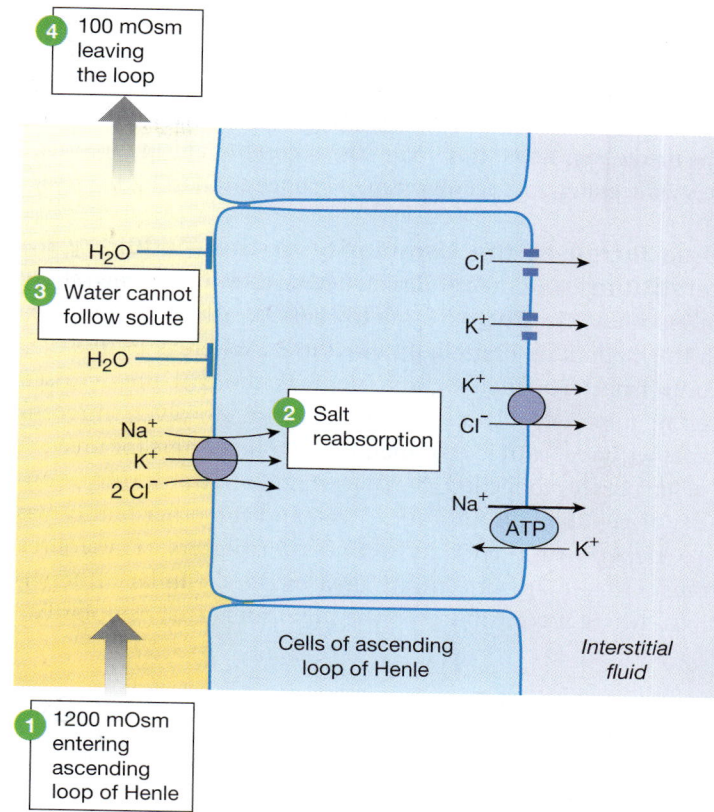

■ **FIGURE 20-11** *Active reabsorption of ions in the thick ascending limb creates a dilute filtrate in the lumen*

the cells on the basolateral side of the epithelium, while K⁺ and Cl⁻ leave the cells together on a cotransport protein or through open channels. NKCC-mediated transport can be inhibited by drugs known as "loop diuretics," such as furosemide (Lasix).

CONCEPT CHECK

6. Explain why a patient taking a loop diuretic that inhibits solute reabsorption will excrete greater-than-normal volumes of urine.

7. Loop diuretics that inhibit the NKCC symporter are sometimes called "potassium-wasting" diuretics. Explain why people who are on loop diuretics must increase their dietary K⁺ intake.

Answers: p. 675

The Vasa Recta Removes Water It is easy to see how transport of solute out of the ascending limb of the loop of Henle dilutes the filtrate and helps concentrate the interstitial fluid in the medulla. Still, why doesn't the water leaving the descending limb of the loop (see Fig. 20-10) *dilute* the interstitial fluid of the medulla? The answer lies in the close anatomical association of the loop of Henle and those peritubular capillaries known as the **vasa recta**.

These capillaries, like the loop of Henle, dip down into the medulla and then go back up to the cortex, forming hairpin loops.

Although textbooks traditionally show a single nephron with a single loop of capillary (as we do in Fig. 20-10), each kidney has thousands of collecting ducts and loops of Henle packed between thousands of vasa recta capillaries, blurring the direct association between a nephron and its vascular supply. Functionally, the direction of blood flow in the vasa recta is opposite the direction of filtrate flow in the loops of Henle, as shown in Figure 20-10.

Water or solutes that leave the tubule move into the vasa recta if an osmotic or concentration gradient exists between the medullary interstitium and the blood in the vasa recta. For example, assume that at the point at which the vasa recta enter the medulla, the blood in the vasa recta is 300 mOsM, isosmotic with the cortex. As the blood flows deeper and deeper into the medulla, it loses water and picks up solutes transported out of the ascending limb of the loop of Henle, carrying these solutes farther into the medulla. By the time the blood reaches the bottom of the vasa recta loop, it has a high osmolarity, similar to that of the surrounding interstitial fluid (1200 mOsM).

Then, as blood in the vasa recta flows back toward the cortex, the high plasma osmolarity attracts the water that is being lost from the descending limb, as Figure 20-10 shows. The movement of this water into the vasa recta decreases the osmolarity of the blood while simultaneously preventing the water from diluting the concentrated medullary interstitial fluid.

The end result of this arrangement is that blood flowing through the vasa recta removes the water reabsorbed from the loop of Henle. Without the vasa recta, water moving out of the descending limb of the loop of Henle would eventually dilute the medullary interstitium. The vasa recta thus are an important part of keeping the medullary solute concentration high.

Urea Increases the Osmolarity of the Medullary Interstitium

The high solute concentration in the medullary interstitium is only partly due to NaCl. Nearly half the solute in the medullary interstitial fluid is urea. Where does this urea come from? For many years scientists thought urea crossed cell membranes only by passive transport. However, in recent years we have learned that membrane transporters for urea are present in the collecting duct. One family of transporters consists of facilitated diffusion carriers, and the other family has Na^+-dependent secondary active transporters. These urea transporters help concentrate urea in the medullary interstitium, where it contributes to the high interstitial osmolarity.

SODIUM BALANCE AND ECF VOLUME

As noted in the introduction to this chapter, we ingest a lot of NaCl—an average of 9 grams per day. This is about 2 teaspoons of salt, or 155 milliosmoles of Na^+ and 155 milliosmoles of Cl^-. Let's see what would happen to our bodies if the kidneys could not get rid of this Na^+. (Remember from Chapter 19 that Cl^- follows the electrical gradient created by Na^+ transport. [⮌ p. 626])

Our normal plasma Na^+ concentration is 135–145 milliosmoles Na^+ per liter of plasma. Because sodium distributes freely between plasma and interstitial fluid, this value also represents our ECF Na^+ concentration. If we add 155 milliosmoles Na^+ to the ECF, how much water would we have to add to keep the ECF Na^+ concentration at 140 mOsM? One form of an equation asking this question is

$$155 \text{ mosmol/X liters} = 140 \text{ mosmol/liter}$$

$$X = 1.1 \text{ liters}$$

We would have to add more than a liter of water to the ECF to compensate for the addition of the Na^+. Normal ECF volume is about 14 liters, and so that increase in volume would represent about an 8% gain! Imagine what that volume increase would do to blood pressure.

Suppose, however, that instead of adding water to keep plasma concentrations constant, we add the NaCl but don't drink any water. What happens to osmolarity now? If we assume that normal total body osmolarity is 300 mOsM and that the volume of fluid in the body is 42 L, the addition of 155 milliosmoles of Na^+ and 155 milliosmoles of Cl^- would increase total body osmolarity to 307 mOsM*—a substantial

---

*$(155 \text{ mosmol } Na^+ + 155 \text{ mosmol } Cl^-)/42 \text{ L} = 7.4 \text{ mosmol/L added}$
300 mOsM initial + 7.4 mOsM added = 307 mOsM final

increase. In addition, because NaCl is a nonpenetrating solute, it would stay in the ECF. Higher osmolarity in the ECF would draw water from the cells, shrinking them and disrupting normal cell function.

Fortunately, our homeostatic mechanisms maintain mass balance: anything extra that comes into the body will be excreted. Figure 20-12 ■ shows a generalized homeostatic pathway for sodium balance in response to salt ingestion. Here's how it works:

The addition of NaCl to the body raises osmolarity. This stimulus triggers two responses: vasopressin secretion and thirst. Vasopressin release causes the kidneys to conserve water (by reabsorbing water from the filtrate) and concentrate the urine. Thirst prompts us to drink water or other fluids. The increased fluid intake decreases osmolarity, but the combination of salt and water intake increases both ECF volume and blood pressure. These increases then trigger another series of control pathways, which bring ECF volume, blood pressure, and total-body osmolarity back into the normal range by excreting extra salt and water.

The kidneys are responsible for most Na^+ excretion, and normally only a small amount of Na^+ leaves the body in feces and perspiration. However, in situations such as vomiting, diarrhea, and heavy sweating, we may lose significant amounts of Na^+ and Cl^- through non-renal routes.

Although we speak of ingesting and losing salt (NaCl), only renal Na^+ absorption is regulated, and the stimuli that set the Na^+ balance pathway in motion are more closely tied to blood volume and blood pressure than to Na^+ levels. Chloride movement usually follows Na^+ movement, either indirectly via the electrochemical gradient created by Na^+ transport or directly via membrane transporters such as the NKCC transporter of the loop of Henle or the Na^+-Cl^- symporter of the distal tubule.

Aldosterone Controls Sodium Balance

The regulation of blood Na^+ levels takes place through one of the most complicated endocrine pathways of the body. The reabsorption of Na^+ in the distal tubules and collecting ducts of the kidney is regulated by the steroid hormone **aldosterone**: the more aldosterone, the more Na^+ reabsorption. Because one target of aldosterone is increased activity of the Na^+-K^+-ATPase, aldosterone also causes K^+ secretion.

Aldosterone is a steroid hormone synthesized in the adrenal cortex, the outer portion of the adrenal gland that sits atop each kidney [⮌ p. 221]. Like other steroid hormones, aldosterone is secreted into the blood and transported on a protein carrier to its target.

The primary site of aldosterone action is the last third of the distal tubule and the portion of the collecting duct that runs through the kidney cortex (the *cortical collecting duct*). The primary target of aldosterone is **principal cells**, or **P cells**

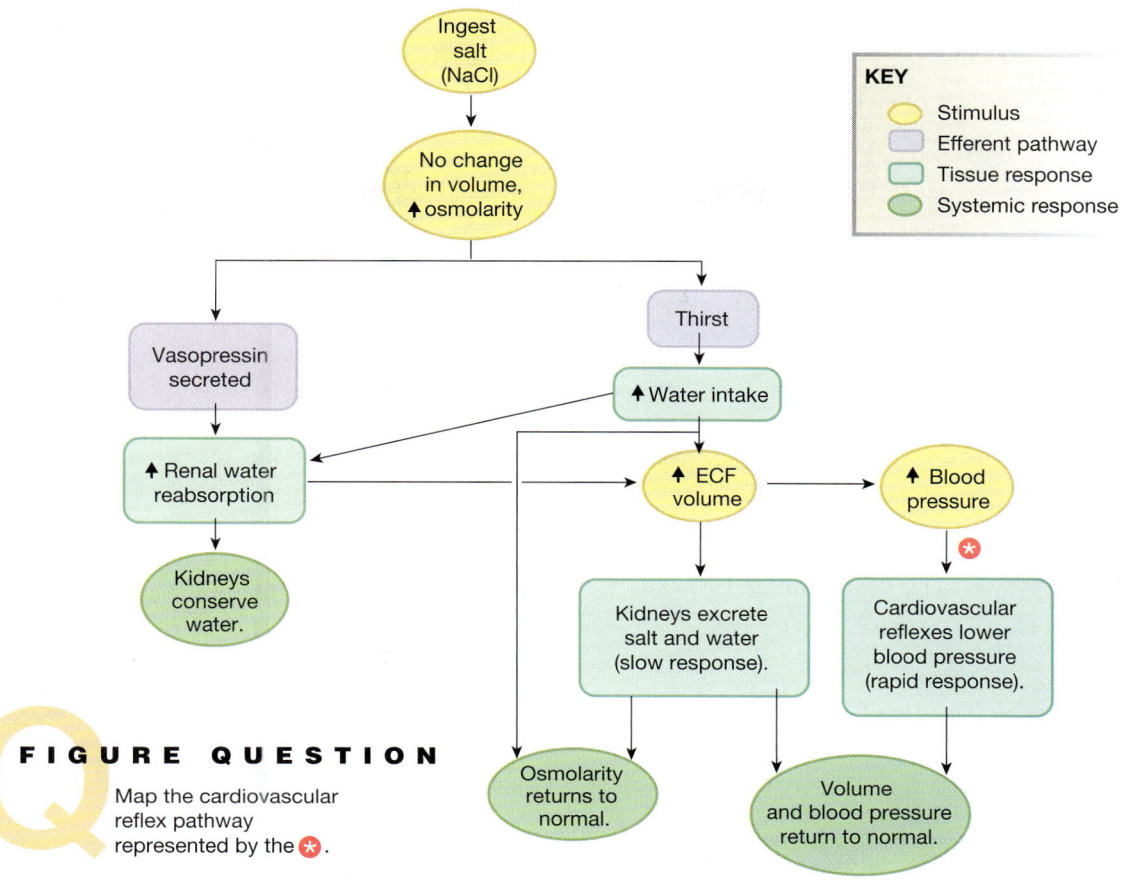

KEY

- ○ Stimulus
- ▢ Efferent pathway
- ▢ Tissue response
- ◯ Systemic response

FIGURE QUESTION

Map the cardiovascular reflex pathway represented by the ✱.

■ **FIGURE 20-12**   *Homeostatic responses to salt ingestion*

(Fig. 20-13 ■). Principal cells are arranged much like other polarized transporting epithelial cells, with Na^+-K^+-ATPase pumps on the basolateral membrane, and various channels and transporters on the apical membrane [🔁 p. 151]. In principal cells, the apical membranes contain leak channels for Na^+ (called ENaC, for *e*pithelial Na^+ *c*hannel) and for K^+ (called ROMK, for *r*enal *o*uter *m*edulla K^+ channel).

Aldosterone enters P cells by simple diffusion. Once inside, it combines with a cytoplasmic receptor (Fig. 20-13 ①). In the early response phase, apical Na^+ and K^+ channels increase their open time under the influence of an as-yet-unidentified signal molecule. As intracellular Na^+ levels rise, the Na^+-K^+-ATPase speeds up, transporting cytoplasmic Na^+ into the ECF and bringing K^+ from the ECF into the P cell. The net result is a rapid increase in Na^+ reabsorption and K^+ secretion that does not require the synthesis of new channel or ATPase proteins. In the slower phase of aldosterone action, newly synthesized channels and pumps are inserted into epithelial cell membranes (Fig. 20-13 ④).

Note that Na^+ and water reabsorption are separately regulated in the distal nephron. Water does not automatically follow Na^+ reabsorption: vasopressin must be present to make the distal-nephron epithelium permeable to water. In contrast, in the proximal tubule, Na^+ reabsorption is automatically fol-

lowed by water reabsorption because the proximal tubule epithelium is always freely permeable to water.

Blood Pressure Is the Primary Stimulus for Aldosterone Secretion

What controls physiological aldosterone secretion from the adrenal cortex? There are two primary stimuli: increased extracellular K^+ concentration and decreased blood pressure. Elevated K^+ concentrations act directly on the adrenal cortex in a reflex that protects the body from hyperkalemia. Decreased blood pressure initiates a complex pathway that results in release of a trophic hormone, **angiotensin II**, that stimulates aldosterone secretion in most situations.

Two additional factors modulate aldosterone release in pathological states: an increase in ECF osmolarity acts directly on adrenal cortex cells to inhibit aldosterone secretion during dehydration, and a severe (10–20 meq/L) decrease in plasma Na^+ can directly stimulate aldosterone secretion.

The Renin-Angiotensin-Aldosterone Pathway Angiotensin II (ANG II) is the usual signal controlling aldosterone release from the adrenal cortex. ANG II is one component of the **renin-angiotensin-aldosterone system (RAAS)**, a complex, multistep pathway for maintaining blood pressure. The

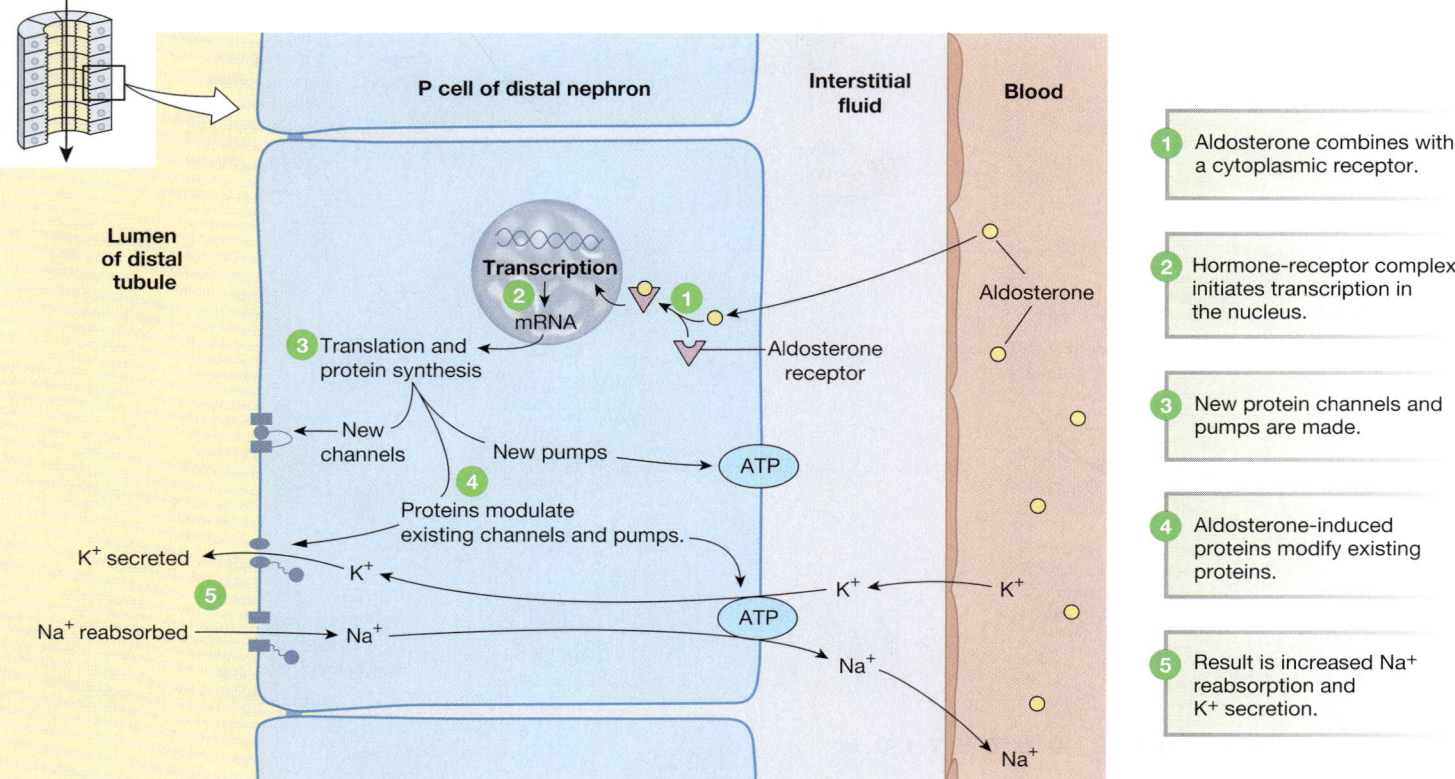

■ **FIGURE 20-13** *Aldosterone action in principal cells*

RAAS pathway begins when juxtaglomerular granular cells in the afferent arterioles of a nephron [🔁p. 618] secrete an enzyme called **renin** (Fig. 20-14 ■). Renin converts an inactive plasma protein, **angiotensinogen**, into **angiotensin I (ANG I)**. (The suffix *-ogen* indicates an inactive precursor.) When ANG I in the blood encounters an enzyme called **angiotensin converting enzyme (ACE)**, ANG I is converted into ANG II.

This conversion was originally thought to take place only in the lungs, but ACE is now known to occur on the endothelium of blood vessels throughout the body. When ANG II in the blood reaches the adrenal gland, it causes synthesis and release of aldosterone. Finally, at the distal nephron, aldosterone initiates a series of intracellular reactions that cause the tubule to reabsorb Na⁺.

The stimuli that begin the RAAS pathway are all related either directly or indirectly to low blood pressure (Fig. 20-15 ■):

1. The *granular cells* are directly sensitive to blood pressure. They respond to low blood pressure in renal arterioles by secreting renin.

2. *Sympathetic neurons,* activated by the cardiovascular control center when blood pressure decreases, terminate on the granular cells and stimulate renin secretion.

3. *Paracrine feedback*—from the macula densa in the distal tubule to the granular cells—stimulates renin release [🔁p. 624]. When fluid flow through the distal tubule is relatively high, the macula densa cells release paracrines,

which inhibit renin release. When fluid flow in the distal tubule decreases, macula densa cells signal the granular cells to secrete renin.

Sodium reabsorption does not directly raise low blood pressure, but retention of Na⁺ increases osmolarity, which stimulates thirst. Fluid intake when the person drinks more water increases ECF volume (see Fig. 20-12). When blood volume increases, blood pressure also increases.

The effects of RAAS pathway are not limited to aldosterone release, however. Angiotensin II is a remarkable hormone with additional effects directed at raising blood pressure. These actions make ANG II an important hormone in its own right, not merely an intermediate step in the aldosterone control pathway.

CONCEPT CHECK

8. In Figure 20-13, what force(s) cause(s) Na⁺ and K⁺ to cross the apical membrane?

Answers: p. 675

Angiotensin II Influences Blood Pressure Through Multiple Pathways

Angiotensin II has significant effects on fluid balance and blood pressure beyond stimulating aldosterone secretion, underscoring the integrated functions of the renal and cardiovascular systems. ANG II increases blood pressure both directly and indirectly through four additional pathways (Fig. 20-14):

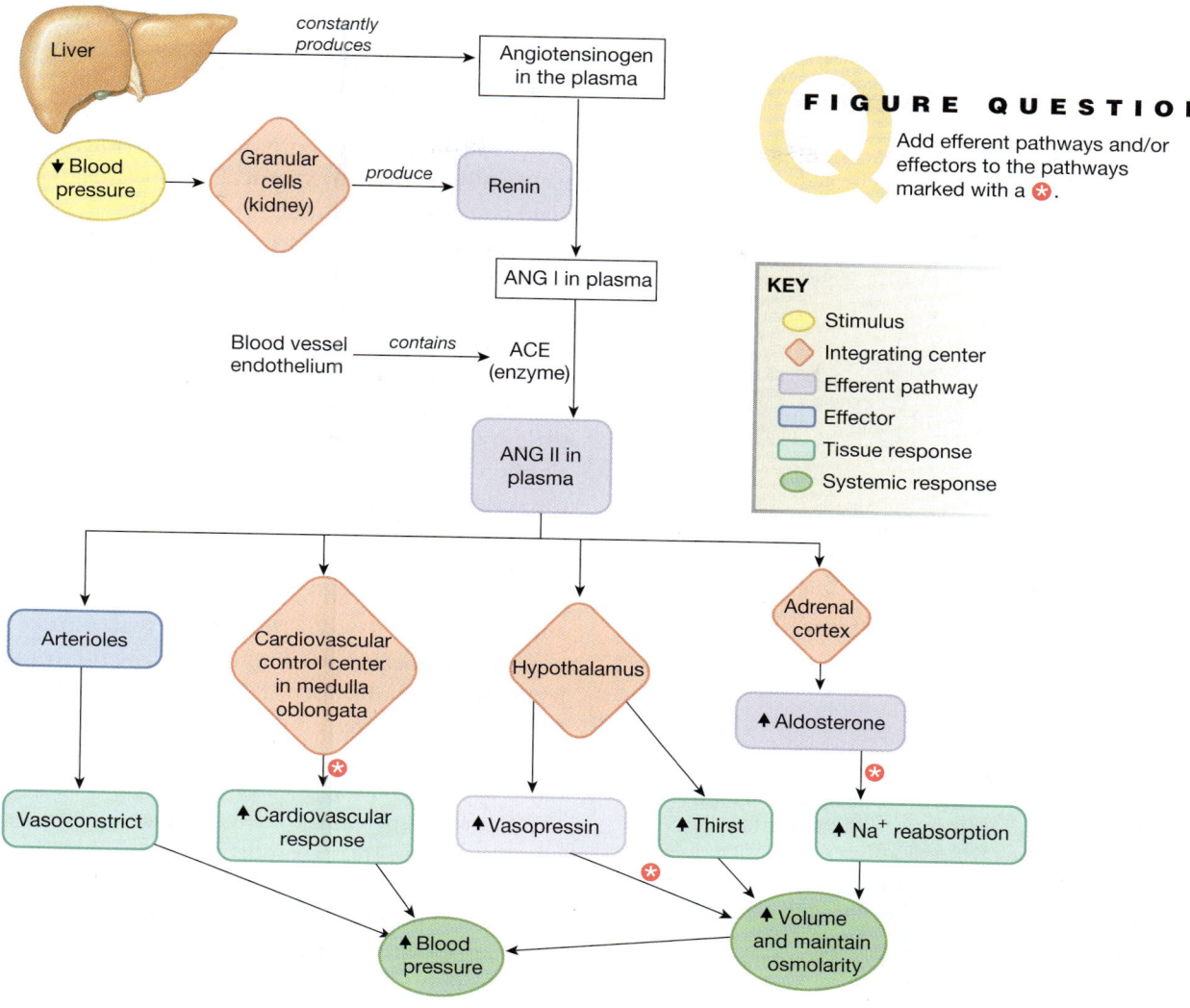

FIGURE QUESTION

Add efferent pathways and/or effectors to the pathways marked with a ✲.

KEY
- ◯ Stimulus
- ◇ Integrating center
- ▢ Efferent pathway
- ▢ Effector
- ▢ Tissue response
- ◯ Systemic response

■ **FIGURE 20-14**   *The renin-angiotensin-aldosterone pathway*

This map outlines the control of aldosterone secretion as well as the blood pressure-raising effects of ANG II.

1. *ANG II increases vasopressin secretion.* ANG II receptors in the hypothalamus initiate this reflex. Fluid retention in the kidney under the influence of vasopressin helps conserve blood volume, thereby maintaining blood pressure.

2. *ANG II stimulates thirst.* Fluid ingestion is a behavioral response that expands blood volume and raises blood pressure.

3. *ANG II is one of the most potent vasoconstrictors* known in humans. Vasoconstriction causes blood pressure to increase without a change in blood volume.

4. *Activation of ANG II receptors in the cardiovascular control center increases sympathetic output to the heart and blood vessels.* Sympathetic stimulation increases cardiac output and vasoconstriction, both of which increase blood pressure.

Once these blood-pressure-raising effects of ANG II became known, it was not surprising that pharmaceutical companies started looking for drugs to block ANG II. Their research produced a new class of antihypertensive drugs called *ACE inhibitors*. These drugs block the ACE-mediated conversion of ANG I to ANG II, thereby helping to relax blood vessels and lower blood pressure. Less ANG II also means less aldosterone release, a decrease in Na$^+$ reabsorption and, ultimately, a decrease in ECF volume. All these responses contribute to lowering blood pressure.

However, the ACE inhibitors are not without side effects. ACE inactivates a cytokine called *bradykinin*. When ACE is inhibited by drugs, bradykinin levels increase, and in some patients this creates a dry, hacking cough. One solution has been the development of drugs called *sartans* that block the blood-pressure-raising effects of ANG II by binding to *AT$_1$ receptors,* a subtype of ANG II receptor.

20

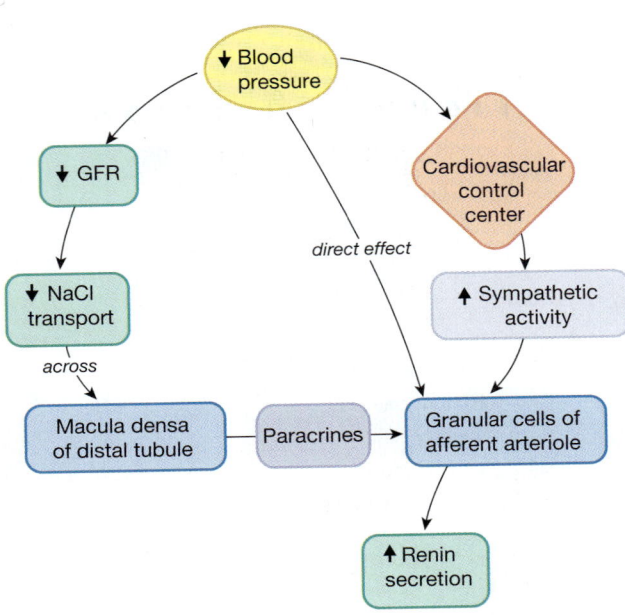

■ FIGURE 20-15 *Decreased blood pressure stimulates renin secretion*

✓ **C O N C E P T   C H E C K**

9. If a person experiences hyperkalemia, what happens to resting membrane potential and the excitability of neurons and the myocardium?

10. A man comes to the doctor with high blood pressure. Tests show that he also has elevated plasma renin levels and atherosclerotic plaques that have nearly blocked blood flow through his renal arteries. How does decreased blood flow in his renal arteries result in elevated renin levels?

11. Describe the pathways through which elevated renin causes high blood pressure in the man mentioned in Concept Check 10.

Answers: p. 675

Atrial Natriuretic Peptide Promotes Na⁺ and Water Excretion

Once it was known that aldosterone and vasopressin increase Na⁺ and water reabsorption, scientists speculated that other hormones might cause Na⁺ loss, or **natriuresis** [*natrium*, sodium + *ourein,* to urinate] and water loss (diuresis) in the urine. If found, these hormones might be used clinically to lower blood volume and blood pressure in patients with essential hypertension [🔁 p. 525]. During years of searching, however, evidence for the other hormones was not forthcoming.

Then, in 1981, a group of Canadian researchers found that injections of homogenized rat atria caused rapid but short-lived excretion of Na⁺ and water in the rats' urine. They hoped they had found the missing hormone, one whose activity would complement that of aldosterone and vasopressin. As it turned out, they had discovered the first natriuretic peptide (NP), one member of a family of hormones that appear to be endogenous RAAS antagonists.

Atrial natriuretic peptide (ANP; also known as *atriopeptin*) is a peptide hormone produced in specialized myocardial cells in the atria of the heart. It is synthesized as part of a large prohormone that is cleaved into several active hormone fragments [🔁 p. 218]. A related hormone, **brain natriuretic peptide** (BNP), is synthesized by ventricular myocardial cells and certain brain neurons. Both natriuretic peptides are released by the heart when myocardial cells stretch more than normal, as would occur with increased blood volume. The peptides bind to membrane receptors that work through a cGMP second messenger system.

At the systemic level, natriuretic peptides enhance Na⁺ and water excretion (Fig. 20-16 ■), but the exact mechanisms by which they do so are not clear. The NPs increase GFR, apparently by making more surface area available for filtration. In addition, they directly decrease NaCl and water reabsorption in the collecting duct. The cellular mechanism by which NPs affect tubular reabsorption is not known.

Natriuretic peptides also act indirectly to increase Na⁺ and water excretion by inhibiting the release of renin, aldosterone, and vasopressin (Fig. 20-16), actions that reinforce the natriuretic-diuretic effect. In addition, natriuretic peptides act directly on the cardiovascular control center of the medulla to lower blood pressure.

Brain natriuretic peptide is now recognized as an important biological marker for heart failure because production of this substance increases with ventricular dilation and increased ventricular pressure. According to one estimate in 2004, more than 70% of U.S. hospitals test for BNP levels to assist in the diagnosis of ventricular failure. This peptide has also been shown to be an independent predictor of heart failure and sudden death from cardiac arrhythmias.

POTASSIUM BALANCE

Aldosterone (but not other factors in the RAAS pathway) also plays a critical role in potassium homeostasis. Only about 2% of the body's K⁺ load is in the ECF, but regulatory mechanisms keep plasma K⁺ concentrations within a narrow range (3.5–5 meq/L). Under normal conditions, mass balance matches K⁺ excretion to K⁺ ingestion. If intake exceeds excretion and plasma K⁺ goes up, aldosterone is released into the blood through the direct effect of hyperkalemia on the adrenal cortex. Aldosterone acting on distal-nephron P cells keeps the cells' apical ion channels open longer and speeds up the Na⁺-K⁺-ATPase, enhancing renal excretion of K⁺.

The regulation of body potassium levels is essential to maintaining a state of well-being. As you learned in Chapter 8, changes in extracellular K⁺ concentration affect the resting membrane potential of all cells [🔁 Fig. 8-19, p. 269]. If plasma (and ECF) K⁺ concentrations decrease (hypokalemia), the concentration gradient between the cell and the ECF becomes larger, more K⁺ leaves the cell, and the resting membrane potential becomes more negative. If ECF K⁺ concentrations increase

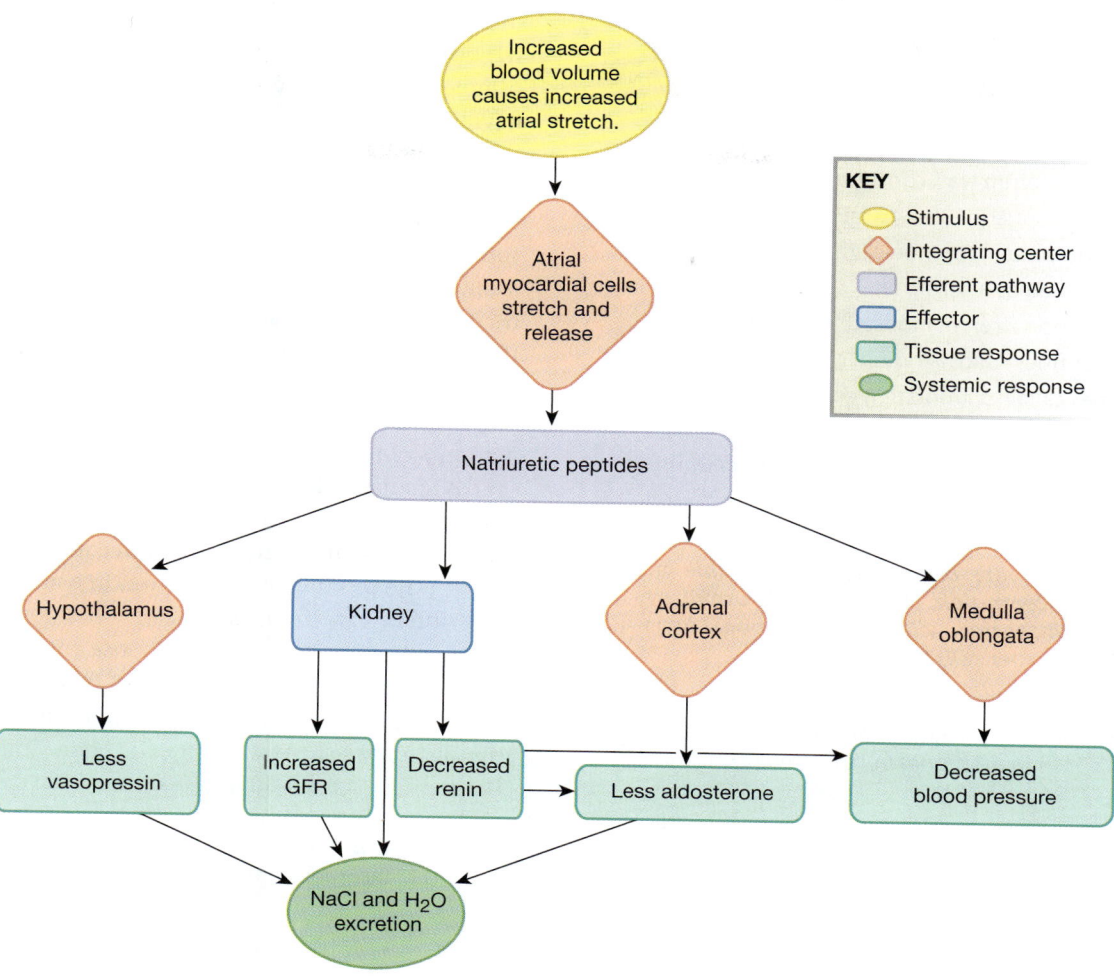

KEY
- Stimulus
- Integrating center
- Efferent pathway
- Effector
- Tissue response
- Systemic response

■ **FIGURE 20-16** *Actions of natriuretic peptides*

(hyperkalemia), the concentration gradient decreases and more K^+ remains in the cell, depolarizing it. (Remember that when plasma K^+ concentrations change, anions such as Cl^- are also added to or subtracted from the ECF in a 1:1 ratio, maintaining overall electrical neutrality.)

Because of the effect of plasma K^+ on excitable tissues, such as the heart, clinicians are always concerned about keeping plasma K^+ within its normal range. If K^+ falls below 3 meq/L or rises above 6 meq/L, the excitable tissues of muscle and nerve begin to show altered function. For example, hypokalemia causes muscle weakness because it is more difficult for hyperpolarized neurons and muscles to fire action potentials. The danger in this condition lies in the failure of respiratory muscles and the heart. Fortunately, skeletal muscle weakness is usually significant enough to lead patients to seek treatment before cardiac problems occur. Mild hypokalemia may be corrected by oral intake of K^+ supplements and K^+-rich foods, such as orange juice and bananas.

Hyperkalemia is a more dangerous potassium disturbance because in this case depolarization of excitable tissues makes them more excitable initially. Subsequently, the cells are unable

to repolarize fully and actually become *less* excitable. In this state, they have action potentials that are either smaller than normal or nonexistent. Cardiac muscle excitability affected by changes in plasma K^+ can lead to life-threatening cardiac arrhythmias.

Disturbances in K^+ balance may result from kidney disease, loss of K^+ in diarrhea, or the use of certain types of diuretics that prevent the kidneys from fully reabsorbing K^+. Inappropriate correction of dehydration can also create K^+ imbalance. Consider a golfer playing a round of golf when the temperature was above 100°F. He was aware of the risk of dehydration, so he drank lots of water to replace fluid lost through sweating. The replacement of lost sweat with pure water kept his ECF volume normal but dropped his total blood osmolarity and his K^+ and Na^+ concentrations. He was unable to finish the round of golf because of muscle weakness, and he required medical attention that included ion replacement therapy. A more suitable replacement fluid would have been one of the sports drinks that include salt and K^+.

Potassium balance is also closely tied to acid-base balance, as you will learn in the final section of this chapter. Correction

RUNNING PROBLEM

The medical staff analyzed Lauren's blood for electrolyte concentrations. Her serum Na^+ concentration was 124 mEq/L. The normal range is 135–145 mEq/L. Lauren's diagnosis was hyponatremia [*hypo-*, below + *natri-*, sodium + *-emia*, blood], defined as a serum Na^+ concentration below 135 mEq/L. Hyponatremia induced by the consumption of large quantities of low-sodium or sodium-free fluid is sometimes called *dilutional hyponatremia*.

Question 3:
Which fluid—ECF or ICF—is being diluted in dilutional hyponatremia?

Question 4:
One way to estimate osmolarity is to double the plasma Na^+ concentration. If normal body osmolarity is 280 mOsM, what effect will Lauren's dilutional hyponatremia have on her cells?

Question 5:
Which organ or tissue are the medical personnel most concerned about in dilutional hyponatremia?

| 642 | 645 | **658** | 661 | 665 | 671 |

of a pH disturbance requires close attention to plasma K^+ levels. Similarly, correction of K^+ imbalance may alter body pH.

BEHAVIORAL MECHANISMS IN SALT AND WATER BALANCE

Although neural, neuroendocrine, and endocrine reflexes play key roles in salt and water homeostasis, behavioral responses are critical in restoring the normal state, especially when ECF volume decreases or osmolarity increases. Drinking water is normally the only way to restore lost water, and eating salt is the only way to raise the body's Na^+ content. Both behaviors are essential for normal salt and water balance. Clinicians must recognize the absence of these behaviors in patients who are unconscious or otherwise unable to obey behavioral urges, and must adjust treatment accordingly. The study of the biological basis for behaviors, including drinking and eating, is a field known as *physiological psychology*.

Drinking Replaces Fluid Loss

Thirst is one of the most powerful urges known in humans. In 1952, the Swedish physiologist Bengt Andersson showed that stimulating certain regions of the hypothalamus triggered drinking behavior. This discovery led to the identification of hypothalamic osmoreceptors that initiate drinking when body osmolarity rises above 280 mOsM. This is an example of a behavior initiated by an internal stimulus.

It is interesting to note that, although increased osmolarity triggers thirst, the act of drinking is sufficient to relieve thirst. The ingested water need not be absorbed in order for thirst to be quenched. As-yet-unidentified receptors in the mouth and pharynx (*oropharynx receptors*) respond to cold water by decreasing thirst and decreasing vasopressin release even though plasma osmolarity remains high. This oropharynx reflex is one reason surgery patients are allowed to suck on ice chips: the ice alleviates their thirst without putting significant amounts of fluid into the digestive system.

A similar reflex exists in camels. When led to water, they will drink just enough to replenish their water deficit. Oropharynx receptors apparently act as a feedforward "metering" system that helps prevent wide swings in osmolarity by matching water intake to water need.

In humans, the reflex response of drinking to counter increased osmolarity is complicated by cultural rituals. Drinking takes place during social events, not just in response to the stimulus of thirst. As a result, our bodies must be capable of eliminating excess fluid ingested in various social situations.

CONCEPT CHECK

12. Incorporate the thirst reflex into Figure 20-7.

Answers: p. 675

Low Na^+ Stimulates Salt Appetite

Thirst is not the only urge associated with fluid balance. **Salt appetite** is a craving for salty foods that occurs when plasma Na^+ concentrations drop. It can be observed in deer and cattle attracted to salt blocks or naturally occurring salt licks. In humans, salt appetite is linked to aldosterone and angiotensin, hormones that regulate Na^+ balance. The centers for salt appetite are in the hypothalamus close to the center for thirst.

Avoidance Behaviors Help Prevent Dehydration

Other behaviors play a role in fluid balance by preventing or promoting dehydration. Desert animals avoid the heat of the day and become active only at night, when environmental temperatures fall and humidity rises. Humans, especially now that we have air conditioning, are not always so wise.

The midday nap, or *siesta,* is a cultural adaptation in tropical countries that keeps people indoors during the hottest part of the day, thereby helping prevent dehydration and overheating. In the United States, we have abandoned this civilized custom and are active continuously during daylight hours, even when the temperature soars during summer in the South and Southwest. Fortunately, our homeostatic mechanisms usually keep us out of trouble.

Osmolarity

	Decrease	No change	Increase
Increase	Drinking large amount of water	Ingestion of isotonic saline	Ingestion of hypertonic saline
No change	Replacement of sweat loss with plain water	Normal volume and osmolarity	Eating salt without drinking water
Decrease	Incomplete compensation for dehydration	Hemorrhage	Dehydration (e.g., sweat loss or diarrhea)

(Volume)

■ **FIGURE 20-17**   *Disturbances in volume and osmolarity*

INTEGRATED CONTROL OF VOLUME AND OSMOLARITY

The body uses an integrated response to correct disruptions of salt and water balance. The cardiovascular system responds to changes in blood volume, and the kidneys respond to changes in blood volume or osmolarity. Maintaining homeostasis throughout the day is a continuous process in which the amounts of salt and water in the body shift, according to whether you just drank a soft drink or sweated through an aerobics class.

In that respect, maintaining fluid balance is like driving a car down the highway: small adjustments keep the car in the center of the lane. However, just as exciting movies feature wild car chases, not sedate driving, the exciting part of fluid homeostasis is the body's response to crisis situations, such as severe dehydration or hemorrhage. In this section we examine challenges to salt and water balance.

Osmolarity and Volume Can Change Independently

Normally, volume and osmolarity are homeostatically maintained within an acceptable range. Under some circumstances, however, fluid loss exceeds fluid gain or vice versa, and the body goes out of balance. Common pathways for fluid loss include excessive sweating, vomiting, diarrhea, and hemorrhage. All of these situations may require medical intervention. In contrast, fluid gain is seldom a medical emergency, unless it is pure water that decreases osmolarity below an acceptable range.

Volume and osmolarity of the ECF can each have three possible states: normal, increased, or decreased. The relation of volume and osmolarity changes can be represented by the matrix in Figure 20-17 ■. The center box represents the normal

state, and the surrounding boxes represent the most common examples of the variations from normal.

In all cases, the appropriate homeostatic compensation for the change acts according to the principle of mass balance: whatever fluid and solute were added to the body must be removed, or whatever was lost must be replaced. However, perfect compensation is not always possible. Let's begin at the upper right corner of Figure 20-17 and move right to left across each row.

1. *Increased volume, increased osmolarity.* A state of increased volume and increased osmolarity might occur if you ate salty food and drank liquids at the same time, such as popcorn and a soft drink at the movies. The net result could be ingestion of hypertonic saline that increases ECF volume and osmolarity. The appropriate homeostatic response is excretion of hypertonic urine. For homeostasis to be maintained, the osmolarity and volume of the urinary output must match the salt and water input from the popcorn and soft drink.

2. *Increased volume, no change in osmolarity.* Moving one cell to the left in the top row, we see that if the proportion of salt and water in ingested food is equivalent to an isotonic NaCl solution, volume will increase but osmolarity will not change. The appropriate response is excretion of isotonic urine whose volume equals that of the ingested fluid.

3. *Increased volume, decreased osmolarity.* This situation would occur if you drank pure water without ingesting any solute. The goal here would be to excrete very dilute urine to maximize water loss while conserving salts. However, because our kidneys cannot excrete pure water, there will always be some loss of solute in the urine. In this situation, urinary output cannot exactly match input, and so compensation is imperfect.

4. *No change in volume, increased osmolarity.* This disturbance (middle row, right cell) might occur if you ate salted popcorn without drinking anything. The ingestion of salt without water increases ECF osmolarity and causes some water to shift from cells to the ECF. The homeostatic response is intense thirst, which prompts ingestion of water to dilute the added solute. The kidneys help by creating highly concentrated urine of minimal volume, conserving water while removing excess NaCl. Once water is ingested, the disturbance becomes that described in situation 1 or situation 2.

5. *No change in volume, decreased osmolarity.* This scenario (middle row, left cell) might occur when a person who is dehydrated replaces lost fluid with pure water, like the golfer described earlier. The decreased volume resulting from the dehydration is corrected, but the replacement fluid has no solutes to replace those lost. Consequently, a new imbalance is created.

This situation led to the development of electrolyte-containing sports beverages. If people working out in hot weather replace lost sweat with pure water, they may restore volume but run the risk of diluting plasma K^+ and Na^+ concentrations to dangerously low levels (*hypokalemia* and *hyponatremia*, respectively).

20

TABLE 20-1   Responses Triggered by Changes in Volume, Blood Pressure, and Osmolarity

STIMULUS	ORGAN OR TISSUE INVOLVED	RESPONSE(S)	FIGURE(S)
Decreased blood pressure			
Direct effects			
	Granular cells	Renin secretion	20-14
	Glomerulus	Decreased GFR	19-7, 20-15
Reflexes			
Carotid and aortic baroreceptors	Cardiovascular control center	Increased sympathetic output, decreased parasympathetic output	15-23, 20-15
Carotid and aortic baroreceptors	Hypothalamus	Thirst stimulation	20-1a
Carotid and aortic baroreceptors	Hypothalamus	Vasopressin secretion	20-7
Atrial baroreceptors	Hypothalamus	Thirst stimulation	20-1a
Atrial baroreceptors	Hypothalamus	Vasopressin secretion	20-7
Increased blood pressure			
Direct effects			
	Glomerulus	Increased GFR (transient)	19-8
	Myocardial cells	Natriuretic peptide secretion	20-16
Reflexes			
Carotid and aortic baroreceptors	Cardiovascular control center	Decreased sympathetic output, increased parasympathetic output	15-22
Carotid and aortic baroreceptors	Hypothalamus	Thirst inhibition	—
Carotid and aortic baroreceptors	Hypothalamus	Vasopressin inhibition	—
Atrial baroreceptors	Hypothalamus	Thirst inhibition	—
Atrial baroreceptors	Hypothalamus	Vasopressin inhibition	—
Increased osmolarity			
Direct effects			
Pathological dehydration	Adrenal cortex	Decreased aldosterone secretion	20-18
Reflexes			
Osmoreceptors	Hypothalamus	Thirst stimulation	20-12
Osmoreceptors	Hypothalamus	Vasopressin secretion	20-7
Decreased osmolarity			
Direct effects			
Pathological hyponatremia	Adrenal cortex	Increased aldosterone secretion	—
Reflexes			
Osmoreceptors	Hypothalamus	Decreased vasopressin secretion	—

RUNNING PROBLEM

During exercise in the heat, sweating rate and sweat composition are quite variable among athletes. Fluid losses can range from less than 0.6 L/h to more than 2.5 L/h, and sweat Na^+ concentrations can range from less than 20 mEq/L to more than 90 mEq/L. The white salt crystals noted on Lauren's face and clothing suggest that she is a "salty sweater" who probably lost a large amount of salt during the race. Follow-up testing revealed that Lauren's sweat Na^+ concentration was 70 mEq/L.

Question 6:
Assuming a sweating rate of 1.0 L/h and a sweat Na^+ level of 70 mEq/L, how much Na^+ did Lauren lose during the 16-hour race?

Question 7:
Total body water for a 60-kg female is approximately 30 L, and her ECF volume is 10 L. Based on the information given in the problem so far, what volume of fluid did Lauren ingest during the race?

642   645   658   **661**   665   671

6. *Decreased volume, increased osmolarity.* Dehydration is a common cause of this disturbance (bottom row, right cell). Dehydration has multiple causes. During prolonged heavy exercise, water loss from the lungs can double while sweat loss may increase from 0.1 liter to as much as 5 liters! Because the fluid secreted by sweat glands is hyposmotic, the fluid left behind in the body becomes hyperosmotic.

    Diarrhea [*diarhein,* to flow through], excessively watery feces, is a pathological condition involving major water and solute loss, this time from the digestive tract. In both sweating and diarrhea, if too much fluid is lost from the circulatory system, blood volume decreases to the point that the heart can no longer pump blood effectively to the brain. In addition, cell shrinkage caused by increased osmolarity disrupts cell function.

7. *Decreased volume, no change in osmolarity.* This situation (bottom row, middle cell) occurs with hemorrhage. Blood loss represents the loss of isosmotic fluid from the extracellular compartment, similar to scooping a cup of seawater out of a large bucketful. If a blood transfusion is not immediately available, the best replacement solution is one that is isosmotic and remains in the ECF, such as isotonic NaCl.

8. *Decreased volume, decreased osmolarity.* This situation (bottom row, left cell) might also result from incomplete compensation of dehydration, but it is uncommon, so we will ignore it.

Dehydration Triggers Renal and Cardiovascular Responses

To understand the body's integrated response to changes in volume and osmolarity, you must first have a clear idea of which pathways become active in response to various stimuli. Table 20-1 ■ is a summary of the many pathways involved in the homeostasis of salt and water balance. For details of individual pathways, refer to the figures cited in Table 20-1.

The homeostatic response to severe dehydration is an excellent example of how the body works to maintain blood volume and cell volume in the face of decreased volume and increased osmolarity. It also illustrates the role of neural and endocrine integrating centers. In dehydration, the adrenal cortex receives two opposing signals. One says, "Secrete aldosterone"; the other says, "Do not secrete aldosterone." The body has multiple mechanisms for dealing with diminished blood volume, but high ECF osmolarity causes cells to shrink and presents a more immediate threat to well-being. Thus, faced with decreased volume and increased osmolarity, the adrenal cortex will not secrete aldosterone. (If secreted, aldosterone would cause Na^+ reabsorption, which could worsen the already-high osmolarity associated with dehydration.)

In severe dehydration, compensatory mechanisms are aimed at restoring normal blood pressure, ECF volume, and osmolarity by (1) conserving fluid to prevent additional loss, (2) triggering cardiovascular reflexes to increase blood pressure, and (3) stimulating thirst so that normal fluid volume and osmolarity can be restored. Figure 20-18 ■ maps the interwoven nature of these responses. This figure is complex and intimidating at first glance, so we will discuss it step by step.

At the top of the map (in yellow) are the two stimuli caused by dehydration: decreased blood volume/pressure, and increased osmolarity. Decreased ECF volume causes decreased blood pressure. Blood pressure acts both directly and as a stimulus for several reflex pathways that are mediated through the carotid and aortic baroreceptors, atrial volume receptors, and the pressure-sensitive granular cells.

1. *The carotid and aortic baroreceptors signal the cardiovascular control center (CVCC) to raise blood pressure.* Sympathetic output from the CVCC increases while parasympathetic output decreases.

    (a) Heart rate goes up as control of the SA node shifts from predominantly parasympathetic to sympathetic.

    (b) The force of ventricular contraction also increases under sympathetic stimulation. The increased force of contraction combines with increased heart rate to increase cardiac output.

    (c) Simultaneously, sympathetic input causes arteriolar vasoconstriction, increasing peripheral resistance.

    (d) Sympathetic vasoconstriction of afferent arterioles in the kidneys decreases GFR, helping conserve fluid.

    (e) Increased sympathetic activity at the granular cells of the kidneys increases renin secretion.

20

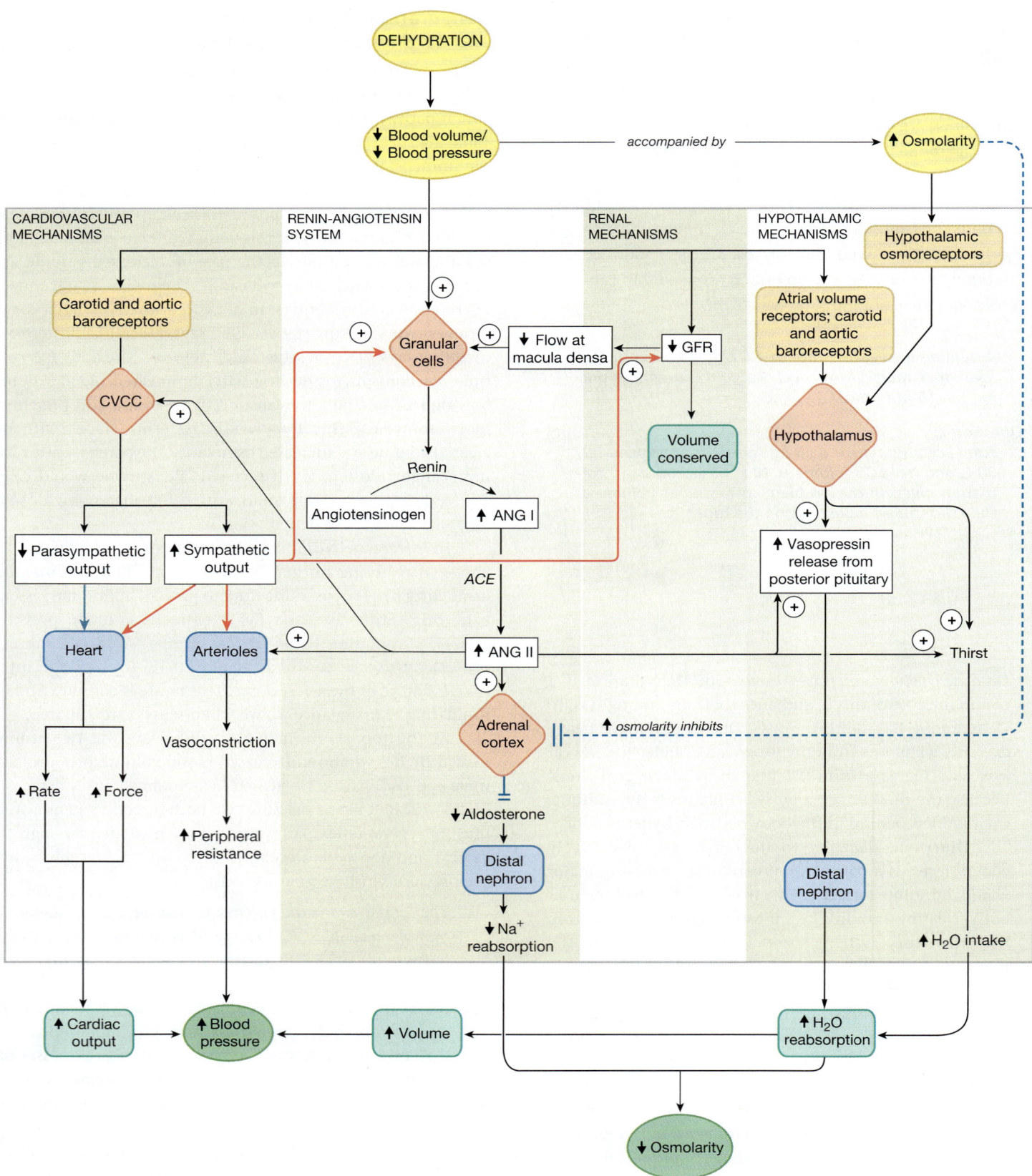

■ **FIGURE 20-18**   *Homeostatic compensation for severe dehydration*

20

2. *Decreased peripheral blood pressure directly decreases GFR.* A lower GFR conserves ECF volume by filtering less fluid into the nephron.

3. *Paracrine feedback causes the granular cells to release renin.* Decreased fluid flow past the macula densa as a result of lower GFR triggers renin release.

4. *Granular cells respond to decreased blood pressure by releasing renin.* The combination of decreased blood pressure, increased sympathetic input onto granular cells, and signals from the macula densa stimulates renin release and ensures increased production of angiotensin II.

5. *Decreased blood pressure, decreased blood volume, increased osmolarity, and increased ANG II production all stimulate vasopressin and the thirst centers of the hypothalamus.*

The redundancy in the control pathways ensures that all four main compensatory mechanisms are activated: cardiovascular responses, ANG II, vasopressin, and thirst.

1. *Cardiovascular responses* combine increased cardiac output and increased peripheral resistance to raise blood pressure. Note, however, that this increase in blood pressure does *not necessarily* mean that blood pressure returns to normal. If dehydration is severe, compensation may be incomplete, and blood pressure may remain below normal.

2. *Angiotensin II* has a variety of effects aimed at raising blood pressure, including stimulation of thirst, vasopressin release, direct vasoconstriction, and reinforcement of cardiovascular control center output. ANG II also reaches the adrenal cortex and attempts to stimulate aldosterone release. In dehydration, however, Na^+ reabsorption worsens the already-high osmolarity. Consequently, high osmolarity at the adrenal cortex directly inhibits aldosterone release, blocking the action of ANG II. The RAAS pathway in dehydration produces the beneficial blood-pressure-enhancing effects of ANG II while avoiding the detrimental effects of Na^+ reabsorption. This is a beautiful example of integrated function.

3. *Vasopressin* increases the water permeability of the renal collecting ducts, allowing water reabsorption to conserve fluid. Without fluid replacement, however, vasopressin cannot bring volume and osmolarity back to normal.

4. *Oral (or intravenous) intake of water* in response to thirst is the only mechanism for replacing lost fluid volume and for restoring ECF osmolarity to normal.

The net result of all four mechanisms is (1) restoration of volume by water conservation and fluid intake, (2) maintenance of blood pressure through increased blood volume, increased cardiac output, and vasoconstriction, and (3) restoration of normal osmolarity by decreased Na^+ reabsorption and increased water reabsorption and intake.

Using the pathways listed in Table 20-1, try to create reflex maps for the seven other disturbances of volume and osmolarity shown in Figure 20-17.

ACID-BASE BALANCE

Acid-base balance (also called pH homeostasis) is one of the essential functions of the body. The pH of a solution is a measure of its H^+ concentration [p. 37]. A normal arterial plasma sample has a H^+ concentration of 0.00004 meq/L, minute compared with the concentrations of other ions. (For example, the plasma concentration of Na^+ is about 135 meq/L.)

Because the body's H^+ concentration is so low, it is commonly expressed on a logarithmic pH scale of 0–14, in which a pH of 7.0 is neutral (neither acidic nor basic). If the pH of a solution is below 7.0, the solution has a H^+ concentration greater than 1×10^{-7} M and is considered acidic. If the pH is above 7.0, the solution has a H^+ concentration lower than 1×10^{-7} M and is considered alkaline (basic).

The normal pH of the body is 7.40, slightly alkaline. A change of 1 pH unit represents a 10-fold change in H^+ concentration. To review the concept of pH and the logarithmic scale on which it is based, see Appendix A.

Enzymes and the Nervous System Are Particularly Sensitive to Changes in pH

The normal pH range of plasma is 7.38–7.42. Extracellular pH usually reflects intracellular pH, and vice versa. Because monitoring intracellular conditions is difficult, plasma values are used clinically as an indicator of ECF and whole body pH. Body fluids that are "outside" the body's internal environment, such as those in the lumen of the gastrointestinal tract or kidney tubule, can have a pH that far exceeds the normal range. Acidic secretions in the stomach, for instance, may create a gastric pH as low as 1, and the pH of urine varies between 4.5 and 8.5, depending on the body's need to excrete H^+ or HCO_3^-.

The concentration of H^+ in the body is closely regulated. Intracellular proteins, such as enzymes and membrane channels, are particularly sensitive to pH because the function of these proteins depends on their three-dimensional shape [p. 31]. Changes in H^+ concentration alter the tertiary structure of proteins by interacting with hydrogen bonds in the molecules, disrupting the proteins' three-dimensional structures and activities.

Abnormal pH may significantly affect the activity of the nervous system. If pH is too low—the condition known as **acidosis**—neurons become less excitable and CNS depression results. Patients become confused and disoriented, then slip into a coma. If CNS depression progresses, the respiratory centers cease to function, causing death.

If pH is too high—the condition known as **alkalosis**—neurons become hyperexcitable, firing action potentials at the slightest signal. This condition shows up first as sensory changes, such as numbness or tingling, then as muscle twitches. If alkalosis is severe, muscle twitches turn into sustained contractions (*tetanus*) that paralyze respiratory muscles.

Disturbances of acid-base balance are associated with disturbances in K^+ balance. This is partly due to a renal transporter

20

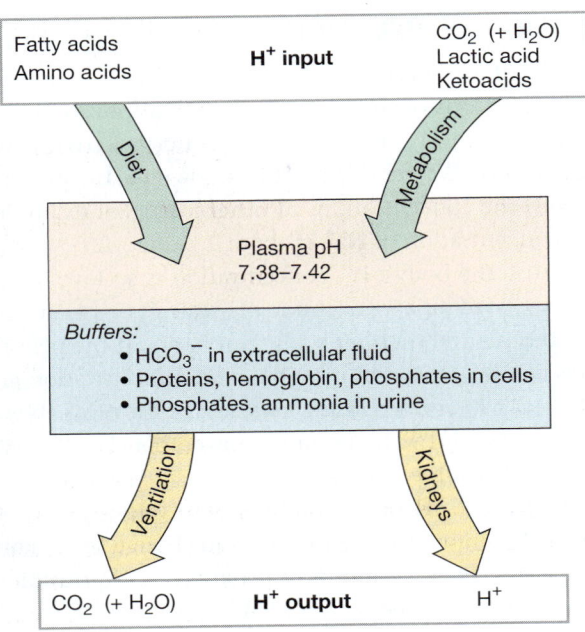

■ FIGURE 20-19   *Hydrogen balance in the body*

that moves K^+ and H^+ ions in an antiport fashion. In acidosis, the kidneys excrete H^+ and reabsorb K^+ using an H^+-K^+-ATPase. In alkalosis, the kidneys reabsorb H^+ and excrete K^+. Potassium imbalance usually shows up as disturbances in excitable tissues, especially the heart.

Acids and Bases in the Body Come from Many Sources

In day-to-day functioning, the body is challenged by intake and production of acids more than bases. Hydrogen ions come both from food and from internal metabolism. Maintaining mass balance requires that acid intake and production be balanced by acid excretion. Hydrogen balance in the body is summarized in Figure 20-19 ■.

Acid Input   Many metabolic intermediates and foods are organic acids that ionize and contribute H^+ to body fluids.* Examples of organic acids include amino acids, fatty acids, intermediates in the citric acid cycle, and lactic acid produced by anaerobic metabolism. Metabolic production of organic acids each day creates a significant amount of H^+ that must be excreted to maintain mass balance.

Under extraordinary circumstances, metabolic organic acid production can increase significantly and create a crisis. For example, severe anaerobic conditions, such as circulatory collapse, produce so much lactic acid that normal homeostatic mechanisms cannot keep pace, resulting in a state of *lactic*

---

*The anion forms of many organic acids end with the suffix –*ate*, such as pyruvate and lactate.

acidosis. In diabetes mellitus, abnormal metabolism of fats and amino acids creates strong acids known as **ketoacids**. These acids cause a state of metabolic acidosis known as *ketoacidosis*.

The biggest source of acid on a daily basis is the production of CO_2 during aerobic respiration. Carbon dioxide is not an acid because it does not contain any hydrogen atoms. However, CO_2 from respiration combines with water to form carbonic acid (H_2CO_3), which dissociates into H^+ and HCO_3^-:

$$CO_2 + H_2O \rightleftharpoons H_2CO_3 \rightleftharpoons H^+ + HCO_3^-$$

This reaction takes place in all cells and in the plasma, but at a slow rate. However, in certain cells of the body, the reaction proceeds very rapidly because of the presence of large amounts of *carbonic anhydrase* [🔁 p. 599]. This enzyme catalyzes the conversion of CO_2 and H_2O to H_2CO_3.

The production of H^+ from CO_2 and H_2O is the single biggest source of acid input under normal conditions. By some estimates, CO_2 from resting metabolism produces 12,500 meq of H^+ each day. If this amount of acid were placed in a volume of water equal to the plasma volume, it would create a H^+ concentration of 4167 meq/L, over one hundred million (10^8) times as concentrated as the normal plasma H^+ concentration of 0.00004 meq/L!

These numbers show that CO_2 from aerobic respiration has the potential to affect pH in the body dramatically. Fortunately, homeostatic mechanisms normally prevent CO_2 from accumulating in the body.

Base Input   Acid-base physiology focuses on acids for good reasons. First, our diet and metabolism have few significant sources of bases. Some fruits and vegetables contain anions that metabolize to HCO_3^-, but the influence of these foods is far outweighed by the contribution of acidic fruits, amino acids, and fatty acids. Second, in acid-base disturbances, those due to excess acid are more common than those due to excess base. For these reasons, the body expends for more resources removing excess acid.

pH Homeostasis Depends on Buffers, the Lungs, and the Kidneys

How does the body cope with minute-to-minute changes in pH? There are three mechanisms: (1) buffers, (2) ventilation, and (3) renal regulation of H^+ and HCO_3^-. Buffers are the first line of defense, always present and waiting to prevent wide swings in pH. Ventilation, the second line of defense, is a rapid, reflexively controlled response that can take care of 75% of most pH disturbances. The final line of defense lies with the kidneys. They are slower than buffers or the lungs but are very effective at coping with any remaining pH disturbance under normal conditions. These three mechanisms help the body balance acid so effectively that normal body pH varies only slightly. Let's take a closer look at each of them.

Buffer Systems Include Proteins, Phosphate Ions, and HCO_3^-

A buffer is a molecule that moderates but does not prevent changes in pH by combining with or releasing H^+ [p. 38]. In the absence of buffers, the addition of acid to a solution causes a sharp change in pH. In the presence of a buffer, the pH change is moderated or may even be unnoticeable. Because acid production is the major challenge to pH homeostasis, most physiological buffers combine with H^+.

Buffers are found both within cells and in the plasma. Intracellular buffers include cellular proteins, phosphate ions (HPO_4^{2-}), and hemoglobin. As we have seen, hemoglobin in red blood cells buffers the H^+ produced by the reaction of CO_2 with H_2O (Fig. 18-14, p. 600).

Each H^+ ion buffered by hemoglobin leaves a matching bicarbonate ion inside the red blood cell. The HCO_3^- then leaves the red blood cell by exchanging with a plasma Cl^-. This exchange is the *chloride shift* described in Chapter 18 [p. 601].

The large amounts of bicarbonate produced from metabolic CO_2 create the most important extracellular buffer system of the body. Plasma HCO_3^- concentration averages 24 meq/L, which is approximately 600,000 times as concentrated as plasma H^+. Although H^+ and HCO_3^- are created in a 1:1 ratio from CO_2 and H_2O, intracellular buffering of H^+ by hemoglobin is a major reason the two ions do not appear in the plasma in the same concentration. The HCO_3^- in plasma can then buffer H^+ from nonrespiratory sources, such as metabolism.

The relationship between CO_2, HCO_3^-, and H^+ in the plasma is expressed by the equation we just looked at:

$$CO_2 + H_2O \rightleftharpoons \underset{\text{carbonic acid}}{H_2CO_3} \rightleftharpoons H^+ + HCO_3^- \quad (1)$$

According to the law of mass action, any change in the amount of CO_2, H^+, or HCO_3^- in the reaction solution will cause the reaction to shift until a new equilibrium is reached. (Water is always in excess in the body and does not contribute to the reaction equilibrium.) For example, if CO_2 increases, the equation shifts to the right, creating one additional H^+ and one additional HCO_3^- from each CO_2 and water:

$$\uparrow CO_2 + H_2O \rightarrow H_2CO_3 \rightarrow \uparrow H^+ + \uparrow HCO_3^- \quad (2)$$

Once a new equilibrium is reached, both H^+ and HCO_3^- levels have increased. The addition of H^+ makes the solution more acidic and therefore lowers its pH. In this reaction, it does not matter that a HCO_3^- buffer molecule has also been produced because HCO_3^- acts as a buffer only when it binds to H^+ and becomes carbonic acid.

Now suppose H^+ is added to the plasma from some metabolic source, such as lactic acid:

$$CO_2 + H_2O \rightleftharpoons H_2CO_3 \rightleftharpoons \uparrow H^+ + HCO_3 \quad (3)$$

In this case, plasma HCO_3^- *can* act as a buffer by combining with some of the added H^+ until the reaction reaches a new equilibrium state. The increase in H^+ shifts the equation to the left:

$$CO_2 + H_2O \leftarrow H_2CO_3 \leftarrow \uparrow H^+ + HCO_3^- \quad (4)$$

Converting some of the added H^+ and bicarbonate buffer to carbonic acid means that at equilibrium, H^+ is still elevated, but not as much as it was initially. The concentration of HCO_3^- is decreased because some has been used as a buffer. The buffered H^+ is converted to CO_2 and H_2O, increasing the amounts of both. At equilibrium, the reaction looks like this:

$$\uparrow CO_2 + \uparrow H_2O \rightleftharpoons H_2CO_3 \rightleftharpoons \uparrow H^+ + \downarrow HCO_3^- \quad (5)$$

The law of mass action is a useful way to think about the relationship between changes in the concentrations of H^+, HCO_3^-, and CO_2, as long as you remember certain qualifications. First, a change in HCO_3^- concentration (as indicated in reaction 5) may not show up clinically as a HCO_3^- concentration outside the normal range. This is because HCO_3^- is 600,000 times as concentrated in the plasma as H^+ is. If both H^+ and HCO_3^- are added to the plasma, you may observe changes in pH (but not in HCO_3^- concentration), because so much HCO_3^- was present initially. Both H^+ and HCO_3^- experience an *absolute* increase in concentration, but because so many HCO_3^- were in the plasma to begin with, the *relative increase* in HCO_3^- goes unnoticed.

As an analogy, think of two football teams playing in a stadium packed with 80,000 fans. If 10 more players (H^+) run out onto the field, everyone notices. But if 10 people (HCO_3^-) come into the stands at the same time, no one pays any attention because there were already so many people watching the game that 10 more make no significant difference.

The second qualification for the law of mass action is that when the reaction shifts to the left and increases plasma CO_2, a nearly instantaneous increase in ventilation takes place (in a normal person). If extra CO_2 is ventilated off, arterial P_{CO_2} may remain normal or even fall below normal as a result of hyperventilation.

RUNNING PROBLEM

The human body attempts to maintain fluid and sodium balance via several hormonal mechanisms. During exercise sessions, increased sympathetic output causes increased production of aldosterone and vasopressin, which promote the retention of Na^+ and water by the kidneys.

Question 8:
What would you expect to happen to vasopressin and aldosterone production in response to dilutional hyponatremia?

| 642 | 645 | 658 | 661 | 665 | 671 |

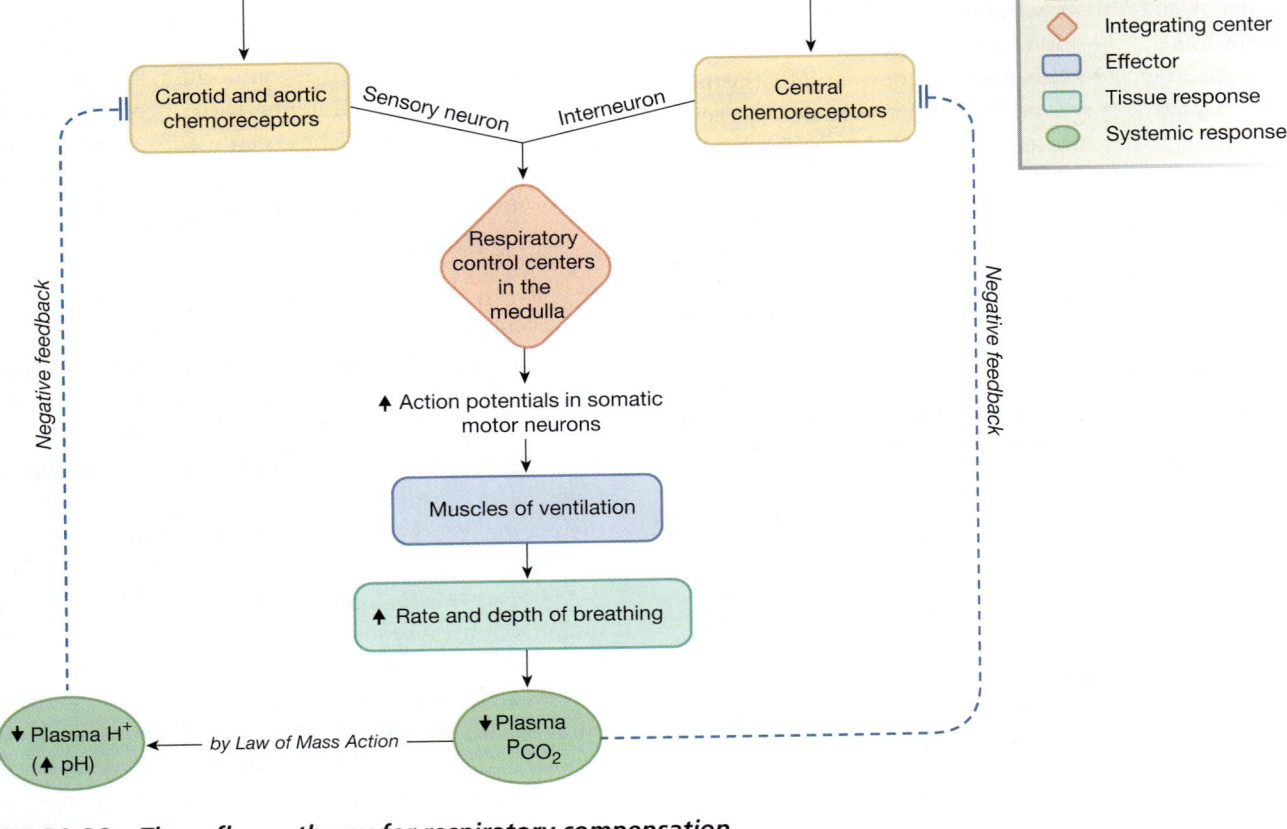

KEY

- ◯ Stimulus
- ▢ Receptor
- ◇ Integrating center
- ▢ Effector
- ◯ Tissue response
- ◯ Systemic response

■ **FIGURE 20-20**   *The reflex pathway for respiratory compensation of metabolic acidosis*

Ventilation Can Compensate for pH Disturbances

The increase in ventilation just described is a *respiratory compensation* for acidosis. Ventilation and acid-base status are intimately linked, as shown by the equation

$$CO_2 + H_2O \rightleftharpoons H_2CO_3 \rightleftharpoons H^+ + HCO_3^-$$

Changes in ventilation can correct disturbances in acid-base balance, but they can also cause them. Because of the dynamic equilibrium between CO_2 and H^+, any change in plasma P_{CO_2} will affect both H^+ and HCO_3^- content of the blood. For example, if a person hypoventilates and P_{CO_2} increases, the equation shifts to the right. More carbonic acid is formed, and H^+ goes up, creating a more acidotic state:

$$\uparrow CO_2 + H_2O \rightarrow H_2CO_3 \rightarrow \uparrow H^+ + \uparrow HCO_3^- \qquad (6)$$

If a person hyperventilates, blowing off CO_2 and thereby decreasing the plasma P_{CO_2}, the equation shifts to the left, which means that H^+ combines with HCO_3^- and becomes carbonic acid, thereby decreasing the H^+ concentration and so raising the pH:

$$\downarrow CO_2 + H_2O \leftarrow H_2CO_3 \leftarrow \downarrow H^+ + \downarrow HCO_3^- \qquad (7)$$

In these two examples, you can see that a change in P_{CO_2} will affect the H^+ concentration and therefore the pH of the plasma. The body uses ventilation as a method for adjusting pH only if a stimulus associated with pH triggers the reflex response. Two stimuli can do so: H^+ and CO_2.

Ventilation is affected directly by plasma H^+ levels through carotid and aortic chemoreceptors (Fig. 20-20 ■). These are located in the aorta and carotid arteries along with the oxygen sensors and blood pressure sensors we discussed previously [⮌ p. 604]. An increase in plasma H^+ stimulates the chemoreceptors, which in turn signal the medullary respiratory control centers to increase ventilation. Increased alveolar ventilation allows the lungs to excrete more CO_2 and convert H^+ to carbonic acid.

The central chemoreceptors of the medulla oblongata cannot respond directly to changes in plasma pH because H^+ does not cross the blood-brain barrier. However, changes in pH change P_{CO_2}, and CO_2 stimulates the central chemoreceptors [⮌ Fig. 18-19, p. 605]. Dual control of ventilation through the central and peripheral chemoreceptors helps the body respond rapidly to changes in either pH or plasma CO_2.

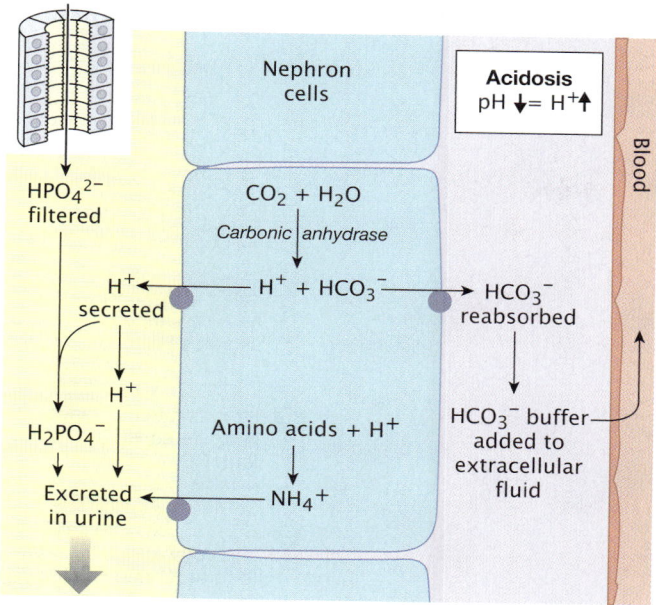

■ FIGURE 20-21 *Overview of renal compensation for acidosis*

The transporters represented in this figure are depicted in greater detail in Figures 20-22 and 20-23.

CONCEPT CHECK

13. Name the muscles of ventilation that might be involved in the reflex represented in Figure 20-20.

14. In equation 6, the amount of HCO_3^- present is increased at equilibrium. Why doesn't this HCO_3^- act as a buffer and prevent acidosis from occurring?

Answers: p. 675

Kidneys Use Ammonia and Phosphate Buffers

The kidneys take care of the 25% of compensation that the lungs cannot handle. They alter pH two ways: (1) directly, by excreting or reabsorbing H^+ and (2) indirectly, by changing the rate at which HCO_3^- buffer is reabsorbed or excreted.

In acidosis, the kidney secretes H^+ into the tubule lumen using direct and indirect active transport (Fig. 20-21 ■). Ammonia from amino acids and phosphate ions (HPO_4^{2-}) in the kidney act as buffers, trapping large amounts of H^+ as NH_4^+ and $H_2PO_4^-$ and allowing more H^+ to be excreted. Phosphate ions are present in filtrate and combine with H^+ secreted into the nephron lumen:

$$HPO_4^{2-} + H^+ \rightleftharpoons H_2PO_4^-$$

Ammonia is made from amino acids, as described in the next section.

Even with these buffers, urine can become quite acidic, down to a pH of about 4.5. While H^+ is being excreted, the kidneys make new HCO_3^- from CO_2 and H_2O. The HCO_3^- is reabsorbed into the blood to act as a buffer and increase pH.

In alkalosis, the kidney reverses the general process just described for acidosis, excreting HCO_3^- and reabsorbing H^+ in an effort to bring pH back into the normal range. Renal compensations are slower than respiratory compensations, and their effect on pH may not be noticed for 24–48 hours. However, once activated, renal compensations effectively handle all but severe acid-base disturbances.

The cellular mechanisms for renal handling of H^+ and HCO_3^- resemble transport processes in other epithelia. However, these mechanisms involve some membrane transporters that you have not encountered before:

1. The **apical Na^+-H^+ antiport** is an indirect active transporter that brings Na^+ into the epithelial cell in exchange for moving H^+ against its concentration gradient into the lumen. This is the same transporter that is active in proximal tubule Na^+ reabsorption [⮂ p. 627].

2. The **basolateral Na^+-HCO_3^- symport** moves Na^+ and HCO_3^- out of the epithelial cell and into the interstitial fluid. This indirect active transporter couples the energy of HCO_3^- diffusing down its concentration gradient to the uphill movement of Na^+ from the cell to the ECF.

3. The **H^+-ATPase** uses energy from ATP to acidify the urine, pushing H^+ against its concentration gradient into the lumen of the distal nephron.

4. The **H^+-K^+-ATPase** puts H^+ into the urine in exchange for reabsorbed K^+. This exchange contributes to the potassium imbalance that sometimes accompanies acid-base disturbances.

5. A **Na^+-NH_4^+ antiport** moves NH_4^+ from the cell to the lumen in exchange for Na^+.

In addition to these transporters, the renal tubule also uses the ubiquitous Na^+-K^+-ATPase and the same HCO_3^--Cl^- antiport protein that is responsible for the chloride shift in red blood cells.

The Proximal Tubule Secretes H^+ and Reabsorbs HCO_3^-

The amount of bicarbonate ion the kidneys filter each day is equivalent to the bicarbonate in a pound of baking soda ($NaHCO_3$)! Most of this HCO_3^- must be reabsorbed to maintain the body's buffer capacity. The proximal tubule reabsorbs most filtered HCO_3^- by indirect methods because there is no apical membrane transporter to bring HCO_3^- into the tubule cell.

Figure 20-22 ■ shows the two pathways by which bicarbonate is reabsorbed in the proximal tubule. (The numbers in the following lists correspond to the steps shown in the figure.) By following this illustration, you will see how the transporters listed in the previous section function together.

The first pathway converts filtered HCO_3^- into CO_2, then back into HCO_3^-:

1. H^+ is secreted from the proximal tubule cell into the lumen in exchange for filtered Na^+, which moves from the lumen into the tubule cell.

20

■ **FIGURE 20-22**   *Proximal tubule H⁺ secretion and the reabsorption of filtered HCO₃⁻*

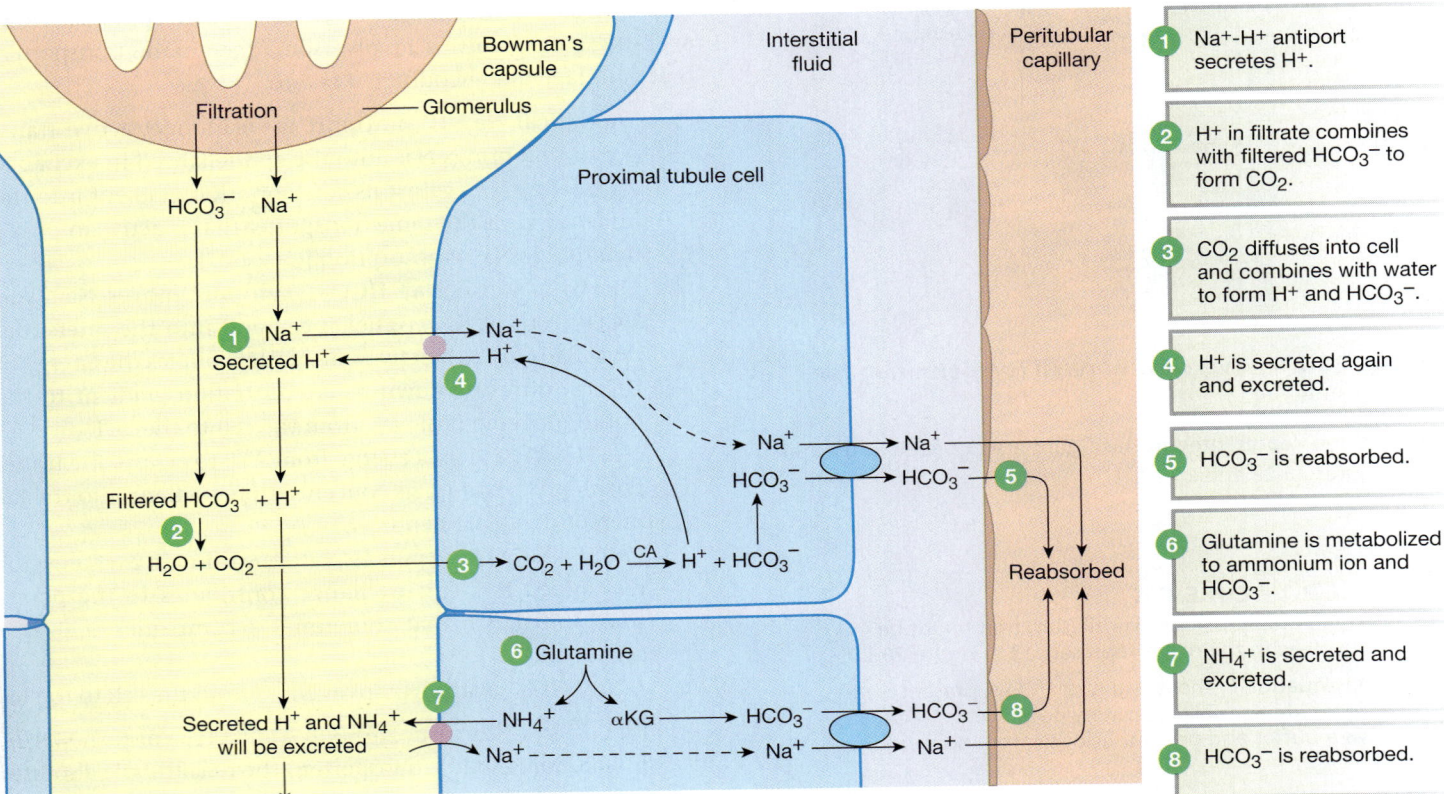

2. The secreted H^+ combines with filtered HCO_3^- to form CO_2 in the lumen.
3. This CO_2 diffuses into the tubule cell and combines with water to form H_2CO_3, which dissociates to H^+ and HCO_3^- in the cytoplasm.
4. The H^+ created in step 3 is secreted, replacing the luminal H^+ that combined with filtered HCO_3^- in step 2.
5. The HCO_3^- created in step 3 is transported out of the cell on the basolateral side by the HCO_3^--Na^+ symporter.

The net result of this process is reabsorption of filtered Na^+ and HCO_3^- and secretion of H^+.

A second way to reabsorb bicarbonate and excrete H^+ comes from metabolism of the amino acid glutamine:

6. Glutamine in the proximal tubule cell loses its two amino groups, which become ammonia (NH_3). The ammonia buffers H^+ to become ammonium ion.
7. The ammonium ion is transported into the lumen in exchange for Na^+.

8. The α-ketoglutarate molecule made from deamination of glutamine is metabolized further to HCO_3^-, which is transported into the blood along with Na^+.

The net result of these pathways is reabsorption of sodium bicarbonate—baking soda, $NaHCO_3$.

The Distal Nephron Controls Acid Excretion

The distal nephron plays a significant role in the fine regulation of acid-base balance. The **intercalated cells** (or **I cells**) interspersed among the principal cells are responsible for acid-base regulation.

Intercalated cells are characterized by high concentrations of carbonic anhydrase in their cytoplasm. This enzyme allows them to rapidly convert CO_2 and water into H^+ and HCO_3^-. The H^+ ions are pumped out of the intercalated cell either by the H^+-ATPase or by an ATPase that exchanges one H^+ for one K^+. Bicarbonate leaves the cell by means of the HCO_3^--Cl^- antiport exchanger.

■ FIGURE 20-23 *Role of intercalated cells in acidosis and alkalosis*

Intercalated cells in the collecting duct secrete or reabsorb H^+ and HCO_3^- according to the needs of the body.

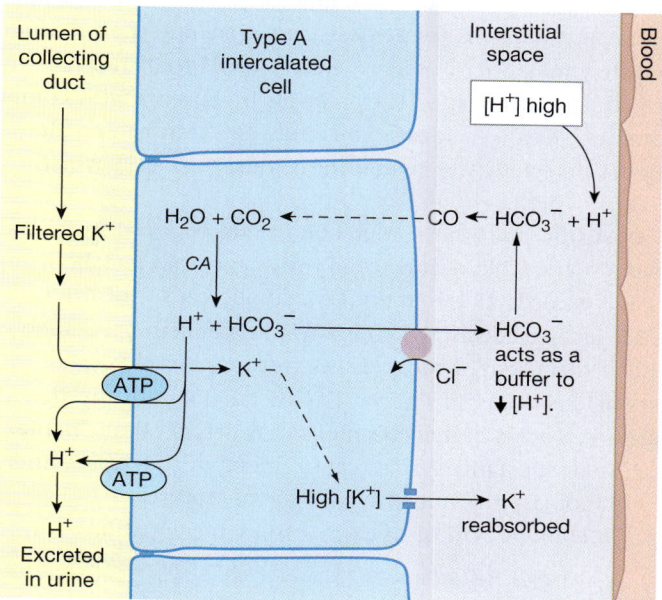

(a) Type A intercalated cells function in acidosis.
H^+ is excreted; HCO_3^- and K^+ are reabsorbed.

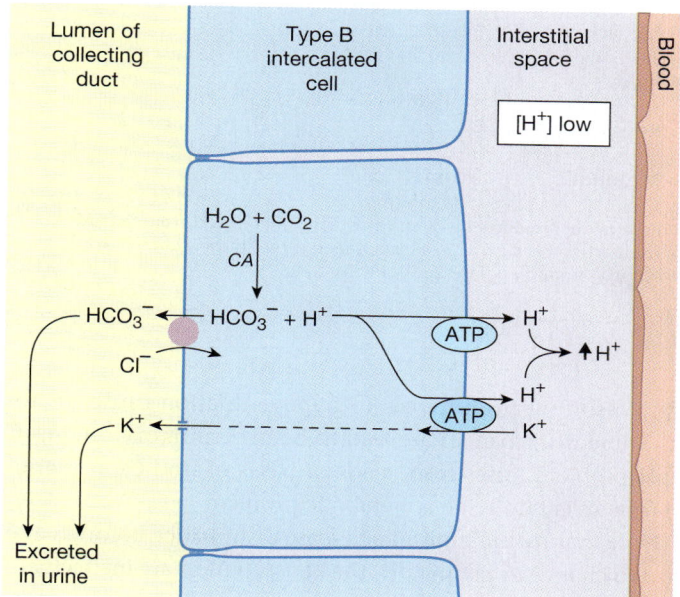

(b) Type B intercalated cells function in alkalosis.
HCO_3^- and K^+ are excreted; H^+ is reabsorbed.

There are two types of intercalated cells, and their transporters are found on different faces of the epithelial cell. During periods of acidosis, type A intercalated cells secrete H^+ and reabsorb bicarbonate. During periods of alkalosis, type B intercalated cells secrete HCO_3^- and reabsorb H^+.

Fig. 20-23a ■ shows how type A intercalated cells work during acidosis, secreting H^+ and reabsorbing HCO_3^-. The process is similar to H^+ secretion in the proximal tubule except for the specific H^+ transporters. The distal nephron uses H^+-ATPase and H^+-K^+-ATPase rather than the Na^+-H^+ antiport protein found in the proximal tubule.

During alkalosis, when the H^+ concentration of the body is too low, H^+ is reabsorbed and HCO_3^- buffer is excreted in the urine (Fig. 20-23b). Once again, the ions are produced by the dissociation of H_2CO_3 formed from H_2O and CO_2. Hydrogen ions are reabsorbed into the ECF on the basolateral side of the nephron cell while HCO_3^- is secreted into the lumen. The polarity of the two types of I cells is reversed, with the transport proteins found on the opposite sides of the cell.

The H^+-K^+-ATPase of the distal nephron helps create parallel disturbances of acid-base balance and K^+ balance. In acidosis, when plasma H^+ is high, the kidney secretes H^+ and reabsorbs K^+. Thus, acidosis is often accompanied by hyperkalemia. (Certain nonrenal events also contribute to elevated ECF K^+ concen-

trations in acidosis.) The reverse is true for alkalosis, when blood H^+ levels are low. The mechanism that allows the distal nephron to reabsorb H^+ simultaneously causes it to secrete K^+, with the result that alkalosis goes hand in hand with hypokalemia.

✓ **CONCEPT CHECK**

15. Why does the K^+-H^+ transporter in the distal nephron require ATP to secrete H^+ but the Na^+-H^+ exchanger in the proximal tubule does not?

16. In hypokalemia, the intercalated cells of the distal nephron reabsorb K^+ from the tubule lumen. What happens to blood pH as a result?
    Answers: p. 675

Acid-Base Disturbances May Be Respiratory or Metabolic in Origin

The three compensatory mechanisms (buffers, ventilation, and renal excretion) take care of most variations in plasma pH. But under some circumstances, the production or loss of H^+ or HCO_3^- is so extreme that compensatory mechanisms fail to maintain pH homeostasis. In these states, the pH of the blood moves out of the normal range of 7.38–7.42. If the body fails to keep pH between 7.00 and 7.70, acidosis or alkalosis can be fatal.

Acid-base problems are classified both by the direction of the pH change (acidosis or alkalosis) and by the underlying cause (metabolic or respiratory). As you learned earlier, changes

20

TABLE 20-2 Plasma P_{CO_2}, Ions, and pH in Acid-Base Disturbances

DISTURBANCE	P_{CO_2}	H^+	pH	HCO_3^-
Acidosis				
Respiratory	↑	↑	↓	↑
Metabolic	Normal* or ↓	↑	↓	↓
Alkalosis				
Respiratory	↓	↓	↑	↓
Metabolic	Normal* or ↑	↓	↑	↓

*These values are different from what you would expect from the law of mass action because almost instantaneous respiratory compensation keeps P_{CO_2} from changing significantly.

in P_{CO_2} resulting from hyper- or hypoventilation cause pH to shift. These disturbances are said to be of respiratory origin. If the pH problem arises from acids or bases of non-CO_2 origin, the problem is said to be a metabolic problem.

Note that by the time an acid-base disturbance becomes evident as a change in plasma pH, the body's buffers are ineffectual. The loss of buffering ability leaves the body with only two options: respiratory compensation or renal compensation. And if the problem is of respiratory origin, only one homeostatic compensation is available—the kidneys. If the problem is of metabolic origin, both respiratory and renal mechanisms can compensate.

The combination of an initial pH disturbance and the resultant compensatory changes is one factor that makes analysis of acid-base disorders in the clinical setting so difficult. In this book we will concentrate on simple scenarios with a single underlying cause. Changes that occur in the four simple acid-base disturbances are listed in Table 20-2 ∎.

Respiratory Acidosis A state of respiratory acidosis occurs when alveolar hypoventilation results in CO_2 retention and elevated plasma P_{CO_2}. Some situations in which this occurs are respiratory depression due to drugs (including alcohol), increased airway resistance in asthma, impaired gas exchange in fibrosis or severe pneumonia, and muscle weakness in muscular dystrophy and other muscle diseases. The most common cause of respiratory acidosis is *chronic obstructive pulmonary disease* (COPD), such as emphysema, in which inadequate gas exchange is compounded by loss of alveolar exchange area.

No matter what the cause of respiratory acidosis, plasma CO_2 levels increase, leading to elevated H^+ and HCO_3^-:

$$\uparrow CO_2 + H_2O \rightarrow H_2CO_3 \rightarrow \uparrow H^+ + \uparrow HCO_3^-$$

The hallmark of respiratory acidosis is decreased pH and elevated bicarbonate levels (Table 20-2). Because the problem is of

respiratory origin, the body cannot carry out respiratory compensation. (However, depending on the problem, mechanical ventilation can be used to assist breathing.)

Any compensation for respiratory acidosis must occur through renal mechanisms that excrete H^+ and reabsorb HCO_3^-. The excretion of H^+ raises plasma pH. Reabsorption of HCO_3^- provides additional buffer that combines with H^+, lowering the H^+ concentration and therefore raising the pH.

In chronic obstructive pulmonary disease, renal compensation mechanisms for acidosis can moderate the pH change, but they may not be able to return the pH to its normal range. If you look at pH and HCO_3^- levels in patients with compensated respiratory acidosis, you find that both those values are higher than they were when the acidosis was at its worst.

Metabolic Acidosis Metabolic acidosis is a disturbance of mass balance that occurs when the dietary and metabolic input of H^+ exceeds H^+ excretion. Metabolic causes of acidosis include lactic acidosis, which is a result of anaerobic metabolism, and ketoacidosis, which results from excessive breakdown of fats or certain amino acids. The metabolic pathway that produces ketoacids in diabetes mellitus is described in Chapter 21. Ingested substances that cause metabolic acidosis include methanol, aspirin, and ethylene glycol (antifreeze).

Metabolic acidosis is expressed by the equation

$$\uparrow CO_2 + H_2O \leftarrow H_2CO_3 \leftarrow \uparrow H^+ + \downarrow HCO_3^- \tag{8}$$

Hydrogen ion concentration increases because of the H^+ contributed by the metabolic acids. This increase shifts the equilibrium represented in the equation to the left, increasing CO_2 levels and using up HCO_3^- buffer.

Metabolic acidosis can also occur if the body loses HCO_3^-. The most common cause of bicarbonate loss is diarrhea, during which HCO_3^- is lost from the intestines. The pancreas produces HCO_3^- from CO_2 and H_2O by a mechanism similar to the renal mechanism illustrated in Figure 20-21. The H^+ made at the same time is released into the blood. Normally, the HCO_3^- is released into the small intestine, then reabsorbed into the blood, buffering the H^+. However, if HCO_3^- is not reabsorbed because a person is experiencing diarrhea, a state of acidosis results.

Whether HCO_3^- concentration is elevated or decreased is an important criterion for distinguishing metabolic acidosis from respiratory acidosis (Table 20-2).

You would think from looking at equation 8 that metabolic acidosis would be accompanied by elevated P_{CO_2}. However, unless the individual also has a lung disease, respiratory compensation takes place almost instantaneously. Both elevated CO_2 and elevated H^+ stimulate ventilation through the pathways described earlier. As a result, P_{CO_2} decreases to normal or even below-normal levels via hyperventilation.

Uncompensated metabolic acidosis is rarely seen clinically. Indeed, a common sign of metabolic acidosis is hyperventilation, evidence of respiratory compensation occurring in response to the acidosis.

The renal compensations discussed for respiratory acidosis also take place in metabolic acidosis: secretion of H^+ and reabsorption of HCO_3^-. Renal compensations take several days to reach full effectiveness, and so they are not usually seen in acute disturbances.

Respiratory Alkalosis

States of alkalosis are much less common than acidotic conditions. Respiratory alkalosis occurs as a result of hyperventilation, when alveolar ventilation increases without a matching increase in metabolic CO_2 production. Consequently, plasma P_{CO_2} falls and alkalosis results when the equation shifts to the left:

$$\downarrow CO_2 + H_2O \leftarrow H_2CO_3 \leftarrow \downarrow H^+ + \downarrow HCO_3^-$$

The decrease in CO_2 shifts the equilibrium to the left, and consequently both plasma H^+ and plasma HCO_3^- decrease. Low plasma HCO_3^- levels in alkalosis indicate a respiratory disorder.

The primary clinical cause of respiratory alkalosis is excessive artificial ventilation. Fortunately, this condition is easily corrected by adjusting the ventilator. The most common physiological cause of respiratory alkalosis is hysterical hyperventilation caused by anxiety. When this is the cause, the neurological symptoms caused by alkalosis can be partially reversed by having the patient breathe into a paper bag. In doing so, the patient rebreathes exhaled CO_2, a process that raises arterial P_{CO_2} and corrects the problem.

Because this alkalosis has respiratory cause, the only compensation available to the body is renal. Filtered bicarbonate, which if reabsorbed could act as a buffer and increase pH even more, is not reabsorbed in the proximal tubule and is secreted in the distal nephron. The combination of HCO_3^- excretion and H^+ reabsorption in the distal nephron decreases the body's HCO_3^- and increases its H^+, both of which help correct the alkalosis.

Metabolic Alkalosis

Metabolic alkalosis has two common causes: excessive vomiting of acidic stomach contents and excessive ingestion of bicarbonate-containing antacids. In both cases, the resulting alkalosis reduces H^+ concentration:

$$\downarrow CO_2 + H_2O \rightarrow H_2CO_3 \rightarrow \downarrow H^+ + \uparrow HCO_3^-$$

The decrease in H^+ shifts the equilibrium to the right, meaning that carbon dioxide (P_{CO_2}) decreases and HCO_3^- goes up.

Just as in metabolic acidosis, respiratory compensation for metabolic alkalosis is rapid. The increase in pH and drop in P_{CO_2} depress ventilation. Less CO_2 is blown off, raising the P_{CO_2} and creating more H^+ and HCO_3^-. This respiratory compensation helps correct the pH problem but elevates HCO_3^- levels even more. However, the compensation is limited because hypoventilation causes hypoxia. Once the arterial P_{O_2} drops below 60 mm Hg, hypoventilation ceases.

The renal response to metabolic alkalosis is the same as that for respiratory alkalosis: HCO_3^- is excreted and H^+ is reabsorbed.

This chapter has used fluid balance and acid-base balance to illustrate functional integration in the cardiovascular, respiratory, and renal systems. Changes in body fluid volume, reflected by changes in blood pressure, trigger both cardiovascular and renal homeostatic responses. Disturbances of acid-base balance are met with compensatory responses from both the respiratory and renal systems. Because of the interwoven responsibilities of these three systems, a disturbance in one system is likely to cause disturbances in the other two. Recognition of this fact is an important aspect of treatment for many clinical conditions.

20

RUNNING PROBLEM CONCLUSION

HYPONATREMIA

In acute cases of dilutional hyponatremia such as Lauren's, the treatment goal is to correct the body's depleted Na^+ load and raise the plasma osmolarity to reduce cerebral swelling. The physicians in the emergency medical tent started a slow intravenous drip of 3% saline and restricted Lauren's oral fluid intake. Over the course of several hours, the combination of Na^+ intake and excretion of dilute urine returned Lauren's plasma Na^+ to normal levels.

Hyponatremia has numerous causes, including inappropriate secretion of antidiuretic hormone (a condition known as SIADH, which stands for syndrome of *inappropriate antidiuretic hormone* secretion). To learn more about medical causes of hyponatremia, Google *hyponatremia*. To learn more about exercise-associated hyponatremia, visit the Gatorade Sports Science Institute at *www.gssiweb.com*.

(This problem was developed by Matt Panke while he was a kinesiology graduate student at the University of Texas.)

	QUESTION	FACTS	INTEGRATION AND ANALYSIS
1.	What are the names of the fluids found in the two major body compartments and the major ions in each fluid?	The two major body compartments are the intracellular space and the extracellular space. Intracellular fluid (ICF) is found in the former, extracellular fluid (ECF) in the latter. The primary ICF ion is K^+, and the major ECF ions are Na^+ and Cl^-.	N/A*

RUNNING PROBLEM CONCLUSION (continued)

QUESTION	FACTS	INTEGRATION AND ANALYSIS
2. Based on Lauren's history, give a reason for why her weight increased during the race.	Lauren reported drinking lots of water and sports drinks. One liter of pure water has a mass of 1 kg.	Lauren's fluid intake was greater than her fluid loss from sweating. A 2-kg increase in body weight means she drank an excess of about 2 L.
3. Which fluid—ECF or ICF—is being diluted in dilutional hyponatremia?	Ingested water distributes itself throughout the ECF and ICF. Sodium is one of the major extracellular cations.	Lauren consumed a large amount of Na-free fluid and therefore diluted her Na^+ stores. Both ECF and ICF have lower osmolarities, but the diluted Na^+ is most noticeable in the ECF.
4. One way to estimate osmolarity is to double the plasma Na^+ concentration. If normal body osmolarity is 280 mOsM, what effect will Lauren's dilutional hyponatremia have on her cells?	Lauren's plasma Na^+ is 124 mEq/L. For Na^+, 1 mEq = 1 milliosmole, and therefore doubling this value tells you that Lauren's estimated plasma osmolarity is 248 mOsM. Water distributes according to tonicity.	At the start of the race, Lauren's cells were at 280 mOsM. The water she ingested distributed according to tonicity, so water entered the ICF from the ECF, resulting in cell swelling.
5. Which organ or tissue are the medical personnel most concerned about in dilutional hyponatremia?	All cells in Lauren's body will swell as a result of excess water ingestion. The brain is encased in the rigid skull.	The bony skull restricts the swelling of brain tissue, causing neurological symptoms, including confusion, headache, and loss of coordination. With lower Na^+ concentrations, death can result.
6. Assuming a sweating rate of 1.0 L/h and a sweat Na^+ level of 70 mEq/L, how much Na^+ did Lauren lose during the 16-hour race?	1.0 L sweat lost/h × 16 h × 70 mEq Na^+/L sweat = 1120 mEq Na^+ lost during the 16-hour race.	Without Na^+ intake to match the amount of Na^+ lost in her sweat, Lauren's Na^+ concentration will fall as she ingests fluid.
7. Total body water for a 60-kg female is approximately 30 L, and her ECF volume is 10 L. Based on the information given so far, what volume of fluid did Lauren ingest during the race?	From the sweating rate given in Question 6, you know that Lauren lost 16 liters of sweat during the race. You also know that she gained 2 kg in weight. One liter of water weighs 1 kg.	Lauren must have ingested at least 18 liters of fluid. You have no information on routes of fluid loss, such as urine and insensible water lost during breathing.
8. What would you expect to happen to vasopressin and aldosterone production in response to dilutional hyponatremia?	Vasopressin secretion is inhibited by a decrease in osmolarity. The usual stimuli for renin or aldosterone release are low blood pressure and hyperkalemia.	Vasopressin secretion increases with hyponatremia. The usual stimuli for aldosterone secretion are absent, but a pathological decrease in plasma Na^+ of 10 mEq/L can stimulate the adrenal cortex to secrete aldosterone. Thus, Lauren's plasma Na^+ may be low enough to increase her aldosterone secretion.

*Not applicable

| 642 | 645 | 658 | 661 | 665 | 671 |

CHAPTER SUMMARY

Homeostasis of body fluid volume, electrolytes, and pH follows the principle of *mass balance*: whatever comes into the body must be excreted. The *control systems* that regulate these parameters are among the most complicated reflexes of the body because of the overlapping functions of the kidneys, lungs, and cardiovascular system. At the cellular level, however, the *movement of molecules across membranes* follows familiar patterns, as transfer of water and solutes from one *compartment* to another depends on osmosis, diffusion, and protein-mediated transport.

Fluid and Electrolyte Homeostasis

1. The renal, respiratory, and cardiovascular systems control fluid and electrolyte balance. Behaviors such as drinking also play an important role. (p. 642; Fig. 20-1)

2. Pulmonary and cardiovascular compensations are more rapid than renal compensation. (p. 643)

Water Balance

IP Urinary: Early Filtrate Processing

3. Most water intake comes from food and drink. The largest water loss is 1.5 liters/day in urine. Smaller amounts are lost in feces, by evaporation from skin, and in exhaled humidified air. (p. 644; Fig. 20-2)

4. Renal water reabsorption conserves water but cannot restore water lost from the body. (p. 644; Fig. 20-3)

5. To produce dilute urine, the nephron must reabsorb solute without reabsorbing water. To concentrate urine, the nephron must reabsorb water without reabsorbing solute. (p. 645)

6. Filtrate leaving the ascending limb of the loop of Henle is dilute. The final concentration of urine depends on the water permeability of the collecting duct. (p. 645; Fig. 20-4)

7. The hypothalamic hormone **vasopressin** controls collecting duct permeability to water in a graded fashion. When vasopressin is absent, water permeability is nearly zero. (p. 646; Fig. 20-5)

8. Vasopressin causes distal nephron cells to insert **aquaporin** water pores in their apical membrane. (p. 647; Fig. 20-6)

9. An increase in ECF osmolarity or a decrease in blood pressure will stimulate vasopressin release from the posterior pituitary. Osmolarity is monitored by hypothalamic **osmoreceptors.** Blood pressure and blood volume are sensed by receptors in the carotid and aortic bodies, and in the atria, respectively. (p. 648; Figs. 20-7, 20-8)

10. The loop of Henle is a **countercurrent multiplier** that creates high osmolarity in the medullary interstitial fluid by actively transporting Na^+, Cl^-, and K^+ out of the nephron. This high medullary osmolarity is necessary for formation of concentrated urine as filtrate flows through the collecting duct. (pp. 650–651; Figs. 20-9, 20-10)

11. The **vasa recta** capillaries carry away water leaving the nephron tubule so that the water does not dilute the medullary interstitium. (p. 651; Fig. 20-10)

12. Urea contributes to the high osmolarity in the renal medulla. (p. 652)

Sodium Balance and the Regulation of ECF Volume

IP Urinary: Late Filtrate Processing

13. The total amount of Na^+ in the body is a primary determinant of ECF volume. (p. 652; Fig. 20-12)

14. The steroid hormone **aldosterone** increases Na^+ reabsorption and K^+ secretion. (p. 652; Fig. 20-13)

15. Aldosterone acts on **principal cells** (P cells) of the distal nephron. This hormone enhances Na^+-K^+-ATPase activity and increases open time of Na^+ and K^+ leak channels. It also stimulates the synthesis of new pumps and channels. (p. 652; Fig. 20-13)

16. Aldosterone secretion can be controlled directly at the adrenal cortex. Increased ECF K^+ stimulates aldosterone secretion, but increased ECF osmolarity inhibits it. (p. 652; Tbl. 20-1)

17. Aldosterone secretion is also stimulated by **angiotensin II.** Granular cells in the kidney secrete **renin**, which converts **angiotensinogen**

in the blood to **angiotensin I. Angiotensin converting enzyme (ACE)** converts ANG I to ANG II. (pp. 653–654; Fig. 20-14)

18. The stimuli for the release of renin are related either directly or indirectly to low blood pressure. (p. 654; Fig. 20-15)

19. ANG II has additional effects that raise blood pressure, including increased vasopressin secretion, stimulation of thirst, vasoconstriction, and activation of the cardiovascular control center. (pp. 654–655; Fig. 20-14)

20. **Atrial natriuretic peptide** (ANP) and **brain natriuretic peptide** (BNP) enhance Na^+ excretion and urinary water loss by increasing GFR, inhibiting tubular reabsorption of NaCl, and inhibiting the release of renin, aldosterone, and vasopressin. (p. 656; Fig. 20-16)

Potassium Balance

21. Potassium homeostasis keeps plasma K^+ concentrations in a narrow range. **Hyperkalemia** and **hypokalemia** cause problems with excitable tissues, especially the heart. (p. 656)

Behavioral Mechanisms in Salt and Water Balance

22. Thirst is triggered by hypothalamic osmoreceptors and relieved by drinking. (p. 658)

23. **Salt appetite** is triggered by aldosterone and angiotensin. (p. 658)

Integrated Control of Volume and Osmolarity

IP Fluids & Electrolytes: Water Homeostasis

24. Homeostatic compensations for changes in salt and water balance follow the law of mass balance. Fluid and solute added to the body must be removed, and fluid and solute lost from the body must be replaced. However, perfect compensation is not always possible. (pp. 659–661; Fig. 20-18; Tbl. 20-2)

Acid-Base Balance

IP Fluids & Electrolytes: Acid/Base Homeostasis

25. The body's pH is closely regulated because pH affects intracellular proteins, such as enzymes and membrane channels. (p. 663)

26. Acid intake from foods and acid production by the body's metabolic processes are the biggest challenge to body pH. The most significant source of acid is CO_2 from respiration, which combines with water to form carbonic acid (H_2CO_3). (p. 664; Fig. 20-19)

27. The body copes with changes in pH by using buffers, ventilation, and renal secretion or reabsorption of H^+ and HCO_3^-. (p. 664; Fig. 20-19)

28. Bicarbonate produced from CO_2 is the most important extracellular buffer of the body. Bicarbonate buffers organic acids produced by metabolism. (p. 665)

29. Ventilation can correct disturbances in acid-base balance because changes in plasma P_{CO_2} affect both the H^+ content and the HCO_3^- content of the blood. An increase in P_{CO_2} stimulates central chemoreceptors. An increase in plasma H^+ stimulates carotid and aortic chemoreceptors. Increased ventilation excretes CO_2 and plasma decreases H^+. (p. 666; Fig. 20-20)

30. In **acidosis**, the kidneys secrete H^+ and reabsorb HCO_3^-. (p. 669; Figs. 20-21, 20-22, 20-23a)

31. In **alkalosis**, the kidneys secrete HCO_3^- and reabsorb H^+. (p. 669; Fig. 20-23b)

32. **Intercalated cells** in the collecting duct are responsible for the fine regulation of acid-base balance. (p. 668; Fig. 20-23)

20

QUESTIONS

(Answers to the Review Questions begin on page A1.)

THE PHYSIOLOGY PLACE

Access more review material online at **The Physiology Place** website. There you'll find review questions, problem-solving activities, case studies, flashcards, and direct links to both *InterActive Physiology*® and PhysioEx™. To access the site, go to *www.physiologyplace.com* and select Human Physiology, Fourth Edition.

LEVEL ONE   REVIEWING FACTS AND TERMS

1. What is an electrolyte? Name five electrolytes whose concentrations must be regulated by the body.

2. List five organs and four hormones important in maintaining fluid and electrolyte balance.

3. Compare the routes by which water enters the body with the routes by which the body loses water.

4. List the receptors that regulate osmolarity, blood volume, blood pressure, ventilation, and pH. Where are they located, what stimulates them, and what compensatory mechanisms are triggered by them?

5. How do the two limbs of the loop of Henle differ in their permeability to water? What makes this difference in permeability possible?

6. Which ion is a primary determinant of ECF volume? Which ion is the determinant of extracellular pH?

7. What happens to the resting membrane potential of excitable cells when plasma K^+ concentrations decrease? Which organ is most likely to be affected by changes in K^+ concentration?

8. Appetite for what two substances is important in regulating fluid volume and osmolarity?

9. Write out the words for the following abbreviations: ADH, ANP, ACE, ANG II, JG apparatus, P cell, I cell.

10. Make a list of all the different membrane transporters in the kidney. For each transporter, tell (a) which section(s) of the nephron contain(s) the transporter; (b) whether the transporter is on the apical membrane only, on the basolateral membrane only, or on both; (c) whether it participates in reabsorption only, in secretion only, or in both of these processes.

11. List and briefly explain three reasons why monitoring and regulating ECF pH are important. What three mechanisms does the body use to cope with changing pH?

12. Which is more likely to accumulate in the body, acids or bases? List some sources of each.

13. What is a buffer? List three intracellular buffers. Name the primary extracellular buffer.

14. Name two ways the kidneys alter plasma pH. Which compounds serve as urinary buffers?

15. Write the equation that shows how CO_2 is related to pH. What enzyme increases the rate of this reaction? Name two cell types that possess high concentrations of this enzyme.

16. When ventilation increases, what happens to arterial P_{CO_2}? To plasma pH? To plasma H^+ concentration?

LEVEL TWO   REVIEWING CONCEPTS

17. **Concept map:** Map the homeostatic reflexes that occur in response to each of the following situations:
    (a) decreased blood volume, normal blood osmolarity
    (b) increased blood volume, increased blood osmolarity
    (c) normal blood volume, increased blood osmolarity

18. Figures 20-20 and 20-21 show the respiratory and renal compensations for acidosis. Draw similar maps for alkalosis.

19. Explain how the loop of Henle and vasa recta work together to create dilute renal filtrate.

20. Diagram the mechanism by which vasopressin alters the composition of urine.

21. Make a table that specifies the following for each substance listed: hormone or enzyme? steroid or peptide? produced by which cell or tissue? target cell or tissue? target has what response?
    (a) ANP   (d) ANG II
    (b) aldosterone   (e) vasopressin
    (c) renin   (f) angiotensin-converting enzyme

22. Name the four main compensatory mechanisms for restoring low blood pressure to normal. Why do you think there are so many homeostatic pathways for raising low blood pressure?

23. Compare and contrast the terms in each set:
    (a) principal cells and intercalated cells
    (b) renin, angiotensin II, aldosterone, ACE
    (c) respiratory acidosis and metabolic acidosis, including causes and compensations
    (d) water reabsorption in proximal tubule, distal tubule, and ascending limb of the loop of Henle
    (e) respiratory alkalosis and metabolic alkalosis, including causes and compensations

LEVEL THREE   PROBLEM SOLVING

24. (a) A 45-year-old man visiting from out of town arrives at the emergency room having an asthma attack caused by pollen. Blood drawn before treatment showed the following: $HCO_3^- = 30$ meq/L (normal: 24), $P_{CO_2} = 70$ mm Hg, pH = 7.24. What is the man's acid-base state? Is this an acute or a chronic situation?

    (b) The man was treated and made a complete recovery. Over the next ten years he continued to smoke a pack of cigarettes a day, and a year ago his family doctor diagnosed chronic obstructive pulmonary disease (emphysema). The man's most recent blood test showed the following: $HCO_3^- = 45$ meq/L, $P_{CO_2} = 85$ mm Hg, pH = 7.34. What is the man's acid-base state now? Is this an acute or a chronic situation?

    (c) Explain why in his second illness his plasma bicarbonate level and P_{CO_2} are higher than in the first illness but his pH is closer to normal.

25. Karen has bulimia, in which she induces vomiting to avoid weight gain. When the doctor sees her, her weight is 89 lb and her respiration rate is 6 breaths/min (normal 12). Her blood HCO_3^- is 62 meq/L (normal: 24–29), arterial blood pH is 7.61, and P_{CO_2} is 61 mm Hg.
    (a) What is her acid-base condition called?
    (b) Explain why her plasma bicarbonate level is so high.
    (c) Why is she hypoventilating? What effect will this have on the pH and total oxygen content of her blood? Explain your answers.

26. Hannah, a 31-year-old woman, decided to have colonic irrigation, a procedure during which large volumes of distilled water were infused into her rectum. During the treatment she absorbed 3000 mL of water. About 12 hours later, her roommate found her in convulsions and took her to the emergency room. Her blood pressure was 140/90, her plasma Na^+ concentration was 106 meq/L (normal: 135 meq/L), and her plasma osmolarity was 270 mOsM. In a concept map or flow chart, diagram all the homeostatic responses her body was using to attempt compensation for the changes in blood pressure and osmolarity.

LEVEL FOUR   QUANTITATIVE PROBLEMS

27. The Henderson-Hasselbalch equation is a mathematical expression of the relationship between pH, HCO_3^- concentration, and dissolved CO_2 concentration. One variant of the equation uses P_{CO_2} instead of dissolved CO_2 concentration:

$$pH = \frac{6.1 + \log\left[HCO_3^-\right]}{0.03 \times P_{CO_2}}$$

(a) If arterial blood has a P_{CO2} of 40 mm Hg and its HCO_3^- concentration is 24 mM, what is its pH? (You will need a log table or calculator with a logarithmic function capability.)

(b) What is the pH of venous blood with the same HCO_3^- concentration but a P_{CO_2} of 46 mm Hg?

28. In extreme dehydration, urine can reach a concentration of 1400 mOsM. If the minimum amount of waste solute that a person must excrete daily is about 600 milliosmoles, what is the minimum urine volume that will be excreted in one day?

29. Hyperglycemia in a diabetic patient leads to osmotic diuresis and dehydration. Given the following information, answer the questions.

Plasma glucose = 400 mg/dL
GFR = 130 mL/min
Glucose T_m = 400 mg/min
Renal plasma flow = 500 mL/min

Normal urine flow = 1 L per day
Normal urine osmolarity = 300 mOsM
Molecular mass of glucose = 180 daltons

(a) How much glucose filters into the nephron each minute?
(b) How much glucose is reabsorbed each minute?
(c) How much glucose is excreted in the urine each day?
(d) Assuming that dehydration causes maximal ADH secretion and allows the urine to concentrate to 1200 mOsM, how much additional urine will this diabetic patient excrete in a day?

ANSWERS

✓ Answers to Concept Check Questions

Page 647

1. Apical membranes will have more water pores when vasopressin is present.
2. The collecting duct cells concentrate organic solutes that counteract the hypertonic effect of the cells' environment.

Page 649

3. Hyperosmotic NaCl is hypertonic and causes the osmoreceptors to shrink, but hyperosmotic urea is hypotonic and causes them to swell. Because only cell shrinkage causes firing, osmoreceptors exposed to urea will not fire.
4. Because vasopressin enhances water reabsorption, vasopressin levels would increase with dehydration.
5. If vasopressin secretion is suppressed, the urine will be dilute.

Page 651

6. Solutes that remain in the lumen when the NKCC symporter is inhibited force water to remain in the lumen with them because urine can be concentrated only to 1200 mOsM. Thus each 12 milliosmoles of un-reabsorbed solute will "hold" an additional 10 mL of water in the urine.
7. Diuretics that inhibit the NKCC symporter leave K^+ in the tubule lumen, where it is likely to be excreted, thus increasing urinary K^+ loss.

Page 653

8. Na^+ and K^+ are moving down their electrochemical gradients.

Page 656

9. In hyperkalemia, resting membrane potential depolarizes. Excitable tissues fire one action potential but are unable to repolarize to fire a second one.
10. Atherosclerotic plaques block blood flow, which decreases GFR and decreases pressure in the afferent arteriole. Both events are stimuli for renin release.

11. Renin secretion begins a cascade that produces angiotensin II. This powerful hormone then causes vasoconstriction, acts on the medullary cardiovascular control center to increase blood pressure, increases production of vasopressin and aldosterone, and increases thirst, resulting in drinking and an increased fluid volume in the body. All these responses may contribute to increased blood pressure.

Page 658

12. On the left side of Figure 20-7, interneurons also lead from hypothalamic osmoreceptors to the hypothalamic thirst centers.

Page 667

13. The muscles of ventilation involved are the diaphragm, the external intercostals, and the scalenes.
14. The bicarbonate level increases as the reaction shifts to the right as a result of added CO_2. Once a new equilibrium state is achieved, bicarbonate cannot act as a buffer because the system is at equilibrium.

Page 669

15. In the distal nephron, both K^+ and H^+ are being moved against their concentration gradients, which requires ATP. In the proximal tubule, Na^+ is moving down its concentration gradient, providing the energy to push H^+ against its gradient.
16. When intercalated cells reabsorb K^+, they secrete H^+, and therefore blood pH increases.

Q Answers to Figure Questions

Page 653

Figure 20-12: See Fig. 15-22, p. 523.

Page 655

Figure 20-14: See Fig. 15-21, p. 522, for the cardiovascular pathway, Fig. 20-13 for the effector cell involved in aldosterone action, and Fig. 20-6 for the effector cell involved in vasopressin action.

20

21

Vibrio cholerae (green), the bacterium that causes the diarrheal disease cholera, occurs in fresh water. Here it is shown attached to the surface of freshwater algae (red).

The Digestive System

BACKGROUND BASICS

A shotgun wound to the stomach seems an unlikely beginning to the scientific study of digestive processes. But in 1822 at Fort Mackinac, a young Canadian trapper named Alexis St. Martin narrowly escaped death when a gun discharged three feet from him, tearing open his chest and abdomen and leaving a hole in his stomach wall. U.S. Army surgeon William Beaumont attended to St. Martin and nursed him back to health over the next two years.

The gaping wound over the stomach failed to heal properly, leaving a *fistula,* or opening, into the lumen. St. Martin was destitute and unable to care for himself, so Beaumont "retained St. Martin in his family for the special purpose of making physiological experiments." In a legal document, St. Martin even agreed to "obey, suffer, and comply with all reasonable and proper experiments of the said William [Beaumont] in relation to . . . the exhibiting . . . of his said stomach and the power and properties . . . and states of the contents thereof."

Beaumont's observations on digestion and on the state of St. Martin's stomach under various conditions created a sensation. In 1832, just before Beaumont's observations were published, the nature of gastric juice [*gaster,* stomach] and digestion in the stomach was a subject of much debate. Beaumont's careful observations went far toward solving the mystery. Like physicians of old who tasted urine when making a diagnosis, Beaumont tasted the mucous lining of the stomach and the

gastric juices. He described them both as "saltish," but mucus was not at all acidic, and gastric fluid was very acidic. Beaumont collected copious amounts of gastric fluid through the fistula, and in controlled experiments he confirmed that gastric fluid digested meat, using a combination of hydrochloric acid and another active factor now known to be the enzyme pepsin.

These observations and others about motility and digestion in the stomach became the foundation of what we know about digestive physiology. Although research today is conducted more at the cellular and molecular level, researchers still create surgical fistulas in experimental animals to observe and sample the contents of the digestive tract.

Why is the digestive system—also referred to as the **gastrointestinal system** [*intestinus,* internal]—of such great interest? The reason is that gastrointestinal diseases today account for nearly one-tenth of the money spent on health care. Many of these conditions, such as heartburn, indigestion, gas, and constipation, are troublesome rather than major health risks, but their significance should not be underestimated. Go into any drugstore and look at the number of over-the-counter medications for digestive disorders to get a feel for the impact digestive diseases have on our society. In this chapter we examine the gastrointestinal system and the remarkable way in which it transforms the food we eat into nutrients for the body's use.

FUNCTION AND PROCESSES OF THE DIGESTIVE SYSTEM

The **gastrointestinal tract**, or **GI tract**, is a long tube passing through the body. The tube has muscular walls lined with epithelium and is closed off by a skeletal-muscle sphincter at each end. Because the GI tract opens to the outside world, the lumen and its contents are actually part of the external environment [Fig. 1-2, p. 3]. (Think of a hole passing through the center of a bead.) The primary function of the GI tract is to move nutrients, water, and electrolytes from the external environment into the body's internal environment.

The food we eat is mostly in the form of macromolecules, such as proteins and complex carbohydrates, so our digestive systems must secrete powerful enzymes to digest food into molecules that are small enough to be absorbed into the body. At the same time, however, these enzymes must not digest the cells of the GI tract itself (*autodigestion*). If protective mechanisms against autodigestion fail, we develop raw patches known as *peptic ulcers* [*peptos,* digested] on the walls of the GI tract.

Another challenge the digestive system faces daily is mass balance: matching input with output. Various exocrine glands and cells secrete about 7 liters of fluid per day into the lumen of the GI tract. These secretions contain digestive enzymes, mucus, electrolytes, and water. They are critical to proper digestive

21

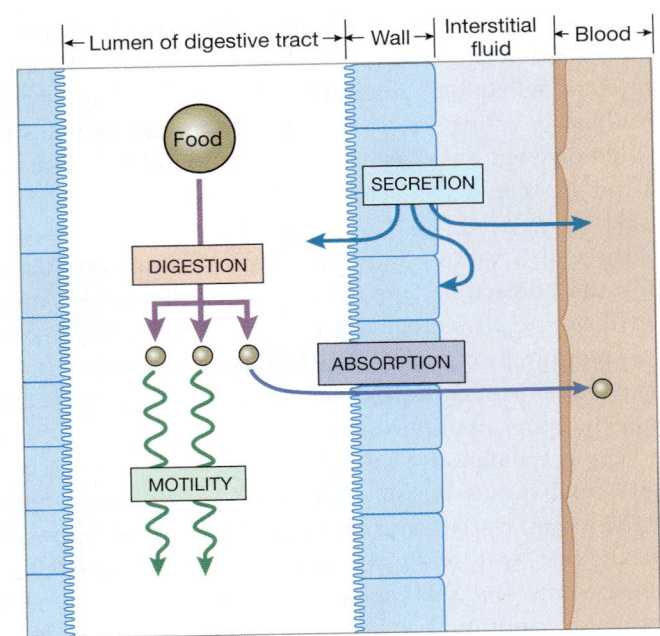

■ **FIGURE 21-1**   *Basic processes of the digestive system*

The four digestive system processes are *digestion* of food into smaller units; *absorption* of substances from the lumen into the ECF; *motility,* the movement of materials through the GI tract; and *secretion* of substances from epithelial cells into the lumen or ECF.

function, but they must be reabsorbed or the body will rapidly dehydrate. Diarrhea (excessively watery feces), for instance, can become an emergency if fluid loss from the GI tract depletes extracellular fluid volume to the point that the circulatory system is unable to maintain adequate blood pressure.

The body meets these physiological challenges by coordinating the four basic processes of the digestive system: digestion, absorption, motility, and secretion (Fig. 21-1 ■). **Digestion** is the chemical and mechanical breakdown of foods into smaller units that can be taken across the intestinal epithelium into the body. **Absorption** is the active or passive transfer of substances from the lumen of the GI tract to the extracellular fluid. **Motility** [*movere*, move + *tillis*, characterized by] is movement of material in the GI tract as a result of muscle contraction. **Secretion** refers to both the transepithelial transfer of water and ions from the ECF to the digestive tract lumen and the release of substances synthesized by GI epithelial cells.

For most nutrients, absorption is not regulated, so "what you eat is what you get." In contrast, motility and secretion are continuously regulated to maximize the availability of absorbable material. Motility is regulated because if food moves through the system too rapidly, there is not enough time for everything in the lumen to be digested and absorbed. Secretion is regulated because if digestive enzymes are not secreted in adequate amounts, food in the GI tract cannot be broken down into an absorbable form.

When digested nutrients have been absorbed and have reached the body's cells, cellular metabolism directs their use or storage (see Chapter 22). Some of the same chemical signal molecules that alter digestive motility and secretion also participate in the control of metabolism, providing an integrating link between the two steps.

Although we tend to think of the digestive system primarily in terms of its digestive function, repelling foreign invaders is another challenge the system faces. Indeed, the greatest contact between the body's internal environment and the outside world occurs in the digestive system, whose total surface area is estimated to be the size of a tennis court. As a result, the GI tract faces daily conflict between the need to absorb water and nutrients, and the need to keep bacteria, viruses, and other pathogens from entering the body. To this end, the transporting epithelium of the GI tract is assisted by an array of physiological defense mechanisms, including mucus, digestive enzymes, acid, and the largest collection of lymphoid tissue in the body, the **gut-associated lymphoid tissue** (GALT). By one estimate, 80% of all lymphocytes [⊜ p. 538] in the body are found in the small intestine.

ANATOMY OF THE DIGESTIVE SYSTEM

The digestive system begins with the oral cavity (mouth and pharynx), which serves as a receptacle for food (Fig. 21-2 ■). In the oral cavity the first stages of digestion begin with chewing and the secretion of saliva by three pairs of **salivary glands:** *sublingual glands* under the tongue, *submandibular glands* under the mandible (jawbone), and *parotid glands* lying near the hinge of the jaw.

Once swallowed, food moves into the GI tract. At intervals along the tract, rings of smooth muscle function as sphincters to separate the tube into segments with distinct functions. Food moves through the tract propelled by waves of muscle contraction. Along the way, secretions are added to the food by secretory epithelium, the liver, and the pancreas, creating a soupy mixture known as **chyme**.

Digestion takes place primarily in the lumen of the tube. The products of digestion are absorbed across the epithelium and pass into the extracellular compartment. From there, they move into the blood or lymph for distribution throughout the body. Any waste remaining in the lumen at the end of the GI tract leaves the body through the opening known as the anus.

CONCEPT CHECK

1. Define digestion.
2. What is the difference between digestion and metabolism [⊜ p. 101]?
3. What is the difference between absorption and secretion?

Answers: p. 714

The Digestive System Consists of Oral Cavity, GI Tract, and Accessory Glandular Organs

A piece of food that enters the mouth and is swallowed passes into the **esophagus**, a narrow tube that travels through the thorax to the abdomen (Fig. 21-2a). The esophageal walls are skeletal muscle initially but transition to smooth muscle about two-thirds of the way down the length. Just below the diaphragm, the esophagus ends at the **stomach**, a baglike organ that can hold as much as 2 liters of food and fluid when fully (if uncomfortably) expanded.

The stomach is divided into three sections: the upper **fundus**, the central **body**, and the lower **antrum** (Fig. 21-2b). The stomach continues digestion, mixing food with acid and enzymes to create chyme. The opening between the stomach and the **small intestine** is guarded by the **pyloric valve**, or **pylorus**. This thickened band of smooth muscle relaxes to allow only small amounts of chyme into the small intestine at any one time. In this way, the stomach acts as an intermediary between the behavioral act of eating and the physiological events of digestion and absorption in the intestine. Integrated signals and feedback loops between the intestine and stomach regulate the rate at which chyme enters the duodenum, ensuring that the intestine will not be overwhelmed with more than it can digest and absorb.

Most digestion takes place in the small intestine, which is also divided into three sections: the **duodenum** (the first 25 cm), **jejunum**, and **ileum** (the latter two together are about 260 cm long). Digestion is carried out by intestinal enzymes, aided by exocrine secretions from two *accessory glandular organs:* the pancreas and the liver (Fig. 21-2a). Secretions from these two organs enter the initial section of the duodenum through ducts. A tonically contracted sphincter (the *sphincter of Oddi*) keeps pancreatic fluid and bile from entering the small intestine except during a meal.

Digestion is essentially completed in the small intestine, and nearly all digested nutrients and secreted fluids are absorbed there, leaving about 1.5 liters of chyme per day to pass into the **large intestine** (Fig. 21-2a). In the **colon**—the proximal section of the large intestine—watery chyme is converted into semisolid **feces** [*faeces*, dregs] as water and electrolytes are absorbed out of the chyme and into the ECF.

When feces are propelled into the terminal section of the large intestine, known as the **rectum**, distension of the rectal wall triggers a *defecation reflex*. Feces leave the GI tract through the **anus**, with its **external anal sphincter** of skeletal muscle, which is under voluntary control. The portion of the GI tract running from the stomach to the anus is collectively called the **gut**.

In a living person, the digestive system from mouth to anus is about 450 cm (nearly 15 feet) long! Of this length, 395 cm (about 13 feet) consists of the large and small intestines. Try to imagine 13 feet of rope ranging from 1 to 3 inches in diameter all coiled up inside your abdomen from the belly button down. The tight arrangement of the abdominal organs helps explain why you feel the need to loosen your belt after consuming a large meal!

Measurements of intestinal length made during autopsies are nearly double those given here because after death, the longitudinal muscles of the intestinal tract relax. This accounts for the wide variation in intestinal length you may encounter in different references.

The GI Tract Wall Has Four Layers

The basic structure of the gastrointestinal wall is similar in the stomach and intestines, although variations exist from one section of the GI tract to another (Fig. 21-2c, d, e). The wall consists of four layers: (1) an inner *mucosa* facing the lumen, (2) a layer known as the *submucosa*, (3) layers of smooth muscle known collectively as the *muscularis externa*, and (4) a covering of connective tissue called the *serosa*.

The Mucosa   The **mucosa**, the inner lining of the gastrointestinal tract, is created from (1) a single layer of epithelial cells; (2) the **lamina propria**, subepithelial connective tissue that holds the epithelium in place; and (3) the **muscularis mucosae**, a thin layer of smooth muscle. Several structural modifications increase the amount of mucosal surface area to enhance absorption.

First, the entire wall is crumpled into folds: *rugae* in the stomach, and *plicae* in the small intestine. The intestinal mucosa also projects into the lumen in small fingerlike extensions known as **villi** (Fig. 21-2d). Additional surface area is added by tubular invaginations of the surface that extend down into the supporting connective tissue. These invaginations are called **gastric glands** in the stomach, and **crypts** in the intestine. Some of the deepest invaginations form secretory **submucosal glands** that open into the lumen through ducts.

Epithelial cells are the most variable feature of the GI tract, changing from section to section. The cells include transporting epithelial cells, endocrine and exocrine secretory cells, and stem cells. Transporting epithelial cells move ions and water into the lumen, and absorb ions, water, and nutrients into the ECF. At the *mucosal* (apical) surface [p. 73], secretory cells release enzymes, mucus, and paracrine molecules into the lumen. At the *serosal* (basolateral) surface, secretory cells secrete hormones into the blood or paracrine messengers into the interstitial fluid, where they act on neighboring cells.

The cell-to-cell junctions that tie GI epithelial cells together vary [p. 69]. In the stomach and colon, the junctions form a tight barrier so that little can pass between the cells. In the small intestine, junctions are not as tight. This intestinal epithelium is considered "leaky" because some water and solutes can be absorbed between the cells (the *paracellular pathway*) instead of through them. We now know that these junctions have plasticity; that is, their "tightness" can be regulated to some extent by the nutritional status of the body.

GI *stem cells* are rapidly dividing, undifferentiated cells that continuously produce new epithelium in the crypts and gastric glands. As stem cells divide, the newly formed cells are

21

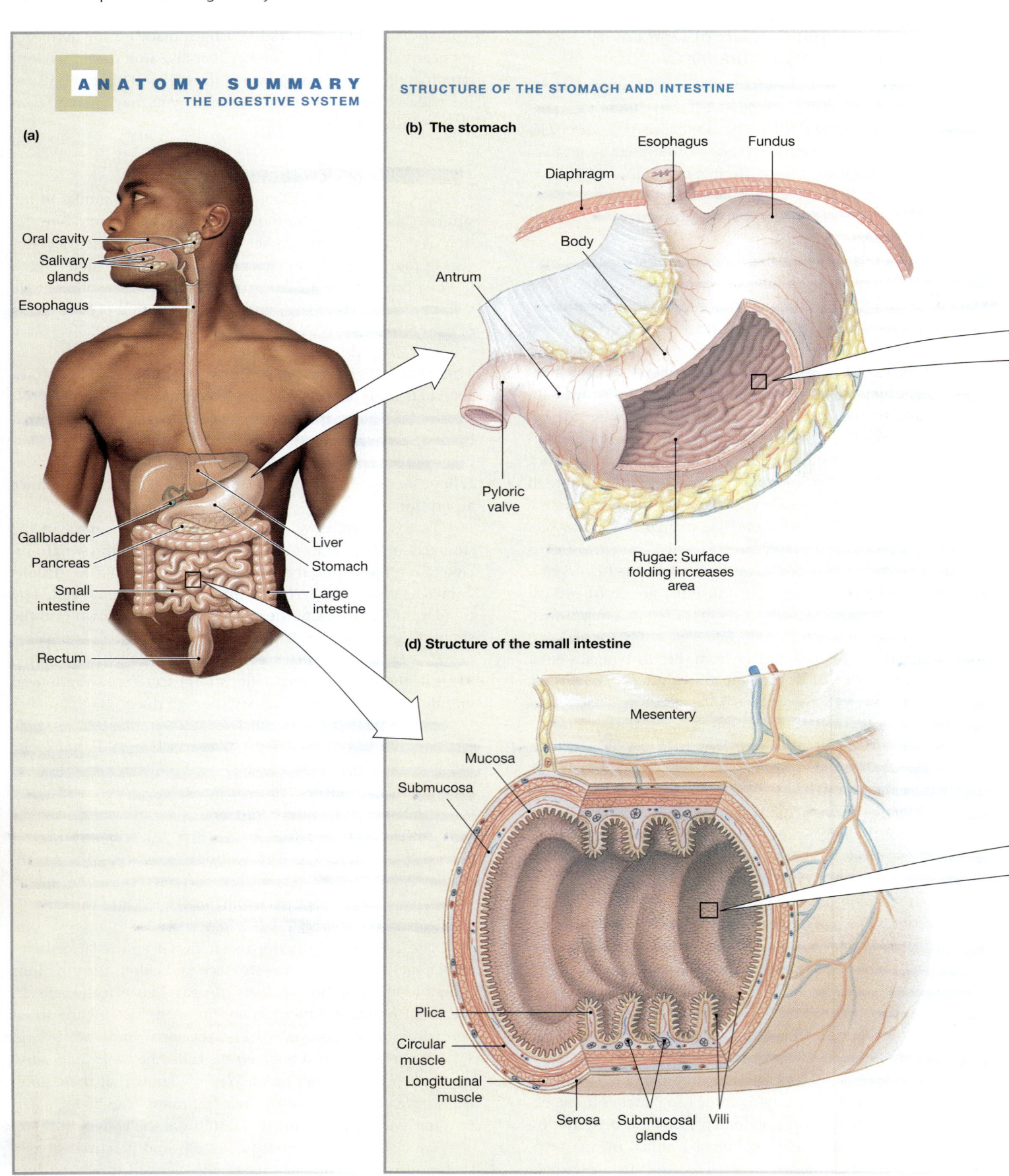

FIGURE 21-2

SECTIONAL VIEWS OF THE STOMACH AND INTESTINE

(c) In the stomach, surface area is increased by invaginations called gastric glands.

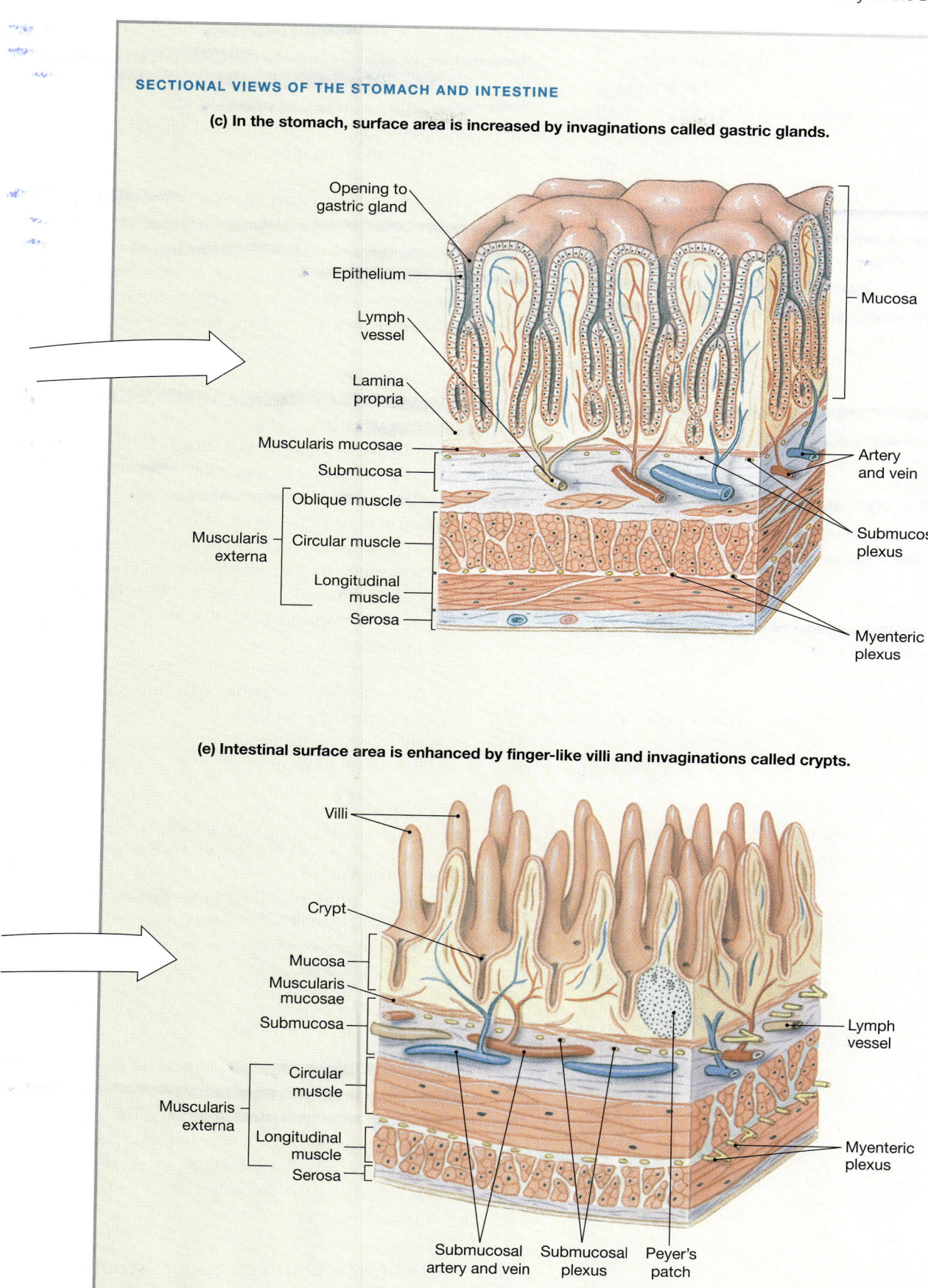

(e) Intestinal surface area is enhanced by finger-like villi and invaginations called crypts.

FIGURE 21-2 *(continued)*

pushed up toward the luminal surface of the epithelium. The average life span of a GI epithelial cell is only a few days, a good indicator of the rough life such cells lead. As with other types of epithelium, the rapid turnover and cell division rate in the GI tract makes these organs susceptible to developing cancers. In 2005, cancers of the colon and rectum (colorectal cancer) were the third leading cause of cancer deaths in the United States.

The **lamina propria** is subepithelial connective tissue that contains nerve fibers and small blood and lymph vessels into which absorbed nutrients pass (Fig. 21-2c). This layer also contains wandering immune cells, such as macrophages and lymphocytes, patrolling for invaders that enter through breaks in the epithelium. In the intestine, collections of lymphoid tissue adjoining the epithelium form small *nodules* and larger **Peyer's patches** (Fig. 21-2e) that create visible bumps in the mucosa. These lymphoid aggregations are a major part of the gut-associated lymphoid tissue (GALT).

The third region of the mucosa, the **muscularis mucosae**, separates the mucosa from the submucosa. The muscularis mucosae is a thin layer of smooth muscle, and contraction of this layer alters the effective surface area for absorption by moving the villi back and forth, like the waving tentacles of a sea anemone.

The Submucosa The layer of the gut wall adjacent to the mucosa, the **submucosa**, is composed of connective tissue with larger blood and lymph vessels (Fig. 21-2c, e). The submucosa

---

RUNNING PROBLEM

A *peptic ulcer* is a raw, inflamed area in the stomach or duodenum that extends from a break in the surface epithelium past the muscularis mucosae. Although most peptic ulcers occur in the duodenum (duodenal ulcers), others are located within the stomach (gastric ulcers). Until 1982, the year *H. pylori* was discovered, peptic ulcers were believed to be caused by high levels of acid and enzyme production in the stomach. However, since the 1980s, scientists have found *H. pylori* in 90% of people with duodenal ulcers and 70% of people with gastric ulcers. To confirm the diagnosis of an ulcer caused by *H. pylori,* Tonya will undergo tests to isolate the bacterium from her stomach.

Question 1:
Why are peptic ulcers found in the duodenum rather than in the jejunum or ileum?

677 **682** 687 691 704 710

---

also contains the **submucosal plexus** [*plexus,* interwoven], one of the two major nerve networks of the **enteric nervous system** [p. 246]. The enteric nervous system helps coordinate digestive function, and the submucosal plexus (also called *Meissner's plexus*) innervates cells in the epithelial layer as well as the smooth muscle of the muscularis mucosae.

The Muscularis Externa and Serosa The outer wall of the gastrointestinal tract, the **muscularis externa**, primarily consists of two layers of smooth muscle: an inner circular layer and an outer longitudinal layer (Fig. 21-2d, e). Contraction of the circular layer decreases the diameter of the lumen; contraction of the longitudinal layer shortens the tube. The stomach has an incomplete third layer of oblique muscle between the circular muscles and the submucosa (Fig. 21-2c).

The second nerve network of the enteric nervous system, the **myenteric plexus** [*myo-,* muscle + *enteron,* intestine], lies between the longitudinal and circular muscle layers. The myenteric plexus (also called *Auerbach's plexus*) controls and coordinates the motor activity of the muscularis externa.

The outer covering of the entire digestive tract, the **serosa**, is a connective tissue membrane that is a continuation of the **peritoneal membrane** (*peritoneum*) lining the abdominal cavity [p. 51]. The peritoneum also forms sheets of **mesentery** that hold the intestines in place so that they do not become tangled as they move.

Now let's take a brief look at the four processes of motility, secretion, digestion, and absorption. Gastrointestinal physiology is a rapidly expanding field, and this textbook does not attempt to be all inclusive. Instead, it focuses on selected broad aspects of digestive physiology.

CONCEPT CHECK

4. Is the lumen of the digestive tract on the apical or basolateral side of the intestinal epithelium? On the serosal or mucosal side?

5. Name the structures a piece of food passes through as it travels from mouth to anus.

6. Why is the digestive system associated with the largest collection of lymphoid tissue in the body?   Answers: p. 714

MOTILITY

Motility in the gastrointestinal tract serves two purposes: moving food from the mouth to the anus and mechanically mixing food to break it into uniformly small particles. This mixing maximizes exposure of the particles to digestive enzymes by increasing their surface area. Gastrointestinal motility is determined by the properties of the tract's smooth muscle and modified by chemical input from nerves, hormones, and paracrine signals.

GI Smooth Muscle Contracts Spontaneously

Most of the gastrointestinal tract is composed of single-unit smooth muscle, with groups of cells electrically connected by

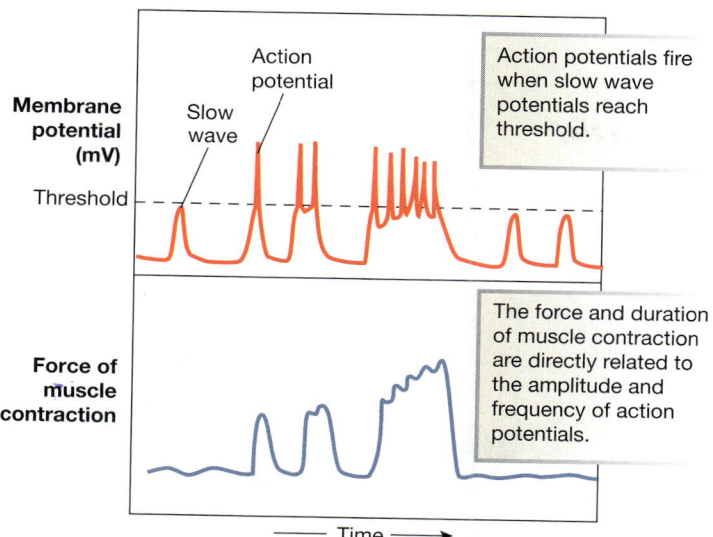

■ **FIGURE 21-3** *Slow wave potentials in GI smooth muscle*

Slow waves occur in spontaneously depolarizing muscle fibers. Slow waves that exceed threshold trigger action potentials (top graph), producing smooth muscle contractions with force and duration that are related to the frequency and amplitude of the action potentials (bottom graph).

gap junctions [🔁 p. 69] to create contracting segments. Different regions exhibit different types of contraction. **Tonic contractions** that are sustained for minutes or hours occur in some smooth muscle sphincters and in the anterior portion of the stomach. **Phasic contractions**, with contraction-relaxation cycles lasting only a few seconds, occur in the posterior region of the stomach and in the small intestine.

Cycles of smooth muscle contraction and relaxation are associated with spontaneous cycles of depolarization and repolarization known as **slow wave potentials** (or simply *slow waves*). Slow wave potentials differ from myocardial pacemaker potentials in that the former have a much slower rate (3–12 waves/min GI versus 60–90 waves/min myocardial) and do not reach threshold with each cycle (Fig. 21-3 ■). A slow wave that does not reach threshold will not cause a contraction in the muscle fiber.

When a slow wave potential does reach threshold, voltage-gated Ca^{2+} channels in the muscle fiber open, Ca^{2+} enters, and the cell fires one or more action potentials. The depolarization phase of the action potential, like that in myocardial autorhythmic cells, is the result of Ca^{2+} entry into the cell. In addition, Ca^{2+} entry initiates muscle contraction [🔁 p. 426].

Contraction of smooth muscle, like that of cardiac muscle, is graded according to the amount of Ca^{2+} that enters the fiber. The longer the duration of the slow wave, the more action potentials fire, and the greater the contraction force in the muscle. Similarly, the longer the duration of the slow wave, the longer the duration

INTERSTITIAL CELLS OF CAJAL

Current experimental evidence suggests that the rhythm of slow wave potentials arises not in the smooth muscle itself but from a network of associated cells known as the *interstitial cells of Cajal* (named for the Spanish neuroanatomist Santiago Ramón y Cajal). These specialized smooth muscle cells are associated with the smooth muscle layers and intrinsic nerve plexus of the gastrointestinal wall. The interstitial cells of Cajal (ICC) apparently act as the pacemakers for smooth muscle slow waves; their depolarizations spread to adjacent muscle fibers through gap junctions. ICCs depend on a tyrosine kinase receptor pathway [🔁 p. 183] called *Kit* to maintain normal activity. Mutant mice with weakly functional *Kit* receptors exhibit abnormal gastrointestinal motility. This observation now has researchers working to establish a link between ICCs and functional bowel disorders, such as irritable bowel syndrome and chronic constipation.

of contraction. Both amplitude and duration can be modified by neurotransmitters, hormones, or paracrine molecules.

Slow wave frequency varies by region of the digestive tract, ranging from 3 waves/min in the stomach to 12 waves/min in the duodenum. Current research indicates that slow waves originate in modified smooth muscle cells called **interstitial cells of Cajal** (see Emerging Concepts box) and pass into smooth muscle cells through gap junctions. The interstitial cells of Cajal are also associated with enteric neurons and may act as an intermediary between the neurons and smooth muscle.

GI Smooth Muscle Exhibits Different Patterns of Contraction

Muscle contractions in the gastrointestinal tract occur in three general patterns. Between meals, when the tract is largely empty, a series of contractions begins in the stomach and passes slowly from section to section, each taking about 90 minutes to reach the large intestine. This pattern, known as the **migrating motor complex**, is a "housekeeping" function that sweeps food remnants and bacteria out of the upper GI tract and into the large intestine.

Muscle contractions during and following a meal fall into one of two other patterns. **Peristalsis** [*peri-*, surrounding + *stalsis*, contraction] is progressive waves of contraction that move from one section of the GI tract to the next, just like the

21

(a) Peristaltic contractions create forward movement.

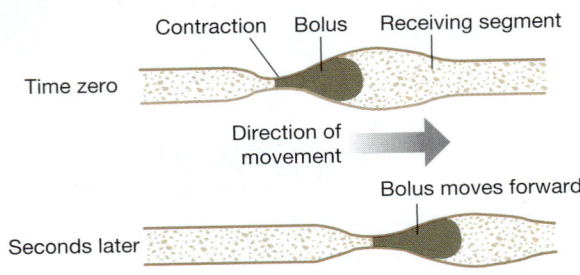

(b) Segmental contractions are responsible for mixing.

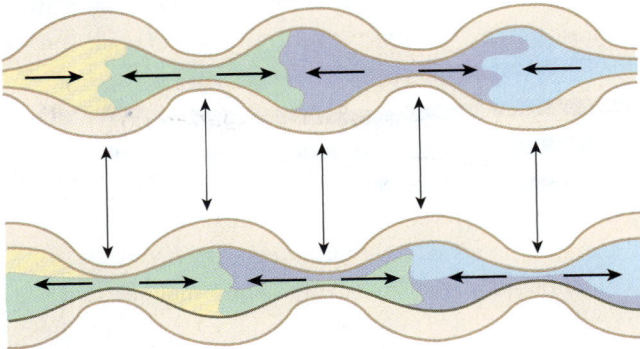

No net forward movement

■ **FIGURE 21-4** *Contractions in the GI tract*

Peristaltic contractions are most important in the esophagus. Segmental contractions are more important in the distal sections of the GI tract.

human "waves" that ripple around a football stadium or basketball arena. In peristalsis, circular muscles contract just behind a mass, or **bolus**, of food. This pushes the bolus forward into a *receiving segment*, where the circular muscles are relaxed (Fig. 21-4a ■). The receiving segment then contracts, continuing the forward movement. Peristaltic contractions push a bolus forward at speeds between 2 and 25 cm/sec.

Peristalsis propels material forward through the esophagus and mixes food as it digests in the stomach. In normal digestion, intestinal peristaltic waves are limited to short distances. Hormones, paracrine signals, and the autonomic nervous system influence peristalsis in all regions of the GI tract.

In **segmental contractions**, short (1–5 cm) segments of intestine alternately contract and relax (Fig. 21-4b). In the contracting segments, circular muscles contract while longitudinal muscles relax. These contractions may occur randomly along the intestine or at regular intervals, as shown in Figure 21-4b. Alternating segmental contractions churn the intestinal contents, mixing them and keeping them in contact with the absorptive epithelium. When segments contract sequentially, in an oral-to-aboral direction [*ab,*- away], digesting material is propelled short distances.

Motility disorders are among the more common gastrointestinal problems. They range from esophageal spasms and

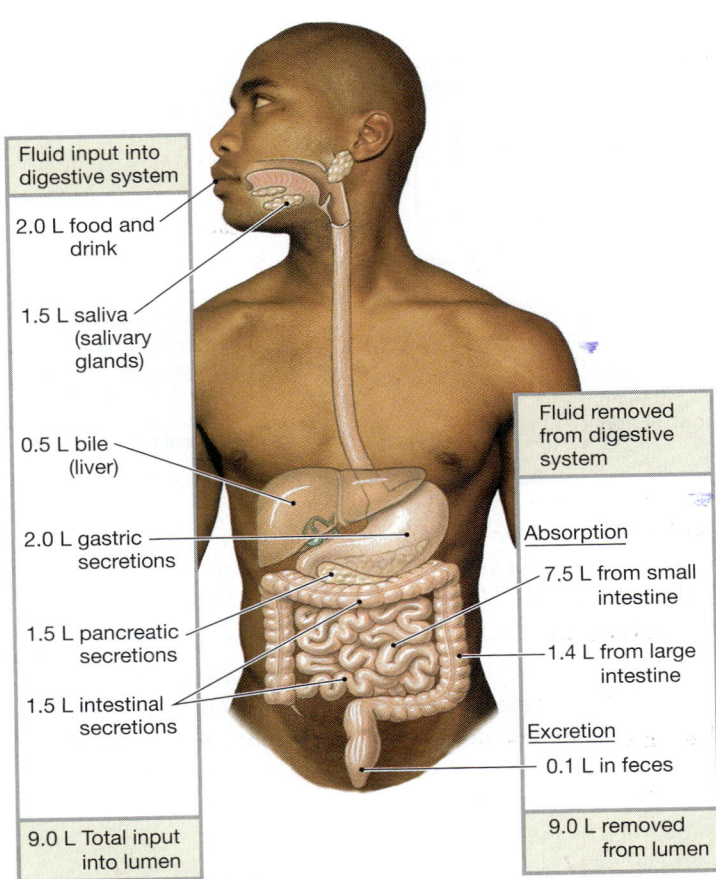

■ **FIGURE 21-5** *Daily mass fluid balance in the digestive system*

To maintain homeostasis, intake must equal output. Thus the volume of fluid entering the GI tract lumen by intake or secretion must equal the volume that leaves the lumen. Of the 2 L of water in food and drink ingested daily, only 0.1 L leaves in the feces. The remaining 1.9 L is absorbed and excreted from the body in urine or through insensible water loss (respiration and sweating).

delayed gastric (stomach) emptying to constipation and diarrhea. *Irritable bowel syndrome* is a chronic functional disorder characterized by altered bowel habits and abdominal pain.

CONCEPT CHECK

7. Why are some sphincters of the digestive system tonically contracted?
   Answers: p. 714

SECRETION

In a typical day, 9 liters of fluid pass through the lumen of an adult's gastrointestinal tract—equal to the contents of three 3-liter soft drink bottles! Only about 2 liters of that volume enter the GI system through the mouth. The remaining 7 liters of fluid come from body water secreted along with enzymes and mucus (Fig. 21-5 ■). About half the secreted fluid comes from accessory organs and glands such as the salivary glands, pancreas, and liver. The remaining 3.5 liters are secreted by epithelial cells of the digestive tract itself.

The significance of the secreted volume becomes more apparent when you realize that the 7 liters of fluid secreted equal one-sixth of total body water volume (42 liters), and more than twice the plasma volume (3 liters). These numbers tell you that if the fluid secreted into the lumen is not reabsorbed as it passes along the GI tract, the body will rapidly dehydrate.

The Digestive System Secretes Ions and Water

A large portion of the 7 liters of fluid secreted by the digestive system each day is composed of water and ions, particularly Na^+, K^+, Cl^-, HCO_3^-, and H^+. The ions are first secreted into the lumen of the tract, then reabsorbed. Water follows osmotic gradients created by the transfer of solutes from one side of the epithelium to the other. Water moves through the epithelial cells via membrane channels or between cells (the paracellular pathway).

Gastrointestinal epithelial cells, like those in the kidney, have distinct apical and basolateral membranes. Each cell surface contains proteins for active transport, facilitated diffusion, and ion movement through open channels. The arrangement of channels and transporters on the apical and basolateral membranes determines the net movement of solutes across the epithelium. In addition, Na^+ may also move through leaky junctions between cells in some parts of the digestive tract.

Many of the membrane transporters of the GI tract are similar to those of the renal tubule. The basolateral membrane contains the ubiquitous Na^+-K^+-ATPase. Cotransporters include the NKCC (Na^+-K^+-$2Cl^-$) symporter, the Cl^--HCO_3^- antiporter, the Na^+-H^+ antiporter, and the H^+-K^+-ATPase. Ion channels include Na^+, K^+, and Cl^- channels, such as the gated Cl^- channel known as the **cystic fibrosis transmembrane conductance regulator**, or **CFTR chloride channel**. Defects in CFTR channel structure or function lead to the disease *cystic fibrosis*.

The sections that follow describe how these membrane proteins are arranged to create acid secretion by the stomach, bicarbonate secretion by the pancreas and duodenum, and isotonic NaCl secretion by the intestines.

Acid Secretion Parietal cells deep in the gastric glands secrete hydrochloric acid into the lumen of the stomach. Acid secretion in the stomach averages 1–3 liters per day and can create a luminal pH as low as 1. The cytoplasmic pH of the parietal cells is about 7.2, which means the cells are pumping H^+ against a gradient that is 2.5 *million* times more concentrated in the lumen.

The parietal cell pathway for acid secretion is depicted in Figure 21-6 ■. The process begins when H^+ from water inside the parietal cell is pumped into the stomach lumen by an

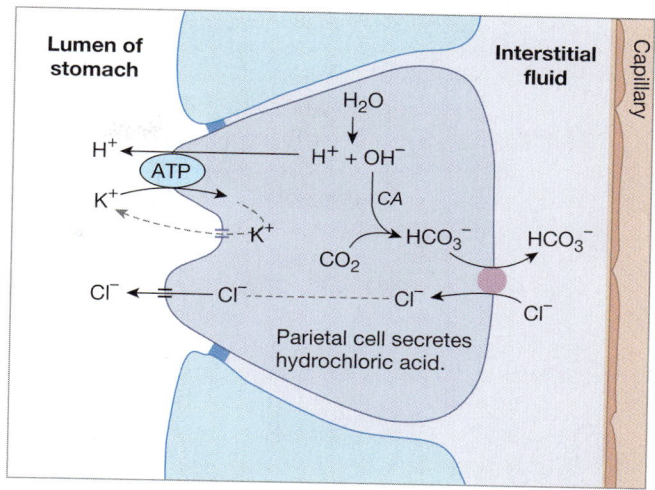

■ **FIGURE 21-6** *Acid secretion by parietal cells*

Hydrochloric acid is secreted into the lumen of the stomach while bicarbonate is absorbed into the gastric blood supply. CA = carbonic anhydrase.

H^+-K^+-ATPase in exchange for K^+ entering the cell. Cl^- then follows H^+ through an open chloride channel, resulting in net secretion of HCl by the cell. (This apical Cl^- channel is not a CFTR channel.) While acid is being secreted into the lumen, bicarbonate made from CO_2 and the OH^- from water is absorbed into the blood. The buffering action of HCO_3^- makes blood leaving the stomach less acidic, creating the *"alkaline tide"* that can be measured as a meal is being digested.

Bicarbonate Secretion Bicarbonate secretion into the duodenum neutralizes acid entering from the stomach. A small amount of bicarbonate comes from duodenal cells, but most comes from the pancreas, which secretes a watery solution of $NaHCO_3$. The exocrine portion of the pancreas consists of lobules called **acini** [*acinus*, grape or berry] that open into ducts whose lumens are part of the body's external environment (Fig. 21-7 ■). The acinar cells secrete digestive enzymes, and the duct cells secrete the $NaHCO_3$ solution. (The pancreas also secretes hormones from *islet cells* tucked among the acinar cells, as illustrated in Figure 21-7.)

Bicarbonate production requires high levels of the enzyme *carbonic anhydrase*, levels similar to those found in renal tubule cells and red blood cells [⊡ p. 599, 668]. Bicarbonate produced from CO_2 and water is secreted by an apical Cl^--HCO_3^- antiport protein (Fig. 21-8 ■). Chloride enters on the basolateral NKCC cotransporter and leaves via an apical CFTR channel. Luminal Cl^- then re-enters the cell in exchange for HCO_3^- entering the lumen. Hydrogen ions produced along with bicarbonate leave the cell on a basolateral Na^+-H^+ antiporter. The H^+ thus reabsorbed into the intestinal circulation helps balance

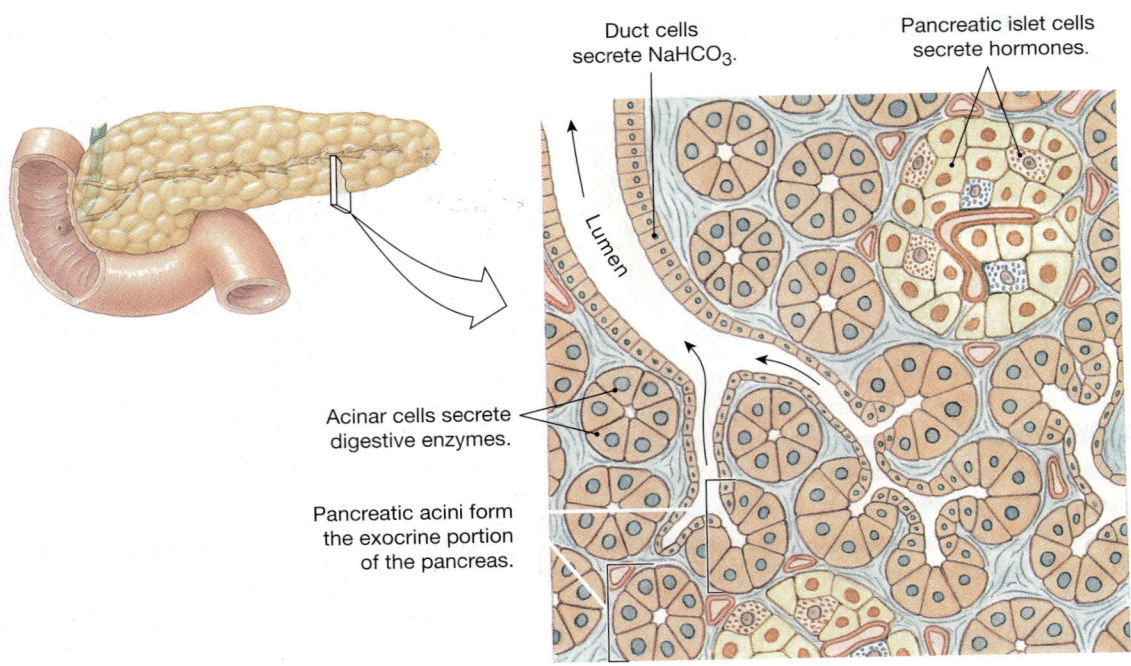

Duct cells secrete NaHCO₃.

Pancreatic islet cells secrete hormones.

Lumen

Acinar cells secrete digestive enzymes.

Pancreatic acini form the exocrine portion of the pancreas.

■ **FIGURE 21-7** *Anatomy of the exocrine and endocrine pancreas*

The secretions of the exocrine pancreas enter the digestive tract through ducts; hormones from the endocrine islet cells enter the blood.

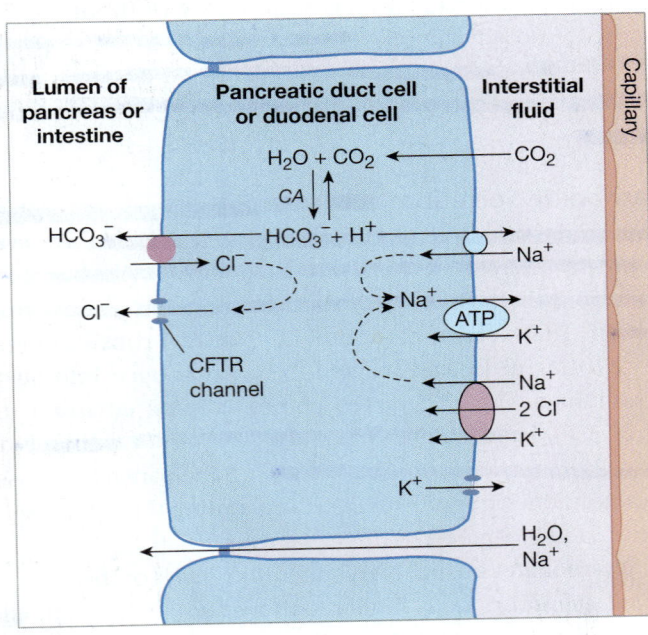

Lumen of pancreas or intestine

Pancreatic duct cell or duodenal cell

Interstitial fluid

Capillary

$H_2O + CO_2 \leftarrow CO_2$

CA

$HCO_3^- \leftarrow$ $HCO_3^- + H^+$

$\rightarrow Na^+$

Cl^-

$Cl^- \leftarrow$ $\rightarrow Na^+$

ATP

K^+

CFTR channel

Na^+

$2\ Cl^-$

K^+

K^+

$H_2O,$
Na^+

■ **FIGURE 21-8** *Bicarbonate secretion*

Cells that produce bicarbonate have high concentrations of carbonic anhydrase (CA). HCO_3^- is exchanged for Cl^- at the apical membrane, and H^+ is exchanged for Na^+ at the basolateral membrane. Bicarbonate secretion occurs in the exocrine pancreas and the duodenum.

HCO_3^- put into the blood when parietal cells secrete H^+ into the stomach (see Fig. 21-6).

Sodium and water movement are passive processes, driven by electrochemical and osmotic gradients. The net movement of negative ions from the ECF to the lumen attracts Na^+, which moves down its electrochemical gradient through leaky junctions between the cells. The secretion of Na^+ and HCO_3^- into the lumen creates an osmotic gradient, and water follows by osmosis. The net result is secretion of a watery sodium bicarbonate solution.

In cystic fibrosis, an inherited defect causes the CFTR channel to be defective or absent. As a result, secretion of chloride and fluid ceases, but goblet cells [p. 75] continue to secrete mucus, resulting in thickened mucus. In the digestive system, the thick mucus clogs small pancreatic ducts and prevents digestive enzyme secretion into the intestine. In airways of the respiratory system, where the CFTR channel is also found, failure to secrete fluid clogs the mucociliary escalator [p. 568] with thick mucus, leading to recurrent lung infections.

NaCl Secretion In addition to ions secreted by the stomach and pancreas, crypt cells in the small intestine and colon secrete an isotonic NaCl solution. The active step is Cl^- secretion, followed by Na^+ and water moving through the paracellular pathway (Fig. 21-9 ■). Chloride enters cells via the NKCC transporter, then exits via an apical CFTR channel. The secreted

RUNNING PROBLEM

Protein digestion in the stomach releases large amounts of urea, the nitrogenous breakdown product of amino acids. *Helicobacter pylori* exploits this sea of urea as protection against the acidic environment of the stomach. The outer membrane of *H. pylori* is studded with ureases, enzymes that convert urea into carbon dioxide and ammonia, which is a base.

Question 2:
   How does the conversion of urea into ammonia protect *H. pylori* from the hostile environment of the stomach?

677   682   **687**   691   704   710

exocytosis [p. 150]. Many intestinal enzymes are not released free into the lumen but remain bound to the apical membranes of intestinal cells, anchored by transmembrane protein "stalks" or lipid anchors [p. 57].

Some digestive enzymes are secreted in an inactive *proenzyme* form known collectively as *zymogens*. Zymogens must be activated in the GI lumen before they can carry out digestion. This late activation allows enzymes to be stockpiled in the cells that make them without damaging those cells. Zymogen names often have the suffix *–ogen* added to the enzyme name, such as *pepsinogen*.

The control pathways for enzyme release vary but include a variety of neural, hormonal, and paracrine signals. Usually, stimulation of parasympathetic neurons in the vagus nerve enhances enzyme secretion.

saline solution mixes with mucus secreted by goblet cells to help lubricate the contents of the gut.

Digestive Enzymes Are Secreted into the Mouth, Stomach, and Intestine

Digestive enzymes are secreted either by exocrine glands (salivary glands and the pancreas) or by epithelial cells in the mucosa of the stomach and small intestine. Enzymes are proteins, which means they are synthesized on the rough endoplasmic reticulum, packaged by the Golgi apparatus into secretory vesicles, and stored in the cell until needed. On demand, they are released by

Specialized Cells Secrete Mucus

Mucus is a viscous secretion composed primarily of glycoproteins collectively called **mucins**. The primary functions of mucus are to form a protective coating over the GI mucosa and to lubricate the contents of the gut. Mucus is made in specialized exocrine cells called *mucous cells* in the stomach and *goblet cells* in the intestine [Fig. 3-27, p. 76]. Goblet cells make up between 10% and 24% of the intestinal cell population. The salivary glands also secrete mucus from specialized cells.

The signals for mucus release include parasympathetic innervation, a variety of neuropeptides found in the enteric nervous system, and cytokines from immunocytes. Parasitic infections and inflammatory processes in the gut both cause substantial increases in mucus output as the body attempts to fortify its protective barrier.

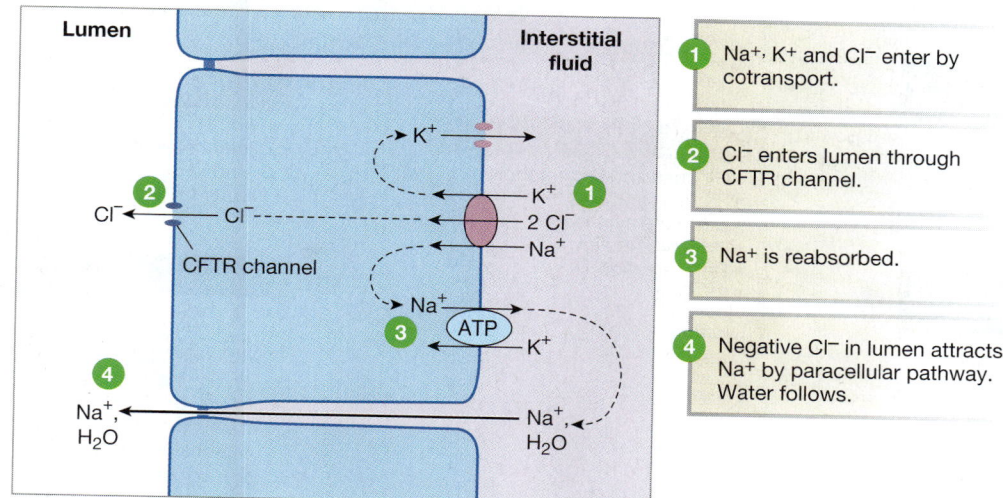

■ **FIGURE 21-9**   *Cl⁻ secretion by intestinal and colonic crypt cells*

Chloride enters cells by indirect active transport and leaves the apical side through a CFTR channel. Sodium and water follow passively by the paracellular pathway.

(a) The liver is the largest of the internal organs, weighing about 1.5 kg (3.3 lb) in an adult. It lies just under the diaphragm, toward the right side of the body.

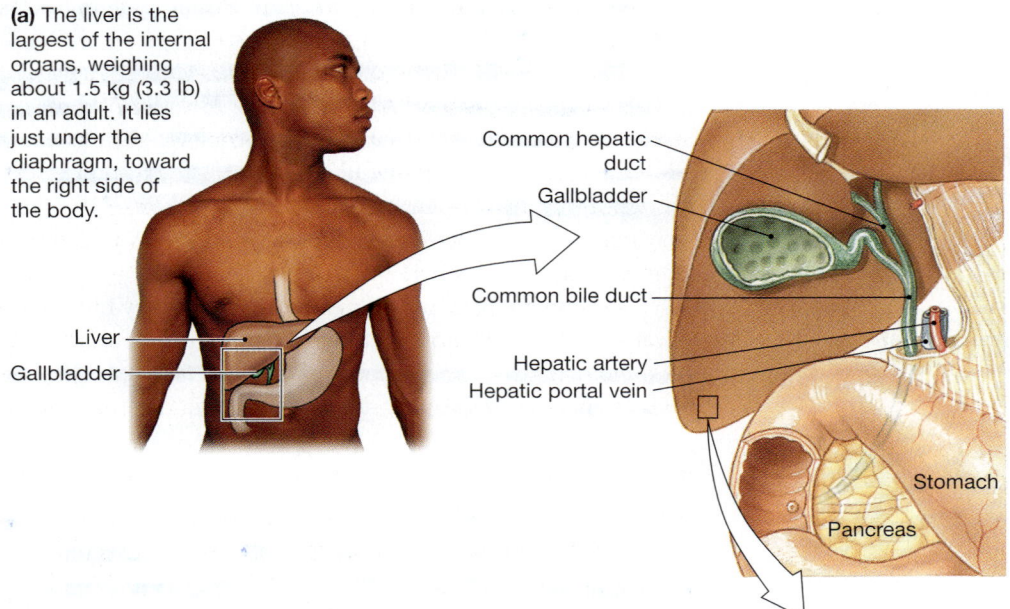

Liver
Gallbladder

Common hepatic duct
Gallbladder
Common bile duct
Hepatic artery
Hepatic portal vein
Stomach
Pancreas

(b) Blood flow to the liver comes from two sources. Oxygenated blood containing metabolites from peripheral tissues reaches the liver via the hepatic artery. Blood to the liver via the hepatic portal vein is rich in absorbed nutrients from the gastro-intestinal tract (Fig. 21-30) and contains hemoglobin breakdown products from the spleen. Blood leaves the liver in the hepatic vein (not shown). Bile synthesized in the liver is secreted into the **common hepatic duct** for storage in the gallbladder. From there, it is secreted into the lumen of the intestine through the **common bile duct**.

(c) The hepatocytes of the liver are organized into irregular hexagonal units called lobules. Each lobule is centered around a central vein that drains blood into the hepatic vein. Along its periphery, a lobule is associated with branches of the hepatic portal vein and hepatic artery. These vessels branch among the hepatocytes, forming **sinusoids** into which the blood flows. About 70% of the surface area of each hepatocyte faces the sinusoids, maximizing the exchange between the blood and the cells. Roughly 15% of the hepatocyte membrane faces the **bile canaliculi**, small channels into which bile is secreted. The canaliculi coalesce into bile ductules that run through the liver alongside the portal veins.

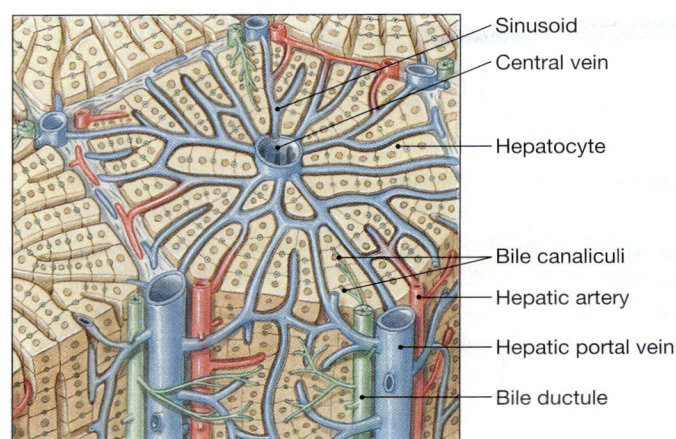

Sinusoid
Central vein
Hepatocyte
Bile canaliculi
Hepatic artery
Hepatic portal vein
Bile ductule

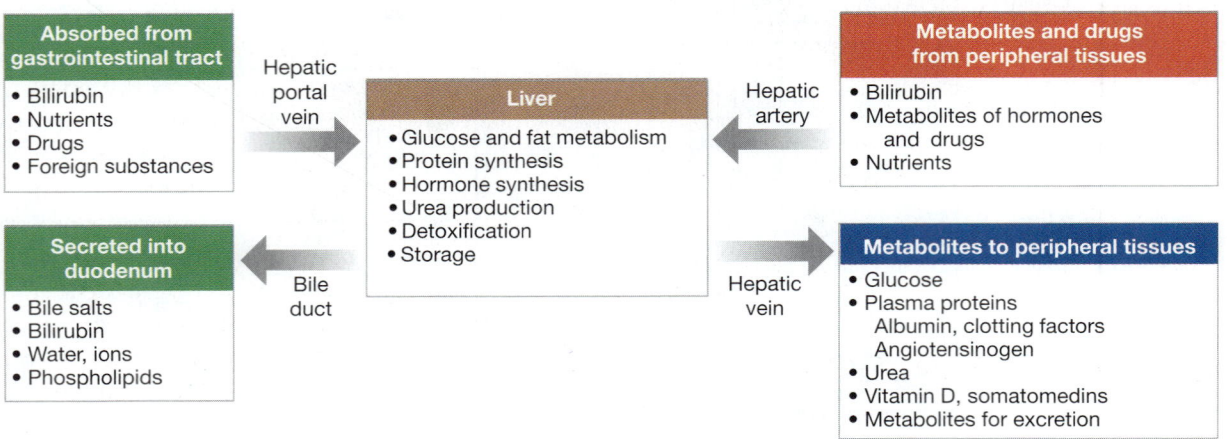

Absorbed from gastrointestinal tract		Metabolites and drugs from peripheral tissues
• Bilirubin • Nutrients • Drugs • Foreign substances	Hepatic portal vein → Liver ← Hepatic artery • Glucose and fat metabolism • Protein synthesis • Hormone synthesis • Urea production • Detoxification • Storage	• Bilirubin • Metabolites of hormones and drugs • Nutrients
Secreted into duodenum		**Metabolites to peripheral tissues**

• Bile salts
• Bilirubin
• Water, ions
• Phospholipids
	← Bile duct / Hepatic vein →	• Glucose
		• Plasma proteins
		  Albumin, clotting factors
		  Angiotensinogen
		• Urea
		• Vitamin D, somatomedins
		• Metabolites for excretion

(d) Blood entering the liver brings nutrients and foreign substances from the digestive tract, bilirubin from hemoglobin breakdown, and metabolites from peripheral tissues of the body. In turn, the liver excretes some of these in the bile and stores or metabolizes others. Some of the liver's products are wastes to be excreted by the kidney; others are essential nutrients, such as glucose. In addition, the liver synthesizes an assortment of plasma proteins.

■ **FIGURE 21-10**

Saliva Is an Exocrine Secretion

Saliva is a complex hyposmotic fluid secreted by the salivary glands of the oral cavity. The components of saliva include water, ions, mucus, and proteins such as enzymes and immunoglobulins. At resting flow rates, saliva is slightly acidic, with a pH of 6–7.

The ionic composition of saliva is determined in two epithelial transport steps. The salivary glands are exocrine glands, with a secretory epithelium (the *acinus*) that opens to the outside environment through a duct (see Fig. 3-28, p. 77). Fluid secreted by the acinar cells resembles extracellular fluid in its ionic composition. As this fluid passes through the duct on its way to the oral cavity, epithelial cells along the duct reabsorb Na^+ and secrete K^+ until the ion ratio in the duct fluid is more like that of intracellular fluid (high K^+ and low Na^+).

Salivation is controlled by the autonomic nervous system. Parasympathetic innervation is the primary stimulus for secretion of saliva, but there is also some sympathetic innervation to the glands. In ancient China, a person under suspicion for a crime might have been given a mouthful of dry rice to chew during questioning. If he could produce enough saliva to moisten the rice and swallow it, he went free. If his nervous state dried up his salivary reflex, however, he was pronounced guilty. Recent research has confirmed that stress, such as that associated with lying, decreases salivary secretion.

The Liver Secretes Bile

Bile is a nonenzymatic solution secreted from hepatocytes, or liver cells (see *Focus on the Liver*, Fig. 21-10 ■). The key components of bile are (1) bile salts that facilitate enzymatic fat digestion, (2) bile pigments, such as bilirubin, that are the waste products of hemoglobin degradation, and (3) cholesterol, which is excreted in the feces. Drugs and other xenobiotics are cleared from the blood by hepatic processing and are also excreted in bile. Bile salts, which act as detergents to solubilize fats during digestion, are made from steroid bile acids combined with amino acids.

Bile is secreted into hepatic ducts that lead to the gallbladder, which stores and concentrates the bile solution. During a meal, contraction of the gallbladder sends bile into the duodenum through the common bile duct, along with a watery solution of bicarbonate and digestive enzymes from the pancreas.

REGULATION OF GI FUNCTION

The digestive system has remarkably complex regulation of motility and secretion. The various control mechanisms, which include neural, endocrine, and local components, include the following:

1. **Long reflexes integrated in the CNS.** A classic neural reflex begins with a stimulus transmitted along a sensory neuron to the CNS, where the stimulus is integrated and acted on. In the digestive system, some classic reflexes originate with sensory receptors in the GI tract, but others originate outside the digestive system (grey arrows in Fig. 21-11 ■). No matter where they originate, digestive reflexes integrated in the CNS are called **long reflexes**.

   Long reflexes that originate completely outside the digestive system include feedforward reflexes [🔁 p. 200] and emotional reflexes. These reflexes are called **cephalic reflexes** because they originate in the cephalic brain [*cephalicus*, head]. *Feedforward reflexes* begin with stimuli such as the sight, smell, sound, or thought of food; they prepare the digestive system for food that the brain is anticipating. For example, if you are hungry and smell dinner cooking, your mouth waters and your stomach growls.

   The influence of emotions on the GI tract illustrates another reflexive link between the brain and the digestive system. Emotional responses that have GI manifestations range from travelers' constipation to "butterflies in the stomach" to psychologically induced diarrhea. Fight-or-flight reactions also influence GI function.

   In long reflexes, the smooth muscle and glands of the GI tract are under autonomic control. In general, we say that parasympathetic neurons to the GI tract, carried mostly in the vagus nerve, are excitatory and enhance GI functions. Sympathetic neurons usually inhibit GI function.

2. **Short reflexes integrated in the enteric nervous system.** Neural control of the GI tract does not rely strictly on the CNS. Instead, the enteric nerve plexus in the gut wall acts as a "little brain," allowing local reflexes to begin, be integrated, and end completely in the GI tract (red arrows in Fig. 21-11). Reflexes that originate within the enteric nervous system (ENS) and are integrated there without outside input are called **short reflexes**. Although the ENS can work in isolation, it also coordinates function with autonomic neurons bringing signals from the CNS.

   The processes controlled by the enteric nervous system are related to motility, secretion, and growth. The submucosal plexus contains sensory neurons that receive signals from the lumen of the gut. The ENS network integrates the sensory information, then initiates responses through submucosal neurons that control secretion by GI epithelial cells and myenteric neurons that influence motility.

3. **Reflexes involving GI peptides.** One particularly challenging part of mastering GI regulation is learning the terminology. Some of the peptide signal molecules involved in digestive regulation were first identified in other body systems, and thus their names have nothing to do with their function in the gastrointestinal system.

   Peptides secreted by cells of the digestive tract may act as hormones or paracrine signals. GI hormones, like all hormones, are secreted into the blood. They act on

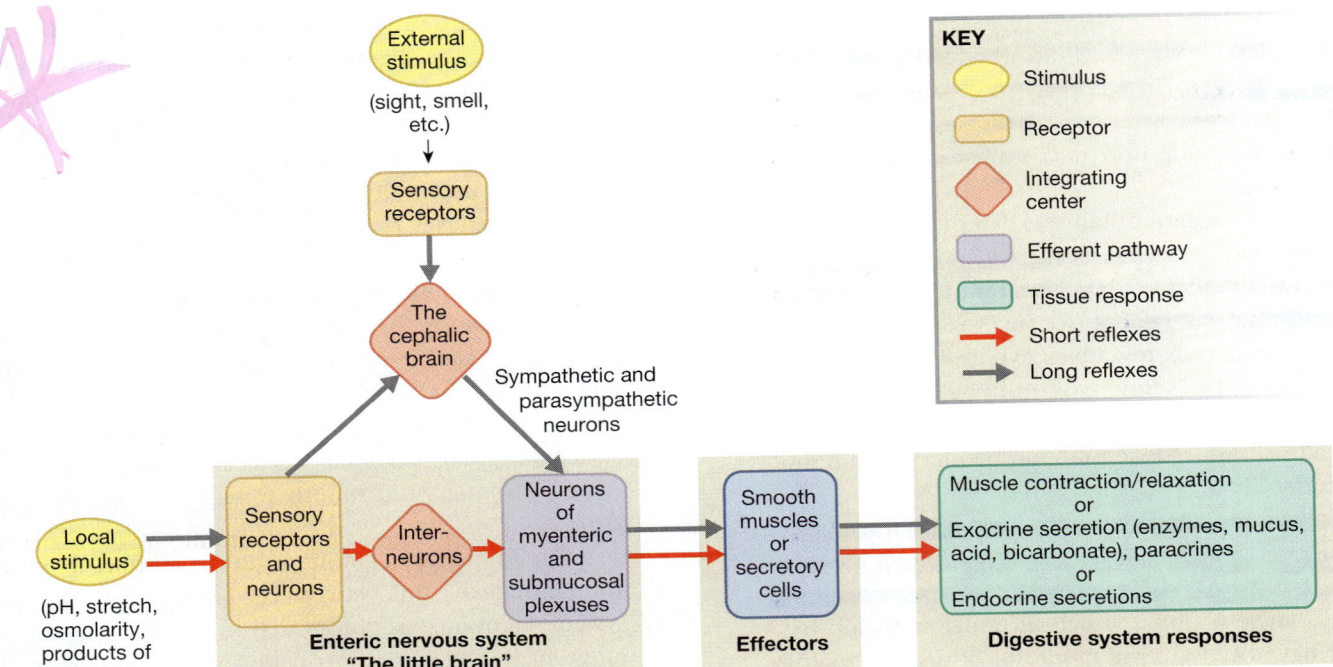

■ FIGURE 21-11 *Integration of long and short reflexes in the digestive system*

Long reflexes are integrated in the CNS; some originate outside the GI tract, but others originate in the enteric nervous system. Short reflexes originate in the enteric nervous system and are carried out by neurons in the wall of the gut.

FIGURE QUESTIONS

- Which effectors and responses are controlled by the myenteric plexus, and which are controlled by the submucosal plexus?
- What type of sensory receptor responds to stretch? to osmolarity? to products of digestion?

the GI tract, on accessory organs such as the pancreas, and on more distant targets, such as the brain. Paracrine molecules are secreted either into the lumen of the GI tract or into the extracellular fluid. Luminal paracrine signals combine with receptors on the apical membrane of the lumen epithelium to elicit a response. Paracrine molecules in the ECF act locally, on cells close to where they were secreted.

In general, the GI peptides excite or inhibit motility and secretion (Fig. 21-12 ■). Motility effects include altered peristaltic activity, contraction of the gallbladder for bile release, and regulated gastric emptying to maximize digestion and absorption. Secretory processes influenced by GI peptides include both endocrine and exocrine functions.

Some of the most interesting digestive pathways are those in which GI peptides act on the brain. For example, in experimental studies the GI hormone cholecystokinin (CCK) was found to enhance *satiety,* the feeling that hunger has been satisfied. However, these findings are complicated by the fact that CCK is also manufactured by neurons and functions as a neurotransmitter in the brain. Another GI peptide, *ghrelin,* is secreted by the stomach and acts on the brain to increase food intake. Peptides and appetite are discussed further in Chapter 22.

The Enteric Nervous System Can Act Independently of the CNS

The enteric nervous system was first recognized more than a century ago, when scientists noted that an increase in GI intraluminal pressure caused a reflex wave of peristaltic contraction to sweep along sections of isolated intestine removed from the body. What they observed was the ability of the ENS to carry out a reflex independent of control by the CNS.

In this respect, the enteric nervous system is much like the nerve networks of jellyfish and sea anemones (phylum Cnidaria) that were introduced in Chapter 9 [p. 293]. You might have seen sea anemones being fed at an aquarium. As a piece of shrimp or fish drifts close to the tentacles, they begin to wave, picking up chemical "odors" through the water. Once the food contacts the tentacles, it is directed toward the mouth, passed from one tentacle to another until it disappears into the digestive cavity.

This purposeful reflex is accomplished without a brain, eyes, or a nose. The anemone's nervous system consists of a network composed of sensory neurons, interneurons, and efferent neurons that control the muscles and secretory cells of the anemone's body. The neurons of the network are linked in a way that allows them to integrate information and act on it. In the

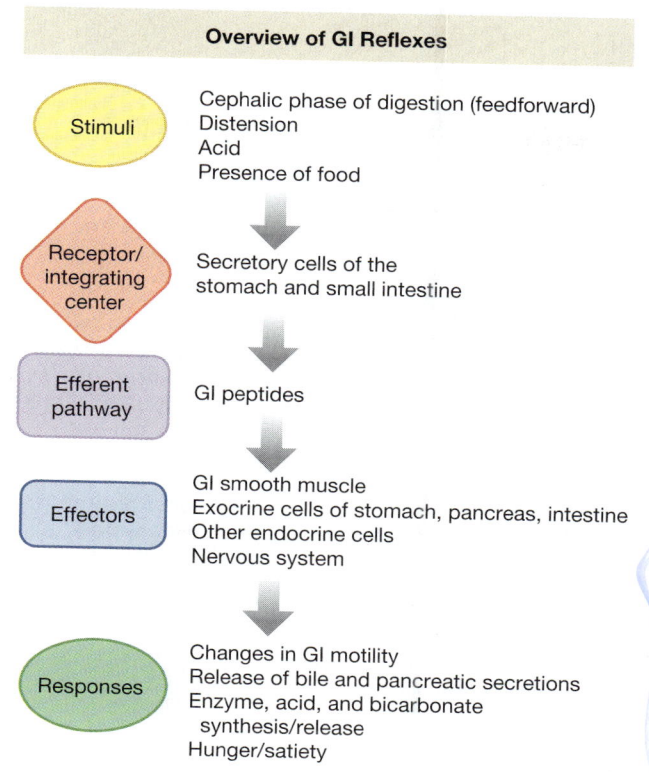

Overview of GI Reflexes

Stimuli
Cephalic phase of digestion (feedforward)
Distension
Acid
Presence of food

Receptor/integrating center
Secretory cells of the stomach and small intestine

Efferent pathway
GI peptides

Effectors
GI smooth muscle
Exocrine cells of stomach, pancreas, intestine
Other endocrine cells
Nervous system

Responses
Changes in GI motility
Release of bile and pancreatic secretions
Enzyme, acid, and bicarbonate
 synthesis/release
Hunger/satiety

■ **FIGURE 21-12**   *Reflexes involving GI peptides*

same way that an anemone captures its food, the ENS receives stimuli and acts on them. Thus, the enteric nervous system is able to function autonomously, independent of efferent signals from the CNS.

RUNNING PROBLEM

Tonya undergoes endoscopy, a procedure in which a fiber-optic scope is inserted into her esophagus and maneuvered into her stomach. Small pincers attached to the end of the scope pluck a piece of her stomach lining, which is sent to the pathology laboratory for study. During the procedure, Tonya is awake but groggy from a tranquilizer. That afternoon she feels fine, but her stomach hurts. She chews an antacid tablet for relief.

Question 3:
  Some antacids contain sodium bicarbonate ($NaHCO_3$), while many others contain aluminum hydroxide ($Al(OH)_3$). Antacids act as buffers that help neutralize hydrochloric acid (HCl) secreted by the stomach. For both types of antacids, write the chemical equation showing how antacids buffer hydrochloric acid.

677   682   687   **691**   704   710

Anatomically and functionally, the enteric nervous system shares many features with the CNS:

1. *Intrinsic neurons.* The **intrinsic neurons** of the two nerve plexuses of the digestive tract are those neurons that lie completely within the wall of the gut. These ENS neurons are thus similar to interneurons of the CNS. (Autonomic neurons that bring signals from the CNS to the digestive system are called **extrinsic neurons**.)

2. *Neurotransmitters and neuromodulators.* ENS neurons release more than 30 neurotransmitters and neuromodulators, most of which are identical to molecules found in the brain. These neurotransmitters are sometimes called *non-adrenergic, non-cholinergic* to distinguish them from the traditional autonomic neurotransmitters norepinephrine and acetylcholine. Among the best-known neurotransmitters and neuromodulators are serotonin, vasoactive intestinal peptide, and nitric oxide.

3. *Support cells.* The support cells of neurons within the ENS are more similar to astroglia of the brain than to Schwann cells of the peripheral nervous system.

4. *Diffusion barrier.* The capillaries that surround ganglia in the ENS are not very permeable and create a diffusion barrier that is similar to the blood-brain barrier of cerebral blood vessels.

5. *Integrating center.* As noted earlier, reflexes that originate in the GI tract can be integrated and acted on without neural signals leaving the ENS. Thus, the neuron network of the ENS is its own integrating center, much like the brain and spinal cord.

It was once thought that if we could explain how the ENS integrates simple behaviors, we could use the system as a model for CNS function. But studying ENS function is difficult because enteric reflexes have no discrete command center. Instead, in an interesting twist, GI physiologists are applying information gleaned from studies of the brain and spinal cord to investigate ENS function. The complex interactions between the enteric and central nervous systems, the endocrine system, and the immune system promise to provide scientists with questions to investigate for many years to come.

GI Peptides Include Hormones, Neuropeptides, and Cytokines

The hormones of the gastrointestinal tract occupy an interesting place in the history of endocrinology. In 1902, two Canadian physiologists, W. M. Bayliss and E. H. Starling, discovered that acidic chyme entering the small intestine from the stomach caused the release of pancreatic juices even when all nerves to the pancreas were cut. Because the only communication remaining between intestine and pancreas was the blood supply that ran between them, Bayliss and Starling postulated the existence of some blood-borne (*humoral*) factor released by the intestine.

When duodenal extracts applied directly to the pancreas stimulated secretion, they knew they were dealing with a

21

TABLE 21-1   The Digestive Hormones

	SECRETED BY	TARGET(S)	EFFECTS ON ENDOCRINE SECRETION	EFFECTS ON EXOCRINE SECRETION
Gastrin	G cells in stomach antrum	ECL cells; parietal cells	None	Stimulates gastric acid
Cholecystokinin (CCK)	Endocrine cells of small intestine; neurons of brain and gut	Gallbladder, pancreas, gastric smooth muscle	None	Stimulates pancreatic enzyme secretion; potentiates bicarbonate secretion; inhibits acid secretion
Secretin	Endocrine cells in small intestine	Pancreas, stomach	None	Stimulates bicarbonate secretion; inhibits gastric acid and gastrin
Gastric inhibitory peptide (GIP)	Endocrine cells in small intestine	Beta cells of endocrine pancreas	Stimulates insulin release (feedforward mechanism)	Inhibits acid secretion
Motilin	Endocrine cells in small intestine	Smooth muscle of antrum and duodenum	None	None
Glucagon-like peptide 1	Endocrine cells in small intestine	Endocrine pancreas	Stimulates insulin release; inhibits glucagon release	Possibly inhibits acid secretion

chemical produced by the duodenum. They named the substance *secretin.* Starling further proposed that the general name *hormone,* from the Greek word meaning "I excite," be given to all humoral agents that act at a site distant from their release.

In 1905, J. S. Edkins postulated the existence of a gastric hormone that stimulated gastric acid secretion. It took more than 30 years for researchers to isolate a relatively pure extract of the gastric hormone, and it was 1964 before the hormone, named *gastrin,* was finally purified.

Why was research on the digestive hormones so slow to develop? A major reason is that GI hormones are secreted by isolated endocrine cells scattered among other cells of the mucosal epithelium. At one time, the only way to obtain these hormones was to make a crude extract of the entire epithelium, a procedure that also liberated digestive enzymes and paracrine molecules made in adjacent cells. Thus, it was very difficult to tell if the physiological effect elicited by the extract came from one hormone, from more than one hormone, or from a paracrine signal such as histamine.

Although researchers have now sequenced more than 30 peptides from the GI mucosa, only some of them are widely accepted as hormones. A few peptides have well-defined paracrine effects, but most fall into a long list of candidate hormones. In addition, we know of nonpeptide regulatory molecules, such as histamine, which functions as a paracrine signal. Because of the uncertainty associated with the field, we will restrict our focus to the major regulatory molecules.

The gastrointestinal hormones are divided into three families. All the members of a family have similar amino acid sequences, and in some cases there is overlap in their ability to bind to receptors.

1. The *gastrin family* includes the hormones **gastrin** and **cholecystokinin** (CCK), plus several variants of each. Their structural similarity means that both gastrin and CCK can bind to and activate the CCK_B receptor found on parietal cells.

2. The *secretin family* includes **secretin**; **vasoactive intestinal peptide** (VIP), a neurocrine [≈ p. 176] molecule; and **GIP**, a hormone known originally as *gastric inhibitory peptide* because it inhibited gastric acid secretion in early experiments. Subsequent studies, however, indicated that GIP administered in lower physiological doses does not block acid secretion. Researchers proposed a new name with the same initials—**glucose-dependent insulinotropic peptide**—that more accurately describes the hormone's action: it stimulates insulin release in response to glucose in

EFFECTS ON MOTILITY	OTHER EFFECTS	STIMULUS FOR RELEASE	RELEASE INHIBITED BY	OTHER INFORMATION
None	Enhanced mucosal cell growth	Peptides and amino acids in lumen; gastrin releasing peptide and ACh in nervous reflexes	pH < 1.5; somatostatin	—
Stimulates gallbladder contraction for bile release; inhibits gastric emptying; promotes intestinal motility	Stimulates satiety	Fatty acids and some amino acids	Somatostatin	Some effects may be due to CCK as a neuropeptide rather than as a hormone
Inhibits gastric emptying	None	Acid in small intestine	Somatostatin	—
None	Satiety and lipid metabolism(?)	Glucose, fatty acids, and amino acids in small intestine	NA	Acid inhibition questionable at physiological concentrations
Stimulates migrating motor complex	Action in brain(?)	Fasting: periodic release every 1.5–2 hours by neural stimulus	NA	Changes associated with both constipation and diarrhea, but relationship is unclear
Slows gastric emptying	Satiety	Mixed meal that includes carbohydrates or fats in the lumen	NA	Related to but not identical to pancreatic glucagon; acts in concert with GIP

the intestinal lumen. However, for the most part *gastric inhibitory peptide* has remained the preferred name.

Another member of the secretin family is the hormone **glucagon-like peptide 1** (GLP-1), which also plays an important role in glucose homeostasis. GIP and GLP-1 will be discussed further in Chapter 22.

3. The third family of peptides contains those that do not fit into the other two families. The primary member of this group is the hormone **motilin**.

The sources, targets, and effects of some GI hormones are summarized in Table 21-1 ■.

DIGESTION AND ABSORPTION

Digestion of macromolecules into absorbable units is accomplished by a combination of mechanical and enzymatic breakdown [🔁 p. 100]. Chewing and churning create smaller pieces of food with more surface area exposed to digestive enzymes. Bile, the complex chemical mixture secreted by the liver, serves a similar purpose by dispersing lipids (more commonly called *fats* in digestive physiology) as fine droplets with greater surface area.

The pH at which different digestive enzymes function best [🔁 p. 97] reflects the location where they are most active. Enzymes that act in the stomach work well at acidic pH; those secreted into the small intestine work best at alkaline pH.

Most absorption takes place in the small intestine, with additional absorption of water and ions in the large intestine. The surface area of the intestine is greatly increased by the presence of finger-like villi (see Fig. 21-2d, e) and by the **brush border** of epithelial cells created from numerous microvilli on the apical surface (Fig. 21-13 ■). Like secretion, absorption of nutrients and ions across the GI epithelium uses many of the same transport proteins as the kidney tubule. Within villi, capillaries pick up most absorbed nutrients except fats, which enter lymph vessels called **lacteals**.

Digestion and absorption are not directly regulated except in a few instances. Instead, they are influenced primarily by motility and secretion in the digestive tract, the two processes that in turn are regulated by hormones, the nervous system, and local control mechanisms.

Carbohydrates Are Absorbed as Monosaccharides

About half the calories the average American ingests are in the form of carbohydrates, mainly *starch* and *sucrose* (table

21

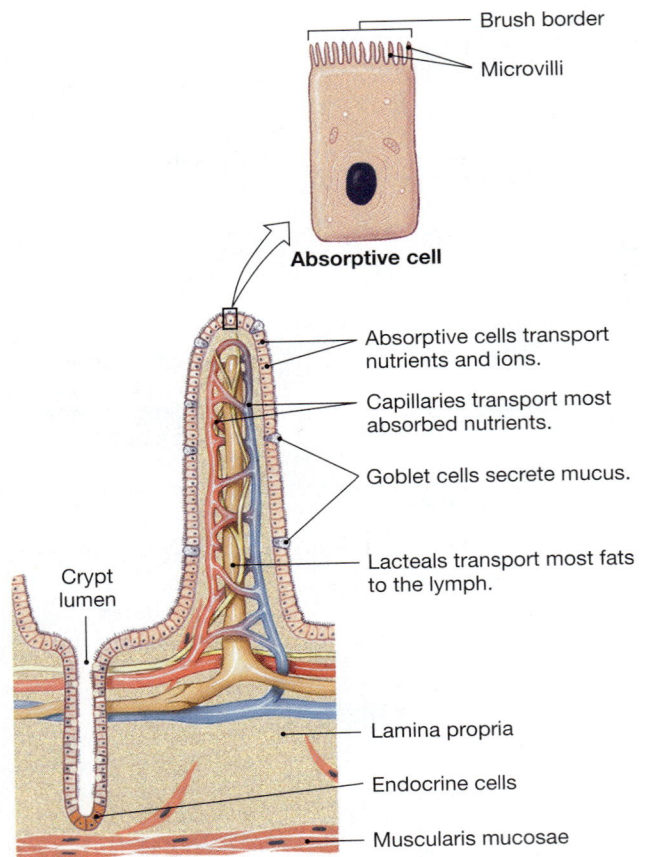

FIGURE 21-13 *A villus and a crypt in the small intestine*

Villi and crypts increase the effective surface area of the small intestine. Stem cells in the crypts produce new epithelial cells to replace those that die or are damaged.

sugar). Other dietary carbohydrates include the glucose polymers *glycogen* and *cellulose,* disaccharides such as *lactose* and *maltose,* and the monosaccharides *glucose* and *fructose* [Fig. 2-7, p. 28].

Intestinal carbohydrate transport is restricted to monosaccharides, which means that all complex carbohydrates and disaccharides must be digested if they are to be absorbed. We are unable to digest cellulose because we lack the necessary enzymes. As a result, the cellulose in plant matter becomes what is known as dietary *fiber,* or *roughage.*

The complex carbohydrates we can digest are starch and glycogen (Fig. 21-14 ■). **Amylase** is the enzyme that breaks long glucose polymers into smaller glucose chains and into the disaccharide maltose. Maltose and other disaccharides are broken down by intestinal brush-border enzymes known as **disaccharidases** (maltase, sucrase, and lactase). The end products of carbohydrate digestion are glucose, galactose, and fructose.

Intestinal glucose and galactose absorption uses transporters identical to those found in the renal proximal tubule: an apical Na^+-glucose SGLT symporter and a basolateral GLUT2 transporter

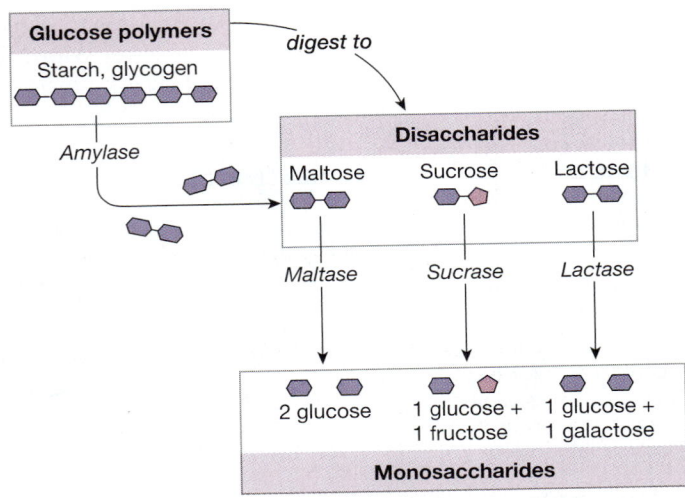

FIGURE 21-14 *Carbohydrate digestion*

(Fig. 21-15 ■). These transporters move galactose as well as glucose. Fructose absorption, however, is not Na^+-dependent. Fructose moves across the apical membrane by facilitated diffusion on the GLUT5 transporter and across the basolateral membrane by GLUT2 [p. 146].

If glucose is the major metabolic substrate for aerobic respiration, why don't intestinal epithelial cells use the glucose they absorb for their own metabolism? How can they keep intracellular glucose concentrations high so that facilitated diffusion will move glucose into the extracellular space? The metabolism of intestinal cells apparently differs from that of most other cells in that intestinal cells do not use glucose as their preferred energy source. Current studies suggest that these cells use the amino acid glutamine as their main source of energy, thus allowing absorbed glucose to pass unchanged into the bloodstream.

Proteins Are Digested into Small Peptides and Amino Acids

Unlike carbohydrates, which are ingested in forms ranging from simple to complex, most ingested proteins are polypeptides or larger [Fig. 2-9, p. 30]. Not all proteins are equally digestible by humans, however. Plant proteins are the least digestible. Among the most digestible is egg protein, 85–90% of which is in a form that can be digested and absorbed. Surprisingly, between 30% and 60% of the protein found in the intestinal lumen comes not from ingested food but from the sloughing of dead cells and from protein secretions such as enzymes and mucus.

The enzymes for protein digestion are classified into two broad groups: endopeptidases (also called *proteases*) and exopeptidases. **Endopeptidases** attack peptide bonds in the interior of the amino acid chain and make smaller peptide fragments from a long chain (Fig. 21-16b ■). These enzymes are secreted as inactive proenzymes from epithelial cells in the stomach, intestine, and pancreas and are activated in the GI

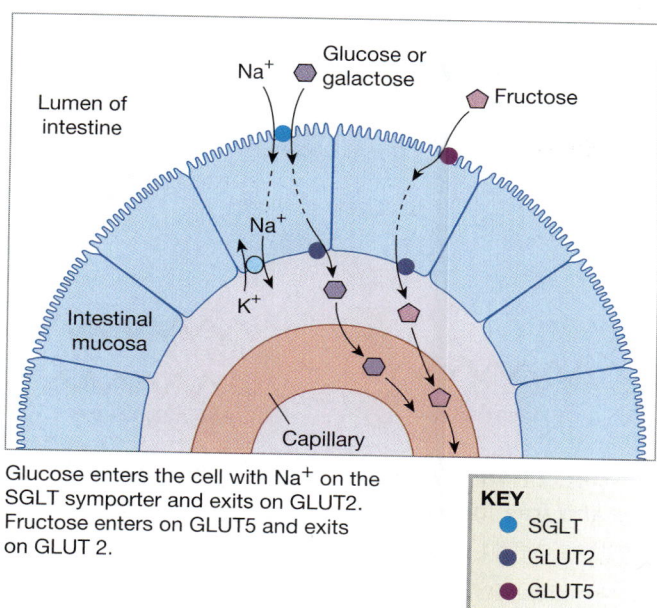

Glucose enters the cell with Na$^+$ on the SGLT symporter and exits on GLUT2. Fructose enters on GLUT5 and exits on GLUT 2.

KEY
- ● SGLT
- ● GLUT2
- ● GLUT5

■ **FIGURE 21-15** *Carbohydrate absorption in the small intestine*

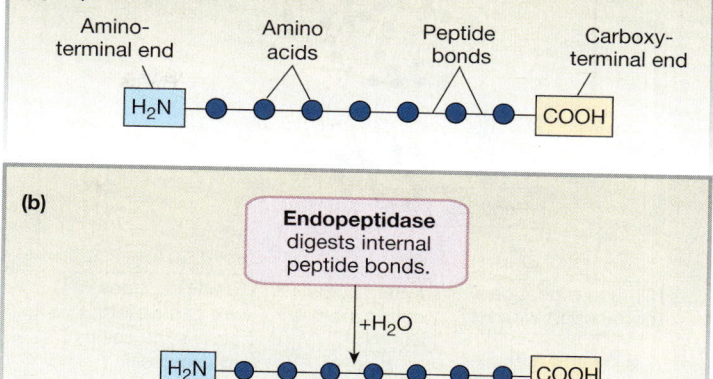

(a) Peptide structure

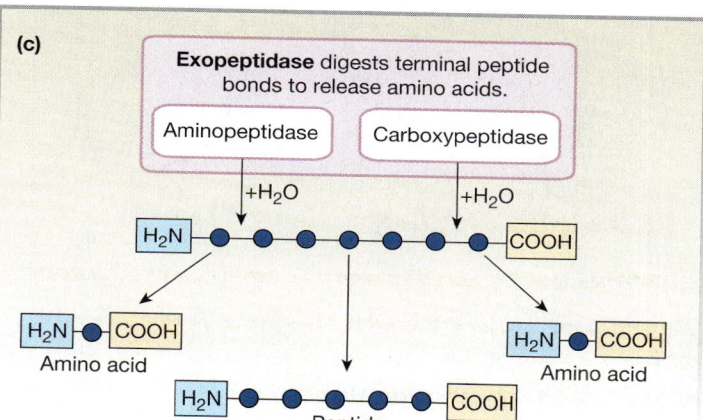

■ **FIGURE 21-16** *Enzymes for protein digestion*

Endopeptidases include pepsin in the stomach, and trypsin and chymotrypsin, which act in the small intestine.

tract lumen. Examples of endopeptidases include **pepsin** secreted in the stomach, and **trypsin** and **chymotrypsin** secreted by the pancreas.

Exopeptidases release single amino acids from peptides by chopping them off the ends, one at a time (Fig. 21-16c). The most important digestive exopeptidases are two isozymes of *carboxypeptidase* secreted by the pancreas. *Aminopeptidases* play a lesser role in digestion.

The primary products of protein digestion are free amino acids, dipeptides, and tripeptides, all of which can be absorbed. Amino acid structure is so variable that multiple amino acid transport systems are found in the intestine. Most free amino acids are carried by Na$^+$-dependent cotransport proteins similar to those in the proximal tubule of the kidney (Fig. 21-17 ■). A few amino acid transporters are H$^+$ dependent.

Dipeptides and tripeptides are carried into the mucosal cell using H$^+$-dependent cotransport. Once inside the epithelial cell, these peptides have two possible fates. Most are digested by cytoplasmic peptidases into amino acids, which are then transported across the basolateral membrane and into the circulation. Those that are not broken down to amino acids are transported intact across the basolateral membrane. The transport system that moves these peptides also is responsible for intestinal uptake of certain drugs, including beta-lactam antibiotics, angiotensin-converting enzyme inhibitors, and thrombin inhibitors.

Some Larger Peptides Can Be Absorbed Intact

Some peptides larger than three amino acids are absorbed by transcytosis [🔲 p. 153] after binding to membrane receptors

on the luminal surface of the intestine. The discovery that ingested proteins can be absorbed as small peptides has implications in medicine because these peptides may act as *antigens*, substances that stimulate antibody formation and result in allergic reactions. Consequently, the intestinal absorption of peptides may be a significant factor in the development of food allergies and food intolerances.

In newborns, peptide absorption takes place primarily in intestinal crypt cells (see Fig. 21-13). At birth, intestinal villi are very small, so the crypts are well exposed to the luminal contents. As the villi grow and the crypts have less access to chyme, the high peptide absorption rates present at birth decline steadily. If parents delay the ingestion of allergy-inducing peptides, the gut has a chance to mature, lessening the likelihood of antibody formation.

21

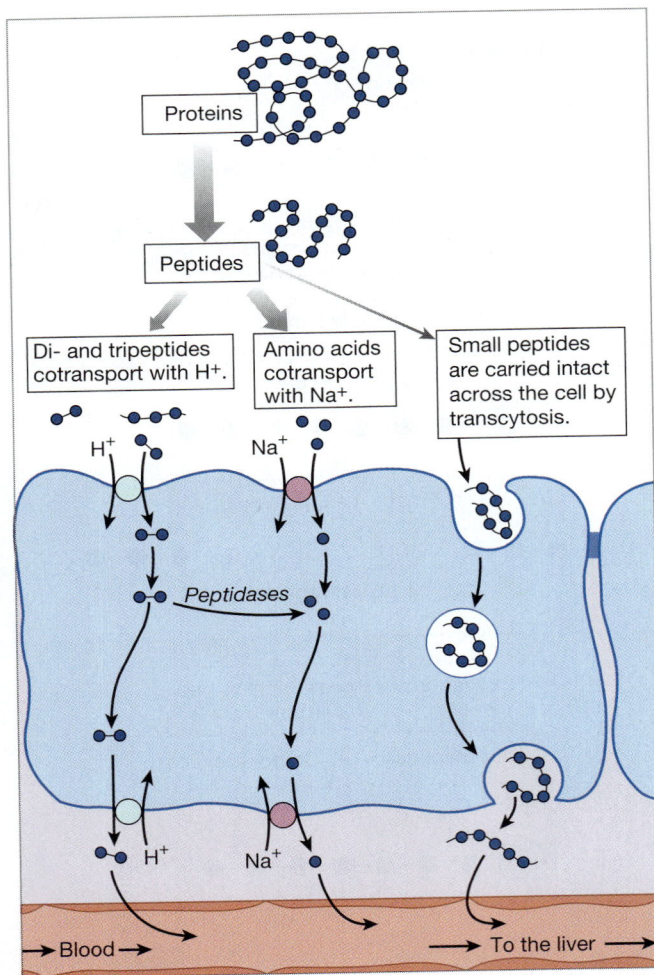

■ **FIGURE 21-17**   *Peptide absorption*

After digestion, proteins are absorbed mostly as free amino acids and as di- and tripeptides. Some peptides larger than tripeptides can be absorbed by transcytosis.

One of the most common antigens responsible for food allergies is gluten, a component of wheat. The incidence of childhood gluten allergies has decreased since the 1970s, when parents were cautioned that infants not be fed gluten-based cereals until they were several months old.

In another medical application, pharmaceutical companies have developed undigestible peptide drugs that can be given orally instead of by injection. Probably the best-known example is DDAVP (1-deamino-8-D-arginine vasopressin), the synthetic analog of vasopressin. If the natural hormone vasopressin is ingested, it is digested rather than absorbed intact. By changing the structure of the hormone slightly, scientists created a synthetic peptide that has the same activity but is absorbed without being digested.

Bile Salts Facilitate Fat Digestion

Fats and related molecules in the Western diet include triglycerides, cholesterol, phospholipids, long-chain fatty acids, and

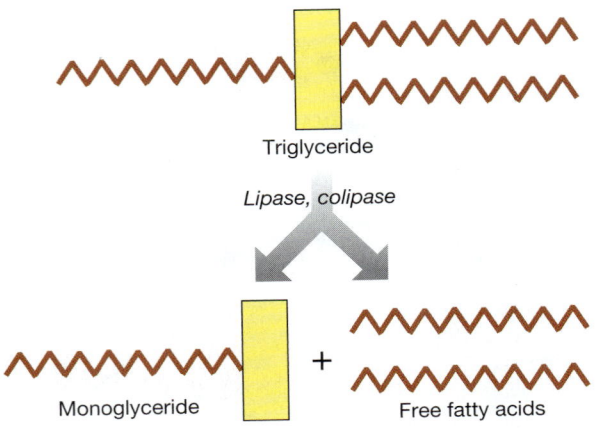

■ **FIGURE 21-18**   *Triglycerides digest into monoglycerides and free fatty acids*

the fat-soluble vitamins [Fig. 2-8, p. 29]. Nearly 90% of our fat calories come from triglycerides because they are the primary form of lipid in both plants and animals.

Enzymatic fat digestion is carried out by **lipases**, enzymes that remove two fatty acids from each triglyceride molecule. The result is one monoglyceride and two free fatty acids (Fig. 21-18 ■). Phospholipids are digested by pancreatic *phospholipase*. Free cholesterol need not be digested before being absorbed.

Fat digestion is complicated by the fact that most lipids are not particularly water soluble and thus form a coarse emulsion of large fat droplets in the aqueous chyme leaving the stomach (Fig. 21-19 ■). In the small intestine, to increase the surface area available for enzymatic fat digestion, bile salts interact with the lipids to break down the coarse emulsion into smaller, more-stable particles. Bile salts, like phospholipids of cell membranes, are *amphipathic* [*amphi-*, on both sides + *pathos*, experience], with a hydrophobic region and a hydrophilic region. The hydrophobic regions of bile salts associate with the surface of lipid droplets (Fig. 21-20a ■) while the polar side chains interact with water, creating a stable emulsion of small water-soluble fat droplets (Fig. 21-19 ①). You can see a similar emulsion when you shake a bottle of salad dressing to combine the oil and aqueous layers.

The bile salt coating of the intestinal emulsion complicates digestion, however, because lipase is unable to penetrate the bile salts. Fat digestion therefore also requires **colipase**, a protein cofactor secreted from the pancreas. Colipase displaces some bile salts, allowing lipase access to fats inside the bile salt coating (Fig. 21-19 ②). As enzymatic and mechanical digestion proceed, fatty acids, bile salts, monoglycerides, phospholipids, and cholesterol form small disk-shaped **micelles** [p. 55]. Micelles then enter the unstirred aqueous layer close to the absorptive cells (*enterocytes*) lining the small intestine lumen.

Because fats are lipophilic, they are absorbed primarily by simple diffusion. Fatty acids and monoglycerides move out of their micelles and diffuse across the apical membrane into the epithelial cells (Fig. 21-19 ③). Cholesterol is transported across

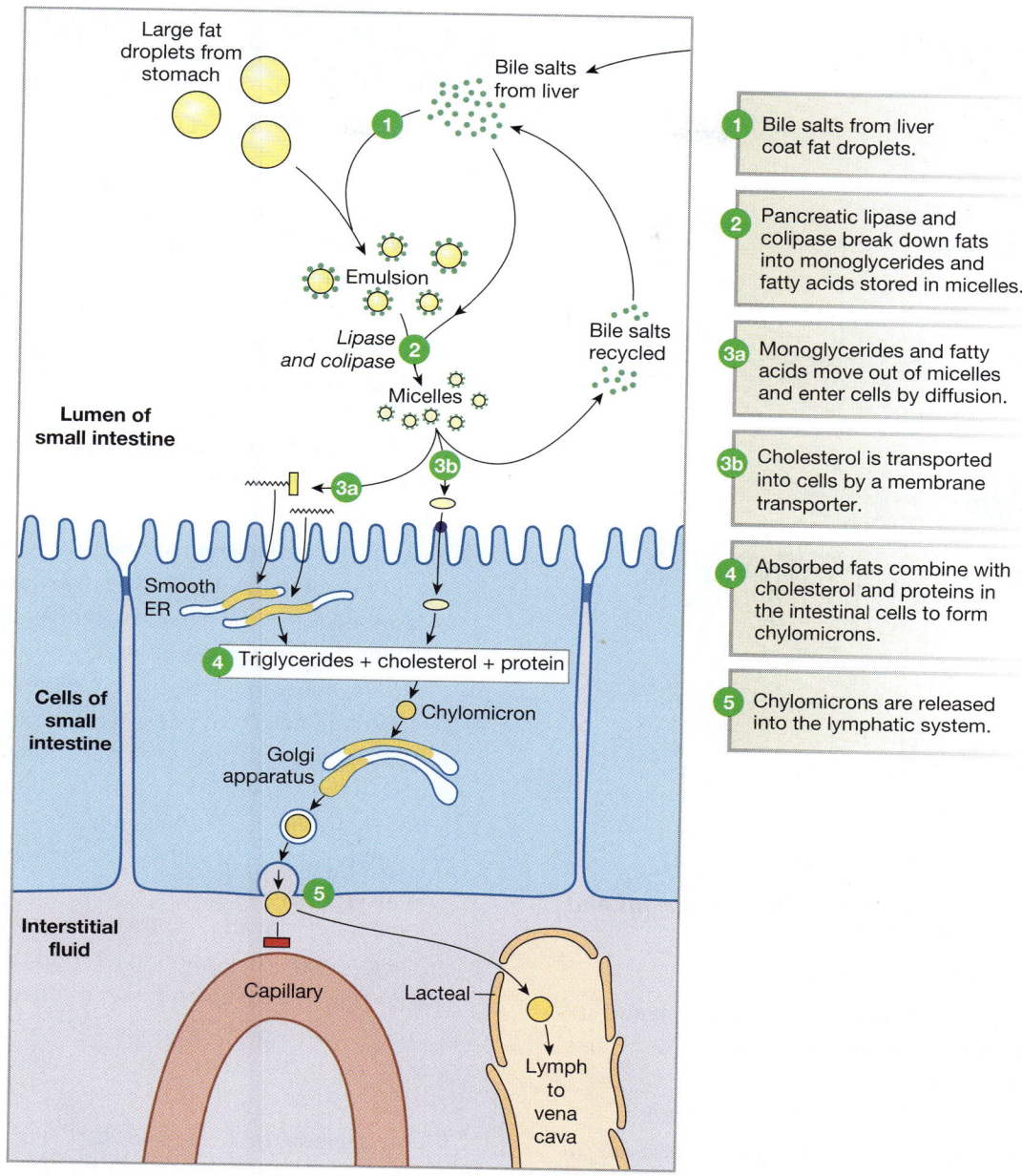

Bile salts from liver

1. Bile salts from liver coat fat droplets.

2. Pancreatic lipase and colipase break down fats into monoglycerides and fatty acids stored in micelles.

3a. Monoglycerides and fatty acids move out of micelles and enter cells by diffusion.

3b. Cholesterol is transported into cells by a membrane transporter.

4. Absorbed fats combine with cholesterol and proteins in the intestinal cells to form chylomicrons.

5. Chylomicrons are released into the lymphatic system.

Large fat droplets from stomach

1

Emulsion

Lipase and colipase

2

Bile salts recycled

Lumen of small intestine

Micelles

3b

3a

Cells of small intestine

Smooth ER

4 Triglycerides + cholesterol + protein

Chylomicron

Golgi apparatus

5

Interstitial fluid

Capillary   Lacteal

Lymph to vena cava

■ **FIGURE 21-19**   *Digestion and absorption of fats*

the apical membrane on a specific, energy-dependent membrane transporter.

Once in the cytoplasm of an intestinal epithelial cell, monoglycerides and fatty acids move to the smooth endoplasmic reticulum, where they are resynthesized into triglycerides (Fig. 21-19 ④). The triglycerides then combine with cholesterol and proteins into large droplets called **chylomicrons**. Because of their size, chylomicrons must be packaged into secretory vesicles and leave the cell by exocytosis.

Once in the extracellular space, the large size of chylomicrons prevents them from crossing the basement membrane surrounding the capillaries (Fig. 21-19 ⑤). Instead, chylomicrons are absorbed into lacteals, the lymph vessels of the villi (see Fig. 21-13).

Chylomicrons pass through the lymphatic system and finally enter the venous blood just before it flows into the heart [p. 519]. Some shorter fatty acids (10 or fewer carbons) are not assembled into chylomicrons. These fatty acids can therefore cross the capillary basement membrane and go directly into the blood.

Nucleic Acids Are Digested into Nitrogenous Bases and Monosaccharides

The nucleic acid polymers DNA and RNA are only a very small part of most diets. They are digested by pancreatic and intestinal enzymes, first into their component nucleotides and then into nitrogenous bases and monosaccharides [Fig. 2-1 p. 32]. The bases are absorbed by active transport, and

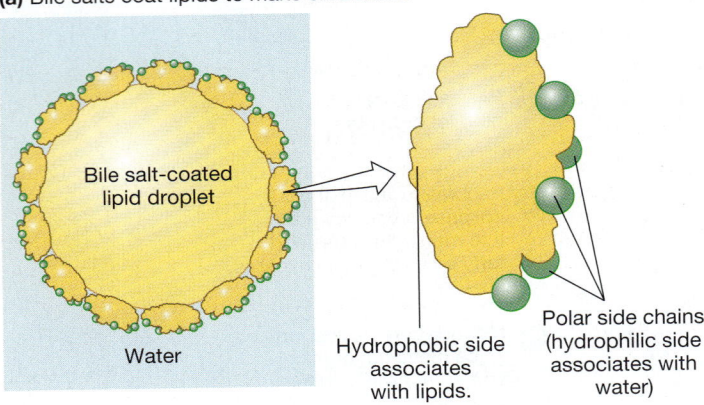

(a) Bile salts coat lipids to make emulsions.

Bile salt-coated lipid droplet

Water

Hydrophobic side associates with lipids.

Polar side chains (hydrophilic side associates with water)

(b) Micelles are small disks with bile salts, phospholipids, fatty acids, cholesterol, and mono- and diglycerides.

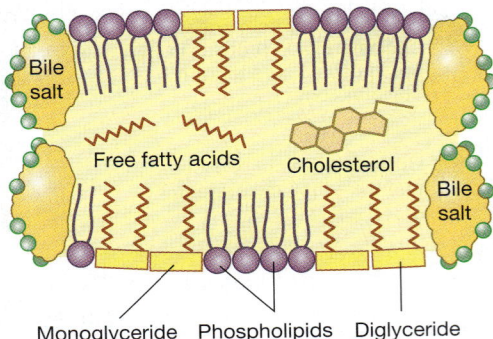

Bile salt

Free fatty acids

Cholesterol

Bile salt

Monoglyceride   Phospholipids   Diglyceride

■ **FIGURE 21-20** *The roles of bile salts in emulsions and micelles*

monosaccharides are absorbed by facilitated diffusion and secondary active transport like other simple sugars.

The Intestine Absorbs Vitamins and Minerals

In general, the fat-soluble vitamins (A, D, E, and K) are absorbed in the small intestine along with fats—one reason that health professionals are concerned about excessive consumption of "fake fats" that are not absorbed (see the box on Olestra). The water-soluble vitamins (C and most B vitamins) are absorbed by mediated transport. The major exception is **vitamin B$_{12}$**, also known as *cobalamin*. This vitamin is made by bacteria, but we obtain most of our dietary supply from seafood, meat, and milk products. The intestinal transporter for B$_{12}$ is found only in the ileum and recognizes B$_{12}$ only when the vitamin is complexed with a protein called **intrinsic factor**, secreted by the stomach.

Mineral absorption usually occurs by active transport. Iron and calcium are two of the few substances whose intestinal absorption is linked to their concentration in the body. For both minerals, a decrease in body concentrations of the mineral [leads] to enhanced uptake at the intestine.

OLESTRA, THE NO-CALORIE FAT SUBSTITUTE

Now that nutritionists have determined that the typical Western diet contains too much fat, many consumers want to reduce the amount of fat in their diet without giving up its flavor and texture. As a result, food manufacturers have been developing low-calorie fat substitutes. In 1996, a fat substitute named Olestra® was approved by the Food and Drug Administration (FDA). Olestra is sucrose polyester: a sucrose molecule with 6–8 fatty acids attached to it. This unusual arrangement of naturally occurring substances tastes like and has the "feel" of dietary fat but can neither be digested by intestinal enzymes nor absorbed across the intestinal epithelium. Foods containing Olestra were introduced with considerable fanfare, but within weeks unhappy consumers were reporting unpleasant side effects—intestinal cramping, gas, and diarrhea—that were significant enough to cause some people to stop using Olestra. In addition, nutritionists are concerned that excessive consumption of Olestra will lead to deficiencies of the fat-soluble vitamins (A, D, E, and K) because these vitamins are normally absorbed along with dietary fats. The FDA requires a warning label on foods made with Olestra and is monitoring its effects.

Iron is ingested as heme iron [⮐ p. 595] in meat and as ionized iron (Fe^{2+}) in some plant products. Heme iron is more readily absorbed, but the mechanism by which it is absorbed is not well understood. Ionized iron is actively absorbed by apical cotransport with H$^+$ or by binding to luminal *transferrin* [⮐ p. 544], forming a complex that is taken into the cell by endocytosis. The mechanism by which Fe^{2+} exits enterocytes is unclear.

Most Ca^{2+} absorption in the gut occurs by passive, unregulated movement through paracellular pathways. Hormonally regulated transepithelial Ca^{2+} transport takes place in the duodenum. Calcium enters the enterocyte through apical Ca^{2+} channels and is actively transported across the basolateral membrane by either a Ca^{2+}-ATPase or by the Na$^+$-Ca^{2+} antiporter. Calcium balance is described in detail in Chapter 22.

The Intestines Absorb Ions and Water

Most water absorption takes place in the small intestine; an additional 0.5 L per day is absorbed in the colon. As with intes-

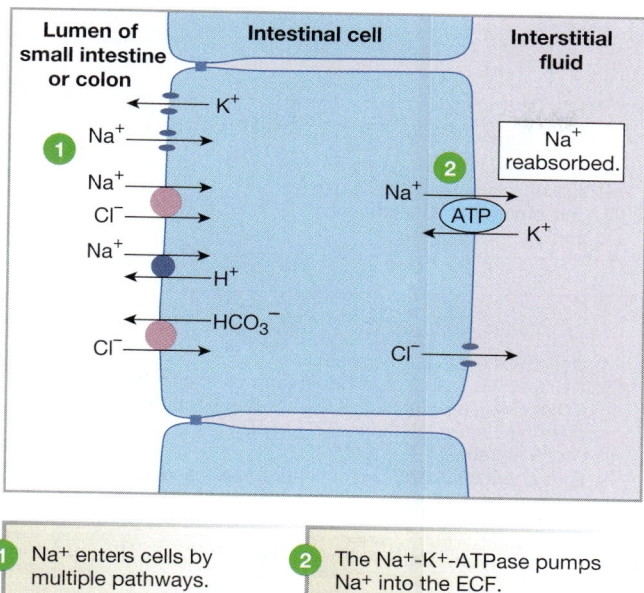

1. Na⁺ enters cells by multiple pathways.

2. The Na⁺-K⁺-ATPase pumps Na⁺ into the ECF.

■ **FIGURE 21-21**   *NaCl reabsorption in the small intestine and colon*

A significant amount of Na^+ reabsorption in the small intestine also occurs through Na^+-coupled organic solute uptake across the apical membrane (not shown in this figure).

tinal fluid secretion, water follows osmotic gradients created by solute absorption. Enterocytes in the small intestine and **colonocytes**, the epithelial cells on the luminal surface of the colon, absorb Na^+ using three membrane proteins (Fig. 21-21 ■): apical Na^+ channels, the Na^+-Cl^- symporter, and the Na^+-H^+ antiporter. In the small intestine, a significant fraction of Na^+ absorption also takes place through Na^+-dependent organic solute uptake, such as the SGLT and the Na^+-amino acid transporters (see Figs. 21-15 and 21-17). On the basolateral side of both enterocytes and colonocytes, the primary transporter for Na^+ is the Na^+-K^+-ATPase. Chloride uptake uses the apical Cl^--HCO_3^- antiporter and a basolateral Cl^- channel. Potassium absorption in the intestine occurs by the paracellular pathway.

In the remainder of this chapter we will follow some food as it passes through the GI tract. As a preview, look at Figure 21-22 ■, a summary of the main events that occur in each section of the GI tract. Food processing is traditionally divided into three phases: a cephalic phase, a gastric phase, and an intestinal phase.

CONCEPT CHECK

8. Recipes for homemade oral rehydration therapy usually include sugar (sucrose) and table salt. Explain how the salt enhances intestinal absorption of glucose.

9. Do bile salts digest triglycerides into monoglycerides and free fatty acids?

Answers: p. 714

THE CEPHALIC PHASE

Digestive processes in the body begin before food ever enters the mouth. Simply smelling, seeing, or even *thinking* about food can make our mouths water and our stomachs rumble. These long reflexes that begin in the brain create a feedforward response known as the **cephalic phase** of digestion (Fig. 21-23 ■). Anticipatory stimuli and the stimulus of food in the oral cavity activate neurons in the medulla oblongata. The medulla in turn sends an efferent signal through autonomic neurons to the salivary glands, and through the vagus nerve to the enteric nervous system. In response to these signals, the stomach, intestine, and accessory glandular organs begin secretion and increase motility in anticipation of the food to come.

Chemical and Mechanical Digestion Begins in the Mouth

When food first enters the mouth, it is met by a flood of the secretion we call saliva. Salivary secretion is under autonomic control and can be triggered by multiple stimuli, including the sight, smell, touch, and even thought of food. The water and mucus in saliva soften and lubricate food to make it easier to swallow. You can appreciate this function if you've ever tried to swallow a dry soda cracker without chewing it thoroughly. Saliva also dissolves food so that we can taste it.

Saliva begins chemical digestion with the secretion of *salivary amylase* and a very small amount of lipase. Amylase breaks starch into maltose after the enzyme is activated by Cl^- in saliva. If you chew on an unsalted soda cracker for a long time, you may be able to detect the conversion of the cracker's flour starch to maltose, which is sweeter.

The final function of saliva is protection. *Lysozyme* is an antibacterial salivary enzyme, and salivary *immunoglobulins* disable bacteria and viruses. In addition, saliva helps wash the teeth and keep the tongue free of food particles.

Mechanical digestion of food begins in the oral cavity with chewing. The lips, tongue, and teeth all contribute to the **mastication** of food, creating a softened, moistened mass (bolus) that can be easily swallowed.

Swallowing Moves Food from the Mouth to the Stomach

Swallowing, or **deglutition**, is a reflex action that pushes a bolus of food or liquid into the esophagus (Fig. 21-24 ■). The stimulus for swallowing is pressure created when the tongue pushes the bolus against the soft palate and the back of the mouth. Sensory input to a swallowing center in the medulla oblongata begins the reflex.

First, the **epiglottis** folds down over the opening of the larynx to prevent food and liquid from entering the airways. At the same time, respiration is inhibited, and the upper esophageal sphincter relaxes as the bolus enters the esophagus. Waves of peristaltic contraction then push the bolus toward the

21

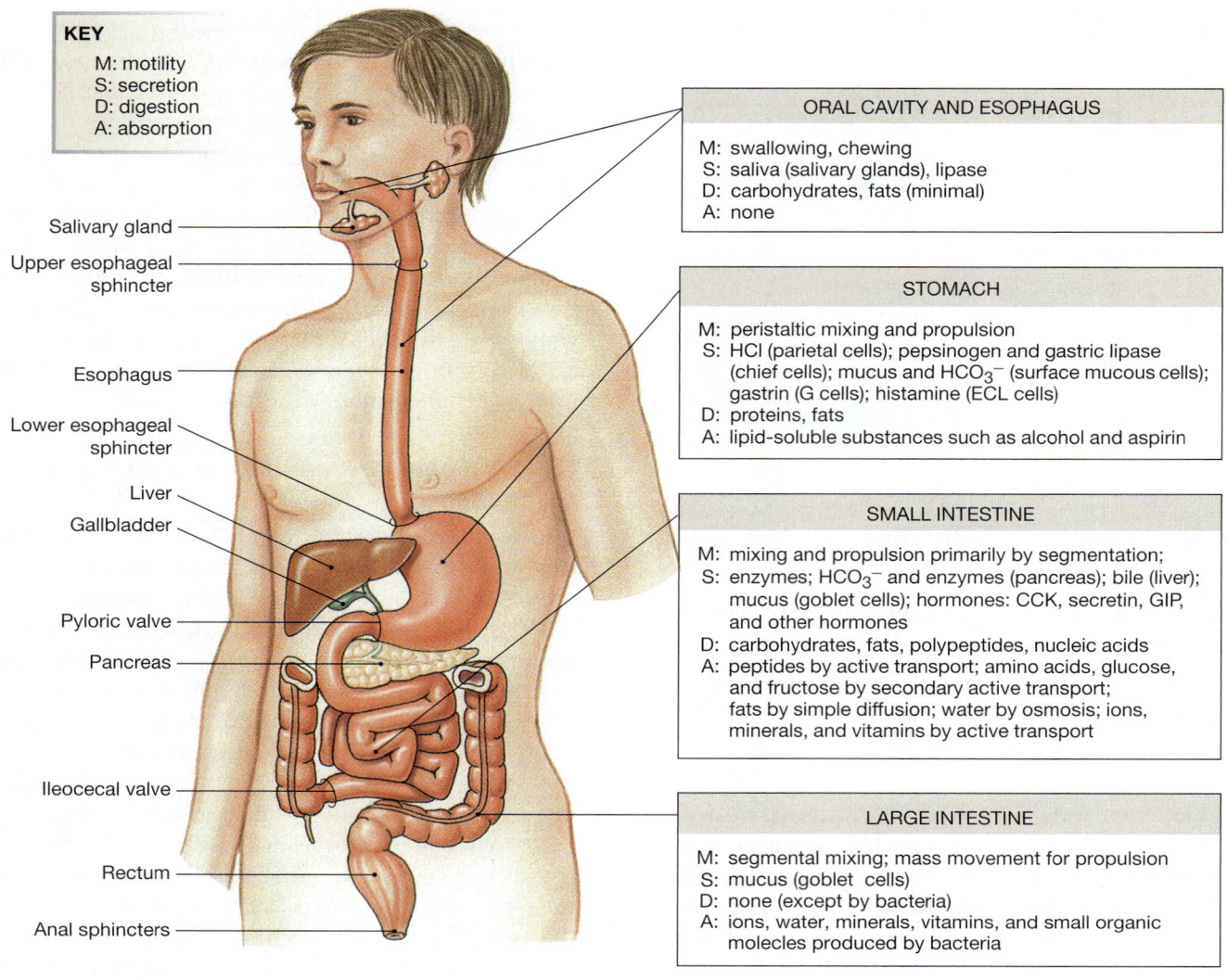

KEY
M: motility
S: secretion
D: digestion
A: absorption

Salivary gland
Upper esophageal sphincter
Esophagus
Lower esophageal sphincter
Liver
Gallbladder
Pyloric valve
Pancreas
Ileocecal valve
Rectum
Anal sphincters

ORAL CAVITY AND ESOPHAGUS

M: swallowing, chewing
S: saliva (salivary glands), lipase
D: carbohydrates, fats (minimal)
A: none

STOMACH

M: peristaltic mixing and propulsion
S: HCl (parietal cells); pepsinogen and gastric lipase (chief cells); mucus and HCO_3^- (surface mucous cells); gastrin (G cells); histamine (ECL cells)
D: proteins, fats
A: lipid-soluble substances such as alcohol and aspirin

SMALL INTESTINE

M: mixing and propulsion primarily by segmentation;
S: enzymes; HCO_3^- and enzymes (pancreas); bile (liver); mucus (goblet cells); hormones: CCK, secretin, GIP, and other hormones
D: carbohydrates, fats, polypeptides, nucleic acids
A: peptides by active transport; amino acids, glucose, and fructose by secondary active transport; fats by simple diffusion; water by osmosis; ions, minerals, and vitamins by active transport

LARGE INTESTINE

M: segmental mixing; mass movement for propulsion
S: mucus (goblet cells)
D: none (except by bacteria)
A: ions, water, minerals, vitamins, and small organic molecles produced by bacteria

■ **FIGURE 21-22** *Summary of motility, secretion, digestion, and absorption in different regions of the digestive system*

stomach, aided by gravity. Gravity is not a requirement, however, as you know if you have ever participated in the party trick of swallowing while standing on your head.

The lower end of the esophagus lies just below the diaphragm and is separated from the stomach by the lower esophageal sphincter. This area is not a true sphincter but a region of relatively high muscle tension that acts as a barrier between the esophagus and the stomach. When food is swallowed, the tension relaxes, allowing the bolus to pass into the stomach.

If the lower esophageal sphincter does not stay contracted, gastric acid and pepsin can irritate the lining of the esophagus, leading to the pain and irritation of heartburn. The walls of the esophagus expand during inspiration, when the intrapleural pressure falls [🔁 p. 572]. Expansion creates subatmospheric pressure in the esophageal lumen and can suck acidic contents out of the stomach if the sphincter is relaxed. The churning action of the stomach when filled with food can also squirt acid back into the esophagus if the sphincter is not fully contracted.

THE GASTRIC PHASE

About 3.5 liters of food, drink, and saliva enter the fundus of the stomach each day. The stomach has three general functions:

1. **Storage.** The stomach stores food and regulates its passage into the small intestine, where most digestion and absorption take place.

2. **Digestion.** The stomach chemically and mechanically digests food into the soupy mixture of uniformly small particles called chyme.

3. **Protection.** The stomach protects the body by destroying many of the bacteria and other pathogens that are swallowed with food or trapped in airway mucus. At the same time, the stomach must protect itself from being damaged by its own secretions.

Before food even arrives, digestive activity in the stomach begins with the long **vagal reflex** of the cephalic phase. Then, when food enters the stomach, stimuli in the gastric lumen

The sight, smell, taste, and thought of food initiate long reflexes that prepare the stomach.

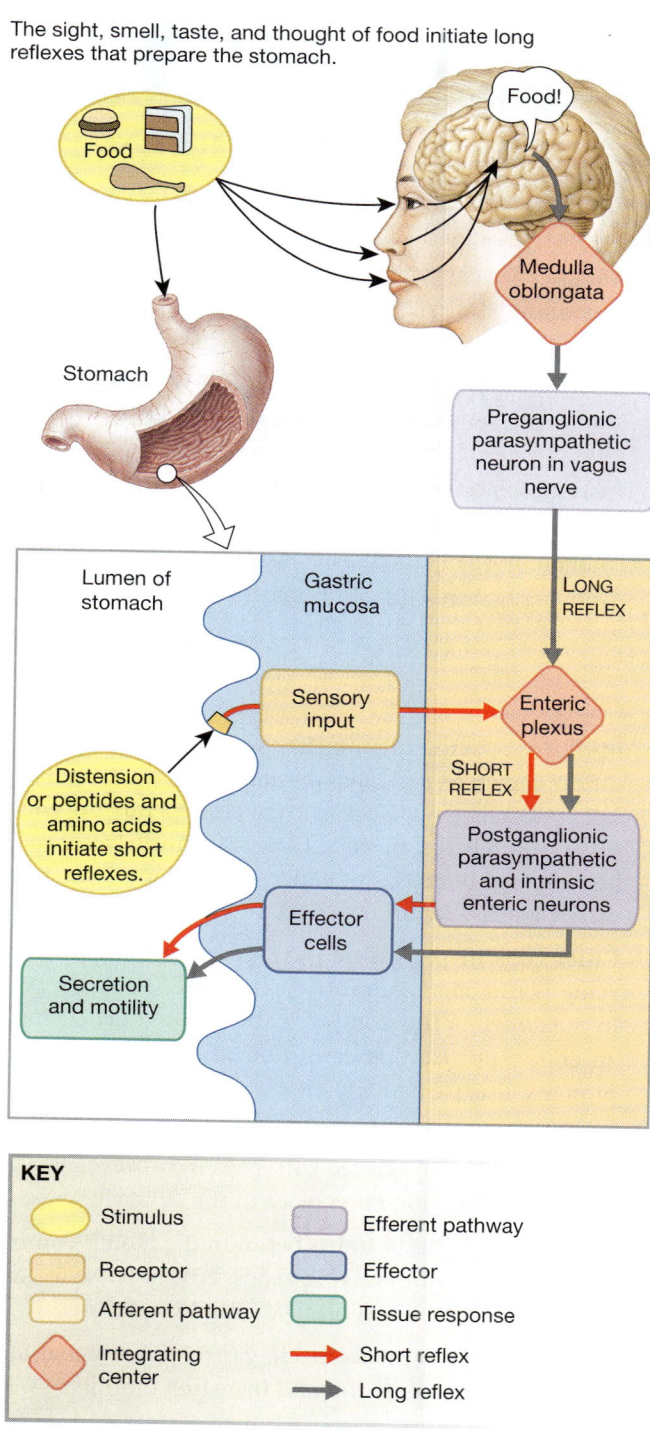

KEY

Stimulus	Efferent pathway
Receptor	Effector
Afferent pathway	Tissue response
Integrating center	Short reflex
	Long reflex

■ **FIGURE 21-23**   *Long and short reflexes of the cephalic and gastric phases of digestion*

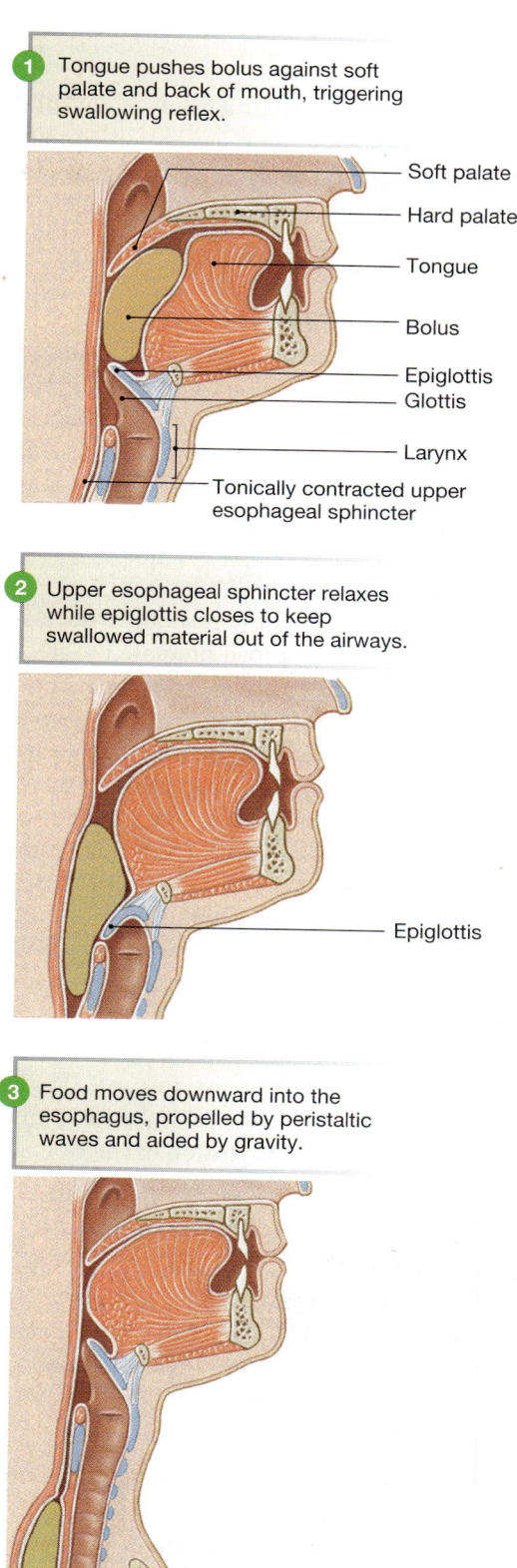

1 Tongue pushes bolus against soft palate and back of mouth, triggering swallowing reflex.

- Soft palate
- Hard palate
- Tongue
- Bolus
- Epiglottis
- Glottis
- Larynx
- Tonically contracted upper esophageal sphincter

2 Upper esophageal sphincter relaxes while epiglottis closes to keep swallowed material out of the airways.

- Epiglottis

3 Food moves downward into the esophagus, propelled by peristaltic waves and aided by gravity.

■ **FIGURE 21-24**   *The swallowing reflex*

21

initiate a series of short reflexes that constitute the **gastric phase** of digestion.

In gastric phase reflexes, distension of the stomach and the presence of peptides or amino acids in the lumen activate endocrine cells and enteric neurons. Hormones, neurocrine secretions, and paracrine molecules then influence motility and secretion.

The Stomach Stores Food

When food arrives, the stomach relaxes and expands to hold the increased volume. This neurally mediated reflex is called *receptive relaxation*. The upper half of the stomach remains relatively quiet, holding food until it is ready to be digested. The storage function of the stomach is perhaps the least obvious aspect of digestion. However, whenever we ingest more than we need from a nutritional standpoint, the stomach must regulate the rate at which food enters the small intestine.

Without such regulation, the small intestine would not be able to digest and absorb the load presented to it, and significant amounts of unabsorbed chyme would pass into the large intestine. The epithelium of the large intestine is not designed for large-scale nutrient absorption, so most of the chyme would pass out in the feces, resulting in diarrhea. This "dumping syndrome" is one of the less pleasant side effects of surgery that removes portions of either the stomach or small intestine.

While the upper stomach is quietly holding food, the lower stomach is busy with digestion. In the distal half of the stomach, a series of peristaltic waves pushes the food down toward the pyloric valve, mixing food with acid and digestive enzymes. As large food particles are digested to the more uniform texture of chyme, each contractile wave squirts a small amount of chyme through the pyloric valve into the duodenum. Enhanced gastric motility during a meal is primarily under neural control and is stimulated by distension of the stomach.

The Stomach Secretes Acid, Enzymes, and Signal Molecules

When food comes into the mouth, the feedforward cephalic vagal reflex begins secretion in the stomach. The various secretions, their stimuli for release, and their functions are summarized in Figure 21-25 ■. Chemical digestion in the stomach is accomplished as follows:

- **Parietal cells** deep in the gastric glands secrete **gastric acid** (HCl), which helps kill bacteria and other ingested microorganisms. Acid also *denatures* proteins [🔁 p. 31], destroying their tertiary structure by breaking disulfide and hydrogen bonds. Unfolding protein chains make the peptide bonds between amino acids accessible to enzymes.

  Parietal cells also secrete *intrinsic factor*. As noted earlier, intrinsic factor complexes with vitamin B_{12} and is essential for B_{12} absorption in the intestine. In the absence

CLINICAL FOCUS

DELAYED GASTRIC EMPTYING

Diabetes mellitus has an impact on almost every organ system. The digestive tract is not exempt. One problem that plagues more than a third of all diabetics is *gastroparesis,* also called delayed gastric emptying. In these patients, the migrating motor complex is absent between meals, and consequently movement of meals from the stomach is slow. Many patients suffer nausea and vomiting as a result. The cause of diabetic gastroparesis is unclear. Some researchers believe it results from diabetic neuropathy, but other evidence suggests that elevated blood glucose levels interfere with the normal hormonal signals for gastric emptying.

of intrinsic factor, vitamin B_{12} deficiency causes the condition known as *pernicious anemia*. In this state, red blood cell synthesis (*erythropoiesis*), which depends on vitamin B_{12}, is severely diminished. Lack of intrinsic factor cannot be remedied directly, but patients with pernicious anemia can be given vitamin B_{12} shots.

- **Chief cells** in the gastric glands secrete the inactive enzyme pepsinogen. Pepsinogen is cleaved to active pepsin in the lumen of the stomach by the action of H^+. Pepsin is an endopeptidase that carries out the initial digestion of proteins. It is particularly effective on collagen and therefore plays an important role in digesting meat.

- *Gastric lipase* is co-secreted with pepsin. However, less than 10% of fat digestion takes place in the stomach.

- Carbohydrate digestion that began in the mouth continues in the stomach until mixing exposes the amylase to gastric acid. Salivary amylase is inactivated at low pH.

Once food enters the stomach, gastric-phase signal molecules promote gastric secretion and increased motility:

- **D cells**, which are closely associated with the parietal cells, secrete the paracrine somatostatin.

- **Enterochromaffin-like** (ECL) **cells** secrete the paracrine histamine.

- **G cells**, found deep in the gastric glands, secrete the hormone gastrin. The release of gastrin is stimulated by the presence of amino acids and peptides in the stomach, by distension of the stomach, and by nervous reflexes mediated by **gastrin-releasing peptide**. Coffee (even if decaffeinated) is also a good stimulant of gastrin release—one reason

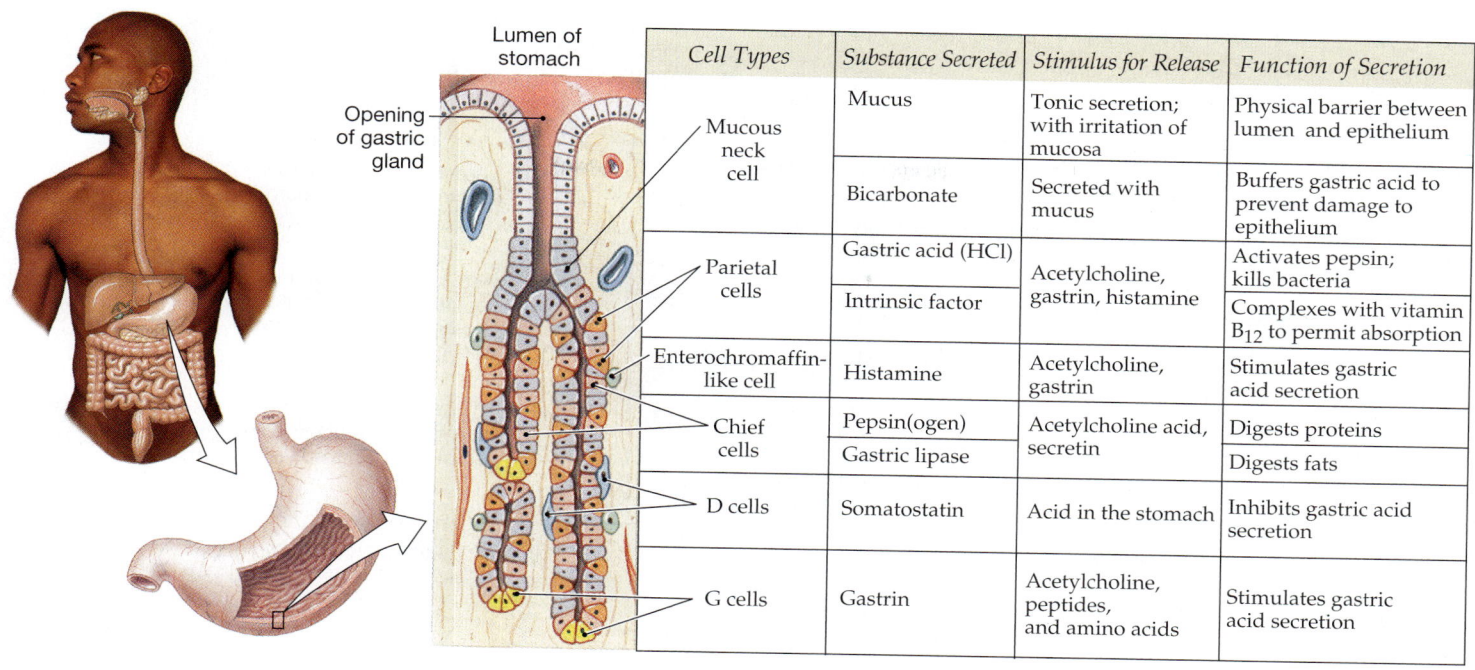

Cell Types	Substance Secreted	Stimulus for Release	Function of Secretion
Mucous neck cell	Mucus	Tonic secretion; with irritation of mucosa	Physical barrier between lumen and epithelium
	Bicarbonate	Secreted with mucus	Buffers gastric acid to prevent damage to epithelium
Parietal cells	Gastric acid (HCl)	Acetylcholine, gastrin, histamine	Activates pepsin; kills bacteria
	Intrinsic factor		Complexes with vitamin B_{12} to permit absorption
Enterochromaffin-like cell	Histamine	Acetylcholine, gastrin	Stimulates gastric acid secretion
Chief cells	Pepsin(ogen)	Acetylcholine acid, secretin	Digests proteins
	Gastric lipase		Digests fats
D cells	Somatostatin	Acid in the stomach	Inhibits gastric acid secretion
G cells	Gastrin	Acetylcholine, peptides, and amino acids	Stimulates gastric acid secretion

■ **FIGURE 21-25** *Activity of secretory cells of the gastric mucosa*

people with excess acid secretion syndromes are advised to avoid coffee. Gastrin release is inhibited by somatostatin and by a luminal pH below 1.5.

The coordinated function of gastric secretion is illustrated in Figure 21-26 ■:

- In a cephalic reflex, parasympathetic neurons from the vagus nerve stimulate G cells to release gastrin into the blood (Fig. 21-26 ①). The presence of amino acids or peptides in the lumen triggers a short reflex for gastrin release.

- Gastrin in turn promotes acid release, both directly and indirectly by stimulating histamine release ②.

- Histamine is released from ECL cells in response to gastrin and acetylcholine from the enteric nervous system ①. Histamine diffuses to its target, the parietal cells, and stimulates acid secretion by combining with H_2 receptors on parietal cells.

- Acid in the stomach lumen stimulates pepsinogen release from chief cells through a short reflex ③. In the lumen, acid converts pepsinogen into pepsin, and protein digestion begins.

- Acid also triggers somatostatin release from D cells ④. Somatostatin acts via negative feedback to inhibit secretion of gastric acid, gastrin, and pepsinogen.

The Stomach Balances Digestion and Protection

Under normal conditions, the gastric mucosa is protected from autodigestion by a mucus-bicarbonate barrier. **Mucous cells** in

the neck of gastric glands secrete both substances. The mucus forms a physical barrier, and the bicarbonate creates a chemical buffer barrier underlying the mucus (Fig. 21-27 ■). Researchers using microelectrodes have shown that the bicarbonate layer just above the cell surface in the stomach has a pH that is close to 7, even when the pH in the lumen is highly acidic at pH 2. Mucus secretion is increased when the stomach is irritated, such as by the ingestion of aspirin (acetylsalicylic acid) or alcohol.

Even the mucus-bicarbonate barrier can fail at times. In *Zollinger-Ellison syndrome,* patients secrete excessive levels of gastrin, usually from gastrin-secreting tumors in the pancreas. As a result, hyperacidity in the stomach overwhelms the normal protective mechanisms and causes a peptic ulcer. In peptic ulcers, acid and pepsin destroy the mucosa, creating holes that extend into the submucosa and muscularis of the stomach and duodenum. (*Acid reflux* [*re-* backward + *fluxus,* flow] into the esophagus can erode the mucosal layer there as well.)

Excess acid secretion is an uncommon cause of peptic ulcers. By far the most common causes are nonsteroidal anti-inflammatory drugs (NSAIDs), such as aspirin, and *Helicobacter pylori,* a bacterium that creates inflammation of the gastric mucosa (see the Running Problem in this chapter).

For many years the primary therapy for excess acid secretion, or *dyspepsia,* was the ingestion of *antacids,* agents that neutralize acid in the gastric lumen. But as molecular biologists explored the mechanism for acid secretion by parietal cells, the potential for new therapies became obvious. Today we have two new classes of drugs to fight hyperacidity: H_2 receptor

21

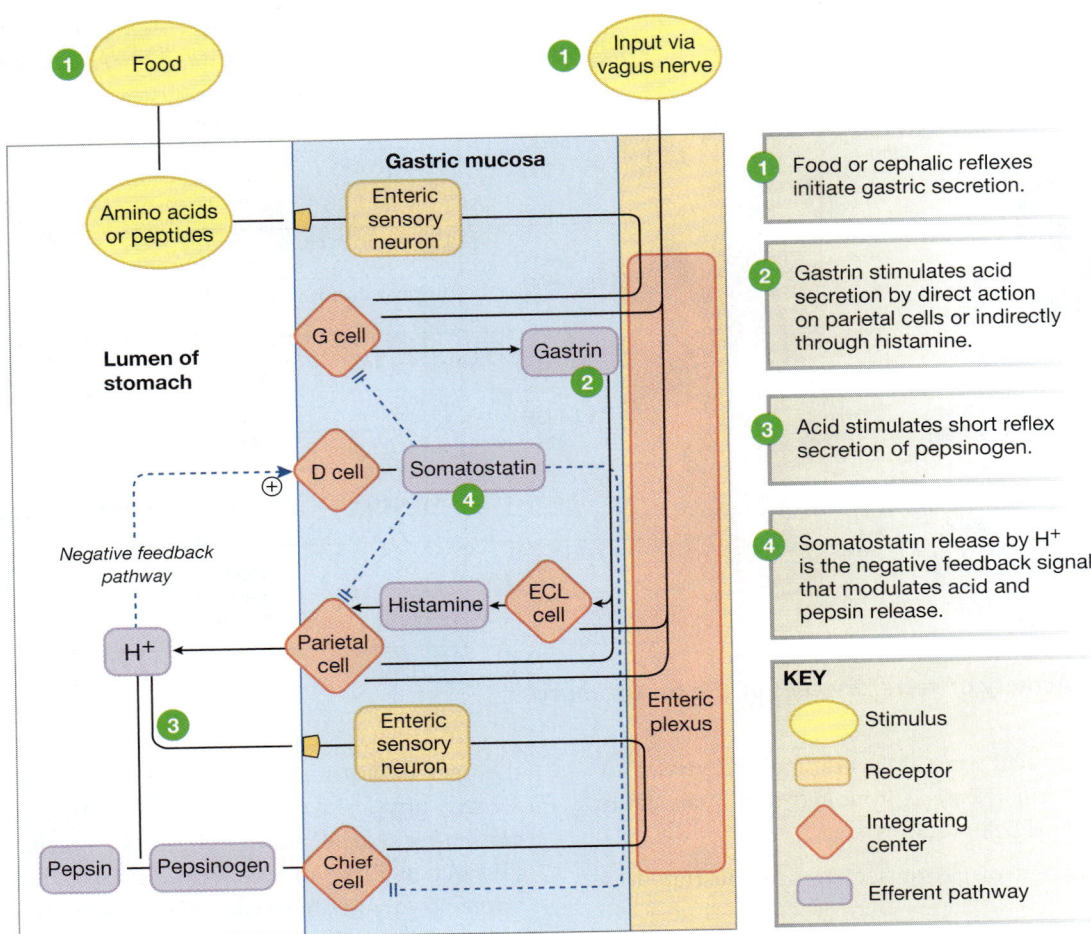

■ FIGURE 21-26 *Integration of cephalic and gastric phase secretion in the stomach*

The cephalic phase is initiated by the sight, smell, sound, or thought of food or by the presence of food in the mouth. The gastric phase is initiated by the arrival of food in the stomach.

antagonists (cimetidine and ranitidine, for example) that block histamine action, and *proton pump inhibitors* (PPIs), which block the H^+-K^+-ATPase (omeprazole and lansoprazole, for example).

THE INTESTINAL PHASE

The net result of the gastric phase is the digestion of proteins in the stomach by pepsin; the formation of chyme by the action of pepsin, acid, and peristaltic contractions; and the controlled entry of chyme into the small intestine, where further digestion and absorption can take place. Once chyme enters the small intestine, the **intestinal phase** of digestion begins. The initiation of the intestinal phase triggers a series of reflexes that feed back to regulate the delivery rate of chyme from the stomach, and feed forward to promote digestion, motility, and utilization of nutrients.

RUNNING PROBLEM

Tonya's stomach still hurts later that night. Her roommate suggests that she try one of the new "acid blockers" she has seen advertised on TV. These drugs are more correctly known as histamine receptor antagonists because they bind competitively to histamine H_2 receptors on parietal cells. Fortunately, Tonya's roommate has just bought a package of these drugs. Two hours after taking the drug, Tonya feels much better.

Question 4:
 How does blocking H_2 receptors stop acid production in the stomach?

677 682 687 691 **704** 710

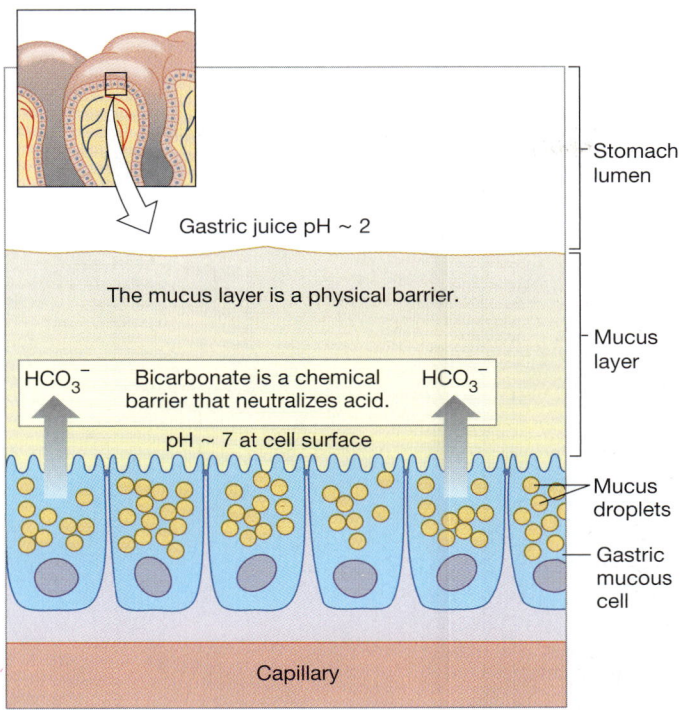

Stomach lumen

Gastric juice pH ~ 2

The mucus layer is a physical barrier.

Mucus layer

HCO_3^- Bicarbonate is a chemical barrier that neutralizes acid. HCO_3^-

pH ~ 7 at cell surface

Mucus droplets

Gastric mucous cell

Capillary

■ **FIGURE 21-27** *The mucus-bicarbonate barrier of the gastric mucosa*

The feedback signals to the stomach are both neural and hormonal (Fig. 21-28 ■):

- Chyme in the intestine activates the enteric nervous system, which then slows gastric motility and secretion. In addition, three hormones reinforce the feedback signal: secretin, cholecystokinin (CCK), and gastric inhibitory peptide (GIP) (see Table 21-1).

- Secretin is released by the presence of acidic chyme in the duodenum. Secretin inhibits acid production and gastric motility, slowing gastric emptying. In addition, secretin stimulates production of pancreatic HCO_3^- to neutralize the acidic chyme that has entered the intestine.

- If a meal contains fats, CCK is secreted into the bloodstream. CCK also slows gastric motility and acid secretion. Because fat digestion proceeds more slowly than either protein or carbohydrate digestion, it is crucial that the stomach allow only small amounts of fat into the intestine at one time.

- If the meal contains carbohydrates, the *incretin hormones* GIP and glucagon-like peptide 1 (GLP-1) are released. In experiments, large doses of GIP inhibit gastric acid secretion, but there is some question whether GIP release reaches these levels with a normal meal. Both GIP and GLP-1 feed forward to promote insulin release by the endocrine pancreas.

- The mixture of acid, enzymes, and digested food in chyme usually forms a hyperosmotic solution. Osmoreceptors in the intestine wall are sensitive to the osmolarity of the entering chyme. When stimulated by high osmolarity, the receptors inhibit gastric emptying in a reflex mediated by some unknown blood-borne substance.

Once food enters the small intestine, it is slowly propelled forward by a combination of slow-wave, segmental, and peristaltic contractions. These actions mix chyme with enzymes and expose digested nutrients to the mucosa for absorption. Forward movement of chyme through the intestine must be slow enough to allow digestion and absorption to go to completion. Parasympathetic innervation, gastrin, and CCK all promote intestinal motility; sympathetic innervation inhibits it.

Bicarbonate in the Small Intestine Neutralizes Gastric Acid

About 5.5 liters of food, fluid, and secretions enter the small intestine each day, and about 3.5 liters of hepatic, pancreatic, and intestinal secretions are added there, making a total input of 9 liters into the lumen (see Fig. 21-5, p. 684). The added secretions include bicarbonate, mucus, bile, and digestive enzymes.

- *Bicarbonate secretion* into the small intestine neutralizes the highly acidic chyme that enters from the stomach. Most bicarbonate comes from the pancreas and is released in response to neural stimuli and secretin.

- Intestinal goblet cells secrete mucus for protection and lubrication.

- *Bile* release into the intestine occurs when gallbladder contraction is stimulated by CCK following ingestion of fats. Bile salts are not altered during fat digestion. When they reach the terminal section of the small intestine (the ileum), they encounter cells that transport them back into the circulation. From there, they return to the liver, are taken back up into the hepatocytes, and resecreted. This recirculation of bile salts is essential to fat digestion because the body's pool of bile salts must cycle from two to five times for each meal. Bilirubin and other wastes secreted in bile are not reabsorbed and pass into the large intestine for excretion.

- *Digestive enzymes* are produced by the intestinal epithelium and acinar cells of the exocrine pancreas. The brush border enzymes, which include peptidases, disaccharidases, and a protease called **enteropeptidase** (previously called *enterokinase*), are anchored to the luminal enterocyte membrane and are not swept out of the small intestine as chyme is propelled forward.

The signals for pancreatic enzyme release include distension of the small intestine, the presence of food in the intestine, neural signals, and the hormone CCK. Pancreatic enzymes enter the intestine in a watery fluid that also contains bicarbonate. Most are secreted as zymogens

21

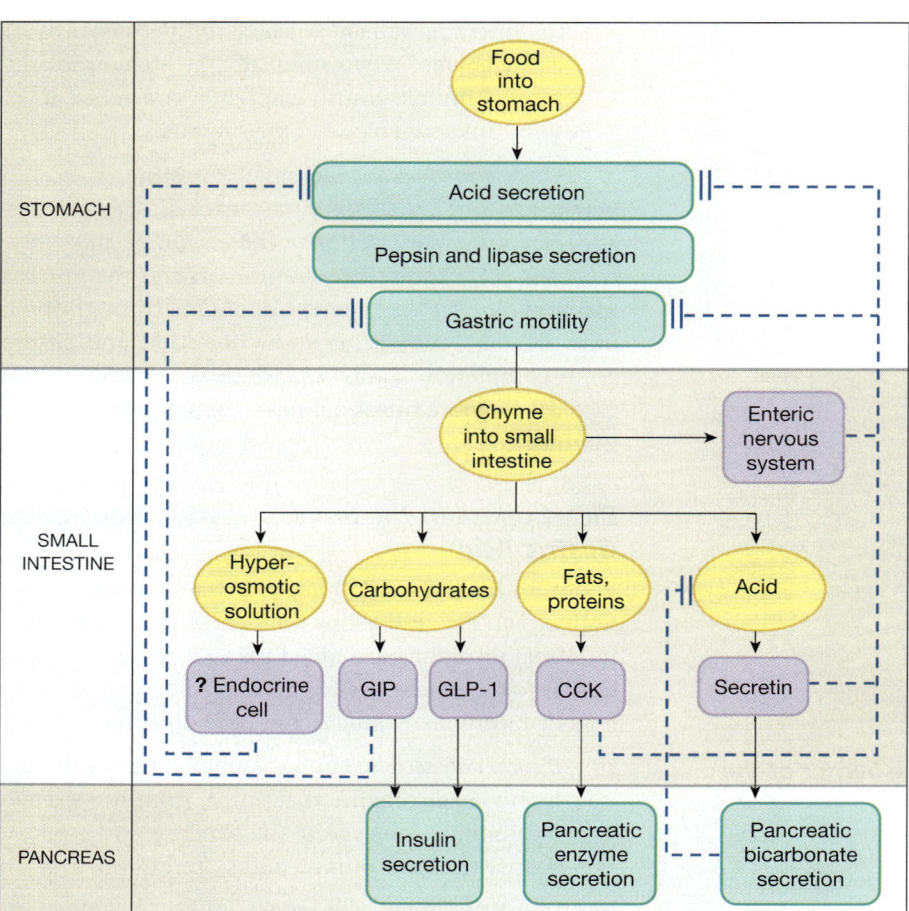

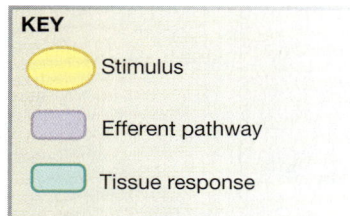

KEY

- Stimulus
- Efferent pathway
- Tissue response

STOMACH

Food into stomach → Acid secretion → Pepsin and lipase secretion → Gastric motility

SMALL INTESTINE

Chyme into small intestine → Enteric nervous system

Hyper-osmotic solution | Carbohydrates | Fats, proteins | Acid

? Endocrine cell | GIP | GLP-1 | CCK | Secretin

PANCREAS

Insulin secretion | Pancreatic enzyme secretion | Pancreatic bicarbonate secretion

■ **FIGURE 21-28**   *The effects of intestinal-phase events on gastric function*

The presence of chyme in the small intestine causes secretion of hormones and neural reflexes that feed back to inhibit gastric motility and secretion.

21

that must be activated upon arrival in the intestine. This activation process is a cascade that begins when brush border enteropeptidase converts inactive trypsinogen to active trypsin (Fig. 21-29 ■). Trypsin then converts the other pancreatic zymogens to their active forms.

Most Fluid Is Absorbed in the Small Intestine

Of the 9 liters that enter the small intestine daily, most (7.5 liters) is reabsorbed there. The transport of organic nutrients and ions, which takes place mostly in the duodenum and jejunum, creates an osmotic gradient for water absorption. Most absorbed nutrients move into capillaries in the villi and from there into the *hepatic portal system* [🔁p. 460]. This specialized region of the circulation has two sets of capillary beds: one that picks up absorbed nutrients at the intestine, and the other that delivers the nutrients directly to the liver (Fig. 21-30 ■). Most digested fats go into the lymphatic system rather than into the blood.

The delivery of absorbed materials directly to the liver underscores the importance of that organ as a biological filter.

Hepatocytes contain a variety of enzymes, such as the *cytochrome P450* isozymes, that metabolize drugs and xenobiotics and clear them from the bloodstream before they reach the systemic circulation. Hepatic clearance is one reason a drug administered orally must often be given in higher doses than the same drug administered by IV infusion.

Most Digestion Occurs in the Small Intestine

Chyme entering the small intestine has undergone relatively little chemical digestion. Protein digestion started in the stomach, but pepsin entering the small intestine is soon inactivated at the higher intestinal pH. Pancreatic proteases and at least 17 additional proteases and peptidases in the brush border continue protein digestion, creating small peptides and free amino acids that can be absorbed.

Carbohydrate digestion in the small intestine finishes converting digestible polysaccharides and disaccharides into monosaccharides that can be absorbed. Pancreatic amylase continues to digest starch into maltose. Maltose and disac-

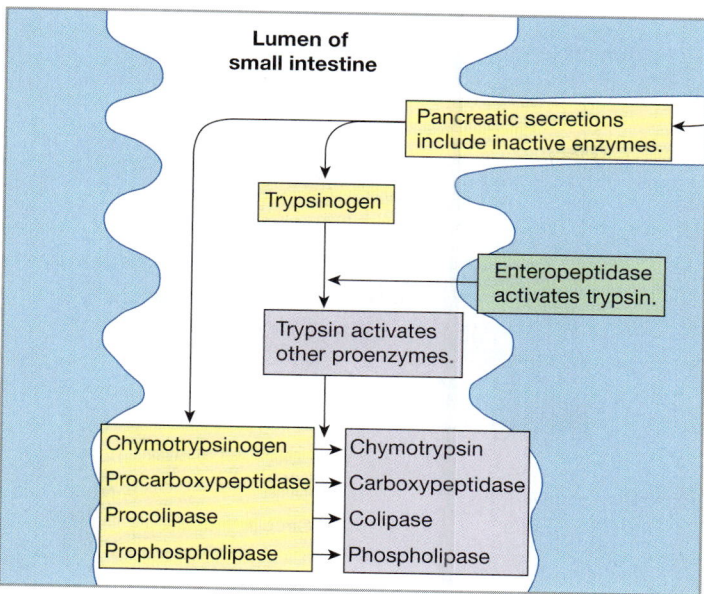

■ **FIGURE 21-29** *Activation of pancreatic zymogens*

Inactive enzymes secreted by the pancreas are activated in a cascade. Trypsinogen is activated to trypsin by brush border enteropeptidase; trypsin then activates other pancreatic enzymes.

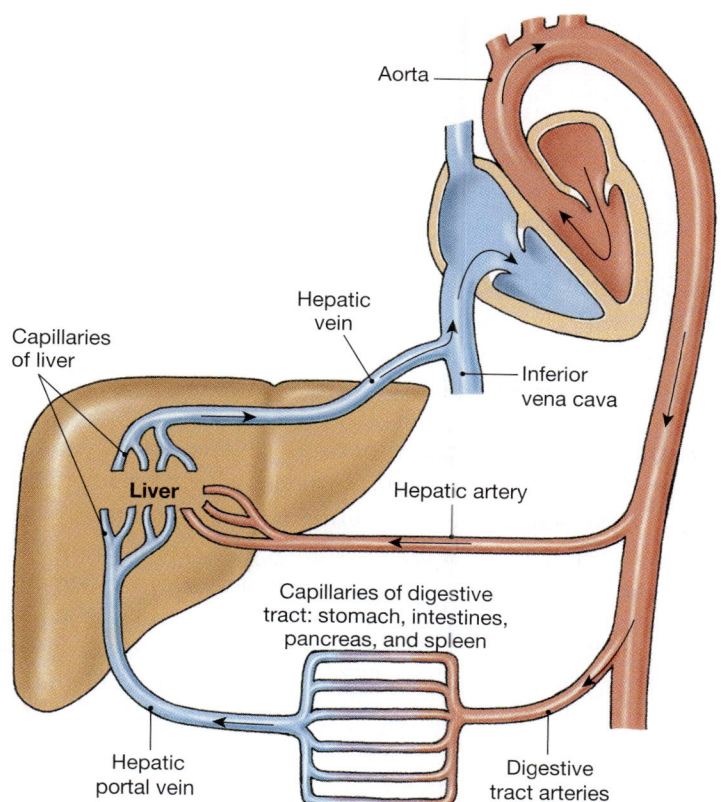

■ **FIGURE 21-30** *The hepatic portal system*

Most nutrients absorbed by the intestine pass through the liver, which serves as a filter that can remove potentially harmful xenobiotics before they get into the systemic circulation.

CLINICAL FOCUS

LACTOSE INTOLERANCE

Lactose, or milk sugar, is a disaccharide composed of glucose and galactose. Ingested lactose must be digested before it can be absorbed, a task accomplished by the intestinal brush border enzyme *lactase*. Generally, lactase is found only in juvenile mammals, except in some humans of European descent. Those people inherit a dominant gene that allows them to produce lactase after childhood. Scientists believe the gene provided a selective advantage to their ancestors, who developed a culture in which milk and milk products played an important role. In non-Western cultures, in which dairy products are not part of the diet after weaning, most adults lack the gene and synthesize less intestinal lactase. Decreased lactase activity is associated with a condition known as *lactose intolerance*. If a person with lactose intolerance drinks milk or eats dairy products, diarrhea may result. In addition, bacteria in the large intestine ferment lactose to gas and organic acids, leading to bloating and flatulence. The simplest remedy is to remove milk from the diet, although milk predigested with lactase is available.

charides from food, such as sucrose and lactose (milk sugar), are digested by the appropriate brush border disaccharidase to their absorbable end products: glucose, galactose, and fructose.

Fats enter the small intestine in the form of a coarse emulsion. In the duodenum, bile salts coat the fat droplets to stabilize them so that digestion can be carried out by pancreatic lipase. Cholesterol and fatty acids are absorbed as described earlier in this chapter.

CONCEPT CHECK

10. List the actions of CCK on the digestive system, and explain how these functions coordinate to promote fat digestion and absorption.

Answers: p. 714

The Large Intestine Concentrates Waste for Excretion

By the end of the ileum, only about 1.5 liters of unabsorbed chyme remain. The colon absorbs most of this volume so that normally only about 0.1 liter of water is lost daily in feces. Chyme enters the large intestine through the **ileocecal valve**. This is a tonically contracted region of muscularis that narrows

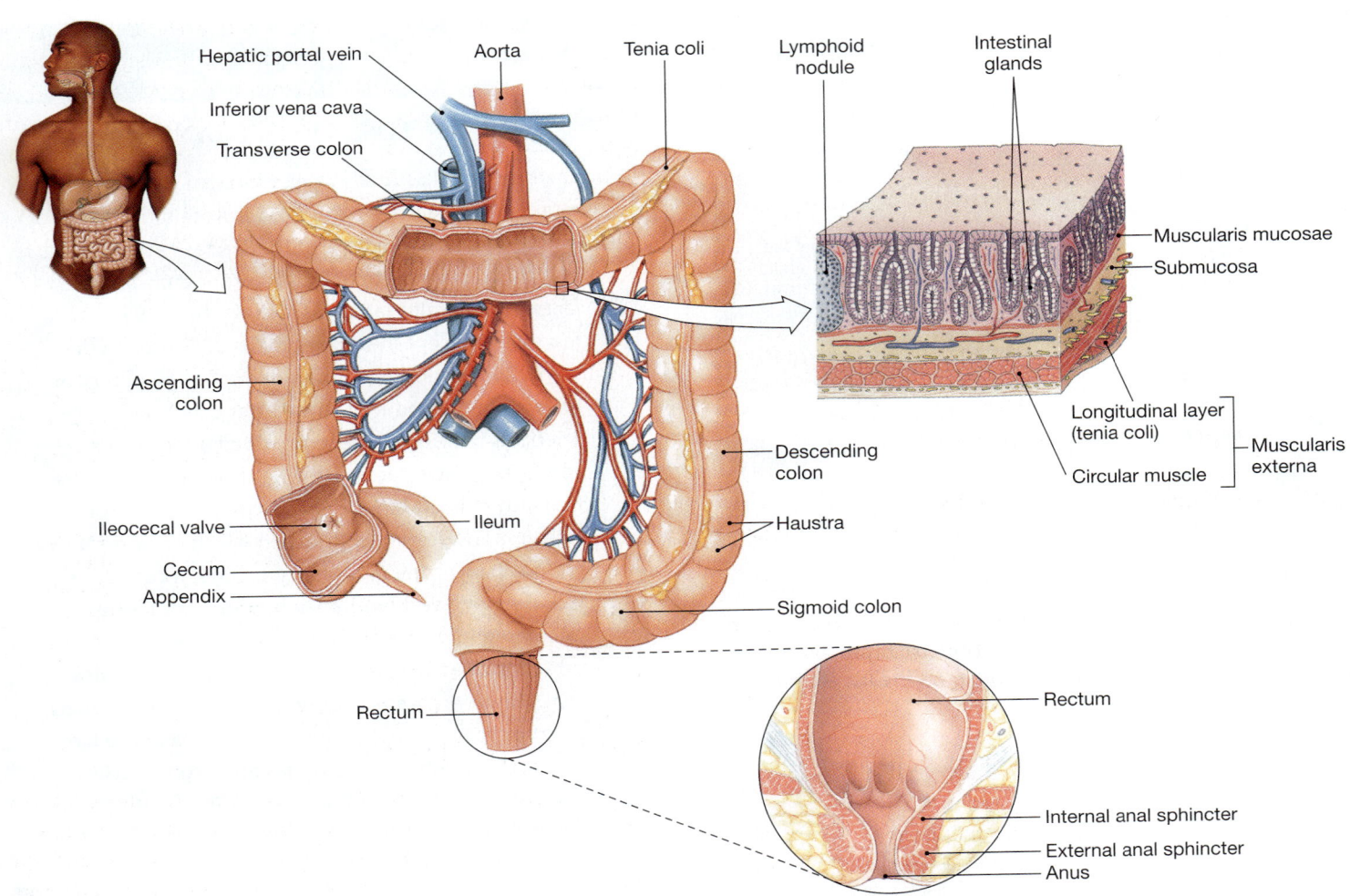

FIGURE 21-31   *Anatomy of the large intestine*

the opening between the ileum and the **cecum**, the initial section of the large intestine (Fig. 21-31 ■). The ileocecal valve relaxes each time a peristaltic wave reaches it. It also relaxes when food leaves the stomach as part of the *gastroileal reflex.*

The large intestine has seven regions. The cecum is a dead-end pouch with the *appendix,* a small fingerlike projection, at its ventral end. Material moves from the cecum upward through the **ascending colon,** horizontally across the body through the **transverse colon,** then down through the **descending colon** and **sigmoid colon** [*sigmoeides,* shaped like a sigma, Σ]. The **rectum** is the short (about 12 cm) terminal section of the large intestine. It is separated from the external environment by the **anus,** an opening closed by two sphincters, an internal smooth muscle sphincter and an external skeletal muscle sphincter.

The wall of the colon differs from that of the small intestine in that the muscularis of the large intestine has an inner circular layer but a discontinuous longitudinal muscle layer concentrated into three bands called the **tenia coli.** Contractions of the tenia pull the wall into bulging pockets called **haustra.**

The mucosa of the colon has two regions, like that of the small intestine. The luminal surface lacks villi and appears

smooth. It is composed of colonocytes and mucus-secreting goblet cells. The crypts contain stem cells that divide to produce new epithelium, as well as goblet cells, endocrine cells, and maturing colonocytes.

Motility in the Large Intestine   Chyme that enters the colon continues to be mixed by segmental contractions. Forward movement is minimal during mixing contractions and depends primarily on a unique colonic contraction known as **mass movement.** A wave of contraction decreases the diameter of a segment of colon and sends a substantial bolus of material forward. These contractions occur 3–4 times a day and are associated with eating and distension of the stomach through the *gastrocolic reflex.* Mass movement is responsible for the sudden distension of the rectum that triggers defecation.

The **defecation reflex** removes undigested feces from the body. Defecation resembles urination in that it is a spinal reflex triggered by distension of the organ wall. The movement of fecal material into the normally empty rectum triggers the reflex. Smooth muscle of the **internal anal sphincter** relaxes, and peristaltic contractions in the rectum push material toward the anus. At the same time, the external anal sphincter, which is under voluntary control, is consciously relaxed if the situation is appropriate. Defecation is often aided by conscious abdominal contractions and forced expiratory movements against a closed glottis (the *Valsalva maneuver*).

Defecation, like urination, is subject to emotional influence. Stress may increase intestinal motility and cause psychosomatic diarrhea in some individuals but decrease motility and cause *constipation* in others. When feces are retained in the colon, either through consciously ignoring a defecation reflex or through decreased motility, continued water absorption creates hard, dry feces that are difficult to expel. One treatment for constipation is glycerin suppositories, small bullet-shaped wads that are inserted through the anus into the rectum. Glycerin attracts water and helps soften the feces to promote defecation.

Digestion and Absorption in the Large Intestine

According to the traditional view of the large intestine, no significant digestion of organic molecules takes place there. However, in recent years this view has been revised. We now know that the numerous bacteria that inhabit the colon break down significant amounts of undigested complex carbohydrates and proteins through fermentation. The end products include lactate and short-chain fatty acids, such as butyric acid. Several of these products are lipophilic and can be absorbed by simple diffusion. The fatty acids, for example, are used by colonocytes as their preferred energy substrate.

Colonic bacteria also produce significant amounts of absorbable vitamins, especially vitamin K. Intestinal gases, such as hydrogen sulfide, that escape from the gastrointestinal tract are a less useful product. Some starchy foods, such as beans, are notorious for their tendency to produce intestinal gas (**flatus**).

Diarrhea Can Cause Dehydration

Diarrhea is a pathological state in which intestinal secretion of fluid is not balanced by absorption, resulting in watery stools. Diarrhea will occur if normal intestinal water absorption mechanisms are disrupted or if there are unabsorbed osmotically active solutes that "hold" water in the lumen. Substances that cause *osmotic diarrhea* include undigested lactose and sorbitol, a sugar alcohol from plants. Sorbitol is used as an "artificial" sweetener in some chewing gums and in foods made for diabetics. Another unabsorbed solute that can cause osmotic diarrhea, intestinal cramping, and gas is Olestra, the "fake fat" made from vegetable oil and sugar.

In clinical settings, patients who need to have their bowels cleaned out before surgery or other procedures are often given four liters of an isotonic solution of polyethylene glycol and electrolytes to drink. Because polyethylene glycol cannot be absorbed, a large volume of unabsorbed solution passes into the colon, where it triggers copious diarrhea that removes all solid waste from the GI tract.

Secretory diarrheas occur when bacterial toxins, such as cholera toxin from *Vibrio cholerae* and *Escherichia coli* entero-

toxin, enhance colonic Cl^- secretion (see Fig. 21-9, p. 687). When excessive fluid secretion is coupled with increased motility, diarrhea results. Secretory diarrhea in response to intestinal infection can be viewed as adaptive because it helps flush pathogens out of the lumen. However, it also has the potential to cause dehydration if fluid loss is excessive.

The World Health Organization estimates that in developing countries, 4 million people die from diarrhea each year. In the United States, diarrhea in children causes about 200,000 hospitalizations a year. Oral replacement fluids for treatment of diarrheal salt and water loss can prevent the *morbidity* (illness) and *mortality* (death) associated with diarrhea. Oral rehydration solutions usually contain glucose or sucrose as well as Na^+, K^+, and Cl^- because the inclusion of a sugar enhances Na^+ absorption. If dehydration is severe, intravenous fluid therapy may be necessary.

✓ CONCEPT CHECK

11. In secretory diarrhea, epithelial cells in the intestinal villi may be damaged or may slough off. In these cases, would it be better to use an oral rehydration solution containing glucose or one containing sucrose? Explain your reasoning.
    Answers: p. 715

IMMUNE FUNCTIONS OF THE GI TRACT

As you learned at the beginning of the chapter, the GI tract is the largest immune organ in the body. Its luminal surface is continuously exposed to disease-causing organisms, and the immune cells of the GALT must prevent these pathogens from entering the body through delicate absorptive tissues. The first lines of defense are the enzymes and immunoglobulins in saliva and the highly acidic environment of the stomach. If pathogens or toxic materials make it into the small intestine, sensory receptors and the immune cells of the GALT respond. Two common responses are diarrhea, just described, and vomiting.

M Cells Sample the Contents of the Gut

The immune system of the intestinal mucosa consists of immune cells scattered throughout the mucosa, clusters of immune cells in Peyer's patches (see Fig. 21-2e, p. 681), and specialized epithelial cells called **M cells** that overlie the Peyer's patches. The M cells provide information about the contents of the lumen to the immune cells of the GALT.

The microvilli of M cells are fewer in number and more widely spaced than in the typical intestinal cell. The apical surface of M cells contains clathrin-coated pits [🔁 p. 148] with embedded membrane receptors. When antigens bind to these receptors, the M cell uses transcytosis to transport them to its basolateral membrane, where they are released into the interstitial fluid. Macrophages and lymphocytes [🔁 p. 538] are waiting in the extracellular compartment for the M cell to present them with antigens.

If the antigens are substances that threaten the body, the immune cells swing into action. They secrete cytokines to attract additional immune cells that can attack the invaders and cytokines that trigger an inflammatory response. A third response to cytokines is increased Cl^-, fluid, and mucus secretion to flush the invaders from the GI tract.

In *inflammatory bowel diseases* (ulcerative colitis and Crohn's disease), the immune response is triggered inappropriately by the normal contents of the gut. One apparently successful experimental therapy for these diseases involves blocking the action of cytokines released by the gut-associated lymphoid tissues.

How certain pathogenic bacteria cross the barrier created by the intestinal epithelium has puzzled scientists for years. The discovery of M cells may provide the answer. It appears that some bacteria, such as *Salmonella* and *Shigella,* have evolved surface molecules that bind to M cell receptors. The M cells then obligingly transport the bacteria across the epithelial barrier and deposit them inside the body, where the immune system immediately reacts. Both bacteria cause diarrhea, and *Salmonella* also causes fever and vomiting.

Vomiting Is a Protective Reflex

Vomiting, or *emesis*, the forceful expulsion of gastric and duodenal contents from the mouth, is a protective reflex that removes toxic materials from the GI tract before they can be absorbed. However, excessive or prolonged vomiting, with its loss of gastric acid, can cause metabolic alkalosis [p. 671].

The vomiting reflex is coordinated through a vomiting center in the medulla. The reflex begins with stimulation of sensory receptors and is often (but not always) accompanied by nausea. A variety of stimuli from all over the body can trigger vomiting. They include chemicals in the blood, such as cytokines and certain drugs; pain; and disturbed equilibrium, such as occurs in a moving car or rocking boat. Tickling the back of the pharynx can also induce vomiting.

Efferent signals from the vomiting center initiate a wave of reverse peristalsis that begins in the small intestine and moves upward. The motility wave is aided by abdominal contraction that increases intra-abdominal pressure. The stomach relaxes so that the increased pressure forces gastric and intestinal contents back into the esophagus and out of the mouth.

During vomiting, respiration is inhibited. The epiglottis and soft palate close off the trachea and nasopharynx to prevent the *vomitus* from being inhaled (*aspirated*). Should acid or small food particles get into the airways, they could damage the respiratory system and cause *aspiration pneumonia.*

In the next chapter we look at the fates of nutrients once they have been absorbed from the digestive tract.

RUNNING PROBLEM CONCLUSION

PEPTIC ULCERS

Tonya's lab results are in. Tests performed on her stomach tissue confirm that she has an *H. pylori* infection that is probably the cause of her duodenal ulcer. Tonya's physician gives her a double antibiotic to take for seven days and changes her H_2 receptor antagonist to a proton pump inhibitor.

"Acid indigestion" and heartburn are two of the most common digestive disorders in the United States. Stroll the aisles of any pharmacy and look at the space devoted to over-the-counter medications for these conditions. Traditional

antacids are buffers, such as calcium carbonate, sodium bicarbonate, aluminum hydroxide, and magnesium hydroxide, that neutralize H^+ in the stomach lumen. Newer therapies have been developed based on physiological studies of gastric acid secretion. To see acid secretion in action, do a Google search for *parietal cell animation.*

Now check your understanding of this running problem by comparing your answers with the information in the following summary table.

	QUESTION	FACTS	INTEGRATION AND ANALYSIS
1	Why are peptic ulcers found in the duodenum rather than in the jejunum or ileum?	Peptic ulcers are caused by gastric acid and stomach enzymes. Chyme moving out of the stomach passes into the duodenum.	If protective mechanisms in the intestine fail, the duodenum is the first part of the small intestine to be exposed to gastric acid and enzymes.
2	How does the conversion of urea into ammonia protect *H. pylori* from the hostile environment of the stomach?	A base is an anion that combines with H^+. Ammonia, a base, combines with H^+ to create ammonium ions, NH_4^+. This ties up free H^+ and increases pH.	*H. pylori* has ammonia-producing enzymes that combine with H^+ to reduce acidity. In this way the bacteria protect themselves from being killed by the acidic environment of the stomach.

21

QUESTION	FACTS	INTEGRATION AND ANALYSIS
3 Write the chemical equations showing how the antacids sodium bicarbonate ($NaHCO_3$) and aluminum hydroxide ($Al(OH)_3$) buffer hydrochloric acid (HCl).	Buffers are molecules that combine with H^+. Bicarbonate (HCO_3^-) and OH^- are buffers.	$NaHCO_3 + HCl \rightarrow NaCl + H_2CO_3$ $Al(OH)_3 + 3\ HCl \rightarrow AlCl_3 + 3\ H_2O$
4 How does blocking H_2 histamine receptors stop acid production in the stomach?	Histamine is a paracrine that stimulates acid secretion by the stomach.	If H_2 receptors are blocked by antagonists, histamine cannot combine with them and stimulate acid secretion.

677   682   687   691   704   **710**

CHAPTER SUMMARY

The digestive system, like the renal system, plays a key role in *mass balance* in the body. Most material that enters the system, whether by mouth or by secretion, is absorbed before it reaches the end of the GI tract. In pathologies such as diarrhea, in which absorption and secretion are unbalanced, the loss of material through the GI tract can seriously disrupt *homeostasis*. Absorption and secretion in the GI tract provide numerous examples of *movement across membranes*, most of which follow patterns you have encountered in other systems. Finally, regulation of GI tract function illustrates the complex interactions that take place between endocrine and neural *control systems* and the immune system.

Function and Processes of the Digestive System

1. The GI tract moves nutrients, water, and electrolytes from the external environment to the internal environment. (p. 677)

2. The four processes of the digestive system are digestion, absorption, motility, and secretion. **Digestion** is chemical and mechanical breakdown of foods into absorbable units. **Absorption** is transfer of substances from the lumen of the GI tract to the ECF. **Motility** is movement of material through the GI tract. **Secretion** refers to the transfer of fluid and electrolytes from ECF to lumen or the release of substances from cells. (p. 678; Fig. 21-1)

3. The GI tract contains the largest collection of lymphoid tissue in the body, the **gut-associated lymphoid tissue (GALT)**. (p. 678)

Anatomy of the Digestive System

IP GI-Anatomy Review

4. **Chyme** is a soupy substance created as ingested food is broken down by mechanical and chemical digestion. (p. 678)

5. Food entering the digestive system passes through the mouth, pharynx, **esophagus, stomach (fundus, body, antrum), small intestine (duodenum, jejunum, ileum), large intestine (colon, rectum)**, and **anus**. (p. 679; Fig. 21-2a)

6. The **salivary glands, pancreas**, and **liver** add exocrine secretions containing enzymes and mucus to the lumen. (p. 679; Fig. 21-2a)

7. The wall of the GI tract consists of four layers: mucosa, submucosa, muscularis externa, and serosa. (p. 679; Fig. 21-2d)

8. The **mucosa** faces the lumen and consists of epithelium, the **lamina propria**, and the **muscularis mucosa**. The lamina propria contains immune cells. Small **villi** and invaginations increase the surface area. (p. 679; Fig. 21-2e)

9. The **submucosa** contains blood vessels and lymph vessels and the **submucosal plexus** of the **enteric nervous system**. (p. 682; Fig. 21-2c, d)

10. The **muscularis externa** consists of a layer of circular muscle and a layer of longitudinal muscle. The **myenteric plexus** lies between these two muscle layers. (p. 682; Fig. 21-2c)

11. The **serosa** is the outer layer of the GI tract wall. It is a connective tissue membrane that is a continuation of the peritoneal membrane. (p. 682)

Motility

IP GI-Motility

12. Motility moves food from mouth to anus and mechanically mixes food. (p. 682)

13. GI smooth muscle cells are electrically connected by gap junctions. Some segments are **tonically contracted**, but others exhibit **phasic contractions**. (p. 683)

14. Intestinal muscle exhibits spontaneous **slow wave potentials**. When a slow wave reaches threshold, it fires action potentials and contracts. (p. 683; Fig. 21-3)

15. Slow waves originate in the **interstitial cells of Cajal**. (p. 683)

16. Between meals, the **migrating motor complex** moves food remnants from the upper GI tract to the lower regions. (p. 683)

17. **Peristaltic contractions** are progressive waves of contraction. Peristalsis is mediated by the enteric nervous system and altered by hormones, paracrine signals, and neuropeptides. (p. 683; Fig. 21-4)

18. **Segmental contractions** are primarily mixing contractions. (p. 684; Fig. 21-4)

21

Secretion

IP GI-Secretion

19. About 2 liters of fluid per day enter the GI tract through the mouth. Another 7 liters of water, ions, and proteins are secreted by the body. Nearly all of this volume is reabsorbed. (p. 684; Fig. 21-5)

20. **Parietal cells** in gastric glands secrete hydrochloric acid. (p. 685; Fig. 21-6)

21. The pancreas secretes a watery $NaHCO_3$ solution from duct cells. (p. 685; Figs. 21-7, 21-8)

22. Intestinal cells secrete Cl^- using the **CFTR chloride channel**. Water and Na^+ follow passively down osmotic and electrochemical gradients. (p. 685; Fig. 21-9)

23. Digestive enzymes are secreted by exocrine glands or by gastric and intestinal epithelium. (p. 687)

24. Mucus from **mucous cells** and **goblet cells** forms a protective coating and lubricates contents of the gut. (p. 687)

25. **Saliva** is an exocrine secretion that contains water, ions, mucus, and proteins. Salivation is under autonomic control. (p. 689)

26. Bile made by **hepatocytes** contains **bile salts**, bilirubin, and cholesterol. Bile is stored and concentrated in the gallbladder (p. 689; Fig. 21-10)

Regulation of GI Function

IP GI-Control of the Digestive System

27. **Short reflexes** originate in the **enteric nervous system** and are integrated there. **Long reflexes** are integrated in the CNS. Long reflexes may originate in the ENS or outside it. (p. 689; Fig. 21-11)

28. Generally, parasympathetic innervation is excitatory for GI function, and sympathetic innervation is inhibitory. (p. 689)

29. GI peptides excite or inhibit motility and secretion. Most stimuli for GI peptide secretion arise from the ingestion of food. (p. 690; Fig. 21-12)

30. GI paracrine molecules, such as histamine, interact with hormones and neural reflexes. (p. 690)

31. The enteric nervous system is called "the little brain" because it can integrate information without input from the CNS. **Intrinsic neurons** lie completely within the ENS (p. 691)

32. Digestive hormones are divided into the gastrin family (**gastrin, cholecystokinin**), secretin family (**secretin, vasoactive intestinal peptide, gastric inhibitory peptide, glucagon-like peptide 1**), and hormones that do not fit into either of those two families (**motilin**). (pp. 692–693; Tbl. 21-1)

Digestion and Absorption

IP GI-Digestion and Absorption

33. Digestion combines mechanical and enzymatic breakdown of food. Digestion is not directly regulated but depends on secretion and motility. (p. 693)

34. Most nutrient absorption takes place in the small intestine. The large intestine absorbs water and ions. (p. 693)

35. **Amylase** digests starch to maltose. **Disaccharidases** digest disaccharides to monosaccharides. (p. 694; Fig. 21-14)

36. Glucose absorption uses the SGLT Na^+-glucose symporter and GLUT2 transporter. Fructose uses the GLUT5 and GLUT2 transporters. (p. 694; Fig. 21-15)

37. **Endopeptidases** (also called proteases) break proteins into smaller peptides. **Exopeptidases** remove amino acids from peptides. (pp. 694–695; Fig. 21-16)

38. Amino acids are absorbed via Na^+-dependent cotransport. Dipeptides and tripeptides are absorbed via H^+-dependent cotransport. Some larger peptides are absorbed intact via transcytosis. (p. 695; Fig. 21-17)

39. Fat digestion requires the enzyme **lipase** and the cofactor **colipase**. (p. 696; Figs. 21-18, 21-19)

40. Fat digestion is facilitated by bile salts, which emulsify fats. As enzymatic and mechanical digestion proceed, fat droplets form **micelles**. (p. 696; Figs. 21-19, 21-20)

41. Fat absorption occurs primarily by simple diffusion. Cholesterol is actively transported. (p. 696)

42. Monoglycerides and fatty acids reassemble into triglycerides in the intestinal cell, then combine with cholesterol and proteins into **chylomicrons**. Chylomicrons are absorbed into the lymph. (p. 697; Fig. 21-19)

43. Nucleic acids are digested and absorbed as nitrogenous bases and monosaccharides. (p. 697)

44. Fat-soluble vitamins are absorbed along with fats. Water-soluble vitamins are absorbed by mediated transport. Vitamin B_{12} absorption requires **intrinsic factor** secreted by the stomach. Mineral absorption usually occurs via active transport. (p. 698)

The Cephalic Phase

45. In the **cephalic phase** of digestion, the sight, smell, or taste of food initiates GI reflexes. (p. 699; Fig. 21-23)

46. Mechanical digestion begins with chewing, or **mastication**. Saliva moistens and lubricates food. Salivary amylase digests carbohydrates. (p. 699)

47. Swallowing, or **deglutition**, is a reflex integrated by a medullary center. (p. 699; Fig. 21-24)

The Gastric Phase

48. The stomach stores food, begins protein and fat digestion, and protects the body from swallowed pathogens. (p. 700)

49. The stomach secretes mucus and bicarbonate from **mucous cells**, gastric acid from parietal cells, pepsinogen from **chief cells**, somatostatin from **D cells**, histamine from **ECL cells**, and gastrin from **G cells**. (p. 702; Fig. 21-25)

50. Gastric function is integrated with the cephalic and intestinal phases of digestion. (p. 703; Figs. 21-26, 21-28)

The Intestinal Phase

51. Acid in the intestine, CCK, and secretin delay gastric emptying. (p. 705; Fig. 21-28)

52. The pancreas secretes bicarbonate to neutralize gastric acid. (p. 705)

53. Intestinal enzymes are part of the **brush border**. Most pancreatic enzymes are secreted as zymogens that need to be activated. (p. 705; Fig. 21-29)

54. Most absorption takes place in the small intestine. (p. 706)

55. Digestion and absorption of nutrients is completed in the small intestine. Most absorbed nutrients go directly to the liver via the hepatic portal system before entering the systemic circulation. (p. 706)

56. The large intestine concentrates the 1.5 liters of chyme that enter it daily. (p. 707)

57. Undigested material in the colon moves forward by **mass movement**. The **defecation reflex**, a spinal reflex subject to higher control, is triggered by sudden distension of the rectum. (p. 708)

58. Colonic bacteria use fermentation to digest organic material. (p. 709)

59. Cells of the colon can both absorb and secrete fluid. Excessive fluid secretion or decreased absorption causes diarrhea. (p. 709)

Immune Functions of the GI Tract

60. Protective mechanisms of the GI tract include acid and mucus production, vomiting, and diarrhea. (p. 709)

61. **M cells** sample gut contents and present antigens to cells of the GALT. (p. 709)

62. **Vomiting** is a protective reflex integrated in the medulla. (p. 710)

QUESTIONS

(Answers to the Review Questions begin on page A1.)

THE PHYSIOLOGY PLACE

Access more review material online at **The Physiology Place** website. There you'll find review questions, problem-solving activities, case studies, flashcards, and direct links to both *InterActive Physiology®* and *PhysioEx™*. To access the site, go to *www. physiologyplace.com* and select Human Physiology, Fourth Edition.

LEVEL ONE REVIEWING FACTS AND TERMS

1. Define the four basic processes of the digestive system and give an example of each.

2. For most nutrients, the processes _____ and _____ are not regulated, whereas _____ and _____ are continuously regulated. Why do you think these differences exist? Defend your answer.

3. Match each of the following descriptions with the appropriate term(s):

   (a) chyme is produced here
   (b) organ where most digestion occurs
   (c) initial section of small intestine
   (d) this adds exocrine secretions to duodenum via a duct
   (e) sphincter between stomach and intestine
   (f) enzymes produced here
   (g) distension of its walls triggers the defecation reflex

   1. colon
   2. stomach
   3. small intestine
   4. duodenum
   5. ileum
   6. jejunum
   7. pancreas
   8. pylorus
   9. rectum
   10. liver

4. List the four layers found in the GI tract walls. What type of tissue predominates in each layer?

5. Describe the functional types of epithelium lining the stomach and intestines.

6. What are Peyer's patches?

7. What purposes does motility serve in the gastrointestinal tract? Which types of tissue contribute to gut motility? Which types of contraction do the tissues undergo?

8. What is a zymogen? What is a proenzyme? List two examples of each.

9. Match each of the following cells with the product(s) it secretes. Items may be used more than once.

   (a) parietal cells
   (b) goblet cells
   (c) brush border cells
   (d) pancreatic cells
   (e) D cells

   1. enzymes
   2. histamine
   3. mucus
   4. pepsinogen
   5. gastrin

   (f) ECL cells
   (g) chief cells
   (h) G cells

   6. somatostatin
   7. HCO_3^-
   8. HCl
   9. intrinsic factor

10. How does each of the following factors affect digestion? Briefly explain how and where each factor exerts its effects.

    (a) emulsification
    (b) neural activity
    (c) pH
    (d) size of food particles

11. Most digested nutrients are absorbed into the _____ of the _____ system, delivering nutrients to the _____ (organ). However, digested fats go into the _____ system because intestinal capillaries have a _____ around them that most lipids are unable to cross.

12. What is the enteric nervous system, and what is its function?

13. What are short reflexes? What types of responses do they regulate? What is meant by the term *long reflex*?

14. What role do paracrines play in digestion? Give specific examples.

LEVEL TWO REVIEWING CONCEPTS

15. **Map exercise:** List the three major groups of biomolecules across the top of a large piece of paper. Down the left side of the paper write *mouth, stomach, small intestine, absorption*. For each biomolecule in each location, fill in the enzymes that digest the biomolecule, the products of digestion for each enzyme, and the location and mechanisms by which these products are absorbed.

16. Define, compare, and contrast the following pairs or sets of terms:

    (a) mastication, deglutition
    (b) microvilli, villi
    (c) peristalsis, segmentation, migrating motor complex, mass movements, defecation, vomiting, diarrhea
    (d) chyme, feces
    (e) short reflexes, long reflexes
    (f) submucosal plexus, myenteric plexus, enteric nervous system, vagus nerve
    (g) cephalic, gastric, and intestinal phases of digestion

17. Diagram the mechanisms by which Na^+, K^+, and Cl^- are transported out of the intestine. Diagram the mechanisms by which H^+ and HCO_3^- are secreted into the lumen.

18. Compare the enteric nervous system with the cephalic brain. Give some specific examples of neurotransmitters, neuromodulators, and supporting cells in the two.

19. List and briefly describe the actions of the members of each of the three groups of GI hormones.

20. Explain how H_2 receptor antagonists and proton pump inhibitors decrease gastric acid secretion.

21

LEVEI THREE PROBLEM SOLVING

21. Draw an intestinal epithelial cell. Label the apical and basolateral membranes, lumen, and ECF. Add and label whatever membrane proteins, organelles, and other structures you need to diagram all the steps needed for the following scenario: An intestinal epithelial cell absorbs some free amino acids and some dipeptides from the intestinal lumen. Inside the cell, the amino acids and dipeptides are reassembled into a digestive enzyme, which is then placed on the apical surface as a brush border enzyme.

22. Draw and label two intestinal cells as you did in question 21. Use the text descriptions of iron and calcium absorption on p. 698 to diagram the absorption of these two minerals. Use arrows labeled with "?" to indicate steps that scientists are unsure about.

23. Erica's baby, Justin, has had a severe bout of diarrhea and is now dehydrated. Is his blood more likely to be acidotic or alkalotic? Why?

24. Mary Littlefeather arrives in her physician's office complaining of severe, steady pain in the upper right quadrant of her abdomen. The pain began shortly after she ate a meal of fried chicken, french fries, and peas. Lab tests and an ultrasound reveal the presence of gallstones in the common bile duct running from the liver, gallbladder, and pancreas into the small intestine.

    (a) Why was Mary's pain precipitated by the meal she ate?
    (b) Which of the following processes will be affected by the gallstones: micelle formation in the intestine, carbohydrate digestion in the intestine, protein absorption in the intestine. Explain your reasoning.

25. Using what you have learned about epithelial transport in this chapter and the renal chapters, draw a picture of the salivary duct cells and lumen. Arrange membrane channels and transporters on the apical and basolateral membranes so that the duct cell absorbs Na^+ and secretes K^+. With neural stimulation, the flow rate of saliva can increase from 0.4 mL/min to 2 mL/min. What do you think happens to the Na^+ and K^+ content of saliva at the higher flow rate?

LEVEL FOUR QUANTITATIVE PROBLEMS

26. Intestinal transport of the amino acid analog MIT (monoiodotyrosine) can be studied using the "everted sac" preparation. A length of intestine is turned inside out, filled with a solution containing MIT, tied at both ends, and then placed in a bath containing nutrients, salts, and an equal concentration of MIT. Changes in the concentration of MIT are monitored in the bath (mucosal or apical side of the inverted intestine), in the intestinal cells (tissue), and within the sac (serosal or basolateral side of the intestine) over a 240-minute period. The results are displayed in the graph shown here. (Data from Nathans et al., *Biochimica et Biophysica Acta* 41:271–282, 1960)

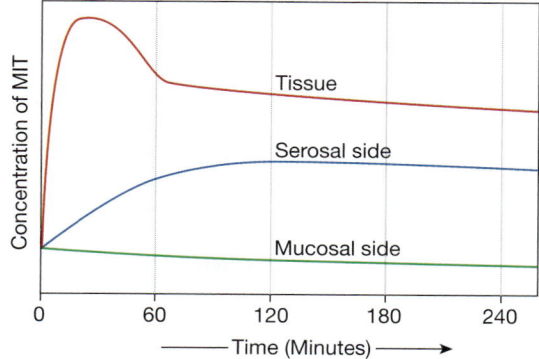

(a) Based on the data shown, is the transepithelial transport of MIT a passive process or an active process?
(b) Which way does MIT move: (1) apical to tissue to basolateral, or (2) basolateral to tissue to apical? Is this movement absorption or secretion?
(c) Is transport across the apical membrane active or passive? Explain your reasoning.
(d) Is transport across the basolateral membrane active or passive? Explain your reasoning.

ANSWERS

✓ Answers to Concept Check Questions

Page 678

1. Digestion is the chemical and mechanical breakdown of food into absorbable units.

2. Digestion takes place in the GI tract lumen, which is external to the body; metabolism takes place in the body's internal environment.

3. Absorption moves material from the GI lumen into the ECF; secretion moves substances from the cells or the ECF into the lumen.

Page 682

4. The lumen of the digestive tract is on the apical or mucosal side of the intestinal epithelium.

5. Mouth → pharynx → esophagus → stomach (fundus, body, antrum) → small intestine (duodenum, jejunum, ileum) → large intestine (colon, rectum) → anus

6. Because the GI tract has a large, vulnerable surface area facing the external environment, it needs the immune cells of lymphoid tissue to combat potential invaders.

Page 684

7. Some sphincters are tonically contracted to close off the GI tract from the outside world and to keep material from passing freely from one section of the tract to another.

Page 699

8. Na^+ from table salt enhances the intestinal absorption of glucose because glucose must be cotransported with Na^+.

9. Bile salts do not digest triglycerides. They simply emulsify them into small particles so that lipase can digest them.

Page 707

10. CCK slows gastric motility and acid secretion, promotes release of pancreatic enzymes and bile, and promotes intestinal motility. Bile from the gallbladder and pancreatic lipase and colipase are necessary for efficient fat digestion. Fat digestion takes longer than digestion of other foodstuffs, and slowing gastric motility gives the intestine more time to digest fats. Decreasing gastric acid secretion also delays movement of chyme from stomach to intestine. Promoting

intestinal motility ensures that bile salts can emulsify fats and that micelles will get adequate exposure to the brush border, where they are absorbed.

Page 709

11. Damage to epithelial cells means that brush border sucrase may be less effective or absent. In these cases, sucrose digestion would be impaired, so it would be better to use a solution containing glucose because this sugar need not be digested before being absorbed.

Q *Answers to Figure Questions*

Page 690

Figure 21-11: (a) The myenteric plexus controls smooth muscle contraction; the submucosal plexus controls smooth muscle contraction and endocrine and exocrine secretions by secretory cells. (b) Stretch receptors respond to stretch, osmoreceptors to osmolarity, and chemoreceptors to products of digestion.

21

22

> Probably no single abnormality has contributed more to our knowledge of the intermediary metabolism of animals than has the disease known as diabetes.
>
> —Helen R. Downes, *The Chemistry of Living Cells, 1955*

Vitamin C crystals viewed under polarized light.

Metabolism and Energy Balance

BACKGROUND BASICS

RUNNING PROBLEM

EATING DISORDERS

Sara Inman and Nicole Baker had been best friends growing up but hadn't seen each other for several semesters. When Sara heard Nicole was in the hospital, she went to visit but wasn't prepared for what she saw. Nicole, who had always been thin, now was so emaciated that Sara hardly recognized her. Nicole had been taken to the emergency room when she fainted in a ballet class. When Dr. Ayani saw Nicole, his concern about her weight was as great as his concern about her broken wrist. At 5' 6'', Nicole weighed only 95 pounds (normal healthy range for that height is 118–155 pounds). Dr. Ayani suspected an eating disorder, probably anorexia nervosa, and he ordered blood tests to confirm his suspicion.

| 717 | 719 | 721 | 722 | 740 | 745 |

Magazine covers at grocery store checkout stands reveal a lot about Americans. Headlines screaming "Lose 10 pounds in a week without dieting" or "CCK: the hormone that makes you thin" vie for attention with glossy photographs of fatty, high-calorie desserts dripping with chocolate and whipped cream. As one magazine article put it, we are a nation obsessed with staying trim and with eating—two mutually exclusive occupations. But what determines when, what, and how much we eat? The factors influencing food intake are an area of intense research because the act of eating is the main point at which our bodies exert control over energy input.

The Brain Controls Food Intake

The control of food intake is a complex process. The digestive system does not regulate energy intake, so we must depend on behavioral mechanisms, such as hunger and satiety [*satis,* enough], to tell us when and how much to eat. Psychological and social aspects of eating, such as parents who say "Clean your plate," complicate the physiological control of food intake. As a result, we still do not fully understand what governs when, what, and how much we eat.

Our current model is based on hypothalamic centers [🖳 Fig. 11-3, p. 379] influenced by neurotransmitters and hormones. Studies using transgenic and knockout mice show that control of these centers involves multiple brain neuropeptides and "brain-gut" hormones. What follows is an overview of this complex and constantly changing field.

The hypothalamus contains two centers that regulate food intake: a **feeding center** that is tonically active and a **satiety**

center that stops food intake by inhibiting the feeding center. Efferent signals from these centers cause changes in eating behavior and create sensations of hunger and fullness. Animals whose feeding center is destroyed will cease to eat. If the satiety center is destroyed, animals overeat and become obese [*obesus,* plump or fat].

There are two classic theories for regulation of food intake: the glucostatic theory and the lipostatic theory. The **glucostatic theory** states that glucose utilization by hypothalamic centers regulates food intake. When blood glucose concentrations decrease, the satiety center is suppressed and the feeding center is dominant. When glucose utilization is up, the satiety center inhibits the feeding center.

The **lipostatic theory** of energy balance proposes that a signal from the body's fat stores to the brain modulates eating behavior so that the body maintains a particular weight. If fat stores are increased, eating decreases. In times of starvation, eating increases. Obesity results from disruption of this pathway.

The discovery of **leptin** [*leptos,* thin], a protein hormone synthesized in adipocytes, provided evidence for a negative-feedback signal between adipose tissue and the brain: as fat stores increase, adipose cells secrete more leptin, and food intake decreases. Leptin is synthesized in adipocytes under control of the *obese (ob) gene*. Mice that lack the *ob* gene (and therefore lack leptin) become obese, as do mice with defective leptin receptors. However, these findings did not translate well to humans, as only a small percentage of obese humans are leptin deficient. The majority of them have *elevated* leptin levels. As we are learning, leptin is only part of the story.

Another important signal molecule is **neuropeptide Y** (NPY), a brain neurotransmitter that seems to be the stimulus for food intake. In normal-weight animals, leptin inhibits NPY in a negative feedback pathway (Fig. 22-1 ■). The peptide **ghrelin**, secreted by the stomach, increases hunger when infused into human subjects. Other neuropeptides and hormones also influence NPY and the hypothalamic centers (Table 22-1 ■). It is interesting to note that many of these peptides affect other functions in addition to food intake. Ghrelin promotes release of growth hormone, for instance, and brain peptides called *orexins* appear to play a role in sleep. Our understanding of how all these factors interact is incomplete.

Appetite and eating are also influenced by sensory input through the nervous system. The simple acts of swallowing and chewing food help create a sensation of fullness. The sight, smell, and taste of food can either stimulate or suppress appetite.

One interesting study tried to determine whether chocolate craving is attributable to psychological factors or to physiological stimuli, such as chemicals in the chocolate.* Subjects were given either dark chocolate, white chocolate (which

---

*"Pharmacological versus sensory factors in the satiation of chocolate craving," *Physiology & Behavior* 56(3):419–422, 1994 Sep.

22

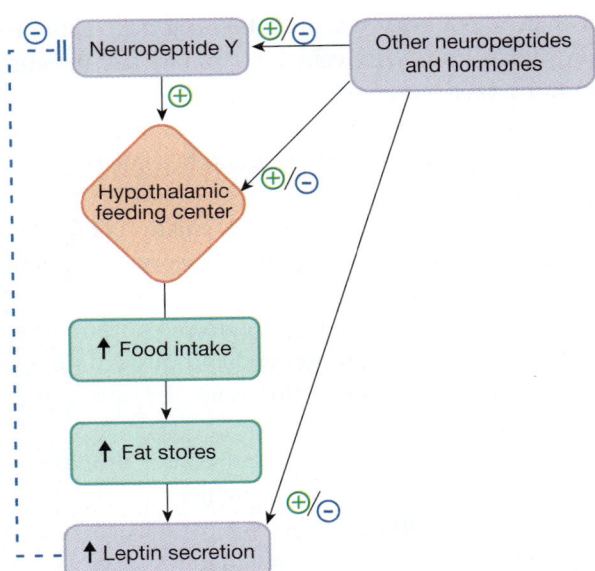

■ FIGURE 22-1   *Roles of peptides in regulation of food intake*

TABLE 22-1	**Some Peptides That Modulate Food Intake**
PEPTIDE	**SOURCE**
Increase food intake	
Neuropeptide Y (NPY)	Hypothalamus
Orexins (also called hypocretins)	Hypothalamus
Galanin	Hypothalamus
Melanin-concentrating hormone (MCH)	Hypothalamus
Ghrelin	Stomach
Decrease food intake	
CCK	Small intestine; neurons
Corticotropin-releasing hormone (CRH)	Hypothalamus
α-Melanocyte-stimulating hormone (α-MSH)	Hypothalamus
CART (cocaine- and amphetamine-regulated transcript)	Hypothalamus
Glucagon-like peptide-1 (GLP-1)	Intestines
PYY$_{3-36}$	Intestines

contains none of the pharmacological agents of cocoa), cocoa capsules, or placebo capsules. The results showed that white chocolate was the best substitute for the real thing, and that aroma plays a significant role in satisfying chocolate craving, suggesting that one aspect of chocolate craving is psychological.

Psychological factors, such as eating or not eating when stressed, are known to play a significant role in regulating food intake. The eating disorder *anorexia nervosa* probably has both psychological and physiological components. The concept of appetite is closely linked to the psychology of eating, which may explain why dieters who crave smooth, cold ice cream cannot be satisfied by a crunchy carrot stick. A considerable amount of money is directed to research on eating behaviors.

C O N C E P T   C H E C K

1. Explain the roles of the satiety and feeding centers. Where are they located?

2. Studies show that most obese humans have elevated leptin levels in their blood. Based on your understanding of endocrine disorders [🔁 p. 232], propose some reasons why leptin is not decreasing food intake in these people.

Answers: p. 748

ENERGY BALANCE

Once food has been digested and absorbed, the body's chemical reactions—known collectively as metabolism—determine what happens to the nutrients contained in the food. Are they destined to burn off as heat? Become muscle? Or turn into extra pounds that make it difficult to zip up blue jeans? In this section we examine energy balance in the body.

Energy Input Equals Energy Output

The *first law of thermodynamics* [🔁 p. 93] states that the total amount of energy in the universe is constant. By extension, this statement means that all energy that goes into a biological system, such as the human body, can be accounted for. In the body, energy is primarily stored in the chemical bonds of molecules, so we can apply the concept of mass balance to energy balance:

$$\text{body energy} = \text{energy intake} - \text{energy output} \qquad (1)$$

Energy intake for humans consists of energy contained in the nutrients we eat, digest, and absorb. *Energy output* is a combination of work performed and energy returned to the environment as heat:

$$\text{energy output} = \text{work} + \text{heat} \qquad (2)$$

The work in equation 2 takes one of three forms [🔁 p. 91]:

1. *Transport work* moves molecules from one side of a membrane to the other. Transport processes bring materials into and out of the body and transfer them between compartments.

2. *Mechanical work* uses intracellular fibers and filaments to create movement. This form of work includes external

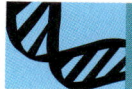

NOVEL APPROACHES FOR DISCOVERING PEPTIDES

In the early days of molecular biology, scientists collected tissues that contained active peptides, isolated and purified the peptides, and then analyzed their amino acid sequence. Now, as we enter the age of proteomics, investigators are using reverse techniques to discover new proteins in the body. One group of researchers, for example, found the *orexin* (or hypocretin) peptides by isolating the mRNA expressed in a particular region of the hypothalamus. The investigators then used the mRNA to create the amino acid sequence of the prepropeptide. The orexin peptides were also discovered simultaneously by a different group of scientists who worked backwards from an "orphan" G protein-coupled receptor [🔁 p. 183] to find its peptide ligand. The endogenous hunger hormone *ghrelin* was discovered by a similar method. Pharmacologists testing synthetic peptides to stimulate the release of growth hormone found that their peptides were binding to a previously unknown receptor, and from that receptor they discovered ghrelin.

work, such as movement created by skeletal muscle contraction, and internal work, such as the movement of cytoplasmic vesicles and the pumping of the heart.

3. *Chemical work* is used for growth, maintenance, and storage of information and energy. Chemical work in the body can be subdivided into synthesis and storage. Storage includes both *short-term energy storage* in high-energy phosphate compounds such as ATP, and *long-term energy storage* in the chemical bonds of glycogen and fat.

In the human body, at least half the energy released in chemical reactions is lost to the environment as unregulated "waste" heat.

Most of the energy-consuming processes in the body are not under conscious control. Body movements, such as walking and exercise, are the only way people can voluntarily increase energy *output*. People can control their energy *intake*, however, by watching what they eat.

Although energy balance is a very simple concept, it is a difficult one for people to accept. Behavior modifications, such as eating less and exercising more, are among the most frequent instructions health care professionals give to patients. These instructions are also the most difficult to follow, and patient compliance is often low.

Energy Use Is Reflected by an Individual's Oxygen Consumption

To compile an energy balance sheet for the human body, we must be able to estimate both the energy content of food (energy intake) and the energy expenditure from heat loss and various types of work (energy output). The most direct way to measure the energy content of food is by **direct calorimetry**. In this procedure, food is burned in an instrument called a *bomb calorimeter,* and the heat released is trapped and measured. The heat released is a direct measure of the energy content of the burned food and is usually measured in kilocalories (kcal). One **kilocalorie** is the amount of heat needed to raise the temperature of 1 liter of water by 1°C. A kilocalorie is the same as a *Calorie* (with a capital C).

Although direct calorimetry is a quick way of measuring the total energy content of food, the *metabolic* energy content of food is slightly less because most foods cannot be fully digested and absorbed. The metabolic energy content of

Sara and Nicole first met in ballet class at age 11. They were both serious about dance and worked hard to maintain the perfect thin ballet-dancer's body. Both girls occasionally took diet pills or laxatives, and it was almost a contest to see who could eat the least food. Sara was always impressed with Nicole's willpower, but then again, Nicole was a perfectionist. Two years ago, Sara found herself less interested in dance and went away to college. Nicole was accepted into a prestigious ballet company and remained focused on her dancing—and on her body image. Nicole's regimented diet became stricter, and if she felt she'd eaten too much, she would make herself throw up. Her weight kept dropping, but each time she looked in the mirror, she saw a fat girl looking back.

Question 1:
If you measured Nicole's leptin level, what would you expect to find?

Question 2:
Would you expect Nicole to have elevated or depressed levels of neuropeptide Y?

717   719   721   722   740   745

22

proteins and carbohydrates is 4 kcal/g. Fats contain more than twice as much energy—9 kcal/g.

The caloric content of any food can be calculated by multiplying the number of grams of each component by its metabolic energy content. For example, if a bagel contains 2 g of fat, 7 g of protein, and 38 g of carbohydrates, its calorie content is (2 g fat × 9 kcal/g) + (7 g protein × 4 kcal/g) + (38 g carbohydrate × 4 kcal/g) = 198 kcal. In the United States, you can find the energy content for various foods on the *Nutrition Facts* label of food packages.

Oxygen Consumption Can Be Equated to Metabolic Rate

Estimating an individual's energy expenditure, or **metabolic rate**, is more complex than figuring the caloric content of ingested food. Heat released by the body can be measured by enclosing a person in a sealed compartment. The difference between heat produced and the person's caloric intake is the energy used for chemical, mechanical, and transport work. Practically speaking, however, measuring total body heat release is not a very easy way to measure energy use.

Probably the most common method for estimating metabolic rate is to measure a person's **oxygen consumption**, the rate at which the body consumes oxygen as it metabolizes nutrients. Recall from Chapter 4 [☞ p. 107] that metabolism of glucose to trap energy in the bonds of ATP is most efficient in the presence of adequate oxygen: $C_6H_{12}O_6 + O_2 + ADP + P_i \rightarrow CO_2 + H_2O + ATP + heat$. Studies have shown that oxygen consumption for different foods is relatively constant at a rate of 1 liter of oxygen consumed for each 4.5–5 kcal of energy released from the food being metabolized. The measurement of oxygen consumption is one form of **indirect calorimetry**.

Another method of estimating metabolic rate is to measure carbon dioxide production, either alone or in combination with oxygen consumption. The equation above shows that aerobic metabolism both consumes O_2 and produces CO_2. However, the ratio of CO_2 produced to O_2 consumed varies with the composition of the diet. This ratio of CO_2 produced to O_2 consumed is known as the **respiratory quotient (RQ)** or the **respiratory exchange ratio (RER)**. RQ varies from a high of 1.0 for a pure carbohydrate diet to 0.8 for pure protein and 0.7 for pure fat. The average American diet has an RQ of about 0.82.

Many Factors Influence Metabolic Rate

Whether measured by O_2 consumption or by CO_2 production, metabolic rate can be highly variable from one person to another or from day to day in a single individual. An individual's lowest metabolic rate is considered the **basal metabolic rate (BMR)**. In reality, metabolic rate would be lowest when an individual is asleep. However, because measuring the BMR of a sleeping person is difficult, metabolic rate is often measured after a 12-hour fast in a person who is awake but resting (a **resting metabolic rate, RMR**).

Metabolic rate (basal, resting, or with activity) is calculated by multiplying oxygen consumption by the number of kilocalories metabolized per liter of oxygen consumed:

metabolic rate (kcal/day) = liters of O_2 consumed/day × kcal/liter of O_2

A mixed diet with an RQ of 0.8 requires one liter of O_2 for each 4.80 kcal metabolized. Thus, for a 70-kg male whose resting oxygen consumption is 430 L/day:

$$resting\ metabolic\ rate = 430\ L\ O_2/day \times 4.80\ kcal/L\ O_2$$
$$= 2064\ kcal/day$$

Factors that affect metabolic rate in humans include age, gender, amount of lean muscle mass, activity level, diet, hormones, and genetics.

1. **Age and gender**. Adult males have an average BMR of 1.0 kcal per hour per kilogram of body weight. Adult females have a lower rate than males: 0.9 kcal/hr/kg. The difference arises because women have a higher percentage of adipose tissue and less lean muscle mass. Metabolic rates in both sexes decline with age. Some of this decline is due to decreases in lean muscle mass.

2. **Amount of lean muscle mass**. Muscle has higher oxygen consumption than adipose tissue, even at rest. (Most of the volume of an adipose tissue cell is occupied by metabolically inactive lipid droplets.) This is one reason weight loss advice often includes weight training in addition to aerobic exercise. Weight training adds muscle mass to the body, which increases basal metabolic rate and results in more calories being burned at rest.

3. **Activity level**. Physical activity and muscle contraction increase metabolic rate over the basal rate. Sitting and lying down consume relatively little energy. Competitive rowing and cycling are among the activities that expend the most energy.

4. **Diet**. Resting metabolic rate increases after a meal, a phenomenon termed **diet-induced thermogenesis**. In other words, there is an energetic cost to the digestion and assimilation of food. Diet-induced thermogenesis is related to the type and amount of food ingested. Fats cause relatively little diet-induced thermogenesis; proteins increase heat production the most. This phenomenon may support the claim of some nutritionists that eating a calorie of fat is different from eating a calorie of protein, although they contain the same amount of energy when measured by direct calorimetry.

5. **Hormones**. Basal metabolic rate is increased by thyroid hormones and by catecholamines (epinephrine and norepinephrine). Some of the peptides that regulate food intake also appear to influence metabolism.

6. **Genetics**. The effect of inherited traits on energy balance can be observed in the variety of normal body types. Some people have very efficient metabolism that converts food

RUNNING PROBLEM

When Nicole's blood test results came back, Dr. Ayani immediately wrote orders to start an intravenous infusion and heart monitoring. The laboratory report showed plasma potassium of 2.5 mEq/L (normal: 3.5–5.0 mEq/L), plasma HCO_3^- of 40 mEq/L (normal: 24–29), and plasma pH of 7.52 (normal: 7.38–7.42). Dr. Ayani admitted Nicole to the hospital for further treatment and evaluation, hoping she could be convinced that she needed help for her anorexia. *Anorexia,* meaning "no appetite," can have both physiological and psychological origins.

Question 3:
What is Nicole's K^+ disturbance called? What effect does her K^+ disturbance have on the resting membrane potential of her cells?

Question 4:
Why does Dr. Ayani want to monitor Nicole's cardiac function?

Question 5:
Based on her clinical values, what is Nicole's acid-base status?

| | | | 717 | 719 | **721** | 722 | 740 | 745 |

energy into energy stored in adipose tissue with little heat loss, while others can eat large amounts of food and never gain weight because their metabolism is less efficient.

Of the factors affecting metabolic rate, a person can voluntarily control only two: energy intake (how much food is eaten) and level of physical activity. If a person's activity includes strength training, which increases lean muscle mass, resting metabolic rate goes up. The addition of lean muscle mass to the body creates additional energy use, which in turn decreases the number of calories that go into storage.

Energy Is Stored in Fat and Glycogen

A person's daily energy requirement, expressed as caloric intake, varies with the needs and activity of the body. For example, large male athletes in an intensive training program may need more than 10,000 kcal per day. On the other hand, a woman engaged in normal activities may require only 2000 kcal/day.

Suppose that our woman's energy requirement could be met by ingesting only glucose. Glucose has an energy content of 4 kcal/g, which means that to get 2000 kcal, she would have to consume 500 g, or 1.1 pounds, of glucose each day. Our

bodies cannot absorb crystalline glucose, however, so those 500 g of glucose would have to be dissolved in water. If the glucose was made as an isosmotic 5% solution, the 500 g would have to be dissolved in 10 liters of water—a substantial volume to drink in a day!

Fortunately, we do not usually ingest glucose as our primary fuel. Proteins, complex carbohydrates, and fats also provide energy. The glucose polymer glycogen [🔁 p. 27] is a more compact form of energy than an equal number of individual glucose molecules. Glycogen also requires less water for hydration.

Our cells therefore convert glucose to glycogen for storage. Normally we keep about 100 g of glycogen in the liver and 200 g in skeletal muscles. But even this 300 g of glycogen can provide only enough energy for 10 to 15 hours. The brain alone requires 150 g of glucose per day.

Consequently, the body keeps most of its energy reserves in compact, high-energy fat molecules. One gram of fat has 9 kcal, more than twice the energy content of an equal amount of carbohydrate or protein. This feature makes adipose tissue a very efficient way for the body to store large amounts of energy in minimal space. Metabolically, however, the energy in fat is harder to access, and the metabolism of fats is slower than that of carbohydrates.

CONCEPT CHECK

3. Name seven factors that can influence a person's metabolic rate.

4. Why does the body store most of its extra energy in fat and not in glycogen?

5. Complete and balance the following equation for aerobic metabolism of one glucose molecule:

$$C_6H_{12}O_6 + O_2 \rightarrow ? + ?$$

6. What is the RQ for the balanced equation in Concept Check 5?

Answers: p. 748

METABOLISM

Chapter 4 introduced the basic pathways of cellular metabolism. **Metabolism** is the sum of all chemical reactions in the body. The reactions making up these pathways (1) extract energy from nutrients, (2) use energy for work, and (3) store excess energy so that it can be used later. Metabolic pathways that synthesize large molecules from smaller ones are called **anabolic pathways** [*ana-,* completion + *metabole,* change]; those that break large molecules into smaller ones are called **catabolic** pathways [*cata-,* down or back].

The classification of a pathway is its net result, not what happens in any individual step of the pathway. For example, in the first step of *glycolysis* [🔁 p. 104], glucose adds a phosphate to become a larger molecule, glucose 6-phosphate. This single reaction is anabolic, but by the end of glycolysis the initial 6-carbon glucose molecule has been converted to two 3-carbon pyruvate molecules. This makes glycolysis a catabolic pathway.

22

When Nicole was admitted to the hospital, her blood pressure was 80/50 and her pulse was a weak and irregular 90 beats per minute. She weighed less than 85% of the minimal healthy weight for a woman of her height and age. She had an intense fear of gaining weight, even though she was underweight. Her menstrual periods were irregular, she had just suffered a fractured wrist from a fall that normally shouldn't have caused a fracture, and her hair was thinning. When Dr. Ayani questioned Nicole, she admitted that she had been feeling weak during dance rehearsals and had been having difficulty concentrating at times.

Question 6:
Based on what you learned in Chapters 14 and 15 about heart rate and blood pressure, speculate on why Nicole has low blood pressure with a rapid pulse.

Question 7:
Would you expect Nicole's renin and aldosterone levels to be normal, elevated, or depressed? How might these levels relate to her K$^+$ disturbance?

Question 8:
Give some possible reasons Nicole had been feeling weak during dance rehearsals.

717 719 721 **722** 740 745

In the human body, we divide metabolism into two states. The period of time following a meal, when the products of digestion are being absorbed, used, and stored, is called the **fed state** or the **absorptive state**. This is an anabolic state in which the energy of nutrient biomolecules is transferred to high-energy compounds or stored in the chemical bonds of other molecules.

Once nutrients from a recent meal are no longer in the bloodstream and available for use by the tissues, the body enters what is called the **fasted state** or the **postabsorptive state**. As the pool of available nutrients in the blood decreases, the body taps into its stored reserves. The postabsorptive state is catabolic because cells break down large molecules into smaller molecules. The energy released by breaking chemical bonds of large molecules is used to do work.

Energy from Ingested Nutrients May Be Used Immediately or Stored

The biomolecules we ingest are destined to meet one of three fates:

1. *Energy.* Biomolecules can be metabolized immediately, with the energy released from broken chemical bonds trapped in ATP, phosphocreatine, and other high-energy compounds. This energy can then be used to do mechanical work.
2. *Synthesis.* Biomolecules entering cells can be used to synthesize basic components needed for growth and maintenance of cells and tissues.
3. *Storage.* If the amount of food ingested exceeds the body's requirements for energy and synthesis, the excess energy goes into storage in the bonds of glycogen and fat. Storage makes energy available for times of fasting.

The fate of an absorbed biomolecule depends on whether it is a carbohydrate, protein, or fat. Figure 22-2 ■ is a schematic diagram that follows these biomolecules from the diet into the three *nutrient pools* of the body: the free fatty acid pool, the glucose pool, and the amino acid pool. **Nutrient pools** are nutrients that are available for immediate use. They are located primarily in the plasma.

Free fatty acids form the primary pool of fats in the blood. They can be used as an energy source by many tissues but are also easily stored as fat (*triglycerides*) in adipose tissue. Carbohydrates are absorbed mostly as glucose. Plasma glucose concentration is the most closely regulated of the three nutrient pools because glucose is the only fuel the brain can metabolize, except in times of starvation. Notice from the locations of the two exit "pipes" on the glucose pool in Figure 22-2 that if the pool falls below a certain level, only the brain has access to glucose. This conservation measure ensures that the brain has an adequate energy supply. Just as the circulatory system gives priority to supplying oxygen to the brain, metabolism also gives priority to the brain.

If the body's glucose pool is within the normal range, most tissues use glucose for their energy source. Excess glucose goes into storage as glycogen. The synthesis of glycogen from glucose is known as **glycogenesis**. Glycogen stores are limited, however, and additional excess glucose is converted to fat (**lipogenesis**).

If plasma glucose concentrations decrease, the body breaks down glycogen (**glycogenolysis**) to glucose. By balancing oxidative metabolism, glycogenesis, glycogenolysis, and lipogenesis, the body maintains plasma glucose concentrations within a narrow range.

If homeostasis fails and plasma glucose exceeds a critical level, as occurs in diabetes mellitus, excess glucose is excreted in the urine. Glucose excretion occurs only when the *renal threshold* for glucose reabsorption is exceeded [p. 629].

The amino acid pool of the body is used primarily for protein synthesis. However, if glucose intake is low, amino acids can be converted into glucose through the pathways known as **gluconeogenesis**. This word literally means "the birth (*genesis*) of new (*neo*) glucose" and refers to the synthesis of glucose from a noncarbohydrate precursor.

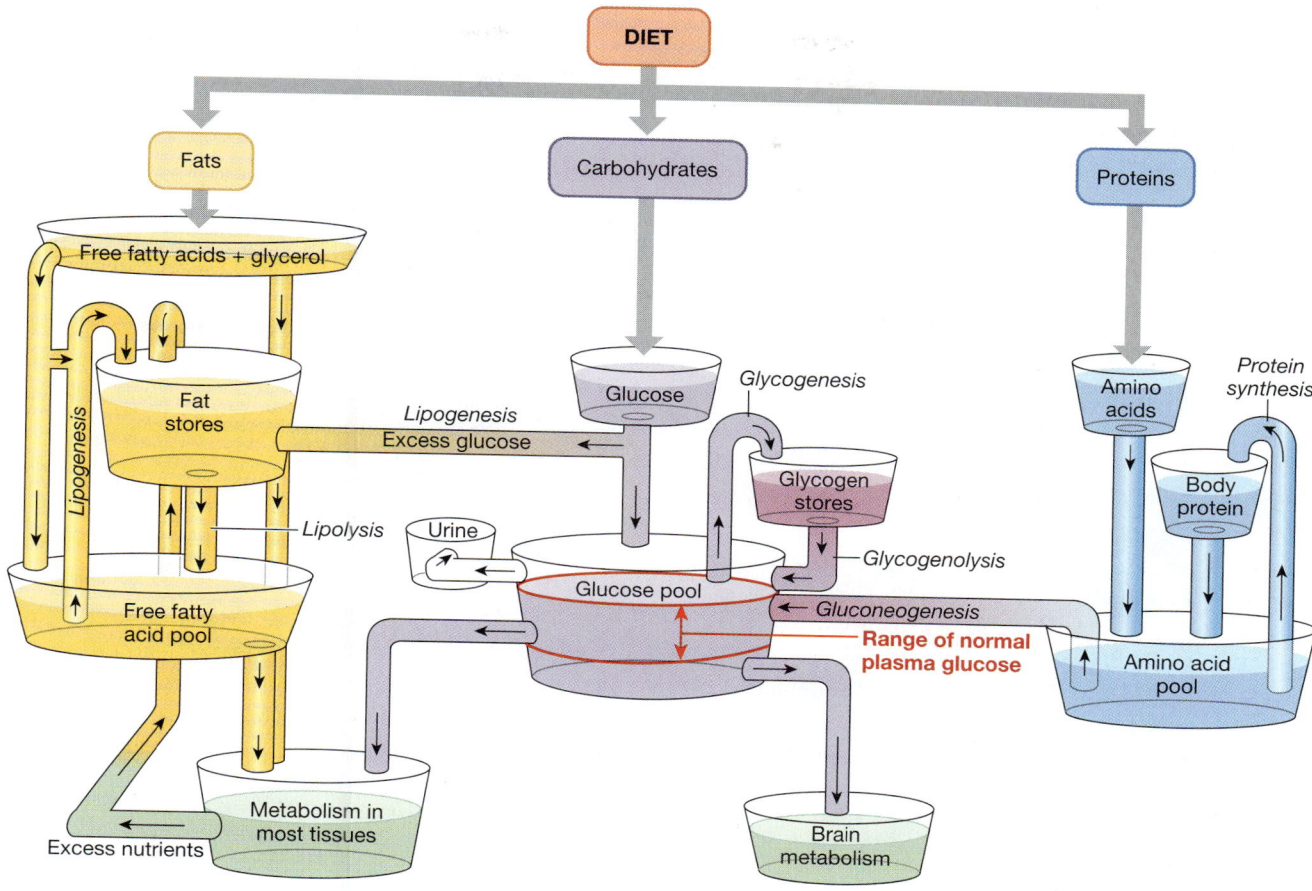

■ **FIGURE 22-2**   *Summary of metabolism*

Adapted from L. L. Langley, *Homeostasis* (New York: Reinhold, 1965).

Amino acids are the main source for glucose through the gluconeogenesis pathways, but glycerol from triglycerides can also be used. Both gluconeogenesis and glycogenolysis are important backup sources for glucose during periods of fasting.

Hormones Control Metabolic Pathways by Changing Enzyme Activity

The most important biochemical pathways for energy production are illustrated in Figure 22-3 ■. This figure does not include all of the metabolic intermediates in each pathway (see Chapter 4 for detailed pathways). Instead, it emphasizes the points at which different pathways intersect, because these intersections are often key points at which metabolism is controlled.

One significant feature of metabolic regulation is the use of different enzymes to catalyze forward and reverse reactions. This dual control, sometimes called *push-pull control,* allows close regulation of a reaction. Figure 22-4 ■ shows how hormones regulate the flow of nutrients through metabolic pathways by altering enzyme activity. In Figure 22-4a, enzyme 1 catalyzes the reaction in which A is converted to B, and enzyme 2 catalyzes the reverse reaction. When the activity of the two enzymes is roughly equal, as soon as A is converted into B,

B is converted back into A. Turnover of the two substrates is rapid, but there is no net production of either A or B.

In Figure 22-4b, the activity of enzyme 1 is enhanced while enzyme 2 catalyzing the reverse reaction is inhibited. This type of dual control is exerted by the pancreatic hormone insulin, which stimulates enzymes for glycogenesis and inhibits enzymes for glycogenolysis. The net result is glycogen synthesis from glucose.

The reverse pattern is shown in Figure 22-4c, in which **glucagon,** another pancreatic hormone, stimulates the enzymes of glycogenolysis while inhibiting the enzymes for glycogenesis. The net result is glucose synthesis from glycogen.

Anabolic Metabolism Dominates in the Fed State

As we've noted, the fed state is anabolic: absorbed nutrients are being used for energy, synthesis, and storage. Table 22-2 ■ summarizes the fates of nutrients in the fed state.

Carbohydrates Provide Energy   Glucose absorbed after a meal enters the circulation of the hepatic portal system and is taken directly to the liver, where about 30% of all ingested glucose is metabolized. The remaining 70% continues in the

22

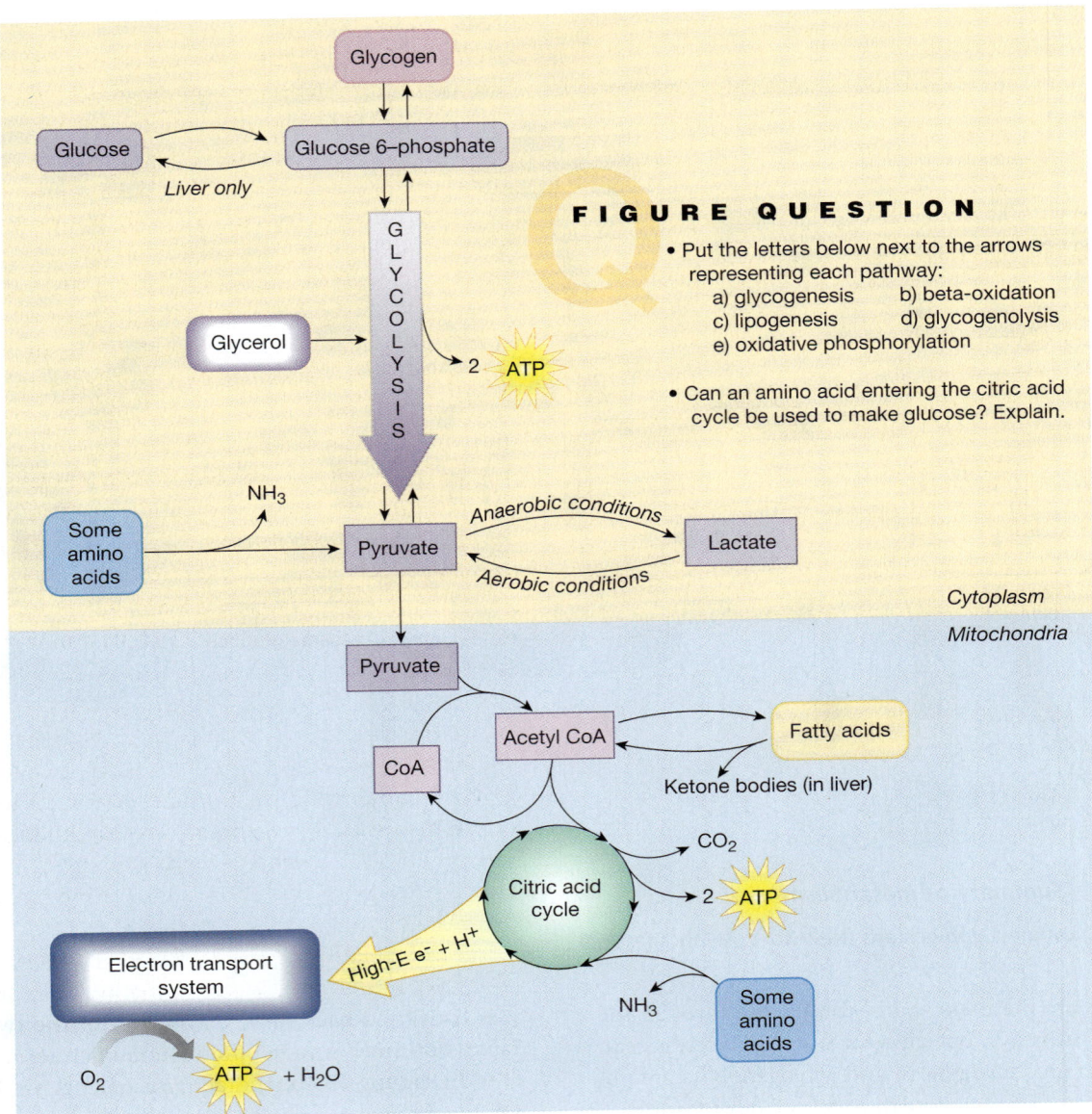

FIGURE QUESTION

• Put the letters below next to the arrows representing each pathway:
   a) glycogenesis   b) beta-oxidation
   c) lipogenesis   d) glycogenolysis
   e) oxidative phosphorylation

• Can an amino acid entering the citric acid cycle be used to make glucose? Explain.

■ **FIGURE 22-3**   *Summary of biochemical pathways for energy production*

bloodstream for distribution to the brain, muscles, and other organs and tissues.

Glucose moves from interstitial fluid into cells via GLUT transporters [🔁 p. 146]. Most glucose absorbed from a meal goes immediately into glycolysis and the citric acid cycle to make ATP. Some glucose is used by the liver for lipoprotein synthesis. Glucose that is not required for energy and synthesis is stored either as glycogen or fat. The body's ability to store glycogen is limited, so most excess glucose is converted to triglycerides and stored in adipose tissue.

CONCEPT CHECK

7. What is the difference between glycogenesis and gluconeogenesis?

8. Are GLUT transporters active or passive transporters?

Answers: p. 748

New Proteins Are Made from Amino Acids   Most amino acids absorbed from a meal go to the tissues for protein synthesis. Like glucose, amino acids are taken first to the liver by the hepatic portal system. The liver uses them to synthesize lipoproteins and plasma proteins, such as albumin, clotting factors, and angiotensinogen.

Amino acids not taken up by the liver are used by cells to create structural or functional proteins, such as cytoskeletal elements, enzymes, and hormones. Amino acids are also incorporated into nonprotein molecules, such as amine hormones and neurotransmitters.

If glucose intake is low, amino acids can be used for energy. However, if more protein is ingested than is needed for synthesis and energy expenditures, excess amino acids are converted to fat. Some bodybuilders spend large amounts of money

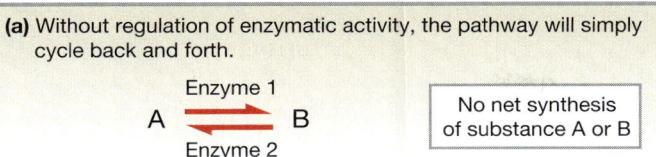

(a) Without regulation of enzymatic activity, the pathway will simply cycle back and forth.

Enzyme 1

A ⇌ B

Enzyme 2

No net synthesis of substance A or B

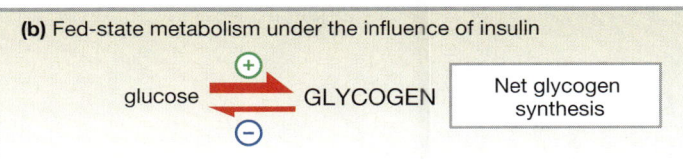

(b) Fed-state metabolism under the influence of insulin

glucose ⇌ GLYCOGEN

Net glycogen synthesis

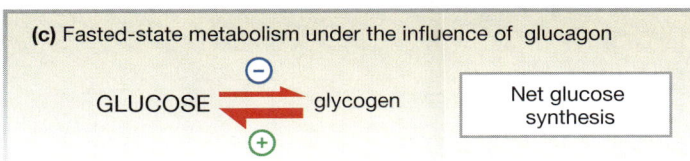

(c) Fasted-state metabolism under the influence of glucagon

GLUCOSE ⇌ glycogen

Net glucose synthesis

■ **FIGURE 22-4** *Push-pull control of metabolism*

In this type of control, different enzymes catalyze the forward and reverse reactions.

on amino acid supplements advertised to build bigger muscles. But these amino acids do not automatically go into protein synthesis. When amino acid intake exceeds the body's need for protein synthesis, excess amino acids are burned for energy or stored as fat.

TABLE 22-2	Fates of Nutrients in the Fed State

Carbohydrates (absorbed primarily as glucose)

1. Used immediately for energy through aerobic pathways*

2. Used for lipoprotein synthesis in liver

3. Stored as glycogen in liver and muscle

4. Excess converted to fat and stored in adipose tissue (glucose → pyruvate → acetyl CoA → fatty acids)

Proteins (absorbed primarily as amino acids)

1. Most amino acids go to tissues for protein synthesis*

2. If needed for energy, amino acids converted in liver to intermediates for aerobic metabolism

3. Excess converted to fat and stored in adipose tissue (amino acids → acetyl CoA → fatty acids)

Fats (absorbed primarily as triglycerides)

1. Stored as fats primarily in liver and adipose tissue*

*Primary fate.

Fats Store Energy   Most ingested fats are assembled into *chylomicrons* in the intestinal epithelium [🔁 p. 697] and enter the venous circulation via the lymphatic vessels (Fig. 22-5 ■). Chylomicrons consist of cholesterol, triglycerides, phospholipids, and lipid-binding proteins called **apoproteins**, or *apolipoproteins* [*apo-*, derived from]. Once these lipid complexes begin to circulate through the blood, **lipoprotein lipase** bound to the capillary endothelium of muscles and adipose tissue converts their triglycerides to free fatty acids and glycerol. These molecules are then used for energy by most cells or reassembled into triglycerides for storage in adipose tissue.

Chylomicron remnants that remain in the circulation are taken up and metabolized by the liver (Fig. 22-5). Cholesterol from the remnants joins the liver's pool of lipids. If cholesterol is in excess, some may be converted to bile salts and excreted in the bile. The remaining cholesterol is added to newly synthesized cholesterol and fatty acids, and packaged into lipoprotein complexes for secretion into the blood.

The lipoprotein complexes that re-enter the blood contain varying amounts of triglycerides, phospholipids, cholesterol, and apoproteins. The more protein a complex contains, the heavier it is, with plasma lipoprotein complexes ranging from *very-low-density lipoprotein* (VLDL) to high-density lipoprotein (HDL). The combination of lipids with proteins makes cholesterol more soluble in plasma, but the complexes are unable to diffuse through cell membranes. Instead, they must be brought into cells by *receptor-mediated endocytosis* [🔁 p. 149]. The apoproteins in the complexes are specific for membrane receptors in different tissues.

Most lipoprotein in the blood is *low-density lipoprotein-cholesterol* [🔁 p. 525]. LDL-C is sometimes known as the "lethal cholesterol" because elevated concentrations of plasma LDL-C are associated with the development of atherosclerosis [🔁 p. 150]. LDL-C complexes contain *apoprotein B* (apoB), which combines with receptors that bring LDL-C into most cells of the body. Several inherited forms of *hypercholesterolemia* (elevated plasma cholesterol levels) have been linked to defective forms of apoB. These abnormal apoproteins may help explain the accelerated development of atherosclerosis in people with hypercholesterolemia.

The second most common lipoprotein in the blood is high-density lipoprotein-cholesterol (HDL-C), sometimes called the "healthy cholesterol" because HDL is the lipoprotein involved in cholesterol transport out of the plasma. HDL-C contains *apoprotein A*, which facilitates cholesterol uptake by the liver and other tissues.

Plasma Cholesterol Levels Are Predictors of Coronary Heart Disease

Of the nutrients in the plasma, lipids and glucose receive the most attention from health professionals. Abnormal glucose metabolism is the hallmark of diabetes mellitus, described later

22

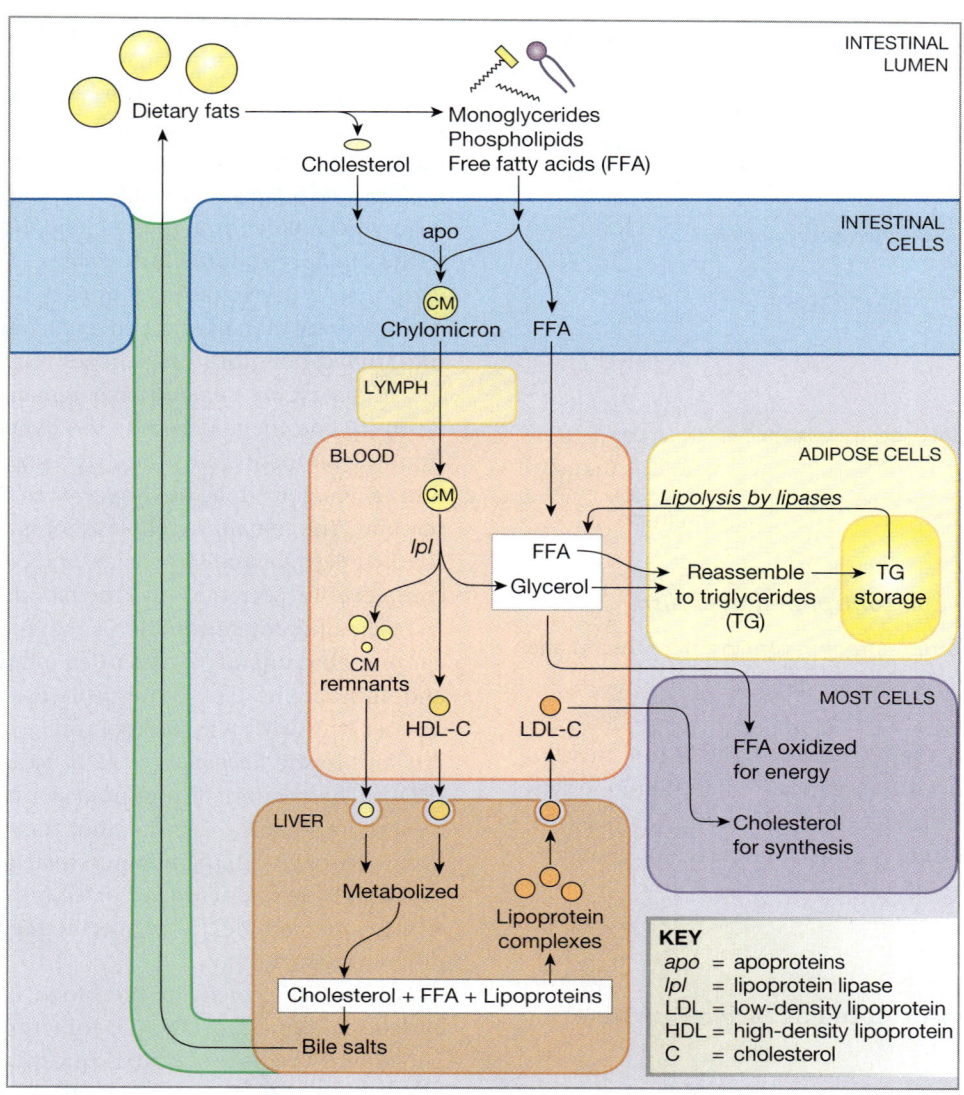

KEY
apo = apoproteins
lpl = lipoprotein lipase
LDL = low-density lipoprotein
HDL = high-density lipoprotein
C = cholesterol

■ **FIGURE 22-5** *Transport and fate of dietary fats*

in this chapter. Abnormal plasma lipids are used as predictors of atherosclerosis and coronary heart disease (CHD) [⮐ p. 525].

Tests to measure blood lipids and assess cardiovascular risk range from simple but less accurate finger-stick blood samples to expensive tests on venous blood that look at all sizes of lipoproteins, from VLDL through HDL. As more epidemiological and treatment data are gathered, experts continue to redefine desirable lipid values. The U.S. National Cholesterol Education Panel issued guidelines in 2001 and updated them in 2004 (*www.nhlbi.nih.gov/guidelines/cholesterol*). Emphasis over the years has shifted from concern about total cholesterol levels (< 200 mg/dL of plasma is recommended) to the absolute amounts and relative proportions of the various subtypes. Some studies indicate that elevated LDL-C is the single largest cholesterol risk factor for CHD (Fig. 22-6 ■), because oxidized LDL-C is taken up by macrophages and leads to the development of atherosclerotic plaques [⮐ Fig. 15-24, p. 527]. Desirable LDL-C values range from < 160 mg/dL for people at low risk for CHD to < 100 mg/dL

for those considered at high risk (including those with diabetes). The HDL-C level in plasma has also been used to predict a person's risk of developing atherosclerosis. Like high LDL-C, low HDL-C (< 40 mg/dL) is associated with a higher risk of developing CHD. More recently, health professionals have started looking at the non-HDL cholesterol value (total cholesterol − HDL-C) as perhaps a better indicator of CHD risk.

Lifestyle modifications (improved diet, smoking cessation, and exercise) can be very effective in improving lipid profiles but can be difficult for patients to implement and sustain. All the pharmacological therapies developed to treat elevated cholesterol target some aspect of cholesterol absorption and metabolism, underscoring the principle of mass balance in cholesterol homeostasis. Decreasing cholesterol uptake or synthesis and increasing cholesterol clearance by metabolism or excretion are effective mechanisms for reducing the amount of cholesterol in the body.

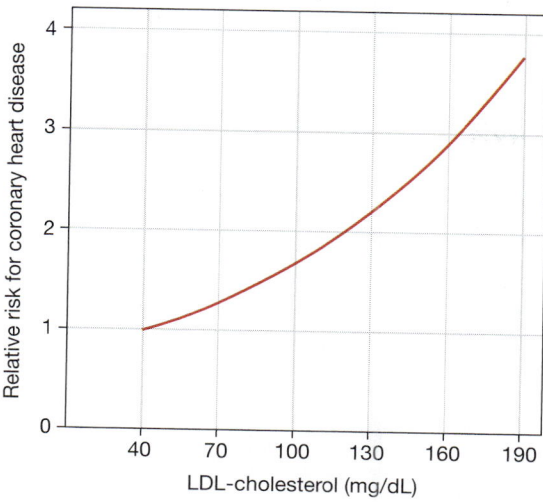

GRAPH QUESTION

Which 30-mg/dL decrease in LDL-C has the biggest effect on decreasing risk of coronary heart disease?

■ **FIGURE 22-6**   *The relationship between LDL-C and risk of developing coronary heart disease*

Data taken from Grundy *et al.*, *Circulation* 110: 227–239, 2004 July 13.

The drugs known as *bile acid sequestrants* [*sequestrare*, to put in the hands of a trustee] bind to bile acids in the intestinal lumen and prevent their reabsorption, thus increasing cholesterol excretion [🔁 p. 696]. Supplied with fewer recycled bile salts, the liver increases bile acid synthesis from cholesterol, thereby decreasing plasma cholesterol when hepatic LDL receptors import cholesterol.

The newest lipid-lowering agent, *ezetimibe,* inhibits intestinal cholesterol transport. Another strategy for reducing intestinal cholesterol uptake is adding plant *sterols* (steroid-alcohols) and *stanols* (saturated sterols [🔁 p. 28]) to the diet. Sterols and stanols displace cholesterol in chylomicrons, which decreases cholesterol absorption.

The remaining lipid-lowering drugs all affect cholesterol metabolism in the liver. The drugs called *statins* inhibit the enzyme *HMG coA reductase,* which mediates cholesterol synthesis in hepatocytes. The *fibrates,* which stimulate a transcription factor called *PPARα* (pronounced *p-par-alpha*), and *niacin* (vitamin B₃, nicotinic acid) decrease LDL-C and increase HDL-C by mechanisms that are not well understood. All the drugs that alter hepatic cholesterol metabolism have a low incidence of liver toxicity and other significant side effects.

✓ CONCEPT CHECK

9. What might you predict the predominant side effects of bile acid sequestrants and ezetimibe to be?

Answers: p. 748

Catabolic Metabolism Dominates in the Fasted State

Once all nutrients from a meal have been digested, absorbed, and distributed to various cells, plasma concentrations of glucose begin to fall. This is the signal for the body to shift from fed- (absorptive) state to fasted- (postabsorptive) state metabolism. The goal of fasted-state metabolism is to maintain plasma glucose concentrations within an acceptable range so that the brain and neurons have adequate fuel.

The liver is the primary source of glucose production during the fasted state (Fig. 22-7 ■). Liver glycogen can provide enough glucose through glycogenolysis to meet 4–5 hours of the body's energy needs.

Skeletal muscle glycogen can be metabolized to glucose in the fasted state but not through direct conversion in the muscle. Muscle cells, like most other cells, lack the enzyme that makes glucose from glucose 6-phosphate. Consequently, glucose 6-phosphate produced from glycogenolysis is metabolized to either pyruvate (aerobic conditions) or lactate (anaerobic conditions). Pyruvate and lactate are then transported to the liver, which uses them to make glucose via gluconeogenesis.

Additional glucose or ATP can be made from amino acids, particularly those in muscle proteins (Fig. 22-7). Enzymes remove amino groups from the amino acids (*deamination*) [🔁 p. 111] and convert the amino groups to urea, which is excreted. Some deaminated amino acids become citric acid cycle intermediates, enter the cycle, and produce ATP. This alternative ATP source thus spares plasma glucose for use by the brain. Other amino acids are processed to pyruvate, which goes to the liver and is made into glucose, as previously described.

In the fasted state, adipose tissue breaks down its stores of triglycerides into fatty acids and glycerol. Glycerol goes to the liver and can be converted to glucose. The fatty acids are released into the blood, from which they can be taken up by many tissues.

Once inside cells, the long carbon chains of fatty acids are chopped into two-carbon acyl units through the process of β-*oxidation* [🔁 p. 111]. In most tissues, these acyl units feed into the citric acid cycle through acetyl CoA and provide a substrate for ATP synthesis through oxidative phosphorylation. If there is excessive fatty acid breakdown, β-oxidation in the liver creates acidic ketone bodies (often called simply *ketones* in physiology and medicine). Ketone bodies enter the blood, creating a state of *ketosis.* The breath of people in ketosis has a fruity odor caused by acetone, a volatile ketone whose odor you might recognize from nail polish remover.

Ketone bodies are the only fuel besides glucose that the brain can use, so in cases of prolonged starvation, these compounds become a significant source of energy for the brain. Ketone bodies can be metabolized by converting them to acetyl CoA, which then enters the citric acid cycle. Unfortunately, the ketone bodies *acetoacetic acid* and *β-hydroxybutyric acid* are

22

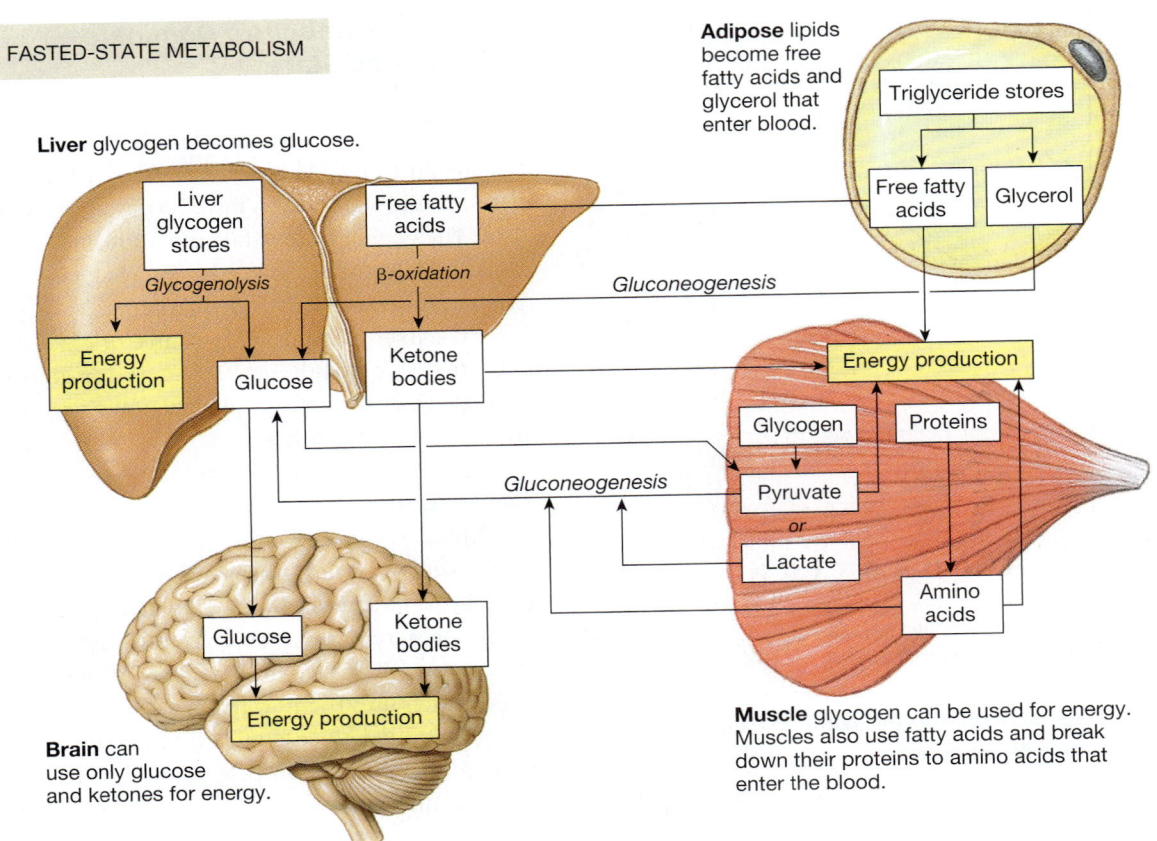

FASTED-STATE METABOLISM

Liver glycogen becomes glucose.

Adipose lipids become free fatty acids and glycerol that enter blood.

Triglyceride stores

Liver glycogen stores

Free fatty acids

Free fatty acids

Glycerol

Glycogenolysis

β-oxidation

Gluconeogenesis

Energy production

Glucose

Ketone bodies

Gluconeogenesis

Energy production

Glycogen

Proteins

Pyruvate

or

Lactate

Amino acids

Glucose

Ketone bodies

Energy production

Brain can use only glucose and ketones for energy.

Muscle glycogen can be used for energy. Muscles also use fatty acids and break down their proteins to amino acids that enter the blood.

■ **FIGURE 22-7** *Fasted-state metabolism*

The function of fasted-state metabolism is maintaining adequate plasma glucose concentrations for the brain.

moderately strong acids, and excessive ketone production leads to a state of *ketoacidosis* [🔁 p 670].

With this summary of metabolic pathways as background, we now turn to the endocrine and neural regulation of metabolism.

CONCEPT CHECK

10. When amino acids are used for energy, which pathways in Figure 22-3 do they follow?

11. Cholesterol is soluble in lipids, so why does plasma cholesterol need the help of a membrane transporter to enter cells?

Answers: p. 748

HOMEOSTATIC CONTROL OF METABOLISM

The endocrine system has primary responsibility for metabolic regulation, although the nervous system does have some influence, particularly in terms of governing food intake. Hour-to-hour regulation depends primarily on the ratio of insulin to glucagon, two hormones secreted by endocrine cells of the pancreas. Both hormones have short half-lives

and must be continuously secreted if they are to have a sustained effect.

The Pancreas Secretes Insulin and Glucagon

The endocrine cells of the pancreas make up less than 2% of the organ's total mass; most pancreatic tissue is devoted to the production and exocrine secretion of digestive enzymes and bicarbonate [🔁 Fig. 21-7, p. 686]. In 1869, the German anatomist Paul Langerhans described small clusters of cells—now known as the **islets of Langerhans**—scattered throughout the body of the pancreas (Figure 22-8b, c ■). The islets contain four distinct cell types, each associated with secretion of one or more peptide hormones.

Nearly three-quarters of the islet cells are **beta cells**, which produce *insulin* and a peptide called *amylin*. Another 20% are **alpha cells**, which secrete **glucagon**. Most of the remaining cells are *somatostatin*-secreting **D cells**. A few rare cells called *PP cells* (or *F cells*) produce *pancreatic polypeptide*.

Like all endocrine glands, the islets are closely associated with capillaries into which the hormones are released. Both sympathetic and parasympathetic neurons terminate on the

KETOGENIC DIETS

Some people intentionally restrict or eliminate their intake of carbohydrates in an effort to lose weight. Diets very low in carbohydrate and high in fat and protein shift metabolism to β-oxidation of fats and production of ketone bodies. These diets are therefore known as *ketogenic diets*. People who go on these diets are often pleased by initial rapid weight loss, but it is due to glycogen breakdown and water loss, not body fat reduction. Sticking with ketogenic diets decreases caloric intake and eventually results in loss of body fat, but any diet that restricts calories will do the same. Among the risks associated with ketogenic diets are dehydration, electrolyte loss, inadequate intake of calcium and vitamins, gout, and kidney problems. The only recommended use of ketogenic diets is for children under 10 who have epilepsy that is not responding fully to drug therapy. For reasons that we do not understand, maintaining a state of ketosis in these children decreases the incidence of seizures.

islets, providing a means by which the nervous system can influence metabolism.

The Insulin-to-Glucagon Ratio Regulates Metabolism

As noted earlier, insulin and glucagon act in antagonistic fashion to keep plasma glucose concentrations within an acceptable range. Both hormones are present in the blood most of the time. It is the ratio of the two hormones that determines which hormone dominates.

In the fed state, when the body is absorbing nutrients, insulin dominates, and the body undergoes net anabolism (Fig. 22-9a ■). Ingested glucose is used for energy production; excess glucose is stored as glycogen or fat. Proteins go primarily to protein synthesis.

In the fasted state, metabolic regulation prevents low plasma glucose concentrations (*hypoglycemia*). When glucagon predominates (Fig. 22-9b), the liver uses glycogen and nonglucose intermediates to synthesize glucose for release into the blood.

Figure 22-10 ■ shows plasma glucagon, glucose, and insulin concentrations over the course of a day. In a normal person, fasting plasma glucose is maintained around 90 mg/dL of plasma. After absorption of nutrients from a meal, plasma glu-

cose rises. The increase in glucose stimulates insulin release, which in turn promotes glucose transfer into cells. Plasma glucose concentrations thus fall back toward the fasting level shortly after each meal.

During an overnight fast, plasma glucose concentrations fall to their lowest values, and insulin secretion also decreases. Glucagon secretion remains relatively steady during the 24-hour period, supporting the theory that it is the ratio of insulin to glucagon that determines the direction of metabolism.

Insulin Is the Dominant Hormone of the Fed State

Insulin is a typical peptide hormone (Table 22-3 ■). It is synthesized as an inactive prohormone and activated prior to secretion

TABLE 22-3	Insulin
Cell of origin	Beta cells of pancreas
Chemical nature	51-amino acid peptide
Biosynthesis	Typical peptide
Transport in the circulation	Dissolved in plasma
Half-life	5 minutes
Factors affecting release	Plasma [glucose] > 100 mg/dL; ↑ blood amino acids; GLP-1 (feedforward reflex); and parasympathetic activity amplifies. Sympathetic activity inhibits.
Target cells or tissues	Liver, muscle, and adipose tissue primarily; brain, kidney, and intestine not insulin dependent
Target receptor	Membrane receptor with tyrosine kinase activity; pathway with insulin-receptor substrates
Whole body or tissue action	↓ Plasma [glucose] by ↑ transport into cells or ↑ metabolic use of glucose
Action at cellular level	↑ Glycogen synthesis; ↑ aerobic metabolism of glucose; ↑ protein and triglyceride synthesis
Action at molecular level	Inserts GLUT transporters in muscle and adipose cells; alters enzyme activity. Complex signal transduction pathway involved.
Feedback regulation	↓ Plasma [glucose] shuts off insulin release.
Other information	Growth hormone and cortisol are antagonistic.

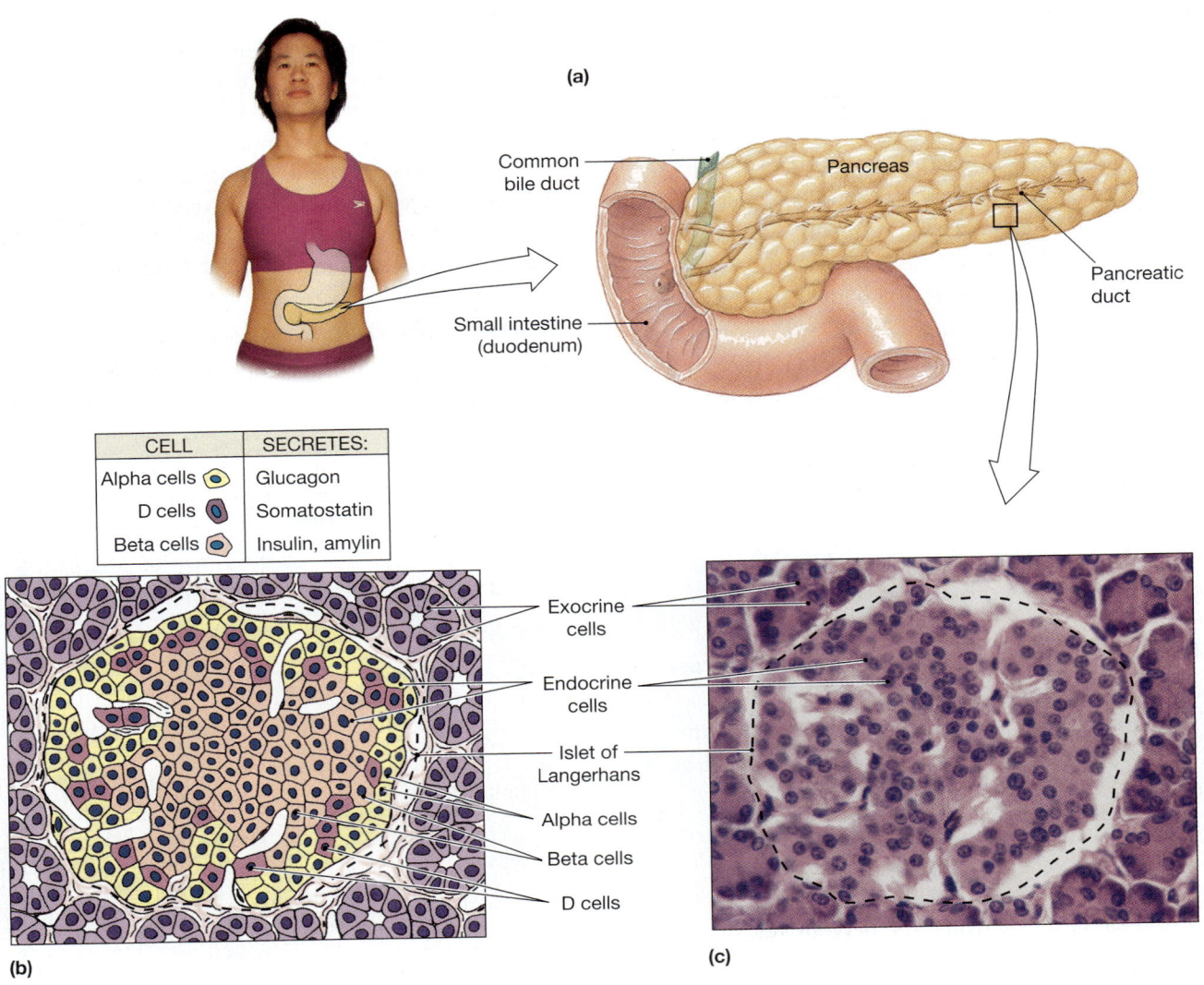

(a)

CELL	SECRETES:
Alpha cells	Glucagon
D cells	Somatostatin
Beta cells	Insulin, amylin

(b)

(c)

■ **FIGURE 22-8** *Anatomy of the pancrease*

(a) Gross anatomy. (b) The cells of the islets of Langerhans, which constitute the
endocrine pancreas.

[⟳ Fig. 7-4c, p. 220]. Glucose is an important stimulus for insulin secretion, but the following factors may stimulate, amplify, or inhibit secretion:

1. **Increased glucose concentrations**. A major stimulus for
   insulin release is plasma glucose concentrations greater
   than 100 mg/dL. Glucose absorbed from the small intestine reaches pancreatic beta cells, where it is taken up by
   GLUT2 transporters [⟳ Fig. 5-38, p. 167]. With more glucose available as substrate, ATP production increases and
   ATP-gated K^+ channels close. The cell depolarizes, voltage-
   gated Ca^{2+} channels open, and Ca^{2+} entry initiates exocytosis of insulin.

2. **Increased amino acid concentrations**. Increased plasma
   amino acid concentrations following a meal also trigger
   insulin secretion.

3. **Feedforward effects of GI hormones**. Recently it has been
   shown that as much as 50% of insulin secretion is stimulated by the hormone *glucagon-like peptide-1 (GLP-1)*. GLP-1
   and GIP (gastric inhibitory peptide) are *incretin* hormones
   produced by cells of the ileum and jejunum in response to
   nutrient ingestion. The incretins travel through the circulation to pancreatic beta cells and may reach them even
   before the first glucose is absorbed. The anticipatory release
   of insulin in response to these hormones prevents a sudden
   surge in plasma glucose concentrations when the meal is
   absorbed. Other GI hormones, such as CCK and gastrin,
   amplify insulin secretion.

4. **Parasympathetic activity**. Parasympathetic activity to
   the GI tract and pancreas increases during and following a
   meal. Parasympathetic input to beta cells stimulates insulin secretion.

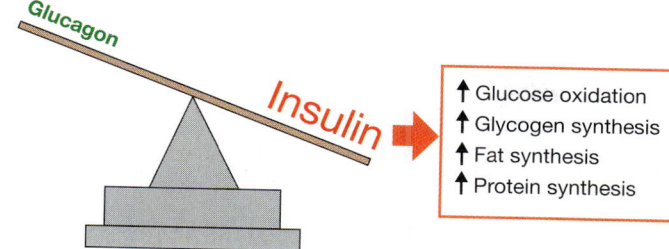

(a) Fed state: insulin dominates

↑ Glucose oxidation
↑ Glycogen synthesis
↑ Fat synthesis
↑ Protein synthesis

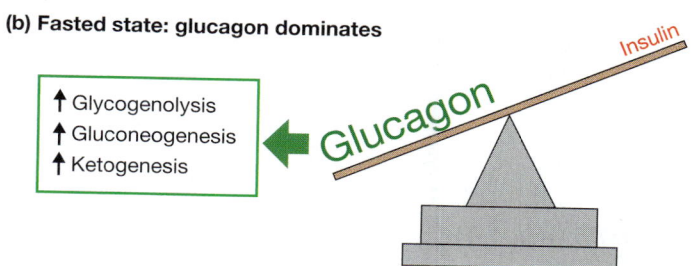

(b) Fasted state: glucagon dominates

↑ Glycogenolysis
↑ Gluconeogenesis
↑ Ketogenesis

■ **FIGURE 22-9**   *Metabolism is controlled by insulin and glucagon*

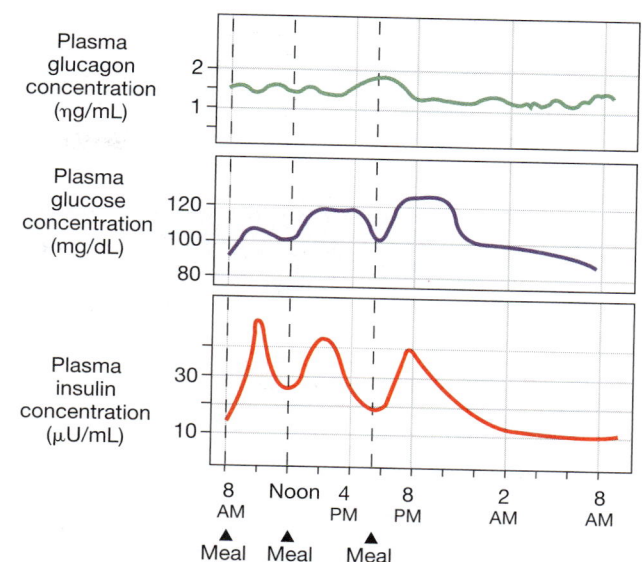

■ **FIGURE 22-10**   *Glucose, glucagon, and insulin levels over a 24-hour period*

Hour-to-hour metabolic regulation depends on the ratio of insulin to glucagon.

5. **Sympathetic activity**. Insulin secretion is inhibited by sympathetic neurons. In times of stress, sympathetic input to the endocrine pancreas increases, reinforced by catecholamine release from the adrenal medulla (Table 22-4 ■). Epinephrine and norepinephrine inhibit insulin secretion and switch metabolism to gluconeogenesis to provide extra fuel for the nervous system and skeletal muscles.

Insulin Promotes Anabolism

Like other peptide hormones, insulin combines with a membrane receptor on its target cells (Fig. 22-11 ■). The insulin receptor has *tyrosine kinase* activity, which initiates complex intracellular cascades whose details are still not completely understood. The activated insulin receptor phosphorylates proteins that include a group known as the **insulin-receptor substrates** (IRS). These proteins act through complicated pathways to influence transport and cellular metabolism. The enzymes that regulate metabolic pathways may be inhibited or activated directly, or their synthesis may be influenced indirectly through transcription factors.

The primary target tissues for insulin are the liver, adipose tissue, and skeletal muscles. The usual target cell response is increased glucose metabolism. In some target tissues, insulin also regulates the GLUT transporters. Other tissues, including the brain and transporting epithelia of the kidney and intestine, do not require insulin for glucose uptake and metabolism.

Insulin lowers plasma glucose in the following four ways:

1. **Insulin increases glucose transport into most, but not all, insulin-sensitive cells**. Adipose tissue and resting skeletal

TABLE 22-4	Adrenal Catecholamines (Epinephrine and Norepinephrine)
Origin	Adrenal medulla (90% epinephrine and 10% norepinephrine)
Chemical nature	Amines made from tyrosine
Biosynthesis	Typical peptide
Transport in the circulation	Some bound to sulfate
Half-life	2 minutes
Factors affecting release	Primarily fight-or-flight reaction through CNS and autonomic nervous system; hypoglycemia
Target cells or tissues	Mainly neurons, pancreatic endocrine cells, heart, blood vessels, adipose tissue
Target receptor	G protein-coupled receptors; α and β subtypes
Second messenger	cAMP (α_2 and all β receptors); IP$_3$ (α_1 receptors)
Whole body or tissue action	↑ Plasma [glucose]; activate fight-or-flight and stress reactions; ↑ glucagon and ↓ insulin secretion
Onset and duration of action	Rapid and brief

22

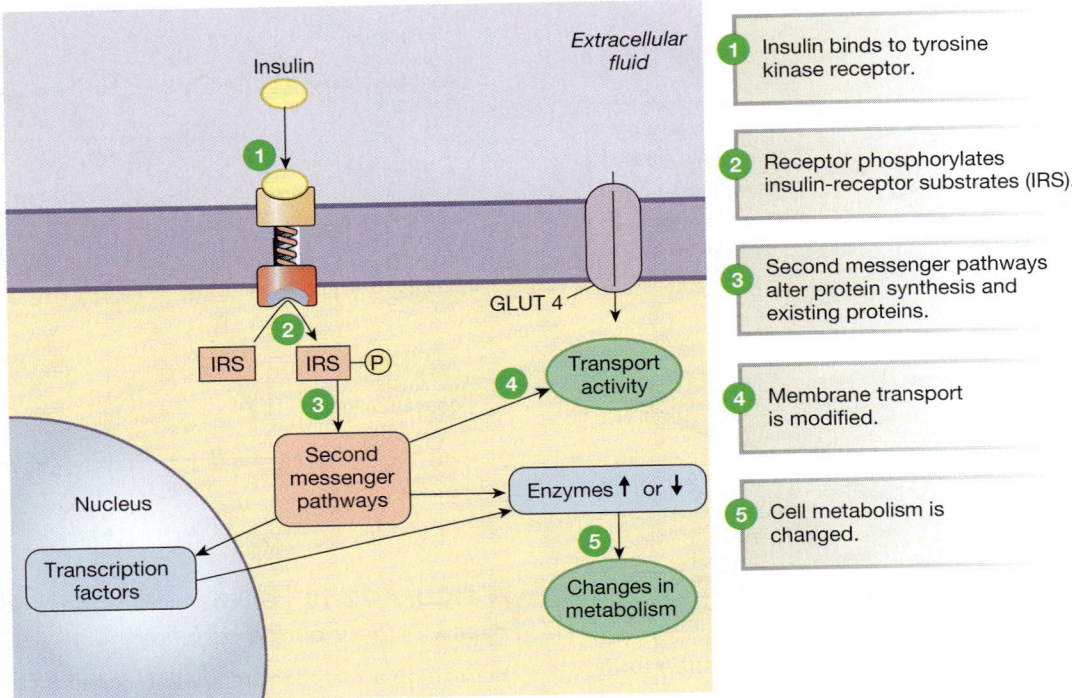

■ **FIGURE 22-11** *Insulin's cellular mechanism of action*

Insulin binds to a tyrosine kinase receptor and activates multiple second messenger pathways mediated by insulin-receptor substrates (IRS).

muscle require insulin for glucose uptake. Without insulin, their GLUT4 transporters are withdrawn from the membrane and stored in cytoplasmic vesicles (Fig. 22-12a ■). When insulin binds to the receptor and activates it, the resulting signal transduction cascade causes the vesicles to move to the cell membrane and insert the GLUT4 transporters by exocytosis (Fig. 22-12b). The cells then take up glucose from the interstitial fluid by facilitated diffusion.

Curiously, *exercising* skeletal muscle is not dependent on insulin activity for its glucose uptake. When muscles

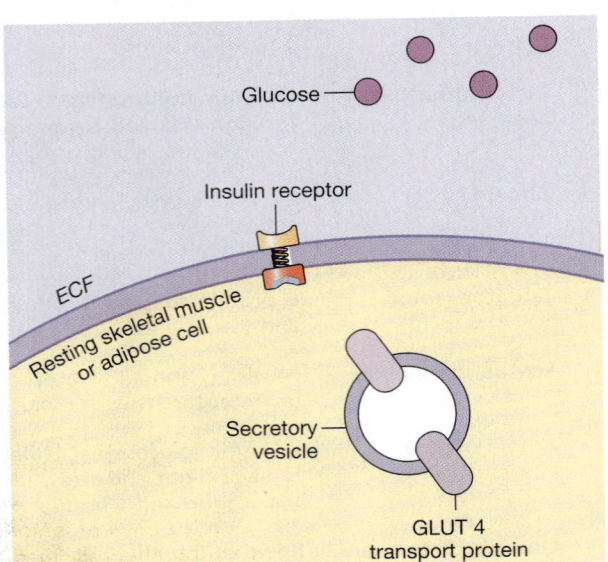

(a) In the absence of insulin, glucose cannot enter the cell.

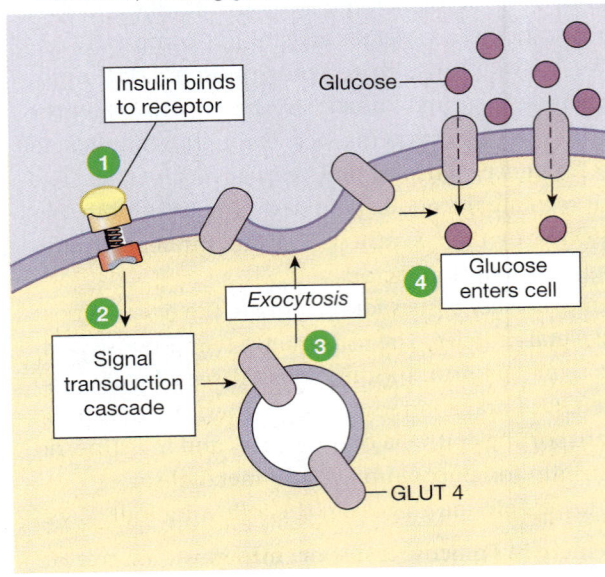

(b) Insulin signals the cell to insert GLUT 4 transporters into the membrane, allowing glucose to enter cell.

■ **FIGURE 22-12** *Insulin enables glucose uptake by adipose tissue and resting skeletal muscle*

(a) Hepatocyte in fed state: liver cell takes up glucose.

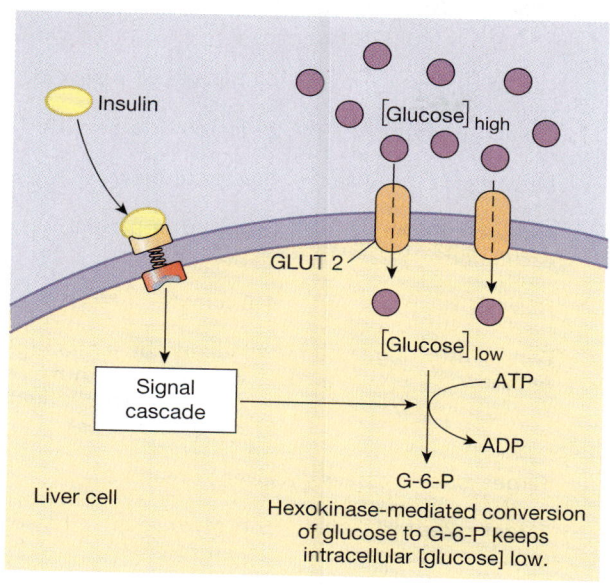

(b) Hepatocyte in fasted state: liver cell makes glucose and transports it out into the blood.

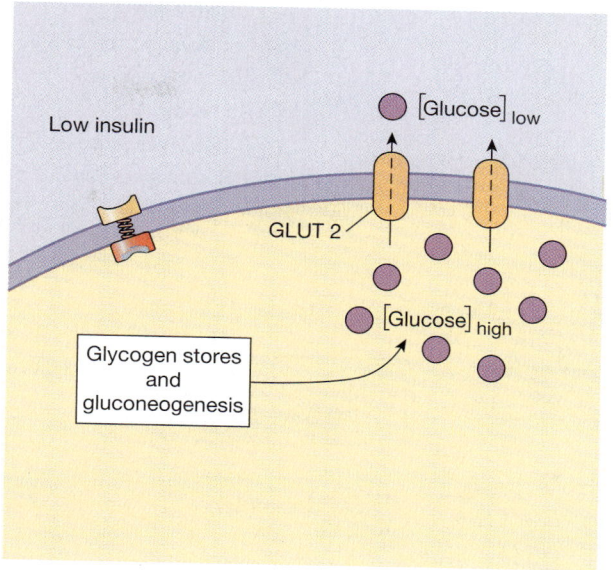

■ **FIGURE 22-13**   *Insulin acts indirectly to alter glucose uptake in hepatocytes*

contract, GLUT4 transporters are inserted into the membrane even in the absence of insulin, and glucose uptake increases. The intracellular signal for this is unclear but appears to involve Ca^{2+} and decreased inorganic phosphate (P_i).

Glucose transport into liver cells (*hepatocytes*) is not *directly* insulin dependent but is influenced by the presence or absence of insulin. Hepatocytes have GLUT2 transporters that are always present in the cell membrane. In the fed state (Fig. 22-13a ■), insulin activates *hexokinase,* the enzyme that phosphorylates glucose to glucose 6-phosphate [🔁 p. 103]. This phosphorylation reaction keeps free intracellular glucose concentrations low relative to plasma concentrations, so that glucose continuously diffuses into the hepatocyte on the GLUT2 transporter.

In the fasted state, when insulin levels are low, glucose moves *out* of the liver and into the blood to help maintain glucose homeostasis. In this process (Fig. 22-13b), hepatocytes are converting glycogen stores and amino acids to glucose. Newly formed glucose moves down its concentration gradient out of the cell using the same GLUT2 transporters operating in the reverse direction. If the GLUT transporters were pulled from the membrane during the fasted state, as they are in muscle and adipose tissue, glucose would have no way to leave the hepatocytes.

2. **Insulin enhances cellular utilization and storage of glucose** (Fig. 22-14 ■). Insulin activates enzymes for glucose utilization (*glycolysis*), and for glycogen and fat synthesis

(*glycogenesis* and *lipogenesis*). Insulin simultaneously inhibits enzymes for glycogen breakdown (*glycogenolysis*), glucose synthesis (*gluconeogenesis*), and fat breakdown (*lipolysis*) to ensure that metabolism moves in the anabolic direction. If more glucose has been ingested than is needed for energy and synthesis, the excess is made into glycogen or fatty acids.

3. **Insulin enhances utilization of amino acids.** Insulin activates enzymes for protein synthesis and inhibits enzymes that promote protein breakdown. If a meal includes protein, amino acids in the ingested food are used for protein synthesis by both the liver and muscle. Excess amino acids are converted to fatty acids.

4. **Insulin promotes fat synthesis.** Insulin inhibits beta-oxidation of fatty acids and promotes conversion of excess glucose or amino acids into triglycerides (*lipogenesis*). Excess triglyceride is stored as lipid droplets in adipose tissue.

Thus, insulin is an anabolic hormone because it promotes glycogen, protein, and fat synthesis. When insulin is absent or deficient, cells go into catabolic metabolism.

CONCEPT CHECK

12. What are the primary target tissues for insulin?

13. Why are glucose metabolism and glucose transport independent of insulin in renal and intestinal epithelium and in neurons?

14. What is the advantage to the body of inhibiting insulin release during a sympathetically mediated fight-or-flight response?

Answers: p. 748

22

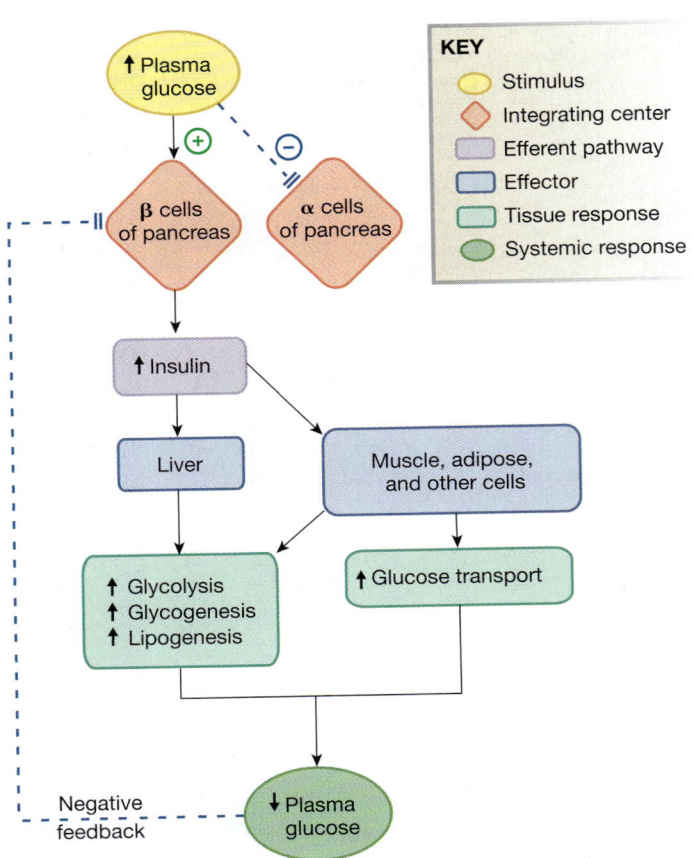

KEY
- ⬭ Stimulus
- ◆ Integrating center
- ▭ Efferent pathway
- ▭ Effector
- ▭ Tissue response
- ⬭ Systemic response

■ **FIGURE 22-14** *Fed-state metabolism under the influence of insulin promotes glucose metabolism by cells*

TABLE 22-5	Glucagon
Cell of origin	Alpha cells of pancreas
Chemical nature	29-amino acid peptide
Biosynthesis	Typical peptide
Transport in the circulation	Dissolved in plasma
Half-life	4–6 minutes
Factors affecting release	Stimulated by plasma [glucose] < 200 mg/dL, with maximum secretion below 50 mg/dL; ↑ blood amino acids
Target cells or tissues	Liver primarily
Target receptor/ second messenger	G protein-coupled receptor linked to cAMP
Whole body or tissue action	↑ Plasma [glucose] by glycogenolysis and gluconeogenesis; ↑ lipolysis leads to ketogenesis in liver
Action at molecular level	Alters existing enzymes and stimulates synthesis of new enzymes
Feedback regulation	↑ Plasma [glucose] shuts off glucagon secretion
Other information	Member of secretin family (along with VIP, GIP, and GLP-1)

Glucagon Is Dominant in the Fasted State

Glucagon, secreted by pancreatic alpha cells, is generally antagonistic to insulin in its effects on metabolism (Table 22-5 ■). When plasma glucose concentrations decline after a meal, insulin secretion slows, and the effects of glucagon on tissue metabolism take on greater significance. As noted earlier, it is the ratio of insulin to glucagon that determines the direction of metabolism rather than an absolute amount of either hormone.

The function of glucagon action is preventing hypoglycemia, and the primary stimulus for glucagon release is plasma glucose concentration. When plasma glucose concentrations fall below 100 mg/dL, glucagon secretion rises dramatically. At glucose concentrations above 100 mg/dL, when insulin is being secreted, glucagon secretion is inhibited and remains at a low but relatively constant level (see Fig. 22-10). The strong relationship between insulin secretion and glucagon inhibition has led to speculation that alpha cells are regulated by some factor linked to insulin rather than by plasma glucose concentrations directly.

The liver is glucagon's primary target tissue (Fig. 22-15 ■). Glucagon stimulates glycogenolysis and gluconeogenesis to increase glucose output. It is estimated that during an overnight fast, 75% of the glucose produced by the liver comes from glycogen stores, and the remaining 25% from gluconeogenesis.

Glucagon release is also stimulated by plasma amino acids. This pathway prevents hypoglycemia after ingestion of a pure protein meal. Let's see how hypoglycemia might occur in the absence of glucagon.

If a meal contains protein but no carbohydrate, amino acids absorbed from the food cause insulin secretion. Even though no glucose has been absorbed, insulin-stimulated glucose uptake increases, and plasma glucose concentrations fall. Unless something counteracts this process, the brain's fuel supply is threatened by hypoglycemia.

Co-secretion of glucagon in this situation prevents hypoglycemia by stimulating hepatic glucose output. Thus, although only amino acids were ingested, both glucose and amino acids are made available to peripheral tissues.

✓ **CONCEPT CHECK**

15. How are glycogenolysis and gluconeogenesis similar, and how are they different?   Answers: p. 748

Diabetes Mellitus Is a Family of Metabolic Diseases

The most common pathology of the pancreatic endocrine system is the family of metabolic disorders known as **diabetes**

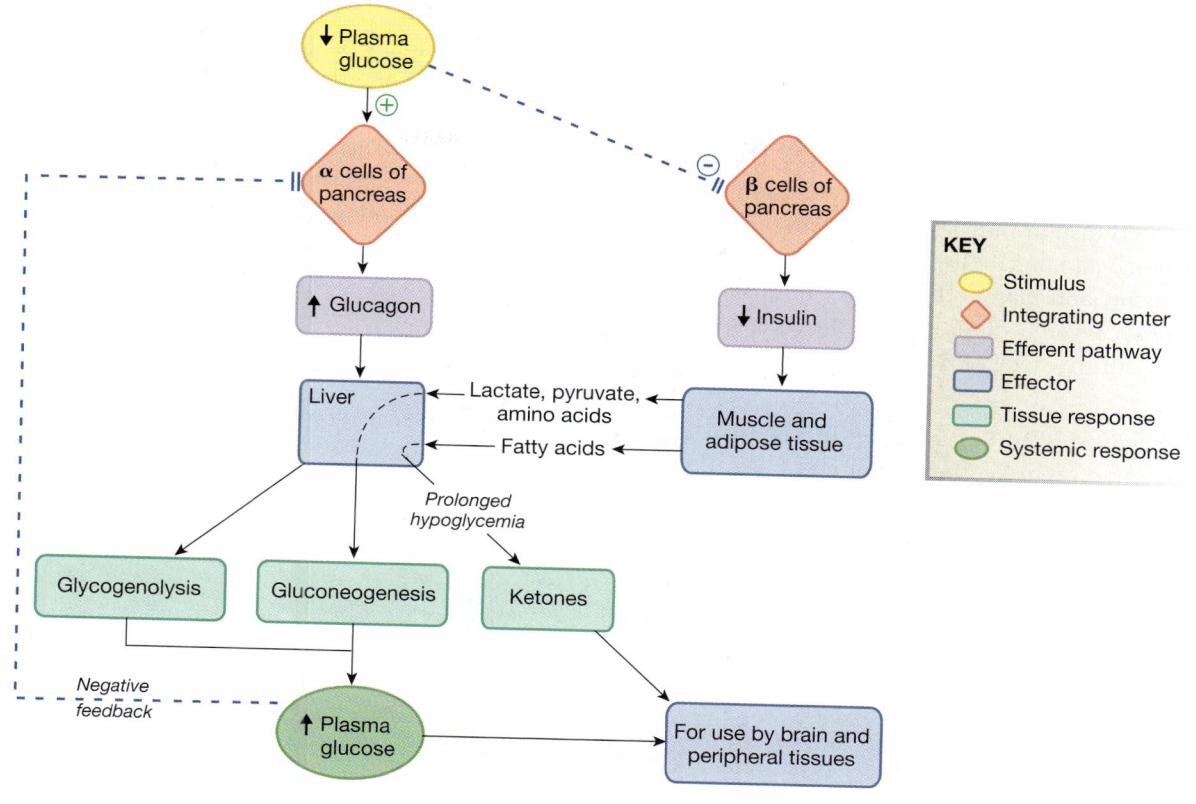

■ FIGURE 22-15   *Endocrine response to hypoglycemia*

Glucagon helps maintain adequate plasma glucose levels by promoting glycogenolysis and gluconeogenesis.

mellitus. Diabetes is characterized by abnormally elevated plasma glucose concentrations (*hyperglycemia*) resulting from inadequate insulin secretion, abnormal target cell responsiveness [🔁 p. 232], or both. Chronic hyperglycemia and its associated metabolic abnormalities cause the many complications of diabetes, including damage to blood vessels, eyes, kidneys, and the nervous system.

Diabetes has been known to affect humans since ancient times, and written accounts of the disorder highlight the calamitous consequences of insulin deficiency. Aretaeus the Cappadocian (A.D. 81–138) wrote of the "wonderful"* nature of this disease that consists of the "melting down of the flesh . . . into urine," accompanied by terrible thirst that cannot be quenched. The copious production of glucose-laden urine gave the disease its name. *Diabetes* refers to the flow of fluid through a siphon, and *mellitus* comes from the word for honey. In the Middle Ages, diabetes was known as "the pissing evil."

The severe type of diabetes described by Aretaeus is **type 1 diabetes mellitus**. It is a condition of insulin deficiency resulting from beta cell destruction. Type 1 diabetes is most commonly an *autoimmune disease* in which the body fails to recognize the beta cells as "self" and destroys them with antibodies and white blood cells.

The other major variant of diabetes mellitus is **type 2 diabetes**. This type of diabetes is also known as *insulin-resistant diabetes* because in most patients, insulin levels in the blood are normal or even elevated initially. Later in the disease process, many type 2 diabetics become insulin deficient and require insulin injections. Type 2 diabetes is actually a whole family of diseases with a variety of causes.

Type 1 Diabetics Are Prone to Ketoacidosis

Type 1 diabetes is a complex disorder whose onset in genetically susceptible individuals is sometimes preceded by a viral infection. Many type 1 diabetics develop their disease in childhood, giving rise to the old name *juvenile-onset diabetes*. About 10% of all diabetics have type 1 diabetes.

Because individuals with type 1 diabetes are insulin deficient, the only treatment is insulin injections. Until the arrival of genetic engineering, most pharmaceutical insulin came from swine, cow, and sheep pancreases. However, once the gene for human insulin was cloned, biotechnology companies began to manufacture artificial human insulin for therapeutic use. In addition, scientists are developing techniques for implanting encapsulated beta cells in the body, in the hope that individuals

---

*In the sense of "causing wonder."

22

with type 1 diabetes will no longer need to rely on regular insulin injections.

The events that follow ingestion of carbohydrate in an insulin-deficient diabetic create a picture of what happens to metabolism in the absence of insulin (Fig. 22-16 ■). Following a meal, nutrient absorption by the intestine proceeds normally because this process is insulin independent. However, nutrient uptake from the blood and metabolism by liver, resting skeletal muscle, adipose tissue, and many other tissues are insulin dependent and therefore severely diminished in the absence of insulin. Lacking nutrients to metabolize, cells go into fasted-state metabolism:

1. *Protein metabolism.* Without glucose for energy and amino acids for protein synthesis, muscles break down their proteins to provide a substrate for ATP production. Amino acids leave the muscles and are transported to the liver.

2. *Fat metabolism.* Adipose tissue in fasted-state metabolism breaks down its fat stores. Fatty acids enter the blood for transport to the liver. The liver uses β-oxidation to break down fatty acids. However, this organ is limited in its ability to send fatty acids through the citric acid cycle, and the excess fatty acids are converted to ketones. Ketone bodies reenter the circulation and can be used by other tissues (such as muscle and brain) for ATP synthesis. (The breakdown of muscle and adipose tissue in the absence of insulin leads to tissue loss and the "melting down of the flesh" described by Aretaeus.)

3. *Glucose metabolism.* In the absence of insulin, glucose remains in the blood, causing hyperglycemia. The liver, unable to take up and metabolize this glucose, initiates fasted-state pathways of glycogenolysis and gluconeogenesis. These pathways produce *additional* glucose from glycogen, amino acids, and glycerol. When the liver sends this glucose into the blood, hyperglycemia worsens.

4. *Brain metabolism.* Tissues that are not insulin dependent, such as most neurons in the brain, carry on metabolism as usual. However, neurons in the brain's satiety center *are* insulin sensitive. Therefore, in the absence of insulin, the satiety center is unable to take up plasma glucose. The center perceives the absence of intracellular glucose as starvation and allows the feeding center to increase food intake. The result is *polyphagia* (excessive eating), a classic symptom associated with untreated type 1 diabetes mellitus.

5. *Osmotic diuresis and polyuria.* If the hyperglycemia of diabetes causes plasma glucose concentrations to exceed the *renal threshold* for glucose, glucose reabsorption in the proximal tubule of the kidney becomes saturated [➋ p. 629]. Consequently, some filtered glucose is not reabsorbed, and it is excreted in the urine (*glucosuria*).

    The presence of additional solute in the collecting duct lumen causes less water to be reabsorbed and more to be excreted (see Ch. 20, question 29, p. 675). This creates large volumes of urine (*polyuria*) and, if unchecked, will cause dehydration. The loss of water in the urine due to unreabsorbed solutes is known as **osmotic diuresis**.

6. *Dehydration.* Dehydration resulting from osmotic diuresis leads to decreased circulating blood volume and decreased blood pressure. Falling blood pressure triggers homeostatic mechanisms for maintaining blood pressure, including secretion of ADH, thirst that causes constant drinking (*polydipsia*), and cardiovascular compensations [➋ Fig. 20-18, p. 662].

7. *Metabolic acidosis.* Metabolic acidosis in diabetes has two potential sources: anaerobic metabolism and ketone body production. Tissues go into anaerobic glycolysis (which creates lactic acid) if cardiovascular compensation fails and blood pressure decreases to the point that perfusion of peripheral tissues becomes inadequate. Lactic acid leaves the cells and enters the blood, contributing to a state of metabolic acidosis.

    Note that anaerobic metabolism is only a secondary cause of metabolic acidosis in uncontrolled insulin-deficient diabetes. The primary cause is the production of acidic *ketone bodies* by the liver. Patients in *diabetic ketoacidosis* exhibit the signs of metabolic acidosis discussed in Chapter 20: increased ventilation, acidification of the urine, and hyperkalemia [➋ p. 670]. If untreated, the combination of ketoacidosis and hypoxia from circulatory collapse can cause coma and even death. The treatment for a patient in diabetic ketoacidosis is insulin replacement, accompanied by fluid and electrolyte therapy to replenish lost volume and ions.

Type 2 Diabetics Often Have Elevated Insulin Levels

Type 2 diabetes is reaching epidemic proportions in the United States and was responsible for 10% of all health care dollars spent in 2002. According to some projections, one of every four Americans will have diabetes in the year 2028. Experts attribute the cause of the epidemic to our sedentary lifestyle, ample food, and overweight and obesity, which affect more than 50% of the population.

Type 2 diabetics account for 90% of all diabetics. A significant genetic predisposition to develop the disease exists among certain ethnic groups. For example, about 25% of Hispanics over age 45 have diabetes. The disease is more common in people over the age of 40, but there is growing concern about the increased diagnosis of type 2 diabetes in children and adolescents. About 80% of type 2 diabetics are obese.

A common hallmark of type 2 diabetes is **insulin resistance**, demonstrated by a delayed response to an ingested glucose load. This response can be detected by a procedure known

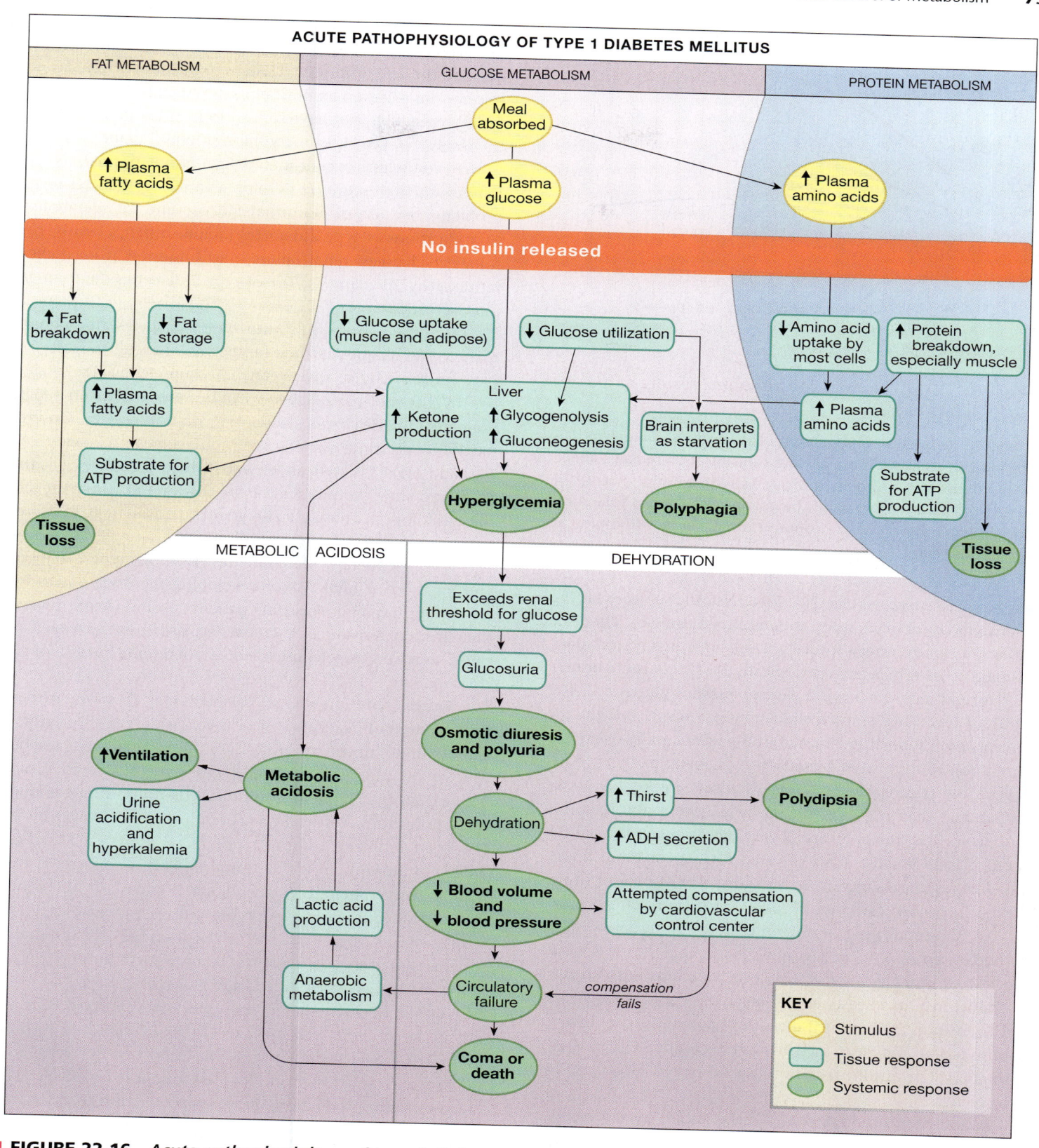

■ **FIGURE 22-16** *Acute pathophysiology of type 1 diabetes mellitus*

Untreated type 1 diabetes is marked by tissue breakdown, glucosuria, polyuria, poly-dipsia, polyphagia, and metabolic ketoacidosis.

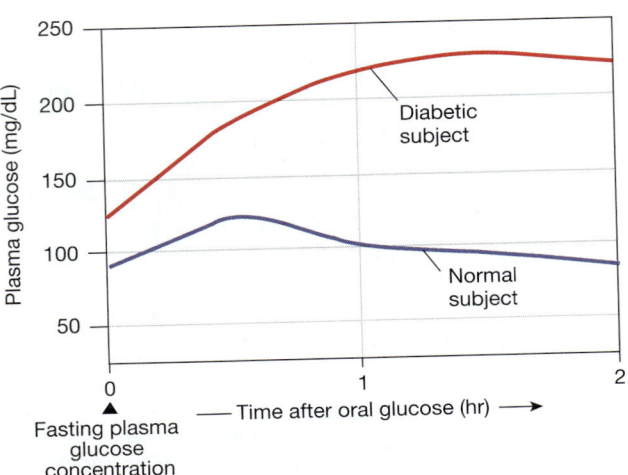

■ FIGURE 22-17   *Normal and abnormal results of glucose tolerance tests*

At time 0, the subject consumes a glucose drink or a meal. In diabetic individuals, the plasma glucose concentration remains above 200 mg/dL after two hours. Individuals with plasma glucose of 140–199 mg/dL after two hours have impaired glucose tolerance. Normal people have glucose concentrations below 140 mg/dL at the two-hour mark.

as a *glucose tolerance test* (Fig. 22-17 ■). First, the subject's fasting plasma glucose concentration is determined (time 0). Then the person consumes either a special glucose drink or a typical meal. Plasma glucose is measured periodically for two or more hours.

Normal subjects have a fasting plasma glucose of 109 mg/dL or less. They show a slight increase in plasma glucose concentration soon after a meal, but the level rapidly returns to normal following insulin secretion. In diabetic patients, however, fasting plasma glucose concentrations are usually above normal, and they rise even higher as glucose is absorbed. Because the cells of the body are slow to remove glucose from the blood, plasma glucose stays elevated for two or more hours. This slow response suggests that insulin, even if present in the blood, is unable to carry out its normal function.

An abnormal glucose tolerance test simply shows that the body's response to an ingested glucose load is not normal. The test cannot distinguish between problems with insulin synthesis, insulin release, or the responsiveness of target tissues to insulin. Some type 2 diabetics have both resistance to insulin action and decreased insulin secretion. Others have normal-to-high insulin secretion but decreased target cell responsiveness.

In addition, although type 2 diabetics are hyperglycemic, they often have elevated glucagon levels as well. This seems contradictory until you realize that the pancreatic alpha cells, like muscle and adipose cells, require insulin for glucose uptake. Thus, in diabetes, the alpha cells do not take up glucose, which prompts them to secrete glucagon. Glucagon then contributes to hyperglycemia by promoting glycogenolysis and gluconeogenesis.

In type 2 diabetes, the acute symptoms of the disease are not nearly as severe as in type 1 because insulin is usually present, and the cells, although resistant to insulin's action, are able to carry out a certain amount of glucose metabolism. The liver, for example, does not have to turn to ketone production, with the result that ketosis is uncommon in type 2 diabetes.

Nevertheless, overall metabolism is not normal, and patients with this condition develop a variety of diabetes-related problems because of abnormal glucose and fat metabolism. Complications of type 2 diabetes include atherosclerosis, neurological changes, renal failure, and blindness from diabetic retinopathy. As many as 70% of type 2 diabetics die from cardiovascular disease.

Because many people with type 2 diabetes are asymptomatic when diagnosed, they can be very difficult to treat. People who come in for their yearly checkup feeling fine, only to be told that they have diabetes, can be very resistant to making dramatic lifestyle changes when they do not feel sick. Unfortunately, by the time diabetic symptoms appear, damage to tissues and organs is well under way. Patient compliance at that point can slow the progress of the disease but cannot reverse the pathological changes. The goal of treatment is to correct hyperglycemia to prevent complications.

The first therapy recommended for most type 2 diabetics and for those at high risk of developing the disease is to lose weight and exercise. For some patients, losing weight reverses their insulin resistance. Exercise decreases hyperglycemia because exercising skeletal muscle does not require insulin for glucose uptake.

Drugs used to treat type 2 diabetes may (1) stimulate beta-cell secretion of insulin, (2) slow the digestion or absorption of carbohydrates in the intestine, (3) inhibit hepatic glucose output, or (4) make target tissues more responsive to insulin (Table 22-6 ■). The newest anti-diabetic drugs mimic endogenous hormones. For example, pramlintide, approved by the U.S. Food and Drug Administration (FDA) in March 2005, is an analog of **amylin**, a peptide hormone that is co-secreted with insulin. Amylin helps regulate glucose homeostasis following a meal by slowing gastric emptying and gastric acid secretion, which delays digestion and absorption of carbohydrates. Amylin also decreases food intake by a central effect on appetite, and it decreases secretion of glucagons.

Other hormone-based therapies still pending FDA approval are incretin *mimetics* (agonists). *Exendin-4* is a GLP-1 mimetic derived from a compound found in the venomous saliva of gila monsters. Exendin-4 has four primary effects: it increases insulin production, decreases production of glucagons, slows gastric emptying, and increases satiety. It has also been associated with weight loss.

In normal physiology, the combined actions of amylin, GIP, and GLP-1 create a self-regulating cycle for glucose absorption and fed-state glucose metabolism. Glucose in the intestine following a meal causes feedforward release of GIP and GLP-1.

22

TABLE 22-6	Drugs for Treating Diabetes	
DRUG	**EFFECT**	**MECHANISM OF ACTION**
Sulfonylureas and meglitinides	Stimulate insulin secretion	Close beta cell K_{ATP} channels and depolarize the cell
α-Glucosidase inhibitors	Decrease intestinal glucose uptake	Block intestinal enzymes that digest complex carbohydrates
Biguanides (e.g., metformin)	Reduce plasma glucose by decreasing hepatic gluconeogennesis	Unclear
PPAR activators ("glitazones")	Increase gene transcription for proteins that promote glucose utilization and fatty acid metabolism	Activate PPAR-γ, a nuclear receptor activator
Amylin analogs (pramlintide)	Reduce plasma glucose	Delay gastric emptying, suppress glucagon secretion, and promote satiety
Incretin (GLP-1) analogs (exendin-4)	Reduce plasma glucose and induce weight loss	Stimulate insulin secretion, reduce glucagon secretion, delay gastric emptying, and promote satiety
Inhaled insulin (Exubera)	Same as endogenous insulin	Same as endogenous insulin

The two incretins travel through the circulation to the pancreas, where they initiate insulin and amylin secretion. Amylin then acts on the GI tract to slow the rate at which food enters the intestine while insulin acts on target tissues to promote glucose uptake and utilization.

Metabolic Syndrome Links Diabetes and Cardiovascular Disease

Clinicians have known for years that people who are overweight are prone to develop type 2 diabetes, atherosclerosis, and high blood pressure. The combination of these three conditions has been formalized into a diagnosis called the **metabolic syndrome**, which highlights the integrative nature of metabolic pathways. People with metabolic syndrome meet at least three of the following five criteria: central (visceral) obesity, blood pressure ≥ 130/85 mm Hg, fasting plasma glucose ≥110 mg/dL, elevated fasting plasma triglyceride levels, and low plasma HDL-C levels. *Central obesity* is defined as a waist circumference greater than 40" in men and 35" in women. Women who have an "apple"-shaped body (widest at the waist) are more prone to developing metabolic syndrome than women who have a pear-shaped body (widest at the hips).

The association between obesity, diabetes, and cardiovascular disease illustrates the fundamental disturbances in cellular metabolism that occur with obesity. One common mechanism known to play a role in both glucose metabolism and lipid metabolism involves the family of nuclear receptors called *peroxisome proliferator-activated receptors* (PPARs). Lipids and lipid-derived molecules bind to PPARs, which turn on a variety of genes. The PPAR subtype called PPARγ (*p-par-gamma*) has been linked to adipocyte differentiation, type 2 diabetes, and foam cells, the endothelial macrophages that have ingested oxidized cholesterol. PPARα, mentioned earlier in the discussion of cholesterol metabolism, is important in hepatic cholesterol metabolism. The PPARs may be important clues to the link between obesity, type 2 diabetes, and atherosclerosis that has evaded scientists for so long.

CONCEPT CHECK

16. Why must insulin be administered as a shot and not as an oral pill?
17. Patients admitted to the hospital with acute diabetic ketoacidosis and dehydration are given insulin and fluids that contain K^+ and other ions. The acidosis is usually accompanied by hyperkalemia, so why is K^+ included in the rehydration fluids? (*Hint:* dehydrated patients may have a high *concentration* of K^+, but their total body fluid volume is low.)   Answers: p. 739

REGULATION OF BODY TEMPERATURE

Type 2 diabetes is an excellent example of the link between body weight and metabolism. The development of obesity may be linked to the efficiency with which the body converts food energy into cell and tissue components. According to one theory, people who are more efficient in transferring energy from food to fat are the ones who put on weight. In contrast, people who are less metabolically efficient can eat the same

22

RUNNING PROBLEM

The medical staff and Nicole's family decided to talk to her about her weight and eating habits. The disorder Dr. Ayani suspects, anorexia nervosa (AN), can have serious physiological consequences. The mortality rate for girls ages 15–24 suffering from AN is 12 times greater than that for the general population. The most common causes of death are cardiac arrest, electrolyte imbalance, and suicide. AN reportedly affects as many as 3% of females in industrialized nations at some point in their lifetime. (While 90% of AN cases are female, the number of male cases is increasing.) Successful treatment of AN includes providing nutrition, psychotherapy, and family therapy. Current research is investigating the usefulness of neuropeptide Y and other brain peptides in treating anorexia.

Question 9:
Why might an NPY agonist help in cases of anorexia?

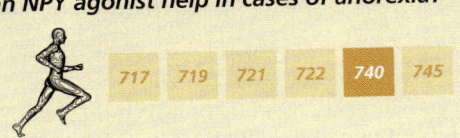

717   719   721   722   **740**   745

number of calories and not gain weight because more food energy is released as heat. Figure 22-18 ■ summarizes the factors contributing to energy balance in the body. Much of what we know about the regulation of energy balance comes from studies on body temperature regulation.

Body Temperature Is a Balance Between Heat Production, Gain, and Loss

Temperature regulation in humans is linked to metabolic heat production (*thermogenesis*). Humans are **homeothermic** animals, which means our bodies regulate internal temperature within a relatively narrow range. Average body temperature is 37° C (98.6° F), with a normal range of 35.5°–37.7° C (96°–99.9° F).

These values are subject to considerable variation, both among individuals and throughout the day in a single individual. The site at which temperature is measured also makes a difference because core body temperature may be higher than temperature at the skin surface. Oral temperatures are about 0.5° C lower than rectal temperatures.

Several factors affect body temperature in a given individual. Body temperature increases with exercise or after a meal (because of *diet-induced thermogenesis*). Temperature also cycles throughout the day: the lowest (basal) body temperature occurs in the early morning, and the highest occurs in the early evening. Women of reproductive age also exhibit a monthly temperature cycle: basal body temperatures are about 0.5° higher in the second half of the menstrual cycle (after ovulation) than before ovulation.

Heat Gain and Loss Are Balanced   Temperature balance in the body, like energy balance, depends on a dynamic equilibrium between heat input and heat output (Fig. 22-19 ■). Heat input has two components: *internal heat production*, which includes heat from normal metabolism and heat released during muscle contraction, and *external heat input* from the environment through either *radiation* or *conduction*.

■ **FIGURE 22-18** *Energy balance in the body*

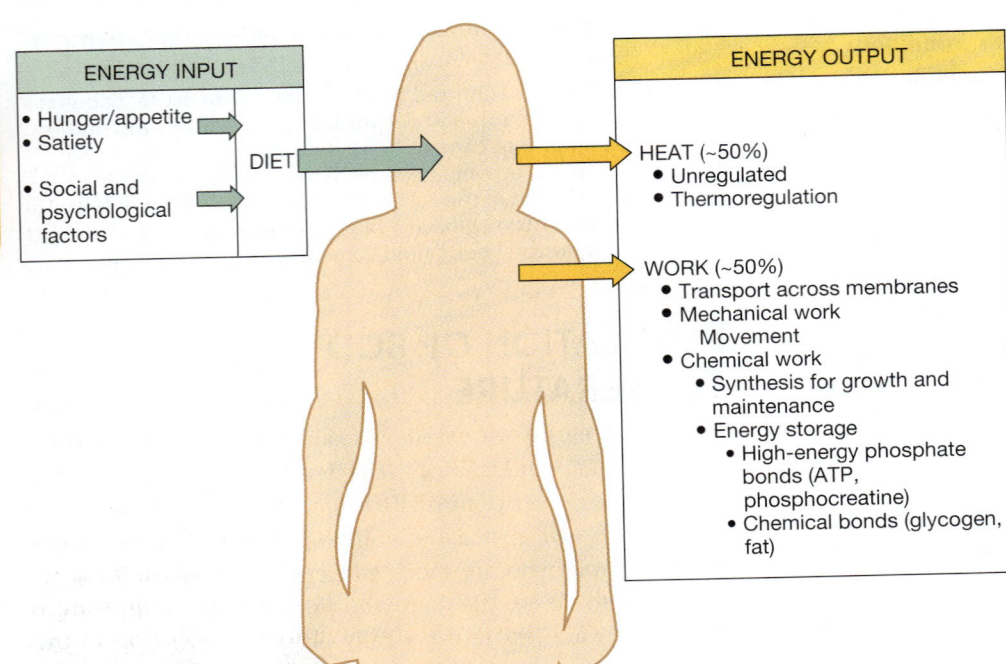

ENERGY INPUT
- Hunger/appetite
- Satiety

DIET

- Social and psychological factors

ENERGY OUTPUT

HEAT (~50%)
- Unregulated
- Thermoregulation

WORK (~50%)
- Transport across membranes
- Mechanical work
  Movement
- Chemical work
  - Synthesis for growth and maintenance
- Energy storage
  - High-energy phosphate bonds (ATP, phosphocreatine)
  - Chemical bonds (glycogen, fat)

EXTERNAL HEAT INPUT + INTERNAL HEAT PRODUCTION = HEAT LOSS

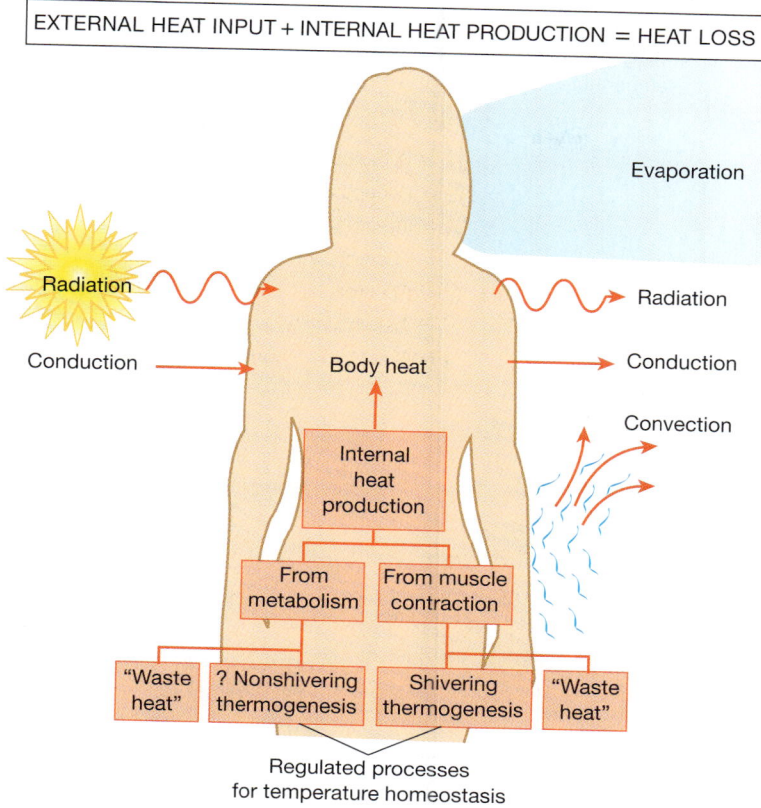

■ **FIGURE 22-19** *Heat balance in the body*

All objects with a temperature above absolute zero give off radiant energy (radiation) with infrared or visible wavelengths. This energy can be absorbed by other objects and constitutes **radiant heat gain** for those objects. You absorb radiant energy each time you sit in the sun or in front of a fire. **Conductive heat gain** is the transfer of heat between objects that are in contact with each other, such as the skin and a heating pad or the skin and hot water.

We lose heat from the body in four ways: conduction, radiation, convection, and evaporation. **Conductive heat loss** is the loss of body heat to a cooler object that is touching the body, such as an icepack or a cold stone bench. **Radiant heat loss** from the human body is estimated to account for nearly half the heat lost from a person at rest in a normal room. *Thermography* is a diagnostic imaging technique that measures radiant heat loss. Some cancerous tumors can be visually identified because they have higher metabolic activity and give off more heat than surrounding tissues.

Radiant and conductive heat loss is enhanced by **convective heat loss**, the process in which heat is carried away by warm air rising from the body's surface. Convective air currents are created wherever a temperature difference in the air exists: hot air rises and is replaced by cooler air. Convection helps move warmed air away from the skin's surface. Clothing,

which traps air and prevents convective air currents, therefore helps retain heat close to the body.

The fourth type of heat loss from the body is **evaporative heat loss**, which takes place as water evaporates at the skin's surface and in the respiratory tract. The conversion of water from the liquid state to the gaseous state requires the input of substantial amounts of heat energy. When water on the body evaporates, it removes heat from the body.

You can demonstrate the effect of evaporative cooling by wetting one arm and letting the water evaporate. As it dries, the wet arm will feel much cooler than the rest of your body because heat is being drawn from the arm to vaporize the water. Similarly, the half-liter of water vapor that leaves the body through the lungs and skin each day takes with it a significant amount of body heat. Evaporative heat loss is affected by the humidity of surrounding air; less evaporation occurs at higher humidities.

Conductive, convective, and evaporative heat loss from the body is enhanced by the *bulk flow* of air across the body, such as air moved across the skin by a fan or breeze. The effect of wind on body temperature regulation in the winter is described by the *wind chill factor*, a combination of absolute environmental temperature and the effect of convective heat loss.

Body Temperature Is Homeostatically Regulated

The human body is usually warmer than its environment and therefore loses heat. However, normal metabolism generates enough heat to maintain body temperature when the environmental temperature stays between 27.8° and 30° C (82° and 86° F). This range is known as the **thermoneutral zone**.

In temperatures above the thermoneutral zone, the body has a net gain of heat because heat production exceeds heat loss. Below the thermoneutral zone, heat loss exceeds heat production. In both cases, the body must use homeostatic compensation to maintain a constant internal temperature.

A human without clothing can thermoregulate at environmental air temperatures between 10° and 55° C (50° and 131° F). Because we are seldom exposed to the higher end of that temperature range, the main physiological challenge in thermal regulation is posed by cold environments.

Humans have been described by some physiologists as tropical animals because we are genetically adapted for life in warm climates. But we have retained a certain amount of genetic flexibility, and the physiological mechanisms by which we thermoregulate have some ability to adapt to changing conditions.

Body temperature regulation is under the control of a *thermoregulatory center* in the hypothalamus. Sensory **thermoreceptors** are located peripherally in the skin and in the anterior hypothalamus. These sensors monitor skin temperature and core body temperature and send that information to the

22

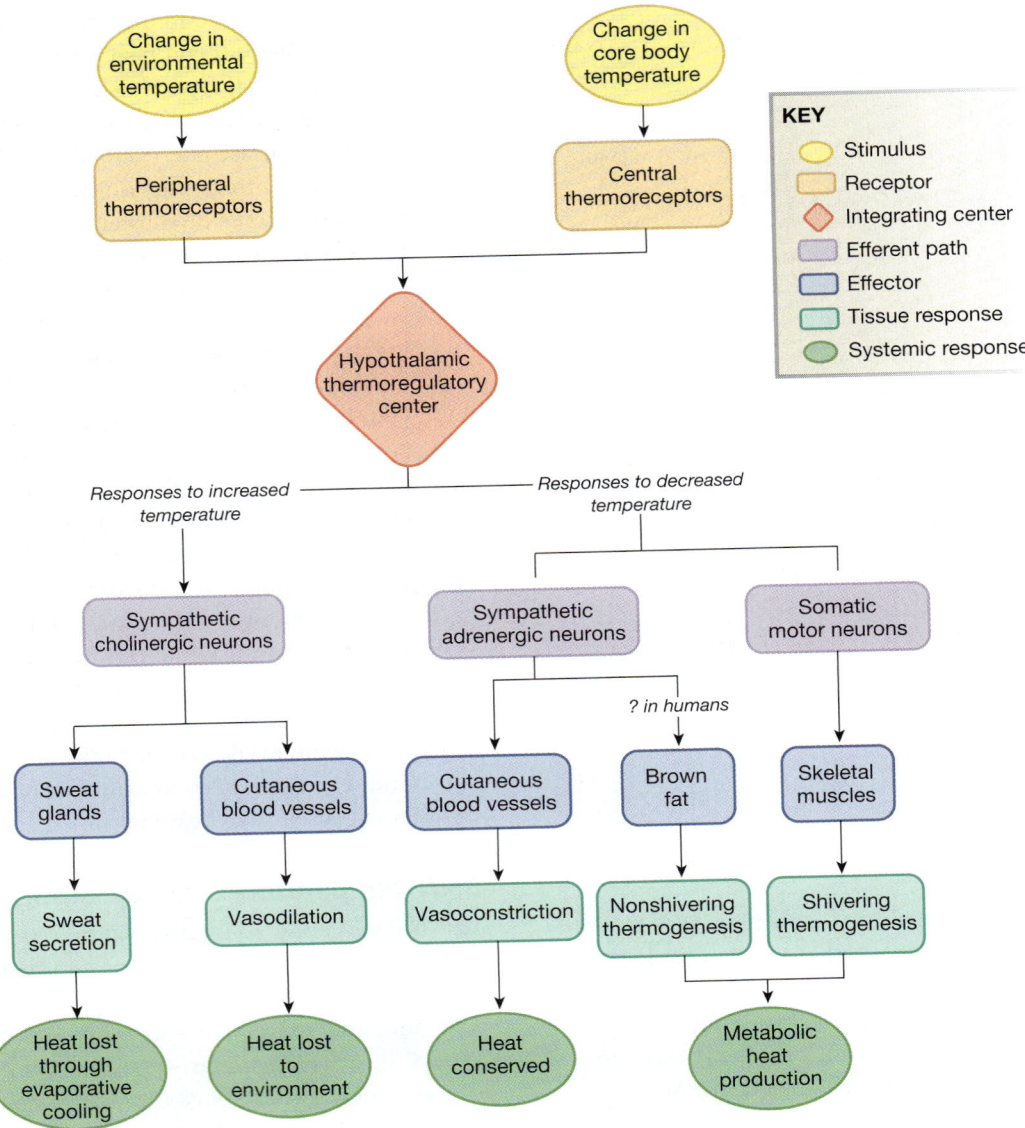

KEY

- ⬭ Stimulus
- ▭ Receptor
- ◇ Integrating center
- ▭ Efferent path
- ▭ Effector
- ▭ Tissue response
- ⬭ Systemic response

■ **FIGURE 22-20** *Thermoregulatory reflexes*

thermoregulatory center. The hypothalamic "thermostat" then compares the input signals with the desired temperature set-point and coordinates an appropriate physiological response to raise or lower the core temperature (Fig. 22-20 ■). Heat loss from the body is promoted by dilation of blood vessels in the skin and by sweating. Heat gain is generated by shivering and possibly by nonshivering thermogenesis.

Alterations in Cutaneous Blood Flow Conserve or Release Heat   Heat loss across the skin surface is regulated by controlling blood flow in *cutaneous* [*cutis*, skin] blood vessels (vessels near the skin's surface). These blood vessels can pick up heat from the environment by convection and transfer it to the body core, or they can lose heat to the surrounding air. Blood flow through cutaneous blood vessels varies from close to zero when heat must be conserved, to nearly one-third of cardiac output when heat

must be released to the environment. Local control influences cutaneous blood flow to a certain degree, possibly through vasodilators produced by the vascular endothelium. However, neural regulation is the primary determining factor.

Most arterioles in the body are under tonic sympathetic control [🔁 p. 512]. If core body temperature falls, the hypothalamus selectively activates sympathetic neurons innervating cutaneous arterioles. The arterioles constrict, increasing their resistance to blood flow and diverting blood to lower-resistance blood vessels in the interior of the body. This response keeps warmer core blood away from the cooler skin surface, thereby reducing heat loss.

In warm temperatures, the opposite happens: cutaneous arterioles dilate to increase blood flow near the skin surface and enhance heat loss. In skin in the extremities, vasodilation results from decreased sympathetic input. In skin away from the extremities, vasodilation is mediated through specialized sympathetic neurons that secrete acetylcholine. It remains unclear whether the effect is direct or mediated by bradykinin or other paracrine vasodilator substances.

✓ CONCEPT CHECK

18. What neurotransmitter and neurotransmitter receptor mediate cutaneous vasoconstriction?

19. What observations might have prompted researchers who discovered sympathetic neurons that secrete ACh to classify them as sympathetic rather than parasympathetic? [*Hint:* 🔁 p. 380]

Answers: p. 748

Sweat Contributes to Heat Loss   Surface heat loss is enhanced by the evaporation of sweat. By some estimates the human integument has 2–3 million sweat glands. The highest concentrations are found on the forehead, scalp, axillae (armpits), palms of the hands, and soles of the feet.

Sweat glands are made of transporting epithelium. Cells deep in the gland secrete an isotonic solution similar to interstitial fluid. As the fluid travels through the duct to the skin, NaCl is reabsorbed, resulting in hypotonic sweat. A typical value for sweat production is 1.5 L/hr. With acclimatization to hot weather, some people sweat at rates of 4–6 L/hr. However, they can maintain this high rate only for short periods unless they are drinking to replace lost fluid volume. Sweat production is regulated by cholinergic sympathetic neurons.

Cooling by evaporative heat loss depends on the evaporation of water from sweat on the skin's surface. Because water evaporates rapidly in dry environments but slowly or not at all in humid ones, the body's ability to withstand high temperatures is directly related to the relative humidity of the air. The combination of heat and humidity is reported by meteorologists as the *heat index* or *humidex*. Air moving across a sweaty skin surface enhances evaporation even with high humidity, which is one reason fans are useful in hot weather.

The Body Produces Heat Through Movement and Metabolism

Heat production by the body falls into two broad categories: (1) unregulated heat production from voluntary muscle contraction and normal metabolic pathways, and (2) regulated heat production for maintaining temperature homeostasis in low environmental temperatures. Regulated heat production is further divided into shivering thermogenesis and nonshivering thermogenesis (Fig. 22-19).

In **shivering thermogenesis**, the body uses shivering (rhythmic tremors caused by skeletal muscle contraction) to generate heat. Signals from the hypothalamic thermoregulatory center initiate these skeletal muscle tremors. Shivering muscle generates five to six times as much heat as resting muscle. Shivering can be partially suppressed by voluntary control.

Nonshivering thermogenesis is metabolic heat production by means other than shivering. In laboratory animals such as the rat, cold exposure significantly increases heat production in brown fat [🔁 p. 77]. The mechanism for brown fat heat production is *mitochondrial uncoupling*. In this process, energy flowing through the electron transport system [🔁 p. 108] is released as heat rather than being trapped in ATP. Mitochondrial uncoupling in response to cold is promoted by thyroid hormones and by increased sympathetic activity.

The importance of nonshivering thermogenesis in adult humans is unclear. Humans are born with significant amounts of brown fat, found primarily in the *interscapular* area between the shoulder blades. In newborns, nonshivering thermogenesis in this brown fat contributes significantly to raising and maintaining body temperature. As children age, however, white fat gradually replaces most brown fat. The brown fat that remains is very difficult to study because the experimental methodology for doing so requires that the fat be extracted from the body.

The body's responses to high and low temperatures are summarized in Figure 22-21 ■. In cold environments, the body tries to reduce heat loss while increasing internal heat production. In hot temperatures, the opposite is true. Notice from Figure 22-21 that voluntary behavioral responses play a significant role in temperature regulation. We reduce activity during hot weather, thereby decreasing muscle heat production. In cold weather, we put on extra clothing, tuck our hands in our armpits, or curl up in a ball to slow heat loss.

The Body's Thermostat Can Be Reset

Variations in body temperature regulation can be either physiological or pathological. Examples of physiological variation include the circadian rhythm of body temperature mentioned earlier, menstrual cycle variations, postmenopausal hot flashes, and fever. These processes share a common mechanism: resetting of the hypothalamic thermostat.

Hot flashes appear to be transient decreases in the thermostat's setpoint caused by the absence of estrogen. When the setpoint is lower, a room temperature that had previously been comfortable suddenly feels too hot. This discomfort triggers the usual thermoregulatory responses to heat, including sweating and cutaneous vasodilation, which leads to flushing of the skin.

For many years *fever* was thought to be a pathological response to infection, but it is now considered part of the body's normal immune response. Toxins from bacteria and other pathogens trigger the release of chemicals known as **pyrogens** [*pyr,* fire] from various immunocytes. Pyrogens are fever-producing cytokines that also have many other effects.

Experimentally, some interleukins (IL-1, IL-6), some interferons, and tumor necrosis factor have all been shown to induce fever. They do so by resetting the hypothalamic thermostat to a higher setpoint. Normal room temperature feels too cold, and the patient begins to shiver, creating additional heat. Pyrogens may also increase nonshivering thermogenesis, causing body temperature to rise.

22

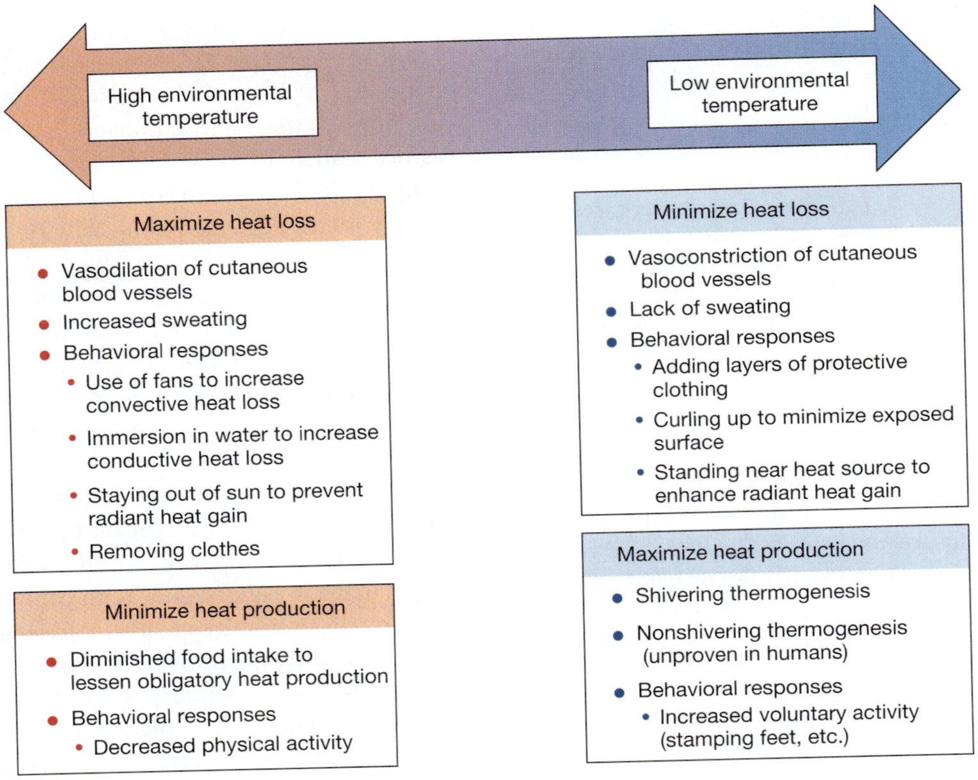

FIGURE 22-21   *Homeostatic responses to environmental extremes*

The adaptive significance of fever is still unclear, but it seems to enhance the activity of white blood cells involved in the immune response. For this reason, some people question whether patients with a fever should be given aspirin and other fever-reducing drugs simply for the sake of comfort. High fever can be dangerous, however; a fever of 41° C (106° F) for more than a brief period causes brain damage.

Pathological conditions in which body temperature strays outside the normal range include different states of *hyperthermia* and *hypothermia*. Heat exhaustion and heat stroke are the most common forms of **hyperthermia**, a condition in which body temperature rises to abnormally high values. **Heat exhaustion** is marked by severe dehydration and core body temperatures of 37.5°–39° C (99.5°–102.2° F). Patients may experience muscle cramps, nausea, and headache. They are usually pale and sweating profusely. Heat exhaustion often occurs in people who are physically active in hot, humid climates to which they are not acclimatized. It also occurs in the elderly, whose ability to thermoregulate is diminished.

Heat stroke is a more severe form of hyperthermia, with higher core body temperatures. The skin is usually flushed and dry. Immediate and rapid cooling of these patients is important, as enzymes and other proteins begin to denature at temperatures above 41° C (106° F). Mortality in heat stroke is nearly 50%.

Malignant hyperthermia, in which body temperature becomes abnormally elevated, is a genetically linked condition. A defective Ca^{2+} channel in skeletal muscle releases too much Ca^{2+} into the cytoplasm. As cell transporters work to move the Ca^{2+} back into mitochondria and the sarcoplasmic reticulum, the heat released from ATP hydrolysis substantially raises body temperature. Some investigators have suggested that a mild version of this process plays a role in nonshivering thermogenesis in mammals.

Hypothermia, a condition in which body temperature falls abnormally low, is also a dangerous condition. As core body temperature falls, enzymatic reactions slow, and the person loses consciousness. When metabolism slows, oxygen consumption also decreases.

Victims of drowning in cold water can sometimes be revived without brain damage if they have gone into a state of hypothermia. This observation led to the development of induced hypothermia for certain surgical procedures, such as heart surgery. The patient is cooled to 21°–24° C (70°–75° F) so that tissue oxygen demand can be met by artificial oxygenation of the blood as it passes through a bypass pump. After surgery is complete, the patient is gradually rewarmed.

✓ CONCEPT CHECK

20. Why must a water bed be heated to allow a person to sleep on it comfortably?

21. Will a person who is exercising outside overheat faster when the air humidity is low or when it is high?

Answers: p. 749

EATING DISORDERS

Nicole finally agreed to undergo counseling and enter a treatment program for anorexia nervosa. She was lucky—her wrist would heal, and her medical complications could have been much worse. After seeing Nicole and discussing her anorexia, Sara realized that she also needed to see a counselor. Even though she was no longer dancing, Sara still used diet pills, diuretics, and laxatives when she became uncomfortable with her weight, and she had started binging and purging—eating much more than normal when she was stressed and then forcing herself to vomit to avoid gaining

any weight. These are the behavioral patterns of *bulimia nervosa* (BN), a condition that is as serious as AN and affects an estimated 4% of females. Its physiological effects and treatments are similar to those for AN.

To learn more about the pathophysiology and treatment of anorexia and bulimia, search for "eating disorders" on the University of California at Davis Health System website, *http://wellness.ucdavis.edu*. For help finding a support group, see the National Association of Anorexia Nervosa and Associated Disorders website at *www.anad.org*.

	QUESTION	FACTS	INTEGRATION AND ANALYSIS
1	If you measured Nicole's leptin level, what would you expect to find?	Leptin is a hormone secreted by adipose tissue.	Nicole has little adipose tissue, so she would have a low leptin level.
2	Would you expect Nicole to have elevated or depressed levels of neuropeptide Y?	NPY is inhibited by leptin. NPY stimulates feeding centers.	Because her leptin level is low, you might predict that NPY would be elevated and feeding stimulated. However, the feeding center is affected by other factors besides NPY (Fig. 22-1). Brain studies of anorexic patients show high levels of CRH (Table 22-1), which opposes NPY and depresses feeding.
3	What is Nicole's K^+ disturbance called? What effect does her K^+ disturbance have on the resting membrane potential of her cells?	Nicole's K^+ is 2.5 mEq/L, and normal is 3.5–5 mEq/L.	Low plasma K^+ is called hypokalemia. Hypokalemia causes the membrane potential to hyperpolarize [🔄 p. 269].
4	Why does Dr. Ayani want to monitor Nicole's cardiac function?	Cardiac muscle is an excitable tissue whose activity depends on changes in membrane potential.	Hypokalemia can alter the membrane potential of cardiac autorhythmic and contractile cells and cause a potentially fatal cardiac arrhythmia.
5	Based on her clinical values, what is Nicole's acid-base status?	Nicole's pH is 7.52, and her plasma HCO_3^- is elevated at 40 mEq/L.	Normal pH is 7.38–7.42, so she is in alkalosis. Her elevated HCO_3^- indicates a metabolic alkalosis. The cause is probably induced vomiting and loss of HCl from her stomach.
6	Based on what you learned in Chapters 14 and 15 about heart rate and blood pressure, speculate on why Nicole has low blood pressure with a rapid pulse.	Her blood pressure is 80/50 (low), and her pulse is 90 (high).	Normally, increasing the heart rate would increase blood pressure. In this case, the increased pulse is a compensatory attempt to raise her low blood pressure. The low blood pressure probably results from dehydration.
7	Would you expect Nicole's renin and aldosterone levels to be normal, elevated, or depressed? How might these levels relate to her K^+ disturbance?	All the primary stimuli for renin secretion are associated with low blood pressure. Renin begins the RAAS pathway that stimulates aldosterone secretion.	Because Nicole's blood pressure is low, you would expect elevated renin and aldosterone levels. Aldosterone promotes renal K^+ secretion, which would lower her body load of K^+. She probably also has low dietary K^+ intake, which contributes to her hypokalemia.

	QUESTION	FACTS	INTEGRATION AND ANALYSIS
8	Give some possible reasons Nicole had been feeling weak during dance rehearsals.	In fasted-state metabolism, the body breaks down skeletal muscle.	Loss of skeletal muscle proteins, hypokalemia, and possibly hypoglycemia could all be causes of Nicole's weakness.
9	Why might an NPY agonist help in cases of anorexia?	NPY stimulates the feeding center.	An NPY agonist might stimulate the feeding center and help Nicole want to eat.

717 719 721 722 740 745

CHAPTER SUMMARY

Energy balance in the body means that the body's energy intake equals its energy output. The same *balance principle* applies to metabolism: the amount of nutrient in each of the body's nutrient pools depends on intake and output. Glucose *homeostasis* is one of the most important goals of regulated metabolism, for without adequate glucose, the brain is unable to function. Flow of material through the biochemical pathways of metabolism depends on the *molecular interactions* of substrates and enzymes.

1. The hypothalamus contains a tonically active **feeding center** and a **satiety center** that inhibits the feeding center. (p. 717)
2. Blood glucose concentrations (the **glucostatic theory**) and body fat content (the **lipostatic theory**) influence food intake. (p. 717)
3. Food intake is influenced by a variety of peptides, including **leptin, neuropeptide Y**, and **ghrelin**. (p. 717; Fig. 22-1)

Energy Balance

4. In energy balance, energy intake equals energy output. (p. 718)
5. The body uses energy for transport, movement, and chemical work. (pp. 718–719)
6. The energy content of food is measured by **direct calorimetry**. (p. 719)
7. Measuring **oxygen consumption** is the most common method of estimating energy expenditure. (p. 720)
8. The **respiratory quotient** (RQ) or **respiratory exchange ratio** (RER) is the ratio of CO_2 produced to O_2 consumed. RQ varies with diet. (p. 720)
9. **Basal metabolic rate** (BMR) is an individual's lowest metabolic rate. Metabolic rate (kcal/day) = L O_2 consumed/day × kcal/L O_2. (p. 720)
10. **Diet-induced thermogenesis** is an increase in heat production after eating. (p. 720)
11. Glycogen and fat are the two primary forms of energy storage in the human body. (p. 721)

Metabolism

12. **Metabolism** is all the chemical reactions that extract, use, or store energy. (p. 721; Figs. 22-2, 22-3)
13. **Anabolic pathways** synthesize small molecules into larger ones. **Catabolic pathways** break large molecules into smaller ones. (p. 721)
14. Metabolism is divided into the **fed** (absorptive) **state** and **fasted** (postabsorptive) **state**. The fed state is anabolic; the fasted state is catabolic. (p. 722)
15. **Glycogenesis** is glycogen synthesis. **Glycogenolysis** is glycogen breakdown. (p. 722; Fig. 22-4)
16. **Gluconeogenesis** is glucose synthesis from noncarbohydrate precursors, especially amino acids. (p. 722)
17. Ingested fats enter the circulation as chylomicrons. **Lipoprotein lipase** removes triglycerides, leaving chylomicron remnants to be taken up and metabolized by the liver. (p. 725; Fig. 22-5)
18. The liver secretes lipoprotein complexes, such as LDL-C and HDL-C. **Apoproteins A** and **B** are the ligands for receptor-mediated endocytosis of lipoprotein complexes. (p. 725; Fig. 22-5)
19. Elevated blood LDL-C and low blood HDL-C are risk factors for coronary heart disease. Therapies for lowering cholesterol decrease cholesterol uptake or synthesis or increase cholesterol clearance. (p. 725)
20. The function of fasted-state metabolism is to maintain adequate plasma glucose concentrations because glucose is normally the only fuel that the brain can metabolize. (p. 727; Fig. 22-7)
21. In the fasted state, the liver produces glucose from glycogen and amino acids. **Beta oxidation** of fatty acids forms acidic **ketone bodies**. (p. 727; Fig. 22-7)

Homeostatic Control of Metabolism

22. Hour-to-hour metabolic regulation depends on the ratio of insulin to glucagon. Insulin dominates the fed state and decreases plasma glucose. Glucagon dominates the fasted state and increases plasma glucose. (p. 728; Figs. 22-9, 22-10)
23. The **islets of Langerhans** secrete insulin and amylin from beta cells, glucagon from alpha cells, and somatostatin from D cells. (p. 728; Fig. 22-8)
24. Increased plasma glucose and amino acid levels stimulate insulin secretion. GI hormones and parasympathetic input amplify it. Sympathetic signals inhibit insulin secretion. (p. 730)

25. Insulin binds to a tyrosine kinase receptor and activates multiple **insulin-receptor substrates**. (p. 731; Fig. 22-11)

26. Major insulin target tissues are the liver, adipose tissue, and skeletal muscles. Some tissues are insulin independent. (p. 731)

27. Insulin increases glucose transport into muscle and adipose tissue, as well as glucose utilization and storage of glucose and fat. (p. 731; Figs. 22-12, 22-14)

28. **Glucagon** stimulates glycogenolysis and gluconeogenesis. (p. 734; Fig. 22-15)

29. **Diabetes mellitus** is a family of disorders marked by abnormal secretion or activity of insulin that causes hyperglycemia. In **type 1 diabetes**, pancreatic beta cells are destroyed by antibodies. In **type 2 diabetes**, target tissues fail to respond normally to insulin. (pp. 734–735)

30. Type 1 diabetes is marked by catabolism of muscle and adipose tissue, glucosuria, polyuria, and metabolic ketoacidosis. Type 2 diabetics have less acute symptoms. In both types, complications include atherosclerosis, neurological changes, and problems with the eyes and kidneys. (p. 736; Fig. 22-16)

31. **Metabolic syndrome** is a condition in which people have central obesity, elevated fasting glucose levels, and elevated lipids. These people are at high risk of cardiovascular disease. (p. 739)

Regulation of Body Temperature

32. Body temperature homeostasis is controlled by the hypothalamus. (p. 741)

33. Heat loss from the body takes place by radiation, conduction, convection, and evaporation. Heat loss is promoted by cutaneous vasodilation and sweating. (p. 743; Fig. 22-19)

34. Heat is generated by **shivering thermogenesis** and by **nonshivering thermogenesis**. (p. 743; Figs. 22-19, 22-20)

QUESTIONS

(Answers to the Review Questions begin on page A1.)

THE PHYSIOLOGY PLACE

Access more review material online at **The Physiology Place** website. There you'll find review questions, problem-solving activities, case studies, flashcards, and direct links to both *InterActive Physiology*® and *PhysioEx*™. To access the site, go to *www.physiologyplace.com* and select Human Physiology, Fourth Edition.

LEVEL ONE   REVIEWING FACTS AND TERMS

1. Define metabolic, anabolic, and catabolic pathways.
2. List and briefly explain the three forms of biological work.
3. Define a kilocalorie. What is direct calorimetry?
4. What is the respiratory quotient (RQ)? What is a typical RQ value for an American diet?
5. Define basal metabolic rate (BMR). Under what conditions is it measured? Why does the average BMR differ in adult males and females? List at least four factors other than gender that may affect BMR in humans.
6. What are the three general fates of biomolecules in the body?
7. What are the main differences between metabolism in the absorptive and postabsorptive states?
8. What is a nutrient pool? What are the three primary nutrient pools of the body?
9. What is the primary goal of fasted-state metabolism?
10. Excess energy in the body is stored in what forms?
11. What are the three possible fates for ingested proteins? For ingested fats?
12. Name the two hormones that regulate glucose metabolism, and explain what effect each hormone has on blood glucose concentrations.
13. What noncarbohydrate molecules can be made into glucose? What are the pathways called through which these molecules are converted to glucose?
14. Under what circumstances are ketone bodies formed? From what biomolecule are ketone bodies formed? How are they used by the body, and why is their formation potentially dangerous?

15. Name two stimuli that increase insulin secretion, and one stimulus that inhibits insulin secretion.
16. What are the two types of diabetes mellitus? How do their causes and basic symptoms differ?
17. What factors release glucagon? What organ is the primary target of glucagon? What effect(s) do(es) glucagon produce?
18. Define the following terms and explain their physiological significance:
    (a) lipoprotein lipase
    (b) amylin
    (c) ghrelin
    (d) neuropeptide Y
    (e) apoprotein
    (f) leptin
    (g) osmotic diuresis
    (h) insulin resistance
19. What effect does insulin have on:
    (a) glycolysis
    (b) gluconeogenesis
    (c) glycogenesis
    (d) lipogenesis
    (e) protein synthesis

LEVEL TWO   REVIEWING CONCEPTS

20. Concept map: Draw a concept map that compares the fed state and the fasted state. For each state, compare metabolism in skeletal muscles, the brain, adipose tissue, and the liver. Indicate which hormones are active in each stage and at what points they exert their influence.
21. Examine the graphs of insulin and glucagon secretion in Figure 22-10. Why have some researchers concluded that the ratio of these two hormones determines whether glucose will be stored or removed from storage?
22. Define, compare, and contrast or relate the terms in each of the following sets:
    (a) glucose, glycogenolysis, glycogenesis, gluconeogenesis, glucagon, glycolysis
    (b) shivering, nonshivering thermogenesis, diet-induced thermogenesis
    (c) lipoproteins, chylomicrons, cholesterol, HDL-C, LDL-C, apoproteins
    (d) direct and indirect calorimetry
    (e) conductive heat loss, radiant heat loss, convective heat loss, evaporative heat loss
    (f) absorptive and postabsorptive states

22

23. Describe (or map) the physiological events that lead to the following signs or symptoms in a type 1 diabetic:
    (a) hyperglycemia
    (b) glucosuria
    (c) polyuria
    (d) ketosis
    (e) dehydration
    (f) severe thirst

24. Both insulin and glucagon are released following ingestion of a protein meal that raises plasma amino acid levels. Why is the secretion of both hormones necessary?

25. Explain the current theory of the control of food intake. Use the following terms in your explanation: hypothalamus, feeding center, satiety center, appetite, leptin, NPY, neuropeptides.

26. Compare human thermoregulation in hot environments and cold environments.

LEVEL THREE   PROBLEM SOLVING

27. Scott is a bodybuilder who consumes large amounts of amino acid supplements in the belief that they will increase his muscle mass. He believes that the amino acids he consumes are stored in his body until he needs them. Is Scott correct? Explain.

28. Draw and label a graph showing the effect of insulin secretion on plasma glucose concentration.

29. One of the debates in fluid therapy for diabetic ketoacidosis (DKA) is whether to administer bicarbonate (bicarb). Although it is generally accepted that bicarb should be given if the patient's blood pH

is < 7.1 (life-threatening), most authorities do not give bicarb otherwise. One reason bicarbonate is not administered relates to the oxygen-binding capacity of hemoglobin. In DKA, patients have low levels of 2,3-DPG [p 598]. When acidosis is corrected rapidly, 2,3-DPG is much slower to recover and may take 24 or more hours to return to normal.

Draw and label a graph of the normal oxygen-dissociation curve [p. 597]. BRIEFLY explain and *draw in lines on the same graph to show:*

(a) what happens to oxygen release during DKA as a result of acidosis and low 2,3-DPG levels.
(b) what happens to oxygen release when the metabolic acidosis is rapidly corrected with bicarbonate.

LEVEL FOUR   QUANTITATIVE PROBLEMS

30. One way to estimate obesity is to calculate a person's **body mass index** (BMI). A body mass index greater than 25 (for men) or 30 (for women) is considered a sign of obesity. Calculate your BMI: divide your body weight in kilograms by the square of your height in meters: kg/m^2. (To convert your weight from pounds to kilograms, use the conversion factor 1 kg/2.2 lb. To convert your height from inches to meters, use the factor 1 m/39.24 in.)

31. What is the calorie content of a serving of spaghetti and meatballs that contains 6 g fat, 30 g carbohydrate, and 8 g protein? What percentage of the calories come from fat?

ANSWERS

✓ Answers to Concept Check Questions

Page 718

1. The feeding center causes an animal to eat, and the satiety center causes an animal to cease eating. Both centers are located in the hypothalamus.

2. The problem might involve abnormal tissue responsiveness—a target cell with no leptin receptors or defective receptors. There might also be a problem with the signal transduction/second messenger pathway.

Page 721

3. These seven factors are age, gender, lean muscle mass, activity, diet, hormones, genetics.

4. One gram of fat contains more than twice the energy of 1 gram of glycogen.

5. $C_6H_{12}O_6 + 6 O_2 \rightarrow 6 CO_2 + 6 H_2O$

6. $RQ = CO_2/O_2 = 6/6 = 1$.

Page 724

7. Glycogenesis is glycogen synthesis; gluconeogenesis is synthesis of glucose from amino acids or glycerol.

8. GLUT transporters are passive facilitated diffusion transporters.

Page 727

9. Bile acid sequestrants and ezetimibe leave bile salts and cholesterol in the intestinal lumen to be excreted, so possible side effects are loose, fatty feces and inadequate absorption of fat-soluble vitamins.

Page 728

10. Amino acids used for energy become pyruvate or enter the citric acid cycle.

11. Plasma cholesterol is bound to a carrier protein and can't diffuse across the cell membrane.

Page 733

12. The primary target tissues for insulin are liver, muscle, and adipose tissue.

13. If glucose uptake depended on insulin, the intestine, kidney tubule, and neurons could not absorb glucose in the fasted state. Neurons use glucose exclusively for metabolism and must always be able to take it up.

14. During fight-or-flight, skeletal muscles need glucose for energy. Inhibiting insulin secretion causes the liver to release glucose into the blood and prevents adipose cells from taking it up, making more glucose available for exercising muscle, which does not require insulin for glucose uptake.

Page 734

15. Glycogenolysis and gluconeogenesis both produce glucose, but glycogenolysis does it via glycogen breakdown, whereas gluconeogenesis uses amino acids or glycerol to make new glucose.

Page 739

16. Insulin is a protein and will be digested if administered orally.

17. Although dehydrated patients may have elevated plasma K^+ concentrations, their total *amount* of K^+ is below normal. If fluid volume is restored to normal with no K^+ added, a below-normal K^+ concentration will result.

Page 743

18. Norepinephrine binds to α-receptors to elicit vasoconstriction.

19. Researchers probably classified the neurons as sympathetic because of where they leave the spinal cord.

Page 744

20. Water cooler than body temperature will draw heat away from the body through conductive heat transfer. If this loss exceeds the body's heat production, the person will feel cold.

21. A person exercising in a humid environment loses the benefit of evaporative cooling and is likely to overheat faster.

Q Answers to Figure and Graph Questions

Page 724

Figure 22-3: 1. (a) next to up arrow from G-6-P to glycogen, (b) next to arrow from fatty acids to acetyl CoA, (c) next to arrow going from acetyl CoA to fatty acids, (d) next to down arrow from glycogen to G-6-P, (e) with electron transport system. 2. No, amino acids entering the citric acid cycle cannot be used to make glucose because the step from pyruvate to acetyl CoA is not reversible.

Page 727

Figure 22-6: The decrease from 190 to 160 mg/dL has the greatest effect.

23

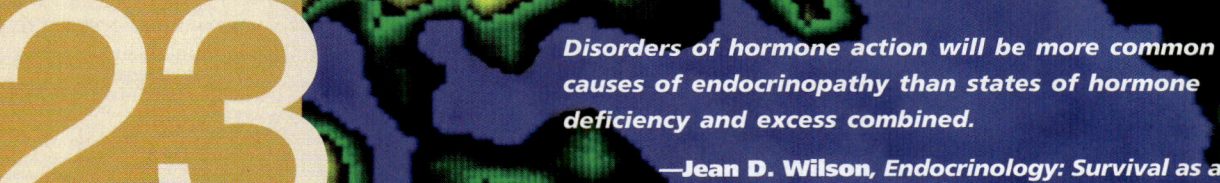

Disorders of hormone action will be more common causes of endocrinopathy than states of hormone deficiency and excess combined.

—**Jean D. Wilson,** *Endocrinology: Survival as a discipline in the 21st century? Annual Reviews of Physiology 62:947–950, 2000*

Radioisotope bone scan of the spine, ribs, and pelvis, using technetium-99m. Purple spots near the top of the spine indicate metastatic prostate cancer.

Endocrine Control of Growth and Metabolism

Review of Endocrine Principles

Adrenal Glucocorticoids

Thyroid Hormones

Growth Hormone

Tissue and Bone Growth

Calcium Balance

BACKGROUND BASICS

RUNNING PROBLEM

HYPERPARATHYROIDISM

"Broken bones, kidney stones, abdominal groans, and psychic moans." Medical students memorize this saying when they learn about *hyperparathyroidism*, a disease in which parathyroid glands [🔁 Fig. 7-2, p. 214] work overtime and produce excess parathyroid hormone (PTH). Dr. Adiaha Spinks suddenly recalls the saying as she examines Prof. Bob Magruder, who has arrived at her office in pain from a kidney stone lodged in his ureter. When questioned about his symptoms, Prof. Magruder also mentions pain in his shin bones, muscle weakness, stomach upset, and a vague feeling of depression. "I thought it was all just the stress of getting my textbook published," he says. To Dr. Spinks, however, Prof. Magruder's combination of symptoms sounds suspiciously like he might be suffering from hyperparathyroidism.

| 751 | 752 | 759 | 765 | 771 | 772 |

In 1998 Mark McGwire made news when he hit 70 home runs, surpassing the single-season home run record Roger Maris established in 1961. McGwire also created a firestorm of controversy when he admitted to taking *androstenedione,* a performance-enhancing steroid prohormone banned by the International Olympic Committee and other groups but not by professional baseball. As a result of the controversy, Congress passed the Anabolic Steroids Act of 2004, which made androstenedione and some other steroid prohormones controlled substances available only by prescription.

What is this prohormone, and why is it so controversial? You will learn more about androstenedione in this chapter as we discuss the hormones that play a role in long-term regulation of metabolism and growth. In normal individuals, these hormones can be difficult to study because their effects are subtle and their interactions with one another complex. As a result, much of what we know about endocrinology comes from studying pathological conditions in which a hormone is either oversecreted or undersecreted. In recent years, however, advances in molecular biology and the use of transgenic animal models have enabled scientists to learn more about hormone action at the cellular level.

REVIEW OF ENDOCRINE PRINCIPLES

Before we delve into the different hormones, let's take a quick review of some basic principles and patterns of endocrinology that were introduced in Chapter 7.

1. **The hypothalamic-pituitary control system** [🔁 p. 226]. Several of the hormones described in this chapter are controlled by hypothalamic and anterior pituitary (*adenohypophyseal*) trophic hormones.

2. **Feedback patterns** [🔁 p. 199]. The negative feedback signal for simple endocrine pathways is the systemic response to the hormone. For example, insulin secretion shuts off when blood glucose concentrations decrease. In complex pathways using the hypothalamic-pituitary control system, the feedback signal may be the hormone itself. In endocrine pathologies, feedback signals may not function correctly.

3. **Hormone receptors** [🔁 p. 217]. Hormone receptors may be on the cell surface or inside the cell.

4. **Cellular responses** [🔁 p. 221]. In general, hormone target cells respond by altering existing proteins or by making new proteins. The historical distinctions between the actions of peptide and steroid hormones are no longer valid. Some steroid hormones exert rapid, nongenomic effects, and some peptide hormones alter transcription and translation.

5. **Magnitude of target cell response** [🔁 p. 233]. The amount of active hormone available to the cell, and the number and activity of target cell receptors, determine the magnitude of target cell response. Cells may up-regulate or down-regulate their receptors to alter their response. Cells that do not have hormone receptors are nonresponsive.

6. **Endocrine pathologies** [🔁 p. 232]. Endocrine pathologies result from (a) excess hormone secretion, (b) inadequate hormone secretion, and (c) abnormal target cell response to the hormone. It now appears that failure of the target cell to respond appropriately to its hormone is a major cause of endocrine disorders.

In the following sections we first examine adrenal corticosteroids and thyroid hormones, two groups of hormones that influence long-term metabolism. We then consider the endocrine control of growth.

ADRENAL GLUCOCORTICOIDS

The adrenal gland, like the pituitary gland, is two embryologically distinct tissues that merged during development. This complex organ secretes multiple hormones, both neurohormones and classic hormones (Fig. 23-1 ■). The *adrenal medulla* occupies a little over a quarter of the inner mass and is composed of modified sympathetic ganglia that secrete catecholamines (mostly epinephrine) to mediate rapid responses in fight-or-flight situations [🔁 p. 377]. The *adrenal cortex* forms the outer three-quarters of the gland and secretes a variety of steroid hormones.

The Adrenal Cortex Secretes Steroid Hormones

The adrenal cortex secretes three major types of steroid hormones: aldosterone (sometimes called a *mineralocorticoid*

RUNNING PROBLEM

Hyperparathyroidism causes breakdown of bone and the release of calcium phosphate into the blood. Elevated plasma Ca^{2+} can affect the function of excitable tissues, such as muscles and neurons. Surprisingly, however, most people with hyperparathyroidism have no symptoms. The condition is usually discovered during blood work performed for a routine health evaluation.

Question 1:
What role does Ca^{2+} play in the normal functioning of muscles and neurons?

Question 2:
What is the technical term for "elevated levels of calcium in the blood"? (Use your knowledge of word roots to construct this term.)

because of its effect on the minerals sodium and potassium) [652], glucocorticoids, and sex hormones. Histologically, the adrenal cortex is divided into three layers, or zones (Fig. 23-1c). The outer *zona glomerulosa* secretes only aldosterone. The inner *zona reticularis* secretes mostly *androgens,* the sex hormones dominant in men. The middle *zona fasciculata* secretes mostly **glucocorticoids,** named for their ability to increase plasma glucose concentrations. **Cortisol** is the main glucocorticoid secreted by the adrenal cortex.

The generalized synthesis pathway for steroid hormones is shown in Figure 23-2 ■. All steroid hormones begin with cholesterol, which is modified by multiple enzymes to end up as aldosterone, glucocorticoids, or sex steroids (androgens and *estrogens* and *progesterone,* the dominant sex hormones in females). The pathways are the same in the adrenal cortex, gonads, and placenta; what differs from tissue to tissue is the distribution of enzymes that catalyze the different reactions. For example, the enzyme that makes aldosterone is found in only one of the three adrenal cortex zones.

This chapter opened with the story of baseball player Mark McGwire and his controversial use of the supplement androstenedione. Figure 23-2 shows that this prohormone is one step in the synthesis of testosterone and dihydrotestosterone. One androstenedione precursor, *dehydroepiandrosterone* (DHEA), is used as a dietary supplement. In the United States, purchase of DHEA is not regulated, despite the fact that this substance is metabolically converted to androstenedione and testosterone, controlled substances whose use is widely banned by sports associations.

The close structural similarity among steroid hormones means that the binding sites on their receptors are also similar, leading to *crossover effects* when one steroid binds to the receptor for a related molecule. For example, *mineralocorticoid receptors* (MRs) for aldosterone are found in the distal nephron. MRs also bind and respond to cortisol, which may be 100 times more concentrated in the blood than aldosterone. What is to keep cortisol from binding to an MR and influencing Na^+ and K^+ excretion? In this example, renal tubule cells with MRs also have an enzyme that converts cortisol to a less-active form with low specificity for the MR, thus blocking crossover. Crossover activity and the tightly linked synthetic pathways for steroid hormones mean that in many endocrine disorders, patients may experience symptoms related to more than one hormone.

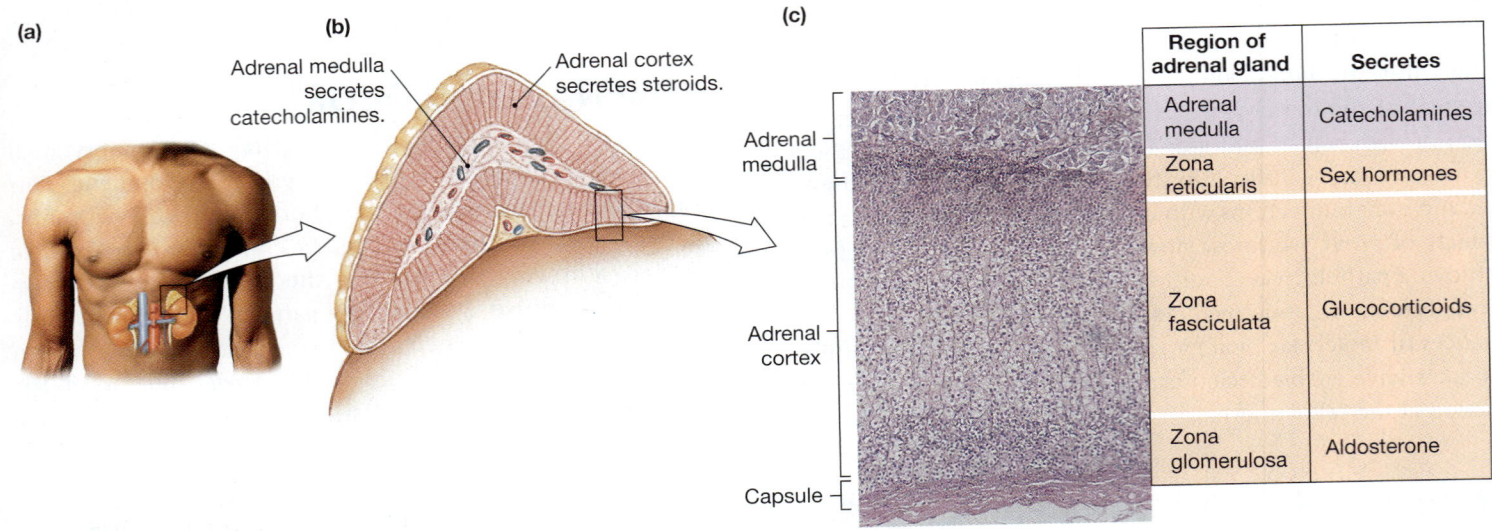

■ **FIGURE 23-1**   *Structure and function of the adrenal gland*

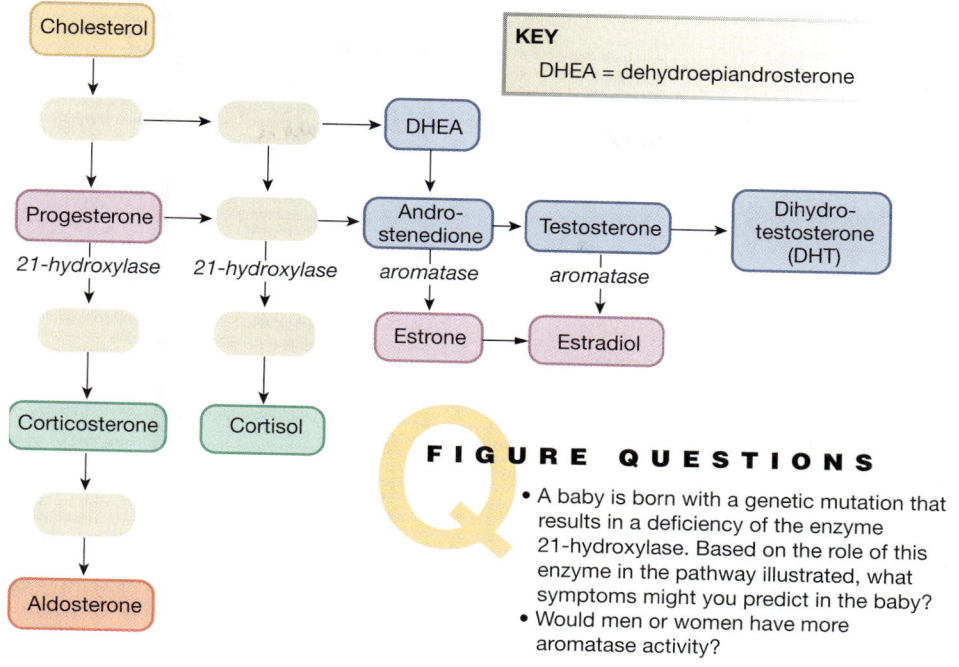

KEY

DHEA = dehydroepiandrosterone

FIGURE QUESTIONS

- A baby is born with a genetic mutation that results in a deficiency of the enzyme 21-hydroxylase. Based on the role of this enzyme in the pathway illustrated, what symptoms might you predict in the baby?
- Would men or women have more aromatase activity?

■ **FIGURE 23-2** *Synthesis pathways for steroid hormones*

All steroid hormones are synthesized from cholesterol. The blank boxes represent intermediate compounds whose names have been omitted for simplicity. Each step is catalyzed by an enzyme, but only two enzymes are shown in this figure.

CONCEPT CHECK

1. Name the two parts of the adrenal gland and the major hormones secreted by each part.
2. For what hormones is androstenedione a prohormone? (See Fig. 23-2.) Why might this prohormone give an athlete an advantage?

Answers: p. 774

Cortisol Secretion Is Controlled by ACTH

The control pathway for cortisol secretion is known as the *hypothalamic-pituitary-adrenal (HPA) pathway* (Fig. 23-3 ■). The HPA pathway begins with hypothalamic **corticotropin-releasing hormone** (CRH), which is secreted into the hypothalamic-hypophyseal portal system and transported to the anterior pituitary. CRH stimulates release of **adrenocorticotropic hormone** (ACTH or *corticotropin*) from the anterior pituitary. ACTH in turn acts on the adrenal cortex to promote synthesis and release of cortisol. Cortisol then acts as a negative feedback signal, inhibiting ACTH and CRH secretion.

Cortisol secretion is continuous and has a strong diurnal rhythm (Fig. 23-4 ■). Secretion normally peaks in the morning and diminishes during the night. Cortisol secretion also increases with stress.

Cortisol follows the typical steroid hormone pattern for synthesis, release, transport, and action [🔁 Fig. 7-7, p. 222]. Once synthesized, it diffuses out of adrenal cells into the plasma, where most of it is transported by a carrier protein,

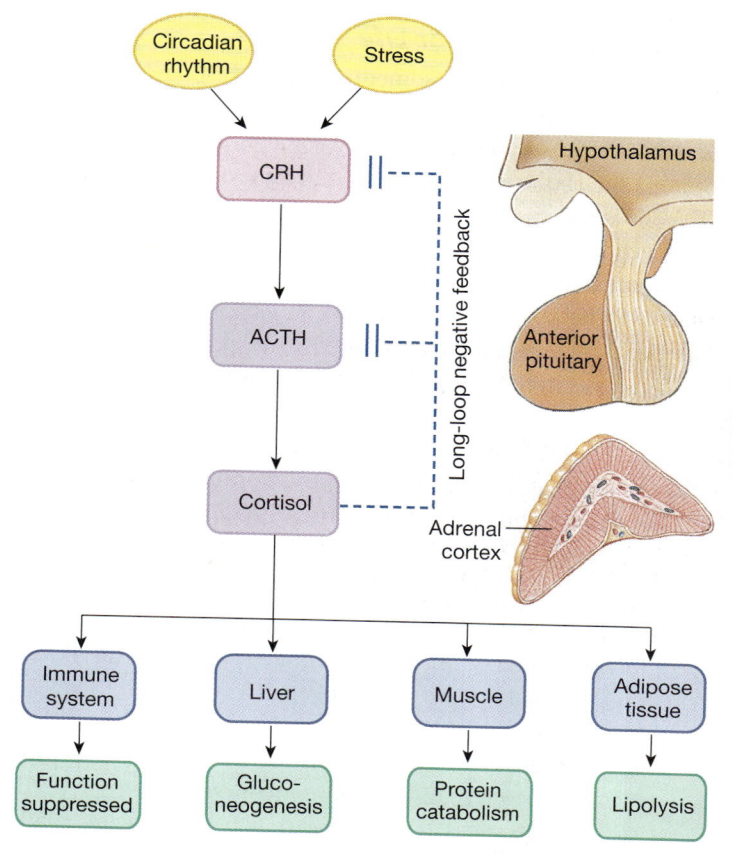

■ **FIGURE 23-3** *The HPA pathway for the control of cortisol secretion*

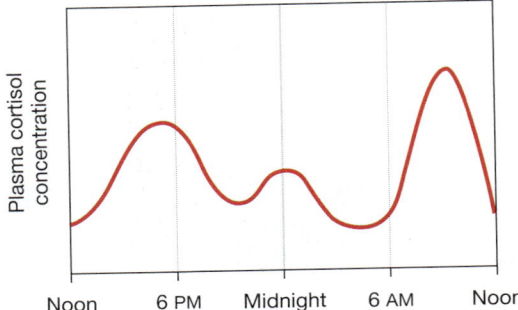

■ FIGURE 23-4 *Circadian rhythm of cortisol secretion*

Blood concentrations of cortisol vary throughout the day, so studies of hormone secretion often measure the cortisol metabolites in over a 24-hour period to obtain an average value for a day's secretion.

corticosteroid-binding globulin (CBG or *transcortin*). Unbound hormone is free to diffuse into target cells.

All nucleated cells of the body have cytoplasmic glucocorticoid receptors. The hormone-receptor complex enters the nucleus, binds to DNA with the aid of a *hormone-response element,* and alters gene expression, transcription, and translation. In general, a tissue's response to glucocorticoid hormones is not evident for 60–90 minutes. However, cortisol's negative feedback effect on ACTH secretion occurs within minutes.

Cortisol Is Essential for Life

Adrenal glucocorticoids are sometimes called the body's stress hormones because of their role in the mediation of long-term stress. Adrenal catecholamines, particularly epinephrine, are responsible for rapid metabolic responses needed in fight-or-flight situations.

Cortisol is essential for life. Animals whose adrenal glands have been removed will die if exposed to any significant environmental stress. The most important metabolic effect of cortisol is its protective effect against *hypoglycemia*. In Chapter 22 you learned that pancreatic glucagon is secreted in response to decreased blood glucose. In the absence of cortisol, however, glucagon is unable to respond adequately to a hypoglycemic challenge. Because cortisol is required for full glucagon and catecholamine activity, it is said to have a *permissive effect* on those hormones [❷ p. 231].

All the metabolic effects of cortisol are directed at preventing hypoglycemia. Overall, cortisol is catabolic (Fig. 23-3, Table 23-1 ■).

1. **Cortisol promotes gluconeogenesis** in the liver. Some glucose produced in the liver is released into the blood, and the rest is stored as glycogen. Thus cortisol increases blood glucose concentrations.
2. **Cortisol causes the breakdown of skeletal muscle proteins** to provide a substrate for gluconeogenesis.
3. **Cortisol enhances lipolysis** so that fatty acids are available to peripheral tissues for energy use. The glycerol from fatty acids can be used for gluconeogenesis.

TABLE 23-1	Cortisol
Origin	Adrenal cortex
Chemical nature	Steroid
Biosynthesis	From cholesterol; made on demand, not stored
Transport in the circulation	On corticosteroid-binding globulin (made in liver)
Half-life	60–90 minutes
Factors affecting release	Circadian rhythm of tonic secretion; stress enhances release
Control pathway	CRH (hypothalamus) → ACTH (anterior pituitary) → cortisol (adrenal cortex)
Target cells or tissues	Most tissues
Target receptor	Intracellular
Whole body or tissue action	↑ Plasma [glucose]; ↓ immune activity; permissive for glucagon and catecholamines
Action at cellular level	↑ Gluconeogenesis and glycogenolysis; ↑ protein catabolism; blocks cytokine production by immune cells
Action at molecular level	Initiates transcription, translation, and new protein synthesis
Feedback regulation	Negative feedback to anterior pituitary and hypothalamus

Cortisol receptors are found in every tissue of the body, but for many targets we do not fully understand the physiological actions of cortisol. However, we can speculate on these actions based on tissue responses to high levels (*pharmacological doses*) of cortisol administered for therapeutic reasons or associated with hypersecretion.

4. **Cortisol suppresses the immune system** through multiple pathways. This effect is discussed in more detail below.
5. **Cortisol causes negative calcium balance**. Cortisol decreases intestinal Ca^{2+} absorption and increases renal Ca^{2+} excretion, resulting in net Ca^{2+} loss from the body. In addition, cortisol is catabolic in bone tissue, causing net breakdown of calcified bone matrix. People who take therapeutic cortisol for extended periods therefore have a higher-than-normal incidence of broken bones.
6. **Cortisol influences brain function**. States of cortisol excess or deficiency cause mood changes and memory and learning alterations. Some of these effects may be mediated by hormones in the cortisol release pathway, such as CRH. This effect of cortisol will be discussed in more detail below.

(a) 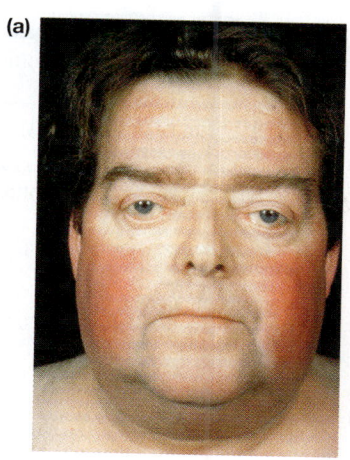   (b)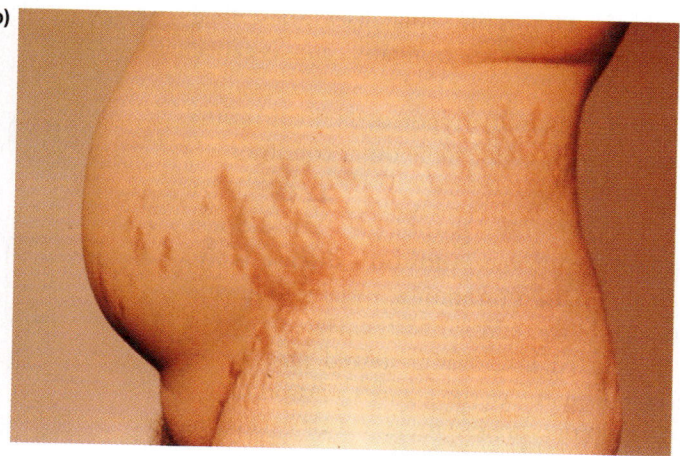

■ **FIGURE 23-5** *Hypercortisolism*

(a) A "moon face" with red cheeks is typical in this condition, also known as Cushing's syndrome. (b) Fat deposits in the trunk. The dark striations result from protein breakdown in the skin.

CONCEPT CHECK

3. What do the abbreviations ACTH, CRH, HPA, and CBG stand for? If there is an alternate name for each term, what is it?

4. You are mountain-biking in Canada and encounter a bear, which chases you up a tree. Is your stress response mediated by cortisol? Explain.

5. The illegal use of anabolic steroids by bodybuilders and athletes periodically receives much attention. Do these illegal steroids include cortisol? Explain.

Answers: p. 774

Cortisol Is a Useful Therapeutic Drug

Cortisol suppresses the immune system by preventing cytokine release and antibody production by white blood cells. It also inhibits the inflammatory response by decreasing leukocyte mobility and migration. These *immunosuppressant effects* of cortisol make it a useful drug for treating a variety of conditions, including bee stings, poison ivy, and pollen allergies. Cortisol also helps prevent rejection of transplanted organs. However, glucocorticoids also have potentially serious side effects because of their metabolic actions. Once *nonsteroidal anti-inflammatory drugs* (NSAIDs), such as ibuprofen, were developed, the use of glucocorticoids for treating minor inflammatory problems was discontinued.

Exogenous administration of glucocorticoids has a negative feedback effect on the anterior pituitary and may shut down ACTH production [Fig. 7-19, p. 232]. Without ACTH stimulation, the adrenal cells that produce cortisol atrophy. For this reason, it is essential that patients taking steroids taper their dose gradually, giving the pituitary and adrenal glands a chance to recover, rather than stopping the drug abruptly.

Cortisol Pathologies Result from Too Much or Too Little Hormone

The most common HPA pathologies result from hormone deficiency and hormone excess. Abnormal tissue responsiveness is an uncommon cause of adrenal steroid disorders.

Hypercortisolism   Excess cortisol in the body is called **hypercortisolism**. It can arise from hormone-secreting tumors or from exogenous administration of the hormone. Cortisol therapy with high doses for more than a week has the potential to cause hypercortisolism—also known as *Cushing's syndrome*, after Dr. Harvey Cushing, who first described the condition in 1932.

Most signs of hypercortisolism can be predicted from the normal actions of the hormone. Excess gluconeogenesis causes hyperglycemia, which mimics diabetes. Muscle protein breakdown and lipolysis cause tissue wasting. Paradoxically, excess cortisol deposits extra fat in the trunk and face, perhaps in part because of increased appetite and food intake. The classic appearance of patients with hypercortisolism is thin arms and legs, obesity in the trunk, and a "moon face" with plump cheeks (Fig. 23-5a ■). CNS effects of too much cortisol include initial mood elevation followed by depression, and difficulty with learning and memory.

Hypercortisolism has three common causes:

1. **An adrenal tumor that autonomously secretes cortisol.** These tumors are not under the control of pituitary ACTH. This condition is an instance of *primary hypercortisolism* [p. 234].

2. **A pituitary tumor that autonomously secretes ACTH.** Excess ACTH prompts the adrenal gland to oversecrete cortisol (*secondary hypercortisolism*). The tumor does not respond to negative feedback. This condition is also called

23

Cushing's disease because it was the actual disease described by Dr. Cushing. (Hypercortisolism from any cause is called Cushing's *syndrome*.)

3. **Iatrogenic (physician-caused) hypercortisolism** occurs secondary to cortisol therapy for some other condition.

Hypocortisolism Hyposecretion pathologies are far less common than Cushing's syndrome. **Addison's disease** is hyposecretion of all adrenal steroid hormones, usually following autoimmune destruction of the adrenal cortex. Hereditary defects in the enzymes needed for adrenal steroid production cause several related syndromes (see the Figure Question in Fig. 23-2). These inherited disorders are often marked by excess androgen secretion because substrate that cannot be made into cortisol or aldosterone is converted to androgens. In newborn girls, excess androgens cause masculinization of the external genitalia, a condition called *adrenogenital syndrome*.

> ✓ **CONCEPT CHECK**
> 6. For primary, secondary, and iatrogenic hypercortisolism, indicate whether the ACTH level is normal, higher than normal, or lower than normal.
> 7. Would someone with Addison's disease have normal, low, or high levels of ACTH in the blood? Answers: p. 774

CRH and ACTH Have Additional Physiological Functions

In recent years, research interest has shifted away from glucocorticoids to CRH and ACTH, the trophic hormones of the HPA pathway. Both peptides are now known to belong to larger families of related molecules, with multiple receptor types found in numerous tissues. Experiments with knockout mice lacking a particular receptor have revealed some of the physiological functions of peptides related to CRH and ACTH.

Two interesting findings of this research are that cytokines secreted by the immune system can stimulate the HPA pathway, and that immune cells have receptors for ACTH and CRH. The association between stress and immune function appears to be mediated through CRH and ACTH, and this association provides one explanation for mind-body interactions in which mental state influences physiological function.

CRH Family The CRH family includes CRH and a related brain neuropeptide called *urocortin*. In addition to its involvement in inflammation and the immune response, CRH is known to decrease food intake [🔁 Tbl. 22-1, p. 718] and has been associated with signals that mark the onset of labor in pregnant women (discussed in Chapter 26). Additional evidence links CRH to anxiety, depression, and other mood disorders.

ACTH and Melanocortins CRH acting on the anterior pituitary stimulates secretion of ACTH. ACTH is synthesized from

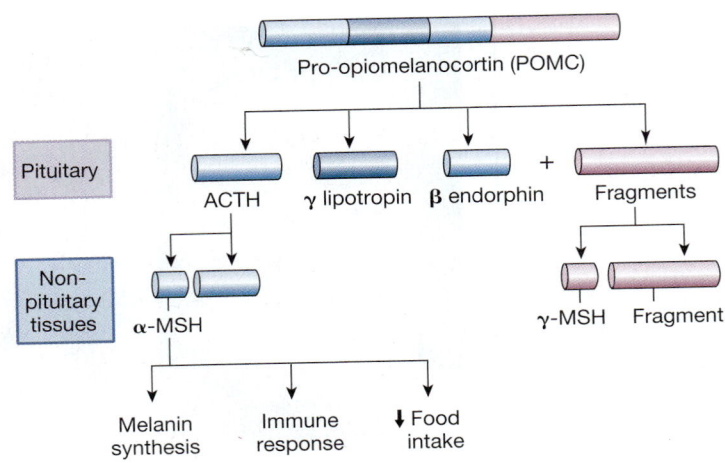

■ **FIGURE 23-6** *Post-translation processing of POMC can create a variety of active peptides*

a large glycoprotein called **pro-opiomelanocortin** (POMC). POMC undergoes post-translational processing to produce a variety of biologically active peptides in addition to ACTH (Fig. 23-6 ■). In the pituitary, POMC products include **β-endorphin**, an endogenous opiate that binds to receptors that block pain perception [🔁 p. 342].

POMC is also processed in tissues outside the pituitary, where its products may be very different from those made in the pituitary. Additional processing of POMC can create *melanocyte-stimulating hormone (MSH)*. **α-MSH** is produced in the brain, where it inhibits food intake, and in the skin, where it acts on **melanocytes**. These cells contain pigments called **melanins** that influence skin color in humans and coat color in rodents (see the Emerging Concepts box).

The MSH hormones plus ACTH have been given the family name **melanocortins**. Five *melanocortin receptors* (MCRs) have been identified. MC2R responds only to ACTH and is the adrenal cortex receptor. MC1R is found in skin melanocytes and responds equally to α-MSH and ACTH. When ACTH is elevated in Addison's disease, the action of ACTH on MC1R leads to increased melanin production and the apparent "tan," or skin darkening, characteristic of this disorder.

> ✓ **CONCEPT CHECK**
> 8. What might be the advantage for the body to co-secrete ACTH and β-endorphin? Answers: p. 774

THYROID HORMONES

The thyroid gland is a butterfly-shaped gland that lies across the trachea at the base of the throat, just below the larynx (Fig. 23-7a ■). It is one of the larger endocrine glands, weighing 15–20 g. The thyroid gland has two distinct endocrine cell types: C ("clear") *cells*, which secrete a calcium-regulating hormone called *calcitonin*, and *follicular cells*, which secrete thyroid hormone. Calcitonin is discussed later with calcium homeostasis.

EMERGING CONCEPTS

MELANOCORTINS AND THE *AGOUTI* MOUSE

Much of what we are learning about the functions of the hypothalamic-pituitary-adrenal (HPA) pathway comes from research on mutant mice. One such animal is the *agouti* mouse, a strain that resulted from a spontaneous mutation first described in 1905. *Agouti* mice with one mutated gene overproduce a protein (called **agouti protein**) that gives them a characteristic yellow coat. Agouti protein is an antagonist to the melanocortin receptor MC1R, which controls melanin synthesis in hair. Of more interest to physiologists, however, is the fact that *agouti* mice overeat and develop adult-onset obesity, hyperglycemia, and insulin resistance—in other words, these mice are a model for type 2 diabetes caused by obesity. Using molecular biology, researchers have learned that agouti protein blocks the receptor MC4R in the brain, the receptor that normally depresses feeding behavior when it binds α-MSH and ACTH. If the receptor is not functional, the animal overeats and becomes obese. This link between melanocortin receptors, eating behavior, and diabetes has opened up a new area of research on treatments to prevent type 2 diabetes. To learn more about the *agouti* mouse models, see the Jackson Laboratory website at *http://www.informatics.jax.org/searches/mlc.cgi?3*.

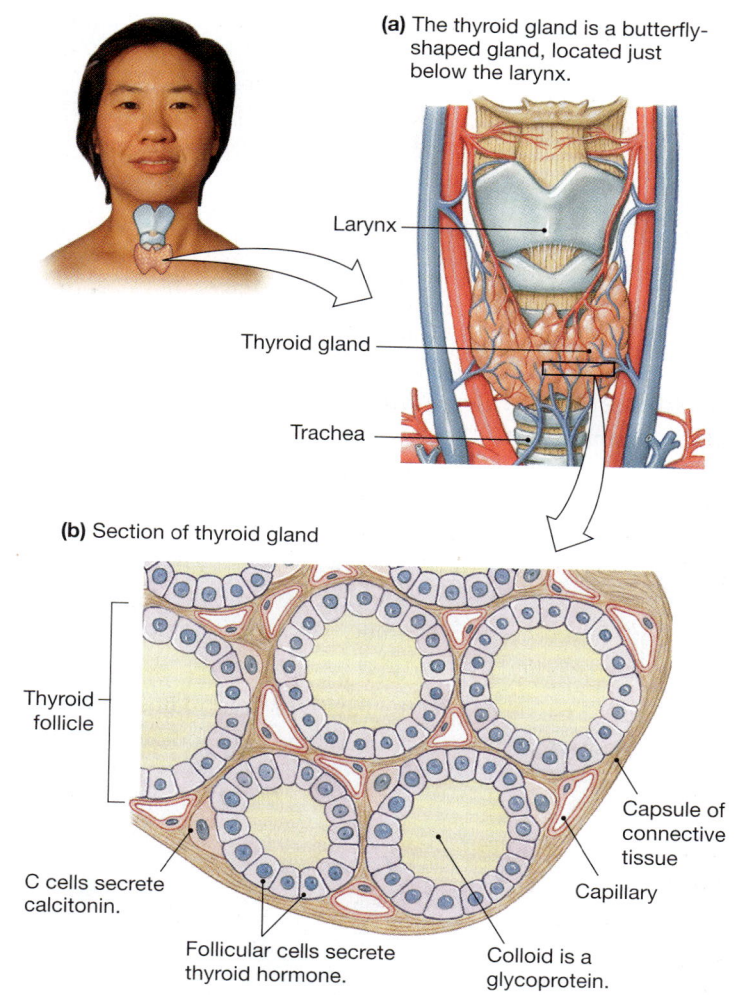

(a) The thyroid gland is a butterfly-shaped gland, located just below the larynx.

Larynx

Thyroid gland

Trachea

(b) Section of thyroid gland

Thyroid follicle

C cells secrete calcitonin.

Follicular cells secrete thyroid hormone.

Colloid is a glycoprotein.

Capsule of connective tissue

Capillary

■ **FIGURE 23-7** *Structure of the thyroid gland*

This endocrine gland secretes both thyroid hormones and calcitonin.

Thyroid Hormones Contain Iodine

Thyroid hormones, like glucocorticoids, have long-term effects on metabolism. Unlike glucocorticoids, however, thyroid hormones are not essential for life. They are essential for normal growth and development in children, however, and infants born with thyroid deficiency will be developmentally delayed unless treated promptly. Because of the importance of thyroid hormones in children, the United States and Canada test all newborns for thyroid deficiency.

Thyroid hormones are amines derived from the amino acid tyrosine, and they are unusual because they contain the element iodine (Fig. 23-8 ■). Currently, thyroid hormones are the only known use for iodine in the body, although a few other tissues also concentrate this mineral.

Synthesis of thyroid hormones takes place in the thyroid **follicles** (also called *acini*), spherical structures whose walls are a single layer of epithelial cells (Fig. 23-7b). The hollow center

of each follicle is filled with a sticky glycoprotein mixture called **colloid**. The colloid holds a 2–3 month supply of thyroid hormones at any one time. Figure 23-9 ■ is an overview of thyroid gland function.

① The follicular cells manufacture enzymes for thyroid-hormone synthesis as well as a glycoprotein called **thyroglobulin**. These proteins are secreted into the center of the follicle.

② Follicular cells absorb dietary iodide, I^-, using the *sodium-iodide symporter* (NIS). I^- movement into the colloid is apparently mediated by an anion transporter.

③ Enzymes combine I^- with tyrosine on the thyroglobulin molecule to make the thyroid hormones **triiodothyronine** (T_3) and **tetraiodothyronine** (T_4, also known as **thyroxine**). (Note the change from *tyrosine* to *thyronine* in the names.) At this point, the hormones are still attached to thyroglobulin.

23

Tyrosine

Thyroxine (T$_4$)

(2 tyrosine + 4 I)

Tyrosine

Triiodothyronine (T$_3$)

(2 tyrosine + 3 I)

■ **FIGURE 23-8**   *Thyroid hormones are made from iodine and tyrosine*

④   The thyroglobulin–T$_3$/T$_4$ complex moves back into the follicular cells by endocytosis.

⑤   Intracellular enzymes free T$_3$ and T$_4$ from the protein.

⑥   Both hormones are lipophilic, so they diffuse out of the follicular cells and into the plasma.

Figure 23-10 ■ depicts the details of thyroid hormone synthesis. The thyroid follicles actively transport iodide across the epithelium. As I$^-$ enters the colloid, the enzyme *thyroid peroxidase* removes an electron from the ion. Other enzymes then add iodine to tyrosine. The addition of one iodine to tyrosine creates **monoiodotyrosine** (MIT); addition of a second iodine creates **diiodotyrosine** (DIT). MIT and DIT then undergo *coupling reactions*. One MIT and one DIT combine to create T$_3$. Two DIT couple to form T$_4$.

Because T$_3$ and T$_4$ are lipophilic molecules, they have limited solubility in plasma. As a result, thyroid hormones bind to plasma proteins, such as **thyroid-binding globulin** (TBG). Most thyroid hormone in the plasma is in the form of T$_4$, and for years it was thought that T$_4$ was the active hormone. However, we now know that T$_3$ is three to five times more active biologically, and that it is the active hormone in target cells.

About 85% of active T$_3$ is made in target cells by enzymatic removal of an iodine from T$_4$. Tissue activation of the hormone adds another layer of control because regulation of the tissue enzyme **deiodinase** allows individual target tissues to control their exposure to active thyroid hormone.

Thyroid receptors, with multiple isoforms, are in the nucleus of target cells. Hormone binding initiates transcription, translation, and synthesis of new proteins.

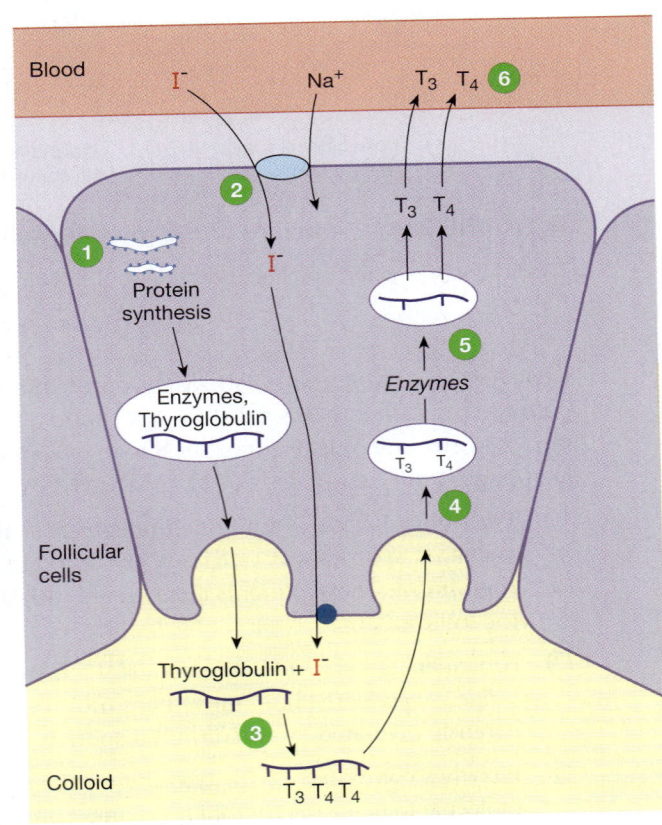

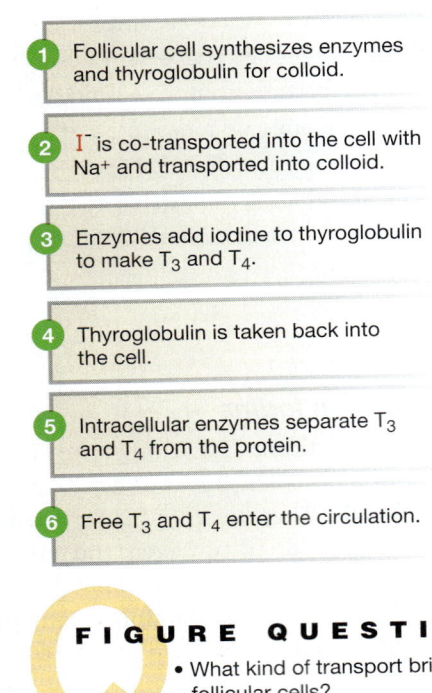

1   Follicular cell synthesizes enzymes and thyroglobulin for colloid.

2   I$^-$ is co-transported into the cell with Na$^+$ and transported into colloid.

3   Enzymes add iodine to thyroglobulin to make T$_3$ and T$_4$.

4   Thyroglobulin is taken back into the cell.

5   Intracellular enzymes separate T$_3$ and T$_4$ from the protein.

6   Free T$_3$ and T$_4$ enter the circulation.

FIGURE QUESTIONS

- What kind of transport brings I$^-$ into follicular cells?
- How does thyroglobulin get into the colloid?
- How does the cell take thyroglobulin back in?
- How do T$_3$ and T$_4$ leave the cell?

■ **FIGURE 23-9**   *Thyroid hormone synthesis takes place in the colloid of the thyroid follicle*

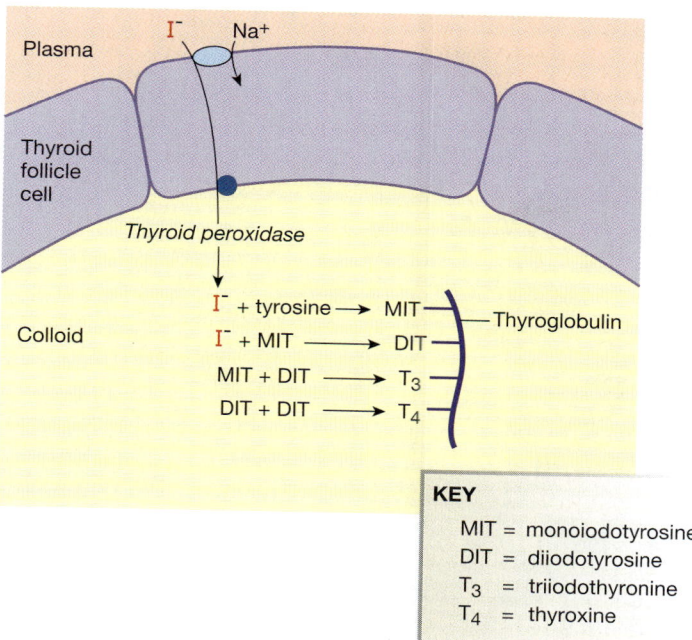

KEY

MIT = monoiodotyrosine
DIT = diiodotyrosine
T_3 = triiodothyronine
T_4 = thyroxine

■ **FIGURE 23-10** *Formation of the thyroid hormones T_3 and T_4*

Iodide is actively concentrated in the thyroid follicle by active transport across the follicle cell membrane. Once in the colloid, iodide combines with the tyrosine to create MIT and DIT. One MIT and one DIT combine to make T_3, and two DITs combine to make T_4. The hormones and their precursors bind to thyroglobulin, a colloid protein.

RUNNING PROBLEM

Elevated blood Ca^{2+} leads to high Ca^{2+} concentrations in the kidney filtrate. Calcium-based kidney stones occur when calcium phosphate or calcium oxalate crystals form and aggregate with organic material in the lumen of the kidney tubule. Once Prof. Magruder's kidney stone passes into the urine, Dr. Spinks sends it for a chemical analysis.

Question 3:
Only free Ca^{2+} in the blood filters into Bowman's capsule at the nephron. A significant portion of plasma Ca^{2+} cannot be filtered. Use what you have learned about filtration at the glomerulus to speculate on why some plasma Ca^{2+} cannot filter [⏩ p. 620].

751 752 **759** 765 771 772

TABLE 23-2	Thyroid Hormones
Cell of origin	Thyroid follicle cells
Chemical nature	Iodinated amine
Biosynthesis	From iodine and tyrosine; formed and stored on parent protein thyroglobulin in follicle colloid
Transport in the circulation	Bound to thyroxine-binding globulin and albumins
Half-life	6–7 days for thyroxine (T_4); about 1 day for triiodothyronine (T_3)
Factors affecting release	Tonic release
Control pathway	TRH (hypothalamus) → TSH (anterior pituitary) → T_3 + T_4 (thyroid) → T_4 deiodinates in tissues to form more T_3
Target cells or tissues	Most cells of the body
Target receptor	Nuclear receptor
Whole body or tissue action	↑ Oxygen consumption (thermogenesis); protein catabolism in adults but anabolism in children; normal development of nervous system
Action at cellular level	Increases activity of metabolic enzymes and Na^+-K^+-ATPase
Action at molecular level	Production of new enzymes
Feedback regulation	T_3 has negative feedback effect on anterior pituitary and hypothalamus

Thyroid Hormones Affect Quality of Life

Thyroid hormones are not essential for life, but they do affect the quality of life. Patients with thyroid excess or deficiency may experience decreased tolerance to heat or cold and mood disturbances, in addition to other symptoms.

The main function of thyroid hormones in adults is to provide substrates for oxidative metabolism (Table 23-2 ■). Thyroid hormones are thermogenic [⏩ p 743] and increase oxygen consumption in most tissues. The exact mechanism is unclear but is at least partly related to changes in ion transport across the cell and mitochondrial membranes. Thyroid hormones also interact with other hormones to modulate protein, carbohydrate, and fat metabolism.

In children, thyroid hormones are necessary for full expression of growth hormone, which means they are essential

23

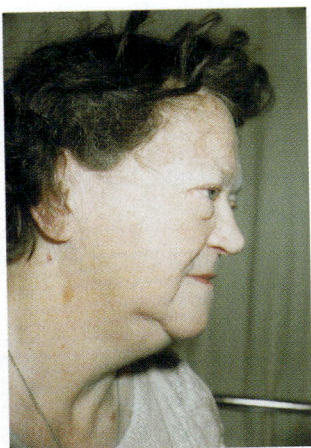

■ FIGURE 23-11   *A woman with myxedema*

In individuals with hypothyroid myxedema, mucopolysaccharide deposits beneath the skin cause bags under the eyes.

for normal growth and development, especially of the nervous system. In the first few years after birth, myelin and synapse formation requires T_3 and T_4. Cytological studies suggest that thyroid hormones regulate microtubule assembly, which is an essential part of neuronal growth.

The actions of thyroid hormones are most observable in people who suffer from either hypersecretion or hyposecretion. Physiological effects that are subtle in normal people often become exaggerated in patients with endocrine disorders.

Hyperthyroidism   A person whose thyroid gland secretes too much hormone suffers from **hyperthyroidism**, which causes changes in metabolism, the nervous system, and the heart.

1. Hyperthyroidism increases oxygen consumption and metabolic heat production. Because of the internal heat generated, these patients have warm, sweaty skin and may complain of being intolerant of heat.
2. Excess thyroid hormone increases protein catabolism and may cause muscle weakness. Patients often report weight loss.
3. The effects of excess thyroid hormone on the nervous system include hyperexcitable reflexes and psychological disturbances ranging from irritability and insomnia to psychosis. The mechanism for psychological disturbances is unclear, but morphological changes in the hippocampus and effects on β-adrenergic receptors have been suggested.
4. Thyroid hormones are known to influence β-adrenergic receptors in the heart, and these effects are exaggerated with hypersecretion. A common sign of hyperthyroidism is rapid heartbeat and increased force of contraction due to up-regulation of β_1-receptors on the myocardium [🔁 p. 491].

Hypothyroidism   Hyposecretion of thyroid hormones affects the same systems altered by hyperthyroidism.

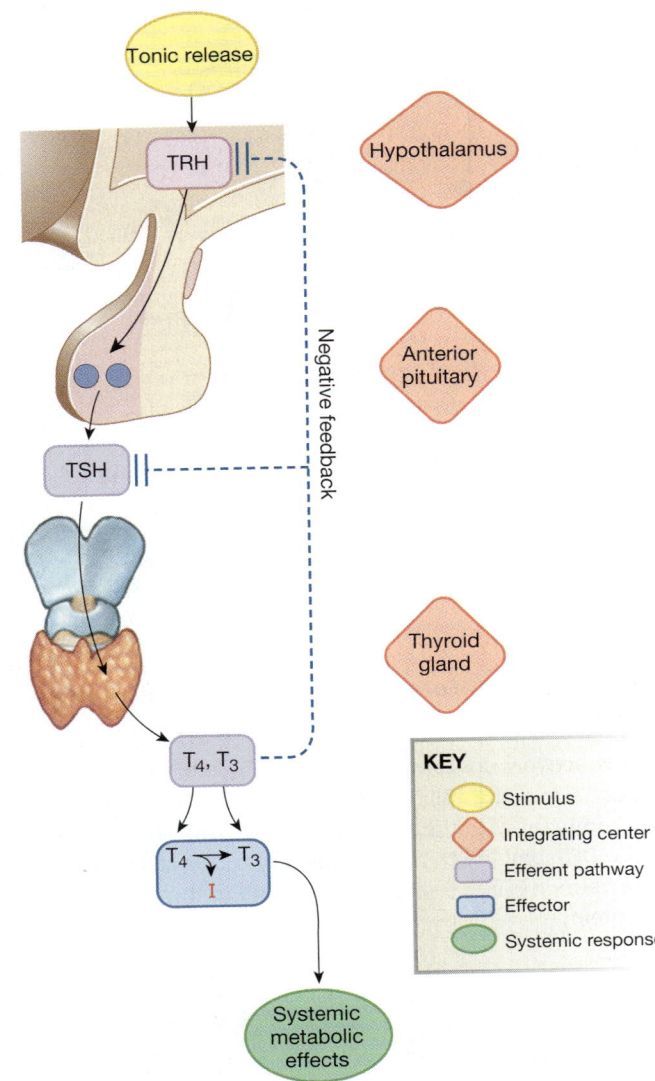

■ FIGURE 23-12   *Pathway of thyroid hormone control*

1. Decreased thyroid hormone secretion slows metabolic rate and oxygen consumption. Patients become intolerant of cold because they are generating less internal heat.
2. Hypothyroidism decreases protein synthesis. In adults, this causes brittle nails, thinning hair, and dry, thin skin. Hypothyroidism also causes accumulation of *mucopolysaccharides* under the skin. These molecules attract water and cause the puffy appearance of *myxedema* (Fig. 23-11 ■). Hypothyroid children have slow bone and tissue growth and are shorter than normal for their age.
3. Nervous system changes in adults include slowed reflexes, slow speech and thought processes, and feelings of fatigue. Deficient thyroid hormone secretion in infancy causes **cretinism**, a condition marked by decreased mental capacity.
4. The primary cardiovascular change in hypothyroidism is bradycardia (slow heart rate [🔁 p. 481]).

TSH Controls the Thyroid Gland

Problems with thyroid hormone secretion can arise either in the thyroid gland or along the control pathway depicted in Figure 23-12 ■. **Thyrotropin-releasing hormone** (TRH) from the hypothalamus controls secretion of the anterior pituitary hormone **thyrotropin**, also known as **thyroid-stimulating hormone** (TSH). This hormone in turns acts on the thyroid gland to promote hormone synthesis. The thyroid hormones normally act as a negative feedback signal to prevent over-secretion.

The trophic action of TSH on the thyroid gland causes enlargement, or *hypertrophy,* of follicular cells. In pathological conditions with elevated TSH levels, the thyroid gland can enlarge to weigh hundreds of grams, forming a huge mass called a **goiter**. A large goiter can almost encircle the neck and will extend downward behind the sternum (Fig. 23-13 ■).

Goiters are the result of excess TSH stimulation of the thyroid gland. Simply knowing that someone has a goiter does not tell you what the pathology is, however. Let's see how both hypothyroidism and hyperthyroidism can be associated with goiter.

Our first example is *primary hypothyroidism* [🔁 p. 232], caused by a lack of iodine in the diet. Without iodine, the thyroid gland cannot make thyroid hormones (Fig. 23-14a ■). Low levels of T_3 and T_4 in the blood remove negative feedback on the hypothalamus and anterior pituitary. TSH secretion rises dramatically, and TSH stimulation enlarges the thyroid gland (goiter). Despite hypertrophy, the gland cannot obtain iodine to make hormone, so the patient remains hypothyroid. These patients exhibit the previously described signs of hypothyroidism.

An enlarged thyroid gland may also be indicative of hypersecretion. In *Graves' disease,* the body produces antibodies called **thyroid-stimulating immunoglobulins,** or **TSI** (Fig. 23-14b). These antibodies mimic the action of TSH by combining with and activating TSH receptors on the thyroid gland, enlarging the gland (goiter) and overstimulating thyroid hormone production. Negative feedback shuts down the body's TRH and TSH secretion but does nothing to block the TSH-like activity of TSI on the thyroid gland. The result is goiter, hypersecretion of T_3 and T_4, and symptoms of hormone excess. Graves' disease is often accompanied by **exophthalmus** (Fig. 23-15 ■), a bug-eyed appearance caused by immune-mediated enlargement of muscles and tissue in the eye socket.

Therapy for thyroid disorders depends on the cause of the problem. Hypothyroidism is treated with oral thyroxine (T_4). Hyperthyroidism can be treated by surgical removal of all or part of the gland, by destruction of thyroid cells with radioactive iodine, or by drugs that block either hormone synthesis (thiourea drugs) or peripheral conversion of T_4 to T_3 (propylthiouracil).

■ **FIGURE 23-13**   *A man with goiter due to excessive TSH stimulation*

CONCEPT CHECK

9. A woman who had her thyroid gland removed because of cancer was given pills containing only T_4. Why was this less-active form of the hormone an effective treatment for her hypothyroidism?

10. Why would excessive production of thyroid hormone, which uncouples mitochondrial ATP production and proton transport, cause a person to become intolerant of heat?

Answers: p. 775

GROWTH HORMONE

Growth in human beings is a continuous process that begins before birth. However, growth rates in children are not steady, with the first two years of life and the adolescent years marked by spurts of rapid growth and development. Normal growth is a complex process that depends on a number of factors:

1. **Growth hormone and other hormones**. Without adequate amounts of growth hormone, children simply will not grow. Thyroid hormones, insulin, and the sex hormones at puberty also play both direct and permissive roles. A deficiency in any one of these hormones will lead to abnormal growth and development.

2. **An adequate diet** that includes protein, sufficient energy (caloric intake), vitamins, and minerals. Many amino acids can be manufactured in the body from other precursors,

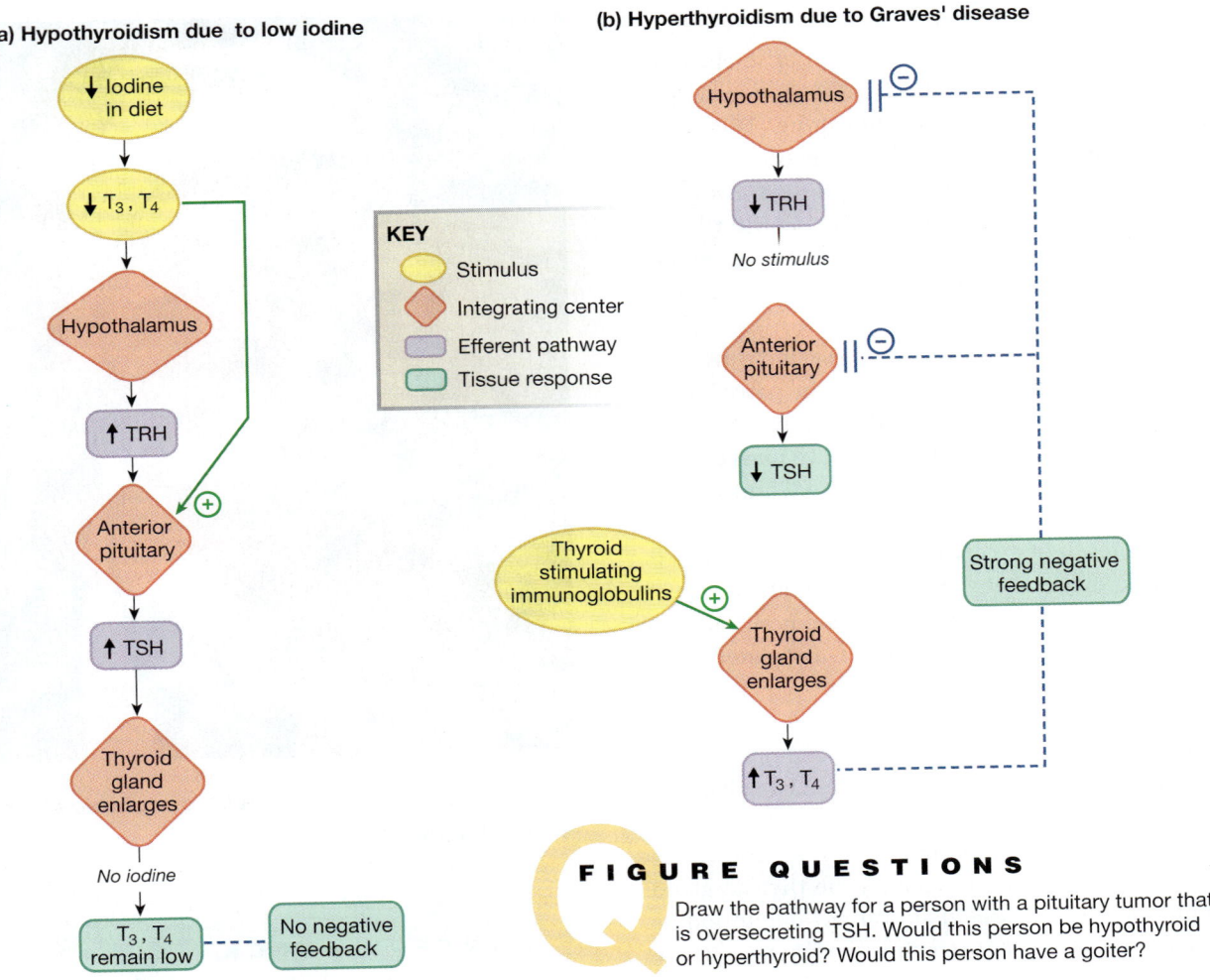

(a) Hypothyroidism due to low iodine

(b) Hyperthyroidism due to Graves' disease

KEY
- Stimulus
- Integrating center
- Efferent pathway
- Tissue response

FIGURE QUESTIONS

Draw the pathway for a person with a pituitary tumor that is oversecreting TSH. Would this person be hypothyroid or hyperthyroid? Would this person have a goiter?

■ **FIGURE 23-14   *Goiter can occur in both hyperthyroidism and hypothyroidism***

(a) In hypothyroidism caused by iodine deficiency, absence of negative feedback increases TSH secretion and results in goiter. (b) In Graves' disease, thyroid-stimulating immune proteins (TSI) bind to thyroid gland TSH receptors and cause the gland to hypertrophy.

but the essential amino acids [🔁 p. 30] must come from dietary sources. Among the minerals, calcium in particular is needed for proper bone formation.

3. **Absence of stress**. Cortisol from the adrenal cortex is released in times of stress and has significant catabolic effects that inhibit growth. Children who are subjected to stressful environments may exhibit a condition known as *failure to thrive* that is marked by abnormally slow growth.

4. **Genetics**. Each human's potential adult size is genetically determined at conception.

Growth Hormone Is Anabolic

Growth hormone (GH or somatotropin [🔁 p. 215]) is released throughout life, although its biggest role is in children. Peak GH secretion occurs during the teenage years. The stimuli for growth hormone release are complex and not well understood,

but they include circulating nutrients, stress, and other hormones interacting with a daily rhythm of secretion (Fig. 23-16 ■).

The stimuli for GH secretion are integrated in the hypothalamus, which secretes two neuropeptides into the hypothalamic-hypophyseal portal system: **growth hormone–releasing hormone** (GHRH) and *growth hormone–inhibiting hormone*, better known as *somatostatin* [🔁 p. 227]. On a daily basis, pulses of GHRH from the hypothalamus stimulate GH release. In adults, the largest pulse of GH release occurs in the first two hours of sleep. It is speculated that GHRH has sleep-inducing properties, but the role of GH in sleep cycles is unclear.

GH is secreted by cells in the anterior pituitary. It is a typical peptide hormone in most respects, except that nearly half the GH in blood is bound to a plasma **growth hormone binding protein**. The binding protein protects plasma GH from being filtered into the urine and extends its half-life by

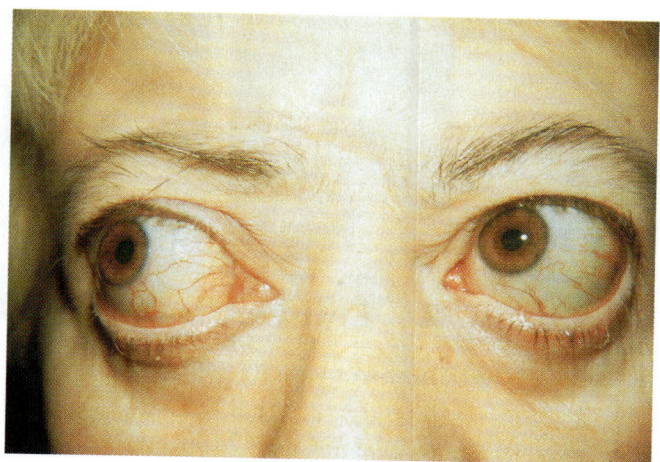

■ FIGURE 23-15 *Exophthalmus*

TABLE 23-3   Growth Hormone

Cell of origin	Anterior pituitary
Chemical nature	191-amino acid peptide; several closely related forms
Biosynthesis	Typical peptide
Transport in the circulation	Half is dissolved in plasma, half is bound to a binding protein whose structure is identical to that of the GH receptor
Half-life	18 minutes
Factors affecting release	Circadian rhythm of tonic secretion; influenced by circulating nutrients, stress, and other hormones in a complex fashion
Target cells or tissues	Trophic on liver for insulin-like growth factor production; also acts directly on many cells
Control pathway	GHRH, somatostatin (hypothalamus); → growth hormone (anterior pituitary)
Target receptor	Membrane receptor with tyrosine kinase activity
Whole body or tissue action (with IGFs)	Bone and cartilage growth; soft tissue growth; ↑ plasma [glucose]
Action at cellular level	Protein synthesis
Action at molecular level	Receptor linked to kinases that phosphorylate proteins to initiate transcription

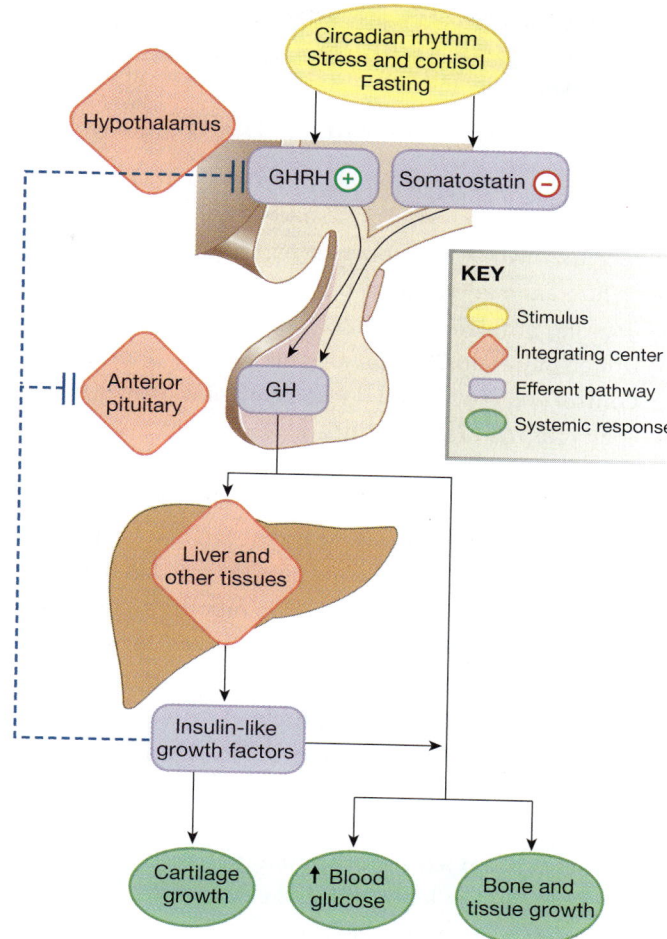

■ FIGURE 23-16 *Pathway of growth hormone control*

12 minutes. Researchers have hypothesized that genetic determination of binding protein concentration plays a role in determining adult height.

The target tissues for GH include both endocrine and nonendocrine cells. GH acts as a trophic hormone to stimulate secretion of **insulin-like growth factors** (IGFs; formerly called *somatomedins*) from the liver and other tissues. IGFs have a negative feedback effect on growth hormone secretion, by acting on the anterior pituitary and on the hypothalamus. IGFs act in concert with growth hormone to stimulate bone and soft tissue growth (Table 23-3 ■).

Metabolically, growth hormone and IGFs are anabolic for proteins and promote protein synthesis, an essential part of tissue growth. Growth hormone also acts with IGFs to stimulate bone growth. IGFs are responsible for cartilage growth. GH increases plasma fatty acid and glucose concentrations by promoting fat breakdown and hepatic glucose output.

CONCEPT CHECK

11. Which pituitary hormone in addition to GH has two hypothalamic factors that regulate its release?

Answers: p. 775

23

CLINICAL FOCUS

NEW GROWTH CHARTS

When you were growing up, did your parents mark your growth each year on a special wall chart? Monitoring growth is an important part of health care for children and adolescents, particularly as we see a growing problem with childhood obesity in the United States. In 2000 the U.S. Centers for Disease Control and Prevention (CDC) issued new growth charts for the first time since 1977. The old charts were based on data gathered from 1929 to 1979 that had surveyed mostly bottle-fed, middle-class white children. We now know that breast-fed and bottle-fed infants grow at different rates: breast-fed babies grow more rapidly in the first two months, then more slowly for the remainder of the first year. There are also socioeconomic differences in growth; babies in lower socioeconomic groups grow more slowly. The new charts take these differences into account. They include body mass index (BMI) information up to age 20 (see Chapter 22, question 30 on p. 748). To see the new charts and learn more about monitoring growth in infants and children, visit the CDC website at *www.cdc.gov/growthcharts*.

■ **FIGURE 23-17**　*Three individuals with acromegaly*
Excess growth hormone secretion in adults causes lengthening of the jaw, coarsening of the features, and growth in hands and feet.

Growth Hormone Is Essential for Normal Growth in Children

The disorders that reflect the actions of growth hormone are most obvious in children. Severe growth hormone deficiency in childhood leads to **dwarfism**, which can result from a problem either with growth hormone synthesis or with defective GH receptors. Unfortunately, neither bovine nor porcine growth hormone is effective as replacement therapy, as only primate growth hormone is active in humans. Prior to 1985, when genetically engineered human growth hormone became available, donated human pituitaries harvested at autopsy were the only source of growth hormone. Fortunately, severe growth hormone deficiency is relatively rare. At the opposite extreme, oversecretion of growth hormone in children leads to **giantism**.

Once bone growth stops in late adolescence, growth hormone cannot further increase height. GH and IGFs can continue to act on cartilage and soft tissues, however. Adults with excessive secretion of growth hormone develop a condition known as **acromegaly**, characterized by lengthening of the jaw, coarsening of facial features, and growth of hands and feet

(Fig. 23-17 ■). Andre the Giant, a French wrestler who also had a role in the classic movie *The Princess Bride*, exhibited signs of both giantism (he grew to 7′4″ tall) and acromegaly before his death at age 47.

Genetically Engineered Human Growth Hormone Raises Ethical Questions

When genetically engineered human growth hormone became available in the mid-1980s, the medical profession was faced with a dilemma. Obviously the hormone should be used to treat children who would otherwise be dwarfs, but what about children with only partial GH deficiency or genetically short children with normal GH secretion levels? This question is complicated by the difficulty of accurately identifying children with partial growth hormone deficiency. And what about children whose parents want them to be taller for athletics? Should these healthy children be given the hormone?

In 2003 the U.S. Food and Drug Administration approved use of a **recombinant human growth hormone** (hGH) for treating children with non-GH-deficient short stature, defined as being more than 2.25 standard deviations below the mean height for their age and sex. (This means children in the bottom 1% of their age-sex group.) In clinical trials, daily injections of the drug for two years resulted in an average height increase of 1.3″ (3.3 cm). At an average cost of $22,000 per year for hGH injections, this comes out to more than $33,000 per inch of height gained. Side effects reported during hGH studies include glucose intolerance [🔗 p. 738] and *pancreatitis* (inflammation of the pancreas). Long-term risks associated with hGH treatment are unknown, and parents must be made aware that hGH therapy has the potential to create psychological problems in children if the results are less than optimum.

TISSUE AND BONE GROWTH

Growth can be divided into two general areas: soft tissue growth and bone growth. In children, bone growth is usually assessed by measuring height, and tissue growth by measuring

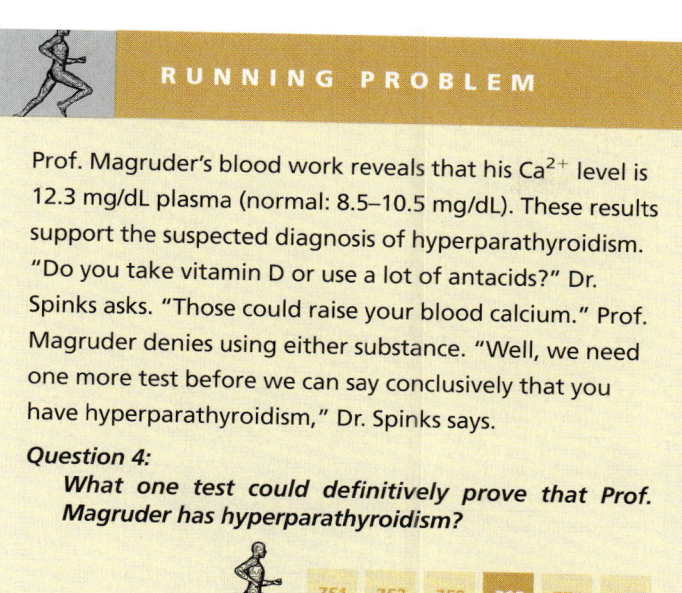

RUNNING PROBLEM

Prof. Magruder's blood work reveals that his Ca^{2+} level is 12.3 mg/dL plasma (normal: 8.5–10.5 mg/dL). These results support the suspected diagnosis of hyperparathyroidism. "Do you take vitamin D or use a lot of antacids?" Dr. Spinks asks. "Those could raise your blood calcium." Prof. Magruder denies using either substance. "Well, we need one more test before we can say conclusively that you have hyperparathyroidism," Dr. Spinks says.

Question 4:
 What one test could definitively prove that Prof. Magruder has hyperparathyroidism?

751   752   759   765   771   772

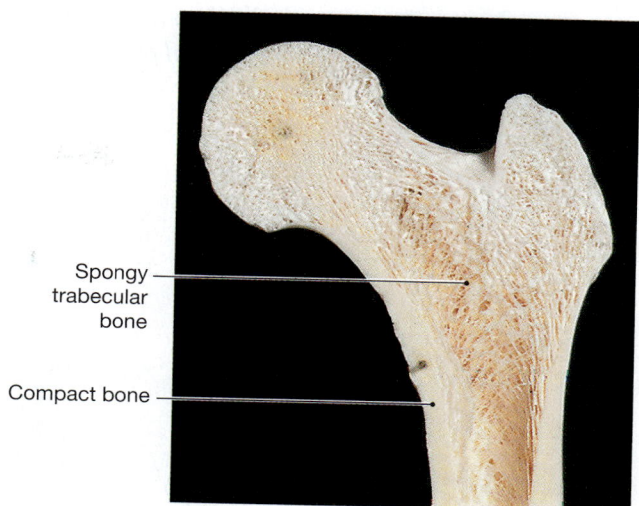

■ **FIGURE 23-18   *Compact and trabecular bone***

weight. Multiple hormones have direct or permissive effects on growth. In addition, we are just beginning to understand how paracrine growth factors interact with classic hormones to influence tissue development and differentiation.

Tissue Growth Requires Hormones and Paracrines

Soft tissue growth requires adequate amounts of growth hormone, thyroid hormone, and insulin. Growth hormone and IGFs are required for tissue protein synthesis and cell division. Under the influence of these hormones, cells undergo both hypertrophy (increased cell size) and **hyperplasia** (increased cell number).

Thyroid hormones play a permissive role in growth and contribute directly to nervous system development. At the target tissue level, thyroid hormone interacts synergistically with growth hormone in protein synthesis and nervous system development. Children with untreated hypothyroidism (cretinism) will not grow to normal height even if they secrete normal amounts of growth hormone.

Insulin supports tissue growth by stimulating protein synthesis and providing energy in the form of glucose. Because insulin is permissive for growth hormone, insulin-deficient children will fail to grow normally even though they may have normal concentrations of growth and thyroid hormones.

Bone Growth Requires Adequate Dietary Calcium

Bone growth, like soft tissue development, requires the proper hormones and adequate amounts of protein and calcium. Bone contains calcified extracellular matrix formed when calcium phosphate crystals precipitate and attach to a collagenous lattice support. The most common form of calcium phosphate is **hydroxyapatite**, $Ca_{10}(PO_4)_6(OH)_2$.

Although the large amount of inorganic matrix in bone makes some people think of it as nonliving, bone is a dynamic tissue, constantly being formed and broken down (*resorbed*). Spaces in the collagen-calcium matrix are occupied by living cells that are well supplied with oxygen and nutrients by blood vessels that run through adjacent channels.

Bones generally have two layers: an outer layer of dense **compact bone** and an inner layer of spongy **trabecular bone** (Fig. 23-18 ■). In some bones, a central cavity is filled with bone marrow [⌇ *Focus on Bone Marrow*, p. 543]. Compact bone provides strength and is thickest where support is needed (such as in the long bones of the legs) or where muscles attach. Trabecular bone is less sturdy and has open, cell-filled spaces between struts of calcified lattice.

Bone growth occurs when matrix is deposited faster than it is resorbed. Specialized bone-forming cells called **osteoblasts** produce enzymes and **osteoid**, a mixture of collagen and other proteins to which hydroxyapatite binds. Recent research has found two other proteins, *osteocalcin* and *osteonectin*, that appear to aid in deposition of the calcified matrix.

Bone diameter increases when matrix deposits on the outer surface of the bone. Linear growth of long bones occurs at specialized regions called **epiphyseal plates**, located at each end of the bone shaft (*diaphysis*) (Fig. 23-19 ■). The side of the plate closer to the end (*epiphysis*) of the bone contains continuously dividing columns of **chondrocytes**, collagen-producing cells of cartilage. As the collagen layer thickens, the older cartilage calcifies and older chondrocytes degenerate, leaving spaces that osteoblasts invade. The osteoblasts then lay down bone matrix on top of the cartilage base. As new bone is added at the ends, the shaft lengthens. Long bone growth continues as long as the epiphyseal plate is active. When osteoblasts complete their work, they revert to a less active form known as **osteocytes**.

23

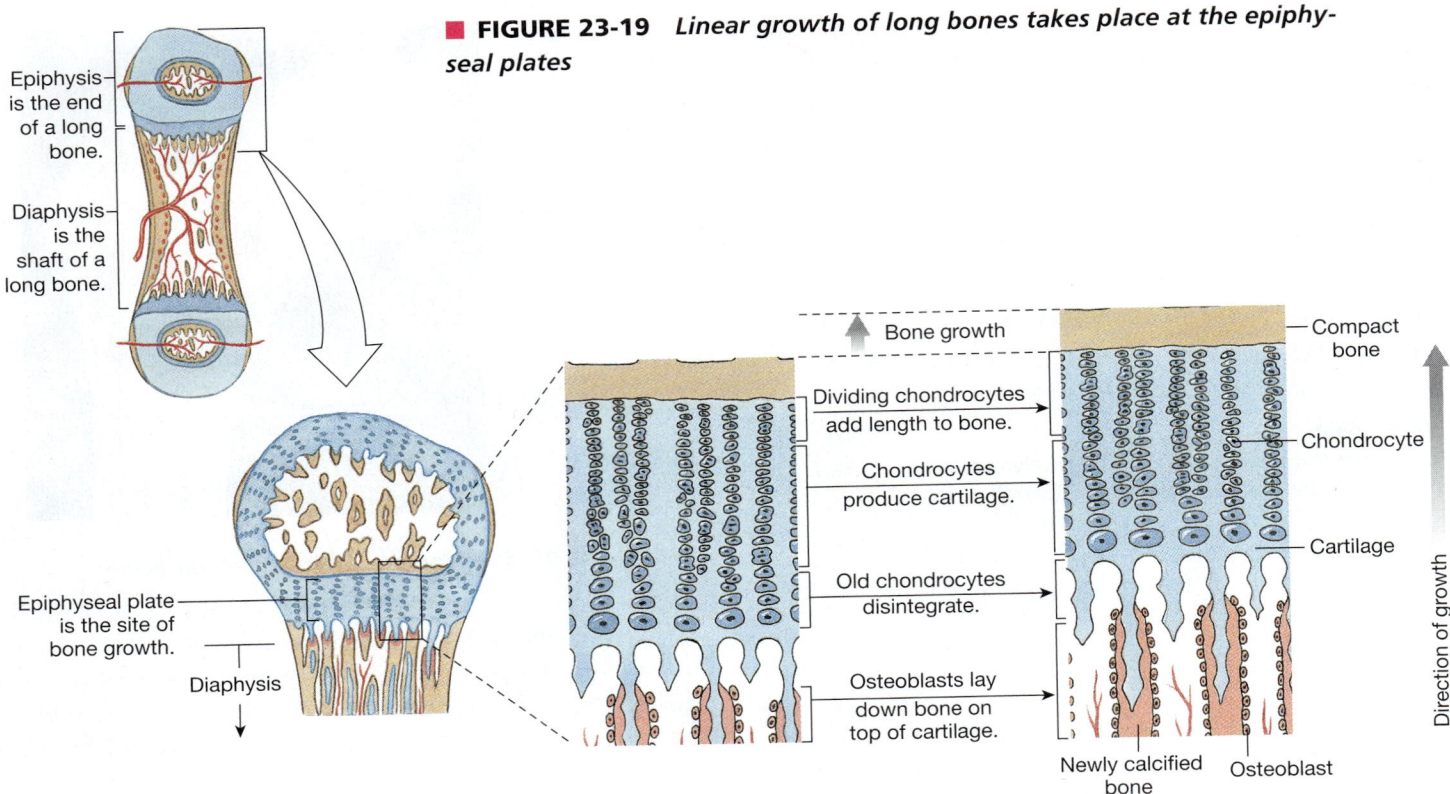

FIGURE 23-19 *Linear growth of long bones takes place at the epiphyseal plates*

Epiphysis is the end of a long bone.

Diaphysis is the shaft of a long bone.

Epiphyseal plate is the site of bone growth.

Diaphysis

Bone growth

Dividing chondrocytes add length to bone.

Chondrocytes produce cartilage.

Old chondrocytes disintegrate.

Osteoblasts lay down bone on top of cartilage.

Compact bone

Chondrocyte

Cartilage

Newly calcified bone

Osteoblast

Direction of growth

Growth of long bone is under the influence of growth hormone and the insulin-like growth factors. In the absence of these hormones, normal bone growth will not occur. Long bone growth is also influenced by steroid sex hormones. The growth spurt of adolescent boys is attributed to increased androgen production. For girls, the picture is less clear-cut. In experiments, estrogens both stimulate and inhibit linear growth.

In all adolescents, the sex hormones eventually inactivate the epiphyseal plate so that long bones no longer grow. Because the epiphyseal plates of various bones close in a regular, ordered sequence, X-rays that show which plates are open and which have closed can be used to calculate a child's "bone age."

Linear bone growth ceases in adults, but bones are dynamic tissues that undergo continual remodeling throughout life under the control of hormones that regulate calcium metabolism in the body.

CONCEPT CHECK

12. Which hormones are essential for normal growth and development?
13. Why don't adults with growth hormone hypersecretion grow taller?

Answers: p. 775

CALCIUM BALANCE

Most calcium in the body—99%, or nearly 2.5 pounds—is found in the bones. This pool is relatively stable, however, and it is the body's small fraction of nonbone calcium that is most

critical to physiological functioning (Table 23-4 ■). As you have learned, Ca^{2+} has several physiological functions:

1. **Ca^{2+} is an important signal molecule**. The movement of Ca^{2+} from one body compartment to another creates Ca^{2+} signals. Calcium entering the cytoplasm initiates exocytosis of synaptic and secretory vesicles, contraction in muscle fibers, or altered activity of enzymes and transporters. Removal of Ca^{2+} from the cytoplasm requires active transport.

2. **Ca^{2+} is part of the intercellular cement that holds cells together at tight junctions**.

3. **Ca^{2+} is a cofactor in the coagulation cascade** [🔁 p. 551]. Although Ca^{2+} is essential for blood coagulation, body Ca^{2+} concentrations never decrease to the point at which coagulation is inhibited. However, removal of Ca^{2+} from a blood sample will prevent the specimen from clotting in the test tube.

4. **Plasma Ca^{2+} concentrations affect the excitability of neurons**. This function of Ca^{2+} has not been introduced before in this text, but it is the function that is most obvious in Ca^{2+}-related disorders. If plasma Ca^{2+} falls too low (**hypocalcemia**), neuronal permeability to Na^{+} increases, neurons depolarize, and the nervous system becomes hyperexcitable. In its most extreme form, hypocalcemia causes sustained contraction (*tetany*) of the respiratory muscles, resulting in asphyxiation. **Hypercalcemia** has the opposite effect, causing depressed neuromuscular activity.

23

TABLE 23-4	Functions of Calcium in the Body	
COMPARTMENT	**PERCENTAGE OF BODY CALCIUM**	**FUNCTION**
Extracellular matrix	99%	Calcified matrix of bone
Extracellular fluid	0.1%	"Cement" for tight junctions; role in myocardial and smooth muscle contraction; neurotransmitter release at synapses; role in excitability of neurons; cofactor in coagulation cascade
Intracellular	0.9%	Signal in second messenger pathways; role in muscle contraction

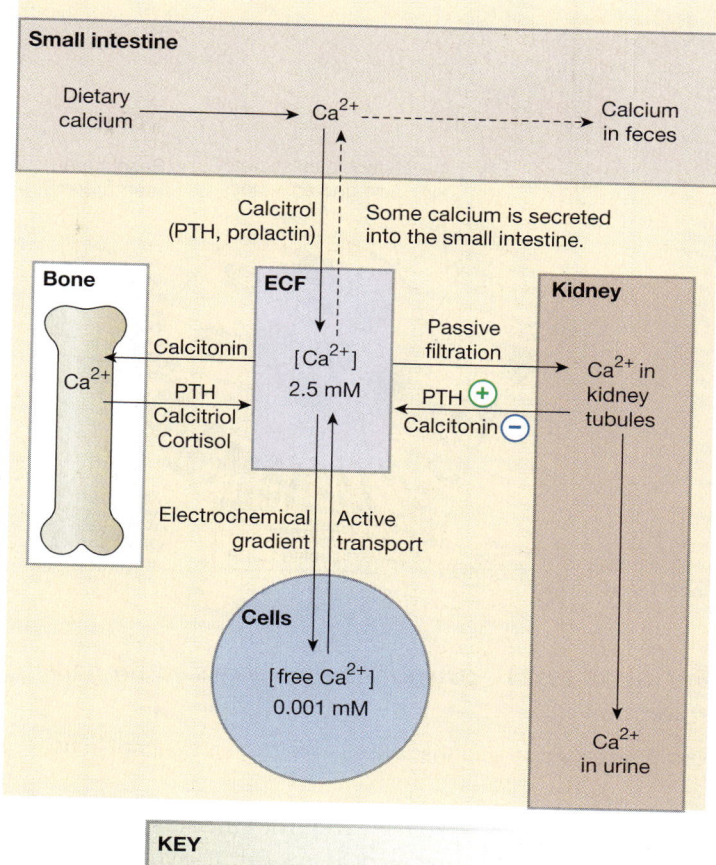

KEY

PTH = parathyroid hormone

■ **FIGURE 23-20** *Calcium balance in the body*

Calcium Concentrations in the Blood Are Closely Regulated

Because calcium is critical to so many physiological functions, the body's plasma Ca^{2+} concentration is very closely regulated. Calcium homeostasis follows the principle of mass balance:

$$\text{Total body calcium} = \text{intake} - \text{output}$$

1. **Total body Ca^{2+}** is all the calcium in the body, distributed among three compartments (Fig. 23-20 ■):
   (a) *Extracellular fluid.* Ionized Ca^{2+} is concentrated in the ECF, with an average Ca^{2+} concentration of 2.5 mM. In the plasma, nearly half the Ca^{2+} is bound to plasma proteins and other molecules. The unbound Ca^{2+} is free to diffuse across membranes through open Ca^{2+} channels.
   (b) *Intracellular Ca^{2+}.* The concentration of free Ca^{2+} in the cytosol is about 0.001 mM. In addition, Ca^{2+} is concentrated inside mitochondria and the sarcoplasmic reticulum. Electrochemical gradients favor movement of Ca^{2+} into the cytosol when Ca^{2+} channels open.
   (c) *Extracellular matrix (bone).* Bone is the largest Ca^{2+} reservoir in the body; most bone Ca^{2+} is in the form of hydroxyapatite crystals. Bone Ca^{2+} forms a reservoir that can be tapped to maintain plasma Ca^{2+} homeostasis. Usually only a small fraction of bone Ca^{2+} is ionized and readily exchangeable, and this pool remains in equilibrium with Ca^{2+} in the interstitial fluid.

2. **Intake** is the Ca^{2+} ingested in the diet and absorbed in the small intestine. Only about one-third of ingested Ca^{2+} is

absorbed, and unlike organic nutrients, Ca^{2+} absorption is hormonally regulated. Most Americans do not eat enough Ca^{2+}-containing foods, however, and intake may not match output.

3. **Output**, or Ca^{2+} loss from the body, occurs primarily through the kidneys, with a small amount excreted in feces. Ionized Ca^{2+} is freely filtered at the glomerulus and then reabsorbed along the length of the nephron. Hormonally regulated reabsorption takes place only in the distal nephron. In that part of the tubule, reabsorption is accomplished by ion movement through an apical Ca^{2+} channel (*ECaC*) and through basolateral Na^{+}-Ca^{2+} antiport and Ca^{2+}-ATPase transporters.

Osteoblasts control the deposition of Ca^{2+} in bone, as previously discussed. The movement of Ca^{2+} out of hydroxyapatite and into the ionized Ca^{2+} pool is controlled by **osteoclasts**, large, mobile, multinucleate cells derived from hematopoietic stem cells [⊟ p. 538]. Osteoclasts are responsible for dissolving bone.

Osteoclasts attach around their periphery to a section of matrix, much like a suction cup (Fig. 23-21 ■). The central

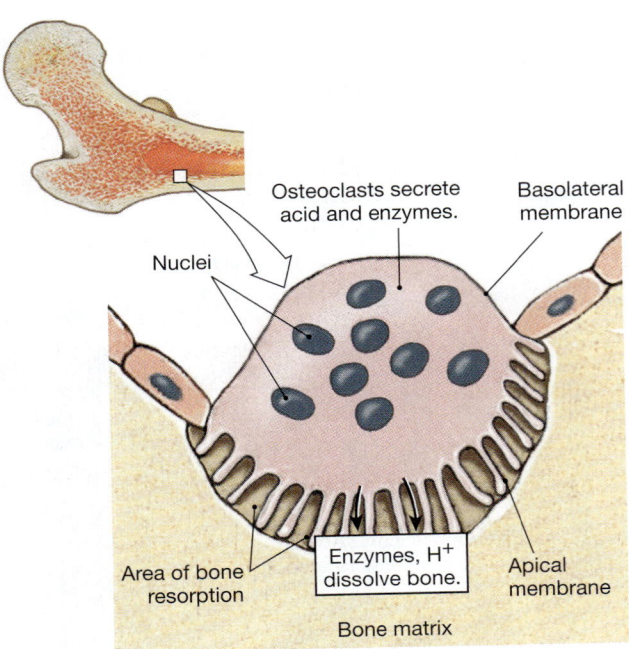

■ **FIGURE 23-21**   *Osteoclasts are responsible for bone resorption*

Osteoclasts secrete acid that dissolves calcium phosphate in bone.

region of the osteoclast secretes acid (with the aid of H^+-ATPase) and *protease* enzymes that work at low pH. The combination of acid and enzymes dissolves the calcified matrix and its collagen support, freeing Ca^{2+}, which can enter the blood.

CONCEPT CHECK

14. What does hypercalcemia do to neuronal membrane potential, and why does that effect depress neuromuscular excitability?

15. Draw a picture of a distal tubule cell and label apical and basolateral membranes, lumen, and ECF. Use the description of distal tubule Ca^{2+} reabsorption in list item 3 (Output) to draw the appropriate transporters.

16. Describe the transport of Ca^{2+} from the tubule cell to the ECF as active, passive, facilitated diffusion, and so on.

Answers: p. 775

Three Hormones Control Calcium Balance

Three hormones regulate the movement of Ca^{2+} between bone, kidney, and intestine: parathyroid hormone, calcitriol (vitamin D_3), and calcitonin (Fig. 23-20). Of these, parathyroid hormone and calcitriol are the most important in adult humans.

Four small **parathyroid glands** in the neck (Fig. 23-22 ■) secrete hormone in response to decreases in plasma Ca^{2+}. The parathyroid glands are found on the dorsal surface of the thyroid gland, and their significance was first recognized in the 1890s by physiologists studying the role of the thyroid gland. These scientists noticed that if they removed all of the thyroid gland from dogs and cats, the animals died in a few days. In

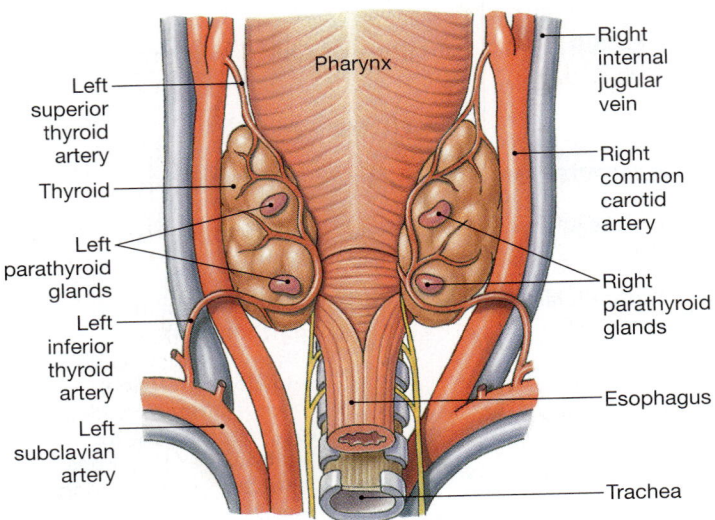

The four parathyroid glands lie hidden behind the thyroid gland.

■ **FIGURE 23-22**   *Parathyroid glands*

This illustration shows the dorsal side of the thyroid gland. (Compare with the ventral view in Fig. 23-7a.)

contrast, rabbits died only if the little parathyroid "glandules" alongside the thyroid were removed. The scientists then looked for parathyroid glands in dogs and cats and found them tucked away behind the larger thyroid gland. If the parathyroids were left behind when the thyroid was surgically removed, the animals lived. The scientists concluded that the parathyroids contained a substance that was essential for life, although the thyroid gland did not.

Parathyroid Hormone   That essential substance is **parathyroid hormone** (PTH, also called *parathormone*), a peptide whose main effect is to increase plasma Ca^{2+} concentrations (Table 23-5 ■). The stimulus for PTH release is a decrease in plasma Ca^{2+}, monitored by a cell membrane **Ca^{2+}-sensing receptor** (CaSR). The CaSR, a G protein-coupled receptor, was the first membrane receptor identified whose ligand was an ion rather than an organic molecule.

By acting on bone, kidney, and intestine, PTH raises plasma Ca^{2+} concentrations. Increased plasma Ca^{2+} acts as negative feedback and shuts off PTH secretion. Parathyroid hormone raises plasma Ca^{2+} three ways:

1. **PTH mobilizes calcium from bone.** Increased bone resorption by osteoclasts takes about 12 hours to become measurable. Curiously, although osteoclasts are responsible for dissolving the calcified matrix and would be logical targets for PTH, they do not have PTH receptors. Instead, PTH effects are mediated by a collection of paracrines, including *osteoprotegerin* (OPG) and an osteoclast differentiation factor called *RANKL*. These paracrine factors are receiving intense scrutiny as potential pharmacological agents.

TABLE 23-5	**Parathyroid Hormone (PTH)**
Cell of origin	Parathyroid glands
Chemical nature	84-amino acid peptide
Biosynthesis	Continuous production, little stored
Transport in the circulation	Dissolved in plasma
Half-life	Less than 20 minutes
Factors affecting release	↓ Plasma Ca^{2+}
Target cells or tissues	Kidney, bone, intestine
Target receptor	Membrane receptor acts via cAMP
Whole body or tissue action	↑ Plasma Ca^{2+}
Action at cellular level	↑ Vitamin D synthesis; ↑ renal reabsorption of Ca^{2+}; ↑ bone resorption
Action at molecular level	Rapidly alters Ca^{2+} transport but also initiates protein synthesis in osteoclasts
Onset of action	2–3 hours for bone, with increased osteoclast activity requiring 1–2 hours; 1–2 days for intestinal absorption; within minutes for kidney transport
Feedback regulation	Negative feedback by ↑ plasma Ca^{2+}
Other information	Osteoclasts have no PTH receptors, so are affected by PTH-induced paracrines. PTH is essential for life; absence causes hypocalcemic tetany.

TABLE 23-6	**Vitamin D_3 (Calcitriol, 1,25-Dihydroxycholecalciferol)**
Cell of origin	Complex biosynthesis; see below
Chemical nature	Steroid
Biosynthesis	Vitamin D formed by sunlight on precursor molecules or ingested in food; converted in two steps (liver and kidney) to $1,25(OH)_2D_3$
Transport in the circulation	Bound to plasma proteins
Stimulus for synthesis	↓ Ca^{2+}; indirectly via PTH; prolactin also stimulates synthesis
Target cells or tissues	Intestine, bone, and kidney
Target receptor	Nuclear
Whole body or tissue action	↑ Plasma Ca^{2+}
Action at molecular level	Stimulates production of calbindin, a Ca^{2+}-binding protein, and of CaSR in parathyroid gland; associated with intestinal transport by unknown mechanism
Feedback regulation	Plasma Ca^{2+} shuts off PTH secretion

2. **PTH enhances renal reabsorption of calcium.** As previously mentioned, regulated Ca^{2+} reabsorption takes place in the distal nephron. PTH simultaneously enhances renal excretion of phosphate by reducing its reabsorption. The opposing effects of PTH on calcium and phosphate are needed to keep their combined concentrations below a critical level. If the concentrations exceed that level, calcium phosphate crystals form and precipitate out of solution. High concentrations of calcium phosphate in the urine are one cause of kidney stones. Additional aspects of phosphate homeostasis are discussed below.

3. **PTH indirectly increases intestinal absorption of calcium** through its influence on vitamin D_3, a process described below.

Calcitriol Intestinal absorption of calcium is enhanced by the action of a hormone known as **1,25-dihydroxycholecalciferol** ($1,25(OH)_2D_3$), also known as **calcitriol** or vitamin D_3 (Table 23-6 ■). The body makes calcitriol from vitamin D that has been obtained through diet or made in the skin by the action of sunlight on precursors made from acetyl CoA (Fig. 23-23 ■). Vitamin D is modified in two steps—first in the liver, then in the kidneys—to make calcitriol.

Calcitriol reinforces the plasma Ca^{2+}-increasing effect of PTH by enhancing Ca^{2+} uptake from the small intestine. In addition, calcitriol facilitates renal reabsorption of Ca^{2+} and helps mobilize Ca^{2+} out of bone.

The production of calcitriol is regulated at the kidney by the action of PTH. Decreased plasma Ca^{2+} increases PTH secretion, which stimulates calcitriol synthesis. Intestinal and renal absorption of Ca^{2+} raises blood Ca^{2+}, turning off PTH in a negative feedback loop that decreases calcitriol synthesis.

23

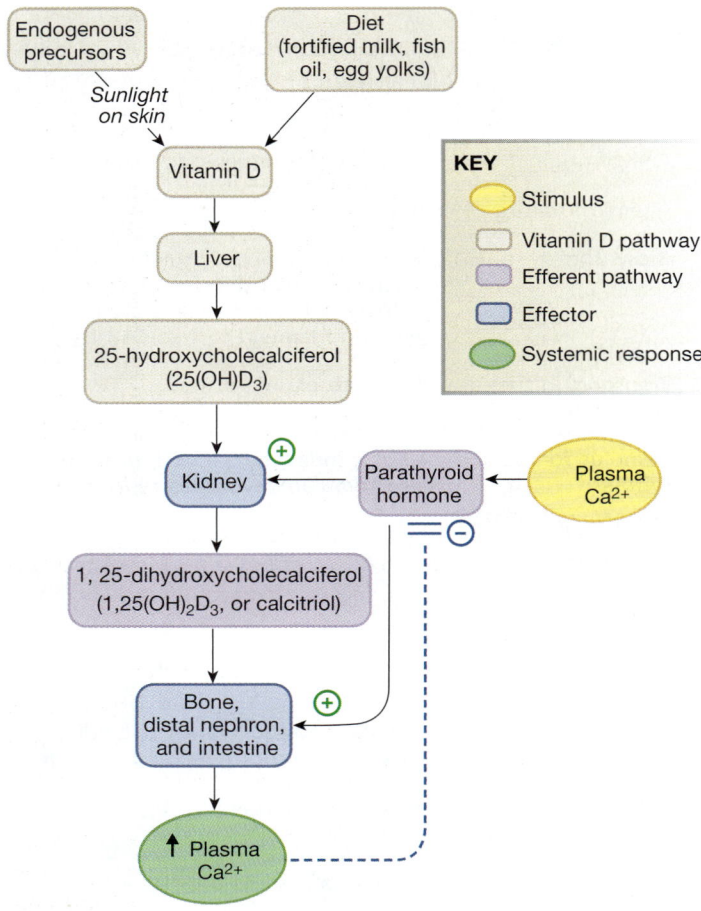

■ FIGURE 23-23 *Endocrine control of calcium balance*

PTH works with calcitriol to promote bone resorption, intestinal Ca^{2+} absorption, and distal nephron Ca^{2+} reabsorption, all of which tend to elevate plasma Ca^{2+} concentrations.

TABLE 23-7	Calcitonin
Cell of origin	C cells of thyroid gland (parafollicular cells)
Chemical nature	32-amino acid peptide
Biosynthesis	Typical peptide
Transport in the circulation	Dissolved in plasma
Half-life	<10 minutes
Factors affecting release	↑ Plasma [Ca^{2+}]
Target cells or tissues	Bone and kidney
Target receptor	G protein-coupled membrane receptor
Whole body or tissue action	Prevents bone resorption; enhances kidney excretion
Action at molecular level	Signal transduction pathways appear to vary during cell cycle
Other information	Experimentally decreases plasma [Ca^{2+}] but has little apparent physiological effect in adult humans; possible effect on skeletal development; possible protection of bone Ca^{2+} stores during pregnancy and lactation

Prolactin, the hormone responsible for milk production in breast-feeding (lactating) women, also stimulates calcitriol synthesis. This action ensures maximal absorption of Ca^{2+} from the diet at a time when metabolic demands for calcium are high.

Calcitonin   The third hormone involved with calcium metabolism is **calcitonin**, a peptide produced by the C cells of the thyroid gland (Table 23-7 ■). Its actions are opposite those of parathyroid hormone. Calcitonin is released when plasma Ca^{2+} increases. Experiments in animals have shown that calcitonin decreases bone resorption and increases renal calcium excretion.

Calcitonin apparently plays only a minor role in daily calcium balance in adult humans. Patients whose thyroid glands have been removed show no disturbance in calcium balance. People with thyroid tumors that secrete large amounts of calcitonin also show no ill effects.

Calcitonin has been used to treat patients with *Paget's disease,* a genetically linked condition in which osteoclasts are

overactive and bone is weakened by resorption. Calcitonin in these patients stabilizes the abnormal bone loss, leading scientists to speculate that this hormone is most important during childhood growth, when net bone deposition is needed, and during pregnancy and lactation, when the mother's body must supply calcium for both herself and her child.

Calcium and Phosphate Homeostasis Are Linked

Phosphate homeostasis is closely linked to calcium homeostasis. Phosphate is the second key ingredient in the hydroxyapatite of bone, $Ca_{10}(PO_4)_6(OH)_2$, and most phosphate in the body is found in bone. However, phosphates have other significant physiological roles, including energy transfer and storage in high-energy phosphate bonds, and activation or deactivation of enzymes, transporters, and ion channels. Phosphates also form part of the DNA and RNA backbone.

Phosphate homeostasis parallels that of Ca^{2+}. Phosphate is absorbed in the intestines, filtered and reabsorbed in the kidneys, and divided between bone, ECF, and intracellular compartments. Vitamin D_3 enhances intestinal absorption of phosphate.

The results of Prof. Magruder's last test confirm that he has hyperparathyroidism. He goes on a low-calcium diet, avoiding milk, cheese, and other dairy products, but several months later he returns to the emergency room with another painful kidney stone. Dr. Spinks sends him to an endocrinologist, who recommends surgical removal of the overactive parathyroid glands. "We can't tell which of the parathyroid glands is most active," the specialist says, "and we'd like to leave you with some parathyroid hormone of your own. So I will take out all four glands, but we'll reimplant two of them in the muscle of your forearm. In many patients, the implanted glands secrete just enough PTH to maintain calcium homeostasis. And if they secrete too much PTH, it is much easier to take them out of your arm than do major surgery on your neck again."

Question 5:

Why can't Prof. Magruder simply take replacement PTH by mouth? (Hint: PTH is a peptide hormone.)

| 751 | 752 | 759 | 765 | 771 | 772 |

Renal excretion is affected by both PTH (which promotes phosphate excretion) and vitamin D_3 (which promotes phosphate reabsorption).

CONCEPT CHECK

17. Name two compounds that store energy in high-energy phosphate bonds.
18. What are the differences between a kinase, a phosphatase, and a phosphorylase?

Answers: p. 775

Osteoporosis Is a Disease of Bone Loss

One of the best-known pathologies of bone function is **osteoporosis**, a metabolic disorder in which bone resorption exceeds bone deposition. The result is fragile, weakened bones that are more easily fractured (Fig. 23-24 ■). Most bone resorption takes place in spongy trabecular bone, particularly in the vertebrae, hips, and wrists.

Osteoporosis is most common in women after menopause, when estrogen concentrations fall. However, older men also develop osteoporosis. Bone loss and small fractures and compression in the spinal column lead to *kyphosis* [hump-back], the

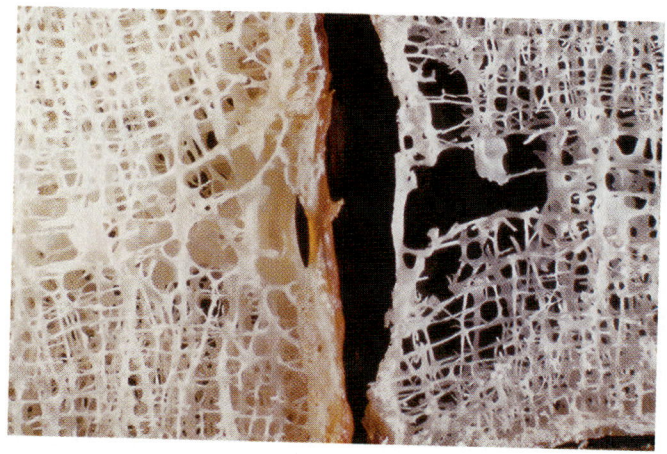

■ **FIGURE 23-24** *Normal bone (left) and bone loss in osteoporosis (right)*

These photographs dramatically illustrate why people with osteoporosis have a high incidence of bone fractures.

stooped, hunchback appearance that is characteristic of advanced osteoporosis in the elderly. Osteoporosis is a complex disease with genetic and environmental components. Risk factors include small, thin body type, postmenopausal age, smoking, and low dietary Ca^{2+} intake.

For many years estrogen or estrogen/progesterone *hormone replacement therapy* (HRT) was used to prevent osteoporosis. However, estrogen therapy alone increases the risk of endometrial and possibly other cancers, and recent studies suggested that estrogen/progesterone HRT might increase risk of heart attacks and strokes. A *selective estrogen receptor modulator (SERM)* called raloxifene has been used to treat osteoporosis, but it is not as effective as drugs that act more directly on bone metabolism. The latter groups include *bisphosphonates*, which induce osteoclast apoptosis and suppress bone resorption, and *teriparatide*, a PTH derivative, which stimulates formation of new bone. Teriparatide consists of the first 34 amino acids of the 84-amino acid PTH molecule and must be injected rather than taken orally. Currently clinical studies are investigating whether some combination of bisphosphonates and teriparatide is more effective in combating osteoporosis than either drug alone.

To avoid osteoporosis in later years, young women need to maintain adequate dietary calcium intake and perform weight-bearing exercises, which increase bone density. Loss of bone mass begins by age 30, long before most women start to think they are at risk, and many women suffer from low bone mass (*osteopenia*) long before they are aware of a problem. Bone mass testing can help with early diagnosis of osteopenia.

RUNNING PROBLEM CONCLUSION

HYPERPARATHYROIDISM

Prof. Magruder had the surgery, and the implanted glands produced an adequate amount of PTH. He must have his plasma Ca^{2+} levels checked regularly for the rest of his life to ensure that the glands continue to function adequately.

To learn more about hyperparathyroidism, try a Google search.

Check your understanding of this running problem by comparing your answers with the information in the summary table.

QUESTION	FACTS	INTEGRATION AND ANALYSIS
1 What role does Ca^{2+} play in the normal functioning of muscles and neurons?	Calcium triggers neurotransmitter release [p. 271] and uncovers the myosin-binding sites on muscle actin filaments [p. 406].	Muscle weakness in hyperparathyroidism is the opposite of what you would predict from knowing the role of Ca^{2+} in muscles and neurons. However, calcium also affects the Na^+ permeability of neurons, and it is this effect that leads to muscle weakness and CNS effects.
2 What is the technical term for "elevated levels of calcium in the blood"?	Prefix for elevated levels: *hyper-*; suffix for "in the blood": *-emia*.	*Hypercalcemia* is the technical term for elevated levels of calcium in the blood.
3 Speculate on why some plasma Ca^{2+} cannot filter into Bowman's capsule.	Filtration at the glomerulus is a selective process that excludes blood cells and most plasma proteins [p. 620].	A significant amount of plasma Ca^{2+} is bound to plasma proteins and therefore cannot filter.
4 What one test could definitively prove that Prof. Magruder has hyperparathyroidism?	Hyperparathyroidism is a condition in which excessive amounts of PTH are secreted.	A test for the amount of PTH in the blood would confirm the diagnosis of hyperparathyroidism.
5 Why can't Prof. Magruder simply take replacement PTH by mouth?	PTH is a peptide hormone.	Ingested peptides will be digested by proteolytic enzymes. Thus PTH taken orally will not be absorbed intact into the body and consequently will not be effective.

751 752 759 765 771 **772**

CHAPTER SUMMARY

Endocrinology is based on the physiological principles of *homeostasis* and *control systems*. Each hormone has stimuli that initiate its secretion, and feedback signals that modulate its release. *Molecular interactions* and *communication across membranes* are also essential to hormone activity. In many instances, such as calcium and phosphate homeostasis, the principle of *mass balance* is the focus of homeostatic regulation.

Review of Endocrine Principles

1. Basic components of endocrine pathways include hormone receptors, feedback loops, and cellular responses. (p. 751)

Adrenal Glucocorticoids

2. The **adrenal cortex** secretes **glucocorticoids**, sex steroids, and aldosterone. (pp. 751–752; Figs. 23-1, 23-2)

3. **Cortisol** secretion is controlled by hypothalamic **CRH** and **ACTH** from the pituitary. Cortisol is the feedback signal. Cortisol is a typical steroid hormone in its synthesis, secretion, transport, and action. (p. 753; Fig. 23-3)

4. Cortisol is catabolic and essential for life. It promotes gluconeogenesis, breakdown of skeletal muscle proteins and adipose tissue, Ca^{2+} excretion, and suppression of the immune system. (p. 754; Tbl. 23-1)

5. **Hypercortisolism** usually results from a tumor or therapeutic administration of the hormone. **Addison's disease** is hyposecretion of all adrenal steroids. (pp. 755–756)

6. CRH and the **melanocortins** have physiological actions in addition to cortisol release. (p. 756; Fig. 23-6)

Thyroid Hormones

7. The thyroid **follicle** has a hollow center filled with **colloid** containing **thyroglobulin** and enzymes. (p. 757; Fig. 23-7)

8. Thyroid hormones are made from tyrosine and iodine. **Tetraiodothyronine** (thyroxine, T_4) is converted in target tissues to the more active hormone **triiodothyronine** (T_3). (p. 757; Figs. 23-8, 23-9, 23-10)

9. Thyroid hormones are not essential for life, but they influence metabolic rate as well as protein, carbohydrate, and fat metabolism. (p. 759; Tbl. 23-2)

10. Thyroid hormone secretion is controlled by **thyrotropin** (thyroid-stimulating hormone, TSH) and **thyrotropin-releasing hormone** (TRH). (p. 761; Figs. 23-12, 23-14)

Growth Hormone

11. Normal growth requires growth hormone, thyroid hormones, insulin, and sex hormones at puberty. Growth also requires adequate diet and absence of stress. (pp. 761–762)

12. Growth hormone is secreted by the anterior pituitary and stimulates secretion of **insulin-like growth factors** (IGFs) from the liver and other tissues. These hormones promote bone and soft tissue growth. (p. 763; Tbl. 23-3)

13. Secretion of growth hormone is controlled by **growth hormone–releasing hormone** (GHRH) and growth hormone–inhibiting hormone (**somatostatin**). (p. 762; Fig. 23-16)

Tissue and Bone Growth

14. **Bone** is composed of **hydroxyapatite** crystals attached to a collagenous support. Bone is a dynamic tissue with living cells. (p. 765; Fig. 23-18)

15. **Osteoblasts** synthesize bone. Long bone growth occurs at **epiphyseal plates**, where **chondrocytes** produce cartilage. (p. 765; Fig. 23-19)

Calcium Balance

IP **Fluids & Electrolytes: Electrolyte Homeostasis**

16. Calcium acts as an intracellular signal for second messenger pathways, exocytosis, and muscle contraction. It also plays a role in cell junctions, coagulation, and neural function (p. 766; Tbl. 23-4)

17. Ca^{2+} homeostasis balances dietary intake, urinary output, and distribution of Ca^{2+} among bone, cells, and the ECF. (p. 767; Fig. 23-20)

18. Decreased plasma Ca^{2+} stimulates **parathyroid hormone** (PTH) secretion by the **parathyroid glands**. (p. 768; Tbl. 23-5)

19. PTH promotes Ca^{2+} resorption from bone, enhances renal Ca^{2+} reabsorption, and increases intestinal Ca^{2+} absorption through its effect on **calcitriol**. (p. 769; Figs. 23-20, 23-23; Tbl. 23-6)

20. Calcitonin from the thyroid gland plays only a minor role in daily calcium balance in adult humans. (p. 770; Tbl. 23-7)

QUESTIONS

(Answers to the Review Questions begin on page A1.)

THE PHYSIOLOGY PLACE

Access more review material online at **The Physiology Place** website. There you'll find review questions, problem-solving activities, case studies, flashcards, and direct links to both *InterActive Physiology®* and PhysioEx™. To access the site, go to *www.physiologyplace.com* and select Human Physiology, Fourth Edition.

LEVEL ONE REVIEWING FACTS AND TERMS

1. Name the zones of the adrenal cortex and the primary hormones secreted in each zone.

2. For (a) cortisol, (b) growth hormone, (c) parathyroid hormone, and (d) T_3 and T_4, draw the full control pathway and show feedback where appropriate. Do not use abbreviations.

3. List four conditions that are necessary for people to achieve their full growth. Include five specific hormones known to exert an effect on growth.

4. Name the thyroid hormones. Which one has the highest activity? How and where is most of it produced?

5. Define each of the following terms and explain its physiological significance:
   (a) melanocortins
   (b) osteoporosis
   (c) hydroxyapatite
   (d) mineralocorticoid
   (e) trabecular bone
   (f) POMC
   (g) epiphyseal plates

6. List seven functions of calcium in the body.

7. Make a table showing the effects of cortisol, thyroid hormones, growth hormone, insulin, and glucagon on protein, carbohydrate, and lipid metabolism.

LEVEL TWO REVIEWING CONCEPTS

8. Mapping exercise: Create a reflex map for each of the following situations:
   (a) Hypercortisolism from an adrenal tumor
   (b) Hypercortisolism from a pituitary tumor
   (c) Hyperthyroidism from a hormone-secreting thyroid tumor
   (d) Hypothyroidism from a pituitary tumor that decreases TSH synthesis

9. Define, compare, and contrast or relate the terms in each set:
   (a) cortisol, glucocorticoids, ACTH, CRH
   (b) thyroid, C cell, follicle, colloid
   (c) thyroglobulin, tyrosine, iodide, TBG, deiodinase, TSH, TRH
   (d) somatotropin, IGF, GHRH, somatostatin, growth hormone binding protein
   (e) giantism, acromegaly, dwarfism
   (f) hyperplasia, hypertrophy
   (g) osteoblast, osteoclast, chondrocyte, osteocyte
   (h) vitamin D, calcitriol, 1,25-dihydroxycholecalciferol, calcitonin, estrogen, PTH

10. Based on what you know about the cellular mechanism of action for T_3, would you expect to see tissue response to this hormone within a few minutes or in more than an hour?

11. If average plasma $[Ca^{2+}]$ is 2.5 mmol/L, what is the concentration in mEq/L?

12. Osteoclasts make acid (H^+) from CO_2 and H_2O. They secrete the acid at their apical membrane and put bicarbonate into the ECF. Draw an osteoclast and diagram this process, including enzymes and the appropriate transporters on each membrane. How many different transporters can you think of that could be used to reabsorb bicarbonate?

LEVEL THREE   PROBLEM SOLVING

13. Diabetic patients who have surgery, become sick, or are under other physiological stress are told to monitor their blood sugar carefully because they may need to increase their insulin dose temporarily. What is the physiological explanation behind this advice?

14. One diagnostic test to determine the cause of hypercortisolism is a *dexamethasone suppression test*. Dexamethasone blocks secretion of ACTH by the pituitary. The following table shows the results from two patients given a dexamethasone suppression test.

	Plasma Cortisol Before Test	Plasma Cortisol After Test
Patient A	High	High
Patient B	High	Low

Can you tell from these results where the patients' pathologies originate? Explain for each patient.

15. When blood test results came back last week, someone in the office spilled a cup of coffee on them, smearing the patient names and some of the numbers. One report shows elevated TSH levels, but the thyroid levels are unreadable. You have three charts waiting for test results on thyroid hormone levels. Your tentative diagnoses, based on physical findings and symptoms, for those three patients are:

Mr. A: primary hypothyroidism
Ms. B: primary hyperthyroidism
Ms. C: secondary hyperthyroidism

(a) Can you tell whose results are on the smeared report, based on the TSH results and the tentative diagnosis?

(b) Can you rule out any of the three people based on those same criteria? Explain.

16. The following graph shows the results of a study done in Boston that compared blood vitamin D levels during summer and winter. Boston is located at 42 degrees north latitude, and weak sunlight in winter there does not allow skin synthesis of vitamin D. (Data from *Am. J. Med.* 112:659–662, 2002 Jun 1)

(a) Summarize the results shown in the graph. How many variables are shown in the graph that you must address in your summary?

(b) Based on what you know, how could you explain the results of the study?

(c) Would taking a multivitamin supplement affect the results?

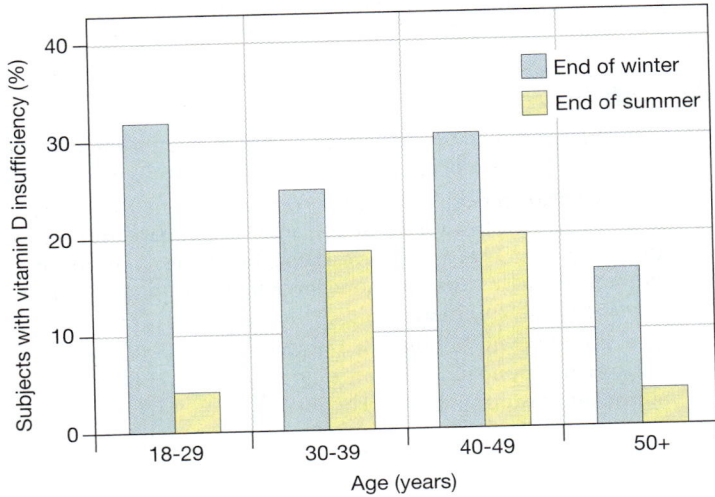

LEVEL FOUR   QUANTITATIVE PROBLEMS

17. Filterable plasma Ca^{2+} is about 5 mg/L. Assume a person has a GFR of 125 mL of plasma filtered per minute.

(a) How much Ca^{2+} will this person filter in a day?

(b) Net dietary Ca^{2+} intake is 170 mg/day. To remain in Ca^{2+} balance, how much Ca^{2+} must he excrete?

(c) What percentage of filtered Ca^{2+} is reabsorbed by the kidney tubule?

ANSWERS

✓ Answers to Concept Check Questions

Page 753

1. The medulla secretes catecholamines (epinephrine, norepinephrine), and the cortex secretes aldosterone, glucocorticoids, and sex hormones.

2. Androstenedione is a prohormone for testosterone. Testosterone is anabolic for skeletal muscle, which might give an athlete a strength advantage.

Page 755

3. ACTH = adrenocorticotropic hormone or corticotropin. CRH = corticotropin-releasing hormone. HPA = hypothalamic-pituitary-adrenal. CBG = corticosteroid-binding globulin or transcortin.

4. This immediate stress response is too rapid to be mediated by cortisol and must be a fight-or-flight response mediated by the sympathetic nervous system and catecholamines.

5. No, because cortisol is catabolic on muscle proteins.

Page 756

6. Primary and iatrogenic hypercortisolism: ACTH is lower than normal due to negative feedback. Secondary hypercortisolism: ACTH is higher because of the ACTH-secreting tumor.

7. Addison's disease: high ACTH due to reduced corticosteroid production and lack of negative feedback.

Page 756

8. ACTH is secreted during stress. If the stress is a physical one caused by an injury, the endogenous opioid β-endorphin can decrease pain and help the person continue functioning.

Page 761

9. In peripheral tissues T_4 is converted to T_3, which is the more active form of the hormone.

10. When mitochondria are uncoupled, energy normally captured in ATP is released as heat. This raises the person's body temperature and causes heat intolerance.

Page 763

11. Prolactin also has two hypothalamic factors that regulate its release [🔁 227].

Page 766

12. Normal growth and development require growth hormone, thyroid hormone, insulin, and insulin-like growth factors.

13. Adults who hypersecrete growth hormone do not grow taller because their epiphyseal plates have closed.

Page 768

14. Hypercalcemia hyperpolarizes the membrane potential, which makes it harder for the neuron to fire an action potential.

15. The figure should resemble Figure 20-13. Instead of a Na^+ channel on the apical membrane, the figure should have a Ca^{2+} channel that allows Ca^{2+} into the cell. The basolateral side should show the ATPase moving Ca^{2+} from cytoplasm to interstitial fluid, and a transporter showing Na^+ entering the cell in exchange for Ca^{2+} leaving.

16. The Na^+-Ca^{2+} antiport performs secondary active transport, and the Ca^{2+}-ATPase performs active transport.

Page 771

17. ATP and phosphocreatine store energy in high-energy phosphate bonds.

18. A kinase transfers a phosphate group from one substrate to another. A phosphatase removes a phosphate group, and a phosphorylase adds one.

Answers to Figure Questions

Page 753

Figure 23-2: (a) A baby born with deficient 21-hydroxylase would have low aldosterone and cortisol levels and an excess of sex steroids, particularly androgens. Low cortisol would decrease the child's ability to respond to stress. Excess androgen would cause masculinization in female infants. (b) Women, who synthesize more estrogens, would have more aromatase.

Page 758

Figure 23-9: I^- comes into the cell by secondary active transport (co-transport with Na^+). Thyroglobulin moves between colloid and cytoplasm by exocytosis and endocytosis. Thyroid hormones leave the cell by simple diffusion.

Page 762

Figure 23-14: A pituitary tumor hypersecreting TSH would cause hyperthyroidism and an enlarged thyroid gland. The pathway would show decreased TRH resulting from short-loop negative feedback from TSH to the hypothalamus, increased TSH caused by the tumor, elevated thyroid hormones, but no negative feedback from the thyroid hormones to the anterior pituitary because the tumor does not respond to feedback signals.

23

24

Although, at first sight, the immune system may appear to be autonomous, it is connected by innumerable structural and functional bridges with the nervous system and the endocrine system, so as to constitute a multisystem.

—Branislav D. Jankovic, *Neuroimmunomodulation: The State of the Art, 1994*

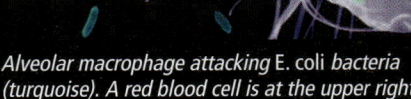

Alveolar macrophage attacking E. coli *bacteria (turquoise). A red blood cell is at the upper right.*

The Immune System

Overview of Immune System Function

Pathogens of the Human Body

The Immune Response

Anatomy of the Immune System

Innate Immunity: Nonspecific Responses

Acquired Immunity: Antigen-Specific Responses

Immune Response Pathways

Neuro-Endocrine-Immune Interactions

BACKGROUND BASICS

specific invaders. The immune system serves three major functions:

RUNNING PROBLEM

TREATMENT FOR AIDS

In 1996, researchers developed a strategy that might offer long-term suppression of HIV, the virus that causes acquired immunodeficiency syndrome (AIDS). In that year, scientists confirmed that a "cocktail" consisting of several drugs, including a *protease inhibitor,* could be effective against HIV, but questions remained. For how long would the new drug combination keep HIV at bay? Would the new drugs work for everyone carrying the infection? George, 28, pushed these questions to the back of his mind. After three years of relatively good health following his positive test for HIV infection, his T lymphocyte count had begun to slide. It was only a matter of time before George became gravely ill.

777   779   793   798   801   803

It seems that since the 1980s human civilization has been visited by new plagues of infectious diseases. Some of them, such as tuberculosis and rabies, are old enemies that we thought we had under control. Others, such as SARS (severe acute respiratory syndrome), monkeypox, and the Hanta and Ebola viruses, seem to be cropping up out of nowhere. The conflict between humans and the invaders that cause disease is literally a fight for survival on both sides. The infectious agents need a host such as the human body to reproduce. But if they kill off all the hosts, they, too, will die out. On the human side of the equation, we must fight off the invaders if humans are to continue to survive as a species.

The body's ability to protect itself from viruses, bacteria, and other disease-causing entities is known as **immunity**, from the Latin word *immunis,* meaning *exempt.* The human immune system consists of the lymphoid tissues of the body, the immune cells, and chemicals (both intracellular and secreted) that coordinate and execute immune functions.

OVERVIEW OF IMMUNE SYSTEM FUNCTION

The key features of the immune system are *specificity* and *memory.* Together these processes enable the body to distinguish "self" from "non-self" and mount a targeted response to specific invaders. The immune system serves three major functions:

1. **It protects the body from disease-causing invaders** known as **pathogens.** Microorganisms (**microbes**) that act as pathogens include bacteria, viruses, fungi, and one-celled protozoans. Larger pathogens include multicellular **parasites,** such as hookworms and tapeworms.

   In addition, virtually any exogenous molecule or cell has the potential to elicit an immune response. Pollens, chemicals, and foreign bodies are examples of substances to which the body may react. Substances that trigger the body's immune response and can react with products of that response are known as **antigens.**

2. **It removes dead or damaged cells,** as well as old red blood cells. Scavenger cells of the immune system patrol the extracellular compartment, gobbling up and digesting dead or dying cells.

3. **It tries to recognize and remove abnormal cells** created when normal cell growth and development go wrong. For example, the diseases we call *cancer* result from abnormal cells that multiply uncontrollably, crowding out normal cells and disrupting function. Scientists believe that cancer cells form on a regular basis but are usually detected and destroyed by the immune system before they get out of control.

Sometimes the body's immune system fails to perform its normal functions. Pathologies of the immune system generally fall into one of three categories: incorrect responses, overactive responses, or lack of response.

1. *Incorrect responses.* If mechanisms for distinguishing self from non-self fail and the immune system attacks the body's normal cells, an *autoimmune disease* results. Type 1 diabetes mellitus, in which proteins made by immune cells destroy pancreatic beta cells, is an example of an autoimmune disease in humans.

2. *Overactive responses.* Allergies are conditions in which the immune system creates a response that is out of proportion to the threat posed by the antigen. In extreme cases, the systemic effects of allergic responses can be life threatening.

3. *Lack of response.* **Immunodeficiency diseases** arise when some component of the immune system fails to work properly. *Primary immunodeficiency* is a family of genetically inherited disorders that range from severe to mild. *Acquired immunodeficiencies* may occur as a result of infection, such as *acquired immunodeficiency syndrome* (AIDS) caused by the *human immunodeficiency virus* (HIV). Acquired immunodeficiencies may also arise as a side effect of drug or radiation therapy, such as those used to treat cancer.

24

PATHOGENS OF THE HUMAN BODY

In the United States, our most prevalent infectious diseases are of viral and bacterial origin. Worldwide, parasites are an additional significant public health concern. For example, malaria, a pathogenic protozoan whose life cycle alternates between human and mosquito hosts, is estimated to infect as many as 100 million people.

Many parasitic organisms, such as the malaria protozoan, are introduced into the body by biting insects. Others enter the body through the digestive tract, brought in by contaminated food and water. Some, such as the fungi that cause valley fever and histoplasmosis, are inhaled. A few, such as the blood fluke *Schistosoma*, burrow through the host's skin. Once in the body, microbes and parasites may enter host cells in an effort to evade the immune response, or they may remain in the extracellular compartment.

Bacteria and Viruses Require Different Defense Mechanisms

Bacteria and viruses differ from each other in several ways (Table 24-1 ■):

1. **Structure**. Bacteria are cells, with a cell membrane that is usually surrounded by a cell wall. Some *encapsulated* bacteria also produce an additional protective outer layer known as a *capsule*. Viruses are not cells. They consist of a nucleic acid (DNA or RNA) core enclosed in a coat of viral proteins called a *capsid*. Some viruses add an *envelope* of phospholipid and protein made from the host's cell membrane and incorporate viral proteins into the envelope (Fig. 24-1a ■).

2. **Living conditions and reproduction**. Most bacteria can survive and reproduce outside a host if they have the required nutrients, temperature, pH, and so on. Viruses *must* use the intracellular machinery of a host cell to replicate.

3. **Susceptibility to drugs**. Most bacteria can be killed by the drugs we call **antibiotics**. These drugs act directly on

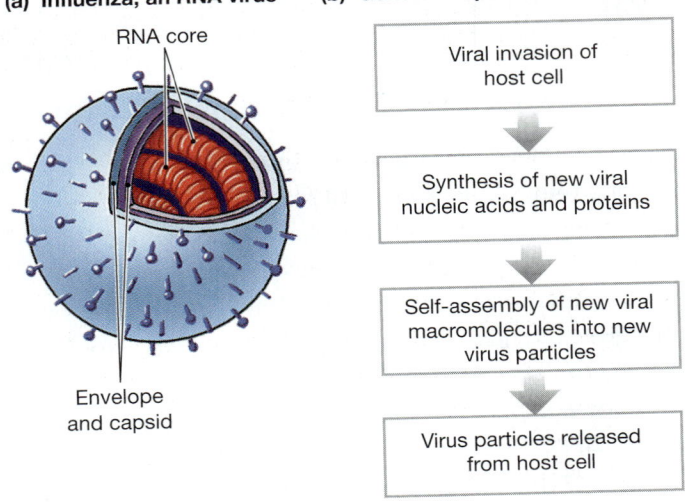

(a) Influenza, an RNA virus

RNA core

Envelope and capsid

(b) General steps of viral replication

Viral invasion of host cell

↓

Synthesis of new viral nucleic acids and proteins

↓

Self-assembly of new viral macromolecules into new virus particles

↓

Virus particles released from host cell

■ **FIGURE 24-1**   *Viral structure and replication*

bacteria and destroy them or inhibit their growth. Viruses cannot be killed by antibiotics. Some viral infections can be treated with *antiviral drugs,* which target viral replication.

Viruses Can Replicate Only Inside Host Cells

The life cycle of a virus begins when the virus invades the host cell (Fig. 24-1b). To cross the human host cell membrane, the virus binds to membrane receptors, triggering endocytosis of the entire virus particle. In an alternative scenario, the virus envelope fuses with the host cell membrane, allowing the core of the virus to enter the cytoplasm.

Once inside the host cell, the virus's nucleic acid takes over the cell's operations. Viruses use their own nucleic acid and the host cell's resources to make new viral nucleic acid and viral proteins. These components assemble into additional virus particles that are released from the host cell to infect other cells.

Viruses can be released from host cells in one of two ways. (1) The virus causes the host cell to rupture, releasing virus particles into the ECF, or (2) virus particles surround themselves with a layer of host cell membrane and then bud off from the surface of the host cell.

Viruses do other kinds of damage to host cells. In taking over a host cell, they may totally disrupt the cell's metabolism so that the cell dies. Some viruses (*Herpes simplex type 1* and varicella-zoster virus, which cause cold sores and chicken pox, respectively) "hide out" in the host cell and replicate only sporadically. Other viruses incorporate their DNA into the host cell DNA. Viruses with this characteristic include HIV and **oncogenic viruses,** which cause cancer.

C O N C E P T   C H E C K

1. Explain the differences between a virus and a bacterium.
2. List the sequence of steps by which a virus replicates itself.

Answers: p. 806

TABLE 24-1	Differences Between Bacteria and Viruses	
	BACTERIA	**VIRUSES**
Structure	Cells; usually surrounded by cell wall	Not cells; nucleic acid core with protein coat
Living conditions	Most can survive and reproduce outside a host	Parasitic; must have a host cell to reproduce
Susceptibility to drugs	Most can be killed or inhibited by antibiotics	Cannot be killed with antibiotics; some can be inhibited with antiviral drugs

EMERGING CONCEPTS

RETROVIRUSES

The RNA viruses known as **retroviruses** replicate in an interesting manner. These viruses do not have DNA that can act as a template for mRNA synthesis; instead they produce an enzyme called **reverse transcriptase**. This enzyme allows the viral RNA to make a complementary strand of DNA. (This is the reverse of the usual process, because normally DNA codes for the synthesis of RNA.) After a period of time (the *latent period*), the virus-infected host cell produces viral mRNA. The viral mRNA then interrupts the host cell's protein synthesis, using host ribosomes to make viral proteins. HIV, the virus that causes AIDS, is a retrovirus.

THE IMMUNE RESPONSE

The body has two lines of defense. Physical and chemical barriers, such as skin, mucus, and stomach acid, first try to keep pathogens out of the body's internal environment. If that line of defense fails, then the internal **immune response** takes over.

Whatever the nature of the foreign material that enters the body, the internal immune response includes the same basic steps: (1) *detection* and *identification* of the foreign substance, (2) *communication* with other immune cells to rally an organized response, (3) *recruitment* of assistance and *coordination* of the response among all participants, and (4) *destruction* or *suppression* of the invader.

Not all invaders can be destroyed by the body's immune system. In some cases the best the body can do is control the damage and keep the invader from spreading. Examples of pathogens that are suppressed by the immune system rather than destroyed include the bacterium that causes tuberculosis; the malaria parasite, which hides inside liver cells; and the herpes viruses, which cause occasional outbreaks of cold sores or genital lesions.

The immune system is distinguished by its extensive use of chemical signaling. Detection, identification, communication, recruitment, coordination, and the attack on the invader all depend on signal molecules such as antibodies and cytokines. **Antibodies**, proteins secreted by certain immune cells, bind antigens and make them more visible to the immune system. **Cytokines** are protein messengers released by one cell that affect the growth or activity of another cell [🔁 p. 176].

The human immune response is generally divided into two categories: nonspecific innate immunity and specific acquired immunity. **Innate immunity** is present from birth [*innatus*, inborn] and is the body's **nonspecific immune response** to invasion. Because the nonspecific innate response does not target a particular pathogen, it begins within minutes to hours. **Inflammation**, apparent on the skin as a red, warm, swollen area, is a hallmark reaction of innate immunity.

Acquired immunity (also called *adaptive immunity*) is directed at specific invaders and thus is the body's **specific immune response**. One characteristic of acquired immunity is that a specific immune response to a first exposure to a pathogen may take days. With repeated exposures, however, the immune system "remembers" prior exposure to the pathogen and reacts more rapidly. Acquired immunity can be divided into *cell-mediated immunity* and *humoral immunity*. (The term *humoral,* referring to the blood, comes from the Hippocratic school of medicine, which classified the body's fluids into four *humors:* blood, phlegm, black bile, and yellow bile.)

As we will see later in this chapter, the nonspecific and specific immune responses overlap. Although we have described them as if they are separate, in reality they are two interconnected parts of a single process. The innate response is the more rapid response, but it does not target a specific invader. It is reinforced by the antigen-specific acquired response, which amplifies the efficacy of the innate response. Communication and coordination among all the different pathways of immunity are vital for maximum protective effect.

Because of the huge numbers of molecules involved in human immunity, and because of the complex interactions between different components of the immune system, the field of immunology is continuously expanding. In this chapter we cover the basics.

RUNNING PROBLEM

HIV is a retrovirus that uses the enzyme *reverse transcriptase* to replicate. Two drugs in the AIDS cocktail, AZT and ddC, are anti-retroviral drugs that work by inhibiting reverse transcriptase. The other drug in the cocktail is a protease inhibitor. Protease is an enzyme needed to cleave key proteins during one step of HIV replication.

Question 1:

Why can't retroviruses reproduce without reverse transcriptase?

24

777 779 793 798 801 803

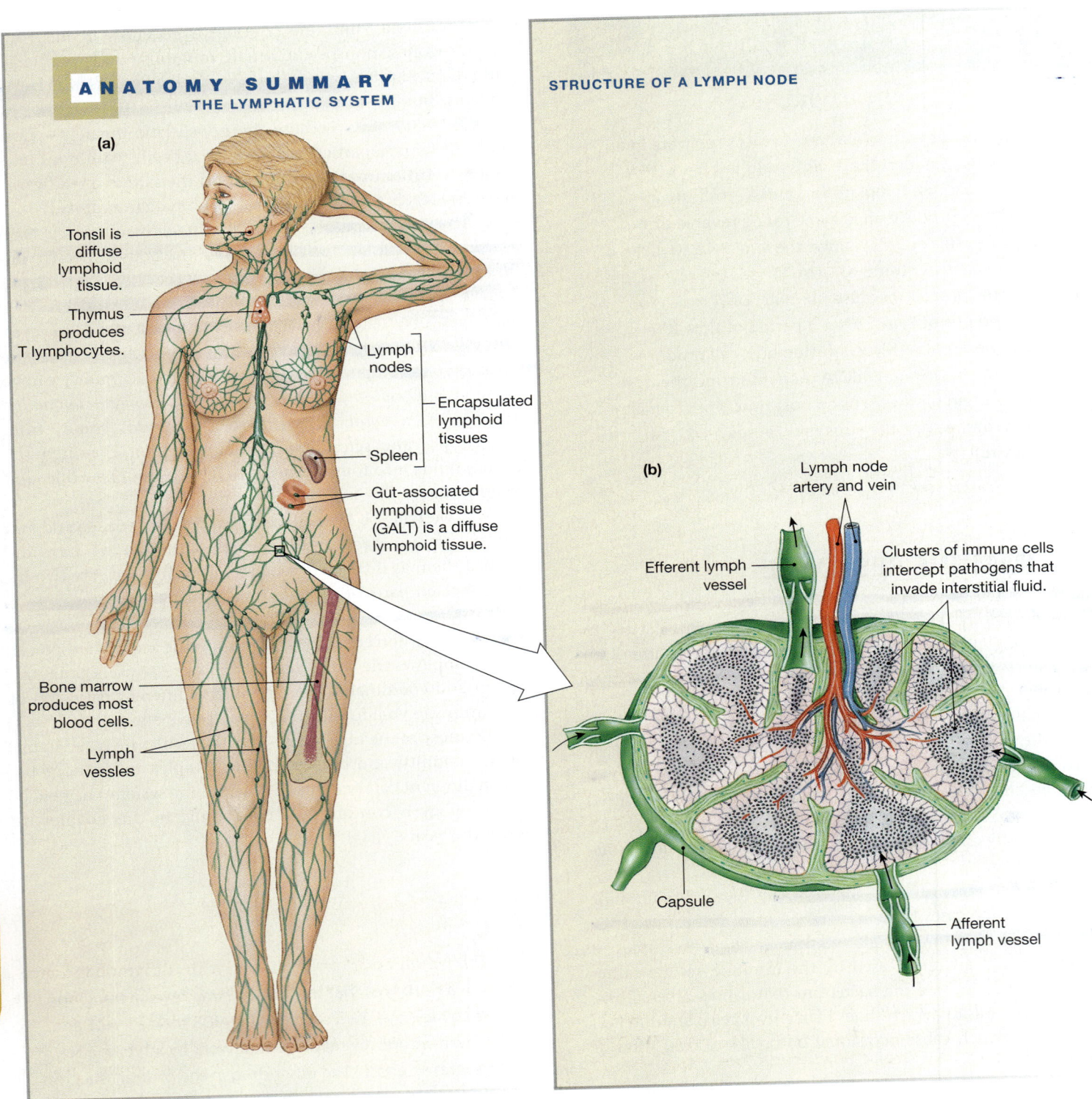

ANATOMY SUMMARY
THE LYMPHATIC SYSTEM

(a)

Tonsil is diffuse lymphoid tissue.

Thymus produces T lymphocytes.

Lymph nodes

Encapsulated lymphoid tissues

Spleen

Gut-associated lymphoid tissue (GALT) is a diffuse lymphoid tissue.

Bone marrow produces most blood cells.

Lymph vessles

STRUCTURE OF A LYMPH NODE

(b)

Lymph node artery and vein

Efferent lymph vessel

Clusters of immune cells intercept pathogens that invade interstitial fluid.

Capsule

Afferent lymph vessel

FIGURE 24-2   *Anatomy of the immune system*

ANATOMY OF THE IMMUNE SYSTEM

The immune system is probably the least anatomically identifiable system of the body because most of it is integrated into the tissues of other organs, such as the skin and gastrointestinal tract. Yet the mass of all immune cells in the body equals the mass of the brain! The immune system has two anatomical components: lymphoid tissues and the cells responsible for the immune response.

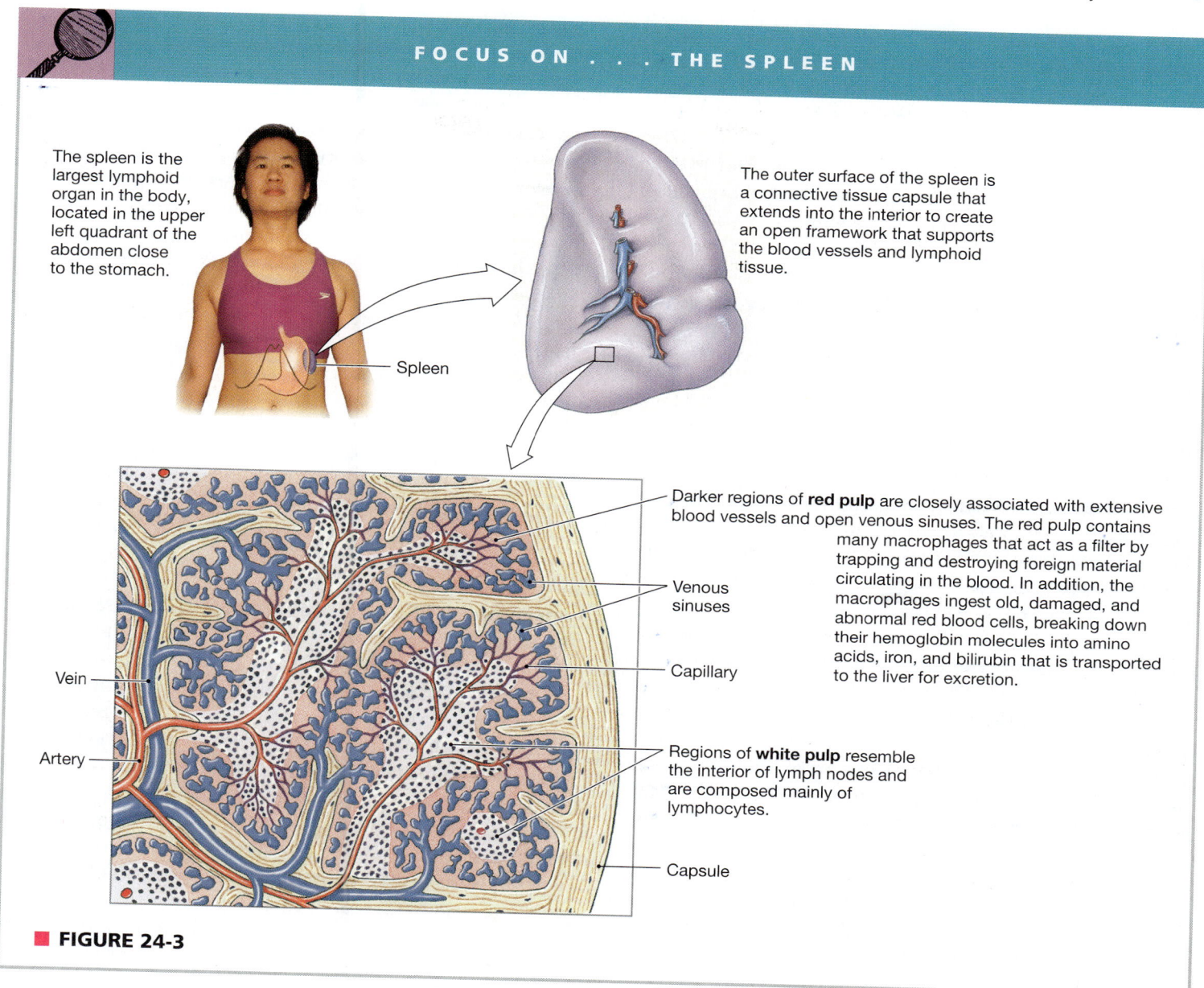

FOCUS ON . . . THE SPLEEN

The spleen is the largest lymphoid organ in the body, located in the upper left quadrant of the abdomen close to the stomach.

Spleen

The outer surface of the spleen is a connective tissue capsule that extends into the interior to create an open framework that supports the blood vessels and lymphoid tissue.

Darker regions of **red pulp** are closely associated with extensive blood vessels and open venous sinuses. The red pulp contains many macrophages that act as a filter by trapping and destroying foreign material circulating in the blood. In addition, the macrophages ingest old, damaged, and abnormal red blood cells, breaking down their hemoglobin molecules into amino acids, iron, and bilirubin that is transported to the liver for excretion.

Venous sinuses

Capillary

Vein

Artery

Regions of **white pulp** resemble the interior of lymph nodes and are composed mainly of lymphocytes.

Capsule

■ **FIGURE 24-3**

Lymphoid Tissues Are Distributed Throughout the Body

The two primary lymphoid tissues are the **thymus gland** [⇄ *Focus on the Thymus Gland*, p. 792] and the **bone marrow**, both sites where cells involved in the immune response form and mature (Fig. 24-2a ■). Secondary lymphoid tissues are sites where mature immune cells interact with pathogens and initiate a response. Secondary tissues are divided into encapsulated tissues and unencapsulated diffuse lymphoid tissues.

The **encapsulated lymphoid tissues** are the **spleen** (Figs. 24-2a and 24-3 ■) and the **lymph nodes** (Fig. 24-2b). Both have an outer wall formed from fibrous collagenous capsules. The spleen contains immune cells positioned so that they monitor the blood for foreign invaders. Phagocytic cells in the spleen also trap and remove aging red blood cells.

The lymph nodes are associated with the lymphatic circulation. Recall from Chapter 15 that there is net flow of fluid out of capillaries and into the interstitial space [⇄ p. 517]. This filtered fluid is picked up by lymph capillaries and passes through the encapsulated lymph nodes on its journey back to the circulation.

Inside lymph nodes, clusters of immune cells intercept pathogens that have entered the interstitial fluid through breaks in the skin or through mucous membranes. Once these microbes have been swept into the lymph, immune cells in the nodes help prevent their spread throughout the body. You have probably noticed that if you have a sinus infection or a sore throat, the lymph nodes in your neck become swollen. These sore, swollen nodes result from the presence of active immune cells that have collected in the nodes to fight the infection.

24

The unencapsulated **diffuse lymphoid tissues** are aggregations of immune cells that appear in other organs of the body (Fig. 24-2). They include the **tonsils** at the posterior nasopharynx; the **gut-associated lymphoid tissue** (GALT), which lies just under the epithelium of the esophagus and intestines [p. 678]; and clusters of lymphoid tissue associated with the skin and the respiratory, urinary, and reproductive tracts. In each case, these tissues contain immune cells positioned to intercept pathogens before they get into the general circulation. Because of the large surface area of the digestive tract epithelium, some authorities consider the GALT to be the largest immune organ in the body. Anatomically, the cells of the immune system can be found in highest concentrations wherever they are most likely to encounter antigens that penetrate the epithelia.

Leukocytes Are the Primary Cells of the Immune System

The primary cells of the immune system are the white blood cells, or **leukocytes**, which were introduced in Chapter 16 [p. 538]. Most leukocytes are much larger than red blood cells, and they are not nearly as numerous. A microliter of whole blood contains about 5 million red blood cells but only about 7000 leukocytes.

Although most leukocytes circulate in the blood, they usually leave the capillaries and function *extravascularly* (outside the vessels). Some types of leukocytes can live out in the tissues for several months, but others may live for only hours or days. Leukocytes are divided into six basic types: (1) *eosinophils*, (2) *basophils* in the blood and the related *mast cells* in the tissues, (3) *neutrophils*, (4) *monocytes* and their derivative *macrophages*, (5) *lymphocytes* and their derivative *plasma cells*, and (6) *dendritic cells*. (Dendritic cells are not usually found in the blood and therefore were not included in our discussion of leukocytes in Chapter 16.)

Leukocytes can be distinguished from one another in stained tissue samples by the shape and size of the nucleus, the staining characteristics of the cytoplasm and cytoplasmic inclusions, and the regularity of the cell border.

Immune Cell Names Reflect Cell Function or Appearance

The terminology associated with immune cells can be very confusing. Some cell types have several variants, and others have been given multiple names (Fig. 24-4 ■). In addition, immune cells can be grouped either functionally or morphologically.

One morphological group is the **granulocytes**, white blood cells whose cytoplasm contains prominent granules. The names of the different cell types come from the staining properties of the granules. Neutrophil granules do not stain darkly with standard blood stains and are therefore "neutral." Basophil granules stain dark blue with basic (alkaline) dye, and eosinophil granules stain dark pink with the acidic dye *eosin*. In all three types of granulocytes, the granule contents are released into the extracellular space, a process known as *degranulation*.

One functional group of leukocytes is the **phagocytes**, white blood cells that engulf and ingest their targets by phagocytosis [p. 148]. This group includes the neutrophils, macrophages,

monocytes (which are macrophage precursors), and eosinophils. A second functional group is the **cytotoxic cells**, so named because they kill the cells they attack. This group includes eosinophils and some types of lymphocytes. A third group is made up of **antigen-presenting cells** (APCs), which display fragments of foreign proteins on their cell surface. This group includes certain lymphocytes, dendritic cells, macrophages, and monocytes.

The terminology associated with macrophages has changed over the history of histology and immunology. For many years, tissue macrophages were known as the *reticuloendothelial system* and were not associated with white blood cells. To confuse matters, the cells were named when they were first described in different tissues, before they were all identified as macrophages. Thus, *histiocytes* in skin, *Kupffer cells* in the liver, *osteoclasts* in bone, *microglia* in the brain, and *reticuloendothelial cells* in the spleen are all names for specialized macrophages. The new name for the reticuloendothelial system is the **mononuclear phagocyte system**, a term that refers both to macrophages in the tissues and to their parent monocytes circulating in the blood.

In the sections that follow, we take a closer look at the six basic types of leukocytes.

Eosinophils Fight Parasites and Contribute to Allergic Reactions
Eosinophils, easily recognized by the bright pink-staining granules in their cytoplasm, are associated with allergic reactions and parasitic diseases. Normally, few eosinophils are found in the peripheral circulation, where they account for only 1–3% of all leukocytes. The life span of an eosinophil in the blood is estimated to be only 6–12 hours.

Most functioning eosinophils are found in the digestive tract, lungs, urinary and genital epithelia, and connective tissue of the skin. These locations reflect their role in defense against parasitic invaders. Eosinophils are known to attach to large antibody-coated parasites, such as the blood fluke *Schistosoma*, and release substances from their granules that damage or kill the parasites. Because eosinophils kill pathogens, they are classified as cytotoxic cells.

Eosinophils also participate in allergic reactions, where they contribute to inflammation and tissue damage by releasing toxic enzymes, oxidative substances, and a protein called *eosinophil-derived neurotoxin*. Although eosinophils have been shown *in vitro* to ingest foreign particles, such as pollen, there is no evidence that they do so *in vivo*.

Basophils Release Histamine and Other Chemicals
Basophils are rare in the circulation but are easily recognized in a stained blood smear by the large, dark blue granules in their cytoplasm. They are very similar to the **mast cells** of tissues, and both cell types release mediators that contribute to inflammation. The granules of these cells contain histamine, **heparin** (an anticoagulant that inhibits blood clotting), cytokines, and other chemicals involved in allergic and immune responses.

Mast cells, like eosinophils, are concentrated in the connective tissue of skin, lungs, and the gastrointestinal tract. In

	Basophils and mast cells	Neutrophils	Eosinophils	Monocytes and macrophages	Lymphocytes and plasma cells	Dendritic cells
% of WBCs in blood	Rare	50–70%	1–3%	1–6%	20–35%	NA
Subtypes and nicknames		Called "polys" or "segs" Immature forms called "bands" or "stabs"		Called the mononuclear phagocyte system	B lymphocytes Plasma cells T lymphocytes Cytotoxic T cells Helper T cells Natural killer cells Memory cells	Also called Langerhans cells, veiled cells
Primary function(s)	Release chemicals that mediate inflammation and allergic responses	Ingest and destroy invaders	Destroy invaders, particularly antibody-coated parasites	Ingest and destroy invaders Antigen presentation	Specific responses to invaders, including antibody production	Recognize pathogens and activate other immune cells by antigen presentation
Classifications		*Phagocytes*				
		Granulocytes				
			Cytotoxic cells		*Cytotoxic cells (some types)*	
					Antigen-presenting cells	

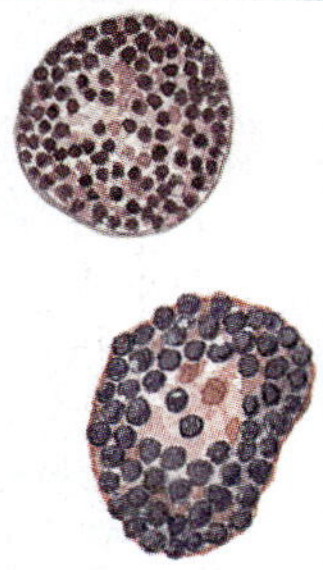

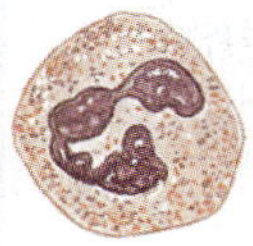

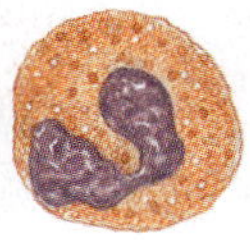

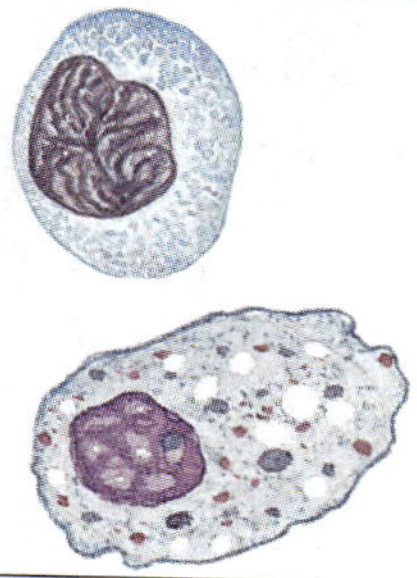

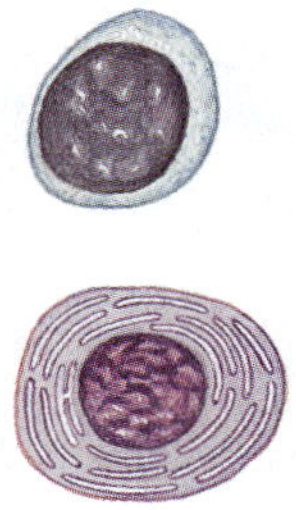

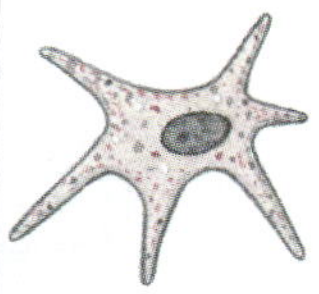

■ **FIGURE 24-4** *Cells of the immune system*

these locations, mast cells are ideally situated to intercept pathogens that are inhaled or ingested or that enter through breaks in the epidermis.

Neutrophils "Eat" Bacteria and Release Cytokines

Neutrophils are the most abundant white blood cells (50–70% of the total) and are most easily identified by a segmented nucleus made up of three to five lobes connected by thin strands of nuclear material. Because of the segmented nucleus, neutrophils are also called *polymorphonuclear leukocytes* ("polys") and *"segs."* Immature neutrophils, occasionally found in the circulation, can be identified by their horseshoe-shaped nucleus. These immature neutrophils go by the nicknames of *"bands"* and *"stabs."*

Neutrophils, like other blood cells, are formed in the bone marrow. They are phagocytic cells that typically ingest and kill 5–20 bacteria during their programmed life span of one or two days. Most neutrophils remain in the blood but can leave the circulation if attracted to an extravascular site of damage or infection. In addition to ingesting bacteria and foreign particles, neutrophils release a variety of cytokines, including fever-causing pyrogens [p. 743] and chemical mediators of the inflammatory response.

Monocytes Mature into Macrophages

Monocytes are the precursor cells of tissue macrophages. Monocytes are not very common in the blood (1–6% of all white blood cells). By some estimates, they spend only eight hours there in transit from the bone marrow to their permanent positions in the tissues.

Once out of the blood, monocytes enlarge and differentiate into **macrophages**. Some tissue macrophages patrol the tissues, creeping along by amoeboid motion. Others find a location and remain fixed in place. In either case, macrophages are the primary scavengers within tissues. They are larger and more effective than neutrophils, ingesting up to 100 bacteria during their life span. Macrophages also remove larger particles, such as old red blood cells and dead neutrophils.

Because they are antigen-presenting cells, macrophages play an important role in the development of acquired immunity. After they ingest and digest molecular or cellular antigens, fragments of processed antigen are inserted into the macrophage membrane as part of surface protein complexes (Fig. 24-5 ■). As noted earlier, dendritic cells and the lymphocytes known as *B lymphocytes* also act as antigen-presenting cells.

24

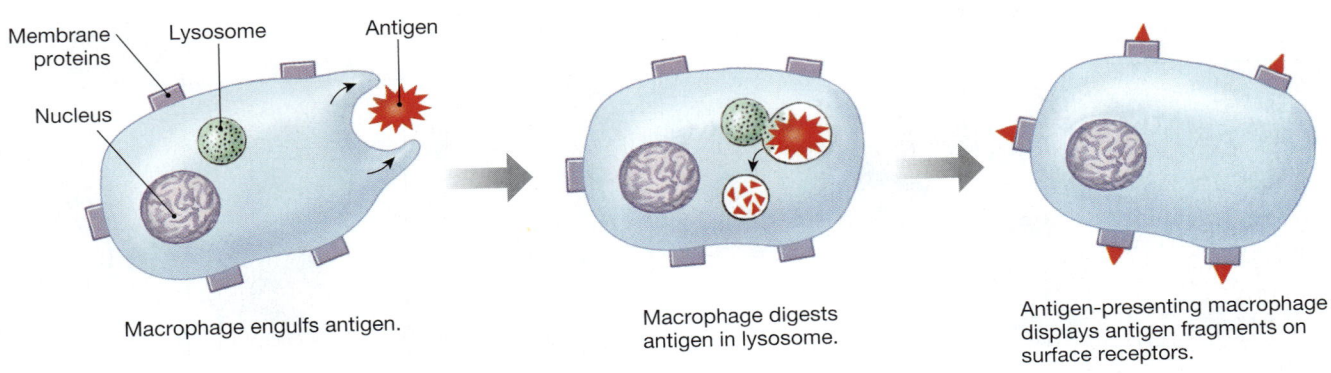

Membrane proteins
Lysosome
Antigen
Nucleus

Macrophage engulfs antigen.

Macrophage digests antigen in lysosome.

Antigen-presenting macrophage displays antigen fragments on surface receptors.

■ **FIGURE 24-5** *Antigen-presenting cells (APCs)*

Lymphocytes Mediate Acquired Immunity Lymphocytes and their derivatives are the key cells that mediate the acquired immune response of the body. Only 5% of all lymphocytes are found in the circulation, where they constitute 20–35% of all white blood cells. Most lymphocytes are found in lymphoid tissues, where they are likely to encounter invaders.

By one estimate, the adult body contains a trillion lymphocytes at any one time. Although most of these cells look alike under the microscope, they have major differences in function and specificity, as you will learn later in the chapter.

Dendritic Cells Activate Lymphocytes Dendritic cells are antigen-presenting cells characterized by long, thin processes that resemble neuronal dendrites. Dendritic cells are found in the skin (where they are called *Langerhans cells*) and in various organs. When dendritic cells recognize and capture antigens, they migrate to secondary lymphoid tissues, where they present the antigens to lymphocytes. Antigen binding activates the lymphocytes.

INNATE IMMUNITY: NONSPECIFIC RESPONSES

The body's first line of defense is to exclude pathogens by physical and chemical barriers. If those barriers fail to keep invaders out, the innate immune system provides the second line of defense. The innate immune response consists of patrolling and stationary leukocytes that attack and destroy invaders. These leukocytes respond in the same fashion to any material they identify as foreign, which is why innate immunity is considered nonspecific. Innate immunity either clears the infection or contains it until the acquired immune response is activated.

Physical and Chemical Barriers Are the Body's First Line of Defense

Physical barriers of the body include the skin [p. 83], the protective mucous linings of the gastrointestinal and genitourinary tracts, and the ciliated epithelium of the respiratory tract.

The digestive and respiratory systems are most vulnerable to microbial invasion because these regions have extensive areas of thin epithelium in direct contact with the external environment. In women, the reproductive tract is also vulnerable, but to a lesser degree. The opening to the uterus is normally sealed by a mucus plug that keeps bacteria in the vagina from ascending into the uterine cavity.

In the respiratory system, inhaled particulate matter is trapped by mucus lining the upper respiratory system. The mucus is then transported upward on the mucociliary escalator to be expelled or swallowed [p. 568]. Swallowed pathogens may be disabled by the acidity of the stomach. In addition, respiratory tract secretions contain **lysozyme**, an enzyme with antibacterial activity. Lysozyme attacks cell wall components of unencapsulated bacteria and breaks them down. However, it cannot digest the capsules of encapsulated bacteria.

Phagocytes Recognize and Ingest Foreign Material

Pathogens that get past the physical barriers of skin and mucus are dealt with by immune cells and the innate immune response. When patrolling and stationary phagocytes in the tissues detect invaders, their response is twofold: they destroy or suppress the invaders by ingesting them (phagocytosis), and they attract additional immune cells by secreting cytokines. Molecules that attract immune cells are called **chemotaxins**. They include bacterial toxins, products of tissue injury (fibrin and collagen fragments), and cytokines released by leukocytes.

The primary phagocytic cells of the immune system are tissue macrophages and neutrophils. If an invader is in a tissue, phagocytes leave the circulation (*extravasation*) by squeezing through pores in the capillary endothelium. Once the phagocytes reach the pathogen, they identify it by chemical cues and then ingest it.

Phagocytosis is similar in macrophages and neutrophils. It is a receptor-mediated event, which ensures that only unwanted particles are ingested. In the simplest reactions, surface molecules on the pathogen act as ligands that bind directly to receptors on the phagocyte membrane (Fig. 24-6a ■). In a sequence

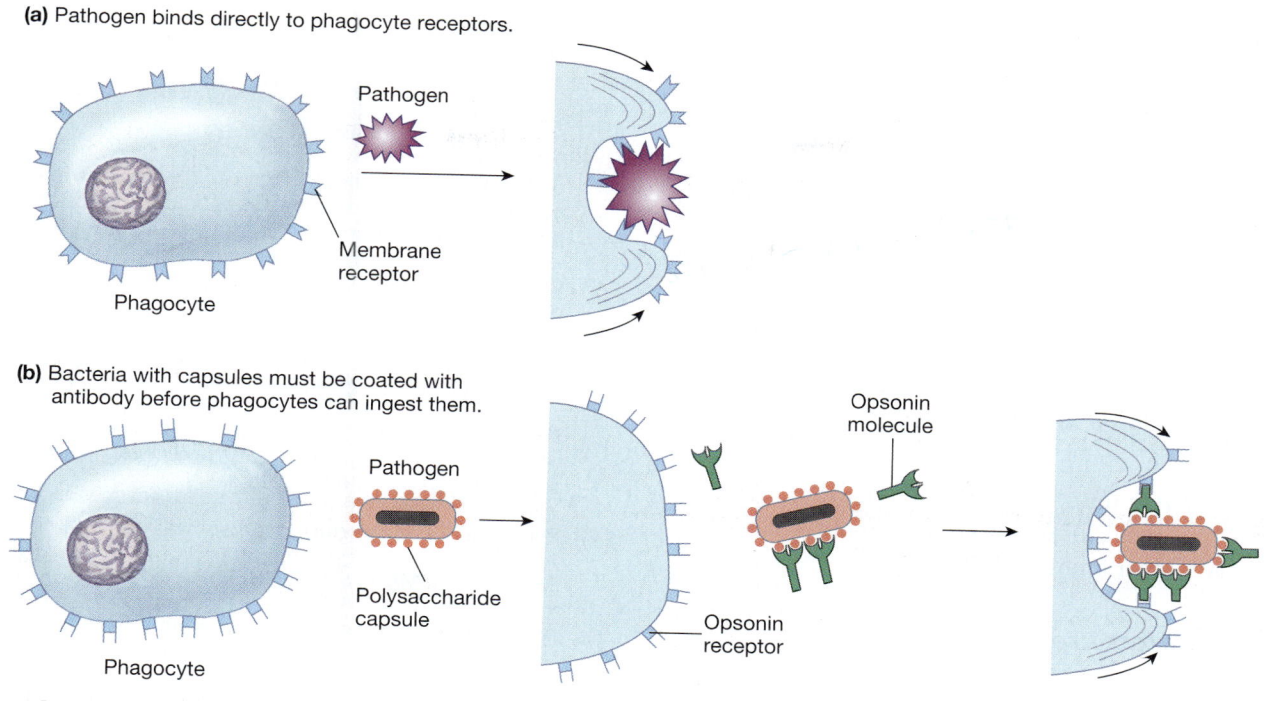

(a) Pathogen binds directly to phagocyte receptors.

Pathogen

Membrane receptor

Phagocyte

(b) Bacteria with capsules must be coated with antibody before phagocytes can ingest them.

Pathogen

Polysaccharide capsule

Phagocyte

Opsonin molecule

Opsonin receptor

■ **FIGURE 24-6** *Phagocytosis*

Some foreign materials activate phagocytes directly (a), but others must be coated with antibodies or other proteins before a phagocyte recognizes them (b).

reminiscent of a zipper closing, the ligands and receptors combine sequentially, so that the phagocyte surrounds the unwanted foreign particle. The process is aided by actin filaments that push arms of the phagocytic cell around the invader.

Phagocyte membranes contain receptors that recognize many different types of foreign particles, both organic and inorganic. Phagocytes ingest unencapsulated bacteria, cell fragments, carbon, and asbestos particles, among other materials. In the laboratory, scientists use the ingestion rate of tiny polystyrene beads as an indicator of phagocytic activity.

Phagocytes cannot instantly recognize all foreign substances, however, because some pathogens lack markers that react with the phagocytes' receptors. For example, certain bacteria have evolved a polysaccharide capsule that masks their surface markers (Fig. 24-6b). These encapsulated bacteria are not as quickly recognized by phagocytes and consequently are more pathogenic because they can grow unchecked until the immune system finally recognizes them and makes antibodies against them. Once that happens, the bacteria are "tagged" with a coat of antibodies so that phagocytes recognize them as something to be ingested.

The antibodies that tag encapsulated bacteria, along with additional plasma proteins, are known collectively as **opsonins** [*opsonin,* "to buy provisions."] In the body, opsonins convert unrecognizable particles into "food" for phagocytes. Opsonins act as bridges between pathogens and phagocytes by binding to receptors on the phagocytes.

Once a pathogen has been ingested by a phagocyte, the ingested particle ends up in a cytoplasmic vesicle called a **phagosome** (Fig. 24-7 ■). Phagosomes fuse with intracellular lysosomes [⊋ p. 66], which contain powerful chemicals that destroy ingested pathogens. Lysosomal contents include enzymes and oxidizing agents, such as hydrogen peroxide (H_2O_2), nitric oxide (NO), and the superoxide anion (O^-).

If an area of infection attracts a large number of phagocytes, the material known as **pus** may form. This thick, whitish to greenish substance is a collection of living and dead neutrophils and macrophages, along with tissue fluid, cell debris, and other remnants of the immune process.

Natural Killer Lymphocytes Eliminate Virally Infected and Tumor Cells

One class of lymphocyte—**natural killer cells,** or NK cells—participates in the innate response against viral infections. NK cells act more rapidly than other lymphocytes in a response that can be measured within hours of a primary viral infection. NK cells induce virus-infected cells to commit suicide (apoptosis [⊋ p. 81]) before the virus can replicate. Complete elimination of the virus requires activation of a specific immune response. NK cells also attack some tumor cells.

NK cells and other lymphocytes secrete various antiviral cytokines, including **interferons,** which were named for their ability to interfere with viral replication. **Interferon-alpha** and

24

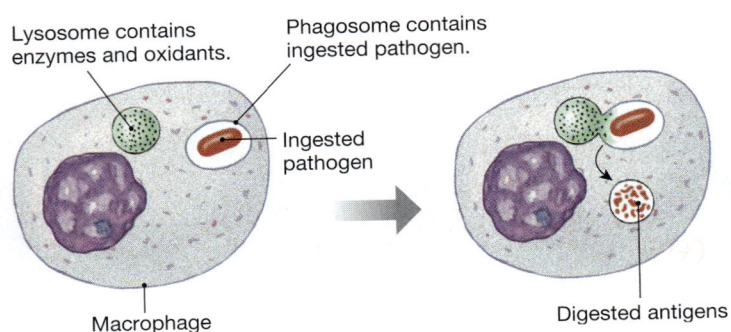

Lysosome contains enzymes and oxidants.

Phagosome contains ingested pathogen.

Ingested pathogen

Macrophage

Digested antigens

■ **FIGURE 24-7** *Lysosomal enzymes digest pathogens that have been enclosed in phagosomes*

interferon-beta target uninfected cells and activate pathways to prevent viral replication. **Interferon-gamma** activates macrophages and other immune cells.

Chemical Mediators Create the Inflammatory Response

Inflammation is a hallmark reaction of innate immunity that has three important roles in fighting infection in damaged tissue: (1) attracting immune cells and chemical mediators to the site, (2) producing a physical barrier to retard the spread of infection, and (3) promoting tissue repair once the infection is under control (a non-immunological function).

The inflammatory response is created when activated tissue macrophages release cytokines. These chemicals attract other immune cells, increase capillary permeability, and cause fever. Immune cells attracted to the site in turn release their own cytokines. Some representative examples of chemicals involved in the innate immune response are described below and listed in Table 24-2 ■, a convenient "scorecard" for following the "players" in these complex processes.

Acute-Phase Proteins Are Released Early in the Immune Response In the time immediately following an injury or pathogen invasion (the acute phase), the body responds by increasing the concentration of various plasma proteins. Some of these proteins, produced mostly by the liver, are given the general name of **acute-phase proteins**. They include molecules that act as opsonins, antiprotease molecules that help prevent tissue damage, and *C-reactive protein* (CRP).

Normally, the levels of acute-phase proteins decline to normal as the immune response proceeds, but in *chronic inflammatory diseases*, such as rheumatoid arthritis, elevated levels of acute-phase proteins may persist. For example, increased levels of C-reactive protein are associated with increased risk of coronary heart disease. This is because atherosclerosis is an inflammatory process that begins when macrophages in blood vessels ingest excess cholesterol and become foam cells [🔁p. 526] that secrete CRP, other cytokines, and growth factors. Elevated CRP is not a specific indicator of atherosclerosis, however,

TABLE 24-2	Chemicals of the Immune Response

FUNCTIONAL CLASSES

Acute phase proteins: Liver proteins released during the acute phase that act as opsonins and enhance the inflammatory response

Chemotaxins: Molecules that attract phagocytes to a site of infection

Cytokines: Proteins released by one cell that affect the growth or activity of another cell

Opsonins: Proteins that coat pathogens so phagocytes recognize and ingest them: antibodies, acute phase proteins, and complement proteins

Pyrogens: Fever-producing substances

SPECIFIC CHEMICALS AND THEIR FUNCTIONS

Antibodies (immunoglobulins, gamma globulins): Proteins secreted by B lymphocytes that fight specific invaders

Bradykinin: Stimulates pain receptors; vasodilator

Complement: Plasma and cell membrane proteins that act as opsonins, cytolytic agents, and mediators of inflammation

C-reactive protein: Opsonin that activates complement cascade

Granzymes: Cytotoxic enzymes that initiate apoptosis

Heparin: An anticoagulant

Histamine: Vasodilator and bronchoconstrictor released by mast cells and basophils

Interferons: Proteins that inhibit viral replication and modulate the immune response

Interleukin-1: Macrophage cytokine that mediates inflammatory response and induces fever

Kinins: Plasma proteins that activate to form bradykinin

Lysozyme: An extracellular enzyme that attacks bacteria

Major histocompatibility complex (MHC): A family of membrane protein complexes involved in cell recognition

Membrane attack complex: A membrane pore protein made in the complement cascade; allows ions and water to enter pathogens

Perforin: A membrane pore protein that allows granzymes to enter the cell; made by NK and cytolytic T cells

Superoxide anion (O^-): Powerful oxidant in phagocyte lysosomes

T-cell receptors: Membrane receptors on T lymphocytes that recognize and bind antigen presented by MHC receptors

24

because CRP levels can also be elevated in other acute and chronic inflammatory conditions.

Histamine Initiates Inflammation
Histamine is found primarily in the granules of mast cells and basophils, and it is the active molecule that helps initiate the inflammatory response when mast cells degranulate. Histamine's actions bring more leukocytes to the injury site to kill bacteria and remove cellular debris.

Histamine opens pores in capillaries, allowing plasma proteins to escape into the interstitial space. This process causes local edema, or swelling. In addition, histamine dilates blood vessels, increasing blood flow to the area. The result of histamine release is a hot, red, swollen area around a wound or infection site.

Mast cell degranulation is triggered by different cytokines in the immune response. Because mast cells are concentrated under mucous membranes that line the airways and digestive tract, the inhalation or ingestion of certain antigens can trigger histamine release. The resultant edema in the nasal passages leads to one annoying symptom of seasonal pollen allergies: the stuffy nose. Fortunately, pharmacologists have developed a variety of drugs called *antihistamines*, which block the action of histamine at its receptor.

Scientists once believed that histamine was the primary chemical responsible for the bronchoconstriction of asthma and for the severe systemic allergic reaction called *anaphylactic shock* [p. 510]. However, we now know that mast cells release powerful lipid mediators in addition to histamine, including leukotrienes, platelet-activating factor, and prostaglandins. These mediators work with histamine to create the bronchoconstriction and hypotension characteristic of anaphylactic shock.

Interleukins Have Widespread Systemic Effects
Interleukins [p. 540] are cytokines initially thought to mediate communication among the body's different types of leukocytes. Scientists have since discovered that many different tissues in the body respond to interleukins.

Interleukin-1 (IL-1) is secreted by activated macrophages and other immune cells. Its main role is to mediate the inflammatory response, but it also has widespread systemic effects on immune function and metabolism. Interleukin-1 modulates the immune response by:

1. *Altering blood vessel endothelium* to ease passage of white blood cells and proteins during the inflammatory response.
2. *Stimulating production of acute-phase proteins* by the liver.
3. *Inducing fever* by acting on the hypothalamic thermostat [p. 743]. IL-1 is a known pyrogen.
4. *Stimulating cytokine and endocrine secretion* by a variety of other cells.

Bradykinin Stimulates Pain Receptors
Kinins are a group of inactive plasma proteins that participate in a cascade similar

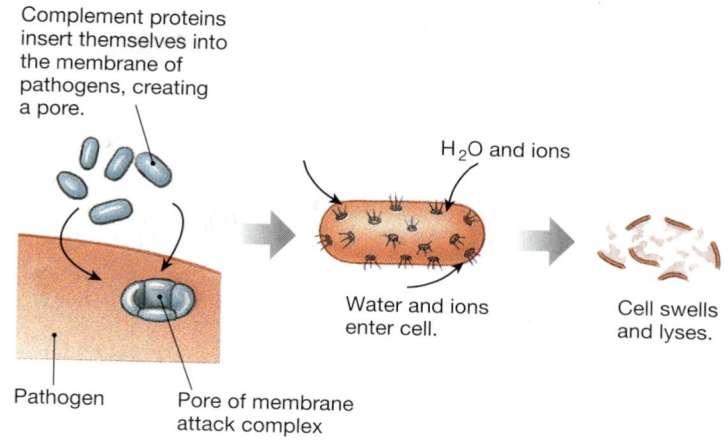

■ **FIGURE 24-8** *Action of a membrane attack complex*

to the coagulation cascade [p. 552]. The end product of the kinin cascade is **bradykinin**, a molecule that has the same vasodilator effects as histamine. Bradykinin also stimulates pain receptors, creating the tenderness associated with inflammation. Pain draws the brain's attention to the injury.

Complement Proteins Are Opsonins and Chemotaxins
Complement is a collective term for a group of more than 25 plasma proteins and cell membrane proteins. The *complement cascade* is similar to the blood coagulation cascade. Various intermediates of the complement cascade act as opsonins, chemical attractants for leukocytes, and agents that cause mast cell degranulation.

The complement cascade terminates with the formation of **membrane attack complex**, a group of lipid-soluble proteins that insert themselves into the cell membranes of pathogens and virus-infected cells and form pores (Fig. 24-8 ■). These pores allow ions to enter the pathogen cells, and water follows by osmosis. As a result, the cells swell and lyse.

CONCEPT CHECK
3. Why does the action of histamine on capillary permeability result in swelling?
Answers: p. 806

ACQUIRED IMMUNITY: ANTIGEN-SPECIFIC RESPONSES
Acquired immune responses are *antigen-specific responses* in which the body recognizes a particular foreign substance and selectively reacts to it. Acquired immunity is mediated primarily by lymphocytes.

There are three main types of lymphocytes: **B lymphocytes**, **T lymphocytes**, and **natural killer (NK) cells**. Activated B lymphocytes develop into **plasma cells**, which secrete antibodies. Activated T lymphocytes develop either into cells that attack and destroy virus-infected cells or into cells that regulate

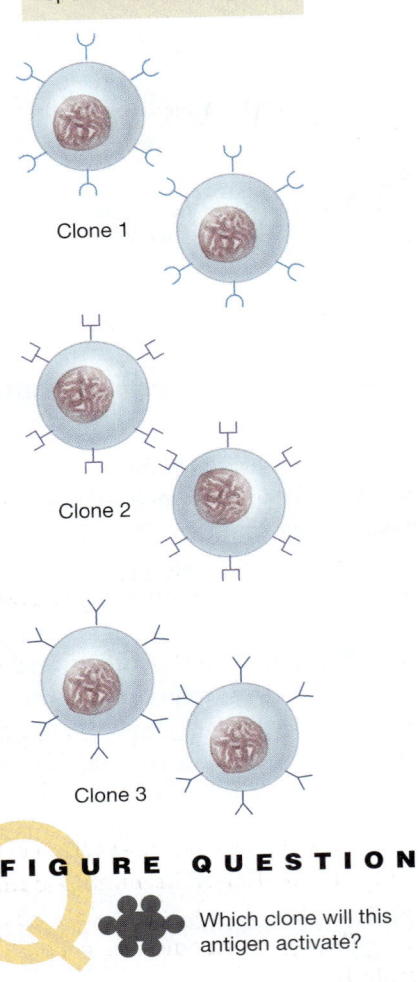

Each cell in a lymphocyte clone has receptors specific to one antigen.

Clone 1

Clone 2

Clone 3

FIGURE QUESTION

Q Which clone will this antigen activate?

■ **FIGURE 24-9** *Lymphocyte clones*

Each of the three pairs of lymphocytes in this figure is from a different clone.

other immune cells. NK cells attack and destroy virus-infected cells and tumor cells. All lymphocytes secrete cytokines that act on immune cells, on non-immune cells, and, sometimes, on pathogens.

The process of acquired immunity overlaps with the process of innate immunity. Cytokines released by the inflammatory response attract lymphocytes to the site of an immune reaction. The lymphocytes release additional cytokines that enhance the inflammatory response.

Acquired immunity can be subdivided into active immunity and passive immunity. **Active immunity** occurs when the body is exposed to a pathogen and produces its own antibodies. Active immunity can occur naturally, when a pathogen invades the body, or artificially, as when we are given vaccinations containing dead or disabled pathogens.

Passive immunity occurs when we acquire antibodies made by another animal. The transfer of antibodies from mother to fetus across the placenta is one example. Injections containing antibodies are another. Travelers going abroad may be injected with *gamma globulin* (antibodies extracted from donated human plasma), but this passive immunity lasts only about three months.

Lymphocytes Are the Primary Cells Involved in the Acquired Immune Response

On a microscopic level, all lymphocytes look alike. At the molecular level, however, the cells can be distinguished from one another by their membrane receptors, which make each lymphocyte specific for a particular ligand. All lymphocytes that are specific for a given ligand form a group known as a **clone** [*klon*, a twig] (Fig. 24-9 ■).

If a specific type of lymphocyte fights each pathogen that enters the body, millions of different types of lymphocytes must be ready to combat millions of different pathogens. But where do all those lymphocytes come from?

At an individual's birth, each clone of lymphocytes is represented by only a few cells, called **naive lymphocytes** [*naif*, natural]. Because the small number of cells in each naive clone is not enough to fight off foreign invaders, the first exposure to an antigen activates the appropriate clone and stimulates it to divide (Fig. 24-10 ■). This process, called **clonal expansion**, creates additional cells in the clone. Naive cells continue to be generated throughout an individual's lifetime.

The newly formed lymphocytes in a clone differentiate into effector cells and memory cells. **Effector cells** carry out the immediate response and then die within a few days. **Memory cells**, in contrast, are long lived and continue reproducing themselves. Second and subsequent exposures to the antigen activate the memory cells, creating a more rapid and stronger secondary response to the antigen.

B Lymphocytes Differentiate into Plasma Cells and Memory Cells

B lymphocytes (also called **B cells**) develop in the bone marrow. (Memory aid: bone starts with B.) Activated B lymphocytes differentiate primarily into specialized cells that secrete antibodies. The word *antibody* describes what the molecules do: work against foreign bodies. Antibodies are also called **immunoglobulins**, and this alternative name describes what the molecules are: globular proteins that participate in the humoral immune response.

Mature B lymphocytes insert antibody molecules into their cell membranes so that the antibodies become surface receptors marking the members of each clone (Fig. 24-9). When a clone of B cells activates in response to antigen exposure, some of the cells differentiate into plasma cells, which are the effector cells for B lymphocytes. Plasma cells do not have antibody proteins

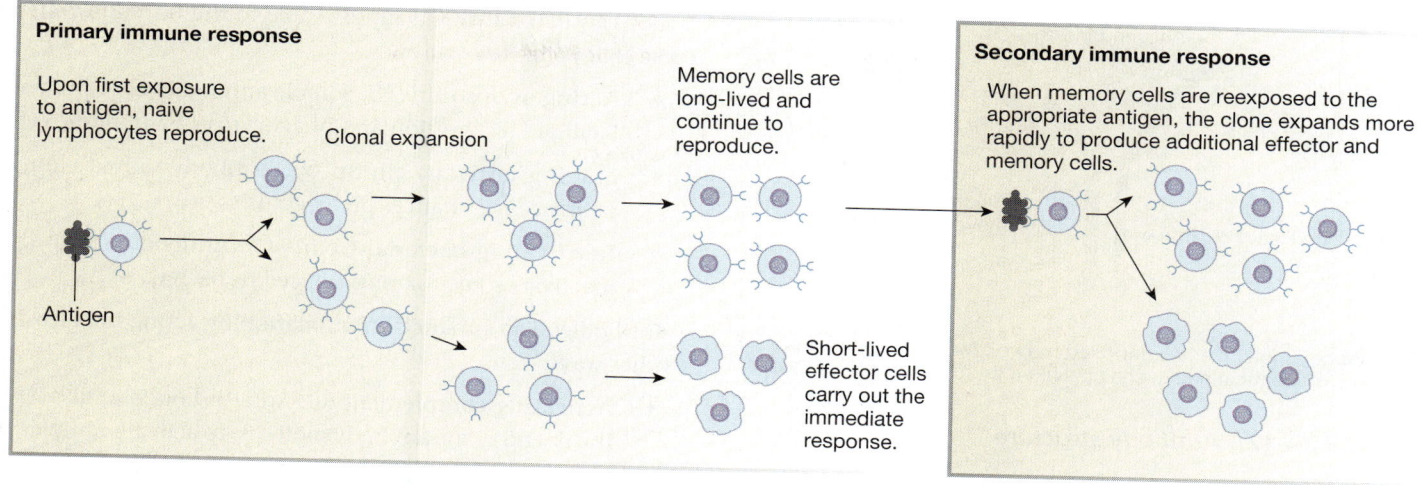

FIGURE 24-10   *How clonal expansion leads to immunologic memory*

bound in their membranes. Instead, they synthesize and secrete additional antibody molecules at incredible rates, estimated to be as high as 2000 molecules per second! Plasma cell antibodies form *humoral immunity,* the soluble antibodies of the plasma.

After each invader has been successfully repulsed, a few memory cells of the clone remain behind, waiting to respond to the next exposure to the same antigen. Figure 24-11 ■ compares the antibody responses—expressed as the concentration of antibodies in the plasma—following the first and second exposures to an antigen. The **primary immune response,** which occurs after the initial exposure, is slower and lower in magnitude because the body has not encountered the antigen previously. The **secondary immune response,** which takes place after second and subsequent exposures, is quicker and larger because it is enhanced by lymphocytes that carry a molecular memory of the first exposure to the antigen.

The existence of a secondary immune response is what makes vaccinations work. A vaccine contains an altered pathogen that no longer harms the host but that can be recognized as foreign by immune cells. The altered pathogen triggers creation of memory cells specific to that particular pathogen. If the vaccinated person is later infected by the pathogen, a stronger and more rapid secondary immune response results. We are now learning, however, that immunity after immunization may not last for a person's lifetime.

Antibodies Are Proteins Secreted by Plasma Cells

Antibodies were among the first aspects of the immune system to be discovered, and traditionally their names—agglutinins, precipitins, hemolysins, and so on—indicated what they do. Today, however, antibodies or immunoglobulins (Ig) are divided into five general classes: IgG, IgA, IgE, IgM, and IgD (pronounced *eye-gee-[letter]*). Antibodies are collectively referred to as **gamma globulins**.

IgGs make up 75% of plasma antibody in adults because they are produced in secondary immune responses. Maternal IgGs cross the placental membrane and give infants immunity in the first few months of life. IgGs activate complement.

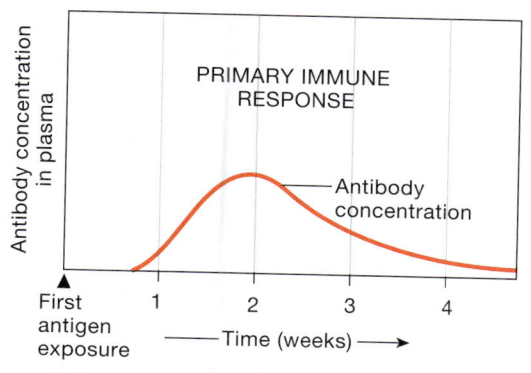

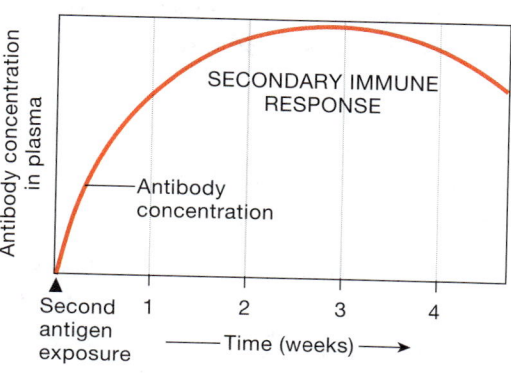

FIGURE 24-11   *Memory in the immune system*

Antibody production in response to the first exposure to an antigen is both slower and weaker than antibody production following subsequent exposures to the same antigen.

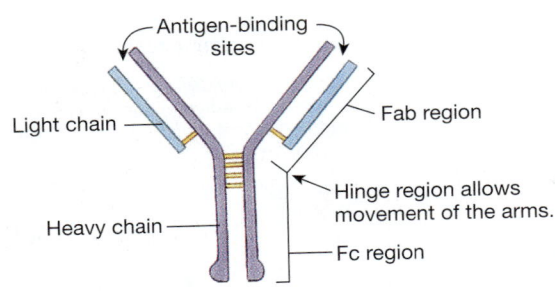

An antibody molecule is composed of two identical light chains and two identical heavy chains, linked by disulfide bonds.

■ **FIGURE 24-12** *Antibody structure*

The two Fab regions contain binding sites for antigens.

IgA antibodies are found in external secretions, such as saliva, tears, intestinal and bronchial mucus, and breast milk, where they disable pathogens before they reach the internal environment.

IgEs are associated with allergic responses. When mast cell receptors bind with IgEs and antigen, the mast cells degranulate and release chemical mediators, such as histamine.

IgM antibodies are associated with primary immune responses and with the antibodies that react to blood group antigens. IgMs also activate complement.

IgD antibody proteins appear on the surface of B lymphocytes along with IgM, but the physiological role of IgDs is unclear.

Antibodies Are Y-Shaped Proteins The basic antibody molecule has four polypeptide chains linked into a Y-shape (Fig. 24-12 ■). Each side of the Y is identical, with one **light chain** attached to one **heavy chain**. The two arms, or **Fab regions**, form antigen-binding sites that confer the antibody's specificity. The stem of the Y-shaped antibody molecule is known as the **Fc region.** The Fc region determines the Ig class to which the antibody belongs. A *hinge region* between the arms and the stem allows flexible positioning of the arms.

In any given antibody molecule, the two light chains are identical and the two heavy chains are identical. However, the chains vary widely among the different antibodies and are specific for the clone from which the antibody comes. Two classes of immunoglobulins (IgM and IgA) are secreted as polymers: IgM is made up of five Y-shaped antibody molecules, and IgA has from one to four antibody molecules.

Antibodies Have Multiple Functions Most antibodies are found in the blood, where they make up about 20% of the plasma proteins in a healthy individual. These antibodies are most effective against extracellular pathogens (such as bacteria), some parasites, antigenic macromolecules, and viruses before they have invaded their host cells.

Antibody function is summarized in Figure 24-13 ■. Although antibodies are not toxic and cannot destroy antigens,

they can make antigens more visible to the immune system in the following three ways:

- **Acting as opsonins** ②. Soluble antibodies coat antigens to facilitate recognition and phagocytosis by immune cells.

- **Making antigens clump** ③. Antibody-caused clumping of antigens enhances phagocytosis.

- **Inactivating bacterial toxins** ③. Antibodies bind to and inactivate some toxins produced by bacteria.

Antibodies also enhance inflammation by acting in at least two other ways:

- **Activating complement** ⑥. Antigen-bound antibodies use the Fc end of the antibody molecule to activate complement.

- **Activating mast cells** ⑤. Mast cells have IgE antibodies attached to their surface. When antigens or complement proteins bind to IgE, the mast cells degranulate, releasing chemicals that mediate the inflammatory response.

The final role of antibodies is to activate immune cells:

- **Antigen-bound antibody activates immune cells.** In the first three functions listed, the antibody binds first to antigen, then to receptors on the immune cell. Phagocytic and some cytotoxic cells have membrane receptors that attract the Fc region of antigen-bound antibodies. The presence of a single Fc receptor eliminates the need to have millions of different receptors to recognize different antigens. Instead, with one type of receptor the immune cells are activated by any antibody-bound antigen.

  If the immune cell is a phagoctye, Fc binding initiates phagocytosis (Fig. 24-13, ③). If the immune cell is a cytotoxic cell (NK cells and eosinophils), Fc binding initiates responses that kill the antibody-bound cell (Fig. 24-13, ④). For example, activated NK cells kill their target cells by triggering apoptosis. This nonspecific response of natural killer cells is called **antibody-dependent cell-mediated cytotoxicity.**

  In the case of B lymphocytes, the antibodies are already an integral part of the lymphocyte (Fig. 24-13, ①). In this case, the binding of antigen activates the cell to produce memory cells and plasma cells.

CONCEPT CHECK

4. Because antibodies are proteins, they are too large to cross cell membranes on transport proteins or through channels. How then do IgAs and other antibodies become part of external secretions such as saliva, tears, and mucus?

5. A child is stung by a bee for the first time. Why should the parent be particularly alert when the child is stung a second time?

Answers: p. 806

Antigen Binding to B Cell Antibodies Activates the Cell

The surface of each B lymphocyte is covered with as many as 100,000 antibody molecules whose Fc ends are inserted into the lymphocyte membrane (Fig. 24-13, ①). The Fab regions of the antibodies are then free to bind to free-floating extracellular

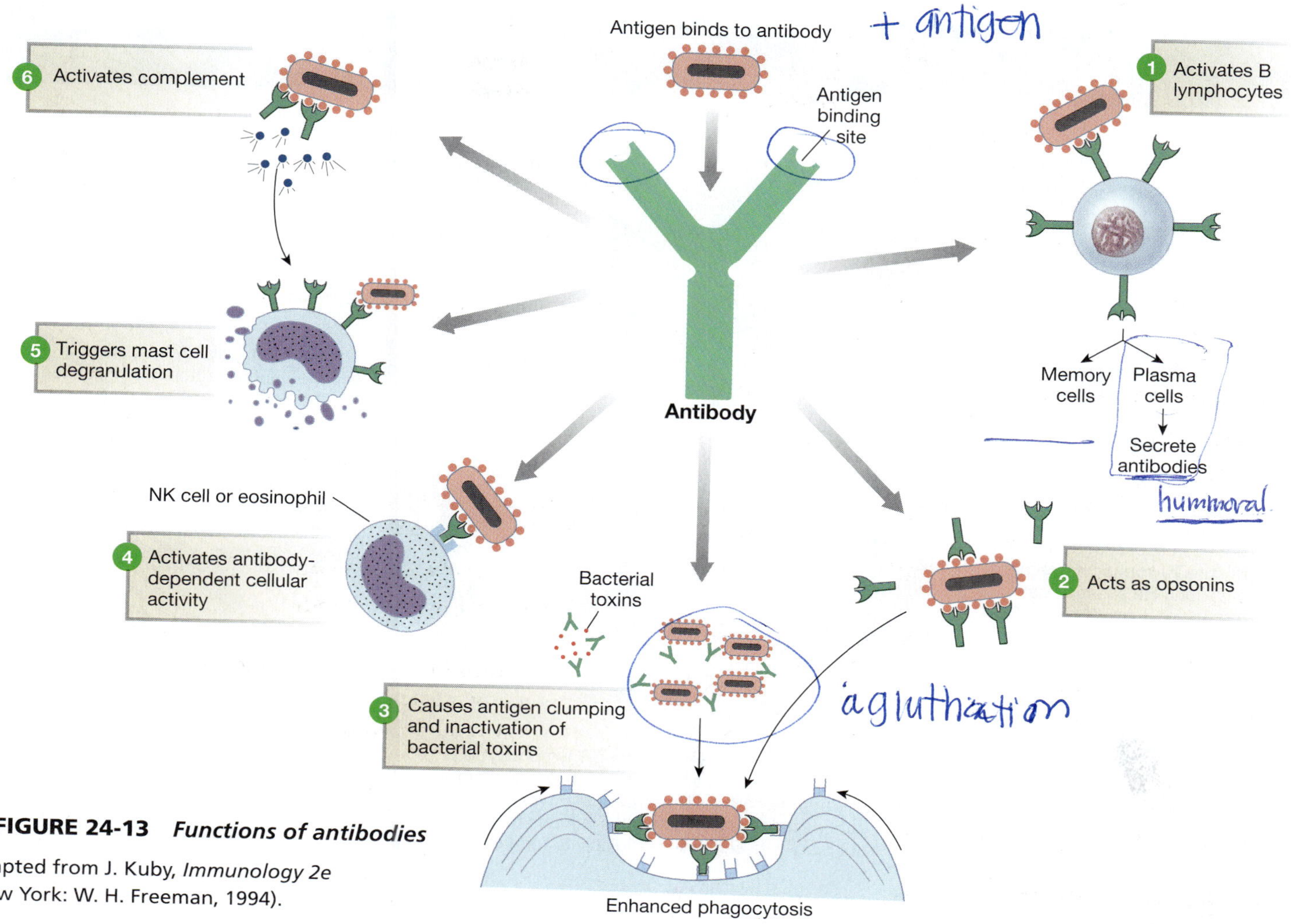

Antigen binds to antibody   *+ antigen*

6 Activates complement

Antigen binding site

5 Triggers mast cell degranulation

Antibody

1 Activates B lymphocytes

Memory cells   Plasma cells

Secrete antibodies

humoral

NK cell or eosinophil

4 Activates antibody-dependent cellular activity

Bacterial toxins

2 Acts as opsonins

aglutination

3 Causes antigen clumping and inactivation of bacterial toxins

Enhanced phagocytosis

■ **FIGURE 24-13**   *Functions of antibodies*

Adapted from J. Kuby, *Immunology 2e* (New York: W. H. Freeman, 1994).

antigens, such as viruses or bacterial toxins. B cell binding neutralizes the pathogen or toxin. Some B cells then differentiate into memory cells that await a subsequent invasion.

For example, *Corynebacterium diphtheriae* is the bacterium that causes diphtheria, an upper respiratory infection. In this disease, the bacterial toxin kills host cells, leaving ulcers that have a characteristic grayish membrane. Natural immunity to the disease occurs when the host develops antibodies that disable the toxin. To develop a vaccine for diphtheria, researchers created an inactivated toxin preparation that did not affect living cells. When administered to a person, the vaccine triggers antibody production without causing any symptoms of the disease. As a result, diphtheria has been almost eliminated in countries with good immunization programs.

T Lymphocytes Must Make Direct Contact with Their Target Cells

Antibodies are effective only against extracellular pathogens. Once a pathogen gets inside a host cell, it can no longer be

"seen" by the humoral immune system. Defending the body against intracellular pathogens is the role of T lymphocytes, which carry out **cell-mediated immunity**. In this process, T cells bind to cells that display foreign antigen fragments as part of a *major histocompatibility complex* (MHC) on their surface.

T lymphocytes develop in the thymus gland from immature precursor cells that migrate there from the bone marrow (Fig. 24-14 ■). Mature T cells bind to MHC-antigen-presenting cells with the help of **T-cell receptors** on the T-cell membrane (Fig. 24-15 ■). T-cell receptors are not antibodies (as B-cell receptors are), although proteins of the two receptor types are closely related. Also, T cells cannot bind to free-floating antigens as B cells do. Instead, T cells can only bind MHC-antigen complexes on the surface of a target cell (Fig. 24-15).

Major Histocompatibility Complexes Incorporates Antigen Fragments   The **major histocompatibility complexes** are a family of membrane protein complexes encoded by a specific set of genes. (These proteins were named when they were discovered to play a role in rejection of foreign tissue

24

FOCUS ON . . . THE THYMUS

The thymus gland is a two-lobed organ located in the thorax just above the heart.

Thyroid gland

Trachea

The thymus gland reaches its greatest size during adolescence. Then it shrinks and is largely replaced by adipose tissue as a person ages. New T lymphocyte production in the thymus is low in adults, but the number of T lymphocytes in the blood does not decrease, suggesting that mature T lymphocytes either have long life spans or can reproduce themselves.

The epithelial cells of the thymus secrete peptide factors that influence the development of T lymphocytes, including interleukins, thymosins, thymopoietin, and thymulin. Although these peptides are sometimes classified as hormones, there is little evidence that they are released into the circulation to act on tissues outside the thymus gland.

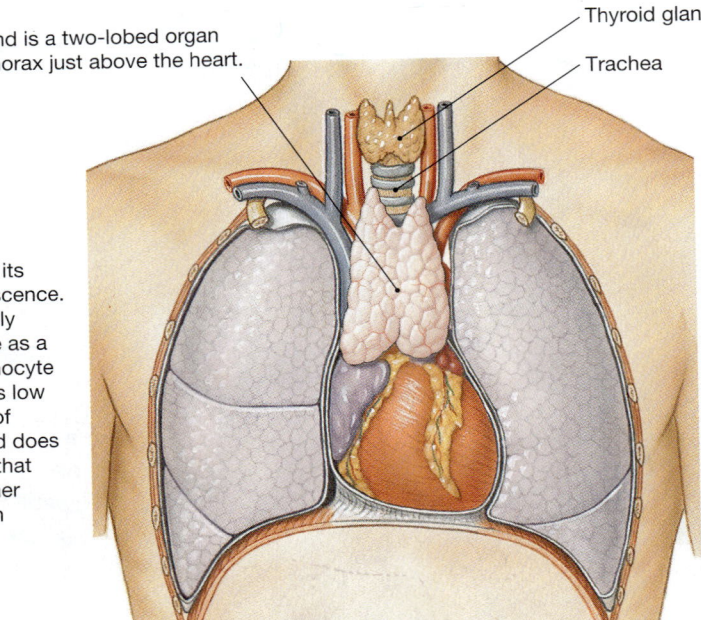

During embryonic development, T lymphocytes synthesize their T-cell receptors and insert them into the membrane. Those cells that would be self-reactive are eliminated, while those that do not react with self tissues multiply to form clones.

■ **FIGURE 24-14**

following organ or tissue transplants.) We know now that every nucleated cell of the body has MHC on its membrane.

MHC proteins combine with fragments of antigens that have been digested within the cell, and the MHC-antigen complex is inserted into the cell membrane by exocytosis (see Fig.

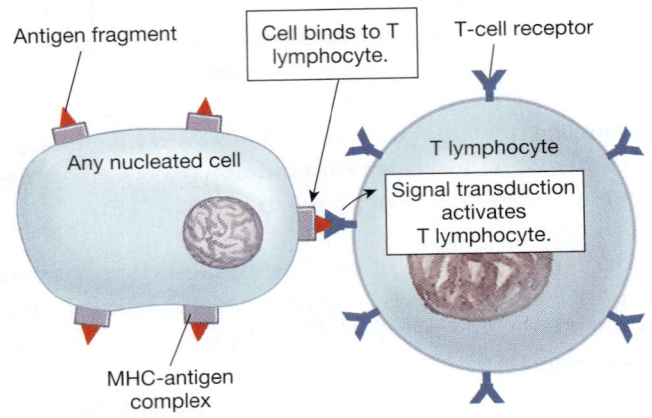

Antigen fragment

Cell binds to T lymphocyte.

T-cell receptor

Any nucleated cell

T lymphocyte

Signal transduction activates T lymphocyte.

MHC-antigen complex

■ **FIGURE 24-15**   *Activation of T lymphocytes*

24-5 for an example). Free antigen in the ECF cannot simply bind to unoccupied MHC receptors on the cell surface.

There are two types of MHC molecules. **MHC class I molecules** are found on all nucleated cells. When viruses and bacteria invade the cell, they are digested into peptide fragments and loaded onto MHC-I "platforms." If a **cytotoxic T cell** (T_C cell) encounters a target cell with foreign antigen fragment on its MHC-I, the T_C cell recognizes the target as either a virus-infected cell or as a tumor cell and kills it to prevent it from reproducing (Fig. 24-16 ■).

MHC class II molecules are found primarily on macrophages, B lymphocytes, and dendritic cells. When an immune cell engulfs and digests an antigen, the fragments are returned to the immune cell membrane combined with MHC class II proteins. If a **helper T cell** (T_H cell) encounters a target cell with a foreign antigen fragment on its MHC-II, the T_H cell responds by secreting cytokines that enhance the immune response.

All MHC proteins are related, but they vary from person to person because of the huge number of alleles (variants of a gene) a person can inherit. There are so many alleles that it is unlikely that any two people other than identical twins will inherit exactly the same set. Major histocompatibility complexes

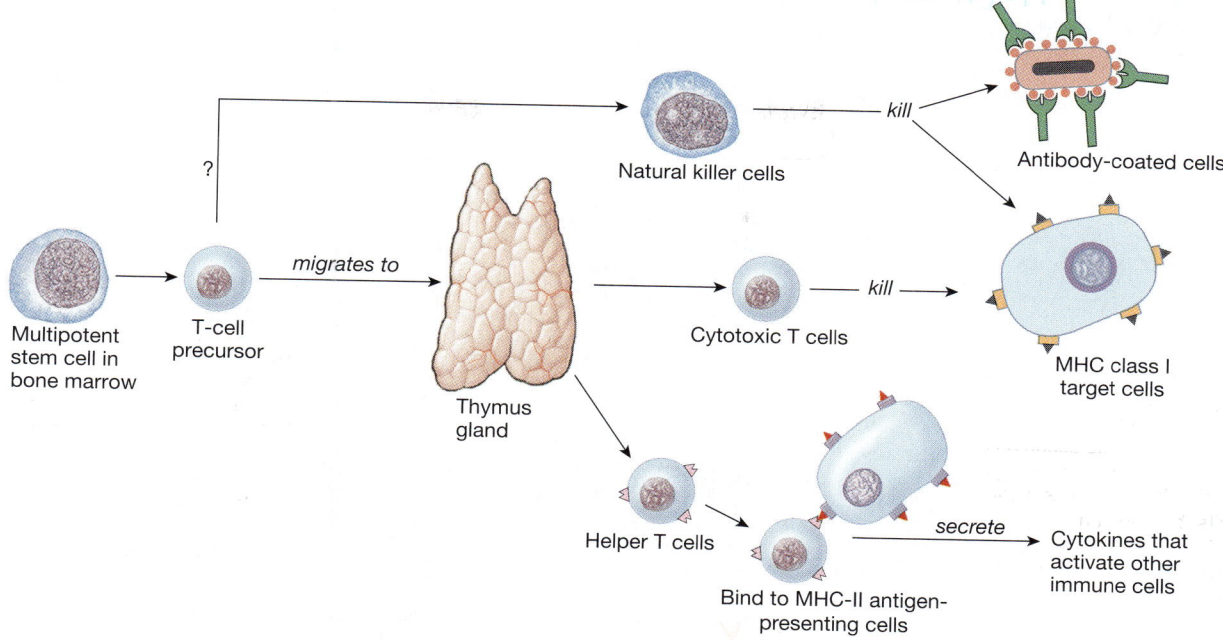

■ **FIGURE 24-16**   *Roles of T lymphocytes and NK cells in cell-mediated immunity*

Cytotoxic and helper T lymphocytes and natural killer cells develop from the same T-cell precursor.

are one reason tissues cannot be transplanted from one person to another without first establishing compatibility. It has recently been discovered that the inheritance of certain MHC alleles is linked to some disorders in which the body fails to recognize its components as "self."

Cytotoxic T Cells Kill Their Targets   Cytotoxic T (T_C) cells attack and destroy cells that display MHC class I-antigen complexes. Although this may be an extreme response, it prevents the reproduction of intracellular invaders such as viruses, some parasites, and some bacteria.

How do cytotoxic T cells kill their targets? They do so in two ways. First they can release a cytotoxic pore-forming molecule called **perforin** along with **granzymes**, enzymes related to the digestive enzymes trypsin and chymotrypsin. When granzymes enter the target cell through perforin channels, they activate an enzyme cascade that induces the cell to commit suicide (*apoptosis*). Second, cytotoxic T cells can induce apoptosis by activating **Fas**, a "death receptor" protein on the target cell membrane that is linked to the enzyme cascade.

Helper T (T_H) cells do not directly attack pathogens and infected cells, but they play an essential role in the immune response by secreting cytokines that influence other cells. The cytokines secreted by helper T cells include (1) *interferon*-gamma (IFN-γ), which activates macrophages; (2) *interleukins* that activate antibody production and cytotoxic T lymphocytes; (3) *colony-stimulating factors* that enhance leukocyte production; and (4) *interleukins* that support the actions of mast cells and

eosinophils. Helper T cells also bind to B cells and promote their differentiation into plasma cells and memory B cells. HIV, the virus that causes AIDS, preferentially infects and destroys helper T cells, leaving the host unable to respond to pathogens that otherwise could be easily suppressed.

RUNNING PROBLEM

Early in the AIDS epidemic, researchers studied the effectiveness of a drug made with CD4—a naturally occurring protein found on the surface of helper T cells, the primary targets of HIV. When HIV binds to CD4, the virus is taken into the cell and begins replicating. Researchers found that adding high amounts of CD4 to test tubes containing HIV and helper T cells reduced the rate of HIV replication. But in human subjects, the results of clinical trials using the CD4 drug were disappointing. The drug did not slow the rate at which HIV infected helper T cells.

Question 2:
   What is the rationale for flooding the body with CD4 protein to prevent HIV replication?

777   779   793   798   801   803

24

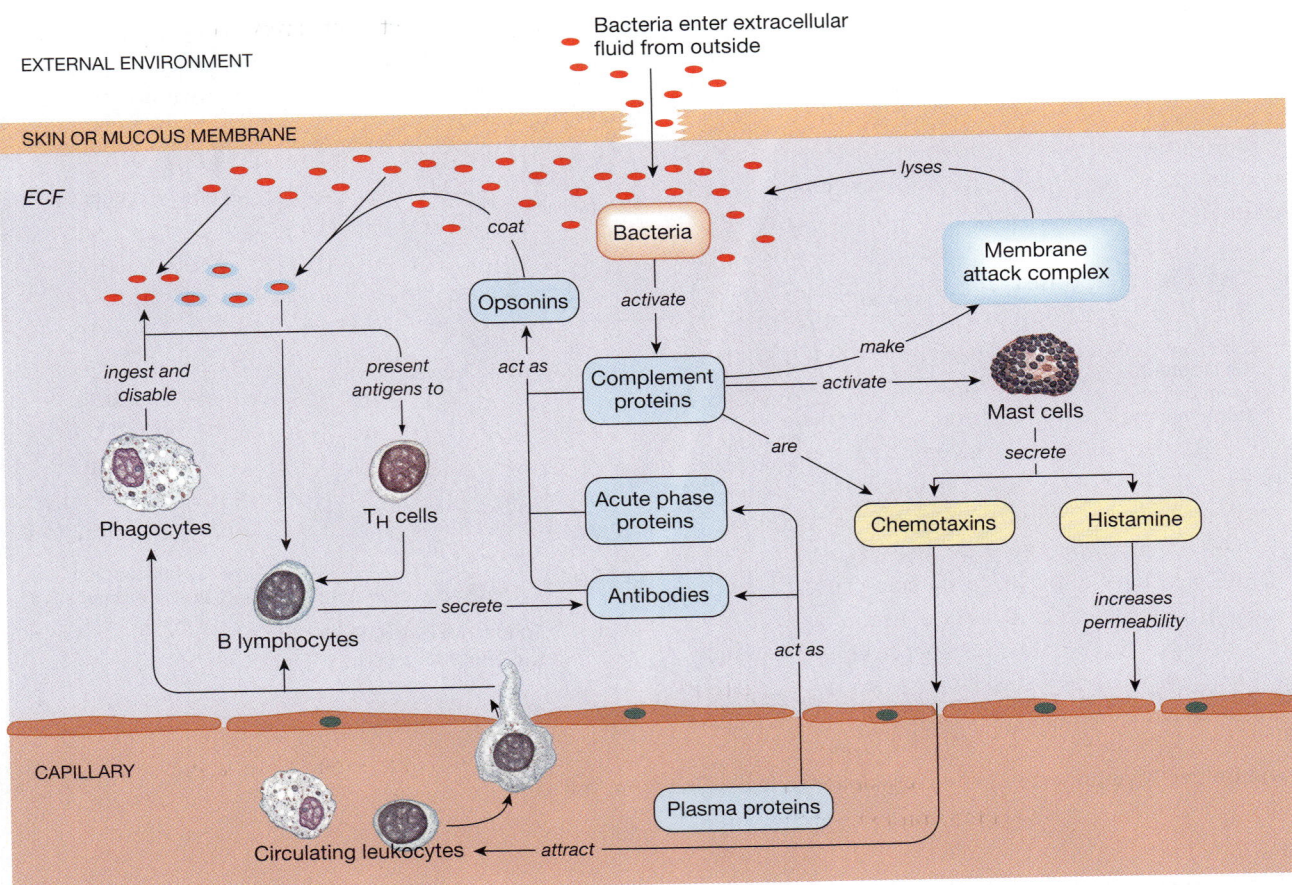

■ **FIGURE 24-17**   *Immune responses to extracellular bacteria*

IMMUNE RESPONSE PATHWAYS

How does the body respond to different kinds of immune challenges? Although the details depend on the particular challenge, the basic pattern is the same. The innate response starts first, and is reinforced by the more specific acquired response. The two pathways are interconnected, so cooperation and communication is essential. In the following sections we examine the body's responses in four situations: an extracellular bacterial infection, a viral infection, an allergic response to pollen, and the transfusion of incompatible blood.

Inflammation Is the Typical Response to Bacterial Invasion

What happens when bacteria invade? As we've seen, if the passive barricades of the skin and mucous membranes fail, bacteria reach the extracellular fluid. There they usually cause an inflammatory response that represents the combined effects of many cells working to get rid of the invader. If bacteria enter the lymph, infection-fighting takes place in lymph nodes as well.

Inflammation is characterized by a red, swollen warm area that is tender or painful. In addition to the nonspecific inflammatory response, lymphocytes attracted to the area produce

antibodies keyed to the specific type of bacterium. The entry of bacteria sets off several interrelated reactions (Fig. 24-17 ■):

1. *Activity of the complement system*. Components of the bacterial cell wall activate the complement system. Some complement proteins are chemical signals (*chemotaxins*) that attract leukocytes from the circulation to help fight the infection. The process in which immune cells leave the circulation and enter the interstitial space is called **extravasation**.

    Complement causes degranulation of mast cells and basophils. Cytokines secreted by mast cells act as additional chemotaxins, attracting more immune cells. Vasoactive chemicals such as histamine dilate blood vessels and increase capillary permeability. The enhanced blood supply to the site creates the redness and warmth of inflammation. Plasma proteins that escape into the interstitial space pull water with them, leading to tissue edema (swelling).

    The complement cascade ends with formation of membrane attack complex molecules that insert themselves into the bacterial wall of unencapsulated bacteria. The subsequent entry of ions and water lyses the bacteria, aided by the enzyme lysozyme. This is a purely chemical process that does not involve immune cells.

2. *Activity of phagocytes.* If the bacteria are not encapsulated, macrophages can begin to ingest the bacteria immediately. But if the bacteria are encapsulated, antibodies must coat the capsule before the bacteria can be ingested by phagocytes. Opsonins enhance the process for bacteria that are not encapsulated. Molecules that act as opsonins include complement, acute phase proteins, and antibodies.

3. *Role of the acquired immune response.* Some elements of the acquired immune response are called into play in bacterial infections. If antibodies against the bacteria are already present, they enhance the innate response by acting as opsonins and neutralizing bacterial toxins.

    Memory B cells attracted to the infection site will be activated if they encounter an antigen they recognize. But this process is much slower than the innate response just described, and it may take several days for plasma cells to begin secreting antibodies against the invader.

    If the infection is a novel one, some of the bacteria will activate naive B cells, and antigen-presenting cells will present bacterial fragments to helper T cells to activate them. This triggers cytokine secretion from the T_H cells, B cell clonal expansion, antibody production by plasma cells, and formation of memory B and T_H cells.

4. *Initiation of repair.* If the initial wound damaged blood vessels underlying the skin, platelets and the proteins of the coagulation cascade will also be recruited to minimize the damage [🔁 p. 548]. Once the bacteria are removed by the immune response, the injured site is repaired under the control of growth factors and other cytokines.

Intracellular Defense Mechanisms Are Needed to Fight Viral Infections

What happens when viruses invade the body? First, they encounter an extracellular phase of immune response similar to that described for bacteria. In the early stages of a viral infection, innate immune responses and antibodies can help control the invasion of the virus.

Once the viruses enter the host's cells, humoral immunity in the form of antibodies is no longer effective. Cytotoxic T lymphocytes (and to a lesser extent, NK cells) are the main defense against intracellular viruses. These cells look for infected host cells, then destroy them.

For years, cell-mediated immunity effected by T lymphocytes, and humoral immunity controlled by B lymphocytes, were considered independent processes. We now know that the two types of immunity are linked. Figure 24-18 ■ depicts how the two types of lymphocytes coordinate to destroy viruses and virus-infected cells. In this figure, we have assumed prior exposure to the virus and the presence of preexisting antibodies in the circulation. Antibodies can play an important defensive role in the early extracellular stages of a viral infection.

1. Antibodies act as opsonins, coating viral particles to make them better targets for macrophages.

2. Antibody-bound viruses are prevented from entering their target cells. However, once the virus is inside the target host cell, antibodies are no longer effective.

3. Macrophages that ingest viruses insert fragments of viral antigen into MHC class II molecules on their membranes. Macrophages also secrete a variety of cytokines. Some of these cytokines initiate the inflammatory response. They produce interferon-α, which causes host cells to make antiviral proteins that keep viruses from replicating. Other macrophage cytokines stimulate NK cells and helper T cells.

4. Helper T cells bind to viral antigen on macrophage MHC class II molecules. Activated helper T cells are the primary mediators for additional antibody production by B lymphocytes.

5. Cytotoxic T cells use viral antigen-MHC class I complexes to recognize the infected host cells. When cytotoxic T cells and NK cells bind to the infected host cell, they secrete the contents of their granules onto the cell surface. Perforin molecules insert pores into the host cell membrane so that granzymes can enter the cell, inducing it to commit suicide and undergo apoptosis. Destruction of infected host cells is a key step in halting the replication of invading viruses.

Antibodies Against a Virus May Not Work in Subsequent Infections   Although Figure 24-18 shows antibodies from a previous infection protecting the body, there is no guarantee that antibodies produced during one infection will be effective against the next invasion by the same virus. Why is this the case? Many viruses mutate constantly, and the protein coat forming the primary antigen may change significantly over time.

The influenza virus is one virus that changes yearly. Consequently, the annual vaccines against influenza must be developed based on virologists' predictions of what mutations will have occurred. If the predictions do not match the mutations or the prevalent strains, the vaccine for that year will not keep someone from catching the flu.

The rapid mutation of viruses is also one reason that researchers have not yet developed an effective vaccine against HIV, the virus that causes **acquired immunodeficiency syndrome** (AIDS). In addition, there is evidence that antibodies coating HIV particles can actually *enhance* virus uptake into target cells. HIV infects cells of the immune system, particularly T lymphocytes, monocytes, and macrophages. When HIV wipes out the helper T cells, cell-mediated immunity against the virus is lost. The general loss of an immune response in AIDS leaves these patients susceptible to a variety of viral, bacterial, fungal, and parasitic infections.

24

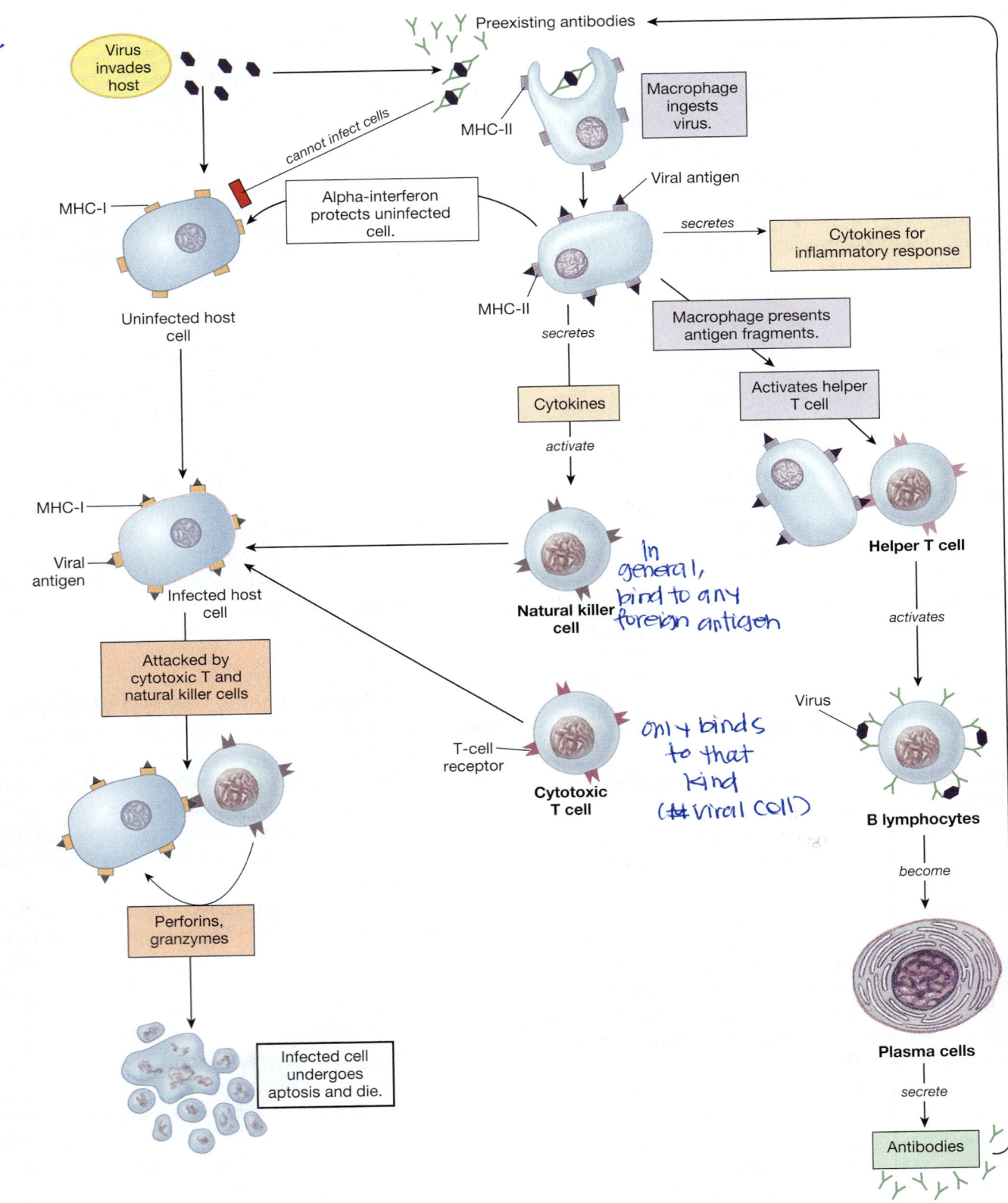

FIGURE 24-18 *Immune responses to viruses*

This figure assumes prior exposure to the virus and preexisting antibodies.

Allergic Responses Are Inflammatory Responses Triggered by Specific Antigens

An **allergy** is an inflammatory immune response to a nonpathogenic antigen. Left alone, the antigen is not harmful to the body. But if an individual is sensitive to the antigen, the body produces an inflammatory response designed to get rid of it. Allergic inflammatory responses can range from mild tissue damage to fatal reactions.

The immune response in allergies is called **sensitivity** or **hypersensitivity** to the antigen. **Immediate hypersensitivity reactions** are mediated by antibodies and occur within minutes of exposure to antigens, which are called **allergens**. **Delayed hypersensitivity reactions** are mediated by helper T cells and macrophages and may take several days to develop.

Allergens can be practically any exogenous molecule: naturally occurring or synthetic, organic or inorganic. Certain foods, insect venoms, and pollen all trigger immediate hypersensitivity reactions. Allergens can be ingested, inhaled, injected, or simply come in contact with the skin.

Delayed hypersensitivity reactions include contact allergies to copper and other base metals. These are common among people who wear costume jewelry. Poison ivy and poison oak also cause common contact allergies.

Allergies have a strong genetic component, so if parents have a ragweed allergy, chances are good that their children will too. The development of allergies requires exposure to the allergen, a factor that is affected by geographical, cultural, and social conditions.

What happens during an immediate hypersensitivity reaction to pollen is depicted in Fig. 24-19 ■.

1. The initial step—first exposure or the sensitization phase ①—is equivalent to the primary immune response discussed previously: the allergen is ingested by and processed by an antigen-presenting cell such as a macrophage, which in turn activates a helper T cell (see the right side of Fig. 24-18).

    The helper T cell activates B lymphocytes that have bound the allergen. This results in plasma cell production of antibodies (IgE and IgG) to the allergen. The IgE antibodies are immediately bound by their Fc ends to the surface of mast cells and basophils. Memory T and memory B cells store the record of the initial allergen exposure.

2. Upon reexposure ②, equivalent to the secondary immune response (see Figs. 24-10 and 24-11), the body reacts more strongly and more rapidly. The allergen binds to IgE already present on mast cells, triggering the immediate release of histamine, cytokines, and other mediators that cause allergic symptoms.

The type of allergic reaction depends on the antibody or immune cell involved. Allergen binding to IgE is the most common response to inhaled, ingested, or injected allergens. When

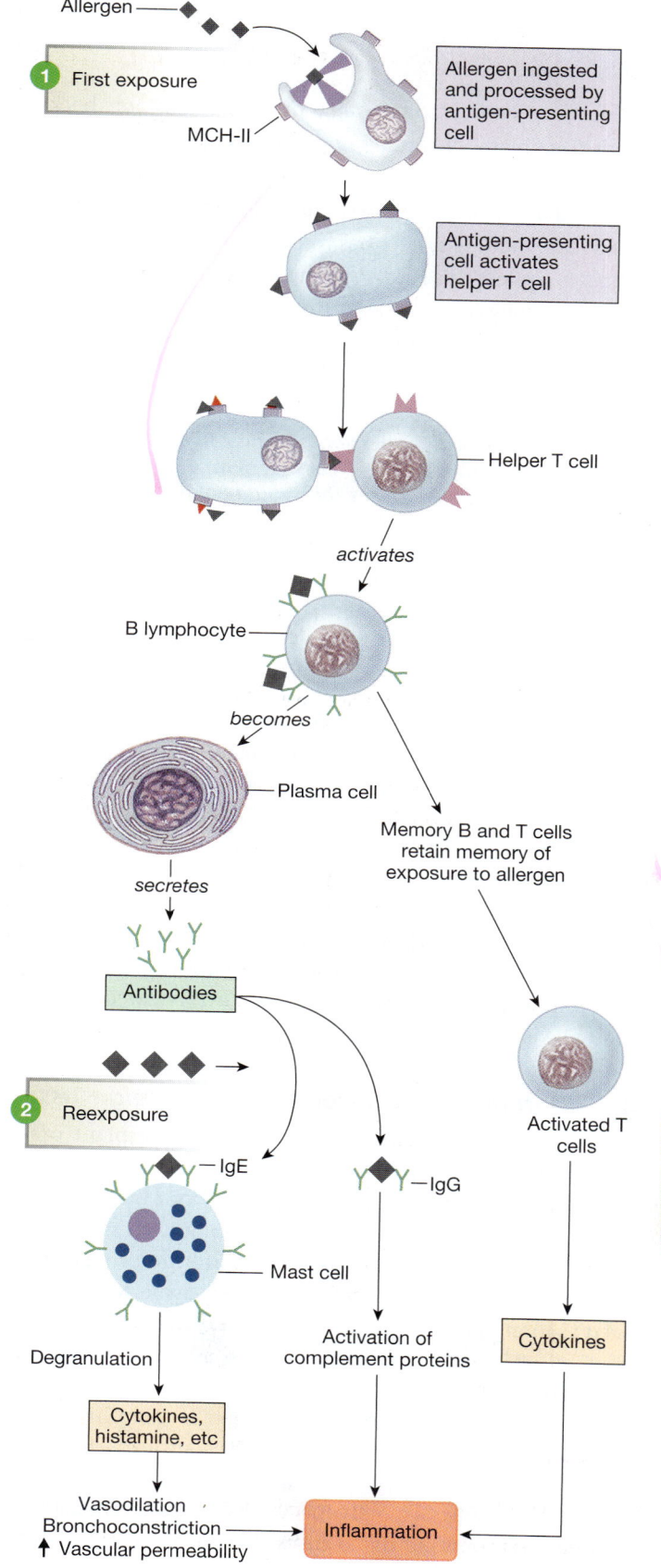

■ **FIGURE 24-19** *Allergic responses*

RUNNING PROBLEM

George remembered the day that his HIV test came back positive. In an HIV test, a sample of blood is tested for the presence of antibodies to HIV. A positive test indicates that HIV particles are circulating in the body, and that the immune system has responded by making antibodies against the virus. After his initial shock, George asked his physician this astute question: "If I'm making antibodies against HIV, why can't these antibodies kill the virus?"

Question 3:
Answer George's question. Hint: recall that HIV eventually destroys helper T cells. Look at Figure 24-18 to see what consequences this would have.

777   779   793   **798**   801   803

allergens bind to IgE antibodies on mast cells, the cells degranulate, releasing histamine and other cytokines. The result is an inflammatory reaction.

The severity of the reaction varies, ranging from localized reactions near the site of allergen entry to systemic reactions such as total body rashes. The most severe IgE-mediated allergic reaction is called **anaphylaxis** [*ana-*, without + *phylax*, guard]. In an anaphylactic reaction, massive release of histamine and other cytokines causes widespread vasodilation, circulatory collapse, and severe bronchoconstriction. Unless treated promptly, anaphylaxis can result in death.

MHC Proteins Allow Recognition of Foreign Tissue

As surgeons have developed techniques to transplant organs between human beings or from animals to humans, physicians have had to wrestle with the problem of donor tissue rejection by the recipient's immune system. The ubiquitous major histocompatibility complex proteins (MHC) are the primary tissue antigens that determine whether donated tissue will be recognized as foreign by the recipient's immune system.

The MHC proteins are also known as **human leukocyte antigens (HLA)** and are classified in an international HLA system. A tissue graft or transplanted organ is more likely to be successful if donor and recipient share HLA antigens. Incompatible matches trigger antibody production and activate cytotoxic T cells and T$_H$ cells. Generally it is T cells that are responsible for acute rejection of solid tissue grafts.

One of the most common examples of tissue donation is a blood transfusion. Human red blood cell (RBC) membranes

(a) Blood type	Antigen on red blood cell	Antibodies in plasma
O	No A or B antigens	"Anti-A" and "anti-B"
A	A antigens	"Anti-B"
B	B antigens	"Anti-A"
AB	A and B antigens	None to A or B

(b)   A mixture of type O and type A blood

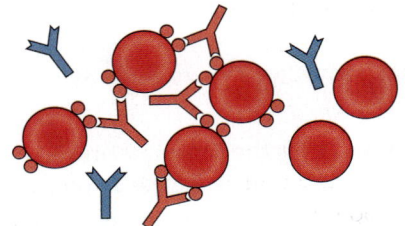

When red blood cells with group A antigens on their membranes are mixed with plasma containing antibodies to group A, the antibodies cause the blood cells to clump, or agglutinate.

FIGURE 24-20   *ABO blood groups*

The antigens and antibodies characteristic of each blood type are shown.

contain antigenic proteins and glycoproteins, but RBCs lack the MHC protein markers for recognizing foreign tissue that are found on nucleated cells. In the absence of MHC proteins, two surface proteins—the ABO blood group antigens and the Rh* antigens—become the most important causes of a rejection reaction after a blood transfusion.

The ABO blood groups consist of four blood types created by combinations of two different glycoprotein antigens (designated A and B) found on the membrane of red blood cells (Fig. 24-20a ■). Each person inherits two alleles (one from each parent) of three possible ABO alleles: A, B, and O (neither A nor B

*Rh stands for Rhesus, a kind of monkey. Rh antigens are also called D antigens.

TABLE 24-3 ABO Blood Group Frequency in the United States

BLOOD GROUP	U.S. POPULATION (%)			
	WHITE	BLACK	ASIAN	NATIVE AMERICAN
O	45	49	40	79
A	40	27	28	16
B	11	20	27	4
AB	4	4	5	<1

antigen will be produced). Because alleles A and B are dominant to allele O, the various combinations of two alleles produces four blood types: A (AA or AO), B (BB or BO), AB, and O (OO) (Table 24-3 ■).

Problems with blood transfusions arise because plasma normally contains antibodies to the ABO group antigens. These antibodies are thought to be stimulated early in life by common substances such as bacteria and food components. They can be measured in the blood of infants as early as age 3–6 months.

People express antibodies to the red blood cell antigen(s) that they do not possess. Thus, people with blood type A have anti-B antibodies in their plasma, people with no antigens on their red blood cells (blood type O) have both anti-A and anti-B antibodies, and people with both antigens on their red blood cells (blood type AB) have no antibodies to A or B antigens.

How does the body respond to a transfusion of incompatible blood? If a person with blood type O is mistakenly given a transfusion of type A blood, for example, an immune reaction takes place. The anti-A antibodies of the type O recipient bind to the transfused type A red blood cells, causing them to clump (*agglutinate*). This reaction is easily seen in a blood sample and forms the basis for the blood-typing tests often performed in student laboratories.

Antibody binding also activates the complement system, resulting in production of membrane attack complexes that cause the transfused cells to swell and leak hemoglobin. Free hemoglobin released into the plasma can result in acute renal failure as the kidneys try to filter the large molecules from the blood. Cross-matching donor and recipient blood is critical prior to giving a blood transfusion.

CONCEPT CHECK

6. A person with AB blood type is transfused with type O blood. What happens and why?

7. A person with O blood type is transfused with type A blood. What happens? Why?

Answers: p. 806

Recognition of Self Is an Important Function of the Immune System

One amazing property of immunity is the ability of the body to distinguish its own cells from foreign cells. The lack of immune response by lymphocytes to cells of the body is known as **self-tolerance** and appears to be due to the elimination of self-reactive lymphocytes. During development of T lymphocytes in the thymus, T lymphocytes develop that can combine with MHC-self-antigen complexes. But by a mechanism that is not understood, these self-reactive T lymphocytes are eliminated through apoptosis. Any self-reactive lymphocytes that arise after the thymus has become inactive are usually either destroyed or rendered ineffective.

When self-tolerance fails, the body makes antibodies against its own components through T cell-activated B lymphocytes. The body's attack on its own cells leads to **autoimmune diseases** [*auto-*, self]. The antibodies produced in autoimmune diseases are specific against a particular antigen and are usually restricted to a particular organ or tissue type. Table 24-4 ■ lists some common autoimmune diseases of humans.

Why does self-tolerance suddenly fail? We still do not know. We have learned, however, that autoimmune diseases often begin in association with an infection. One potential trigger for autoimmune diseases is foreign antigens that are similar to human antigens. When the body makes antibodies to the foreign antigen, those antibodies have enough cross-reactivity with human tissues to do some damage.

TABLE 24-4 Some Common Autoimmune Diseases in Humans

DISEASE	ANTIBODIES PRODUCED AGAINST
Graves' disease (hyperthyroidism)	Thyroid-stimulating hormone receptor on thyroid cells
Insulin-dependent diabetes mellitus	Pancreatic beta cell antigens
Multiple sclerosis	Myelin of CNS neurons
Myasthenia gravis	Acetylcholine receptor of motor endplate
Rheumatoid arthritis	Collagen
Systemic lupus erythematosus	Intracellular nucleic acid protein complexes (antinuclear antibodies)
Guillain-Barré syndrome (acute inflammatory demyelinating polyneuropathy)	Myelin of peripheral nerves

24

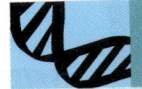

BIOTECHNOLOGY

ENGINEERED ANTIBODIES

One goal of cancer treatments is to destroy cancerous cells while minimizing damage to normal body cells. Scientists have used both chemotherapy and radiation for this approach but with mixed success. The newest cancer treatments are based on the genetic engineering of antibodies against tumor surface proteins, such as the growth factor receptor HER2 that occurs in high density on some breast cancer tumors. The engineered antibody then binds to its antigen on the tumor cell. In the case of the HER2 receptor, the antibody blocks the action of a growth factor the tumor cell needs to reproduce. Other antibodies have been engineered to carry chemotherapy drugs or radioactive molecules that specifically target cancer cells.

One strategy for developing these antibody treatments involves injecting the antigen into mice, which then produce antibodies against the antigen. Murine (mouse) antibodies injected into humans normally would trigger an immune response against the mouse protein, so scientists have used genetic engineering to create "humanized antibodies." In this procedure, the antigen-binding portion of the mouse Fab region is integrated into the Fab region of a human antibody, resulting in a protein that is less likely to be rejected in the human body.

One example of an autoimmune disease is type 1 diabetes mellitus, in which the body makes *islet cell antibodies* that destroy pancreatic beta cells but leave other endocrine cells untouched [🔁 p. 735]. Another autoimmune condition, discussed in Chapter 23, is the hyperthyroidism of Graves' disease [🔁 p. 761]. The body makes *thyroid-stimulating immunoglobulins* that mimic thyroid-stimulating hormone and cause the thyroid gland to oversecrete hormone.

Severe autoimmune diseases are sometimes treated by administering glucocorticosteroids, such as cortisol and its derivatives. Glucocorticoids depress immune system function by suppressing bone marrow production of leukocytes and decreasing the activity of circulating immune cells. Although helpful in suppressing symptoms of the autoimmune disease, large doses of steroids may trigger the same signs and symptoms as hormone oversecretion, causing patients to develop the central obesity and moon-face characteristic of Cushing's syndrome [🔁 p. 755].

Immune Surveillance Allows the Body to Remove Abnormal Cells

What is the role of the human immune system in protecting the body against cancer? This question holds considerable interest for scientists. One theory, known as **immune surveillance**, proposes that although cancerous cells develop on a regular basis, they are detected and destroyed before they can spread. Some evidence supports that theory, but it does not explain why so many cancerous tumors develop every year, or why people with depressed immune function do not develop cancerous tumors in all types of tissues. However, immune surveillance does appear to recognize and control virus-associated tumors, presumably because of the viral antigen presented to NK and cytotoxic T cells.

Many types of endogenous tumors lack surface antigens that allow them to be recognized as abnormal by the immune system. One active area of cancer research is investigating ways to activate the immune system to fight cancerous tumor cells.

NEURO-ENDOCRINE-IMMUNE INTERACTIONS

One of the most fascinating and rapidly developing areas in physiology involves the link between the mind and the body. Although serious scientists scoffed at the topic for many years, the interaction of emotions and somatic illnesses has been described for centuries. The Old Testament says, "A merry heart doeth good like a medicine; but a broken spirit drieth the bones" (Proverbs 17:22). Most societies have tales of people who lost the will to live and subsequently died without obvious illness, or of people given up for dead who made remarkable recoveries.

Today, the medical field is beginning to recognize the reality of the mind-body connection. Psychosomatic illnesses and the placebo effect have been accepted for many years. Why should we not investigate the possibility that cancer patients can enhance their immune function by visualizing their immune cells gobbling up the abnormal cancer cells?

The study of brain-immune interactions is now a recognized field known as **neuroimmunomodulation**. Currently the field is still in its infancy, and the results of many studies raise more questions than they answer. In one experiment, for example, mice were repeatedly injected with a chemical that suppressed the activity of their lymphocytes. During each injection, they were also exposed to the odor of camphor, a chemical that does not affect the immune system. After a period of conditioning, the mice were exposed only to the camphor odor. When the researchers checked the mice's lymphocyte function, they found that the odor of camphor had suppressed the activity of the immune cells, just as the chemical suppressant had done previously. The pathways through which this conditioning occurred are still largely unexplained.

What do we know at this point about the relationship between the immune, nervous, and endocrine systems?

1. **The three systems share common signal molecules and common receptors for those molecules.** Chemicals once thought to be exclusive to a single cell or tissue are being discovered all over the body. Immune cells secrete hormones, and leukocyte cytokines are produced by non-immune cells. For example, lymphocytes secrete thyrotropin (TSH), ACTH, growth hormone, and prolactin, as well as hypothalamic corticotropin-releasing hormone (CRH).

   Receptors for hormones, neurotransmitters, and cytokines are being discovered everywhere. Neurons in the brain have receptors for cytokines produced by immune cells. Natural killer cells have opiate receptors and beta adrenergic receptors. Everywhere we turn, the nervous, endocrine, and immune systems seem to share chemical signal molecules and their receptors.

2. **Hormones and neuropeptides can alter the function of immune cells.** It has been known for years that increased cortisol levels due to stress were associated with decreased antibody production, depressed lymphocyte proliferation, and diminished activity of natural killer cells. Substance P, a neuropeptide, has been shown to induce degranulation of mast cells in the mucosa of the intestines and the respiratory tract. And sympathetic innervation of the bone marrow increases antibody synthesis and production of cytotoxic T cells.

3. **Cytokines from the immune system can affect neuroendocrine function.** Stressors such as bacterial and viral infections or tumors can induce stress responses in the central nervous system through cytokines released by immune cells. Interleukin-1 is probably the best-studied cytokine in this response, but the induction of cortisol release by lymphocyte ACTH is receiving considerable attention.

   It was once believed that cortisol secretion depended on neural signals translated through the hypothalamic CRH-anterior pituitary ACTH pathway. Now it appears that pathogenic stressors can activate the cortisol pathway by causing immune cells to secrete ACTH.

The interaction of the three systems is summarized in the model shown in Fig. 24-21 ■. The nervous, endocrine, and immune systems are closely linked through bidirectional communication carried out using cytokines, hormones, and neuropeptides. The brain is linked to the immune system by autonomic neurons, CNS neuropeptides, and leukocyte cytokines. The nervous system controls endocrine glands by secretion of hypothalamic-releasing hormones, but immune cells secrete some of the same trophic hormones. Hormones from endocrine

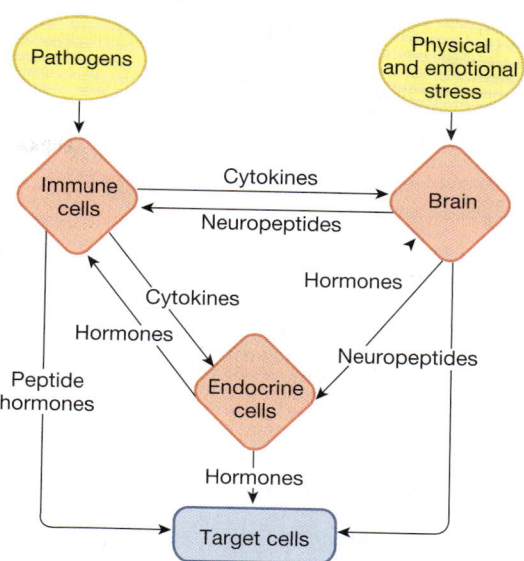

■ **FIGURE 24-21** *Model for interaction between nervous, endocrine, and immune systems*

glands feed back to influence both the nervous and immune systems.

The result is a complex network of chemical signals subject to modulation by outside factors. These outside factors include physical and emotional stimuli integrated through the brain, pathogenic stressors integrated through the immune system, and a variety of miscellaneous factors, including magnetic fields, chemical factors from brown adipose tissue, and melatonin from the pineal gland. It will probably take years for us to decipher these complex pathways.

RUNNING PROBLEM

George began taking the AIDS drug cocktail. On the day he filled the prescription, his T-cell count was 338. (A normal T-cell count is 1000.) Within four weeks, his T-cell count was 489. Within two months, it was inching toward 900. The amount of virus in George's blood (called the "viral load") had also decreased significantly. George was thrilled. "I don't know if this will last forever," he told his doctor. "But I'm going to enjoy this reprieve while I can."

Question 4:
What correlation can be made between T-cell count and the viral load of HIV in George's blood?

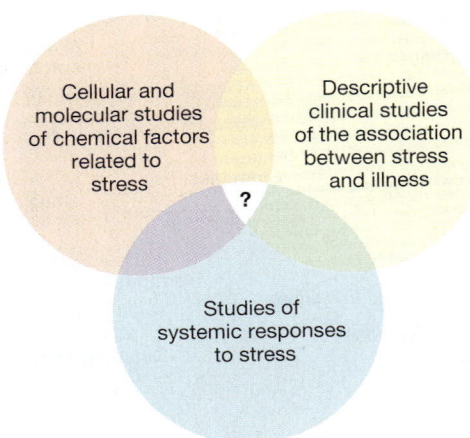

■ **FIGURE 24-22**   *Neuroimmunomodulation*

The field of neuroimmunomodulation is still in its infancy, and we do not fully understand how the nervous, immune, and endocrine systems interact to affect health and well-being.

Stress Alters Immune System Function

One area of interest is the link between inability to cope with stress and the development of illnesses. The modern study of stress is attributed to Hans Selye, beginning in 1936. He defined **stress** as nonspecific stimuli that disturb homeostasis and elicit an invariable stress response that he termed the *general adaptation syndrome*.

Selye's stress response consisted of stimulation of the adrenal glands, followed by suppression of the immune system due to high levels of circulating glucocorticoids [🔁 p. 751]. For many years, Selye's experiment was the benchmark for defining stress. However, since the 1970s, the definition of stress and the stress response has broadened.

Stressors—the events or items that create stress—are highly variable and difficult to define in experimental settings. Acute stress is different from chronic stress. A person's reaction to stress is affected by loneliness and by whether the person feels in control of the stressful situation. Many stressors are sensed and interpreted by the brain, leading to modulation of the stressor by experience and expectations. A stressor to one person may not affect another.

Most physical and emotional stressors are integrated through the central nervous system. The two classic stress responses are (1) the rapid, neurally mediated **fight-or-flight reaction**, and (2) the elevation of adrenal cortisol levels with associated suppression of the immune response. Fight-or-flight is a rapid reaction to acute stress. The cortisol response is a better indicator of chronic or repetitive stress.

One difficulty in studying the body's response to stress is the complexity involved in integrating information from three levels of response (Fig. 24-22 ■). Scientists have gathered a good deal of information at each of three levels: cellular and molecular factors related to stress, systemic responses to stress, and clinical studies describing the relationships between stress and illness. In some cases, we have linked two different levels together, although in other cases the experimental evidence is scanty or even contradictory. What we still do not understand is the big picture, the way that all three levels of response overlap.

Experimental progress is slow because many traditional study animals, such as rodents, are not good models for studying stress in humans. At the same time, controlled experiments that create stress in humans are difficult to design and execute for both practical and ethical reasons. Much of our research is observational. One study, for example, followed natural killer cell activity over an 18-month period in two groups of people: spouse caregivers of Alzheimer's disease patients (a group that reports high psychosocial stress) and a matched group of controls. For a subset of the caregivers, high perceived stress at the beginning of the study was associated with low NK cell activity at the end. However, many variables—including age, gender, exercise, and use of alcohol and prescription drugs—must be taken into account before we can conclude that stress was the causative factor.

Modern Medicine Incorporates Mind-Body Therapeutics

The role of mind-body interactions in traditional medicine is receiving more attention in recent years. The U.S. National Center for Complementary and Alternative Medicine (*http://nccam.nih.gov*) supports rigorously designed research studies on the efficacy of complementary medical therapies, such as meditation, yoga, hypnosis, visual imagery, and biofeedback. Although the placebo and nocebo effects [🔁 p. 11] are widely accepted in the traditional medical community, the role of complementary therapies such as laughter therapy is still being studied. Can watching a funny movie really help your immune system? Recent studies suggest that it can. In these studies, researchers measured the activity of immune cells in subjects before and after watching a humorous movie or video. In both studies, immune cells became more active. You can learn more about *humor therapy* from the Association for Applied and Therapeutic Humor (*www.aath.org*).

In spite of the difficulties of studying the integration of the immune, endocrine, and nervous systems, scientists are beginning to piece together the intricate tapestry of mind-body interactions. When we can explain at the molecular level why students tend to get sick just before exam time, and why relaxation therapy enhances the immune response, we will be well on the way to understanding the complex connections between the brain and the immune system.

RUNNING PROBLEM CONCLUSION

TREATMENT FOR AIDS

In this running problem you learned about a treatment for AIDS that consists of two anti-retroviral drugs and a protease inhibitor. The combination of these drugs seems to stem the tide of HIV replication more effectively than either of the drugs used alone. Other drugs have been developed, and tests for an HIV vaccine are still in progress. To learn more about current treatments, see *www.aidsinfo.nih.gov* or

AIDS.org at *www.aids.org*. For statistics on HIV and AIDS in the United States, go to *www.cdc.gov/hiv/stats.htm*. For international statistics, go to the World Health Organization at *www.who.int/emc-hiv/*

Now check your understanding of this running problem by comparing your answers to the information in the summary table.

	QUESTION	FACTS	INTEGRATION AND ANALYSIS
1	Why can't retroviruses reproduce without reverse transcriptase?	Retroviruses have no DNA to use as a template to make viral mRNA. Reverse transcriptase is an enzyme that allows retroviruses to make complementary DNA from their own viral RNA.	Without reverse transcriptase, retroviruses cannot make viral DNA. Without viral DNA, the virus cannot make viral mRNA or viral proteins. Therefore, without reverse transcriptase, a retrovirus cannot replicate.
2	What is the rationale for flooding the body with CD4 protein to prevent HIV replication?	CD4 is one receptor on helper T cells to which HIV binds. Binding causes the virus to be taken into the T cell.	Introducing large amounts of CD4 protein into the body might "trick" HIV into binding to these free proteins instead of to CD4 on helper T cells. This would reduce the number of virus particles that could invade T cells.
3	George asks, "If I'm making antibodies against HIV, why can't these antibodies kill the virus"? Answer George's question.	Viruses can only be affected by antibodies when they are outside host cells. HIV bound to antibody is taken up into helper T cells.	HIV enters helper T cells and, over time, destroys them.
4	What correlation can be made between T-cell count and the viral load?	George's T-cell count increases while the viral load decreases.	Apparently, the drug cocktail is effective in decreasing the rate of T-cell infection by HIV. This has two results. As fewer T cells are killed by the virus, the T-cell count rises. As the drugs suppress viral replication, viral load decreases.

777 779 793 798 801 **803**

CHAPTER SUMMARY

In this chapter you learned how the immune system protects the body from pathogens and how it removes damaged tissue and abnormal cells. An immune response requires detection of the foreign substance, communication with other immune cells, and coordination of the response. These functions depend on *chemical communication* and *molecular interactions* between receptors, antibodies, and antigens. The *compartmentation* of the body affects how the immune system fights pathogens: some pathogens hide inside cells while others remain in the extracellular compartment.

Overview of Immune System Function

1. **Immunity** is the ability of the body to protect itself from pathogens. The human immune system consists of lymphoid tissues, immune cells, and cytokines. (p. 777)

2. **Antigens** are substances that trigger an immune response and react with products of that response. (p. 777)

Pathogens of the Human Body

3. Bacteria are cells that can survive and reproduce outside of a living host. Antibiotic drugs kill bacteria. (p. 778)

4. Viruses are DNA or RNA cores enclosed in protective proteins. They invade host cells and reproduce using the host's intracellular

machinery. The immune system must destroy infected host cells to kill a virus. (p. 778; Fig. 24-1)

The Immune Response

5. The immune response includes: (1) *detection* and *identification*, (2) *communication* with other immune cells, (3) *recruitment* of assistance and *coordination* of the response, and (4) *destruction* or *suppression* of the invader. (p. 779)

6. **Innate immunity** is the rapid, **nonspecific immune response**. **Acquired immunity** is slower and is an **antigen-specific response**. The two types of immunity overlap. (p. 779)

Anatomy of the Immune System

7. Immune cells form and mature in the **thymus gland** and **bone marrow**. The **spleen** and **lymph nodes** are encapsulated lymphoid tissues. Unencapsulated **diffuse lymphoid tissues** include **tonsils** and the **gut-associated lymphoid tissue (GALT)**. (p. 781; Figs. 24-2, 24-3, 24-14)

8. **Eosinophils** are **cytotoxic cells** that kill parasites. **Basophils** and **mast cells** release histamine, **heparin**, and other cytokines. (p. 782; Fig. 24-4)

9. **Neutrophils** are phagocytic cells that release pyrogens and cytokines that mediate inflammation. (p. 783; Fig. 24-4)

10. **Monocytes** are the precursors of tissue **macrophages**. **Antigen-presenting cells (APCs)** display antigen fragments in proteins on their membrane. (p. 783; Figs. 24-4, 24-5)

11. **Lymphocytes** and **plasma cells** mediate antigen-specific immune responses. (p. 784)

12. **Dendritic cells** capture antigens and present them to lymphocytes. (p. 784)

Innate Immunity: Nonspecific Responses

13. Physical and chemical barriers, such as skin and mucus, are the first line of defense. (p. 784)

14. **Chemotaxins** attract immune cells to pathogens. (p. 784)

15. Phagocytosis requires that surface molecules on the pathogen bind to membrane receptors on the phagocyte. **Opsonins** are antibodies or complement proteins that coat foreign material. (p. 785; Figs. 24-6, 24-7)

16. **Natural killer cells** kill certain tumor cells and virus-infected cells. They bind to the Fc stem of antibodies that are bound to a target cell. This is called **antibody-dependent cell-mediated cytotoxicity**. (pp. 785, 790)

17. **Inflammation** results from cytokines released by activated tissue macrophages. (p. 787)

18. Inflammatory cytokines include **acute phase proteins**, histamine, **interleukins, bradykinin**, and **complement proteins**. (pp. 786–787)

19. The complement cascade forms **membrane attack complex**, pore-forming molecules that cause pathogens to rupture. (p. 787; Fig. 24-8)

Acquired Immunity: Antigen-Specific Responses

20. Lymphocytes mediate the acquired immune response. Lymphocytes that react to one type of antigen form a **clone**. (p. 788; Fig. 24-9)

21. First exposure to an antigen activates a **naive lymphocyte** clone and causes it to divide (**clonal expansion**). Newly formed lymphocytes differentiate into **effector cells** or **memory cells**. Additional exposure to the antigen creates a faster and stronger **secondary immune response**. (p. 788; Figs. 24-10, 24-11)

22. Lymphocyte cytokines include **interleukins** and **interferons**, which interfere with viral replication. **Interferon-gamma** activates macrophages and other immune cells. (p. 793)

23. **B lymphocytes** secrete antibodies. **T lymphocytes** and natural killer cells attack and destroy infected cells. (p. 788)

24. **Antibodies** bind to pathogens. Mature B lymphocytes are covered with antibodies that act as surface receptors. Activated B lymphocytes differentiate into **plasma cells**, which secrete soluble antibodies. (p. 788)

25. The five classes of immunoglobulins (Ig), or antibodies, are IgG, IgA, IgE, IgM, and IgD, collectively known as **gamma globulins**. (p. 789)

26. The antibody molecule is Y shaped, with four polypeptide chains. The arms, or **Fab regions**, contain antigen-binding sites. The stem, or **Fc region**, can bind to immune cells. Some immunoglobulins are polymers. (p. 790; Fig. 24-12)

27. Soluble antibodies act as opsonins, serve as a bridge between antigens and leukocytes, activate complement proteins and mast cells, and disable viruses and toxins. (p. 790; Fig. 24-13)

28. B cell antibodies bind antigens to activate B lymphocytes and initiate the production of additional antibodies. (p. 790; Fig. 24-13)

29. **T lymphocytes** are responsible for **cell-mediated immunity**, in which the lymphocytes bind to target cells using **T-cell receptors**. (p. 791; Fig. 24-15)

30. T lymphocytes react to antigen bound to **major histocompatibility complex** proteins on the target cell. **MHC class I molecules** displaying antigen cause cytotoxic T (T_C) cells to kill the target cell. **MHC class II molecules** with antigen activate helper T (T_H) cells to secrete cytokines. (pp. 791–792; Fig. 24-16)

31. **Cytotoxic T (T_C) cells** release **perforin**, a pore-forming molecule that allows **granzymes** to enter the target cell and trigger it into committing suicide (**apoptosis**). (p. 793)

32. **Helper T (T_H) cells** secrete cytokines that influence other cells. (p. 793; Fig. 24-16)

Immune Response Pathways

33. Bacteria usually cause a nonspecific inflammatory response. In addition, lymphocytes produce antibodies keyed to the specific type of bacterium. (p. 794; Fig. 24-17)

34. Innate immune responses and antibodies help control the early stages of a viral infection. Once the viruses enter the host's cells, cytotoxic T cells must destroy infected host cells. (p. 795; Fig. 24-18)

35. An **allergy** is an inflammatory immune response to a nonpathogenic **allergen**. The body gets rid of the allergen through the inflammatory response. (p. 797; Fig. 24-19)

36. **Immediate hypersensitivity reactions** to allergens are mediated by antibodies and occur within minutes of exposure to antigen. **Delayed hypersensitivity reactions** are mediated by T lymphocytes and may take several days to develop. (p. 797)

37. The major histocompatibility complex (MHC) proteins found on all nucleated cells determine tissue compatibility. Lymphocytes will attack any MHC complexes they do not recognize as self proteins. (p. 798)

38. Human red blood cells lack MHC proteins but contain other antigenic membrane proteins and glycoproteins, such as the ABO and Rh antigens. (p. 798; Fig. 24-20)

24

39. **Self-tolerance** is the body's ability to distinguish its own cells and not produce an immune response. When self-tolerance fails, the body makes antibodies against its own components, creating an **autoimmune disease**. (p. 799)

40. The theory of **immune surveillance** proposes that cancerous cells develop on a regular basis but are detected and destroyed by the immune system before they can spread. (p. 800)

Neuro-Endocrine-Immune Interactions

41. The nervous, endocrine, and immune systems are linked together by signal molecules and receptors. The study of these interactions is a field known as **neuroimmunomodulation**. (p. 800; Fig. 24-21)

42. The link between inability to cope with stress and the development of illnesses is believed to result from neuroimmunomodulation. (p. 802)

QUESTIONS

(Answers to the Review Questions begin on page A1.)

LEVEL ONE REVIEWING FACTS AND TERMS

1. Define immunity. What is meant by memory and specificity in the immune system?
2. Name the anatomical components of the immune system.
3. List and briefly summarize the three main functions of the immune system.
4. Name two ways viruses are released from a host cell. Give three ways in which viruses damage host cells or disrupt their function.
5. Suppose a pathogen has invaded the body. What four steps occur if the body is able to destroy the pathogen? What compromise may be made if the pathogen cannot be destroyed?
6. Define the following terms and explain their significance:
   (a) anaphylaxis
   (b) agglutinate
   (c) extravascular
   (d) degranulation
   (e) acute phase protein
   (f) clonal expansion
   (g) immune surveillance
7. How are histiocytes, Kupffer cells, osteoclasts, and microglia related?
8. What is the mononuclear phagocytic system, and what role does it play in the immune system?
9. Match each of the following cell types with its description:
   (a) lymphocyte
   (b) neutrophil
   (c) monocyte
   (d) dendritic cell
   (e) eosinophil
   (f) basophil
   1. most abundant leukocyte; phagocytic; lives 1–2 days
   2. cytotoxic; is associated with allergic reactions and parasitic infestations
   3. precursor of macrophages; relatively rare in blood smears
   4. related to mast cells; releases cytokines such as histamine and heparin
   5. most of these cells reside in the lymphoid tissues
   6. these cells capture antigens, then migrate to the lymphoid tissues
10. Give examples of physical and chemical barriers to infection.
11. Name the three main types of lymphocytes and their subtypes. Explain the functions and interactions of each group.
12. Define self-tolerance. Which cells are responsible? Relate self-tolerance to autoimmune disease.
13. What is meant by the term *neuroimmunomodulation*?
14. Explain the terms *stress, stressor,* and *general adaptation syndrome*.

LEVEL TWO REVIEWING CONCEPTS

15. **Concept map:** Map the following terms. You may add additional terms.
   - acquired immunity
   - antibody
   - antigen
   - antigen-presenting cell
   - B lymphocyte
   - bacteria
   - basophil
   - chemotaxin
   - cytokine
   - cytotoxic cell
   - cytotoxic T cell
   - dendritic cell
   - eosinophil
   - Fab region
   - Fc region
   - helper T cell
   - innate immunity
   - memory cell
   - MHC class I
   - MHC class II
   - monocyte
   - natural killer cell
   - neutrophil
   - opsonin
   - phagocyte
   - phagosome
   - plasma cell
   - T lymphocyte
   - viruses
16. Why do lymph nodes often swell and become tender or even painful when you are sick?
17. Summarize the effects of histamine, interleukin-1, acute phase proteins, bradykinin, complement, and interferon-γ. Are these chemicals antagonistic or synergistic to one another?
18. Compare and contrast the terms in the following sets of terms:
   (a) pathogen, microbe, pyrogen, antigen, antibody, antibiotic
   (b) infection, inflammation, allergy, autoimmune disease
   (c) virus, bacteria, allergen, retrovirus
   (d) chemotaxin, opsonin, cytokine, interleukin, bradykinin, interferon
   (e) acquired immunity, innate immunity, humoral immunity, cell-mediated immunity, specific immunity, nonspecific immunity
   (f) immediate and delayed hypersensitivity
   (g) membrane attack complex, granzymes, perforin
19. Diagram and label the parts of an antibody molecule. What is the significance of each part?
20. Diagram the steps in the process of inflammation.
21. Diagram the steps in the process of fighting viral infections.
22. Diagram the steps in the process of an allergic reaction.

LEVEL THREE PROBLEM SOLVING

23. One ABO blood type is called the "universal donor," and one is called the "universal recipient." Which types are these, and why are they given those names?

24

24. Maxie Moshpit claims that Snidley Superstar is the father of her child. Maxie has blood type O. Snidley has blood type B. Baby Moshpit has blood type O. Can you rule out Snidley as the father of the baby with only this information? Explain.

25. Every semester around exam time more students visit the Student Health Center with colds and viral infections. The physicians attribute the increased incidence of illness to stress. Can you explain the biological basis for that relationship? What are some other possible explanations?

26. Barbara is troubled with rheumatoid arthritis, characterized by painful, swollen joints from inflamed connective tissue. She notices that it flares up when she is tired and stressed. This condition is regarded as an autoimmune disease. Outline the physiological reactions that would lead to this condition.

27. The differential white cell count [Fig. 16-3, p. 541] is an important diagnostic tool in medicine. If a patient's "diff" shows an increase in the number of immature neutrophils ("bands"), do you think the cause is more likely to be a bacterial or a viral infection? If the count shows an increase in eosinophils (*eosinophilia*), is the cause more likely to be a viral infection or a parasitic infection? Explain your reasoning.

ANSWERS

✓ Answers to Concept Check Questions

Page 778

1. Bacteria are cells; viruses are not. Bacteria can reproduce outside a host; viruses cannot. Bacteria can be killed by antibiotics; viruses cannot.

2. Virus enters host cell and uses its viral nucleic acids to tell host cell to make viral nucleic acids and proteins. These assemble into new virus particles, which are then released from the host cell.

Page 787

3. When capillary permeability increases, proteins move from plasma to interstitial fluid. This decreases the colloid osmotic force opposing capillary filtration, and additional fluid accumulates in the interstitial space (swelling or edema).

Page 790

4. Antibodies can be moved across cells by transcytosis or released from cells by exocytosis.

5. The child developed antibodies to bee venom on first exposure, so if the child will have a severe allergic reaction to bee venom, it will occur on the second exposure. Allergic reactions are mediated by IgE. After the first bee string, IgE antibodies secreted in response to the venom are bound to the surface of mast cells. At the second exposure, bee venom binding to the IgE causes the mast cells to degranulate, causing an allergic reaction that may be severe enough to cause anaphylaxis.

Page 799

6. Type O blood has neither A nor B antigens, but even if it did, type AB blood does not produce anti-A or anti-B antibodies. Therefore, no immune reaction will occur.

7. If type A blood is given to a type O recipient, the recipient's anti-A antibodies will cause agglutination.

Q Answers to Figure Questions

Page 788
Figure 24-9: The antigen will activate clone 1.

One fascinating aspect of the physiology of the human being at work is that it provides basic information about the nature and the range of the functional capacity of different organ systems.

—Per-Olof Åstrand and Kaare Rodahl, *Textbook of Work Physiology, 1977*

Integrative Physiology III: Exercise

BACKGROUND BASICS

HEAT STROKE

"Dangerous heat continues for fifth straight day," reads the morning headline. Had Colleen Warren, 19, taken time to scan the story, she might have avoided the life-threatening health emergency she faced a few hours later. But Colleen overslept and rushed off to field hockey practice. The weather this week had been unbearably hot and humid, just as it had been all summer. The forecast today was for a high of 105° F, with high humidity. In her haste, Colleen forgot her water bottle. "No big deal," she thought as she hopped on her bike to peddle the two miles to the practice field.

| 808 | 811 | 812 | 814 | 816 | 818 |

One of the most common challenges to body homeostasis is not a disease but an everyday activity: exercise. Exercise comes in many forms. Distance running and cycling are examples of a dynamic endurance exercise. Weight lifting and strength training are examples of resistance training. In this chapter we examine dynamic exercise as a challenge to homeostasis that is met with an integrated response from multiple body systems. In many ways, exercise is the ideal example for teaching physiological integration. Everyone is familiar with it, and unlike high-altitude mountaineering or deep-sea diving, it involves no special environmental conditions. Moreover, exercise is a normal physiological state, not a pathological one—although it can be affected by disease and (if excessive) can result in injury.

Note that this chapter is not intended to be a complete discussion of exercise physiology. For further information, readers should consult an exercise physiology textbook.

METABOLISM AND EXERCISE

Exercise begins with skeletal muscle contraction, an active process that requires ATP for energy. But where does the ATP for muscle contraction come from? A small amount is in the muscle fiber when contraction begins (Fig. 25-1 ■). As this ATP is transformed into ADP, another phosphate compound, **phosphocreatine** (PCr), transfers energy from its high-energy phosphate bond to ADP. The transfer replenishes the muscle's supply of ATP [🔁 p. 410].

Together, the combination of muscle ATP and phosphocreatine is adequate to support only about 15 seconds of intense exercise, such as sprinting or power lifting. Subsequently, muscle fibers must manufacture additional ATP from energy stored in nutrients. Some of these molecules are contained within the muscle fiber itself. Others must be mobilized from the liver and adipose tissue, then transported to muscles through the circulation.

The primary substrates for energy production are carbohydrates and fats. Recall from Chapter 4 that the most efficient production of ATP occurs through aerobic pathways such as the glycolysis-citric acid cycle pathway [🔁 p. 107]. If the cell has adequate amounts of oxygen for oxidative phosphorylation, then both glucose and fatty acids can be metabolized to provide ATP (Fig. 25-1).

If the oxygen requirement of a muscle fiber exceeds its oxygen supply, energy production from fatty acids decreases dramatically, and glucose metabolism shifts to anaerobic pathways. Anaerobic metabolism is also known as **glycolytic metabolism**, because in low oxygen conditions, anaerobic glycolysis is the primary pathway for ATP production. When a cell lacks oxygen for oxidative phosphorylation, the final product of glycolysis—pyruvate—is converted to lactic acid [🔁 p. 106]. In general, exercise that depends on anaerobic metabolism cannot be sustained for an extended period.

Anaerobic metabolism has the advantage of speed, producing ATP 2.5 times as rapidly as aerobic pathways do (Fig. 25-2 ■). But this advantage comes with two distinct disadvantages: (1) anaerobic metabolism provides only 2 ATP per glucose, compared with an average of 30–32 ATP per glucose for oxidative metabolism, and (2) anaerobic metabolism contributes to a state of metabolic acidosis by producing lactic acid. (However, the CO_2 generated during metabolism is a more significant source of acid.)

Where does glucose for aerobic and anaerobic ATP production come from? The body has three sources: the plasma glucose pool, intracellular stores of glycogen in muscles and liver, and "new" glucose made in the liver through *gluconeogenesis* [🔁 p. 112]. Muscle and liver glycogen stores provide enough energy substrate to release about 2000 kcal (equivalent to about 20 miles of running in the average person), more than adequate for the exercise that most of us do. However, glucose alone cannot provide sufficient ATP for endurance athletes such as marathon runners. To meet their energy demands, they rely on the energy stored in fats, estimated at 70,000 kcal per person.

In reality, aerobic exercise of any duration uses both fatty acids and glucose as substrates for ATP production. About 30 minutes after aerobic exercise begins, the concentration of free fatty acids in the blood increases significantly, indicating that fats are being mobilized from adipose tissue. However, the breakdown of fatty acids through the process of β-oxidation [🔁 p. 111] is slower than glucose metabolism through glycolysis, so muscle fibers use a combination of fatty acids and glucose to meet their energy needs.

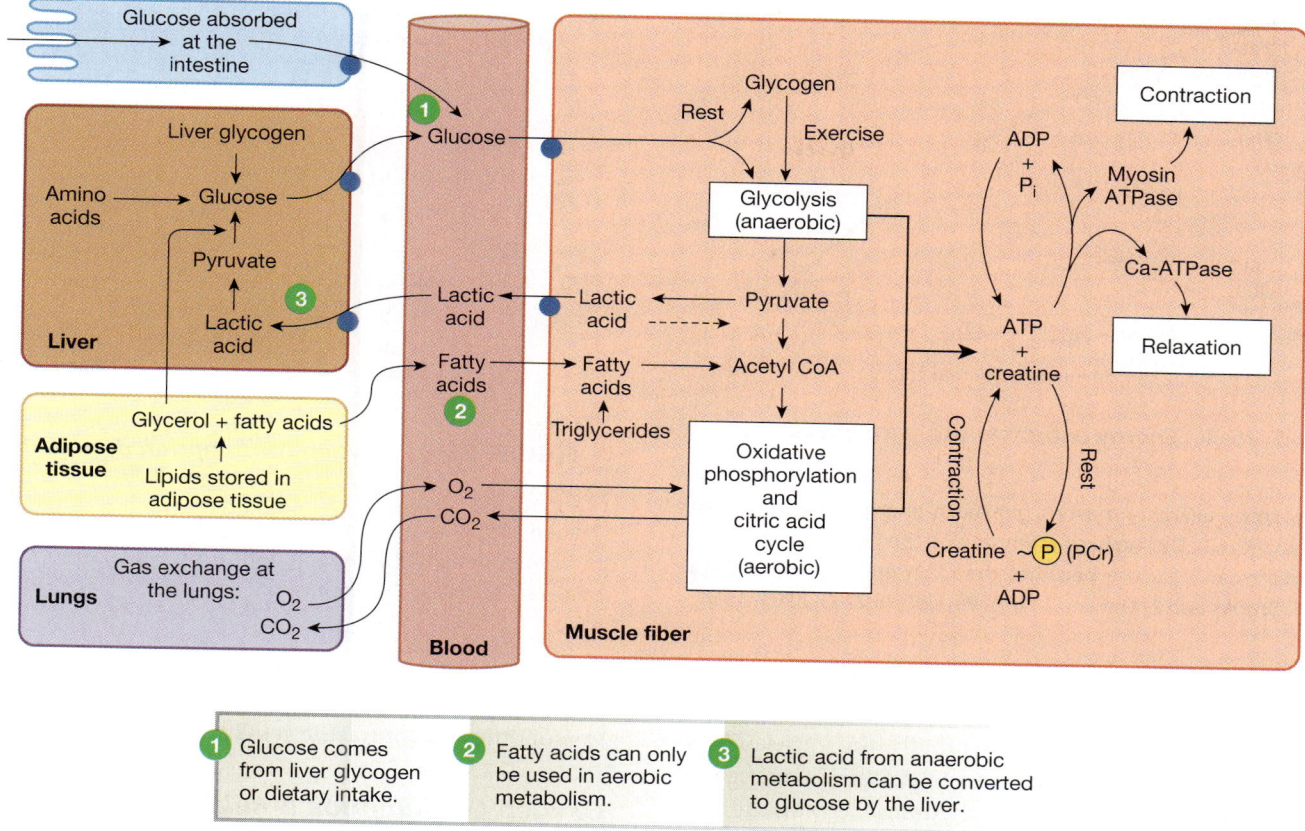

■ FIGURE 25-1   *Overview of muscle metabolism*

ATP for muscle contraction is continuously produced by aerobic metabolism of glucose and fatty acids. During short bursts of activity, when ATP demand exceeds the rate of aerobic ATP production, anaerobic glycolysis produces ATP and lactic acid.

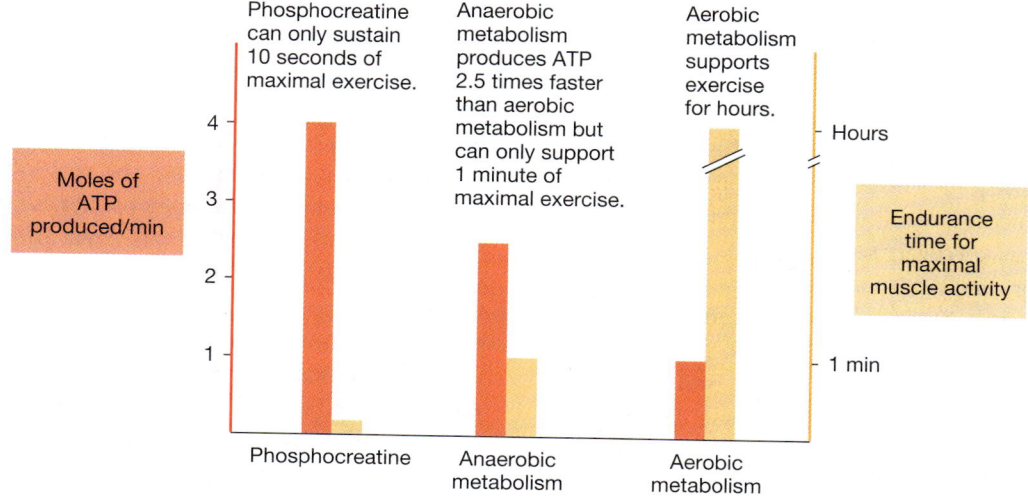

■ FIGURE 25-2   *Speed of ATP production compared with ability to sustain maximal muscle activity*

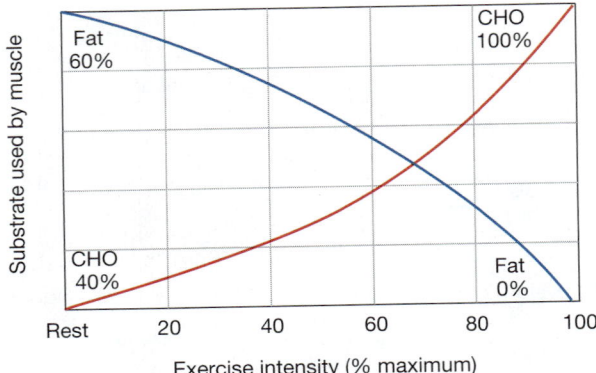

■ FIGURE 25-3 *Energy substrate use with increasing exercise*

At low intensity exercise, muscles get more energy from fats than from glucose (CHO). During high-intensity exercise (levels greater than 70% of maximum), glucose becomes the main energy source. Data from G. A. Brooks and J. Mercier, *J. App. Physiol.* 76:2253–2261, 1994.

At lower exercise intensities, most of the energy for ATP production comes from fats (Fig. 25-3 ■). As exercise intensity increases and ATP is consumed more rapidly, the muscle fibers begin to use a larger proportion of glucose. When exercise exceeds about 70% of maximum, carbohydrates become the primary source of energy.

Aerobic training increases both fat and glycogen stores within muscle fibers. Endurance training also increases the activity of enzymes for β-oxidation and converts muscle fibers from fast-twitch glycolytic to fast twitch oxidative-glycolytic [🔁 p. 412].

Hormones Regulate Metabolism During Exercise

Several hormones that affect glucose and fat metabolism change their pattern of secretion during exercise. Plasma concentrations of glucagon, cortisol, the catecholamines (epinephrine and norepinephrine), and growth hormone all increase during exercise. Cortisol and the catecholamines, along with growth hormone, promote the conversion of triglycerides to glycerol and fatty acids. Glucagon, catecholamines, and cortisol also mobilize liver glycogen and raise plasma glucose levels. A hormonal environment that favors the conversion of glycogen into glucose is desirable, because glucose is a major energy substrate for exercising muscle.

Curiously, although plasma glucose concentrations rise with exercise, the secretion of insulin decreases. This response is contrary to what you might predict, because normally an increase in plasma glucose stimulates insulin release. During exercise, however, insulin secretion is suppressed, probably by sympathetic input onto the beta cells of the pancreas.

What could be the advantage of lower insulin levels during exercise? For one thing, less insulin means that cells other than muscle fibers reduce their glucose uptake, sparing blood glucose for use by muscles. Actively contracting muscle cells, on the other hand, are not affected by low levels of insulin because they do not require insulin for glucose uptake. Contraction

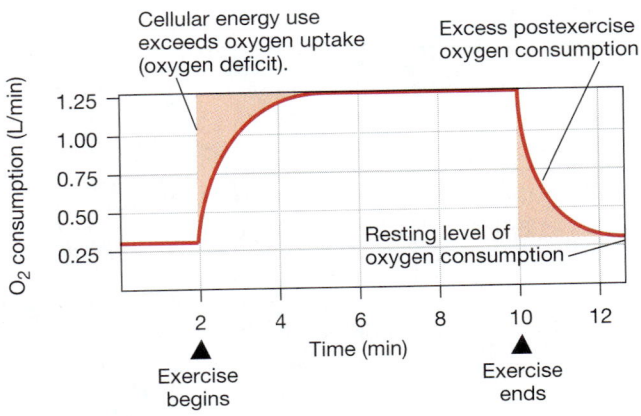

■ FIGURE 25-4 *Oxygen consumption during and after exercise*

Oxygen supply to exercising cells lags behind energy use, creating an oxygen deficit. The oxygen deficit is compensated for by excess postexercise oxygen consumption.

stimulates the insulin-independent translocation of GLUT4 transporters to the muscle membrane, increasing glucose uptake in proportion to contractile activity.

Oxygen Consumption Is Related to Exercise Intensity

The activities we call exercise range widely in intensity and duration, from the rapid and relatively brief burst of energy exerted by a sprinter or power lifter, to the sustained effort of a marathoner. Physiologists traditionally quantify the intensity of a period of exercise by measuring oxygen consumption ($\dot{V}_{O_2}$). **Oxygen consumption** refers to the fact that oxygen is used up, or consumed, during oxidative phosphorylation, when it combines with hydrogen in the mitochondria to form water [🔁 p. 108].

Oxygen consumption is a measure of cellular respiration and is usually measured in liters of oxygen consumed per minute. A person's *maximal rate of oxygen consumption* ($\dot{V}_{O_2max}$) is an indicator of the ability to perform endurance exercise. The greater the $\dot{V}_{O_2max}$, the greater the person's predicted ability to do work. A metabolic hallmark of exercise is an increase in oxygen consumption that persists even after the activity ceases (Fig. 25-4 ■). When exercise commences, oxygen consumption increases so rapidly that it is not immediately matched by the oxygen supplied to the muscles. During this lag time, ATP is provided by muscle ATP reserves, phosphocreatine, and aerobic metabolism supported by oxygen stored on muscle myoglobin and blood hemoglobin [🔁 p. 412].

The use of these muscle stores creates an *oxygen deficit* because their replacement requires aerobic metabolism and oxygen uptake. Once exercise stops, oxygen consumption is slow to resume its resting level. The **excess postexercise oxygen consumption** (EPOC; formerly called the oxygen debt) represents oxygen being used to metabolize lactate, restore ATP and phosphocreatine levels, and replenish the oxygen bound to myoglobin. Other factors that play a role in elevating postexercise oxygen consumption include increased body temperature and circulating catecholamines.

Several Factors Limit Exercise

What are the factors that limit exercise capacity? To some extent, the answer depends on the type of exercise. Resistance training such as strength training depends heavily on anaerobic metabolism to meet energy needs. The situation is more complex with aerobic or endurance exercise. Is the limiting factor for aerobic exercise the ability of the exercising muscle to use oxygen efficiently? The ability of the cardiovascular system to deliver oxygen to the tissues? The ability of the respiratory system to provide oxygen to the blood?

One possible limiting factor in exercise is the ability of muscle fibers to obtain and use oxygen. If muscle mitochondria are limited in number, or if they have insufficient oxygen supply, the muscle fibers are unable to produce ATP rapidly. Although data suggest that muscle metabolism is not the limiting factor for maximum exercise capacity, muscle metabolism has been shown to influence submaximal exercise capacity—a finding that explains the increase in numbers of muscle mitochondria and capillaries with endurance training.

The question of whether the pulmonary system or the cardiovascular system limits maximal exercise was resolved when research showed that ventilation is only 65% of its maximum when cardiac output has reached 90% of its maximum. From that information, exercise physiologists concluded that the ability of the cardiovascular system to deliver oxygen and nutrients to the muscle at a rate that supports aerobic metabolism is a major factor in determining maximum oxygen consumption.

Next we will examine the reflexes that integrate breathing and cardiovascular function during exercise.

RUNNING PROBLEM

By the time Colleen reached the practice field, she was already sweating and her face was flushed. At 9 A.M. the air temperature was 80° F and the humidity was 54%. Colleen took a quick drink of water from the large water container and ran out to the field.

Question 1:
 "Humidity" is the percentage of water vapor present in air. Why is the thermoregulatory mechanism of sweating less efficient in humid environments?

808　**811**　812　814　816　818

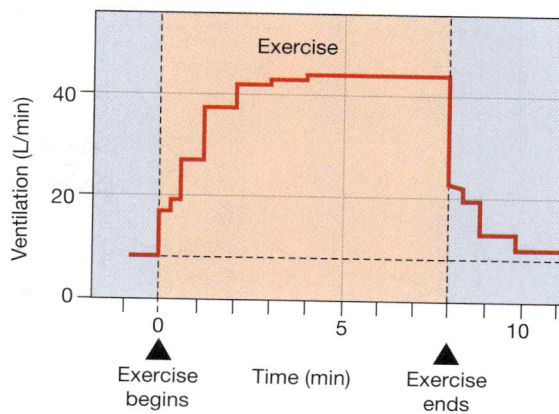

■ **FIGURE 25-5** *Changes in ventilation with submaximal exercise*

Ventilation rate jumps as soon as exercise begins, despite the fact that neither arterial P_{CO_2} nor P_{O_2} has changed. Modified from P. Dejours, *Handbook of Physiology* (Washington, D.C.: American Physiological Society, 1964).

VENTILATORY RESPONSES TO EXERCISE

Think about what happens to your breathing when you exercise. Exercise is associated with both increased rate and increased depth of breathing, resulting in enhanced alveolar ventilation [⊜ p. 577]. **Exercise hyperventilation**, or **hyperpnea**, results from a combination of feedforward signals from central command neurons in the motor cortex and sensory feedback from peripheral receptors.

When exercise begins, mechanoreceptors and proprioceptors [⊜ p. 439] in muscles and joints send information about movement to the motor cortex. Descending pathways from the motor cortex to the respiratory control center of the medulla oblongata then immediately increase ventilation (Fig. 25-5 ■).

As muscle contraction continues, sensory information feeds back to the respiratory control center to ensure that ventilation and tissue oxygen use remain closely matched. Sensory receptors involved in the secondary response probably include central, carotid, and aortic chemoreceptors that monitor P_{CO_2}, pH, and P_{O_2} [⊜ p. 604]; proprioceptors in the joints; and possibly receptors located within the exercising muscle itself. Pulmonary stretch receptors [⊜ p. 607] were once thought to play a role, but recipients of heart-lung transplants display a normal ventilatory response to exercise even though the neural connections between lung and brain are absent.

Exercise hyperventilation maintains nearly normal arterial P_{O_2} and P_{CO_2} by steadily increasing alveolar ventilation in proportion to the level of exercise. The compensation is so effective that when arterial P_{O_2}, P_{CO_2}, and pH are monitored during mild to moderate exercise, they show no significant change (Fig. 25-6 ■). This observation means that the once-accepted causes of increased ventilation during mild to moderate

25

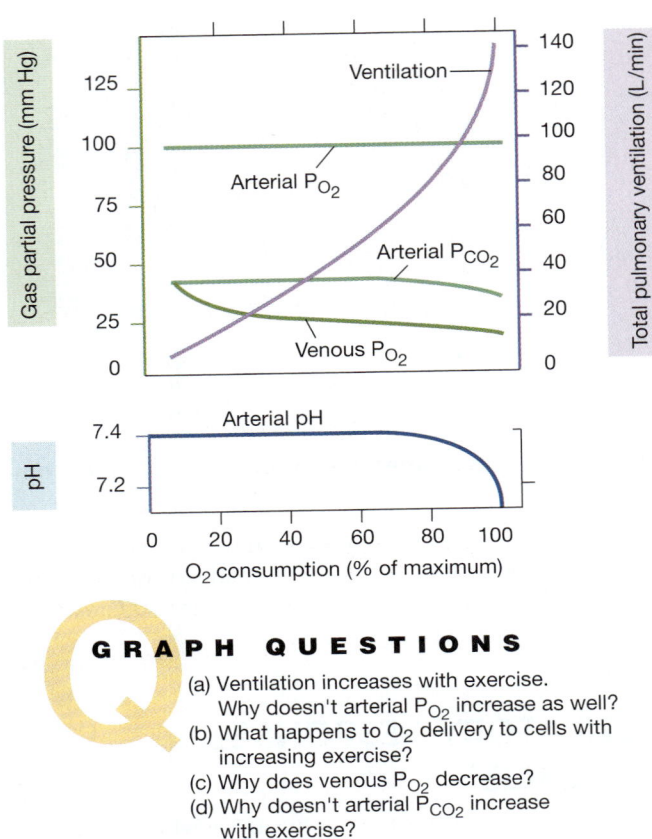

GRAPH QUESTIONS

(a) Ventilation increases with exercise. Why doesn't arterial P_{O_2} increase as well?
(b) What happens to O_2 delivery to cells with increasing exercise?
(c) Why does venous P_{O_2} decrease?
(d) Why doesn't arterial P_{CO_2} increase with exercise?
(e) Why does arterial P_{CO_2} decrease with maximum exercise?

■ **FIGURE 25-6** *Changes in blood gas, partial pressures, and arterial pH with exercise*

exercise—reduced arterial P_{O_2}, elevated arterial P_{CO_2}, and decreased plasma pH—must not be correct. Thus the chemoreceptors or medullary respiratory control center, or both, must be responding to other exercise-induced signals.

Several factors have been postulated to be these signals, including sympathetic input to the carotid body and changes in plasma K^+ concentration. During even mild exercise, extracellular K^+ increases as repeated action potentials in the muscle fibers allow K^+ to move out of cells. Carotid chemoreceptors are known to respond to increased K^+ by increasing ventilation. However, because K^+ concentration changes slowly, this mechanism does not explain the sharp initial rise in ventilation at the onset of activity.

It appears likely that the initial change in ventilation is caused by sensory input from muscle mechanoreceptors combined with parallel descending pathways from the motor cortex. Once exercise is under way, sensory input keeps ventilation matched to metabolic needs.

CONCEPT CHECK

1. If venous P_{O_2} decreases as exercise intensity increases, what do you know about the P_{O_2} of muscle cells as exercise intensity increases?

Answers: p. 820

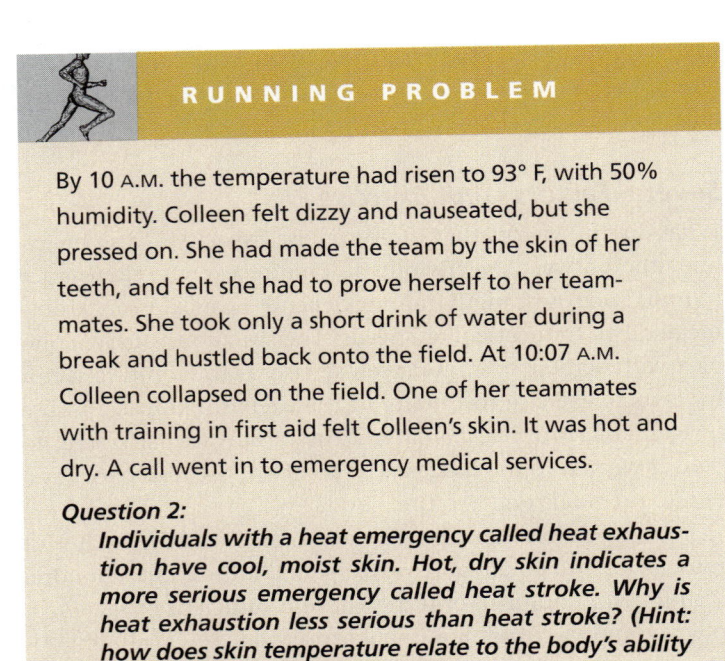

RUNNING PROBLEM

By 10 A.M. the temperature had risen to 93° F, with 50% humidity. Colleen felt dizzy and nauseated, but she pressed on. She had made the team by the skin of her teeth, and felt she had to prove herself to her teammates. She took only a short drink of water during a break and hustled back onto the field. At 10:07 A.M. Colleen collapsed on the field. One of her teammates with training in first aid felt Colleen's skin. It was hot and dry. A call went in to emergency medical services.

Question 2:
Individuals with a heat emergency called heat exhaustion have cool, moist skin. Hot, dry skin indicates a more serious emergency called heat stroke. Why is heat exhaustion less serious than heat stroke? (Hint: how does skin temperature relate to the body's ability to regulate body temperature?)

808   811   812   814   816   818

CARDIOVASCULAR RESPONSES TO EXERCISE

When exercise begins, mechanosensory input from working limbs combines with descending pathways from the motor cortex to activate the cardiovascular control center in the medulla oblongata. The center responds with sympathetic discharge that increases cardiac output and causes vasoconstriction in many peripheral arterioles.

Cardiac Output Increases During Exercise

During strenuous exercise, cardiac output rises dramatically. In untrained individuals, cardiac output goes up fourfold, from 5 L/min to 20 L/min. In trained athletes, it may go up six to eight times, reaching as much as 40 L/min. As you learned in Chapter 14 [🔁 Fig. 14-31, p. 494], cardiac output is related to heart rate and stroke volume:

$$\text{cardiac output (CO)} = \text{heart rate} \times \text{stroke volume}$$

If the factors that influence heart rate and stroke volume are considered, then

$$CO = (\text{SA node rate} + \text{autonomic nervous system input})$$
$$\times (\text{venous return} + \text{force of contraction})$$

Let's examine which of those factors has the greatest effect on cardiac output during exercise in a healthy heart. During exercise, venous return is enhanced by skeletal muscle contraction and deep inspiratory movements [🔁 p. 491]. It is therefore

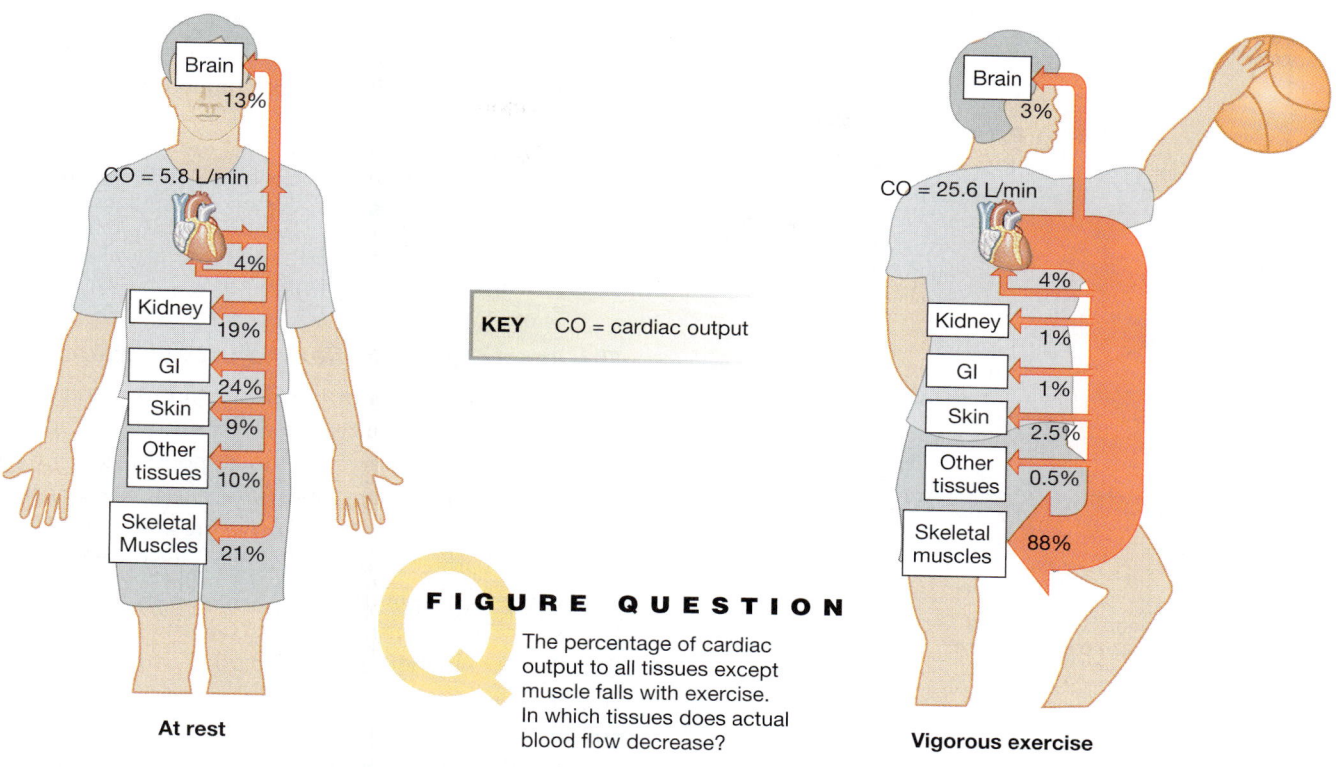

KEY   CO = cardiac output

FIGURE QUESTION

The percentage of cardiac output to all tissues except muscle falls with exercise. In which tissues does actual blood flow decrease?

At rest

Vigorous exercise

■ **FIGURE 25-7**   *Distribution of cardiac output at rest and during exercise*

Vasoconstriction in nonexercising tissues combined with vasodilation in exercising muscle shunts blood to muscles.

tempting to postulate that, owing to Starling's law of the heart, the cardiac muscle fibers simply stretch in response to increased venous return, thereby increasing contractility.

However, overfilling of the ventricles is potentially dangerous, because overstretching may damage the fibers. One way the body offsets increased venous return is by speeding up heart rate. If the interval between contractions is shorter, the heart will have less time to fill and is less likely to be damaged by excessive stretch.

The initial change in heart rate at the onset of exercise is due to decreased parasympathetic activity at the sinoatrial (SA) node [🔁 p. 489]. As cholinergic inhibition lessens, heart rate rises from its resting rate to around 100 beats per minute, the intrinsic pacemaker rate of the SA node. At that point, sympathetic output from the cardiovascular control center escalates.

Sympathetic stimulation has two effects on the heart. First, it increases contractility so that the heart squeezes out more blood per stroke (increased stroke volume). Second, sympathetic innervation increases heart rate so that the heart has less time to relax, protecting it from overfilling. The combination of faster heart rate and greater stroke volume increases cardiac output during exercise.

Peripheral Blood Flow Redistributes to Muscle During Exercise

At rest, skeletal muscles receive less than a fourth of the cardiac output, or about 1.2 L/min. During exercise, a significant shift

in peripheral blood flow takes place because of local and reflex reactions. During strenuous exercise in highly trained athletes, the combination of increased cardiac output and vasodilation can increase blood flow through exercising muscle to more than 22 L/min! The relative distribution of blood flow to tissues also shifts. About 88% of cardiac output is diverted to the exercising muscle, up from 21% at rest (Fig. 25-7 ■).

The redistribution of blood flow during exercise results from a combination of vasodilation in skeletal muscle arterioles and vasoconstriction in other tissues. At the onset of exercise, sympathetic signals from the cardiovascular control center cause vasoconstriction in peripheral tissues. As muscles become active, changes in the microenvironment of muscle tissue take place: tissue O_2 concentrations decrease, while temperature, CO_2, and acid in the interstitial fluid around muscle fibers increase. All of these factors act as paracrines that cause local vasodilation that overrides the sympathetic signal for vasoconstriction. The net result is shunting of blood flow from inactive tissues to the exercising muscles, where it is needed.

Blood Pressure Rises Slightly During Exercise

What happens to blood pressure during exercise? You learned in Chapter 15 that peripheral blood pressure is determined by a combination of cardiac output and peripheral resistance [🔁 p. 508]. Cardiac output increases during exercise, thereby contributing to

25

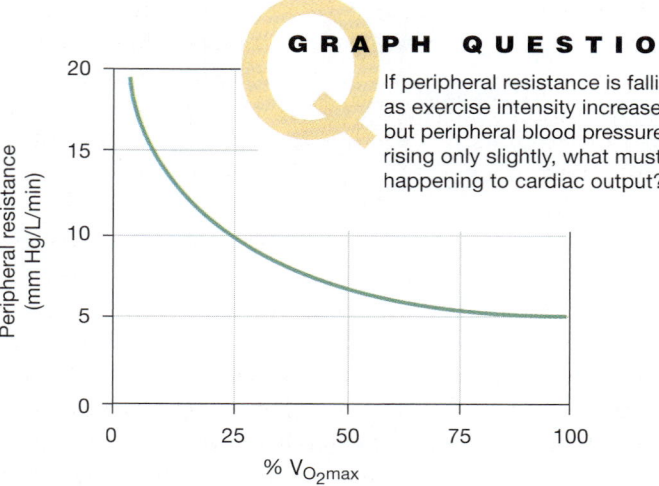

(a) Peripheral resistance decreases due to vasodilation in exercising muscle.

GRAPH QUESTION

If peripheral resistance is falling as exercise intensity increases, but peripheral blood pressure is rising only slightly, what must be happening to cardiac output?

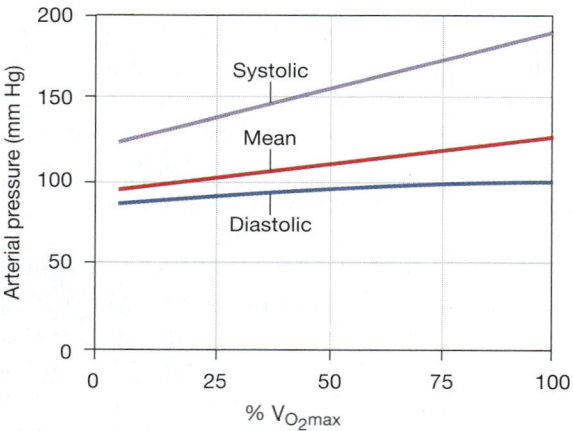

(b) Mean arterial blood pressure rises slightly despite a drop in resistance.

■ **FIGURE 25-8** *Changes in peripheral resistance and arterial blood pressure during exercise*

increased blood pressure. The changes resulting from peripheral resistance are harder to predict, however, because some peripheral arterioles are constricting while others are dilating:

$$\text{mean arterial blood pressure} = \text{cardiac output} \times \text{peripheral resistance}$$

Skeletal muscle vasodilation decreases peripheral resistance to blood flow. At the same time, sympathetically induced vasoconstriction in nonexercising tissues offsets the vasodilation, but only partially. Consequently, total peripheral resistance to blood flow falls dramatically as exercise commences, reaching a minimum at about 75% of $\dot{V}_{O_2max}$ (Fig. 25-8a ■).

If no other compensation occurred, this decrease in peripheral resistance would dramatically lower arterial blood pressure. However, increased cardiac output cancels out decreased peripheral resistance. When blood pressure is monitored during exercise, mean arterial blood pressure actually increases slightly as

exercise intensity increases (Fig. 25-8b). The fact that it increases at all, however, suggests that the normal baroreceptor reflexes that control blood pressure are not functioning during exercise.

CONCEPT CHECK

2. In Figure 25-8b, why does the line for mean blood pressure lie closer to diastolic pressure instead of being evenly centered between systolic and diastolic pressures? (*Hint*: what is the equation for calculating mean blood pressure?)

Answers: p. 820

The Baroreceptor Reflex Adjusts to Exercise

Normally, homeostasis of blood pressure is regulated through peripheral baroreceptors in the carotid and aortic bodies: an increase in blood pressure initiates responses that return blood pressure to normal. But during exercise, blood pressure increases without activating homeostatic compensation. Why is the normal baroreceptor reflex absent during exercise?

There are several theories. According to one, signals from the motor cortex during exercise reset the arterial baroreceptor threshold to a higher pressure. Blood pressure can then increase slightly during exercise without triggering the homeostatic counter-regulatory responses.

Another theory suggests that signals in baroreceptor afferent neurons are blocked in the spinal cord by presynaptic inhibition [🔁 p. 279] at some point before the afferent neurons synapse with central nervous system neurons. This central inhibition inactivates the baroreceptor reflex during exercise.

A third theory is based on the postulated existence of muscle chemoreceptors that are sensitive to metabolites (probably H^+) produced during strenuous exercise. When stimulated, these chemoreceptors signal the CNS that tissue blood flow is not adequate to remove muscle metabolites or keep the muscle in aerobic metabolism. The chemoreceptor input is reinforced by sensory input from mechanoreceptors in the working limbs. The CNS response to this sensory input is to override the baroreceptor reflex and raise blood pressure to enhance muscle perfusion. The same hypothetical muscle chemoreceptors may play a role in ventilatory responses to exercise.

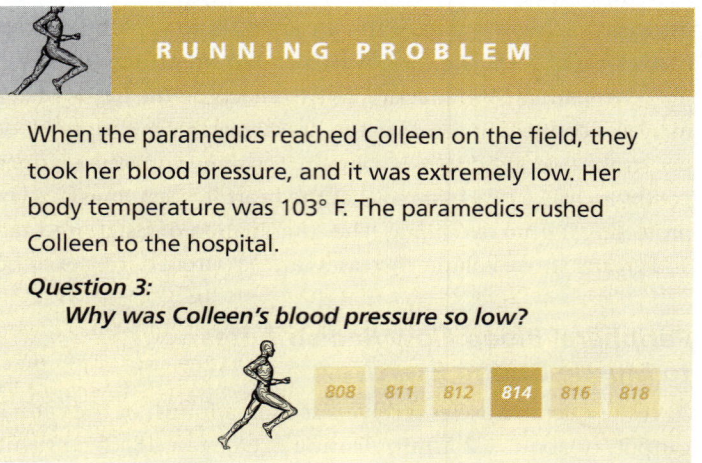

RUNNING PROBLEM

When the paramedics reached Colleen on the field, they took her blood pressure, and it was extremely low. Her body temperature was 103° F. The paramedics rushed Colleen to the hospital.

Question 3:
Why was Colleen's blood pressure so low?

808 811 812 **814** 816 818

FEEDFORWARD RESPONSES TO EXERCISE

Interestingly, there is a significant *feedforward* element [🔁 p. 200] in the physiological responses to exercise. It is easy to explain physiological changes that occur with exercise as reactions to the disruption of homeostasis. However, many of these changes occur in the absence of the normal stimuli or before the stimuli are present. For example, ventilation rates jump as soon as exercise begins (Fig. 25-5), even though experiments have shown that arterial P_{O_2} and P_{CO_2} do not change (Fig. 25-6).

How does the feedforward response work? As exercise begins, proprioceptors in the muscles and joints send information to the motor cortex of the brain. Descending signals from the motor cortex go not only to the exercising muscles but also along parallel pathways to the cardiovascular and respiratory control centers and to the limbic system of the brain.

Output from the limbic system and cardiovascular control center triggers generalized sympathetic discharge. As a result, an immediate slight increase in blood pressure marks the beginning of exercise. Sympathetic discharge causes widespread vasoconstriction, increasing blood pressure. Once exercise has begun, this increase in blood pressure compensates for decreases in blood pressure resulting from muscle vasodilation.

As exercise proceeds, reactive compensations become superimposed on the feedforward changes. For example, when exercise reaches 50% of aerobic capacity, muscle chemoreceptors detect the buildup of lactic acid and metabolites and send this information to central command centers in the brain. The command centers then maintain changes in ventilation and circulation that were initiated in a feedforward manner. Thus, the integration of systems in exercise involves both common reflex pathways and some unique centrally mediated reflex pathways.

TEMPERATURE REGULATION DURING EXERCISE

As exercise continues, heat released through metabolism creates an additional challenge to homeostasis. Most of the energy released during metabolism is not converted into ATP but instead is released as heat. (Efficiency of energy conversion from organic substrates to ATP is only 20–25%.) With continued exercise, heat production exceeds heat loss, and core body temperature rises. In endurance events, body temperature can reach 40°–42° C (104°–108° F), which we would normally consider a fever.

This rise in body temperature during exercise triggers two thermoregulatory mechanisms: sweating and increased cutaneous blood flow [🔁 p. 742]. Both mechanisms help regulate body temperature, but both can also disrupt homeostasis in other ways. While sweating lowers body temperature

through evaporative cooling, the loss of fluid from the extracellular compartment can cause dehydration and significantly reduce circulating blood volume. Because sweat is a hypotonic fluid, the extra water loss increases body osmolarity. The combination of decreased ECF volume and increased osmolarity during extended exercise sets in motion the complex homeostatic pathway for dehydration described in Chapter 20, including thirst and renal conservation of water [🔁 Fig. 20-18, p. 662].

The other thermoregulatory mechanism—increased blood flow to the skin—causes body heat loss to the environment through convection [🔁 p. 741]. However, increased sympathetic output during exercise tends to vasoconstrict cutaneous blood vessels, which opposes the thermoregulatory response. The primary control of vasodilation in the skin during exercise appears to come from a sympathetic vasodilator system. Activation of these neurons as body core temperature rises dilates cutaneous blood vessels without altering sympathetic vasoconstriction in other body tissues.

Although cutaneous vasodilation is essential for thermoregulation, it disrupts homeostasis by decreasing peripheral resistance and diverting blood flow from the muscles. In the face of these contradictory demands, the body initially gives preference to thermoregulation. However, if central venous pressure falls below a critical minimum, the body abandons thermoregulation in the interest of maintaining blood flow to the brain.

The degree to which the body can adjust to both demands depends on the type of exercise being performed and its intensity and duration. Strenuous exercise in hot, humid environments can severely impair normal thermoregulatory mechanisms and cause *heat stroke*, a potentially fatal condition. Unless prompt measures are taken to cool the body, core temperatures can go as high as 43° C (109° F).

It is possible for the body to adapt to repeated exercise in hot environments, however, through **acclimatization**. In this process, physiological mechanisms shift to fit a change in environmental conditions. As the body adjusts to exercise in the heat, sweating begins sooner and doubles or triples in volume, enhancing evaporative cooling. With acclimatization, sweat also becomes more dilute, as salt is reabsorbed from the sweat glands under the influence of increased aldosterone. Salt loss in an unacclimatized person exercising in the heat may reach 30 g NaCl per day, but that value decreases to as little as 3 g after a month of acclimatization.

EXERCISE AND HEALTH

Physical activity has many positive effects on the human body. The lifestyles of humans have changed dramatically since we were hunter-gatherers, but our bodies still seem to work best with a certain level of physical activity. Several common pathological conditions—including high blood pressure,

25

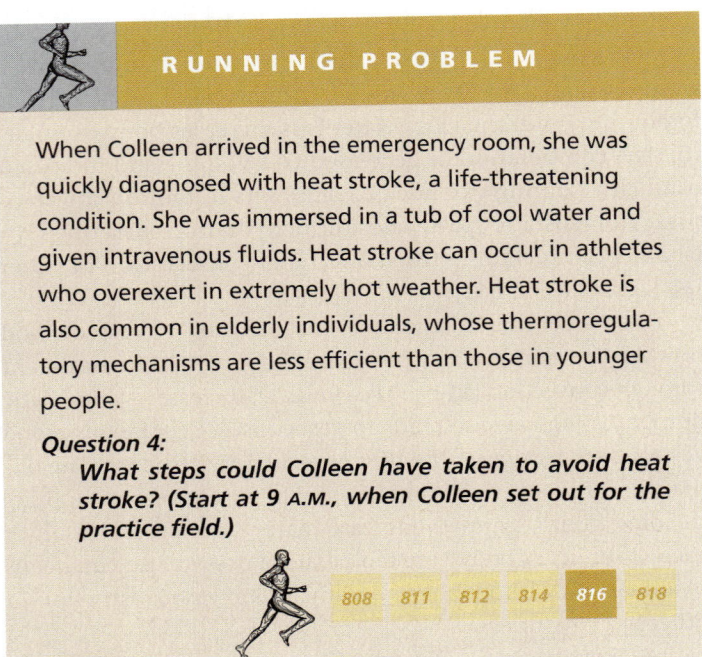

RUNNING PROBLEM

When Colleen arrived in the emergency room, she was quickly diagnosed with heat stroke, a life-threatening condition. She was immersed in a tub of cool water and given intravenous fluids. Heat stroke can occur in athletes who overexert in extremely hot weather. Heat stroke is also common in elderly individuals, whose thermoregulatory mechanisms are less efficient than those in younger people.

Question 4:
What steps could Colleen have taken to avoid heat stroke? (Start at 9 A.M., when Colleen set out for the practice field.)

808 811 812 814 **816** 818

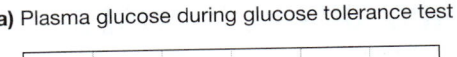

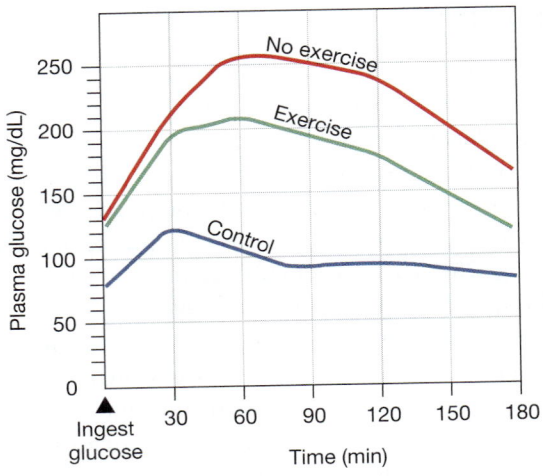

strokes, and diabetes mellitus—can be improved by physical activity. Even so, developing regular exercise habits is one lifestyle change that many people find difficult to make. In this section we look at the effects exercise has on several common health conditions.

Exercise Lowers the Risk of Cardiovascular Disease

As early as the 1950s, scientists showed that physically active men have a lower rate of heart attacks than do men who lead a sedentary lifestyle. These studies started many investigations into the exact relationship between cardiovascular disease and exercise. Scientists have subsequently demonstrated that exercise has positive benefits for both men and women. These benefits include lowering blood pressure, decreasing plasma triglyceride levels, and raising plasma HDL-cholesterol levels. High blood pressure is a major risk factor for strokes, while elevated triglycerides and low HDL-cholesterol levels are associated with development of atherosclerosis and increased risk of heart attack.

Overall, exercise reduces the risk of death or illness from a variety of cardiovascular diseases. Even such mild exercise as walking has significant health benefits that could reduce the risk of developing cardiovascular diseases or diabetes and the complications of obesity in the estimated 40 million adult Americans with sedentary lifestyles.

Type 2 Diabetes Mellitus May Improve with Exercise

Regular exercise is now widely accepted as effective in preventing and alleviating type 2 diabetes mellitus. With regular exercise, skeletal muscle fibers up-regulate both the number of glucose transporters and the number of insulin receptors on their mem-

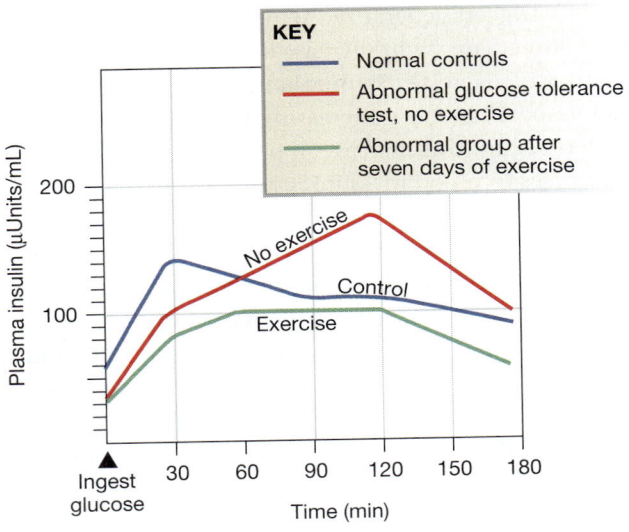

■ **FIGURE 25-9** *The effects of exercise on glucose tolerance and insulin secretion*

(a) Plasma glucose levels in normal men (blue line), in men with type 2 diabetes who had not been exercising (red line), and in those same diabetic men after seven days of exercise (green line). (b) Plasma insulin levels in groups of men in the categories described in part (a). Data from B. R. Seals et al., *J. App. Physiol.* 56(6):1521–1525, 1984; and M. A. Rogers et al., *Diabetes Care* 11:613–618, 1988.

brane. The addition of glucose transporters decreases the muscle's dependence on insulin for glucose uptake.

Increasing the number of insulin receptors makes the muscle fibers more sensitive to insulin. A smaller amount of insulin then can achieve a response that previously required more insulin. Because the cells are responding to lower insulin levels, the endocrine pancreas secretes less insulin. This lessens the stress on the pancreas, resulting in a lower incidence of type 2 diabetes mellitus.

Figure 25-9 ■ shows the effects of seven days of exercise on glucose utilization and insulin secretion in men with mild type 2 diabetes. Individuals in the experiment underwent *glucose tolerance tests,* in which they ingested 100 g of glucose after an overnight fast. Their plasma glucose levels were assessed before and for 120 minutes after ingesting the glucose. Simultaneous measurements were made of plasma insulin.

The graph in Fig. 25-9a shows glucose tolerance tests in control subjects (blue line) and in the diabetic men before and after exercise (red and green lines, respectively). Fig. 25-9b shows concurrent insulin secretion in the three groups. After only seven days of exercise, both the glucose tolerance test and insulin secretion in exercising diabetic subjects had shifted to a pattern that was more like that of the normal control subjects. These results demonstrate the beneficial effect of exercise on glucose transport and metabolism, and support the recommendation that patients with type 2 diabetes maintain a regular exercise program.

Stress and the Immune System May Be Influenced by Exercise

Another topic that is receiving much attention is the interaction of exercise with the immune system. Epidemiological evidence looking at large populations of people suggests that exercise is associated with a reduced incidence of disease and with increased longevity. Moreover, many people believe that exercise boosts immunity, prevents cancer, and helps HIV-infected patients combat AIDS.

To date, however, the results of rigidly controlled research studies have not supported those viewpoints. Indeed, good evidence suggests that strenuous exercise is a form of stress that suppresses the immune response. Immune suppression may be due to corticosteroid release, or it may be due to release of interferon-γ during strenuous exercise.

Researchers have proposed that the relationship between exercise and immunity can be represented by a J-shaped curve (Fig. 25-10 ■). People who exercise moderately have more effective immune systems than those who are sedentary, but people who exercise strenuously may experience a decrease in immune function because of the stress of the exercise.

Another area of exercise physiology filled with interesting though contradictory results is the effect of exercise on stress, depression, and other psychological parameters. Research has shown an inverse relationship between exercise and depression: people who exercise regularly are significantly less likely to be clinically depressed than are people who do not exercise

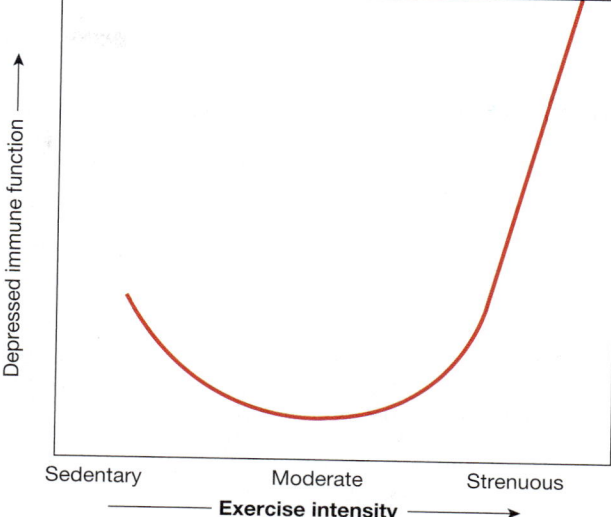

Moderate exercise enhances immunity, but strenuous exercise is a form of stress that depresses immunity.

■ **FIGURE 25-10**   *The relationship between immune function and exercise*

This J-shaped curve indicates that individuals who exercise in moderation have greater immune function than either sedentary individuals or those who exercise strenuously.

regularly. Although the association exists, assigning cause and effect to the two parameters is difficult. Are the exercisers less depressed because they exercise? Or do depressed individuals exercise less because they are depressed?

Many published studies appear to show that regular exercise is effective in reducing depression. But a careful analysis of experimental design suggests that the conclusions of those studies may be overstated. The subjects in many of the experiments were being treated concurrently with drugs or psychotherapy, so it is difficult to attribute improvement in their condition solely to exercise.

In addition, participation in exercise studies gives subjects a period of social interaction, another factor that might play a role in the reduction of stress and depression. The assertion that exercise reduces depression would be better supported by a demonstration of the effects of exercise on biological parameters, such as disturbances of monoamine neurotransmitters [➔ p. 317]. To date, however, little research of that type has been conducted.

25

RUNNING PROBLEM CONCLUSION

HEAT STROKE

Colleen was kept overnight in the hospital for observation. She was unable to practice with the team for the remainder of the season because victims of heat stroke are more sensitive to high temperatures for some time after the episode.

To learn more about the symptoms and treatment of heat stroke, try a Google search. Then check your understanding of this problem by comparing your answers to the information in the summary table.

	QUESTION	FACTS	INTEGRATION AND ANALYSIS
1	Why is the thermoregulatory mechanism of sweating less efficient in humid environments?	"Humidity" is the percentage of water vapor present in air. Thermoregulation in warm environments includes sweating and evaporative cooling.	Evaporation is slower in humid air, so evaporative cooling is less effective in high humidity.
2	Individuals with a heat emergency called heat exhaustion have cool, moist skin. Hot, dry skin indicates a more serious emergency called heat stroke. Why is heat exhaustion less serious than heat stroke?	Sweating helps the body regulate temperature through evaporative cooling. As evaporation occurs, the skin surface cools.	Cool, moist skin indicates that the sweating mechanism is still functioning. Hot, dry skin indicates that the sweating mechanism has failed. Subjects with hot, dry skin are likely to have higher internal temperatures.
3	Why was Colleen's blood pressure so low?	Blood volume decreases when the body loses large amounts of water through sweating.	Colleen had been sweating but not replacing the fluid she lost. This caused her blood volume to decrease, with a corresponding decrease in blood pressure.
4	What steps could Colleen have taken to avoid heat stroke? (Start at 9 A.M., when Colleen set out for the practice field.)	Heat stroke is caused by dehydration resulting from excessive sweating. Dehydration results in a drop in blood pressure. Peripheral blood vessels constrict in an effort to maintain pressure and blood flow to the brain. Constricted blood vessels in the skin cannot release excess heat. In addition, the sweating response is inhibited to prevent further fluid loss.	Colleen could have avoided heat stroke by: (1) consuming large amounts of fluid (water plus electrolytes) before and during practice; (2) avoiding unnecessary exercise such as bicycling to practice; (3) stopping activity at the first signs of heat emergency (dizziness and nausea); (4) seeking shade and drinking large amounts of fluid to replenish lost fluids; (5) applying cool wet towels or ice to bring body temperature lower.

808 811 812 814 816 **818**

25

CHAPTER SUMMARY

In this chapter you learned about exercise and the physiological challenges it presents. *Integration* and *coordination* between the body's *physiological* *control systems* allow the internal environment to remain relatively constant, despite the challenges to *homeostasis* that exercise presents.

Metabolism and Exercise

1. Exercising muscle requires a steady supply of ATP from metabolism or from conversion of phosphocreatine. (p. 808; Fig. 25-1)

2. Carbohydrates and fats are the primary energy substrates. Glucose can be metabolized through both oxidative and anaerobic pathways, but fatty acid metabolism requires oxygen. (p. 808; Fig. 25-1)

3. Anaerobic **glycolytic metabolism** converts glucose to lactic acid. Glycolytic metabolism is 2.5 times more rapid than aerobic pathways but is not as efficient at ATP production. (p. 808; Fig. 25-2)

4. Glucagon, cortisol, catecholamines, and growth hormone influence glucose and fatty acid metabolism during exercise. These hormones favor the conversion of glycogen to glucose. (p. 810)

5. Although plasma glucose concentrations rise with exercise, the secretion of insulin decreases. This response reduces glucose uptake by most cells, making more glucose available for exercising muscle. (p. 810)

6. The intensity of exercise is indicated by **oxygen consumption** ($\dot{V}_{O_2}$). A person's maximal rate of oxygen consumption ($\dot{V}_{O_2max}$) is an indicator of that person's ability to perform endurance exercise. (p. 810)

7. Oxygen consumption increases rapidly at the onset of exercise. **Excess postexercise oxygen consumption** is due to ongoing metabolism, increased body temperature, and circulating catecholamines. (p. 810; Fig. 25-4)

8. Muscle mitochondria increase in size and number with endurance training. (p. 811)

9. At maximal exertion, the ability of the cardiovascular system to deliver oxygen and nutrients appears to be the primary limiting factor. (p. 811)

Ventilatory Responses to Exercise

10. Exercise hyperventilation results from feedforward signals from the motor cortex, and sensory feedback from peripheral sensory receptors. (p. 811; Fig. 25-5)

11. Arterial P_{O_2}, P_{CO_2}, and pH do not change significantly during mild to moderate exercise. Exercise hyperventilation is due to some other factor such as increased plasma K^+ levels. (p. 812; Fig. 25-6)

Cardiovascular Responses to Exercise

12. Cardiac output increases with exercise because of increased venous return and sympathetic stimulation of heart rate and contractility. (p. 812; Fig. 25-7)

13. Blood flow through exercising muscle increases dramatically when skeletal muscle arterioles dilate. Arterioles in other tissues constrict. (p. 813; Fig. 25-7)

14. Decreased tissue O_2 and glucose or increased muscle temperature, CO_2, and acid act as paracrines and cause local vasodilation. (p. 813)

15. Mean arterial blood pressure increases slightly as exercise intensity increases. The baroreceptors that control blood pressure change their set-points during exercise. (p. 814; Fig. 25-8)

Feedforward Responses to Exercise

16. When exercise begins, feedforward responses prevent significant disruption of homeostasis. (p. 815)

Temperature Regulation During Exercise

17. Heat released during exercise is dissipated by sweating and increased cutaneous blood flow. (p. 815)

Exercise and Health

18. Physical activity can help prevent or decrease the risk of developing high blood pressure, strokes, and type 2 diabetes mellitus. (p. 816)

QUESTIONS

(Answers to the Review Questions begin on page A1.)

THE PHYSIOLOGY PLACE

Access more review material online at **The Physiology Place** website. There you'll find review questions, problem-solving activities, case studies, flashcards, and direct links to both *InterActive Physiology*® and PhysioEx™. To access the site, go to *www. physiologyplace.com* and select Human Physiology, Fourth Edition.

LEVEL ONE REVIEWING FACTS AND TERMS

1. Name the two muscle compounds that store energy in the form of high-energy phosphate bonds.

2. The most efficient ATP production is through *aerobic/anaerobic* pathways. When these pathways are being used, then *glucose/fatty acids/both/neither* can be metabolized to provide ATP.

3. What are the differences between aerobic and anaerobic metabolism?

4. List three sources of glucose that can be metabolized to ATP, either directly or indirectly.

5. List four hormones that promote the conversion of triglycerides into fatty acids. What effects do these hormones have on plasma glucose levels?

6. What is meant by the term *oxygen deficit,* and how is it related to excessive postexercise oxygen consumption?

7. What organ system is the limiting factor for maximal exertion?

8. In endurance events, body temperature can reach 40°–42° C. What is normal body temperature? What two thermoregulatory mechanisms are triggered by this change in temperature during exercise?

LEVEL TWO REVIEWING CONCEPTS

9. **Concept map:** Map the metabolic, cardiovascular, and respiratory changes that occur during exercise. Include the signals to and from the nervous system, and show what specific areas signal and coordinate the exercise response.

10. What causes insulin secretion to decrease during exercise, and why is this decrease adaptive?

11. State two advantages and two disadvantages of anaerobic glycolysis.

12. Compare and contrast each of the terms in the following sets of terms, especially as they relate to exercise:
    (a) ATP, ADP, PCr
    (b) myoglobin, hemoglobin

13. Match the following responses to the brain area that controls it. Brain areas may be used once, more than once, or not at all. Some responses may have more than one answer.
    (a) pons
    (b) medulla oblongata
    (c) midbrain
    (d) motor cortex
    (e) hypothalamus
    (f) cerebellum
    (g) brain not involved; involves local control

    1. changes in cardiac output
    2. vasoconstriction
    3. exercise hyperventilation
    4. increased stroke volume
    5. increased heart rate
    6. coordination of skeletal muscle movement

14. Specify whether each of the following parameters stays the same, increases, or decreases when a person becomes better conditioned for athletic activities:
    (a) heart rate during exercise
    (b) resting heart rate

(c) cardiac output during exercise

(d) resting cardiac output

(e) breathing rate during exercise

(f) blood flow to muscles during exercise

(g) blood pressure during exercise

(h) total peripheral resistance during exercise

15. Why doesn't increased venous return during exercise overstretch the heart muscle?

16. Diagram the three theories that explain why the normal baroreceptor reflex is absent during exercise.

17. List and briefly discuss the benefits of a lifestyle that includes regular exercise.

18. Explain how exercise decreases blood glucose in type 2 diabetes mellitus.

LEVEL THREE   PROBLEM SOLVING

19. You have decided to manufacture a new sports drink that will help athletes, from football players to gymnasts. List at least five different ingredients you would include in your drink, and indicate why each is important for the athlete.

LEVEL FOUR   QUANTITATIVE PROBLEMS

20. You are a well-conditioned athlete. At rest, your heart rate is 60 beats per minute and your stroke volume is 70 mL/beat. What is your cardiac output? At one point during exercise, your heart rate goes up to 120 beats/min. Does your cardiac output increase proportionately? Explain.

ANSWERS

✓ Answers to Concept Check Questions

Page 812

1. If venous P_{O_2} decreases, P_{O_2} in the cells is also decreasing.

Page 814

2. The mean blood pressure line lies closer to the diastolic pressure line because the heart spends more time in diastole than systole.

Q Answers to Figure and Graph Questions

Page 812

Figure 25-6: (a) Arterial P_{O_2} remains constant because pulmonary ventilation is matched to blood flow through the lungs. (b) Although arterial P_{O_2} is constant, oxygen delivery to cells increases due to increased cardiac output (not shown). (c) Venous P_{O_2} drops as exercise increases because cells are removing more oxygen from hemoglobin as oxygen consumption increases. (d) Arterial P_{CO_2} does not change until oxygen consumption reaches 80% of maximum. (e) As the person begins to hyperventilate, arterial (and alveolar) P_{CO_2} declines.

Page 813

Figure 25-7: Blood flow to an organ is calculated by multiplying cardiac output (L/min) times the percentage of flow to that organ. When rest and exercise values are compared, actual blood flow decreases only in the kidneys, GI tract, and "other tissues."

Page 814

Figure 25-8: Mean arterial pressure is cardiac output times resistance. If resistance is falling but MAP is increasing, then cardiac output must be increasing.

26

Birth, and copulation, and death. That's all the facts when you come to brass tacks.

—T. S. Eliot, *Sweeney Agonistes*

Mouse spermatozoa showing DNA (blue) in the heads and microtubules (red) in the flagellated tails.

Reproduction and Development

BACKGROUND BASICS

RUNNING PROBLEM

INFERTILITY

Peggy and Larry have just about everything to make them happy: successful careers, a loving marriage, a comfortable home. But one thing is missing: after five years of marriage, they have been unable to have a child. Today, Peggy and Larry have their first appointment with Charles Coddington, an infertility specialist. "Finding the cause of your infertility is going to require some painstaking detective work," Dr. Coddington explains. He will begin his workup of Peggy and Larry by asking detailed questions about their reproductive histories. Based on the answers to these questions, he will then order tests to pinpoint the problem.

| 822 | 834 | 843 | 844 | 849 | 853 | 855 |

Imagine growing up as a girl, then at the age of 12 or so, finding that your voice is deepening and your genitals are developing into those of a man. This scenario actually happens to a small number of men who have a condition known as *pseudohermaphroditism* [*pseudes,* false + *hermaphrodites,* the dual-sex offspring of Hermes and Aphrodite]. These men have the internal sex organs of a male but inherit a gene that causes a deficiency in one of the male hormones. Consequently, they are born with external genitalia that appear feminine, and they are raised as girls. At **puberty** [*pubertas,* adulthood], the period when a person makes the transition from being nonreproductive to being reproductive, pseudohermaphrodites begin to secrete more male hormones. As a result, they develop some, but not all, of the characteristics of men. Not surprisingly, a conflict arises: should these individuals change gender or remain female? Most choose to change and continue life as men.

Reproduction is one area of physiology in which we humans like to think of ourselves as significantly advanced over other animals. We mate for pleasure as well as procreation, and women are always sexually receptive (that is, not only during fertile periods), but just how different are we?

Like other terrestrial animals, we have internal fertilization that allows motile flagellated sperm to remain in an aqueous environment. To facilitate the process, we have mating and courtship rituals, as do other animals. Development is also internal, protecting the growing embryo from dehydration and cushioning it with a layer of fluid.

Humans are *sexually dimorphic;* that is, males and females are physically distinct. This distinction is sometimes blurred by dress and hairstyle, but these are cultural acquisitions.

Although everyone agrees that male and female humans are physically dimorphic, we are still debating whether we are behaviorally and psychologically dimorphic as well.

Sex hormones play a significant role in the behavior of other mammals, acting on adults as well as influencing the brain of the developing embryo.* Their role in humans is more controversial. Human fetuses are exposed to sex hormones while in the uterus, but it is unclear how much gender imprinting occurs. Does the preference of little girls for dolls and of little boys for toy guns have a biological basis or a cultural basis? We have no answer yet, but growing evidence suggests that at least part of our brain structure is influenced by sex hormones before we ever leave the womb.

This chapter addresses the biology of human reproduction and development. We will begin with the fertilized egg, or **zygote**, and examine sex determination and sexual differentiation.

SEX DETERMINATION

The male and female sex organs consist of three sets of structures: the gonads, the internal genitalia, and the external genitalia. **Gonads** [*gonos,* seed] are the organs that produce **gametes** [*gamein,* to marry], the reproductive cells (eggs and sperm) that unite to form a new individual. The male gonads are the **testes** (singular *testis*), which produce **sperm** (*spermatozoa*). The female gonads are the **ovaries**, which produce eggs, or **ova** (singular *ovum*). The undifferentiated gonadal cells destined to produce eggs and sperm are called **germ cells**. The **internal genitalia** consist of accessory glands and ducts that connect the gonads with the outside environment. The **external genitalia** include all external reproductive structures.

Sexual development is programmed in the human genome [Appendix C]. Each nucleated cell of the body except eggs and sperm contains 46 chromosomes (the *diploid number*): 22 matched (*homologous*) pairs of **autosomes** and one pair of **sex chromosomes** (Fig. 26-1 ■). The 22 pairs of autosomal chromosomes direct development of the human body form and of variable characteristics such as hair color and blood type. The two sex chromosomes, designated as either X or Y, contain genes that direct development of internal and external sex organs. The X chromosome is larger than the Y chromosome and includes many genes that are missing from the Y chromosome.

Eggs and sperm each have a half set of 23 chromosomes (the *haploid number*) because of *meiotic* divisions during their creation. When egg and sperm unite, the resulting zygote contains a unique set of 46 chromosomes. Thus, humans inherit one chromosome of each homologous pair from the mother, and the other from the father.

---

*Once the fertilized egg begins to divide (2-cell stage, 4-cell stage, etc.) it becomes first an embryo (weeks 0–8 of development), then a fetus (8 weeks until birth).

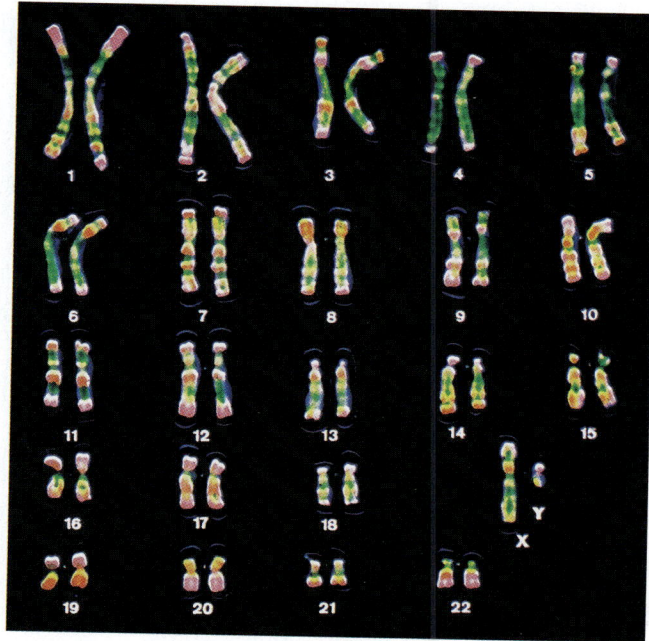

■ FIGURE 26-1   *Human chromosomes*

Humans have 23 pairs of chromosomes, which in this figure have been arranged in homologous pairs. The presence of an X chromosome and a Y chromosome (lower right) means that these chromosomes came from a male.

The Sex Chromosomes Determine Genetic Sex

The sex chromosomes a person inherits determine the genetic sex of that individual. Genetic females are XX, and genetic males are XY (Fig. 26-2 ■). Females inherit one X chromosome from each parent. Males inherit a Y chromosome from their father and an X chromosome from their mother. The Y chromosome is essential for development of the male reproductive organs.

If sex chromosomes are abnormally distributed at fertilization, the presence or absence of a Y chromosome determines whether development proceeds along male or female lines. The presence of a Y chromosome means the embryo will become male, even if the zygote also has multiple X chromosomes. For instance, an XXY zygote will become male. A zygote that inherits only a Y chromosome (YO) will die because the larger X chromosome contains essential genes that are missing from the Y chromosome.

In the absence of a Y chromosome, an embryo will develop into a female. Therefore, a zygote that gets only one X chromosome (XO; Turner's syndrome) will develop into a female. Two X chromosomes are needed for normal female reproductive function, however.

Once the ovaries develop in a female fetus, one X chromosome in each cell of her body inactivates and condenses into a clump of nuclear chromatin known as a *Barr body*. (Barr bodies in females can be seen in stained cheek epithelium.) The selection of the X chromosome that becomes inactive during development is random: some cells will have an active maternal

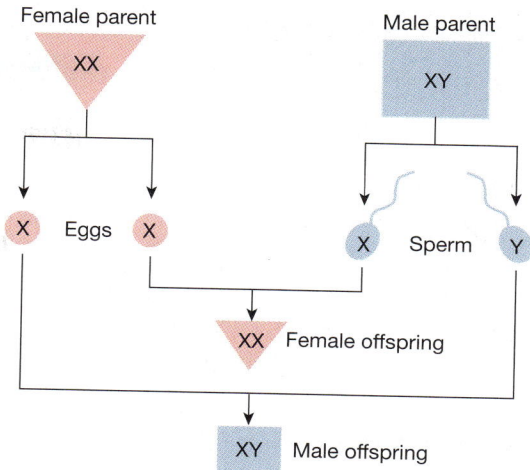

■ FIGURE 26-2   *Inheritance of X and Y chromosomes*

Each egg produced by a female (XX) has an X chromosome. Sperm produced by a male (XY) will have either an X chromosome or a Y chromosome.

X chromosome and others have an active paternal X chromosome. Because inactivation occurs early in development—before cell division is complete—all cells of a given tissue will usually have the same active X chromosome, either maternal or paternal.

CONCEPT CHECK

1. Name the male and female gonads and gametes.

Answers: p. 859

CLINICAL FOCUS

X-LINKED INHERITED DISORDERS

Normally, a person inherits two copies of the gene for a given trait: one copy from the mother and one from the father. However, many genes found on the X chromosome (*X-linked genes*) have no matching gene on the much smaller Y chromosome. Females inherit two X chromosomes and always get two copies of X-linked genes, so the expression of X-linked traits follows the usual pattern of gene dominance and recession. Males, however, receive only one copy of an X-linked gene—on the X chromosome, received from the mother—so males *always* exhibit the traits associated with an X-linked gene. If the maternally inherited X-linked gene is defective, male offspring will exhibit the mutation. Among the identified X-linked diseases are Duchenne muscular dystrophy [🔁 p. 420], hemophilia [🔁 p. 554], and color-blindness.

26

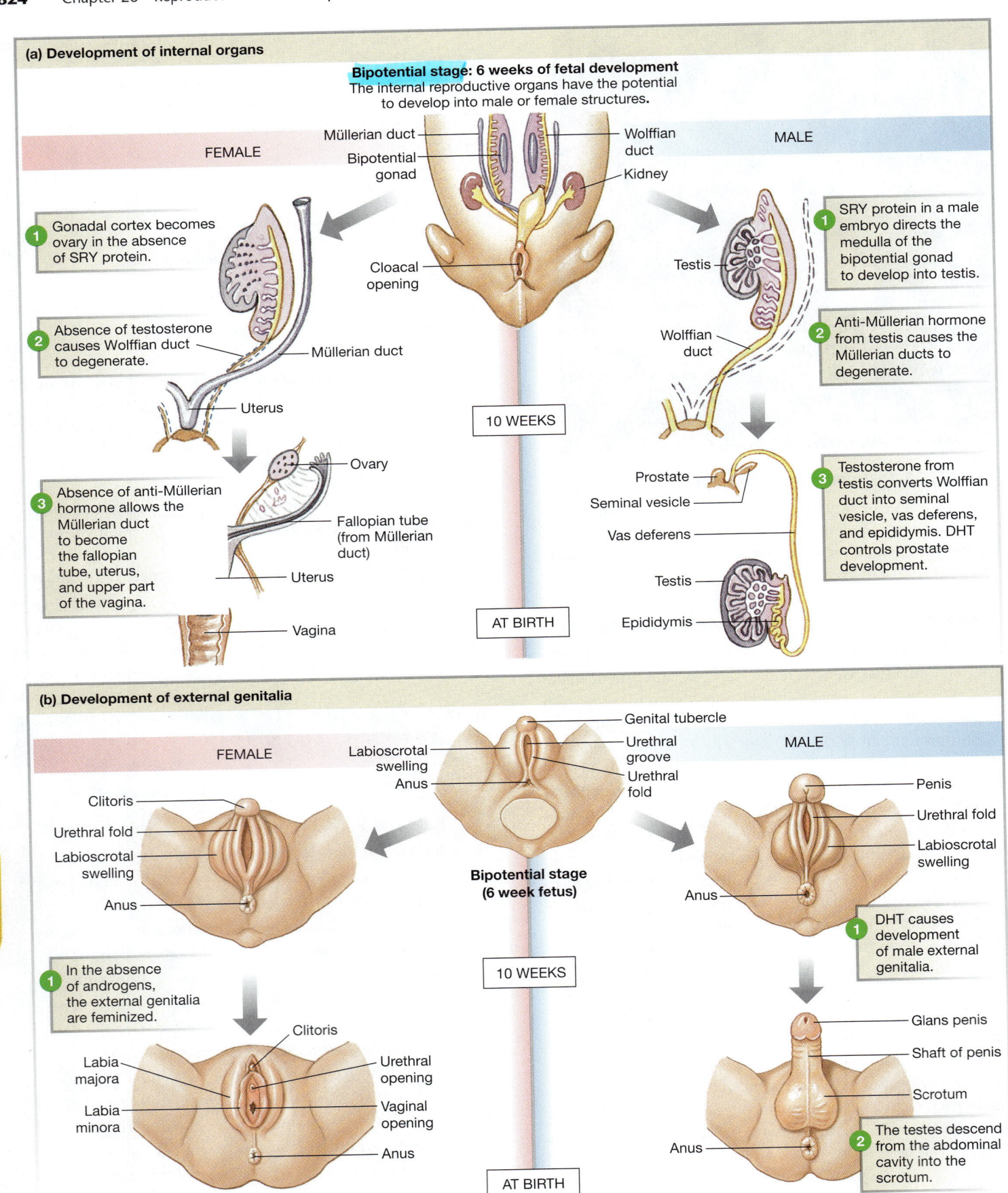

(a) Development of internal organs

Bipotential stage: 6 weeks of fetal development
The internal reproductive organs have the potential
to develop into male or female structures.

FEMALE

Müllerian duct
Bipotential gonad
Wolffian duct
Kidney

MALE

1 Gonadal cortex becomes ovary in the absence of SRY protein.

Cloacal opening

Testis

1 SRY protein in a male embryo directs the medulla of the bipotential gonad to develop into testis.

2 Absence of testosterone causes Wolffian duct to degenerate.

Müllerian duct

Wolffian duct

2 Anti-Müllerian hormone from testis causes the Müllerian ducts to degenerate.

Uterus

10 WEEKS

3 Absence of anti-Müllerian hormone allows the Müllerian duct to become the fallopian tube, uterus, and upper part of the vagina.

Ovary

Fallopian tube (from Müllerian duct)

Uterus

Prostate
Seminal vesicle

Vas deferens

Testis

3 Testosterone from testis converts Wolffian duct into seminal vesicle, vas deferens, and epididymis. DHT controls prostate development.

Vagina

AT BIRTH

Epididymis

(b) Development of external genitalia

Genital tubercle
Urethral groove
Urethral fold

FEMALE

Labioscrotal swelling
Anus

MALE

Clitoris
Urethral fold
Labioscrotal swelling

Penis
Urethral fold
Labioscrotal swelling

Anus

Anus

1 In the absence of androgens, the external genitalia are feminized.

Bipotential stage (6 week fetus)

1 DHT causes development of male external genitalia.

10 WEEKS

Clitoris

Labia majora

Labia minora

Anus

Urethral opening

Vaginal opening

Anus

Glans penis
Shaft of penis
Scrotum

2 The testes descend from the abdominal cavity into the scrotum.

AT BIRTH

26

FIGURE 26-3 *Sexual development in the human embryo*

Sexual Differentiation Occurs in the Second Month of Development

The sex of an early embryo is difficult to determine because reproductive structures do not begin to differentiate until the seventh week of development. Before differentiation, the embryonic tissues are considered *bipotential* because they cannot be morphologically identified as male or female.

The bipotential gonad has an outer cortex and an inner medulla (Fig. 26-3a ■). Under the influence of the appropriate developmental signal (described below), the medulla will develop into a testis ①; in the absence of that signal, the cortex will differentiate into ovarian tissue.

The bipotential internal genitalia consist of two pairs of accessory ducts: **Wolffian ducts** derived from the embryonic kidney, and **Müllerian ducts**. As development proceeds along either male or female lines, one pair of ducts develops while the other degenerates (Fig. 26-3a ②).

The bipotential external genitalia consist of a *genital tubercle, urethral folds, urethral groove,* and *labioscrotal swellings* (Fig. 26-3b). These structures differentiate into the male and female reproductive structures as development progresses (Table 26-1 ■).

What directs some single-cell zygotes to become males, and others to become females? Gender determination depends on the presence or absence of the *sex-determining region of the Y chromosome,* or **SRY gene**. In the absence of the SRY gene and its products, the gonads develop into ovaries. In the presence of a functional SRY gene, the bipotential gonads develop into testes.

Male Embryonic Development The SRY gene produces a protein (**SRY protein** or *testis-determining factor*) that binds to DNA and activates additional genes, including SOX9, WT1

(Wilms' tumor protein), and SF1 (steroidogenic factor). The protein products of these genes direct development of the gonadal medulla into a testis (Fig. 26-4 ■). Note that testicular development does *not* require male sex hormones such as **testosterone**. The developing embryo cannot secrete testosterone until after the gonads differentiate into testes.

Once the testes differentiate, they begin to secrete three hormones that influence development of the male internal and external genitalia. Testicular **Sertoli cells** secrete glycoprotein **anti-Müllerian hormone** (AMH; also called *Müllerian inhibiting substance*). Testicular **Leydig cells** secrete testosterone and its derivative **dihydrotestosterone (DHT)**. These two **androgens** [*andro-*, male] are the dominant steroid hormones in males. Testosterone and DHT both bind to the same *androgen receptor,* but the two ligands elicit different responses.

In the developing fetus, anti-Müllerian hormone causes the embryonic Müllerian ducts to regress (Fig. 26-3a, ② male). Testosterone converts the Wolffian ducts into male accessory structures: epididymis, vas deferens, and seminal vesicle (③ male and Table 26-1). Later in fetal development, testosterone controls migration of the testes from the abdomen into the *scrotum,* or scrotal sac. The remaining male sex characteristics, such as differentiation of the external genitalia, are controlled primarily by DHT.

The importance of DHT in male development came to light in studies of the male pseudohermaphrodites described in the opening of this chapter. These men inherit a defective gene for **5α-reductase**, the enzyme that catalyzes the conversion of testosterone to DHT. Despite normal testosterone secretion, these men have inadequate levels of DHT, and as a result the male external genitalia and prostate gland fail to develop fully during fetal development. At birth, the infants appear to

TABLE 26-1	**Sexual Differentiation**	
MALE	**BIPOTENTIAL STRUCTURE**	**FEMALE**
Glans penis	← Genital tubercle →	Clitoris
Shaft of penis	← Urethral folds and groove →	Labia minora, opening of vagina and urethra
Shaft of penis and scrotum	← Labioscrotal swellings →	Labia majora
Regresses	← Gonad (cortex) →	Forms ovary
Forms testis	← Gonad (medulla) →	Regresses
Becomes epididymis, vas deferens, and seminal vesicle (testosterone present)	← Wolffian duct →	Regresses (testosterone absent)
Regresses (anti-Müllerian hormone present)	← Müllerian duct →	Becomes Fallopian tube, uterus, cervix, and upper 1/3 of vagina (anti-Müllerian hormone absent)

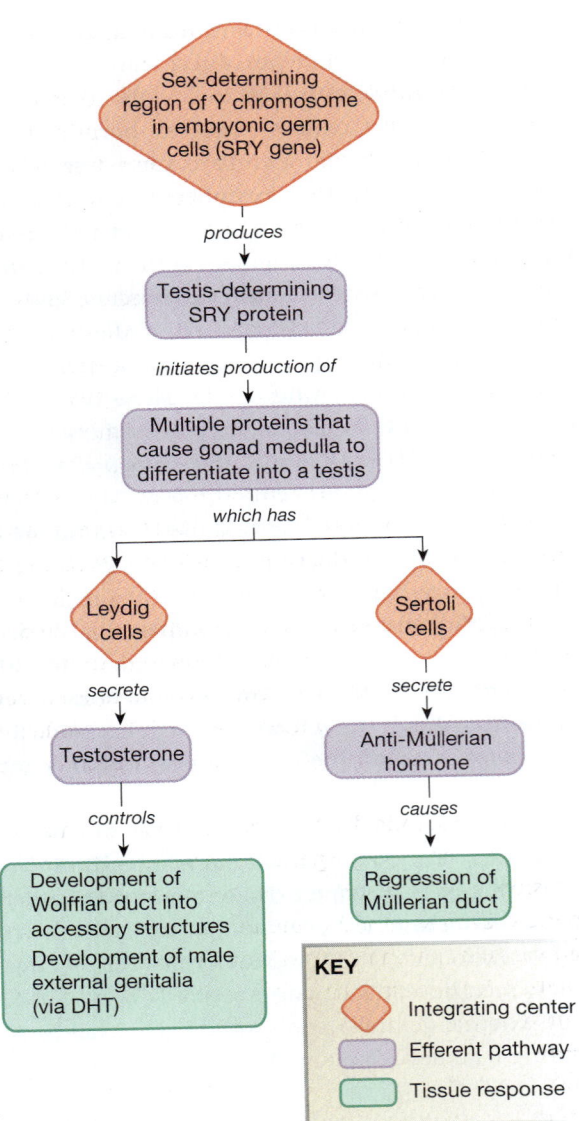

■ FIGURE 26-4 *Role of the SRY gene in male development*

be female and are raised as such. However, at puberty, the testes again begin to secrete testosterone, causing masculinization of the external genitalia, pubic hair growth (although scanty facial and body hair), and deepening voice. By studying the 5α-reductase defect in these individuals, scientists have been able to separate the effects of testosterone from those of DHT.

CONCEPT CHECK

2. Where in a target cell would you expect to find receptors for androgens? Where would you expect to find receptors for AMH?

Answers: p. 859

Exposure of nongenital tissues to testosterone during embryonic development is known to have masculinizing effects, such as altering the brain's responsiveness to certain hormones. One controversial aspect of the masculinizing effects of testosterone is its influence on human sexual behavior and gender

identity. It is well documented that in many nonhuman mammals, adult sexual behavior depends on the absence or presence of testosterone during critical periods of brain development. However, a similar cause/effect relationship has never been proved in humans. In human behavior, it is very difficult to separate biological influences from environmental factors, and it will probably be years before this question is resolved.

Female Embryonic Development In female embryos, which have no SRY gene, the cortex of the bipotential gonad develops into ovarian tissue (Fig. 26-3a, ① female). Without inhibition from testicular AMH, the Müllerian ducts develop into the upper portion of the **vagina**, the **uterus**, and the **Fallopian tubes** (after the anatomist Fallopius, who first described them; also known as **oviducts**). In the absence of testosterone, the Wolffian ducts degenerate, and without DHT the external genitalia take on female characteristics. Thus, absence of the SRY gene and testicular hormones creates a female.

CONCEPT CHECK

3. Why was King Henry VIII of England wrong to blame his wives when they were unable to produce a male heir to the throne?

4. Which sex will a zygote become if it inherits only one X chromosome (XO)?

5. If the testes are removed from an early male embryo, why does it develop a uterus and Fallopian tubes rather than the normal male accessory structures? Will the embryo have male or female external genitalia? Explain.

Answers: p. 859

BASIC PATTERNS OF REPRODUCTION

The male testis and female ovary both produce hormones and gametes, and they share other similarities, as might be expected of organs having the same origin. However, male and female gametes are very different from each other. Eggs are some of the largest cells in the body [Fig. 3-10, p. 59]. They are nonmotile and must be moved through the reproductive tract on currents created by smooth muscle contraction or the beating of cilia. Sperm, in contrast, are quite small. They are the only flagellated cells in the body and are highly motile so that they can swim up the female reproductive tract in their search for an egg to fertilize.

The timing of gamete production, or **gametogenesis**, is also very different in males and females. Women are born with all the eggs, or **oocytes**, they will ever have. During the reproductive years, the eggs mature. They are released from the ovaries roughly once a month for about 40 years. Thereafter female reproductive cycles cease (*menopause*).

Men, on the other hand, manufacture sperm continuously from the time they reach reproductive maturity. Sperm and testosterone production diminishes with age but does not cease the way women's reproductive cycles do.

DETERMINING GENDER

The first question new parents typically ask about their child is, "Is it a boy or a girl?" Sometimes the answer is not obvious because in approximately 1 in 3000 births, the gender of the child cannot easily be determined. Multiple criteria might be used to establish an individual's gender: genetic, chromosomal, gonadal, morphological, or even psychological characteristics. For example, presence of a Y chromosome with a functional SRY gene results in testes development and could be one criterion for "maleness." However, it is possible for an infant to have a Y chromosome and not appear to be male because of a defect in the complicated cascade of developmental steps that occur after testicular differentiation.

Currently there is ongoing debate about how best to decide gender in cases where there is doubt. Traditionally, gender determination has been based on appearance of the external genitalia at birth, but the idea that individuals should be allowed to choose their gender when they become old enough is gaining ground. The gender a person considers himself or herself to be is called the person's *gender identity*. You can read more about causes of ambiguous genitalia and the current criteria used to decide a child's gender in the American Academy of Pediatrics policy statement "Evaluation of the Newborn with Developmental Anomalies of the External Genitalia," *Pediatrics* 106(1):138–142, 2000 (July) (available online at *http://pediatrics.aappublications.org/*).

Gametogenesis Begins in Utero and Resumes During Puberty

Figure 26-5 ■ compares the male and female patterns of gametogenesis. In both sexes, germ cells of the embryonic gonads first undergo a series of mitotic divisions to increase their numbers. After that, the germ cells are ready to undergo **meiosis**, the cell division process through which gametes are formed.

When meiosis begins, the cell's DNA replicates, creating the equivalent of 46 duplicated chromosomes. However, cell and chromosomal division do not take place. Each duplicated chromosome is in the form of two identical **sister chromatids** linked together at a region known as the **centromere**. The resultant cells, called *primary gametes,* still have 46 chromosomes, but each chromosome is duplicated and contains twice the normal amount of DNA.

In the **first meiotic division**, the duplicated chromosomes remain paired as sister chromatids when the primary gamete divides into two *secondary gametes*. Each secondary gamete thus has 23 doubled chromosomes. In the **second meiotic division**, secondary gametes divide and the sister chromatids separate. If division of the secondary gamete goes to completion (as it does in males), the result is two cells each containing 23 single chromosomes. Theoretically gametogenesis is identical in males and females, but the timing of mitotic and meiotic divisions is very different.

Male Gametogenesis At birth, the testes of a newborn boy have not progressed beyond the mitosis stage and contain only immature germ cells. After birth, the gonads become inactive until puberty, the period in the early teen years when the gonads mature.

At puberty, germ cell mitosis resumes (Fig. 26-5). From that point onward, the germ cells, known as **spermatogonia** (singular *spermatogonium*), have two possible fates. Some continue to undergo mitosis throughout the male's reproductive life. Others are destined to undergo meiosis and become **primary spermatocytes**.

In the first meiotic division, one primary spermatocyte divides into two **secondary spermatocytes**. In the second meiotic division, each secondary spermatocyte divides into two **spermatids**. Each spermatid contains 23 single chromosomes, the haploid number characteristic of a gamete. The spermatids then mature into sperm. Thus, one primary spermatocyte creates four sperm.

Female Gametogenesis In the embryonic ovary, germ cells are called **oögonia** (singular *oögonium*) (Fig. 26-5). Oögonia complete mitotic replication and the first stage of meiosis by the fifth month of fetal development. At this time, germ cell mitosis ceases and no further oocytes can be formed. At birth the ovary contains about half a million primary gametes, or **primary oocytes**.

The oocyte's first meiotic division takes place following puberty. Each primary oocyte divides into two cells, a large **egg** (**secondary oocyte**) and a tiny **first polar body**. Despite the size difference, each cell contains 23 duplicated chromosomes. The first polar body disintegrates, either with or without undergoing a second meiotic division.

Meanwhile, the egg begins the second meiotic division. After the sister chromatids separate from each other, meiosis pauses. The final step of meiosis, in which sister chromatids in the egg go to separate cells, does not take place unless the egg is fertilized.

The ovary releases the mature egg during a process known as **ovulation**. If the egg is not fertilized, meiosis never goes to completion, and the egg disintegrates or passes out of the body.

26

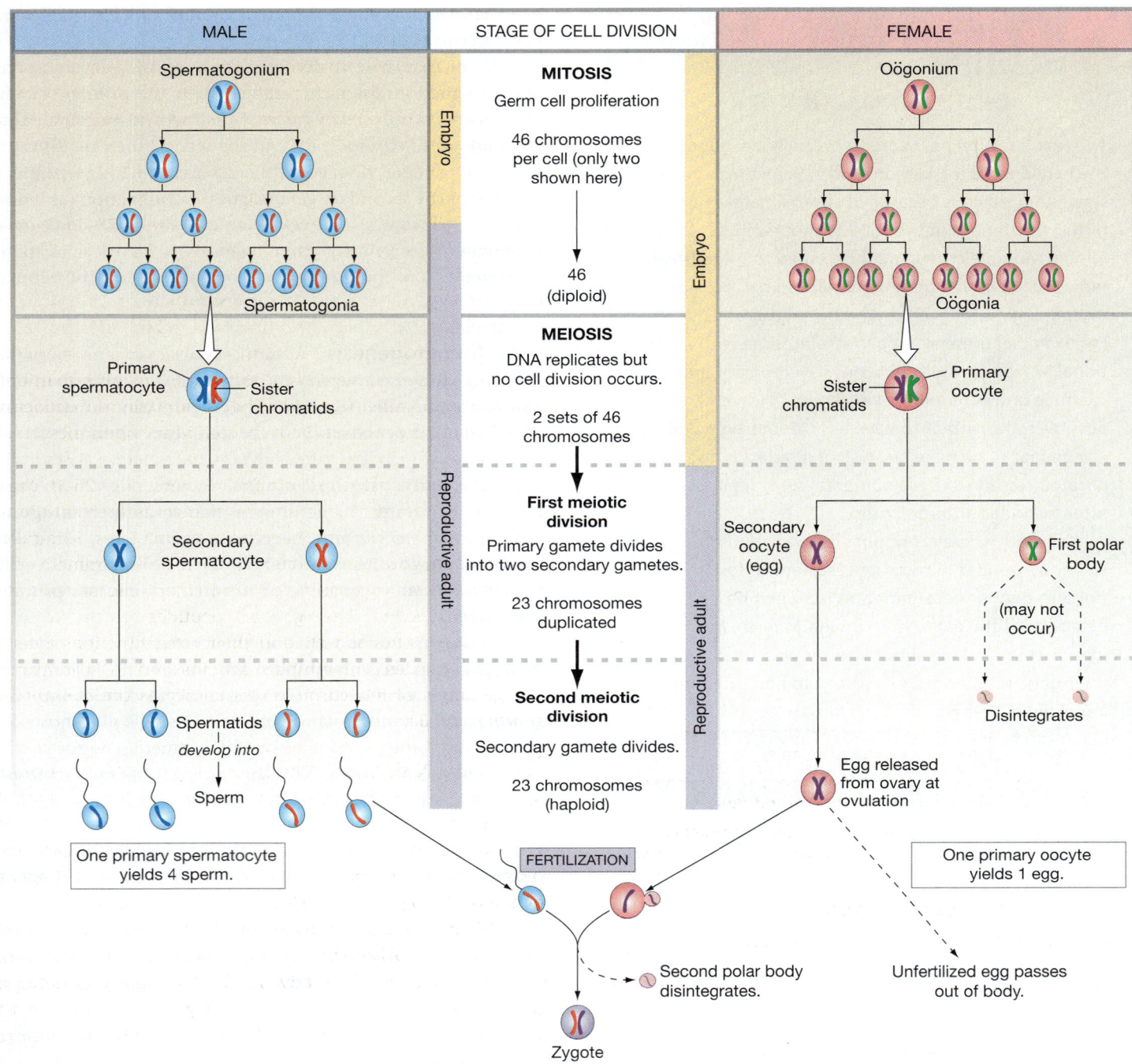

MALE | **STAGE OF CELL DIVISION** | **FEMALE**

MALE

Spermatogonium

Spermatogonia

Primary spermatocyte — Sister chromatids

Secondary spermatocyte

Spermatids
develop into
Sperm

One primary spermatocyte yields 4 sperm.

STAGE OF CELL DIVISION

Embryo

MITOSIS

Germ cell proliferation

46 chromosomes per cell (only two shown here)

46 (diploid)

MEIOSIS

DNA replicates but no cell division occurs.

2 sets of 46 chromosomes

First meiotic division

Primary gamete divides into two secondary gametes.

23 chromosomes duplicated

Second meiotic division

Secondary gamete divides.

23 chromosomes (haploid)

FERTILIZATION

Second polar body disintegrates.

Zygote

Reproductive adult

FEMALE

Oögonium

Oögonia

Sister chromatids — Primary oocyte

Secondary oocyte (egg)   First polar body

(may not occur)

Disintegrates

Egg released from ovary at ovulation

One primary oocyte yields 1 egg.

Unfertilized egg passes out of body.

Embryo

Reproductive adult

■ **FIGURE 26-5** *Gametogenesis*

For simplicity, this figure shows only one of the body's 22 pairs of autosomes. In mitosis, all chromosomes replicate and go to new cells. In meiosis, the chromosomes duplicate themselves, but each chromosome and its duplicate remain linked as sister chromatids joined by a centromere. Mitosis is completed *in utero* in females but continues after puberty in males. At fertilization, the male and female gametes each contribute one autosome of each pair to the zygote.

If fertilization by a sperm occurs, the final step of meiosis takes place. Half the sister chromatids remain in the fertilized egg (zygote), while the other half are released in a **second polar body**. The second polar body, like the first, degenerates. In females, each primary oocyte thus gives rise to only one egg.

Gametogenesis in both males and females is under the control of hormones from the brain and from endocrine cells in the gonads. Some of these hormones are identical in males and females, but others are different.

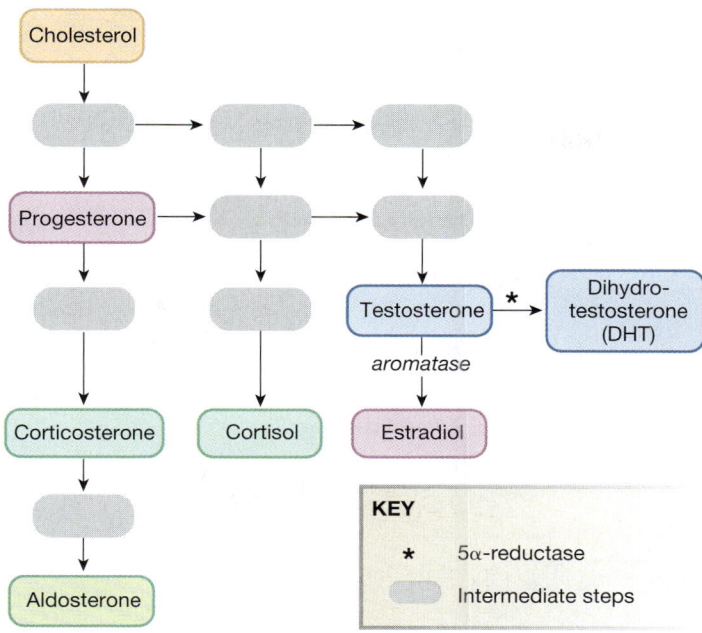

■ FIGURE 26-6   *Synthesis pathways for steroid hormones*

The blank boxes represent intermediate compounds whose names have been omitted for simplicity.

KEY

*   5α-reductase

    Intermediate steps

✔ **CONCEPT CHECK**

6. At what stage of development is the gamete in a newborn male? In a newborn female?

7. Compare the amount of DNA in the first polar body with the amount of DNA in the second polar body.

8. How many gametes are formed from one primary oocyte? From one primary spermatocyte?   Answers: p. 859

The Brain Directs Reproduction

The pathways that regulate reproduction begin with secretion of peptide hormones by the hypothalamus and anterior pituitary. These trophic hormones control gonadal secretion of the steroid sex hormones, including androgens, **estrogens**, and **progesterone**. The steroid hormones are closely related to one another and arise from the same steroid precursors (Fig. 26-6 ■). Both sexes produce both androgens and estrogens, but androgens predominate in males, and estrogens are dominant in females.

In men, most testosterone is secreted by the testes, but about 5% comes from the adrenal cortex. Testosterone is converted in peripheral tissues to its more potent derivative DHT. Some of the physiological effects attributed to testosterone are actually the result of DHT activity.

Males synthesize some estrogens, but the feminizing effects of these compounds are usually not obvious in males. Both testes and ovaries contain the enzyme **aromatase**, which converts testosterone to the female sex hormone **estradiol**, the

---

main estrogen in humans. A small amount of estrogen is made in peripheral tissues.

In women, the ovary produces estrogens (particularly estradiol and *estrone*) and *progestins*, particularly progesterone. The ovary and the adrenal cortex produce small amounts of androgens.

Control Pathways   The hormonal control of reproduction in both sexes follows the basic hypothalamus-anterior pituitary-peripheral gland pattern (Fig. 26-7 ■). **Gonadotropin-releasing hormone** (GnRH*) from the hypothalamus controls secretion of two anterior pituitary **gonadotropins: follicle stimulating hormone (FSH)** and **luteinizing hormone (LH)**. FSH and LH in turn act trophically on the gonads.

FSH, along with steroid sex hormones, is required to initiate and maintain gametogenesis. LH acts primarily on endocrine cells, stimulating production of the steroid sex hormones.

Although primary control of gonadal function arises in the brain, the gonads also influence their own function. Both ovary and testis secrete peptide hormones that act directly on the pituitary. **Inhibins** inhibit FSH secretion; related peptides called **activins** stimulate FSH secretion. Activins also promote spermatogenesis, oocyte maturation, and embryonic nervous system development. These gonadal peptides are produced in nongonadal tissues as well, and their other functions are still being investigated. The inhibins, activins, and AMH are part of a large family of related molecules known as the *transforming growth factor-β* family.

Feedback Pathways   The feedback pathways for trophic hormones follow the general pattern of long-loop and short-loop feedback described in Chapter 7 [🌐 p. 227]. Gonadal steroids alter secretion of GnRH, FSH, and LH in a long-loop response, and the pituitary gonadotropins inhibit GnRH release by a short-loop path (Fig. 26-7).

When circulating levels of gonadal steroids are low, the pituitary secretes FSH and LH (Table 26-2 ■). Once steroid secretion reaches a certain level, negative feedback inhibits gonadotropin release. Androgens always maintain negative feedback on gonadotropin release: as androgen levels go up, FSH and LH secretion diminishes.

On the other hand, only lower concentrations of estrogen have a negative feedback effect. If estrogen rises rapidly to above a threshold level for at least 36 hours, feedback changes from negative to positive, and gonadotropin release (particularly LH) is *stimulated*. The paradoxical effects of estrogen on gonadotropin release play a significant role in the female reproductive cycle, as you will learn later in this chapter.

We still do not fully understand the mechanism underlying the change from negative to positive feedback with estrogen.

26

---

*GnRH is sometimes called *luteinizing hormone releasing hormone* (LHRH) because it was first thought to have its primary effect on LH.

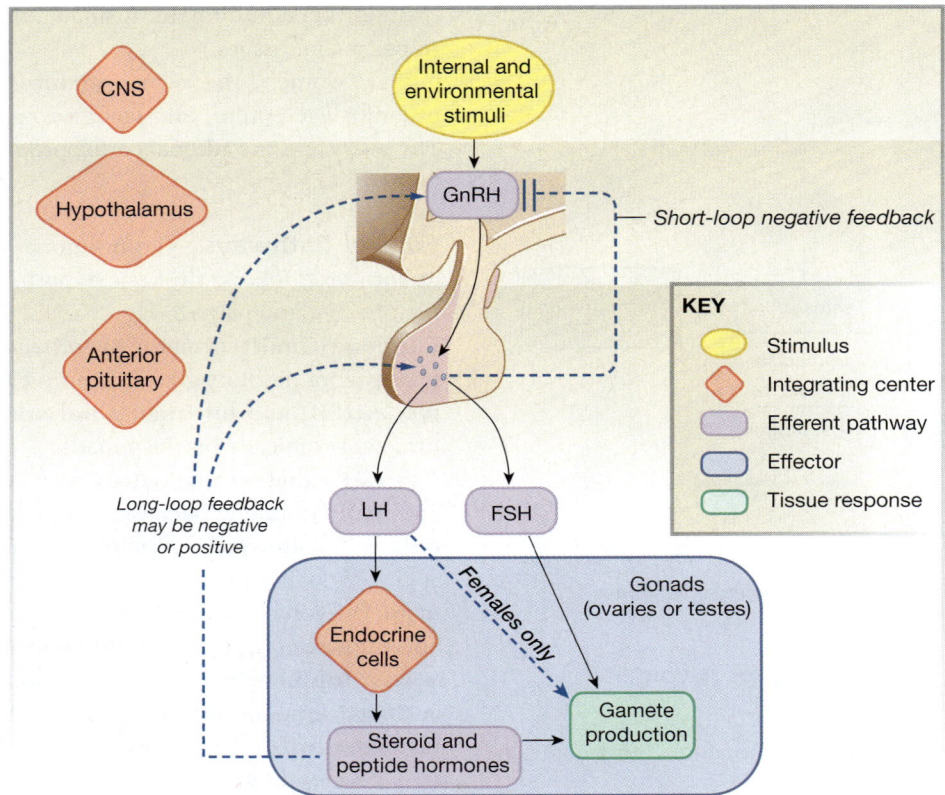

■ **FIGURE 26-7**   *General pattern of hormonal control of reproduction*

Some evidence suggests that high levels of estrogen increase the number of GnRH receptors, making the anterior pituitary more sensitive to GnRH (up-regulation of receptors [📖 p. 190]). Other evidence points to a direct effect of estrogen on hypothalamic release of GnRH.

Pulsatile GnRH Release   Starting at puberty, tonic GnRH release from the hypothalamus occurs in small pulses every 1–3 hours in both males and females. The region of the hypothalamus that contains the GnRH neuron cell bodies has been called a **pulse generator** because it coordinates the periodic pulsatile secretion of GnRH.

Scientists wondered why tonic GnRH release occurred in pulses rather than in a steady fashion, but several studies have shown the significance of the pulses. Children who suffer from a GnRH deficiency will not mature sexually in the absence of go-nadotropin stimulation of the gonads. If treated with steady infu-sions of GnRH through drug-delivery pumps, these children still fail to mature sexually. But if the pumps are adjusted to deliver GnRH in pulses similar to those that occur naturally, the children will go through puberty. Apparently, steady high levels of GnRH cause down-regulation of the GnRH receptors on gonadotropin cells, making the pituitary unable to respond to GnRH.

This down-regulation is the basis for the therapeutic use of GnRH in treating certain disorders. For example, patients with prostate and breast cancers stimulated by androgens or estro-gens may be given GnRH agonists to slow the growth of the cancer cells. It seems paradoxical to give these patients a drug that stimulates secretion of androgens and estrogens, but after a brief increase in FSH and LH, the pituitary becomes insensi-tive to GnRH. Then FSH and LH secretion decreases, and gonadal output of steroid hormones also falls. In essence, the GnRH agonist creates chemical castration that reverses when the drug is no longer administered.

TABLE 26-2	**Effects of Sex Steroids on Gonadotropin Release**	
STEROID HORMONE	**EFFECT**	**GONADOTROPIN LEVEL**
Low estrogen or androgen	Absence of negative feedback	Increases
Moderate estrogen or androgen	Negative feedback	Decreases
High androgen	Negative feedback	Decreases
Sustained high estrogen	Positive feedback	Increases

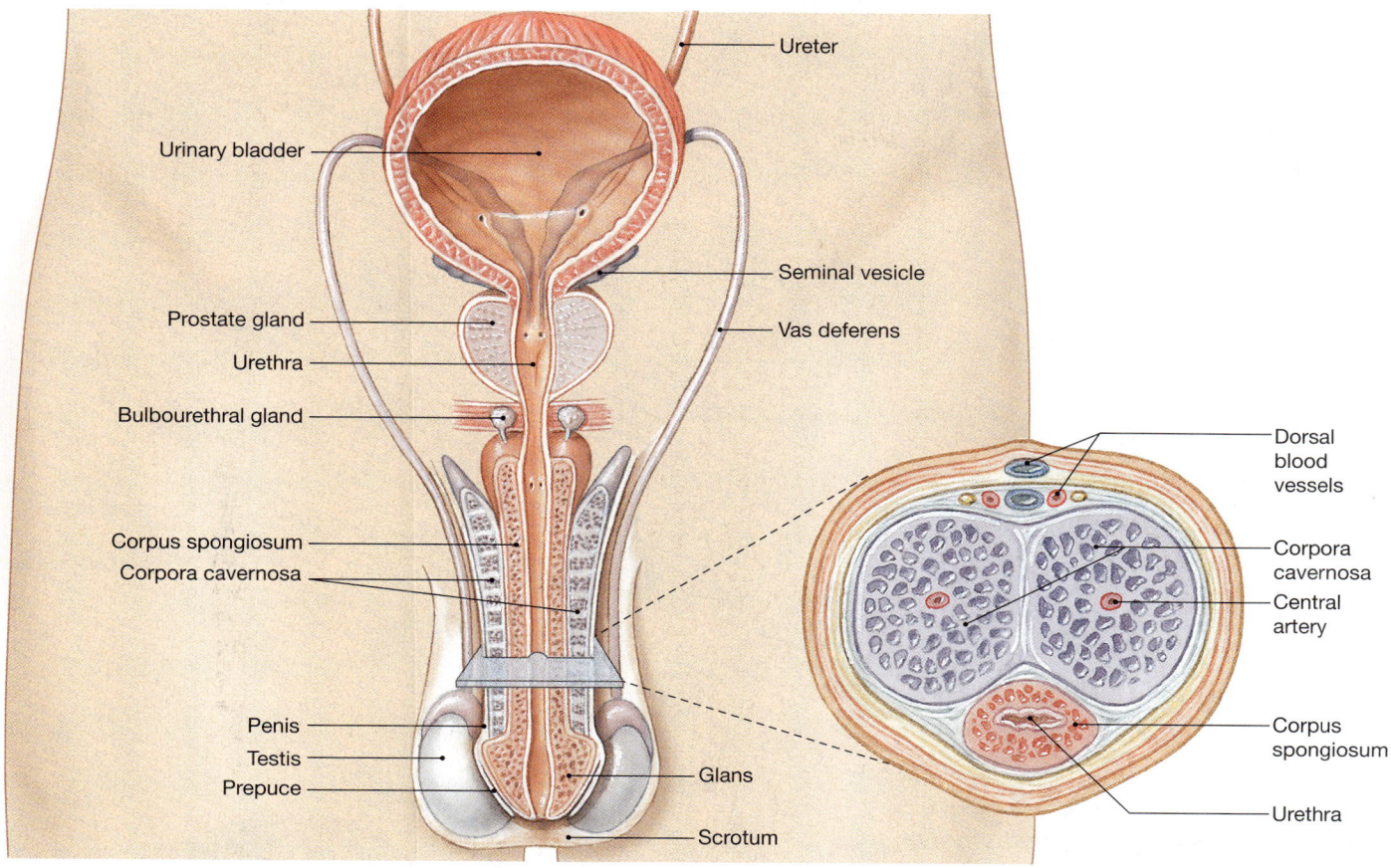

■ FIGURE 26-8   *Reproductive structures in the male*

Reproduction Is Influenced by Environmental Factors

Among the least-understood influences on reproductive hormones and gametogenesis are environmental effects. In men, factors that influence gametogenesis are difficult to monitor short of requesting periodic sperm counts. Disruption of the normal reproductive cycle in women is easier to study because physiological uterine bleeding in the menstrual cycle is easily monitored.

Factors that affect reproductive function in women include stress, nutritional status, and changes in the day-night cycle, such as those which occur with travel across time zones or with shift work. The hormone **melatonin** from the pineal gland [🔁 p. 237] mediates reproduction in seasonally breeding animals, such as birds and deer, and researchers are investigating whether melatonin also plays a role in seasonal and daily rhythms in humans.

Environmental estrogens are also receiving a lot of attention. These are naturally occurring compounds, such as the *phytoestrogens* of plants, or synthetic compounds that have been released into the environment. Some of these compounds bind to estrogen receptors and mimic estrogen's effects. Others are anti-estrogens that block estrogen receptors or interfere with second messenger pathways or protein synthesis. Growing evidence suggests that some of these endocrine disruptors can adversely influence developing embryos and even have their effects passed down to subsequent generations.

Now that you have learned the basic patterns of hormone secretion and gamete development, let's look in detail at the male and female reproductive systems.

CONCEPT CHECK

9. What does aromatase do?
10. What do the following abbreviations stand for? (Spelling counts!) FSH, DHT, SRY, LH, GnRH, AMH
11. Name the hypothalamic and anterior pituitary hormones that control reproduction.

Answers: p. 859

MALE REPRODUCTION

The male reproductive system consists of the testes, the internal genitalia (accessory glands and ducts), and the external genitalia. The external genitalia consist of the **penis** and the **scrotum**, a saclike structure that contains the testes. The **urethra** serves as a common passageway for sperm and urine, although not simultaneously. It runs through the ventral aspect of the shaft of the penis (Fig. 26-8 ■) and is surrounded by a

26

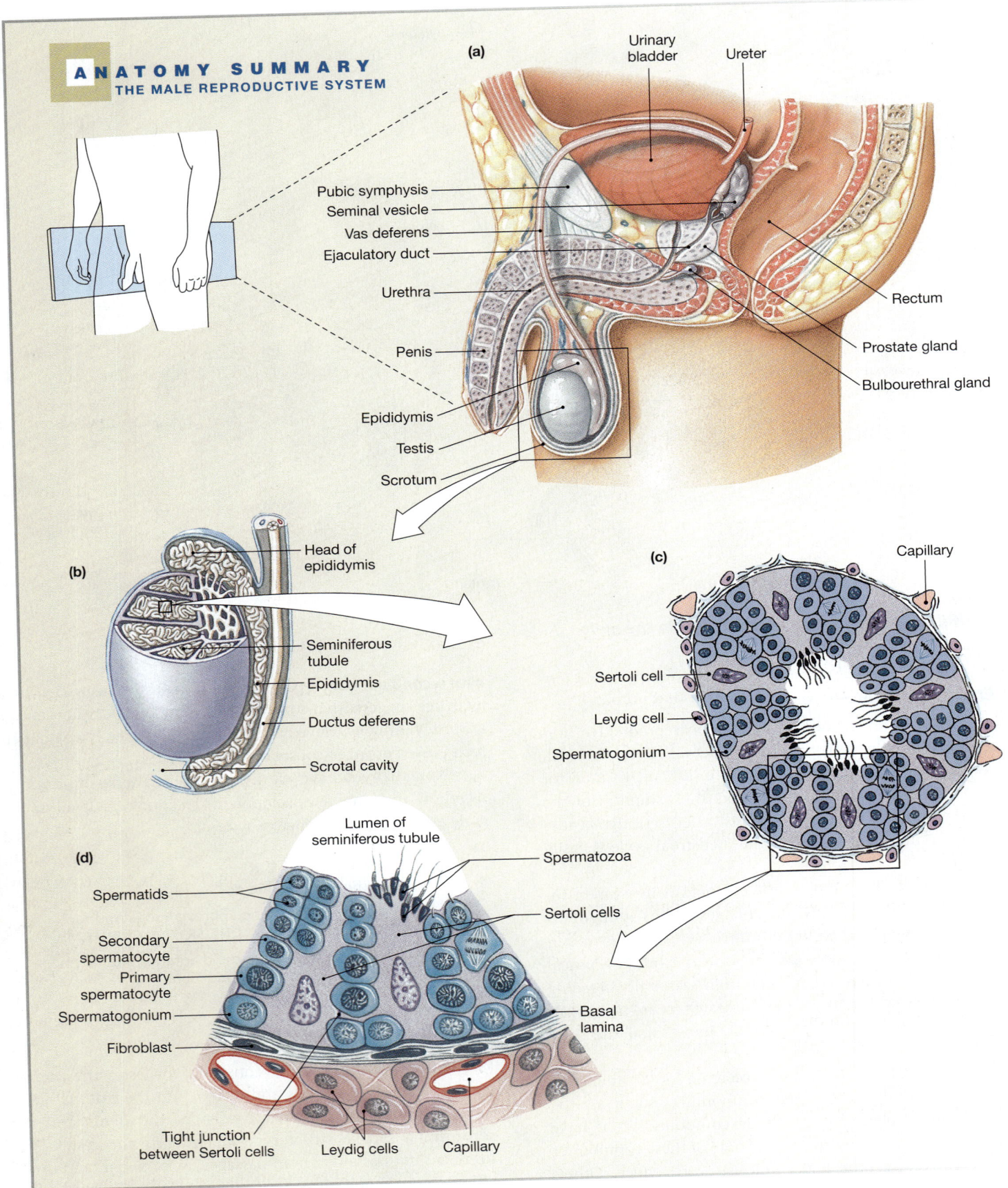

ANATOMY SUMMARY
THE MALE REPRODUCTIVE SYSTEM

(a)

Urinary bladder

Ureter

Pubic symphysis
Seminal vesicle
Vas deferens
Ejaculatory duct
Urethra

Penis

Epididymis

Testis

Scrotum

Rectum

Prostate gland

Bulbourethral gland

(b)

Head of epididymis

Seminiferous tubule

Epididymis

Ductus deferens

Scrotal cavity

(c)

Capillary

Sertoli cell

Leydig cell

Spermatogonium

(d)

Lumen of seminiferous tubule

Spermatids

Secondary spermatocyte

Primary spermatocyte

Spermatogonium

Fibroblast

Spermatozoa

Sertoli cells

Basal lamina

Tight junction between Sertoli cells

Leydig cells

Capillary

26

■ **FIGURE 26-9**

spongy column of tissue known as the **corpus spongiosum** [*corpus*, body; plural *corpora*]. The corpus spongiosum and two columns of tissue called the **corpora cavernosa** constitute the erectile tissue of the penis.

The tip of the penis is enlarged into a region called the **glans** that at birth is covered by a layer of skin called the **foreskin**, or **prepuce**. In some cultures, the foreskin is removed surgically in a procedure called **circumcision**. In the United States, this practice goes through cycles of popularity. Proponents of the procedure claim that it is necessary for good hygiene, and they cite evidence suggesting that the incidence of penile cancer, sexually transmitted diseases, and urinary tract infections is lower in circumcised men. Opponents claim that it is cruel to subject newborn boys to an unnecessary and possibly painful procedure.

The scrotum is an external sac into which the testes migrate during fetal development. This location outside the abdominal cavity is necessary because normal sperm development requires a temperature that is 2°–3° F lower than core body temperature. Men who have borderline or low sperm counts are advised to switch from jockey-style underwear, which keeps the scrotum close to the body, to boxer shorts, which allow the testes to stay cooler.

The failure of one or both testes to descend is known as **cryptorchidism** [*crypto*, hidden + *orchis*, testicle] and occurs in 1–3% of newborn males. If left alone, about 80% of cryptorchid testes spontaneously descend later. Those that remain in the abdomen through puberty become sterile and are unable to produce sperm.

Although cryptorchid testes lose their gametogenic potential, they can produce androgens, indicating that hormone production is not as temperature sensitive as sperm production. Because undescended testes are prone to become cancerous, authorities recommend that they be moved to the scrotum with testosterone treatment or, if necessary, surgically.

The male accessory glands and ducts include the **prostate gland**, the **seminal vesicles**, and the **bulbourethral (Cowper's) glands**. The bulbourethral glands and seminal vesicles empty their secretions into the urethra through ducts. The individual glands of the prostate open directly into the urethral lumen.

The prostate gland is the best known of the three accessory glands because of its medical significance. Cancer of the prostate is the second most common form of cancer in men (next to lung cancer), and *benign prostatic hypertrophy* (enlargement) creates problems for many men after age 50. Because the prostate gland completely encircles the urethra, its enlargement causes difficulty in urinating by narrowing the passageway.

Fetal development of the prostate gland, like that of the external genitalia, is under the control of dihydrotestosterone. Discovery of the role of DHT in prostate growth led to the development of *finasteride*, a 5α-reductase inhibitor that blocks DHT production. This drug was the first nonsurgical treatment for benign prostatic hypertrophy. Studies are currently under way to determine if lowering (but not eliminating) DHT levels will decrease the incidence of cancer of the prostate gland.

The Testes Produce Sperm and Testosterone

The human testes are paired ovoid structures about 5 cm by 2.5 cm (Fig. 26-9a ■). The word *testis* means "witness" in Latin, and its application to the male gonad comes from the fact that in ancient Rome, men taking an oath placed one hand on their genitals.

The testes have a tough outer fibrous capsule that encloses masses of coiled **seminiferous tubules** clustered into 250–300 compartments (Fig. 26-9b). Between tubules is interstitial tissue consisting primarily of blood vessels and the testosterone-producing Leydig cells (Fig. 26-9c). The seminiferous tubules constitute nearly 80% of the testicular mass in an adult. Each individual tubule is 0.3–1 meter long, and, if stretched out and laid end to end, the entire mass would extend for about the length of two and a half football fields.

The seminiferous tubules leave the testis and join the **epididymis** [*epi-*, upon + *didymos*, twin], a single duct that forms a tightly coiled cord on the surface of the testicular capsule (Fig. 26-9b). The epididymis becomes the **vas deferens** [*vas*, vessel + *deferre*, to carry away from], also known as the **ductus deferens**. This duct passes into the abdomen, where it eventually empties into the urethra, the passageway from the urinary bladder to the external environment (see Fig. 26-8).

Seminiferous Tubules

Each seminiferous tubule contains two types of cells: spermatogonia in various stages of becoming sperm and **Sertoli cells** (Fig. 26-9c, d). The developing spermatocytes stack in columns from the outer edge of the tubule to the lumen. Between each column is a single Sertoli cell that extends from the outer edge of the tubule to the lumen. Surrounding the outside of the tubule is a basal lamina (Fig. 26-9d) that acts as a barrier, preventing certain large molecules in the interstitial fluid from entering the tubule but allowing testosterone to enter easily.

Adjacent Sertoli cells in a tubule are linked to each other by tight junctions that form an additional barrier between the lumen of the tubule and the interstitial fluid outside the tubule. These tight junctions are sometimes called the **blood-testis barrier** because functionally they behave much like the impermeable capillaries of the blood-brain barrier, restricting movement of molecules between two compartments. The basal lamina and tight junctions create three compartments: the tubule lumen, a *basal compartment* on the basolateral side of the Sertoli cells, and the interstitial fluid. Because of the barriers between these compartments, the luminal fluid has a composition different from that of interstitial fluid, with low concentrations of glucose and high concentrations of K^+ and steroid hormones.

26

RUNNING PROBLEM

Infertility can be caused by problems in either the man or the woman. Sometimes, however, both partners have problems that contribute to their infertility. In general, male infertility is caused by low sperm counts, abnormalities in sperm morphology, or abnormalities in the reproductive structures that carry sperm. Female infertility may be caused by problems in hormonal pathways that govern maturation and release of eggs or by abnormalities of the reproductive structures (cervix, uterus, ovaries, oviducts). Because tests of male fertility are simple to perform, Dr. Coddington first analyzes Larry's sperm. In this test, trained technicians examine a fresh sperm sample under a microscope. They note the shape and motility of the sperm and estimate the concentration of sperm in the sample.

Question 1:
  Name (in order) the male reproductive structures that carry sperm from the testes to the external environment.

Question 2:
  A new technique for the treatment of male infertility involves retrieval of sperm from the epididymis. The retrieved sperm can then be used to fertilize an egg, which is then implanted in the uterus. Which causes of male infertility might make this treatment necessary?

| 822 | **834** | 843 | 844 | 849 | 853 | 855 |

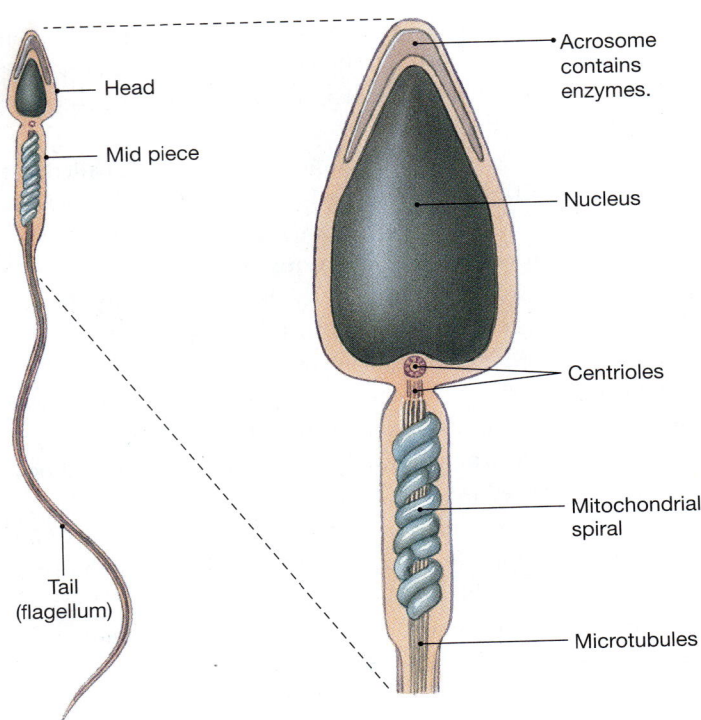

FIGURE 26-10 *Sperm structure*

dense structure, while a lysosome-like vesicle called an **acrosome** flattens out to form a cap over the tip of the nucleus. The acrosome contains enzymes essential for fertilization. Mitochondria to produce energy for sperm movement concentrate in the midpiece of the sperm body, along with microtubules that extend into the tail [⟳ p. 62]. The result is a small, motile gamete that bears little resemblance to the parent spermatid.

Sperm are released into the lumen of the seminiferous tubule, along with secreted fluid. From there, they are free to move out of the testis. The entire development process—from spermatogonium division until sperm release—takes about 64 days. At any given time, different regions of the tubule contain spermatocytes in different stages of development. The staggering of developmental stages allows sperm production to remain nearly constant at a rate of *200 million* sperm per day. That may sound like an extraordinarily high number, but it is about the number of sperm released in a single ejaculation.

Sperm just released from Sertoli cells are not yet mature and are incapable of swimming. They are pushed out of the tubule lumen by other sperm and by bulk flow of fluid secreted by Sertoli cells. Sperm entering the epididymis complete their maturation during the 12 or so days of their transit time, aided by protein secretions from epididymal cells.

Sperm Production Spermatogonia, the germ cells that undergo meiotic division to become sperm, are found clustered near the basal ends of the Sertoli cells, just inside the basal lamina of the seminiferous tubules (Fig. 26-9c, d). In this basal compartment, they undergo mitotic divisions that produce additional germ cells. Some of the spermatogonia remain near the outer edge of the tubule to produce future spermatogonia; others enter meiosis and become primary spermatocytes.

As spermatocytes differentiate into sperm, they move inward toward the tubule lumen, continuously surrounded by Sertoli cells. The tight junctions of the blood-testis barrier break and reform around the migrating cells, ensuring that the barrier remains intact. By the time spermatocytes reach the luminal ends of Sertoli cells, they have divided twice and become spermatids.

Spermatids remain embedded in the apical membrane of Sertoli cells while they complete the transformation into sperm, losing most of their cytoplasm and developing a flagellated tail (Fig. 26-10 ■). The chromatin of the nucleus condenses into a

Sertoli Cells The function of Sertoli cells is to regulate sperm development. Another name for Sertoli cells is *sustentacular cells* because they provide sustenance, or nourishment, for the developing spermatogonia. Sertoli cells manufacture and secrete

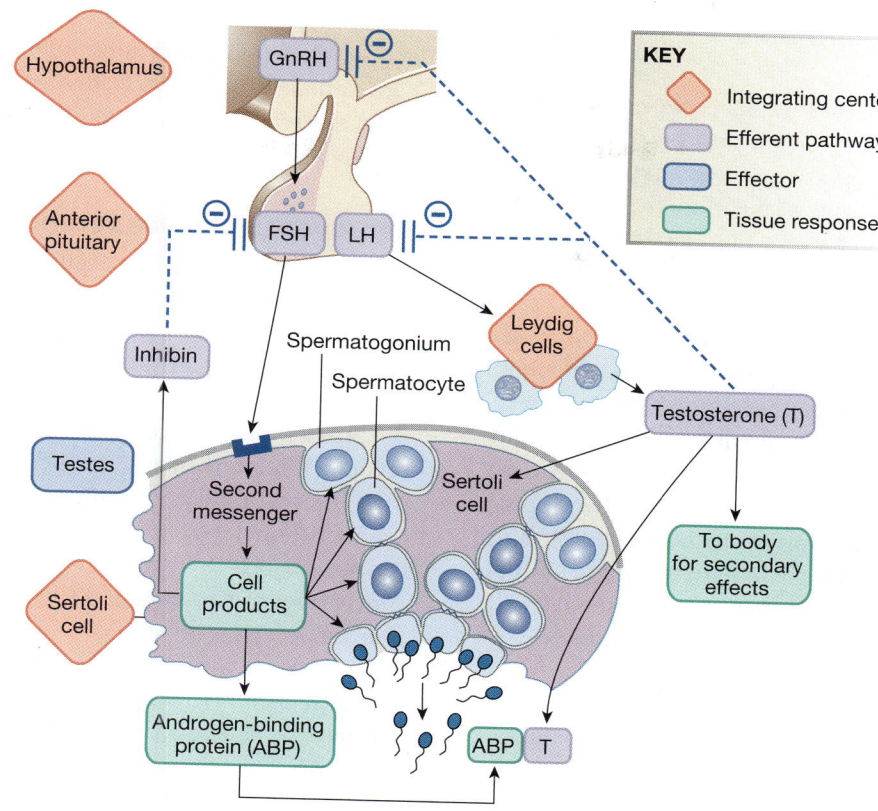

KEY

◇	Integrating center
▭	Efferent pathway
▭	Effector
▭	Tissue response

■ **FIGURE 26-11**   *Hormonal control of spermatogenesis*

proteins that range from the hormones inhibin and activin to growth factors, enzymes, and **androgen-binding protein** (ABP). ABP is secreted into the seminiferous tubule lumen, where it binds to testosterone (Fig. 26-11 ■). Protein binding makes testosterone less lipophilic and concentrates it in the luminal fluid.

Leydig Cells   Leydig cells, located in the interstitial tissue between seminiferous tubules (Fig. 26-9c, d), secrete testosterone. They are first active in the fetus, when testosterone is needed to direct development of male characteristics. After birth, the cells become inactive until puberty, when they resume testosterone production. The Leydig cells also convert some testosterone to estradiol.

Spermatogenesis Requires Gonadotropins and Testosterone

The hormonal control of spermatogenesis follows the general pattern described previously: hypothalamic GnRH promotes release of LH and FSH from the anterior pituitary (Fig. 26-11). FSH and LH in turn stimulate the testes. The gonadotropins were named originally for their effect on the female ovary, but the same names have been retained in the male.

GnRH release is pulsatile, peaking every 1.5 hours, and LH release follows the same pattern. FSH levels are not as obviously related to GnRH secretion because FSH secretion is also influenced by inhibin and activin.

The target of FSH is Sertoli cells because (unlike oocytes) male germ cells do not have FSH receptors. FSH stimulates Sertoli synthesis of paracrine molecules needed for spermatogonia mitosis and spermatogenesis. In addition, FSH stimulates production of androgen-binding protein and inhibin.

The primary target of LH is the Leydig cells, which produce testosterone. In turn, testosterone feeds back to inhibit LH release. Testosterone is essential for spermatogenesis, but its actions also appear to be mediated by Sertoli cells. Spermatocytes have no androgen receptors and cannot respond directly to testosterone, but they do have receptors for androgen-binding protein. Sertoli cells, which produce androgen-binding protein, have androgen receptors.

Spermatogenesis is a very difficult process to study *in vivo* or *in vitro,* and the available animal models may not accurately reflect the situation in the human testis. Therefore, it may be some time before we can say with certainty how testosterone and FSH regulate spermatogenesis.

CONCEPT CHECK

12. What do Sertoli cells secrete? What do Leydig cells secrete?

13. Because GnRH agonists cause down-regulation of GnRH receptors, what would be the advantages and disadvantages of using these drugs as a male contraceptive?

Answers: p. 859

26

Male Accessory Glands Contribute Secretions to Semen

The male reproductive tract has three accessory glands—bulbourethral glands, seminal vesicles, and prostate—whose primary function is to secrete various fluid mixtures. When sperm leave the vas deferens during ejaculation, they are joined by these secretions, resulting in a sperm-fluid mixture known as **semen**. About 99% of the volume of semen is fluid added from the accessory glands.

Accessory gland contributions to the composition of semen are listed in Table 26-3 ■. Semen provides a liquid medium for sperm delivery, including mucus for lubrication, buffers to neutralize the usually acidic environment of the vagina, and nutrients for sperm metabolism. Seminal vesicles contribute prostaglandins [↺ p. 189], lipid-related molecules that appear to influence sperm motility and transport in both the male and female reproductive tracts. Prostaglandins were originally believed to come from the prostate gland, and the name was well established by the time their true source was discovered.

In addition to providing a medium for sperm, accessory gland secretions help protect the male reproductive tract from pathogens that might ascend the urethra from the external environment. The secretions physically flush out the urethra and supply immunoglobulins, lysozyme, and other compounds with antibacterial action. One interesting component of semen is zinc. Its role in reproduction is unclear, but concentrations of zinc below a certain level are associated with male infertility.

Androgens Influence Secondary Sex Characteristics

Androgens have a number of effects on the body in addition to gametogenesis. These effects are divided into primary and secondary sex characteristics. **Primary sex characteristics** are the internal sexual organs and external genitalia that distinguish males from females. As you have already learned, androgens are responsible for the differentiation of male genitalia during embryonic development and for their growth during puberty.

The **secondary sex characteristics** are other traits that distinguish males from females. The male body shape is sometimes described as an inverted triangle, with broad shoulders and narrow waist and hips. The female body is usually more pear shaped, with broad hips and narrow shoulders. Androgens are responsible for such typically male traits as beard and body hair growth, muscular development, thickening of the vocal chords with subsequent lowering of the voice, and behavioral effects, such as the sex drive (**libido**).

Androgens are anabolic hormones that promote protein synthesis, which gives them their street name of *anabolic steroids*. The illicit use of these drugs by athletes has been widespread despite possible adverse side effects such as liver tumors and infertility. One of the more interesting side effects is the apparent addictiveness of anabolic steroids. Withdrawal from taking the drugs may be associated with behavioral changes that include excessive aggression ('roid rage), depression, or psychosis. These psychiatric disturbances suggest that human brain function can be modulated by sex steroids, just as the brain function of other animals can. Fortunately, many side effects of anabolic steroids are reversible once their use is discontinued.

CONCEPT CHECK

14. Explain why the use of exogenous anabolic steroids might shrink a man's testes and make him temporarily infertile.

Answers: p. 859

FEMALE REPRODUCTION

Female reproduction is a more complicated topic than male reproduction because of the cyclic nature of gamete production in the ovary.

The Female Reproductive Tract Includes Ovaries and Uterus

The female external genitalia are known collectively as either the **vulva** or the **pudendum** [*vulva*, womb; *pudere*, to be ashamed]. They are shown in Figure 26-12a ■, the view seen by a health care worker who is about to do a pelvic exam or take a Pap smear [↺ p. 51].

TABLE 26-3	Composition of Semen	
COMPONENT	**FUNCTION**	**SOURCE**
Sperm	Gametes	Seminiferous tubules
Mucus	Lubricant	Bulbourethral glands
Water	Provides liquid medium	All accessory glands
Buffers	Neutralize acidic environment of the vagina	Prostate, bulbourethral glands
Nutrients Fructose Citric acid Vitamin C Carnitine	Nourish sperm	Seminal vesicles Prostate Seminal vesicles Epididymis
Enzymes	Clot semen in vagina, then liquefy the clot	Seminal vesicles and prostate
Zinc	Unknown; possible association with fertility	Unknown
Prostaglandins	Smooth muscle contraction; may aid sperm transport	Seminal vesicles

Starting at the periphery are the **labia majora** [*labium*, lip], folds of skin that arise from the same embryonic tissue as the scrotum. Within the labia majora are the **labia minora**, derived from embryonic tissues that in the male give rise to the shaft of the penis (see Fig. 26-3). The **clitoris** is a small bud of erectile, sensory tissue at the anterior end of the vulva, enclosed by the labia minora and an additional fold of tissue equivalent to the foreskin of the penis.

In females, the urethra opens to the external environment between the clitoris and the vagina [*vagina*, sheath], the cavity that acts as receptacle for the penis during copulation. At birth, the external opening of the vagina is partially closed by a thin ring of tissue called the **hymen**, or *maidenhead*. The hymen is external to the vagina, not within it, so the normal use of tampons during menstruation will not rupture the hymen. However, it can be stretched by normal activities such as horseback riding and therefore is not an accurate indicator of a woman's virginity.

Now let's follow the path of sperm deposited in the vagina during intercourse. To continue into the female reproductive tract, sperm must pass through the narrow opening of the **cervix**, the neck of the uterus that protrudes slightly into the upper end of the vagina (Fig. 26-12b, c). The cervical canal is lined with mucous glands whose secretions create a barrier between the vagina and uterus. Sperm that make it through the cervical canal find themselves in the lumen of the uterus, or *womb*, a hollow, muscular organ slightly smaller than a woman's clenched fist.

The uterus is the structure in which fertilized eggs implant and develop during pregnancy. It is composed of three tissue layers (Fig. 26-12f): a thin outer connective tissue covering, a thick middle layer of smooth muscle known as the **myometrium**, and an inner layer known as the **endometrium** [*metra*, womb]. The endometrium consists of an epithelium with glands that dip into a connective tissue layer below. The thickness and character of the endometrium vary during the menstrual cycle. Cells of the epithelial lining alternately proliferate and slough off, accompanied by a small amount of bleeding in the process known as **menstruation** [*menstruus*, monthly].

Sperm swimming upward through the uterus leave its cavity through openings into the two Fallopian tubes (Fig. 26-12c). The Fallopian tubes are 20–25 cm long and about the diameter of a drinking straw. Their walls have two layers of smooth muscle, longitudinal and circular, similar to the walls of the intestine. A ciliated epithelium lines the inside of the tubes. Fluid movement created by the cilia and aided by muscular contractions transports an egg along the Fallopian tube toward the uterus. If sperm moving up the tube encounter an egg moving down the tube, fertilization may occur. Pathological conditions in which ciliary function is absent are associated with female infertility and with pregnancies in which the embryo implants in the Fallopian tube rather than the uterus.

The flared open end of the Fallopian tube divides into fingerlike projections called **fimbriae** [*fimbriae*, fringe]. The fimbriae (Fig. 26-12c) are held close to the adjacent ovary by connective tissue, which helps ensure that eggs released from the surface of the ovary will be swept into the tube rather than floating off into the abdominal cavity.

The Ovary Produces Eggs and Hormones

The ovary is an elliptical structure, about 2–4 cm long (Fig. 26-12d). It has an outer connective tissue layer and an inner connective tissue framework known as the **stroma** [*stroma*, mattress]. Most of the ovary consists of a thick outer *cortex* filled with ovarian follicles in various stages of development or decline. The small central *medulla* contains nerves and blood vessels.

The ovary, like the testis, produces both gametes and hormones. As discussed earlier, about 7 million oögonia in the embryonic ovary develop into half a million primary oocytes. Each primary oocyte is enclosed in a **primary follicle** with a single layer of **granulosa cells** separated by a basement membrane from an outer layer of cells known as the **theca** [*theke*, case or cover] (Fig. 26-12e).

A Menstrual Cycle Lasts About One Month

Female humans produce gametes in monthly cycles (average 28 days; normal range 24–35 days). These cycles are commonly called **menstrual cycles** because they are marked by a 3–7 day period of bloody uterine discharge known as the **menses** [*menses*, months], or **menstruation**. The menstrual cycle can be described by changes that occur in follicles of the ovary (**ovarian cycle**) and in the endometrial lining of the uterus (**uterine cycle**). Figure 26-13 ■ is a summary figure showing a typical menstrual cycle.

The ovarian cycle is divided into three phases:

1. **Follicular phase.** The first part of the ovarian cycle, known as the **follicular phase**, is a period of follicular growth in the ovary. This phase is the most variable in length and lasts from 10 days to 3 weeks.
2. **Ovulation.** Once one or more follicles have ripened, the ovary releases the oocyte(s) during **ovulation**.
3. **Luteal phase.** The phase of the ovarian cycle following ovulation is known as the *postovulatory* or **luteal phase**. The second name comes from the transformation of a ruptured follicle into a **corpus luteum** [*corpus*, body + *luteus*, yellow], named for its yellow pigment and lipid deposits. The corpus luteum secretes hormones that continue the preparations for pregnancy. If a pregnancy does not occur, the corpus luteum ceases to function after about two weeks, and the ovarian cycle begins again.

The endometrial lining of the uterus goes through its own cycle—the uterine cycle—regulated by ovarian hormones:

1. **Menses.** The beginning of the follicular phase in the ovary corresponds to menstrual bleeding from the uterus.
2. **Proliferative phase.** The latter part of the ovary's follicular phase corresponds to the **proliferative phase** in the uterus, during which the endometrium adds a new layer of cells in anticipation of pregnancy.

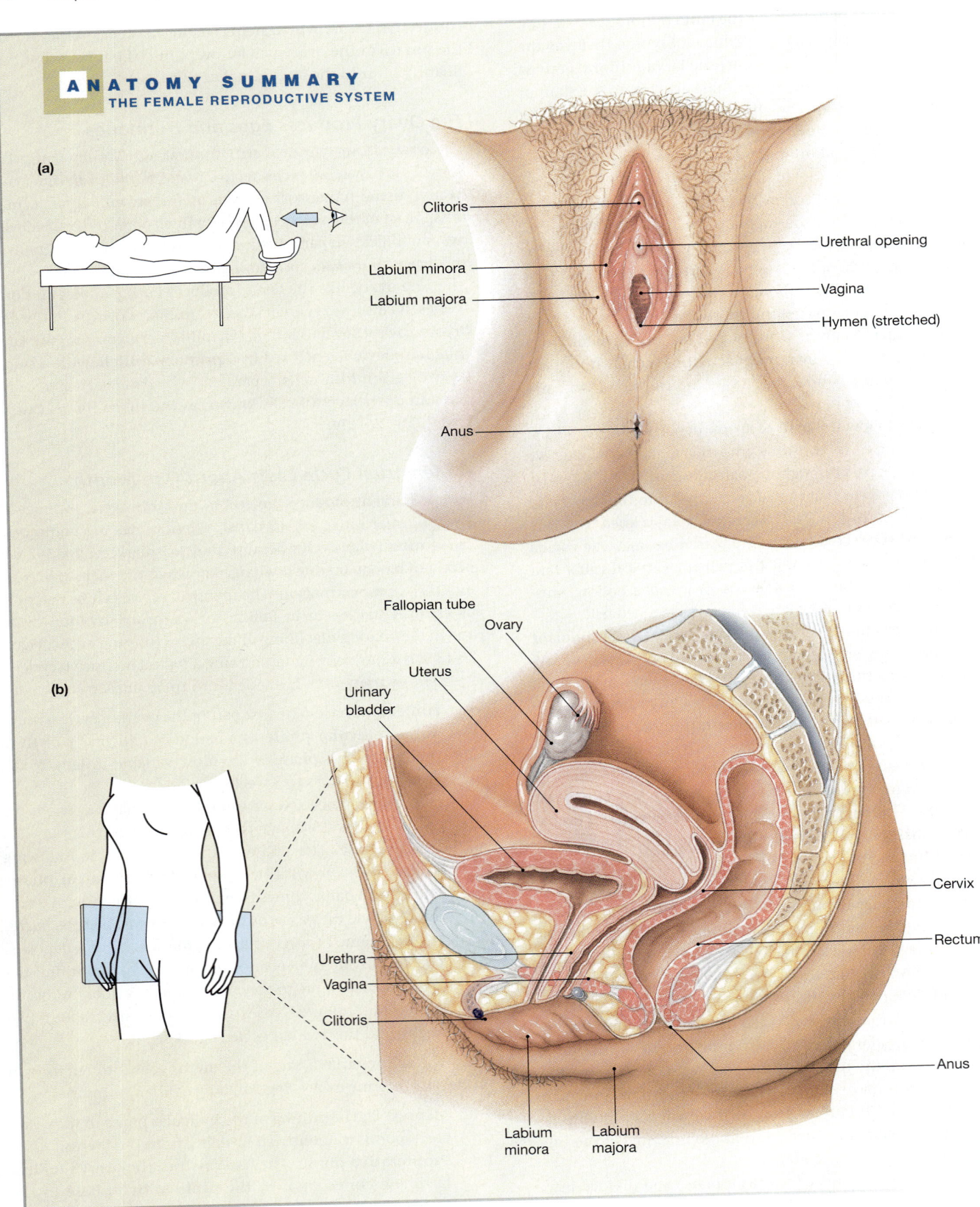

ANATOMY SUMMARY
THE FEMALE REPRODUCTIVE SYSTEM

(a)

Clitoris

Labium minora

Labium majora

Anus

Urethral opening

Vagina

Hymen (stretched)

(b)

Fallopian tube

Ovary

Uterus

Urinary
bladder

Urethra

Vagina

Clitoris

Cervix

Rectum

Anus

Labium
minora

Labium
majora

FIGURE 26-12

STRUCTURE OF THE UTERUS AND OVARY

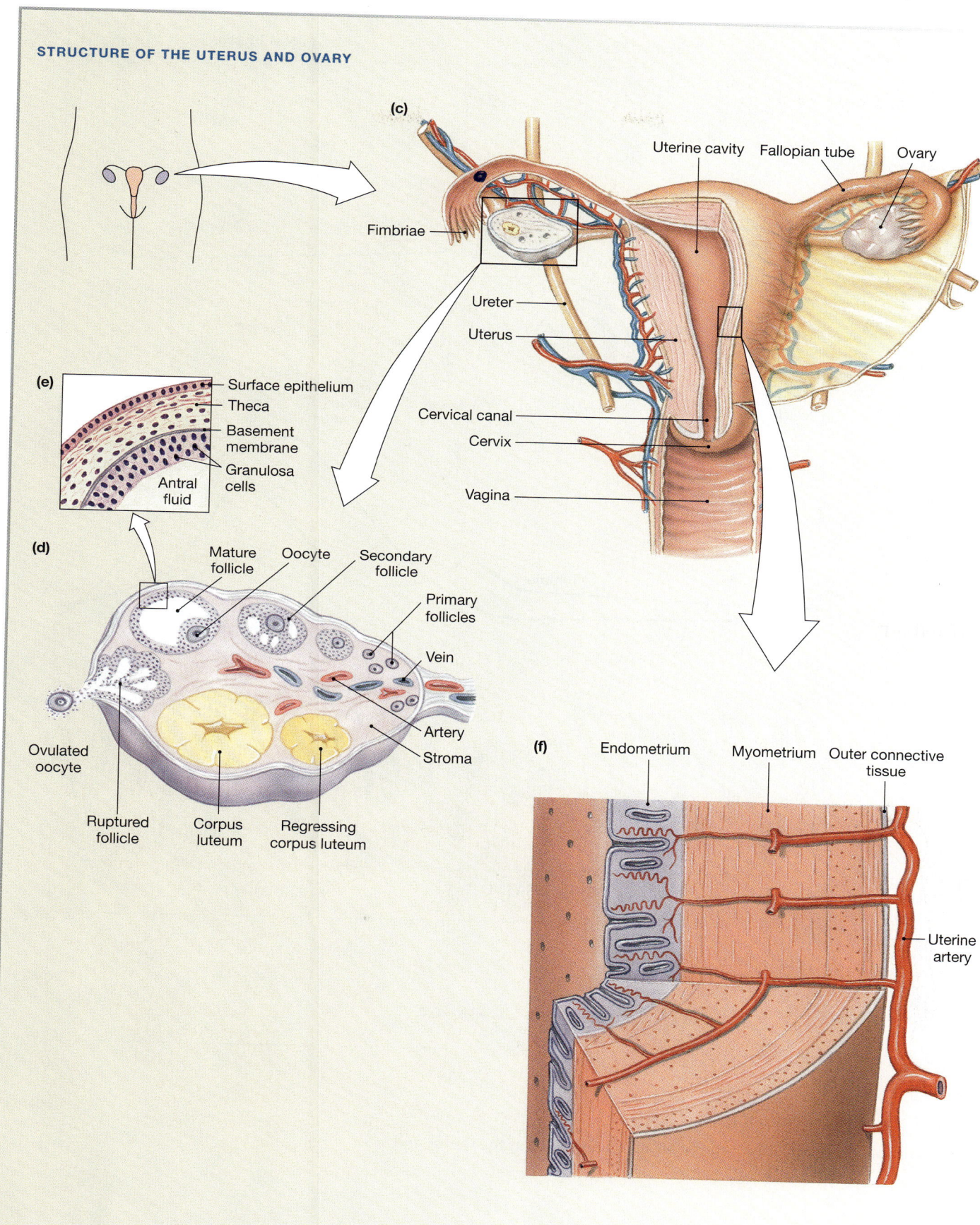

(c)

Uterine cavity
Fallopian tube
Ovary

Fimbriae

Ureter

Uterus

Cervical canal

Cervix

Vagina

(e)
Surface epithelium
Theca
Basement membrane
Granulosa cells
Antral fluid

(d)
Mature follicle
Oocyte
Secondary follicle
Primary follicles
Vein
Artery
Stroma
Ovulated oocyte
Ruptured follicle
Corpus luteum
Regressing corpus luteum

(f)
Endometrium
Myometrium
Outer connective tissue
Uterine artery

26

■ **FIGURE 26-12** **(continued)**

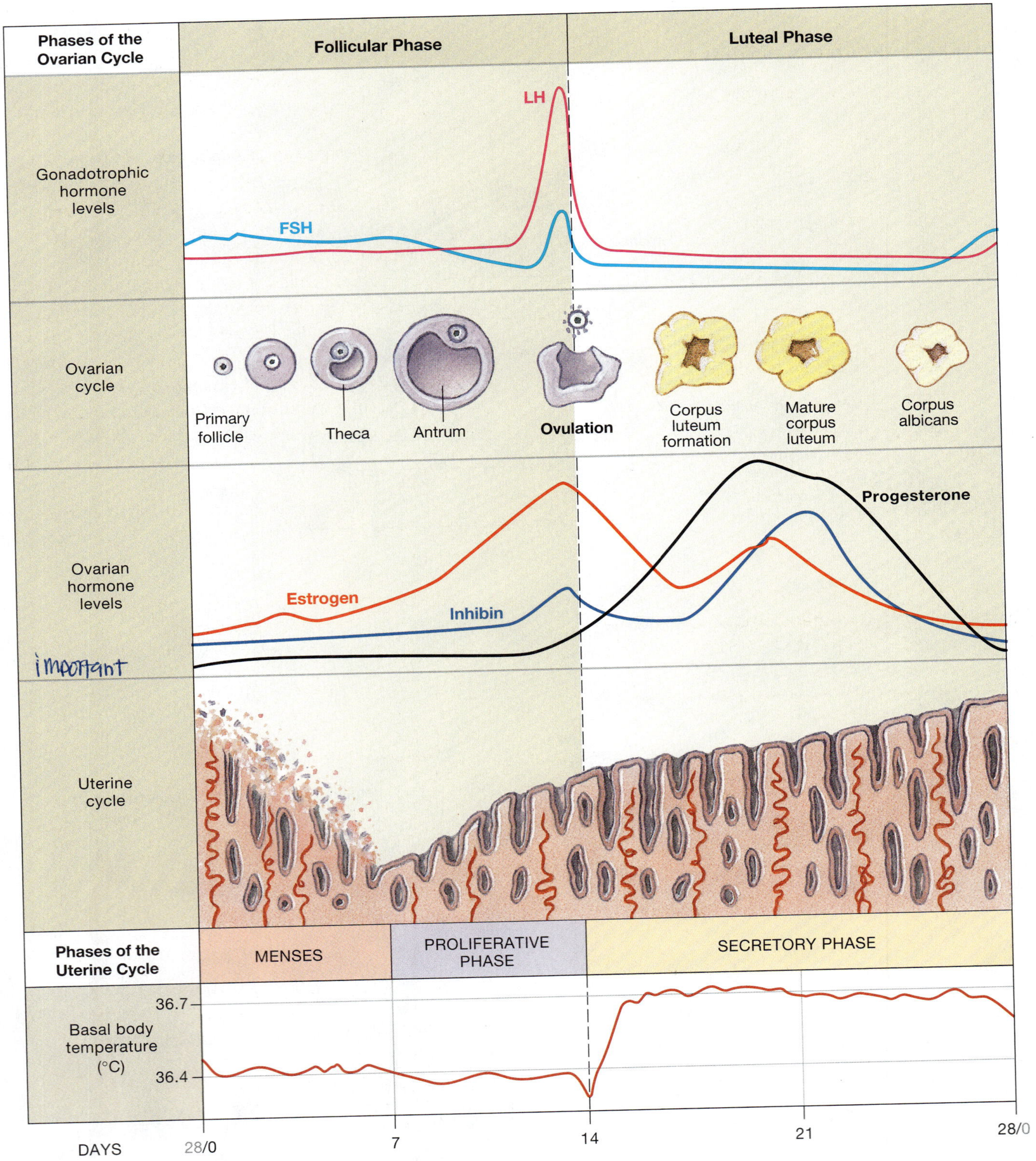

Phases of the Ovarian Cycle	Follicular Phase		Luteal Phase

Gonadotrophic hormone levels

LH

FSH

Ovarian cycle

Primary follicle — Theca — Antrum — **Ovulation** — Corpus luteum formation — Mature corpus luteum — Corpus albicans

Ovarian hormone levels

Progesterone

Estrogen

Inhibin

important

Uterine cycle

Phases of the Uterine Cycle	MENSES	PROLIFERATIVE PHASE	SECRETORY PHASE

Basal body temperature (°C)

36.7

36.4

DAYS   28/0   7   14   21   28/0

■ **FIGURE 26-13** *The menstrual cycle*

This 28-day menstrual cycle is divided into phases based on events in the uterus and ovary.

26

3. **Secretory phase**. After ovulation, hormones from the corpus luteum convert the thickened endometrium into a secretory structure. Thus, the luteal phase of the ovarian cycle corresponds to the **secretory phase** of the uterine cycle. If no pregnancy occurs, the superficial layers of the secretory endometrium are lost during menstruation as the uterine cycle begins again.

Hormonal Control of the Menstrual Cycle Is Complex

The ovarian and uterine cycles are under the primary control of various hormones:

- GnRH from the hypothalamus
- FSH and LH from the anterior pituitary
- Estrogen, progesterone, and inhibin from the ovary

During the follicular phase, the dominant steroid hormone is estrogen (Fig. 26-13). In the luteal phase, progesterone is dominant, although estrogen is still present.

Now let's go through an ovarian cycle in detail.

Early Follicular Phase   The first day of menstruation is day 1 of a cycle. This point was chosen to start the cycle because the bleeding of menstruation is an easily observed physical sign. Just before the beginning of each cycle, gonadotropin secretion from the anterior pituitary increases. Under the influence of FSH, several follicles in the ovaries begin to mature (Table 26-4 ■).

As the follicles grow, their granulosa cells (under the influence of FSH) and their thecal cells (under the influence of LH) start to produce steroid hormones. Thecal cells synthesize androgens that diffuse into the neighboring granulosa cells, where aromatase converts them to estrogens (Fig. 26-14a ■).

Gradually increasing estrogen levels in the circulation have several effects. Estrogen exerts negative feedback on pituitary FSH and LH secretion, thus preventing the development of additional follicles in the same cycle. At the same time, estrogen stimulates additional estrogen production by the granulosa cells. This positive feedback loop allows the follicles to continue estrogen production even though FSH and LH levels decrease.

As the follicles enlarge, granulosa cells secrete fluid that collects in a central cavity in the follicle known as the **antrum**

TABLE 26-4	Development of Ovarian Follicles

The contents of the follicle are given starting at the center with the ovum and moving outward. The zona pellucida (see Fig. 26-16) is a glycoprotein coat that protects the ovum.

STAGE	OVUM	ZONA PELLUCIDA	GRANULOSA CELLS	ANTRUM	BASAL LAMINA	THECA
Primary follicle (before FSH stimulation)	Primary oocyte	Minimal	Single layer	None	Separates granulosa and theca	Single cell layer plus blood vessels
Secondary follicle (early follicular phase)	Primary oocyte	Increased in width	2–6 cell layer	None	Present	Single cell layer
Tertiary follicle (late follicular phase)	Primary oocyte, then secondary oocyte with division arrested	Present	3–4 cell layer	Develops within granulosa layer and fills with fluid; swells to 15–20 mm in diameter	Present	*Inner layer:* secretory and small blood vessels *Outer layer:* connective tissue, smooth muscle cells, large blood vessels
Corpus luteum (luteal phase)	None	None	Converted to luteal cells	Fills with migrating cells	Disappears	Converted to luteal cells
Corpus albicans (post-luteal phase)	None	None	Cells degenerate	None	None	Cells degenerate

26

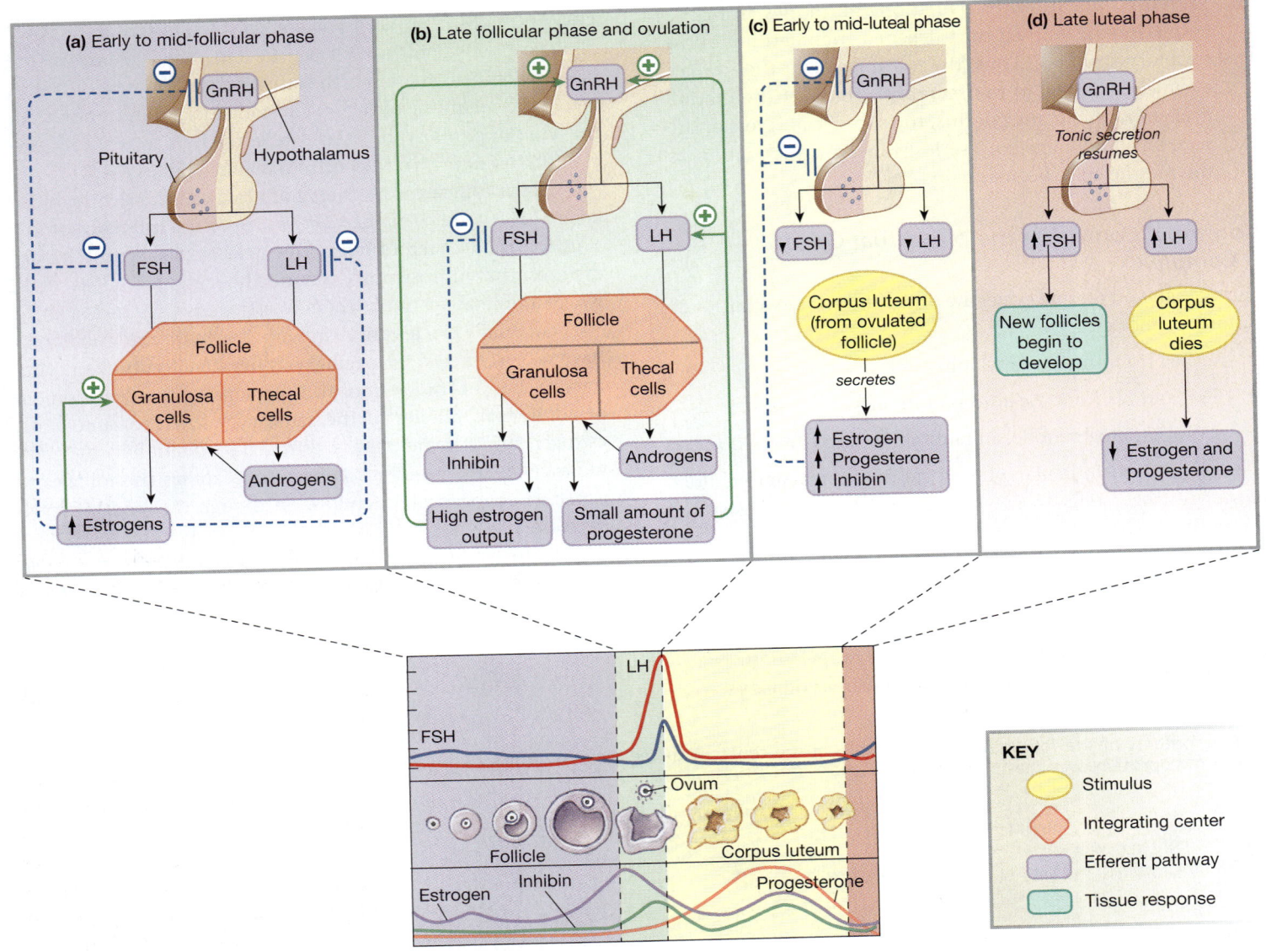

■ **FIGURE 26-14** *Hormonal control of the menstrual cycle*

[*antron,* cave]. Antral fluid (see Fig. 26-12e) contains hormones and enzymes needed for ovulation. At each stage of follicular development, some follicles undergo *atresia* (hormonally regulated cell death). Only a few follicles reach the final stage, and usually only one dominant follicle develops until ovulation.

In the uterus, menstruation ends during the early follicular phase. Under the influence of estrogen from developing follicles, the endometrium begins to grow, or proliferate. This period is characterized by an increase in cell number and by enhanced blood supply to bring nutrients and oxygen to the thickening endometrium. Estrogen also causes mucous glands of the cervix to produce clear, watery mucus.

Late Follicular Phase As the follicular phase nears its end, ovarian estrogen secretion peaks (Fig. 26-13). By this point of the cycle, only one follicle is still developing. As the follicular

phase ends, granulosa cells of this dominant follicle begin to secrete inhibin and progesterone in addition to estrogen (Fig. 26-14b). Estrogen, which had exerted a negative feedback effect on GnRH earlier in the follicular phase, changes to positive feedback.

Immediately before ovulation, the persistently high levels of estrogen, aided by rising levels of progesterone, enhance pituitary responsiveness to GnRH. As a result, LH secretion increases dramatically, a phenomenon known as the *LH surge.* FSH surges also, but to a lesser degree, presumably because it is being suppressed by inhibin and estrogen.

The LH surge is an essential part of ovulation. Without it, the final steps of oocyte maturation cannot take place. Meiosis resumes in the developing follicle with the first meiotic division, which converts the primary oocyte into a secondary oocyte (egg) and a polar body, which is extruded (see Fig. 26-5).

RUNNING PROBLEM

The results of Larry's sperm analysis are normal. Dr. Coddington is therefore able to rule out sperm abnormalities as a cause of Peggy and Larry's infertility. Peggy is instructed to take her body temperature daily and record the results on a chart. This temperature tracking is intended to determine whether or not she is ovulating. Following ovulation, body temperature rises slightly and remains elevated through the remainder of the menstrual cycle.

Question 3:
  For which causes of female infertility is temperature tracking useful? For which causes is it not useful?

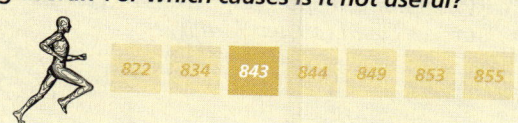

| 822 | 834 | **843** | 844 | 849 | 853 | 855 |

While this division is taking place, antral fluid collects and the follicle grows to its greatest size, preparing to release the egg.

High levels of estrogen in the late follicular phase prepare the uterus for a possible pregnancy. The endometrium grows to a final thickness of 3–4 mm. Just before ovulation, the cervical glands produce copious amounts of thin, stringy mucus to facilitate sperm entry. The stage is set for ovulation.

Ovulation   About 16–24 hours after LH peaks, ovulation occurs. The mature follicle secretes *collagenase*, which dissolves collagen in the connective tissue holding the follicular cells together. The breakdown products of collagen create an inflammatory reaction, attracting leukocytes that secrete prostaglandins into the follicle. The prostaglandins may cause smooth muscle cells in the outer theca to contract, rupturing the follicle wall at its weakest point. Antral fluid spurts out along with the egg, which is surrounded by two to three layers of granulosa cells. The egg is swept into the Fallopian tube and carried away to be fertilized or to die.

In addition to promoting follicular rupture, the LH surge causes follicular thecal cells to migrate into the antral space, mingling with the former granulosa cells and filling the cavity. Both cell types then transform into *luteal cells* of the corpus luteum. This process, known as *luteinization,* involves biochemical and morphological changes. Newly formed luteal cells accumulate lipid droplets and glycogen granules in their cytoplasm and begin to secrete progesterone. Estrogen synthesis diminishes.

Early to Mid-Luteal Phase   After ovulation, the corpus luteum produces steadily increasing amounts of progesterone and estrogen. Progesterone is the dominant hormone of the luteal phase. Estrogen levels increase but never reach the peak seen before ovulation.

The combination of estrogen and progesterone exerts negative feedback on the hypothalamus and anterior pituitary. Gonadotropin secretion, further suppressed by luteal inhibin production, remains shut down throughout most of the luteal phase (Fig. 26-14c).

Under the influence of progesterone, the endometrium continues its preparation for pregnancy. Endometrial glands coil, and additional blood vessels grow into the connective tissue layer. Endometrial cells deposit lipids and glycogen in their cytoplasm. These deposits will provide nourishment for a developing embryo while the **placenta**, the fetal-maternal connection, is developing.

Progesterone also causes cervical mucus to thicken. Thicker mucus creates a plug that blocks the cervical opening, preventing bacteria as well as sperm from entering the uterus.

One interesting effect of progesterone is its thermogenic ability. During the luteal phase of an ovulatory cycle, a woman's basal body temperature, taken immediately upon awakening and before getting out of bed, jumps 0.3°–0.5° F and remains elevated until menstruation. Because this change in the temperature setpoint occurs after ovulation, it cannot be used effectively to predict ovulation. However, it is a simple way to assess whether a woman is having ovulatory or *anovulatory* (non-ovulating) cycles.

Late Luteal Phase and Menstruation   The corpus luteum has an intrinsic life span of approximately 12 days. If pregnancy does not occur, the corpus luteum spontaneously undergoes *apoptosis* [⊞ p. 81] to become an inactive structure called a **corpus albicans** [*albus,* white]. As the luteal cells degenerate, progesterone and estrogen production decrease (Fig. 26-14d). This decrease removes the negative feedback signal to the pituitary and hypothalamus, and secretion of FSH and LH increases.

Maintenance of a secretory endometrium depends on the presence of progesterone. When the corpus luteum degenerates and hormone production falls, blood vessels in the surface layer of the endometrium contract. Without oxygen and nutrients, the surface cells die. About two days after the corpus luteum ceases to function, or 14 days after ovulation, the endometrium begins to slough its surface layer, and menstruation begins.

Menstrual discharge from the uterus totals about 40 mL of blood and 35 mL of serous fluid and cellular debris. There are usually few clots of blood in the menstrual flow because of the presence of *plasmin* [⊞ p. 551], which breaks up clots. Menstruation continues for 3–7 days, well into the follicular phase of the next ovulatory cycle.

Estrogens and Androgens Influence Female Secondary Sex Characteristics

Estrogens control the development of primary sex characteristics in females, just as androgens control them in males. Estrogens also control the most prominent female secondary sex traits: breast development and the pattern of fat distribution (hips and upper thighs). Other female secondary sex characteristics, however, are governed by androgens produced in the adrenal cortex. Pubic and axillary (armpit) hair growth and libido (sex drive) are under the control of adrenal androgens.

✓ **CONCEPT CHECK**

15. Name the phases of the ovarian cycle and the corresponding phases of the uterine cycle.

16. What side effects would you predict in female athletes who take anabolic steroids to build muscles?

17. Aromatase converts testosterone to estrogen. What would happen to the ovarian cycle of a woman given an aromatase inhibitor?

Answers: p. 859

PROCREATION

Reproduction throughout the animal kingdom is marked by species-specific behaviors designed to ensure that egg and sperm meet. For aquatic animals that release gametes into the water, coordinated timing is everything. Interaction between males and females of these species may be limited to chemical communication by pheromones.

In terrestrial vertebrates, internal fertilization requires interactive behaviors and specialized adaptations of the genitalia. The female must have an internal receptacle for sperm (the vagina) and the male must possess an organ (the penis) capable of placing sperm in the receptacle. The penis of human males is *flaccid* (soft and limp) in its resting state, not capable of penetrating the narrow opening of the vagina. The processes through which the penis first stiffens and enlarges (**erection**) and then releases sperm from the ducts of the reproductive tract (**ejaculation**) make up the main components of the male sex act. Without these events, fertilization cannot take place.

The Human Sexual Response Has Four Phases

The human sex act—also known as sexual intercourse, copulation, or **coitus** [*coitio*, a coming together]—is highly variable in some ways and highly stereotypical in other ways.

Human sexual response in both sexes is divided into four phases: (1) excitement, (2) plateau, (3) orgasm, and (4) resolution. In the excitement phase, various erotic stimuli prepare the genitalia for the act of copulation. For the male, excitement involves erection of the penis. For the female, it includes erection of the clitoris and vaginal lubrication. In both sexes, erection is a state of vasocongestion in which arterial blood flow into spongy erectile tissue exceeds venous outflow.

Erotic stimuli include sexually arousing tactile stimuli as well as psychological stimuli. Because the latter vary widely among individuals and among cultures, what is erotic to one person or in one culture may be considered disgusting by another individual or in another culture. Regions of the body that possess receptors for sexually arousing tactile stimuli are called

RUNNING PROBLEM

Results of the temperature tracking for several months reveal that Peggy is ovulating regularly. Dr. Coddington therefore believes that her ovaries are functioning normally. Other possible causes for this couple's infertility include abnormalities in Peggy's cervix, Fallopian tubes, or uterus. Dr. Coddington next decides to order a post-coital test. In this test, the couple is instructed to have intercourse 12 hours before the physician visit. Cervical mucus is then analyzed. This test will also analyze the interaction between sperm and mucus.

Question 4:
What abnormalities in the cervix, Fallopian tubes, and uterus could cause infertility?

822 834 843 **844** 849 853 855

erogenous zones and include the genitalia as well as the lips, tongue, nipples, and ear lobes.

In the plateau phase, changes that started during excitement intensify and peak in an **orgasm** (climax). In both sexes, orgasm is a series of muscular contractions accompanied by intense pleasurable sensations and increased blood pressure, heart rate, and respiration rate. In females, the uterus and walls of the vagina contract. In males, the contractions usually result in the ejaculation of semen from the penis. Female orgasm is not required for pregnancy.

The final phase of the sexual response is resolution, a period during which the physiological parameters that changed in the first three phases slowly return to normal.

The Male Sex Act Includes Erection and Ejaculation

A key element to successful copulation is the ability of the male to achieve and sustain an erection. Sexual excitement from either tactile or psychological stimuli triggers the **erection reflex**, a spinal reflex that is subject to control from higher centers in the brain. The urination and defecation reflexes are similar types of reflexes [🔁 pp. 635, 708].

In its simplest form, the erection reflex begins with tactile stimuli sensed by mechanoreceptors in the glans penis or other erogenous zones (Fig. 26-15 ■). Sensory neurons signal the spinal integration center, which inhibits vasoconstrictive sympathetic input on penile arterioles. Simultaneously, nitric oxide produced by increased parasympathetic input actively dilates the penile arterioles. As arterial blood flows into the open spaces of the erectile tissue, it passively compresses the veins

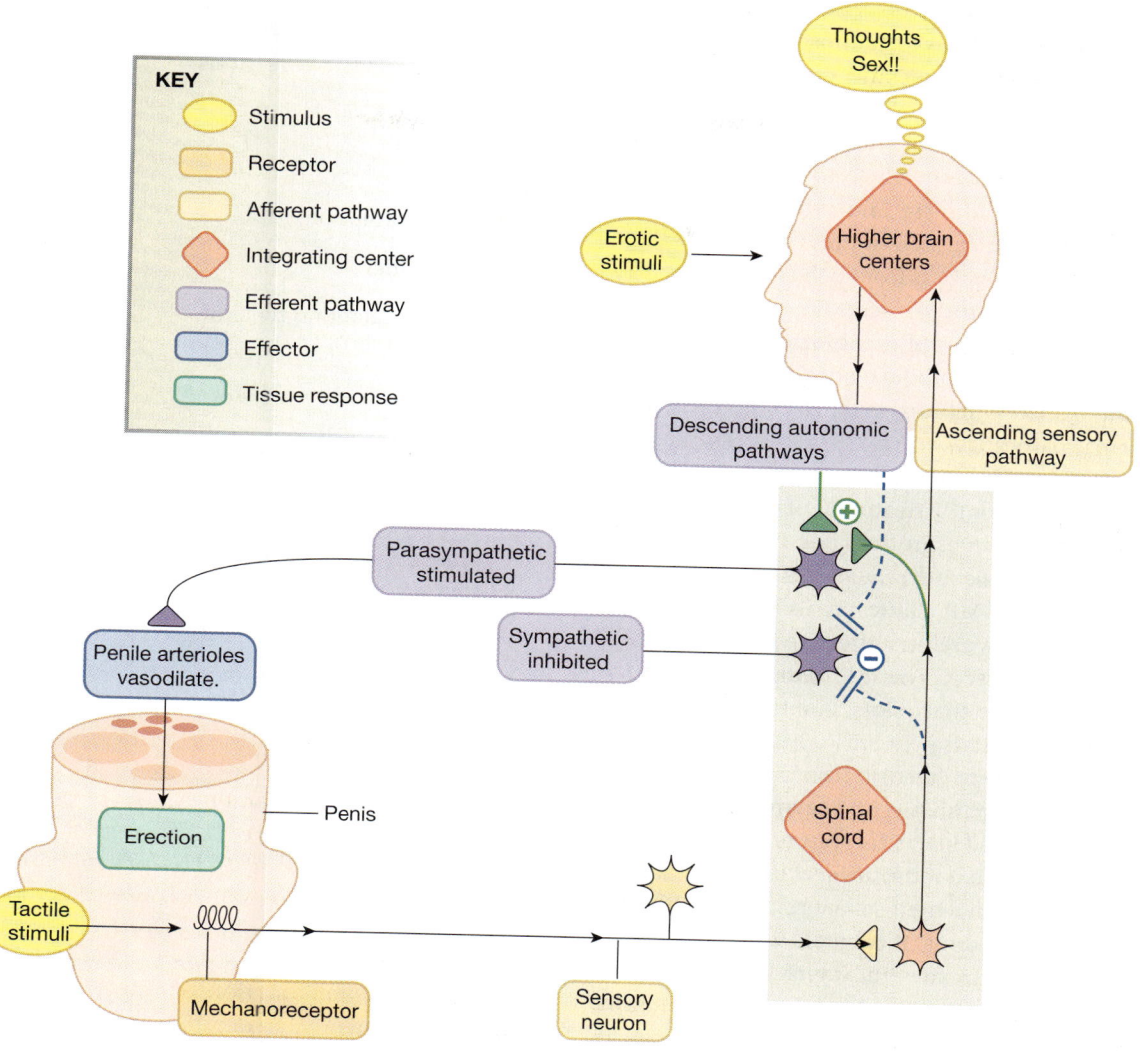

KEY

Stimulus	
Receptor	
Afferent pathway	
Integrating center	
Efferent pathway	
Effector	
Tissue response	

■ **FIGURE 26-15** *The erection reflex*

Erection can take place without input from higher brain centers. It can also be stimulated (and inhibited) by descending pathways from the cerebral cortex.

and traps blood. The erectile tissue becomes engorged, stiffening and lengthening the penis within 5–10 seconds.

The climax of the male sexual act coincides with emission and ejaculation. **Emission** is the movement of sperm out of the vas deferens and into the urethra, where they are joined by secretions from the accessory glands to make semen. The average semen volume is 3 mL (range 2–6 mL), of which less than 10% is sperm.

During ejaculation, semen in the urethra is expelled to the exterior by a series of rapid muscular contractions accompanied by sensations of intense pleasure—the orgasm. A sphincter at the base of the bladder contracts to prevent sperm from entering the bladder and urine from joining the semen.

Both erection and ejaculation can occur in the absence of mechanical stimulation. Sexually arousing thoughts, sights, sounds, emotions, and dreams can all initiate sexual arousal and even lead to orgasm in both men and women. In addition, nonsexual penile erection accompanies rapid eye movement (REM) sleep.

Sexual Dysfunction Affects Males and Females

The inability to obtain or sustain a penile erection is known as **erectile dysfunction** (ED) or *impotence*. Erectile dysfunction is a matter of global concern because inability to achieve and sustain an erection disrupts the sex act for both men and women. Organic (physiological and anatomical) causes of ED include neural and hormonal problems, vascular insufficiency, and drug-induced ED. A variety of psychological causes can also contribute to ED.

Alcohol inhibits sexual performance in both men and women, as noted by Shakespeare in *Macbeth* (II, iii). When

26

MacDuff asks, "What three things does drink especially provoke?" the porter answers, "Marry, sir, nose-painting, sleep, and urine. Lechery, sir, it provokes and unprovokes: it provokes the desire, but it takes away the performance." Several antidepressant drugs list loss of libido among their side effects.

Erectile dysfunction in men over age 40 is now considered a marker for cardiovascular disease and atherosclerosis, and sometimes ED is the first clinical sign of these conditions. Erections occur when neurotransmitters from pelvic nerves increase endothelial production of nitric oxide (NO), which increases cGMP and results in vasodilation of penile arterioles. Endothelial dysfunction and failure to produce adequate NO occur in atherosclerosis and diabetes mellitus, making ED an early manifestation of vascular pathology.

In 1998 the U.S. Food and Drug Administration (FDA) approved sildenafil (Viagra®) for the treatment of erectile dysfunction. Sildenafil and similar drugs in the same class prolong the effects of nitric oxide by blocking *phosphodiesterase*-5 (PDE-5), the enzyme that degrades cGMP. Clinical trials have shown that phosphodiesterase inhibitors are very effective in correcting ED but are not without side effects. The U.S. Federal Aviation Administration issued an order that pilots should not take sildenafil within six hours of flying because 3% of men report impaired color vision (a blue or greenish haze). This impairment occurs because sildenafil also inhibits an enzyme in the retina.

When the FDA approved PDE-5 inhibitors for male erectile dysfunction, women wondered if the drug, which promotes the erection reflex, would improve their sexual response. Although women do have clitoral erections, the female sexual response is more complicated. Studies on the efficacy of PDE-5 inhibitors for orgasmic dysfunction in women have had mixed results. Instead, pharmaceutical companies are testing other drugs for female sexual dysfunction. The most promising candidates, now in late clinical trials, are based on testosterone, the androgen that creates libido in both sexes.

Contraceptives Are Designed to Prevent Pregnancy

One disadvantage of sexual intercourse for pleasure rather than reproduction is the possibility of an unplanned pregnancy. On average, 85% of young women who have sexual intercourse without using any form of birth control will get pregnant within a year. Many women, however, get pregnant after just a single unprotected encounter. Couples who hope to avoid unwanted pregnancies generally use some form of birth control, or **contraception**.

Contraceptive practices fall into several broad groups. **Abstinence**, the total avoidance of sexual intercourse, is the surest method to avoid pregnancy (and sexually transmitted diseases). Some couples practice abstinence only during times of suspected fertility.

Interventional methods of contraception include (1) barrier methods, which prevent union of eggs and sperm; (2) methods

TABLE 26-5	Efficacy of Various Contraceptive Methods
METHOD	**PREGNANCY RATE WITH TYPICAL USE***
No contraception	85%
Spermicides	29%
Abstinence during times of predicted fertility	25%
Diaphragm, cervical cap, or sponge	16–32%†
Oral contraceptive pills	8%
Intrauterine devices (IUDs)	< 1%
Implanted hormonal contraceptives (Norplant™)	< 1%
Male condom	15%
Female condom	21%
Sterilization	< 1%

*Rates reflect unintentional pregnancies in the first year of using the method. Data are from *www.contraceptivetechnology.org/table.html* (Accessed 6/14/05).

†Lower rates are in women who have never delivered a child.

that prevent implantation of the fertilized egg; and (3) hormonal treatments that decrease or stop gamete production. The efficacy of interventional contraceptives depends in part on how consistently and correctly they are used (Table 26-5 ■).

Barrier Methods   Contraceptive methods based on chemical or physical barriers are among the earliest recorded means of birth control. Once people made the association between pregnancy and semen, they concocted a variety of physical barriers and *spermicides* [*cida*, killer] to kill sperm. An ancient Egyptian papyrus with the earliest known references to birth control describes the use of vaginal plugs made of leaves, feathers, figs, and alum held together with crocodile and elephant dung. Sea sponges soaked in vinegar and disks of oiled silk have also been used at one time or another. In subsequent centuries women used douches of garlic, turpentine, and rose petals to rinse the vagina after intercourse. As you can imagine, many of these methods also caused vaginal or uterine infections.

Modern versions of the female barrier include the **diaphragm**, introduced into the United States in 1916. These rubber domes and a smaller version called a *cervical cap* are usually filled with a spermicidal cream, then inserted into the top of the vagina so they cover the cervix. One advantage to the diaphragm is that it is nonhormonal. When used properly and regularly, diaphragms are highly effective (97–99%). However, they are not always used because they must be inserted close to the time of intercourse, and consequently about 20% of women who depend on diaphragms for contraception are pregnant within the first year. Another female barrier contraceptive that was recently re-introduced is the **contraceptive sponge**, which contains a spermicidal chemical.

The male barrier contraceptive is the **condom**, a closed sheath that fits closely over the penis to catch ejaculated semen. Males have used condoms made from animal bladders and intestines for centuries. Condoms lost popularity when oral contraceptives came into widespread use in the 1960s and 1970s, but in recent years they have regained favor because they combine pregnancy protection with protection from many sexually transmitted diseases. However, latex condoms may cause allergic reactions, and there is evidence that HIV can pass through pores in some condoms currently produced. A female version of the condom is also commercially available. It covers the cervix and completely lines the vagina, providing more protection from sexually transmitted diseases.

Sterilization is the most effective contraceptive method for sexually active people, but it is a surgical procedure and is not easily reversed. Female sterilization is called **tubal ligation** and consists of tying off and cutting the Fallopian tubes. A woman with a tubal ligation will still ovulate, but the eggs remain in the abdomen. The male form of sterilization is the **vasectomy**, in which the vas deferens is tied and clipped. Sperm are still made in the seminiferous tubules, but because they cannot leave the reproductive tract, they are reabsorbed.

Implantation Prevention   Some contraceptive methods do not prevent fertilization but do keep a fertilized egg from establishing itself in the endometrium. They include **intrauterine devices (IUDs)** as well as chemicals that change the properties of the endometrium. IUDs are plastic devices that are inserted into the uterine cavity, where they create a mild inflammatory reaction that prevents implantation. They have low failure rates (0.5% per year) but side effects that range from pain and bleeding to infertility caused by pelvic inflammatory disease and blockage of the Fallopian tubes.

Hormonal Treatments   Techniques for decreasing gamete production depend on altering the hormonal milieu of the body. In centuries past, women would eat or drink various plant concoctions for contraception. Some of these substances actually worked because the plants contained estrogen-like compounds. Modern pharmacology has improved on this method, and now women can choose between oral contraceptive pills, injections lasting three months, or capsules inserted under the skin.

The **oral contraceptives**, also known as *birth control pills*, were first made available in 1960. They rely on various combinations of estrogen and progesterone that inhibit gonadotropin secretion from the pituitary. Without adequate FSH and LH, ovulation is suppressed. In addition, progesterones in the contraceptive pills thicken the cervical mucus and help prevent sperm penetration. These hormonal methods of contraception are highly effective when used correctly but also carry some risks, including an increased incidence of blood clots and strokes (especially in women who smoke).

Development of a male hormonal contraceptive has been slow because of undesirable side effects. Contraceptives that block testosterone secretion or action are also likely to decrease the male libido or even cause impotence. Both side effects are unacceptable to men who would be most interested in using the contraceptive. Some early male oral contraceptives irreversibly suppressed sperm production, which was also unacceptable. It now appears, however, that a combination of oral progestin to suppress sperm production plus injected testosterone to maintain libido is a promising candidate for a male hormonal contraceptive.

Contraceptive vaccines are based on antibodies against various components of the male and female reproductive systems, such as antisperm or antiovum antibodies. These contraceptives appear to be long lasting and reversible, and they can be administered as shots. Some contraceptive vaccines are being tested in clinical trials.

Infertility Is the Inability to Conceive

While some couples are trying to prevent pregnancy, others are spending thousands of dollars trying to get pregnant using **assisted reproductive technology** (ART). *Infertility* is the inability of a couple to conceive a child after a year of unprotected intercourse. For years, infertile couples had no choice but adoption if they wanted to have a child, but incredible strides have been made in this field since the 1970s. As a result, many infertile couples today are able to have children.

Infertility can arise from a problem in the male, the female, or both. Male infertility usually results from a low sperm count or an abnormally high number of defective sperm. Female infertility can be mechanical (blocked Fallopian tubes or other structural problems) or hormonal, leading to decreased or absent ovulation. One problem involving both partners is that the woman may produce antibodies to her partner's sperm. In addition, not all pregnancies go to a successful conclusion. By some estimates, as many as a third of all pregnancies spontaneously terminate—many within the first weeks, before the woman is even aware that she was pregnant.

Some of the most dramatic advances have been made in the field of *in vitro* fertilization, in which a woman's ovaries are hormonally manipulated to ovulate multiple eggs at one time. The eggs are collected surgically and fertilized outside the body. The developing embryos are then placed in the woman's uterus, which has been primed for pregnancy by hormonal

26

(a)

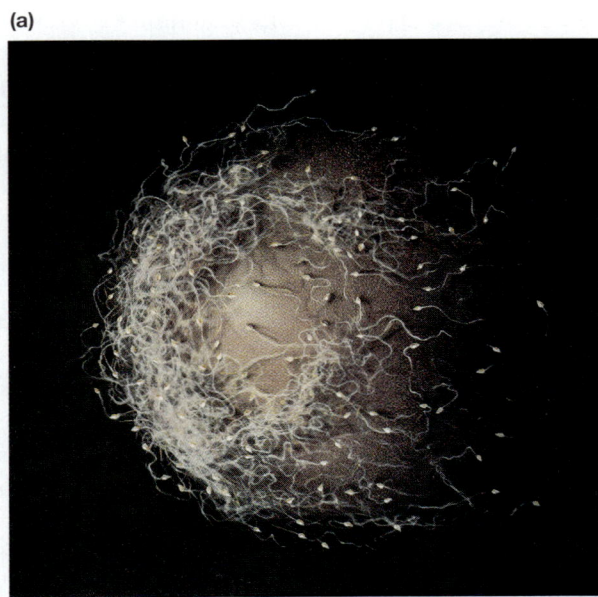

(b) Capacitated sperm release enzymes from their acrosomes in order to penetrate the cells and zona pellucida surrounding the egg.

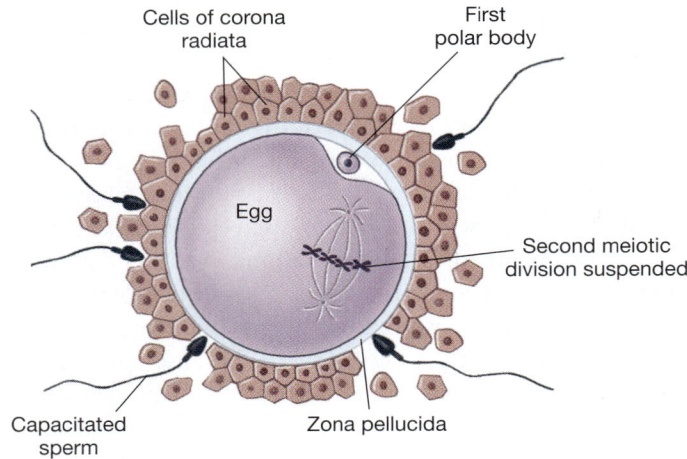

Cells of corona radiata

First polar body

Egg

Second meiotic division suspended

Capacitated sperm

Zona pellucida

■ **FIGURE 26-16**   *Fertilization*

The photograph shows the tremendous difference in the size of human sperm and egg.

therapy. Because of the expense and complicated nature of the procedure, multiple embryos are usually placed in the uterus at one time, which may result in multiple births. *In vitro* fertilization has allowed some infertile couples to have children, with a 2002 success rate in the United States of about 34%.

PREGNANCY AND PARTURITION

Now let's return to a recently ovulated egg and some sperm deposited in the vagina and follow them through fertilization, pregnancy, and **parturition**, the birth process.

Fertilization Requires Capacitation

Once an egg is released from the ruptured follicle, it is swept into the Fallopian tube by beating cilia. Meanwhile, sperm deposited in the vagina must go through their final maturation step, **capacitation**, which enables the sperm to swim rapidly and fertilize an egg. The process apparently involves the reorganization of molecules in the outer membrane of the sperm head.

Normally, capacitation takes place in the female reproductive tract, which presents a problem for *in vitro* fertilization. Those sperm must be artificially capacitated by placing them in physiological saline supplemented with human serum. Much of what we know about human fertilization has come from infertility research aimed at improving the success rate of *in vitro* fertilization.

Fertilization of an egg by a sperm is the result of a chance encounter, possibly aided by chemical attractants produced by

the egg. An egg can be fertilized for only about 12–24 hours after ovulation. Sperm in the female reproductive tract remain viable for four to six days.

Fertilization normally takes place in the distal part of the Fallopian tube. Of the millions of sperm in a single ejaculation, only about 100 reach this point. To fertilize the egg, a sperm must penetrate both an outer layer of loosely connected granulosa cells (the *corona radiata*) and a protective glycoprotein coat called the **zona pellucida** (Fig. 26-16 ■). To get past these barriers, capacitated sperm release powerful enzymes from the acrosome in the sperm head, a process known as the **acrosomal reaction**. The enzymes dissolve cell junctions and the zona pellucida, allowing the sperm to wiggle their way toward the egg. The first sperm to reach the egg quickly finds sperm-binding receptors on the oocyte membrane and fuses its membrane to the egg membrane (Fig. 26-17 ■). The fused section of membrane opens, and the sperm nucleus sinks into the egg's cytoplasm. Fusion of the egg and sperm membranes signals the egg to resume meiosis and complete its second division. The final meiotic division creates a second polar body, which is ejected. At this point, the 23 chromosomes of the sperm join the 23 chromosomes of the egg, creating a zygote nucleus with a full set of genetic material.

The fusion of sperm and oocyte membrane prevents **polyspermy**, in which more than one sperm fertilizes an egg, by triggering a chemical reaction called the **cortical reaction**. Membrane-bound **cortical granules** in the peripheral cytoplasm of the egg release their contents into the space just outside the egg membrane. These chemicals rapidly alter the membrane and

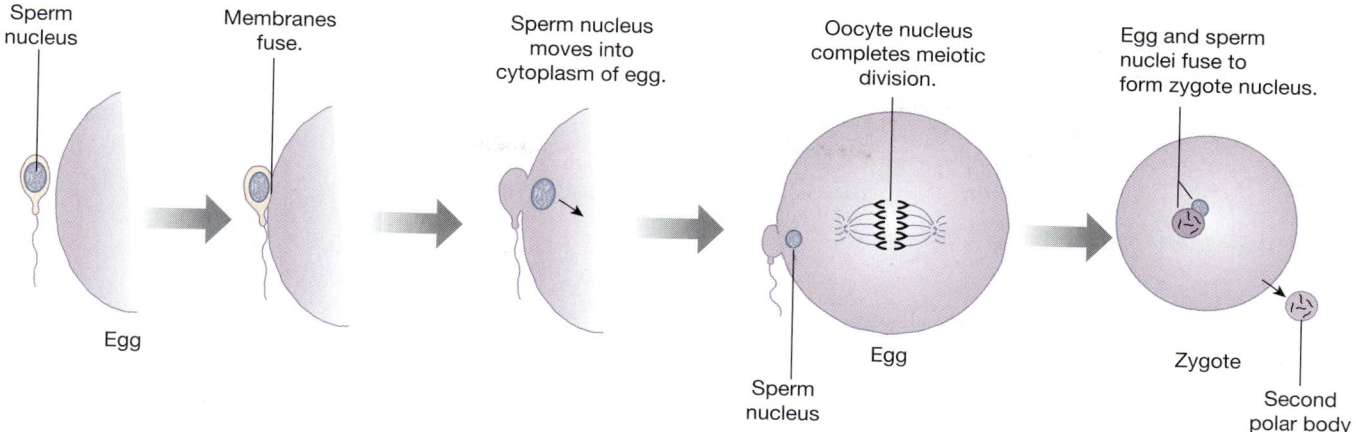

Sperm
nucleus

Membranes
fuse.

Sperm nucleus
moves into
cytoplasm of egg.

Oocyte nucleus
completes meiotic
division.

Egg and sperm
nuclei fuse to
form zygote nucleus.

Egg

Sperm
nucleus

Egg

Zygote

Second
polar body

■ **FIGURE 26-17**   *Fusion of sperm and egg to form a zygote*

surrounding zona pellucida so that additional sperm cannot penetrate or bind.

Once the egg is fertilized and becomes a zygote, it begins mitotic division as it slowly makes its way along the Fallopian tube to the uterus, where it will settle for the remainder of the **gestation** period [*gestare*, to carry in the womb].

The Developing Embryo Implants in the Endometrium

The dividing embryo takes four or five days to move through the Fallopian tube into the uterine cavity (Fig. 26-18 ■). Under the influence of progesterone, smooth muscle of the tube relaxes, and transport proceeds slowly. By the time the developing embryo reaches the uterus, it consists of a hollow ball of about 100 cells called a **blastocyst**.

Some of the outer layer of blastocyst cells will become the **chorion**, an *extraembryonic membrane* that will enclose the embryo and form the placenta (Fig. 26-19a ■). The inner cell mass of the blastocyst will develop into the embryo and into other extraembryonic membranes. These membranes include the

amnion, which secretes *amniotic fluid* in which the developing embryo floats; the **allantois**, which becomes part of the umbilical cord that links the embryo to the mother; and the **yolk sac**, which degenerates early in human development.

Implantation of the blastocyst into the uterine wall normally takes place about 7 days after fertilization. The blastocyst secretes enzymes that allow it to invade the endometrium, like a parasite burrowing into its host. As it does so, endometrial cells grow out around the blastocyst until it is completely engulfed.

As the blastocyst continues dividing and becomes an embryo, cells that will become the placenta form fingerlike **chorionic villi** that penetrate into the vascularized endometrium. Enzymes from the villi break down the walls of maternal blood vessels until the villi are surrounded by pools of maternal blood (Fig. 26-19b). The blood of the embryo and that of the mother do not mix, but nutrients, gases, and wastes are exchanged across the membranes of the villi. Many of these substances move by simple diffusion, but some, such as maternal antibodies, are transported across the membrane.

The placenta continues to grow during pregnancy until, by delivery, it is about 20 cm in diameter (the size of a small dinner plate). The placenta receives as much as 10% of the total maternal cardiac output. The tremendous blood flow to the placenta is one reason sudden abnormal separation of the placenta from the uterine wall is a medical emergency.

The Placenta Secretes Hormones During Pregnancy

As the blastocyst implants in the uterine wall and the placenta begins to form, the corpus luteum is nearing the end of its preprogrammed 12-day life span. Unless the developing embryo sends a hormonal signal, the corpus luteum will disintegrate, progesterone and estrogen levels will drop, and the embryo will be flushed from the body along with the surface layers of endometrium during menstruation. Several hormones that prevent menstruation during pregnancy are secreted by the

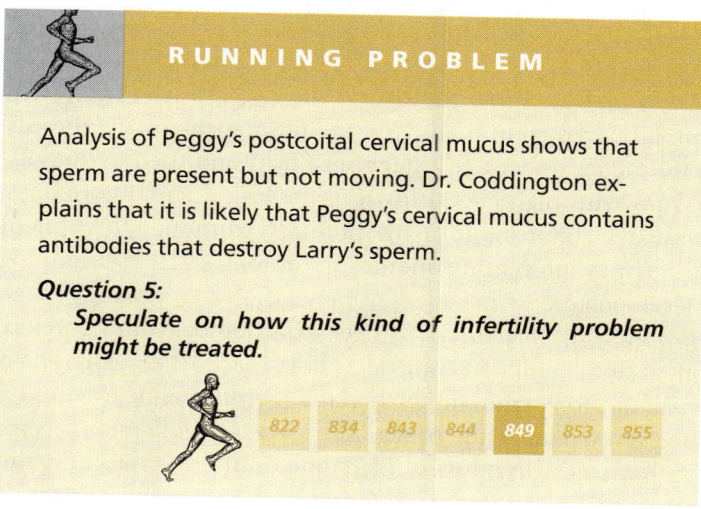

RUNNING PROBLEM

Analysis of Peggy's postcoital cervical mucus shows that sperm are present but not moving. Dr. Coddington explains that it is likely that Peggy's cervical mucus contains antibodies that destroy Larry's sperm.

Question 5:
   Speculate on how this kind of infertility problem might be treated.

822   834   843   844   **849**   853   855

26

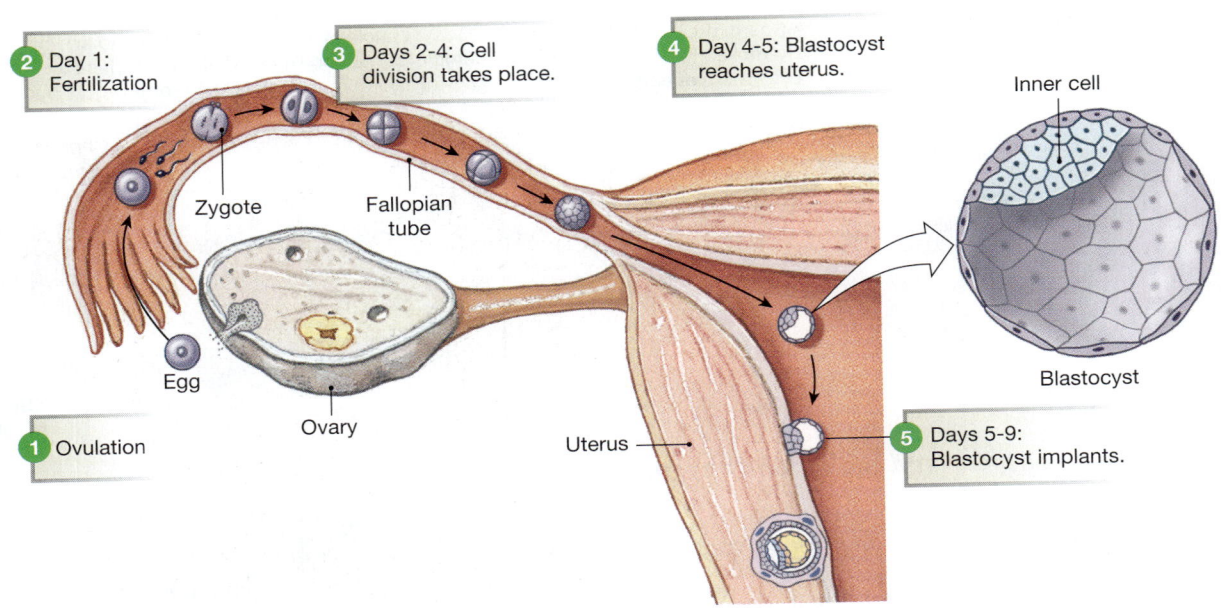

2 Day 1: Fertilization

3 Days 2-4: Cell division takes place.

4 Day 4-5: Blastocyst reaches uterus.

Inner cell

Zygote

Fallopian tube

Egg

Ovary

1 Ovulation

Uterus

5 Days 5-9: Blastocyst implants.

Blastocyst

■ **FIGURE 26-18** *Ovulation, fertilization, and implantation*

placenta, including human chorionic gonadotropin, human placental lactogen, estrogen, and progesterone.

Human Chorionic Gonadotropin The corpus luteum remains active during early pregnancy because of **human chorionic gonadotropin** (hCG), a peptide hormone secreted by the chorionic villi and developing placenta. hCG is structurally related to LH, and it binds to LH receptors. Under the influence of hCG, the corpus luteum keeps producing progesterone to keep the endometrium intact.

By the seventh week of development, however, the placenta has taken over progesterone production, and the corpus luteum is no longer needed. At that point, it finally degenerates. Human chorionic gonadotropin production by the placenta peaks at three months of development, then diminishes.

A second function of hCG is stimulation of testosterone production by the developing testes in male fetuses. As you learned in the opening sections of this chapter, fetal testosterone and its metabolite DHT are essential for expression of male characteristics and for descent of the testes into the scrotum before birth.

Human chorionic gonadotropin is the chemical detected by pregnancy tests. Because hCG can induce ovulation in rabbits, years ago the urine from a woman who suspected she was pregnant was injected into a rabbit. The rabbit's ovaries were then inspected for signs of ovulation. It took several days for the women to learn the results of this test. Today, with modern biochemical techniques, women can perform their own pregnancy tests in a few minutes in the privacy of their home.

Human Placental Lactogen (hPL) Another peptide hormone produced by the placenta is **human placental lactogen**

(hPL), also known as *human chorionic somatomammotropin* (hCS). This hormone, structurally related to growth hormone and prolactin, was initially believed to be necessary for breast development during pregnancy and for milk production (**lactation**). Although hPL probably does contribute to lactation, women who do not make hPL during pregnancy because of a genetic defect still have adequate breast development and milk production.

A second role for hPL is alteration of the mother's glucose and fatty acid metabolism to support fetal growth. Maternal glucose moves across the membranes of the placenta by facilitated diffusion and enters the fetal circulation. During pregnancy, about 4% of women develop *gestational diabetes mellitus*, with elevated blood glucose levels caused by insulin resistance, similar to type 2 diabetes. After delivery, glucose metabolism in most of these women returns to normal, but these mothers and their babies are at higher risk of developing type 2 diabetes later in life.

Estrogen and Progesterone Estrogen and progesterone are produced continuously during pregnancy, first by the corpus luteum under the influence of hCG and then by the placenta. With high circulating levels of these steroid hormones, feedback suppression of the pituitary continues throughout pregnancy, preventing another set of follicles from beginning development.

During pregnancy, estrogen contributes to the development of the milk-secreting ducts of the breasts. Progesterone is essential for maintaining the endometrium and in addition helps suppress uterine contractions. The placenta makes a variety of other hormones, including inhibin, prolactin, and prorenin, but the function of most of them remains unclear.

26

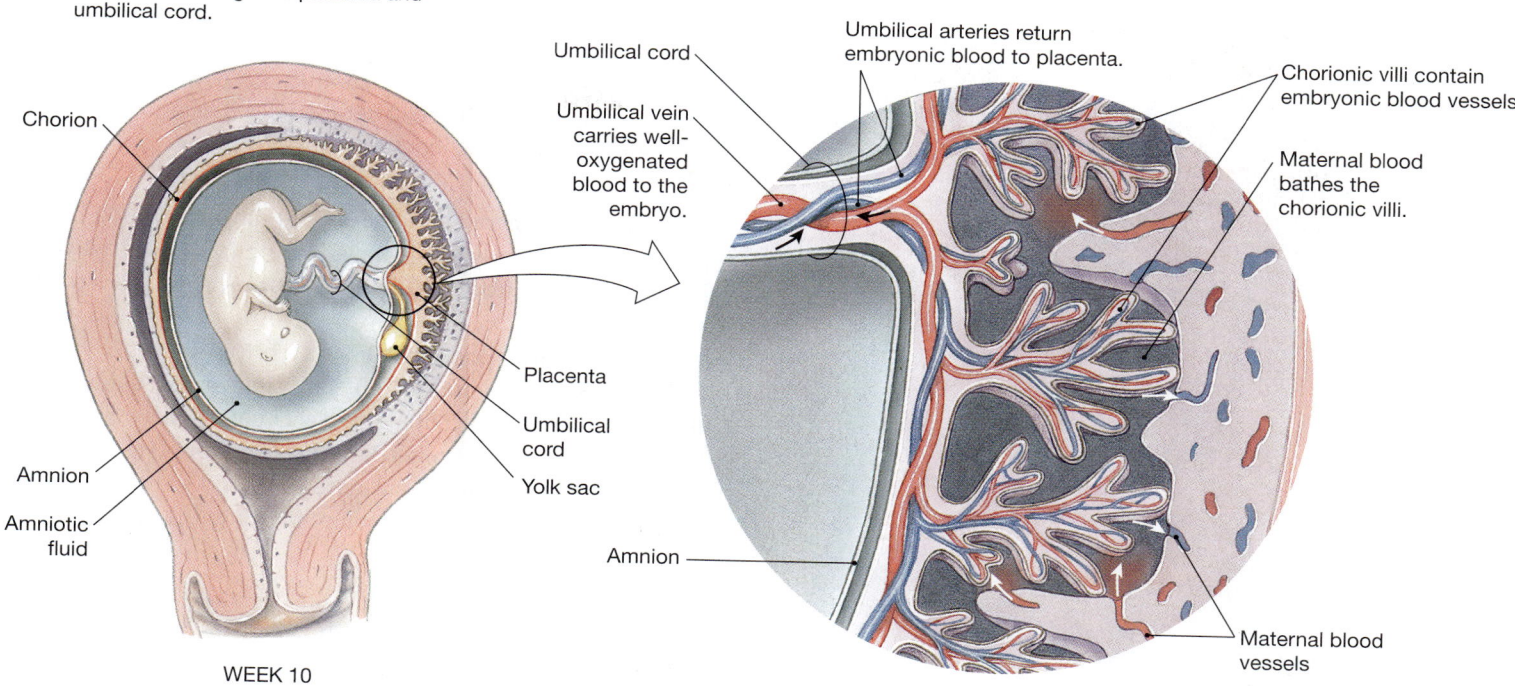

(a) The developing embryo floats in amniotic fluid. It obtains oxygen and nutrients from the mother through the placenta and umbilical cord.

Chorion

Amnion

Amniotic fluid

WEEK 10

Umbilical cord

Umbilical vein carries well-oxygenated blood to the embryo.

Placenta

Umbilical cord

Yolk sac

(b) Some material is exchanged across placental membranes by diffusion, but other material must be transported.

Umbilical arteries return embryonic blood to placenta.

Chorionic villi contain embryonic blood vessels.

Maternal blood bathes the chorionic villi.

Amnion

Maternal blood vessels

■ **FIGURE 26-19**   *The placenta*

Pregnancy Ends with Labor and Delivery

Parturition normally occurs between the 38th and 40th weeks of gestation. What triggers this process? For many years, researchers developed animal models of the signals that initiate parturition, only to discover recently that many of those models do not apply to humans. Parturition begins with **labor**, the rhythmic contractions of the uterus that push the fetus out into the world. Signals that initiate these contractions could begin with either the mother or the fetus, or they could be a combination of signals from both.

In many nonhuman mammals, a decrease in estrogen and progesterone levels marks the beginning of parturition. A decrease in progesterone levels is logical, as progesterone inhibits uterine contractions. In humans, however, levels of these hormones do not decrease until labor is well under way.

Another possible labor trigger is oxytocin, the peptide hormone that causes uterine muscle contraction. As a pregnancy nears full term, the number of uterine oxytocin receptors increases. However, studies have shown that oxytocin secretion does not increase until after labor begins. Synthetic oxytocin is often used to induce labor in pregnant women, but it is not always effective. Apparently, the start of labor requires something more than adequate amounts of oxytocin.

Another possibility for the induction of labor is that the fetus somehow signals that it has completed development. One theory supported by clinical evidence is that corticotropin-releasing hormone (CRH) secreted by the placenta is the signal

to begin labor. (CRH is also a hypothalamic releasing factor that controls release of ACTH from the anterior pituitary.) In the weeks prior to delivery, maternal blood CRH levels increase rapidly. In addition, women with elevated CRH levels as early as 15 weeks of gestation are more likely to go into premature labor.

Although we do not know for certain what initiates parturition, we do understand the sequence of events. Once the contractions of labor begin, a positive feedback loop consisting of mechanical and hormonal factors is set into motion. The fetus is normally oriented head down (Fig. 26-20a ■). At the beginning of labor it repositions itself lower in the abdomen ("the baby has dropped") and begins to push on the softened cervix (Fig. 26-20b). The peptide hormone **relaxin**, secreted by the ovaries and placenta, apparently helps soften ("ripen") the cervix and also loosens ligaments holding the pelvic bones together.

Cervical stretch triggers uterine contractions that move in a wave from the top of the uterus down, pushing the fetus farther into the pelvis. The lower portion of the uterus stays relaxed, and the cervix stretches and dilates. Cervical stretch starts a positive feedback cycle of escalating contractions (Fig. 26-21 ■). The contractions are reinforced by secretion of oxytocin from the posterior pituitary [🗟 p. 226], with continued cervical stretch reinforcing oxytocin secretion.

Prostaglandins are produced in the uterus in response to CRH and oxytocin secretion. Prostaglandins are very effective at causing uterine muscle contractions at any time. They are the

26

(a) Fully developed fetus

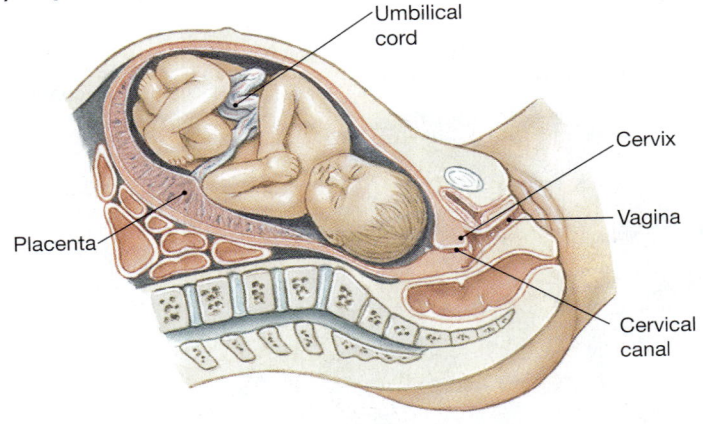

(b) Cervical dilation

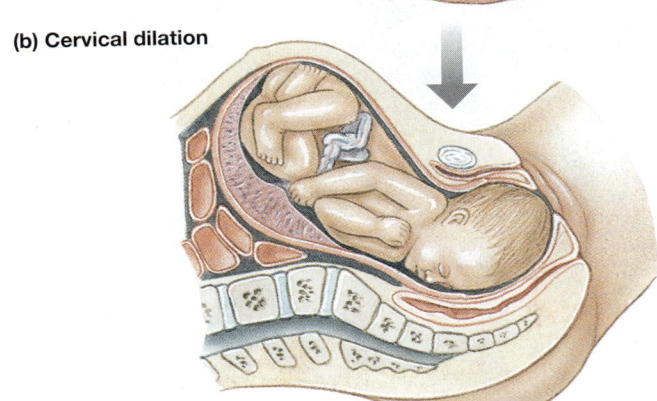

(c) Delivery

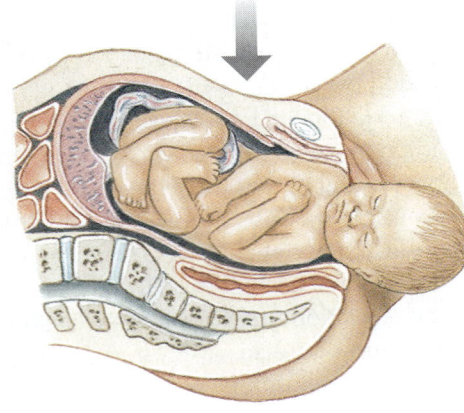

(d) Expulsion of the placenta

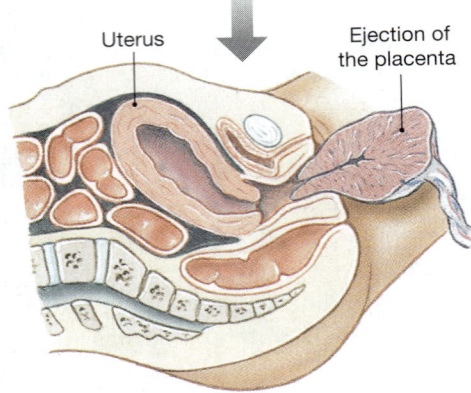

■ **FIGURE 26-20** *Parturition: the birth process*

(a) As labor begins, the fetus is normally head down in the uterus. (b) Rhythmic uterine contractions push the head against the softened cervix, stretching and dilating it. (c) Once the cervix is fully dilated and stretched, the uterine contractions push the fetus out through the vagina. (d) Shortly after the fetus is delivered, the placenta detaches from the uterine wall and is expelled.

primary cause of menstrual cramps and have been used to induce abortion in early pregnancy. During labor and delivery, prostaglandins reinforce the uterine contractions induced by oxytocin (Fig. 26-21).

As the contractions of labor intensify, the fetus moves down though the vagina and out into the world (Fig. 26-20c), still attached to the placenta. The placenta then detaches from the uterine wall and is expelled a short time later (Fig. 26-20d). Uterine contractions clamp the maternal blood vessels and help prevent excessive bleeding, although typically the mother loses about 240 mL of blood.

The Mammary Glands Secrete Milk During Lactation

A newborn has lost its source of maternal nourishment through the placenta and must rely on an external source of food instead. Primates, who normally have only one or two offspring at a time, have two functional mammary glands. A mammary gland is composed of about 20 milk-secreting lobules, each made of branched hollow ducts of secretory epithelium surrounded by *myoepithelial* contractile cells (Fig. 26-22 ■). Interestingly, the

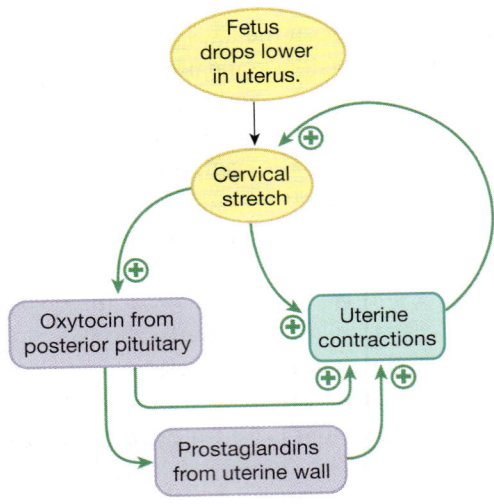

■ **FIGURE 26-21** *The positive feedback loop of parturition*

Initiation of this loop requires a uterus that is ready to respond to these stimuli. The loop stops when the fetus is delivered and the cervix is no longer being stretched.

mammary gland epithelium is closely related to the secretory epithelium of sweat glands, so milk secretion and sweat secretion share some common features.

During puberty, the breasts begin to develop under the influence of estrogen. The milk ducts grow and branch, and fat is deposited behind the glandular tissue. During pregnancy, the glands develop further under the direction of estrogen, growth hormone, and cortisol. The final development step also requires progesterone, which converts the duct epithelium into a secretory structure. This process is similar to progesterone's effect on the uterus, in which progesterone makes the endometrium into a secretory tissue during the luteal phase.

Although estrogen and progesterone stimulate mammary development, they inhibit secretion of milk. Milk production is stimulated by prolactin from the anterior pituitary [p. 230]. Prolactin is an unusual pituitary hormone in that its secretion is primarily controlled by **prolactin-inhibiting hormone** (PIH) from the hypothalamus. Good evidence suggests that PIH is actually dopamine, an amine neurohormone related to epinephrine and norepinephrine [p. 274].

During the later stages of pregnancy, PIH secretion falls, and prolactin reaches levels 10 or more times those found in nonpregnant women. Prior to delivery, when estrogen and progesterone are also high, the mammary glands produce only

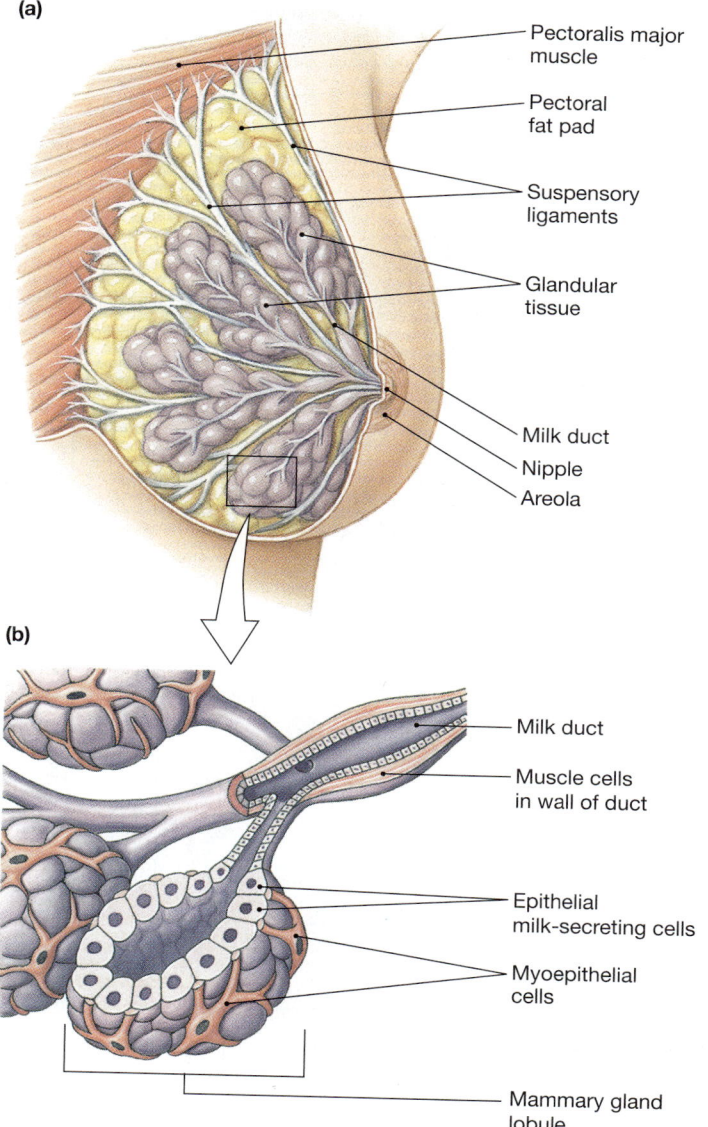

FIGURE 26-22 *Mammary glands*
Epithelial cells of the mammary glands secrete milk into the lumen of the gland.

small amounts of a thin, low-fat secretion called **colostrum**. After delivery, when estrogen and progesterone decrease, the glands produce greater amounts of milk that contains 4% fat and substantial amounts of calcium. Proteins in colostrum and milk include maternal immunoglobulins, secreted into the duct and absorbed intact by the infant's intestinal epithelium [p. 695]. This process transfers some of the mother's immunity to the infant during its first weeks of life.

Suckling, the mechanical stimulus of the infant nursing at the breast, reinforces the inhibition of PIH begun in the last weeks of pregnancy (Fig. 26-23). In the absence of PIH, prolactin secretion increases, resulting in milk production. Pregnancy is not a

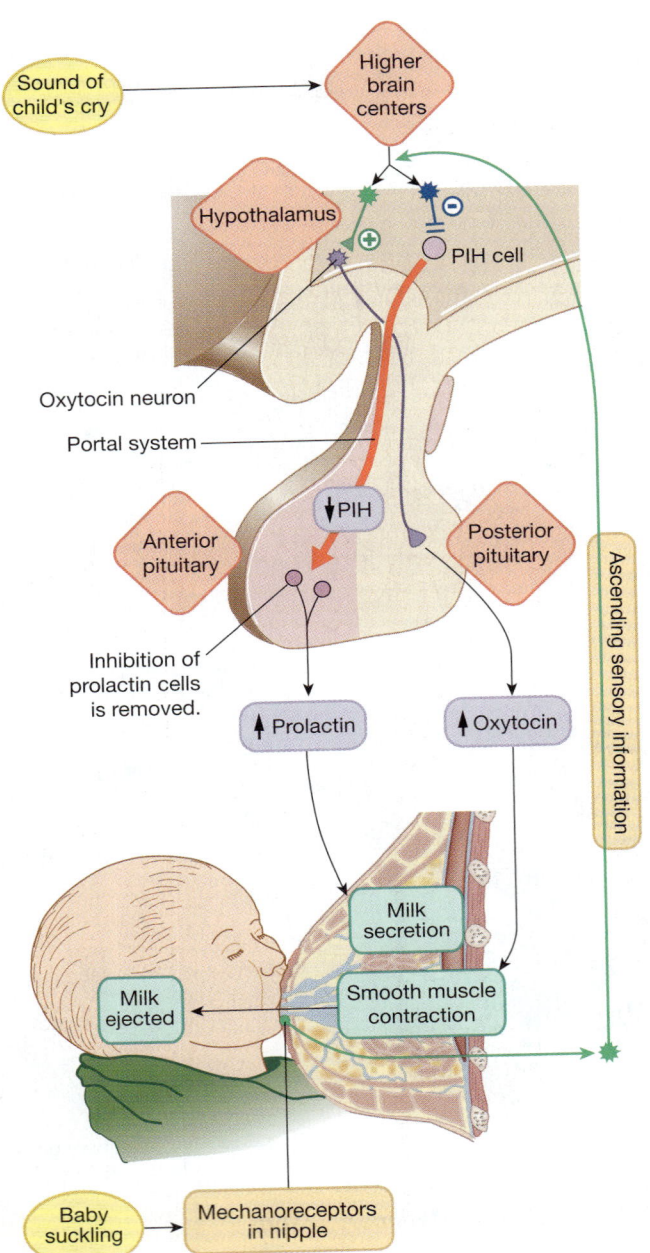

FIGURE 26-23 *The hormonal control of milk secretion and release*

the infant's mouth. Although prolactin release requires the mechanical stimulus of suckling, oxytocin release can be stimulated by various cerebral stimuli, including the thought of the child. Many nursing mothers experience inappropriate milk release triggered by hearing someone else's child cry.

Prolactin Has Other Physiological Roles

Although we discussed prolactin in the context of nursing mothers, all non-nursing women and men have tonic prolactin secretion that exhibits a diurnal cycle, peaking during sleep. Prolactin is related to growth hormone and plays a role in other reproductive and nonreproductive processes. For example, prolactin is synthesized in the uterine endometrium during normal menstrual cycles. Male knockout mice who lack prolactin or a prolactin receptor have decreased fertility.

Some interesting research has established a role for prolactin in neuroimmunomodulation [p. 800]. Both prolactin and growth hormone appear to be necessary for normal differentiation of T lymphocytes in the thymus gland, an observation supported by impaired immune function in animals with *hypoprolactinemia*. In contrast, several autoimmune diseases, including multiple sclerosis, systemic lupus erythematosus, and autoimmune thyroiditis, have been linked to elevated levels of prolactin.

GROWTH AND AGING

The reproductive years begin with the events surrounding puberty and end with decreasing gonadal hormone production.

Puberty Marks the Beginning of the Reproductive Years

In girls, the onset of puberty is marked by budding breasts and the first menstrual period, called **menarche**, a time of ritual significance in many cultures. In the United States, the average age at menarche is 12 years (normal range is considered 8 to 13 years).

In boys, the signs of puberty are more subtle. They include growth and maturation of the external genitalia; development of secondary sex characteristics, such as pubic and facial hair and lowering of voice pitch; change in body shape; and growth in height. The age range for male puberty is 9 to 14 years.

Puberty requires maturation of the hypothalamic-pituitary control pathway. Before puberty, the child has low levels of both steroid sex hormones and gonadotropins. Because low sex hormone levels normally enhance gonadotropin release, the combination of low steroids and low gonadotropins indicates that the pituitary is not yet sensitive to steroid levels in the blood.

At puberty, the hypothalamic GnRH-secreting neurons increase their pulsatile secretion of GnRH, which in turn increases gonadotropin release. The signal responsible for the onset of

requirement for lactation, and some women who have adopted babies have been successful in breast-feeding.

The ejection of milk from the glands, known as the **let-down reflex**, requires the presence of oxytocin from the posterior pituitary. Oxytocin initiates smooth muscle contraction in the uterus and breasts. In the *postpartum* (after delivery) uterus, oxytocin-induced contractions help return the uterus to its prepregnancy size.

In the lactating breast, oxytocin causes contraction of myoepithelial cells surrounding the mammary glands. This contraction creates high pressure that sends the milk squirting into puberty is still unclear. One theory says that it is the genetically programmed maturation of hypothalamic neurons. We know that puberty has a genetic basis because inherited patterns of maturation are common. If a woman did not start her menstrual periods until she was 16, for example, it is likely that her daughters will also have late menarche.

The hormone *leptin* [p. 717] may also contribute to the onset of puberty. Undernourished women with little adipose tissue and low leptin levels often stop having menstrual periods (*amenorrhea*), and knockout mice without leptin are infertile. Presumably improved nutrition over the last century increased individuals' prepubertal fat stores and leptin secretion, which could interact with other factors to initiate puberty.

Menopause and Andropause Are a Consequence of Aging

Several centuries ago in America, many people died of acute illnesses while still reproductively active. Now modern medicine has overcome most acute illnesses, and we are living well past the time that most of us are likely to have children.

Women's reproductive cycles stop completely at the time known as **menopause**. The physiology of menopause has been well studied. After about 40 years of menstrual cycles, a woman's periods become irregular and finally cease. The cessation of reproductive cycles is due not to the pituitary but to the ovaries, which can no longer respond to gonadotropins. In the absence of negative feedback, gonadotropin levels increase dramatically in an effort to stimulate the ovaries into maturing more follicles.

The absence of estrogen in postmenopausal women leads to symptoms of varying severity. These may include hot flashes [p. 743], atrophy of genitalia and breasts, and osteoporosis as calcium is lost from bones [p. 771]. Hormone replacement therapy (HRT) for women in menopause traditionally consists of estrogen or a combination of estrogen and progesterone. This treatment has become controversial, however, because of studies that suggest that its risks outweigh its benefits.

A newer drug therapy for menopause uses *selective estrogen receptor modulators* (SERMs). These drugs bind with different affinities to the two estrogen receptor subtypes, which allows the drugs to mimic the beneficial effects of estrogen on bone while avoiding the potentially detrimental effects on breasts and uterus.

In men, testosterone production decreases with age, and about half of men over the age of 50 have symptoms of **andropause**. The existence of physiological andropause in men is still somewhat controversial because the physical and psychological symptoms of aging men are not clearly linked to a decline in testosterone. Many men remain reproductively active as they age, and it is not uncommon for men in their fifties or sixties to have children with younger women. Postmenopausal women also remain sexually active, although not reproductively active, and some report a more fulfilling sex life once the fear of unwanted pregnancy has been removed.

RUNNING PROBLEM CONCLUSION

INFERTILITY

In this running problem you learned how the cause of infertility is diagnosed in a typical couple. To learn more about infertility, see literature from the American Society for Reproductive Medicine at *www.asrm.org* or go to Medline Plus (*www.nlm.nih.gov/medlineplus*) and look under Health Topics.

Now test your understanding of the running problem by checking your answers against the information in this summary table.

	QUESTION	FACTS	INTEGRATION AND ANALYSIS
1	Name (in order) the male reproductive structures that carry sperm from the testes to the external environment.	The male reproductive structures include the testes, accessory glandular organs, a series of ducts, and the external genitalia.	Sperm leaving the testes pass into the epididymis, then into the vas deferens, and finally exit the body via the urethra.
2	Which causes of male infertility might make retrieval of sperm from the epididymis necessary?	The epididymis is the first duct the sperm enter upon leaving the testes.	If the infertility problem is due to blockage or congenital defects in the vas deferens or urethra, removal of sperm from the epididymis might be useful. If the problem is caused by low sperm count or abnormal sperm morphology, this technique would probably not be useful.

26

3	For which causes of female infertility is temperature tracking useful? For which causes is it not useful?	Basal body temperature rises slightly following ovulation.	Temperature tracking is a useful way to tell if a woman is ovulating, but it cannot reveal structural problems in the female reproductive tract.
4	What abnormalities in the cervix, Fallopian tubes, and uterus could cause infertility?	The cervix, Fallopian tubes, and uterus are hollow structures through which sperm must pass.	Any blockage of these organs resulting from disease or congenital defects would prevent normal movement of sperm and cause infertility. Hormonal problems might cause the endometrium to develop incompletely, preventing implantation of the embryo.
5	Speculate on how infertility due to cervical mucus antibodies against sperm might be treated.	Antibodies in the cervical mucus react with antigenic material in the semen or on the sperm, causing the sperm to become immobile.	If the antigenic material can be removed from the semen, this might help the problem of sperm immobilization. A semen sample can be washed to remove nonsperm components. If the antigens are part of the sperm, this method will not work.
6	Should ART or intrauterine insemination be recommended for Peggy and Larry? Why?	ART is used when ovulation is abnormal. Intrauterine insemination is used when ovulation is normal.	Intrauterine insemination can be used to overcome cervical factors, as this technique bypasses the cervix. Because Peggy can ovulate, intrauterine insemination should be recommended for Peggy and Larry.

822   834   843   844   849   853   **855**

CHAPTER SUMMARY

In this chapter you learned how the human race perpetuates itself through reproduction. The reproductive system has some of the most complex *control systems* of the body, in which multiple hormones interact in an ever-changing fashion. *Homeostasis* in the adult reproductive system is anything but steady state, particularly during the female menstrual cy-cle, when the *feedback effects* of estrogen change from negative to positive and back again. An example of *positive feedback* occurs with oxytocin secretion during labor and delivery. The testis provides a nice example of *compartmentation*, with the lumen of the seminiferous tubules, where sperm develop, isolated from the rest of the extracellular compartment.

Sex Determination

1. The sex organs consist of **gonads, internal genitalia**, and **external genitalia**. (p. 822)

2. Male **testes** produce **sperm**. Female **ovaries** produce eggs, or **ova**. Embryonic cells that will produce **gametes** (eggs and sperm) are called **germ cells**. (p. 822)

3. Humans have 46 chromosomes. (p. 822; Fig. 26-1)

4. The genetic sex of an individual depends on the **sex chromosomes**: females are XX, and males are XY. In the absence of a Y chromosome, an embryo will develop into a female. (p. 823; Fig. 26-2)

5. The **SRY gene** on the Y chromosomes produces **SRY protein**, a *testis determining factor* that converts the bipotential gonad into a testis. In the absence of SRY protein, the gonad becomes an ovary. (p. 825)

6. Testicular **Sertoli cells** secrete **anti-Müllerian hormone**, which causes the **Müllerian ducts** to regress. **Leydig cells** secrete testosterone, which converts **Wolffian ducts** to male accessory structures. **Dihydrotestosterone** promotes development of the prostate gland and external genitalia. (p. 825; Figs. 26-3, 26-4)

7. Absence of testosterone and AMH causes Müllerian ducts to develop into **Fallopian tubes (oviducts), uterus**, and **vagina**. In females, the Wolffian ducts regress. (p. 826; Figs. 26-3, 26-4)

Basic Patterns of Reproduction

8. **Gametogenesis** begins with mitotic divisions of **spermatogonia** and **oögonia**. The first step of meiosis creates **primary spermatocytes** and **primary oocytes**. The first meiotic division creates two identical **secondary spermatocytes** in males or a large secondary oocyte (egg) and a tiny **first polar body** in females. (p. 827; Fig. 26-5)

9. The second meiotic division in males creates haploid **spermatids** that mature into sperm. In females, the second meiotic division does not take place unless the egg is fertilized. (p. 827; Fig. 26-5)

10. In both sexes, **gonadotropin-releasing hormone** (GnRH) controls the secretion of **follicle stimulating hormone** (FSH) and **luteinizing hormone** (LH) from the anterior pituitary. FSH and steroid sex hormones regulate gametogenesis in gonadal gamete-producing cells. LH stimulates production of steroid sex hormones. (p. 829; Fig. 26-7)

11. The steroid sex hormones include **androgens**, **estrogens**, and **progesterone**. **Aromatase** converts androgens to estrogens. **Inhibin** inhibits secretion of FSH, and **activin** stimulates FSH secretion. (p. 829; Fig. 26-6)

12. Gonadal steroids generally suppress secretion of GnRH, FSH, and LH. However, if estrogen rises rapidly above a threshold level for at least 36 hours, its feedback changes to positive and stimulates gonadotropin release. (p. 829)

13. After puberty, tonic GnRH release occurs in small pulses every 1–3 hours from a region of the hypothalamus called a **pulse generator**. (p. 830)

Male Reproduction

14. The **corpus spongiosum** and **corpora cavernosa** make up the erectile tissue of the penis. The **glans** is covered by the **foreskin**. The urethra runs though the **penis**. (p. 833; Fig. 26-8)

15. The testes migrate into the scrotum during fetal development. Failure of one or both testes to descend is known as **cryptorchidism**. (p. 833)

16. The testes consist of **seminiferous tubules** and interstitial tissue containing blood vessels and Leydig cells. The seminiferous tubules join the **epididymis**, which becomes the **vas deferens**. The vas deferens empties into the urethra. (p. 833; Fig. 26-9)

17. A seminiferous tubule contains spermatogonia, spermatocytes, and Sertoli cells. Tight junctions between Sertoli cells form a **blood-testis barrier**. (p. 833; Fig. 26-9)

18. Spermatogonia in the tubule enter meiosis, becoming primary spermatocytes, spermatids, and finally sperm in about 64 days. (p. 834; Fig. 26-9)

19. Sertoli cells regulate sperm development. They also produce inhibin, activin, growth factors, enzymes, and **androgen-binding protein**. (p. 835; Fig. 26-11)

20. Leydig cells produce 95% of a male's testosterone. The other 5% comes from the adrenal cortex. (p. 835)

21. FSH stimulates Sertoli cell production of androgen-binding protein, inhibin, and paracrine molecules. Leydig cells produce testosterone under the direction of LH. (p. 835; Fig. 26-11)

22. The **prostate gland**, **seminal vesicles**, and **bulbourethral glands** secrete the fluid component of **semen**. (p. 836)

23. **Primary sex characteristics** are the internal sexual organs and external genitalia. **Secondary sex characteristics** are other features of the body, such as body shape. (p. 836)

Female Reproduction

24. Female external genitalia, called the **vulva** or **pudendum**, are the **labia majora**, **labia minora**, and **clitoris**. The urethra opening is between the clitoris and the vagina. (p. 837; Fig. 26-12)

25. The uterine tissue layers are outer connective tissue, **myometrium**, and **endometrium**. (p. 837; Fig. 26-12)

26. Fallopian tubes are lined with ciliated epithelium. The bulk of an ovary consists of ovarian follicles. (p. 837; Fig. 26-12)

27. Eggs are produced in monthly **menstrual cycles**. (p. 837; Fig. 26-13)

28. In the **ovarian cycle**, the **follicular phase** is a period of follicular growth. **Ovulation** is the release of an egg from its follicle. In the **luteal phase**, the ruptured follicle becomes a **corpus luteum**. (p. 837; Fig. 26-13)

29. The **menses** begin the **uterine cycle**. This is followed by a **proliferative phase**, with endometrial thickening. Following ovulation, the endometrium goes into a **secretory phase**. (p. 837; Fig. 26-13)

30. Follicular **granulosa cells** secrete estrogen. As the follicular phase ends, a surge in LH is necessary for oocyte maturation. (p. 841; Fig. 26-13)

31. The corpus luteum secretes progesterone and some estrogen, which exert negative feedback on the hypothalamus-anterior pituitary. (p. 843; Fig. 26-14)

32. Estrogens and androgen control primary and secondary sex characteristics in females. (p. 843)

Procreation

33. The human sex act is divided into four phases; (1) excitement, (2) plateau, (3) orgasm, and (4) resolution. (p. 844)

34. The male **erection reflex** is a spinal reflex that can be influenced by higher brain centers. Parasympathetic input mediated by nitric oxide actively vasodilates the penile arterioles. (p. 844; Fig. 26-15)

35. **Emission** is the movement of sperm out of the vas deferens and into the urethra. **Ejaculation** is the expulsion of semen to the external environment. (p. 845)

36. Contraceptive methods include **abstinence**, **barrier methods**, **implantation prevention**, and **hormonal treatments**. (p. 846)

37. Infertility can arise from a problem in the male, the female, or both. *In vitro* fertilization has allowed some infertile couples to have children. (p. 847)

Pregnancy and Parturition

38. Sperm must go through **capacitation** before they can fertilize an egg. (p. 848)

39. Fertilization normally takes place in the Fallopian tube. Capacitated sperm release acrosomal enzymes (the **acrosomal reaction**) to dissolve cell junctions and the **zona pellucida** of the egg. The first sperm to reach the egg fertilizes it. (p. 848; Figs. 26-16, 26-17)

40. Fusion of egg and sperm membranes initiates a **cortical reaction** that prevents **polyspermy**. (p. 848)

41. The developing embryo is a hollow **blastocyst** when it reaches the uterus. Once the blastocyst implants, it develops **chorionic**, **amniotic**, and **allantoic membranes**. (p. 849; Figs. 26-18, 26-19a)

42. The **chorionic villi** of the placenta are surrounded by pools of maternal blood where nutrients, gases, and wastes are exchanged between mother and embryo. (p. 849; Fig. 26-19b)

43. The corpus luteum remains active during early pregnancy because of **human chorionic gonadotropin** (hCG) produced by the developing embryo. (p. 850)

44. The placenta secretes hCG, estrogen, progesterone, and **human placental lactogen**. This last hormone plays a role in maternal metabolism. (p. 850)

45. Estrogen during pregnancy contributes to development of milk-secreting ducts in the breasts. Progesterone is essential for maintaining the endometrium and, along with **relaxin**, helps suppress uterine contractions. (pp. 850–851)

26

46. **Parturition** normally occurs in the 38th–40th weeks of gestation. It begins with **labor** and ends with delivery of the fetus and placenta. A positive feedback loop of oxytocin secretion causes uterine muscle contraction. (p. 851; Figs. 26-20, 26-21)

47. Following delivery, the mammary glands produce milk under the influence of prolactin. Milk is released during nursing by oxytocin, causing mammary gland myoepithelial cells to contract. (p. 852; Figs. 26-22, 26-23)

48. Prolactin plays a role in immune function in both sexes. (p. 853)

Growth and Aging

49. In girls, puberty begins with **menarche**, the first menstrual period, at age 8–13 years. The age range for the onset of puberty in boys is 9 to 14 years. (p. 854)

50. The cessation of reproductive cycles in women is known as the **menopause**. Some men exhibit symptoms of **andropause**. (p. 855)

QUESTIONS

(Answers to the Review Questions begin on page A1.)

THE PHYSIOLOGY PLACE

Access more review material online at **The Physiology Place** website. There you'll find review questions, problem-solving activities, case studies, flashcards, and direct links to both *InterActive Physiology®* and *PhysioEx™*. To access the site, go to *www.physiologyplace.com* and select Human Physiology, Fourth Edition.

LEVEL ONE   REVIEWING FACTS AND TERMS

1. Match each of the following items with all the terms it applies to:
   (a) X or Y
   (b) inactivated X chromosome
   (c) XX
   (d) XY
   (e) XX or XY
   (f) autosomes

   1. chromosomes other than sex chromosomes
   2. fertilized egg
   3. sperm or ova
   4. sex chromosomes
   5. germ cells
   6. male chromosomes
   7. female chromosomes
   8. Barr body

2. The Y chromosome contains a region for male sex determination that is known as the _____ gene.

3. List the functions of the gonads. How do the products of gonadal function differ in males and females?

4. Trace the anatomical routes to the external environment followed by a newly formed sperm and by an ovulated egg. Name all structures the gametes pass through on their journey.

5. Define each of the following terms and describe its significance to reproductive physiology:
   (a) aromatase
   (b) blood-testis barrier
   (c) androgen binding protein
   (d) first polar body
   (e) acrosome

6. Decide if each of the following statements is true or false, and defend your answer.
   (a) All testosterone is produced in the testes.
   (b) Each sex hormone is produced only by members of one sex.
   (c) Anabolic steroid use appears to be addictive, and withdrawal symptoms include psychological disturbances.
   (d) High levels of estrogen in the late follicular phase help prepare the uterus for menstruation.
   (e) Progesterone is the dominant hormone of the luteal phase of the ovarian cycle.

7. What is semen? What are its main components, and where are they produced?

8. List and give a specific example of the various methods of contraception. Which is/are most effective? Least effective?

LEVEL TWO   REVIEWING CONCEPTS

9. **Concept maps:** Map the following groups of terms. You may add additional terms.

List 1	List 2
anti-Müllerian hormone	antrum
DHT	corpus luteum
Leydig cells	endometrium
Müllerian ducts	follicle
Sertoli cells	granulosa cells
sperm	myometrium
spermatids	ovum
spermatocytes	thecal cells
spermatogonia	
SRY	
testosterone	
Wolffian ducts	

10. Diagram the hormonal control of gametogenesis in males.

11. Diagram the menstrual cycle, distinguishing between the ovarian cycle and the uterine cycle. Include all relevant hormones.

12. Why are X-linked traits exhibited more frequently by males than females?

13. Define and relate each of the following terms in each group:
   (a) gamete, zygote, germ cell, embryo, fetus
   (b) coitus, erection, ejaculation, orgasm, emission, erogenous zones
   (c) capacitation, zona pellucida, acrosomal reaction, cortical reaction, cortical granules
   (d) puberty, menarche, menopause, andropause

14. Compare the actions of each of the following hormones in males and females:
   (a) FSH
   (b) inhibin
   (c) activin
   (d) GnRH
   (e) LH
   (f) DHT
   (g) estrogen
   (h) testosterone
   (i) progesterone

15. Compare and contrast the events of the four phases of sexual intercourse in males and in females.

16. Discuss the roles of each of the following hormones in pregnancy, labor and delivery, and mammary gland development and lactation:
   (a) human chorionic gonadotropin
   (b) luteinizing hormone
   (c) human placental lactogen
   (d) estrogen
   (e) progesterone
   (f) relaxin
   (g) prolactin

LEVEL THREE   PROBLEM SOLVING

17. Down syndrome is a chromosomal defect known as "trisomy" (three copies instead of two) of chromosome 21. The extra chromosome

usually comes from the mother. Speculate what causes trisomy, using what you have learned about the events surrounding fertilization.

18. Sometimes the follicle fails to rupture at ovulation, even though it appears to have gone through all stages of development. This condition results in benign ovarian cysts, and the unruptured follicles can be palpated as bumps on the surface of the ovary. If the cysts persist, symptoms of this condition often mimic pregnancy, with missed menstrual periods and tender breasts. Explain how these symptoms occur, using diagrams as needed.

19. An XY individual inherits a mutation that results in completely nonfunctional androgen receptors.

    (a) Is this person genetically male or female?
    (b) Will this person have functional ovaries, functional testes, or incompletely developed or nonfunctional gonads?
    (c) Will this person have Wolffian ducts or their derivatives? Müllerian ducts or their derivatives?
    (d) Will this person have the external appearance of a male or a female?

20. The babies of mothers with gestational diabetes mellitus tend to weigh more at birth. They are also at risk of developing hypoglycemia immediately following birth. Use what you have learned about diabetes and insulin to explain these two observations. *Hint:* these babies have normal insulin responses.

LEVEL FOUR   QUANTITATIVE PROBLEMS

21. The following graph shows the results of an experiment in which normal men were given testosterone over a period of months (indicated by the beige bar from A to E). Control values of hormones

were measured prior to the start of the experiment. From time B to time C, the men were also given FSH. From time D to time E, they were also given LH. Based on the information given, answer the following questions.

(a) Why did testosterone level increase beginning at point A?
(b) Why did LH and FSH levels decrease beginning at point A?
(c) Predict what happened to the men's sperm production in the A–B interval, the B–C interval, and the D–E interval.

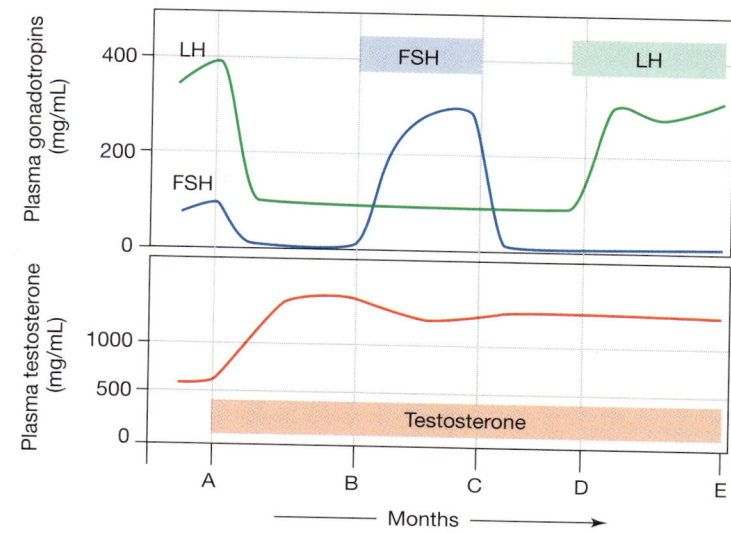

ANSWERS

✓ Answers to Concept Check Questions

Page 823

1. Female gonad: ovary; female gamete: egg, or ovum. Male gonad: testis; male gametes: sperm.

Page 826

2. Primary androgen receptors are in the cytoplasm or nucleus of the target cell. AMH has membrane receptors.

Page 826

3. The male parent donates the chromosome that determines sex of the zygote; therefore, the wives were not at fault.

4. An XO fetus will be a female because she lacks a Y chromosome.

5. Lack of AMH from the testes allows Müllerian ducts to develop into uterus and Fallopian tubes. External genitalia will be female because there is no DHT for development of male genitalia.

Page 829

6. A newborn male's gametes are spermatogonia; a newborn female's gametes are primary oocytes.

7. The first polar body has twice as much DNA as the second polar body.

8. Each primary oocyte forms one egg; each primary spermatocyte forms four sperm.

Page 831

9. Aromatase converts testosterone to estradiol.

10. FSH = follicle-stimulating hormone, DHT = dihydrotestosterone, SRY = sex-determining region of Y chromosome, LH = luteinizing hormone, GnRH = gonadotropin releasing hormone, AMH = anti-Müllerian hormone.

11. Hypothalamic GnRH, and FSH and LH from the anterior pituitary, control reproduction.

Page 835

12. Sertoli cells secrete inhibin, activin, androgen-binding protein, enzymes, and growth factors. Leydig cells secrete testosterone.

13. The advantage is that GnRH agonists decrease FSH and LH, so the testes stop producing sperm. The disadvantage is that the testes also stop producing testosterone, which causes decreased sex drive.

Page 836

14. Exogenous anabolic steroids (androgens) shut down FSH and LH secretion, which in turn makes the testes shrink and stop producing sperm.

Page 844

15. Ovarian cycle: follicular phase, ovulation, and luteal phase. The menses and proliferative phases of the uterine cycle correspond to the follicular phase and ovulation; the secretory uterine phase corresponds to the luteal phase.

16. Women who take anabolic steroids may experience growth of facial and body hair, deepening of the voice, increased libido, and irregular menstrual cycles.

17. A woman given an aromatase inhibitor would have decreased estrogen production.

26

Appendix A — Answers to End-of-Chapter Review Questions

CHAPTER 1

LEVEL ONE REVIEWING FACTS AND TERMS

1. Physiology is the study of the normal functioning of a living organism and its component parts. Anatomy is the study of structure.
2. See Figure 1-1.
3. See Table 1-1.
4. Physiology integrates body function across all levels of organization, from cells to organs, and emphasizes the coordinated function of body systems.
5. Homeostasis is the maintenance of internal stability. Body temperature and water balance are two homeostatically maintained functions.
6. Four major themes are homeostasis and control systems, structure-function relationships, biological energy, and communication.

LEVEL TWO REVIEWING CONCEPTS

7. Mapping: Maps are highly individual. The best way to evaluate your map is to compare it with one done by a classmate or to ask your instructor for comments.
8. (a) Tissues are collections of cells that carry out related functions. Organs are collections of tissues that form structural and functional units.
   (b) The *x*-axis is the independent variable, and the *y*-axis is the dependent variable.
   (c) The dependent variable changes when the independent variable is manipulated or altered during the experiment.
   (d) The teleological approach is a functional approach that is concerned with the "why" of a physiological system. The mechanistic approach is concerned with the physiological mechanisms of physiology (the "hows" of a physiological system).
   (e) The internal environment of the human body is the extracellular fluid. The external environment is the world outside the body.
   (f) In a blind study, the subjects do not know the treatment they are receiving. In a double-blind study, neither the subject nor the administrator of the experiment knows which treatment is the control and which is the active treatment. In a crossover study, one subject serves as both the control and the experimental subject.
9. Lumens that are in contact with the external environment include nasal and oral cavities, external ear, lacrimal ducts, ducts of the integumentary system (sweat, sebaceous, and mammary glands), urethra, anus, and vagina. Lumens that are not in contact with the external environment include organs of the digestive tract (esophagus, stomach, small and large intestines), accessory digestive organs (ducts of the salivary glands, pancreas, liver and gall bladder, and the gall bladder itself), urinary organs (renal tubules and renal pelvis of the kidneys, ureters), reproductive organs (seminiferous tubules, epididymis, ductus deferens, ejaculatory duct, ducts of prostate, bulbourethral glands, and seminal vesicles), cardiovascular organs (heart chambers, blood vessels), lymphatic vessels, deeper respiratory organs (larynx, trachea, bronchi, bronchioles, alveoli), canals and ventricles in central nervous system, cavities in bones, and hollow organelles.

10. The endocrine and nervous systems coordinate functions in all organ systems. Protection of the body is provided by integumentary, digestive, cardiovascular, and immune systems. The respiratory system exchanges oxygen and carbon dioxide with the external environment. The digestive system takes in nutrients, and the digestive and urinary systems eliminate digestive and metabolic waste products and water. The integumentary system loses water and solutes to the environment.

LEVEL THREE PROBLEM SOLVING

11. (a) "Because of gravity" is an incorrect mechanistic answer.
    (b) "To bring oxygen and nutrients to the cells" is a correct teleological answer.
    (c) "Because if it didn't, we would die" is a concrete teleological answer.
    (d) "Because of the pumping action of the heart" is a correct mechanistic answer.
12. Other problems that terrestrial animals have had to overcome include the requirement of an aqueous environment for fertilization (internal fertilization in mammals; many other terrestrial animals return to water to breed); requirement for aqueous environment for embryonic development (eggs in birds, some reptiles and insects); internal development in mammals, some reptiles, and insects; physical support (exoskeletons in insects, internal skeletons in vertebrates).

LEVEL FOUR QUANTITATIVE PROBLEMS

13. (a) The independent variable was time (*x*-axis). The dependent variable was body length (*y*-axis).
    (b) There was no control. A proper control would have been a similar group of fish treated identically except for being fed a diet with normal levels of vitamin D.
    (c) A graph should have time in days on the *x*-axis and body length on the *y*-axis. A line graph is most appropriate for these data.
    (d) Growth was slowest from days 0–3 (1 mm/3 days) and most rapid for days 6–9 and days 18–21 (each 3 mm/3 days).
14. (a) The independent variable was the concentration of the soaking solution. The dependent variable was volume change of the potato slices.
    (b) The volume measurements before soaking provide a baseline but there is no control in this test.
    (c) A bar graph or a scatter plot with best-fit line could be used to graph these data. A best-fit-line graph would allow you to estimate volume change at intermediate salt concentrations, such as 5%.
15. (a) This graph is a scatter plot.
    (b) The investigators were asking if there is a relationship between midarm muscle circumference and aerobic fitness.
    (c) There appears to be no relationship between midarm muscle circumference and aerobic fitness.

16. (a) This question has no "correct" answer. For peer critiques of the study, read the editorials published in the same issue: *New England Journal of Medicine* 347(2):132–33 and 137–39, 2002, July 11.
  (b) The placebo group could have reported decreased pain because they believed that the surgery had helped (a placebo effect) or because other interventions (the anesthesia and incision of the surgery, pain medication, physical therapy) had helped.
  (c) The study is directly applicable to a limited population: male veterans, under age 76, predominantly white, with osteoarthritis or degenerative joint disease.
  (d) This was a blind study.
  (e) The investigators were trying to determine whether a placebo effect could account for post-surgical improvement.

CHAPTER 2

LEVEL ONE   REVIEWING FACTS AND TERMS

1. Three major essential elements in the human body are carbon, hydrogen, and oxygen.
2. Atoms bind to form molecules.
3. (a) proton   (b) neutron   (c) electron
4. The number of protons determines the atomic number.
5. Calcium, carbon, oxygen, sodium, nitrogen, potassium, hydrogen, and phosphorous.
6. Isotopes have the same number of *protons* and *electrons* but variable numbers of *neutrons*.
7. Unstable isotopes emit *radiation*. *Nuclear medicine* uses radiation for diagnosis and treatment of disease.
8. Paired electrons are stable. Atoms with unpaired electrons have a higher probability of gaining or losing electrons to other atoms.
9. An atom that gains or loses one or more electrons is called an *ion*.
10. (a) 2   (b) 4   (c) 1   (d) 3
11. *Nonpolar* compounds have even distribution of electrons; *polar* compounds have their electrons distributed unevenly. Polar compounds dissolve more readily in water and are said to be hydrophilic (water loving).
12. pH indicates the H^+ concentration. *Acidic* solutions have a pH less than 7; *basic* or *alkaline* solutions have a pH greater than 7.
13. *Buffers* help prevent changes in pH.
14. The four kinds of biomolecules are proteins (collagen, keratin, hemoglobin), carbohydrates (glucose, sucrose, fructose), lipids (cholesterol, phospholipids), and nucleic acids (ATP, DNA, RNA).
15. (a) 4   (b) 5   (c) 6   (d) 1   (e) 3
16. *Lipoproteins* are proteins plus fatty components. *Glycoproteins* are proteins plus carbohydrates.
17. (a) 1   (b) 5   (c) 4   (d) 2   (e) 3
18. (a) 3   (b) 1   (c) 5   (d) 2   (e) 4, 6
19. A nucleotide is composed of one or more phosphate groups, a five-carbon sugar, and a carbon-nitrogen ring called a base.
20. Any molecule that binds to another molecule is called a *ligand*.
21. (a) 4   (b) 3   (c) 2
22. A *cofactor* is an ion or small organic function group that must be present in order for an enzyme to work.
23. A protein which loses its conformation is said to be *denatured*.

LEVEL TWO   REVIEWING CONCEPTS

24. Maps will be different. Figure 2-1 is a good starting map for List 1. Check your map against your classmate's, or ask your instructor to look for omissions and errors.

25. (a) Na has 11 electrons.
  (b) The net electrical charge on a sodium atom is zero.
  (c) Na has 12 neutrons.
  (d) If the atom loses one electron, it would be called an ion, or more specifically, a cation.
  (e) The Na ion has a charge of +1.
  (f) Na^+
  (g) If Na loses a proton, it becomes a neon atom.
  (h) Ne
26. $[H^+] = 10^{-3}$ M = *pH 3; acidic.* $[H^+] = 10^{-10}$ M = *pH 10; basic*
27. ATP has usable energy in a high-energy bond. DNA stores genetic information in cells. RNA translates the genetic information of DNA into proteins. cAMP assists transfer of signals from outside of cells to the inside of cells.
28. Primary structure is the sequence of amino acids in a peptide chain. The secondary structure folds the chain into either an α-helix or β-pleated sheet. Tertiary structure is the final three-dimensional shape of the protein (globular or fibrous). Quaternary structure is the combination of two or more protein subunits to form a larger molecule.
29. DNA contains the bases adenine, guanine, cytosine, and thymine. RNA substitutes uracil for thymine. DNA has the sugar deoxyribose; RNA has the sugar ribose. DNA is a doubled stranded molecule with two chains of bases linked in an alpha helix. RNA is a single stranded molecule.
30. Purines (adenine and guanine) are larger and consist of two carbon rings. Pyrimidines are smaller and have only one carbon ring.
31. Isoforms are structurally related proteins whose functions are similar but whose affinities for ligands differ.
32. (a) 4, 5   (b) 3   (c) 2, 1

LEVEL THREE   PROBLEM SOLVING

33. Nucleotides contain all of the elements listed. Pure carbohydrates would have a C:H:O ratio of 1:2:1. Fats would have mostly carbon and hydrogen and little oxygen. Proteins would not have phosphorus and would have less nitrogen relative to carbon. The nitrogenous bases have larger proportions of N than amino acids do; the phosphate groups of nucleotides are mostly oxygen and phosphorus.
34. According to the equation, carbon dioxide combines with water to ultimately produce H^+ and HCO_3^- (bicarbonate ion). An increase in H^+ results in a decrease in the pH, as the blood becomes more acidic. Any negatively charged ion (anion) that attracts and binds H^+ has the potential to act as a buffer.
35. The atomic mass would remain the same while the atomic number would increase to 54. The resulting atom would be xenon.

LEVEL FOUR   QUANTITATIVE PROBLEMS

36. (a) $C_6H_{12}O_6$ (glucose); m.w. 180
  (b) CO_2; m.w. 44
  (c) H_2O; m.w. 18
  (d) $C_3H_7O_2N$ (alanine); m.w. 89
  (e) $C_5H_{10}O_5$ (a pentose); m.w. 150
37. 0.9% solution = 0.9 g/100 mL. Weigh out 9 g NaCl, place into a 1 liter flask, and dissolve in enough water to yield 1 L of solution.
38. (a) 1 M = 6.02×10^{23} molecules of NaCl
  (b) A 1 M solution contains 1000 millimoles.
  (c) 1 M contains one equivalent of Na^+.
  (d) 58.5 g/liter = 5.85 g/100 mL = 5.85% solution.

39. 5% glucose = 5 g/100 mL or 10 g in 200 mL of solution. Molarity: 5 g/100 mL = 50 g/L × 1 mole/180 g = 0.278 moles/L or 278 millimoles/L (278 mM). 500 mL of 5% glucose would have 25 g glucose × 1 mole/180 g = 139 millimoles glucose in 500 mL.

40. Myoglobin has a higher affinity for oxygen than hemoglobin because at lower oxygen concentrations, more myoglobin has oxygen bound to it.

CHAPTER 3

LEVEL ONE REVIEWING FACTS AND TERMS

1. Four general functions of cell membranes are (1) to act as a barrier between cell and extracellular environment, (2) to regulate the exchange of material between cell and its environment, (3) to act as a point of transfer of information between the cell and other cells, and (4) to provide structural support.

2. The *fluid mosaic model* has a bilayer of *phospholipids* with embedded *proteins* and *carbohydrates* on the extracellular surface.

3. The cell membrane is mostly phospholipids and proteins.

4. Inclusions are particles of insoluble material in the cytosol, such as glycogen granules, protein fibers and ribosomes. Membranous organelles, such as mitochondria and the Golgi complex, are separated from the cytosol by one or more phospholipid membranes.

5. The cytoskeleton is a flexible, changeable, three-dimensional scaffolding of actin, microfilaments, intermediate filaments, and microtubules. The cytoskeleton (1) provides mechanical strength, (2) stabilizes the positions of organelles and membrane proteins, (3) transports material into the cell and throughout the cytoplasm, (4) links cells together and supports material outside cell, and (5) is responsible for various kinds of movement.

6. (a) 2   (b) 3   (c) 1   (d) 4

7. (a) 3   (b) 5   (c) 4   (d) 1   (e) 2

8. Lysosomal enzymes are activated by very acidic conditions.

9. Exocrine glands produce watery *serous secretions* and stickier *mucous secretions*.

10. *Endocrine* glands secrete hormones.

11. The four tissue types are connective tissue (such as tendons that hold muscles to bones); epithelial tissues (such as the protective epithelium of the skin); neural tissue (such as the brain); and muscular tissue (such as the heart and skeletal muscles).

12. The largest organ of the body is the skin.

13. (a) 1   (b) 1   (c) 4   (d) 3   (e) 4   (f) 4
    (g) 4   (h) 1   (i) 1

14. Exocrine glands in the skin include sweat glands that secrete watery sweat, apocrine glands that secrete waxy or milky secretions, and sebaceous glands that secrete a mixture of lipids.

15. The matrix of a mitochondrion is the internal compartment of the organelle. The matrix of tissues is noncellular material found outside the cells.

LEVEL TWO REVIEWING CONCEPTS

16. Map: Check your map against Figures 3-11, 3-12, 3-13, and 3-14.

17. The three types of cell junctions are adhesive junctions, tight junctions, and gap junctions. They are all alike in that they link cells together and have proteins that are connected to the cell membrane. Adhesive junctions are designed to allow twisting and stretching of the tissues in which they are found. Tight junctions are designed to prevent movement of materials between the cells they link. Gap junctions are designed to allow material to pass from the cytoplasm of one cell directly into the cytoplasm of the connecting cell.

18. Pancreatic cells that manufacture insulin have more rough endoplasmic reticulum (RER); proteins are made on ribosomes bound to the RER membranes. Steroid hormones, such as cortisol, are lipids and are made on the smooth endoplasmic reticulum (SER), which lacks ribosomes.

19. Vesicles are bubbles or spheres made from phospholipid membranes similar to the cell membrane. Many vesicles are formed when ends of the Golgi apparatus are pinched off. Lysosomes and peroxisomes with enzymes remain in the cytoplasm, while the contents of secretory vesicles are released from the cell.

20. A stratified epithelium has many cell layers that offer more protection than the single-cell layer of a simple epithelium.

21. See Figure 3-25. Tight junctions prevent movement of material between the cells; leaky junctions allow some material to pass between the cells.

22. The glucose would enter the *intracellular fluid* of the intestinal cell, then move from the cell into *interstitial fluid* first, then into the *plasma*.

23. Cholesterol decreases membrane permeability to water by filling space between the interior tails of phospholipids.

24. Bone is a rigid connective tissue due to calcification; cartilage is firm but elastic. Bones are the primary support structure for the body; cartilage forms the ear, nose, larynx, and spine and helps hold bones together at the joints.

25. (a) A lumen is the hollow inside of an organ or tube. The wall is the layer of cells that make up the organ or tube.

    (b) Cytoplasm is everything inside the cell membrane except for the nucleus; cytosol is the semi-gelatinous, intracellular fluid that contains dissolved nutrients, ions, and waste products.

    (c) Myosin is a motor protein filament found in the cytoplasm. Keratin is a structural protein fiber made in connective tissue by keratinocytes.

26. This is an example of apoptosis because it is a normal part of development.

27. (a) cell junctions: 1 (flow through gap junctions), 2 (fusion of tight junction proteins), 4 (strength of desmosomes)

    (b) cell membrane: 1 (membrane receptors), 2 (membrane enzymes), 3 (phospholipid bilayer forms barrier), 4 (fluidity), 5 (ATP-dependent membrane protein transporters)

    (c) cytoskeleton: 2 (microtubules direct movement of chromosomes), 4 (strength), 5 (ATP required for actin-myosin interaction)

    (d) organelles: 2 (mRNA binds to ribosomes), 3 (membrane-bounded organelles), 5 (ATP-dependent processes like protein synthesis)

    (e) cilia: 2 (microtubules and dynein), 4 (strength and flexibility), 5 (ATP-dependent movement)

28. Though the extracellular matrix may be rigid, it is a dynamic structure. The proteins that comprise this matrix can be disassembled and re-assembled as necessary, to allow the tissues to change size and shape.

LEVEL THREE PROBLEM SOLVING

29. Cilia of the respiratory passages constantly sweep mucus and trapped particles away from the lungs, to the pharynx where they can be swallowed harmlessly. When they fail to beat, inhaled particulate matter and pathogens are more likely to reach the delicate lungs. Possible consequences include infections, inflammation due

to presence of irritants, cancer, and increased effort of breathing due to accumulation of mucus. The smoker's cough is necessary to remove the mucus that would otherwise be swept away by the cilia.

30. Epithelial cells have one of the highest rates of mitosis of any cell type in the body, because many are in vulnerable locations and need to be replaced frequently. For example, the epithelial lining of the stomach is frequently exposed to acids and enzymes which kill the cells after a few days. Tissues whose cells undergo mitosis more often are more likely to develop abnormal cell division.

31. Matrix metalloproteinases are enzymes secreted by cancer cells that dissolve the extracellular matrix, enabling cancer cells to escape from the original tumor and spread to other tissues. Blocking these enzymes may prevent metastasis (spread) of cancer cells from the organ where they arise, thus preventing them from interfering with normal physiology of other organs.

CHAPTER 4

LEVEL ONE   REVIEWING FACTS AND TERMS

1. The three forms of work are transport work (moving substances across cell membranes); chemical work (making proteins for cell structures); and mechanical work, (muscle contraction).

2. Potential energy is stored energy, such as the energy stored in chemical bonds or concentration gradients. Kinetic energy is the energy of motion, such as muscles contracting.

3. The First Law says that there is a fixed amount of energy in the universe. The Second Law says that without input of energy, a system will become progressively less organized or ordered.

4. *Metabolism* is the sum of chemical processes in the cell.

5. Water and CO_2 are the *reactants* or *substrates*. The speed of a reaction is called the *reaction rate*.

6. *Enzymes* are proteins that speed up chemical reactions by *decreasing* their activation energy.

7. 1. (d)   2. (a)   3. (f)   4. (c)

8. *-ase* is the suffix added to most enzyme names.

9. Organic molecules that must be present for an enzyme to function are called *coenzymes*. Many coenzymes require dietary *vitamins* for their synthesis.

10. Molecules that gain electrons are said to be *reduced*; molecules that lose electrons are said to be *oxidized*.

11. Removal of water is *dehydration;* use of water to break down polymers is *hydrolysis*.

12. Removal of an amino acid group is *deamination*; transfer of an amino group to another molecule is *transamination*. Amino acids that are removed are broken down into urea and uric acid.

13. *Catabolic* reactions release energy; *anabolic* reactions synthesize large biomolecules. Metabolic energy is measured in kilocalories.

14. Accumulation of metabolic end products may result in *feedback inhibition*.

15. Transport of H^+ into the inner mitochondrial membrane creates a concentration gradient that stores energy. When the ions move back across the membrane, the energy that is released is trapped in the high-energy bond of ATP.

16. The two carrier molecules that bring electrons to the electron transport system are NADH and $FADH_2$.

17. The breakdown of lipids is *lipolysis*. Fatty acids are broken down through *beta-oxidation*. The acetyl CoA formed from lipids then can go into the citric acid cycle.

18. Carbohydrates and proteins have 4 kilocalories per gram; fat has 9.

19. The smooth endoplasmic reticulum is the primary organelle involved in lipid synthesis.

LEVEL TWO   REVIEWING CONCEPTS

20. For map 1, use Figure 4-13 as a starting point. Use Figures 4-25 and 4-26 to help create map 2.

21. Potential energy stored in chemical bonds can be used to perform work, can be transferred to another molecule, or can be released as heat to the environment.

22. 1. (b)   2. (a)   3. (b)   4. (a)   5. (c)   6. (c)

23. It is better to store enzymes in an inactive form so that they cannot harm the cell if they are accidentally released.

24. Aerobic breakdown of one glucose yields 30–32 ATP, while anaerobic breakdown yields only 2 ATP. Anaerobic breakdown is much faster and does not require oxygen, but the energy yield is much less.

25. Conversion of glycogen to glucose to glucose 6-phosphate requires the energy of one ATP, but conversion of glycogen directly to glucose 6-phosphate does not use an ATP.

26. Transcription, which takes place in the nucleus, is the synthesis of mRNA from the sense strand of DNA. Translation is the conversion of information coded in mRNA into a string of amino acids; it takes place on cytoplasmic or RER ribosomes.

27. Anticodons are part of a tRNA molecule. Amino acids attach to tRNA at the opposite end of the molecule from the anticodon.

28. The energy in ATP is primarily within the high-energy bond holding the third phosphate group onto the rest of the molecule, and chemical bonds contain potential energy.

29. If the reaction cannot proceed without the contribution of energy from ATP, the activation energy must be large compared to a reaction that does not require ATP.

LEVEL THREE   PROBLEM SOLVING

30. Cells that do not use glucose as their primary energy source could use certain amino acids or fatty acids instead. For example, the transporting epithelia of the intestine use the amino acid glutamate.

31. The mRNA sequence would be GCGAUGUUCAGUCCAUGGCAU-UGC. The anticodons would be UAC (tyrosine), AGU (serine), AGG (arginine), GUA (valine).

LEVEL FOUR   QUANTITATIVE PROBLEMS

32. Exergonic.

33. The polypeptide will have 149 amino acids.

CHAPTER 5

LEVEL ONE   REVIEWING FACTS AND TERMS

1. Membrane proteins act as structural proteins (link the cytoskeleton to fibers in the matrix); transporters (channels that allow water to cross the membrane); receptors (receptors for hormones); and enzymes (digestive enzymes in the intestine).

2. Active transport requires the direct or indirect use of energy. Passive transport uses only the energy stored in a concentration gradient.

3. Simple and facilitated diffusion and osmosis are examples of passive transport. Phagocytosis, exocytosis, and endocytosis are examples of active transport.

4. Four factors that speed up the rate of diffusion include a greater concentration difference, smaller distance, higher temperature, and smaller molecular size.

5. (a) 4   (b) 1, 6   (c) 3   (d) 5

6. Materials enter cells by simple diffusion through the phospholipid bilayer, by protein-mediated transport, or in vesicles (endocytosis/phagocytosis).

7. Cotransporters of molecules in the same direction are called *symport* carriers, in the opposite direction are called *antiport* carriers. A transport protein that moves only one molecule is a *uniport* carrier.

8. The two types of active transport are *direct* and *indirect*.

9. A molecule that moves freely between body compartments is said to be a *penetrating* solute. A molecule that is not able to enter cells is called a *nonpenetrating* solute.

10. The toddler has the highest percentage of body water (d), followed by the 25-year-old male (a), the 25-year-old female (b), and the 65-year-old female (c). Body weight has no direct effect on the percentage of body water, although people with more fat will have less water.

11. Osmolarity is the concentration of osmotically active particles in solution and is usually given in osmoles or milliosmoles per liter.

12. A hypotonic solution will cause net gain of water by the cell, whereas a hypertonic solution will cause net loss of water by the cell. Tonicity is determined by relative concentrations of nonpenetrating solutes in cell versus solution.

13. The four principles of electricity are: (1) like charges repel while opposite charges attract; (2) every positive ion has a matching negative ion; (3) energy must be used to separate ions or electrons and protons; and (4) conductors allow ions to move through them, while insulators keep ions separated.

14. (a) 7   (b) 1, 4   (c) 6   (d) 5   (e) 8   (f) 3   (g) 2

15. The equilibrium potential is the electrical gradient that is equal and opposite to a given concentration gradient.

16. Conductors allow free movement of electrical charge; insulators prevent movement of charge.

LEVEL TWO   REVIEWING CONCEPTS

17. Use Figures 5-4, 5-7, 5-24, and 5-26 to help create your map.

Cell (67% of volume)	Interstitial fluid (25%)	Plasma (8%)
An^-, $Proteins^-$ $K^+ > Na^+$ $HCO_3^- > Cl^-$	$Na^+ > K^+$ $Cl^- > HCO_3^- >>>$ Proteins	$Na^+ > K^+$ $Cl^- > HCO_3^-$ Proteins

Cell membrane ↑   Endothelium ↑

19. The three factors that influence diffusion across a membrane are lipid solubility, thickness of the membrane, and surface area of the membrane. Diffusion across the membrane requires that a molecule pass through the lipid core of the membrane, so the more lipophilic a substance is, the faster it will diffuse. Diffusion is slower across thicker membranes and faster when there is more surface area.

20. Specificity is the ability of an enzyme or transporter to work on one specific molecule or class of molecules. Competition refers to the fact that the different substrates will compete for the binding site on the enzyme or transport protein. Saturation means that the protein will work at a maximum rate when all the binding sites are filled. Facilitated diffusion uses protein transporters, so each transporter can be described by its specificity (glucose transporter, amino acid transporter, etc.). If two or more different substrates are present, the substrates will compete for transport. And if enough substrate is present, transport rate will reach a maximum rate that cannot be exceeded.

21. (a) hypotonic   (b) into the cells

22. The system started at equilibrium, so to disrupt the equilibrium requires input of energy. Therefore, some type of active transport has occurred.

23. (a) False. A 2 M NaCl solution = 4 OsM, but a 2 M glucose solution = 2 OsM.
    (b) True. See reasoning for 23a.
    (c) False. Water moves to dilute the more concentrated solute, therefore, water will move from the 2 OsM glucose to the 4 OsM NaCl.
    (d) False. As water moves out of the glucose solution, its volume decreases.

24. (a) A 1 M NaCl solution would be 2 OsM and isosmotic. (c) The glucose solution would have to have a molarity greater than 4 M for this statement to be true. (d) Once (c) is true, (d) becomes true.

25. An electrical gradient means the separation of electrical charge and is not concerned with what kinds of molecules are carrying those charges. An electrochemical gradient takes into account both the concentration gradient of a molecule and the electrical gradient.

LEVEL THREE   PROBLEM SOLVING

26. Sodium leak channels on the apical side will allow Na^+ into the cell down its electrochemical gradient. There must be no water pores on the apical side or water will follow the Na^+. The removal of Na^+ but not water from the lumen makes the remaining fluid hypotonic. Na^+ entering the cell on the apical side is pumped out of the cell into the ECF on the basolateral side by the Na^+-K^+-ATPase.

27. Glucose moves into cells by facilitated diffusion using a protein carrier. Insulin could increase glucose uptake in a variety of ways. (1) It could increase the number of carrier proteins available for transport. (2) It could increase the affinity of the carrier for glucose. (3) Because diffusion depends on maintaining a concentration gradient for glucose, insulin could act on the cell's metabolism to keep the intracellular glucose concentration low. In reality, insulin uses methods 1 and 3.

28. The three terms apply to both enzymes and transporters. In one instance, the substrates are altered by the protein; and in the other, the substrates are simply moved unchanged across a membrane. In both instances, the substrate(s) binds to the protein at a specific binding site.

29. For all parts, note that the concentrations are given in mM rather than mOsM. 1 mM urea = 1 mOsM urea, but (assuming complete dissociation) 1mM NaCl = 2 mOsM NaCl.
    (a) Hyperosmotic (450 mOsM total) but isotonic (300 mOsM NaCl)
    (b) Hyposmotic (250 mOsM total) and hypotonic (200 mOsM NaCl)
    (c) Isosmotic (300 mOsM total) and hypotonic (200 mOsM NaCl)
    (d) Hyperosmotic (400 mOsM total) and isotonic (300 mOsM NaCl)
    (e) Hyperosmotic (350 mOsM total) and hypotonic (200 mOsM NaCl)

30. When sugars are added to membrane proteins in the ER and Golgi, they are added to proteins facing the lumen of these organelles, which will become the lumen of the secretory vesicles that bud off the Golgi. In Figure 5-24 on p. 149, you can see that whatever faces the inside of a vesicle will face the outside of the cell after being inserted into the membrane. Therefore the sugar tails of the membrane proteins will face the extracellular side once the proteins have been inserted into the membrane.

LEVEL FOUR   QUANTITATIVE PROBLEMS

31. Plasma osmolarity = 296 mOsM
32. (a) ICF volume = 29.5 L; interstitial fluid volume = 9.8
    (b) Total solute = 12.432 osmoles; ECF solute = 3.7 osmoles; ICF solute = 8.732 osmoles; plasma solute = 0.799 osmoles.
33. Half-normal saline is approximately 154 mOsM.
34. (a) increases   (b) decreases   (c) increase   (d) decrease
35. The graph shows simple diffusion (a). You know that it is not active transport because the concentration inside the cell does not increase past equilibrium (concentration in = concentration out). The graph is the same shape as the saturation graph in Figure 5-22 but the axes are not the same. There is only one solute being transported, so there is no evidence for competition.

CHAPTER 6

LEVEL ONE   REVIEWING FACTS AND TERMS

1. The two routes for long-distance signal delivery are electrical signals carried by neurons and chemical signals that are distributed through the bloodstream.
2. The nervous and endocrine systems maintain homeostasis.
3. Chemical and electrical signals are sent throughout the body. Chemical signals are available to all cells because they are distributed through the blood.
4. Homeostasis is the process of maintaining a relatively stable internal environment.
5. Parameters that are maintained homeostatically include blood pressure, body temperature, heart rate, and blood glucose concentrations.
6. A sensory receptor receives information about changes in the environment and relays that information to an integrating center. A membrane receptor allows a chemical ligand to bind to a cell and start a reaction. Any cell with the receptor for a chemical will be a target of that chemical, but without the proper receptor, a cell cannot respond.
7. The signal ligand or first messenger binds to a receptor which activates and changes intracellular effectors.
8. The three amplifier enzymes are (a) adenylyl cyclase, (b) guanylyl cyclase, and (c) phospholipase C.
9. Protein kinases add phosphates from ATP to the substrate.
10. Stimulus, receptor, afferent pathway, integrating center, efferent pathway, effector, response.
11. Central receptors are located within the central nervous system; peripheral receptors are associated with sensory neurons of the peripheral nervous system.
12. (a) 3   (b) 1   (c) 4   (d) 5   (e) 2
13. Receptors for signal pathways are found in the nucleus, cytosol, or cell membrane.
14. Daily fluctuations in a parameter are called circadian rhythms. These cycles arise in special cells in the brain.

15. Down-regulation results in a decreased sensitivity to a prolonged signal.
16. Down-regulation occurs when receptor number or receptor affinity for the substrate decreases.
17. In negative feedback, the effector moves the system in the opposite direction from the stimulus.

LEVEL TWO   REVIEWING CONCEPTS

18. (a) **Gap junctions** are formed when protein channels, called **connexons**, connect the cytoplasm of two cells. **Connexins** are the proteins that form the channels. When gap junctions are open, the cytoplasm of adjoining cells is continuous, forming a functional **syncytium**, or a unit that behaves like a single cell with multiple nuclei. Gap junctions can be found in the heart.
    (b) These are all chemicals secreted by cells into the extracellular space. **Paracrines** act on cells close to the cell that secreted them; **autocrines** act on the same cell that secretes them. Histamine is an example of a paracrine. **Cytokines** are peptides that can act as autocrines and paracrines or that can be distributed in the blood like **hormones**. Many growth factors are cytokines. **Neurocrines** are chemicals secreted by neurons, or nerve cells. Neurotransmitters, neurohormones, and neuromodulators are all neurocrines.
    (c) **Agonists** are molecules that have the same action as another molecule. Estrogens in birth control pills are examples of agonists. **Antagonists** are molecules whose action opposes that of another molecule. Antagonists to estrogens and androgens are used to treat hormone-dependent cancers.
    (d) **Transduction** is the process by which a signal molecule transfers its information from the extracellular fluid to the cytoplasm. Transduction often involves a **cascade**, or series, of steps, with **amplification** taking place at each step so that one signal molecule can result in a larger signal.
19. The four classes of membrane receptors are: (1) ligand-gated channels such as the ATP-gated K+ channel (2) integrin receptors, e.g., receptors of platelets (3) receptor enzymes such as tyrosine kinase receptors, and (4) G-protein-coupled receptors such as those that activate adenylyl cyclase and subsequently cAMP.
20 Walter Cannon was the father of American physiology. His four postulates are:
    (a) The nervous system keeps body functions such as blood pressure and body temperature within normal limits.
    (b) Some body functions are regulated in an up-and-down (or tonic) fashion rather than in an on-off fashion.
    (c) Some functions are controlled by two signals that act in opposition: one signal enhances the function while the other signal inhibits it.
    (d) The response of a cell to a particular chemical signal depends on the receptor that the cell has for that signal. Different cells can have different receptors and therefore exhibit different responses.
21. Stimulus: a change that begins a response (example: touching a hot stove).
    Receptor: a cell or structure that senses the change of the stimulus (example: temperature receptor).
    Afferent pathway: the means by which information about the stimulus is conveyed to an integrating center, an input signal (example: a sensory nerve).
    Integrating center: a cell or group of cells that receives incoming information, decides whether and how it should be acted upon, and sends a signal that begins a response (example: the brain).

Efferent pathway: the signal that goes from the integrating center to the cell or tissue that will carry out the response (example: a nerve or a hormone).

Effector: the cell or tissue that will carry out the response (example: a muscle).

Response: what the cell or tissue does to react to the change caused by the stimulus (example: pull your hand away from the hot stove).

22. In negative feedback, the signal tells the response to turn off. Positive feedback reinforces the stimulus and keeps the response going. Both positive and negative feedback happen after the response has taken place; a feedforward response gets the response loop started before the stimulus does. Positive feedback helps keep a parameter within a range for homeostasis. Feedforward helps prevent big changes in the regulated function. Positive feedback makes a change bigger and keeps it going once started by a stimulus.

23. The primary advantage of neural control over endocrine control is speed. The primary advantage of endocrine control is the ability to affect widely separated tissues with a single signal. Neural is better for short-acting responses, endocrine for long-acting.

24. (a) negative feedback
    (b) positive feedback
    (c) negative feedback
    (d) negative feedback

25. (a) Tissues respond to glucagon by increasing blood glucose.
    (b) Breast tissues secrete and release milk.
    (c) the urinary bladder
    (d) sweat glands

26. (a) endocrine cells that secrete glucagon (the endocrine pancreas)
    (b) insufficient information given in question. The hormone oxytocin controls milk release (letdown) and the hormone prolactin regulates milk secretion.
    (c) the nervous system
    (d) the nervous system

LEVEL THREE   PROBLEM SOLVING

27. (a) stimulus = cold wind causes body temperature to decrease; receptor = temperature receptors; afferent pathway = sensory nerve cells; integrating center = CNS; efferent pathway = efferent neurons; effectors = muscles used to pull an afghan around you; response = afghan conserves heat and causes body temperature to increase.
    (b) stimulus = smell of sticky buns; receptor = odor receptors in the nose; afferent path = sensory neurons; integrating center = CNS; efferent pathway = skeletal muscles; response = walk to bakery, buy, and devour buns.

LEVEL FOUR   QUANTITATIVE PROBLEMS

28. (a) This paragraph describes amplification and a cascade.
    (b) One rhodopsin creates (1000 × 4000) or 4,000,000 GMP.

CHAPTER 7

LEVEL ONE   REVIEWING FACTS AND TERMS

1. *Endocrinology* is the study of hormones.

2. Hormones alter the rate of enzymatic reactions, control transport of molecules into and out of the cell, or change gene expression and protein synthesis in their target cells.

3. See Figure 7-2 to check your answers.

4. (a) 4   (b) 5   (c) 1   (d) 2   (e) 3

5. d - b - c - a

6. Hormones are secreted into the *blood* for transport to a *distant target* and take effect at *very low* concentrations.

7. The half-life of a hormone is the amount of time required for half of a dose of hormone to disappear from the blood.

8. Hormones are broken down into their metabolites by the *kidneys* and *liver,* and the metabolites are excreted in the *urine* and *bile,* respectively.

9. Candidate hormones are often called *factors*.

10. Peptide hormones are made of three or more amino acids. Most hormones are peptides (see Figure 7-2). Steroid hormones are derived from cholesterol; examples include hormones made by the gonads, adrenal cortex, and placenta. Amine hormones are made from single amino acids; thyroid hormones, epinephrine, norepinephrine, and dopamine are examples.

11. (a) peptide hormones (some of the amines are lipophobic but some are lipophilic)
    (b) peptides and some amines
    (c) steroids
    (d) peptides and some amines
    (e) peptides
    (f) steroids
    (g) peptides
    (h) all classes
    (i) steroids and some amines
    (j) steroids

12. Steroids are not very water-soluble, or the carrier protects the hormone from being broken down and extends its half-life.

13. In the nucleus, steroid hormones act as a *transcription* factor, activating *genes* that control synthesis of new *proteins*.

14. Some steroid hormones also have receptors on the *cell membrane*.

15. Melatonin is made from *tryptophan,* and the other amines are made from *tyrosine*.

16. *Trophic* hormones control the secretion of other hormones.

17. In complex endocrine pathways, the hormones themselves may act as the *feedback* signals.

18. Neurosecretory hormones (or neurohormones) are synthesized and secreted from nerve cells or neurons.

19. The posterior pituitary secretes oxytocin and vasopressin, both peptide neurohormones.

20. The hypothalamic-hypophyseal portal system is composed of capillaries that pick up hormones secreted in the hypothalamus and deliver them directly to a set of capillaries in the adjacent anterior pituitary. The direct connection allows very small amounts of hypothalamic hormone to control the endocrine cells of the anterior pituitary.

21. Prolactin controls milk production by the breast and is not trophic. The remaining hormones are all trophic hormones. Growth hormone, or somatotropin, alters tissue metabolism and controls secretion of somatomedins. Corticotropin, or ACTH, controls cortisol secretion. Thyrotropin, or TSH, controls thyroid hormone secretion. Follicle-stimulating hormone (FSH) and luteinizing hormone (LH) control hormone secretion by the gonads as well as gamete production.

22. Long-loop negative feedback goes from a peripheral endocrine gland back to the pituitary and hypothalamus to shut off trophic hormone secretion. In short-loop negative feedback, the trophic hormone of the anterior pituitary has a negative feedback effect on the hypothalamus.

23. *Synergism* occurs when the effects of two hormones combined is greater than their individual effects added together. If two hormones must be present for a full effect to be demonstrated, the relationship is called *permissiveness*. If one hormone opposes the actions of another, the hormone is *antagonistic*.

LEVEL TWO   REVIEWING CONCEPTS

24. Use Figure 7-3 to help with the map for List 1 and use Figures 7-12 and 7-13 for List 2.

25. (a) All these chemicals are secreted by cells for actions on other cells. Paracrines act on adjacent cells; cytokines act on adjacent or distant cells; hormones act on distant cells. Cytokines are always peptides; hormones may be peptides, steroids, or amines. Cytokines are made on demand; peptide hormones are made in advance and stored in the endocrine cell.
    (b) A primary endocrine pathology arises in the last endocrine gland of a complex endocrine reflex. Secondary pathologies arise in a gland secreting a trophic hormone. Either pathology may result in too much or too little hormone being secreted.
    (c) Hypersecretion is excess secretion of hormone; hyposecretion is secretion of less-than-normal amounts of hormone.
    (d) Both are part of the pituitary gland and both secrete peptide hormones. The anterior pituitary gland is a true endocrine gland, but the posterior pituitary is neural tissue. The signal for posterior pituitary neurohormone release is an action potential. Anterior pituitary hormones are released in response to hypothalamic trophic hormones.

26. See Table 7-1.

LEVEL THREE   PROBLEM SOLVING

27. The meanings of the words given have not changed significantly. Endocrine receptors are membrane or intracellular proteins. Endocrine target cells may up- or down-regulate their responses. Specificity refers to the selective interaction between a protein and its ligand. Enzymes, hormone receptors, and transport proteins are all proteins that have a binding site in their structure.

28. In patient A, cortisol hypersecretion is a result of ACTH hypersecretion (a secondary pathology). When dexamethasone suppresses secretion of ACTH, the adrenal gland is no longer being stimulated and cortisol secretion decreases. Patient B has primary hypercortisolism, with the problem arising in the adrenal gland. His normal negative feedback pathways do not operate, and the adrenal gland continues oversecreting cortisol even though ACTH secretion has been suppressed by dexamethasone.

29. (a) See Figure 26-11.
    (b) Both LH and testosterone are necessary for gamete formation. Administering testosterone would not directly suppress gamete formation for birth control. However, testosterone does have a negative feedback effect on the anterior pituitary and would shut off LH secretion. Because LH is needed for gamete production, its absence would suppress gamete synthesis. Thus, administration of testosterone could be an effective means of birth control.

LEVEL FOUR   QUANTITATIVE PROBLEMS

30. Half-life is 3 hours.
31. (a) Group A   (b) Group B   (c) Group A
32. *x*-axis is plasma glucose concentration. *y*-axis is insulin secretion. As *X* increases, *Y* increases.

CHAPTER 8
LEVEL ONE   REVIEWING FACTS AND TERMS

1. Afferent neurons carry messages from sensory receptors to the CNS. Their cell bodies are located close to the CNS and long processes extend from the cell bodies to the receptors. Interneurons are completely contained within the CNS and are often extensively branched. Efferent neurons carry signals from the CNS to effectors. They tend to have short, branched dendrites and extended axons.

2. Somatic motor neurons control *skeletal muscles*. *Autonomic* neurons control smooth and cardiac muscles, glands, and some adipose tissue.

3. Autonomic neurons are classified as either *sympathetic* or *parasympathetic*.

4. (a) 3   (b) 1   (c) 2   (d) 5   (e) 4

5. Neurons and glial cells are the two primary cell types of the nervous system.

6. See Figures 8-2 and 8-4.

7. (c). Answer (b) is not correct because not all axonal transport uses microtubules and not all substances moved will be secreted.

8. (a) 1, 4   (b) 2, 3, 5, 6

9. d

10. The four major ion channels are (1) Na^+ channels, voltage-gated along the axon and ligand-gated or mechanically-gated Na^+ channels on dendrites, (2) voltage-gated K^+ channels along the axon, (3) voltage-gated Ca^{2+} channels in the axon terminal, and (4) chemically-gated Cl^- channels.

11. e - b - d - a - c

12. Answers b and d are correct.

13. The resting cell is more permeable to K^+ than Na^+. Although Na^+ contributes little to the resting potential, it is key to the action potential.

14. Na^+ is more concentrated outside the cell than inside.

15. K^+ is more concentrated inside the cell.

16. An action potential occurs when Na^+ enters the cell.

17. The resting membrane potential is due to the high K^+ permeability of the cell.

18. The myelin sheath is formed from cells that create multiple wraps of insulating membrane around neurons to prevents current from leaking out of the axon.

19. Factors that enhance conduction speed are the diameter of the axon and the presence or absence of insulating myelin.

20. Neurotransmitters are removed by enzymatic degradation, reabsorption and diffusion.

21. See Figures 8-16 and 8-18.

LEVEL TWO   REVIEWING CONCEPTS

22. See Figures 8-1 and 8-5.
23. Answer d is correct.
24. The PNS has two types of glial cells: Schwann cells and satellite cells, and the CNS has four types: oligodendrocytes, microglia, astrocytes, and ependymal cells. Oligodendrocytes and Schwann cells form myelin. Satellite cells form supportive capsules around nerve cell bodies located in ganglia. Astrocytes transfer nutrients between the blood and neurons. Microglia are immune cells. Ependymal cells separate the fluid compartments of the CNS.

25. f - c - g - e - b - k - c - a - h - i - d.
26. (a) depolarize
    (b) hyperpolarize
    (c) repolarize
    (d) depolarize. See Figure 5-37 on p. 166.
27. (a) Na^+ entry depolarizes;
    (b) K^+ leaving hyperpolarizes;
    (c) Cl^- entry hyperpolarizes;
    (d) Ca^{2+} entry depolarizes
28. (a) Threshold level signals trigger action potentials. Suprathreshold will also trigger action potentials but subthreshold will not unless they are summed through either spatial or temporal summation. Action potentials are all-or-none events that cannot be summed. Overshoot is the portion of the action potential above 0 mV. Undershoot is the after-hyperpolarization portion of the action potential (see Figure 8-12).
    (b) Graded potentials may be depolarizing or hyperpolarizing. If the graded potential occurs in a postsynaptic cell, the depolarizing graded potential is an excitatory postsynaptic potential (EPSP) and a hyperpolarizing graded potential is inhibitory, or an IPSP. Refractory periods are the periods of time following an action potential during which an additional action potential cannot be fired. No stimulus can trigger another action potential during the absolute refractory period but a suprathreshold stimulus can trigger an action potential during the relative refractory period.
    (c) See answer to question 1.
    (d) Sensory neurons are afferent neurons; all others are efferent neurons. For classification and comparison, see answer to question 1.
    (e) Fast synaptic potentials result from the opening of ion channels by neurotransmitters; they occur rapidly and are short-lived. Slow synaptic potentials are mediated through second messengers and last longer. They may include changes in the open states of ion channels but may also alter cellular proteins.
    (f) Temporal summation occurs when multiple stimuli arrive at the trigger zone close together in time. Spatial summation occurs when multiple stimuli arriving at the neuron at different locations arrive simultaneously at the trigger zone.
    (g) Divergence is the pattern where a single neuron branches and its collaterals synapse on multiple target neurons. Convergence is the pattern where many presynaptic neurons provide input to a smaller number of postsynaptic neurons.
29. Information about strength is coded by the frequency of action potentials; duration is coded by the duration of a train of repeated action potentials.
30. b

LEVEL THREE   PROBLEM SOLVING

31. Although the neurons are fully developed, all the necessary synapses have not yet been made between neurons and between neurons and effectors.
32. Closure of the voltage-gated Na^+ channels is also initiated by depolarization, but the inactivation gates close slower than the activation gates open, allowing ions to flow for a short period of time.
33. The correct answers are b, d, and h.
34. (a) thermal   (b) chemical   (c) chemical   (d) chemical
    (e) chemical   (f) chemical   (g) mechanical

LEVEL FOUR   QUANTITATIVE PROBLEMS

35. (a) An increase in sodium permeability will make alpha larger. That in turn will increase the numerator and make the membrane potential more positive.
    (b) Membrane potential is −78.5 mV.
36. (a) $(12 \times 2\ mV = +24) + (3 \times -3\ mV = -9)$ = signal strength of +15. Potential −70 + 15 = −55. Threshold is −50, so no action potential. (Potential must be equal to or more positive than threshold.)
    (b) $(11 \times 2 = +22) + (3 \times -3 = -9) = +13$. Potential −70 + 13 = −57. Threshold is −60, so action potential will fire.
    (c) $(14 \times 2 = +28) + (-9) = +19$. Potential −70 + 19 = −51. Threshold is −50, so no action potential.

CHAPTER 9

LEVEL ONE   REVIEWING FACTS AND TERMS

1. *Plasticity* is the ability to change connections in response to sensory input and experience.
2. *Cognitive* functions deal with thought, and *affective* functions deal with emotion.
3. The *cerebrum* allows human reasoning and cognition.
4. The brain is inside the *skull*, and the spinal cord is inside the *vertebral column*.
5. From the bones inward, the meninges are the dura mater, the arachnoid membrane, and the pia mater.
6. CSF physically protects the brain: its buoyancy reduces the brain's weight, and the fluid provides a cushion between the brain and the bones. CSF also provides chemical protection by creating a closely regulated external environment for the cells of the brain.
7. (a) HCO_3^- is lower in CSF
   (b) Ca^{2+} is lower in CSF
   (c) glucose is lower in CSF
   (d) H^+ is higher in CSF
   (e) Na^+ is the same in CSF and plasma
   (f) K^+ is lower in CSF
8. Neurons normally metabolize *glucose*. Low glucose in the blood is *hypoglycemia*. Neurons have high *oxygen* consumption, and they receive about *15%* of the blood pumped by the heart.
9. (a) 5   (b) 7   (c) 9   (d) 3   (e) 1   (f) 2   (g) 6
   (h) 8   (i) 4
10. The blood-brain barrier consists of capillaries that are much less leaky and much more selective than capillaries elsewhere in the body due to tight junctions between endothelial cells. The function of this barrier is to closely regulate which substances are allowed out of the blood and into brain tissue.
11. Gray matter has nerve cell bodies, dendrites, and axon terminals; it forms clusters (nuclei) or layers in the brain and spinal cord. Gray matter is where information passes from neuron to neuron. White matter is mostly myelinated axons; tracts of white matter carry information up and down the spinal cord to the brain.
12. (a) Sensory fields direct perception. (b) The motor cortex directs movement. (c) Association areas integrate information and direct voluntary behavior.
13. Cerebral lateralization is the asymmetrical distribution of function between the two lobes of the cerebrum. The left brain is the center for language and verbal functions, while the right brain is the center for spatial skills.

14. Gamma-aminobutyric acid (GABA) in the brain and glycine in the spinal cord are inhibitory amino acid neurotransmitters that open Cl^- channels and hyperpolarize their target cells.

15. REM, or rapid eye movement sleep, is the period of sleep when most dreaming takes place. It is marked by rapid, low-amplitude EEG waves, flaccid paralysis, and depression of homeostatic function. Slow-wave sleep, or deep sleep, is marked by high amplitude, low frequency waves on the EEG and unconscious body movements.

16. The hypothalamus contains centers for homeostasis. Reflexes and behaviors controlled by the hypothalamus include body temperature, thirst, reproductive functions, hunger and satiety, and influence of cardiovascular function. Emotional input to the hypothalamus comes from the limbic system.

17. The *amygdala* appears to be the center of basic instincts.

18. Learning can be categorized into associative and nonassociative learning. In habituation, a person responds less and less to a repeated stimulus; in sensitization, a person experiences an enhanced response to a dangerous or unpleasant stimulus.

19. Integration of spoken language takes place in Broca's area and Wernicke's area of the cortex.

LEVEL TWO   REVIEWING CONCEPTS

20. The map should include information contained in Table 9-1 and Figures 9-15 and 9-16.

21. Cerebrospinal fluid is secreted into the ventricles and flows into the subarachnoid space around the brain and spinal cord before being reabsorbed by special regions of the cerebral arachnoid membrane.

22. The sensory system, behavioral state system, and cognitive system regulate motor output by the CNS.

23. The receptor for the molecule determines whether it acts as a neurotransmitter (opens ion channels) or a neuromodulator (acts through a second messenger).

24. (a) The diffuse modulatory systems in the brain stem influence attention, motivation, wakefulness, memory, motor control, mood, and metabolism. The reticular formation in the brain stem influences arousal and sleep, muscle tone, breathing, blood pressure, and pain. The limbic system is the interior region of the cerebrum and links higher cognitive functions with more primitive emotions such as fear.

    (b) Memory is divided into short-term memory that disappears unless consolidated, and long-term memory that is stored for recall. Long-term memory includes reflexive, or unconscious, memory, and declarative, or conscious, memory.

    (c) Nuclei are clusters of nerve cell bodies in the CNS; ganglia (in vertebrates) are clusters of nerve cell bodies outside the CNS.

    (d) Tracts are bundles of axons within the CNS; nerves are bundles of axons outside the CNS. Horns are extensions of spinal cord gray matter that connect to peripheral nerves. Nerve fibers are bundles of axons. Roots are the two branches of a peripheral nerve that enter the spinal cord. Sensory neurons go to the dorsal root and efferent neurons exit the spinal cord through the ventral root.

25. Primary somatic sensory cortex is in the parietal lobe. The visual cortex processes information from the eyes. The auditory cortex processes information from the ears. The olfactory cortex processes information from the nose. The motor cortices in the frontal lobes control skeletal muscle movements. The association areas integrate sensory information into perception.

26. (a) Lower frequency waves would have peaks farther apart.
    (b) Larger amplitude waves would have taller peaks.
    (c) Higher frequency waves would have peaks closer together.

27. Those motivational states known as drives all create increased state of arousal, goal-oriented behavior, and coordination of disparate behaviors to achieve goals.

28. Changes that take place at synapses as memories are formed include up-regulation of receptors and formation of new synapses.

LEVEL THREE   PROBLEM SOLVING

29. With expressive aphasia Mr. Anderson could understand people but was unable to communicate in any way that made sense. He might be able to select the correct words but cannot put them together appropriately. If he had receptive aphasia, he would be unable to understand communication from other people. Speech centers are in the left brain. If music centers are in the right brain, then perhaps information from Wernicke's area can be integrated by the right brain so that Mr. Anderson can musically string together words so that they make sense.

30. Learning probably occurred, but learning need not be translated into behavioral responses. The participants who didn't buckle their seat belts learned that wearing seat belts was important but did not consider this knowledge important enough to act on.

31. One conclusion might be that sleep-deprived dogs are producing a substance that induces sleep. Possible controls would be putting CSF from normal dogs into sleep-deprived dogs, CSF from normal dogs into normal dogs, and CSF from sleep-deprived dogs into other sleep-deprived dogs.

32. (1) No, other information that should be taken into consideration include genetics, age, and general health. (2) The application of this study would be limited to women of similar age, background, and health. Other factors you would be interested in would include the ethnicity of the participants, factors as listed in (1), and geographical location.

CHAPTER 10

LEVEL ONE   REVIEWING FACTS AND TERMS

1. The afferent division carries information from sensory receptors to the CNS.

2. Proprioception is the ability to tell where our body is in space and to sense the relative locations of different body parts.

3. All sensory pathways include a stimulus that acts on a sensory receptor and a sensory neuron that begins at the receptor and ends in the CNS.

4. Mechanoreceptors respond when they are physically deformed by pressure, sound, stretch, etc. Chemoreceptors respond to the presence of or changes in the concentration of specific chemicals, such as glucose or oxygen. Photoreceptors respond to photons of light. Thermoreceptors respond to varying degrees of heat.

5. Information is sensed within the *sensory field* of a neuron.

6. (a) 3   (b) 2   (c) 1, 2   (d) 2, 3   (e) 4

7. *Transduction* is the process by which stimulus energy is converted into a change in membrane potential. The form of energy to which a receptor responds best is called its *adequate stimulus*. The minimum stimulus required to activate a receptor is known as the *threshold*.

8. When a sensory receptor membrane depolarizes (or hyperpolarizes in a few cases), the change in membrane potential is called the *receptor potential*. These are graded potentials.

9. The adequate stimulus for a receptor is the form of energy to which the receptor is most sensitive.

10. The organization of sensory regions in the *cortex* preserves the topographical organization of sensory receptors so that perceptions are localized to the area where the stimulus occurred. The two exceptions to this rule are olfaction and hearing, where the brain uses timing of stimulus reception to compute a location.

11. Lateral inhibition occurs when the sensory neurons surrounding a sensory field are inhibited, which enhances the contrast between the stimulus and surrounding unstimulated areas.

12. Tonic receptors adapt slowly to stimuli and respond to stimuli that need to be constantly monitored, such as noxious stimuli. Phasic receptors adapt rapidly to stimuli and stop responding unless the stimulus changes. An example is smell. Phasic receptors allow new stimuli to be sensed while ignoring old stimuli that are presumably not important or a threat.

13. Pain in the arm from cardiac ischemia is an example of *referred pain*.

14. Sweet, salty, bitter, sour, and umami are the five basic taste sensations. Nutritious foods taste sweet (e.g., fruit) or umami (e.g., meat) or salty (e.g., salted popcorn); sour foods may be either nutritious (e.g., juice) or unsafe due to spoilage; bitter foods may contain toxins or may be nutritious (unsweetened chocolate).

15. The frequency of the sound waves per second is measured in *hertz (Hz)*. The loudness or intensity of sound is a function of the *wave amplitude* and is measured in *decibels (dB)*. The range of hearing for the average human ear is over the range of *20–20,000 Hz*, with the most acute hearing in the range of *1000–3000 Hz*.

16. The basilar membrane codes for pitch. Spatial coding refers to the translation of different wave frequencies into stimulation of different areas of the membrane.

17. Correct answer is a. (Action potentials have constant amplitude and refractory period).

18. Once sound waves have been transformed into electrical signals in the cochlea, sensory neurons transfer information to the *medulla,* with collaterals then taking information to the *reticular formation* and *cerebellum*. The main auditory pathway synapses in the *midbrain* and *thalamus* before finally projecting to the *auditory cortex* in the *cerebrum*.

19. The *semicircular canals* sense rotation, and the *otolith organs* respond to linear forces. Movement is the *dynamic* component, and head position while standing is the *static* component.

20. b, a, d, c, e

21. The three primary colors of vision are *red, blue,* and *green*. Seeing these colors stimulate photoreceptors called *cones*. Lack of the ability to distinguish some colors is called *color-blindness*.

22. The six types of retinal cells are rods and cones (the photoreceptors), bipolar cells, ganglion cells, horizontal cells, and amacrine cells. Photoreceptors transduce light energy into electrical signals. Signal processing takes place in the bipolar neurons and ganglion cells, modulated by input from horizontal and amacrine cells.

LEVEL TWO   REVIEWING CONCEPTS

23. (a) Special senses (hearing, vision, smell, taste, and equilibrium) have receptors localized in the head. Somatic senses are more general senses (touch, pressure, vibration, temperature) with receptors are located all over the body.
    (b) See Table 10-4.
    (c) Sharp pain is transmitted by small, myelinated Aδ fibers. Dull, diffuse pain is transmitted through small, unmyelinated C fibers.
    (d) In conductive hearing loss, sound cannot be transmitted through the external or middle ear. In sensorineural hearing loss, the inner ear has been damaged. In central hearing loss, the auditory pathways from the inner ear or the auditory cortex have sustained damage.

24. The brain can distinguish up to seven distant sensory areas: 1, 2, 3, 1 + 2, 1 + 3, 2 + 3, and 1 + 2 + 3.

25. Ascending pathways for pain go to the limbic system and hypothalamus, which explains the link between pain, emotional distress, and symptoms such as nausea and vomiting.

26. Olfaction begins with the nasal olfactory receptors whose axons project through the cribriform plate of the ethmoid bone to the olfactory bulbs of the brain, where they synapse with secondary sensory neurons. Secondary and higher-order neurons project to the olfactory cortex. Parallel pathways go to the amygdala and hippocampus. G_{olf} is a special G protein in olfactory receptors, which, when activated by an odorant, increases intracellular cAMP. cAMP opens ion channels, leading to depolarization of the cell, and initiating the olfactory signal into the brain.

27. Bitter, sweet, and umami flavor chemicals bind to membrane receptors on taste buds, activating second messenger systems. Different G protein-linked receptors are involved for the three ligands, activating several different signal transduction pathways. Some trigger release of intracellular calcium while others open calcium channels. Salty and sour tastes are really just the ions Na^+ and H^+; these flavor ions enter the receptor through ion channels. Entry of these ions into the cell depolarizes the membrane potential, which in turn opens voltage-gated calcium channels. Calcium triggers exocytosis of neurotransmitter, which passes the signal to the next neuron in the pathway.

28. a, g, j, h, c, e, i, b, f, d

29. See Figures 10-23, 10-24, and 10-25.

30. In accommodation, the lens changes shape due to contraction/relaxation of the ciliary muscles. The change in lens shape keeps light focused on the retina as objects move toward and away from the eye. Loss of this reflex is called presbyopia.

31. Presbyopia is loss of accommodation due to stiffening of the lens with age. Myopia or near-sightedness is due to a longer-than-normal distance between lens and retina; hyperopia or far-sightedness is due to a shorter-than-normal distance. Color blindness is caused by a defect in the visual pigments of some of the cones.

32. Intensity can be coded by the frequency of action potentials, and duration by the duration of a train of action potentials.

33. Start with Figure 10-28 and the basic components of vision. Work in details and related terms from the text.

34. See Table 10-1 and the section for each special sense.

LEVEL THREE   PROBLEM SOLVING

35. You are testing the somatic sense of touch-pressure, mediated through free nerve endings and Merkel receptors. When she only feels one probe, it is because both needles are within the same receptive field.

36. Some well-known tests include having people try to walk a straight line or stand on one leg with the eyes closed. Other tests include higher brain functions, such as reciting the alphabet backwards.

37. The first sense to test for a child with poor speech is hearing. If children cannot hear well, they cannot hear the sounds to imitate for speech.

38. Normally, shining a light in either eye should induce pupillary constriction in both eyes (the consensual reflex). Given the normal reflex upon shining light into the right eye, it appears that the motor components of the reflex to both eyes are intact. Absence of the reflex upon stimulating the left eye suggests damage to the left retina and/or to the left optic nerve.

CHAPTER 11
LEVEL ONE   REVIEWING FACTS AND TERMS

1. The two divisions of the efferent part of the autonomic nervous system are the somatic motor division, which controls skeletal muscles, and the autonomic division, which controls smooth muscle, cardiac muscle, glands, and some adipose tissue.

2. The autonomic nervous system can also be called the *visceral* nervous system because it controls the internal organs, or viscera. Some functions controlled by the autonomic nervous system include heart rate, blood pressure, and digestive function.

3. The two divisions of the autonomic nervous system are the sympathetic and parasympathetic divisions. Anatomically, sympathetic neurons exit the spinal cord in the thoracic and lumbar regions and have ganglia close to the spinal cord; parasympathetic neurons exit from the brain stem or sacral region of the spinal cord and have ganglia on or close to their targets. The neurotransmitter released onto the target is norepinephrine for the sympathetic and acetylcholine for the parasympathetic divisions. Physiologically, sympathetic innervation controls fight-or-flight reactions while parasympathetic is in charge of rest-and-digest functions.

4. The adrenal medulla is a neuroendocrine gland that is considered to be a modified sympathetic ganglion. It secretes epinephrine primarily, along with a small amount of norepinephrine.

5. Neurons that secrete acetylcholine are described as *cholinergic;* those that secrete norepinephrine are called *adrenergic* or *noradrenergic* neurons.

6. Four things that can happen to autonomic neurotransmitters after they are released include diffusion away from the synapse, breakdown by enzymes in the synapse, reuptake into the presynaptic neuron, or combination with a membrane receptor.

7. The main enzyme responsible for catecholamine degradation is *monoamine oxidase,* abbreviated *MAO.*

8. Somatic motor pathways are excitatory, composed of a single neuron, and synapse with skeletal muscles.

9. Acetylcholinesterase is the enzyme that breaks down ACh in the synapse.

10. Nicotinic cholinergic receptors are found on the postsynaptic cell of the neuromuscular junction.

LEVEL TWO   REVIEWING CONCEPTS

11. Use the information in Figures 11-10 and 11-11 to create this map.

12. Divergence of neural pathways in the autonomic nervous system allows a single signal to have effects on multiple targets.

13. (a) Neuroeffector junctions occur along the length of the autonomic axon anywhere that there is a varicosity; neuromuscular junctions (NMJs) of the somatic nervous system occur at the axon terminals of the somatic motor neuron. Transmitter release and activity in the synapse are similar at both types of junctions.
    (b) Alpha and beta receptors are adrenergic; nicotinic and muscarinic are cholinergic. Nicotinic are the only receptors found in the somatic motor division; autonomic divisions all have

nicotinic receptors on the postganglionic neuron. Adrenergic and muscarinic receptors are found on autonomic targets.

14. (a) Autonomic ganglia contain the nerve cell bodies of postganglionic autonomic neurons. CNS nuclei are clusters of nerve cell bodies in the brain and spinal cord.
    (b) The adrenal and pituitary glands are complex glands composed of both true endocrine tissue (adrenal cortex and anterior pituitary) and neuroendocrine tissue (adrenal medulla and posterior pituitary).
    (c) Boutons are found at the ends of axons; varicosities are areas of neurotransmitter release strung out like beads along the length of the branched ends of autonomic neurons.

15. (a) 1, 2   (b) 3   (c) 4   (d) 3

16. (d), (e)

LEVEL THREE   PROBLEM SOLVING

17. The electrochemical gradient for Na^+ is greater than that for K^+.

18. (a) Endocytosis probably brings the marker into the axon terminal.
    (b) Parasympathetic autonomic ganglia are located close to the target organ.
    (c) Postganglionic parasympathetic neurons secrete acetylcholine onto their targets.

19. Any animal poisoned by curare will experience a loss in function of skeletal muscles. The animal would become paralyzed and unable to move its limbs, so it could not flee. Respiratory muscles will also be affected, and if the dose of curare is sufficient, the animal will die by suffocation.

LEVEL FOUR   QUANTITATIVE PROBLEMS

20. Cigarette smoking among high school students increased from 1991 to 1997, then began to decrease. (b) Males are more likely to be smokers than females, and whites smoke more than Hispanics, who smoke more than blacks. Therefore, white males are the most likely to smoke and black females are the least likely to smoke.

CHAPTER 12
LEVEL ONE   REVIEWING FACTS AND TERMS

1. The three types of muscle are smooth, cardiac, and skeletal. Skeletal muscles are attached to bones.

2. The two striated muscles are cardiac and skeletal muscle.

3. Skeletal muscle is controlled strictly by somatic motor neurons.

4. (a) False, muscles comprise about 40% of body weight.
   (b) true
   (c) true
   (d) true

5. Skeletal muscle components are connective tissue sheath, sarcolemma, myofibrils, and myofilaments.

6. Modified endoplasmic reticulum of skeletal muscle is called *sarcoplasmic reticulum.* It stores and concentrates Ca^{2+} ions.

7. T-tubules allow *action potentials* to travel to the interior of the muscle fiber.

8. Myofibrils are made of actin, myosin, troponin, tropomyosin, titin and nebulin. Myosin produces the power stroke, by binding to and pulling on actin.

9. Z disk forms the boundaries of a sarcomere. I band has a Z disk in the middle; A band consists of the thick filaments; H zone is the lighter region of the A band; M line, which divides the A band in half, is where thick filaments link to each other.

10. Titin and nebulin keep myofilaments in alignment. In addition, titin is elastic and helps stretched muscles return to their resting length.

11. During muscle contraction, the *A band* (myosin) remains a constant length. The H zone and I band shorten during contraction, and the Z disks approach each other.

12. The sliding filament theory states that a muscle cell contracts by the increasing overlap of thin and thick filaments in each sarcomere of each myofibril. Myosin heads on the thick filament bind to actin. When myosin heads bend, they pull the actin filaments toward the center of the sarcomere.

13. Fast-twitch glycolytic fibers: a, b, e; Fast-twitch oxidative fibers: g, f, d, e; Slow-twitch oxidative fibers: c, d, f, h

14. An increase in intracellular free Ca^{2+} activates troponin, which repositions tropomyosin, uncovering actin's myosin-binding sites.

15. Somatic motor neurons release acetylcholine.

16. The motor end plate is the region of a muscle fiber where the synapse occurs. It contains ACh receptors. Influx of Na^+ at the motor end plate causes the cell to depolarize and fire an action potential.

17. A single contraction/relaxation cycle in a skeletal muscle fiber is known as a *twitch.*

18. ATP binding to myosin causes myosin to dissociate from actin. ATP hydrolysis causes the myosin head to swing and bind to a new actin molecule. Release of the inorganic phosphate initiates the power stroke.

19. The basic unit of contraction in an intact skeletal muscle is the *motor unit.* The force of contraction within a skeletal muscle is increased by *recruitment* of additional motor units.

20. The two types are single-unit (visceral) and multi-unit smooth muscle.

LEVEL TWO   REVIEWING CONCEPTS

21. Use Figures 12-3 to 12-6 to help construct this map.

22. Use Figures 12-9 to 12-11 to help construct this map.

23. The action potential activates DHP receptors that open Ca^{2+} channels in the sarcoplasmic reticulum membrane. Calcium then diffuses out of the sarcoplasmic reticulum.

24. Muscle cells can make some ATP by energy transfer from phosphocreatine. Oxidative fibers that require oxygen can use glucose and fatty acids to make ATP; glycolytic fibers get their ATP primarily from anaerobic glycolysis.

25. Muscle fatigue is a state in which a muscle can no longer generate or sustain the expected force. The physical cause of fatigue is unclear and may involve changes in ion concentrations, depletion of nutrients, or effects of these on excitation-contraction coupling. Central fatigue is a subjective feeling of fatigue and may be a protective mechanism. Muscle cells can become more fatigue-resistant by increasing the size and number of mitochondria and by increasing blood supply to the muscle.

26. To vary force, the body uses different types of motor units and recruits different numbers of motor units. Small movements use motor units with fewer muscle fibers; gross movements use motor units with more fibers.

27. See Table 12-3.

28. The sarcoplasmic reticulum stores Ca^{2+} and releases it upon command. The Ca^{2+} for smooth muscle contraction enters the cell from the extracellular fluid.

29. (a) Fast-twitch oxidative-glycolytic are smaller, contain some myoglobin, use both oxidative and glycolytic metabolism for ATP production, and are more fatigue-resistant. Fast-twitch glycolytic fibers are largest in diameter, rely primarily on anaerobic glycolysis for ATP synthesis, and are least fatigue-resistant. Slow-twitch fibers develop tension more slowly and maintain tension longer, are the most fatigue-resistant, depend primarily on oxidative phosphorylation for ATP production, have more mitochondria, greater vascularity, large amounts of myoglobin, and are smallest in diameter.

(b) A single contraction-relaxation cycle is a twitch. Tetanus is a contraction with little to no relaxation, which occurs at high stimulus rates.

(c) Action potentials in both motor neurons and skeletal muscle cells are the result of inward sodium current and outward potassium current through voltage-gated ion channels. In a motor neuron the action potential triggers ACh release, which ultimately leads to muscle contraction. The muscle action potential triggers Ca^{2+} release from the sarcoplasmic reticulum; Ca^{2+} in turn triggers contraction.

(d) In a motor neuron, temporal summation determines whether or not the neuron will produce an action potential. Summation in a muscle cell occurs when the peak force increases as the stimulation rate increases. This summation results as the concentration of free calcium inside the cell increases, thus more cross bridges can form and more force can be produced.

(e) Isotonic contraction moves a load. Isometric contraction creates tension without moving a load. In most movements, there is an isotonic phase and an isometric phase. Concentric action occurs when a muscle shortens in a controlled movement of a load. Eccentric action occurs when a muscle lengthens in a controlled movement of a load. Doing a biceps curl would involve alternating concentric and eccentric actions.

(f) Slow-wave potentials are cycles of depolarization and repolarization of membrane potential that occur in smooth muscle cells. Pacemaker potentials are repetitive slow depolarizations to threshold that occur in some types of smooth muscle cells and in cardiac muscle.

(g) In skeletal muscle, Ca^{2+} from internal stores regulates contraction. In smooth muscle, Ca^{2+} for contraction comes from both internal stores and extracellular fluid.

30. Release of Ca^{2+} from the SR of smooth muscle involves an IP_3-activated channel. Ca^{2+} influx from the ECF occurs through channels that open in response to depolarization, stretch, or chemical signals. Depolarization results from slow-wave or pacemaker potentials. Stretch results from increasing volume of contents of the hollow organ whose wall contains smooth muscle, for example, the urinary bladder. Autonomic neurotransmitters, hormones, and paracrines are important chemical factors.

LEVEL THREE   PROBLEM SOLVING

31. (a) Adding ATP will allow the cross bridges to detach, and myosin will be able to unbind from actin. If insufficient Ca^{2+} is available, the muscle will relax.

(b) If both ATP and Ca^{2+} are available, the muscle will continue in the contraction cycle until it is completely contracted. The muscle will remain contracted as long as free calcium is present in sufficient concentrations.

32. Because curare does not interfere with secretion of acetylcholine but does prevent contraction of skeletal muscle, it must interfere with a process that follows ACh release. These include diffusion of

ACh across the synaptic cleft, ACh binding to muscle fiber receptors, opening of the receptor-linked ion channels, and ensuing ion flux to produce the muscle action potential. To interfere with ion flow, curare would either have to bind to a significant number of ions, or block ion channels. Curare has been shown to bind to the acetylcholine receptor, thereby preventing ACh activation of the associated channels.

33. The muscle characteristic most dependent upon an individual's height is muscle length, because it is related to bone length. Assuming these athletes are lean, differences in weight and therefore muscle mass are correlated with muscle strength, so heavier athletes should have stronger muscles. More important factors are the relative endurance and strength required for a given sport. Any given muscle will have a combination of three fiber types, with the exact ratios depending upon genetics and specific type of athletic training.

    (a) Running up and down the basketball court requires both endurance, because of the duration of the game, and strength, because players must run fast. Precisely passing or shooting the ball requires variable strength dependent upon the distance the ball must move, as the ball itself represents a minimal load. Jumping up to the basket while shooting the ball requires strength, as the load is the entire mass of the athlete, but the activity is not sustained, therefore endurance is not an issue here. The leg muscles would have a greater proportion of fast-twitch glycolytic fibers, which generate the most strength for jumping, and fast-twitch oxidative, which have more endurance for running. The arm and shoulder muscles will have a higher proportion of fast-twitch glycolytic, because shooting requires fast and precise contraction, and great force when shooting over a distance.

    (b) A steer wrestler requires great strength in both the upper and lower limbs and trunk, but the duration of the activity is relatively short. The muscles will have a greater proportion of fast-twitch glycolytic fibers.

    (c) Female figure skaters require both strength and endurance. The legs do most of the work, producing great force during jumps, and sustaining activity while traversing the ice. The trunk muscles must hold the skater in precise positions. The trunk muscles will have a higher proportion of slow-twitch oxidative fibers for endurance. The leg muscles will have a higher proportion of fast-twitch oxidative, for moving across the ice, and fast-twitch glycolytic, for powering jumps.

    (d) A gymnastics routine involves great strength in arms and legs, and great endurance in trunk and limb muscles. Arm and leg muscles will have higher proportions of fast-twitch glycolytic fibers, to generate strength on the balance beam, uneven parallel bars, and floor exercises. Slow-twitch oxidative fibers in limb and trunk muscles will maintain the precise positioning of the body during all routines.

LEVEL FOUR   QUANTITATIVE PROBLEMS

34. The data suggest that muscle fatigue may result from lactate accumulation or loss of PCr. To read the original paper, go to *http://jap.physiology.org* and use the SEARCH function to find the paper.

35. (a) The biceps would need to exert 7.5 kg of force to hold the arm stationary, a 125% increase over a biceps inserted at 5 cm.

    (b) With a 7-kg weight at 20 cm, the biceps needs to exert an additional 28 kg of force. This is less than if the weight is placed in the hand.

CHAPTER 13

LEVEL ONE   REVIEWING FACTS AND TERMS

1. All neural reflexes begin with a *stimulus*.

2. Somatic reflexes involve *skeletal* muscles; *autonomic*, or visceral, reflexes are controlled by autonomic neurons.

3. The pathway pattern that brings information from many neurons into a smaller number of neurons is *convergence*.

4. If a presynaptic modulatory neuron decreases the amount of neurotransmitter released by its target neuron, the modulation is termed *presynatic inhibition*.

5. Autonomic reflexes are also called *visceral* reflexes because many of them involve the internal organs known collectively as the viscera.

6. Spinal reflexes include urination and defecation. Cranial reflexes include control of heart rate, blood pressure, and body temperature.

7. The limbic system is involved in the conversion of emotion into visceral reflexes. Some reflexes that may be influenced by emotion include heart rate, gastrointestinal function, and breathing rate.

8. The simplest autonomic reflex has two neuron-neuron synapses and one neuron-target synapse. The neuron-neuron synapses are in the spinal cord between the sensory neuron and the preganglionic efferent neuron, and in the autonomic ganglia, between the pre- and postganglionic neurons.

9. The three types of sensory receptors for muscle reflexes are the Golgi tendon organ, the muscle spindle, and the proprioreceptors called joint capsule mechanoreceptors.

10. Even at rest, muscles maintain *tone*.

11. Increase. This reflex is useful because it prevents damage from overstretching.

12. (a) 2, 3, 5, 6   (b) 1, 2, 6   (c) 1, 2, 4

13. The Golgi tendon organ responds to both *stretch* and *contraction* but more strongly to contraction. Activation *decreases* muscle contraction via inhibition of the *alpha motor* neuron.

14. The simplest reflexes are skeletal muscle reflexes, with one synapse between two neurons (monosynaptic). The knee jerk (patellar tendon) reflex is an example.

15. The three types of movement are reflex movements, such as the knee jerk; voluntary movements, such as playing the piano; and rhythmic movements, such as walking. Reflex movements are integrated in the spinal cord; the other two types must involve the brain at some point. Reflex movements are involuntary; rhythmic movements are initiated and terminated by signals from the cerebral cortex but in between are involuntary.

LEVEL TWO   REVIEWING CONCEPTS

16. Alpha-gamma coactivation allows muscle spindles to continue functioning when the muscle contracts and would otherwise release tension on the spindle. When the muscle contracts (under control of alpha motor neurons), the ends of the spindles also contract (under the control of gamma motor neurons) to maintain stretch on the central portion of the spindle.

17. If M is an inhibitory neuron, neurotransmitter release by P will decrease when M's neurotransmitter hyperpolarizes the postsynaptic membrane potential of P.

18. (a) The patellar reflex is used to assess function of the local reflex arc and the central nervous system components that regulate limb movement. A reduced or absent reflex could indicate damage to the quadriceps muscle, the nerves that control it, or

the area of the spinal cord where the nerves originate. A hyperactive reflex could indicate damage to the spinal cord at a level above the reflex arc, or damage to motor areas of the brain.

(b) This reflex is fairly easy to inhibit by conscious intent or if nervous, so the reflex would probably be less apparent. The origin of this inhibition is the primary motor cortex. The inhibitory cells will produce IPSPs in the spinal motor neuron.

(c) If the brain is distracted by some other task, the inhibitory signals will presumably stop, and the physician can more easily assess the reflex.

LEVEL THREE   PROBLEM SOLVING

19. (a) Ca^{2+}-activated transmitter release is prevented.
(b) Cell would hyperpolarize due to Cl^- entry, and voltage-gated Ca^{2+} channels in terminal would not open.
(c) Same answer as (b), but because K^+ would leave cell.

20. See Figures 13-12 and 13-13. Parts of the brain include the brain stem, cerebellum, basal ganglia, thalamus, cerebral cortex (visual cortex, association areas, motor cortex).

21. (a) Fright activates the sympathetic nervous system, which regulates a multitude of physiological functions, including the heart rate, piloerector muscles of the skin, and other startle reflexes.

(b) The limbic system is involved in emotions, including fear. Functions of the limbic system include regulating other primitive drives such as sex, rage, aggression and hunger, and reflexes including urination, defecation, blushing, blanching, and piloerection. The limbic system influences primarily autonomic motor output. Heart, blood vessels, respiratory muscles, smooth muscle, and glands are some of the target organs involved.

(c) Piloerection occurs because there are smooth muscles called arrector pili muscles that attach to the base of each hair. Smooth muscles are controlled by the autonomic division.

22. Both toxins are produced by bacteria of the Clostridium genus. *Clostridium tetani* are in soil, and can enter the body through a cut. *Clostridium botulini* can be present in improperly canned food, and enter the body upon ingestion. Both toxins produce paralysis (lack of control) of skeletal muscles. Tetanus toxin inhibits secretion of glycine from interneurons that normally inhibit somatic motor neurons. This releases the neurons from inhibition, so they trigger prolonged contractions in skeletal muscles; this is called spastic paralysis. Botulinum toxin blocks secretion of acetylcholine from somatic motor neurons. Skeletal muscles do not contract unless acetylcholine is present, so this toxin produces flaccid paralysis.

CHAPTER 14

LEVEL ONE   REVIEWING FACTS AND TERMS

1. (a) William Harvey was the first European to describe the closed circulatory system.
(b) Frank and Starling described the relationship between stretch of ventricular muscle and the force of its contraction.
(c) Malpighi described capillaries.

2. Three functions of the cardiovascular system are transport of materials entering and leaving the body, defense, and cell-to-cell communication.

3. a, e, d, b, f, c

4. The primary reason blood flows is a *pressure* gradient. In humans, this value is highest at the *left ventricle* and in the *aorta*; it is lowest

in the *right atrium*. In a system in which fluid is flowing, pressure decreases over distance due to *friction*.

5. If vasodilation occurs in a blood vessel, pressure *decreases*.

6. Cell junctions between myocardial cells are *intercalated disks* that contain *gap junctions*.

7. SA node to internodal pathways of atrial conducting system to AV node to bundle of His (left and right branches) to Purkinje fibers to ventricular myocardium.

8. (a) End-systolic volume is the volume of blood left in ventricle after heart contracts; end-diastolic is volume in the ventricle as heart begin to contract.
(b) Sympathetic control (epinephrine or norepinephrine on β_1-adrenergic receptors) increases heart rate. Parasympathetic control (ACh on muscarinic receptors) decreases heart rate.
(c) Diastole is the relaxation phase of the heart; systole is the contraction phase.
(d) Pulmonary circulation goes to the lungs; systemic circulation goes to the rest of the body.
(e) Action potentials pass from the SA node, a group of autorhythmic cells at the "top" of the right atrium, to the AV node, which are autorhythmic cells in the floor of the right atrium.

9. (a) 11   (b) 12   (c) 3   (d) 14   (e) 8   (f) 1
(g) 10   (h) 2   (i) 6   (j) 4.
See chapter for definitions of unused terms.

10. Vibrations from AV closure cause the "lub" sound and semilunar valve closure causes the "dup" sound.

11. (a) heart rate
(b) end-diastolic volume
(c) stroke volume
(d) cardiac output
(e) blood volume

LEVEL TWO   REVIEWING CONCEPTS

12. (a) Refer to Figure 14-1.
(b) Use Figures 14-31 and 14-27 as a starting point for your map.

13. See Figures 14-24 and 14-26.

14. See Table 12-3. The unique properties of cardiac muscle that are essential include the strong cell-to-cell junctions, gap junctions for electrical conduction, and the modification of some muscle cells into autorhythmic cells for rapid conduction and acting as a pacemaker

15. Contractions in cardiac muscle cannot sum or exhibit tetanus because of the long refractory period that prevents a new action potential until the heart muscle has relaxed.

16. See Figure 14-21. Atrial relaxation and ventricular contraction overlap during the QRS complex. The mechanical events following the waves are either contraction or relaxation of the heart muscle.

17. (a) 3; 5 in the last part   (b) 5   (c) 3   (d) 5   (e) 2
(f) 2   (g) 5   (h) 6

18. An ECG provides information about heart rate, heart rhythm (regular or irregular), conduction velocity, and the electrical condition of heart tissue. By implication only, it suggests what mechanical events are occurring. An ECG does not give any direct information about force of contraction.

19. An inotropic effect is an effect on force of cardiac contraction. Norepinephrine and cardiac glycosides such as digitalis have a positive inotropic effect.

LEVEL THREE   PROBLEM SOLVING

20. Calcium channel blockers decrease cardiac output by slowing heart rate and decrease force of contraction by blocking Ca^{2+} entry and decreasing Ca^{2+}-induced Ca^{2+} release. Beta blockers decrease cardiac output by blocking increased heart rate and force of contraction by norepinephrine and epinephrine.

21. (a) Cpt. Jeffers' heart muscle has been damaged by lack of oxygen and the cells are unable to contract as strongly. Thus, less blood is being pumped out of the ventricle each time the ventricle contracts.
    (b) Leads are two recording electrodes placed on the surface of the body to measure electrical activity.
    (c) Leads are effective because electricity is conducted through body fluids to the skin surface.

22. A longer than normal P-R interval in an ECG might result from a conduction problem at the AV node or in the ventricular conduction system.

23. A rapid atrial depolarization rate is dangerous because it can cause a rapid ventricular rate. Also, if the rate is too fast, not all action potentials can be conducted to the ventricles due to the refractory period of the heart muscle. This can result in an arrhythmia. The AV node is destroyed to prevent the rapid atrial signals from being passed to the ventricles, and the ventricular pacemaker is implanted so that the ventricles have an electrical signal that tells them to contract at an appropriate rate.

LEVEL FOUR   QUANTITATIVE PROBLEMS

24. SV/EDV = 0.25. If SV = 40 mL, EDV = 160 mL. SV = EDV − ESV, so 40 mL = 160 mL − ESV, therefore ESV = 120 mL. CO = HR × SV = 100 bpm × 40 mL/beat = 4000 mL/min or 4 L/min.

25. (a) 162.2 cm H_2O (b) 108.1 cm H_2O

26. 5200 mL/min or 5.2 L/min

27. 85 mL

28. (a) 1 min   (b) 12 sec

CHAPTER 15

LEVEL ONE   REVIEWING FACTS AND TERMS

1. The first priority of blood pressure homeostasis is to maintain adequate perfusion to the brain and heart.

2. (a) 6, 9   (b) 1, 2   (c) 4, 7   (d) 3, 5, 6, 8   (e) 3, 4

3. Blood vessel walls have endothelium, where exchange takes place in capillaries and where important paracrines are secreted; elastic tissue, which allows the vessel walls to recoil after being stretched; smooth muscle, which provides resistance to stretch and which can also contract; fibrous connective tissue, which also provides resistance to stretch.

4. Arterioles regulates blood flow to individual tissues.

5. Aortic pressure reaches *120 mm Hg* during systole. During *diastole,* aortic pressure declines to a typical low value of *80 mm Hg*. This blood pressure reading would be written as *120/80*.

6. The rapid pressure increase can be felt as a *pulse*. The equation for pulse pressure is systolic pressure minus diastolic pressure.

7. Venous return is aided by one-way valves in the veins, the skeletal muscle pump, and low pressure in the thorax during breathing.

8. Hypertension, or elevated blood pressure, threatens well-being because high pressure in the blood vessels can cause a weakened blood vessel to rupture and bleed.

9. Korotkoff sounds result when cuff pressure is lower than systolic pressure but higher than diastolic pressure.

10. See Table 15-2 for paracrines that cause vasodilation. Two other ways to control arterioles are contraction by sympathetic neurons (α-receptors) and vasodilation by epinephrine (β₂-receptors in certain organs).

11. Hyperemia is a localized region of increased blood flow. In active hyperemia, the increased blood flow is in response to an increase in metabolism. Reactive hyperemia is an increase in flow that follows a period of decreased blood flow.

12. Most systemic arterioles have *sympathetic* innervation. Exceptions are brain arterioles (no neural regulation), and arterioles in the penis and clitoris, controlled by the parasympathetic division.

13. (a) 1, 5   (b) 2, 6   (c) 1, 2, 4   (d) 3, 8   (e) none of the above

14. The digestive tract, liver, kidneys, and skeletal muscles receive over two-thirds of the cardiac output at rest. The kidneys have the highest flow of blood, on a per unit weight basis.

15. Capillary density in a tissue is directly proportional to the tissue's metabolic rate. Cartilage has one of the lowest capillary densities, while muscles and glands have the highest.

16. (a) diffusion   (b) diffusion or transcytosis
    (c) facilitated diffusion   (d) osmosis

17. The lymphatic vessels participate in immune, circulatory, and digestive system functions.

18. Edema is excess fluid in the interstitial space. It arises when capillary filtration exceeds the ability of the lymphatic system to remove the filtered fluid. Causes for this might be lower capillary oncotic pressure due to decreased plasma proteins or blockage of the lymphatic vessels by a tumor or other pathology.

19. (a) Perfusion is blood flow though a tissue.
    (b) Colloid osmotic pressure is the contribution of plasma proteins to the osmotic pressure of the plasma.
    (c) Vasoconstriction is a decrease in blood vessel diameter (usually an arteriole or a vein).
    (d) Angiogenesis is the growth of new blood vessels, especially capillaries, into a tissue.
    (e) Metarterioles are small vessels that go between arterioles and venules. They can act as bypass channels when precapillary sphincters constrict to prevent blood flow into capillaries.
    (f) Pericytes are cells surrounding the capillary endothelium that regulate the capillary leakiness.

20. The two major lipoprotein carriers of cholesterol are *HDL* and *LDL*. LDL-C is bad in elevated amounts; elevated HDL-C is good.

LEVEL TWO   REVIEWING CONCEPTS

21. Concept Map: Use Figure 15-10 as the starting point and add additional detail wherever possible.

22. (a) Lymphatic capillaries and systemic capillaries both have thin walls but the pores of lymphatic capillaries are larger. Lymphatic capillaries dead-end and have contractile fibers to help fluid flow, whereas systemic capillaries depend on systemic blood pressure for flow.
    (b) In general, the sympathetic division raises blood pressure by increasing cardiac output (heart rate and stroke volume) and causing vasoconstriction. The parasympathetic division can decrease blood pressure by decreasing heart rate, but has no effect on vasodilation/constriction.
    (c) Lymph fluid is similar to blood plasma minus most plasma proteins. Blood also has nearly half its volume occupied by blood cells.

(d) Continuous capillaries have small pores and regulate the movement of substances better than fenestrated capillaries, which have larger pores. The latter can also open large gaps to allow proteins and blood cells to pass into the circulation.

(e) Hydraulic (hydrostatic) pressure forces fluid out of capillaries; colloid osmotic pressure of plasma proteins draws fluid into capillaries. In systemic capillaries, hydraulic pressure over the length of the capillary exceeds colloid osmotic pressure, leading to net filtration of fluid out of the capillary.

23. Preventing Ca^{2+} entry into cardiac and smooth muscle decreases the ability of those muscles to contract. Decreasing Ca^{2+} entry into autorhythmic cells decreases heart rate and cardiac output. Neurons and other cells are unaffected by these drugs because they have different types of calcium channels with which the drugs do not interact.

24. Myogenic autoregulation refers to the ability of vascular smooth muscle to regulate its own contraction. It may result from Ca^{2+} influx when the muscle is stretched, which causes a reflex contraction, or it may be due to paracrine release from the endothelium.

25. Left ventricular failure will cause blood to pool in the lungs, increasing hydrostatic pressure in pulmonary capillaries. This may cause edema in the lungs, which results in shortness of breath when oxygen has trouble diffusing into the body. If the pressure increases back into the systemic venous circulation, the increased hydrostatic pressure will cause peripheral edema and increased venous pressure.

LEVEL THREE   PROBLEM SOLVING

26. The sight of blood is integrated in cerebral cortex, which sends output to the CVCC in the medulla oblongata, which results in increased parasympathetic output and decreased sympathetic output, resulting in decreased heart rate and decreased blood pressure.

27. Cells in the intact arteriole wall (in the endothelium) detect changes in oxygen and communicate these changes via a paracrine to the smooth muscle.

28. (a) Robert's uncontrollable risk factors include being male, being middle-aged, and having a family history of cardiovascular disease on both sides of his family. His only controllable risk factor is his elevated blood pressure (hypertension).

(b) Robert has hypertension because he had systolic blood pressure greater than 140 or diastolic pressure greater than 90 on several occasions. It would be useful to confirm that this was not doctor-anxiety by having him take his blood pressure for a week or so at locations away from the doctor's office, such as at a drug store.

(c) Beta blockers may help lower blood pressure by preventing norepinephrine and epinephrine from combining with β1 receptors in the heart, thus lowering cardiac output and MAP.

29. (a) MAP will increase (MAP ∝ R), flow through vessels 1 and 2 will decrease (F ∝ 1/R), and flow through vessels 3 and 4 will increase (F ∝ MAP).

(b) Stimulus = increase in pressure, receptor = baroreceptor in artery, control center = vasomotor center, response = systemic vasodilation to decrease pressure.

(c) Filtration pressure downstream from the constriction decreases, because flow is reduced and flow affects pressure.

30. Atropine increases heart rate by acting as a competitive inhibitor and binding to the muscarinic receptor, preventing ACh from binding.

31. (a) increases afterload
(b) resistance increases and pressure increases

32. (a) In Figure 14-1 you could draw the connection from pulmonary artery to aorta. In Figure 14-7d you can see the remnant of the closed ductus connecting the aorta and pulmonary artery.

(b) The lungs are not functioning.
(c) systemic
(d) Left side must generate more pressure
(e) Blood will flow from the aorta into the pulmonary artery.

LEVEL FOUR   QUANTITATIVE PROBLEMS

33. If diameter goes from 2 to 4, flow increases 16-fold.

34. Answers will vary. For a 50-kg individual with a resting pulse of 70 bpm, her weight in blood will be pumped in approximately 10 minutes.

35. Mean arterial pressure (MAP) = 1/3 (115 − 73) + 73 = 87 mm Hg. Pulse pressure = 115 − 73 = 42 mm Hg.

36. 250 mL oxygen/min = CO × (200 − 160 mL oxygen/L blood). His cardiac output is 6250 mL/min or 6.25 L/min.

CHAPTER 16
LEVEL ONE   REVIEWING FACTS AND TERMS

1. The fluid portion of blood, *plasma*, is composed mainly of *water*.

2. The three types of plasma proteins are albumins, globulins, and fibrinogen. For functions, see Table 16-1. Albumins are the most prevalent.

3. The cellular elements of blood include erythrocytes or red blood cells (transport O_2 and CO_2); leukocytes or white blood cells (defense); and platelets (clotting).

4. Blood cell production is *hematopoiesis*. It begins in the embryo in the yolk sac, liver, spleen, and bone marrow. By birth, hematopoiesis is restricted to the bone marrow, and by adulthood it occurs only in the axial skeleton and the proximal ends of the long bones.

5. Colony-stimulating factors stimulate hematopoiesis. Cytokines are released by one cell to act on another cell. Interleukins are cytokines released by leukocytes to act on other leukocytes. These three types of chemical signals influence growth and differentiation of blood cells. For examples, see Table 16-2.

6. Red blood cell production is erythropoiesis, white blood cell production is leukopoiesis and platelet production is thrombopoiesis.

7. The hormone that directs red blood cell synthesis is *erythropoietin*. It is primarily produced in the kidney when there is hypoxia, or low oxygen in the tissues.

8. The hematocrit is the percent of total blood volume occupied by packed (centrifuged) red cells. Normal hematocrit for men is 40–54% and for women, 37–47%.

9. An erythroblast is the immature large, nucleated precursor of the mature erythrocyte, which is smaller and lacks a nucleus. Three distinct characteristics of erythrocytes are the biconcave disk shape, the lack of a nucleus, and the red color created by hemoglobin.

10. Iron in the diet is important for hemoglobin synthesis.

11. (a) Jaundice is a yellow color to the skin and sclera of the eyes due to elevated levels of bilirubin, a hemoglobin metabolite, in the blood.

(b) Anemia is a low level of hemoglobin, usually reflected by a decreased hematocrit.

(c) Transferrin is the plasma protein that acts as a carrier for iron.

(d) Hemophilia is a group of inherited diseases in which some aspect of the coagulation cascade is abnormal, resulting in decreased ability of the blood to clot.

12. Chemicals that prevent blood clotting are called anticoagulants.

LEVEL TWO   REVIEWING CONCEPTS

13. List 1: see Figure 16-11. List 2: see Figure 16-13. List 3: see Figure 16-7.

14. The intrinsic pathway is initiated when exposed collagen and other triggers, such as a glass test tube wall, activate plasma protein factor XII. The extrinsic pathway begins when damaged tissue exposes tissue factor (III) that activates factor VII. Factors III and VII work directly on the common pathway and also interact with the intrinsic pathway. The two pathways unite at the common pathway to initiate the formation of thrombin. See Figure 16-12.

15. Activated platelets cannot stick to undamaged regions of endothelium that release prostacyclin and nitric oxide (NO).

LEVEL THREE   PROBLEM SOLVING

16. Rachel is pale and tired at the first test because she has a low red blood cell count and low hemoglobin content in her blood. Jen looks for bruising as a sign that Rachel's platelet count is too low. Jen recommends proteins and vitamins for hemoglobin synthesis and the production of new blood cell components. Iron is also necessary for hemoglobin synthesis. Rachel is advised to avoid crowds to prevent being exposed to infections such as viruses; her WBC count is low and her ability to fight infection is decreased. By 20 days post-chemotherapy, all of Rachel's blood counts are back into the low-normal range.

17. (a) The plasma protein transferrin is elevated in this disease.
    (b) The liver, which stores iron, may be damaged in hemochromatosis.
    (c) One simple way to decrease the body's overload of iron in hemochromatosis is to withdraw blood. This illustrates the principle of mass balance: if intake has exceeded output, restore the body load by increasing output.

18. Some other factor that is essential for red blood cell synthesis must be lacking. These would include iron for hemoglobin production, folic acid, and vitamin B_{12}.

LEVEL FOUR   QUANTITATIVE PROBLEMS

19. Total blood volume in a 200-lb man is about 6.4 L; his plasma volume is about 3.1 L. A 130-lb woman has total blood volume of 4.1 L and plasma volume of about 2.4 L.

20. Her total blood volume is 3.5 L and her total erythrocyte volume is 1.4 L.

CHAPTER 17

LEVEL ONE   REVIEWING FACTS AND TERMS

1. Four functions of the respiratory system are gas exchange, vocalization, pH regulation, and protection.

2. Cellular respiration is the metabolic process in which oxygen and nutrients are used for energy production. External respiration is the exchange of gases between the atmosphere and cells.

3. Quiet inspiration uses the external intercostals, scalenes, and diaphragm. Normal quiet expiration is generally passive and involves no significant muscle contraction. In active expiration, the internal intercostals and abdominal muscles contract. These are all skeletal muscles.

4. Pleural fluid allows the lungs to slide freely over the internal surface of the thoracic cage.

5. Atmosphere to nose and mouth, pharynx, larynx, trachea, main bronchus, secondary bronchi, bronchioles, epithelium of the alveoli, interstitial fluid, and capillary endothelium.

6. See Figure 17-2g for structure of an alveolus. Type I cells are for gas exchange; type II produce surfactant. Macrophages ingest foreign material. Capillary endothelium is almost fused to the alveolar epithelium to minimize diffusion distance, and the surface of an alveolus is almost covered with capillaries.

7. Pulmonary circulation: right ventricle to pulmonary trunk, to left and right pulmonary arteries, smaller arteries, arterioles, capillaries, venules, and small veins. The two pulmonary veins empty into the left atrium. The pulmonary circuit contains about 0.5 liter of blood at any given moment. Typical pulmonary arterial blood pressure is 25/8, compared with 120/80 for the systemic arterial pressure.

8. Inspired air is warmed, humidified, and cleaned (filtered) during its passage through the airways.

9. During inspiration, most of the volume change of the chest is due to movement of the diaphragm.

10. Intrapleural pressure is always subatmospheric; it is lower at the end of inspiration and highest near the end of expiration. Intraalveolar is equal to atmospheric at the beginning and end of inspiration. It is below atmospheric during inspiration and above atmospheric during expiration. See Figure 17-11.

11. Surfactant lowers the surface tension of water. In the lungs, it makes it easier for the lungs to inflate and stay inflated.

12. Radius of the airways, especially the bronchioles, plays the largest role in changing resistance in the human respiratory system.

13. (a) 1   (b) 2   (c) 1   (d) 2

14. (a) V_T = the part that varies from 2.0 to 2.5 L, IRV = from 2.5 to 3.75 L, ERV = from 2.0 to 1.0 L, RV = from 0 to 1.0 L, VC = from 1.0 to 3.75 L, and TLC = from 0 to 3.75 L.
    (b) V_T = 0.5 L, IRV = 1.25 L, ERV = 1.0 L.
    (c) Ventilation rate is 3 breaths/15 seconds = 0.2 breath per second × 60 seconds/minute = 12 breaths per minute.

LEVEL TWO   REVIEWING CONCEPTS

15. See Figures 17-9 and 17-11.

16. (a) Compliance refers to the ability of a substance to stretch; elastance is a measure of how well a stretched substance returns to its unstretched state.
    (b) Ventilation is the exchange of air between the atmosphere and the lungs. Inspiration is movement of air into the lungs. Expiration is movement of air out of the lungs.
    (c) Intrapleural pressure is always subatmospheric; intra-alveolar pressures vary from subatmospheric to above atmospheric.
    (d) Total pulmonary ventilation refers to how much air enters and leaves the airways in a given period of time. Alveolar ventilation refers to how much air enters and leaves the alveoli in a given period of time. Alveolar ventilation is a better indicator of the efficiency of breathing.
    (e) Type 1 alveolar cells refer to the thin alveolar cells for gas exchange while Type II alveolar cells are the cells that synthesize and secrete surfactants.
    (f) Pulmonary circulation carries the blood from the heart to the lung. Systemic circulation carries blood to and from most tissues of the body.

17. Bronchoconstrictors: histamine, leukotrienes, acetylcholine (muscarinic); bronchodilators: carbon dioxide, epinephrine (β_2)

18. (a) decrease   (b) decrease   (c) decrease   (d) increase
    (e) decrease   (f) increase

19. **Pneumothorax** is presence of air in the pleural cavity. **Spirometer** is a device used to measure the volume of inhaled and/or exhaled air. **Auscultation** is listening for body sounds. **Hypoventilation** is decreased pulmonary ventilation. **Bronchoconstriction** is a decrease in the radius of the bronchioles. **Minute volume** is the total pulmonary ventilation. **Partial pressure** of gas is that portion of total pressure in a mixture of gases that is contributed by a specific gas.

20. (a) The volume of air in the balloon is equal to the coyote's *vital capacity*. This volume can be measured by adding tidal volume with expiratory and inspiratory reserve volumes.
    (b) In ten years, the coyote will not be able to put as much air in the balloon because lung function tends to decrease with age as elasticity and compliance diminish.

21. b – b – d – d

22. The x-axis should be time, and the y-axis should be P_{O_2}. During inspiration, the P_{O_2} of the primary bronchi will increase, as fresh air ($P_{O_2} = 160$ mm Hg) pushes the stale air ($P_{O_2} = 100$ mm Hg) into the alveoli. During expiration, the P_{O_2} will decrease, as the oxygen-depleted air exits. The curve will vary from 100 mm Hg to 160 mm Hg.

23. (a) Decreased compliance (decreased ability to expand) requires more work to inflate the lungs.
    (b) Lungs will inflate less readily.
    (c) Elastance is not directly affected by changes in compliance.
    (d) Airway resistance is not affected by changes in compliance.

24. (a) decreases   (b) increases   (c) decreases

LEVEL THREE   PROBLEM SOLVING

25. Total pulmonary ventilation = ventilation rate × tidal volume. Her resting pulmonary ventilation is 400 mL/breath × 13 breaths/min = 5200 mL/min.

    (a) 400 mL/breath × 20 breaths/min = 8000 mL/min
    (b) 550 mL/breath × 13 breaths/min = 7150 mL/min
    (c) 500 mL/breath × 15 breaths/min = 7500 mL/min

    Increasing her respiratory rate to 20 breaths/min has the largest effect. In real life, both respiratory rate and tidal volume increase during exercise, so she would have an even higher total pulmonary ventilation.

26. (a) 600 mL/breath × 16 breaths/min = 9600 mL/min.
    (b) Dilating the bronchioles reduces the resistance of the airways, allowing greater volume of air flow into and out of the lungs. The patient is able to force more air out of the lungs on expiration, which increases her ERV and decreases her RV.
    (c) From Figure 17-7, normal values for a female are 12–20 breaths/min, V_T = 500 mL, ERV = 700 mL, RV = 1100 mL, IC = 2400 mL, VC = 3100 mL, IRV = 1900, FRC = 1800 mL, and TLC = 4200 mL. Her respiratory rate is normal, and her lung volumes are abnormal. Her high RV is confirmed by the X-ray data showing overinflation. In obstructive lung diseases such as asthma, the bronchioles tend to collapse on expiration, trapping air in the lungs and resulting in hyperinflation. Her IRV = VC − V_T − ERV = 2800 − 600 − 1000 = 1200 mL. Her low IRV accounts for most of the low vital capacity, and is to be expected in someone with asthma, where airway resistance is high, air flow is low, and the lungs are already overinflated at the beginning of inspiration. Her higher tidal volume may be the result of the constant effort she must exert to breathe.

LEVEL FOUR   QUANTITATIVE PROBLEMS

27. $P_1 V_1 = P_2 V_2$ New volume = 200 mL

28. (a) 21% O_2 = 160 mm Hg, 78% nitrogen = 593 mm Hg, 0.3% CO_2 = 3 mm Hg
    (b) 40% O_2 = 304 mm Hg, 13% nitrogen = 99 mm Hg, 45% CO_2 = 342 mm Hg, 2% H_2 = 15 mm Hg
    (c) 10% O_2 = 76 mm Hg, 15% nitrogen = 114 mm Hg, 1% argon = 8 mm Hg, 25% CO_2 = 190 mm Hg

29. 400 mL/breath × 12 breaths/min = total pulmonary ventilation of 4800 mL/min. Before an exam, ventilation is 7200 mL/min. Alveolar ventilation is (400–120) × rate, or 3360 (at rest) and 5040 mL/min (before exam).

30. Tidal volume = 417 mL/breath. IRV = 3383 mL

31. Lung volume is 1.1 L. (Did you forget to subtract the volume of the spirometer?)

CHAPTER 18

LEVEL ONE   REVIEWING FACTS AND TERMS

1. Pressure gradients for oxygen and carbon dioxide, solubility of gases in water, alveolar capillary perfusion, pH of blood, temperature.

2. 98% of oxygen is bound to hemoglobin. Remaining oxygen is dissolved in the plasma.

3. Factors that alter oxygen-hemoglobin binding are P_{O_2}, temperature, pH, and the amount of hemoglobin available for binding. Of these, amount of hemoglobin is the most important.

4. Hemoglobin has four globular protein chains, each wrapped around a central heme group that contains iron, the essential element.

5. The respiratory control centers are found in the *medulla* and *pons*. The dorsal respiratory group contains neurons for inspiration; the ventral respiratory group contains neurons for inspiration and for active expiration. A central pattern generator is a group of neurons that interact spontaneously to control rhythmic contraction of certain muscle groups.

6. The central chemoreceptors increase ventilation when P_{CO_2} increases. The peripheral chemoreceptors in the carotid and aortic bodies increase ventilation primarily in response to P_{O_2} less than 60 mm Hg. P_{CO_2} is the most important chemical signal for ventilation.

7. The protective reflexes of the respiratory system include an irritant-mediated bronchoconstriction and the Hering-Breuer reflex that prevents overinflation. The latter is initiated when tidal volume exceeds one liter.

8. Pressure gradients drive the exchange of respiratory gases.

9. Decreased atmospheric P_{O_2}, decreased alveolar perfusion, anemia and blood loss, decreased pulmonary ventilation, toxicity from carbon monoxide or cyanide, increased thickness of respiratory membrane as with pulmonary edema, decreased surface area of respiratory membrane as with emphysema.

LEVEL TWO   REVIEWING CONCEPTS

10. Start with Figure 18-13.

11. Most oxygen is bound to hemoglobin, not dissolved in the plasma, (P_{O_2}).

12. (a) 98% of arterial O_2 is bound to hemoglobin, while the remaining 2% is dissolved in plasma. 70% of CO_2 is converted to bicarbonate and H^+, 23% is bound to Hb, and the remaining 7% is dissolved in plasma.

(b) Partial pressure of a gas in solution refers to the amount that dissolves at equilibrium when the solution is exposed to the same partial pressure. Concentration is amount of gas per volume of solution, measured in units such as moles per liter. While solution partial pressure and concentration are proportional, the concentration is affected by the gas solubility, and therefore is not the same as the gas concentration in air.

13. Decreased pH (increased acidity) promotes oxygen unloading from hemoglobin.

14. Hypoxia is low oxygen inside cells. COPD is chronic obstructive pulmonary disease, such as asthma, in which increased airway resistance results in decreased pulmonary ventilation. Hypercapnia is elevated carbon dioxide.

15. Oxygen is not very soluble in water, and the metabolic requirement for oxygen in most multicellular animals would not be met without an oxygen-transport molecule.

16. (a) Graph should have an x-axis of ventilation in liters/minute and a y-axis of arterial P_{O_2}, in mm Hg, resembling Figure 18-9. Relationship is direct, with an increase in ventilation producing an increase in arterial P_{O_2}. Because maximum arterial P_{O_2} will depend upon oxygen solubility in water, the slope of the graph should decrease as dissolved oxygen approaches saturation.
    (b) Graph should have an x-axis of arterial P_{CO_2} in mm Hg, and a y-axis of ventilation in liters/minute. Relationship is direct, with an increase in arterial P_{CO_2} producing a reflexive increase in ventilation, mediated by both the central and peripheral chemoreceptors. Because there is a maximum ventilation determined by properties of the neurons and muscles involved in ventilation, slope of curve should decrease as it approaches this maximum.

17. (a) Amount of dissolved gas increases with P_{O_2}.
    (b) As P_{O_2} increases, amount of oxygen bound to hemoglobin increases, until 100% saturation is reached.

18. Arterial P_{O_2} is not changed because P_{O_2} depends on the P_{O_2} of the alveoli, not on how much Hb is available for oxygen transport.

19. (a) See Figure 18-20.   (b) See Figure 18-15.

LEVEL THREE   PROBLEM SOLVING

20. The increase in his dead space decreases alveolar ventilation.
    (a) increases   (b) decreases   (c) increases   (d) decreases

21. Person (a) has slightly reduced oxygen pressure but at $P_{O_2} = 80$, saturation of hemoglobin will be about 95%. If oxygen content with normal hemoglobin at 100 mm Hg P_{O_2} is 197 mL O_2/L at 98% saturation, then oxygen content at $P_{O_2} = 80$ mm Hg is 190 mL O_2/L blood (197 × (0.95/0.98)). Person (b) has reduced hemoglobin of 12 g/dL at 98% saturation, so oxygen content would be 157.6 mL O_2/L blood (197 × (12/15)). The increased P_{O_2} did not compensate for the decreased hemoglobin content and (b) has less total oxygen.

22. (a) decrease   (b) decrease   (c) decrease

23. (a) Respiratory movements depend upon signals originating above the level of the cut, which could include any area of the brain.
    (b) Ventilation depends upon signals originating in the medulla and/or pons.
    (c) Respiratory rhythm is controlled by the medulla alone, but other important aspects of normal respiration such as depth depend upon signals originating in the pons or higher.

24. With chronic elevated P_{CO_2}, the central chemoreceptors adapt and CO_2 is no longer a chemical drive for ventilation. The primary chemical signal for ventilation becomes low oxygen (below 60 mm Hg).

Thus, when the patient is given oxygen and the P_{O_2} goes above the threshold, there is no chemical drive for ventilation, so the patient stops breathing.

25. (a) Hemoglobin at the alveoli is about 96% saturated. At an exercising cell, it is about 23% saturated.
    (b) Bzork does not consume much oxygen because he only uses about 20% of the oxygen that his hemoglobin can carry. With exercise, his oxygen consumption increases dramatically, and his hemoglobin releases more than 3/4 of the oxygen it can carry.

26. All three lines show that as P_{CO_2} increases, ventilation increases. Line A shows that a decrease in P_{CO_2} potentiates this increase in ventilation (when compared to Line B). Line C shows that ingestion of alcohol diminishes the effect of increasing P_{CO_2} on ventilation. Because alcohol is a CNS-depressant, we can hypothesize that the pathway that links increased P_{CO_2} and increased ventilation is integrated in the CNS.

LEVEL FOUR   QUANTITATIVE PROBLEMS

27. The maximum O_2 carrying capacity is 1.65 mL O_2/gm Hb.

28. Oxygen consumption is 202.5 mL O_2/min.

29. Nothing happens to the curve. The percent of available Hb bound to O_2 is unchanged at any given P_{O_2}. However, with less Hb available, less oxygen will be transported.

CHAPTER 19
LEVEL ONE   REVIEWING FACTS AND TERMS

1. Five characteristics are color (concentration), odor (infection or excreted substances), clarity (presence of cells), taste (presence of glucose), and froth (presence of proteins)

2. Kidney functions are regulation of extracellular fluid volume (to maintain adequate blood pressure), regulation of osmolarity (so that cells don't shrink or swell too much), maintenance of ion balance (especially Ca^{2+} and K^+, both of which influence the function of neurons), regulation of pH (cells malfunction if pH homeostasis is not maintained), excretion of wastes and foreign substances (to prevent toxic effects), and production of hormones (that regulate red blood cell production, Ca^{2+} balance, and Na^+ balance).

3. At any given time, 20–25% of the cardiac output goes to the kidneys.

4. Urine flows into the paired ureters, urinary bladder, and urethra.

5. (a), (e), (b), (g), (f), (d), (c), (h)

6. The filtration barriers are the glomerular capillary endothelium, basal lamina, and the epithelium of Bowman's capsule. Blood cells and most plasma proteins are usually excluded by these layers.

7. Hydraulic pressure of flowing blood promotes glomerular filtration. The fluid pressure (hydrostatic pressure) of fluid in Bowman's capsule and the oncotic (osmotic) pressure created by plasma proteins oppose it. Net driving force is the arithmetic sum of these pressures.

8. GFR stands for glomerular filtration rate. A typical GFR is 125 mL/min or 180 L/day.

9. (a) Juxtaglomerular apparatus is a specialized group of cells found where the distal tubule passes between the afferent and efferent arterioles. It is composed of the macula densa cells in the distal tubule wall and granular cells in the arteriole wall. (b) Paracrine signals from the macula densa play a role in autoregulation of GFR and in the secretion of renin.

(c) Mesangial cells in the basal lamina of the glomerulus alter the size of the filtration slits formed by the interlacing "fingers" of

(d) podocytes, the specialized epithelial cells that surround the glomerular capillaries. Changes in slit size alter GFR.

(e) The urinary bladder has an internal smooth muscle sphincter that is passively contracted and an external skeletal muscle sphincter that is tonically (actively) contracted.

(f) The renal cortex is the outer layer of the kidney; it contains renal corpuscles, proximal and distal tubules, and parts of the loop of Henle and collecting ducts.

10. Most reabsorption (80%) occurs in the proximal tubule. When a molecule or ion is reabsorbed from the lumen of the nephron, it goes into the peritubular capillaries and from there into the systemic venous circulation, back to the right atrium. If a substance is filtered and not reabsorbed, it is excreted in the urine.

11. (a) 2, 3   (b) 3, 4   (c) 3, 5   (d) 1   (e) 5

12. Penicillin, K^+, and H^+ are secreted into the tubule lumen.

13. Creatinine is used to estimate GFR in humans.

14. Micturition is urination.

LEVEL TWO   REVIEWING CONCEPTS

15. The map of factors that influence or control GFR should include information from Figures 19-4, 19-6, 19-7, 19-8, and 19-9.

16. (a) Filtration and secretion both involve movement of material from blood to tubule lumen, but filtration is a bulk flow process created by pressure gradients while secretion is usually a mediated transport process. Excretion is also bulk flow but involves movement from the lumen of the kidney to the outside world.

(b) Transport maximum for a transport system is the maximum rate at which the carriers can work due to saturation of available binding sites by substrate. The renal threshold is the plasma concentration at which saturation occurs; at this point, the solute begins to appear in the urine.

(c) These substances are used as examples of renal transport. Creatinine is an endogenous compound but the other three are xenobiotics (foreign substances). All but inulin are filtered, not reabsorbed, and are secreted. Inulin is filtered but neither reabsorbed nor secreted.

(d) Clearance refers to the rate at which plasma is cleared of a substance (mL plasma cleared of substance X/min). GFR is expressed as the filtration rate of plasma (mL plasma filtered/min). Excretion is removal of urine, expressed as mL urine/min.

17. This kidney allows rapid removal of foreign substances that are filtered but not reabsorbed. The clearance rate for these substances is equal to the GFR if the substances are not also secreted.

18. If the afferent arteriole of a nephron constricts, the GFR in that nephron decreases. If the efferent arteriole of a nephron constricts, the GFR in that nephron increases.

19. The micturition reflex is shown in Figure 19-18. With toilet training, higher brain centers monitor sensory input from the expanding bladder and inhibit the reflex until an appropriate time. Higher brain centers can inhibit micturition either consciously or unconsciously or can be used to initiate the reflex.

20. Bladder smooth muscle contracts under parasympathetic control, so blocking muscarinic receptors decreases bladder contraction.

LEVEL THREE   PROBLEM SOLVING

21. (a) Filtration rate is proportional to plasma concentration of inulin. Filtration is a passive, bulk flow process that does not show saturation. The excretion rate of inulin exceeds the filtra-

tion rate, indicating that inulin is also secreted into the tubule lumen. The slope of the excretion rate is higher initially, then it decreases to match the slope of the filtration rate. The point where the slope changes represents saturation of transporters that secrete inulin. Inulin is thus filtered, secreted (transported from the peritubular capillaries into the lumen after filtration), and excreted (eliminated from the body). No evidence for reabsorption is presented.

(b) The line indicating net secretion will be approximately superimposed on the filtration line until the slope change, after which the secretion line is horizontal (no further increase in rate due to saturation).

22. Drawing should resemble Figure 19-12. Place transporters as described on apical and basolateral membranes. Cl^- moves between the cells.

23. The dialysis fluid can resemble plasma without any waste substances, such as urea. This will allow diffusion of solutes and water from the blood into the dialysis fluid but diffusion will stop at the desired concentration. To remove excess water from the blood, the dialysis fluid can be made more concentrated.

LEVEL FOUR   QUANTITATIVE PROBLEMS

24. Renal blood flow (25%) is 1 L/min.

25. The first specimen contains 20,000 mg creatinine. Clearance = 1000 L plasma/day. Normally creatinine clearance = GFR. However, this value of 1000 L/day is not all realistic for GFR (normal average is 180 L/day). The repeat test has 4000 mg of creatinine and gives a clearance of 200 L/day, which is within normal limits. The abnormal values on the first test were probably a laboratory error. Dwight's kidney function is normal.

26. For any solute that filters: plasma concentration × GFR = filtration rate. At the transport maximum: filtration rate = reabsorption rate of T_m

By substitution: plasma concentration × GFR = T_m

The renal threshold represents the plasma concentration at which the transporters working at their maximum (T_m). By substitution:

$$\text{renal threshold} \times \text{GFR} = T_m$$

Mermaid's GFR is 250 mL/min and T_m is 50 mg/min, so renal threshold is 0.2 mg glucose/mL plasma.

Clearance = excretion rate/plasma concentration. At 15 mg glucose/mL plasma, 3750 mg/min filter and 50 mg/min reabsorb, so 3700 mg/min are excreted.

$$\text{Clearance} = \frac{3700 \text{ mg glucose/min}}{15 \text{ mg glucose/mL plasma}} =$$

246.7 mL plasma cleared/min

CHAPTER 20

LEVEL ONE   REVIEWING FACTS AND TERMS

1. Electrolytes are ions, which can conduct electric current through a solution. Some electrolytes whose concentrations the body must regulate are Na^+, K^+, Ca^{2+}, H^+, HPO_4^{2-} and HCO_3^-.

2. Organs include the kidneys, lungs, heart, blood vessels, and organs of the digestive tract. Four important hormones are vasopressin (ADH), aldosterone, atrial natriuretic peptides (ANP), and the renin-angiotensin pathway.

3. Water enters the body only through ingested food and drink. A small amount of water is manufactured by the body as a product of metabolism. The body loses water in exhaled air, by evaporation and perspiration from the skin, through the kidneys and in the f

4. See Table 20-2.

5. The descending limb of the loop of Henle is permeable to water but does not transport salts because its cells lack the appropriate membrane transporters. The ascending limb is impermeable to water; however, it does have the membrane transporters to transport NaCl from the lumen into interstitial fluid.

6. Na^+ is a primary determinant of extracellular fluid volume. H^+ is the determinant of extracellular pH.

7. When plasma concentrations of K^+ decrease, more K^+ leaves the cell, and the membrane potential becomes more negative (hyperpolarizes). The heart is most likely to be affected by changes in K^+ concentration.

8. Salt appetite and thirst are important in regulating fluid volume and osmolarity.

9. ADH = antidiuretic hormone; ANP = atrial natriuretic peptide; ACE = angiotensin-converting enzyme; ANGII = angiotensin II; JG (apparatus) = juxtaglomerular; P cell = principal cell; I cell = intercalated cell.

10. Use the information in the following figures to compile your list: Figures 19-10 and 19-12, 20-7, 20-8, 20-12, 20-19, 20-20, and 20-21.

11. pH homeostasis is important because pH alters protein structure and therefore enzyme activity, membrane transporters, and the nervous system. The body copes with changing pH by using buffers, renal compensatory mechanisms, and respiratory compensation.

12. Acids are more likely to accumulate in the body. Acid sources include CO_2 from cellular respiration, metabolic acids, and acids ingested as food. Sources of bases include some foods.

13. A buffer is a molecule that helps prevent or slow changes in pH. Three intracellular buffers are proteins, phosphate ions, and hemoglobin. The primary extracellular buffer is HCO_3^-.

14. The kidneys excrete or reabsorb H^+ or HCO_3^-. Ammonia and phosphates are urinary buffers.

15. $CO_2 + H_2O \rightleftharpoons H_2CO_3 \rightleftharpoons H^+ + HCO_3^-$. Carbonic anhydrase increases the rate of this reaction. Renal tubule cells and red blood cells have high concentrations of carbonic anhydrase.

16. Arterial P_{CO_2} decreases, pH increases, and plasma H^+ concentration decreases.

LEVEL TWO REVIEWING CONCEPTS

17. To map the homeostatic reflexes that occur in response to each of the situations, use the information in Table 20-2 as the starting point and compile all the different pathways into a single map similar to Figure 20-16. Be sure to include all the steps of the reflex pathways on your map (stimulus, receptor, etc.).

18. This map will combine information from Figures 20-18 through 20-21.

19. See Figure 20-4.

20. This diagram should combine Figures 20-5 and 20-6.

21. (a) ANP—peptide hormone from atrial myocardial cells. Causes excretion of Na^+ and water by its actions on the kidney and by inhibiting ADH release from hypothalamus.
    (b) Aldosterone—steroid from adrenal cortex. Causes increased Na^+ reabsorption and increased K^+ excretion by its action on the distal tubule.
    (c) Renin—enzyme/hormone (peptide) from JG cells. Acts on blood angiotensinogen, converting it to ANGI.
    (d) ANGII—peptide hormone made from precursor molecule made by liver. Increases blood pressure and tries to maintain or increase blood volume by actions on arterioles, brain, and adrenal cortex
    (e) ADH—peptide hormone from hypothalamus. Increases water reabsorption by the distal nephron.
    (f) Angiotensin-converting enzyme—enzyme (protein) found on vascular endothelium. Converts ANGI to ANGII (see Figure 20-14).

22. Compensatory mechanisms for restoring low blood pressure are vasoconstriction, increased cardiac output, conservation of water by the kidneys, and increased thirst to increase fluid volume. If blood pressure falls too low, blood supply to the brain will diminish, resulting in damage or death.

23. (a) Principal and intercalated cells are both in the distal tubule. P cells are associated with aldosterone-mediated Na^+ reabsorption; I cells are involved with acid-base regulation.
    (b) Renin, angiotensin II, aldosterone, and ACE are all parts of the RAAS system. Renin and ACE function primarily as enzymes, while ANGII and aldosterone are hormones. For more details, see Figure 20-14.
    (c) In both respiratory and metabolic acidosis, body pH falls below 7.38. Respiratory acidosis results from CO_2 retention (from any number of causes), while metabolic acidosis results from excessive production of metabolic acids. Respiratory acidosis is moderated by buffers and can be compensated for by renal excretion of H^+ and retention of HCO_3^-. Metabolic acidosis is moderated by buffers and can be compensated by increased ventilation (decreasing arterial P_{CO_2}) and increased renal excretion of H^+ and retention of HCO_3^-. In respiratory acidosis, arterial P_{CO_2} will be elevated; in metabolic acidosis, it is usually decreased.
    (d) Water reabsorption in the proximal tubule and ascending limb of the loop of Henle is not regulated by ADH; water reabsorption in the distal tubule is altered by ADH. The proximal and distal tubule cells have water pores, but the apical membrane of the cells of the ascending limb of the loop of Henle does not.
    (e) Respiratory and metabolic alkalosis are both conditions in which body pH goes above 7.42. Metabolic alkalosis may be caused by excessive ingestion of bicarbonate-containing antacids or by loss of acidic stomach contents due to vomiting; respiratory alkalosis may be caused by hyperventilation. Metabolic alkalosis can be compensated by decreased ventilation (increasing arterial P_{CO_2}) and by decreased renal excretion of H^+ and increased excretion of HCO_3^-. Respiratory alkalosis can only be compensated by the renal mechanisms.

LEVEL THREE PROBLEM SOLVING

24. (a) The man is in acidosis. Plasma bicarbonate ion is elevated, probably because of his elevated P_{CO_2}, which also causes elevated H^+. This is an acute situation, because the bronchoconstriction is a short-term event.
    (b) The man is in acidosis again, with greater elevation of HCO_3^- and P_{CO_2}. As he now has a chronic disease that contributes to his chemical imbalance, the imbalance is chronic.
    (c) Renal compensation has partially increased his pH by excretion of H^+ and reabsorption of HCO_3^-, which also acts as a buffer. His P_{CO_2} is elevated because of his emphysema.

25. (a) Karen is in metabolic alkalosis, for which her body is trying to compensate.
    (b) Her plasma bicarbonate is high because after vomiting acid (H^+), her body was left with bicarbonate ions.

(c) Hypoventilation increases P_{CO_2} and also increases HCO_3^- and H^+. The increase in H^+ decreases her pH (compensation) but her HCO_3^- goes up even more. Hypoventilation will also decrease her arterial P_{O_2}, and therefore decrease the total oxygen content of her blood (for review, see Figure 18-20).

26. Hannah's blood pressure is high. Absorption of 3 L water will expand total body volume and therefore blood volume. Her plasma Na^+ concentration and plasma osmolarity are low. Use Table 20-2 to select the reflex pathways for your map.

LEVEL FOUR   QUANTITATIVE PROBLEMS

27. (a) pH = 6.1 + log 24/(.03 × 40) = 7.40   (b) 7.34
28. 428.6 mL (600 mosml/? L = 1400 mosmol/L)
29. (a) 400 mg glucose/100 mL × 130 mL/min = 520 mg glucose/min filters.
    (b) Can reabsorb up to T_m therefore 400 mg/min reabsorbed.
    (c) Excreted = filtered − reabsorbed or 120 mg/min × 1440 min/day = 172.8 g/day excreted.
    (d) Must convert grams of glucose to milliosmoles: 172.8 g × 180 g/mole × 1000 mosmol/mole = 960 mosmol glucose excreted/day. Concentration = amount/volume. 1200 mOsM = 960 mosmol/? liters. To excrete the glucose will require 0.8 additional liter of urine.

CHAPTER 21
LEVEL ONE   REVIEWING FACTS AND TERMS

1. Digestion is the chemical or mechanical breakdown of ingested nutrients (e.g., proteins) into a form that can be absorbed by the body. Absorption is movement from lumen to extracellular fluid (water); secretion is movement from ECF to lumen (enzymes). Motility is the movement of material through the digestive tract.
2. *Absorption* and *digestion* are not regulated, whereas *secretion* and *motility* are continuously regulated. Secretion and motility are necessary for proper digestion, without which the body cannot obtain food. By not regulating absorption and digestion, the body ensures that it will always absorb the maximum available nutrients.
3. (a) 2   (b) 3   (c) 4   (d) 7, 10   (e) 8   (f) 2, 3, 7   (g) 2
4. The layers (lumen outward) are mucosa (epithelium, connective tissue and smooth muscle), submucosa (connective tissue), musculature (smooth muscle), and serosa (connective tissue).
5. Secretory epithelium (endocrine and exocrine) lines the stomach, while absorptive epithelium with a few secretory cells lines the intestines.
6. Peyer's patches are clusters or nodes of lymphoid tissue in the mucosal layer.
7. Motility moves food through the gastrointestinal tract (usually from mouth to anus) and helps mix food with GI secretions. Gut motility results from contraction of longitudinal and circular muscle layers to create either propulsive peristaltic movements or mixing segmental movements.
8. Zymogens are inactive proenzymes in the digestive system. They must have a segment of protein chain removed before the enzyme becomes active. Examples are pepsinogen-pepsin and trypsinogen-trypsin.
9. (a) 8, 9   (b) 3, 7   (c) 1, 3, 7   (d) 1, 3, 7   (e) 6
   (f) 2   (g) 4   (h) 5
10. Temperature is the only factor that does not significantly affect digestion in humans. Emulsification of fats takes place in the stomach and small intestine with the aid of lipase, mechanical

mixing, and bile salts. Neural activity plays an important role in motility and secretion along the length of the digestive tract. Acidic pH in the stomach helps break down food and digest microorganisms. Surface area of food is increased by mechanical digestion and increases the area upon which enzymes can act.

11. Most digested nutrients are absorbed into *capillaries* of the *hepatic portal* system, delivering nutrients to the *liver*. However, digested fats go into the *lymphatic* system because intestinal capillaries have a *basement membrane (basal lamina)* around them that most lipids are unable to cross.

12. The ENS is a network of neurons contained within the digestive tract that can sense a stimulus, integrate information, and create an appropriate response without integration or input from the CNS. Although the ENS can work independently, it also interacts with the CNS through sensory and autonomic neurons.

13. Short reflexes are reflexes mediated entirely within the enteric nervous system. They regulate endocrine and exocrine secretion and GI motility. Long reflexes are GI reflexes integrated in the CNS.

14. Paracrines in the GI tract help mediate secretion and motility. Two examples are serotonin (5-HT) and histamine.

LEVEL TWO   REVIEWING CONCEPTS

15. Use Figure 21-22 as the basis for the map and add details to it.
16. (a) Mastication is chewing, deglutition is swallowing.
    (b) Villi are created by folds of the intestine, while microvilli are created by folds of the cell membrane on individual cells. Both increase the surface area.
    (c) Migrating motor complex, peristaltic contraction, and segmentation are all patterns of GI muscle contraction. Migrating motor complex contractions move food remnants and bacteria from the stomach to the large intestine between meals. Peristaltic contractions are progressive waves of contraction that move from one section of the GI tract to another. Segmental contraction is the contraction and relaxation of short segments of the intestine. Mass movements push a large bolus of material into the rectum, triggering defecation (excretion of feces). Vomiting, a protective reflex integrated in the medulla, is the forceful expulsion of gastric and duodenal contents from the mouth. Diarrhea is excessive amounts of watery stool.
    (d) Chyme is the watery combination of semidigested food and GI secretions in the GI tract. Feces are the usually-solid waste material that remains after digestion and absorption are complete. Chyme is produced in the stomach; feces are produced in the large intestine.
    (e) Short reflexes originate within and are integrated within the enteric nervous system. Long reflexes may originate in the GI tract or elsewhere and are integrated within the CNS.
    (f) The ENS is a unique division of the nervous system that controls digestive function. It consists of the submucosal plexus in the submucosal layer of the gut wall, and the myenteric plexis, which lies between the two muscle layers of the digestive tract wall. The vagus nerve carries sensory information and efferent signals between the brain and many internal organs, including the ENS.
    (g) Cephalic phase consists of digestive reflexes triggered by stimuli received in the brain, such as the smell, sight, or taste of food. Gastric phase begins with food entering the stomach, triggering a series of short reflexes. Intestinal phase begins when chyme enters the small intestine.
17. See Figures 21-6, 21-8, 21-9, and 21-21.

18. The ENS and cephalic brain use similar neurotransmitters and neuromodulators, including serotonin, VIP, and NO. Enteric support cells are similar to astroglia in the CNS. The GI capillaries are not very permeable, like the blood-brain barrier surrounding those in the brain. Both divisions act as integrating centers.

19. See Table 21-1 for specific hormones.

20. See Figure 21-26.

LEVEL THREE PROBLEM SOLVING

21. See Figure 21-17.

22. Severe diarrhea may be associated with loss of small intestine contents and bicarbonate, which results in metabolic acidosis.

23. (a) The liver and gall bladder produce and store bile salts needed for fat digestion. Ingestion of a fatty meal triggers reflex contraction of the gall bladder, but the blocked bile duct prevents secretion of the bile, causing pain.

    (b) Micelle formation will be decreased due to lack of bile salts, and carbohydrate digestion will be decreased due to lack of pancreatic secretions containing amylase. Protein absorption will be decreased some because of low pancreatic protease secretion, but protein digestion begins in the stomach and is continued by enzymes bound to the brush border, so it does not stop completely when the bile duct is blocked. Therefore, some digested proteins will be absorbed.

25. Apical membrane has Na^+ and K^+ leak channels. Basolateral membrane has the Na^+-K^+-ATPase. At high flow, saliva has more Na^+ and less K^+.

LEVEL FOUR QUANTITATIVE PROBLEMS

26. (a) MIT started out with equal concentrations in both solutions but by the end of the experiment, it was more concentrated on the serosal side. Therefore, MIT must be moving by active transport.

    (b) MIT moves apical to basolateral. This movement is absorption.

    (c) Transport across the apical membrane goes from the bath into the tissue. The tissue has more MIT than the bath. Therefore, this must be active transport.

    (d) Transport across the basolateral membrane goes from the tissue into the sac of intestine. The tissue has more MIT than fluid on the basolateral side. Therefore, this must be passive transport.

CHAPTER 22

LEVEL ONE REVIEWING FACTS AND TERMS

1. Metabolic pathways are all pathways used for synthesis or for energy production, use, or storage. Anabolic pathways are primarily synthetic; catabolic pathways break down large molecules into smaller ones.

2. Biological work includes transport (such as moving molecules across membranes), mechanical work (such as movement of muscles), and chemical work (such as protein synthesis).

3. A kilocalorie is the amount of heat required to raise the temperature of 1 liter of water by 1° C. In direct calorimetry, food is burned to see how much energy it contains.

4. The respiratory quotient is the ratio of CO_2 produced to O_2 used in cellular metabolism. A typical RQ value for an American diet is about 0.82.

5. BMR is an individual's lowest metabolic rate, measured at rest after sleep and a 12-hour fast. Average BMR is higher in adult males than

in females because females have more adipose tissue with a lower respiration rate. Other factors that affect BMR are age, physical activity, amount of lean muscle mass, diet, hormones, and genetics.

6. Biomolecules will be broken down for energy, used for synthesis, or stored.

7. Absorptive state metabolism is dominated by anabolic reactions and nutrient storage. Postabsorptive state metabolism mobilizes stored nutrients and uses them for energy and synthesis.

8. A nutrient pool is a group of nutrients that are available for the cells to use; most are found in the blood. The three primary pools are glucose, free fatty acids, and amino acids.

9. The primary goal of metabolism during the fasted state is to maintain adequate levels of glucose for the brain.

10. Excess energy is stored in small amounts of glycogen and in larger amounts of adipose tissue fat.

11. Three fates of ingested proteins include protein synthesis, energy, and conversion to fat or glucose for storage. The fate of ingested fats includes lipid synthesis (such as for cell membranes), energy, and storage as fats.

12. Insulin decreases blood glucose and glucagon increases it.

13. Amino acids and the glycerol portion of fatty acids can be made into glucose through gluconeogenesis.

14. Ketone bodies are produced by excessive breakdown of fatty acids, as occurs in starvation. Ketones can be burned as fuel by the brain and many peripheral tissues. Many ketone bodies are strong acids, so high concentrations can create metabolic acidosis.

15. Two stimuli for insulin secretion are increased plasma glucose and parasympathetic input onto pancreatic beta cells. Sympathetic stimulation inhibits insulin secretion.

16. Type 1 diabetes mellitus results from an absolute lack of insulin production. In type 2 diabetes, the body produces insulin but the cells do not respond normally to the hormone. Both types are characterized by elevated fasting blood glucose levels. In type 1, the body uses fats and proteins for fuel, leading to muscle wasting, ketosis, polyuria, polydipsia, and osmotic diuresis. Type 2 is not as severe because the cells can use some glucose.

17. Glucagon release is stimulated by low plasma glucose or increased plasma amino acids. The liver is its primary target, and glucagon increases glycogenolysis and gluconeogenesis to increase plasma glucose.

18. (a) Lipoprotein lipase is a capillary endothelium enzyme that converts blood triglycerides into free fatty acids and monoglycerides.

    (b) Amylin is co-secreted with insulin and slows gastric emptying and gastric acid secretion.

    (c) Ghrelin is a "hunger hormone" secreted by the stomach.

    (d) NPY is produced by the hypothalamus and causes an increase in food intake.

    (e) Apoproteins are the protein components of lipoproteins. Apoprotein B on LDL-C facilitates transport into most cells. Defective apoproteins have been linked to some inherited hypercholesterolemia.

    (f) Leptin is a "satiety hormone" produced by adipocytes.

    (g) Osmotic diuresis is loss of water in the urine due to high amounts of urine solutes. Hyperglycemia, as in diabetes mellitus, causes dehydration through osmotic diuresis.

    (h) Insulin resistance is seen in type 2 diabetes when target cells fail to response to insulin.

19. (a) stimulates   (b) inhibits   (c) stimulates
    (d) stimulates   (e) stimulates

LEVEL TWO   REVIEWING CONCEPTS

20. Hints for your concept maps: Try to base your maps on a figure in the book, such as Figures 22-2, 22-5, or 22-6. Use different colors for each organ or hormone. These will probably be the most complex maps you've done yet!

21. The concentration of glucagon varies only slightly during the day, while the concentration of insulin cycles according to food intake. Thus it appears that it is the ratio rather than an absolute amount of hormone that determines the direction of metabolism.

22. (a) Glucose is a monosaccharide. Glycogenolysis is glycogen breakdown. Glycogenesis is glycogen production from glucose. Gluconeogenesis is glucose synthesis from amino acids and fats. Glucagon is a hormone that increases plasma glucose. Glycolysis is the first pathway in glucose metabolism for ATP production.

   (b) Thermogenesis is heat production by cells. Shivering thermogenesis occurs when muscles twitch, producing heat as a by-product. Non-shivering thermogenesis occurs in all cells, but especially brown fat, proportional to their metabolic activity. Diet-induced thermogenesis represents the heat generated by the digestive and anabolic reactions that occur during the absorptive state.

   (c) Lipoproteins are transport molecules. Chylomicrons are lipoprotein complexes assembled in the intestinal epithelium and absorbed into the lymphatic system. Cholesterol is a steroid component of cell membranes and precursor to steroid hormones. HDL-C contains high-density lipoprotein and takes cholesterol into liver cells, where it is metabolized or excreted. LDL-C is the most prevalent form of cholesterol-based lipoprotein and elevated concentrations are associated with atherosclerosis. Apoprotein is the protein component of lipoproteins.

   (d) Calorimetry is measurement of energy content and is a means of determining metabolic rate. Direct calorimetry involves measuring actual heat production when food is burned. Indirect calorimetry measures oxygen consumption or CO_2 production.

   (e) Conductive heat loss is loss of body heat to a cooler object. Radiant heat loss results from production of infrared electromagnetic waves by any object above absolute zero temperature. Convective heat loss results from upward movement of warm air and its replacement by cooler air. Evaporative heat loss occurs when water evaporates.

   (f) The absorptive metabolic state exists when nutrients are absorbed and anabolic processes exceed catabolic processes. Post-absorptive state exists when catabolic processes exceed anabolic processes to maintain plasma glucose.

23. (a) Hyperglycemia results from the lack of insulin production.

   (b) Glucosuria results when the glucose concentration in filtrate exceeds the kidney's capacity to reabsorb glucose.

   (c) Polyuria results from osmotic diuresis caused by glucosuria.

   (d) Ketosis results from increased metabolism of fatty acids, in the absence of cells' ability to utilize glucose.

   (e) Dehydration is a consequence of polyuria due to osmotic diuresis.

   (f) Severe thirst is a consequence of dehydration.

24. If a person ingests a pure protein meal and only insulin is released, the person's blood glucose concentrations might fall too low due to insulin action. Glucagon secretion by the same stimulus ensures that glucose concentrations in the blood will remain within normal levels.

25. See Figure 22-1.

26. See Figures 22-19 and 22-20.

LEVEL THREE   PROBLEM SOLVING

27. Once Scott's new muscle proteins are made and once the limited pool of free amino acids in the blood is filled, excess amino acids will be stored as glycogen or fat.

28. As insulin secretion (x-axis) increases, plasma glucose concentration (y-axis) decreases.

29. Graph should be very similar to Figure 18-9 on p. 597. (a) Acidosis promotes O_2 dissociation from hemoglobin, shifting the curve to the right (like Fig. 18-10 on p. 598). The effect of low DPG, however, would shift the curve to the left (like Fig. 18-11 on p. 599). The net effect of both conditions would be somewhere between the right shift expected from acidosis (increased unloading) and the left shift expected from low DPG (decreased unloading). (b) Bicarbonate would bind some of the free H^+, bringing pH back up. As pH approached normal, the dissociation curve would shift back to the left. With DPG still remaining low, the curve would be between the left shift for low DPG and normal. Oxygen release after treatment would therefore be depressed, and cells would be deprived, compared to the untreated condition.

LEVEL FOUR   QUANTITATIVE PROBLEMS

30. Answers will vary. A 64-inch woman weighing 50 kg will have a BMI of approximately 19.

31. Fat: 6 g × 9 kcal/g = 54 kcal. Carbohydrate: 30 g × 4 kcal/g =120 kcal. Protein: 8 g × 4 kcal/g = 32 kcal. Total = 206 kcal. 54/206 = 26% of calories from fat.

CHAPTER 23

LEVEL ONE   REVIEWING FACTS AND TERMS

1. The zones are zona glomerulosa (aldosterone), zona fasciculata (glucocorticoids), and zona reticularis (sex steroids, primarily androgens).

2. (a) corticotropin releasing hormone (hypothalamus) → adrenocorticotropic hormone (anterior pituitary) → cortisol (adrenal cortex) feeds back to directly inhibit secretion of both CRH and ACTH.

   (b) growth hormone releasing hormone and growth hormone inhibiting hormone (hypothalamus) → growth hormone (anterior pituitary)

   (c) decreased blood Ca^{2+} → parathyroid (parathyroid glands) → increase in blood Ca^{2+} by increasing reabsorption of bone, among other effects → negative feedback inhibits secretion of PTH.

   (d) Thyrotropin releasing hormone (hypothalamus) → thyroid stimulating hormone (thyrotropin) (anterior pituitary) → triiodothyronine (T_3) and thyroxine (T_4) (thyroid gland) → negative feedback to hypothalamus and anterior pituitary

3. Conditions: adequate diet, absence of stress, and adequate amounts of thyroid and growth hormones. Other important hormones: insulin, somatomedins, and sex hormones at puberty.

4. Triiodothyronine (T_3) and tetraiodothyronine (T_4 or thyroxine). T_3 is the most active thyroid hormone; most of it is made from T_4 in peripheral tissues.

5. (a) Melanocortins include ACTH and melanocyte-stimulating hormone (MSH).

   (b) Osteoporosis is bone loss that occurs when bone reabsorption exceeds bone deposition.

   (c) Hydroxyapatite is the inorganic portion of the bone matrix, consisting mainly of calcium salts.

(d) Mineralocorticoids are steroid hormones that regulate minerals, i.e., aldosterone.

(e) Trabecular bone is small spicules of bone that surround red marrow cavities, in areas known as spongy bone.

(f) POMC is pro-opiomelanocortin, the inactive precursor to ACTH and other active molecules.

(g) Epiphyseal plates are growth zones in long bones, comprised of cartilage.

6. Functions: blood clotting, cardiac muscle excitability and contraction, skeletal and smooth muscle contraction, second messenger systems, exocytosis, tight junctions, strength of bones and teeth.

7. In this table, A indicates promotion of anabolism, C indicates promotion of catabolism.

META-BOLIC TARGET	CORTI-SOL	THYROID HORM-ONES	GROWTH HORM-ONE	INS-ULIN	GLU-CAGON
PROTEIN	C (in skeletal muscle)	A (children) C (adults)	A	A	C
CARBO-HYDRATE	C	C	C	A	C
LIPID	C	C	C	A	C

LEVEL TWO   REVIEWING CONCEPTS

8. (a) See Figure 7-20c, p. 234.
   (b) See Figure 7-20b, p. 234.
   (c) Substitute hormones of thyroid pathway into Figure 7-20c.
   (d) Substitute hormones of thyroid pathway into Figure 7-21a, p. 235.

9. (a) CRH from hypothalamus stimulates secretion of ACTH from anterior pituitary which stimulates secretion of glucocorticoids such as cortisol from adrenal cortex, zona fasciculata.
   (b) Follicle cells of thyroid gland secrete colloid from which thyroid hormones are produced; C cells secrete calcitonin.
   (c) Thyroid hormone synthesis is controlled by TSH, whose release is controlled by TRH. In the thyroid gland tyrosine and iodine combine on thyroglobulin to make thyroid hormones. Thyroid-binding globulin (TBG) carries lipophilic thyroid hormones during transport in the blood. Deiodinase in target removes the iodine from T_4 to create T_3.
   (d) Growth hormone releasing hormone (GHRH) stimulates secretion of growth hormone (GH or somatotropin) from the anterior pituitary. Somatostatin (also known as growth hormone inhibiting hormone) inhibits production of GH. Growth hormone binding protein binds about half the GH in the blood. Insulin-like growth factors (IGFs) from the liver act with GH to promote growth.
   (e) Dwarfism results from severe GH deficiency in childhood. Giantism results from hypersecretion of GH during childhood. Acromegaly is lengthening of jaw and growth in hands and feet, caused by hypersecretion of GH in adults.
   (f) Hyperplasia is increased cell number. Hypertrophy is increased cell size.
   (g) Osteoblasts are immature osteocytes, bone cells that secrete organic bone matrix. Chondrocytes are cartilage cells. Osteoclasts are bone-destroying cells.

(h) PTH increases blood calcium by stimulating bone reabsorption, renal reabsorption, and intestinal absorption of calcium. Calcitriol, also know as 1,25-dihydroxycholecalciferol, is a vitamin D derivative that mediates the PTH effect on intestinal absorption of calcium. Calcitonin decreases bone reabsorption of calcium. Estrogen promotes bone deposition.

10. Thyroid hormone effects on metabolic rate are apparent within a few minutes and are thought to be related to changes in ion transport across cell and mitochondrial membranes. Thyroid effects on growth in children would require more than an hour to be apparent.

11. An ion equivalent is its molarity times the number of charges/ion. For Ca^{2+}: 2.5 mmol/L $\times$ 2 = 5 mEq/L.

12. An osteoclast is illustrated in Fig. 23-21. The cell makes acid with the help of carbonic anhydrase: $CO_2 + H_2O \rightarrow H_2CO_3 \rightarrow HCO_3^- + H^+$. The apical membrane secretes HCl using a proton pump and the basolateral membrane secretes HCO_3^- with either a chloride-bicarbonate antiporter or a sodium-bicarbonate symporter

LEVEL THREE   PROBLEM SOLVING

13. Physiological stress stimulates secretion of cortisol, increases blood glucose. An increase in insulin would oppose this effect.

14. A normal response to ACTH suppression by dexamethasone would be a decrease in plasma cortisol. Patient A shows no response to ACTH suppression, suggesting there may be an adrenal tumor that is insensitive to ACTH. Patient B does show a decrease in cortisol production following ACTH suppression, suggesting that the problem may be a pituitary tumor.

15. Mr. A would have elevated TSH. Ms. B would have low levels of TSH. Mrs. C would have elevated TSH. (a) It is not possible to determine if the lab slip has the results of Mr. A or Mrs. C, without being able to read the thyroid hormone levels. (b) Ms. B can be ruled out, because her TSH would be low, if the tentative diagnosis is correct.

16. (a) People in all age groups showed vitamin D insufficiency when measured at the end of winter. This deficiency was most pronounced in the 18–29 age group and least pronounced in the 50+ age group. At the end of summer, far fewer subjects were deficient in vitamin D. The variables are season when blood was collected, age group, and % of people with vitamin D insufficiency.
   (b) Energy from the sun is required for precursors in the skin to be converted to vitamin D. Days are shorter in the winter, and at northern latitudes like Boston, people spend less time outside during the winter. This explains the difference in data between the two seasons. Fewer than half the people tested were deficient, however, suggesting that most people consumed enough vitamin D. The biggest seasonal difference occurred in the 18–29 age group. These younger people probably spent significantly more time outside in the summer than members of the other groups.
   (c) Vitamin D is present in multivitamin supplements, therefore consumption of these supplements should reduce vitamin D insufficiency.

LEVEL FOUR   QUANTITATIVE PROBLEMS

17. (a) 5 mg Ca^{2+}/L plasma $\times$ 125 mL plasma filtered/min $\times$ 1440 min/day = 900 mg Ca^{2+} filtered/day
   (b) To remain in Ca^{2+} balance, he must excrete 170 mg/day.
   (c) 900 mg filtered $-$ 170 mg excreted = 730 mg reabsorbed. 730/900 = 81%.

CHAPTER 24

LEVEL ONE   REVIEWING FACTS AND TERMS

1. Immunity is the body's ability to defend itself against disease-causing pathogens. Memory refers to cells that activate upon repeat exposure to an antigen, producing a heightened immune response. Specificity refers to antibody production that is tailored to bind specific antigens.

2. Anatomical components are thymus gland, bone marrow, spleen, lymph nodes, and diffuse lymphoid tissues.

3. Functions are to protect the body against foreign pathogens; to remove dead or damaged tissues and cells; to recognize and remove abnormal "self" cells.

4. Viruses bud off the host cell or kill and rupture the host cell. Viruses damage host cells by killing them, by taking over their metabolism, or by causing them to become cancerous and to reproduce uncontrollably.

5. The body must detect the pathogen, recognize it as foreign, organize a response, recruit assistance from other cells, and destroy the pathogen. If the pathogen cannot be destroyed, the body may suppress it to keep it from spreading.

6. (a) Anaphylaxis is a severe IgE-mediated allergic reaction characterized by widespread vasodilation, circulatory collapse, and bronchoconstriction.
    (b) Agglutinate means to clump together. When blood cells with one blood group antigen are exposed to the matching antibody, the antibody-antigen reaction causes the blood cells to agglutinate.
    (c) Many steps of the immune reaction take place extravascularly (outside the blood vessels).
    (d) Degranulation refers to the loss of intracellular granules when a cell releases chemicals stored in cytoplasmic granules.
    (e) Acute phase proteins are released in the early stages of injury or infection; they act as opsonins that coat pathogens.
    (f) Clonal expansion is the process by which one cell of a clone divides to make many identical cells.
    (g) Immune surveillance is the ability of the immune system to scan the body for abnormal cells (especially cancerous) and destroy these cells before they cause harm.

7. Histiocytes, Kupffer cells, osteoclasts, and microglia are the tissue-specific names given to specialized macrophages in the tissues.

8. The mononuclear phagocytic system includes monocytes and macrophages, which ingest and destroy invaders and abnormal cells.

9. (a) 5   (b) 1   (c) 3   (d) 6   (e) 2   (f) 4

10. Physical barriers include the skin, mucous membranes, and the mucociliary escalator of the respiratory tract. Chemical barriers include nonspecific chemicals such as lysozymes, opsonins, and enzymes, and specific chemicals such as antibodies.

11. B-lymphocytes secrete antibodies; T-lymphocytes, and natural killer cells kill infected cells either directly or indirectly. T-lymphocytes bind to antigen presented by MHC-complexes; natural killer cells can also bind to antibodies coating foreign cells.

12. Self-tolerance is the ability of the body's immune system to ignore the body's own cells. Self-tolerance occurs because normally T lymphocytes that would react with "self" cells die. If self-tolerance fails and the body makes antibodies against itself, autoimmune disease results.

13. Neuroimmunomodulation refers to the ability of the nervous system to influence immune function, either positively or negatively.

14. Stress is a nonspecific stimulus that disturbs homeostasis; a stressor is a stimulus that causes stress. The general adaptation syndrome is a response to stress that includes activation of the adrenal glands (both the fight-or-flight response of the adrenal medulla and the less dramatic response of cortisol secretion), followed by suppression of the immune system.

LEVEL TWO   REVIEWING CONCEPTS

15. Use the figures and the tables of the chapter to help create your map.

16. Lymph nodes trap bacteria and other pathogens and immune cells then create a localized inflammatory response within the nodes as they fight off the pathogens. This response includes swelling and the secretion of substances that activate nociceptors, resulting in swollen, sore nodes.

17. Histamine is a vasodilator that opens pores in capillaries, making it easier for immune cells and proteins in the blood to get into the extracellular space. Interleukin-1 also increases capillary permeability, stimulates acute phase proteins, causes fever, and stimulates cytokine and endocrine secretions. Acute phase proteins act as opsonins and help prevent tissue damage. Bradykinin is another vasodilator and stimulates pain receptors. Complement is a group of proteins that act as opsonins and attractants for leukocytes, signal mast cells to release histamine, and form membrane attack complex that puts pores in pathogens. Gamma interferon activates macrophages. These molecules all work together so they are not antagonistic. If the effect of them working together is greater than the sum of them working alone, then they are considered synergistic.

18. (a) Pathogens are any organisms that cause disease. Pyrogens are fever-causing chemicals. Antigens are substances that trigger an immune response and react with the products of the response. Antibodies are disease-fighting chemicals produced by the body, but antibiotics are drugs that destroy bacteria and fungi.
    (b) An infection is an illness caused by a pathogen, especially a virus or bacterium; inflammation is a nonspecific response to cell damage or extracellular invaders, including nonpathogens such as a splinter. An allergy is an inflammatory response to a nonpathogenic invader, such as plant pollen. In autoimmune diseases, the body creates a specific antibody-mediated response to its own cells.
    (c) Allergens are nonpathogenic substances that create allergic reactions; the other substances are all pathogens. Bacteria are cellular organisms that reproduce in the body's extracellular space. Viruses and retroviruses are acellular parasites that must invade the host's cells to reproduce. Retroviruses cause the host cell to make viral DNA so that the virus can reproduce.
    (d) Chemotaxins are any chemicals that attract immune cells to a specific location. Cytokines are peptides made on demand and secreted by cells for action on other cells. Opsonins are proteins that coat and tag foreign material so that it can be recognized by the immune system. Interleukins are cytokines that initially were thought to act only on leukocytes. Interferons are lymphocyte cytokines that aid in the immune response. Bradykinin is a paracrine vasodilator.
    (e) The human immune response is divided into nonspecific (innate) immunity and specific (acquired) immunity. Innate immunity is present from birth. Acquired immunity, also called adaptive immunity, is directed at specific invaders. Acquired immunity can be divided into cell-mediated immunity and humoral immunity (presence of antibodies in the blood).

(f) Immediate hypersensitivity responses are mediated by antibodies and occur within minutes of exposure to the allergens. Delayed hypersensitivity reactions may take several days to develop and are mediated by helper T cells and macrophages.

(g) These are all chemicals of the immune response. Membrane attack complex and perforin are membrane pore proteins. Perforins specifically allow granzymes, which are cytotoxic enzymes, to enter the cell.

19. The diagram should resemble Figure 24-12. The Fc region determines which of the five antibody classes a given antibody belongs to. The Fab region contains the antigen-binding sites that confer the antibody's specificity.

20. See Figure 24-17.

21. See Figure 24-18.

22. See Figure 24-19.

LEVEL THREE   PROBLEM SOLVING

23. Blood type O is the universal donor because these erythrocytes lack A or B surface antigens and therefore do not trigger an immune response. Blood type AB is the universal recipient because these erythrocytes have both the A and the B antigens and therefore no antibodies to A or B antigens.

24. Maxie's genotype must be OO, as must the baby's. Snidley's genotype can be either BB or BO. Baby received an O gene from Maxie, and could have received the other O gene from Snidely. Thus, it is possible that Snidley is the father of Maxie's baby.

25. Emotional stress causes increased secretion of cortisol, which suppresses the immune system. It is also likely that students are spending more time inside because they are studying more and they may be having closer contact with fellow students in study groups.

26. Autoimmune diseases occur when immune cells treat a normal self protein as if it were an antigen, and thus a specific tissue is attacked. Autoimmune diseases often begin in association with an infection, and are thought to represent cross-reactivity of the antibodies that developed because of the infection. Because stress suppresses the immune system, the associated infection may be prolonged.

27. An increase in neutrophils is associated with bacterial infection. An increase in eosinophils is associated with parasitic infection.

CHAPTER 25

LEVEL ONE   REVIEWING FACTS AND TERMS

1. ATP and phosphocreatine store energy in muscles.

2. The most efficient ATP production comes through *aerobic* pathways. When these pathways are being used then *both glucose and fatty acids* can be metabolized to provide ATP.

3. Aerobic metabolism: oxygen required, glucose goes through glycolysis and the citric acid cycle. Energy transferred to ATP through mitochondrial oxidative phosphorylation in the mitochondria. Produces 30–32 molecules of ATP/glucose. Anaerobic metabolism: no oxygen used, glucose undergoes glycolysis to lactic acid. Produces only 2 ATP/glucose molecule consumed.

4. Three sources of glucose for cells are glycogen, plasma glucose, and glucose produced through gluconeogenesis.

5. Cortisol, growth hormone, epinephrine, and norepinephrine all promote conversion of triglycerides into fatty acids and increase plasma glucose levels.

6. At the beginning of exercise, muscle ATP use exceeds aerobic ATP production so cellular stores of ATP are used. This creates an oxygen deficit reflected by increased oxygen consumption after exercise ceases.

7. The cardiovascular system is thought to be the major limiting factor.

8. Normal body temperature is 37° C. Increases in body temperature trigger sweating and cutaneous vasodilation.

LEVEL TWO   REVIEWING CONCEPTS

9. Figures whose information might be incorporated into the map can be found in Chapters 4, 15, 17, 18, 23, and 25.

10. Insulin secretion decreases during exercise because of sympathetic input on pancreatic beta cells. This decrease causes the liver to produce glucose and keeps insulin-sensitive tissues from taking up glucose, thus sparing glucose for the brain and exercising muscle, whose glucose uptake does not require insulin.

11. Two advantages of anaerobic glycolysis are that it is fast and it uses glucose, which is readily available. Two disadvantages are that it has a low ATP yield per glucose and that it contributes to metabolic acidosis.

12. (a) ATP is the energy source for muscle contraction. ADP can accept a high-energy phosphate from PCr and become ATP.

(b) Myoglobin is an intracellular muscle protein that aids diffusion of oxygen from the blood to the mitochondria. Hemoglobin is an oxygen-binding pigment in red blood cells that transports oxygen from lungs to cells.

13. (a) 3   (b) 1, 2, 3, 4, 5   (c) 1, 2, 4, 5, 6   (d) 6   (e) no match   (f) 6   (g) 1 (venous return), 4

14. (a) increases   (b) decreases   (c) increases   (d) increases   (e) increases   (f) increases   (g) stays the same   (h) decreases

15. Increased heart rate during exercise shortens the filling time and helps offset the increased end diastolic volume that might be expected from increased venous return.

16. Baroreceptor reflex may be absent in exercise because (1) the baroreceptor reflex resets to a higher setpoint, (2) afferent signals from the baroreceptors are being blocked in transit up the spinal cord, or (3) chemo- and mechanoreceptor input from exercising tissues overrides the baroreceptor input.

17. People with a lifestyle that includes regular exercise have lower risk of heart attacks, lower blood pressure, better lipid profiles, and lower risk of developing type 2 diabetes.

18. Exercising muscle does not require insulin for glucose uptake, so regular exercise can help keep blood glucose levels normal.

LEVEL THREE   PROBLEM SOLVING

19. A sports drink should contain water, NaCl, and K^+ to replace fluid and ions lost in sweat. In addition, a sports drink should contain a carbohydrate such as glucose that is easily absorbed and readily metabolized to form ATP.

LEVEL FOUR   QUANTITATIVE PROBLEMS

20. 60 beats/minute × 70 ml/beat = 4200 ml/min cardiac output. If heart rate doubles to 120 beats/min, CO goes to 8400 ml/min, or doubles also.

CHAPTER 26

LEVEL ONE   REVIEWING FACTS AND TERMS

1. (a) 3, 4, 5   (b) 8   (c) 2, 7   (d) 2, 6   (e) 2   (f) 1

2. The region for male sex determination is the *SRY* gene.

3. The gonads produce gametes and secrete sex hormones. The female gamete is the egg or ovum; the male is the sperm. Female gonadal hormones are estrogen, progesterone, androgens, and inhibin. Male gonadal hormones are androgens and inhibin.

4. Newly formed sperm: lumen of seminiferous tubule → epididymis → ductus (vas) deferens → ejaculatory duct (passing the seminal vesicles, prostate gland, and bulbourethral glands) → urethra. Ovulated egg: fallopian tube → uterine cavity → cervix → vagina

5. (a) Aromatase converts androgens to estrogens.
   (b) The blood-testis barrier is formed by tight junctions that prevent free movement of substances between the blood and lumen of the seminiferous tubule.
   (c) Androgen binding protein is made by Sertoli cells and secreted into the seminiferous tubule lumen, where it binds and concentrates androgens.
   (d) The first polar body is formed by the first meiotic division of a primary oocyte. It disintegrates and has no further function.
   (e) The acrosome is a lysosome-like structure in the head of a sperm whose enzymes are essential for fertilization.

6. (a) False. Some testosterone is produced in the adrenal glands of both genders.
   (b) False. Men produce estrogen, women produce androgens, and both genders produce FSH, LH, and inhibin.
   (c) True.
   (d) False. High levels of late follicular estrogen help prepare the uterus for implantation of a fertilized ovum and act as a negative feedback signal to the hypothalamus/pituitary.
   (e) True.

7. Semen is a sperm-fluid mixture made mostly by the accessory glands. See Table 26-3 for a list of components and their sources.

8. The most effective form of contraception is abstinence. The least effective forms of contraception rely on avoiding intercourse during times when the female thinks she might be fertile.

LEVEL TWO   REVIEWING CONCEPTS

9. List 1: use Figures 26-3, 26-4, and 26-5 to create your map. List 2: use Figures 26-12, 26-13, and 26-14.

10. See Figure 26-11.

11. See Figure 26-14.

12. Males have one X chromosome and one Y chromosome, which often does not have a gene to match one found on the X chromosome. Thus, a male may inherit a recessive X trait and will exhibit it, while a female who inherits the same recessive trait will not exhibit it if her second X chromosome has the dominant gene for the trait.

13. (a) A gamete is a haploid germ cell: eggs and sperm. A diploid zygote is formed from the fusion of egg and sperm; as it undergoes mitosis, the zygote becomes an embryo. By the 8th week of pregnancy, the embryo becomes a fetus.
    (b) Coitus is the sex act, or intercourse. In erection, the penis becomes stiff and engorged with blood. During the male orgasm, sperm move into the urethra during emission, then move out of the body with the other components of semen during ejac-

ulation. Erogenous zones are portions of the body that have receptors for sexually arousing stimuli.
    (c) Before sperm can fertilize an egg, they must undergo capacitation. When sperm reach an egg, they undergo an acrosomal reaction that helps the sperm penetrate the protective zona pellucida around the egg. Once a sperm fuses with the oocyte membrane, cortical granules in the egg cytoplasm release their contents to change the properties of the egg membrane (the cortical reaction).
    (d) Puberty is the period of time in which a person becomes sexually mature. In females, its onset is marked by menarche, the first menstrual period. In females, reproductive cycles cease during the time known as the menopause. In men, testosterone production decreases with age but the existence of andropause is still controversial.

14. (a) FSH in both sexes stimulates gamete production.
    (b) Inhibin in both sexes inhibits FSH secretion.
    (c) Activin in both sexes stimulates FSH secretion.
    (d) GnRH in both sexes acts on the anterior pituitary to stimulate release of FSH and LH.
    (e) LH in both sexes stimulates production of sex hormones by the gonads; in females, LH is also necessary for gamete maturation.
    (f) DHT is a metabolite of testosterone that is responsible for the development of the male genitalia in the fetus.
    (g) Estrogen is present in both sexes but has its predominant effect in females, where it is involved in gamete formation and some secondary characteristics.
    (h) Testosterone in males plays a role in gamete formation. In both sexes it is responsible for some secondary sex traits such as hair growth.
    (i) Progesterone is found in females only; it helps prepare the uterus for pregnancy.

15. The four phases are very similar in both genders. In the excitement phase, the male penis and the female clitoris become erect due to increased blood flow into the organ. The vagina also secretes fluids for lubrication. In male orgasm, ejaculation takes place, while in female orgasm the uterus and vaginal walls contract.

16. (a) hCG is secreted by the developing placenta and keeps the corpus luteum from dying.
    (b) LH is very similar to hCG but otherwise has no direct role in pregnancy and subsequent events.
    (c) HPL apparently plays a role in regulation of maternal metabolism during pregnancy.
    (d) Estrogen is needed for breast development and to act as a negative feedback to prevent new follicles from developing.
    (e) Progesterone is used for maintenance of the uterine lining and to prevent uterine contractions. It also has a role in mammary gland development.
    (f) Relaxin helps prevent uterine contractions.
    (g) PIH levels decrease so that prolactin levels will increase, allowing milk production.

LEVEL THREE   PROBLEM SOLVING

17. Normally after fertilization the second polar body, containing a haploid set of chromosomes, is released from the zygote. If all or some of the second polar body chromosomes are retained, the embryo will have three copies of a chromosome instead of just two.

18. If the unovulated cysts continue to secrete estrogen and do not develop into corpora lutea, the uterine lining will continue to grow and the breasts will develop, just as during pregnancy.

19. (a) Male
    (b) nonfunctional testes
    (c) no ducts of either type
    (d) female

20. During pregnancy the mother's high blood glucose concentration is available to the fetus, which metabolizes the extra energy and gains weight. The fetus also up-regulates insulin secretion to handle the glucose coming across the placenta. After birth, when insulin is still high but glucose drops to normal, the baby may become hypoglycemic.

LEVEL FOUR   QUANTITATIVE PROBLEMS

21. (a) Testosterone increased at point A because it was being administered to the subjects.
    (b) LH and FSH decreased at point A because of the negative feedback effect of testosterone.
    (c) Sperm production decreased in the A-B interval because FSH and LH decreased. It increased toward the end of the B-C interval because FSH allowed sperm production to resume. Sperm production did not increase significantly during the D-E interval.

Physics and Math

Richard D. Hill and Daniel Biller, *University of Texas*

INTRODUCTION

This appendix discusses selected aspects of **biophysics**, the study of physics as it applies to biological systems. Because living systems are in a continual exchange of force and energy, it is necessary to define these important concepts. According to the seventeenth-century scientist Sir Isaac Newton, a body at rest tends to stay at rest, and a body in motion tends to continue moving in a straight line unless the body is acted upon by some force (Newton's First Law). Newton further defined **force** as an influence, measurable in both intensity and direction, that operates on a body in such a manner as to produce an alteration of its state of rest or motion. Put another way, force gives **energy** to a quantity, or mass, thereby enabling it to do work. In general, a driving force multiplied by a quantity yields energy, or work. Some relevant examples of this principle include:

Mechanical force × distance =
$$\text{mechanical energy or mechanical work}$$

Gas pressure × volume of gas =
$$\text{mechanical energy or mechanical work}$$

Osmotic pressure × molar value =
$$\text{osmotic energy or osmotic work}$$

Electrical potential × charge =
$$\text{electrical energy or electrical work}$$

Temperature × entropy = heat energy or mechanical work

Chemical potential × concentration =
$$\text{chemical energy or chemical work}$$

Energy exists in two general forms: kinetic energy and potential energy. **Kinetic energy** [*kinein*, to move] is the energy possessed by a mass in motion. **Potential energy** is energy possessed by a mass because of its position. Kinetic energy (*KE*) is equal to one-half the mass (*m*) of a body in motion multiplied by the square of the velocity (*v*) of the body:

$$KE = 1/2\ mv^2$$

Potential energy (*PE*) is equal to the mass (*m*) of a body multiplied by acceleration due to gravity (*g*) times the height (*h*) of the body above the earth's surface:

$$PE = mgh \qquad \text{where } g = 10 \text{ m/s}^2$$

Both kinetic and potential energy are measured in joules.

BASIC UNITS OF MEASUREMENT

For physical concepts to be useful in scientific endeavors, they must be measurable and should be expressed in standard units of measurement. Some fundamental units of measure include the following:

Length (*l*): Length is measured in meters (m).

Time (*t*): Time is measured in seconds (s).

Mass (*m*): Mass is measured in kilograms (kg), and is defined as the weight of a body in a gravitational field.

Temperature (*T*): Absolute temperature is measured on the Kelvin (K) scale,

where K = degrees Celsius (°C) + 273.15

and °C = (degrees Fahrenheit −32)/1.8

Electric current (*I*): Electric current is measured in amperes (A).

Amount of substance (*n*): The amount of a substance is measured in moles (mol).

Using these fundamental units of measure, we can now establish standard units for physical concepts (Table B-1 ■). Although these are the standard units for these concepts at this time, they are not the only units ever used to describe them. For instance, force can also be measured in dynes, energy can be measured in calories, pressure can be measured in torr or mm Hg, and power can be measured in horsepower. However, all of these units can be converted into a standard unit counterpart, and vice versa.

The remainder of this appendix will discuss some biologically relevant applications of physical concepts. This discussion will

TABLE B-1	Standard Units for Physical Concepts	
MEASURED CONCEPT	**STANDARD (SI*) UNIT**	**MATHEMATICAL DERIVATION/ DEFINITION**
Force	Newton (N)	$1 \text{ N} = 1 \text{ kg m/s}^2$
Energy/Work/Heat	Joule (J)	$1 \text{ J} = 1 \text{ N} \cdot \text{m}$
Power	Watt (W)	$1 \text{ W} = 1 \text{ J/s}$
Electrical charge	Coulomb (C)	$1 \text{ C} = 1 \text{ A} \cdot \text{s}$
Potential	Volt (V)	$1 \text{ V} = 1 \text{ J/C}$
Resistance	Ohm (Ω)	$1\ \Omega = 1 \text{ V/A}$
Capacitance	Farad (F)	$1 \text{ F} = 1 \text{ C/V}$
Pressure	Pascal (Pa)	$1 \text{ Pa} = 1 \text{ N/m}^2$

*SI = Système International d'Unités

include topics such as bioelectrical principles, osmotic principles, and behaviors of gases and liquids relevant to living organisms.

BIOELECTRICAL PRINCIPLES

Living systems are composed of different molecules, many of which exist in a charged state. Cells are filled with charged particles such as proteins and organic acids, and ions are in continual flux across the cell membrane. Therefore, electrical forces are important to life.

When molecules gain or lose electrons, they develop positive or negative charges. A basic principle of electricity is that opposite charges attract and like charges repel. A force must act on a charged particle (a mass) to bring about changes in its position. Therefore, there must be a force acting on charged particles to cause attraction or repulsion, and this electrical force can be measured. Electrical force increases as the strength (number) of charges increases, and it decreases as the distance between the charges increases. This observation has been called **Coulomb's law**, and can be written:

$$F = \frac{q_1 q_2}{\varepsilon d^2}$$

where q_1 and q_2 are the electrical charges (coulombs), d is the distance between the charges (meters), ε is the dielectric constant, and F is the force of attraction or repulsion, depending on the type of charge on the particles.

When opposite charges are separated, a force acts over a distance to draw them together. As the charges move together, work is being done by the charged particles and energy is being released (Fig. B-1 ■). Conversely, to separate the united charges, energy must be added and work done. If charges are separated and kept apart, they have the potential to do work. This electrical potential is called **voltage**. Voltage is measured in **volts (V)**.

If electrical charges are separated and there is a potential difference between them, then the force between the charges will allow electrons to flow. Electron flow is called an electric **current**. The **Faraday constant (F)** is an expression of the electrical charge carried by one mole of electrons and is equal to 96,485 coulombs/mole.

The amount of current that flows depends on the nature of the material between the charges. If that material hinders the electron flow, then it is said to offer **resistance (R)**, measured in ohms. Current is inversely proportional to resistance, such that current decreases as resistance increases. If a material offers a high resistance, then that material is called an **insulator**. If resistance is low, and current flows relatively freely, then the material is called a **conductor**. Current, voltage, and resistance are related by **Ohm's law**, which states:

$$V = IR$$

where V = potential difference in volts
I = current in amperes
R = resistance in ohms

If you separate two opposite charges, there will be an electric force between them.

If you increase the number of charges that are separated, the force increases.

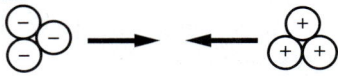

If you increase the distance between the charges, the force decreases.

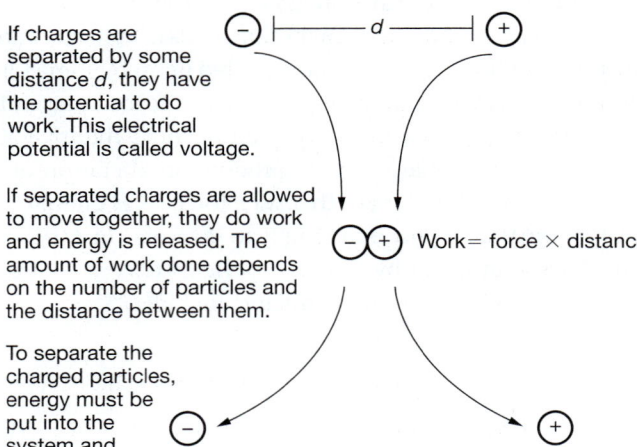

If charges are separated by some distance d, they have the potential to do work. This electrical potential is called voltage.

If separated charges are allowed to move together, they do work and energy is released. The amount of work done depends on the number of particles and the distance between them.

Work= force × distance

To separate the charged particles, energy must be put into the system and work is done.

■ **FIGURE B-1   Electrical force**

In biological systems, pure water is not a good conductor, but water containing dissolved NaCl is a fairly good conductor because ions provide charges to carry the current. In biological membranes, the lipids have few or no charged groups, so they offer high resistance to current flow across them. Thus, different cells can have different electrical properties depending on their membrane lipid composition and the permeability of their membranes to ions.

OSMOTIC PRINCIPLES

Freezing point, vapor pressure, boiling point, and osmotic pressure are properties of solutions collectively called **colligative properties**. These properties depend on the number of solute particles present in a solution. **Osmotic pressure** is the force that drives the diffusion of water across a membrane. Because there are no solutes in pure water, it has no osmotic pressure. However, if one adds a solute like NaCl, the greater the concentration (c) of a solute dissolved in water, the greater the osmotic pressure. The osmotic pressure (π) varies directly with the concentration of solute (number of particles (n) per volume (V)):

$$\pi = (n/V)RT$$

$$\pi = cRT$$

where R is the ideal gas constant (8.3145 joules/K mol) and T is the absolute temperature in Kelvin.

Osmotic pressure can be measured by determining the mechanical pressure that must be applied to a solution so that osmosis ceases.

Water balance in the body is under the control of osmotic pressure gradients (concentration gradients). Most cellular membranes allow water to pass freely because water is a small, uncharged molecule. Therefore, by making cellular membranes selectively permeable to solutes, the body can control the movement of water by controlling the movement of solutes.

RELEVANT BEHAVIORS OF GASES AND LIQUIDS

The respiratory and circulatory systems of the human body have developed according to the physical laws that govern the behaviors of gases and liquids. This section will discuss some of the important laws that govern these behaviors and how our body systems utilize these laws.

Gases   The **ideal gas law** states:

$$PV = nRT$$

where P = pressure of gases in the system
V = volume of the system
n = number of moles in gas
T = temperature
R = ideal gas constant (8.3145 J/K mol)

If n and T are kept constant for all pressures and volumes in a system, then any two pressures and volumes in that system are related by Boyle's Law,

$$P_1V_1 = P_2V_2$$

where P represents pressure and V represents volume.

This principle is relevant to the human lungs because the concentration of gas in the lungs is relatively equal to that in the atmosphere. In addition, body temperature is maintained at a constant temperature by homeostatic mechanisms. Therefore, if the volume of the lungs is changed, then the pressure in the lungs will change inversely. For example, an increase in pressure causes a decrease in volume, and vice versa.

Liquids   **Fluid pressure** (or hydrostatic pressure) is the pressure exerted by a fluid on a real or hypothetical body. In other words, the pressure exists whether or not there is a body submerged in the fluid. Fluid exerts a pressure (P) on an object submerged in it at a certain depth from the surface (h). **Pascal's law** allows us to find the fluid pressure at a specified depth for any given fluid. It states:

$$P = \rho gh$$

where: P = fluid pressure (measured in pascals, Pa)
ρ = density of the fluid
g = acceleration due to gravity (10 m/s^2)
h = depth below the surface of the fluid

Fluid pressure is unrelated to the shape of the container in which the fluid is situated.

REVIEW OF LOGARITHMS

Understanding logarithms ("logs") is important in biology because of the definition of pH:

$$pH = -\log_{10}[H^+]$$

This equation is read as "pH is equal to the negative log to the base 10 of the hydrogen ion concentration." But what is a logarithm?

A logarithm is the exponent to which you would have to raise the base (10) to get the number in which you are interested. For example, to get the number 100, you would have to square the base (10):

$$10^2 = 100$$

The base 10 was raised to the second power; therefore, the log of 100 is 2:

$$\log 100 = 2$$

Some other simple examples include:

$10^1 = 10$ The log of 10 is 1.
$10^0 = 1$ The log of 1 is 0.
$10^{-1} = 0.1$ The log of 0.1 is -1.

What about numbers that fall between the powers of 10? If the log of 10 is 1 and the log of 100 is 2, the log of 70 would be between 1 and 2. The actual value can be looked up on a log table or ascertained with most calculators.

To calculate pH, you need to know another rule of logs that says:

$$-\log x = \log(1/x)$$

and a rule of exponents that says:

$$1/10^x = 10^{-x}$$

Suppose you have a solution whose hydrogen ion concentration [H$^+$] is 10^{-7} mEq/L. What is the pH of this solution?

$$pH = -\log[H^+]$$

$$pH = -\log(10^{-7})$$

Using the rule of logs, this can be rewritten as

$$pH = \log(1/10^{-7})$$

Using the rule of exponents, this can be rewritten as

$$pH = \log 10^7$$

The log of 10^7 is 7, so the solution has a pH of 7.

Natural logarithms (ln) are logs in the base e. The mathematical constant e is approximately equal to 2.7183.

Genetics

Richard D. Hill, *University of Texas*

WHAT IS DNA?

Deoxyribonucleic acid (DNA) is the macromolecule that stores the information necessary to build structural and functional cellular components. It also provides the basis for inheritance when DNA is passed from parent to offspring. The union of these concepts about DNA allows us to devise a working definition of a gene. A **gene** (1) is a segment of DNA that codes for the synthesis of a protein, and (2) acts as a unit of inheritance that can be transmitted from generation to generation. The external appearance (**phenotype**) of an organism is determined to a large extent by the genes it inherits (**genotype**). Thus, one can begin to see how variation at the DNA level can cause variation at the level of the entire organism. These concepts form the basis of **genetics** and evolutionary theory.

NUCLEOTIDES AND BASE-PAIRING

DNA belongs to a group of macromolecules called **nucleic acids.** **Ribonucleic acid (RNA)** is also a nucleic acid, but it has different functions in the cell that are not discussed here (see Chapter 4, p. 115). Nucleic acids are polymers made from monomers [*mono-*, one] called **nucleotides**. Each nucleotide consists of a *nucleoside* (a pentose, or 5-carbon, sugar covalently bound to a nitrogenous base) and a phosphoric acid with at least one phosphate group (Fig. C-1a■). Nitrogenous bases in nucleic acids are classified as either **purines** or **pyrimidines**. The purine bases are **guanine (G)** and **adenine (A)**; the pyrimidine bases are **cytosine (C)**, **thymine (T)**, found in DNA only, and **uracil (U)**, found in RNA only. To remember which DNA bases are pyrimidines, look at the first syllable. The word "pyrimidine" and names of the DNA pyrimidine bases all have a "y" in the first syllable.

When nucleotides link together to form polymers such as DNA and RNA, the phosphate group of one nucleotide bonds covalently to the sugar group of the adjacent nucleotide (Fig. C-1b, c). The end of the polymer that has an unbound sugar is called the 3' ("three prime") end. The end of the polymer with the unbound phosphate is called the 5' end.

DNA STRUCTURE

In humans, many millions of nucleotides are joined together to form DNA. Eukaryotic DNA is commonly in the form of a double-stranded double helix (Fig. C-2■) that looks like a twisted ladder or twisted zipper. The sugar-phosphate sides, or backbone, are the same for every DNA molecule, but the sequence of the nucleotides is unique for each individual organism.

The backbone of the double helix is formed by covalent **phosphodiester bonds** that link a deoxyribose sugar from one nucleotide to the phosphate group on the adjacent nucleotide. The "rungs" of the double helix are created when the nitrogenous bases on one DNA strand form hydrogen bonds with nitrogenous bases on the adjoining DNA strand. This phenomenon is called **base-pairing**. The base-pairing rules are as follows:

1. Purines pair only with pyrimidines.
2. Guanine (G) forms three hydrogen bonds with cytosine (C) in both DNA and RNA.
3. Adenine (A) forms with two hydrogen bonds with thymine (T) in DNA or with uracil (U) in RNA.

The number of hydrogen bonds is directly related to the amount of energy necessary to break the base pair. Thus, more energy is required to break G:::C bonds (each ":" represents a hydrogen bond) than A::T bonds. This fact can be useful experimentally to make general conjectures about the similarity of two DNA samples.

The two strands of DNA are bound in **antiparallel** orientation, so that the 3' end of one strand is bound to the 5' end of the second strand (see Fig. C-1c). This organization has important implications for DNA replication.

DNA Replication Is Semi-Conservative

In order to be transmitted from one generation to the next, DNA must be replicated. Furthermore, the process of replication must be accurate and fast enough for a living system. The base-pairing rules for nitrogenous bases provide a means for making an appropriate replication system.

In DNA replication, special proteins unzip the DNA double helix and build new DNA by pairing new nucleotide molecules to the two existing DNA strands. The result of this replication is two double-stranded DNA molecules, such that each DNA molecule contains one DNA strand from the template and one newly synthesized DNA strand. This form of replication is called **semi-conservative replication**.

Replication of DNA is bidirectional. A portion of DNA that is "unzipped" and has enzymes performing replication is called a **replication fork** (Fig. C-2). Replication begins at many points (**replicons**), and it continues along both parent strands simultaneously until all the replication forks join.

Nucleotides bond together to form new strands of DNA with the help of an enzyme called **DNA polymerase**. DNA polymerase can only add nucleotides to the 3' end of a growing strand of DNA. For this reason, DNA is said to replicate in a 5' to 3' direction.

(a)

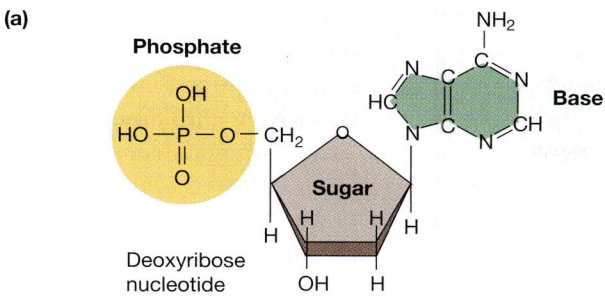

Phosphate
Base
Sugar
Deoxyribose nucleotide

(b)

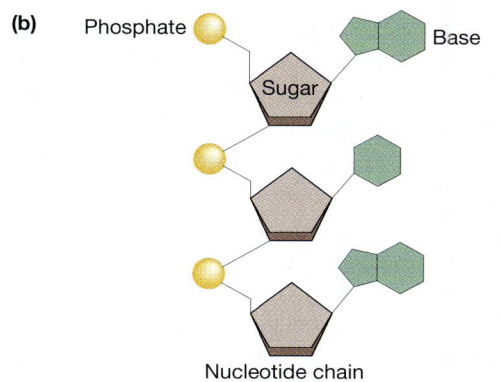

Phosphate
Base
Sugar
Nucleotide chain

(c)

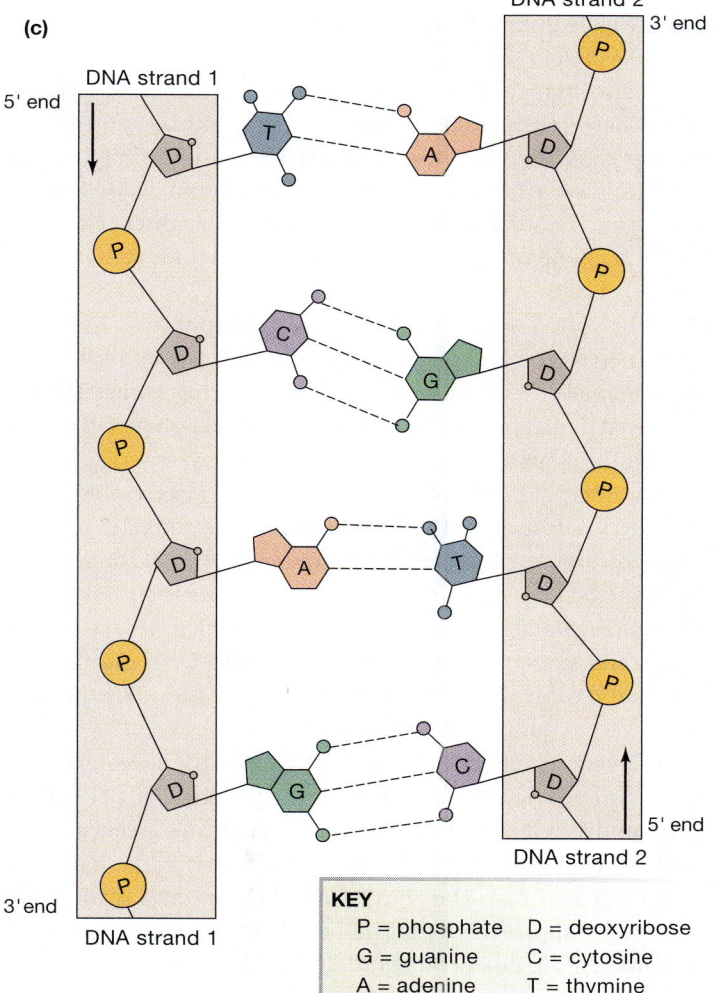

DNA strand 2
3' end
DNA strand 1
5' end
3' end
DNA strand 1
5' end
DNA strand 2

KEY
P = phosphate   D = deoxyribose
G = guanine   C = cytosine
A = adenine   T = thymine

■ **FIGURE C-1** *Nucleotides and DNA*

(a) A nucleotide is composed of a pentose sugar, a nitrogenous base, and a phosphate group. **(b)** Nucleotides link sugar to phosphate to form nucleic acids. The end of the nucleic acid with an unbound sugar is designated the 3' end; the end with the unbound phosphate is the 5' end. **(c)** Two nucleotide strands link by hydrogen binding between complementary base pairs to form the double helix of DNA. Adenine always pairs to thymine, and guanine pairs to cytosine.

The antiparallel orientation of the DNA strands and the directionality of DNA polymerase force replication into two different modes: **leading strand replication** and **lagging strand replication**. The DNA polymerase can replicate continuously along only one parent strand of DNA: the parent strand in the 3' to 5' orientation. The DNA replicated continuously is called the **leading strand**.

The DNA replication along the other parent strand is discontinuous because of the strand's 5' to 3' orientation. DNA replication on this strand occurs in short fragments called **Okazaki fragments** that are synthesized in the direction away from the replication fork. Another enzyme known as **DNA ligase** later connects these fragments into a continuous strand. The DNA replicated in this way is called the **lagging strand**. Because the 5' ends of the lagging strand of DNA cannot be replicated by DNA polymerase, a specialized enzyme called **telomerase** has arisen to replicate the 5' ends.

Much of the accuracy of DNA replication comes from base pairing, but on occasion, mistakes in replication happen. However, several quality control mechanisms are in place to keep the error rate at 1 error/10^9 to 10^{12} base pairs. **Genome** (the entire amount of DNA in an organism) sizes in eukaryotes range from 10^9 to 10^{11} base pairs per genome, so this error rate is low enough to prevent many lethal mutations, yet still allows genetic variation to arise.

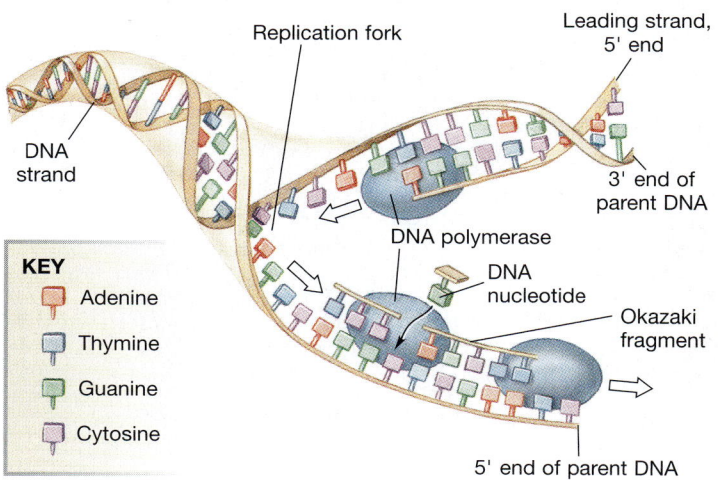

Replication fork
Leading strand, 5' end
DNA strand
3' end of parent DNA
DNA polymerase
DNA nucleotide
Okazaki fragment
5' end of parent DNA

KEY
Adenine
Thymine
Guanine
Cytosine

■ **FIGURE C-2** *DNA replication*

FUNCTIONS OF DNA

One of the key functions of DNA is its ability to code for the synthesis of proteins that participate in structural or functional aspects of a cell. The information coded in DNA is *transcribed* into mRNA [🔁 p. 115], whose code then is *translated* into proteins with the help of tRNAs attached to specific amino acids (Table C-1■). The

TABLE C-1	The Amino Acids of the Human Body

Each amino acid has both a three-letter abbreviation of its name and a single-letter abbreviation, used to spell out the sequence of amino acids in large proteins. The essential amino acids are marked with a star. These nine amino acids cannot be made by the body and must be obtained from the diet.

AMINO ACID	THREE-LETTER ABBREVIATION	ONE-LETTER SYMBOL
Alanine	Ala	A
Arginine	Arg	R
Asparagine	Asn	N
Asparagine or aspartic acid	Asx	B
Aspartic acid	Asp	D
Cysteine	Cys	C
Glutamic acid	Glu	E
Glutamine	Gln	Q
Glutamine or glutamic acid	Glx	Z
Glycine	Gly	G
*Histidine	His	H
*Isoleucine	Ile	I
*Leucine	Leu	L
*Lysine	Lys	K
*Methionine	Met	M
*Phenylalanine	Phe	F
Proline	Pro	P
Serine	Ser	S
*Threonine	Thr	T
*Tryptophan	Trp	W
*Tyrosine	Tyr	Y
*Valine	Val	V

*Essential amino acids

other key function of DNA is its ability to act as a unit of inheritance when transmitted across generations.

Before we begin our discussion of DNA as a unit of inheritance, a few terms need to be introduced. A **chromosome** is one complete molecule of DNA, and each chromosome contains many genes. An **allele** is a form of a gene; interactions between the cell products of alleles determine how that gene will be expressed in the phenotype of an individual. **Somatic** [*soma,* body] **cells** are those cells that comprise the majority of the body (e.g., a skin cell, a liver cell); they are not directly involved with passing on genetic information to future generations. Each somatic cell in a human contains two alleles of each gene, one allele inherited from each parent. For this reason, human somatic cells are called **diploid** ("two chromosome sets"), meaning that they have two complete sets of all their chromosomes. In contrast, **germ cells** pass genetic information directly to the next generation. In human males, the germ cells are the **spermatozoa** (sperm), and in human females, the germ cells are the **oocytes** (eggs). Human germ cells are called **haploid** ("half of the chromosome sets") because each germ cell only contains one chromosome set, which is equal to half of the number of chromosome sets in somatic cells. When a human male germ cell joins with a human female germ cell, the result is a fertilized egg (zygote) containing the diploid number of chromosomes. If this zygote eventually develops into a healthy adult, that adult will have diploid somatic cells and haploid germ cells.

Cells alternate between periods of cell growth and cell division. There are two types of cell division: mitosis and meiosis. **Mitosis** is cell division by somatic cells that results in two daughter cells, each with a diploid set of chromosomes. **Meiosis**, in contrast, is cell division that results in four daughter cells, each with a haploid set of chromosomes. The daughter cells develop into germ cells.

The period of cell growth is called **interphase**. It is divided into three stages: G_1, a period of cell growth, protein synthesis, and organelle production; **S**, the period during which DNA is replicated in preparation for cell division; and G_2, a period of protein synthesis and final preparations for cell division (Fig. C-3■). During interphase, the DNA in the nucleus is not visible under the light microscope without dyes because it is uncoiled and diffuse. However, as a cell prepares for division, it must condense all its DNA to form more manageable packages. Each eukaryotic DNA molecule has millions of base pairs, which, if laid end-to-end, could stretch out to about 6 cm. If this DNA did not coil tightly and condense for cell division, imagine how difficult moving it around during cell division would be.

There is a hierarchy of DNA packaging in the cell (Fig. C-4■). Each chromosome begins with a linear molecule of DNA about 2 nm in diameter. Then proteins called **histones** associate with the DNA to form a fiber about 10 nm in diameter that looks like "beads on a string." The "beads" are **nucleosomes** made up of histones wrapped in DNA. The beaded string can twist into a **solenoid** [*solen,* pipe] coil about 30 nm in diameter, consisting

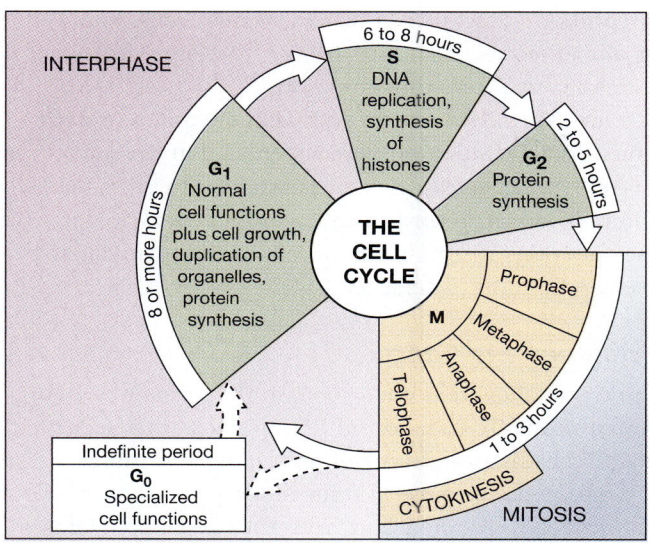

■ FIGURE C-3 *The cell cycle*

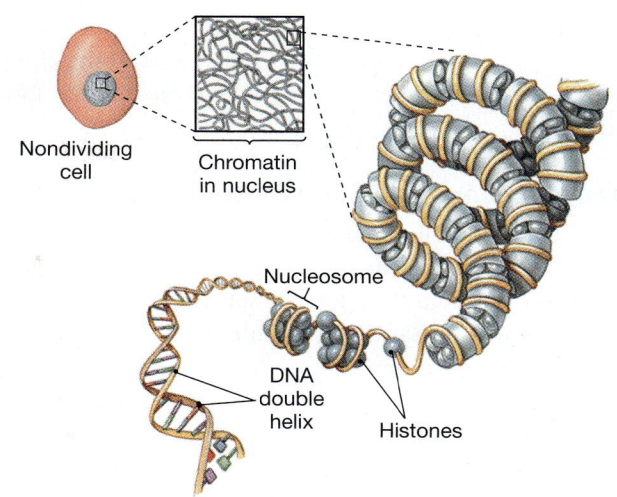

■ FIGURE C-4 *Levels of organization of DNA*

of about 6 nucleosomes per turn. The 30-nm solenoid structure can form looped domains, and these looped domains can attach to **nonhistone** (structural) **proteins** to form a 250–300-nm fiber called **chromatid fiber**. The chromatid fiber then coils to form the **chromosome fiber** (about 700 nm in diameter) that is visible during cell division. In this state of packaging, the cell is ready for division.

Mitosis Creates Two Identical Daughter Cells

As stated earlier, mitosis is the cellular division of a somatic cell that results in two diploid daughter cells. The steps of mitosis are **prophase, metaphase, anaphase**, and **telophase** (Fig. C-5 ■). The entire somatic cell cycle can be remembered by the acronym, IPMAT, in which the "I" stands for interphase and the other letters stand for the steps of mitosis that follow.

Prophase   During prophase, chromatin becomes condensed and microscopically visible as duplicate chromosomes. The duplicated chromosomes form **sister chromatids**, which are joined

to each other at the **centromere**. The centriole pair [⮂ p. 61] duplicates and the centriole pairs move to opposing ends of the cell. The **mitotic spindle**, composed of microtubules, assembles between them. The nuclear membrane begins to break down and disappears by the end of prophase.

Metaphase   In metaphase, mitotic spindle fibers extending from the centrioles attach to the centromere of each chromosome. The forty-six chromosomes, each consisting of a pair of sister chromatids, line up at the "equator" of the cell.

Anaphase   During anaphase, the spindle fibers pull the sister chromatids apart, so that an identical copy of each chromosome moves toward each pole of the cell. By the end of anaphase, an identical set of forty-six chromosomes is present at each pole. At this point, the cell has a total of ninety-two chromosomes.

Telophase   The actual division of the parent cell into two daughter cells takes place during telophase. In **cytokinesis**, the cytoplasm divides when an actin contractile ring tightens at the midline of the cell. The result is two separate daughter cells,

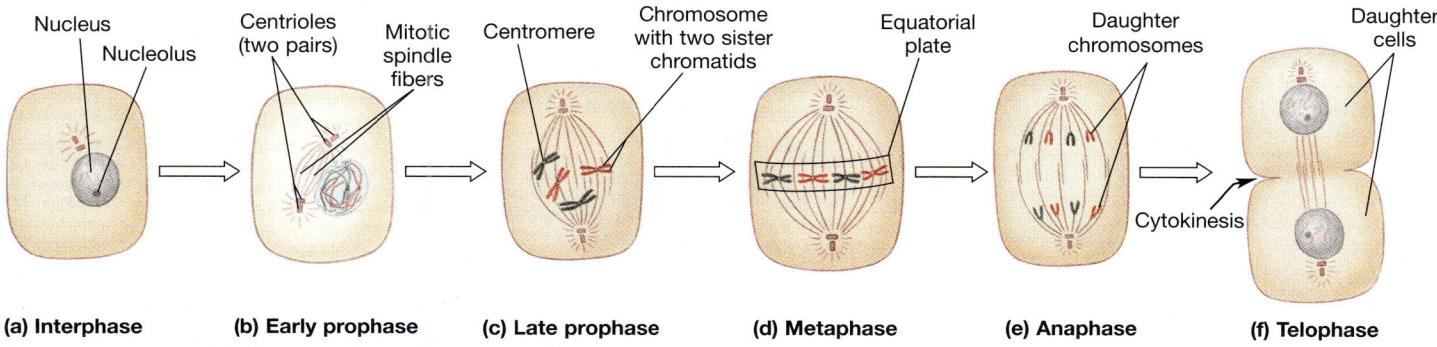

(a) Interphase   **(b) Early prophase**   **(c) Late prophase**   **(d) Metaphase**   **(e) Anaphase**   **(f) Telophase**

■ FIGURE C-5 *Mitosis*

each with a full diploid set of chromosomes. The spindle fibers disintegrate, nuclear envelopes are formed around the chromosomes in each cell, and the chromatin returns to its loosely coiled state.

Mutations Change the Sequence of DNA

Over the course of a lifetime, there are countless opportunities for mistakes to arise in the replication of DNA. A change in a DNA sequence, such as the addition, substitution, or deletion of a base, is a **point mutation**. If a mutation is not corrected, it may cause a change in the gene product. These changes may be relatively minor, or they may result in dysfunctional gene products that could kill the cell or the organism. Only rarely does a mutation result in a beneficial change in a gene product. Fortunately, our cells contain enzymes that can detect and repair damage to DNA.

Some mutations are caused by **mutagens**, which are factors that increase the rate of mutation. Various chemicals, ionizing radiation such as X-rays and atomic radiation, ultraviolet light, and other factors can behave as mutagens. Mutagens either alter the base code of DNA or interfere with repair enzymes, thereby promoting mutation.

Mutations that occur in body cells are called **somatic mutations**. Somatic mutations are perpetuated in the somatic cells of an individual, but they are not passed on to subsequent generations. However, **germ-line mutations** can also occur. Because these mutations arise in the germ cells of an individual, they are passed on to future generations.

Oncogenes and Cancer

Proto-oncogenes are normal genes in the genome of an organism that primarily code for protein products that regulate cell growth, cell division, and cell adhesion. Mutations in these proto-oncogenes give rise to **oncogenes**, genes that induce uncontrolled cell proliferation and the condition known as **cancer.** The mutations in proto-oncogenes that give rise to cancer-causing oncogenes are often the result of viral activity.

ANTERIOR (situated in front of): in humans, toward the front of the body (see VENTRAL) D-1■

POSTERIOR (situated behind): in humans, toward the back of the body (see DORSAL) D-2■

MEDIAL (middle, as in *median strip*): located nearer to the midline of the body (the line that divides the body into mirror-image halves)

LATERAL (side, as in a *football lateral*): located toward the sides of the body

DISTAL (distant): farther away from the point of reference or from the center of the body

PROXIMAL (closer, as in *proximity*): closer to the center of the body

SUPERIOR (higher): located toward the head or the upper part of the body

INFERIOR (lower): located away from the head or from the upper part of the body

PRONE: lying on the stomach, face downward

SUPINE: lying on the back, face up

DORSAL: refers to the back of the body

VENTRAL: refers to the front of the body

IPSILATERAL: on the same side as

CONTRALATERAL: on the opposite side from

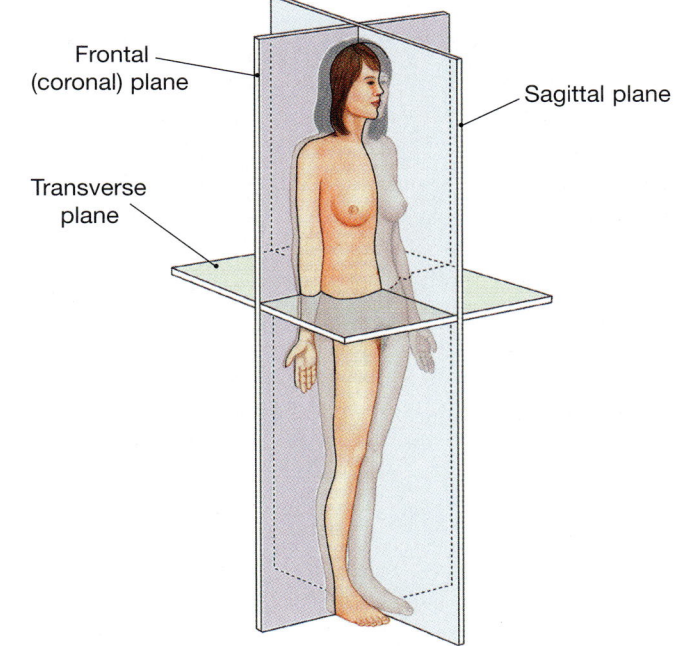

Frontal (coronal) plane

Sagittal plane

Transverse plane

■ **FIGURE D-1** *Sectional planes*

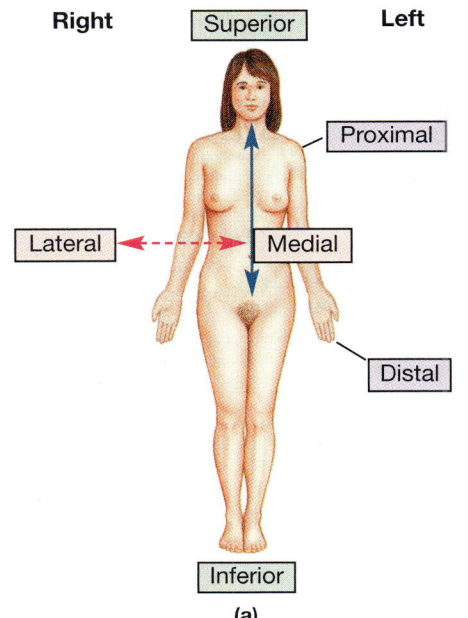

Right   Superior   Left

Proximal

Lateral   Medial

Distal

Inferior

(a)

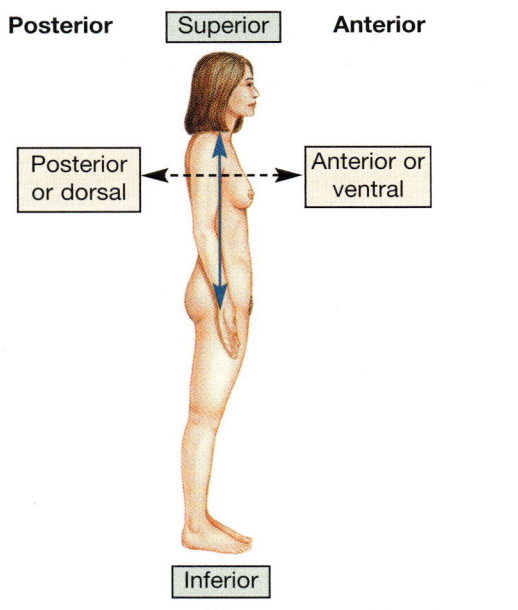

Posterior   Superior   Anterior

Posterior or dorsal ←-→ Anterior or ventral

Inferior

(b)   ■ **FIGURE D-2** *Body directions*

NOTE: A *t* following a page number indicates tabular material and an *f* indicates a figure.

functions of, 568–569, 569f
smooth muscle of, in asthma, 429
airway resistance, 576, 577t
alanine, A-36t
albumin Plasma protein made in the liver
(Ch 2, 7, 15, 16), 220, 536, 538t
denaturation of, 42
alcohol, 273t, 282
alcohol consumption, erectile dysfunction and,
845–846
aldosterone A steroid hormone that stimulates
Na^+ reabsorption and K^+ secretion in the
kidney (Ch 7, 20, 23), 215f, 221f, 652–653,
653–654, 654f, 655f, 656f, 665, 829f
adrenal secretion of, 751–752, 752f, 753f
algorithm, 525
alkaline phosphatase, 96t
alkaline solution, 37, 663
alkaline tide, 685
alkalosis Extracellular pH greater than 7.42
(Ch 20), 663, 670t, 671
intercalated cells in, 669, 669f
renal compensation for, 667
allantois Extraembryonic membrane that
becomes part of the umbilical cord
(Ch 26), 849
allele, A-36
allergen Any substance capable of triggering
an allergic reaction (Ch 24), 797, 797f
allergy (allergic response), 777, 797–798, 797f
eosinophils in, 782
all-or-none phenomenon, 258
allosteric activator, 41
allosteric inhibitor, 41
allosteric modulation, 41
allosteric modulator Bind to an enzyme away
from the binding site and change the shape
of the active site (Ch 2), 41, 52f
alpha-adrenergic receptor (α receptor)
Membrane receptor that binds to
norepinephrine and epinephrine (Ch 6, 11),
189, 190, 190f, 193, 273t, 276, 383–384,
384t, 385t
agonists and antagonists of, 386t
arteriole control and, 513, 514f
alpha-blockers, 386
alpha-bungarotoxin Snake toxin that is a
nicotinic receptor antagonist (Ch 8), 270,
273t, 386t, 391
alpha cell pancreatic cell that secretes glucagon.
Synonym: A cell (Ch 22), 730f, 734, 735f
alpha (α) chain, 595, 596f
alpha-gamma coactivation Simultaneous
activation of alpha and gamma motor
neurons so that during muscle contraction,
the intrafusal fibers continue to monitor
tension in the muscle (Ch 13), 440–442, 442f
alpha- (α) glucosidase inhibitors, 739t
alpha helix Spiral configuration formed by
some amino acid chains (Ch 2), 30f,
31, 122
alpha interferon, 785–786, 795, 796f
alpha- (α) ketoglutarate, 668
alpha- (α) melanocyte-stimulating hormone
(MSH), 718t, 756, 756f
alpha motor neuron Neurons that innervate
extrafusal muscle fibers and cause muscle
contraction (Ch 13), 439, 441f
alpha₁- (α₁) receptor, 384, 385t
alpha₂- (α₂) receptor, 384, 385t
5-alpha-reductase Enzyme that converts
testosterone to DHT (Ch 26), 825–826

alpha wave Low amplitude, high-frequency
brain waves characteristic of the awake-
resting state (Ch 9), 314f
alternative splicing The processing of mRNA to
make different proteins from a single
strand of DNA (Ch 4), 116f, 118, 119f
alternative therapy, 802
altitude, 588, 591, 592, 593, 596
altitude hypoxia, 592, 593, 596, 601
altitude sickness, 592, 593
alveolar blood flow, 579–580, 580t, 581f
alveolar cells, types I and II, 561, 563f
alveolar epithelium, 561, 563f
alveolar macrophage Immune cells that patrol
the alveoli (Ch 17), 563f, 574
alveolar membrane, gas exchange affected by
changes in, 592–593, 592f
alveolar P_{CO_2}, 578–579, 579f, 590, 590f
alveolar P_{O_2}, 578–579, 579f, 590, 590f
decreased, 591–592, 591f
alveolar pressure, 570
expiration and, 572
inspiration and, 570–572, 571f
alveolar ventilation The volume of air that
reaches the alveoli each minute (Ch 17,
18), 577, 578t, 579t
breathing pattern and, 577, 578–580, 578t,
579f, 580t, 581f
exercise and, 811
pathological conditions affecting, 591–592,
591f, 592–593
alveoli The exchange surface of the lungs, where
oxygen and carbon dioxide transfer
between air and the blood (Ch 17), 560,
561, 563f, 564f
Alzheimer's disease (Ch 9), 319–320
amacrine cells, in retina, 364, 365f
AMAN. See acute motor axonal polyneuropathy
ambiguous genitalia, 827
amenorrhea, 855
aminase, 100t
amination Addition of an amino group to a
molecule (Ch 4), 101
amine hormone Derivatives of single amino
acids, either tyrosine or tryptophan (Ch 7),
214–215f, 218t, 222, 223f
amine neurotransmitter Neurotransmitters
made from amino acids, including the
catecholamines, histamine, and serotonin
(Ch 8), 273t, 274
amino acid Molecule with a central carbon atom
linked to a hydrogen atom, an amino
group, a carboxyl group, and a variable
group of atoms designated "R." The
building blocks of proteins (Ch 2, 4, 8, 21,
22, Appendix C), 30-31, 30f, 695, 695f,
724–725, A-36, A-36t
amine hormone synthesis and, 222, 223f
glucagon release and, 734, 735f
glucose synthesis and, 111, 112–113, 113f,
722–723, 723f, 727, 728f
insulin secretion and, 730, 733
mRNA translation in production
of, 118–119, 120f
as neurotransmitter, 273t, 274–275
protein catabolism and, 103, 111, 111f, 695, 695f
protein synthesis and, 30, 30f, 104f, 722, 723f,
724–725
structure of, 30–31, 30f
substitution of, inherited diseases and, 114
amino acid pool, 722, 723f
amino acid supplements, 724–725

amino group Functional group whose
composition is - NH_2 (Ch 2), 27t, 30f, 31,
111, 119
in exchange reaction, 101
aminopeptidase Digestive enzyme that
removes amino acids from the NH_2
terminal end of a peptide (Ch 2, 21), 39,
695, 695f
ammonia, 111, 111f
ammonia buffer, 667
ammonium ion, 111, 111f, 668, 668f
amnesia, anterograde, 318
amnion Extraembryonic membrane that secretes
amniotic fluid (Ch 26), 849, 851f
amniotic fluid, 849, 851f
AMPA receptor Glutamate receptor-channel that
allows net Na^+ influx (Ch 8), 273t, 277
in long-term potentiation, 282, 282f
ampere (A), A-31
amphipathic (Ch 21), 696
amplification, 180, 180f
in sound transduction, 350
amplifier enzyme A membrane enzyme that
creates two or more second messengers
during signal transduction (Ch 6), 180,
181f, 181t
amplitude
of graded potential, 255, 256, 256f
of sound wave, 349, 349f
ampulla (vestibular apparatus) (Ch 10),
355, 356f
amu. See atomic mass unit
amygdala Portion of the brain linked to
emotion and memory (Ch 9), 305f, 308,
308f, 315–316
amylase Enzyme that digests starch to maltose
(Ch 4, 21), 97t, 694, 694f, 699
salivary, 699
amylin Peptide cosecreted with insulin (Ch 22),
728, 738–739
amylin analogs, 739t
amyloid protein (Ch 9), 319
anabolic metabolism/pathway, 101, 112–122, 721,
723–727. See also anabolism
anabolic steroids, 751, 836
anabolism Metabolic pathways that require a net
input of energy and that synthesize small
molecules into larger ones (Ch 4, 22), 101,
112–122, 721, 723–727
insulin and, 731–733, 732f, 733f, 734f
anaerobic Adjective pertaining to a process that
does not require oxygen (Ch 4), 103
anaerobic glycolysis, 410
anaerobic metabolism, 103, 106–107, 106t, 107f
exercise and, 808, 809f
analgesia, 275, 342
analgesic drugs Drugs that relieve pain
(Ch 10), 342
opiates/opioid peptides, 275, 342
anal sphincters, 679, 708, 708–709, 708f
anaphase, A-37, A-37f
anaphylactic shock (anaphylaxis) Allergic
reaction marked by bronchoconstriction
and vasodilation with hypotension (Ch 15),
189, 510, 787, 798
anatomical positions of human body, A-39
anatomic dead space The portions of the
airways that do not exchange gases with
the blood (Ch 17), 577
anatomy The study of structure (Ch 1, 9), 2.
See also specific organ, structure, system
anaxonic neuron (Ch 8), 246, 248f

bilirubin Breakdown product of heme groups from hemoglobin (Ch 16, 21), 545, 545*f*, 688*f*, 689, 705

binding protein, 38

binding site, 39
  receptor, 189
    diseases/drugs associated with changes in, 191, 192*t*

binocular vision Three-dimensional vision from overlapping visual fields of two eyes (Ch 10), 369–370, 370*f*

binocular zone, 369, 370*f*

bioelectrical principles, A-32

bioenergetics, 93

biological energy, 9

biological rhythm The cyclic variation of a biological process (Ch 6), 199, 200–201, 201*f*, 370
  body temperature and, 201, 740, 743
  of cortisol secretion, 201, 201*f*, 753, 753*f*, 754*f*
  of prolactin secretion, 854

biological systems, energy in, 9, 90–93, 91*f*, 92*f*

biome, 6

biomolecule Organic molecules associated with living organisms: carbohydrates, lipids, proteins and nucleotides (Ch 2), 27–34. *See also specific type*

biophysics, A-31 to A-33

biorhythm. *See* biological rhythm

biosphere, 2, 2*f*

biotin, 97

bipolar neuron Neuron with a single axon and single dendrite (Ch 8, 10), 246, 248*f*
  in retina, 364, 365*f*, 369

bipotential gonad Embryonic tissue that has the potential to develop into either testis or ovary (Ch 26), 824*f*, 825, 825*t*

birth control, 846–847, 846*t*

birth control pill, 846*t*, 847

2,3-bisphosphoglycerate. *See* 2,3-diphosphoglycerate

bisphosphonates, 771

bitter taste, sensation of, 345, 347*f*

blackwater fever, 614

bladder. *See* urinary bladder

bladder sphincters, 421, 635, 635*f*

-blast (suffix), 76

blastocyst Early embryo, consisting of a hollow ball of cells (Ch 26), 849, 850*f*

bleaching of visual pigment Process in which light activation causes opsin to release retinal (Ch 10), 367, 368*f*

bleb Weakened section of the lung tissue (Ch 17), 81, 573

bleeding (venesection), 536

bleeding/blood loss. *See* hemorrhage

blind spot Region of the retina with no photoreceptors because the optic nerve and blood vessels exit the eye. Synonym: optic disk (Ch 10), 359, 359*f*, 364

blind study An experiment in which the subject does not know if he or she is receiving the experimental treatment or a placebo (Ch 1), 11

blister, 71

blood The circulating portion of the extracellular fluid (Ch 3, 14, 15, 16, 18), 78, 78*t*, 80*f*, 535–557
  composition of, 536–538, 537*f*. *See also* blood cells; plasma
  deoxygenated, 459–460
  distribution of, 514, 515*f*, 516*f*

blood pressure and, 509, 510*f*
  to brain, 300
  exercise and, 813, 813*f*
gas transport in, 588*f*, 593–602, 602*f*. *See also* gas transport
  hormone secretion into, 216
  osmotic changes in, red cell shape affected by, 543, 544*f*
  oxygenation of, 459*f*, 460

blood-brain barrier Tight junctions in the brain capillaries that prevent free exchange of many substances between the blood and the cerebrospinal fluid (Ch 9, 15), 69, 299–300, 299*f*, 503

blood cells, 78*t*, 537*f*, 538. *See also specific type*
  production of, 538–542, 539*f*, 540*t*, 543*f*

blood clot, 77, 548, 551, 552*f*. *See also* coagulation
  integrin receptor defects and, 184
  myocardial infarction and, 457, 461, 527–528
  plaque rupture and, 526, 527*f*

blood count, 540–541, 541*f*

blood doping, 536, 542, 547, 551, 555

blood flow. *See also specific type of blood vessel*
  alveolar, 579–580, 580*t*, 581*f*
  body temperature and, 742–743, 815
  distribution of, 514, 515*f*, 516*f*
    exercise and, 813, 813*f*
  myogenic autoregulation of, 510–511
  pressure, volume, resistance relationships and, 460–464, 461*f*, 462*f*, 463*f*, 465*f*, 505, 505*t*, 507–508, 508*f*, 510*f*, 514, 515*f*
    glomerular filtration rate and, 623–624, 623*f*
  rate of, 464, 465*f*
  resistance to, 462–463, 463*f*, 464, 507–508, 508*f*, 509–513, 510*f*, 514, 515*f*
  valves regulating
    cardiac, 459, 465, 467*f*, 468–470, 469*f*
    venous, 459, 459*f*, 506, 506*f*
  velocity of, 464, 465*f*, 516–517, 517*f*
    in capillaries, 516–517, 517*f*

blood fluke, 778

blood glucose test, 175

blood groups, 58, 798–799, 798*f*, 799*t*

blood loss. *See* hemorrhage

blood pressure The pressure exerted by blood on the walls of the blood vessels. Usually measured in the systemic arteries (Ch 15, 17, 20, 25), 504–509, 504*f*, 505*f*, 505*t*
  afterload and, 493
  aldosterone secretion and, 653–654, 655*f*, 656*f*
  baroreceptor reflex and, 521–525, 523*f*, 524*f*
  blood distribution and, 509, 510*f*
  blood vessels and, 460–461, 461*f*
  blood volume change and, 508–509, 509*f*, 510*f*
  cardiac output and, 507–508, 508*f*, 510*f*
  cardiovascular disease and, 528, 528*f*
  elevated. *See* hypertension
  exercise and, 813–814, 814*f*
  glomerular filtration rate (GFR) and, 622–624, 623*f*
  integrated response to changes in, 643, 643*f*, 661–663, 662*f*
  mean arterial, 464, 506
  measurement of, 506–507, 507*f*
  peripheral resistance and, 507–508, 508*f*, 510*f*
  pulmonary arterial, 564
  regulation of, 507–509, 508*f*, 509*f*, 510*f*, 511*t*, 521–525, 522*f*, 523*f*, 524*f*, 614, 660*t*
    angiotensin II in, 654–655, 655*f*
    renin-angiotensin-aldosterone system in, 654, 655*f*, 656*f*
  responses to changes in, 660*t*, 661–663, 662*f*

in shock, 510
  systemic, 460–461, 460*f*, 505–506, 505*f*, 506*f*
  volume, flow, resistance relationships and, 460–464, 461*f*, 462*f*, 463*f*, 465*f*, 505, 505*t*, 507–509, 508*f*, 509*f*, 510*f*
  water balance and, 648, 649*f*, 661–663, 662*f*

blood substitutes, 594

blood sugar. *See* diabetes mellitus

blood-testis barrier Tight junctions between Sertoli cells that prevent free exchange between the extracellular fluid and the lumen of the seminiferous tubules (Ch 26), 833

blood transfusion, 798–799
  blood doping and, 547, 551

blood type (blood groups), 58, 798–799, 798*f*, 799*t*

blood vessel, 458, 459*f*, 468*t*, 502–504, 502*f*. *See also specific type or organ or system supplied*
  angiogenesis and, 503–504
  cutaneous, thermoregulation and, 742–743
  damage to, 548–549, 549*f*
  epithelia of, 73, 73*f*, 502
  local control of, 194
  pressure gradient in, 460–461, 460*f*, 505–506, 505*f*
  size of
    resistance and, 463, 463*f*
    velocity and, 464, 465*f*, 516–517, 517*f*
  sympathetic control of, 380*f*
  tonic control of diameter of, 192, 193*f*
  walls of, 502, 502*f*

blood volume
  disturbances of, 659–661, 659*f*
    integrated response to, 643, 643*f*, 660*t*, 661–663, 662*f*
  pressure, flow, resistance relationships and, 460–464, 461*f*, 462*f*, 463*f*, 465*f*, 508–509, 509*f*, 510*f*
  renal regulation of, 508, 509*f*, 614
  water balance and, 648, 649*f*, 661–663, 662*f*

"blue bloater," 559. *See also* chronic bronchitis

B lymphocyte (B cell) White blood cell that secretes antibodies, 783*f*, 784, 787–788, 788–789, 789*f*
  in allergic reaction, 797, 797*f*
  in bacterial infection, 794*f*, 795
  in viral infection, 795, 796*f*

BMR. *See* basal metabolic rate

BNP. *See* brain natriuretic peptide

body cavities, 51, 52*f*

body compartments, 9, 51, 52*f*. *See also* compartmentation

body fluid compartments, 52, 52*f*

body load, in mass balance, 129, 129*f*

body mass index (BMI), 764

body movement
  control of, 435–455
    cerebellum in, 305–306
    integrated, 445–451
    neural, 448–451, 448*t*, 449*f*, 450*f*, 451*f*
  heat production and, 743
  mechanics of, 417–421, 418*f*, 419*f*, 420*f*, 421*f*
  types of, 446–448, 447*t*
  vestibular apparatus providing information about, 355, 447

body positioning (equilibrium), 328*t*, 355–357, 356*f*, 357*f*, 358*f*, 447
  receptors in, 196, 196*f*, 355, 355–357, 356*f*, 357*f*

body positions, anatomic, A-39

body of stomach, 679, 680*f*

hyperparathyroidism and, 752, 759, 765, 772
phosphate homeostasis and, 770–771
calcium channel Ion channel that allows movement of Ca²⁺ across a membrane (Ch 5, 6, 10, 11, 12, 14, 15), 138, 186, 254, 271, 346, 391, 408, 471, 491, 511
ECaC, 767
in smooth muscle contraction, 427, 427f, 683
stretch-activated, 427
voltage-gated
in insulin secretion, 166, 167f, 730
neurotoxins blocking, 270
calcium channel blocker Drugs that block Ca²⁺ channels; used to treat high blood pressure (Ch 15), 191, 521, 528, 529
calcium equilibrium potential, 253t
calcium-induced Ca²⁺ release Process in which Ca²⁺ entry into a muscle fiber triggers release of additional Ca²⁺ from the sarcoplasmic reticulum (Ch 14), 471, 472f
calcium ion (Ca²⁺), 22t. *See also* calcium
calcium phosphate, 752, 765
calcium release channel, in excitation-contraction coupling, 408, 408f
in cardiac muscle, 471, 472f
in smooth muscle, 423–424
calcium-sensing receptor (CaSR) Receptor found on parathyroid cells that responds to changes in plasma Ca²⁺ concentrations (Ch 23), 224, 768
abnormal, 192t
calcium signal, 92, 766. *See also* calcium, as intracellular signal
calcium spark, 186
caldesmon A regulatory protein on smooth muscle actin (Ch 12), 427
calmodulin Intracellular second messenger that binds Ca²⁺ (Ch 6, 12), 182t, 186, 425, 425f, 426
Calorie. *See* kilocalorie
calorie, 101
calorimetry, 719–720
calponin Actin-associated regulatory protein of smooth muscle (Ch 12), 427
CAM. *See* cell adhesion molecule
cAMP. *See* cyclic AMP
canal of Schlemm, 358, 359f
cancer, 58, 75, 84–85, A-38
anchoring junctions and, 71
angiogenesis and, 504
cervical, 51, 67, 75, 81, 84–85
immune system and, 777, 800
oncogenes/oncogenic viruses and, 778, A-38
prostate, 833
cancer treatment
angiogenesis and, 504
engineered antibodies in, 800
guided imagery in, 438
candidate hormone Molecules that have not been shown to fulfill all the qualifications of a hormone (Ch 7), 216
cannabinoid receptors (Ch 8), 275
Cannon, Walter B. The father of American physiology (Ch 1), 5, 192
Cannon's postulates, 192–193
capacitation Changes in sperm that confer the ability to swim rapidly and fertilize an egg (Ch 26), 848–849, 848f
capacity Sum of two or more lung volumes (Ch 17), 567–568, 568f
CAPD. *See* continuous ambulatory peritoneal dialysis

capillary Smallest blood vessels where blood exchanges material with the interstitial fluid (Ch 5, 14, 15), 457, 459f, 460, 501f, 502, 502f, 503, 514–519, 516f
blood-brain barrier and, 299–300, 299f, 503
blood flow velocity in, 516–517, 517f
epithelia of, 72, 73, 74f, 502
lymph, 519
pressure in, 460f, 505f
pulmonary, 564, 580, 581f
capillary absorption, 517
capillary exchange Movement of fluid between the plasma and interstitial fluid across the capillary endothelium (Ch 15), 514–519, 518f
edema and, 519–520
capillary filtration, 517–519, 518f
capillary hydrostatic pressure, 518, 518f
edema and, 519–520
glomerular filtration and, 621–622, 622f
reabsorption and, 629–630
capsaicin, 340
capsid, viral, 778, 778f
capsule, bacterial Polysaccharide wall around a bacterium that prevents phagocytes from binding to the bacterium (Ch 24), 778
capsule, muscle spindle, 439
carbaminohemoglobin Hemoglobin with bound carbon dioxide (Ch 18), 601
carbohydrate Biomolecule whose basic structure is (CH₂Oₙ). Includes sugars, starch, glycogen (Ch 2, 5, 12, 21, 22), 27, 28f, 693–694, 723–724. *See also* glucose
in combined molecules, 32
digestion and absorption of, 693–694, 694f, 695f, 706–707
as energy source, 103, 104–112, 104f, 106t, 723–724, 724f
for skeletal muscle, 113, 410
membrane, 54f, 54t, 58
metabolism of, 722, 723–724, 723f, 724f, 725t
exercise and, 808, 809f, 810, 810f
insulin deficiency and, 736, 737f. *See also* diabetes mellitus
carbon, 21
carbon dioxide (CO₂) Gaseous product of aerobic respiration (Ch 4, 17, 18, 20), 107, 599
acid production and, 664
bicarbonate ions and, 599–601
exchange of, 588f
hemoglobin binding of, 601
partial pressure of, 566t, 589, 589f, 590, 590t
in acid-base disturbances, 670, 670t, 671
alveolar, 578–579, 579f, 590, 590f
arterial and venous, 590, 590t
arteriolar and bronchiolar control and, 580, 580t, 581f
central chemoreceptor monitoring of, 604, 605–606, 605f, 606f
exercise hyperventilation and, 811–812, 812f
oxygen-hemoglobin dissociation curve affected by, 598, 598f
peripheral chemoreceptor monitoring of, 604
pH affected by, 666, 666f
in pulmonary edema, 593
pH affected by, 664, 666, 666f, 670, 670t, 671
production of, metabolic rate and, 720
removal of at lungs, 601
solubility of, 589, 589f
transport of, 458, 458t, 588f, 599–602, 600f, 602f
vasoactive effects of, 511t
ventilation affected by, 604–606, 605f, 606f

carbonic acid (H₂CO₃) Made from the reaction of CO₂ and water (Ch 18, 20), 37, 600f, 664
carbonic anhydrase Enzyme that catalyzes the conversion of carbon dioxide and water into carbonic acid (Ch 4, 18, 20, 21), 98, 599–600, 600f, 664
in bicarbonate secretion, 685, 686f
carbon monoxide (CO), 188, 275
carboxyl group, 27t, 30f, 31, 119
carboxypeptidase Enzyme that breaks peptide bonds at the carboxy terminal end of a peptide (Ch 21), 695, 695f, 707f
cardiac arrhythmia, 254, 483, 484
cardiac chambers, 465–468, 466–467f
cardiac cycle The period of time from the end of one heart beat through the end of the next beat (Ch 14), 480–487, 488f
electrical events of, 480, 481, 481f, 482f, 488f
mechanical events of, 480, 482f, 484–486, 485f, 488f
pressure-volume curves representing, 486–487, 487f
cardiac enzymes, 96, 97t, 471
cardiac glycoside Drugs such as ouabain and digitalis that block the Na⁺-K⁺-ATPase (Ch 14), 492, 525
cardiac hypertrophy, hypertension and, 528
cardiac murmurs, 486
cardiac muscle Striated muscle of the heart (Ch 3, 11, 12, 14), 397, 397f, 429, 429t, 465, 467f, 470–477, 470f
action potential in, 472–475, 473f, 474f, 475f, 476t
ECG compared with, 481, 482f
autonomic control of, 382–383, 475–476, 476f
contraction of, 397, 429, 429t, 470–477, 470f
electrical conduction in coordination of, 477–479, 477f, 478f
force of, 472, 473f
graded, 472
neural and endocrine control of, 491–492, 491f, 492f
excitation-contraction coupling in, 471, 472f
length-tension relationships in, 472, 473f
movement control in, 451–452
refractory period in, 473–474, 474f
cardiac notch, 562f
cardiac output (CO) The amount of blood pumped per ventricle per unit time (Ch 14, 15, 19, 25), 464, 489, 493, 494f, 502, 507–508, 510f
exercise affecting, 812–813, 813f
kidneys receiving, 615
in myocardial infarction, 495
in shock, 510
cardiac valve, 459, 465, 467f, 468–470, 469f
cardiogenic shock, 510
cardiovascular control center (CVCC) Neurons in the medulla oblongata that integrate sensory information and direct autonomic responses aimed at maintaining adequate blood pressure (Ch 15, 20), 521–522, 522f, 661, 662f
angiotensin II affecting, 655, 655f
cardiovascular disease, 525–528
diabetes and, 526, 738
exercise affecting risk for, 816
hypertension and, 528, 528f
inflammatory markers for, 526
risk factors for, 525

(continues)

pulmonary capillaries, 564, 580, 581*f*
pulmonary circulation That portion of the circulation that carries blood to and from the lungs (Ch 14, 17), 459*f*, 460, 561–564
pulmonary edema Excessive interstitial fluid volume in the lungs (Ch 18), 528, 592–593
alveolar ventilation/gas exchange and, 591*f*, 592–593
high-altitude (HAPE), 593
pulmonary function tests, 567, 567*f*, 580–581
pulmonary trunk The single artery that receives blood from the right ventricle; splits into left and right pulmonary arteries (Ch 14, 16), 465, 561
pulmonary valve The semilunar valve between the right ventricle and the pulmonary trunk (Ch 14), 467*f*, 469*f*, 470
pulmonary vein Vessel that carries well-oxygenated blood from the lung to the left heart (Ch 14), 459*f*, 460, 465, 466*f*, 561–564
pulmonary ventilation, total, 577, 578*f*
pulse Pressure wave that is transmitted through the fluid of the cardiovascular system (Ch 15), 505
pulse generator Region of the hypothalamus that coordinates the pulsatile secretion of GnRH (Ch 26), 830
pulse oximeter, 593
pulse pressure The strength of the pulse wave, defined as the systolic pressure minus the diastolic pressure (Ch 15), 505, 505*f*
pumps. *See* ATPase
pupil, 358*f*, 359*f*, 360
autonomic control of, 380*f*
pupillary reflex Constriction of the pupil in response to light (Ch 10), 360, 360*f*
purine, 33, A-34
as neurotransmitter, 273*t*, 275
receptors for, 273*t*, 275
purinergic receptor A receptor that binds to purines, such as AMP or ATP (Ch 8), 273*t*, 275
Purkinje fiber Specialized myocardial cell that rapidly conducts electrical signals to the apex of the heart (Ch 14), 477, 478*f*
pus, 785
push-pull control, 723, 725*f*
P wave Wave of the ECG that represents atrial depolarization (Ch 14), 480, 481*f*, 482*f*, 488*f*
pyloric sphincter. *See* pylorus
pyloric valve. *See* pylorus
pylorus The region of increased muscle tone separating the stomach and small intestine (Ch 21), 679, 680*f*
pyramid Region of the medulla where neurons from one side of the body cross to the other (Ch 9, 13), 305, 450, 451*f*
pyramidal cell, 311
pyramidal tract Descending pathways from basal nuclei that do not pass through the pyramids (Ch 13), 450, 451*f*
pyrimidine, 33, A-34
pyrogen Fever-causing substances (Ch 22, 24), 743, 786*t*
pyruvate, 104–106, 104*f*, 105*f*, 106*t*, 727, 728*f*
aerobic metabolism of, citric acid cycle and, 107–108, 108*f*
anaerobic metabolism of, lactate production and, 106–107, 106*t*, 107*f*
PYY₃₋₃₆, 718*t*

QRS complex Wave complex that represents ventricular depolarization and atrial repolarization (Ch 14), 480, 481*f*, 488*f*
QT interval From the beginning of the Q wave to the end of the T wave. Corresponds to ventricular contraction (Ch 12), 481*f*
quaternary structure, of protein Arrangement of a protein with multiple peptide chains (Ch 2), 30*f*, 31
quiet breathing, 567, 570–571, 570*f*, 571*f*, 603–604, 604*f*
Q wave First wave of ventricular depolarization (Ch 14), 482*f*

RAAS. *See* renin-angiotensin-aldosterone system
rabies, 777
Rabs, 150
radiant heat gain, 741, 741*f*
radiant heat loss, 741, 741*f*
radiation Energy emitted by unstable isotopes (Ch 2), 21–22, 741
radioactive iodine, 231, 761
radioisotope Unstable isotopes that emit energy (Ch 2), 21–22
radio waves, 364, 364*f*
raloxifene, 771
ramping, 603, 604*f*
ranitidine, 704
RANKL, 768
Ranvier, nodes of Unmyelinated regions on myelinated axons (Ch 8), 250, 251*f*, 267, 267*f*
rapid eye movement sleep. *See* REM sleep
Ray, Felix, 328
reabsorption Movement of filtered material from the lumen of the nephron to the blood (Ch 19), 618, 619, 619*f*, 620, 620*f*, 626–630, 626*f*
active, 627–628, 627*f*
passive, 628
peritubular capillary pressures and, 629–630
saturation and, 628–629, 629*f*, 630*f*
urine concentration and, 645
vasopressin affecting, 646–647, 647*f*, 648*f*
reactant, 93, 93*t*
reaction rate, 43–44, 43*f*, 93
enzymes affecting, 98, 98*f*
substrate concentration affecting, 39, 98, 99, 99*f*
reactive hyperemia An increase in tissue blood flow following a period of low perfusion (Ch 15), 512, 512*f*
receiving segment, 684, 684*f*
receptive aphasia Inability to understand spoken or visual information due to damage to Wernicke's area (Ch 9), 320
receptive field The region within which a sensory neuron can sense a stimulus (Ch 10), 330–331, 331*f*
receptive relaxation, 702
receptor (1) A cellular protein that binds to a ligand; (2) A cell or group of cells that continually monitor changes in the internal or external environment (Ch 2, 5, 6, 10, 11), 38, 137–138, 138*f*
agonists and antagonists and, 189, 190*f*
conversion of stimuli by, 330
diseases/drugs associated with abnormalities of, 191, 192*t*, 233
down-regulation of, 190–191
hormone, 176, 177*f*, 217, 219, 221, 221*f*, 751
abnormal, 233
isoforms of, 189–190, 190*f*

membrane, 137–138, 138*f*, 179, 179*f*
multiple, for one ligand, 189–190, 190*f*, 193
multiple ligands for, 189
multiple meanings for, 196, 196*f*
nociceptors, 340
odorant, 344*f*, 345
phasic, 334–335, 336*f*
in reflex control pathway, 195, 195*f*, 196–197, 196*f*, 198, 198*f*
sensitivity of, 329–330
sensory, 83*f*, 195, 195*f*, 196–197, 196*f*, 198, 198*f*, 204*f*, 205*t*
for somatic senses, 337–342, 337*f*, 338*f*, 338*t*, 340*t*
specificity and competition and, 189
taste, 345, 346*f*
temperature, 196*f*, 330, 330*t*, 339–340
termination of activity of, 191, 217
tonic, 334, 336*f*
touch, 333, 338–339, 339*f*
up-regulation of, 190–191
receptor adaptation A repeated stimulus loses its ability to stimulate a receptor (Ch 10), 334, 335
receptor channel, 179, 185
receptor-enzyme A class of membrane proteins that bind ligands on the extracellular side and activate enzymes (usually tyrosine kinase or guanylyl cyclase) on the intracellular side (Ch 6), 179, 179*f*, 181*t*, 182–183, 183*f*
receptor-mediated endocytosis A ligand binds to a membrane protein, which triggers endocytosis of the membrane-receptor complex (Ch 5, 22), 149, 149*f*, 725
receptor potential Graded potential in a special senses receptor (Ch 10), 330
receptor proteins, 38, 177, 178–180, 178*f*, 179*f*. *See also* receptor
diseases/drugs associated with alterations in, 191, 192*t*
recessive genetic disorder, 98
reciprocal inhibition The relaxation of antagonistic muscles during a contraction reflex (Ch 13), 443, 444
recognition of self Ability of the immune system to recognize self and not create an immune response. This ability is lost in autoimmune diseases (Ch 24), 777, 799–800, 799*t*
recombinant human EPO, blood doping and, 551, 555
recombinant human growth hormone (hGH), 764
recording electrode, 162, 163*f*
recruitment Addition of motor units to increase the force of contraction in a muscle (Ch 12), 416–417
rectal temperature, 740
rectum The distal segment of the large intestine (Ch 21), 679, 680*f*, 708, 708*f*
red blood cell (RBC). *See* erythrocyte
red blood cell antigens, 798–799, 798*f*, 799*t*
red blood cell membrane, composition of, 54*t*
red bone marrow, 540
red cell count, 541*f*
"red muscle" Muscle that has lots of mitochondria and good blood supply so that it can carry out oxidative metabolism (Ch 12), 412
red pulp, 781*f*
reduced molecule, 100, 100*t*
reductase, 100*t*
5-α-reductase Enzyme that converts testosterone to DHT (Ch 26), 825–826, 829*f*

Photo Credits

About the Author
Author: Dee Silverthorn

Chapter 1
Page 1: Niki Sianni/Photonica/Getty Images.

Chapter 2
Page 19: Edy Kieser, *www.edykieser.ch*

Chapter 3
Page 50: Todd Derksen; **3-8:** D. E. Saslowsky, J. Lawrence, X. Ren, D. A. Brown, R. M. Henderson, J. M. Edwardson, Placental alkaline phosphatase is efficiently targeted to rafts in supported lipid bilayers, Journal of Biological Chemistry, Pages 26966-70, Vol. 277:30, 26 July 2002. c.2002 Journal of Biological Chemistry, American Society for Biochemistry and Molecular Biology, Inc., Bethesda, MD, 20814. JBC Online, *http://www.jbc.org;* **3-13c:** Robert W. Riess; **3-14a:** Fawcett/Hirokawa/Heuser/Science Researchers, Inc.; **3-16:** CNRI/Science Source/Photo Researchers, Inc.; **3-17:** Robert W. Riess; **3-17:** Robert W. Riess; **3-18b:** Brad J. Marsh, David N. Mastronarde, Karolyn F. Buttle, Kathryn E. Howell, and J. Richard McIntosh, Inaugural Article: Organellar relationships in the Golgi region of the pancreatic beta cell line, HIT-T15, visualized by high resolution electron tomography. Proceedings of the National Academy of Sciences of the United States of America. Fig. 2, Pages 2399-2406, Vol. 98:5, 27 February 2001. c.2001 by the National Academy of Sciences; **3-19:** Dr. Don Fawcett/Photo Researchers, Inc.; 3-20 Biophoto Associates/Photo Researchers, Inc.; **3-21 (top):** Prof. H. Wartenberg/Dr. H. Jastrow's EM Atlas, *www.drjastrow.de;* **3-21 (bottom):** Photo Researchers, Inc.; **3-23:** Todd Derksen; **3-26b:** Custom Medical Stock Photo, Inc.; **3-27:** Todd Derksen; **3-29:** Ward's Natural Science Establishment, Inc.; **3-30c:** John D. Cunningham/Visuals Unlimited; **3-31:** Frederic H. Martini; **3-33a:** SPL/Photo Researchers; **3-33b:** SPL/Photo Researchers, Photo Researchers, Inc.

Chapter 4
Page 89: Photo Researchers, Inc.

Chapter 5
Page 128: Alex Gray/The Welcome Trust Medical Photographic Library.

Chapter 7
Page 211: CNRI/Phototake NYC; **17.01:** Dee Silverthorn.

Chapter 8
Page 243: Peter Brophy/The Welcome Trust Medical Photographic Library; **8-17a:** Todd Derksen, **8-17b:** David M. Phillips/ Visuals Unlimited; **8-31:** Timothy M. Gomez, University of Wisconsin-Madison.

Chapter 9
Page 291: McGill University/CNRI/Phototake NYC; **9-12b:** Robert Brons/BPS/Stone; **9-17:** Marcus Raichle, Washington University School of Medicine.

Chapter 10
Page 327: E.S. Lein, X. Zhao, F.H. Gage Defining a molecular atlas of the hippocampus using DNA microarrays and high-throughput in situ hybridization. Journal of Neuroscience 2004 Apr 14; 24 (15):3879-89, (i); **10-15b:** Todd Derksen; **10-28:** Webvision, John Moran Eye Center, University of Utah.

Chapter 11
Page 376: Dr. Flora Love.

Chapter 12
Page 396: University of Texas at Austin; **12-1a:** Ward's Natural Science Establishment; **12-1b:** Phototake NYC; **12-1c:** Todd Derksen; **12-15 (left):** D. Comack, ed. Ham's Histology 9th ed. Philadelphia: J.B. Lippincott, 1987. By permission; **12-15 (top right):** Frederic H. Martini; **12-15 (bottom right):** Frederic H. Martini; **12-26:** Biophoto Associates/Photo Researchers, Inc.

Chapter 14
Page 456: Carole L. Moncman, Ph.D.

Chapter 15
Page 500: Amy Brock and Don Ingber, Harvard Medical School/ Children's Hospital; **15-20:** World Health Organization.

Chapter 16
Page 535: Diane Gray Molecular Probes, Inc. Eugene, Oregon, USA; **16-5a:** Todd Derksen; **16-6a:** Visuals Unlimited; **16-6b:** Visuals Unlimited; **16-6c:** Visuals Unlimited; **16-8:** Photo Researchers, Inc.; **16-9b:** Todd Derksen; **16-9c:** Todd Derksen.

Chapter 17
Page 558: Barbara A. Danowski Department of Biology Union College Schenectady, New York, USA; **17-8:** Frederic H. Martini.

Chapter 18
Page 587: Phototake NYC.

Chapter 19
Page 613: Ruben M. Sandoval, Indiana Center for Biological Microscopy, Indiana University School of Medicine, Indianapolis Indiana, USA; **19-1f:** Todd Derksen; **19-4b:** Todd Derksen.

Chapter 21
Page 676: Mark L. Tamplin, Anne L. Gauzens, and Rita R. Colwell.

Chapter 22
Page 716: Roland O. Marsh, Jr. Madigan Army Medical Center Tacoma, Washington, USA; **22-8c:** Ward's Natural Science Establishment, Inc.

Chapter 23
Page 750: CNRI/SPL/Photo Researchers, Inc.; **23-1c:** Ward's Natural Science Establishment, Inc.; **23-5a:** Biophoto Associates/ Science Source/Photo Researchers, Inc.; **23-5b:** Biophoto Associates/Photo Researchers, Inc.; **23-11:** Martin Rotker/Phototake NYC; **23-13:** Alison Wright/CORBIS; **23-15:** Ralph Eagle/Science Source/Photo Researchers, Inc.; **23-17:** American Journal of Medicine; **23-17:** American Journal of Medicine; **23-17:** American Journal of Medicine; **23-18:** Ralph T. Hutchings; **23-24:** Dr. Michael Klein/Peter Arnold, Inc.

Chapter 24
Page 776: Dr. Dennis Kunkel/Phototake NYC.

Chapter 26
Page 821: Vanessa Rawe, Ph.D., Laboratory of Reproductive Biology and Special Studies Center of Studies in Gynecology and Reproduction, Viamonte, Buenos Aires, ARGENTINA. Photographed at Pittsburgh Development Center, MWRI, University of Pittsburgh School of Medicine, Pittsburgh, PA; **26-01:** CNRI/ Science Photo Library/Photo Researchers, Inc. **26-16a:** Francis Leroy/Biocosmos/ Science Photo Library/Custom Medical Stock Photo, Inc.

Measurements and Conversions

PREFIXES

deci-	(d)	1/10	0.1	1×10^{-1}
centi-	(c)	1/100	0.01	1×10^{-2}
milli-	(m)	1/1000	0.001	1×10^{-3}
micro-	(μ)	1/1,000,000	0.000001	1×10^{-6}
nano-	(n)	1/1,000,000,000	0.000000001	1×10^{-9}
pico-	(p)	1/1,000,000,000,000	0.000000000001	1×10^{-12}
kilo-	(k)		1000.	1×10^{3}

METRIC SYSTEM

1 meter (m)	=	100 centimeters (cm)	=	1000 millimeters (mm)
1 centimeter (cm)	=	10 millimeters (mm)	=	0.01 meters (m)
1 millimeter (mm)	=	1000 micrometers (μm; also called micron, μ)		
1 angstrom (Å)	=	1/10,000 micrometer	=	1×10^{-7} millimeters
1 liter (L)	=	1000 milliliters (mL)		
1 deciliter (dL)	=	100 milliliters (mL)	=	0.1 liters (L)
1 cubic centimeter (cc)	=	1 milliliter (mL)		
1 milliliter (mL)	=	1000 microliters (μL)		
1 kilogram (kg)	=	1000 grams (g)		
1 gram (g)	=	1000 milligrams (mg)		
1 milligram (mg)	=	1000 micrograms (μg)		

CONVERSIONS

1 yard (yd)	= 0.92 meter
1 inch (in)	= 2.54 centimeters
1 meter	= 1.09 yards
1 centimeter	= 0.39 inch
1 liquid quart (qt)	= 946 milliliters
1 fluid ounce (oz)	= 8 fluid drams = 29.57 milliliters (mL)
1 liter	= 1.05 liquid quarts
1 pound (lb)	= 453.6 grams
1 kilogram	= 2.2 pounds

TEMPERATURE

FREEZING

0 degrees Celsius (°C) = 32 degrees Fahrenheit (°F) = 273 Kelvin (K)

To convert degrees Celsius (°C) to degrees Fahrenheit (°F):
(°C × 9/5) + 32

To convert degrees Fahrenheit (°F) to degrees Celsius (°C):
(°F − 32) × 5/9

NORMAL VALUES OF BLOOD COMPONENTS

SUBSTANCE OR PARAMETER	NORMAL RANGE	MEASURED IN
Calcium (Ca^{2+})	4.3–5.3 meq/L	Serum
Chloride (Cl^{-})	100–108 meq/L	Serum
Potassium (K^{+})	3.5–5.0 meq/L	Serum
Sodium (Na^{+})	135–145 meq/L	Serum
pH	7.35–7.45	Whole blood
P_{O_2}	75–100 mm Hg	Arterial blood
P_{CO_2}	34–45 mm Hg	Arterial blood
Osmolality	280–296 mosmol/kg water	Serum
Glucose, fasting	70–110 mg/dL	Plasma
Creatinine	0.6–1.5 mg/dL	Serum
Protein, total	6.0–8.0 g/dL	Serum

Modified from W. F. Ganong, *Review of Medical Physiology* (Norwalk: Appleton & Lange), 1995.